GRAY'S
ANATOMY

GRAY'S ANATOMY

HENRY GRAY F.R.S.

DRAWINGS BY
H. V. CARTER M.D.

EDITED BY
T. PICKERING PICK F.R.C.S.
&
ROBERT HOWDEN M.A., M.B., C.M.

This sixteenth edition first published in 1905 by
Longmans, Green and Company

This edition published in 2003 by Senate an
Imprint of Merchant Book Co. Ltd.
P.O. Box 10, East Molesey, KT8 2WZ

ISBN 1 85958 018 1

Printed and bound in Finland

THE FIRST EDITION OF THIS WORK

WAS DEDICATED TO

SIR BENJAMIN COLLINS BRODIE, BART.

F.R.S., D.C.L.

IN ADMIRATION OF

HIS GREAT TALENTS

AND IN REMEMBRANCE OF

MANY ACTS OF KINDNESS SHOWN TO THE ORIGINAL AUTHOR OF THE BOOK

FROM AN

EARLY PERIOD OF HIS PROFESSIONAL CAREER

PREFACE

TO

THE SIXTEENTH EDITION

THE text of this edition has been carefully revised and in part rewritten. Many additional drawings have been introduced, more especially in connection with the chapters on Embryology, the Nervous System, and the Organs of Sense, while at the same time care has been taken to keep the size of the book within reasonable limits. A further use of colour has been employed in order to render the drawings more effective and instructive.

The half-tone illustrations which were utilised to a considerable extent in recent editions, and which required for their successful reproduction a highly glazed heavy paper, have been replaced by woodcuts and line drawings. These not only harmonise more closely with Carter's original figures, but have permitted of a lighter paper being used for printing purposes, with the result that the present volume weighs considerably less than the last.

The Editors desire to express their indebtedness to Dr. R. A. BOLAM, Lecturer on Physiology and Histology in the University of Durham College of Medicine, Newcastle-upon-Tyne, who has revised the chapter on Histology and furnished it with several new illustrations.

It is hoped that the present edition will maintain and enhance the reputation which the work has enjoyed during the past forty-seven years.

CONTENTS

GENERAL ANATOMY OR HISTOLOGY

EMBRYOLOGY

OSTEOLOGY

THE ARTICULATIONS

MUSCLES AND FASCIÆ

THE BLOOD VASCULAR SYSTEM

a

THE LYMPH VASCULAR SYSTEM

THE NERVOUS SYSTEM

a 2

CONTENTS

ORGANS OF SPECIAL SENSE

ORGANS OF VOICE AND RESPIRATION

ORGANS OF DIGESTION

CONTENTS

SURGICAL ANATOMY OF FEMORAL HERNIA

SURGICAL ANATOMY OF PERINÆUM AND ISCHIO-RECTAL REGION

ERRATA

In the inscription to figure 566, page 798, for 17. *Eminentia acustici* read 17. *Eminentia acustica.* On lines 18 and 25, page 799, for *area acustici* read *area acustica* ; and on line 20 for *striæ acustici* read *striæ acusticæ*

LIST OF ILLUSTRATIONS

The illustrations, when copied from any other work, have the author's name affixed. When no such acknowledgment is made the drawing is to be considered original.

GENERAL ANATOMY OR HISTOLOGY

EMBRYOLOGY

OSTEOLOGY

ARTICULATIONS

MUSCLES AND FASCIÆ

BLOOD VASCULAR SYSTEM

VEINS

LYMPHATIC SYSTEM

NERVOUS SYSTEM

SPINAL NERVES

ORGANS OF SENSE

ORGANS OF VOICE AND RESPIRATION

ORGANS OF DIGESTION AND THEIR APPENDAGES

THE URINARY AND GENERATIVE ORGANS

DUCTLESS GLANDS

SURGICAL ANATOMY OF HERNIA

SURGICAL ANATOMY OF PERINÆUM AND
ISCHIO-RECTAL REGION

GENERAL ANATOMY OR HISTOLOGY

THE ANIMAL CELL (fig. 1)

ALL the tissues and organs of which the body is composed were originally developed from a microscopic structure (the *ovum*), consisting of a soft jelly-like granular material enclosed in a membrane, and containing a vesicle, or small spherical body, inside which are one or more solid spots. This may be regarded as a perfect cell. Moreover, all the solid tissues can be shown to consist largely of similar cells, differing, it is true, in external form, but essentially similar to an ovum.

In the higher organisms all such cells may be defined as 'nucleated masses of protoplasm of microscopic size.' The two essentials, therefore, of an animal cell in the higher organisms are : the presence of a soft jelly-like granular material, similar to that found in the ovum, and which is usually styled *protoplasm* ; and

FIG. 1.—Diagram of a cell. (Modified from Wilson.)

Cell wall — Attraction sphere (centro-sphere) enclosing two cen-trosomes

Nucleolus

Nuclear membrane — Net-knot of chromatin form-ing a pseudo-nucleus

Chromatin network

Vacuoles — Cell-inclusions (paraplasm)

a small spherical body embedded in it, and termed a *nucleus* ; * the remaining constituents of the ovum, viz. its limiting membrane and the solid spot contained in the nucleus, called the *nucleolus*, are not considered essential to the cell, and in fact many cells exist without them.

Protoplasm (*cytoplasm*) is a material probably of variable constitution, but yielding to the chemist on its disintegration bodies chiefly of proteid nature. Lecithin and cholesterin are constantly found in it, as well as inorganic salts, chief among which are the phosphates and chlorides of the alkali metals and calcium. It is of a semi-fluid, viscid consistence, and appears either as a hyaline substance, homogeneous and clear, or else it exhibits a granular appearance. This granular appearance, under a high power of the microscope, is seen to be due to the fact that protoplasm consists of an elastic and refractile network or honeycombed

---

* In certain lower forms of life, masses of protoplasm without any nuclei have been described by Huxley and others, as cells.

B

reticulum, containing in its meshes a clear, semi-fluid homogeneous substance. The former is known as *spongioplasm*, the latter as *hyaloplasm*. The granular appearance is often caused by the knots of the network being mistaken for granules; but, in addition to this, protoplasm often contains true granules, some of which are proteid in nature and probably essential constituents; others are fat, glycogen or pigment granules, and are regarded as adventitious material taken in from without, and hence are styled cell-inclusions or *paraplasm*. The size and shape of the meshes of the spongioplasm vary in different cells and in different parts of the same cell. In many fixed cells, e.g. epithelial cells, the external layer becomes denser than the rest, and often altered by the deposition in it of some chemical substance, so as to constitute a membrane which encloses the rest of the protoplasm and forms the *cell wall*. The relative amount of spongioplasm and hyaloplasm varies in different cells; the latter preponderating in the young cell and the former increasing in amount, at the expense of the hyaloplasm, as the cell grows.

The most striking characteristics of protoplasm are its vital properties of *motion* and *nutrition*. By motion is meant the property which protoplasm has of changing its shape and position by some intrinsic power, which enables it to thrust out from its main body an irregular process, into which the whole of the protoplasmic substance is gradually drawn, so that the mass comes to occupy a new position. This, on account of its resemblance to the movements observed in the Amœba or Proteus animalcule, has been termed 'amœboid movement.' Ciliary movement, or the vibration of hair-like processes from the surface of any structure, may also be regarded as a variety of the motion with which protoplasm is endowed.

Nutrition is the power which protoplasm has of attracting to itself the materials necessary for its growth and maintenance from surrounding matter. When any foreign particle comes in contact with the protoplasmic substance, it becomes incorporated in it, being enwrapped by one or more processes projected from the parent mass which enclose it. When thus taken up, it may remain in the substance of the protoplasm for some time without change, or may be again extruded.

The **Nucleus** is a minute body, embedded in the protoplasm, and usually of a spherical or oval form, its size having little relation to the size of the cell. It is surrounded by a well-defined wall, the *nuclear membrane*, which encloses the nuclear contents. These are known as the *nuclear substance* (nuclear matrix), which is composed of a homogeneous material and a stroma or network. The former is probably of the same nature as the hyaloplasm of the cell; but the latter, which forms also the wall of the nucleus, differs from the spongioplasm of the cell substance. It is sometimes known as the *chromoplasm* or *intranuclear network*, and consists of a network of fibres or filaments arranged in a reticular manner. These filaments stain very readily with certain dyes; they are therefore named *chromatin*; while the interstitial substance does not stain readily, and is hence called *achromatin*. In some resting nuclei, i.e. nuclei which are not undergoing subdivision, the nuclear filaments do not form a network, but present the appearance of a convoluted skein, similar to that found in a nucleus about to undergo division, which will be immediately described.

Within the nuclear matrix are one or more highly refracting bodies, termed *nucleoli*, connected with the nuclear membrane by the nuclear filaments. They are regarded as being of two kinds. Some are mere local condensations ('net-knots') of the chromoplasm; these are irregular in shape and are termed *pseudo-nucleoli*; others are distinct bodies differing from the pseudo-nucleoli both in nature and chemical composition; they may be termed *true nucleoli*, and are usually found in resting cells.

The nuclear substance differs chemically from ordinary protoplasm in containing *nuclein*, in its power of resisting the action of acids and alkalies, in its imbibing more intensely the stain of carmine, hæmatoxylin, &c., and in its remaining unstained by some reagents which colour ordinary protoplasm.

Recent investigations tend to show that most living cells contain, in addition to their protoplasm and nucleus, a minute particle which, on account of the power it appears to possess of attracting the surrounding protoplasmic granules, is termed the *attraction particle* or *centrosome*; it usually lies near the nucleus. The spherical arrangement of fibrillar rows of granules surrounding the central particle is termed the *attraction-sphere* or *centrosphere*. These spheres are as a rule double, and are connected by a spindle-shaped system of delicate fibrils (*achromatic spindle*). They are best seen in young cells which are about to undergo the process of division, a process believed to commence in these bodies.

The process of reproduction of cells is described as being brought about by *indirect* or by *direct division*. *Indirect division* or *karyokinesis* (*karyomitosis*) has been observed in all the tissues—generative cells, epithelial tissue, connective tissue, muscular tissue, and nerve tissue, and probably it will ultimately be shown that the division of cells always takes place in this way, and that the process of reproduction of cells by direct division is, as some observers believe, merely a sort of imperfect or abnormal karyokinesis.

The process of indirect cell division is characterised by a series of complex changes in the nucleus, leading to its subdivision ; this being followed by cleavage of the cell protoplasm. Starting with the nucleus in the quiescent or *resting* stage, these changes may be briefly grouped under the four following phases :

1. *Prophase.*—The nuclear network of chromatin filaments assumes the form of a twisted *skein* or *spirem*, while the nuclear membrane and nucleolus disappear. The convoluted skein of chromatin divides into a definite number of V-shaped loops or *chromosomes*. The number of chromosomes varies in different animals—in man it is believed to be always sixteen. Coincident with or preceding these changes the centrosome, or attraction particle, which usually lies by the side of the nucleus, undergoes subdivision, and the two resulting centrosomes, each surrounded by a centrosphere, are seen to be connected by a spindle of delicate achromatic fibres, the *achromatic spindle*. These centrosomes move away from each other—one towards each extremity of the nucleus—and the fibrils of the achromatic spindle are correspondingly lengthened. The centrosomes are now situated one at either extremity or pole of the elongated spindle, and each is surrounded by a centrosphere, from which fibrils radiate into the investing protoplasm. A line encircling the spindle midway between its poles is named the *equator*, and around this the V-shaped chromosomes arrange themselves in the form of a star, thus constituting the *mother star* or *monaster*.

2. *Metaphase.*—Each V-shaped chromosome now undergoes longitudinal cleavage into two equal halves or *daughter chromosomes*, the cleavage commencing at the apex of the V and extending along its divergent limbs. The daughter chromosomes, thus separated, travel in opposite directions along the fibrils of the achromatic spindle towards the centrosomes, around which they group themselves, and thus two star-like figures are formed, one at either pole of the achromatic spindle. This is termed the *diaster*.

3. *Anaphase.*—The V-shaped daughter chromosomes now arrange themselves into a *skein* or *spirem*, and eventually form the network of chromatin which is characteristic of the resting nucleus. The nuclear membrane and nucleolus are also differentiated during this phase. The cell protoplasm begins to appear constricted around the equator of the achromatic spindle, where double rows of granules are also sometimes seen. The constriction deepens and the original cell gradually becomes divided.

4. *Telophase.*—In this stage the cell is completely divided into two new cells, each with its own nucleus, centrosome and centrosphere, which assume the ordinary positions occupied by such structures in the resting stage.

The series of diagrams (fig. 2), by Professor S. Delépine, is intended to explain the formation of some of the most important changes observed in nuclei of cells during karyokinesis (mitosis) ; it is based chiefly on the work of Flemming, Strasburger, E. van Beneden, Rabl, O. Hertwig, Henneguy, &c. A. *Resting nucleus.* Nucleolus and nuclear membrane visible. A centrosome is represented near the nucleus. B and C. *Skein* or *spirem*. Chromatic filaments much convoluted. Evidence of longitudinal splitting begins to be distinct in several parts. The centrosome has divided ; the nuclear membrane is becoming indistinct. The two centrosomes are widely separated, and the space between them is occupied by the achromatic spindle. (Two arrows point to the positions which the centrosomes will ultimately occupy ; during their passage to these points the achromatic spindle seems to be within the nucleus.) The nuclear membrane has disappeared. D. *Mother star, monaster.* The nuclear segments (chromosomes) resulting from the breaking-up of the chromatic filament into fragments of nearly equal length have moved towards the equator of the spindle, where they now form an equatorial

FIG. 2.—Karyokinesis : or indirect cell-division.

SIDE OR EQUATORIAL   END OR POLAR   SIDE OR EQUATORIAL.
VIEW.   VIEW.   VIEW.

A. Resting nucleus. B. Skein or spirem, close. C. Skein or spirem, open. D. Mother star, monaster. E. Metaphase. F. Daughter stars or diaster. G. Daughter skeins or dispirem, beginning to form. H. Daughter skeins or dispirem, formed. I. Resting daughter nuclei.

plate. These segments are all split longitudinally. E. *Metaphase*. One half of each chromosome moves in the direction of one pole and the other half in that of the other pole, being guided towards the centrosomes by the achromatic filaments. F. *Daughter stars* or *diaster*. G. *Daughter skeins* or *dispirem*, beginning to form. Segments in the form of thick loops not closely packed. H. *Daughter skeins* or *dispirem*, formed. Segments more closely packed and less distinct, owing to the formation of anastomoses. I. *Resting daughter nuclei.* Cell completely divided into two, but bridges remain between them in the region previously occupied by the achromatic filaments, these being specially distinct in certain cells (e.g. prickle cells). The nucleus has a distinct nuclear membrane and a nucleolus.

In the reproduction of cells by *direct* division the process is brought about either by segmentation or by gemmation. In reproduction by *segmentation* or *fission*, the nucleus becomes constricted in its centre, assuming an hour-glass shape, and then divides into two. This leads to a cleavage or division of the whole protoplasmic mass of the cell ; and thus two daughter cells are formed, each containing a nucleus. These daughter cells are at first smaller than the original mother cell ; but they grow, and the process may be repeated in them, so that multiplication may take place rapidly. In reproduction by *gemmation*, a budding-off or separation of a portion of the nucleus and parent cell takes place, and, becoming separated, forms a new organism.

The **cell-wall,** which is not an essential constituent, and in fact is often absent, is merely the external layer of the protoplasm, firmer than the rest of the cell, and often thickened by the deposit in it of certain chemical substances. It forms a flexible, transparent, finely striated membrane, sometimes furnished with minute pores, so as to be permeable to fluids.

THE NUTRITIVE FLUIDS

The **circulating fluids** of the body, which subserve its nutrition, are the blood, the lymph, and the chyle.

THE BLOOD

The blood is an opaque, rather viscid fluid, of a bright red or scarlet colour when it flows from the arteries, of a dark red or purple colour when it flows from the veins. It is salt to the taste, and has a peculiar faint odour and an alkaline reaction. Its specific gravity is about 1·060, and its temperature is generally about 99° F., though varying slightly in different parts of the body.

General Composition of the Blood.—Blood consists of a faintly yellow fluid, the *plasma* or *liquor sanguinis*, in which are suspended numerous minute particles, the *blood corpuscles*, the majority of which are coloured and give to the blood its red tint. If a drop of blood is placed in a thin layer on a glass slide and examined under the microscope, a number of these corpuscles will be seen immersed in the clear fluid plasma.

The **Blood Corpuscles** are chiefly of two kinds : (1) Coloured corpuscles or Erythrocytes, (2) Colourless corpuscles or Leucocytes. A third variety, the Blood platelets, are of subsidiary importance.

1. **Coloured or Red Corpuscles** (*erythrocytes*), when examined under the microscope, are seen to be circular discs, biconcave in profile. They have no nuclei, but, in consequence of their biconcave shape, present, according to the alteration of focus under an ordinary high power, a central part, sometimes bright, sometimes dark, which has the appearance of a nucleus (fig. 3, *a*). It is to their aggregation that the blood owes its red hue, although when examined by transmitted light their colour appears to be only a faint reddish-yellow. Their size varies slightly even in the same drop of blood, but it may be stated that their ordinary diameter is about $\frac{1}{3200}$ of an inch, while their thickness is about $\frac{1}{12000}$ of an inch or nearly one quarter of their diameter. Besides these there are found, especially in disease (e.g. anæmic conditions), certain smaller corpuscles of about one-half or one-third of the size just indicated ; these are termed *microcytes*, and are very scarce in normal blood. The number of red corpuscles in the blood is enormous ; between 4,000,000 and 5,000,000 are contained in a cubic millimetre. Power states that the red corpuscles of an adult would present an aggregate surface of about 3,000 square yards. Each corpuscle consists of

a colourless elastic spongework or stroma, condensed at the periphery to form an investing membrane, and uniformly diffused throughout this are the coloured fluid contents.   The stroma is composed mainly of *nucleo-proteid* and of the fatty substances, *lecithin* and *cholesterin*, while the coloured material consists chiefly of the respiratory proteid, *hæmoglobin*, which contains a proportion of iron in addition to the ordinary proteid elements.   This proteid has a great affinity for oxygen, and, when removed from the body, crystallises readily under certain circumstances.   It is very soluble in water, the addition of which to a drop of blood speedily dissolves out the hæmoglobin from the corpuscles.

If the web of a frog's foot be spread out and examined under the microscope, the blood is seen to flow in a continuous stream through the vessels, and the corpuscles show no tendency to adhere to each other or to the wall of the vessel. Doubtless the same is the case in the human body; but when the blood is drawn and examined on a slide without reagents, the corpuscles often collect into heaps like rouleaux of coins (fig. 3, *b*). It has been suggested that this phenomenon may be explained by alteration in surface tension. During life the red corpuscles may be seen to change their shape under pressure so as to adapt themselves, to some extent, to the size of the vessel.

FIG. 3.—Human red blood-corpuscles.
Highly magnified.

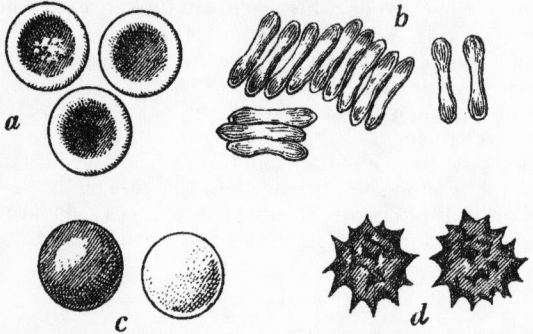

a. Seen from the surface. *b.* Seen in profile and forming rouleaux.
c. Rendered spherical by water. *d.* Rendered crenate by salt solution.

They are, however, highly elastic, and speedily recover their shape when the pressure is removed.   They are readily influenced by the medium in which they are placed and by the specific gravity of the medium.   In water they swell up, lose their shape, and become globular (fig. 3, *c*).   Subsequently the hæmoglobin becomes dissolved out, and the envelope can barely be distinguished as a faint circular outline.   Solutions of salt or sugar, denser than the plasma, give them a stellate or crenated appearance (fig. 3, *d*), but the usual shape may be restored by diluting the solution to the same specific gravity as the plasma.   The crenated outline may be produced as the first effect of the passage of an electric shock : subsequently, if sufficiently strong, the shock ruptures the envelope.   A solution of salt or sugar, of the same specific gravity as the plasma, merely separates the blood corpuscles mechanically, without changing their shape.

The **Colourless Corpuscles** or *leucocytes* are of various sizes, some no larger, others even smaller, than the red corpuscles.   In human blood, however, the majority are rather larger than the red corpuscles, and measure about $\frac{1}{2000}$ to $\frac{1}{2500}$ of an inch in diameter.   On the average from 10,000 to 12,000 leucocytes are found in each cubic millimetre of blood.

They consist of minute masses of nucleated protoplasm, and exhibit several varieties, which are differentiated from each other chiefly by the occurrence or non-occurrence of granules in their protoplasm and by the staining reactions of these granules when present (fig. 4).   (1) The most numerous and important are spherical in shape, and are characterised by a nucleus, which often consists of two or three parts (multipartite) connected together by fine threads of chromatin. The protoplasm is clear, and contains a number of very fine granules, which stain with acid dyes, as eosin (fig. 4, P).   These cells are termed the *polymorphonuclear* leucocytes.   (2) A second variety comprises from 2 to 4 per cent. of the leucocytes ; they are larger than the previous kind, and are made up of a coarsely granular protoplasm, the granules being highly refractile and grouped round a single nucleus of horse-shoe shape (fig. 4, E).   These granules stain deeply with eosin, and the cells are therefore often termed *eosinophile* corpuscles. (3) The third variety is called the *hyaline* cell (fig. 4, H).   This is usually about the same size as the eosinophile cell, and, when at rest, is spherical in shape and contains a single round or oval nucleus.   The protoplasm is free from granules, but is not quite transparent, having the appearance of ground glass. (4) The fourth kind of colourless corpuscle is designated the *lymphocyte* (fig. 4, L), because it is identical with the lymphoid cell derived from the lymphatic glands, the spleen, tonsil, and thymus. It is the smallest of the leucocytes, and consists chiefly of a spheroidal nucleus with very little surrounding protoplasm of a

FIG. 4.—Varieties of leucocytes found in human blood.

homogeneous nature; it is regarded as the immature form of the hyaline cell. The third and fourth varieties together constitute from 20 to 30 per cent. of the colourless cells, but of these two varieties the lymphocytes are by far the more numerous. Leucocytes having in their protoplasm granules which stain with basic dyes (basophile) have been described as occurring in human blood, but they are rarely found except in disease.

The white corpuscles are very various in shape in living blood (fig. 5), because many of them have the power of constantly changing their form by protruding finger-shaped or filamentous processes of their own substance, by which they move, and take up granules from the surrounding medium. In locomotion the corpuscle pushes out a process of its substance—a *pseudopodium*, as it is called—and then shifts the rest of the body into it. In the same way when any granule or particle

FIG. 5.—Human colourless blood-corpuscle, showing its successive changes of outline within ten minutes when kept moist on a warm stage. (Schofield.)

comes in its way it wraps a pseudopodium round it, and then withdrawing it, lodges the particle in its own substance. By means of these amœboid properties the cells have the power of wandering or emigrating from the blood-vessels by penetrating their walls and thus finding their way into the extra-vascular spaces. A chemical investigation of the protoplasm of the leucocytes shows the presence of nucleo-proteid and of a globulin. The occurrence of small amounts of fat and glycogen may also be demonstrated.

The **Blood platelets** are discoid or irregularly shaped, colourless, refractile bodies, much smaller than the red cells. Considerable discussion has arisen as to their significance. Recent observers have shown that under the action of certain stains the centrally situated portion of the blood platelet takes on an appearance suggestive of a nucleus. In spite of this, and of the fact that they have been observed in the blood-vessels during life, there is still a tendency to regard them as products of disintegration of the white cells, or as precipitates, possibly of nucleo-proteid, and not as living elements of the blood.

Origin of the Blood Corpuscles.—In the embryo the red corpuscles are developed from mesoblastic cells in the vascular area of the blastoderm. These cells unite with one another to form a network, their nuclei multiply in number, and around some of the nuclei an aggregation of coloured protoplasm takes place. After a time the network becomes hollowed out by an accumulation of fluid, and forms capillary blood-vessels, and in the fluid those nuclei which are surrounded by coloured protoplasm float as the first red blood cells.* The embryonic corpuscles are thus nucleated, and, further, they have the power of amœboid movement. These cells disappear in later embryonic life, to be replaced by smaller non-nucleated corpuscles, having all the characters of the adult erythrocytes, which, according to Schäfer, are formed within certain cells of the connective tissue. Small globules of reddish colouring matter appear in the protoplasm of these cells, and these eventually becoming larger, more uniform in size and disc-shaped, float in a cavity which results from the coalescence of numerous vacuoles. The cells, becoming more hollowed, join with neighbouring cells to form new blood-vessels, and these become connected with previously existing vessels. In post-embryonic life the important source of the red corpuscles is the red marrow in the ends of the long bones and especially in the ribs and sternum. Here are found special, nucleated, coloured cells, termed *erythroblasts*, which are probably direct descendants of the nucleated, embryonic red cells. These erythroblasts by atrophy and disappearance of their nuclei (or, as some observers maintain, by their extrusion) and by assumption of the biconcave form are transformed into the adult red corpuscles. Of the white corpuscles of the blood, the lymphocytes are derived from lymphatic tissue generally, and from the lymphatic glands especially, and enter the blood by way of the lymph stream; the hyaline cells probably develop from the lymphocytes, while the eosinophile cells are believed to originate mainly in the bone marrow and possibly also in the connective tissues.

The **Plasma or Liquor Sanguinis,** the fluid portion of the blood, has a yellowish tint, is alkaline in reaction, and of a specific gravity of 1·028. It contains in solution about 10 per cent. of solids, of which four-fifths are proteid in nature; the remainder being salts, chiefly chlorides, phosphates, and sulphates of the alkali metals; carbohydrates, chiefly sugar; fats and soaps; cholesterin, urea, and other nitrogenous extractives. The proteids are three in number, *serum albumin, serum globulin,* and *fibrinogen.* Fibrinogen is a body of the globulin class, but differs from serum globulin in several respects. It is the substance from which the *fibrin,* which plays so important a part in the clotting of the blood, is derived.

Coagulation of the Blood.—When blood is drawn from the body and allowed to stand, it solidifies in the course of a very few minutes into a jelly-like mass or *clot,* which has the same appearance and volume as the fluid blood, and, like it, looks quite uniform. Soon, however, drops of a transparent yellowish fluid, the *serum,* begin to ooze from the surface of the mass and to collect around it. Coincidently the clot begins to contract, so that in the course of about twenty-four hours, having become considerably smaller and firmer than the first formed jelly-like mass, it floats in a quantity of yellowish serum. The clotting of the blood is due to the formation of a fine meshwork of the insoluble material, *fibrin,* which entangles and encloses the blood corpuscles. It is supposed that when blood is drawn a nucleo-proteid, termed *prothrombin,* appears in the plasma, probably as the result of disintegration of some of the white cells and perhaps also the blood platelets. This substance interacts with soluble lime salts in the blood, and a fresh body, *thrombin* or *fibrin ferment,* is the result. The thrombin then acts on the fibrinogen in solution in the plasma, converting it into insoluble fibrin, while at the same time a very small amount of a new proteid of the globulin type passes into solution.

Fibrin may be obtained, practically free from corpuscles, by whipping the blood, after it has been withdrawn from the body, with a bundle of twigs, to which the fibrin adheres as it is formed. By various means the clotting of the blood may be retarded so that the plasma may be obtained free from corpuscles; from this plasma there may be derived fibrin and serum, without the cellular elements. Fibrin thus obtained is a white or buff-coloured stringy substance, and when observed, in the course of formation, under the microscope, shows a meshwork of fine fibrils. After exposure to the air for some time it becomes

---

* Recent observations tend to show that the endothelial lining of the vessels and the blood corpuscles are of hypoblastic origin.

hard, dry, brown and brittle. It is one of the class of coagulated proteids, insoluble in hot or cold water, saline solution, alcohol, or ether. Under the action of dilute hydrochloric acid it swells up but does not dissolve, but, when thus swollen, is readily dissolved by a solution of pepsin.

Serum, with the exception of its proteids, has a composition identical with that of plasma. The fibrinogen, characteristic of plasma, has disappeared, and the fibrin ferment or thrombin is found instead, together with the serum albumin and serum globulin which are not involved in the process of coagulation.

The relation of the various constituents of the blood to each other may be easily understood by a reference to the subjoined plan.

$$\text{Blood}\begin{cases} \text{Corpuscles} \\ \text{Plasma} \begin{cases} \text{Fibrin} \\ \text{Serum} \end{cases} \end{cases} \text{Clot}$$

Gases of the Blood.—When blood is exposed to the vacuum of an air-pump, 100 volumes are found to yield about 60 volumes of gas. The gases present are carbon dioxide, oxygen and nitrogen, and they occur in the following proportions in arterial and venous blood :

	Carbon dioxide	Oxygen	Nitrogen
Arterial blood . .	40 vols.	20 vols.	1 to 2 vols.
Venous blood . .	46 to 50 vols.	10 to 12 vols.	1 to 2 vols.

The greater quantity of the oxygen is in loose chemical combination with the hæmoglobin of the red corpuscles. The carbon dioxide exists in combination for the most part as sodium bicarbonate and carbonate. The nitrogen is in simple solution in the plasma.

Blood Crystals.—Hæmoglobin, as already stated, readily crystallises when separated from the blood corpuscles. In human blood the crystals are elongated prisms (fig. 6, A), and in the majority of animals belong to the rhombic system, though in the squirrel hexagonal plates are met with. Small brown prismatic

FIG. 6.—Blood-crystals.

A. Hæmoglobin crystals from human blood. B. Hæmin crystals from blood treated with acetic acid. C. Hæmatoidin crystals from an old apoplectic clot.

crystals of *hæmin* (fig. 6, B) may be obtained by mixing dried blood with common salt and boiling with a few drops of glacial acetic acid. A drop of the mixture on a slide will show the characteristic crystals on cooling. *Hæmatoidin* crystals (fig. 6, C) occur sometimes in old blood clots.

LYMPH AND CHYLE

Lymph is a transparent, colourless or slightly yellow fluid, which is conveyed by a set of vessels, named *lymphatics*, into the blood. These vessels arise in nearly all parts of the body as *lymph capillaries*. They take up the blood plasma which has exuded from the blood capillaries into the tissue spaces where it has nourished the tissue elements, and return it into the veins close to the heart, there to be mixed with the mass of blood. The greater number of these lymphatics empty themselves into one main duct, the *thoracic duct*, which passes upwards along the front of the spine and opens into the large veins on the left side of the root of the neck. The remainder empty themselves into a smaller duct which terminates in the corresponding veins on the right side of the neck.

Lymph, as its name implies, is a watery fluid of sp. gr. about 1·015, closely resembling the blood plasma, but more dilute, containing only about 5 per cent. of proteids and 1 per cent. of salt and extractives. When examined under the microscope, leucocytes of the lymphocyte class are found floating in the transparent fluid. They are always increased in number after the passage of the lymph through lymphoid tissue, as in lymphatic glands. They are constantly furnishing a fresh supply of colourless corpuscles to the blood.

Chyle is an opaque, milky-white fluid, absorbed by the villi of the small intestine from the food, and carried by a set of vessels similar to the lymphatics, named *lacteals*, to the commencement of the thoracic duct, where it is intermingled with the lymph and poured into the circulation through the same channels. It must be borne in mind that these two sets of vessels, lymphatics and lacteals, though differing in name, are identical in structure, and that the character of the fluid they convey is different only while digestion is going on. At other times the lacteals convey a transparent, nearly colourless lymph.

Chyle exactly resembles lymph in its physical and chemical properties, except that it has, in addition to the other constituents of lymph, a quantity of finely divided fatty particles, the so-called 'molecular basis of chyle' to which the milky appearance is due. It contains a little more proteid than lymph, but the chief difference lies in the large quantity of fats, soaps, lecithin and cholesterin which are found in the chyle. Lymph and chyle, containing, as they do, fibrinogen in solution and leucocytes, clot on removal from the body, the coagulum being free from red cells, and presenting a clear or whitish jelly-like appearance.

EPITHELIUM

All the surfaces of the body—the external surface of the skin, the internal surface of the digestive, respiratory, and genito-urinary tracts, the closed serous cavities, the inner coat of the vessels, the acini and ducts of all secreting and excreting glands, the ventricles of the brain and the central canal of the spinal cord—are covered by one or more layers of simple cells, called *epithelium* or *epithelial cells*. These cells are also present in the terminal parts of the organs of special sense, and in some other structures, as the pituitary and thyroid bodies. They serve various purposes, forming in some cases a protective layer, in others acting as agents in secretion and excretion, and again in others being concerned in the elaboration of the organs of special sense. Thus, in the skin, the main purpose served by the epithelium (here called the *epidermis*) is that of protection. As the surface is worn away by the agency of friction or change of temperature new cells are supplied, and thus the surface of the true skin and the vessels and nerves which it contains are defended from damage. In the gastro-intestinal mucous membrane and in the glands, the epithelial cells appear to be the principal agents in separating the secretion from the blood or from the alimentary fluids. In other situations (as the nose, fauces, and respiratory passages) the chief office of the epithelial cells appears to be to maintain an equable temperature by the moisture with which they keep the surface always slightly lubricated. In the serous cavities they also keep the opposed layers moist, and thus facilitate their movements on each other. Finally, in all internal parts they insure a perfectly smooth surface.

FIG. 7.—Simple pavement epithelium.

Epithelium consists of one or more layers of cells, united together by an interstitial cement substance, supported on a basement membrane, and is naturally grouped into two classes according as to whether there is a single layer of cells (*simple epithelium*) or more than one (*stratified epithelium*). A third variety (*transitional epithelium*) is that in which cells, in three or four layers, are so fitted together that the appearance is not one of distinct stratification. The different varieties of simple epithelium are generally spoken of as squamous or pavement, columnar, glandular or spheroidal, and ciliated.

The *pavement* epithelium (fig. 7) is composed of flat, nucleated scales of different shapes, usually polygonal, and varying in size. These cells fit together by their edges, like the tiles of a mosaic pavement. The nucleus is generally flattened, but may be spheroidal. The flattening depends upon the thinness of the .cell. The protoplasm of the cell presents a fine reticulum or honeycombed network, which gives to the cell the appearance of granulation. This kind of epithelium forms the lining of the air-cells of the lungs. The *endothelium*, which covers the serous membranes, and which lines the heart, blood-vessels, lymphatics, and the anterior chamber of the eye, is also of the pavement type, being composed of a single layer of flattened transparent squamous cells, joined edge to edge in such a manner as to form a membrane of cells. Endothelium differs from epithelium in that it originates from the embryonic mesoblast, while epithelium arises, as a rule, from the epiblast or hypoblast.

The *columnar* or *cylindrical* epithelium (fig. 8) is formed of cylindrical or rod-shaped cells set together so as to form a complete layer, resembling, when viewed in profile, a palisade. The cells have a prismatic figure, more or less

FIG. 8.—Columnar epithelium from an intestinal villus.

Striated free borders of cells

Basement membrane

FIG. 9.—Goblet cells. (From Kirke's 'Physiology.')

flattened from mutual pressure, and are set upright on the surface on which they are supported. Their protoplasm is always more or less reticulated, and fine longitudinal striæ may be seen in it. They possess a nucleus which is oval in shape and contains an intranuclear network.

This form of epithelium covers the mucous membrane of nearly the whole gastro-intestinal tract and the glands of that part, the greater part of the urethra, the vas deferens, the prostate, Cowper's glands, Bartholini's glands, and a portion of the uterine mucous membrane. In a modified form it also covers the ovary.

Goblet or *chalice* cells are a modification of the columnar cell. They appear to be formed by an alteration in shape of the columnar epithelium (ciliated or otherwise) consequent on the formation of granules which consist of a substance called *mucigen* in the interior of the cell. This distends the upper part of the cell, while the nucleus is pressed down towards its deep part, until the cell bursts and the mucus is discharged on to the surface of the mucous membrane (as shown in fig. 9), the cell then assuming the shape of an open cup or chalice.

The *glandular* or *spheroidal* epithelium (fig. 10) is composed of spheroidal or polyhedral cells, but the cells may be columnar or cubical in shape in some situations. Like other forms of epithelial cells, the protoplasm is a fine reticulum, which gives to the cells the appearance of granulation. They are found in the terminal recesses of secreting glands, and the protoplasm of the cells usually contains the materials which the cells secrete.

FIG. 10.—Spheroidal epithelium. Magnified 250 times.

Ciliated epithelium (fig. 11) may be of any of the preceding forms, but generally inclines to the columnar shape. It is distinguished by the presence of minute processes, which are direct prolongations of the cell-protoplasm, like hairs or eyelashes (cilia) standing up from the free surface. If the cells are examined during life or immediately on removal from the living body (for which in the human subject the removal of a nasal polypus offers a convenient opportunity) in a weak solution of salt, the cilia will be seen in lashing motion; and if the cells are separate, they will often be seen to be moved about in the field by this motion.

FIG. 11.—Ciliated epithelium from the human trachea. Magnified 350 times.

a. Innermost layers of the elastic longitudinal fibres. b. Homogeneous innermost layer of the mucous membrane. c. Deepest round cells. d. Middle elongated cells. e. Superficial cells, bearing cilia.

The situations in which ciliated epithelium is found in the human body are : the respiratory tract from the nose downwards to the smallest ramifications of the bronchial tubes, except a part of the pharynx and the surface of the vocal cords; the tympanum and Eustachian tube ; the Fallopian tube and upper portion of the uterus; the vasa efferentia, coni vasculosi and the first part of the excretory duct of the testicle; and the ventricles of the brain and central canal of the spinal cord.

Stratified epithelium (fig. 13) consists of several layers of cells superimposed one on the top of the other and varying greatly in shape. The cells of the deepest layer are for the most part columnar

FIG. 12.—Epithelial cells from the oral cavity of man. Magnified 350 times.

a. Large. b. Middle sized. c. The same with two nuclei.

FIG. 13. — Stratified epithelium from the œsophagus.

FIG. 14.—Portion of epidermis from a section of the skin of the finger. (Ranvier.) From Schäfer's 'Essentials of Histology.')

Stratum corneum

Stratum lucidum

Stratum granulosum

Prickle-cells of stratum Malpighii

in shape, and as a rule form a single layer, placed vertically on the supporting membrane ; above these are several layers of spheroidal cells, which as they approach the surface become more and more compressed, until the superficial layers are found to consist of flattened scales (fig. 12), the margins of which overlap one another so as to present an imbricated appearance. They here undergo a chemical change from the conversion of their protoplasm into a horny substance (keratin).

Certain cells found in the deeper layers of stratified epithelium, and termed *prickle cells* (fig. 14), constitute a variety of squamous epithelium. These cells possess short, fine fibrils, which pass from their margins to those of neighbouring cells, serving to connect them together. They are not closely connected together by cement-substance, but are separated from each other by intercellular channels, across which these fine fibrils may be seen bridging, and this gives to the cell, when isolated, the appearance of being covered over with a number of short spines, in consequence of the fibrils being broken through. They were first described by Max Schultze and Virchow, and it was believed by them that the cells were dovetailed together. Subsequently this was shown not to be so by Martyn, who pointed out that the prickles were attached to each other by their apices; and recently Delépine has stated that he believes the prickles of prickle cells are parts of fibrils forming internuclear bundles between the nuclei of the cells of an epithelium in a state of active growth (see page 5, and fig. 2).

FIG. 15.—Transitional epithelium.

Transitional epithelium occurs in the ureters and urinary bladder. Here the cells of the most superficial layer are cubical, with depressions on their under surfaces, which fit on to the rounded ends of the cells of the second layer, which are pear-shaped, the apices touching the basement-membrane. Between their tapering points is a third variety of cells, filling in the intervals between them, and of smaller size than those of the other two layers (fig. 15).

CONNECTIVE TISSUES

The term **connective tissue** includes a number of tissues which possess this feature in common, viz. that they serve the general purpose in the animal economy of supporting and connecting the tissues of the body. These tissues may differ considerably from each other in appearance, but they present nevertheless many points of relationship, and are moreover developed from the same layer of the embryo, the mesoblast. They are divided into three great groups : (1) the connective tissues proper, (2) cartilage, and (3) bone. Blood, which has already been described, is, strictly speaking, a form of connective tissue, and is so dealt with by many histologists.

THE CONNECTIVE TISSUES PROPER

Several forms or varieties of connective tissue are recognised : (1) Areolar tissue. (2) White fibrous tissue. (3) Yellow elastic tissue. (4) Mucous tissue. (5) Retiform tissue. They are all composed of a homogeneous matrix, in which are embedded cells and fibres—the latter of two kinds, white and yellow or elastic. The distinction between the different forms of tissue depends upon the relative preponderance of one or other kind of fibre, of cells, or of matrix.

Areolar tissue (fig. 16) is so called because its meshes are easily distended, and thus separated into areolæ or spaces, which open freely into each other, and are consequently easily blown up with air, or permeated by fluid, when injected into any part of the tissue. Such spaces, however, do not exist in the natural condition of the body, but the whole tissue forms one unbroken membrane composed of a number of interlacing fibres, variously superimposed. Hence the term 'the cellular membrane' is in many parts of the body more appropriate than its more modern equivalent. The chief use of the areolar tissue is to bind parts together ; while by the laxity of its fibres, and the permeability of its areolæ, it allows them to move on each other, and affords a ready exit for inflammatory and other effused fluids. It is one of the most extensively distributed of all the tissues.

It is found beneath the skin, in a continuous layer all over the body, connecting it to the subjacent parts. In the same way it is situated beneath the mucous and serous membranes. It is also found between muscles, vessels, and nerves, forming investing sheaths for them, and connecting them with surrounding structures. In addition to this, it is found in the interior of organs, binding together the various lobes and lobules of the compound glands, the various coats of the hollow viscera, and the fibres of muscles &c., and thus forms one of the most important connecting media of the various structures or organs of which the body is made up. In many parts the areolæ or interspaces of areolar tissue are occupied by fat-cells, constituting *adipose tissue*, which will presently be described.

Areolar tissue presents to the naked eye a flocculent appearance, somewhat like spun-silk. When stretched out, it is seen to consist of delicate soft elastic threads interlacing with each other in every direction, and forming a network of extreme delicacy. When examined under the microscope (fig. 16) it is found to be

Fig. 16.—Subcutaneous tissue from a young rabbit. Highly magnified. (Schäfer.)

Plasma cell

White fibres

Elastic fibres

Fibrillated cell

Lamellar cell

composed of white fibres and yellow elastic fibres intercrossing in all directions, and united together by a homogeneous cement or ground substance, the *matrix*, showing cell-spaces wherein lie many cellular elements, the *connective tissue corpuscles*; these contain the protoplasm out of which the whole is developed and regenerated.

The *white fibres* are arranged in waving bands or bundles of minute transparent homogeneous filaments or fibrillæ. The bundles have a tendency to split up longitudinally or send off slips to join neighbouring bundles and receive others in return, but the individual fibres are unbranched and never join other fibres; the *yellow elastic fibres* have a well-defined outline and are considerably larger in size than the white fibrillæ. They vary much, being from the $\frac{1}{24000}$ to the $\frac{1}{4000}$ of an inch in diameter. The fibres form bold and wide curves, branch, and freely anastomose with each other. They are homogeneous in appearance, and tend to curl up, especially at their broken ends.

Connective tissue corpuscles.—The cells of areolar tissue are of three principal kinds: (1) Flattened lamellar cells, which may be either branched or unbranched. The branched lamellar cells are composed of clear cell substance, in which is contained an oval nucleus. The processes of these cells unite so as to form an open network, as in the cornea. The unbranched cells are joined edge to edge like the cells of an epithelium. The 'tendon cells,' presently to be described, are an example of this variety. (2) Granule cells, which are ovoid or spheroidal in shape, and formed of a soft protoplasm, containing granules which are albuminous in character and stain deeply with eosin. (3) Plasma cells of Waldeyer, varying greatly in size and form, but always to be distinguished from the other two varieties by containing a largely vacuolated protoplasm. The vacuoles are filled with fluid, and the protoplasm between the spaces is clear, with occasionally a few scattered granules.

In addition to these three typical forms of connective tissue corpuscles, areolar tissue may be seen to possess *wandering cells*, i.e. leucocytes which have emigrated from the neighbouring vessels, and in some instances, as in the choroid coat of the eye, cells filled with granules of pigment (*pigment-cells*).

Fig. 17.—White fibrous tissue. High power.

The connective tissue corpuscles lie in spaces in the ground substance between the bundles of fibres, and these spaces may be brought into view by treating the tissue with nitrate of silver and exposing it to the light. This will colour the ground substance and leave the cell-spaces unstained.

White fibrous tissue (fig. 17) is a true connecting structure, and serves three purposes in the animal economy. In the form of ligaments it serves to bind bones together; in the form of tendons it serves to connect muscles to bones or other structures; and it forms an investing or protecting structure to various organs in the form of membranes. Examples of where it serves this latter office are to be found in the muscular fasciæ or sheaths, the periosteum, and perichondrium; the investments of the various glands (such as the tunica albuginea testis, the capsule of the kidney, &c.), the investing sheath of the nerves (epineurium), and of various organs, as the penis and the eye (sheath of the corpora cavernosa and corpus spongiosum and of the sclerotic). In white fibrous tissue, as its name implies, the white fibres predominate, the matrix being apparent only as a cement substance, the yellow elastic fibres comparatively few, while the tissue-cells are arranged in a special manner. It presents to the naked eye the appearance of silvery white glistening fibres, covered over with a quantity of loose flocculent tissue which binds the fibres together and carries the blood-vessels (fig. 18). It is not possessed of any elasticity, and only the very slightest extensibility; it is exceedingly strong, so that upon the application of any external violence the bone with which it is connected will fracture before the fibrous tissue will give way. In ligaments and tendons the bundles of fibres run parallel with each other; in membranes they intersect one another in different places. The cells found in white fibrous tissue are often called 'tendon-cells.' They are situated on the surface of groups of bundles and are quadrangular in shape, arranged in rows, in single file, each cell being separated from its neighbours by a narrow line of cement-substance. The nucleus is generally situated at one end of the cell, the nucleus of the adjoining

Fig. 18.—Connective tissue. (Klein and Noble Smith.)

a, The white fibrous element—a layer of more or less sharply outlined, parallel, wavy bundles of connective tissue fibrils. On the surface of this layer is *b*, a network of fine elastic fibres.

cell being in close proximity to it (fig. 19). Upon the addition of acetic acid to white fibrous tissue it swells up into a glassy-looking indistinguishable mass. When boiled in water it is converted almost completely into gelatin, the white fibres being composed of the albuminoid *collagen*, which is often regarded as the anhydride of gelatin.

Yellow elastic tissue.—In certain parts of the body a tissue is found which when viewed in mass is of a yellowish colour, and is possessed of great elasticity; so that it is capable of considerable extension, and when the extending force is withdrawn returns at once to its original condition. This is *yellow elastic tissue*, which may be regarded as a connective tissue in which the yellow elastic fibres have developed to the practical exclusion of the other elements. It is found in the ligamenta subflava, in the vocal cords, in the longitudinal coat of the trachea and bronchi, in the inner coats of the blood-vessels, especially the larger arteries, and to a very considerable extent in the thyro-hyoid, crico-thyroid, and stylo-hyoid

FIG. 19.—Tendon of mouse's tail, stained with logwood, showing chains of cells between the tendon-bundles. (From Quain's 'Anatomy.' E. A. Schäfer.)

FIG. 20.—Yellow elastic tissue. High power.

ligaments. It is also found in the ligamentum nuchæ of the lower animals (fig. 20). In some parts where the fibres are broad and large and the network close, the tissue presents the appearance of a membrane, with gaps or perforations corresponding to the intervening spaces. This is to be found in the inner coat of the arteries, and to it the name of *fenestrated membrane* has been given by Henle. The yellow elastic fibres remain unaltered by acetic acid. Chemically they are composed of the albuminoid body, *elastin*.

Vessels and Nerves of Connective Tissue.—The *blood-vessels* of connective tissue are very few—that is to say, there are few actually destined for the tissue itself, although many vessels may permeate one of its forms, the areolar tissue, carrying blood to other structures. In white fibrous tissue the blood-vessels usually run parallel to the longitudinal bundles and between them, sending transverse communicating branches across; in some forms, as in the periosteum and dura mater, they are fairly numerous. In yellow elastic tissue, the blood-vessels also run between the fibres, and do not penetrate them. *Lymphatic* vessels are very numerous in most forms of connective tissue, especially in the areolar tissue beneath the skin and the mucous and serous surfaces. They are also found in abundance in the sheaths of tendons, as well as in the tendons themselves.

Nerves are to be found in the white fibrous tissue, where they terminate in a special manner; but it is doubtful whether any nerves terminate in areolar tissue; at all events, they have not yet been demonstrated, and the tissue is possessed of very little sensibility.

Development of Connective Tissue.—Connective tissue is developed from embryonic connective-tissue cells derived from the mesoblast. These cells, at first rounded, become fusiform and branched, and ultimately connective-tissue corpuscles. A mucinous intercellular substance or matrix, partly derived from the cells themselves and partly from the lymph exuded by the neighbouring blood-vessels, gradually separates the cells. In the matrix the fibres are deposited, probably under the influence of the cells, but not by any transformation of the cell protoplasm. In the case of yellow elastic fibres, rows of granules of elastin are first laid down; these eventually fuse into the fully developed fibre.

Mucous tissue exists chiefly in the 'jelly of Wharton,' which forms the bulk of the umbilical cord, but is also found in other situations in the fœtus, chiefly as a stage in the development of connective tissue. It consists of a matrix, largely made up of mucin, in which are nucleated cells with branching

Fig. 21.—Mucous tissue from the umbilical cord of the human fœtus (four months).

and anastomosing processes (fig. 21). Few fibres are seen in typical mucous tissue, though, at birth, the umbilical cord shows considerable development of fibres. In the adult the vitreous humour of the eye is a persistent form of mucous tissue, in which there are no fibres, and from which the cells have disappeared, leaving only the mucinous ground substance.

Retiform or **reticular** connective tissue (fig. 22) is found extensively in many parts of the body, constituting the framework of some organs and entering into the construction of many mucous membranes. It is a variety of connective tissue, in which the intercellular or ground substance has, in a great measure, disappeared, and has been replaced by fluid. It is apparently composed almost entirely of extremely fine bundles of white fibrous tissue, forming an intricate network, and chemically it yields gelatin. The fibres are covered and concealed in places by flattened branched connective-tissue cells. In many situations the interstices of the network are filled with rounded lymph-corpuscles, and the tissue is then termed **lymphoid** or **adenoid tissue** (see fig. 78).

Basement membranes, formerly described as homogeneous membranes, are really a form of connective tissue. They constitute the supporting membrane, or *membrana propria*, on which is placed the epithelium of mucous membranes or

C

secreting glands, and are also found in other situations. By means of staining with nitrate of silver they may be shown to consist of flattened cells in close apposition, and joined together by their edges, thus forming an example of an epithelioid arrangement of connective-tissue cells. In some situations the cells,

FIG. 22.—Retiform connective tissue, from a lymphatic gland.

instead of adhering by their edges, give off branching processes, which join with similar processes of other cells and so form a network, rather than a continuous membrane. In other instances basement membranes are composed of elastic tissue, as in the cornea, or, again, in other cases, of condensed ground substance.

ADIPOSE TISSUE

In almost all parts of the body the ordinary areolar tissue contains a variable quantity of fat. The principal situations where it is not found are the sub-cutaneous tissue of the eyelids, the penis and scrotum, the nymphæ, within the cavity of the cranium, and in the lungs, except near their roots. Nevertheless its

FIG. 23.—Adipose tissue. High power.

a. Starlike appearance, from crystallisation of fatty acids.

distribution is not uniform; in some parts it is collected in great abundance, as in the subcutaneous tissue, especially of the abdomen; around the kidneys; on the surface of the heart between the furrows, and in some other situations. Lastly, fat enters largely into the formation of the marrow of bones. A distinction must be made between fat and adipose tissue; the latter being a distinct tissue, the former an oily matter, which in addition to forming adipose tissue is also widely present in the body, as in the fat of the brain and liver, and in the blood, chyle, &c.

Adipose tissue consists of small vesicles, *fat-cells*, lodged in the meshes of areolar tissue. The fat-cells (fig. 23) vary in size, but are of about the average diameter of $\frac{1}{500}$ of an inch. They are formed of an exceedingly delicate protoplasmic membrane, filled with fatty matter, which is liquid during life, but becomes solidified after death. They are round or spherical where they have not been subjected to pressure; otherwise they assume a more or less angular outline. A nucleus is always present and can be easily demonstrated by staining with hæmatoxylin; in the natural condition it is so compressed by the contained oily matter as to be scarcely recognisable. These fat-cells are contained in clusters in the areolæ of fine connective tissue, and are held together mainly by a network of capillary blood-vessels, which are distributed to them.

Chemically the oily material in the cells is composed of the fats, olein, palmitin, and stearin, which are glycerin compounds with fatty acids. Sometimes fat crystals form in the cells after death (fig. 23, *a*). By boiling the tissue in ether or strong alcohol, the fat may be extracted from the vesicle, which is then seen empty and shrunken.

FIG. 24.—Development of fat. (Klein and Noble Smith.)

a. Minute artery. *v*. Minute vein. *c*. Capillary blood-vessels in the course of formation; they are not yet completely hollowed out, there being still left in them protoplasmic septa. *d*. The ground substance, containing numerous nucleated cells, some of which are more distinctly branched and flattened than others, and appear therefore more spindle-shaped.

Fat is said to be first detected in the human embryo about the fourteenth week. The fat-cells are formed by the transformation of connective-tissue corpuscles, in which small droplets of oil are formed; these coalesce to produce a larger drop, and this increases until it distends the corpuscle, the remaining protoplasm and the nucleus being displaced towards the periphery of the cell (fig. 24).

PIGMENT

In various parts of the body **pigment** is found; most frequently in epithelial cells and in the cells of connective tissue. Pigmented *epithelial cells* are found in the external layer of the retina and on the posterior surface of the iris. Pigment is likewise found in the epithelial cells of the deeper layers of the cuticle in some parts of the body—such as the areola of the nipple and in coloured patches of skin and especially in the skin of the coloured races, and in hair. It is also found in the epithelial cells of the olfactory region, and of the membranous labyrinth of the ear.

In the *connective-tissue cells* pigment is frequently met with in the lower vertebrates. In man it is found in the choroid coat of the eye (fig. 25), and in the iris of all but the light blue eyes and the albino. It is also occasionally met with in the cells of retiform tissue and in the pia mater of the upper part of the spinal cord. These cells are characterised by their large size and branched processes, which, as well as the body of the cells, are filled with

FIG. 25.—Pigment-cells from the choroid coat of the eyeball.

C 2

granules. The pigment consists of dark brown or black granules of very small size closely packed together within the cells, but not invading the nucleus. Occasionally the pigment is yellow, and when occurring in the cells of the cuticle constitutes 'freckles.'

CARTILAGE

Cartilage is a non-vascular structure which is found in various parts of the body—in adult life chiefly in the joints, in the parietes of the thorax, and in various tubes, such as the air-passages, nostrils, and ears, which are to be kept permanently open. In the foetus, at an early period, the greater part of the skeleton is cartilaginous. As this cartilage is afterwards replaced by bone, it is called *temporary*, in contradistinction to that which remains unossified during the whole of life, and which is called *permanent*.

Cartilage is divided, according to its minute structure, into hyaline cartilage, white fibro-cartilage, and yellow or elastic fibro-cartilage. Besides these varieties met with in the adult human subject, there is a variety called *cellular cartilage*, which consists entirely, or almost entirely, of cells, united in some cases by a network of very fine fibres, in other cases apparently destitute of any intercellular substance. This is found in the external ear of rats, mice, and some other animals, and is present in the chorda dorsalis of the human embryo, but is not found in any other human structure. The various cartilages in the body are also classified, according to their function and position, into articular, interarticular, costal, and membraniform.

Hyaline cartilage, which may be taken as the type of this tissue, consists of a gristly mass of a firm consistence, but of considerable elasticity and of a pearly-bluish colour. Except where it coats the articular ends of bones, it is covered externally by a fibrous membrane, the *perichondrium*, from the vessels of which it imbibes its nutritive fluids, being itself destitute of blood-vessels. It contains no nerves. Its intimate structure is very simple. If a thin slice is examined under the microscope, it will be found to consist of cells of a rounded or bluntly angular form, lying in groups of two or more in a granular or almost homogeneous

FIG. 26.—Human cartilage-cells from the cricoid cartilage.   × 350.

matrix (fig. 26). The cells, when arranged in groups of two or more, have generally a straight outline where they are in contact with each other, and in the rest of their circumference are rounded. The cell-contents consist of clear translucent protoplasm, in which fine interlacing filaments and minute granules may sometimes be seen ; embedded in this are one or two round nuclei, having the usual intranuclear network. The cells are embedded in cavities in the matrix, called *cartilage lacunæ*; around these the matrix is arranged in concentric lines, as if it had been formed in successive portions around the cartilage cells. This constitutes the so-called *capsule* of the space. Each lacuna is generally occupied by a single cell, but during the division of the cells it may contain two, four, or eight cartilage-cells. By exposure to the action of an electric shock the cell assumes a jagged outline and shrinks away from the interior of the capsule.

The matrix is transparent and apparently without structure, or else presents a dimly granular appearance, like ground glass. Some observers have shown that the matrix of hyaline cartilage, and especially the articular variety, after prolonged maceration, can be broken up into fine fibrils. These fibrils are probably of the same nature, chemically, as the white fibres of connective tissue. It is believed by some histologists that the matrix is permeated by a number of fine channels, which connect the lacunæ with each other, and that these canals communicate with the lymphatics of the perichondrium, and thus the structure is permeated with a current of nutrient fluid. This, however, is somewhat doubtful.

Articular cartilage, costal cartilage, and temporary cartilage are all of the hyaline variety. They present minute differences in the size and shape of their cells and in the arrangement of their matrix. In **articular cartilage**, which shows no tendency to ossification, the matrix is finely granular under a high power; the cells and nuclei are small, and are disposed parallel to the surface in the superficial part, while nearer to the bone they become vertical. Articular cartilages have a tendency to split in a vertical direction; in disease this tendency becomes very manifest. Articular cartilage is not covered by perichondrium, on its free surface, where it is exposed to friction, though a layer of connective tissue can be traced in the adult over a small part of its circumference continuous with that of the synovial membrane, and here the cartilage-cells are more or less branched and pass insensibly into the branched connective-tissue corpuscles of the synovial membrane.

Articular cartilage forms a thin incrustation upon the joint-surfaces of the bones, and its elasticity enables it to break the force of any concussion, while its smoothness affords ease and freedom of movement. It varies in thickness according to the shape of the articular surface on which it lies; where this is convex the cartilage is thickest at the centre, where the greatest pressure is received; and the reverse is the case on the concave articular surfaces. Articular cartilage appears to derive its nutriment partly from the vessels of the neighbouring synovial membrane, partly from those of the bone upon which it is implanted. Toynbee has shown that the minute vessels of the cancellous tissue as they approach the articular lamella dilate and form arches, and then return into the substance of the bone.

In **costal cartilage** the cells and nuclei are large, and the matrix has a tendency to fibrous striation, especially in old age (fig. 27). In the thickest parts of the costal cartilages a few large vascular channels may be detected. This appears, at first sight, to be an exception to the statement that cartilage is a non-vascular tissue, but is not so really, for the vessels give no branches to the cartilage substance itself, and the channels may rather be looked upon as involutions of the perichondrium.

Fig. 27.—Costal cartilage from a man seventy-six years of age, showing the development of fibrous structure in the matrix. In several portions of the specimen two or three generations of cells are seen enclosed in a parent cell-wall. High power.

The ensiform cartilage may be regarded as one of the costal cartilages, and the cartilages of the nose and of the larynx and trachea (except the epiglottis and cornicula laryngis, which are composed of elastic fibro-cartilage) resemble them in microscopical characters.

Temporary cartilage and the process of its ossification will be described with bone.

The hyaline cartilages, especially in adult and advanced life, are prone to calcify—that is to say, to have their matrix permeated by the salts of lime without any appearance of true bone. The process of calcification occurs also and still more frequently, according to Rollett, in such cartilages as those of the trachea and in the costal cartilages, which are prone afterwards to conversion into true bone.

White fibro-cartilage consists of a mixture of white fibrous tissue and cartilaginous tissue in various proportions; it is to the first of these two constituents that its flexibility and toughness are chiefly owing, and to the latter its elasticity. When examined under the microscope it is found to be made up of fibrous connective tissue arranged in bundles, with cartilage-cells between the bundles; these to a certain extent resemble tendon-cells, but may be distinguished from them by being surrounded by a concentrically striated area of cartilage matrix and by their being less flattened (fig. 28). The fibro-cartilages admit of arrangement into four groups—interarticular, connecting, circumferential, and stratiform.

1. The *interarticular fibro-cartilages (menisci)* are flattened fibro-cartilaginous plates, of a round, oval, triangular, or sickle-like form, interposed between the articular cartilages of certain joints. They are free on both surfaces, thinner towards their centre than at their circumference, and held in position by the attachment of their margins and extremities to the surrounding ligaments. The synovial membrane of the joint is prolonged over them a short distance from their attached margins. They are found in the temporo-mandibular, sterno-clavicu-lar, acromio-clavicular, wrist and knee joints. These car-tilages are usually found in those joints which are most exposed to violent concus-sion and subject to frequent movement. Their use is— to maintain the apposition of the opposed surfaces in their various motions ; to increase the depth of the articular surfaces and give ease to the gliding move-ment ; to moderate the effects of great pressure and deaden the intensity of the shocks to which the parts may be subjected.

Fig. 28.—White fibro-cartilage from an intervertebral disc.

Humphry has pointed out that these interarticular fibro-cartilages serve an important purpose in increasing the variety of movements in a joint. Thus, in the knee-joint, there are two kinds of motion, viz. angular movement and rotation, although it is a hinge joint, in which, as a rule, only one variety of motion is permitted ; the former movement takes place between the condyles of the femur and the interarticular cartilage, the latter between the cartilage and the head of the tibia. So, also, in the temporo-mandibular joint, the up-ward and downward move-ment of opening and shut-ting the mouth takes place between the fibro-cartilage and the jaw-bone, the grind-ing movement between the glenoid cavity and the fibro-cartilage, the latter moving with the jaw-bone.

Fig. 29.—Yellow cartilage, ear of horse. High power.

2. The *connecting fibro-cartilages* are interposed be-tween the bony surfaces of those joints which admit of only slight mobility, as between the bodies of the vertebræ and between the pubic bones. They form discs, which adhere closely to both of the opposed sur-faces, and are composed of concentric rings of fibrous tissue, with cartilaginous laminæ interposed, the former tissue predominating towards the circumference, the latter towards the centre.

3. The *circumferential fibro-cartilages* consist of a rim of fibro-cartilage, which surrounds the margin of some of the articular cavities, as the cotyloid cavity of the hip, and the glenoid cavity of the shoulder ; they serve to deepen the articular surface, and to protect its edges.

4. The *stratiform fibro-cartilages* are those which form a thin coating to osseous grooves through which the tendons of certain muscles glide. Small masses of fibro-cartilages are also developed in the tendons of some muscles, where they glide over bones, as in the tendons of the Peroneus longus and the Tibialis posticus.

Yellow or **elastic fibro-cartilage** is found in the human body in the auricle of the external ear, the Eustachian tubes, the cornicula laryngis, and the epiglottis. It consists of cartilage-cells and a matrix, the latter being pervaded in every direction, except immediately around each cell, where there is a variable amount of non-fibrillated hyaline, intercellular substance, by a network of yellow elastic fibres, branching and anastomosing in all directions (fig. 29). The fibres resemble those of yellow elastic tissue, both in appearance and in being unaffected by acetic acid; and according to Rollett their continuity with the elastic fibres of the neighbouring tissue admits of being demonstrated.

The distinguishing feature of cartilage as to its chemical composition, is that it yields on boiling a substance called *chondrin*, very similar to gelatin, but differing from it in several of its reactions. It is now believed that chondrin is not a simple body, but a mixture of gelatin with mucinoid substances, chief among which, perhaps, is a compound termed *chondro-mucoid*.

BONE

Structure and Physical Properties of Bone.—Bone is one of the hardest structures of the animal body; it possesses also a certain degree of toughness and elasticity. Its colour, in a fresh state, is pinkish white externally, and deep red within. On examining a section of any bone, it is seen to be composed of two kinds of tissue, one of which is dense in texture, like ivory, and is termed *compact tissue*; the other consists of slender fibres and lamellæ, which join to form a reticular structure; this, from its resemblance to lattice-work, is called *cancellous tissue*. The compact tissue is always placed on the exterior of the bone; the cancellous is always internal. The relative quantity of these two kinds of tissue varies in different bones, and in different parts of the same bone, as strength or lightness is requisite. Close examination of the compact tissue shows it to be extremely porous, so that the difference in structure between it and the cancellous tissue depends merely upon the different amount of solid matter, and the size and number of spaces in each; the cavities being small in the compact tissue and the solid matter between them abundant, while in the cancellous tissue the spaces are large and the solid matter is in smaller quantity.

Bone during life is permeated by vessels and is enclosed, except where it is coated with articular cartilage, in a fibrous membrane, the *periosteum*, by means of which many of these vessels reach the hard tissue. If the periosteum is stripped from the surface of the living bone small bleeding points are seen, which mark the entrance of the periosteal vessels; and on section during life, every part of the bone will be seen to exude blood from the minute vessels which ramify in it. The interior of the bones of the limbs presents a cylindrical cavity filled with marrow and lined by a highly vascular areolar structure, called the *medullary membrane* or *internal periosteum*, which, however, is rather the areolar envelope of the cells of the marrow, than a definite membrane.

The **periosteum** adheres to the surface of the bones in nearly every part, excepting at their cartilaginous extremities. When strong tendons or ligaments are attached to the bone, the periosteum is incorporated with them. It consists of two layers closely united together, the outer one formed chiefly of connective tissue, containing occasionally a few fat-cells; the inner one, of elastic fibres of the finer kind, forming dense membranous networks, which can be again separated into several layers. In young bones the periosteum is thick and very vascular, and is intimately connected at either end of the bone with the epiphysial cartilage, but less closely with the shaft, from which it is separated by a layer of soft tissue, containing a number of granular corpuscles or 'osteoblasts,' in which ossification proceeds on the exterior of the young bone. Later in life the periosteum is thinner, less vascular, and the osteoblasts have become converted into an epithelioid layer, separated from the rest of the periosteum in many places by cleft-like spaces, which are supposed to serve for the transmission of lymph. The periosteum serves as a nidus for the ramification of the vessels previous to their distribution in the bone; hence the liability of bone to exfoliation or necrosis, when denuded of this membrane by injury or disease. Fine nerves and lymphatics, which generally accompany the arteries, may also be demonstrated in the periosteum.

The **marrow** not only fills up the cylindrical cavity in the shafts of the long bones, but also occupies the spaces of the cancellous tissue and extends into the larger bony canals (Haversian canals) which contain the blood-vessels. It differs in composition in different bones. In the shafts of adult long bones the marrow is of a *yellow* colour, and contains, in 100 parts, 96 of fat, 1 of areolar tissue and vessels, and 3 of fluid, with extractive matter, and consists of a basis of connective tissue, supporting numerous blood-vessels and cells, most of which are fat-cells, but some are 'marrow-cells,' such as occur in the red marrow, to be immediately described. In the flat and short bones, in the articular ends of the long bones, in the bodies of the vertebræ, in the cranial diploë, and in the sternum and ribs, it is of a *red* colour, and contains, in 100 parts, 75 of water and 25 of solid matter, consisting of cell-globulin, nucleo-proteid, extractives, salts, and only a small proportion of fat. The red marrow consists of a small quantity of connective tissue, blood-vessels, and numerous cells (fig. 30), some few of which are fat-cells, but the great majority are roundish nucleated cells, the true 'marrow-cells' of Kölliker. These marrow-cells proper, or *myelocytes*, resemble in appearance lymphoid corpuscles, and like them are amœboid ; they generally have a hyaline protoplasm, though some show granules either oxyphile or basophile in reaction.

Fig. 30.—Human bone-marrow.

Normoblast with dividing nucleus

Myelocyte
Erythrocyte
Myeloplaxe
Myelocyte
Fat

Eosinophile cell
Normoblasts
Myelocyte dividing
Myelocyte
Fat

A number of *eosinophile* cells are also present. Among the marrow-cells may be seen smaller cells, which possess a slightly pinkish hue ; these are the *erythroblasts* or *normoblasts*, from which, as we have seen, the red corpuscles of the adult are derived, and which may be regarded as descendants of the nucleated coloured corpuscles of the embryo. *Giant-cells* (*myeloplaxes, osteoclasts*), large, multi-nucleated, protoplasmic masses, are also to be found in both sorts of adult marrow, but more particularly in red marrow. They were believed by Kölliker to be concerned in the absorption of bone-matrix, and hence the name which he gave to them—*osteoclasts*. They excavate small shallow pits or cavities in the bone, which are named *Howship's lacunæ*, in which they are found lying.

Vessels of Bone.—The blood-vessels of bone are very numerous. Those of the compact tissue are derived from a close and dense network of vessels ramifying in the periosteum. From this membrane, vessels pass into the minute orifices in the compact tissue, and run through the canals which traverse its substance. The cancellous tissue is supplied in a similar way, but by less numerous and larger vessels, which, perforating the outer compact tissue, are distributed to the cavities of the spongy portion of the bone. In the long bones, numerous apertures may be seen at the ends near the articular surfaces, some of which give passage to the arteries of the larger set of vessels referred to ; but the most numerous and largest apertures are for the veins of the cancellous tissue, which run separately from the arteries. The medullary canal in the shafts of the long bones is supplied by one large artery (or sometimes more), which enters the bone at the nutrient foramen (situated in most cases near the centre of the shaft), and perforates obliquely the compact structure. The *medullary* or *nutrient* artery, usually accompanied by one or two veins, sends branches upwards and downwards, to supply the medullary membrane, which lines the central cavity and the adjoining canals. The ramifications of this vessel anastomose with the arteries both of the cancellous and compact tissues. In most of the flat, and in many of the short spongy bones, one or more large apertures are observed, which transmit, to the central parts of the bone, vessels corresponding to the medullary arteries and veins. The veins emerge from the long bones in three places (Kölliker)—(1) by one or two large veins, which accompany the artery ; (2) by numerous large and small veins at the articular extremities ; (3) by many small veins which arise in the compact substance. In the flat cranial bones the veins are large, very numerous, and run in tortuous canals in the diploic tissue, the sides of the canals being formed by a thin lamella of bone, perforated here and there for the passage of branches from the adjacent cancelli. The same condition is also found in all cancellous tissue, the veins being enclosed and supported by osseous structure, and having exceedingly thin coats. When the bony structure is divided, the vessels remain patulous, and do not contract in the canals in which they are contained. Hence the occurrence of purulent absorption after amputation, in those cases where the stump becomes inflamed, and the cancellous tissue is infiltrated and bathed in pus.

Lymphatic vessels, in addition to those found in the periosteum, have been traced by Cruikshank into the substance of bone, and Klein describes them as running in the Haversian canals.

Nerves are distributed freely to the periosteum, and accompany the nutrient arteries into the interior of the bone. They are said by Kölliker to be most numerous in the articular extremities of the long bones, in the vertebræ, and the larger flat bones.

Minute Anatomy.—The intimate structure of bone, which in all essential particulars is identical in the compact and in the cancellous tissue, is most easily studied in a transverse section from the compact wall of one of the long bones after maceration, such as is shown in fig. 31.

FIG. 31.—From a transverse section of the diaphysis of the humerus. Magnified 350 times.

If this is examined with a rather low power the bone will be seen to be mapped out into a number of circular districts : each one of which consists of a central hole surrounded by a number of concentric rings. These districts are termed *Haversian systems* ; the central hole is an *Haversian canal*, and the rings are layers of bone-tissue arranged concentrically around the central canal, and termed *lamellæ*. Moreover, on closer examination, it will be found that between these lamellæ, and therefore also arranged concentrically around the central canal, are a number of little dark specks, the *lacunæ*, and that these lacunæ are connected with each other and with the central Haversian canal by a number of fine dark lines, which radiate like the spokes of a wheel and are called *canaliculi*. All

a. Haversian canals. *b.* Lacunæ, with their canaliculi in the lamellæ of these canals. *c.* Lacunæ, of the interstitial lamellæ. *d.* Others at the surface of the Haversian systems, with canaliculi given off from one side.

these structures—the concentric lamellæ, the lacunæ, and the canaliculi—may be seen in any single Haversian system, forming a circular district round a central, Haversian, canal. Between these circular systems, filling in the irregular intervals which are left between them, are other lamellæ, with their lacunæ and canaliculi, running in various directions, but more or less curved (fig. 32). These are termed *interstitial* lamellæ. Again, other lamellæ, for the most part found

FIG. 32.—Transverse section of compact tissue of bone.
Magnified about 150 diameters. (Sharpey.)

on the surface of the bone, are arranged concentrically to the circumference of bone, constituting, as it were, a single Haversian system of the whole bone, of which the medullary cavity would represent the Haversian canal. These latter lamellæ are termed *circumferential*, or by some authors *primary* or *fundamental* lamellæ, to distinguish them from those laid down around the axis of the Haversian canals, which are then termed *secondary* or *special* lamellæ.

FIG. 33.—Section parallel to the surface from the shaft of the femur. Magnified 100 times.

a. Haversian canals. *b.* Lacunæ seen from the side. *c.* Others seen from the surface in lamellæ which are cut horizontally.

The *Haversian canals*, seen as round holes in a transverse section of bone at or about the centre of each Haversian system, may be demonstrated to be true canals, if a longitudinal section is made, as in fig. 33. It will then be seen that these round holes are tubes cut across, which run parallel with the longitudinal axis of the bone for a short distance, and then branch and communicate. They vary considerably in size, some being as large as $\frac{1}{200}$ of an inch in diameter; the average size being, however, about $\frac{1}{500}$ of an inch. Near the medullary cavity the canals are larger than those near the surface of the bone. Each canal contains one or two blood-vessels, with a small quantity of delicate connective tissue and some nerve filaments. In the larger ones there are also lymphatic spaces, and branched cells, the processes of which communicate, through the canaliculi, with the branched processes of certain bone cells in the substance of the bone. Those canals near the surface of the bone open upon it by minute orifices, and those near the medullary cavity open in the same way into this space, so that the whole of the bone is permeated by a system of blood-vessels running through the bony canals in the centre of the Haversian systems.

The *lamellæ* are thin plates of bone-tissue encircling the central canal, and may be compared, for the sake of illustration, to a number of sheets of paper pasted one over another around a central hollow cylinder. After macerating a piece of bone in dilute mineral acid, these lamellæ may be stripped off in a longitudinal direction as thin films. If one of these is examined with a high

FIG. 34.—Lamellæ torn from a decalcified human parietal bone to show the perforating fibres of Sharpey. (Copied from a drawing by Allen Thomson.)

Perforating fibres of Sharpey

Apertures from which the fibres have been withdrawn

Decussating fibres in lamella

power under the microscope it will be found to be composed of a finely reticular structure, presenting the appearance of lattice-work made up of very slender, transparent fibres, decussating obliquely, and coalescing at the points of intersection so as to form an exceedingly delicate network. These fibres are composed of fine fibrils, identical with those of white connective tissue. The intercellular matrix between the fibres has been replaced by calcareous deposit which the acid dissolves. In many places the various lamellæ may be seen to be held together by tapering fibres, which run obliquely through them, pinning or bolting them together. These fibres were first described by Sharpey, and were named by him *perforating fibres* (fig. 34).

The *lacunæ* are situated between the lamellæ, and consist of a number of oblong spaces. In an ordinary microscopic section, viewed by transmitted light, they appear as dark, oblong, opaque spots, and were formerly believed to be solid cells. Subsequently, when it was seen that the Haversian canals were channels which lodge the vessels of the part, and the canaliculi minute tubes by which the plasma of the blood circulates through the tissue, the theory was formulated that the lacunæ were hollow spaces filled during life with the same fluid, and only lined (if lined at all) by a delicate membrane. But this view was eventually proved to be erroneous, for examination of the structure of bone, when recent, led Virchow to believe that the lacunæ are

FIG. 35.—Nucleated bone-cells and their processes, contained in the bone-lacunæ and their canaliculi respectively. From a section through the vertebra of an adult mouse. (Klein and Noble Smith.)

occupied during life by a branched cell, termed a bone-cell or bone-corpuscle, the processes from which pass down the canaliculi—a view which is now universally accepted (fig. 35). It is by means of these cells that the fluids necessary for nutrition are brought into contact with the ultimate tissue of bone.

The *canaliculi* are exceedingly minute channels, which pass *across* the lamellæ and connect the lacunæ with neighbouring lacunæ and also with the Haversian canal. From this central canal a number of canaliculi are given off, which radiate from it, and open into the first set of lacunæ arranged around the Haversian canal, between the first and second lamellæ. From these lacunæ a second set of canaliculi are given off, which pass outwards to the next series of lacunæ, and so on until they reach the periphery of the Haversian system ; here the canaliculi given off from the last series of lacunæ do not communicate with the lacunæ of neighbouring Haversian systems, but after passing outwards for a short distance form loops and return to their own lacuna. Thus every part of an Haversian system is supplied with nutrient fluids derived from the vessels in the Haversian canal and traversing the canaliculi and lacunæ.

The *bone-cells* are contained in the lacunæ, which, however, they do not completely fill. They are flattened nucleated cells, which Virchow has shown are homologous with those of connective tissue. The cells are branched, and the branches, especially in young bones, pass into the canaliculi from the lacunæ.

If a longitudinal section is examined, as in fig. 33, the structure is seen to be the same. The appearance of concentric rings is replaced by that of lamellæ or rows of lacunæ, parallel to the course of the Haversian canals, and these canals appear like half-tubes instead of circular spaces. The tubes are seen to branch and communicate, so that each separate Haversian canal runs only a short distance. In other respects the structure has much the same appearance as in transverse sections.

In sections of thin plates of bone (as in the walls of the cells which form the cancellous tissue) the Haversian canals are absent, and the canaliculi open into the spaces of the cancellous tissue (medullary spaces), which thus have the same function as the Haversian canals in the more compact bone.

Chemical Composition.—Bone consists of an animal and an earthy part intimately combined together.

The animal part may be obtained by immersing the bone for a considerable time in dilute mineral acid, after which process the bone comes out exactly

the same shape as before, but perfectly flexible, so that a long bone (one of the ribs, for example) can easily be tied in a knot. If now a transverse section is made (fig. 36), the same general arrangement of the Haversian canals, lamellæ, lacunæ, and canaliculi is seen, though not so plainly as in the ordinary section.

The earthy part may be separately obtained by calcination, by which the animal matter is completely burnt out. The bone will still retain its original form, but it will be white and brittle, will have lost about one-third of its original weight, and will crumble down with the slightest force. The earthy matter confers on bone its hardness and rigidity, and the animal matter its tenacity.

The animal basis is largely composed of *ossein*, which is identical with the collagen of white fibrous tissue, so that when boiled with water, especially under pressure, it is almost entirely resolved into gelatin.

The organic matter of bone forms about *one-third*, or 33·3 per cent. ; the inorganic matter, *two-thirds*, or 66·7 per cent. Of the earthy matter, five-sixths is calcium phosphate, the remainder consisting of calcium carbonate, calcium fluoride, calcium chloride, and magnesium phosphate, with small amounts of sodium chloride and sulphate. Even after the removal of all the marrow, a small percentage of fat is still found in bone.

Some of the diseases to which bones are liable mainly depend on the disproportion between the two constituents of bone. Thus in the disease called rickets, so common in the children of the poor, the bones become bent and curved, either from the superincumbent weight of the body, or under the action of certain muscles. This depends upon some defect of nutrition by which bone becomes deprived of its normal proportion of earthy matter, while the animal matter is of unhealthy quality. In the vertebræ of a rickety subject, Bostock found in 100 parts 79·75 animal and 20·25 earthy matter.

Development of Bone.—In the fœtal skeleton some bones are preceded by membrane, such as those forming the roof and sides of the skull ; others, such as the bones of the limbs, are preceded by rods of cartilage. Hence two kinds of ossification are described : the *intramembranous* and the *intracartilaginous*.

Intramembranous Ossification.—In the case of bones which are developed in membrane no cartilaginous mould precedes the appearance of the bone tissue. The membrane, which occupies the place of the future bone, is of the nature of connective tissue, and ultimately forms the periosteum. At this stage it is seen to be composed of fibres and granular cells in a matrix. The outer portion is more fibrous, while, internally, the cells or *osteoblasts* predominate ; the whole tissue is richly supplied with blood-vessels. At the outset of the process of bone formation a little network of bony spiculæ is first noticed radiating from the point or centre of ossific tion. When these rays of growing bone are examined by the microscope they are found to consist at their growing point of a network of fine clear fibres and granular corpuscles with an intervening ground substance (fig. 37).

FIG. 37.—Part of the growing edge of the developing parietal bone of a fœtal cat. (After J. Lawrence.)

The fibres are termed *osteogenetic* fibres, and are made up of fine fibrils differing little from those of white fibrous tissue. Like them they are probably deposited in the matrix through the influence of the cells—in this case the osteoblasts. The osteogenetic fibres soon assume a dark and granular appearance from the deposition of calcareous granules in the fibres and in the intervening matrix, and as they calcify they are found to enclose some of the granular corpuscles or osteoblasts. By the fusion of the calcareous granules the bony tissue again assumes a more transparent appearance, but the fibres are no longer so distinctly seen. The involved osteoblasts form the corpuscles of the future bone, the spaces in which they are enclosed constituting the lacunæ. As the osteogenetic fibres grow out to the periphery they continue to calcify, and give rise to fresh bone spicules. Thus a network of bone is formed, the meshes of which contain the blood-vessels and a delicate connective tissue crowded with osteoblasts. The bony trabeculæ thicken by the addition of fresh layers of bone formed by the osteoblasts on their surface, and the meshes are correspondingly encroached upon. Subsequently successive layers of bony tissue are deposited under the periosteum and round the larger vascular channels which become the Haversian canals, so that the bone increases much in thickness.

Intracartilaginous Ossification.—Just before ossification begins the bone is entirely cartilaginous, and in a long bone, which may be taken as an example, the process commences in the centre and proceeds towards the extremities, which for some time remain cartilaginous. Subsequently a similar process commences in one or more places in those extremities and gradually extends through them. The extremities do not, however, become joined to the shaft by bony tissue until growth has ceased, but are attached to it by a layer of cartilaginous tissue termed the *epiphysial cartilage*.

The first step in the ossification of the cartilage is that the cartilage-cells, at the point where ossification is commencing and which is termed a *centre of ossification*, enlarge and arrange themselves in rows (fig. 38). The matrix in which they are embedded increases in quantity, so that the cells become further separated from each other. A deposit of calcareous material now takes place in this matrix, between the rows of cells, so that they become separated from each

[FIG. 38.—Section of fœtal bone of cat.

FIG. 39.—Part of a longitudinal section of the developing femur of a rabbit.

ir. Irruption of the subperiosteal tissue. p. Fibrous layer of the periosteum. o. Layer of osteoblasts. im. Subperiosteal bony deposit. (From Quain's 'Anatomy,' E. A. Schäfer).

a. Flattened cartilage-cells. b. Enlarged cartilage-cells. c, d. Newly formed bone. e. Osteoblasts. f. Giant-cells or osteoclasts. g, h. Shrunken cartilage-cells. (From 'Atlas of Histology,' Klein and Noble Smith.)

other by longitudinal columns of calcified matrix, presenting a granular and opaque appearance. Here and there the matrix between two cells of the same row also becomes calcified, and transverse bars of calcified substance stretch across from one calcareous column to another. Thus there are longitudinal groups of the cartilage-cells enclosed in oblong cavities, the walls of which are formed of calcified matrix, which cuts off all nutrition from the cells, and they, in consequence, waste, leaving spaces called the *primary areolæ* (Sharpey).

At the same time that this process is going on in the centre of the solid bar of cartilage of which the fœtal bone consists, certain changes are taking place on its surface. This is covered by a very vascular membrane, the *perichondrium*, entirely similar to the embryonic connective tissue already described as constituting the basis of membrane bone, on the inner surface of which—that is to say, on the surface in contact with the cartilage—are gathered the formative cells, the *osteoblasts*. By the agency of these cells a thin layer of bony tissue is being formed between the perichondrium and the cartilage, by the *intramembranous* mode of ossification just described. There are then, in this first stage of ossification, two processes going on simultaneously : in the centre of the cartilage the formation of a number of oblong spaces, formed of calcified matrix and containing the withered cartilage-cells, and on the surface of the cartilage the formation of a layer of true membrane-bone. The second stage consists in the prolongation into the cartilage of processes of the deeper or osteogenetic layer of the perichondrium, which has now become periosteum (fig. 38, *ir*). The processes consist of blood-vessels and cells—*osteoblasts*, or bone-formers, and *osteoclasts*, or bone-destroyers. The latter are similar to the giant-cells (myeloplaxes) found in marrow, and they excavate passages through the new-formed bony layer by absorption, and pass through it into the calcified matrix (fig. 38). Wherever these processes come in contact with the calcified walls of the primary areolæ they absorb it, and thus cause a fusion of the original cavities and the formation of larger spaces, which are termed the *secondary areolæ* (Sharpey) or *medullary spaces* (Müller). These secondary spaces, the original cartilage-cells having disappeared, become filled with embryonic marrow, consisting of osteoblasts and vessels, and derived, in the manner described above, from the osteogenetic layer of the periosteum (fig. 39).

Thus far there has been traced the formation of enlarged spaces (secondary areolæ), the perforated walls of which are still formed by calcified cartilage-matrix, containing an embryonic marrow, derived from the processes sent in from the osteogenetic layer of the periosteum, and consisting of blood-vessels and round cells, osteoblasts (fig. 39). The walls of these secondary areolæ are at this time of only inconsiderable thickness, but they become thickened by the deposition of layers of new bone on their interior. This process takes place in the following manner. Some of the osteoblasts of the embryonic marrow, after undergoing rapid division, arrange themselves as an epithelioid layer on the surface of the wall of the space (fig. 40).

FIG. 40.—Transverse section from the femur of a human embryo about eleven weeks old.

a. A medullary sinus cut transversely : and *b.* another longitudinally. *c.* Osteoblasts. *d.* Newly formed osseous substance of a lighter colour. *e.* That of greater age. *f.* Lacunæ with their cells. *g.* A cell still united to an osteoblast.

This layer of osteoblasts forms a bony stratum, and thus the wall of the space becomes gradually covered with a layer of true osseous substance. On this a second layer of osteoblasts arrange themselves, and in their turn form an osseous layer. By the repetition of this process the original cavity becomes very much reduced in size, and at last only remains as a small circular hole in the centre, containing the remains of the embryonic marrow—that is, a blood-vessel and a few osteoblasts. This small cavity constitutes the Haversian canal of the perfectly ossified bone. The successive layers of osseous matter which have been laid down, and which encircle this central canal, constitute the lamellæ, of which, as we have seen, each Haversian system is made up. As the successive layers of osteoblasts form osseous tissue, certain of the osteoblastic cells remain included between the various bony layers. These persist as the corpuscles of the future bone, the spaces enclosing them forming the lacunæ (fig. 39). The canaliculi, at first extremely short, are supposed to be extended by absorption, so as to meet those of neighbouring lacunæ.

Such are the changes which may be observed at one particular point, the centre of ossification. While they have been going on a similar process has been set up in the surrounding parts and has been gradually proceeding towards the end of the shaft, so that in the ossifying bone all the changes described above may be seen in different parts, from the true bone in the centre of the shaft to the hyaline cartilage at the extremities. The bone thus formed differs from the bone of the adult in being more spongy and less regularly lamellated.

In this way the steps of a process have been described by which a solid bony mass is produced, having vessels running into it from the periosteum ; Haversian canals in which these vessels run ; medullary spaces filled with fœtal marrow ; lacunæ with their contained bone-cells ; and canaliculi growing out of these lacunæ.

This process of ossification, however, is not the origin of the whole of the skeleton, for even in those bones in which the ossification proceeds in a great measure from a single centre, situated in the cartilaginous shaft of a long bone, a considerable part of the original bone is formed by intramembranous ossification beneath the perichondrium or periosteum ; so that the girth of the bone is increased by bony deposit from the deeper layer of this membrane.

The shaft of the bone is at first solid, but a tube is hollowed out in it by absorption around the vessels passing into it, which becomes the medullary canal. This absorption is supposed to be brought about by large giant-cells, the so-called osteoclasts of Kölliker (fig. 39, f). They vary in shape and size, and are known by containing a large number of clear nuclei, sometimes as many as twenty. The occurrence of similar cells in some tumours of bones has led to such tumours being denominated ' myeloid.'

As more and more bone is removed by this process of absorption from the interior of the bone to form the medullary canal, so more and more bone is deposited on the exterior from the periosteum, until at length it has attained the shape and size which it is destined to retain during adult life. As the ossification of the cartilaginous shaft extends towards the articular ends it carries with it, as it were, a layer of cartilage, or the cartilage grows as it ossifies, and thus the bone is increased in length. During this period of growth the articular end, or epiphysis, remains for some time entirely cartilaginous, then a bony centre appears in it, and it commences the same process of intracartilaginous ossification ; but this process never extends to any great distance. The epiphyses remain separated from the shaft by a narrow cartilaginous layer for a definite time. This layer ultimately ossifies, the distinction between shaft and epiphysis is obliterated, and the bone assumes its completed form and shape. The same remarks also apply to the processes of bone which are separately ossified, such as the trochanters of the femur. The bones, having been formed, continue to grow until the body has acquired its full stature. They increase in length by ossification continuing to extend in the epiphysial cartilage, which goes on growing in advance of the ossifying process. They increase in circumference by deposition of new bone, from the deeper layer of the periosteum, on their external surface, and at the same time an absorption takes place from within, by which the medullary cavity is increased.

The medullary spaces which characterise the cancellous tissue are produced by the absorption of the original fœtal bone in the same way as that by which the original medullary canal is formed. The distinction between the cancellous and the compact tissue appears to depend essentially upon the extent to which this process of absorption has been carried ; and we may perhaps remind the reader that in morbid states of the bone inflammatory absorption produces exactly the same change, and converts portions of bone, naturally compact, into cancellous tissue.

The number of ossific centres varies in different bones. In most of the short bones ossification commences by a single point in the centre, and proceeds towards the circumference. In the long bones there is a central point of ossification for the shaft or diaphysis : and one or more for each extremity, the epiphysis. That for the shaft is the first to appear. The union of the epiphyses with the shaft takes place in the reverse order to that in which their ossification began, with the exception of the fibula, and appears to be regulated by the direction of the nutrient artery of the bone. Thus, the nutrient arteries of the bones of the arm and fore-arm are directed towards the elbow, and the epiphyses of the bones forming this joint become united to the shaft before those at the opposite extremity. In the lower limb, on the other hand, the nutrient arteries pass in a direction from the knee : that is, upwards in the femur, downwards in the tibia and fibula ; and in them it is observed that the upper epiphysis of the femur, and the lower epiphysis of the tibia and fibula, become first united to the shaft.

Where there is only one epiphysis, the medullary artery is directed towards that end of the bone where there is no additional centre ; as towards the acromial end of the clavicle, towards the distal end of the metacarpal bone of the thumb and great toe, and towards the proximal end of the other metacarpal and meta-tarsal bones.

Besides these epiphyses for the articular ends, there are others for projecting parts or processes, which are formed separately from the bulk of the bone. For an account of these, the reader must be referred to the description of the individual bones.

A knowledge of the exact periods when the epiphyses become joined to the shaft is often of great importance in medico-legal inquiries. It also aids the surgeon in the diagnosis of many of the injuries to which the joints are liable ; for it not infrequently happens that, on the application of severe force to a joint, the epiphysis becomes separated from the shaft, and such injuries may be mistaken for fracture or dislocation.

MUSCULAR TISSUE

The **muscles** are formed of bundles of reddish fibres, endowed with the property of contractility. The two principal kinds of muscular tissue found in the body are the voluntary and involuntary. The former of these, from the characteristic appearances which its fibres exhibit under the microscope, is known as ' striped ' muscle, and from the fact that it is capable of being put into action and controlled by the will, as ' voluntary ' muscle. The fibres of the latter do not present any cross-striped appearance, and for the most part are not under the control of the will; hence they are known as the 'unstriped' or 'involuntary' muscles. The muscular fibres of the heart differ in certain particulars from both these groups, and they are therefore separately described as ' cardiac ' muscular fibres.

Thus it will be seen that there are three varieties of muscular fibres : (1) trans-versely striated muscular fibres, which are for the most part voluntary and under the control of the will, but some of which are not so, such as the muscles of the pharynx and upper part of the œsophagus. This variety of muscle is sometimes called *skeletal* ; (2) transversely striated muscular fibres, which are not under the control of the will, i.e. the cardiac muscle ; (3) plain or unstriped muscular fibres, which are involuntary and controlled by a differ-ent part of the nervous system from that which controls the activity of the voluntary muscles. Such are the muscular walls of the stomach and intestine, of the uterus and bladder, of the blood-vessels, &c.

FIG. 41.— Transverse section from the sterno-mastoid in man. Magnified 50 times.

a. Epimysium. *b.* Fasciculus. *c.* Peri-mysium. *d.* Fibre.

The **striped** or **voluntary muscles** are composed of bundles of fibres enclosed in a delicate web called the ' perimysium,' in contradistinction to the sheath of areolar tissue which invests the entire muscle, the ' epimysium ' (fig. 41). The bundles are termed ' fasciculi ;' they are prismatic in shape, of different sizes in different muscles, and for the most part placed parallel to one another, though they have a tendency to converge towards their tendinous attachments. Each fasciculus is made up of a strand of *fibres*, which also run parallel with each other, and which are separated from one another by a delicate connective tissue derived from the perimysium and termed *endomysium*. This does not form the sheath of the fibres, but serves to support the blood-vessels and nerves ramifying between them. The fibres are enclosed in a separate and distinct sheath of their own, but it is not areolar tissue and is therefore not derived from the perimysium.

D

A *muscular fibre* may be said to consist of a soft contractile substance enclosed in a tubular sheath, named by Bowman the *sarcolemma*. The fibres are cylindrical or prismatic in shape, and are of no great length, not exceeding, it is said, an inch and a half. They end either by blending with the tendon or apo-neurosis, or else by rounded or tapering extremities which are connected to the neighbouring fibres by means of the sarcolemma. Their breadth varies in man from $\frac{1}{200}$ to $\frac{1}{600}$ of an inch. As a rule, the fibres do not divide or anastomose; but occasionally, especially in the tongue and facial muscles, the fibres may be seen to divide into several branches. The precise mode in which the muscular fibre joins the tendon has been variously described by different observers. It may, perhaps, be sufficient to say that the sarcolemma, or membranous investment of the muscular fibre, appears to become blended with a small bundle of fibres, into which the tendon becomes subdivided, while the muscular substance termi-nates abruptly and can be readily made to retract from the point of junction. The areolar tissue between the fibres appears to be prolonged more or less into the tendon, so as to form a kind of sheath around the tendon bundles for a longer or

Fig. 42.—Two human mus-cular fibres. Magnified 350 times.

Fig. 43.—Fragments of striped muscular fibres, showing a cleavage in opposite directions. Magnified 300 diameters.

A. Longitudinal cleavage. The longitudinal and transverse lines are both seen. Some longitudinal lines are darker and wider than the rest, and are not continuous from end to end. This results from partial separation of the fibrillæ. *c*. Fibrillæ separated from one another by violence at the broken end of the fibre, and marked by transverse lines equal in width to those on the fibre. *c' c''* represent two appearances commonly presented by the separated single fibrillæ (more highly magnified). At *c'* the borders and transverse lines are all perfectly rectilinear, and the included spaces perfectly rectangular. At *c''* the borders are scalloped and the spaces bead-like. When most distinct and definite the fibrilla presents the former of these appearances. B. Transverse cleavage. The longitudinal lines are scarcely visible. *a*. Incomplete fracture following the opposite sur-faces of a disc, which stretches across the interval, and retains the two fragments in connection. The edge and surfaces of this disc are seen to be minutely granular, the granules corresponding in size to the thickness of the disc, and to the distance between the faint longitudinal lines. *b*. Another disc nearly detached. *b'*. Detached disc, more highly magnified, showing the sarcous elements.

In the one, the bundle of fibrillæ (*b*) is torn, and the sarcolemma (*a*) is seen as an empty tube.

shorter distance. When muscular fibres are attached to the skin or mucous membranes, their fibres are described by Hyde Salter as becoming continuous with those of the areolar tissue.

The *sarcolemma*, or tubular sheath of the fibre, is a transparent, elastic, and apparently homogeneous membrane of considerable toughness, so that it will sometimes remain entire when the included substance is ruptured (see fig. 42), On the internal surface of the sarcolemma in mammalia, and also in the substance of the fibre in the lower animals, elongated nuclei are seen, and in connection with these a row of granules, apparently fatty, is sometimes observed.

Upon examination of a voluntary muscular fibre by transmitted light, it is found to be apparently marked by alternate light and dark bands or striæ, which pass transversely, or somewhat obliquely, round the fibre (fig. 42). The dark and light bands are of nearly equal breadth, and alternate with great regularity. They vary in breadth from about $\frac{1}{1200}$ to $\frac{1}{1700}$ of an inch. If the surface is carefully focussed, rows of granules will be detected at the point of junction of the dark and light bands, and very fine longitudinal lines may be seen running through the dark bands and joining these granules together. By treating the specimen with certain reagents (e.g. chloride of gold) fine lines may be seen running transversely between the granules uniting them together. This appearance is believed to be due to a reticulum or network of interstitial substance lying between the contractile portions of the muscle. The longitudinal striation gives the fibre the appearance of being made up of a bundle of fibrillæ, which have been termed *sarcostyles* or *muscle columns*, and if the fibre is hardened in alcohol, it can be broken up longitudinally and the sarcostyles separated from each other (fig. 43, A). The reticulum, with its longitudinal and transverse meshes, is called *sarcoplasm*.

If now a transverse section of a muscular fibre is made, it is seen to be divided into a number of areas, called the *areas of Cohnheim*, more or less polyhedral in shape, and consisting of the transversely divided sarcostyles, surrounded by transparent series of sarcoplasm (fig. 43, B, b').

Upon closer examination, and by somewhat altering the focus, the appearances become more complicated, and are susceptible of various interpretations. The

FIG. 44.—A. Portion of a medium-sized human muscular fibre. Magnified nearly 800 diameters. B. Separated bundles of fibrils, equally magnified.

FIG. 45.—Part of a striped muscular fibre of the water-beetle, prepared with absolute alcohol. Magnified 300 diameters. (Klein and Noble Smith.)

a. Sarcolemma. *b.* Membrane of Krause: owing to contraction during hardening, the sarcolemma shows regular bulgings. At the side of Krause's membrane is the transparent lateral disc.
Several nuclei of muscle-corpuscles are shown, and in them a minute network.

a, a. Larger, and *b, b,* smaller collections. *c.* Still smaller.
d, d. The smallest which could be detached.

transverse striation, which in figs. 42 and 43 appears as a mere alternation of dark and light bands, is resolved into the appearance seen in fig. 44, which shows a series of broad dark bands, separated by light bands, which are divided into two by a dark dotted line. This line is termed *Krause's membrane* (fig. 46, *k*), because it was believed by Krause to be an actual membrane, continuous with the sarcolemma, and dividing the light band into two compartments. It is now more usually regarded as being due to an optical phenomenon, from the light being reflected between discs of different refrangibility. In addition to the membrane of Krause, fine clear lines may be made out, with a sufficiently high power, crossing the centre of the dark band ; these are known as the *lines of Hensen* (fig. 46, H).

Formerly it was supposed by Bowman that a muscular fibre was made up of a number of quadrangular particles, which he named sarcous elements, joined together like so many bricks forming a column, and he came to this conclusion because he found that under the influence of certain reagents the fibre could be broken up transversely into discs, as well as longitudinally into fibrillæ (fig. 43, B). But it is now believed that this cross cleavage is purely artificial, and that a muscular fibre is built up of fibrillæ and not of small quadrangular particles.

Assuming that this is so, we have now to consider a little more in detail the minute structure of these longitudinal fibrillæ, or sarcostyles, as they are termed. Perhaps there are few subjects in histology which have received more attention, and in which the appearances seen under the microscope have been more differently interpreted, than the minute anatomy of muscular fibre. Schäfer has worked out this subject, particularly in the wing muscles of insects, which are peculiarly adapted for this purpose on account of the large amount for interstitial sarcoplasm which separates the sarcostyles. In the following description that given by Schäfer will be closely followed (fig. 46).

Each sarcostyle may be said to be made up of successive portions, each of which Schäfer terms a *sarcomere*. This is the portion situated between two membranes of Krause, which transversely divides the light band. Each sarcomere consists of a central dark part, which forms a portion of the dark band of the whole fibre, and is named by Schäfer a *sarcous element.** This sarcous element really consists of two parts, superimposed one on the top of the other, and when the fibre is stretched these two parts become separated from each other at the line of Hensen (fig. 46, A). On either side of this central dark portion is a clear layer, most visible when the fibre is extended; this is situated between the dark centre and the membrane of Krause, and when the sarcomeres are joined together to form the sarcostyle, constitutes the light band of the striated muscular fibre.

When the sarcostyle is extended, the clear intervals are well marked and plainly to be seen; when, on the other hand, the sarcostyle is contracted,

FIG. 46.—Diagram of a sarcomere. (After Schäfer.)
A. In moderately extended condition.   B. In a contracted condition.

k. k. Membranes of Krause.   H. Line or plane of Hensen.   S.E. Poriferous sarcous element.

that is to say, the muscle is in a state of contraction, these clear portions are very small or they may have disappeared altogether (fig. 46, B). When the sarcostyle is stretched to its full extent, not only is the clear portion very well marked, but the dark portion—the sarcous element—will be seen to be separated into its two constituents along the line of Hensen.

The sarcous element does not lie free in the sarcomere, for when the sarcostyle is stretched, so as to render the clear portion visible, very fine lines, which are probably septa, may be seen running through it from the sarcous element to the membrane of Krause.

Schäfer explains these phenomena in the following way. He considers that each sarcous element is made up of a number of longitudinal channels, which open into the clear part towards the membrane of Krause but are closed at the line of Hensen. When the muscular fibre is contracted the *clear* part of the muscular substance finds its way into these channels or tubes and is therefore hidden from sight, but at the same time it swells up the sarcous element and widens and shortens the sarcomere. When, on the other hand, the fibre is extended, this clear substance finds its way out of the tubes and collects between the sarcous element and the membrane of Krause, and gives the appearance of the light part between these two structures; by this means it elongates and narrows the sarcomere.

If this view is true it is a matter of great interest, and, as Schäfer has shown, harmonises the contraction of muscle with the amœboid action of protoplasm.

* This must not be confused with the 'sarcous element of Bowman' (see above).

In an amœboid cell there is a framework of spongioplasm, which stains with hæmatoxylin and similar reagents, enclosing in its meshes a clear substance, hyaloplasm, which will not stain with these reagents. Under stimulation the hyaloplasm passes into the pores of the spongioplasm; without stimulation it tends to pass out as in the formation of pseudopodia. In muscle there is the same thing, viz. a framework of spongioplasm staining with hæmatoxylin—the substance of the sarcous element—and this encloses a clear hyaloplasm, the clear substance of the sarcomere, which resists staining with this reagent. During contraction of the muscle—i.e. stimulation—this clear substance passes into the pores of the spongioplasm; while during extension of the muscle—i.e. when there is no stimulation—it tends to pass out of the spongioplasm.

In this way the contraction is brought about: under stimulation the protoplasmic material (the clear substance of the sarcomere) recedes into the sarcous element, causing the sarcomere to widen out and shorten. The contraction of the muscle is merely the sum total of this widening out and shortening of these bodies.

The *capillaries* of striped muscle are very abundant, and form a sort of rectangular network, the branches of which run longitudinally in the endomysium between the muscular fibres, and are joined at short intervals by transverse

FIG. 47.—Non-striated muscular fibre. (From Kirke's 'Physiology.')

anastomosing branches. The larger vascular channels, arteries and veins, are found only in the perimysium, between the muscular fasciculi.

Nerves are profusely distributed to striped muscle. Their mode of termination will be described on a subsequent page.

The existence of *lymphatic* vessels in striped muscle has not been ascertained, though they have been found in tendons and in the sheath of the muscle.

The **unstriped, plain,** or **involuntary muscle** is found in the walls of the hollow viscera—viz. the lower half of the œsophagus and the whole of the remainder of the gastro-intestinal tube; in the trachea and bronchi, and the alveoli and infundibula of the lungs; in the gall-bladder and ductus communis choledochus; in the large ducts of the salivary and pancreatic glands; in the pelvis and calyces of the kidney, the ureter, bladder, and urethra; in the female sexual organs—viz. the ovary, the Fallopian tubes, the uterus (enormously developed in pregnancy), the vagina, the broad ligaments, and the erectile tissue of the clitoris; in the male sexual organs—viz. the dartos of the scrotum, the vas deferens and epididymis, the vesiculæ seminales, the prostate gland, and the corpora cavernosa and corpus spongiosum; in the ducts of certain glands, as in Wharton's duct; in the capsule and trabeculæ of the spleen; in the mucous membranes, forming the muscularis mucosæ; in the skin, forming the arrectores pilorum, and also in the sweat-glands; in the arteries, veins, and lymphatics; in the iris and the ciliary muscle.

Plain or unstriped muscle is made up of spindle-shaped cells, called *contractile fibre-cells*, collected into bundles and held together by a cement substance (fig. 47)

These bundles are further aggregated into larger bundles, or flattened bands, and bound together by ordinary connective tissue.

The *contractile fibre-cells* are elongated, spindle-shaped, nucleated cells of various sizes, averaging from $\frac{1}{600}$ to $\frac{1}{300}$ of an inch in length, and $\frac{1}{4500}$ to $\frac{1}{3300}$ of an inch in breadth. On transverse section they are more or less polyhedral in shape, from mutual pressure. They present a faintly longitudinal striated appearance, and consist of an elastic cell-wall containing a central bundle of fibrillæ, representing the contractile substance, and an oval or rod-like nucleus, which includes, within a membrane, a fine network communicating at the poles of the nucleus with the contractile fibres (Klein). The adhesive interstitial cement substance, which connects the fibre-cells together, represents the endomysium, or delicate connective tissue which binds the fibres of striped muscular tissue into fasciculi ; while the tissue connecting the individual bundles together represents the perimysium. The unstriped muscle, as a rule, is not under the control of the will, neither is the contraction rapid nor does it involve the whole muscle, as is the case with the voluntary muscles. The membranes which are composed of the unstriped muscle slowly contract in a part of their extent, generally under the influence of a mechanical stimulus, as that of distension or of cold ; and then the contracted part slowly relaxes while another portion of the membrane takes up the contraction. This peculiarity of action is most strongly marked in the intestines, constituting their *vermicular motion*.

Fig. 48. — Anastomosing muscular fibres of the heart seen in a longitudinal section. On the right the limits of the separate cells with their nuclei are exhibited somewhat diagrammatically.

Cardiac Muscular Tissue.—The fibres of the heart differ very remarkably from those of other striped muscles. They are smaller by one-third, and their transverse striæ are by no means so well marked. The fibres are made up of distinct quadrangular cells joined end to end (fig. 48). Each cell contains a clear oval nucleus, situated near the centre of the cell. The extremities of the cells have a tendency to branch or divide, the subdivisions uniting with offsets from other cells, and thus producing an anastomosis of the fibres. The connective tissue between the bundles of fibres is much less than in ordinary striped muscle, and no sarcolemma has been proved to exist.

Development of Muscle Fibres.—Voluntary muscular fibres are developed from the mesoblast, the embryonic cells of which elongate, show multiplication of nuclei, and eventually become striated ; the striation is first obvious at the side of the fibre, spreads around the circumference, and ultimately extends to the centre. The nuclei, at first situated centrally, gradually pass out to assume their final position immediately beneath the sarcolemma. In the case of plain muscle the mesoblastic cells assume a pointed shape at the extremities and become flattened, the nucleus also lengthening out to its permanent rod-like form.

Chemical Composition of Muscle.—In chemical composition the muscular fibres may be said, in round numbers, to consist of 75 per cent. of water, about 20 per cent. of proteids, 2 per cent. of fat, 1 per cent. of nitrogenous extractives and carbohydrates, and 2 per cent. of salts, which are mainly potassium phosphate and carbonate.

NERVOUS TISSUE

The **nervous tissues** of the body are comprised in two great systems—the *cerebro-spinal* and the *sympathetic*.

The *cerebro-spinal* system comprises the brain (including the medulla oblongata), the spinal cord, the cranial nerves, the spinal nerves, and the ganglia connected with these nerves. The *sympathetic* system consists of a double chain of ganglia, with the nerves which go to and come from them. It is not directly connected with the brain or spinal cord, though it is so indirectly by means of its numerous communications with the cranial and spinal nerves.

The nervous tissues are found microscopically to be composed of *nerve-cells* and their various processes, together with a supporting tissue called *neuroglia*, which, however, is found only in the brain and spinal cord. Certain long processes of the nerve-cells are of special importance, and it is convenient to consider them apart from the cells; they are known as *nerve-fibres*.

To the naked eye a difference is obvious between certain portions of the brain and spinal cord, viz. the *grey matter* and the *white matter*. This is found to be due to the fact that the grey matter is largely composed of nerve-cells, while the white matter contains only their long processes, the nerve-fibres. It is in the former, as is generally supposed, that nervous impressions and impulses originate, and by the latter that they are conducted. Hence the grey matter forms the essential constituent of all the ganglionic centres, both those in the isolated ganglia and those aggregated in the cerebro-spinal axis; while the white matter is found in all the commissural portions of the nerve-centres and in all the cerebro-spinal nerves.

Neuroglia, the peculiar ground substance in which are embedded the true nervous constituents of the brain and spinal cord, consists of fibres and cells.

FIG. 49.—Neuroglia cells of brain shown by Golgi's method. (After Andriezen.) (Copied from Schäfer's 'Essentials of Histology.')

A. Cell with branched processes. B. Spider cell with unbranched processes.

Some of the cells are stellate in shape, and their fine processes become neuroglia fibres, which extend radially and unbranched (fig. 49, B) among the nerve-cells and fibres which they aid in supporting. Other cells give off fibres which branch repeatedly (fig. 49, A). In addition to these fibres there are others which do not appear to be connected with the neuroglia cells. They start from the epithelial cells lining the ventricles of the brain and central canal of the spinal cord, and pass through the nervous tissue, branching repeatedly to terminate in slight enlargements on the pia mater. Thus, neuroglia is evidently a connective tissue in function but is not so in development; it is epiblastic in origin, whereas all connective tissues are mesoblastic.

Nerve-cells or **ganglion-cells** are largely aggregated in the grey substance of the brain and spinal cord, but smaller collections of these cells also form the swellings seen on many nerves, which are called *ganglia*. These are found chiefly upon the spinal and cranial nerve-roots and in connection with the sympathetic nerves. Each nerve-cell consists of a finely fibrillated protoplasmic material, of a reddish or yellowish-brown colour, which occasionally presents patches of a deeper tint, caused by the aggregation of pigment-granules at one side of the nucleus, as in the substantia nigra and locus cœruleus. The protoplasm also sometimes contains peculiar angular granules, which stain deeply with basic dyes, such as methylene blue; these are known as *Nissl's granules* (fig. 52). The nucleus is, as a rule, a large, well-defined, round, vesicular body, often presenting an

FIG. 50.—Various forms of nerve-cells.

A. Pyramidal cell. B. Small multipolar cell, in which the axon quickly divides into numerous branches. C. Small fusiform cell. D and E. Ganglion cells (E shows T-shaped division of axon). *ax.* Axon. *c.* Capsule.

FIG. 51.—Bipolar nerve-cell from the spinal ganglion of the pike. (After Kölliker.)

FIG. 52.—Motor nerve-cell from ventral horn of spinal cord of rabbit. (After Nissl.) The angular and spindle-shaped Nissl bodies are well shown.

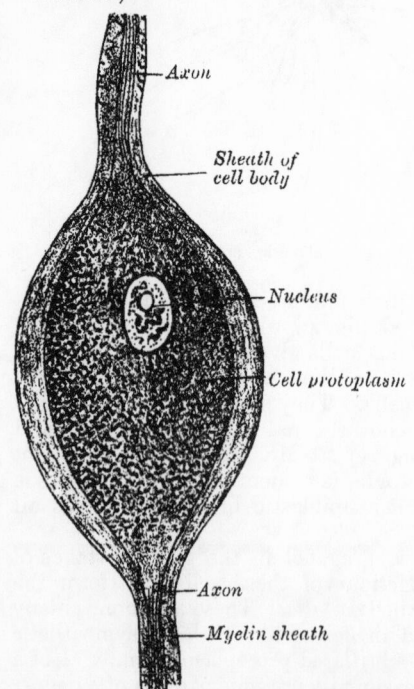

Axon

Sheath of cell body

Nucleus

Cell protoplasm

Axon

Myelin sheath

Axon

Nucleolus

intranuclear network, and containing a nucleolus which is peculiarly clear and brilliant. The nerve-cells vary in shape and size, and have one or more processes. They may be divided for purposes of description into three groups, according to the number of processes which they possess: (1) Unipolar cells, which are found in the spinal ganglia; their single process, after a short course, divides in a T-shaped manner. (2) Bipolar cells, also found in the spinal ganglia (fig. 51), when the cells are in an embryonic condition. They are best demonstrated in the sympathetic ganglion-cells of a frog. Sometimes the processes com-off from opposite poles of the cell, and the cell then assumes a spindle shape; at others they both emerge at the same point. In some cases where two fibres are apparently connected with a cell, one of the fibres is really derived from an adjoining nerve-cell and is passing to end in a ramification around the ganglion-cell, or, again, it may be coiled spirally round the nerve process which is issuing from the cell. (3) Multipolar cells, which are caudate or stellate in shape, and characterised by their large size and by the tail-like processes which issue from them. The processes are of two kinds: one of them is termed the *axis-cylinder process* or *axon* because it becomes the axis cylinder of a nerve-fibre (figs. 52, 53, 54). The others are termed the *protoplasmic processes* or *dendrons*; they begin to divide and subdivide as soon as they emerge from the cell, and finally end in minute twigs and become lost among the other elements of the nervous tissue.

FIG. 53.—Pyramidal cell from the cerebral cortex of a mouse. (After Ramón y Cajal.)

FIG. 54.—Cell of Purkinje from the cerebellum of a cat. (After Ramón y Cajal.)

Axon

Nerve-fibres are found universally in the peripheral nerves, and in the white substance of the brain and spinal cord. The fibres are of two kinds, the *medullated* or *white* fibres, and the *non-medullated* or *grey* fibres.

The *medullated* fibres form the white part of the brain and spinal cord, and also the greater part of the cerebro-spinal nerves, and give to these structures their opaque, white aspect. When perfectly fresh they appear to be homogeneous; but soon after removal from the body they present, when examined by transmitted light, a double outline or contour, as if consisting of two parts (fig. 55). The central portion is named the *axis cylinder of Purkinje*; around this is a sort of sheath of fatty material, staining black with osmic acid, named the *white substance of Schwann*, which gives to the fibre its double contour, and the whole is enclosed in a delicate membrane, the *neurilemma, primitive sheath*, or *nucleated sheath of Schwann* (fig. 55).

The *axis cylinder* is the essential part of the nerve-fibre, and is always present; the other parts, the medullary sheath and the neurilemma, being occasionally absent, especially at the origin and termination of the nerve-fibre. It undergoes no interruption from its origin in the nerve-centre to its peripheral termination, and must be regarded as a direct prolongation of a nerve-cell. It constitutes about one-half or one-third of the nerve-tube, the white substance being greater in proportion in the nerves thàn in the central organs. It is quite transparent, and is therefore indistinguishable in a perfectly fresh and natural state of the nerve. It is made up of exceedingly fine fibrils, which stain darkly with gold chloride (fig. 56). At its termination the axis cylinder of a nerve-fibre may be seen to break up into fibrillæ, confirming the view of its structure. These

Fig. 55.—White or medullated nerve-fibres, showing the sinuous outline and double contours. (After Schäfer.)

Fig. 56.—Longitudinal section through a nerve-fibre from the sciatic nerve of a frog. ×830. (After Böhm and Davidoff.)

Fibrils of axis cylinder

Neurilemma

Segment of Lantermann

fibrillæ have been termed the *primitive fibrillæ* of Schultze. The axis cylinder is said by some to be enveloped in a special, reticular sheath, which separates it from the white matter of Schwann, and is composed of a substance called *neurokeratin*. The more common opinion is that this network or reticulum is contained in the white matter of Schwann, and by some it is believed to be produced by the action of the reagents employed to show it. The *medullary sheath* or *white matter of Schwann* (fig. 56) is regarded as being a fatty matter in a fluid state, which insulates and protects the essential part of the nerve—the axis cylinder.

Fig. 57.—A node of Ranvier of a medullated nerve-fibre, viewed from above, magnified about 750 diameters. The medullary sheath is discontinuous at the node, whereas the axis cylinder passes from one segment into the other. At the node the sheath of Schwann appears thickened. (Klein and Noble Smith.)

The white matter varies in thickness to a very considerable extent, in some forming a layer of extreme thinness, so as to be scarcely distinguishable, in others forming about one-half the nerve-tube. The size of the nerve-fibres, which varies from $\frac{1}{1200}$ to $\frac{1}{2000}$ of an inch, depends mainly upon the amount of the white substance, though

the axis cylinder also varies in size within certain limits. The white matter of Schwann does not always form a continuous sheath to the axis cylinder, but undergoes interruptions in its continuity at regular intervals, giving to the fibre the appearance of constriction at these points. These were first described by Ranvier, and are known as the *nodes of Ranvier* (fig. 57). The portion of nerve-fibre between two nodes is called an *internodal segment*. The neurilemma or primitive sheath is not interrupted at the nodes, but passes over them as a continuous membrane. In addition to these interruptions oblique clefts may be seen in the medullary sheath, subdividing it into irregular portions, which are termed *medullary segments*, or *segments of Lantermann* (fig. 56). There is reason to believe that these clefts are artificially produced in the preparation of the specimens. Medullated nerve-fibres, when examined, frequently present a beaded or varicose appearance : this is due to manipulation and pressure causing the oily matter to collect into drops, and in consequence of the extreme delicacy of the primitive sheath, even slight pressure will cause the transudation of the fatty matter, which collects as drops of oil outside the membrane. This is, of course, promoted by the action of certain reagents.

The *neurilemma* or *primitive sheath* (sometimes called the *tubular membrane* or *sheath of Schwann*) presents the appearance of a delicate, structureless membrane. Here and there beneath it, and situated in depressions in the white matter of Schwann, are nuclei surrounded by a small amount of protoplasm. The nuclei are oval and somewhat flattened, and bear a definite relation to the nodes of Ranvier ; one nucleus generally lying in the centre of each internode. The primitive sheath is not present in all medullated nerve-fibres, being absent in those fibres which are found in the brain and spinal cord.

Non-medullated Fibres.—Most of the nerves of the sympathetic system, and some of the cerebro-spinal, consist of another variety of nervous fibres, which are called the *grey* or *gelatinous* nerve-fibres—*fibres of Remak* (fig. 58). These consist of a central core or axis cylinder enclosed in a nucleated sheath, which tends to split into fibrillæ, and is probably of the nature of neurokeratin. In external appearance the gelatinous nerves are semi-transparent and grey or yellowish-grey. The individual fibres vary in size, generally averaging about half the size of the medullated fibres.

Fig. 58.—A small nervous branch from the sympathetic of a mammal.

Development of Nerve-cells and Fibres.—The nerve-cells are developed from certain of the cells which line the neural canal or form the neural crest of the embryo (see section on development). Some of these cells assume a rounded form and are termed *neuroblasts*, and from each neuroblast there grows out a process, the axis-cylinder process or axon, and subsequently the branching processes or dendrons. The axis cylinders, at first naked, acquire their medullary sheath, possibly by some metamorphosis of their outer layer. The neurilemma is thought to be derived from mesoblastic cells which become flattened and wrapped round the fibre, the cement substance at their apposed ends forming the material which stains with silver nitrate at the nodes of Ranvier. Nerve-cells in the sympathetic and peripheral ganglia take their origin from small collections of neuroblasts, which are split off from the rudimentary spinal ganglia.

a. Two medullated nerve-fibres among a number of grey nerve-fibres, *b.*

Cells which are, originally, similar to neuroblasts seem to give rise to neuroglia cells, numerous processes sprouting from the cell to form the neuroglial fibres.

Chemical Composition.—The amount of water in nervous tissue varies with the situation. Thus in the grey matter of the cerebrum it constitutes about 83 per cent., in the white matter from the same region about 70 per cent., while in the peripheral nerves, such as the sciatic, it may fall to 60 per cent. The solids consist of proteids (in the grey matter they form half the total solids), neurokeratin, nuclein, protagon, lecithin, cerebrosides, cholesterin, nitrogenous extractives, and salts with some gelatin and fat from the adherent connective tissue.

The nervous structures are divided, as before mentioned, into two great systems—viz. the *cerebro-spinal*, comprising the brain and spinal cord, the nerves connected with these structures, and the ganglia situated on them ; and the *sympathetic*, consisting of a double chain of ganglia and the nerves connected with them. All these structures require separate consideration ; they are composed of the two kinds of nervous tissue above described, intermingled in various proportions, and having, in some parts, a very intricate arrangement.

The **brain** or **encephalon** is that part of the cerebro-spinal system which is contained in the cavity of the skull. It is divided into several parts, which will be described subsequently. In these parts the grey or vesicular nervous matter is found on the surface of the brain, forming the convolutions of the cerebrum, and the laminæ of the cerebellum. Again, grey matter is found in the interior of the brain, collected into large and distinct masses or ganglionic bodies, such as the corpus striatum, optic thalamus, and corpora quadrigemina. Finally, grey matter is found intermingled intimately with the white, but without definite arrangement, as in the grey matter in the pons Varolii and the floor of the fourth ventricle.

The white matter of the brain is divisible into three distinct classes of fibres: (1) Diverging or peduncular fibres, which connect the hemispheres with the medulla oblongata and the spinal cord. (2) Commissural fibres, which connect together the two hemispheres. (3) Association fibres, which connect different parts of the same hemisphere.

The manner in which these fibres are intermingled with each other and with the grey matter in the brain and spinal cord is very intricate, and can only be fully understood by a careful study of the details of its descriptive anatomy. The further consideration of this subject will therefore be deferred until after the description of the various divisions of which the cerebro-spinal system is made up.

FIG. 59.—Transverse section through a microscopic nerve, representing a compound nerve-bundle, surrounded by perineurium. Magnified 120 diameters.

The medullated fibres are seen as circles with a central dot, viz. medullary sheath and axis cylinder, in transverse section. They are embedded in endoneurium, containing numerous nuclei, which belong to the connective-tissue cells of the latter. (Klein and Noble Smith.)

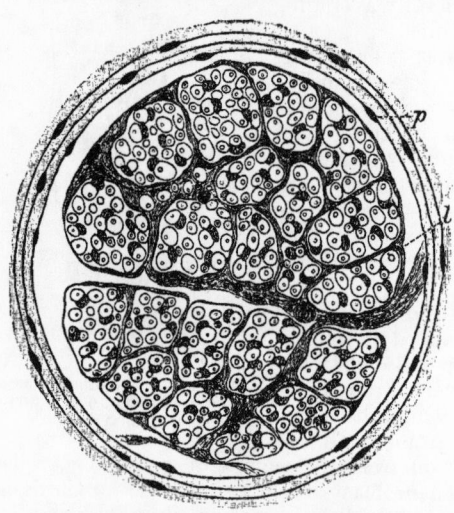

Perineurium, consisting of laminæ of fibrous connective-tissue, alternating with flattened nucleated connective-tissue cells. *l*. Lymph-space between perineurium and surface of nerve-bundle.

The **nerves** are round or flattened cords, formed of the nerve-fibres already described. They are connected at one end with the cerebro-spinal centre or with the ganglia, and are distributed at the other end to the various textures of the body; they are subdivided into two great classes—the *cerebro-spinal* nerves, which proceed from the cerebro-spinal axis, and the *sympathetic* or *ganglionic* nerves, which proceed from the ganglia of the sympathetic.

The **cerebro-spinal nerves** consist of numerous nerve-fibres collected together and enclosed in a membranous sheath (fig. 59). A small bundle of primitive fibres, enclosed in a tubular sheath, is called a *funiculus*; if the nerve is of small size, it may consist only of a single funiculus; but if large, the funiculi are collected together into larger bundles or fasciculi, which are bound together in a common membranous investment, and constitute the nerve.

In structure, the common membranous investment, or sheath of the whole nerve, which is called the *epineurium*, as well as the septa given off from it, and which separate the fasciculi, consists of connective tissue, composed of white and yellow elastic fibres, the latter existing in great abundance. The tubular sheath of the funiculi, called the *perineurium*, is a fine, smooth, transparent membrane, which may be easily separated, in the form of a tube, from the fibres it encloses; in structure it is made up of connective tissue, which has a distinctly lamellar arrangement, consisting of several lamellæ, separated from each other by spaces containing lymph. The nerve-fibres are held together and supported within the funiculus by delicate connective tissue, called the *endoneurium*. It is continuous with septa which pass inwards from the innermost layer of the perineurium, and shows a ground substance in which are embedded fine bundles of fibrous connective tissue which run for the most part longitudinally.

It serves to support capillary vessels, which are arranged so as to form a net-work with elongated meshes. The cerebro-spinal nerves consist almost exclusively of the medullated nerve-fibres, the non-medullated existing in very small proportions.

The blood-vessels supplying a nerve terminate in a minute capillary plexus, the vessels composing which pierce the perineurium, and run, for the most part, parallel with the fibres; they are connected together by short, transverse vessels, forming narrow, oblong meshes, similar to the capillary system of muscle. Fine non-medullated nerve-fibres accompany these capillary vessels, *vaso-motor fibres,* and break up into elementary fibrils, which form a network around the vessel. Horsley has also demonstrated certain medullated fibres as running in the epineurium and terminating in small spheroidal tactile corpuscles or end-bulbs of Krause. These nerve-fibres, which Marshall believes to be sensory, and which he has termed *nervi nervorum,* are considered by him to have an important bearing upon certain neuralgic pains.

The nerve-fibres, as far as is at present known, do not coalesce, but pursue an uninterrupted course from the centre to the periphery. In separating a nerve, however, into its component funiculi, it may be seen that these do not pursue a perfectly insulated course, but occasionally join at a very acute angle with other funiculi proceeding in the same direction; from this, branches are given off, to join again in like manner with other funiculi. It must be distinctly understood, however, that in these communications the nerve-fibres do not coalesce, but merely pass into the sheath of the adjacent nerve, become intermixed with its nerve-fibres, and again pass on, to become blended with the nerve-fibres in some adjoining funiculus.

Nerves, in their course, subdivide into branches, and these frequently communicate with branches of a neighbouring nerve. The communications which thus take place form what is called a *plexus.* Sometimes a plexus is formed by the primary branches of the trunks of the nerves—as the cervical, brachial, lumbar, and sacral plexuses—and occasionally by the terminal funiculi, as in the plexuses formed at the periphery of the body. In the formation of a plexus, the component nerves divide, then join, and again subdivide in such a complex manner that the individual funiculi become interlaced most intricately; so that each branch leaving a plexus may contain filaments from each of the primary nervous trunks which form it. In the formation also of smaller plexuses at the periphery of the body there is a free interchange of the funiculi and primitive fibres. In each case, however, the individual filaments remain separate and distinct, and do not inosculate with one another.

It is probable that through this interchange of fibres the different branches passing off from a plexus have a more extensive connection with the spinal cord than if they each had proceeded to be distributed without such connection with other nerves. Consequently the parts supplied by these nerves have more extended relations with the nervous centres; by this means, also, groups of muscles may be associated for combined action.

The **sympathetic** nerves are constructed in the same manner as the cerebro-spinal nerves, but consist mainly of non-medullated fibres, collected in funiculi, and enclosed in a sheath of connective tissue. There is, however, in these nerves, a certain admixture of medullated fibres, and the amount varies in different nerves, and may be known by their colour. Those branches of the sympathetic which present a well-marked grey colour are composed more especially of gelatinous nerve-fibres, intermixed with a few medullated fibres; while those of a white colour contain more of the latter fibres, and a few of the former. Occasionally, the grey and white cords run together in a single nerve, without any intermixture, as in the branches of communication between the sympathetic ganglia and the spinal nerves, or in the communicating cords between the ganglia.

The nerve-fibres, both of the cerebro-spinal and sympathetic system, convey impressions of a twofold kind. The *sensory* nerves, called also *centripetal* or *afferent* nerves, transmit to the nervous centres impressions made upon the peripheral extremities of the nerves, and in this way the mind, through the medium of the brain, becomes conscious of external objects. The *motor* nerves, called also *centrifugal* or *efferent* nerves, transmit impressions from the nervous centres to the parts to which the nerves are distributed, these impressions either exciting muscular contraction, or influencing the processes of nutrition, growth, and secretion.

Origin and Termination of Nerves.—By the expression ' the termination of nerve-fibres ' is signified their connections with the nerve-centres, and with the parts they supply. The former are sometimes called their *origin*, or *central* termination; the latter their *peripheral* termination. The origin in some cases is single—that is to say, the whole nerve emerges from the nervous centre by a single root; in other instances the nerve arises by two or more roots, which come off from different parts of the nerve-centre, sometimes widely apart from each other, and it often happens, when a nerve arises in this way by two roots, that the functions of these two roots are different; as, for example, in the spinal nerves, each of which arises by two roots, the anterior of which is motor, and the posterior sensory. The point where the nerve root or roots emerge from the nervous centre is named the *superficial* or *apparent* origin, but the fibres of which the nerve consists can be traced for a certain distance into the nervous centre to some portion of the grey substance, which constitutes the *deep* or *real* origin of the nerve. The manner in which these fibres arise at their deep origin varies with their functions. The centrifugal or efferent nerve-fibres originate in the nerve-cells of the grey substance, the axis-cylinder processes of these cells being prolonged to form the fibres. In the case of the centripetal or afferent nerves the fibres grow inwards either from nerve-cells in the organs of special sense (e.g. the retina) or from nerve-cells in the ganglia. Having entered the nerve-centre they branch and send their ultimate twigs among the cells, without, however, uniting with them.

Peripheral Terminations of Nerves.—Nerve-fibres terminate peripherally in various ways, and these may be conveniently studied in the sensory and motor nerves respectively. **Sensory** nerves would appear to terminate either in *minute primitive fibrillæ* or networks of these; or else in special terminal organs, which have been termed *peripheral end-organs*, and of which there are several principal varieties, viz. the end-bulbs of Krause, the tactile corpuscles of Wagner, the Pacinian corpuscles, and the neuro-tendinous and neuro-muscular spindles.

Termination in Fibrillæ.—When a medullated nerve-fibre approaches its termination, the white matter of Schwann suddenly disappears, leaving only the axis cylinder, surrounded by the neurilemma, and forming a non-medullated fibre. This, after a time, loses its neurilemma, and consists only of an axis cylinder, which can be seen, in preparations stained with chloride of gold, to be made up of fine varicose fibrils. Finally, the axis cylinder breaks up into its constituent primitive nerve-fibrillæ, which often present regular varicosities and anastomose with one another, thus forming a network. This network passes between the elements of the tissue to which the nerves are distributed, which is always epithelial, the nerve-fibrils lying in the interstitial substance between the epithelial cells, and there terminating, though some observers maintain that the actual terminations are within the cells. In this way nerve-fibres have been found to terminate in the epithelium of the skin and mucous membranes, and in the anterior epithelium of the cornea.

FIG. 60.—End-bulb of Krause.

a. Medullated nerve-fibre.   *b.* Capsule of corpuscle.   (From Klein's 'Elements of Histology.')

The **end-bulbs of Krause** (fig. 60) are minute cylindrical or oval bodies, consisting of a capsule formed by the expansion of the connective-tissue sheath of a medullated fibre, and containing a soft semi-fluid core in which the termination of the axis cylinder is situated, ending either as a bulbous extremity, or in a coiled-up plexiform mass. End-bulbs are found in the conjunctiva of the eye, where they are spheroidal in shape in man, but cylindrical in most other animals, in the mucous membrane of the lips and tongue, and in the epineurium of nerve-trunks. They are also found in the genital organs of both sexes, the penis in the male and the clitoris in the female. In this situation they have a mulberry-like appearance, from being constricted by connective-tissue septa into from two to six knob-like masses, and have received the name of *genital corpuscles*. Very similar corpuscles are found in the epineurium of nerve-trunks. In the synovial membrane of certain joints (e.g. those of the fingers), rounded or oval end-bulbs have been found ; these are designated *articular end-bulbs*.

Tactile corpuscles have been described by Grandry as occurring in the papillæ of the beak and tongue of birds, and by Merkel as occurring in the papillæ and epithelium of the skin of man and animals, especially in those parts of the skin devoid of hair. They consist of a capsule composed of a very delicate, nucleated membrane, and contain two or more granular, somewhat flattened cells, between which the medullated nerve-fibre, which enters the capsule by piercing its investing membrane, is supposed to terminate.

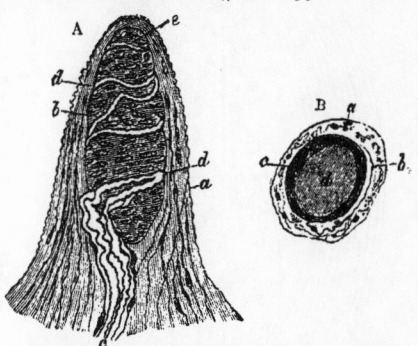

FIG. 61.—Papilla of the hand, treated with acetic acid. Magnified 350 times.

The **tactile corpuscles** (fig. 61), described by Wagner and Meissner, are oval-shaped bodies, made up of connective tissue, and consisting of a capsule, and imperfect membranous septa, derived from it, which penetrate its interior. The axis cylinder of the medullated fibres passes through the capsule, and having entered the corpuscle terminates in a small globular or pyriform enlargement, near the inner surface of the capsule. These tactile corpuscles have been described as occurring in the

A. Side view of a papilla of the hand. *a.* Cortical layer. *b.* Tactile corpuscle, with transverse nuclei. *c.* Small nerve of the papilla, with neurilemma. *d.* Its two nervous fibres running with spiral coils around the tactile corpuscle. *e.* Apparent termination of one of these fibres. B. A tactile papilla seen from above so as to show its transverse section. *a.* Cortical layer. *b.* Nerve-fibre. *c.* Outer layer of the tactile body, with nuclei. *d.* Clear interior substance.

papillæ of the corium of the hand and foot, the front of the forearm, the skin of the lips, and the mucous membrane of the tip of the tongue, the palpebral conjunctiva, and the skin of the nipple. They are not found in all the papillæ ; but from their existence in those parts in which the skin is highly sensitive, it is probable that they are specially concerned in the sense of touch, though their absence from the papillæ of other tactile parts shows that they are not essential to this sense.

Ruffini has described a special variety of nerve-ending in the subcutaneous tissue of the human finger (fig. 62). These are usually known as *Ruffini's endings*.

FIG. 62.—Nerve-ending of Ruffini.
(After A. Ruffini, ' Arch. ital. de Biol. Turin,' t. xxi. 1894.)

They are principally situated at the junction of the corium with the subcutaneous tissue ; they are oval in shape, and consist of a strong connective-tissue sheath, inside which the nerve-fibre divides into numerous branches, which show varicosities and end in small free knobs. They resemble the organs of Golgi.

The **Pacinian corpuscles** * (fig. 63) are found in the human subject chiefly on the nerves of the palm of the hand and sole of the foot and in the genital organs

* Often called in German anatomical works ' corpuscles of Vater.'

of both sexes lying in the subcutaneous tissue ; but they have also been described as connected with the nerves of the joints, and in some other situations, as the mesentery of the cat and along the tibia of the rabbit. Each of these corpuscles is attached to and encloses the termination of a single nerve-fibre. The corpuscle, which is perfectly visible to the naked eye (and which can be most easily demonstrated in the mesentery of a cat), consists of a number of lamellæ or capsules arranged more or less concentrically around a central clear space, in which the nerve-fibre is contained. Each lamella is composed of bundles of fine connective-tissue fibres, and is lined on its inner surface by a single layer of flattened epithelioid cells. The central clear space, which is elongated or cylindrical in shape, is filled with a transparent material, in the middle of which is the single medullated fibre, which traverses the space to near its distal extremity. Here it terminates in a rounded knob or end, sometimes bifurcating previously, in which case each branch has a similar arrangement. Todd and Bowman have described minute arteries as entering by the sides of the nerves and forming capillary loops in the intercapsular spaces, and even penetrating into the central space. Other authors describe the artery as entering the corpuscle at the pole opposite to the nerve-fibre.

FIG. 63.—Pacinian corpuscle, with its system of capsules and central cavity.

a. Arterial twig, ending in capillaries, which form loops in some of the intercapsular spaces, and one penetrates to the central capsule. *b.* The fibrous tissue of the stalk prolonged from the neurilemma. *n.* Nerve-tube advancing to the central capsule, there losing its white matter, and stretching along the axis to the opposite end, where it is fixed by a tubercular enlargement.

Herbst has described a somewhat similar ' nerve-ending ' to the Pacinian corpuscle, as being found in the mucous membrane of the tongue of the duck, and in some other situations. It differs, however, from the Pacinian corpuscles, in being smaller, its capsule thinner and more closely approximated, and especially in the fact that the axis cylinder in the central clear space is coated with a continuous row of nuclei. These bodies are known as the *corpuscles of Herbst.*

Neuro-tendinous spindles.—The nerves supplying tendons have a special modification of the terminal fibres, especially numerous at the point where the

FIG. 64.—Organ of Golgi (neuro-tendinous spindle) from the human tendo Achillis. (After Ciaccio.)

Nerve-fibre

Tendon bundles

Organ of Golgi, showing ramification of nerve-fibrils

Muscular fibres

tendon is becoming muscular. The tendon bundles become enlarged, and the nerve-fibres—one, two, or more in number—penetrate between the fasciculi of the tendon and spread out between the fibres to end in irregular discs or varicosities. A spindle-shaped body is thus formed, composed of tendon bundles and nerve-fibres, which is known as the *organ of Golgi* (fig. 64).

Neuro-muscular spindles.—In the majority of voluntary muscles there have been found special end-organs consisting of a small bundle of peculiar muscular fibres (intrafusal fibres), embryonic in type, invested by a capsule, within which nerve-fibres, experimentally shown to be sensory in origin, terminate. These neuro-muscular spindles vary in length from $\frac{1}{30}$ to $\frac{1}{6}$ of an inch and have a distinctly fusiform appearance. The large medullated nerve-fibres passing to the end-organ are from one to three or four in number; entering the fibrous capsule, they divide several times, and, losing their medulla, ultimately end in naked axis cylinders encircling the intrafusal fibres by flattened expansions, or irregular ovoid or rounded discs (fig. 65). Neuro-muscular spindles have not yet been demonstrated in the tongue or eye muscles.

In the organs of **special sense** the nerves appear to terminate in cells, which belong to the epithelial class, and have received the name of *sensory* or *nerve-epithelium* cells. This is not, however, the real state of the case; the nerve-fibre is in reality a process from the epithelial cell, and terminates by branching around a ganglion-cell. The stimulus carried by it is continued onwards by an axis cylinder, derived from the ganglion, to the brain. These nerve-

FIG. 65.—Middle third of a terminal plaque in the muscle spindle of an adult cat. (After Ruffini.)

epithelium cells must therefore be regarded as modified forms of nerve-cells. They will be more particularly described in the chapter on the organs of special sense.

Motor nerves are to be traced either into unstriped or striped muscular fibres. In the **unstriped** or **involuntary** muscles the nerves are derived from the sympathetic, and are composed mainly of the non-medullated fibres. Near their termination they divide into a number of branches, which communicate and form an intimate plexus. At the junction of the branches small triangular nuclear bodies (ganglion-cells) are situated. From these plexuses minute branches are given off, which divide and break up into the ultimate fibrillæ of which the nerve is composed. These fibrillæ course between the involuntary muscle-cells, and, according to Elischer, terminate on the surface of the cell, opposite the nucleus, in a minute swelling. Arnold and Frankenhäuser believed that these ultimate fibrillæ penetrated the muscular cell, and ended in the nucleus. More recent observation has, however, tended to disprove this.

In the **striped** or **voluntary** muscle, the nerves supplying the muscular fibres are derived from the cerebro-spinal nerves, and are composed mainly of medullated fibres. The nerve, after entering the sheath of the muscle, breaks up into fibres, or bundles of fibres, which form plexuses, and gradually divide until, as a rule, a single nerve-fibre enters a single muscular fibre. Sometimes, however, if the muscular fibre is long, more than one nerve-fibre enters it. Within the muscular fibre the nerve terminates in a special expansion, called by Kühne,

E

who first accurately described them, *motorial end-plates* (fig. 66).* The nerve-fibre, on approaching the muscular fibre, suddenly loses its white matter of Schwann, which abruptly terminates; the neurilemma becomes continuous with

FIG. 66.—Muscular fibres of *Lacerta viridis* with the terminations of nerves.

a. Seen in profile. P P. The nerve end-plates. S S. The base of the plate, consisting of a granular mass with nuclei. *b.* The same as seen in looking at a perfectly fresh fibre, the nervous ends being probably still excitable. (The forms of the variously divided plate can hardly be represented in a woodcut by sufficiently delicate and pale contours to reproduce correctly what is seen in nature.) *c.* The same as seen two hours after death from poisoning by curare.

the sarcolemma of the muscle, and only the axis cylinder enters the muscular fibre, where it at once spreads out, ramifying like the roots of a tree, imme-diately beneath the sarcolemma, and

FIG. 67.—Section through a microscopic ganglion. Magnified 300 diameters. (Klein and Noble Smith.)

c. Capsule of the ganglion. *n.* Nerve-fibres passing out of the ganglion. The nerve-fibres which entered the ganglion are not represented. The nerve-fibres are ordinary medullated fibres, but the details of their structure are not shown, owing to the low magnifying power. The ganglion-cells are invested by a special capsule, lined by a few nuclei, which are here repre-sented as if contained *in* the capsule.

is embedded in a layer of granular matter, containing a number of clear, oblong nuclei, the whole constituting an end-plate from which the con-tractile wave of the muscular fibre is said to start.

The **Ganglia** may be regarded as separate small aggregations of nerve-cells, connected with each other, with the cerebro-spinal axis, and with the nerves in various situations. They are found on the posterior root of each of the spinal nerves; on the posterior or sensory root of the fifth cranial nerve; on the facial and auditory nerves; on the glosso-pharyngeal and pneumo-gastric nerves. They are also found in a connected series along each side of the vertebral column, forming the trunk of the sympathetic; and on the branches of that nerve, generally in the plexuses or at the point of junction of two or more nerves with each other, or with branches of the cerebro-spinal system. On section they are seen to consist of a reddish-grey substance, traversed by nume-rous white nerve-fibres; they vary considerably in form and size; the largest are found in the cavity of the abdomen; the smallest, not visible to the naked eye, exist in considerable numbers upon the nerves distributed to the different viscera. The ganglia are invested by a smooth and firm, closely

* They had, however, previously been noticed, though not accurately described, by Doyère, who named them 'nerve-hillocks.'

adhering, membranous envelope, consisting of dense areolar tissue; this sheath is continuous with the perineurium of the nerves, and sends numerous processes into the interior of the ganglion, which support the blood-vessels supplying its substance.

In structure all ganglia are essentially similar (fig. 67), consisting of the same structural elements—viz. a collection of nerve-cells and nerve-fibres. Each nerve-cell has a nucleated sheath, which is continuous with the sheath of the nerve-fibre with which the cell is connected. The nerve-cells in the ganglia of the spinal nerves are pyriform in shape, and have only one process, the axis cylinder or axon. A short distance from the cell and while still within the ganglion this process divides in a T-shaped manner, one limb of the cross-bar passing centrally and forming the central portion of a sensory nerve-fibre; the other limb passing peripherally to form the axis-cylinder process of the peripheral nerve-fibre. In the sympathetic ganglia the nerve-cells are multipolar and have one axis-cylinder process or axon and several protoplasmic processes or dendrons. The former of these emerges from the ganglion as a non-medullated nerve-fibre. Similar cells are found in the ganglia connected with the fifth cranial nerve, and these ganglia are therefore regarded by some as the cranial portions of the sympathetic system. The nerve-cells are disposed in the ganglia in groups of varying size, and these groups are separated from each other by bundles of nerve-fibres, some of which traverse the ganglia without being connected with the cells.

THE VASCULAR SYSTEM

The **Vascular System**, exclusive of its central organ, the heart, is divided into four classes of vessels: the arteries, capillaries, veins, and lymphatics. The minute structure of these vessels will be briefly described here, the reader being referred to the body of the work for the details of their ordinary anatomy.

Structure of Arteries (fig. 68).—The arteries are composed of three coats: internal or endothelial coat (*tunica intima* of Kölliker); middle muscular coat (*tunica media*); and external connective-tissue coat (*tunica adventitia*).

FIG. 68.—Transverse section through a small artery and vein of the mucous membrane of the epiglottis of a child. Magnified about 350 diameters. (Klein and Noble Smith.)

A. Artery, showing the nucleated endothelium, *e*, which lines it: the vessel being contracted, the endothelial cells appear very thick. Underneath the endothelium is the wavy elastic intima. The chief part of the wall of the vessel is occupied by the circular muscle-coat *m*: the staff-shaped nuclei of the muscle-cells are well seen. Outside this is *a*, part of the adventitia. This is composed of bundles of connective-tissue fibres, shown in section, with the nuclei of the connective-tissue corpuscles. The adventitia gradually merges into the surrounding connective tissue. V. Vein showing a thin endothelial membrane, *e*, raised accidentally from the intima, which on account of its delicacy is seen as a mere line on the media *m*. This latter is composed of a few circular unstriped muscle-cells. *a*. The adventitia, similar in structure to that of an artery.

The two inner coats together are very easily separated from the external, as by the ordinary operation of tying a ligature on an artery. If a fine string be tied forcibly upon an artery and then taken off, the external coat will be found undivided, but the internal coats are divided in the track of the ligature and can easily be further dissected from the outer coat. The *inner coat* can be separated from the middle by a little maceration, or it may be stripped off in small pieces; but, on account of its friability, it cannot be separated as a complete membrane. It is a fine, transparent, colourless structure which is highly elastic, and is commonly corrugated into longitudinal wrinkles. The inner coat consists of: (1) A layer of pavement-endothelium, the cells of which are polygonal, oval, or fusiform, and have very distinct round or oval nuclei. This endothelium is brought into view most distinctly by staining with nitrate of silver. (2) A sub-endothelial layer, consisting of delicate connective tissue with branched cells

E 2

lying in the interspaces of the tissue. In arteries of less than a line in diameter the subendothelial layer consists of a single stratum of stellate cells, and the connective tissue is only largely developed in vessels of a considerable size. (3) An elastic or fenestrated layer, which consists of a membrane containing a network of elastic fibres, having principally a longitudinal direction, and in which, under the microscope, small elongated apertures or perforations may be seen, giving it a fenestrated appearance. It was therefore called by Henle the *fenestrated membrane*. This membrane forms the chief thickness of the inner coat, and can be separated into several layers, some of which present the appearance of a network of longitudinal elastic fibres, and others present a more membranous character, marked by pale lines having a longitudinal direction. The fenestrated membrane in microscopic arteries is a very thin layer; but in the larger arteries, and especially in the aorta, it has a very considerable thickness.

The *middle coat* (*tunica media*) is distinguished from the inner by its colour and by the transverse arrangement of its fibres, in contradistinction to the longitudinal direction of those of the inner coat. In the smaller arteries it consists principally of muscular tissue, being made up of plain muscle-fibres in fine bundles, arranged in lamellæ and disposed circularly around the vessel. These lamellæ vary in number according to the size of the vessel; the very small arteries having only a single layer, and those not larger than one-tenth of a line in diameter three or four layers. It is to this coat that the great thickness of the walls of the artery is mainly due (fig. 68, A, *m*). In the larger vessels, as the iliac, femoral, and carotid, elastic fibres unite to form lamellæ, which alternate with the layers of muscular fibre and are united by elastic fibres, which pass between the muscular bundles, and are connected

FIG. 69.—Longitudinal section of artery and vein.

a. An artery from the mesentery of a child, ·062‴, and *b*, vein ·067‴ in diameter, treated with acetic acid and magnified 350 times. *a*. Tunica adventitia, with elongated nuclei. β. Nuclei of the contractile fibre-cells of the tunica media, seen partly from the surface, partly apparent in transverse section. γ. Nuclei of the endothelial cells. δ. Elastic longitudinal fibrous coat.

with the fenestrated membrane of the inner coat (fig. 70). In the largest arteries, as the aorta and innominate, the amount of elastic tissue is very considerable. In these vessels also bundles of white connective tissue have been found in small quantities in the middle coat. The muscle-fibre cells of which the middle coat is made up are about $\frac{1}{500}$ of an inch in length and contain well-marked, rod-shaped nuclei, which are often slightly curved.

The *external coat* (*tunica adventitia*) consists mainly of fine and closely felted bundles of white connective tissue, but also contains elastic fibres in all but the smallest arteries. The elastic tissue is much more abundant next the tunica media, and it is sometimes described as forming here, between the adventitia and media, a special layer, the *tunica elastica externa* of Henle. This layer is most marked in arteries of medium size. In the largest vessels the external coat is relatively thin; but in small arteries it is of greater proportionate thickness. In the smaller arteries it consists of a single layer of white connective tissue and elastic fibres; while in the smallest arteries, just above the capillaries, the elastic fibres are wanting, and the connective tissue, of which the coat is composed, becomes more homogeneous the nearer it approaches the capillaries, and is gradually reduced to a thin membranous envelope, which finally disappears.

Some arteries have extremely thin coats in proportion to their size; this is especially the case in those situated in the cavity of the cranium and spinal canal, the difference depending on the greater thinness of the external and middle coats.

The arteries, in their distribution throughout the body, are included in a thin fibro-areolar investment, which forms what is called their *sheath*. In the limbs this is usually formed by a prolongation of the deep fascia; in the upper part of the thigh it consists of a continuation downwards of the transversalis and iliac fasciæ of the abdomen; in the neck, of a prolongation of the deep cervical fascia. The included vessel is loosely connected with its sheath by delicate areolar tissue; and the sheath usually encloses the accompanying veins, and sometimes a nerve. Some arteries, as those in the cranium, are not included in sheaths.

All the larger arteries are supplied with blood-vessels like the other organs of the body; they are called the *vasa vasorum*. These nutrient vessels arise from a branch of the artery or from a neighbouring vessel, at some considerable distance from the point at which they are distributed; they ramify in the loose areolar tissue connecting the artery with its sheath, and are distributed to the external coat, but do not, in man, penetrate the other coats; though in some of the larger mammals some few vessels have been traced into the middle coat. Minute veins serve to return the blood from these vessels; they empty themselves into the vein or veins accompanying the artery. Lymphatic vessels and lymphatic spaces are also present in the outer coat.

Arteries are also supplied with nerves, which are derived chiefly from the sympathetic, but partly from the cerebrospinal system. They form intricate plexuses upon the

FIG. 70.—Section of a medium-sized artery.
(After Grünstein.)

Endothelial and subendothelial layer of inner coat
Elastic layer
Innermost layers of middle coat

Outermost layers of middle coat
Innermost part of outer coat

Outermost part of outer coat

surfaces of the larger trunks, and run along the smaller arteries as single filaments or bundles of filaments, which twist around the vessel and unite with each other in a plexiform manner. The branches derived from these plexuses penetrate the external coat and are principally distributed to the muscular tissue of the middle coat, and thus regulate, by causing the contraction and relaxation of this tissue, the amount of blood sent to any part.

The Capillaries.—The smaller arterial branches (excepting those of the cavernous structure of the sexual organs, of the spleen, and in the uterine placenta) terminate in a network of vessels which pervade nearly every tissue of the body. These vessels, from their minute size, are termed capillaries (*capillus*, a 'hair'). They are interposed between the smallest branches of the arteries and the commencing veins, constituting a network, the branches of which maintain the same diameter throughout; the meshes of the network being more uniform in shape and size than those formed by the anastomoses of the small arteries and veins.

The *diameter* of the capillaries varies in the different tissues of the body, their usual size being about $\frac{1}{3000}$ of an inch. The smallest are those of the brain, and the mucous membrane of the intestines ; and the largest those of the skin and the marrow of bone, where they are stated to be as large as $\frac{1}{1200}$ of an inch. The *form* of the capillary net varies in the different tissues, the meshes being generally rounded or elongated.

The *rounded form of mesh* is most common, and prevails where there is a dense network, as in the lungs, in most glands and mucous membranes, and in the cutis ; here the meshes are more or less angular, sometimes nearly quadrangular, or polygonal, or more often irregular and not of an absolutely circular outline.

Elongated meshes are observed in the muscles and nerves, the meshes being usually of a parallelogram form, the long axis of the mesh running parallel with the long axis of the nerve and fibre. Sometimes the capillaries have a *looped* arrangement ;· a single vessel projecting from the common network and

FIG. 71.—Capillaries from the mesentery of a guinea-pig after treatment with solution of nitrate of silver.

FIG. 72.—Finest vessels on the arterial side. From the human brain. Magnified 300 times.

a. Cells. b. Their nuclei.

returning after forming one or more loops, as in the papillæ of the tongue and skin.

The number of the capillaries, and the size

1. Smallest artery. 2. Transition vessel. 3. Coarser capillaries. 4. Finer capillaries. a. Structureless membrane still with some nuclei, representative of the tunica adventitia. b. Nuclei of the muscular fibre-cells. c. Nuclei within the small artery, perhaps appertaining to an endothelium. c̓. Nuclei in the transition vessels.

of the meshes, determine the degree of vascularity of a part. The closest network and the smallest interspaces are found in the lungs and in the choroid coat of the eye. In these situations the interspaces are smaller than the capillary vessels themselves. In the kidney, in the conjunctiva, and in the cutis the interspaces are from three to four times as large as the capillaries which form them ; and in the brain from eight to ten times as large as the capillaries in their long diameter, and from four to six times as large in their transverse diameter. In the adventitia of arteries the width of the meshes is ten times that of the capillary vessels. As a general rule, the more active the function of the organ, the closer is its capillary net and the larger its supply of blood ; the meshes of the network are very narrow in all growing parts, in the glands and in the mucous membranes ; wider in bones and ligaments, which are comparatively inactive ; and nearly altogether absent in tendons, in which very little organic change occurs after their formation.

Structure.—The walls of the capillaries consist of a fine transparent endothelial layer, composed of cells joined edge to edge by an interstitial cement substance,

and continuous with the endothelial cells which line the arteries and veins.
When stained with nitrate of silver the edges which bound the epithelial cells are
brought into view (fig. 71). These cells are of large size and of an irregular
polygonal or lanceolate shape, each containing an oval nucleus which may be
brought into view by carmine or hæmatoxylin. Between their edges, at various
points of their meeting, roundish dark spots are sometimes seen, which have
been described as stomata, though they are closed by intercellular substance.
They have been believed to be the situation through which the white corpuscles
of the blood, when migrating from the blood-vessels, emerge; but this view,
though probable, is not universally accepted.

Kolossow, a Russian observer, describes these cells as having a rather more
complex structure. He states that they consist of two parts : of hyaline ground-
plates, and of a protoplasmic granular part, in which is embedded the nucleus, on
the outside of the ground-plates. The hyaline internal coat of the capillaries
does not form a complete membrane, but consists of 'plates' which are inelastic,

FIG. 73.—Section of a medium-sized vein.

Endothelium

Elastic layer

Middle coat

Outer coat

and though in contact with each other, are not continuous; when therefore the
capillaries are subjected to intravascular pressure, the plates become separated
from each other; the protoplasmic portions of the cells, on the other hand, are
united together.

In many situations a delicate sheath or envelope of branched nucleated
connective-tissue cells is found around the simple capillary tube, particularly in
the larger ones ; and in other places, especially in the glands, the capillaries are
invested with retiform connective tissue.

In the largest capillaries (which ought, perhaps, to be described rather as the
smallest arteries or pre-capillaries) there is, outside the epithelial layer, a muscular
layer, consisting of contractile fibre-cells, arranged tranversely, as in the tunica
media of the arteries (fig. 72).

The veins, like the arteries, are composed of three coats—internal, middle, and
external ; and these coats are, with the necessary modifications, analogous to the
coats of the arteries ; the internal being the endothelial, the middle the muscular,
and the external the connective or areolar (fig. 73). The main difference between

the veins and the arteries is in the comparative weakness of the middle coat of the former, and to this is due the fact that the veins do not stand open when divided, as the arteries do, and that they are passive rather than active organs of the circulation.

In the veins immediately above the capillaries the three coats are hardly to be distinguished. The endothelium is supported on an outer membrane separable into two layers, the outer of which is the thicker, and consists of a delicate, nucleated membrane (adventitia), while the inner is composed of a network of longitudinal elastic fibres (media). In the veins next above these in size (one-fifth of a line according to Kölliker) a connective-tissue layer containing numerous muscle-fibres circularly disposed can be traced, forming the middle coat, while the elastic and connective-tissue elements of the outer coat become more distinctly perceptible. In the middle-sized veins the typical structure of these vessels becomes clear. The endothelium is of the same character as in the arteries, but its cells are more oval and less fusiform. It is supported by a connective-tissue layer, consisting of a delicate network of branched cells, and external to this is a layer of elastic fibres disposed in the form of a network in place of the definite fenestrated membrane seen in arteries. This constitutes the *internal* coat. The *middle* coat is composed of a thick layer of connective tissue with elastic fibres, intermixed, in some veins, with a transverse layer of muscular tissue. The white fibrous element is in considerable excess, and the elastic fibres are in much smaller proportion in the veins than in the arteries. The *outer* coat consists of areolar tissue, as in the arteries, with longitudinal elastic fibres. In the largest veins the outer coat is from two to five times thicker than the middle coat, and contains a large number of longitudinal muscular fibres. This is most distinct in the inferior vena cava, especially at the termination of this vein in the heart, in the trunks of the hepatic veins, in all the large trunks of the vena portæ, in the splenic, superior mesenteric, external iliac, renal and azygos veins. In the renal and portal veins it extends through the whole thickness of the outer coat, but in the other veins mentioned a layer of connective and elastic tissue is found external to the muscular fibres. All the large veins which open into the heart are covered for a short distance with a layer of striped muscular tissue continued on to them from the heart. Muscular tissue is wanting—(1) in the veins of the maternal part of the placenta; (2) in the venous sinuses of the dura mater and the veins of the pia mater of the brain and spinal cord; (3) in the veins of the retina; (4) in the veins of the cancellous tissue of bones; (5) in the venous spaces of the corpora cavernosa. The veins of the above-mentioned parts consist of an internal endothelial lining supported on one or more layers of areolar tissue.

Most veins are provided with valves, which serve to prevent the reflux of the blood. They are formed by a reduplication of the inner coat, strengthened by connective tissue and elastic fibres, and are covered on both surfaces with endothelium, the arrangement of which differs on the two surfaces. On the surface of the valve next the wall of the vein, the cells are arranged transversely; while on the other surface, over which the current of blood flows, the cells are arranged vertically in the direction of the current. The valves are semilunar. They are attached by their convex edge to the wall of the vein; the concave margin is free, directed in the course of the venous current, and lies in close apposition with the wall of the vein as long as the current of blood takes its natural course; if, however, any regurgitation takes place, the valves become distended, their opposed edges are brought into contact, and the current is interrupted. Most commonly two such valves are found placed opposite one another, more especially in the smaller veins or in the larger trunks at the point where they are joined by smaller branches; occasionally there are three and sometimes only one. The wall of the vein on the cardiac side of the point of attachment of each segment of the valve is expanded into a pouch or sinus, which gives to the vessel, when injected or distended with blood, a knotted appearance. The valves are very numerous in the veins of the extremities, especially of the lower extremities; these vessels having to conduct the blood against the force of gravity. They are absent in the very small veins, i.e. those less than $\frac{1}{12}$ of an inch in diameter, also in the venæ cavæ, the hepatic veins, portal vein and most of its branches, the renal, uterine, and ovarian veins. A few valves are found in the spermatic veins, and one also at their point of junction with the renal vein and inferior vena cava respectively. The cerebral and spinal veins, the veins of the cancellated tissue of bone, the pulmonary veins, and the umbilical vein and its branches, are also destitute of valves. They are occasionally found, few in number, in the venæ azygos and intercostal veins.

The veins are supplied with nutrient vessels, *vasa vasorum*, like the arteries. Nerves also are distributed to them in the same manner as to the arteries, but in much less abundance.

The **lymphatic** vessels, including in this term the lacteal vessels, which are identical in structure with them, are composed of three coats. The *internal* is an endothelial and elastic coat. It is thin, transparent, slightly elastic, and ruptures sooner than the other coats. It is composed of a layer of elongated endothelial cells with serrated margins, by which the adjacent cells are dovetailed into one another. These are supported on an elastic membrane. The *middle* coat is composed of smooth muscular and fine elastic fibres, disposed in a transverse direction. The *external* coat consists of connective tissue, intermixed with smooth muscular fibres, longitudinally or obliquely disposed. It forms a protective covering to the other coats, and serves to connect the vessel with the neighbouring structures. The above description applies only to the larger lymphatics ; in the smaller vessels there is no muscular or elastic coat, and their structure consists only of a connective-tissue coat, lined by endothelium. The thoracic duct is a somewhat more complex structure than the other lymphatics ; it presents a distinct subendothelial layer of branched corpuscles, similar to that found in the arteries, and in the middle coat is a layer of connective tissue with its fibres arranged longitudinally. The lymphatics are supplied by nutrient vessels, which are distributed to their outer and middle coats ; and here also have been traced many non-medullated nerve-fibres in the form of a fine plexus of fibrils.

The lymphatics are very generally provided with valves, which assist materially in effecting the circulation of the fluid they contain. These valves are formed of a thin layer of fibrous tissue, lined on both surfaces by endothelium, which presents the same arrangement upon the two surfaces as was described in connection with the valves of veins. Their form is semilunar ; they are attached by their convex edge to the sides of the vessel, the concave edge being free and directed along the course of the contained current. Usually two such valves, of equal size, are found opposite one another ; but occasionally exceptions occur, especially at or near the anastomoses of lymphatic vessels. Thus, one valve may be of very rudimentary size and the other increased in proportion.

The valves in the lymphatic vessels are placed at much shorter intervals than in the veins. They are most numerous near the lymphatic glands, and they are found more frequently in the lymphatics of the neck and upper extremity than in the lower. The wall of the lymphatics immediately above the point of attachment of each segment of a valve is expanded into a pouch or sinus, which gives to these vessels, when distended, the knotted or beaded appearance which they present. Valves are wanting in the vessels composing the plexiform network in which the lymphatics usually originate on the surface of the body.

Origin of Lymphatics.—The finest visible lymphatic vessels (lymphatic capillaries) form a plexiform network in the tissues and organs, and they consist of a single layer of endothelial plates, with more or less sinuous margins. These vessels commence in an intercommunicating system of clefts or spaces, which have no complete endothelial lining, in the connective tissue of the different organs. They have been named the *rootlets* of the lymphatics, and are identical with the spaces in which the connective-tissue corpuscles are contained. This then is properly regarded as one method of their commencement, where the lymphatic vessels are apparently continuous with spaces in the connective tissue, and Klein has described and figured a direct communication between these spaces and the lymphatic vessel.* But the lymphatics have also other modes of origin, for the intestinal lacteals commence by closed extremities, though some observers believe that the closed extremity is continuous with a minute network contained in the substance of the villus, through which the lacteal is connected with the endothelial cells covering it. Again, it seems now to be conclusively proved that the serous membranes present stomata or openings between the endothelial cells

* *Atlas of Histology*, pl. viii. fig. xiv.

(fig. 74), by which there is an open communication with the lymphatic system and through which the lymph is thought to be pumped by the alternate dilatation and contraction of the serous surface, due to the movements of respiration and circulation, so that the serous and synovial sacs may be regarded, in a certain sense, as large lymph cavities or sinuses.* Von Recklinghausen was the first to observe the passage of milk and other coloured fluids through these stomata on the peritoneal surface of the central tendon of the diaphragm. Again, in most glandular structures the lymphatic capillaries have a lacunar origin. Here they begin in irregular clefts or spaces in the tissue of the part ; occupying the penetrating connective tissue and surrounding the lacunæ or tubules of the gland, and in many places separating the capillary network from the alveolus or tubule, so that the interchange between the blood and the secreting cells of the part must be carried on through these lymph-spaces or lacunæ. Closely allied to this is the mode of origin of lymphatics in perivascular and perineural spaces. Sometimes a minute artery may be seen to be ensheathed for a certain distance by a lymphatic capillary vessel, which is often many times wider than a blood capillary. These are known as perivascular lymphatics.

FIG. 74.—Stomata of serous membranes.

1. Endothelium from the under surface of the *centrum tendineum* of the rabbit. *a*. Stomata. 2. Endothelium of the mediastinum of the dog. *a*. Stomata. 3. Section through the pleura of the same animal. *b*. Free orifices of short lateral passages of the lymph-canals. (Copied from Ludwig, Schweigger-Seidel and Dybkowsky.)

Terminations of Lymphatics.—The lymphatics, including the lacteals, discharge their contents into the veins at two points : namely, at the angles of junction of the subclavian and internal jugular veins : on the left side by means of the thoracic duct, and on the right side by the right lymphatic duct. (See description of lymphatics.)

Lymphatic Glands (*lymph glands*) are small oval or bean-shaped bodies, situated in the course of lymphatic and lacteal vessels so that the lymph and chyle pass through them on their way to the blood. They generally present on one side a slight depression—the *hilum*—through which the blood-vessels enter and leave the interior. The efferent lymphatic vessel also emerges from the gland at this spot, while the afferent vessels enter the organ at different parts of the periphery. On section (fig. 75), a lymphatic gland displays two different structures : an external, of lighter colour—the *cortical* ; and an internal, darker—the *medullary*. The cortical structure does not form a complete investment, but is deficient at the hilum, where the medullary portion reaches the surface of the gland ; so that the efferent vessel is derived directly from the medullary structure, while the afferent vessels empty themselves into the cortical substance.

Lymphatic glands consist of (1) a fibrous envelope, or *capsule*, from which a framework of processes (*trabeculæ*) proceeds inwards, imperfectly dividing the gland into open spaces freely communicating with each other ; (2) a quantity of lymphoid tissue occupying these spaces without completely filling them ; (3) a free supply of blood-vessels, which are supported on the trabeculæ ; and (4) the *afferent* and *efferent* vessels. The nerves passing into the hilum are few in number and are chiefly distributed to the blood-vessels supplying the gland.

The *capsule* is composed of connective tissue with some plain muscle-fibres, and from its internal surface are given off a number of membranous processes or trabeculæ, consisting, in man, of connective tissue, with a small admixture of plain muscle-fibres ; but in many of the lower animals composed almost entirely of involuntary muscle. They pass inwards, radiating towards the centre of the

---

* The resemblance between lymph and serum led Hewson long ago to regard the serous cavities as sacs into which the lymphatics open. Recent microscopic discoveries confirm this opinion in a very interesting manner.

gland, for a certain distance—that is to say, for about one-third or one-fourth of the space between the circumference and the centre of the gland. In some animals

FIG. 75.—Section of human lymphatic gland.

a. Capsule. *b.* Blood-vessels. *c.* Cortical portion of gland. *g.* Germ-centres. *h.* Hilum. *l.* Efferent lymph-vessels. *m.* Medullary portion of gland. *n.* Nodules. *t.* Trabeculæ.

they are sufficiently well marked as to apparently divide the peripheral or cortical portion of the gland into a number of compartments (so-called follicles), but in

FIG. 76.—Follicle from a lymphatic gland of the dog, in vertical section.

FIG. 77.—From the medullary substance of an inguinal gland of the ox. (After His.)

a. Lymph-tube, with its complicated system of vessels. *b.* Retinacula stretched between the tube and the septa. *c.* Portion of another lymph-tube. *d.* Septa.

a. Reticular sustentacular substance of the more external portion, *b,* of the more internal, and *c,* of the most external and most finely webbed part on the surface of the follicle. *d.* Origin of a large lymph-tube. *e.* Of a smaller one. *f.* Capsule. *g.* Septa. *h.* Vas afferens. *i.* Investing space of the follicle, with its retinacula. *k.* One of the divisions of the septa. *l.* Attachment of the lymph-tubes to the septa.

man this arrangement is not apparent. The larger trabeculæ springing from the capsule break up into finer bands, and these interlace to form a meshwork in the central or medullary portion of the gland. In these spaces formed by the interlacing trabeculæ (fig. 76) is contained the proper gland-substance or lymphoid tissue. The gland-pulp does not, however, completely fill the spaces, but leaves, between its outer margin and the enclosing trabeculæ, a channel or space of uniform width throughout. This is termed the *lymph-path* or *lymph-sinus* (fig. 78). Running across it are a number of finer trabeculæ of retiform connective tissue, the fibres of which are, for the most part, covered by ramified cells.

On account of the peculiar arrangement of the framework of the organ, the gland-pulp in the cortical portion is disposed in the form of nodules, and in the medullary part in the form of rounded cords. It consists of ordinary lymphoid tissue, being made up of a delicate reticulum of retiform tissue, which is continuous with that in the lymph-paths, but marked off from it by a closer reticulation; in its meshes are closely packed lymph-corpuscles, traversed by a dense plexus of capillary blood-vessels. It is probable, moreover, that the reticular tissue of the gland-pulp and the lymph-paths is continuous with that of the trabeculæ, and ultimately with that of the capsule of the gland. The nodules or follicles in the cortical portion of the gland frequently show, in their centres, areas where karyokinetic figures indicate the division of the lymph-corpuscles. These areas are termed *germ-centres* (fig. 75).

FIG. 78.—Section of lymphatic-gland tissue.

a. Trabeculæ. *b.* Small artery in substance of same. *c.* Lymph-paths.
d. Lymph-corpuscles. *e.* Capillary plexus.

The *afferent vessels*, as above stated, enter at all parts of the periphery of the gland, and after branching and forming a dense plexus in the substance of the capsule open into the lymph-sinuses of the cortical part. In doing this, they lose all their coats except their endothelial lining, which is continuous with a layer of similar cells lining the lymph-paths. In like manner the *efferent* vessel commences from the lymph-sinuses of the medullary portion. The stream of lymph carried to the gland by the afferent vessel thus passes through the plexus in the capsule to the lymph-paths of the cortical portion, where it is exposed to the action of the gland-pulp; flowing through these it enters the paths or sinuses of the medullary portion, and finally emerges from the hilum by means of the efferent vessel. The stream of lymph in its passage through the lymph-sinuses is much retarded by the presence of the reticulum. Hence morphological elements, either normal or morbid, are easily arrested and deposited in the sinuses. This is a matter of considerable importance in connection with the subject of poisoned wounds and the absorption of the poison by the lymphatic system, since by this means septic organisms carried along the lymphatic vessels may be arrested in the lymph-sinuses of the gland-tissue, and thus be prevented from entering the general circulation. Many lymph-corpuscles pass with the efferent lymph-stream to join the general blood-stream. The arteries of the gland enter at the hilum, and either pass at once to the gland-pulp, to break up into a capillary plexus, or else run along the trabeculæ, partly to supply them and partly running across the lymph-paths to assist in forming the capillary plexus of the gland-pulp. This plexus traverses the lymphoid tissue, but does not pass into the lymph-sinuses. From it the veins commence and emerge from the organ at the same place as that at which the artery enters.

THE SKIN AND ITS APPENDAGES

The **skin** (fig. 79) is the principal seat of the sense of touch, and may be regarded as a covering for the protection of the deeper tissues; it plays an important part in the regulation of the body temperature, and is also an excretory and absorbing organ. It consists principally of a layer of vascular connective tissue, named the *dermis*, *corium*, or *cutis vera*, and an external covering of epithelium, termed the *epidermis* or *cuticle*. On the surface of the former layer are the sensitive *papillæ*; and within, or embedded beneath it, are certain organs with special functions: namely, the *sweat-glands*, *hair-follicles*, and *sebaceous glands*.

FIG. 79.—A diagrammatic sectional view of the skin (magnified).

Duct of sweat-gland

Tactile corpuscle

Duct of sweat-gland

Adipose tissue

Artery

Pacinian corpuscle

Stratum corneum

Stratum lucidum

Stratum granulosum

Stratum Malpighii

Stratum germinativum

Dermis

Sweat-gland

Nerve

The **epidermis** or **cuticle** (*scarf-skin*) is non-vascular, and consists of stratified epithelium (fig. 80). It is accurately moulded on the papillary layer of the dermis. It forms a defensive covering to the surface of the true skin, and limits the evaporation of watery vapour from its free surface. It varies in thickness in different parts. In some situations, as in the palms of the hands and soles of the feet, it is thick, hard, and horny in texture. This may be in a measure due to the fact that these parts are exposed to intermittent pressure, but that this is not the only cause is proved by the fact that the condition exists to a very considerable extent at birth. The more superficial layers of cells, called the *horny* layer (*stratum corneum*), may be separated by maceration from the deeper layers, which are called the *rete mucosum* or *stratum Malpighii*, and which consist of several layers of differently shaped cells. The free surface of the epidermis is marked by a network of linear furrows of variable size, marking out the surface into a number of spaces of polygonal or lozenge-shaped form. Some of these furrows are large, as opposite the flexures of the joints, and correspond to the folds in the dermis produced by their movements. In other situations, as upon the back of the hand, they are exceedingly fine, and intersect one another at various angles. Upon the palmar surface of the hand and fingers, and upon the sole of the foot, these lines are very distinct, and are disposed in curves ; they depend upon the large size and peculiar arrangement of the papillæ upon which the epidermis is placed. The deep surface of the epidermis is accurately moulded upon the papillary layer of the dermis, each papilla being invested by its epidermic sheath ; so that when this layer is removed by maceration, it presents on its under surface a number of pits or depressions corresponding to the elevations in the papillæ, as well as the ridges left in the intervals between them. Fine tubular prolongations are continued from this layer into the ducts of the sudoriferous and sebaceous glands.

FIG. 80.—Section of epidermis. (Ranvier.)

Stratum corneum

Stratum lucidum
Stratum granulosum

Stratum Malpighii

- - *Nerve-fibrils*

Stratum germinativum

In structure, the epidermis consists of several layers of epithelial cells, agglutinated together and having a laminated arrangement. These several layers may be described as composed of four different strata from within outwards : (1) the *stratum Malpighii*, composed of several layers of epithelial cells, of which the deepest layer is columnar in shape and placed perpendicularly on the surface of the corium, their lower ends being denticulate, to fit into corresponding denticulations of the true skin ; this deepest layer is sometimes termed the basilar layer or *stratum germinativum* ; the succeeding laminæ consist of cells of a more rounded or polyhedral form, the contents of which are soft, opaque, granular, and soluble in acetic acid. They are often marked on their surfaces with ridges and furrows, and are covered with numerous fibrils, which connect the surfaces of the cells : these are known as *prickle* cells (see page 13). They contain numerous epidermic fibrils, which are stained violet with hæmatoxylin and red by carmine, and form threads of union connecting adjacent cells. Between the cells are fine intercellular clefts which serve for the passage of lymph and in which lymph-corpuscles or pigment-granules may be found. (2) Immediately superficial to these are two or three layers of flattened, spindle-shaped cells, the *stratum granulosum*, which contain granules that become deeply stained in hæmatoxylin ; the granules consist of a material named *eleidin*, an intermediate substance in the formation of keratin. They are supposed to be cells in a transitional stage between the proto-plasmic cells of the stratum Malpighii and the horny cells of the superficial layers. (3) Above this layer, the cells become indistinct, and appear, in sections, to form a homogeneous or dimly striated membrane, composed of closely packed scales, in which traces of a flattened nucleus may be found. It is called the *stratum lucidum*. (4) As these cells successively approach the surface by the development of fresh layers from beneath, they assume a flattened form, from the evaporation of their fluid contents, and consist of many layers of horny epithelial scales in which no nucleus is discernible, forming the *stratum corneum*. These cells are unaffected by acetic acid, the protoplasm having become changed into horny material or *keratin*. According to Ranvier they contain granules of a material

FIG. 81.—Section of skin, showing the epidermis and dermis: a hair in its follicle: the arrector pili muscle : sebaceous glands.

which has the characters of beeswax. The deepest layer of the stratum Malpighii is separated from the papillæ by an apparently homogeneous basement-membrane, which is most distinctly brought into view in specimens prepared with chloride of gold. This, according to Klein, is merely the deepest portion of the epithelium, and is 'made up of the basis of the individual cells, which have undergone a chemical and morphological alteration.' The black colour of the skin in the negro, and the tawny colour among some of the white races, is due to the presence of pigment in the cells of the cuticle. This pigment is more especially distinct in the cells of the deeper layer, or stratum Malpighii, and is similar to that found in the cells of the pigmentary layer of the retina. As the cells approach the surface and desiccate, the colour becomes partially lost ; the disappearance of the pigment from the superficial layers of the epidermis is, however, difficult to explain.

The **dermis, corium,** or **cutis vera** is tough, flexible, and highly elastic, in order to defend the parts beneath from violence.

It varies in thickness, from a quarter of a line to a line and a half, in different parts of the body. Thus it is very thick in the palms of the hands and soles of the feet; thicker on the posterior aspect of the body than the front, and on the outer than the inner side of the limbs. In the eyelids, scrotum, and penis it is exceedingly thin and delicate. The skin generally is thicker in the male than in the female, and in the adult than in the child.

The corium consists of felted connective tissue, with a varying amount of elastic fibres and numerous blood-vessels, lymphatics, and nerves. The fibro-areolar tissue forms the framework of the cutis, and is differently arranged in different parts, so that it is usual to describe it as consisting of two layers: the deeper or *reticular* layer, and the superficial or *papillary* layer. Unstriped muscular fibres are found in the superficial layers of the corium, wherever hairs are present; and in the subcutaneous areolar tissue of the scrotum, penis, labia majora of the female, and the nipples. In the last situation the fibres are disposed in bands, closely reticulated and arranged in superimposed laminæ.

The *reticular* layer consists of strong interlacing fibrous bands, composed chiefly of the white variety of fibrous tissue, but containing, also, some fibres of the yellow elastic tissue, which vary in number in different parts; and connective-tissue corpuscles, which are often to be found flattened against the white fibrous tissue bundles. Towards the attached surface the fasciculi are large and coarse, and the areolæ which are left by their interlacement are large, and occupied by

Fig. 82.—Longitudinal section through human nail and its nail groove (sulcus).

adipose tissue and sweat-glands. Below this the elements of the skin become gradually blended with the subcutaneous areolar tissue, which, except in a few situations, contains fat. Towards the free surface the fasciculi are much finer, and their mode of interlacing close and intricate.

The *papillary layer* is situated upon the free surface of the *reticular layer*; it consists of numerous small, highly sensitive, and vascular eminences, the *papillæ*, which rise perpendicularly from its surface. The papillæ are conical-shaped eminences, having a round or blunted extremity, occasionally divided into two or more parts, and are received into corresponding pits on the under surface of the cuticle. Their average length is about $\frac{1}{100}$ of an inch, and they measure at their base $\frac{1}{250}$ of an inch in diameter. On the general surface of the body, more especially in those parts which are endowed with slight sensibility, they are few in number, short, exceedingly minute, and irregularly scattered over the surface; but in some situations, as upon the palmar surface of the hands and fingers, upon the plantar surface of the feet and toes, and around the nipple, they are long, of large size, closely-aggregated together, and arranged in parallel curved lines, forming the elevated ridges seen on the free surface of the epidermis. Each ridge contains two rows of papillæ, and between the two rows the ducts of the sweat-glands pass outwards to open on the summit of the ridges. In structure the papillæ consist of very small and closely interlacing bundles of finely fibrillated tissue, with a few elastic fibres; within this tissue is a capillary loop, and in some papillæ, especially in the palms of the hands and the fingers, there are tactile corpuscles.

The *arteries* supplying the skin form a network in the subcutaneous tissue, from which branches are given off to supply the sweat-glands, the hair-follicles, and the fat. Other branches are given off which constitute a plexus immediately beneath the corium; from this, fine capillary vessels pass into the papillæ, forming, in the smaller ones, a single capillary loop, but in the larger, a more or less convoluted vessel. There are numerous *lymphatics* supplied to the skin, which form two networks, superficial and deep, communicating with each other and with those of the subcutaneous tissue by oblique branches. They originate in the cell-spaces of the tissue.

The *nerves* of the skin terminate partly in the epidermis and partly in the cutis vera. The former are prolonged into the epidermis from a dense plexus in the superficial layer of the corium and terminate between the cells in bulbous extremities; or, according to some observers, in the deep epithelial cells themselves. The latter terminate in end-bulbs, touch-corpuscles, or Pacinian bodies, in the manner already described; and, in addition to these, a considerable number of fibrils are distributed to the hair-follicles, which are said to entwine the follicle in a circular manner. Other nerve-fibres are supplied to the plain muscular fibres of the hair-follicles (arrectores pili) and to the muscular coat of the blood-vessels. These are probably non-medullated fibres.

The **appendages of the skin** are the nails, the hairs, the sudoriferous and sebaceous glands, and their ducts.

The nails and hairs are peculiar modifications of the epidermis, consisting essentially of the same cellular structure as that tissue.

The **nails** (figs. 82 and 83) are flattened, elastic structures of a horny texture, placed upon the dorsal surface of the terminal phalanges of the fingers and toes.

FIG. 83.—Transverse section through human nail and its sulcus.

Each nail is convex on its outer surface, concave within, and is implanted by a portion, called the *root*, into a groove in the skin; the exposed portion is called the *body*, and the anterior extremity the *free edge*. The nail is firmly adherent to the dermis, being accurately moulded upon its surface, as the epidermis is in other parts. The part of the dermis beneath the body and root of the nail is called the *matrix*, because it is the part from which the nail is produced. Corresponding to the body of the nail, the matrix is thick, and raised into a series of longitudinal ridges which are very vascular, and the colour is seen through the transparent tissue. Behind this, near the root of the nail, there are papillæ which are small, less vascular, and have no regular arrangement, and here the tissue of the nail is somewhat more opaque; hence this portion is of a whiter colour, and is called the *lunula* on account of its shape.

The cuticle, as it passes forwards on the dorsal surface of the finger or toe, is attached to the surface of the nail, a little in advance of its root; at the extremity of the finger it is connected with the under surface of the nail a little behind its free edge. The cuticle and horny substance of the nail (both epidermic structures) are thus directly continuous with each other. The nails consist of a greatly thickened stratum lucidum, the stratum corneum forming merely the thin cuticular fold (*eponychium*) which overlaps the lunula. The cells have a laminated arrangement, and are essentially similar to those composing the epidermis. The cells of the deepest layer, which lie in contact with the papillæ of the matrix, are columnar in form and arranged perpendicularly to the surface; those which succeed them are of a rounded or polygonal form, the more superficial ones becoming broad, thin, and flattened, and so closely packed as to make the limits of each cell very

F

indistinct. It is by the successive growth of new cells at the root and under surface of the body of the nail that it advances forwards and maintains a due thickness, while, at the same time, the growth of the nail in the proper direction is secured. As these cells in their turn become displaced by the growth of new ones, they assume a flattened form, and finally become compacted together into a firm, dense, horny texture. In *chemical composition* the nails resemble the upper layers of the epidermis. According to Mulder, they contain a somewhat larger proportion of carbon and sulphur.

The **hairs** are peculiar modifications of the epidermis, and consist essentially of the same structure as that membrane. They are found on nearly every part of the surface of the body, excepting the palms of the hands, soles of the feet, and the glans penis. They vary much in length, thickness, and colour in different parts of the body and in different races of mankind. In some parts, as in the skin of the eyelids, they are so short as not to project beyond the follicles containing them; in others, as upon the scalp, they are of considerable length; again, in other parts, as the eyelashes, the hairs of the pubic region, and the whiskers and beard, they are remarkable for their thickness. Straight hairs are stronger than curly hairs, and present on transverse section a cylindrical or oval outline; curly hairs, on the other hand, are flattened.

Fig. 84.—Transverse section of hair-follicle.

Hyaline layer

Fibrous part
of hair

Medulla of
hair

Huxley's
layer

Henle's layer

Outer or
dermic coat

A hair consists of a *root*, the part implanted in the skin; the *shaft* or *stem*, the portion projecting from its surface; and the *point*.

The *root of the hair* presents at its extremity a bulbous enlargement, which is whiter in colour and softer in texture than the shaft, and is lodged in a follicular involution of the epidermis called the *hair - follicle* (fig. 81). When the hair is of considerable length the follicle extends into the subcutaneous cellular tissue. The hair-follicle commences on the surface of the skin with a funnel-shaped opening, and passes inwards in an oblique or curved direction—the latter in curly hairs—to become dilated at its deep extremity, where it corresponds with the bulbous condition of the hair which it contains. It has opening into it, near its free extremity, the orifices of the ducts of one or more sebaceous glands. At the bottom of each hair-follicle is a small conical, vascular eminence or papilla, similar in every respect to those found upon the surface of the skin; it is continuous with the dermic layer of the follicle, is highly vascular, and probably supplied with nervous fibrils. In structure the hair-follicle consists of two coats—an outer or *dermic*, and an inner or *epidermic*.

The *outer* or *dermic* coat is formed mainly of fibrous tissue; it is continuous with the corium, is highly vascular, and supplied by numerous minute nervous filaments. It consists of three layers (fig. 84). The most internal, next the cuticular lining of the follicle, consists of a hyaline basement-membrane having a glassy, transparent appearance, which is well marked in the larger hair-follicles, but is not very distinct in the follicles of minute hairs. It is continuous with the basement-membrane of the surface of the corium. External to this is a compact layer of fibres and spindle-shaped cells arranged circularly around the follicle. This layer extends from the bottom of the follicle as high as the entrance of the ducts of the sebaceous glands. Externally is a thick layer of connective tissue, arranged in longitudinal bundles, forming a more open texture and corresponding to the reticular part of the corium. In this are contained the blood-vessels and nerves.

The *inner* or *epidermic* layer is closely adherent to the root of the hair, so that when the hair is plucked from its follicle this layer most commonly adheres to it and forms what is called the *root-sheath*. It consists of two strata named respectively the *outer* and *inner root-sheaths*; the former of these corresponds with the Malpighian layer of the epidermis, and resembles it in the rounded form and soft character of its cells; at the bottom of the hair-follicle these cells become continuous with those of the root of the hair. The *inner* root-sheath consists of a delicate cuticle next the hair, composed of a thin layer of imbricated scales having a downward direction, so that they fit accurately over the upwardly directed imbricated scales of the hair itself; then of one or two layers of horny, flattened, nucleated cells, known as *Huxley's layer*; and finally of a single layer of horny oblong cells without visible nuclei, called *Henle's layer*.

The hair-follicle contains the root of the hair, which terminates in a bulbous extremity, and is excavated so as to exactly fit the papilla from which it grows. The bulb is composed of polyhedral epithelial cells, which as they pass upwards into the root of the hair become elongated and spindle-shaped, except some in the centre which remain polyhedral. Some of these latter cells contain pigment-granules, which give rise to the colour of the hair. It occasionally happens that these pigment-granules completely fill the cells in the centre of the bulb; this gives rise to the dark tract of pigment often found, of greater or less length, in the axis of the hair.

The *shaft of the hair* consists of a central pith or medulla, the fibrous part of the hair, and the cuticle externally. The *medulla* occupies the centre of the shaft and ceases towards the point of the hair. It is usually wanting in the fine hairs covering the surface of the body, and commonly in those of the head. It is more opaque and deeper coloured when viewed by transmitted light than the fibrous part; but when viewed by reflected light it is white. It is composed of rows of polyhedral cells, which contain granules of eleidin and frequently air-bubbles. The *fibrous* portion of the hair constitutes the chief part of the shaft; its cells are elongated and unite to form flattened fusiform fibres. Between the fibres are found minute spaces which contain either pigment-granules in dark hair, or minute air-bubbles in white hair. In addition to this there is also a diffused pigment contained in the fibres. The cells which form the *hair-cuticle* consist of a single layer which surrounds those of the fibrous part; they are converted into thin, flat scales, having an imbricated arrangement.

Connected with the hair-follicles are minute bundles of involuntary muscular fibres, termed the *arrectores pili*. They arise from the superficial layer of the corium, and are inserted into a thickened portion of the outer surface of the hair-follicle, below the entrance of the duct of the sebaceous gland. They are placed on the side towards which the hair slopes, and by their action elevate the hair (fig. 81).* The sebaceous gland is situated in the angle which the arrector muscle forms with the superficial portion of the hair-follicle, and contraction of the muscle thus tends to squeeze the sebaceous secretion out from the duct of the gland.

The **sebaceous glands** are small, sacculated, glandular organs, lodged in the substance of the corium. They are found in most parts of the skin, but are especially abundant in the scalp and face: they are also very numerous around the apertures of the anus, nose, mouth, and external ear, but are wanting in the palms of the hands and soles of the feet. Each gland consists of a single duct, more or less capacious, which terminates in a cluster of small secreting pouches or saccules. The sacculi connected with each duct vary, as a rule, in number from two to five, but, in some instances, may be as many as twenty. They are

---

* Arthur Thomson suggests that the contraction of these muscles on follicles which contain weak, flat hairs will tend to produce a permanent curve in the follicle, and this curve will be impressed on the hair which is moulded within it, so that the hair, on emerging through the skin, will be curled. Curved hair-follicles are characteristic of the scalp of the Bushman.

composed of a transparent, colourless membrane, enclosing a number of epithelial cells. Those of the outer or marginal layer are small and polyhedral, and are continuous with the lining cells of the duct. The remainder of the sac is filled with larger cells, containing fat, except in the centre, where the cells have become broken up, leaving a cavity filled with their débris and a mass of fatty matter, which constitutes the sebaceous secretion. The orifices of the ducts open most frequently into the hair-follicles, but occasionally upon the general surface, as in the labia minora and the free margin of the lips. On the nose and face the glands are of large size, distinctly lobulated, and often become much enlarged from the accumulation of pent-up secretion. The largest sebaceous glands are those found in the eyelids—the Meibomian glands.

The **sudoriferous** or **sweat glands** are the organs by which a large portion of the aqueous and gaseous materials are excreted by the skin. They are found in almost every part of this structure, and are situated in small pits on the under surface of the corium, or, more frequently, in the subcutaneous areolar tissue, surrounded by a quantity of adipose tissue. They are small, lobular, reddish bodies, consisting of a single convoluted tube, from which the efferent duct proceeds upwards through the corium and cuticle, becomes somewhat dilated at its extremity, and opens on the surface of the cuticle by an oblique valve-like aperture. The duct, as it passes through the epidermis, presents a spiral arrangement, being twisted like a corkscrew, in those parts where the epidermis is thick; where, however, it is thin, the spiral arrangement does not exist. In the superficial layers of the corium the duct is straight, but in the deeper layers it is convoluted or even twisted. The spiral course of these ducts is particularly distinct in the thick cuticle of the palm of the hand and sole of the foot. The size of the glands varies. They are especially large in those regions where the amount of perspiration is great, as in the axillæ, where they form a thin, mammillated layer of a reddish colour, which corresponds exactly to the situation of the hair in this region; they are large also in the groin. Their number varies. They are most numerous on the palm of

FIG. 85.—Coiled tube of a sweat-gland cut in various directions.

a. Longitudinal section of the proximal part of the coiled tube. *b.* Transverse section of the same. *c.* Longitudinal section of the distal part of the coiled tube. *d.* Transverse section of the same. (From Klein and Noble Smith's 'Atlas of Histology.')

the hand, presenting, according to Krause, 2,800 orifices on a square inch of the integument, and are rather less numerous on the sole of the foot. In both of these situations the orifices of the ducts are exceedingly regular, and open on the curved ridges. In other situations they are more irregularly scattered, but the number in a given extent of surface presents a fairly uniform average. In the neck and back they are least numerous, their number amounting to 417 on the square inch (Krause). Their total number is estimated by the same writer at 2,381,248, and, supposing the aperture of each gland to represent a surface of $\frac{1}{516}$ of a line in diameter, he calculates that the whole of these glands would present an evaporating surface of about eight square inches. Each gland consists of a single tube intricately convoluted, terminating at one end by a blind extremity, and opening at the other end upon the surface of the skin. In the larger glands this single duct usually divides and subdivides dichotomously; the smaller ducts ultimately terminating in short cæcal pouches, rarely anastomosing. The wall of the duct is thick, the width of the canal seldom exceeding one-third of its diameter. The tube, both in the gland and where it forms the excretory duct, consists of two layers—an outer, formed by fine areolar tissue, and an inner layer of epithelium (fig. 85). The external or fibro-cellular coat is thin, continuous with the superficial layer of the corium, and extends only as high as the surface of the true skin. The epithelial lining in the distal part of the coiled tube of the gland proper consists of a single layer of cubical epithelium, supported on a basement-membrane, and beneath it, between the epithelium and the fibro-cellular coat, a layer of longitudinally or obliquely arranged fibres, which are usually regarded as muscular, though the evidence that this is so is not con-clusive. In the duct and the proximal part of the coiled tube of the gland proper there are two or more layers of polyhedral cells, lined on their internal surface, i.e. next the lumen of the tube, by a delicate membrane or cuticle, and on their outer surface by a limiting membrana propria, but there are no muscular fibres. The epithelium is continuous with the epidermis and with the delicate internal cuticle of the epidermic portion of the tube. When the cuticle is care-fully removed from the surface of the cutis, these convoluted tubes of epithelium may be drawn out and form short, thread-like processes on its under surface.

The contents of the smaller sweat-glands are quite fluid; but in the larger glands the contents are semi-fluid and opaque, and contain a number of coloured granules and cells which appear analogous to epithelial cells.

SEROUS MEMBRANES

The **serous membranes** form shut sacs and may be regarded as lymph-sacs, from which lymphatic vessels arise by stomata or openings between the endo-thelial cells (see page 57). The sac consists of one portion which is applied to the walls of the cavity which it lines—the *parietal* portion; and another reflected over the surface of the organ or organs contained in the cavity—the *visceral* portion. Sometimes the sac is arranged quite simply, as the tunica vaginalis testis; at others with nume-rous involutions or recesses, as the peritoneum, in which, nevertheless, the membrane can always be traced con-tinuously around the whole circumference. The sac is completely closed, so that no communication exists between the serous cavity and the parts in its neighbourhood. An ap-parent exception exists in the peritoneum of the female; for the Fallopian tube opens freely into the peritoneal cavity in the dead subject so that a bristle can be passed from the one into the other. But this communication is closed during life, except at the moment of the passage of the ovum out of the ovary into the tube, as is proved by the fact that no interchange of fluids ever takes place between the two cavities in dropsy of the peritoneum, or in accumulation of fluid in the Fallopian tubes.* The serous membrane is sometimes supported by a firm, fibrous layer, as is the case with the pericardium, and such membranes may be spoken of as 'fibro-serous.'

FIG. 86.—Part of peritoneal surface of the central tendon of diaphragm of rabbit, prepared with nitrate of silver.

s. Stomata. *l.* Lymph-channels. *t.* Tendon-bundles. The stomata are surrounded by germinating epithelial cells. (From 'Handbook for the Physiological Laboratory.' Klein.)

The various serous membranes are the peritoneum, lining the cavity of the abdomen; the two pleuræ and the pericardium, covering the lungs and heart respectively; and the tunicæ vaginales, surrounding the testicles in the scrotum.†

* The communication between the uterine cavity and the peritoneal sac is not only apparent in the dead subject, but is an anatomical fact, which is established by the con-tinuity of its epithelium with that covering the uterus, Fallopian tubes, and fimbriæ.

† The arachnoid membrane covering the brain and spinal cord was formerly regarded as a serous membrane, but is now no longer classed with them, as it differs from them in structure, and does not form a shut sac as do the other serous membranes.

Serous membranes are thin, transparent, glistening structures, consisting of homogeneous basement-membranes lined on their inner surface by a single layer of polygonal or pavement endothelial cells, supported on a matrix of fibrous connective tissue, with networks of fine elastic fibres, in which are contained numerous capillaries and lymphatics.   On the surface of the endothelium between the cells numerous apertures or interruptions are to be seen.   Some of these are stomata, surrounded by a ring of cubical endothelium (see fig. 86), and communicate with a lymphatic capillary; others (*pseudostomata*) are mere interruptions in the endothelial layer, and are occupied by processes of the branched connective-tissue corpuscle of the subjacent tissue or by accumulations of the intercellular cement substance.

The amount of fluid contained in these closed sacs is, in most cases, only sufficient to moisten the surface, but not to furnish any appreciable quantity of fluid.   When a small quantity can be collected, it is found to resemble lymph, and like that fluid coagulates spontaneously; but when secreted in large quantities, as in dropsy, it is a more watery fluid, yet still contains a considerable amount of proteid which is coagulated on boiling.

FIG. 87.—Villus of synovial membrane.   (After Hammar.)

SYNOVIAL MEMBRANES

Synovial membranes, like serous membranes, are connective-tissue membranes placed between two movable tissues, so as to diminish friction, as in movable joints; or between a tendon and a bone, where the former glides over the latter; and between the skin and various subcutaneous bony prominences.

The synovial membranes are composed essentially of connective tissue, with the cells and fibres of that structure, containing numerous vessels and nerves.   It was formerly supposed that these membranes were analogous in structure to the serous membranes, and consisted of a layer of flattened cells on a basement-membrane.   No such continuous layer, however, exists, although here and there are patches of cells probably epithelial in nature. They are surrounded and held together by an albuminous ground-substance.   Long villus-like processes (fig. 87) are often found projecting from the surface of synovial membranes; they are covered by small rounded cells, and are supposed to extend the surface for the secretion of the fluid which moistens the membranes and which is named *synovia*.   It is a rich lymph, plus a mucin-like substance, and to the latter constituent it owes its viscidity.   A further description of the synovial membranes will be given with the anatomy of the joints.

MUCOUS MEMBRANES

Mucous membranes line all those passages by which the internal parts communicate with the exterior, and are continuous with the skin at the various orifices of the surface of the body.   They are soft and velvety, and very vascular, and their surface is coated over by their secretion, *mucus*, which is of a tenacious consistence, and serves to protect them from the foreign substances introduced into the body with which they are brought in contact.

They are described as lining the two tracts—the gastro-pulmonary and the genito-urinary; and all, or almost all, mucous membranes may be classed as belonging to and continuous with the one or the other of these tracts.

The deep surfaces of these membranes are attached to the parts which they line by means of connective tissue, which is sometimes very abundant, forming a loose and lax bed, so as to allow considerable movement of the opposed surfaces on each other. It is then termed the *submucous tissue*. At other times it is exceedingly scanty, and the membrane is closely connected to the tissue beneath; sometimes, for example, to muscle, as in the tongue; sometimes to cartilage, as in the larynx; and sometimes to bone, as in the nasal fossæ and sinuses of the skull.

In structure a mucous membrane is composed of *corium* and *epithelium*. The epithelium is of various forms, including the squamous, columnar, and ciliated, and is often arranged in several layers. This epithelial layer is supported by the corium, which is analogous to the dermis of the skin, and consists of connective tissue, either simply areolar, or containing a greater or less quantity of lymphoid tissue. This tissue is usually covered on its external surface by a transparent basement-membrane generally composed of clear flattened cells, placed edge to edge; on this the epithelium rests. It is only in some situations that the basement-membrane can be demonstrated. The corium is an exceedingly vascular membrane, containing a dense network of capillaries, which lie immediately beneath the epithelium and are derived from small arteries in the submucous tissue.

The fibro-vascular layer of the corium contains, besides the areolar tissue and vessels, unstriped muscle-cells, which form in many situations a definite layer called the *muscularis mucosæ*. These are situated in the deepest part of the membrane, and are plentifully supplied with nerves. Other nerves pass to the epithelium and terminate between the cells. Lymphatic vessels are found in great abundance, commencing either by blind extremities or in networks, and communicating with plexuses in the submucous tissue.

Embedded in the mucous membrane are found numerous glands, and projecting from it are processes (villi and papillæ) analogous to the papillæ of the skin. These glands and processes, however, exist only at certain parts, and it will be more convenient to defer their description so that they may be described as they occur.

SECRETING GLANDS

The **secreting glands** are organs whose cells produce, by the metabolism of their protoplasm, certain substances, called 'secretions,' of a more or less definite composition; the material for the secretion being primarily selected from the blood. The essential parts therefore of a secreting gland are *cells*, which have the power of extracting from the blood certain matters, and in some cases converting them into new chemical compounds; and *blood-vessels*, by which the blood is brought into close relationship with these cells. The general arrangement in all secreting structures—that is to say, not only in secreting glands, but also in secreting membranes—is that the cells are arranged on one surface of an extra-vascular basement-membrane, which supports them, and a minute plexus of capillary vessels ramifies on the other surface of the membrane. The cells then extract from the blood certain constituents which pass through the membrane into the cells, where they are prepared and elaborated. The basement-membrane does not, however, always exist, and any free surface would appear to answer the same purpose in some cases.

By the various modifications of this secreting surface the different glands are formed. This is generally effected by an invagination of the membrane in different ways, the object being to increase the extent of secreting surface within a given bulk.

In the simplest form a single invagination takes place, constituting a *simple* gland; this may be either in the form of an open tube (fig. 88, A), or the walls of the tube may be dilated so as to form a saccule (fig. 88, B). These are named the *simple tubular* or *saccular* glands. Or, instead of a short tube, the invagination may be lengthened to a considerable extent, and then coiled up to occupy less space. This constitutes the *simple convoluted tubular* gland, an example of which may be seen in the sweat-glands of the skin (fig. 88, C).

If, instead of a single invagination, secondary invaginations take place from the primary one, as in fig. 88, D and E, the gland is then termed a compound one. These secondary invaginations may assume either a saccular or tubular form, and so constitute the two subdivisions—the *compound saccular* or *racemose* gland, and the *compound tubular*. The racemose gland in its simplest form consists of a primary invagination which forms a sort of duct, upon the extremity of which are found a number of secondary invaginations, called saccules or alveoli, as in Brunner's glands (fig. 88, D). But, again, in other instances, the duct, instead of being simple, may divide into branches, and these again into other branches, and so on ; each ultimate ramification terminating in a dilated cluster of saccules, and thus we may have the secreting surface almost indefinitely extended, as in the salivary glands (fig. 88, E). In the compound tubular glands the division of the primary duct takes place in the same way as in the racemose glands, but the branches retain their tubular form, and do not terminate in saccular recesses, but become greatly lengthened out (fig. 88, F). The best example of this form of gland is to be found in the kidney. All these varieties of glands are produced by a more or less complicated invagination of a secreting membrane, and they are

FIG. 88.—Diagrammatic plan of the varieties of secreting glands.

A. Simple gland.   B. Sacculated simple gland.   C. Simple convoluted tubular gland.   D, E. Racemose gland.
F. Compound tubular gland.

all identical in structure : that is to say, the saccules or tubes, as the case may be, are lined with cells, generally spheroidal or columnar in figure, and on their outer surface is an intimate plexus of capillary vessels. The secretion, whatever it may be, is eliminated by the cells from the blood, and is poured into the saccule or tube, and so finds its way out through the primary invagination on to the free surface of the secreting membrane. In addition, however, to these glands, which are formed by an *invagination* of the secreting membrane, there are some few others which are formed by a *protrusion* of the same structure, as in the vascular fringes of synovial membranes. This form of secreting structure is not nearly so frequently met with.

There are also certain glands which are described as capable of *internal secretion*, wherein are no ducts leading to any free surface, the secretion being carried either directly into the blood stream, or indirectly through the medium of lymphatics. Such are the thyroid and suprarenal glands.

EMBRYOLOGY

THE whole body is developed out of the female germ-cell or **ovum** after it has been fertilised by the male germ-cell or **spermatozoon**. The ovum is a simple nucleated cell, and all the complicated changes by which the various tissues and organs of the body are formed from it, after fertilisation, are the result of two general processes, viz. the *segmentation* or *cleavage* of cells, and their *differentiation*. The former consists of the division of the nucleus and surrounding cell-substance, whereby the original cell is represented by two, a process which is again and again repeated. The differentiation of cells is a term used to describe that unknown power or tendency impressed upon cells, apparently identical in structure, whereby they assume different shapes and become associated, it may be, with the formation of widely differing tissues.

The earliest stages of the development of the human embryo have not yet been observed, and therefore many of the statements which are accepted in human embryology regarding these early stages are made on the strength of what has been observed to occur in the lower animals.

Having regard to the main purpose of this work, it is impossible, in the space available in this chapter, to describe fully, or illustrate adequately, all the phenomena which occur in the different stages of the development of the human body. The principal facts only will be given— the student being referred for further details to one or other of the text-books on embryology.

FIG. 89.—Human ovum from a middle-sized follicle. Magnified 350 times.

a. Zona pellucida or zona striata. *b*. Yolk or vitellus. *c*. Germinal vesicle and germinal spot.

STRUCTURE OF THE OVUM

The human ovum is extremely minute, measuring only from $\frac{1}{150}$ to $\frac{1}{125}$ of an inch in diameter, and is situated within a Graafian follicle of the ovary.* By the rupture of the Graafian follicle on to the surface of the ovary the ovum is liberated, and is then conveyed by the Fallopian tube or oviduct to the cavity of the uterus. Unless it be fertilised by the spermatozoon, it undergoes no further development, and is discharged from the uterus. If, on the other hand, fertilisation occurs, the fertilised ovum is retained within the uterus, and is developed into a new being.

In appearance and structure the ovum (figs. 89, 90) differs little from an ordinary cell, but distinctive names have been applied to its several parts ; thus, the body of the cell is known as the *vitellus* or *yolk*, the nucleus as the *germinal vesicle*, and the nucleolus as the *germinal spot*. The ovum is enclosed within a thick, transparent envelope, the *zona pellucida* or *zona striata*, adhering to the outer surface of which there are several layers of cells, derived from those of the Graafian follicle, which collectively constitute the *corona radiata*.

The **vitellus** or **yolk** comprises (1) the *cytoplasm* of the ordinary animal cell, with its network or cyto-reticulum filled with cell-sap or cytolymph. The

* See description of the ovary on a future page.

cytoplasm of the ovum is frequently termed the *formative yolk*, in order to distinguish it from (2) the *nutritive yolk* or *deutoplasm*, which consists of numerous rounded granules of fatty and albuminoid substances embedded in the cytoplasm. In the mammalian ovum the nutritive yolk is extremely small in amount, and is only of service in nourishing the ovum in the earliest stages of its development, whereas in the egg of the bird there is sufficient to supply the chick with nutriment throughout the whole period of incubation. The nutritive yolk not only varies in amount, but in its mode of distribution within the egg: thus, in some animals it is nearly uniformly distributed throughout the cytoplasm; in others it is centrally placed and is surrounded by the cytoplasm; in still others it is accumulated at the lower pole of the ovum, while the cytoplasm, being lighter, occupies the upper pole. An *attraction sphere* with its *centrosome* is present in the ova of the lower animals, and is probably represented in the ova of mammals by the body of Balbiani. This body is not visible during all the stages

FIG. 90.—Ovum of rabbit. Highly magnified. (After Waldeyer.)

Part of corona radiata

Germinal spot

Germinal vesicle

Zona pellucida or striata

Yolk granule

of the development of the ovum, and is 'most readily seen before the space appears in the Graafian follicle. Then the body in question lies in the immediate neighbourhood of the nucleus. It consists of a lighter central sphere enclosing one or two smaller spheres, and surrounded by a more darkly staining protoplasm.' *

The **germinal vesicle** or nucleus is a large spherical body which at first occupies a nearly central position, but becomes eccentric as the growth of the ovum proceeds. Its structure is that of an ordinary cell-nucleus, and consists of an intranuclear network of achromatin filled with nuclear sap or karyoplasm; while connected with, or embedded in, the achromatic reticulum are a number of chromatin masses or chromosomes, which may present the appearance of a skein or may assume the form of rods or loops. It is now very generally admitted that in every species of animal the nucleus contains a fixed and definite number of chromosomes, a number which varies from two in certain threadworms to as many as a hundred and sixty-eight in the crustacea; in man, as well as in the ox and guinea pig, the number is said to be sixteen. The nucleus is enclosed by a delicate nuclear membrane, and contains in its interior a well-defined nucleolus or germinal spot.

* Robinson, 'Hunterian Lectures on the Mammalian Ovum and Placenta.' *Journal of Anatomy and Physiology*, vol. xxxviii.

COVERINGS OF THE OVUM

The **zona pellucida** or **zona striata** (fig. 90) is probably secreted by the cells of the corona radiata. It consists of a thick, clear membrane, which, under the higher powers of the microscope, is seen to be perforated by numerous fine radially arranged channels. These give to it a striated appearance, and may suffice for the passage of nutritive materials to the ovum ; they may also provide an entrance for the spermatozoa at the time of fertilisation. In some animals (e.g. insects) the zona pellucida presents one, sometimes several small perforations or *micropyles*, by which the spermatozoa are believed to enter. The zona pellucida persists for some time after fertilisation has occurred, and may act as a source of protection during the earlier stages of segmentation.*

The **corona radiata** (fig. 90) consists of two or three strata of cells ; they are derived from the cells of the Graafian follicle, and adhere to the outer surface of the zona pellucida when the ovum is set free from the follicle. Its cells are radially arranged around the zona, those of the innermost layer being columnar in shape, and sending, according to some observers, delicate processes from their deep ends into the channels of the zona. The cells of the corona radiata soon disappear ; though in some animals they secrete, or are replaced by, a layer of adhesive albumen, which may not only assist in protecting and nourishing the ovum, but may help to fix it temporarily to the uterine mucous membrane.

The phenomena attending the discharge of the ova from the Graafian follicles belong more to the ordinary functions of the ovary than to the general subject of the development of the body, and are therefore described with the anatomy of the ovaries on a subsequent page.

MATURATION OF THE OVUM

The ovum is incapable of being fertilised by the spermatozoon until after it has undergone a process of maturation. This takes place either before or immediately after its escape from the Graafian follicle, and consists essentially of

Fig. 91.—Formation of polar bodies in asterias glacialis.
(Slightly modified from Hertwig.)

In fig. I the polar spindle (*sp*) has advanced to the surface of the egg. In fig. II a small elevation (*pb'*) is formed which receives half of the spindle. In fig. III the elevation is constricted off, forming the first polar body (*pb'*), and a second spindle is formed. In fig. IV is seen a second elevation which in fig. V has been constricted off as the second polar body (*pb²*). Out of the remainder of the spindle (*f.pn* in fig. VI) the female pronucleus is developed.

an unequal subdivision of the ovum into four cells, three of which are minute and are incapable of further development ; these are named the *polar bodies*, while the

* Robinson (*op. cit.*) suggests that ' it probably prevents the contact and union of the chorionic cells on the outer surface of the developing ovum with the decidua until the ovum has attained a sufficient size. It also probably serves to prevent contact of the embryonic ectoderm with the uterine wall until the differentiation of the embryonic ectoderm cells has taken place to such an extent that they are no longer disposed to fuse with the decidua.'

fourth and much larger cell constitutes the *mature ovum*. The phenomena of maturation (fig. 91) may be shortly described as follows : the germinal vesicle approaches the surface of the ovum, and, having undergone the usual mitotic changes, divides ; one-half of its contained chromatin, surrounded by a small amount of cytoplasm, is then separated from the main mass of the ovum to form the first polar body. The process is repeated, and a second polar body, containing one-half of the remaining chromatin, is separated, while the first, in many cases, becomes divided into two, thus forming three in all. The remainder of the germinal vesicle recedes towards the centre of the ovum, and is now known as the *female pronucleus* ; from what has been said, it will be seen that it contains only one-fourth of the chromatin originally present in the germinal vesicle.

The number of polar bodies varies in the ova of different animals ; in some only one is formed, in others two—the latter condition being probably explained by the fact that the first polar body has not undergone subdivision by the time the second is separated from the ovum. Coincident with the extrusion of the polar bodies a shrinking of the yolk or vitellus takes place, and a fluid, the *perivitelline fluid*, collects between it and the zona pellucida. In this the polar bodies are situated, and in it also spermatozoa may be subsequently seen (e.g. in the ovum of the rabbit).

THE SPERMATOZOON

The spermatozoa or male germ-cells are developed within the tubuli seminiferi of the testicles. They are present in enormous numbers in the seminal fluid, and consist of small but greatly modified cells. The human spermatozoon possesses a head, a neck, a connecting piece or body, and a tail (fig. 92).

The **head** is oval or elliptical, but flattened, so that when viewed in profile it is pear-shaped. Its anterior two-thirds are covered by a layer of modified protoplasm, which is named the *head-cap*. This, in some animals (e.g. the salamander), is prolonged into a barbed spearlike process or *perforator*, which probably facilitates the entrance of the spermatozoon into the ovum. Bardeleben and E. Nelson have described spearlike perforators as being present in the human spermatozoa, but other observers deny their presence. Waldeyer inclines to the view that in man the perforator consists of the anterior sharp margin of the head-cap, and acts as a cutting rather than a boring apparatus. The posterior part of the head exhibits an affinity for certain reagents, and presents a transversely striated appearance, being crossed by three or four dark bands. In some animals a central rodlike filament extends forwards for about two-thirds of the length of the head, while in others a rounded body is seen near its centre. The head contains a mass of chromatin, and is generally regarded as the nucleus of the cell surrounded by a thin envelope.

The **neck** immediately succeeds the head, and in some animals appears as a marked constriction, which, however, is not very clearly differentiated in the human spermatozoon. At the junction of the head and neck two or three rounded nodules, the *anterior centrosome bodies*, are seen ; behind these is a band of homogeneous substance.

The **connecting piece** or **body** is rodlike and about the same length as the head. At its anterior end are two or three rounded nodules, the *posterior centrosome bodies* : fine threads connect these with the anterior centrosome bodies. An axial filament runs through the connecting piece, and is continued backwards into the tail. It is probably a derivative of the centrosome bodies. Outside the axial filament is a layer of protoplasm which contains a spirally arranged fibril.

The **tail** is of great length, and contains an axial thread or filament which is composed of delicate fibrillæ. It is surrounded by a protoplasmic envelope, which may contain a spiral thread or may present a striated appearance. Further, in some animals there is attached to the connecting piece and tail a thin undulatory membrane, along the free edge of which there is a marginal filament. The terminal portion of the tail is named the *end-piece*, and consists of the axial filament only.

It is interesting to note that in the development of the spermatozoa a nuclear reduction occurs similar to that seen during the maturation of the ovum. In the testicle are certain cells identical with primitive ova. These are termed *spermatogonia*, and each divides into a couple of *spermatocytes*, which again undergo subdivision, giving rise to four *spermatids* or young spermatozoa. From this it will be seen that the spermatozoon contains only one-fourth of the chromatin which originally existed in the nucleus of the spermatogone. The matured ovum and the spermatozoon may therefore be looked upon as of the same morphological value. It must, however, be kept in mind that all four

FIG. 92.—Human spermatozoon. (Diagrammatic.)

A. Surface view. B. Profile view. In c the head, neck, and connecting piece are highly magnified to show the centrosome bodies and their uniting threads, the spirally arranged fibril, and the axial filament.

spermatozoa are functionally active, whereas the polar bodies given off during the maturation of the ovum are apparently of no functional value.

In virtue of their tails, which act as propellers, the spermatozoa, in the fresh condition, are capable of free movement, and if placed in favourable surroundings (e.g. in the female passages) may retain their vitality for several days.

FERTILISATION OF THE OVUM

Fertilisation consists in the union of the spermatozoon with the mature ovum (fig. 93). This usually takes place in the upper part of the Fallopian tube, and the ovum is then conveyed to the cavity of the uterus—a journey which probably occupies two or three days. Should the passage of the fertilised ovum along the tube be arrested, a tubal pregnancy is the result. Numerous spermatozoa may pierce the zona pellucida (e.g. in the rabbit as many as sixty have been seen in its interior), but only one, under normal conditions, enters the vitellus and takes part in the process of fertilisation. At the point where the spermatozoon is about to pierce the vitellus the latter is drawn out into a conical elevation, termed the *cone of attraction.* As soon as the spermatozoon has entered at this point, a membrane appears to be secreted over the aperture of its entrance to prevent the passage of additional spermatozoa.   Occasionally a second spermatozoon may enter the vitellus, thus giving rise to a condition of *polyspermy* : when this occurs the ovum usually develops in an abnormal manner and gives rise to a monstrosity. Having pierced the vitellus, the tail of the spermatozoon disappears, while its head assumes the form of a nucleus containing a cluster of chromosomes.

FIG. 93.—Fertilisation of the ovum of an echinoderm.

s. Spermatozoon. *m.pr.* Male pronucleus. *f.pr.* Female pronucleus. 1. Accession of a spermatozoon to the periphery of the vitellus. 2. Its penetration. 3. Transformation of the head of the spermatozoon into the male pronucleus. 4, 5. Blending of the male and female pronuclei. (From Quain's 'Anatomy,' Selenka.)

This constitutes the *male pronucleus,* and associated with it there is seen a centrosome and attraction sphere.   The male pronucleus passes more deeply into the vitellus, and coincident with this the granules of the surrounding cytoplasm become radially arranged.   The male and female pronuclei migrate towards each other, and, meeting near the centre of the vitellus, fuse to form a new nucleus, the *first segmentation nucleus,* which therefore contains both male and female nuclear substance; the former transmits the individualities of the father, the latter those of the mother, to the future embryo.

SEGMENTATION OF THE FERTILISED OVUM

After it has been fertilised the ovum undergoes repeated subdivision into a number of small cells (figs. 94, 95).   The first segmentation nucleus exhibits the usual mitotic changes, and these are succeeded by a division of the ovum into two cells of equal or unequal size.   Whether the two cells may differ in quality as well as in quantity is as yet a matter of some doubt.   (Consult Hertwig's 'Handbuch der vergleichenden und experimentellen Entwickelungslehre der Wirbeltiere.')   The process is repeated again and again, so that the two cells are succeeded by four, eight, sixteen, thirty-two, and so on, with the result that a mass of cells is found within the zona pellucida (which itself takes no share in the process, but ultimately disappears), and to this mass the term *morula* is applied. The segmentation of the mammalian ovum may not take place in the regular sequence of two, four, eight, &c., since one of the two first-formed cells may subdivide more rapidly than the other, giving rise to a three- or a five-cell stage. The cells of the morula are at first closely aggregated inside the zona pellucida ; but soon they become arranged into an outer or peripheral layer, the *primitive ectoderm* or *trophoblast,* which does not contribute to the formation of the embryo proper, and an *inner cell-mass,* from which the embryo is developed (fig. 96). Fluid collects between the trophoblast and the greater part of the inner cell-mass, and thus the morula is converted into a vesicle, the *blastodermic vesicle.*   The inner cell-mass remains in contact, however, with the trophoblast at one pole of the ovum ; this is named the *embryonic pole,* since it indicates the situation where the future embryo will be developed.   The cells of the trophoblast become differentiated into two strata : an outer, termed the *syncytium* or *plasmodioblast,* so named because it consists of a layer of protoplasm studded with nuclei, but

FIG. 94.—First stages of segmentation of a mammalian ovum : semi-diagrammatic.
(From a drawing by Allen Thomson.)

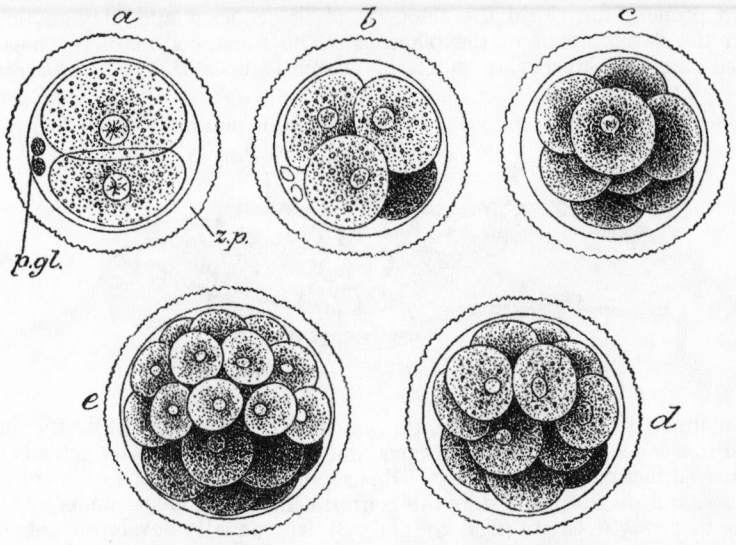

z.p. Zona pellucida.   *p.gl.* Polar bodies.   *a.* Two-cell stage.   *b.* Four-cell stage.   *c.* Eight-cell stage.
d, e. Morula stage.

FIG. 95.—Ovum of the rabbit at the end of the process of segmentation.

ep. Primitive ectoderm.   *hy.* Primitive entoderm.   *bp.* Place where the ectoderm has not yet grown over the entoderm.
(From Balfour, after Ed. van Beneden.)

FIG. 96.—Blastodermic vesicle of Vespertilio murinus.   (After Van Beneden.)

showing no evidence of subdivision into cells; and an inner layer of prismatic epithelium, which is named the *cytoblast* or *layer of Langhans*. As already stated, the cells of the trophoblast do not contribute to the formation of the embryo proper; they form the ectoderm of the chorion and play an important part in the development of the placenta. The inner cell-mass becomes differentiated into two layers: an outer, of prismatic cells, the *embryonic ectoderm*,

FIG. 97.—Section through embryonic area of Vespertilio murinus. (After Van Beneden.)

and an inner, of flattened cells, the *entoderm* (figs. 96 to 98); the latter is applied to the deep aspect of the former, and from there gradually extends around the inner surface of the trophoblast. Spaces appear between the cells of the inner cell-mass, and at the same time the central cells of the mass undergo atrophy, and in this way a cavity (figs. 97 and 98) is gradually developed between the

FIG. 98.—Section through embryonic area of Vespertilio murinus (after Van Beneden) to show the formation of the amniotic cavity.

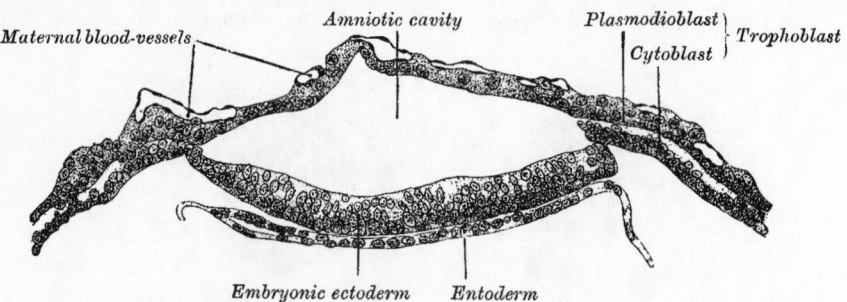

embryonic ectoderm and the overlying trophoblast: this is termed the *primitive amniotic cavity*; it persists in certain of the bats, and probably also in man and monkeys, and forms in them the permanent amniotic cavity. In other animals the trophoblastic roof of this primitive amniotic cavity disappears, and the embryonic ectoderm assumes a superficial position, its margin being continuous with the trophoblast. The formation of the amnion will be again referred to.

THE EMBRYONIC AREA

In reptiles, birds, and mammals, only a part of the ovum is utilised in the development of the embryo proper, the remainder being used up in the formation of membranes and other appendages which are concerned with its protection and nutrition; the blastodermic membrane therefore may be divided into *embryonic* and *extra-embryonic areas* (figs. 99, 100). If the ovum, at this stage, be viewed from the surface it will be seen to exhibit a centrally placed, circular, opaque area surrounded by a more transparent portion. The central opaque part is the embryonic area (area embryonalis), and is equal in extent with the embryonic ectoderm already referred to; the peripheral clearer portion is the extra-embryonic area. The circumference of the embryonic region ' remains as a relatively slow-growing area, while the embryonic and extra-embryonic portions of the wall of the ovum rapidly increase in extent. Under these circumstances, it follows that the margin of the embryonic area will soon appear

as a ring between the upper embryonic and the lower or extra-embryonic parts of the ovum, both of which have expanded beyond it in all directions' (Robinson).*
The circumference of the embryonic area corresponds with the future umbilicus.

The primitive streak; formation of the mesoderm.—The embryonic area becomes oval and then pear-shaped, the wider end being directed forwards. At the narrow, posterior end an opaque, crescentic or anchor-shaped patch makes its appearance, and gradually extends forwards as a dark streak along the middle line of the area for about one-half of its length; this is termed the *primitive streak* (figs. 99, 100), and indicates the commencing development of a third layer of cells, which constitutes the *mesoderm* or *mesoblast*. A shallow groove, the *primitive groove*, makes its appearance on the surface of the streak, and the anterior end of this groove communicates by means of an aperture, the *blastopore*, with that part of the blastodermic vesicle which ultimately forms the primitive alimentary canal. The cells of the axial part of the embryonic ectoderm multiply rapidly and, growing downwards, blend with those of the underlying entoderm; this thickening

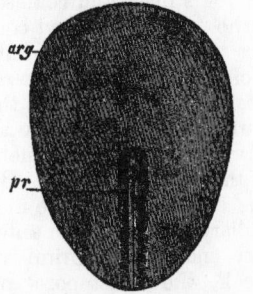

FIG. 99.—Embryo of a rabbit.
(After Kölliker.)

arg. Embryonic area. *pr.* Primitive
streak.

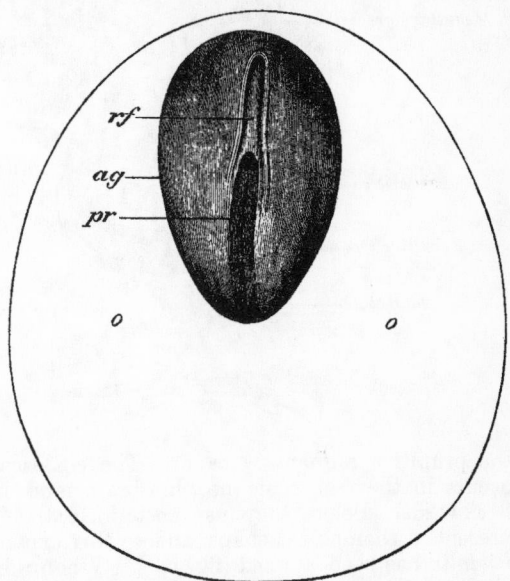

FIG. 100.—Embryonic area of the ovum of
rabbit at the seventh day.

ag. Embryonic area *o, o.* Region of the blastodermic vesicle
immediately surrounding the embryonic area. *pr.* Primitive
streak. *rf.* Medullary groove. (From Kölliker.)

of the ectoderm gives rise to a linear opacity, the primitive streak. From the sides and posterior extremity of the primitive streak a layer of cells, the mesoderm or mesoblast, extends outwards between the ectoderm and entoderm, and so converts the bilaminar into a trilaminar blastoderm. Although the mesoderm is mainly derived from the cells of the ectoderm, possibly those of the entoderm also contribute to it.

The extension of the mesoderm takes place throughout the whole of the embryonic and extra-embryonic areas of the ovum, except in certain regions. One of these is seen immediately in front of the neural tube. Here the mesoderm extends forward in the form of two crescentic masses, which curve inwards and meet in the middle line so as to enclose behind them an area which is devoid of mesoderm and is named the *oral plate* or *bucco-pharyngeal area*, since it afterwards forms the septum between the primitive mouth and primitive pharynx. In front of the bucco-pharyngeal area, where the lateral crescents of mesoderm have fused in the middle line, the pericardium is afterwards developed, and this region is therefore designated the *pericardial area*. A second region where the mesoderm is absent, at least for a time, is that immediately in front of the pericardial area. This is the region where the *pro-amnion* is developed; in man, however, a pro-amnion is apparently never formed.

* 'The Early Stages of the Development of the Pericardium,' by Professor Arthur Robinson. *Journal of Anatomy and Physiology*, vol. xxxvii.

THE NEURAL GROOVE AND TUBE

In front of the primitive streak two longitudinal ridges, caused by a looping or folding up of the ectoderm, make their appearance, one on either side of the middle line. These are named the *medullary folds* or *laminæ*; they commence some distance behind the anterior end of the embryonic area, where they are continuous with each other, and from there gradually extend backwards, one on either side of the primitive streak. Between these folds is seen a mesial, longitudinal furrow, the *neural* or *medullary groove* (figs. 100, 101), which gradually deepens as the medullary folds become elevated. Ultimately the folds meet and coalesce in the middle line and convert the neural groove into a closed tube, the *neural tube* or *canal* (fig. 104), the surrounding ectodermal wall of which forms the rudiment of the nervous system. By the coalescence of the medullary folds over the anterior end of the primitive streak, the blastopore no longer opens on the surface but into the closed canal of the neural tube, and thus a communication, the *neurenteric canal*, is established between the neural tube and

FIG. 101.—Human embryo—length 2 mm. Dorsal view, with the amnion laid open. × 30. (After Graf Spee; reconstruction.)

Yolk-sac

Amnion

Medullary groove

Neurenteric canal

Primitive streak

Body-stalk

the primitive alimentary canal. The coalescence of the medullary folds first occurs in the region of the hind-brain, and from there extends forwards and backwards. Before they fuse posteriorly, the hinder part of the neural groove presents a rhomboidal shape, and to this expanded portion the term *sinus rhomboidalis* has been applied (fig. 103). When the medullary folds are in the act of closing in to form the neural tube, a crest or ridge of cells is separated from their

FIG. 102.—Section across the anterior part of the medullary groove.
(By Schäfer. Modified from Quain's 'Anatomy.')

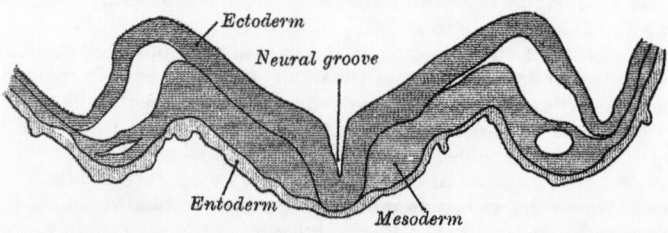

Ectoderm

Neural groove

Entoderm

Mesoderm

edges. This is termed the *neural crest* or *ganglion ridge* (fig. 138), and is at first situated between the dorsal aspect of the neural tube and the overlying ectoderm: it forms the rudiment of the ganglia of the spinal and cranial nerves.

The cephalic end of the neural groove exhibits several dilatations, which, when the tube is closed, assume the form of three vesicles; these constitute the three primary cerebral vesicles, and correspond respectively to the future *fore-brain*, *mid-brain*, and *hind-brain* (fig. 103). Their walls are developed into the nervous tissue and neuroglia of the brain, while their cavities become modified to form the brain ventricles. The remainder of the tube forms the central canal of the spinal cord, the neuroglial and nervous elements of which are formed from the ectodermal wall of the tube.

THE NOTOCHORD OR CHORDA DORSALIS

The notochord (figs. 104 and 106, 9, 10) consists of a rod of cells which is situated immediately below the neural tube and which constitutes the foundation of the axial skeleton, since around it are developed the bodies of the vertebræ and the intervertebral discs; it entirely disappears in the situation of the former, but persists in the centre of the latter as the *nucleus pulposus* (figs. 116 and 117). Its development is coincident with that of the neural tube. On the ventral aspect of the neural groove an axial thickening of the entoderm takes place; this thickening assumes the appearance of a furrow—the *chordal furrow*—the margins of which come into contact, and so convert it into a solid rod of cells — the *notochord* — which is then separated from the entoderm. It lies at first between the neural tube and the entoderm of the primitive alimentary canal, but soon becomes surrounded by the ingrowing mesoderm, which encloses it in a sheath, the *notochordal sheath*. It extends throughout the entire length of the future vertebral column, and its cephalic extremity reaches as far as the anterior end of the mid-brain. The ce-

Fig. 103.—Chick embryo of thirty-three hours' incubation, viewed from the dorsal aspect. × 30. (From Duval's 'Atlas d'Embryologie.')

Head fold of amnion partly covering the fore-brain

Mid-brain

Hind-brain

Nerve ganglion
Otic vesicle

Heart

Vitelline vein

Fourteenth protovertebral somite

Paraxial mesoderm

Medullary ridge

Sinus rhomboidalis

Remains of primitive streak

phalic portion is at first in contact with the entoderm of the fore-gut, and its anterior end lies immediately behind the pituitary invagination of the stomatodœum. It becomes partly surrounded by mesoderm, and is then seen to end in a hook-like extremity in the region of the future dorsum sellæ of the sphenoid bone.

FORMATION OF THE BODY CAVITY OR CŒLOM

As the mesoderm extends between the ectoderm and entoderm, it splits, except in the immediate neighbourhood of the neural tube, into an upper or

FIG. 104.—Section across the dorsal part of a chick embryo of forty-five hours' incubation. (Balfour.)

Neural canal   *Proto-vertebral somite*   *Wolffian duct*   *Body cavity or cœlom*   Somatic mesoderm

Ectoderm

Entoderm

Notochord   Aorta   Splanchnic mesoderm

somatic, and a lower or *splanchnic* layer. The former becomes applied to the inner surface of the ectoderm, and with it forms the *somatopleure*; while the latter adheres to the entoderm, and with it forms the *splanchnopleure* (figs. 104, 105, 106). The space between the two mesodermic layers is termed the *body cavity* or *cœlom*. The portion of this space which is enclosed within the embryo (embryonic cœlom) is developed into the cavities of the peritoneum, pleuræ, and pericardium. The portion outside the embryo (extra-embryonic cœlom) envelops the yolk-sac. That part of the mesoderm which remains undivided on either side of the neural tube is termed the *paraxial mesoderm*, to distinguish it from the more laterally placed portions which constitute the *lateral mesoderm*.

FIG. 105.—Transverse sections through the embryo chick before and some time after the closure of the medullary canal. (After Remak.)

A. At the end of first day. 1. Notochord. 2. Neural groove. 3. Edge of medullary fold. 4. Ectoderm. 5. Mesoderm. 6. Entoderm. 7. Protovertebral somite. B. On third day in lumbar region. 1. Notochord in its sheath. 2. Neural canal now closed. 3. Wall of neural canal. 4. Ectoderm. 5. Somatic mesoderm. 5'. Splanchnic mesoderm. 4×5. Fold of amnion. 6. Alimentary canal.

THE PROTOVERTEBRAL SOMITES, OR PRIMITIVE SEGMENTS

About the third week the paraxial mesoderm becomes transversely segmented and converted into a series of well-defined, more or less cubical areas, the *protovertebral* or *mesodermic somites* (figs. 103 and 104), which extend from the region of the hind-brain along the entire length of the trunk on either side of the middle line.

They lie immediately under the ectoderm on the lateral aspect of the neural tube and notochord, and are connected to the lateral mesoblast by a narrow isthmus of cells, the *intermediate cell-mass*, in which the genito-urinary organs are developed. The cells of each somite encircle a central cavity—the *myocœl*—which, however, soon becomes filled with angular and spindle-shaped cells. The somites of the trunk may be arranged in the following groups, viz. : cervical 8, thoracic 12, lumbar 5, sacral 5, and coccygeal

Description of fig. 106.—1–5 are diagrammatic representations of cross and longitudinal sections through the hen's egg at different stages of incubation. 6 represents a cross-section through an embryo fish. 7. A longitudinal section through a Selachian embryo. 8 and 9. Half of a cross-section through an embryo chick of two and three days respectively (after Kölliker). 10. Cross-section through a five days' chick embryo (after Remak); and 11. Longitudinal section through a chick embryo. The ectoderm is coloured blue: the mesoderm red, the entoderm green, and the yolk-sac yellow. The reference letters apply to all the diagrams. (From Hertwig's 'Entwickelungsgeschichte.')
ep. Ectoderm. *me.* Mesoderm. *hy.* Entoderm. *mr.* Medullary ridges. *N.* Neural canal. *af.* Amniotic fold. *aaf, paf, laf.* Anterior, posterior, and lateral amniotic folds. *A.C.* Amniotic cavity. *Am.* Amnion. *Ch.* Chorion. *du.* Dermal umbilicus. *lf.* Lateral fold. *hf.* Head fold. *vd.* Vitello-intestinal duct. *ig.* Intestinal groove. *al.* Allantois.

FIG. 106.—Diagrams to illustrate the development of the embryo with its yolk-sac,
amnion, and allantois.   (From Hertwig's 'Embryology.')

ys. Yolk-sac.   iu. Intestinal umbilicus.   me¹. Somatic mesoderm.   me². Splanchnic mesoderm.   dm. Dorsal
mesentery.   vm. Ventral mesentery.   pc′, pc². Pleuro-peritoneal cavity.   va. Vascular area.   va′. Limit of vascular
area.   ya. Yolk area.   ya′. Limit of yolk area.   m. Mouth.   nc. Notochord.   ao. Aorta.   a. Anus.   y. Yolk mass.
In 10, y is placed in the primitive alimentary canal.   W.b. Wolffian body.   fg. Fore-gut.   hg. Hind-gut.   tf. Tail
fold.   nc. Neurenteric canal.

from 5 to 8. Those of the head are three or four in number, and extend forward as far as the primitive ear-capsule.

The three layers of the blastoderm are therefore named, from without inwards, the ectoderm, mesoderm, and entoderm (figs. 102 and 104). Each has distinctive characters and gives rise to certain tissues of the body.

The **ectoderm** consists of columnar cells, which are, however, somewhat flattened or cubical towards the margin of the embryonic area. It forms the whole of the nervous system, the epidermis of the skin, the lining cells of the sebaceous, sweat and mammary glands, the hairs and nails, the epithelium of the nose and adjacent air-sinuses, and that of the cheeks and roof of the mouth. From it also are derived the enamel of the teeth, and the anterior lobe of the pituitary body, the epithelium of the cornea, conjunctiva, and lachrymal glands, and the neuro-epithelium of the sense organs.

The **entoderm** consists at first of flattened cells, which subsequently become columnar. It forms the epithelial lining of the whole of the alimentary canal

FIG. 107.—Section through the embryo which is represented in fig. 101.
(After Graf Spee.)

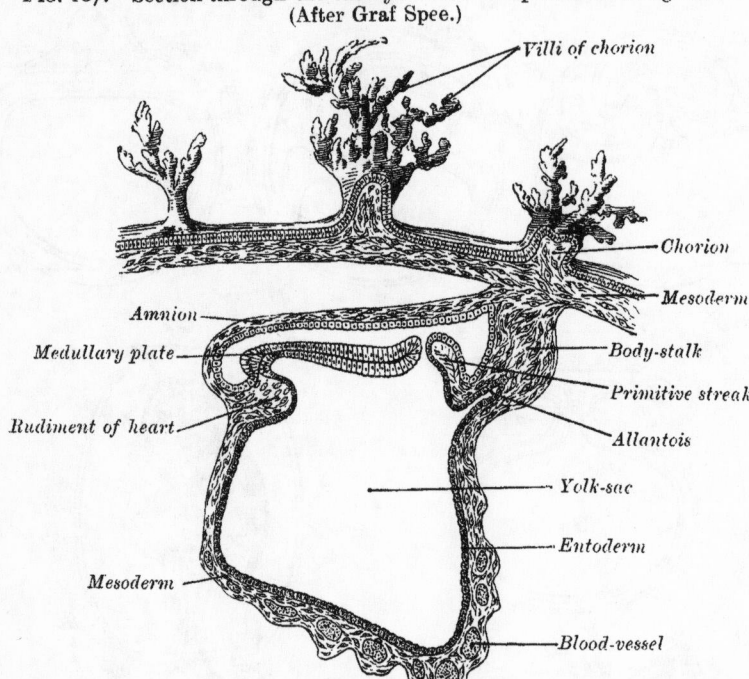

excepting part of the mouth and pharynx and the terminal part of the rectum (which are lined by involutions of the ectoderm), the lining cells of all the glands which open into the alimentary canal, including those of the liver and pancreas, the epithelium of the Eustachian tube and tympanic cavity, of the trachea, bronchi, and air-cells of the lungs, of the urinary bladder and part of the urethra, and that which lines the follicles of the thyroid and thymus glands.

The **mesoderm** consists of loosely arranged branched cells surrounded by a considerable amount of intercellular fluid. From it the remaining tissues of the body are developed. The epithelial lining of the heart and blood-vessels is, however, regarded by some as being of entodermal origin.

DELIMITATION OF THE EMBRYO

As has been pointed out, the margin of the embryonic area is of relatively slow growth compared with the rapidly developing embryonic and extra-embryonic parts of the ovum, and thus it forms a ring of constriction between these two areas, which expand above and below it. The embryonic area is in this manner constricted off from the blastodermic vesicle to form the *embryo* ; and as this takes place, a small part of the blastodermic vesicle is enclosed within the embryo and

FIG. 108.—Diagram of a transverse section of an embryo, showing the mode of formation of the amnion. The amniotic folds have nearly united in the middle line. (From Quain's 'Anatomy,' vol. i. pt. 1, 1890.)

constitutes the *primitive alimentary canal*, while the larger portion of the vesicle is left outside the embryo, and is termed the *yolk-sac* or *umbilical vesicle* (fig. 108). The passage between the primitive alimentary canal and the yolk-sac is named the *vitelline* or *omphalo-mesenteric duct* (fig. 166). Although the embryo grows in all directions, it increases much more rapidly in length than in width ; and coincident with this, the cephalic and caudal extremities are bent downwards to form the *cephalic* and *caudal folds* respectively (figs. 109 and 166). With the forward growth of the head the pericardial area is folded backwards under it, and comes to lie in the ventral wall of the primitive alimentary canal, its original upper surface, in this manner, becoming its under surface. When the cephalic and caudal flexures have been formed, the primitive alimentary canal presents the appearance of a nearly straight tube, closed at its two extremities. This tube is divided into three portions, viz.: (*a*) the *fore-gut*, between the pericardium and the notochord ; (*b*) the *mid-gut*, opening directly into the yolk-sac ; and (*c*) the *hind-gut*, contained within the caudal fold (fig. 166).

FIG. 109.—Diagrammatic section through the ovum of a mammal in the long axis of the embryo.

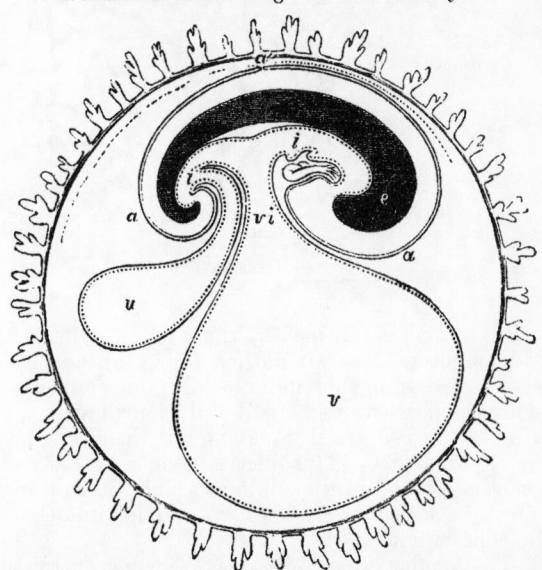

e. The cranio-vertebral axis. *i, i*. The primitive alimentary canal. *a*. The amnion. *a'*. The point of reflection into the false amnion. *v*. Yolk-sac, communicating with the mid-gut. *vi*. The vitello-intestinal duct. *u*. The allantois. The ovum is surrounded externally by the villous chorion.

MEMBRANES AND APPENDAGES OF THE FŒTUS

These are concerned with the protection and nourishment of the embryo, and comprise (1) the yolk-sac, the amnion, the chorion, the allantois, and the umbilical cord, which are of fœtal origin ; (2) the uterine decidua, which is produced by a modification of the mucous membrane of the uterus ; and (3) the placenta, which is derived partly from fœtal and partly from maternal tissues.

The Yolk-sac (figs. 101, 107 and 108).—In the human embryo this is of small size, and only supplies nourishment to the embryo during the earlier stages of its existence. It is an appendage of the alimentary canal, and consists, as has been explained, of that portion of the blastodermic vesicle which is not enclosed within the body of the embryo to form the primitive alimentary canal, and like that tube it is lined by entoderm, outside of which is a layer of mesoderm. It assumes the form of an oval sac, which is filled with fluid, the *vitelline fluid* ; it is situated on the ventral aspect of the embryo, and communicates with the mid-gut by means of the vitelline duct. Blood is conveyed to the wall of the sac by the primitive aortæ, and after circulating through a wide-meshed capillary plexus, termed the

FIG. 110.—Human embryo of about fourteen days old with yolk-sac. (After His.) (From Kollmann's 'Entwickelungsgeschichte.')

Dorsal aorta

Primitive jugular vein

Amnion

Cardinal vein

Dorsal aorta

Body-stalk

Chorionic villi

vascular area, is returned by the vitelline veins to the tubular heart. This constitutes the *vitelline circulation* (fig. 110), and by means of it nutritive material is absorbed from the vitelline fluid and conveyed to the embryo. The yolk-sac increases in size up to the end of the fourth week, and then presents the appearance of a pear-shaped vesicle, opening into the alimentary canal by a long narrow tube, the vitelline duct. This duct is sooner or later obliterated, but a trace of it is sometimes found in the adult as a diverticulum from the small intestine ; this is known as *Meckel's diverticulum*, and is situated from two to four feet above the ilio-colic junction.

The Amnion is a membranous sac which surrounds and protects the embryo. It is at first of small size, but increases considerably towards the middle of pregnancy as the fœtus acquires the power of independent movement. It exists only in reptiles, birds, and mammals, which are hence called 'Amniota' ; being absent in amphibia and fishes, which are consequently termed 'Anamnia.' In man, monkeys, and some of the bats the primitive amniotic cavity probably persists (fig. 98). As already explained (page 80), Van Beneden regards the amniotic cavity as being produced by a hollowing-out of the inner cell-mass, owing to the atrophy and disappearance of some of its cells.   Mall, on the other hand, suggests that the amnion is formed by an inversion of the blastoderm.

In reptiles, birds, and certain mammals it is developed in the following manner. At the point of constriction where the primitive alimentary canal of the embryo joins the yolk-sac—i.e. in the region of the future umbilicus—a reflection or folding upwards of the somatopleure takes place.   This fold first makes its appearance at the cephalic extremity, and subsequently at the caudal end and sides of the embryo, and, gradually rising more and more, its different parts meet over the dorsal aspect of the embryo.   After coming into contact they fuse, and the septum between them disappears, so that the inner layer of the cephalic part becomes continuous with the inner layer of the caudal and lateral parts, and the outer with the outer.   Thus there are two membranes formed, one produced by the inner layer of the fold—the amnion—which encloses a space over the back of the embryo—the amniotic cavity ; the outer membrane is termed the *serosa, false amnion,* or *chorion* (figs. 106 and 108).   The space be-tween the amnion and the chorion constitutes the extra-embryonic cœlom, already re-ferred to, and is at first in communication with the em-bryonic cœlom or primitive pleuro - peritoneal   cavity. When first formed the amnion is in close contact with the body of the embryo, but about the fourth or fifth week fluid begins to accumulate within it.   This fluid constitutes the *liquor amnii,* and, increasing in quantity, causes the am-nion to expand and ultimately to adhere to the inner surface of the chorion, so that the extra-embryonic part of the cœlom is obliterated.   The amnion therefore covers the inner surface of the chorion and the fœtal aspect of the placenta.   The liquor amnii increases in quantity up to the sixth or seventh month of   pregnancy,   after   which it diminishes somewhat in amount.   It allows of the free movements of the fœtus during the later stages of

FIG. 111.—Transverse section through the dorsal region of an embryo chick, end of third day. (From Foster and Balfour.)

Am. Amnion.   *mp.* Muscle-plate.   *cv.* Cardinal vein.   *Ao.* Dorsal aorta, at the point where its two roots begin to join.   *Ch.* Notochord. *Wd.* Wolffian duct.   *Wb.* Commencement of formation of Wolffian body.   *ep.* Entoderm.   *So.* Somatopleure.   *hy.* Ectoderm.   The section passes through the place where the alimentary canal communicates with the yolk-sac.   *Sp.* Splanchnopleure.   *mc.* Medullary canal.

pregnancy, and also protects it by diminishing the risk of injury from without.   It contains less than 2 per cent. of solids, which consist of urea and other extractives, inorganic salts, a small amount of albumen, and frequently a trace of sugar.   That part of the fluid is swallowed by the fœtus is proved by the fact that epidermal débris and hairs have been found among the contents of the fœtal alimentary canal.

The Chorion (figs. 109 and 112).—The chorion is developed from the wall of the blastodermic vesicle, and consists of two layers : an outer formed by the primitive ectoderm or trophoblast, and an inner by the somatic mesoblast (figs. 107 and 114).   The trophoblast is made up of an internal layer of cubical or prismatic cells, the *cytoblast* or *layer of Langhans,* and an external layer of richly nucleated protoplasm devoid of cell boundaries, the *syncytium.*   Numerous villous processes, the *chorionic villi,* project into the decidua from the entire outer surface of the chorion.   These are at first small and non-vascular, and consist of tropho-blast only, but they increase in size and undergo ramification ; while the mesoderm, carrying branches of the umbilical vessels, grows into them, and in this way they are vascularised. These villi invade and destroy the decidua of the uterus, and at the same time absorb from it nutritive materials for the growth of the embryo. Until about the end of the second month of pregnancy the villi which cover the chorion are of an almost uniform size (fig. 109), but after this they develop unequally. The greater part of the chorion is in contact with the decidua reflexa (fig. 115), and over this portion the villi, with their contained vessels, undergo atrophy, so that by the fourth month scarcely a trace of them is left, and hence this part of the chorion becomes smooth, and is named the *chorion læve*; as it takes no share in the formation of the placenta, it is also named the non-placental part of the chorion. On the other hand, the villi on that part of the chorion which is in contact with the decidua serotina increase greatly in size and complexity, and hence this part is named the *chorion frondosum*. Since it forms the fœtal

FIG. 112.—Magnified view of the human embryo of four weeks, with the membranes opened. (From Leishmann, after Coste.)

y. The umbilical vesicle with the vitelline vessels, *v*, and its long tubular attachment to the intestine.   *c*. The villi of the chorion.   *m*. The amnion opened.   *u*. Cul-de-sac of the allantois, and on each side of this the umbilical vessels passing out to the chorion.   In the embryo : *a*. The eye.   *e*. The otic vesicle.   *h*. The heart.   *l*. The liver.   *o*. The upper ; *p*, the lower limb.   *w*. Wolffian body, in front of which are the mesentery and intestine. The Wolffian duct and tubes are not represented.

portion of the placenta, it is appropriately named the placental part of the chorion, and its villi, the placental villi (fig. 115).

The Allantois (figs. 166 and 167).—The allantois arises as a diverticulum from that part of the hind-gut which later forms the cloaca : it grows out into a mass of mesoderm which lies below and around the tail end of the embryo. The diverticulum is lined by entoderm and covered by mesoderm, and in the latter are carried the allantoic or umbilical vessels.

In reptiles, birds, and many mammals the allantois becomes expanded into a vesicle (figs. 106 and 109) which projects into the extra-embryonic cœlom, i.e. into the space between the amnion and the chorion. If its further development be traced in the bird, it is seen to project to the right side of the embryo, and, gradually expanding, it spreads over its dorsal surface as a flattened sac between the amnion and the chorion, and, extending in all directions, ultimately surrounds the yolk. Its outer wall becomes applied to, and fuses with, the chorion, which lies immediately inside the shell membrane. Blood is carried to the allantoic sac by the two allantoic or umbilical arteries which are continuous with the primitive aortæ, and, after circulating through the allantoic capillaries, is returned to the primitive heart by the two umbilical veins. In this way the allantoic circulation, which is of the utmost importance in connection with the respiration and nutrition of the chick, is established. Oxygen is taken from, and carbonic acid is given up to, the atmosphere through the egg-shell, while nutritive materials are at the same time absorbed by the blood from the yolk.

In man and other primates the nature of the allantois is entirely different from that just described. Here it exists merely as a narrow, tubular diverticulum of the hind-gut, and never assumes the form of a vesicle outside the embryo. With the formation of the amnion the embryo is, in most animals, entirely separated from the chorion, and is only again united to it when the allantoic mesoderm spreads over and becomes applied to its inner surface. The human embryo, on the other hand, as was pointed out by His, is never wholly separated from the chorion, its tail end being from the first connected with the chorion by means of a thick band of mesoderm, named the *body-stalk* (*Bauchstiel* of His—figs. 107 and 167) ; into this stalk the tube of the allantois extends (fig. 167). Moreover, in the human embryo the allantoic diverticulum is seen before the hind-gut is formed, and appears as a tubular protrusion from the blastodermic cavity (fig. 107). The body-stalk is at first attached to the hind-end of the embryo, but with the growth of the tail and the formation of the caudal flexure it assumes a ventral position, and the tubular allantois is then seen to open from the cloacal part of the hind-gut. The intra-abdominal part of the allantois becomes the fibrous cord of the *urachus*, and the body-stalk is elongated to form the umbilical cord.

The allantoic mesoderm gradually spreads itself over the inner surface of the chorion, and, invading the chorionic villi, supplies them with blood-vessels. The chorionic villi are embedded in the uterine

FIG. 113.—Diagrammatic sections of the uterine mucous membrane : (A) of the non-pregnant uterus ; (B) of the pregnant uterus, showing the thickened mucous membrane and the altered condition of the uterine glands.

Mucous membrane

A

Muscular fibres

Stratum compactum

B

Stratum spongiosum

Basal layer

Muscular fibres

decidua, and thus, by means of the allantoic circulation, nutritive substances are absorbed from the maternal blood and carried through the allantoic veins to the fœtus ; at the same time, the waste products of the fœtus are passed into the maternal blood. Coincident with the atrophy of the non-placental villi of the chorion, the corresponding branches of the allantoic vessels become obliterated. Those vessels, on the other hand, which supply the placental villi undergo great enlargement, and constitute the fœtal blood-vessels of the placenta.

The decidua.—Some four or five days before each menstrual period, changes are initiated in the mucous membrane of the uterus in order to render it fit for the reception of a fertilised ovum. It becomes congested and hypertrophied, its connective-tissue cells are increased in number, and its glands are expanded. If the ovum be not fertilised, the superficial layer of the mucous membrane rapidly degenerates and is cast off, while the superficial vessels are at the same time ruptured, causing the hæmorrhage of menstruation. After the cessation of the menstrual flow, the mucous membrane is regenerated, and the epithelium of the glands spreads over its surface and forms a new epithelial covering. If the ovum be fertilised, these degenerative processes do not occur : the mucous membrane is retained *in situ* and undergoes further changes, which result in the formation of the *decidua*.

These further changes are as follows : Its thickness and its vascularity are greatly increased ; its glands are elongated and open on its free surface by funnel-shaped orifices, while their deeper portions are tortuous and dilated into irregular spaces. The interglandular tissue is also increased in quantity, and is crowded with large round, oval, or polygonal cells, termed *decidual cells*. These changes are well advanced by the second month of pregnancy, and at this time the mucous membrane consists of the following strata (fig. 113) : (1) *Stratum compactum*, next the free surface ; in this the uterine glands are only slightly expanded, and are lined by columnar cells. (2) *Stratum spongiosum*, in which the gland tubes are greatly dilated and very tortuous, and ultimately come to be separated by only a small amount of interglandular tissue, while their lining cells are flattened or cubical. It is through the deep part of this layer that the placenta is separated after the birth of the child. (3) A thin *basal layer*, next the uterine muscular fibres, containing the deepest parts of the uterine glands, which are not dilated, and are seen to be lined with columnar epithelium. It is from this epithelium that the epithelial lining of the uterus is regenerated after pregnancy. The decidua lines the whole of the *body* of the uterus, without, however, occluding the orifices of the Fallopian tubes.

When the fertilised ovum reaches the interior of the uterus it attaches itself to the decidua, generally in the region of the fundus uteri. It has been customary to describe the decidua as growing up around the ovum so as ultimately to cover it in—the part which grows over and covers it in being named the *decidua reflexa* or *capsularis*. The portion of the decidua which intervenes between the ovum and the uterine muscular fibres is known as the *decidua serotina*, while that which lines the remainder of the body of the uterus is termed the *decidua vera* (fig. 115). In view, however, of the observations of Hubert Peters, who has investigated the placental development in a human ovum estimated to be less than a week old, this description must be somewhat modified. In Peters' specimen (fig. 114) the entire ovum was lenticular in shape, its longest diameter being parallel to the decidua serotina. The whole ovum was embedded in the decidua on the posterior wall of the uterus, forming a slight projection towards the uterine cavity. At the central part of this prominence the uterine epithelium was absent over an area measuring 1 mm. in diameter, and this aperture was closed by a blood-clot. Peters' observations therefore lead to the conclusion that as soon as the fertilised ovum enters the uterus it adheres to the decidua, and, destroying the surface epithelium, rapidly sinks into the substance of the stratum compactum, in which it becomes embedded. The ovum quickly expands, and excavates the stratum compactum both in width and depth, and the overhanging part of the decidua forms the decidua reflexa or capsularis ; the original aperture by which the ovum entered being closed by blood-clot, which ultimately undergoes organisation. No remains of the surface epithelium of the uterus were found on the inner surface of the decidua reflexa, nor over the decidua serotina. The decidua reflexa contained no uterine glands, nor were any glands seen opening on the surface of the decidua serotina. Remains of dilated glands were present in the deeper part of the decidua serotina, but their communications with the surface no longer existed. Villous processes covered the whole surface of the ovum and projected into the surrounding decidua. Many of these villi consisted of trophoblast only, but some contained cores of mesoderm. Spaces filled with maternal blood were found in the trophoblastic syncytium, and the processes of the latter were seen to embrace the maternal capillaries. The amnion was quite closed over the rudiment of the embryo, and the latter consisted of a patch of columnar ectodermal cells, on the ventral aspect of which there existed a small closed yolk-sac. The amnion and yolk-sac were enveloped by mesoderm, and the latter was continuous with the chorionic mesoderm by a cord of cells which represented the body-stalk.

FIG. 114.—Section through early ovum embedded in the uterine decidua. Semi-diagrammatic. (Modified from Peters.)

a. Amniotic cavity. b. Maternal blood-vessels. e. Embryonic ectoderm. en. Entoderm. f. Mass of fibrin. g. Uterine glands. m. Mesoderm. mm. Uterine decidua. s. Body-stalk. t. Trophoblast. u. Uterine epithelium. v. Yolk-sac.

Coincident with the growth of the embryo, the decidua reflexa is thinned and expanded (fig. 115). Its vascular supply is diminished, and it undergoes degeneration, and at the same time the non-placental villi atrophy and disappear. The space between the decidua reflexa and the decidua vera is gradually obliterated, so that by the third month of pregnancy the two are in contact. By the fifth month the decidua reflexa has practically disappeared, and, as a consequence, the chorion læve comes directly into contact with the decidua vera. During the succeeding months of pregnancy the decidua vera also undergoes atrophy, owing to the increased pressure. The glands of the stratum compactum are obliterated, and their epithelium is lost. In the stratum spongiosum the glands are compressed and present the appearance of slit-like fissures, while their epithelium undergoes degeneration. In the deepest layer, however, the glandular epithelium retains a columnar or cubical form.

The Placenta.—The placenta connects the fœtus to the uterine wall, and is the organ by means of which the nutritive, respiratory, and excretory functions of the fœtus are carried on. It consists of two parts—fœtal and maternal—the former being derived from the villi of the chorion frondosum with their contained blood-vessels, the latter from the decidua serotina, i.e. the part of the uterine decidua which intervenes between the ovum and the muscular wall of the uterus (fig. 115). It has been pointed out that at first the chorionic villi consist of trophoblast only—viz. the layer of Langhans and the syncytium—and that later they are invaded by the allantoic mesoderm, which conveys to them branches of the umbilical arteries. The syncytium proliferates rapidly, and burrows into and

FIG. 115.—Sectional plan of the gravid uterus in the third and fourth month. (Modified from Wagner.)

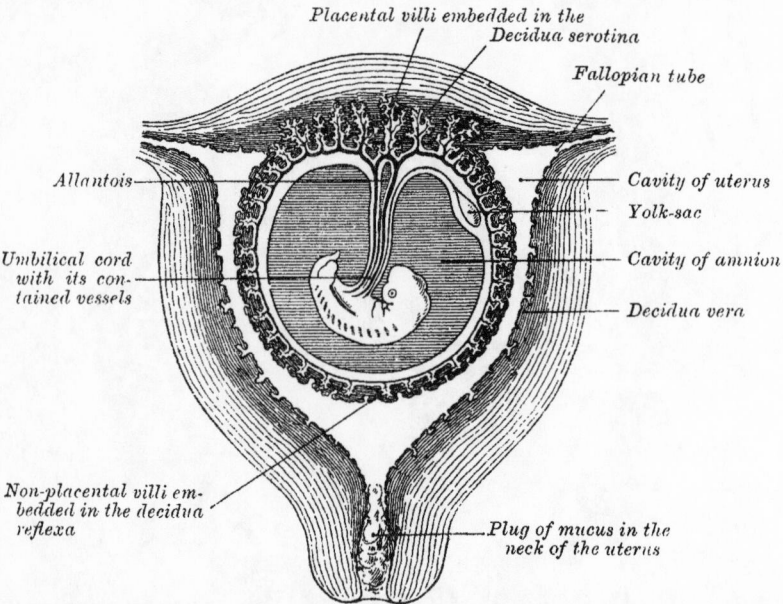

Placental villi embedded in the Decidua serotina

Fallopian tube

Allantois

Cavity of uterus

Yolk-sac

Umbilical cord with its contained vessels

Cavity of amnion

Decidua vera

Non-placental villi embedded in the decidua reflexa

Plug of mucus in the neck of the uterus

absorbs the tissues of the decidua. It also becomes vacuolated, and the extremities of its processes surround the maternal capillaries, which are seen to open directly into the spaces in the syncytium ; and thus the maternal blood comes to circulate through these spaces (fig. 114). Subsequently the non-placental villi of the chorion, which project into the decidua reflexa, become atrophied, and this part of the chorion is rendered smooth (chorion læve). On the other hand, the villi of the chorion frondosum, which project into the decidua serotina, increase enormously in size and complexity, and practically replace the whole of the superficial layer of this part of the decidua ; and the greatly ramified villi come to be suspended in a series of freely communicating spaces, the *maternal blood-sinuses*, which are filled with maternal blood. The maternal blood is conveyed to these spaces by the uterine arteries, and, after circulating slowly through them, is drained away by the uterine veins.

The endothelial lining of the uterine vessels can only be traced for a short distance from the mouths of the vessels on to the walls of the spaces, and is then lost. Whether the spaces beyond this point are lined by modified decidual cells, or by an expansion of the trophoblastic syncytium, is not yet determined ; it must, however, be clearly understood that they are not lined by uterine epithelium.

Each chorionic villus contains a branch of an umbilical artery and a tributary of the umbilical vein, with an intervening capillary plexus.   These are surrounded by thin mesodermal walls, which, in turn, are covered by two strata of cells. These two strata are probably both of fœtal origin, and may represent the layer of Langhans and the modified syncytium.   As to their nature there is, however, a marked difference of opinion—some observers regarding one (that next the maternal blood), others both, as being of maternal origin.   The fœtal blood is separated by these thin layers from the maternal blood in the sinuses, and hence there is no intermingling of the fœtal and maternal blood-currents.   Nevertheless, the fœtal blood is able to absorb, through the thin walls of the villi, oxygen and nutritive materials from the maternal blood, and give up to the latter its waste products.   The blood, thus purified, is carried back to the fœtus by the umbilical vein.   It will thus be seen that the placenta not only establishes a mechanical connection between the mother and the fœtus, but subserves for the latter the purposes of nutrition, respiration, and excretion.   In favour of the view that the placenta possesses certain selective powers may be mentioned the fact that glucose is more plentiful in the maternal than in the fœtal blood.   It is interesting to note also that the proportion of iron, and of lime and potash, is increased during the last months of pregnancy.   Further, there seems evidence that the maternal leucocytes may migrate into the fœtal blood, since leucocytes are much more numerous in the blood of the umbilical vein than in that of the umbilical arteries.

Separation of the placenta.—After the child is born, the placenta and membranes are expelled from the uterus as the *after-birth*.   The separation of the placenta from the uterine wall takes place through the deepest part of the stratum spongiosum, and necessarily causes rupture of the uterine vessels.   The orifices of the torn vessels are, however, closed by the firm contraction of the uterine muscular fibres, and this, together with the formation of a blood-clot over the placental site, prevents *post partum*-hæmorrhage.   The epithelial lining of the uterus is regenerated by the proliferation of the epithelium which lines the persistent portions of the uterine glands in the basal layer of the decidua.

After its separation the placenta appears as a flattened disc, which weighs about a pound, and has a diameter of from six to eight inches.   Its average thickness is about an inch and a quarter, but diminishes rapidly towards the circumference of the disc.   Its outer or decidual surface is comparatively smooth, but on inspection is seen to be incompletely divided into a number of lobules or *cotyledons* ; the fissures between the lobules correspond with septa which pass inwards towards the chorionic surface of the placenta.   Most of these septa end in irregular or pointed processes after extending for a short distance ; but others, especially those near the edge of the placenta, extend through its thickness and are attached to the chorion.   In the early months these septa convey branches of the uterine arteries which open into the maternal sinuses on the surfaces of the septa.   The inner or chorionic surface of the placenta is smooth, being closely invested by the amnion.   Seen through the latter, the chorion presents a mottled appearance, consisting of grey, purple, or yellowish areas.   The umbilical cord is usually attached near the centre of the placenta, but may be inserted anywhere between the centre and the margin.   In some cases it is inserted into the membranes, i.e. the *velamentous* insertion.   From the attachment of the cord the larger branches of the umbilical vessels radiate under the amnion, the veins being deeper than the arteries, and of a greater size.   Occasionally remains of the vitelline duct and yolk-sac may be observed beneath the amnion, close to the cord, the former as an attenuated thread, the latter as a minute sac.

On section, the placenta presents a soft, spongy appearance, caused by the greatly branched villi, which are surrounded by a varying amount of maternal blood—the latter giving rise to the dark red colour of the placenta.   Many of the larger villi extend from the chorionic to the decidual surface, while others are attached to the septa which extend inwards from the decidual surface of the placenta ; but the great majority hang free in the maternal sinuses, like the branches of a tree.   Owing to the rapid thinning of the placenta at the periphery of the disc, its decidual and chorionic surfaces come into contact.

The umbilical cord.—The rudiment of the umbilical cord is represented by the constricting ring which separates the rapidly growing embryo from the extra-embryonic area of the ovum.   Included in this ring are the body-stalk and the vitelline duct—the former containing the allantoic diverticulum and the umbilical vessels, the latter forming the communication between the alimentary canal and the yolk-sac. The body-stalk is a continuation of the tail end of the embryo, connecting the latter with the chorion. On transverse section, the amnion is seen to arch over its dorsal aspect, which is covered by thickened ectoderm and shows a trace of a medullary groove. Running through its mesoderm are the two umbilical arteries and the two umbilical veins, together with the canal of the allantois—the last being lined by entodermal epithelium. Its ventral surface is bounded by the extra-embryonic cœlom, and is in contact with the vitelline duct and yolk-sac. With the rapid elongation of the embryo and the formation of the tail fold, the body-stalk is carried forwards to the ventral surface of the embryo, where its mesoderm blends with that of the yolk-sac and the vitelline duct. The lateral leaves of somatopleure then grow round on each side, and, meeting on the ventral aspect of the allantois, enclose the vitelline duct and its vessels, together with a part of the embryonic cœlom; the latter is ultimately obliterated. The cord, therefore, is not covered by the amnion, but by a layer of ectoderm which is continuous with that of the amnion at the placental end of the cord, around which the amnion is attached. The various constituents of the cord are enveloped by embryonic gelatinous tissue (jelly of Wharton). The vitelline vessels and duct, together with the right umbilical vein, undergo atrophy, and disappear; and thus the cord, at birth, contains a pair of umbilical arteries and only one umbilical vein.

DEVELOPMENT OF THE EMBRYO

The further development of the embryo will, perhaps, be best understood by giving an outline of the principal facts relating to the development of the chief parts of which the body consists, viz. the skeleton, the pharyngeal cavity, mouth &c., the nervous centres, the organs of the senses, the circulatory system, the alimentary canal and its appendages, the respiratory and the genito-urinary organs. The reader is also referred to the chronological table of the development of the fœtus at the end of this section.

DEVELOPMENT OF THE SKELETON

The entire skeleton is of mesoblastic origin; it may be divided into (a) the skeleton of the trunk (axial skeleton), comprising the vertebral column, skull, ribs, and sternum, and (b) that of the limbs (appendicular skeleton).

The vertebral column.—The notochord (fig. 104) is the first rudiment of the vertebral column, and forms a central axis, around which its segments are developed. It is derived from the entoderm, and consists of a rod of cells, which lies on the ventral aspect of the neural tube and reaches from the anterior end of the mid-brain to the extremity of the tail. The paraxial mesoderm divides from before backwards into a number of segments, the protovertebral somites or primitive segments (fig. 103), which are arranged symmetrically on either side of the neural tube; and to each pair of somites a pair of spinal nerves is distributed. Each somite at first contains a central cavity or myocœl, which, however, is soon filled with a core or nucleus of angular and spindle-shaped cells. The cells of each somite become arranged into three groups: viz. (1) an external or dorsal, the *cutis plate*; (2) an internal, the *muscle plate* or *myotome*; and (3) a ventral, the *sclerotome*. The cells of the sclerotome are largely derived from those which constitute the core of the myocœl. Fusion of the individual sclerotomes in an antero-posterior direction soon occurs, and all trace of their originally segmented condition is lost; they then form a continuous strand of cells, the *scleratogenous layer*, which extends along the ventral and lateral aspects of the neural tube. The cells of this layer proliferate rapidly, and, gradually extending inwards, enclose the notochord in a continuous investment. From this investment the base of the skull, the vertebræ and their ligaments, and the membranes of the spinal cord are developed. It becomes thicker and more condensed opposite each protovertebral somite, and then extends outwards and backwards in the form of an arch, which is named the *primitive vertebral bow*. The mesial part of this bow lies below the notochord, and is named the *hypochordal bar* or *brace*. The lateral portions of the bow extend both dorsally and ventrally—the dorsal extension surrounds the neural tube, and forms the future neural arch of the vertebra; while the ventral becomes the costal process. The body of the vertebra is developed from that part of the investing mass which lies on the dorsal aspect, the hypochordal bar, and immediately surrounds the notochord. It makes its appearance as two small cartilaginous nodules, one on each side of the notochord, around which they ultimately fuse to form the cartilaginous vertebral body.

Except in the case of the first cervical vertebra, the hypochordal bar of the primitive vertebral bow disappears by fusing with the intervertebral disc. In the first cervical vertebra, on the other hand, the entire bow persists, the hypochordal bar being developed into the anterior arch of this bone. The cartilaginous body of this vertebra fuses with that of the second cervical to form its odontoid process.

The portions of the notochord which are surrounded by the bodies of the vertebræ atrophy, and ultimately disappear, while the parts which lie in the centre of the intervertebral discs undergo enlargement and persist throughout life as the nucleus pulposus in the centre of these discs (figs. 116 and 117).

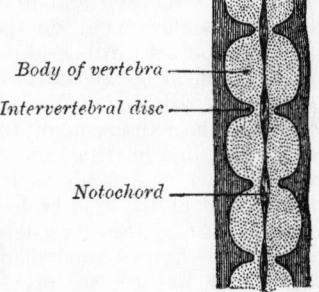

FIG. 116.—Longitudinal section of vertebral column of an eight weeks' human fœtus. (Kölliker.)

Body of vertebra

Intervertebral disc

Notochord

FIG. 117.—Sagittal section through the intervertebral disc and adjacent parts of two vertebræ of an advanced sheep's embryo. (Kölliker.)

Anterior common ligament

Posterior common ligament

Cartilaginous end of vertebral body

Enlargement of notochord in disc

Intervertebral disc

Slight enlargement of notochord in the cartilaginous end

Development of the ribs.—The ribs are formed from the ventral or costal processes of the primitive vertebral bows already referred to. Each of these processes grows ventrally into the body wall, inclining at first backwards and outwards, so as to pass on the caudal aspect of the corresponding muscle-plate. They are at first continuous with the primitive vertebræ; but when the latter become cartilaginous, the direct continuity is lost at the costo-vertebral articulations. Normally, there are only twelve pairs of ribs, but the costal processes of the last cervical or of the first lumbar vertebra are sometimes developed into movable ribs.

Development of the sternum.—The view generally held regarding the development of the sternum is that it arises as a paired structure from the ventral ends of the ribs, and that the two halves fuse in the middle line. The anterior ends of the nine upper costal bars join to form a cartilaginous strip on each side of the middle line—the two strips being at first connected by membrane. They then become united in the middle line from before backwards, and so give rise to a longitudinal piece of cartilage, which represents the manubrium and gladiolus. The ventral ends of the eighth and ninth cartilages fuse to form the xiphoid

H

appendix, and subsequently lose their connection with the sternum. Sometimes the eighth retains its original connection, and constitutes an eighth true rib, which occurs more frequently on the right than on the left side. This bilateral condition of the primitive sternum would serve to explain the occurrence of fissures or holes in the bone, as well as that rare anomaly, a completely divided sternum.

Paterson,* on the other hand, after reviewing the literature on this subject, and giving the results of his own observations on the embryonic and adult conditions of the sternum in man and some of the lower animals, says : 'The weight of evidence is all on the side of the primary association of the sternum with the shoulder girdle and its secondary connection with the ribs.'

Development of the skull—Up to a certain stage the development of the skull corresponds with that of the vertebral column ; but it is modified later in association with the expansion of the brain-vesicles, the formation of the organs of smell, sight, and hearing, and the development of the mouth and pharynx. The notochord, which extends as far forwards in the base of the future skull as the anterior end of the mid-brain, becomes partly surrounded by mesoderm. The posterior part of this mesodermal investment corresponds with the future basi-occiput, and shows a subdivision into four segments, which are separated by the roots of the hypoglossal nerve. The mesoderm then extends over the brain-

FIG. 118.—Diagrams of the cartilaginous cranium. (Wiedersheim.)

vesicles, and thus the entire brain is enclosed by a mesodermal investment, which is termed the *membranous primordial cranium*. From this are developed the bones of the skull and the membranes of the brain, together with the muscles, blood-vessels, true skin, and subcutaneous tissues of the scalp. In the shark and dog-fish this membranous cranium undergoes complete chondrification, and forms the cartilaginous skull or *chondro-cranium* of these animals. In mammals, on the other hand, the process of chondrification is limited to the base of the skull—the roof and sides being covered in by membrane. Thus, the bones of the base of the skull are preceded by cartilage, those of the roof and sides by membrane. The posterior part of the base of the skull is developed around the notochord, and exhibits a segmented condition analogous to that of the vertebral column, while the anterior part arises in front of the notochord. The base of the skull may therefore be divided into (a) *chordal* or *vertebral*, and (b) *prechordal* or *prevertebral* portions.

In the lower vertebrates two pairs of cartilages are developed : viz. a pair of parachordal cartilages, one on either side of the notochord ; and a pair of prechordal cartilages, the *trabeculæ cranii*, in front of the notochord. The *parachordal cartilages* (fig. 118) unite to form a cartilaginous plate. from which the basi-occiput and basi-sphenoid are developed. On the lateral aspect of the parachordal cartilages the *otic* or *auditory vesicles* are situated, and the mesoderm

---

* *The Human Sternum*, by Professor A. Melville Paterson, 1904.

enclosing them is soon converted into cartilage, forming the *cartilaginous ear-capsules*. These cartilaginous ear-capsules, which are of an oval shape, fuse with the lateral aspects of the basilar plate, and from them arise the petro-mastoid portions of the temporal bones. The *trabeculæ cranii* (fig. 118) are two curved bars of cartilage which embrace the region of the future pituitary fossa; their posterior ends soon unite with the basilar plate, while their anterior ends join to form the *ethmoidal plate*, which extends forwards between the fore-brain and the olfactory pits. Later, the trabeculæ extend across and fuse below the pituitary body, forming the floor of the pituitary fossa, and so cutting off the anterior lobe of the latter from the mouth cavity. The mesial part of the ethmoidal plate forms the bony and cartilaginous parts of the nasal septum, while its lateral prolongations form the roof and upper part of the outer wall of the nose. From the posterior parts of the trabeculæ cranii the sphenoid bone and its external pterygoid plates are developed. Thus, it will be seen that the bones which form the base of the skull are preceded by the cartilaginous chondro-cranium. Those of the vault, on the other hand, are of membranous formation, and are termed *dermal* or *covering*

FIG. 119.—Profile view of the head of a human embryo, estimated as twenty-seven days old. (After His.)

Lens

Maxillary process

Mandibular arch

Olfactory pit

Hyomandibular cleft

Auditory vesicle

Hyoid arch

Thyro-hyoid arch

Sinus præcervicalis

bones. They are partly developed from the mesoderm which lies superficial to the primordial cranium, and partly from that which lies outside the entoderm of the fore-gut. They comprise the upper part of the tabular portion of the occipital (interparietal), the squamous-temporals and tympanic rings, the parietals, the frontal, the vomer, the internal pterygoid plates, and the bones of the face. Some of them remain distinct throughout life (e.g. parietal and frontal), while others join with the bones of the chondro-cranium (e.g. interparietal, squamous-temporals, and internal pterygoid plates).

Recent observations have shown that, in mammals, the basi-cranial cartilage, both in the chordal and prechordal regions of the base of the skull, is developed as a single plate, which extends from behind forwards. In man, however, its posterior part shows an indication of its being developed from two chondrifying centres which fuse rapidly in front and below. The relation of this cartilaginous plate to the notochord differs in different animals. In the rat embryo it lies beneath the notochord (Robinson); in the sheep, pig, calf, and ferret, the cranial part of the notochord is enclosed within it; in man, the anterior and posterior thirds of the cartilage surround the notochord, but its middle third lies

H 2

on the dorsal aspect of the latter, which in this region is placed between the cartilage and the wall of the pharynx.

Development of the limbs.—The limbs begin to make their appearance in the third week as small elevations or buds at the side of the trunk. Prolongations from several protovertebral somites extend into each bud, and carry with them

FIG. 120.—Under surface of the head of a human embryo about twenty-nine days old. (After His.)

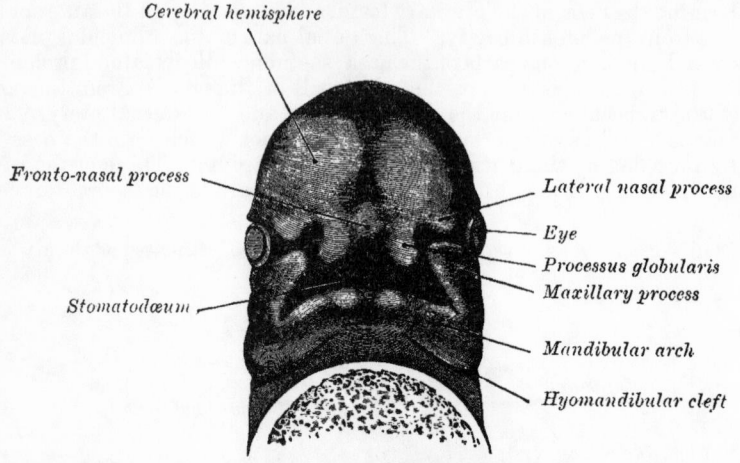

the anterior or ventral divisions of the corresponding spinal nerves. By the sixth week the three chief divisions of the limb are marked off by furrows—the upper into arm, forearm, and hand; the lower into thigh, leg, and foot. The limbs are at first directed backwards nearly parallel to the long axis of the trunk, and each presents two surfaces and two borders. Of the surfaces, one—the future *flexor* surface of the limb—is directed ventrally; the other, the

FIG. 121.—Under surface of the head of a human embryo about thirty days old. (After His.)

extensor surface, dorsally; while one border, the *pre-axial*, looks forward towards the cephalic end of the embryo, and the other, the *post-axial*, backwards towards the caudal end. The external condyle of the humerus, the radius and the thumb lie along the pre-axial border in the case of the upper limb; and the internal condyle of the femur, the tibia and the great toe along the corresponding border of the lower limb. The pre-axial part is derived from the anterior somites, the post-axial from the posterior somites of the limb bud ; and this explains, to a large extent, the innervation of the adult limb, the nerves of the more anterior somites being distributed along the pre-axial (radial or tibial), and those of the more posterior segments along the post-axial (ulnar or fibular) border of the limb. The limbs next undergo

FIG. 122.—The head and neck of a human embryo thirty-two days old, seen from the ventral surface. The floor of the mouth and pharynx have been removed. (His.) (From Marshall's 'Vertebrate Embryology.')

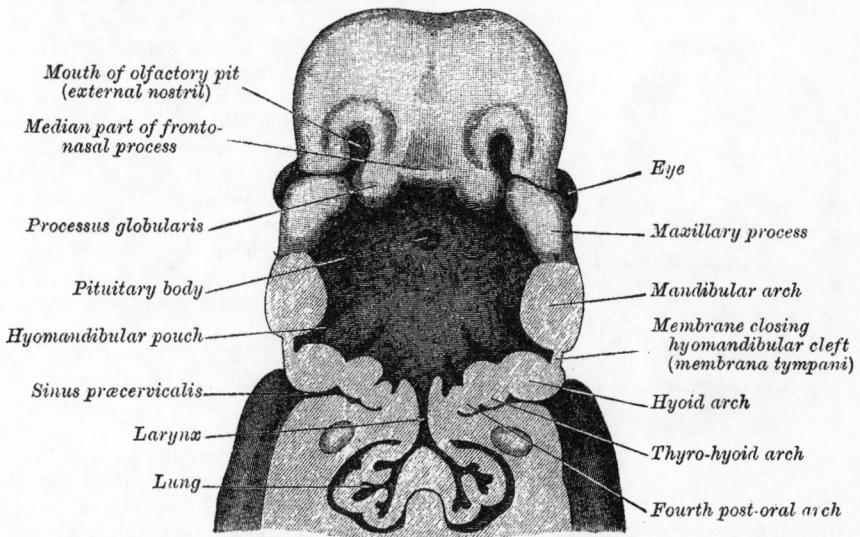

a rotation or torsion through an angle of 90° around their long axes. This rotation takes place in an opposite direction in the fore as compared with the hind limb, being outwards and forwards in the former, and inwards and backwards in the latter. In this manner the pre-axial (radial) border of the fore-limb is directed outwards, while the pre-axial (tibial) border of the hind-limb is directed

FIG. 123.—The roof of the mouth of a human embryo about two and a half months old, showing the mode of formation of the palate. (His.) (From Marshall's 'Vertebrate Embryology.')

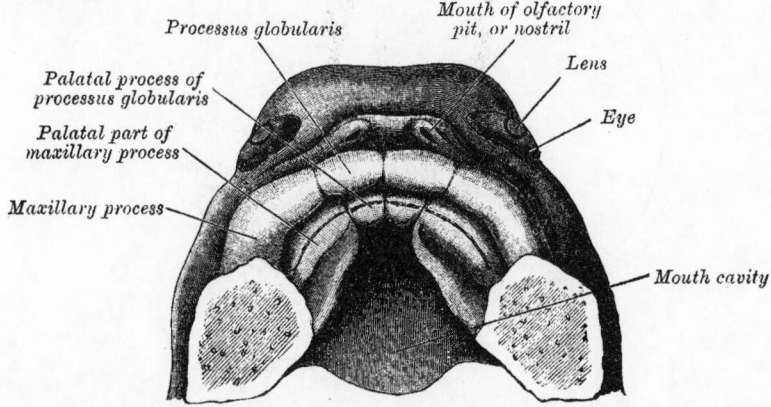

inwards ; and thus the flexor surface of the fore-limb is turned forwards, and that of the hind-limb backwards.

Development of the pharynx.—The pharynx is developed from the anterior part of the fore-gut, in the lateral walls of which a series of furrows or incomplete clefts appear (figs. 122 and 160). These are named the *visceral clefts*, and take origin as paired grooves or pouches from the side of the fore-gut ; over each groove a corresponding indentation of the ectoderm occurs, so that the latter comes into contact with the entodermal lining of the fore-gut, and the two layers unite to form thin septa along the bottom of the grooves between the fore-gut and the exterior. In gill-bearing animals these septa disappear, and the grooves become complete clefts—the gill-clefts—opening from

FIG. 124.—Head of a human embryo of about eight weeks, in which the nose and mouth are formed. (His.)

the pharynx on to the exterior; perforation, however, does not occur in birds and mammals. By the thickening of the mesoderm in front of and behind each cleft, a series of rounded bars or arches, the *visceral arches*, is formed. The dorsal ends of these arches are attached to the sides of the head, while their ventral extremities ultimately meet in the middle line of the neck. Six arches in all make their appearance, of which the first four are visible externally, while the last two are never elevated above the surface. In each arch there is developed a cartilaginous bar, which gives to it firmness and stability; and in each there is also found one of the primitive aortic arches. The first arch is named the *mandibular*; the second, the *hyoid*; the third, the *thyro-hyoid* (fig. 122); while the others have no distinctive names.

The mandibular arch lies between the first visceral cleft and the stomatodœum, and is developed into the lower lip and mandible. Its cartilaginous bar is known as *Meckel's cartilage* (fig. 125). The upper and lower ends of this cartilage are ossified: the upper to form the malleus, and probably the incus;* the lower to form the part of the mandible adjacent to the symphysis menti. The intervening part of the cartilage disappears, but the membrane surrounding its lower part becomes ossified to form a large part of the jaw, while the upper part of its enveloping membrane is represented

FIG. 125.—Head and neck of a human embryo eighteen weeks old, with Meckel's cartilage and hyoid bar exposed. (After Kölliker.)

by the internal lateral or spheno-mandibular ligament. The second or hyoid arch assists in forming the side and front of the neck. From its cartilage are developed the styloid process, stylo-hyoid ligament, and lesser cornu of the hyoid

* Some regard the incus as arising from the proximal end of the hyoid bar, while Gadow (*Phil. Trans.* vol. clxxix.) inclines to the view that the malleus, incus, and stapes arise from a cartilaginous plate, the hyo-mandibula, which binds the proximal ends of the mandibular and hyoid bars together.

bone. The cartilage of the third or thyro-hyoid arch gives origin to the great cornu of the hyoid bone. The lower ends of the second and third arches unite with those of the opposite side, and form a transverse band between the furcula behind and the tuberculum impar in front (fig. 126). This band soon becomes V-shaped, its two limbs extending forward, one on either side of the tuberculum impar; from it are developed the body of the hyoid bone, and the posterior third of the tongue. The lower portions of the cartilages of the fourth and fifth arches unite to form the thyroid cartilage.

FIG. 126.—The floor of the pharynx of a human embryo about fifteen days old. × 50. (From His.)

Mandibular arch

Hyoid arch

Tuberculum impar

Furcula

Third arch

Fourth arch

The first and second arches grow more rapidly than those behind them, with the result that the latter become, to a certain extent, telescoped within the former, and a deep depression, the *sinus præcervicalis*, (fig. 122) is formed on the side of the neck. This sinus is bounded in front by the hyoid arch, and ultimately becomes obliterated by the fusion of its anterior and posterior walls. The outer part of the first cleft becomes the external auditory meatus, while the inner part of the same cleft forms the Eustachian tube and tympanic cavity. The septum between the outer and inner parts of this cleft becomes invaded by mesoderm, and forms the membrana tympani. No traces of the outer parts of the second, third, and fourth clefts persist. The inner part of the second cleft is subdivided into an upper and a lower portion by the palate. The former persists as the fossa of Rosenmüller, or lateral recess of the naso-pharynx; in the latter the tonsil is developed, above which a trace of the cleft persists as the supratonsillar fossa. From the pharyngeal aspect of the third cleft the thymus gland arises as an entodermal diverticulum on each side, and from the corresponding part of the fourth cleft similar diverticula give origin to the lateral parts of the thyroid body.

Development of the tongue (figs. 126 to 128).—The tongue is developed in the floor of the pharynx. The rudiment of the anterior or buccal portion

FIG. 127.—The floor of the pharynx of a human embryo about twenty-three days old. × 30. (From His.)

Tuberculum impar

Posterior part of tongue

Furcula

Mandibular arch

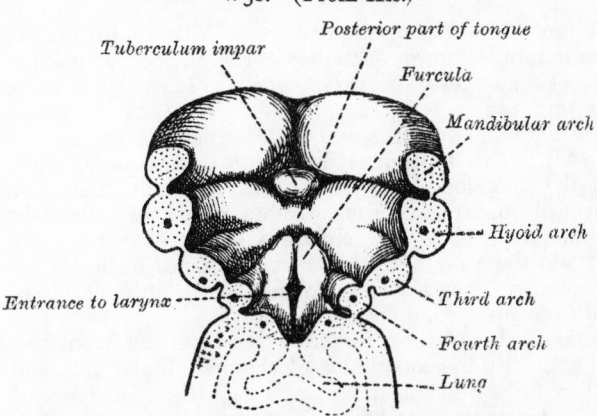

Entrance to larynx

Hyoid arch

Third arch

Fourth arch

Lung

appears during the third week as a rounded elevation, immediately behind the ventral ends of the mandibular arches. This elevation is named the *tuberculum impar* (figs. 126 and 127); it extends forwards on the oral surface of the mandibular arch, and increases markedly in size by the development of a pair of lateral tongue-elevations, which raise themselves from the inner surface of the maxillary arch, and, blending with the tuberculum impar, form the tip and greater portion of the buccal part of the tongue. These lateral growths correspond with similar structures which were described by E. Kallius in the development of the tongue of the lizard. (Consult article by E. Göppert in Hertwig's 'Entwickelungslehre.') From the ventral ends of the fourth arch there arises a second and larger elevation, in the centre of which is a median groove or furrow. This second elevation is named the *furcula* (fig. 126), and at first is merely separated from the tuberculum impar by a depression. Soon, however, they become separated by a ridge formed by the forward growth and fusion of the ventral ends of the second and third arches. The posterior or pharyngeal part of the tongue is developed from this ridge, which extends forwards in the form of a **V**, so as to embrace between its two limbs the tuberculum impar (figs. 127 and 128). At the apex of the **V** a pit-like invagination occurs, to form the middle thyroid rudiment, and this depression persists in the adult as the *foramen cæcum* of the tongue. The union of the anterior and posterior parts of the tongue is marked by a **V**-shaped depression (sulcus terminalis), the apex of which is at the foramen cæcum, while the two limbs run outwards and forwards, parallel to, but a little behind, the circumvallate papillæ. The prominent anterior part of the furcula forms the epiglottis; the furrow behind it is the entrance to the larynx; and the anterior parts of its lateral margins constitute the aryteno-epiglottidean folds.

Development of the mouth.—The mouth is developed partly from the stomatodœum, and partly from the floor of the anterior portion of the fore-gut. By the growth of the head end of the embryo, and the formation of the cephalic flexure, the pericardial area and the oral plate or bucco-pharyngeal area are

FIG. 128.—Floor of mouth of an embryo slightly older than that shown in fig. 127.
× 16. (From His.)

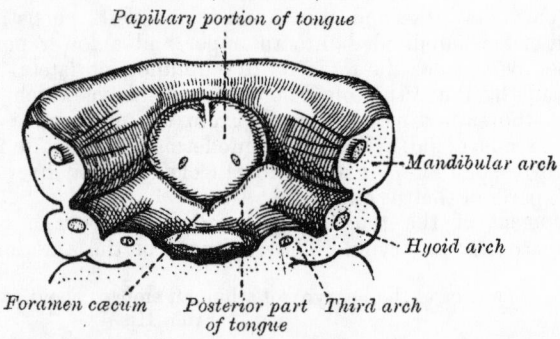

Papillary portion of tongue

Mandibular arch

Hyoid arch

Foramen cæcum　　*Posterior part*　*Third arch*
　　　　　　　　　of tongue

carried to the ventral surface of the embryo. With the further expansion of the brain, and the bulging forwards of the pericardium, the oral plate comes to occupy a depression between these two prominences. This depression constitutes the *stomatodœum* (fig. 120). It is lined by ectoderm, and is separated from the anterior end of the fore-gut by the oral plate, which is now named the *pharyngeal septum* (fig. 167). This septum is devoid of mesoderm, being formed by the apposition of the stomatodœal ectoderm with the fore-gut entoderm; at the end of a fortnight it disappears, and thus a communication is established between the mouth and the future pharynx. No trace of the pharyngeal septum is found in the adult; and the communication just mentioned must not be confused with the isthmus faucium, since His has shown that the anterior pillars of the fauces are developed from the second visceral arches.

The visceral arches extend forwards between the stomatodœum and the pericardium; and with the completion of the mandibular arch and the formation of the maxillary processes, the mouth assumes the appearance of a pentagonal orifice, which is bounded in front by the fronto-nasal process which covers the fore-brain and contains the anterior part of the coalesced trabeculæ cranii, behind by the mandibular arch, and laterally by the maxillary processes (fig. 120). With the inward growth and fusion of the palatal processes, the upper portion of the stomatodœum is shut off to form the nasal cavities, while from its lower or buccal portion the roof and anterior part of the mouth, together with the teeth, are developed.

The **salivary glands** arise as diverticula from the epithelial lining of the mouth, and their rudiments appear in the following order, viz.: the parotid during the fourth week, the submaxillary in the sixth week, and the sublingual during the ninth week (Hammar).

The **tonsils** are developed from the lower part of the second visceral cleft, immediately behind the anterior pillars of the fauces. The entoderm which lines this shallow pouch grows in the form of a number of solid buds into the surrounding mesoderm. These buds become hollowed out by the degeneration and casting off of their central cells, and by this means the tonsillar crypts are formed. Lymphoid cells accumulate around the crypts, and become grouped to form the lymphoid follicles; the latter, however, are not well defined until after birth.

The **thymus gland** first appears in the form of two flask - shaped entodermal diverticula, which arise, one on either side, from the third visceral cleft. Each diverticulum extends outwards and backwards into the surrounding mesoderm, and, although its opening into the pharynx is soon obliterated, the neck of the flask persists for some time as a cellular cord. By further proliferation of the cells which line the flask, buds of cells are formed, which become surrounded and isolated by the invading mesoderm. In the latter, numerous lymphoid cells make their appearance, and are aggregated to form lymphoid follicles. These lymphoid cells are probably derivates of the entodermal cells which lined the original diverticulum and its subdivisions.

The **thyroid body** is developed from a median and two lateral diverticula. The median diverticulum appears about the fourth week, immediately behind the tuberculum impar of the tongue, between the

FIG. 129.—Vertical section of the head in early embryos of the rabbit. Magnified. (From Mihalkovics.)

A. From an embryo of five millimetres in length. B. From an embryo of six millimetres in length. C. Vertical section of the anterior end of the notochord and pituitary body, &c., from an embryo sixteen millimetres long. In A the bucco-pharyngeal membrane is still present. In B it is in the process of disappearing, and the stomatodœum now communicates with the primitive pharynx. c. Fore-brain. mc. Mid-brain. mo. Hind-brain. m. Wall of brain cavity. if. Infundibulum. am. Amnion. sp.e. Spheno-ethmoidal; bc. Central; and sp.o. Spheno-occipital parts of basis cranii. h. Heart. f. Anterior extremity of foregut, i. tha. Thalamus. p. Original position of pituitary diverticulum, py. ch. Notochord. ph. Pharynx.

mandibular and hyoid arches. It grows downwards and backwards as a tubular duct, which bifurcates and subsequently subdivides to form a series of cellular cords. The lateral diverticula arise from the inner aspect of the fourth visceral cleft, and fuse with the median portion. The connection of the lateral diverticula with the pharynx disappears early. That of the median rudiment is termed the *thyro-glossal duct*; its continuity is subsequently interrupted by the development of the body of the hyoid bone, above which it is represented by the foramen cæcum of the tongue, and below by the pyramidal lobe of the thyroid body.

The **pituitary body, or hypophysis cerebri.**—This consists of a large anterior, and a small posterior, lobe: the former is derived from the ectoderm of the stomatodœum, the latter from the floor of the fore-brain. About the fourth week there appears a pouch-like diverticulum of the ectodermal lining of the roof of the stomatodœum. This, the pituitary involution or *pouch of Rathke* (fig. 129), is the rudiment of the anterior lobe of the pituitary body, and extends upwards in front of the cephalic end of the notochord and the remnant of the pharyngeal septum, and comes into contact with the under surface of the fore-brain. It is then constricted off to form a closed vesicle, but remains for a time connected to the ectoderm of the stomatodœum by a solid cord of cells. The vesicle sends out hollow processes into the surrounding mesoderm, and is gradually converted into a mass of small, tortuous tubules lined with columnar or cubical cells. The upwardly directed pituitary involution becomes applied to the antero-lateral aspect of a downwardly directed diverticulum from the base of the fore-brain. This diverticulum constitutes the future infundibulum in the floor of the third ventricle, while its lower extremity becomes modified to form the posterior lobe of the pituitary body. In some of the lower animals this lobe contains nerve-cells and nerve-fibres, but in man and the higher vertebrates these are replaced by connective tissue. A canal (*cranio-pharyngeal canal*) is sometimes found extending from the pituitary fossa to the under surface of the skull, and marks the original position of Rathke's pouch.

Development of the nose and face (figs. 120 to 124).—The nasal cavities are formed from the stomatodœum, while the outer nose is derived from its antero-lateral boundaries. Two areas of thickened ectoderm, the *olfactory areas*, appear immediately under the fore-brain in the anterior wall of the stomatodœum, one on either side of the *fronto-nasal process*. By the upgrowth of the surrounding mesoderm and ectoderm these areas are converted into pits, the *olfactory pits* (fig. 119), which divide the fronto-nasal process into a *mesial* and two *lateral nasal processes* (fig. 120). The rounded lateral angles of the mesial nasal process constitute the *globular processes* of His (figs. 121, 122). The olfactory pits form the rudiments of the nasal cavities, and extend backwards between the mesial and lateral nasal processes into the roof of the stomatodœum. From their ecto-dermal lining the olfactory epithelium and part of the olfactory bulb are derived. The globular processes are prolonged backwards as plates, termed the *nasal laminæ*: these laminæ are at first some distance apart, but, gradually approach-ing, they ultimately fuse, and form the nasal septum; while the processes them-selves meet in the middle line, and form the premaxillæ and central part of the upper lip (fig. 123).

The depressed part of the fronto-nasal process between the globular processes forms the lower part of the nasal septum or *columella*; while above this is seen a prominent angle, which becomes the future point, and still higher a flat area, the future bridge, of the nose (fig. 124). The lateral nasal processes form the alæ of the nose. Continuous with the dorsal end of the mandibular arch, and growing forwards from its cephalic border, is a triangular process—the *maxillary process*—the ventral extremity of which is separated from the mandibular arch by a >-shaped notch (fig. 120). The maxillary process grows forwards to form the outer wall and floor of the orbit, and meets with the lateral nasal process, from which, however, it is separated for a time by a groove—the *oculo-nasal sulcus*—which extends from the furrow encircling the eyeball to the olfactory pit. The maxillary processes ultimately fuse with the lateral nasal and globular processes, converting the oculo-nasal sulci into the lachrymal sacs and nasal ducts, and at the same time forming the lateral parts of the upper lip and the posterior boundary of the anterior nares. The maxillary processes also give rise to the lower portion of the lateral wall of the nasal cavity—the upper part of this wall, together with the roof, being developed from the ethmoid plate of the cartilaginous chondro-cranium. The nasal cavity is shut off from the buccal part of the stomatodœum by the development of the palate, the greater part of which is formed by a pair of shelf-like *palatal processes* which extend inwards from the maxillary processes (fig. 123); these coalesce with each other in the middle line, and constitute the entire palate, except a small part in front which is formed by the premaxillary bones. The union of the palatal processes with the premaxillæ is deficient in the middle line, where an aperture remains—the naso-palatine canal. The union of the parts which form the palate commences in front, the premaxillary and palatal processes joining in the eighth week, while the region of the future hard palate is completed by the ninth, and that of the soft palate by the eleventh week. The deformity known as cleft palate results from a non-union of the palatal processes, and that of hare-lip through a non-union of the maxillary and globular processes. The nasal cavity becomes divided into the two nasal fossæ by a vertical septum, which extends downwards and backwards from the fronto-nasal process and nasal laminæ, and which unites below with the palatal processes. Into this a plate of cartilage extends from the under aspect of the ethmoid plate of the chondro-cranium. The anterior part of this persists as the septal cartilage of the nose, but the posterior and upper parts are replaced by the vomer and mesethmoid. On each side of the nasal septum, at its lower and anterior part, the ectoderm is invaginated to form a blind pouch or diverticulum, which extends backwards and upwards into the nasal septum. This forms the rudiment of *Jacobson's organ*, which opens below, close to the junction of the premaxillary and maxillary bones.

DEVELOPMENT OF THE NERVOUS SYSTEM

The entire nervous system is of ectodermal origin, and its first rudiment is seen in the medullary or neural groove which extends along the dorsal aspect of the embryo (fig. 100). By the elevation and ultimate fusion of the medullary ridges, the groove is converted into the neural tube (fig. 104). Along the line

FIG. 130.—Transverse section of the spinal cord of a human embryo at the beginning of the fourth week. (After His.) The left edge of the figure corresponds to the lining of the central canal.

Germinal cell
Neuroblast
Nuclei of spongioblasts
Myelospongium network

of fusion of the medullary ridges a flattened crest of ectodermal cells—the *neural crest* or *ganglion ridge*—attaches, for a time, the neural tube to the overlying ectoderm (fig. 138). This crest is, however, soon cut off from the ectoderm, and, adhering to the neural tube, forms the rudiment of the ganglia of the cranial and spinal nerves and sympathetic system. The anterior end of the neural tube becomes expanded to form the three primary brain-vesicles which are subsequently modified to form the ventricular cavities of the brain (except the fifth); the remainder of the tube forms the central canal of the spinal cord (fig. 103). From its surrounding wall the nervous elements and the neuroglia of the brain and spinal cord are developed. It will be convenient to study (1) the development of the spinal cord, (2) that of the brain.

Development of the Spinal Cord.—At first the wall of the neural tube is composed of a single layer of columnar ectodermal cells. Soon the lateral parts of the wall become thickened, while the dorsal and ventral parts remain thin, and are named the *roof* and *floor plates*, or the *dorsal* and *ventral laminæ* (figs. 131 and 132). A transverse section of the tube at this stage presents an oval outline, while its lumen has the appearance of a slit. The cells which constitute the wall of the tube are differentiated into two sets : viz. (a) *spongioblasts* or young neuroglia cells, and (b) *germinal cells*, which are the parents of the *neuroblasts* or young nerve-cells (fig. 130). The spongioblasts are elongated and columnar, and

extend from the lumen of the tube to its peripheral wall—their inner and outer
ends being modified to form the inner and outer limiting membranes of the cord.
The parts of the spongioblasts abutting against the central canal retain their
columnar character, and ultimately form the layer of ciliated columnar epithelium
which lines this canal. Their outer parts, on the other hand, undergo ramification and form a sponge-like network, termed the *myelospongium*, from which the
neuroglia or sustentacular tissue of the cord is developed. The branching of the
spongioblasts is most marked near the periphery of the cord, and this outer part,
in consequence, assumes the appearance of a fine reticulum.

The germinal cells are large, round or oval, and first make their appearance
between the inner ends of the neuroglia cells on the lateral aspects of the central
canal. They increase rapidly in number, so that by the fourth week they form
an almost continuous layer on each side of the tube. No germinal cells are
found in the roof or floor plates; the roof-plate retains, in certain regions of the
brain, its epithelial character; elsewhere, its cells become spongioblasts. The
nuclei of many of the germinal cells exhibit mitotic changes, indicating that the
cells are undergoing rapid subdivision. By such subdivision they give rise to
the neuroblasts or young nerve-cells. The neuroblasts migrate outwards from

FIG. 131.—Section of spinal cord of a four weeks' embryo. (His.)

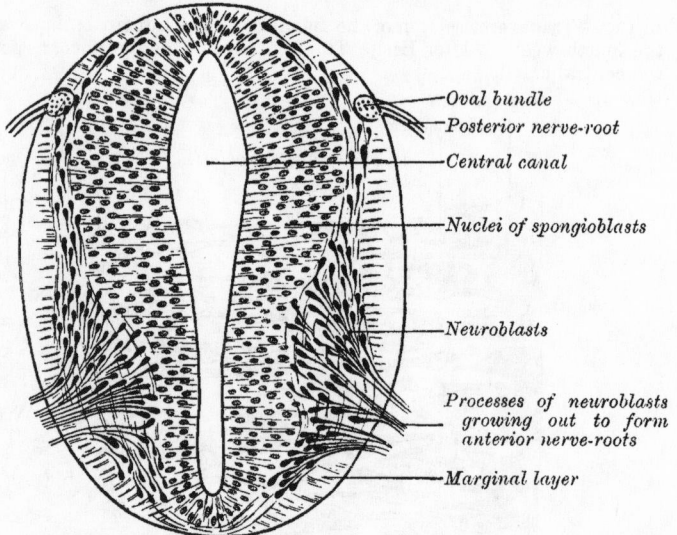

the sides of the central canal, and at the same time they become pear-shaped;
the tapering part of the cell undergoes still further elongation, and forms the
axis-cylinder or axon of the cell.

A transverse section of the cord exhibits three layers, viz.: (1) a *marginal
layer* or *marginal veil*, consisting of a fine neuroglia network, in which the future
white matter of the cord is developed. (2) An intermediate layer, the representative of the future grey matter of the cord. This is crowded with neuroblasts, and
is sometimes termed the *mantle layer*. (3) An internal or *ependymal layer*, next the
central canal, in which the germinal cells were first seen, but which, after their
subdivision and migration, becomes the epithelium of the central canal. The
lateral walls of the slit-like canal increase in thickness, and the canal itself widens
out near its dorsal extremity, and assumes a somewhat lozenge-shaped appearance. The widest part of the canal serves to subdivide the lateral wall of the
neural tube into a *dorsal* or *alar*, and a *ventral* or *basal lamina* (fig. 132), a subdivision which extends forwards into the brain.

The ventral part of the mantle layer becomes thickened, and on cross-section
appears as a triangular patch between the marginal and ependymal layers. This
thickening is the rudiment of the anterior horn of grey matter, and contains many
neuroblasts, the axis-cylinders of which pass out through the marginal layer and
form the anterior root of the spinal nerve (fig. 131). The thickening of the
mantle layer gradually extends backwards, and forms the posterior horn of grey

matter. The axons of many of the neuroblasts in the alar lamina pass forward, and cross in the floor-plate to the opposite side of the cord; these form the rudiment of the anterior white commissure of the cord. The cells of the neural crest having lost their original connection with the overlying ectoderm, become differentiated into a series of oval masses which constitute the rudiments of the spinal ganglia and, except in the region of the tail, are equal in number to the protovertebral somites. They are arranged symmetrically on the two sides of the cord, and consist at first of round or oval cells. These cells, however, soon become bipolar or spindle-shaped, and are drawn out into central and peripheral processes. The former grow backwards and inwards, and, becoming connected with the spinal cord, constitute the posterior or dorsal roots of the spinal nerves; while the latter grow outwards, and, joining the fibres of the anterior nerve-root, form, together with them, the spinal nerve. In the spinal cord of a six weeks' embryo the central processes form a well-defined *oval bundle* in the marginal layer of the alar lamina (fig. 131); but with the subsequent development of the posterior horn of grey matter, this bundle is displaced inwards, and forms the rudiment of the posterior white column of the cord. The two processes of the ganglion cell become gradually more and more approximated, and ultimately arise from a single stem in a T-shaped manner; the original bipolar condition of the cells is, however, retained in the ganglia of the auditory nerve. The central processes of the ganglion cells are generally regarded as axons, while the peripheral are by many classed as dendrites. The anterior and lateral white columns consist at first of the axons of the neuroblasts; they are, however, at a later stage, largely augmented by the pyramidal tracts which descend from the cerebral cortex. By the growth of the anterior horns of grey matter, and by the increase in size of the anterior columns, a furrow is formed between the lateral halves of the cord anteriorly; this gradually deepens to form the anterior median fissure. The mode of formation of the posterior fissure is somewhat uncertain. Many believe that it is produced by a growing together

FIG. 132.—Section of the medulla in the cervical region, at six weeks. Magnified 50 diameters.

1. Central canal. 2. Alar lamina. 3. Anterior grey matter. 4. Posterior nerve-root. 5. Floor-plate. 6. Roof-plate. 7. Anterior column. 8. Basal lamina. 9. Posterior column. 10. Anterior roots. 11. Posterior roots.

of the walls of the posterior part of the central canal. Robinson * traverses this view, and points out that the so-called posterior fissure is occupied by a fibrillated tissue, which is probably of a spongioblastic origin, since its fibrils can be traced directly into the posterior grey commissure.

Up to the fourth month of fœtal life the spinal cord occupies the entire length of the spinal canal, and the spinal nerves pass outwards at right angles to the cord. From this time onwards, the spinal column grows more rapidly than the cord, and the latter, being fixed above through its continuity with the brain, gradually assumes a higher position within the canal. By the sixth month its lower end reaches only as far as the upper end of the sacral canal; at birth it is on a level with the third lumbar vertebra, and in the adult it terminates at the lower border of the first or upper border of the second lumbar segment. A delicate filament, the *filum terminale*, extends from its lower end as far as the coccyx.

The ganglia of the sympathetic system are generally regarded as being developed as off-shoots from the ganglia on the roots of the cranial and spinal nerves.

Development of the brain.—The brain is developed from the anterior end of the neural tube, which at an early period becomes expanded into three vesicles, the primary cerebral vesicles (figs. 103 and 133). These are marked off from each other by intervening constrictions, and are named the *fore-brain* or *prosencephalon*, the *mid-brain* or *mesencephalon*, and the *hind-brain* or *rhombencephalon*—the last

* *Studies in Anatomy*, Owens College, 1891.

being continuous with the spinal cord. Soon after the appearance of these vesicles the embryonic brain becomes bent upon itself in a somewhat zigzag fashion. This results from the unequal growth of its different parts, and gives rise to three flexures. The first of these to appear is seen in the region of the mid-brain, and is named the *primary cephalic* or *mid-brain flexure* (fig. 133, A, above the letters *Pm*). By means of it, the fore-brain is bent in a ventral direction around the anterior end of the notochord and fore-gut, with the result that the floor of the fore-brain comes to lie almost parallel with that of the hind-brain (fig. 133, C). This flexure causes the mid-brain to become, for a time, the most prominent part of the brain, since its dorsal surface corresponds with the convexity of the curve. The second bend to appear is at the junction of the hind-brain and spinal cord. This is termed the *cervical* or *nuchal flexure* (fig. 133, B, *NK*), and increases from the third to the end of the fifth week, when the hind-brain forms nearly a right angle with the spinal cord. These two flexures affect the whole head, but the cervical bend is diminished when the erection of the head takes place—a process which commences after the fifth week. The third bend is named the *Varolian* or *pontine flexure* (fig. 133, A and B, *Br*), because it is found in the region of the future pons Varolii. It differs from the other two as follows : (*a*) its convexity is forwards, and (*b*) it does not affect the head. The lateral walls of the brain-tube, like those of the spinal cord, are divided by an internal furrow into alar or dorsal and basal or ventral laminæ.

The *hind-brain or rhombencephalon.*—The cavity of the hind-brain becomes the fourth ventricle. If measured at the time when the primary cephalic flexure makes its appearance, the length of the hind-brain will be found to exceed the combined length of the other two vesicles. Immediately behind the mid-brain it exhibits a marked constriction, which is named the *isthmus rhombencephali* (fig. 133, J), and is best seen when the brain is viewed from the dorsal aspect. It is customary to divide the hind-brain into two parts : viz. an upper, called the *metencephalon*, and a lower, the *myelencephalon*. The cerebellum is developed by a thickening of the roof, the pons by a similar change in the floor and lateral walls of the metencephalon. The floor and lateral walls of the myelencephalon are thickened to form the medulla oblongata, while its roof, on the other hand, remains thin, and, retaining to a great extent its epithelial nature, is expanded in a lateral direction ; moreover, by the growth and backward extension of the cerebellum, it is folded inwards towards the cavity of the fourth ventricle. It assists in forming the dorsal wall of this cavity, and is also invaginated as the epithelial covering of its choroid plexuses. From its upper part is developed the inferior medullary velum ; below, it is continuous with the obex and ligulæ.

The development of the medulla oblongata resembles that of the spinal cord, but at the same time exhibits one or two interesting modifications. On transverse section the myelencephalon at this stage is seen to consist of two lateral walls, connected across the middle line by floor and roof plates, as in the cord (figs. 136 and 137). Each lateral wall consists of an alar and a basal lamina, separated by an internal furrow, the remains of which are represented in the adult brain by the foveæ on the floor of the fourth ventricle. The contained cavity is more or less triangular in outline, its base being formed by the roof-plate, which is thin and greatly expanded transversely. Neuroblasts are developed in the alar and basal laminæ ; they become pear-shaped, and their narrow stalks are elongated to form the axis-cylinders of the nerve-fibres. Opposite the furrow or boundary between the alar and basal laminæ a bundle of nerve-fibres attaches itself to the outer surface of the alar lamina. This is named the *tractus solitarius* (fig. 137), and is formed by the sensory fibres of the glosso-pharyngeal and vagus nerves. It is the homologue of the *oval bundle* seen in the cord, and, like it, is developed by an ingrowth of fibres from the ganglia of the neural crest. At first it is applied to the outer surface of the alar lamina, but it soon becomes buried, owing to the growth over it of the neighbouring parts. By the fifth week the dorsal part of the alar lamina bends in an outward direction along its entire length, to form what is termed the *rhombic lip* (fig. 137). Within a few days this lip becomes applied to, and unites with, the outer surface of the main part of the alar lamina, and so covers in the tractus solitarius and also the spinal root of the fifth nerve.

Neuroblasts accumulate in the mantle layer : those in the basal lamina correspond with the cells in the anterior horn of the spinal cord, and, like them, give

FIG. 133.—Profile views of the brain of human embryos at three several stages, reconstructed from sections. (His.) (Copied from Quain's 'Anatomy.')

A. Brain of an embryo of about fifteen days, magnified 35 diameters. B. Brain of an embryo about three and a half weeks old. The optic vesicle has been cut away. C. Brain of an embryo about seven and a half weeks old. The optic stalk is cut through. A. Optic vesicle. H. Telencephalon. Z. Thalamencephalon. M. Mid-brain. J. Isthmus rhombencephali. Hh. Metencephalon. N. Myelencephalon. Gb. Otic vesicle. Rf. Roof of fourth ventricle. NK. Cervical flexure. Br. Pontine flexure. Pm. Mammillary process. Tr. Infundibulum. Hp (in B). Outline of hypophysis-fold of buccal epiblast. Rl. Rhinencephalon. In C the basilar artery is represented along its whole course.

origin to motor nerve-fibres; in the medulla they are, however, arranged in groups or nuclei, instead of forming a continuous column.   From the alar lamina and its rhombic lip, neuroblasts migrate into the basal lamina, and become aggregated to form the olivary nuclei, while many send their axis-cylinders through the floor-plate to the opposite side of the medulla, and thus constitute the rudiment of the raphé of the medulla.   By means of this thickening of the ventral portion of the medulla the motor nuclei are buried deeply from the surface, and, in the adult, are found close to the floor of the fourth ventricle.   This is still further accentuated : (*a*) by the development of the anterior pyramids, which are formed about the fourth month by the downward growth of the motor fibres from the cerebral cortex ; and (*b*) by the fibres which pass to and from the cerebellum. The isthmus rhombencephali represents the future valve of Vieussens, or superior medullary velum, and the superior peduncles of the cerebellum.

The *pons Varolii* is developed from the ventro-lateral wall of the metencephalon by a process similar to that which has been described for the medulla.

The *cerebellum* is developed in the roof of the anterior part of the hind-brain (fig. 135).   The alar laminæ become thickened to form two lateral plates, which fuse in the middle line.   The outer surface of the cerebellum is at first smooth and convex, but is soon divided by two fissures into three primary lobes, anterior, middle, and posterior.   The fissure between the anterior and middle lobes is named the *fissura prima*, and represents the preclival fissure of the adult cerebellum ; that separating the middle and posterior lobes is termed the *fissura secunda*, and is the rudiment of the post-pyramidal fissure.   The lingula, the central lobe and its alæ, the culmen monticuli, and anterior crescentic lobes are developed from the anterior primary lobe ; while the uvula, nodule, and flocculus arise from the posterior primary lobe.   The remainder of the cerebellum is developed from the middle primary lobe, which undergoes great lateral expansion, and forms the larger portions of the cerebellar hemispheres.   These are subsequently subdivided by the great horizontal fissure, which, although an important feature in the adult, is developmentally of secondary interest.   The rudiment of the cerebellum at first projects in a dorsal direction ; but, by the backward growth of the cerebrum, it is folded downwards and somewhat flattened, and the thin roof-plate of the fourth ventricle, originally continuous with the posterior border of the cerebellum, is projected inwards towards the cavity of the ventricle.

The mid-brain or mesencephalon.—The mid-brain (fig. 135) exists for a time as a thin-walled cavity of some size, and is separated from the isthmus rhombencephali behind, and from the fore-brain in front, by slight constrictions.   Its cavity is reduced in diameter, and forms the Sylvian aqueduct of the adult brain. Its basal laminæ become thickened to form the crura cerebri, which are at first of small size, but rapidly increase after the fourth month.   The neuroblasts of these laminæ are grouped in relation to the sides and floor of the Sylvian aqueduct, and constitute the nuclei of the third and fourth nerves, and of the descending root of the fifth nerve.   By a similar thickening process its alar laminæ are developed into the corpora quadrigemina.   The dorsal part of the wall for a time undergoes expansion, and presents an internal median furrow and a corresponding external ridge ; these, however, disappear, and the latter is replaced by a groove.   Subsequently two oblique furrows extend inwards and backwards, and the thickened lamina is thus subdivided into the quadrigeminal bodies.

The fore-brain.—At a very early stage of its existence—in some animals, even before the fusion of the medullary laminæ—the fore-brain expands on each side in the form of a hollow diverticulum ; these diverticula are termed the *optic vesicles*.   The central part of the fore-brain grows forwards and upwards between these vesicles, while at the same time the outer parts of the vesicles become enlarged ; and thus the entire fore-brain exhibits an early subdivision into three parts—an intermediate portion, or fore-brain proper, and two lateral portions, the optic vesicles (fig. 153).   These vesicles may therefore be looked upon as hollow appendages of the lower and outer aspect of the fore-brain.   With the subsequent expansion of the fore-brain and of the outer parts of the optic vesicles, the latter communicate with the former by narrow tubular stalks, the *optic stalks*.   The optic stalk represents the future optic nerve, while the optic vesicle gives rise to the retina and portions of the ciliary body and iris.   The fore-brain proper next grows in a downward and forward direction ; and from the lateral walls of this part, the cerebral hemispheres arise as hollow diverticula, which rapidly expand to form two large pouches, one on each side; these diverticula form the rudiments of the lateral ventricles, and communicate with the cavity of the fore-brain proper by relatively wide openings which ultimately form the *foramen of Monro* in the adult brain (fig. 134).

This anterior part of the fore-brain, together with the cerebral hemispheres, constitutes the *telencephalon*, while the hinder part of the fore-brain is named the *thalamencephalon* or *diencephalon*; both of these contribute to the formation of the third ventricle. In the middle line between the cerebral hemispheres the anterior wall of the brain-vesicle is formed by a thin lamina, the *lamina terminalis* (fig. 134), which reaches from the optic stalks to the foramen of Monro. A transverse section of the thalamencephalon at the fifth week shows the same parts as the spinal cord and medulla, viz. a pair of lateral walls connected across the middle line by floor and roof plates. Moreover, each lateral wall shows a division into alar and basal laminæ, which are separated internally by a furrow which was named by Reichert the *sulcus of Monro*. The roof-plate is thin and longitudinally folded, while the floor-plate is bulged downwards. The alar lamina of the thalamencephalon becomes thickened to form the optic thalamus of its own side, while from the corresponding part of the telencephalon the cerebral hemisphere is derived. At first the optic thalami are seen on the outer surface of the brain, but are subsequently hidden by the growth over them of the cerebral

FIG. 134.—Median section of fore-brain of a human embryo of 12·5 mm. in length. (After His.) (From Kollmann's 'Entwickelungsgeschichte.')

Roof of fore-brain

Cerebral hemisphere

Foramen of Monro

Rhinencephalon

Lamina terminalis

Optic recess

Optic tract

Sulcus of Monro

Tegmentum

Cephalic flexure

Mammillary elevation

Recessus infundibuli

hemispheres. The thalami grow inwards, and thus gradually narrow the cavity between them into a slit-like aperture which forms the greater part of the third ventricle. The mesial surfaces of the optic thalami ultimately adhere, in part, to each other across the cavity, and by this means the middle or grey commissure takes its rise. From the basal laminæ and floor the parts which occupy the interpeduncular space in the adult brain are developed, and their early condition can be recognised in a mesial section of the fore-brain at the fifth week. On examining the floor of such a section (fig. 134) from behind forwards, the following structures are seen: (1) an elevation, convex inferiorly, which, by the fourth month, is subdivided by a median groove into two tubercles, the corpora mammillaria; (2) a slight swelling, the tuber cinereum; the lower part of this is ultimately narrowed into a funnel-shaped diverticulum—the infundibulum—while the most dependent part of it is modified to form the posterior lobe of the pituitary body; the anterior lobe of this body has been seen to arise as a diverticulum of the ectoderm of the stomatodœum. The anterior wall of the vesicle is formed by the lamina terminalis, and at the angle of union of the floor and anterior wall is the recessus opticus, leading into the optic stalk (fig. 135); this angle corresponds with the optic chiasma in the adult brain. Above the lamina terminalis is the foramen of Monro, or passage leading into the cavity of the cerebral hemisphere. From the optic recess a groove, the sulcus of Monro, already referred to, passes backwards on the lateral wall of the thalamencephalon, and indicates its division between the alar and basal laminæ. A trace of this sulcus is seen in the adult

I

brain in the form of a slight groove extending from the foramen of Monro towards the Sylvian aqueduct.

The greater part of the roof-plate of the thalamencephalon remains thin and epithelial, and is subsequently invaginated by the choroid plexuses of the third ventricle. The posterior part of the roof, however, is modified to form the pineal gland and posterior commissure.

The pineal gland arises about the sixth week as an upward evagination of the roof-plate, immediately in front of the mid-brain (fig. 135). This ultimately becomes solid, except its proximal part, which persists as the recessus pinealis. In lizards the pineal evagination is elongated into a stalk, resembling the optic stalk, and its peripheral extremity is expanded into a hollow sac in which a rudimentary lens and retina are formed. The stalk becomes solid, and in it nerve-fibres can be recognised; in these animals, therefore, the pineal body forms a

FIG. 135.—Median section of brain of human fœtus during the third month. (After His.)

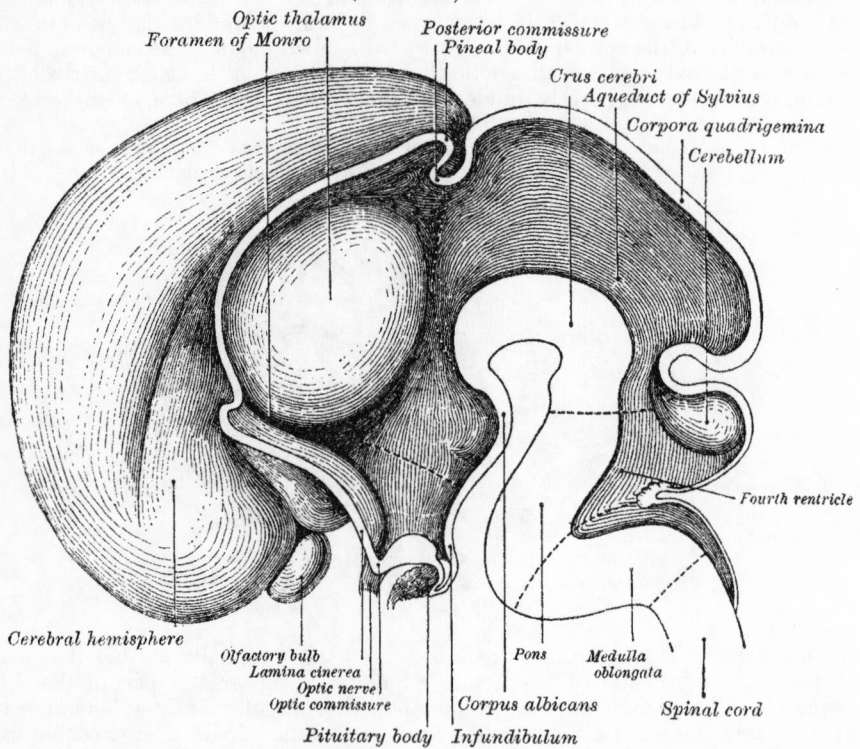

Optic thalamus
Foramen of Monro
Posterior commissure
Pineal body
Crus cerebri
Aqueduct of Sylvius
Corpora quadrigemina
Cerebellum
Fourth ventricle
Cerebral hemisphere
Olfactory bulb
Lamina cinerea
Optic nerve
Optic commissure
Pons
Medulla oblongata
Corpus albicans
Spinal cord
Pituitary body
Infundibulum

rudimentary eye. The posterior commissure first appears as a thickening of the dorsal wall of the fore-brain immediately behind the pineal evagination.

The cerebral hemispheres (figs. 134, 135).—It has been already stated that the cerebral hemispheres arise as lateral diverticula of the alar laminæ of the anterior part of the fore-brain; these diverticula expand rapidly and soon attain a large size. In the fourth week a longitudinal ridge appears on the under surface of each hemisphere, close to the lamina terminalis. This ridge is named the *olfactory area*, or *rhinencephalon* (fig. 134), and corresponding with it, there is an internal groove. Subsequently it is divided by a furrow or constriction into an anterior and a posterior part; the anterior part grows forwards as a hollow stalk continuous with the anterior part of the ventricular cavity. The stalk becomes solid, and forms the rudiment of the olfactory peduncle or tract and the trigonum olfactorium, while its free extremity is enlarged to form the olfactory bulb; a strand of gelatinous substance in the interior of the latter indicates the site of the original cavity. The posterior part forms the locus perforatus anticus and the peduncles of the corpus callosum (gyrus subcallosus). The rhinencephalon remains rudimentary in man, but forms a conspicuous part of the brain in osmatic animals.*

The cerebral hemispheres and their olfactory lobes grow forward on either side of the lamina terminalis. The hemispheres also expand upwards and backwards, their mesial surfaces being separated from each other by a septum of mesodermic tissue continuous with the membranous cranium ; this septum is the rudiment of the falx cerebri, and is continuous behind with a similar septum which passes between the occipital part of the cerebrum and the cerebellum and constitutes the rudimentary tentorium cerebelli. The hemispheres also grow backwards and downwards, covering first the thalamencephalon, and next the mid-brain, while by the seventh month they have extended so as to overlap the upper surface of the cerebellum ; this great expansion of the cerebral hemispheres is characteristic of the brains of mammals, and attains its maximum in the brain of man. The floor of the hemisphere is thickened to form the corpus striatum. This thickening begins to appear about the second month, and extends to the posterior end of the primitive hemisphere ; and, as a consequence, when this part of the hemisphere grows downwards to form the temporal lobe, the posterior part of the corpus striatum is carried into the roof of the descending horn of the ventricle, where it is seen as the tail of the caudate nucleus in the adult brain. Corresponding with this internal thickening, a shallow depression makes its appearance on the lateral surface of the hemisphere ; this constitutes the rudiment of the Sylvian fissure, and is named the *Sylvian fossa*. The floor of the fossa represents the future island of Reil, which is gradually submerged by the growth over it of the surrounding parts of the vesicle ; and thus, in the adult, the island lies deeply in the bottom of the Sylvian fissure. The parts of the cerebral vesicle which grow over it constitute the temporal, fronto-parietal, and orbital opercula of the adult brain. The cavity of the hemisphere forms the lateral ventricle, and is at first of a hemispherical or semilunar shape ; but, coincident with the growth of the hemisphere, it becomes extended into three prolongations, the anterior, posterior, and descending cornua of the adult cavity, the last representing the original posterior extremity of the vesicle.

The roof-plate which connects the hemispheres above the foramen of Monro remains thin, and by the fifth week is invaginated as lateral folds into the ventricular cavities. The vascular mesoderm which lies between the hemispheres extends into these folds, and forms the rudiment of the choroid plexuses ; these at the fifth month are of large size, and nearly fill the cavities of the ventricles.

Fissures of the cerebrum.—The surface of the cerebral hemisphere is at first smooth, but ultimately becomes indented by depressions or furrows, which constitute its fissures or sulci. These were formerly described as being of three kinds, viz. : *transitory, complete*, and *partial*.

Certain furrows or infoldings of the thin cerebral wall have been described as occurring during the second and third months, and subsequently undergoing, with one or two exceptions, complete obliteration. These, from their temporary nature, have been named *transitory fissures*. It has been pointed out, however, that if the brain be hardened *in situ* before post-mortem changes have taken place, its surface is found to be destitute of any such furrows, and the balance of evidence is in favour of the view that the so-called transitory fissures are artificially produced after death.

A *complete fissure* is one which indents the whole thickness of the wall of the cerebral hemisphere, and which, as a consequence, gives rise to a corresponding elevation in the ventricular cavity. To this category belong the hippocampal and calcarine and collateral fissures, which were regarded as owing their existence to the persistence of certain of the so-called transitory fissures just referred to. The Sylvian fissure is sometimes described as a complete fissure, but strictly speaking this is not correct ; its development has already been referred to.

The *incomplete fissures* are superficial foldings of the grey matter, and produce no corresponding elevations in the interior of the ventricle.

---

* In addition to the parts mentioned, the rhinencephalon includes the septum lucidum, fornix, gyrus supracallosus, gyrus dentatus, hippocampus, and uncus.

The Nerves

1. *The spinal nerves.*—Each spinal nerve arises from the spinal cord by a dorsal or afferent, and a ventral or efferent root, and, as already stated (pages 108, 109), the fibres of the former root are developed as outgrowths of the cells of the spinal ganglia, while those of the latter consist of the axons of the neuroblasts of the anterior horn of the cord. The afferent and efferent roots join, immediately beyond the spinal ganglion, to form the spinal nerve, which then divides into its anterior and posterior primary divisions, fibres from both roots being carried into each of the two divisions.

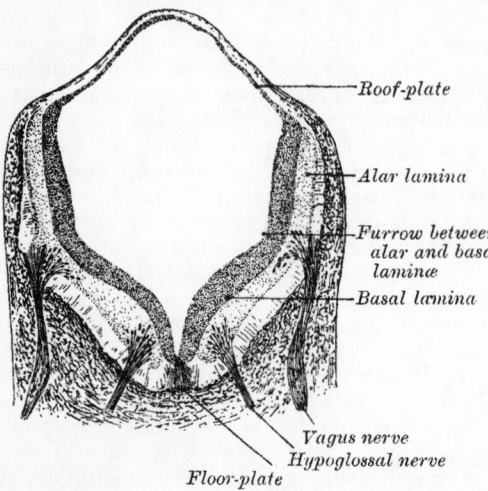

Fig. 136.—Transverse section of medulla oblongata of human embryo. × 32. (From Kollmann's 'Entwickelungsgeschichte.')

Roof-plate

Alar lamina

Furrow between alar and basal laminæ

Basal lamina

Vagus nerve
Hypoglossal nerve
Floor-plate

2. *The cranial nerves.*—With the exception of the olfactory and optic nerves, which will be specially considered, the cranial nerves are developed in a similar manner to the spinal nerves. The sensory or afferent nerves are derived from the cells of the ganglion rudiments of the neural crest. The central processes of these cells grow into the brain and form the roots of the nerves, while the peripheral processes extend outwards and constitute their fibres of distribution. It has been seen, in considering the development of the medulla oblongata (page 110), that the *tractus solitarius* (fig. 137), derived from the fibres which grow inwards from the ganglion rudiments of the glossopharyngeal and vagus nerves, is the homologue of the *oval bundle* in the cord which had its origin in the posterior nerve-roots. The efferent or motor nerves arise as outgrowths of the neuroblasts situated in the basal laminæ of the mid-

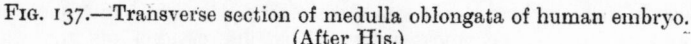

Fig. 137.—Transverse section of medulla oblongata of human embryo. (After His.)

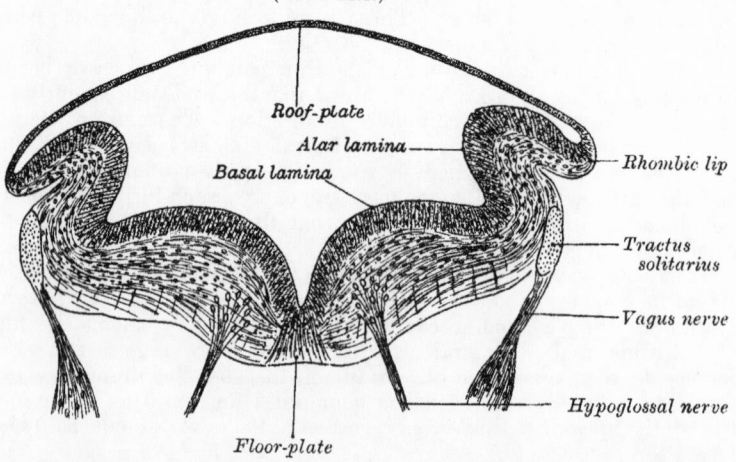

Roof-plate
Alar lamina
Basal lamina
Rhombic lip
Tractus solitarius
Vagus nerve
Hypoglossal nerve
Floor-plate

and hind-brain. While, however, the anterior spinal nerve-roots arise in one series from the basal lamina, the cranial motor nerves are grouped into two sets, according as they arise from the mesial or lateral parts of the basal lamina. To the former set belong the third, fourth, sixth, and twelfth nerves ; to the latter, the spinal accessory and the motor fibres of the fifth, seventh, ninth, and tenth nerves.

The *olfactory nerves* are outgrowths of neuroblasts which are developed from the ectoderm of the olfactory pit (page 106). These neuroblasts pass out from the ectoderm, and form a ganglion which subsequently fuses with the olfactory bulb. From the neuroblasts of this ganglion processes grow, centripetally to form the nerve-roots, and centrifugally to form the olfactory nerves.

The development of the optic nerve will be considered with that of the eye.

Development of the Eye.—The optic nerve and retina are developed as an outgrowth from the rudimentary brain, which extends towards the side of the head,

FIG. 138.—Transverse section of a chick embryo of twenty-nine hours' incubation. (From Duval's 'Atlas d'Embryologie.')

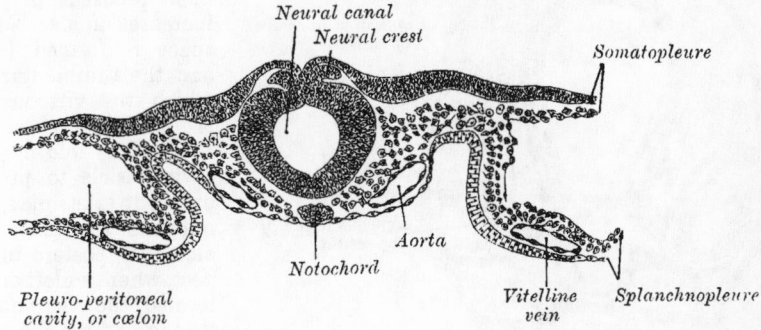

and is there met by an ingrowth from the ectoderm, out of which the fibres of the crystalline lens and the epithelium of the conjunctiva, cornea, and lachrymal gland are developed.

The first appearance of the eye consists in a hollow, laterally directed protrusion of the fore-brain ; this is called the *optic vesicle* (fig. 139). It is at first an open cavity communicating by a hollow stalk with that of the cerebral vesicle. As it is prolonged outwards, the ectoderm lying immediately over it becomes thickened, and then forms a depression which gradually encroaches on the most prominent part of the optic vesicle ; this in its turn appears to recede before it,

FIG. 139.—Transverse section of head of chick embryo of forty-eight hours' incubation. (From Duval's 'Atlas d'Embryologie.')

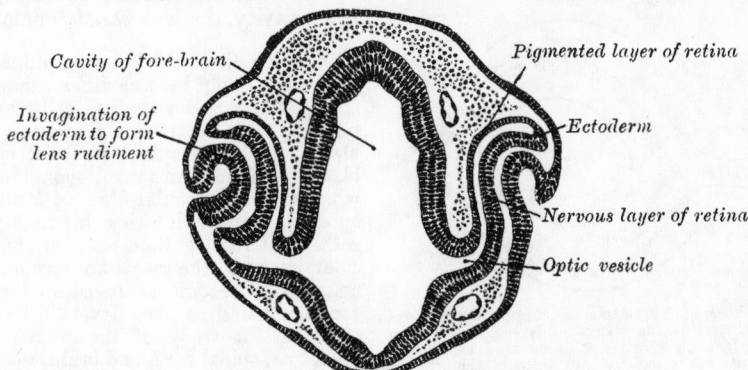

so as to become at first depressed and then inverted in the manner indicated in figs. 139 and 140, so that the cavity of the vesicle is almost obliterated by the folding back of its anterior half, and the original vesicle converted into a cup, the *optic cup*, in which the involuted ectodermal layer, the rudiment of the lens, is received (fig. 140) ; at the same time the proximal part of the vesicle becomes elongated and narrowed into a hollow stalk, the *optic stalk*. This cup-shaped cavity consists therefore of two layers which are continuous with each other at the cup-margin ; the outer is thin, and eventually forms the pigmented layer of the retina ; the inner is thick, and is converted into the nervous layers of the retina (fig. 140).* Between the two is the remnant of the cavity of the original optic vesicle, which finally becomes obliterated by the union of its two layers. The mouth of the optic cup overlaps the equator of the lens as far as the future aperture of the pupil. In this region the inner or retinal layer of the cup is not differentiated into nervous elements, but remains as a single layer of columnar cells, which becomes applied to the cells of the pigmented layer, and the conjoined strata form the *pars ciliaris* and *pars iridica retinæ* of the adult (fig. 143). As development proceeds the optic cup increases in size, and thus a space is formed between it and the rudimentary lens, in which the vitreous humour is developed (figs. 141 and 143). The folding in of the optic vesicle to produce the optic cup takes place not only opposite the lens, but also along its postero-inferior aspect, where a cleft or fissure is formed, the *choroidal fissure*, through which the mesoderm extends to form the vitreous humour. This gap or cleft is continued for some distance into the stalk of the optic vesicle, and thus allows a process of mesoderm, with the arteria centralis retinæ, to extend along the cleft (fig. 142). After a time the gap or fissure becomes closed, by a coalescence of its margins, and the arteria centralis retinæ assumes its permanent position in the centre of the optic nerve.

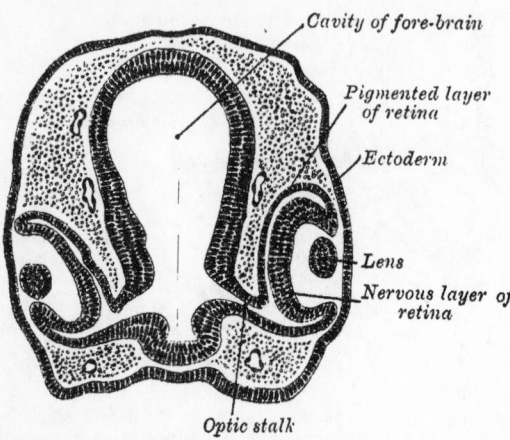

FIG. 140.—Transverse section of head of chick embryo of fifty-two hours' incubation. (From Duval's 'Atlas d'Embryologie.')

Cavity of fore-brain

Pigmented layer of retina

Ectoderm

Lens

Nervous layer of retina

Optic stalk

The lens is at first a thickening of the ectoderm, then a depression or involution takes place, thus forming an open follicle, the margins of which gradually approach each other and coalesce, forming a cavity, the *lens vesicle*, enclosed by

FIG. 141.—Diagrammatic sketch of a vertical longitudinal section through the eyeball of a human fœtus of four weeks. (After Kölliker.) Magnified 100 diameters. The section is a little to the side, so as to avoid passing through the ocular cleft.

vp

P

vp

vs

l

c

op

c. The ectoderm. *l.* The lens. *op.* Optic stalk with its cavity, *vp.* *p.* The pigmented layer of the retina. *r.* The inner layer from which the nervous elements of the retina are formed. *vs.* Rudiment of the vitreous humour.

* The cells of the inner or retinal layer of the optic cup become differentiated into spongioblasts and germinal cells, and the latter by their subdivision give rise to neuroblasts. As in the spinal cord, the spongioblasts ramify to form a myelospongium, from which the sustentacular fibres of Müller, the outer and inner limiting membranes, together with the ground-work of the molecular layers of the retina are formed. The neuroblasts become arranged to form the ganglionic and nuclear layers. Cameron, after careful study of the neuroblasts in the retina, spinal cord, and brain, maintains that they consist of nuclei only, and that they possess no cytoplasmic investment; the 'clear protoplasm,' which has been described as surrounding them during mitotic division, being merely the achromatic nuclear substance set free owing to the disappearance of the nuclear membrane. He further maintains that the nuclei of the nuclear layers undergo subsequent multiplication by *direct* division. The layer of rods and cones is first developed in the central part of the optic cup, and from there gradually extends towards the cup-margin. The rods and cones first appear as clear globules, which, after being protruded through the inner limiting membrane, rapidly increase in size—a process which would seem to depend on their power of digesting and absorbing the pigment from the cells of the pigmented layer. ('Development of the Retina in Amphibia,' *Journal of Anatomy and Physiology*, vol. xxxix., 1905.)

ectodermal cells. At the point of involution the external layer of ectoderm separates from the lens and passes freely over the surface, so that the lens becomes disconnected from the general ectoderm, and recedes into the optic cup,

FIG. 142.—Optic cup and choroidal fissure seen from below, from a human embryo of about four weeks. (Kollmann.)

while the cuticular layer covering it is developed into the corneal epithelium. The cells forming the posterior or inner wall of the lens vesicle rapidly increase in size, becoming elongated and developed into the lens fibres, and, filling up the

FIG. 143.—Horizontal section through the eye of an eighteen days' embryo rabbit. × 30. (Kölliker.)

cavity, convert it into a solid body. The cells on the anterior wall retain their cellular character, and form the anterior lens epithelium of the adult. The optic cup contains a quantity of mesodermal tissue continuous with the general mesoderm through the choroidal fissure. This tissue becomes converted into the vitreous humour, and surrounds the lens with a vascular membrane—the *vascular capsule of the lens*. From the central artery of the retina several branches are prolonged forwards through the vitreous body to the capsule of the lens, but by the sixth month these have all undergone atrophy except one, which persists till the ninth month as the *arteria hyaloidea*. It disappears, however, before birth, and its position is indicated in the adult by the *canalis hyaloideus of Stilling*. The front part of the vascular capsule of the crystalline lens forms the *membrana pupillaris*, and also attaches the iris to the capsule of the lens. It disappears about the seventh month.

The optic stalk becomes solid by the union of the margins of the choroidal fissure and by the obliteration of the stalk cavity. By far the greater number of the optic nerve-fibres arise as the axis-cylinder processes of the nerve-cells in the retina, and pass along the optic stalk to the brain. Some of the fibres, however, are centrifugal, and arise from nerve-cells in the brain. The outer layer of the optic cup forms the pigmentary layer of the retina, while its inner layer becomes differentiated into (*a*) spongioblasts, which are the rudiments of the supporting tissues of the retina, and (*b*) neuroblasts, which form its nervous elements (see page 118).

The sclera, cornea, and choroid are developed from the mesoderm surrounding the optic vesicle.

The eyelids are formed as small cutaneous folds (fig. 143), which at the end of the third month come together and unite in front of the globe and cornea. This union is broken up and the eyelids separate before the end of foetal life.

The lachrymal sac and nasal duct result from a thickening of the ectoderm in the groove between the lateral nasal and maxillary processes. This thickening becomes hollowed out into a channel, and the lips of the groove meet over it, enclose it, and convert it into a duct, which eventually opens into the nasal fossa. The epithelium of the cornea and conjunctiva, and that which lines the ducts and alveoli of the lachrymal gland, are of ectodermal origin, as are also the eyelashes and the lining cells of the glands which open on the lid-margins.

Development of the Ear.—The first rudiment of the ear appears shortly after that of the eye, in the form of a thickening of the ectoderm covering the hind-brain. The thickening is followed by an involution of the ectoderm to form the *auditory pit* (fig. 144), which becomes deeper and deeper, and sinking towards the base of the skull, forms a flask-shaped cavity. The mouth of the flask is then closed, and thus a shut sac is formed, the *otic vesicle* (fig. 145), which by its

FIG. 144.—Section through the head of a human embryo, about twelve days old, in the region of the hind-brain. (Kollmann.)

FIG. 145.—Section through hind-brain and otic vesicle of an embryo more advanced than that of fig. 144. (After His.)

Cavity of hind-brain

Auditory pit

Epiblast

Notochord

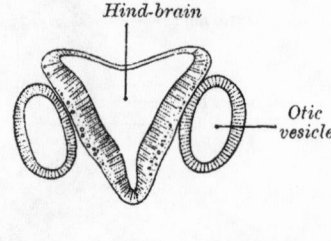

Hind-brain

Otic vesicle

sinking inwards comes to be placed between the ali-sphenoid and basi-occipital matrices. From it the epithelial lining of the labyrinth is formed. The otic vesicle is embedded in a mass of mesoblastic tissue, which rapidly undergoes chondrification and ossification to form the bony labyrinth. The vesicle becomes pear-shaped; and the neck of the flask, or *recessus labyrinthi*, prolonged upwards, forms the aquæductus vestibuli. From it are given off certain diverticula, from which the various parts of the labyrinth are formed. One from the anterior end gradually elongates, and, forming a tube, bends on itself and becomes the cochlea. Three others, which appear on the surface of the vesicle, form the semicircular

canals, of which the external canal is the last to be developed (figs. 147 and 148).
Subsequently, a constriction takes place in the original vesicle, which nearly
divides it into two, and from these are formed the utricle and saccule (fig. 148).
Finally, the auditory nerve, which has been developed from the 'neural crest'

FIG. 146.—Left auditory vesicle of a human embryo
of four weeks, seen from
the outer surface. (W.
His, jun.)

FIG. 147.—Left auditory vesicle of a human embryo of
five weeks, seen from the outer surface. (W. His, jun.)

in the manner above de-
scribed (page 116), pierces
the auditory capsule in
two main divisions—one
for the vestibule, the other
for the cochlea. The middle
ear and Eustachian tube are the remains of the inner part of the first branchial
cleft (hyomandibular), and are closed externally by the membrana tympani, which
originally consists of a layer of epiblast externally, and a layer of hypoblast
internally ; between these two layers the mesoblast extends to form the substantia

FIG. 148.—Transverse section through head of fœtal sheep, in the
region of the labyrinth.   × 30.   (After Boettcher.)

propria of the membrane.   With regard to the exact mode of development of
the ossicles of the middle ear there is some difference of opinion.   The view
generally maintained is that the *incus* and *malleus* are developed from the
proximal end of the mandibular (Meckel's) cartilage (fig. 125) ; that the base of
the *stapes* is formed by the ossification of the mesoderm which fills in the foramen
ovale, while its arch is developed around a small vessel, the stapedial artery, which

subsequently undergoes atrophy. As already stated (footnote, page 102), Gadow regards all three ossicles as being derived from the hyomandibula.

The external auditory meatus is formed from the outer part of the hyomandibular cleft, while the pinna is developed by the gradual differentiation of six tubercles which appear around the outer margin of the cleft. Two tubercles

FIG. 149.—Transverse section of the canal of the cochlea of a fœtal cat.
(After Boettcher and Ayres.)

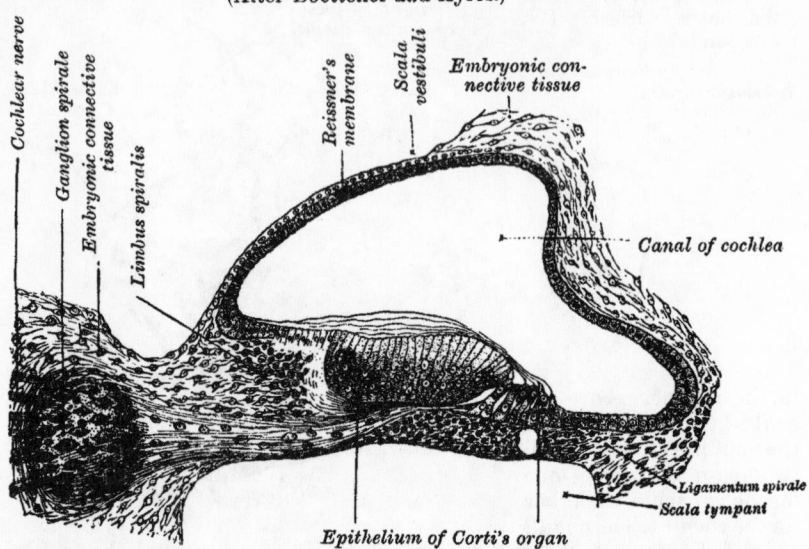

appear on the posterior edge of the mandibular arch; these represent the rudiments of the tragus and crus helicis. Three are found on the hyoid arch, and indicate, from below upwards, the lobule, antitragus, and antihelix. One

FIG. 150.—Left ears of human embryos estimated at thirty-five and thirty-eight days respectively. (After His.)

arises above the cleft, and grows downwards behind the antitragus and antihelix; from it and its downward prolongation the upper and posterior parts of the helix are developed (fig. 150).

Development of the Skin, Glands, and Soft Parts.—The epidermis and its appendages, consisting of the hairs, nails, sebaceous and sweat glands, are developed from the ectoderm, while the corium or true skin is of mesodermic origin, being derived from the cutis plates of the protovertebral somites. About the fifth week the epidermis consists of two layers of cells, the deeper one corresponding to the rete mucosum. The subcutaneous fat appears about the fourth month, and the papillæ of the true skin about the sixth. A considerable desquamation of epidermis takes place during fœtal life, and this desquamated epidermis, mixed with a sebaceous secretion, constitutes the *vernix caseosa*, with which the skin is smeared during the last three months of fœtal life. The nails are formed at the third month, and begin to project from the epidermis about the sixth. The hairs appear between the third and fourth months in the form of solid downgrowths of the deeper layer of the epidermis, which then become inverted by papillary projections from the corium. About the fifth month, the fœtal hairs (*lanugo*) appear, first on the head and then on the other parts ; they drop off after birth, and give place to the permanent hairs. The cellular structures of the sudoriferous and sebaceous glands are formed from the ectoderm, while the connective tissue and blood-vessels are derived from the mesoderm. The mammary gland is also formed partly from mesoderm and partly from ectoderm—its blood-vessels and connective tissue being derived from the former, its cellular elements from the latter. Its first rudiment is seen about the third month, in the form of a number of small inward projections of the ectoderm which invade the mesoderm ; from these, secondary tracts of cellular elements radiate and subsequently give rise to the glandular follicles and ducts. The development of the former, however, remains imperfect, except in the adult female.

Development of the Muscles.—The voluntary muscles are developed from the muscle-plates or myotomes of the protovertebral somites. By the end of the third week the cells of the muscle-plates commence to elongate in an antero-posterior direction, and, losing their cellular character, are converted into striped muscular fibres. By the fusion of the myotomes in an antero-posterior direction, the longitudinal muscles of the back are developed; and from extensions of the myotomes into the body wall and limb buds, the muscles of these parts arise. The involuntary muscles are derived from the splanchnopleure mesoblast, and are therefore not connected in any way with the protovertebral somites.

Development of the Blood-vascular System.—There are three distinct stages in the development of the circulatory system, each in accordance with the manner in which nourishment is provided for at different periods of the existence of the individual. In the first stage there is the *vitelline circulation*, during which nutriment is extracted from the *vitellus* or contents of the yolk-sac. In the second stage there is the *placental circulation*, during which nutrition is obtained by means of the placenta from the blood of the mother. In the third stage there is the *complete circulation of the adult*, commencing after birth, during which nutrition is provided for by the organs of the individual itself.

Blood-vessels first make their appearance in the mesodermal wall of the yolk-sac, i.e. outside the body of the embryo. Here the mesodermal cells become arranged into solid strands or cords which join to form a close-meshed network. The peripheral cells of these strands become flattened and joined to each other by their edges to form the walls of the primitive blood-vessels, which at this stage are simple endothelial tubes. Fluid collects within these tubes, and the more centrally situated cells of the cell-cords are thus pushed to the sides of the vessels and appear as masses of loosely arranged cells which project towards the lumen of the tube. These masses are termed *blood islands*, and their cells acquire colouring matter (hæmoglobin), and are then detached to form the blood-corpuscles.[*] The earliest blood-corpuscles are all nucleated : they are also capable of subdivision and of executing amœboid movements, and in these respects they resemble white blood-corpuscles. Soon, however, true white blood-corpuscles make their appearance, and, according to Beard,[†] are first derived from the rudiments of the thyroid gland.

Coincident with the development of the blood-vessels in the vascular area, the first rudiment of the heart appears within the body of the embryo as a pair of tubular vessels which are developed in the splanchnopleure of the pericardial area. These are named the *primitive aortæ*, and a direct continuity is soon

---

[*] Some observers incline to the view that the blood-corpuscles are of entodermal origin, being developed from the endothelium of the vessels, the sequence of the development of the different structures being : first the heart, then the blood-vessels, and lastly the blood-corpuscles. (Consult Dr. E. Mehnert's *Biomechanik*, Jena, 1898.)

[†] *Anatomischer Anzeiger*, December 1900.

established between them and the vessels of the vascular area. Each receives anteriorly a vein—the vitelline vein—from the yolk-sac, and is prolonged backwards on the lateral aspect of the notochord under the name of the dorsal aorta. The dorsal aortæ end at first on the yolk-sac; but with the development of the allantois, they are continued backwards through the body-stalk as the umbilical arteries to the villi of the chorion.

By the forward growth and flexure of the head the pericardial area and the anterior portions of the primitive aortæ are folded backwards on the ventral aspect of the fore-gut, and the original position of the somatopleure and splanchnopleure layers of the pericardial area is reversed, the latter being placed on the dorsal aspect of the former. Each primitive aorta now consists of a ventral and a dorsal part connected anteriorly by an arch. These three parts are named respectively the anterior ventral aorta, the dorsal aorta, and the first cephalic arch. The vitelline veins which enter the embryo through the anterior wall of the umbilical orifice are now continuous with the posterior ends of the anterior ventral aortæ. With the formation of the tail-fold the posterior parts of the primitive aortæ are carried forward in a ventral direction to form the posterior ventral aortæ and primary caudal arches.* In the pericardial region the two primitive vessels grow together and fuse to form a single tubular heart (fig. 151), the posterior end of which receives the two vitelline veins, while from its anterior

FIG. 151.—Diagram to illustrate the simple tubular condition of the heart. (Drawn from Ecker-Ziegler model.)

FIG. 152.—Heart further advanced than in fig. 151. (Drawn from Ecker-Ziegler model.)

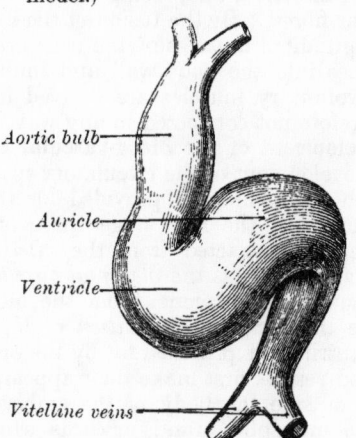

Aortic bulb
Ventricle
Auricle
Sinus venosus
Vitelline veins

Aortic bulb
Auricle
Ventricle
Vitelline veins

end there arise the two anterior ventral aortæ.† By the rhythmical contraction of the tubular heart the blood is forced through the aortæ and blood-vessels of the vascular area, from which it is returned to the heart by the vitelline veins. This constitutes the vitelline circulation (fig. 110), and by means of it nutriment is absorbed from the vitellus.

The vitelline veins at first open separately into the posterior end of the tubular heart, but after a time their terminal portions fuse, and the two vessels communicate with the heart through a common orifice. The vitelline veins ultimately drain the blood from the alimentary canal, and are modified to form the portal vein. This is caused by the growth of the liver, which interrupts their direct continuity with the heart; and the blood returned by them circulates through the liver before reaching the heart.

Coincident with the atrophy of the yolk-sac the vitelline circulation diminishes and ultimately ceases, while more and more blood is simultaneously carried through the umbilical arteries to the villi of the chorion. Subsequently, as the non-placental chorionic villi atrophy, their vessels disappear; and then the umbilical arteries convey the whole of their contents to the placenta, whence it is returned to the heart by the umbilical veins. In this manner the placental

* Young and Robinson, *Journal of Anatomy and Physiology*, vol. xxxii.
† In most fishes and in the amphibia the heart originates as a single median tube.

circulation is established, and by means of it nutritive materials are absorbed from, and waste products given up to, the maternal blood.

The umbilical veins, like the vitelline, become interrupted by the liver, and the blood returned by them passes through this organ before reaching the heart. Ultimately the right umbilical vein shrivels up and disappears, as will be explained later on.

During the occurrence of these changes great alterations take place in the primitive heart and blood-vessels, and now require description.

Further Development of the Heart.—The following is an outline of the changes which take place during the further development of the heart.

The simple tubular heart, already described, becomes elongated and bent on itself so as to form an **S**-shaped loop (figs. 151 and 152), the anterior part bending to the right and the posterior part to the left. The intermediate portion

FIG. 153.—Head of chick embryo of about thirty-eight hours' incubation, viewed from the ventral surface. × 26. (From Duval's 'Atlas d'Embryologie.')

arches transversely from right to left, and then turns sharply forwards into the anterior part of the loop. Slight constrictions make their appearance in the tube and divide it from behind forwards into four parts, viz.: (1) the *sinus venosus* (*sinus reuniens* of His); (2) the common auricle; (3) the common ventricle; (4) the aortic bulb (figs. 151 to 153). From the aortic bulb the two ventral

FIG. 154.—Heart of human embryo of about fifteen days. (Reconstruction by His.)

aortæ extend forwards in the floor of the pharynx. The constriction between the auricle and ventricle is well marked and constitutes the *auricular canal*, while that between the ventricle and the aortic bulb is less distinct and is termed the *fretum Halleri*. The former indicates the site of the future auriculo-ventricular valves, the latter that of the semilunar valves.

The sinus venosus is at first situated in the septum transversum (a layer of mesoderm in which the liver and the central tendon of the Diaphragm are developed) behind the common auricle, and is formed by the union of three pairs of veins, viz.: (1) the veins or ducts of Cuvier from the body of the embryo; (2) the omphalo-mesenteric veins from the yolk-sac; (3) the umbilical veins from the placenta (fig. 156). The sinus is at first placed transversely, and opens by a median aperture into the common auricle. Soon, however, it assumes an oblique position, and its right half or horn becomes larger than the left, while the

FIG. 155.—Heart showing expansion of auricles.
(Drawn from Ecker-Zeigler model.)

opening into the auricle now communicates with the right portion of the auricular cavity. The right horn ultimately becomes incorporated with and forms a part of the right auricle, the line of union between it and the auricle proper being indicated in the interior of the adult auricle by a vertical crest, the *crista terminalis* of His. The left horn, which ultimately receives only the left duct of Cuvier, persists as the coronary sinus (fig. 162). The omphalo-mesenteric and umbilical veins are soon replaced by a single vessel, the inferior vena cava, and

FIG. 156.—Heart of human embryo, 4·2 mm. long, seen from behind. (His.)

the three veins (inferior vena cava and right and left Cuvierian ducts) open into the dorsal aspect of the auricle by a common slit-like aperture. The upper part of this aperture represents the opening of the permanent superior vena cava, the lower that of the inferior vena cava, and the intermediate part the orifice of the coronary sinus. The slit-like aperture lies obliquely, and is guarded by two valve-like folds of endocardium, the *right* and *left venous valves*. The left venous valve disappears, while the right is subsequently divided to form the Eustachian and Thebesian valves. At the lower extremity of the slit is a triangular thickening, the *spina vestibuli* of His, which partly closes the aperture between the two auricles, and which, according to His, takes a part in the formation of both the interauricular and interventricular septa.

The common auricle and ventricle are each subdivided into two cavities, and the aortic bulb is divided into the pulmonary artery and aorta as follows :

The auricular canal is at first a short straight tube connecting the auricular with the ventricular portion of the heart, but it becomes overlapped by the growing

FIG. 157.—Diagrams to show the development of the septum of the aortic bulb and of the ventricles. (Born.)

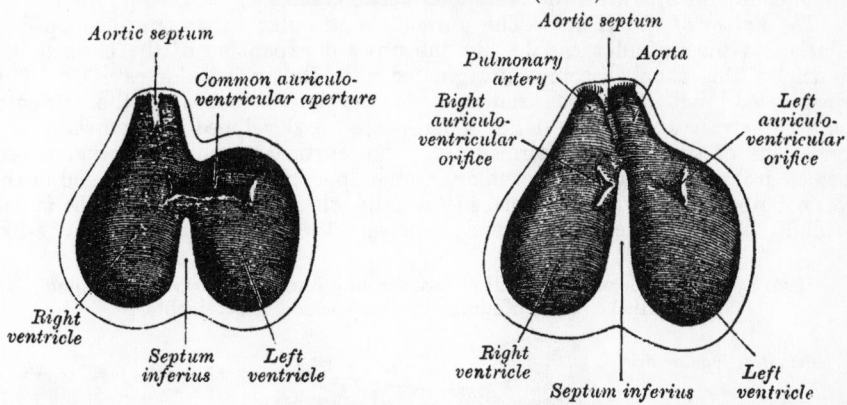

auricles and ventricles so that its position on the surface of the heart is only indicated by an annular constriction (fig. 155). Its lumen is reduced to a transverse slit, and two thickenings appear, one on its dorsal and another on its ventral wall. These thickenings, or *endocardial cushions* (fig. 159) as they are termed, project into the canal, and, meeting in the middle line, unite to form the *septum intermedium* which divides the canal into two channels, the future right and left auriculo-ventricular orifices.

The common auricular cavity becomes subdivided into right and left auricles by a septum, the *septum superius*, which grows downwards into the auricular cavity so that the two auricles communicate with each other only below the margin of the septum. This communication (ostium primum of Born) does not, however, represent the foramen ovale, for the septum grows downwards and blends with that of the auricular canal formed by the fusion of the endocardial cushions. The foramen ovale (ostium secundum of Born) results from a perforation of the primary septum. A second septum grows downwards from the upper wall of the auricle to the right of the primary septum, and forms the annulus ovalis. Shortly after birth it fuses with the primary septum, and by this means the foramen ovale is closed. Sometimes this fusion is incomplete and the upper part of the foramen remains patent.

The common ventricle becomes divided by a septum, the *septum inferius* (fig. 157), which grows upwards from the lower part of the ventricle, its position

FIG. 158.—Transverse sections through the aortic bulb to show the growth of the aortic septum. The lowest section is on the left, the highest on the right of the figure. (After His.)

being indicated on the surface of the heart by a furrow. It extends upwards almost as far as the auricular canal, but for some time an interventricular foramen exists between it and the septum of the auricular canal (septum intermedium).

The aortic bulb is divided by the *aortic septum*. This makes its appearance at the distal end of the bulb as two ridge-like thickenings of its endothelial

lining ; these increase in size, and, projecting into the lumen, ultimately fuse
to form the septum, and thus the aortic bulb is divided into the pulmonary artery
and the aorta.   The aortic septum takes a spiral course towards the proximal end
of the bulb, so that the two vessels lie side by side above ; but near the heart the
pulmonary artery is in front of the aorta (fig. 158).   The septum grows down
into the ventricle as an oblique partition, which ultimately blends with the septum
inferius of the ventricles in such a way as to bring the right ventricle into communication with the pulmonary artery, and through the latter with the sixth pair
of aortic arches ; while the left ventricle is brought into continuity with the aorta
which communicates with the remaining aortic arches.

The Valves of the Heart.—The auriculo-ventricular valves are developed in
relation to the auricular canal.   By the upward expansion of the bases of the
ventricles this canal becomes invaginated into the ventricular cavities.   This
invaginated part forms the rudiment of the lateral cusps of the auriculo-
ventricular valves ; their mesial or septal cusps are developed as downward prolongations of the septum intermedium.   The aortic and pulmonary valves are
formed from four endocardial cushions which appear within the lower end of the
aortic bulb, two of which are placed laterally, the third in front and the fourth
behind.   The lateral cushions merely represent the lower parts of the ridge-like

Fig. 159.—Interior of dorsal half of heart from a human embryo 10 mm. long.
      (After His.)   (From Kollmann's ' Entwickelungsgeschichte.')

thickenings which grow in to form the aortic septum.   When the aortic septum
is completed these lateral cushions are each subdivided into two, thus giving rise
to six thickenings—the rudiments of the semilunar valves—three at the aortic
and three at the pulmonary orifice.

Peculiarities of the fœtal heart.—In early fœtal life the heart is placed
directly under the head and is relatively of large size.   Later it assumes its
position in the thorax, but lies at first in the middle line ; towards the end of
pregnancy it gradually becomes oblique in direction.   Its auricular portion is at
first larger than the ventricular part, and the two auricles communicate freely
through the foramen ovale.   In consequence of the communication between the
pulmonary artery and the aorta, through the ductus arteriosus, the contents of
the right ventricle are mainly carried into the latter vessel instead of to the
lungs, and hence the wall of the right ventricle is as thick as that of the left.
At the end of fœtal life, however, the left ventricle is thicker than the right,
a difference which becomes more and more emphasised after birth.

Further Development of the Arteries.—It has been seen (page 124) that the
primitive aortæ arise from the aortic bulb, and that each consists of a ventral and
a dorsal part which are continuous through the first aortic arch.   The dorsal aortæ
run backwards at first on either side of the notochord, but after a time they fuse
to form a single trunk, which constitutes the greater part of the future descending
aorta.   The first aortic arches pass through the mandibular arches, and behind

them five additional pairs are developed within the corresponding visceral arches; so that, in all, six pairs of aortic arches are formed (fig. 161). The first and second arches pass between the ventral and dorsal aortæ, while the others arise at first by a common trunk from the aortic bulb. As the neck elongates, the ventral aortæ are drawn out, and the third and fourth arches arise directly from these vessels.

In fishes these arches persist and give off branches to the gills, in which the blood is oxygenated. In mammals some of them remain as permanent structures, while others disappear or become obliterated (fig. 161).

The ventral aortæ.—These persist on both sides. The right forms (*a*) the innominate artery, (*b*) the right common and external carotid arteries. The left

FIG. 160.—Profile view of a human embryo estimated at twenty or twenty-one days old. (After His.)

gives rise to (*a*) the short portion of the aortic arch, which reaches from the origin of the innominate artery to that of the left common carotid artery; (*b*) the left common and external carotid trunks.

The aortic arches.—The first and second disappear; the third constitutes the commencement of the internal carotid artery, and is therefore named the *carotid arch*. The fourth right arch forms the right subclavian as far as the origin of its internal mammary branch; while the fourth left arch constitutes the arch of the aorta between the origin of the left carotid artery and the termination of the ductus arteriosus. The fifth arch disappears on both sides. The sixth right arch disappears; that on the left side gives off the pulmonary arteries and persists as the ductus arteriosus. This duct remains pervious during the whole of fœtal life, but after birth becomes obliterated. His found that in the

K

early embryo the right and left arches each gave a branch to the lungs, but that later both pulmonary arteries took origin from the left arch.

The dorsal aortæ.—In front of the third aortic arches the dorsal aortæ persist and form the forward continuation of the internal carotid arteries. Behind the third arch the right dorsal aorta disappears as far as the point where the two dorsal aortæ fuse to form the descending aorta. The part of the left dorsal aorta which intervenes between the third and fourth arches disappears, while the remainder persists to form the descending part of the arch of the aorta. A constriction, the *aortic isthmus*, is sometimes seen in the aorta between the origin of the left subclavian and the attachment of the ductus arteriosus.

Sometimes the right subclavian artery arises from the aortic arch beyond the origin of the left subclavian and passes upwards and to the right behind the trachea and œsophagus. This condition may be explained by the persistence of the right dorsal aorta and the obliteration of the fourth right arch.

Fig. 161.—Scheme of the aortic arches and their destination.
(Modified from Kollmann.)

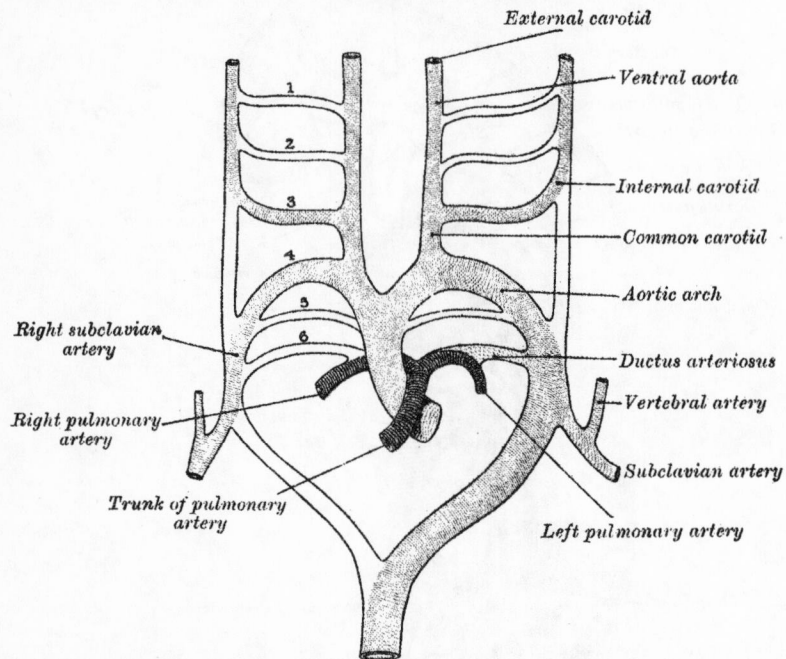

In birds the fourth right arch forms the permanent arch of the aorta; in reptiles the fourth arch on both sides persists and gives rise to the double aortic arch in these animals.

The heart originally lies on the ventral aspect of the pharynx, immediately behind the stomatodœum. With the elongation of the neck and development of the lungs it recedes within the thorax, and, as a consequence, the ventral aortæ are drawn out and the original position of the fourth and fifth arches is greatly modified. Thus, on the right side the fourth recedes to the root of the neck, while on the left side it is withdrawn within the thorax. The recurrent laryngeal nerves originally passed to their distribution on the outer sides of the sixth pair of arches, and therefore become pulled backwards with the descent of these structures, so that in the adult the left hooks round the ductus arteriosus; owing to the disappearance of the fifth and sixth right arches the right nerve hooks round that immediately above them, i.e. the commencement of the subclavian artery. A series of segmental arteries arises from the primitive dorsal aortæ, those in the neck alternating with the cervical segments of the spine. The segmental artery which lies between the sixth and seventh cervical segments is of special interest, since it forms the lower part of the vertebral artery and, when the fore-limb bud appears, sends a branch to it (i.e. the subclavian artery); the upper part of the vertebral artery is formed by an antero-posterior anastomosis between the higher segmental arteries. From this segmental artery the entire left subclavian and the greater part of the right subclavian are formed.

The subclavian artery is prolonged into the limb under the names of the axillary and brachial arteries, and these together constitute the arterial stem for the upper arm. The direct continuation of this stem in the forearm is formed by the anterior interosseous artery; while the radial and ulnar vessels, which ultimately exceed this vessel in size, are in reality lateral branches of the main stem.

The formation of the primary caudal arches has already been referred to (page 124), and the fusion of the dorsal aortæ to form the greater part of the systemic aorta has also been pointed out (page 128). The middle sacral artery of the adult was formerly regarded as the direct continuation of the adult aorta, but

Fig. 162.—Human embryo with heart and anterior body wall removed to show the sinus venosus and its tributaries. (After His.) (From Kollmann's ' Entwicke. lungsgeschichte.')

Right primitive jugular vein   *Left primitive jugular vein*

Right cardinal vein   *Left cardinal vein*

Right duct of Cuvier

Sinus venosus   *Left duct of Cuvier*

Right hepatic vein   *Left hepatic vein*

Portal vein   *Left umbilical vein*

Portal vein

Right umbilical vein   *Left umbilical vein*

Umbilical cord

Young and Robinson (*op. cit.*) maintain that it ' is a secondary branch, probably representing fused segmental arteries.' They have further pointed out that while the dorsal and ventral extremities of the primary caudal arches remain, their middle portions ' disappear and are replaced by " secondary " caudal arches which lie to the outer sides of the Wolffian ducts.' ' The vessels which are to be looked upon as the posterior continuations of the primitive aorta in the adult in man, rodents, &c., are the common iliac, internal iliac, and hypogastric arteries.'

The hypogastric arteries are continued into the umbilical cord as the umbilical arteries. After birth they become obliterated from the umbilicus as far as the origin of the superior vesical arteries.

The primary arterial stem for the lower limb is formed by the sciatic artery, which accompanies the great sciatic nerve along the posterior aspect of the thigh to the back of the knee, whence it is continued as the peroneal artery. This arrangement exists in reptiles and amphibians. The femoral artery arises later

as a branch of the common iliac, and, passing down the front and inner side of
the thigh to the bend of the knee, joins the sciatic artery.   The femoral quickly
enlarges, and, coincident with this, the part of the sciatic immediately above the
knee undergoes atrophy.   The anterior and posterior tibial arteries, like the radial
and ulnar in the forearm, are branches of the main arterial stem.

   Development of the Veins.—The formation of the great veins of the embryo
may be best considered under two groups, visceral and parietal.

   The *visceral veins* are the two vitelline or omphalo-mesenteric veins bringing
the blood from the yolk-sac, and the two umbilical or allantoic veins returning
the blood from the placenta ; these four veins open close together into the sinus
venosus (fig. 156).

   The vitelline veins run upwards at first in front, and subsequently on either
side of the intestinal canal.   They unite on the ventral aspect of the canal, and
then encircle the duodenal portion of the intestinal tube by forming around it
two venous rings, the first on its dorsal, the second on its ventral aspect.   The
portions of the veins above the upper ring become invaded by the developing

FIG. 163.—The liver, and the veins in connection with it, of a human embryo, twenty-four
   or twenty-five days old, as seen from the ventral surface.   (After His.)   (Copied from
   Milnes Marshall's ' Embryology.')

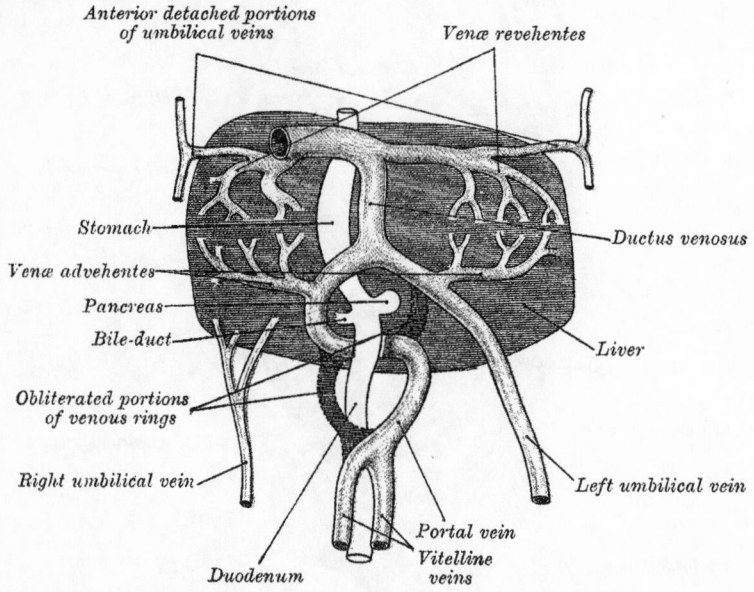

liver and broken up by it into a network of smaller vessels, the central part of
the network consisting of a capillary plexus.   The branches which convey the
blood to this plexus are named the *venæ advehentes*, and become the branches of
the portal vein ; while the vessels which drain the plexus into the sinus venosus
are termed the *venæ revehentes*, and form the future hepatic veins (figs. 162 and
163).   Ultimately the left vena revehens no longer communicates directly with
the sinus venosus, but opens into the right vena revehens.

   The lower part of the *portal vein* is formed from the fused vitelline veins
which receive the veins from the alimentary canal; its upper part is derived
from the venous rings by the persistence of the left half of the lower and the
right half of the upper ring, so that the vessel forms a spiral turn round the
duodenum (fig. 163).

   The two umbilical veins fuse early to form a single trunk in the allantois, but
remain double for some time within the embryo and pass forwards to the sinus
venosus in the side walls of the body.   Like the vitelline veins, their direct
connection with the sinus venosus becomes interrupted by the invasion of the
liver, and thus at this stage the whole of the blood from the yolk-sac and
placenta passes through the substance of the liver before it reaches the heart.
The right umbilical vein shrivels up and almost entirely disappears ; the left, on

the other hand, becomes enlarged and opens into the upper venous ring of the vitelline veins.   Finally a direct branch is established between this ring and the right hepatic vein; this branch is named the *ductus venosus*, and, enlarging rapidly, it forms a wide channel through which most of the blood, returned from the placenta, is carried direct to the heart without passing through the liver. A small proportion of the blood from the placenta is, however, conveyed from the left umbilical vein to the liver through the left vena advehens.   The left umbilical vein and the ductus venosus undergo atrophy and obliteration after birth, and form respectively the ligamentum teres and ligamentum venosum of the liver (fig. 163).

The *Parietal Veins.*—The first indication of a parietal system consists in the appearance of two short transverse veins (the *ducts of Cuvier*), which open, one on either side, into the auricular portion of the heart.   Each of these ducts is formed by an ascending and descending vein.   The ascending veins return the blood from the parietes of the trunk and from the Wolffian bodies, and are called *cardinal* veins.   The descending veins return the blood from the head, and are called *primitive jugular* veins (fig. 160).   The blood from the lower limbs is collected by the right and left iliac veins, which, in the earlier stages

FIG. 164.—Diagram to illustrate the development of the principal systemic veins. (Hertwig.)

dc. Duct of Cuvier. *je, ji.* External and internal jugular veins. *s.* Subclavian. *vh.* Hepatic veins. *U.* Umbilical. *ci, ci².* Vena cava inferior. *ca.* Cardinal veins. *ilcd, ilcs.* Right and left common iliac veins. *ad, as.* Right and left innominate veins. *cs.* Vena cava superior. *css.* Rudimentary portion of left superior vena cava. *cc.* Coronary sinus. *az.* Azygos major. *hz, hz¹.* Azygos minor. *ile.* External iliac. *ili.* Internal iliac. *r.* Renal vein.

of development, open into the corresponding right and left cardinal veins (fig. 164); later on, a transverse branch (the left common iliac vein) connects the lower ends of the two cardinal veins, and through this the blood is carried into the right cardinal vein.   The portion of the left cardinal vein above the left common iliac vein becomes atrophied as high as the level of the left renal vein, above which it persists as the superior and inferior azygos minor veins and the lower part of the left superior intercostal vein.   The right cardinal vein, which now receives the blood from both lower extremities, forms a large venous trunk along the posterior abdominal wall; up to the level of the renal veins it forms the lower part of the inferior vena cava.   Above the level of the renal veins the right cardinal vein persists as the vena azygos major, and receives the right intercostal veins, while the azygos minor veins are brought into communication with it by the development of transverse branches in front of the spinal column (fig. 164).

Inferior vena cava (fig. 164).—As just stated, the lower part of the inferior vena cava is formed from the post-renal part of the right cardinal vein.   The upper part of the inferior vena cava arises as a small vein which lies between the two primitive kidneys and which is continuous superiorly with the conjoined ductus venosus and right hepatic vein.   This small vein divides below into two

branches which anastomose with the right and left cardinal veins at the level of
the renal veins.   Through the anastomosis thus formed with the right cardinal
vein the blood from the latter is carried into the small vein, which then
becomes enlarged to form the pre-renal part of the vena cava inferior.   Coincident
with the atrophy of the post-renal part of the left cardinal vein, the blood from
the left renal vein and its tributaries is also carried into the vena cava inferior.

In consequence of the atrophy of the Wolffian bodies the cardinal veins
diminish in size; the primitive jugular veins, on the other hand, become
enlarged, owing to the rapid development of the head and brain.   They are
further augmented by receiving the vein (*subclavian*) from the upper extremity,
and so come to form the chief veins of the Cuvierian ducts; these ducts gradually
assume an almost vertical position in consequence of the descent of the heart
into the thorax.   The right and left Cuvierian ducts are originally of the same
diameter, and are frequently termed the *right* and *left superior venæ cavæ*.   By the
development of a transverse branch (the future *left innominate vein*) between the
two primitive jugular veins, the blood is carried across from the left to the right
internal jugular, which thus becomes enlarged to form the upper part of the
superior vena cava of the adult; the lower part of this vessel is formed by the
right Cuvierian duct.   Below the origin of this transverse branch the left primitive
jugular vein and left Cuvierian duct atrophy, the former constituting the upper
part of the left superior intercostal vein, while the latter is represented by the
vestigial fold and oblique vein of Marshall (fig. 164).   Both right and left
superior venæ cavæ are present in some animals, and are occasionally found in
the adult human being.   The oblique vein of Marshall passes downwards across
the back of the left auricle to open into the coronary sinus, which, as already
indicated, represents the persistent left horn of the sinus venosus.

The primitive jugular or anterior cardinal veins are situated on the ventral
surface of the brain, on the mesial side of the cranial nerve-roots.   A considerable
portion of each of these veins disappears and is replaced by a vein which
is developed on the lateral aspect of the cranial nerves from the fifth to the
twelfth inclusive.   This new vein (*vena capitis lateralis*) drains the blood from
the mid and hind brains, and leaves the skull in company with the seventh nerve.
The blood from the hind-brain is collected into a vein (the future lateral sinus)
which passes through the foramen jugulare on the lateral aspect of the vagus
nerve; here the two vessels join to form the internal jugular vein.   On the
dorsal aspect of the ear-capsule an anastomotic channel is opened up between the
vena capitis lateralis and the lateral sinus; and, coincident with this, the portion of
the former vein which extends from the fifth to the tenth cranial nerve becomes
obliterated, and thus the whole of the blood from the brain is ultimately drained
away by the lateral sinuses.   The primitive jugular vein is therefore represented
in the adult by the internal jugular, and not by the external jugular as is usually
stated—the latter being a vessel of later formation.   (Consult 'Die Entwickelung
des Blutgefäss-systems,' by Hochstetter, in Hertwig's 'Entwickelungslehre;' and
also an article by Mall in the 'American Journal of Anatomy,' vol. iv. December
1904.)

The fœtal circulation is described on a future page.

The pericardium.—As already pointed out (page 81), the anterior portion of
the embryonic area in front of the oral plate or bucco-pharyngeal membrane
is named the pericardial area.   Previous to the formation of the head-fold, the
mesoderm has divided into its somatic and splanchnic layers, and these two
layers, together with the intervening cœlomic space, extend forwards on either
side of the bucco-pharyngeal membrane into the pericardial area; the part of the
cœlom contained within this area becomes the cavity of the pericardium.   This
is, at first, in the shape of a crescent, the lateral horns of which extend backwards
on either side of the bucco-pharyngeal membrane and are continuous with the
peritoneal part of the cœlomic space.   The primitive blood-vessels, which, in
the pericardial region, fuse to form the primitive heart, are developed in the
splanchnic mesoderm of the pericardial area.   By the rapid elongation of the
embryo, and the formation of the head-fold, the pericardial area and its contained
blood-vessels are folded backwards to form the ventral wall of the fore-gut.   By
means of this process the surfaces of the pericardial area are consequently
reversed, its splanchnic layer being now situated on the dorsal aspect of its
somatic layer, while its original anterior limit comes to form the front boundary

of the umbilicus. The vitelline veins, bringing the blood from the yolk-sac, enter the embryo through the anterior wall of the umbilicus and pass upwards and forwards to open into the tubular heart, which is, for a time, suspended along its entire length, from the ventral aspect of the fore-gut, by a dorsal mesentery (dorsal mesocardium) (fig. 183). By the absorption of the middle part of this dorsal mesocardium the great transverse sinus of the pericardium, behind the ascending aorta and pulmonary artery, is formed.

In amphibians and birds the pericardium is developed by the fusion of the lateral halves of the cœlom in the middle line beneath the fore-gut, and therefore in these animals there exists, for a period, a ventral mesocardium ; but Professor Robinson has shown that the pericardial cavity in mammals is from the first a single cavity, and that there is never at any time a ventral mesocardium.

The mesoderm immediately in front of the umbilicus becomes thickened to form the septum transversum, above which are situated the lateral horns of the pericardial cavity. These assume the form of tubular passages on the sides of the fore-gut, and constitute the communications between the pericardial and peritoneal parts of the cœlom (fig. 183). The lung buds grow out behind the ducts of Cuvier into these passages, and push their way outwards and forwards into the tissue of the septum transversum. The expansion of the pleural cavities therefore takes place in the septum, which by this means is differentiated into the central part of the Diaphragm and the posterior wall of the pericardium. The anterior limit of the septum transversum is indicated by the Cuvierian ducts (superior venæ cavæ), by the growth of which the passages between the pericardium and pleuræ are closed.

Development of the Alimentary Canal.—As already indicated (page 87), the primitive alimentary canal is formed, at an early stage, by the enclosure within the embryo of a portion of the blastodermic vesicle, and is seen to consist of

FIG. 165.—Diagrammatic outline of a longitudinal vertical section of the chick on the fourth day.

ep. Ectoderm. *sm.* Somatic mesoderm. *hy.* Entoderm. *vm.* Splanchnic mesoderm. *af.* Cephalic fold. *pf.* Caudal fold. *am.* Cavity of true amnion. *ys.* Yolk-sac. *i.* Intestine. *s.* Fore-gut. *a.* Future anus, still closed. *m.* The mouth. *me.* The mesentery. *al.* The allantoic vesicle. *pp.* Space between inner and outer folds of amnion. *h.* Heart. *ri.* Vitelline duct. (From Quain's 'Anatomy,' Allen Thomson.)

three parts, viz. : (1) the *fore-gut*, within the cephalic flexure and dorsal to the heart ; (2) the *mid-gut*, opening freely into the yolk-sac ; and (3) the *hind-gut*, within the caudal flexure. At first the fore-gut and hind-gut end blindly (figs. 165 and 166). The anterior end of the fore-gut is separated from the stomatodœum by the pharyngeal septum ; the hind-gut terminates posteriorly in the cloaca which is closed externally by the cloacal membrane. The formation of the mouth, and the subsequent communication between it and the cephalic end of the fore-gut, have already been considered ; the manner in which the anus is formed will be discussed presently.

From the fore-gut are developed the pharynx, œsophagus, stomach, and greater part of the duodenum,* and further, as diverticula from the duodenum, the liver and pancreas ; from the hind-gut, the rectum (except the anus) and a considerable part of the colon, and as a tubular outgrowth from it the hollow stalk of the allantois ; the mid-gut gives origin to the remainder of the alimentary tube.

* The level of the opening of the common bile-duct is sometimes given as the junction of the fore-gut with the mid-gut.

The upper part of the fore-gut becomes dilated to form the pharynx (fig. 167), in relation to which the branchial arches are developed (fig. 122) (see page 101);

FIG. 166.—Diagram of a longitudinal section of a mammalian embryo. Very early. (After Quain.)

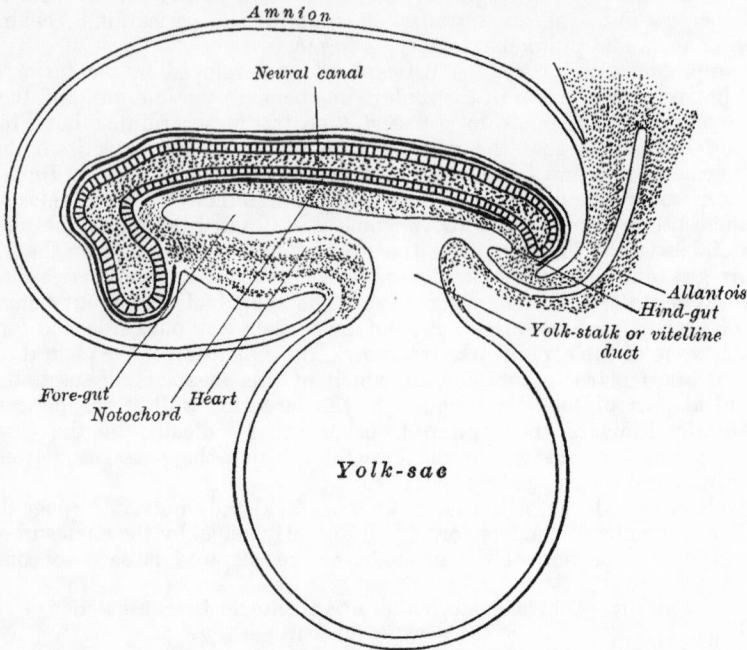

FIG. 167.—Human embryo about fifteen days old. Brain and heart represented from right side. Alimentary canal and yolk-sac in mesial section. (After His.)

the succeeding part remains tubular, and with the descent of the stomach is elongated to form the œsophagus. About the fourth week a fusiform dilatation, the future stomach, makes its appearance, and beyond this the mid-gut opens freely into the yolk-sac (figs. 167 and 168). The opening is at first wide, but is

FIG. 168.—Sketches in profile of two stages in the development of the human alimentary canal. (His.) Fig. A × 30. Fig. B × 20.

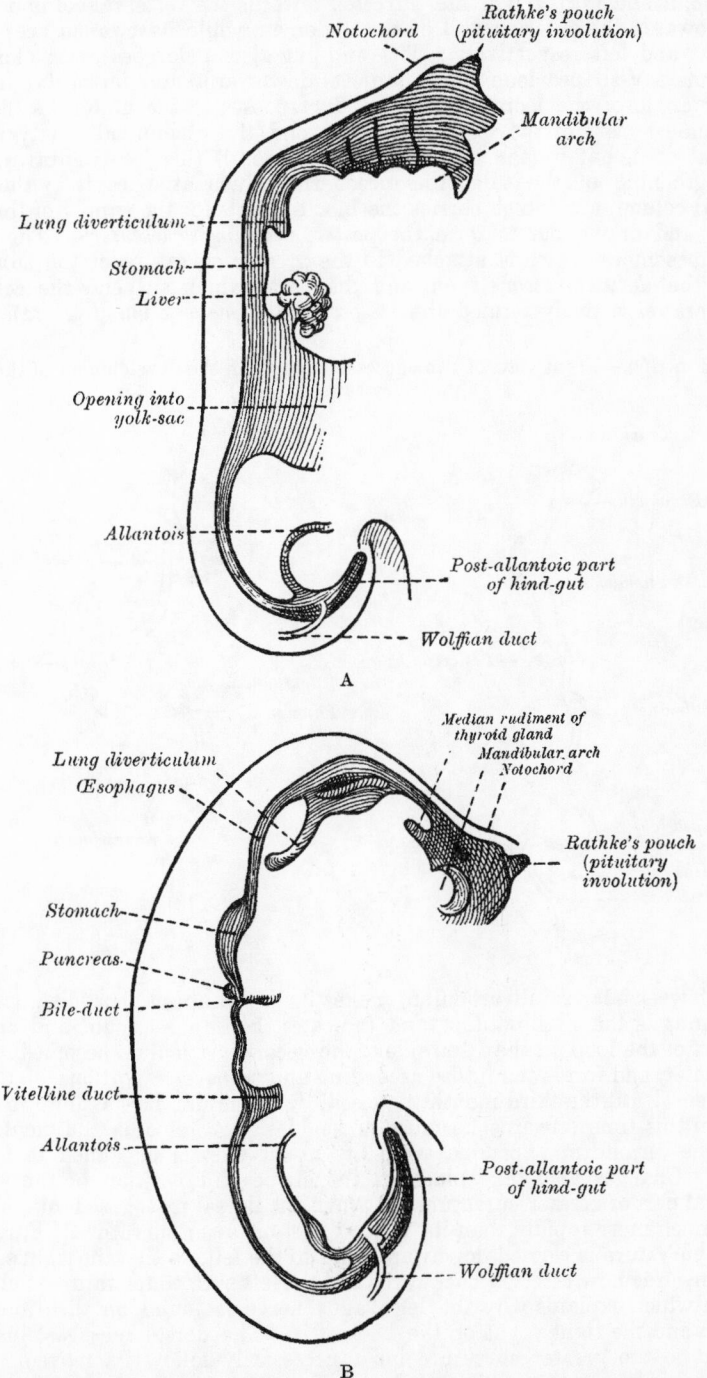

gradually narrowed into a tubular stalk, the *yolk-stalk* or *vitelline duct*. The proximal part of this duct persists in a small percentage of subjects and constitutes *Meckel's diverticulum*, which is found about three or four feet above the ileo-cæcal valve. At this stage, therefore, the alimentary canal forms a nearly straight tube in front of the notochord and primitive aortæ (fig. 166). From the stomach to the rectum it is attached to the notochord by a band of mesoderm, from which the common mesentery of the gut is subsequently developed. The stomach undergoes a further dilatation, and its two curvatures can be recognised (figs. 168, B, and 172), the greater directed towards the vertebral column and the lesser towards the anterior wall of the abdomen, while its two surfaces look to the right and left respectively. The mid-gut also undergoes great elongation, and forms a **V**-shaped loop which projects downwards and forwards ; from the bend or angle of the loop the vitelline duct passes to the umbilicus (fig. 172). For a time a part of the loop extends beyond the abdominal cavity into the umbilical cord, but by the end of the third month this is withdrawn. With the lengthening of the tube, the mesoderm, which attaches it to the future vertebral column and which carries the blood-vessels for the supply of the gut, is thinned and drawn out to form the *posterior common mesentery*. The portion of this mesentery which is attached to the greater curvature of the stomach is named the *dorsal mesogastrium*, and the parts which suspend the colon and rectum are respectively termed the *mesocolon* and *mesorectum* (fig. 172). About

Fig. 169.—Front view of two successive stages in the development of the alimentary canal. (His.)

the sixth week a lateral diverticulum makes its appearance a short distance behind the opening of the vitelline duct, and indicates the future cæcum and appendix. The part of the loop on the distal side of the cæcal diverticulum becomes increased in diameter, and forms the future ascending and transverse portions of the large intestine. Until the third month the cæcal diverticulum has a uniform calibre, but from this time onwards its most dependent part remains rudimentary and forms the vermiform appendix, while its upper part is expanded to form the cæcum. Changes also take place in the shape and position of the stomach. Its dorsal part or greater curvature, to which the dorsal mesogastrium is attached, grows much more rapidly than its ventral part or lesser curvature. Further, the greater curvature is carried downwards and to the left, so that the right surface of the stomach is now directed backwards and the left surface forwards—a change in position which explains why the left vagus nerve is found on the front of the stomach and the right vagus on the back of it. The dorsal mesogastrium being attached to the greater curvature must necessarily follow its movements, and hence it becomes greatly elongated and drawn outwards from the vertebral column, and, like the stomach, what was originally its right surface is now directed backwards and its left forwards. In this way a pouch, the *bursa omentalis*, is formed behind the stomach ; this pouch is the future lesser sac of the peritoneum, and it increases in size as the alimentary tube undergoes further development; the entrance to the pouch constitutes the future *foramen of Winslow* (figs. 173, 176). The duodenum is developed from that part of the tube which immediately succeeds the stomach. This undergoes little elongation, being more or less fixed in position by the liver and pancreas, which arise as diverticula from it. The duodenum is at first suspended by a mesentery, and projects forwards in the form of a loop. The loop and its mesentery are subsequently displaced by the transverse colon, so that the right surface of the duodenal mesentery is directed backwards, and, adhering to the parietal peritoneum, is lost. The remainder of the canal becomes greatly elongated, and as a consequence the tube is coiled on itself, and this elongation demands a corresponding increase in the width of the intestinal attachment of the mesentery, which becomes plaited or folded.

At this stage the small and large intestine are attached to the vertebral column by a common mesentery, the coils of the small intestine falling to the right of the middle line, while the large intestine lies on the left side.*

The gut is now rotated upon itself, so that the large intestine is carried over in front of the small intestine, and the cæcum is placed immediately below the liver; about the sixth month the cæcum descends into the right iliac fossa, and the large intestine forms an arch consisting of the ascending, transverse, and descending portions of the colon—the transverse portion crossing in front of the duodenum and lying just below the greater curvature of the stomach; within this arch the coils of the small intestine are disposed (figs. 173, 177). Sometimes the downward

FIG. 170.—Schematic and enlarged cross section through the body of a human embryo in the region of the mesogastrium. Beginning of third month. (Toldt.)

Left suprarenal capsule   *Right suprarenal capsule*

FIG. 171.—Same section as in fig. 170, at end of third month.   (Toldt.)

progress of the cæcum is arrested, so that in the adult it may be found lying immediately below the liver instead of in the right iliac region.

Further changes take place in the bursa omentalis and in the common mesentery, and give rise to the peritoneal relations seen in the adult. The bursa omentalis, which at first reaches only as far as the greater curvature of the stomach, grows downwards to form the great omentum, and this downward extension lies in front of the transverse colon and the coils of the small intestine. The anterior layer of the transverse mesocolon is at first quite distinct from the posterior layer of the great omentum, but ultimately the two blend, and hence the great omentum appears as if attached to the transverse colon (fig. 174). The

* Sometimes this condition persists throughout life, and it is then found that the duodenum does not cross from the right to the left side of the vertebral column, but lies entirely on the right side of the mesial plane, where it is continued into the jejunum; the arteries to the small intestine (rami intestini tenuis) also arise from the right instead of the left side of the superior mesenteric artery.

mesenteries of the ascending and descending parts of the colon disappear in the majority of cases, while that of the small intestine assumes the oblique attachment characteristic of its adult condition.

FIG. 172.—Abdominal part of alimentary canal and its attachment to the primitive or common mesentery. Human embryo of six weeks. (After Toldt.) (From Kollmann's 'Entwickelungsgeschichte.')

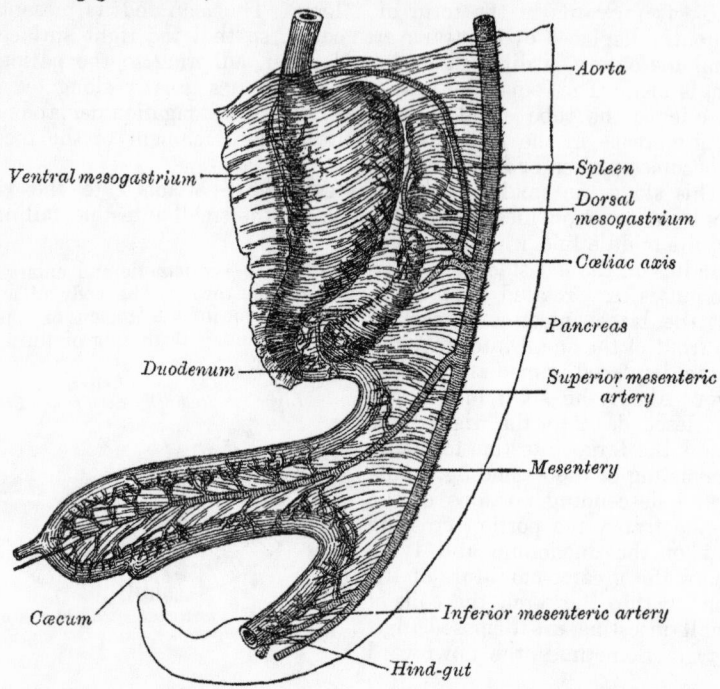

FIG. 173.—Diagrams to illustrate two stages in the development of the human alimentary canal and its mesentery. The arrow indicates the entrance to the bursa omentalis.

The small omentum is formed by a thinning of the mesoderm or *ventral mesogastrium*, which attaches the stomach and duodenum to the anterior

abdominal wall.   By the subsequent growth of the liver this leaf of mesoderm is
divided into two parts, viz. : the small omentum between the stomach and liver,

FIG. 174.—Diagrams to illustrate the development of the great omentum
and transverse mesocolon.

and the falciform ligament between the liver and the abdominal wall and
Diaphragm (fig. 175).

FIG. 175.—The primitive mesentery of a six weeks' human embryo, half schematic.
(Kollmann.)

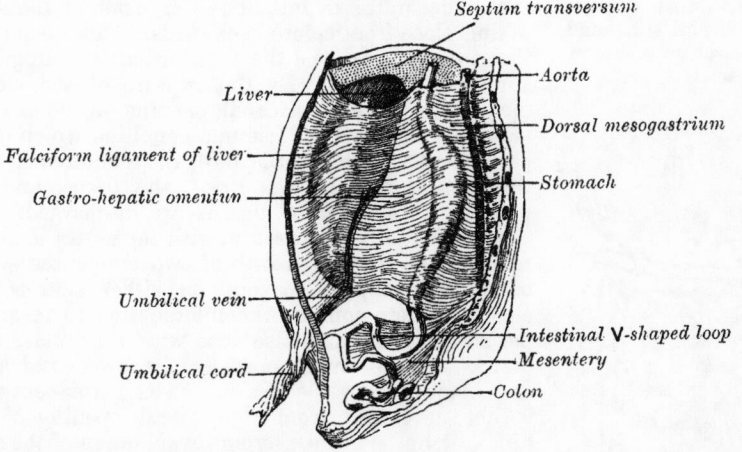

Formation of the rectum and anus.—The hind-gut is at first prolonged backwards into the body-stalk as the tube of the allantois ; but, with the growth and
flexure of the tail-end of the embryo, the body-stalk, with its contained allantoic
tube, is carried forwards to the ventral aspect of the body, and consequently a
bend is formed at the junction of the hind-gut and allantois.   This bend becomes
dilated into a pouch, which constitutes the *entodermal cloaca* ; into its dorsal part
the hind-gut opens, and from its ventral part the allantois passes forwards.   At a
later stage the Wolffian and Müllerian ducts also open into its ventral portion.
The cloaca is, for a time, shut off from the exterior by a membrane, the *cloacal
membrane*, formed by the apposition of the ectoderm and entoderm, and reaching,
at first, as far forwards as the future umbilicus.   Behind the latter, however, the
mesoderm subsequently extends inwards to form the lower part of the abdominal
wall and symphysis pubis.   By the growth of the surrounding tissues the cloacal
membrane comes to lie at the bottom of a depression which is lined by ectoderm
and named the *ectodermal cloaca*.

The entodermal cloaca is divided into a dorsal and a ventral part by means of a partition consisting of two lateral septa which grow inwards and unite with

FIG. 176.—Schematic figure of the bursa omentalis, &c.   Human embryo of eight weeks.   (Kollmann.)

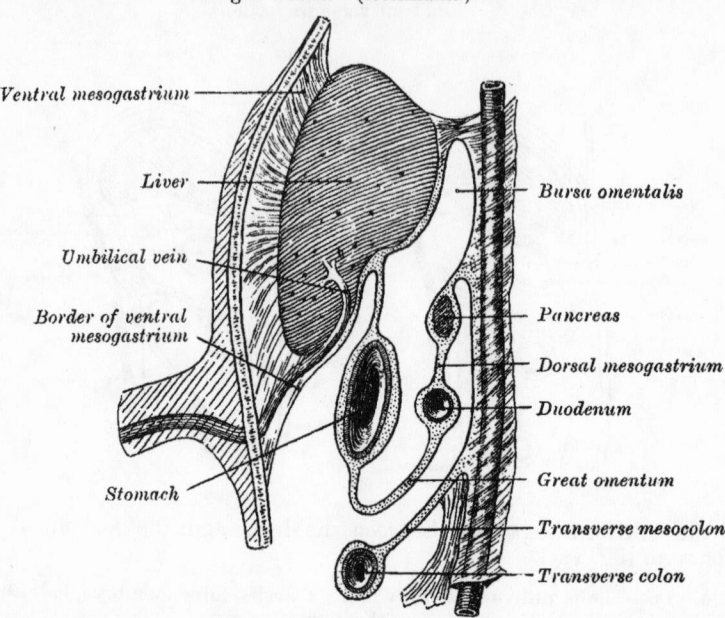

Ventral mesogastrium

Liver

Umbilical vein

Border of ventral mesogastrium

Stomach

Bursa omentalis

Pancreas

Dorsal mesogastrium

Duodenum

Great omentum

Transverse mesocolon

Transverse colon

FIG. 177.—Final disposition of the intestines and their vascular relations. (Jonnesco.)

A. Aorta. H. Hepatic artery. S. Splenic artery. M, Col. Branches of superior mesenteric artery. m, m'. Branches of inferior mesenteric artery.

each other in the middle line—the fusion of the septa taking place from before backwards.   The dorsal part forms the rectum, and the anterior part the urogenital sinus and bladder.   By the rupture of the cloacal membrane the cloacal canal, opening on the exterior, is formed, thus giving rise to a condition which exists permanently in the reptile, bird, and monotreme.   Into this cloacal channel the urine, the fæces, and the products of the genital organs are discharged.   The communication of the rectum with the cloaca is obliterated by the inward growth of two eminences, which make their appearance one on either side of the cloaca.   These join in the middle line to form the perineal septum, and also fuse with the hinder edge· of the septum which separates the urogenital sinus and bladder from the rectum.   The permanent anus is not developed from the cloacal opening of the hind-gut, but is formed by an invagination of the ectoderm behind the perineal septum.   This invagination is termed the proctodæum, and it meets with the ventral aspect of the hind-gut and forms with it the anal membrane.   By the absorption of this membrane the anus is formed.   A small part of the hind-gut projects backwards beyond the anal orifice; this portion is named the post-anal gut, and usually becomes obliterated and disappears.[*]

F. Wood-Jones [†] gives a different account from

[*] Consult, in this connection, the following article: 'A Contribution to the Morphology of the Human Urino-genital Tract,' by D. Berry Hart, M.D., F.R.C.P.E.   Journal of Anatomy and Physiology, April 1901, vol. xxxv.

[†] 'The Nature of the Malformations of the Rectum and Urogenital Passages,' by F. Wood-Jones, M.B., B.Sc., M.R.C.S.   British Medical Journal, December 17, 1904.

the above as to the manner in which the rectum is separated from the cloaca. He maintains that the growth of the hind-gut keeps pace with that of the hind-end of the embryo, and 'buds backwards past its cloacal orifice, past its old termination in the allantois, and forms the portion of the hind-gut distal to the allantois'; this portion he terms the *post-allantoic gut* (figs. 178, 179, and 186). 'The cloacal opening of the hind-gut is now normally lost; originally a small opening in the embryo of 12 somites (which is about 2 mm. in total length), the rapid growth of the hind-gut, the post-allantoic gut, and of the allantois itself, together with the lateral infolding of the wall described by Keibel, serve to close the opening of the hind-gut into the cloaca.' This view, which affords a satisfactory explanation of the varieties of imperforate rectum and anus which are

FIG. 178.—Diagram to illustrate the development of the post-allantoic gut. The hind-gut opens freely into the cloaca. (After Wood-Jones.)

Hind-gut

Allantois

Cloaca

Post-allantoic gut

sometimes found, leads to the conclusion that the cloaca does not contribute to the formation of the rectum, and that the septa which have been described as fusing in the middle line to form the perinæum are non-existent.

The peritoneal cavity is formed by the abdominal portion of the cœlom, which becomes separated from the thoracic portion by the development of the Diaphragm.

The *liver* arises in the form of a diverticulum or hollow outgrowth from the ventral surface of the duodenum (figs. 167, 168, and 183). This diverticulum is lined by entoderm, and grows upwards and forwards into the septum transversum, and there gives off two solid buds of cells which represent the right and the left lobes of the liver. The solid buds of cells grow into columns or cylinders, termed the *hepatic cylinders*, which branch and anastomose to form a close meshwork. This network invades the vitelline and umbilical veins, and breaks up these vessels into a series of capillaries, which ramify in the meshes of the cellular network and ultimately form the venous capillaries of the liver.

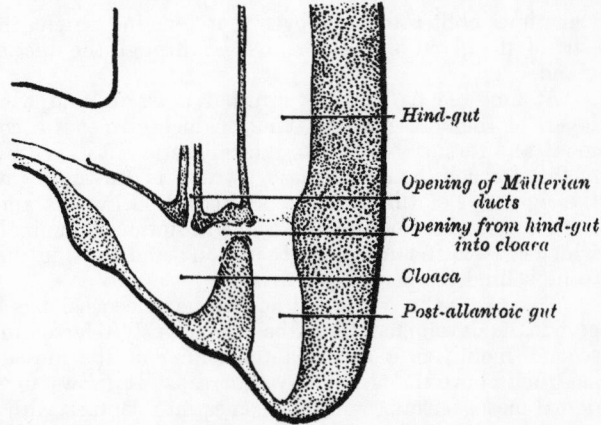

FIG. 179.—Diagram to illustrate the further development of the post-allantoic gut. The hind-gut still opens into the cloaca. The opening of the Müllerian ducts is also seen. (After Wood-Jones.)

Hind-gut

Opening of Müllerian ducts

Opening from hind-gut into cloaca

Cloaca

Post-allantoic gut

By the continued growth and ramification of the hepatic cylinders the mass of the liver is gradually formed. The original diverticulum from the duodenum forms the common bile-duct, and from this the cystic duct and gall-bladder arise as a hollow evagination.

As the liver undergoes enlargement, both it and the ventral mesogastrium of the fore-gut are gradually differentiated from the septum transversum; and from the under surface of the latter the liver projects downwards into the abdominal cavity. By the growth of the liver the ventral mesogastrium is divided into two parts, of which the anterior forms the falciform ligament, and the posterior the gastro-hepatic omentum. About the third month the liver almost fills the abdominal cavity. From this period the relative development of the liver is less active, more especially that of the left lobe, which now becomes smaller than the right; but up to the end of fœtal life the liver remains relatively larger than in the adult.

The pancreas.—During the fourth week the pancreas arises as two hollow buds, a larger dorsal and a smaller ventral bud (fig. 180). The former springs as a diverticulum from the dorsal aspect of the duodenum, a short distance above the hepatic diverticulum, and, growing upwards and backwards into the dorsal mesogastrium, forms the main part of the pancreas. The ventral part grows into the ventral mesogastrium and gives origin to the lower part of the head of the gland. It arises as an evagination from the commencement of the hepatic diverticulum, and thus, from the first, the duct of this part of the gland opens into the duodenum through an orifice common to it and the bile-duct. By the seventh week the dorsal and ventral parts have become united and their ducts communicate with each other. From this time onward the duct of the ventral part of the gland increases greatly in size, and forms the continuation of the main duct (canal of Wirsung). On the other hand, that portion of the duct of the dorsal part of the gland which extends from the point of union of the two ducts to the duodenum undergoes little enlargement, and forms the duct of Santorini, which opens about three-quarters of an inch above the orifice of the common bile-duct. The opening of the duct of Santorini into the duodenum is

FIG. 180.—The rudiments of the pancreas in a six weeks' human embryo. (After Hamburger.) (From Kollmann's 'Entwickelungsgeschichte.')

Duct of Wirsung

Bile-duct

Ventral pancreas — — *Dorsal pancreas*

Duct of Santorini — — *Stomach*

sometimes obliterated, and even if it remains patent the secretion of the dorsal part of the gland is mostly conveyed through the duct of the ventral part of the gland.

At first the pancreas is directed upwards and backwards between the two layers of the dorsal mesogastrium, which give to it a complete peritoneal investment, and its surfaces look to the right and left. With the change in the position of the stomach the dorsal mesogastrium is drawn downwards and to the left, and the right side of the pancreas is directed backwards and the left forwards. The right surface becomes applied to the posterior abdominal wall, and the peritoneum, which covered it, undergoes absorption; and thus, in the adult, the gland appears to lie behind the peritoneal cavity.

The spleen (fig. 172).—Although the spleen belongs to the group of ductless glands, its development may be conveniently referred to here. It appears in the second month as a localised thickening of the mesoderm in the dorsal mesogastrium above the tail of the pancreas. It grows towards the left side of the dorsal mesogastrium, and thus comes into contact with the right surface of the stomach. With the change in position of this viscus the spleen is carried to the left, and comes to lie behind the cardiac part of the stomach and in contact with the left kidney. The part of the dorsal mesogastrium which intervened between the spleen and the greater curvature of the stomach forms the gastro-splenic omentum.

Development of the Respiratory Organs.—Towards the end of the third week a deep longitudinal furrow (figs. 126 and 127) appears in the ventral wall of the fore-gut, commencing at the level of the fourth visceral arch and reaching backwards nearly as far as the stomach. It is bounded in front by an elevation termed the *furcula*, and laterally by two ridges. By the union of the posterior parts of the two ridges the groove is converted into a tube-like diverticulum, which is lined by entoderm and which grows backwards on the ventral aspect of the œsophagus. The upper end of this diverticulum is expanded to form the larynx; the furcula is the future epiglottis, and the upper parts of the lateral ridges constitute the aryteno-epiglottidean folds. The thyroid cartilage is developed from the cartilages of the fourth and fifth visceral arches, while that of the sixth visceral arch appears to be modified to form the cricoid and arytenoid cartilages and the cartilages of the trachea.

The lower end of the tube-like diverticulum bifurcates into a larger right and a smaller left bud, the *right* and *left lung buds*, and each of these is further subdivided—the right into three and the left into two parts; these subdivisions are the early indications of the corresponding lobes of the lungs (figs. 122 and 169). These lung buds undergo further subdivision and ramification, and ultimately end in minute expanded extremities—the infundibula of the lung. After the sixth month the air-vesicles begin to make their appearance on the infundibula in the form of minute pouches.

Development of the Diaphragm (figs. 181 and 182). The following description is based on that given by Keith.* The *central tendon* of the Diaphragm is

FIG. 181.—The thoracic aspect of the Diaphragm of a newly born child in which the communication between the peritoneum and pleura has not been closed on the left side; the position of the opening is marked on the right side by the spino-costal hiatus. (After Keith.)

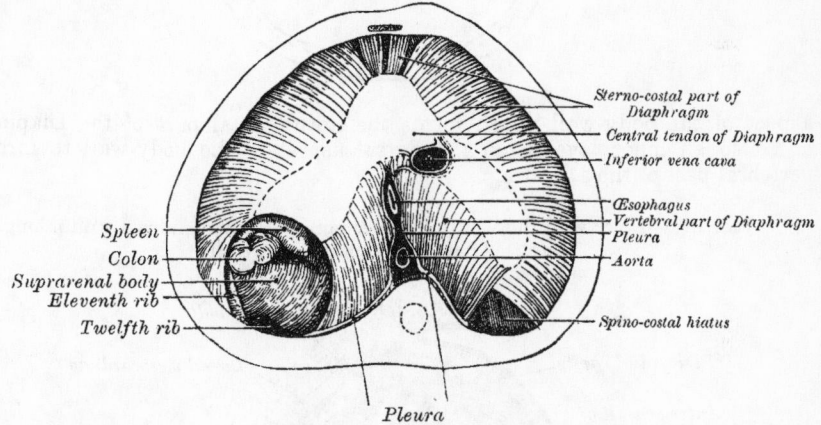

Sterno-costal part of Diaphragm
Central tendon of Diaphragm
Inferior vena cava
Œsophagus
Vertebral part of Diaphragm
Pleura
Aorta
Spino-costal hiatus
Spleen
Colon
Suprarenal body
Eleventh rib
Twelfth rib
Pleura

derived from the septum transversum; each half of its muscular portion is developed in two parts, viz.: (*a*) a *sterno-costal* portion, which is derived from the ventral longitudinal musculature of the embryonic neck; and (*b*) a *spinal* or *vertebral* portion, which arises from the bodies of the vertebræ and arcuate ligaments, and is derived from the cervical part of the transversalis muscle. The pleuro-peritoneal opening is closed by the approximation of the sterno-costal and vertebral parts; and the spino-costal fibrous hiatus, best seen on the left half of the Diaphragm, marks its position after closure. Sometimes the opening remains patent, giving rise to a congenital diaphragmatic hernia.

The formation or separation of the Diaphragm—for at first it forms part of the wall of the body-cavity—results from the development of the pleural cavities and lungs. The lung buds appear in the cervical region of the embryo, and they, together with the parts of the cœlom in which they are contained, undergo a rapid development, growing forwards and outwards into the tissue of the dorsal part of the septum transversum and of the body-wall; within that tissue the pleural cavities are excavated. The pleural cavities also develop within the body-wall towards the ventral median line, thus separating the pericardium from the thoracic wall (see arrows in fig. 182). In this manner the

* *Human Embryology and Morphology*, by Arthur Keith, M.D., F.R.C.S., 2nd edition, 1904. Consult also an article on the development of the Diaphragm, by the same author, in vol.xxxix. of the *Journal of Anatomy and Physiology*.

pleural cavities are excavated within the body-wall, dorsal to and on each side of the pericardium. The formation of the pleural cavities not only separates the pericardium from the body-wall, but also an inner layer from the ventro-lateral

Fig. 182.—Diagram of transverse section through rabbit embryo. (After Keith.)

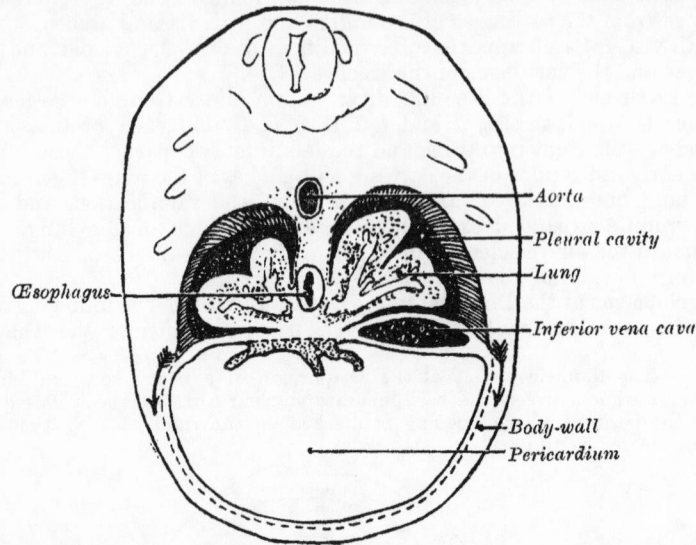

aspect of the body-wall which forms the sterno-costal part of the Diaphragm, and also an inner layer from the dorsal aspect of the body-wall to form the vertebral part of the Diaphragm.

Fig. 183.—Liver with the septum transversum. Human embryo 3 mm. long. × 16. (After model and figure by His.)

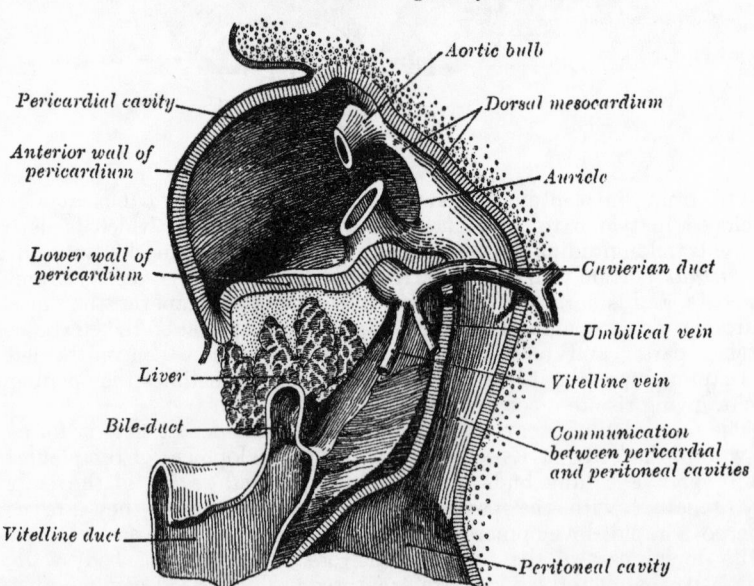

Development of the Urinary and Generative Organs.—The urinary and generative organs are developed from the intermediate cell-mass which is situated between the protovertebral somites and the lateral plates of mesoderm.

Wolffian ducts and Wolffian bodies.—In the outer part of the intermediate cell-mass, immediately under the ectoderm, a longitudinal cord of cells makes its appearance. This cellular cord sinks into the subjacent mesoderm, and, acquiring a lumen, constitutes the *Wolffian duct,** which passes backwards and opens into the urogenital sinus. On the inner side of this duct a series of tubules, the *Wolffian tubules,* are developed. Each tubule opens externally into the Wolffian duct, while its opposite end is invaginated by a tuft of capillary blood-vessels to form a glomerulus. These tubules increase in number, and collectively constitute

FIG. 184.—Section of the urogenital area of a chick embryo of the fourth day. (Waldeyer.)

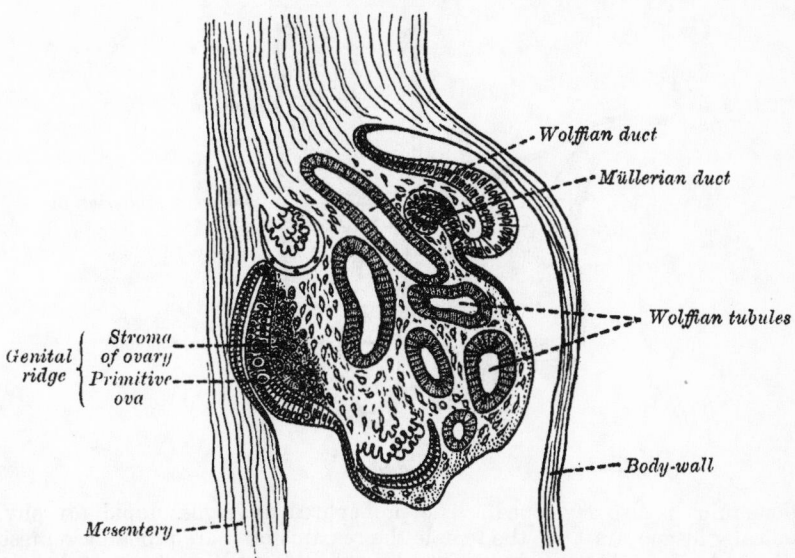

the *Wolffian body* or *mesonephros* (fig. 185). At the beginning of the second month this body forms an elongated spindle-shaped structure, which projects into the cœlomic cavity on either side of the dorsal mesentery and reaches from the septum transversum in front to the fifth lumbar somite behind. The Wolffian body persists and forms the permanent kidney in fishes and amphibians, but in reptiles, birds, and mammals it is superseded by the *metanephros*, which forms

FIG. 185.—Enlarged view from the front of the left Wolffian body before the establishment of the distinction of sex.   (From Farre, after Kobelt.)

a, a, b, c, d. Tubular structure of the Wolffian body. *e.* Wolffian duct. *f.* Its upper extremity. *g.* Its termination in *x,* the urogenital sinus. *h.* The duct of Müller. *i.* Its upper, funnel-shaped extremity. *k.* Its lower end, terminating in the urogenital sinus. *l.* The genital ridge, ovary or testicle.

the permanent kidney in these animals. The anterior tubules of the Wolffian body become attached to the sexual eminence or genital ridge from which the ovary in the female, and the testicle in the male, are developed. Coincident with the development of the permanent kidneys, the Wolffian bodies atrophy, and this process proceeds to a much greater extent in the female than in the male.

* The Wolffian duct is by many embryologists regarded as being of ectodermal origin, formed by a longitudinal invagination of the ectoderm which overlies the intermediate cell-mass.

The pronephros.—In front of the Wolffian body there are developed a number
of tubules which form the *pronephros* or *head-kidney*—an early embryonic structure
in all vertebrates.   This consists of a series of transverse tubules which open into
a duct termed the *pronephric duct*; this duct is continuous posteriorly with
the Wolffian duct.   Each pronephric tubule communicates by means of a funnel-
shaped, ciliated opening with the cœlomic cavity, and in the course of each duct

FIG. 186.—Diagram to illustrate the separation of the hind-gut from the cloaca.   The
hind-gut has now lost its cloacal opening.   The post-allantoic gut is about to meet the
proctodœal depression.   (After Wood-Jones.)

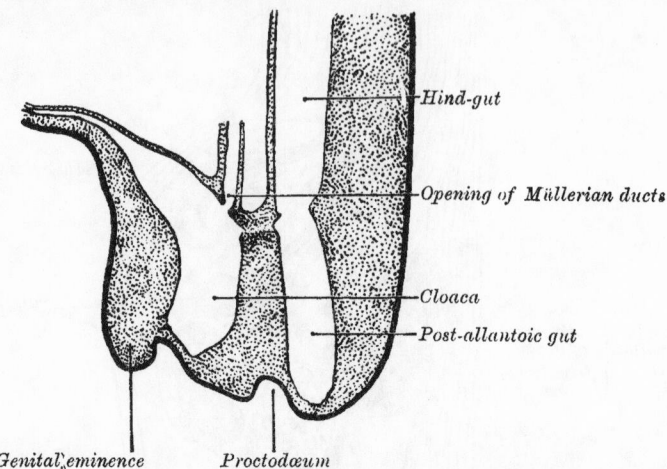

a glomerulus is also developed.   The pronephros undergoes rapid atrophy and
practically disappears.   In the female the remains of it are probably represented
by the hydatids of Morgagni at the fimbriated end of the Fallopian tube ; in the
male, by the stalked hydatid at the upper end of the testicle.

In the male the Wolffian duct persists, and forms the tube of the epididymis,
the vas deferens, and common ejaculatory duct, while the seminal vesicle arises

FIG. 187.—Diagram to illustrate the formation of the vagina, bladder, and urethra.
The Müllerian ducts have lost their opening into the urogenital sinus, and the new
solid vagina has grown down and later becomes canalised.   (After Wood-Jones.)

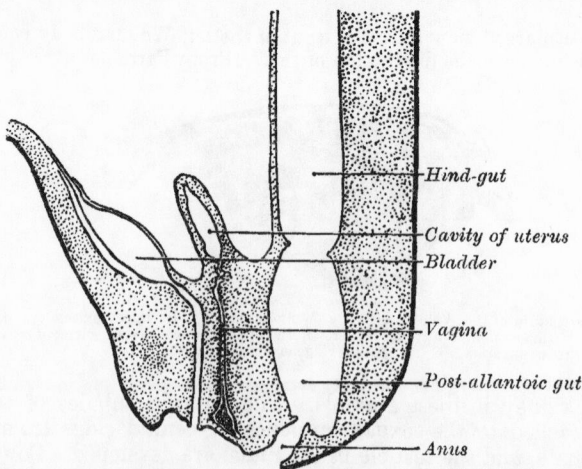

as a lateral diverticulum from its hinder end.   The anterior Wolffian tubules
form the rete testis, vasà efferentia, and coni vasculosi of the testicle ; while the
posterior tubules are represented by the vasa aberrantia of the globus minor, and
by the organ of Giraldès, which is sometimes found in front of the spermatic
cord above the globus major (fig. 188, c).

FIG. 188.—Diagrams to show the development of male and female generative organs from a common type. (Allen Thomson.)

A.—Diagram of the primitive urogenital organs in the embryo previous to sexual distinction.

3. Ureter. 4. Urinary bladder. 5. Urachus. *ot.* The genital ridge from which either the ovary or testicle is formed. *W.* Left Wolffian body. *w, w.* Right and left Wolffian ducts. *m, m.* Right and left Müllerian ducts uniting together and running with the Wolffian ducts in *gc,* the genital cord. *ug.* Sinus urogenitalis. *i.* Lower part of the intestine. *cl.* Cloaca. *cp.* Elevation which becomes clitoris or penis. *ls.* Fold of integument from which the labia majora or scrotum are formed.

B.—Diagram of the female type of sexual organs.

o. The left ovary. *po.* Parovarium (epoöphoron of Waldeyer). *W.* Scattered remains of Wolffian tubes near it (paroöphoron of Waldeyer). *dG.* Remains of the left Wolffian duct, such as give rise to the duct of Gärtner, represented by dotted lines; that of the right side is marked *w. f.* The abdominal opening of the left Fallopian tube. *u.* Uterus. The Fallopian tube of the right side is marked *m. g.* Round ligament, corresponding to gubernaculum. *i.* Lower part of the intestine. *ra.* Vagina. *h.* Situation of the hymen. *C.* Gland of Bartholin (Cowper's gland), and immediately above it the urethra. *cc.* Corpus cavernosum clitoridis. *sc.* Vascular bulb or corpus spongiosum. *n.* Nympha. *l.* Labium. *v.* Vulva.

C.—Diagram of the male type of sexual organs.

t. Testicle in the place of its original formation. *e.* Caput epididymis. *vd.* Vas deferens. *W.* Scattered remains of the Wolffian body, constituting the organ of Giraldès, or the paradidymis of Waldeyer. *vh.* Vas aberrans. *m.* Müllerian duct, the upper part of which remains as the hydatid of Morgagni; the lower part, represented by a dotted line descending to the prostatic vesicle, constitutes the occasionally existing cornu and tube of the uterus masculinus. *g.* The gubernaculum. *i.* Lower part of the intestine. *vs.* The vesicula seminalis. *pr.* The prostate gland. *C.* Cowper's gland of one side. *cp.* Corpora cavernosa penis cut short. *sp.* Corpus spongiosum urethrae. *s.* Scrotum. *t',* together with the dotted lines above, indicates the direction in which the testicle and epididymis descend from the abdomen into the scrotum.

In the female the Wolffian bodies and ducts become atrophied. The remains of the Wolffian tubules are represented by the epoöphoron or organ of Rosen-müller, and by the paroöphoron, two small collections of rudimentary, blind tubules which are situated in the mesosalpinx. The lower part of the Wolffian duct disappears, while the upper part persists as the functionless duct of Gärtner * (figs. 188, B, and 190).

The permanent kidney, or metanephros.—The rudiments of the permanent kidneys make their appearance about the end of the first month. Each arises as a diverticulum from the hind-end of the Wolffian duct, close to where the latter opens into the urogenital sinus. This diverticulum grows upwards and forwards into the posterior part of the intermediate cell-mass, where its blind or anterior extremity becomes dilated and subsequently divides into several buds, which form the rudiments of the pelvis and calyces of the kidney. By further subdivision it gives rise to the collecting tubules of the kidney; whether the secretory tubules are developed from the renal diverticulum or from the surrounding mesoderm is not as yet determined. The mesoderm around the subdivisions of the diverticulum becomes condensed to form the connective tissue and vessels of the kidney, while the diverticulum is elongated to form the ureter, the posterior extremity of which opens at first into the hind-end of the Wolffian duct; after the sixth week, it separates from the Wolffian duct and opens independently into the part of the cloaca which ultimately becomes the bladder. The manner in which this separation is brought about is not fully known.†

The secretory tubules of the kidney become arranged into pyramidal masses or lobules, and the lobulated condition of the kidney exists for some time after birth, while traces of it may even be found in the adult. The kidney of the ox and many other animals remains lobulated throughout life.

The Müllerian ducts.—Shortly after the formation of the Wolffian ducts a second pair of ducts is developed; these are named the *Müllerian ducts.* Each arises on the outer aspect of the corresponding Wolffian body as a tubular invagi-nation of the cells lining the cœlom (fig. 184). The orifice of the invagination remains patent, and undergoes enlargement and modification to form the abdominal ostium of the Fallopian tube. The ducts pass backwards on the outer aspects of the Wolffian bodies, but towards the posterior end of the embryo they cross to the inner side of the Wolffian ducts, and thus come to lie side by side between and behind the latter—the four ducts forming what is termed the *genital cord.*

Ultimately, the Müllerian ducts open into the urogenital sinus between the orifices of the Wolffian ducts, and terminate on an elevation named the *Müllerian eminence.* Berry Hart describes them as ending blindly on this eminence.

In the male the Müllerian ducts atrophy, but a trace of their anterior ends is represented by the sessile hydatids of the epididymis, while their terminal fused portions form the uterus masculinus or sinus pocularis in the floor of the prostatic portion of the urethra (fig. 188, c).

In the female the Müllerian ducts persist and undergo further development.

---

* Berry Hart (*op. cit.*) has described the Wolffian ducts as ending at the site of the future hymen in bulbous enlargements, which he has named the *Wolffian bulbs*; and states that the hymen is formed by these bulbs, 'aided by a special involution from below of the cells lining the urogenital sinus.' He further believes that 'the lower third of the vagina is due to the coalescence of the upper portion of the urogenital sinus and the lower ends of the Wolffian ducts,' and 'that the epithelial lining of the vagina is derived from the Wolffian bulbs.' He also regards the colliculus seminalis of the male urethra as being formed from the lower part of the Wolffian ducts.

† The separation of the ureter from the Wolffian duct may be brought about by the absorption of the hinder end of the latter into the genito-urinary chamber, and by the growth of the wall of this chamber between the openings. Robinson (*Proceedings of the Anatomical Society of Great Britain and Ireland*, May 1903, page lxiii) states, regarding an embryo of about seven weeks, that 'from the posterior or lower opening of the Wolffian duct a grooved ridge, the *Wolffian ledge*, runs caudally on the wall of the genito-urinary chamber and gradually disappears at the junction of the Wolffian angle with the body of the chamber. The lateral margins of the groove are continuous anteriorly with the lateral margins of the Wolffian duct, and apparently fuse together to form the ventral wall of the lower part of the duct. . . . Obviously, if the lateral margins of the groove were to fuse from before backwards, the aperture of the Wolffian duct would be carried further backward in the chamber, and its distance from the opening of the ureter increased.'

The portions which lie in the genital cord fuse to form the uterus and vagina ; the parts in front of this cord remain separate, and each forms the corresponding Fallopian tube—the abdominal ostium of which is developed from the anterior extremity of the original tubular invagination from the cœlom. The fusion of the Müllerian ducts begins in the third month, and the septum formed by their fused mesial walls disappears from below upwards, and thus the cavities of the vagina and uterus are produced. About the fifth month an annular constriction marks the position of the neck of the uterus, and after the sixth month the walls of the uterus begin to thicken. The development of the vagina in the manner just described would necessitate the growth of a septum between it and the urethra ; but Wood-Jones (*op. cit.*) maintains that no such septum exists, and that ' the vagina is, for a great part of fœtal life, a solid rod, and not an open canal at all ' (fig. 187). He says : ' Early in the history of the embryo the Müllerian ducts open into the urogenital sinus at its upper part (fig. 186) ; late in its history they open at the hind-end of the vagina, and for a considerable interval they have no opening at all—the old one being lost and the new one not yet formed. No septal division is employed in this change ; but as the hind-gut, when its cloacal opening is lost, re-establishes communication with the exterior by a new downgrowth, so the Müllerian ducts, when their cloacal opening becomes obliterated, tunnel a new passage to the hind-end. The active agents in this strange growth are two epithelial masses that have been described by Berry Hart as the Wolffian bulbs, but to give this name to them is to give a definite idea as to their origin, and this seems to be by no means clear.'

The urethra.—In the female the urethra is formed from the upper part of the urogenital sinus, viz. that part which lies above the openings of the Wolffian and Müllerian ducts. The portion of the sinus below these openings becomes gradually shortened, and is ultimately opened out to form the vestibule, and in this manner the urethra and vagina come to open separately on the surface. Wood-Jones regards the female urethra as ' the cloacal remnant in its simplest form,' and points out that it does not remain tubular throughout fœtal life, but is for a time ' obliterated more or less completely by the proliferation of the vaginal bulbs.' Developmentally considered, the male urethra consists of three parts : 1. The prostatic and membranous portions, which are derived from the urogenital sinus. 2. The penile portion as far as the base of the glans penis, which is formed by the fusion from behind forwards, in the middle line, of the inner genital folds. This fusion is apparently at first solid, but the epidermis extends backwards along the line of union, and by central desquamation forms the lumen of the urethra. 3. The part contained within the glans penis ; this is formed by the ingrowth and subsequent central desquamation of a solid plug of epidermis which perforates the glans and meets the canal behind (Berry Hart).

The union of the inner genital folds may be arrested at any stage of their growth, giving rise to partial or complete hypospadias.

The *prostate gland* originally consists of two separate portions, each of which arises as a series of diverticular buds from the epithelial lining of the urogenital sinus, and which appear between the third and fourth months. These buds become tubular, and form the glandular substance of the two lobes, which ultimately meet and fuse behind the urethra and also extend on to its ventral aspect. The third or middle lobe is formed as an extension of the lateral lobes between the common ejaculatory ducts and the bladder. Skene's ducts in the female urethra are regarded as the analogues of the male prostatic glands.

The glands of Cowper in the male, and of Bartholin in the female, also arise as diverticula from the epithelial lining of the urogenital sinus.

The urinary bladder.—The trigone of the bladder is formed from the upper part of the urogenital sinus ; the remainder of the hollow viscus being developed from the part of the cloaca which lies above the sinus. The allantois extends from the summit of the bladder to the umbilicus, and usually undergoes complete obliteration to form the fibrous cord of the urachus. In some cases, however, the allantoic canal remains patent, and urine may escape by it at the umbilicus. If the urethra is looked upon as the remnant of the cloaca, then the bladder, with the exception of the trigone, must be regarded as being developed by a dilatation of the proximal part of the allantois.

Ovaries and Testicles.—The first appearance of the ovaries and testicles is essentially the same in the two sexes, and consists in a thickening of the epithelial layer which lines the peritoneal or body cavity on the inner side of the Wolffian ridge. Beneath this thickened epithelium an increase in the mesoblast takes place, forming a distinct projection. This is termed the *genital ridge* (fig. 184), and from it the testicle in the male, and the ovary in the female, are developed. At first the Wolffian body and genital ridge are suspended by a common mesentery, but as the embryo grows the genital ridge gradually becomes pinched off from the Wolffian body, with which it is at first continuous, though it still remains connected to the remnant of this body by a fold of peritoneum, the *mesorchium* or *mesovarium*. About the seventh week the distinction of sex in the genital ridge begins to be perceptible.

The *ovary*, thus formed from the genital ridge, consists of a central part of connective tissue covered by a layer of epithelium, the *germinal epithelium*. Between the cells of the germinal epithelium a number of larger cells, the *primitive ova*, are found, and these are carried into the subjacent stroma by budlike ingrowths of the germinal epithelium, the cells of which surround the primitive

FIG. 189.—Section of the ovary of a newly born child.   (Waldeyer.)

ova ; and in this manner the primitive Graafian follicles are formed. The remains of the germinal epithelium on the surface of the ovary form the permanent epithelial covering of this organ. According to Beard the primitive ova are early set apart during the segmentation of the ovum and migrate into the germinal ridge.

Waldeyer taught, and for many years his views have been accepted, that the primitive germ-cells were derived from the 'germinal epithelium' covering the genital ridge. Beard,* on the other hand, maintains that in the skate they are not derived from this epithelium, but are probably formed during the later stages of cell-cleavage, before there is any trace of an embryo ; and a similar view was advanced by Nussbaum as to their origin in amphibia. Beard says : 'At the close of segmentation many of the future germ-cells lie in the segmentation cavity just beneath the site of the future embryo, and there is no doubt they subsequently wander into it.' The germ-cells, 'after they enter the resting phase, are sharply marked off from the cells of the embryo by entire absence of mitoses among them.' They can be further recognised by their irregular form and amœboid-like processes, and by the fact that their cytoplasm has no affinity for ordinary stains, but assumes a brownish tinge when treated by osmic acid. The path along which they travel into the embryo is a very definite one—viz. 'from the yolk-sac upwards between the splanchnopleure and gut in the hinder portion of the embryo.' This pathway, named by Beard the *germinal path*, 'leads them directly to the position which they ought finally to take up in the " germinal ridge " or nidus.' A considerable number apparently never reach their proper destination, since 'vagrant germ-cells are found in all sorts of places, but more particularly on the mesentery.' Some of these may possibly find their way into the germinal ridge ; some probably undergo atrophy, while others may persist and become the seat of dermoid tumours.

The *testicle* is developed in a very similar way to the ovary. Like the ovary, in its earliest stages it consists of a central mass of connective tissue covered by germinal epithelium, among which larger cells, the *primitive sperm-cells*, are seen. These are carried into the subjacent stroma by tubes of germinal epithelium, which form the lining of the seminiferous tubules, while the primitive sperm-cells form the spermatogonia. The seminiferous tubules become connected with outgrowths from the Wolffian body, which, as before mentioned, form the rete testis and vasa efferentia.

The development of the other parts of the internal male and female organs has already been described.

Descent of the Testes.—The testes, at an early period of fœtal life, are placed at the back part of the abdominal cavity, behind the peritoneum and a little below the kidneys. The anterior surface and sides are invested by peritoneum. At about the third month of intra-uterine life a peculiar structure, the *gubernaculum testis*, makes its appearance. This structure is at first a slender band, extending from that part of the skin of the groin which afterwards forms the scrotum through the inguinal canal to the body and epididymis of the testicle, and is then continued upwards in front of the kidney towards the Diaphragm. As development advances, the peritoneum covering the testicle

Fig. 190.—Adult ovary, parovarium, and Fallopian tube.
(From Farre, after Kobelt.)

a, a. Epoöphoron formed from the upper part of the Wolffian body. b. Remains of the uppermost tubes sometimes forming hydatids. c. Middle set of tubes. d. Some lower atrophied tubes. e. Atrophied remains of the Wolffian duct. f. The terminal bulb or hydatid. h. The Fallopian tube, originally the duct of Müller. i. Hydatid attached to the extremity. l. The ovary.

encloses it and forms a mesentery, the *mesorchium*, which also encloses the gubernaculum and forms two folds, one above the testicle and the other below it. The one above the testicle is the *plica vascularis*, and contains ultimately the spermatic vessels ; the one below, the *plica gubernatrix*, contains the lower part of the gubernaculum, which has now grown into a thick cord ; it terminates below at the internal ring in a tube of peritoneum, the processus vaginalis, which protrudes itself down the inguinal canal. The lower part of the gubernaculum by the fifth month has become a thick cord, while the upper part has disappeared. The lower part can now be seen to consist of a central core of unstriped muscle-fibre, and outside this of a firm layer of striped elements, connected, behind the peritoneum, with the abdominal wall. As the scrotum develops, the lower end of the gubernaculum is carried with the skin to which it is attached to the bottom of this pouch. The fold of peritoneum, constituting the processus vaginalis, projects itself downwards into the inguinal canal, and emerges at the external abdominal ring, pushing before it a part of the internal oblique muscle and the aponeurosis of the external oblique, which form respectively the cremaster muscle and the external spermatic fascia. It forms a gradually elongating depression or *cul-de-sac*, which eventually reaches the bottom of the scrotum, and into this the testicle is drawn by the growth of the body of the fœtus, for the gubernaculum does not grow commensurately with the growth of other parts, and therefore the testicle, being attached by the gubernaculum to the bottom of the scrotum, is prevented from rising as the body grows, and is drawn first into the inguinal canal and eventually into the scrotum. It seems certain also that the gubernacular cord becomes shortened as development proceeds, and this assists in causing the testicle to reach the bottom of the scrotum. By the eighth month the testicle has reached the scrotum, preceded by the lengthened pouch of peritoneum, the processus vaginalis, which communicates by its upper extremity with the peritoneal cavity. Just before birth the upper part of the pouch usually becomes closed, and this obliteration extends gradually downwards to within a short distance of the testis. The process of peritoneum surrounding the testis, which is now entirely cut off from the general peritoneal cavity, constitutes the *tunica vaginalis.**

In the female there is also a gubernaculum, which effects a considerable change in the position of the ovary, though not so extensive a change as that of the testicle in the male. The gubernaculum in the female, as it lies on either side in contact with the fundus of the uterus formed by the union of the Müllerian ducts, con- tracts adhesions to this organ, and thus the ovary is prevented from descending below this level. The remains of the gubernaculum—that is to say, the part below the attachment of the cord to the uterus to its termination in the labia majora—ultimately forms the round ligament of the uterus. A pouch of peritoneum accompanies it along the inguinal canal, analogous to the processus vaginalis in the male: it is called the *canal of Nuck*. In rare cases the gubernaculum may fail to contract adhesions to the uterus, and then the ovary descends through the inguinal canal into the labia majora, extending down the canal of Nuck, and under these circumstances resembles in position the testicle in the male.

Surgical Anatomy.—Abnormalities in the formation and in the descent of the testicle may occur. The testicle may fail to be developed: or the testicle may be fully developed, and the vas deferens may be undeveloped in whole or part; or again, both testicle and vas deferens may be fully developed, but the duct may not become connected to the gland. The testicle may fail in its descent, or it may descend into some abnormal position. Thus it may be retained in the position where it was primarily developed, below the kidney; or it may descend to the internal abdominal ring, but fail to pass through this opening; it may be retained in the inguinal canal, which is perhaps the most common position; or it may pass through the external abdominal ring and remain just outside it, failing to pass to the bottom of the scrotum. On the other hand, it may get into some abnormal position: it may pass the scrotum and reach the perinæum, or it may fail to enter the inguinal canal, and may find its way through the femoral ring into the crural canal, and present itself on the thigh at the saphenous opening. There is still a third class of cases of abnormality of the testicle, where the organ has descended in due course into the scrotum, but is malplaced. The most common form of this is where the testicle is *inverted*: that is to say, the organ is rotated so that the epididymis is connected to the front of the scrotum, and the body, surrounded by the tunica vaginalis, is directed backwards. In these cases the vas deferens is to be felt in the front of the cord. The condition is of importance in connection with hydrocele and hæmatocele, and the position of the testicle should always be carefully ascertained before performing any operation for these affections. Again, more rarely, the testicle may be *reversed*. This is a condition in which the top of the testicle, indicated by the globus major of the epididymis, is at the bottom of the scrotum, and the vas deferens comes off from the summit of the organ. Cases sometimes occur, generally in the young adult, in which the spermatic cord becomes twisted. In consequence of this the circulation through it is partially or completely arrested; if the latter, the testicle becomes gangrenous; if the former, it may undergo atrophy.

The **external organs of generation** (fig. 191), like the internal, pass through a stage in which there is no distinction of sex. It is therefore necessary to describe this stage, and then follow the development of the female and male organs respectively.

In front of the cloacal membrane an elevation appears during the sixth week. This is named the *genital eminence*, and represents the rudiment of the penis or clitoris. It is bounded laterally and below by two prominent folds, termed the *external genital folds*. By the growth of the surrounding parts the ecto- dermal cloaca assumes the form of a cleft, the *urogenital cleft*, into which the urogenital sinus opens after the disappearance of the cloacal membrane; the

---

* The obliteration of the process of peritoneum which accompanies the cord, and is hence called the *funicular process*, is often incomplete. See section on Inguinal Hernia.

FIG. 191.—Stages in the development of the external sexual organs in the male and female.   (Drawn from the Ecker-Ziegler models.)

A

Hind-limb
Cloaca
Tail

B

Genital eminence
Outer genital fold
Inner genital fold
Genital groove

C

Glans penis
Outer genital fold
Inner genital fold
Genital groove
Perinæum
Anus

D

Glans clitoridis
Outer genital fold
Urogenital cleft
Inner genital fold
Perinæum
Anus

E

Prepuce
Glans clitoridis
Labia majora
Labia minora
Vestibule
Vagina
Anus

F

Glans penis
Genital groove
Scrotum
Raphe
Anus

margins of the cleft constitute the *inner genital folds*. On the lower aspect of the genital eminence there is seen a furrow or groove, the *genital groove,* which is continuous behind with the urogenital fissure. All these parts are well developed by the second month, yet no distinction of sex is possible.

Female Organs.—The female organs are developed by an easy transition from the above. The lower part of the urogenital sinus persists as the vestibule of the vagina and the urethra. The genital eminence forms the clitoris, the external genital folds the labia majora, and the inner genital folds the labia minora and the prepuce and frænum of the clitoris.

Male Organs.—In the male the changes are greater. The genital eminence is developed into the glans and corpora cavernosa of the penis, the glans appearing in the third month, and the corpora cavernosa in the fourth. The genital groove closes and thus forms the spongy portion of the urethra (see page 151). The bulb and corpus spongiosum of the urethra are formed by the fusion of the inner genital folds. The prepuce is generally described as growing forward from the root of the glans penis, its inner surface at first adhering to the glans and then becoming free. It is formed, however, by the epidermis passing in round the glans, so that on coronal section the epidermis presents a horse-shoe arrangement. By the breaking down of the more centrally situated cells of this epidermal invagination the prepuce becomes movable. ' Adherent prepuce is not an adhesion really, but a hindered central desquamation ' (Berry Hart, *op. cit.*). The external genital folds unite in the middle line to form the scrotum, at about the same time as the genital groove closes—viz. between the third and fourth month.

The *suprarenal bodies* are developed from two different sources. The medullary part of the organ is of epiblastic origin, and is derived from the tissues forming the sympathetic ganglia of the abdomen, while the cortical portion is of mesoblastic origin, and originates as an outgrowth from the upper part of the Wolffian body. The two parts are at first quite distinct, but become combined in the process of development. The suprarenal bodies are at first larger than the kidneys ; about the tenth week they equal them in size, and from that time decrease relatively to the kidney, though they remain, throughout fœtal life, proportionately much larger than in the adult.

The following table is translated from the work of Beaunis and Bouchard, with some alterations, especially in the earlier weeks. It will serve to present a *résumé* of the above facts in an easily accessible form.*

* It will be noticed that the time assigned in this table for the appearance of the first rudiment of some of the bones varies in some cases from that assigned in the description of the various bones. This is a point on which anatomists differ, and which probably varies in different cases.

CHRONOLOGICAL TABLE

OF

THE DEVELOPMENT OF THE FŒTUS

(Modified from Beaunis and Bouchard)

First Week.—During the early part of this period the ovum is in the Fallopian tube. Having been fertilised in the upper part of the tube, it slowly passes down, undergoing segmentation, and reaches the uterus before the end of the first week. In Peter's specimen, the age of which was probably from five to six days, the ovum was found embedded in the decidua on the posterior wall of the uterus and enveloped by a decidua reflexa, the central part of which, however, consisted merely of a layer of fibrin. The ovum was in the form of a sac, the outer wall of which consisted of a layer of trophoblast; inside this was a thin layer of mesoderm composed of round, oval, and spindle-shaped cells. Numerous villous processes—some consisting of trophoblast only, others possessing a core of mesoderm—projected from the surface of the ovum into the surrounding decidua. Inside this sac the rudiment of the embryo was found in the form of a patch of ectoderm, covered by a small but completely closed amnion. It possessed a minute yolk-sac and was surrounded by mesoderm, which was connected by a band to that lining the trophoblast.

Second Week.—By the end of this week the ovum has increased considerably in size, and the decidua reflexa forms a complete fold over it. Its villi consist of a syncytial layer, together with the layer of Langhans, and the majority of them are vascularised. The embryo has assumed a definite form : its cephalic and caudal extremities are easily distinguished and are folded towards the yolk-sac. The primitive groove and streak are evident—the former terminating anteriorly in the neurenteric canal. The neural groove and ridges have appeared, and the rudiment of the notochord is being differentiated from the entoderm. The embryo is more completely separated from the yolk-sac ; the rudimentary heart has made its appearance ; the allantoic diverticulum extends into the body-stalk ; all three layers of the blastoderm are present, and the paraxial mesoderm is being segmented into the protovertebral somites.

Third Week.—By the end of the third week the flexures of the embryo have taken place, so that it is strongly curved. The protovertebral somites number about thirty. In the nervous system the primary divisions of the brain are visible, and the primitive ocular and auditory vesicles are already formed. The primary circulation is established. The alimentary canal presents a nearly straight tube communicating with the yolk-sac. The branchial arches are formed and the bucco-pharyngeal membrane has disappeared. The limbs have appeared as short buds. The Wolffian bodies are visible.

Fourth Week.—The umbilical vesicle has attained its full development. The caudal extremity projects. The upper and the lower limbs appear. The heart separates into a right and left heart. The special ganglia and anterior roots of the spinal nerves, the olfactory fossæ, the lungs and the pancreas can be made out. The cerebral hemispheres appear as hollow buds, and the ganglia of some of the cranial nerves are visible.

Fifth Week.—The allantois is vascular in its whole extent. The first traces of the hands and feet can be seen. The aortic bulb divides into aorta and pulmonary artery. The duct of Müller and genital gland are visible. The ossification of the clavicle and the lower jaw commences. The cartilage of Meckel occupies the first post-oral arch. The tubercles which form the rudiment of the pinna are visible. The rudiment of the penis or clitoris, in the form of the genital tubercle, is present.

Sixth Week.—The activity of the umbilical vesicle ceases. The pharyngeal clefts disappear. The vertebral column, primitive cranium, and ribs assume the cartilaginous condition. The posterior roots of the nerves, the membranes of the nervous centres, the bladder, kidney, tongue, larynx, thyroid body, and the germs of teeth are apparent.

Seventh Week.—The muscles begin to be perceptible. The points of ossification of the ribs, scapula, shaft of humerus, femur, tibia, palate, and upper jaw appear.

Eighth Week.—The distinction of arm and forearm, and of thigh and leg, is apparent, as well as the interdigital clefts. The capsule of the lens and pupillary membrane, the interventricular and commencement of the interauricular septum, the salivary glands, the spleen and suprarenal capsules are distinguishable. The larynx begins to become cartilaginous. All the vertebral bodies are cartilaginous. The points of ossification for the ulna, radius, fibula, and ilium make their appearance. The two halves of the hard palate unite. The sympathetic nerves are now for the first time to be discerned.

Ninth Week.—The corpus striatum and the pericardium are first apparent. The ovary and testicle can be distinguished from each other. The genital furrow appears. The osseous nuclei of the bodies and arches of the vertebræ, of the frontal, vomer, and malar bones, of the shafts of the metacarpal and metatarsal bones, and of the phalanges appear. The union of the hard palate is completed. The gall-bladder is seen.

Third Month.— The formation of the foetal placenta advances rapidly. The projection of the caudal extremity disappears. It is possible to distinguish the male and female organs from each other. The Cloacal aperture is divided into two parts. The cartilaginous arches on the dorsal region of the spine close. The points of ossification for the lachrymal, nasal, squamous portion of temporal and ischium appear, as well as the orbital centre of the superior maxillary. The pons Varolii and fissure of Sylvius can be made out. The eyelids, the hair, and the nails begin to form. The mammary gland, the epiglottis, and prostate are beginning to develop. The union of the testicle with the canals of the Wolffian body takes place. By the end of this month the length of the foetus is about 7 cm., but if the legs be included it is from 9 to 10 cm.

Fourth Month.—The closure of the cartilaginous arches of the spine is complete. Osseous points for the first sacral vertebra and os pubis appear. The ossification of the malleus and incus takes place. The corpus callosum, the membrana lamina spiralis, the cartilage of the Eustachian tube, and the tympanic ring are seen. Fat is first developed in the subcutaneous cellular tissue. The tonsils are seen, and the closure of the genital furrow and the formation of the scrotum and prepuce take place. By the end of the fourth month the foetus is from 12 to 13 cm. in length, but if the legs be included it is from 16 to 20 cm.

Fifth Month.—The two layers of the decidua begin to coalesce. Osseous nuclei of the axis and odontoid process, of the lateral points of the first sacral vertebra, of the median points of the second, and of the lateral masses of the ethmoid make their appearance. Ossification of the stapes and the petrous bone, and ossification of the germs of the teeth, take place. The germs of the permanent teeth and the organ of Corti appear, The eruption of hair on the head commences. The sudoriferous glands, Brunner's glands, the follicles of the tonsil and base of the tongue, and the lymphatic glands appear at this period. The differentiation between the uterus and vagina becomes apparent. By the end of this month the total length of the foetus, including the legs, is from 25 to 27 cm.

Sixth Month.—The points of ossification for the anterior root of the transverse process of the seventh cervical vertebra, the lateral points of the second cervical vertebra, the median points of the third, the manubrium sterni and the os calcis appear. The sacro-vertebral angle forms. The cerebral hemispheres cover the cerebellum. The papillæ of the skin, the sebaceous glands, and Peyer's patches make their appearance. The free border of the nail projects from the corium of the dermis. The walls of the uterus thicken. Measured from vertex to heels, the total length of the foetus at the end of this month is from 30 to 32 cm.

Seventh Month.—The additional points of the first sacral vertebra, the lateral points of the third, the median point of the fourth, the first osseous point of the body of the sternum, and the osseous point for the astragalus appear. Meckel's cartilage disappears. The cerebral convolutions, the island of Reil, and the tubercula quadrigemina are apparent. The pupillary membrane atrophies. The testicle passes into the vaginal process of the peritoneum. From vertex to heels the total length at the end of the seventh month is from 35 to 36 cm., i.e. about 14 in. Its weight is a little over 3 lbs.

Eighth Month.—Additional points for the second sacral vertebra, lateral points for the fourth, and median points for the fifth sacral vertebræ can be seen. The total length, i.e. from head to heels, at the end of the eighth month is about 40 cm. (16 in.), and its weight varies between 4½ and 5½ lbs.

Ninth Month.—Additional points for the third sacral vertebra, lateral points for the fifth, osseous points for the middle turbinated bone, for the body and great cornu of the hyoid, for the second and third pieces of the body of the sternum, and for the lower end of the femur appear. Ossification of the bony lamina spiralis and axis of the cochlea takes place. The eyelids open, and the testicles are in the scrotum. At full time the foetus weighs from 6½ to 8 lbs., and measures from head to heels about 50 cm. (20 in.).

DESCRIPTIVE AND SURGICAL
ANATOMY

---------※---------

OSTEOLOGY

THE SKELETON

THE term *skeleton*, in its widest application, includes the bones, cartilages, and ligaments of the body; it is, however, generally used to indicate only the bony framework.

The entire skeleton in the adult consists of 200 distinct bones. These are—

The Spine or vertebral column (sacrum and coccyx included) .	26
Cranium	8
Face	14
Hyoid bone, sternum, and ribs	26
Upper extremities	64
Lower extremities	62

200

In this emuneration the patellæ are included as separate bones, but the smaller sesamoid bones, and the ossicula auditûs, are not reckoned. The teeth belong to the tegumentary system.

These bones are divisible into four classes : *Long, Short, Flat,* and *Irregular.*

The **Long Bones** are found in the limbs, where they form a system of levers which have to sustain the weight of the trunk, and to confer the power of locomotion. A long bone consists of a shaft and two extremities. The *shaft* is a hollow cylinder, contracted and narrowed to afford greater space for the bellies of the muscles ; the walls consist of dense, compact tissue of great thickness in the middle, but becoming thinner towards the extremities ; the cancellous tissue is scanty, and the bone is hollowed out in its interior to form the *medullary canal.* The *extremities* are generally somewhat expanded, for greater convenience of mutual connection, for the purposes of articulation, and to afford a broad surface for muscular attachment. Here the bone is made up of cancellous tissue with only a thin coating of compact substance. The medullary canal and the spaces in the cancellous tissue are filled with marrow. The long bones are not straight, but curved ; the curve generally taking place in two directions, thus affording greater strength to the bone. The bones belonging to this class are : the *clavicle, humerus, radius, ulna, femur, tibia, fibula, metacarpal* and *metatarsal* bones, and the *phalanges.*

Short Bones.—Where a part of the skeleton is intended for strength and compactness, and its motion is at the same time slight and limited, it is divided into a number of small pieces united together by ligaments, and the separate bones are short and compressed, such as those of the *carpus* and *tarsus.* These bones, in their structure, are spongy throughout, excepting at their surface, where there is a thin crust of compact substance. The *patellæ,* together with the other sesamoid bones, are by some regarded as short bones.

Flat Bones.—Where the principal requirement is either extensive protection or the provision of broad surfaces for muscular attachment, the osseous structure is expanded into broad, flat plates, as in the bones of the skull and the shoulder-blade. These bones are composed of two thin layers of compact tissue enclosing between them a variable quantity of cancellous tissue. In the cranial bones, these layers of compact tissue are familiarly known as the *tables* of the skull; the outer one is thick and tough; the inner one thinner, denser, and more brittle, and hence termed the *vitreous table*. The intervening cancellous tissue is called the *diploë*, and this, in certain regions of the skull, becomes absorbed so as to leave spaces filled with air (*air-sinuses*) between the two tables. The flat bones are: the *occipital, parietal, frontal, nasal, lachrymal, vomer, scapula, os innominatum, sternum, ribs*, and, according to some, the *patella*.

The **Irregular** or **Mixed Bones** are such as, from their peculiar form, cannot be grouped under either of the preceding heads. Their structure is similar to that of other bones, consisting of a layer of compact tissue externally, and of cancellous tissue within. The irregular bones are: the *vertebræ, sacrum, coccyx, temporal, sphenoid, ethmoid, malar, superior maxillary, inferior maxillary, palate, inferior turbinated*, and *hyoid*.

Surfaces of Bones.—If the surface of any bone is examined, certain eminences and depressions are seen, to which descriptive anatomists have given the following names.

These eminences and depressions are of two kinds: *articular* and *non-articular*. Well-marked examples of articular eminences are found in the heads of the humerus and femur; and of articular depressions, in the glenoid cavity of the scapula, and the acetabulum. Non-articular eminences are designated according to their form. Thus, a broad, rough, uneven elevation is called a *tuberosity, protuberance*, or *process*; a small, rough prominence, a *tubercle*; a sharp, slender, pointed eminence, a *spine*; a narrow, rough elevation, running some way along the surface, a *ridge, crest*, or *line*.

The non-articular depressions are also of very variable form, and are described as *fossæ, pits, depressions, grooves, furrows, fissures, notches*, &c. These non-articular eminences and depressions serve to increase the extent of surface for the attachment of ligaments and muscles, and are usually well marked in proportion to the muscularity of the subject.

A prominent process projecting from the surface of a bone, which it has never been separate from or movable upon, is termed an *apophysis* (from ἀπόφυσις, *an excrescence*); but if such process is developed as a separate piece from the rest of the bone, to which it is afterwards joined, it is termed an *epiphysis* (from ἐπίφυσις, *an accretion*). The main part of the bone, or shaft, which is formed from the primary centre of ossification, is termed the *diaphysis*, and is separated, during growth, from the epiphysis by a layer of cartilage, at which growth in length of the bone takes place.

Bones are often traversed by canals or foramina, not only for the entrance of nutrient vessels into their interior, but also for the transmission of vessels and nerves from one part to another.

In describing the various surfaces or aspects of a bone, or indeed of any other structure, the body is supposed to be in the erect position. Any surface, extremity, or other part directed upwards towards the head is termed *superior*, while those directed downwards towards the feet are termed *inferior*. Surfaces directed forwards towards the front of the body are termed *anterior* or *ventral*, while those directed backwards are *posterior* or *dorsal*. Those surfaces which are directed towards a median, antero-posterior, vertical plane are termed *internal* or *mesial*, while those directed away from this plane are *external* or *lateral*.

The minute structure, growth, and composition of bone are described on pages 25 and 28.

THE SPINE

The **Spine** is a flexuous and flexible column, formed of a series of bones called *vertebræ* (from *vertere*, to turn).

The **Vertebræ** are thirty-three in number, and have received the names *cervical, thoracic* or *dorsal, lumbar, sacral*, and *coccygeal*, according to the position which they occupy; seven being found in the cervical region, twelve in the thoracic, five in the lumbar, five in the sacral, and four in the coccygeal.

This number is sometimes increased by an additional vertebra in one region, or the number may be diminished in one region, the deficiency being supplied by an additional vertebra in another. These observations do not apply to the cervical portion of the spine, the number of bones forming which is seldom increased or diminished.

The Vertebræ in the upper three regions of the spine are separate throughout the whole of life; but those found in the sacral and coccygeal regions are in the adult firmly united, so as to form two bones—five entering into the formation of the upper bone or *sacrum*, and four into the terminal bone of the spine or *coccyx*.

General Characters of a Vertebra

Each **vertebra** consists of two essential parts—an anterior solid segment, the *body* or *centrum*, and a posterior segment, the *neural arch*. The arch is formed of two *pedicles*, and two *laminæ*, which with the body form the *vertebral foramen*; they support seven *processes*—viz. four *articular*, two *transverse*, and one *spinous*.

The bodies of the vertebræ are piled one upon the other, forming a strong pillar, for the support of the cranium and trunk; the arches form a hollow cylinder behind the bodies for the protection of the spinal cord. The different vertebræ are connected together by means of the articular processes and the intervertebral fibro-cartilages; while the transverse and spinous processes serve as levers for the attachment of muscles which move the different parts of the spine. Lastly, between each pair of vertebræ apertures exist through which the spinal nerves pass from the cord. Each of these constituent parts must now be separately examined.

The **Body** or **centrum** is the largest part of a vertebra. It forms a segment of a solid column for the support of the trunk. Above and below, it is flattened; its upper and lower surfaces are rough, for the attachment of the intervertebral fibro-cartilages, and present a rim around their circumference. In front, it is convex from side to side, concave from above downwards. Behind, it is flat from above downwards and slightly concave from side to side. Its anterior surface is perforated by a few small apertures, for the passage of nutrient vessels; while on the posterior surface is a single large, irregular aperture, or occasionally more than one, for the exit of veins from the body of the vertebra—the *venæ basis vertebræ*.

The **Pedicles** are two short, thick processes of bone, which project backwards, one on each side, from the upper part of the body of the vertebra, at the line of junction of its posterior and lateral surfaces. The concavities above and below the pedicles are the *intervertebral notches*; they are four in number, two on each side, the inferior ones being generally the deeper. When the vertebræ are articulated, the notches of each contiguous pair of bones form the *intervertebral foramina*, which communicate with the spinal canal and transmit the spinal nerves and blood-vessels.

The **Laminæ** are two broad plates of bone which are connected to the body by the pedicles. They fuse in the middle line posteriorly, and so complete the vertebral foramen, which serves for the lodgment and protection of the spinal cord. Their upper borders and the lower part of their anterior surfaces are rough for the attachment of the ligamenta subflava.

The **Spinous Process** projects backwards from the junction of the two laminæ, and serves for the attachment of muscles and ligaments.

The **Articular Processes,** four in number, two on each side, spring from the junction of the pedicles with the laminæ. The two superior project upwards, their articular surfaces being directed more or less backwards; the two inferior project downwards, their articular surfaces looking more or less forwards.[*]

The **Transverse Processes,** two in number, project one at each side from the point where the lamina joins the pedicle, between the superior and inferior articular processes. They serve for the attachment of muscles and ligaments.

Characters of the Cervical Vertebræ (fig. 192)

The **Cervical Vertebræ** are smaller than those in the other regions of the spine, and may readily be distinguished by a foramen in the transverse process,

[*] It may, perhaps, be as well to remind the reader that the direction of a surface is determined by that of a line drawn at right angles to it.

M

which does not exist in the transverse process of either the dorsal or lumbar vertebræ.

The **Body** is small, comparatively dense, and broader from side to side than from before backwards. The anterior and posterior surfaces are flattened and of equal depth ; the former is placed on a lower level than the latter, and its inferior border is prolonged downwards, so as to overlap the upper and fore part of the vertebra below. Its upper surface is concave transversely, and presents a projecting lip on each side ; its lower surface is convex from side to side, concave from before backwards, and presents laterally a shallow concavity, which receives the corresponding projecting lip of the adjacent vertebra. The *pedicles* are directed outwards and backwards, and are attached to the body midway between the upper and lower borders, so that the superior intervertebral notch is as deep as the inferior, but it is, at the same time, narrower. The *laminæ* are narrow, long, thinner above than below ; they enclose the spinal foramen, which is large, and of a triangular form. The *spinous process* is short, and bifid at the extremity to afford greater extent of surface for the attachment of muscles, the two divisions being often of unequal size. They increase in length from the fourth to the seventh. The *articular processes* are flat, oblique, and of an oval form : the superior are directed backwards and upwards ; the inferior forwards and downwards. The *transverse processes* are short, directed obliquely outwards and forwards, bifid at their extremity, and marked by a groove along their

FIG. 192.—A cervical vertebra.

upper surface, which runs downwards and outwards from the superior intervertebral notch, and serves for the transmission of one of the cervical nerves. They are situated in front of the articular processes and on the outer side of the pedicles, and are pierced at their base by a foramen, the *foramen transversarium* or *vertebrarterial foramen*, for the transmission of the vertebral artery, vein, and plexus of nerves. Each process is formed by two roots : the anterior root, sometimes called the *costal process*, arises from the side of the body, and is the homologue of the rib in the dorsal region of the spine ; the posterior root springs from the junction of the pedicle with the lamina, and corresponds with the transverse process in the dorsal region. It is by the junction of the two that the foramen for the vertebral vessels is formed. The extremity of each of these roots forms the *anterior* and *posterior tubercles* of the transverse processes.*

The peculiar vertebræ in the cervical region are the first, or *Atlas* ; the second, or *Axis* ; and the seventh, or *Vertebra prominens*. The great modifications in the form of the atlas and axis are designed to admit of the nodding and rotatory movements of the head.

The **Atlas** (fig. 193) is so named because it supports the globe of the head. The chief peculiarities of this bone are that it has neither body nor spinous

* The anterior tubercle of the transverse process of the sixth cervical vertebra is of large size, and is sometimes known as ' Chassaignac's ' or the ' carotid tubercle.' It is in close relation with the carotid artery, which lies in front and a little external to it ; so that, as was first pointed out by Chassaignac, the vessel can with ease be compressed against it.

process. The body is detached from the rest of the bone, and forms the odontoid process of the second vertebra. The atlas is ring-like, and consists of an anterior arch, a posterior arch, and two lateral masses. The *anterior* arch forms about one-fifth of the ring : its anterior surface is convex, and presents about its centre a *tubercle*, for the attachment of the Longus colli muscle ; posteriorly it is concave, and marked by a smooth, oval or circular facet, for articulation with the odontoid process of the axis. The upper and lower borders respectively give attachment to the anterior occipito-atlantal and the anterior atlanto-axial ligaments ; the former connects it with the occipital bone above and the latter with the axis below. The *posterior* arch forms about two-fifths of the circumference of the bone : it terminates behind in a *tubercle*, which is the rudiment of a spinous process, and gives origin to the Rectus capitis posticus minor. The diminutive size of this process prevents any interference in the movements between the atlas and the cranium. The posterior part of the arch presents above and behind a rounded edge for the attachment of the posterior occipito-atlantal ligament, while in front, immediately behind each superior articular process, is a groove, sometimes converted into a foramen by a delicate bony spiculum which arches backwards from the posterior extremity of the superior articular process. These grooves represent the superior intervertebral notches, and are peculiar from being situated behind the articular processes, instead of in front of them. They serve for the transmission of the vertebral artery, which,

FIG. 193.—First cervical vertebra, or Atlas.

ascending through the foramen in the transverse process, winds round the lateral mass in a direction backwards and inwards. They also transmit the suboccipital (first spinal) nerve. On the under surface of the posterior arch, in the same situation, are two other grooves, placed behind the lateral masses, and representing the inferior intervertebral notches of other vertebræ. They are much less marked than the superior. The lower border gives attachment to the posterior atlanto-axial ligament, which connects it with the axis. The *lateral masses* are the most bulky and solid parts of the atlas, in order to support the weight of the head : they present two articulating processes above, and two below. The two superior are of large size, oval, concave, and approach each other in front, but diverge behind : they are directed upwards, inwards, and a little backwards, each forming a kind of cup for the corresponding condyle of the occipital bone, and are admirably adapted to the nodding movements of the head. Not infrequently they are partially subdivided by a more or less deep indentation which encroaches upon each lateral margin. The inferior articular processes are circular in form, flattened or slightly concave and directed downwards and inwards, articulating with the axis, and permitting the rotatory movements of the head. Just below the inner margin of each superior articular surface is a small tubercle, for the attachment of the transverse ligament which, stretching across the ring of the atlas, divides it into two unequal parts—the anterior or smaller segment receiving the odontoid process of the axis, the posterior allowing the transmission of the spinal cord and its membranes. This part of the spinal canal is of considerable size, to afford space for the spinal cord ; and hence lateral displacement of the

M 2

atlas may occur without compression of this structure. The *transverse processes* are of large size, project directly outwards and downwards from the lateral masses, and serve for the attachment of special muscles which assist in rotating the head. They are long, and do not present an anterior and posterior tubercle, which have become fused into one mass; they are perforated at their base by a foramen for the vertebral artery, which is directed from below, upwards and backwards.

The **Axis** (fig. 194) is so named from forming the pivot upon which the first vertebra, carrying the head, rotates. The most distinctive character of this bone is the strong, prominent process, tooth-like in form (hence the name *odontoid*), which rises perpendicularly from the upper surface of the body. The *body* is deeper in front than behind, and prolonged downwards anteriorly so as to overlap the upper and fore part of the next vertebra. It presents in front a median longitudinal ridge, separating two lateral depressions for the attachment of the Longus colli muscle of either side. The *odontoid process* presents two articulating surfaces: one in front, of an oval form, for articulation with the atlas; another behind for the transverse ligament—the latter frequently encroaching on the sides of the process. The apex is pointed, and gives attachment to the middle odontoid or check ligament. Below the apex, the process is somewhat enlarged and presents on either side a rough impression for the attachment of the lateral

FIG. 194.—Second cervical vertebra, or Axis.

odontoid or check ligaments, which connect it to the occipital bone; the base of the process, where it is attached to the body, is constricted, so as to prevent displacement from the transverse ligament, which binds it in this situation to the anterior arch of the atlas. Sometimes, however, this process does become displaced, especially in children, in whom the ligaments are more relaxed: instant death is the result of this accident. The internal structure of the odontoid process is more compact than that of the body. The *pedicles* are broad and strong, especially their anterior extremities, which coalesce with the sides of the body and the root of the odontoid process. They are covered above by the superior articulating surfaces. The *laminæ* are thick and strong, and the spinal foramen large, but smaller than that of the atlas. The *transverse processes* are very small, not bifid, but terminating in a single tubercle; they are perforated by the foramen for the vertebral artery, which is directed obliquely upwards and outwards. The *superior articular surfaces* are round, slightly convex, directed upwards and outwards, and are peculiar in being supported on the body, pedicles, and transverse processes. The *inferior articular surfaces* have the same direction as those of the other cervical vertebræ. The *superior intervertebral notches* are very shallow, and lie behind the articular processes; the *inferior* in front of them, as in the other cervical vertebræ. The *spinous process* is of large size, very strong, deeply channelled on its under surface, and presents a bifid, tubercular extremity for the attachment of muscles which serve to rotate the head upon the spine.

Seventh Cervical (fig. 195).—The most distinctive character of this vertebra is the existence of a long and prominent spinous process; hence the name 'vertebra prominens.' This process is thick, nearly horizontal in direction, not bifurcated, but terminating in a tubercle, which has attached to it the lower end of the ligamentum nuchæ. The *transverse process* is of considerable size, its posterior tubercles are large and prominent, while the anterior are small and faintly marked; its upper surface has usually a shallow groove, and it seldom presents more than a trace of bifurcation at its extremity. The foramen in the transverse process is sometimes as large as in the other cervical vertebræ, but is generally smaller on one or both sides, and sometimes wanting. On the left side it occasionally gives passage to the vertebral artery; more frequently the vertebral vein traverses it on both sides; but the usual arrangement is for both artery and vein to pass in front of the transverse process, and not through the foramen. Sometimes the anterior root of the transverse process exists as a separate bone, and attains a large size. It is then known as a 'cervical rib.'

FIG. 195.—Seventh cervical vertebra, or vertebra prominens.

CHARACTERS OF THE THORACIC OR DORSAL VERTEBRÆ

The **Dorsal Vertebræ** (fig. 196) are intermediate in size between those in the cervical and those in the lumbar region, and increase in size from above

FIG. 196.—A dorsal vertebra.

downwards, the upper vertebræ being much smaller than those in the lower part of the region. The dorsal vertebræ may be at once recognised by the presence

of one or more facets or half-facets on the sides of the body for the heads of
the ribs.

The **bodies** of the dorsal vertebræ resemble those in the cervical and lumbar
regions at the respective ends of this portion of the spine; but in the middle of
the dorsal region their form is very characteristic, being heart-shaped, and as
broad in the antero-posterior as in the lateral direction. They are slightly thicker
behind than in front, flat above and below, convex and prominent in front, deeply

FIG. 197.—Peculiar dorsal vertebræ.

{ *An entire facet above ;*
  a demi-facet below

– *A demi-facet above*

– *One entire facet*

{ *An entire facet*
– *No facet on trans. proc.*
  which is rudimentary

{ *An entire facet*
  No facet on trans-
    verse process
– *Infer. artic. process*
  convex and turned
  outward

concave behind, slightly constricted laterally and in front, and marked on each
side, near the root of the pedicle, by two demi-facets, one above, the other below.
These are covered with cartilage in the recent state, and, when the vertebræ are
articulated with one another, form, with the intervening fibro-cartilages, oval
surfaces for the reception of the heads of the corresponding ribs. The *pedicles* are
directed backwards and slightly upwards, and the inferior intervertebral notches
are of large size, and deeper than in any other region of the spine. The *laminæ*

are broad, thick, and imbricated—that is to say, they overlap one another like tiles on a roof. The spinal foramen is small, and of a circular form. The *spinous process* is long, triangular on transverse section, directed obliquely downwards, and terminates in a tubercular extremity. These processes overlap one another from the fifth to the eighth, but are less oblique in direction above and below. The *articular processes* are flat, nearly vertical in direction, and project from the upper and lower parts of the pedicles; the superior being directed backwards and a little outwards and upwards, the inferior forwards and a little inwards and downwards. The *transverse processes* arise from the same parts of the arch as the posterior roots of the transverse processes in the neck, and are situated behind the articular processes and pedicles; they are thick, strong, and of considerable length, directed obliquely backwards and outwards, and present a clubbed extremity, which is tipped on its anterior part by a small, concave surface, for articulation with the tubercle of a rib. Besides the articular facet for the rib, three indistinct tubercles may be seen rising from the transverse processes of the lower three or four dorsal vertebræ, one at the upper border, one at the lower border, and one externally. In man, they are comparatively small in size, and serve only for the attachment of muscles. But in some animals they attain considerable magnitude, either for the purpose of more closely connecting the segments of this portion of the spine, or for muscular and ligamentous attachment.

The peculiar dorsal vertebræ are the *first, ninth, tenth, eleventh,* and *twelfth* (fig. 197).

The **First Dorsal Vertebra** presents, on each side of the *body*, a single entire articular facet for the head of the first rib, and a half-facet for the upper half of the second. The body is like that of a cervical vertebra, being broad transversely; its upper surface is concave, and lipped on each side. The *superior articular surfaces* are oblique, directed upwards and backwards, but not outwards; the *spinous process* thick, long, and almost horizontal. The *transverse processes* are long, and the upper intervertebral notch deeper than in the other vertebræ of this series.

The **Ninth Dorsal** has no demi-facet below. In some subjects, however, the ninth has two demi-facets on each side; when this occurs the tenth has only a demi-facet at the upper part.

The **Tenth Dorsal** has (except in the cases just mentioned) an entire articular facet on each side above, which is partly placed on the outer surface of the pedicle. It has no demi-facet below.

In the **Eleventh Dorsal,** the body approaches in its form and size to the lumbar. The articular facets for the heads of the ribs, one on each side, are of large size, and placed chiefly on the pedicles, which are thicker and stronger in this and the next vertebra than in any other part of the dorsal region. The *spinous process* is short, and nearly horizontal in direction. The *transverse processes* are very short, tubercular at their extremities, and have no articular facets for the tubercles of the ribs.

The **Twelfth Dorsal** has the same general characters as the eleventh, but may be distinguished from it by the inferior articular processes being convex and turned outwards, like those of the lumbar vertebræ; by the general form of the body, laminæ, and spinous process, approaching to that of the lumbar vertebræ; and by the transverse processes being subdivided into three elevations, the superior, inferior, and external tubercles, which correspond to the mammillary, accessory, and transverse processes of the lumbar vertebræ. Traces of similar elevations are found on the tenth and eleventh dorsal vertebræ (*vide ut supra*).

CHARACTERS OF THE LUMBAR VERTEBRÆ

The **Lumbar Vertebræ** (fig. 198) are the largest segments of the movable part of the vertebral column, and can at once be distinguished by the absence of the foramen in the transverse process, the characteristic feature of the cervical vertebræ, and by their not having any articulating facet on the side of the body, the distinguishing mark of the dorsal vertebræ.

The **body** is large, and has a greater diameter from side to side than from before backwards, a little thicker in front than behind, flattened or slightly concave above and below, concave behind, and deeply constricted in front and at the sides, presenting prominent margins, which afford a broad basis for the support of the superincumbent weight. The *pedicles* are very strong, directed backwards from the upper part of the bodies; consequently, the inferior intervertebral notches are of considerable depth. The *laminæ* are broad, short, and strong; and the spinal foramen triangular, larger than in the dorsal, but smaller than in the

FIG. 198.—Lumbar vertebra.

cervical, region. The *spinous process* is thick and broad, somewhat quadrilateral, projecting backwards, thicker below than above, and terminating in a rough, uneven border. The *superior articular processes* are concave, and look backwards and inwards; the *inferior*, convex, look forwards and outwards; the former are separated by a much wider interval than the latter, and embrace the lower

FIG. 199.—Lumbar vertebra.

articulating processes of the vertebra above. The *transverse processes* are long, slender, directed transversely outwards in the upper three lumbar vertebræ, slanting a little upwards in the lower two. In the upper vertebræ of the series they arise from the junction of the pedicle and the lamina, but as they descend they appear to advance and arise from the pedicle and posterior part of the body of the vertebra. They are situated in front of the articular processes instead of behind them as in the dorsal vertebræ, and are homologous with the ribs. Of the three tubercles noticed in connection with the transverse processes of the lower dorsal vertebræ, the *superior* one becomes connected in this region with the back part of the superior articular process, and has received the name of *mammillary* process; the *inferior* is represented by a small process pointing downwards, situated at the back part of the base of the transverse process, and called the *accessory* process—this is the true transverse process, which is rudimental in this region of the spine; the *external* one, the so-called transverse process, is the homologue of the rib, and hence is sometimes called the *costal* process (fig. 199). Although in man these are comparatively small, in some animals they attain considerable size, and serve to lock the vertebræ more closely together.

The **Fifth Lumbar** vertebra is characterised by having the body much thicker in front than behind, which accords with the prominence of the sacro-vertebral articulation; by the smaller size of its spinous process; by the wide interval between the inferior articulating processes; and by the greater size and thickness of its transverse processes, which spring from the body as well as from the pedicles.

Structure of the Vertebræ.—The body is composed of light, spongy, cancellous tissue, having a thin coating of compact tissue on its external surface perforated by numerous orifices, some of large size, for the passage of vessels; its interior is traversed by one or two large canals, for the reception of veins, which converge towards a single large, irregular aperture, or several small apertures, at the posterior part of the body of each bone. The arch and processes projecting from it have, on the contrary, a thick covering of compact tissue.

Development.—Each vertebra is formed of four primary centres of ossification (fig. 200), one for each lamina and its processes, and two for the body.* Ossification commences in the laminæ of the upper cervical vertebræ about the sixth week of fœtal life, the date becoming gradually later as one passes down the column. The ossific granules first appear in the situation where the transverse processes afterwards project, and spread backwards to the spine, forwards into the pedicles, and outwards into the transverse and articular processes. Ossification in the body commences in the centre of the cartilage about the eighth week in the mid-dorsal region, and from there subsequently extends upwards and downwards along the column. The ossific granules form two closely approximated centres, which speedily coalesce to form one ossific point. At birth these three pieces are perfectly separate. It is important to note that a portion of the postero-lateral aspect of the body is formed from the centre of the neural arch, and the line of junction of the body and arch (*neuro-central synchondrosis*) encroaches on the former. It therefore follows that, in the dorsal region, the facets for the heads of the ribs lie behind the neuro-central synchondrosis and are ossified from the centres for the neural arch. During the first year the laminæ become united behind, the union taking place first in the lumbar region and then extending upwards through the dorsal and lower cervical regions. About the third year the bodies of the upper cervical vertebræ become joined to the arch on each side, and the junction takes place from above downwards, but is not completed in the lower lumbar vertebræ until the sixth year. Before puberty, no other changes occur, excepting a gradual increase in the growth of these primary centres; the upper and under surfaces of the bodies, and the ends of the transverse and spinous processes, being tipped with cartilage, in which ossific granules are not as yet deposited. About the sixteenth year (fig. 201), five secondary centres appear, one for the tip of each transverse process, one for the extremity of the spinous process; and two epiphysial plates, one for the upper and the other for the lower surface of the body (fig. 202). These fuse with the rest of the bone about the age of five-and-twenty.

Exceptions to this mode of development occur in the first, second, and seventh cervical vertebræ, and in those of the lumbar region.

* By many observers it is asserted that the bodies of the vertebræ are developed from a single centre which speedily becomes bilobed, so as to give the appearance of two nuclei; but that there are two centres, at all events sometimes, is evidenced by the fact that the two halves of the body of the vertebra may remain distinct throughout life, and be separated by a fissure through which a protrusion of the spinal membranes may take place, constituting an anterior spina bifida.

The Atlas (fig. 203).—The atlas is usually ossified from three centres. Two of these are destined for the lateral masses, the ossification of which commences about the seventh week of fœtal life near the articular processes, and extends backwards ; at birth, these portions of bone are separated from one another behind by a narrow interval filled with cartilage. Between the third and fourth years they unite either directly or through the medium of a separate centre, developed in the cartilage in the middle line. The anterior arch, at birth, is altogether cartilaginous and consists of the hypochordal brace or bar (see page 96), which persists in the case of the atlas ; in this a separate nucleus appears about the end of the first year after birth, and joins the lateral masses from the sixth to the eighth year after birth— their line of union extending across the anterior portions of the superior articular facets. Sometimes there are two nuclei developed in the cartilage, one on either side of the median line, which join to form a single mass. Occasionally there is no separate centre, the anterior arch being formed by the gradual extension forwards and ultimate junction of the two lateral masses.

The **Axis** (fig. 204) is developed by *five* primary and *two* secondary centres. The body and arch of this bone are formed in the same manner as the corresponding parts in the other vertebræ ; one centre (or two, which speedily coalesce) for the lower part of the body, and one for each lamina. The centres for the laminæ appear about the seventh or eighth week, that for the body about the fourth month. The odontoid process consists originally of an extension upwards of the cartilaginous mass, in which the lower part of the body is formed. At about the sixth month of fœtal life, two osseous nuclei make their appearance in the base of this process : they are placed laterally, and join before birth to form a conical bilobed mass deeply cleft above ; the interval between the cleft and the summit of the process is formed by a wedge-shaped piece of cartilage. The base of the process is separated from the body by a cartilaginous interval, which gradually becomes ossified at its circumference, but remains cartilaginous in its centre until advanced age.* In this cartilage,

FIG. 200.—Development of a vertebra.

By 4 primary centres

2 for body (8th week)

1 for each lamina (6th week)

FIG. 201.

By 3 secondary centres

1 for each trans. process 16 years

1 for spinous process (16 years)

FIG. 202.

By 2 additional plates

1 for upper surface of body

16 years

1 for under surface of body

FIG. 203.—Atlas.

By 3 centres

1 for anter. arch (1st yr.) not constant

1 for each lateral mass } 7th week

FIG. 204.—Axis.

By 7 centres

2nd year

6th month
1 for each lateral mass

1 for body (4th month)
1 for under surface of body

FIG. 205.—Lumbar vertebra.

2 additional centres for tubercles on superior articular process

Exceptional cases

* See Cunningham, *Journ. Anat.* vol. xx. p. 238.

rudiments of the lower epiphysial lamella of the atlas and the upper epiphysial lamella of the axis may sometimes be found. Finally, as Humphry has demonstrated, the apex of the odontoid process has a separate nucleus, which appears in the second year and joins about the twelfth year. This is the upper epiphysial lamella of the atlas. In addition to these there is a secondary centre for a thin epiphysial plate on the under surface of the body of the bone.

The Seventh Cervical.—The anterior or costal part of the transverse process of the seventh cervical is sometimes developed from a separate osseous centre at about the sixth month of fœtal life, and joins the body and posterior division of the transverse process between the fifth and sixth years. Occasionally this process continues as a separate piece, and, becoming lengthened outwards and forwards, constitutes what is known as a cervical rib. This separate ossific centre for the costal process has also been found in the fourth, fifth, and sixth cervical vertebræ.

The **Lumbar Vertebræ** (fig. 205) have *two additional centres* for the mammillary tubercles, which project from the back part of the superior articular processes. The transverse process of the first lumbar is sometimes developed as a separate piece, which may remain permanently unconnected with the rest of the bone, thus forming a lumbar rib—a peculiarity which is rarely met with.

Progress of Ossification in the Spine generally.—Ossification of the laminæ of the vertebræ commences in the cervical region of the spine, and proceeds gradually downwards. Ossification of the bodies, on the other hand, commences about the mid-dorsal region, and extends both upwards and downwards Although, however, the ossific nuclei for the bodies first make their appearance in the dorsal vertebræ, the lumbar and first sacral are those in which these nuclei are largest at birth.

Attachment of Muscles.—To the *Atlas* are attached nine pairs : the Longus colli, Rectus capitis anticus minor, Rectus lateralis, Obliquus capitis superior and inferior, Splenius colli, Levator anguli scapulæ, First Intertransverse, and Rectus capitis posticus minor.

To the *Axis* are attached eleven pairs : the Longus colli, Levator anguli scapulæ, Splenius colli, Scalenus medius, Transversalis cervicis, Intertransversales, Obliquus capitis inferior, Rectus capitis posticus major, Semispinalis colli, Multifidus spinæ, Interspinales.

To the remaining vertebræ, generally, are attached thirty-five pairs and a single muscle : *anteriorly*, the Rectus capitis anticus major, Longus colli, Scalenus anticus, medius, and posticus, Psoas magnus and parvus, Quadratus lumborum, Diaphragm, Obliquus abdominis internus, and Transversalis abdominis ; *posteriorly*, the Trapezius, Latissimus dorsi, Levator anguli scapulæ, Rhomboideus major and minor, Serratus posticus superior and inferior, Splenius, Erector spinæ, Ilio-costalis, Longissimus dorsi, Spinalis dorsi, Cervicalis ascendens, Transversalis colli, Trachelo-mastoid, Complexus, Biventer cervicis, Semispinalis dorsi et colli, Multifidus spinæ, Rotatores spinæ, Interspinales, Supraspinales, Intertransversales, Levatores costarum.

SACRAL AND COCCYGEAL VERTEBRÆ

The **Sacral** and **Coccygeal Vertebræ** consist at an early period of life of nine separate pieces which are united in the adult, so as to form two bones, five entering into the formation of the sacrum, four into that of the coccyx. Occasionally, the coccyx consists of five bones.[*]

THE SACRUM

The **Sacrum** (sacer, *sacred*) is a large, triangular bone (fig. 206), situated at the lower part of the vertebral column, and at the upper and back part of the pelvic cavity, where it is inserted like a wedge between the two innominate bones ; its upper part or base articulates with the last lumbar vertebra, its apex with the coccyx. The sacrum is curved upon itself, and placed very obliquely, its upper extremity projecting forwards, and forming, with the last lumbar

---

[*] Sir George Humphry describes this as the usual composition of the coccyx.—*On the Skeleton*, p. 456.

vertebra, a very prominent angle, called the *promontory* or *sacro-vertebral angle*; while its central part is projected backwards, so as to give increased capacity to the pelvic cavity. It presents for examination an anterior and posterior surface, two lateral surfaces, a base, an apex, and a central canal.

The **Anterior Surface** is concave from above downwards, and slightly so from side to side. In the middle are seen four transverse ridges, indicating the original division of the bone into five separate pieces. The portions of bone intervening between the ridges correspond to the bodies of the vertebræ. The body of the first segment is of large size, and in form resembles that of a lumbar vertebra; the succeeding ones diminish in size from above downwards, are flattened from before backwards, and curved so as to accommodate themselves to the form of the sacrum, being concave in front, convex behind. At each end of the ridges above mentioned are seen the *anterior sacral foramina*, analogous to the intervertebral foramina, four in number on each side, somewhat rounded in

Fig. 206.—Sacrum, anterior surface.

form, diminishing in size from above downwards, and directed outwards and forwards : they give exit to the anterior divisions of the sacral nerves and entrance to the lateral sacral arteries. External to these foramina is the *lateral mass*, consisting, at an early period of life, of separate segments; in the adult, these become blended with the bodies and with each other. Each lateral mass is traversed by four broad, shallow grooves, which lodge the anterior divisions of the sacral nerves as they pass outwards, the grooves being separated by prominent ridges of bone which give attachment to the slips of the Pyriformis muscle.

If a vertical section is made through the centre of the sacrum (fig. 207), the bodies are seen to be united at their circumference by bone, a wide interval being left centrally, which, in the recent state, is filled by intervertebral substance. In some bones, this union is more complete between the lower segments than between the upper ones.

The **Posterior Surface** (fig. 208) is convex and much narrower than the anterior. In the middle line are three or four tubercles, which represent the rudimentary spinous processes of the sacral vertebræ. Of these tubercles, the first is generally prominent, and distinct from the rest; the second and third are either separate or united into a tubercular ridge, which diminishes in size from above downwards; the fourth usually, and the fifth always remaining undeveloped. External to the spinous processes on each side are the *laminæ*, broad and well marked in the first three pieces; sometimes the fourth, and generally the fifth, are only partially developed and fail to meet in the middle line. These partially developed laminæ are prolonged downwards, as rounded processes, the *sacral cornua*, and are connected to the cornua of the coccyx. Between them the bony wall of the lower end of the sacral canal is imperfect, and is liable to be opened in the sloughing of bed-sores. External to the laminæ is a linear series of indistinct tubercles representing the *articular processes*; the upper pair are large, well developed, and correspond in shape and direction to the superior articulating processes of a lumbar vertebra; the second and third are small; the fourth and fifth (usually blended together) are situated on each side of the sacral canal and assist in forming the sacral cornua. External to the articular processes are the four *posterior sacral foramina*; they are smaller in size and less regular in form than the anterior, and transmit the posterior divisions of the sacral nerves. On the outer side of the posterior sacral foramina is a series of tubercles, the rudimentary *transverse processes* of the sacral vertebræ. The first pair of transverse tubercles are of large size, very distinct, and correspond with each superior angle of the bone; they, together with the second pair, which are of small size, give attachment to the horizontal part of the sacro-iliac ligament; the third give attachment to the oblique fasciculi of the posterior sacro-iliac ligaments; and the fourth and fifth to the great sacro-sciatic ligaments. The interspace between the spinous and transverse processes on the back of the sacrum presents a wide, shallow concavity, called the *sacral groove*; it is continuous above with the vertebral groove, and lodges the origin of the Multifidus spinæ.

The **Lateral Surface**, broad above, becomes narrowed into a thin edge below. Its upper half presents in front a broad, ear-shaped surface for articulation with the ilium. This is called the *auricular* surface, and in the fresh state is coated with fibro-cartilage. It is bounded posteriorly by deep and uneven impressions, for the attachment of the posterior sacro-iliac ligaments. The lower half is thin and sharp, and terminates in a projection called the *inferior lateral angle*; below this angle is a notch, which is converted into a foramen by articulation with the transverse process of the upper piece of the coccyx, and transmits the anterior division of the fifth sacral nerve. This lower, sharp border gives attachment to the greater and lesser sacro-sciatic ligaments, and to some fibres of the Gluteus maximus posteriorly, and to the Coccygeus in front.

The **Base** of the sacrum, which is broad and expanded, is directed upwards and forwards. In the middle is seen a large oval articular surface, the upper aspect of the body of the first sacral vertebra, which is connected with the under surface of the body of the last lumbar vertebra by a fibro-cartilaginous disc. It

FIG. 207.—Vertical section of the sacrum.

is bounded behind by the large, triangular orifice of the sacral canal. The orifice
is formed behind by the laminæ and spinous process of the first sacral vertebra :
the superior articular processes project from it on each side ; they are oval, concave, directed backwards and inwards, like the superior articular processes of a
lumbar vertebra ; and in front of each articular process is an intervertebral notch,
which forms the lower part of the foramen between the last lumbar and first
sacral vertebra. Lastly, on each side of the large oval articular plate is a broad
and flat triangular surface of bone, which extends outwards, supports the Psoas
magnus muscle and lumbo-sacral cord, and is continuous on each side with the
iliac fossa. This is called the *ala* of the sacrum, and gives attachment to a few

FIG. 208.—Sacrum, posterior surface.

of the fibres of the Iliacus muscle. The posterior part of the ala represents the
transverse process, and its anterior part the costal process of the first sacral
segment.

The **Apex**, directed downwards and slightly forwards, presents a small, oval,
concave surface for articulation with the coccyx.

The **Spinal Canal** (fig. 207) runs throughout the greater part of the bone ; it
is large and triangular in form above ; small and flattened, from before backwards, below. In this situation its posterior wall is incomplete, from the non-
development of the laminæ and spinous processes. It lodges the sacral nerves,
and is perforated by the anterior and posterior sacral foramina, through which
these pass out.

Structure.—It consists of much loose, spongy tissue within, invested externally
by a thin layer of compact tissue.

Differences in the Sacrum of the Male and Female.—The sacrum in the
female is shorter and wider than in the male ; the lower half forms a greater
angle with the upper ; the upper half of the bone being nearly straight, the lower
half presenting the greatest amount of curvature. The bone is also directed
more obliquely backwards, which increases the size of the pelvic cavity and
renders the sacro-vertebral angle more projecting. In the male the curvature is

more evenly distributed over the whole length of the bone, and is altogether greater than in the female.

Peculiarities of the Sacrum.—This bone, in some cases, consists of six pieces; occasionally, the number is reduced to four. Occasionally the bodies of the first and second segments are not joined, or the laminæ and spinous processes have not coalesced. Sometimes the upper pair of transverse tubercles are not joined to the rest of the bone on one or both sides; and, lastly, the sacral canal may be open for nearly the lower half of the bone, in consequence of the imperfect development of the laminæ and spinous processes. The sacrum, also, varies considerably with respect to its degree of curvature.

FIG. 209.—Development of the sacrum.

Additional centres for the first three pieces *

At birth

Development (figs. 209, 210, 211).— The sacrum, formed by the union of five vertebræ, has usually *thirty-five* centres of ossification, but this number may be increased.

The *bodies* of the sacral vertebræ have each three ossific centres: one for the central part, one for the epiphysial plate on its upper and another for that on its under surface. Occasionally the primary centres for the bodies of the first and second piece of the sacrum are double.

FIG. 210.

At 4½ yrs.

The arch of each sacral vertebra is developed by two centres, one for each lamina. These unite with each other behind, and subsequently join the body.

The *lateral masses* have six additional centres, two for each of the first three vertebræ.* These centres make their appearance above and to the outer side of the anterior sacral foramina (fig. 209), and are developed into separate segments (fig. 210); they are subsequently blended with each other, and with the bodies and transverse processes, to form the lateral masses.

FIG. 211.

Two epiphysial laminæ for each lateral surface *

At 25th year

Lastly, each *lateral surface* of the sacrum is developed by two epiphysial plates (fig. 211): one for the auricular surface, and one for the remaining part of the thin lateral edge of the bone.

Period of Development.—At about the eighth or ninth week of fœtal life, ossification of the central part of the body of the first sacral vertebra commences and is rapidly followed by deposit of ossific matter in the second and third; but ossification does not commence in the lower two segments until between the fifth and eighth months of fœtal life. Between the sixth and eighth months ossification of the laminæ takes place; and about the same time, the centres for the lateral masses of the first three sacral vertebræ make their appearance. The period at which the arch becomes completed by the junction of the laminæ with the bodies in front, and with each other behind, varies in different segments. The junction between the laminæ and the bodies takes place first in the lower vertebræ as early as the second year, but is not effected in the uppermost until the fifth or sixth year. About the sixteenth year the epiphysial plates for the upper and under surfaces of the bodies are formed; and between the eighteenth and twentieth years, those for each lateral surface of the sacrum make their appearance. The

---

* Occasionally the fourth vertebra has also these two additional centres.

bodies of the sacral vertebræ are, during early life, separated from each other by intervertebral discs. But about the eighteenth year the two lowest segments become joined together by bone. This process gradually extends upwards until all the segments become united; and the bone is completely formed from the twenty-fifth to the thirtieth year of life.

Articulations.—With four bones: the last lumbar vertebra, coccyx, and the two innominate bones.

Attachment of Muscles.—To eight pairs: in front, the Pyriformis and Coccygeus, and a portion of the Iliacus to the base of the bone; behind, the Gluteus maximus, Latissimus dorsi, Multifidus spinæ, and Erector spinæ, and sometimes the Extensor coccygis.

THE COCCYX

The **Coccyx** (κόκκυξ, *cuckoo*), so called from having been compared to a cuckoo's beak (fig. 212), is usually formed of four small segments of bone, and is the most rudimentary part of the vertebral column. In each of the first three segments may be traced a rudimentary body, articular and transverse processes; the last piece (sometimes the third) is a mere nodule of bone, without distinct processes. All the segments are destitute of pedicles, laminæ, and spinous processes; and, consequently, of intervertebral foramina and spinal canal. The first segment is the largest; it resembles the lowermost sacral vertebra, and often exists as a separate piece; the last three, diminishing in size from above downwards, are usually blended together so as to form a single bone. The gradual diminution in the size of the pieces gives this bone a triangular form, the base of the triangle joining the end of the sacrum. It presents for examination an anterior and posterior surface, two borders, a base, and an apex. The *anterior surface* is slightly concave, and marked with three transverse grooves, indicating the points of junction of the different pieces. It has attached to it the anterior sacro-coccygeal ligament and Levator ani muscle, and supports the lower end of the rectum. The *posterior surface* is convex, marked by transverse grooves similar to those on the anterior surface; and presents on each side a lineal row of tubercles, the rudimentary articular processes of the coccygeal vertebræ. Of these, the superior pair are large, and are called the *cornua of the coccyx*: they project upwards, and articulate with the cornua of the sacrum, the junction between these two bones completing the fifth posterior sacral foramen for the transmission of the fifth sacral nerve. The *lateral borders* are thin, and present a series of small eminences, which represent the transverse processes of the coccygeal vertebræ. Of these, the first on each side is the largest, flattened from before backwards, and often ascends to join the lower part of the thin lateral edge of the sacrum, thus completing the fifth anterior sacral foramen for the transmission of the anterior division of the fifth sacral nerve; the others diminish in size from above downwards, and are often wanting. The *borders* of the coccyx are narrow, and give attachment on each side to the sacro-sciatic ligaments, to the Coccygeus muscle in front of the ligaments, and to the Gluteus maximus behind them. The *base* presents an oval surface for articulation with the sacrum. The *apex* is rounded, and has attached to it the tendon of the external Sphincter muscle of the anus. It may be bifid, and is sometimes deflected to one or other side.

Development.—The coccyx is developed by *four* centres, one for each piece. Occasionally one of the first three pieces of this bone is developed by two centres,

FIG. 212.—Coccyx.

Cornua

Rudim. trans. process

Anterior surface

Posterior surface

placed side by side. The ossific nuclei make their appearance in the following order: in the first segment, shortly after birth; in the second piece, at from five to ten years; in the third, from ten to fifteen years; in the fourth, from fifteen to twenty years. As age advances, these various segments become united with each other from below upwards, the union between the first and second segments being frequently delayed until after the age of twenty-five or thirty. At a late period of life, especially in females, the coccyx often becomes joined to the end of the sacrum.

Articulation.—With the sacrum.

Attachment of Muscles. — To four pairs and one single muscle: on either side, the Coccygeus; behind, the Gluteus maximus and Extensor coccygis, when present; at the apex, the Sphincter ani; and in front, the Levator ani.

OF THE SPINE IN GENERAL

The **Spinal Column,** formed by the junction of the vertebræ, is situated in the median line, at the posterior part of the trunk; its average length in the male is about two feet four inches. Of this length the cervical part measures about five inches, the dorsal about eleven, the lumbar about seven, and the sacrum and coccyx about five. The female spine is about two feet in length.

Viewed from in front, the width of the bodies of the vertebræ will be found to increase from the second cervical to the first dorsal; there is then a slight diminution in the next three vertebræ; below this there is again a gradual and progressive increase in width as low as the sacro-vertebral angle. From this point there is a rapid diminution, terminating below at the apex of the coccyx.

Viewed laterally (fig. 213), the spinal column presents several curves, which correspond to the different regions of the column, and are called *cervical, dorsal, lumbar,* and *pelvic.* The *cervical* curve commences at the apex of the odontoid process, and terminates at the middle of the second dorsal vertebra; it is convex in front, and is the least marked of all the curves. The *dorsal* curve, which is concave forwards, commences at the middle of the second and terminates at the middle of the twelfth dorsal vertebra. Its most prominent point behind corresponds to the spine of the seventh dorsal. The *lumbar* curve is more marked in the female than in the male;

FIG. 213.—Lateral view of the spine.

1st cervical or Atlas
2nd cervical or Axis
3
4
5
6
7
1st dorsal
2
3
4
5
6
7
8
9
10
11
12
1st lumbar
2
3
4
5

N

it commences at the middle of the last dorsal vertebra, and terminates at the sacro-vertebral angle. It is convex anteriorly; the convexity of the lower three vertebræ being much greater than that of the upper two. The *pelvic* curve commences at the sacro-vertebral articulation, and terminates at the point of the coccyx. It is concave anteriorly. The dorsal and pelvic curves are the primary curves, and begin to be formed about the fourth month of fœtal life. The cervical and lumbar curves are compensatory or secondary, and are developed after birth in order to maintain the erect position.

If a body is enveloped in plaster of Paris and a vertical section in the median line is made, it will be found, as pointed out by Humphry, that a plumb-line dropped from the middle of the odontoid process of the axis will pass through the middle of the body of the second dorsal vertebra, through the middle of the body of the last dorsal, through the middle and anterior-inferior edge of the last lumbar, and bisects a line drawn transversely through the heads of the thigh bones. It is known from experiment that the line of gravity of the head passes through the middle of the odontoid process; it therefore follows that the line of gravity of the head passes through the points of confluence of the three superior curves of the spine and through the heads of the thigh bones, so that the weight of the skull and its contents is directly transmitted to the pelvis and lower extremities when the body is in the erect position.

The spine has also a slight *lateral* curvature, the convexity of which is directed towards the right side. This may be produced, as Bichat first explained, chiefly by muscular action; most persons using the right arm in preference to the left, especially in making long-continued efforts, when the body is curved to the right side. In support of this explanation, it has been found, by Béclard, that in one or two individuals who were left-handed, the lateral curvature was directed to the left side. Others regard this curvature as being produced by the aortic arch and upper part of the thoracic aorta—a view which is supported by the fact that in cases where the viscera are transposed and the aorta is on the right side, the convexity of the lateral curvature is directed to the left side.

The movable part of the spinal column presents for examination an anterior, a posterior, and two lateral surfaces: a base, summit, and spinal canal.

The **anterior surface** presents the bodies of the vertebræ joined to each other in the recent state by the intervertebral discs. The bodies are broad in the cervical region, narrow in the upper part of the dorsal, and broadest in the lumbar region. The whole of this surface is convex transversely, concave from above downwards in the dorsal region, and convex in the same direction in the cervical and lumbar regions.

The **posterior surface** presents in the median line the spinous processes. In the cervical region these are short, horizontal, with bifid extremities. In the dorsal region, they are directed obliquely above, assume almost a vertical direction in the middle, and are horizontal, with a slight inclination downwards, below, as are also the spines of the lumbar vertebræ. They are separated by considerable intervals in the loins, by narrower intervals in the neck, and are closely approximated in the middle of the dorsal region. Occasionally one of these processes deviates a little from the median line—a fact to be remembered in practice, as irregularities of this sort are attendant also on fractures or displacements of the spine. On either side of the spinous processes, extending the whole length of the column, is the *vertebral groove* formed by the laminæ in the cervical and lumbar regions, where it is shallow, and by the laminæ and transverse processes in the dorsal region, where it is deep and broad. In the recent state, these grooves lodge the deep muscles of the back. External to the vertebral grooves are the articular processes, and still more externally the transverse processes. In the dorsal region, the latter processes stand backwards, on a plane considerably posterior to the same processes in the cervical and lumbar regions. In the cervical region, the transverse processes are placed in front of the articular processes, and on the outer side of the pedicles, between the intervertebral foramina. In the dorsal region they are posterior to the pedicles, intervertebral foramina, and articular processes. In the lumbar, they are placed in front of the articular processes, but behind the intervertebral foramina.

The **lateral surfaces** are separated from the posterior by the articular processes in the cervical and lumbar regions, and by the transverse processes in the dorsal. These surfaces present in front the sides of the bodies of the vertebræ, marked in the dorsal region by the facets for articulation with the heads of the ribs. More posteriorly are the intervertebral foramina, formed by the juxtaposition of the intervertebral notches, oval in shape, smallest in the cervical and upper part of the dorsal regions, and gradually increasing in size to the last lumbar. They are situated between the transverse processes in the neck, and in front of them in the thoracic and lumbar regions, and transmit the spinal nerves.

The **base** of that portion of the vertebral column which is formed by the twenty-four movable vertebræ consists of the under surface of the body of the fifth lumbar vertebra ; and the **summit**, of the upper surface of the atlas.

The **vertebral** or **spinal canal** follows the different curves of the spine ; it is largest in those regions in which the spine enjoys the greatest freedom of movement, as in the neck and loins, where it is wide and triangular ; and narrow and rounded in the thoracic region, where motion is more limited.

Surface Form.—The only parts of the vertebral column which lie closely under the skin, and so directly influence surface form, are the apices of the spinous processes. These are always distinguishable at the bottom of a furrow, which, more or less evident, runs down the mesial line of the back from the external occipital protuberance above to the middle of the sacrum below. In the neck the furrow is broad, and terminates in a conspicuous projection, which is caused by the spinous processes of the seventh cervical vertebra (vertebra prominens) and the first dorsal, the latter being the more prominent of the two. Above this the spinous process of the sixth cervical vertebra may sometimes be seen to form a projection ; the other cervical spines are sunken, and are not visible, though the spine of the axis can be felt, and generally also the spines of the third, fourth, and fifth cervical vertebræ. In the dorsal region, the furrow is shallow, and during stooping disappears, and then the spinous processes become more or less visible. The markings produced by these spines are small and close together. In the lumbar region the furrow is deep, and the situation of the lumbar spines is frequently indicated by little pits or depressions, especially if the muscles in the loins are well developed and the spine incurved. They are much larger and farther apart than in the dorsal region. In the sacral region the furrow is shallower, presenting a flattened area which terminates below at the most prominent part of the posterior surface of the sacrum, formed by the spinous process of the third sacral vertebra. At the bottom of the furrow may be felt the irregular posterior surface of the bone. Below this, in the deep groove leading to the anus, the coccyx may be felt. In order to identify any particular spine it is customary to count from the prominence caused by the seventh cervical and first dorsal spines ; of these two, the spine of the first dorsal is the more prominent. It is useful, however, to bear in mind that the root of the spine of the scapula is on a level with the interval between the spines of the third and fourth dorsal vertebræ ; the apex or inferior angle of the scapula is on a level with the interval between the spines of the seventh and eighth dorsal ; the highest point of the crest of the ilium corresponds to the fourth lumbar spine, and the posterior superior spine of the ilium corresponds to the second sacral spine. The only other portions of the vertebral column which can be felt from the surface are the transverse processes of three of the cervical vertebræ—viz. the first, the sixth, and the seventh. The transverse process of the atlas can be felt as a rounded nodule of bone just below and in front of the apex of the mastoid process, along the anterior border of the Sterno-mastoid. The transverse process of the sixth cervical vertebra is of surgical importance. If deep pressure be made in the neck, in the course of the carotid artery, opposite the cricoid cartilage, the prominent anterior tubercle of the transverse process of the sixth cervical vertebra can be felt. This has been named *Chassaignac's tubercle,* and against it the carotid artery may be most conveniently compressed by the finger. The transverse process of the seventh cervical vertebra can also be often felt. Sometimes the anterior root, or costal process, is large and segmented off, forming a cervical rib.

Surgical Anatomy.—Occasionally the coalescence of the laminæ is not completed, and consequently a cleft is left in the arches of the vertebræ, through which a protrusion of the spinal membranes (dura mater and arachnoid), and generally of the spinal cord itself, takes place, constituting the malformation known as *spina bifida.* This condition is most common in the lumbo-sacral region, but it may occur in the dorsal or cervical region, or the arches throughout the whole length of the canal may remain unapproximated. In some rare cases, in consequence of the non-coalescence of the two primary centres from which the body is formed, a similar condition may occur in front of the canal, the bodies of the vertebræ being found cleft and the tumour projecting into the thorax, abdomen, or pelvis, between the lateral halves of the bodies affected.

The construction of the spinal column of a number of pieces, securely connected together and enjoying only a slight degree of movement between any two individual pieces, but permitting of a very considerable range of movement as a whole, allows a sufficient degree of mobility without any material diminution of strength. The many joints of which the spine is composed, together with the very varied movements to which it is subjected, render it liable to sprains ; but so closely are the individual vertebræ articulated that these sprains are rarely or ever severe, and an amount of violence sufficiently great to produce tearing of the ligaments would tend rather to cause a dislocation or fracture. The further safety of the column and its slight liability to injury is provided for by its disposition in curves, instead of in one straight line. For it is an elastic column, and must first bend before it breaks ; under these circumstances, being made up of three curves, it represents three columns, and greater force is required to produce bending of a short column, than of a longer one that is equal to it in breadth and material. Again, the safety of the column is provided for by the interposition of the intervertebral discs between the bodies of the vertebræ, which act as admirable buffers in counteracting the effects of violent jars or shocks. Fracture-dislocation of the spine may be caused by direct or indirect violence or by a combination of the two, as when a person, falling from a height, strikes against some prominence and is doubled over it. The fractures from indirect violence are the more common, and here the bodies of the vertebræ are compressed, while the arches are torn asunder ; in fracture from direct violence, on the other hand, the arches are compressed and the bodies of the vertebræ separated from each other. It will therefore be seen that in both classes of injury the spinal marrow is the part least likely to be injured, and may escape damage even where there has been considerable lesion of the bony framework. For. as Mr. Jacobson states, ' being lodged in the centre of the column, it occupies neutral ground in respect to forces which might cause fracture. For it is a law in mechanics that when a beam, as of timber, is exposed to breakage and the force does not exceed the limits of the strength of the material, one division resists compression, another laceration of the particles, while the third, between the two, is in a negative condition.' [*] Applying this principle to the spine, it will be seen that, whether the fracture-dislocation be produced by direct or indirect violence, one segment, either the anterior or posterior, will be exposed to compression, the other to laceration, and the intermediate part, where the cord is situated, will be in a neutral state. When a fracture-dislocation is produced by indirect violence, the displacement is almost always the same ; the upper segment being driven forwards on the lower, so that the cord is compressed between the body of the vertebra below and the arch of the vertebra above.

The parts of the spine most liable to be injured are : (1) the junction of the dorsal with the lumbar region, including the last two dorsal and the upper two lumbar vertebræ, for this part is near the middle of the column and there is therefore a greater amount of leverage, and moreover the portion above is comparatively fixed, and the vertebræ which form this portion of the column, though smaller than those below, have nevertheless to bear almost as great a weight; (2) the cervico-dorsal region, because here the flexible cervical portion of the spine joins the more fixed dorsal region ; and (3) the atlanto-axial region, because it enjoys an extensive range of movement, and, being near the skull, is influenced by violence applied to the head.

In fracture-dislocation it has been proposed to cut down and expose the injured bones, in order to facilitate replacement and remove pressure from the spinal cord; and much controversy has arisen as to the value of this operation. The rule for the guidance of the surgeon in these cases appears to be as follows : If there is evidence of complete destruction or division of the spinal cord at the seat of lesion, the operation is practically use-less ; but if there seems to be any probability that a portion of the cord has escaped, an operation is indicated, on the chance that when pressure has been removed, a sufficient quantity of the substance of the cord may remain intact to allow of its carrying on its functions to a certain extent. Laminectomy is also performed in other classes of cases : (1) where the laminæ have been driven in by direct violence ; (2) where paraplegia super-venes after an appreciable interval since the injury, and is therefore probably due to the effusion of blood or inflammatory exudation ; (3) where paraplegia occurs in cases of spinal caries, if the paraplegic symptoms persist for a long period, or threaten life from intractable cystitis or bed-sores.

THE SKULL

The **Skull** or **Cranium** (κράνος, *a helmet*) is supported on the summit of the vertebral column, and is of an oval shape, wider behind than in front. It is composed of a series of flattened or irregular bones which, with one exception (the lower jaw), are immovably jointed together. It consists of two parts : (1) the Brain-case (cranium cerebrale), which encloses and protects the brain ; and (2) the skeleton of the Face (cranium viscerale). The Brain-case is composed of *eight* bones—viz. the *occipital*, two *parietal*, *frontal*, two *temporal*, *sphenoid*, and *ethmoid*. The Face is composed of *fourteen* bones—viz. the two *nasal*, two *superior maxillary*, two *lachrymal*, two *malar*, two *palate*, two *inferior turbinated*, *vomer*,

[*] Holmes's *System of Surgery*, vol. i. p. 529. 1883.

and *inferior maxillary*. The *ossicula auditûs*, the *teeth*, and *Wormian bones* are not included in this enumeration.

Skull, 22 bones
{
  Brain-case, 8 bones
  {
    Occipital.
    Two Parietal.
    Frontal.
    Two Temporal.
    Sphenoid.
    Ethmoid.
  }
  Face, 14 bones
  {
    Two Nasal.
    Two Superior Maxillary.
    Two Lachrymal.
    Two Malar.
    Two Palate.
    Two Inferior Turbinated.
    Vomer.
    Inferior Maxillary.
  }
}

The Hyoid bone, situated at the root of the tongue and attached to the base of the skull by ligaments, has also to be considered in this section.

The Occipital Bone

The **Occipital Bone** (ob, caput, *against the head*) is situated at the back part and base of the cranium. It is trapezoid in shape and is much curved on itself (fig. 214). It presents at its front and lower part a large oval aperture, the

FIG. 214.—Occipital bone. Outer surface.

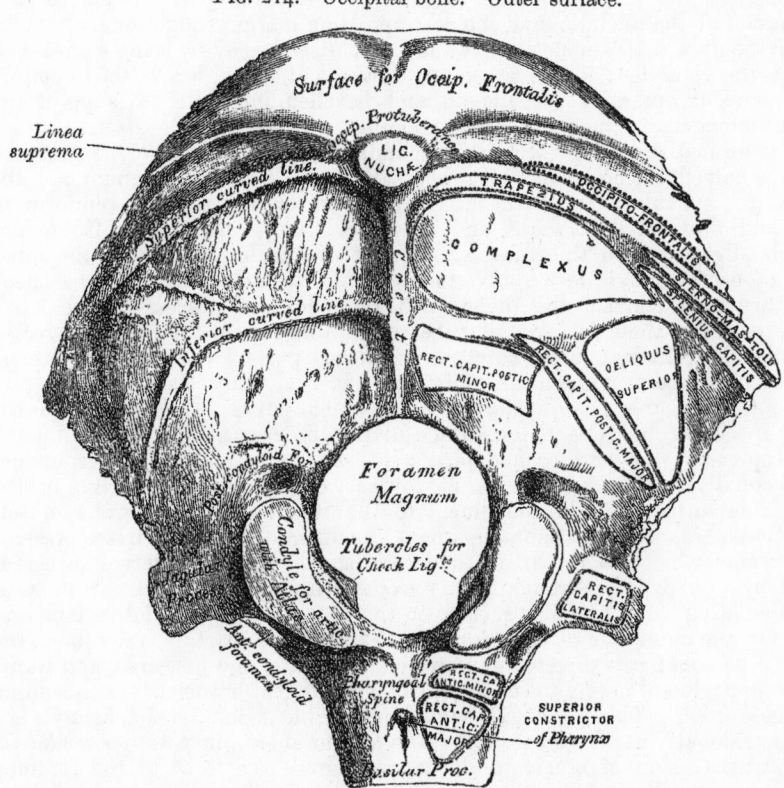

foramen magnum, by which the cranial cavity communicates with the spinal canal. The portion of bone behind this opening is curved and expanded and forms the *tabula*; the portion in front is a thick, elongated mass of bone, the *basilar process*; while on either side of the foramen are situated the *lateral* or *condylic portions*, bearing the condyles, by which the bone articulates with the atlas. It presents for examination two surfaces, four borders, and four angles.

The **external surface** is convex. Midway between the summit of the bone and the posterior margin of the foramen magnum is a prominent tubercle, the *inion* or *external occipital protuberance*, and descending from it as far as the foramen, a vertical ridge, the *external occipital crest*. This protuberance and crest give attachment to the Ligamentum nuchæ, and vary in prominence in different skulls. Passing outwards from the occipital protuberance is a semicircular ridge on each side, the *superior curved line*. Above this line there is often a second less distinctly marked ridge, called the *highest curved line* (*linea suprema*), to which the epicranial aponeurosis is attached. The bone between these two lines is smoother and denser than the rest of the surface. Running parallel with these from the middle of the crest is another semicircular ridge on each side, the *inferior curved line*. The surface of the bone above the linea suprema is rough and porous, and, in the recent state, is covered by the Occipito-frontalis muscle, while the superior and inferior curved lines, together with the surfaces of bone between and below them, serve for the attachment of several muscles. The superior curved line gives attachment internally to the Trapezius, externally to the muscular origin of the Occipito-frontalis, and to the Sterno-cleido-mastoid. to the extent shown in fig. 214; the depressions between the curved lines to the Complexus internally, the Splenius capitis and Obliquus capitis superior externally. The inferior curved line, and the depressions below it, afford insertion to the Rectus capitis posticus, major and minor.

The **foramen magnum** is a large, oval aperture, its long diameter extending from before backwards. It transmits the medulla oblongata and its membranes, the spinal accessory nerves, the vertebral arteries, the anterior and posterior spinal arteries, and the occipito-axial ligaments. Its back part is wide for the transmission of the medulla, and the corresponding margin rough for the attachment of the dura mater enclosing it; the fore part is narrower, being encroached upon by the condyles; it has projecting towards it, from below, the odontoid process, and its margins are smooth and bevelled internally to support the medulla oblongata. The point corresponding with the middle of its anterior margin is termed the *basion*; that corresponding with the middle of its posterior margin is named the *opisthion*. On each side of the foramen magnum are the *condyles*, for articulation with the atlas; they are convex, oval or reniform in shape, and directed downwards and outwards; they converge in front, and encroach slightly upon the anterior segment of the foramen. On the inner border of each condyle is a rough tubercle for the attachment of the check ligaments which connect this bone with the odontoid process of the axis; while external to them is a rough tubercular prominence, the *jugular process*, channelled in front by a deep notch, which forms part of the jugular foramen, or foramen lacerum posterius. This notch is frequently bisected by a projection of bone, the *intrajugular process*, which curves outwards from the margin of the notch. The under surface of this process presents an eminence, which represents the *paramastoid process* of some mammals. This eminence is occasionally large, and extends as low as the transverse process of the atlas. This surface affords attachment to the Rectus capitis lateralis muscle and to the lateral occipito-atlantal ligament; its upper or cerebral surface presents a deep groove which lodges part of the lateral sinus, while its external surface is marked by a quadrilateral rough facet, covered with cartilage in the fresh state, and articulating with a similar surface on the petrous portion of the temporal bone. On the outer side of each condyle, near its fore part, is a foramen, the *anterior condyloid*; it is directed downwards, outwards, and forwards, and transmits the hypoglossal nerve, and occasionally a meningeal branch of the ascending pharyngeal artery. This foramen is sometimes double. Behind each condyle is a fossa,* occasionally perforated at the bottom by a foramen, the *posterior condyloid*, for the transmission of a vein to the lateral sinus. In front of the foramen magnum is a strong quadrilateral plate of bone, the *basilar process*, wider behind than in front; its under surface, which is rough, presenting in the median line

---

* This fossa presents many variations in size. It is usually shallow, and the foramen small; occasionally wanting on one or both sides. Sometimes both fossa and foramen are large, but confined to one side only; more rarely, the fossa and foramen are very large on both sides.

a tubercular ridge, the *pharyngeal spine*, for the attachment of the tendinous raphe and Superior constrictor of the pharynx ; and on each side of it, rough depressions for the attachment of the Rectus capitis anticus, major and minor.

The **Internal** or **Cerebral Surface** (fig. 215) is deeply concave.   The posterior part or tabula is divided by a crucial ridge into four fossæ.   The two superior fossæ receive the occipital lobes of the cerebrum, and present slight depressions corresponding to their convolutions.   The two inferior, which receive the hemispheres of the cerebellum, are larger than the former, and comparatively smooth ; both are marked by slight grooves for the lodgment of arteries.   At the point of meeting of the four divisions of the crucial ridge is an eminence, the *internal occipital protuberance*.   It nearly corresponds to that on the outer surface, and is perforated by one or more large vascular foramina.   From this eminence, the

FIG. 215.—Occipital bone.   Inner surface.

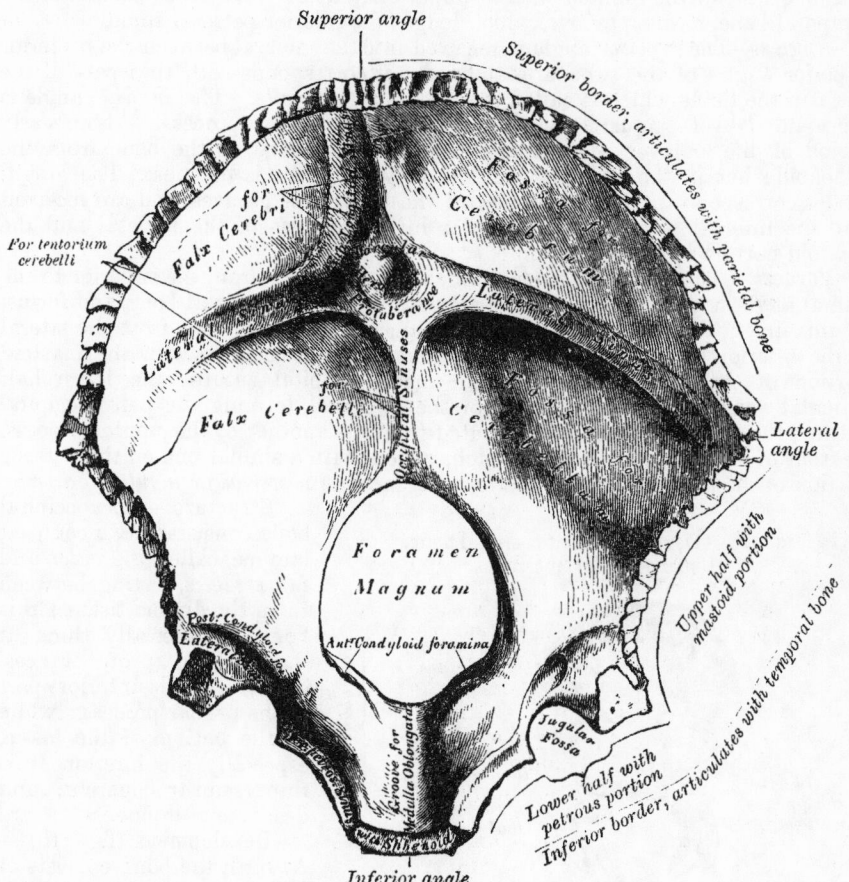

superior division of the crucial ridge runs upwards to the superior angle of the bone ; on one side, generally the right, it presents a deep groove for the superior longitudinal sinus, the margins of which give attachment to the falx cerebri. The inferior division, the *internal occipital crest*, runs to the posterior margin of the foramen magnum, on the edge of which it becomes gradually lost ; this ridge, which is bifurcated below, serves for the attachment of the falx cerebelli.   It is usually marked by a single groove, which commences at the back part of the foramen magnum and lodges the occipital sinus.   Occasionally two sinuses exist, and the groove is then double.   The transverse grooves pass outwards to the lateral angles ; they are deeply channelled, for the lodgment of the lateral sinuses, their prominent margins affording attachment to the tentorium cerebelli.*   At

---

* Usually one of the transverse grooves is deeper and broader than the other : occasionally, both grooves are of equal depth and breadth, or both equally indistinct.   The

the point of meeting of these grooves is a depression, the *torcular Herophili*,* placed a little to one or the other side of the internal occipital protuberance. More anteriorly is the foramen magnum, and on each side of it, but nearer its anterior than its posterior part, the internal opening of the anterior condyloid foramen ; the internal openings of the posterior condyloid foramina are a little external and posterior, protected by a small arch of bone. At this part of the internal surface there is a very deep groove, in which the posterior condyloid foramen, when it exists, has its termination. This groove is continuous, in the complete skull, with the transverse groove on the posterior part of the bone, and lodges the end of the same sinus, the lateral. In front of the foramen magnum is the basilar process, presenting a shallow depression, the *basilar groove*, which slopes from behind, upwards and forwards, and supports the medulla oblongata and part of the pons Varolii, and on each side of the basilar process is a narrow channel, which, when united with a similar channel on the petrous portion of the temporal bone, forms a groove, which lodges the inferior petrosal sinus.

Angles.—The *superior* angle is received into the interval between the posterior superior angles of the two parietal bones : it corresponds with that part of the skull in the fœtus which is called the *posterior fontanelle*. The *inferior* angle is represented by the square-shaped surface of the basilar process. At an early period of life, a layer of cartilage separates this part of the bone from the sphenoid ; but in the adult, the union between them is osseous. The *lateral* angles correspond to the outer ends of the transverse grooves, and are received into the interval between the posterior inferior angles of the parietal and the mastoid portion of the temporal.

Borders.—The *superior* border extends on each side from the superior to the lateral angle, is deeply serrated for articulation with the parietal bone, and forms, by this union, the lambdoid suture ; the *inferior* border extends from the lateral to the inferior angle ; its upper half is rough, and articulates with the mastoid portion of the temporal, forming the masto-occipital suture ; the lower half articulates with the petrous portion of the temporal, forming the petro-occipital suture ; these two portions are separated from one another by the jugular process. In front of this process is a deep notch, which, with a similar one on the petrous portion of the temporal, forms the *foramen lacerum posterius* or *jugular foramen*.

FIG. 216.—Development of occipital bone.
Usually by seven centres.

At birth
the 4 pieces
are separate

4 *for interparietal part of tabular surface*

2 *for cartilaginous part of tabular surface*

1 *for each condyloid portion*

2 *for basilar portion*

Join about 4 yr.

Join 5–6 yr.

Structure.—The occipital bone consists of two compact laminæ, called the *outer* and *inner tables*, having between them the diploic tissue ; this bone is especially thick at the ridges, protuberances, condyles, and anterior part of the basilar process ; while at the bottom of the fossæ, especially the inferior, it is thin, semi-transparent, and destitute of diploë.

Development (fig. 216).— At birth the bone consists of four distinct parts : a *tabular* or expanded portion, which lies behind the foramen magnum ; two *condylar* parts, which form the sides of the foramen ; and a *basilar* part, which lies in front of the foramen. The upper portion of the tabular surface—that is to say, the portion above the level of the external occipital protuberance—is developed from membrane, and may remain separated from the rest of the bone throughout life when it constitutes the *interparietal* bone ; the rest of the bone is developed from cartilage. The number of nuclei for the interparietal part of the tabular portion, which is developed from membrane, varies from two to four. They appear about the thirteenth week and coalesce about the fifteenth. The

broader of the two transverse grooves is nearly always continuous with the vertical groove for the superior longitudinal sinus.

* The columns of blood coming in different directions were supposed to be *pressed* together at this point (*torcular*, a wine-press).

cartilaginous portion of the tabular part begins to ossify from two centres, which appear about the seventh week of fœtal life and soon unite to form a single piece. Union of the upper and lower portions takes place about the third or fourth month. Each of the condylar parts begins to ossify from a single centre about the end of the eighth week. The basilar portion is ossified from two centres, one placed in front of the other. These appear about the sixth week and rapidly coalesce, so that the ossification from one centre is frequently described. About the fourth year, the tabular and the two condyloid pieces join; and about the sixth year, the bone consists of a single piece. At a later period, between the eighteenth and twenty-fifth years, the occipital and sphenoid become united, forming a single bone.

Articulations.—With six bones : two parietal, two temporal, sphenoid, and atlas.

Attachment of Muscles.—To twelve pairs ; to the superior curved line are attached the Occipito-frontalis, Trapezius, and Sterno-cleido-mastoid. To the space between the curved lines, the Complexus,* Splenius capitis, and Obliquus capitis superior ; to the inferior curved line, and the space between it and the foramen magnum, the Rectus capitis posticus, major and minor ; to the transverse process, the Rectus capitis lateralis ; and to the basilar process, the Rectus capitis anticus, major and minor, and the Superior constrictor of the pharynx.

THE PARIETAL BONES

The **Parietal Bones** (paries, *a wall*) form, by their union, the sides and roof of the skull. Each bone is of an irregular quadrilateral form, and presents for examination two surfaces, four borders, and four angles.

FIG. 217.—Left parietal bone. External surface.

Articulates with opposite parietal bone

Parietal foramen

Parietal eminence

Articulates with frontal bone

Articulates with occipital bone

Upper Temporal Ridge

Lower Temporal Ridge

TEMPORAL MUSCLE

Squamous portion of temporal bone

Sphenoid

Mastoid portion

Surfaces.—The *external surface* (fig. 217) is convex, smooth, and marked about its centre by an eminence, called the *parietal eminence*, which indicates the

* To these the Biventer cervicis should be added, if it is regarded as a separate muscle.

point where ossification commenced. Crossing the middle of the bone in an arched direction are two well-marked curved lines or ridges, the *upper* and *lower temporal ridges*; the former gives attachment to the temporal fascia, while the latter indicates the upper limit of the origin of the temporal muscle. Above these ridges, the surface of the bone is rough and porous, and covered by the aponeurosis of the Occipito-frontalis; between them the bone is smoother and more polished than the rest; below them the bone forms part of the temporal fossa, and affords attachment to the Temporal muscle. At the back part of the superior border, close to the sagittal suture, is a small foramen, the *parietal foramen*, which transmits a vein to the superior longitudinal sinus, and sometimes a small branch of the occipital artery. Its existence is not constant, and its size varies considerably.

The **internal surface** (fig. 218) is concave; it presents depressions for lodging the convolutions of the cerebrum, and numerous furrows for the ramifications of

FIG. 218.—Left parietal bone. Internal surface.

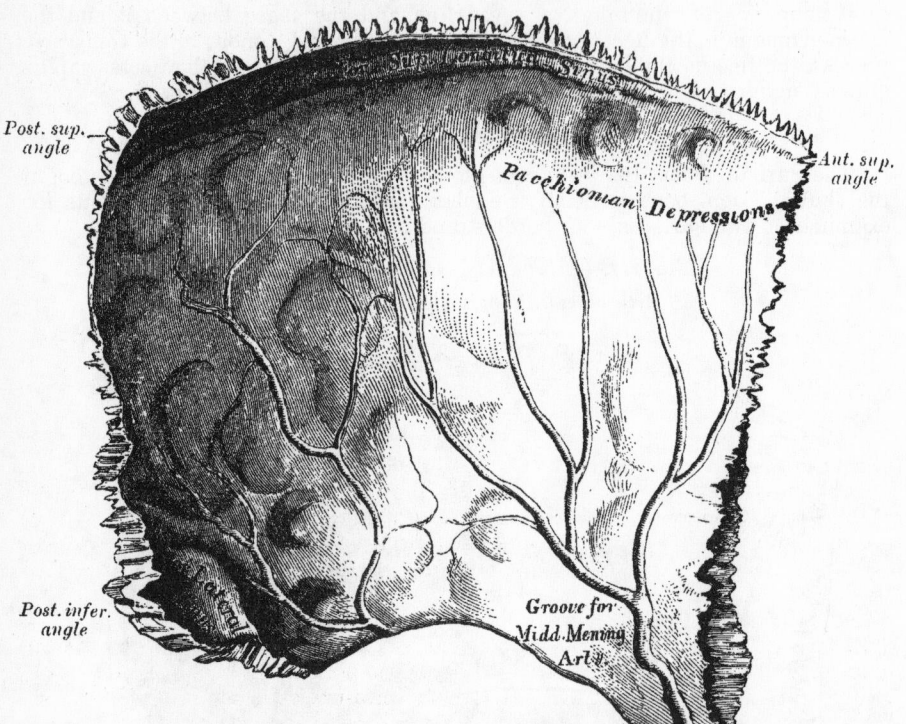

the middle meningeal artery; the latter run upwards and backwards from the anterior inferior angle, and from the central and posterior part of the lower border of the bone. Along the upper margin is part of a shallow groove, which, when joined to the opposite parietal, forms a channel for the superior longitudinal sinus, the elevated edges of which afford attachment to the falx cerebri. Near the groove are seen several depressions, especially in the skulls of old persons; they lodge the Pacchionian bodies. The internal opening of the parietal foramen is also seen when that aperture exists.

Borders.—The *superior*, the longest and thickest, is dentated to articulate with its fellow of the opposite side, forming the sagittal suture. The *inferior* is divided into three parts: of these, the anterior is thin and pointed, bevelled at the expense of the outer surface, and overlapped by the tip of the great wing of the sphenoid; the middle portion is arched, bevelled at the expense of the outer surface, and overlapped by the squamous portion of the temporal; the posterior portion is thick and serrated for articulation with the mastoid portion of the temporal. The *anterior* border, deeply serrated, is bevelled at the expense of the outer surface above and of the inner below; it articulates with the frontal bone, forming the *coronal suture*. The *posterior* border, deeply denticulated, articulates with the occipital, forming the *lambdoid suture*.

Angles.—The *anterior superior* angle, thin and pointed, corresponds with the union of the sagittal and coronal sutures ; this point is named the *bregma*. This region in the fœtal skull is membranous, and is called the *anterior fontanelle*. The *anterior inferior* angle is thin and lengthened, being received in the interval between the great wing of the sphenoid and the frontal. Its inner surface is marked by a deep groove, sometimes a canal, for the anterior branch of the middle meningeal artery. The *posterior superior* angle corresponds with the point of junction of the sagittal and lambdoid sutures—a point which is termed the *lambda*. In the fœtus this part of the skull is membranous, and is called the *posterior fontanelle*. The *posterior inferior* angle articulates with the mastoid portion of the temporal bone, and generally presents on its inner surface a broad, shallow groove which lodges part of the lateral sinus. The point of union of the posterior inferior angle of the parietal with the occipital and mastoid part of the temporal is named the *asterion*.

Development.—The parietal bone is formed in membrane, being developed by one centre, which appears at the parietal eminence about the seventh or eighth week of fœtal life. Ossification gradually extends from the centre to the circumference of the bone : the angles are consequently the parts last formed, and it is here that the fontanelles exist previous to the completion of the growth of the bone. Occasionally the parietal bone is divided into two parts, upper and lower, by an antero-posterior suture.

Articulations.—With five bones : the opposite parietal, the occipital, frontal, temporal, and sphenoid.

Attachment of Muscles.—One only, the Temporal.

The Frontal Bone

The **Frontal Bone** (frons, *the forehead*) resembles a cockle-shell in form, and consists of two portions—a *vertical* or *frontal* portion situated at the anterior part of the cranium, forming the forehead ; and a *horizontal* or *orbito-nasal* portion which enters into the formation of the roof of the orbits and nasal fossæ.

Vertical Portion.—*External Surface* (fig. 219).—In the median line, traversing the lower part of the bone, may occasionally be seen the remains of the suture which represents the line of union of the two parts of which the bone consists at an early period of life. This is well marked in young subjects, but usually becomes obliterated in adult life, except at the lower part, where traces of it may sometimes remain. It is termed the *frontal* (metopic) suture. On either side of this suture, a little below the centre of the bone, is a rounded eminence, the *frontal eminence*. These eminences vary in size in different individuals, are occasionally unsymmetrical, and are especially prominent in cases of well-marked cerebral development. The whole surface of the bone above this part is smooth, and covered by the aponeurosis of the Occipito-frontalis muscle. Below the frontal eminence, and separated from it by a shallow groove, is the *superciliary ridge*, broad internally, where it is continuous with the nasal eminence, but less distinct as it arches outwards. These ridges are, in most cases, caused by the projection outwards of the frontal air sinuses,* are usually larger and more rounded in the male than the female, and give attachment to the Orbicularis palpebrarum and Corrugator supercilii. The degree of prominence of the superciliary ridges is not necessarily an index of the size of the frontal air sinuses, since in some cases the latter are large where the former are faintly marked. On the other hand, the ridges may be prominent, and the sinuses rudimentary or absent, as in the skull of the Maori. Between the two superciliary ridges is a smooth surface, the *glabella* or *nasal eminence*. Beneath the superciliary ridge is the *supra-orbital*

---

* Some confusion is occasioned to students commencing the study of anatomy by the name 'sinus' having been given to two perfectly different kinds of space connected with the skull. It may be as well, therefore, to state here, at the outset, that the 'sinuses' in the interior of the cranium which produce the grooves on the inner surface of the bones are venous channels along which the blood runs in its passage back from the brain, while the 'sinuses' external to the cranial cavity (the frontal, sphenoidal, ethmoidal, and maxillary) are hollow spaces in the bones themselves which communicate with the nostrils, and contain air.

arch, a curved and prominent margin, which forms the upper boundary of the orbit, and separates the vertical from the horizontal portion of the bone. The outer part of the arch is sharp and prominent, affording to the eye, in that situation, considerable protection from injury; the inner part is less prominent. At the junction of the internal and middle thirds of this arch is a notch, some-times converted into a foramen, and called the *supra-orbital notch* or *foramen* It transmits the supra-orbital artery, vein, and nerve. A small aperture is seen in the upper part of the notch, which transmits a vein from the diploë to join the supra-orbital vein. The supra-orbital arch terminates externally in the *external angular process*, and internally in the *internal angular process*. The external angular process is strong, prominent, and articulates with the malar bone; running upwards and backwards from it are two well-marked lines, which, commencing together at the external angular process, soon diverge from each other and run in a curved direction across the bone. These are the *upper* and *lower temporal ridges*; the upper gives attachment to the temporal fascia, the

FIG. 219.—Frontal bone. Outer surface.

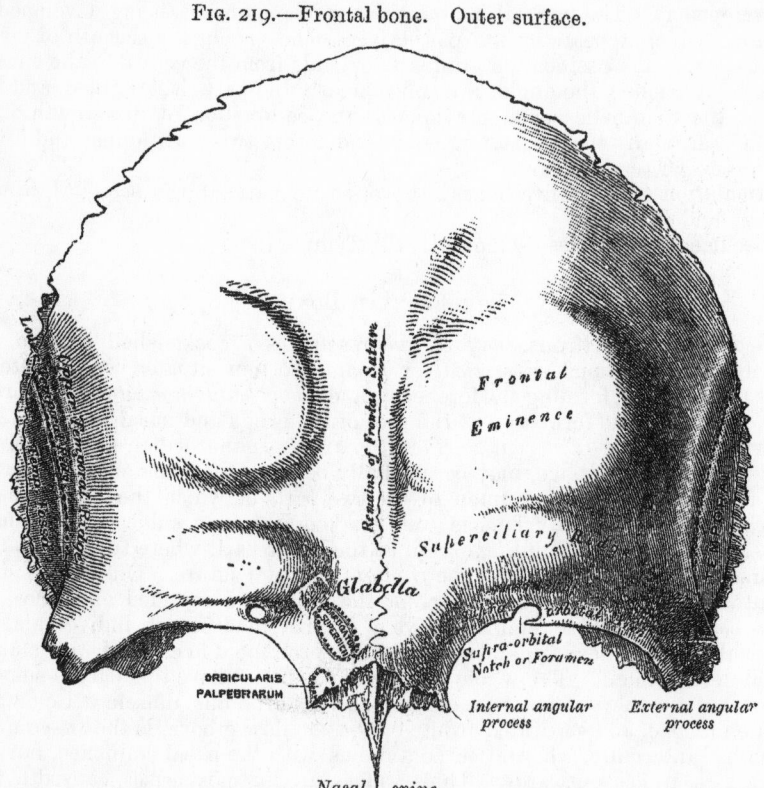

lower to the temporal muscle. Beneath them is a slight concavity, which forms the anterior part of the temporal fossa, and gives origin to the Temporal muscle. The internal angular processes are less marked than the external, but descend to a lower level, and articulate with the lachrymal bones. Between the internal angular processes is a rough, uneven interval, the *nasal notch*, which articulates in the middle line with the nasal bones, and on either side with the nasal process of the superior maxillary bone. The term *nasion* is applied to the middle of the fronto-nasal suture. From the centre of this notch projects a process, the *nasal process*, which extends beneath the nasal bones and nasal processes of the superior maxillary bones, and supports the bridge of the nose. The nasal process termi-nates below in a long pointed process, the *nasal spine*, and on either side of this is a small grooved surface which enters into the formation of the roof of the nasal fossa. The nasal spine forms part of the septum of the nose, articu-lating in front with the nasal bones and behind with the perpendicular plate of the ethmoid.

Internal Surface (fig. 220).—Along the middle line in the upper part of the bone is a vertical groove, the edges of which unite below to form a ridge, the *frontal crest*; the groove lodges the superior longitudinal sinus, while its margins and the crest afford attachment to the falx cerebri. The crest terminates below in a small notch which is converted into a foramen by articulation with the ethmoid. It is called the *foramen cæcum*, and varies in size in different subjects; it is sometimes impervious, and lodges a process of the falx cerebri; when open, it transmits a vein from the lining membrane of the nose to the superior longitudinal sinus. On either side of the groove the bone is deeply concave, presenting depressions for the convolutions of the brain, and numerous small furrows for lodging the ramifications of the anterior branches of the middle meningeal arteries. Several small, irregular fossæ are also seen on either side of the groove, for the reception of the Pacchionian bodies.

FIG. 220.—Frontal bone. Inner surface.

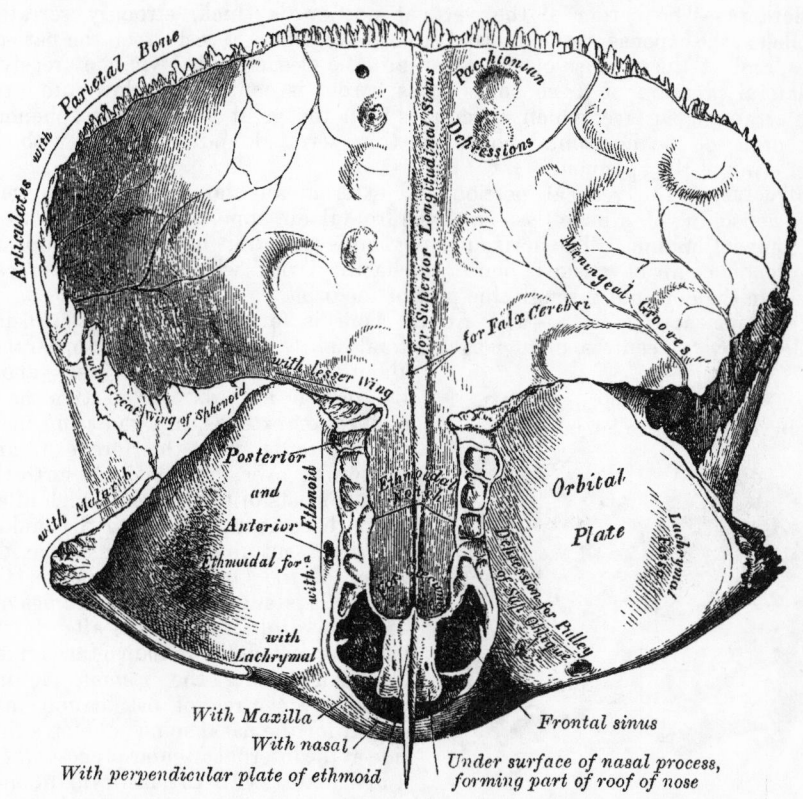

Horizontal Portion—This portion of the bone consists of two thin plates, the *orbital plates*, which form the vaults of the orbits, and are separated from one another by a median gap, the *ethmoidal notch*.

The *external surface* of each orbital plate consists of a smooth, concave, triangular lamina of bone, marked at its anterior and external part (immediately beneath the external angular process) by a shallow depression, the *lachrymal fossa*, for lodging the lachrymal gland; and at its anterior and internal part, by a depression (sometimes a small tubercle), the *trochlear fossa*, for the attachment of the cartilaginous pulley of the Superior oblique muscle of the eyeball. The ethmoidal notch separates the two orbital plates; it is quadrilateral, and filled up, in the articulated skull, by the cribriform plate of the ethmoid. The margins of this notch present several half-cells, which, when united with corresponding half-cells on the upper surface of the ethmoid, complete the ethmoidal cells; two grooves are also seen crossing these edges transversely; they are converted into canals by articulation with the ethmoid, and are called the *anterior* and *posterior ethmoidal* canals : they open on the inner wall of the orbit. The anterior one transmits the nasal nerve and anterior ethmoidal vessels, the posterior one the posterior ethmoidal vessels. In front of the ethmoidal notch, on either side of the nasal spine, are the openings of the frontal air sinuses. These are two irregular cavities, which extend backwards, upwards and outwards, a variable distance, between the two tables of the skull, and are separated from one another by a thin, bony septum, which is often displaced to one side. Absent at birth, they are usually fairly well developed between the ninth and twelfth years. These cavities vary in size in different persons, are larger in men than in women, and are frequently of unequal size on the two sides, the right being commonly the larger.* They are lined by mucous membrane, and communicate with the nose by the infundibulum. They rarely communicate with each other.

The *internal surface* presents the convex upper surfaces of the orbital plates, separated from each other in the middle line by the ethmoidal notch, and marked by depressions for the convolutions of the frontal lobes of the brain, and faint grooves for meningeal arteries derived from the ophthalmic.

Borders.—The border of the vertical portion is thick, strongly serrated, bevelled at the expense of the internal table above, where it rests upon the parietal bones, and at the expense of the external table at each side, where it receives the lateral pressure of those bones ; this border is continued below into a triangular, rough surface, which articulates with the great wing of the sphenoid. The border of the horizontal portion is thin, serrated, and articulates with the lesser wing of the sphenoid.

Structure.—The vertical portion and external angular processes are very thick, consisting of diploic tissue and the frontal air sinuses contained between two compact laminæ. The horizontal portion is thin, translucent, and composed entirely of compact tissue ; hence the facility with which instruments can penetrate the cranium through this part of the orbit.

Development (fig. 221).—The frontal bone is formed in membrane, being developed by *two* centres, one for each lateral half, which make their appearance about the seventh or eighth week, above the orbital arches. From this point ossification extends, in a radiating manner, upwards into the forehead, and backwards over the orbit. At birth the bone consists of two pieces, which afterwards become united, along the middle line, by a suture which runs from the vertex of the bone to the root of the nose. This suture usually becomes obliterated within a few years after birth ; but it occasionally remains throughout life, constituting the *metopic* suture. Secondary centres of ossification may appear for the nasal spine ; one on either side at the internal angular process where it articulates with the lachrymal bone ;

Fig. 221.—Frontal bone at birth. Developed by two lateral halves.

and sometimes there is one on either side at the lower end of the coronal suture. This latter centre may remain ununited, and is known as the *pterion ossicle* ; or it may join with the parietal, sphenoid, or temporal bone.

Articulations.—With twelve bones : two parietal, the sphenoid, the ethmoid, two nasal, two superior maxillary, two lachrymal, and two malar.

Attachment of Muscles.—To three pairs : the Corrugator supercilii, Orbicularis palpebrarum, and Temporal on each side.

THE TEMPORAL BONES

The **Temporal Bones** (tempus, *time*) are situated at the sides and base of the skull. Each consists of four parts, viz. . the *squamous* or *squamo-zygomatic*, the *petro-mastoid*, the *tympanic plate*, and the *styloid process*.

* Aldren Turner (*The Accessory Sinuses of the Nose*, 1901) gives the following measurements for a sinus of average size : height, 1¼ in. ; breadth, 1 in. ; depth from before backwards, 1 in.

The **squamous portion** (squama, *a scale*), the anterior and upper part of the bone, is scale-like in form, and thin and translucent in texture (fig. 222). Its *outer surface* is smooth, convex, and grooved at its back part for the deep temporal arteries; it affords attachment to the Temporal muscle, and forms part of the temporal fossa. At its back part may be seen a curved ridge—part of the *temporal ridge*; it serves for the attachment of the temporal fascia, and limits the origin of the Temporal muscle. The boundary between the squamous and mastoid portions of the bone, as indicated by traces of the original suture, lies fully half an inch below this ridge. Projecting from the lower part of the squamous portion is a long, arched process of bone, the *zygoma* or zygomatic process. This process is at first directed outwards, its two surfaces looking upwards and downwards; it then appears as if twisted upon itself, and runs forwards, its surfaces now looking inwards and outwards. The superior border of the process is long, thin, and sharp, and serves for the attachment of the temporal fascia. The inferior, short, thick, and arched, has attached to it some fibres of the Masseter muscle. Its outer surface is convex and subcutaneous; its inner is concave, and affords attachment to the Masseter muscle. The extremity, broad and deeply

FIG. 222.—Left temporal bone. Outer surface.

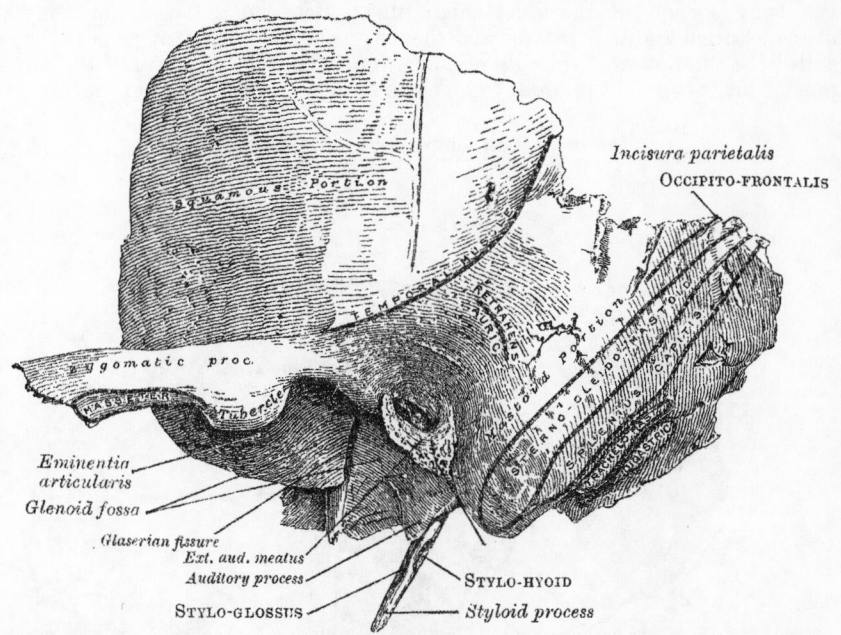

serrated, articulates with the malar bone. The zygomatic process is connected to the temporal bone by three divisions, called its *roots*—an anterior, middle, and posterior. The anterior, which is short but broad and strong, is directed inwards, to terminate in a rounded eminence, the *eminentia articularis*. This eminence forms the front boundary of the glenoid fossa, and in the recent state is covered with cartilage. The middle root is known as the *post-glenoid process*, and is prominent in young bones. It separates the mandibular portion of the glenoid fossa from the external auditory meatus, and terminates at the commencement of a well-marked fissure, the *Glaserian fissure*. The posterior root, which is strongly marked, runs from the upper border of the zygoma, backwards above the external auditory meatus. It is termed the *supra-mastoid crest*, and forms the posterior part of the lower temporal ridge. At the junction of the anterior root with the zygoma is a projection, called the *tubercle*, for the attachment of the external lateral ligament of the lower jaw; and between the anterior and middle roots is an oval depression, forming part of the glenoid fossa (γλήνη, *a socket*), for the reception of the condyle of the lower jaw. The glenoid fossa is bounded, in front, by the eminentia articularis; behind by the *tympanic plate*, which separates it from the external auditory meatus; it is divided into two parts by a narrow slit, the *Glaserian fissure*. The anterior or mandibular part, formed by the squamous portion of the bone, is smooth, covered in the recent state with cartilage, and articulates with the condyle of the lower jaw. This part of the fossa presents posteriorly a small conical eminence, the *post-glenoid process*, already referred to. This process is the representative of a prominent tubercle which, in some of the mammalia, descends behind the condyle of the jaw, and prevents it being displaced backwards during mastication (Humphry). The posterior part of the glenoid fossa, which lodges a portion of the parotid gland, is formed by the *tympanic plate*, which constitutes the anterior, the inferior, and part of the posterior wall of the external auditory meatus. This plate of bone terminates above in the Glaserian fissure, and below forms a sharp edge, the *vaginal* process, which encircles the root of the styloid process. The Glaserian fissure, which leads into the tympanum, lodges the processus gracilis of the malleus, and transmits the tympanic branch of the internal maxillary artery. The chorda tympani nerve passes through a canal (*canal of Huguier*), separated from the anterior edge of the Glaserian fissure by a thin scale of bone and situated on the outer side of the Eustachian tube, in the retiring angle between the squamous and petrous portions of the temporal bone. This thin scale of bone is derived from the roof of the tympanum, and forms the greater part of the outer wall of the bony portion of the Eustachian tube. Between the posterior bony wall of the external auditory meatus and the posterior root of the zygoma is the area called the *suprameatal triangle* of Macewen. Through this space the surgeon pushes his instrument in order to reach the antrum of the mastoid process.

FIG. 223.—Left temporal bone. Inner surface.

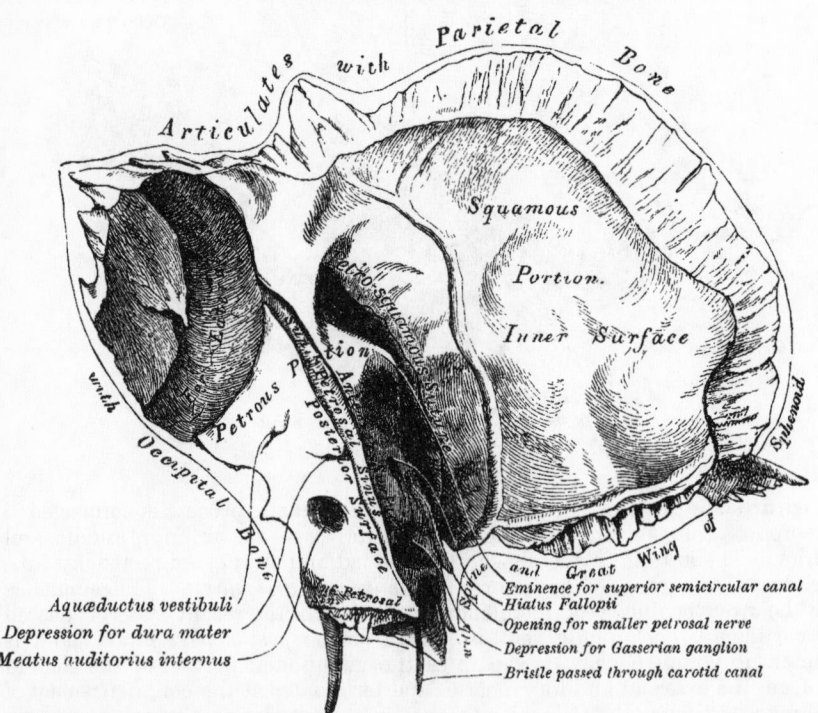

Aquæductus vestibuli
Depression for dura mater
Meatus auditorius internus

The **internal surface** of the squamous portion (fig. 223) is concave, presents numerous depressions for the convolutions of the cerebrum, and two well-marked grooves for the branches of the middle meningeal artery.

Borders.—The superior border is thin, bevelled at the expense of the internal surface, so as to overlap the lower border of the parietal bone, forming the squamous suture. The anterior inferior border is thick, serrated, and bevelled, alternately at the expense of the inner and outer surfaces, for articulation with the great wing of the sphenoid. Posteriorly, the superior border ends by forming an angle with the mastoid portion of the bone, called the *incisura parietalis*.

The **Petro-mastoid Portion** consists of (a) the *mastoid process*, which forms a prominent, nipple-like mass behind the external auditory meatus; and (b) the *petrous portion*, which is pyramidal in shape, and projects inwards and forwards to form part of the floor of the skull.

The **Mastoid Portion** (μαστός, *a nipple* or *teat*) is situated at the posterior part of the bone; its outer surface is rough, and gives attachment to the Occipito-frontalis and Retrahens auriculam muscles. It is perforated by numerous foramina; one of these, of large size, situated near the posterior border of the bone, is termed the *mastoid foramen*; it transmits a vein to the lateral sinus and a small artery from the occipital, to supply the dura mater. The position and size of this foramen are very variable. It is not always present; sometimes it is situated in the occipital bone, or in the suture between the temporal and the occipital. The mastoid portion is continued below into a conical projection, the *mastoid process*, the size and form of which vary somewhat; it is larger in the male than in the female. This process serves for the attachment of the Sterno-mastoid, Splenius capitis, and Trachelo-mastoid muscles. On the inner side of the mastoid process is a deep groove, the *digastric fossa* or *incisura mastoidea*, for

FIG. 224.—Section through the temporal bone, showing the communication of the cavity of the tympanum with the mastoid antrum.

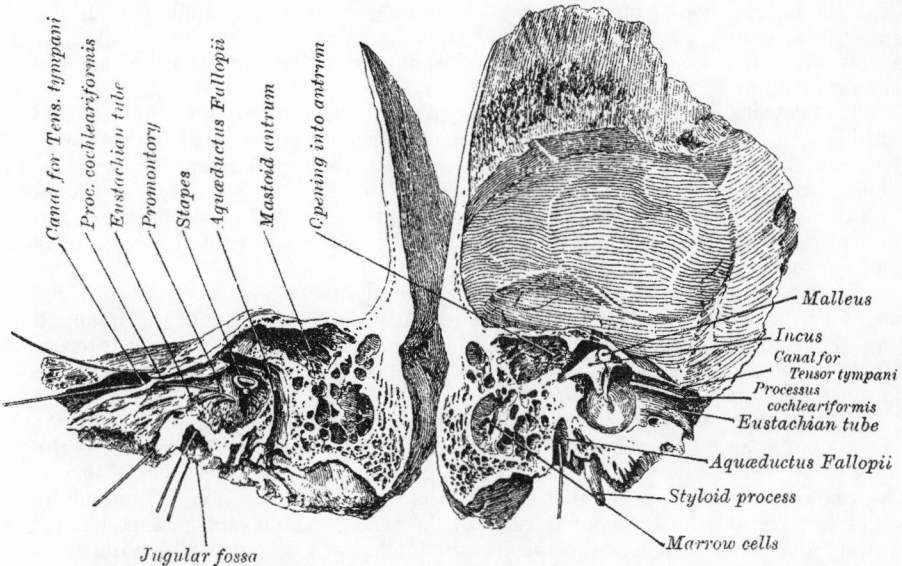

the attachment of the Digastric muscle; and, running parallel with it, but more internal, the *occipital groove*, which lodges the occipital artery. The internal surface of the mastoid portion presents a deep, curved groove, the *fossa sigmoidea*, which lodges part of the lateral sinus; and into it may be seen opening the mastoid foramen. The groove for the lateral sinus is separated from the inner-most of the mastoid air-cells by only a thin lamina of bone, and even this may be partly deficient. A section of the mastoid process shows it to be hollowed out into a number of cellular spaces, communicating with each other, called the *mastoid cells*, which exhibit the greatest possible variety as to their size and number. At the upper and front part of the bone these cells are large and irregular and contain air, but towards the lower part of the bone they diminish in size, while those at the apex of the mastoid process are quite small and usually contain marrow. Occasionally they are entirely absent, and the mastoid is solid throughout. In addition to these a large irregular cavity (fig. 224) is situated at the upper and front part of the bone. It is called the *mastoid antrum*, and must be distinguished from the mastoid cells, though it communicates with them. Like the mastoid cells it is filled with air and lined by a prolongation of the mucous membrane of the tympanic cavity, with which it communicates. The mastoid antrum is bounded above by a thin plate of bone, the *tegmen tympani*, which

O

separates it from the middle fossa of the base of the skull; below by the mastoid process; externally by the squamous portion of the bone just below the supra-mastoid crest, and internally by the external semicircular canal of the internal ear which projects into its cavity. The opening by which it communicates with the tympanum is situated at the superior internal angle of the posterior wall of that cavity, and opens into that portion of the tympanic cavity which is known as the *attic* or *epitympanic recess*; that is to say, the portion of the tympanic cavity which is above the level of the membrana tympani.

The mastoid cells are not developed until towards puberty; hence the prominence of the mastoid process in the adult: the mastoid antrum, on the other hand, is of large size at birth.

In consequence of the communication which exists between the tympanic cavity and mastoid antrum, inflammation of the lining membrane of the former cavity may easily travel backwards to that of the antrum and mastoid air-cells, leading to caries and necrosis of their walls and the risk of transference of the inflammation to the lateral sinus or encephalon.

Borders.—The superior border of the mastoid portion is broad and rough, its serrated edge sloping outwards, for articulation with the posterior inferior angle of the parietal bone. The posterior border, also uneven and serrated, articulates with the inferior border of the occipital bone between its lateral angle and jugular process. Anteriorly it is fused with the squamous portion of the bone above, and below enters into the formation of the external auditory meatus and the cavity of the tympanum.

The **Petrous Portion** (πέτρος, *a stone*), so named from its extreme density and hardness, is a pyramidal process of bone, wedged in at the base of the skull between the sphenoid and occipital bones. It is directed inwards, forwards, and a little upwards. It presents for examination a base, an apex, three surfaces, and three borders; and contains, in its interior, the essential parts of the organ of hearing. The *base* is applied against the internal surface of the squamous and mastoid portions.

The **apex** of the petrous portion, rough and uneven, is received into the angular interval between the posterior border of the greater wing of the sphenoid and the basilar process of the occipital; it presents the anterior or internal orifice of the carotid canal, and forms the posterior and external boundary of the foramen lacerum medium.

The **anterior surface** of the petrous portion (fig. 223) forms the posterior part of the middle fossa of the base of the skull. This surface is continuous with the squamous portion, to which it is united by a suture, the *petro-squamous suture*, the remains of which are distinct even at a late period of life. It is marked by digital depressions for the convolutions of the brain, and presents six points for examination: 1. an eminence (*eminentia arcuata*) near the centre, which indicates the situation of the superior semicircular canal: 2. in front and a little to the outer side of this eminence a depression, indicating the position of the tympanum; here the layer of bone which separates the tympanum from the cranial cavity is extremely thin, and is known as the *tegmen tympani*: 3. a shallow groove, sometimes double, leading outwards and backwards to an oblique opening, the *hiatus Fallopii*, for the passage of the large superficial petrosal nerve and the petrosal branch of the middle meningeal artery: 4. a smaller opening, occasionally seen external to the latter, for the passage of the small superficial petrosal nerve: 5. near the apex of the bone, the termination of the carotid canal, the wall of which in this situation is deficient in front: 6. above this canal a shallow depression for the reception of the Gasserian ganglion.

The **posterior surface** forms the front part of the posterior fossa of the base of the skull, and is continuous with the inner surface of the mastoid portion of the bone. It presents three points for examination: 1. about its centre is a large orifice, the *meatus acusticus internus*, the size of which varies considerably; its margins are smooth and rounded; and it leads into a short canal, about four lines in length, which runs directly outwards. This canal is closed externally by a vertical plate, the *lamina cribrosa*, which is divided by a horizontal crest, the *crista falciformis*, into two unequal portions (fig. 225). Each portion is further subdivided by a little vertical ridge into anterior and posterior parts. The *lower* portion presents three sets of foramina; one group, just below the posterior part of the crest, forms the *area cribrosa media*, and consists of a number of small openings for the nerves to the saccule; below and posterior to this, the *foramen singulare*, or opening for the nerve to the posterior semicircular canal; in front and below the first, the *tractus spiralis foraminosus*, consisting of a number of small spirally arranged openings, which terminate in the *canalis centralis cochleæ* and transmit the nerves to the cochlea. The *upper* portion, that above the crista, presents behind a series of small openings, the *area cribrosa superior*, for the passage of filaments to the utricle and superior and external semicircular canals, and, in front, the *area facialis*, with one large opening, the commencement of the aquæductus Fallopii, for the passage of the facial nerve : 2. behind the meatus auditorius is a small slit almost hidden by a thin plate of bone, leading to a canal, the *aquæductus vestibuli*, which transmits the *ductus endolymphaticus* together with a small artery and vein : 3. in the interval between these two openings, but above them, is an angular depression which lodges a process of the dura mater, and transmits a small vein into the cancellous tissue of the bone.

FIG. 225.—Diagrammatic view of the fundus of the internal auditory meatus. (Testut.)

1. Crista falciformis.   2. Area facialis, with (2′) Internal opening of the aquæductus Fallopii.   3. Ridge separating the area facialis from the area cribrosa superior.   4. Area cribrosa superior, with (4′) Openings for nerve filaments.   5. Anterior inferior cribriform area, with (5′) the tractus spiralis foraminosus, and (5″) the canalis centralis of the cochlea.   6. Ridge separating the tractus spiralis foraminosus from the area cribrosa media.   7. Area cribrosa media, with (7′) orifices for nerves to saccule.   8. Foramen singulare.

In the child this depression is represented by a large fossa, the *fossa subarcuata*, which extends backwards as a blind tunnel under the superior semicircular canal.

The **inferior** or **basilar surface** (fig. 226) is rough and irregular, and forms part of the base of the skull. Passing from the apex to the base, this surface presents eleven points for examination : 1. a rough surface, quadrilateral in form, which serves partly for the attachment of the Levator palati muscle and cartilaginous portion of the Eustachian tube ; and partly for connection with the basilar process of the occipital bone through the intervention of dense fibrous tissue : 2. the large, circular aperture of the carotid canal, which ascends at first vertically, and then, making a bend, runs horizontally forwards and inwards ; it transmits the internal carotid artery, and the carotid plexus of nerves : 3. to the inner side of the opening for the carotid canal and close to the posterior border, in front of the jugular fossa, is a triangular depression, on the floor of which is a small opening, the *aquæductus cochleæ*, which lodges a tubular prolongation of the dura mater and transmits a vein from the cochlea to join the internal jugular : 4. behind these openings a deep depression, the *jugular fossa*, which varies in depth and size in different skulls ; it lodges the bulb of the internal jugular vein, and, with a similar depression on the margin of the jugular process of the occipital bone, forms the foramen lacerum posterius, or jugular foramen : 5. a small foramen for the passage of Jacobson's nerve (the tympanic branch of the glossopharyngeal) ; this foramen is seen in the bony ridge dividing the carotid canal from the jugular fossa : 6. a small foramen in the outer part of the jugular fossa, for the entrance of the auricular branch of the pneumogastric (Arnold's nerve) : 7. behind the jugular fossa, a quadrilateral area, the *jugular surface* ; it is covered with cartilage in the recent state, and articulates with the jugular process of the occipital bone :. 8. the *vaginal process*, a very broad, sheath-like plate of bone, which extends backwards from the carotid canal and divides behind into two laminæ, the outer of which is continuous with the tympanic plate, the inner with the jugular process ; between these laminæ is the 9th point for examination, the *styloid process*, a sharp spine, about an inch in length : 10. the *stylo-mastoid foramen*, a rather large orifice, placed between the styloid and mastoid processes ; it is the termination of the aquæductus Fallopii, and transmits the facial nerve and stylo-mastoid artery : 11. the *auricular fissure*, situated between the tympanic plate and mastoid processes, for the exit of the auricular branch of the pneumo-gastric nerve.

Borders.—The *superior*, the longest, is grooved for the superior petrosal sinus, and has attached to it the tentorium cerebelli; at its inner extremity is a semilunar notch, upon which the fifth nerve lies. The *posterior* border is intermediate in length between the superior and the anterior. Its inner half is marked by a groove, which, when completed by its articulation with the occipital bone, forms the channel for the inferior petrosal sinus. Its outer half presents an excavation—the *jugular fossa*—which, with a similar notch on the occipital, forms the foramen lacerum posterius. A projecting eminence of bone occasionally stands out from the centre of the notch, and divides the foramen into two. The *anterior* border is divided into two parts—an outer joined to the squamous portion by a suture (*petro-squamous*), the remains of which are distinct; an inner, free, articulating with the spinous process of the sphenoid. At the angle of junction of the petrous and squamous portions are seen two canals, separated from

Fig. 226.—Left temporal bone. Inferior surface.

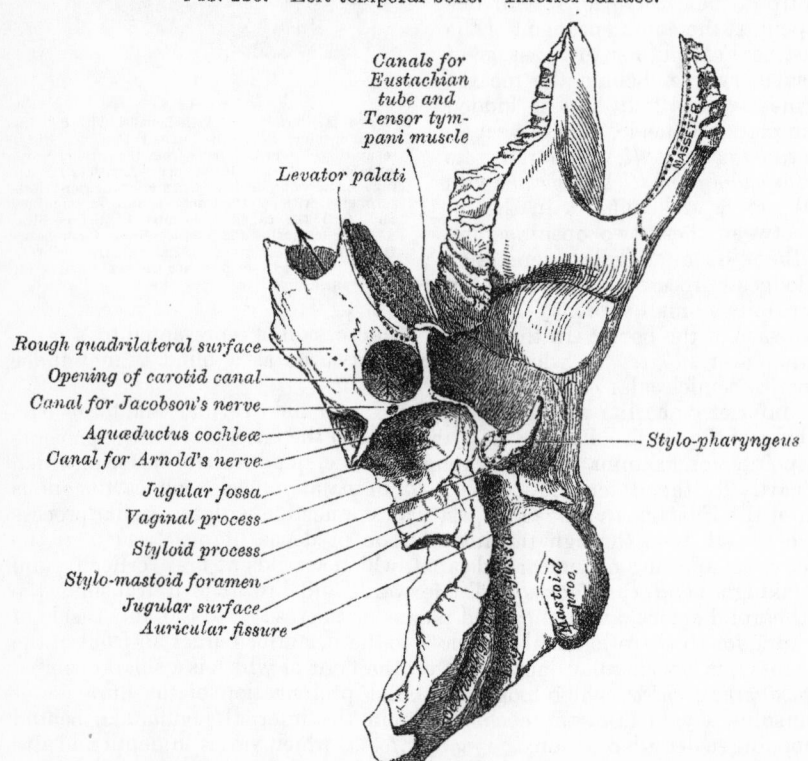

Canals for
Eustachian
tube and
Tensor tym-
pani muscle

Levator palati

Rough quadrilateral surface
Opening of carotid canal
Canal for Jacobson's nerve
Aquæductus cochleæ
Canal for Arnold's nerve
Jugular fossa
Vaginal process
Styloid process
Stylo-mastoid foramen
Jugular surface
Auricular fissure

Stylo-pharyngeus

one another by a thin plate of bone, the *processus cochleariformis*; they both lead into the tympanum, the upper one transmitting the Tensor tympani muscle, the lower one forming the bony part of the Eustachian tube.

The tympanic cavity and auditory ossicles, and the internal ear, are described with the organ of hearing.

The **Tympanic Plate** consists of a curved plate of bone lying below the squamous portion and in front of the mastoid process. Its postero-superior surface is concave, and forms the anterior wall, the floor, and part of the posterior wall of the bony external auditory meatus. Internally, it presents a narrow furrow, the *sulcus tympanicus*, for the attachment of the membrana tympani. Its antero-inferior surface is quadrilateral and slightly concave; it constitutes the posterior boundary of the glenoid cavity, and is in contact with the retro-mandibular part of the parotid gland. Its outer border is free and rough; it is named the *external auditory process*, and gives attachment to the cartilaginous part of the external auditory meatus. Internally, the tympanic plate is fused with the petrous portion, and appears in the retreating angle between it and the squamous portion, where it lies below and to the outer side of the orifice of the Eustachian tube. Posteriorly, it blends with the squamous and mastoid parts, and forms the anterior boundary of the auricular fissure. Its antero-superior border fuses externally with the back of the post-glenoid process, while internally it bounds the Glaserian fissure. The lower border is thin and sharp at its inner part; its outer part splits to enclose the root of the styloid process, and is therefore named the *vaginal process*. The central portion of the tympanic plate is thin, and in a considerable percentage of skulls is perforated by a hole, the *foramen of Huschke*.

The **external auditory meatus** is directed inwards and slightly forwards : at the same time it forms a slight curve, so that the floor of the canal is convex upwards. It measures about three-quarters of an inch in length, and presents an oval or elliptical shape—its long axis being directed downwards and slightly backwards. As has been pointed out, its anterior wall, its floor, and the lower part of its posterior wall are formed by the tympanic plate ; the roof and upper part of the posterior wall belong to the squamous portion. Its inner end is closed, in the recent state, by the membrana tympani : the upper limit of its outer orifice is formed by the posterior root of the zygoma, immediately below which there is sometimes seen a small spine, the *suprameatal* spine, situated at the upper and posterior part of the orifice.

The **Styloid process** is slender, pointed, and of varying length ; it projects downwards and forwards, from the under aspect of the temporal bone, beyond the tympanic plate. Its proximal part (*tympano-hyal*) is ensheathed by the vaginal process, while its projecting portion (*stylo-hyal*) gives attachment to the stylo-hyoid and stylo-mandibular ligaments, and to the Stylo-glossus, Stylo-hyoid, and Stylo-pharyngeus muscles. The stylo-hyoid ligament extends from the apex of the process to the lesser cornu of the hyoid bone, and may undergo partial or complete ossification.

Structure.—The squamous portion is like that of the other cranial bones : the mastoid portion cellular, and the petrous portion dense and hard.

Development (fig. 227).—The temporal bone is developed by *eight* centres, exclusive of those for the internal ear and the ossicula—viz. one for the squamous portion including the zygoma, one for the tympanic plate, four for the petro-mastoid part, and two for the styloid process. Just before the close of fœtal life the temporal bone consists of four parts : 1. The *squamo-zygomatic*, which is ossified in membrane from a single nucleus ; this appears near the root of the zygoma about the second month. 2. The *petro-mastoid*, which is developed from four centres ; these make their appearance in the cartilaginous ear-capsule about the fifth or sixth month. One (*prootic*) appears in the neighbourhood of the eminentia arcuata, spreads in front and above the internal auditory meatus and extends to the apex of the bone ; it forms part of the cochlea, vestibule, superior semicircular canal,

FIG. 227.—Development of the temporal bone. By eight centres.

1 for squamous portion including zygoma : 2nd month

1 for tympanic ring

4 for petrous and mastoid portions

2 for styloid process

Unite about time of birth

At birth 3 pieces separate ; unite during 1st year

Bristle passed through carotid canal

and inner wall of the tympanic cavity. A second (*opisthotic*) appears at the promontory on the inner wall of the tympanum and surrounds the fenestra rotunda ; it forms the floor of the tympanum and vestibule, surrounds the carotid canal, invests the outer and lower part of the cochlea, and spreads inwards below the internal auditory meatus. A third (*pterotic*) roofs in the antrum and tympanic cavity ; while the fourth (*epiotic*) appears near the posterior semicircular canal and extends to form the mastoid process (Vrolik). 3. The *tympanic ring*, an incomplete circle in the concavity of which is a groove, the *sulcus tympanicus,* for the attachment of the circumference of the tympanic membrane. This is also ossified from a single centre which appears about the third month. 4. The *styloid process,* which is developed from the proximal part of the cartilage of the second visceral arch by two centres : one for the base, which appears before birth, and is termed the *tympano-hyal;* the other, comprising the rest of the process, is named the *stylo-hyal,* and does not appear until after birth. Shortly before birth the tympanic plate joins with the squamous. The petro-mastoid joins with the squamous during the first year, and the tympano-hyal portion of the styloid process about the same time. The stylo-hyal does not join the rest of the bone until after puberty, and in some skulls never becomes united. The chief subsequent changes in this bone are : (1) The tympanic ring extends outwards and backwards so as to form the tympanic plate. This extension does not, however, take place at an equal rate all round the circumference of the ring, but occurs most rapidly on its anterior and posterior portions, and these outgrowths meet and blend, and thus, for a time, there exists in the floor of the meatus a foramen, the *foramen of Huschke*: this foramen may persist throughout life. (2) The glenoid cavity is at first extremely shallow, and looks outwards as well as downwards ; it becomes deeper and is ultimately directed downwards. Its change in direction is accounted for as follows : The part of the squamous temporal which supports it lies at first *below* the level of the zygoma. As, however, the base of the skull increases in width, this lower part of the squama is directed horizontally inwards to contribute to the middle fossa of the skull, and its surfaces therefore come to look upwards and downwards ; the attached portion of the zygoma becomes everted, and projects like a shelf at right angles to the squama. (3) The mastoid portion is at first quite flat, and the stylo-mastoid foramen and rudimentary styloid process lie immediately behind the tympanic ring. With the development of the air-cells the outer part of the mastoid portion grows downwards and forwards to form the mastoid process, and the styloid process and stylo-mastoid foramen now come to lie on the under surface. The descent of the foramen is necessarily accompanied by a corresponding lengthening of the aqueduct of Fallopius. (4) The downward and forward growth of the mastoid process also pushes forward the tympanic plate, so that the portion of it which formed the original floor of the meatus and contained the foramen of Huschke is ultimately found in the anterior wall. (5) With the gradual increase in size of the petrous portion, the *fossa subarcuata* becomes filled up and almost obliterated.

Articulations.—With five bones : occipital, parietal, sphenoid, inferior maxillary, and malar.

Attachment of Muscles.—To fourteen : to the squamous portion, the Temporal ; to the zygoma, the Masseter ; to the mastoid portion, the Occipito-frontalis, Sterno-mastoid, Splenius capitis, Trachelo-mastoid, Digastricus, and Retrahens auriculam ; to the styloid process, the Stylo-pharyngeus, Stylo-hyoid, and Styloglossus ; and to the petrous portion, the Levator palati, Tensor tympani, and Stapedius.

THE SPHENOID BONE

The **Sphenoid Bone** (σφήν, *a wedge*) is situated at the anterior part of the base of the skull, articulating with all the other cranial bones, and binding them firmly and solidly together. In its form it somewhat resembles a bat with its wings extended ; and is divided into a central portion or body, two greater and two lesser wings extending outwards on each side of the body, and two processes— the pterygoid processes—which project from it below.

The **body** is of large size, and hollowed out in its interior so as to form a mere shell of bone. It presents for examination *four* surfaces—a superior, an inferior, an anterior, and a posterior.

The **Superior Surface** (fig. 228).—In front is seen a prominent spine, the *ethmoidal spine,* for articulation with the cribriform plate of the ethmoid ; behind this a smooth surface presenting, in the median line, a slight longitudinal eminence, with a depression on each side, for lodging the olfactory lobes of the brain. This surface is bounded behind by a ridge, which forms the anterior border of a narrow, transverse groove, the *optic groove,* above and behind which lies the optic commissure ; the groove terminates on either side in the *optic foramen,* for the passage of the optic nerve and ophthalmic artery. Behind the optic groove is a small eminence, olive-like in shape, the *olivary process*; and still more posteriorly, a deep depression, the *pituitary fossa*, or *sella turcica*, which lodges the pituitary body. This fossa is perforated by numerous foramina, for the transmission of nutrient vessels into the substance of the bone. It is bounded in front by two small eminences, one on either side, called the *middle clinoid processes* (κλίνη, *a bed*), which are sometimes connected by a spiculum of bone to the anterior clinoid processes (see below), and behind by a square-shaped plate of bone, the *dorsum ephippii* or *dorsum sellæ*, terminating at its superior angles in two tubercles, the *posterior clinoid processes*, the size and form of which vary considerably in different individuals. These processes deepen the pituitary fossa, and serve for the attachment of prolongations from the tentorium cerebelli. The sides of the dorsum ephippii are notched for the passage of the sixth pair of nerves, and below present a sharp process, the *petrosal process*, which is joined to the apex of the petrous portion of the temporal bone, forming the inner boundary of the middle lacerated foramen. Behind the dorsum sellæ, the bone presents a shallow depression, which slopes obliquely backwards, and is continuous with the basilar groove of the occipital bone; it is called the *clivus*, and supports the upper part of the pons Varolii. On either side of the body is a

FIG. 228.—Sphenoid bone. Superior surface.

broad groove, curved something like the italic letter *f*; it lodges the internal carotid artery and the cavernous sinus, and is called the *carotid* or *cavernous groove*. Along the outer margin of this groove, at its posterior part, is a ridge of bone, in the angle between the body and greater wing, called the *lingula*. The **posterior surface**, quadrilateral in form, is joined to the basilar process of the occipital bone. During childhood these bones are separated by a layer of cartilage; but in after-life (between the eighteenth and twenty-fifth years) this becomes ossified, ossification commencing above and extending downwards; and the two bones then form one piece. The **anterior surface** (fig. 229) presents, in the middle line, a vertical ridge of bone, the *sphenoidal crest*, which articulates with the perpendicular plate of the ethmoid, forming part of the septum of the nose. On either side of it are irregular openings leading into the *sphenoidal air-cells* or *sinuses*. These are two large, irregular cavities hollowed out of the interior of the body of the sphenoid bone, and separated from one another by a more or less complete perpendicular bony septum. Their form and size vary considerably; * they are seldom symmetrical, and are often partially subdivided by irregular osseous laminæ. Occasionally, they extend into the basilar process of the occipital nearly as far as the foramen magnum. The septum is seldom

---

* Aldren Turner (*op. cit.*) gives the following as their average measurements: vertical height, $\frac{7}{8}$ in.; antero-posterior depth, $\frac{7}{8}$ in.; transverse breadth, $\frac{3}{4}$ in.

quite vertical, being commonly bent to one or the other side. These sinuses begin to be developed in the third year, and are of a considerable size by the age of six. They are partially closed, in front and below, by two thin, curved plates of bone, the *sphenoidal turbinated* bones, leaving a round opening at their upper parts, by which they communicate with the upper and back part of the nose and occasionally with the posterior ethmoidal cells or sinuses. The lateral margin of this surface presents a serrated edge, which articulates with the os planum of the ethmoid, completing the posterior ethmoidal cells; the lower margin, also rough and serrated, articulates with the orbital process of the palate bone; and the upper margin with the orbital plate of the frontal bone. The **inferior surface** presents, in the middle line, a triangular spine, the *rostrum*, which is continuous with the sphenoidal crest on the anterior surface, and is received into a deep fissure between the alæ of the vomer. On each side may be seen a projecting lamina of bone, which runs horizontally inwards from near the base of the pterygoid process: these laminæ are termed the *vaginal processes*, and articulate with the edges of the vomer. Close to the root of the pterygoid process is a groove, formed into a complete canal when articulated with the

Fig. 229.—Sphenoid bone.   Anterior surface.*

sphenoidal process of the palate bone; it is called the *pterygo-palatine canal*, and transmits the pterygo-palatine vessels and pharyngeal nerve.

The **Greater Wings** are two strong processes of bone, which arise from the sides of the body, and are curved in a direction upwards, outwards and backwards and prolonged behind into a sharp-pointed extremity, the *spinous process* of the sphenoid. Each wing presents three surfaces and a circumference. The **superior** or **cerebral surface** (fig. 228) forms part of the middle fossa of the skull; it is deeply concave, and presents depressions for the convolutions of the brain. At its anterior and internal part is seen a circular aperture, the *foramen rotundum*, for the transmission of the second division of the fifth nerve. Behind and external to this is a large, oval foramen, the *foramen ovale*, for the transmission of the third division of the fifth nerve, the small meningeal artery, and sometimes the small superficial petrosal nerve.† At the inner side of the foramen ovale, a small aperture, the *foramen Vesalii*, may occasionally be seen opposite the root of the pterygoid process; it opens near the scaphoid fossa, and transmits a small vein from the cavernous sinus. Lastly, in the posterior angle, near to the spine of the sphenoid, is a short canal, sometimes double,

* In this figure, both the anterior and inferior surfaces of the body of the sphenoid bone are shown, the bone being held with the pterygoid processes almost horizontal.

† The small petrosal nerve sometimes passes through a special canal (*canaliculus innominatus* of Arnold) situated on the inner side of the foramen spinosum, the foramen ovale.

the *foramen spinosum*; it transmits the middle meningeal artery and vein and a recurrent branch from the third division of the fifth cranial nerve. The **external surface** (fig. 229) is convex, and divided by a transverse ridge, the *pterygoid ridge* or *infratemporal crest*, into two portions. The superior or larger, convex from above downwards, concave from before backwards, enters into the formation of the temporal fossa, and gives attachment to part of the Temporal muscle. The inferior portion, smaller in size and concave, enters into the formation of the zygomatic fossa, and affords attachment to the External pterygoid muscle. It presents the opening of the foramen ovale, and, at its posterior part, a sharp-pointed eminence of bone, the *spinous process*, which is perforated by the foramen spinosum and frequently grooved on its inner aspect for the chorda tympani nerve. To the spinous process are connected the internal lateral ligament of the mandible and the Tensor palati muscle. The cartilaginous part of the Eustachian tube is supported by the inner surface of the spine of the sphenoid. The pterygoid ridge, dividing the temporal and zygomatic portions, gives attachment to part of the External pterygoid muscle. At its inner and anterior extremity is a triangular spine of bone, which serves to increase the extent of origin of this muscle. The **anterior** or **orbital surface**, smooth, and quadrilateral in shape, forms the greater part of the outer wall of the orbit. It is bounded above by a serrated edge, for articulation with the frontal bone; below, by a rounded border, which enters into the formation of the spheno-maxillary fissure. Internally, it presents a sharp border, which forms the lower boundary of the sphenoidal fissure and has projecting from about its centre a little tubercle of bone, which gives attachment to the inferior common ligament of origin of the muscles of the eyeball; and at its upper part is a notch for the transmission of a recurrent branch of the lachrymal artery; externally, it presents a serrated margin for articulation with the malar bone. At the inner and lower part of the anterior surface, immediately below the inner end of the sphenoidal fissure, is a grooved surface, which presents the opening of the foramen rotundum and forms the posterior wall of the spheno-maxillary fossa. *Circumference of the great wing* (fig. 228): commencing from behind; that portion of the circumference which extends from the body of the sphenoid to the spine is serrated and articulates by its outer half with the petrous portion of the temporal bone; while the inner half forms the anterior boundary of the foramen lacerum medium, and presents the posterior aperture of the Vidian canal for the passage of the Vidian nerve and artery. In front of the spine the circumference of the great wing presents a serrated edge, bevelled at the expense of the inner table below, and of the outer table above, which articulates with the squamous portion of the temporal bone. At the tip of the great wing a triangular portion is seen, bevelled at the expense of the internal surface, for articulation with the anterior inferior angle of the parietal bone; this region is named the *pterion*. Internal to this is a triangular, serrated surface, for articulation with the frontal bone: this surface is continuous internally with the sharp inner edge of the orbital plate, which assists in the formation of the sphenoidal fissure, and externally with the serrated margin for articulation with the malar bone.

The **Lesser Wings** (*processes of Ingrassias*) are two thin, triangular plates of bone, which arise from the upper and lateral parts of the body of the sphenoid, and, projecting transversely outwards, terminate in sharp points (fig. 228). The superior surface of each is smooth, flat, broader internally than externally, and supports part of the frontal lobe of the brain. The inferior surface forms the back part of the roof of the orbit, and the upper boundary of the sphenoidal fissure or foramen lacerum anterius. It gives attachment, close to the margin of the optic foramen, to the Levator palpebræ superioris muscle. The *sphenoidal fissure* is of a triangular form, and leads from the cavity of the cranium into the orbit: it is bounded internally by the body of the sphenoid; above, by the lesser wing; below, by the internal margin of the orbital surface of the great wing; and is converted into a foramen by the articulation of this bone with the frontal. It transmits the third, the fourth and the sixth nerves, the three branches of the ophthalmic division of the fifth nerve, some filaments from the cavernous plexus of the sympathetic, the orbital branch of the middle meningeal artery, a recurrent branch from the lachrymal artery to the dura mater, and the ophthalmic vein. The anterior border of the lesser wing is serrated for articulation with the frontal bone ; the posterior, smooth and rounded, is received into the Sylvian fissure of the brain. The inner extremity of this border forms the *anterior clinoid process*, to which the tentorium cerebelli is attached. The lesser wing is connected to the side of the body by two roots, the upper thin and flat, the lower thicker, obliquely directed, and presenting on its outer side, near its junction with the body, a small tubercle, for the attachment of the superior common tendon of origin of three of the muscles of the eye. Between the two roots is the *optic foramen*, for the transmission of the optic nerve and ophthalmic artery.

The **Pterygoid Processes** (πτέρυξ, *a wing* ; εἶδος, *likeness*), one on each side, descend perpendicularly from the point where the body and greater wing unite (fig. 230). Each process consists of an external and an internal plate, which are joined together by their anterior borders above, but are separated below, leaving an angular cleft, the *pterygoid notch*, in which the pterygoid process or tuberosity of the palate bone is received. The two plates diverge from each other from their line of connection in front, so as to form a **V**-shaped fossa, the *pterygoid fossa*. The *external pterygoid plate* is broad and thin, and turned a little outwards ; its outer surface forms part of the inner wall of the zygomatic fossa, and gives attachment to the External pterygoid ; its inner surface forms part of the pterygoid fossa, and gives attachment to the Internal pterygoid. The *internal pterygoid plate* is much narrower and longer, curving outwards, at its extremity, into a hook-like process of bone, the *hamular process*, around which turns the tendon of the Tensor palati muscle. The outer surface of this plate forms part of the pterygoid fossa, the inner surface constitutes the outer boundary of the posterior aperture of the nares. On the posterior surface of the base of the process, above the pterygoid fossa, is a small, oval, shallow depression, the *scaphoid fossa*, from which

FIG. 230.—Sphenoid bone. Posterior surface.

arises the Tensor palati, and above which is seen the posterior orifice of the Vidian canal. Below and to the inner side of the Vidian canal, on the posterior surface of the base of the internal plate, is a little prominence, which is known by the name of the *pterygoid tubercle*. The pharyngeal aponeurosis is attached to the posterior edge of the internal plate, and to its lower third the Superior constrictor of the pharynx. The anterior surface of the pterygoid process is very broad at its base, and forms the posterior wall of the spheno-maxillary fossa. It supports Meckel's ganglion. It presents, above, the anterior orifice of the Vidian canal ; and below, a rough margin, which articulates with the perpendicular plate of the palate bone.

The **Sphenoidal Spongy Bones** are two thin, curved plates of bones, which exist as separate pieces until puberty, and occasionally are not joined to the sphenoid in the adult. They are situated at the anterior and inferior part of the body of the sphenoid, an aperture of variable size being left in their anterior wall, through which the sphenoidal sinuses open into the nasal fossæ. They are irregular in form, and taper to a point behind, being broader and thinner in front. Their upper surface, which looks towards the cavity of the sinus, is concave ; their under surface is convex, and forms part of the roof of the nasal fossæ. Each bone articulates in front with the ethmoid, externally with the palate ; its pointed posterior extremity is placed above the vomer, and is received between the root of the pterygoid process on the outer side and the rostrum of the sphenoid on the inner.*

* A small portion of sphenoidal turbinated bone sometimes enters into the formation of the inner wall of the orbit, between the os planum of the ethmoid in front, the orbital plate of the palate below, and the frontal above. Cleland, *Roy. Soc. Trans.* 1862.

Development.—Up to about the eighth month of fœtal life the sphenoid bone consists of two distinct parts : a posterior or *post-sphenoid* part, which comprises the pituitary fossa, the greater wings, and the pterygoid processes ; and an anterior or *pre-sphenoid* part, to which the anterior part of the body and lesser wings belong. It is developed by fourteen centres : eight for the post-sphenoid division, and six for the pre-sphenoid (fig. 231). The eight centres for the post-sphenoid are—one for each greater wing and external pterygoid plate, one for each internal pterygoid plate, two for the posterior part of the body, and one on each side for the lingula. The six for the pre-sphenoid are one for each lesser wing, two for the anterior part of the body, and one for each sphenoidal turbinated bone.*

FIG. 231.—Plan of the development of sphenoid. By fourteen centres.

one for each lesser wing — *two for anterior part of body*

one for each internal pterygoid plate — *one for each lingula*

one for each great wing and external pterygoid plate

one for each Sphenoidal turbinated bone

Post-sphenoid Division.—The first nuclei to appear are those for the greater wings (*ali-sphenoids*). They make their appearance between the foramen rotundum and foramen ovale about the eighth week, and from them the external pterygoid plates are also formed.† Soon after, the nuclei for the posterior part of the body appear, one on either side of the sella turcica, and become blended together about the middle of fœtal life. The internal pterygoid plates are ossified in membrane, and their centres probably appear about the ninth or tenth week (Fawcett, *op. cit.*) ; they become joined to the external pterygoid plate about the sixth month. The remaining centres appear about the fourth month, and those for the lingulæ speedily become joined to the rest of the bone.

Pre-sphenoid Division.—The first nuclei to appear are those for the lesser wings (*orbito-sphenoids*). They make their appearance about the ninth week, at the outer borders of the optic foramina. A second pair of nuclei appear on the inner side of the foramina shortly after, and becoming united, form the front part of the body of the bone. The remaining two centres for the sphenoidal turbinated bones make their appearance about the fifth month. At birth they consist of small triangular laminæ, and it is not till the third year that they become hollowed out and cone-shaped. About the fourth year they become fused with the lateral masses of the ethmoid, i.e. several years before they unite with the sphenoid, and hence, from an embryological point of view, may be regarded as belonging to the ethmoid.

The pre-sphenoid is united to the post-sphenoid about the eighth month, so that at birth the bone consists of three pieces—viz. the body in the centre, and on each side the great wings with the pterygoid processes. The lesser wings become joined to the body at about the time of birth. In the first year after birth the greater wings and body are united. From the tenth to the twelfth year the spongy bones are partially united to the sphenoid, their junction being completed by the twentieth year. Lastly, the sphenoid joins the occipital from the eighteenth to the twenty-fifth year. Between the pre- and post-sphenoid divisions there are occasionally seen the remains of a canal (cranio-pharyngeal canal), through which, in early fœtal life, the pituitary diverticulum (or pouch of Rathke) of the buccal ectoderm is transmitted.

The sphenoid has attached to it certain intrinsic ligaments, which have received special names. The most important of these are : the *pterygo-spinous*, which is attached to the spinous process and the internal pterygoid plate (see *cervical fascia*) ; the *spheno-mandibular*, or internal lateral ligament of the lower jaw (see *temporo-mandibular joint*) ; the *interclinoid*, a fibrous process which passes from the anterior to the posterior clinoid process ; and the *carotico-clinoid*,

---

* According to Cleland, the sphenoidal turbinated bones are ossified from four distinct centres.

† Fawcett (*Anatomischer Anzeiger*, March 1905) states that the external pterygoid plate is ossified in membrane, and is not a downward continuation of the cartilaginous great wing.

which passes from the anterior to the middle clinoid process. These ligaments occasionally ossify, and form adventitious foramina.

Articulations.—The sphenoid articulates with *all* the bones of the cranium, and five of the face—the two malar, two palate, and vomer: the exact extent of articulation with each bone is shown in the accompanying figures.*

Attachment of Muscles.—To eleven pairs: the Temporal, External pterygoid, Internal pterygoid, Superior constrictor, Tensor palati, Levator palpebræ, and Superior oblique, Superior rectus, Internal rectus, Inferior rectus, External rectus of the eyeball.

The Ethmoid Bone

The **Ethmoid** (ἠθμός, *a sieve*) is an exceedingly light, spongy bone, of a cubical shape, situated at the anterior part of the base of the cranium, between the two orbits, at the root of the nose, and contributing to each of these cavities. It consists of three parts: a horizontal plate, which forms part of the base of the cranium; a perpendicular plate, which constitutes part of the septum nasi; and two lateral masses of cells.

The **Horizontal** or **Cribriform Plate** (fig. 232) forms part of the anterior fossa of the base of the skull, and is received into the ethmoid notch of the frontal bone between the two orbital plates, roofing in the nasal fossæ below. Projecting upwards from the middle line of this plate is a thick, smooth, triangular process of bone, the *crista galli*, so called from its resemblance to a cock's comb. Its base joins the cribriform plate. Its posterior border, long, thin, and slightly curved, serves for the attachment of the falx cerebri. Its anterior border, short and thick, articulates with the frontal bone, and presents two small projecting alæ, which are received into corresponding depressions in the frontal, completing the foramen cæcum behind. Its sides are smooth and sometimes bulging; in which case it is found to enclose a small air-sinus.† On each side of the crista galli, the cribriform plate is narrow, and deeply grooved, to support the bulb of the olfactory lobe of the cerebrum, and perforated by foramina for the passage of the olfactory nerves. These foramina are arranged in three rows: the innermost, which are the largest and least numerous, are lost in grooves on the upper part of the septum; the foramina of the outer row are continued on to the surface of the upper spongy bone. The foramina of the middle row are the smallest; they perforate the bone, and transmit nerves to the roof of the nose. At the front part of the cribriform plate, on each side of the crista galli, is a small fissure, which transmits the nasal branch of the ophthalmic nerve; and at its posterior part a triangular notch, which receives the ethmoidal spine of the sphenoid.

FIG. 232.—Ethmoid bone.
Outer surface of right lateral mass (enlarged).

The **Perpendicular Plate** (fig. 233) is a thin, flattened lamella of bone, which descends from the under surface of the cribriform plate, and assists in forming the septum of the nose. It is much thinner in the middle than at the circumference, and is generally deflected a little to one side. Its anterior border articulates with the nasal spine of the frontal bone and the crest of the nasal bones. Its posterior border, divided into two parts, articulates by its upper half

---

* It also sometimes articulates with the tuberosity of the maxilla (see page 210).

† Sir George Humphry states that the crista galli is commonly inclined to one side, usually the opposite to that towards which the lower part of the perpendicular plate is bent.—*The Human Skeleton*, 1858, p. 277.

with the sphenoidal crest of the sphenoid, by its lower half with the vomer. The inferior border serves for the attachment of the septal cartilage of the nose. On each side of the perpendicular plate numerous grooves and canals are seen, leading from foramina on the cribriform plate; they lodge filaments of the olfactory nerves.

FIG. 233.—Perpendicular plate of ethmoid (enlarged). Shown by removing the right lateral mass.

The **Lateral Masses** of the ethmoid consist of a number of thin-walled cellular cavities, the *ethmoidal cells*, interposed between two vertical plates of bone, the outer one of which forms part of the orbit, and the inner one part of the nasal fossa of the corresponding side. In the disarticulated bone many of these cells appear to be broken; but when the bones are articulated, they are closed in at every part, except where they open into the nasal fossæ. The upper surface of each lateral mass presents a number of apparently half-broken cellular spaces; these are closed in, when articulated, by the edges of the ethmoidal notch of the frontal bone. Crossing this surface are two grooves on each side, converted into canals by articulation with the frontal; they are the *anterior* and *posterior ethmoidal* canals, and open on the inner wall of the orbit. The posterior surface also presents large, irregular cellular cavities, which are closed in by articulation with the sphenoidal turbinated bones and orbital process of the palate. The cells at the anterior surface are completed by the lachrymal bone and nasal process of the superior maxillary, and those below by the superior maxillary. The outer surface of each lateral mass is formed of a thin, smooth, oblong plate of bone, called the *os planum*; it forms part of the inner wall of the orbit, and articulates, above, with the orbital plate of the frontal; below, with the superior maxillary; in front, with the lachrymal; and behind, with the sphenoid and orbital process of the palate.

From the inferior part of each lateral mass, immediately behind the os planum, there projects downwards and backwards an irregular lamina of bone, called the *unciform process*, from its hook-like form; it serves to close in the upper part of the orifice of the antrum of Highmore, and articulates with the ethmoidal process of the inferior turbinated bone. It is often broken in disarticulating the bones.

FIG. 234.—Ethmoid bone. Inner surface of right lateral mass (enlarged).

The inner surface of each lateral mass (fig. 234) forms part of the outer wall of the nasal fossa of the corresponding side. It is formed of a thin lamella of bone, which descends from the under surface of the cribriform plate, and terminates below in a free, convoluted margin, the *middle turbinated* bone. The whole of this surface is rough, and marked above by numerous grooves, which run nearly vertically downwards from the cribriform plate; they lodge branches of the olfactory nerve, which are distributed on the mucous membrane covering the superior turbinated bone. The back part of this surface is subdivided by a narrow oblique fissure, the *superior meatus* of the nose, bounded above by a thin, curved plate of bone, the *superior turbinated* bone. By means of an orifice at the upper part of this fissure, the posterior ethmoidal cells open into the nose. Below, and in front of the superior meatus, is seen the convex surface of the middle turbinated bone. It extends along the whole length of the inner surface of each lateral mass; its lower margin is free and thick, and its concavity, directed outwards, assists in forming the middle meatus of the nose. It is by a large orifice at the upper and front part of the middle meatus that the anterior ethmoidal cells, and through them the frontal sinuses, communicate with the nose, by means of a funnel-shaped canal, the *infundibulum*. The cellular cavities of each lateral mass, thus walled in by the os planum on the outer side, and by the other bones already mentioned, are divided by transverse bony partitions into three sets, which do not communicate with each other; they are termed the *anterior, middle,* and *posterior ethmoidal air-cells*. The anterior cells communicate with the frontal sinuses above, and the middle meatus below, by means of a long, flexuous canal, the *infundibulum*; the middle also open into the middle meatus; the posterior open into the superior meatus, and communicate (occasionally) with the sphenoidal sinuses.

Development.—The ethmoid ossifies in the cartilage of the nasal capsule by *three* centres : one for the perpendicular lamella, and one for each lateral mass.

The lateral masses are first developed, ossific granules making their appearance in the os planum between the fourth and fifth months of foetal life, and extending into the spongy bones. At birth, the bone consists of the two lateral masses, which are small and ill developed. During the first year after birth, the perpendicular plate and crista galli begin to ossify, from a single nucleus, and become joined to the lateral masses about the beginning of the second year. The cribriform plate is ossified partly from the perpendicular plate and partly from the lateral masses. The formation of the ethmoidal cells does not commence until about the fourth or fifth year.

Articulations.—With fifteen bones : the sphenoid, two sphenoidal turbinated, the frontal, and eleven of the face—the two nasal, two superior maxillary, two lachrymal, two palate, two inferior turbinated, and the vomer.

No muscles are attached to this bone.

The Fontanelles (figs. 235, 236)

The early stages of the development of the cranium have already been described in the section on development. Before birth, the bones at the vertex and sides of the cranium are separated from each other by membranous intervals, in which bone is deficient. These intervals are principally found at the four angles of the parietal bones; hence there are six of them. Their formation is due to the wave of ossification in the parietal bones being

Fig. 235.—Skull at birth, showing the anterior and posterior fontanelles.

Fig. 236.—The lateral fontanelles.

circular and the bones quadrilateral; the ossific matter first meets at the margins of the bones, at the points nearest to their centres of ossification, and spaces are left at the angles, which are called *fontanelles*, so named from the pulsations of the brain, which are perceptible at the anterior fontanelle, and were likened to the rising of water in a fountain.

The anterior fontanelle is the largest; it is lozenge-shaped, and corresponds to the junction of the sagittal and coronal sutures; the posterior fontanelle, of smaller size, is triangular, and is situated at the junction of the sagittal and lambdoid sutures; the remaining ones are situated at the inferior angles of each parietal bone. The latter are closed soon after birth; the two at the two superior angles remain open longer : the posterior being closed in a few months after birth; the anterior remaining open until the first or second year. These spaces are gradually filled in by an extension of the ossifying process, or by the development of a Wormian bone. Sometimes the anterior fontanelle remains open beyond two years, and is rarely persistent throughout life.

SUPERNUMERARY OR WORMIAN* BONES

In addition to the constant centres of ossification of the cranium, additional ones may be found in the course of the sutures. These form irregular, isolated bones, interposed between the cranial bones, and have been termed *Wormian bones* or *ossa triquetra*. They are most frequently found in the course of the lambdoid suture, but occasionally also occupy the situation of the fontanelles, especially the posterior and, more rarely, the anterior. One, the *pterion ossicle*, is often found between the anterior inferior angle of the parietal bone and the greater wing of the sphenoid (fig. 236). They have a tendency to be more or less symmetrical on the two sides of the skull, and they vary much in size, being in some cases not larger than a pin's head, and confined to the outer table ; in other cases so large that one pair of these bones may form the whole of the occipital bone above the superior curved lines, as described by Béclard and Ward. Their number is generally limited to two or three ; but more than a hundred have been found in the skull of an adult hydrocephalic skeleton. In their development, structure, and mode of articulation, they resemble the other cranial bones.

CONGENITAL FISSURES AND GAPS

An arrest in the ossifying process may give rise to deficiencies, or gaps ; or to fissures which are of importance from a medico-legal point of view, as they are liable to be mistaken for fractures. The fissures generally extend from the margins towards the centre of the bone, but the gaps may be found in the middle as well as at the edges. In course of time they may become filled with a thin lamina of bone.

BONES OF THE FACE

The Facial Bones are fourteen in number—viz. the

Two Nasal.	Two Palate.
Two Superior Maxillary.	Two Inferior Turbinated.
Two Lachrymal.	Vomer.
Two Malar.	Inferior Maxillary or Mandible.

' Of these, the upper and lower jaws are the fundamental bones for mastication, and the others are accessories ; for the chief function of the facial bones is to provide an apparatus for mastication, while subsidiary functions are to provide for the sense-organs (eye, nose, tongue) and a vestibule to the respiratory and vocal organs. Hence the variations in the shape of the face in man and the lower animals depend chiefly on the question of the character of their food and their mode of obtaining it.' †

NASAL BONES

The **Nasal** (nasus, *the nose*) are two small oblong bones, varying in size and form in different individuals ; they are placed side by side at the middle and upper part of the face, forming, by their junction, ' the bridge' of the nose (fig. 237). Each bone presents for examination two surfaces and four borders. The *outer* surface (fig. 238) is concavo-convex from above downwards, convex from side to side; it is covered by the Pyramidalis and Compressor nasi muscles, and gives attachment at its upper part to a few fibres of the Occipito-frontalis muscle (Theile). It is marked by numerous small arterial furrows, and perforated about its centre by a foramen, sometimes double, for the transmission of

---

* Wormius, a physician in Copenhagen, is said to have given the first detailed description of these bones.

† W. W. Keen. American edition.

a small vein. The *inner* surface (fig. 239) is concave from side to side, and is traversed from above downwards by a longitudinal groove (sometimes a canal), for the passage of a branch of the nasal nerve. The *superior* border is narrow, thick, and serrated for articulation with the nasal notch of the frontal bone. The *inferior* border is broad, thin, sharp, inclined obliquely downwards, outwards, and backwards, and serves for the attachment of the upper lateral cartilage of the nose. It presents, about its middle, a notch, through which

FIG. 237.—Nasal bone and Lachrymal bone *in situ*.

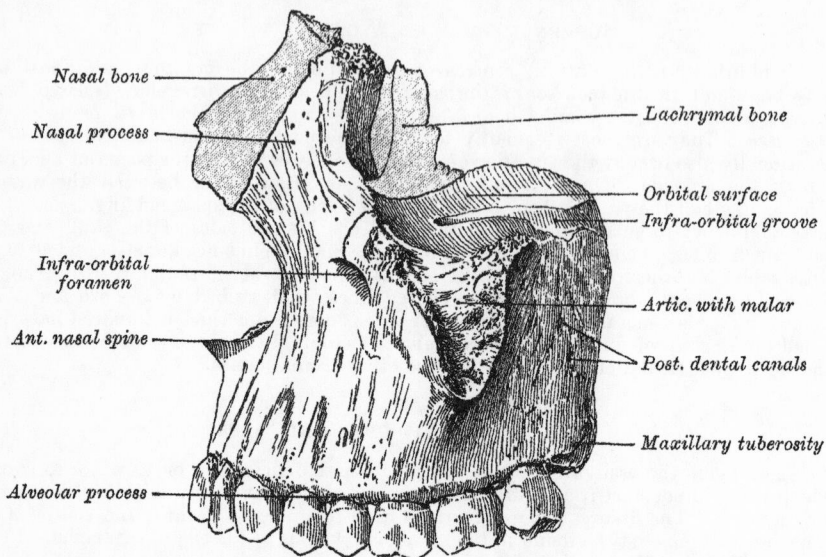

passes the branch of the nasal nerve above referred to ; and is prolonged at its inner extremity into a sharp spine, which, when articulated with the opposite bone, forms the *nasal angle*. The *external* border is serrated, bevelled at the expense of the internal surface above, and of the external below, to articulate with the nasal process of the superior maxillary. The *internal* border, thicker above than below, articulates with its fellow of the opposite side, and is prolonged behind into a vertical crest, which forms part of the septum of the nose :

FIG. 238.—Right nasal bone.   FIG. 239.—Left nasal bone.
Outer surface.   Inner surface.

this crest articulates, from above downwards, with the nasal spine of the frontal, the perpendicular plate of the ethmoid, and the septal cartilage of the nose.

Development.—By *one* centre for each bone, which appears about the eighth week.

Articulations.—With four bones : two of the cranium, the frontal and ethmoid, and two of the face, the opposite nasal and the superior maxillary.

Attachment of Muscles.—A few fibres of the Occipito-frontalis muscle.

SUPERIOR MAXILLARY BONES OR MAXILLÆ

The **Superior Maxillary Bones** (maxilla, *the jaw-bone*) are the most important bones of the face from a surgical point of view, on account of the number of diseases to which some of their parts are liable. Their careful examination is, therefore, a matter of considerable interest. They are the largest bones of the face, excepting the mandible, and form, by their union, the whole of the upper jaw. Each bone assists in the formation of three cavities, viz.: the roof of the mouth, the floor and outer wall of the nasal fossæ, and the floor of the orbit; it also enters into the formation of two fossæ, the zygomatic and spheno-maxillary, and two fissures, the spheno-maxillary and pterygo-maxillary.

The bone presents for examination a body and four processes—malar, nasal, alveolar, and palate.

The **body** is somewhat pyramidal in shape, and is hollowed out in its interior to form a large cavity, the *antrum of Highmore*, or *maxillary sinus*. Its surfaces are four—an external or facial, a posterior or zygomatic, a superior or orbital, and an internal or nasal.

The **external** or **facial surface** (fig. 240) is directed forwards and outwards. It presents at its lower part a series of eminences corresponding to the position

FIG. 240.—Left superior maxillary bone. Outer surface.

of the fangs of the teeth. Just above those of the incisor teeth is a depression, the *incisive* or *myrtiform fossa*, which gives origin to the Depressor alæ nasi; and, below it, to the alveolar border is attached a slip of the Orbicularis oris. Above and a little external to it, the Compressor nasi arises. More external is another depression, the *canine fossa*, larger and deeper than the incisive fossa, from which it is separated by a vertical ridge, the *canine eminence*, corresponding to the socket of the canine tooth. The canine fossa gives origin to the Levator anguli oris. Above the canine fossa is the *infra-orbital foramen*, the termination of the infra-orbital canal; it transmits the infra-orbital vessels and nerve. Sometimes the infra-orbital canal opens by two, very rarely by three, orifices on the face. Above the infra-orbital foramen is the margin of the orbit, which affords partial attachment to the Levator labii superioris proprius. Internally, the facial surface is limited by a deep concavity, the *nasal notch*, to the sharp margin of which is attached the Dilatator naris posterior.

P

The **posterior** or **zygomatic surface** is convex, directed backwards and outwards, and forms part of the zygomatic fossa. It is separated from the facial surface by the malar process and by a strong ridge of bone, which extends upwards from the socket of the second molar tooth. It presents about its centre several apertures leading to canals in the substance of the bone ; they are termed the *posterior dental canals*, and transmit the posterior dental vessels and nerves. At the lower part of this surface is a rounded eminence, the *maxillary tuberosity*, especially prominent after the growth of the wisdom-tooth, rough on its inner side for articulation with the tuberosity of the palate-bone and sometimes with the external pterygoid plate. It gives attachment to a few fibres of origin of the Internal pterygoid muscle. Immediately above this is a smooth surface, which forms the anterior boundary of the spheno-maxillary fossa ; from this fossa a groove, for the lodgment of the second division of the fifth nerve, extends outwards and becomes continuous with the infra-orbital groove on the orbital surface of the bone ; a second groove runs obliquely downwards, and is converted into a canal by articulation with the palate-bone, forming the *posterior palatine* canal for the large palatine nerve and descending palatine artery.

The **superior** or **orbital surface** is thin, smooth, triangular, and forms part of the floor of the orbit. It is bounded internally by an irregular margin which in front presents a notch, the *lachrymal notch*, for the reception of the lachrymal bone ; in the middle it articulates with the os planum of the ethmoid, and behind with the orbital process of the palate-bone ; it is bounded behind by a smooth, rounded edge which enters into the formation of the spheno-maxillary fissure, and which sometimes articulates at its outer extremity with the orbital plate of the sphenoid ; and it is bounded in front by part of the circumference of the orbit, which is continuous, on the inner side with the nasal, on the outer side with the malar process of the bone. Near the middle line of the orbital surface is a deep groove, the *infra-orbital*, for the passage of the infra-orbital vessels and nerve. The groove commences at the middle of the outer border of this surface, where it is continuous with that near the upper edge of the posterior surface, and, passing forwards, terminates in a canal, which subdivides into two branches. One of the canals, the *infra-orbital*, opens just below the margin of the orbit ; the other, which is smaller, runs downwards in the substance of the anterior wall of the antrum ; it is called the *anterior dental canal*, and transmits the anterior dental vessels and nerve to the front teeth of the upper jaw. From the back part of the infra-orbital canal, a second small canal is sometimes given off, which runs downwards in the outer wall of the antrum, and conveys the middle dental nerve to the bicuspid teeth. Occasionally, this canal is derived from the anterior dental. At the inner and fore part of the orbital surface, just external to the lachrymal groove for the nasal duct, is a depression, which gives origin to the Inferior oblique muscle of the eye.

The **internal surface** (fig. 241) is unequally divided into two parts by a horizontal projection of bone, the *palate process* : the portion above the palate process forms part of the outer wall of the nasal fossæ ; that below it forms part of the cavity of the mouth. The superior division of this surface presents a large, irregular opening leading into the *antrum of Highmore*. At the upper border of this aperture are numerous broken cellular cavities, which, in the articulated skull, are closed in by the ethmoid and lachrymal bones. Below the aperture is a smooth concavity which forms part of the inferior meatus of the nasal fossæ, and behind it is a rough surface for articulation with the perpendicular plate of the palate bone and is traversed by a groove which, commencing near the middle of the posterior border, runs obliquely downwards and forwards, and forms, when completed by its articulation with the palate-bone, the *posterior palatine canal*. In front of the opening of the antrum is a deep groove, converted into a canal by the lachrymal and inferior turbinated bones. It is called the *lachrymal groove*, and lodges the nasal duct. More anteriorly is a well-marked rough ridge, the *inferior turbinated crest*, for articulation with the inferior turbinated bone. The shallow concavity above this ridge forms part of the middle meatus of the nose ; while that below it forms part of the inferior meatus. The portion of this surface below the palate process is concave, rough and uneven, and perforated by numerous small foramina for the passage of nutrient vessels. It enters into the formation of the roof of the mouth.

The **Antrum of Highmore,** or **Maxillary Sinus,** is a large pyramidal cavity, hollowed out of the body of the maxillary bone : its apex, directed outwards, is formed by the malar process ; its base, by the outer wall of the nose. Its walls are everywhere exceedingly thin, and correspond to the orbital, facial, and zygomatic surfaces of the body of the bone. Its inner wall, or base, presents, in the disarticulated bone, a large, irregular aperture, which communicates with the nasal fossa. The margins of this aperture are thin and ragged, and the aperture itself is much contracted by the articulation of its margins with the ethmoid above, the inferior turbinated below, and the palate-bone behind.* In the articulated skull, this cavity communicates with the middle meatus of the nasal fossa, generally by two small apertures left between the above-mentioned bones. In the recent state, usually only one small opening exists, near the upper part of the cavity, sufficiently large to admit the end of a probe ; the other being closed by mucous membrane.

The infra-orbital canal usually projects into the cavity of the antrum as a well-marked ridge, which passes downwards and inwards from the roof to the

FIG. 241.—Left superior maxillary bone.   Inner surface.

anterior wall, and additional ridges are sometimes seen in the cavity. On its posterior wall are the *posterior dental canals*, transmitting the posterior dental vessels and nerves to the teeth. The floor is formed by the alveolar process of the jaw, and, in a cavity of average size, is on a level with the floor of the nose ; where the cavity is large it reaches below this level. Projecting into it are several conical processes, corresponding to the roots of the first and second molar teeth ;† in some cases it is perforated by the teeth in this situation. The size of the cavity varies in different skulls, and even on the two sides of the same skull.‡

* In some cases, at any rate, the lachrymal bone also encroaches slightly on the anterior superior portion of the opening, and assists in forming the inner wall of the antrum.

† The number of teeth whose fangs are in relation with the floor of the antrum is variable. The antrum 'may extend so as to be in relation to all the teeth of the true maxilla, from the canine to the *dens sapientiæ*.'—See Salter on Abscess of the Antrum, in *A System of Surgery*, edited by T. Holmes, 2nd edit. vol. iv. p. 356.

‡ Aldren Turner (*op. cit.*) gives the following measurements as those of an average-sized antrum : vertical height opposite first molar tooth, $1\frac{1}{2}$ in. ; transverse breadth, I in. ; and antero-posterior depth, $1\frac{1}{4}$ in.

P 2

The extreme thinness of the walls of this cavity enables one to explain how a tumour growing from the antrum encroaches upon the adjacent parts, pushing up the floor of the orbit, and displacing the eyeball, projecting inwards into the nose, protruding forwards on to the cheek, and making its way backwards into the zygomatic fossa, and downwards into the mouth.

The **Malar Process** is a rough triangular eminence, situated at the angle of separation of the facial from the zygomatic surface. In front it forms part of the facial surface; behind, it is concave, and forms part of the zygomatic fossa; above, it is rough and serrated for articulation with the malar bone; while below, a prominent ridge marks the division between the facial and zygomatic surfaces. A small part of the Masseter muscle arises from this process.

The **Nasal Process** is a strong, triangular plate of bone, which projects upwards, inwards, and backwards, by the side of the nose, forming part of its lateral boundary. Its *external* surface is concave, smooth, perforated by numerous foramina, and gives attachment to the Levator labii superioris alæque nasi, the Orbicularis palpebrarum, and Tendo oculi. Its *internal* surface forms part of the outer wall of the nasal fossa: at its upper part it presents a rough, uneven surface, which articulates with the ethmoid bone, closing in the anterior ethmoidal cells; below this is a transverse ridge, the *superior turbinated crest*, for articulation with the middle turbinated bone of the ethmoid, bounded below by a shallow smooth concavity which forms part of the middle meatus; below this again is the inferior turbinated crest (already described), where the nasal process joins the body of the bone. Its *upper* border articulates with the frontal bone. Its *anterior* border is thin, directed obliquely upwards and forwards, and presents a serrated edge for articulation with the nasal bone; its *posterior* border is thick, and hollowed into a groove, the *lachrymal groove*, for the nasal duct: of the two margins of this groove, the inner articulates with the lachrymal bone, the outer forms part of the circumference of the orbit. Where the latter joins the orbital surface is a small tubercle, the *lachrymal tubercle*; this serves as a guide to the position of the lachrymal sac in the operation for fistula lachrymalis. The lachrymal groove in the articulated skull is converted into a canal by the lachrymal bone and lachrymal process of the inferior turbinated bone; it is directed downwards, and a little backwards and outwards, is about the diameter of a goose-quill, slightly narrower in the middle than at either extremity, and terminates below in the inferior meatus. It lodges the nasal duct.

The **Alveolar Process** is the thickest and most spongy part of the bone, broader behind than in front, and excavated into deep cavities for the reception of the teeth. These cavities are eight in number, and vary in size and depth according to the teeth they contain. That for the canine tooth is the deepest; those for the molars are the widest, and subdivided into minor cavities by septa; those for the incisors are single, but deep and narrow. The Buccinator muscle arises from the outer surface of this process, as far forward as the first molar tooth. When the superior maxillary bones are articulated with each other, their alveolar processes together form the *alveolar arch*; the centre of the anterior margin of this arch is named the *alveolar point*.

The **Palate Process**, thick and strong, projects horizontally inwards from the inner surface of the bone. It is much thicker in front than behind, and forms a considerable part of the floor of the nose and the roof of the mouth.

Its *inferior* surface (fig. 242) is concave, rough and uneven, and forms, when articulated with the palatal process of the opposite bone, the anterior three-fourths of the hard palate. This surface is perforated by numerous foramina for the passage of the nutrient vessels; channelled at the back part of its alveolar border by a longitudinal groove, sometimes a canal, for the transmission of the posterior palatine vessels, and the anterior or large palatine nerve from Meckel's ganglion; and presents little depressions for the lodgment of the palatine glands. When the two superior maxillary bones are articulated together, a large orifice may be seen in the middle line, immediately behind the incisor teeth. This is the *anterior palatine canal* or *fossa*. On examining the bottom of this fossa four canals are seen: two branch off laterally to the right and left nasal fossæ, and two lie in the middle line. The lateral canals are named the *foramina of Stenson*, and through each of them passes the anterior or terminal branch of the descending or posterior palatine arteries, which ascends from the mouth to the nasal fossæ, and in them is lodged the remains of the organ of Jacobson. The canals in the middle line are termed the *foramina of Scarpa*, and transmit the naso-palatine nerves, the left passing through the anterior, and the right through the posterior canal. On the palatal surface of the process, a delicate linear suture, well seen in young skulls, may sometimes be noticed extending from the anterior palatine fossa to the interval between the lateral incisor and the canine tooth. This marks out the *intermaxillary*, or *incisive bone*, which in some animals exists permanently as a separate piece. It includes the whole thickness of the alveolus, the corresponding part of the floor of the nose and the anterior nasal spine, and contains the sockets of the incisor teeth. The *upper* surface is concave from side to side, smooth, and forms part of the floor of the nose. It presents the upper orifices of the foramina of Stenson and Scarpa, the former being on either side of the middle line, the latter in the intermaxillary suture, and therefore not visible unless the two bones are placed in apposition. The *outer* border of the palate process is incorporated with the rest of the bone. The *inner* border is thicker in front than

FIG. 242.—The palate and alveolar arch.

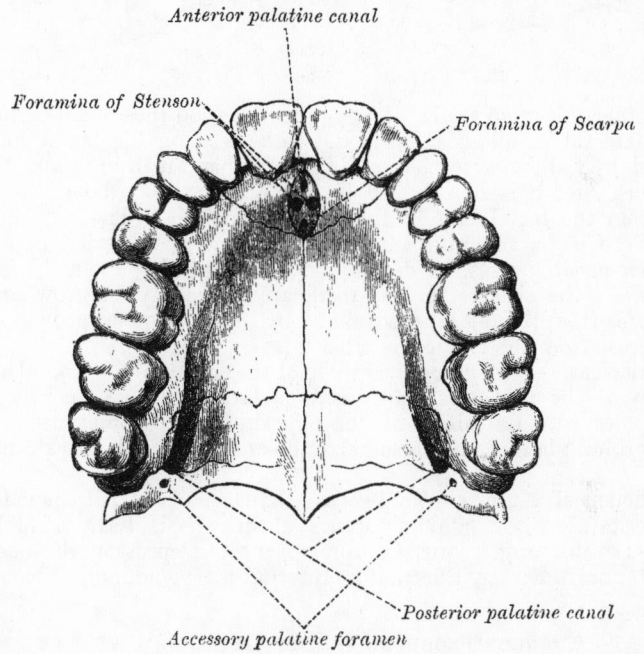

Anterior palatine canal

Foramina of Stenson

Foramina of Scarpa

Posterior palatine canal

Accessory palatine foramen

behind, and is raised above into a ridge, the *nasal crest*, which, with the corresponding ridge in the opposite bone, forms a groove for the reception of the vomer. In front, this ridge rises to a considerable height, and this portion is named the *incisor crest*. The *anterior* margin is bounded by the thin, concave border of the opening of the nose prolonged forwards internally into a sharp process, forming, with a similar process of the opposite bone, the *anterior nasal spine*. The *posterior* border is serrated for articulation with the horizontal plate of the palate-bone. The middle of the inferior border of the anterior nasal aperture at the base of the nasal spine is named the *subnasal point*.

Development.—This bone commences to ossify at a very early period, and ossification proceeds in it with such rapidity that it is difficult to ascertain with certainty its precise number of centres. It appears probable, however, that it is ossified from six centres, which are deposited in membrane. One, the *orbito-nasal*, forms that portion of the body of the bone which lies internal to the infra-orbital canal, including the floor of the orbit and the outer wall of the nasal fossa; a second, the *malar*, gives origin to that portion of the bone which lies external to the infra-orbital canal and the malar process; from a third, the *palatine*, is developed the palatine process posterior to Stenson's canal and the adjoining part of the nasal wall ; a fourth, the *pre-maxillary*, forms the front part of the alveolus which carries the incisor teeth and corresponds to the pre-maxillary bone of the lower animals ;* a fifth, the *nasal*, gives rise to the nasal process and the portion of the bone above the canine tooth ; and a sixth, the *infravomerine*, lies between the palatine and pre-maxillary centres and beneath the vomer ; this centre, together with the corresponding centre of the opposite bone, separates the foramina of Stenson from each other. These centres appear about the eighth week, and by the tenth week the three first-named have fused together and the bone consists of two portions, one the maxilla proper,

FIG. 243.—Development of superior maxillary bone.   At birth.

ANTERIOR SURFACE   INFERIOR SURFACE

and the other the pre-maxilla. The suture between these two portions persists on the palate till middle life, but is not to be seen on the facial surface. This is believed by Callender to be due to the fact that the front wall of the sockets of the incisor teeth is not formed by the pre-maxillary bone, but by an out-growth from the facial part of the superior maxilla. The antrum appears as a shallow groove on the inner surface of the bone at an earlier period than any of the other nasal sinuses, its development commencing about the fourth month of fœtal life. The sockets for the teeth are formed by the growing downwards of two plates from the dental groove, and by the subsequent growth of partitions jutting across from the one to the other.

Articulations.—With *nine* bones : two of the cranium, the frontal and ethmoid, and seven of the face—viz. the nasal, malar, lachrymal, inferior turbinated, palate, vomer, and its fellow of the opposite side. Sometimes it articulates with the orbital plate of the sphenoid, and sometimes with its external pterygoid plate.

Attachment of Muscles.—To twelve : the Orbicularis palpebrarum, Obliquus oculi inferior, Levator labii superioris alæque nasi, Levator labii superioris proprius, Levator anguli oris, Compressor nasi, Depressor alæ nasi, Dilatator naris posterior, Masseter, Buccinator, Internal pterygoid, and Orbicularis oris.

CHANGES PRODUCED IN THE UPPER JAW BY AGE

At birth the transverse and antero-posterior diameters of the bone are greater than the vertical. The nasal process is well marked and the body of the bone consists of little more than the alveolar process, while the teeth-sockets reach almost to the floor of the orbit. The antrum of Highmore presents the appearance of a slit-like furrow on the outer wall of the nose. In the adult the vertical diameter is the greater, owing to the development of the alveolar process and the increase in size of the antrum. In old age the bone approaches again in character to the infantile condition : its height is diminished, and after the loss of the teeth the alveolar process is absorbed, and the lower part of the bone contracted and diminished in thickness.

THE LACHRYMAL BONES

The **Lachrymal** (lachryma, *a tear*) are the smallest and most fragile bones of the face. They are situated at the front part of the inner wall of the orbit (fig. 237), and resemble somewhat in form, thinness, and size, a finger-nail ; hence they are termed the *ossa unguis*. Each bone presents for examination

---

* Some anatomists believe that the pre-maxillary bone is ossified by two centres (see page 250).

two surfaces and four borders. The *external* or *orbital* surface (fig. 244) is divided by a vertical ridge, the *lachrymal crest*, into two parts. The portion of bone in front of this ridge presents a smooth, concave, longitudinal groove, the free margin of which unites with the nasal process of the superior maxillary bone, completing the lachrymal groove. The upper part of this groove lodges the lachrymal sac; the lower part the nasal duct. The portion of bone behind the ridge is smooth, slightly concave, and forms part of the inner wall of the orbit. The ridge, with a part of the orbital surface immediately behind it, affords attachment to the Tensor tarsi muscle: it terminates below in a small, hook-like projection, the *hamular process*, which articulates with the lachrymal tubercle of the superior maxillary bone, and completes the upper orifice of the lachrymal groove. It sometimes exists as a separate piece, which is then called the *lesser lachrymal bone*. The *internal* or *nasal* surface presents a depressed furrow, corresponding to the ridge on its outer surface. The surface of bone in front of this forms part of the middle meatus of the nose; and that behind it articulates with the ethmoid bone, filling in the anterior ethmoidal cells. Of the *four borders* the *anterior* is the longest, and articulates with the nasal process of the superior maxillary bone. The *posterior*, thin and uneven, articulates with the os planum of the ethmoid. The *superior*, the shortest and thickest, articulates with the internal angular process of the frontal bone. The *inferior* is divided by the lower edge of the vertical crest into two parts: the posterior part articulates with the orbital plate of the superior maxillary bone; the anterior portion is prolonged downwards into a pointed process (the *descending* or *turbinal process*), which articulates with the lachrymal process of the inferior turbinated bone, and assists in the formation of the lachrymal groove.

Fig. 244.—Left lachrymal bone. External surface. (Slightly enlarged.)

Development.—By a single centre, which makes its appearance soon after ossification of the vertebræ has commenced.

Articulations.—With four bones: two of the cranium, the frontal and ethmoid, and two of the face, the superior maxillary and the inferior turbinated.

Attachment of Muscles.—To one muscle, the Tensor tarsi.

The Malar Bones

The **Malar** (mala, *the cheek*) are two small, quadrangular bones, situated at the upper and outer part of the face: they form the prominence of the cheek, part of the outer wall and floor of the orbit, and part of the temporal and zygomatic fossæ (fig. 245). Each bone presents for examination an external and an internal surface; four processes, the frontal, orbital, maxillary, and zygomatic; and four borders.

The **external surface** (fig. 246) is smooth, convex, and perforated near its centre by one or two small apertures, the *malar foramina*, for the passage of the malar nerves and vessels; it is covered by the Orbicularis palpebrarum muscle, and affords attachment to the Zygomaticus major and minor muscles.

The **internal surface** (fig. 247), directed backwards and inwards, is concave, presenting internally a rough, triangular surface, for articulation with the superior maxillary bone, and externally a smooth, concave surface, which above forms the anterior boundary of the temporal fossa, and below, where it is wider, a part of the zygomatic fossa. This surface presents, a little above its centre, the aperture of one or two malar canals, and affords attachment to a portion of the Masseter muscle at its lower part.

Of the four processes, the **frontal** is thick and serrated, and articulates with the external angular process of the frontal bone. To its orbital margin is attached the external tarsal ligament. The **orbital** process is a thick, strong plate, which projects backwards from the orbital margin of the bone. Its *supero-internal* surface, smooth and concave, forms, by its junction with the orbital surface of the superior maxillary bone and with the great wing of the sphenoid, part of the floor and outer wall of the orbit. Its *infero-external* surface, smooth and convex, forms part of the zygomatic and temporal fossæ. Its *anterior* margin, smooth and rounded, forms part of the circumference of the orbit. Its *superior* margin, rough, and directed horizontally, articulates

FIG. 245.—Malar bone *in situ*.

with the frontal bone behind the external angular process. Its *posterior* margin is rough, and serrated for articulation with the great wing of the sphenoid;

FIG. 246.—Left malar bone.
Outer surface.

FIG. 247.—Left malar bone.
Inner surface.

internally it is also serrated for articulation with the orbital surface of the superior maxillary. At the angle of junction of the sphenoidal and maxillary portions, a short, rounded, non-articular margin is generally seen; this forms the anterior boundary of the spheno-maxillary fissure : occasionally, this non-articular margin does not exist, the fissure being completed by the direct junction of the maxillary and sphenoid bones, or by the interposition of a small Wormian bone in the angular interval between them. On the *inner* surface of the orbital process are seen the orifices of one or two temporo-malar canals ; one of these usually opens on the posterior surface, the other (occasionally there are two) on the facial surface : they transmit filaments (temporo-malar) of the orbital branch of the superior maxillary nerve. The **maxillary** process is a rough, triangular surface which articulates with the superior maxillary bone. The **zygomatic** process, long, narrow, and serrated, articulates with the zygomatic process of the temporal bone. Of the **four borders,** the *antero-superior* or *orbital* is smooth, concave, and forms a considerable part of the circumference of the orbit. The *antero-inferior* or *maxillary* border is rough, and bevelled at the expense of its inner table, to articulate with the superior maxillary bone ; affording attachment by its margin to the Levator labii superioris proprius, just at its point of junction with the superior maxillary. The *postero-superior* or *temporal* border, curved like an italic letter *f*, is continuous above with the commencement of the temporal ridge ; below, with the upper border of the zygomatic arch : it affords attachment to the temporal fascia. The *postero-inferior* or *zygomatic* border is continuous with the lower border of the zygomatic arch, affording attachment by its rough edge to the Masseter muscle.

Development.—The malar bone ossifies generally from three centres, which appear about the eighth week—one for the zygomatic and two for the orbital portion—and fuse about the fifth month of fœtal life. After birth, the bone is sometimes seen to be divided, by a horizontal suture, into an upper and larger and a lower and smaller division. In some quadrumana the malar bone consists of two parts, an orbital and a malar.

Articulations.—With four bones : three of the cranium, frontal, sphenoid, and temporal ; and one of the face, the superior maxillary.

Attachment of Muscles.—To four : the Levator labii superioris proprius, Zygomaticus major and minor, and Masseter.

THE PALATE BONES

The **Palate Bones** (palatum, *the palate*) are situated at the back part of the nasal fossæ : they are wedged in between the superior maxillary bones and the

FIG. 248.—Palate bone *in situ.*

pterygoid processes of the sphenoid (fig. 248). Each bone assists in the formation of three cavities : the floor and outer wall of the nose, the roof of the mouth, and the floor of the orbit; it enters into the formation of two fossæ : the spheno-maxillary and pterygoid, and one fissure, the spheno-maxillary. In form the palate-bone somewhat resembles the letter L, and may be divided into an inferior or horizontal plate and a superior or vertical plate.

The **Horizontal Plate** is quadrilateral, and presents two surfaces and four borders. The *superior* surface, concave from side to side, forms the back part of the floor of the nose. The *inferior* surface, slightly concave and rough, forms the posterior fourth of the hard palate. At its posterior part may be seen a transverse ridge, more or less marked, for the attachment of part of the aponeurosis of the Tensor palati muscle. At the outer extremity of this ridge is a deep groove converted into a canal by its articulation with the tuberosity of the superior maxillary bone, and forming the *posterior palatine canal.* Near this groove, the orifices of one or two small canals, *accessory posterior palatine,* may be seen. The *anterior* border is serrated, bevelled at the expense of its inferior surface, and articulates with the palate process of the superior maxillary bone. The *posterior* border is concave, free, and serves for the attachment of the soft palate. Its inner extremity is sharp and pointed, and, when united with the opposite bone, forms a projecting process, the *posterior nasal spine,* for the attachment of the Azygos uvulæ muscle. The *external* border is united with the lower part of the perpendicular plate almost at right angles. The *internal* border, the thickest, is serrated for articulation with its fellow of the opposite side ; its superior edge is raised into a ridge, which, united with the opposite bone, forms a crest for articulation with the posterior part of the lower edge of the vomer.

FIG. 249.—Left palate bone. Internal view. (Enlarged.)

Orbital Process
Orbital Surface
Maxillary surface
Superior meatus
Spheno-palatine foramen
VERTICAL PLATE
Maxillary process
HORIZONTAL PLATE

The **Vertical Plate** (fig. 249) is thin, of an oblong form, and directed upwards and a little inwards. It presents two surfaces, an external and an internal, and four borders.

The **internal surface** exhibits at its lower part a broad, shallow depression, which forms part of the inferior meatus of the nose. Immediately above this is a well-marked, horizontal ridge, the *inferior turbinated crest,* for articulation with the inferior turbinated bone ; superior to this, a second broad, shallow depression, which forms part of the middle meatus, surmounted above by a horizontal ridge less prominent than the inferior, the *superior turbinated crest,* for articulation with the middle turbinated bone. Above the superior turbinated crest is a narrow, horizontal groove, which forms part of the superior meatus.

The **external surface** is rough and irregular throughout the greater part of its extent, for articulation with the inner surface of the superior maxillary bone, its upper and back part being smooth where it enters into the formation of the spheno-maxillary fossa ; it is also smooth in front, where it covers the posterior part of the orifice of the antrum. Towards the back part of this surface is a deep groove, converted into a canal, the *posterior palatine,* by its articulation with the superior maxillary bone. It transmits the posterior or descending palatine vessels, and the large palatine nerve from Meckel's ganglion.

The *anterior* border is thin, irregular, and presents, opposite the inferior turbinated crest, a pointed, projecting lamina, the *maxillary process,* which is directed forwards, and closes in the lower and back part of the opening of the antrum. The *posterior* border (fig. 250) presents a deep groove, the edges of which are serrated for articulation with the pterygoid process of the sphenoid.

At the lower part of this border is seen a pyramidal process of bone, the *pterygoid process* or *tuberosity* of the palate, which is received into the angular interval between the two pterygoid plates of the sphenoid at their inferior extremity. This process presents at its back part a median groove and two lateral surfaces. The groove is smooth, and forms part of the pterygoid fossa affording attachment to the Internal pterygoid muscle ; while the lateral surfaces are rough and uneven, for articulation with the anterior border of each pterygoid plate. The base of this process, continuous with the horizontal portion of the bone, presents the apertures of the *posterior* and *external accessory palatine canals*, through which pass the posterior and external palatine branches of Meckel's ganglion ; while its outer surface is rough, for articulation with the inner surface of the body of the superior maxillary bone.

FIG. 250.—Left palate bone.   Posterior view.
(Enlarged.)

The *superior* border of the vertical plate presents two well-marked processes separated by an intervening notch or foramen. The anterior, or larger, is called the *orbital process* ; the posterior, the *sphenoidal*.

The **Orbital Process**, directed upwards and outwards, is placed on a higher level than the sphenoidal. It presents five surfaces, which enclose a hollow cellular cavity, and is connected to the perpendicular plate by a narrow, constricted neck. Of these five surfaces, three are articular, two non-articular or free. The three articular are : (1) the *anterior* or *maxillary* surface, directed forwards, outwards, and downwards, of an oblong form, and rough for articulation with the superior maxillary bone. (2) The *posterior* or *sphenoidal* surface is directed backwards, upwards, and inwards. It ordinarily presents a small, open cell, which communicates with the sphenoidal cell, and the margins of which are serrated for articulation with the vertical part of the sphenoidal turbinated bone. (3) The *internal* or *ethmoidal* surface is directed inwards, upwards, and forwards, and articulates with the lateral mass of the ethmoid bone. In some cases, the cellular cavity above mentioned opens on this surface of the bone ; it then communicates with the posterior ethmoidal cells. More rarely it opens on both surfaces, and then communicates with the posterior ethmoidal and the sphenoidal cells. The non-articular or free surfaces are the *superior* or *orbital*, directed upwards and outwards, of a triangular shape, concave, smooth, and forming the back part of the floor of the orbit ; and the *external* or *zygomatic* surface, smooth and of an oblong form, is directed outwards, backwards, and downwards towards the spheno-maxillary fossa, and is separated from the orbital by a rounded border, which enters into the formation of the spheno-maxillary fissure.

The **Sphenoidal Process** of the palate-bone is a thin, compressed plate, much smaller than the orbital, and directed upwards and inwards. It presents three surfaces and two borders. The *superior* surface, the smallest of the three, articulates with the under surface of the sphenoidal turbinated bone, and reaches as far as the ala of the vomer ; it presents a groove, which contributes to the formation of the pterygo-palatine canal. The *internal* surface is concave, and forms part of the outer wall of the nasal fossa. The *external* surface is divided into an articular and non-articular portion : the former is rough, for articulation with the inner surface of the internal pterygoid plate of the sphenoid ; the latter is smooth, and forms part of the spheno-maxillary fossa. The *anterior* border forms the posterior boundary of the spheno-palatine foramen. The *posterior* border, serrated at the expense of the outer table, articulates with the inner surface of the internal pterygoid plate.

The orbital and sphenoidal processes are separated from one another by a deep notch, which is converted into a foramen, the *spheno-palatine*, by articulation with the sphenoidal turbinated bone. Sometimes the two processes are united above, and form between them a complete foramen, or the notch is crossed by one or more spiculæ of bone, giving rise to two or more foramina. In the articulated skull, this foramen opens into the back part of the outer wall of the superior meatus, and transmits the spheno-palatine vessels and the superior nasal and naso-palatine nerves.

Development.—From a single centre, which makes its appearance about the second month at the angle of junction of the two plates of the bone. From this point ossification spreads inwards to the horizontal plate, downwards into the tuberosity, and upwards into the vertical plate. Some authorities describe the bone as ossifying from four centres: one for the tuberosity and portion of the vertical plate behind the posterior palatine groove; a second for the rest of the vertical and the horizontal plates; a third for the orbital, and a fourth for the sphenoidal process. In the fœtus, the horizontal plate is much longer than the vertical; and even after it is fully ossified, the whole bone is at first remarkable for its shortness.

Articulations.—With six bones: the sphenoid, ethmoid, superior maxillary, inferior turbinated, vomer, and opposite palate.

Attachment of Muscles.—To three: the Tensor palati, Azygos uvulæ, and Internal pterygoid.

The Inferior Turbinated Bones

The **Inferior Turbinated Bones** (turbo, *a whirl*) are situated one on each side of the outer wall of the nasal fossæ. Each consists of a layer of thin, spongy bone, curled upon itself like a scroll—hence its name 'turbinated'—and extends horizontally along the outer wall of the nasal fossa, immediately below the orifice of the antrum (fig. 251). Each bone presents two surfaces, two borders, and two extremities.

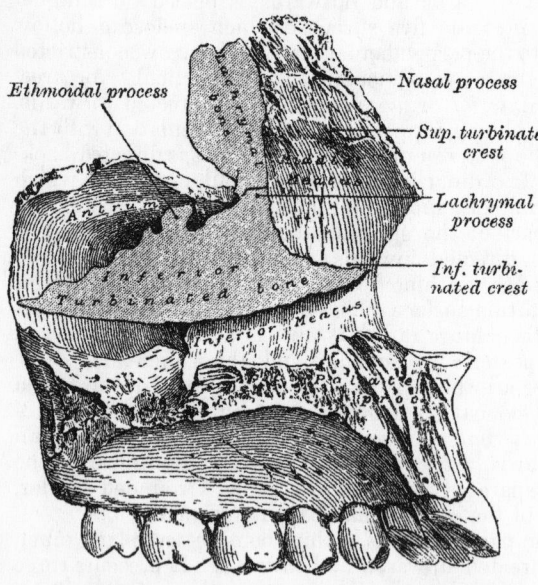

Fig. 251.—Inferior turbinated bone and Lachrymal bone *in situ*.

Ethmoidal process

Nasal process

Sup. turbinated crest

Lachrymal process

Inf. turbinated crest

Its *internal surface* (fig. 252) is convex, perforated by numerous apertures, and traversed by longitudinal grooves and canals for the lodgment of arteries and veins. In the recent state it is covered by the lining membrane of the nose. Its *external surface* is concave (fig. 253), and forms part of the inferior meatus. Its *upper border* is thin, irregular, and connected to various bones along the outer wall of the nose. It may be divided into three portions: of these, the anterior articulates with the inferior turbinated crest of the superior maxillary bone; the posterior with the inferior turbinated crest of the palate-bone; the middle portion of the superior border presents three well-marked processes, which vary much in their size and form. Of these, the anterior and smallest is situated at the junction of the anterior fourth with the posterior three-fourths of the bone: it is small and pointed, and is called the *lachrymal process*; it articulates, by its apex, with the descending or turbinal process of the lachrymal bone, and, by its margins, with the groove on the back of the nasal process of the superior maxillary, and thus assists in forming the canal for the nasal duct. At the junction of the two middle fourths of the bone, but encroaching on its posterior fourth, a broad, thin plate, the *ethmoidal process*, ascends to join the unciform process of the ethmoid ; from the lower border of this process a thin lamina of bone curves downwards and outwards, hooking over the lower edge of the orifice of the antrum, which it narrows below: it is called the *maxillary process*, and fixes the bone firmly to

FIG. 252.—Right inferior turbinated bone. Internal surface.　　FIG. 253.—Right inferior turbinated bone. External surface.

the outer wall of the nasal fossa.　The *inferior border* is free, thick, and cellular in structure, more especially in the middle of the bone.　Both *extremities* are more or less narrow and pointed, the posterior being the more tapering.　If the bone is held so that its outer concave surface is directed backwards (i.e. towards

FIG. 254.—Vomer *in situ*.

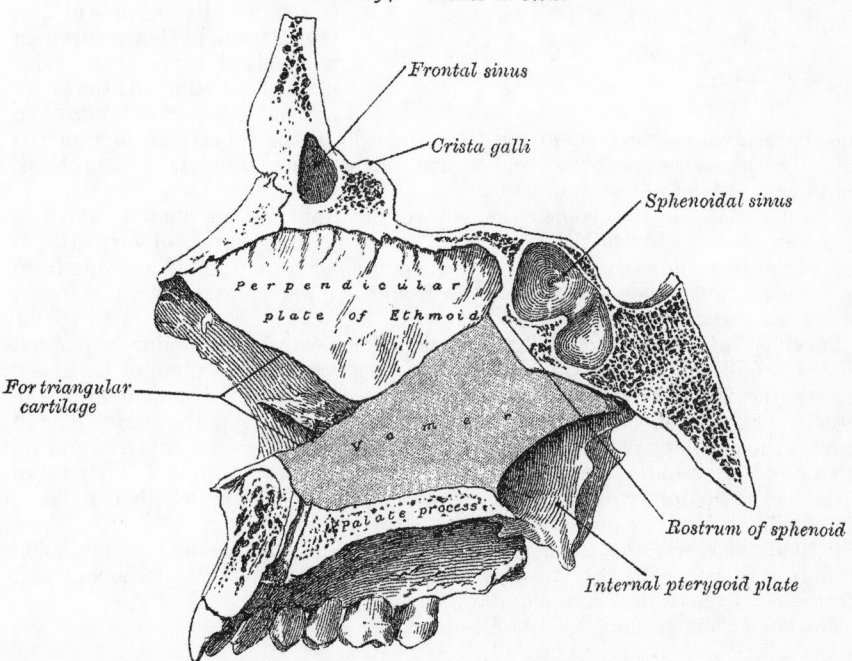

the holder), and its superior border, from which the lachrymal and ethmoidal processes project, upwards, the lachrymal process will be directed to the side to which the bone belongs.*

Development.—By a single centre, which makes its appearance about the middle of foetal life.

Articulations.—With four bones : one of the cranium, the ethmoid, and three of the face, the superior maxillary, lachrymal, and palate.

No muscles are attached to this bone.

THE VOMER

The **Vomer** (vomer, *a ploughshare*) is a single bone, situated vertically at the back part of the nasal fossæ, forming part of the septum of the nose (fig. 254).

---

* If the lachrymal process is broken off, as is often the case, the side to which the bone belongs may be known by recollecting that the maxillary process is nearer the back than the front of the bone.

It is thin, somewhat like a ploughshare in form ; but it varies in different individuals, being frequently bent to one or the other side ; it presents for examination two surfaces and four borders. The *lateral surfaces* (fig. 255) are smooth, marked by small furrows for the lodgment of blood-vessels, and by a groove, sometimes a canal, the *naso-palatine*, which runs obliquely downwards and forwards to the intermaxillary suture ; it transmits the naso-palatine nerve. The *superior border*, the thickest, presents a deep furrow,

FIG. 255.—The vomer.

bounded on each side by a horizontal projecting ala of bone ; it receives the rostrum of the sphenoid, while the margins of the alæ articulate with the vaginal processes of the internal pterygoid plates behind, and with the sphenoidal processes of the palate bones in front. The *inferior border* articulates with the crest formed by the superior maxillary and palate bones. The upper half of the *anterior border* usually consists of two laminæ of bone, between which is received the perpendicular plate of the ethmoid ; the lower half, also separated into two lamellæ, receives between them the lower margin of the septal cartilage of the nose. The *posterior border* is free, concave, and separates the nasal fossæ behind. It is thick and bifid above, thin below.

The surfaces of the vomer are covered by mucous membrane, which is intimately connected with the periosteum, with the intervention of very little, if any, submucous connective tissue. Hence polypi are rarely found growing from this surface, though they frequently grow from the outer walls of the nasal fossæ, where the submucous tissue is abundant.

Development.—The vomer at an early period consists of two laminæ, separated by a very considerable interval, and enclosing between them a plate of cartilage, the *vomerine cartilage*, which is prolonged forwards to form the remainder of the septum. Ossification commences in the membrane overlying the cartilage, and not from the cartilage itself, at its postero-inferior part by two centres, one on each side of the middle line, which extend to form the two laminæ. They begin to coalesce at the lower part, the cartilage becoming absorbed, but their union is not complete until after puberty.

Articulations.—With six bones : two of the cranium, the sphenoid and ethmoid ; and four of the face, the two superior maxillary and the two palate-bones ; and with the cartilage of the nasal septum.

The vomer has no muscles attached to it.

The Inferior Maxillary Bone, or Mandible

The **Inferior Maxillary Bone**, or **Mandible**, the largest and strongest bone of the face, serves for the reception of the lower teeth. It consists of a curved, horizontal portion, the *body*, and two perpendicular portions, the *rami*, which join the back part of the body nearly at right angles.

The **Horizontal Portion**, or **Body** (fig. 256), is convex in its general outline, and curved somewhat like a horseshoe. It presents for examination two surfaces and two borders. The **external surface** is convex from side to side, concave from above downwards. In the median line is a faint ridge, the *symphysis*, which extends from the upper to the lower border of the bone, and indicates the line of junction of the two pieces of which the bone is composed at an early period of life. The lower part of the ridge terminates in a prominent triangular eminence, the *mental process*. This eminence is rounded below, and often presents a median depression separating two processes, the *mental tubercles*. It forms the chin, a feature peculiar to the human skull. On either side of the symphysis, just below the cavities for the incisor teeth, is a depression, the *incisive fossa*, for the attachment of the Levator menti (Levator labii inferioris); more externally is attached a small portion of the Orbicularis oris, and, still more externally, is a foramen, the *mental foramen*, for the passage of the mental vessels and nerve. This foramen is placed below the interval between the two bicuspid teeth.

FIG. 256.—Inferior maxillary bone. Outer surface. Side view.

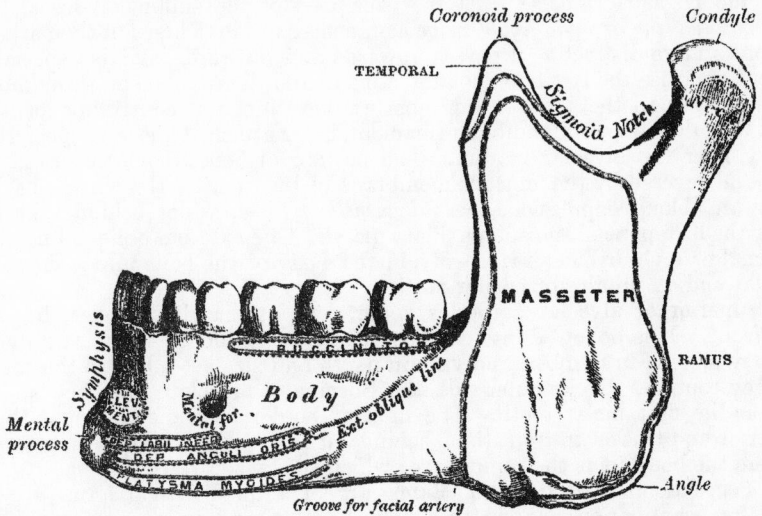

Running outwards from the base of the mental process on each side is a ridge, the *external oblique* line. The ridge is at first nearly horizontal, but afterwards inclines upwards and backwards, and is continuous with the anterior border of the ramus: it affords attachment to the Depressor labii inferioris and Depressor anguli oris; below it the Platysma myoides is attached.

FIG. 257.—Inferior maxillary bone. Inner surface. Side view.

The **internal surface** (fig. 257) is concave from side to side, convex from above downwards. In the middle line is an indistinct linear depression, corresponding to the symphysis externally; on either side of this depression, just below its centre, are four prominent tubercles, placed in pairs, two above and two below;

they are called the *genial tubercles*, and afford attachment, the upper pair to the Genio-hyo-glossi muscles, the lower pair to the Genio-hyoidei muscles. Some-times the tubercles on each side are blended into one, at others they all unite into an irregular eminence, or again, nothing but an irregularity may be seen on the surface of the bone at this part. On either side of the genial tubercles is an oval depression, the *sublingual fossa*, for lodging the sublingual gland ; and beneath the fossa, a rough depression, which gives attachment to the anterior belly of the Digastric muscle. At the back part of the sublingual fossa, the *internal oblique line* or *mylo-hyoid ridge* commences ; it is at first faintly marked, but becomes more distinct as it passes upwards and outwards, and is especially prominent opposite the last two molar teeth ; it affords attachment throughout its whole extent to the Mylo-hyoid muscle : the Superior constrictor of the pharynx with the pterygo-mandibular ligament, being attached above its posterior extremity, near the alveolar margin. The portion of bone above this ridge is smooth, and covered by the mucous membrane of the mouth ; the portion below presents an oblong depression, the *submaxillary fossa*, wider behind than in front, for the lodgment of the submaxillary gland. The external oblique line and the internal or mylo-hyoidean ridge divide the body of the bone into a superior or alveolar and an inferior or basilar portion.

The **superior** or **alveolar** border is wider, and its margins thicker, behind than in front. It is hollowed into cavities, for the reception of the teeth ; these cavities are sixteen in number, and vary in depth and size according to the teeth which they contain. To its outer side, the Buccinator muscle is attached as far forward as the first molar tooth. The **inferior** border is rounded, longer than the superior and thicker in front than behind ; it may present a shallow groove, just where the body joins the ramus, over which the facial artery turns.

The **Perpendicular Portions,** or **Rami,** are of a quadrilateral form. Each presents for examination two surfaces, four borders, and two processes. The *external surface* is flat, marked with ridges, and gives attachment throughout nearly the whole of its extent to the Masseter muscle. The *internal surface* presents about its centre the oblique aperture of the inferior dental canal, for the passage of the inferior dental vessels and nerve. The margin of this opening is irregular ; it presents in front a prominent ridge, surmounted by a sharp spine, the *lingula*, which gives attachment to the internal lateral ligament of the lower jaw, and at its lower and back part a notch leading to a groove, the *mylo-hyoidean*, which runs obliquely downwards to the back part of the submaxillary fossa, and lodges the mylo-hyoid vessels and nerve. Behind the groove is a rough surface, for the insertion of the Internal pterygoid muscle. The inferior dental canal runs obliquely downwards and forwards in the substance of the ramus, and then horizontally forwards in the body ; it is here placed under the alveoli, with which it communicates by small openings. On arriving at the incisor teeth, it turns back to communicate with the mental foramen, giving off two small canals, which run forward, to be lost in the cancellous tissue of the bone beneath the incisor teeth. This canal, in the posterior two-thirds of the bone, is situated nearer the internal surface of the jaw ; and in the anterior third, nearer its external surface. Its walls are composed of compact tissue at either extremity, and of cancellous in the centre. It contains the inferior dental vessels and nerve, from which branches are distributed to the teeth through small apertures at the bases of the alveoli. The *lower border* of the ramus is thick, straight, and con-tinuous with the body of the bone. At its junction with the posterior border is the *angle of the jaw*, which is either inverted or everted, and marked by rough, oblique ridges on each side, for the attachment of the Masseter externally, and the Internal pterygoid internally ; the stylo-mandibular ligament is attached to the angle between these muscles. The *anterior border* is thin above, thicker below, and continuous with the external oblique line. The *posterior border* is thick, smooth, rounded, and covered by the parotid gland. The *upper border* of the ramus is thin, and presents two processes, separated by a deep concavity, the *sigmoid notch*. Of these processes, the anterior is the *coronoid*, the posterior the *condyloid*.

The **Coronoid Process** is a thin, flattened triangular eminence of bone, which varies in shape and size in different subjects, and serves chiefly for the attachment of the Temporal muscle. Its *external surface* is smooth, and affords insertion to the Temporal and Masseter muscles. Its *internal surface* gives attachment to the Temporal muscle, and presents the commencement of a longitudinal ridge, which is continued to the posterior part of the alveolar process. On the outer side of this ridge is a deep groove, continued below on the outer side of the alveolar process; this ridge and part of the groove afford attachment, above, to the Temporal; below, to the Buccinator muscle.

The **Condyloid Process**, shorter but thicker than the coronoid, consists of two portions: the *condyle*, and the constricted portion which supports the condyle, the *neck*. The *condyle* is of an oblong form, its long axis being transverse, and set obliquely on the neck in such a manner that its outer end is a little more forward and a little higher than its inner. It is convex from before backwards, and from side to side, the articular surface extending farther on the posterior than on the anterior aspect. At its outer extremity is a small tubercle for the attachment of the external lateral ligament of the temporo-mandibular joint. The *neck* of the condyle is flattened from before backwards, and strengthened by ridges which descend from the fore part and sides of the condyle. Its lateral margins are narrow, the external one giving attachment to part of the external lateral ligament. Its posterior surface is convex; its anterior is hollowed out on its inner side by a depression (the *pterygoid fossa*) for the attachment of the External pterygoid muscle.

The **Sigmoid Notch**, separating the two processes, is a deep semilunar depression, crossed by the masseteric vessels and nerve.

Development.—The lower jaw is developed principally from membrane, but partly from cartilage. The process of ossification commences early—earlier than any other bone except the clavicle. The greater part of the bone is formed from a centre of ossification (*dentary*), which appears between the fifth and sixth weeks in the membrane on the outer surface of Meckel's cartilage. A second centre (*splenial*) arises in the membrane on the inner surface of the cartilage, and from this centre the inner wall of the sockets of the teeth is formed; this terminates above in the lingula. The anterior extremity of Meckel's cartilage becomes ossified, forming the portion of the body of the bone which lies on each side of the symphysis. The central portion of the cartilage becomes absorbed and the upper or posterior part is converted into fibrous tissue and remains as a permanent structure, the internal lateral ligament of the temporo-mandibular joint; so that the upper part of the ramus and its processes have no connection with Meckel's cartilage, but are ossified from two supplemental patches of cartilage, which appear one in the condyle, the other in the coronoid process. Some authorities also describe a separate centre, preceded by a cartilaginous matrix for the angle. At birth, the bone consists of two halves, united by a fibrous symphysis, in which ossification takes place during the first year.

Articulation.—With the glenoid fossæ of the two temporal bones.

Attachment of Muscles.—To fifteen pairs; to its external surface, commencing at the symphysis, and proceeding backwards: Levator menti, Depressor labii inferioris, Depressor anguli oris, Platysma myoides, Buccinator, Masseter; a portion of the Orbicularis oris (Accessorii Orbicularis inferioris) is also attached to this surface. To its internal surface, commencing at the same point: Genio-hyo-glossus, Genio-hyoideus, Mylo-hyoideus, Digastric, Superior constrictor of the pharynx, Temporal, Internal pterygoid, External pterygoid.

CHANGES PRODUCED IN THE LOWER JAW BY AGE

The changes which the lower jaw undergoes after birth, relate (1) to the alterations effected in the body of the bone by the first and second dentitions, the loss of the teeth in the aged, and the subsequent absorption of the alveoli; (2) to the size and situation of the dental canal; and (3) to the angle at which the ramus joins with the body.

At birth (fig. 258), the bone consists of lateral halves, united by fibrous tissue. The body is a mere shell of bone, containing the sockets of the two incisor, the canine, and the two temporary molar teeth, imperfectly partitioned from one another. The dental canal is of large size, and runs near the lower border of the bone, the mental foramen opening beneath the socket of the first molar. The angle is obtuse (175°), and the condyloid portion nearly in the same horizontal line with the body; the neck of the condyle is short, and bent backwards. The coronoid process is of comparatively large size, and situated at right angles with the rest of the bone.

After birth (fig. 259), the two segments of the bone become joined at the symphysis, from below upwards, in the first year; but a trace of separation may be visible in the

Side View of the Lower Jaw at Different Periods of Life

Fig. 258.—At birth.

Fig. 259.—In childhood.

Fig. 260.—In the adult.

Fig. 261.—In old age.

beginning of the second year, near the alveolar margin. The body becomes elongated in its whole length, but more especially behind the mental foramen, to provide space for the three additional teeth developed in this part. The depth of the body becomes greater, owing to increased growth of the alveolar part, to afford room for the fangs of the teeth, and by thickening of the subdental portion which enables the jaw to withstand the powerful action of the masticatory muscles; but the alveolar portion is the deeper of the two, and, consequently, the chief part of the body lies above the oblique line. The dental canal, after the second dentition, is situated just above the level of the mylo-hyoid ridge; and the mental foramen occupies the position usual to it in the adult. The angle becomes less obtuse, owing to the separation of the jaws by the teeth. (About the fourth year it is 140°.)

In the adult (fig. 260), the alveolar and basilar portions of the body are usually of equal depth. The mental foramen opens midway between the upper and lower border of the bone, and the dental canal runs nearly parallel with the mylo-hyoid line. The ramus is almost vertical in direction, and joins the body at a little more than a right angle.

In old age (fig. 261), the bone becomes greatly reduced in size: for with the loss of the teeth the alveolar process is absorbed, and the basilar part of the bone alone remains; consequently, the chief part of the bone is below the oblique line. The dental canal, with the mental foramen opening from it, is close to the alveolar border. The ramus is oblique in direction, the angle obtuse, and the neck of the condyle more or less bent backwards.

HYOID BONE

The **Hyoid bone** is named from its resemblance to the Greek upsilon; it is also called the *lingual bone*, because it supports the tongue, and gives attachment to its numerous muscles. It is a bony arch, shaped like a horse-shoe, and consisting of five segments, a body, two greater cornua, and two lesser cornua. It is suspended from the tips of the styloid processes of the temporal bones by ligamentous bands, the *stylo-hyoid* ligaments.

The **Body** (*basi-hyal*) forms the central part of the bone, and is of a quadrilateral form; its *anterior surface* (fig. 262), convex, directed forwards and upwards, is divided into two parts by a vertical ridge which descends along the median line, and is crossed at right angles by a horizontal ridge, so that this surface is divided into four spaces or depressions. At the point of meeting of these two lines is a prominent elevation, the *tubercle*. The portion above the horizontal ridge is directed upwards, and is sometimes described as the superior border. The anterior surface gives attachment to the Genio-hyoid in the greater part of its extent; above, to the Genio-hyo-glossus; below, to the Mylo-hyoid, Stylo-hyoid, and aponeurosis of the Digastric (suprahyoid aponeurosis); and

FIG. 262.—Hyoid bone. Anterior surface (enlarged).

Greater cornu
MIDDLE CONSTRICTOR *of Pharynx*
HYO-GLOSSUS
CHONDRO-GLOSSUS
Lesser cornu
GENIO-HYO-GLOSSUS
THYRO-HYOID
Body
STYLO-HYOID
OMO-HYOID
STERNO-HYOID   MYLO-HYOID
GENIO-HYOID

between these to part of the Hyo-glossus. The *posterior surface* is smooth, concave, directed backwards and downwards, and separated from the epiglottis by the thyro-hyoid membrane and a quantity of loose areolar tissue; a bursa intervenes between it and the thyro-hyoid membrane. The *superior border* is rounded, and gives attachment to the thyro-hyoid membrane, part of the Genio-hyo-glossi, and Chondro-glossi muscles. The *inferior border* gives attachment, in front, to the Sterno-hyoid; behind, to the Omo-hyoid and to part of the Thyro-hyoid, at its junction with the great cornu. It also gives attachment to the Levator glandulæ thyroideæ, when this muscle is present. The *lateral surfaces*, after middle life, are joined to the greater cornua. In early life they are connected to the cornua by cartilaginous surfaces, and held together by ligaments, and occasionally a synovial membrane is found between them.

The **Greater Cornua** (*thyro-hyal*) project backwards from the lateral surfaces of the body; they are flattened from above downwards, diminish in size from before backwards, and terminate posteriorly in a tubercle for the attachment of the lateral thyro-hyoid ligament. The outer surface gives attachment to the Hyo-glossus; the upper border to the Middle constrictor of the pharynx; the lower to part of the Thyro-hyoid muscle.

The **Lesser Cornua** (*cerato-hyal*) are two small, conical-shaped eminences, attached by their bases to the angles of junction between the body and greater cornua, and giving attachment by their apices to the stylo-hyoid ligaments.* The smaller cornua are connected to the body of the bone by a distinct diarthrodial joint, which usually persists throughout life, but occasionally becomes ankylosed.

Development.—By *five* centres : one for the body, and one for each cornu. Ossification commences in the body about the eighth month, and in the greater cornua towards the end of fœtal life. Ossification of the lesser cornua commences during the first or second year after birth. According to some authorities there are two centres for the body.

Attachment of Muscles.— Sterno-hyoid, Thyro-hyoid; Omo-hyoid, aponeurosis of the Digastric, Stylo-hyoid, Mylo-hyoid, Genio-hyoid, Genio-hyo-glossus, Chondro-glossus, Hyo-glossus, Middle constrictor of the pharynx, and occasionally a few fibres of the Inferior lingualis. It also gives attachment to the thyro-hyoid membrane, and the stylo-hyoid, thyro-hyoid, and hyo-epiglottic ligaments.

Surface Form.—The hyoid bone can be felt in the receding angle below the chin, and the finger can be carried along the whole length of the bone to the greater cornu, which is situated just below the angle of the jaw. This process of bone is best perceived by making pressure on one cornu and so pushing the bone over to the opposite side, when the cornu of this side will be distinctly felt immediately beneath the skin. It is an important landmark in ligature of the lingual artery.

Surgical Anatomy.—The hyoid bone is occasionally fractured, generally from direct violence, as in hanging ; forcible grasping of the throat in garotting or throttling, or by a blow ; but cases are also recorded where it has arisen from muscular action. The fracture generally occurs about the junction of the greater cornu with the body of the bone, but sometimes takes place through the latter. In consequence of the muscles of the tongue having important connections with this bone, there is great pain upon any attempt being made to move the tongue, as in speaking or swallowing.

THE SUTURES

The bones of the Skull, with the exception of the Mandible, are connected to each other by means of *Sutures*. That is, the articulating surfaces or edges of the bones are more or less roughened or uneven, and are closely adapted to each other, a small amount of intervening fibrous tissue, the *sutural ligament*, fastening them together. The *Sutures of the Brain-case* may be divided into three sets : 1. those at the vertex ; 2. those at the side ; 3. those at the base.

The sutures at the vertex of the skull are four : the *metopic, sagittal, coronal,* and *lambdoid*.

The **Metopic** or **Frontal Suture** is the suture formed by the junction of the two halves of the frontal bone. At birth the two halves of the frontal bone are separated by the suture, but the greater part of it usually disappears about the fifth or sixth year. In the adult it may not exist at all, or may be represented by a slight fissure, just above the glabella ; but according to Cunningham it exists in about 8 per cent. of Europeans, in whom it is of more frequent occurrence than in the lower races of man.

The **Sagittal Suture** (*interparietal*) is formed by the junction of the two parietal bones, and extends from the middle of the frontal bone, backwards to the superior angle of the occipital. In childhood, and sometimes in the adult, when the two halves of the frontal bone are not united, it is continued forwards to the root of the nose. This suture is sometimes perforated, near its posterior extremity, by the parietal foramen ; and in front, where it joins the coronal suture, a space is occasionally left, which encloses a large Wormian bone.

---

* These ligaments in many animals are distinct bones, and in man may undergo partial ossification.

The **Coronal Suture** (*fronto-parietal*) extends transversely across the vertex of the skull, and connects the frontal with the parietal bones. It commences at the extremity of the greater wing of the sphenoid on one side, and terminates at the same point on the opposite. The dentations of the suture are more marked at the sides than at the summit, and are so constructed that the frontal rests on the parietal above, while laterally the frontal supports the parietal.

The **Lambdoid Suture** (*occipito-parietal*), so called from its resemblance to the Greek letter Λ, connects the occipital with the parietal bones. It commences on each side at the mastoid portion of the temporal bone, and inclines upwards and backwards to the end of the sagittal suture. The dentations of this suture are very deep and distinct, and are often interrupted by several small Wormian bones.

The sutures at the side of the skull extend from the external angular process of the frontal bone to the lower end of the lambdoid suture. From before backwards, they consist of the *fronto-malar, spheno-malar, spheno-frontal, spheno-parietal, squamo-sphenoidal, squamo-parietal*, and *masto-parietal*.

The **Fronto-malar** connects the external angular process of the frontal bone with the parietal.

The **Spheno-malar** is situated between the anterior border of the great wing of the sphenoid and the orbital process of the malar.

The **Spheno-frontal** joins the great wing of the sphenoid to the frontal behind its external angular process.

The **Spheno-parietal** is placed between the anterior inferior angle of the parietal and the great wing of the sphenoid. This suture varies greatly in length in different skulls, and is sometimes absent owing to the articulation of the squamous-temporal with the frontal.

The **Squamo-sphenoidal** connects the posterior border of the greater wing of the sphenoid with the anterior border of the squamous-temporal.

The **Squamo-parietal**, or **Squamous Suture**, is arched. It is formed by the squamous portion of the temporal bone overlapping the middle division of the lower border of the parietal.

The **Masto-parietal** is a short suture, deeply dentated, situated between the posterior inferior angle of the parietal and the superior border of the mastoid portion of the temporal.

The sutures at the base of the skull are, the *basilar* in the centre, and on each side, the *petro-occipital*, the *masto-occipital*, the *petro-sphenoidal*, and the *squamo-sphenoidal*.

The **Basilar Suture** is formed by the junction of the basilar surface of the occipital bone with the posterior surface of the body of the sphenoid. At an early period of life a thin plate of cartilage exists between these bones; but in the adult they become fused into one. From the outer extremity of the basilar suture to the termination of the lambdoid, an irregular suture exists, which is subdivided into two portions. The inner portion, formed by the union of the petrous part of the temporal with the occipital bone, is termed the **petro-occipital**. The outer portion, formed by the junction of the mastoid part of the temporal with the occipital, is called the **masto-occipital**. In the petro-occipital suture, a thin plate of cartilage exists; in the masto-occipital is occasionally found the opening of the mastoid foramen. Between the outer extremity of the basilar suture and the spheno-parietal, an irregular suture is seen uniting the sphenoid and temporal bones. The inner and smaller portion of this suture is termed the **petro-sphenoidal**; it runs between the petrous portion of the temporal and the great wing of the sphenoid : the outer portion, of greater length, and arched, connects the squamous portion of the temporal with the great wing of the sphenoid ; it is called the **squamo-sphenoidal**.

The cranial bones are connected with those of the face, and the facial bones with each other, by numerous sutures, which, though distinctly marked, have received no special names. The only remaining suture deserving especial consideration is the **transverse**. This extends across the upper part of the face, and is formed by the junction of the frontal with the bones which lie below it; it extends from the external angular process of one side to the same point on the opposite side, and connects the frontal with the malar, the sphenoid, the ethmoid, the lachrymal, the superior maxillary, and the nasal bones.

The sutures remain separate for a considerable period after the complete formation of the skull. It is probable that they serve the purpose of permitting the growth of the bones at their margins ; while their peculiar formation, together with the interposition of the sutural ligament between the bones forming them, prevents the dispersion of blows or jars received upon the skull. Humphry remarks 'that, as a general rule, the sutures are first obliterated at the parts in which the ossification of the skull was last completed—viz. in the neighbourhood of the fontanelles ; and the cranial bones seem in this respect to observe a similar law to that which regulates the union of the epiphyses to the shafts of the long bones.' The same author remarks that the time of their disappearance is extremely variable : they are sometimes found well marked in skulls edentulous with age, while in others which have only just reached maturity they can hardly be traced. The obliteration of the sutures takes place sooner on the inner than on the outer surface of the skull. The sagittal and coronal sutures are as a rule the first to become ossified—the process starting near the posterior extremity of the former and the lower ends of the latter.

THE SKULL AS A WHOLE

The Skull, as a whole, varies much in size and in shape. As regards the size, it is important to note that it forms a good index of the development of the brain, and is most conveniently estimated by its capacity. It is filled with shot, and the amount required to fill it ascertained by means of a graduated measure. Skulls which contain from 1·350 to 1·450 cubic centimetres are termed *mesocephalic* ; those above 1·450 c.c., *megacephalic* ; and those below 1·350 c.c., *microcephalic*. Alterations in shape depend to a great extent upon the relation of the length to the breadth. On viewing a series of skulls from above, it will be found that the shape may vary from nearly circular to an elongated oval. In order to express these differences in figures, the length and the breadth of the skull are taken : the length, by measuring from the most prominent part of the glabella to the point of the occipital bone, in the median line farthest removed from the glabella ; the breadth, by taking the greatest transverse diameter of the skull above the supramastoid ridges. The proportion of the breadth to the length $\left(\dfrac{\text{breadth} \times 100}{\text{length}}\right)$ is the cephalic index. Where this is over 80, the skull is said to be *brachycephalic* ; where it is between 75 and 80, it is said to be *mesaticephalic* ; and where it is below 75, it is *dolichocephalic*. The shape of the head and face also largely depends upon the size of the jaws and the extent to which they project forwards. The degree of projection is indicated by comparing the length of a line drawn from the centre of the anterior margin of the upper alveolar arch to the middle of the anterior margin of the foramen magnum, with a line drawn from the middle of the naso-frontal suture to the same point. This constitutes the *gnathic* or *alveolar index*, and may be expressed thus :

$$\frac{\text{Basi-alveolar length} \times 100}{\text{Basi-nasal length}} = \text{gnathic index.}$$

All skulls with a gnathic index below 98 are *orthognathous* ; between 98 and 103, *mesognathous* ; and above 103, *prognathous*. The skull may also vary in form from want of symmetry.

The Skull, formed by the union of the several cranial and facial bones already described, when considered as a whole, is divisible into five regions : a superior region or vertex, an inferior region or base, two lateral regions, and an anterior region, the face.

VERTEX OF THE SKULL

The **Superior Region,** or **Vertex,** presents two surfaces, an external and an internal.

The **external surface** is bounded, in front, by the glabella and supra-orbital ridges ; behind, by the occipital protuberance and superior curved lines of the occipital bone ; laterally, by an imaginary line extending from the outer end of the superior curved line, along the temporal ridge, to the external angular process of the frontal bone. This surface includes the greater part of the vertical portion of the frontal, the greater part of the parietal, and the superior third of the occipital bone ; it is smooth, convex, of an elongated oval form, crossed transversely by the coronal suture, and from before backwards by the sagittal, which terminates behind in the lambdoid, while immediately above the glabella is the remnant of the metopic suture. The point of junction of the coronal and sagittal sutures is named the *bregma*, and is represented by a line drawn vertically upwards from the external auditory meatus, the head being in its normal position. The point of junction of the sagittal and lambdoid sutures is called the *lambda*, and is about two inches and three-quarters above the external occipital protuberance. From before backwards may be seen the superciliary ridges and frontal eminences ; on each side of the sagittal suture are the parietal foramen and parietal eminence, and still more posteriorly the convex surface of the occipital bone. In the neighbourhood of the parietal foramen the skull is often flattened, and the name of *obelion* is sometimes given to that point of the sagittal suture which lies in line with the two parietal foramina.

The **internal surface** is concave, presents depressions for the convolutions of the cerebrum, and numerous furrows for the lodgment of branches of the meningeal arteries. Along the middle line of this surface is a longitudinal groove, narrow in front, where it commences at the frontal crest, but broader behind ; it lodges the superior longitudinal sinus, and by its margins affords attachment to the falx cerebri. On either side of it are several depressions for the Pacchionian bodies, and at its back part, the internal openings of the parietal foramina. This surface is crossed, in front, by the coronal suture ; from before backwards, by the sagittal ; behind by the lambdoid.

Base of the Skull

The **Inferior Region,** or **Base of the Skull,** presents two surfaces—an internal or cerebral, and an external or basilar.

The **internal** or **cerebral surface** (fig. 263) presents three fossæ, called the *anterior, middle,* and *posterior* fossæ of the cranium.

The **Anterior Fossa** is formed by the orbital plates of the frontal, the cribriform plate of the ethmoid, the anterior third of the superior surface of the body, and the upper surface of the lesser wings of the sphenoid, and is defined posteriorly by the posterior border of the lesser wings of the sphenoid. It is the most elevated of the three fossæ, convex externally where it corresponds to the roof of the orbit, concave in the situation of the cribriform plate of the ethmoid. It is traversed by three sutures, the *ethmo-frontal, ethmo-sphenoidal,* and *fronto-sphenoidal* ; and lodges the frontal lobes of the cerebrum. It presents, in the median line, from before backwards, the commencement of the *groove* for the superior longitudinal sinus, and the *frontal crest* for the attachment of the falx cerebri ; the *foramen cæcum,* an aperture formed between the frontal bone and the crista galli of the ethmoid, which, if pervious, transmits a small vein from the nose to the superior longitudinal sinus ; behind the foramen cæcum, the *crista galli,* the posterior margin of which affords attachment to the falx cerebri ; on either side of the crista galli, the *olfactory groove* formed by the cribriform plate, which supports the olfactory bulb, and presents three rows of foramina for the transmission of its nervous filaments to the nose, and in front a slit-like opening for the nasal branch of the ophthalmic division of the fifth nerve. On the outer side of each olfactory groove are the internal openings of the *anterior* and *posterior ethmoidal foramina* ; the former, situated about the middle of the outer margin of the olfactory groove, transmits the anterior ethmoidal vessels and the nasal nerve ; the nerve runs in a groove along the outer edge of the cribriform plate, to the slit-like opening above mentioned ; while the posterior ethmoidal foramen opens at the back part of this margin under cover of the projecting lamina of the sphenoid, and transmits a meningeal branch from the posterior ethmoidal artery. Farther back in the middle line is the *ethmoidal spine,* bounded behind by a slight elevation, separating two shallow longitudinal grooves which support the olfactory lobes. Behind this is a transverse sharp ridge, running outwards on either side to the anterior margin of the optic foramen, and separating the anterior from the middle fossa of the base of the skull. The anterior fossa presents, laterally, depressions for the convolutions of the brain, and grooves for the lodgment of the anterior meningeal arteries.

The **Middle Fossa**, deeper than the preceding, is narrow in the middle line, but becomes wider at the side of the skull. It is bounded in front by the posterior

FIG. 263.—Base of the skull. Inner or cerebral surface.

margin of the lesser wing of the sphenoid, the anterior clinoid process, and the ridge forming the anterior margin of the optic groove; behind, by the superior border of the petrous portion of the temporal, and the dorsum ephippii;

externally by the squamous portion of the temporal, anterior inferior angle of the parietal bone, and greater wing of the sphenoid. It is traversed by four sutures, the *squamo-parietal, spheno-parietal, squamo-sphenoidal,* and *petro-sphenoidal.*

In the middle line, from before backwards, is the *optic groove,* behind which lies the optic commissure; the groove terminates on each side in the optic foramen, for the passage of the optic nerve and ophthalmic artery; behind the optic groove is the *olivary process,* and laterally the *anterior clinoid processes,* to which are attached processes of the tentorium cerebelli. Farther back is the *sella turcica,* a deep depression, which lodges the pituitary gland, bounded in front by a small eminence on either side, the *middle clinoid process,* and behind by a broad square plate of bone, the *dorsum ephippii,* surmounted at each superior angle by a tubercle, the *posterior clinoid process;* beneath the latter process is a notch, for the sixth nerve. On each side of the sella turcica is the *cavernous groove:* it is broad, shallow, and curved somewhat like the italic letter *f;* it commences behind at the foramen lacerum medium, and terminates on the inner side of the anterior clinoid process, and presents along its outer margin a ridge of bone. This groove lodges the cavernous sinus, the internal carotid artery, and the nerves of the orbit. The sides of the middle fossa are of considerable depth: they present depressions for the convolutions of the brain, and grooves for the branches of the middle meningeal arteries; the latter commence on the outer side of the foramen spinosum, and consist of two large branches, an anterior and a posterior; the former passing upwards and forwards to the anterior inferior angle of the parietal bone, the latter passing upwards and backwards. The following foramina are also to be seen from before backwards. Most anteriorly is the *foramen lacerum anterius,* or *sphenoidal fissure,* formed above by the lesser wing of the sphenoid; below, by the greater wing; internally, by the body of the sphenoid; and sometimes completed externally by the orbital plate of the frontal bone. It transmits to the orbital cavity the third, the fourth, the three branches of the ophthalmic division of the fifth, the sixth nerve, some filaments from the cavernous plexus of the sympathetic, the orbital branch of the middle meningeal artery, and from the orbital cavity a recurrent branch from the lachrymal artery to the dura mater, and the ophthalmic vein. Behind the inner extremity of the sphenoidal fissure is the *foramen rotundum,* for the passage of the second division of the fifth or superior maxillary nerve; still more posteriorly is the *foramen Vesalii,* which varies in size in different individuals, and is often absent. When present, it opens below at the outer side of the scaphoid fossa, and transmits a small vein. Behind and external to the latter opening is the *foramen ovale,* which transmits the third division of the fifth or inferior maxillary nerve, the small meningeal artery, and the small superficial petrosal nerve.* On the outer side of the foramen ovale is the *foramen spinosum,* for the passage of the middle meningeal artery and veins, and on the inner side of the foramen ovale, the *foramen lacerum medium.* The lower part of this aperture is filled up by a layer of fibro-cartilage in the recent state, while its upper and inner parts transmit the internal carotid artery and its sympathetic plexus of nerves. The Vidian nerve and a meningeal branch from the ascending pharyngeal artery pierce the layer of fibro-cartilage. On the anterior surface of the petrous portion of the temporal bone is seen, from without inwards, the eminence caused by the projection of the superior semicircular canal; in front of and a little outside this a depression corresponding to the roof of the tympanum; the groove leading to the hiatus Fallopii, for the transmission of the great superficial petrosal nerve and the petrosal branch of the middle meningeal artery; beneath it, the smaller groove, for the passage of the small superficial petrosal nerve; and, near the apex of the bone, the depression for the Gasserian ganglion and the orifice of the carotid canal, for the passage of the internal carotid artery and carotid plexus of nerves.

The **Posterior Fossa,** deeply concave, is the largest of the three, and situated on a lower level than either of the preceding. It is formed by the dorsum sellæ and clivus of the sphenoid, by the occipital, the petrous and mastoid portions of the temporal, and the posterior inferior angle of the parietal bone; it is crossed by four sutures, the *petro-occipital,* the *masto-occipital,* the *masto-parietal,* and

* See footnote, page 200.

(in the young skull) the *basilar*; and lodges the cerebellum, pons Varolii, and medulla oblongata. It is separated from the middle fossa in and near the median line by the dorsum ephippii, and on each side by the superior border of the petrous portion of the temporal bone. This border serves for the attachment of the tentorium cerebelli, is grooved for the superior petrosal sinus, and at its inner extremity presents a notch, upon which the fifth nerve rests. The circumference of the fossa is bounded posteriorly by the grooves for the lateral sinuses. In the centre of this fossa is the *foramen magnum*, bounded on either side by a rough tubercle, which gives attachment to the odontoid or check ligaments; and a little above these are seen the internal openings of the *anterior condyloid foramina*, through which pass the hypoglossal nerve and a meningeal branch from the ascending pharyngeal artery. In front of the foramen magnum is a grooved surface, formed by the basilar process of the occipital bone and by the posterior third of the superior surface of the body of the sphenoid, which supports the medulla oblongata and pons Varolii, and articulates on each side with the petrous portion of the temporal bone, forming the *petro-occipital suture*, the anterior half of which is grooved for the inferior petrosal sinus, the posterior half being encroached upon by the *foramen lacerum posterius*, or *jugular foramen*. Through the anterior portion of this foramen passes the inferior petrosal sinus; through the posterior, the lateral sinus and some meningeal branches from the occipital and ascending pharyngeal arteries; and through the middle, the glosso-pharyngeal, pneumogastric, and spinal accessory nerves. Above the jugular foramen is the *internal auditory meatus*, for the facial and auditory nerves and auditory artery; behind and external to this is the slit-like opening leading into the aquæductus vestibuli, which lodges the ductus endolymphaticus; while between these, and near the superior border of the petrous portion, is a small, triangular depression, the remains of the fossa subarcuata, which lodges a process of the dura mater and occasionally transmits a small vein. Behind the foramen magnum are the *inferior occipital fossæ*, which lodge the hemispheres of the cerebellum, separated from one another by the *internal occipital .crest*, which serves for the attachment of the falx cerebelli, and lodges the occipital sinus. The posterior fossæ are surmounted by the deep transverse grooves for the lodgment of the *lateral sinuses*. These channels, in their passage to the jugular foramina, groove the occipital bone, the posterior inferior angle of the parietal, the mastoid portion of the temporal, and the jugular process of the occipital, and terminate at the back part of the jugular foramen. Where this sinus grooves the mastoid portion of the temporal bone, the orifice of the *mastoid foramen* may be seen; and, just previous to its termination, the *posterior condyloid foramen* opens into it. Neither foramen is constant.

The **External Surface** of the Base of the Skull (*norma basalis*) is extremely irregular (fig. 264). It is bounded in front by the incisor teeth in the upper jaw; behind, by the superior curved lines of the occipital bone; and laterally by the alveolar arch, the lower border of the malar bone, the zygoma, and an imaginary line, extending from the zygoma to the mastoid process and extremity of the superior curved line of the occiput. It is formed by the palate processes of the superior maxillary and palate bones, the vomer, the pterygoid processes, under surface of the great wing, spinous processes and part of the body of the sphenoid, the under surface of the squamous, mastoid, and petrous portions of the temporal, and the under surface of the occipital bone. The anterior part of the base of the skull is raised above the level of the rest of this surface (when the skull is turned over for the purpose of examination), and is bounded in front and laterally by the alveolar process, which is thicker behind than in front, and excavated by sixteen depressions for lodging the teeth of the upper jaw, the cavities varying in depth and size according to the teeth they contain. Immediately behind the incisor teeth is the *anterior palatine fossa*. At the bottom of this fossa may usually be seen four apertures: two placed laterally, the *foramina of Stenson*, which open above, one in the floor of each nostril, and transmit the anterior branches of the posterior palatine vessels, and two in the median line in the intermaxillary suture, the *foramina of Scarpa*, one in front of the other, the anterior transmitting the left, and the posterior (the larger) the right, naso-palatine nerve. These two latter canals are sometimes wanting, or they may join to form a single one, or one of them may open into one of the lateral canals above referred to. The palatine vault is concave, uneven, perforated by numerous foramina, marked by depressions for the palatine glands, and crossed by a crucial suture, formed by the junction of the four bones of which it is composed. At the front part of this surface, a delicate

FIG. 264.—Base of the skull. External surface.

linear suture may be seen in the young skull, passing outwards and forwards from the anterior palatine fossa to the interval between the lateral incisor and canine teeth, and marking off the pre-maxillary portion of the bone (fig. 242). At each posterior angle of the hard palate is the *posterior palatine foramen*, for the transmission of the posterior palatine vessels and large descending palatine nerve ; and running forwards and inwards from it a groove, for the same vessels and nerve. Behind the posterior palatine foramen is the *tuberosity of the palate-bone*, perforated by one or more accessory posterior palatine canals, and marked by the commencement of a ridge, which runs transversely inwards, and serves for the attachment of the tendinous expansion of the Tensor palati muscle. Projecting backwards from the centre of the posterior border of the hard palate is the *posterior nasal spine*, for the attachment of the Azygos uvulæ muscle. Behind and above the hard palate are the posterior apertures of the nares (*choanæ*), separated by the vomer, bounded above by the body of the sphenoid, below by the horizontal plate of the palate-bone, and laterally by the internal pterygoid plate of the sphenoid. Each aperture measures about an inch in the vertical, and about half an inch in the transverse direction. At the base of the vomer may be seen the expanded alæ of this bone, receiving between them the rostrum of the sphenoid. Near the lateral margins of the vomer, at the root of the pterygoid processes, are the *pterygo-palatine canals*. The pterygoid process, which bounds the posterior nares on each side, presents near its base the *pterygoid* or *Vidian canal*, for the Vidian nerve and artery. Each process consists of two plates, which bifurcate at the extremity to receive the tuberosity of the palate-bone, and are separated behind by the pterygoid fossa, which lodges the Internal pterygoid muscle. The internal plate is long and narrow, presenting on the outer side of its base the *scaphoid fossa*, for the origin of the Tensor palati muscle, and at its extremity the *hamular process*, around which the tendon of this muscle turns. The external pterygoid plate is broad, forms the inner boundary of the zygomatic fossa, and affords attachment, by its outer surface, to the External pterygoid muscle.

Behind the nasal fossæ in the middle line is the basilar surface of the occipital bone, presenting in its centre the *pharyngeal spine* for the attachment of the Superior constrictor muscle of the pharynx, with depressions on each side for the insertion of the Rectus capitis anticus major and minor. At the base of the external pterygoid plate is the *foramen ovale*, for the transmission of the third division of the fifth nerve, the small meningeal artery, and some-times the small petrosal nerve ; behind this, the *foramen spinosum*, which transmits the middle meningeal artery and veins, and the prominent spinous process of the sphenoid, which gives attachment to the internal lateral ligament of the lower jaw and the Tensor palati muscle. External to the spinous process is the *glenoid fossa*, divided into two parts by the Glaserian fissure (page 191), the anterior portion concave, smooth, bounded in front by the eminentia articularis, and serving for the articulation of the condyle of the lower jaw ; the posterior portion rough, bounded behind by the tympanic plate, and serving for the reception of part of the parotid gland. Emerging from between the laminæ of the vaginal process of the tympanic plate is the *styloid process* ; and at the base of this process is the *stylo-mastoid foramen*, for the exit of the facial nerve, and entrance of the stylo-mastoid artery. External to the stylo-mastoid foramen is the *auricular fissure*, for the auricular branch of the pneumogastric, between the tympanic plate and the mastoid process. Upon the inner side of the mastoid process is a deep groove, the *digastric fossa* ; and a little more internally, the *occipital groove*, for the occipital artery. At the base of the internal pterygoid plate is a large and somewhat triangular aperture, the *foramen lacerum medium*, bounded in front by the great wing of the sphenoid, behind by the apex of the petrous portion of the temporal bone, and internally by the body of the sphenoid and basilar process of the occipital bone ; it presents in front the posterior orifice of the Vidian canal ; behind, the aperture of the carotid canal. The basilar surface of this opening is filled up in the recent state by a fibro-cartilaginous plate, across the upper or cerebral aspect of which the internal carotid artery passes. External to this aperture, the *petro-sphenoidal suture* is observed, at the outer termination of which are the orifice of the bony part of the Eustachian tube and the orifice of the canal for the Tensor tympani muscle. Behind this suture is the under surface of the petrous portion of the temporal bone, presenting, from within outwards, the quadrilateral, rough surface, part of which affords attachment to the Levator palati muscle ; external to this surface the orifice of the carotid canal and, to its inner side, the depression leading to the aquæductus cochleæ, the former transmitting the internal carotid artery and the carotid plexus of the sympathetic, the latter serving for the passage of a vein from the cochlea. Behind the carotid canal is a large aperture, the *jugula.· foramen*, formed in front by the petrous portion of the temporal, and behind by the occipital ; it is generally larger on the right than on the left side, and is partly subdivided into three compartments. The anterior is for the passage of the inferior petrosal sinus ; the posterior for the lateral sinus and some meningeal branches from the occipital and ascending pharyngeal arteries ; the central one for the glosso-pharyngeal, pneumogastric, and spinal accessory nerves. On the ridge of bone dividing the carotid canal from the jugular foramen is the small foramen for the transmission of Jacobson's nerve (tympanic branch of the glosso-pharyngeal) ; and on the wall of the jugular foramen, near the root of the styloid process, is the small aperture for the transmission of Arnold's nerve (auricular branch of the vagus). Behind the basilar surface of the occipital bone is the *foramen magnum*, bounded on each side by the condyles, rough internally for the attachment of the check or odontoid ligaments, and presenting externally a rough surface, the *jugular process*, which serves for the attachment of the Rectus capitis lateralis muscle and the lateral occipito-atlantal ligament. The foramen magnum transmits the medulla oblongata and its membranes, the spinal accessory nerves, the vertebral arteries, the anterior and posterior spinal arteries, and the occipito-axial ligaments. On either side of each condyle anteriorly is the *anterior condyloid fossa*, perforated by the anterior condyloid foramen, for the passage of the hypoglossal nerve and a meningeal artery. Behind each condyle is the *posterior condyloid fossa*, perforated on one or both sides by the posterior condyloid foramen, for the transmission of a vein to the lateral sinus. Behind the foramen magnum is the *external occipital crest*, terminating above at the *external occipital protuberance*, while on each side are seen the *superior* and *inferior curved lines* ; these, as well as the surfaces of bone between them, are rough for the attachment of the muscles, which are enumerated on page 185.

LATERAL REGION OF THE SKULL

The **Lateral Region** of the Skull (fig. 265) is of a somewhat triangular form, the base of the triangle being formed by a line extending from the external angular process of the frontal bone along the temporal ridge backwards to the outer extremity of the superior curved line of the occiput : and the sides by two lines, the one drawn downwards and backwards from the external angular process of the frontal bone to the angle of the lower jaw, the other from the angle of the jaw upwards and backwards to the outer extremity of the superior curved line. This region is divisible into three portions—temporal fossa, mastoid portion, and zygomatic fossa.

THE TEMPORAL FOSSA

The **Temporal Fossa** is bounded above and behind by the temporal ridges, which extend from the external angular process of the frontal upwards and backwards across the frontal and parietal bones, and then curve downwards and forwards to terminate in the posterior root of the zygomatic process (*supra-mastoid crest*). In front, it is bounded by the frontal, malar, and great wing of the sphenoid ; externally, by the zygomatic arch, formed conjointly by the malar and temporal bones ; below, it is separated from the zygomatic fossa by the pterygoid ridge, or infratemporal crest, on the outer surface of the great wing of the sphenoid. This fossa is formed by five bones, part of the frontal, great wing of the sphenoid, parietal, squamous portion of the temporal, and malar bones, and is traversed by six sutures, part of the *transverse facial, spheno-malar, coronal, spheno-parietal, squamo-parietal,* and *squamo-sphenoidal.* The point where the coronal suture crosses the superior temporal ridge is named the *stephanion* ; and the region where the parietal, frontal, squamous temporal, and greater wing of the sphenoid meet at the anterior inferior angle of the parietal bone is named the *pterion*.   This point is a little above the level of the external angular process of the frontal bone and about an inch and a quarter behind it.   This fossa is deeply concave in front, convex behind, traversed by grooves which lodge branches of the deep temporal arteries, and filled by the Temporal muscle.

The Mastoid Portion

The **Mastoid Portion** of the side of the skull is bounded in front by the tubercle of the zygoma ; above, by a line which runs from the posterior root of the zygoma to the end of the mastoid-parietal suture ; behind and below by the

Fig. 265.—Side view of the skull.

masto-occipital suture.   It is formed by the mastoid and part of the squamous and tympanic portions of the temporal bone ; its surface is convex and rough for the attachment of muscles, and presents, from behind forwards, the mastoid foramen, the mastoid process, the suprameatal triangle, the external auditory meatus, bounded behind, below and in front by the tympanic plate, and, most anteriorly, the temporo-maxillary articulation.

The point where the posterior inferior angle of the parietal meets the occipital bone and mastoid portion of the temporal is named the *asterion*.

The Zygomatic Fossa

The **Zygomatic Fossa** (fig. 266) is an irregularly shaped cavity, situated below and on the inner side of the zygoma ; bounded, in front, by the zygomatic surface of the superior maxillary bone and the ridge which descends from its malar process; behind, by the eminentia articularis and spine of sphenoid; above, by the great wing of the sphenoid below the pterygoid ridge and by the under part of the squamous temporal; below, by the alveolar border of the superior maxilla; internally, by the external pterygoid plate; and externally, by the zygomatic arch and by the ramus and coronoid process of the lower jaw. It contains the lower part of the Temporal, the External and Internal pterygoid muscles, together with the internal maxillary artery and vein, and inferior maxillary nerve, and their branches. The foramen ovale and foramen spinosum open on its roof, and the posterior dental canals on its anterior wall. At its upper

FIG. 266.—Zygomatic fossa.

and inner part may be observed two fissures, which together form a **F**-shaped fissure, the horizontal limb being named the spheno-maxillary, and the vertical one the pterygo-maxillary fissure.

The **Spheno-maxillary Fissure,** horizontal in direction, opens into the outer and back part of the orbit. It is formed above by the lower border of the orbital surface of the great wing of the sphenoid; below, by the external border of the orbital surface of the superior maxilla and the orbital process of the palate-bone; externally, by a small part of the malar bone: * internally, it joins at right angles with the pterygo-maxillary fissure. This fissure opens a communication from the orbit into three fossæ—the temporal, zygomatic, and spheno-maxillary; it transmits the superior maxillary nerve and its temporo-malar branch, the infra-orbital vessels, the ascending branches from Meckel's ganglion, and a vein which connects the ophthalmic vein with the pterygoid venous plexus.

The **Pterygo-maxillary Fissure** is vertical, and descends at right angles from the inner extremity of the preceding; it is a triangular interval, formed by the divergence of the superior maxillary bone from the pterygoid process of the sphenoid. It serves to connect the spheno-maxillary fossa with the zygomatic fossa, and transmits branches of the internal maxillary artery.

The Spheno-maxillary Fossa

The **Spheno-maxillary Fossa** is a small, triangular space situated at the angle of junction of the spheno-maxillary and pterygo-maxillary fissures, and placed beneath the apex of the orbit. It is formed above by the under surface of the body of the sphenoid and by the orbital process of the palate-bone; in front, by

* Occasionally the superior maxillary bone and the sphenoid articulate with each other at the anterior extremity of this fissure; the malar is then excluded from it.

the superior maxillary bone ; behind, by the anterior surface of the base of the pterygoid process and lower part of the anterior surface of the great wing of the sphenoid ; internally, by the vertical plate of the palate. This fossa has three fissures terminating in it, the *sphenoidal, spheno-maxillary,* and *pterygo-maxillary*; it communicates with the orbit by the spheno-maxillary fissure ; with the nasal fossæ by the spheno-palatine foramen, and with the zygomatic fossa by the pterygo-maxillary fissure. It also communicates with the cavity of the cranium, and has five foramina opening into it. Of these there are three on the posterior wall : the *foramen rotundum* above ; below, and internal to this, the *Vidian canal*; and still more inferiorly and internally, the *pterygo-palatine canal*. On the inner wall is the *spheno-palatine foramen* by which the spheno-maxillary communicates with the nasal fossa ; and below is the superior orifice of the *posterior palatine canal,* besides occasionally the orifices of the *accessory posterior palatine* canals. The fossa contains the superior maxillary nerve and Meckel's ganglion, and the termination of the internal maxillary artery.

ANTERIOR REGION OF THE SKULL

The **Anterior Region** of the Skull (*norma frontalis*), which forms the face, is of an oval form, presents an irregular surface, and is excavated for the reception of two of the organs of sense, the eye and the nose. It is bounded above by the glabella and margins of the orbit; below, by the lower border of the body of the mandible; on each side, by the malar bone, and ramus of the jaw. In the median line are seen from above downwards the *glabella,* and diverging from it are the *superciliary ridges,* which indicate the situation of the frontal sinuses and support the eyebrows. Beneath the glabella is the *fronto-nasal suture,* the mid-point of which is termed the *nasion,* and below this is the arch of the nose, formed by the nasal bones, and the nasal processes of the superior maxillary. The nasal arch is convex from side to side, concavo-convex from above downwards, presenting in the median line the internasal suture between the nasal bones, and laterally the naso-maxillary suture between the nasal bone and the nasal process of the superior maxillary bone. Below the nasal bones is seen the opening of the anterior nares, which is heart-shaped, with the narrow end upwards, and presents laterally the thin, sharp margins serving for the attachment of the lateral cartilages of the nose, and in the middle line, a prominent process, the *anterior nasal spine.** Still lower is the *intermaxillary suture,* and on each side of it the *incisive fossa.* Beneath this fossa are the alveolar processes of the upper and lower jaws, containing the incisor teeth, and at the lower part of the median line, the *symphysis of the chin,* the *mental process,* with its two *mental tubercles,* separated by a median groove, and the *incisive fossa* of the lower jaw.

On each side, proceeding from above downwards, is the *supra-orbital ridge,* terminating externally in the external angular process at its junction with the malar, and internally in the internal angular process ; towards the inner third of this ridge is the *supra-orbital notch* or *foramen,* for the passage of the supra-orbital vessels and nerve. Beneath the supra-orbital ridge is the opening of the orbit, bounded externally by the orbital ridge of the malar bone ; below, by the orbital ridge formed by the malar and superior maxillary bones ; internally, by the nasal process of the superior maxillary, and the internal angular process of the frontal bone. On the outer side of the orbit is the quadrilateral anterior surface of the malar bone perforated by one or two small malar foramina. Below the inferior margin of the orbit is the *infra-orbital foramen,* the termination of the infra-orbital canal, and beneath this the *canine fossa,* which gives attachment to the Levator anguli oris ; still lower are the alveolar processes, containing the teeth of the upper and lower jaws. Beneath the alveolar arch of the lower jaw

---

* Skulls may be classified according to the shape of the anterior opening of the nose. This is expressed by the *nasal index* thus: $\dfrac{\text{nasal width} \times 100}{\text{nasal height}}$ = the nasal index. Skulls are termed: (1) *leptorhine,* where the nasal index is below 48, as in the mixed Europeans, Eskimos, and American Indians ; (2) *mesorhine,* where the index is between 48 and 53, as in the Chinese, Burmese, and Malays ; (3) *platyrhine,* where the index is above 53, as in the Negro, Kaffir, and native Australian.

is the *mental foramen* for the passage of the mental vessels and nerve, the *external oblique line*, and at the lower border of the bone, at the point of junction of the body with the ramus, a shallow groove for the passage of the facial artery.

THE ORBITS

The **Orbits** (fig. 267) are two quadrilateral pyramidal cavities, situated at the upper and anterior part of the face, their bases being directed forwards and outwards, and their apices backwards and inwards, so that the axes of the two, if continued backwards, would meet over the body of the sphenoid bone. Each orbit is formed of *seven* bones, the frontal, sphenoid, ethmoid, superior maxillary, malar, lachrymal, and palate; but three of these, the frontal, ethmoid, and

FIG. 267.—Anterior region of the skull.

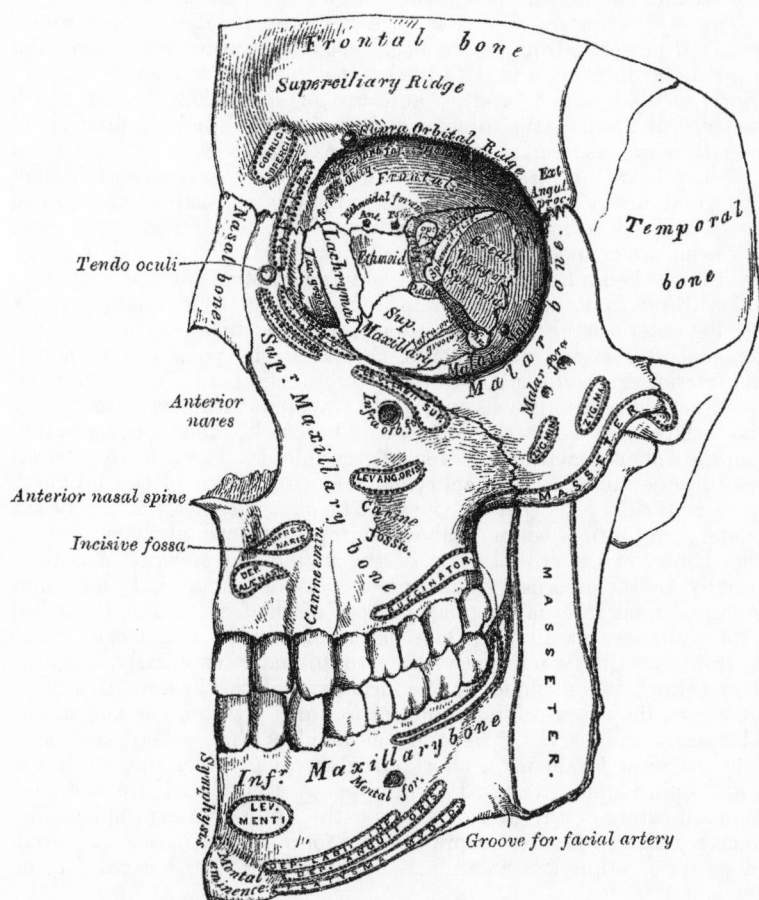

sphenoid, enter into the formation of *both* orbits, so that the two cavities are formed of *eleven* bones only. Each cavity presents for examination a roof, a floor, an inner and an outer wall, four angles, a circumference or base, and an apex. The **roof** is concave, directed downwards and slightly forwards, and formed in front by the orbital plate of the frontal; behind by the lesser wing of the sphenoid. This surface presents internally the depression for the cartilaginous pulley of the Superior oblique muscle; externally, the depression for the lachrymal gland; and posteriorly, the suture connecting the frontal and lesser wing of the sphenoid.

The **floor** is directed upwards and outwards, and is of less extent than the roof; it is formed chiefly by the orbital process of the superior maxillary bone; in front, to a small extent, by the orbital process of the malar, and behind, by

R

the superior surface of the orbital process of the palate. The floor presents at its anterior and internal part, just external to the lachrymal groove, a depression for the attachment of the Inferior oblique muscle; externally, the suture between the malar and superior maxillary bones; near its middle, the infra-orbital groove; and posteriorly, the suture between the maxillary and palate bones.

The **inner wall** is flattened, nearly vertical, and formed from before backwards by the nasal process of the superior maxilla, the lachrymal, os planum of the ethmoid, and a small part of the body of the sphenoid. This surface presents the lachrymal groove, the crest of the lachrymal bone, and the sutures connecting the lachrymal with the superior maxillary, the ethmoid with the lachrymal in front, and the ethmoid with the sphenoid behind.

The **outer wall** is directed forwards and inwards, and is formed in front by the orbital process of the malar bone; behind, by the orbital surface of the greater wing of the sphenoid. On it are seen the orifices of one or two malar canals, and the suture connecting the sphenoid and malar bones.

Angles.—The *superior external angle* is formed by the junction of the upper and outer walls; it presents, from before backwards, the suture connecting the frontal with the malar in front, and with the great wing of the sphenoid behind; quite posteriorly is the foramen lacerum anterius, or sphenoidal fissure, which transmits the third, the fourth, the three branches of the ophthalmic division of the fifth, the sixth nerve, some filaments from the cavernous plexus of the sympathetic, the orbital branch of the middle meningeal artery, a recurrent branch from the lachrymal artery to the dura mater, and the ophthalmic vein. The *superior internal angle* is formed by the junction of the upper and inner wall, and presents the suture connecting the frontal bone with the lachrymal in front, and with the ethmoid behind. The point of junction of the *anterior* border of the lachrymal with the frontal has been named the *dacryon*. This angle presents two foramina, the anterior and posterior ethmoidal, the former transmitting the anterior ethmoidal vessels and nasal nerve, the latter the posterior ethmoidal vessels. The *inferior external angle,* formed by the junction of the outer wall and floor, presents the spheno-maxillary fissure, which transmits the superior maxillary nerve and its orbital branches, the infra-orbital vessels, and the ascending branches from the spheno-palatine or Meckel's ganglion. The *inferior internal angle* is formed by the union of the lachrymal and os planum of the ethmoid, with the superior maxillary and palate bones. The *circumference*, or base, of the orbit, quadrilateral in form, is bounded above by the supra-orbital ridge; below, by the anterior border of the orbital plate of the malar and superior maxillary bones; externally, by the external angular process of the frontal and the malar bones; internally, by the internal angular process of the frontal, and the nasal process of the superior maxillary. The circumference is marked by three sutures, the fronto-maxillary internally, the fronto-malar externally, and the malo-maxillary below; it contributes to the formation of the lachrymal groove, and presents, above, the supra-orbital notch (or foramen), for the passage of the supra-orbital vessels and nerve. The *apex*, situated at the back of the orbit, corresponds to the optic foramen,[*] a short, circular canal, which transmits the optic nerve and ophthalmic artery. It will thus be seen that there are *nine* openings communicating with each orbit—viz. the optic foramen, sphenoidal fissure, spheno-maxillary fissure, supra-orbital foramen, infra-orbital canal, anterior and posterior ethmoidal foramina, malar foramina, and canal for the nasal duct.

The Nasal Fossæ

The **Nasal Fossæ** are two large, irregular cavities, situated one on either side of the middle line of the face, extending from the base of the cranium to the roof of the mouth, and separated from each other by a thin vertical septum. They communicate by two large apertures, the *anterior nares*, with the front of the face; and by the two *posterior nares* with the naso-pharynx behind. These fossæ are much narrower above than below, and in the middle than at the

---

[*] Quain, Testut, and others give the apex of the orbit as corresponding with the inner end of the sphenoidal fissure. It seems better, however, to adopt the statement in the text, since the muscles of the eyeball take origin around the optic foramen, and diverge from it to the globe of the eye.

anterior or posterior openings: their depth, which is considerable, is much greater in the middle than at either extremity. Each nasal fossa communicates with four sinuses, the frontal above, the sphenoidal behind, and the maxillary and ethmoidal on the outer wall. Each fossa also communicates with four cavities: with the orbit by the lachrymal groove, with the mouth by the anterior palatine canal, with the cranium by the olfactory foramina, and with the spheno-maxillary fossa by the spheno-palatine foramen; and they occasionally communicate with each other by an aperture in the septum. The bones entering into their formation are fourteen in number: three of the cranium, the frontal, sphenoid, and ethmoid, and all the bones of the face, excepting the malar and lower jaw. Each cavity is bounded by a roof, a floor, an inner and an outer wall.

The **upper wall**, or **roof** (fig. 268), is long, narrow, and horizontal in its centre, but sloped downwards at its anterior and posterior extremities; it is formed in

FIG. 268.—Roof, floor, and outer wall of left nasal fossa.

front by the nasal bones and nasal spine of the frontal, which are directed downwards and forwards; in the middle, by the cribriform plate of the ethmoid, which is horizontal; and behind, by the under surface of the body of the sphenoid, and sphenoidal turbinated bones, the ala of the vomer and the sphenoidal process of the palate-bone, which are directed downwards and backwards. This surface presents, from before backwards, the internal aspect of the nasal bones; on their outer side, the suture formed between the nasal bone and the nasal process of the superior maxillary; on their inner side, the elevated crest which receives the nasal spine of the frontal and the perpendicular plate of the ethmoid, and articulates with its fellow of the opposite side; while the surface of the bones is perforated by a few small vascular apertures, and presents the longitudinal groove for the nasal nerve: farther back is the suture, connecting the frontal with the nasal in front, and the ethmoid behind, the olfactory foramina and nasal slit on the under surface of the cribriform plate, and the suture between it and the sphenoid behind: quite posteriorly are seen the sphenoidal turbinated bones, the orifices of the sphenoidal sinuses, and the articulation of the alæ of the vomer with the under surface of the body of the sphenoid.

R 2

The **floor** is flattened from before backwards, concave from side to side, and wider in the middle than at either extremity. It is formed in front by the palate process of the superior maxillary; behind, by the palate process of the palate-bone. This surface presents, from before backwards, the anterior nasal spine; behind this, the upper orifices of the anterior palatine canal; internally, the elevated crest which articulates with the vomer; and behind, the suture between the palate and superior maxillary bones, and the posterior nasal spine.

The **inner wall**, or septum (fig. 269), is a thin vertical partition, which separates the nasal fossæ from one another; it is occasionally perforated, so that the fossæ communicate, and it is frequently deflected considerably to one side.* It is formed, in front, by the crest of the nasal bones and nasal spine of the frontal; in the middle, by the perpendicular plate of the ethmoid; behind, by the vomer and rostrum of the sphenoid; below, by the crest of the superior maxillary and palate bones. It presents, in front, a large, triangular notch, which receives the septal cartilage of the nose; and behind, the grooved edge of the vomer. Its

Fig. 269.—Inner wall of nasal fossæ, or septum of nose.

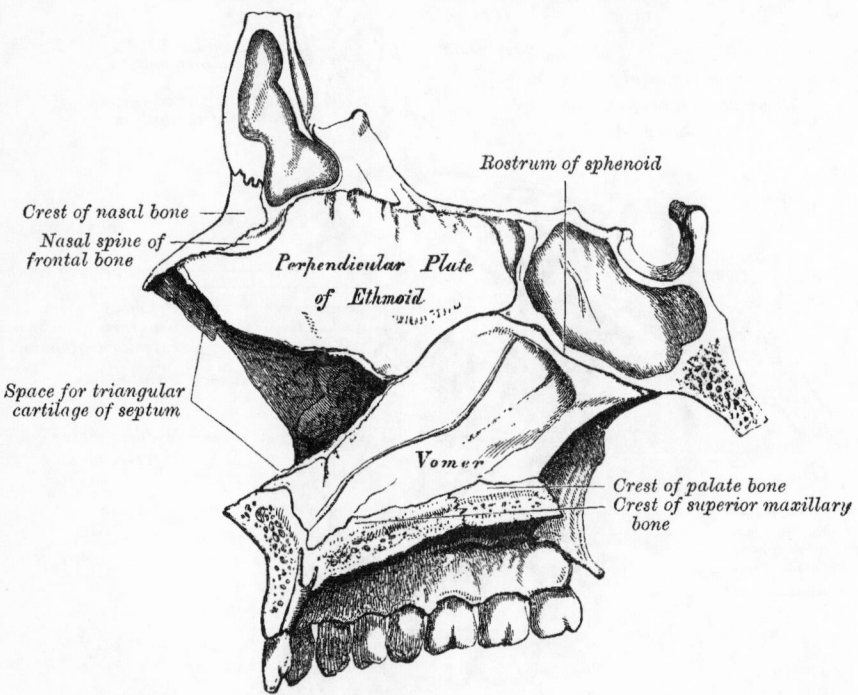

surface is marked by numerous vascular and nervous canals and the groove for the naso-palatine nerve, and is traversed by sutures connecting the bones of which it is formed.

The **outer wall** (fig. 268) is formed, in front, by the nasal process of the superior maxillary and by the lachrymal bone; in the middle, by the ethmoid and inner surface of the superior maxillary and inferior turbinated bones; behind, by the vertical plate of the palate-bone, and the internal pterygoid plate of the sphenoid. This surface presents three irregular longitudinal passages, or *meatuses*, termed the superior, middle, and inferior meatuses of the nose. The *superior meatus*, the smallest of the three, is situated at the upper part of each nasal fossa, occupying the middle third of the outer wall. It lies between the superior and middle turbinated bones, and has opening into it two foramina, the *spheno-palatine* at the back of its outer wall, and the *posterior ethmoidal cells* at the front part of the outer wall. The sphenoidal sinus opens into a recess, the *spheno-ethmoidal recess*, which is placed above and behind the superior turbinated bone. The *middle meatus* is situated between the middle and inferior

* See footnote, page 204.

turbinated bones, and extends from the anterior to the posterior end of the inferior turbinated bone. It presents in front the orifice of the *infundibulum*, by which the middle meatus communicates with the frontal sinus and the anterior ethmoidal cells. The middle ethmoidal cells also open into this meatus, while at the centre of the outer wall is the *orifice of the antrum of Highmore*, which in a considerable percentage of skulls is duplicated. The *inferior meatus*, the largest of the three, is the space between the inferior turbinated bone and the floor of the nasal fossa. It extends almost the entire length of the outer wall of the nose, is broader in front than behind, and presents anteriorly the lower *orifice of the canal for the nasal duct*.

The **anterior nares** present a heart-shaped or pyriform opening, whose long axis is vertical, and narrow extremity upwards. This opening in the recent state is much contracted by the cartilages of the nose. It is bounded above by the inferior border of the nasal bone ; laterally by the thin, sharp margin which separates the facial from the nasal surface of the superior maxillary bone ; and below by the same border, where it slopes inwards to join its fellow of the opposite side at the anterior nasal spine.

The **posterior nares** or **choanæ** are the two posterior oval openings of the nasal fossæ, by which they communicate with the upper part of the pharynx. They are situated below and in front of the basilar process of the occipital bone, and are bounded above by the under surface of the body of the sphenoid and alæ of the vomer ; below, by the posterior border of the horizontal plate of the palate-bone ; externally, by the inner surface of the internal pterygoid plate ; internally, they are separated from each other by the posetrior border of the vomer.

Surface Form.—The various bony prominences or landmarks which are to be easily felt and recognised in the head and face, and which afford the means of mapping out the important structures comprised in this region, are as follows :

1. Supra-orbital arch.	8. Parietal eminence.
2. Internal angular process.	9. Temporal ridge.
3. External angular process.	10. Frontal eminence.
4. Zygomatic arch.	11. Superciliary ridge.
5. Mastoid process.	12. Nasal bone.
6. External occipital protuberance.	13. Lower margin of orbit.
7. Superior curved line of occipital bone.	14. Lower jaw.

1. The *supra-orbital arch* is to be felt throughout its entire extent, covered by the eyebrow. It forms the upper boundary of the circumference or base of the orbit and separates the face from the forehead. It is strong and arched, and terminates internally at the root of the nose, in the *internal angular process*, which articulates with the lachrymal bone. Externally it terminates in the *external angular process* which articulates with the malar bone. This arched ridge is sharper and more defined in its outer than in its inner half, and forms an overhanging process which protects and shields the lachrymal gland. It also protects the eye in its most exposed situation and in the direction from which blows are more likely to descend. At the junction of the inner and middle third of the arch, a slight interruption in the outline may sometimes be felt ; this is the *supra-orbital notch*. Generally the notch is converted into a foramen, and the interruption cannot be felt. A line carried from this notch or foramen downwards over the face, between the two bicuspid teeth of the lower jaw, passes over the infra-orbital and the mental foramina, and thus constitutes a guide to the point of exit from the bones of the face of the three largest cutaneous branches of the fifth cranial nerve. The supra-orbital arch varies in prominence in different individuals. It is more marked in the male than in the female, and in some races of mankind than others. In the less civilised races, as the forehead recedes backwards, the supra-orbital arch becomes more prominent and approaches more to the characters of the monkey tribe, in which the supra-orbital arches are very largely developed and acquire additional prominence from the oblique direction of the frontal bone. 2. The *internal angular process* is scarcely to be felt. Its position is indicated by the angle formed by the supra-orbital arch with the nasal process of the superior maxillary bone and the lachrymal bone at the inner side of the orbit. Between the internal angular processes of the two sides is a broad surface which assists in forming the root of the nose, and immediately above this a broad, smooth, somewhat triangular area, the *glabella*, situated between and connecting the superciliary ridges. 3. The *external angular process* is much more strongly marked than the internal, and is plainly to be felt. It is formed by the junction or confluence of the supra-orbital and temporal ridges, and, articulating with the malar bone, it serves to a very considerable extent to support the bones of the face. In carnivorous animals the external angular process does

not articulate with the malar, and therefore this lateral support to the bones of the face is not present. 4. The *zygomatic arch* can be plainly felt throughout its entire length, being situated almost immediately under the skin. It is formed by the malar bone and the zygomatic process of the temporal bone. At its anterior extremity, it is broad and constitutes the prominence of the cheek ; the posterior part is narrower, and terminates just in front and a little above the tragus of the external ear : its upper border may be traced backwards, as the posterior root, above the tragus and the external auditory meatus to join the posterior part of the lower temporal ridge, forming the *supramastoid crest*. A spot in this line, immediately in front of the upper border of the tragus and between it and the condyle of the jaw, is known as the *pre-auricular point*. This is an important landmark, since the temporal vessels and the auriculo-temporal nerve cross it, and two inches vertically above it is the lower end of the fissure of Rolando. The lower border is more plainly to be felt than the upper, in consequence of the dense temporal fascia being attached to the latter, which somewhat obscures its outline. Its shape differs very much in individuals and in different races of mankind. In the most degraded type of skull— as, for instance, in the skull of the negro of the Guinea Coast—the malar bones project forwards and not outwards, and the zygoma at its posterior extremity extends farther outwards before it is twisted on itself to be prolonged forwards. This makes the zygomatic arch stand out in bold relief, and affords greater space for the Temporal muscle. In skulls which have a more pyramidal shape, as in the Esquimaux or Greenlander, the malar bones do not project forwards and downwards under the eyes, as in the preceding form, but take a direction outwards, forming with the zygoma a large, rounded sweep or segment of a circle. Thus it happens that if two lines are drawn from the zygomatic arches, touching the temporal ridges, they meet above the top of the head, instead of being parallel, or nearly so, as in the European skull, in which the zygomatic arches are not nearly so prominent. This gives to the face a more or less oval type. 5. Behind the ear is the *mastoid portion of the temporal bone*, plainly to be felt, and terminating below in a nipple-shaped process. Its anterior border can be traced immediately behind the concha, and its apex is about on a level with the lobule of the ear. It is rudimentary in infancy, but gradually develops in childhood, and is more marked in the negro than in the European. 6. The *external occipital protuberance (inion)* is always plainly to be felt just at the level where the skin of the neck joins that of the head. At this point the skull is thick for the purposes of safety, while radiating from it are numerous curved arches or buttresses of bone which give to this portion of the skull further security. 7. Running outwards on either side from the external occipital protuberance is an arched ridge of bone, which can be more or less plainly perceived. This is the *superior curved line* of the occipital bone, and gives attachment to some of the muscles which keep the head erect on the spine ; accordingly, we find it more developed in the negro tribes, in whom the jaws are much more massive, and therefore require stronger muscles to prevent their extra weight carrying the head forwards. Below this line the surface of bone at the back of the head is obscured by the overlying muscles, except in the middle line, where the external occipital crest can generally be felt at the bottom of the nuchal furrow. Above it, the vault of the cranium is thinly covered with soft structures, so that the form of this part of the head is almost exactly that of the upper portion of the occipital, the parietal, and the frontal bones themselves ; and, in bald persons, even the lines of junction of the bones, especially the junction of the occipital and parietal at the lambdoid suture, may be defined as a slight depression, caused by the thickening of the borders of the bones in this situation. 8. In the line of the greatest transverse diameter of the head are generally to be found the *parietal eminences*, one on each side of the middle line ; though sometimes these eminences are not situated at the point of the greatest transverse diameter, which is at some other prominent part of the parietal region. Each denotes the point where ossification of the parietal bone began. They are much more prominent and better marked in early life, in consequence of the sharper curve of the bone at this period, so that it describes the segment of a smaller circle. Later in life, as the bone grows, the curve spreads out and forms the segment of a larger circle, and the eminence becomes less distinguishable. In consequence of this sharp curve of the bone in early life, the whole of the vault of the skull has a squarer shape than it has in later life, and this appearance may persist in some rickety skulls. The eminence is more apparent in the negro's skull than in that of the European. This is due to greater flattening of the temporal fossa in the former skull to accommodate the larger Temporal muscle which exists in these races. The parietal eminence is particularly exposed to injury from blows or falls on the head, but fracture is to a certain extent prevented by the shape of the bone, which forms an arch, so that the force of the blow is diffused over the bone in every direction. 9. At the side of the head is the *temporal ridge*. Commencing at the external angular process, it may be felt as a curved ridge, passing upwards and then curving backwards, on the frontal bone, separating the forehead from the temporal fossa. It may then be traced, passing backwards in a curved direction, over the parietal bone, and, though less marked, still generally to be recognised. Finally, the ridge curves downwards and forwards, and terminates in the posterior root of the zygoma. Victor Horsley has shown, in an article on the ' Topography of the Cerebral Cortex,' that the second temporal ridge (see page 186) can be made out on the living body. 10. The *frontal eminences* vary in prominence in different individuals, and are often not symmetrical on the two sides of the body, one being more pronounced than the other. This is especially noticeable in the skull of the young child or infant, and becomes less marked as age advances. The prominence of the frontal eminences depends more upon the general shape of the whole bone than upon the size of the protuberances themselves. As the skull is more highly developed in consequence of increased intellectual capacity, so the frontal bone becomes more upright, and the frontal eminences stand out in bolder relief. Thus they may be considered as affording, to a certain extent, an indication of the development of the hemispheres of the brain beneath, and of the mental powers of the individual. They are not so much exposed to injury as the parietal eminences. In falls forward the upper extremities are involuntarily thrown out, and break the force of the fall, and thus shield the frontal bone from injury. 11. Below the frontal eminences on the forehead are the *superciliary ridges*, which denote the position of the frontal sinuses, and vary in different individuals, being, as a rule, small in the female, absent in children, and sometimes unusually prominent in the male, when the frontal sinuses are largely developed. The degree of prominence of the superciliary ridges is not, however, necessarily related to the size of the sinuses, for sometimes large sinuses are present in cases where there is but little elevation of the ridges, and in other cases strongly marked ridges may exist, with little formation of air-spaces. They commence on either side of the glabella, and at first present a rounded form, which gradually fades away at their outer ends. 12. The *nasal bones* form the prominence of the nose. They vary much in size and shape, and to them is due the varieties in the contour of this organ and much of the character of the face. Thus, in the Mongolian or Ethiopian they are flat, broad, and thick at their base, giving to these tribes the flattened nose by which they are characterised, and differing very decidedly from the Caucasian, in whom the nose, owing to the shape of the nasal bones, is narrow, elevated at the bridge, and elongated downwards. At the root of the nose, and a little below the glabella, is the fronto-nasal suture (*nasion*). Beneath it, the nasal bones are thin and connected with the cartilages of the nose, and the angle or arch formed by their union serves to throw out the bridge of the nose, and is much more marked in some individuals than others. 13. The *lower margin of the orbit*, formed by the superior maxillary bone and the malar bone, is plainly to be felt throughout its entire length. It is continuous internally with the nasal process of the superior maxillary bone, which forms the inner boundary of the orbit. At the point of junction of the lower margin of the orbit with the nasal process is to be felt a little tubercle, which can be plainly perceived by running the finger along the bone in this situation. This tubercle serves as a guide to the position of the lachrymal sac, which is situated above and behind it. 14. The outline of the *lower jaw* is to be felt throughout its entire length. Just in front of the tragus of the external ear, and below the zygomatic arch, the condyle can be made out. When the mouth is opened, this prominence of bone can be perceived advancing out of the glenoid fossa on to the eminentia articularis, and receding again when the mouth is closed. From the condyle the posterior border of the ramus can be felt extending down to the angle. A line drawn from the condyle to the angle would indicate the exact position of this border. From the angle to the symphysis of the chin, the lower, rounded border of the body of the bone is plainly to be felt. At the point of junction of the two halves of the bone is a well-marked, triangular eminence, the *mental process*, which forms the prominence of the chin.

Surgical Anatomy.—An arrest in the ossifying process may give rise to deficiencies or gaps; or to fissures, which are of importance from a medico-legal point of view, as they are liable to be mistaken for fractures. The fissures generally extend from the margin towards the centre of the bone, but gaps may be found in the middle, as well as at the edges. In course of time they may become covered with a thin lamina of bone.

Occasionally a protrusion of the brain or its membranes may take place through one of these gaps in an imperfectly developed skull, or through one of the sutures, owing to non-closure. When the protrusion consists of membranes only, and is filled with cerebro-spinal fluid, it is called a *meningocele*; when the protrusion consists of brain as well as membranes, it is termed an *encephalocele*; and when the protruded brain is a prolongation from one of the ventricles, and is distended by a collection of fluid from an accumulation in the ventricle, it is termed an *hydrencephalocele*. This latter condition is most commonly met with at the root of the nose, where a protrusion of the anterior horn of the lateral ventricle takes place through a deficiency of the fronto-nasal suture. These malformations are usually found in the middle line, and most frequently at the back of the head, the protrusion taking place through the fissures which separate the centres of ossification from which the tabular portion of the occipital bone is originally developed (see page 185). They generally occur through the upper part of the vertical fissure, which is the last to ossify, but not uncommonly through the lower part, when the foramen magnum may be incomplete. More rarely these protrusions have been met with in other situations than those two above mentioned, both through normal fissures, as the sagittal, lambdoid, and other sutures, and also through abnormal gaps and deficiencies at the sides, and even at the base of the skull.

The chief function of the skull is to protect the brain and its appendages from any form of violence to which they may be subjected. We find, therefore, that those portions of the skull which are most exposed to external violence are thicker than others, which are shielded from injury by overlying muscles. Thus, the skull-cap is thick and dense, whereas the squamous portion of the temporal bone, being protected by the temporal muscle and the inferior occipital fossæ, being shielded by the muscles at the back of the neck, are thin and fragile. Fracture of the skull is further prevented by its elasticity, by its rounded shape, and its construction of a number of secondary elastic arches, each made up of a single bone. The manner in which vibrations are transmitted through the bones of the skull is also of importance as regards its protective mechanism, at all events as far as the base is concerned. In the vault, the bones being of a fairly equal thickness and density, vibrations are transmitted in a uniform manner in all directions. But, owing to the varying thickness and density of the bones forming the base, this is not so ; and therefore we find in this situation special buttresses which serve to carry the vibrations in certain definite directions. At the front of the skull, on either side is the ridge which separates the anterior from the middle fossa of the base ; and behind, the ridge or buttress which separates the middle from the posterior fossa ; and if any violence is applied to the vault, the vibrations would be carried along these buttresses to the sella turcica, where they meet. This part has been termed the 'centre of resistance,' and here there is a special protective mechanism to guard the brain. The subarachnoid space is dilated, and an increased quantity of cerebro-spinal fluid acts as a water-cushion to shield the brain from injury. In like manner, when violence is applied to the base of the skull, as in falls upon the feet, the vibrations are carried backwards through the occipital crest, and forwards through the basilar process and body of the sphenoid to the vault of the skull.

Fractures of the skull are best considered as affecting either the vault or the base. Fractures of the vault may, and generally do, involve the whole thickness of the bone ; but sometimes one table may be fractured without any corresponding injury to the other. Thus, the outer table of the skull may be splintered and driven into the diploë, or in the frontal or mastoid regions into the frontal or mastoid cells, without any injury to the internal table. And, on the other hand, the internal table has been fractured, and portions of it depressed and driven inwards, without any fracture of the outer table. As a rule, in fractures of the skull, the inner table is more splintered and comminuted than the outer, and this is due to several causes. It is thinner and more brittle ; the force of the violence as it passes inwards becomes broken up, and is more diffused by the time it reaches the inner table ; the bone being in the form of an arch bends as a whole and spreads out, and thus presses the particles together on the convex surface of the arch, i.e. the outer table, and forces them asunder on the concave surface or inner table ; and, lastly, there is nothing firm under the inner table to support it and oppose the force. Fractures of the vault may be simple fissures, or starred and comminuted fractures, and these may be depressed or elevated. These latter cases of fracture with elevation of the fractured portion are uncommon, and can only be produced by direct wound. In comminuted fracture, a portion of the skull is broken into several pieces, the lines of fracture radiating from a centre where the chief impact of the blow was felt ; if depressed, a fissure circumscribes the radiating lines, enclosing a portion of skull. If this area is circular it is termed a 'pond' fracture, and would in all probability have been caused by a round instrument, as a life preserver or hammer ; if elliptical in shape it is termed a 'gutter' fracture, and would owe its shape to the instrument which had produced it, as a poker.

Fractures of the base of the skull may be produced by *indirect* or *direct* violence. I. In the former class of cases the violence is applied to the vertex or some part of the cranial convexity, as when a person falls from a height on to his head and a fracture of the base results. The mechanism of this form of fracture was formerly explained by the doctrine of *contre-coup*, i.e. that the force was transmitted from one side of the skull to the other ; but this idea is now completely exploded, and there are at the present day two theories as the mode of causation of these fractures, both probably being responsible for a certain number of cases. (*a*) Aran's theory of *irradiation* holds that all fractures of the base are produced by a fissure, which starts from the point of injury and radiates to the base. There can be little doubt that many cases of fracture of the base, especially of the middle fossæ, are caused in this way ; but it is insufficient to explain all, since instances have been met with of fracture of the base of the skull in which there has been no fracture of the vault. (*b*) To explain these cases, another theory, known as the *compression* or *bursting* theory, has been suggested. If a hollow, elastic sphere is compressed, from above downwards, it will bulge laterally, and, if the compression is sufficient, it will eventually burst in the situation where it bulges. Now, the skull is an elastic sphere, and when compression is applied to it, its diameter will be reduced along the line of greatest pressure and will therefore be increased in other directions, and may increase to such an extent that bursting may result. In a hollow elastic sphere of uniform thickness, the bulging and subsequent bursting take place at the equatorial line midway between the two points of compression ; but the skull is not of uniform thickness, and therefore the bulging and subsequent bursting take place at the weakest part.

II. Direct violence applied to the base of the skull may cause fracture in several different ways: by the impact of the vertebral column against the condyles of the occipital bone, in falls on the buttocks or feet; by the condyle of the lower jaw being driven against the glenoid fossa, in blows or falls on the chin; by the thrust of a pointed instrument through the orbit or nose; by gunshot wounds through the mouth; and by falls or a stab on the back of the head, fracturing the occipital bone.

In the majority of cases of fracture of the base, the fracture is compound. In the anterior fossa, if the fissure extends across the cribriform plate, the nasal mucous membrane is usually torn and the fracture rendered compound into the nose; or the fissure may extend into the sphenoidal or ethmoid sinuses and thus open up a communication between the fracture and the external air. In the middle fossa, the fracture often opens up the tympanic cavity, and if the membrana tympani is torn, the fracture is compound; or if it runs across the pituitary fossa, the mucous membrane of the roof of the naso-pharynx is generally torn and the fracture rendered compound.

The most common place for fracture of the base to occur is through the middle fossa, and here the fissure usually takes a fairly definite course. Starting from the point struck, which is generally somewhere in the neighbourhood of the parietal eminence, it runs downwards through the parietal and squamous portion of the temporal bone and across the petrous portion of this bone, frequently traversing and implicating the internal auditory meatus, to the middle lacerated foramen. From this it may pass across the body of the sphenoid, through the pituitary fossa, to the middle lacerated foramen of the other side, and may indeed travel round the whole cranium, so as to completely separate the anterior from the posterior part. The course of the fracture should be borne in mind, as it explains the symptoms to which fracture in this region may give rise: thus, if the fissure pass across the internal auditory meatus, injury to the facial and auditory nerves may result, with consequent facial paralysis and deafness; or the tubular prolongation of the arachnoid around these nerves in the meatus may be torn and thus permit of the escape of the cerebro-spinal fluid should there be a communication between the internal ear and the tympanum together with rupture of the membrana tympani, as is frequently the case: again, if the fissure passes across the pituitary fossa and the muco-periosteum covering the under surface of the body of the sphenoid is torn, blood will find its way into the pharynx and be swallowed, and after a time vomiting of blood will result. Fractures of the anterior fossa, involving the bones forming the roof of the orbit and nasal fossa, are generally the result of blows on the forehead; but fracture of the cribriform plate of the ethmoid may be a complication of fracture of the nasal bone. When the fracture implicates the roof of the orbit, the blood finds its way into this cavity, and, travelling forwards, appears as a subconjunctival ecchymosis. If the roof of the nasal fossa be fractured, the blood escapes from the nose. In rare cases there may be also escape of cerebro-spinal fluid from the nose, where the dura mater and arachnoid have been torn. In fractures of the posterior fossa, extravasation of blood may appear at the nape of the neck.

Diseases of the Skull.—Owing to the fact that the pericranium does not play an important part in the nutrition and regeneration of the skull, as does the periosteum in the long bones, it is advisable to discuss *inflammatory* diseases of the skull, as affecting the bones and the membrane covering them together, under the term *osteomyelitis.* This is generally caused by septic infection of a scalp wound, exposing and bruising the bone, or in compound fractures, and is termed *septic osteomyelitis.* Occasionally it may occur independently of injury, and then follows the same course, and is due to the same causes, as acute infective osteomyelitis in the long bones.

The most common chronic diseases of the skull are due to syphilis and tubercle. In *acquired syphilis* the disease usually occurs as *nodes*, which arise most commonly in the pericranium, but may also arise in the diploë, or more rarely on the inner surface of the skull. The formation of gummata under the periosteum generally leads to *caries*, which may be either limited if the gumma is localised, or widespread if the gumma is diffuse. The caries is often complicated by *necrosis*, for a condition of sclerosis is frequently set up in the surrounding bone, and the vessels in the Haversian canals become compressed and the vitality of the bone interfered with; hence we often find a central necrosing area surrounded by a zone of caries. A common result of syphilitic disease of the skull is the production of large, hard masses of bone on its surface, which gives it a tuberculated appearance; in other cases, the skull presents a curious worm-eaten appearance: this is due to the fact that the osteogenetic powers of the periosteum are small and the formation of bone on the surface slight. In *hereditary syphilis*, in addition to the formation of gummata, which are usually of the subperiosteal variety, atrophic or hypertrophic changes may take place. In the atrophic cases, the bone becomes abnormally thin, or even perforated, generally where there is pressure, as from the pillow or nurse's arm. Hence they are usually met with in the parietal bones or vertical plate of the occiput. In the hypertrophic cases, a deposit of porous bone takes place around the anterior fontanelle in the parietal and frontal bones; these deposits are separated by the coronal and sagittal sutures, and give to the skull an appearance like a 'hot cross bun.' They are known as *Parrot's nodes*, and such a skull has received the name of *natiform*, from its fancied resemblance to the buttocks.

Tuberculous disease sometimes occurs in the bones of the skull, and is plainly differentiated from syphilitic disease by the fact that it always commences in the bone itself and only affects the pericranium and dura mater secondarily; and by the fact that it is never accompanied by the formation of bosses or osteophytes, as in syphilis. It may involve the whole thickness of the bone and cause perforation, or it may destroy only the external table and part of the diploë and produce a non-perforating ulcer of the bone. Or, in a third variety, the tuberculous granulation tissue may perforate the internal table of the skull and spread between the bone and dura mater.

The most common tumours of the skull are the osteomata and the sarcomata; but in addition a few cases of hydatid cyst, originating in the diploë, have been recorded. The osteoma is generally the ivory exostosis, though cases of spongy exostosis do occur. Sarcomata of the skull may arise either from the pericranium or the diploë, but it is usually impossible to distinguish clinically between the two. Carcinoma, if it occurs in the skull, is always secondary to cancer in some other part of the body.

Hypertrophic changes occur in the skull in ostitis deformans, acromegaly, leontiasis ossea, and in rickets. In these latter cases the skull becomes enlarged from the formation of a soft, spongy bone, and its shape peculiar: the forehead is high, square and projecting, and the antero-posterior diameter is long, in relation to its transverse diameter. The bones of the face are small and ill-developed, and this gives the appearance of a larger head than actually exists. The bones may become the seat of cranio-tabes, and the anterior fontanelle is long in closing, sometimes not being obliterated till the fifth or sixth year of life.

The mastoid antrum, situated in the mastoid portion of the temporal bone, is sometimes the seat of suppuration from a direct extension backwards from the tympanic cavity. In such cases, the surgeon has to open the antrum in order to give exit to the pus. This he does by introducing his gouge in the suprameatal triangle (see page 192). A line is drawn horizontally through the upper border of the bony external auditory meatus, and a second vertically through the posterior wall of the meatus, and the gouge is applied in the angle where these two lines intersect each other. If the instrument is introduced at a higher level, it will open the cavity of the skull. It is to be carried in the direction of the external auditory canal—inwards, forwards, and a little upwards—for the distance of from 1 to 1½ cm., when the antrum will be reached. In some cases of middle-ear trouble, septic thrombosis of the lateral sinus takes place, and it becomes necessary to trephine and explore the sinus. This is done by applying the trephine at a point an inch behind the centre of the external auditory meatus, and a quarter of an inch above Reid's base-line: that is to say, a line drawn backwards from the lower margin of the orbit through the centre of the external auditory meatus. In cases of abscess of the brain from middle-ear disease, when the abscess is presumably either in the temporo-sphenoidal lobe or the cerebellum, the operation devised by Dean should be performed. The pin of an average-sized trephine is to be introduced an inch and a quarter behind and a quarter of an inch above the centre of the external auditory meatus, and a crown of bone removed. This will expose a part of the lateral sinus, and the dura mater just above it. By slightly enlarging the opening in an upward direction with gouge forceps, the temporo-sphenoidal lobe of the cerebrum can be explored. Should no pus be found here, the opening must be enlarged in a similar manner downwards and backwards. This will expose the whole of the lateral sinus and the dura mater below it and by incising the membrane in this situation, the cerebellum can be explored.

In connection with the bones of the face a common malformation is *cleft palate*. The cleft usually starts posteriorly, and in its most elementary form may consist simply of a bifid uvula; or the cleft may extend through the soft palate; or the posterior part or the whole of the hard palate may be involved, the cleft extending as far forwards as the anterior palatine canal. In the severest forms, the cleft extends through the alveolus and may pass either between the two maxillary bones, or laterally between the pre-maxillary bone and the rest of the upper jaw; that is to say, between the lateral incisor and canine teeth. In some instances, the cleft has been noticed to pass outwards between the central and lateral incisor teeth; and this has induced some anatomists to believe that the premaxillary bone is developed from two centres, and not from one, as was stated in the description of the bone. Thus, Albrecht believes that, instead of there being two premaxillary bones, each supporting two incisor teeth, there are four bones, two on either side of the median line, and each one carrying one tooth. The mesial segment, bearing a central incisor, is called an *endo-gnathion*; the lateral segment, bearing the lateral incisor, is called a *meso-gnathion*; and the rest of the maxilla is termed the *exo-gnathion*. When a cleft palate is lateral, it may affect one or both sides. If the latter, the central part is frequently displaced forwards and may be united to the septum of the nose, the deficiency in the alveolus being complicated with a cleft in the lip (hare-lip). On examining a cleft palate in which the alveolus is not implicated, the cleft will generally appear to be in the mesial line, but occasionally is unilateral and in some cases bilateral. To understand this it must be borne in mind that three processes are concerned in the formation of the palate—the two palatal processes of the superior maxilla, which grow in horizontally and unite in the middle line; and the ethmo-vomerine process, which grows downwards from the base of the skull and fronto-nasal process to unite with the palatal process in the mesial line. In those cases where the palatal processes fail to unite with each other and with the mesial process, the cleft of the palate is median : where one palatal process unites with the mesial septum, the other failing to do so, the appearance of the cleft in the palate is unilateral. The right process is the one which usually joins, and the cleft is therefore on the left side. In some cases where the palatal processes fail to meet in the middle, the ethmo-vomerine process grows downwards into the cleft and thus gives the appearance of a bilateral cleft. Occasionally there may be a hole in the middle line of the hard palate, the anterior part of the hard and the soft palate being perfect. But this is rare, because, as a rule, the union of the various processes progresses from before backwards ; and therefore the posterior part of the palate is more frequently involved than the anterior.

The bones of the face are sometimes fractured as the result of direct violence. The two most commonly broken are the nasal bone and the inferior maxilla, and of these, the latter is by far the most frequently fractured of all the bones of the face. Fracture of the *nasal* bone is for the most part transverse, and takes place about half an inch from the free margin. The broken portion may be displaced backwards or more generally to one side by the force which produced the lesion, as there are no muscles here which can cause displacement. The *malar* bone is probably never broken alone—that is to say, unconnected with a fracture of the other bones of the face. The *zygomatic arch* is occasionally fractured, and when this occurs from direct violence, as is usually the case, the fragments may be displaced inwards. Fractures of the *superior maxilla* may vary much in degree, from the chipping off of a portion of the alveolar arch, to an extensive comminution of the whole bone from severe violence, as the kick of a horse. The most common situation for a fracture of the *inferior maxillary bone* is in the neighbourhood of the canine tooth, as at this spot the jaw is weakened by the deep socket for the fang of this tooth ; it is next most frequently fractured at the angle ; then at the symphysis, and finally the neck of the condyle or the coronoid process may be broken. Occasionally a double fracture may occur, one in either half of the bone. The fractures are usually compound, from laceration of the mucous membrane covering the gums. The displacement is mainly the result of the same violence as produced the injury, but may be further increased by the action of the muscles passing from the neighbourhood of the symphysis to the hyoid bone.

The superior and inferior maxillary bones are both of them frequently the seat of necrosis ; though the disease more often affects the lower than the upper jaw, probably on account of the greater supply of blood to the latter. It may be the result of periostitis, from tooth irritation, injury, or the action of some specific poison, as syphilis, or from salivation by mercury ; it sometimes occurs in children after attacks of the exanthematous fevers, and a special form occurs from the action of the fumes of phosphorus in persons engaged in the manufacture of matches.

Tumours attack the jaw-bones not infrequently, and these may be either innocent or malignant : in the upper jaw cysts may occur in the antrum, constituting the so-called dropsy of the antrum ; or, again, cysts may form in either jaw in connection with the teeth : either cysts connected with the roots of fully developed teeth, the 'dental cyst ;' or cysts connected with imperfectly developed teeth, the 'dentigerous cyst.' Solid innocent tumours include the fibroma, the chondroma, and the osteoma. Of malignant tumours there are two classes, the sarcoma and the epithelioma. The sarcomata are of various kinds, the spindle-celled and round-celled of a very malignant character, and the myeloid sarcoma principally affecting the alveolar margin of the bone. Of the epitheliomata we find the squamous variety spreading to the bone from the palate or gum, and the cylindrical epithelioma originating in the antrum or nasal fossæ.

Both superior and inferior maxillary bones occasionally require removal for tumours and for some other conditions. In order to remove the upper jaw the patient should be placed in the recumbent position, in a good light, with the head and shoulders just raised. The central incisor tooth, on the affected side, is then extracted. An incision is made, commencing just below the inner canthus of the eye, along the side of the nose, round the ala, and down the middle line of the upper lip into the mouth. A second incision is made from the commencement of the first, along the lower border of the orbit as far as the prominence of the malar bone. The flap thus formed is reflected outwards, so as to expose the bone. All bleeding is now arrested, before proceeding with the operation. The periosteum attached along the lower margin of the orbit is now to be incised, and with the handle of the scalpel the periosteum covering the floor of the orbit is to be raised from the bone ; for in all cases it is essential that this fibrous layer should be left intact. The mouth is now widely opened with a gag, and the mucous membrane covering the hard palate incised down to the bone in the middle line, and the soft palate separated from the hard. The surgeon now proceeds to divide the connections of the bone with the other bones of the face, having first separated the ala of the nose from its bony attachment. This is done partly with a narrow saw and partly with cutting bone forceps. They are (1) the junction with the malar bone, the line of section being carried into the sphenomaxillary fissure : (2) the nasal process of the superior maxillary bone ; a small portion of its upper extremity, connected with the nasal bone in front, the lachrymal bone behind, and the frontal bone above, being left : (3) the connection with the bone on the opposite side and with the palate bone in the roof of the mouth. The bone is now firmly grasped with lion-forceps; and by means of a rocking movement upwards and downwards, the remaining attachments of the orbital plate with the ethmoid, and of back of the bone with the palate, broken through. Occasionally, in removing the upper jaw, it will be found that the orbital plate can be saved, and this should always be done if possible. A horizontal saw-cut is to be made just below the infra-orbital foramen, and the bone cut through in this situation.

Removal of one half of the lower jaw is sometimes required. The patient should be placed in a recumbent position, in a good light, with the head and shoulders well raised. The central incisor tooth on the affected side is first extracted. An incision is then made in the middle line, through the lower lip and over the chin, to just below the symphysis of the jaw. From the termination of this the incision is carried outwards, along the lower margin of the bone to its angle. This will necessarily divide the facial artery and vein, which must be at once secured. The incision is now to be carried upwards along the posterior border of the ramus of the bone, to a point a little below the lobule of the ear. The flap thus formed is now raised by dividing the reflection of the mucous membrane from the lip on to it and the muscles attached to its external surface, including the Masseter. The jaw is then divided by sawing the bone through the socket of the extracted tooth. By dividing the bone in this situation, and not through the symphysis, the attachment of the Genio-hyo-glossus to the genial tubercle is preserved, and all danger of the tongue falling backwards and occluding the larynx is prevented. The bone being divided and pulled outwards, the knife is inserted along the inner side of the bone, and the muscles attached to this surface separated from it; in doing this, care must be taken to keep close to the bone, to avoid injuring the submaxillary gland. The anterior part of the bone being thus freed, the surgeon seizes it in his left hand, and by strongly everting it, allows space to divide the Internal pterygoid muscle and the inferior dental vessels and nerve. He now depresses it, so as to bring into view the coronoid process, from which he separates the Temporal muscle. When the tendon of this muscle has been completely divided, the bone can be still further depressed; and this should be done without any eversion, for fear of danger to the internal maxillary artery, which would be brought up into the wound by everting the bone. When the bone is sufficiently depressed, the condyle of the jaw will come into view. The capsular ligament is opened in front with the point of the knife, and the ligamentous tissues on the other side of the joint carefully divided. It is wiser now to abandon the use of the knife, and tear through the External pterygoid muscle and other structures by twisting out the bone. If the knife is used it should be kept closely in contact with the bone, for the danger of wounding the internal maxillary artery in this stage of the operation is very considerable.

The antrum of Highmore occasionally requires tapping for suppuration. This may be done through the socket of a tooth, preferably the first molar, the fangs of which are most intimately connected with the antrum, or through the facial aspect of the bone above the alveolar process. This latter method does not perhaps afford such efficient drainage, but there is less chance of food finding its way into the cavity. The operation may be performed by incising the mucous membrane above the second molar tooth, and driving a trocar or any sharp-pointed instrument into the cavity.

THE THORAX

The skeleton of the **Thorax**, or **Chest**, is an osseo-cartilaginous cage, containing and protecting the principal organs of respiration and circulation. It is conical in shape, being narrow above and broad below, flattened from before backwards, and longer behind than in front. It is somewhat reniform on transverse section.

Boundaries.—The *posterior* surface is formed by the twelve dorsal or thoracic vertebræ and the posterior part of the ribs. It is concave from above downwards, and presents on each side of the middle line a deep groove, in consequence of the direction backwards and outwards which the ribs take from their vertebral extremities to their angles. The *anterior* surface is flattened or slightly convex, and inclined forwards from above downwards. It is formed by the sternum and costal cartilages. The *lateral* surfaces are convex; they are formed by the ribs, separated from each other by spaces, the *intercostal spaces*. These spaces are eleven in number, and are occupied by the intercostal muscles and membranes.

The *upper opening* of the thorax is reniform in shape, being broader from side to side than from before backwards. It is formed by the first dorsal vertebra behind, the upper margin of the sternum in front, and the first rib on each side. It slopes downwards and forwards, so that the anterior part of the ring is on a lower level than the posterior. The antero-posterior diameter is about two inches and the transverse about four. The *lower* opening is formed by the twelfth dorsal vertebra behind, by the eleventh and twelfth ribs at the sides, and in front by the cartilages of the tenth, ninth, eighth, and seventh ribs, which ascend on either side and form an angle, the *subcostal angle*, from the apex of which the ensiform cartilage projects. It is wider transversely than from before backwards. It

FIG. 270.—Anterior surface of sternum and costal cartilages.

slopes obliquely downwards and backwards; so that the cavity of the thorax is much deeper behind than in front. The Diaphragm closes in the opening, forming the floor of the thorax.

In the female, the thorax differs from that of the male as follows: 1. Its general capacity is less. 2. The sternum is shorter. 3. The upper margin of the sternum is on a level with the lower part of the body of the third dorsal vertebra, whereas in the male it is on a level with the lower part of the body of the second dorsal vertebra. 4. The upper ribs are more movable, and so allow a greater enlargement of the upper part of the thorax than in the male.

The Sternum

The **Sternum** (στέρνον, the chest) (figs. 270, 271) is a flat, narrow bone, situated in the median line of the front of the chest, and consisting, in the adult, of three portions. It has been likened to an ancient sword : the upper piece, representing the handle, is termed the *manubrium ;* the middle and largest piece, which represents the chief part of the blade, is termed the *gladiolus ;* and the inferior piece, which is likened to the point of the sword, is termed the *ensiform process* or *xiphoid appendix.* In its natural position its inclination is oblique from above, downwards and forwards. It is slightly convex in front and concave behind, broad above, becoming narrowed at the point where the first and second pieces are connected ; after which it again widens a little, and is pointed at its extremity. Its average length in the adult is about seven inches, being rather longer in the male than in the female.

Fig. 271.—Posterior surface of sternum.

The **Manubrium** (*presternum*) is of a somewhat triangular form, broad and thick above, narrow below at its junction with the middle piece. Its *anterior surface,* convex from side to side, concave from above downwards, is smooth, and affords attachment on each side to the Pectoralis major and sternal origin of the Sterno-cleido-mastoid muscles. In well-marked bones the ridges limiting the attachment of these muscles are very distinct. Its *posterior surface,* concave and smooth, affords attachment on each side to the Sterno-hyoid and Sterno-thyroid muscles. The *superior border,* the thickest, presents at its centre the *presternal notch ;* and on each side an oval articular surface, directed upwards, backwards, and outwards, for articulation with the sternal end of the clavicle. The *inferior border* presents an oval, rough surface, covered in the recent state with a thin layer of cartilage, for articulation with the second portion of the bone. The *lateral borders* are marked above by a depression for the first costal cartilage, and below by a small facet, which, with a similar facet on the upper angle of the middle portion of the bone, forms a notch for the reception of the costal cartilage of the second rib. These articular surfaces are separated by a narrow, curved edge, which slopes from above downwards and inwards.

The **Gladiolus** (*mesosternum*), considerably longer, narrower, and thinner than the first piece, is broader below than above. Its *anterior surface* is nearly flat, directed upwards and forwards, and marked by three transverse ridges which cross the bone opposite the third, fourth, and fifth articular depressions.* At the junction of the third and fourth pieces is occasionally seen an orifice, the *sternal foramen,* of varying size and form. This surface affords attachment on each side to the sternal origin of the Pectoralis major. The *posterior surface,* slightly concave, is also marked by three transverse lines ; but they are less distinct than those in front : this surface affords attachment below, on each side, to the Triangularis sterni muscle, and occasionally presents the posterior opening of the sternal foramen. The *superior border* has an oval surface

---

* Patterson (*The Human Sternum,* 1904), who examined 524 specimens, points out that these ridges are altogether absent in 26·7 per cent.; that in 69 per cent. a ridge exists opposite the third costal attachment; in 39 per cent., opposite the fourth; and in 4 per cent. only, opposite the fifth.

for articulation with the manubrium. The *inferior border* is narrow, and articulates with the ensiform appendix. Each *lateral border*, at its superior angle, has a small facet, which, with a similar facet on the manubrium, forms a cavity for the cartilage of the second rib ; below this are four angular depressions which receive the cartilages of the third, fourth, fifth, and sixth ribs, while the inferior angle has a small facet, which, with a corresponding one on the ensiform appendix, forms a notch for the cartilage of the seventh rib. These articular depressions are separated by a series of curved interarticular intervals, which diminish in length from above downwards, and correspond to the intercostal spaces. Most of the cartilages belonging to the true ribs, as will be seen from the foregoing description, articulate with the sternum at the lines of junction of its primitive component segments. This is well seen in many of the lower animals, where the separate parts of the bone remain ununited longer than in man. In this respect a striking analogy exists between the mode of connection of the ribs with the vertebral column, and the connection of the costal cartilages with the sternum.

The **Ensiform Process** or **Xiphoid Appendix** (*metasternum*) is the smallest of the three pieces : it is thin and elongated in form, cartilaginous in structure in youth, but more or less ossified at its upper part in the adult. Its *anterior surface* affords attachment to the chondro-xiphoid ligament and a small part of the Rectus abdominis ; its *posterior surface*, to some of the fibres of the Diaphragm and Triangularis sterni muscles ; its *lateral borders*, to the aponeurosis of the abdominal muscles. Above, it articulates with the lower end of the gladiolus, and at each superior angle presents a facet for the lower half of the cartilage of the seventh rib ; below, by its pointed extremity, it gives attachment to the linea alba. This portion of the sternum varies much in appearance, being sometimes pointed, broad and thin, sometimes bifid, or perforated by a round hole, occasionally curved, or deflected considerably to one or the other side.

Structure.—The bone is composed of delicate, highly vascular cancellous tissue, covered by a thin layer of compact bone, which is thickest in the manubrium, between the articular facets for the clavicles.

Development.—The cartilaginous sternum originally consists of two bars, situated one on either side of the mesial plane and connected with the cartilages of the nine upper ribs of its own side. These two bars fuse with each other along the middle line, and the bone, including the ensiform appendix, is developed by *six* centres : one for the first piece or manubrium, four for the second piece or gladiolus, and one for the ensiform appendix (fig. 272). Up to the middle of foetal life the sternum is entirely cartilaginous, and when ossification takes place the ossific granules are deposited in the middle of the intervals between the articular depressions for the costal cartilages, in the following order : in the manubrium and first piece of the gladiolus, during the sixth month ; in the second and third pieces of the gladiolus, during the seventh month ; in its fourth piece, within the first year, or between the first and second years after birth ; and in the ensiform appendix, between the fifth and eighteenth years, by a single centre which makes its appearance at the upper part, and proceeds gradually downwards. To these may be added the occasional existence, as described by Breschet, of two small episternal centres, which make their appearance one on each side of the presternal notch. They are probably vestiges of the episternal bone of the monotremata and lizards.[*] It occasionally happens that some of the segments are formed from more than one centre, the number and position of which vary (fig. 274). Thus, the first piece may have two, three, or even six centres. When two are present, they are generally situated one above the other, the upper one being the larger ; [†] the second piece has seldom more than one ; the third, fourth, and fifth pieces are often formed from two centres placed laterally, the irregular union of which will serve to explain the rare occurrence of the sternal foramen (fig. 275), or of the vertical fissure which occasionally intersects this part of the bone,

---

[*] Out of 141 sterna between the ages of birth and sixteen years, Patterson (*op. cit.*) found the fourth or lowest centre for the gladiolus present only in 38 cases, i.e. 26·9 per cent.

[†] Sir George Humphry states that this is ' probably the more complete condition.'

FIG. 272.—Development of the sternum, by six centres.

Time
of
appearance

1 for manubrium   } 6th month

2 }
4 for gladiolus { 3 } 7th month
4 }
5 1st year after birth

1 for ensiform
cartilage 5th to 18th year

FIG. 273.

Time
of
union

Rarely unite, except in old age

Between puberty and the 25th year

Soon after puberty

Partly cartilaginous to advanced life

FIG. 274.—Peculiarities.

In
number of
centres

for first piece, two or more centres

for second piece, usually one

for third
for fourth } 2, placed laterally
for fifth

FIG. 275.

And in
mode of
union

Arrest of development of lateral pieces,
producing:

Sternal fissure, and

Sternal foramen

and which is further explained by the manner in which the cartilaginous matrix, in which ossification takes place, is formed. Union of the various centres

Fig. 276.—A central rib of left side.

of the gladiolus commences about puberty, from below, and proceeds upwards, so that by the age of twenty-five they are all united, and this portion of bone consists of one piece (fig. 273). The ensiform cartilage may become joined to the gladiolus before thirty, but this occurs more frequently after forty; on the other hand, it sometimes remains ununited in old age. The manubrium is occasionally joined to the gladiolus in advanced life by bone. When this union takes place, however, it is generally only superficial, the central portion of the sutural cartilage remaining unossified.

Articulations.—With the clavicles and seven costal cartilages on each side.

Attachment of Muscles.—To nine pairs and one single muscle: the Pectoralis major, Sterno-cleido-mastoid, Sterno-hyoid, Sterno-thyroid, Triangularis sterni, the aponeuroses of the Obliquus externus, Obliquus internus, Transversalis, and Rectus muscles, and Diaphragm.

THE RIBS

The **Ribs** are elastic arches of bone, which form the chief part of the thoracic walls. They are twelve in number on each side; but this number may be increased by the development of a cervical or lumbar rib, or may be diminished to eleven. The first seven are connected behind with the spine, and in front with the sternum, through the intervention of the costal cartilages; they are called *true* ribs (*vertebro-sternal*).* The remaining five are *false* ribs; of these, the first three have their cartilages attached to the cartilage of the rib above (*vertebro-chondral*): the last two are free at their anterior extremities and are termed *floating* (*vertebral*) ribs. The ribs vary in their direction, the upper ones being less oblique than the lower. The extent of obliquity reaches its maximum at the ninth rib, and gradually decreases from that rib to the twelfth. The ribs are situated one below the other in such a manner that spaces are left between them, which are called *intercostal spaces*. The length of these spaces corresponds to the length of the ribs and their cartilages; their breadth is greater in front than behind, and between the upper than between the lower ribs. The ribs increase in length from the first to the seventh, below which they diminish to the twelfth.

* Sometimes the eighth rib cartilage articulates with the sternum; this condition occurs more frequently on the right than on the left side.

S

In breadth they decrease from above downwards ; in the upper ten the greatest breadth is at the sternal extremity.

Common Characters of the Ribs (fig. 276).—A rib from the middle of the series should be taken in order to study the common characters of these bones.

Each rib presents two extremities, a posterior or vertebral, an anterior or sternal, and an intervening portion—the body or shaft.

The **posterior** or **vertebral extremity** presents for examination a head, neck, and tuberosity. The **head** (fig. 277) is marked by a kidney-shaped articular surface, divided by a horizontal ridge into two facets for articulation with the depression formed by the junction of the bodies of two contiguous dorsal vertebræ ; the upper facet is small, the inferior one of larger size ; the ridge separating them serves for the attachment of the interarticular ligament. The **neck** is the flattened portion of the rib which extends outwards from the head ; it is about an inch long, and is placed in front of the transverse process of the lower of the two vertebræ with which the head articulates. Its *anterior surface* is flat and smooth, its *posterior* rough, for the attachment of the middle costo-transverse ligament, and perforated by numerous foramina, the direction of which is less constant than those found on the inner surface of the shaft. Of its two borders the *superior* presents a rough crest for the attachment of the anterior costo-transverse ligament ; its *inferior border* is rounded. On the posterior surface of the neck, just where it joins the shaft, and nearer the lower than the upper border, is an eminence—the **tubercle** ; it consists of an articular and a non-articular portion. The *articular portion*, the more internal and inferior of

FIG. 277.—Vertebral extremity of a rib.   External surface.

The two, presents a small, oval surface for articulation with the extremity of the transverse process of the lower of the two vertebræ to which the head is connected. The *non-articular portion* is a rough elevation, which affords attachment to the posterior costo-transverse ligament. The tubercle is much more prominent in the upper than in the lower ribs.

The **shaft** is thin and flat, so as to present two surfaces, an external and an internal ; and two borders, a superior and an inferior. The *external surface* is convex, smooth, and marked, at its back part, a little in front of the tuberosity, by a prominent line, directed obliquely from above downwards and outwards ; this gives attachment to a tendon of the Ilio-costalis muscle, or of one of its accessory portions, and is called the *angle*. At this point the rib is bent in two directions. If the rib is laid upon its lower border, it will be seen that the portion of the shaft in front of the angle rests upon this border, while the portion of the shaft behind the angle is bent inwards and at the same time tilted upwards. The interval between the angle and the tuberosity increases gradually from the second to the tenth rib. The portion of bone between these two parts is rounded, rough, and irregular, and serves for the attachment of the Longissimus dorsi muscle. The portion of bone between the tubercle and sternal extremity is also slightly twisted upon its own axis, the external surface looking downwards behind the angle and a little upwards in front of it. This surface presents, towards its sternal extremity, an oblique line, the *anterior angle*. The *internal surface* is concave, smooth, directed a little upwards behind the angle, a little downwards in front of it, and is marked by a ridge which commences at the lower extremity of the head ; this ridge is strongly marked as far as the inner side of the angle, and gradually becomes lost at the junction of the anterior with the middle third of the bone. The interval between it and the inferior border presents a groove, *subcostal*, for the intercostal vessels and nerve. At the back part of the bone, this groove belongs to the inferior border, but just in front of the angle, where it is deepest and broadest, it corresponds to the internal surface. The superior edge of the groove is rounded; it serves for the attachment of the Internal intercostal muscle. The inferior edge corresponds to the lower margin of the rib, and gives attachment to the External intercostal muscle. Within the groove are seen the orifices of numerous small foramina, which traverse the wall of the shaft obliquely from before backwards. The *superior border*, thick and rounded, is marked by an external and an internal lip, more distinct behind than in front, which serve for the attachment of the External and Internal intercostal muscles. The *inferior border*, thin and sharp, has attached to it the External intercostal muscle. The **anterior** or **sternal extremity** is flattened, and presents a porous, oval, concave depression, into which the costal cartilage is received.

PECULIAR RIBS

The ribs which require especial consideration are five in number—viz. the first, second, tenth, eleventh, and twelfth.

The **first rib** (fig. 278) is one of the shortest and the most curved of all the ribs; it is broad and flat, its surfaces looking upwards and downwards, and its borders inwards and outwards. The *head* is of small size, rounded, and presents only a single articular facet for articulation with the body of the first dorsal vertebra. The *neck* is narrow and rounded. The *tuberosity*, thick and prominent, rests on the outer border. There is no angle, but in this situation the rib is slightly bent, with the convexity of the bend upwards, so that the head of the bone is directed downwards. The *upper surface* of the shaft is marked by two shallow depressions, separated from each other at the inner border of the rib by a small rough surface, the *scalene tubercle*, for the attachment of the Scalenus anticus muscle—the groove in front of it transmitting the subclavian vein, that behind it the subclavian artery. Between the groove for the subclavian artery and the tuberosity is a second rough surface for the attachment of the Scalenus medius muscle. The *under surface* is smooth, and destitute of a subcostal groove. The *outer border* is convex, thick, and rounded, and at its posterior part gives attachment to the first serration of the Serratus magnus; the *inner* is concave, thin, and sharp, and marked about its centre by the scalene tubercle. The *anterior extremity* is larger and thicker than any of the other ribs.

The **second rib** (fig. 279) is much longer than the first, but bears a very considerable resemblance to it in the direction of its curvature. The non-articular portion of the tuberosity is occasionally only feebly marked. The *angle* is slight, and situated close to the tuberosity, and the shaft is not twisted, so that both ends touch any plane surface upon which it may be laid; but there is a similar though smaller bend, with its convexity upwards, to that found in the first rib. The shaft is not horizontal, like that of the first rib; its *outer surface*, which is convex, looking upwards and a little outwards. It presents, near the middle, a rough eminence for the attachment of the lower part of the first and the whole of the second digitation of the Serratus magnus; behind and above which is attached the Scalenus posticus. The *inner surface*, smooth and concave, is directed downwards and a little inwards: it presents a short subcostal groove towards its posterior part.

The **tenth rib** (fig. 280) has only a single articular facet on its head.

The **eleventh** and **twelfth ribs** (figs. 281 and 282) have each a single articular facet on the head, which is of rather large size; they have no neck or tuberosity, and are pointed at the extremity. The eleventh has a slight angle and a shallow subcostal groove. The twelfth has neither, and is much shorter than the eleventh, and the head has a little inclination downwards. Sometimes the twelfth rib is even shorter than the first.

Structure.—The ribs consist of highly vascular cancellous tissue, enclosed in a thin, compact layer.

Development.—Each rib, with the exception of the last two, is developed by *three* centres: one each for the shaft, the head, and the tubercle. The last

two ribs have only *two* centres, that for the tubercle being wanting.   Ossification commences in the shaft at a very early period, before its appearance
in the vertebræ.   The epiphysis of the head, which is of a slightly angular shape,
and that for the tubercle, of a lenticular form, make their appearance between the
sixteenth and twentieth years, and are not united to the rest of the bone until
about the twenty-fifth year.

Peculiar ribs.

FIG. 278.

1st
Shortest
Surfaces horizontal

has a single artic. facet

SCALENUS MEDIUS

Groove for Subclavian Vein

Groove for Subclavian Artery

no Angle

First serration
of Serratus magnus

FIG. 279.

Angle
slightly marked
and close to
tuberosity

2.nd

Second
Serration

Third Serration

Rough muscular eminence

FIG. 280.

Single articular facet —

Tuberosity   10th

Angle

FIG. 281.

Single articular facet -

no
Tuberosity   11th

Angle

FIG. 282.

Single articular facet —

Tuberosity   no

12 th   no
Angle

Attachment of Muscles.—To nineteen : the Internal and External intercostals,
Scalenus anticus, Scalenus medius, Scalenus posticus, Pectoralis minor, Serratus
magnus, Obliquus externus, Quadratus lumborum, Diaphragm, Latissimus dorsi,
Serratus posticus superior, Serratus posticus inferior, Ilio-costalis, Musculus
accessorius ad ilio-costalem, Longissimus dorsi, Cervicalis ascendens, Levatores
costarum, and Infracostales.

The Costal Cartilages

The **Costal Cartilages** (fig. 270, page 253) are bars of white, hyaline cartilage, which serve to prolong the ribs forward to the front of the chest, and contribute very materially to the elasticity of its walls. The first seven are connected with the sternum, the next three with the lower border of the cartilage of the preceding rib. The cartilages of the last two ribs have pointed extremities, which terminate in free ends in the walls of the abdomen. Like the ribs, the costal cartilages vary in their length, breadth, and direction. They increase in length from the first to the seventh, then gradually diminish to the last. Their breadth, as well as that of the intervals between them, diminishes from the first to the last. They are broad at their attachment to the ribs, and taper towards their sternal extremities, excepting the first two, which are of the same breadth throughout, and the sixth, seventh, and eighth, which are enlarged where their margins are in contact. In direction they also vary: the first descends a little, the second is horizontal, the third ascends slightly, while all the rest follow the course of the ribs for a short extent, and then ascend to the sternum or preceding cartilage. Each costal cartilage presents two surfaces, two borders, and two extremities. The *anterior surface* is convex, and looks forwards and upwards : that of the first gives attachment to the costo-clavicular ligament and the Subclavius muscle ; that of the second, third, fourth, fifth, and sixth, at their sternal ends, to the Pectoralis major.* The others are covered by, and give partial attachment to, some of the great flat muscles of the abdomen. The *posterior surface* is concave, and directed backwards and downwards, the first giving attachment to the Sterno-thyroid, the third to the sixth inclusive to the Triangularis sterni, and the six or seven inferior ones to the Transversalis muscle and the Diaphragm. Of the two borders, the *superior* is concave, the *inferior* convex : they afford attachment to the Internal intercostal muscles, the upper border of the sixth giving attachment to the Pectoralis major muscle. The contiguous borders of the sixth, seventh, and eighth, and sometimes the ninth and tenth, costal cartilages present small, smooth, oblong-shaped facets at the points where they articulate. Of the two extremities, the *outer* one is continuous with the osseous tissue of the rib to which it belongs. The *inner* extremity of the first is continuous with the sternum ; the six succeeding ones have rounded extremities, which are received into shallow concavities on the lateral margins of the sternum. The inner extremities of the eighth, ninth, and tenth costal cartilages are pointed, and are connected with the cartilage above. Those of the eleventh and twelfth are free and pointed.

The costal cartilages are most elastic in youth, those of the false ribs being more so than those of the true. In old age they become of a deep yellow colour, and are prone to calcify.

Attachment of Muscles.—To nine : the Subclavius, Sterno-thyroid, Pectoralis major, Internal oblique, Transversalis, Rectus, Diaphragm, Triangularis sterni, and Internal intercostals.

Surface Form.—The bones of the chest are to a very considerable extent covered by muscles, so that in the strongly developed muscular subject they are for the most part concealed. In the emaciated subject, on the other hand, the ribs, especially in the lower and lateral regions, stand out as prominent ridges with the sunken, intercostal spaces between them.

In the middle line, in front, the superficial surface of the sternum is to be felt throughout its entire length, at the bottom of a deep median furrow situated between the two great Pectoral muscles and called the *sternal furrow*. These muscles overlap the anterior surface somewhat, so that the whole of the sternum in its entire width is not subcutaneous, and this overlapping is greater opposite the centre of the bone than above and below, so that the furrow is wide at its upper and lower part, but narrower in the middle. The centre of the upper border of the sternum is visible, constituting the *presternal notch*, and is in the same horizontal plane as the lower border of the body of the second dorsal vertebra ; the lateral parts of this border are obscured by the tendinous origins of the Sterno-mastoid muscles, which present themselves as oblique tendinous cords, which narrow and deepen the notch. Lower down on the subcutaneous surface a well-defined transverse ridge, the *angle of Ludovic*, is always to be felt. This denotes the line of junction of the manubrium and body of the bone ; it lies in the same horizontal plane as

---

* The first and seventh occasionally give origin to the same muscle.

the body of the fifth dorsal vertebra, and is a useful guide to the second costal cartilage, and thus to the identity of any given rib. The second rib being found, through its costal cartilage, it is easy to count downwards and find any other. From the middle of the sternum the furrow spreads out, and, exposing more of the surface of the bone, terminates at the junction of the body of the bone with the ensiform cartilage, on the same level as the intervertebral cartilage between the ninth and tenth dorsal vertebræ. Immediately below this is the *infrasternal notch*; between the points of junction of the seventh costal cartilages to the sternum, and below the notch, is a triangular depression, the *epigastric fossa*, or pit of the stomach (scrobiculus cordis), bounded laterally by the cartilages of the seventh ribs; in it the ensiform cartilage can be felt. The sternum in its vertical diameter presents a general convexity forwards, the most prominent point of which is at the joint between the manubrium and gladiolus.

On each side of the sternum the costal cartilages and ribs on the front of the chest are partially obscured by the great Pectoral muscle; through which, however, they are to be felt as ridges, with yielding intervals between them, corresponding to the intercostal spaces. Of these spaces, the one between the second and third ribs is the widest, the next two somewhat narrower, and the remainder, with the exception of the last two, comparatively narrow.

The lower border of the Pectoralis major muscle corresponds to the fifth rib, and below this, on the front of the chest, the broad, flat outline of the ribs, as they begin to ascend, and the more rounded outline of the costal cartilages, are often visible. The lower boundary of the front of the thorax, the *abdomino-thoracic arch*, which is most plainly seen by bending the body backwards, is formed by the ensiform cartilage and the cartilages of the seventh, eighth, ninth, and tenth ribs, and the extremities of the eleventh and twelfth ribs or their cartilages.

On each side of the chest, from the axilla downwards, the flattened external surfaces of the ribs may be defined in the form of oblique ridges, separated by depressions corresponding to the intercostal spaces. They are, however, covered by muscles, which, when strongly developed, obscure their outline to a certain extent. Nevertheless, the ribs, with the exception of the first, can generally be followed over the front and sides of the chest without difficulty. The first rib, being almost completely covered by the clavicle and scapula, can only be distinguished in a small portion of its extent. At the back, the angles of the ribs form a slightly marked oblique line, on each side of and some distance from the vertebral spines. This line diverges somewhat as it descends, and external to it is a broad, convex surface, caused by the projection of the ribs beyond their angles. Over this surface, except where covered by the scapula, the individual ribs can be distinguished.

For clinical purposes and convenience of description, the surface of the chest has been mapped out by arbitrary lines into certain definite areas. Of these lines, some are vertical and others transverse. On the front of the chest the most important vertical lines are the *mid-sternal*, which run down the median line of the sternum and the *mammary*, which runs vertically downwards from a point midway between the centre of the presternal notch and the tip of the acromion process. This line, if prolonged, will cross Poupart's ligament at a point midway between the anterior superior spine of the ilium and the symphysis pubis. The transverse lines are also two in number, and are drawn across the front of the chest at the levels of the junction of the third and sixth costal cartilages to the sternum. These lines divide the front of the thorax on either side into the following regions. Above the upper transverse line, the space to the inner side of the mid-clavicular line is the *upper sternal* region, and to the outer side the *infraclavicular* region. The space between the two transverse lines to the inner side of the mid-clavicular line is the *lower sternal* region; and that to the outer side of the same line, the *mammary* region. The space below the lower transverse line is the *inframammary* region. The lower transverse line, if continued on to the lateral area of the thorax, divides this surface into an *axillary* region above the line and an *infra-axillary* region below the line. This lateral area is bounded in front and behind by two vertical lines—that in front, the *anterior axillary line*, being drawn from the anterior fold of the axilla; and that behind, the *posterior axillary line*, from the posterior fold of the axilla. By some, this lateral area of the thorax is further divided by a *mid-axillary* line, drawn downwards from the apex of the axilla. On the posterior aspect of the thorax the *scapular line* is drawn vertically through the inferior aspect of the scapula.

Surgical Anatomy.—Malformations of the sternum present nothing of surgical importance beyond the fact that abscesses of the mediastinum may sometimes escape through the sternal foramen. Fracture of the sternum is by no means common, owing, no doubt, to the elasticity of the ribs and their cartilages which support it like so many springs. When broken, it is frequently associated with fracture of the spine, and may be caused by forcibly bending the body either backwards; or forwards until the chin becomes impacted against the top of the sternum. It may also be fractured by direct violence or by muscular action. The fracture usually occurs in the upper half of the body of the bone. Dislocation of the gladiolus from the manubrium also takes place, and is sometimes described as a fracture.

The bone being subcutaneous is frequently the seat of gummatous tumours and not uncommonly is affected with caries. Occasionally the bone, and especially its ensiform appendix, becomes altered in shape and driven inwards by the pressure, in workmen, of tools against their chest.

The ribs are frequently broken, though from their connections and shape they are able to withstand great force, yielding under the injury and recovering themselves like a spring. The middle ones of the series are the most liable to fracture. The first and to a less extent the second, being protected by the clavicle, are rarely fractured ; and the eleventh and twelfth on account of their loose and floating condition enjoy a like immunity. The fracture generally occurs from indirect violence, from forcible compression of the chest wall, and the bone then gives way at its weakest part, i.e. just in front of the angle. But the ribs may also be broken by direct violence, when the bone gives way and is driven inwards at the point struck, or they may be broken by muscular action. It seems probable, however, that in these latter cases the bone has undergone some atrophic changes. Fracture of the ribs is frequently complicated with some injury to the viscera contained within the thorax or upper part of the abdominal cavity, and this is most likely to occur in fractures from direct violence.

Fracture of the costal cartilages may also take place, though it is a comparatively rare injury.

The thorax is frequently found to be altered in shape in certain diseases.

In *rickets*, the ends of the ribs, where they join the costal cartilages, become enlarged, giving rise to the so-called 'rickety rosary.' Outside these enlargements the softened ribs sink in, so as to present a groove passing downwards and outwards on each side of the sternum. This bone is forced forwards by the bending of the ribs, and the antero-posterior diameter of the chest is increased. The ribs from the second to the eighth are the ones affected, the lower ones being prevented from falling in by the presence of the liver, stomach, and spleen ; and when the liver and spleen are enlarged, as they sometimes are in rickets, the lower ribs may be pushed outwards. causing a transverse groove (Harrison's sulcus) just above the costal arch. This deformity is known under the name of 'pigeon-breast,' and is primarily due to some chronic obstruction to the entry of air into the thorax, though it is more prone to occur in the softened bones of rickety children than in the healthy, where the resistance of the thoracic walls is greater. The *phthisical chest* is often long and narrow, flattened from before backwards, and with great obliquity of the ribs and projection of the scapulæ. In *pulmonary emphysema* the chest is enlarged in all its diameters, and presents on section an almost circular outline. It has received the name of the 'barrel-shaped chest.' In severe cases of *lateral curvature of the spine*, the thorax becomes much distorted. In consequence of the rotation of the bodies of the vertebræ, which takes place in this disease, the ribs opposite the convexity of the dorsal curve become extremely convex behind, being thrown out and bulging, and at the same time flattened in front, so that the two ends of the same rib are almost parallel. Coincident with this the ribs on the opposite side, on the concavity of the curve, are sunk and depressed behind, and bulging and convex in front. In addition to this the ribs often become welded together by bony material.

The chest walls are sometimes unduly rigid, owing to excessive deposit of calcareous matter in the bones and costal cartilages. This is usually found in old age, but occasionally in the young, and interferes more or less with the respiratory movements. Not infrequently the ribs are the seat of a tuberculous periostitis, leading to necrosis. The disease is usually situated on the internal surface of the rib ; and hence the best treatment is to excise the portion of the rib which is the seat of the disease, when a carious cavity containing a sequestrum will be found on its internal surface.

In cases of empyema the chest requires opening to evacuate the pus. There is considerable difference of opinion as to the best position for doing this. Probably the best place in most cases will be found to be between the fifth and sixth ribs, in or a little in front of the mid-axillary bone. This is the last part of the cavity to be closed by the expansion of the lung ; it is not thickly covered by soft parts ; the space between the two ribs is sufficiently great to allow of the introduction of a fair-sized drainage tube, and the opening is in a dependent position, when the patient is confined to bed, as he usually inclines towards the affected side, so as to allow the sound lung the freest possible play, and this position permits of efficient drainage.

THE EXTREMITIES

The extremities, or limbs, are those long, jointed appendages of the body which are connected to the trunk by one end, and free in the rest of their extent. They are *four* in number : an *upper* or *thoracic pair*, connected with the thorax through the intervention of the shoulder, and subservient mainly to prehension ; and a *lower pair*, connected with the pelvis, intended for support and locomotion. Both pairs of limbs are constructed after one common type. so that they present numerous analogies ; while, at the same time, certain differences are observed between the upper and lower pair, dependent on the peculiar offices they have to perform.

The bones by which the upper and lower limbs are attached to the trunk are named respectively the *shoulder* and *pelvic girdles*, and they are constructed on the same general type, though presenting certain modifications relating to the different uses to which the upper and lower limbs are respectively applied. The *shoulder girdle* is formed by the scapulæ and clavicles, and is imperfect in front and behind. In front, however, the girdle is completed by the upper end of the sternum, with which the inner extremities of the clavicles articulate. Behind, the girdle is widely imperfect, and the scapulæ are connected to the trunk by muscles only. The *pelvic girdle* is formed by the innominate bones, and is completed in front, through the symphysis pubis, at which the two innominate bones articulate with each other. It is imperfect behind, but the intervening gap is filled in by the upper part of the sacrum. The pelvic girdle therefore presents, with the sacrum, a complete ring, comparatively fixed, and presenting an arched form which confers upon it a solidity manifestly intended for the support of the trunk, and in marked contrast to the lightness and mobility of the shoulder girdle.

With regard to the morphology of these girdles, it will be noticed that they each consist of a dorsal and a ventral section : the dorsal section of the upper extremity is the scapula ; of the lower, the ilium, and these two parts coincide. The ventral section, in the lower extremity, divides into two branches : one, anterior, is the os pubis ; the other, posterior, is the ischium ; in the upper extremity the arrangement is not quite so clear, the posterior branch of the ventral section is generally believed to be the coracoid process and corresponds to the ischium, and the anterior division by some is regarded as concurring with the precoracoid of reptiles and amphibia : which would thus be the homologue of the os pubis. The clavicle is looked upon as of secondary or dermic origin, and as having no representative in the lower limb. Others believe that the clavicle is analogous to the ventral anterior segment of the shoulder girdle, and therefore is the representative of the os pubis.

THE UPPER EXTREMITY

The bones of the upper extremity consist of those of the shoulder girdle, of the arm, the forearm, and the hand.

THE SHOULDER GIRDLE

The Shoulder Girdle consists of two bones, the clavicle and the scapula.

THE CLAVICLE (figs. 283, 284)

The **Clavicle** (clavis, *a key*), or collar-bone, forms the anterior portion of the shoulder girdle. It is a long bone, curved somewhat like the italic letter *f*, and placed nearly horizontally at the upper and anterior part of the thorax, immediately above the first rib. It articulates by its inner extremity with the upper border of the sternum, and by its outer extremity with the acromion process of the scapula ; serving to sustain the upper extremity in the various positions which it assumes, while, at the same time, it allows of great latitude of motion in the arm.* It presents a double curvature, when looked at in front ; the convexity being forwards at the sternal end, and the concavity at the scapular end. Its outer third is flattened from above downwards, and extends, in the natural position of the bone, from a point opposite the coracoid process to the acromion. Its inner two-thirds are of a prismatic form, and extend from the sternum to a point opposite the coracoid process of the scapula.

* The clavicle acts especially as a fulcrum to enable the muscles to give lateral motion to the arm. It is accordingly absent in those animals whose fore limbs are used only for progression, but is present for the most part in those animals whose anterior extremities are clawed and used for prehension, though in some of them—as, for instance, in a large number of the carnivora—it is merely a rudimentary bone suspended among the muscles, and not articulating either with the scapula or sternum.

External or **Flattened Portion**.—The *outer third* is flattened from above down-wards, so as to present two surfaces, an upper and a lower ; and two borders, an anterior and a posterior. The *upper surface* is flattened, rough, marked by impressions for the attachment of the Deltoid in front, and the Trapezius behind ; between these two impressions, externally, a small portion of the bone is sub-cutaneous. The *under surface* is flattened. At its posterior border, a little external to the point where the prismatic joins with the flattened portion, is a rough eminence, the *conoid tubercle* ; this, in the natural position of the bone, surmounts the coracoid process of the scapula, and gives attachment to the conoid ligament. From this tubercle, an oblique line, rarely a depression, passes forwards and outwards to near the outer end of the anterior border ; it is called the *oblique line*, or *trapezoid ridge*, and affords attachment to the trapezoid ligament. The *anterior border* is concave, thin, and rough, and gives attach-ment to the Deltoid ; it occasionally presents, at its inner end, at the commence-ment of the deltoid impression, a tubercle, the *deltoid tubercle*, which is sometimes to be felt in the living subject. The *posterior border* is convex, rough, broader than the anterior, and gives attachment to the Trapezius.

FIG. 283.—Left clavicle. Superior surface.

Sternal extremity   *Acromial extremity*

FIG. 284.—Left clavicle. Inferior surface.

Internal or **Prismatic Portion**.—The prismatic portion forms the *inner two-thirds* of the bone. It is curved so as to be convex in front, concave behind, and is marked by three borders, separating three surfaces. The *anterior border* is continuous with the anterior margin of the flat portion. Its outer portion is smooth, and corresponds to the interval between the attachment of the Pectoralis major and Deltoid muscles ; at the inner half of the clavicle it forms the lower boundary of an elliptical space for the attachment of the clavicular portion of the Pectoralis major, and approaches the posterior border of the bone. The *superior border* is continuous with the posterior margin of the flat portion, and separates the anterior from the posterior surface. Smooth and rounded externally, it becomes rough towards the inner third for the attach-ment of the Sterno-mastoid muscle, and terminates at the upper angle of the sternal extremity. The *posterior* or *subclavian border* separates the posterior from the inferior surface, and extends from the conoid tubercle to the rhomboid impression. It forms the posterior boundary of the groove for the Subclavius muscle, and gives attachment to a layer of cervical fascia, enveloping the Omo-hyoid muscle. The *anterior surface* is included between the superior and anterior borders. Its inner part is directed forwards and upwards ; its outer part looks still more upwards, and becomes continuous with the superior surface of the flat portion. Externally, it is smooth, convex, nearly subcutaneous, being covered only by the Platysma; but, corresponding to the inner half of the bone, it is divided by a more or less prominent line into two parts: a lower portion, elliptical in form, rough, and slightly convex, for the attachment of the Pectoralis major; and an upper part, which is rough, for the attachment of the Sterno-cleido-mastoid. Between the two muscular impressions is a small, subcutaneous interval. The *posterior* or *cervical surface* is smooth, flat, and looks backwards towards the root of the neck. It is limited, above, by the superior border; below, by the subclavian border; internally, by the margin of the sternal extremity; externally, it is continuous with the posterior border of the flat portion. It is concave from within outwards, and is in relation, by its lower part, with the suprascapular vessels. This surface, at the junction of the inner and outer curves, is also in close relation with the brachial plexus and subclavian vessels. It gives attachment, near the sternal extremity, to part of the Sterno-hyoid muscle; and presents, at or near the middle, a foramen, directed obliquely outwards, which transmits the chief nutrient artery of the bone. Sometimes there are two foramina on the posterior surface, or one on the posterior, and another on the inferior surface. The *inferior* or *subclavian surface* is bounded, in front, by the anterior border; behind, by the subclavian border. It is narrow internally, but gradually increases in width externally, and is continuous with the under surface of the flat portion. Commencing at the sternal extremity may be seen a small facet for articulation with the cartilage of the first rib. This is continuous with the articular surface at the sternal end of the bone. External to this is a broad, rough surface, the *rhomboid impression*, rather more than an inch in length, for the attachment of the costo-clavicular (rhomboid) ligament. The remaining part of this surface is occupied by a longitudinal groove, broad and smooth externally, narrow and more uneven internally; it gives attachment to the Subclavius muscle, and, by its margins, to the costo-coracoid membrane, which splits to enclose the muscle. Not infrequently this groove is subdivided into two parts by a longitudinal line, which gives attachment to the intermuscular septum of the Subclavius muscle.

The **inner** or **sternal extremity** of the clavicle is triangular in form, directed inwards, and a little downwards and forwards; and presents an articular facet, concave from before backwards, convex from above downwards, which articulates with the sternum through the intervention of an interarticular fibro-cartilage; the circumference of the articular surface is rough, for the attachment of numerous ligaments. The posterior border of this surface is prolonged backwards so as to increase the size of the articular facet; the upper border gives attachment to the interarticular fibro-cartilage, and the lower border is continuous with the costal facet on the inner end of the inferior or subclavian surface, which articulates with the cartilage of the first rib.

The **outer** or **acromial extremity,** directed outwards and forwards, presents a small, flattened, oval facet, which looks obliquely downwards, for articulation with the acromion process of the scapula. The circumference of the articular facet is rough, especially above, for the attachment of the acromio-clavicular ligaments.

Peculiarities of the Bone in the Sexes and in Individuals.—In the female, the clavicle is generally shorter, thinner, less curved, and smoother than in the male. In those persons who perform considerable manual labour, which brings into constant action the muscles connected with this bone, it becomes thicker and more curved, and its ridges for muscular attachment become prominently marked. The right clavicle is generally longer, thicker, and rougher than the left.

Structure.—The shaft, as well as the extremities, consists of cancellous tissue, invested in a compact layer much thicker in the middle than at either end. The clavicle is highly elastic, by reason of its curves. From the experiments of Ward, it has been shown that it possesses sufficient longitudinal elastic force to project its own weight nearly two feet on a level surface, when a smart blow is struck on it; and sufficient transverse elastic force, opposite the centre of its anterior convexity, to throw its own weight about a foot. This extent of elastic power must serve to moderate very considerably the effect of concussions received upon the point of the shoulder.

Development.—By *two* centres: one for the shaft, and one for the sternal extremity. The centre for the shaft appears very early, before any other bone;

according to Béclard, as early as the thirtieth day. The centre for the sternal end makes its appearance about the eighteenth or twentieth year, and unites with the rest of the bone about the twenty-fifth year.

Articulations.—With the sternum, scapula, and cartilage of the first rib.

Attachment of Muscles. — To six : the Sterno-cleido-mastoid, Trapezius, Pectoralis major, Deltoid, Subclavius, and Sterno-hyoid.

Surface Form.—The clavicle can be felt throughout its entire length, even in persons who are very fat. Commencing at the inner end, the enlarged sternal extremity, where the bone projects above the upper margin of the sternum, can be felt, forming with the sternum and the rounded tendon of the Sterno-mastoid a V-shaped notch, the *presternal notch*. Passing outwards, the shaft of the bone can be defined immediately under the skin, with its convexity forwards in the inner two-thirds, the surface partially obscured above and below by the attachments of the Sterno-mastoid and Pectoralis major muscles. In the outer third it forms a gentle curve backwards, and terminates at the outer end in a somewhat enlarged extremity which articulates with the acromial process of the scapula. The direction of the clavicle is almost, if not quite, horizontal when the arm is lying quietly by the side, though in well-developed subjects it may incline a little upwards at its outer end. Its direction is, however, very changeable with the varying movements of the shoulder-joint. The clavicle inclines backwards, so that its outer or acromial extremity is on a plane posterior to its sternal end. This causes the shoulder to be thrown backwards away from the thorax ; and if a solution of continuity of the bone takes place, the point of the shoulder is carried forwards and inwards, having lost the buttress-like action of the bone.

Surgical Anatomy.—The clavicle is more frequently broken than any other bone in the body. This is due to the fact that it is much exposed to violence, and is the only bony connection between the upper limb and the trunk, acting as a buttress, which keeps the point of the shoulder away from the thorax. The bone, moreover, is slender, and is very superficial. It may be broken by direct or indirect violence, or by muscular action. The most common cause is, however, from indirect violence, as the result of force applied to the hand or shoulder, and the bone then gives way at the junction of its outer with its inner two-thirds : that is to say, at the junction of the two curves, for this is its weakest part. The fracture is generally oblique, and the displacement of the outer fragment is downwards, forwards, and inwards. The deformity is mainly due to the weight of the arm acting upon the fragment, when the buttress-like action of the bone is gone, assisted by the muscles which pass from the thorax to the upper extremity. The inner fragment, as a rule, is little displaced. Beneath the bone the main vessels of the upper limb and the great nerve-cords of the brachial plexus lie on the first rib and are liable to be wounded in fracture ; especially in fracture from direct violence, when the force of the blow drives the broken ends inwards. Fortunately the Subclavius muscle is interposed between the structures and the clavicle, and this often protects them from injury.

The clavicle is not uncommonly the seat of sarcomatous tumours, rendering the operation of excision of the entire bone necessary. This is an operation of considerable difficulty and danger. It is best performed by exposing the bone freely, disarticulating at the acromial end, and turning it inwards. The removal of the outer part is comparatively easy, but resection of the inner part is fraught with difficulty, the main danger being the risk of wounding the great veins which are in relation with its under surface.

The Scapula

The **Scapula** (σκαπάνη, *a spade*) forms the back part of the shoulder girdle. It is a large, flat bone, triangular in shape, situated at the posterior aspect and side of the thorax, between the second and seventh or sometimes the eighth ribs, its internal border or base being about an inch from, and nearly but not quite parallel with, the spinous processes of the vertebræ, so that it is rather closer to them above than below. It presents for examination two surfaces, three borders, and three angles.

The **anterior surface, or venter** (fig. 285), presents a broad concavity, the *subscapular fossa*. It is marked, in the inner two-thirds, by several oblique ridges, which pass from behind outwards and upwards ; the outer third is smooth. The oblique ridges give attachment to the tendinous intersections, and the surfaces between them to the fleshy fibres, of the Subscapularis muscle. The outer third of the fossa, which is smooth, is covered by, but does not afford attachment to, the fibres of this muscle. The venter is separated from the internal border by smooth, triangular areas at the superior and inferior angles, and in the interval between these by a narrow ridge which is often deficient. These triangular areas and the intervening ridge afford attachment to the Serratus magnus muscle. The subscapular fossa presents a transverse depression at its upper part, where the bone appears to be bent on itself, forming a considerable angle, called the *subscapular angle*, thus giving greater strength to the body of the bone from its arched form; while the summit of the arch serves to support the spine and acromion process. It is in this situation that the fossa is deepest; so that the

FIG. 285.—Left scapula. Anterior surface, or venter.

thickest part of the Subscapularis muscle lies in a line perpendicular to the plane of the glenoid cavity, and must consequently operate most effectively on the head of the humerus, which is contained in that cavity.

The **posterior surface**, or **dorsum** (fig. 286), is arched from above downwards, alternately concave and convex from side to side. It is subdivided unequally into two parts by the *spine*; the portion above the spine is called the *supraspinous fossa*, and that below it the *infraspinous fossa*.

The *supraspinous fossa*, the smaller of the two, is concave, smooth, and broader at the vertebral than at the humeral extremity. It affords attachment by its inner two-thirds to the Supraspinatus muscle.

The *infraspinous fossa* is much larger than the preceding; towards its vertebral margin a shallow concavity is seen at its upper part; its centre presents a prominent convexity, while towards the axillary border is a deep groove which runs from the upper towards the lower part. The inner two-thirds of this surface affords attachment to the Infraspinatus muscle; the outer third is only covered by it, without giving origin to its fibres. This surface is marked near the axillary border by an elevated ridge, which runs from the lower part of the

FIG. 286.—Left scapula. Posterior surface, or dorsum.

glenoid cavity, downwards and backwards to the posterior border, about an inch above the inferior angle. The ridge serves for the attachment of a strong aponeurosis, which separates the Infraspinatus from the two Teres muscles. The surface of bone between this line and the axillary border is narrow in the upper two-thirds of its extent, and is crossed near its centre by a groove for the passage of the dorsalis scapulæ vessels; it affords attachment to the Teres minor. Its lower third presents a broader, somewhat triangular surface, which gives origin to the Teres major, and over which the Latissimus dorsi glides; frequently the latter muscle takes origin by a few fibres from this part. The broad and narrow portions of bone above alluded to are separated by an oblique line, which runs from the axillary border, downwards and backwards, to meet the elevated ridge : to it is attached the aponeurosis which separates the two Teres muscles from each other.

The **Spine** is a prominent plate of bone, which crosses obliquely the inner four-fifths of the dorsum of the scapula at its upper part, and separates the supra- from the infra-spinous fossa : it commences at the vertebral border by a smooth, triangular surface over which the tendon of insertion of the lower part of the Trapezius glides, separated from the bone by a bursa ; and, gradually becoming more elevated as it passes outwards, terminates in the acromion process, which overhangs the shoulder-joint. The spine is triangular, and flattened from above downwards, its apex corresponding to the vertebral border, its base (which is directed outwards) to the neck of the scapula. It presents two surfaces and three borders. Its *superior surface* is concave, assists in forming the supraspinous fossa, and affords attachment to part of the Supraspinatus muscle. Its *inferior surface* forms part of the infraspinous fossa, gives origin to part of the Infraspinatus muscle, and presents near its centre the orifice of a nutrient canal. Of the three borders, the *anterior* is attached to the dorsum of the bone ; the *posterior*, or *crest of the spine*, is broad, and presents two lips and an intervening rough interval. To the superior lip is attached the Trapezius, to the extent shown in the figure. A rough tubercle is generally seen occupying that portion of the spine which receives the tendon of insertion of the lower part of this muscle. To the inferior lip, throughout its whole length, is attached the Deltoid. The interval between the lips is also partly covered by the tendinous fibres of these muscles. The *external border*, or *base*, the shortest of the three, is slightly concave ; its edge, thick and round, is continuous above with the under surface of the acromion process, below with the neck of the scapula. The narrow portion of bone external to this border, and separating it from the glenoid cavity, is called the *great scapular notch*, and serves to connect the supra- and infra-spinous fossæ.

The **Acromion Process**, so called from forming the summit of the shoulder (ἄκρον, *a summit* ; ὦμος, *the shoulder*), is a large and somewhat triangular or oblong process, flattened from behind forwards, directed at first a little outwards, and then curving forwards and upwards, so as to overhang the glenoid cavity. Its *upper surface*, directed upwards, backwards, and outwards, is convex, rough, and gives attachment to some fibres of the Deltoid, and in the rest of its extent it is subcutaneous. Its *under surface* is smooth and concave. Its *outer border* is thick and irregular, and presents three or four tubercles for the tendinous origins of the Deltoid muscle. Its *inner margin*, shorter than the outer, is concave, gives attachment to a portion of the Trapezius muscle, and presents about its centre a small, oval surface for articulation with the acromial end of the clavicle. Its *apex*, which corresponds to the point of meeting of these two borders in front, is thin, and has attached to it the coraco-acromial ligament.

Borders.—Of the three borders of the scapula, the *superior* is the shortest and thinnest ; it is concave, and extends from the superior angle to the base of the coracoid process. At its outer part is a deep, semicircular notch, the *suprascapular*, formed partly by the base of the coracoid process. This notch is converted into a foramen by the transverse ligament, and serves for the passage of the supra-scapular nerve ; sometimes this foramen is entirely surrounded by bone. The adjacent margin of the superior border affords attachment to the Omo-hyoid muscle. The *external*, or *axillary*, *border* is the thickest of the three. It com-mences above at the lower margin of the glenoid cavity, and inclines obliquely downwards and backwards to the inferior angle. Immediately below the glenoid cavity is a rough impression (the *infraglenoid tubercle*), about an inch in length, which affords attachment to the long head of the Triceps muscle ; in front of this is a longitudinal groove, which extends as far as its lower third, and affords origin to part of the Subscapularis muscle. The inferior third of this border, which is thin and sharp, serves for the attachment of a few fibres of the Teres major behind and of the Subscapularis in front. The *internal*, or *vertebral*, *border*, also named the *base*, is the longest of the three, and extends from the superior to the inferior angle of the bone. It is arched, intermediate in thickness between the superior and the external borders, and the portion of it above the spine is bent considerably outwards, so as to form an obtuse angle with the lower part. The vertebral border presents an anterior lip, a posterior lip, and an intermediate space. The *anterior lip* affords attachment to the Serratus magnus; the *posterior lip*, to the Supraspinatus above the spine, the Infraspinatus below; the interval between the two lips, to the Levator anguli scapulæ above the triangular surface at the commencement of the spine, the Rhomboideus minor to the edge of that surface; the Rhomboideus major being attached by means of a fibrous arch, connected above to the lower part of the triangular surface at the base of the spine, and below to the lower part of the posterior border.

Angles.—Of the three angles, the *superior*, formed by the junction of the superior and internal borders, is thin, smooth, rounded, somewhat inclined outwards, and gives attachment to a few fibres of the Levator anguli scapulæ muscle. The *inferior* angle, thick and rough, is formed by the union of the vertebral and axillary borders, its outer surface affording attachment to the Teres major and frequently to a few fibres of the Latissimus dorsi. The *anterior* angle is the thickest part of the bone, and forms what is called the *head* of the scapula. The head presents a shallow, pyriform, articular surface, the *glenoid cavity* (γλήνη, *a socket*), whose longest diameter is from above downwards, and its direction outwards and forwards. It is broader below than above; at its apex is a slight impression (the *supraglenoid tubercle*), to which is attached the long tendon of the Biceps muscle. It is covered with cartilage in the recent state; and its margins, slightly raised, give attachment to a fibro-cartilaginous structure, the *glenoid ligament*, by which its cavity is deepened. The *neck* of the scapula is the slightly depressed surface which surrounds the head; it is more distinct on the posterior than on the anterior surface, and below than above. In the latter situation it has, arising from it, a thick prominence, the coracoid process.

The **Coracoid Process**, so called from its fancied resemblance to a crow's beak (κόραξ, *a crow*), is a thick, curved process of bone, which arises by a broad base from the upper part of the neck of the scapula; it is directed at first upwards and inwards; then, becoming smaller, it changes its direction, and passes forwards and outwards. The ascending portion, flattened from before backwards, presents in front a smooth, concave surface, over which passes the Subscapularis muscle. The horizontal portion is flattened from above downwards; its upper surface is convex and irregular, and gives attachment to the Pectoralis minor; its under surface is smooth; its inner border is rough, and gives attachment to the Pectoralis minor; its outer border is also rough for the coraco-acromial ligament, while the apex is embraced by the conjoined tendon of origin of the short head of the Biceps and of the Coraco-brachialis and gives attachment to the costo-coracoid ligament. At the inner side of the root of the coracoid process is a rough impression for the attachment of the conoid ligament; and running from it obliquely forwards and outwards, on to the upper surface of the horizontal portion, an elevated ridge for the attachment of the trapezoid ligament.

Structure.—In the head, processes, and all the thickened parts of the bone, the scapula is composed of cancellous tissue, while in the rest of its extent it is composed of a thin layer of dense, compact tissue. The central part of the supraspinous fossa and the upper part of the infraspinous fossa, but especially the former, are usually so thin as to be semi-transparent; occasionally the bone is found wanting in this situation, and the adjacent muscles come into contact.

Development (fig. 287).—By *seven* or more centres: one for the body, two for the coracoid process, two for the acromion, one for the vertebral border, and one for the inferior angle.

Ossification of the body of the scapula commences about the second month of fœtal life, by the formation of an irregular quadrilateral plate of bone, immediately behind the glenoid cavity. This plate extends so as to form the chief part of the bone, the spine growing up from its posterior surface about the third month. At birth, a large part of the scapula is osseous, but the glenoid cavity, coracoid and acromion processes, the posterior border, and inferior angle are cartilaginous. From the fifteenth to the eighteenth month after birth, ossification takes place in the middle of the coracoid process, which as a rule becomes joined with the rest of the bone about the fifteenth year. Between the fourteenth and twentieth years, ossification of the remaining centres takes place in quick succession, and usually in the following order: first, in the root of the coracoid process, in the form of a broad scale; secondly, near the base of the acromion process; thirdly, in the inferior angle and contiguous part of the posterior

border; fourthly, near the extremity of the acromion; fifthly, in the posterior
border. The acromion process, besides being ossified from two separate nuclei,
has its base formed by an extension into it of the centre of ossification which
belongs to the spine, the extent of which varies in different cases. The two
separate nuclei unite, and then join with the extension from the spine. The
upper third of the glenoid cavity is ossified from a separate centre (subcoracoid),
which makes its appearance between the tenth and eleventh years and joins
between sixteen and eighteen. Further, an epiphysial plate appears for the
lower part of the glenoid cavity, while the tip of the coracoid process frequently
presents a separate nucleus. These various epiphyses are joined to the bone
by the twenty-fifth year. Sometimes failure of union between the acromion
process and spine occurs, the junction being effected by fibrous tissue, or by an

FIG. 287.—Plan of the development of the scapula. By seven centres.

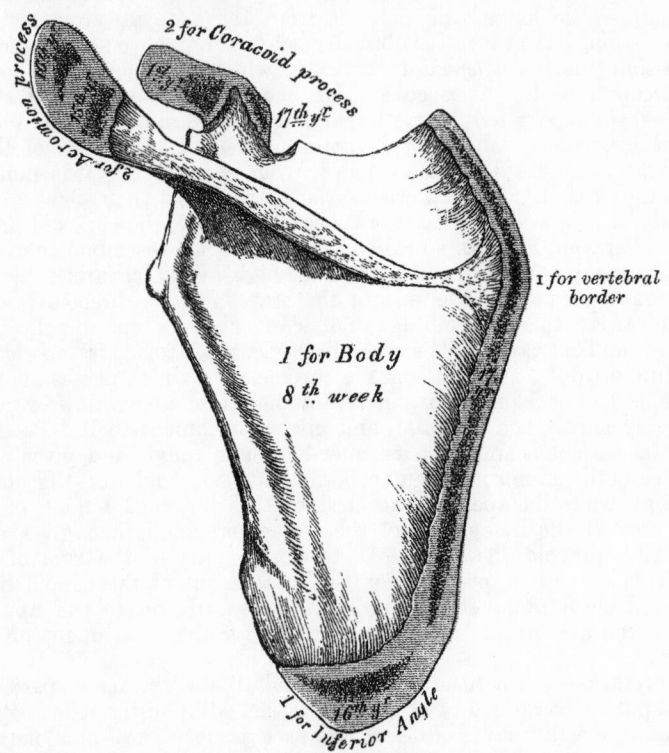

imperfect articulation; in some cases of supposed fracture of the acromion
with ligamentous union, it is probable that the detached segment was never
united to the rest of the bone.

Articulations.—With the humerus and clavicle.

Attachment of Muscles.—To seventeen: to the anterior surface, the Subscapularis; posterior surface, Supraspinatus, Infraspinatus, Teres minor, Teres
major; spine, Trapezius, Deltoid; superior border, Omo-hyoid; vertebral
border, Serratus magnus, Levator anguli scapulæ, Rhomboideus minor and
major; axillary border, Triceps; apex of glenoid cavity, long head of the
Biceps; coracoid process, short head of the Biceps, Coraco-brachialis, Pectoralis
minor; and to the inferior angle occasionally a few fibres of the Latissimus
dorsi.

Surface Form.—The only parts of the scapula which are truly subcutaneous are the
spine and acromion process, but, in addition to these, the coracoid process, the internal or
vertebral border and inferior angle, and, to a less extent, the axillary border, may be defined.
The acromion process and spine of the scapula are easily felt throughout their entire length,
forming, with the clavicle, the arch of the shoulder. The acromion can be ascertained to
be connected to the clavicle at the acromio-clavicular joint by running the finger along

it, its position being often indicated by an irregularity or bony outgrowth from the clavicle close to the joint. The acromion can be felt forming the point of the shoulder, and from this can be traced backwards to join the spine of the scapula. The place of junction is denoted by a prominence, which is sometimes called the acromial angle. From here the spine can be felt as a distinct ridge of bone, marked on the surface as an oblique depression, which becomes less and less distinct and terminates a little external to the spinous processes of the vertebræ. Its termination is indicated by a slight dimple in the skin, on a level with the interval between the third and fourth dorsal spines. Below this point the vertebral border of the scapula may be traced, running downwards and outwards, and thus diverging from the vertebral spines, to the inferior angle of the bone, which can be recognised, although covered by the Latissimus dorsi muscle. From this angle the axillary border can usually be traced through its thick muscular covering, forming, with the muscles, the posterior fold of the axilla. The coracoid process may be felt about an inch below the junction of the middle and outer third of the clavicle. Here it is covered by the anterior border of the Deltoid, and lies a little to the outer side of a slight depression, which corresponds to the interval between the Pectoralis major and Deltoid muscles. When the arms are hanging by the side, the upper angle of the scapula corresponds to the upper border of the second rib or the interval between the first and second dorsal spines, the inferior angle to the upper border of the eighth rib or the interval between the seventh and eighth dorsal spines.

Surgical Anatomy.—Fractures of the body of the scapula are rare, owing to the mobility of the bone, the thick layer of muscles by which it is encased on both surfaces, and the elasticity of the ribs on which it rests. Fracture of the neck of the bone is also uncommon. The most frequent course of the fracture is from the suprascapular notch to the infra-glenoid tubercle, and it derives its principal interest from its simulation to a subglenoid dislocation of the humerus. The diagnosis can be made by noting the alteration in the position of the coracoid process. A fracture of the neck external to and not including the coracoid process is said to occur, but it is exceedingly doubtful whether such an accident ever takes place. The acromion process is more frequently broken than any other part of the bone, and there is sometimes, in young subjects, a separation of the epiphysis. It is believed that many of the cases of supposed fracture of the acromion, with fibrous union, which have been found on post-mortem examination are really cases of imperfectly united epiphysis. The coracoid process is occasionally broken off, either from direct violence or, perhaps rarely, from muscular action.

Tumours of various kinds grow from the scapula. Of the innocent form of tumours probably the osteomata are the most common. When this tumour grows from the venter of the scapula, as it sometimes does, it is of the compact variety, such as usually grows from membrane-formed bones, as the bones of the skull. This would appear to afford evidence that this portion of the bone is formed from membrane, and not like the rest of the bone from cartilage. Sarcomatous tumours sometimes grow from the scapula and may necessitate removal of the bone, with or without amputation of the upper limb. The bone may be excised by a T-shaped incision, and the flaps being reflected, the removal is commenced from the posterior or vertebral border, so that the subscapular vessels which lie along the axillary border are among the last structures divided, and can be at once secured.

THE ARM

The **Arm** is that portion of the upper extremity which is situated between the shoulder and the elbow. Its skeleton consists of a single bone, the humerus.

The Humerus

The **Humerus** is the longest and largest bone of the upper extremity; it presents for examination a shaft and two extremities.

The **Upper Extremity** consists of a large, rounded *head*, joined to the shaft by a constricted portion, called the *neck*, and two other eminences, the *greater* and *lesser tuberosities* (fig. 288).

The **head**, nearly hemispherical in form,* is directed upwards, inwards, and a little backwards, and articulates with the glenoid cavity of the scapula; its surface is smooth, and coated with cartilage in the recent state. The circumference of its articular surface is slightly constricted, and is termed the *anatomical neck*, in contradistinction to the constriction which exists below the tuberosities. The

---

* Though the head is nearly hemispherical in form, its margin, as Humphry has shown, is by no means a true circle. Its greatest measurement is, from the top of the bicipital groove in a direction downwards, inwards, and backwards. Hence it follows that the greatest elevation of the arm can be obtained by rolling the articular surface in this direction—that is to say, obliquely upwards, outwards, and forwards.

T

FIG. 288.—Left humerus.   Anterior view

Surgical neck—

Brachio-radialis or
Supinator radii longus

Extensor carpi radialis
longior

Common origin of
Flexor carpi radialis
Palmaris longus
Flexor sublinis digitorum
Flexor carpi ulnaris

Common origin of
Extensor carpi rad. brev.
  „   communis digitorum
  „   minimi digiti
  „   carpi ulnaris
Supinator brevis

latter is called the *surgical neck*, from its being frequently the seat of fracture. It should be remembered, however, that fracture of the *anatomical neck* does sometimes, though rarely, occur.

The **anatomical neck** is obliquely directed, forming an obtuse angle with the shaft. It is more distinctly marked in the lower half of its circumference than in the upper half, where it presents a narrow groove, separating the head from the tuberosities. Its circumference affords attachment to the capsular ligament, and is perforated by numerous vascular foramina.

The **greater tuberosity** is situated on the outer side of the head and lesser tuberosity. Its upper surface is rounded and marked by three flat impressions, separated by two slight ridges: the superior impression gives insertion to the tendon of the Supraspinatus; the middle to the Infraspinatus; the inferior one, and the shaft of the bone below it, to the Teres minor. The outer surface of the great tuberosity is convex, rough, and continuous with the outer side of the shaft.

The **lesser tuberosity** is more prominent, although smaller, than the greater: it is situated in front of the head, and is directed inwards and forwards. Its summit presents a prominent impression for the insertion of the tendon of the Subscapularis muscle. The tuberosities are separated from one another by a deep groove, the *bicipital groove*, which lodges the long tendon of the Biceps muscle, and transmits a branch of the anterior circumflex artery to the shoulder-joint. It commences above between the two tuberosities, passes obliquely downwards and a little inwards, and terminates at the junction of the upper with the middle third of the bone. It is deep and narrow at the commencement, and becomes shallow and a little broader as it descends. Its borders are called, respectively, the anterior and posterior *bicipital ridges*, and form the upper part of the anterior and internal borders of the shaft of the bone. In the recent state it is covered with a thin layer of cartilage, lined by a prolongation of the synovial membrane of the shoulder-joint, and gives insertion to the tendon of the Latissimus dorsi muscle.

The **Shaft** of the humerus is almost cylindrical in the upper half of its extent, prismatic and flattened below, and presents three borders and three surfaces for examination.

The **anterior border** runs from the front of the great tuberosity above to the coronoid depression below, separating the internal from the external surface. Its upper part is very prominent and rough, and forms the outer lip of the bicipital groove. It is sometimes called the *anterior bicipital* or *pectoral ridge*, and serves for the insertion of the tendon of the Pectoralis major. About its centre it forms the anterior boundary of the rough deltoid impression; below, it is smooth and rounded, affording attachment to the Brachialis anticus.

The **external border** runs from the back part of the greater tuberosity to the external epicondyle, and separates the external from the posterior surface. It is rounded and indistinctly marked in its upper half, serving for the attachment of the lower part of the insertion of the Teres minor, and below this giving origin to the external head of the Triceps muscle; its centre is traversed by a broad but shallow, oblique depression, the *musculo-spiral groove*; its lower part is marked by a prominent, rough margin, a little curved from behind forwards, the *external supracondylar ridge*, which presents an anterior lip for the origin of the Supinator longus (Brachio-radialis) above, and Extensor carpi radialis longior below, a posterior lip for the Triceps, and an intermediate space for the attachment of the external intermuscular septum.

The **internal border** extends from the lesser tuberosity to the internal epicondyle. Its upper third is marked by a prominent ridge, the posterior lip of the bicipital groove, which gives insertion to the tendon of the Teres major. About its centre is a slight impression for the insertion of the Coraco-brachialis, and just below this is seen the entrance of the nutrient canal, directed downwards. Sometimes there is a second canal situated at the commencement of the musculo-spiral groove for a nutrient artery derived from the superior profunda branch of the brachial artery. The inferior third of this border is raised into a slight ridge, the *internal supracondylar ridge*, which becomes very prominent below; it presents an anterior lip for the origin of the Brachialis anticus, a posterior lip for the internal head of the Triceps, and an intermediate space for the attachment of the internal intermuscular septum.

T 2

FIG. 289.—Left humerus.
Posterior surface.

The **external surface** is directed outwards above, where it is smooth, rounded, and covered by the Deltoid muscle; forwards and outwards below, where it is slightly concave from above downwards, and gives origin to part of the Brachialis anticus muscle. About the middle of this surface is seen a rough, triangular impression for the insertion of the Deltoid muscle; and below it the musculo-spiral groove, directed obliquely from behind, forwards, and downwards, and transmitting the musculo-spiral nerve and superior profunda artery.

The **internal surface**, less extensive than the external, is directed inwards above, forwards and inwards below; at its upper part it is narrow, and forms the floor of the bicipital groove which gives insertion to the tendon of the Latissimus dorsi. The middle part of this surface is slightly rough for the attachment of some of the fibres of the tendon of insertion of the Coracobrachialis; its lower part is smooth, concave from above downwards, and gives origin to the Brachialis anticus muscle.*

The **posterior surface** (fig. 289) appears somewhat twisted, so that its upper part is directed a little inwards, its lower part backwards and a little outwards. Nearly the whole of this surface is covered by the external and internal heads of the Triceps, the former

---

* A small, hook-shaped process of bone, the *supracondylar process*, varying from $\frac{1}{10}$ to $\frac{3}{4}$ of an inch in length, is not infrequently found projecting from the inner surface of the shaft of the humerus two inches above the internal epicondyle. It is curved downwards, forwards, and inwards, and its pointed extremity is connected to the internal border, just above the inner epicondyle, by a ligament or fibrous band, which gives origin to a portion of the Pronator radii teres; through the arch completed by this fibrous band the median nerve and brachial artery pass, when these structures deviate from their usual course. Sometimes the nerve alone is transmitted through it, or the nerve may be accompanied by the ulnar artery, in cases of high division of the brachial. A well-marked groove is usually found behind the process, in which the nerve and artery are lodged. This space is analogous to the supracondyloid foramen in many animals, and probably serves in them to protect the nerve and artery from compression during the contraction of the muscles in this region. A detailed account of this process is given by Struthers, in his *Anatomical and Physiological Observations*, p. 202. According to J. Wood, an accessory portion of the Coraco-brachialis muscle is frequently connected with this process (*Journal of Anat. and Phys.* No. 1. Nov. 1866, p. 47).

of which arises from its upper and outer part, the latter from its inner and back part, the two being separated by the musculo-spiral groove.

The **Lower Extremity** is flattened from before backwards, and curved slightly forwards; it terminates below in a broad, articular surface, which is divided into two parts by a slight ridge. Projecting on either side are the external and internal epicondyles. The articular surface extends a little lower than the epicondyles, and is curved slightly forwards, so as to occupy the more anterior part of the bone; its greatest breadth is in the transverse diameter, and it is obliquely directed, so that its inner extremity occupies a lower level than the outer. The outer portion of the articular surface presents a smooth, rounded eminence, which has received the name of the *capitellum*, or *radial head* of the humerus; it articulates with the cup-shaped depression on the head of the radius, and is limited to the front and lower part of the bone, not extending as far back as the other portion of the articular surface. On the inner side of this eminence is a shallow groove, in which is received the inner margin of the head of the radius. Above the front part of the capitellum is a slight depression, the *radial fossa*, which receives the anterior border of the head of the radius, when the forearm is flexed. The inner portion of the articular surface, the *trochlea*, presents a deep depression between two well-marked borders. This surface is convex from before backwards, concave from side to side, and occupies the anterior, lower, and posterior parts of the bone. The external border, less prominent than the internal, separates it from the groove which articulates with the margin of the head of the radius. The internal border is thicker, more prominent, and consequently of greater length, than the external. The grooved portion of the articular surface fits accurately within the greater sigmoid cavity of the ulna; it is broader and deeper on the posterior than on the anterior aspect of the bone, and is inclined obliquely from behind forwards, and from without inwards. Above the front part of the trochlear surface is a smaller depression, the *coronoid fossa*, which receives the coronoid process of the ulna during flexion of the forearm. Above the back part of the trochlear surface is a deep, triangular depression, the *olecranon fossa*, in which the summit of the olecranon process is received in extension of the forearm. These fossæ are separated from one another by a thin, transparent lamina of bone, which is sometimes perforated, forming the *supratrochlear foramen*; their margins afford attachment to the anterior and posterior ligaments of the elbow-joint, and they are lined, in the recent state, by the synovial membrane of this articulation. The articular surfaces, in the recent state, are covered with a thin layer of cartilage. The external epicondyle is a small, tubercular eminence, less prominent than the internal, curved a little forwards, and giving attachment to the external lateral ligament of the elbow-joint, and to a tendon common to the origin of some of the Extensor and Supinator muscles. The internal epicondyle, larger and more prominent, and therefore more liable to fracture, than the external, is directed a little backwards: it gives attachment to the internal lateral ligament, to the Pronator radii teres, and to a tendon common to the origin of some of the Flexor muscles of the forearm. The ulnar nerve runs in a groove at the back of the internal condyle. These epicondyles are directly continuous above with the external and internal supra-condylar ridges.

Structure.—The extremities consist of cancellous tissue, covered with a thin, compact layer; the shaft is composed of a cylinder of compact tissue, thicker at the centre than at the extremities, and hollowed out by a large medullary canal, which extends along its whole length.

Development (fig. 290).—By *eight* centres: one for the shaft, one for the head, one for each tuberosity, one for the radial head, one for the trochlear portion of the articular surface, and one for each epicondyle. The nucleus for the shaft appears near the centre of the bone in the eighth week, and soon extends towards the extremities. At birth the humerus is ossified nearly in its whole length, only the extremities remaining cartilaginous. During the first year, sometimes before birth, ossification commences in the head of the bone, and during the third year the centre for the great tuberosity, and during the fifth that for the lesser tuberosity, make their appearance. By the sixth year the centres for the head and tuberosities have increased in size and become joined, so as to form a single large epiphysis, which fuses with the shaft about the twentieth year.

The lower end of the humerus is developed in the following manner: At the end of the second year ossification commences in the capitellum, and from this point extends inwards, so as to form the chief part of the articular end of the bone, the centre for the inner part of the trochlea not appearing until about the age of twelve. Ossification commences in the internal epicondyle about the fifth year, and in the external one about the thirteenth or fourteenth year. About the sixteenth or seventeenth year, the outer epicondyle and both portions of the articulating surface (having already joined) unite with the shaft, and at the eighteenth year the inner epicondyle becomes joined to it.

FIG. 290.—Plan of the development of the humerus. By seven centres.

Epiphyses of head and tuberosities blend at fifth year, and unite with shaft at twentieth year

Unites with shaft at eighteenth year

Articulations.—With the glenoid cavity of the scapula, and with the ulna and radius.

Attachment of Muscles.—To twenty-four: to the greater tuberosity, the Supraspinatus, Infraspinatus, and Teres minor; to the lesser tuberosity, the Subscapularis; to the anterior bicipital ridge, the Pectoralis major; to the posterior bicipital ridge, the Teres major; to the bicipital groove, the Latissimus dorsi; to the shaft, the Deltoid, Coraco-brachialis, Brachialis anticus, external and internal heads of the Triceps; to the internal epicondyle, the Pronator radii teres, and common tendon of the Flexor carpi radialis, Palmaris longus, Flexor sublimis digitorum, and Flexor carpi ulnaris; to the external supracondylar ridge, the Supinator longus and Extensor carpi radialis longior; to the external epicondyle, the common tendon of the Extensor carpi radialis brevior, Extensor communis digitorum, Extensor minimi digiti, Extensor carpi ulnaris, and Supinator brevis; to the back of the external epicondyle, the Anconeus.

Surface Form.—The humerus is almost entirely clothed by the muscles which surround it, and the only parts of this bone which are strictly subcutaneous are small portions of the internal and external epicondyles. In addition to these, the tuberosities and a part of the head of the bone can be felt under the skin and muscles by which they are covered. Of these the greater tuberosity forms the most prominent bony point of the shoulder, extending beyond the acromion process and covered by the Deltoid muscle. It influences materially the surface form of the shoulder. It is best felt while the arm is lying loosely by the side; if the arm be raised, it recedes from under the finger. The lesser tuberosity, directed forwards and inwards, is to be felt to the inner side of the greater tuberosity, just below the acromio-clavicular joint. Between the two tuberosities lies the bicipital groove. This can be defined by placing the finger, and making firm pressure, just internal to the greater tuberosity; then, by rotating the humerus, the groove will be felt to pass under the finger as the bone is rotated. With the arm abducted from the side the lower part of the head of the bone is to be felt by pressing deeply in the axilla. On each side of the elbow-joint, and just above it, the internal and external epicondyles of the bone are to be felt. Of these the internal is the more prominent, but the internal supracondylar ridge, passing upwards from it, is much less marked than the external, and, as a rule, is not to be felt. Occasionally, however, we find along this border the hook-shaped process mentioned above. The external epicondyle is to be seen most plainly during semiflexion of the forearm, and its position is indicated by a depression between the attachment of the adjacent muscles. From it is to be felt a strong bony ridge running up the outer border of the shaft of the bone. This is the external supracondylar ridge; it is concave forwards, and corresponds with the curved direction of the lower extremity of the humerus.

Surgical Anatomy.—There are several points of surgical interest connected with the humerus. First, as regards its development. The upper end, though the first to ossify, is the last to join the shaft, and the length of the bone is mainly due to growth from the upper epiphysial line. Hence, in cases of amputation of the arm in young subjects, the humerus continues to grow considerably, and the end of the bone which immediately after the operation was covered with a thick cushion of soft tissue begins to project, thinning the soft parts and rendering the stump conical. This may necessitate the removal of a couple of inches or so of the bone, and even after this operation a recurrence of the conical stump may take place.

There are several points of surgical interest in connection with fractures. First, as regards their causation ; the bone may be broken by direct or indirect violence like the other long bones, but, in addition to this, it is probably more frequently fractured by muscular action than any other of this class of bone in the body. It is usually the shaft, just below the insertion of the Deltoid, which is thus broken. The accident has been known to happen from throwing a stone, and again in an apparently healthy adult, from cutting a piece of hard ‘cake tobacco’ on a table. In this latter case there was no disease of the bone that could be discovered. Fractures of the upper end may take place through the anatomical neck, through the surgical neck, or separation of the greater tuberosity may occur. Fracture of the anatomical neck is a very rare accident ; in fact, it is doubted by some whether it ever occurs. These fractures are usually considered to be intracapsular, but they are probably partly within and partly without the capsule, as the lower part of the capsule is inserted some little distance below the anatomical neck, while the upper part is attached to it. They may be impacted or non-impacted. In most cases there is little or no displacement on account of the capsule, in whole or in part, remaining attached to the lower fragment. But occasionally a very remarkable alteration in position takes place ; the upper fragment turns on its own axis, so that the cartilaginous surface of the head rests against the upper end of the lower fragment. When the fractured end is entirely separated from all its surroundings, its vascular supply must be entirely cut off, and one would expect it, theoretically, to necrose. But this must be exceedingly rare, for Gurlt was unable to find a single authenticated case recorded. Separation of the upper epiphysis of the humerus sometimes occurs in the young subject, and is marked by a characteristic deformity, by which the lesion may be at once recognised. This consists in the presence of an abrupt projection at the front of the joint some short distance below the coracoid process, caused by the upper end of the lower fragment. In fractures of the shaft of the humerus the lesion may take place at any point, but appears to be more common in the lower than the upper part of the bone. The points of interest in connection with these fractures are : (1) that the musculo-spiral nerve may be injured as it lies in the groove on the bone, or may become involved in the callus which is subsequently thrown out ; and (2) the frequency of non-union. The latter is believed to be more common in the humerus than in any other bone, and various causes have been assigned for it. It would seem most probably to be due to the difficulty that there is in fixing the shoulder-joint and the upper fragment, and possibly the elbow-joint and lower fragment also. Other causes which have been assigned for the non-union are : (1) that in attempting passive motion of the elbow-joint to overcome any rigidity which may exist, the movement does not take place at the articulation, but at the seat of fracture ; or that the patient, in consequence of the rigidity of the elbow, in attempting to flex or extend the forearm, moves the fragment and not the joint. (2) The presence of small portions of muscular tissue between the broken ends. (3) Want of support to the elbow, so that the weight of the arm tends to drag the lower fragment away from the upper. An important distinction to make in fractures of the lower end of the humerus, is between those that involve the joint and those which do not ; the former are always serious, as they may lead to impairment of the utility of the limb. They include the T-shaped fracture and oblique fractures which involve the articular surface. The fractures which do not involve the joint are the transverse above the epicondyles, and the so-called epitrochlear fracture, where the tip of the internal epicondyle is broken off, generally from direct violence.

Under the head of separation of the epiphysis two separate injuries have been described. One where the whole of the four ossific centres which form the lower extremity of the bone are separated from the shaft ; and secondly, where the articular portion is alone separated, the two epicondyles remaining attached to the shaft of the bone. The epiphysial line between the shaft and lower end runs across the bone just above the tips of the epicondyles, a point to be borne in mind in performing the operation of excision.

Tumours originating from the humerus are of frequent occurrence. A not uncommon place for a chondroma to grow from is the shaft of the bone somewhere in the neighbourhood of the insertion of the Deltoid. Sarcomata frequently grow from this bone.

FIG. 291.—Bones of the left forearm.   Anterior surface.

ULNA

RADIUS

Flexor sublimis digitorum

Pronator radii teres

Occasional origin of Flexor longus pollicis

Artic. with Humerus

Head

Neck

Bicipital tuberosity

Radial origin of Flexor sublimis digitorum

Styloid process

Supinator longus

Groove for Extensor ossis metacarpi pollicis and Extensor brevis pollicis

Styloid process

THE FOREARM

The **Forearm** is that portion of the upper extremity which is situated between the elbow and wrist. Its skeleton is composed of two bones, the ulna and the radius.

THE ULNA

The **Ulna** (figs. 291, 292), so called from its forming the elbow (ὠλένη), is a long bone, prismatic in form, placed at the inner side of the forearm, parallel with the radius. It is the larger and longer of the two bones. Its upper extremity, of great thickness and strength, forms a large part of the articulation of the elbow-joint; it diminishes in size from above downwards, its lower extremity being very small, and excluded from the wrist-joint by the interposition of an interarticular fibro-cartilage. It is divisible into a shaft and two extremities.

The **Upper Extremity,** the strongest part of the bone, presents for examination two large, curved processes, the olecranon process and the coronoid process; and two concave, articular cavities, the greater and lesser sigmoid cavities.

The **Olecranon Process** (ὠλένη, elbow; κρανίον, head) is a large, thick, curved eminence, situated at the upper and back part of the ulna. It is curved forwards at the summit so as to present a prominent tip which is received into the olecranon fossa in extension of the forearm; its base being contracted where it joins the shaft. This is the narrowest part of the upper end of the ulna, and, consequently, the most usual seat of fracture. The posterior surface of the olecranon, directed backwards, is triangular, smooth, subcutaneous, and covered by a bursa. Its upper surface is of quadrilateral form, marked behind by a rough impression for the insertion of the Triceps muscle; and in front, near the margin, by a slight transverse groove for the attachment of part of the posterior ligament of the elbow-joint. Its anterior surface is smooth, concave, covered with cartilage in the recent state, and forms the upper and back part of the great sigmoid cavity. The lateral borders present a continuation of the same groove that was seen on the margin of the superior surface; they serve for the attachment of ligaments —viz. the back part of the internal lateral ligament internally, the posterior ligament externally. From its inner border a part of the Flexor carpi ulnaris arises; while to the outer border is attached the Anconeus.

The **Coronoid Process** (κορώνη, anything hooked like a crow's beak) is a triangular eminence of bone which projects horizontally forwards from the upper and front part of the ulna. Its base is continuous with the shaft, and of considerable strength; so much so that fracture of it is an accident of rare occurrence. Its apex is pointed, slightly curved upwards, and received into the coronoid depression of the humerus in flexion of the forearm. Its upper surface is smooth, concave, and forms the lower part of the greater sigmoid cavity. The under surface is concave, and marked internally by a rough impression for the insertion of the Brachialis anticus. At the junction of this surface with the shaft is a rough eminence, the *tubercle of the ulna*, for the attachment of the oblique ligament. Its outer surface presents a narrow, oblong, articular depression, the *lesser sigmoid* cavity. The inner surface, by its prominent, free margin, serves for the attachment of part of the internal lateral ligament. At the front part of this surface is a small, rounded eminence for the origin of one head of the Flexor sublimis digitorum; behind the eminence, a depression for part of the origin of the Flexor profundus digitorum; and, descending from the eminence, a ridge, which gives origin to one head of the Pronator radii teres. Frequently, the Flexor longus pollicis arises from the lower part of the coronoid process by a rounded bundle of muscular fibres.

The **Greater Sigmoid Cavity,** so called from its resemblance to the old shape of the Greek letter Σ, is a semilunar depression of large size, formed by the olecranon and coronoid processes, and serving for articulation with the trochlear surface of the humerus. About the middle of either lateral border of this cavity is a notch, which contracts it somewhat, and serves to indicate the junction of the two processes of which it is formed. The cavity is concave from above downwards, and divided into two lateral parts by a smooth, elevated ridge which runs from the summit of the olecranon to the tip of the coronoid process. Of these two portions, the internal is the larger, and is slightly concave transversely;

FIG. 292.—Bones of the left forearm.   Posterior surface.

ULNA

RADIUS

Subcutaneous & covered by Bursa

Flexor sublimis digitorum

SUPINATOR BREVIS

EXTENSOR OSSIS METACARPI POLLICIS

Oblique

FLEXOR DIGITORUM PROFUNDUS

Posterior border of ulna : giving attachment to an aponeurosis common to { *Extensor carpi ulnaris Flexor carpi ulnaris Flexor profundus digitorum* }

EXTENSOR BREVIS POLLICIS

For Extensor carpi ulnaris

For Extensor minimi digiti

For { *Extensor indicis Extensor communis digitorum* }

For Ext. carpi radialis longior } *Extensor carpi radialis brevior* } *Extensor longus pollicis* the external portion is convex above, slightly concave below. The articular surface, in the recent state, is covered with a thin layer of cartilage.

The **Lesser Sigmoid Cavity** is a narrow, oblong, articular depression on the outer side of the coronoid process, and receives the lateral articular surface of the head of the radius. It is concave from before backwards; and its extremities, which are prominent, serve for the attachment of the orbicular ligament. In the recent state, it is covered with a thin layer of cartilage.

The **Shaft**, at its upper part, is prismatic in form, and curved from behind forwards, and from without inwards, so as to be convex behind and externally; its central part is quite straight; its lower part is rounded, smooth, and bent a little outwards; it tapers gradually from above downwards, and presents for examination three borders, and three surfaces.

The **anterior border** commences above at the prominent inner angle of the coronoid process, and terminates below in front of the styloid process. It is well marked above, smooth and rounded in the middle of its extent, and affords attachment to the Flexor profundus digitorum: its lower fourth, marked off from the rest of the border by the commencement of an oblique ridge on the anterior surface, serves for the origin of the Pronator quadratus. It separates the anterior from the internal surface.

The **posterior border** commences above at the apex of the triangular subcutaneous surface at the back part of the olecranon, and terminates below at the back part of the styloid process; it is well marked in the upper three-fourths, and gives attachment to an aponeurosis which affords a common origin to the Flexor carpi ulnaris, the Extensor carpi ulnaris, and the Flexor profundus digitorum muscles; its lower fourth is smooth and rounded. This border separates the internal from the posterior surface.

The **external** or **interosseous border** commences above by the union of two lines, which converge one from each extremity of the lesser sigmoid cavity, enclosing between them a triangular space for the origin of part of the Supinator brevis; it terminates below at the middle of the head of the ulna. Its two middle fourths are very prominent, its lower fourth is smooth and rounded. This border gives attachment to the interosseous membrane, and separates the anterior from the posterior surface.

The **anterior surface**, much broader above than below, is concave in its upper three-fourths, and affords attachment to the Flexor profundus digitorum; its lower fourth, also concave, is covered by the Pronator quadratus. The lower fourth is separated from the remaining portion of the bone by a prominent ridge, directed obliquely from above downwards and inwards; this ridge (the *oblique* or *Pronator ridge*) marks the extent of origin of the Pronator quadratus. At the junction of the upper with the middle third of the bone is the nutrient canal, directed obliquely upwards and inwards.

The **posterior surface**, directed backwards and outwards, is broad and concave above; somewhat narrower and convex in the middle of its course; narrow, smooth, and rounded below. It presents, above, an oblique ridge, which runs from the posterior extremity of the lesser sigmoid cavity, downwards to the posterior border; the triangular surface above this ridge receives the insertion of the Anconeus muscle, while the upper part of the ridge itself affords attachment to the Supinator brevis. The surface of bone below this is subdivided by a longitudinal ridge, sometimes called the *perpendicular line*, into two parts: the internal part is smooth, and covered by the Extensor carpi ulnaris; the external portion, wider and rougher, gives origin from above downwards to part of the Supinator brevis, the Extensor ossis metacarpi pollicis, the Extensor longus pollicis, and the Extensor indicis muscles.

The **internal surface** is broad and concave above, narrow and convex below. It gives origin by its upper three-fourths to the Flexor profundus digitorum muscle: its lower fourth is subcutaneous.

The **Lower Extremity** of the ulna is of small size, and excluded from the articulation of the wrist-joint. It presents for examination two eminences, the outer and larger of which is a rounded, articular eminence, termed the *head* of the ulna; the inner, narrower and more projecting, is a non-articular eminence, the *styloid process*. The *head* presents an articular facet, part of which, of an oval or semilunar form, is directed downwards, and articulates with the upper surface of the interarticular fibro-cartilage which separates it from the wrist-joint; the remaining portion, directed outwards, is narrow, convex, and received into the sigmoid cavity of the radius. The *styloid process* projects from the inner and back part of the bone, and descends a little lower than the head, terminating in a rounded summit, which affords attachment to the internal lateral ligament of the wrist. The head is separated from the styloid process by a depression for the attachment of the apex of the triangular interarticular fibro-cartilage; and behind, by a shallow groove for the passage of the tendon of the Extensor carpi ulnaris.

Fig. 293.—Plan of the development of the ulna. By three centres.

Olecranon

Appears at tenth year

Joins shaft at sixteenth year

Shaft 8th

Appears at fourth year

Joins shaft at twentieth year

Inferior extremity

Structure.—Similar to that of the other long bones.

Development.—By *three* centres : one for the shaft, one for the inferior extremity, and one for the olecranon (fig. 293). Ossification commences near the middle of the shaft about the eighth week, and soon extends through the greater part of the bone. At birth the ends are cartilaginous. About the fourth year, a separate osseous nucleus appears in the middle of the head, and soon extends into the styloid process. About the tenth year, ossific matter is deposited in the olecranon near its extremity, the chief part of this process being formed from an extension of the shaft of the bone into it. The upper epiphysis is joined to the shaft about the sixteenth year, the lower about the twentieth year.

Articulations. — With the humerus and radius.

Attachment of Muscles.—To fourteen : to the olecranon, the Triceps, Anconeus, and one head of the Flexor carpi ulnaris. To the coronoid process, the Brachialis anticus, Pronator radii teres, Flexor sublimis digitorum, and Flexor profundus digitorum ; generally also the Flexor longus pollicis. To the shaft, the Flexor profundus digitorum, Pronator quadratus, Flexor carpi ulnaris, Extensor carpi ulnaris, Anconeus, Supinator brevis, Extensor ossis metacarpi pollicis, Extensor longus pollicis, and Extensor indicis.

Surface Form.—The most prominent part of the ulna on the surface of the body is the olecranon process, which can always be identified at the back of the elbow-joint. When the forearm is flexed, the upper quadrilateral surface can be felt, directed backwards ; during extension it recedes into the olecranon fossa, and the contracting fibres of the Triceps prevent its being perceived. At the back of the olecranon is the smooth, triangular, subcutaneous surface, continuous below with the posterior border of the shaft of the bone, and felt in every position of the forearm. During extension, the upper border of the olecranon is slightly above the level of the internal epicondyle, and the process itself is nearer to this epicondyle than the outer one. Running down the back of the forearm, from the apex of the triangular surface which forms the posterior surface of the olecranon, is a prominent ridge of bone, the posterior border of the ulna. This can be made out throughout the entire length of the shaft of the bone, from the olecranon above to the styloid process below. As it passes down the forearm it pursues a sinuous course and inclines to the inner side, so that, though it is situated in the middle of the back of the limb above, it is on the inner side of the wrist at its termination. It becomes rounded off in its lower third, and may be traced below to the small, subcutaneous surface of the styloid process. Internal to this border the lower fourth of the inner surface is to be felt. The styloid process presents itself as a prominent tubercle of bone, continuous above with the posterior subcutaneous border of the ulna, and terminating below in a blunt apex, which lies a little internal, and behind, but on a level with, the wrist-joint. The styloid process is best perceived when the hand is in the same line as the bones of the forearm, and in a position midway between supination and pronation. If the forearm is pronated while the finger is placed on the process, it will be felt to recede, and another prominence of bone will appear just behind and above it. This is the head of the ulna, which articulates with the lower end of the radius and the triangular interarticular fibro-cartilage, and now projects between the tendons of the Extensor carpi ulnaris and the Extensor minimi digiti muscles.

The Radius

The **Radius** (radius, *a ray*, or *spoke of a wheel*) is situated on the outer side of the forearm, lying side by side with the ulna, which exceeds it in length and size (figs. 291, 292). Its upper end is small, and forms only a slight part of the elbow-joint; but its lower end is large, and forms the chief part of the wrist. It is a long bone, prismatic in form, slightly curved longitudinally, and, like other long bones, has a shaft and two extremities.

The **Upper Extremity** presents a head, neck, and tuberosity. The *head* is of a cylindrical form, depressed on its upper surface into a shallow cup which articulates with the capitellum or radial head of the humerus. In the recent state it is covered with a layer of cartilage, which is thinnest at its centre. Around the circumference of the head is a smooth, articular surface, broad internally where it articulates with the lesser sigmoid cavity of the ulna; narrow in the rest of its circumference, where it rotates within the orbicular ligament. It is coated with cartilage in the recent state. The head is supported on a round, smooth, and constricted portion of bone, called the *neck*, which presents, behind, a slight ridge, for the insertion of part of the Supinator brevis. Beneath the neck, at the inner and front aspect of the bone, is a rough eminence, the *bicipital tuberosity*. Its surface is divided into two parts by a vertical line—a posterior, rough portion, for the insertion of the tendon of the Biceps muscle; and an anterior, smooth portion, on which a bursa is interposed between the tendon and the bone.

The **Shaft** of the bone is prismoid in form, narrower above than below, and slightly curved, so as to be convex outwards. It presents three surfaces, separated by three borders.

The **anterior border** extends from the lower part of the tuberosity above, to the anterior part of the base of the styloid process below. It separates the anterior from the external surface. Its upper third is very prominent; and from its oblique direction, downwards and outwards, has received the name of the *oblique line* of the radius. It gives insertion, externally, to the Supinator brevis; internally, there arises from it the Flexor longus pollicis, and between these the Flexor sublimis digitorum. The middle third of the anterior border is indistinct and rounded. Its lower fourth is sharp, prominent, affords insertion to the Pronator quadratus and gives attachment to the posterior annular ligament of the wrist; it terminates in a small tubercle, into which is inserted the tendon of the Supinator longus.

The **posterior border** commences above, at the back part of the neck of the radius, and terminates below, at the posterior part of the base of the styloid process; it separates the posterior from the external surface. It is indistinct above and below, but well marked in the middle third of the bone.

The **internal** or **interosseous border** commences above, at the back part of the tuberosity, where it is rounded and indistinct; it becomes sharp and prominent as it descends, and at its lower part divides into two ridges, which pass to the anterior and posterior margins of the sigmoid cavity. To the posterior of the two ridges the lower part of the interosseous membrane is attached, while the triangular surface between the ridges gives insertion to part of the Pronator quadratus. This border separates the anterior from the posterior surface, and has the interosseous membrane attached to it throughout the greater part of its extent.

The **anterior surface** is concave in its upper three-fourths, and gives origin to the Flexor longus pollicis muscle; it is broad and flat in its lower fourth, and gives insertion to the Pronator quadratus. A prominent ridge limits the attachment of the Pronator quadratus below, and between this and the inferior border is a triangular rough surface for the attachment of the anterior ligament of the wrist-joint. At the junction of the upper and middle third of this surface is the nutrient foramen, which is directed obliquely upwards.

The **posterior surface** is round, convex, and smooth in the upper third of its extent, and covered by the Supinator brevis muscle. Its middle third is broad, slightly concave, and gives origin to the Extensor ossis metacarpi pollicis above, the Extensor brevis pollicis below. Its lower third is broad, convex, and covered by the tendons of the muscles, which subsequently run in the grooves on the lower end of the bone.

The **external surface** is round and convex throughout its entire extent. Its upper third gives insertion to the Supinator brevis muscle. About its centre is seen a rough ridge, for the insertion of the Pronator radii teres muscle. Its lower part is narrow, and covered by the tendons of the Extensor ossis metacarpi pollicis and Extensor brevis pollicis muscles.

The **Lower Extremity** of the radius is large, of quadrilateral form, and provided with two articular surfaces—one at the extremity, for articulation with the carpus, and one at the inner side of the bone, for articulation with the ulna. The carpal articular surface is triangular, concave, smooth, and divided by a slight antero-posterior ridge into two parts. Of these, the external is of a triangular form, and articulates with the scaphoid bone; the inner, quadrilateral, articulates with the semilunar. The articular surface for the ulna is called the *sigmoid cavity* of the radius; it is narrow, concave, smooth, and articulates with the head of the ulna. These two articular surfaces are separated from each other by a prominent ridge, to which is attached the base of the triangular fibro-cartilage; a structure which serves to separate the carpal joint from the inferior radio-ulnar articulation. The circumference of this end of the bone presents three surfaces—an anterior, external, and posterior. The *anterior surface*, rough and irregular, affords attachment to the anterior ligament of the wrist-joint. The *external surface* is prolonged obliquely downwards into a strong, conical projection, the *styloid process*, which gives attachment by its base to the tendon of the Supinator longus, and by its apex to the external lateral ligament of the wrist-joint. The outer surface of this process is marked by a flat groove, which runs obliquely downwards and forwards, and gives passage to the tendons of the Extensor ossis metacarpi pollicis, and the Extensor brevis pollicis. The *posterior surface* is convex, affords attachment to the posterior ligament of the wrist, and is marked by three grooves. Proceeding from without inwards, the first groove is broad, but shallow, and subdivided into two by a slight ridge: the outer of these two transmits the tendon of the Extensor carpi radialis longior, the inner the tendon of the Extensor carpi radialis brevior. The second, which is near the centre of the bone, is a deep but narrow groove, bounded on its outer side by a sharply defined ridge; it is directed obliquely from above, downwards and outwards, and transmits the tendon of the Extensor longus pollicis. The third, lying most internally, is a broad groove, for the passage of the tendons of the Extensor indicis and Extensor communis digitorum.

Structure.—Similar to that of the other long bones.

Development (fig. 294). — By *three* centres: one for the shaft, and one for each extremity. That for the shaft makes its appearance near the centre of the bone, during the eighth week of fœtal life. About the end of the second year, ossification commences in the lower end; and at the fifth year, in the upper end. At the age of seventeen or eighteen the upper epiphysis becomes joined to the shaft; the lower epiphysis becoming united during the twentieth year.

FIG. 294.—Plan of the development of the radius. By three centres.

Head

Appears at fifth year   *Unites with shaft about puberty*

Appears at second year   *Unites with shaft about twentieth year*

Lower extremity

Articulations.—With four bones: the humerus, ulna, scaphoid, and semilunar.

Attachment of Muscles.—To nine: to the tuberosity, the Biceps; to the oblique ridge, the Supinator brevis, Flexor sublimis digitorum, and Flexor longus pollicis; to the shaft (its anterior surface), the Flexor longus pollicis and Pronator quadratus; (its posterior surface), the Extensor ossis metacarpi pollicis and Extensor brevis pollicis; (its outer surface), the Pronator radii teres; and to the styloid process, the Supinator longus.

Surface Form.—Below, and a little in front of, the posterior surface of the external condyle a part of the head of the radius may be felt, covered by the orbicular and external lateral ligaments.   There is in this situation a little dimple in the skin, which is most visible when the arm is extended, and which marks the position of the head of the bone.   If the finger is placed on this dimple, and the semiflexed forearm pronated and supinated, the head of the bone will be distinctly perceived rotating in the lesser sigmoid cavity.   The upper half of the shaft of the radius cannot be felt, as it is surrounded by the fleshy bellies of the muscles arising from the external condyle.   The lower half of the shaft can be readily examined, though covered by tendons and muscles and not strictly subcutaneous.   If traced downwards, the shaft will be felt to terminate in a lozenge-shaped, convex surface on the outer side of the base of the styloid process.   This is the only subcutaneous part of the bone, and from its lower extremity the apex of the styloid process will be felt bending inwards towards the wrist.   About the middle of the posterior aspect of the lower extremity of the bone is a well-marked ridge, the *dorsal radial tubercle*, best perceived when the hand is slightly flexed on the wrist.   It forms the outer boundary of the oblique groove on the posterior surface of the bone, through which the tendon of the Extensor longus pollicis runs, and serves to keep that tendon in its place.

Surgical Anatomy.—Cases of congenital absence, either partial or complete, of both radius and ulna, have been recorded.   Absence of the radius is not uncommonly accompanied by absence of the thumb.   The two bones of the forearm are more often broken together, than is either the radius or ulna separately.   It is therefore convenient to consider the fractures of both bones in the first instance and subsequently to mention the principal fractures which take place in each bone.   These fractures may be produced by either direct or indirect violence, though more commonly by direct violence.   When indirect force is applied to the forearm the radius as a rule alone gives way, though both bones may suffer.   The fracture from indirect force generally takes place somewhere about the middle of the bones, fracture from direct violence may occur at any part, more often, however, in the lower half of the bones.   The fracture is usually transverse, but may be more or less oblique.   A point of interest in connection with these fractures is the tendency that there is for the two bones to unite across the interosseous membrane ; the limb should therefore be put up in a position midway between supination and pronation, which is not only the most comfortable position, but also separates the bones most widely from each other and therefore diminishes the risk of the bones becoming united across the interosseous membrane.   The splints, anterior and posterior, which are applied in these cases should be rather wider than the limb, so as to prevent any lateral pressure on the bones.   In these cases there is a greater liability to gangrene from the pressure of the splints than in other parts of the body.   This is no doubt due principally to two causes : (1) the flexion of the forearm compressing to a certain extent the brachial artery and retarding the flow of blood to the limb ; and (2) the superficial position of the two main arteries of the forearm in a part of their course, and their liability to be compressed by the splints.   The special fractures of the ulna are : (1) fracture of the olecranon.   This is usually caused by direct violence, as in falls on the elbow with the forearm flexed, but occasionally by muscular action by the sudden contraction of the Triceps.   The most common place for the fracture to occur is at the constricted portion where the olecranon joins the shaft of the bone, and the fracture may be either transverse or oblique ; but any part may be broken, and even a thin shell may be torn off.   Fractures from direct violence are occasionally comminuted.   The displacement is sometimes very slight, owing to the fibrous structures around the process not being torn.   (2) Fracture of the coronoid process may occur as a complication of dislocation backwards of the bones of the forearm, but it is doubtful if it ever occurs as an uncomplicated injury.   (3) Fractures of the shaft of the ulna may occur at any part, but usually take place at the middle of the bone or a little below it.   They are generally the result of direct violence, but may occur as a complication of dislocation of the radius.   (4) The styloid process may be knocked off by direct violence.   Fractures of the radius consist of (1) fracture of the head of the bone ; this for the most part takes place in conjunction with some other lesion, but may occur as an uncomplicated injury.   (2) Fracture of the neck may also take place, but is usually complicated with other injury.   (3) Fractures of the shaft of the radius are very common, and may take place at any part of the bone.   They may be caused either by direct or indirect violence.   In fractures of the upper third of the shaft of the bone—that is to say, above the insertion of the Pronator radii teres—the displacement is very great.   The upper fragment is strongly supinated by the Biceps and Supinator brevis, and flexed by the Biceps ; while the lower fragment is pronated and drawn towards the ulna by the two pronators. If such a fracture is put up in the ordinary position, midway between supination and pronation, the fracture will unite with the upper fragment in a position of supination, and the lower one in the mid-position, and thus considerable impairment of the movements of the hand will result.   The limb should be put up with the forearm supinated.   (4) The most important fracture of the radius is that of the lower end (Colles's fracture).   The fracture is transverse, and generally takes place about an inch from the lower extremity.   It is caused by falls on the palm of the hand, and is an injury of advanced life, occurring more

Fig. 295.—Bones of the left hand. Dorsal surface.

FIG. 296.—Bones of the left hand. Palmar surface.

frequently in the female than in the male. In consequence of the manner in which the fracture is caused, the upper fragment becomes driven into the lower, and impaction is the result; excess of violence may, however, disimpact; the lower fragment becoming split up into two or more pieces, so that no fixation occurs. Separation of the lower epiphysis of the radius may take place in the young. This injury and Colles's fracture may be distinguished from other injuries in this neighbourhood—especially dislocation, with which it is liable to be confounded—by observing the relative positions of the styloid processes of the ulna and radius. In the natural condition of parts, with the arm hanging by the side, the styloid process of the radius is on a lower level than that of the ulna: that is to say, nearer the ground. After fracture or separation of the epiphysis this process is on the same or on a higher level than that of the ulna, whereas it would be unaltered in position in dislocation.

THE HAND

The skeleton of the Hand is subdivided into three segments—the Carpus or wrist bones; the Metacarpus or bones of the palm; and the Phalanges or bones of the digits.

THE CARPUS

The bones of the **Carpus** (καρπός, *the wrist*), eight in number, are arranged in two rows. Those of the upper row, enumerated from the radial to the ulnar side, are the scaphoid, semilunar, cuneiform, and pisiform; those of the lower row, enumerated in the same order, are the trapezium, trapezoid, os magnum, and unciform.

COMMON CHARACTERS OF THE CARPAL BONES

Each bone (excepting the pisiform) presents six surfaces. Of these the *anterior* or *palmar* and the *posterior* or *dorsal* are rough, for ligamentous attachment; the dorsal surface being the broader, except in the scaphoid and semilunar. The *superior* or *proximal* and *inferior* or *distal* are articular, the superior generally convex, the inferior concave; and the *internal* and *external* are also articular when in contact with contiguous bones, otherwise rough and tubercular. The structure in all is similar, consisting of cancellous tissue enclosed in a layer of compact bone. Each bone is also developed from a single centre of ossification.

BONES OF THE UPPER ROW

SCAPHOID (fig. 297)

The **Scaphoid** is the largest bone of the first row. It has received its name from its fancied resemblance to a boat, being broad at one end, and narrowed, like a prow, at the opposite (σκάφη, *a boat*; εἶδος, *like*). It is situated at the upper and outer part of the carpus, its long axis being from above downwards, outwards, and forwards. The *superior surface* is convex, smooth, of triangular shape, and articulates with the lower end of the radius. The *inferior surface*,

FIG. 297.—The left scaphoid.

directed downwards, outwards, and backwards, is smooth, convex, also triangular, and divided by a slight ridge into two parts, the external of which articulates with the trapezium, the inner with the trapezoid. The *posterior* or *dorsal surface* presents a narrow, rough groove, which runs the entire length of the bone, and serves for the attachment of ligaments. The *anterior* or *palmar surface* is concave above, and elevated at its lower and outer part into a prominent, rounded projection, called the *tuberosity*, which projects forwards from the front of the carpus and gives attachment to the anterior annular ligament of the wrist and sometimes origin to a few fibres of the Abductor pollicis. The *external surface* is rough and narrow, and gives attachment to the external lateral ligament of the wrist. The *internal surface* presents two articular facets ; of these, the superior or smaller is flattened, of semilunar form, and articulates with the semilunar ; the inferior or larger is concave, forming, with the semilunar bone, a concavity for the head of the os magnum.

To ascertain to which hand the bone belongs, hold it with the superior or radial, convex, articular surface upwards, and the posterior surface—i.e. the narrow, non-articular, grooved surface—towards you. The tubercle on the outer surface points to the side to which the bone belongs.*

Articulations.—With five bones : the radius above, trapezium and trapezoid below, os magnum and semilunar internally.

Attachment of Muscles.—Occasionally a few fibres of the Abductor pollicis.

SEMILUNAR (fig. 298)

The **Semilunar** bone may be distinguished by its deep concavity and crescentic outline (*semi*, half ; *luna*, moon). It is situated in the centre of the upper row of the carpus, between the scaphoid and cuneiform. The *superior surface*, convex, smooth, and bounded by four edges, articulates with the radius. The *inferior surface* is deeply concave, and of greater extent from before backwards than transversely : it articulates with the head of the os magnum, and, by a long, narrow facet (separated by a ridge from the general surface), with the unciform bone. The *anterior* or *palmar* and *posterior* or *dorsal surfaces* are rough, for the attachment of ligaments, the former being the broader, and of a somewhat rounded form. The *external surface* presents a narrow, flattened, semilunar facet for articulation with the scaphoid. The *internal surface* is marked by a smooth, quadrilateral facet, for articulation with the cuneiform.

FIG. 298.—The left semilunar.

For cuneiform   *For radius*

For uniform   *For os magnum*   *For scaphoid*

Hold it with the convex articular surface for the radius upwards, and the narrowest non-articular surface towards you. The semilunar facet for the scaphoid will be on the side to which the bone belongs.

Articulations.—With five bones : the radius above, os magnum and unciform below, scaphoid and cuneiform on either side.

CUNEIFORM (fig. 299)

The **Cuneiform** (cuneus, *a wedge* ; forma, *likeness*) may be distinguished by its pyramidal shape (*os triquetrum*), and by its having an oval, isolated facet for articulation with the pisiform bone. It is situated at the upper and inner side of the carpus. The *superior surface* presents an internal, rough, non-articular portion, and an external or articular portion, which is convex, smooth, and articulates with the triangular interarticular fibro-cartilage of the wrist. The *inferior surface*, directed outwards, is concave, sinuously curved, and smooth for articulation with the unciform. The *posterior* or *dorsal surface* is rough for the attachment of ligaments. The *anterior* or *palmar surface* presents, at its inner side, an oval facet, for articulation with the pisiform ; and is rough externally,

FIG. 299.
The left cuneiform.

For pisiform   *For semilunar*

For unciform

* In these directions each bone is supposed to be placed in its natural position—that is, such a position as it would occupy when the arm is hanging by the side, the forearm in a position of supination, the thumb being directed outwards and the palm of the hand looking forwards.

for ligamentous attachment. The *external surface*, the base of the pyramid, is marked by a flat, quadrilateral, smooth facet, for articulation with the semi-lunar. The *internal surface*, the summit of the pyramid, is pointed and roughened, for the attachment of the internal lateral ligament of the wrist.

Hold the bone with the surface supporting the pisiform facet away from you, and the concavo-convex surface for the unciform downwards. The base of the wedge (i.e. the broad end of the bone) will be on the side to which it belongs.

Articulations.—With three bones : the semilunar externally, the pisiform in front, the unciform below ; and with the triangular, interarticular fibro-cartilage which separates it from the lower end of the ulna.

PISIFORM (fig. 300)

The **Pisiform** (pisum, *a pea* ; forma, *likeness*) may be known by its small size, and by its presenting a single articular facet. It is situated on a plane anterior to the other bones of the carpus ; it is spheroidal in form, with its long diameter directed vertically. Its *posterior* surface presents a smooth, oval facet, for articulation with the cuneiform : this facet approaches the superior, but not the inferior, border of the bone. The *anterior* or *palmar surface* is rounded and rough, and gives attachment to the anterior annular ligament and to the Flexor carpi ulnaris and Abductor minimi digiti muscles. The *outer* and *inner surfaces* are also rough, the former being concave, the latter usually convex.

FIG. 300.
The left pisiform.

For cuneiform

Hold the bone with the posterior surface—that which presents the articular facet—towards you, in such a manner that the faceted portion of the surface is uppermost. The outer, concave surface will point to the side to which it belongs.

Articulation.—With one bone, the cuneiform.

Attachment of Muscles.—To two : the Flexor carpi ulnaris and Abductor minimi digiti ; and to the anterior annular ligament.

BONES OF THE LOWER ROW

TRAPEZIUM (fig. 301)

The **Trapezium** (τράπεζα, *a table*) is of very irregular form. It may be distinguished by a deep groove, for the tendon of the Flexor carpi radialis muscle. It is situated at the external and inferior part of the carpus, between the scaphoid and first metacarpal bone. The *superior surface* is directed upwards and inwards ; internally, it is smooth, and articulates with the scaphoid ; externally, it is

FIG. 301.—The left trapezium.

For trapezoid   *Groove*   *For scaphoid*

Ridge

For 2nd metacarpal   *For trapezoid*

For 1st metacarpal

For 2nd metacarpal

rough, and becomes continuous with the external surface. The *inferior surface*, directed downwards and inwards, is oval, concave from side to side, convex from before backwards, so as to form a saddle-shaped surface, for articulation with the base of the first metacarpal bone. The *anterior* or *palmar surface* is narrow and rough. At its upper part is a deep groove, running from above obliquely downwards and inwards : it transmits the tendon of the Flexor carpi radialis, and is bounded externally by a prominent ridge, the *oblique ridge* of the trapezium. This surface gives origin to the Abductor pollicis, Opponens pollicis, and Flexor brevis pollicis muscles, and also affords attachment to the anterior annular ligament. The *posterior* or *dorsal surface* is rough. The *external surface* is also broad and rough, for the attachment of ligaments. The *internal surface* presents two articular facets : the upper one, large and concave, articulates with the trapezoid ; the lower one, small and oval, with the base of the second metacarpal bone.

Hold the bone with the saddle-shaped surface downwards and the grooved surface away from you. The prominent, rough, non-articular surface points to the side to which the bone belongs.

Articulations.—With four bones : the scaphoid above, the trapezoid and second metacarpal bones internally, the first metacarpal below.

Attachment of Muscles.—Abductor pollicis, Opponens pollicis, and part of the Flexor brevis pollicis.

TRAPEZOID (fig. 302)

The **Trapezoid** is the smallest bone in the second row. It may be known by its wedge-shaped form, the broad end of the wedge forming the dorsal, the narrow end the palmar surface ; and by its having four articular surfaces touching each other, and separated by sharp edges.

The *superior surface*, quadrilateral in form, smooth, and slightly concave, articulates with the scaphoid. The *inferior surface* articulates with the upper end of the second metacarpal bone ; it is convex from side to side, concave from before backwards, and subdivided, by an elevated ridge, into two unequal lateral facets. The *posterior* or *dorsal* and

FIG. 302.—The left trapezoid.

For scaphoid   *Ant. surface*   *For trapezium*

For os magnum

Post. surface   *For 2nd metacarpal*

anterior or *palmar surfaces* are rough, for the attachment of ligaments, the former being the larger of the two. The *external surface*, convex and smooth, articulates with the trapezium. The *internal surface* is concave and smooth in front, for articulation with the os magnum ; rough behind, for the attachment of an interosseous ligament.

Hold the bone with the larger, non-articular surface towards you, and the smooth, quadrilateral articular surface upwards. The convex, articular surface will point to the side to which the bone belongs.[*]

Articulations.—With four bones : the scaphoid above, second metacarpal bone below, trapezium externally, os magnum internally.

OS MAGNUM (fig. 303)

The **Os Magnum** is the largest bone of the carpus, and occupies the centre of the wrist. It presents, above, a rounded portion or head, which is received into

FIG. 303.—The left os magnum.

For semilunar

For scaphoid

For unciform

For trapezoid

For 2nd metacarpal   *For 3rd metacarpal*   *For 4th metacarpal*   *Anterior surface*

the concavity formed by the scaphoid and semilunar bones ; a constricted portion or neck ; and below this, the body. The *superior surface* is round, smooth, and

---

[*] Occasionally in a badly marked bone there is some difficulty in ascertaining to which side this bone belongs ; the following method will sometimes be found useful. Hold the bone with its broader, non-articular surface upwards, so that its sloping border is directed towards you. The border will slope to the side to which the bone belongs.

articulates with the semilunar. The *inferior surface* is divided by two ridges into three facets, for articulation with the second, third, and fourth metacarpal bones, that for the third (the middle facet) being the largest of the three. The *posterior* or *dorsal surface* is broad and rough; the *anterior* or *palmar*, narrow, rounded, and also rough, for the attachment of ligaments and a part of the Adductor obliquus pollicis. The *external surface* articulates with the trapezoid by a small facet at its anterior inferior angle, behind which is a rough depression for the attachment of an interosseous ligament. Above this is a deep and rough groove, which forms part of the neck, and serves for the attachment of ligaments, bounded superiorly by a smooth, convex surface, for articulation with the scaphoid. The *internal surface* articulates with the unciform by a smooth, concave, oblong facet, which occupies its posterior and superior parts; and is rough in front, for the attachment of an interosseous ligament.

Hold the bone with the broader, non-articular surface towards you, and the head upwards. The small, articular facet at the anterior inferior angle of the external surface will point to the side to which the bone belongs.

Articulations.—With seven bones: the scaphoid and semilunar above; the second, third, and fourth metacarpal bones below; the trapezoid on the radial side; and the unciform on the ulnar side.

Attachment of Muscles.—Part of the Adductor obliquus pollicis.

UNCIFORM (fig. 304)

The **Unciform** (uncus, *a hook*; forma, *likeness*) may be readily distinguished by its wedge-shaped form, and the hook-like process which projects from its palmar surface. It is situated at the inner and lower angle of the carpus, with its base downwards, resting on the two inner metacarpal bones, and its apex directed upwards and outwards. The *superior surface*, the apex of the wedge, is narrow, convex, smooth, and articulates with the semilunar. The *inferior surface* articulates with the fourth and fifth metacarpal bones, the concave surface for each being separated by a ridge, which runs from before backwards. The *posterior* or *dorsal surface* is triangular and rough, for ligamentous attachment. The *anterior* or *palmar surface* presents, at its lower and inner side, a curved, hook-like process of bone, the *unciform process*, directed from the palmar surface forwards and outwards. It gives attachment, by its apex, to the annular ligament and the Flexor carpi ulnaris; by its inner surface to the Flexor brevis minimi digiti and the Opponens minimi digiti; and is grooved on its outer side, for the passage

FIG. 304.—The left unciform.

For semilunar
For cuneiform
Unciform process

For os magnum
For 4th metatarsal
For 5th metatarsal

of the Flexor tendons into the palm of the hand. This is one of the four eminences on the front of the carpus to which the anterior annular ligament is attached; the others being the pisiform internally, the oblique ridge of the trapezium and the tuberosity of the scaphoid externally. The *internal surface* articulates with the cuneiform by an oblong facet, cut obliquely from above, downwards and inwards. The *external surface* articulates with the os magnum by its upper and posterior part, the remaining portion being rough, for the attachment of ligaments.

Hold the bone with the hooked process away from you, and the articular surface, divided into two parts for the metacarpal bones, downwards. The concavity of the process will be on the side to which the bone belongs.

Articulations.—With five bones: the semilunar above, the fourth and fifth metacarpal below, the cuneiform internally, the os magnum externally.

Attachment of Muscles.—To three: the Flexor brevis minimi digiti, the Opponens minimi digiti, the Flexor carpi ulnaris.

THE METACARPUS

The **Metacarpal Bones** are five in number: they are long cylindrical bones, presenting for examination a shaft and two extremities.

COMMON CHARACTERS OF THE METACARPAL BONES

The **shaft** is prismoid in form, and curved longitudinally, so as to be convex in the longitudinal direction behind, concave in front. It presents three surfaces: two lateral and one posterior. The *lateral surfaces* are concave, for the attachment of the Interossei muscles, and separated from one another by a prominent anterior ridge. The *posterior* or *dorsal surface* presents in its distal half a smooth, triangular, flattened area which is covered, in the recent state, by the tendons of the Extensor muscles. This triangular surface is bounded by two lines, which commence in small tubercles situated on the dorsal aspect on either side of the digital extremity, and, running backwards, converge to meet some distance behind the centre of the bone and form a ridge which runs along the rest of the dorsal surface to the carpal extremity. This ridge separates two lateral, sloping surfaces for the attachment of the Dorsal interossei muscles.* To the tubercles on the digital extremities are attached the lateral ligaments of the metacarpo-phalangeal joints.

The **carpal extremity, or base,** is of a cuboidal form, and broader behind than in front: it articulates above with the carpus, and on each side with the adjoining metacarpal bones; its *dorsal* and *palmar surfaces* are rough, for the attachment of tendons and ligaments.

The **digital extremity, or head,** presents an oblong surface markedly convex from before backwards, less so from side to side, and flattened laterally; it articulates with the proximal phalanx; it is broader, and extends farther forwards, on the palmar than on the dorsal aspect. It is longer in the antero-posterior than in the transverse diameter. On either side of the head is a tubercle for the attachment of the lateral ligament of the metacarpo-phalangeal joint. The *posterior surface*, broad and flat, supports the Extensor tendons; the *anterior surface* is grooved in the middle line for the passage of the Flexor tendons and marked on each side by an articular eminence continuous with the terminal articular surface.

PECULIAR CHARACTERS OF THE METACARPAL BONES

The **metacarpal bone of the thumb** (fig. 305) is shorter and stouter than the rest, diverges to a greater degree from the carpus, and its *palmar surface* is directed inwards towards the palm. The *shaft* is flattened and broad on its dorsal aspect, and does not present the ridge which is found on the other metacarpal bones; it is concave from above downwards, on its palmar surface. On its outer border is inserted the Opponens pollicis muscle, while its inner border gives origin to the outer head of the First dorsal interosseous muscle. The *carpal extremity*, or *base*, presents a concavo-convex surface, for articulation with the trapezium; it has no lateral facets, but presents externally a tubercle for the insertion of the Extensor ossis metacarpi pollicis. The *digital extremity* is less convex than that of the other metacarpal bones, broader from side to side than from before backwards. It presents on its palmar aspect two distinct articular eminences for the two sesamoid bones in the tendons of the Flexor brevis pollicis; the outer one being the larger of the two.

FIG. 305.—The first metacarpal. (Left.)

For trapezium   For trapezium

The side to which this bone belongs may be known by holding it in the position it occupies in the hand, with the carpal extremity upwards and the

---

* By these sloping surfaces the metacarpal bones of the hand may be at once distinguished from the metatarsal bones of the foot.

dorsal surface backwards; the tubercle for the Extensor ossis metacarpi pollicis
will point to the side to which it belongs.

Attachment of Muscles.—To four: the Opponens pollicis, the Extensor ossis
metacarpi pollicis, the Flexor brevis pollicis, and the First dorsal interosseous.

The **metacarpal bone of the index finger** (fig. 306) is the longest, and its base
the largest, of the four remaining bones. Its *carpal extremity* is prolonged upwards
and inwards, forming a prominent ridge. The dorsal and palmar surfaces of this
extremity are rough, for the attachment of tendons and ligaments. It presents
four articular facets: three on the upper aspect of the base; the middle one of the
three is the largest, concave from side to side, convex from before backwards
for articulation with the trapezoid; the external one is a small, flat, oval facet,
for articulation with the trapezium; the internal one, on the summit of the ridge,
is long and narrow, for articulation with the os magnum. The fourth facet is on
the inner or ulnar side of the extremity of the bone, and is for articulation with
the third metacarpal bone.

The side to which this bone belongs is indicated by the absence of the lateral
facet on the outer (radial) side of its base, so that if the bone is placed with its

Fig. 306.—The second metacarpal.　　　　　　　Fig. 307.—The third metacarpal.
(Left.)　　　　　　　　　　　　　　　　　　　(Left.)

For trapezoid　For trapezium　For 3rd metacarpal　　　Styloid　For 2nd metacarpal　For os magnum　For 4th metacarpal
For os magnum　　　　　　　　　　　　　　　　process

base towards the student, and the palmar surface upwards, the side on which
there is no lateral facet will be that to which it belongs.

Attachment of Muscles.—To six: Flexor carpi radialis, Extensor carpi radialis
longior, Adductor obliquus pollicis, First and Second dorsal interossei, and First
palmar interosseous.

The **metacarpal bone of the middle finger** (fig. 307) is a little smaller than the
preceding; it presents a pyramidal eminence, the *styloid process*, on the radial
side of its base (dorsal aspect), which extends upwards behind the os magnum;
immediately below this, on the dorsal aspect, is a rough surface for the attachment of the Extensor carpi radialis brevior. The carpal articular facet is
concave behind, flat in front, and articulates with the os magnum. On the
radial side is a smooth, concave facet for articulation with the second metacarpal
bone, and on the ulnar side two small, oval facets, for articulation with the
fourth metacarpal.

The side to which this bone belongs is easily recognised by the styloid
process on the radial side of its base. With the palmar surface uppermost, and
the base towards the student, this process points towards the side to which the
bone belongs.

Attachment of Muscles.—To six: Extensor carpi radialis brevior, Flexor carpi radialis, Adductor transversus pollicis, Adductor obliquus pollicis, and Second and Third dorsal interossei.

The **metacarpal bone of the ring-finger** (fig. 308) is shorter and smaller than the preceding, and its base small and quadrilateral; the carpal surface of the base presenting two facets, a large one externally for articulation with the unciform, and a small one internally for the os magnum. On the radial side are two oval facets, for articulation with the third metacarpal bone; and on the ulnar side, a single concave facet, for the fifth metacarpal.

If this bone is placed with the base towards the student, and the palmar surface upwards, the radial side of the base, which has two facets for articulation with the third metacarpal bone, will be on the side to which it belongs. If, as sometimes happens in badly marked bones, one of these facets is indistinguishable, the side may be known by selecting the surface on which the larger articular facet is present. This facet is for the fifth metacarpal bone, and would therefore be situated on the ulnar side—that is, the one to which the bone does *not* belong.

Attachment of Muscles.—To three: the Third and Fourth dorsal and Second palmar interossei.

FIG. 308.—The fourth metacarpal. (Left.)　　　　FIG. 309.—The fifth metacarpal. (Left.)

For os magnum　*For 3rd metacarpal*　*For unciform*　*For 5th metacarpal*　*For 4th metacarpal*　*For cuneiform*

The **metacarpal bone of the little finger** (fig. 309) presents on its base one facet, which is concavo-convex, and articulates with the unciform bone, and one lateral, articular facet on its radial side, which articulates with the fourth metacarpal bone. On its ulnar side is a prominent tubercle, for the insertion of the tendon of the Extensor carpi ulnaris. The dorsal surface of the shaft is marked by an oblique ridge, which extends from near the ulnar side of the upper extremity to the radial side of the lower. The outer division of this surface serves for the attachment of the Fourth dorsal interosseous muscle; the inner division is smooth, triangular, and covered by the Extensor tendons of the little finger.

If this bone is placed with its base towards the student, and its palmar surface upwards, the side of the head which has a lateral facet will be that to which the bone belongs.

Attachment of Muscles.—To five: the Extensor carpi ulnaris, Flexor carpi ulnaris, Opponens minimi digiti, Fourth dorsal, and Third palmar interossei.

Articulations.—Besides the phalangeal articulations, the first metacarpal bone articulates with the trapezium; the second with the trapezium, trapezoid, os magnum, and third metacarpal bone; the third with the os magnum, and second and fourth metacarpal bones; the fourth with the os magnum, unciform, and third and fifth metacarpal bones; and the fifth with the unciform and fourth metacarpal bone.

The *first* has no lateral facets on its carpal extremity; the *second* has no lateral facet on its radial side, but one on its ulnar side; the third has one on its radial and two on its ulnar side; the fourth has two on its radial and one on its ulnar side; and the fifth has one on its radial side only.

THE PHALANGES

The **Phalanges** (*internodia*) are the bones of the digits; they are fourteen in number, three for each finger, and two for the thumb. They are long bones, and present for examination a shaft and two extremities. The *shaft* tapers from above downwards, is convex posteriorly, concave in front from above downwards, flat from side to side, and marked laterally by rough ridges, which give attachment to the fibrous sheaths of the Flexor tendons. The *metacarpal extremity* or *base*, in the first row presents an oval, concave, articular surface, broader from side to side than from before backwards; and the same extremity in the other two rows, a double concavity, separated by a longitudinal median ridge, extending from before backwards. The *digital extremities* are smaller than the bases, and terminate, in the first and second rows, in two small, lateral condyles, separated by a slight groove; the articular surface being prolonged farther on the palmar than on the dorsal surface, especially in the first row.

The **Ungual Phalanges** are convex on their dorsal, flat on their palmar surfaces; they are recognised by their small size, and by a roughened, elevated surface of a horse-shoe form on the palmar aspect of their ungual extremity, which serves to support the sensitive pulp of the finger.

Articulations.—The first row with the metacarpal bones and the second row of phalanges; the second row with the first and third; the third with the second row.

Attachment of Muscles.—To the base of the first phalanx of the thumb, five muscles: the Extensor brevis pollicis, Flexor brevis pollicis, Abductor pollicis, Adductor transversus and obliquus pollicis. To the second phalanx, two: the Flexor longus pollicis and the Extensor longus pollicis. To the base of the first phalanx of the index finger, the First dorsal and the First palmar interosseous; to that of the middle finger, the Second and Third dorsal interosseous; to that of the ring-finger, the Fourth dorsal and the Second palmar interosseous; and to that of the little finger, the Third palmar interosseous, the Flexor brevis minimi digiti, and Abductor minimi digiti. To the second phalanges, the Flexor sublimis digitorum, Extensor communis digitorum, and, in addition, the Extensor indicis to the index finger, the Extensor minimi digiti to the little finger. To the third phalanges, the Flexor profundus digitorum and Extensor communis digitorum.

DEVELOPMENT OF THE BONES OF THE HAND

The **Carpal Bones** are each developed by a *single* centre. At birth, they are all cartilaginous. Ossification proceeds in the following order (fig. 310): In the os magnum and unciform an ossific point appears during the first year, the former preceding the latter; in the cuneiform, at the third year; in the semilunar and trapezium, at the fifth year, the former preceding the latter; in the scaphoid, at the sixth year; in the trapezoid, during the eighth year; and in the pisiform, about the twelfth year.

Occasionally an additional bone, the *os centrale*, is found on the back of the carpus, lying between the scaphoid, trapezoid, and os magnum. During the second month of fœtal life it is represented by a small cartilaginous nodule, which usually fuses with the cartilaginous scaphoid. Sometimes the styloid process of the third metacarpal is detached and forms an additional ossicle.

The **Metacarpal Bones** are each developed by *two* centres: one for the shaft and one for the digital extremity of each of the four inner metacarpal bones; one for the shaft and one for the base of the metacarpal bone of the thumb, which in this respect resembles the phalanges.* It will be seen, therefore, that the so-called

---

* Allen Thomson demonstrated the fact that the first metacarpal bone is often developed from three centres: that is to say, there is a separate nucleus for the distal end, forming a distinct epiphysis visible at the age of seven or eight years. He also states that there are traces of a proximal epiphysis in the second metacarpal bone. *Journal of Anatomy*, 1869.

metacarpal bone of the thumb is ossified in the same manner as the phalanges, and this has led some anatomists to regard this digit as being made up of three phalanges, and not of a metacarpal bone and two phalanges. Ossification commences in the centre of the shaft about the eighth or ninth week, the centre for the first metacarpal bone being the last to appear, and gradually proceeds to either end of the bone ; about the third year the digital extremities of the four inner metacarpal bones, and the base of the first metacarpal, begin to ossify, and they unite about the twentieth year.

The **Phalanges** are each developed by *two* centres : one for the shaft, and one for the base. Ossification commences · in the shaft, in all three rows, at about the eighth week, and gradually involves the whole of the bone excepting the upper extremity. Ossification of the base commences in the first row

FIG. 310.—Plan of the development of the hand.

CARPUS
One centre for each bone :
All cartilaginous at birth

METACARPUS
Two centres for each bone :
One for shaft
One for digital extremity,
except first

PHALANGES
Two centres for each bone :
One for shaft
One for metacarpal
extremity

between the third and fourth years, and a year later in those of the second and third rows. The two centres become united, in each row, between the eighteenth and twentieth years.

In the ungual phalanges the centre for the shaft appears at the distal extremity of the phalanx, instead of at the middle of the shaft as is the case with the other phalanges. Moreover, of all the bones of the hand they are the first to begin to ossify.

Surface Form.—On the front of the wrist are two subcutaneous eminences, one on the radial side, the larger and flatter, produced by the tuberosity of the scaphoid and the ridge on the trapezium ; the other on the ulnar side, caused by the pisiform bone. The tubercle of the scaphoid is to be felt just below and internal to the apex of the styloid process of the radius, between the tendons of the Extensor ossis metacarpi pollicis and the Flexor carpi radialis. It is best perceived by extending the hand on the forearm. Half an inch below this tubercle is to be felt another prominence, better marked than the tubercle ; this is the ridge on the trapezium, which gives attachment to some of the short muscles of the thumb. On the inner side of the front of the wrist the pisiform bone is to be felt, forming a small but prominent projection in this situation. It is some distance below the styloid process of the ulna, and may be said to be just below the level of the styloid process of the radius, and is crossed by the crease which separates the front of the forearm from the palm of the hand. The rest of the front of the carpus is covered by tendons and the annular ligament, and entirely concealed, with the exception of the hooked process of the unciform, which can only be made out with difficulty. The back of the carpus is convex and covered by the Extensor tendons, so that the posterior surface of the cuneiform is the only bone which can be felt. Below the carpus the dorsal surfaces of the metacarpal bones, except the fifth, are covered by tendons, and are only visible in very thin hands. The dorsal surface of the fifth is, however, subcutaneous throughout almost its whole length, and is plainly to be perceived and felt. In addition to this, slightly external to the middle line of the hand, is a prominence, frequently well marked, but occasionally indistinct, formed by the styloid process of the metacarpal bone of the middle finger. This prominence is in the same line as the dorsal radial tubercle, and is an inch and a half vertically above it. The heads of the metacarpal bones are plainly to be felt and seen, rounded in contour and standing out in bold relief under the skin, when the fist is clenched. It should be borne in mind that when the fingers are flexed on the hand, the articular surfaces of the first phalanges glide off the heads of the metacarpal bones on to their anterior surfaces; so that the heads of these bones form the prominence of the knuckles and receive the force of any blow which may be given. The head of the third metacarpal bone is the most prominent, and receives the greater part of the shock of the blow. This bone articulates with the os magnum, so that the concussion is carried through this bone to the scaphoid and semilunar, with which the head of the os magnum articulates, and by these bones is transferred to the radius, along which it may be carried to the capitellum of the humerus. The enlarged extremities of the phalanges are to be plainly felt : they constitute the joints of the fingers. When the digits are bent, the proximal phalanges of the joints form prominences, which in the joint between the first and second phalanges are slightly hollowed, in accordance with the grooved shape of their articular surfaces, while in the last row the prominence is flattened and square-shaped. In the palm of the hand the four inner metacarpal bones are covered by muscles, tendons, and the palmar fascia, and no part of them but their heads is to be distinguished. With regard to the thumb, on the dorsal aspect, the base of the metacarpal bone forms a prominence, below the styloid process of the radius; the shaft is to be felt, covered by tendons, terminating at its head in a flattened prominence, in front of which can be felt the sesamoid bones.

Surgical Anatomy.—The carpal bones are little liable to fracture, except from extreme violence, when the parts are so comminuted as to necessitate amputation. Occasionally they are the seat of tuberculous disease. The metacarpal bones and the phalanges are sometimes broken from direct violence. The first metacarpal bone is the one most commonly fractured, then the second, the fourth and the fifth, the third being the one least frequently broken. There are two diseases of the metacarpal bones and phalanges which require special mention on account of their constant occurrence. One is tuberculous dactylitis, consisting in a deposit of tuberculous material in the medullary canal, expansion of the bone, with subsequent caseation and resulting necrosis. The other is chondroma, which is perhaps more commonly found in connection with the metacarpal bones and phalanges than with any other bones. The tumours are usually multiple, and may spring either from the medullary canal, or from the periosteum.

THE LOWER EXTREMITY

The bones of the lower extremity consist of those of the pelvic girdle, of the thigh, of the leg, and of the foot.

THE PELVIC GIRDLE

The **Pelvic Girdle** consists of a single bone, the *os innominatum*, by which the thigh is connected to the trunk.

The Os Innominatum

The **Os Innominatum** (in, *not*; nomino, *I name*), or *nameless bone*, so called from bearing no resemblance to any known object, is a large, irregularly shaped, flat bone, constricted in the centre and expanded above and below. With its fellow of the opposite side, it forms the sides and anterior wall of the pelvic cavity. In young subjects it consists of three separate parts, which meet and form a large, cup-like cavity, the *acetabulum*, situated near the middle of the outer surface of the bone : and, although in the adult these have become united, it is usual to describe the bone as divisible into three portions—the ilium, the ischium, and the os pubis.

The **ilium**, so called from its supporting the flank (*ilia*), is the superior, broad and expanded portion which runs upwards from the acetabulum, and forms the prominence of the hip.

FIG. 311.—Right os innominatum. External surface.

The **ischium** (ἰσχίον, *the hip*) is the inferior and strongest portion of the bone ; it proceeds downwards from the acetabulum, expands into a large tuberosity, and then, curving forwards, forms, with the descending ramus of the os pubis, a large aperture, the obturator foramen.

The **os pubis** is that portion which extends inwards and downwards from the acetabulum to articulate in the middle line with the bone of the opposite side : it forms the front of the pelvis, supports the external organs of generation, and has received its name from the skin over it being covered with hair.

The **Ilium** presents for examination two surfaces—an external and an internal
—a crest, and two borders—an anterior and a posterior.

The **External Surface** (fig. 311) is divided into two parts: an upper or gluteal,
and a lower or acetabular.   The *upper portion*, known as the *dorsum of the ilium*,
is directed backwards and outwards behind ; downwards and outwards in front.
It is smooth, convex in front, deeply concave behind ; bounded above by the
crest, below by the upper border of the acetabulum ; in front and behind, by the
anterior and posterior borders.   This surface is crossed in an arched direction by
three semicircular ridges—the superior, middle, and inferior curved lines.   The
superior curved line, the shortest of the three, commences at the crest, about
two inches in front of its posterior extremity; it is at first distinctly marked, but
as it passes downwards and backwards to the upper part of the great sacro-sciatic
notch, where it terminates, it becomes less marked, and is often altogether lost.
Behind this line is a narrow semilunar surface, the upper part of which is rough
and gives origin to part of the Gluteus maximus ; the lower part is smooth and
has no muscular fibres attached to it.   The *middle curved line*, the longest of the
three, commences at the crest, about an inch behind its anterior extremity, and,
taking a curved direction downwards and backwards, terminates at the upper
part of the great sacro-sciatic notch.   The space between the superior and middle
curved lines and the crest is concave, and gives origin to the Gluteus medius
muscle.   Near the central part of this line the orifice of a nutrient foramen may
often be observed.   The *inferior curved line*, the least distinct of the three, commences in front at the notch on the anterior border, and, taking a curved direction
backwards and downwards, terminates at the middle of the great sacro-sciatic
notch.   The surface of bone included between the middle and inferior curved
lines is concave from above downwards, convex from before backwards, and gives
origin to the Gluteus minimus muscle.   Beneath the inferior curved line, and
corresponding to the upper part of the acetabulum, is a roughened surface (sometimes a depression), from which arises the reflected tendon of the Rectus femoris
muscle.   The *lower* or *acetabular part* of the external surface enters into the
formation of the acetabulum, of which it forms rather less than two-fifths.   It is
separated from the upper gluteal portion of this surface by a prominent rim,
which forms part of the margin of the acetabular cavity.

The **Internal Surface** (fig. 312) of the ilium is bounded above by the crest ;
below, it is continuous with the pelvic surfaces of the os pubis and ischium, a
faint line only indicating the place of union ; in front and behind, it is bounded
by the anterior and posterior borders.   It presents a large, smooth, concave
surface, called the *iliac fossa*, or *venter ilii*, which gives origin to the Iliacus
muscle, and presents at its lower part the orifice of a nutrient canal ; and below
this a smooth, rounded border, the *linea ilio-pectinea*, which separates the iliac
fossa from that portion of the internal surface which enters into the formation of
the true pelvis, and which gives origin to part of the Obturator internus muscle.
Behind the iliac fossa is a rough surface, divided into two portions, an anterior
and a posterior.   The anterior or *auricular surface*, so called from its resemblance
in shape to the ear, is coated with cartilage in the recent state, and articulates
with a surface of similar shape on the side of the sacrum.   The posterior portion
is rough, for the attachment of the posterior sacro-iliac ligaments and for a part
of the origin of the Erector and Multifidus spinæ muscles.

The **crest** of the ilium is convex in its general outline and sinuously curved,
being concave inwards in front, concave outwards behind.   It is longer in the
female than in the male, very thick behind, and thinner at the centre than at the
extremities.   It terminates at either end in a prominent eminence, the *anterior
superior* and *posterior superior iliac spines*.   The surface of the crest is broad,
and divided into an external lip, an internal lip, and an intermediate space.
About two inches behind the anterior superior spinous process there is a prominent tubercle on the outer lip.   To the external lip are attached the Tensor
fasciæ femoris, Obliquus externus abdominis, and Latissimus dorsi, and along its
whole length the fascia lata ; to the space between the lips, the Internal oblique ;
to the internal lip, the Transversalis abdominis, Quadratus lumborum, and
Erector spinæ, the Iliacus, and the fascia iliaca.

The **anterior border** of the ilium is concave.   It presents two projections,
separated by a notch.   Of these, the uppermost, situated at the junction of the
crest and anterior border, is called the *anterior superior iliac spine*, the

outer border of which gives attachment to the fascia lata, and the origin of the Tensor fasciæ femoris ; its inner border, to the Iliacus ; while its extremity affords attachment to Poupart's ligament and the origin of the Sartorius muscle. Beneath this eminence is a notch which gives origin to the Sartorius, and across which passes the external cutaneous nerve. Below the notch is the *anterior inferior iliac spine*, which terminates in the upper lip of the acetabulum ; it gives attachment to the straight tendon of the Rectus femoris muscle and the iliofemoral ligament. On the inner side of the anterior inferior spine is a broad,

FIG. 312.—Right os innominatum. Internal surface.

shallow groove, over which passes the Ilio-psoas muscle. This groove is bounded internally by an eminence, the *ilio-pectineal*, which marks the point of union of the ilium and os pubis.

The **posterior border** of the ilium, shorter than the anterior, also presents two projections separated by a notch, the *posterior superior* and the *posterior inferior iliac spines*. The former corresponds with that portion of the inner surface of the ilium which serves for the attachment of the oblique portion of the sacro-iliac ligaments and the Multifidus spinæ ; the latter to the auricular portion which articulates with the sacrum.   Below the posterior inferior spine is a deep notch, the *great sacro-sciatic*.

The **Ischium** forms the lower and back part of the os innominatum.   It is divisible into a thick and solid portion, the *body* ; a large, rough eminence, on which the trunk rests in sitting, the *tuberosity* ; and a thin part, which passes forwards and slightly upwards, the *ramus*.

The **body,** somewhat triangular in form, presents three surfaces, external, internal, and posterior ; and three borders, external, internal, and posterior.   The *external surface* corresponds to that portion of the acetabulum which is formed by the ischium ; it is smooth and concave, and constitutes a little more than two-fifths of the acetabular cavity ; its outer margin is bounded by a prominent rim or lip, the external border, to which the cotyloid fibro-cartilage is attached.   Below the acetabulum, between it and the tuberosity, is a deep groove, along which the tendon of the Obturator externus muscle runs, as it passes outwards to be inserted into the digital fossa of the femur.   The *internal surface* is smooth, concave, and enters into the formation of the lateral boundary of the true pelvic cavity. This surface is perforated by two or three large, vascular foramina, and gives origin to part of the Obturator internus muscle.   The *posterior surface* is quadri-lateral in form, broad and smooth.   Below, where it joins the tuberosity, it presents a groove continuous with that on the external surface for the tendon of the Obturator externus muscle.   The lower edge of this groove is formed by the tuberosity of the ischium, and gives origin to the Gemellus inferior muscle.   This surface is limited, externally, by the margin of the acetabulum ; behind by the posterior border ; it supports the Pyriformis, the two Gemelli, and the Obturator internus muscles, in their passage outwards to the great trochanter.   The *external border* forms the prominent rim of the acetabulum, and separates the posterior from the external surface.   To it is attached the cotyloid fibro-cartilage. The *internal border* is thin, and forms the outer circumference of the obturator foramen.   The *posterior border* of the body of the ischium presents, a little below the centre, a thin and pointed triangular eminence, the *spine of the ischium*, more or less elongated in different subjects ; its external surface gives attach-ment to the Gemellus superior, its internal surface to the Coccygeus, Levator ani, and white line of the pelvic fascia ; while to the pointed extremity is connected the lesser sacro-sciatic ligament.   Above the spine is a notch of large size, the *great sacro-sciatic*, converted into a foramen by the lesser sacro-sciatic ligament ; it transmits the Pyriformis muscle, the gluteal vessels, the superior and inferior gluteal nerves, the sciatic vessels, the greater and lesser sciatic nerves, the internal pudic vessels and nerve, and the nerves to the Obturator internus and Quadratus femoris.   Of these, the gluteal vessels and superior gluteal nerve pass out above the Pyriformis muscle, the other structures below it. Below the spine is a smaller notch, the *lesser sacro-sciatic* ; it is smooth, coated in the recent state with cartilage, the surface of which presents two or three ridges corresponding to the subdivisions of the tendon of the Obturator internus, which winds over it.   It is converted into a foramen by the sacro-sciatic liga-ments, and transmits the tendon of the Obturator internus, the nerve which supplies that muscle, and the internal pudic vessels and nerve.

The **tuberosity** presents for examination three surfaces : external, internal, and posterior.   The *external surface* is quadrilateral in shape and rough for the attachment of muscles.   It is bounded above by the groove for the tendon of the Obturator externus ; in front it is limited by the posterior margin of the obturator foramen, and below it is continuous with the ramus of the bone ; behind, it is bounded by a prominent margin which separates it from the posterior surface. In front of this margin the surface gives origin to the Quadratus femoris, and anterior to this to some of the fibres of origin of the Obturator externus.   The lower part of the surface gives origin to part of the Adductor magnus.   The *internal surface* forms part of the bony wall of the true pelvis.   In front, it is limited by the posterior margin of the obturator foramen.   Behind, it is bounded by a sharp ridge, for the attachment of a falciform prolongation of the great sacro-sciatic ligament, and, more anteriorly, gives origin to the Transversus perinæi and Erector penis muscles.   The *posterior surface* is divided into two portions : a lower, rough, somewhat triangular part, and an upper, smooth, quadrilateral portion.   The anterior portion is subdivided by a prominent vertical ridge, passing from base to apex, into two parts ; the outer one gives attachment to the Adductor magnus, the inner to the great sacro-sciatic ligament. The upper portion is subdivided into two facets by an oblique ridge, which runs downwards and outwards; from the upper and outer facet arises the Semi-membranosus; from the lower and inner, the long head of the Biceps and the Semitendinosus.

The **ramus** is the thin, flattened part of the ischium, which ascends from the tuberosity upwards and inwards, and joins the descending ramus of the os pubis—their point of junction being indicated in the adult by a raised line. The outer surface of the ramus is uneven, for the origin of the Obturator externus muscle, and also some fibres of the Adductor magnus; its inner surface forms part of the anterior wall of the pelvis. Its inner border is thick, rough, slightly everted, forms part of the outlet of the pelvis, and presents two ridges and an intervening space. The ridges are continuous with similar ones on the descending ramus of the os pubis: to the outer one is attached the deep layer of the superficial perineal fascia (fascia of Colles), and to the inner the superficial layer of the triangular ligament of the urethra. If these two ridges are traced downwards, they will be found to join with each other just behind the point of origin of the Transversus perinæi muscle; here the two layers of fascia are continuous behind the posterior border of the muscle. To the intervening space, just in front of the point of junction of the ridges, is attached the Transversus perinæi muscle, and in front of this a portion of the crus penis vel clitoridis and the Erector penis vel clitoridis muscle. Its outer border is thin and sharp, and forms part of the inner margin of the obturator foramen.

The **Os Pubis** is the anterior part of the os innominatum, and, with the bone of the opposite side, constitutes the front boundary of the true pelvic cavity. It is divisible into a body, an ascending and a descending ramus.

The **body** is somewhat quadrilateral in shape, and presents for examination two surfaces and three borders. The *anterior surface* is rough, directed downwards and outwards, and serves for the origin of various muscles. To the upper and inner angle, immediately below the crest, is attached the Adductor longus; lower down, from without inwards, are attached the Obturator externus, the Adductor brevis, and the upper part of the Gracilis. The *posterior surface*, convex from above downwards, concave from side to side, is smooth, and forms part of the anterior wall of the pelvis. It gives origin to the Levator ani, Obturator internus, a few muscular fibres prolonged from the bladder, and the pubo-prostatic ligaments. The *upper border* presents for examination a prominent tubercle, which projects forwards and is called the *spine*; to it is attached the outer pillar of the external abdominal ring and Poupart's ligament. Passing upwards and outwards from this is a prominent ridge, forming part of the *ilio-pectineal line* which marks the brim of the true pelvis: to it is attached a portion of the conjoined tendon of the Internal oblique and Transversalis muscles, Gimbernat's ligament and the triangular fascia of the abdomen. Internal to the spine of the os pubis is the *crest*, which extends from this process to the inner extremity of the bone. It affords attachment, anteriorly, to the conjoined tendon of the Internal oblique and Transversalis; and posteriorly, to the Rectus and Pyramidalis muscles. The point of junction of the crest with the inner border of the bone is called the *angle*; to it, as well as to the symphysis, is attached the internal pillar of the external abdominal ring. The *internal border* is articular; it is oval, covered by eight or nine transverse ridges, or a series of nipple-like processes arranged in rows, separated by grooves; they serve for the attachment of a thin layer of cartilage, placed between it and the central fibro-cartilage. The *outer border* presents a sharp margin, which forms part of the circumference of the obturator foramen and affords attachment to the obturator membrane.

The **ascending** or **superior ramus** extends from the body to the point of junction of the os pubis with the ilium, and forms the upper part of the circumference of the obturator foramen. It presents for examination a superior, inferior, and posterior surface, and an outer extremity. The *superior surface* presents a continuation of the ilio-pectineal line, already mentioned as commencing at the pubic spine. In front of this ridge, the surface of bone is triangular in form, wider externally than internally, smooth, and is covered by the Pectineus muscle. The surface is bounded externally by a rough eminence, the *ilio-pectineal*, which serves to indicate the point of junction

X

of the ilium and os pubis. The triangular surface is bounded below by a prominent ridge, the *obturator crest*, which extends from the cotyloid notch to the spine of the os pubis. The *inferior surface* forms the upper boundary of the obturator foramen, and presents, externally, a broad and deep, oblique groove, for the passage of the obturator vessels and nerve; and internally, a sharp margin which is part of the circumference of the obturator foramen, and to which the obturator membrane is attached. The *posterior surface* constitutes part of the anterior boundary of the true pelvis. It is smooth, convex from above downwards, and affords origin to some fibres of the Obturator internus. The *outer extremity*, the thickest part of the ramus, forms one-fifth of the cavity of the acetabulum.

The **descending** or **inferior ramus** of the os pubis is thin and flattened. It passes outwards and downwards, becoming narrower as it descends and joins with the ramus of the ischium. Its *anterior surface* is rough, for the origin of muscles—the Gracilis along its inner border; a portion of the Obturator externus where it enters into the formation of the foramen of that name; and between these two muscles, the Adductores brevis and magnus from within outwards. The *posterior surface* is smooth, and gives origin to the Obturator internus, and, close to the inner margin, to the Compressor urethræ. The *inner border* is thick, rough, and everted, especially in females. It presents two ridges, separated by an intervening space. The ridges extend downwards, and are continuous with similar ridges on the ramus of the ischium; to the external one is attached the fascia of Colles, and to the internal one the superficial layer of the triangular ligament of the urethra. The *outer border* is thin and sharp, forms part of the circumference of the obturator foramen, and gives attachment to the obturator membrane.

The **cotyloid cavity,** or **acetabulum,** is a deep, cup-shaped, hemispherical depression, directed downwards, outwards, and forwards; formed internally by the os pubis, above by the ilium, behind and below by the ischium; a little less than two-fifths being formed by the ilium, a little more than two-fifths by the ischium, and the remaining fifth by the pubic bone. It is bounded by a prominent, uneven rim, which is thick and strong above, and serves for the attachment of the *cotyloid ligament*, which contracts its orifice, and deepens the surface for articulation. It presents below a deep notch, the *cotyloid notch*, which is continuous with a circular depression, the *fossa acetabuli*, at the bottom of the cavity: this depression is perforated by numerous apertures, lodges a mass of fat, and its margins, as well as those of the notch, serve for the attachment of the ligamentum teres. The notch is converted into a foramen by the transverse ligament which passes across it; through this foramen the nutrient vessels and nerves enter the joint.

The **obturator** or **thyroid foramen** is a large aperture, situated between the ischium and os pubis. In the male it is large, of an oval form, its longest diameter being obliquely from before backwards; in the female it is smaller, and more triangular. It is bounded by a thin, uneven margin, to which a strong membrane is attached; and presents, anteriorly, a deep groove, which runs from the pelvis obliquely inwards and downwards. This groove is converted into a foramen by a ligamentous band, a specialised part of the obturator membrane. This band is attached on either side to a tubercle: one on the internal border of the ischium, just in front of the cotyloid notch; the other on the inferior margin of the posterior surface of the ascending ramus of the os pubis. Through the foramen the obturator vessels and nerve pass out of the pelvis.

Structure.—This bone consists of much cancellous tissue, especially where it is thick, enclosed between two layers of dense, compact tissue. In the thinner parts of the bone, as at the bottom of the acetabulum and centre of the iliac fossa, it is usually semi-transparent, and composed entirely of compact tissue.

Development (fig. 313).—By *eight* centres: three primary—one for the ilium, one for the ischium, and one for the os pubis; and *five* secondary—one for the crest of the ilium, one for the anterior inferior spine (said to occur more frequently in the male than in the female), one for the tuberosity of the ischium, one for the symphysis pubis (more frequent in the female than in the male), and one or more for the Y-shaped piece at the bottom of the acetabulum. These various centres appear in the following order: in the lower part of the ilium, immediately above the sciatic notch, about the eighth or ninth week of foetal life; in the body of the ischium, about the third month; in the body of the os pubis, between the fourth and fifth months. At birth, the three primary centres are quite separate, the crest, the bottom of the acetabulum, the ischial tuberosity, and the rami of the ischium and pubes being still cartilaginous. About the seventh or eighth year, the rami of the os pubis and ischium are almost completely united by bone. About the thirteenth or fourteenth year, the three divisions of the bone have extended their growth into the bottom of the acetabulum, being separated from each other by a Y-shaped portion of cartilage, which now presents traces of ossification, often by two or more centres. One of these, the *os acetabuli*, appears about the age of twelve, between the ilium and os pubis, and fuses with them about the age of eighteen. It forms the pubic part of the acetabulum. The ilium and ischium then become joined, and lastly the os pubis to the ischium, through the intervention of this Y-shaped portion. At about the age of puberty, ossification takes place in each of the

FIG. 313.—Plan of the development of the os innominatum.

By eight centres { *Three primary (Ilium, Ischium, and Os Pubis)* / *Five secondary* }

The three primary centres unite through Y-shaped piece about puberty.
Epiphyses appear about puberty, and unite about 25th year.

remaining portions, and they become joined to the rest of the bone between the twentieth and twenty-fifth years. Separate centres are frequently found for the pubic and ischial spines, and another for the crest and angle of the os pubis.

Articulations.—With its fellow of the opposite side, the sacrum and femur.

Attachment of Muscles.—To the *ilium*, sixteen. To the outer lip of the crest, the Tensor fasciæ femoris, Obliquus externus abdominis, and Latissimus dorsi; to the internal lip, the Iliacus, Transversalis, Quadratus lumborum, and Erector spinæ; to the interspace between the lips, the Obliquus internus. To the outer surface of the ilium, the Gluteus maximus, Gluteus medius, Gluteus minimus, reflected tendon of the Rectus; to the upper part of the great sacro-sciatic notch, a portion of the Pyriformis; to the internal surface, the Iliacus; to that portion of the internal surface below the linea ilio-pectinea, the Obturator internus; to the internal surface of the posterior superior spine, the Multifidus spinæ; to the anterior border, the Sartorius and straight tendon of the Rectus. To the *ischium*,

X 2

thirteen. To the outer surface of the ramus, the Obturator externus and Adductor magnus ; to the internal surface, the Obturator internus and Erector penis. To the spine, the Gemellus superior, Levator ani, and Coccygeus. To the tuberosity, the Biceps, Semitendinosus, Semimembranosus, Quadratus femoris, Adductor magnus, Gemellus inferior, Transversus perinæi, Erector penis. To the os *pubis*, sixteen : Obliquus externus, Obliquus internus, Transversalis, Rectus, Pyramidalis, Psoas parvus, Pectineus, Adductor magnus, Adductor longus, Adductor brevis, Gracilis, Obturator externus and internus, Levator ani, Compressor urethræ, and occasionally a few fibres of the Accelerator urinæ.

THE PELVIS (figs. 314, 315)

The **Pelvis,** so called from its resemblance to a basin (L. *pelvis*), is stronger and more massively constructed than either the cranial or the thoracic cavity ; it is a bony ring, interposed between the movable vertebræ of the spinal column, which it supports, and the lower extremities, upon which it rests. It is composed of four bones : the two ossa innominata, which bound it on either side and in front ; and the sacrum and coccyx, which complete it behind.

FIG. 314.—Male pelvis (adult).

The pelvis is divided by an oblique plane passing through the prominence of the sacrum, the linea ilio-pectinea, and the upper margin of the symphysis pubis, into the false and true pelvis. The circumference of this plane is termed the *pelvic brim.*

The **false pelvis** is the expanded portion of the cavity which is situated above and in front of this plane. It is bounded on each side by the ossa ilii ; in front it is incomplete, presenting a wide interval between the spinous processes of the ilia on either side, which is filled up in the recent state by the parietes of the abdomen ; behind, in the middle line, is a deep notch. This broad, shallow cavity is fitted to support the intestines, and to transmit part of their weight to the anterior wall of the abdomen, and is, in fact, really a portion of the abdominal cavity. The term false pelvis is incorrect, and this space ought more properly to be regarded as part of the hypogastric and iliac regions of the abdomen.

The **true pelvis** is that part of the pelvic cavity which is situated below and behind the pelvic brim. It is smaller than the false pelvis, but its bony walls are more perfect. For convenience of description, it is divided into a superior circumference or inlet, an inferior circumference or outlet, and a cavity.

The **superior circumference** forms the brim of the pelvis, the included space being called the *inlet* (fig. 316). It is formed by the linea ilio-pectinea, completed in front by the crests of the pubic bones, and behind by the anterior margin of the base of the sacrum and sacro-vertebral angle. The inlet of the pelvis is somewhat heart-shaped, obtusely pointed in front, diverging on either side, and

FIG. 315.—Female pelvis (adult).

encroached upon behind by the projection forwards of the promontory of the sacrum. It has three principal diameters: antero-posterior or conjugate, transverse, and oblique. The antero-posterior extends from the sacro-vertebral angle to the symphysis pubis; its average measurement is four inches in the male, four and three-quarters in the female. The transverse extends across the greatest

FIG. 316.—Diameters of the pelvic inlet.

width of the inlet, from the middle of the brim on one side to the same point on the opposite; its average measurement is four and a half inches in the male, five and a quarter in the female. The oblique extends from the margin of the brim, corresponding to the ilio-pectineal eminence on one side, to the sacro-iliac articulation on the opposite side; its average measurement is four and a quarter inches in the male, and five in the female.

The **cavity** of the true pelvis is bounded in front and below by the symphysis pubis and the bodies of the pubic bones ; above and behind, by the concavity of the sacrum and coccyx, which, curving forwards above and below, contracts the inlet and outlet of the canal ; and laterally it is bounded by a broad, smooth, quadrangular surface of bone, corresponding to the inner surface of the body of the ischium and that part of the ilium which is below the ilio-pectineal line. The cavity is shallow in front, measuring at the symphysis an inch and a half in depth, three inches and a half in the middle, and four inches and a half posteriorly. From this description, it will be seen that the cavity of the pelvis is a short, curved canal, considerably deeper on its posterior than on its anterior wall. It contains, in the recent subject, the rectum, bladder, and part of the organs of generation. The rectum is placed at the back of the pelvis, and corresponds to the curve of the sacro-coccygeal column ; the bladder in front, behind the symphysis pubis. In the female, the uterus and vagina occupy the interval between these viscera.

The **lower circumference** of the pelvis is very irregular, and forms its *outlet* (fig. 317). It is bounded by three prominent eminences : one posterior, formed by the point of the coccyx ; and one on each side, the tuberosities of the ischia. These eminences are separated by three notches : one in front, the *pubic arch*, formed by the convergence of the rami of the ischia and pubic bones on each

FIG. 317.—Diameters of the pelvic outlet.

side. The other notches, one on each side, are formed by the sacrum and coccyx behind, the ischium in front, and the ilium above : they are called the *sacro-sciatic notches* ; in the natural state they are converted into foramina by the lesser and greater sacro-sciatic ligaments. In the recent state, when the ligaments are *in situ*, the outlet of the pelvis is lozenge-shaped, bounded, in front, by the subpubic ligament and the rami of the ossa pubis and ischia ; on each side by the tuberosities of the ischia ; and behind, by the great sacro-sciatic ligaments and the tip of the coccyx.

The diameters of the outlet of the pelvis are two, antero-posterior and transverse. The *antero-posterior (conjugate) diameter* extends from the tip of the coccyx to the lower part of the symphysis pubis ; its average measurement is three inches and three-quarters in the male, and five inches in the female. The antero-posterior diameter varies with the length of the coccyx, and is capable of increase or diminution, on account of the mobility of that bone. The *transverse diameter* extends from the posterior part of one ischiatic tuberosity to the same point on the opposite side ; the average measurement is three and a half inches in the male, and four and three-quarters in the female.*

* The measurements of the pelvis given above are fairly accurate, but different measurements are given by various authors, no doubt due in a great measure to differences in the physique and stature of the population from whom the measurements have been taken.

Position of the Pelvis.—In the erect posture, the pelvis is placed obliquely with regard to the trunk of the body: the plane of the bony ring, which is the brim of the true pelvis, forms an angle of from 50° to 60° with the horizontal plane. The pelvic surface of the symphysis pubis looks upwards and backwards, the concavity of the sacrum and coccyx downwards and forwards; the base of the sacrum in well-formed female bodies being nearly four inches above the upper border of the symphysis pubis, and the apex of the coccyx a little more than half an inch above its lower border. In consequence of this obliquity of the pelvis, the line of gravity of the head, which passes through the middle of the odontoid process of the axis and through the points of junction of the curves of the vertebral column to the sacro-vertebral angle, descends towards the front of the cavity, so that it bisects a line drawn transversely through the middle of the heads of the thigh-bones (see page 178). And thus the centre of gravity of the head is placed immediately over the heads of the thigh-bones on which the trunk is supported.

Axes of the Pelvis (fig. 318).—A line carried at right angles to the plane of the inlet at its middle, would correspond at one extremity with the umbilicus and at the other with the middle of the coccyx: the axis of the inlet is therefore directed downwards and backwards. The axis of the outlet, produced upwards, would touch the base of the sacrum, and is therefore directed downwards and forwards. The axis of the cavity is curved like the cavity itself: this curve corresponds to the concavity of the sacrum and coccyx, the extremities being indicated by the central points of the inlet and outlet. A knowledge of the direction of these axes serves to explain the course of the foetus in its passage through the pelvis during parturition. It is also important to the surgeon, as indicating the direction of the force required in the removal of calculi from the bladder, by the perineal operation, and as determining the direction in which instruments should be used in operations upon the pelvic viscera.

Differences between the Male and Female Pelvis.—The *female* pelvis, looked at as a whole, is distinguished from the *male* by the bones being more delicate and by its depth being smaller.

FIG. 318.—Vertical section of the pelvis, with lines indicating the axes of the pelvis.

The whole pelvis is less massive, and its bones are lighter and more slender, and its muscular impressions are slightly marked. The ilia are less sloped, and the anterior iliac spines more widely separated; hence the greater prominence of the hips. The *inlet* in the female is larger than in the male; it is more nearly circular, and its obliquity is greater. The *cavity* is shallower and wider; the sacrum is shorter, wider, and its upper part is less curved; the obturator foramina are triangular and smaller in size than in the male. The *outlet* is larger and the coccyx more movable. The sacro-sciatic notches are shallower, and the spines of the ischia project less inwards. The tuberosities of the ischia and the acetabula are wider apart, and the former are more everted. The pubic symphysis is less deep, and the pubic arch is wider and more rounded than in the male, where it is an angle rather than an arch. In consequence of this the width of the fore part of the pelvic outlet is much increased and the passage of the foetal head facilitated.

The size of the pelvis varies, not only in the two sexes, but also in different members of the same sex. This does not appear to be influenced in any way by the height of the individual. Women of short stature, as a rule, have broad pelves. Occasionally the pelvis is equally contracted in all its dimensions, so much so that all its diameters measure an inch less than the average, and this even in well-formed women of average height. The principal divergences, however, are found at the inlet, and affect the relation of the antero-posterior to the transverse diameter. Thus the inlet of the pelvis may be elliptical either in a transverse or antero-posterior direction ; the transverse diameter in the former, and the antero-posterior in the latter, greatly exceeding the other diameters. Again, the inlet of the pelvis in some instances is seen to be almost circular.

The same differences are found in various races. European women are said to have the most roomy pelves. That of the negress is smaller, circular in shape, and with a narrow pubic arch. The Hottentots and Bushwomen possess the smallest pelves.

In the *fœtus*, and for several years after birth, the pelvis is small in proportion to that of the adult, and the projection of the sacro-vertebral angle less marked. The generally accepted opinion that the female pelvis does not acquire its sexual characters until after puberty has been shown by recent observations * to be erroneous, the characteristic differences between the male and female pelvis being distinctly indicated as early as the fourth month of fœtal life.

Surface Form.—The pelvic bones are so thickly covered with muscles that it is only at certain points that they approach the surface and can be felt through the skin. In front, the anterior superior spine of the ilium is easily to be recognised ; a portion of it is subcutaneous, and in thin subjects may be seen to stand out as a prominence at the outer extremity of the fold of the groin. In fat subjects its position is marked by an oblique depression among the surrounding fat, at the bottom of which the bony process may be felt. Proceeding upwards and outwards from this process, the sinuously curved crest of the ilium may be traced throughout its whole length. Its highest point is on a level with the spine of the fourth lumbar vertebra : upon its outer lip, about two and a half inches behind the anterior superior spine, is a prominence, known as the *tubercular point of the ilium.* The crest is represented, in muscular subjects, on the surface, by a groove or furrow, the *iliac furrow,* caused by the projection of the fleshy fibres of the External oblique muscle of the abdomen ; the iliac furrow lies slightly below the level of the crest. It terminates behind in the posterior superior spine, the position of which is indicated by a slight depression on a level with the spinous process of the second sacral vertebra. Between the two posterior superior spines, but at a lower level, is to be felt the spinous process of the third sacral vertebra (see page 179). Another part of the bony pelvis which is easily accessible to the touch is the tuberosity of the ischium, situated beneath the gluteal fold, and, when the hip is flexed, easily to be felt, as it is then uncovered by muscle. Finally, the spine of the os pubis can always be readily felt, and constitutes an important surgical guide, especially in connection with the subject of hernia. It is nearly in the same horizontal plane as the upper edge of the great trochanter. In thin subjects it is very apparent, but in the obese it is obscured by the pubic fat. It can, however, be detected by following up the tendon of origin of the Adductor longus muscle. A line drawn from the anterior superior spine of the ilium to the most prominent part of the tuberosity of the ischium passes across the hip at a level with the upper border of the great trochanter. It is known as *Nélaton's line,* and is of service in detecting any displacement of the trochanter in fractures or dislocation in this situation. If a line be drawn from the posterior superior spine of the ilium to the outer part of the tuberosity of the ischium, it will cross the spine of the ischium about four inches below the posterior superior iliac spine. The great sciatic foramen will lie above, and the lesser foramen below this point.

Surgical Anatomy.—There is arrest of development in the bones of the pelvis in cases of extroversion of the bladder ; the anterior part of the pelvic girdle being deficient, the bodies of the pubic bones imperfectly developed, and the symphysis absent. ' The pubic bones are separated to the extent of from two to four inches, the superior rami shortened and directed forwards, and the obturator foramen diminished in size, narrowed, and turned outwards. The iliac bones are straightened out more than normal. The sacrum is very peculiar. The lateral curve, instead of being concave, is flattened out or even convex, with the ilio-sacral facets turned more outward than normal, while the vertical curve is straightened.' †

Fractures of the pelvis are divided into fractures of the false pelvis and of the true pelvis. Fractures of the false pelvis vary in extent ; a small portion of the crest may be broken, or one of the spinous processes may be torn off, and this may be the result of muscular action ; or the bone may be extensively comminuted. This latter accident is the result of some crushing violence, and may be complicated with fracture of the true pelvis. These cases may be accompanied by injury to the intestine as it lies in the hollow of the bone, or to

---

* Fehling, *Zeitschr. für Geburt. u. Gynäk.* Bd. ix. und x. ; and Arthur Thomson, *Journal of Anatomy and Physiology,* vol. xxxiii.

† Wood. Heath's *Dictionary of Practical Surgery,* i. 426.

the iliac vessels as they course along the margin of the true pelvis. Fractures of the true pelvis generally occur through the ascending ramus of the os pubis and the ramus of the ischium, as this is the weakest part of the bony ring, and may be caused either by crushing violence applied in an antero-posterior direction, when the fracture occurs from direct force, or by compression laterally, when the acetabula are pressed together and the bone gives way in the same place from indirect violence. It is frequently accompanied by a fracture in the neighbourhood of the sacro-iliac joint, either through the lateral mass of the sacrum, or through the ilium of the same side, parallel and close to the articulation. The cause of this is that, after the pelvic ring has given way in front, the continued violence falls on this part, and the bone yields on one side of the joint, being weaker than the strong unyielding ligaments of the joint itself. Sometimes the fracture may be double, occurring on both sides of the body. It is in these cases that the contained viscera are likely to be injured: the urethra, the bladder, the rectum, the vagina in the female, the small intestines, and even the uterus, have all been lacerated by a displaced fragment. Fractures of the acetabulum are occasionally met with : either a portion of the rim may be broken off, or a fracture may take place through the bottom of the cavity and the head of the femur be driven inwards and project into the pelvic cavity. Separation of the Y-shaped cartilage at the bottom of the acetabulum may also occur in the young subject, separating the bone into its three anatomical portions.

The sacrum is rarely broken by direct violence—i.e. blows, kicks, or falls on the part. The lesion may be complicated with injury to the nerves of the sacral plexus, leading to paralysis and loss of sensation in the lower extremity, or to incontinence of fæces from paralysis of the Sphincter ani.

The coccyx is not infrequently fractured from blows or kicks. The fracture is attended with great pain in walking and on making any expiratory effort, such as coughing, defæcation, &c., because the Coccygeus muscle, which is attached to this bone, forms part of the lower diaphragm of the abdomen. Falls or blows on the coccyx, unaccompanied by fracture, sometimes give rise to a severe neuralgic pain, which is exceedingly intractable and difficult of cure. The condition is known as *coccygodynia*, and for its relief removal of the coccyx has been practised of late years, but the results have not been very satisfactory.

The pelvic bones often undergo important deformity in rickets, the effects of which in the adult woman may interfere seriously with child-bearing. The deformity is due mainly to the weight of the spine and trunk, which presses on the sacro-vertebral angle and greatly increases it, so that the antero-posterior diameter of the pelvis is diminished. But, in addition to this, the weight of the viscera on the venter ilii causes them to expand and the tuberosities of the ischia to be incurved. In osteomalacia also great deformity may occur. The weight of the trunk causes an increase in the sacro-vertebral angle, and a lessening of the antero-posterior diameter of the inlet, and at the same time the pressure of the acetabula on the heads of the thigh-bones causes these cavities, with the adjacent bone, to be pushed upwards and backwards, so that the oblique diameters of the pelvis are also diminished, and the cavity of the pelvis assumes a triradiate shape, with the symphysis pubis pushed forwards.

THE THIGH

The **Thigh** is that portion of the lower extremity which is situated between the pelvis and the knee. Its skeleton consists of a single bone, the femur.

The Femur

The **Femur** (femur, the *thigh*) is the longest,* largest, and strongest bone in the skeleton, and almost perfectly cylindrical in the greater part of its extent. In the erect posture it is not vertical, being separated from its fellow above by a considerable interval, which corresponds to the breadth of the pelvis, but inclining gradually downwards and inwards, so as to approach its fellow towards its lower part, for the purpose of bringing the knee-joint near the line of gravity of the body. The degree of this inclination varies in different persons, and is greater in the female than in the male, on account of the greater breadth of the pelvis. The femur, like other long bones, is divisible into a shaft and two extremities.

The **Upper Extremity** presents for examination a head, a neck, a great and a small trochanter.

The **head**, which is globular, and forms rather more than a hemisphere, is directed upwards, inwards, and a little forwards, the greater part of its convexity being above and in front. Its surface is smooth, coated with cartilage in the

---

* In a man six feet high, it measures eighteen inches—one-fourth of the whole body.

FIG. 319.—Right femur. Anterior surface.

Obturator internus
and Gemelli

Pyriformis

Great Trochanter

Head

Artic. with Acetabulum

Neck

Depression for ligamentum teres

Tubercle

Lesser Trochanter

CRURÆUS

Shaft

SUB-CRURÆUS

Outer tuberosity

Inner tuberosity

Artic. with Patella

Outer Condyle

Inner Condyle

recent state, with the exception of an ovoid depression, which is situated a little below and behind its centre, and gives attachment to the ligamentum teres. The **neck** is a flattened pyramidal process of bone, which connects the head with the shaft, and forms with the latter a wide angle. The angle is widest in infancy, and becomes lessened during growth, so that at puberty it forms a gentle curve from the axis of the shaft. In the adult, the neck forms an angle of about 125° with the shaft, but this varies in inverse proportion to the development of the pelvis and the stature. In consequence of the prominence of the hips and widening of the pelvis in the female, the neck of the thigh-bone forms more nearly a right angle with the shaft than it does in man. It has been stated that the angle diminishes in old age and the direction of the neck becomes horizontal, but this statement is founded on insufficient evidence. Humphry stated that the angle decreases during the period of growth, but after full growth has been attained it does not usually undergo any change, even in old age. He further states that the angle varies considerably in different persons of the same age. It is smaller in short than in long bones, and when the pelvis is wide.* The neck is flattened from before backwards, contracted in the middle, and broader at its outer extremity, where it is connected with the shaft, than at its summit, where it is attached to the head. The vertical diameter of the outer half is increased by the thickening of the lower edge, which slopes downwards to join the shaft at the small trochanter, so that the outer half of the neck is flattened from before backwards, and its vertical diameter measures one-third more than the antero-posterior. The inner half is smaller, and of a more circular shape. The *anterior surface* of the neck is perforated by numerous vascular foramina. The *posterior surface* is smooth, and is broader and more concave than the anterior; it gives attachment to the posterior part of the capsular ligament of the hip-joint, about half an inch above the posterior intertrochanteric line. The *superior border* is short and thick, and terminates externally at the great trochanter; its surface is perforated by large foramina. The *inferior border*, long and narrow, curves a little backwards, to terminate at the small trochanter.

* *Journal of Anatomy and Physiology*, vol. xxiii.

The **Trochanters** (τροχάω, *to run* or *roll*) are prominent processes of bone which afford leverage to the muscles which rotate the thigh on its axis. They are two in number, great and small.

The **Great Trochanter** is a large, irregular, quadrilateral eminence, situated at the outer side of the neck, at its junction with the upper part of the shaft. It is directed a little outwards and backwards, and, in the adult, is about three-quarters of an inch lower than the head. It presents for examination two surfaces and four borders. The *external surface*, quadrilateral in form, is broad, rough, convex, and marked by a prominent diagonal impression, which extends from the posterior superior to the anterior inferior angle, and serves for the insertion of the tendon of the Gluteus medius. Above the impression is a triangular surface, sometimes rough for part of the tendon of the same muscle, sometimes smooth for the interposition of a bursa between that tendon and the bone. Below and behind the diagonal line is a smooth, triangular surface, over which the tendon of the Gluteus maximus muscle plays, a bursa being interposed. The *internal surface* is of much less extent than the external, and presents at its base a deep depression, the *digital* or *trochanteric fossa*, for the insertion of the tendon of the Obturator externus muscle, and above and in front of this an impression for the insertion of the Obturator internus and Gemelli. The *superior border* is free; it is thick and irregular, and marked near the centre by an impression for the insertion of the Pyriformis. The *inferior border* corresponds to the line of junction of the base of the trochanter with the outer surface of the shaft; it is marked by a rough, prominent, slightly curved ridge, which gives origin to the upper part of the Vastus externus muscle. The *anterior border* is prominent, somewhat irregular, as well as the surface of bone immediately below it; it affords insertion at its outer part to the Gluteus minimus. The *posterior border* is very prominent, and appears as a free, rounded edge, which forms the back part of the digital fossa.

The **Small Trochanter** is a conical eminence, which varies in size in different subjects; it projects from the lower and back part of the base of the neck. Its base is triangular, and connected with the adjacent parts of the bone by three well-marked borders: two of these are above—the *internal* continuous with the lower border of the neck, the *external* with the posterior intertrochanteric line—while the *inferior border* is continuous with the middle division of the linea aspera. Its summit, which is directed inwards and backwards, is rough, and gives insertion to the tendon of the Ilio-psoas. The Iliacus is also inserted into the shaft below the small trochanter, between the Vastus internus in front and the Pectineus behind.

A well-marked prominence, of variable size, which projects at the junction of the upper part of the neck with the great trochanter, is called the *tubercle of the femur*; it is the point of meeting of five muscles: the Gluteus minimus externally, the Vastus externus below, and the tendon of the Obturator internus and Gemelli above. Running obliquely downwards and inwards from the tubercle is the *spiral line* of the femur, or *anterior intertrochanteric line*; it winds round the inner side of the shaft, below the lesser trochanter, and terminates in the linea aspera, about two inches below this eminence. Its upper half is rough, and affords attachment to the ilio-femoral ligament of the hip-joint; its lower half is less prominent, and gives origin to the upper part of the Vastus internus. Running obliquely downwards and inwards from the summit of the great trochanter on the posterior surface of the neck is a very prominent well-marked ridge, the *posterior intertrochanteric line*. Its upper half forms the posterior border of the great trochanter, and its lower half runs downwards and inwards to the upper and back part of the lesser trochanter. A slight ridge sometimes commences about the middle of the posterior intertrochanteric line, and passes vertically downwards for about two inches along the back part of the shaft: it is called the *linea quadrata*, and gives attachment to the Quadratus femoris and a few fibres of the Adductor magnus muscles.*

The **Shaft**, almost cylindrical in form, is a little broader above than in the centre, and somewhat flattened from before backwards below. It is slightly

---

* Generally there is merely a slight thickening about the middle of the intertrochanteric line, marking the attachment of the upper part of the Quadratus femoris. This is termed by some anatomists the *tubercle of the Quadratus*.

FIG. 320.—Right femur. Posterior surface.

arched, so as to be convex in front and
concave behind, where it is strengthened
by a prominent longitudinal ridge, the
linea aspera. It presents for examination three borders, separating three surfaces. Of the three borders, one, the
linea aspera, is posterior ; the other two
are placed laterally.

The **linea aspera** (fig. 320) is a prominent longitudinal ridge or crest, on the
middle third of the bone, presenting an
external lip, an internal lip, and a rough
intermediate space. Above, the linea
aspera is prolonged by three ridges. The
most external one is very rough, and is
continued almost vertically upwards to
the base of the great trochanter. It is
sometimes termed the *gluteal ridge*, and
gives attachment to part of the Gluteus
maximus muscle : its upper part is often
elongated into a roughened crest, on
which is a more or less well-marked,
rounded tubercle, a rudimental third
trochanter. The middle ridge, the least
distinct, is continued to the base of the
trochanter minor, and gives attachment
to the Pectineus muscle ; and the internal one is lost above in the spiral line
of the femur. Below, the linea aspera
is prolonged by two ridges, which enclose
between them a triangular space, the
popliteal surface, upon which rests the
popliteal artery. Of these two ridges,
the outer one is the more prominent,
and descends to the summit of the outer
condyle. The inner one is less marked,
especially at its upper part, where it is
crossed by the femoral artery. It terminates, below, at the summit of the internal condyle, in a small tubercle, the
Adductor tubercle, which affords insertion
to the tendon of the Adductor magnus.

From the inner lip of the linea aspera
and its inner prolongation above and
below, the Vastus internus arises ; and
from the outer lip and its outer prolongation above, the Vastus externus takes
origin. The Adductor magnus is inserted into the linea aspera, to its outer
prolongation above, and its inner prolongation below. Between the Vastus
externus and the Adductor magnus are
attached two muscles—viz. the insertion
of the Gluteus maximus above and the
origin of the short head of the Biceps
below. Between the Adductor magnus and the Vastus internus four
muscles are inserted : the Iliacus and
Pectineus above ; the Adductor brevis
and Adductor longus below. The linea
aspera is perforated a little below its
centre by the nutrient canal, which is
directed obliquely upwards.

The two *lateral borders* of the femur are only slightly marked, the outer one extending from the anterior inferior angle of the great trochanter to the anterior extremity of the external condyle ; the inner one from the spiral line, at a point opposite the trochanter minor, to the anterior extremity of the internal condyle. The internal border marks the limit of attachment of the Crureus muscle internally.

The *anterior surface* includes that portion of the shaft which is situated between the two lateral borders. It is smooth, convex, broader above and below than in the centre, slightly twisted, so that its upper part is directed forwards and a little outwards, its lower part forwards and a little inwards. From the upper three-fourths of this surface the Crureus arises ; the lower fourth is separated from the muscle by the intervention of the synovial membrane of the knee-joint and a bursa, and affords attachment to the Subcrureus to a small extent. The *external surface* includes the portion of bone between the external border and the outer lip of the linea aspera ; it is continuous above with the outer surface of the great trochanter, below with the outer surface of the external condyle : to its upper three-fourths is attached the outer portion of the Crureus muscle. The *internal surface* includes the portion of bone between the internal border and the inner lip of the linea aspera ; it is continuous above with the lower border of the neck, below with the inner side of the internal condyle : it is covered by the Vastus internus muscle.

The **Lower Extremity,** larger than the upper, is of a cuboid form, flattened from before backwards, and divided into two large eminences, the *condyles* (κόνδυλος, *a knuckle*), by an interval which presents a smooth depression in front called the *trochlea,* and a notch of considerable size behind—the *intercondyloid notch.* The *external condyle* is the more prominent anteriorly, and is the broader both in the antero-posterior and transverse diameters. The *internal condyle* is the narrower, longer, and more prominent inferiorly. The difference in the length of the two condyles is only observed when the bone is perpendicular, and depends upon the obliquity of the thigh-bones, in consequence of their separation above at the articulation with the pelvis. If the femur is held obliquely, the surfaces of the two condyles will be seen to be nearly horizontal. The two condyles are directly continuous in front, and form a smooth trochlear surface, which articulates with the patella. This surface presents a median groove, which extends downwards and backwards to the intercondyloid notch ; and two lateral convexities, of which the external is the broader, more prominent, and prolonged farther upwards upon the front of the outer condyle. The external border of this articular surface is also more prominent, and ascends higher than the internal one. This trochlear portion of the articular surface is separated from the remainder, which articulates with the tibia and the semilunar cartilages, by two faintly marked grooves, which pass obliquely forwards and outwards and forwards and inwards respectively from the anterior margin of the intercondyloid notch. They receive the anterior edges of the two semilunar fibro-cartilages when the knee is extended. The intercondyloid notch lodges the crucial ligaments ; it is bounded laterally by the opposed surfaces of the two condyles, and in front by the lower end of the shaft.

Outer Condyle.—The *outer surface* of the external condyle presents, a little behind its centre, an eminence, the *outer tuberosity* ; it is less prominent than the inner tuberosity, and gives attachment to the external lateral ligaments of the knee. Immediately beneath it is a groove which commences at a depression a little behind the centre of the lower border of this surface : the front part of this depression gives origin to the Popliteus muscle, the tendon of which is lodged in the groove during flexion of the knee. The groove is smooth, covered with cartilage in the recent state, and runs upwards and backwards to the posterior extremity of the condyle. The *inner surface* of the outer condyle forms one of the lateral boundaries of the intercondyloid notch, and gives attachment, by its posterior part, to the anterior crucial ligament. The *inferior surface* is convex, smooth, and broader than that of the internal condyle. The posterior extremity is convex and smooth : just above and to the outer side of the articular surface is a depression for the origin of the outer head of the Gastrocnemius, above which is the origin of the Plantaris.

Inner Condyle.—The *inner surface* of the inner condyle presents a convex eminence, the *inner tuberosity,* rough, for the attachment of the internal lateral ligament. The *outer side* of the inner condyle forms one of the lateral boundaries of the intercondyloid notch, and gives attachment, by its anterior part, to the posterior crucial ligament. Its *inferior* or *articular surface* is convex, and presents a less extensive surface than the external condyle. Just above the articular surface of the condyle, behind, is a depression for the tendon of origin of the inner head of the Gastrocnemius.

Structure.—The shaft of the femur is a cylinder of compact tissue, hollowed by a large medullary canal. The cylinder is of great thickness and density in the middle third of the shaft, where the bone is narrowest and the medullary canal well formed; but above and below this the cylinder gradually becomes thinner, owing to a separation of the layers of the bone into cancelli, which project into the medullary canal and finally obliterate it, so that the upper and lower ends of the shaft, and the articular extremities more especially, consist of cancellated tissue, invested by a thin compact layer.

The arrangement of the cancelli in the ends of the femur is remarkable. In the upper end they are arranged in two sets. One, starting from the top of the head, the upper surface of the neck, and the great trochanter, converge to the inner part of the circumference of the shaft (fig. 321); these are placed in the direction of greatest pressure, and serve to support the vertical weight of the body. The second set are planes of lamellæ intersecting the former nearly at right angles, and are situated in the line of the greatest tension—that is to say, along the lines in which the muscles and ligaments exert their traction. In the head of the bone these planes are arranged in a curved form, in order to strengthen the bone when exposed to pressure in all directions. In the midst of the cancellous tissue of the neck is a vertical plane of compact bone, the *femoral spur* (*calcar femorale*) which commences at the point where the neck joins the shaft midway between the small trochanter and the internal border of the shaft of the bone, and extends in the direction of the digital fossa (fig. 322). This materially strengthens this portion of the bone. Another point in connection with the structure of the neck of the femur requires mention, especially on account of its influence on the production of fracture in this situation. It will be noticed that a considerable portion of the great trochanter lies behind the level of the posterior surface of the neck, and if a section be made through the trochanter at this level, it will be seen that the posterior wall of the neck is prolonged into the trochanter. This prolongation is termed by Bigelow the 'true neck,'* and forms a thin, dense plate of bone, which passes beneath the posterior intertrochanteric ridge towards the outer surface of the bone.

Fig. 321.—Diagram showing the arrangement of the cancelli of the neck of the femur.

In the lower end, the cancelli spring on all sides from the inner surface of the cylinder, and descend in a perpendicular direction to the articular surface, the cancelli being strongest and having a more accurately perpendicular course above the condyles. In addition to this, however, horizontal planes of cancellous tissue are to be seen, so that the spongy tissue in this situation presents an appearance of being mapped out into a series of rectangular areas.

Articulations.—With three bones: the os innominatum, tibia, and patella.

Development (fig. 323).—The femur is developed by *five* centres: one for the shaft, one for each extremity, and one for each trochanter. Of all the long bones, except the clavicle, it is the first to show traces of ossification; this commences in the shaft, at about the seventh week of fœtal life; the centres of ossification

* *Bigelow on the Hip*, p. 121.

in the epiphyses appearing in the following order : In the lower end of the bone, at the ninth month of fœtal life * (from this the condyles and tuberosities are formed) ; in the head, at the end of the first year after birth ; in the great trochanter, during the fourth year ; and in the small trochanter, between the thirteenth and fourteenth. The order in which the epiphyses are joined to the shaft is the reverse of that of their appearance : their junction does not commence until after puberty, the small trochanter being first joined, then the great, then the head, and, lastly, the inferior extremity (the first in which ossification commenced), which is not united until the twentieth year.

FIG. 322.—Calcar femorale.

FIG. 323.—Plan of the development of the femur. By five centres.

Attachment of Muscles.—To twenty-three. To the great trochanter : the Gluteus medius, Gluteus minimus, Pyriformis, Obturator internus, Obturator externus, Gemellus superior, Gemellus inferior, and Quadratus femoris. To the lesser trochanter : the Psoas magnus and the Iliacus below it. To the shaft : the Vastus externus, Gluteus maximus, short head of the Biceps, Adductor magnus, Pectineus, Adductor brevis, Adductor longus, Vastus internus, Crureus, and Subcrureus. To the condyles : the Gastrocnemius, Plantaris, and Popliteus.

Surface Form.—The femur is covered with muscles, so that in fairly muscular subjects the shaft is not to be detected through its fleshy covering, and the only parts always accessible to the touch are the outer surface of the great trochanter and the lower expanded end of the bone. In very thin subjects, the head of the bone may be felt through the overlying muscles, immediately below and external to the mid-point of Poupart's ligament. The external surface of the great trochanter is to be felt, especially in certain positions of the limb. Its situation is generally indicated by a depression, owing to the thickness of the Gluteus medius and minimus, which project above it. When, however, the thigh is flexed, and especially if crossed over the opposite one, the trochanter produces a blunt eminence on the surface. The upper border is about on a level with the centre of the hip-joint, and its exact level is indicated by a line drawn from the anterior superior spinous

---

* This is said to be the only epiphysis in which ossification begins before birth ; though according to some observers the centre for the upper epiphysis of the tibia also appears before birth.

process of the ilium, over the outer side of the hip, to the most prominent point of the tuberosity of the ischium. This is known as Nélaton's line. The outer and inner condyles of the lower extremity are accessible to the touch. The outer one is more subcutaneous than the inner one, and readily felt. The tuberosity on it is comparatively little developed, but can be more or less easily recognised. The inner condyle is more thickly covered, and this gives a general convex outline to this part, especially when the knee is flexed. The tuberosity on it is easily felt, and at the upper part of the condyle the sharp tubercle for the insertion of the tendon of the Adductor magnus can be recognised without difficulty. When the knee is flexed, and the patella situated in the interval between the condyles and the upper end of the tibia, a part of the trochlear surface of the femur can be made out above the patella.

Surgical Anatomy.—There are one or two points about the ossification of the femur bearing on practice to which allusion must be made. It has been stated that the lower end of the femur is the only epiphysis in which ossification has commenced at the time of birth. The presence of this ossific centre is, therefore, a proof, in newly born children found dead, that the child has arrived at the full period of utero-gestation, and is always relied upon in medico-legal investigations. The position of the epiphysial line should be carefully noted. It is on a level with the Adductor tubercle, and the epiphysis does not, therefore, form the whole of the cartilage-clad portion of the lower end of the bone. It is essential to bear this point in mind in performing excision of the knee, since growth in length of the femur takes place chiefly from the lower epiphysis, and any interference with the epiphysial cartilage in a young child would involve such ultimate shortening of the limb, from want of growth, as to render the limb almost useless. Separation of the lower epiphysis may take place up to the age of twenty, at which time it becomes completely joined to the shaft of the bone; but, as a matter of fact, few cases occur after the age of sixteen or seventeen. The epiphysis of the head of the femur is of interest principally on account of its being the seat of origin in a large number of cases of tuberculous disease of the hip-joint. The disease commences in the majority of cases in the highly vascular and growing tissue at the end of the shaft in the neighbourhood of the epiphysial cartilage, and from here extends into the joint.

Fractures of the femur are divided, like those of the other long bones, into fractures of the upper end; of the shaft; and of the lower end. The fractures of the upper end may be classified into (1) fracture of the neck; (2) fracture at the junction of the neck with the great trochanter; (3) fracture of the great trochanter; and (4) separation of the epiphysis, either of the head or of the great trochanter. The first of these, fracture of the neck, is usually termed intracapsular fracture, but this is scarcely a correct designation, as, owing to the attachment of the capsular ligament, the fracture is partly within and partly without the capsule when the fracture occurs at the lower part of the neck. It generally takes place in old people, principally women, and usually from a very slight degree of indirect violence. Probably the main cause of the fracture occurring in old people is in consequence of the degenerative changes which the bone has undergone. Merkel believes that it is mainly due to the absorption of the calcar femorale. These fractures are occasionally impacted. As a rule they unite by fibrous tissue, but frequently no union takes place, and the surfaces of the fracture become smooth and eburnated.

Fractures at the junction of the neck with the great trochanter are usually termed extra-capsular, but this designation is also incorrect, as the fracture is partly within the capsule, owing to its attachment in front to the anterior intertrochanteric line, which is situated below the line of fracture. These fractures are produced by direct violence to the great trochanter, as from a blow or fall laterally on the hip. From the manner in which the accident is caused, the neck of the bone is driven into the trochanter, where it may remain impacted, or the trochanter may be split up into two or more fragments, and thus no fixation takes place.

Fractures of the great trochanter may be either ' oblique fracture through the trochanter major, without implicating the neck of the bone' (Astley Cooper), or separation of the great trochanter. Most of the recorded cases of this latter injury occurred in young persons, and were probably cases of separation of the epiphysis of the great trochanter. Separation of the epiphysis of the head of the femur has been said to occur, but has never been verified by post-mortem examination.

Fractures of the shaft may occur at any part, but the most usual situation is at or near the centre of the bone. They may be caused by direct or indirect violence or by muscular action. Fractures of the upper third of the shaft are almost always the result of indirect violence, while those of the lower third are the result, for the most part, of direct violence. In the middle third fractures occur from both forms of injury in about equal proportions. Fractures of the shaft are generally oblique, but they may be transverse, longitudinal, or spiral. The transverse fracture occurs most frequently in children. The fractures of the lower end of the femur include transverse fracture above the condyles, the most common; and this may be complicated by a vertical fracture between the condyles, constituting the T-shaped fracture. In these cases the popliteal artery is in danger of being wounded. Oblique fracture separating either the internal or external condyle, and a longitudinal incomplete fracture between the condyles, may also take place.

The femur as well as the other bones of the leg are frequently the seat of acute necrosis in young children. This is no doubt due to their greater exposure to injury, which is often the exciting cause of this disease. Tumours are not infrequently found growing from the femur : the most common forms being sarcoma, which may grow either from the periosteum or from the medullary tissue within the interior of the bone ; and exostosis, which is commonly found originating in the neighbourhood of the epiphysial cartilage of the lower end.

THE LEG

The skeleton of the Leg consists of three bones : the Patella, a large, sesamoid bone, placed in front of the knee ; the Tibia ; and the Fibula.

THE PATELLA (figs. 324, 325)

The **Patella** (patella, *a small pan*) is a flat, triangular bone, situated at the anterior part of the knee-joint. It is usually regarded as a sesamoid bone, developed in the tendon of the Quadriceps extensor. It resembles these bones (1) in its being developed in a tendon ; (2) in its centre of ossification presenting a knotty or tubercular outline ; (3) in its structure being composed mainly of dense cancellous tissue, as in the other sesamoid bones. It serves to protect the front of the joint, and increases the leverage of the Quadriceps extensor by making it act at a greater angle. It presents an anterior and posterior surface, three borders, and an apex.

The **anterior surface** is convex, perforated by small apertures, for the passage of nutrient vessels, and marked by numerous rough, longitudinal striæ. This surface is covered, in the recent state, by an expansion from the tendon of the Quadriceps extensor, which is continuous below with the superficial fibres of the ligamentum patellæ. It is separated from the integument by a bursa. The **posterior surface** presents a smooth, oval-shaped, articular surface, covered with cartilage in the recent state, and divided into two facets by a vertical ridge,

Fig. 324.—Right patella. Anterior surface.

Fig. 325.—Right patella. Posterior surface.

which descends from the superior border towards the inferior angle of the bone. The ridge corresponds to the groove on the trochlear surface of the femur, and the two facets to the inner and outer parts of the same surface ; the outer facet being the broader and deeper. This character serves to indicate the side to which the bone belongs. Below the articular surface is a rough, convex, non-articular depression, the lower half of which gives attachment to the ligamentum patellæ ; the upper half being separated from the head of the tibia by adipose tissue.

The **superior border** is thick, and sloped from behind, downwards and forwards : it gives attachment to that portion of the Quadriceps extensor which is derived from the Rectus and Crureus muscles. The **lateral borders** are thinner, converging below : they give attachment to that portion of the Quadriceps extensor derived from the external and internal Vasti muscles.

The **apex** is pointed, and gives attachment to the ligamentum patellæ.

Structure.—The patella consists of a nearly uniform dense cancellous tissue, covered by a thin compact lamina. The cancelli immediately beneath the anterior surface are arranged parallel with it. In the rest of the bone they radiate from the posterior articular surface towards the other parts of the bone.

Development.—By a single centre, which makes its appearance in the second or third year, but which may be delayed until the sixth year. More rarely, the bone is developed by two centres, placed side by side. Ossification is completed about the age of puberty.

Articulations.—With the two condyles of the femur.

Y

FIG. 326.—Bones of the right leg. Anterior surface.

Head
Styloid process
External lateral ligament
FIBULA
TIBIA
Spine
Internal Tuberosity
External Tuberosity
Tubercle
LIGAMENT PATELLÆ
SARTORIUS
SEMI TENDINOSUS GRACILIS
EXTENSOR LONGUS DIGITORUM EXTENSOR PROPRIUS HALLUCIS PERONEUS TERTIUS
Attachment of Interosseous Membrane
Crest
Internal Surface, Subcutaneous
Face covered by Ext. Tendon
Internal malleolus
External malleolus

Attachment of Muscles.—To four: the Rectus, Crureus, Vastus internus, and Vastus externus. These muscles, joined at their insertion, constitute the Quadriceps extensor cruris.

Surface Form.—The external surface of the patella can be seen and felt in front of the knee. In the extended position of the limb the internal border is a little more prominent than the outer, and if the Quadriceps extensor is relaxed, the bone can be moved from side to side and appears to be loosely fixed. If the joint is flexed, the patella recedes into the hollow between the condyles of the femur and the upper end of the tibia, and becomes firmly fixed against the femur.

Surgical Anatomy.—The main surgical interest about the patella is in connection with fractures, which are of common occurrence. They may be produced by muscular action—that is to say, by violent contraction of the Quadriceps extensor, while the limb is in a position of semiflexion, so that the bone is snapped across the condyles; or by direct violence, such as falls on the knee. In the former class of cases the fracture is transverse; in the latter it may be oblique, longitudinal, stellate, or the bone variously comminuted. The principal interest in these cases attaches to their treatment. Owing to the displacement of the fragments, and the difficulty there is in maintaining them in apposition, union takes place by fibrous tissue, and this may subsequently stretch, producing wide separation of the fragments and permanent lameness. Various plans, including opening the joint and suturing the fragments, have been advocated for overcoming this difficulty.

In the larger number of cases of fracture of the patella, the knee-joint is involved, the cartilage which covers its posterior surface being also torn. In some cases of fracture from direct violence, however, this need not necessarily happen, the lesion involving only the superficial part of the bone; and, as Morris has pointed out, it is an anatomical possibility, in complete fracture, if the lesion involve only the lower and non-articular part of the bone, for it to take place without injury to the synovial membrane.

THE TIBIA (figs. 326, 328)

The **Tibia** (tibia, *a flute* or *pipe*) is situated at the front and inner side of the leg, and, excepting the femur, is the longest and largest bone in the skeleton. It is prismoid in form, expanded above, where it enters into the knee-joint, more slightly enlarged below. In the male, its direction is vertical, and parallel with the bone of the opposite side; but in the female it has a slightly oblique direction downwards and outwards, to compensate for the oblique direction of the femur inwards. It presents for examination a shaft and two extremities.

The **Upper Extremity,** or **Head,** is large, and expanded on each side into two lateral eminences, the *tuberosities. Superiorly,* the tuberosities present two smooth articular surfaces, one on either side of the middle line. The inner of these is oval in shape, and slightly concave from side to side and from before backwards. The outer, nearly circular, is concave from side to side, but slightly convex from before backwards, especially at its posterior part, where it is prolonged on to the posterior surface for a short distance. The central portions of these facets articulate with the condyles of the femur, while their peripheral portions support the semilunar cartilages of the knee, which here intervene between the two bones. In the middle line, but nearer the posterior than the anterior aspect of the bone, is an eminence, the *spinous process* of the tibia, surmounted on each side by a prominent tubercle, on to the lateral aspect of which the facets, just described, are prolonged; in front and behind the spinous process is a rough depression for the attachment of the anterior and posterior crucial ligaments and the semilunar fibro-cartilages (fig. 327). The *anterior surfaces* of the tuberosities are continuous with one another, forming a single large surface, which is somewhat flattened: it is triangular, broad above, and perforated by large vascular foramina: narrow below where it terminates in a prominent oblong elevation

FIG. 327.—Upper surface of tibia, showing attachment of crucial ligament and fibro-cartilages.

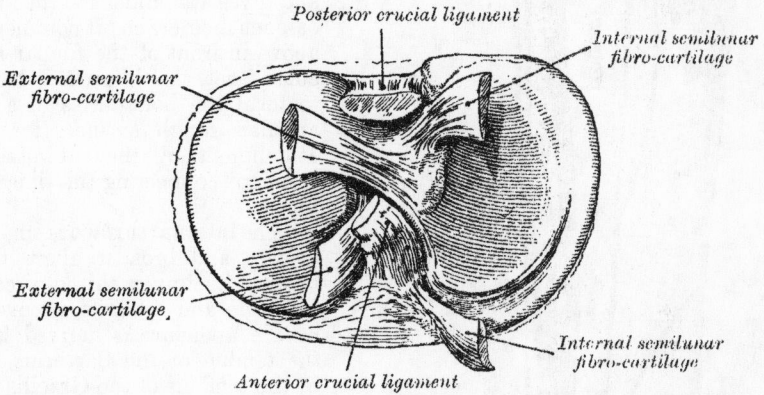

Posterior crucial ligament

Internal semilunar fibro-cartilage

External semilunar fibro-cartilage

External semilunar fibro-cartilage

Internal semilunar fibro-cartilage

Anterior crucial ligament

of large size, the *tubercle* of the tibia, which gives attachment to the ligamentum patellæ; a bursa intervenes between the deep surface of the ligament and the part of the bone immediately above the tubercle. *Posteriorly,* the tuberosities are separated from each other by a shallow depression, the *popliteal notch,* which gives attachment to part of the posterior crucial ligament, and part of the posterior ligament of the knee-joint. The *inner tuberosity* presents posteriorly a deep transverse groove, for the insertion of the tendon of the Semi-membranosus. Its *lateral surface* is convex, rough and prominent: it gives attachment to the internal lateral ligament. The *outer tuberosity* presents posteriorly a flat articular facet, nearly circular in form, directed downwards, backwards, and outwards, for articulation with the head of the fibula. Its *lateral surface* is convex and rough, more prominent in front than the internal: it presents an eminence, situated on a level with the upper border of the tubercle of the tibia at the junction of its anterior and outer surfaces, for the attachment of the ilio-tibial band. Just below this the Extensor longus digitorum and a slip from the Biceps are attached.

The **Shaft** of the tibia is of a triangular prismoid form, broad above, gradually decreasing in size to its most slender part, at the commencement of its lower fourth, where fracture most frequently occurs. It then enlarges again towards its lower extremity. It presents for examination three borders and three surfaces.

The **anterior border,** the most prominent of the three, is called the *crest of the* *tibia*, or, in popular language, the *shin*; it commences above at the tubercle, and terminates below at the anterior margin of the inner malleolus. This border is sinuous, prominent in the upper two-thirds of its extent, smooth and rounded below; it gives attachment to the deep fascia of the leg.

Fig. 328.—Bones of the right leg. Posterior surface.

Articulates with astragalus

The **internal border** is smooth and rounded above and below, but more prominent in the centre; it commences at the back part of the inner tuberosity, and terminates at the posterior border of the internal malleolus; its upper part gives attachment to the internal lateral ligament of the knee to the extent of about two inches, and insertion to some fibres of the Popliteus muscle; from its middle third some fibres of the Soleus and Flexor longus digitorum muscles take origin.

The **external border, or interosseous ridge,** is thin and prominent, especially its central part, and gives attachment to the interosseous membrane; it commences above in front of the fibular articular facet, and bifurcates below, to form the boundaries of a triangular rough surface, for the attachment of the interosseous ligament connecting the tibia and fibula.

The **internal surface** is smooth, convex, and broader above than below; its upper third, directed forwards and inwards, is covered by the aponeurosis derived from the tendon of the Sartorius, and by the tendons of the Gracilis and Semitendinosus, all of which are inserted nearly as far forwards as the anterior border; in the rest of its extent it is subcutaneous.

The **external surface** is narrower than the internal; its upper two-thirds presents a shallow groove for the origin of the Tibialis anticus muscle; its lower third is smooth, convex, curves gradually forwards to the anterior aspect of the bone, and is covered from within outwards by the tendons of the following muscles: Tibialis anticus, Extensor proprius hallucis, Extensor longus digitorum.

The **posterior surface** (fig. 328) presents, at its upper part, a prominent ridge, the *oblique line* of the tibia, which extends from the back part of the articular facet for the fibula obliquely downwards, to the internal border, at the junction of its upper and middle thirds. It marks the lower limit of the insertion of the Popliteus muscle and serves for the attachment of the popliteal fascia; it gives origin to part of the Soleus, Flexor longus digitorum, and Tibialis posticus muscles; the triangular area, above and to the inner side of this line, gives insertion to the Popliteus muscle. The middle third of the posterior surface is divided by a vertical ridge into lateral halves: the ridge is well marked at its commencement at the oblique line, but becomes gradually indistinct below; the inner and broader half gives origin to the Flexor longus digitorum, the outer and narrower to part of the Tibialis posticus. The remaining part of the bone presents a smooth surface covered by the Tibialis posticus, Flexor longus digitorum, and Flexor longus hallucis muscles. Immediately below the oblique line is the medullary foramen, which is large and directed obliquely downwards.

The **Lower Extremity**, much smaller than the upper, presents five surfaces; it is prolonged downwards, on its inner side into a strong process, the *internal malleolus*. The *inferior surface* of the bone is quadrilateral, and smooth for articulation with the astragalus. It is concave from before backwards, broader in front than behind, and traversed from before backwards by a slight elevation, separating two lateral depressions. It is narrow internally, where the articular surface becomes continuous with that on the inner malleolus. The *anterior surface* of the lower extremity is smooth and rounded above, and covered by the tendons of the Extensor muscles of the toes; its lower margin presents a rough transverse depression, for the attachment of the anterior ligament of the ankle-joint; the *posterior surface* presents a superficial groove directed obliquely downwards and inwards, continuous with a similar groove on the posterior surface of the astragalus, and serving for the passage of the tendon of the Flexor longus hallucis; the *external surface* presents a triangular rough depression for the attachment of the inferior interosseous ligament connecting it with the fibula; the lower part of this depression is smooth, covered with cartilage in the recent state, and articulates with the fibula. The surface is bounded by two prominent borders, continuous above with the interosseous ridge; they afford attachment to the anterior and posterior inferior tibio-fibular ligaments. The *internal surface* of the lower extremity is prolonged downwards to form a strong pyramidal process, flattened from without inwards—the *internal malleolus*. The *inner surface* of this process is convex and subcutaneous; its *outer surface* is smooth and slightly concave, and articulates with the astragalus; its *anterior border* is rough, for the attachment of the anterior fibres of the internal lateral or Deltoid ligament; its *posterior border* presents a broad and deep groove, directed obliquely downwards and inwards, which is occasionally double; this groove transmits the tendons of the Tibialis posticus and Flexor longus digitorum muscles. The *summit* of the internal malleolus is marked by a rough depression behind, for the attachment of the internal lateral ligament of the ankle-joint.

FIG. 329.—Plan of the development of the tibia. By three centres.

Upper extremity

Appears at birth

Joins shaft about 20th year

Shaft
7th

Appears at 2nd year

Joins shaft about 18th year

Lower extremity

Structure.—Like that of the other long bones. At the junction of the middle and lower thirds, where the bone is smallest, the wall of the shaft is thicker than in other parts, in order to compensate for the smallness of the calibre of the bone.

Development.—By *three* centres (fig. 329): one for the shaft, and one for each extremity. Ossification commences in the centre of the shaft about the seventh week, and gradually extends towards either extremity. The centre for the upper epiphysis appears before or shortly after birth; it is flattened in form, and has a thin tongue-shaped process in front, which forms the tubercle. That for the lower epiphysis appears in the second year. The lower epiphysis joins the shaft at about the eighteenth, and the upper one about the twentieth year. Two additional centres occasionally exist, one for the tongue-shaped process of the upper epiphysis, which forms the tubercle, and one for the inner malleolus.

Articulations.—With three bones : the femur, fibula, and astragalus.

Attachment of Muscles.—To twelve : to the inner tuberosity, the Semimembranosus ; to the outer tuberosity, the Tibialis anticus and Extensor longus digitorum and Biceps ; to the internal surface of the shaft, the Sartorius, Gracilis, and Semitendinosus ; to its external surface, the Tibialis anticus ; to its posterior surface, the Popliteus, Soleus, Flexor longus digitorum, and Tibialis posticus ; to the tubercle, the ligamentum patellæ, by which the Quadriceps extensor muscle is inserted into the tibia. In addition to these muscles, the Tensor fasciæ femoris is inserted indirectly into the tibia, through the ilio-tibial band, and the Peroneus longus occasionally derives a few fibres of origin from the outer tuberosity.

Surface Form.—A considerable portion of the tibia is subcutaneous and easily to be felt. At the upper extremity the tuberosities are to be recognised just below the knee. The internal one is broad and smooth, and merges into the subcutaneous surface of the shaft below. The external one is narrower and more prominent, and on it, about midway between the apex of the patella and the head of the fibula, may be felt a prominent tubercle for the insertion of the ilio-tibial band. In front of the upper end of the bone, between the tuberosities, is the tubercle of the tibia, forming an oval eminence, which is continuous below with the anterior border or crest of the bone. This border can be felt, forming the prominence of the shin, in the upper two-thirds of its extent as a sharp and flexuous ridge. In the lower third of the leg the border disappears and the bone is concealed by the tendons of the muscles on the front of the leg. Internal to the anterior border is to be felt the broad internal surface of the tibia, slightly encroached upon by the muscles in front and behind. It commences above at the wide expanded inner tuberosity and terminates below at the internal malleolus. The internal malleolus is a broad prominence situated on a higher level and somewhat farther forwards than the external malleolus. It overhangs the inner border of the arch of the foot. Its anterior border is nearly straight, its posterior border presents a sharp edge, which forms the inner margin of the groove for the tendon of the Tibialis posticus muscle.

The Fibula (figs. 326, 328)

The **Fibula** (fibula, *a clasp*) is situated at the outer side of the leg. It is the smaller of the two bones, and, in proportion to its length, the most slender of all the long bones ; it is placed on the outer side of the tibia, with which it is connected above and below. Its upper extremity is small, placed towards the back of the head of the tibia, below the level of the knee-joint, and excluded from its formation ; the lower extremity inclines a little forwards, so as to be on a plane anterior to that of the upper end ; it projects below the tibia, and forms the outer ankle. The bone presents for examination a shaft and two extremities.

The **Upper Extremity, or Head,** is of an irregular quadrate form, presenting above a flattened articular facet, directed upwards, forwards and inwards, for articulation with a corresponding facet on the external tuberosity of the tibia. On the outer side is a thick and rough prominence, continued behind into a pointed eminence, the *styloid process*, which projects upwards from the posterior part of the head. The prominence, at its upper and outer part, gives attachment to the tendon of the Biceps muscle, and to the long external lateral ligament of the knee, the ligament dividing the tendon into two parts. The summit of the styloid process gives attachment to the short external lateral ligament. The remaining part of the circumference of the head is rough, for the attachment of muscles and ligaments. It presents in front a tubercle for the origin of the upper and anterior part of the Peroneus longus, and the adjacent surface gives attachment to the anterior superior tibio-fibular ligament ; and behind, another tubercle for the attachment of the posterior superior tibio-fibular ligament, and the origin of the upper fibres of the Soleus muscle.

The **Shaft** presents four borders—the antero-external, the antero-internal, the postero-external, and the postero-internal; and four surfaces—anterior, posterior, internal, and external.

The **antero-external border** commences above in front of the head, runs vertically downwards to a little below the middle of the bone, and then curving somewhat outwards, bifurcates so as to embrace a triangular subcutaneous surface immediately above the outer surface of the external malleolus. This border gives attachment to an intermuscular septum, which separates the Extensor muscles on the anterior surface of the leg from the Peroneus longus and brevis muscles on the outer surface.

The **antero-internal border**, or **interosseous ridge**, is situated close to the inner side of the preceding, and runs nearly parallel with it in the upper third of its extent, but diverges from it so as to include a broader space in the lower two-thirds. It commences above just beneath the head of the bone (sometimes it is quite indistinct for about an inch below the head), and terminates at the apex of a rough triangular surface immediately above the articular facet of the external malleolus. It serves for the attachment of the interosseous membrane, which separates the Extensor muscles in front from the Flexor muscles behind.

The **postero-external border** is prominent; it commences above at the base of the styloid process, and terminates below in the posterior border of the outer malleolus. It is directed outwards above, backwards in the middle of its course, backwards and a little inwards below, and gives attachment to an aponeurosis which separates the Peronei muscles on the outer surface of the shaft from the Flexor muscles on its posterior surface.

The **postero-internal border,** sometimes called the *oblique line*, commences above at the inner side of the head, and terminates by becoming continuous with the antero-internal border or interosseous ridge at the lower fourth of the bone. It is well marked and prominent at the upper and middle parts of the bone. It gives attachment to an aponeurosis which separates the Tibialis posticus from the Soleus above and the Flexor longus hallucis below.

The **anterior surface** is the interval between the antero-external and antero-internal borders. It is extremely narrow and flat in the upper third of its extent; broader and grooved longitudinally in its lower third; it serves for the origin of three muscles, the Extensor longus digitorum, Extensor proprius hallucis, and Peroneus tertius.

The **posterior surface** is the space included between the postero-external and the postero-internal borders; it is continuous below with the triangular surface above the articular facet of the outer malleolus; it is directed backwards above, backwards and inwards at its middle, directly inwards below. Its upper third is rough, for the origin of the Soleus muscle; its lower part presents a triangular surface, connected to the tibia by a strong interosseous ligament, and between these two points the entire surface is covered by the fibres of origin of the Flexor longus hallucis muscle. At about the middle of this surface is the nutrient foramen, which is directed downwards.

The **internal surface** is the interval included between the antero-internal and the postero-internal borders. It is directed inwards, and is grooved for the origin of the Tibialis posticus muscle.

The **external surface** is the space between the antero-external and postero-external borders. It is much broader than the preceding, and often deeply grooved, is directed outwards in the upper two-thirds of its course, backwards in the lower third, where it is continuous with the posterior border of the external malleolus. This surface is completely occupied by the Peroneus longus and brevis muscles.

The **Lower Extremity**, or **external malleolus**, is of a pyramidal form, somewhat flattened from without inwards, and is longer, and descends lower, than the internal malleolus. Its *external surface* is convex, subcutaneous, and continuous with the triangular, subcutaneous surface on the outer side of the shaft. The *internal surface* presents in front a smooth triangular facet, broader above than below, and convex from above downwards, which articulates with a corresponding surface on the outer side of the astragalus. Behind and beneath the articular surface is a rough depression, which gives attachment to the posterior fasciculus of the external lateral ligament of the ankle. The *anterior border* is thick and rough, and marked below by a depression for the attachment of the anterior fasciculus of the external lateral ligament. The *posterior border* is broad and marked by a shallow groove, for the passage of the tendons of the Peroneus longus and brevis muscles. The *summit* is rounded, and gives attachment to the middle fasciculus of the external lateral ligament.

In order to distinguish the side to which the bone belongs, hold it with the lower extremity downwards, and the broad groove for the Peronei tendons backwards, i.e. towards the holder: the triangular subcutaneous surface will then be directed to the side to which the bone belongs.

Articulations.—With two bones: the tibia and astragalus.

Development.—By *three* centres (fig. 330): one for the shaft, and one for each extremity. Ossification commences in the shaft about the eighth week of fœtal life, a little later than in the tibia, and extends gradually towards the extremities. At birth both ends are cartilaginous. Ossification commences in the lower end in the second year, and in the upper one about the fourth year. The lower epiphysis, the first in which ossification commences, becomes united to the shaft about the twentieth year; the upper epiphysis joins about the twenty-fifth year. Ossification appearing first in the lower epiphysis is contrary to the rule which prevails with regard to the commencement of ossification in epiphyses, viz. that the epiphysis towards which the nutrient artery is directed commences to ossify last; but it follows the rule which prevails with regard to the union of epiphyses, by uniting first.

Attachment of Muscles.—To nine: to the head, the Biceps, Soleus, and Peroneus longus; to the anterior surface of the shaft, the Extensor longus digitorum, Peroneus tertius, and Extensor proprius hallucis; to the internal surface, the Tibialis posticus; to the posterior surface, the Soleus and Flexor longus hallucis; to the external surface, the Peroneus longus and brevis.

Fig. 330.—Plan of the development of the fibula. By three centres.

Upper extremity

Appears about 4th year — Unites about 25th year

Appears at 2nd year — Unites about 20th year

Lower extremity

Surface Form.—The only parts of the fibula which are to be felt are the head, the lower part of the external surface of the shaft, and the external malleolus. The head is to be seen and felt behind and to the outer side of the outer tuberosity of the tibia. It presents a small, prominent, triangular eminence slightly above the level of the tubercle of the tibia. The external malleolus presents a narrow elongated prominence, situated on a plane posterior to the internal malleolus and reaching to a lower level. From it may be traced the lower third or half of the external surface of the shaft of the bone in the interval between the Peroneus tertius in front and the other two Peronei tendons behind.

Surgical Anatomy.—In fractures of the bones of the leg, both bones are generally fractured, but each bone may be broken separately, the fibula more frequently than the tibia. Fracture of both bones may be caused either by direct or indirect violence. When it occurs from indirect force, the fracture in the tibia is at the junction of the middle and lower third of the bone. Many causes conduce to render this the weakest part of the bone. The fracture of the fibula is usually at rather a higher level. These fractures present great variety, both as regards their direction and condition. They may be oblique, transverse, longitudinal, or spiral. When oblique, they are for the most part the result of indirect violence, and the direction of the fracture is from behind, downwards, forwards, and inwards in many cases, but may be downwards and outwards, or downwards and backwards. When transverse, the fracture is often at the upper part of the bone, and is the result of direct violence. The spiral fracture of the tibia generally commences as a vertical fissure, involving the ankle-joint, and is associated with fracture of the fibula higher up. It is the result of torsion, from twisting of the body while the foot is fixed.

Fig. 331.—Bones of the right foot.   Dorsal surface.

FIG. 332.—Bones of the right foot. Plantar surface.

Fractures of the tibia alone are almost always the result of direct violence, except where the malleolus is broken off by twists of the foot. Fractures of the fibula alone may arise from indirect or direct force, those of the lower end being usually the result of the former, and those higher up being caused by a direct blow on the part.

The tibia and fibula, like the femur, are often the seat of acute necrosis. Chronic abscess is more frequently met with in the cancellous tissue of the head and lower end of the tibia than in any other bone of the body. The abscess is of small size, very chronic, and probably the result of tuberculous osteitis in the highly vascular growing tissue at the end of the shaft near the epiphysial cartilage in the young subject.

The tibia is the bone which is most commonly and most extensively distorted in rickets. It gives way at the junction of the middle and lower third, its weakest part, and presents a curve forwards with generally some lateral displacement.

THE FOOT (figs. 331, 332)

The skeleton of the Foot consists of three divisions: the Tarsus, Metatarsus, and Phalanges.

THE TARSUS

The bones of the **Tarsus** are seven in number: viz. the calcaneum or os calcis, astragalus, cuboid, navicular, internal, middle, and external cuneiform bones.

THE CALCANEUM (fig. 333)

The **Calcaneum**, or **Os Calcis** (calx, *the heel*), is the largest and strongest of the tarsal bones. It is situated at the lower and back part of the foot, serving to transmit the weight of the body to the ground, and forming a strong lever for the

FIG. 333.—The left os calcis. A. Postero-external view. B. Antero-internal view.

For ext. calcaneo-astragaloid ligament
Groove for Peroneus brevis
Groove for Peroneus longus
Peroneal tubercle
For tendo Achillis

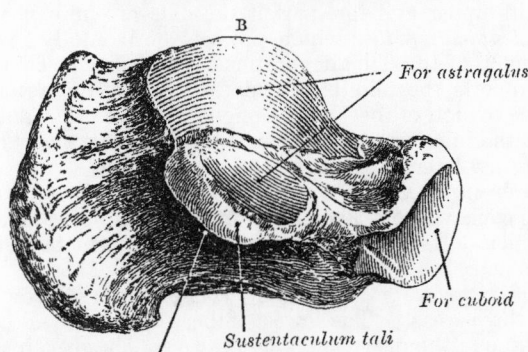

For astragalus
For cuboid
Sustentaculum tali
Groove for Flexor longus hallucis

muscles of the calf. It is irregularly cuboidal in form, having its long axis directed forwards and outwards; it presents for examination six surfaces.

The *superior surface* is formed behind by the upper aspect of that part of the bone which projects backwards to form the heel. This varies in length in different individuals; is convex from side to side, concave from before backwards, and upon it rests a mass of fat placed in front of the tendo Achillis. In front of this area is a large, usually somewhat oval-shaped facet, which looks upwards and forwards; it is convex from behind forwards, and articulates with the posterior calcanean facet on the under aspect of the astragalus. It is bounded anteriorly by a deep depression which is continued backwards and inwards in the form of a groove. In the articulated foot this groove lies below a similar one on the under aspect of the astragalus, and the two form a canal (*sinus tarsi*) for the lodgment of the interosseous calcaneo-astragaloid ligament. In front and to the inner side of this groove is an elongated facet, concave from behind forwards, and with its long axis directed forwards and outwards. This facet is frequently divided into two by a notch: of the two, the posterior, and larger, is supported on a projecting process of bone, termed the *sustentaculum tali*, and articulates with the middle calcanean facet on the under aspect of the astragalus; the anterior, and smaller, is placed on the anterior part of the body of the bone, and articulates with the anterior calcanean facet on the astragalus. The outer portion of the upper surface of the anterior part of the bone is rough, for the attachment of ligaments and for the origin of the Extensor brevis digitorum muscle.

The *inferior surface* is uneven, wider behind than in front, and convex from side to side; it is bounded posteriorly by two tubercles, separated by a depression; the *external*, small, prominent, and rounded, gives origin to part of the Abductor minimi digiti; the *internal*, broader and larger, for the support of the heel, gives attachment, by its prominent inner margin, to the Abductor hallucis, and in front to the Flexor brevis digitorum muscles and plantar fascia; the depression between the tubercles gives origin to the Abductor minimi digiti. The rough surface in front of the tubercles gives attachment to the long plantar ligament, and to the outer head of the Flexor accessorius muscle; while to a prominent tubercle nearer the anterior part of this surface, as well as to a transverse groove in front of it, is attached the short plantar ligament.

The *external surface* is broad behind and narrow in front, flat, and almost subcutaneous; it presents near its centre a tubercle, for the attachment of the middle fasciculus of the external lateral ligament. At its upper and anterior part, this surface gives attachment to the external calcaneo-astragaloid ligament; and in front of the tubercle it presents a narrow surface marked by two oblique grooves, separated by an elevated ridge, or tubercle, which varies much in size in different bones; it is named the *peroneal tubercle*, and gives attachment to a fibrous process from the external annular ligament. The *superior groove* transmits the tendon of the Peroneus brevis; the *inferior*, the tendon of the Peroneus longus.

The *internal surface* is deeply concave; it is directed obliquely downwards and forwards, and serves for the transmission of the plantar vessels and nerves into the sole of the foot; it affords attachment to part of the Flexor accessorius muscle. At its upper and fore part it presents an eminence of bone, the *lesser process* or *sustentaculum tali*, which projects horizontally inwards, and to it a slip of the tendon of the Tibialis posticus is attached. This process is concave above, and supports the middle articular surface of the astragalus; below, it is grooved for the tendon of the Flexor longus hallucis. Its anterior margin gives attachment to the inferior calcaneo-navicular ligament, and its inner margin to a part of the internal lateral ligament of the ankle-joint.

The *anterior surface*, of a somewhat triangular form, articulates with the cuboid. It is concave from above, downwards and outwards, and convex in the opposite direction. Its inner border gives attachment to the inferior calcaneonavicular ligament.

The *posterior surface* is prominent, convex, wider below than above, and divisible into three areas. The lowest of these is rough, and covered by the fatty and fibrous tissue of the heel; the middle, also rough, gives insertion to the tendo Achillis and Plantaris; while the highest is smooth, and is covered by a bursa which intervenes between it and the tendo Achillis.

Articulations.—With two bones: the astragalus and cuboid.

Attachment of Muscles.—To eight; part of the Tibialis posticus, the tendo Achillis, Plantaris, Abductor hallucis, Abductor minimi digiti, Flexor brevis digitorum, Flexor accessorius, and Extensor brevis digitorum.

The Astragalus (fig. 334)

The **Astragalus** (ἀστράγαλος, *a die*) is the largest of the tarsal bones, next to the os calcis. It occupies the middle and upper part of the tarsus, supporting the tibia above, articulating with the malleoli on either side, resting below upon

Fig. 334.—The left astragalus. A. Superior and external view. B. Inferior and internal view. C. Viewed from below.

A

For navicular　　*Neck*　　*Sup. surface for tibia*

For ext. malleolus　　*Groove for Flexor longus hallucis*

B

For inner malleolus

For navicular

Groove for Flexor longus hallucis

For os calcis

C

For navicular

For inferior calcaneo-navicular ligament

Anterior calcanean facet

Groove for interosseous ligament

Middle and posterior calcanean facets

Groove for tendon of Flexor longus hallucis

the os calcis, and articulating in front with the navicular. This bone may easily be recognised by its large oval head, by the broad articular facet on its upper convex surface, or by the two articular facets separated by a deep groove on its under concave surface. It consists of a *head, neck,* and *body.*

The *superior surface* of the **body** presents, behind, a broad smooth trochlear surface, for articulation with the tibia. The trochlea is broader in front than behind, convex from before backwards, slightly concave from side to side: in front of it is the upper surface of the neck of the bone. The *inferior surface* presents two articular facets separated by a deep groove. The groove runs obliquely forwards and outwards, becoming gradually broader and deeper in front: it corresponds with a similar groove upon the upper surface of the os calcis, and forms, when articulated with that bone, a canal (*sinus tarsi*) filled up in the recent state by the interosseous calcaneo-astragaloid ligament. Of the two articular facets, the posterior (*posterior calcanean facet*) is the larger, of an oblong form, and deeply concave from without inwards and forwards: it articulates with the corresponding facet on the upper surface of the os calcis; the anterior (*middle calcanean facet*) is shorter and narrower, of an elongated oval form, convex longitudinally, and articulates with the upper surface of the sustentaculum tali of the os calcis. The *internal surface* presents at its upper part a pear-shaped articular facet for the inner malleolus, continuous above with the trochlear surface; below the articular surface is a rough depression, for the attachment of the deep portion of the internal lateral ligament of the ankle-joint. The *external surface* presents a large triangular facet, concave from above downwards, for articulation with the external malleolus; its anterior half is continuous above with the trochlear surface; and in front of it is a rough depression for the attachment of the anterior fasciculus of the external lateral ligament of the ankle-joint. Professor E. Fawcett * has directed attention to a triangular facet which comes into contact with the inferior tibio-fibular ligament during flexion of the ankle. It is situated between the posterior half of the outer border of the trochlea and the corresponding part of the base of the triangular facet for the external malleolus of the fibula. It is triangular in shape, the apex being directed forwards and the base backwards. Its length from base to apex is from three-quarters of an inch to an inch; its breadth at the base, about a quarter of an inch. The *posterior surface* is narrow, and traversed by a groove, which runs obliquely downwards and inwards, and transmits the tendon of the Flexor longus hallucis, external to which is a prominent tubercle, to which the posterior fasciculus of the external lateral ligament is attached. This tubercle is sometimes separated from the rest of the astragalus, and is then known as the *os trigonum*. To the inner side of the groove is a second, but less marked tubercle.

The **neck** is directed forwards and inwards, and comprises the constricted portion of the bone between the body and the oval head. Its upper and inner surfaces are rough, for the attachment of ligaments; its external surface is concave, directed downwards and outwards, and is continuous below with the deep groove for the interosseous calcaneo-astragaloid ligament.

The **head** looks forwards and inwards, and its anterior surface presents a large, oval, convex facet, for articulation with the navicular. Its inferior surface has two facets, which are best seen in the recent condition. The inner of these, situated immediately in front of the middle calcaneal facet, is convex, triangular or semi-oval in shape, and rests on the inferior calcaneo-navicular ligament; the outer (*anterior calcanean facet*) is somewhat flattened, and articulates with the upper surface of the anterior part of the os calcis.

To ascertain to which foot the bone belongs, hold it with the broad articular surface upwards, and the rounded head forwards; the lateral triangular articular surface for the external malleolus will then point to the side to which the bone belongs.

Articulations.—With four bones: tibia, fibula, os calcis, and navicular.

THE CUBOID (fig. 335)

The **Cuboid** (κύβος, *a cube*; εἶδος, *like*) bone is placed on the outer side of the foot, in front of the os calcis, and behind the fourth and fifth metatarsal bones. It is of a pyramidal shape, its base being directed inwards, its apex outwards. It may be distinguished from the other tarsal bones by the existence of a deep groove on its under surface, for the tendon of the Peroneus longus muscle. It presents for examination six surfaces: three articular, and three non-articular.

* *Edinburgh Medical Journal,* 1895.

The **non-articular surfaces** are the superior, inferior, and external. The *superior* or *dorsal surface*, directed upwards and outwards, is rough, for the attachment of numerous ligaments. The *inferior* or *plantar surface* presents in front a deep groove, which runs obliquely from without, forwards and inwards; it lodges the tendon of the Peroneus longus, and is bounded behind by a prominent ridge, to which is attached the long calcaneo-cuboid ligament. The ridge terminates externally in an eminence, the *tuberosity of the cuboid*, the surface of which presents a convex facet, over which the sesamoid bone or cartilage, frequently found in the tendon of the Peroneus longus, glides. The surface of bone behind the groove is rough, for the attachment of the short plantar ligament, a few fibres of the Flexor brevis hallucis and a fasciculus from the tendon of the Tibialis posticus. The *external surface*, the smallest and narrowest of the three, presents a deep notch formed by the commencement of the peroneal groove.

The **articular surfaces** are the posterior, anterior, and internal. The *posterior surface* is smooth, triangular, and concavo-convex, for articulation with the anterior surface of the os calcis. The *anterior surface*, of smaller size, but also irregularly

FIG. 335.—The left cuboid. A. Antero-internal view. B. Postero-external view.

triangular, is divided by a vertical ridge into two facets : the inner, quadrilateral in form, articulates with the fourth metatarsal bone ; the outer, larger and more triangular, articulates with the fifth metatarsal. The *internal surface* is broad, rough, irregularly quadrilateral, presenting at its middle and upper part a smooth oval facet, for articulation with the external cuneiform bone ; and behind this (occasionally) a smaller facet, for articulation with the navicular ; it is rough in the rest of its extent, for the attachment of strong interosseous ligaments.

To ascertain to which foot the bone belongs, hold it so that its under surface, marked by the peroneal groove, looks downwards, and the large concavo-convex articular surface backwards, towards the holder: the narrow non-articular surface, marked by the commencement of the peroneal groove, will point to the side to which the bone belongs.

Articulations.—With four bones : the os calcis, external cuneiform, and the fourth and fifth metatarsal bones ; occasionally with the navicular.

Attachment of Muscles.—Part of the Flexor brevis hallucis and a slip from the tendon of the Tibialis posticus.

The Navicular (fig. 336)

The **Navicular** or **Scaphoid bone** is situated at the inner side of the tarsus, between the astragalus behind and the three cuneiform bones in front. It may be distinguished by its form, being concave behind, convex and subdivided into three facets in front.

The *anterior surface* is convex from side to side, and subdivided by two ridges into three facets, for articulation with the three cuneiform bones. The *posterior surface* is oval, concave, broader externally than internally, and articulates with the rounded head of the astragalus. The *superior surface* is convex from side to side, and rough for the attachment of ligaments. The *inferior surface* is irregular, and also rough for the attachment of ligaments. The *internal surface* presents a rounded tubercular eminence, the *tuberosity* of the navicular, the lower part of which projects, and gives attachment to part of the tendon of the Tibialis posticus. The *external surface* is rough, and irregular, for the attachment of ligamentous fibres, and occasionally presents a small facet for articulation with the cuboid bone.

To ascertain to which foot the bone belongs, hold it with the concave articular surface backwards, and the convex dorsal surface upwards; the external surface,

FIG. 336.—The left navicular.　A. Antero-external view.　B. Postero-internal view.

i.e. the surface opposite the tubercle, will point to the side to which the bone belongs.

Articulations.—With four bones : astragalus and three cuneiform ; occasionally also with the cuboid.

Attachment of Muscles.—Part of the Tibialis posticus.

THE CUNEIFORM BONES

The **Cuneiform Bones** have received their name from their wedge-like shape (cuneus, *a wedge*; forma, *likeness*). They are three in number, and form, with the cuboid, the anterior row of the tarsus, being placed between the navicular behind, the three innermost metatarsal bones in front, and the cuboid externally. They are named, from their position, *internal, middle,* and *external.*

THE INTERNAL CUNEIFORM (fig. 337)

The **Internal Cuneiform** is the largest of the three. It is situated at the inner side of the foot, between the navicular behind and the base of the first metatarsal in front. It may be distinguished from the other two by its large size, and by its not presenting such a distinct wedge-like form. Without the others, it may be known by the large kidney-shaped anterior articulating surface, and by the prominence on the inferior or plantar surface for the attachment of the Tibialis posticus. It presents for examination six surfaces.

FIG. 337.—The left internal cuneiform.
A. Antero-internal view.　B. Postero-external view.

The *internal surface* is subcutaneous, and forms part of the inner border of the foot; it is broad, quadrilateral, and presents at its anterior inferior angle a smooth oval impression, into which the tendon of the Tibialis anticus is partially inserted; in the rest of its extent it is rough, for the attachment of ligaments. The *external surface* is concave, presenting, along its superior and posterior borders, a narrow reversed L-shaped surface for articulation with the middle cuneiform behind, and second metatarsal bone in front: in the rest of its extent it is rough for the attachment of ligaments and part of the tendon of the Peroneus longus. The *anterior surface*, kidney-shaped, much larger than the posterior, articulates with the metatarsal bone of the great toe. The *posterior surface* is triangular, concave, and articulates with the innermost and largest of the three facets on the anterior surface of the navicular. The *inferior* or *plantar surface* is rough, and forms the base of the wedge; it presents a prominent tuberosity at its back part for the insertion of part of the tendon of the Tibialis posticus. It also gives insertion in front to part of the tendon of the Tibialis anticus. The *superior surface* is the narrow pointed end of the wedge, which is directed upwards and outwards; it is rough for the attachment of ligaments.

To ascertain to which side the bone belongs, hold it so that its superior narrow edge looks upwards, and the long, kidney-shaped, articular surface forwards; the external surface, marked by its vertical and horizontal articular facets, will point to the side to which it belongs.

Articulations.—With four bones: navicular, middle cuneiform, first and second metatarsal bones.

Attachment of Muscles.—To three: the Tibialis anticus and posticus, and Peroneus longus.

THE MIDDLE CUNEIFORM (fig. 338)

The **Middle Cuneiform**, the smallest of the three, is of very regular wedge-like form, the broad extremity being placed upwards, the narrow end downwards. It is situated between the other two bones of the same name, and articulates with the navicular behind, and the second metatarsal in front. It is smaller than the external cuneiform bone, from which it may be further distinguished by the L-shaped articular facet, which runs round the upper and back part of its inner surface.

The *anterior surface*, triangular in form, and narrower than the posterior, articulates with the base of the second metatarsal bone. The *posterior surface*, also triangular, articulates with the middle facet on the anterior surface of the navicular. The *internal surface* presents a reversed L-shaped articular facet, running along the superior and posterior borders, for articulation with the internal cuneiform,

FIG. 338.—The left middle cuneiform,
A. Antero-internal view.
B. Postero-external view.

For int. cuneiform   *For navicular*

For 2nd metatarsal   *For ext. cuneiform*

and is rough in the rest of its extent for the attachment of ligaments. The *external surface* presents posteriorly a smooth facet for articulation with the external cuneiform bone. The *superior surface* forms the base of the wedge; it is quadrilateral, broader behind than in front, and rough for the attachment of ligaments. The *inferior surface*, sharp and tubercular, is also rough for ligamentous attachment, and for the insertion of a slip from the tendon of the Tibialis posticus.

To ascertain to which foot the bone belongs, hold its superior or dorsal surface upwards, the larger, triangular surface being towards the holder: the smooth facet (limited to the posterior border) will then point to the side to which it belongs.

Articulations.—With four bones: navicular, internal and external cuneiform, and second metatarsal bone.

Attachment of Muscle.—A slip from the tendon of the Tibialis posticus is attached to this bone.

THE EXTERNAL CUNEIFORM (fig. 339)

The **External Cuneiform**, intermediate in size between the two preceding, is of a very regular wedge-like form, the broad extremity being placed upwards, the narrow end downwards. It occupies the centre of the front row of the tarsus between the middle cuneiform internally, the cuboid externally, the navicular behind, and the third metatarsal in front. It is distinguished from

Z

the internal cuneiform bone by its more regular wedge-like shape, and by the absence of the kidney-shaped articular surface: from the middle cuneiform, by the absence of the reversed L-shaped facet, and by the two articular facets which are present on both its inner and outer surfaces. It presents six surfaces for examination.

The *anterior surface*, triangular in form, articulates with the third metatarsal bone. The *posterior surface* articulates with the most external facet of the navicular, and is rough below for the attachment of ligamentous fibres. The *internal surface* presents two articular facets, separated by a rough depression: the anterior one, sometimes divided into two, articulates with the outer side of the base of the second metatarsal bone; the posterior one skirts the posterior border, and

FIG. 339.—The left external cuneiform.
A. Postero-internal view. B. Antero-external view.

For navicular

For mid. cuneiform

For 4th metatarsal   For cuboid

A   B

For 2nd metatarsal   For 3rd metatarsal

articulates with the middle cuneiform; the rough depression between the two gives attachment to an interosseous ligament. The *external surface* also presents two articular facets, separated by a rough non-articular surface; the anterior facet, situated at the superior angle of the bone, is small and semi-oval in shape, and articulates with the inner side of the base of the fourth metatarsal bone; the posterior and larger one is triangular or oval, and articulates with the cuboid; the rough, non-articular surface serves for the attachment of an interosseous ligament. The three facets for articulation with the three metatarsal bones are continuous with one another; those for articulation with the middle cuneiform and navicular are also continuous, but that for articulation with the cuboid is usually separate. The *superior* or *dorsal surface* is of an oblong form; its postero-external angle being prolonged backwards. The *inferior* or *plantar surface* is a rounded margin, and serves for the attachment of part of the tendon of the Tibialis posticus, part of the Flexor brevis hallucis, and ligaments.

To ascertain to which side the bone belongs, hold it with the broad dorsal surface upwards, the prolonged edge backwards; the separate articular facet for the cuboid will point to the side to which it belongs.

Articulations.—With six bones: the navicular, middle cuneiform, cuboid, and second, third, and fourth metatarsal bones.

Attachment of Muscles.—To two: part of the Tibialis posticus, and Flexor brevis hallucis.

The number of tarsal bones may be reduced owing to congenital ankylosis, which may occur between the os calcis and cuboid, the os calcis and navicular, the os calcis and astragalus, or the astragalus and navicular.

THE METATARSAL BONES

The **Metatarsal Bones** are five in all, and are numbered one to five, in accordance with their position from within outwards; they are long bones, and present for examination a shaft and two extremities.

COMMON CHARACTERS OF THE METATARSAL BONES

The *Shaft* is prismoid in form, tapers gradually from the tarsal to the phalangeal extremity, and is curved longitudinally, so as to be concave below, slightly convex above. The *posterior extremity*, or *base*, is wedge-shaped, articulating by its terminal surface with the tarsal bones, and by its lateral surfaces with the contiguous metatarsal bones: its dorsal and plantar surfaces are rough for the attachment of ligaments. The *anterior extremity*, or *head*, presents a terminal convex articular surface, oblong from above downwards, and extending farther backwards below than above. Its sides are flattened, and each presents a depression, surmounted by a tubercle, for ligamentous attachment. Its under surface is grooved in the middle line for the passage of the Flexor tendon, and marked on each side by an articular eminence continuous with the terminal articular surface.

PECULIAR CHARACTERS OF THE METATARSAL BONES

The **First** (fig. 340) is remarkable for its great thickness, and is the shortest of all the metatarsal bones. The *shaft* is strong, and of well-marked prismoid form. The *posterior extremity* presents, as a rule, no lateral articular facets, but occasionally on the outer side there is an oval facet, by which it articulates with the second metatarsal bones. Its terminal articular surface is of large size and kidney-shaped; its circumference is grooved, for the tarso-metatarsal ligaments, and internally gives insertion to part of the tendon of the Tibialis anticus; its inferior angle presents a rough oval prominence for the insertion of the tendon of the Peroneus longus. The *head* is of large size; on its plantar surface are two grooved facets, over which glide sesamoid bones; the facets are separated by a smooth elevated ridge.

This bone is known by the single kidney-shaped articular surface on its base; the deeply grooved appearance of the plantar surface of its head; and its

FIG. 340.—The first metatarsal.
(Left.)

FIG. 341.—The second metatarsal.
(Left.)

For internal cuneiform   *For Peroneus longus*

For 3rd metatarsal

For internal cuneiform   *For middle cuneiform*   *For external cuneiform*

great thickness, relatively to its length. When it is placed in its natural position, the concave border of the kidney-shaped articular surface on its base points to the foot to which the bone belongs.

Attachment of Muscles.—To three: part of the Tibialis anticus, the Peroneus longus, and the First dorsal interosseous.

The **Second** (fig. 341) is the longest and largest of the remaining metatarsal bones, being prolonged backwards into the recess formed between the three cuneiform bones. Its *tarsal extremity* is broad above, narrow and rough below. It presents four articular surfaces: one behind, of a triangular form, for articulation with the middle cuneiform; one at the upper part of its internal lateral surface, for articulation with the internal cuneiform; and two on its external lateral surface, an upper and lower, separated by a rough non-articular interval. Each of these articular surfaces is divided by a vertical ridge into two, thus making four facets; the two anterior of these articulate with the third metatarsal; the two posterior (sometimes continuous) with the external cuneiform.

In addition to these articular surfaces, there is occasionally a fifth when this bone articulates with the first metatarsal bone. It is oval in shape, and is situated on the inner side of the shaft near the base.

The facets on the tarsal extremity of the second metatarsal bone serve at once to distinguish it from the rest, and to indicate the foot to which it belongs ; there being one facet at the upper angle of the internal surface, and two facets, each subdivided into two parts, on the external surface, pointing to the foot to which the bone belongs. The fact that the two posterior subdivisions of these external facets sometimes run into one should not be forgotten.

Attachment of Muscles.—To four : the Adductor obliquus hallucis, First and Second dorsal interossei, a slip from the tendon of the Tibialis posticus, and occasionally a slip from the Peroneus longus.

The **Third** (fig. 342) articulates behind, by means of a triangular smooth surface, with the external cuneiform ; on its inner side, by two facets, with the second metatarsal ; and on its outer side, by a single facet, with the fourth metatarsal bone. The latter facet is situated at the upper angle of the base.

Fig. 342.—The third metatarsal.　　　　Fig. 343.—The fourth metatarsal.
(Left.)　　　　　　　　　　　　　　　　　　(Left.)

For 2nd　　　For 2nd　　　　For 4th
metatarsal　For ext.　metatarsal　metatarsal
　　　　　cuneiform

For 3rd meta-
tarsal

For cuboid

For external cuneiform　　　For 5th metatarsal

The third metatarsal bone is known by its having at its tarsal end two undivided facets on the inner side, and a single facet on the outer. This distinguishes it from the second metatarsal, in which the two facets, found on one side of its tarsal end, are each subdivided into two. The single facet (when the bone is put in its natural position) is on the side to which the bone belongs.

Attachment of Muscles.—To five : Adductor obliquus hallucis, Second and Third dorsal and First plantar interossei, and a slip from the tendon of the Tibialis posticus.

The **Fourth** (fig. 343) is smaller in size than the preceding ; its *tarsal extremity* presents an oblique quadrilateral surface for articulation with the cuboid ; a smooth facet on the inner side, divided by a ridge into an anterior portion for articulation with the third metatarsal, and a posterior portion for articulation with the external cuneiform ; on the outer side a single facet, for articulation with the fifth metatarsal.

The fourth metatarsal is known by its having a single facet on either side of the tarsal extremity, that on the inner side being divided into two parts. If this subdivision be not recognisable, the fact that its tarsal end is bent somewhat outwards will indicate the side to which it belongs.

Attachment of Muscles.—To five: Adductor obliquus hallucis, Third and Fourth dorsal and Second plantar interossei, and a slip from the tendon of the Tibialis posticus.

The **Fifth** (fig. 344) is recognised by the tubercular eminence, the *tuberosity*, on the outer side of its base. It articulates behind, by a triangular surface cut obliquely from without inwards, with the cuboid; and internally, with the fourth metatarsal. On the inner part of the dorsal aspect of the base is inserted the tendon of the Peroneus tertius, and on the dorsal aspect of the tuberosity that of the Peroneus brevis. A strong band of the plantar fascia connects the projecting part of the tuberosity with the outer tuberosity on the under surface of the os calcis. The plantar aspect of the base is grooved for the tendon of the Abductor minimi digiti, and gives origin to the Flexor brevis minimi digiti.

FIG. 344.—The fifth metatarsal. (Left.)

For 4th metatarsal   For cuboid   Tuberosity

The projection on the outer surface of this bone at its tarsal end at once distinguishes it from the others, and points to the side to which it belongs.

Attachment of Muscles.—To six: the Peroneus brevis, Peroneus tertius, Flexor brevis minimi digiti, Adductor transversus hallucis, Fourth dorsal and Third plantar interossei.

Articulations.—Each bone articulates with one or more of the tarsal bones by its proximal extremity, and by the other with one of the first row of phalanges. The number of tarsal bones with which each metatarsal articulates is one for the first, three for the second, one for the third, two for the fourth, and one for the fifth.

PHALANGES

The **Phalanges** of the foot, both in number and general arrangement, resemble those of the hand; there being two in the great toe, and three in each of the other toes. They differ from them, however, in their size, the shafts being much reduced in length and, especially in the first row, laterally compressed.

The phalanges of the *first row* closely resemble those of the hand. The *shaft* is compressed from side to side, convex above, concave below. The *posterior extremity* is concave; and the *anterior extremity* presents a trochlear surface, for articulation with the second phalanges.

The phalanges of the *second row* are remarkably small and short, but rather broader than those of the first row.

The *ungual* phalanges, in form, resemble those of the fingers; but they are smaller, flattened from above downwards, presenting a broad base for articulation with the second row, and an expanded extremity for the support of the nail and end of the toe.

Articulations.—The first row, with the metatarsal bones behind and second phalanges in front; the second row, of the four outer toes, with the first and third phalanges; of the great toe, with the first phalanx; the third row, of the four outer toes, with the second phalanges.

Attachment of Muscles.—To the first phalanges. Great toe, five muscles: innermost tendon of Extensor brevis digitorum, Abductor hallucis, Adductor obliquus hallucis, Flexor brevis hallucis, Adductor transversus hallucis. Second toe, three muscles: First and Second dorsal interossei, and First lumbrical. Third toe, three muscles: Third dorsal and First plantar interossei, and Second lumbrical. Fourth toe, three muscles: Fourth dorsal and Second plantar interossei, and Third lumbrical. Fifth toe, four muscles: Flexor brevis minimi digiti, Abductor minimi digiti, Third plantar interosseous, and Fourth lumbrical.—Second phalanges. Great toe: Extensor longus hallucis, Flexor longus hallucis. Other toes: Flexor brevis digitorum, and one slip of the common tendon of the Extensor longus and Extensor brevis digitorum.*—Third phalanges: Flexor longus digitorum, and two slips of the common tendon of the Extensor longus and Extensor brevis digitorum.

DEVELOPMENT OF THE BONES OF THE FOOT (fig. 345)

The **Tarsal bones** are each developed by a single centre, excepting the os calcis, which has an epiphysis for its posterior extremity. The centres make their appearance in the following order: os calcis, at the sixth month of fœtal life; astragalus, about the seventh month; cuboid, at the ninth month; external

FIG. 345.—Plan of the development of the foot.

cuneiform, during the first year; internal cuneiform, in the third year; middle cuneiform and navicular, in the fourth year. The epiphysis for the posterior extremity of the os calcis appears at the tenth year, and unites with the rest of the bone soon after puberty. The tubercle on the posterior surface of the astragalus is sometimes ossified from a separate centre, and may remain distinct from the main mass of the bone, when it is named the *os trigonum*.

The **Metatarsal bones** are each developed by *two* centres: one for the shaft, and one for the digital extremity, in the four outer metatarsals; one for the shaft, and one for the proximal extremity, in the metatarsal bone of the great toe.† Ossification commences in the centre of the shaft about the ninth week, and

---

* Except the second phalanx of the fifth toe, which receives no slip from the Extensor brevis digitorum.

† As was noted in the first metacarpal bone, so in the first metatarsal, there is often a second epiphysis for its distal extremity (see footnote, page 298).

extends towards either extremity. The centre in the proximal end of the first metatarsal bone appears about the third year; the centres in the distal ends of the other bones between the fifth and eighth years; they become joined between the eighteenth and twentieth years.

The **Phalanges** are developed by *two* centres for each bone: one for the shaft, and one for the metatarsal extremity. The centre for the shaft appears about the tenth week, that for the epiphysis between the fourth and tenth years; they join the shaft about the eighteenth year.

SKELETON OF THE FOOT AS A WHOLE

The skeleton of the foot is constructed on the same principles as the hand, but modified to form a firm basis of support for the rest of the body when in the erect position. It is more solidly built up, and its component parts are less movable on each other than in the hand. This is especially the case with the great toe, which has to assist in supporting the body, and is therefore constructed with greater solidity; it lies parallel with the other toes, and has a very limited degree of mobility, whereas the thumb, which is occupied in numerous and varied movements, is constructed in such a manner as to permit of great mobility; its metacarpal bone is directed away from the others, so as to form an acute angle with the second, and it enjoys a considerable range of motion at its articulation with the carpus. The foot is placed at right angles to the leg—a position which is almost peculiar to man, and has relation to the erect position which he maintains. In order to allow of its supporting the weight of the whole body in this position with the least expenditure of material, it is constructed in the form of an arch. This arch is not, however, made up of two equal limbs. The hinder one, which is made up of the os calcis and the posterior part of the astragalus, is about half the length of the anterior limb, and measures about three inches. The anterior limb consists of the rest of the tarsal and the metatarsal bones, and measures about seven inches. It may be said to consist of two parts, an inner segment made up of the head of the astragalus, the navicular, the three cuneiforms, and the three inner metatarsal bones; and an outer segment composed of the os calcis, the cuboid and the two outer metatarsal bones. The summit of the arch is at the superior articular surface of the astragalus; and its two extremities—that is to say the two piers on which the arch rests in standing—are the tubercles on the under surface of the os calcis posteriorly, and the heads of the metatarsal bones anteriorly. The weakest part of the arch is the joint between the astragalus and scaphoid, and here it is more liable to yield in those who are overweighted, and in those in whom the ligaments which complete and preserve the arch are relaxed. This weak point in the arch is braced on its concave surface by the inferior calcaneo-navicular ligament, which is more elastic than most other ligaments, and thus allows the arch to yield from jars or shocks applied to the anterior portion of the foot, and quickly restores it to its pristine condition. This ligament is supported internally by blending with the Deltoid or internal lateral ligament of the ankle, and inferiorly by the tendon of the Tibialis posticus muscle, which is spread out into a fan-shaped insertion, and prevents undue tension of the ligament or such an amount of stretching as would permanently elongate it.

In addition to this longitudinal arch the foot presents a transverse arch, at the anterior part of the tarsus and hinder part of the metatarsus. This, however, can scarcely be described as a true arch, but presents more the character of a half-dome. The inner border of the central portion of the longitudinal arch is elevated from the ground, and from this point the bones arch over to the outer border, which is in contact with the ground, and, assisted by the longitudinal arch, produce a sort of rounded niche on the inner side of the foot, which gives the appearance of a transverse as well as a longitudinal arch.

The line of the foot, from the point of the heel to the toes, is not quite straight, but is directed slightly outwards, so that the inner border is a little convex and the outer border concave. This disposition of the bones becomes more marked when the longitudinal arch of the foot is lost, as in the disease known under the name of ‘flat-foot.’

Surface Form.—On the dorsum of the foot the individual bones are not to be distinguished, with the exception of the head of the astragalus, which forms a rounded projection in front of the ankle-joint when the foot is forcibly extended. The whole surface forms a smooth convex outline, the summit of which is the ridge formed by the head of the astragalus, the navicular, the middle cuneiform, and the second metatarsal bones; from this it gradually inclines outwards and more rapidly inwards. On the inner side of the foot, the internal tuberosity of the os calcis and the ridge separating the inner from the posterior surface of the bone may be felt most posteriorly. In front of this, and below the internal malleolus, may be felt the projection of the sustentaculum tali. Passing forwards is the well-marked tuberosity of the navicular bone, situated about an inch or an inch and a quarter in front of the internal malleolus. Farther towards the front, the ridge formed by the base of the first metatarsal bone can be obscurely felt, and from this the shaft of the bone can be traced to the expanded head articulating with the base of the first phalanx of the great toe. Immediately beneath the base of this phalanx, the internal sesamoid bone is to be felt. Lastly, the expanded ends of the bones forming the last joint of the great toe are to be felt. On the outer side of the foot the most posterior bony point is the outer tuberosity of the os calcis, with the ridge separating the posterior from the outer surface of the bone. In front of this the greater part of the external surface of the os calcis is subcutaneous; on it, below and in front of the external malleolus, may be felt the peroneal ridge, when this process is present. Farther forwards, the base of the fifth metatarsal bone forms a prominent and well-defined landmark, and in front of this the shaft of the bone, with its expanded head, and the base of the first phalanx, may be made out. The sole of the foot is almost entirely covered by soft parts, so that but few bony parts are to be made out, and these somewhat obscurely. The hinder part of the under surface of the os calcis and the heads of the metatarsal bones, with the exception of the first, which is concealed by the sesamoid bones, may be recognised.

Surgical Anatomy.—Considering the injuries to which the foot is subjected, it is surprising how seldom the tarsal bones are fractured. This is no doubt due to the fact that the tarsus is composed of a number of bones, articulated by a considerable extent of surface and joined together by very strong ligaments, which serve to break the force of violence applied to this part of the body. When fracture does occur, these bones being composed for the most part of a soft cancellous structure, covered only by a thin shell of compact tissue, are often extensively comminuted, especially as most of the fractures are produced by direct violence; and, having only a very scanty amount of soft parts over them, the fractures are very often compound, and amputation is often necessary.

When fracture occurs in the anterior group of tarsal bones, it is almost invariably the result of direct violence; but fractures of the posterior group—that is, of the calcaneum and astragalus—are usually produced by falls from a height on to the feet; though fracture of the os calcis may be caused by direct violence or by muscular action. The posterior part of the bone—that is, the part behind the articular surfaces—is the most common seat of the fracture, though some few cases of fracture of the sustentaculum tali and of vertical fracture between the two articulating facets have been recorded. The neck of the astragalus, being the weakest part of the bone, is most frequently fractured, though fractures may occur in any part and almost in any direction, either associated or not with fracture of other bones.

In cases of club-foot, especially in congenital cases, the bones of the tarsus become altered in shape and size, and displaced from their proper positions. This is principally the case in congenital equino-varus, in which the astragalus, particularly about the head, becomes twisted and atrophied, and a similar condition may be present in the other bones, more especially the navicular. The tarsal bones are peculiarly liable to become the seat of tuberculous caries from comparatively trivial injuries. There are several reasons to account for this. They are composed of a delicate cancellated structure, surrounded by intricate synovial membranes. They are situated at the farthest point from the central organ of the circulation and exposed to vicissitudes of temperature; and, moreover, on their dorsal surface are thinly clad with soft parts which have but a scanty blood-supply. And finally, after slight injuries, they are not maintained in a condition of rest to the same extent as some other parts of the body after similar injuries. Caries of the calcaneum or astragalus may remain limited to the one bone for a long period, but when one of the other bones is affected, the remainder frequently become involved, in consequence of the disease spreading through the large and complicated synovial membrane which is more or less common to these bones.

Amputation of the whole or a part of the foot is often required either for injury or disease. The principal amputations are as follows: (1) Syme's: amputation at the ankle-joint by a heel-flap, with removal of the malleoli and sometimes a thin slice from the lower end of the tibia. (2) Roux's: amputation at the ankle-joint by a large internal flap. (3) Pirogoff,s amputation: removal of the whole of the tarsal bones, except the posterior part of the os calcis, and a thin slice from the tibia and fibula including the two malleoli. The sawn surface of the os calcis is then turned up and united to the similar surface of the tibia. (4) Subastragaloid amputation: removal of the foot below the astragalus through the joint between it and the os calcis. This operation has been modified by Hancock, who leaves the posterior third of the os calcis and turns it up against the denuded surface of the astragalus. This latter operation is of doubtful utility, and is rarely performed.

(5) Chopart's or medio-tarsal: removal of the anterior part of the foot with all the tarsal bones except the os calcis and astragalus; disarticulation being effected through the astragalo-scaphoid and calcaneo-cuboid joints. (6) Lisfranc's: amputation of the anterior part of the foot through the tarso-metatarsal joints. This has been modified by Hey, who disarticulated through the joints of the four outer metatarsal bones with the tarsus, and sawed off the projecting internal cuneiform; and by Skey, who sawed off the base of the second metatarsal bone and disarticulated the others.

The bones of the tarsus occasionally require removal individually. . This is especially the case with the astragalus and os calcis for disease limited to the one bone, or again the astragalus may require excision in cases of subastragaloid dislocation, or, as recommended by Lund, in cases of inveterate talipes. The cuboid has been removed for the same reason by Solly. But both these two latter operations have fallen very much into disuse, and have been superseded by resection of a wedge-shaped piece of bone from the outer side of the tarsus. Finally, Mickulicz and Watson have devised operations for the removal of more extensive portions of the tarsus. Mickulicz's operation consists in the removal of the os calcis and astragalus, along with the articular surfaces of the tibia and fibula, and also of the scaphoid and cuboid. The remaining portion of the tarsus is then brought into contact with the sawn surfaces of the tibia and fibula, and fixed there. The result is a position of the shortened foot resembling talipes equinus. Watson's operation is adapted to those cases where the disease is confined to the anterior tarsal bones. By two lateral incisions he saws through the bases of the metatarsal bones in front and opens up the joints between the scaphoid and astragalus, and the cuboid and os calcis, and removes the intervening bones.

The metatarsal bones and phalanges are nearly always broken by direct violence, and in the majority of cases the injury is the result of severe crushing accidents, necessitating amputation. The metatarsal bones, and especially that of the great toe, are frequently diseased, either in tuberculous subjects or in perforating ulcer of the foot.

SESAMOID BONES

The sesamoid bones are small, more or less rounded masses, cartilaginous in early life, which are developed in those tendons which exert a great amount of pressure upon the parts over which they glide. It is said that they are more commonly found in the male than in the female, and in persons of an active muscular habit than in those who are weak and debilitated. They are invested throughout their whole surface by the fibrous tissue of the tendon in which they are found, excepting upon that side which lies in contact with the part over which they play, where they present a free articular facet. They may be divided into two kinds: those which glide over the articular surfaces of joints, and those which play over the cartilaginous facets found on the surfaces of certain bones.

The sesamoid bones of the joints in the upper extremity are two on the palmar surface of the metacarpo-phalangeal joint in the thumb, developed in the tendons of the Flexor brevis pollicis; of these the internal is the larger; sometimes one or two opposite the metacarpo-phalangeal articulations of the index and little fingers; and, still more rarely, one opposite the same joints of the third and fourth fingers, at the interphalangeal joint of the thumb and at the articulation between the second and third phalanges of the index finger. In the lower extremity they consist of the patella, which is developed in the tendon of the Quadriceps extensor muscle; two small sesamoid bones, of which the inner is the larger, found in the tendons of the Flexor brevis hallucis, opposite the metatarso-phalangeal joint of the great toe; occasionally one at the metatarso-phalangeal joints of the second and fifth toes, and, still more rarely, at the corresponding joints of the third and fourth toes, or at the interphalangeal joint of the great toe.

Those found in the tendons which glide over certain bones occupy the following situations: one sometimes found in the tendon of the Biceps cubiti, opposite the tuberosity of the radius; one in the tendon of the Peroneus longus, where it glides through the groove in the cuboid bone; one, appearing late in life, in the tendon of the Tibialis anticus, opposite the smooth facet of the internal cuneiform bone; one is found in the tendon of the Tibialis posticus, opposite the inner side of the head of the astragalus; one in the outer head of the Gastrocnemius, behind the outer condyle of the femur; and one in the conjoined tendon of the Psoas and Iliacus, where it glides over the os pubis. Sesamoid bones are found occasionally in the tendon of the Gluteus maximus, as it passes over the great trochanter; and in the tendons which wind round the inner and outer malleoli.

THE ARTICULATIONS

THE various bones of the Skeleton are connected together at different parts of their surfaces, and such a connection is designated by the name of *Joint* or *Articulation*. If the joint is *immovable*, as between the cranial and most of the facial bones, the adjacent margins of the bones are applied in almost close contact, a thin layer of fibrous membrane, the *sutural ligament*, and, at the base of the skull, in certain situations, a thin layer of cartilage being interposed. Where slight movement is required, combined with great strength, the osseous surfaces are united by tough and elastic fibro-cartilages, as in the joints between the vertebral bodies and in the interpubic articulation ; but in the *movable* joints, the bones forming the articulation are generally expanded for greater convenience of mutual connection, covered by *cartilage*, held together by strong bands or capsules of fibrous tissue, called *ligaments*, and partially lined by a membrane, the *synovial membrane*, which secretes a fluid to lubricate the various parts of which the joint is formed : so that the structures which enter into the formation of a movable joint are bone, cartilage, fibro-cartilage, ligament, and synovial membrane.

Bone constitutes the fundamental element of all the joints. In the long bones, the extremities are the parts which form the articulations ; they are generally somewhat enlarged, and consist of spongy cancellous tissue with a thin coating of compact substance. In the flat bones, the articulations usually take place at the edges ; and in the short bones at various parts of their surface. The layer of compact bone which forms the joint surface, and to which the articular cartilage is attached, is called the *articular lamella*. It is white, extremely dense, and varies in thickness. Its structure differs from ordinary bone-tissue in this respect, that it contains no Haversian canals, and its lacunæ are larger than in ordinary bone, and have no canaliculi. The vessels of the cancellous tissue, as they approach the articular lamella, turn back in loops, and do not perforate it ; this layer is consequently denser and firmer than ordinary bone, and is evidently designed to form an unyielding support for the articular cartilage.

The **articular cartilage**, which covers the articular surfaces of bone, is of the hyaline variety, and has been described in the section on General Anatomy.

Ligaments, properly so called, are peculiar to the movable joints, and serve to connect together the articular surfaces of bones. They are composed mainly of bundles of *white fibrous tissue* placed parallel with, or closely interlaced with one another, and presenting a white, shining, silvery aspect. In the freely movable joints, or diarthrodia, they form envelopes or capsules connecting together the articular extremities of bones. These capsules in particular places undergo thickening and form strong ligamentous bands, which, though strictly speaking constituent parts of the capsule, are usually described as distinct ligaments. A ligament is pliant and flexible, so as to allow of the most perfect freedom of movement, but strong, tough, and inextensile, so as not to yield readily under the most severely applied force ; it is consequently well adapted to serve as the connecting medium between the bones. Some ligaments consist entirely of *yellow elastic tissue*, as the ligamenta subflava, which connect together the laminæ of adjacent vertebræ, and the ligamentum nuchæ in the lower animals.

In these cases it will be observed that the elasticity of the ligament is intended to act as a substitute for muscular power.

Synovial membrane is composed of a thin, delicate connective tissue, with branched connective-tissue corpuscles. Its secretion is thick, viscid, and glairy, like the white of egg; and is hence termed *synovia*. The synovial membranes in the body admit of subdivision into three kinds—articular, bursal, and vaginal.

The *articular synovial membranes* are found in all the freely movable joints. The membrane invests the inner surface of the capsule enclosing the joint, and is reflected over the surface of any tendons passing through its cavity, as the tendon of the Popliteus in the knee, and the tendon of the Biceps in the shoulder. Hence the articular synovial membrane may be regarded as a short wide tube, attached by its open ends to the margins of the articular cartilages and covering the inner surface of the various ligaments which connect the articular surfaces, so that along with the cartilages it completely encloses the joint cavity. In the fœtus this membrane is said, by Toynbee, to be continued over the surface of the cartilages; but in the adult it is wanting, excepting at their circumference, upon which it encroaches for a short distance and to which it is firmly attached. In some of the joints the synovial membrane is thrown into folds, which pass across the cavity. They are called *synovial ligaments*, and are especially distinct in the knee. In other joints there are flattened folds, subdivided at their margins into fringe-like processes, the vessels of which have a convoluted arrangement. These latter generally project from the synovial membrane near the margin of the cartilage, and lie flat upon its surface. They consist of connective tissue, covered with endothelium, and contain fat-cells in variable quantities, and, more rarely, isolated cartilage-cells. The larger folds often contain considerable quantities of fat. They were described, by Clopton Havers, as *mucilaginous glands*, and as the source of the synovial secretion. Under certain diseased conditions, similar processes are found covering the entire surface of the synovial membrane, forming a mass of pedunculated fibro-fatty growths, which projects into the joint. Similar structures are also found in some of the bursal and vaginal synovial membranes.

The *bursal synovial membranes* are found interposed between surfaces which move upon each other, as in the gliding of a tendon, or of the integument over projecting bony surfaces. They admit of subdivision into two kinds, the *bursæ mucosæ* and the *bursæ synoviæ*. The *bursæ mucosæ* are large, simple, or irregular cavities in the subcutaneous areolar tissue, enclosing a clear viscid fluid. They are found in various situations, as between the integument and the front of the patella, over the olecranon, the malleoli, and other prominent parts. The *bursæ synoviæ* are found interposed between muscles or tendons as they play over projecting bony surfaces, as between the Glutei muscles and the surface of the great trochanter. They consist of a thin wall of connective tissue, partially covered by patches of cells, and contain a viscid fluid. Where one of these exists in the neighbourhood of a joint, it may communicate with its cavity, as in the case of the bursa between the tendon of the Psoas and Iliacus and the capsular ligament of the hip, or that interposed between the deep surface of the Subscapularis and the capsular ligament of the shoulder-joint.

The *vaginal synovial membranes* (*synovial sheaths*) serve to facilitate the gliding of tendons in the osseo-fibrous canals through which they pass. The membrane is here arranged in the form of a sheath, one layer of which adheres to the wall of the canal, and the other is reflected upon the surface of the enclosed tendon; the space between the two free surfaces of the membrane containing synovia. These sheaths are chiefly found surrounding the tendons of the flexor and extensor muscles of the fingers and toes, as they pass through the osseo-fibrous canals in the hand or foot.

Synovia is a transparent, yellowish-white, or slightly reddish fluid, viscid like the white of egg, having an alkaline reaction and slightly saline taste. It consists, according to Frerichs, in the ox, of 94·85 water, 0·56 mucus and epithelium, 0·07 fat, 3·51 albumen and extractive matter, and 0·99 salts.

CLASSIFICATION OF JOINTS

The articulations are divided into three classes : *synarthrosis*, or immovable ; *amphiarthrosis*, or mixed ; and *diarthrosis*, or movable joints.

1. Synarthrosis. Immovable Articulations

Synarthrosis includes all those articulations in which the surfaces of the bones are in almost direct contact, fastened together by an intervening mass of connective tissue or hyaline cartilage, and in which there is no appreciable motion, as the joints between the bones of the cranium and face, excepting those of the lower jaw. The varieties of synarthrosis are four in number : Sutura, Schindylesis, Gomphosis, and Synchrondrosis.

Sutura (*a seam*) is that form of articulation where the contiguous margins of flat bones are united by a thin layer of fibrous tissue. It is met with only in the skull. When the articulating surfaces are connected by a series of processes and indentations interlocked together, it is termed a *true suture* (*sutura vera*) ; of which there are three varieties : sutura dentata, serrata, and limbosa. The surfaces of the bones are not in direct contact, being separated by a layer of membrane, continuous externally with the pericranium, internally with the dura mater. The *sutura dentata* (dens, *a tooth*) is so called from the tooth-like form of the projecting articular processes, as in the suture between the parietal bones. In the *sutura serrata* (serra, *a saw*) the edges of the two bones forming the articulation are serrated like the teeth of a fine saw, as between the two portions of the frontal bone. In the *sutura limbosa* (limbus, *a selvage*), besides the dentated processes, there is a certain degree of bevelling of the articular surfaces, so that the bones overlap one another, as in the suture between the parietal and frontal bones. When the articulation is formed by roughened surfaces placed in apposition with one another, it is termed a *false suture* (*sutura notha*), of which there are two kinds, the *sutura squamosa* (squama, *a scale*), formed by the overlapping of two contiguous bones by broad bevelled margins, as in the squamo-parietal (squamous) suture ; and the *sutura harmonia* (ἁρμονία, *a joining together*), where there is simple apposition of two contiguous rough bony surfaces, as in the articulation between the two superior maxillary bones, or of the horizontal plates of the palate bones.

Schindylesis (σχινδύλησις, *a fissure*) is that form of articulation in which a thin plate of bone is received into a cleft or fissure formed by the separation of two laminæ in another bone, as in the articulation of the rostrum of the sphenoid and perpendicular plate of the ethmoid with the vomer, or in the reception of the latter in the fissure between the superior maxillary and palate bones.

Gomphosis (γόμφος, *a nail*) is an articulation formed by the insertion of a conical process into a socket, as a nail is driven into a board ; this is not illustrated by any articulation between bones, properly so called, but is seen in the articulation of the teeth with the alveoli of the maxillary bones.

Synchondrosis.—Where the connecting medium is cartilage the joint is termed a synchondrosis. This is a temporary form of joint, for the cartilage becomes converted into bone before adult life. Such a joint is found between the epiphyses and shafts of long bones, and in the junction between the occipital bone and the sphenoid and between the petrous portion of the temporal bone and the jugular process of the occipital bone.

2. Amphiarthrosis. Mixed Articulations

In this form of articulation only a slight amount of movement is possible, the contiguous osseous surfaces being either connected together by broad flattened discs of fibro-cartilage, of a more or less complex structure, which adhere to each bone, as in the articulation between the bodies of the vertebræ, and in the pubic symphysis. This is termed **Symphysis.** Or, secondly, the bony surfaces are united by an interosseous ligament, as in the inferior tibio-fibular articulation. To this the term **Syndesmosis** is applied.

3. DIARTHROSIS. MOVABLE ARTICULATIONS

This form of articulation includes the greater number of the joints in the body, mobility being their distinguishing character. They are formed by the approximation of two contiguous bony surfaces, covered with cartilage, connected by ligaments, and lined by synovial membrane. The varieties of joints in this class have been determined by the kind of motion permitted in each. There are two varieties in which the movement is uniaxial; that is to say, all movements take place around one axis. In one form, the Ginglymus, this axis is practically speaking transverse; in the other, the Trochoid or pivot-joint, it is longitudinal. There are two varieties where the movement is biaxial, or around two horizontal axes at right angles to each other, or at any intervening axis between the two. These are the Condyloid and saddle-joint. There is one form of joint where the movement is polyaxial, the Enarthrosis or ball-and-socket joint. And finally there are the Arthrodia or gliding joints.

Ginglymus or **Hinge-joint** (γίγγλυμος, *a hinge*).—In this form of joint the articular surfaces are moulded to each other in such a manner as to permit motion only in one plane, forwards and backwards; the extent of motion at the same time being considerable. The direction which the distal bone takes in this motion is never in the same plane as that of the axis of the proximal bone, but there is always a certain amount of alteration from the straight line during flexion. The articular surfaces are connected together by strong lateral ligaments, which form their chief bond of union. The most perfect forms of ginglymus are the interphalangeal joints and the joint between the humerus and ulna; the knee and ankle are less perfect, as they allow a slight degree of rotation or lateral movement in certain positions of the limb.

Trochoid (*pivot-joint*).—Where the movement is limited to rotation, the joint is formed by a pivot-like process turning within a ring, or the ring on the pivot, the ring being formed partially of bone, partly of ligament. In the superior radio-ulnar articulation, the ring is formed partly by the lesser sigmoid cavity of the ulna; in the rest of its extent, by the orbicular ligament; here, the head of the radius rotates within the ring. In the articulation of the odontoid process of the axis with the atlas, the ring is formed in front by the anterior arch of the atlas; behind, by the transverse ligament; here, the ring rotates round the odontoid process.

Condyloid Articulation.—In this form of joint, an ovoid articular head, or condyle, is received into an elliptical cavity in such a manner as to permit of flexion and extension, adduction and abduction and circumduction, but no axial rotation. The articular surfaces are connected together by anterior, posterior, and lateral ligaments. An example of this form of joint is found in the wrist.

Articulation by Reciprocal Reception (*saddle-joint*).—In this variety the articular surfaces are concavo-convex; that is to say, they are inversely convex in one direction and concave in the other. The movements are the same as in the preceding form; that is to say, there is flexion, extension, adduction, abduction, and circumduction, but no axial rotation. The articular surfaces are connected by a capsular ligament. The best example of this form of joint is the carpo-metacarpal joint of the thumb.

Enarthrosis is that form of joint in which the distal bone is capable of motion around an indefinite number of axes, which have one common centre. It is formed by the reception of a globular head into a deep cup-like cavity (hence the name 'ball-and-socket'), the parts being kept in apposition by a capsular ligament strengthened by accessory ligamentous bands. Examples of this form of articulation are found in the hip and shoulder.

Arthrodia is that form of joint which admits of a gliding movement; it is formed by the approximation of plane surfaces, or one slightly concave, the other slightly convex; the amount of motion between them being limited by the ligaments, or osseous processes, surrounding the articulation; as in the articular processes of the vertebræ, the carpal joints, except that of the os magnum with the scaphoid and semilunar bones, and the tarsal joints with the exception of the joint between the astragalus and the navicular.

On the next page, in a tabular form, are the names, distinctive characters, and examples of the different kinds of articulations.

Synarthrosis, or Immovable Joint. Surfaces separated by fibrous membrane or cartilage, without any intervening synovial cavity, and immovably connected with each other. As in joints of cranium and face (except the lower jaw).

Sutura. Articulation by processes and indentations interlocked together.

Sutura vera (true) articulate by indented borders.

Dentata, having tooth-like processes. As in interparietal suture.

Serrata, having serrated edges like the teeth of a saw. As in interfrontal suture.

Limbosa, having bevelled margins, and dentated processes. As in fronto-parietal suture.

Sutura notha (false) articulate by rough surfaces.

Squamosa, formed by thin bevelled margins, overlapping each other. As in squamo-parietal suture.

Harmonia, formed by the apposition of contiguous rough surfaces. As in intermaxillary suture.

Schindylesis.—Articulation formed by the reception of a thin plate of one bone into a fissure of another. As in articulation of rostrum of sphenoid with vomer.

Gomphosis.—Articulation formed by the insertion of a conical process into a socket. As in the teeth.

Synchondrosis.—When the connecting medium is cartilage, as in the occipito-sphenoid joint.

Amphiarthrosis, Mixed Articulation.

Symphysis.—Surfaces connected by fibro-cartilage, not separated by synovial membrane, and having limited motion. As in joints between bodies of vertebræ.

Syndesmosis.—Surfaces united by an interosseous ligament. As in the inferior tibio-fibular articulation.

Diarthrosis, Movable Joint.

Ginglymus.—Hinge-joint; motion limited to two directions, forwards and backwards. Articular surfaces fitted together so as to permit of movement in one plane. As in the interphalangeal joints and the joint between the humerus and the ulna.

Trochoides or Pivot-joint.—Articulation by a pivot process turning within a ring, or ring around a pivot. As in superior radio-ulnar articulation, and atlanto-axial joint.

Condyloid.—Ovoid head received into elliptical cavity. Movements in every direction except axial rotation. As the wrist-joint.

Reciprocal Reception (Saddle-joint).—Articular surfaces inversely convex in one direction and concave in the other. Movement in every direction except axial rotation. As in the carpo-metacarpal joint of the thumb.

Enarthrosis.—Ball-and-socket joint; capable of motion in all directions. Articulations by a globular head received into a cup-like cavity. As in hip- and shoulder-joints.

Arthrodia.—Gliding-joint; articulations by plane surfaces, which glide upon each other. As in carpal and tarsal articulations.

The Kinds of Movement admitted in Joints

The movements admissible in joints may be divided into four kinds : gliding, angular, circumduction, and rotation. These movements are often, however, more or less combined in the various joints, so as to produce an infinite variety, and it is seldom that we find only one kind of motion in any particular joint.

Gliding movement is the simplest kind of motion that can take place in a joint, one surface gliding or moving over another without any angular or rotatory movement. It is common to all movable joints; but in some, as in most of the articulations of the carpus and tarsus, it is the only motion permitted. This movement is not confined to plane surfaces, but may exist between any two contiguous surfaces, of whatever form, limited by the ligaments which enclose the articulation.

Angular movement occurs only between the long bones, and by it the angle between the two bones is increased or diminished. It may take place in four directions: forwards and backwards, constituting flexion and extension, or inwards and outwards, from the mesial line of the body (or in the fingers and toes from the middle line of the hand or foot), constituting adduction and abduction. The strictly ginglymoid or hinge-joints admit of flexion and extension only. Abduction and adduction, combined with flexion and extension, are met with in the more movable joints; as in the hip, shoulder, and metacarpal joint of the thumb, and partially in the wrist.

Circumduction is that degree of motion which takes place between the head of a bone and its articular cavity, while the limb is made to circumscribe a conical space, the base of which corresponds with the inferior extremity of the limb, the apex with the articular cavity; this kind of motion is best seen in the shoulder- and hip-joints.

Rotation is the movement of a bone around an axis; the latter may be formed by a separate bone, as in the case of the pivot formed by the odontoid process of the axis around which the atlas turns; or a bone may rotate around its own longitudinal axis, as in the rotation of the humerus and femur at the shoulder- and hip-joints respectively; or the axis of rotation may not be quite parallel to the long axis of the bone, as in the movement of the radius on the ulna during pronation and supination of the hand, where the axis of rotation is represented by a line connecting the centre of the head of the radius above with the centre of the head of the ulna below.

Ligamentous Action of Muscles.—The movements of the different joints of a limb are combined by means of the long muscles which pass over more than one joint, and which, when relaxed and stretched to their greatest extent, act as elastic ligaments in restraining certain movements of one joint, except when combined with corresponding movements of the other—these latter movements being usually in the opposite direction. Thus the shortness of the hamstring muscles prevents complete flexion of the hip, unless the knee-joint is also flexed, so as to bring their attachments nearer together. The uses of this arrangement are threefold. 1. It co-ordinates the kinds of movement which are the most habitual and necessary, and enables them to be performed with the least expenditure of power. 'Thus in the usual gesture of the arms, whether in grasping or rejecting, the shoulder and the elbow are flexed simultaneously, and simultaneously extended,' in consequence of the passage of the Biceps and Triceps cubiti over both joints. 2. It enables the short muscles which pass over only one joint to act upon more than one. 'Thus, if the Rectus femoris remain tonically of such length that, when stretched over the extended hip, it compels extension of the knee, then the Gluteus maximus becomes, not only an extensor of the hip, but an extensor of the knee as well.' 3. It provides the joints with ligaments which, while they are of very great power in resisting movements to an extent incompatible with the mechanism of the joint, at the same time spontaneously yield when necessary. 'Taxed beyond its strength a ligament will be ruptured, whereas a contracted muscle is easily relaxed; also, if neighbouring joints be united by ligaments, the amount of flexion or extension of each must remain in constant proportion to that of the other; while, if the union be by muscles, the separation of the points of attachment of those muscles may vary considerably in different varieties of movement, the muscles adapting themselves tonically to the length required.' The quotations are from a very interesting paper by Cleland in the 'Journal of Anatomy and Physiology,' No. 1, 1866, p. 85; by whom this important fact in the mechanism of joints was first clearly pointed out, though it was independently observed afterwards by other anatomists. W. W. Keen points out how important it is 'that the surgeon should remember this ligamentous action of muscles in making passive motion— for instance, at the wrist after Colles's fracture. If the fingers be extended, the wrist can be flexed to a right angle. If, however, they be first flexed as in "making a fist," flexion at the wrist is quickly limited to from forty to fifty degrees in different persons, and is very painful beyond that point. Hence passive motion here should be made with the fingers extended. In the leg, when flexing the hip, the knee should be flexed.' Keen further points out that 'a beautiful illustration of this is seen in the perching of birds, whose toes are forced to clasp the perch by just such a passive ligamentous action so soon as they stoop. Hence they can go to sleep and not fall off the perch.'

The articulations may be grouped into those of the trunk, those of the upper extremity, and those of the lower extremity.

ARTICULATIONS OF THE TRUNK

These may be divided into the following groups, viz.:

I. Of the vertebral column.
II. Of the atlas with the axis.
III. Of the spine with the cranium.
IV. Of the lower jaw.
V. Of the ribs with the vertebræ.

VI. Of the cartilages of the ribs with the sternum, and with each other.
VII. Of the sternum.
VIII. Of the vertebral column with the pelvis.
IX. Of the pelvis.

I. ARTICULATIONS OF THE VERTEBRAL COLUMN

There are two distinct varieties of articulation in the vertebral column: 1. Those between the bodies of the vertebræ, which form a series of amphiarthrodial joints; these are termed the *intercentral joints*. 2. Those between the articular process, which form a series of arthrodial joints, and are termed *interneural*.

1. INTERCENTRAL ARTICULATIONS

The intercentral articulations, or articulations between the bodies of the vertebræ, belong to the class of amphiarthrodial joints, and the individual vertebræ enjoy very little movement on each other. When, however, this slight degree of movement between the pairs of bones takes place in all the joints of the spine, the range of movement is very considerable. The ligaments of this articulation are the following:

Anterior Common Ligament.   Posterior Common Ligament.
Intervertebral Substance.

The **Anterior Common Ligament** (figs. 346, 347, 354, 358) is a broad and strong band of fibres, which extends along the anterior surface of the bodies of the vertebræ, from the axis to the sacrum. It is broader below than above, thicker in the dorsal than in the cervical or lumbar regions, and somewhat thicker opposite the front of the body of each vertebra than opposite the intervertebral substance. It is attached, above, to the body of the axis, where it is continuous with the anterior atlanto-axial ligament, and from this is continued upwards as a rounded cord, superficial to the anterior atlanto-axial ligament, to the under surface of the basilar process of the occipital bone. Here it is connected with the tendon of insertion of the Longus colli muscle. It extends down as far as the upper bone of the sacrum. It consists of dense longitudinal fibres, which are intimately adherent to the intervertebral substance, and the prominent margins of the vertebræ; but less closely to the middle parts of the bodies. In the latter situation the fibres are exceedingly thick, and serve to fill up the concavities on their front surface, and to make the anterior surface of the spine more even. This ligament is composed of several layers of fibres, which vary in length, but are closely interlaced with each other. The most superficial or longest fibres extend between four or five vertebræ. A second, subjacent set extend between two or three vertebræ; while a third set, the shortest and deepest, extend from one vertebra to the next. At the sides of the bodies the ligament consists of a few short fibres, which pass from one vertebra to the next, separated from the median portion by large oval apertures, for the passage of vessels.

The **Posterior Common Ligament** (figs. 346, 350) is situated within the spinal canal, and extends along the posterior surface of the bodies of the vertebræ, from the body of the axis, where it is continuous with the occipito-axial ligament, to the sacrum. It is broader above than below, and thicker in the dorsal than in the cervical or lumbar regions. In the situation of the intervertebral substance and contiguous margins of the vertebræ, where the ligament is more intimately adherent, it is broad, and presents a series of dentations with intervening concave margins; but it is narrow and thick over the centre of the bodies, from which it is separated by the *venæ basis vertebræ*. This ligament is composed of smooth, shining, longitudinal fibres, denser and more compact than those of the anterior ligament, and consists of a superficial layer occupying the interval between three or four vertebræ, and of a deeper layer which extends between one vertebra and the next adjacent to it. It is separated from the dura mater of the spinal cord by some loose connective tissue which is very liable to serous infiltration.

The **Intervertebral Substance** (figs. 346, 355) is a lenticular disc of composite structure interposed between the adjacent surfaces of the bodies of the vertebræ, from the axis to the sacrum, and forming the chief bond of connection between

FIG. 346.—Vertical section of two vertebræ and their ligaments, from the lumbar region.

these bones. The discs vary in shape, size, and thickness, in different parts of the spine. In *shape* they correspond with the surfaces of the bodies between which they are placed, except in the cervical region, where they are slightly smaller from side to side than the corresponding bodies. They are oval in the cervical and lumbar regions, and circular in the dorsal. Their *size* is greatest in the lumbar region. In *thickness* they vary not only in the different regions of the spine, but in different parts of the same disc: they are thicker in front than behind in the cervical and lumbar regions, and thus contribute to the anterior convexity of the column in these regions; while they are of nearly uniform thickness in the dorsal region, the anterior concavity in this region being almost entirely owing to the shape of the vertebral bodies. The intervertebral discs form about one-fourth of the spinal column, exclusive of the first two vertebræ; but this amount is not equally distributed between the various bones; the dorsal portion of the spine having, in proportion to its length, a much smaller amount than the cervical and lumbar regions, which necessarily gives to the latter parts greater pliancy and freedom of movement. The intervertebral discs are adherent, by their surfaces, to a thin layer of hyaline cartilage which covers

A A

the upper and under surfaces of the bodies of the vertebræ, and in which, in early life, the epiphysial plate develops; and by their circumference are closely connected in front to the anterior, and behind to the posterior common ligament; while in the dorsal region they are joined laterally, by means of the interarticular ligament, to the heads of those ribs which articulate with two vertebræ: they consequently form part of the articular cavities in which the heads of these bones are received.

Structure of the Intervertebral Substance.—The intervertebral substance is composed, at its circumference, of laminæ of fibrous tissue and fibro-cartilage, forming the *annulus fibrosus*; and, at its centre, of a soft, pulpy, highly elastic substance, of a yellowish colour, which rises up considerably above the surrounding level when the disc is divided horizontally. This pulpy substance (*nucleus pulposus*) is especially well developed in the lumbar region, and is the remains of the chorda dorsalis. The laminæ are arranged concentrically one within the other, the outermost consisting of ordinary fibrous tissue, but the others and more numerous consist of white fibro-cartilage. These plates are not quite vertical in their direction, those near the circumference being curved outwards and closely approximated; while those nearest the centre curve in the opposite direction, and are somewhat more widely separated. The fibres of which each plate is composed are directed, for the most part, obliquely from above downwards; the fibres of adjacent plates passing in opposite directions and varying in every layer: so that the fibres of one layer are directed across those of another, like the limbs of the letter X. This laminar arrangement belongs to about the outer half of each disc. The pulpy substance presents no concentric arrangement, and consists of a fine fibrous matrix, containing angular cells, united to form a reticular structure.

2. Interneural Articulations

The interneural articulations, or articulations between the articular processes of the vertebræ, belong to the arthrodial variety of movable joints. The processes are connected together by capsular ligaments, and the joints are lined by synovial membranes; but in addition to these are a number of accessory ligaments, which connect together the laminæ, spinous and transverse processes. The ligaments of the interneural articulations are as follows:

Capsular.	Supraspinous.
Ligamenta subflava.	Interspinous.
Intertransverse.	

The **Capsular Ligaments** (fig. 348) are thin and loose ligamentous sacs, attached to the contiguous margins of the articulating processes of each vertebra, through the greater part of their circumference, and completed internally by the ligamenta subflava. They are longer and looser in the cervical than in the dorsal or lumbar regions. The capsular ligaments are lined on their inner surface by **synovial membrane**.

The **Ligamenta Subflava** (fig. 346) are interposed between the laminæ of the vertebræ, from the axis to the first segment of the sacrum. They are most distinct when seen from the interior of the spinal canal: when viewed from the outer surface they appear short, being overlapped by the laminæ. Each ligament consists of two lateral portions, which commence on each side at the root of the articular process, and pass backwards to the point where the laminæ converge to form the spinous process, where their margins are in contact and to a certain extent united; slight intervals being left for the passage of small vessels. These ligaments consist of yellow elastic tissue, the fibres of which, almost perpendicular in direction, are attached to the anterior surface of the lamina above, some distance from its inferior margin, and to the posterior surface, as well as to the upper margin of the lamina below. In the cervical region they are thin in texture, but very broad and long; they become thicker in the dorsal region; and in the lumbar acquire very considerable thickness. Their highly elastic property serves to preserve the upright posture, and to assist the spine in resuming it, after it has been flexed. These ligaments do not exist between the occiput and atlas, or between the atlas and axis.

The **Supraspinous Ligament** (fig. 346) is a strong fibrous cord, which connects together the apices of the spinous processes from the seventh cervical to the spinous processes of the sacrum. It is thicker and broader in the lumbar than in the dorsal region, and intimately blended, in both situations, with the neighbouring aponeurosis. The most superficial fibres of this ligament connect three or four vertebræ; those more deeply seated pass between two or three vertebræ; while the deepest connect the contiguous extremities of neighbouring vertebræ. It is continued upwards to the external occipital protuberance and occipital crest, as the ligamentum nuchæ, which, in the human subject, is thin, and forms merely an intermuscular septum.

The **Ligamentum nuchæ** is a fibrous membrane, which, in the neck, represents the supraspinous ligaments of the lower vertebræ. It extends from the external occipital protuberance and crest to the spinous process of the seventh cervical vertebra. From its anterior border a fibrous lamina is given off, which is attached to the posterior tubercle of the atlas, and the spinous process of each of the cervical vertebræ, so as to form a septum between the muscles on each side of the neck. In man it is merely the rudiment of an important elastic ligament, which, in some of the lower animals, serves to sustain the weight of the head.

The **Interspinous Ligaments** (fig. 346), thin and membranous, are interposed between the spinous processes. These ligaments extend from the root to the summit of each spinous process, connecting together their adjacent margins. They meet the ligamenta subflava in front and the supraspinous ligament behind. They are narrow and elongated in the dorsal region; broader, quadrilateral in form, and thicker in the lumbar region; and only slightly developed in the neck.

The **Intertransverse Ligaments** consist of bundles of fibres, interposed between the transverse processes. In the cervical region they consist of a few irregular, scattered fibres; in the dorsal, they are rounded cords intimately connected with the deep muscles of the back; in the lumbar region they are thin and membranous.

Actions.—The movements permitted in the spinal column are, Flexion, Extension, Lateral movement, Circumduction, and Rotation.

In *Flexion*, or movement of the spine forwards, the anterior common ligament is relaxed, and the intervertebral substances are compressed in front; while the posterior common ligament, the ligamenta subflava, and the inter- and supraspinous ligaments are stretched, as well as the posterior fibres of the intervertebral discs. The interspaces between the laminæ are widened, and the inferior articular processes of the vertebræ above glide upwards, upon the articular processes of the vertebræ below. Flexion is the most extensive of all the movements of the spine.

In *Extension*, or movement of the spine backwards, an exactly opposite disposition of the parts takes place. This movement is not extensive, being limited by the anterior common ligament, and by the approximation of the spinous processes.

Flexion and extension are free in the lower part of the lumbar region between the third and fourth and fourth and fifth lumbar vertebræ; above the third they are much diminished, and reach their minimum in the middle and upper part of the back. They increase again in the neck, the capability of motion backwards from the upright position being in this region greater than that of the motion forwards, whereas in the lumbar region the reverse is the case.

In *Lateral Movement*, the sides of the intervertebral discs are compressed, the extent of motion being limited by the resistance offered by the surrounding ligaments, and by the approximation of the transverse processes. This movement may take place in any part of the spine, but is freest in the neck and loins.

Circumduction is very limited, and is produced merely by a succession of the preceding movements.

Rotation is produced by the twisting of the intervertebral substances; this, although only slight between any two vertebræ, produces a considerable extent of movement, when it takes place in the whole length of the spine, the front of the upper part of the column being turned to one or the other side. This movement occurs only to a slight extent in the neck, but is freer in the upper part of the dorsal region, and is altogether absent in the lumbar region.

It is thus seen that the *cervical region* enjoys the greatest extent of each variety of movement, flexion and extension especially being very free. In the *dorsal region*, the three movements of flexion, extension, and circumduction are only permitted to a slight extent; while rotation is very free in the upper part and ceases below. In the lumbar region there is free flexion, extension, and lateral movement, but no rotation.

As Humphry has pointed out, the movements permitted are mainly due to the shape and position of the articular processes. In the loins the inferior articular processes are turned outwards and embraced by the superior; this renders rotation in this region of the spine impossible, while there is nothing to prevent a sliding upwards and downwards of the surfaces on each other so as to allow of flexion and extension. In the dorsal region, on the other hand, the articulating processes, by their direction and mutual adaptation, especially at the upper part of the series, permit of rotation, but prevent extension and flexion; while in the cervical region the greater obliquity and lateral slant of the articular processes allow not only flexion and extension, but also rotation.

The principal muscles which produce *flexion* are the Sterno-mastoid, Rectus capitis anticus major, and Longus colli; the Scaleni; the abdominal muscles and the Psoas magnus. *Extension* is produced by the fourth layer of the muscles of the back, assisted in the neck by the Splenius, Semispinalis dorsi et colli, and the Multifidus spinæ. *Lateral* motion is produced by the fourth layer of the muscles of the back, by the Splenius and the Scaleni, the muscles of one side only acting; and *rotation* by the action of the following muscles of one side only, viz. the Sterno-mastoid, the Rectus capitis anticus major, the Scaleni, the Multifidus spinæ, the Complexus, and the abdominal muscles.

II. ARTICULATION OF THE ATLAS WITH THE AXIS

The articulation of the Atlas with the Axis is of a complicated nature, comprising no fewer than four distinct joints. There is a pivot articulation between the odontoid process of the axis and the ring formed between the anterior arch of the atlas and the transverse ligament (see fig. 349). Here there are two joints: one in front between the posterior surface of the anterior arch of the atlas and the front of the odontoid process; the other between the anterior surface of the transverse ligament and the back of the process. Between the articular processes of the two bones there is on either side an arthrodial or gliding joint. The ligaments which connect these bones are, the

Two Capsular.	Posterior Atlanto-axial.
Anterior Atlanto-axial.	Transverse.

The **Capsular Ligaments** are two thin and loose capsules, connecting the lateral masses of the atlas with the superior articular surfaces of the axis, the fibres being strengthened at the posterior and inner part of the articulation by a distinct ligamentous band, the *accessory ligament*, which is attached below to the body of the axis near the base of the odontoid process.

The **Anterior Atlanto-axial Ligament** (fig. 347) is a strong membranous layer, fixed, above, to the lower border of the anterior arch of the atlas; below, to the front of the body of the axis. It is strengthened in the middle line by a rounded cord, which is connected, above, to the tubercle on the anterior arch of the atlas, and below to the body of the axis, being a continuation upwards of the anterior common ligament of the spine. These ligaments are in relation, in front, with the Recti antici majores.

The **Posterior Atlanto-axial Ligament** (fig. 348) is a broad and thin membranous layer, attached, above, to the lower border of the posterior arch of the atlas; below, to the upper edge of the laminæ of the axis. This ligament supplies the place of the ligamenta subflava, and is in relation, behind, with the Inferior oblique muscles.

The **Transverse Ligament** * (figs. 349, 350) is a thick, strong band, which arches across the ring of the atlas, and serves to retain the odontoid process in firm connection with its anterior arch. This ligament is flattened from before backwards, broader and thicker in the middle than at either extremity, and

---

* It has been found necessary to describe the transverse ligament with those of the atlas and axis; but the student must remember that it is really a portion of the mechanism by which the movements of the head on the spine are regulated; so that the connections between the atlas and axis ought always to be studied in association with those between the latter bones and the skull.

firmly attached on each side to a small tubercle on the inner surface of the lateral mass of the atlas. As it crosses the odontoid process, a small fasciculus (*superior crus*) is prolonged upwards, and another (*inferior crus*) downwards, from the superficial or posterior fibres of the ligament: the former is inserted into the basilar process of the occipital bone, in close relation with the occipito-axial ligament; the latter descends, to be attached to the posterior surface of the body of the axis; hence, the whole ligament has received the name of *cruciform*. The transverse ligament divides the ring of the atlas into two unequal parts: of these, the posterior and larger serves for the transmission of the spinal cord and its membranes and the spinal accessory nerves; the anterior and smaller contains the odontoid process. Since the space between the anterior arch of the atlas and the transverse ligament is smaller at the lower part than the upper (because the transverse ligament embraces firmly the narrow neck of the odontoid process), this process is retained in firm connection with the atlas after all the other ligaments have been divided.

FIG. 347.— Occipito-atlantal and atlanto-axial ligaments.   Front view.

There are *four* **Synovial Membranes** in this articulation: one lining the inner surface of each of the capsular ligaments; one between the anterior surface of the odontoid process and the anterior arch of the atlas, and one between the posterior surface of the odontoid process and the transverse ligament. The latter often communicates with those between the condyles of the occipital bone and the articular surfaces of the atlas.

Actions.—This joint allows the rotation of the atlas (and, with it, of the cranium) upon the axis, the extent of rotation being limited by the odontoid ligaments.

The principal muscles by which this action is produced are the Sterno-mastoid and Complexus of one side, acting with the Rectus capitis anticus major, Splenius, Trachelo-mastoid, Rectus capitis posticus major, and Inferior oblique of the other side.

III. ARTICULATIONS OF THE SPINE WITH THE CRANIUM

The ligaments connecting the spine with the cranium may be divided into two sets, those connecting the occipital bone with the atlas, and those connecting the occipital bone with the axis.

LIGAMENTS CONNECTING THE ATLAS WITH THE OCCIPITAL BONE

The articulation between the atlas and the occipital bone is a double condyloid joint. Its ligaments are, the

Two Capsular.	Posterior Occipito-atlantal.
Anterior Occipito-atlantal.	Two Lateral Occipito-atlantal.

The **Capsular Ligaments** surround the condyles of the occipital bone, and connect them with the articular processes of the atlas ; they consist of thin and loose capsules, which are lined by synovial membrane.

The **Anterior Occipito-atlantal Ligament** (fig. 347) is a broad membranous layer, composed of densely woven fibres, which passes between the anterior margin of the foramen magnum above, and the whole length of the upper border of the anterior arch of the atlas below. Laterally, it is continuous with the capsular ligaments. In front, it is strengthened in the middle line by a strong, narrow, rounded cord, which is attached, above, to the basilar process of the occiput, and, below, to the tubercle on the anterior arch of the atlas. This ligament is in relation, in front, with the Recti antici minores ; behind, with the odontoid ligaments.

The **Posterior Occipito-atlantal Ligament** (fig. 348) is a very broad but thin membranous lamina, intimately blended with the dura mater. It is connected, above, to the posterior margin of the foramen magnum ; below, to the upper border of the posterior arch of the atlas. At each side this ligament is not attached to the bone, but presents a rounded inferior border, which forms, with

FIG. 348.—Occipito-atlantal and atlanto-axial ligaments. Posterior view.

the superior intervertebral notch, an opening for the passage of the vertebral artery and suboccipital nerve. This free border, which arches over the artery and nerve, sometimes becomes ossified. It is in relation, behind, with the Recti postici minores and Obliqui superiores ; in front, with the dura mater of the spinal canal, to which it is intimately adherent.

The **Lateral Ligaments** are thickened portions of the capsular ligament, reinforced by bundles of fibrous tissue, which are directed obliquely upwards and inwards, attached above to the jugular process of the occipital bone ; below, to the base of the transverse process of the atlas.

Synovial Membranes.—There are two synovial membranes in this articulation : one lining the inner surface of each of the capsular ligaments. These occasionally communicate with that between the posterior surface of the odontoid process and the transverse ligament.

Actions.—The movements permitted in this joint are flexion and extension, which give rise to the ordinary forward and backward nodding of the head, besides slight lateral motion to one or the other side. When either of these actions is carried beyond a slight extent, the whole of the cervical portion of

Fig. 349.—Articulation between odontoid process and atlas.

the spine assists in its production. Flexion is mainly produced by the action of the Rectus capitis anticus major and minor and the Sterno-mastoid muscles ; extension by the Rectus capitis posticus major and minor, the Superior oblique, the Complexus, Splenius, and upper fibres of the Trapezius. The Recti laterales are concerned in the lateral movement, assisted by the Trapezius, Splenius, Complexus, and the Sterno-mastoid of the same side, all acting together. According to Cruveilhier, there is a slight motion of rotation in this joint.

LIGAMENTS CONNECTING THE AXIS WITH THE OCCIPITAL BONE

Occipito-axial.   Three Odontoid.

To expose these ligaments, the spinal canal should be laid open by removing the posterior arch of the atlas, the laminæ and spinous process of the axis, and the portion of the occipital bone behind the foramen magnum, as seen in fig. 350.

The **Occipito-axial Ligament** (*membrana tectoria*) is situated within the spinal canal. It is a broad, strong band which covers the odontoid process and its ligaments, and appears to be a prolongation upwards of the posterior common ligament of the spinal column. It is attached, below, to the posterior surface of the body of the axis, and, becoming expanded as it ascends, is inserted into the basilar groove of the occipital bone, in front of the foramen magnum, where it blends with the dura mater of the skull.

Relations.—By its anterior surface with the transverse ligament, by its posterior surface with the dura mater.

The **Odontoid** or **Check Ligaments** (*alar ligaments*) are strong, rounded, fibrous cords, which arise one on either side of the upper part of the odontoid process, and, passing obliquely upwards and outwards, are inserted into the rough depressions on the inner side of the condyles of the occipital bone. In the triangular interval left between these ligaments another fibrous cord (*ligamentum apicis dentis* or *middle odontoid ligament*) may be seen, which passes almost perpendicularly from the apex of the odontoid process to the anterior margin of the foramen magnum, being intimately blended with the deep portion of the anterior occipito-atlantal ligament, and upper fasciculus of the transverse ligament of the atlas. It is regarded as a rudimental intervertebral disc, and in it traces of the notochord may persist.

Actions.—The odontoid ligaments serve to limit the extent to which rotation of the cranium may be carried ; hence they have received the name of *check ligaments*.

In addition to these ligaments which connect the atlas and axis to the skull, the ligamentum nuchæ must be regarded as one of the ligaments by which the spine is connected with the cranium. It has been described on page 355.

FIG. 350.—Occipito-axial and atlanto-axial ligaments. Posterior view, obtained by removing the arches of the vertebræ and the posterior part of the skull.

Surgical Anatomy.—The ligaments which unite the component parts of the vertebræ together are so strong, and these bones are so interlocked by the arrangement of their articulating processes, that dislocation is very uncommon, and, indeed, unless accompanied by fracture, rarely occurs, except in the upper part of the neck. Dislocation of the occiput from the atlas has only been recorded in one or two cases; but dislocation of the atlas from the axis, with rupture of the transverse ligament, is much more common: it is the mode in which death is produced in many cases of execution by hanging. In the lower part of the neck—that is, below the third cervical vertebra—dislocation unattended by fracture occasionally takes place.

IV. ARTICULATION OF THE LOWER JAW (TEMPORO-MANDIBULAR)

This is a ginglymo-arthrodial joint; the parts entering into its formation on each side are, above, the anterior part of the glenoid cavity of the temporal bone and the eminentia articularis; and, below, the condyle of the lower jaw. The ligaments are the following :

Capsular.   Internal Lateral.
External Lateral.   Interarticular Fibro-cartilage.
                 Stylo-mandibular.

The **Capsular Ligament** forms a thin and loose, but distinct capsule, attached above to the circumference of the glenoid cavity and the articular surface immediately in front; below, to the neck of the condyle of the lower jaw. It is thinnest on the inner side.

The **External Lateral Ligament** (fig. 351) is an accessory band of the capsular ligament, and is not separable from it. It consists of two short, narrow fasciculi, one in front of the other, attached, above, to the outer surface of the zygoma and to the tubercle on its lower border; below, to the outer surface and posterior border of the neck of the lower jaw. It is broader above than below; its fibres are parallel with one another, and directed obliquely downwards and backwards. Externally, it is covered by the parotid gland, and by the integument.

FIG. 351.—Temporo-mandibular articulation. External view.

The **Internal Lateral Ligament** (*spheno-mandibular*) (fig. 352) is not a true ligament, as it is developed from the fibrous covering of a part of Meckel's cartilage. It is a flat thin band which is attached above to the spinous process of the sphenoid bone, and, becoming broader as it descends, is inserted into the lingula of the dental foramen. Its outer surface is in relation, above, with the External pterygoid muscle; lower down, it is separated from the neck of the condyle by the internal maxillary artery; and still more inferiorly, the inferior dental vessels and nerve and a lobule of the parotid gland separate it from the ramus of the jaw. The inner surface is in relation with the Internal pterygoid muscle.

FIG. 352.—Temporo-mandibular articulation. Internal view.

The **Interarticular Fibro-cartilage** (fig. 353) is a thin plate of an oval form, placed horizontally between the condyle of the jaw and the glenoid cavity. Its upper surface is concavo-convex from before backwards, and a little convex transversely, to accommodate itself to the form of the glenoid cavity. Its under surface, where it is in contact with the condyle, is concave. Its circumference is connected to the capsular ligament; and in front to the tendon of the External pterygoid muscle. It is thicker at its circumference, especially behind, than at its centre. The fibres of which it is composed have a concentric arrangement, more apparent at the circumference than at the centre. Its surfaces are smooth. It divides the joint into two cavities, each of which is furnished with a synovial membrane.

The **Synovial Membranes,** two in number, are placed one above, and the other below, the fibro-cartilage. The upper one, the larger and looser of the two,

FIG. 353.—Vertical section of temporomandibular articulation.

is continued from the margin of the cartilage covering the glenoid cavity and eminentia articularis on to the upper surface of the fibro-cartilage. The lower one passes from the under surface of the fibro-cartilage to the neck of the condyle of the jaw, being prolonged downwards a little farther behind than in front. The interarticular cartilage is sometimes perforated in its centre, and the two synovial sacs then communicate with each other.

The **Stylo-mandibular Ligament** is a specialised band of the cervical fascia, which extends from near the apex of the styloid process of the temporal bone to the angle and posterior border of the ramus of the lower jaw, between the Masseter and Internal pterygoid muscles. This ligament separates the parotid from the submaxillary gland, and has attached to its inner side part of the fibres of origin of the Stylo-glossus muscle. Although usually classed among the ligaments of the jaw, it can only be considered as an accessory to the articulation.

The *nerves* of this joint are derived from the auriculo-temporal and masseteric branches of the inferior maxillary. The *arteries* are derived from the temporal branch of the external carotid.

Actions.—The movements permitted in this articulation are very extensive. Thus, the jaw may be depressed or elevated, or it may be carried forwards or backwards. It must be borne in mind that there are two distinct joints in this articulation—that is to say, one between the condyle of the jaw and the interarticular fibro-cartilage, and another between the fibro-cartilage and the glenoid fossa ; when the jaw is depressed, as in opening the mouth, the movements which take place in these two joints are not the same. In the lower compartment, that between the condyle and the fibro-cartilage, the movement is of a ginglymoid or hinge-like character, the condyle rotating on a transverse axis on the fibro-cartilage ; while in the upper compartment the movement is of a gliding character, the fibro-cartilage, together with the condyle, gliding forwards on to the eminentia articularis. These two movements take place simultaneously, the condyle and fibro-cartilage move forwards on the eminence, and at the same time the condyle revolves on the fibro-cartilage. In the opposite movement of shutting the mouth the reverse action takes place ; the fibro-cartilage glides back, carrying the condyle with it, and this at the same time revolves back to its former position. When the jaw is carried horizontally forwards, as in protruding the lower incisors in front of the upper, the movement takes place principally in the upper compartment of the joint, the fibro-cartilage, carrying with it the condyle, gliding forwards on the glenoid fossa. This is because the movement is mainly effected by the External pterygoid muscles, which are inserted into the condyles and interarticular fibro-cartilages. The grinding or chewing movement is produced by the alternate movement of one condyle, with its fibro-cartilage, forwards and backwards, while the other condyle moves simultaneously in the opposite direction ; at the same time the condyle undergoes a vertical rotation on its own axis on the fibro-cartilage in the lower compartment. One condyle advances and rotates, the other condyle recedes and rotates, in alternate succession.

The lower jaw is *depressed* by its own weight, assisted by the Platysma, the Digastric, the Mylo-hyoid, and the Genio-hyoid. It is *elevated* by the anterior part of the Temporal, Masseter, and Internal pterygoid. It is drawn *forwards* by the simultaneous action of the External pterygoid, and the superficial fibres of the Masseter ; and it is drawn *backwards* by the deep fibres of the Masseter and the posterior fibres of the Temporal muscle. The grinding movement is caused by the alternate action of the two External pterygoids.

Surface Form.—The temporo-mandibular articulation is quite superficial, situated below the base of the zygoma, in front of the tragus and external auditory meatus, and behind the posterior border of the upper part of the Masseter muscle. Its exact position can be at once ascertained by feeling for the condyle of the jaw, the working of which can be distinctly felt in the movements of the lower jaw in opening and shutting the mouth. When the mouth is opened wide, the condyle advances out of the glenoid fossa on to the eminentia articularis, and a depression is felt in the situation of the joint.

Surgical Anatomy.—The lower jaw is dislocated only in one direction—viz. forwards. The accident is caused by violence or muscular action. When the mouth is open, the condyle is situated on the eminentia articularis, and any sudden violence, or even a sudden muscular spasm, as during a convulsive yawn, may displace the condyle forwards into the zygomatic fossa. The displacement may be unilateral or bilateral, according as one or both of the condyles are displaced. The latter of the two is the more common.

Sir Astley Cooper described a condition which he termed ' subluxation.' It occurs principally in delicate women, and is believed by some to be due to relaxation of the ligaments, permitting too free movement of the bone, and possibly some displacement of the fibro-cartilage. Others have believed that it is due to gouty or rheumatic changes in the joint.

In close relation to the condyle of the jaw is the external auditory meatus and the tympanum ; any force, therefore, applied to the bone is liable to be attended with damage to these parts, or inflammation in the joint may extend to the ear, or on the other hand inflammation of the middle ear may involve the articulation and cause its destruction, thus leading to ankylosis of the joint. In children, arthritis of this joint may follow the exanthemata, and in adults occurs as the result of some constitutional conditions, as rheumatism or gout. The temporo-mandibular joint is also occasionally the seat of osteo-arthritis, leading to great suffering during efforts of mastication. A peculiar affection sometimes attacks the neck and condyle of the lower jaw, consisting in hypertrophy and elongation of these parts and consequent protrusion of the chin to the opposite side.

V. ARTICULATIONS OF THE RIBS WITH THE VERTEBRÆ

The articulations of the ribs with the vertebral column may be divided into two sets : 1. Those which connect the heads of the ribs with the bodies of the vertebræ (*costo-central*). 2. Those which connect the necks and tubercles of the ribs with the transverse processes (*costo-transverse*).

1. COSTO-CENTRAL ARTICULATION (fig. 354)

These constitute a series of gliding joints, in which there is a certain hinge-like action, and hence may be characterised as ginglymo-arthrodial joints. They are formed by the articulation of the heads of the ribs with the cavities on the contiguous margins of the bodies of the dorsal vertebræ and the intervertebral substance between them, except in the case of the first, tenth, eleventh, and twelfth ribs, where the cavity is formed by a single vertebra. The bones are connected by the following ligaments :

Capsular. Anterior Costo-vertebral or Stellate.
Interarticular.

The **Capsular Ligament** surrounds and encloses the joint, being composed of short, strong fibres, which pass between the head of the rib and the circumference of the articular cavity formed by the intervertebral disc and the adjacent vertebræ. It is most distinct at the upper and lower parts of the articulation ; and at its upper part some of its fibres pass through the intervertebral foramen to the back of the intervertebral disc, and, behind, it is continuous with the middle costo-transverse ligament.

The **Anterior Costo-vertebral** or **Stellate Ligament** is a specialised part of the capsule, and connects the anterior part of the head of each rib with the sides of the bodies of two vertebræ, and the intervertebral disc between them. It consists of three flat bundles of ligamentous fibres, which are attached to the anterior part of the head of the rib, just beyond the articular surface. The superior bundle passes upwards to be connected with the body of the vertebra above; the inferior one descends to the body of the vertebra below; and the middle one, the smallest and least distinct, passes horizontally inwards, to be attached to the intervertebral substance.

Relations.—In front, with the thoracic ganglia of the sympathetic, the pleura, and, on the right side, with the vena azygos major; behind, with the interarticular ligament and synovial membranes.

In the first rib, which articulates with a single vertebra, this ligament does not present a distinct division into three fasciculi; its fibres, however, radiate, and are attached to the body of the last cervical vertebra, as well as to the body of the vertebra with which the rib articulates. In the tenth, eleventh, and twelfth ribs, which also articulate with a single vertebra, the division does not exist; but the fibres of the ligament in each case radiate and are connected with the vertebra above, as well as that with which the rib articulates.

FIG. 354.—Costo-vertebral and costo-transverse articulations.
Anterior view.

Upper synovial cavity
Interarticular ligament
Lower synovial cavity

The **Interarticular Ligament** is situated in the interior of the joint. It consists of a short band of fibres, flattened from above downwards, attached by one extremity to the sharp crest which separates the two articular facets on the head of the rib, and by the other to the intervertebral disc. It divides the joint into two cavities, which have no communication with each other. In the first, tenth, eleventh, and twelfth ribs, the interarticular ligament does not exist; consequently, there is but one cavity. This ligament is the analogue of the *ligamentum conjugale* of some mammals, which unites the heads of opposite ribs across the back of the intervertebral disc.

The Synovial Membranes.—There are two synovial membranes in each of the articulations in which there is an interarticular ligament, one above and one below this structure: one only in those joints where there is a single cavity.

2. COSTO-TRANSVERSE ARTICULATION (fig. 355)

The articular portion of the tubercle of the rib and adjacent transverse process form an arthrodial joint.

In the *eleventh* and *twelfth ribs* this articulation is wanting.

The ligaments connecting these parts are, the

Capsular.   Middle Costo-transverse (Interosseous).
Anterior Costo-transverse.   Posterior Costo-transverse.

The **Capsular Ligament** is a thin, membranous sac attached to the circumference of the articular surfaces, and enclosing a small synovial membrane.

In the *eleventh* and *twelfth ribs* this ligament is absent.

The **Anterior** or **Superior Costo-transverse Ligament** consists of two sets of fibres : one (anterior) is attached below to the sharp crest on the upper border of the neck of each rib, and passes obliquely upwards and outwards, to the lower border of the transverse process immediately above ; the other (posterior) is attached below to the neck of the rib, and passes upwards and inwards to the base of the transverse process and outer border of the lower articular process of the vertebra above. This ligament is in relation, in front, with the intercostal vessels and nerves ; behind, with the Longissimus dorsi. Its *internal border* is thickened and free, and bounds an aperture which transmits the posterior branches of the intercostal vessels and nerves. Its *external border* is continuous with a thin aponeurosis, which covers the External intercostal muscle.

The *first rib* has no anterior costo transverse ligament.

The **Middle Costo-transverse** or **Interosseous Ligament** consists of short but strong fibres, which pass between the rough surface on the posterior part of the neck of each rib and the anterior surface of the adjacent transverse process. In

FIG. 355.—Costo-transverse articulation. Seen from above.

order fully to expose this ligament a horizontal section should be made across the transverse process and corresponding part of the rib ; or the rib may be forcibly separated from the transverse process, and the fibres of the ligament put on the stretch.

In the *eleventh* and *twelfth ribs* this ligament is quite rudimentary or wanting.

The **Posterior Costo-transverse Ligament** is a short but thick and strong fasciculus, which passes obliquely from the summit of the transverse process to the rough non-articular portion of the tubercle of the rib. This ligament is shorter and more oblique in the upper than in the lower ribs. Those corresponding to the superior ribs ascend, while those of the inferior ribs descend slightly.

In the *eleventh* and *twelfth ribs* this ligament is wanting.

Actions.—The heads of the ribs are so closely connected to the bodies of the vertebræ by the stellate and interarticular ligaments, and the necks and tubercles of the ribs to the transverse processes, that only a slight gliding movement of the articular surfaces on each other can take place in these articulations. The result of this gliding movement with respect to the six upper ribs, consists in an elevation of the front and middle portion of the rib, the hinder part being prevented from performing any upward movement by its close connection with the spine. In this gliding movement the rib rotates on an axis corresponding with a line drawn through the two articulations, Costo-central and Costo-transverse, which the rib forms with the spine. With respect to the seventh, eighth, ninth, and tenth ribs, each one, besides rotating in a similar manner to the upper six, also rotates on an axis corresponding with a line drawn from the head of the rib to the sternum. By the first movement—that of rotation of the rib on an axis corresponding with a line drawn through the two articulations which this bone forms with the spine—an elevation of the anterior part of the rib takes place, and a consequent enlargement of the antero-posterior diameter of the chest. None of the ribs lie in the horizontal plane : they are all directed more or less obliquely, so that their anterior extremities lie on a lower level than their posterior, and this obliquity increases from the first to the seventh, and then again decreases. If any one rib is examined—say, that in which there is the greatest obliquity—it is obvious that as its sternal extremity is carried upwards, it must also be thrown forwards ; so that the rib may be regarded as a radius, moving on the vertebral joint as a centre, and causing the sternal attachment to describe an arc of a circle in the vertical plane of the body. Since all the ribs are oblique and connected in front to the sternum by the elastic costal cartilages, they must have a tendency to thrust the sternum forwards, and so increase the antero-posterior diameter of the chest. By the second movement—that of the rotation of the rib on an axis corresponding with a line drawn from the head of the rib to the sternum—an elevation of the middle portion of the rib takes place, and consequently an increase in the transverse diameter of the chest. For the ribs not only slant downwards and forwards from their vertebral attachment, but they are also oblique in relation to their transverse plane—that is to say, their middle is on a lower level than either their vertebral or sternal extremities. It results from this that when the ribs are raised, the centre portion is thrust outwards, somewhat after the fashion in which the handle of a bucket is thrust away from the side when raised to a horizontal position, and the lateral diameter of the chest is increased (see fig. 356). The mobility of the different ribs varies very much. The first rib is more fixed than the others, on account of the weight of the upper extremity and the strain of the ribs beneath ; but on the freshly dissected thorax it moves as freely as the rest. From the same causes the movement of the second rib is also not very extensive. In the other ribs, this mobility increases successively down to the last two, which are very movable. The ribs are generally more movable in the female than in the male.

FIG. 356.—Diagrams showing the axis of rotation of the ribs in the movements of respiration. The one axis of rotation corresponds with a line drawn through the two articulations which the rib forms with the spine (*a*, *b*), and the other with a line drawn from the head of the rib to the sternum (A, B). (From Kirke's 'Handbook of Physiology.')

VI. ARTICULATION OF THE CARTILAGES OF THE RIBS WITH THE STERNUM (CHONDRO-STERNAL) (fig. 357)

The articulations of the cartilages of the true ribs with the sternum are arthrodial joints, with the exception of the first, in which the cartilage is almost always directly united with the sternum, and which must, therefore, be regarded as a synarthrodial articulation. The ligaments connecting them are—

Capsular.
Anterior Chondro-sternal.
Posterior Chondro-sternal.

Interarticular Chondro-sternal.
Anterior Chondro-xiphoid.
Posterior Chondro-xiphoid.

The **Capsular Ligaments** surround the joints formed between the cartilages of the true ribs and the sternum. They are very thin, intimately blended with the anterior and posterior ligaments, and strengthened at the upper and lower parts of the articulations by a few fibres, which pass from the cartilage to the side of the sternum. These ligaments protect the synovial membranes.

The **Anterior Chondro-sternal Ligament** is an accessory part of the capsular ligament, and consists of a broad and thin membranous band that radiates from the front of the inner extremity of the cartilages of the true ribs to the anterior surface of the sternum. It is composed of fasciculi which pass in different directions. The *superior fasciculi* ascend obliquely, the *inferior* pass obliquely downwards, and the *middle fasciculi* horizontally. The superficial fibres of this ligament are the longest; they intermingle with the fibres of the ligaments above and below them, with those of the opposite side, and with the tendinous fibres of origin of the Pectoralis major, forming a thick fibrous membrane, which covers the surface of the sternum. This is more distinct at the lower than at the upper part.

The **Posterior Chondro-sternal Ligament** is also a part of the capsular ligament, but is less thick and distinct than the anterior; it is composed of fibres which radiate from the posterior surface of the sternal end of the cartilages of the true ribs to the posterior surface of the sternum, becoming blended with the periosteum.

The **Interarticular Chondro-sternal Ligaments.**—These are only found constantly between the second costal cartilages and the sternum. The cartilage of the *second rib* is connected with the sternum by means of an interarticular ligament, attached by one extremity to the cartilage of the second rib, and by the other extremity to the cartilage which unites the first and second pieces of the sternum. This articulation is provided with two synovial membranes. Occasionally the cartilage of the *third rib* is connected with the sternum by means of an interarticular ligament which is attached by one extremity to the cartilage of the third rib, and by the other extremity to the point of junction of the second and third pieces of the sternum. Still more rarely a similar condition is found in the other four joints of the series. In the lower two the ligament sometimes completely obliterates the cavity, so as to convert it into a synarthrosis.

The **Anterior Chondro-xiphoid.**—This is a band of ligamentous fibres, which connects the anterior surface of the seventh costal cartilage, and occasionally also that of the sixth, to the anterior surface of the ensiform appendix. It varies in length and breadth in different subjects.

The **Posterior Chondro-xiphoid** is a similar band of fibres on the internal or posterior surface, though less thick and distinct.

Synovial Membranes.—There is no synovial membrane between the first costal cartilage and the sternum, as this cartilage is directly continuous with the sternum. There are two synovial membranes in the articulation of the second costal cartilages to the sternum. There is generally one synovial membrane in each of the joints between the third, fourth, fifth, sixth, and seventh costal cartilages to the sternum; but it is sometimes absent in the sixth and seventh chondro-sternal joints. If an interarticular ligament exists in any of these joints, there are two synovial cavities. After middle life the articular surfaces lose their polish, become roughened, and the synovial membranes appear to be wanting. In old age, the articulations do not exist, the cartilages of most of the ribs becoming continuous with the sternum.

Actions.—The movements which are permitted in the chondro-sternal articulations are limited to elevation and depression, and these only to a slight extent.

ARTICULATIONS OF THE CARTILAGES OF THE RIBS WITH EACH OTHER (INTERCHONDRAL) (fig. 357)

The contiguous borders of the sixth, seventh, and eighth, and sometimes the ninth and tenth, costal cartilages articulate with each other by small, smooth,

FIG. 357.—Chondro-sternal, chondro-xiphoid, and interchondral articulations. Anterior view.

oblong-shaped facets. Each articulation is enclosed in a thin *capsular ligament* lined by *synovial membrane*, and strengthened externally and internally by ligamentous fibres (*interchondral ligaments*) which pass from one cartilage to the other. Sometimes the fifth costal cartilage, more rarely that of the ninth, articulates, by its lower border, with the adjoining cartilage by a small oval facet; more frequently they are connected together by a few ligamentous fibres. Occasionally, the articular surfaces above mentioned are wanting.

The outer extremity of each costal cartilage is received into a depression in the sternal end of the rib, and the two are held together by the periosteum.

VII. ARTICULATIONS OF THE STERNUM

The first piece of the sternum is united to the second either by an amphi-arthrodial joint—a single piece of true fibro-cartilage uniting the segments—or by a diarthrodial joint, in which each bone is clothed with a distinct lamina of cartilage, adherent on one side, free on the other. In the latter case, the cartilage covering the gladiolus is continued without interruption on to the cartilages of the second ribs. Rivington has found the diarthrodial form of joint in about one-third of the specimens examined by him, Maisonneuve more frequently. It appears to be rare in childhood, and is formed, in Rivington's opinion, from the amphiarthrodial form, by absorption. The diarthrodial joint seems to have no tendency to ossify at any age, while the amphiarthrodial is more liable to do so, and has been found ossified as early as thirty-four years of age. The two segments are further connected by an

Anterior Intersternal Ligament.
Posterior Intersternal Ligament.

The **Anterior Intersternal Ligament** consists of a layer of fibres, having a longitudinal direction; it blends with the fibres of the anterior chondro-sternal ligaments on both sides, and with the tendinous fibres of origin of the Pectoralis major. This ligament is rough, irregular, and much thicker below than above.

The **Posterior Intersternal Ligament** is disposed in a somewhat similar manner on the posterior surface of the articulation.

VIII. ARTICULATION OF THE VERTEBRAL COLUMN WITH THE PELVIS

The ligaments connecting the last lumbar vertebra with the sacrum are similar to those which connect the movable segments of the spine with each other—viz.: 1. The continuation downwards of the anterior and posterior common ligaments. 2. The intervertebral substance connecting the flattened oval surfaces of the two bones and forming an amphiarthrodial joint. 3. Ligamenta subflava, connecting the arch of the last lumbar vertebra with the posterior border of the sacral canal. 4. Capsular ligaments connecting the articulating processes and forming a double arthrodia. 5. Inter- and supra-spinous ligaments.

Two additional ligaments connect the pelvis with the spine; these are the lumbo-sacral and ilio-lumbar.

The **Lumbo-sacral Ligament** (fig. 358) is a short, thick, triangular fasciculus, which is connected above to the lower and front part of the transverse process of the last lumbar vertebra, passes obliquely outwards, and is attached below to the lateral surface of the base of the sacrum, becoming blended with the anterior sacro-iliac ligament. The ligament is in relation, in front, with the Psoas muscle.

The **Ilio-lumbar Ligament** (fig. 358), the thickened lower edge of the anterior lamella of the lumbar fascia, passes horizontally outwards from the apex of the transverse process of the last lumbar vertebra to the crest of the ilium immediately in front of the sacro-iliac articulation. It is of a triangular form, thick and narrow internally, broad and thinner externally. It is in relation, in front, with the Psoas muscle; behind, with the muscles occupying the vertebral groove; above, with the Quadratus lumborum.

IX. ARTICULATIONS OF THE PELVIS

The ligaments connecting the bones of the pelvis with each other may be divided into four groups : 1. Those connecting the sacrum and ilium. 2. Those passing between the sacrum and ischium. 3. Those connecting the sacrum and coccyx. 4. Those between the two pubic bones.

B B

FIG. 358.—Articulations of pelvis and hip. Anterior view.

Aperture of communication
with
Bursa under Psoas and
Iliacus

FIG. 359.—Articulations of pelvis and hip. Posterior view.

Anterior or lesser
sacro-sciatic
ligament

1. Articulation of the Sacrum and Ilium (Sacro-iliac joint)

The sacro-iliac articulation is an amphiarthrodial joint, formed between the lateral surfaces of the sacrum and the ilium. The articular, ear-shaped surface of each bone is covered with a thin plate of cartilage, thicker on the sacrum than on the ilium. These are in close contact with each other, and to a certain extent united together by irregular patches of softer fibro-cartilage, and at their upper and posterior part by fine fibres of interosseous fibrous tissue. In a considerable part of their extent, especially in advanced life, they are not connected together, but are separated by a space containing a synovial-like fluid, and hence the joint presents the characters of a diarthrosis.

The ligaments connecting these surfaces are the anterior and posterior sacro-iliac.

The **Anterior Sacro-iliac Ligament** (fig. 358) consists of numerous thin bands, which connect the anterior surfaces of the sacrum and ilium.

The **Posterior Sacro-iliac** (fig. 359) is a strong interosseous ligament, situated in a deep depression between the sacrum and ilium behind, and forming the

Fig. 360.—Side view of pelvis, showing the great and lesser sacro-sciatic ligaments.

Anterior sacro-iliac ligament

Poupart's ligament

Great sacro-sciatic ligament

Lesser sacro-sciatic ligament

Great sacro-sciatic ligament

Obturator membrane

chief bond of connection between those bones. It consists of numerous strong fasciculi, which pass between the bones in various directions. Three of these are of large size; the *two superior*, nearly horizontal in direction, arise from the first and second transverse tubercles on the posterior surface of the sacrum, and are inserted into the rough, uneven surface at the posterior part of the inner surface of the ilium. The third fasciculus, oblique in direction, is attached by one extremity to the third transverse tubercle on the posterior surface of the sacrum, and by the other to the posterior superior spine of the ilium; it is sometimes called the *oblique sacro-iliac ligament*.

The position of the sacro-iliac joint is indicated by the posterior superior spine of the ilium. This process is immediately behind the centre of the articulation.

2. Ligaments passing between the Sacrum and Ischium (fig. 360)

The Great Sacro-sciatic (Posterior).
The Lesser Sacro-sciatic (Anterior).

The **Great** or **Posterior Sacro-sciatic Ligament** is situated at the lower and back part of the pelvis. It is flat, and triangular in form; narrower in the middle than at the extremities ; attached by its broad base to the posterior inferior spine of the ilium, to the fourth and fifth transverse tubercles of the sacrum, and to the lower part of the lateral margin of that bone and the coccyx.   Passing obliquely downwards, outwards, and forwards, it becomes narrow and thick, but at its insertion into the inner margin of the tuberosity of the ischium, it increases in breadth, and is prolonged forwards along the inner margin of the ramus, forming what is known as the *falciform ligament*, the free concave edge of which gives attachment to the obturator fascia.   One of its surfaces is turned towards the perinæum, the other towards the Obturator internus muscle.   The lower border of the ligament is directly continuous with the tendon of origin of the long head of the Biceps muscle, and by many is believed to be the proximal end of this muscle, cut off by the projection of the tuberosity of the ischium.

The *posterior surface* of this ligament gives origin, by its whole extent, to fibres of the Gluteus maximus.   Its *anterior surface* is united to the lesser sacro-sciatic ligament.   Its *external border* forms, above, the posterior boundary of the great sacro-sciatic foramen, and, below, the posterior boundary of the lesser sacro-sciatic foramen.   Its *lower border* forms part of the boundary of the perinæum. It is pierced by the coccygeal branch of the sciatic artery and coccygeal nerve.

The **Lesser** or **Anterior Sacro-sciatic Ligament,** much shorter and smaller than the preceding, is thin, triangular in form, attached by its apex to the spine of the ischium, and internally, by its broad base, to the lateral margin of the sacrum and coccyx, anterior to the attachment of the great sacro-sciatic ligament with which its fibres are intermingled.

It is in relation, *anteriorly*, with the Coccygeus muscle, to which it is closely connected ; *posteriorly*, it is covered by the great sacro-sciatic ligament, and crossed by the internal pudic vessels and nerve.   Its *superior border* forms the lower boundary of the great sacro-sciatic foramen ; its *inferior border*, part of the lesser sacro-sciatic foramen.

These two ligaments convert the sacro-sciatic notches into foramina.   The *superior* or *great* sacro-sciatic foramen is bounded, in front and above, by the posterior border of the os innominatum ; behind, by the great sacro-sciatic ligament ; and below, by the lesser sacro-sciatic ligament.   It is partially filled up, in the recent state, by the Pyriformis muscle which passes through it. Above this muscle, the gluteal vessels and superior gluteal nerve emerge from the pelvis ; and below it, the sciatic vessels and nerves, the internal pudic vessels and nerve, the inferior gluteal nerve, and the nerves to the obturator internus and quadratus femoris also make their exit from the pelvis.   The *inferior* or *lesser* sacro-sciatic foramen is bounded, in front, by the tuber ischii ; above, by the spine and lesser sacro-sciatic ligament ; behind, by the greater sacro-sciatic ligament.   It transmits the tendon of the Obturator internus muscle, its nerve, and the internal pudic vessels and nerve.

3. Articulation of the Sacrum and Coccyx

This articulation is an amphiarthrodial joint, formed between the oval surface at the apex of the sacrum, and the base of the coccyx.   It is analogous to the joints between the bodies of the vertebræ, and is connected by similar ligaments. They are the

Anterior Sacro-coccygeal.	Lateral Sacro-coccygeal.
Posterior Sacro-coccygeal.	Interposed Fibro-cartilage.

Interarticular.

The **Anterior Sacro-coccygeal Ligament** consists of a few irregular fibres, which descend from the anterior surface of the sacrum to the front of the coccyx, becoming blended with the periosteum.

The **Posterior Sacro-coccygeal Ligament** is a flat band, of a pearly tint, which arises from the margin of the lower orifice of the sacral canal, and descends to be inserted into the posterior surface of the coccyx.   This ligament completes the lower and back part of the sacral canal.   Its superficial fibres are much longer than the more deeply seated.   This ligament is in relation, behind, with the Gluteus maximus.

The **Lateral Sacro-coccygeal Ligaments** connect the transverse processes of the coccyx to the lower lateral angles of the sacrum.

A **Fibro-cartilage** is interposed between the contiguous surfaces of the sacrum and coccyx; it differs from that interposed between the bodies of the vertebræ in being thinner, and its central part firmer in texture. It is somewhat thicker in front and behind than at the sides. Occasionally, a synovial membrane is found when the coccyx is freely movable, which is more especially the case during pregnancy.

The **Interarticular Ligaments** are thin bands of ligamentous tissue, which unite the cornua of the two bones.

The different segments of the coccyx are connected together by an extension downwards of the anterior and posterior sacro-coccygeal ligaments, a thin annular disc of fibro-cartilage being interposed between each segment. In the adult male, all the pieces become ossified; but in the female, this does not commonly occur until a later period of life. The separate segments of the coccyx are first united, and at a more advanced age the joint between the sacrum and coccyx is obliterated.

Actions.—The movements which take place between the sacrum and coccyx, and between the different pieces of the latter bone, are forwards and backwards; they are very limited. Their extent increases during pregnancy.

4. Articulation of the Ossa Pubis (Symphysis Pubis)

The articulation between the pubic bones is an amphiarthrodial joint, formed by the junction of the two oval articular surfaces of the ossa pubis. The ligaments of this articulation are, the

Anterior Pubic.	Superior Pubic.
Posterior Pubic.	Subpubic.

Interpubic Disc.

The **Anterior Pubic Ligament** (fig. 358) consists of several superimposed layers, which pass across the front of the articulation. The superficial fibres pass obliquely from one bone to the other, decussating and forming an interlacement with the fibres of the aponeurosis of the External oblique and the tendon of the Rectus muscles. The deep fibres pass transversely across the symphysis, and are blended with the fibro-cartilage.

The **Posterior Pubic Ligament** consists of a few thin, scattered fibres, which unite the two pubic bones posteriorly.

The **Superior Pubic Ligament** is a band of fibres, which connects together the two pubic bones superiorly.

The **Subpubic Ligament** is a thick, triangular arch of ligamentous fibres, connecting together the two pubic bones below, and forming the upper boundary of the pubic arch. Above, it is blended with the interarticular

FIG. 361.—Vertical section of the symphysis pubis. Made near its posterior surface.

fibro-cartilage; laterally, it is united with the descending rami of the ossa pubis; below, it is free, and is separated from the triangular ligament of the perinæum by the opening through which the dorsal vein of the penis passes into the pelvis. Its fibres are closely connected, and have an arched direction.

The **Interpubic Disc** connects the pubic bones in front. Each of the two bony surfaces is covered by a thin layer of hyaline cartilage, which is firmly connected to the bone by a series of nipple-like processes which accurately fit within corresponding depressions on the osseous surfaces. These opposed cartilaginous surfaces are connected together by an intermediate lamina of fibro-cartilage which varies in thickness in different subjects. It often contains a cavity in its centre, probably formed by the softening and absorption of the fibro-cartilage, since it rarely appears before the tenth year of life, and is not lined by synovial membrane. It is larger in the female than in the male, but it is very doubtful whether it enlarges, as was formerly supposed, during pregnancy. It is most frequently limited to the upper and back part of the joint; but it occasionally reaches to the front, and may extend the entire length of the cartilage. This cavity may be easily demonstrated by making a transverse vertical section of the symphysis pubis near its posterior surface (fig. 361).

The **Obturator Membrane** is more properly regarded as analogous to the muscular fasciæ, with which it will be described.

ARTICULATIONS OF THE UPPER EXTREMITY

The articulations of the Upper Extremity may be arranged as follows :

I. Sterno-clavicular.	VI. Radio-ulnar.
II. Acromio-clavicular.	VII. Wrist.
III. Scapular ligaments.	VIII. Carpal.
IV. Shoulder.	IX. Carpo-metacarpal.
V. Elbow.	X. Metacarpo-phalangeal.

XI. Phalangeal.

I. STERNO-CLAVICULAR ARTICULATION (fig. 362).

The **Sterno-clavicular** is regarded by most anatomists as an arthrodial joint; but Cruveilhier considers it to be an articulation by reciprocal reception. Probably the former opinion is the correct one; the varied movements, which the joint enjoys, being due to the interposition of an interarticular fibro-cartilage between

Fig. 362.—Sterno-clavicular articulation. Anterior view.

the joint surfaces. The parts entering into its formation are the sternal end of the clavicle, the upper and lateral part of the first piece of the sternum, and the cartilage of the first rib. The articular surface of the clavicle is much larger than that of the sternum, and is invested with a layer of cartilage,* which is considerably thicker than that on the latter bone. The ligaments of this joint are, the

Capsular.	Interclavicular.
Anterior Sterno-clavicular.	Costo-clavicular (rhomboid).
Posterior Sterno-clavicular.	Interarticular fibro-cartilage.

The **Capsular Ligament** completely surrounds the articulation, consisting of fibres of varying degrees of thickness and strength. Those in front and behind are of considerable thickness, and form the anterior and posterior sterno-clavicular ligaments; but those above and below, especially in the latter situation, are thin and scanty, and partake more of the character of connective tissue than true fibrous tissue.

* According to Bruch, the sternal end of the clavicle is covered by a tissue, which is rather fibrous than cartilaginous in structure.

The **Anterior Sterno-clavicular Ligament** is a part of the capsular ligament. It is a broad band of fibres, which covers the anterior surface of the articulation, being attached, above, to the upper and front part of the inner extremity of the clavicle ; and, passing obliquely downwards and inwards, is attached, below, to the front of the upper part of the first piece of the sternum. This ligament is covered by the sternal portion of the Sterno-cleido-mastoid and the integument ; behind, it is in relation with the interarticular fibro-cartilage and the two synovial membranes.

The **Posterior Sterno-clavicular Ligament,** also a part of the capsular ligament, is a similar band of fibres, which covers the posterior surface of the articulation, being attached, above, to the upper and back part of the inner extremity of the clavicle ; and, passing obliquely downwards and inwards, is attached, below, to the back of the upper part of the first piece of the sternum. It is in relation, in front, with the interarticular fibro-cartilage and synovial membranes ; behind, with the Sterno-hyoid and Sterno-thyroid muscles.

The **Interclavicular Ligament** is a flattened band, which varies considerably in form and size in different individuals ; it passes in a curved direction from the upper part of the inner extremity of one clavicle to the other, and is also attached to the upper margin of the sternum. It is in relation, in front, with the integument ; behind, with the Sterno-thyroid muscles.

The **Costo-clavicular Ligament** (rhomboid) is short, flat, and strong : it is of a rhomboid form, attached, below, to the upper and inner part of the cartilage of the first rib, it ascends obliquely backwards and outwards, and is attached, above, to the rhomboid depression on the under surface of the clavicle. It is in relation, in front, with the tendon of origin of the Subclavius ; behind, with the subclavian vein.

The **Interarticular Fibro-cartilage** is a flat and nearly circular disc, interposed between the articulating surfaces of the sternum and clavicle. It is attached, above, to the upper and posterior border of the articular surface of the clavicle ; below, to the cartilage of the first rib, at its junction with the sternum ; and by its circumference to the anterior and posterior sterno-clavicular and interclavicular ligaments. It is thicker at the circumference, especially its upper and back part, than at its centre, or below. It divides the joint into two cavities, each of which is furnished with a separate synovial membrane.

Of the two **Synovial Membranes** found in this articulation, one is reflected from the sternal end of the clavicle, over the adjacent surface of the fibro-cartilage, and cartilage of the first rib ; the other is placed between the articular surface of the sternum and adjacent surface of the fibro-cartilage ; the latter is the larger of the two.

Actions.—This articulation is the centre of the movements of the shoulder, and admits of a limited amount of motion in nearly every direction—upwards, downwards, backwards, forwards, as well as circumduction. When these movements take place in the joint, the clavicle in its motion carries the scapula with it, this bone gliding on the outer surface of the chest. This joint therefore forms the centre from which all movements of the supporting arch of the shoulder originate, and is the only point of articulation of this part of the skeleton with the trunk. 'The movements attendant on elevation and depression of the shoulder take place between the clavicle and the interarticular fibro-cartilage, the bone rotating upon the ligament on an axis drawn from before backwards through its own articular facet. When the shoulder is moved forwards and backwards, the clavicle, with the interarticular fibro-cartilage, rolls to and fro on the articular surface of the sternum, revolving, with a sliding movement, round an axis drawn nearly vertically through the sternum. In the circumduction of the shoulder, which is compounded of these two movements, the clavicle revolves upon the interarticular fibro-cartilage, and the latter, with the clavicle, rolls upon the sternum.'* Elevation of the clavicle is principally limited by the costo-clavicular ligament ; depression, by the interclavicular ligament and interarticular fibro-cartilage. The muscles which *raise* the clavicle, as in shrugging the shoulders, are the upper fibres of the Trapezius, the Levator anguli scapulæ, the clavicular head of the Sterno-mastoid, assisted to a certain extent by the two Rhomboids, which pull the vertebral border of the Scapula backwards and

* Humphry, *On the Human Skeleton*, page 402.

upwards and so raise the clavicle. The *depression* of the clavicle is principally effected by gravity, assisted by the Subclavius, Pectoralis minor, and lower fibres of the Trapezius. It is drawn *backwards* by the Rhomboids and the middle and lower fibres of the Trapezius, and *forwards* by the Serratus magnus and Pectoralis minor.

Surface Form.—The position of the sterno-clavicular joint may be easily ascertained by feeling the enlarged sternal end of the collar-bone just external to the long, cord-like, sternal origin of the Sterno-mastoid muscle. If this muscle is relaxed by bending the head forwards, a depression just internal to the end of the clavicle, and between it and the sternum, can be felt, indicating the exact position of the joint, which is subcutaneous. When the arm hangs by the side, the cavity of the joint is V-shaped. If the arm is raised, the bones become more closely approximated, and the cavity becomes a mere slit.

Surgical Anatomy.—The strength of this joint mainly depends upon its ligaments, and it is owing to this, and to the fact that the force of the blow is usually transmitted along the long axis of the clavicle, that dislocation rarely occurs, and that the bone is broken rather than displaced. When dislocation does occur, the course which the displaced bone takes depends more upon the direction in which the violence is applied than upon the anatomical construction of the joint; it may be either forwards, backwards, or upwards. The chief point worthy of note, as regards the construction of the joint, in connection with dislocation, is the fact that, owing to the shape of the articular surfaces, and the strength of the joint mainly depending upon the ligaments, the displacement when reduced is very liable to recur, and hence it is extremely difficult to keep the end of the bone in its proper place.

II. ACROMIO-CLAVICULAR ARTICULATION (fig. 363)

The **Acromio-clavicular** is an arthrodial joint, formed between the outer extremity of the clavicle and the inner margin of the acromion process of the scapula. Its ligaments are, the

Capsular.
Superior Acromio-clavicular.
Inferior Acromio-clavicular.

Interarticular Fibro-cartilage.
Coraco-clavicular { Trapezoid and Conoid.

The **Capsular Ligament** completely surrounds the articular margins, and is specially strong above and below, where it forms the superior and inferior acromio-clavicular ligaments. It consists of fibres arranged parallel to each other and passing between the adjacent borders of the two bones.

The **Superior Acromio-clavicular Ligament** is a quadrilateral band, a part of the capsular ligament, which covers the superior part of the articulation, extending between the upper part of the outer end of the clavicle and the adjoining part of the upper surface of the acromion. It is composed of parallel fibres, which interlace with the aponeurosis of the Trapezius and Deltoid muscles; below, it is in contact with the interarticular fibro-cartilage (when it exists) and the synovial membranes.

The **Inferior Acromio-clavicular Ligament**, also a part of the capsular ligament, is somewhat thinner than the preceding; it covers the under part of the articulation, and is attached to the adjoining surfaces of the two bones. It is in relation, above, with the synovial membranes, and in rare cases with the interarticular fibro-cartilage; below, with the tendon of the Supraspinatus.

The **Interarticular Fibro-cartilage** is frequently absent in this articulation. When it exists, it generally only partially separates the articular surfaces, and occupies the upper part of the articulation. More rarely, it completely separates the joint into two cavities.

The Synovial Membrane.—There is usually only one synovial membrane in this articulation, but when a complete interarticular fibro-cartilage exists, there are two.

The **Coraco-clavicular Ligament** serves to connect the clavicle with the coracoid process of the scapula. It does not properly belong to this articulation, but as it forms a most efficient means in retaining the clavicle in contact with the acromial process, it is usually described with it. It consists of two fasciculi, called the *trapezoid* and *conoid ligaments*.

The *Trapezoid Ligament*, the anterior and external fasciculus, is broad, thin, and quadrilateral: it is placed obliquely between the coracoid process and the clavicle. It is attached, below, to the upper surface of the coracoid process; above, to the oblique line on the under surface of the clavicle. Its anterior border is free; its posterior border is joined with the conoid ligament: the two forming, by their junction, a projecting angle.

The *Conoid Ligament*, the posterior and internal fasciculus, is a dense band of fibres, conical in form, the base being directed upwards, the summit downwards. It is attached by its apex to a rough impression at the base of the coracoid process, internal to the preceding; above, by its expanded base, to the conoid tubercle on the under surface of the clavicle, and to a line proceeding internally from it for half an inch. These ligaments are in relation, in front, with the Subclavius and Deltoid; behind, with the Trapezius. They serve to limit rotation of the scapula; the Trapezoid limiting rotation forwards, and the Conoid backwards.

Actions.—The movements of this articulation are of two kinds. 1. A gliding motion of the articular end of the clavicle on the acromium. 2. Rotation of the

FIG. 363.—The left shoulder-joint, acromio-clavicular articulation, and proper ligaments of scapula.

scapula forwards and backwards upon the clavicle, the extent of this rotation being limited by the two portions of the coraco-clavicular ligament.

The acromio-clavicular joint has important functions in the movements of the upper extremity. It has been well pointed out by Humphry, that if there had been no joint between the clavicle and scapula, the circular movement of the scapula on the ribs (as in throwing the shoulders backwards or forwards) would have been attended with a greater alteration in the direction of the shoulder than is consistent with the free use of the arm in such positions, and it would have been impossible to give a blow straight forwards with the full force of the arm; that is to say, with the combined force of the scapula, arm, and forearm. 'This joint,' as he happily says, 'is so adjusted as to enable either bone to turn in a hinge-like manner upon a vertical axis drawn through the other, and it permits the surfaces of the scapula, like the baskets in a roundabout swing, to look the same way in every position, or nearly so.' Again, when the whole arch formed by the clavicle and scapula rises and falls (in elevation or depression of the shoulders), the joint between these two bones enables the scapula still to maintain its lower part in contact with the ribs.

Surface Form.—The position of the acromio-clavicular joint can generally be ascertained by the slightly enlarged extremity of the outer end of the clavicle, which causes it to project above the level of the acromion process of the scapula. Sometimes this enlargement is so considerable as to form a rounded eminence, which is easily to be felt. The joint lies in the plane of a vertical line passing up the middle of the front of the arm.

Surgical Anatomy.—The acromio-clavicular joint owes its security mainly to the coraco-clavicular ligament, which limits the amount of movement of the outer end of the clavicle either upwards, backwards, or forwards. Owing to the slanting shape of the articular surfaces of this joint, dislocation generally occurs downwards: that is to say, the acromion process of the scapula is dislocated under the outer end of the clavicle; but dislocation in the opposite direction has been described. The displacement is often incomplete, on account of the strong coraco-clavicular ligaments, which remain untorn. The same difficulty exists, as in the sterno-clavicular dislocation, in maintaining the ends of the bone in position after reduction.

III. PROPER LIGAMENTS OF THE SCAPULA (fig. 363)

The proper ligaments of the scapula are, the

Coraco-acromial.   Transverse.
                        Spino-glenoid.

The **Coraco-acromial Ligament** is a strong triangular band, extending between the coracoid and acromial processes. It is attached, by its apex, to the summit of the acromion just in front of the articular surface for the clavicle; and by its broad base to the whole length of the outer border of the coracoid process. Its posterior fibres are directed inwards, its anterior fibres forwards and inwards. This ligament completes the vault formed by the coracoid and acromion processes for the protection of the head of the humerus. It is in relation, above, with the clavicle and under surface of the Deltoid; below, with the tendon of the Supra-spinatus muscle, a bursa being interposed. Its outer border is continuous with a dense lamina that passes beneath the Deltoid upon the tendons of the Supra- and Infra-spinatus muscles. This ligament is sometimes described as consisting of two marginal bands and a thinner intervening portion, the two bands being attached respectively to the apex and base of the coracoid process, and joining together at their attachment into the acromion process. When the Pectoralis minor is inserted, as occasionally is the case, into the capsule of the shoulder-joint instead of into the coracoid process, it passes between these two bands, and the intervening portion is then deficient.

The **Transverse** or **Suprascapular Ligament** converts the suprascapular notch into a foramen. It is a thin and flat fasciculus, narrower at the middle than at the extremities, attached by one end to the base of the coracoid process, and by the other to the inner extremity of the scapular notch. The suprascapular nerve passes through the foramen; the suprascapular vessels pass over the ligament. This ligament is sometimes ossified.

The **Spino-glenoid Ligament** consists of a band of fibres, situated on the posterior surface of the neck of the scapula and stretching from the outer border of the spine to the margin of the glenoid cavity. It forms an arch under which the suprascapular vessels and nerve pass as they enter the infraspinous fossa.

Movements of Scapula.—The scapula is capable of being moved upwards and downwards, forwards and backwards, or, by a combination of these movements, circumducted on the wall of the chest. The muscles which *raise* the scapula are the upper fibres of the Trapezius, the Levator anguli scapulæ, and the two Rhomboids; those which *depress* it are the lower fibres of the Trapezius, the Pectoralis minor, and, through the clavicle, the Subclavius. The scapula is drawn *backwards* by the Rhomboids and the middle and lower fibres of the Trapezius, and *forwards* by the Serratus magnus and Pectoralis minor, assisted, when the arm is fixed, by the Pectoralis major. The mobility of the scapula is very considerable, and greatly assists the movements of the arm at the shoulder-joint. Thus, in raising the arm from the side, the Deltoid and Supraspinatus can only lift it to a right angle with the trunk, the further elevation of the limb being effected by the Trapezius and Serratus magnus moving the scapula on the wall of the chest. This mobility is of special importance in ankylosis of the shoulder-joint, the movements of this bone compensating to a very great extent for the immobility of the joint.

IV. SHOULDER-JOINT (fig. 363)

The **Shoulder** is an enarthrodial or ball-and-socket joint. The bones entering into its formation are the large globular head of the humerus, which is received into the shallow glenoid cavity of the scapula, an arrangement which permits of very considerable movement, while the joint itself is protected against displacement by the tendons which surround it and by atmospheric pressure. The ligaments do not maintain the joint surfaces in apposition, because when they alone remain the humerus can be separated to a considerable extent from the glenoid cavity; their use, therefore, is to limit the amount of movement. Above, the joint is protected by an arch, formed by the under surface of the coracoid and acromion processes, and the coraco-acromial ligament. The articular surfaces are covered by a layer of cartilage: that on the head of the humerus is thicker at the centre than at the circumference, the reverse being the case in the glenoid cavity. The ligaments of the shoulder are, the

Capsular.	Gleno-humeral.
Coraco-humeral.	Transverse Humeral.

Glenoid.*

The **Capsular Ligament** completely encircles the articulation, being attached, above, to the circumference of the glenoid cavity beyond the glenoid ligament; below, to the anatomical neck of the humerus, approaching nearer to the articular cartilage above than in the rest of its extent. It is thicker above and below than elsewhere, and is remarkably loose and lax, and much larger and longer than is necessary to keep the bones in contact, allowing them to be separated from each other more than an inch, an evident provision for that extreme freedom of movement which is peculiar to this articulation. It is strengthened, above, by the Supraspinatus; below, by the long head of the Triceps; behind, by the tendons of the Infraspinatus and Teres minor; and in front, by the tendon of the Subscapularis. The capsular ligament usually presents three openings: one anteriorly, below the coracoid process, establishes a communication between the synovial membrane of the joint and a bursa beneath the tendon of the Subscapularis muscle. The second, which is not constant, is at the posterior part, where a communication sometimes exists between the joint and a bursal sac under the tendon of the Infraspinatus muscle. The third is seen between the tuberosities of the humerus, for the passage of the long tendon of the Biceps muscle.

The **Coraco-humeral** is a broad band which strengthens the upper part of the capsular ligament. It arises from the outer border of the coracoid process, and passes obliquely downwards and outwards to the front of the great tuberosity of the humerus, being blended with the tendon of the Supraspinatus muscle. This ligament is intimately united to the capsular ligament by its hinder and lower border; but its superior and anterior border presents a free edge, which overlaps the capsular ligament.

Gleno-humeral Ligaments.—In addition to the coraco-humeral ligament, the capsular ligament is strengthened by three supplemental bands in the interior of the joint, which are named the *gleno-humeral ligaments*. These may be best seen by opening the capsule at the back of the joint and removing the head of the humerus. One of them is situated on the inner side of the joint, and passes from the inner edge of the glenoid cavity to the lower part of the lesser tuberosity of the humerus. This is sometimes known as *Flood's ligament*. A second is situated at the lower part of the joint, and passes from the under edge of the glenoid cavity to the under part of the anatomical neck of the humerus, and is known as *Schlemm's ligament*. A third is situated at the upper part of the joint, and is fixed above to the apex of the glenoid

---

* The long tendon of origin of the Biceps muscle also acts as one of the ligaments of this joint. See the observations on page 351, on the function of the muscles passing over more than one joint.

cavity close to the root of the coracoid process, and passing downwards along
the inner edge of the tendon of the Biceps, is attached below to the lesser
tuberosity of the humerus. In addition to these, the capsule is strengthened
externally on its anterior aspect by two bands derived from the tendons of
the Pectoralis major and Teres major respectively.

The Transverse Humeral Ligament.—This is a broad band of fibrous tissue
passing from the lesser to the greater tuberosity of the humerus, and always
limited to that portion of the bone which lies above the epiphysial line. It converts the bicipital groove into an osseo-aponeurotic canal, and is the analogue of
the strong process of bone which connects the summits of the two tuberosities
in the musk ox.

The **Glenoid Ligament** is a fibro-cartilaginous rim attached round the margin
of the glenoid cavity. It is triangular on section, the thickest portion being
fixed to the circumference of the cavity, the free edge being thin and sharp.
It is continuous above with the long tendon of the Biceps muscle, which
gives off two fasciculi, to blend with the fibrous tissue of the ligament.
This ligament deepens the cavity for articulation, and protects the edges of the
bone.

The **Synovial Membrane** is reflected from the margin of the glenoid cavity
over the glenoid ligament which surrounds it; it is then reflected over the

FIG. 364.—Vertical sections through the shoulder-joint, the arm being vertical
and horizontal. (After Henle.)

internal surface of the capsular ligament, and covers the lower part and sides of
the anatomical neck of the humerus as far as the cartilage covering the head
of the bone. The long tendon of the Biceps muscle which passes through the
capsular ligament is enclosed in a tubular sheath of synovial membrane, which
is reflected upon it at the point where it perforates the capsule, and is continued
around it as far as the summit of the glenoid cavity. The tendon of the Biceps
thus traverses the articulation, but it is not contained in the interior of the
synovial cavity.

Bursæ.—The bursæ in the neighbourhood of the shoulder-joint are the
following: (1) A constant bursa is situated between the tendon of the Subscapularis muscle and the capsule of the joint, which communicates with the
synovial membrane through an opening in the front of the capsular ligament:
(2) a bursa is sometimes found between the tendon of the Infraspinatus and
the capsule, which occasionally communicates with the joint: (3) a large bursa
exists between the under surface of the Deltoid muscle and the capsule, which
does not communicate with the joint; this bursa is prolonged under the
acromion process and coraco-acromial ligament, and intervenes between these
structures and the capsule of the joint: (4) a large bursa mucosa is situated
on the summit of the acromion: (5) a bursa is frequently found between the
coracoid process and the capsule of the joint: (6) there is a bursa beneath
the Coraco-brachialis muscle; (7) one between the Teres major and the long

head of the Triceps ; (8) one in front of, and another behind the tendon of the Latissimus dorsi.

The *Muscles* in relation with the joint are, above, the Supraspinatus ; below, the long head of the Triceps ; in front, the Subscapularis ; behind, the Infraspinatus and Teres minor ; within, the long tendon of the Biceps. The Deltoid is placed most externally, and covers the articulation on its outer side, as well as in front and behind.

The *Arteries* supplying the joint are articular branches of the anterior and posterior circumflex, and suprascapular.

The *Nerves* are derived from the circumflex and suprascapular.

Actions.—The shoulder-joint is capable of movement in every direction, forwards, backwards, abduction, adduction, circumduction, and rotation. The humerus is drawn *forwards* by the Pectoralis major, anterior fibres of the Deltoid, Coraco-brachialis, and by the Biceps, when the forearm is flexed ; *backwards* by the Latissimus dorsi, Teres major, posterior fibres of the Deltoid, and by the Triceps when the forearm is extended ; it is *abducted* (elevated) by the Deltoid and Supraspinatus ; it is *adducted* (depressed) by the Subscapularis, Pectoralis major, Latissimus dorsi, and Teres major, and by the weight of the limb ; it is *rotated outwards* by the Infraspinatus and Teres minor ; and it is *rotated inwards* by the Subscapularis, Latissimus dorsi, Teres major, and Pectoralis major.

The most striking peculiarities in this joint are : 1. The large size of the head of the humerus in comparison with the depth of the glenoid cavity, even when supplemented by the glenoid ligament. 2. The looseness of the capsule of the joint. 3. The intimate connection of the capsule with the muscles attached to the head of the humerus. 4. The peculiar relation of the Biceps tendon to the joint.

It is in consequence of the relative size of the two articular surfaces, and the looseness of the capsular ligament, that the joint enjoys such free movement in every possible direction. When these movements of the arm are arrested in the shoulder-joint by the contact of the bony surfaces, and by the tension of the corresponding fibres of the capsule, together with that of the muscles acting as accessory ligaments, they can be carried considerably farther by the movements of the scapula, involving, of course, motion at the acromio- and sterno-clavicular joints. These joints are therefore to be regarded as accessory structures to the shoulder-joint (see page 377). The extent of these movements of the scapula is very considerable, especially in extreme elevation of the arm, which movement is best accomplished when the arm is thrown somewhat forwards and outwards, because the margin of the head of the humerus is by no means a true circle ; its greatest diameter is from the bicipital groove, downwards, inwards, and backwards, and the greatest elevation of the arm can be obtained by rolling its articular surface in the direction of this measurement. The great width of the central portion of the humeral head also allows of very free horizontal movement when the arm is raised to a right angle, in which movement the arch formed by the acromion, the coracoid process, and the coraco-acromial ligament, constitutes a sort of supplemental articular cavity for the head of the bone.

The looseness of the capsule is so great that the arm will fall about an inch from the scapula when the muscles are dissected from the capsular ligament, and an opening made in it to counteract the atmospheric pressure. The movements of the joint, therefore, are not regulated by the capsule so much as by the surrounding muscles and by the pressure of the atmosphere, an arrangement which 'renders the movements of the joint much more easy than they would otherwise have been, and permits a swinging, pendulum-like vibration of the limb when the muscles are at rest' (Humphry). The fact, also, that in all ordinary positions of the joint the capsule is not put on the stretch, enables the arm to move freely in all directions. Extreme movements are checked by the tension of appropriate portions of the capsule, as well as by the interlocking of the bones. Thus it is said that 'abduction is checked by the contact of the great tuberosity with the upper edge of the glenoid cavity ; adduction by the tension of the coraco-humeral ligament' (Beaunis et Bouchard). Cleland * maintains that the

* *Journ. of Anat. and Phys.* No. 1, 1886, p. 85.

limitations of movement at the shoulder-joint are due to the structure of the joint itself, the glenoid ligament fitting, in different positions of the elevated arm, into the anatomical neck of the humerus.

Cathcart * has pointed out that in abducting the arm and raising it above the head, the scapula rotates throughout the whole movement with the exception of a short space at the beginning and at the end; that the humerus moves on the scapula not only from the hanging to the horizontal position but also in passing upwards as it approaches the vertical above; that the clavicle moves not only during the second half of the movement but in the first as well, though to a less extent—i.e. the scapula and clavicle are concerned in the first stage as well as in the second; and that the humerus is partly involved in the second as well as chiefly in the first.

The intimate union of the tendons of the four short muscles with the capsule converts these muscles into elastic and spontaneously acting ligaments of the joint, and it is regarded as being also intended to prevent the folds into which all portions of the capsule would alternately fall in the varying positions of the joint from being driven between the bones by the pressure of the atmosphere.

The peculiar relations of the Biceps tendon to the shoulder-joint appear to subserve various purposes. In the first place, by its connection with both the shoulder and elbow the muscle harmonises the action of the two joints, and acts as an elastic ligament in all positions, in the manner previously adverted to (see page 351). Next, it strengthens the upper part of the articular cavity, and prevents the head of the humerus from being pressed up against the acromion process, when the Deltoid contracts, instead of forming the centre of motion in the glenoid cavity. By its passage along the bicipital groove it assists in rendering the head of the humerus steady in the various movements of the arm. When the arm is raised from the side it assists the Supra- and Infra-spinatus in rotating the head of the humerus in the glenoid cavity. It also holds the head of the bone firmly in contact with the glenoid cavity, and prevents its slipping over its lower edge, or being displaced by the action of the Latissimus dorsi and Pectoralis major, as in climbing and many other movements.

Surface Form.—The direction and position of the shoulder-joint may be indicated by a line drawn from the middle of the coraco-acromial ligament, in a curved direction, with its convexity inwards, to the innermost part of that portion of the head of the humerus which can be felt in the axilla when the arm is forcibly abducted from the side. When the arm hangs by the side, not more than one-third of the head of the bone is in contact with the glenoid cavity, and three-quarters of its circumference is in front of a vertical line drawn from the anterior border of the acromion process.

Surgical Anatomy.—Owing to the construction of the shoulder-joint and the freedom of movement which it enjoys, as well as in consequence of its exposed situation, it is more frequently dislocated than any other joint in the body. Dislocation occurs when the arm is abducted, and when, therefore, the head of the humerus presses against the lower and front part of the capsule, which is the thinnest and least supported part of the ligament. The rent in the capsule almost invariably takes place in this situation, and through it the head of the bone escapes, so that the dislocation in most instances is primarily subglenoid. The head of the bone does not usually remain in this situation, between the tendons of the Subscapularis and the Triceps, but generally assumes some other position, which varies according to the direction and amount of force producing the dislocation and the relative strength of the muscles in front and behind the joint. In consequence of the muscles at the back being stronger than those in front, and especially on account of the long head of the Triceps preventing the bone passing backwards, dislocation forwards is much more common than backwards. The most frequent position which the head of the humerus ultimately assumes is on the front of the neck of the scapula, beneath the coracoid process, and hence named subcoracoid dislocation. Occasionally, in consequence probably of a greater amount of force being brought to bear on the limb, the head is driven farther inwards, and rests on the upper part of the front of the chest, beneath the clavicle (subclavicular). Sometimes it remains in the position in which it was primarily displaced, resting on the axillary border of the scapula (subglenoid), and rarely it passes backwards and remains in the infraspinatus fossa, beneath the spine (subspinous).

The shoulder-joint is sometimes the seat of all those inflammatory affections, both acute and chronic, which attack joints, though perhaps less frequently than some other articulations of equal size and importance. Acute synovitis may result from injury, rheumatism, or pyæmia, or may follow secondarily on acute epiphysitis of infants. It is attended with

* *Journ. of Anat. and Phys.* vol. xviii. 1884.

effusion into the joint, and when this occurs the capsule is evenly distended, and the contour of the joint rounded. Special projections may occur at the site of the openings in the capsular ligament. Thus a swelling may appear just in front of the joint, internal to the lesser tuberosity, from effusion into the bursa beneath the Subscapularis muscle ; or, again, a swelling which is sometimes bilobed may be seen in the interval between the Deltoid and Pectoralis major muscles, from effusion into the diverticulum, which runs down the bicipital groove with the tendon of the Biceps. The effusion into the synovial membrane can be best ascertained by examination from the axilla, where a soft, elastic, fluctuating swelling can usually be felt. In cases of septic synovitis, where incision is required, the opening should be made in front, over the most prominent point of the swelling. After the pus has been evacuated, a counter-opening should be made behind, so as to ensure efficient drainage.

Tuberculous arthritis not infrequently attacks the shoulder-joint, and may lead to total destruction of the articulation, when ankylosis may result, or long-protracted suppuration may necessitate excision. This joint is also one of those which is most liable to be the seat of osteo-arthritis, and may also be affected in gout and rheumatism ; or in locomotor ataxy, when it becomes the seat of Charcot's disease.

Ankylosis is occasionally met with in the shoulder-joint, as the result of destructive changes. The ankylosis usually takes place with the arm in a dependent position, and any attempt to raise the arm is attended by a rotation of the scapula on the wall of the chest.

Excision of the shoulder-joint may be required in cases of arthritis (especially the tuberculous form) which have gone on to destruction of the articulation ; in compound dislocations and fractures, particularly those arising from gunshot injuries, in which there has been extensive injury to the head of the bone ; in some cases of old unreduced dislocation, where there is much pain ; and possibly in some few cases of growth connected with the upper end of the bone. The operation is best performed by making an incision from the middle of the coraco-acromial ligament down the arm for about three inches : this will expose the bicipital groove and the tendon of the Biceps, which may be either divided or hooked out of the way, according as to whether it is implicated in the disease or not. The capsule is freely opened, and the muscles attached to the greater and lesser tuberosities of the humerus divided. The head of the bone can then be thrust out of the wound and sawn off ; or divided with a narrow saw *in situ* and subsequently removed. The section should be made, if possible, just below the articular surface, so as to leave the bone as long as possible. The glenoid cavity must then be examined, and gouged if carious.

V. ELBOW-JOINT (figs. 365, 366)

The **Elbow** is a *ginglymus* or hinge-joint. The bones entering into its formation are the trochlea of the humerus, which is received into the greater sigmoid cavity of the ulna, and admits of the movements peculiar to this joint, viz. flexion and extension; while the lesser, or radial, head of the humerus articulates with the cup-shaped depression on the head of the radius; the circumference of the head of the radius articulates with the lesser sigmoid cavity of the ulna, allowing of the movement of rotation of the radius on the ulna, the chief action of the superior radio-ulnar articulation. The articular surfaces are covered with a thin layer of cartilage, and connected together by a capsular ligament of unequal thickness, being especially thickened on its two sides, and, to a less extent, in front and behind. These thickened portions are usually described as distinct ligaments under the following names :

Anterior.   Internal Lateral.
Posterior.   External Lateral.

The **Anterior Ligament** (fig. 365) is a broad and thin fibrous layer, which covers the anterior surface of the joint. It is attached to the front of the internal epicondyle and to the front of the humerus immediately above the coronoid and radial fossæ; below, to the anterior surface of the coronoid process of the ulna and orbicular ligament, being continuous on each side with the lateral ligaments. Its superficial fibres pass obliquely from the inner condyle of the humerus outwards to the orbicular ligament. The middle fibres, vertical in direction, pass from the upper part of the coronoid depression and become partly blended with the preceding, but mainly inserted into the anterior surface of the coronoid process. The deep or transverse set intersects these at right angles. This ligament is in relation, in front, with the Brachialis anticus, except at its outermost part; behind, with the synovial membrane.

The **Posterior Ligament** (fig. 366) is thin and membranous, and consists of transverse and oblique fibres. Above, it is attached to the humerus immediately behind the capitellum and close to the inner margin of the trochlear surface, to the lateral margins of the olecranon fossa, and to the back of the external condyle some little distance from the trochlear surface. Below, it is fixed to the upper and outer aspects of the olecranon process, to the posterior part of the orbicular ligament, and to the ulna behind the lesser sigmoid cavity. The transverse fibres form a strong band which bridges across the olecranon fossa ; under cover of this band a pouch of synovial membrane and a pad of fat project into the upper part of the fossa when the joint is extended In the fat are a few scattered fibrous bundles, which pass from the deep aspect of the transverse band to the upper part of the fossa. This ligament is in relation, behind, with the tendon of the Triceps and the Anconeus ; in front, with the synovial membrane.

Fig. 365.—Left elbow-joint, showing anterior and internal ligaments.

Fig. 366.—Left elbow-joint, showing posterior and external ligaments.

The **Internal Lateral Ligament** (fig. 365) is a thick triangular band consisting of two portions, an anterior and posterior, united by a thinner intermediate portion. The *anterior portion*, directed obliquely forwards, is attached, above, by its apex, to the front part of the internal epicondyle of the humerus ; and, below, by its broad base, to the inner margin of the coronoid process. The *posterior portion*, also of triangular form, is attached, above, by its apex, to the lower and back part of the internal epicondyle ; below, to the inner margin of the olecranon. Between these two bands a few intermediate fibres descend from the internal epicondyle to blend with a *transverse band* of ligamentous tissue which bridges across the notch between the olecranon and coronoid processes. This ligament is in relation, internally, with the Triceps and Flexor carpi ulnaris muscles, and the ulnar nerve, and gives origin to part of the Flexor sublimis digitorum.

The **External Lateral Ligament** (fig. 366) is a short and narrow fibrous band, less distinct than the internal, attached, above, to a depression below the external epicondyle of the humerus ; below, to the orbicular ligament, some of its most posterior fibres passing over that ligament, to be inserted into the outer margin of the ulna. This ligament is intimately blended with the tendon of origin of the Supinator brevis muscle.

The **Synovial Membrane** is very extensive. It extends from the margin of the articular surface of the humerus, and lines the coronoid and olecranon fossæ on that bone : from these points, it is reflected over the anterior, posterior, and lateral ligaments ; and forms a pouch between the lesser sigmoid cavity, the internal surface of the orbicular ligament, and the circumference of the head of the radius. Projecting into the cavity is a crescentic fold of synovial membrane, between the radius and ulna, suggesting the division of the joint into two : one the humero-radial, the other the humero-ulnar.

Between the capsular ligament and the synovial membrane are three masses of fat : one, the largest, over the olecranon fossa, which is pressed into the fossa by the Triceps during flexion ; a second, over the coronoid fossa ; and a third over the radial fossa. These are pressed into their respective fossæ during extension.

The *Muscles* in relation with the joint are, in front, the Brachialis anticus ; behind, the Triceps and Anconeus ; externally, the Supinator brevis, and the common tendon of origin of the Extensor muscles ; internally, the common tendon of origin of the Flexor muscles, and the Flexor carpi ulnaris, with the ulnar nerve (fig. 367).

The *Arteries* supplying the joint are derived from the anastomosis between the superior profunda, inferior profunda, and anastomotic branches of the brachial, with the anterior, posterior, and interosseous recurrent branches of the ulnar, and the recurrent branch of the radial. These vessels form a complete chain of inosculation around the joint.

The *Nerves* are derived from the ulnar, as it passes between the internal condyle and the olecranon ;

FIG. 367.—Sagittal section of the right elbow-joint, taken somewhat obliquely and seen from the radial aspect. (After Braune.)

Median cephalic vein

Head of radius

Radial nerve

a filament from the musculo-cutaneous (Rüdinger), and two from the median (Macalister).

Actions.—The elbow-joint comprises three different portions, viz. : the joint between the ulna and humerus, that between the head of the radius and the humerus, and the superior radio-ulnar articulation, described below. All these articular surfaces are invested by a common synovial membrane, and the movements of the whole joint should be studied together. The combination of the movements of flexion and extension of the forearm with those of pronation and supination of the hand, which is ensured by the two being performed at the same joint, is essential to the accuracy of the various minute movements of the hand.

The portion of the joint between the ulna and humerus is a simple hinge-joint, and allows of movements of flexion and extension only. Owing to the obliquity of the trochlear surface of the humerus, this movement does not take place in a straight line ; so that when the forearm is extended and supinated, the axes of the arm and forearm are not in the same line, the upper portion of the

C C

limb forming an angle with the lower, and the hand, with the forearm, being directed outwards. During flexion, on the other hand, the forearm and the hand tend to approach the middle line of the body, and thus enable the hand to be easily carried to the face. The shape of the trochlear surface of the humerus, with its prominences and depressions, accurately adapted to the great sigmoid cavity, prevents any lateral movement. *Flexion* is produced by the action of the Biceps and Brachialis anticus, assisted by the muscles arising from the internal condyle of the humerus and the Supinator longus ; *extension*, by the Triceps and Anconeus, assisted by the Extensors of the wrist and by the Extensor communis digitorum and Extensor minimi digiti.

The joint between the head of the radius and the capitellum or radial head of the humerus is an arthrodial joint. The bony surfaces would of themselves constitute an enarthrosis and allow of movement in all directions, were it not for the orbicular ligament by which the head of the radius is bound down firmly to the sigmoid cavity of the ulna, and which prevents any separation of the two bones laterally. It is to the same ligament that the head of the radius owes its security from dislocation, which would otherwise occur, as a consequence of the shallowness of the cup-like surface on the head of the radius. In fact, but for this ligament, the tendon of the Biceps would be liable to pull the head of the radius out of the joint.* In complete extension, the head of the radius glides so far backwards that its edge is plainly felt at the back of the articulation. Flexion and extension of the elbow-joint are limited by the tension of the structures on the front and back of the joint ; the limitation of flexion being also aided by the soft structures of the arm and forearm coming into contact.

In combination with any position of flexion or extension, the head of the radius can be rotated in the upper radio-ulnar joint, carrying the hand with it. The hand is directly articulated to the lower surface of the radius only, and the concave or sigmoid surface on the lower end of the radius travels round the lower end of the ulna. The latter bone is excluded from the wrist-joint by the inter-articular fibro-cartilage. Thus, rotation of the head of the radius round an axis which passes through the centre of the radial head of the humerus imparts circular movement to the hand through a very considerable arc.

Surface Form.—If the forearm be slightly flexed on the arm, a curved crease or fold with its convexity downwards may be seen running across the front of the elbow, extending from one condyle to the other. The centre of this fold is some slight distance above the line of the joint. The position of the radio-humeral portion of the joint can be at once ascertained by feeling for a slight groove or depression between the head of the radius and the capitellum of the humerus at the back of the articulation.

Surgical Anatomy.—From the great breadth of the joint, and the manner in which the articular surfaces are interlocked, and also on account of the strong lateral ligaments and the support which the joint derives from the mass of muscles attached to each condyle of the humerus, lateral displacement of the bones is very uncommon; whereas antero-posterior dislocation, on account of the shortness of the antero-posterior diameter, the weakness of the anterior and posterior ligaments, and the want of support of muscles, occurs much more frequently. Dislocation backwards takes place when the forearm is in a position of extension, and forwards when in a position of flexion. For, in the former position, that of extension, the coronoid process is not interlocked into the coronoid fossa, and loses its grip to a certain extent, whereas the olecranon process is in the olecranon fossa, and entirely prevents displacement forwards. On the other hand, during flexion, the coronoid process is in the coronoid fossa, and prevents dislocation backwards, while the olecranon loses its grip and is not so efficient, as during extension, in preventing a forward displacement. When lateral dislocation does take place it is generally incomplete.

Dislocation of the elbow-joint is of common occurrence in children, far more common than dislocation of any other articulation, for, as a rule, fracture of a bone more frequently takes place, under the application of any severe violence, in young persons, than dislocation. In lesions of this joint there is often very great difficulty in ascertaining the exact nature of the injury.

The elbow-joint is occasionally the seat of acute synovitis. The joint cavity then becomes distended with fluid, the bulging showing itself principally around the olecranon process, that is to say, on its inner and outer sides and above, in consequence of the laxness of the posterior ligament. Sometimes a well-marked triangular projection may be seen on the outer side of the olecranon, from bulging of the synovial membrane

---

* Humphry, *op. cit.* p. 419.

beneath the Anconeus muscle. Again, there is often some swelling just above the head of the radius, in the line of the radio-humeral joint. There is not generally much swelling at the front of the joint, though sometimes deep-seated fulness beneath the Brachialis anticus may be noted. When suppuration occurs the abscess usually points at one or other border of the Triceps muscle; occasionally the pus discharges itself in front, near the insertion of the Brachialis anticus muscle. In cases of suppurative synovitis, incisions should be made into the joint on either side of the olecranon, care being observed on the inner side to avoid wounding the ulnar nerve. Chronic synovitis, usually of tuberculous origin, is of common occurrence in the elbow-joint: under these circumstances the forearm tends to assume the position of semiflexion, which is that of greatest ease and relaxation of ligaments. It should be borne in mind, that should ankylosis occur in this or the extended position, the limb will not be nearly so useful as if ankylosed in a position of rather less than a right angle. Loose cartilages are sometimes met with in the elbow-joint, not so commonly, however, as in the knee; nor do they, as a rule, give rise to such urgent symptoms, and rarely require operative interference. The elbow-joint is also sometimes affected with osteo-arthritis, but this affection is less common in this articulation than in some other of the larger joints.

Excision of the elbow is principally required for three conditions—viz. tuberculous arthritis, injury and its results, and faulty ankylosis—but may be necessary for some other rarer conditions, such as disorganising arthritis after pyæmia, unreduced dislocations, and osteo-arthritis. The results of the operation are, as a rule, more favourable than those of excision of any other joint, and it is one, therefore, that the surgeon should never hesitate to perform, especially in the first three of the conditions mentioned above. The operation is best performed by a single vertical incision down the back of the joint; a transverse incision, over the outer condyle, being added if the parts are much thickened and fixed. A straight incision is made about four inches long, the mid-point of which is on a level with and a little to the inner side of the tip of the olecranon. This incision is made down to the bone, through the substance of the Triceps muscle. The operator with the point of his knife, and guarding the soft parts with his thumb-nail, separates them from the bone. In doing this there are two structures which he should carefully avoid: the ulnar nerve, which lies parallel to his incision, but a little internal, as it courses down between the internal condyle and the olecranon process; and the prolongation of the Triceps into the deep fascia of the forearm over the Anconeus muscle. Having cleared the bones and divided the lateral and posterior ligaments, the forearm is strongly flexed and the ends of the bones turned out and sawn off. The turning out of the ends of the bones is rendered easier by first cutting off the olecranon process with a pair of cutting bone forceps. The section of the humerus should be through the base of the condyles, that of the ulna and radius should be just below the level of the lesser sigmoid cavity of the ulna and the neck of the radius. In this operation the object is to obtain such union as shall allow free motion of the bones of the forearm; and, therefore, passive motion must be commenced early—that is to say, about the tenth day.

VI. RADIO-ULNAR ARTICULATIONS

The articulation of the radius with the ulna is effected by ligaments, which connect together both extremities as well as the shafts of these bones. They may, consequently, be subdivided into three sets: 1. The superior radio-ulnar articulation; 2. the middle radio-ulnar union; and, 3. the inferior radio-ulnar articulation.

1. SUPERIOR RADIO-ULNAR ARTICULATION

This articulation is a trochoid or pivot-joint. The parts entering into its formation are the inner side of the circumference of the head of the radius and the lesser sigmoid cavity of the ulna. Its only ligament is the *annular* or *orbicular*.

The **Orbicular Ligament** (fig. 366) is a strong band of ligamentous fibres, which surrounds the head of the radius, and retains it in firm connection with the lesser sigmoid cavity of the ulna. It forms about four-fifths of an osseo-fibrous ring, attached by each end to the extremities of the lesser sigmoid cavity, and is smaller at the lower part of its circumference than above, by which means the head of the radius is more securely held in its position. Its upper border blends with the anterior and posterior ligaments of the elbow, while its *outer surface* is strengthened by the external lateral ligament of the elbow, and affords origin to part of the Supinator brevis muscle. Its *inner surface* is smooth, and lined by synovial membrane. The synovial membrane is continuous with that which lines the elbow-joint.

Actions.—The movement which takes place in this articulation is limited to rotation of the head of the radius within the ring formed by the orbicular ligament, and the lesser sigmoid cavity of the ulna; rotation forwards being called *pronation*; rotation backwards, *supination*. Supination is performed by the Biceps and Supinator brevis, assisted to a slight extent by the Extensor muscles of the thumb and, in certain positions, by the Supinator longus. Pronation is performed by the Pronator radii teres and the Pronator quadratus, assisted, in some positions, by the Supinator longus.

Surface Form.—The position of the superior radio-ulnar joint is marked on the surface of the body by the little dimple on the back of the elbow which indicates the position of the head of the radius.

Surgical Anatomy.—Dislocation of the head of the radius alone is not an uncommon accident, and occurs most frequently in young persons from falls on the hand when the forearm is extended and supinated, the head of the bone being displaced forward. It is attended by rupture of the orbicular ligament. Occasionally a peculiar injury, which is supposed to be a subluxation, occurs in young children in lifting them from the ground by the hand or forearm. It is believed that the head of the radius is displaced downwards in the orbicular ligament, the upper border of which becomes folded over the head of the radius, between it and the capitellum of the humerus. The forearm becomes fixed in a position of semiflexion, midway between supination and pronation, and great pain is complained of upon any attempt to move the joint.

2. Middle Radio-ulnar Union

The shafts of the radius and ulna are connected by the

Oblique ligament and the Interosseous membrane.

The **Oblique** or **Round Ligament** (fig. 365) is a small, flattened, fibrous band, which extends obliquely downwards and outwards, from the tubercle of the ulna at the base of the coronoid process to the radius a little below the bicipital tuberosity. Its fibres run in the opposite direction to those of the interosseous ligament; and it appears to be placed as a substitute for it in the upper part of the interosseous interval. This ligament is sometimes wanting.

The **Interosseous Membrane** is a broad and thin plane of fibrous tissue descending obliquely downwards and inwards, from the interosseous ridge on the radius to that on the ulna. It is deficient above, commencing about an inch beneath the tubercle of the radius; is broader in the middle than at either extremity; and presents an oval aperture just above its lower margin for the passage of the anterior interosseous vessels to the back of the forearm. This ligament serves to connect the bones, and to increase the extent of surface for the attachment of the deep muscles. Between its upper border and the oblique ligament an interval exists, through which the posterior interosseous vessels pass. Two or three fibrous bands are occasionally found on the posterior surface of this membrane, which descend obliquely from the ulna towards the radius, and which have consequently a direction contrary to that of the other fibres. It is in relation, *in front*, by its upper three-fourths, with the Flexor longus pollicis on the outer side, and with the Flexor profundus digitorum on the inner, lying in the interval between which are the anterior interosseous vessels and nerve; by its lower fourth with the Pronator quadratus; *behind*, with the Supinator brevis, Extensor ossis metacarpi pollicis, Extensor brevis pollicis, Extensor longus pollicis, Extensor indicis; and, near the wrist, with the anterior interosseous artery and posterior interosseous nerve.

3. Inferior Radio-ulnar Articulation

This is a pivot-joint, formed by the head of the ulna received into the sigmoid cavity at the inner side of the lower end of the radius. The articular surfaces are covered by a thin layer of cartilage, and connected together by the following ligaments:

Anterior Radio-ulnar.   Posterior Radio-ulnar.
Interarticular Fibro-cartilage.

The **Anterior Radio-ulnar Ligament** (fig. 368) is a narrow band of fibres extending from the anterior margin of the sigmoid cavity of the radius to the anterior surface of the head of the ulna.

FIG. 368.—Ligaments of wrist and hand.   Anterior view.

The **Posterior Radio-ulnar Ligament** (fig. 369) extends between similar points on the posterior surface of the articulation.

The **Interarticular Fibro-cartilage** (fig. 371) is triangular in shape, and is placed transversely beneath the head of the ulna, binding the lower end of this

FIG. 369.—Ligaments of wrist and hand.   Posterior view.

bone and the radius firmly together.   Its periphery is thicker than its centre, which is thin and occasionally perforated.   It is attached by its apex to a depression which separates the styloid process of the ulna from the head of that bone; and by its base, which is thin, to the prominent edge of the radius, which separates the sigmoid cavity from the carpal articulating surface. Its margins are united to the ligaments of the wrist-joint. Its *upper surface*, smooth and concave, articulates with the head of the ulna, forming an arthrodial joint; its *under surface*, also concave and smooth, forms part of the wrist-joint and articulates with the cuneiform and inner part of the semilunar bones. Both surfaces are lined by a synovial membrane : the upper surface, by one peculiar to the radio-ulnar articulation; the under surface, by the synovial membrane of the wrist.

The **Synovial Membrane** (fig. 371) of this articulation has been called, from its extreme looseness, the *membrana sacciformis*; it extends horizontally inwards between the head of the ulna and the interarticular fibro-cartilage, and upwards between the radius and the ulna, forming here a very loose *cul-de-sac*. The quantity of synovia which it contains is usually considerable.

Actions.—The movement in the inferior radio-ulnar articulation is just the reverse of that in the superior radio-ulnar joint. It consists of a movement of rotation of the lower end of the radius round an axis which corresponds to the centre of the head of the ulna. When the radius rotates forwards, *pronation* of the forearm and hand is the result; and when backwards, *supination*. It will thus be seen that in pronation and supination of the forearm and hand the radius describes a segment of a cone, the axis of which extends from the centre of the head of the radius to the middle of the head of the ulna. In this movement, however, the ulna is not quite stationary, but rotates a little in the opposite direction. So that it also describes the segment of a cone, though of smaller size than that described by the radius. The movement which causes this alteration in the position of the head of the ulna takes place principally at the shoulder-joint by a rotation of the humerus, but possibly also to a slight extent at the elbow-joint.*

Surface Form.—The position of the inferior radio-ulnar joint may be ascertained by feeling for a slight groove at the back of the wrist, between the prominent head of the ulna and the lower end of the radius, when the forearm is in a state of almost complete pronation.

VII. RADIO-CARPAL OR WRIST-JOINT

The **Wrist** is a condyloid articulation. The parts entering into its formation are the lower end of the radius and under surface of the interarticular fibro-cartilage on the one hand; and the scaphoid, semilunar, and cuneiform bones on the other. The articular surface of the radius and the under surface of the interarticular fibro-cartilage form together a transversely elliptical concave surface, the receiving cavity. The articular surfaces of the scaphoid, semilunar, and cuneiform bones present a smooth, convex surface, the *condyle*, which is received into the concavity above mentioned. All the bony surfaces of the articulation are covered with cartilage, and connected together by a capsule, which is divided into the following ligaments :

| External Lateral. | Anterior. |
| Internal Lateral. | Posterior. |

The **External Lateral Ligament** (*radio-carpal*) (fig. 368) extends from the summit of the styloid process of the radius to the outer side of the scaphoid, some of its fibres being prolonged to the trapezium and annular ligament. It has in relation with it the radial artery, which separates the ligament from the tendons of the Extensor ossis metacarpi and Extensor brevis pollicis.

The **Internal Lateral Ligament** (*ulno-carpal*) is a rounded cord, attached, above, to the extremity of the styloid process of the ulna; and dividing below into two fasciculi, one of which is attached to the inner side of the cuneiform bone, the other to the pisiform bone and annular ligament.

The **Anterior Ligament** is a broad membranous band, attached, above, to the anterior margin of the lower end of the radius, to its styloid process and to the ulna; its fibres pass downwards and inwards to be inserted into the palmar surface of the scaphoid, semilunar, and cuneiform bones, some of the fibres being continued to the os magnum. In addition to this broad membrane, there

---

* See *Journ. of Anat. and Phys.* vol. xix. parts ii., iii. and iv.

is a distinct rounded fasciculus, superficial to the rest, which passes from the base of the styloid process of the ulna to the semilunar and cuneiform bones. This ligament is perforated by numerous apertures for the passage of vessels, and is in relation, in front, with the tendons of the Flexor profundus digitorum and Flexor longus pollicis; behind, with the synovial membrane of the wrist-joint, where it is closely adherent to the anterior border of the triangular fibro-cartilage of the inferior radio-ulnar articulation.

The **Posterior Ligament** (fig. 369), less thick and strong than the anterior, is attached, above, to the posterior border of the lower end of the radius; its fibres pass obliquely downwards and inwards, to be attached to the dorsal surface of the scaphoid, semilunar, and cuneiform bones, being continuous with those of the dorsal carpal ligaments. This ligament is in relation, behind, with the Extensor tendons of the fingers; in front, with the synovial membrane of the wrist, where it is blended with the triangular fibro-cartilage.

The **Synovial Membrane** (fig. 371) lines the inner surface of the ligaments above described, extending from the lower end of the radius and interarticular fibro-cartilage above to the articular surfaces of the carpal bones below. It is loose and lax, and presents numerous folds, especially behind.

Relations.—The wrist-joint is covered in front by the Flexor, and behind by the Extensor tendons; it is also in relation with the radial and ulnar arteries.

The *Arteries* supplying the joint are the anterior and posterior carpal branches of the radial and ulnar, the anterior and posterior interosseous, and some ascending branches from the deep palmar arch.

The *Nerves* are derived from the ulnar and posterior interosseous.

Actions.—The movements permitted in this joint are flexion, extension, abduction, adduction, and circumduction. Its actions will be further studied with those of the carpus, with which they are combined.

Surface Form.—On the front of the lower part of the forearm three transverse furrows may be generally seen: the uppermost is on a level with the styloid process of the ulna; the middle corresponds fairly accurately with the wrist-joint; and the lowest indicates the position of the mid-carpal articulation.

Surgical Anatomy.—The wrist-joint is rarely dislocated, its strength depending mainly upon the numerous strong tendons which surround the articulation. Its security is further provided for by the number of small bones of which the carpus is made up, and which are united by very strong ligaments. The slight movement which takes place between the several bones serves to break the jars that result from falls or blows on the hand. Dislocation backwards, which is the more common, simulates to a considerable extent Colles's fracture of the radius, and is liable to be mistaken for it. The diagnosis can be easily made out by observing the relative position of the styloid processes of the radius and the ulna. In the natural condition the styloid process of the radius is on a lower level, i.e. nearer the ground, when the arm hangs by the side, than that of the ulna, and the same would be the case in dislocation. In Colles's fracture, on the other hand, the styloid process of the radius is on the same, or even a higher level than that of the ulna.

The wrist-joint is occasionally the seat of acute synovitis, the result of traumatism, or

Fig. 370.—Longitudinal section of the right forearm, hand, and third finger, viewed from the ulnar aspect. (After Braune.)

arising in the rheumatic or pyæmic state.   When the synovial sac is distended with fluid,
the swelling is greatest on the dorsal aspect of the wrist, showing a general fulness, with
some bulging between the tendons.   The inflammation is prone to extend to the intercarpal joints and to attack also the sheaths of the tendons in the neighbourhood.   Chronic
inflammation of the wrist is generally tuberculous, and often leads to similar disease in
the synovial sheaths of adjacent tendons and of the intercarpal joints.   The disease, therefore, when progressive, frequently leads to necrosis of the carpal bones, and the result is
often unsatisfactory.

VIII. ARTICULATIONS OF THE CARPUS

These articulations may be subdivided into three sets :

> 1. The Articulations of the First Row of Carpal Bones.
> 2. The Articulations of the Second Row of Carpal Bones.
> 3. The Articulations of the Two Rows with each other.

1. ARTICULATIONS OF THE FIRST OR PROXIMAL ROW OF CARPAL BONES

These are arthrodial joints.   The ligaments connecting the scaphoid, semilunar, and cuneiform bones are—

Dorsal.   Palmar.
Two Interosseous.

The **Dorsal Ligaments,** two in number, are placed transversely behind the
bones of the first row; they connect the scaphoid and semilunar, and the semilunar and cuneiform.

The **Palmar Ligaments,** also two, connect the scaphoid and semilunar, and
the semilunar and cuneiform bones; they are less strong than the dorsal, and
placed very deeply below the Flexor tendons and the anterior ligament of the wrist.

The **Interosseous Ligaments** (fig. 371) are two narrow bundles of fibrous
tissue, connecting the semilunar bone, on one side with the scaphoid, and on the
other with the cuneiform.   They are on a level with the superior surfaces of
these bones, and their upper surfaces are smooth, and form with the bones
which they connect, the convex articular surface of the wrist-joint.

The ligaments connecting the pisiform bone are—

Capsular.   Two Palmar.

The **Capsular Ligament** is a thin membrane which connects the pisiform
bone to the cuneiform.   It is lined with a separate synovial membrane.

The **two Palmar Ligaments** are strong fibrous bands which connect the
pisiform to the unciform, the *piso-unciform,* and to the base of the fifth metacarpal bone, the *piso-metacarpal ligament* (fig. 368).   These ligaments are, in
reality, prolongations of the tendon of the Flexor carpi ulnaris.

2. ARTICULATIONS OF THE SECOND OR DISTAL ROW OF CARPAL BONES

These also are arthrodial joints.   The articular surfaces are covered with
cartilage, and connected by the following ligaments :

Dorsal.   Palmar.
Three Interosseous.

The **Dorsal Ligaments,** three in number, extend transversely from one bone
to another on the dorsal surface, connecting the trapezium with the trapezoid, the
trapezoid with the os magnum, and the os magnum with the unciform.

The **Palmar Ligaments,** also three, have a similar arrangement on the palmar
surface.

The **three Interosseous Ligaments,** much thicker than those of the first row,
are placed one between the os magnum and the unciform, a second between the
os magnum and the trapezoid, and a third between the trapezium and trapezoid.
The first of these is much the strongest, and the third is sometimes wanting.

3. ARTICULATIONS OF THE TWO ROWS OF CARPAL BONES WITH EACH OTHER

The joint between the scaphoid, semilunar, and cuneiform on the one hand, and the second row of the carpus on the other, is named the *mid-carpal joint*, and is made up of three distinct portions; in the centre the head of the os magnum and the superior surface of the unciform articulate with the deep cup-shaped cavity formed by the scaphoid and semilunar bones, and constitute a sort of ball-and-socket joint. On the outer side the trapezium and trapezoid articulate with the scaphoid, and on the inner side the unciform articulates with the cuneiform, forming gliding joints.

The ligaments are—

Anterior or Palmar. External Lateral.
Posterior or Dorsal. Internal Lateral.

The **Anterior** or **Palmar Ligament** consists of short fibres, which pass, for the most part, from the palmar surface of the bones of the first row to the front of the os magnum.

The **Posterior** or **Dorsal Ligament** consists of short, irregular bundles of fibres passing between the bones of the first and second row on the dorsal surface of the carpus.

The **Lateral Ligaments** are very short: one is placed on the radial, the other on the ulnar side of the carpus; the former, the stronger and more distinct, connecting the scaphoid and trapezium, the latter the cuneiform and unciform; they are continuous with the lateral ligaments of the wrist-joint. In addition to these ligaments, a slender interosseous band sometimes connects the os magnum and the scaphoid.

The **Synovial Membrane of the Carpus** is very extensive; it passes from the under surface of the scaphoid, semilunar, and cuneiform bones to the upper surface of the bones of the second row, sending upwards two prolongations— between the scaphoid and semilunar, and the semilunar and cuneiform—sending downwards three prolongations between the four bones of the second row. The prolongation between the trapezium and trapezoid, or that between the trapezoid and os magnum, is often continued onwards, through the absence of the interosseous ligament, into the carpo-metacarpal joints, sometimes of the four inner metacarpal bones; sometimes of the second and third metacarpal bones only. In the latter condition, the carpo-metacarpal joints of the fourth and fifth metacarpal bones have a separate synovial membrane. The synovial membranes of these joints are prolonged for a short distance between the meta-carpal bones. There is a separate synovial membrane between the pisiform and cuneiform bones.

Actions.—The articulation of the hand and wrist, considered as a whole, is divided into three parts: (1) the radius and the interarticular fibro-cartilage; (2) the meniscus, formed by the scaphoid, semilunar, and cuneiform—the pisiform bone having no essential part in the movement of the hand; (3) the hand proper, the metacarpal bones with the four carpal bones on which they are supported, viz. the trapezium, trapezoid, os magnum, and unciform. These three elements form two joints: (1) the superior (wrist-joint proper), between the meniscus and bones of the forearm; (2) the inferior, between the hand and meniscus (transverse or mid-carpal joint).

(1) The articulation between the forearm and carpus is a true condyloid articulation, and therefore all movements but rotation are permitted. Flexion and extension are the most free, and of these a greater amount of extension than flexion is permitted on account of the articulating surfaces extending farther on the dorsal than on the palmar aspect of the carpal bones. In this movement the carpal bones rotate on a transverse axis drawn between the tips of the styloid processes of the radius and ulna. A certain amount of adduction (or ulnar flexion) and abduction (or radial flexion) is also permitted. Of these the former is considerably greater in extent than the latter, on account of the shorter length of the styloid process of the ulna, abduction being limited by the contact of the styloid process of the radius with the trapezium. In this movement the carpus revolves upon an antero-posterior axis drawn through the centre of the wrist. Finally, circumduction is permitted by the consecutive movements of adduction, extension, abduction, and flexion, with intermediate movements

between them.   There is no rotation, but this is provided for by the supination and pronation of the radius on the ulna.   The movement of *flexion* is performed by the Flexor carpi radialis, the Flexor carpi ulnaris, and the Palmaris longus ; *extension* by the Extensor carpi radialis longior et brevior and the Extensor carpi ulnaris ; *adduction* (ulnar flexion) by the Flexor carpi ulnaris and the Extensor carpi ulnaris ; and *abduction* (radial flexion) by the Extensors of the thumb, and the Extensor carpi radialis longior and brevior and the Flexor carpi radialis.

(2) The chief movements permitted in the transverse or mid-carpal joint are flexion and extension and a slight amount of rotation.   In flexion and extension, which is the movement most freely enjoyed, the trapezium and trapezoid on the radial side and the unciform on the ulnar side glide forwards and backwards on the scaphoid and cuneiform respectively, while the head of the os magnum and the superior surface of the unciform rotate in the cup-shaped cavity of the scaphoid and semilunar.   Flexion at this joint is freer than extension.   A very trifling amount of rotation is also permitted, the head of the os magnum rotating round a vertical axis drawn through its own centre, while at the same time a slight gliding movement takes place in the lateral portions of the joint.

IX.   CARPO-METACARPAL ARTICULATIONS

1.   ARTICULATION OF THE METACARPAL BONE OF THE THUMB WITH THE TRAPEZIUM

This is a joint of reciprocal reception, and enjoys great freedom of movement, on account of the configuration of its articular surfaces, which are saddle-shaped, so that, on section, each bone appears to be received into a cavity in the other, according to the direction in which they are cut.   The joint is surrounded by a capsular ligament.

The **Capsular Ligament** is thick but loose, and passes from the circumference of the upper extremity of the metacarpal bone to the rough edge bounding the articular surface of the trapezium ; it is thickest externally and behind, and lined by a separate *synovial membrane*.

Movements.—In the articulation of the metacarpal bone of the thumb with the trapezium, the movements permitted are flexion, extension, adduction, abduction, and circumduction.   When the joint is flexed, the metacarpal bone is moved inwards and forwards, and the thumb is opposed to the fingers.   It is by this peculiar movement of *opposition* that the tip of the thumb is brought into contact with the palmar surfaces of the slightly flexed fingers one after another.

2.   ARTICULATIONS OF THE METACARPAL BONES OF THE FOUR INNER FINGERS WITH THE CARPUS

The joints formed between the carpus and four inner metacarpal bones are arthrodial joints.   The ligaments are—

Dorsal.   Palmar.
Interosseous.

The **Dorsal Ligaments,** the strongest and most distinct, connect the carpal and metacarpal bones on their dorsal surface.   The second metacarpal bone receives two fasciculi, one from the trapezium, the other from the trapezoid ; the third metacarpal receives two, one from the trapezoid, and one from the os magnum ; the fourth two, one from the os magnum, and one from the unciform ; the fifth receives a single fasciculus from the unciform bone, which is continuous with a similar ligament on the palmar surface, forming an incomplete capsule.

The **Palmar Ligaments** have a somewhat similar arrangement on the palmar surface, with the exception of the third metacarpal, which has three ligaments, an external one from the trapezium, situated above the sheath of the tendon of the Flexor carpi radialis ; a middle one from the os magnum ; and an internal one from the unciform.

The **Interosseous Ligaments** consist of short, thick fibres, which are limited to one part of the carpo-metacarpal articulation ; they connect the contiguous inferior angles of the os magnum and unciform with the adjacent surfaces of the third and fourth metacarpal bones.

The **Synovial Membrane** is frequently a continuation of that between the two rows of carpal bones. Occasionally, the articulation of the unciform with the fourth and fifth metacarpal bones has a separate synovial membrane.

The synovial membranes of the wrist and carpus (fig. 371) are thus seen to be five in number. The *first*, the *membrana sacciformis*, passes from the lower end of the ulna to the sigmoid cavity of the radius, and lines the upper surface of the interarticular fibro-cartilage. The *second* passes from the lower end of the radius and interarticular fibro-cartilage above, to the bones of the first row below. The *third*, the most extensive, passes between the contiguous margins of the two rows of carpal bones, and sometimes, in the event of one of the interosseous ligaments being absent, between the bones of the second row to the carpal extremities of the four inner metacarpal bones. The *fourth*, from the margin of the trapezium to the metacarpal bone of the thumb. The *fifth*, between the adjacent margins of the cuneiform and pisiform bones. Occasionally the carpo-metacarpal joints have a separate synovial membrane (see page 393).

FIG. 371.—Vertical section through the articulations at the wrist, showing the five synovial membranes.

Actions.—The movement permitted in the carpo-metacarpal articulations of the four inner fingers is limited to a slight gliding of the articular surfaces upon each other, the extent of which varies in the different joints. Thus the articulation of the metacarpal bone of the little finger is most movable, then that of the ring-finger. The metacarpal bones of the index and middle fingers are almost immovable.

3. ARTICULATIONS OF THE METACARPAL BONES WITH EACH OTHER

The carpal extremities of the four inner metacarpal bones articulate with one another at each side by small surfaces covered with cartilage, and connected together by dorsal, palmar, and interosseous ligaments.

The **Dorsal** and **Palmar Ligaments** pass transversely from one bone to another on the dorsal and palmar surfaces. The *Interosseous Ligaments* pass between their contiguous surfaces, just beneath their lateral articular facets.

The **Synovial Membrane** between the lateral facets is a reflection from that between the two rows of carpal bones.

The **Transverse Metacarpal Ligament** (fig. 372) is a narrow fibrous band, which passes transversely across the anterior surfaces of the digital extremities of the four inner metacarpal bones, connecting them together. It is blended anteriorly with the anterior (glenoid) ligament of the metacarpo-phalangeal articulations. To its posterior border is connected the fascia which covers the Interossei muscles. Its anterior surface is concave where the Flexor tendons pass over it. Behind it the tendons of the Interossei muscles pass to their insertion.

X. METACARPO-PHALANGEAL ARTICULATIONS (fig. 372)

These articulations are of the condyloid kind, formed by the reception of the rounded head of the metacarpal bone into a shallow cavity in the extremity of the first phalanx, with the exception of that of the thumb, which presents more of the characters of a ginglymoid joint. The ligaments are—

Anterior.
Two Lateral.

FIG. 372.—Articulations of the phalanges.

Metacarpo-phalangeal articulation

Phalangeal articulations

The **Anterior Ligaments** (*Glenoid Ligaments* of Cruveilhier) are thick, dense, fibro-cartilaginous structures, placed upon the palmar surface of the joints in the intervals between the lateral ligaments, to which they are connected; they are loosely united to the metacarpal bone, but are very firmly attached to the base of the first phalanx. Their palmar surfaces are intimately blended with the transverse metacarpal ligament, and present a groove for the passage of the Flexor tendons, the sheath surrounding which is connected to each side of the groove. By their deep surfaces, they form part of the articular facets for the heads of the meta-carpal bones, and are lined by synovial membranes.

The **Lateral Ligaments** are strong, rounded cords, placed one on each side of the joint, and attached by one extremity to the posterior tubercle and adjacent depression on the side of the head of the metacarpal bone, and by the other to the contiguous extremity of the phalanx.

Actions.—The movements which occur in these joints are flexion, extension, adduction, abduction, and circumduction; the lateral movements are very limited.

Surface Form.—The prominences of the knuckles do not correspond to the position of the joints either of the metacarpo-phalangeal or interphalangeal articulations. These prominences are invariably formed by the distal ends of the proximal bone of each joint, and the line indicating the position of the joint must be sought considerably in front of the middle of the knuckle. The usual rule for finding these joints is to flex the distal phalanx on the proximal one to a right angle; the position of the joint is then indicated by an imaginary line drawn along the middle of the lateral aspect of the proximal phalanx.

XI. ARTICULATIONS OF THE PHALANGES

These are hinge joints. The ligaments are—

Anterior.   Two Lateral.

The arrangement of these ligaments is similar to those in the metacarpo-phalangeal articulations; the Extensor tendon supplies the place of a posterior ligament.

Actions.—The only movements permitted in the phalangeal joints are flexion and extension; these movements are more extensive between the first and second phalanges than between the second and third. The movement of flexion is very considerable, but extension is limited by the anterior and lateral ligaments.

ARTICULATIONS OF THE LOWER EXTREMITY

The articulations of the Lower Extremity comprise the following:

I. Hip.   V. Tarsal.
II. Knee.   VI. Tarso-metatarsal.
III. Tibia with Fibula.   VII. Metatarsal.
IV. Ankle.   VIII. Metatarso-phalangeal.
IX. Phalangeal.

I. HIP-JOINT (fig. 373)

This articulation is an enarthrodial, or ball-and-socket joint, formed by the reception of the head of the femur into the cup-shaped cavity of the acetabulum. The articulating surfaces are covered with cartilage, that on the head of the femur being thicker at the centre than at the circumference, and covering the entire surface with the exception of a depression just below its centre for the ligamentum teres; that covering the acetabulum is much thinner at the centre than at the circumference. It forms an incomplete cartilaginous ring, of a horse-shoe shape, being deficient below, where there is a circular depression, which is occupied in the recent state by a mass of fat, covered by synovial membrane. The ligaments of the joint are, the

Capsular.   Teres.
Ilio-femoral.   Cotyloid.
Transverse.

The **Capsular Ligament** is a strong, dense, ligamentous capsule, embracing the margin of the acetabulum above, and surrounding the neck of the femur below. Its *upper circumference* is attached to the acetabulum, two or three lines external to the cotyloid ligament, above and behind; but in front, it is attached to the outer margin of the ligament, and opposite to the notch where the margin of the cavity is deficient, it is connected to the transverse ligament, and by a few fibres to the edge of the obturator foramen. Its *lower circumference* surrounds the neck of the femur, being attached, in front, to the spiral or anterior intertrochanteric line; above, to the base of the neck; behind, to the neck of the bone, about half an inch above the posterior intertrochanteric line; below, to the lower part of the neck of the femur, close to the lesser trochanter. From this insertion some of the fibres are reflected upwards over the neck of the femur as longitudinal bands, termed *retinacula*. The capsule is much thicker at the upper and fore part of the joint, where the greatest amount of resistance is required, than below and internally, where it is thin, loose, and longer than in any other part. It consists of two sets of fibres, circular and longitudinal. The circular fibres (*zona orbicularis*) are most abundant at the lower and back part of the capsule, and form a sling or collar around the neck of the femur. Anteriorly they blend with the deep surface of the ilio-femoral ligament, and through its medium reach the anterior inferior spine of the ilium. The longitudinal fibres are greatest in amount at the upper and front part of the capsule, where they form distinct bands, or accessory ligaments, of which the most important is the

ilio-femoral. The other accessory bands are known as the *pubo-femoral*, passing
from the ilio-pectineal eminence and obturator crest to the front of the capsule,
where some of its fibres blend with the lower segment of the ilio-femoral ligament; and *ischio-capsular*, passing from the ischium, just below the acetabulum,
to blend with the circular fibres at the lower part of the joint. The external

FIG. 373.—Left hip-joint laid open.

surface (fig. 358, page 370) is rough, covered by numerous muscles, and separated
in front from the Psoas and Iliacus by a synovial bursa, which not infrequently
communicates, by a circular aperture, with the cavity of the joint. It differs
from the capsular ligament of the shoulder in being much less loose and lax, and
in not being perforated for the passage of a tendon.

The **Ilio-femoral Ligament** (figs. 358 and 374) is a band of great strength
which lies in front of the joint; it is intimately connected with the capsular
ligament, and serves to strengthen it in this situation. It is attached, above, to
the lower part of the anterior inferior spine of the ilium; and, diverging below,
forms two bands, of which one passes downwards to be inserted into the lower
part of the anterior intertrochanteric line; the other passes downwards and
outwards to be inserted into the upper part of the same line. Between the two
bands is a thinner part of the capsule. In some joints there is no division, and
the ligament spreads out into a flat triangular band which is attached below
into the whole length of the anterior intertrochanteric line. This ligament is
frequently called the Y-shaped ligament of Bigelow; and the outer or upper of
the two bands is sometimes described as a separate ligament, under the name
of the *ilio-trochanteric ligament.*

The **Ligamentum Teres** is a triangular, somewhat flattened band implanted
by its apex into the depression a little behind and below the centre of the head
of the femur, and by its broad base into the margins of the cotyloid notch,
becoming blended with the transverse ligament. It is formed of connective
tissue, surrounded by a tubular sheath of synovial membrane. Occasionally only
the synovial fold exists, or the ligament may be altogether absent. The ligament
is made tense when the hip is semiflexed, and the limb then adducted or rotated
outwards; it is, on the other hand, relaxed when the limb is abducted. It has,

however, but little influence as a ligament, though it may to a certain extent limit movement.

The **Cotyloid Ligament** is a fibro-cartilaginous rim attached to the margin of the acetabulum, the cavity of which it deepens; at the same time it protects the edge of the bone, and fills up the inequalities on its surface. It bridges over the notch as the *transverse ligament*, and thus forms a complete circle, which closely surrounds the head of the femur, and assists in holding it in its place. It is prismoid on section, its base being attached to the margin of the acetabulum, and its opposite edge being free and sharp; while its two surfaces are invested by synovial membrane, the external one being in contact with the capsular ligament, the internal one being inclined inwards so as to narrow the acetabulum, and embrace the cartilaginous surface of the head of the femur. It is much thicker above and behind than below and in front, and consists of close compact fibres, which arise from different points of the circumference of the acetabulum, and interlace with each other at very acute angles.

FIG. 374.—Hip-joint, showing the ilio-femoral ligament. (After Bigelow.)

FIG. 375.—Vertical section through hip-joint. (Henle.)

The **Transverse Ligament** is in reality a portion of the cotyloid ligament, though differing from it in having no cartilage cells among its fibres. It consists of strong, flattened fibres, which cross the notch at the lower part of the acetabulum, and convert it into a foramen. Thus an interval is left beneath the ligament for the passage of nutrient vessels to the joint.

The **Synovial Membrane** is very extensive. Commencing at the margin of the cartilaginous surface of the head of the femur, it covers all that portion of the neck which is contained within the joint; from the neck it is reflected on the internal surface of the capsular ligament, covers both surfaces of the cotyloid ligament and the mass of fat contained in the depression at the bottom of the acetabulum, and is prolonged in the form of a tubular sheath around the ligamentum teres, as far as the head of the femur. It sometimes communicates through a hole in the capsular ligament between the inner band of the Y-shaped ligament and the pubo-femoral ligament with a bursa situated on the under surface of the Ilio-psoas muscle.

The muscles in relation with the joint are, in front, the Psoas and Iliacus, separated from the capsular ligament by a synovial bursa; above, the reflected head of the Rectus and Gluteus minimus, the latter being closely adherent to the capsule; internally, the Obturator externus and Pectineus; behind, the Pyriformis, Gemellus superior, Obturator internus, Gemellus inferior, Obturator externus, and Quadratus femoris (fig. 376).

The arteries supplying the joint are derived from the obturator, sciatic, internal circumflex, and gluteal.

The nerves are articular branches from the sacral plexus, great sciatic, obturator, accessory obturator, and a filament from the branch of the anterior crural supplying the Rectus.

Actions.—The movements of the hip are very extensive, and consist of flexion, extension, adduction, abduction, circumduction, and rotation.

The hip-joint presents a very striking contrast to the shoulder-joint in the much more complete mechanical arrangements for its security and for the limitation of its movements. In the shoulder, as has been seen, the head of the humerus is not adapted at all in size to the glenoid cavity, and is hardly

Fig. 376.—Relation of muscles to the capsule of the hip-joint.
From a drawing by F. A. Barton.

restrained in any of its ordinary movements by the capsular ligament. In the hip-joint, on the contrary, the head of the femur is closely fitted to the acetabulum for a distance extending over nearly half a sphere, and at the margin of the bony cup it is still more closely embraced by the cotyloid ligament, so that the head of the femur is held in its place by that ligament even when the fibres of the capsule have been quite divided (Humphry). The ilio-femoral ligament is the strongest of all the ligaments in the body, and is put on the stretch by any attempt to extend the femur beyond a straight line with the trunk. That is to say, this ligament is the chief agent in maintaining the erect position without muscular fatigue; for a vertical line passing through the centre of gravity of the trunk falls behind the centres of rotation in the hip-joints, and therefore the pelvis tends to fall backwards, but is prevented by the tension of the ilio-femoral and capsular ligaments. The security of the joint may be also provided for by the two bones being directly united through the ligamentum teres; but it is doubtful whether this so-called ligament can have much influence upon the mechanism of the joint. Flexion of the hip-joint is arrested by the soft parts of the thigh and abdomen being brought into contact, when the leg is flexed on the thigh ; and by the action of the hamstring muscles when the leg is extended ; * extension by the tension of the ilio-femoral ligament ; adduction by the thighs coming into contact ; adduction with flexion by the outer band of the ilio-femoral ligament, the outer part of the capsular ligament ; abduction by the inner band of the ilio-femoral ligament and the pubo-femoral band ; rotation outwards by the outer band of the ilio-femoral ligament ; and rotation inwards by the ischio-capsular ligament and the hinder part of the capsule. The muscles which *flex* the femur on the pelvis are, the Psoas, Iliacus, Rectus, Sartorius, Pectineus, Adductor longus and brevis, and the anterior fibres of the Gluteus medius and minimus. *Extension* is mainly performed by the Gluteus maximus, assisted by the hamstring muscles. The thigh is *adducted* by the Adductor magnus, longus and brevis, the Pectineus, the Gracilis, and lower part of the Gluteus maximus, and *abducted* by the Gluteus medius and minimus, and upper part of the Gluteus maximus. The muscles which *rotate* the thigh *inwards* are the anterior fibres of the Gluteus medius, the Gluteus minimus, and the Tensor fasciæ femoris ; while those which rotate it *outwards* are the posterior fibres of the Gluteus medius, the Pyriformis, Obturator externus and internus, Gemellus superior and inferior, Quadratus femoris, Iliacus, Gluteus maximus, the three Adductors, the Pectineus, and the Sartorius.

Surface Form.—A line drawn from the anterior superior spinous process of the ilium to the most prominent part of the tuberosity of the ischium (Nélaton's line) runs through the centre of the acetabulum, and would, therefore, indicate the level of the hip-joint ; or, in other words, the upper border of the great trochanter, which lies on Nélaton's line, is on a level with the centre of the hip-joint.

Surgical Anatomy.—In dislocation of the hip, 'the head of the thigh-bone may rest at any point around its socket ' (Bryant) ; but whatever position the head ultimately assumes, the primary displacement is generally downwards and inwards, the capsule giving way at its weakest—that is, its lower and inner—part. The situation that the head of the bone subsequently assumes is determined by the degree of flexion or extension, and of outward or inward rotation of the thigh at the moment of luxation, influenced, no doubt, by the ilio-femoral ligament, which is not easily ruptured. When, for instance, the head is forced backwards, this ligament forms a fixed axis, round which the head of the bone rotates, and is thus driven on to the dorsum of the ilium. The ilio-femoral ligament also influences the position of the thigh in the various dislocations : in the dislocations backwards it is tense, and produces inversion of the limb ; in the dislocation on to the pubes, it is relaxed, and therefore allows the external rotators to evert the thigh ; while in the thyroid dislocation it is tense, and produces flexion. The muscles inserted into the upper part of the femur, with the exception of the Obturator internus, have very little direct influence in determining the position of the bone. But Bigelow has endeavoured to show that the Obturator internus is the principal agent in deciding whether, in the backward dislocations, the head of the bone shall be ultimately lodged on the dorsum of the ilium, or in or near the sciatic notch. In both dislocations the head passes, in the first instance, in the same direction ; but, as Bigelow asserts, in the displacement on to the dorsum, the head of the bone travels up behind the acetabulum, in front of the muscle ; while in the dislocation into the sciatic notch the head passes behind the muscle, and is prevented from reaching the dorsum, in consequence of the tendon of the muscle arching over the neck of the bone, and it therefore remains in the neighbourhood of the sciatic notch. Bigelow distinguishes these two forms of dislocation by describing them as dislocations backwards, ' above and below ' the Obturator internus.

The ilio-femoral ligament is rarely torn in dislocations of the hip, and this fact is taken advantage of by the surgeon in reducing these dislocations by manipulation. It is made to act as the fulcrum to a lever, of which the long arm is the shaft of the femur, and the short arm the neck of the bone.

The hip-joint is rarely the seat of acute synovitis from injury, on account of its deep position and its thick covering of soft parts. Acute inflammation may, and does, frequently occur as the result of constitutional conditions, as rheumatism, pyæmia, &c. When, in these cases, effusion takes place, and the joint becomes distended with fluid, the swelling is not very easy to detect on account of the thickness of the capsule and the depth of the articulation. It is principally to be found on the front of the joint, just internal to the ilio-femoral ligament ; or behind, at the lower and back part. In these two places the capsule is thinner than elsewhere. Disease of the hip-joint is much oftener of a chronic character, and is usually of tuberculous origin. It begins either in the bones or in

---

* The hip-joint cannot be completely flexed, in most persons, without at the same time flexing the knee, on account of the shortness of the hamstring muscles.—Cleland, *Journ. of Anat. and Phys.* No. I. Old Series, p. 87.

the synovial membrane; usually in the former, and probably, in most cases, in the growing, highly vascular tissue in the neighbourhood of the epiphysial cartilage. In this respect it differs very materially from tuberculous arthritis of the knee, where the disease usually commences in the synovial membrane. The reasons why the disease so commonly begins in this situation are twofold: first, this part being the centre of rapid growth, its nutrition is unstable and apt to pass into inflammatory action; and, secondly, great strain is thrown upon it, from the frequency of falls and blows upon the hip, which causes crushing of the epiphysial cartilage or the cancellous tissue in its neighbourhood, with the results likely to follow such an injury. In addition to these, the depth of the joint protects it from the causes of synovitis.

In chronic hip-disease the affected limb assumes an altered position, the cause of which it is important to understand. In the early stage of a typical case, the limb is flexed, abducted, and rotated outwards. In this position all the ligaments of the joint are relaxed: the front of the capsule by flexion; the outer band of the ilio-femoral ligament by abduction; and the inner band of this ligament and the back of the capsule by rotation outwards. It is, therefore, the position of greatest ease. The condition is not quite obvious at first, upon examining a patient. If the patient is laid in the supine position, the affected limb will be found to be extended and parallel with the other. But it will be seen that the pelvis is tilted downwards on the diseased side and the limb apparently longer than its fellow, and that the lumbar spine is arched forwards (lordosis). If now the thigh is abducted and flexed, the tilting downwards and the arching forwards of the pelvis disappear. The condition is thus explained. A limb which is flexed and abducted is obviously useless for progression, and, in order to overcome the difficulty, the patient depresses the affected side of his pelvis in order to produce parallelism of his limbs, and at the same time rotates his pelvis on its transverse horizontal axis, so as to direct the limb downwards, instead of forwards. In the later stages of the disease the limb becomes flexed and adducted and inverted. This position probably depends upon muscular action, at all events as regards the adduction. The Adductor muscles are supplied by the obturator nerve, which also largely supplies the joint. These muscles are therefore thrown into reflex action by the irritation of the peripheral terminations of this nerve in the inflamed articulation.

Osteo-arthritis is not uncommon in the hip-joint, and is said to be more common in the male than in the female, in whom the knee-joint is more frequently affected. It is a disease of middle age or advanced life.

Congenital dislocation is more commonly met with in the hip-joint than in any other articulation. The displacement usually takes place on to the dorsum ilii. It gives rise to extreme lordosis, and a waddling gait is noticed as soon as the child commences to walk.

Excision of the hip may be required for disease or for injury, especially gunshot. It may be performed either by an anterior incision or a posterior one. The former entails less interference with important structures, especially muscles, than the posterior one, but permits of less efficient drainage. In these days, however, when the surgeon aims at securing healing of the wound without suppuration, this second desideratum is not of so much importance. In the operation from the front, an incision is made three to four inches in length, starting immediately below and external to the anterior superior spinous process of the ilium, downwards and inwards between the Sartorius and Tensor fasciæ femoris, to the neck of the bone, dividing the capsule at its upper part. A narrow-bladed saw now divides the neck of the femur, and the head of the bone is extracted with sequestrum forceps. All diseased tissue is carefully removed with a sharp spoon or scissors, and the cavity thoroughly flushed out with a hot antiseptic fluid.

The posterior method consists in making an incision three or four inches long, commencing midway between the top of the great trochanter and the anterior superior spine, and ending over the shaft, just below the trochanter. The muscles are detached from the great trochanter, and the capsule opened freely. The head and neck are freed from the soft parts and the bone sawn through just below the top of the trochanter with a narrow saw. The head of the bone is then levered out of the acetabulum. In both operations, if the acetabulum is eroded, it must be freely gouged.

II. KNEE-JOINT

The knee-joint was formerly described as a ginglymus or hinge-joint, but is really of a much more complicated character. It must be regarded as consisting of three articulations in one: one between each condyle of the femur and the corresponding semilunar cartilage and tuberosity of the tibia, which are condyloid joints, and one between the patella and the femur, which is partly arthrodial, but not completely so, since the articular surfaces are not mutually adapted to each other, so that the movement is not a simple gliding one. This view of the construction of the knee-joint receives confirmation from the study of the articulation in some of the lower mammals, where three synovial membranes are sometimes found, corresponding to these three subdivisions, either entirely distinct or only connected together by small communications. This view is further rendered probable by the existence of the two crucial ligaments within the joint, which must be regarded as the external and internal lateral ligaments of the inner and outer joints respectively. The existence of the ligamentum mucosum would further indicate a tendency to separation of the synovial cavity into two minor sacs, one corresponding to each joint.

The bones entering into the formation of the knee-joint are the condyles of the femur above, the head of the tibia below, and the patella in front. They are connected together by ligaments, some of which are placed on the exterior of the joint, while others occupy its interior.

External Ligaments	*Interior Ligaments*
Capsular.	Anterior, or External Crucial.
Anterior, or Ligamentum Patellæ.	Posterior, or Internal Crucial.
	Two Semilunar Fibro-cartilages.
Posterior, or Ligamentum Posticum Winslowii.	Transverse.
	Coronary.
Internal Lateral.	Ligamentum mucosum) Processes of Syn-
Two External Lateral.	Ligamenta alaria) ovial membrane.

The **Capsular Ligament** consists of an exceedingly thin, but strong fibrous membrane which fills in the intervals left between the specially named bands, which are constituents of the capsule and inseparably connected with it. Unlike other diarthrodial joints, the capsular ligament does not form a complete investment for the joint, for it is deficient in front and above, beneath the tendon of the Quadriceps extensor. It is largely made up of expansions from the fascia lata and from the tendons surrounding the joint. In front, expansions from the Vasti muscles and fascia lata fill in the interval between the anterior and lateral ligaments, constituting the *lateral patellar ligaments,* and form the anterior part of the capsular ligament. Behind, it consists of vertical fibres which arise from the condyles and intercondyloid notch of the femur, and are augmented by fibres derived from the tendon of the Semimembranosus muscle. Here it forms part of the posterior ligament. On the outer side of the joint, a prolongation from the ilio-tibial band fills in the interval between the posterior and the external lateral ligaments, and partly covers the latter. On the inner side, expansions from the Sartorius and Semimembranosus pass upwards to the internal lateral ligament and strengthen the capsule on the inner side of the joint.

The **Anterior Ligament,** or **Ligamentum Patellæ** (fig. 377), is the central portion of the common tendon of the Extensor muscles of the thigh, which is continued from the patella to the tubercle of the tibia, supplying the place of an anterior ligament. It is a strong, flat, ligamentous band, about three inches in length, attached,

FIG. 377.—Right knee-joint. Anterior view.

above, to the apex and adjoining margins of the patella and the rough depression on its posterior surface; below, to the tubercle of the tibia; its superficial fibres being continuous over the front of the patella with those of the tendon of the Quadriceps extensor. The lateral portions of the tendon of the Extensor muscles pass down on either side of the patella, attached to the borders of this bone and its ligament, to be inserted into the upper extremity of the tibia on each side of the tubercle; externally, these portions merge into the capsular ligament, as stated above, forming the lateral patellar ligaments. The posterior surface of the ligamentum patellæ is separated from the synovial membrane of the joint by a large pad (*infrapatellar*) of fat, and from the tibia by a synovial bursa.

The **Posterior Ligament (Ligamentum Posticum Winslowii)** (fig. 378) is a broad, flat, fibrous band, formed of fasciculi separated from one another by apertures for the passage of vessels and nerves. It is attached above to the upper margin of the intercondyloid notch and posterior surface of the femur close to the articular margins of the condyles, and below to the posterior margin of the head of the tibia. Superficial to the main part of the ligament is a strong fasciculus derived from the tendon of the Semimembranosus, and passing from the back part of the inner tuberosity of the tibia obliquely upwards and

FIG. 378.—Right knee-joint.   FIG. 379.—Right knee-joint.
          Posterior view.   Showing internal ligaments.

outwards to the back part of the outer condyle of the femur, where it becomes lost in the general ligament.* The posterior ligament forms part of the floor of the popliteal space, and the popliteal artery rests upon it.

The **Internal Lateral Ligament** is a broad, flat, membranous band, situated nearer to the back than the front of the joint. It is attached, above, to the inner tuberosity of the femur immediately below the Adductor tubercle; below, to the inner tuberosity and inner surface of the shaft of the tibia. The fibres of the posterior part of the ligament are short, and are inserted into the inner part of the tuberosity of the tibia above the groove for the Semimembranosus muscle. They incline backwards as they descend. The anterior part of the ligament is a flattened band, about four inches in length, which inclines forwards as it descends. It is inserted into the inner surface of the shaft of the tibia about

---

* This oblique band is by some anatomists termed the Ligamentum Posticum Winslowii, in contradistinction to the Posterior ligament, which is regarded as part of the capsular ligament.

an inch and a half below the level of the tuberosity. It is crossed, at its lower part, by the tendons of the Sartorius, Gracilis, and Semitendinosus muscles, a synovial bursa being interposed. Its *deep surface* covers the anterior portion of the tendon of the Semimembranosus, with which it is connected by a few fibres, the synovial membrane of the joint, and the inferior internal articular vessels and nerve; it is intimately adherent to the internal semilunar fibro-cartilage.

The **Long External Lateral Ligament** is a strong, rounded fibrous cord, situated nearer to the back than the front of the joint. It is attached, above, to the back part of the outer tuberosity of the femur, immediately above the groove for the Popliteus muscle; below, to the outer side of the head of the fibula, in front of the styloid process. Its *outer surface* is covered by the tendon of the Biceps, which divides at its insertion into two parts, separated by the ligament. Passing beneath the ligament are the tendon of the Popliteus muscle, and the inferior external articular vessels and nerve. It has no attachment to the external semilunar fibro-cartilage.

The **Short External Lateral Ligament** is an accessory bundle of fibres placed behind and parallel with the preceding, attached, above, to the lower and back part of the outer tuberosity of the femur; below, to the summit of the styloid process of the fibula. This ligament is intimately connected with the capsular

FIG. 380.—Head of tibia, with semilunar fibro-cartilages. Seen from above. Right side.

ligament, while passing beneath it are the tendon of the Popliteus muscle, and the inferior external articular vessels and nerve.

The **Crucial** are two interosseous ligaments of considerable strength, situated in the interior of the joint, nearer its posterior than its anterior part. They are called *crucial* because they cross each other somewhat like the lines of the letter **X**; and have received the names *anterior* and *posterior*, from the position of their attachment to the tibia.

The **Anterior** or **External Crucial Ligament** (fig. 379) is attached to the depression in front of the spine of the tibia, being blended with the anterior extremity of the external semilunar fibro-cartilage, and passing obliquely upwards, backwards, and outwards, is inserted into the inner and back part of the outer condyle of the femur.

The **Posterior** or **Internal Crucial Ligament** is stronger, but shorter and less oblique in its direction, than the anterior. It is attached to the back part of the depression behind the spine of the tibia, to the popliteal notch, and to the posterior extremity of the external semilunar fibro-cartilage; and passes upwards, forwards, and inwards, to be inserted into the outer and fore part of the inner condyle of the femur. It is in relation, in front, with the anterior crucial ligament; behind, with the capsular ligament.

The **Semilunar Fibro-cartilages** (fig. 380) are two crescentic lamellæ, which serve to deepen the surfaces of the head of the tibia, for articulation with the condyles of the femur. The circumferential border of each cartilage is thick, convex, and attached to the inside of the capsule of the knee; the border which is directed towards the centre of the joint is thin, concave, and free. Their upper surfaces are concave, and in relation with the condyles of the femur; their lower surfaces are flat, and rest upon the head of the tibia. Each cartilage covers nearly the peripheral two-thirds of the corresponding articular surface of the tibia; both surfaces of the cartilages are smooth, and invested by synovial membrane.

The **Internal Semilunar Fibro-cartilage** is nearly semicircular in form, a little elongated from before backwards, and broader behind than in front; its anterior extremity, thin and pointed, is attached to a depression on the anterior margin of the head of the tibia, in front of the anterior crucial ligament; its posterior extremity is attached to the depression behind the spine, between the attachments of the external semilunar fibro-cartilage and the posterior crucial ligaments.

The **External Semilunar Fibro-cartilage** forms nearly an entire circle, covering a larger portion of the articular surface than the internal one. It is grooved on its outer side for the tendon of the Popliteus muscle, which separates it from the external lateral ligament. Its extremities, at their insertion, are interposed between the two extremities of the internal semilunar fibro-cartilage; the anterior extremity being attached in front of the spine of the tibia to the outer side of, and behind, the anterior crucial ligament, with which it blends; the posterior extremity being attached behind the spine of the tibia and in front of the posterior extremity of the internal semilunar fibro-cartilage. Just before its insertion posteriorly it gives off a strong fasciculus, the *ligament of Wrisberg*, which passes obliquely upwards and inwards, to be inserted into the inner condyle of the femur, close to and behind the attachment of the posterior crucial ligament. Occasionally a small fasciculus is given off, which passes forwards to be inserted into the back part of the anterior crucial ligament. The external semilunar fibro-cartilage gives off from its anterior convex margin a fasciculus, which forms the transverse ligament.

The **Transverse Ligament** is a band of fibres which passes transversely from the anterior convex margin of the external semilunar fibro-cartilage to the anterior convex margin of the internal semilunar fibro-cartilage; its thickness varies considerably in different subjects, and it is sometimes absent altogether.

The **Coronary Ligaments** are merely portions of the capsular ligament, which connect the circumference of each of the semilunar fibro-cartilages with the margin of the head of the tibia.

The **Synovial Membrane** of the knee-joint is the largest and most extensive in the body. Commencing at the upper border of the patella, it forms a short *cul-de-sac* beneath the Quadriceps extensor tendon of the thigh, on the lower part of the front of the shaft of the femur: this very frequently communicates with a synovial bursa interposed between the tendon and the front of the femur, by an orifice of variable size. On each side of the patella, the synovial membrane extends beneath the aponeurosis of the Vasti muscles, and more especially beneath that of the Vastus internus. Below the patella it is separated from the anterior ligament by a considerable quantity of adipose tissue, known as the *infrapatellar pad.* In this situation it sends off a triangular prolongation, which extends from the anterior part of the joint below the patella, to the front of the intercondyloid notch. This fold has been termed the **ligamentum mucosum.** It also sends off two fringe-like folds, called the **ligamenta alaria,** which extend from the sides of the ligamentum mucosum, upwards and laterally between the patella and femur. On either side of the joint, it passes downwards from the femur, lining the capsule to its point of attachment to the semilunar cartilages; it may then be traced over the upper surfaces of these cartilages to their free borders, and thence along their under surfaces to the tibia. At the back part of the external one it forms a *cul-de-sac* between the groove on its surface and the tendon of the Popliteus; it surrounds the crucial ligaments, and lines the inner surface of the ligaments which enclose the joint. The pouch of synovial membrane between the Extensor tendon and front of the femur is supported, during the movements of the knee, by a small muscle, the Subcrureus, which is inserted into it.

The folds of synovial membrane and the fatty processes contained in them act, as it seems, mainly as padding to fill up interspaces and obviate concussion. Sometimes the bursa beneath the Quadriceps extensor is completely shut off from the rest of the synovial cavity, thus forming a closed sac between the Quadriceps and the lower part of the front of the femur ; or it may communicate with the synovial cavity by a minute aperture.

The bursæ about the knee-joint are the following :

In front there are four bursæ : one is interposed between the patella and the skin ; another of small size between the upper part of the tibia and the ligamentum patellæ ; a third between the lower part of the tuberosity of the tibia and the skin ; and a fourth between the anterior surface of the lower end of the femur and the under surface of the Quadriceps extensor cruris, which usually communicates with the synovial membrane of the knee-joint. On the outer side there are four bursæ : (1) one beneath the outer head of the Gastrocnemius (which sometimes communicates with the joint) ; (2) one above the external lateral ligament, between it and the tendon of the Biceps ; (3) one beneath the external lateral ligament, between it and the tendon of the Popliteus (this is sometimes only an expansion from the next bursa) ; (4) one beneath the tendon of the Popliteus, between it and the condyle of the femur, which is almost always an extension from the synovial membrane. On the inner side there are five bursæ : (1) one beneath the inner head of the Gastrocnemius, which sends a prolongation between the tendons of the Gastrocnemius

FIG. 381.—Longitudinal section through the middle of the right knee-joint. (After Braune.)

and Semimembranosus. This bursa often communicates with the joint ; (2) one above the internal lateral ligament between it and the tendons of the Sartorius, Gracilis, and Semitendinosus ; (3) one beneath the internal lateral ligament between it and the tendon of the Semimembranosus (this is sometimes only an expansion from the next bursa) ; (4) one beneath the tendon of the Semimembranosus, between it and the head of the tibia ; (5) occasionally there is a bursa between the tendons of the Semimembranosus and Semitendinosus.

Structures around the Joint.—In front, and at the sides, the Quadriceps extensor ; on the outer side, the tendons of the Biceps and the Popliteus and the external popliteal nerve ; on the inner side, the Sartorius, Gracilis, Semitendinosus and Semimembranosus ; behind, an expansion from the tendon of the Semimembranosus, the popliteal vessels, and the internal popliteal nerve, Popliteus, Plantaris, and inner and outer heads of the Gastrocnemius, some lymphatic glands, and fat.

The *Arteries* supplying the joint are the anastomotica magna, a branch of the femoral, the articular branches of the popliteal, the anterior and posterior recurrent branches of the anterior tibial, and the descending branch from the external circumflex of the profunda.

The *Nerves* are derived from the obturator, anterior crural, and external and internal popliteal.

Actions.—The knee-joint permits of movements of flexion and extension, and, in certain positions, of slight rotation inwards and outwards. The movement of flexion and extension does not, however, take place in a simple hinge-like manner, as in other joints, but is a complicated movement, consisting of a certain amount of gliding and rotation ; so that the same part of one articular surface is not always applied to the same part of the other articular surface, and the axis of motion is not a fixed one. If the joint is examined while in a condition of extreme flexion, the posterior part of the articular surfaces of the tibia will be found to be in contact with the posterior rounded extremities of the condyles of the femur ; and if a simple hinge-like movement were to take place, the axis, round which the revolving movement of the tibia occurs, would be in the back part of the condyle. If the leg be now brought forwards into a position of semi-flexion, the upper surface of the tibia will be seen to glide over the condyles of the femur, so that the middle part of the articular facets are in contact, and the axis of rotation must therefore have shifted forwards to nearer the centre of the condyles. If the leg be now fully extended, a still further gliding takes place, and a further shifting forwards of the axis of rotation. This is not, however, a simple movement, but is accompanied by a certain amount of rotation outwards round a vertical axis drawn through the centre of the head of the tibia. This rotation is due to the greater length of the internal condyle and to the fact that the anterior portion of its articular surface is inclined obliquely outwards. In consequence of this, it will be seen that towards the close of the movement of extension—that is to say, just before complete extension is effected— the tibia glides upwards and outwards over this oblique surface on the inner condyle, and the leg is therefore necessarily rotated outwards. In flexion of the joint the converse of these movements takes place : the tibia glides backwards round the end of the femur, and at the commencement of the movement the tibia is directed downwards and inwards along the oblique curve of the inner condyle, thus causing an inward rotation to the leg.

During flexion and extension, the patella moves on the lower end of the femur, but this movement is not a simple gliding one ; for if the articular surface of this bone is examined, it will be found to present on each side of the central vertical ridge two less marked transverse ridges, which divide the surface, except a small portion along the inner border, which is cut off by a slight vertical ridge, into six facets (see fig. 382), and therefore does not present a uniform curved surface, as would be the case if a simple gliding movement took place. These six facets—three on each side of the median vertical ridge—correspond to and denote the parts of the bone respectively in contact with the condyles of the femur during flexion, semiflexion, and extension. In flexion, only the upper facets on the patella are in contact with the condyles of the femur ; the lower two-thirds of the bone rests upon the mass of fat which occupies the space between the femur and tibia. In the semiflexed position of the joint, the middle facets on the patella rest upon the most prominent portion of the condyles, and thus afford greater leverage to the Quadriceps by increasing its distance from the centre of motion. In complete extension the patella is drawn up so that only the lower facets are in contact with the articular surfaces of the condyles. The narrow strip along the inner border is in contact with the outer aspect of the internal condyle when the leg is fully flexed at the knee-joint. As in the elbow, so it is in the knee—the axis of rotation in flexion and extension is not precisely at right angles to the axis of the bone, but during flexion there is a certain amount of alteration of plane ; so that, whereas in flexion the femur and tibia are in the same plane, in extension the one bone

FIG. 382.—View of the posterior surface of the right patella. Showing diagrammatically the areas of contact with the femur in different positions of the knee.

forms an angle with the other. There is, however, this difference between the two extremities—that in the upper, during extension, the humeri are parallel, and the bones of the forearm diverge ; in the lower, the femora converge below, and the tibiæ are parallel.

In addition to the slight rotation during flexion and extension, the tibia enjoys an independent rotation on the condyles of the femur, in certain positions of the joint. The movement takes place between the interarticular fibro-cartilages and the tibia, whereas the movement of flexion and extension takes place between the interarticular fibro-cartilages and the femur. So that the knee may be said to consist of two joints, separated by the fibro-cartilages : an upper (menisco-femoral), in which flexion and extension take place ; and a lower (menisco-tibial), allowing of a certain amount of rotation. The latter movement can only take place in the semiflexed position of the limb, when all the ligaments are relaxed.

During *flexion* the ligamentum patellæ is put upon the stretch, as is also the posterior crucial ligament in extreme flexion. The other ligaments are all relaxed by flexion of the joint, though the relaxation of the anterior crucial ligament is very trifling. Flexion is only checked during life by the contact of the leg with the thigh. In the act of extending the leg upon the thigh the ligamentum patellæ is tightened by the Quadriceps extensor ; but when the leg is fully extended, as in the erect posture, the ligament becomes relaxed, so as to allow free lateral movement to the patella, which then rests on the front of the lower end of the femur. The other ligaments, with the exception of the posterior crucial, which is partly relaxed, are all on the stretch. When the limb has been brought into a straight line, extension is checked mainly by the tension of all the ligaments except the posterior crucial and ligamentum patellæ. The movements of *rotation*, of which the knee is capable, are permitted in the semiflexed condition by the partial relaxation of both crucial ligaments, as well as the lateral ligaments. Rotation inwards appears to be limited by the tension of the anterior crucial ligament, and by the interlocking of the two ligaments ; but rotation outwards does not appear to be checked by either crucial ligament, since they uncross during the execution of this movement, but by the lateral ligaments, especially the internal. The main function of the crucial ligaments is to act as a direct bond of union between the tibia and femur, preventing the former bone from being carried too far backwards or forwards. Thus the anterior crucial ligament prevents the tibia from being carried too far forwards by the Extensor tendons, and the posterior crucial checks too great movement backwards by the Flexors. They also assist the lateral ligaments in resisting any lateral bending of the joint. The interarticular cartilages are intended, as it seems, to adapt the surface of the tibia to the shape of the femur to a certain extent, so as to fill up the intervals which would otherwise be left in the varying positions of the joint, and to interrupt the jars which would be so frequently transmitted up the limb in jumping or falls on the feet ; also to permit of the two varieties of motion, flexion and extension, and rotation, as explained above. The patella is a great defence to the knee-joint from any injury inflicted in front, and it distributes upon a large and tolerably even surface, during kneeling, the pressure which would otherwise fall upon the prominent ridges of the condyles ; it also affords leverage to the Quadriceps extensor muscle to act upon the tibia, and Ward has pointed out * how this leverage varies in the various positions of the joint, so that the action of the muscle produces velocity at the expense of force in the commencement of extension, and, on the contrary, at the close of extension tends to diminish velocity, and therefore the shock to the ligaments at the moment tension of these structures takes place.

Extension of the leg on the thigh is performed by the Quadriceps extensor. *Flexion* by the hamstring muscles, assisted by the Gracilis and Sartorius, and, indirectly, by the Gastrocnemius, Popliteus, and Plantaris. *Rotation outwards* by the Biceps, and *rotation inwards* by the Popliteus, Semitendinosus, and, to a slight extent, the Semimembranosus, the Sartorius, and the Gracilis.

Surface Form.—The interval between the two bones entering into the formation of the knee-joint can always easily be felt. If the limb is extended, it is situated on a slightly higher level than the apex of the patella ; but if the limb is slightly flexed, a knife carried

---

* *Human Osteology*, p. 405.

horizontally backwards immediately below the apex of the patella would pass directly into the joint. When the knee is semiflexed, the internal border of the inner condyle of the femur, the upper border of the inner tuberosity of the tibia, and the inner margin of the patella form a triangular depressed area, which coincides with the level of the joint and with the internal semilunar fibro-cartilage. If this cartilage is displaced inwards, a gap will be felt in this situation. When the knee-joint is distended with fluid, the outline of the synovial membrane at the front of the knee may be fairly well mapped out.

Surgical Anatomy.—From a consideration of the construction of the knee-joint, it would at first sight appear to be one of the least secure of any of the joints in the body. It is formed between the two longest bones, and therefore the amount of leverage which can be brought to bear upon it is considerable; the articular surfaces are but ill adapted to each other, and the range and variety of motion which it enjoys are great. All these circumstances tend to render the articulation insecure; but, nevertheless, on account of the powerful ligaments which bind the bones together, the joint is one of the strongest in the body, and dislocation from traumatism is a rare occurrence. When, on the other hand, the ligaments have been softened or destroyed by disease, partial displacement is liable to occur, and is frequently brought about by the mere action of the muscles displacing the articular surfaces from each other. The tibia may be dislocated in any direction from the femur—forwards, backwards, inwards, or outwards; or a combination of two of these dislocations may occur—that is, the tibia may be dislocated forwards and laterally, or backwards and laterally; and any of these dislocations may be complete or incomplete. As a rule, however, the antero-posterior dislocations are complete, the lateral ones incomplete.

One or other of the semilunar cartilages may become displaced and nipped between the femur and tibia. The accident is produced by a twist of the leg when the knee is flexed, and is accompanied by a sudden pain and fixation of the knee in a flexed position. The cartilage may be displaced either inwards or outwards: that is to say, either inwards towards the tibial spine, so that the cartilage becomes lodged in the intercondyloid notch; or outwards, so that the cartilage projects beyond the margin of the two articular surfaces.

Acute synovitis, the result of traumatism or exposure to cold, is very common in the knee, on account of its superficial position. When distended with fluid, the swelling shows itself above and at the sides of the patella, reaching about an inch, occasionally two inches or more, above the trochlear surface of the femur, and extending a little higher under the Vastus internus than the Vastus externus. The swelling extends lower at the inner side of the patella than it does on its outer side. The lower level of the synovial membrane is just above the level of the upper part of the head of the fibula. In the middle line it covers the upper third of the ligamentum patellæ, being separated from it, however, by the infrapatellar or pad fat. Chronic synovitis principally shows itself in the form of pulpy degeneration of the synovial membrane, leading to tuberculous arthritis. The reasons why tuberculous disease of the knee usually commences in the synovial membrane appear to be the complex and extensive nature of this sac; the extensive vascular supply to it; and the fact that injuries are generally diffused and applied to the front of the joint rather than to the ends of the bones. Syphilitic disease not infrequently attacks the knee-joint. In the hereditary form of the disease it is usually symmetrical, attacking both joints, which become filled with synovial effusion, and is very intractable and difficult of cure. In the tertiary form of the disease, gummatous infiltration of the synovial membrane may take place. The knee is one of the joints most commonly affected with osteo-arthritis, and is said to be more frequently the seat of this disease in women than in men. The occurrence of the so-called loose cartilages is almost confined to the knee, though they are occasionally met with in the elbow, and, rarely, in some other joints. Many of them occur in cases of osteo-arthritis, in which calcareous or cartilaginous material is formed in one of the synovial fringes and constitutes the foreign body, and may or may not become detached, in the former case only meriting the usual term, 'loose' cartilage. In other cases they have their origin in the exudation of inflammatory lymph, and possibly, in some rare instances, a portion of the articular cartilage or one of the semilunar cartilages becomes detached and constitutes the foreign body.

In inflammatory affections of the knee-joint, the position of greatest ease, and therefore the one which is always assumed, is that of slight flexion. In this position there is the most complete relaxation of ligamentous structures, and therefore the greatest diminution in the tension caused by the effusion. If this flexed position is maintained for any length of time, it becomes permanent from fibrous adhesions taking place, and the utility of the limb is materially impaired. Attention should therefore be paid by the surgeon to the position of the limb; and by carefully applied splints, with the leg in an extended position, this untoward result should be prevented. In cases of septic synovitis, incisions to evacuate the pus should be made vertically on either side of the patella, between it and the condyles of the femur: a drainage tube can then be passed across the joint from one side to the other.

Genu valgum, or knock knee, is a common deformity of childhood, in which, owing to changes in and about the joint, the angle between the outer border of the tibia and femur is diminished, so that as the patient stands the two internal condyles of the femora are in contact, but the two internal malleoli of the tibiæ are more or less widely separated from each other. When, however, the knees are flexed to a right angle, the two legs are practically parallel with each other. At the commencement of the disease there is a yielding of the internal lateral ligament and other fibrous structures on the inner side of the joint; as a result of this there is a constant undue pressure of the outer tuberosity of the tibia against the outer condyle of the femur. This extra pressure causes arrest of growth and, possibly, wasting of the outer condyle, and a consequent tendency for the tibia to become separated from the internal condyle. To prevent this the internal condyle becomes depressed; probably, as was first pointed out by Mickulicz, by an increased growth of the lower end of the diaphysis on its inner side, so that the line of the epiphysis becomes oblique instead of transverse to the axis of the bone, with a direction downwards and inwards.

Genu varum is a deformity in which the knees are widely separated, the femur and tibia being curved outwards. In early life, the disease is due to rickets; in advanced life, it may be due to osteitis deformans.

Excision of the knee-joint is most frequently required for tuberculous disease of this articulation, but is also practised in cases of disorganisation of the knee after rheumatic fever, pyæmia, &c., in osteo-arthritis, and in old neglected infantile paralysis, when there is a flail-like, distorted limb. It is also occasionally called for in cases of injury, gunshot or otherwise. The operation is best performed either by a horse-shoe incision, starting from one condyle, descending as low as the tubercle of the tibia, where it crosses the front of the leg, and then carried upward to the other condyle; or by a transverse incision across the patella. In this latter incision the patella is either removed or sawn across, and the halves subsequently sutured together. The bone having been cleared, and in those cases where the operation is performed for tuberculous disease, all pulpy tissue having been carefully removed, the section of the femur is first made. This should never include, in children, more than, at the most, two-thirds of the articular surface, otherwise the epiphysial cartilage will be involved, with disastrous results as regards the growth of the limb. Afterwards a thin slice should be removed from the upper end of the tibia, not more than half an inch. If any diseased tissue still appears to be left in the bones, it should be removed with the gouge, rather than that a further section of the bones should be made.

The bursæ about the knee-joint are sometimes the seat of enlargement. The prepatellar bursa—i.e. the bursa between the front of the patella and the skin—is frequently affected in individuals who are in the habit of constantly kneeling, and it is then known as 'housemaid's knee.' The bursa beneath the Semimembranosus tendon also constantly becomes enlarged, and forms a fluctuating swelling at the back of the knee. During extension, the swelling is firm and tense; but during flexion it becomes soft, and, as the bursa often communicates with the synovial membrane, the fluid it contains, when enlarged, can be made to disappear by pressure when the knee is flexed.

III. ARTICULATIONS BETWEEN THE TIBIA AND FIBULA

The articulations between the tibia and fibula are affected by ligaments which connect both extremities, as well as the shafts of the bones. They may, consequently, be subdivided into three sets: 1. The Superior Tibio-fibular articulation. 2. The Middle Tibio-fibular ligament or interosseous membrane. 3. The Inferior Tibio-fibular articulation.

1. SUPERIOR TIBIO-FIBULAR ARTICULATION

This articulation is an arthrodial joint. The contiguous surfaces of the bones present two flat, oval facets covered with cartilage, and connected together by the following ligaments:

Capsular.

Anterior Superior Tibio-fibular.   Posterior Superior Tibio-fibular.

The **Capsular Ligament** surrounds the articulation, being attached round the margins of the articular facets on the tibia and fibula, and is much thicker in front than behind.

The **Anterior Superior Ligament** (fig. 379) consists of two or three broad and flat bands, which pass obliquely upwards and inwards from the front of the head of the fibula to the front of the outer tuberosity of the tibia.

The **Posterior Superior Ligament** (fig. 378) is a single thick and broad band, which passes upwards and inwards from the back part of the head of the fibula to the back part of the outer tuberosity of the tibia. It is covered by the tendon of the Popliteus muscle.

A **Synovial Membrane** lines this articulation, which at its upper and back part is occasionally continuous with that of the knee-joint.

2. Middle Tibio-fibular Ligament or Interosseous Membrane

An interosseous membrane extends between the contiguous margins of the tibia and fibula, and separates the muscles on the front from those on the back of the leg. It consists of a thin, aponeurotic lamina composed of oblique fibres, which for the most part pass downwards and outwards between the interosseous ridges on the two bones; some few fibres, however, pass in the opposite direction, downwards and inwards. It is broader above than below. Its upper margin does not quite reach the superior tibio-fibular joint, but presents a free concave border, above which is a large, oval aperture for the passage of the anterior tibial vessels forwards to the anterior aspect of the leg. At its lower part is an opening for the passage of the anterior peroneal vessels. It is continuous below with the inferior interosseous ligament; and is perforated in numerous parts for the passage of small vessels. It is in relation, in front, with the Tibialis anticus, Extensor longus digitorum, Extensor proprius hallucis, Peroneus tertius, and the anterior tibial vessels and nerve; behind, with the Tibialis posticus and Flexor longus hallucis.

3. Inferior Tibio-fibular Articulation

This articulation is formed by the rough, convex surface of the inner side of the lower end of the fibula, connected with a concave rough surface on the outer side of the tibia. Below, to the extent of about two lines, these surfaces are smooth, and covered with cartilage, which is continuous with that of the ankle-joint. The ligaments of this joint are—

Anterior Inferior Tibio-fibular.	Transverse or Inferior.
Posterior Inferior Tibio-fibular.	Inferior Interosseous.

The **Anterior Inferior Ligament** (fig. 384) is a flat, triangular band of fibres, broader below than above, which extends obliquely downwards and outwards between the adjacent margins of the tibia and fibula, on the front aspect of the articulation. It is in relation, in front, with the Peroneus tertius, the aponeurosis of the leg, and the integument; behind, with the inferior interosseous ligament; and lies in contact with the cartilage covering the astragalus.

The **Posterior Inferior Ligament,** smaller than the preceding, is disposed in a similar manner on the posterior surface of the articulation.

The **Transverse** or **Inferior Ligament** lies under cover of the posterior ligament, and is a strong, thick band of yellowish fibres which passes transversely across the back of the joint, from the external malleolus to the posterior border of the articular surface of the tibia, almost as far as its malleolar process. This ligament projects below the margin of the bones, and forms part of the articulating surface for the astragalus.

The **Inferior Interosseous Ligament** consists of numerous short, strong, fibrous bands, which pass between the contiguous rough surfaces of the tibia and fibula, and constitute the chief bond of union between the bones. This ligament is continuous, above, with the interosseous membrane.

The **Synovial Membrane** lining the articular surface is derived from that of the ankle-joint.

Actions.—The movement permitted in these articulations is limited to a very slight gliding of the articular surfaces one upon another.

IV. ANKLE-JOINT

The **Ankle** is a ginglymus, or hinge-joint. The bones entering into its formation are the lower extremity of the tibia and its malleolus, and the external malleolus of the fibula, which form a mortise to receive the upper convex surface of the astragalus and its two lateral facets. The bony surfaces are covered with cartilage, and connected by a capsule, which in places forms thickened bands constituting the following ligaments :

Capsular.

| Anterior. | Internal Lateral. |
| Posterior. | External Lateral. |

The **Capsular Ligament** is an imperfect ligamentous structure which surrounds the joint, and is attached, above, to the borders of the articular surface of the tibia ; and below, to the astragalus around its upper articular surface. The parts of which it is composed vary considerably in strength, and are described under separate names as distinct ligaments. These are :

The **Anterior Ligament** (fig. 383), which is a broad, thin, membranous layer, attached, above, to the anterior margin of the lower extremity of the tibia ; below, to the margin of the astragalus, in front of its articular surface. It is in relation, in front, with the Extensor tendons of the toes, with the tendons of the Tibialis anticus and Peroneus tertius, and the anterior tibial vessels and nerve ; behind, it lies in contact with the synovial membrane.

FIG. 383.—Ankle-joint : tarsal and tarso-metatarsal articulations. Internal view. Right side.

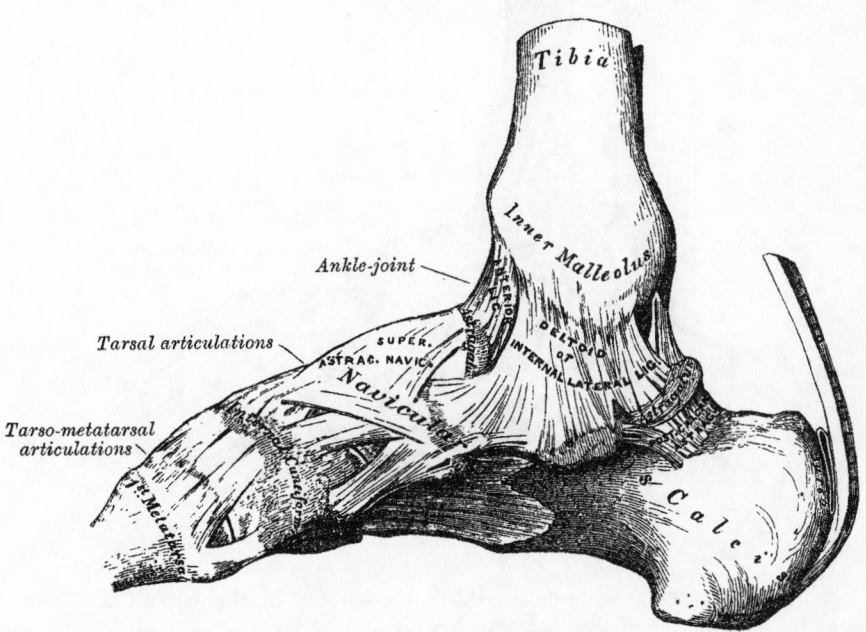

The **Posterior Ligament** is very thin, and consists principally of transverse fibres. It is attached, above, to the margin of the articular surface of the tibia, blending with the transverse tibio-fibular ligament ; below, to the astragalus behind its superior articular facet. Externally, where a somewhat thickened band of transverse fibres is attached to the hollow on the inner surface of the external malleolus, it is thicker than internally.

The **Internal Lateral** or **Deltoid Ligament** is a strong, flat, triangular band, attached, above, to the apex and anterior and posterior borders of the inner malleolus. The most anterior fibres pass forwards to be inserted into the navicular bone and the inner margin of the inferior calcaneo-navicular ligament ; the middle descend almost perpendicularly to be inserted into the sustentaculum tali of the os calcis ; and the posterior fibres pass backwards and outwards to be attached to the inner side of the astragalus. This ligament is sometimes described as consisting of five parts, viz. : *anterior astragalo-tibial, posterior astragalo-tibial, tibio-navicular, calcaneo-tibial,* and *deep astragalo-tibial.* They are only, however, differentiated bands of the one ligament made

in the process of dissection.   This ligament is covered by the tendons of the
Tibialis posticus and Flexor longus digitorum muscles.

The **External Lateral Ligament** (fig. 384) consists of three distinctly specialised
fasciculi of the capsule, taking different directions, and separated by distinct
intervals, for which reason it is described by some anatomists as three distinct
ligaments.*

The *anterior fasciculus* (anterior astragalo-fibular), the shortest of the three,
passes from the anterior margin of the external malleolus, forwards and inwards,
to the astragalus, in front of its external articular facet.

The *posterior fasciculus* (posterior astragalo-fibular), the strongest and most
deeply seated, passes inwards from the depression at the inner and back part of
the external malleolus to a prominent tubercle on the posterior surface of the
astragalus.   Its fibres are almost horizontal in direction.

The *middle fasciculus* (calcaneo-fibular), the longest of the three, is a narrow,
rounded cord, passing from the apex of the external malleolus downwards and

FIG. 384.—Ankle-joint : tarsal and tarso-metatarsal articulations.
External view.   Right side.

slightly backwards to a tubercle on the outer surface of the os calcis.   It is
covered by the tendons of the Peroneus longus and brevis.

The **Synovial Membrane** invests the inner surface of the ligaments, and sends
a duplicature upwards between the lower extremities of the tibia and fibula for a
short distance.

Relations.—The tendons, vessels, and nerves in connection with the joint are,
in front, from within outwards, the Tibialis anticus, Extensor proprius hallucis,
anterior tibial vessels, anterior tibial nerve, Extensor longus digitorum, and
Peroneus tertius ; behind, from within outwards, the Tibialis posticus, Flexor
longus digitorum, posterior tibial vessels, posterior tibial nerve, Flexor longus
hallucis ; and, in the groove behind the external malleolus, the tendons of the
Peroneus longus and brevis.

The *Arteries* supplying the joint are derived from the malleolar branches of
the anterior tibial and the peroneal.

The *Nerves* are derived from the anterior and posterior tibial.

Actions.—It must be borne in mind that when the body is in the erect
position, the foot is at right angles to the leg, and therefore the normal position
of the ankle-joint is one of flexion.   The movements of the joint are those of

* Humphry *On the Skeleton*, p. 559.

dorsiflexion and extension. Dorsiflexion consists in the approximation of the dorsum of the foot to the front of the leg, while in extension the heel is drawn up and the toes pointed downwards. The malleoli tightly embrace the astragalus in all positions of the joint, so that any slight degree of lateral movement which may exist is simply due to stretching of the inferior tibio-fibular ligaments, and slight bending of the shaft of the fibula. Of the ligaments, the internal, or deltoid, is of very great power—so much so, that it usually resists a force which fractures the process of bone to which it is attached. Its middle portion, together with the middle fasciculus of the external lateral ligament, binds the bones of the leg firmly to the foot, and resists displacement in every direction. Its anterior and posterior fibres limit extension and flexion of the foot respectively, and the anterior fibres also limit abduction. The posterior portion of the external lateral ligament assists the middle portion in resisting the displacement of the foot backwards, and deepens the cavity for the reception of the astragalus. The anterior fasciculus is a security against the displacement of the foot forwards, and limits extension of the joint. The movements of inversion and eversion of the foot, together with the minute changes in form by which it is applied to the ground or takes hold of an object in climbing, &c., are mainly

FIG. 385.—Section of the right foot near its inner border, dividing the tibia, astragalus, navicular, internal cuneiform, and first metatarsal bone, and the first phalanx of the great toe. (After Braune.)

effected in the tarsal joints; the one which enjoys the greatest amount of motion being that between the astragalus and os calcis behind, and the navicular and cuboid in front. This is often called the *transverse* or *medio-tarsal joint*, and it can, with the subordinate joints of the tarsus, replace the ankle-joint in a great measure when the latter has become ankylosed.

Extension of the foot upon the tibia and fibula is produced by the Gastrocnemius, Soleus, Plantaris, Tibialis posticus, Peroneus longus and brevis, Flexor longus digitorum, and Flexor longus hallucis. *Dorsiflexion*, by the Tibialis anticus, Peroneus tertius, Extensor longus digitorum, and Extensor proprius hallucis.* *Inversion*, in the extended position, is produced by the Tibialis anticus and posticus; and *eversion* by the Peronei.

Surface Form.—The line of the ankle-joint may be indicated by a transverse line drawn across the front of the lower part of the leg, about half an inch above the level of the tip of the internal malleolus. It can be felt on either side of the Extensor tendons; and during extension of the foot, the superior saddle-shaped facet of the astragalus can be plainly perceived below the anterior border of the lower end of the tibia.

Surgical Anatomy.—As the ankle-joint is a very strong and powerful articulation, displacement of the trochlear surface of the astragalus from the tibio-fibular mortise is not of

---

* The student must bear in mind that the Extensor longus digitorum and Extensor proprius hallucis are *extensors* of the toes, but *flexors* of the ankle; and that the Flexor longus digitorum and Flexor longus hallucis are *flexors* of the toes, but *extensors* of the ankle.

common occurrence, and great force is required to produce it. Nevertheless, dislocation does occasionally occur, both in an antero-posterior and a lateral direction. In the latter, which is the more common, fracture is a necessary accompaniment of the injury. The dislocation in these cases is somewhat peculiar, and is not a displacement in a horizontally lateral direction, such as usually occurs in lateral dislocations of ginglymoid joints, but the astragalus undergoes a partial rotation round an antero-posterior axis drawn through its own centre, so that the superior surface, instead of being directed upwards, is inclined more or less inwards or outwards according to the variety of the displacement.

The ankle-joint is more frequently sprained than any joint in the body, and this may lead to acute synovitis. In these cases, when the synovial sac is distended with fluid, the bulging appears principally in the front of the joint, beneath the anterior tendons, and on either side, between the Tibialis anticus and the internal lateral ligament on the inner side, and between the Peroneus tertius and the external lateral ligament on the outer side. In addition to this, bulging often occurs posteriorly, and a fluctuating swelling may be detected on either side of the tendo Achillis.

Chronic synovitis may result from constant sprains, and when once this joint has been sprained it is more liable to a recurrence of the injury than it was before; or the synovitis may be tuberculous in its origin, the disease usually commencing in the astragalus and extending to the joint, though it may commence as a synovitis the result probably of some slight strain in a tuberculous subject.

Excision of the ankle-joint is not often performed, for two reasons. In the first place, disease of the articulation for which this operation is indicated is frequently associated with disease of the tarsal bones, which prevents its performance; and, secondly, the foot after excision is often of very little use; far less, in fact, than after a Symes's amputation, which is, therefore, a preferable operation in these cases. Excision, may, however, be attempted in cases of tuberculous arthritis, in a young and otherwise healthy subject, where the disease is limited to the bones forming the joint. It may also be required after injury, where the vessels and nerves have not been damaged and the patient is young and free from visceral disease. The excision is best performed by two lateral incisions. One commencing two and a half inches above the external malleolus, carried down the posterior border of the fibula, round the end of this bone, and then forwards and downwards as far as the calcaneo-cuboid joint, midway between the tip of the external malleolus and the tuberosity on the fifth metatarsal bone. Through this incision the fibula is cleared, the external lateral ligament is divided, and the bone sawn through about half an inch above the level of the ankle-joint and removed. A similar curved incision is now made on the inner side of the foot, commencing two and a half inches above the lower end of the tibia, carried down the posterior border of the bone, round the internal malleolus and forwards and downwards to the tuberosity of the navicular bone. Through this incision the tibia is cleared in front and behind, the internal lateral, the anterior and posterior ligaments divided, and the end of the tibia protruded through the wound by displacing the foot outwards, and sawn off sufficiently high to secure a healthy section of bone. The articular surface of the astragalus is now to be sawn off or the whole bone removed. In cases where the operation is performed for tuberculous arthritis, the latter course is probably preferable, as the injury done by the saw is frequently the starting point of fresh caries; and after removal of the whole bone the shortening is not appreciably increased, and the result as regards union appears to be as good as when two sawn surfaces of bone are brought into apposition.

V. ARTICULATIONS OF THE TARSUS

1. ARTICULATIONS OF THE Os Calcis and Astragalus

The articulations between the os calcis and astragalus are two in number—anterior and posterior. Of these, the anterior forms part of the joint between the os calcis, astragalus, and navicular bone, and will be described hereafter as the astragalo-calcaneo-navicular articulation. The posterior or astragalo-calcanean articulation is formed between the posterior and larger facet on the inferior surface of the astragalus, and the external facet on the superior surface of the os calcis. It is an arthrodial joint, and the two bones are connected together by the following ligaments:

Capsular.	Posterior Calcaneo-astragaloid.
External Calcaneo-astragaloid.	Anterior Calcaneo-astragaloid.
Internal Calcaneo-astragaloid.	Interosseous.

The **Capsular Ligament** surrounds the two articular surfaces, and consists for the most part of short fibres, which are split up into distinct slips, which form the specially named ligaments of the articulation; between them there is only a weak fibrous investment.

The **External Calcaneo-astragaloid Ligament** (fig. 384) is a short, strong fasciculus, passing from the outer surface of the astragalus, immediately beneath its external malleolar facet, to the outer surface of the os calcis. It is placed in front of the middle fasciculus of the external lateral ligament of the ankle-joint, with the fibres of which it is parallel.

The **Internal Calcaneo-astragaloid Ligament** is a band of fibres connecting the internal tubercle of the back of the astragalus with the back of the sustentaculum tali. Its fibres blend with those of the inferior calcaneo-navicular ligament.

The **Posterior Calcaneo-astragaloid Ligament** (fig. 383) connects the external tubercle of the astragalus with the upper and inner part of the os calcis; it is a short band, the fibres of which radiate from their narrow attachment to the astragalus.

The **Anterior Calcaneo-astragaloid Ligament** extends from the front and outer surface of the neck of the astragalus to the superior surface of the os calcis. It forms the posterior boundary of the anterior calcaneo-astragaloid joint, and is sometimes described as the *anterior interosseous ligament*.

The **Interosseous Ligament** forms the chief bond of union between the bones. It is, in fact, the united capsular ligaments of the two joints mentioned above, the astragalo-calcaneo-navicular and the astragalo-calcanean, and consists of two partially united layers of fibres, one belonging to the anterior and the other to the posterior joint. It is attached by one extremity to the groove between the articulating facets on the under surface of the astragalus; by the other, to a corresponding depression on the upper surface of the os calcis. It is very thick and strong, being at least an inch in breadth from side to side, and serves to unite the os calcis and astragalus solidly together.

The **Synovial Membrane** (fig. 387) lines the capsule of the joint, and is distinct from the other synovial membranes of the tarsus.

Actions.—The movements permitted between the astragalus and os calcis are limited to a gliding of the one bone on the other in a direction from before backwards, and from side to side.

2. ARTICULATION OF THE ASTRAGALUS WITH THE OS CALCIS AND NAVICULAR BONE

The articulation between the astragalus and navicular is an arthrodial joint: the rounded head of the astragalus being received into the concavity formed by the posterior surface of the navicular, the anterior articulating surface of the calcaneum, and the upper surface of the inferior calcaneo-navicular ligament, which are all directly continuous with each other. There are two ligaments in this joint:

Capsular.   Superior Astragalo-navicular.

The **Capsular Ligament** consists of an imperfectly developed layer of fibres, except posteriorly, where they become greatly increased, and form, with a part of the capsule of the astragalo-calcanean joint, the strong interosseous ligament which fills in the canal formed by the opposing grooves on the os calcis and astragalus, as above mentioned.

The **Superior Astragalo-navicular Ligament** is a broad band, which passes obliquely forwards from the neck of the astragalus to the superior surface of the navicular bone. It is thin, and weak in texture, and covered by the Extensor tendons. The inferior calcaneo-navicular supplies the place of an inferior ligament.

The **Synovial Membrane** lines all parts of the capsule of the joint.

Actions.—This articulation permits of considerable mobility; but its feebleness is such as to allow occasionally of dislocation of the other bones of the tarsus from the astragalus.

3. ARTICULATIONS OF THE OS CALCIS WITH THE CUBOID

The ligaments connecting the os calcis with the cuboid are five in number:

Capsular

Dorsal { Superior Calcaneo-cuboid.
{ Internal Calcaneo-cuboid (Interosseous).

Plantar { Long Calcaneo-cuboid.
{ Short Calcaneo-cuboid.

E E

The **Capsular Ligament** is an imperfectly developed layer, in which are certain strengthened bands, which form the other named ligaments of the joint.

The **Superior** or **Dorsal Calcaneo-cuboid Ligament** (fig. 384) is a thin but broad fasciculus, which passes between the contiguous surfaces of the os calcis and cuboid, on the dorsal surface of the joint.

The **Internal Calcaneo-cuboid (Interosseous) Ligament** (fig. 384) is a short, but thick and strong, band of fibres, arising from the os calcis, in the deep hollow which intervenes between it and the astragalus, and closely blended, at its origin, with the superior calcaneo-navicular ligament, so as to form with it a **V**-shaped structure. It is inserted into the inner side of the cuboid bone. This ligament forms one of the chief bonds of union between the first and second rows of the tarsus.

The **Long Calcaneo-cuboid (Long Plantar) Ligament** (fig. 386), the more superficial of the two plantar ligaments, is the longest of all the ligaments of the tarsus : it is attached to the under surface of the os calcis, from near the tuberosities to the anterior tubercle ; its fibres pass forwards to be attached to the ridge on the under surface of the cuboid bone, the more superficial fibres being continued onwards to the bases of the second, third, and fourth metatarsal bones. This ligament crosses the groove on the under surface of the cuboid bone, converting it into a canal for the passage of the tendon of the Peroneus longus.

FIG. 386.—Ligaments of plantar surface of the foot.

The **Short Calcaneo - cuboid (Short Plantar) Ligament** lies nearer to the bones than the preceding, from which it is separated by a little areolar tissue. It is exceedingly broad, about an inch in length, and extends from the tubercle, and the depression in front of it, on the fore part of the under surface of the os calcis, to the inferior surface of the cuboid bone behind the peroneal groove.

Synovial Membrane. — The synovial membrane in this joint is distinct. It lines the inner surface of the ligaments.

Actions.—The movements permitted between the os calcis and cuboid are limited to a slight gliding upon each other.

The *transverse tarsal* or *medio-tarsal joint* is formed by the articulation of the os calcis with the cuboid, and by the articulation of the astragalus with the navicular. The movement which takes place in this joint is more extensive than that in the other tarsal joints, and consists of a sort of rotation by means of which the sole of the foot may be slightly flexed and extended or carried inwards (inverted) and outwards (everted).

4. THE LIGAMENTS CONNECTING THE OS CALCIS AND NAVICULAR

Though these two bones do not directly articulate, they are connected by two ligaments :

> Superior or External Calcaneo-navicular.
> Inferior or Internal Calcaneo-navicular.

The **Superior** or **External Calcaneo-navicular** (fig. 384) arises, as already mentioned, with the internal calcaneo-cuboid in the deep hollow between the astragalus and os calcis; it passes forwards from the upper surface of the anterior extremity of the os calcis to the outer side of the navicular bone. These two ligaments resemble the letter Y, being blended together behind, but separated in front.

The **Inferior** or **Internal Calcaneo-navicular** (fig. 386) is by far the larger and stronger of the two ligaments between these bones; it is a broad and thick band of fibres, which passes forwards and inwards from the anterior margin of the sustentaculum tali of the os calcis to the under surface of the navicular bone. This ligament not only serves to connect the os calcis and navicular, but supports the head of the astragalus, forming part of the articular cavity in which it is received. The *upper surface* presents a fibro-cartilaginous facet, lined by the synovial membrane continued from the anterior calcaneo-astragaloid articulation, upon which a portion of the head of the astragalus rests. Its *under surface* is in contact with the tendon of the Tibialis posticus muscle; * its inner border is blended with the fore part of the Deltoid ligament, thus completing the socket for the head of the astragalus.

Surgical Anatomy.—The inferior calcaneo-navicular ligament, by supporting the head of the astragalus, is principally concerned in maintaining the arch of the foot, and, when it yields, the head of the astragalus is pressed downwards, inwards, and forwards by the weight of the body, and the foot becomes flattened, expanded, and turned outwards, constituting the disease known as *flat-foot*. This ligament contains a considerable amount of elastic fibres, so as to give elasticity to the arch and spring to the foot; hence it is sometimes called the 'spring' ligament. It is supported, on its under surface, by the tendon of the Tibialis posticus, which spreads out at its insertion into a number of fasciculi, which are attached to most of the tarsal and metatarsal bones; this prevents undue stretching of the ligament, and is a protection against the occurrence of flat-foot.

5. The Articulation of the Navicular with the Cuneiform Bones

The navicular is connected to the three cuneiform bones by

Dorsal and Plantar ligaments.

The **Dorsal Ligaments** are small, longitudinal bands, arranged as three bundles, one to each of the cuneiform bones. That bundle of fibres which connects the navicular with the internal cuneiform is continuous round the inner side of the articulation with the plantar ligament which connects these two bones.

The **Plantar Ligaments** have a similar arrangement to those on the dorsum. They are strengthened by processes given off by the tendon of the Tibialis posticus.

The **Synovial Membrane** of these joints is part of the great tarsal synovial membrane.

Actions.—The movements permitted between the navicular and cuneiform bones are limited to a slight gliding upon each other.

6. The Articulation of the Navicular with the Cuboid

The navicular bone is connected with the cuboid by

Dorsal, Plantar, and Interosseous ligaments.

The **Dorsal Ligament** passes obliquely forwards and outwards from the navicular to the cuboid bone.

The **Plantar Ligament** passes nearly transversely between these two bones.

The **Interosseous Ligament** consists of strong transverse fibres which pass between the rough non-articular portions of the lateral surfaces of these two bones.

The **Synovial Membrane** of this joint is part of the great tarsal synovial membrane.

Actions.—The movements permitted between the navicular and cuboid bones are limited to a slight gliding upon each other.

* Hancock describes an extension of this ligament upwards on the inner side of the foot, which completes the socket of the joint in that direction. *Lancet*, 1866, vol. i. p. 618.

7. The Articulation of the Cuneiform Bones with each other

These bones are connected together by

Dorsal, Plantar, and Interosseous ligaments.

The **Dorsal Ligaments** consist of two bands which pass transversely : one connecting the internal with the middle cuneiform, and the other connecting the middle with the external cuneiform.

The **Plantar Ligaments** have a similar arrangement to those on the dorsum. They are strengthened by the processes given off from the tendon of the Tibialis posticus.

The **Interosseous Ligaments** consist of strong transverse fibres which pass between the rough non-articular portions of the lateral surfaces of the adjacent cuneiform bones.

The **Synovial Membrane** of these joints is part of the great tarsal synovial membrane.

Actions.—The movements permitted between the cuneiform bones are limited to a slight gliding upon each other.

8. The Articulation of the External Cuneiform Bone with the Cuboid

These bones are connected together by

Dorsal, Plantar, and Interosseous ligaments.

The **Dorsal Ligament** passes transversely between these two bones.

The **Plantar Ligament** has a similar arrangement. It is strengthened by a process given off from the tendon of the Tibialis posticus.

The **Interosseous Ligament** consists of strong transverse fibres which pass between the rough non-articular portions of the lateral surfaces of the adjacent sides of these two bones.

The **Synovial Membrane** of this joint is part of the great tarsal synovial membrane.

Actions.—The movements permitted between the external cuneiform and cuboid are limited to a slight gliding upon each other.

Nerve-supply.—All the joints of the tarsus are supplied by the anterior tibial nerve.

Surgical Anatomy.—In spite of the great strength of the ligaments which connect the tarsal bones together, dislocation at some of the tarsal joints does occasionally occur; though, on account of the spongy character of the bones, they are more frequently broken than dislocated, as the result of violence. When dislocation does take place, it is most commonly in connection with the astragalus; for not only may this bone be dislocated from the tibia and fibula at the ankle-joint, but the other bones may be dislocated from it, the trochlear surface of the bone remaining *in situ* in the tibio-fibular mortise. This constitutes what is known as the *sub-astragaloid* dislocation. Or, again, the astragalus may be dislocated from all its connections—from the tibia and fibula above, the os calcis below, and the navicular in front—and may even undergo a rotation, on either a vertical or a horizontal axis. In the former case the long axis of the bone becoming directed across the joint, so that the head faces the articular surface on one or other malleolus ; or, in the latter, the lateral surfaces becoming directed upwards and downwards, so that the trochlear surface faces to one or the other side. Finally, dislocation may occur at the medio-tarsal joint, the anterior tarsal bones being luxated from the astragalus and calcaneum. The other tarsal bones are also, occasionally, though rarely, dislocated from their connections.

VI. TARSO-METATARSAL ARTICULATIONS

These are arthrodial joints. The bones entering into their formation are four tarsal bones, viz. the internal, middle and external cuneiform, and the cuboid, which articulate with the bases of the metatarsal bones of the five toes. The metatarsal bone of the great toe articulates with the internal cuneiform ; that of the second is deeply wedged in between the internal and external cuneiform, resting against the middle cuneiform, and being the most strongly articulated of all the metatarsal bones ; the third metatarsal articulates with the extremity of the external cuneiform ; the fourth, with the cuboid and external cuneiform ; and the fifth, with the cuboid. The articular surfaces are covered with cartilage and connected together by the following ligaments :

<table>
<tr><td>Dorsal.</td><td>Plantar.</td><td>Interosseous.</td></tr>
</table>

The **Dorsal Ligaments** consist of strong, flat bands, which connect the tarsal with the metatarsal bones. The first metatarsal is connected to the internal cuneiform by a single broad, thin band ; the second has three dorsal ligaments, one from each cuneiform bone ; the third has one from the external cuneiform ; the fourth has two, one from the external cuneiform and one from the cuboid ; and the fifth, one from the cuboid.

The **Plantar Ligaments** consist of longitudinal and oblique bands connecting the tarsal and metatarsal bones, but disposed with less regularity than on the dorsal surface. Those for the first and second metatarsal are the most strongly marked ; the second and third metatarsal receive strong bands, which pass obliquely across from the internal cuneiform ; the plantar ligaments of the fourth and fifth metatarsal consist of a few scanty fibres derived from the cuboid.

The **Interosseous Ligaments** are three in number: internal, middle, and external. The *internal* one is the strongest of the three, and passes from the outer surface of the internal cuneiform to the adjacent angle of the second metatarsal. The *middle* one, less strong than the preceding, connects the external cuneiform with the adjacent angle of the second metatarsal. The *external* interosseous ligament connects the outer angle of the external cuneiform with the adjacent side of the third metatarsal.

The **Synovial Membrane** between the internal cuneiform bone and the first metatarsal bone is a distinct sac. The synovial membrane between the middle and external cuneiform behind, and the second and third metatarsal bones in front, is part of the great tarsal synovial membrane. Two prolongations are sent forwards from it, one between the adjacent sides of the second and third metatarsal bones, and one between the third and fourth metatarsal bones. The synovial membrane between the cuboid and the fourth and fifth metatarsal bones is a distinct sac. From it a prolongation is sent forwards between the fourth and fifth metatarsal bones.

Actions.—The movements permitted between the tarsal and metatarsal bones are limited to a slight gliding upon each other.

VII. ARTICULATIONS OF THE METATARSAL BONES WITH EACH OTHER

The base of the first metatarsal bone is not connected with the second metatarsal bone by any ligaments ; in this respect it resembles the thumb.

The bases of the four outer metatarsal bones are connected by dorsal, plantar, and interosseous ligaments.

The **Dorsal Ligaments** pass transversely between the adjacent metatarsal bones.

The **Plantar Ligaments** have a similar arrangement to those on the dorsum.

The **Interosseous Ligaments** consist of strong transverse fibres which pass between the rough non-articular portions of the lateral surfaces.

The **Synovial Membrane** between the second and third, and the third and fourth, metatarsal bones is part of the great tarsal synovial membrane.

The synovial membrane between the fourth and fifth metatarsal bones is a prolongation of the synovial membrane of the cubo-metatarsal joint.

Actions.—The movement permitted in the tarsal ends of the metatarsal bones is limited to a slight gliding of the articular surfaces upon one another.

THE SYNOVIAL MEMBRANES IN THE TARSAL AND METATARSAL JOINTS

The **Synovial Membranes** (fig. 387) found in the articulations of the tarsus and metatarsus are six in number: one for the posterior calcaneo-astragaloid articulation; a second for the anterior calcaneo-astragaloid and astragalo-navicular articulations; a third for the calcaneo-cuboid articulation ; and a fourth for the articulations of the navicular with the three cuneiforms, the three cuneiforms with each other, the external cuneiform with the cuboid, and the middle and external cuneiforms with the bases of the second and third metatarsal bones, and the lateral surfaces of the second, third, and fourth metatarsal bones with each other ; a fifth for the internal cuneiform with the metatarsal bone of the great toe ; and

FIG. 387.—Oblique section of the articulations of the tarsus and metatarsus. Showing the six synovial membranes.

a sixth for the articulation of the cuboid with the fourth and fifth metatarsal bones. A small synovial membrane is sometimes found between the contiguous surfaces of the navicular and cuboid bones.

Nerve-supply.—The nerves supplying the tarso-metatarsal joints are derived from the anterior tibial.

The *digital extremities* of all the metatarsal bones are connected together by the *transverse metatarsal ligament.*

The **Transverse Metatarsal Ligament** is a narrow band which passes transversely across the anterior extremities of all the metatarsal bones, connecting them together. It is blended anteriorly with the plantar (glenoid) ligament of the metatarso-phalangeal articulations. To its posterior border is connected the fascia covering the Interossei muscles. Its inferior surface is concave where the Flexor tendons pass over it. Above it the tendons of the Interossei muscles pass to their insertion. It differs from the transverse metacarpal ligament in that it connects the metatarsal bone of the great toe with the rest of the metatarsal bones.

VIII.   METATARSO-PHALANGEAL ARTICULATIONS

The metatarso-phalangeal articulations are of the condyloid kind, formed by the reception of the rounded head of the metatarsal bone into a superficial cavity in the extremity of the first phalanx.

The ligaments are—

Plantar.   Two Lateral.

The **Plantar Ligaments** (Glenoid ligaments of Cruveilhier) are thick, dense, fibrous structures. Each is placed on the plantar surface of the joint in the interval between the lateral ligaments, to which they are connected ; they are loosely united to the metatarsal bones, but very firmly to the bases of the first phalanges. Their plantar surface is intimately blended with the transverse metatarsal ligament, and presents a groove for the passage of the Flexor tendons, the sheath surrounding which is connected to each side of the groove. By their deep surface, they form part of the articular surface for the head of the metatarsal bone, and are lined by a synovial membrane.

The **Lateral Ligaments** are strong, rounded cords, placed one on each side of the joint, and attached, by one extremity, to the posterior tubercle on the side of the head of the metatarsal bone ; and, by the other, to the contiguous extremity of the phalanx.

The **Posterior Ligament** is supplied by the Extensor tendon placed over the back of the joint.

Actions.—The movements permitted in the metatarso-phalangeal articulations are flexion, extension, abduction, and adduction.

IX. ARTICULATIONS OF THE PHALANGES

The articulations of the phalanges are ginglymoid joints.

The ligaments are—

Plantar.   Two Lateral.

The arrangement of these ligaments is similar to those in the metatarso-phalangeal articulations : the Extensor tendon supplies the place of a posterior ligament.

Actions.—The only movements permitted in the phalangeal joints are flexion and extension ; these movements are more extensive between the first and second phalanges than between the second and third. The movement of flexion is very considerable, but extension is limited by the plantar and lateral ligaments.

Surface Form.—The principal joints which it is necessary to distinguish, with regard to the surgery of the foot, are the medio-tarsal and the tarso-metatarsal joints. The joint between the astragalus and the navicular is best found by means of the tubercle of the navicular bone, for the line of the joint is immediately behind this process. If the foot is grasped and forcibly extended, a rounded prominence, the head of the astragalus, will appear on the inner side of the dorsum in front of the ankle-joint, and if a knife is carried downwards, just in front of this prominence and behind the line of the navicular tubercle, it will enter the astragalo-navicular joint. The calcaneo-cuboid joint is situated midway between the external malleolus and the prominent end of the fifth metatarsal bone. The plane of the joint is in the same line as that of the astragalo-navicular. The position of the joint between the fifth metatarsal bone and the cuboid is easily found by the projection of the fifth metatarsal bone, which is the guide to it. The direction of the line of the joint is very oblique, so that, if continued inwards, it would pass through the inner side of the head of the first metatarsal bone. The joints between the third and fourth metatarsal bones and the cuboid and external cuneiform are the direct continuation inwards of the previous joint, but their planes are less oblique. The tarso-metatarsal articulation of the great toe corresponds to a groove which can be felt by making firm pressure on the inner side of the foot one inch in front of the tubercle on the navicular bone ; and the joint between the second metatarsal bone and the middle cuneiform is to be found on the dorsum of the foot, half an inch behind the level of the tarso-metatarsal joint of the great toe. The line of the joints between the metatarsal bones and the first phalanges is about an inch behind the webs of the corresponding toes.

THE MUSCLES AND FASCIÆ *

THE Muscles are connected with the bones, cartilages, ligaments, and skin, either directly or through the intervention of fibrous structures, called tendons or aponeuroses. Where a muscle is attached to bone or cartilage, the fibres terminate in blunt extremities upon the periosteum or perichondrium, and do not come into direct relation with the osseous or cartilaginous tissue. Where muscles are connected with the skin, they either lie as a flattened layer beneath it, or are connected with its areolar tissue by larger or smaller bundles of fibres, as in the muscles of the face.

The muscles vary extremely in their form. In the limbs, they are of considerable length, especially the more superficial ones; they surround the bones, and constitute an important protection to the various joints. In the trunk, they are broad, flattened, and expanded, forming the parietes of the cavities which they enclose; hence the reason of the terms, *long*, *broad*, *short*, &c., used in the description of a muscle.

There is a considerable variation in the arrangement of the fibres of certain muscles with reference to the tendons to which they are attached. In some, the fibres are parallel and run directly from their origin to their insertion; these are quadrilateral muscles, such as the Thyro-hyoid. A modification of these is found in the fusiform muscles, in which the fibres are not quite parallel, but slightly curved, so that the muscle tapers at each end; in their action, however, they resemble the quadrilateral muscles. Secondly, in other muscles the fibres are convergent; arising by a broad origin, they converge to a narrow or pointed insertion. This arrangement of fibres is found in the triangular muscles—e.g. the Temporal. In some muscles, which otherwise would belong to the quadrilateral or triangular type, the origin and insertion are not in the same plane, but the plane of the line of origin intersects that of their insertion: such is the case in the Pectineus muscle. Thirdly, in some muscles the fibres are oblique and converge, like the plumes of a quill-pen, to one side of a tendon which runs the entire length of the muscle. Such a muscle is rhomboidal or penniform, as the Peronei. A modification of these rhomboidal muscles is found in those cases where oblique fibres converge to both sides of a central tendon which runs down the middle of the muscle; these are called bipenniform, and an example is afforded in the Rectus femoris. Finally, we have muscles in which the fibres are arranged in curved bundles in one or more planes, as in the Sphincter muscles. The arrangement of the muscular fibres is of considerable importance in respect to their relative strength and range of movement. Those muscles where the fibres are long and few in number have great range, but diminished strength; where, on the other hand, the fibres are short and more numerous, there is great power, but lessened range.

Muscles differ much in size: the Gastrocnemius forms the chief bulk of the back of the leg, and the fibres of the Sartorius are nearly two feet in length,

* The Muscles and Fasciæ are described conjointly, in order that the student may consider the arrangement of the latter in his dissection of the former. It is rare for the student of anatomy in this country to have the opportunity of dissecting the fasciæ separately; and it is for this reason, as well as from the close connection that exists between the muscles and their investing sheaths, that they are considered together. Some general observations are first made on the anatomy of the muscles and fasciæ, the special description being given in connection with the different regions.

while the Stapedius, a small muscle of the internal ear, weighs about a grain, and its fibres are not more than two lines in length.

The names applied to the various muscles have been derived : 1, from their situation, as the Tibialis, Radialis, Ulnaris, Peroneus ; 2, from their direction, as the Rectus abdominis, Obliqui capitis, Transversalis ; 3, from their uses, as Flexors, Extensors, Abductors, &c. ; 4, from their shape, as the Deltoid, Trapezius, Rhomboideus ; 5, from the number of their divisions, as the Biceps, the Triceps ; 6, from their points of attachment, as the Sterno-cleido-mastoid, Sterno-hyoid, Sterno-thyroid.

In the description of a muscle, the term *origin* is meant to imply its more fixed or central attachment ; and the term *insertion* the movable point to which the force of the muscle is directed ; but the origin is absolutely fixed in only a very small number of muscles, such as those of the face, which are attached by one extremity to the bone, and by the other to the movable integument ; in the greater number, the muscle can be made to act from either extremity.

In the dissection of the muscles, the student should pay especial attention to the exact *origin, insertion,* and *actions* of each, and its more important *relations* with surrounding parts. An accurate knowledge of the points of attachment of the muscles is of great importance in the determination of their action. By a consideration of the action of the muscles, the surgeon is able to explain the causes of displacement in various forms of fracture, and the causes which produce distortion in various deformities, and, consequently, to adopt appropriate treatment in each case. The relations, also, of some of the muscles, especially those in immediate apposition with the larger blood-vessels, and the surface-markings they produce, should be carefully remembered, as they form useful guides in the application of ligatures to those vessels.

Tendons are white, glistening, fibrous cords, varying in length and thickness, sometimes round, sometimes flattened, of considerable strength, and devoid of elasticity. They consist almost entirely of white fibrous tissue, the fibrils of which have an undulating course parallel with each other, and are firmly united together. They are very sparingly supplied with blood-vessels, the smaller tendons presenting in their interior no trace of them. Nerves supplying tendons have a special modification of their terminal fibres, named organ of Golgi (see page 48). The tendons consist principally of a substance which yields gelatin.

Aponeuroses are flattened or ribbon-shaped tendons, of a pearly-white colour, iridescent, glistening, and similar in structure to the tendons. They are only sparingly supplied with blood-vessels.

The tendons and aponeuroses are connected, on the one hand, with the muscles ; and, on the other hand, with the movable structures, as the bones, cartilages, ligaments, and fibrous membranes (for instance, the sclerotic). Where the muscular fibres are in a direct line with those of the tendon or aponeurosis, the two are directly continuous, the muscular fibre being distinguishable from that of the tendon only by its striation. But where the muscular fibre joins the tendon or aponeurosis at an oblique angle, the former terminates, according to Kölliker, in rounded extremities, which are received into corresponding depressions on the surface of the latter, the connective tissue between the fibres being continuous with that of the tendon. The latter mode of attachment occurs in all the penniform and bipenniform muscles, and in those muscles the tendons of which commence in a membranous form, as the Gastrocnemius and Soleus.

The fasciæ (fascia, *a bandage*) are fibro-areolar or aponeurotic laminæ, of variable thickness and strength, found in all regions of the body, investing the softer and more delicate organs. The fasciæ have been subdivided, from the situation in which they are found, into two groups, superficial and deep.

The *superficial fascia* is found immediately beneath the integument over almost the entire surface of the body. It connects the skin with the deep or aponeurotic fascia, and consists of fibro-areolar tissue, containing in its meshes pellicles of fat in varying quantity. In the eyelids and scrotum, where fat is rarely deposited, this tissue is very liable to serous infiltration. The superficial fascia varies in thickness in different parts of the body : in the groin it is so thick as to be capable of being subdivided into several laminæ. Beneath the fatty layer of the superficial fascia, which is immediately subcutaneous, there is generally another layer of the same structure, comparatively devoid of adipose tissue, in which the trunks of the subcutaneous vessels and nerves are found, as the superficial epigastric vessels in the abdominal region, the radial and ulnar veins in the forearm, the saphenous veins in the leg and thigh, and the superficial lymphatic glands; certain cutaneous muscles also are situated in the superficial fascia, as the Platysma myoides in the neck, and the Orbicularis palpebrarum around the eyelids. This fascia is most distinct at the lower part of the abdomen, the scrotum, perinæum, and extremities; is very thin in those regions where muscular fibres are inserted into the integument, as on the side of the neck, the face, and around the margin of the anus. It is very dense in the scalp, in the palms of the hands, and soles of the feet, forming a fibro-fatty layer, which binds the integument firmly to the subjacent structure.

The superficial fascia connects the skin to the subjacent parts, facilitates the movement of the skin, serves as a soft nidus for the passage of vessels and nerves to the integument, and retains the warmth of the body, since the fat contained in its areolæ is a bad conductor of heat.

The *deep fascia* is a dense, inelastic, unyielding fibrous membrane, forming sheaths for the muscles, and in some cases affording them broad surfaces for attachment. It consists of shining tendinous fibres, placed parallel with one another, and connected together by other fibres disposed in a rectilinear manner. It is usually exposed on the removal of the superficial fascia, forming a strong investment, which not only binds down collectively the muscles in each region, but gives a separate sheath to each, as well as to the vessels and nerves. The fasciæ are thick in unprotected situations, as on the outer side of a limb, and thinner on the inner side. The deep fasciæ assist the muscles in their action, by the degree of tension and pressure they make upon their surface: and, in certain situations, this is increased and regulated by muscular action, as, for instance, by the Tensor fasciæ femoris and Gluteus maximus in the thigh, by the Biceps in the upper and lower extremities, and Palmaris longus in the hand. In the limbs, the fasciæ not only invest the entire limb, but give off septa which separate the various muscles, and are attached beneath to the periosteum: these prolongations of fasciæ are usually spoken of as intermuscular septa.

The Muscles and Fasciæ may be arranged, according to the general division of the body, into those of the cranium, face, and neck; those of the trunk; those of the upper extremity; and those of the lower extremity.

MUSCLES AND FASCIÆ OF THE CRANIUM AND FACE

The Muscles of the Cranium and Face consist of ten groups, arranged according to the region in which they are situated.

I. Cranial Region.	VI. Maxillary Region.
II. Auricular Region.	VII. Mandibular Region.
III. Palpebral Region.	VIII. Intermaxillary Region.
IV. Orbital Region.	IX. Temporo-mandibular Region.
V. Nasal Region.	X. Pterygo-mandibular Region.

The muscles contained in each of these groups are the following:

I. Cranial Region.
Occipito-frontalis.

II. Auricular Region.
Attrahens auriculam.
Attollens auriculam.
Retrahens auriculam.

III. Palpebral Region.
Orbicularis palpebrarum.
Tensor tarsi.
Corrugator supercilii.

IV. Orbital Region.
Levator palpebræ.
Rectus superior.
Rectus inferior.
Rectus internus.
Rectus externus.
Obliquus superior.
Obliquus inferior.

V. Nasal Region.
Pyramidalis nasi.
Levator labii superioris alæque nasi.

Dilatator naris posterior.
Dilatator naris anterior.
Compressor naris.
Compressor narium minor.
Depressor alæ nasi.

VI. *Maxillary Region.*
Levator labii superioris.
Levator anguli oris.
Zygomaticus major.
Zygomaticus minor.

VII. *Mandibular Region.*
Levator labii inferioris.

Depressor labii inferioris.
Depressor anguli oris.

VIII. *Intermaxillary Region.*
Orbicularis oris.
Buccinator.
Risorius.

IX. *Temporo-mandibular Region.*
Masseter.
Temporal.

X. *Pterygo-mandibular Region.*
Pterygoideus externus.
Pterygoideus internus.

I. Cranial Region—Occipito-frontalis

Dissection (fig. 388).—The head being shaved, and a block placed beneath the back of the neck, make a vertical incision through the skin from before backwards, commencing at the root of the nose in front, and terminating behind at the occipital protuberance; make a second incision in a horizontal direction along the forehead and round the side of

Fig. 388.—Dissection of the head, face, and neck.

1, *Dissection of scalp*

2, 3, *of auricular region*

4, 5, 6, *of face*

7, 8, *of neck*

the head, from the anterior to the posterior extremity of the preceding. Raise the skin in front, from the subjacent muscle, from below upwards; this must be done with great care, in order to avoid cutting through the vessels and nerves which lie immediately beneath the skin.

The Skin of the Scalp.—This is thicker than in any other part of the body, but its thickness varies in different situations: it is thickest in the occipital region, whereas in the temporal and mastoid regions it is moderately thin. It is intimately adherent to the superficial fascia. The hair-follicles are very closely set together, and extend throughout the whole thickness of the skin. It also contains a number of sebaceous glands.

The **superficial fascia** in the cranial region is a firm, dense, fibro-fatty layer, intimately adherent to the integument, and to the Occipito-frontalis and its tendinous aponeurosis; it is continuous, behind, with the superficial fascia at the back part of the neck; and, laterally, is continued over the temporal fascia. It contains between its layers the superficial vessels and nerves and much granular fat.

The **Occipito-frontalis** (fig. 389) is a broad musculo-fibrous layer, which covers the whole of one side of the vertex of the skull, from the occiput to the eyebrow.

It consists of two muscular slips, connected by an intervening tendinous aponeurosis. The *occipital portion* (sometimes called the *occipitalis* muscle) is thin, quadrilateral in form, and about an inch and a half in length; it arises from the outer two-thirds of the superior curved line of the occipital bone, and from the mastoid portion of the temporal. Its fibres of origin are tendinous, but they soon become muscular, and ascend in a parallel direction to terminate in the tendinous aponeurosis. The *frontal portion* (sometimes called the *frontalis*

FIG. 389.—Muscles of the head, face, and neck.

muscle) is thin, of a quadrilateral form, and intimately adherent to the superficial fascia. It is broader, its fibres are longer, and their structure paler than the occipital portion. It has no bony attachments. Its internal fibres are continuous with those of the Pyramidalis nasi. Its middle fibres become blended with the Corrugator supercilii and Orbicularis palpebrarum; and the outer fibres are also blended with the latter muscle over the external angular process. From these attachments, the fibres are directed upwards and join the aponeurosis below the coronal suture. The inner margins of the frontal portions of the two muscles are joined together for some distance above the root of the nose; but between the occipital portions there is a considerable, though variable, interval, which is occupied by the aponeurosis.

The **aponeurosis** covers the upper part of the vertex of the skull, being continuous across the middle line with the aponeurosis of the opposite muscle. Behind, it is attached, in the interval between the occipital origins, to the occipital protuberance and highest curved lines of the occipital bone; in front, it forms a short and narrow prolongation between the frontal portions; and on each side it has connected with it the Attollens and Attrahens muscles of the pinna; in this situation it loses its aponeurotic character, and is continued over the temporal fascia to the zygoma as a layer of laminated areolar tissue. This aponeurosis is closely connected to the integument by the firm, dense, fibro-fatty layer which forms the superficial fascia: it is connected with the pericranium by loose cellular tissue, which allows of a considerable degree of movement of the integument.

Nerves.—The frontal portion of the Occipito-frontalis is supplied by the facial nerve; its occipital portion, by the posterior auricular branch of the same nerve.

Actions.—The frontal portion of the muscle raises the eyebrows and the skin over the root of the nose, and at the same time draws the scalp forwards, throwing the integument of the forehead into transverse wrinkles. The posterior portion draws the scalp backwards. By bringing alternately into action the frontal and occipital portions the entire scalp may be moved forwards and backwards. In the ordinary action of the muscles, the eyebrows are elevated, and at the same time the aponeurosis is fixed by the posterior portion, thus giving to the face the expression of surprise: if the action is more exaggerated, the eyebrows are still further raised, and the skin of the forehead thrown into transverse wrinkles, as in the expression of fright or horror.

Surgical Anatomy.—From an anatomical point of view, the scalp consists of five layers, viz. the skin, subcutaneous tissue, Occipito-frontalis muscle and its aponeurosis, subaponeurotic connective tissue, and pericranium. But from a surgical standpoint it is better to regard the first three of these structures as a single layer, since they are all intimately fused together, and when torn off in an accident, or turned down as a flap in a surgical operation, remain firmly connected to each other. In consequence of the dense character of the subcutaneous tissue, the amount of swelling which occurs as the result of inflammation is slight; and the edges of a wound which does not involve the Occipito-frontalis muscle or its aponeurosis do not gape. The blood-vessels, also, which lie in this tissue, when wounded, are unable to freely contract and retract; and therefore the hæmorrhage from scalp wounds is often very considerable, but can always be arrested by pressure—a matter of great importance, as it is often very difficult or impossible to pick up with forceps a wounded vessel in the scalp.

The skin of the scalp is abundantly supplied with sebaceous and sudoriparous glands. The former are sometimes the seat of cystic enlargement, constituting the so-called *atheromatous cysts* or *wens*.

The subaponeurotic connective tissue is, from a surgical point of view, of considerable importance. It is loose and lax, and is easily torn through; and hence, when a flap wound occurs in the scalp, this is the tissue which is torn when the flap is separated from the parts beneath. The vessels are therefore torn down with the flap, and there is little risk of sloughing, unless the vitality of the part has been actually destroyed by the injury. In consequence of its loose nature and feeble vitality, any septic inflammation is apt to assume a very diffuse form and spread all over the skull, and, unless relieved by timely incisions, lead to extensive destruction of the integument, from the tension to which it is subjected.

II. Auricular Region (fig. 389)

Attrahens auriculam.   Attollens auriculam.
                        Retrahens auriculam.

These three small muscles are placed immediately beneath the skin around the pinna. In man, in whom the pinna is almost immovable, they are rudimentary. They are the analogues of large and important muscles in some of the mammalia.

Dissection.—This requires considerable care, and should be performed in the following manner: To expose the Attollens auriculam, draw the pinna or broad part of the ear downwards, when a tense band will be felt beneath the skin, passing from the side of the head to the upper part of the concha; by dividing the skin over this band, in a direction from below upwards, and then reflecting it on each side, the muscle is exposed. To bring into view the Attrahens auriculam, draw the helix backwards by means of a hook, when the muscle will be made tense, and may be exposed in a similar manner to the preceding. To expose the Retrahens auriculam, draw the pinna forwards, when the muscle, being made tense, may be felt beneath the skin, at its insertion into the back part of the concha, and may be exposed in the same manner as·the other muscles.

The **Attrahens auriculam** (*Auricularis anterior*), the smallest of the three, is thin, fan-shaped, and its fibres pale and indistinct; they arise from the lateral edge of the aponeurosis of the Occipito-frontalis, and converge to be inserted into a projection on the front of the helix.

Relations.—*Superficially*, with the skin; *deeply*, with the areolar tissue derived from the aponeurosis of the Occipito-frontalis, beneath which are the superficial temporal artery and vein and the temporal fascia.

The **Attollens auriculam** (*Auricularis superior*), the largest of the three, is thin and fan-shaped: its fibres arise from the aponeurosis of the Occipito-frontalis, and converge to be inserted by a thin, flattened tendon into the upper part of the cranial surface of the pinna.

Relations.—*Superficially*, with the integument; *deeply*, with the areolar tissue derived from the aponeurosis of the Occipito-frontalis, beneath which is the temporal fascia.

The **Retrahens auriculam** (*Auricularis posterior*) consists of two or three fleshy fasciculi, which arise from the mastoid portion of the temporal bone by short aponeurotic fibres. They are inserted into the lower part of the cranial surface of the concha.

Relations.—*Superficially*, with the integument; *deeply*, with the mastoid portion of the temporal bone and the posterior auricular artery and nerve.

Nerves.—The Attrahens and Attollens auriculam are supplied by the temporal branch of the facial nerve; the Retrahens auriculam is supplied by the posterior auricular branch of the same nerve.

Actions.—In man, these muscles possess very little action: the Attrahens auriculam draws the ear forwards and upwards; the Attollens auriculam slightly raises it; and the Retrahens auriculam draws it backwards.

III. Palpebral Region (fig. 389)

Orbicularis palpebrarum.   Corrugator supercilii.
Tensor tarsi.   Levator palpebræ.

Dissection (fig. 388).—In order to expose the muscles of the face, continue the longitudinal incision, made in the dissection of the Occipito-frontalis, down the median line of the face to the tip of the nose, and from this point onwards to the upper lip; and carry another incision along the margin of the lip to the angle·of the mouth, and transversely across the face to the angle of the jaw. Then make an incision in front of the external ear, from the angle of the jaw upwards, to join the transverse incision made in exposing the Occipito-frontalis. These incisions include a square-shaped flap, which should be removed with care, in the direction marked in the figure, as the muscles at some points are intimately adherent to the integument.

The **Orbicularis palpebrarum** is a sphincter muscle, which surrounds the circumference of the orbit and eyelids. It arises from the internal angular process of the frontal bone, from the nasal process of the superior maxillary bone in front of the lachrymal groove for the nasal duct, and from the anterior surface and borders of a short tendon, the *tendo oculi*, or *internal tarsal ligament*, placed at the inner angle of the orbit. From this origin, the fibres are directed outwards, forming a broad, thin, and flat layer, which covers the eyelids, surrounds the circumference of the orbit, and spreads out over the temple, and downwards on the cheek. The palpebral portion (pars ciliaris) of the Orbicularis is thin and pale; it arises from the bifurcation of the tendo oculi, and forms a series of concentric curves, which are inserted on the outer side of the eyelids into the external tarsal ligament. The orbital portion (pars orbicularis) is thicker and of a reddish colour: its fibres are well developed, and form complete ellipses. The upper fibres of this portion blend with the Occipito-frontalis and Corrugator supercilii.

Relations.—By its *superficial surface*, with the integument. By its *deep surface*, above, with the Occipito-frontalis and Corrugator supercilii, with which it is intimately blended, and with the supra-orbital vessels and nerve ; below, it covers the lachrymal sac, and the origin of the Levator labii superioris alæque nasi, the Levator labii superioris, and the Zygomaticus minor muscles. *Internally*, it is occasionally blended with the Pyramidalis nasi. *Externally*, it lies on the temporal fascia. On the eyelids, it is separated from the conjunctiva by the tarsal ligaments, the tarsal plates, and the Meibomian glands ; and in the case of the upper lid, by the tendon of the Levator palpebræ muscle.

The *tendo oculi*, or *internal tarsal ligament*, is a short tendon, about two lines in length and one in breadth, attached to the nasal process of the superior maxillary bone in front of the lachrymal groove. Crossing the lachrymal sac, it divides into two parts, each division being attached to the inner extremity of the corresponding tarsal plate. As the tendon crosses the lachrymal sac, a strong aponeurotic lamina is given off from the posterior surface, which expands over the sac, and is attached to the ridge on the lachrymal bone. This is the reflected aponeurosis of the tendo oculi.

The *external tarsal ligament* is a much weaker structure than the tendo oculi. It is attached to the margin of the frontal process of the malar bone, and passes inwards to the outer commissure of the eyelid, where it divides into two slips, which are attached to the margins of the respective tarsal plates.

Use of Tendo oculi.—Besides giving attachment to part of the Orbicularis palpebrarum, and to the tarsal plates, it serves to suck the tears into the lachrymal sac, by its attachment to the sac. Thus, each time the eyelids are closed, the tendo oculi becomes tightened, through the action of the Orbicularis, and draws the wall of the lachrymal sac outwards and forwards, so that a vacuum is made in it, and the tears are sucked along the lachrymal canals into it.

FIG. 390.—Horner's muscle. (From a preparation in the Museum of the Royal College of Surgeons of England.)

Horner's muscle

Orbicularis palpebrarum

Puncta lachrymalia

Lev. lab. sup. alæque nasi

The **Tensor tarsi** (Horner's muscle) is a small thin muscle, about three lines in breadth and six in length, situated at the inner side of the orbit, behind the tendo oculi and lachrymal sac (fig. 390). It arises from the crest and adjacent part of the orbital surface of the lachrymal bone, and, passing across the lachrymal sac, divides into two slips, which cover the lachrymal canals, and are inserted into the tarsal plates internal to the puncta lachrymalia. Its fibres appear to be continuous with those of the palpebral portion of the Orbicularis palpebrarum, from which they are usually considered to be derived ; it is occasionally very indistinct.

The **Corrugator supercilii** is a small, narrow, pyramidal muscle, placed at the inner extremity of the eyebrow, beneath the Occipito-frontalis and Orbicularis palpebrarum muscles. It arises from the inner extremity of the superciliary ridge ; whence its fibres pass upwards and outwards, between the palpebral and orbital portions of the Orbicularis palpebrarum, and are inserted into the deep surface of the skin, opposite the middle of the orbital arch.

Relations.—By its *anterior surface*, with the Occipito-frontalis and Orbicularis palpebrarum muscles. By its *posterior surface*, with the frontal bone and supratrochlear nerve.

Nerves.—The Orbicularis palpebrarum, Corrugator supercilii, and Tensor tarsi are supplied by the facial nerve. Recent investigations tend to show that the Orbicularis palpebrarum, Corrugator supercilii, and frontal part of the Occipito-frontalis, are in reality supplied by fibres of the third nerve, which descend through the pons Varolii to join the facial nerve.

Actions.—The Orbicularis palpebrarum is the sphincter muscle of the eyelids. The palpebral portion acts involuntarily, closing the lids gently, as in sleep or in blinking; the orbicular portion is subject to the will. When the entire muscle is brought into action, the skin of the forehead, temple, and cheek is drawn inwards towards the inner angle of the orbit, and the eyelids are firmly closed, as in photophobia. When the skin of the forehead, temple, and cheek is thus drawn inwards by the action of the muscle it is thrown into folds, especially radiating from the outer angle of the eyelids, which give rise in old age to the so-called 'crow's feet.' The Levator palpebræ is the direct antagonist of this muscle; it raises the upper eyelid and exposes the globe of the eye. The Corrugator supercilii draws the eyebrow downwards and inwards, producing the vertical wrinkles of the forehead. It is the 'frowning' muscle, and may be regarded as the principal agent in the expression of suffering. The Tensor tarsi draws the eyelids and the extremities of the lachrymal canals inwards and compresses them against the surface of the globe of the eye; thus placing them in the most favourable situation for receiving the tears. It serves, also, to compress the lachrymal sac.

IV. Orbital Region (fig. 391)

Levator palpebræ superioris.	Rectus internus.
Rectus superior.	Rectus externus.
Rectus inferior.	Obliquus oculi superior.

Obliquus oculi inferior.

Dissection.—To open the cavity of the orbit, remove the skull-cap and brain; then saw through the frontal bone at the inner extremity of the supra-orbital ridge, and externally at its junction with the malar. Break in pieces the thin roof of the orbit by a few slight blows of the hammer, and take it away; drive forward the superciliary portion of the frontal bone by a smart stroke, but do not remove it, as that would destroy the pulley of the Obliquus superior. When the fragments are cleared away, the periosteum of the orbit will be exposed; this being removed, together with the fat which fills the cavity of the orbit, the several muscles of this region can be examined. The dissection will be facilitated by distending the globe of the eye. In order to effect this, puncture the optic nerve near the eyeball with a curved needle, and push the needle onwards into the globe; insert the point of a blowpipe through this aperture, and force a little air into the cavity of the eyeball; then apply a ligature round the nerve so as to prevent the air escaping. The globe being now drawn forwards, the muscles will be put upon the stretch.

The **Levator palpebræ superioris** is thin, flat, and triangular in shape. It arises from the under surface of the lesser wing of the sphenoid, above and in front of the optic foramen, from which it is separated by the origin of the Superior rectus. At its origin, it is narrow and tendinous, but soon becomes broad and fleshy, and finally terminates in a wide aponeurosis, which is inserted into the upper margin of the superior tarsal plate. From this aponeurosis a thin expansion is continued onwards, passing between the fibres of the Orbicularis to be inserted into the skin of the lid, and some deeper fibres blend with an expansion from the sheath of the Superior rectus muscle, and are with it prolonged into the conjunctiva. At each angle of the lid, the aponeurosis blends with the palpebral ligaments; and, at the inner angle, is also attached to the pulley of the Superior oblique muscle of the eye, which it serves to render tense.

Relations.—By its *upper surface*, with the frontal nerve, the supra-orbital artery, and the periosteum of the orbit and lachrymal gland; and, in the lid, with the inner surface of the tarsal ligament. By its *under surface*, with the Superior rectus; and, in the lid, with the conjunctiva. A small branch of the third nerve enters its under surface.

The **Four Recti muscles** arise from a fibrous ring which surrounds the optic foramen and encloses the optic nerve. From this ring a little tendinous bridge is prolonged over the lower and inner part of the sphenoidal fissure, to be inserted into a tubercle of bone on the margin of the greater wing of the sphenoid, bounding the sphenoidal fissure. Two specialised parts of this fibrous ring may be made out: a lower, the more distinct, the *ligament* or *tendon of Zinn*, which gives origin to the Inferior rectus, part of the Internal rectus, and the lower head of origin of the External rectus; and an upper, which gives origin to the Superior rectus, the rest of the Internal rectus, and the upper head of the

FIG. 391.—Muscles of the right orbit.

External rectus. This upper band is sometimes termed the *superior tendon of Lockwood*. Each muscle passes forward in the position implied by its name, to be inserted by a tendinous expansion into the sclerotic coat, about three or four lines from the margin of the cornea. Between the two heads of the External rectus is a narrow interval, through which pass the two divisions of the third nerve, the nasal branch of the ophthalmic division of the fifth nerve, the sixth nerve, and the ophthalmic vein. Although these muscles present a common origin and are inserted in a similar manner into the sclerotic coat, there are certain differences to be observed in them, as regards their length and breadth. The Internal rectus is the broadest; the External is the longest; and the Superior is the thinnest and narrowest.

FIG. 392.—The relative position and attachment of the muscles of the left eyeball.

The **Superior oblique** is a fusiform muscle, placed at the upper and inner side of the orbit, internal to the Levator palpebræ. It arises about a line above the inner margin of the optic foramen, above and internal to the origin of the Superior rectus, and, passing forwards to the inner angle of the orbit, terminates in a rounded tendon, which plays in a fibro-cartilaginous ring or pulley attached to a depression beneath the internal angular process of the frontal bone, the contiguous surfaces of the tendon and ring being lined by a delicate synovial membrane, and enclosed in a thin fibrous investment. The tendon is reflected backwards, outwards, and downwards beneath the Superior rectus to the outer part of the globe of the eye, and is inserted into the sclerotic coat, behind the equator of the eyeball, the insertion of the muscle lying between the Superior and External recti.

Relations.—By its *upper surface*, with the periosteum lining the roof of the

F F

orbit, and the fourth nerve. The tendon, where it lies on the globe of the eye, is covered by the Superior rectus. By its *under surface,* with the nasal nerve, ethmoidal arteries, and the upper border of the Internal rectus.

The **Inferior oblique** is a thin, narrow muscle, placed near the anterior margin of the orbit. It arises from the orbital plate of the superior maxillary bone, external to the lachrymal groove. Passing outwards, backwards, and upwards, between the Inferior rectus and the floor of the orbit, and then between the eyeball and the External rectus, it is inserted into the outer part of the sclerotic coat between the Superior and External rectus, near to, but somewhat behind, the tendon of insertion of the Superior oblique.

Relations.—By its *ocular surface,* with the globe of the eye, and with the Inferior rectus. By its *orbital surface,* with the periosteum covering the floor of the orbit, and with the External rectus. Its borders look forwards and backwards ; the posterior one receives a branch of the third nerve.

Nerves.—The Levator palpebræ, Inferior oblique, and all the Recti excepting the External, are supplied by the third nerve ; the Superior oblique, by the fourth ; the External rectus, by the sixth.

Actions.—The Levator palpebræ raises the upper eyelid, and is the direct antagonist of the Orbicularis palpebrarum. The four Recti muscles are attached in such a manner to the globe of the eye that, acting singly, they will turn it either upwards, downwards, inwards, or outwards, as expressed by their names. The movement produced by the Superior or Inferior rectus is not quite a simple one, for inasmuch as they pass obliquely outwards and forwards to the eyeball, the elevation or depression of the cornea is accompanied by a certain deviation inwards, with a slight amount of rotation, which, however, is corrected by the Oblique muscles ; the Inferior oblique correcting the deviation inwards of the Superior rectus, and the Superior oblique that of the Inferior rectus. The contraction of the External and Internal recti, on the other hand, produces a purely horizontal movement. If any two contiguous Recti of one eye act together they carry the globe of the eye in the diagonal of these directions, viz. upwards and inwards, upwards and outwards, downwards and inwards, or downwards and outwards. A little consideration will show that sometimes the corresponding Recti of the two eyes act in unison, and at others the opposite Recti act together. Thus, in turning the eyes to the right, the External rectus of the right eye will act in unison with the Internal rectus of the left eye ; but if both eyes are directed to an object in the middle line at a short distance, the two Internal recti will act in unison. The movement of circumduction, as in looking round a room, is performed by the alternate action of the four Recti. The oblique muscles rotate the eyeball on its antero-posterior axis, this kind of movement being required for the correct viewing of an object when the head is moved laterally, as from shoulder to shoulder, in order that the picture may fall in all respects on the same part of the retina of each eye.*

Fasciæ of the Orbit.—The connective tissue of the orbit is in various places condensed into thin membranous layers, which may be conveniently described as (1) the orbital fascia ; (2) the sheath of the muscles ; and (3) the covering of the eyeball.

(1) The *orbital fascia.* This forms the periosteum of the orbit. It is loosely connected to the bones, from which it can be readily separated. Behind, it is connected with the dura mater by processes which pass through the optic foramen and sphenoidal fissure, and with the sheath of the optic nerve. In front, it is connected with the periosteum at the margin of the orbit, and sends off a process which assists in forming the palpebral fascia. From its internal surface two processes are given off : one to enclose the lachrymal gland, the other to hold the pulley of the Superior oblique muscle in position. A layer of non-striped muscle, the *orbitalis muscle* of H. Müller, may be seen bridging across the spheno-maxillary fissure.

(2) The sheaths of the muscles give off expansions to the margins of the orbit which limit the action of the muscles.

---

* 'On the Oblique Muscles of the Eye in Man and Vertebrate Animals,' by John Struthers, M.D., in *Anatomical and Physiological Observations.* For a fuller account than our space allows of the various co-ordinated actions of the muscles of a single eye and of both eyes, the reader may be referred to M. Foster's *Text-book of Physiology.*

(3) The fascia covering the eyeball—Tenon's capsule—will be described with the anatomy of the eyeball.

Surgical Anatomy.—The position and exact point of insertion of the tendons of the Internal and External recti muscles into the globe should be carefully examined from the front of the eyeball, as the surgeon is often required to divide the one or the other muscle for the cure of strabismus. In convergent strabismus, which is the more common form of the disease, the eye is turned inwards, requiring the division of the Internal rectus. In the divergent form, which is more rare, the eye is turned outwards, the External rectus being especially implicated. The deformity produced in either case is to be remedied by division of one or the other muscle. The operation is thus performed: the lids are to be well separated; the eyeball is rotated outwards or inwards, and the conjunctiva raised by a pair of forceps, and incised immediately beneath the lower border of the tendon of the muscle to be divided, a little behind its insertion into the sclerotic; the submucous areolar tissue is then divided, and into the small aperture thus made, a blunt hook is passed upwards between the muscle and the globe, and the tendon of the muscle and conjunctiva covering it, divided by a pair of blunt-pointed scissors. Or the tendon may be divided by a subconjunctival incision, one blade of the scissors being passed upwards between the tendon and the conjunctiva, and the other between the tendon and the sclerotic. The student, when dissecting these muscles, should remove on one side of the subject the conjunctiva from the front of the eye, in order to see more accurately the position of the tendons, while on the opposite side the operation may be performed.

V. NASAL REGION (fig. 389)

Pyramidalis nasi.	Dilatator naris anterior.
Levator labii superioris alæque nasi.	Compressor naris.
Dilatator naris posterior.	Compressor narium minor.
Depressor alæ nasi.	

The **Pyramidalis nasi** is a small pyramidal slip placed over the nasal bone. Its origin is by tendinous fibres from the fascia covering the lower part of the nasal bone and upper part of the cartilage, where it blends with the Compressor naris, and it is inserted into the skin over the lower part of the forehead between the two eyebrows, its fibres decussating with those of the Occipito-frontalis.

Relations.—By its *upper surface*, with the skin. By its *under surface*, with the frontal and nasal bones.

The **Levator labii superioris alæque nasi** is a thin triangular muscle, placed by the side of the nose, and extending between the inner margin of the orbit and upper lip. It arises by a pointed extremity from the upper part of the nasal process of the superior maxillary bone, and, passing obliquely downwards and outwards, divides into two slips, one of which is inserted into the cartilage of the ala of the nose; the other is prolonged into the upper lip, becoming blended with the Orbicularis oris and Levator labii superioris proprius.

Relations.—In front, with the integument; and with a small part of the Orbicularis palpebrarum above.

The **Dilatator naris posterior** is a small muscle, which is placed partly beneath the elevator of the nose and lip. It arises from the margin of the nasal notch of the superior maxilla, and from the sesamoid cartilages, and is inserted into the skin near the margin of the nostril.

The **Dilatator naris anterior** is a thin delicate fasciculus, passing from the cartilage of the ala of the nose to the integument near its margin. This muscle is situated in front of the preceding.

The **Compressor naris** is a small, thin, triangular muscle, arising by its apex from the superior maxillary bone, above and a little external to the incisive fossa; its fibres proceed upwards and inwards, expanding into a thin aponeurosis which is continuous on the bridge of the nose with that of the muscle of the opposite side, and with the aponeurosis of the Pyramidalis nasi.

The **Compressor narium minor** is a small muscle, attached by one end to the alar cartilage, and by the other to the integument at the end of the nose.

The **Depressor alæ nasi** is a short, radiated muscle, arising from the incisive fossa of the superior maxilla; its fibres ascend to be inserted into the septum, and back part of the ala of the nose. This muscle lies between the mucous membrane and muscular structure of the lip.

Nerves.—All the muscles of this group are supplied by the facial nerve.

F F 2

Actions.—The Pyramidalis nasi draws down the inner angle of the eyebrows and produces transverse wrinkles over the bridge of the nose. The Levator labii superioris alæque nasi draws upwards the upper. lip and ala of the nose; its most important action is upon the nose, which it dilates to a considerable extent. The action of this muscle produces a marked influence over the countenance, and it is the principal agent in the expression of contempt and disdain. The two Dilatatores nasi enlarge the aperture of the nose. Their action in ordinary breathing is to resist the tendency of the nostrils to close from atmospheric pressure, but in difficult breathing they may be noticed to be in violent action, as well as in some emotions, as anger. The Depressor alæ nasi is a direct antagonist of the other muscles of the nose, drawing the ala of the nose downwards, and thereby constricting the aperture of the nares. The Compressor naris depresses the cartilaginous part of the nose and draws the alæ together.

VI. Superior Maxillary Region (fig. 389)

Levator labii superioris.	Zygomaticus major.
Levator anguli oris.	Zygomaticus minor.

The **Levator labii superioris (proprius)** is a thin muscle, of a quadrilateral form. It arises from the lower margin of the orbit immediately above the infra-orbital foramen, some of its fibres being attached to the superior maxilla, others to the malar bone; its fibres converge to be inserted into the muscular substance of the upper lip, between the attachment of the Levator labii superioris alæque nasi and the Levator anguli oris.

Relations.—By its *superficial surface* above, with the lower segment of the Orbicularis palpebrarum; below, it is subcutaneous. By its *deep surface*, it conceals the origin of the Compressor naris and Levator anguli oris muscles; also, the infra-orbital vessels and nerve, as they escape from the infra-orbital foramen.

The **Levator anguli oris** arises from the canine fossa, immediately below the infra-orbital foramen; its fibres incline downwards and a little outwards, to be inserted into the angle of the mouth, intermingling with those of the Zygomaticus major, the Depressor anguli oris, and the Orbicularis.

Relations.—By its *superficial surface*, with the Levator labii superioris and the infra-orbital vessels and nerves. By its *deep surface*, with the superior maxilla, the Buccinator, and the mucous membrane.

The **Zygomaticus major** is a slender fasciculus, which arises from the malar bone, in front of the zygomatic suture, and descending obliquely downwards and inwards, is inserted into the angle of the mouth, where it blends with the fibres of the Levator anguli oris, the Orbicularis oris, and the Depressor anguli oris.

Relations.—By its *superficial surface*, with the subcutaneous adipose tissue. By its *deep surface*, with the Masseter and Buccinator muscles and the facial artery and vein.

The **Zygomaticus minor** arises from the malar bone, immediately behind the maxillary suture, and passing downwards and inwards, is continuous with the Orbicularis oris at the outer margin of the Levator labii superioris. It lies in front of the preceding.

Relations.—By its *superficial surface*, with the integument and the Orbicularis palpebrarum above. By its *deep surface*, with the Masseter, Buccinator, and Levator anguli oris, and the facial artery and vein.

Nerves.—This group of muscles is supplied by the facial nerve.

Actions.—The Levator labii superioris is the proper elevator of the upper lip, carrying it at the same time a little forwards. It assists in forming the naso-labial ridge, which passes from the side of the nose to the upper lip and gives to the face an expression of sadness. The Levator anguli oris raises the angle of the mouth and assists the Levator labii superioris in producing the naso-labial ridge. The Zygomaticus major draws the angle of the mouth backwards and upwards, as in laughing; while the Zygomaticus minor, being inserted into the outer part of the upper lip and not into the angle of the mouth, draws it backwards, upwards, and outwards, and thus gives to the face an expression of sadness.

VII. INFERIOR MAXILLARY REGION (fig. 389)

Levator labii inferioris (Levator menti).
Depressor labii inferioris (Quadratus menti).
Depressor anguli oris (Triangularis menti).

Dissection.—The two principal muscles in this region may be dissected by making a vertical incision through the integument from the margin of the lower lip to the chin; a second incision should then be carried along the margin of the lower jaw as far as the angle, and the integument carefully removed in the direction shown in fig. 388. The Levator labii inferioris is to be dissected by everting the lower lip and raising the mucous membrane.

The **Levator labii inferioris (Levator menti)** is a small conical fasciculus, placed on the side of the frænum of the lower lip. It arises from the incisive fossa, external to the symphysis of the lower jaw; its fibres descend to be inserted into the integument of the chin.

Relations.—On its *inner surface*, with the mucous membrane; in the *median line*, it is blended with the muscle of the opposite side; and on its *outer side*, with the Depressor labii inferioris.

The **Depressor labii inferioris (Quadratus menti)** is a small quadrilateral muscle. It arises from the external oblique line of the lower jaw, between the symphysis and mental foramen, and passes obliquely upwards and inwards, to be inserted into the integument of the lower lip, its fibres blending with the Orbicularis oris, and with those of its fellow of the opposite side. It is continuous with the fibres of the Platysma at its origin. This muscle contains much yellow fat intermingled with its fibres.

Relations.—By its *superficial surface*, with part of the Depressor anguli oris, and with the integument, to which it is closely connected. By its *deep surface*, with the mental vessels and nerves, the mucous membrane of the lower lip, the labial glands, and the Levator menti, with which it is intimately united.

The **Depressor anguli oris (Triangularis menti)** is triangular in shape, arising, by its broad base, from the external oblique line of the lower jaw, whence its fibres pass upwards, to be inserted, by a narrow fasciculus, into the angle of the mouth. It is continuous with the Platysma at its origin, and with the Orbicularis oris and Risorius at its insertion, and some of its fibres are directly continuous with those of the Levator anguli oris.

Relations.—By its *superficial surface*, with the integument. By its *deep surface*, with the Depressor labii inferioris and Buccinator.

Nerves.—This group of muscles is supplied by the facial nerve.

Actions.—The Levator labii inferioris raises the lower lip, and protrudes it forwards, and at the same time wrinkles the integument of the chin, expressing doubt or disdain. The Depressor labii inferioris draws the lower lip directly downwards and a little outwards, as in the expression of irony. The Depressor anguli oris depresses the angle of the mouth, being the antagonist of the Levator anguli oris and Zygomaticus major; acting with these muscles, it will draw the angle of the mouth directly backwards.

VIII. INTERMAXILLARY REGION

Orbicularis oris.   Buccinator.   Risorius.

Dissection.—The dissection of these muscles may be considerably facilitated by filling the cavity of the mouth with tow, so as to distend the cheeks and lips; the mouth should then be closed by a few stitches, and the integument carefully removed from the surface.

The **Orbicularis oris** (fig. 389) is not a sphincter muscle like the Orbicularis palpebrarum, but consists of numerous strata of muscular fibres, having different directions, which surround the orifice of the mouth. These fibres are partially derived from the other facial muscles which are inserted into the lips, and are partly fibres proper to the lips themselves. Of the former, a considerable number are derived from the Buccinator and form the deeper stratum of the Orbicularis. Some of them—namely, those near the middle of the Buccinator—decussate at the angle of the mouth, those arising from the upper jaw passing to the lower lip, and those from the lower jaw to the upper lip. Other fibres of the muscle, situated at its upper and lower part, pass across the lips from side to side without decussation.   Superficial to this stratum is a second, formed by the Levator and Depressor anguli oris, which cross each other at the angle of the mouth; those from the Depressor passing to the upper lip, and those from the Levator to the lower lip, along which they run to be inserted into the skin near the median line.   In addition to these there are fibres from other muscles inserted into the lips, the Levator labii superioris, the Levator labii superioris alæque nasi, the Zygomatici, and the Depressor labii inferioris; these intermingle with the transverse fibres above described, and have principally an oblique direction.   The proper fibres of the lips are oblique, and pass from the under surface of the skin to the mucous membrane, through the thickness of the lip.   And in addition to these are fibres by which the muscle is connected directly with the maxillary bones and the septum of the nose.   These consist, in the upper lip, of four bands, two of which (*Musculus incisivus superior*) arise from the alveolar border of the superior maxilla, opposite the lateral incisor tooth, and arching outwards on each side are continuous at the angles of the mouth with the other muscles inserted into this part.   The two remaining muscular slips, called the *Naso-labialis*, connect the upper lip to the back of the septum of the nose : as they descend from the septum, an interval is left between them.   It is this interval which forms the depression seen on the surface of the skin beneath the septum of the nose, which is called the *philtrum*.   The additional fibres for the lower segment (*Musculus incisivus inferior*) arise from the inferior maxilla, externally to the Levator labii inferioris, and arch outwards to the angles of the mouth, to join the Buccinator and the other muscles attached to this part.

Relations.—By its *superficial surface*, with the integument, to which it is closely connected.   By its *deep surface*, with the buccal mucous membrane, the labial glands, and coronary vessels.   By its *outer circumference*, it is blended with the numerous muscles which converge to the mouth from various parts of the face.   Its *inner circumference* is free, and covered by mucous membrane.

The **Buccinator** (fig. 401) is a broad, thin muscle, quadrilateral in form, which occupies the interval between the jaws at the side of the face.   It arises from the outer surface of the alveolar processes of the upper and lower jaws, corresponding to the three molar teeth; and behind, from the anterior border of the pterygo-mandibular ligament, through which it is continuous with the Superior constrictor of the pharynx.   The fibres converge towards the angle of the mouth, where the central fibres intersect each other, those from below being continuous with the upper segment of the Orbicularis oris, and those from above with the inferior segment; the highest and lowest fibres continue forward uninterruptedly into the corresponding segment of the lip without decussation.

Relations.—By its *superficial surface*, behind, with a large mass of fat, which separates it from the ramus of the lower jaw, the Masseter, and a small portion of the Temporal muscle, and which has been named the *suctorial pad*, because it is supposed to assist in the act of sucking.   By its outer surface, the Buccinator muscle is in relation with the Zygomatici, Risorius, Levator anguli oris, Depressor anguli oris, and Stenson's duct, which pierces it opposite the second molar tooth of the upper jaw; the facial artery and vein cross it from below upwards; it is also crossed by the branches of the facial and buccal nerves.   By its *internal surface*, with the buccal glands and mucous membrane of the mouth.

The *pterygo-mandibular ligament* separates the Buccinator muscle from the Superior constrictor of the pharynx.   It is a tendinous band, derived from the deep cervical fascia, attached by one extremity to the apex of the internal pterygoid plate, and by the other to the posterior extremity of the internal oblique line of the lower jaw.   Its *inner surface* corresponds to the cavity of the mouth, and is lined by mucous membrane.   Its *outer surface* is separated from the ramus of the jaw by a quantity of adipose tissue.   Its *posterior border* gives attachment to the Superior constrictor of the pharynx; its *anterior border*, to the fibres of the Buccinator (see fig. 401).

The **Risorius** (*Risorius Santorini*) consists of a narrow bundle of fibres, which arises in the fascia over the Masseter muscle and, passing horizontally forwards, is inserted into the skin at the angle of the mouth (fig. 389).   It is placed superficial to the Platysma, and is broadest at its outer extremity. This muscle varies much in its size and form.

Nerves.—The muscles in this group are all supplied by the facial nerve. The buccal branch of the inferior maxillary nerve pierces the Buccinator muscle, and by some anatomists is regarded as partly supplying this muscle. Probably it merely pierces it on its way to the mucous membrane of the cheek.

Actions.—The Orbicularis oris in its ordinary action produces the direct closure of the lips; by its deep fibres, assisted by the oblique ones, it closely applies the lips to the alveolar arch. The superficial part, consisting principally of the decussating fibres, brings the lips together and also protrudes them forwards. The Buccinators contract and compress the cheeks, so that, during the process of mastication, the food is kept under the immediate pressure of the teeth. When the cheeks have been previously distended with air, the Buccinator muscles expel it from between the lips, as in blowing a trumpet. Hence the name (*buccina*, a trumpet). The Risorius retracts the angles of the mouth, and produces the unpleasant expression which is sometimes seen in tetanus, and is known as ' risus sardonicus.'

IX. TEMPORO-MANDIBULAR REGION

Masseter.   Temporal.

Masseteric Fascia.—Covering the Masseter muscle, and firmly connected with it, is a strong layer of fascia, derived from the deep cervical fascia. Above, this fascia is attached to the lower border of the zygoma, and behind it covers the parotid gland, constituting the *parotid fascia.*

The **Masseter** is exposed by the removal of this fascia (fig. 389); it is a short, thick muscle, somewhat quadrilateral in form, consisting of two portions, superficial and deep. The *superficial portion*, the larger, arises by a thick, tendinous aponeurosis from the malar process of the superior maxilla, and from the anterior two-thirds of the lower border of the zygomatic arch: its fibres pass downwards and backwards, to be inserted into the angle and lower half of the outer surface of the ramus of the jaw. The *deep portion* is much smaller, and more muscular in texture; it arises from the posterior third of the lower border and the whole of the inner surface of the zygomatic arch; its fibres pass downwards and forwards, to be inserted into the upper half of the ramus and outer surface of the coronoid process of the jaw. The deep portion of the muscle is partly concealed, in front, by the superficial portion; behind, it is covered by the parotid gland. The fibres of the two portions are united at their insertion.

Relations.—By its *superficial surface*, with the Zygomatici, the parotid gland and socia parotidis, and Stenson's duct; the branches of the facial nerve and the transverse facial vessels, which cross it; the masseteric fascia; the Risorius, Platysma myoides, and the integument. By its *deep surface*, with the Temporal muscle at its insertion, the ramus of the jaw, the Buccinator and the long buccal nerve, from which it is separated by a mass of fat. The masseteric nerve and artery enter it on its deep surface. Its *posterior margin* is overlapped by the parotid gland. Its *anterior margin* projects over the Buccinator muscle; and the facial vein lies on it below.

The **Temporal fascia** is seen, at this stage of the dissection, covering in the Temporal muscle. It is a strong, fibrous investment, covered, on its outer surface, by the Attrahens and Attollens auriculam muscles, the aponeurosis of the Occipito-frontalis, and by part of the Orbicularis palpebrarum. The temporal vessels and the auriculo-temporal nerve cross it from below upwards. Above, it is a single layer, attached to the entire extent of the upper temporal ridge; but below, where it is attached to the zygoma, it consists of two layers, one of which is inserted into the outer, and the other into the inner border of the zygomatic arch. A small quantity of fat, the orbital branch of the temporal artery, and a filament from the orbital, or temporo-malar, branch of the superior maxillary nerve, are contained between these two layers. It affords attachment by its inner surface to the superficial fibres of the Temporal muscle.

Dissection.—In order to expose the Temporal muscle, remove the temporal fascia, which may be effected by separating it at its attachment along the upper border of the zygoma, and dissecting it upwards from the surface of the muscle. The zygomatic arch should then be divided, in front, at its junction with the malar bone; and, behind, near the external auditory meatus, and drawn downwards with the Masseter, which should be detached from its insertion into the ramus and angle of the jaw. The whole extent of the Temporal muscle is then exposed.

The **Temporal** (fig. 393) is a broad, radiating muscle, situated at the side of the head, and occupying the entire extent of the temporal fossa. It arises from the whole of the temporal fossa except that portion of it which is formed by the malar bone. Its attachment extends from the external angular process of the frontal in front, to the mastoid portion of the temporal behind; and from the curved line on the frontal and parietal bones above, to the pterygoid ridge on the great wing of the sphenoid below. It also arises from the inner surface of the temporal fascia. Its fibres converge as they descend, and terminate in an aponeurosis, the fibres of which, radiated at its commencement, converge into a thick and flat tendon, which is inserted into the inner surface, apex, and anterior border of the coronoid process of the jaw, nearly as far forwards as the last molar tooth.

FIG. 393.—The Temporal muscle; the zygoma and Masseter having been removed.

Relations.—By its *superficial surface*, with the integument, the temporal fascia, the Attrahens and Attollens auriculam muscles, the temporal vessels and nerves, the aponeurosis of the Occipito-frontalis, the temporal fascia, the zygoma, and Masseter. By its *deep surface*, with the temporal fossa, the External pterygoid and part of the Buccinator muscles, the internal maxillary artery, its deep temporal branches, the deep temporal nerves, and the buccal vessels and nerve. Behind the tendon are the masseteric vessels and nerve. Its anterior border is separated from the malar bone by a mass of fat.

Nerves.—Both muscles are supplied by the inferior maxillary nerve.

X. PTERYGO-MANDIBULAR REGION (fig. 394).

External pterygoid.   Internal pterygoid.

Dissection.—The Temporal muscle having been examined, saw through the base of the coronoid process, and draw it upwards, together with the Temporal muscle, which should be detached from the surface of the temporal fossa. Divide the ramus of the jaw just below the condyle, and also, by a transverse incision extending across the middle, just above the dental foramen; remove the fragment of bone, and the Pterygoid muscles will be exposed.

The **External pterygoid** is a short, thick muscle, somewhat conical in form, which extends almost horizontally between the zygomatic fossa and the condyle of the jaw. It arises by two heads, separated by a slight interval: the *upper* arises from the inferior portion of the external surface of the greater wing of the sphenoid and from the pterygoid ridge, which separates the zygomatic from the temporal fossa; the *lower* from the outer surface of the external pterygoid plate. Its fibres pass horizontally backwards and outwards, to be inserted into a depression in front of the neck of the condyle of the lower jaw, and into the corresponding part of the interarticular fibro-cartilage of the temporo-mandibular articulation.

Relations.—By its *external surface*, with the ramus of the lower jaw, the internal maxillary artery, which crosses it,* the tendon of the Temporal muscle and the Masseter. By its *internal surface*, it rests against the upper part of the Internal pterygoid, the internal lateral ligament, the middle meningeal artery, and inferior maxillary nerve; by its *upper border* it is in relation with the temporal and masseteric branches of the inferior maxillary nerve; by its lower border it is in relation with the inferior dental and gustatory nerves.

FIG. 394.—The Pterygoid muscles; the zygomatic arch and a portion of the ramus of the jaw having been removed.

Through the interval between the two portions of the muscle, the buccal nerve emerges and the internal maxillary artery passes, when the trunk of this vessel lies on the lower part of the muscle (see fig. 394).

The **Internal pterygoid** is a thick quadrilateral muscle, and resembles the Masseter in form. It arises from the pterygoid fossa, being attached to the inner surface of the external pterygoid plate, and to the grooved surface of the tuberosity of the palate bone, and by a second slip from the outer surface of the tuberosities of the palate and superior maxillary bones; its fibres pass downwards, outwards, and backwards, to be inserted, by a strong tendinous lamina, into the lower and back part of the inner side of the ramus and angle of the lower jaw, as high as the dental foramen.

Relations.—By its *external surface*, with the ramus of the lower jaw, from which it is separated, at its upper part, by the External pterygoid, the internal lateral ligament, the internal maxillary artery, the dental vessels and nerves, the lingual nerve, and a process of the parotid gland. By its *internal surface*, with the Tensor palati, being separated from the Superior constrictor of the pharynx by a cellular interval.

Nerves.—These muscles are supplied by the inferior maxillary nerve.

* In many cases the artery will be found under cover of the muscle.

Actions.—The Temporal, Masseter, and Internal pterygoid raise the lower jaw against the upper with great force. The superficial portion of the Masseter assists the External pterygoid in drawing the lower jaw forwards upon the upper ; the jaw being drawn back again by the deep fibres of the Masseter, and posterior fibres of the Temporal. The External pterygoid muscles are the direct agents in the trituration of the food, drawing the lower jaw directly forwards, so as to make the lower teeth project beyond the upper. If the muscle of one side acts, the corresponding side of the jaw is drawn forwards, and the other condyle remaining comparatively fixed, the symphysis deviates to the opposite side. The alternation of these movements on the two sides produces trituration.

Surface Form.—The outline of the muscles of the head and face cannot be traced on the surface of the body, except in the case of two of the masticatory muscles. Those of the head are thin, so that the outline of the bone is perceptible beneath them. Those of the face are small, covered by soft skin, and often by a considerable layer of fat, so that their outline is concealed ; but they serve to round off and smooth prominent borders, and to fill up what would be otherwise unsightly angular depressions. Thus, the Orbicularis palpebrarum rounds off the prominent margin of the orbit, and the Pyramidalis nasi fills in the sharp depression beneath the glabella, and thus softens and tones down the abrupt depression which is seen on the unclothed bone. In like manner, the labial muscles, converging to the lips, and assisted by the superimposed fat, fill in the sunken hollow of the lower part of the face. Although the muscles of the face are usually described as arising from the bones, and inserted into the nose, lips, and corners of the mouth, they have fibres inserted into the skin of the face along their whole extent, so that almost every point of the skin of the face has its muscular fibre to move it; hence it is that when in action the facial muscles produce alterations in the skin-surface, giving rise to the formation of various folds or wrinkles, or otherwise altering the relative position of parts, so as to produce the varied expressions with which the face is endowed; these muscles are therefore termed the ' Muscles of expression.' The only two muscles in this region which greatly influence surface-form are the Masseter and the Temporal. The Masseter is a quadrilateral muscle, which imparts fulness to the hinder part of the cheek. When the muscle is firmly contracted, as when the teeth are clenched, its outline is plainly visible ; the anterior border forms a prominent vertical ridge, behind which is a considerable fulness, especially marked at the lower part of the muscle ; this fulness is entirely lost when the mouth is opened, and the muscle no longer in a state of contraction. The Temporal muscle is fan-shaped, and fills the Temporal fossa, substituting for it a somewhat convex form, the anterior part of which, on account of the absence of hair over the temple, is more marked than the posterior, and stands out in strong relief when the muscle is in a state of contraction.

MUSCLES AND FASCIÆ OF THE NECK

The Muscles of the Neck may be arranged into groups, corresponding with the region in which they are situated.

These groups are nine in number :

I. Superficial Region.
II. Infrahyoid Region.
III. Suprahyoid Region.
IV. Lingual Region.
V. Pharyngeal Region.
VI. Palatal Region.
VII. Anterior Vertebral Region.
VIII. Lateral Vertebral Region.
IX. Muscles of the Larynx.

The muscles contained in each of these groups are the following :

I. Superficial Region.

Platysma myoides.   Sterno-cleido-mastoid.

II. Infrahyoid Region.

Sterno-hyoid.
Sterno-thyroid.
Thyro-hyoid.
Omo-hyoid.

III. Suprahyoid Region.

Digastric.
Stylo-hyoid.
Mylo-hyoid.
Genio-hyoid.

IV. *Lingual Region.*

Extrinsic.
- Genio-hyo-glossus.
- Hyo-glossus.
- Chondro-glossus.
- Stylo-glossus.
- Palato-glossus.

Intrinsic.
- Superior lingualis.
- Inferior lingualis.
- Transverse lingualis.
- Vertical lingualis.

V. *Pharyngeal Region.*
Inferior constrictor.
Middle constrictor.
Superior constrictor.
Stylo-pharyngeus.
Palato-pharyngeus.
Salpingo-pharyngeus.

VI. *Palatal Region.*
Levator palati.
Tensor palati.
Azygos uvulæ.
Palato-glossus.
Palato-pharyngeus.
Salpingo-pharyngeus.

VII. *Anterior Vertebral Region.*
Rectus capitis anticus major.
Rectus capitis anticus minor.
Rectus capitis lateralis.
Longus colli.

VIII. *Lateral Vertebral Region.*
Scalenus anticus.
Scalenus medius.
Scalenus posticus.

IX. *Muscles of the Larynx.*
Included in the description of the Larynx.

I. SUPERFICIAL CERVICAL REGION

Platysma myoides.   Sterno-cleido-mastoid.

Dissection.—A block having been placed at the back of the neck, and the face turned to the side opposite that to be dissected, so as to place the parts upon the stretch, make two transverse incisions: one from the chin, along the margin of the lower jaw, to the mastoid process; and the other along the upper border of the clavicle. Connect these by an oblique incision made in the course of the Sterno-mastoid muscle, from the mastoid process to the sternum; the two flaps of integument having been removed in the direction shown in fig. 388, the superficial fascia will be exposed.

The **Superficial cervical fascia** is a thin aponeurotic lamina, which is hardly demonstrable as a separate membrane. Beneath it is found the Platysma myoides muscle.

The **Platysma myoides** (fig. 389) is a broad, thin plane of muscular fibres, placed immediately beneath the superficial fascia on each side of the neck. It arises by thin, fibrous bands from the fascia covering the upper part of the Pectoral and Deltoid muscles; its fibres pass over the clavicle, and proceed obliquely upwards and inwards along the side of the neck. The anterior fibres interlace, below and behind the symphysis menti, with the fibres of the muscle of the opposite side; the posterior fibres pass over the lower jaw, some of them being attached to the bone below the external oblique line, others passing on to be inserted into the skin and subcutaneous tissue of the lower part of the face, many of these fibres blending with the muscles about the angle and lower part of the mouth. Sometimes fibres can be traced to the Zygomatic muscles, or to the margin of the Orbicularis oris. Beneath the Platysma, the external jugular vein may be seen descending from the angle of the jaw to the clavicle.

Relations.—By its *external surface*, with the integument, to which it is united more closely below than above. By its *internal surface*, with the Pectoralis major and Deltoid, and with the clavicle. In the *neck*, with the external and anterior jugular veins, the deep cervical fascia, the superficial branches of the cervical plexus, the Sterno-mastoid, Sterno-hyoid, Omo-hyoid, and Digastric muscles. Behind the Sterno-mastoid muscle, it covers in the posterior triangle of the neck. On the *face*, it is in relation with the parotid gland, the facial artery and vein, and the Masseter and Buccinator muscles.

Actions.—The Platysma myoides produces a slight wrinkling of the surface of the skin of the neck, in an oblique direction, when the entire muscle is brought into action. Its anterior portion, the thickest part of the muscle, depresses the lower jaw; it also serves to draw down the lower lip and angle of the mouth on each side, being one of the chief agents in the expression of melancholy.

The **Deep cervical fascia** (fig. 395) lies under cover of the Platysma myoides muscle and constitutes a complete investment for the neck. It also forms a sheath for the carotid vessels, and, in addition, is prolonged deeply in the shape of certain processes or lamellæ, which come into close relation with the structures situated in front of the vertebral column.

The investing portion of the fascia is attached behind to the ligamentum nuchæ and to the spine of the seventh cervical vertebra. It forms a thin investment to the Trapezius muscle, and at its anterior border is continued forwards as a rather loose areolar layer, covering the posterior triangle of the neck, to the posterior border of the Sterno-mastoid muscle, where it begins to assume the appearance of a true fascia. Along the hinder edge of this muscle it divides to

FIG. 395.—Section of the neck at about the level of the sixth cervical vertebra. Showing the arrangement of the deep cervical fascia.

enclose it, and at the anterior margin again forms a single lamella, which roofs in the anterior triangle of the neck, and, reaching forwards to the middle line, is continuous with the corresponding part from the opposite side of the neck. In the middle line of the neck it is attached to the symphysis menti and body of the hyoid bone.

Above, the fascia is attached to the superior curved line of the occiput, to the mastoid process of the temporal, and to the whole length of the body of the jaw. Opposite the angle of the jaw the fascia is very strong, and binds the anterior edge of the Sterno-mastoid firmly to that bone. Between the jaw and the mastoid process it ensheathes the parotid gland—the layer which covers the gland extending upwards under the name of the *parotid fascia* to be fixed to the zygomatic arch.   From the layer which passes under the parotid a strong band extends upwards to the styloid process, forming the *stylo-mandibular ligament.* From this three prolongations form : the *internal lateral ligament of the lower jaw,* the *pterygo-mandibular* and the *pterygo-spinous ligaments.*   The latter of these three, the **pterygo-spinous ligament**, is a membranous band which stretches across from the upper half of the posterior free border of the external pterygoid plate to the spinous process of the sphenoid bone.   It occasionally ossifies, and produces an adventitious foramen between its upper border and the base of the skull, through which pass the branches of the third division of the fifth nerve, which supply the muscles of mastication.

Below, the fascia is attached to the acromion process, the clavicle, and manubrium sterni.   Some little distance above the last, however, it splits into two layers, superficial and deep.   The former is attached to the anterior border of the manubrium, the latter to its posterior border and to the interclavicular ligament. Between these two layers is a slit-like interval, the *suprasternal space,* or *space of Burns.*   It contains a small quantity of areolar tissue, and sometimes a lymphatic gland ; the lower portions of the anterior jugular veins and their transverse connecting branch ; and also the sternal heads of the Sterno-mastoid muscles.

The fascia which lines the deep aspect of the Sterno-mastoid gives off certain important processes, viz. : (1) A process to envelop the tendon of the Omo-hyoid, and bind it down to the sternum and first costal cartilage.   (2) A strong sheath, the *carotid sheath,* which encloses the carotid artery, internal jugular vein, and the vagus nerve.   (3) The *prevertebral fascia,* which extends inwards behind the carotid vessels, where it assists in forming their sheath, and passes in front of the prevertebral muscles.   It thus forms the posterior limit of a fibrous compartment, which contains the larynx and trachea, the thyroid gland, and the pharynx and œsophagus.   The prevertebral fascia is fixed above to the base of the skull, while below it is continued into the thorax in front of the Longus colli muscles. Parallel to the carotid vessels and along their inner aspect the prevertebral fascia gives off a thin lamina, the *bucco-pharyngeal fascia,* which closely invests the Constrictor muscles of the pharynx, and is continued forward from the Superior constrictor on to the Buccinator.   It is attached to the prevertebral layer by loose connective tissue only, and thus an easily distended space, the *retro-pharyngeal space,* is found between them.   This space is limited above by the base of the skull, while below it extends behind the œsophagus into the thorax, where it is continued into the posterior mediastinum.   The prevertebral fascia is prolonged downwards and outwards behind the carotid vessels and in front of the Scaleni muscles, and forms a sheath for the brachial nerves and subclavian vessels in the posterior triangle of the neck, and, continued under the clavicle as the axillary sheath, is attached to the deep surface of the costo-coracoid membrane. Immediately above the clavicle an areolar space exists between the investing layer and the sheath of the subclavian vessels, and in it are found the lower part of the external jugular vein, the descending clavicular nerves, the suprascapular and transversalis colli vessels, and the posterior belly of the Omo-hyoid muscle. This space extends downwards behind the clavicle, and is limited below by the fusion of the costo-coracoid membrane with the anterior wall of the axillary sheath.   (4) The *pretracheal fascia,* which extends inwards in front of the carotid vessels, and assists in forming the carotid sheath.   It is further continued behind the depressor muscles of the hyoid bone, and, after enveloping the thyroid body, is prolonged in front of the trachea to meet the corresponding layer of the opposite side.   Above, it is fixed to the hyoid bone, while below it is carried downwards in front of the trachea and large vessels at the root of the neck, and ultimately blends with the fibrous pericardium.

Surgical Anatomy.—The cervical fascia is of considerable importance from a surgical point of view.   As will be seen from the foregoing description, it may be divided into three layers : (1) a superficial layer ; (2) a layer passing in front of the trachea, and forming with the superficial layer a sheath for the depressors of the hyoid bone ; (3) a prevertebral layer passing in front of the bodies of the cervical vertebræ, and forming with the second layer a space in which is contained the trachea, œsophagus, &c.   The superficial layer forms a complete investment for the neck.   It is attached behind to the ligamentum nuchæ and the spine of the seventh cervical vertebra : above, it is fixed to the external occipital protuberance, to the superior curved line of the occiput, to the mastoid process, to the zygoma and the lower jaw; below, it is attached to the manubrium sterni, the clavicle, the acromion process, and the spine of the scapula; in front, it blends with the fascia of the opposite side. This layer would oppose the extension of abscesses or new growths towards the surface, and pus forming beneath it would have a tendency to extend laterally. If the pus is in the posterior triangle, it might extend backwards under the Trapezius, forwards under the Sterno-mastoid, or downwards under the clavicle for some distance, until stopped by the junction of the cervical fascia to the costo-coracoid membrane. If it is contained in the anterior triangle, it might find its way into the anterior mediastinum, being situated in front of the layer of fascia which passes down into the thorax to become continuous with the pericardium; but owing to the less density and thickness of the fascia in this situation it more frequently finds its way through it and points above the sternum. The second layer of fascia is connected above with the hyoid bone. It passes down beneath the depressors and in front of the thyroid body and trachea to become continuous with the fibrous layer of the pericardium. Laterally it invests the great vessels of the neck and is connected with the superficial layer beneath the Sterno-mastoid. Pus forming beneath this layer would in all probability find its way into the posterior mediastinum. The third layer (the prevertebral fascia) is connected above to the base of the skull. Pus forming beneath this layer, in cases, for instance, of caries of the bodies of the cervical vertebræ, might extend towards the posterior and lateral part of the neck and point in this situation, or might perforate this layer of fascia and the pharyngeal fascia and point into the pharynx (retro-pharyngeal abscess).

In cases of cut throat the cervical fascia is of considerable importance. When the wound involves only the superficial layer the injury is usually trivial, the special danger being injury to the external jugular vein, and the special complication diffuse cellulitis. But where the second of the two layers is opened up, important structures may be injured, and lead to serious results.

It may be worth while mentioning that in Burns's space is contained the sternal head of origin of the Sterno-mastoid muscle, so that this space is opened in division of this tendon. The anterior jugular vein is also contained in the same space.

The **Sterno-mastoid** or **Sterno-cleido-mastoid** (fig. 396) is a large, thick muscle, which passes obliquely across the side of the neck, being enclosed between the two laminæ of the investing layer of the deep cervical fascia. It is thick and narrow at its central part, but is broader and thinner at each extremity. It arises, by two heads, from the sternum and clavicle. The *sternal portion* is a rounded fasciculus, tendinous in front, fleshy behind, which arises from the upper and anterior part of the first piece of the sternum, and is directed upwards, outwards, and backwards. The *clavicular portion* arises from the inner third of the superior border and anterior surface of the clavicle, being composed of fleshy and aponeurotic fibres; it is directed almost vertically upwards. These two portions are separated from one another, at their origin, by a triangular interval, but become gradually blended, below the middle of the neck, into a thick, rounded muscle, which is inserted, by a strong tendon, into the outer surface of the mastoid process, from its apex to its superior border, and by a thin aponeurosis into the outer half of the superior curved line of the occipital bone. The Sterno-mastoid varies much in its extent of attachment to the clavicle: in one case the clavicular may be as narrow as the sternal portion; in another, as much as three inches in breadth. When the clavicular origin is broad, it is occasionally sub-divided into numerous slips, separated by narrow intervals. More rarely, the corresponding margins of the Sterno-mastoid and Trapezius have been found in contact. In the application of a ligature to the third part of the subclavian artery, it will be necessary, where the muscles come close together, to divide a portion of one or of both.

This muscle divides the quadrilateral space at the side of the neck into two triangles, an anterior and a posterior. The boundaries of the *anterior triangle* are, in front, the median line of the neck; above, the lower border of the body of the jaw, and an imaginary line drawn from the angle of the jaw to the Sterno-mastoid; behind, the anterior border of the Sterno-mastoid muscle. The apex of the triangle is at the upper border of the sternum. The boundaries of the *posterior triangle* are, in front, the posterior border of the Sterno-mastoid; below, the middle third of the clavicle; behind, the anterior margin of the Trapezius.* The apex corresponds with the meeting of the Sterno-mastoid and Trapezius on the occipital bone.

---

* The anatomy of these triangles will be more exactly described with that of the vessels of the neck.

Relations.—By its *superficial surface*, with the integument and Platysma, from which it is separated by the external jugular vein, several of the superficial branches of the cervical plexus, and the anterior layer of the deep cervical fascia. By its *deep surface*, it is in relation with the Sterno-clavicular articulation ; a process of the deep cervical fascia ; the Sterno-hyoid, Sterno-thyroid, Omo-hyoid, posterior belly of the Digastric, Levator anguli scapulæ, Splenius and Scaleni muscles ; common carotid artery, internal and anterior jugular veins, commencement of the internal and external carotid arteries, the occipital, subclavian, transversalis colli, and suprascapular arteries and veins ; the phrenic, pneumogastric, hypoglossal, descendens and communicantes hypoglossi nerves ; the spinal accessory nerve, which pierces its upper third ; the cervical plexus, parts of the thyroid and parotid glands and deep lymphatic glands.

FIG. 396.—Muscles of the neck, and boundaries of the triangles.

Nerves.—The Platysma myoides is supplied by the facial nerve ; the Sternocleido-mastoid by the spinal accessory and branches from the anterior primary divisions of the second and third cervical nerves.

Actions.—When only one Sterno-mastoid muscle acts, it draws the head towards the shoulder of the same side, assisted by the Splenius and the Obliquus capitis inferior of the opposite side. At the same time it rotates the head so as to carry the face towards the opposite side. If the head is fixed, the two muscles assist in elevating the thorax in forced inspiration.

Surface Form.—The anterior edge of the muscle forms a very prominent ridge beneath the skin, which it is important to notice, as it constitutes a guide to the surgeon in making the necessary incisions for ligature of the common carotid artery, and for œsophagotomy.

Surgical Anatomy.—The surgical anatomy of the Sterno-mastoid muscle is of importance in connection with the deformity known as *wry-neck*, which is due to a paralytic or spastic condition of this muscle. The wry-neck may be temporary, as the result of direct irritation of the muscle or of the nerves which supply it, and may occur in acute glandular enlargement, cellulitis of the neck, myositis of the muscle, or cervical caries.

Or it may be permanent, and is then very often due to injury to the muscle during birth, especially in breech presentations; rupture of the muscle and subsequent cicatricial contraction taking place.

In these cases, division of the muscle is often necessary to effect a cure, and this may be done either subcutaneously or through an open wound. The subcutaneous method is thus performed : The external jugular and anterior jugular veins having been, if possible, defined, a tenotomy knife is introduced close to the margin of one tendon of origin of the muscle, about half an inch above the clavicle, and the tenotome passed flat behind the tendon and then turned forwards, and the tendon divided from behind forwards while the muscle is put well upon the stretch by an assistant. The other tendon is then divided in a similar manner. In dividing the clavicular origin, it is always desirable to introduce the tenotome along the posterior border, in order to avoid the external jugular vein. The open method is, however, much to be preferred, as being the more effectual and the less dangerous, if precautions are taken to keep the wound aseptic. The tendons of origin are freely exposed by a horizontal incision across the root of the neck and carefully divided ; any tense bands of fascia which can be felt should also be divided. The wound is now sutured and dressed, and the head fixed in a perfectly straight position by a plaster-of-Paris bandage.

II. INFRAHYOID REGION (figs. 396, 397)

Sterno-hyoid.	Thyro-hyoid.
Sterno-thyroid.	Omo-hyoid.

Dissection.—The muscles in this region may be exposed by removing the deep fascia from the front of the neck. In order to see the entire extent of the Omo-hyoid, it is necessary to divide the Sterno-mastoid at its centre, and turn its ends aside, and to detach the Trapezius from the clavicle and scapula. This, however, should not be done until the Trapezius has been dissected.

The **Sterno-hyoid** is a thin, narrow, riband-like muscle, which arises from the inner extremity of the clavicle, the posterior sterno-clavicular ligament, and the upper and posterior part of the manubrium sterni ; passing upwards and inwards, it is inserted, by short, tendinous fibres, into the lower border of the body of the os hyoides. This muscle is separated, below, from its fellow by a considerable interval ; but the two muscles come into contact with one another in the middle of their course, and from this upwards, lie side by side. It some-times presents, immediately above its origin, a transverse tendinous intersection, like those in the Rectus abdominis.

Relations.—By its *superficial surface*, below, with the sternum, the sternal end of the clavicle, and the Sterno-mastoid ; and above, with the Platysma and deep cervical fascia. By its *deep surface*, with the Sterno-thyroid, Crico-thyroid, and Thyro-hyoid muscles, the thyroid gland, the superior thyroid vessels, the thyroid cartilage, the crico-thyroid and thyro-hyoid membranes. Usually two bursæ are found between the crico-thyroid membrane and the under surface of this muscle. Occasionally there is only one, and in some instances no bursa at all. They are called the Sterno-hyoid bursæ.

The **Sterno-thyroid** is shorter and wider than the preceding muscle, beneath which it is situated. It arises from the posterior surface of the manubrium sterni, below the origin of the Sterno-hyoid, and from the edge of the cartilage of the first rib, and sometimes of the second rib also ; and is inserted into the oblique line on the side of the ala of the thyroid cartilage. This muscle is in close contact with its fellow at the lower part of the neck ; and is occasionally traversed by a transverse or oblique tendinous intersection like those in the Rectus abdominis.

Relations.—By its *anterior surface*, with the Sterno-hyoid, Omo-hyoid, and Sterno-mastoid. By its *posterior surface*, from below upwards, with the trachea, vena innominata, common carotid (and on the right side the arteria innominata), the thyroid gland and its vessels, and the lower parts of the larynx and pharynx. The inferior thyroid vein lies along its inner border, a relation which it is important to remember in the operation of tracheotomy. On the left side the deep surface of the muscle is in relation to the œsophagus.

The **Thyro-hyoid** is a small, quadrilateral muscle appearing like a continuation of the Sterno-thyroid. It arises from the oblique line on the side of the thyroid cartilage, and passes vertically upwards to be inserted into the lower border of the body and greater cornu of the hyoid bone.

Relations.—By its *external surface*, with the Sterno-hyoid and Omo-hyoid muscles. By its *internal surface*, with the thyroid cartilage, the thyro-hyoid membrane, and the superior laryngeal vessels and nerve.

The **Omo-hyoid** passes across the side of the neck, from the scapula to the hyoid bone. It consists of two fleshy bellies, united by a central tendon. It arises from the upper border of the scapula, and occasionally from the transverse ligament which crosses the suprascapular notch; its extent of attachment to the scapula varying from a few lines to an inch. From this origin, the posterior belly forms a flat, narrow fasciculus, which inclines forwards and slightly upwards across the lower part of the neck, being bound down to the clavicle by a fibrous expansion, and passing behind the Sterno-mastoid muscle, it becomes tendinous; it then changes its direction, forming an obtuse angle, and terminates in the anterior belly, which passes almost vertically upwards, close to the outer border

Fig. 397.—Muscles of the neck. Anterior view.

of the Sterno-hyoid, to be inserted into the lower border of the body of the os hyoides, just external to the insertion of the Sterno-hyoid. The central tendon of this muscle, which varies much in length and form, is held in position by a process of the deep cervical fascia, which sheaths it. This process is prolonged down, to be attached to the clavicle and first rib. It is by this means that the angular form of the muscle is maintained.

This muscle subdivides each of the two large triangles at the side of the neck into two smaller triangles; the two posterior ones being the *posterior superior* or *occipital*, and the *posterior inferior* or *subclavian triangles*; the two anterior, the *anterior superior* or *carotid*, and the *anterior inferior* or *muscular triangles*.

Relations.—By its *superficial surface*, with the Trapezius, the Sterno-mastoid, deep cervical fascia, Platysma, and integument. By its *deep surface*, with the Scaleni muscles, phrenic nerve, lower cervical nerves, which go to form the brachial plexus, the suprascapular vessels and nerve, sheath of the common carotid artery and internal jugular vein, the Sterno-thyroid and Thyro-hyoid muscles.

G G

Nerves.—The Thyro-hyoid and anterior belly of the Omo-hyoid are supplied by the descendens hypoglossi; the Sterno-hyoid, Sterno-thyroid, and posterior belly of the Omo-hyoid, by branches from the loop of communication between the descendens and communicantes hypoglossi.

Actions.—These muscles depress the larynx and hyoid bone, after they have been drawn up with the pharynx in the act of deglutition. The Omo-hyoid muscles not only depress the hyoid bone, but carry it backwards, and to one or the other side. It is concerned especially in prolonged inspiratory efforts; for by tensing the lower part of the cervical fascia it lessens the inward suction of the soft parts, which would otherwise compress the great vessels and the apices of the lungs. The Thyro-hyoid may act as an elevator of the thyroid cartilage, when the hyoid bone ascends, drawing upwards the thyroid cartilage, behind the os hyoides. The Sterno-thyroid acts as a depressor of the thyroid cartilage.

III. Suprahyoid Region (figs. 396, 397)

Digastric.   Mylo-hyoid.
Stylo-hyoid.   Genio-hyoid.

Dissection.—To dissect these muscles, a block should be placed beneath the back of the neck, and the head drawn backwards, and retained in that position. On the removal of the deep fascia, the muscles are at once exposed.

The **Digastric** consists of two fleshy bellies united by an intermediate, rounded tendon. It is a small muscle, situated below the side of the body of the lower jaw, and extending, in a curved form, from the side of the head to the symphysis of the jaw. The *posterior belly*, longer than the anterior, arises from the digastric groove on the inner side of the mastoid process of the temporal bone, and passes downwards, forwards, and inwards. The *anterior belly* arises from a depression on the inner side of the lower border of the jaw, close to the symphysis, and passes downwards and backwards. The two bellies terminate in the central tendon which perforates the Stylo-hyoid muscle, and is held in connection with the side of the body and the greater cornu of the hyoid bone by a fibrous loop, lined by a synovial membrane. A broad aponeurotic layer is given off from the tendon of the Digastric on each side, which is attached to the body and great cornu of the hyoid bone; this is termed the *suprahyoid aponeurosis*. It forms a strong layer of fascia between the anterior portion of the two muscles, and a firm investment for the other muscles of the suprahyoid region which lie deeper.

The Digastric muscle divides the anterior superior triangle of the neck into two smaller triangles; the upper, or *submaxillary*, being bounded, above, by the lower border of the body of the jaw, and a line drawn from its angle to the mastoid process; below, by the posterior belly of the Digastric and the Stylo-hyoid muscles; in front, by the anterior belly of the Digastric: the lower or *carotid triangle* being bounded above by the posterior belly of the Digastric, behind by the Sterno-mastoid, below by the Omo-hyoid.

Relations.—By its *superficial surface*, with the Mastoid process, the Platysma, Sterno-mastoid, part of the Splenius, Trachelo-mastoid, and Stylo-hyoid muscles, and the parotid gland. By its *deep surface*, the anterior belly lies on the Mylo-hyoid; the posterior belly on the Stylo-glossus, Stylo-pharyngeus, and Hyo-glossus muscles, the external carotid artery and its occipital, lingual, facial, and ascending pharyngeal branches, the internal carotid artery, internal jugular vein, and hypoglossal nerve.

The **Stylo-hyoid** is a small, slender muscle, lying in front of, and above, the posterior belly of the Digastric. It arises from the back and outer surface of the styloid process, near the base; and, passing downwards and forwards, is inserted into the body of the hyoid bone, just at its junction with the greater cornu, and just above the Omo-hyoid. This muscle is perforated, near its insertion, by the tendon of the Digastric.

Relations.—By its *superficial surface*, above with the parotid gland and deep cervical fascia; below it is superficial, being situated immediately beneath the deep cervical fascia. By its *deep surface*, with the posterior belly of the Digastric, the external carotid artery, with its lingual and facial branches, the Hyo-glossus muscle, and the hypoglossal nerve.

The Stylo-hyoid ligament.—In connection with the Stylo-hyoid muscle may be described a ligamentous band, the *Stylo-hyoid ligament*. It is a fibrous cord, often containing a little cartilage in its centre, which continues the styloid process down to the hyoid bone, being attached to the tip of the former and the small cornu of the latter. It is often more or less ossified, and in many animals forms a distinct bone, the *epihyal*.

The anterior belly of the Digastric should be removed, in order to expose the next muscle.

The **Mylo-hyoid** is a flat, triangular muscle, situated immediately beneath the anterior belly of the Digastric, and forming, with its fellow of the opposite side, a muscular floor for the cavity of the mouth. It arises from the whole length of the mylo-hyoid ridge of the lower jaw, extending from the symphysis in front to the last molar tooth behind. The posterior fibres pass inwards and slightly downwards, to be inserted into the body of the os hyoides. The middle and anterior fibres are inserted into a median fibrous raphé, extending from the symphysis of the lower jaw to the hyoid bone, where they join at an angle with the fibres of the opposite muscle. This median raphé is sometimes wanting; the muscular fibres of the two sides are then directly continuous with one another.

Relations.—By its *cutaneous* or *under surface*, with the Platysma, the anterior belly of the Digastric, the suprahyoid aponeurosis, the submaxillary gland, sub-mental vessels, and mylo-hyoid vessels and nerve. By its *deep* or *superior surface*, with the Genio-hyoid, part of the Hyo-glossus, and Stylo-glossus muscles, the hypoglossal and lingual nerves, the submaxillary ganglion, the sublingual gland, the deep portion of the submaxillary gland and Wharton's duct; the sublingual and ranine vessels, and the buccal mucous membrane.

Dissection.—The Mylo-hyoid should now be removed, in order to expose the muscles which lie beneath; this is effected by reflecting it from its attachments to the hyoid bone and jaw, and separating it by a vertical incision from its fellow of the opposite side.

The **Genio-hyoid** is a narrow, slender muscle, situated immediately beneath* the inner border of the preceding. It arises from the inferior genial tubercle on the inner side of the symphysis of the jaw, and passes downwards and backwards, to be inserted into the anterior surface of the body of the os hyoides. This muscle lies in close contact with its fellow of the opposite side, and increases slightly in breadth as it descends.

Relations.—It is covered by the Mylo-hyoid, and lies along the lower border of the Genio-hyo-glossus.

Nerves.—The Mylo-hyoid and anterior belly of the Digastric are supplied by the mylo-hyoid branch of the inferior dental; the Stylo-hyoid and posterior belly of the Digastric, by the facial; the Genio-hyoid, by the hypoglossal.

Actions.—This group of muscles performs two very important actions. They raise the hyoid bone, and with it the base of the tongue, during the act of degluti-tion; or, when the hyoid bone is fixed by its depressors and those of the larynx, they depress the lower jaw. During the first act of deglutition, when the mass of food is being driven from the mouth into the pharynx, the hyoid bone, and with it the tongue, is carried upwards and forwards by the anterior belly of the Digastric, the Mylo-hyoid, and Genio-hyoid muscles. In the second act, when the mass is passing through the pharynx, the direct elevation of the hyoid bone takes place by the combined action of all the muscles; and after the food has passed, the hyoid bone is carried upwards and backwards by the posterior belly of the Digastric and Stylo-hyoid muscles, which assist in preventing its return into the mouth.

* This refers to the depth of the muscles from the skin in the order of dissection. In the erect position of the body each of these muscles lies above the preceding.

IV. Lingual Region

Genio-hyo-glossus.   Chondro-glossus.
Hyo-glossus.   Stylo-glossus.
                    Palato-glossus.

Dissection.—After completing the dissection of the preceding muscles, saw through the lower jaw just external to the symphysis. Then draw the tongue forwards, and attach it, by a stitch, to the nose; when its muscles, which are thus put on the stretch, may be examined.

The **Genio-hyo-glossus** has received its name from its triple attachment to the jaw, hyoid bone, and tongue, but its connection with the hyoid bone is very slight or altogether absent. It is a flat, triangular muscle, placed vertically close to the middle line, its apex corresponding with its point of origin from the lower jaw, its base with its insertion into the tongue and hyoid bone. It arises by a short tendon from the superior genial tubercle on the inner side of the symphysis

Fig. 398.—Muscles of the tongue. Left side.

of the jaw, immediately above the Genio-hyoid; from this point the muscle spreads out in a fan-like form, a few of the inferior fibres extending downwards, to be attached by a thin aponeurosis into the upper part of the body of the hyoid bone, a few fibres passing between the Hyo-glossus and Chondro-glossus to blend with the Constrictor muscles of the pharynx; the middle fibres passing backwards, and the superior ones upwards and forwards, to enter the whole length of the under surface of the tongue, from the base to the apex. The muscles lie one on either side of the median plane; behind, they are quite distinct from each other, and are separated at their insertion into the under surface of the tongue by a tendinous raphé, which extends through the middle of the organ; in front, the two muscles are more or less blended: distinct fasciculi are to be seen passing off from one muscle, crossing the middle line, and inter-secting with bundles of fibres derived from the muscle on the other side.

Relations.—By its *internal surface* it is in contact with its fellow of the opposite side. By its *external surface*, with the Inferior lingualis, the Hyo-glossus, the lingual artery and hypoglossal nerve, the lingual nerve, and sublingual gland.

By its *upper border*, with the mucous membrane of the floor of the mouth (frænum linguæ). By its *lower border*, with the Genio-hyoid.

The **Hyo-glossus** is a thin, flat, quadrilateral muscle, which arises from the side of the body, and whole length of the greater cornu of the hyoid bone, and passes almost vertically upwards to enter the side of the tongue, between the Stylo-glossus and Inferior lingualis. Those fibres of this muscle which arise from the body of the hyoid bone are directed upwards and backwards, overlapping those arising from the greater cornu, which are directed upwards and forwards.

Relations.—By its *external surface*, with the Digastric, the Stylo-hyoid, Styloglossus, and Mylo-hyoid muscles, the submaxillary ganglion, the lingual and hypoglossal nerves, Wharton's duct, the ranine vein, the sublingual gland, and the deep portion of the submaxillary gland. By its *deep surface*, with the Stylohyoid ligament, the Genio-hyo-glossus, Inferior lingualis, and Middle constrictor, the lingual vessels, and the glosso-pharyngeal nerve.

The **Chondro-glossus** (fig. 399) is a distinct muscular slip, though it is sometimes described as a part of the Hyo-glossus, from which, however, it is separated by the fibres of the Genio-hyo-glossus, which pass to the side of the pharynx. It is about three-quarters to an inch in length, and arises from the inner side and base of the lesser cornu and contiguous portion of the body of the hyoid bone, and passes directly upwards to blend with the intrinsic muscular fibres of the tongue, between the Hyo-glossus and Geniohyo-glossus. A small slip of muscular fibres is occasionally found, arising from the cartilago triticea in the thyro-hyoid ligament, and passing upwards and forwards, enters the tongue with the hindermost fibres of the Hyo-glossus.

The **Stylo-glossus,** the shortest and smallest of the three styloid muscles, arises from the anterior and outer side of the styloid process, near its apex, and from the stylo-mandibular ligament, to which its fibres, in most cases, are attached by a thin aponeurosis. Passing downwards and forwards between the internal and external carotid arteries, and becoming nearly horizontal in its direction, it divides upon the side of the tongue into two portions : one longitudinal, which enters the side of the tongue near its dorsal surface, blending with the fibres of the Inferior lingualis in front of the Hyo-glossus ; the other oblique, which overlaps the Hyo-glossus muscle and decussates with its fibres.

FIG. 399.—Muscles of the tongue from below. (From a preparation in the Museum of the Royal College of Surgeons of England.)

Chondro-glossus

Relations.—By its *external surface*, from above downwards, with the parotid gland, the Internal pterygoid muscle, the lingual nerve, and the mucous membrane of the mouth. By its *internal surface*, with the tonsil, the Superior constrictor, and the Hyo-glossus muscle.

The **Palato-glossus,** or **Constrictor isthmi faucium,** although it is one of the muscles of the tongue, serving to draw its base upwards during the act of deglutition, is more nearly associated with the soft palate, both in its situation and function ; it will, consequently, be described with that group of muscles.

Nerves.—The Palato-glossus is probably innervated by the spinal accessory nerve, through the pharyngeal plexus; the remaining muscles of this group by the hypoglossal.

Muscular substance of tongue.—The muscular fibres of the tongue run in various directions. These fibres are divided into two sets—Extrinsic and Intrinsic. The extrinsic muscles of the tongue are those which have their origin external to, and only their terminal fibres contained in, the substance of the organ. They are: the Stylo-glossus, the Hyo-glossus, the Palato-glossus, the Genio-hyo-glossus, and part of the Superior constrictor of the pharynx (Pharyngo-glossus). The intrinsic are those which are contained entirely within the tongue, and form the greater part of its muscular structure.

The tongue consists of symmetrical halves separated from each other in the middle line by a fibrous septum. Each half is composed of muscular fibres arranged in various directions, containing much interposed fat, and supplied by vessels and nerves.

To demonstrate the various fibres of the tongue, the organ should be subjected to prolonged boiling, in order to soften the connective tissue; the dissection may then be commenced from the dorsum.

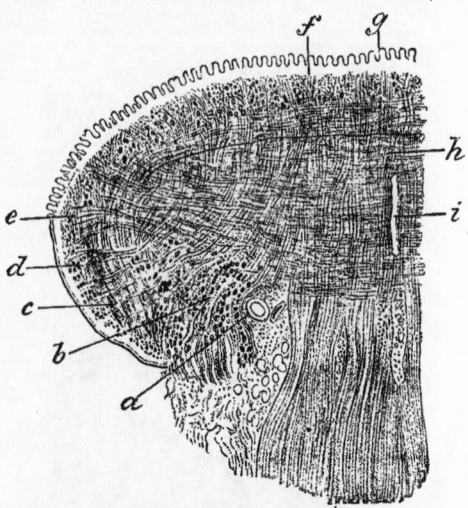

FIG. 400.—Coronal section of tongue. Showing intrinsic muscles. (Altered from Krause.)

a. Lingual artery. *b.* Inferior lingualis, cut through. *c.* Fibres of Hyo-glossus. *d.* Oblique fibres of Stylo-glossus. *e.* Insertion of Transverse lingualis. *f.* Superior lingualis. *g.* Papillæ of tongue. *h.* Vertical fibres of Genio-hyo-glossus intersecting Transverse lingualis. *i.* Septum.

Immediately beneath the mucous membrane is a submucous, fibrous layer, into which the muscular fibres which terminate on the surface of the tongue are inserted. Upon removing this, with the mucous membrane, the first stratum of muscular fibres is exposed. This belongs to the group of intrinsic muscles, and has been named the *Superior lingualis*. It consists of a thin layer of oblique and longitudinal fibres, which arise from the submucous fibrous layer, close to the Epiglottis, and from the fibrous septum, and pass forwards and outwards to the edges of the tongue. Between its fibres pass some vertical fibres derived from the Genio-hyo-glossus and from the vertical intrinsic muscle, which will be described later on. Beneath this layer is the second stratum of muscular fibres, derived principally from the extrinsic muscles. In front, it is formed by the fibres derived from the Stylo-glossus, running along the side of the tongue, and sending one set of fibres over the dorsum which runs obliquely forwards and inwards to the middle line, and another set of fibres, seen at a later period of the dissection, on to the under surface of the sides of the anterior part of the tongue, which runs forwards and inwards, between the fibres of the Hyo-glossus, to the middle line. Behind this layer of fibres, derived from the Stylo-glossus, are fibres derived from the Hyo-glossus, together with some few fibres of the Palato-glossus. The Hyo-glossus, entering the side of the under surface of the tongue, between the Stylo-glossus and Inferior lingualis, passes round its margin and spreads out into a layer on the dorsum, which occupies the middle third of the organ, and runs almost transversely inwards to the septum. It is reinforced by some fibres from the Palato-glossus; other fibres of this muscle pass more deeply and intermingle with the next layer. The posterior part of the second layer of the muscular fibres of the tongue is derived from those fibres of the Hyo-glossus which arise from the lesser cornu of the hyoid bone, and are here described as a separate muscle—the Chondro-glossus. The fibres of this muscle are arranged in a fan-shaped manner, and spread out over the posterior third of the tongue. Beneath this layer is the great mass of the intrinsic muscles of the tongue, intersected at right angles by the terminal fibres of one of the extrinsic muscles—

the Genio-hyo-glòssus. This portion of the tongue is paler in colour and softer in texture than that already described, and is sometimes designated the medullary portion in contradistiction to the firmer superficial part, which is termed the cortical portion. It consists largely of transverse fibres, the *Transverse lingualis*, and of vertical fibres, the *Vertical lingualis*. The Transverse lingualis forms the largest portion of the third layer of muscular fibres of the tongue. The fibres arise from the median septum, and pass outwards to be inserted into the submucous 'fibrous layer at the sides of the tongue. Intermingled with these transverse intrinsic fibres are transverse extrinsic fibres derived from the Palato-glossus and the Superior constrictor of the pharynx. These Transverse extrinsic fibres, however, run in the opposite direction, passing inwards towards the septum. Intersecting the transverse fibres are a large number of vertical fibres derived partly from the Genio-hyo-glossus and partly from intrinsic fibres, the *Vertical lingualis*. The fibres derived from the Genio-hyo-glossus enter the under surface of the tongue on each side of the median septum from base to apex. They ascend in a radiating manner to the dorsum, being inserted into the submucous fibrous layer covering the tongue on each side of the middle line. The Vertical lingualis is found only at the borders of the fore part of the tongue, external to the fibres of the Genio-hyo-glossus. Its fibres extend from the upper to the under surface of the organ, decussating with the fibres of the other muscles, and especially with the Transverse lingualis. The fourth layer of muscular fibres of the tongue consists partly of extrinsic fibres derived from the Stylo-glossus, and partly of intrinsic fibres, the *Inferior lingualis*. At the sides of the under surface of the organ are some fibres derived from the Stylo-glossus ; as this muscle runs forwards at the side of the tongue, it gives off fibres which pass forwards and inwards between the fibres of the Hyo-glossus and form an inferior oblique stratum which joins in front with the anterior fibres of the Inferior lingualis. The Inferior lingualis is a longitudinal band, situated on the under surface of the tongue, and extending from the base to the apex of the organ. Behind, some of its fibres are connected with the body of the hyoid bone. It lies between the Hyo-glossus and the Genio-hyo-glossus; in front of the Hyo-glossus it comes into relation with the Stylo-glossus, with the fibres of which it blends. It is in relation by its under surface with the ranine artery.

Surgical Anatomy.—The fibrous septum which exists between the two halves of the tongue is very complete, so that the anastomosis between the two lingual arteries is not very free, a fact often illustrated by injecting one-half of the tongue with coloured size, while the other half is left uninjected, or is injected with size of a different colour.

This is a point of considerable importance in connection with removal of one-half of the tongue for cancer, an operation which is now frequently resorted to when the disease is strictly confined to one side of the tongue. If the mucous membrane is divided longitudinally exactly in the middle line, the tongue can be split into halves along the median raphé, without any appreciable hæmorrhage, and the diseased half can then be removed.

Actions.—The movements of the tongue, although numerous and complicated, may be understood by carefully considering the direction of the fibres of its muscles. The *Genio-hyo-glossi* muscles, by means of their posterior fibres, draw the base of the tongue forwards, so as to protrude the apex from the mouth. The anterior fibres draw the tongue back into the mouth. The whole length of these two muscles acting along the middle line of the tongue draw it downwards, so as to make it concave from side to side, forming a channel along which fluids may pass toward the pharynx, as in sucking. The *Hyo-glossi* muscles depress the tongue, and draw down its sides, so as to render it convex from side to side. The *Stylo-glossi* muscles draw the tongue upwards and backwards. The *Palato-glossi* muscles draw the base of the tongue upwards. The intrinsic muscles are mainly concerned in altering the shape of the tongue, whereby it becomes shortened, narrowed from side to side, or curved in different directions; thus, the Superior and Inferior linguales tend to shorten the tongue, but the former, in addition, turn the tip and sides upwards so as to render the dorsum concave, while the latter pull the tip downwards and cause the dorsum to become convex. The Transverse lingualis narrows and elongates the tongue, and the Vertical lingualis flattens and broadens it. The complex arrangement of the muscular fibres of the tongue, and the various directions in which they run, give to this organ the power of assuming the various forms necessary for the enunciation of the different consonantal sounds; and Macalister states 'there is reason to believe that the musculature of the tongue varies in different races owing to the hereditary practice and habitual use of certain motions required for enunciating the several vernacular languages.'

V. Pharyngeal Region

Inferior constrictor.
Middle constrictor.
  Palato-pharyngeus. }
  Salpingo-pharyngeus. } (See next section.)
Superior constrictor.
Stylo-pharyngeus.

Dissection (fig. 401).—In order to examine the muscles of the pharynx, cut through the trachea and œsophagus just above the sternum, and draw them upwards by dividing the loose areolar tissue connecting the pharynx with the front of the vertebral column. The parts being drawn well forwards, apply the edge of the saw immediately behind the styloid processes, and saw the base of the skull through from below upwards. The pharynx and mouth should then be stuffed with tow, in order to distend its cavity and render the muscles tense and easier of dissection.

The **Inferior constrictor,** the most superficial and thickest of the three constrictors, arises from the sides of the cricoid and thyroid cartilages. To the cricoid cartilage it is attached in the interval between the Crico-thyroid muscle in front, and the articular facet for the inferior cornu of the thyroid cartilage behind. To the thyroid cartilage it is attached to the oblique line on the side of the great ala, to the cartilaginous surface behind it, nearly as far as its posterior border, and to the inferior cornu. From these attachments the fibres spread backwards and inwards, to be inserted into the fibrous raphé in the posterior median line of the pharynx. The inferior fibres are horizontal, and continuous with the circular fibres of the œsophagus; the rest ascend, increasing in obliquity, and overlap the Middle constrictor.

Fig. 401.—Muscles of the pharynx. External view.

Relations.—The Inferior constrictor is covered by a thin membrane which surrounds the entire pharynx (bucco-pharyngeal fascia). *Behind,* it is in relation with the vertebral column and the prevertebral fascia and muscles; *laterally,* with the thyroid gland, the common carotid artery, and the Sterno-thyroid muscle; by its *internal surface,* with the Middle constrictor, the Stylo-pharyngeus, Palato-pharyngeus, the pharyngeal aponeurosis and mucous membrane of the pharynx. The internal laryngeal nerve and the laryngeal branch of the Superior thyroid artery pass near the upper border, and the inferior, or recurrent laryngeal nerve, and the laryngeal branch of the inferior thyroid artery, beneath the lower border of this muscle, previous to their entering the larynx.

The **Middle constrictor** is a flattened, fan-shaped muscle, smaller than the preceding. It arises from the whole length of the upper border of the greater cornu of the hyoid bone, from the lesser cornu, and from the stylo-hyoid ligament. The fibres diverge from their origin: the lower ones descending beneath the Inferior constrictor, the middle fibres passing transversely, and the upper fibres ascending and overlapping the Superior constrictor. The muscle is inserted into the posterior median fibrous raphé, blending in the middle line with the one of the opposite side.

Relations.—This muscle is separated from the Superior constrictor by the glosso-pharyngeal nerve and the Stylo-pharyngeus muscle and Stylo-hyoid ligament; and from the Inferior constrictor by the internal laryngeal nerve and laryngeal branch of the superior thyroid artery. *Behind*, it lies on the prevertebral fascia, the Longus colli, and the Rectus capitis anticus major. *On each side* it is in relation with the carotid vessels, the pharyngeal plexus, and some lymphatic glands. Near its origin it is covered by the Hyo-glossus, from which it is separated by the lingual vessels. It lies upon the Superior constrictor, the Stylo-pharyngeus, the Palato-pharyngeus, the pharyngeal aponeurosis, and the mucous membrane of the pharynx.

The **Superior constrictor** is a quadrilateral muscle, thinner and paler than the other Constrictors, and situated at the upper part of the pharynx. It arises from the lower third of the posterior margin of the internal pterygoid plate and its hamular process, from the pterygo-mandibular ligament, from the alveolar process of the mandible above the posterior extremity of the mylo-hyoid ridge, and by a few fibres from the side of the tongue. From these points the fibres curve backwards, to be inserted into the median raphé, being also prolonged by means of a fibrous aponeurosis to the pharyngeal spine on the basilar process of the occipital bone. The superior fibres arch beneath the Levator palati and the Eustachian tube, the interval between the upper border of the muscle and the base of the skull being deficient in muscular fibres, and closed by the pharyngeal aponeurosis. This interval is known as the *sinus of Morgagni*.

Relations.—By its *outer surface*, with the prevertebral fascia and muscles, the vertebral column, the internal carotid and ascending pharyngeal arteries, the internal jugular vein and pharyngeal venous plexus, the glosso-pharyngeal, pneumogastric, spinal accessory, hypoglossal, lingual, and sympathetic nerves, the Middle constrictor and internal pterygoid muscles, the Styloid process, the Stylo-hyoid ligament, and the Stylo-pharyngeus. By its *internal surface*, with the Palato-pharyngeus, the tonsil, the pharyngeal aponeurosis and mucous membrane of the pharynx. Its *lower border* is separated from the Middle constrictor of the pharynx by the Stylo-pharyngeus muscle and the glosso-pharyngeal nerve.

The **Stylo-pharyngeus** is a long, slender muscle, round above, broad and thin below. It arises from the inner side of the base of the styloid process, passes downwards along the side of the pharynx between the Superior and Middle constrictors, and spreads out beneath the mucous membrane, where some of its fibres are lost in the Constrictor muscles; and others, joining with the Palato-pharyngeus, are inserted into the posterior border of the thyroid cartilage. The glosso-pharyngeal nerve runs on the outer side of this muscle, and crosses over it in passing forward to the tongue.

Relations.—*Externally*, with the Stylo-glossus muscle, the parotid gland, the external carotid artery, and the Middle constrictor. *Internally*, with the internal carotid, the internal jugular vein, the Superior constrictor, Palato-pharyngeus and mucous membrane.

Nerves.—The Constrictors are supplied by branches from the pharyngeal plexus, the Stylo-pharyngeus by the glosso-pharyngeal nerve, and the Inferior constrictor by additional branches from the external and recurrent laryngeal nerves.

Actions.—When deglutition is about to be performed, the pharynx is drawn upwards and dilated in different directions, to receive the food propelled into it from the mouth. The Stylo-pharyngei, which are much farther removed from one another at their origin than at their insertion, draw the sides of the pharynx upwards and outwards, and so increase its transverse diameter; its breadth in the antero-posterior direction being increased by the larynx and tongue being carried forwards in their ascent. As soon as the morsel is received in the pharynx, the Elevator muscles relax, the bag descends, and the Constrictors contract upon it, and convey it gradually downwards into the œsophagus. Besides its action in deglutition, the pharynx also exerts an important influence in the modulation of the voice, especially in the production of the higher tones.

VI. PALATAL REGION

Levator palati.	Palato-glossus.
Tensor palati.	Palato-pharyngeus.
Azygos uvulæ.	Salpingo-pharyngeus.

Dissection (fig. 402).—Lay open the pharynx from behind, by a vertical incision extending from its upper to its lower part, and partially divide the occipital attachment by a transverse incision on each side of the vertical one ; the posterior surface of the soft palate is then exposed. Having fixed the uvula so as to make it tense, the mucous membrane and glands should be carefully removed from the posterior surface of the soft palate and the muscles of this part are at once exposed.

The **Levator palati** is a long, thick, rounded muscle, placed on the outer side of the posterior nares. It arises from the under surface of the apex of the petrous portion of the temporal bone, and from the inner surface of the cartilaginous

Fig. 402.—Muscles of the soft palate. The pharynx being laid open from behind.

portion of the Eustachian tube ; after passing above the upper concave margin of the Superior constrictor, it extends obliquely downwards and inwards, its fibres spreading out in the soft palate as far as the middle line, where they blend with those of the opposite side.

Relations.—*Externally*, with the Tensor palati, Superior constrictor, and Eustachian tube. *Internally*, with the mucous membrane of the pharynx. *Posteriorly*, with the posterior fasciculus of the Palato-pharyngeus, the Azygos uvulæ, and the mucous lining of the soft palate.

The **Tensor** or **Circumflexus palati** is a broad, thin, riband-like muscle, placed on the outer side of the Levator palati, and consisting of a vertical and a horizontal portion. The vertical portion arises by a flat lamella from the scaphoid fossa at the base of the internal pterygoid plate ; from the spine of the sphenoid and the outer side of the cartilaginous portion of the Eustachian tube : descending vertically between the internal pterygoid plate and the inner surface of the Internal pterygoid muscle, it terminates in a tendon, which winds round the hamular process, being retained in this situation by some of the fibres of origin of the Internal pterygoid muscle. Between the hamular process and the tendon is a small bursa. The tendon or horizontal portion then passes horizontally inwards, and is inserted into a broad aponeurosis, the *palatine aponeurosis*, and into the transverse ridge on the horizontal portion of the palate bone.

Relations.—*Externally*, with the Internal pterygoid. *Internally*, with the Levator palati, from which it is separated by the Eustachian tube and Superior constrictor, and with the internal pterygoid plate. In the soft palate, its tendon and the palatine aponeurosis is anterior to that of the Levator palati, being covered by the Palato-glossus and the mucous membrane.

Palatine Aponeurosis.—Attached to the posterior border of the hard palate is a thin, firm, fibrous lamella which supports the muscles and gives strength to the soft palate. It is thicker above than below, where it becomes very thin and difficult to define. Laterally, it is continuous with the pharyngeal aponeurosis.

The **Azygos uvulæ** is not a single muscle, as would be inferred from its name, but a pair of narrow cylindrical fleshy fasciculi, placed one on either side of the median line of the soft palate. Each muscle arises from the posterior nasal spine of the palate bone, and from the contiguous tendinous aponeurosis of the soft palate, and descends to be inserted into the uvula.

Relations.—*Anteriorly*, with the tendinous expansion of the Levatores palati ; *behind*, with the posterior fasciculus of the Palato-pharyngeus and the mucous membrane.

The next two muscles are exposed by removing the mucous membrane from the pillars of the fauces throughout nearly their whole extent.

The **Palato-glossus (Constrictor isthmi faucium)** is a small fleshy fasciculus, narrower in the middle than at either extremity, forming, with the mucous membrane covering its surface, the anterior pillar of the fauces. It arises from the anterior surface of the soft palate, where it is continuous with the muscle of the opposite side, and, passing downwards, forwards, and outwards in front of the tonsil, is inserted into the side of the tongue, some of its fibres spreading over the dorsum, and others passing deeply into the substance of the organ to intermingle with the Transverse lingualis.

The **Palato-pharyngeus** is a long, fleshy fasciculus, narrower in the middle than at either extremity, forming, with the mucous membrane covering its surface, the posterior pillar of the fauces. It is separated from the Palato-glossus by an angular interval, in which the tonsil is lodged. It arises from the soft palate by an expanded fasciculus, which is divided into two parts by the Levator palati and Azygos uvulæ. The *posterior fasciculus* lies in contact with the mucous membrane, and joins with the corresponding muscle in the middle line; the *anterior fasciculus*, the thicker, lies in the soft palate between the Levator and Tensor, and joins in the middle line the corresponding part of the opposite muscle. Passing outwards and downwards behind the tonsil, the Palato-pharyngeus joins the Stylo-pharyngeus, and is inserted with that muscle into the posterior border of the thyroid cartilage, some of its fibres being lost on the side of the pharynx, and others passing across the middle line posteriorly, to decussate with the muscle of the opposite side.

Relations.—In the soft palate, its *posterior surface* is covered by mucous membrane, from which it is separated by a layer of palatine glands. By its *anterior surface* it is in relation with the Tensor palati. Where it forms the posterior pillars of the fauces, it is covered by mucous membrane, excepting on its outer surface. In the *pharynx* it lies between the mucous membrane and the Constrictor muscles.

The **Salpingo-pharyngeus.**—This muscle arises from the inferior part of the Eustachian tube near its orifice ; it passes downwards and blends with the posterior fasciculus of the Palato-pharyngeus.

In a dissection of the soft palate from its posterior or nasal surface to its anterior or oral surface, the muscles would be exposed in the following order : viz. the posterior fasciculus of the Palato-pharyngeus, covered by the mucous membrane reflected from the floor of the nasal fossæ; the Azygos uvulæ; the Levator palati; the anterior fasciculus of the Palato-pharyngeus; the aponeurosis of the Tensor palati, and the Palato-glossus covered by a reflection from the oral mucous membrane.

Nerves.—The Tensor palati is supplied by a branch from the otic ganglion; the remaining muscles of this group are in all probability supplied by the bulbar portion of the spinal accessory through the pharyngeal plexus.*

Actions.—During the *first stage* of deglutition, the morsel of food is driven back into the fauces by the pressure of the tongue against the hard palate: the base of the tongue being, at the same time, retracted, and the larynx raised with the pharynx, and carried forwards under it. During the *second stage* the entrance to the larynx is closed, not, as was formerly supposed, by the folding backwards of the epiglottis over it, but, as Anderson Stuart has shown, by the drawing forward of the arytenoid cartilages towards the cushion of the epiglottis —a movement produced by the contraction of the external thyro-arytenoid, the arytenoid and aryteno-epiglottidean muscles.

The morsel of food after leaving the tongue passes on to the posterior or laryngeal surface of the epiglottis, and glides along this for a certain distance; † then the Palato-glossi muscles, the constrictors of the fauces, contract behind the food; the soft palate is slightly raised by the Levator palati, and made tense by the Tensor palati; and the Palato-pharyngei, by their contraction, pull the pharynx upwards over the morsel of food, and come nearly together, the uvula filling up the slight interval between them. By these means the food is prevented from passing into the naso-pharynx; at the same time, the latter muscles form an inclined plane, directed obliquely downwards and backwards, along the under surface of which the morsel descends into the lower part of the pharynx. The Salpingo-pharyngeus raises the upper and lateral part of the pharynx—i.e. that part which is above the point where the Stylo-pharyngeus is attached to the pharynx.

Surgical Anatomy.—After the operation for the closure of a cleft in the palate, the palate muscles, especially the Tensor and Levator palati, have a tendency to retard the healing process by active traction upon the line of suture. To obviate this, it is necessary to divide them. This is best done by making longitudinal incisions, on either side, parallel to the cleft and just internal to the hamular process, in such a position as to avoid the posterior palatine artery.

VII. Anterior Vertebral Region

Rectus capitis anticus major.   Rectus capitis lateralis.
Rectus capitis anticus minor.   Longus colli.

The **Rectus capitis anticus major** (fig. 403), broad and thick above, narrow below, appears like a continuation upwards of the Scalenus anticus. It arises by four tendinous slips from the anterior tubercles of the transverse processes of the third, fourth, fifth, and sixth cervical vertebræ, and ascends, converging towards its fellow of the opposite side, to be inserted into the basilar process of the occipital bone.

Relations.—By its *anterior surface*, with the pharynx, the sympathetic nerve, and the sheath enclosing the internal and common carotid artery, internal jugular vein, and the pneumogastric nerve. By its *posterior surface*, with the Longus colli, the Rectus capitis anticus minor, and the upper cervical vertebræ.

The **Rectus capitis anticus minor** is a short, flat muscle, situated immediately behind the upper part of the preceding. It arises from the anterior surface of the lateral mass of the atlas, and from the root of its transverse process, and passing obliquely upwards and inwards, is inserted into the basilar process immediately behind the preceding muscle.

Relations.—By its *anterior surface*, with the Rectus capitis anticus major. By its *posterior surface*, with the front of the occipito-atlantal articulation.

---

* 'The Innervation of the Soft Palate,' by William Aldren Turner, M.B. *Journal of Anatomy and Physiology*, vol. xxiii. p. 523.

† Walton (quoted by A. Stuart) maintains that the epiglottis is not essential to the deglutition even of liquids.

The **Rectus capitis lateralis** is a short, flat muscle, which arises from the upper surface of the transverse process of the atlas, and is inserted into the under surface of the jugular process of the occipital bone.

Relations.—By its *anterior surface*, with the internal jugular vein. By its *posterior surface*, with the vertebral artery. On its *outer side* lies the occipital artery ; on its *inner side* the suboccipital nerve.

The **Longus colli** is a long, flat muscle, situated on the anterior surface of the spine, between the atlas and the third dorsal vertebra. It is broad in the middle, narrow and pointed at each extremity, and consists of three portions : a superior oblique, an inferior oblique, and a vertical portion. The *superior oblique portion* arises from the anterior tubercles of the transverse processes of the third, fourth, and fifth cervical vertebræ ; and, ascending obliquely inwards, is inserted by a

FIG. 403.—The prevertebral muscles.

narrow tendon into the tubercle on the anterior arch of the atlas. The *inferior oblique portion*, the smallest part of the muscle, arises from the front of the bodies of the first two or three dorsal vertebræ ; and, ascending obliquely outwards, is inserted into the anterior tubercles of the transverse processes of the fifth and sixth cervical vertebræ. The *vertical portion* lies directly on the front of the spine ; it arises, below, from the front of the bodies of the upper three dorsal and lower three cervical vertebræ, and is inserted into the front of the bodies of the second, third, and fourth cervical vertebræ above.

Relations.—By its *anterior surface*, with the prevertebral fascia, the pharynx, the œsophagus, the sympathetic nerve, the sheath of the great vessels of the neck, the inferior thyroid artery, and recurrent laryngeal nerve. By its *posterior surface*, with the cervical and dorsal portions of the spine. Its *inner border* is separated from the opposite muscle by a considerable interval below ; but they approach each other above.

VIII. LATERAL VERTEBRAL REGION

Scalenus anticus.   Scalenus medius.

Scalenus posticus.

The **Scalenus anticus** is a conical-shaped muscle, situated deeply at the side of the neck, behind the Sterno-mastoid.   It arises from the anterior tubercles of

FIG. 404.—Muscles of the neck.   (From a preparation in the Museum of the Royal College of Surgeons of England.)

the transverse processes of the third, fourth, fifth, and sixth cervical vertebræ, and descending, almost vertically, is inserted by a narrow, flat tendon into the scalene tubercle on the inner border of the first rib, and also into the upper surface of the rib between the grooves for the subclavian artery and vein.   The lower part of this muscle separates the subclavian artery and vein : the latter being in front, and the former, with the brachial plexus, behind.

Relations.—*In front*, with the clavicle, the Subclavius, Sterno-mastoid, and Omo-hyoid muscles, the transversalis colli, the suprascapular and ascending cervical arteries, the subclavian vein, and the phrenic nerve.   By its *posterior surface*, with the cords of the brachial plexus, the subclavian artery and the pleura, which separate it from the Scalenus medius. It is separated from the Longus colli, on the inner side, by the vertebral artery. On the anterior tubercles of the transverse processes of the cervical vertebræ, between the attachments of the Scalenus anticus and Longus colli, lies the ascending cervical branch of the inferior thyroid artery.

The **Scalenus medius,** the largest and longest of the three Scaleni, arises from the posterior tubercles of the transverse processes of the lower six cervical vertebræ, and descending along the side of the vertebral column is inserted by a broad attachment into the upper surface of the first rib, behind the groove for the subclavian artery, as far back as the tubercle. It is separated from the Scalenus anticus by the subclavian artery below and the cervical nerves above. The posterior thoracic, or nerve of Bell, is formed in the substance of the Scalenus medius and emerges from it. The nerve to the Rhomboids also pierces it.

Relations.—By its *anterior surface*, with the Sterno-mastoid ; it is crossed by the clavicle, the Omo-hyoid muscle, subclavian artery, and the cervical nerves. To its *outer side* is the Levator anguli scapulæ, and the Scalenus posticus muscle.

The **Scalenus posticus,** the smallest of the three Scaleni, arises, by two or three separate tendons, from the posterior tubercles of the transverse processes of the lower two or three cervical vertebræ, and, diminishing as it descends, is inserted by a thin tendon into the outer surface of the second rib, behind the attachment of the Serratus magnus. This is the most deeply placed of the three Scaleni, and is occasionally blended with the Scalenus medius.

Nerves.—The Rectus capitis anticus major and minor and the Rectus lateralis are supplied by the first cervical nerve, and from the loop formed between it and the second; the Longus colli and Scaleni, by branches from the anterior divisions of the lower cervical nerves (fifth, sixth, seventh, and eighth) before they form the brachial plexus. The Scalenus medius also receives a filament from the deep external branches of the cervical plexus.

Actions.—The Rectus capitis anticus major and minor are the direct antagonists of the muscles at the back of the neck, serving to restore the head to its natural position after it has been drawn backwards. These muscles also serve to flex the head, and from their obliquity, rotate it, so as to turn the face to one or the other side. The Longus colli flexes and slightly rotates the cervical portion of the spine. The Scaleni muscles, when they take their fixed point from above, elevate the first and second ribs, and are, therefore, inspiratory muscles. When they take their fixed point from below, they bend the spinal column to one or the other side. If the muscles of both sides act, lateral movement is prevented, but the spine is slightly flexed. The Rectus lateralis, acting on one side, bends the head laterally.

Surface Form.—The muscles in the neck, with the exception of the Platysma myoides, are invested by the deep cervical fascia, which softens down their form, and is of considerable importance in connection with deep cervical abscesses and tumours, modifying the direction of their growth and causing them to extend laterally instead of towards the surface. The *Platysma myoides* does not influence surface form except it is in action, when it produces wrinkling of the skin of the neck, which is thrown into oblique ridges parallel with the fasciculi of the muscle. Sometimes this contraction takes place suddenly and repeatedly, as a sort of spasmodic twitching, the result of a nervous habit. The *Sterno-cleido-mastoid* is the most important muscle of the neck as regards its surface form. If the muscle is put into action by drawing the chin downwards and to the opposite shoulder, its surface form will be plainly outlined. The sternal origin will stand out as a sharply defined ridge, while the clavicular origin will present a flatter and less prominent outline. The fleshy middle portion will appear as an oblique roll or elevation, with a thick rounded anterior border gradually becoming less marked above. On the opposite side, i.e. on the side to which the head is turned, the outline is lost, its place being occupied by an oblique groove in the integument. When the muscle is at rest its anterior border is still visible, forming an oblique rounded ridge, terminating below in the sharp outline of the sternal head. The posterior border of the muscle does not show above the clavicular head. The anterior border is defined by drawing a line from the tip of the mastoid process to the sterno-clavicular joint. It is an important surface-marking in the operation of ligature of the common carotid artery and some other operations. Between the sternal and clavicular heads is a slight depression, most marked when the muscle is in action, which overlies the lower part of the internal jugular vein. This is bounded below by the prominent sternal extremity of the clavicle. Between the sternal origins of the two muscles is a V-shaped space, the *suprasternal notch*, more pronounced below, and becoming toned down above, where the Sterno-hyoid and Sterno-thyroid muscles, lying upon the trachea, become more prominent. Above the hyoid bone, in the middle line, the anterior belly of the *Digastric* to a certain extent influences surface form. It corresponds to a line drawn from the symphysis of the lower jaw to the side of the body of the hyoid bone, and renders this part of the hyo-mental region convex. In the posterior triangle of the neck, the posterior belly of the *Omo-hyoid*, when in action, forms a conspicuous object, especially in thin necks, presenting a cord-like form running across this region, almost parallel with, and a little above, the clavicle.

MUSCLES AND FASCIÆ OF THE TRUNK

The Muscles of the Trunk may be arranged in four groups, corresponding with the region in which they are situated.

I. The Back.
II. The Thorax.
III. The Abdomen.
IV. The Pelvic outlet.

I. Muscles of the Back

The muscles of the back are very numerous, and may be subdivided into five layers.

First Layer

Trapezius.
Latissimus dorsi.

Second Layer

Levator anguli scapulæ.
Rhomboideus minor.
Rhomboideus major.

Third Layer

Serratus posticus superior.
Serratus posticus inferior.
Splenius capitis.
Splenius colli.

Fourth Layer

Sacral and Lumbar Regions
Erector spinæ.

Dorsal Region
Ilio-costalis.
Musculus accessorius ad ilio-costalem.

Longissimus dorsi.
Spinalis dorsi.

Cervical Region
Cervicalis ascendens.
Transversalis cervicis.
Trachelo-mastoid.
Complexus.
Biventer cervicis.
Spinalis colli.

Fifth Layer

Semispinalis dorsi.
Semispinalis colli.
Multifidus spinæ.
Rotatores spinæ.
Interspinales.
Extensor coccygis.
Intertransversales.
Rectus capitis posticus major.
Rectus capitis posticus minor.
Obliquus capitis inferior.
Obliquus capitis superior.

First Layer

Trapezius.   Latissimus dorsi.

Dissection (fig. 405).—Place the body in the prone position, with the arms extended over the sides of the table, and the chest and abdomen supported by several blocks, so as to render the muscles tense. Then make an incision along the middle line of the back from the occipital protuberance to the coccyx. Make a transverse incision from the upper end of this to the mastoid process; and a third incision from its lower end, along the crest of the ilium to about its middle. This large intervening space should, for convenience of dissection, be subdivided by a fourth incision, extending obliquely from the spinous process of the last dorsal vertebra, upwards and outwards, to the acromion process. This incision corresponds with the lower border of the Trapezius muscle. The flaps of integument are then to be removed in the direction shown in the figure.

The **superficial fascia** is exposed upon removing the skin from the back. It forms a layer of considerable thickness and strength, in which a quantity of granular pinkish fat is contained. It is continuous with the superficial fascia in other parts of the body. The **deep fascia** is a dense fibrous layer, attached above

to the superior curved line of the occipital bone : in the middle line it is attached
to the ligamentum nuchæ, and to the spinous processes and supraspinous ligaments of all the vertebræ below the seventh cervical : laterally, in the neck it is
continuous with the deep cervical fascia ; over the shoulder it is attached to the
spine of the scapula and the acromion process,
and is continued downwards over the Deltoid
muscle to the arm ; on the thorax it is continuous with the deep fascia of the axilla and
chest, and on the abdomen with that covering
the abdominal muscles : below, it is attached
to the crest of the ilium.

FIG. 405.—Dissection of the
muscles of the back.

The **Trapezius** (fig. 406) is a broad, flat,
triangular muscle, placed immediately beneath
the skin and fascia, and covering the upper
and back part of the neck and shoulders.   It
arises from the external occipital protuberance
and the inner third of the superior curved line
of the occipital bone ; from the ligamentum
nuchæ, the spinous process of the seventh
cervical, and those of all the dorsal vertebræ ;
and from the corresponding portion of the
supraspinous ligament.   From this origin, the
superior fibres proceed downwards and outwards, the inferior upwards and outwards ;
and the middle horizontally ; the superior
fibres are inserted into the outer third of the
posterior border of the clavicle ; the middle
fibres into the inner margin of the acromion
process, and into the superior lip of the
posterior border or crest of the spine of the
scapula ; the inferior fibres converge near the
scapula, and terminate in a triangular aponeurosis, which glides over a smooth surface
at the inner extremity of the spine, to be
inserted into a tubercle at the outer part of
this smooth surface.   The Trapezius is fleshy
in the greater part of its extent, but tendinous
at its origin and insertion.   At its occipital origin, it is connected to the bone
by a thin fibrous lamina, firmly adherent to the skin, and wanting the lustrous,
shining appearance of aponeuroses.   At its origin from the spines of the vertebræ,
it is connected to the bones by means of a broad semi-elliptical aponeurosis,
which occupies the space between the sixth cervical and the third dorsal
vertebræ, and forms, with the aponeurosis of the opposite muscle, a tendinous
ellipse.   The rest of the muscle arises by numerous short tendinous fibres.   If
the Trapezius is dissected on both sides, the two muscles resemble a trapezium,
or diamond-shaped quadrangle : two angles corresponding to the shoulders ; a
third to the occipital protuberance ; and the fourth to the spinous process of the
last dorsal vertebra.

The clavicular insertion of this muscle varies as to the extent of its attachment : it sometimes advances as far as the middle of the clavicle, and may
occasionally become blended with the posterior edge of the Sterno-mastoid, or
overlap it.   This should be borne in mind in the operation for tying the third
part of the subclavian artery.

Relations.—By its *superficial surface*, with the integument.   By its *deep
surface*, in the neck, with the Complexus, Splenius, Levator anguli scapulæ, and
Rhomboideus minor ; in the back, with the Rhomboideus major, Supraspinatus,
Infraspinatus, the Vertebral aponeurosis (which separates it from the prolongations of the Erector spinæ), and the Latissimus dorsi.   The superficial cervical
artery, the spinal accessory nerve, and branches from the third and fourth cervical
nerves, pass beneath the anterior border of this muscle.   The anterior margin of
its cervical portion forms the posterior boundary of the posterior triangle of the
neck, the other boundaries being the Sterno-mastoid in front, and the clavicle
below.

H H

FIG. 406.—Muscles of the back. On the left side, the first layer is exposed ; on the right side, the second layer and part of the third.

The **Latissimus dorsi** is a broad flat muscle, which covers the lumbar and the lower half of the dorsal regions, and is gradually contracted into a narrow fasciculus at its insertion into the humerus. It arises by tendinous fibres from the spinous processes of the six inferior dorsal vertebræ and from the posterior layer of the lumbar fascia (see page 469), by which it is attached to the spines of the lumbar and sacral vertebræ, and to the supraspinous ligament. It also arises from the posterior third of the external lip of the crest of the ilium, behind the insertion of the External oblique, and by fleshy digitations from the three or four lower ribs, which are interposed between similar processes of the External oblique muscle (fig. 411, page 484). From this extensive origin the fibres pass in different directions, the upper ones horizontally, the middle obliquely upwards, and the lower vertically upwards, so as to converge and form a thick fasciculus, which crosses the inferior angle of the scapula, and usually receives a few fibres from it. The muscle curves around the lower border of the Teres major, and is twisted upon itself, so that the superior fibres become at first posterior and then inferior, and the vertical fibres at first anterior and then superior. It terminates in a short quadrilateral tendon, about three inches in length, which passes in front of the tendon of the Teres major, and is inserted into the bottom of the bicipital groove of the humerus; its insertion extending higher on the humerus than that of the tendon of the Pectoralis major. The lower border of the tendon of this muscle is united with that of the Teres major, the surfaces of the two being separated near their insertion by a bursa; another bursa is sometimes interposed between the muscle and the inferior angle of the scapula. This muscle at its insertion gives off an expansion to the deep fascia of the arm.

A muscular slip, the *axillary arch*, varying from 3 to 4 inches in length, and from ¼ to ¾ of an inch in breadth, occasionally arises from the upper edge of the Latissimus dorsi about the middle of the posterior fold of the axilla, and crosses the axilla in front of the axillary vessels and nerves, to join the under surface of the tendon of the Pectoralis major, the Coraco-brachialis, or the fascia over the Biceps. The position of this abnormal slip is a point of interest in its relation to the axillary artery, as it crosses the vessel just above the spot usually selected for the application of a ligature, and may mislead the surgeon during the operation. It may be easily recognised by the transverse direction of its fibres. Struthers found it, in 8 out of 105 subjects, occurring seven times on both sides.

There is usually a fibrous slip which passes from the lower border of the tendon of the Latissimus dorsi, near its insertion, to the long head of the Triceps. This is occasionally muscular, and is the representative of the *dorso-epitrochlearis* muscle of apes.

Relations.—Its *superficial surface* is subcutaneous, excepting at its upper part, where it is covered by the Trapezius, and at its insertion, where its tendon is crossed by the axillary vessels and the brachial plexus of nerves. By its *deep surface*, it is in relation with the Lumbar fascia, the Serratus posticus inferior, the lower external intercostal muscles and ribs, inferior angle of the scapula, Rhomboideus major, Infraspinatus, and Teres major. Its outer margin is separated below from the External oblique by a small triangular interval, the base of which is formed by the crest of the ilium, and its floor by the Internal oblique muscle of the abdomen. This is known as the *triangle of Petit*, and is sometimes the site of a lumbar hernia. Another triangle of practical importance is situated behind the scapula. It is bounded above by the Trapezius, below by the Latissimus dorsi, and externally by the vertebral border of the scapula; the floor is partly formed by the Rhomboideus. If the scapula is drawn forwards by folding the arms across the chest, and the trunk bent forwards, a part of the sixth and seventh ribs and the interspace between them become subcutaneous and available for auscultation. The space is therefore known as the *triangle of auscultation*.

Nerves.—The Trapezius is supplied by the spinal accessory, and by branches from the anterior divisions of the third and fourth cervical nerves; the Latissimus dorsi by the sixth, seventh, and eighth cervical nerves through the middle or long subscapular nerve.

Second Layer

Levator anguli scapulæ.   Rhomboideus minor
                        Rhomboideus major.

Dissection.—The Trapezius must be removed, in order to expose the next layer; to effect this, detach the muscle from its attachment to the clavicle and spine of the scapula, and turn it back towards the spine.

The **Levator anguli scapulæ** is situated at the back part and side of the neck. It arises by tendinous slips from the transverse process of the atlas and from the posterior tubercles of the transverse processes of the second, third, and fourth cervical vertebræ; these, becoming fleshy, are united so as to form a flat muscle, which passes downwards and backwards, and is inserted into the posterior border of the scapula, between the superior angle and the triangular smooth surface at the root of the spine.

Relations.—By its *superficial surface*, with the integument, Trapezius, and Sterno-mastoid. By its *deep surface*, with the Splenius colli, Transversalis cervicis, Cervicalis ascendens, and Serratus posticus superior muscles, and with the posterior scapular artery and the nerve to the Rhomboids.

The **Rhomboideus minor** arises from the lower part of the ligamentum nuchæ and the spinous processes of the seventh cervical and first dorsal vertebræ. Passing downwards and outwards, it is inserted into the margin of the triangular smooth surface at the root of the spine of the scapula. This small muscle is usually separated from the Rhomboideus major by a slight cellular interval.

Relations.—By its *superficial (posterior) surface*, with the Trapezius. By its *deep surface*, with the same structures as the Rhomboideus major.

The **Rhomboideus major** is situated immediately below the preceding, the adjacent margins of the two being occasionally united. It arises by tendinous fibres from the spinous processes of the second, third, fourth, and fifth dorsal vertebræ and the supraspinous ligament, and is inserted into a narrow tendinous arch, attached above to the lower part of the triangular surface at the root of the spine; below, to the inferior angle, the arch being connected to the border of the scapula by a thin membrane. When the arch extends, as it occasionally does, only a short distance, the muscular fibres are inserted into the scapula itself.

Relations.—By its *superficial surface*, with the Latissimus dorsi. By its *deep surface*, with the Serratus posticus superior, the posterior scapular artery, and the nerve to the muscle, the vertebral aponeurosis which separates it from the prolongations of the Erector spinæ, the Intercostal muscles, and ribs.

Nerves.—The Rhomboid muscles are supplied by branches from the anterior division of the fifth cervical nerve; the Levator anguli scapulæ by the anterior division of the third and fourth cervical nerves, and frequently by a branch from the nerve to the Rhomboids.

Actions.—The movements effected by the preceding muscles are numerous, as may be conceived from their extensive attachment. The whole of the *Trapezius* when in action retracts the scapula and braces back the shoulder; if the head is fixed, the upper part of the Trapezius will elevate the point of the shoulder, as in supporting weights; when the lower fibres are brought into action they assist in depressing the bone. The middle and lower fibres of the muscle rotate the scapula, causing elevation of the acromion process. If the shoulders are fixed, both Trapezii, acting together, will draw the head directly backwards; or if only one acts, the head is drawn to the corresponding side.

The *Latissimus dorsi*, when it acts upon the humerus, depresses it, draws it backwards, and at the same time rotates it inwards. It is the muscle which is principally employed in giving a downward blow, as in felling a tree or in sabre practice. If the arm is fixed, the muscle may act in various ways upon the trunk: thus, it may raise the lower ribs and assist in forcible inspiration; or, if both arms are fixed, the two muscles may assist the Abdominal and great Pectoral muscles in suspending and drawing the whole trunk forwards, as in climbing or walking on crutches.

The *Levator anguli scapulæ* raises the superior angle of the scapula, assisting the Trapezius in bearing weights or in shrugging the shoulders. If the shoulder is fixed, the Levator anguli scapulæ inclines the neck to the corresponding side and rotates it in the same direction. The *Rhomboid* muscles carry the inferior angle backwards and upwards, thus producing a slight rotation of the scapula upon the side of the chest, the Rhomboideus major acting especially on the lower angle of the scapula, through the tendinous arch by which it is inserted. The Rhomboid muscles acting together with the middle and inferior fibres of the Trapezius will draw the scapula directly backwards towards the spine

THIRD LAYER

Serratus posticus superior.   Serratus posticus inferior

Splenius { Splenius capitis.
          { Splenius colli.

Dissection.—To bring into view the third layer of muscles, remove the whole of the second, together with the Latissimus dorsi, by cutting through the Levator anguli scapulæ and Rhomboid muscles near their origin, and reflecting them downwards, and by dividing the Latissimus dorsi in the middle by a vertical incision carried from its upper to its lower part, and reflecting the two halves of the muscle.

The **Serratus posticus superior** is a thin, flat, quadrilateral muscle, situated at the upper and back part of the thorax. It arises by a thin and broad aponeurosis from the lower part of the ligamentum nuchæ, and from the spinous processes of the last cervical and two or three upper dorsal vertebræ and from the supraspinous ligament. Inclining downwards and outwards, it becomes muscular, and is inserted, by four fleshy digitations, into the upper borders of the second, third, fourth, and fifth ribs, a little beyond their angles.

Relations.—By its *superficial surface*, with the Trapezius, Rhomboidei, and Levator anguli scapulæ. By its *deep surface*, with the Splenius, and the vertebral aponeurosis, which separates it from the prolongations of the Erector spinæ, and with the Intercostal muscles and ribs.

The **Serratus posticus inferior** is situated at the junction of the dorsal and lumbar regions: it is of an irregularly quadrilateral form, broader than the preceding, and separated from it by a wide interval. It arises by a thin aponeurosis from the spinous processes of the last two dorsal and two or three upper lumbar vertebræ, and from the supraspinous ligament. Passing obliquely upwards and outwards, it becomes fleshy, and divides into four flat digitations, which are inserted into the lower borders of the four lower ribs, a little beyond their angles. The thin aponeurosis of origin is intimately blended with the lumbar fascia.

Relations.—By its *superficial surface,* with the Latissimus dorsi. By its *deep surface*, with the Erector spinæ, ribs, and Intercostal muscles. Its upper margin is continuous with the vertebral aponeurosis.

The *Vertebral aponeurosis* is a thin, fibrous lamina, extending along the whole length of the back part of the thoracic region, serving to bind down the long Extensor muscles of the back which support the spine and head, and separate them from those muscles which connect the spine to the upper extremity. It consists of longitudinal and transverse fibres blended together, forming a thin lamella, which is attached in the median line to the spinous processes of the dorsal vertebræ; externally, to the angles of the ribs; and below, to the upper border of the Serratus posticus inferior and portion of the lumbar fascia, which gives origin to the Latissimus dorsi; above, it passes beneath the Serratus posticus superior and the Splenius, and blends with the deep fascia of the neck.

The *Lumbar fascia* or *aponeurosis* (fig. 406), which may be regarded as the posterior aponeurosis of the Transversalis abdominis muscle, consists of three laminæ, which are attached as follows: the posterior layer, to the spines of the lumbar and sacral vertebræ and the supraspinous ligament; the middle, to the tips of the transverse processes of the lumbar vertebræ and the intertransverse ligaments; the anterior, to the roots of the lumbar transverse processes. The posterior layer is continued above as the vertebral aponeurosis, while inferiorly it is fixed to the outer lip of the iliac crest. With this layer are blended the aponeurotic origin of the Serratus posticus inferior and part of that of the Latissimus dorsi. The middle layer is attached above to the last rib, and below to the iliac crest; the anterior layer is fixed below to the ilio-lumbar ligament and iliac crest ; while above it is thickened to form the external arcuate ligament of the Diaphragm, and stretches from the tip of the last rib to the transverse process of the first or second lumbar vertebra. These three layers, together with the vertebral column, enclose two spaces, the posterior of which is occupied by the Erector spinæ muscle, and the anterior by the Quadratus lumborum.

Now detach the Serratus posticus superior from its origin, and turn it outwards, when the Splenius muscle will be brought into view.

The **Splenius** is a broad sheet situated at the back of the neck and upper part of the dorsal region. At its origin, it is a single muscle, which arises, by tendinous fibres, from the lower half of the ligamentum nuchæ, from the spinous processes of the last cervical and of the six upper dorsal vertebræ, and from the supraspinous ligament. From this origin, the fleshy fibres proceed obliquely upwards and outwards, forming a broad flat muscle, which divides as it ascends into two portions, the Splenius capitis and Splenius colli.

The **Splenius capitis** is inserted, under cover of the Sterno-mastoid, into the mastoid process of the Temporal bone, and into the rough surface on the occipital bone just beneath the superior curved line.

The **Splenius colli** is inserted, by tendinous fasciculi, into the posterior tubercles of the transverse processes of the two or three upper cervical vertebræ.

The Splenius is separated from its fellow of the opposite side by a triangular interval, in which is seen the Complexus.

Relations.—By its *superficial surface*, with the Trapezius, from which it is separated below by the Rhomboidei and the Serratus posticus superior. It is covered at its insertion by the Sterno-mastoid, and at the lower and back part of the neck by the Levator anguli scapulæ. By its *deep surface*, with the Spinalis dorsi, Longissimus dorsi, Semispinalis colli, Complexus, Trachelo-mastoid, and Transversalis cervicis.

Nerves.—The Splenius is supplied from the external branches of the posterior divisions of the middle and lower cervical nerves ; the Serratus posticus superior is supplied by branches from the upper three or four intercostal nerves ; the Serratus posticus inferior, by branches from the ninth, tenth, and eleventh intercostal nerves.

Actions.—The Serrati are respiratory muscles. The Serratus posticus superior elevates the ribs ; it is therefore an inspiratory muscle ; while the Serratus posticus inferior draws the lower ribs downwards and backwards, and thus elongates the thorax. It also fixes the lower ribs, thus assisting the inspiratory action of the Diaphragm and resisting the tendency which it has to draw the lower ribs upwards and forwards. It must therefore be regarded as a muscle of inspiration. This muscle is also probably a tensor of the vertebral aponeurosis. The Splenii muscles of the two sides, acting together, draw the head directly backwards, assisting the Trapezius and Complexus ; acting separately, they draw the head to one or the other side, and slightly rotate it, turning the face to the same side. They also assist in supporting the head in the erect position.

FOURTH LAYER

1. Erector spinæ.

a. Outer Column.	*b. Middle Column.*
Ilio-costalis.	Longissimus dorsi.
Musculus accessorius.	Transversalis cervicis.
Cervicalis ascendens.	Trachelo-mastoid.

c. Inner Column.
Spinalis dorsi.

2. Complexus.

Dissection.—To expose the muscles of the fourth layer, remove entirely the Serrati and the vertebral and lumbar fasciæ. Then detach the Splenius by separating its attachment to the spinous processes and reflecting it outwards.

The **Erector spinæ** (fig. 407), and its prolongations in the dorsal and cervical regions, fill up the vertebral groove on each side of the spine. It is covered in

FIG. 407.—Muscles of the back. Deep layers.

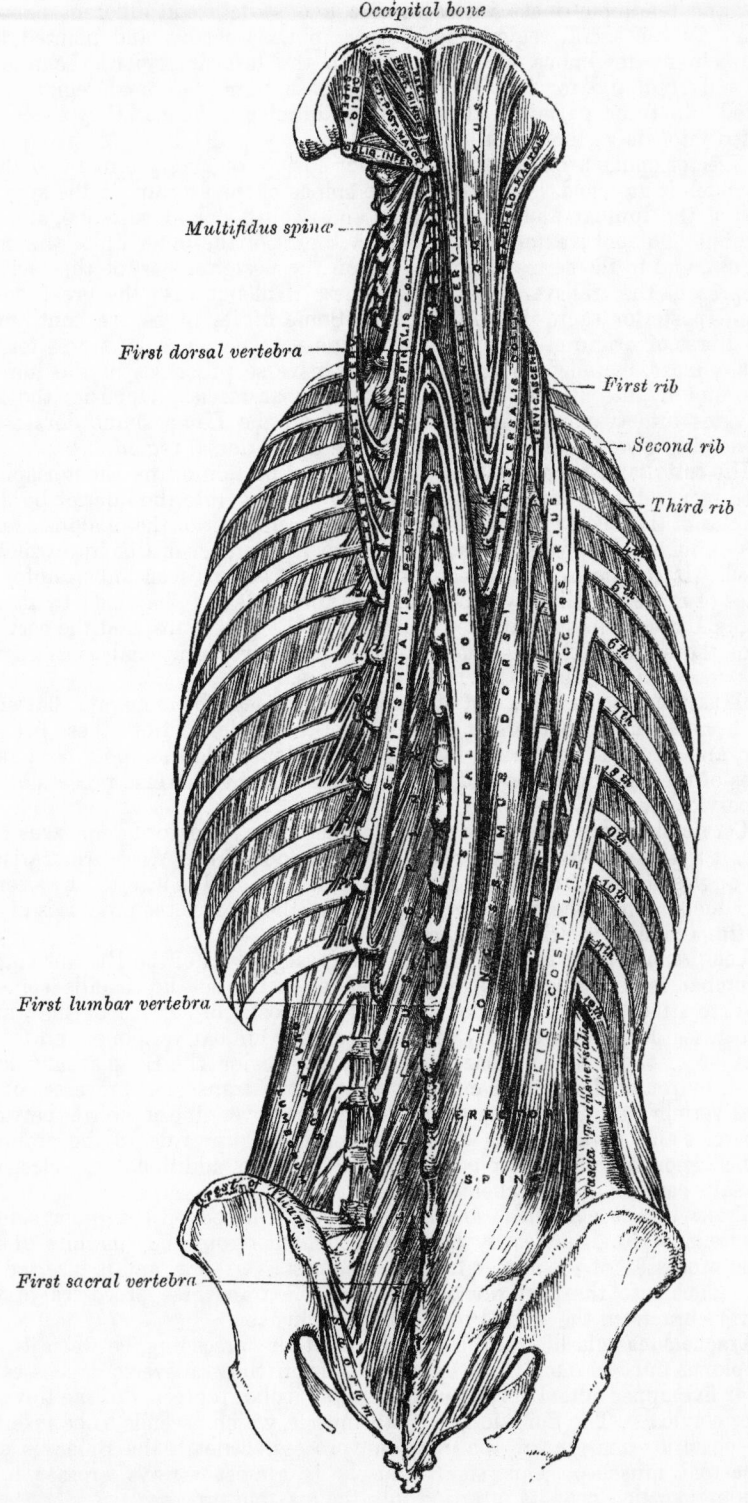

the lumbar region by the lumbar fascia; in the dorsal region by the Serrati muscles and the vertebral aponeurosis; and in the cervical region by a layer of cervical fascia continued beneath the Trapezius and the Splenius. This large muscular and tendinous mass varies in size and structure at different parts of the spine. In the sacral region, the Erector spinæ is narrow and pointed, and its origin chiefly tendinous in structure. In the lumbar region, the muscle becomes enlarged, and forms a thick fleshy mass. In the dorsal region, it is subdivided into three parts, which gradually diminish in size as they ascend to be inserted into the vertebræ and ribs.

The Erector spinæ arises from the anterior surface of a very broad and thick tendon, which is attached, internally, to the spines of the sacrum, to the spinous processes of the lumbar and the eleventh and twelfth dorsal vertebræ, and the supraspinous ligament; externally, to the back part of the inner lip of the crest of the ilium, and to the series of eminences on the posterior part of the sacrum, which represent the transverse processes, where it blends with the great sacro-sciatic and posterior sacro-iliac ligaments. Some of its fibres are continuous with the fibres of origin of the Gluteus maximus. The muscular fibres form a large fleshy mass, bounded in front by the transverse processes of the lumbar vertebræ, and by the middle lamella of the lumbar fascia. Opposite the last rib it divides into two parts, the Ilio-costalis and the Longissimus dorsi; the Spinalis dorsi is given off from the latter in the upper dorsal region.

The **Ilio-costalis** or **Sacro-lumbalis**, the external portion of the Erector spinæ, is inserted, generally, by six or seven flattened tendons, into the inferior borders of the angles of the six or seven lower ribs. The number of the tendons of this muscle is, however, very variable, and therefore the number of ribs into which it is inserted. If this muscle is reflected outwards, it will be seen to be reinforced by a series of muscular slips, which arise from the angles of the ribs; by means of these the Ilio-costalis is continued upwards to the upper ribs, and the cervical portion of the spine. The accessory portions form two additional muscles, the Musculus accessorius and the Cervicalis ascendens.

The **Musculus accessorius ad ilio-costalem** arises, by separate flattened tendons, from the upper borders of the angles of the six lower ribs: these become muscular, and are finally inserted, by separate tendons, into the upper borders of the angles of the six upper ribs and into the back of the transverse process of the seventh cervical vertebra.

The **Cervicalis ascendens** is the continuation of the Accessorius upwards into the neck; it is situated on the inner side of the tendons of the Accessorius, arising from the angles of the third, fourth, fifth, and sixth ribs, and is inserted by a series of slender tendons into the posterior tubercles of the transverse processes of the fourth, fifth, and sixth cervical vertebræ.

The **Longissimus dorsi** is the middle and largest portion of the Erector spinæ. In the lumbar region, where it is as yet blended with the Ilio-costalis, some of the fibres are attached to the whole length of the posterior surface of the transverse processes and the accessory processes of the lumbar vertebræ, and to the middle layer of the lumbar fascia. In the dorsal region, the Longissimus dorsi is inserted, by rounded tendons, into the tips of the transverse processes of all the dorsal vertebræ, and by fleshy processes into the lower nine or ten ribs between their tubercles and angles. This muscle is continued upwards, to the cranium and cervical portion of the spine, by means of two additional muscles, the Transversalis cervicis and Trachelo-mastoid.

The **Transversalis cervicis** (*transversalis capitis*), placed on the inner side of the Longissimus dorsi, arises by long thin tendons from the summits of the transverse processes of the upper four or five dorsal vertebræ, and is inserted by similar tendons into the posterior tubercles of the transverse processes of the cervical vertebræ from the second to the sixth inclusive.

The **Trachelo-mastoid** lies on the inner side of the preceding, between it and the Complexus muscle. It arises, by tendons, from the transverse processes of the four or five upper dorsal vertebræ, and the articular processes of the three or four lower cervical. The fibres form a small muscle, which ascends to be inserted into the posterior margin of the mastoid process, beneath the Splenius and Sterno-mastoid muscles. This small muscle is almost always crossed by a tendinous intersection near its insertion into the mastoid process.

The **Spinalis dorsi** is situated at the inner side of the Longissimus dorsi, with which it is intimately blended, and arises by three or four tendons, from the spinous processes of the first two lumbar and the last two dorsal vertebræ : these, uniting, form a small muscle, which is inserted, by separate tendons, into the spinous processes of the dorsal vertebræ, the number varying from four to eight. It is intimately united with the Semispinalis dorsi, which lies beneath it.

The *Spinalis colli* is an inconstant muscle, which arises from the lower part of the ligamentum nuchæ, the spine of the seventh cervical, and sometimes from those of the first and second dorsal vertebræ, and is inserted into the spinous process of the axis, and occasionally into the spinous processes of the two vertebræ below it. This muscle was found absent in five cases out of twenty-four.

Relations.—The Erector spinæ and its prolongations are bound down to the vertebræ and ribs in the lumbar and dorsal regions by the lumbar fascia and the vertebral aponeurosis. The inner part of these muscles covers the muscles of the fifth layer. In the neck, they are in relation, by their *superficial surface*, with the Trapezius and Splenius ; by their *deep surface*, with the Semispinalis dorsi and colli, and the Recti and Obliqui.

The **Complexus** is a broad thick muscle, situated at the upper and back part of the neck, beneath the Splenius, and internal to the Transversalis cervicis and Trachelo-mastoid. It arises, by a series of tendons, from the tips of the transverse processes of the upper six or seven dorsal and the last cervical vertebræ, and from the articular processes of the three cervical above this. The tendons, uniting, form a broad muscle, which passes obliquely upwards and inwards, and is inserted into the innermost depression between the two curved lines of the occipital bone. This muscle, about its middle, is traversed by an imperfect tendinous intersection. The term *Biventer cervicis* is given to the inner portion of the Complexus ; this portion is usually separated from the rest of the muscle, and consists of two fleshy bellies connected by an intervening tendon.

Relations.—The Complexus is covered by the Splenius and the Trapezius. It lies on the Rectus capitis posticus major and minor, the Obliquus capitis superior and inferior, and on the Semispinalis colli, from which it is separated by the profunda cervicis artery, the princeps cervicis artery, and branches of the posterior primary divisions of the cervical nerves. The Biventer cervicis is separated from its fellow of the opposite side by the ligamentum nuchæ.

FIFTH LAYER

The fifth layer, or rather group, of muscles comprises the

Semispinalis dorsi.	Extensor coccygis.
Semispinalis colli.	Intertransversales.
Multifidus spinæ.	Rectus capitis posticus major.
Rotatores spinæ.	Rectus capitis posticus minor.
Interspinales.	Obliquus capitis inferior.

Obliquus capitis superior.

Dissection.—Remove the muscles of the preceding layer by dividing and turning aside the Complexus ; then detach the Spinalis and Longissimus dorsi from their attachments, divide the Erector spinæ at its connection below to the sacral and lumbar spines, and turn it outwards. The muscles filling up the interval between the spinous and transverse processes are then exposed.

The **Semispinalis dorsi** (fig. 407) consists of thin, narrow, fleshy fasciculi, interposed between tendons of considerable length. It arises by a series of small tendons from the transverse processes of the lower dorsal vertebræ, from the sixth to the tenth inclusive ; and is inserted, by five or six tendons, into the spinous processes of the upper four dorsal and lower two cervical vertebræ.

The **Semispinalis colli**, thicker than the preceding, arises by a series of tendinous and fleshy fibres from the transverse processes of the upper five or six dorsal vertebræ, and is inserted into the cervical spinous processes, from the axis to the fifth inclusive. The fasciculus connected with the axis is the largest and is chiefly muscular in structure.

Relations.—By their *superficial surface*, from below upwards, with the Spinalis dorsi, Longissimus dorsi, Splenius, Complexus, the profunda cervicis artery, the princeps cervicis artery, and the internal branches of the posterior divisions of the first, second, and third cervical nerves. By their *deep surface*, with the Multifidus spinæ.

The **Multifidus spinæ** consists of a number of fleshy and tendinous fasciculi, which fill up the groove on either side of the spinous processes of the vertebræ, from the sacrum to the axis. In the sacral region, these fasciculi arise from the back of the sacrum, as low as the fourth sacral foramen, and from the aponeurosis of origin of the Erector spinæ; from the inner surface of the posterior superior spine of the ilium, and posterior sacro-iliac ligaments; in the lumbar regions, from the mammillary processes; in the dorsal region, from the transverse processes; and in the cervical region, from the articular processes of the four lower vertebræ. Each fasciculus, passing obliquely upwards and inwards, is inserted into the whole length of the ·spinous process of one of the vertebræ above. These fasciculi vary in length: the most superficial, ·the longest, pass from one vertebra to the third or fourth above; those next in order pass from one vertebra to the second or third above; while the deepest connect two contiguous vertebræ.

Relations.—By its *superficial surface*, with the Longissimus dorsi, Spinalis dorsi, Semispinalis dorsi, and Semispinalis colli. By its *deep surface*, with the laminæ and spinous processes of the vertebræ, and with the Rotatores spinæ in the dorsal region.

The **Rotatores spinæ** are found only in the dorsal region of the spine, beneath the Multifidus spinæ; they are eleven in number on each side. Each muscle is small and somewhat quadrilateral in form; it arises from the upper and back part of the transverse process, and is inserted into the lower border and outer surface of the lamina of the vertebra above, the fibres extending as far inwards as the root of the spinous process. The first is found between the first and second dorsal; the last, between the eleventh and twelfth. Sometimes the number of these muscles is diminished by the absence of one or more from the upper or lower end.

The **Interspinales** are short muscular fasciculi, placed in pairs between the spinous processes of the contiguous vertebræ, one on each side of the interspinous ligament. In the *cervical region*, they are most distinct, and consist of six pairs, the first being situated between the axis and third ·vertebra, and the last between the last cervical and the first dorsal. They are small narrow bundles, attached, above and below, to the apices of the spinous processes. In the *dorsal region*, they are found between the first and second vertebræ, and sometimes between the second and third; and below, between the eleventh and twelfth. In the *lumbar region*, there are four pairs of these muscles in the intervals between the five lumbar vertebræ. There is also occasionally one in the interspinous space, between the last dorsal and first lumbar, and between the fifth lumbar and the sacrum.

The **Extensor coccygis** is a slender muscular fasciculus, which is not always present; it extends over the lower part of the posterior surface of the sacrum and coccyx. It arises by tendinous fibres from the last bone of the sacrum, or first piece of the coccyx, and passes downwards to be inserted into the lower part of the coccyx. It is a rudiment of the Extensor muscle of the caudal vertebræ of the lower animals.

The **Intertransversales** are small muscles placed between the transverse processes of the vertebræ. In the *cervical region* they are most developed, consisting of rounded muscular and tendinous fasciculi, which are placed in pairs, passing between the anterior and the posterior tubercles respectively of the transverse processes of two contiguous vertebræ, separated from one another by the anterior division of the cervical nerve, which lies in the groove between them. In this region there are seven pairs of these muscles, the first pair being between the atlas and axis, and the last pair between the seventh cervical and first dorsal vertebræ. In the *dorsal region* they are least developed, consisting chiefly of rounded tendinous cords in the intertransverse spaces of the upper three dorsal vertebræ; but between the transverse processes of the lower three dorsal vertebræ, and between the transverse processes of the last dorsal and the first lumbar, they are muscular in structure. In the *lumbar region* they are arranged in pairs, on either side of the spine; one set occupying the entire interspace between the transverse processes of the lumbar vertebræ, the *intertransversales* *laterales*; the other set, *intertransversales mediales*, passing from the accessory process of one vertebra to the mammillary process of the vertebra below.*

The **Rectus capitis posticus major** arises by a pointed tendinous origin from the spinous process of the axis, and, becoming broader as it ascends, is inserted into the outer part of the inferior curved line of the occipital bone and the surface of bone immediately below it. As the muscles of the two sides pass upwards and outwards, they leave between them a triangular space, in which are seen the Recti capitis postici minores muscles.

Relations.—By its *superficial surface*, with the Complexus; and, at its insertion, with the Superior oblique. By its *deep surface*, with part of the Rectus capitis posticus minor, the posterior arch of the atlas, the posterior occipito-atlantal ligament, and part of the occipital bone.

The **Rectus capitis posticus minor,** the smallest of the four muscles in this region, is of a triangular shape; it arises by a narrow pointed tendon from the tubercle on the posterior arch of the atlas, and, becoming broader as it ascends, is inserted into the inner part of the inferior curved line of the occipital bone and the surface between it and the foramen magnum.

Relations.—By its *superficial surface*, with the Complexus and the Rectus capitis posticus major. By its *deep surface*, with the posterior occipito-atlantal ligament.

The **Obliquus capitis inferior,** the larger of the two Oblique muscles, arises from the apex of the spinous process of the axis, and passes outwards and slightly upwards, to be inserted into the lower and back part of the transverse process of the atlas.

Relations.—By its *superficial surface*, with the Complexus and with the posterior division of the second cervical nerve which turns round its lower border. By its *deep surface*, with the vertebral artery, and posterior atlanto-axial ligament.

The **Obliquus capitis superior,** narrow below, wide and expanded above, arises by tendinous fibres from the upper surface of the transverse process of the atlas, joining with the insertion of the preceding, and, passing obliquely upwards and inwards, is inserted into the occipital bone, between the two curved lines, external to the Complexus.

Relations.—By its *superficial surface*, with the Complexus and Trachelo-mastoid and occipital artery. By its *deep surface*, with the posterior occipito-atlantal ligament.

The **Suboccipital triangle.**—Between the two oblique muscles and the Rectus capitis posticus major a triangular interval exists, the *suboccipital triangle*. This triangle is bounded, above and internally, by the Rectus capitis posticus major; above and externally, by the Obliquus capitis superior; below and externally, by the Obliquus capitis inferior. It is covered in by a layer of dense fibro-fatty tissue, situated beneath the Complexus muscle. The floor is formed by the posterior occipito-atlantal ligament and the posterior arch of the atlas. It contains the vertebral artery, as it runs in a deep groove on the upper surface of the posterior arch of the atlas; and the posterior division of the suboccipital nerve.

Nerves.—The fourth and fifth layers of the muscles of the back are supplied by the posterior primary divisions of the spinal nerves.

Actions.—When both the Spinales dorsi contract, they extend the dorsal region of the spine; when only one contracts, it helps to bend the dorsal portion of the spine to one side. The Erector spinæ, comprising the Ilio-costalis and the Longissimus dorsi with their accessory muscles, serves, as its name implies, to maintain the spine in the erect posture; it also serves to bend the trunk backwards when it is required to counterbalance the influence of any weight at the front of the body—as, for instance, when a heavy weight is suspended from the neck, or when there is any great abdominal distension, as in pregnancy or dropsy; the peculiar gait under such circumstances depends upon the spine being drawn backwards, by the counterbalancing action of the Erector spinæ muscles. The muscles which form the continuation of the Erector spinæ upwards steady the head and neck, and fix them in the upright position. If the

---

* The student is referred to an article on the morphology of the human intertransverse muscles, by J. Dunlop Lickley; *Journal of Anatomy and Physiology*, vol. xxxix. 1904.

Ilio-costalis and Longissimus dorsi of one side act, they serve to draw down the chest and spine to the corresponding side. The Cervicales ascendentes, taking their fixed points from the cervical vertebræ, elevate those ribs to which they are attached; taking their fixed points from the ribs, both muscles help to extend the neck; while one muscle bends the neck to its own side. The Transversales cervicis, when both muscles act, taking their fixed points from below, bend the neck backwards. The Trachelo-mastoid, when both muscles act, taking their fixed points from below, bend the head backwards; while, if only one muscle acts, the face is turned to the side on which the muscle is acting, and then the head is bent to the shoulder. The two Recti muscles draw the head backwards. The Rectus capitis posticus major, owing to its obliquity, rotates the cranium, with the atlas, round the odontoid process, turning the face to the same side. The Multifidus spinæ acts successively upon the different parts of the spine: thus, the sacrum furnishes a fixed point from which the fasciculi of this muscle act upon the lumbar region; these then become the fixed points for the fasciculi moving the dorsal region, and so on throughout the entire length of the spine; it is by the successive contraction and relaxation of the separate fasciculi of this and other muscles that the spine preserves the erect posture without the fatigue that would necessarily have been produced had this position been maintained by the action of a single muscle. The Multifidus spinæ, besides preserving the erect position of the spine, serves to rotate it, so that the front of the trunk is turned to the side opposite to that from which the muscle acts, this muscle being assisted in its action by the Obliquus externus abdominis. The Complexi draw the head directly backwards; if one muscle acts, it draws the head to one side, and rotates it so that the face is turned to the opposite side. The Superior oblique draws the head backwards; and, from the obliquity in the direction of its fibres, will slightly rotate the cranium, turning the face to the opposite side. The Obliquus capitis inferior rotates the atlas, and with it the cranium, round the odontoid process, turning the face to the same side. The Semispinales, when the muscles of the two sides act together, help to extend the spine; when the muscles of one side only act, they rotate the dorsal and cervical parts of the spine, turning the body to the opposite side. The Interspinales by approximating the spinous process help to extend the spine. The Intertransversales approximate the transverse processes, and help to bend the spine to one side. The Rotatores spinæ assist the Multifidus spinæ to rotate the spine, so that the front of the trunk is turned to the side opposite to that from which the muscle acts.

Surface Form.—The surface forms produced by the muscles of the back are numerous and difficult to analyse unless they are considered in systematic order. The most superficial layer, consisting of large strata of muscular substance, influences to a certain extent the surface form, and at the same time reveals the forms of the layers beneath. The *Trapezius* at the upper part of the back, and in the neck, covers over and softens down the outline of the underlying muscles. Its anterior border forms the posterior boundary of the posterior triangle of the neck. It presents a slight undulating ridge which passes downwards and forwards from the occiput to the junction of the middle and outer third of the clavicle. The tendinous ellipse formed by a part of the origin of the two muscles at the back of the neck is always to be seen as an oval depression, more marked when the muscle is in action. A slight dimple on the skin opposite the interval between the spinous processes of the third and fourth dorsal vertebræ marks the triangular aponeurosis by which the inferior fibres are inserted into the root of the spine of the scapula. From this point the inferior border of the muscle may be traced as an undulating ridge to the spinous process of the twelfth dorsal vertebra. In like manner, the *Latissimus dorsi* softens down and modulates the underlying structures at the lower part of the back and lower part of the side of the chest. In this way it modulates the outline of the Erector spinæ; of the Serratus posticus inferior, which is sometimes to be discerned through it, and is sometimes entirely obscured by it; of part of the Serratus magnus and Superior oblique, which it covers, and of the convex oblique ridges formed by the ribs with the intervening intercostal spaces. The anterior border of the muscle is the only part which gives a distinct surface form. This border may be traced, when the muscle is in action, as a rounded edge, starting from the crest of the ilium, and passing obliquely forwards and upwards to the posterior border of the axilla, where it combines with the Teres major in forming a thick rounded fold, the posterior boundary of the axillary space. The muscles in the second layer influence to a very considerable extent the surface form of the back of the neck and upper part of the trunk. The *Levator anguli scapulæ* reveals itself as a prominent divergent line, running downwards and outwards, from the transverse processes of the upper cervical vertebræ to the angle of the scapula, covered over and toned down by the overlying Trapezius. The *Rhomboïdei* produce, when in action, a vertical eminence between the vertebral border of the scapula and the spinal furrow, varying in intensity according to the condition of contraction or relaxation of the Trapezius muscle, by which they are for the most part covered. The lowermost part of the Rhomboideus major is uncovered by the Trapezius and forms on the surface an oblique ridge running upwards and inwards from the inferior angle of the scapula. Of the muscles of the third layer of the back, the *Serratus posticus superior* does not in any way influence surface form. The *Serratus posticus inferior,* when in strong action, may occasionally be revealed as an elevation beneath the Latissimus dorsi. The *Splenii* by their divergence serve to broaden out the upper part of the back of the neck and produce a local fulness in this situation, but do not otherwise influence surface form. Beneath all these muscles those of the fourth layer—the *Erector spinæ* and its continuations—influence the surface form in a decided manner. In the loins, the Erector spinæ, bound down by the lumbar fascia, forms a rounded vertical eminence, which determines the depth of the spinal furrow, and which below tapers to a point on the posterior surface of the sacrum and becomes lost there. In the back it forms a flattened plane which gradually becomes lost. In the neck the only part of this group of muscles which influences surface form is the *Trachelo-mastoid,* which produces a short convergent line across the upper part of the posterior triangle of the neck, appearing from under cover of the posterior border of the Sterno-mastoid and being lost below beneath the Trapezius.

II. Muscles and Fasciæ of the Thorax

The muscles belonging exclusively to this region are few in number. They are, the

Intercostales externi.	Triangularis sterni.
Intercostales interni.	Levatores costarum.
Infracostales.	Diaphragm.

Intercostal fasciæ.—A thin but firm layer of fascia covers the outer surface of the External intercostal and the inner surface of the Internal intercostal muscles; and a third layer, the *middle intercostal fascia,* more delicate, is interposed between the two planes of muscular fibres. These are the intercostal fasciæ: they are best marked in those situations where the muscular fibres are deficient, as between the External intercostal muscles and sternum, in front; and between the Internal intercostals and spine, behind.

The **Intercostal muscles** (fig. 422) are two thin planes of muscular and tendinous fibres, placed one over the other, filling up the intercostal spaces, and being directed obliquely between the margins of the adjacent ribs. They have received the name 'external' and 'internal,' from the position they bear to one another. The tendinous fibres are longer and more numerous than the muscular: hence the walls of the intercostal spaces possess very considerable strength, to which the crossing of the muscular fibres materially contributes.

The **External intercostals** are eleven in number on each side. They extend from the tubercles of the ribs, behind, to the commencement of the cartilages of the ribs, in front, where they terminate in a thin membrane, the *anterior intercostal membrane,* which is continued forwards to the sternum. They arise from the lower border of each rib, and are inserted into the upper border of the rib below. In the two lowest spaces they extend to the end of the cartilages, and in the upper two or three spaces they do not quite extend to the ends of the ribs. Their fibres are directed obliquely downwards and forwards, in a similar direction to those of the External oblique muscle of the abdomen. They are thicker than the Internal intercostals.

Relations.—By their *outer surface,* with the muscles which immediately invest the chest, viz. the Pectoralis major and minor, Serratus magnus, and Rhomboideus major, Serratus posticus superior and inferior, Scalenus posticus, Ílio-costalis, Longissimus dorsi, Cervicalis ascendens, Transversalis cervicis, Levatores costarum, Obliquus externus abdominis, and the Latissimus dorsi. By their *internal surface,* with the intercostal vessels and nerve, the Internal intercostal muscles, and the posterior intercostal membrane.

The **Internal intercostals** are also eleven in number on each side. They commence anteriorly at the sternum, in the interspaces between the cartilages of the true ribs, and from the anterior extremities of the cartilages of the false ribs, and extend backwards as far as the angles of the ribs; whence they are continued to the vertebral column by a thin aponeurosis, the *posterior intercostal membrane*. They arise from the ridge on the inner surface of each rib, as well as from the corresponding costal cartilage, and are inserted into the upper border of the rib below. Their fibres are directed obliquely downwards and backwards, passing in the opposite direction to the fibres of the External intercostal muscles.

Relations.—By their *external surface*, with the intercostal vessels and nerves, and the External intercostal muscles; near the sternum, with the anterior intercostal membrane, and the Pectoralis major. By their *internal surface*, with the pleura costalis, Triangularis sterni, and Diaphragm.

The **Infracostales (subcostales)** consist of muscular and aponeurotic fasciculi, which vary in number and length: they are placed on the inner surface of the ribs, where the Internal intercostal muscles cease; they arise from the inner surface of one rib, and are inserted into the inner surface of the first, second, or third rib below. Their fibres run in the same direction as those of the Internal intercostals. They are most frequent between the lower ribs.

FIG. 408.—Posterior surface of sternum and costal cartilages, showing Triangularis sterni muscle. (From a preparation in the Museum of the Royal College of Surgeons of England.)

The **Triangularis sterni** (fig. 408) is a thin plane of muscular and tendinous fibres, situated upon the inner wall of the front of the chest. It arises from the lower third of the posterior surface of the sternum, from the posterior surface of the ensiform cartilage, and from the sternal ends of the costal cartilages of the three or four lower true ribs. Its fibres diverge upwards and outwards, to be inserted by digitations into the lower border and inner surfaces of the costal cartilages of the second, third, fourth, fifth, and sixth ribs. The lowest fibres of this muscle are horizontal in their direction, and are continuous with those of the Transversalis: those which succeed are oblique, while the superior fibres are almost vertical. This muscle varies much in its attachment, not only in different bodies, but on opposite sides of the same body.

Relations.—*In front*, with the sternum, ensiform cartilage, costal cartilages, Internal intercostal muscles, and internal mammary vessels. *Behind*, with the pleura, pericardium, and anterior mediastinum.

The **Levatores costarum** (fig. 407), twelve in number on each side, are small tendinous and fleshy bundles, which arise from the extremities of the transverse processes of the seventh cervical and eleven upper dorsal vertebræ, and, passing obliquely downwards and outwards like the fibres of the External intercostals, are inserted into the upper border of the rib below them, between the tubercle and the angle. The inferior Levatores divide into two fasciculi, one of which is inserted as above described; the other fasciculus passes down to the second rib below its origin; thus, each of the lower ribs receives fibres from the transverse processes of two vertebræ.

Nerves.—The muscles of this group are supplied by the intercostal nerves.

The **Diaphragm** (διάφραγμα, *a partition wall*) (fig. 409) is a dome-shaped musculo-fibrous septum, consisting of muscular fibres externally, which arise from the circumference of the thoracic cavity and pass upwards and inwards to converge to a central tendon. It separates the thorax from the abdomen, forming the floor of the former cavity and the roof of the latter. It arises from the whole of the internal circumference of the thorax: being attached, in front, by two fleshy slips to the ensiform cartilage; on either side, to the inner surface of the cartilages and bony portions of the six inferior ribs, interdigitating with the Transversalis abdominis; and behind, to two aponeurotic arches, named the *ligamenta arcuata*; and by the *crura*, to the lumbar vertebræ. The fibres from these sources vary in length; those arising from the ensiform appendix are very

FIG. 409.—The Diaphragm. Under surface.

short and occasionally aponeurotic; those from the ligamenta arcuata, and more especially those from the cartilages of the ribs at the side of the chest, are longer, describe well-marked curves as they ascend, and finally converge to be inserted into the circumference of the central tendon. Between the origin from the ensiform appendix and that from the cartilages of the adjoining ribs, the fibres of the Diaphragm are deficient, the interval being filled by areolar tissue; this interval transmits the superior epigastric branch of the internal mammary artery, covered on the thoracic side by the pleuræ; on the abdominal, by the peritoneum. A triangular gap is sometimes seen between the fibres springing from the internal and those arising from the external arcuate ligament. When it exists the kidney is only separated from the pleura by fatty and areolar tissue.

The *ligamentum arcuatum internum* is a tendinous arch, thrown across the upper part of the Psoas magnus muscle, on each side of the spine. It is connected, by one end, to the outer side of the body of the first or second lumbar vertebra, being continuous with the outer side of the tendon of the corresponding crus ; and, by the other end, to the front of the transverse process of the first, and sometimes also to that of the second, lumbar vertebra.

The *ligamentum arcuatum externum* is the thickened upper margin of the anterior lamella of the lumbar fascia ; it arches across the upper part of the Quadratus lumborum, being attached, by one extremity, to the front of the transverse process of the first lumbar vertebra ; and, by the other, to the apex and lower margin of the last rib.

The Crura.—The Diaphragm is connected to the spine by two *crura* or *pillars*, which are situated on the bodies of the lumbar vertebræ, on each side of the aorta. The crura, at their origin, are tendinous in structure : the right crus, larger and longer than the left, arising from the anterior surface of the bodies and intervertebral substances of the three upper lumbar vertebræ ; the left, from the two upper ; both blending with the anterior common ligament of the spine. These tendinous portions of the crura pass forwards and inwards and gradually converge to meet in the middle line, forming an arch, the *ligamentum arcuatum medium,* beneath which pass the aorta, vena azygos major, and thoracic duct. From this tendinous arch muscular fibres arise, which diverge, the outermost being directed upwards and outwards to the central tendon ; the innermost decussating in front of the aorta, and then diverging, so as to surround the œsophagus before ending in the central tendon. The fibres derived from the right crus are the most numerous, and pass in front of those derived from the left.

The *Central* or *Trefoil Tendon* of the Diaphragm is a thin but strong tendinous aponeurosis, situated near the centre of the vault formed by the muscle, but somewhat closer to the front of the chest than the back, so that the posterior muscular fibres are the longer. It is situated immediately below the pericardium, with which it is partly blended. It is shaped somewhat like a trefoil leaf, consisting of three divisions, or leaflets, separated from one another by slight indentations. The right leaflet is the largest ; the middle one, directed towards the ensiform cartilage, the next in size ; and the left, the smallest. In structure, the tendon is composed of several planes of fibres, which intersect one another at various angles, and unite into straight or curved bundles—an arrangement which affords it additional strength.

The *Openings* in the Diaphragm consist of three large and several smaller apertures. The former are the aortic, the œsophageal, and the vena caval.

The *aortic opening* is the lowest and the most posterior of the three large apertures connected with this muscle, being at the level of the twelfth dorsal vertebra. It is situated slightly to the left of the middle line, immediately in front of the spinal column, and is, therefore, *behind* the Diaphragm. It is an osseo-aponeurotic aperture, formed by the ligamentum arcuatum medium thrown across the front of the bodies of the vertebræ, from the crus on one side to that on the other, and transmits the aorta, vena azygos major, and thoracic duct. Sometimes the vena azygos major is transmitted upwards through the right crus. Occasionally some tendinous fibres are prolonged across the bodies of the vertebræ from the inner part of the lower end of the crura, passing behind the aorta, and thus converting the opening into a fibrous ring.

The *œsophageal opening* is situated at the level of the tenth dorsal vertebra ; it is elliptical in form, muscular in structure, and formed by the decussating fibres of the two crura, which act as a sort of sphincter. It is placed above, in front, and a little to the left of the preceding. It transmits the œsophagus and pneumogastric nerves and some small œsophageal arteries. The anterior margin of this aperture is occasionally tendinous, being formed by the margin of the central tendon.

The *opening for the vena cava (foramen quadratum)* is the highest, and is situated about the level of the disc between the eighth and ninth dorsal vertebræ ; it is quadrilateral in form, tendinous in structure, and placed at the junction of the right and middle leaflets of the central tendon, its margins being adherent to the wall of the inferior vena cava.

The *right crus* transmits the greater and lesser splanchnic nerves of the right side ; the *left crus* transmits the greater and lesser splanchnic nerves of the left side, and the vena azygos minor. The gangliated cords of the sympathetic usually enter the abdominal cavity by passing behind the internal arcuate ligaments. The superior epigastric branch of the internal mammary artery, and some lymphatics from the convex surface of the liver and the abdominal wall, pass through the triangular interspace behind the ensiform cartilage, where the muscular fibres of the Diaphragm are deficient.

The *Serous Membranes* in relation with the Diaphragm are four in number: three lining its upper or thoracic surface; one, its abdominal. The three serous membranes on its upper surface are the pleura on either side, and the serous layer of the pericardium, which covers the middle portion of the tendinous centre. The serous membrane covering the greater part of its under surface is a portion of the general peritoneal membrane of the abdominal cavity.

The Diaphragm is arched, being convex towards the chest, and concave to the abdomen. The *right portion* is accurately moulded over the convex surface of the liver; resting upon its upper surface is the concave base of the right lung. The *left portion* is lower than the right by about three-quarters of an inch. It supports the base of the left lung, and covers the great end of the stomach, the spleen, and left kidney. The *central portion*, which supports the heart, is on a slightly lower level than the two lateral portions.

Nerves.—The Diaphragm is supplied by the phrenic and lower intercostal nerves and the phrenic plexus of the sympathetic.

Actions.—The Intercostals are the chief agents in the movement of the ribs in ordinary respiration. When the first rib is elevated and fixed by the Scaleni, the External intercostals raise the other ribs, especially their fore part, and so increase the capacity of the chest from before backwards; at the same time they evert their lower borders, and so enlarge the thoracic cavity transversely. The Internal intercostals, at the side of the thorax, depress the ribs, and invert their lower borders, and so diminish the thoracic cavity; but at the fore part of the chest these muscles assist the External intercostals in raising the cartilages.* The Levatores costarum assist the External intercostals in raising the ribs. The Triangularis sterni draws down the costal cartilages; it is therefore an expiratory muscle.

The Diaphragm is the principal muscle of inspiration. When in a condition of rest the muscle presents a domed surface, concave towards the abdomen; and consists of a circumferential muscular and a central tendinous part. When the muscular fibres contract, they become less arched, or nearly straight, and thus cause the central tendon to descend, and in consequence the level of the floor of the chest is lowered, the vertical diameter of the chest being proportionately increased. In this descent the different parts of the tendon move unequally. The left leaflet descends to the greatest extent; the right to a less extent, on account of the liver; and the central leaflet the least, because of its connection to the pericardium. In descending, the Diaphragm presses on the abdominal viscera, and so to a certain extent causes a projection of the abdominal wall; but in consequence of these viscera not yielding completely, the central tendon becomes a fixed point, and enables the circumferential muscular fibres to act *from* it, and so elevate the lower ribs and expand the lower part of the thoracic cavity; and Duchenne has shown that the Diaphragm has the power of

---

* The view of the action of the Intercostal muscles given in the text is that which is taught by Hutchinson (*Cycl. of Anat. and Phys.* art. 'Thorax'), and is usually adopted in our schools. It is, however, much disputed. Hamberger believed that the External intercostals act as elevators of the ribs, or muscles of inspiration, while the Internal act in expiration. Haller taught that both sets of muscles act in common—viz. as muscles of inspiration—and this view is adopted by many of the best anatomists of the Continent, and appears supported by many observations made on the human subject under various conditions of disease, and on living animals after the muscles have been exposed under chloroform. The reader may consult an interesting paper by Cleland in the *Journal of Anat. and Phys.* No. II., May 1867, p. 209, 'On the Hutchinsonian Theory of the Action of the Intercostal Muscles.' He refers to Henle, Luschka, Budge, and Bäumler, *Observations on the Action of the Intercostal Muscles*, Erlangen, 1860. (In *New Syd. Soc.'s Year-Book* for 1861, p. 69.) W. W. Keen has come to the conclusion, from experiments made upon a criminal executed by hanging, that the External intercostals are muscles of expiration, as they pulled the ribs down, while the Internal intercostals pulled the ribs up, and are muscles of inspiration. (*Trans. Coll. Phys. Philadelphia*, Third Series, vol. i., 1875, p. 97.)

I I

elevating the ribs, to which it is attached, by its contraction, if the abdominal viscera are *in situ*, but that if these organs are removed, this power is lost. When at the end of inspiration, the Diaphragm relaxes, the thoracic walls return to their natural position in consequence of their elastic reaction and of the elasticity and weight of the displaced viscera.*

In all expulsive acts the Diaphragm is called into action, to give additional power to each expulsive effort. Thus, before sneezing, coughing, laughing, and crying; before vomiting; previous to the expulsion of the urine and fæces, or of the fœtus from the womb, a deep inspiration takes place.

The height of the Diaphragm is constantly varying during respiration; it also varies according to the degree of distension of the stomach and intestines, and the size of the liver. After a forced expiration, the right arch is on a level, in front, with the fourth costal cartilage; at the side, with the fifth, sixth, and seventh ribs; and behind, with the eighth rib; the left arch being usually from one to two ribs' breadth below the level of the right one. In a forced inspiration, it descends from one to two inches; its slope would then be represented by a line drawn from the ensiform cartilage towards the tenth rib.

Muscles of Inspiration and Expiration.—The muscles which assist the action of the Diaphragm in ordinary tranquil inspiration are the Intercostals and the Levatores costarum, as above stated, and the Scaleni. When the need for more forcible action exists, the shoulders and the base of the scapula are fixed, and then the powerful muscles of forced inspiration come into play; the chief of these are the Trapezius, the Pectoralis minor, the Serratus posticus superior and inferior, and the Rhomboidei. The lower fibres of the Serratus magnus may possibly assist slightly in dilating the chest by raising and everting the ribs. The Sterno-mastoid also, when the head is fixed, assists in forced inspiration, by drawing up the sternum, and by fixing the clavicle, and thus affording a fixed point for the action of the muscles of the chest. The Ilio-costalis and Quadratus lumborum assist in forced inspiration by fixing the last rib (see page 493).

The ordinary action of expiration is hardly effected by muscular force, but results from a return of the walls of the thorax to a condition of rest, owing to their elasticity and to that of the lungs. Forced expiratory actions are performed mainly by the flat muscles (Obliqui and Transversalis) of the abdomen, assisted also by the Rectus. Other muscles of forced expiration are the Internal intercostals and the Triangularis sterni (as above mentioned).

III. Muscles of the Abdomen

The muscles of the abdomen may be divided into two groups: 1. The superficial muscles of the abdomen; 2. The deep muscles of the abdomen.

1. Superficial Muscles

The muscles of this group are, the

Obliquus Externus.	Transversalis.
Obliquus Internus.	Rectus.

Pyramidalis.

Dissection (fig. 410).—To dissect the abdominal muscles, make a vertical incision from the ensiform cartilage to the symphysis pubis; a second incision from the umbilicus obliquely upwards and outwards to the outer surface of the chest, as high as the lower border of the fifth or sixth rib; and a third, commencing midway between the umbilicus and pubes, transversely outwards to the anterior superior iliac spine, and along the crest of the ilium as far as its posterior third. Then reflect the three flaps included between these incisions from within outwards, in the lines of direction of the muscular fibres. If necessary, the abdominal muscle may be made tense by inflating the peritoneal cavity through the umbilicus.

The **superficial fascia** of the abdomen consists over the greater part of the abdominal wall of a single layer of fascia, which contains a variable amount of fat; but as this layer approaches the groin it is easily divisible into two, between

* For a detailed description of the general relations of the Diaphragm, and its action, refer to Sibson's *Medical Anatomy*.

which are found the superficial vessels and nerves and the superficial inguinal lymphatic glands. The superficial layer (*fascia of Camper*) is thick, areolar in texture, containing adipose tissue in its meshes, the quantity of which varies in different subjects. Below, it passes over Poupart's ligament, and is continuous with the superficial fascia of the thigh. In the male, this fascia is continued over the penis and outer surface of the cord to the scrotum, where it helps to form the dartos. As it passes to the scrotum it changes its character, becoming thin, destitute of adipose tissue, and of a pale reddish colour, and in the scrotum it acquires some involuntary muscular fibres. From the scrotum it may be traced backwards to be continuous with the superficial fascia of the perinæum. In the female, this fascia is continued into the labia majora. The deeper layer (*fascia of Scarpa*) is thinner and more membranous in character than the superficial layer, and contains a considerable quantity of yellow elastic fibres. It is loosely connected by areolar tissue to the aponeurosis of the External oblique, but in the middle line it is more intimately adherent to the linea alba and to the symphysis pubis, and is prolonged on to the dorsum of the penis, forming the suspensory ligament; above, it is continuous with the superficial fascia over the rest of the trunk; below, it blends with the fascia lata of the thigh a little below Poupart's ligament; internally and below, it is continued over the penis and spermatic cord to the scrotum, where it helps to form the dartos. From the scrotum it may be traced backwards to be continuous with the deep layer of the superficial fascia of the perinæum (fascia of Colles). In the female, it is continued into the labia majora.

Fig. 410.—Dissection of abdomen.

3. *Dissection of inguinal hernia*

The **Obliquus externus abdominis** (fig. 411) is situated on the side and fore part of the abdomen; being the largest and the most superficial of the three flat muscles in this region. It is broad, thin, and irregularly quadrilateral, its muscular portion occupying the side, its aponeurosis the anterior wall of the abdomen. It arises, by eight fleshy digitations, from the external surfaces and lower borders of the eight inferior ribs; these digitations are arranged in an oblique line which runs downwards and backwards; the upper ones being attached close to the cartilages of the corresponding ribs; the lowest to the apex of the cartilage of the last rib; the intermediate ones, to the ribs at some distance from their cartilages. The five superior serrations increase in size from above downwards, and are received between corresponding processes of the Serratus magnus; the three lower ones diminish in size from above downwards, receiving between them corresponding processes from the Latissimus dorsi. From these attachments the fleshy fibres proceed in various directions. Those from the lowest ribs pass nearly vertically downwards, to be inserted into the anterior half of the outer lip of the crest of the ilium; the middle and upper fibres, directed downwards and forwards, terminate in an aponeurosis, opposite a line drawn from the prominence of the ninth costal cartilage to the anterior superior spinous process of the ilium.

The **Aponeurosis of the External oblique** is a thin but strong membranous structure, the fibres of which are directed obliquely downwards and inwards. It is joined with that of the opposite muscle along the median line, and covers the whole of the front of the abdomen; above, it is connected with the lower border of the Pectoralis major; below, its fibres are closely aggregated together, and extend obliquely across from the anterior superior spine of the ilium to the spine of the os pubis and the linea ilio-pectinea. In the median line, it interlaces with the aponeurosis of the opposite muscle, forming the linea alba, which extends from the ensiform cartilage to the symphysis pubis.

That portion of the aponeurosis which extends between the anterior superior spine of the ilium and the spine of the os pubis is a thick band, folded inwards, and continuous below with the fascia lata ; it is called *Poupart's ligament*. The portion which is reflected from Poupart's ligament at the spine of the os pubis along the pectineal line is called *Gimbernat's ligament*. From the point of attachment of the latter to the pectineal line, a few fibres pass upwards and

FIG. 411.—The External oblique muscle.

External
abdominal ring
Gimbernat's
ligament

Pubes

inwards, behind the inner pillar of the ring, to the linea alba. They diverge as they ascend, and form a thin triangular fibrous layer, which is called the *triangular fascia*.

In the aponeurosis of the External oblique, immediately above the crest of the os pubis, is a triangular opening, the *external abdominal ring*, formed by a separation of the fibres of the aponeurosis in this situation.

Relations.—By its *external surface*, with the superficial fascia, superficial epigastric and circumflex iliac vessels, and some cutaneous nerves. By its *internal surface*, with the Internal oblique, the lower part of the eight inferior ribs, and Intercostal muscles, the Cremaster, the spermatic cord in the male, and round ligament in the female. Its *posterior border*, extending from the last rib to the crest of the ilium, is fleshy throughout and free ; it is occasionally overlapped by the Latissimus dorsi, though generally a triangular interval exists between the two muscles near the crest of the ilium, in which is seen a portion of the Internal oblique. This triangle, *Petit's triangle*, is therefore bounded in front by the External oblique, behind by the Latissimus dorsi, below by the crest of the ilium, while its floor is formed by the Internal oblique (fig. 406).

The following parts of the aponeurosis of the External oblique muscle require to be further described, viz. the external abdominal ring, the intercolumnar fibres and fascia, Poupart's ligament, Gimbernat's ligament, and the triangular fascia.

The External or Superficial abdominal ring.—Just above, and to the outer side of the crest of the os pubis, an interval is seen in the aponeurosis of the External oblique, called the *External abdominal ring.* The aperture is oblique in direction, somewhat triangular in form, and corresponds with the course of the fibres of the aponeurosis. It usually measures from base to apex about an inch, and transversely about half an inch. It is bounded below by the crest of the os pubis ; above, by a series of curved fibres, the *intercolumnar*, which pass across the upper angle of the ring, so as to increase its strength ; and on each side, by the margins of the opening in the aponeurosis, which are called the *columns* or *pillars of the ring.*

The external pillar, which is at the same time inferior from the obliquity of its direction, is the stronger ; it is formed by that portion of Poupart's ligament which is inserted into the spine of the os pubis ; it is curved so as to form a kind of groove, upon which the spermatic cord rests. The internal or superior pillar is a broad, thin, flat band which is attached to the front of the symphysis pubis, interlacing with its fellow of the opposite side, that of the right side being superficial.

The external abdominal ring gives passage to the spermatic cord and ilio-inguinal nerve in the male, and to the round ligament of the uterus and the ilio-inguinal nerve in the female : it is much larger in men than in women, on account of the large size of the spermatic cord, and hence the greater frequency of inguinal hernia in men.

The **intercolumnar fibres** are a series of curved tendinous fibres, which arch across the lower part of the aponeurosis of the External oblique. They have received their name from stretching across between the two pillars of the external ring, describing a curve with the convexity downwards. They are much thicker and stronger at the outer margin of the external ring, where they are connected to Poupart's ligament, than internally, where they are inserted into the linea alba. They are more strongly developed in the male than in the female. The intercolumnar fibres increase the strength of the lower part of the aponeurosis, and prevent the divergence of the pillars from one another.

These intercolumnar fibres, as they pass across the external abdominal ring, are themselves connected together by delicate fibrous tissue, thus forming a fascia which, as it is attached to the pillars of the ring, covers it in, and is called the *intercolumnar fascia.* This **intercolumnar fascia** is continued down as a tubular prolongation around the spermatic cord and testis, and encloses them in a distinct sheath ; hence it is also called the *external spermatic fascia.* The external abdominal ring is only seen as a distinct aperture after the external spermatic fascia has been removed from its pillars.

The sac of an inguinal hernia, in passing through the external abdominal ring, receives an investment from the intercolumnar fascia.

Poupart's ligament is the lower border of the aponeurosis of the External oblique muscle, which extends from the anterior superior spine of the ilium to the spine of the os pubis. From this latter point it is reflected backwards and outwards to be attached to the pectineal line for about half an inch, forming Gimbernat's ligament. Its general direction is curved downwards towards the thigh, where it is continuous with the fascia lata. Its outer half is rounded and oblique in direction. Its inner half gradually widens at its attachment to the os pubis, is more horizontal in direction, and lies beneath the spermatic cord.

Nearly the whole of the space included between the crural arch and the innominate bone is filled in by the parts which descend from the abdomen into the thigh. These will be referred to again on a subsequent page.

Gimbernat's ligament is that part of the aponeurosis of the External oblique muscle which is reflected backwards and outwards from the spine of the os pubis to be inserted into the pectineal line. It is about half an inch in length, larger in the male than in the female, almost horizontal in direction in the erect posture, and of a triangular form with the base directed outwards. Its base, or outer margin, is concave, thin, and sharp, and forms the inner boundary of the crural ring. Its apex corresponds to the spine of the os pubis. Its posterior margin is attached to the pectineal line, and is continuous with the pubic portion of the fascia lata. Its anterior margin is attached to Poupart's ligament. Its surfaces are directed upwards and downwards.

The **triangular fascia** of the abdomen is a layer of tendinous fibres of a triangular shape, formed by an expansion from Gimbernat's ligament and the outer pillar of the ring. It passes inwards beneath the spermatic cord, and expands into a somewhat fan-shaped fascia, lying behind the inner pillar of the external abdominal ring, and in front of the conjoined tendon, and interlaces with the ligament of the other side at the linea alba.

Ligament of Cooper.—This is a strong ligamentous band, which was first described by Sir Astley Cooper. It extends upwards and backwards from the base of Gimbernat's ligament along the ilio-pectineal line, to which it is attached. It is strengthened by the fascia transversalis, by the pectineal aponeurosis, and by a lateral expansion from the lower attachment of the linea alba (adminiculum lineæ albæ).

Dissection.—Detach the External oblique by dividing it across, just in front of its attachment to the ribs, as far as its posterior border, and separate it below from the crest of the ilium as far as the anterior superior spine ; then detach the muscle carefully from the Internal oblique, which lies beneath it, and turn it towards the opposite side.

The **Obliquus internus abdominis** (fig. 412), thinner and smaller than the preceding, beneath which it lies, is of an irregularly quadrilateral form, and situated at the side and fore part of the abdomen. It arises, by fleshy fibres, from the outer half of Poupart's ligament, being attached to the groove on its upper surface ; from the anterior two-thirds of the middle lip of the crest of the ilium, and from the posterior lamella of the lumbar fascia. From this origin the fibres diverge: those from Poupart's ligament, few in number, and paler in colour than the rest, arch downwards and inwards across the spermatic cord in the male and the round ligament in the female, and, becoming tendinous, are inserted, conjointly with those of the Transversalis, into the crest of the os pubis and inner part of the ilio-pectineal line behind Gimbernat's ligament, forming what is known as the conjoined tendon of the Internal oblique and Transversalis ; those from the anterior third of the iliac origin are horizontal in their direction, and, becoming tendinous along the lower fourth of the linea semilunaris, pass in front of the Rectus muscle to be inserted into the linea alba ; those which arise from the middle third of the origin from the crest of the ilium pass obliquely upwards and inwards, and terminate in an aponeurosis, which divides at the outer border of the Rectus muscle into two lamellæ, which are continued forwards, one in front and the other behind this muscle, to the linea alba : the posterior lamella being also connected to the cartilages of the seventh, eighth, and ninth ribs ; the most posterior fibres pass almost vertically upwards, to be inserted into the lower borders of the cartilages of the three lower ribs, being continuous with the Internal intercostal muscles.

The *conjoined tendon* of the Internal oblique and Transversalis is mainly formed by the lower part of the tendon of the Transversalis, and is inserted into the crest of the os pubis and ilio-pectineal line, immediately behind the external abdominal ring, serving to protect what would otherwise be a weak point in the abdominal wall. Sometimes this tendon is insufficient to resist the pressure from within, and is carried forward in front of the protrusion through the external abdominal ring, forming one of the coverings of direct inguinal hernia ; or the hernia forces its way through the fibres of the conjoined tendon. The conjoined tendon is sometimes divided into an outer and an inner portion— the former being termed the *ligament of Hesselbach* ; the latter, the *ligament of Henle.*

The aponeurosis of the Internal oblique is continued forward to the middle line of the abdomen, where it joins with the aponeurosis of the opposite muscle at the linea alba, and extends from the margin of the thorax to the os pubis. At the outer margin of the Rectus muscle, this aponeurosis, for the upper three-fourths of its extent, divides into two lamellæ, which pass, one in front and the other behind the muscle, enclosing it in a sheath, and reuniting on its inner border at the linea alba; the anterior layer is blended with the aponeurosis of the External oblique muscle; the posterior layer with that of the Transversalis. Along the lower fourth, the aponeurosis ceases to split, and passes altogether in front of the Rectus without any separation, and thus a deficiency is left in

FIG. 412.—The Internal oblique muscle.

the sheath of the muscle behind; this is marked above by a sharp lunated margin, having its concavity downwards, and known as the *semilunar fold of Douglas*.

Relations.—By its *external surface*, with the External oblique, Latissimus dorsi, spermatic cord, and external ring. By its *internal surface*, with the Transversalis muscle, the lower intercostal vessels and nerves, the ilio-hypogastric and the ilio-inguinal nerves. Near Poupart's ligament, it lies on the fascia transversalis, internal ring, and spermatic cord.

The **Cremaster muscle** is a thin muscular layer, composed of a number of fasciculi which arise from the middle of Poupart's ligament where its fibres are continuous with those of the Internal oblique and also occasionally with the Transversalis. It passes along the outer side of the spermatic cord, descends with it through the external abdominal ring upon the front and sides of the cord, and forms a series of loops which differ in thickness and length in different subjects. Those at the upper part of the cord are exceedingly short, but they become in succession longer and longer, the longest reaching down as low as the testicle, where a few are inserted into the tunica vaginalis. These loops are united together by areolar tissue, and form a thin covering over the cord and testis, the *cremasteric fascia*. The fibres ascend along the inner side of the cord, and are inserted by a small pointed tendon into the spine and crest of the os pubis and front of the sheath of the Rectus muscle.

It will be observed that the origin and insertion of the Cremaster is precisely similar to that of the lower fibres of the Internal oblique. This fact affords an easy explanation of the manner in which the testicle and cord are invested by this muscle. At an early period of fœtal life the testis is placed at the lower and back part of the abdominal cavity, but during its descent towards the scrotum, which takes place before birth, it carries on it some fibres from the lower part of the muscle, which accompany the testicle and cord into the scrotum.

Dissection.—Detach the Internal oblique in order to expose the Transversalis beneath. This may be effected by dividing the muscle above, at its attachment to the ribs; below, at its connection with Poupart's ligament and the crest of the ilium; and behind, by a vertical incision extending from the last rib to the crest of the ilium. The muscle should previously be made tense by drawing upon it with the fingers of the left hand, and if its division is carefully effected, the cellular interval between it and the Transversalis, as well as the direction of the fibres of the latter muscle, will afford a clear guide to their separation; above the crest of the ilium the ascending branches of the deep circumflex iliac vessels are interposed between them, and form an important guide in separating them. The muscle should then be thrown inwards towards the linea alba.

The **Transversalis muscle** (fig. 413), so called from the direction of its fibres, is the most internal of the flat muscles of the abdomen, being placed immediately beneath the Internal oblique. It arises by fleshy fibres from the outer third of Poupart's ligament; from the anterior three-fourths of the inner lip of the crest of the ilium; from the inner surface of the cartilages of the six lower ribs, interdigitating with the Diaphragm; and from the lumbar aponeurosis, which may be regarded as the posterior aponeurosis of the muscle, and which has been seen to divide into three lamellæ (see page 469). The muscle terminates in front in a broad aponeurosis, the lower fibres of which curve downwards and inwards, and are inserted, together with those of the Internal oblique, into the lower part of the linea alba, the crest of the os pubis and pectineal line, forming what is known as the conjoined tendon of the Internal oblique and Transversalis. Throughout the rest of its extent the aponeurosis passes horizontally inwards, and is inserted into the linea alba; its upper three-fourths passing behind the Rectus muscle, blending with the posterior lamella of the Internal oblique; its lower fourth passing in front of the Rectus.

Relations.—By its *external surface*, with the Internal oblique, the lower intercostal nerves, and the inner surface of the cartilages of the lower ribs. By its *internal surface*, with the fascia transversalis, which separates it from the extra-peritoneal fat and parietal layer of the peritoneum. Its lower border forms the upper boundary of the inguinal canal.

Dissection.—To expose the Rectus muscle, open its sheath by a vertical incision extending from the margin of the thorax to the os pubis, and then reflect the two portions from the surface of the muscle, which is easily done, excepting at the lineæ transversæ, where so close an adhesion exists that the greatest care is requisite in separating them. Now raise the outer edge of the muscle, in order to examine the posterior layer of the sheath. By dividing the muscle in the centre, and turning its lower part downwards, the point where the posterior wall of the sheath terminates in the fold of Douglas will be seen.

The **Rectus abdominis** is a long flat muscle, which extends along the whole length of the front of the abdomen, being separated from its fellow of the opposite side by the linea alba. It is much broader, but thinner, above than below, and arises by two tendons, the external or larger being attached to the crest of the os pubis; the internal, smaller portion, interlacing with its fellow of the opposite side, and being connected with the ligaments covering the front of the symphysis pubis. The fibres ascend, and the muscle is inserted by three portions of unequal size into the cartilages of the fifth, sixth, and seventh ribs.

The upper portion, attached principally to the cartilage of the fifth rib, usually has some fibres of insertion into the anterior extremity of the rib itself. Some fibres are occasionally connected with the costo-xiphoid ligaments, and side of the ensiform cartilage.

The Rectus muscle is traversed by tendinous intersections, three in number, which have received the name of *lineæ transversæ*. One of these is usually situated opposite the umbilicus, and two above that point; of the latter, one

FIG. 413.—The Transversalis, Rectus, and Pyramidalis muscles.

corresponds to the extremity of the ensiform cartilage, and the other is about midway between the ensiform cartilage and the umbilicus. These intersections pass transversely or obliquely across the muscle in a zigzag course; they rarely extend completely through its substance; they may pass only halfway across it, and are intimately adherent in front to the sheath in which the muscle is enclosed. Sometimes one or two additional lines may be seen, one usually below the umbilicus; the position of the other, when it exists, is variable. These additional lines are for the most part incomplete.

The Rectus is enclosed in a sheath (fig. 414) formed by the aponeuroses of the Oblique and Transversalis muscles, which are arranged in the following manner. When the aponeurosis of the Internal oblique arrives at the outer margin of the Rectus, it divides into two lamellæ, one of which passes in front of the Rectus, blending with the aponeurosis of the External oblique; the other, behind it, blending with the aponeurosis of the Transversalis; and these, joining again at its inner border, are inserted into the linea alba. This arrangement of the aponeurosis exists from the costal margin to midway between the umbilicus and symphysis pubis, where the posterior wall of the sheath terminates in a thin curved margin, the *semilunar fold of Douglas*, the concavity of which looks downwards towards the pubes : below this level the aponeuroses of all three muscles pass in front of the Rectus without any separation. The extremities of the fold of Douglas descend as pillars to the os pubis. The inner pillar is attached to the symphysis pubis ; the outer pillar, which was named by Braune the ligament of Hesselbach, passes downwards as a distinct band on the inner side of the internal abdominal ring, and there its fibres divide into two sets, internal and external : the internal fibres are attached to the ascending ramus of the os pubis and the pectineal fascia ; the external ones pass to the Psoas fascia, to the deep surface of Poupart's ligament, and to the tendon of the Transversalis on the outer side of the ring. The Rectus muscle, in the situation

FIG. 414.—A transverse section of the abdomen in the lumbar region.

where its sheath is deficient, is separated from the peritoneum by the transversalis fascia. Since the tendon of the Internal oblique muscle only reaches as high as the costal margin, it follows that above this level the sheath of the Rectus is deficient behind, the muscle resting directly on the cartilages of the ribs, and being covered merely by the tendon of the External oblique.

The **Pyramidalis** is a small muscle, triangular in shape, placed at the lower part of the abdomen, in front of the Rectus, and contained in the same sheath with that muscle. It arises by tendinous fibres from the front of the os pubis and the anterior pubic ligament ; the fleshy portion of the muscle passes upwards, diminishing in size as it ascends, and terminates by a pointed extremity, which is inserted into the linea alba, midway between the umbilicus and the os pubis. This muscle may be wanting on one or both sides ; the lower end of the Rectus then becomes proportionately increased in size. Occasionally it has been found double on one side, or the muscles of the two sides are of unequal size. Sometimes it extends higher than what is stated above.

Besides the Rectus and Pyramidalis muscles, the sheath of the Rectus contains the superior and deep epigastric arteries, the terminations of the lumbar arteries and of the lower intercostal arteries and nerves.

Nerves.—The abdominal muscles are supplied by the lower intercostal nerves. The Transversalis and Internal oblique also receive filaments from the hypogastric branch of the ilio-hypogastric and sometimes from the ilio-inguinal. The Cremaster is supplied by the genital branch of the genito-crural.

In the description of the abdominal muscles, mention has frequently been made of the linea alba, lineæ semilunares, and lineæ transversæ; when the dissection of the muscles is completed, these structures should be examined.

The **linea alba** is a tendinous raphé seen along the middle line of the abdomen, extending from the ensiform cartilage to the symphysis pubis, to which it is attached. It is placed between the inner borders of the Recti muscles, and is formed by the blending of the aponeuroses of the Obliqui and Transversales muscles. It is narrow below, corresponding to the linear interval existing between the Recti; but broader above, as these muscles diverge from one another in their ascent, becoming of considerable breadth after great distension of the abdomen from pregnancy or ascites. At its lower end the linea alba has a double insertion—its superficial fibres passing in front of the inner heads of the Recti to the symphysis pubis, while its deeper fibres form a triangular lamella, which is attached behind the Recti to the posterior lip of the crest of the pubis, and which is named the *adminiculum lineæ albæ*. It presents numerous apertures for the passage of vessels and nerves; the largest of these is the umbilicus, which in the fœtus transmits the umbilical vessels, but in the adult is obliterated, the cicatrix being stronger than the neighbouring parts; hence umbilical hernia occurs in the adult *near* the umbilicus, while in the fœtus it occurs *at* the umbilicus. The linea alba is in relation, in front, with the integument, to which it is adherent, especially at the umbilicus; behind, it is separated from the peritoneum by the transversalis fascia, which above the umbilicus is adherent to it, but below is separated from it by a certain amount of fat, which must not be mistaken for extra-peritoneal fat.

The **lineæ semilunares** are two curved tendinous lines placed one on each side of the linea alba. Each corresponds with the outer border of the Rectus muscle, extends from the cartilage of the ninth rib to the pubic spine, and is formed by the aponeurosis of the Internal oblique at its point of division to enclose the Rectus, where it is reinforced in front by the External oblique, and behind by the Transversalis.

The **lineæ transversæ** are narrow transverse lines which intersect the Recti muscles, as already mentioned; they connect the lineæ semilunares with the linea alba.

Actions.—The abdominal muscles perform a threefold action.

When the pelvis and thorax are fixed, they compress the abdominal viscera, by constricting the cavity of the abdomen, in which action they are materially assisted by the descent of the Diaphragm. By these means assistance is given in expelling the fœtus from the uterus, the fæces from the rectum, the urine from the bladder, and the contents of the stomach in vomiting.

If the pelvis and spine are fixed, these muscles compress the lower part of the thorax, materially assisting expiration. If the pelvis alone is fixed, the thorax is bent directly forward, when the muscles of both sides act, or to either side when those of the two sides act alternately, rotation of the trunk at the same time taking place to the opposite side.

If the thorax is fixed, these muscles, acting together, draw the pelvis upwards, as in climbing; or, acting singly, they draw the pelvis upwards, and bend the vertebral column to one side or the other. The Recti muscles, acting from below, depress the thorax, and consequently flex the vertebral column; when acting from above, they flex the pelvis upon the vertebral column. The Pyramidales are tensors of the linea alba.

The **transversalis fascia** is a thin aponeurotic membrane which lies between the inner surface of the Transversalis muscle and the extra-peritoneal fat. It forms part of the general layer of fascia which lines the abdominal parietes, and is directly continuous with the iliac and pelvic fasciæ. In the inguinal region, the transversalis fascia is thick and dense in structure and joined by fibres from the aponeurosis of the Transversalis muscle, but it becomes thin and cellular as it ascends to the Diaphragm, and blends with the fascia covering the under aspect of this muscle. In front, it unites across the middle line with the fascia on the opposite side of the body; and behind, it becomes lost in the fat which covers the posterior surfaces of the kidneys. Below, it has the following attachments:

posteriorly, it is connected to the whole length of the crest of the ilium, between the attachments of the Transversalis and Iliacus muscles ; between the anterior superior spine of the ilium and the femoral vessels it is connected to the posterior margin of Poupart's ligament, and is there continuous with the iliac fascia. Internal to the femoral vessels it is thin and attached to the os pubis and pectineal line, behind the conjoined tendon, with which it is united ; corresponding to the region where the femoral vessels pass into the thigh, this fascia descends in front of them, forming the anterior wall of the femoral sheath. Beneath Poupart's ligament it is strengthened by a band of fibrous tissue, which is only loosely connected to Poupart's ligament, and is specialised as the *deep crural arch.* The spermatic cord in the male and the round ligament in the female pass through this fascia ; the point where they pass through is called the *internal abdominal ring.* This opening is not visible externally owing to a prolongation of the transversalis fascia on these structures, forming the *infundibuliform fascia.*

The **Internal** or **deep abdominal ring** is situated in the transversalis fascia, midway between the anterior superior spine of the ilium and the symphysis pubis, and about half an inch above Poupart's ligament. It is of an oval form, the extremities of the oval directed upwards and downwards, varies in size in different subjects, and is much larger in the male than in the female. It is bounded, above and externally, by the arched fibres of the Transversalis ; below and internally, by the deep epigastric vessels. It transmits the spermatic cord in the male and the round ligament in the female. From its circumference a thin funnel-shaped membrane, the *infundibuliform fascia,* is continued round the cord and testis, enclosing them in a distinct covering.

When the sac of an oblique inguinal hernia passes through the internal or deep abdominal ring, the infundibuliform process of the transversalis fascia forms one of its coverings.

Extra-peritoneal Connective Tissue.—Between the inner surface of the general layer of the fascia which lines the interior of the abdominal and pelvic cavities, and the peritoneum, there is a considerable amount of connective tissue, which is termed the *extra-peritoneal* or *subperitoneal connective tissue,* and which, according to Anderson and Makins, is 'a portion of a widespread system of mesoblastic connective tissue, which surrounds the great vessels of the trunk.' It may be described as consisting of two portions, parietal and visceral.

The *parietal* portion lines the cavity in varying quantities in different situations. It is especially abundant on the posterior wall of the abdomen, and particularly around the kidneys, where it contains much fat. On the anterior wall of the abdomen, with the exception of the pubic region, and on the roof of the abdomen, it is scanty, and here the transversalis fascia is more closely connected with the peritoneum. In the pelvis, there is a considerable amount of extra-peritoneal connective tissue.

The *visceral* portion follows the course of the branches of the abdominal aorta between the layers of the mesenteries and other folds of peritoneum which connect the various viscera to the abdominal wall, and assists in fixing them. The two portions are directly continuous with each other.

The deep crural arch.—Curving over the vessels, just at the point where they become femoral, on the abdominal side of Poupart's ligament and loosely connected with it, is a thickened band of fibres called the deep crural arch. It is apparently a thickening of the fascia transversalis, joining externally the centre of Poupart's ligament, and arching across the front of the crural sheath to be inserted by a broad attachment into the spine of the os pubis and ilio-pectineal line, behind the conjoined tendon. In some subjects this structure is not very prominently marked, and not infrequently it is altogether wanting.

Surface Form.—The skin of the abdomen is thin and sensitive. In the male, it is often thickly hair-clad, especially towards the lower part of the middle line. In the female, the hairs are confined to the pubes. After distension from pregnancy or other causes, the skin presents transverse white lines, which are quite smooth, being destitute of papillæ. These are known as *striæ gravidarum.*

The only muscles of the abdomen which have any considerable influence on surface form are the External oblique and Rectus muscles. With regard to the External oblique, the upper digitations of its origin from the ribs are well marked in a muscular subject, intermingled with the serrations of the Serratus magnus ; the lower digitations are not visible, being covered by the thick border of the Latissimus dorsi. Its attachment to the crest of the ilium, in conjunction with the Internal oblique, forms a thick oblique roll, which determines the iliac furrow. Sometimes on the front of the lateral region of the abdomen an undulating outline marks the spot where the muscular fibres terminate and the aponeurosis commences. The outer border of the Rectus is defined by the *linea semi-lunaris*, which may be exactly determined by putting the muscle into action. It corresponds with a curved line, with its convexity outwards, drawn from the end of the cartilage of the ninth rib to the spine of the os pubis, so that the centre of the line, at or near the umbilicus, is three inches from the median line. The inner border of the Rectus corresponds to the *linea alba*, marked on the surface of the body by a groove, the *abdominal furrow*, which extends from the infrasternal fossa to, or to a little below, the umbilicus, where it gradually becomes lost. The surface of the Rectus presents three transverse furrows, the *lineæ transversæ*. The upper two of these, one opposite or a little below the tip of the ensiform cartilage, and another, midway between this point and the umbilicus, are usually well marked; the third, opposite the umbilicus, is not so distinct. The umbilicus, situated in the linea alba, varies in position as regards its height. It is always situated above a zone drawn round the body opposite the highest point of the crest of the ilium, generally being about three-quarters of an inch to an inch above this line. It usually corresponds, therefore, to the fibro-cartilage between the third and fourth lumbar vertebræ.

2. DEEP MUSCLES OF THE ABDOMEN

Psoas magnus.	Iliacus.
Psoas parvus.	Quadratus lumborum.

The Psoas magnus, the Psoas parvus, and the Iliacus muscles, with the fascia covering them, will be described with the Muscles of the Lower Extremity (see page 541).

The fascia covering the Quadratus lumborum.—This is the most anterior of the three layers of the lumbar aponeurosis. It is a thin layer of fascia which, passing over the anterior surface of the Quadratus lumborum, is attached, internally, to the bases of the transverse processes of the lumbar vertebræ; below, to the ilio-lumbar ligament; and above, to the apex and lower border of the last rib.

The portion of this fascia which extends from the transverse process of the first lumbar vertebra to the apex and lower border of the last rib, constitutes the ligamentum arcuatum externum.

The **Quadratus lumborum** (fig. 407, page 471) is situated in the lumbar region. It is irregularly quadrilateral in shape, and broader below than above. It arises by aponeurotic fibres from the ilio-lumbar ligament and the adjacent portion of the crest of the ilium for about two inches, and is inserted into the lower border of the last rib for about half its length and, by four small tendons, into the apices of the transverse processes of the four upper lumbar vertebræ. Occasionally a second portion of this muscle is found situated in front of the preceding. It arises from the upper borders of the transverse processes of three or four of the lower lumbar vertebræ, and is inserted into the lower margin of the last rib. The Quadratus lumborum is contained in a sheath formed by the anterior and middle lamellæ of the lumbar aponeurosis.

Relations.—Its *anterior surface* (or rather the fascia which covers its anterior surface) is in relation with the colon, the kidney, the Psoas muscle, and the Diaphragm. Between the fascia and the muscle are the last dorsal, ilio-hypogastric and ilio-inguinal nerves. Its *posterior surface* is in relation with the middle lamella of the lumbar fascia, which separates it from the Erector spinæ. The Quadratus lumborum extends, however, beyond the outer border of the Erector spinæ.

Nerve-supply.—The anterior divisions of the last dorsal and first and second lumbar nerves.

Actions.—The Quadratus lumborum draws down the last rib, and acts as a muscle of inspiration by helping to fix the origin of the Diaphragm. If the thorax and spine are fixed, it may act upon the pelvis, raising it towards its own side when only one muscle is put in action; and when both muscles act together, either from below or above, they flex the trunk.

IV. Muscles of the Pelvic Outlet

The muscles of this region are situated at the pelvic outlet in the ischio-rectal region and the perinæum. They include the following :

1. Muscles of the ischio-rectal region.
2. Muscles of the perinæum : A. In the Male ; B. In the Female.

1. Muscles of the Ischio-rectal Region

Corrugator cutis ani.	Internal sphincter ani.
External sphincter ani.	Levator ani.

Coccygeus.

The Corrugator cutis ani.—Around the anus is a thin stratum of involuntary muscular fibre, which radiates from the orifice. Internally, the fibres fade off into the submucous tissue, while externally they blend with the true skin. By its contraction it raises the skin into ridges around the margin of the anus.

The **External sphincter ani** is a thin, flat plane of muscular fibres, elliptical in shape and intimately adherent to the integument surrounding the margin of the anus. It measures about three or four inches in length, from its anterior to its posterior extremity, being about an inch in breadth, opposite the anus. It consists of two strata, superficial and deep. The *superficial,* constituting the main portion of the muscle, arises from a narrow tendinous band, the *ano-coccygeal raphé,* which stretches from the tip of the coccyx to the posterior margin of the anus ; it forms two flattened planes of muscular tissue, which encircle the anus and meet in front to be inserted into the central tendinous point of the perinæum, joining with the Transversus perinæi, the Levator ani, and the Accelerator urinæ. The *deeper portion* forms a complete sphincter to the anal canal. Its fibres surround the canal, closely applied to the deep Sphincter, and in front blend with the other muscles at the central point of the perinæum. In a considerable proportion of cases the fibres decussate in front of the anus, and are continuous with the Transversus perinæi. Posteriorly, they are not attached to the coccyx, the fibres of opposite sides being continuous behind the anal canal. The upper edge of the muscle is ill-defined, since fibres are given off from it to join the Levator ani.

Nerve-supply.—A branch from the anterior division of the fourth sacral and the inferior hæmorrhoidal branch of the internal pudic.

Actions.—The action of this muscle is peculiar : 1. It is, like other muscles, always in a state of tonic contraction, and having no antagonistic muscle it keeps the anal canal and orifice closed. 2. It can be put into a condition of greater contraction under the influence of the will, so as to more firmly occlude the anal aperture in expiratory efforts, unconnected with defæcation. 3. Taking its fixed point at the coccyx, it helps to fix the central point of the perinæum, so that the Accelerator urinæ may act from this fixed point.

The **Internal sphincter ani** is a muscular ring which surrounds the anal canal for about an inch ; its inferior border being contiguous with, but quite separate from, the External sphincter. This muscle is about two lines in thickness, and is formed by an aggregation of the involuntary circular fibres of the intestine. It surrounds the canal for about an inch, and terminates about a quarter of an inch from the external orifice. It is paler in colour and less coarse in texture than the External sphincter.

Actions.—Its action is entirely involuntary. It helps the External sphincter to occlude the anal aperture.

The **Levator ani** (fig. 415) is a broad, thin muscle, situated on each side of the pelvis. It is attached to the inner surface of the sides of the true pelvis, and descending unites with its fellow of the opposite side to form the greater part of the floor of the pelvic cavity. It supports the viscera in this cavity and surrounds the various structures which pass through it. It arises, in front, from the posterior surface of the body of the os pubis on the outer side of the symphysis ; behind, from the inner surface of the spine of the ischium ; and between these two points, from the pelvic fascia. Posteriorly, this fascial origin corresponds, more or less closely, with the 'white line'—i.e. with the angle of division between the obturator and recto-vesical layers of the pelvic fascia—but in front, the muscle arises from the fascia at a varying distance above the 'white line,' in some cases reaching nearly as high as the canal for the obturator vessels and nerve. The fibres pass downwards to the middle line of the floor of the pelvis, and are inserted, the most posterior into the side of the last two segments of the coccyx; those placed more anteriorly unite with the muscle of the opposite side, in a median fibrous raphé (*ano-coccygeal raphé*), which extends between the coccyx and the margin of the anus. The middle fibres, which form the larger portion of the muscle, are inserted into the side of the rectum, blending with the fibres of the Sphincter muscles; lastly, the anterior fibres, the longest, descend upon the side of the prostate gland to unite beneath it with the muscle of the opposite side, blending with the fibres of the External sphincter and Transversus perinæi muscles at the central tendinous point of the perinæum.

FIG. 415.—Side view of pelvis, showing Levator ani.
(From a preparation in the Museum of the Royal College of Surgeons of England.)

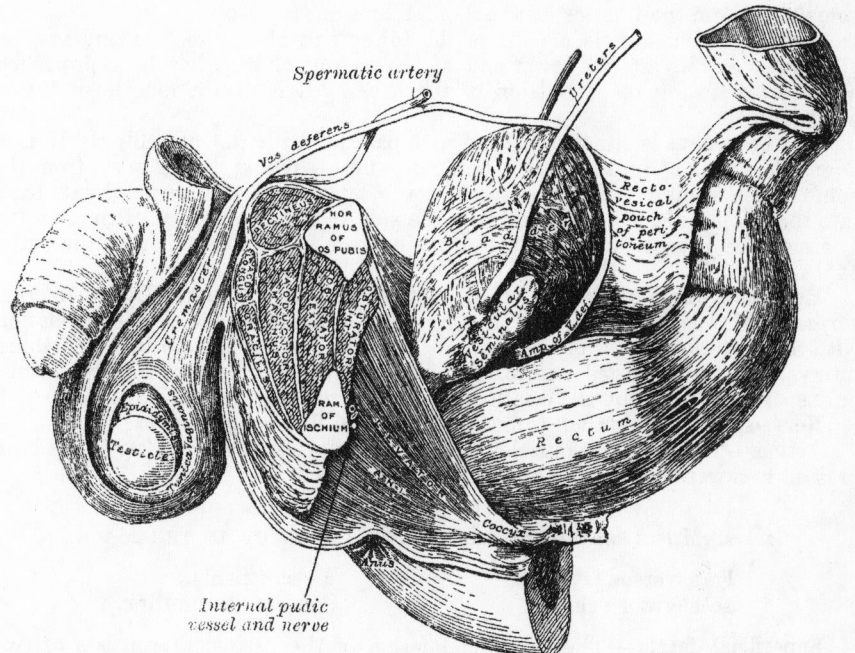

Peter Thompson * has not been able to satisfy himself that any of the fibres of the Levator ani terminate in the rectal walls.

The anterior portion is occasionally separated from the rest of the muscle by connective tissue. From this circumstance, as well as from its peculiar relation with the prostate gland, descending by its side, and surrounding it as in a sling, it has been described by Santorini and others as a distinct muscle, under the name of *Levator prostatæ*. In the female, the anterior fibres of the Levator ani descend upon the side of the vagina.

Relations.—By its *upper* or *pelvic surface*, with the recto-vesical fascia, which separates it from the prostate, rectum, and peritoneum. By its *lower* or *perineal surface*, it forms the inner boundary of the ischio-rectal fossa, and is covered by a thin layer of fascia, the *ischio-rectal* or *anal fascia*, given off from the obturator fascia. Its *posterior border* is free and separated from the Coccygeus muscle by a cellular interspace. Its *anterior border* is separated from the muscle of the opposite side by a triangular space, through which the urethra, and in the female the vagina, passes from the pelvis.

The Levator ani may be divided into ilio-coccygeal and pubo-coccygeal parts.

The *ilio-coccygeus* arises from the ischial spine and from the posterior part of the pelvic fascia, and is attached to the coccyx and ano-coccygeal raphé: it is usually thin,

---

* *The Myology of the Pelvic Floor*, 1899.

and may fail entirely, or be largely replaced by fibrous tissue. An accessory slip at its posterior part is sometimes named the *ilio-sacralis*. The *pubo-coccygeus* arises from the back of the pubis and from the anterior part of the pelvic fascia, and 'is directed backwards almost horizontally along the side of the anal canal towards the coccyx and sacrum, to which it finds attachment. Between the termination of the vertebral column and the anus, the two pubo-coccygei muscles come together and form a thick, fibro-muscular layer lying on the raphé formed by the ilio-coccygei' (Thompson). The greater part of this muscle is inserted into the coccyx and into the last one or two pieces of the sacrum. This insertion into the vertebral column is, however, not admitted by all observers. The fibres which form a sling for the rectum are named the *pubo-rectalis* or *sphincter recti*. They arise from the lower part of the symphysis pubis, and from the upper layer of the triangular ligament. They meet with the corresponding fibres of the opposite side around the lower part of the rectum, and form for it a strong sling. The pre-anal fibres of the Levator ani form the Levator prostatæ muscle.

Nerve-supply.—A branch from the anterior division of the fourth sacral nerve and a branch from the pudic nerve, which is sometimes derived from the perineal, sometimes from the inferior hæmorrhoidal division.

Actions.—This muscle constricts the lower end of the rectum and vagina. It elevates and inverts the lower end of the rectum after it has been protruded and everted during the expulsion of the fæces. It is also a muscle of forced expiration.

The **Coccygeus** is situated behind and parallel with the preceding. It is a triangular plane of muscular and tendinous fibres, arising, by its apex, from the spine of the ischium and lesser sacro-sciatic ligament, and inserted, by its base, into the margin of the coccyx and into the side of the lower piece of the sacrum. It assists the Levator ani and Pyriformis in closing in the back part of the outlet of the pelvis.

Relations.—By its *inner* or *pelvic surface*, with the rectum. By its *external surface*, with the lesser sacro-sciatic ligament. The *lower border* is in relation with the posterior border of the Levator ani, but separated from it by a cellular interval : its *upper border* is in relation with the lower border of the Pyriformis, but separated from it by the sciatic and internal pudic vessels and nerve.

Nerve-supply.—A branch from the fourth and fifth sacral nerves.

Action.—The Coccygei muscles raise and support the coccyx, after it has been pressed backwards during defæcation or parturition.

2. A. MUSCLES AND FASCIÆ OF THE PERINÆUM IN THE MALE

Transversus perinæi.   Erector penis.
Accelerator urinæ.   Compressor urethræ.

Superficial fascia.—The superficial fascia of the perinæum consists of two layers, superficial and deep, as in other regions of the body.

The *superficial layer* is thick, loose, areolar in texture, and contains much adipose tissue in its meshes, the amount of which varies in different subjects. In front, it is continuous with the dartos of the scrotum ; behind, it is continuous with the subcutaneous areolar tissue surrounding the anus ; and, on either side, with the same fascia on the inner side of the thighs. In the middle line, it is adherent to the skin of the raphé and to the deep layer of the superficial fascia. This layer should be carefully removed after it has been examined, when the deep layer will be exposed.

The *deep layer of superficial fascia* (Fascia of Colles) is thin, aponeurotic in structure, and of considerable strength, serving to bind down the muscles of the root of the penis. It is continuous, in front, with the dartos of the scrotum, the deep fascia of the penis, the fascia of the spermatic cord, and Scarpa's fascia upon the anterior wall of the abdomen ; on either side it is firmly attached to the margins of the rami of the os pubis and ischium, external to the crus penis, and as far back as the tuberosity of the ischium ; posteriorly, it curves round the Transversus perinæi muscles to join the lower margin of the triangular ligament. In the middle line, it is connected with the superficial fascia and the median septum of the Accelerator urinæ muscle. This fascia not only covers the muscles in this region, but sends upwards a vertical septum from its deep surface, which separates the back part of the subjacent space into two, the septum being incomplete in front.

The central tendinous point of the Perinæum.—This is a fibrous point in the middle line of the perinæum, between the urethra and the rectum, and about half an inch in front of the anus. At this point five muscles converge and are attached: viz. the External sphincter ani, the Accelerator urinæ, the two Transversi perinæi, and the anterior fibres of the Levator ani; so that by the contraction of these muscles, which extend in different directions, it serves as a fixed point of support.

The **Transversus perinæi** is a narrow muscular slip, which passes more or less transversely across the perineal space in front of the anus. It arises by tendinous fibres from the inner and fore part of the tuberosity of the ischium, and, passing inwards, is inserted into the central tendinous point of the perinæum, joining in this situation with the muscle of the opposite side, the External sphincter ani behind, and the Accelerator urinæ in front. In some cases, the fibres of the

FIG. 416.—The perinæum.
The integument and superficial layer of superficial fascia reflected.

deeper layer of the Sphincter ani decussate in front of the anus and are continued into this muscle. Occasionally it gives off fibres, which join with the Accelerator urinæ of the same side.

Nerve-supply.—The perineal branch of the internal pudic.

Actions.—The simultaneous contraction of the two muscles serves to fix the central tendinous point of the perinæum.

The **Accelerator urinæ** (*Bulbo-cavernosus*) is placed in the middle line of the perinæum, in front of the anus. It consists of two symmetrical halves, united along the median line by a tendinous raphé. It arises from the central tendon of the perinæum, and from the median raphé in front. From this point its fibres diverge like the plumes of a quill-pen; the most posterior form a thin layer, which is lost on the superficial surface of the triangular ligament; the middle fibres encircle the bulb and adjacent parts of the corpus spongiosum, and join with the fibres of the opposite side, on the upper part of the corpus spongiosum, in a strong aponeurosis; the anterior fibres, the longest and most distinct, spread out over the sides of the corpus cavernosum, to be inserted partly into that body, anterior to the Erector penis, and occasionally extending to the os pubis; and partly

K K

terminating in a tendinous expansion, which covers the dorsal vessels of the penis. The latter fibres are best seen by dividing the muscle longitudinally, and dissecting it outwards from the surface of the urethra.

Actions.—This muscle serves to empty the canal of the urethra, after the bladder has expelled its contents; during the greater part of the act of micturition its fibres are relaxed, and it only comes into action at the end of the process. The middle fibres are supposed, by Krause, to assist in the erection of the corpus spongiosum, by compressing the erectile tissue of the bulb. The anterior fibres, according to Tyrrel, also contribute to the erection of the penis, as they are inserted into, and continuous with, the fascia of the penis, compressing the dorsal vein during the contraction of the muscle.

The **Erector penis** (*Ischio-cavernosus*) covers the crus penis. It is an elongated muscle, broader in the middle than at either extremity, and situated on either side of the lateral boundary of the perinæum. It arises by tendinous and fleshy fibres from the inner surface of the tuberosity of the ischium, behind

FIG. 417.—The superficial muscles and vessels of the perinæum.

the crus penis; and from the rami of the pubis and ischium on each side of the crus. From these points fleshy fibres succeed, which end in an aponeurosis which is inserted into the sides and under surface of the crus penis.

Nerve-supply.—The perineal branch of the internal pudic.

Action.—The Erector penis compresses the crus penis, and retards the return of the blood through the veins, and thus serves to maintain the organ erect.

Between the muscles just examined a triangular space exists, bounded internally by the Accelerator urinæ, externally by the Erector penis, and behind by the Transversus perinæi. The floor of this space is formed by the triangular ligament of the urethra (deep perineal fascia), and running from behind forwards in it are the superficial perineal vessels and nerves, and the long pudendal nerve. The transverse perineal artery courses along the posterior boundary of the space on the Transversus perinæi muscle.

The **triangular ligament of the urethra** (*deep perineal fascia*) is stretched almost horizontally across the pubic arch, so as to close in the front part of the outlet of the pelvis. It consists of two dense membranous laminæ, which are united along their posterior borders, but are separated in front by intervening structures. The superficial of these two layers, the *inferior layer of the tri-angular ligament*, is triangular in shape, about an inch and a half in depth. Its apex is directed forwards, and is separated from the subpubic ligament by an oval opening for the transmission of the dorsal vein of the penis. Its lateral margins are attached on each side to the rami of the os pubis and ischium, above the crura penis. Its base is directed towards the rectum, and connected to the central tendinous point of the perinæum. It is continuous with the deep layer of the superficial fascia behind the Transversus perinæi muscle, and with a thin fascia which covers the cutaneous surface of the Levator ani muscle (*anal fascia*).

This layer of the triangular ligament is perforated, about an inch below the symphysis pubis, by the urethra, the aperture for which is circular in form and about three or four lines in diameter; by the arteries to the bulb and the ducts of Cowper's glands close to the urethral orifice; by the arteries to the corpora cavernosa—one on each side close to the pubic arch and about halfway along

Fig. 418.—Triangular ligament or deep perineal fascia.
On the left side the superficial layer has been removed.

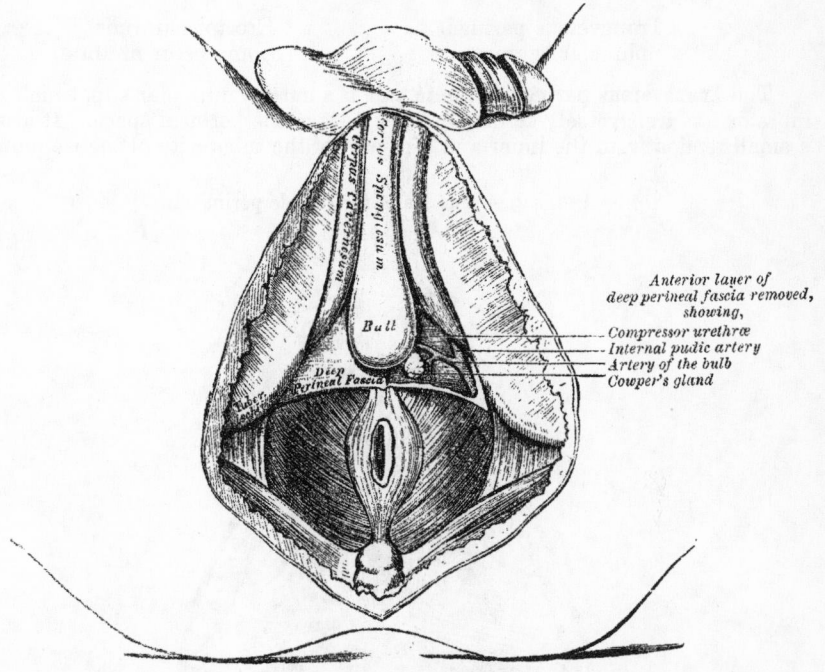

the attached margin of the ligament; by the dorsal arteries and nerves of the penis near the apex of the ligament. Its base is also perforated by the superficial perineal vessels and nerves, while between its apex and the subpubic ligament the dorsal vein of the penis passes upwards into the pelvis.

If the superficial or inferior layer of the triangular ligament is detached on either side, the following structures will be seen between it and the deep layer: the dorsal vein of the penis; the membranous portion of the urethra, and the Compressor urethræ muscle; Cowper's glands and their ducts; the pudic vessels and dorsal nerve of penis; the artery and nerve of the bulb, and a plexus of veins.

The deep layer of the ligament (*superior layer of the triangular ligament*) is derived from the obturator fascia and stretches across the pubic arch. If the obturator fascia is traced inwards after covering the Obturator internus muscle, it will be found to be attached by some of its deeper or anterior fibres to the inner margin of the ischio-pubic ramus, while its superficial or posterior fibres pass over this attachment to form the superior layer of the triangular ligament. Behind, this layer of the fascia is continuous with the inferior layer and with the fascia of Colles, and in front it is continuous with the recto-vesical fascia on the prostate gland. It is pierced by the urethra, or rather consists of two halves which are separated in the middle line by the urethra passing between them.

The **Compressor urethræ** (*Constrictor urethræ*) surrounds the whole length of the membranous portion of the urethra, and is contained between the two layers of the triangular ligament. It arises, by aponeurotic fibres, from the junction of the rami of the os pubis and ischium, to the extent of half or three-quarters of an inch: each segment of the muscle passes inwards, and divides into two fasciculi, which surround the urethra from the prostate gland behind, to the bulbous portion of the urethra in front; and unite, at the upper and lower surfaces of this tube, with the muscle of the opposite side, by means of a tendinous raphé.

Nerve-supply.—The perineal branch of the internal pudic.

Actions.—The muscles of both sides act together as a sphincter, compressing the membranous portion of the urethra. During the transmission of fluids they, like the Acceleratores urinæ, are relaxed, and only come into action at the end of the process to eject the last drops of the fluid.

2. B. MUSCLES OF THE PERINÆUM IN THE FEMALE (fig. 419)

Transversus perinæi.	Erector clitoridis.
Sphincter vaginæ.	Compressor urethræ.

The **Transversus perinæi** in the **female** is a narrow muscular slip, which passes more or less transversely across the back part of the perineal space. It arises by a small tendon from the inner and fore part of the tuberosity of the ischium, and,

FIG. 419.—Muscles of the female perinæum.

passing inwards, is inserted into the central point of the perinæum, joining in this situation with the muscle of the opposite side, the External sphincter ani behind, and the Sphincter vaginæ in front.

Nerve-supply.—The perineal branch of the internal pudic.

Action.—The simultaneous contraction of the two muscles serves to fix the central tendinous point of the perinæum.

The **Sphincter vaginæ** surrounds the orifice of the vagina, and is analogous to the Accelerator urinæ in the male. It covers the cuter aspect of the vestibular bulbs, and is attached posteriorly to the central tendinous point of the perinæum, where it blends with the External sphincter ani. Its fibres pass forwards on each side of the vagina, to be inserted into the corpora cavernosa of the clitoris, a fasciculus crossing over the body of the organ so as to compress the dorsal vein.

Nerve-supply.—The perineal branch of the internal pudic.

Actions.—The Sphincter vaginæ diminishes the orifice of the vagina. The anterior fibres contribute to the erection of the clitoris, as they are inserted into and are continuous with the fascia of the clitoris, compressing the dorsal vein during the contraction of the muscle.

The **Erector clitoridis** (*Ischio-cavernosus*) corresponds with the Erector penis in the male, but is smaller. It covers the unattached surface of the crus clitoridis. It is an elongated muscle, broader at the middle than at either extremity, and situated on either side of the lateral boundary of the perinæum. It arises by tendinous and fleshy fibres from the inner surface of the tuberosity of the ischium, behind the crus clitoridis; from the surface of the crus; and from the adjacent portion of the ramus of the ischium. From these points fleshy fibres succeed, and end in an aponeurosis, which is inserted into the sides and under surface of the crus clitoridis.

Nerve-supply.—The perineal branch of the internal pudic.

Actions.—The Erector clitoridis compresses the crus clitoridis and retards the return of blood through the veins, and thus serves to maintain the organ erect.

The **triangular ligament** (*deep perineal fascia*) in the female is not so strong as in the male. It is attached to the pubic arch, its apex being connected with the subpubic ligament. It is divided in the middle line by the aperture of the vagina, with the external coat of which it becomes blended, and in front of this is perforated by the urethra. Its posterior border is continuous, as in the male, with the deep layer of the superficial fascia around the Transversus perinæi muscle.

Like the triangular ligament in the male it consists of two layers, between which are to be found the following structures: the dorsal vein of the clitoris, a portion of the urethra and the Compressor urethræ muscle, the glands of Bartholin and their ducts; the pudic vessels and the dorsal nerve of the clitoris; the arteries of the bulbi vestibuli, and a plexus of veins.

The **Compressor urethræ** (*Constrictor urethræ*) arises on each side from the margin of the descending ramus of the os pubis. The fibres, passing inwards, divide into two sets: those of the fore part of the muscle are directed across the subpubic arch in front of the urethra to blend with the muscular fibres of the opposite side; while those of the hinder and larger part pass inwards to blend with the wall of the vagina behind the urethra.

Nerve-supply.—The perineal branch of the internal pudic.

MUSCLES AND FASCIÆ OF THE UPPER EXTREMITY

The Muscles of the Upper Extremity are divisible into groups, corresponding with the different regions of the limb.

I. OF THE THORACIC REGION

1. *Anterior Thoracic Region*

Pectoralis major.   Pectoralis minor.
Subclavius.

2. *Lateral Thoracic Region*

Serratus magnus.

II. OF THE SHOULDER AND ARM

3. *Acromial Region*

Deltoid.

4. *Anterior Scapular Region*

Subscapularis.

5. *Posterior Scapular Region*

Supraspinatus.   Teres minor.
Infraspinatus.   Teres major.

6. *Anterior Humeral Region*

Coraco-brachialis.   Biceps.
Brachialis anticus.

7. *Posterior Humeral Region*

Triceps.   Subanconeus.

III. OF THE FOREARM

8. *Anterior Radio-ulnar Region*

Superficial Layer.
Pronator radii teres.
Flexor carpi radialis.
Palmaris longus.
Flexor carpi ulnaris.
Flexor sublimis digitorum.

Deep Layer.
Flexor profundus digitorum.
Flexor longus pollicis.
Pronator quadratus.

9. *Radial Region*

Supinator longus (Brachio-radialis).
Extensor carpi radialis longior.
Extensor carpi radialis brevior.

10. *Posterior Radio-ulnar Region*

Superficial Layer.
Extensor communis digitorum.
Extensor minimi digiti.
Extensor carpi ulnaris.
Anconeus.

Deep Layer.
Supinator brevis.
Extensor ossis metacarpi pollicis.
Extensor brevis pollicis.
Extensor longus pollicis.
Extensor indicis.

IV. OF THE HAND

11. *Radial Region*

Abductor pollicis.
Opponens pollicis.
Flexor brevis pollicis.
Adductor obliquus pollicis.
Adductor transversus pollicis.

12. *Ulnar Region*

Palmaris brevis.
Abductor minimi digiti.
Flexor brevis minimi digiti.
Opponens minimi digiti.

13. *Middle Palmar Region*

Lumbricales.   Interossei palmares.
Interossei dorsales.

Dissection of Pectoral Region and Axilla (fig. 420).—The arm being drawn away from the side nearly at right angles with the trunk, and rotated outwards, make a vertical incision through the integument in the median line of the chest, from the upper to the lower part of the sternum; a second incision along the lower border of the Pectoral muscle, from the ensiform cartilage to the inner side of the axilla; a third, from the sternum along the clavicle, as far as its centre; and a fourth, from the middle of the clavicle obliquely downwards, along the interspace between the Pectoral and Deltoid muscles, as low as the fold of the armpit. The flap of integument is then to be dissected off in the direction indicated in the figure, but not entirely removed, as it should be replaced on completing the dissection. If a transverse incision is now made from the lower end of the sternum to the side of the chest, as far as the posterior fold of the armpit, and the integument reflected outwards, the axillary space will be more completely exposed.

I. MUSCLES AND FASCIÆ OF THE THORACIC REGION

1. *Anterior Thoracic Region*

Pectoralis major.   Pectoralis minor.
Subclavius.

The **superficial fascia** of the thoracic region is a loose cellulo-fibrous layer, enclosing masses of fat in its spaces. It is continuous with the superficial fascia of the neck and upper extremity above, and of the abdomen below. Opposite the mamma, it divides into two layers, one of which passes in front, the other behind that gland; and from both of these layers numerous septa pass into its substance, supporting its various lobes: from the anterior layer, fibrous processes pass orwards to the integument and nipple. These processes were called by Sir A. Cooper the *ligamenta suspensoria*, from the support they afford to the gland in this situation.

The **deep fascia** of the thoracic region is a thin aponeurotic lamina, covering the surface of the great Pectoral muscle, and sending numerous prolongations between its fasciculi: it is attached, in the middle line, to the front of the sternum; and, above, to the clavicle; externally and below it becomes continuous with the fascia over the shoulder, axilla, and thorax. It is very thin over the upper part of the muscle, thicker in the interval between the Pectoralis major and Latissimus dorsi, where it closes in the axillary space, and divides at the outer margin of the latter muscle into two layers, one of which passes in front, and the other behind it; these proceed as far as the spinous processes of the dorsal vertebræ, to which they are attached.   As the fascia leaves the lower edge of the Pectoralis major to pass across the floor of the axilla it sends a layer upwards under cover of the muscle : this lamina splits to envelop the Pectoralis minor, at the upper edge of which it becomes continuous with the costo-coracoid membrane.

The hollow of the armpit, seen when the arm is abducted, is mainly produced by the traction of this fascia on the axillary floor, and hence it is sometimes named the *suspensory ligament* of the axilla.   At the lower part of the thoracic region the deep fascia is well developed, and is continuous with the fibrous sheath of the Recti muscles.

Surgical Anatomy.—The axillary fascia prevents the extension of the pus, in cases of suppuration in the axilla, in a downward direction, and the matter has a tendency to spread upwards, beneath the Pectoral muscles, towards the root of the neck.   Early evacuation is therefore necessary.   The incision should be made midway between the anterior and posterior axillary folds, so as to avoid the long thoracic and subscapular vessels.

The **Pectoralis major** (fig. 421) is a broad, thick, triangular muscle, situated at the upper and fore part of the chest and in front of the axilla.   It arises from the anterior surface of the sternal half of the clavicle ; from half the breadth of the anterior surface of the sternum, as low down as the attachment of the cartilage of the sixth or seventh rib ; this portion of its origin consists of aponeurotic fibres, which intersect those of the opposite muscle ; it also arises from the cartilages of all the true ribs, with the exception, frequently, of the first, or of the seventh, or both ; and from the aponeurosis of the External oblique muscle of the abdomen.   The fibres from this extensive origin converge towards its insertion, giving to the muscle a radiated appearance.   Those fibres which arise from the clavicle pass obliquely outwards and downwards, and are usually separated from the rest by a slight interval : those from the lower part of the sternum, and the cartilages of the lower true ribs, pass upwards and outwards ; while the middle fibres pass horizontally.   They all terminate in a flat tendon, about two inches broad, which is inserted into the outer bicipital ridge of the humerus.   This tendon consists of two laminæ, placed one in front of the other, and usually blended together below.   The anterior, the thicker, receives the clavicular and upper half of the sternal portion of the muscle ; and its fibres are inserted in the same order as that in which they arise : that is to say, the outermost fibres of origin from the clavicle are inserted at the uppermost part of the bicipital ridge ; the upper fibres of origin from the sternum pass down to the lowermost part of this anterior lamina of the tendon and extend as low as the tendon of the Deltoid and join with it.   The posterior lamina of the tendon receives the attachment of the lower half of the sternal portion and the deeper part of the muscle from the costal cartilages.   These deep fibres, and particularly those from the lower costal cartilages, ascend the higher, turning backwards successively behind the superficial and upper ones, so that the tendon appears to be twisted.   The posterior lamina reaches higher on the humerus than the anterior one, and from it an expansion is given off which covers the bicipital groove and blends with the capsule of the shoulder-joint.   From the deepest fibres of this lamina at its

FIG. 420.—Dissection of upper extremity.

3. *Dissection of shoulder and arm*

1. *Dissection of pectoral region and axilla*

2. *Bend of elbow*

4. *Forearm*

5. *Palm of hand*

insertion an expansion is given off which lines the bicipital groove of the humerus, while from the lower border of the tendon a third expansion passes downwards to the fascia of the arm.

Relations.—By its *anterior surface*, with the integument, the superficial fascia, the Platysma, some of the branches of the descending cervical nerves, the mammary gland, and the deep fascia. By its *posterior surface*: its *thoracic portion*, with the sternum, the ribs and costal cartilages, the costo-coracoid membrane, the Subclavius, Pectoralis minor, Serratus magnus, and the Intercostals; its *axillary portion* forms the anterior wall of the axillary space, and

Fig. 421.—Muscles of the chest and front of the arm. Superficial view.

covers the axillary vessels and nerves, the Biceps and Coraco-brachialis muscles. Its *upper border* lies parallel with the Deltoid, from which it is separated by a slight interspace in which lie the cephalic vein and humeral branch of the acromio-thoracic artery. Its *lower border* forms the anterior margin of the axilla, being at first separated from the Latissimus dorsi by a considerable interval; but both muscles gradually converge towards the outer part of the space.

Dissection.—Detach the Pectoralis major by dividing the muscle along its attachment to the clavicle, and by making a vertical incision through its substance a little external to its line of attachment to the sternum and costal cartilages. The muscle should then be reflected outwards, and its tendon carefully examined. The Pectoralis minor is now exposed, and immediately above it, in the interval between its upper border and the clavicle, a strong fascia, the *costo-coracoid membrane*.

The **costo-coracoid membrane** is a strong fascia, situated under cover of the clavicular portion of the Pectoralis major muscle. It occupies the interval between the Pectoralis minor and Subclavius muscles, and protects the axillary vessels and nerves. Traced upwards, it splits to enclose the Subclavius muscle, and its two layers are attached to the clavicle, one in front of and the other behind the muscle ; the latter layer fuses with the deep cervical fascia and with the sheath of the axillary vessels. Internally, it blends with the fascia covering

Fig. 422.—Muscles of the chest and front of the arm, with the boundaries of the axilla.

the first two intercostal spaces, and is attached also to the first rib internal to the origin of the Subclavius muscle. Externally, it is very thick and dense, and is attached to the coracoid process. The portion extending from its attachment to the first rib to the coracoid process is often whiter and denser than the rest ; this is sometimes called the *costo-coracoid ligament*. Below this, it is thin, and at the upper border of the Pectoralis minor it splits into two layers to invest the muscle ; from the lower border of the Pectoralis minor it is continued downwards to join the axillary fascia, and outwards to join the fascia over the short head of the Biceps. The costo-coracoid membrane is pierced by the cephalic vein, the acromio-thoracic artery and vein, superior thoracic artery, and external anterior thoracic nerve.

The **Pectoralis minor** (fig. 422) is a thin, flat, triangular muscle, situated at the upper part of the thorax, beneath the Pectoralis major. It arises by three tendinous digitations, from the upper margin and outer surface of the third, fourth, and fifth ribs, near their cartilages, and from the aponeurosis covering the Intercostal muscles; the fibres pass upwards and outwards, and converge to form a flat tendon, which is inserted into the inner border and upper surface of the coracoid process of the scapula.

Relations.—By its *anterior surface*, with the Pectoralis major, the external anterior thoracic nerve, and the thoracic branches of the acromio-thoracic artery. By its *posterior surface*, with the ribs, Intercostal muscles, Serratus magnus, the axillary space, and the axillary vessels and brachial plexus of nerves. Its upper border is separated from the clavicle by a narrow triangular interval which is occupied by the costo-coracoid membrane. In this space are the first part of the axillary vessels and nerves. Running parallel to the lower border of the muscle is the long thoracic artery.

The costo-coracoid membrane should now be removed, when the Subclavius muscle will be seen.

The **Subclavius** is a small triangular muscle, placed in the interval between the clavicle and the first rib. It arises by a short, thick tendon from the first rib and its cartilage at their junction, in front of the rhomboid ligament; the fleshy fibres proceed obliquely upwards and outwards, to be inserted into the groove on the under surface of the clavicle between the rhomboid and conoid ligaments.

Relations.—By its *upper surface*, with the clavicle. By its *deep surface*, it is separated from the first rib by the subclavian vessels and brachial plexus of nerves. Its *anterior surface* is separated from the Pectoralis major by the costo-coracoid membrane, which, with the clavicle, forms an osseo-fibrous sheath in which the muscle is enclosed.

If the costal attachment of the Pectoralis minor is divided across, and the muscle reflected outwards, the axillary vessels and nerves are brought fully into view, and should be examined.

Nerves.—The Pectoralis major is supplied by the external and internal anterior thoracic nerves; through these nerves the muscle receives filaments from all the spinal nerves entering into the formation of the brachial plexus; the Pectoralis minor receives its fibres from the eighth cervical and first dorsal nerves through the internal anterior thoracic nerve. The Subclavius is supplied by a filament derived from the fifth and sixth cervical nerves.

Actions.—If the arm has been raised by the Deltoid, the Pectoralis major will, conjointly with the Latissimus dorsi and Teres major, depress it to the side of the chest. If acting alone, it adducts and draws forwards the arm, bringing it across the front of the chest, and at the same time rotates it inwards. The Pectoralis minor depresses the point of the shoulder, drawing the scapula downwards and inwards to the thorax, and throwing the inferior angle backwards. The Subclavius depresses the shoulder, drawing the clavicle downwards and forwards. When the arms are fixed, all three muscles act upon the ribs, drawing them upwards and expanding the chest, and thus becoming very important agents in forced inspiration. Asthmatic patients always assume an attitude which fixes the shoulders, so that all these muscles may be brought into action to assist in dilating the cavity of the chest.

2. Lateral Thoracic Region

Serratus magnus.

The **Serratus magnus** (fig. 423) is a thin, irregularly quadrilateral muscle, situated between the ribs and the scapula at the upper and lateral part of the chest. It arises by fleshy digitations from the outer surfaces and upper borders of the eight or nine upper ribs, and from the aponeurosis covering the intercostal muscles between them. Each digitation (except the first) arises from the corresponding rib; the first digitation arises from the first and second ribs, and from the fascia covering the first intercostal space. From this extensive attachment the fibres pass backwards, closely applied to the chest-wall, and reach the vertebral border of the scapula, and are inserted into its ventral aspect in the following manner. The first digitation, arising from the first and second ribs, is inserted into a triangular area on the ventral aspect of the superior angle. The next two digitations (i.e. from the second and third ribs) spread out to form a thin triangular sheet, the base of which is directed backwards and is inserted into nearly the whole length of the ventral aspect of the vertebral border. The lower five or six digitations converge to form a fan-shaped mass, the apex of which is inserted, by muscular and by tendinous fibres, into a triangular impression on the ventral aspect of the inferior angle. The lower four slips interdigitate at their origin with the upper five slips of the External oblique muscle of the abdomen.

Relations. — This muscle is partly covered, in front, by the Pectoral muscles and by the mammary gland; behind, by the Subscapularis. The axillary vessels and nerves lie upon its upper part, while its *deep surface* rests upon the ribs and Intercostal muscles.

Nerve.—The Serratus magnus is supplied by the posterior thoracic nerve, which is derived from the fifth, sixth, and seventh cervical nerves.

Actions.—The Serratus magnus, as a whole, carries the scapula forwards, and at the same time raises the vertebral border of the bone. It is therefore concerned in the action of pushing. Its lower and stronger fibres move forwards the lower angle and assist the Trapezius in rotating the bone at the sterno-clavicular joint, and thus assist this muscle in raising the acromion and supporting weights upon the shoulder. It is also an assistant to the Deltoid in raising the arm, inasmuch as during the action of this latter muscle it fixes the scapula and so steadies the glenoid cavity on which the head of the humerus rotates. After the Deltoid has raised the arm to a right angle with the trunk, the Serratus magnus and the Trapezius, by rotating the scapula, raise the arm into an almost vertical position. It is possible that when the shoulders are fixed the lower fibres of the Serratus magnus may assist in raising and everting the ribs; but it is not the important inspiratory muscle which it was formerly believed to be.

Fig. 423.—Serratus magnus. (From a preparation in the Museum of the Royal College of Surgeons of England.)

Spine of scapula

Ext. oblique

Eighth rib

Surgical Anatomy.—When the muscle is paralysed, the vertebral border, and especially the lower angle of the scapula, leave the ribs and stand out prominently on the surface, giving a peculiar 'winged' appearance to the back. The patient is unable to raise the arm, and an attempt to do so is followed by a further projection of the lower angle of the scapula from the back of the thorax.

Dissection.—After completing the dissection of the axilla, if the muscles of the back have been dissected, the upper extremity should be separated from the trunk. Saw through the clavicle at its centre, and then cut through the muscles which connect the scapula and arm with the trunk, viz. the Pectoralis minor in front, Serratus magnus at the side, and the Levator anguli scapulæ, the Rhomboids, Trapezius, and Latissimus dorsi behind. These muscles should be cleaned and traced to their respective insertions. Then make an incision through the integument, commencing at the outer third of the clavicle, and extending along the margin of that bone, the acromion process, and spine of the scapula; the integument should be dissected from above downwards and outwards, when the fascia covering the Deltoid is exposed (fig. 420, 3).

II. Muscles and Fasciæ of the Shoulder and Arm

The **superficial fascia** of the upper extremity is a thin fibro-cellular layer, containing the superficial veins and lymphatics, and the cutaneous nerves. It is most distinct in front of the elbow, where it contains very large superficial veins and nerves; in the hand it is hardly demonstrable, the integument being closely adherent to the deep fascia by dense fibrous bands. Subcutaneous bursæ are found in this fascia over the acromion, the olecranon, and the knuckles.

The deep fascia of the upper extremity comprises the aponeurosis of the shoulder, arm, and forearm, the anterior and posterior annular ligaments of the carpus, and the palmar fascia. These will be considered in the description of the muscles of the several regions.

3. Acromial Region

Deltoid.

The **deep fascia** covering the Deltoid (deltoid aponeurosis) is a fibrous layer, which covers the outer surface of the muscle, thick and strong behind, where it is continuous with the Infraspinatus fascia, thinner over the rest of its extent. It sends numerous prolongations between the fasciculi of the muscle. In front, it is continuous with the fascia covering the great Pectoral muscle; behind, with that covering the Infraspinatus; above, it is attached to the clavicle, the acromion, and spine of the scapula; below, it is continuous with the deep fascia of the arm.

The **Deltoid** (fig. 421) is a large, thick, triangular muscle, which gives the rounded outline to the shoulder, and has received its name from its resemblance to the Greek letter Δ reversed. It covers the shoulder-joint in front, behind, and on its outer side. It arises from the outer third of the anterior border and upper surface of the clavicle; from the outer margin and upper surface of the acromion process; and from the lower lip of the posterior border of the spine of the scapula, as far back as the triangular surface at its inner end. From this extensive origin the fibres converge towards their insertion, the middle passing vertically, the anterior obliquely backwards, the posterior obliquely forwards; they unite to form a thick tendon, which is inserted into a rough triangular prominence on the middle of the outer side of the shaft of the humerus. At its insertion the muscle gives off an expansion to the deep fascia of the arm. This muscle is remarkably coarse in texture, and the arrangement of its muscular fibres is somewhat peculiar; the central portion of the muscle—that is to say, the part arising from the acromion process—consists of oblique fibres, which arise in a bipenniform manner from the sides of tendinous intersections, generally four in number, which are attached above to the acromion process and pass downwards parallel to one another in the substance of the muscle. The oblique muscular fibres thus formed are inserted into similar tendinous intersections, generally three in number, which pass upwards from the insertion of the muscle into the humerus and alternate with the descending septa. The portions of the muscle which arise from the clavicle and spine of the scapula are not arranged in this manner, but pass from their origin above, to be inserted into the margins of the inferior tendon.

Relations.—By its *superficial surface*, with the integument, the superficial and deep fasciæ, Platysma, and supra-acromial nerves. Its *deep surface* is separated from the capsule of the shoulder-joint by a large synovial bursa, and covers the coracoid process, coraco-acromial ligament, Pectoralis minor, Coraco-brachialis, both heads of the Biceps, the tendon of the Pectoralis major, the insertions of the Supraspinatus, Infraspinatus, and Teres minor, the scapular and external heads of the Triceps, the circumflex vessels and nerve, and the upper part of the shaft of the humerus. Its *anterior border* is separated at its upper part from the Pectoralis major by a cellular interspace, which lodges the cephalic vein and humeral branch of the acromio-thoracic artery: lower down the two muscles are in close contact. Its *posterior border* rests on the Infraspinatus and Triceps muscles.

Nerves.—The Deltoid is supplied by the fifth and sixth cervical through the circumflex nerve.

Actions.—The Deltoid raises the arm directly from the side, so as to bring it at right angles with the trunk. Its anterior fibres, assisted by the Pectoralis major, draw the arm forwards; and its posterior fibres, aided by the Teres major and Latissimus dorsi, draw it backwards.

Surgical Anatomy.—The Deltoid is very liable to atrophy, and when in this condition simulates dislocation of the shoulder-joint, as there is flattening of the shoulder and apparent prominence of the acromion process; the distance also between the acromion process and the head of the bone is increased, and the tips of the fingers can be inserted between them. Atrophy of the Deltoid may be due to disuse or loss of trophic influence, either from injury to the circumflex nerve or cord lesions, as in infantile paralysis.

Dissection.—Divide the Deltoid across, near its upper part, by an incision carried along the margin of the clavicle, the acromion process, and spine of the scapula, and reflect it downwards, when the structures under cover of it will be seen.

4. *Anterior Scapular Region*

Subscapularis.

The **subscapular fascia** is a thin membrane attached to the entire circumference of the subscapular fossa, and affording attachment by its inner surface to some of the fibres of the Subscapularis muscle: when this is removed, the Subscapularis muscle is exposed.

The **Subscapularis** (fig. 422) is a large triangular muscle, which fills up the subscapular fossa, arising from its internal two-thirds, with the exception of a narrow margin along the posterior border, and the surfaces at the superior and inferior angles which afford attachment to the Serratus magnus: it also arises from the lower two-thirds of the groove on the axillary border of the bone. Some fibres arise from tendinous laminæ, which intersect the muscle, and are attached to ridges on the bone; and others from an aponeurosis, which separates the muscle from the Teres major and the long head of the Triceps. The fibres pass outwards, and, gradually converging, terminate in a tendon, which is inserted into the lesser tuberosity of the humerus and the front of the capsular ligament of the shoulder-joint. The tendon of the muscle is separated from the neck of the scapula and anterior part of the capsular ligament of the shoulder-joint by a large bursa, which communicates with the cavity of the joint by an aperture in the capsular ligament.

Relations.—The *anterior surface* of this muscle forms a considerable part of the posterior wall of the axilla, and is in relation with the Serratus magnus, Coraco-brachialis, and Biceps, the axillary vessels and brachial plexus of nerves, and the subscapular vessels and nerves. Its *posterior surface* is in relation with the scapula and the capsular ligament of the shoulder-joint. Its *lower border* is contiguous with the Teres major and Latissimus dorsi.

Nerves.—The Subscapularis is supplied by the fifth and sixth cervical nerves through the upper and lower subscapular nerves.

Actions.—The Subscapularis rotates the head of the humerus inwards; when the arm is raised, it draws the humerus forwards and downwards. It is a powerful defence to the front of the shoulder-joint, preventing displacement of the head of the bone.

5. *Posterior Scapular Region* (fig. 424)

Supraspinatus.	Teres minor.
Infraspinatus.	Teres major.

Dissection.—To expose these muscles, and to examine their mode of insertion into the humerus, detach the Deltoid and Trapezius from their attachment to the spine of the scapula and acromion process. Remove the clavicle by dividing the ligaments connecting it with the coracoid process, and separate it at its articulation with the scapula: divide the acromion process near its root with a saw. The fragments being removed, the tendons of the posterior Scapular muscles will be fully exposed, and can be examined. A block should be placed beneath the shoulder-joint, so as to make the muscles tense.

The **supraspinous fascia** is a thick and dense membranous layer, which completes the osseo-fibrous case in which the Supraspinatus muscle is contained; affording attachment, by its deep surface, to some of the fibres of the muscle. It is thick internally, but thinner externally under the coraco-acromial ligament. When this fascia is removed, the Supraspinatus muscle is exposed.

The **Supraspinatus muscle** occupies the whole of the supraspinous fossa, arising from its internal two-thirds, and from the strong fascia which covers its surface. The muscular fibres converge to a tendon, which passes across the upper part of the capsular ligament of the shoulder-joint, to which it is intimately adherent, and is inserted into the highest of the three impressions on the great tuberosity of the humerus.

Relations.—By its *upper surface*, with the Trapezius, the clavicle, the acromion, the coraco-acromial ligament, and the Deltoid. By its *under surface*, with the scapula, the suprascapular vessels and nerve, and upper part of the shoulder-joint.

The **infraspinous fascia** is a dense fibrous membrane, covering in the Infra-spinatus muscle and fixed to the circumference of the infraspinous fossa; it affords attachment, by its inner surface, to some fibres of that muscle. At the

FIG. 424.—Muscles on the dorsum of the scapula and the Triceps.

point where the Infraspinatus commences to be covered by the Deltoid, this fascia divides into two layers : one layer passes over the Deltoid muscle, helping to form the Deltoid fascia already described ; the other passes beneath the Deltoid to the shoulder-joint.

The **Infraspinatus** is a thick triangular muscle, which occupies the chief part of the infraspinous fossa, arising by fleshy fibres from its internal two-thirds ; and by tendinous fibres from the ridges on its surface : it also arises from a strong fascia which covers it externally, and separates it from the Teres major and minor. The fibres converge to a tendon, which glides over the external border of the spine of the scapula, and, passing across the posterior part of the capsular ligament of the shoulder-joint, is inserted into the middle impression on the great tuberosity of the humerus. The tendon of this muscle is sometimes separated from the capsule of the shoulder-joint by a synovial bursa, which may communicate with the joint cavity.

Relations.—By its *posterior surface*, with the Deltoid, the Trapezius, Latissimus dorsi, and the integument. By its *anterior surface*, with the scapula, the suprascapular nerve, the suprascapular and dorsalis scapulæ vessels, and the capsular ligament of the shoulder-joint. Its *lower border* is in contact with the Teres major and minor, and is occasionally united with the latter.

The **Teres minor** is a narrow, elongated muscle, which arises from the dorsal surface of the axillary border of the scapula for the upper two-thirds of its extent, and from two aponeurotic laminæ, one of which separates this muscle from the Infraspinatus, the other from the Teres major; its fibres pass obliquely upwards and outwards, and terminate in a tendon, which is inserted into the lowest of the three impressions on the great tuberosity of the humerus, and, by fleshy fibres, into the humerus immediately below it. The tendon of this muscle passes across, and is united with, the posterior part of the capsular ligament of the shoulder-joint.

Relations.—By its *posterior surface*, with the Deltoid, and the integument. By its *anterior surface*, with the scapula, the dorsal branch of the subscapular artery, the long head of the Triceps, and the shoulder-joint. By its *upper border*, with the Infraspinatus. By its *lower border*, with the Teres major, from which it is separated anteriorly by the long head of the Triceps.

The **Teres major** is a thick but somewhat flattened muscle, which arises from the oval surface on the dorsal aspect of the inferior angle of the scapula, and from the fibrous septa interposed between it and the Teres minor and Infraspinatus; the fibres are directed upwards and outwards, and terminate in a flat tendon, about two inches in length, which is inserted into the inner bicipital ridge of the humerus. The tendon of this muscle, at its insertion into the humerus, lies behind that of the Latissimus dorsi, from which it is separated by a synovial bursa, the two tendons being, however, united along their lower borders for a short distance.

Relations.—By its *posterior surface*, with the Latissimus dorsi below, and the long head of the Triceps above. By its *anterior surface*, with the Subscapularis, Latissimus dorsi, Coraco-brachialis, short head of the Biceps, the axillary vessels, and brachial plexus of nerves. Its *upper border* is at first in relation with the Teres minor, from which it is afterwards separated by the long head of the Triceps. Its *lower border* forms, in conjunction with the Latissimus dorsi, part of the posterior boundary of the axilla. The Latissimus dorsi at first covers the origin of the Teres major, then wraps itself obliquely round its lower border, so that its tendon ultimately comes to lie in front of that of the Teres major.

Nerves.—The Supra- and Infra-spinatus muscles are supplied by the fifth and sixth cervical nerves through the suprascapular nerve; the Teres minor, by the fifth cervical, through the circumflex; and the Teres major, by the fifth and sixth cervical, through the lower subscapular.

Actions.—The Supraspinatus assists the Deltoid in raising the arm from the side, and fixes the head of the humerus in the glenoid cavity. The Infraspinatus and Teres minor rotate the head of the humerus outwards: they also assist in carrying the arm backwards. One of the most important uses of these three muscles is the great protection they afford to the shoulder-joint, the Supraspinatus supporting it above, and preventing displacement of the head of the humerus upwards, while the Infraspinatus and Teres minor protect it behind, and prevent dislocation backwards. The Teres major assists the Latissimus dorsi in drawing the humerus downwards and backwards, when previously raised, and rotating it inwards; when the arm is fixed it may assist the Pectoral and Latissimus dorsi muscles in drawing the trunk forwards.

6. *Anterior Humeral Region* (fig. 422)

Coraco-brachialis.   Biceps.   Brachialis anticus.

Dissection.—The arm being placed on the table, with the front surface uppermost, make a vertical incision through the integument along the middle line, from the outer extremity of the anterior fold of the axilla, to about two inches below the elbow-joint, where it should be joined by a transverse incision, extending from the inner to the outer side of the forearm; the two flaps being reflected on either side, the fascia should be examined (fig. 420).

The **deep fascia** of the arm is continuous with that covering the Deltoid and the great Pectoral muscle, by means of which it is attached, above, to the clavicle, acromion, and spine of the scapula ; it forms a thin, loose, membranous sheath investing the muscles of the arm, and sends inwards septa between them ; it is composed of fibres disposed in a circular or spiral direction, and connected together by vertical and oblique fibres. It differs in thickness at different parts, being thin over the Biceps, but thicker where it covers the Triceps, and over the condyles of the humerus : it is strengthened by fibrous aponeuroses, derived from the Pectoralis major and Latissimus dorsi on the inner side, and from the Deltoid externally. On either side it gives off a strong intermuscular septum, which is attached to the corresponding supracondylar ridge and condyle of the humerus. These septa serve to separate the muscles of the anterior from those of the posterior brachial region. The *external intermuscular septum* extends from the lower part of the anterior bicipital ridge, along the external supracondylar ridge, to the outer epicondyle ; it is blended with the tendon of the Deltoid, gives attachment to the Triceps behind, to the Brachialis anticus, Supinator longus, and Extensor carpi radialis longior, in front ; and is perforated by the musculo-spiral nerve and superior profunda artery. The *internal intermuscular septum*, thicker than the preceding, extends from the lower part of the posterior lip of the bicipital groove below the Teres major, along the internal supracondylar ridge to the inner epicondyle ; it is blended with the tendon of the Coraco-brachialis, and affords attachment to the Triceps behind and the Brachialis anticus in front. It is perforated by the ulnar nerve, the inferior profunda artery, and the posterior branch of the anastomotic artery. At the elbow, the deep fascia is attached to all the prominent points round the joint, viz. the epicondyles of the humerus and the olecranon process of the ulna, and is continuous with the deep fascia of the forearm. Just below the middle of the arm, on its inner side, in front of the internal intermuscular septum, is an oval opening in the deep fascia, which transmits the basilic vein and some lymphatic vessels. On the removal of this fascia, the muscles, vessels, and nerves of the anterior humeral region are exposed.

The **Coraco-brachialis**, the smallest of the three muscles in this region, is situated at the upper and inner part of the arm. It arises by fleshy fibres from the apex of the coracoid process, in common with the short head of the Biceps, and from the intermuscular septum, between the two muscles ; the fibres pass downwards, backwards, and a little outwards, to be inserted by means of a flat tendon into an impression at the middle of the inner surface and internal border of the shaft of the humerus between the origins of the Triceps and Brachialis anticus. It is perforated by the musculo-cutaneous nerve. The inner border of the muscle forms a guide to the position of the terminal portion of the axillary and upper part of the brachial arteries.

Relations.—By its *anterior surface*, with the Pectoralis major above, and at its insertion with the brachial vessels and median nerve which cross it. By its *posterior surface*, with the tendons of the Subscapularis, Latissimus dorsi, and Teres major, the inner head of the Triceps, the humerus, and the anterior circumflex vessels. By its *inner border*, with the brachial artery and the median and musculo-cutaneous nerves. By its *outer border*, with the short head of the Biceps and Brachialis anticus.

The **Biceps** (*Biceps flexor cubiti*) is a long fusiform muscle, occupying the whole of the anterior surface of the arm, and divided above into two portions or heads, from which circumstance it has received its name. The short head arises by a thick flattened tendon from the apex of the coracoid process, in common with the Coraco-brachialis. The long head arises from the supraglenoid tubercle at the upper margin of the glenoid cavity, and is continuous with the glenoid ligament. This tendon arches over the head of the humerus, being enclosed in a special sheath of the synovial membrane of the shoulder-joint ; it then passes through an opening in the capsular ligament at its attachment to the humerus, and descends in the bicipital groove, in which it is retained by the transverse humeral ligament and by a fibrous prolongation from the tendon of the Pectoralis major. Each tendon is succeeded by an elongated muscular belly, and the two bellies, although closely applied to each other, can readily be separated until within about three inches of the elbow-joint. Here they end in a flattened tendon, which is inserted into the rough posterior portion of the tuberosity of the radius, a synovial bursa being interposed between the tendon and the front part of the tuberosity. As the tendon of the muscle approaches the radius it is twisted upon itself, so that its anterior surface becomes external and is applied to the tuberosity of the radius at its insertion : opposite the bend of the elbow the tendon gives off, from its inner side, a broad aponeurosis, the *bicipital fascia* (*semilunar fascia*), which passes obliquely downwards and inwards across the brachial artery, and is continuous with the deep fascia covering the origins of the Flexor muscles of the forearm (fig. 421). The inner border of this muscle forms a guide to the position of the brachial artery, in tying the vessel in the middle of the arm.*

Relations.—Its *anterior surface* is overlapped above by the Pectoralis major and Deltoid ; in the rest of its extent it is covered by the superficial and deep fasciæ and the integument. Its *posterior surface* rests above on the shoulder-joint and upper part of the humerus ; below, it lies on the Brachialis anticus, with the musculo-cutaneous nerve intervening between the two, and on the Supinator brevis. Its *inner border* is in relation with the Coraco-brachialis, and overlaps the brachial vessels, and median nerve ; its *outer border*, with the Deltoid and Supinator longus.

The **Brachialis anticus** is a broad muscle, which covers the elbow-joint and the lower half of the front of the humerus. It is somewhat compressed from before backwards, and is broader in the middle than at either extremity. It arises from the lower half of the front of the humerus ; and commences above at the insertion of the Deltoid, which it embraces by two angular processes. Its origin extends below, to within an inch of the margin of the articular surface. It also arises from the intermuscular septa, but more extensively from the inner than the outer, from which it is separated below by the Brachio-radialis (Supinator longus) and Extensor carpi radialis longior. Its fibres converge to a thick tendon, which is inserted into a rough depression on the anterior surface of the coronoid process of the ulna, being received into an interval between two fleshy slips of the Flexor profundus digitorum.

Relations.—By its *anterior surface*, with the Biceps, the brachial vessels, musculo-cutaneous and median nerves. By its *posterior surface*, with the humerus and front of the elbow-joint. By its *inner border*, with the Triceps, ulnar nerve, and Pronator radii teres, from which it is separated by the intermuscular septum. By its *outer border*, with the musculo-spiral nerve, radial recurrent artery, the Brachio-radialis, and Extensor carpi radialis longior.

Nerves.—The muscles of this group are supplied by the musculo-cutaneous nerve. The Brachialis anticus usually receives an additional filament from the musculo-spiral. The Coraco-brachialis receives its supply primarily from the seventh cervical, the Biceps and Brachialis anticus from the fifth and sixth cervical nerves.

Actions.—The Coraco-brachialis draws the humerus forwards and inwards, and at the same time assists in elevating it towards the scapula. The Biceps is a flexor of the forearm : it is also a powerful supinator, and serves to render tense the deep fascia of the forearm by means of the bicipital fascia given off from its tendon. The Brachialis anticus is a flexor of the forearm, and forms an important defence to the elbow-joint. When the forearm is fixed, the Biceps and Brachialis anticus flex the arm upon the forearm, as is seen in efforts of climbing.

Surgical Anatomy.—The long tendon of the Biceps is sometimes dislocated from the Bicipital groove. When this takes place, the arm becomes fixed in a position of abduction, but the head of the humerus can be felt in its proper position. It can generally be replaced by flexing the forearm on the arm and rotating the limb. Rupture of the long tendon of the Biceps may also take place.

---

* A third head to the Biceps is occasionally found (Theile says as often as once in eight or nine subjects), arising at the upper and inner part of the Brachialis anticus, with the fibres of which it is continuous, and inserted into the bicipital fascia and inner side of the tendon of the Biceps. In most cases this additional slip passes behind the brachial artery in its course down the arm. In some instances the third head consists of two slips, which pass down, one in front, the other behind the artery, concealing the vessel in the lower half of the arm.

7. Posterior Humeral Region

Triceps.   Subanconeus.

The **Triceps** (*Triceps extensor cubiti*) (fig. 424) is situated on the back of the arm, extending the entire length of the posterior surface of the humerus. It is of large size, and divided above into three parts : hence its name. These three portions have been named (1) the middle, scapular, or long head ; (2) the external, or long humeral head ; and (3) the internal, or short humeral head.

The *middle* or *scapular head* arises, by a flattened tendon, from a rough triangular depression on the scapula, immediately below the glenoid cavity, being blended at its upper part with the capsular ligament ; the muscular fibres pass downwards between the two other portions of the muscle, and join with them in the common tendon of insertion.

The *external head* arises from the posterior surface of the shaft of the humerus, between the insertion of the Teres minor and the upper part of the musculo-spiral groove ; from the external border of the humerus and the external intermuscular septum : the fibres from this origin converge towards the common tendon of insertion.

The *internal head* arises from the posterior surface of the shaft of the humerus, below the musculo-spiral groove : it is narrow and pointed above, and extends below the insertion of the Teres major to within an inch of the trochlear surface : it also arises from the internal border of the humerus and from the back of the whole length of the internal and lower part of the external intermuscular septa. The fibres of this portion of the muscle are directed, some downwards to the olecranon, while others converge to the common tendon of insertion.

The *common tendon* of the Triceps commences about the middle of the back part of the muscle : it consists of two aponeurotic laminæ, one of which is subcutaneous and covers the posterior surface of the muscle for the lower half of its extent ; the other is more deeply seated in the substance of the muscle ; after receiving the attachment of the muscular fibres, they join together above the elbow, and are inserted, for the most part, into the posterior portion of the upper surface of the olecranon process ; a band of fibres is, however, continued downwards, on the outer side, over the Anconeus, to blend with the deep fascia of the forearm.

The long head of the Triceps descends between the Teres minor and Teres major, dividing the triangular space between these two muscles and the humerus into two smaller spaces, one triangular, the other quadrangular (fig. 424). The triangular space contains the dorsalis scapulæ vessels ; it is bounded by the Teres minor above, the Teres major below, and the scapular head of the Triceps externally : the quadrangular space transmits the posterior circumflex vessels and the circumflex nerve ; it is bounded by the Teres minor above, the Teres major below, the scapular head of the Triceps internally, and the humerus externally.

Relations.—By its *posterior surface*, with the Deltoid above : in the rest of its extent it is subcutaneous. By its *anterior surface*, with the humerus, musculo-spiral nerve, superior profunda vessels, and back part of the elbow-joint. Its *middle* or *long head* is in relation, behind, with the Deltoid and Teres minor ; in front, with the Subscapularis, Latissimus dorsi, and Teres major.

The **Subanconeus** is a name given to a few fibres from the under surface of the lower part of the Triceps muscle, which are inserted into the posterior ligament and synovial membrane of the elbow-joint. By some authors it is regarded as the analogue of the Subcrureus in the lower limb, but it is not a separate muscle.

Nerves.—The Triceps is supplied by the seventh and eighth cervical nerves through the musculo-spiral nerve.

Actions.—The Triceps is the great extensor muscle of the forearm, serving, when the forearm is flexed, to extend the elbow-joint. It is the direct antagonist of the Biceps and Brachialis anticus. When the arm is extended, the long head of the muscle may assist the Teres major and Latissimus dorsi in drawing the humerus backwards and in adducting it to the thorax. The long head of the Triceps protects the under part of the shoulder-joint, and prevents displacement of the head of the humerus downwards and backwards. The Subanconeus draws up the posterior ligament during extension of the forearm.

Surgical Anatomy.—The existence of the band of fibres from the Triceps to the fascia of the forearm is of importance in excision of the elbow, and should always be carefully preserved from injury by the operator, as by means of these fibres the patient is enabled to extend the forearm, a movement which would otherwise mainly be accomplished by gravity—that is to say, by allowing the forearm to drop from its own weight.

III. MUSCLES AND FASCIÆ OF THE FOREARM

Dissection.—To dissect the forearm, place the limb in the position indicated in fig. 420, make a vertical incision along the middle line from the elbow to the wrist, and a transverse incision at the extremity of this; the superficial structures being removed, the deep fascia of the forearm is exposed.

The **deep fascia** of the forearm, continuous above with that enclosing the arm, is a dense, highly glistening aponeurotic investment, which forms a general sheath enclosing the muscles in this region; it is attached, behind, to the olecranon and posterior border of the ulna, and gives off from its deep surface numerous intermuscular septa, which enclose each muscle separately. Below, it is continuous in front with the anterior annular ligament of the wrist, and forms a sheath for the tendon of the Palmaris longus muscle, which passes over the annular ligament to be inserted into the palmar fascia. Behind, near the wrist-joint, it becomes much thickened by the addition of many transverse fibres, and forms the posterior annular ligament. It consists of circular and oblique fibres, connected together by numerous vertical fibres. It is much thicker on the posterior than on the anterior surface, and at the lower than at the upper part of the forearm, and is strengthened above by tendinous fibres derived from the Biceps in front, and from the Triceps behind. Its deep surface gives origin to muscular fibres, especially at the upper part of the inner and outer sides of the forearm, and forms the boundaries of a series of conical-shaped cavities, in which the muscles are contained. Besides the vertical septa separating the individual muscles, transverse septa are given off both on the anterior and posterior surfaces of the forearm, separating the deep from the superficial layer of muscles. Numerous apertures exist in the fascia for the passage of vessels and nerves; one of these, of large size, situated at the front of the elbow, serves for the passage of a communicating branch between the superficial and deep veins. Near the wrist, it is perforated on its anterior surface by the ulnar artery and nerve.

The muscles of the forearm may be subdivided into groups corresponding to the region they occupy. One group occupies the inner and anterior aspect of the forearm, and comprises the Flexor and Pronator muscles. Another group occupies its outer side; and a third its posterior aspect. The two latter groups include all the Extensor and Supinator muscles.

8. *Anterior Radio-ulnar Region*

The muscles in this region are divided for convenience of description into two groups or layers, superficial and deep.

Superficial Layer

Pronator radii teres.   Palmaris longus.
Flexor carpi radialis.   Flexor carpi ulnaris.
                    Flexor sublimis digitorum.

These muscles take origin from the internal epicondyle of the humerus by a common tendon; they receive additional fibres from the deep fascia of the forearm near the elbow, and from the septa which pass inwards from this fascia between the individual muscles.

The **Pronator radii teres** has two heads of origin. One, the larger and more superficial, arises from the humerus, immediately above the internal epicondyle, and from the tendon common to the origin of the other muscles; also from the fascia of the forearm, and intermuscular septum between it and the Flexor carpi radialis. The other head is a thin fasciculus, which arises from the inner side of the coronoid process of the ulna, and joins the preceding at an acute angle. Between the two heads the median nerve enters the forearm. The muscle passes obliquely across the forearm from the inner to the outer side, and terminates in a flat tendon, which turns over the outer margin of the radius, and is inserted into a rough impression at the middle of the outer surface of the shaft of that bone.

Relations.—By its *anterior surface* throughout the greater part of its extent with the deep fascia; at its insertion it is crossed by the radial vessels and nerve and covered by the Brachio-radialis. By its *posterior surface*, with the Brachialis anticus, Flexor sublimis digitorum, the median nerve, and ulnar artery: the small, or deep, head being interposed between the two latter structures. Its *outer border* forms the inner boundary of a triangular space (*antecubital fossa*), situated in front of the elbow-joint, which contains the brachial artery, median nerve, and tendon of the Biceps muscle. Its *inner border* is in contact with the Flexor carpi radialis.

Surgical Anatomy.—This muscle, when suddenly brought into very active use, as in the game of lawn tennis, is apt to be strained, producing slight swelling, tenderness, and pain on putting the muscle into action. This is known as 'lawn-tennis arm.'

The **Flexor carpi radialis** lies on the inner side of the preceding muscle. It arises from the internal epicondyle by the common tendon, from the fascia of the forearm, and from the intermuscular septa between it and the Pronator radii teres externally, the Palmaris longus internally, and the Flexor sublimis digitorum beneath. Slender and aponeurotic in structure at its commencement, it increases in size, and terminates in a tendon, which forms rather more than the lower half of its length. This tendon passes through a canal in the outer part of the annular ligament, runs through a groove in the os trapezium (which is converted into a canal by a fibrous sheath, and lined by a synovial membrane), and is inserted into the base of the metacarpal bone of the index finger, and by a slip into the base of the metacarpal bone of the middle finger. The radial artery, in the lower part of the forearm, lies between the tendon of this muscle and the Supinator longus, and may easily be tied in this situation.

Relations.—By its *superficial surface*, with the deep fascia and the integument. By its *deep surface*, with the Flexor sublimis digitorum, Flexor longus pollicis, and wrist-joint. By its *outer border*, with the Pronator radii teres and the radial vessels. By its *inner border*, with the Palmaris longus above, and the median nerve below.

The **Palmaris longus** is a slender, fusiform muscle, lying on the inner side of the preceding. It arises from the inner epicondyle of the humerus by the common tendon, from the deep fascia, and the intermuscular septa between it and the adjacent muscles. It terminates in a slender, flattened tendon, which passes over the upper part of the annular ligament, to end in the central part of the palmar fascia and lower part of the annular ligament, frequently sending a tendinous slip to the short muscles of the thumb. This muscle is often absent, and is subject to very considerable variations: it may be tendinous above and muscular below; or it may be muscular in the centre, with a tendon above and below; or it may present two muscular bundles with a central tendon; or finally it may consist simply of a mere tendinous band.

Relations.—By its *superficial surface*, with the deep fascia. By its *deep surface*, with the Flexor sublimis digitorum. *Internally*, with the Flexor carpi ulnaris. *Externally*, with the Flexor carpi radialis. Just above the wrist the median nerve lies close to the tendon, on its outer and posterior side.

The **Flexor carpi ulnaris** lies along the ulnar side of the forearm. It arises by two heads, connected by a tendinous arch, beneath which pass the ulnar nerve and posterior ulnar recurrent artery. One head arises from the inner epicondyle of the humerus by the common tendon; the other from the inner margin of the olecranon and from the upper two-thirds of the posterior border of the ulna by an aponeurosis, common to it and the Extensor carpi ulnaris and Flexor profundus digitorum; and from the intermuscular septum between it and the Flexor sublimis digitorum. The fibres terminate in a tendon, which occupies the anterior part of the lower half of the muscle, and is inserted into the pisiform bone, and is prolonged from this to the fifth metacarpal and unciform bones, by the piso-metacarpal and piso-unciform ligaments: it is also attached by a few fibres to the annular ligament. The ulnar artery lies on the outer side of the tendon of this muscle, in the lower two-thirds of the forearm; the tendon forming a guide in tying the vessel in this situation.

Relations.—By its *superficial surface*, with the deep fascia, with which it is intimately connected for a considerable extent. By its *deep surface*, with the Flexor sublimis digitorum, the Flexor profundus digitorum, the Pronator quadratus, and the ulnar vessels and nerve. By its *outer* or *radial border*, with the Palmaris longus above, and the ulnar vessels and nerve below.

The **Flexor sublimis digitorum** (*perforatus*) is placed beneath the preceding muscles, which therefore must be removed in order to bring its attachment into view. It is the largest of the muscles of the superficial layer, and arises by three heads. One head arises from the internal epicondyle of the humerus by the common tendon, from the internal lateral ligament of the elbow-joint, and from the intermuscular septum common to it and the preceding muscles. The second head arises from the inner side of the coronoid process of the ulna, above the ulnar origin of the Pronator radii teres (fig. 291, page 280). The third head arises from the oblique line of the radius, extending from the tubercle to the insertion of the Pronator radii teres. The fibres pass vertically downwards, forming a broad and thick muscle, which speedily separates into two planes of muscular fibres, superficial and deep: the superficial plane divides into two parts which end in tendons for the middle and ring fingers; the deep plane also divides into two parts, which end in tendons for the index and little fingers, but previously to having done so, it gives off a muscular slip, which joins that part of the superficial plane which is intended for the ring finger. As the four tendons thus formed pass beneath the annular ligament into the palm of the hand, they are arranged in pairs, the superficial pair corresponding to the middle and ring fingers, the deep pair to the index and little fingers. The tendons diverge from one another as they pass onwards. Opposite the bases of the first phalanges each tendon divides into two slips, to allow of the passage of the corresponding tendon of the Flexor profundus digitorum; the two portions of the tendon then unite and form a grooved channel for the reception of the accompanying deep Flexor tendon. Finally they subdivide a second time, to be inserted into the sides of the second phalanges about their middle. After leaving the palm, these tendons accompanied by the deep Flexor tendons lie in osseo-aponeurotic canals (fig. 426). The canals are formed by strong fibrous bands, which arch across the tendons, and are attached on each side to the margins of the phalanges. Opposite the middle of the proximal and second phalanges the sheath is very strong, and the fibres pass transversely; but opposite the joints it is much thinner, and the fibres pass obliquely. Each sheath is lined by a synovial membrane, which is reflected on the contained tendons.

Fig. 425.—Front of the left forearm. Superficial muscles.

Relations.—In the forearm, by its *superficial surface*, with the deep fascia and all the preceding superficial muscles; by its *deep surface*, with the Flexor profundus digitorum, Flexor longus pollicis, the ulnar vessels and nerve, and the median nerve. In the hand, its tendons are in relation, *in front*, with the palmar fascia, superficial palmar arch, and the branches of the median nerve; *behind* with the tendons of the deep Flexor and the Lumbricales.

Fig. 426.—Front of the left forearm.
Deep muscles.

Deep Layer

Flexor profundus digitorum.
Flexor longus pollicis.
Pronator quadratus.

Dissection.—Divide each of the superficial muscles at its centre, and turn either end aside; the deep layer of muscles, together with the median nerve and ulnar vessels, will then be exposed.

The **Flexor profundus digitorum** (*perforans*) (fig. 426) is situated on the ulnar side of the forearm, immediately beneath the superficial Flexors. It arises from the upper three-fourths of the anterior and inner surfaces of the shaft of the ulna, embracing the insertion of the Brachialis anticus above, and extending, below, to within a short distance of the Pronator quadratus. It also arises from a depression on the inner side of the coronoid process; by an aponeurosis from the upper three-fourths of the posterior border of the ulna, in common with the Flexor and Extensor carpi ulnaris; and from the ulnar half of the interosseous membrane. The fibres form a fleshy belly of considerable size, which divides into four tendons: these pass under the annular ligament beneath the tendons of the Flexor sublimis digitorum. Opposite the first phalanges, the tendons pass through the openings in the tendons of the Flexor sublimis digitorum, and are finally inserted into the bases of the last phalanges. The portion of the muscle for the index finger is usually distinct throughout, but the tendons for the three inner fingers are connected together by cellular tissue and tendinous slips, as far as the palm of the hand. The tendons of this muscle and those of the Flexor sublimis digitorum, while contained in the osseo-aponeurotic canals of the fingers, are invested in a synovial sheath, and are connected to each other, and to the phalanges, by slender, tendinous filaments, called *vincula accessoria tendinum*. There are two sets of these: (*a*) the *ligamenta brevia*, which are two in number in each finger, and consist of triangular bands of fibres connecting the tendon of the Flexor sublimis digitorum to the front of the first interphalangeal joint and head of the first phalanx, and the tendon of the Flexor profundus digitorum to the front of the second interphalangeal joint and head of the second phalanx ; (b) the *ligamenta longa*, which connect the under surface of the tendons of the Flexor profundus digitorum to the subjacent Flexor sublimis after the tendons of the former have passed through the latter. The ligamentum breve of the deep Flexor tendons consists largely of yellow elastic tissue, and may assist in drawing down the tendon after flexion of the finger.*

Four small muscles, the Lumbricales, are connected with the tendons of the Flexor profundus in the palm. They will be described with the muscles of the hand.

Relations.—By its *superficial surface*, in the forearm, with the Flexor sublimis digitorum, the Flexor carpi ulnaris, the ulnar vessels and nerve, and the median nerve ; and in the hand, with the tendons of the superficial Flexor. By its *deep surface*, in the forearm, with the ulna, the interosseous membrane, the Pronator quadratus ; and in the hand, with the Interossei, Adductores pollicis, and deep palmar arch. By its *ulnar border*, with the Flexor carpi ulnaris. By its *radial border*, with the Flexor longus pollicis, the anterior interosseous vessels and nerve being interposed.

The **Flexor longus pollicis** is situated on the radial side of the forearm, lying on the same plane as the preceding. It arises from the grooved anterior surface of the shaft of the radius : commencing, above, immediately below the tuberosity and oblique line, and extending, below, to within a short distance of the Pronator quadratus. It also arises from the adjacent part of the interosseous membrane, and generally by a fleshy slip from the inner border of the coronoid process, or from the internal condyle of the humerus. The fibres proceed downwards, and terminate in a flattened tendon, which passes beneath the annular ligament, is then lodged in the interspace between the outer head of the Flexor brevis pollicis and the Adductor obliquus pollicis, and, entering an osseo-aponeurotic canal similar to those for the other Flexor tendons, is inserted into the base of the last phalanx of the thumb.

Relations.—By its *superficial surface*, with the Flexor sublimis digitorum, Flexor carpi radialis, Brachio-radialis, and radial vessels. By its *deep surface*, with the radius, interosseous membrane, and Pronator quadratus. By its *ulnar border*, with the Flexor profundus digitorum, from which it is separated by the anterior interosseous vessels and nerve.

The **Pronator quadratus** is a small, flat, quadrilateral muscle, extending transversely across the front of the radius and ulna, above their carpal extremities. It arises from the pronator ridge on the lower part of the anterior surface of the shaft of the ulna ; from the inner part of the anterior surface of the lower fourth of the ulna ; and from a strong aponeurosis which covers the inner third of the muscle. The fibres pass outwards and slightly downwards, to be inserted into the lower fourth of the outer border and the anterior surface of the shaft of the radius. The deeper fibres of the muscle are inserted into the triangular area above the sigmoid cavity of the radius—an attachment comparable with the ulnar origin of the Supinator brevis from the triangular area below the lesser sigmoid cavity of that bone.

Relations.—By its *superficial surface*, with the Flexor profundus digitorum, the Flexor longus pollicis, Flexor carpi radialis, and the radial vessels. By its *deep surface*, with the radius, ulna, and interosseous membrane.

Nerves.—All the muscles of the superficial layer are supplied by the median nerve, excepting the Flexor carpi ulnaris, which is supplied by the ulnar. The Pronator radii teres, the Flexor carpi radialis, and the Palmaris longus derive their supply primarily from the sixth cervical ; the Flexor sublimis digitorum from the seventh and eighth cervical and first dorsal, and the Flexor carpi ulnaris from the eighth cervical and first dorsal nerves. Of the deep layer, the Flexor profundus digitorum is supplied by the eighth cervical and first dorsal through the ulnar and anterior interosseous branch of the median. The remaining two muscles, Flexor longus pollicis and Pronator quadratus, are also supplied by the eighth cervical and first dorsal through the anterior interosseous branch of the median.

Actions.—These muscles act upon the forearm, the wrist, and hand. The

---

* Marshall, *Brit. and For. Med.-Chir. Rev.* 1853.

Pronator radii teres helps to rotate the radius upon the ulna, rendering the hand prone; when the radius is fixed, it assists the other muscles in flexing the forearm. The Flexor carpi radialis is one of the flexors of the wrist; when acting alone, it flexes the wrist, inclining it to the radial side. It can also assist in pronating the forearm and hand, and, by continuing its action, in bending the elbow. The Flexor carpi ulnaris is one of the flexors of the wrist; when acting alone, it flexes the wrist, inclining it to the ulnar side; and by continuing to contract, it bends the elbow. The Palmaris longus is a tensor of the palmar fascia. It also assists in flexing the wrist and elbow. The Flexor sublimis digitorum flexes first the middle, and then the proximal phalanx. It assists in flexing the wrist and elbow. The Flexor profundus digitorum is one of the flexors of the phalanges. After the Flexor sublimis has bent the second phalanx, the Flexor profundus flexes the terminal one; but it cannot do so until after the contraction of the superficial muscle. It also assists in flexing the wrist. The Flexor longus pollicis is a flexor of the phalanges of the thumb. When the thumb is fixed, it also assists in flexing the wrist. The Pronator quadratus helps to rotate the radius upon the ulna, rendering the hand prone.

9. *Radial Region* (fig. 427)

Brachio-radialis (Supinator longus).   Extensor carpi radialis longior.
Extensor carpi radialis brevior.

Dissection.—Divide the integument in the same manner as in the dissection of the anterior brachial region; and, after having examined the cutaneous vessels and nerves and deep fascia, remove all these structures. The muscles will then be exposed. The removal of the fascia will be considerably facilitated by detaching it from below upwards. Great care should be taken to avoid cutting across the tendons of the muscles of the thumb, which cross obliquely the larger tendons running down the back of the radius.

The **Brachio-radialis** (Supinator longus) is the most superficial muscle on the radial side of the forearm: it is fleshy for the upper two-thirds of its extent, tendinous below. It arises from the upper two-thirds of the external supra-condylar ridge of the humerus, and from the external intermuscular septum, being limited above by the musculo-spiral groove. The fibres terminate above the middle of the forearm in a flat tendon, which is inserted into the outer side of the base of the styloid process of the radius.
Relations.—By its *superficial surface*, with the integument and fascia for the greater part of its extent; near its insertion it is crossed by the Extensor ossis metacarpi pollicis and the Extensor brevis pollicis. By its *deep surface*, with the humerus, the Extensor carpi radialis longior and brevior, the insertion of the Pronator radii teres, and the Supinator brevis. By its *inner border*, above the elbow, with the Brachialis anticus, the musculo-spiral nerve, and radial recurrent artery; and in the forearm with the radial vessels and nerve.
The **Extensor carpi radialis longior** is placed partly beneath the preceding muscle. It arises from the lower third of the external supracondylar ridge of the humerus, from the external intermuscular septum, and by a few fibres from the common tendon of origin of the Extensor muscles of the forearm. The fibres terminate at the upper third of the forearm in a flat tendon, which runs along the outer border of the radius, beneath the Extensor tendons of the thumb; it then passes beneath the posterior annular ligament of the wrist, where it lies in a groove on the back of the radius common to it and the Extensor carpi radialis brevior, immediately behind the styloid process, and is inserted into the posterior surface of the base of the metacarpal bone of the index finger, on its radial side.
Relations.—By its *superficial surface*, with the Supinator longus, and fascia of the forearm. Its *outer side* is crossed obliquely by the Extensor tendons of the thumb. By its *deep surface*, with the elbow-joint, the Extensor carpi radialis brevior, and back part of the wrist.
The **Extensor carpi radialis brevior** is shorter, as its name implies, and thicker than the preceding muscle, beneath which it is placed. It arises from the external epicondyle of the humerus, by a tendon common to it and the three following muscles: from the external lateral ligament of the elbow-joint; from a strong aponeurosis which covers its surface; and from the intermuscular

septa between it and the adjacent muscles. The fibres terminate about the middle of the forearm in a flat tendon, which is closely connected with that of the preceding muscle, and accompanies it to the wrist; it passes beneath the Extensor tendons of the thumb, then beneath the annular ligament, and, diverging somewhat from its fellow, is inserted into the posterior surface of the base of the metacarpal bone of the middle finger, on its radial side. Under the posterior annular ligament of the wrist the tendon lies in a shallow groove on the back of the radius, to the ulnar side of the groove which lodges the tendon of the Extensor carpi radialis longior, and separated from it by an indistinct ridge.

The tendons of the two preceding muscles pass through the same compartment of the annular ligament, and are lubricated by a single synovial membrane.

Relations. — By its *superficial surface*, with the Extensor carpi radialis longior, and with the Extensor muscles of the thumb which cross it. By its *deep surface*, with the Supinator brevis, tendon of the Pronator radii teres, radius, and wrist-joint. By its *ulnar border*, with the Extensor communis digitorum.

10. *Posterior Radio-ulnar Region* (fig. 427)

The muscles in this region are divided for purposes of description into two groups or layers, superficial and deep.

Superficial Layer

Extensor communis digitorum.
Extensor minimi digiti.
Extensor carpi ulnaris.
Anconeus.

The **Extensor communis digitorum** is situated at the back part of the forearm. It arises from the external epicondyle of the humerus, by the common tendon; from the deep fascia, and the intermuscular septa between it and the adjacent muscles. It divides below into four tendons, which pass, together with that of the Extensor indicis,

FIG. 427.—Posterior surface of the forearm. Superficial muscles.

Ext. ossis metacarpi pollicis

Ext. brevis pollicis

Ext. longus pollicis

Tendon of Ext. indicis

through a separate compartment of the annular ligament, lubricated by a synovial membrane. The tendons then diverge, and, after passing across the back of the hand, are inserted into the second and third phalanges of the fingers in the following manner: Opposite the metacarpo-phalangeal articulation each tendon is bound by fasciculi to the lateral ligaments and serves as the posterior ligament; after having passed the joint, it spreads out into a broad aponeurosis, which covers the dorsal surface of the first phalanx and is reinforced, in this situation, by the tendons of the Interossei and Lumbricales. Opposite the first phalangeal joint this aponeurosis divides into three slips, a middle and two lateral: the former is inserted into the base of the second phalanx; and the two lateral, which are continued onwards along the sides of the second phalanx, unite by their contiguous margins, and are inserted into the dorsal surface of the last phalanx. As the tendons cross the phalangeal joints, they furnish them with posterior ligaments. The tendon to the index finger is accompanied by the Extensor indicis, which lies on its inner side. On the back of the hand, the three inner tendons—those to the middle, ring, and little fingers— are connected by two obliquely placed bands, one from the third tendon passing downwards and outwards to the second tendon, and the other passing from the same tendon downwards and inwards to the fourth. Occasionally the first tendon is connected to the second by a thin transverse band.

Relations.—By its *superficial surface*, with the fascia of the forearm and hand, the posterior annular ligament, and integument. By its *deep surface*, with the Supinator brevis, the Extensor muscles of the thumb and index finger, the posterior interosseous vessels and nerve, the wrist-joint, carpus, metacarpus, and phalanges. By its *radial border*, with the Extensor carpi radialis brevior. By its *ulnar border*, with the Extensor minimi digiti and Extensor carpi ulnaris.

The **Extensor minimi digiti** is a slender muscle placed on the inner side of the Extensor communis, with which it is generally connected. It arises from the common Extensor tendon by a thin, tendinous slip; and from the intermuscular septa between it and the adjacent muscles. Its tendon runs through a separate compartment in the annular ligament behind the inferior radio-ulnar joint, then divides into two as it crosses the hand, and is finally inserted into the expansion of the Extensor tendon on the dorsum of the first phalanx of the little finger.

The **Extensor carpi ulnaris** is the most superficial muscle on the ulnar side of the forearm. It arises from the external epicondyle of the humerus, by the common tendon; by an aponeurosis from the posterior border of the ulna in common with the Flexor carpi ulnaris and the Flexor profundus digitorum; and from the deep fascia of the forearm. This muscle terminates in a tendon, which runs through a groove behind the styloid process of the ulna, passing through a separate compartment in the annular ligament, and is inserted into the prominent tubercle on the ulnar side of the base of the metacarpal bone of the little finger.

Relations.—By its *superficial surface*, with the deep fascia of the forearm. By its *deep surface*, with the ulna, and the muscles of the deep layer.

The **Anconeus** is a small, triangular muscle, placed behind and below the elbow-joint, and appears to be a continuation of the external portion of the Triceps. It arises by a separate tendon from the back part of the external epicondyle of the humerus, and is inserted into the side of the olecranon, and upper fourth of the posterior surface of the shaft of the ulna; its fibres diverge from their origin, the upper ones being directed transversely, the lower obliquely inwards.

Relations.—By its *superficial surface*, with a strong fascia derived from the Triceps. By its *deep surface*, with the elbow-joint, the orbicular ligament, the ulna, a small portion of the Supinator brevis, and the posterior interosseous recurrent artery.

Deep Layer (fig. 429)

Supinator brevis.	Extensor brevis pollicis.
Extensor ossis metacarpi pollicis.	Extensor longus pollicis.
	Extensor indicis.

The **Supinator brevis** (fig. 428) is a broad muscle, of a hollow cylindrical form, curved round the upper third of the radius. It consists of two distinct planes of muscular fibres, between which lies the posterior interosseous nerve.

The two planes arise in common : the superficial one by tendinous, and the deeper by muscular fibres from the external epicondyle of the humerus ; from the external lateral ligament of the elbow-joint, and the orbicular ligament of the radius ; from the ridge on the ulna, which runs obliquely downwards from the posterior extremity of the lesser sigmoid cavity ; from the triangular depression below that cavity ; and from a tendinous expansion which covers the surface of the muscle. The superficial fibres surround the upper part of the radius, and are inserted into the outer edge of the bicipital tuberosity and to the oblique line of the radius, as low down as the insertion of the Pronator radii teres. The upper fibres of the deeper plane form a sling-like fasciculus, which encircles the neck of the radius above the tuberosity and is attached to the back part of its inner surface : the greater part of this portion of the muscle is inserted into the posterior and external surface of the shaft, midway between the oblique line and the head of the bone. Between the insertion of the two planes, the posterior interosseous nerve passes to the back of the forearm.

FIG. 428.—Supinator brevis. (From a preparation in the Museum of the Royal College of Surgeons of England.)

Relations.—By its *superficial surface*, with the superficial Extensors and the Brachio-radialis, and the radial vessels and nerve. By its *deep surface*, with the elbow - joint, the interosseous membrane, and the radius.

The **Extensor ossis metacarpi pollicis** is the most external and the largest of the deep Extensor muscles ; it lies immediately below the Supinator brevis, with which it is sometimes united. It arises from the outer part of the posterior surface of the shaft of the ulna below the insertion of the Anconeus, from the interosseous membrane, and from the middle third of the posterior surface of the shaft of the radius. Passing obliquely downwards and outwards, it terminates in a tendon, which runs through a groove on the outer side of the lower end of the radius, accompanied by the tendon of the Extensor brevis pollicis, and is inserted into the outer side of the base of the metacarpal bone of the thumb. It occasionally gives off two slips, near its insertion : one to the Trapezium, and the other to blend with the origin of the Abductor pollicis.

Relations.—By its *superficial surface*, with the Extensor communis digitorum, Extensor minimi digiti, and fascia of the forearm ; and with the branches of the posterior interosseous artery and nerve which cross it. By its *deep surface*, with the ulna, interosseous membrane, radius, the tendons of the Extensor carpi radialis longior and brevior, which it crosses obliquely ; and, at the outer side of the wrist, with the radial vessels. By its *upper border*, with the Supinator brevis. By its *lower border*, with the Extensor brevis pollicis.

The **Extensor brevis pollicis** (*Extensor primi internodii pollicis*), the smallest muscle of this group, lies on the inner side of the preceding It arises from the posterior surface of the shaft of the radius, below the Extensor ossis metacarpi pollicis and from the interosseous membrane. Its direction is similar to that of the Extensor ossis metacarpi pollicis, its tendon passing through the same groove on the outer side of the lower end of the radius, to be inserted into the base of the first phalanx of the thumb. It is closely connected with the Extensor ossis metacarpi pollicis, of which it is usually regarded as a segmentation.

Relations. — The same as those of the Extensor ossis metacarpi pollicis.

The **Extensor longus pollicis** (*Extensor secundi internodii pollicis*) is much larger than the preceding muscle, the origin of which it partly covers. It arises from the outer part of the posterior surface of the middle third of the shaft of the ulna, below the origin of the Extensor ossis metacarpi pollicis, and from the interosseous membrane. It terminates in a tendon, which passes through a separate compartment in the annular ligament, lying in a narrow, oblique groove at the back part of the lower end of the radius. It then crosses obliquely the tendons of the Extensor carpi radialis longior and brevior, being separated from the other Extensor tendons of the thumb by a triangular interval, in which the radial artery is found; and is finally inserted into the base of the last phalanx of the thumb.

Relations.—By its *superficial surface*, with the same parts as the Extensor ossis metacarpi pollicis. By its *deep surface*, with the ulna, interosseous membrane, the posterior interosseous nerve, radius, the wrist, the radial vessels, and metacarpal bone of the thumb.

The **Extensor indicis** is a narrow, elongated muscle, placed on the inner side of, and parallel with, the preceding. It arises from the posterior surface of the shaft of the ulna, below the origin of the Extensor longus pollicis and from the interosseous membrane. Its tendon passes under the annular ligament in the same compartment as that which transmits the Extensor communis digitorum, and subsequently joins the ulnar side of the tendon of the Extensor communis which belongs to the index finger, opposite the lower end of the corresponding metacarpal bone.

Relations.—The relations are similar to those of the preceding muscles.

Nerves.—The Brachio-radialis is supplied by the fifth and sixth, the Extensor carpi radialis longior and brevior by the sixth and seventh, and the Anconeus by the seventh and eighth cervical nerves, all through the musculo-spiral nerve ; the remaining muscles are enervated through the posterior interosseous nerve, the Supinator brevis being supplied by the sixth cervical, and all the other muscles by the seventh cervical.

Actions.—The muscles of the radial and posterior aspects of the forearm, which comprise all the Extensor and Supinator muscles, act upon the forearm, wrist, and hand ; they are the direct antagonists of the Pronator and Flexor muscles. The Anconeus assists the Triceps in extending the forearm. The chief action of the Brachio-radialis is that of a flexor of the elbow-joint, but in addition to this it may act both as a supinator and as a pronator : that is to say, if the forearm is forcibly pronated, it will act as a supinator, and bring the bones into a position midway between supination and pronation ; and, *vice versâ*, if the arm is forcibly supinated, it will act as a pronator, and bring the bones into the same position, midway between supination and pronation. The action of the muscle is therefore to throw the forearm and hand into the position they naturally occupy when placed across the chest. The Supinator brevis is a supinator : that is to say, when the radius has been carried across the ulna in pronation, and the back of the hand is directed forwards, this muscle carries the radius back again to its normal position on the outer side of the ulna, and the palm of the hand is again directed forwards. The Extensor carpi radialis longior extends the wrist and abducts the hand. It may also assist in bending the elbow-joint ; at all events it serves to fix or steady this articulation. The Extensor carpi radialis brevior assists the Extensor carpi radialis longior in extending the wrist, and may also act slightly as an abductor of the hand. The Extensor carpi ulnaris helps to extend the hand, but when acting alone inclines it towards the ulnar side : by its continued action it extends the elbow-joint. The Extensor communis digitorum extends the phalanges, then the wrist, and finally the elbow. It acts principally on the proximal phalanges, the middle and terminal phalanges being acted upon mainly by the Interossei and Lumbricales. It has also a tendency to separate the fingers as it extends them. The Extensor minimi digiti extends the little finger, and by its continued action assists in extending the wrist. It is owing to this muscle that the little finger can be extended or pointed while the others are flexed. The chief action of the Extensor ossis metacarpi pollicis is to carry the thumb outwards and backwards from the palm of the hand, and hence it has been called the *Abductor pollicis longus.* By its continued action it helps to extend and abduct the wrist. The Extensor brevis pollicis extends the proximal phalanx of the thumb. By its continued action it helps to extend and abduct the wrist. The Extensor longus pollicis extends the terminal phalanx of the thumb. By its continued action it helps to extend and abduct the wrist. The Extensor indicis extends the index finger, and by its continued action assists in extending the wrist. It is owing to this muscle that the index finger can be extended or pointed while the others are flexed.

Surgical Anatomy.—The tendons of the Extensor muscles of the thumb are liable to become strained, and their sheaths inflamed, after excessive exercise, producing a sausage-shaped swelling along the course of the tendons and giving a peculiar creaking sensation to the finger when the muscles act. In consequence of its often being caused by such movements as wringing clothes, it is known as ' washerwoman's sprain.'

IV. MUSCLES AND FASCIÆ OF THE HAND

The **Muscles of the Hand** are subdivided into three groups : 1. Those of the thumb, which occupy the radial side and produce the *thenar* eminence ; 2. Those of the little finger, which occupy the ulnar side and give rise to the *hypothenar* eminence ; 3. Those in the middle of the palm and within the interosseous spaces.

Dissection (fig. 420).—Make a transverse incision across the front of the wrist, and a second across the heads of the metacarpal bones : connect the two by a vertical incision in the middle line, and continue it through the centre of the middle finger. The anterior and posterior annular ligaments, and the palmar fascia, should then be dissected.

The **Anterior Annular Ligament** is a strong, fibrous band, which arches over the carpus, converting the deep groove on the front of the carpal bones into a canal, beneath which pass the Flexor tendons of the digits. It is attached, internally, to the pisiform bone and the hook of the unciform bone; and externally, to the tuberosity of the scaphoid, and to the inner part of the anterior surface and the ridge on the trapezium. It is continuous, above, with the deep fascia of the forearm, of which it may be regarded as a thickened portion; and

FIG. 430.—Transverse section through the wrist,
showing the annular ligaments and the canals for the passage of the tendons.

below, with the palmar fascia. It is crossed by the ulnar vessels and nerve, and the cutaneous branches of the median and ulnar nerves. At its outer extremity is the tendon of the Flexor carpi radialis, which lies in the groove on the trapezium between the attachments of the annular ligament to the bone. It has inserted into its anterior surface a part of the tendon of the Palmaris longus and part of the tendon of the Flexor carpi ulnaris, and has arising from

FIG. 431.—Transverse section through the carpus,
showing the relative positions of the tendons, vessels, and nerves. (Henle.)

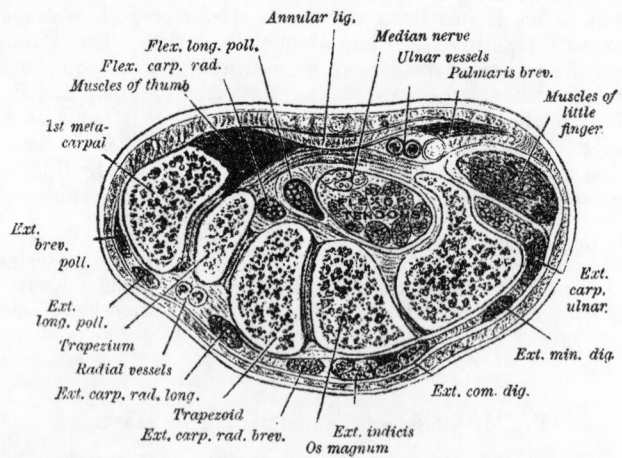

it, below, the small muscles of the thumb and little finger. Beneath it pass the tendons of the Flexor sublimis and profundus digitorum, the tendon of the Flexor longus pollicis, and the median nerve.

The **Synovial Membranes of the Flexor Tendons at the Wrist.**—There are two synovial membranes which enclose all the tendons as they pass beneath this ligament, one for the Flexor sublimis and profundus digitorum, the other for the Flexor longus pollicis. They extend up into the forearm for about an inch above the annular ligament, and occasionally communicate with each other under the ligament. The sheath which surrounds the Flexor tendons of the fingers extends downwards about halfway along the metacarpal bones, where it terminates in blind diverticula around the tendons to the index, middle, and ring fingers. In the case of the little finger, it is prolonged on its tendons, and usually communicates with the synovial sheath of that digit. The sheath which envelops the tendon of the Flexor longus pollicis is continued along the thumb as far as the insertion of the tendon.

The **Posterior Annular Ligament** is a strong, fibrous band, extending obliquely downwards and inwards across the back of the wrist, and consisting of the deep fascia of the back of the forearm, strengthened by the addition of some transverse fibres. It binds down the Extensor tendons in their passage to the fingers, being attached, internally, to the styloid process of the ulna, the cuneiform and pisiform bones; externally, to the outer margin of the radius; and, in its passage across the wrist, to the elevated ridges on the posterior surface of the radius. Between it and the bones are formed six compartments for the passage of tendons, each of which is lined by a separate synovial membrane. These are, from without inwards: 1. On the outer side of the styloid process, for the tendons of the Extensor ossis metacarpi and Extensor brevis pollicis. 2. Behind the styloid process, for the tendons of the Extensor carpi radialis longior and brevior. 3. About the middle of the posterior surface of the radius, for the tendon of the Extensor longus pollicis. 4. To the inner side of the latter, for the tendons of the Extensor communis digitorum and Extensor indicis. 5. Opposite the interval between the radius and ulna, for the Extensor minimi digiti. 6. Grooving the back of the ulna, for the tendon of the Extensor carpi

Fig. 432.—Diagram showing the arrangement of the synovial sheaths of the palm and fingers.

ulnaris. The synovial membranes lining these sheaths are usually very extensive, reaching from above the annular ligament, down upon the tendons for a variable distance on the back of the hand.

The **deep palmar fascia** (fig. 433) forms a common sheath which invests the muscles of the hand. It consists of a central and two lateral portions.

The *central portion* occupies the middle of the palm, is triangular in shape, of great strength and thickness, and binds down the tendons and protects the vessels and nerves in this situation. It is narrow above, where it is attached to the lower margin of the annular ligament, and receives the expanded tendon of the Palmaris longus muscle. Below, it is broad and expanded, and divides into four slips, one for each finger. Each slip gives off superficial fibres, which are inserted into the skin of the palm and finger, those to the palm joining the skin at the furrow corresponding to the metacarpo-phalangeal articulation, and those to the fingers passing into the skin at the transverse fold at the base of the fingers. The deeper part of each slip subdivides into two processes, which are inserted into the fibrous sheaths which confine the Flexor tendons. From the sides of these processes, offsets are sent backwards, to be attached to the transverse metacarpal ligament. By this arrangement short channels are formed on the front of the lower ends of the metacarpal bones, through which the Flexor tendons pass. Between the two processes into which each slip divides is attached the *digital sheath*. The intervals left in the fascia, between the four fibrous slips, transmit the digital vessels and nerves, and the tendons of the Lumbricales. At the points of division of the palmar fascia into the slips above mentioned, numerous strong, transverse fibres bind the separate processes together and form the *superficial transverse metacarpal ligament*. The palmar fascia is intimately adherent to the integument by dense fibro-areolar tissue forming the superficial palmar fascia, and gives origin by its inner margin to the Palmaris brevis; it covers the superficial palmar arch, the tendons of the Flexor muscles, and the branches of the median and ulnar nerves; and on each side it gives off a vertical septum, which is continuous with the interosseous aponeurosis, and separates the lateral from the middle palmar group of muscles.

FIG. 433.—Palmar fascia.
(From a preparation in the Museum of the Royal College of Surgeons of England.)

The *lateral portions* of the palmar fascia are thin, fibrous layers, which cover, on the radial side, the muscles of the ball of the thumb, and, on the ulnar side, the muscles of the little finger: they are continuous with the dorsal fascia, and in the palm with the central portion of the palmar fascia.

The **Superficial Transverse Ligament of the Fingers** is a thin, fibrous band, which stretches across the roots of the four fingers, and is closely attached to the skin of the clefts, and internally to the fifth metacarpal bone, forming a sort of rudimentary web. Beneath it the digital vessels and nerves pass onwards to their destination.

Surgical Anatomy.—The palmar fascia is liable to undergo contraction, producing a very inconvenient deformity, known as 'Dupuytren's contraction.' The ring and little fingers are most frequently implicated, but the middle, the index, and the thumb may be involved. The proximal phalanx is drawn down and cannot be straightened, and the two distal phalanges become similarly flexed as the disease advances.

Owing to their constant exposure to injury and septic influences, the fingers are very liable to become the seat of serious inflammatory mischief. To this inflammation the term *paronychia* is given, and this affection may assume various degrees of severity. In the mildest cases the disease is confined to the superficial layer of the skin, and suppuration takes place beneath it. This is known as *subcuticular paronychia*, and is a comparatively simple condition, and an incision through the epidermis will at once relieve it. The only complication is that the pus may burrow under the nail, causing increased pain. A more severe condition is the *paronychia cellulosa*, in which the pulp of the end of the finger is involved. This is attended with intense throbbing pain, owing to the fact that the inflamed area is covered by thick and often horny epithelium, when the disease occurs in the labouring classes, as it so frequently does. In these cases, unless a timely incision is made, the inflammation is liable to involve the periosteum covering the phalanx, as there is least resistance in this direction; and *subperiosteal paronychia* is set up, which is followed by necrosis of a part or the whole of the ungual phalanx. In other cases, the inflammation may involve the theca of the Flexor tendons, and a *thecal paronychia* may be set up. The inflammation then rapidly spreads up the sheath; but the extent will depend upon the particular digit involved. From the description of the Flexor sheaths given above (page 527), it will be evident that inflammation of the sheath of the thumb and little finger may prove a far more formidable affection than that of the other three digits, because the sheaths of these two digits communicate with the large synovial sheaths which surround the Flexor tendons, and the inflammation may extend into the palm of the hand and beneath the annular ligament into the forearm.

In order to relieve these conditions, free and early incisions are necessary, and must be made with discrimination, in order to avoid wounding important structures. In the pulp of the finger—i.e. over the distal phalanx—the incision should be made in the middle line and down to the bone. In the rest of the finger, the incision should be made in the middle line over the phalanges, and not over the interphalangeal joints. In the palm of the hand, incisions may be made either on the distal or proximal side of the superficial palmar arch. On the distal sides, the incisions should be made over the metacarpal bones, preferably those of the index and middle finger. On the proximal side, the safest line of incision is along the radial side of the hypothenar eminence, between the ulnar artery and nerve internally, and the median nerve externally. When suppuration has extended under the annular ligament, and incisions are required in the forearm, the positions in which they should be made are over the tendons of the Flexor sublimis digitorum, between the median nerve and the ulnar artery, and over the tendon of the Flexor longus pollicis, between the radial artery and the tendon of the Flexor carpi radialis muscle.

Chronic inflammation of the Flexor tendons of the fingers, probably tuberculous, is occasionally met with, constituting a disease known as 'compound palmar ganglion:' it presents an hourglass outline, with a swelling in front of the wrist and in the palm of the hand, and a constriction, corresponding to the annular ligament, between the two. The fluid can be forced from the one swelling to the other under the ligament, and when this is done, a crackling sensation is sometimes perceived, from the presence of 'melon-seed' bodies in the interior of the ganglion.

11. *Radial Region* (figs. 434, 435)

Abductor pollicis.	Flexor brevis pollicis.
Opponens pollicis.	Adductor obliquus pollicis.

Adductor transversus pollicis.

The **Abductor pollicis** is a thin, flat muscle, placed immediately beneath the integument. It arises from the annular ligament, the tuberosity of the scaphoid, and the ridge of the trapezium, frequently by two distinct slips; and, passing outwards and downwards, is inserted by a thin, flat tendon into the radial side of the base of the first phalanx of the thumb and the capsule of the metacarpo-phalangeal articulation.

Relations.—By its *superficial surface*, with the palmar fascia and superficialis volæ artery, which frequently perforates it. By its *deep surface*, with the Opponens pollicis, from which it is separated by a thin aponeurosis. Between its *inner border* and the Flexor brevis pollicis is a narrow cellular interval.

The **Opponens pollicis** is a small, triangular muscle, placed beneath the preceding. It arises from the palmar surface of the ridge on the trapezium and from the annular ligament, passes downwards and outwards, and is inserted into the whole length of the metacarpal bone of the thumb on its radial side.

Relations.—By its *superficial surface*, with the Abductor and Flexor brevis pollicis. By its *deep surface*, with the trapezio-metacarpal articulation. By its *inner border*, with the Adductor obliquus pollicis.

M M

The **Flexor brevis pollicis** consists of two portions, outer and inner. The *outer and more superficial portion* arises from the outer two-thirds of the lower border of the annular ligament and the lower part of the ridge on the trapezium; it passes along the outer side of the tendon of the Flexor longus pollicis, and, becoming tendinous, has a sesamoid bone developed in its tendon, and is inserted into the outer side of the base of the first phalanx of the thumb. The *inner and deeper portion* of the muscle is very small, and arises from the ulnar side of the first metacarpal bone between the Adductor obliquus pollicis and the outer head of the First dorsal interosseous, and is inserted into the inner side of the base of the first phalanx with the Adductor obliquus pollicis.

Relations.—By its *superficial surface*, with the palmar fascia. By its *deep surface*, with the tendon of the Flexor longus pollicis. By its *external surface*, with the Opponens pollicis. *Behind*, with the Adductor obliquus pollicis.

The **Adductor obliquus pollicis** arises by several slips from the os magnum, the bases of the second and third metacarpal bones, the anterior carpal ligaments,

FIG. 434.—Muscles of the thumb.
(From a preparation in the Museum of the Royal College of Surgeons of England.)

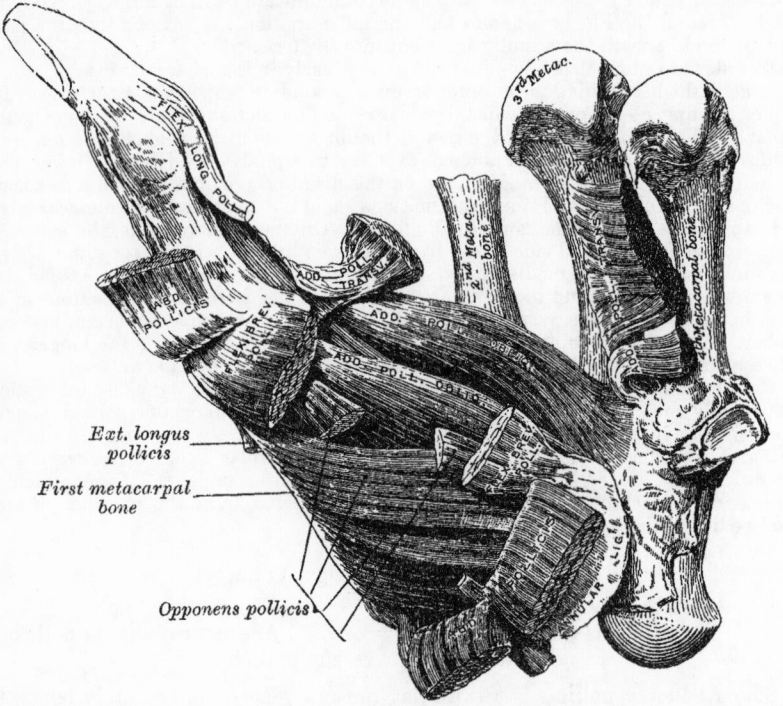

and the sheath of the tendon of the Flexor carpi radialis. From this origin the greater number of fibres pass obliquely downwards and converge to a tendon, which, uniting with the tendons of the deeper portion of the Flexor brevis pollicis and the Adductor transversus, is inserted into the inner side of the base of the first phalanx of the thumb, a sesamoid bone being developed in the tendon of insertion. A considerable fasciculus, however, passes more obliquely outwards beneath the tendon of the long flexor to join the superficial portion of the short flexor and the Abductor pollicis.

Relations.—By its *superficial surface*, with the Flexor longus pollicis and the outer head of the Flexor brevis pollicis. Its *deep surface* is in relation with the deep palmar arch, which passes between the two adductors.

The **Adductor transversus pollicis** (fig. 434) is the most deeply seated of this group of muscles. It is of a triangular form, arising by its broad base from the lower two-thirds of the metacarpal bone of the middle finger on its palmar surface; the fibres, proceeding outwards, converge, to be inserted, with the inner part of the Flexor brevis pollicis, and the Adductor obliquus pollicis, into the ulnar side of the base of the first phalanx of the thumb.

Relations.—By its *superficial surface*, with the Adductor obliquus pollicis, the tendons of the Flexor profundus, and the Lumbricales. Its *deep surface* covers the first two interosseous spaces, from which it is separated by a strong aponeurosis.

Nerves.—The Abductor, Opponens, and outer head of the Flexor brevis pollicis are supplied by the sixth cervical through the median nerve; the inner head of the Flexor brevis, and the Adductors, by the eighth cervical through the ulnar nerve.

Actions.—The actions of the muscles of the thumb are almost sufficiently indicated by their names. This segment of the hand is provided with three extensors—an extensor of the metacarpal bone, an extensor of the first, and an extensor of the second phalanx; these occupy the dorsal surface of the forearm and hand. There are also three flexors on the palmar surface—a flexor of the metacarpal bone, a flexor of the proximal, and a flexor of the terminal phalanx; there is also an abductor and two adductors. The Abductor pollicis moves the metacarpal bone of the thumb outwards: that is, away from the index finger. The Opponens pollicis flexes the metacarpal bone: that is, draws it inwards over the palm, and at the same time rotates the bone, so as to turn the ball of the thumb towards the fingers, thus producing the movement of opposition. The Flexor brevis pollicis flexes and adducts the proximal phalanx of the thumb. The Adductores pollicis move the metacarpal bone of the thumb inwards: that is, towards the index finger. These muscles give to the thumb its extensive range of motion. It will be noticed, however, that in consequence of the position of the first metacarpal bone, these movements differ from the corresponding movements of the metacarpal bones of the other fingers. Thus extension of the thumb more nearly corresponds to the motion of abduction in the other fingers, and flexion to adduction.

12. *Ulnar Region* (fig. 435)

Palmaris brevis.   Flexor brevis minimi digiti.
Abductor minimi digiti.   Opponens minimi digiti.

The **Palmaris brevis** is a thin quadrilateral muscle, placed beneath the integument on the ulnar side of the hand. It arises, by tendinous fasciculi, from the annular ligament and palmar fascia; the fleshy fibres pass inwards, to be inserted into the skin on the inner border of the palm of the hand.

Relations.—By its *superficial surface*, with the integument, to which it is intimately adherent, especially by its inner border. By its *deep surface*, with the inner portion of the palmar fascia, which separates it from the ulnar vessels and nerve, and from the muscles of the ulnar side of the hand.

The **Abductor minimi digiti** is situated on the ulnar border of the palm of the hand. It arises from the pisiform bone and from the tendon of the Flexor carpi ulnaris, and terminates in a flat tendon, which divides into two slips; one is inserted into the ulnar side of the base of the first phalanx of the little finger. The other slip is inserted into the ulnar border of the aponeurosis of the Extensor minimi digiti.

Relations.—By its *superficial surface*, with the inner portion of the palmar fascia, and the Palmaris brevis. By its *deep surface*, with the Opponens minimi digiti. By its *outer border*, with the Flexor brevis minimi digiti.

The **Flexor brevis minimi digiti** lies on the same plane as the preceding muscle, on its radial side. It arises from the convex aspect of the hook of the unciform bone, and anterior surface of the annular ligament, and is inserted into the inner side of the base of the first phalanx of the little finger. It is separated from the Abductor at its origin, by the deep branches of the ulnar artery and nerve. This muscle is sometimes wanting; the Abductor is then, usually, of large size.

Relations.—By its *superficial surface*, with the internal portion of the palmar fascia, and the Palmaris brevis. By its *deep surface*, with the Opponens. The deep branch of the ulnar artery and the corresponding branch of the ulnar nerve pass between the Abductor and Flexor brevis minimi digiti muscles.

The **Opponens minimi digiti** (fig. 426) is of a triangular form, and placed immediately beneath the preceding muscles. It arises from the convexity of the hook of the unciform bone, and contiguous portion of the annular ligament ; its fibres pass downwards and inwards, to be inserted into the whole length of the metacarpal bone of the little finger, along its ulnar margin.

FIG. 435.—Muscles of the left hand. Palmar surface.

Relations.—By its *superficial surface*, with the Flexor brevis, and Abductor minimi digiti. By its *deep surface*, with the Interossei muscles in the fourth metacarpal space, the metacarpal bone, and the Flexor tendons of the little finger.

Nerves.—All the muscles of this group are supplied by the eighth cervical nerve through the ulnar nerve.

Actions.—The Abductor minimi digiti abducts the little finger from the ring finger. It corresponds to a dorsal interosseous muscle. It also assists in flexing the proximal phalanx. The Flexor brevis minimi digiti abducts the little finger from the middle line of the hand. It also assists in flexing the proximal phalanx. The Opponens minimi digiti draws forwards the fifth metacarpal bone, so as to deepen the hollow of the palm. The Palmaris brevis corrugates the skin on the inner side of the palm of the hand.

13. *Middle Palmar Region*

Lumbricales.   Interossei dorsales.
Interossei palmares.

The **Lumbricales** (fig. 435) are four small fleshy fasciculi, accessories to the Flexor profundus digitorum muscle. They arise from the tendons of the deep Flexor : the first and second, from the radial side and palmar surface of the tendons of the index and middle fingers respectively ; the third, from the contiguous sides of the tendons of the middle and ring fingers ; and the fourth,

FIG. 436.—The Dorsal interossei   FIG. 437.—The Palmar interossei
of left hand.   of left hand.

from the contiguous sides of the tendons of the ring and little fingers. They pass to the radial side of the corresponding fingers, and opposite the metacarpo-phalangeal articulation each tendon is inserted into the tendinous expansion of the Extensor communis digitorum, covering the dorsal aspect of each finger.

The **Interossei muscles** (figs. 436, 437) are so named from occupying the intervals between the metacarpal bones, and are divided into two sets, a dorsal and palmar.

The **Dorsal interossei** are four in number, larger than the palmar, and occupy the intervals between the metacarpal bones. They are bipenniform muscles, arising by two heads from the adjacent sides of the metacarpal bones, but more extensively from the metacarpal bone of the finger into which the muscle is inserted. They are inserted into the bases of the first phalanges and into the aponeurosis of the common Extensor tendon. Between the double origin of each of these muscles is a narrow triangular interval, through the first of which the radial artery passes ; through each of the other three a perforating branch from the deep palmar arch is transmitted.

The *First dorsal interosseous muscle*, or *Abductor indicis*, is larger than the others. It is flat, triangular in form, and arises by two heads, separated by a fibrous arch, for the passage of the radial artery from the dorsum to the palm of the hand. The outer head arises from the upper half of the ulnar border of the first metacarpal bone ; the inner head, from almost the entire length of the radial border of the second metacarpal bone ; the tendon is inserted into the radial side of the index finger. The *second* and *third dorsal interossei* are inserted into the middle finger, the former into its radial, the latter into its ulnar side. The *fourth* is inserted into the ulnar side of the ring finger.

The **Palmar interossei,** three in number, are smaller than the Dorsal, and placed upon the palmar surface of the metacarpal bones, rather than between them. They arise from the entire length of the metacarpal bone of one finger, and are inserted into the side of the base of the first phalanx and aponeurotic expansion of the common Extensor tendon of the same finger.

The *first* arises from the ulnar side of the second metacarpal bone, and is inserted into the same side of the first phalanx of the index finger. The *second* arises from the radial side of the fourth metacarpal bone, and is inserted into the same side of the ring finger. The *third* arises from the radial side of the fifth metacarpal bone, and is inserted into the same side of the little finger. From this account it may be seen that each finger is provided with two Interossei muscles, with the exception of the little finger, in which the Abductor muscle takes the place of one of the pair.

Nerves.—The two outer Lumbricales are supplied by the sixth cervical nerve, through the third and fourth digital branches of the median nerve : the two inner Lumbricales and all the Interossei are supplied by the eighth cervical nerve, through the deep palmar branch of the ulnar nerve. Brooks states that the third lumbrical received a twig from the median in twelve out of twenty-one cases.

Actions.—The Palmar interossei muscles adduct the fingers to an imaginary line drawn longitudinally through the centre of the middle finger ; and the Dorsal interossei abduct the fingers from that line. In addition to this the Interossei, in conjunction with the Lumbricales, flex the first phalanges at the metacarpo-phalangeal joints, and extend the second and third phalanges in consequence of their insertion into the expansion of the Extensor tendons. The Extensor communis digitorum is believed to act almost entirely on the first phalanges.

Surface Form.—The *Pectoralis major* largely influences surface form and conceals a considerable part of the thoracic wall in front. Its sternal origin presents a festooned border, which bounds and determines the width of the sternal furrow Its clavicular origin is somewhat depressed and flattened, and between the two portions of the muscle is often an oblique depression, which differentiates the one from the other. The outer margin of the muscle is generally well marked above, and bounds the infraclavicular fossa, a triangular interval which separates the Pectoralis major from the Deltoid. It gradually becomes less marked as it approaches the tendon of insertion, and is more closely blended with the Deltoid muscle. The lower border of the Pectoralis major forms the rounded anterior axillary fold, and corresponds with the direction of the fifth rib. The *Pectoralis minor* influences surface form. When the arm is raised, its lowest slip of origin produces a local fulness just below the border of the anterior fold of the axilla, and so serves to break the sharp line of the lower border of the Pectoralis major muscle which is produced when the arm is in this position. The origin of the *Serratus magnus* causes a very characteristic surface marking. When the arm is raised from the side, in a well-developed subject, the five or six lower serrations are plainly discernible, forming a zigzag line, caused by the series of digitations, which diminish in size from above downwards, and have their apices arranged in the form of a curve. When the arm is lying by the side, the first serration to appear, at the lower margin of the Pectoralis major, is the one attached to the fifth rib. The *Deltoid*, with the prominence of the upper extremity of the humerus, produces the rounded outline of the shoulder. It is rounder and fuller in front than behind, where it presents a somewhat flattened form. Above, its anterior border presents a rounded, slightly curved eminence, which bounds externally the infraclavicular fossa ; below, it is closely united with the Pectoralis major. Its posterior border is thin, flattened, and scarcely marked above ; below, it is thicker and more prominent. When the muscle is in action, the middle portion becomes irregular, presenting alternate longitudinal elevations and depressions : the elevations corresponding to the fleshy portions ; the depressions, to the tendinous intersections of the muscle. The insertion of the Deltoid is marked by a depression on the outer side of the middle of the arm. Of the scapular muscles, the only one which materially influences surface form is the *Teres major*, which assists the Latissimus dorsi in forming the thick, rounded, posterior fold of the axilla. When the arm is raised, the Coraco-brachialis reveals itself as a long, narrow elevation, which emerges from under cover of the anterior fold of the axilla and runs downwards, internal to the shaft of the humerus. When the arm is hanging by the side, its front and inner part presents the prominence of the Biceps, bounded on either side by an intermuscular depression. This muscle determines the contour of the front of the arm, and extends from the anterior margin of the axilla to the bend of the elbow. Its upper tendons are concealed by the Pectoralis major and the Deltoid, and its lower tendon sinks into the space at the bend of the elbow. When the muscle is in a state of complete contraction—that is to say, when the forearm has been flexed and supinated—it presents a rounded convex form, bulged out laterally, and its length is diminished. On each side of the Biceps, at the lower part of the arm, the *Brachialis anticus* is discernible. On the outer side it forms a narrow eminence, which extends some distance up the arm along the border of the Biceps. On the inner side it shows itself only as a little fulness just above the elbow. On the back of the arm the long head of the *Triceps* may be seen as a longitudinal eminence emerging from under cover of the Deltoid, and gradually merging into the longitudinal flattened plane of the tendon of the muscle on the lower part of the back of the arm. The tendon of insertion of the muscle extends about halfway up the back of the arm, where it forms an elongated flattened plane when the muscle is in action. Under similar conditions the surface forms produced by the three heads of the muscle are well seen. On the anterior aspect of the elbow are to be seen two muscular elevations, one on each side, separated above, and converging below so as to form a triangular space. Of these, the inner elevation, consisting of the flexors and pronator, forms the prominence along the inner side and front of the forearm. It is a fusiform mass, pointed above at the internal condyle, and gradually tapering off below. The *Pronator radii teres*, the innermost muscle of the group, forms the boundary of the triangular space at the bend of the elbow. It is shorter, less prominent, and more oblique than the outer boundary. The most prominent part of the eminence is produced by the *Flexor carpi radialis*, the muscle next in order on the inner side of the preceding one. It forms a rounded prominence above, and may be traced downwards to its tendon, which can be felt lying on the front of the wrist, nearer to the radial than to the ulnar border and to the inner side of the radial artery. The *Palmaris longus* presents no surface marking above, but, below, is the most prominent tendon on the front of the wrist, standing out, when the muscle is in action, as a sharp tense cord beneath the skin. The *Flexor sublimis digitorum* does not directly influence surface form. The position of its four tendons on the front of the lower part of the forearm is indicated by an elongated depression between the tendons of the Palmaris longus and the Flexor carpi ulnaris. The *Flexor carpi ulnaris* occupies a small part of the posterior surface of the forearm, and is separated from the extensor and supinator group, which occupies the greater part of this surface, by the ulnar furrow, produced by the subcutaneous posterior border of the ulna. Its tendon can be perceived along the ulnar border of the front of the forearm, and is most marked when the hand is flexed and adducted. The deep muscles of the front of the forearm have no direct influence on surface form. The external group of muscles of the forearm, consisting of the extensors and supinators, occupy the outer, and a considerable portion of the posterior, surface of this region. It has a fusiform outline, which is altogether on a higher level than the pronato-flexor group. Its apex emerges from between the Triceps and Brachialis anticus muscles some distance above the elbow-joint, and acquires its greatest breadth opposite the external condyle, and thence gradually shades off into a flattened surface. About the middle of the forearm it divides into two longitudinal eminences which diverge from each other, leaving a triangular interval between them. The outer of these two groups of muscles consists of the Supinator longus and the Extensor carpi radialis longior and brevior, which forms a longitudinal eminence descending from the external condylar ridge in the direction of the styloid process of the radius. The other and more posterior group consists of the Extensor communis digitorum, the Extensor minimi digiti, and the Extensor carpi ulnaris. It commences above as a tapering form at the external condyle of the humerus, and is separated behind at its upper part from the Anconeus by a well-marked furrow; and below, from the pronato-flexor mass, by the ulnar furrow. In the triangular interval left between these two groups, the Extensor muscles of the thumb and index finger are seen. The only two muscles of this region which require special mention, as independently influencing surface form, are the Supinator longus and the Anconeus. The inner border of the *Supinator longus* forms the outer boundary of the triangular space at the bend of the elbow. It commences as a rounded border above the condyle, and is longer, less oblique, and more prominent than the inner boundary. Lower down, the muscle forms a full fleshy mass on the outer side of the upper part of the forearm, and below tapers into a tendon, which may be traced down to the styloid process of the radius. The *Anconeus* presents a distinct and characteristic surface form in the shape of a triangular, slightly elevated surface, immediately external to the subcutaneous posterior surface of the olecranon, and differentiated from the common extensor group by a well-marked oblique longitudinal depression. The upper angle of the triangle corresponds to the external condyle, and is marked by a depression of dimple in this situation. In the interval, caused by the divergence from each other or the two groups of muscles into which the extensor and supinator group is divided at the lower part of the forearm, an oblique elongated eminence is seen, caused by the emergence of two of the extensors of the thumb from their deep origin at the back of the forearm, This eminence, full above, and becoming flattened out and partially subdivided below, runs downwards and outwards over the back and outer surface of the radius to the outer side of the wrist-joint, where it forms a ridge, especially marked when the thumb is extended, which passes onwards to the posterior aspect of the thumb. The tendons of most of the Extensor muscles are to be seen and felt at the level of the wrist-joint. Most externally are the tendons of the Extensor ossis metacarpi pollicis and the Extensor brevis pollicis, forming a vertical ridge over the outer side of the joint from the styloid process of the radius to the thumb. Internal to this is the oblique ridge produced by the tendon of the Extensor longus pollicis, very noticeable when the muscle is in action. The Extensor carpi radialis longior is scarcely to be felt, but the Extensor carpi radialis brevior can be distinctly perceived, as a vertical ridge emerging from under the ulnar border of the tendon of the Extensor longus pollicis, when the hand is forcibly extended at the wrist. Internal to this, again, can be felt the tendons of the Extensor indicis, Extensor communis digitorum, and Extensor minimi digiti; the latter tendon being separated from those of the common extensor by a slight furrow. The muscles of the hand are principally concerned, as far as regards surface form, in producing the thenar and hypothenar eminences, and individually are not to be distinguished, on the surface, from each other. The *Adductor transversus pollicis* is, however, an exception to this; its anterior border gives rise to a ridge across the web of skin connecting the thumb to the rest of the hand. The thenar eminence is much larger and rounder than the hypo-thenar one, which presents a longer and narrower eminence along the ulnar side of the hand. When the *Palmaris brevis* is in action it produces a wrinkling of the skin over the hypothenar eminence, and a deep dimple on the ulnar border of the hand. On the back of the hand the *Dorsal interossei* produce elongated swellings between the meta-carpal bones. When the thumb is closely adducted to the hand, the first dorsal inter-osseous (Abductor indicis) forms a prominent fusiform bulging; the other interossei are not so marked.

The skin over the inner side and front of the forearm is thin, smooth and sensitive; it contains few hairs and many sweat-glands. Over the outer side and back of the arm and forearm it is thicker, denser, not so sensitive, and contains more hairs and fewer sweat-glands. Over the olecranon the cuticle is thick and rough; the skin loosely connected to the underlying tissues, and transversely wrinkled when the forearm is extended. At the front of the wrist, the skin presents three transverse wrinkles, which correspond to the position of the styloid process of the ulna, the wrist-joint, and the mid-carpal joint respectively. The skin of the palm of the hand differs considerably from that of the forearm. At the wrist, it suddenly becomes hard and dense and covered with a thick layer of cuticle. The skin in the thenar region presents these characteristics less than elsewhere. In spite of this hardness and density, the skin of the palm is exceedingly sensitive and very vascular. It is destitute of hair, and contains no sebaceous follicles. The skin of the palm is tied down by fibrous bands along the lines of flexion, producing certain furrows of a permanent character. One of these (*linea vitalis*), starting from about the tubercle of the scaphoid, extends downwards and inwards, and then downwards and outwards, curving round the thenar eminence, and terminates on the radial border of the hand, a little above the metacarpo-phalangeal joint of the index finger. It corresponds to the outer border of the central portion of the palmar fascia, and is produced by the special movement of opposition of the thumb to the other fingers. A second line (*linea cephalica*) commences at the termination of the first, and extends obliquely across the palm, upwards and inwards, to the ulnar margin about the middle of the fifth metacarpal bone. A third (*linea mensalis*) commences at the ulnar border of the hand, about an inch below the termination of the linea cephalica, and extends outwards across the palm over the heads of the third, fourth, and fifth metacarpal bones. The last two lines are caused by flexion of the fingers at the metacarpo-phalangeal joint. From the linea mensalis, furrows pass vertically downwards along each finger, corresponding to the lines of attachment of the skin to the sheaths of the Flexor tendons. These are called *valleculæ*, and between them are small projections or hillocks, caused by bulging of the subcutaneous tissues when the tendons are stretched. They are named, from the thumb to the little finger, *eminentia Veneris, Jovis, Saturni, Solis,* and *Mercurii.* Over the fingers the skin again becomes thinner, especially at the flexures of the joints; and over the terminal phalanges it is thrown into numerous ridges, in consequence of the arrange-ment of the papillæ in it. These ridges form, in different individuals, distinctive and permanent patterns, which may be used for purposes of identification. The superficial fascia in the palm is made up of dense fibro-fatty tissue. This tissue binds down the skin so firmly to the deep palmar fascia that very little movement is permitted between the two.

Surgical Anatomy.—The student, having completed the dissection of the muscles of the upper extremity, should consider the effects likely to be produced by the action of the various muscles in fracture of the bones.

In considering the actions of the various muscles upon fractures of the upper extremity, the most common forms of injury have been selected both for illustration and description.

Fracture of the *middle of the clavicle* (fig. 438) is always attended with considerable displacement of the outer fragment, which is drawn downwards and inwards, and at the same time rotated, so that its outer end is carried forwards and its inner end backwards.

The displacement is produced as follows: the outer fragment is drawn *downwards* by the weight of the arm, the Trapezius not being able to support the weight of the limb. It is drawn *inwards* by the Subclavius and Pecto-ralis minor, possibly assisted by the Pectoralis major and Latissimus dorsi; and is rotated on an axis drawn through it own centre by the Serratus magnus, which causes the scapula to rotate on the wall of the chest, and carries the acromion and outer end of the outer fragment of the clavicle forwards, and so carries the inner end of the outer portion backwards. The depression of the whole outer fragment is produced by the weight of the arm and by the contraction of the Deltoid. The causes of displacement having been ascertained, it is easy to apply the appropriate treatment. The outer fragment is to be drawn outwards, and, together with the scapula, raised upwards to a level with the inner fragment, and retained in that position.

FIG. 438.—Fracture of the middle of the clavicle.

In fracture of the *acromial end of the clavicle*, between the conoid and trapezoid ligaments, only slight displacement occurs, as these ligaments, from their oblique insertion, serve to hold both portions of the bone in apposition. Fracture, also, of the *sternal end*, internal to the costo-clavicular liga-ment, is attended with only slight displacement, this ligament serving to retain the fragments in close apposition.

Fracture of the *acromion process* outside the ligaments usually arises from violence applied to the upper and outer part of the shoulder. There is great displacement; the outer fragment being drawn downwards by the weight of the arm, and rotated forwards and inwards, so that it forms a right angle with the rest of the bone.

Fracture of the *coracoid process* is an extremely rare accident, and is usually caused by a sharp blow on the point of the shoulder. In many cases there appears to be little or no displacement, from the fact that the coraco-clavicular ligament remains intact, and keeps the separated fragment from displacement. But in some cases displacement does occur, the fragment being drawn downwards and forwards by the Coraco-brachialis and Biceps, and inwards by the Pectoralis minor. In order to relax these muscles and replace the fragments in close apposition, the forearm should be flexed so as to relax the Biceps, and the arm drawn forwards and inwards across the chest so as to relax the Coraco-brachialis; the humerus should then be pushed upwards against the coraco-acromial ligament, and the arm retained in that position.

FIG. 439.—Fracture of the surgical neck of the humerus.

Fracture of the *surgical neck of the humerus* (fig. 439) is very common, is attended with con-siderable displacement, and its appearances corre-spond somewhat with those of dislocation of the head of the humerus into the axilla. The upper fragment is slightly elevated under the coraco-acromial ligament by the muscles attached to the greater and lesser tuberosities; the lower fragment is drawn inwards by the Pectoralis major, Latissi-mus dorsi, and Teres major; and the humerus is thrown obliquely outwards from the side by the Deltoid, and occasionally elevated so as to cause the upper end of the lower fragment to project beneath and in front of the coracoid process. The deformity is reduced by fixing the shoulder, and drawing the arm outwards and down-wards. To counteract the opposing muscles, and to keep the fragments in position, a conical-shaped pad should be placed in the axilla, and the arm bandaged to the side by a broad roller passed round the chest, in such a manner that the elbow is carried slightly forwards, so as to throw the upper end of the lower fragment backwards and outwards towards the head of the bone. The whole is then covered with a carefully moulded guttapercha or poroplastic shoulder-cap.

In fracture of the *shaft of the humerus* below the insertion of the Pectoralis major, Latissimus dorsi, and Teres major, and above the insertion of the Deltoid, there is also considerable deformity, the upper fragment being drawn inwards by the first-mentioned muscles, and the lower fragment upwards and outwards by the Deltoid, producing shortening of the limb, and a considerable prominence at the seat of fracture, from the fractured ends of the bone riding over one another, especially if the fracture takes place in an oblique direction. The fragments may be brought into apposition by extension from the elbow, and retained in that position by adopting the same means as in the preceding injury.

FIG. 440.—Fracture of the humerus above the condyles.

In fractures of the *shaft of the humerus* immediately below the insertion of the Deltoid, the amount of deformity depends greatly upon the direction of the fracture. If it occurs in a transverse direction, only slight displacement takes place, the upper fragment being drawn a little forwards; but in oblique fracture, the combined actions of the Biceps and Brachialis anticus muscles in front, and the Triceps behind, draw upwards the lower fragment, causing it to glide over the upper fragment, either backwards or forwards, according to the direction of the fracture. Simple extension reduces the deformity, and the application of a shoulder cap and splints to the arm will retain the fragments in apposition (see page 279). Care should be taken not to raise the elbow; but the forearm and hand may be supported in a sling.

Fracture of the *humerus* (fig. 440) immediately above the condyles deserves very attentive consideration, as the general appearances correspond somewhat with those produced by separation of the epiphysis of the humerus, and with those of dislocation of the radius and ulna backwards. If the direction of the fracture is oblique, from above, downwards and forwards, the lower fragment is drawn upwards by the Brachialis anticus and Biceps in front, and the Triceps behind; and at the same time is drawn backwards behind the upper fragment by the Triceps. This injury may be diagnosed from dislocation by the increased mobility in fracture, the existence of crepitus, and the fact of the deformity being remedied by extension, on the discontinuance of which it is reproduced. The age of the patient is of importance in distinguishing this form of injury from separation of the epiphysis. In some cases where the injury has been produced by falls on the elbow, the lower fragment is drawn upwards and forwards, causing a considerable prominence in front; and the upper fragment projects backwards beneath the tendon of the Triceps muscle.

FIG. 441.—Fracture of the olecranon.

Fracture of the *olecranon process* (fig. 441) is a frequent accident. The detached fragment is displaced upwards, by the action of the Triceps muscle, from half an inch to two inches; the prominence of the elbow is consequently lost, and a deep hollow is felt at the back part of the joint, which is much increased on flexing the limb. The patient at the same time loses, more or less, the power of extending the forearm. The treatment consists in wiring the fragments together; but if for some reason this operation is not desirable, the fragments should be approximated by strapping or a figure-of-8 bandage, and the arm put up in an extended position in order to relax the Triceps. Massage and passive movements must be employed, for fear of ankylosis. Union is generally ligamentous.

Fracture of the *neck of the radius* is an exceedingly rare accident, and is usually caused by direct violence. Its diagnosis is somewhat obscure, on account of the slight deformity visible, the injured part being surrounded by a large number of muscles; but the movements of pronation and supination are entirely lost. The lower fragment is drawn forwards and slightly upwards by the Biceps, and inwards by the Pronator radii teres, its displacement forwards and upwards being counteracted in some degree by the Supinator brevis. The treatment essentially consists in relaxing the Biceps, Supinator brevis, and Pronator radii teres muscles, by flexing the forearm, and placing it in a position midway between pronation and supination, extension having been previously made so as to bring the parts in apposition.

In fracture of the *radius* below the insertion of the Biceps, but above the insertion of the Pronator radii teres, the upper fragment is strongly supinated by the Biceps and Supinator brevis, and at the same time drawn forwards and flexed by the Biceps ; the lower fragment is pronated and drawn inwards towards the ulna by the pronators. Thus there is extreme displacement with very little deformity. In treating such a fracture the arm must be put up in a position of supination, otherwise union will take place with great impairment of the movements of the hand. In fractures of the radius below the insertion of the Pronator radii teres (fig. 442), the upper fragment is drawn upwards by the Biceps, and inwards by the Pronator radii teres, holding a position midway between pronation and supination, and a degree of fulness in the upper half of the forearm is thus produced: the lower fragment is drawn downwards

FIG. 442.—Fracture of the shaft of the radius.

and inwards towards the ulna by the Pronator quadratus, and thrown into a state of pronation by the same muscle ; at the same time, the Supinator longus, by elevating the styloid process, into which it is inserted, will serve to depress the upper end of the lower fragment still more towards the ulna. In order to relax the opposing muscles the forearm should be bent, and the limb placed in a position midway between pronation and supination ; the fracture is then easily reduced by extension from the wrist and elbow : well-padded splints should be applied on both sides of the forearm from the elbow to the wrist ; the hand being allowed to fall will, by its own weight, counteract the action of the Pronator quadratus and Supinator longus, and elevate the lower fragment to the level of the upper one.

In fracture of the *shaft of the ulna* the upper fragment retains its usual position, but the lower fragment is drawn outwards towards the radius by the Pronator quadratus, producing a well-marked depression at the seat of fracture, and some fulness on the dorsal and palmar surfaces of the forearm. The fracture is easily reduced by extension from the wrist and forearm. The forearm should be flexed, and placed in a position midway between pronation and supination, and well-padded splints applied from the elbow to the ends of the fingers.

In fracture of the *shafts of the radius and ulna together*, the lower fragments are drawn upwards, sometimes forwards, sometimes backwards, according to the direction of the fracture, by the combined actions of the Flexor and Extensor muscles, producing a

FIG. 443.—Fracture of the lower end of the radius.

degree of fulness on the dorsal or palmar surface of the forearm ; at the same time the lower fragments are drawn into contact by the Pronator quadratus, the radius being in a state of pronation : the upper fragment of the radius is drawn upwards and inwards by the Biceps and Pronator radii teres to a higher level than the ulna ; the upper portion of the ulna is slightly elevated by the Brachialis anticus. The fracture may be reduced by extension from the wrist and elbow, and the forearm should be placed in the same position as in fracture of the ulna.

In fracture of the *lower end of the radius* (fig. 443) the displacement which is produced is very considerable, and bears some resemblance to dislocation of the carpus backwards, from which it should be carefully distinguished. The lower fragment is displaced backwards and upwards, but this displacement is probably due to the force of the blow driving the portion of the bone into this position and not to any muscular influence. The upper fragment projects forwards, often lacerating the substance of the Pronator quadratus, and is drawn by this muscle into close contact with the lower end of the ulna, causing a projection on the anterior surface of the forearm, immediately above the carpus, from the Flexor tendons being thrust forwards. This fracture may be distinguished from dislocation by the deformity being removed on making sufficient extension, when crepitus may be occasionally detected; at the same time, on extension being discontinued, the parts immediately resume their deformed appearance (see also page 290). The age of the patient will assist in determining whether the injury is fracture or separation of the epiphysis. The treatment consists in flexing the forearm, and making powerful extension from the wrist and elbow, depressing at the same time the radial side of the hand, and retaining the parts in that position by well-padded *pistol-shaped* splints.

MUSCLES AND FASCIÆ OF THE LOWER EXTREMITY

The Muscles of the Lower Extremity are subdivided into groups, corresponding with the different regions of the limb.

I. Iliac Region
Psoas magnus.
Psoas parvus.
Iliacus.

II. Thigh
1. *Anterior Femoral Region*
Tensor fasciæ femoris.
Sartorius.
Rectus femoris.
Quadriceps extensor { Vastus externus.
Vastus internus.
Crureus.
Subcrureus.

2. *Internal Femoral Region*
Gracilis.
Pectineus.
Adductor longus.
Adductor brevis.
Adductor magnus.

3. *Gluteal Region*
Gluteus maximus.
Gluteus medius.
Gluteus minimus.
Pyriformis.
Obturator internus.
Gemellus superior.
Gemellus inferior.
Quadratus femoris.
Obturator externus.

4. *Posterior Femoral Region*
Biceps.
Semitendinosus.
Semimembranosus.

III. Leg
5. *Anterior Tibio-fibular Region*
Tibialis anticus.
Extensor proprius hallucis.

Extensor longus digitorum.
Peroneus tertius.

6. *Posterior Tibio-fibular Region*
Superficial Layer
Gastrocnemius.
Soleus.
Plantaris.

Deep Layer
Popliteus.
Flexor longus hallucis.
Flexor longus digitorum.
Tibialis posticus.

7. *Fibular Region*
Peroneus longus.
Peroneus brevis.

IV. Foot
8. *Dorsal Region*
Extensor brevis digitorum.

9. *Plantar Region*
First Layer
Abductor hallucis.
Flexor brevis digitorum.
Abductor minimi digiti.

Second Layer
Flexor accessorius.
Lumbricales.

Third Layer
Flexor brevis hallucis.
Adductor obliquus hallucis.
Flexor brevis minimi digiti.
Adductor transversus hallucis

Fourth Layer
The Interossei.

I. Muscles and Fasciæ of the Iliac Region

Psoas magnus.   Psoas parvus.   Iliacus.

Dissection.—No detailed description is required for the dissection of these muscles. On the removal of the viscera from the abdomen, they are exposed, covered by the peritoneum and a thin layer of fascia, the iliac fascia.

The **Fascia covering the Psoas and Iliacus** is the aponeurotic layer which lines the back part of the abdominal cavity, and covers the Psoas and Iliacus muscles throughout their whole extent. It is thin above, and becomes gradually thicker below as it approaches Poupart's ligament.

The *portion covering the Psoas* is thickened above to form the ligamentum arcuatum internum, which stretches from the transverse process of the first lumbar vertebra to the body of the second; internally, it is attached by a series of arched processes to the intervertebral discs, and prominent margins of the bodies of the vertebræ, and to the upper part of the sacrum; the intervals so left, opposite the constricted portions of the bodies, transmitting the lumbar arteries and veins and filaments of the sympathetic cord. Externally, above the crest of the ilium, this portion of the fascia is continuous with the anterior lamella of the lumbar fascia covering the front of the Quadratus lumborum (see page 470), but below the crest of the ilium it is continuous with the fascia covering the Iliacus.

The *portion investing the Iliacus* (Fascia iliaca*) is connected, externally, to the whole length of the inner border of the crest of the ilium; and internally, to the brim of the true pelvis, where it is continuous with the periosteum; and at the ilio-pectineal eminence it receives the tendon of insertion of the Psoas parvus, when that muscle exists. External to the femoral vessels, this fascia is intimately connected to the posterior margin of Poupart's ligament, and is continuous with the fascia transversalis. Immediately to the outer side of the femoral vessels the fascia iliaca is prolonged backwards and inwards from Poupart's ligament as a band, the *ilio-pectineal ligament*, which is attached to the ilio-pectineal eminence. This ligament divides the space between Poupart's ligament and the innominate bone into two parts, the inner of which transmits the femoral vessels, the outer the ilio-psoas and the anterior crural nerve (fig. 360). Internal to the vessels the iliac fascia is attached to the ilio-pectineal line behind the conjoined tendon, where it is again continuous with the transversalis fascia; and corresponding to the point where the femoral vessels pass into the thigh, this fascia descends behind them forming the posterior wall of the femoral sheath. This portion of the iliac fascia which passes behind the femoral vessels is also attached to the ilio-pectineal line beyond the limits of the attachment of the conjoined tendon; at this part it is continuous with the pubic portion of the fascia lata of the thigh. The external iliac vessels lie in front of the iliac fascia, but all the branches of the lumbar plexus behind it; it is separated from the peritoneum by a quantity of loose areolar tissue.

The **Psoas magnus** (fig. 445) is a long fusiform muscle placed on the side of the lumbar region of the spine and margin of the pelvis. It arises from the front of the bases and lower borders of the transverse processes of the lumbar vertebræ by five fleshy slips; also from the sides of the bodies, and the corresponding intervertebral substances of the last dorsal and all the lumbar vertebræ. The muscle is connected to the bodies of the vertebræ by five slips; each slip is attached to the upper and lower margins of two vertebræ, and to the intervertebral substance between them; the slips themselves being connected by tendinous arches which extend across the constricted part of the bodies, and beneath which pass the lumbar arteries and veins and filaments of the sympathetic cord. These tendinous arches also give origin to muscular fibres, and protect the blood-vessels and nerves from pressure during the action of the muscle. The first slip is attached to the contiguous margins of the last dorsal and first lumbar vertebræ; the last to the contiguous margins of the fourth and fifth lumbar vertebræ, and to the intervertebral substance. From these points the muscle proceeds downward across the brim of the pelvis, and, diminishing gradually in size, passes beneath Poupart's ligament, and

* The student must not confound this fascia with the *iliac portion of the fascia lata* (see page 545).

terminates in a tendon, which, after receiving nearly the whole of the fibres of the Iliacus, is inserted into the lesser trochanter of the femur.

Relations.—In the lumbar region: By its *anterior surface*, with the ligamentum arcuatum internum, the iliac fascia, the extra-peritoneal fat and peritoneum, the kidney, Psoas parvus, renal vessels, ureter, spermatic vessels, genito-crural nerve, and the colon. By its *posterior surface*, with the transverse processes of the lumbar vertebræ, and the Quadratus lumborum, from which it is separated by the anterior lamella of the lumbar fascia. The lumbar plexus is situated in the posterior part of the substance of the muscle. By its *inner side*, the muscle is in relation with the bodies of the lumbar vertebræ, the lumbar arteries, the ganglia of the sympathetic nerve, and their branches of communication with the spinal nerves; the lumbar glands; the vena cava inferior on the right, and the aorta on the left side, and along the brim of the pelvis with the external iliac artery. In the thigh it is in relation, in front, with the fascia lata; behind, with the capsular ligament of the hip, from which it is separated by a synovial bursa, which frequently communicates with the cavity of the joint through an opening of variable size; by its *inner border*, with the Pectineus and internal circumflex artery, and also with the femoral artery, which slightly overlaps it; by its *outer border*, with the anterior crural nerve and Iliacus muscle.

The **Psoas parvus** is a long slender muscle, placed in front of the Psoas magnus. It arises from the sides of the bodies of the last dorsal and first lumbar vertebræ and from the intervertebral substance between them. It forms a small muscular bundle, which terminates in a long flat tendon, inserted into the ilio-pectineal line and eminence, and, by its outer border, into the fascia iliaca. This muscle is often absent, and, according to Cruveilhier, sometimes double.

Relations.—The Psoas parvus is covered by the peritoneum, and, at its origin, by the ligamentum arcuatum internum; it rests on the Psoas magnus.

The **Iliacus** is a flat, triangular muscle, which fills up the whole of the iliac fossa. It arises from the upper two-thirds of this fossa, and from the inner margin of the crest of the ilium; behind, from the anterior sacro-iliac and the ilio-lumbar ligaments, and base of the sacrum; in front, it reaches as far as the anterior superior and anterior inferior spinous processes of the ilium, and the notch between them. The fibres converge to be inserted into the outer side of the tendon of the Psoas, some of them being prolonged on to the shaft of the femur for about an inch below and in front of the lesser trochanter.*

Relations.—Within the abdomen: By its *anterior surface*, with the iliac fascia, which separates the muscle from the extra-peritoneal fat and peritoneum, and with the external cutaneous nerve; on the right side, with the cæcum; on the left side, with the sigmoid flexure of the colon. By its *posterior surface*, with the iliac fossa. By its *inner border*, with the Psoas magnus, and anterior crural nerve. In the thigh, it is in relation, by its *anterior surface*, with the fascia lata, Rectus, Sartorius, and profunda femoris artery; behind, with the capsule of the hip-joint, a synovial bursa common to it and the Psoas magnus being interposed.

Nerves.—The Psoas magnus is supplied by the anterior branches of the second and third lumbar nerves; the Psoas parvus, when it exists, is supplied by the anterior branch of the first lumbar nerve; and the Iliacus by the anterior branches of the second and third lumbar nerves through the anterior crural.

Actions.—The Psoas and Iliacus muscles, acting from above, flex the thigh upon the pelvis: acting from below, the femur being fixed, the muscles of both sides bend the lumbar portion of the spine and pelvis forwards. They also serve to maintain the erect position, by supporting the spine and pelvis upon the femur, and assist in raising the trunk when the body is in the recumbent posture.

The *Psoas parvus* is a tensor of the iliac fascia.

Surgical Anatomy.—In the iliac fascia there is no definite septum between the portions of fascia covering the Psoas and Iliacus respectively, and the fascia is only connected to the subjacent muscles by a quantity of loose connective tissue. When an abscess forms beneath this fascia, as it is very apt to do, the matter is contained in an osseo-fibrous cavity

---

* The Psoas and Iliacus are sometimes regarded as a single muscle, the *Ilio-psoas*, having two heads of origin and a single insertion.

which is closed on all sides within the abdomen, and is open only at its lower part, where the fascia is prolonged over the muscle into the thigh.

Abscess within the sheath of the Psoas muscle (*Psoas abscess*) is generally due to tuberculous caries of the bodies of the lower dorsal and lumbar vertebræ. When the disease is in the dorsal region, the matter tracks down the posterior mediastinum, in front of the bodies of the vertebræ, and, passing beneath the Ligamentum arcuatum internum, enters the sheath of the Psoas muscle, down which it passes as far as the pelvic brim; it then gets beneath the iliac portion of the fascia, and fills up the iliac fossa. In consequence of the attachment of the fascia to the pelvic brim, it rarely finds its way into the pelvis, but passes by a narrow opening under Poupart's ligament into the thigh, to the outer side of the femoral vessels. It thus follows that a Psoas abscess may be described as consisting of four parts: (1) a somewhat narrow channel at its upper part, in the Psoas sheath; (2) a dilated sac in the iliac fossa; (3) a constricted neck under Poupart's ligament; and (4) a dilated sac in the upper part of the thigh. When the lumbar vertebræ are the seat of the disease, the matter finds its way directly into the substance of the muscle. The muscular fibres are destroyed, and the nervous cords contained in the abscess are isolated and exposed in its interior; the femoral vessels which lie in front of the fascia remain intact, and the peritoneum seldom becomes implicated. All Psoas abscesses do not, however, pursue this course: the matter may leave the sheath of the muscle above the crest of the ilium, and tracking backwards may point in the loin (*lumbar abscess*); or it may point above Poupart's ligament in the inguinal region; or it may follow the course of the iliac vessels into the pelvis, and, passing through the great sacro-sciatic notch, discharge itself on the back of the thigh; or it may open into the bladder, or find its way into the perinæum.

II. Muscles and Fasciæ of the Thigh

1. *Anterior Femoral Region*

Tensor fasciæ femoris.	Quadriceps	Rectus femoris.
Sartorius.	extensor	Vastus externus.
		Vastus internus.
		Crureus.

Subcrureus.

Dissection.—To expose the muscles and fasciæ in this region, make an incision along Poupart's ligament, from the anterior superior spine of the ilium to the spine of the os pubis; a vertical incision from the centre of this, along the middle of the thigh to below the knee-joint; and a transverse incision from the inner to the outer side of the leg, at the lower end of the vertical incision. The flaps of integument having been removed, the superficial and deep fasciæ should be examined. The more advanced student should commence the study of this region by an examination of the anatomy of femoral hernia, and Scarpa's triangle, the incisions for the dissection of which are marked out in fig. 444.

The **superficial fascia** forms a continuous layer over the whole of the thigh, consisting of areolar tissue, containing in its meshes much fat, and capable of being separated into two or more layers, between which are found the superficial vessels and nerves. It varies in thickness in different parts of the limb; in the groin it is thick, and the two layers are separated from one another by the superficial inguinal lymphatic glands, the internal saphenous vein, and several smaller vessels. One of these two layers, the superficial, is continuous above with the superficial fascia of the abdomen. The deep layer of the superficial fascia is a very thin, fibrous layer, best marked on the inner side of the long saphenous vein and below Poupart's ligament. It is placed beneath the sub-cutaneous vessels and nerves and upon the surface of the fascia lata. It is intimately adherent to the fascia lata a little below Poupart's ligament. It covers the saphenous opening in the fascia lata, being closely united to its circumference, and is connected to the sheath of the femoral vessels, corresponding to its under surface. The portion of fascia covering this aperture is perforated by the internal saphenous vein and by numerous blood and lymphatic vessels, hence it has been termed the *cribriform fascia*, the openings for these vessels having been likened to the holes in a sieve. The cribriform fascia adheres closely both to the superficial fascia and to the fascia lata, so that it is described by some anatomists as part of the fascia lata, but is usually considered as belonging to the superficial fascia. It is not until the cribriform fascia has been cleared away that the saphenous opening is seen, so that this opening does not in ordinary cases exist naturally, but is the result of dissection. Callender, however, speaks of cases in which, probably as the result of pressure from enlarged inguinal lymphatic glands, the fascia has become atrophied, and a saphenous opening exists independent of dissection. A femoral hernia in passing through the saphenous opening receives the cribriform fascia as one of its coverings. A large subcutaneous bursa is found in the superficial fascia over the patella.

The **deep fascia** of the thigh is exposed on the removal of the superficial fascia, and is named, from its great extent, the *fascia lata*; it constitutes a uniform investment for the whole of this region of the limb, but varies in thickness in different parts; thus, it is thicker in the upper and outer part of the thigh, where it receives a fibrous expansion from the Gluteus maximus muscle, and where the Tensor fasciæ femoris is inserted between its layers: it is very thin behind and at the upper and inner part, where it covers the Adductor muscles, and again becomes stronger around the knee, receiving fibrous expansions from the tendon of the Biceps externally, and from the Sartorius internally, and Quadriceps extensor cruris in front. The fascia lata is attached, above and behind, to the back of the sacrum and coccyx; externally, to the crest of the ilium; in front, to Poupart's ligament, and to the body of the os pubis; and internally, to the descending ramus of the os pubis, to the ramus and tuberosity of the ischium, and to the lower border of the great sacrosciatic ligament. From its attachment to the crest of the ilium it passes down over the Gluteus medius muscle to the upper border of the Gluteus maximus, where it splits into two layers, one passing superficial to and the other beneath this muscle. At the lower border of the muscle the two layers reunite. Externally, the fascia lata receives the greater part of the tendon of insertion of the Gluteus maximus, and becomes proportionally thickened. The portion of the fascia lata attached to the front part of the crest of the ilium, and corresponding to the origin of the Tensor fasciæ femoris, passes down the outer side of the thigh as two layers, one superficial to and the other beneath this muscle; these at the lower end of the muscle become blended together into a thick and strong band, having first received the insertion of the muscle. This band is continued downwards, under the name of the *ilio-tibial band*, to be inserted into the external tuberosity of the tibia. The part of the ilio-tibial band which lies beneath the Tensor fasciæ femoris is prolonged upwards to the capsule of the hip, with the outer part of which it becomes continuous. Below, the fascia lata is attached to all the prominent points around the knee-joint, viz. the condyles of the femur, tuberosities of the tibia, and head of the fibula. On each side of the patella it is strengthened by transverse fibres given off from the lower part of the Vasti muscles, which are attached to and support this bone. Of these the outer are the stronger, and are continuous with the ilio-tibial band. From the inner surface of the fascia lata are given off two strong intermuscular septa, which are attached to the whole length of the linea aspera and its prolongations above and below: the external and stronger one, which extends from the insertion of the Gluteus maximus to the outer condyle, separates the Vastus externus in front from the short head of the Biceps behind, and gives partial origin to these muscles; the inner one, the thinner of the two, separates the Vastus internus from the Adductor and Pectineus muscles. Besides these

Fig. 444.—Dissection of lower extremity. Front view.

1. *Dissection of femoral hernia, and Scarpa's triangle*

2. *Front of thigh*

3. *Front of leg*

4. *Dorsum of foot*

there are numerous smaller septa, separating the individual muscles, and enclosing each in a distinct sheath.

The Saphenous Opening.—At the upper and inner part of the thigh, a little below the inner end of Poupart's ligament, a large oval-shaped aperture is observed after the superficial fascia has been cleared off: it transmits the internal saphenous vein, and other smaller vessels, and is termed the *saphenous opening.* In order to consider more correctly the mode of formation of this aperture, the fascia lata in this part of the thigh is described as consisting of two portions, an iliac portion and a pubic portion.

The *iliac portion* is the part of the fascia lata on the outer side of the saphenous opening. It is attached, externally, to the crest of the ilium and its anterior superior spine, to the whole length of Poupart's ligament, and to the pectineal line in conjunction with Gimbernat's ligament. From the spine of the os pubis it is reflected downwards and outwards, forming an arched margin, the *falciform process* or boundary of the saphenous opening; this margin overlies and is adherent to the anterior layer of the sheath of the femoral vessels: to its edge is attached the cribriform fascia. The upward and inward prolongation of the falciform process is named the *superior cornu*; its downward and inward prolongation, the *inferior cornu.* The latter is well defined, and is continuous behind the saphenous vein with the pubic portion of the fascia lata.

The *pubic portion* is situated at the inner side of the saphenous opening; at the lower margin of this aperture it is continuous with the iliac portion; traced upwards, it covers the surface of the Pectineus, Adductor longus, and Gracilis muscles, and, passing behind the sheath of the femoral vessels, to which it is closely united, is continuous with the fascia iliaca, and is attached above to the ilio-pectineal line. From this description it may be observed that the iliac portion of the fascia lata passes in front of the femoral vessels, and the pubic portion behind them, so that an apparent aperture exists between the two, through which the internal saphenous passes to join the femoral vein.

The fascia should now be removed from the surface of the muscles. This may be effected by pinching it up between the forceps, dividing it, and separating it from each muscle in the course of its fibres.

The **Tensor fasciæ femoris** arises from the anterior part of the outer lip of the crest of the ilium, from the outer surface

FIG. 445.—Muscles of the iliac and anterior femoral regions.

of the anterior superior spinous process, and part of the outer border of the notch below it, between the Gluteus medius and Sartorius, and from the inner surface of the fascia lata. It is inserted between two layers of the fascia lata about the junction of the middle and upper thirds of the thigh. From the point of insertion the fascia is continued downwards to the external tuberosity of the tibia as a thickened band, the *ilio-tibial band.*

Relations.—By its *superficial surface*, with the fascia lata and the integument. By its *deep surface*, with the Gluteus medius, Rectus femoris, Vastus externus, and the ascending branches of the external circumflex artery. By its *anterior border*, with the Sartorius, from which it is separated below by a triangular space, in which is seen the Rectus femoris. By its *posterior border*, with the Gluteus medius.

The **Sartorius,** the longest muscle in the body, is flat, narrow, and ribbon-like; it arises by tendinous fibres from the anterior superior spinous process of the ilium and the upper half of the notch below it, passes obliquely across the upper and anterior part of the thigh, from the outer to the inner side of the limb, then descends vertically, as far as the inner side of the knee, passing behind the inner condyle of the femur, and terminates in a tendon; this curves obliquely forwards and expands into a broad aponeurosis, which is inserted, in front of the Gracilis and Semitendinosus, into the upper part of the inner surface of the shaft of the tibia, nearly as far forwards as the crest. The upper part of the aponeurosis is curved backwards over the upper edge of the tendon of the Gracilis so as to be inserted behind it. An offset, derived from its upper margin, blends with the fibrous capsule of the knee-joint, and another, given off from its lower border, blends with the fascia on the inner side of the leg.

The relations of this muscle to the femoral artery should be carefully examined, as it constitutes the chief guide in tying the vessel. In the upper third of the thigh it forms the outer side of a triangular space, *Scarpa's triangle*, the inner side of which is formed by the inner border of the Adductor longus, and the base, turned upwards, by Poupart's ligament; the femoral artery passes perpendicularly through the middle of this space from its base to its apex. In the middle third of the thigh, the femoral artery is contained in Hunter's canal, on the roof of which lies the Sartorius.

Relations.—By its *superficial surface*, with the fascia lata and integument. By its *deep surface*, with the Rectus, Iliacus, Vastus internus, anterior crural nerve, sheath of the femoral vessels, Adductor longus, Adductor magnus, Gracilis, Semitendinosus, long saphenous nerve, and internal lateral ligament of the knee-joint.

The **Quadriceps extensor** includes the four remaining muscles on the front of the thigh. It is the great Extensor muscle of the leg, forming a large fleshy mass, which covers the front and sides of the femur, being united below into a tendon, attached to the patella, and above, subdivided into separate portions, which have received distinct names. Of these, one occupying the middle of the thigh, connected above with the ilium, is called the *Rectus femoris*, from its straight course. The other divisions lie in immediate connection with the shaft of the femur, which they cover from the trochanters to the condyles. The portion on the outer side of the femur is termed the *Vastus externus*; that covering the inner side, the *Vastus internus*; and that in front, the *Crureus*.

The **Rectus femoris** is situated in the middle of the anterior region of the thigh; it is fusiform in shape, and its superficial fibres are arranged in a bipenniform manner, the deep fibres running straight down to the deep aponeurosis. It arises by two tendons: one, the anterior or straight, from the anterior inferior spinous process of the ilium; the other, the posterior or reflected tendon, from a groove above the brim of the acetabulum; the two unite at an acute angle, and spread into an aponeurosis, which is prolonged downwards on the anterior surface of the muscle and from which the muscular fibres arise.* The muscle terminates in a broad and thick aponeurosis, which occupies the lower two-thirds of its

---

* W. R. Williams, in an interesting paper in the *Journ. of Anat. and Phys.* vol. xiii. p. 204, points out that the reflected tendon is the real origin of the muscle, and is alone present in early fœtal life. The direct tendon is merely an accessory band of condensed fascia. The paper will well repay perusal, though in some particulars the description in the text is more generally accurate.

posterior surface, and, gradually becoming narrowed into a flattened tendon, is inserted into the patella in common with the Vasti and Crureus.

Relations.—By its *superficial surface*, with the anterior fibres of the Gluteus minimus, the Tensor fasciæ femoris, the Sartorius, and the Iliacus ; by its lower three-fourths, with the fascia lata. By its *posterior surface*, with the hip-joint, the external circumflex vessels, branches of the anterior crural nerve, and the Crureus and Vasti muscles.

The **Vastus externus** is the largest part of the Quadriceps extensor. It arises by a broad aponeurosis, which is attached to the upper half of the anterior inter-trochanteric line, to the anterior and inferior borders of the root of the great trochanter, to the outer lip of the gluteal ridge, and to the upper half of the outer lip of the linea aspera : this aponeurosis covers the upper three-fourths of the muscle, and from its inner surface many fibres take origin. A few additional fibres arise from the tendon of the Gluteus maximus, and from the external intermuscular septum between the Vastus externus and short head of the Biceps. The fibres form a large fleshy mass, which is attached to a strong aponeurosis, placed on the under surface of the muscle at its lower part : this becomes contracted and thickened into a flat tendon, which is inserted into the outer border of the patella, blending with the great Extensor tendon, and giving an expansion to the capsule of the knee-joint.

Relations.—By its *superficial surface*, with the Rectus, the Tensor fasciæ femoris, the fascia lata, and the tendon of the Gluteus maximus, from which it is separated by a synovial bursa. By its *deep surface*, with the Crureus, some large branches of the external circumflex artery and anterior crural nerve being interposed.

The **Vastus internus** and **Crureus** appear to be inseparably united, but when the Rectus femoris has been reflected a narrow interval will be observed extending upwards from the inner border of the patella between the two muscles. Here they can be separated, and the separation should be continued upwards as far as the lower part of the anterior intertrochanteric line, where, however, the two muscles are frequently continuous.

The **Vastus internus** arises from the lower half of the anterior intertrochanteric line, the spiral line, the inner lip of the linea aspera, the upper part of the internal supracondylar line, and the tendons of the Adductor longus and Adductor magnus and internal intermuscular septum. Its fibres are directed downwards and forwards, and are chiefly attached to an aponeurosis which lies on the deep surface of the muscle and is inserted into the inner border of the patella and the Quadriceps extensor tendon, an expansion being sent to the capsule of the knee-joint.

The **Crureus** arises from the front and outer aspect of the shaft of the femur in its upper two-thirds and from the lower part of the external intermuscular septum. Its fibres end in a superficial aponeurosis, which forms the deep part of the Quadriceps extensor tendon.

Relations.—The inner edge of the Crureus is in contact with the anterior edge of the Vastus internus, but when separated from each other, as directed above, the latter muscle is seen merely to overlap the inner aspect of the femoral shaft without taking any fibres of origin from it. The Vastus internus is partly covered by the Rectus and Sartorius, but where these separate near the knee it becomes superficial, and produces a well-marked prominence above the inner aspect of the knee. In the middle third of the thigh it forms the outer wall of Hunter's canal, which contains the femoral vessels and the long saphenous nerve—the roof of the canal being formed by a strong fascia (derived from the membranous expansion of the fascia lata, which lies beneath the Sartorius) which extends from the Vastus internus to the Adductores longus and magnus. The Crureus is almost completely hidden by the Rectus femoris and Vastus externus. The deep surface of the two muscles is in relation to the femur and Subcrureus muscle. A synovial bursa is situated between the femur and the portion of the Quadriceps extensor tendon above the patella ; in the adult it usually communicates with the synovial cavity of the knee-joint.

The *tendons* of the different portions of the Quadriceps extensor unite at the lower part of the thigh, so as to form a single strong tendon, which is inserted into the upper part of the patella, some few fibres passing over it to blend with the Ligamentum patellæ. More properly, the patella may be regarded as a sesamoid bone, developed in the tendon of the Quadriceps; and the Ligamentum patellæ, which is continued from the lower part of the patella to the tuberosity of the tibia, as the proper tendon of insertion of the muscle, the lateral patellar ligaments (see page 403) being fascial expansions from its borders. A synovial bursa is interposed between the tendon and the upper part of the front of the tibia; and another, the *pre-patellar bursa*, is placed over the patella itself. This latter bursa often becomes enlarged, constituting 'housemaid's knee.'

The **Subcrureus** is a small muscle, usually distinct from the Crureus, but occasionally blended with it, which arises from the anterior surface of the lower part of the shaft of the femur, and is inserted into the upper part of the synovial membrane of the knee-joint. It sometimes consists of several separate muscular bundles.

Nerves.—The Tensor fasciæ femoris is supplied by the fourth and fifth lumbar and first sacral nerves through the superior gluteal nerve; the other muscles of this region, by the second, third, and fourth lumbar nerves, through branches of the anterior crural.

Actions.—The Tensor fasciæ femoris is a tensor of the fascia lata; continuing its action, the oblique direction of its fibres enables it to abduct and to rotate the thigh inwards. In the erect posture, acting from below, it will serve to steady the pelvis upon the head of the femur; and by means of the ilio-tibial band it steadies the condyles of the femur on the articular surfaces of the tibia, and assists the Gluteus maximus in supporting the knee in the extended position. The Sartorius flexes the leg upon the thigh, and, continuing to act, flexes the thigh upon the pelvis; it next abducts and rotates the thigh outwards. It was formerly supposed to adduct the thigh, so as to cross one leg over the other, and hence received its name of Sartorius, or tailor's muscle (*sartor*, a tailor), because it was supposed to assist in crossing the legs in the squatting position. When the knee is bent, the Sartorius assists the Semitendinosus, Semimembranosus, and Popliteus in rotating the tibia inwards. Taking its fixed point from the leg, it flexes the pelvis upon the thigh, and, if one muscle acts, assists in rotating the pelvis. The Quadriceps extensor extends the leg upon the thigh. The Rectus muscle assists the Psoas and Iliacus in supporting the pelvis and trunk upon the femur. It also assists in flexing the thigh on the pelvis, or if the thigh is fixed it will flex the pelvis. The Vastus internus draws the patella inwards as well as upwards.

Surgical Anatomy.—A few fibres of the Rectus muscle are occasionally ruptured from severe strain. This accident is especially liable to occur during the games of football and cricket, and is sometimes known as 'cricket thigh.' The patient experiences a sudden pain in the part, as if he had been struck, and the Rectus muscle stands out and is felt to be tense and rigid. The accident is often followed by considerable swelling from inflammatory effusion. Occasionally the Quadriceps extensor may be torn away from its insertion into the patella; or the tendon of the patella may be ruptured about an inch above the bone. This accident is caused in the same manner as fracture of the patella by muscular action, viz. by a violent muscular effort to prevent falling while the knee is in a position of semiflexion. A distinct gap can be felt above the patella, and, owing to the retraction of the muscular fibres, union may fail to take place.

2. *Internal Femoral Region*

Gracilis.	Adductor longus.
Pectineus.	Adductor brevis.

Adductor magnus.

Dissection.—These muscles are at once exposed by removing the fascia from the fore part and inner side of the thigh. The limb should be abducted, so as to render the muscles tense and easier of dissection.

The **Gracilis** (figs. 445, 448) is the most superficial muscle on the inner side of the thigh. It is thin and flattened, broad above, narrow and tapering below. It arises by a thin aponeurosis from the lower half of the margin of the symphysis and the anterior half of the pubic arch. The fibres pass vertically downwards, and terminate in a rounded tendon, which passes behind the internal condyle of the femur, and, curving round the inner tuberosity of the tibia, becomes flattened, and is inserted into the upper part of the inner surface of the shaft of the tibia, below the tuberosity. A few of the fibres of the lower part of the tendon are prolonged into the deep fascia of the leg. The tendon of this muscle is situated immediately above that of the Semitendinosus, and its upper edge is overlapped by the tendon of the Sartorius with which it is in part blended. As it passes across the internal lateral ligament of the knee-joint, it is separated from it by a synovial bursa common to it and the Semitendinosus muscle.

Relations.—By its *superficial surface,* with the fascia lata and the Sartorius below; the internal saphenous vein crosses it obliquely near its lower part, lying superficial to the fascia lata. The internal saphenous nerve emerges between its tendon and that of the Sartorius. By its *deep surface,* with the Adductor brevis and the Adductor magnus, and the internal lateral ligament of the knee-joint.

The **Pectineus** (fig. 445) is a flat, quadrangular muscle, situated at the anterior part of the upper and inner aspect of the thigh. It arises from the linea ilio-pectinea, and to a slight extent from the surface of bone in front of it, between the pectineal eminence and spine of the os pubis, and from the fascia covering the anterior surface of the muscle; the fibres pass downwards, backwards, and outwards, to be inserted into a rough line leading from the small trochanter to the linea aspera.

Relations.—By its *anterior surface,* with the pubic portion of the fascia lata, which separates it from the femoral vessels and internal saphenous vein. By its *posterior surface,* with the capsular ligament of the hip-joint, the Adductor brevis and Obturator externus muscles, the obturator vessels and nerve being interposed. By its *outer border,* with the Psoas, a cellular interval separating them, through which pass the internal circumflex vessels. By its *inner border,* with the margin of the Adductor longus.

The **Adductor longus,** the most superficial of the three Adductors, is a flat triangular muscle, lying on the same plane as the Pectineus. It arises, by a flat, narrow tendon, from the front of the os pubis, at the angle of junction of the crest with the symphysis; and soon expands into a broad fleshy belly, which, passing downwards, backwards, and outwards, is inserted, by an aponeurosis, into the linea aspera, between the Vastus internus and the Adductor magnus, with both of which it is usually blended.

Relations.—By its *anterior surface,* with the fascia lata, the Sartorius, and, near its insertion, with the femoral artery and vein. By its *posterior surface,* with the Adductor brevis and magnus, the anterior division of the obturator nerve, and with the profunda artery and vein

Fig. 446.—Deep muscles of the internal femoral region.

near its insertion. By its *outer border*, with the Pectineus. By its *inner border*, with the Gracilis.

The Pectineus and Adductor longus should now be divided near their origin, and turned downwards, when the Adductor brevis and Obturator externus will be exposed.

The **Adductor brevis** is situated immediately behind the two preceding muscles. It is somewhat triangular in form, and arises by a narrow origin from the outer surface of the body and descending ramus of the os pubis, between the Gracilis and Obturator externus. Its fibres, passing backwards, outwards, and downwards, are inserted, by an aponeurosis, into the line leading from the small trochanter to the linea aspera and into the upper part of the linea aspera, immediately behind the Pectineus and upper part of the Adductor longus.

Relations.—By its *anterior surface*, with the Pectineus, Adductor longus, profunda femoris artery, and anterior division of the obturator nerve. By its *posterior surface*, with the Adductor magnus, and posterior division of the obturator nerve. By its *outer border*, with the internal circumflex artery, the Obturator externus, and conjoined tendon of the Psoas and Iliacus. By its *inner border*, with the Gracilis and Adductor magnus. This muscle is pierced, near its insertion, by the second or first and second perforating branches of the profunda femoris artery.

The Adductor brevis should now be cut away near its origin, and turned outwards when the entire extent of the Adductor magnus will be exposed.

The **Adductor magnus** is a large triangular muscle, forming a septum between the muscles on the inner and those on the back of the thigh. It arises from a small part of the descending ramus of the os pubis, from the ramus of the ischium, and from the outer margin of the inferior part of the tuberosity of the ischium. Those fibres which arise from the ramus of the os pubis are very short, horizontal in direction, and are inserted into the rough line leading from the great trochanter to the linea aspera, internal to the Gluteus maximus ; those from the ramus of the ischium are directed downwards and outwards with different degrees of obliquity, to be inserted, by means of a broad aponeurosis, into the linea aspera and the upper part of its internal prolongation below. The internal portion of the muscle, composed principally of those fibres which arise from the tuberosity of the ischium, forms a thick fleshy mass consisting of coarse bundles which descend almost vertically, and terminate about the lower third of the thigh in a rounded tendon, which is inserted into the adductor tubercle on the inner condyle of the femur, being connected by a fibrous expansion to the line leading upwards from the tubercle to the linea aspera. At the insertion of the muscle, a series of osseo-aponeurotic openings, formed by tendinous arches attached to the bone, is seen. The upper four openings are small, and give passage to the perforating branches of the profunda femoris artery. The lowest is of large size, and transmits the femoral vessels from Hunter's canal to the popliteal space.

Relations.—By its *anterior surface*, with the Pectineus, Adductor brevis, Adductor longus, and the femoral and profunda vessels and obturator nerve. By its *posterior surface*, with the great sciatic nerve, the Gluteus maximus, Biceps, Semitendinosus, and Semimembranosus. By its *superior* or *shortest border*, it lies parallel with the Quadratus femoris, the internal circumflex artery passing between them. By its *internal* or *longest border*, with the Gracilis, Sartorius, and fascia lata. By its *external* or *attached border*, it is inserted into the femur behind the Adductor brevis and Adductor longus, which separate it from the Vastus internus ; and in front of the Gluteus maximus and short head of the Biceps, which separate it from the Vastus externus.

Nerves.—The three Adductor muscles and the Gracilis are supplied by the third and fourth lumbar nerves through the obturator nerve ; the Adductor magnus receiving an additional branch from the sacral plexus through the great sciatic. The Pectineus is supplied by the second, third, and fourth lumbar nerves through the anterior crural, and by the accessory obturator, from the third lumbar, when it exists. Occasionally it receives a branch from the obturator nerve.*

* Paterson describes the Pectineus as consisting of two incompletely separated strata, of which the outer or dorsal stratum, which is constant, is supplied by the anterior

Actions.—The Pectineus and three Adductors adduct the thigh powerfully; they are especially used in horse exercise, the sides of the horse being grasped between the knees by the contraction of these muscles. In consequence of the obliquity of their insertion into the linea aspera, they rotate the thigh outwards, assisting the external Rotators, and when the limb has been abducted, they draw it inwards, carrying the thigh across that of the opposite side. The Pectineus and Adductor brevis and longus assist the Psoas and Iliacus in flexing the thigh upon the pelvis. In progression, also, all these muscles assist in drawing forwards the hinder limb. The Gracilis assists the Sartorius in flexing the leg and rotating it inwards; it is also an adductor of the thigh. If the lower extremities are fixed, these muscles may take their fixed point from below and act upon the pelvis, serving to maintain the body in an erect posture; or, if their action is continued, to flex the pelvis forwards upon the femur.

Surgical Anatomy.—The Adductor longus is liable to be severely strained in those who ride much on horseback, or its tendon to be ruptured by suddenly gripping the saddle. And, occasionally, especially in cavalry soldiers, the tendon may become ossified, constituting the 'rider's bone.'

3. *Gluteal Region*

Gluteus maximus.	Obturator internus.
Gluteus medius.	Gemellus superior.
Gluteus minimus.	Gemellus inferior.
Pyriformis.	Quadratus femoris.

Obturator externus.

Dissection (fig. 447).—The subject should be turned on its face, a block placed beneath the pelvis to make the buttocks tense, and the limbs allowed to hang over the end of the table, with the foot inverted, and the thigh abducted. Make an incision through the integument along the crest of the ilium to the middle of the sacrum and thence downwards to the tip of the coccyx, and carry a second incision from that point obliquely downwards and outwards to the outer side of the thigh, four inches below the great trochanter. The portion of integument included between these incisions is to be removed in the direction shown in the figure.

The **Gluteus maximus** (fig. 448), the most superficial muscle in the gluteal region, is a very broad and thick fleshy mass, of a quadrilateral shape, which forms the prominence of the nates. Its large size is one of the most characteristic points in the muscular system in man, connected as it is with the power he has of maintaining the trunk in the erect posture. In structure the muscle is remarkably coarse, being made up of muscular fasciculi lying parallel with one another, and collected together into large bundles, separated by deep cellular intervals. It arises from the superior curved line of the ilium, and the portion of bone, including the crest, immediately above and behind it; from the posterior surface of the lower part of the sacrum, the side of the coccyx; the aponeurosis of the Erector spinæ muscle, the great sacro-sciatic ligament, and the fascia (gluteal aponeurosis) covering the Gluteus medius. The fibres are directed obliquely downwards and outwards; those forming the upper and larger portion of the muscle, together with the superficial fibres of the lower portion, terminate in a thick tendinous lamina, which passes across the great trochanter, and is inserted into the fascia lata covering the outer side of the thigh; the deeper fibres of the lower portion of the muscle are inserted into the rough line leading from the great trochanter to the linea aspera between the Vastus externus and Adductor magnus.

Three *synovial bursæ* are usually found in relation with the deep surface of this muscle. One of these, of large size, and generally multilocular, separates it from the great trochanter. A second, often wanting, is situated on the tuberosity of the ischium. A third is found between the tendon of the muscle and the Vastus externus.

Relations.—By its *superficial surface*, with a thin fascia, which separates it from the subcutaneous tissue. By its *deep surface*, from above downwards, with the ilium, sacrum, coccyx, and great sacro-sciatic ligament, part of the Gluteus

crural nerve, or in its absence by the accessory obturator, with which it is intimately related; while the inner or ventral stratum, when present, is supplied by the obturator nerve.—*Journal of Anatomy and Physiology*, vol. xxvi. p. 43.

medius; Pyriformis, Gemelli, Obturator internus, Quadratus femoris, the tuberosity of the ischium, great trochanter, the origin of the Biceps, Semitendinosus, Semimembranosus, and Adductor magnus muscles. The superficial part of the gluteal artery reaches the deep surface of the muscle by passing between the Pyriformis and the Gluteus medius; the sciatic and internal pudic vessels and nerves, and muscular branches from the sacral plexus issue from the pelvis below the Pyriformis. The first perforating artery and the terminal branches of the internal circumflex artery are also found under cover of the muscle. Its *upper border* is thin, and connected with the Gluteus medius by the fascia lata. Its *lower border* is free and prominent and is crossed by the fold of the nates.

Fig. 447.—Dissection of lower extremity. Posterior view.

1. Dissection of gluteal region

3. Back of thigh

2. Popliteal space

4. Back of leg

5. Sole of foot

Dissection.—Divide the Gluteus maximus near its origin by a vertical incision carried from its upper to its lower border; a cellular interval will be exposed, separating it from the Gluteus medius and External rotator muscles beneath. The upper portion of the muscle is to be altogether detached, and the lower portion turned outwards; the loose areolar tissue filling up the interspace between the trochanter major and tuberosity of the ischium being removed, the parts already enumerated as exposed by the removal of this muscle will be seen.

The **Gluteus medius** is a broad, thick, radiating muscle, situated on the outer surface of the pelvis. Its posterior third is covered by the Gluteus maximus; its anterior two-thirds by the fascia lata, which separates it from the superficial fascia and integument. It arises from the outer surface of the ilium, between the superior and middle curved lines, and from the outer lip of that portion of the crest which is between them; it also arises from the dense fascia (gluteal aponeurosis) covering its outer surface. The fibres converge to a strong flattened tendon, which is inserted into the oblique ridge which runs downwards and forwards on the outer surface of the great trochanter. A synovial bursa separates the tendon of the muscle from the surface of the trochanter in front of its insertion.

Relations.—By its *superficial surface*, with the Gluteus maximus behind, the Tensor fasciæ femoris and deep fascia in front. By its *deep surface*, with the Gluteus minimus, deep branches of the gluteal vessels, and the superior gluteal nerve. It *anterior border* is blended with the Gluteus minimus. Its *posterior border* lies parallel with the Pyriformis, the superficial branches of the gluteal vessels intervening.

This muscle should now be divided near its insertion and turned upwards, when the Gluteus minimus will be exposed.

The **Gluteus minimus**, the smallest of the three Glutei, is placed immediately beneath the preceding. It is fan-shaped, arising from the outer surface of the ilium, between the middle and inferior curved lines, and behind, from the margin of the great sacro-sciatic notch: the fibres converge to the deep surface of a radiated aponeurosis, which, terminating in a tendon, is inserted into an impression on the anterior border of the great trochanter, and gives an expansion to the capsule of the hip-joint. A synovial bursa is interposed between the tendon and the great trochanter.

Relations.—By its *superficial surface*, with the Gluteus medius, and the gluteal vessels and superior gluteal nerve. By its *deep surface*, with the ilium, the reflected tendon of the Rectus femoris, and capsular ligament of the hip-joint. Its *anterior margin* is blended with the Gluteus medius. Its *posterior margin* is in contact and sometimes joined with the tendon of the Pyriformis.

The **Pyriformis** is a flat muscle, pyramidal in shape, lying almost parallel with the posterior margin of the Gluteus medius. It is situated partly within the pelvis against its posterior wall, and partly at the back of the hip-joint. It arises from the front of the sacrum by three fleshy digitations, attached to the portions of bone between the first, second, third, and fourth anterior sacral foramina, and also from the grooves leading from the foramina : a few fibres also arise from the margin of the great sacro-sciatic foramen, and from the anterior surface of the great sacro-sciatic ligament. The muscle passes out of the pelvis through the great sacro-sciatic foramen, the upper part of which it fills, and is inserted by a rounded tendon into the upper border of the great trochanter, behind, but often partly blended with, the tendon of the Obturator internus and Gemelli muscles.

Relations.—By its *anterior surface, within the pelvis*, with the Rectum (especially on the left side), the sacral plexus of nerves, and the branches of the internal iliac vessels ; *external to the pelvis*, with the posterior surface of the ischium and capsular ligament of the hip-joint. By its *posterior surface, within the pelvis*, with the sacrum ; and *external to it*, with the Gluteus maximus. By its *upper border*, with the Gluteus medius, and the gluteal vessels and superior gluteal nerve. By its *lower border*, with the Gemellus superior and Coccygeus ; the sciatic vessels and nerves, the internal pudic vessels and nerve, and muscular branches from

FIG. 448.—Muscles of the hip and thigh.

Inner hamstring tendons

Sartorius

Gracilis

Semitendinosus

Semimembranosus

Outer hamstring tendon

Biceps

the sacral plexus, passing from the pelvis in the interval between the two muscles. The muscle is frequently pierced by the external popliteal nerve.

The **obturator membrane** (fig. 360) is a thin layer of interlacing fibres, which closes the obturator foramen. It is attached, externally, to the margin of the foramen; internally, to the posterior surface of the ischio-pubic ramus, below and internal to the margin of the foramen. It is occasionally incomplete, and presents at its upper and outer part a small canal, which is bounded below by a thickened band of fibres, for the passage of the obturator vessels and nerve (see page 306). Both Obturator muscles are connected with this membrane.

Dissection.—The next muscle, as well as the origin of the Pyriformis, can only be seen when the pelvis is divided and the viscera removed.

The **Obturator internus,** like the preceding muscle, is situated partly within the cavity of the pelvis, and partly at the back of the hip-joint. It arises from the inner surface of the anterior and external wall of the pelvis, where it surrounds the greater part of the obturator foramen, being attached to the descending ramus of the os pubis and the ramus of the ischium, and at the side to the inner surface of the innominate bone below and behind the pelvic brim, reaching from the upper part of the great sacro-sciatic foramen above and behind to the thyroid foramen below and in front. It also arises from the inner surface of the obturator membrane except at its posterior part, from the tendinous arch which completes the canal for the passage of the obturator vessels and nerve and to a slight extent from the obturator layer of the pelvic fascia, which covers it. The fibres converge rapidly, and are directed backwards and downwards, and terminate in four or five tendinous bands, which are found on its deep surface; these bands are reflected at a right angle over the grooved surface of the ischium between its spine and tuberosity. This bony surface is covered by smooth cartilage, which is separated from the tendon by a synovial bursa, and presents one or more ridges which correspond with the furrows between the tendinous bands. These leave the pelvis by the lesser sacro-sciatic foramen and unite into a single flattened tendon, which passes horizontally outwards, and, after receiving the attachment of the Gemelli, is inserted into the fore part of the inner surface of the great trochanter in front of the Pyriformis. A synovial bursa, narrow and elongated in form, is usually found between the tendon of this muscle and the capsular ligament of the hip: it occasionally communicates with the bursa between the tendon and the tuberosity of the ischium, the two forming a single sac.

In order to display the peculiar appearances presented by the tendon of this muscle, it must be divided near its insertion and reflected inwards.

Relations.—*Within the pelvis,* this muscle is in relation, by its *anterior surface,* with the obturator membrane and inner surface of the anterior wall of the pelvis; by its *posterior surface,* with the pelvic and obturator fasciæ, which separate it from the Levator ani; and it is crossed by the internal pudic vessels and nerve. This surface forms the outer boundary of the ischio-rectal fossa. *External to the pelvis,* it is covered by the Gluteus maximus, crossed by the great sciatic nerve, and rests on the back part of the hip-joint. As the tendon of the Obturator internus emerges from the lesser sacro-sciatic foramen it is overlapped by the two Gemelli, while nearer its insertion the Gemelli pass in front of it and form a groove in which the tendon lies.

The **Gemelli** are two small muscular fasciculi, accessories to the tendon of the Obturator internus, which is received into a groove between them. They are called *superior* and *inferior.*

The **Gemellus superior,** the smaller of the two, arises from the outer surface of the spine of the ischium, and passing horizontally outwards becomes blended with the upper part of the tendon of the Obturator internus, and is inserted with it into the inner surface of the great trochanter. This muscle is sometimes wanting.

Relations.—By its *superficial surface,* with the Gluteus maximus and the sciatic vessels and nerves. By its *deep surface,* with the capsule of the hip-joint. By its *upper border,* with the lower margin of the Pyriformis. By its *lower border,* with the tendon of the Obturator internus.

The **Gemellus inferior** arises from the upper part of the tuberosity of the ischium, where it forms the lower edge of the groove for the Obturator internus tendon, and, passing horizontally outwards, is blended with the lower part of the tendon of the Obturator internus, and inserted with it into the inner surface of the great trochanter.

Relations.—By its *superficial surface*, with the Gluteus maximus and the sciatic vessels and nerves. By its *deep surface*, with the capsular ligament of the hip-joint. By its *upper border*, with the tendon of the Obturator internus. By its *lower border*, with the tendon of the Obturator externus and Quadratus femoris.

The **Quadratus femoris** is a short, flat muscle, quadrilateral in shape (hence its name), situated between the Gemellus inferior and the upper margin of the Adductor magnus. It arises from the upper part of the external lip of the tuberosity of the ischium, and, proceeding horizontally outwards, is inserted

FIG. 449.—Obturator externus muscle.
(From a preparation in the museum of the Royal College of Surgeons of England.)

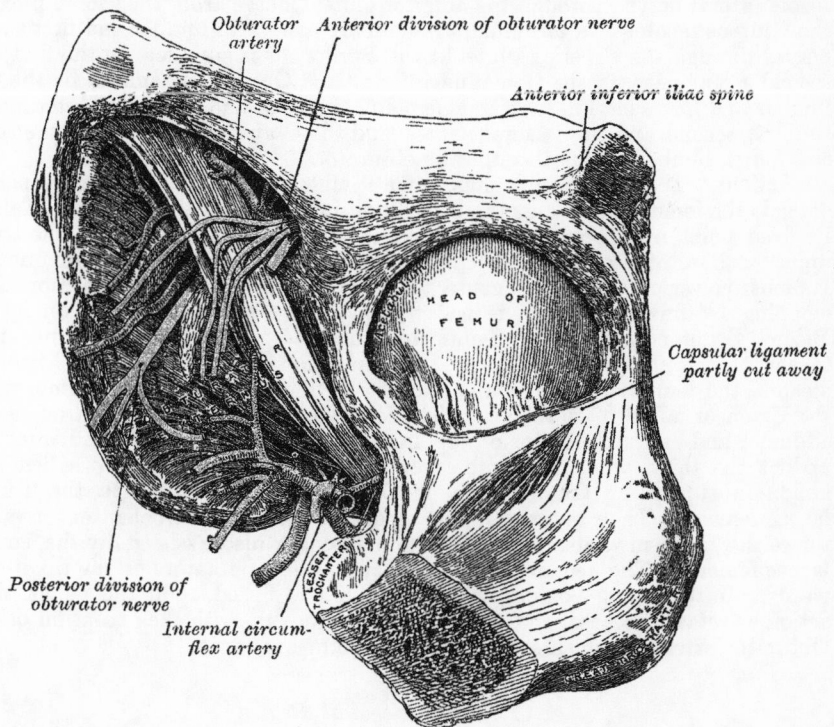

into the upper part of the linea quadrata—that is, the line which crosses the posterior intertrochanteric line. A synovial bursa is often found between the under surface of this muscle and the small trochanter, which it covers.

Relations.—By its *posterior surface*, with the Gluteus maximus and the sciatic vessels and nerves. By its *anterior surface*, with the tendon of the Obturator externus and small trochanter, and with the capsule of the hip-joint. By its *upper border*, with the Gemellus inferior. Its *lower border* is separated from the Adductor magnus by the terminal branches of the internal circumflex vessels.

Dissection.—In order to expose the next muscle (the Obturator externus), it is necessary to remove the Psoas, Iliacus, Pectineus, and Adductores brevis and longus muscles from the front and inner side of the thigh; and the Gluteus maximus and Quadratus femoris from the back part. Its dissection should, consequently, be postponed until the muscles of the anterior and internal femoral regions have been examined.

The **Obturator externus** (fig. 449) is a flat, triangular muscle, which covers the outer surface of the anterior wall of the pelvis. It arises from the margin of bone immediately around the inner side of the obturator foramen, viz. from the body and ramus of the os pubis, and the ramus of the ischium ; it also arises from the inner two-thirds of the outer surface of the obturator membrane, and from the tendinous arch which completes the canal for the passage of the obturator vessels and nerves. The fibres from the pubic arch extend on to the inner surface of the bone, from which they obtain a narrow origin between the margin of the foramen and the attachment of the membrane. The fibres converging pass backwards, outwards, and upwards, and terminate in a tendon which runs across the back part of the hip-joint, and is inserted into the digital fossa of the femur.

Relations.—By its *anterior surface*, with the Psoas, Iliacus, Pectineus, Adductor magnus, and Adductor brevis ; and more externally, with the neck of the femur and capsule of the hip-joint. The obturator artery and vein lie between this muscle and the obturator membrane ; the superficial part of the obturator nerve lies above the muscle, and the deep branch perforates it. By its *posterior surface*, with the obturator membrane and Quadratus femoris.

Nerves.—The Gluteus maximus is supplied by the fifth lumbar and first and second sacral nerves through the inferior gluteal nerve from the sacral plexus ; the Gluteus medius and minimus, by the fourth and fifth lumbar and first sacral nerves through the superior gluteal ; the Pyriformis is supplied by the first and second sacral nerves ; the Gemellus inferior and Quadratus femoris by the last lumbar and first sacral nerve ; the Gemellus superior and Obturator internus by the first, second, and third sacral nerves, and the Obturator externus by the third and fourth lumbar nerves through the obturator.

Actions.—The Gluteus maximus, when it takes its fixed point from the pelvis, extends the femur and brings the bent thigh into a line with the body. Taking its fixed point from below, it acts upon the pelvis, supporting it and the trunk upon the head of the femur, which is especially obvious in standing on one leg. Its most powerful action is to cause the body to regain the erect position after stooping, by drawing the pelvis backwards, being assisted in this action by the Biceps, Semitendinosus, and Semimembranosus. The Gluteus maximus is a tensor of the fascia lata, and by its connection with the ilio-tibial band it steadies the femur on the articular surface of the tibia during standing, when the Extensor muscles are relaxed. The lower part of the muscle also acts as an adductor and external rotator of the limb. The Gluteus medius and minimus abduct the thigh, when the limb is extended, and are principally called into action in supporting the body on one limb, in conjunction with the Tensor fasciæ femoris. Their anterior fibres, by drawing the great trochanter forwards, rotate the thigh inwards, in which action they are also assisted by the Tensor fasciæ femoris. The remaining muscles are powerful rotators of the thigh outwards. In the sitting posture, when the thigh is flexed upon the pelvis, their action as rotators ceases, and they become abductors, with the exception of the Obturator externus, which still rotates the femur outwards.

4. *Posterior Femoral Region*

Biceps.   Semitendinosus.   Semimembranosus.
(*Hamstring muscles.*)

Dissection (fig. 447).—Make a vertical incision along the middle of the back of the thigh, from the lower fold of the nates to about three inches below the back of the knee-joint, and there connect it with a transverse incision, carried from the inner to the outer side of the leg. Make a third incision transversely at the junction of the middle with the lower third of the thigh. The integument having been removed from the back of the knee, and the boundaries of the popliteal space examined, the removal of the integument from the remaining part of the thigh should be continued, when the fascia and muscles of this region will be exposed.

The **Biceps** (*Biceps flexor cruris*) is a large muscle, of considerable length, situated on the posterior and outer aspect of the thigh (fig. 448). It has two heads of origin : one, the long head, from the lower and inner impression on the back part of the tuberosity of the ischium, by a tendon common to it and the Semitendinosus, and from the lower part of the great sacro-sciatic ligament ; the other, or short head, from the outer lip of the linea aspera, between the

Adductor magnus and Vastus externus, extending up almost as high as the insertion of the Gluteus maximus; from the outer prolongation of the linea aspera to within two inches of the outer condyle; and from the external intermuscular septum. The fibres of the long head form a fusiform belly, which, passing obliquely downwards and a little outwards, terminates in an aponeurosis which covers the posterior surface of the muscle, and receives the fibres of the short head: this aponeurosis becomes gradually contracted into a tendon, which is inserted into the outer side of the head of the fibula, and by a small slip into the lateral surface of the external tuberosity of the tibia. At its insertion the tendon divides into two portions, which embrace the long external lateral ligament of the knee-joint. From the posterior border of the tendon a thin expansion is given off to the fascia of the leg. The tendon of this muscle forms the outer hamstring.

Relations.—By its *superficial surface*, with the Gluteus maximus, and the small sciatic nerve, the fascia lata, and integument. By its *deep surface*, with the Semimembranosus, Adductor magnus, and Vastus externus, the great sciatic nerve, and, near its insertion, with the external head of the Gastrocnemius, Plantaris, the superior external articular artery, and the external popliteal nerve.

The **Semitendinosus,** remarkable for the great length of its tendon, is situated at the posterior and inner aspect of the thigh. It arises from the lower and inner impression on the tuberosity of the ischium, by a tendon common to it and the long head of the Biceps; it also arises from an aponeurosis which connects the adjacent surfaces of the two muscles to the extent of about three inches after their origin. It forms a fusiform muscle, which, passing downwards and inwards, terminates a little below the middle of the thigh in a long round tendon which lies along the inner side of the popliteal space, then curves around the inner tuberosity of the tibia, and is inserted into the upper part of the inner surface of the shaft of that bone, nearly as far forwards as its anterior border. At its insertion it gives off from its lower border a prolongation to the deep fascia of the leg. This tendon lies behind the tendon of the Sartorius, and below that of the Gracilis, to which it is united. A tendinous intersection is usually observed about the middle of the muscle.

Relations.—By its *superficial surface*, with the Gluteus maximus and fascia lata. By its *deep surface*, with the Semimembranosus, Adductor magnus, inner head of the Gastrocnemius, and internal lateral ligament of the knee-joint, the last being separated from the tendon by a bursa.

The **Semimembranosus,** so called from its membranous tendon of origin, is situated at the back part and inner side of the thigh. It arises by a thick tendon from the upper and outer impression on the back part of the tuberosity of the ischium, above and to the outer side of the Biceps and Semitendinosus, and is inserted into the groove on the inner and back part of the inner tuberosity of the tibia. The tendon of the muscle at its origin expands into an aponeurosis, which covers the upper part of its anterior surface: from this aponeurosis muscular fibres arise, and converge to another aponeurosis which covers the lower part of its posterior surface and contracts into the tendon of insertion. The tendon of the muscle at its insertion gives off certain fibrous expansions: one of these, of considerable size, passes upwards and outwards to be inserted into the back part of the outer condyle of the femur, forming part of the posterior ligament of the knee-joint; a second is continued downwards to the fascia which covers the Popliteus muscle; while a few fibres join the internal lateral ligament of the joint.

The tendons of the two preceding muscles form the inner hamstrings.

Relations.—By its *superficial surface*, with the Gluteus maximus, Semitendinosus, Biceps, and fascia lata. By its *deep surface*, with the origin of the Quadratus femoris, popliteal vessels, Adductor magnus, and inner head of the Gastrocnemius, from which it is separated by a synovial bursa. By its *inner border*, with the Gracilis. By its *outer border*, with the great sciatic nerve, and its internal popliteal branch.

Nerves.—The muscles of this region are supplied by the fourth and fifth lumbar and the first, second, and third sacral nerves through the great sciatic nerve.

Actions.—The hamstring muscles flex the leg upon the thigh. When the knee is semiflexed, the Biceps, in consequence of its oblique direction downwards

and outwards, rotates the leg slightly outwards; and the Semitendinosus, and
to a slight extent the Semimembranosus, rotate the leg inwards, assisting the
Popliteus.   Taking their fixed point from below, these muscles serve to support
the pelvis upon the head of the femur, and to draw the trunk directly backwards,
as in raising it from the stooping position or in feats of strength, when the body
is thrown backwards in the form of an arch.   When the leg is extended on the
thigh they limit the amount of flexion of the trunk on the' lower limbs.

Surgical Anatomy.—In disease of the knee-joint, contraction of the hamstring tendons
is a frequent complication; this causes flexion of the leg, and a partial dislocation of the
tibia backwards, with a slight degree of rotation outwards, probably due to the action of
the Biceps muscle.   The hamstring tendons occasionally require subcutaneous division in
some forms of spurious ankylosis of the knee-joint dependent upon permanent contraction
and rigidity of the Flexor muscles, or from stiffening of the ligamentous and other tissues
surrounding the joint, the result of disease.   This is effected by putting the tendon upon
the stretch, and inserting a narrow, sharp-pointed knife between it and the skin: the
cutting edge being then turned towards the tendon, it should be divided, taking care that
the wound in the skin is not at the same time enlarged.   The relation of the external
popliteal nerve, which lies in close apposition to the inner border of the tendon of the
Biceps, must always be borne in mind in dividing this tendon; and probably, in unskilled
hands, a free incision and exposure of the tendon, before division, is a safer proceeding.

III.   MUSCLES AND FASCIÆ OF THE LEG

The muscles of the leg may be divided into three groups: those on the
anterior, those on the posterior, and those on the outer side.

5. *Anterior Tibio-fibular Region*

Tibialis anticus.	Extensor longus digitorum.
Extensor proprius hallucis.	Peroneus tertius.

Dissection (fig. 444).—The knee should be bent, a block placed beneath it, and the
foot kept in an extended position; then make an incision through the integument in the
middle line of the leg to the ankle, and continue it along the dorsum of the foot to the
toes.   Make a second incision transversely across the ankle, and a third in the same
direction across the bases of the toes; remove the flaps of integument included between
these incisions in order to examine the deep fascia of the leg.

The **Deep Fascia of the Leg** forms a complete investment to the muscles, but
is not continued over the subcutaneous surfaces of the bones.   It is continuous
above with the fascia lata, and is attached around the knee to the patella, the
ligamentum patellæ, the tubercle and tuberosities of the tibia, and the head
of the fibula.   Behind, it forms the popliteal fascia, covering in the popliteal
space; here it is specially thick, being strengthened by transverse fibres, and is
perforated by the external saphenous vein.   It receives an expansion from the
tendon of the Biceps on the outer side, and from the tendons of the Sartorius,
Gracilis, and Semitendinosus on the inner side; in front, it blends with the
periosteum covering the subcutaneous surface of the tibia, and with that covering
the head and external malleolus of the fibula; below, it is continuous with the
annular ligaments of the ankle.   It is thick and dense in the upper and anterior
part of the leg, and gives attachment, by its deep surface, to the Tibialis anticus
and Extensor longus digitorum muscles; but thinner behind, where it covers
the Gastrocnemius and Soleus muscles.   Its deep surface gives off, on the outer
side of the leg, two strong intermuscular septa, which enclose the Peronei
muscles, and separate them from the muscles on the anterior and posterior
tibial regions and several smaller and more slender processes, which enclose the
individual muscles in each region; at the same time a broad transverse intermuscular septum, called the *deep transverse fascia of the leg*, intervenes between
the superficial and deep muscles in the posterior tibio-fibular region.

Remove the fascia by dividing it in the same direction as the integument, excepting
opposite the ankle, where it should be left intact.   Commence the removal of the fascia
from below, opposite the tendons, and detach it in the line of direction of the muscular
fibres.

The **Tibialis anticus** is situated on the outer side of the tibia; it is thick and fleshy at its upper part, tendinous below. It arises from the outer tuberosity and upper half or two-thirds of the external surface of the shaft of the tibia; from the adjoining part of the interosseous membrane; from the deep surface of the fascia; and from the intermuscular septum between it and the Extensor longus digitorum: the fibres pass vertically downwards, and terminate in a tendon, which is apparent on the anterior surface of the muscle at the lower third of the leg. After passing through the innermost compartment of the anterior annular ligament, it is inserted into the inner and under surface of the internal cuneiform bone, and base of the metatarsal bone of the great toe.

Relations.—By its *anterior surface*, with the fascia, and with the annular ligament. By its *posterior surface*, with the interosseous membrane, tibia, ankle-joint, and inner side of the tarsus: this surface also overlaps the anterior tibial vessels and nerve in the upper part of the leg. By its *inner surface*, with the tibia. By its *outer surface*, with the Extensor longus digitorum, and Extensor proprius hallucis, and the anterior tibial vessels and nerve.

The **Extensor proprius hallucis** is a thin, elongated, and flattened muscle, situated between the Tibialis anticus and Extensor longus digitorum. It arises from the anterior surface of the fibula for about the middle two-fourths of its extent, its origin being internal to that of the Extensor longus digitorum; it also arises from the interosseous membrane to a similar extent. The fibres pass downwards, and terminate in a tendon, which occupies the anterior border of the muscle, passes through a distinct compartment in the lower portion of the annular ligament, crosses the anterior tibial vessels near the bend of the ankle, and is inserted into the base of the last phalanx of the great toe. Opposite the metatarsophalangeal articulation, the tendon gives off a thin prolongation on each side, which covers the surface of the joint. It usually sends an expansion, from the inner side of the tendon, to be inserted into the base of the first phalanx.

Relations.—By its *anterior surface*, with the fascia, and the anterior annular ligament. By its *posterior surface*, with the interosseous membrane, fibula, tibia, and ankle-joint. By its *outer side*, with the Extensor longus digitorum above, the dorsalis pedis vessels, anterior tibial nerve, and Extensor brevis digitorum below. By its *inner side*, with the Tibialis anticus and the anterior tibial vessels above. The muscle is external to the anterior tibial vessels in the upper part of the leg; but in the lower third its tendon crosses over them, so that it lies internal to them on the dorsum of the foot.

Fig. 450.—Muscles of the front of the leg.

The **Extensor longus digitorum** is an elongated, flattened, penniform muscle, situated at the outer part of the front of the leg. It arises from the outer tuberosity of the tibia; from the upper three-fourths of the anterior surface of the shaft of the fibula; from the interosseous membrane; from the deep surface of the fascia; and from the intermuscular septa between it and the Tibialis anticus on the inner, and the Peronei on the outer side. The tendon passes under the annular ligament in company with the Peroneus tertius, and divides into four slips, which run forward on the dorsum of the foot, and are inserted into the second and third phalanges of the four lesser toes. The mode in which the tendons are inserted is the following: each of the three inner tendons, opposite the metatarso-phalangeal articulation, is joined, on its outer side, by a tendon of the Extensor brevis digitorum. Each receives a fibrous expansion from the Interossei and Lumbricales, and then spreads out into a broad aponeurosis, which covers the dorsal surface of the first phalanx: this aponeurosis, at the articulation of the first with the second phalanx, divides into three slips, a middle one, which is inserted into the base of the second phalanx; and two lateral slips, which, after uniting on the dorsal surface of the second phalanx, are continued onwards, to be inserted into the base of the third.

Relations.—By its *anterior surface*, with the fascia and the annular ligament. By its *posterior surface*, with the fibula, interosseous membrane, ankle-joint, and Extensor brevis digitorum. By its *inner side*, with the Tibialis anticus, Extensor proprius hallucis, and anterior tibial vessels and nerve. By its *outer side*, with the Peroneus longus and brevis.

The **Peroneus tertius** is a part of the Extensor longus digitorum, and might be described as its fifth tendon. The fibres belonging to this tendon arise from the lower third or more of the anterior surface of the fibula; from the lower part of the interosseous membrane; and from an intermuscular septum between it and the Peroneus brevis. The tendon, after passing through the same canal in the annular ligament as the Extensor longus digitorum, is inserted into the dorsal surface of the base of the metatarsal bone of the little toe. This muscle is sometimes wanting.

Nerves.—These muscles are supplied by the fourth and fifth lumbar and first sacral nerves through the anterior tibial nerve.

Actions.—The Tibialis anticus and Peroneus tertius are the direct flexors of the foot at the ankle-joint; the former muscle, when acting in conjunction with the Tibialis posticus, raises the inner border of the foot (i.e. inverts the foot); and the latter, acting with the Peroneus brevis and longus, raises the outer border of the foot (i.e. everts the foot). The Extensor longus digitorum and Extensor proprius hallucis extend the phalanges of the toes, and, continuing their action, flex the foot upon the leg. Taking their fixed point from below, in the erect posture, all these muscles serve to fix the bones of the leg in the perpendicular position, and give increased strength to the ankle-joint.

6. *Posterior Tibio-fibular Region*

Dissection (fig. 447).—Make a vertical incision along the middle line of the back of the leg, from the lower part of the popliteal space to the heel, connecting it below by a transverse incision extending between the two malleoli; the flaps of integument being removed, the fascia and muscles should be examined.

The muscles in this region of the leg are subdivided into two layers—superficial and deep. Those of the superficial layer constitute a powerful muscular mass, forming the calf of the leg. Their large size is one of the most characteristic features of the muscular apparatus in man, and bears a direct relation to his ordinary attitude and mode of progression.

Superficial Layer

Gastrocnemius.   Soleus.   Plantaris.

The **Gastrocnemius** is the most superficial muscle, and forms the greater part of the calf. It arises by two heads, which are connected to the condyles of the femur by two strong flat tendons. The inner and larger head takes its origin from a depression at the upper and back part of the inner condyle and from the adjacent part of the femur. The outer head arises from an impression on the outer side of the external condyle and from the posterior surface of the femur immediately above the outer part of the condyle. Both heads, also, arise from the subjacent part of the capsular ligament of the knee. Each tendon spreads out into an aponeurosis, which covers the posterior surface of that portion of the muscle to which it belongs. From the anterior surface of these tendinous expansions, muscular fibres are given off; those of the inner head being thicker and extending lower than those of the outer. The fibres in the median line unite at an angle in a median tendinous raphé, which expands into a broad aponeurosis on the anterior surface of the muscle, and into this the remaining fibres are inserted. The aponeurosis, gradually contracting, unites with the tendon of the Soleus, and forms with it the tendo Achillis.

FIG. 451.—Muscles of the back of the leg Superficial layer.

Relations.—By its *superficial surface*, with the fascia of the leg, which separates it from the external saphenous vein and nerve. By its *deep surface*, with the posterior ligament of the knee-joint, the Popliteus, Soleus, Plantaris, popliteal vessels, and internal popliteal nerve. Beneath the tendon of the inner head is a synovial bursa, which, in some cases, communicates with the cavity of the knee-joint. The tendon of the outer head sometimes contains a sesamoid fibro-cartilage or bone, where it plays over the corresponding outer condyle; and one is occasionally found in the tendon of the inner head.

The Gastrocnemius should be divided across, just below its origin, and turned downwards, in order to expose the next two muscles.

The **Soleus** is a broad flat muscle situated immediately beneath the Gastrocnemius. It has received its name from its resemblance in shape to a sole-fish. It arises by tendinous fibres from the back part of the head of the fibula, and from the upper third of the posterior surface of its shaft; from the oblique line of the tibia, and from the middle third of its internal border; some fibres also arise from a tendinous arch placed between the tibial and fibular origins of the muscle, beneath which the popliteal vessels and internal popliteal nerve run. The fibres pass backwards to an aponeurosis which covers the posterior surface of the muscle, and this gradually becoming thicker and narrower, joins with the tendon of the Gastrocnemius, and forms with it the tendo Achillis.

Relations.—By its *superficial surface*, with the Gastrocnemius and Plantaris. By its *deep surface*, with the Flexor longus digitorum, Flexor longus hallucis,

O O

Tibialis posticus, and posterior tibial vessels and nerve, from which it is separated by the transverse intermuscular septum or deep transverse fascia of the leg.

The **tendo Achillis,** the common tendon of the Gastrocnemius and Soleus, is the thickest and strongest tendon in the body. It is about six inches in length, and commences near the middle of the leg, but receives fleshy fibres on its anterior surface, almost to its lower end. Gradually becoming contracted below, it is inserted into the middle part of the posterior surface of the os calcis, a synovial bursa being interposed between the tendon and the upper part of this surface. The tendon spreads out somewhat at its lower end, so that its narrowest part is about an inch and a half above its insertion. The tendon is covered by the fascia and the integument, and is separated from the deep muscles and vessels by a considerable interval filled up with areolar and adipose tissue. Along its outer side, but superficial to it, is the external saphenous vein.

The **Plantaris** is an extremely diminutive muscle, placed between the Gastrocnemius and Soleus, and remarkable for its long and delicate tendon. It arises from the lower part of the outer prolongation of the linea aspera, and from the posterior ligament of the knee-joint. It forms a small fusiform belly, about three or four inches in length, terminating in a long slender tendon which crosses obliquely between the two muscles of the calf, and, running along the inner border of the tendo Achillis, is inserted with it into the posterior part of the os calcis. This muscle is sometimes double, and at other times wanting. Occasionally, its tendon is lost in the internal annular ligament, or in the fascia of the leg.

Nerves.—The Gastrocnemius is supplied by the first and second sacral nerves, and the Plantaris by the fourth and fifth lumbar and first sacral nerves through the internal popliteal. The Soleus is supplied by the first and second sacral nerves through the internal popliteal and posterior tibial.

Actions.—The muscles of the calf are the chief extensors of the foot at the ankle-joint. They possess considerable power, and are constantly called into use in standing, walking, dancing, and leaping; hence the large size they usually present. In walking, these muscles draw powerfully upon the os calcis, raising the heel from the ground; the body being thus supported on the raised foot, the opposite limb can be carried forwards. In standing, the Soleus, taking its fixed point from below, steadies the leg upon the foot and prevents the body from falling forwards, to which there is a constant tendency from the superincumbent weight. The Gastrocnemius, acting from below, serves to flex the femur upon the tibia, assisted by the Popliteus. The Plantaris is the rudiment of a large muscle which exists in some of the lower animals, and is continued over the os calcis to be inserted into the plantar fascia. In man it is an accessory to the Gastrocnemius, extending the ankle if the foot is free, or bending the knee if the foot is fixed. Possibly, acting from below, by its attachment to the posterior ligament of the knee-joint, it may pull that ligament backwards during flexion, and so protect it from being compressed between the two articular surfaces.

Deep Layer (fig. 452)

Popliteus.   Flexor longus digitorum.
Flexor longus hallucis.   Tibialis posticus.

Dissection.—Detach the Soleus from its attachment to the fibula and tibia, and turn it downwards, and the deep layer of muscles is exposed, covered by the deep transverse fascia of the leg.

The **Deep Transverse Fascia** of the leg is a transversely placed, intermuscular septum, between the superficial and deep muscles in the posterior tibio-fibular region. On either side it is connected to the margins of the tibia and fibula. Above, where it covers the Popliteus, it is thick and dense, and receives an expansion from the tendon of the Semimembranosus; it is thinner in the middle of the leg; but below, where it covers the tendons passing behind the malleoli, it is thickened and continuous with the internal annular ligament.

This fascia should now be removed, commencing from below opposite the tendons, and detaching it from the muscles in the direction of their fibres.

The **Popliteus** is a thin, flat, triangular muscle, which forms the lower part of the floor of the popliteal space. It arises by a strong tendon about an inch in length, from a depression at the anterior part of the popliteal groove on the outer side of the external condyle of the femur, and to a small extent from the posterior ligament of the knee-joint; and is inserted into the inner two-thirds of the triangular surface above the oblique line on the posterior surface of the shaft of the tibia, and into the tendinous expansion covering the surface of the muscle.

Fig. 452.—Muscles of the back of the leg. Deep layer.

Relations.—The tendon of the muscle is covered by that of the Biceps and by the external lateral ligament of the knee-joint; it grooves the posterior border of the external semilunar fibro-cartilage, and is invested by the synovial membrane of the knee-joint. The fascia covering the muscle separates it from the Gastrocnemius, Plantaris, popliteal vessels, and internal popliteal nerve. The deep surface of the muscle is in contact with the knee-joint and back of the tibia.

The **Flexor longus hallucis** is situated on the fibular side of the leg, and is the largest and most superficial of the three next muscles. It arises from the lower two-thirds of the posterior surface of the shaft of the fibula, with the exception of an inch at its lowest part; from the lower part of the interosseous membrane; from an intermuscular septum between it and the Peronei, externally; and from the fascia covering the Tibialis posticus internally. The fibres pass obliquely downwards and backwards, and terminate in a tendon which occupies nearly the whole length of the posterior surface of the muscle. This tendon lies in a groove on the posterior surface of the lower end of the tibia; it then grooves the posterior surface of the astragalus, and the under aspect of the sustentaculum tali of the os calcis, and passes into the sole of the foot, where it runs forwards between the two heads of the Flexor brevis hallucis, and is inserted into the base of the last phalanx of the great toe. The grooves on the astragalus and os calcis, which contain the tendon of the muscle, are converted by tendinous fibres into distinct canals, lined by synovial membrane. As the tendon passes forwards in the sole of the foot, it is situated above, and crosses from without inwards, the tendon of the Flexor longus digitorum, to which it is connected by a tendinous slip.

Relations.—By its *superficial surface*, with the Soleus and tendo Achillis, from which it is separated by the deep transverse fascia. By its *deep surface*, with the fibula, Tibialis posticus, the peroneal vessels, the lower part of the interosseous membrane, and the ankle-joint. By its *outer border*, with the Peronei. By its *inner border*, with the Tibialis posticus, and posterior tibial vessels and nerve.

The **Flexor longus digitorum** (*flexor perforans*) is situated on the tibial side of the leg. At its origin it is thin and pointed, but gradually increases in size as it descends. It arises from the posterior surface of the shaft of the tibia, from immediately below the oblique line to within three inches of its extremity, internal to the tibial origin of the Tibialis posticus; some fibres also arise from the fascia covering the Tibialis posticus. The fibres terminate in a tendon, which runs nearly the whole length of the posterior surface of the muscle. This tendon passes behind the internal malleolus, in a groove, common to it and the Tibialis posticus, but separated from the latter by a fibrous septum; each tendon being contained in a special sheath lined by a separate synovial membrane. It then passes obliquely forwards and outwards, superficial to the internal lateral ligament into the sole of the foot (fig. 454), where it crosses the tendon of the Flexor longus hallucis,* lying on its plantar aspect and receiving from it a strong tendinous slip; it then becomes expanded and is joined by the Flexor accessorius, and finally divides into four tendons, which are inserted into the bases of the last phalanges of the four lesser toes, each tendon passing through an opening in the tendon of the Flexor brevis digitorum opposite the base of the first phalanx.

Relations.—In the leg: By its *superficial surface*, with the posterior tibial vessels and nerve, and the deep transverse fascia, which separates it from the Soleus muscle; by its *deep surface*, with the tibia and Tibialis posticus. In the foot, it is covered by the Abductor hallucis and Flexor brevis digitorum, and crosses superficial to the Flexor longus hallucis.

The **Tibialis posticus** lies between the two preceding muscles, and is the most deeply seated of all the muscles in the leg. It commences above by two pointed processes, separated by an angular interval, through which the anterior tibial vessels pass forwards to the front of the leg. It arises from the whole of the posterior surface of the interosseous membrane, excepting its lowest part; from the outer portion of the posterior surface of the shaft of the tibia, between the commencement of the oblique line above and the junction of the middle and lower third of the shaft below; and from the upper two-thirds of the internal surface of the fibula; some fibres also arise from the deep transverse fascia, and from the intermuscular septa separating it from the adjacent muscles on either side. This muscle, in the lower fourth of the leg, passes in front of the Flexor longus digitorum, and terminates in a tendon, which lies in a groove behind the inner malleolus, with the tendon of that muscle, but enclosed in a separate sheath; it then passes through another sheath, over the internal lateral ligament into the foot, and then beneath the inferior calcaneo-navicular ligament, and is inserted into the tuberosity of the navicular bone. The tendon of this muscle contains a sesamoid fibro-cartilage, as it runs under the inferior calcaneo-navicular ligament, and gives off fibrous expansions, one of which passes backwards to the sustentaculum tali of the os calcis, others forwards and outwards to the three cuneiforms, the cuboid, and the bases of the second, third, and fourth metatarsal bones (fig. 455).

Relations.—By its *superficial surface*, with the soleus, from which it is separated by the deep transverse fascia, the Flexor longus digitorum, the posterior tibial vessels and nerve, and the peroneal vessels. By its *deep surface*, with the interosseous ligament, the tibia, fibula, and ankle-joint.

Nerves.—The Popliteus is supplied by the fourth and fifth lumbar and first sacral nerves, through the internal popliteal; the Flexor longus digitorum and Tibialis posticus by the fifth lumbar and first sacral; and the Flexor longus hallucis by the fifth lumbar, and first and second sacral nerves through the posterior tibial.

Actions.—The Popliteus assists in flexing the leg upon the thigh; when the leg is flexed, it will rotate the tibia inwards. It is especially called into action at the commencement of the act of bending the knee, inasmuch as it produces a slight inward rotation of the tibia, which is essential in the early stage of this movement. The Tibialis posticus is a direct extensor of the foot at the ankle-joint; acting in conjunction with the Tibialis anticus, it turns the sole of the foot inwards (i.e. inverts the foot), antagonising the Peronei, which turn it outwards (evert it). In the sole of the foot the tendon of the Tibialis posticus lies directly below the inferior calcaneo-navicular ligament, and is therefore an important factor in maintaining the arch of the foot. The Flexor longus digitorum and Flexor longus hallucis are the direct flexors of the phalanges, and, continuing

---

* That is, in the order of dissection of the sole of the foot.

their action, extend the foot upon the leg; they assist the Gastrocnemius and Soleus in extending the foot, as in the act of walking, or in standing on tiptoe. In consequence of the oblique direction of the tendon of the long flexor, the toes would be drawn inwards, were it not for the Flexor accessorius muscle, which is inserted into the outer side of its tendon, and draws it to the middle line of the foot during its action. Taking their fixed point from the foot, these muscles serve to maintain the upright posture by steadying the tibia and fibula, perpendicularly, upon the ankle-joint. They also serve to raise these bones from the oblique position they assume in the stooping posture.

7. Fibular Region

Peroneus longus.   Peroneus brevis.

Dissection.—The muscles are readily exposed, by removing the fascia covering their surface, from below upwards, in the line of direction of their fibres.

The **Peroneus longus** is situated at the upper part of the outer side of the leg, and is the more superficial of the two muscles. It arises from the head and upper two-thirds of the outer surface of the shaft of the fibula, from the deep surface of the fascia, and from the intermuscular septa, between it and the muscles on the front, and those on the back of the leg: occasionally also by a few fibres from the outer tuberosity of the tibia. Between its attachment to the head and to the shaft of the fibula there is a small area of bone from which no muscular fibres arise; here the external popliteal nerve passes beneath the muscle. It terminates in a long tendon, which runs behind the outer malleolus, in a groove common to it and the tendon of the Peroneus brevis, behind which it lies, the groove being converted into a canal by a fibrous band, and the tendons invested by a common synovial membrane; it then extends obliquely forwards across the outer side of the os calcis, below its peroneal tubercle, being contained in a separate fibrous sheath, lined by a prolongation of the synovial membrane which lines the groove behind the malleolus. Having reached the outer side of the cuboid bone, it runs in a groove on the under surface of that bone, which is converted into a canal by the long calcaneo-cuboid ligament, and is lined by a synovial membrane: the tendon then crosses the sole of the foot obliquely, and is inserted into the outer side of the base of the metatarsal bone of the great toe and the internal cuneiform bone. Occasionally it sends a slip to the base of the second metatarsal bone. The tendon changes its direction at two points: first, behind the external malleolus; secondly, on the outer side of the cuboid bone; in both of these situations the tendon is thickened, and, in the latter, a sesamoid fibro-cartilage, or sometimes a bone, is usually developed in its substance.

Relations.—By its *superficial surface*, with the fascia and integument; by its *deep surface*, with the fibula, external popliteal nerve, the Peroneus brevis, os calcis and cuboid bones; by its *anterior border*, with an intermuscular septum, which intervenes between it and the Extensor longus digitorum; by its *posterior border*, with an intermuscular septum, which separates it from the Soleus above and the Flexor longus hallucis below.

The **Peroneus brevis** lies under cover of the Peroneus longus, and is a shorter and smaller muscle. It arises from the lower two-thirds of the external surface of the shaft of the fibula, internal to the Peroneus longus; and from the intermuscular septa, separating it from the adjacent muscles on the front and back part of the leg. The fibres pass vertically downwards, and terminate in a tendon, which runs in front of that of the preceding muscle, through the same groove behind the external malleolus, being contained in the same fibrous sheath, and lubricated by the same synovial membrane. It then passes through a separate sheath on the outer side of the os calcis, above that for the tendon of the Peroneus longus, the two tendons being here separated by the peroneal tubercle, and is finally inserted into the tuberosity at the base of the metatarsal bone of the little toe, on its outer side.

Relations.—By its *superficial surface*, with the Peroneus longus and the fascia of the leg and foot. By its *deep surface*, with the fibula and outer side of the os calcis.

Nerves.—The Peroneus longus and brevis are supplied by the fourth and fifth lumbar and first sacral nerves through the musculo-cutaneous branch of the external popliteal nerve.

Actions.—The Peroneus longus and brevis extend the foot upon the leg, in conjunction with the Tibialis posticus, antagonising the Tibialis anticus and Peroneus tertius, which are flexors of the foot. The Peroneus longus also everts the sole of the foot; hence the extreme eversion occasionally observed in fracture of the lower end of the fibula, where that bone offers no resistance to the action of this muscle. From the oblique direction of the Peroneus longus tendon across the sole of the foot it is an important agent in the maintenance of the transverse arch of the foot. Taking their fixed point below, the Peronei serve to steady the leg upon the foot. This is especially the case in standing upon one leg, when the tendency of the superincumbent weight is to throw the leg inwards : the Peroneus longus overcomes this tendency by drawing on the outer side of the leg, and thus maintains the perpendicular direction of the limb.

Surgical Anatomy.—The student should now consider the position of the tendons of the various muscles of the leg, their relation with the ankle-joint and surrounding blood-vessels, and especially their action upon the foot, as their rigidity and contraction give rise to one or other of the kinds of deformity known as *club foot.* The most simple and common deformity, and one that is rarely, if ever, congenital, is *talipes equinus,* the heel being raised by rigidity and contraction of the Gastrocnemius muscle, and the patient walking upon the ball of the foot. In *talipes varus,* the foot is forcibly adducted and the inner side of the sole raised, sometimes to a right angle with the ground, by the action of the Tibialis anticus and posticus. In *talipes valgus,* the outer edge of the foot is raised by the Peronei muscles, and the patient walks on the inner ankle. In *talipes calcaneus* the toes are raised by the Extensor muscles, the heel is depressed and the patient walks upon it. Other varieties of deformity are met with, as *talipes equino-varus, equino-valgus,* and *calcaneo-valgus,* whose names sufficiently indicate their nature. Of these, talipes equino-varus is the most common congenital form ; the heel is raised by the tendo Achillis, the inner border of the foot drawn upwards by the Tibialis anticus, the anterior two-thirds twisted inwards by the Tibialis posticus, and the arch increased by the contraction of the plantar fascia, so that the patient walks on the middle of the outer border of the foot. Each of these deformities may sometimes be successfully relieved by division of the opposing tendons and fascia : by this means the foot regains its proper position, and the tendons heal by the organisation of lymph thrown out between the divided ends. The operation is easily performed by putting the contracted tendon upon the stretch, and dividing it by means of a narrow, sharp-pointed knife inserted beneath it.

Rupture of a few of the fibres of the Gastrocnemius, or rupture of the Plantaris tendon, not uncommonly occurs, especially in men somewhat advanced in life, from some sudden exertion, and frequently occurs during the game of lawn tennis, and is hence known as ' lawn-tennis leg.' The accident is accompanied by a sudden pain, and produces a sensation as if the individual had been struck a violent blow on the part. The tendo Achillis is also sometimes ruptured. It is stated that John Hunter ruptured his tendo Achillis while dancing, at the age of forty. The bursa beneath the tendo Achillis, between it and the posterior surface of the os calcis, sometimes becomes inflamed, especially in pedestrians and ' long-distance ' walkers. It causes great and disabling pain, and entirely prevents the sufferer from continuing his walk.

IV. MUSCLES AND FASCIÆ OF THE FOOT

The fibrous bands, or thickened portions of the fascia of the leg, which bind down the tendons in front of and behind the ankle in their passage to the foot, should now be examined; they are termed the *annular ligaments,* and are three in number—anterior internal, and external.

The **anterior annular ligament** consists of a superior or transverse portion, which binds down the Extensor tendons as they descend on the front of the tibia and fibula ; and an inferior or Y-shaped portion, which retains them in connection with the tarsus, the two portions being connected by a thin intervening layer of fascia. The transverse portion is attached externally to the lower end of the fibula and internally to the tibia ; above, it is continuous with the fascia of the leg ; it contains one synovial sheath for the tendon of the Tibialis anticus ; the other tendons and the anterior tibial vessels and nerve passing beneath it, but without any synovial sheath. The Y-shaped portion is placed in front of the ankle-joint, the stem of the Y being attached externally to the upper surface of the os calcis, in front of the depression for the interosseous ligament; it is directed inwards, as a double layer, one lamina passing in front, and the other behind, the tendons of the Peroneus tertius and Extensor longus digitorum. At the inner border of the latter tendon these two layers join together, forming a sheath in which the tendons are enclosed, surrounded by a synovial membrane. From the inner extremity of this sheath the two limbs of the Y diverge: one passes upwards and inwards, to be attached to the internal malleolus, passing over the Extensor proprius hallucis and the vessels and nerves, but enclosing the Tibialis anticus and its synovial sheath by a splitting of its fibres. The other limb extends downwards and inwards to be attached to the inner border of the plantar fascia, and passes over the tendons of the Extensor proprius hallucis and Tibialis anticus and also the vessels and nerves. These two tendons are contained in separate synovial sheaths situated beneath the ligament.

The **internal annular ligament** is a strong fibrous band, which extends from the inner malleolus above to the internal margin of the os calcis below, converting a series of grooves in this situation into canals, for the passage of the tendons of the Flexor muscles and vessels into the sole of the foot. It is continuous by its upper border with the deep fascia of the leg, and by its lower border with the plantar fascia and the fibres of origin of the Abductor hallucis muscle. The four canals which it forms transmit, from within outwards, first, the tendon of the Tibialis posticus; second, the tendon of the Flexor longus digitorum; third, the posterior tibial vessels and nerve, which run through a broad space beneath the ligament; lastly, in a canal formed partly by the astragalus, the tendon of the Flexor longus hallucis. The canals for the tendons are lined by separate synovial membranes.

The **external annular ligament** extends from the extremity of the outer malleolus to the outer surface of the os calcis: it binds down the tendons of the Peroneus longus and brevis muscles in their passage beneath the outer ankle. The two tendons are enclosed in one synovial sheath.

8. Dorsal Region

Extensor brevis digitorum.

The **fascia** on the dorsum of the foot is a thin membranous layer, continuous above with the anterior margin of the lower part of the annular ligament; it becomes gradually lost opposite the heads of the metatarsal bones, and on each side blends with the lateral portions of the plantar fascia; it forms a sheath for the tendons placed on the dorsum of the foot. On the removal of this fascia, the muscle and tendons of the dorsal region of the foot are exposed.

The **Extensor brevis digitorum** (fig. 450) is a broad thin muscle, which arises from the fore part of the upper and outer surfaces of the os calcis, in front of the groove for the Peroneus brevis; from the external calcaneo-astragaloid ligament; and from the common limb of the Y-shaped portion of the anterior annular ligament. It passes obliquely across the dorsum of the foot, and terminates in four tendons. The innermost, which is the largest, is inserted into the dorsal surface of the base of the first phalanx of the great toe, crossing the dorsalis pedis artery; the other three, into the outer sides of the long Extensor tendons of the second, third, and fourth toes.

Relations.—By its *superficial surface*, with the fascia of the foot, the tendons of the Extensor longus digitorum and Peroneus tertius. By its *deep surface*, with the tarsal and metatarsal arteries and bones, and the Dorsal interossei muscles.

Nerves.—It is supplied by the anterior tibial nerve.

Actions.—The Extensor brevis digitorum is an accessory to the long Extensor, extending the phalanges of the four inner toes, but acting only on the first phalanx of the great toe. The obliquity of its direction counteracts the oblique movement given to the toes by the long Extensor, so that, both muscles acting together, the toes are evenly extended.

9. Plantar Region

Dissection of the Sole of the Foot.—The foot should be placed on a high block with the sole uppermost, and firmly secured in that position. Carry an incision round the heel and along the inner and outer borders of the foot to the great and little toes. This incision should divide the integument and thick layer of granular fat beneath, until the fascia is visible ; the skin and fat should then be removed from the fascia in a direction from behind forwards, as seen in fig. 447.

The **plantar fascia** is of great strength, and consists of pearly-white glistening fibres, disposed, for the most part, longitudinally : it is divided into a central and two lateral portions.

The *central portion,* the thickest, is narrow behind and attached to the inner tubercle of the os calcis, posterior to the origin of the Flexor brevis digitorum ; and becoming broader and thinner in front, divides near the heads of the metatarsal bones into five processes, one for each of the toes. Each of these processes divides opposite the metatarso-phalangeal articulation into two strata, superficial and deep. The superficial stratum is inserted into the skin of the transverse sulcus which separates the toes from the sole. The deeper stratum divides into two slips which embrace the sides of the Flexor tendons of the toes, and blend with the sheaths of the tendons, and laterally with the transverse metatarsal ligament, thus forming a series of arches through which the tendons of the short and long flexors pass to the toes. The intervals left between the five processes allow the digital vessels and nerves, and the tendons of the Lumbricales muscles, to become superficial. At the point of division of the fascia, numerous transverse fibres are superadded, which serve to increase the strength of the fascia at this part by binding the processes together, and connecting them with the integument. The central portion of the plantar fascia is continuous with the lateral portions at each side, and sends upwards into the foot, at their point of junction, two strong vertical intermuscular septa, broader in front than behind, which separate the middle from the external and internal plantar group of muscles ; from these again thinner transverse septa are derived, which separate the various layers of muscles in this region. The upper surface of this fascia gives attachment behind to the Flexor brevis digitorum muscle.

The *lateral portions* of the plantar fascia are thinner than the central piece, and cover the sides of the foot.

The *outer portion* covers the under surface of the Abductor minimi digiti ; it is thin in front and thick behind, where it forms a strong band between the outer tubercle of the os calcis and the base of the fifth metatarsal bone ; it is continuous internally with the middle portion of the plantar fascia, and externally with the dorsal fascia.

The *inner portion* is thin, and covers the under aspect of the Abductor hallucis muscle; it is attached behind to the internal annular ligament, and is continuous around the side of the foot with the dorsal fascia, and externally with the middle portion of the plantar fascia.

The muscles in the plantar region of the foot may be divided into three groups, in a similar manner to those in the hand. Those of the internal plantar region are connected with the great toe, and correspond with those of the thumb ; those of the external plantar region are connected with the little toe, and correspond with those of the little finger ; and those of the middle plantar region are connected with the tendons intervening between the two former groups. But in order to facilitate the dissection of these muscles, it will be found more convenient to divide them into four layers, as they present themselves, in the order in which they are successively exposed.

First Layer

Abductor hallucis. Flexor brevis digitorum.
Abductor minimi digiti.

Dissection.—Remove the fascia on the inner and outer sides of the foot, commencing in front over the tendons, and proceeding backwards. The central portion should be divided transversely in the middle of the foot, and the two flaps dissected forwards and backwards.

The **Abductor hallucis** lies along the inner border of the foot. It arises from the inner tubercle on the under surface of the os calcis ; from the internal annular ligament ; from the plantar fascia covering it ; and from the intermuscular septum between it and the Flexor brevis digitorum. The fibres terminate in a tendon, which is inserted, together with the innermost tendon of the Flexor brevis hallucis, into the inner side of the base of the first phalanx of the great toe.

Relations.—By its *superficial surface*, with the plantar fascia. By its *deep surface*, with the Flexor brevis hallucis, the Flexor accessorius, and the tendons of the Flexor longus digitorum and Flexor longus hallucis, the Tibialis anticus and posticus, the plantar vessels and nerves. Its outer border is in relation to the Flexor brevis digitorum.

The **Flexor brevis digitorum** (*flexor perforatus*) lies in the middle of the sole of the foot, immediately beneath* the central part of the plantar fascia, with which it is firmly united. It arises by a narrow tendinous process, from the inner tubercle of the os calcis, from the central part of the plantar fascia, and from the intermuscular septa between it and the adjacent muscles. It passes forwards, and divides into four tendons, one for each of the four outer toes. Opposite the bases of the first phalanges, each tendon divides into two slips, to allow of the passage of the corresponding tendon of the Flexor longus digitorum; the two portions of the tendon then unite and form a grooved channel for the reception of the accompanying long Flexor tendon. Finally, they divide a second time, to be inserted into the sides of the second phalanges about their middle. The mode of division of the tendons of the Flexor brevis digitorum, and their insertion into the phalanges, is analogous to the Flexor sublimis digitorum in the hand.

Relations.—By its *superficial surface*, with the plantar fascia. By its *deep surface*, with the Flexor accessorius, the Lumbricales, the tendons of the Flexor longus digitorum, and the external plantar vessels and nerve, from which it is separated by a thin layer of fascia. The *outer* and *inner borders* are separated from the adjacent muscles by means of vertical prolongations of the plantar fascia.

Fibrous sheaths of the Flexor tendons.—These are not so well marked as in the fingers. The Flexor tendons of the toes as they run along the phalanges are retained against the bones by fibrous sheaths, forming osseo-aponeurotic canals. These sheaths are formed by strong fibrous bands, which arch across the tendons, and are attached on each side to the margins of the phalanges. Opposite the middle of the proximal and second phalanges the sheath is very strong, and the fibres pass transversely; but opposite the joints it is much thinner, and the fibres are directed obliquely. Each sheath is lined by a synovial membrane, which is reflected on the contained tendon.

The **Abductor minimi digiti** lies along the outer border of the foot. It arises, by a very broad origin, from the outer tubercle of the os calcis, from the under surface of the os calcis between the two tubercles, from the fore part of the inner

Fig. 453.—Muscles of the sole of the foot. First layer.

---

* That is, in the order of dissection of the sole of the foot.

tubercle, from the plantar fascia, and the intermuscular septum between it and
the Flexor brevis digitorum. Its tendon, after gliding over a smooth facet on
the under surface of the base of the fifth metatarsal bone, is inserted, with the
short Flexor of the little toe, into the outer side of the base of the first phalanx
of this toe.

Relations.—By its *superficial surface*, with the plantar fascia. By its *deep
surface*, with the Flexor accessorius, the Flexor brevis minimi digiti, the long
plantar ligament, and the tendon of the Peroneus longus. On its *inner side* are
the external plantar vessels and nerve,
and it is separated from the Flexor brevis
digitorum by a vertical septum of the
plantar fascia.

Fig. 454.—Muscles of the sole of the foot.
Second layer.

Dissection.—The muscles of the superficial layer should be divided at their origin,
by inserting the knife beneath each, and
cutting obliquely backwards, so as to detach
them from the bone; they should then be
drawn forwards, in order to expose the second
layer, but not cut away at their insertion.
The two layers are separated by a thin membrane, the *deep plantar fascia*, on the removal of which is seen the tendon of the Flexor
longus digitorum, the Flexor accessorius, the
tendon of the Flexor longus hallucis, and
the Lumbricales. The long Flexor tendons
diverge from each other at an acute angle:
the Flexor longus hallucis runs along the
inner side of the foot, on a plane superior to
that of the Flexor longus digitorum, the
direction of which is obliquely outwards.

Second Layer

Flexor accessorius.
Lumbricales.

The **Flexor accessorius** arises by two
heads, which are separated from each
other by the long plantar ligament:
the inner or larger, which is muscular,
being attached to the inner concave surface of the os calcis, below the groove
which lodges the tendon of the Flexor
longus hallucis; the outer head, flat and
tendinous, to the outer border of the
inferior surface of the os calcis, in front
of its lesser tubercle, and to the long
plantar ligament: the two portions join
at an acute angle, and are inserted into
the outer margin and upper and under
surfaces of the tendon of the Flexor
longus digitorum, forming a kind of
groove, in which the tendon is lodged.[*]

Relations.—By its *superficial surface*,
with the muscles of the superficial layer,
from which it is separated by the external plantar vessels and nerves. By its
deep surface, with the os calcis, and long calcaneo-cuboid ligament.

The **Lumbricales** are four small muscles, accessory to the tendons of the
Flexor longus digitorum: they arise from the tendons of the long Flexor, as far
back as their angles of division, each arising from two tendons, except the
internal one. Each muscle terminates in a tendon, which passes forwards on
the inner side of the four lesser toes, and is inserted into the expansion of the
long Extensor tendon on the dorsum of the first phalanx of the corresponding toe.

[*] According to Turner, the fibres of the Flexor accessorius end in aponeurotic bands,
which contribute slips to the second, third, and fourth digits.

Dissection.—The Flexor tendons should be divided at the back part of the foot, and the Flexor accessorius at its origin, and drawn forwards, in order to expose the third layer.

Third Layer

Flexor brevis hallucis.	Flexor brevis minimi digiti.
Adductor obliquus hallucis.	Adductor transversus hallucis.

The **Flexor brevis hallucis** arises, by a pointed tendinous process, from the inner part of the under surface of the cuboid bone, from the contiguous portion of the external cuneiform, and from the prolongation of the tendon of the Tibialis posticus, which is attached to that bone. The muscle divides, in front, into two portions, which are inserted into the inner and outer sides of the base of the first phalanx of the great toe, a sesamoid bone being developed in each tendon at its insertion. The inner portion of this muscle is blended with the Abductor hallucis previous to its insertion; the outer with the Adductor obliquus hallucis; the tendon of the Flexor longus hallucis lies in a groove between them.

Relations.—By its *superficial surface*, with the Abductor hallucis and the tendon of the Flexor longus hallucis. By its *deep surface*, with the tendon of the Peroneus longus, and metatarsal bone of the great toe. By its *inner border*, with the Abductor hallucis. By its *outer border*, with the Adductor obliquus hallucis.

The **Adductor obliquus hallucis** is a large, thick, fleshy mass, passing obliquely across the foot, and occupying the hollow space between the four inner metatarsal bones. It arises from the tarsal extremities of the second, third, and fourth metatarsal bones, and from the sheath of the tendon of the Peroneus longus, and is inserted, together with the outer portion of the Flexor brevis hallucis, into the outer side of the base of the first phalanx of the great toe.

The small muscles of the great toe, the Abductor, Flexor brevis, Adductor obliquus, and Adductor transversus, like the similar muscles of the thumb, give off fibrous expansions, at their insertions, to blend with the long Extensor tendons.

The **Flexor brevis minimi digiti** lies on the metatarsal bone of the little toe, and much resembles one of the Interossei. It arises from the base of the metatarsal bone of the little toe, and from the sheath of the Peroneus longus; its tendon is inserted into the base of the first phalanx of the little toe on its outer side. Occasionally some of the deeper fibres of the muscle are inserted into the outer part of the distal half of the fifth metatarsal bone; these are described by some as a distinct muscle, the Opponens minimi digiti.

Relations.—By its *superficial surface*, with the plantar fascia and tendon of the Abductor minimi digiti. By its *deep surface*, with the fifth metatarsal bone.

The **Adductor transversus hallucis** (*Transversus pedis*) is a narrow, flat, muscular fasciculus, stretched transversely across the heads of the metatarsal bones,

FIG. 455.—Muscles of the sole of the foot. Third layer.

between them and the Flexor tendons. It arises from the inferior metatarso-phalangeal ligaments of the three outer toes, sometimes only from the third and fourth, and from the transverse ligament of the metatarsus; and is inserted into the outer side of the first phalanx of the great toe; its fibres being blended with the tendon of insertion of the Adductor obliquus hallucis.

Relations.—By its *superficial surface*, with the tendons of the long and short Flexors and Lumbricales. By its *deep surface*, with the Interossei.

Fourth Layer
The Interossei.

The **Interossei muscles** in the foot are similar to those in the hand, with this exception, that they are grouped around the middle line of the *second* toe, instead of the middle line of the *third* finger. They are seven in number, and consist of two groups, dorsal and plantar.

The **Dorsal interossei,** four in number, are situated between the metatarsal bones. They are bipenniform muscles, arising by two heads from the adjacent

FIG. 456.—The Dorsal interossei. | FIG. 457.—The Palmar interossei.
Left foot. | Left foot.

sides of the metatarsal bones between which they are placed; their tendons are inserted into the bases of the first phalanges, and into the aponeurosis of the common Extensor tendon. In the angular interval left between the heads of each muscle at its posterior extremity, one of the perforating arteries passes to the dorsum of the foot; except in the First interosseous muscle, where the interval allows the passage of the communicating branch of the dorsalis pedis artery. The First dorsal interosseous muscle is inserted into the inner side of the second toe; the other three are inserted into the outer sides of the second, third, and fourth toes.

The **Plantar interossei,** three in number, lie beneath, rather than between, the metatarsal bones. They are single muscles, and are each connected with but one metatarsal bone. They arise from the bases and inner sides of the shaft of the third, fourth, and fifth metatarsal bones, and are inserted into the inner sides of the bases of the first phalanges of the same toes, and into the aponeurosis of the common Extensor tendon.

Nerves.—The Flexor brevis digitorum, the Flexor brevis and Abductor hallucis, and the innermost Lumbrical * are supplied by the internal plantar nerve. All the other muscles in the sole of the foot by the external plantar. The first dorsal interosseous muscle frequently receives an extra filament from the internal branch of the anterior tibial nerve on the dorsum of the foot, and the second dorsal interosseous a twig from the external branch of the same nerve.

Actions.—All the muscles of the foot act upon the toes, and in describing their action they may be grouped as Abductors, Adductors, Flexors, or Extensors. The *Abductors* are the Dorsal interossei, the Abductor hallucis, and the Abductor minimi digiti. The Dorsal interossei are abductors from an imaginary line passing through the axis of the second toe, so that the first muscle draws the second toe inwards, towards the great toe ; the second muscle draws the same toe outwards ; the third draws the third toe, and the fourth draws the fourth toe in the same direction. Like the interossei in the hand, each assists in flexing the proximal phalanx and extending the two terminal phalanges. The Abductor hallucis abducts the great toe from the others, and also flexes the proximal phalanx of this toe. And in the same way the action of the Abductor minimi digiti is twofold, as an abductor of this toe from the others, and also as a flexor of the proximal phalanx. The *Adductors* are the Plantar interossei, the Adductor obliquus hallucis, and the Adductor transversus hallucis. The plantar interosseous muscles adduct the third, fourth, and fifth toes, towards the imaginary line passing through the second toe, and by means of their insertion into the aponeurosis of the Extensor tendon they assist in flexing the proximal phalanx and extending the two terminal phalanges. The Adductor obliquus hallucis is chiefly concerned in adducting the great toe towards the second one, but also assists in flexing this toe. The Adductor transversus hallucis approximates all the toes and thus increases the curve of the transverse arch of the metatarsus. The *Flexors* are the Flexor brevis digitorum, the Flexor accessorius, the Flexor brevis hallucis, the Flexor brevis minimi digiti, and the Lumbricales. The Flexor brevis digitorum flexes the second phalanges upon the first, and, continuing its action, flexes the first phalanges also, and brings the toes together. The Flexor accessorius assists the long Flexor of the toes and converts the oblique pull of the tendons of that muscle into a direct backward pull upon the toes. The Flexor brevis minimi digiti flexes the little toe and draws its metatarsal bone downwards and inwards. The Lumbricales, like the corresponding muscles in the hand, assist in flexing the proximal phalanx, and by their insertion into the long Extensor tendon aid that muscle in straightening the two terminal phalanges. The only muscle in the *Extensor* group is the Extensor brevis digitorum. It extends the first phalanx of the great toe and assists the long Extensor in extending the next three toes, and at the same time gives to the toes an outward direction when they are extended.

Surface Form.—The skin of the thigh, especially above in the hollow of the groin and on the inner side, is thin, smooth, and elastic, and contains few hairs, except in the neighbourhood of the pubes. Towards the outer side it becomes thicker, and the hairs more numerous. The skin over the buttock is also fairly thick, with low sensibility and vascularity. As a rule, it is destitute of conspicuous hairs, except towards the post-anal furrow, where, in some males, an abundant development of hair is present. The skin over the front of the knee is covered by thickened epidermis ; it is loose and thrown into transverse wrinkles when the leg is extended: that of the leg is thin, especially on the inner side, and covered with numerous large hairs. On the dorsum of the foot, the skin is thin, loosely connected to subjacent parts, and contains few hairs. On the sole, and especially the heel, the epidermis is of great thickness, and here, as in the palm of the hand, there are no hairs or sebaceous follicles.

Of the muscles of the thigh, those of the iliac region have no influence on surface form, while those of the anterior femoral region, being to a great extent superficial, largely contribute to the surface form of this part of the body. The *Tensor fasciæ femoris* produces a broad elevation immediately below the anterior portion of the crest of the ilium and behind the anterior superior spinous process. From its lower border, a longitudinal groove, corresponding to the ilio-tibial band, may be seen running down the outer side of the thigh to the outer side of the knee-joint. The *Sartorius* muscle,

---

* Formerly the two inner Lumbricales were described as being supplied by the internal plantar nerve. Brooks, however (*Journal of Anatomy*, vol. xxi. p. 575), in ten dissections found that in nine of them only the inner Lumbrical obtained its nerve supply from this source. In the tenth instance the first and second Lumbricales were supplied by both external and internal plantar.

when it is brought into action by flexing the leg on the thigh, and the thigh on the pelvis, and rotating the thigh outwards, presents a well-marked surface form. At its upper part, where it constitutes the outer boundary of Scarpa's triangle, it forms a prominent oblique ridge, which becomes changed into a flattened plane below, and this gradually merges in a general fulness on the inner side of the knee-joint. When the Sartorius is not in action, a depression exists between the Extensor quadriceps and the Adductor muscles, running obliquely downwards and inwards from the apex of Scarpa's triangle to the inner side of the knee, which corresponds to this muscle. In the depressed angle formed by the divergence of the Sartorius and Tensor fasciæ femoris muscles, just below the anterior superior spinous process of the ilium, the *Rectus femoris* muscle appears, and, below this, determines to a great extent the convex form of the front of the thigh. In a well-developed subject, the borders of the muscle, when in action, can be clearly defined. The *Vastus externus* forms a long flattened plane on the outer side of the thigh, traversed by the longitudinal groove formed by the ilio-tibial band. The *Vastus internus*, on the inner side of the lower half of the thigh, gives rise to a considerable prominence, which increases towards the knee and terminates somewhat abruptly in this situation with a full, curved outline. The *Crureus* and *Subcrureus* are completely hidden, and do not directly influence surface form. The *Adductor muscles*, constituting the internal femoral group, are not to be individually distinguished from each other, with the exception of the upper tendon of the Adductor longus and the lower tendon of the Adductor magnus. The upper tendon of the *Adductor longus*, when the muscle is in action, stands out as a prominent ridge, which runs obliquely downwards and outwards from the neighbourhood of the pubic spine, and forms the inner boundary of a flattened triangular space on the upper part of the front of the thigh, known as Scarpa's triangle. The lower tendon of the *Adductor magnus* can be distinctly felt as a short ridge extending down to the Adductor tubercle on the internal condyle, between the Sartorius and Vastus internus. The Adductor group of muscles fills in the triangular space at the upper part of the thigh, formed between the oblique femur and the pelvic wall, and to them is due the contour of the inner border of the thigh, the *Gracilis* largely contributing to the smoothness of the outline. These muscles are not marked off on the surface from those of the posterior femoral region by any intermuscular depression; but on the outer side of the thigh these latter muscles are defined from the Vastus externus by a distinct marking, corresponding to the external intermuscular septum. The *Gluteus maximus* and a part of the *Gluteus medius* are the only muscles of the buttock which influence surface form. The other part of the Gluteus medius, the Gluteus minimus, and the External rotators are completely hidden. The *Gluteus maximus* forms the full rounded outline of the buttock; it is more prominent behind, compressed in front, and terminates at its tendinous insertion in a depression immediately behind the great trochanter. Its lower border does not correspond to the gluteal fold, but is much more oblique, being marked by a line drawn from the side of the coccyx to the junction of the upper with the lower two-thirds of the thigh on the outer side. From beneath the lower margin of this muscle the *Hamstring muscles* appear, at first narrow and not well marked, but, as they descend, becoming more prominent and widened out, and eventually dividing into two well-marked ridges, which constitute the upper boundaries of the popliteal space, and are formed by the tendons of the inner and outer Hamstring muscles respectively. In the upper part of the thigh these muscles are not to be individually distinguished from each other; but lower down, the separation between the Semitendinosus and Semimembranosus is denoted by a slight intermuscular marking. The external hamstring tendon formed by the *Biceps* is seen as a thick cord running down to the head of the fibula. The inner hamstring tendons comprise the Semitendinosus, the Semimembranosus, and the Gracilis. The *Semitendinosus* is the most internal of these, and can be felt, in certain positions of the limb, as a sharp cord; the *Semimembranosus* is thick, and the *Gracilis* is situated a little farther forwards than the other two.

All the muscles on the front of the leg appear to a certain extent somewhere on the surface, but the form of this region is mainly dependent upon the Tibialis anticus and the Extensor longus digitorum. The *Tibialis anticus* is well marked, and presents a fusiform enlargement at the outer side of the tibia, and projects beyond the crest of the shin bone. From the muscular mass, its tendon may be traced downward, standing out boldly, when the muscle is in action, on the front of the tibia and ankle-joint, and coursing down to its insertion along the inner border of the foot. A well-marked groove separates this muscle externally from the *Extensor longus digitorum*, which fills up the rest of the space between the upper part of the shaft of the tibia and fibula. This muscle does not present so bold an outline as the Tibialis anticus, and its tendon below, diverging from the tendon of the Tibialis anticus, forms with the latter a sort of plane, in which may be seen the tendon of the Extensor proprius hallucis. A groove on the outer side of the Extensor longus digitorum, seen most plainly when the muscle is in action, separates the tendon from a slight eminence corresponding to the *Peroneus tertius*. The fleshy fibres of the *Peroneus longus* are strongly marked at the upper part of the outer side of the leg, especially when the muscle is in action. It forms a bold swelling, separated by furrows from the Extensor longus digitorum in front and the Soleus behind. Below, the fleshy fibres terminate abruptly in a tendon which overlaps the more flattened form of the *Peroneus brevis*. At the External malleolus the tendon of the Peroneus brevis is more marked than that of the Peroneus longus. On the dorsum of the foot the tendons of the Extensor muscles, emerging from beneath the anterior annular ligament, spread out, and can be distinguished in the following order: the most internal and largest is the Tibialis anticus, then the Extensor proprius hallucis; next comes the Extensor longus digitorum, dividing into four tendons to the four outer toes; and lastly, most externally, is the Peroneus tertius. The flattened form of the dorsum of the foot is relieved by the rounded outline of the fleshy belly of the *Extensor brevis digitorum*, which forms a soft fulness on the outer side of the tarsus in front of the external malleolus, and by the Dorsal interossei, which bulge between the metatarsal bones. At the back of the knee is the popliteal space, bounded above by the tendons of the Hamstring muscles; below, by the two heads of the Gastrocnemius. Below this space is the prominent fleshy mass of the calf of the leg, produced by the *Gastrocnemius* and *Soleus*. When these muscles are in action, as in standing on tiptoe, the borders of the Gastrocnemius are well defined, presenting two curved lines, which converge to the tendon of insertion. Of these borders, the inner is more prominent than the outer. The fleshy mass of the calf terminates somewhat abruptly below in the tendo Achillis, which stands out prominently on the lower part of the back of the leg. It presents a somewhat

FIG. 458.—Fracture of the neck of the femur within the capsular ligament.

Pyriformis
Gemellus superior
Obturator internus
Gemellus inferior
Obturator externus
Quadratus femoris

tapering form in the upper three-fourths of its extent, but widens out slightly below. When the muscles of the calf are in action, the lateral portions of the *Soleus* may be seen, forming curved eminences, of which the outer is the longer, on either side of the Gastrocnemius. Behind the inner border of the lower part of the shaft of the tibia, a well-marked ridge, produced by the tendon of the Tibialis posticus, is visible when this muscle is in a state of contraction.

On the sole of the foot the superficial layer of muscles influences surface form; the *Abductor minimi digiti* most markedly. This muscle forms a narrow rounded elevation along the outer border of the foot, while the *Abductor hallucis* does the same, though to a less extent, on the inner side. The *Flexor brevis digitorum*, bound down by the plantar fascia, is not very apparent; it produces a flattened form, covered by the thickened skin of the sole, which is here thrown into numerous wrinkles.

Surgical Anatomy.—The student should now consider the effects produced by the action of the various muscles in fractures of the bones of the lower extremity. The more common forms of fracture are selected for illustration and description.

In fracture of the *neck of the femur internal to the capsular ligament* (fig. 458), the characteristic marks are slight shortening of the limb, and eversion of the foot, neither of which symptoms occurs, however, in some cases until some time after the injury. The eversion is caused by the weight of the limb rotating it outwards. The shortening is produced by the action of the Glutei, and by the Rectus femoris in front, and the Biceps, Semitendinosus and Semimembranosus behind.

In fracture of the *femur just below the trochanters* (fig. 459), the upper fragment, the portion chiefly displaced, is tilted forwards almost at right angles with the pelvis, by the combined action of the Psoas and Iliacus ; and, at the same time, everted and drawn out- wards by the external rotator and Glutei muscles, causing a marked prominence at the upper and outer side of the thigh, and much pain from the bruising and laceration of the muscles. The limb is shortened, in consequence of the lower fragment being drawn upwards by the Rectus in front, and the Biceps, Semimembranosus and Semitendinosus behind ; and is, at the same time, everted. This fracture may be reduced by two different methods : either by direct relaxation of all the opposing muscles, to effect which the limb should be put up in such a manner that the thigh is flexed on the pelvis and the leg on the thigh ; or by overcoming the contraction of the muscles, by continued extension, which may be effected by means of the long splint.

Oblique fracture of the femur *immediately above the condyles* (fig. 460) is a formidable injury, and attended with considerable displacement. On examination of the limb, the

FIG. 459.—Fracture of the femur below the trochanters.

FIG. 460.—Fracture of the femur above the condyles.

FIG. 461.—Fracture of the patella.

Semimembranosus
Semitendinosus

lower fragment may be felt deep in the popliteal space, being drawn backwards by the Gastrocnemius and Plantaris muscles ; and upwards by the hamstring and Rectus muscles. The pointed end of the upper fragment is drawn inwards by the Pectineus and Adductor muscles, and tilted forwards by the Psoas and Iliacus, piercing the Rectus muscle, and occasionally the integument. Relaxation of these muscles, and direct approximation of the broken fragments, are effected by placing the limb on a double inclined plane. The greatest care is requisite in keeping the pointed extremity of the upper fragment in proper position ; otherwise, after union of the fracture, the power of extension of the limb is partially destroyed, from the Rectus muscle being held down by the fractured end of the bone, and from the patella, when elevated, being drawn upwards against the projecting fragment.

In fracture of the *patella* (fig. 461) the fragments are separated by the effusion which takes place into the joint, and possibly by the action of the Quadriceps extensor ; the extent of separation of the two fragments depending upon the degree of laceration of the ligamentous structures around the bone.

In oblique fracture of the *shaft of the tibia* (fig. 462), if the fracture has taken place obliquely from above, downwards and forwards, the fragments ride over one another, the lower fragments being drawn backwards and upwards by the powerful action of the muscles of the calf ; the pointed extremity of the upper fragment projects forwards immediately beneath the integument, often protruding through it, and rendering the fracture a com- pound one. If the direction of the fracture is the reverse of that shown in the figure, the pointed extremity of the lower fragment projects forwards, riding upon the lower end of the upper one. By bending the knee, which relaxes the opposing muscles, and making extension from the ankle and counter-extension at the knee, the fragments may be brought into apposition. It is often necessary, however, in compound fracture, to remove a portion of the projecting bone with the saw before complete adaptation can be effected.

FIG. 462.—Oblique fracture of the shaft of the tibia.

FIG. 463.—Fracture of the fibula, with dislocation of the foot outwards—'Pott's fracture.'

Fracture of the *fibula with dislocation of the foot outwards* (fig. 463), commonly known as 'Pott's fracture,' is one of the most frequent injuries of the ankle-joint. The fibula is fractured about three inches above the ankle; in addition to this the internal malleolus is broken off, or the deltoid ligament torn through, and the end of the tibia displaced from the corresponding surface of the astragalus. The foot is markedly everted, and the sharp edge of the upper end of the fractured malleolus presses strongly against the skin; at the same time, the heel is drawn up by the muscles of the calf. This injury can generally be reduced by flexing the leg at right angles with the thigh, which relaxes all the opposing muscles, and by making extension from the ankle and counter-extension at the knee.

THE BLOOD VASCULAR SYSTEM

THE blood vascular system comprises the heart and blood-vessels with their contained fluid, the blood. The composition of the blood and the minute anatomy of the blood-vessels have already been considered in the section on Histology.

The Heart is the central organ of the entire system, and consists of a hollow muscle; by its contraction the blood is pumped to all parts of the body through a complicated series of tubes, termed *arteries*. The arteries undergo enormous ramification in their course throughout the body, and end in very minute vessels, called *arterioles*, which in their turn open into a close-meshed network of microscopic vessels, termed *capillaries*. After the blood has passed through the capillaries it is collected into a series of larger vessels, called *veins*, by which it is again returned to the heart. The passage of the blood through the heart and blood-vessels constitutes what is termed the *circulation* of the blood, of which the following is an outline.

The human heart is divided by a septum into two halves, right and left, each half being further divided into two cavities, the upper of the two being termed the *auricle* and the lower the *ventricle*. The heart therefore consists of four chambers or cavities, two forming the right half, the right auricle and right ventricle, and two the left half, the left auricle and left ventricle. The auricles are receiving chambers, and the ventricles distributing ones. The right half of the heart contains venous or impure blood; the left, arterial or pure blood. From the cavity of the left ventricle the pure blood is carried into a large artery, the *aorta*, through the numerous branches of which it is distributed to all parts of the body, with the exception of the lungs. In its passage through the capillaries of the body the blood gives up to the tissues the materials necessary for their growth and nourishment, and at the same time receives from the tissues the waste products resulting from their metabolism, and in doing so becomes changed from arterial or pure blood into venous or impure blood, which is collected by the veins and through them returned to the right auricle of the heart. From this cavity the impure blood passes into the right ventricle, from which it is conveyed through the *pulmonary arteries* to the lungs. In the capillaries of the lungs it again becomes arterialised, and is then carried to the left auricle by the *pulmonary veins*. From this cavity it passes into that of the left ventricle, from which the cycle once more begins.

The course of the blood from the left ventricle through the body generally to the right side of the heart constitutes the greater or *systemic* circulation, while its passage from the right ventricle through the lungs to the left side of the heart is termed the lesser or *pulmonary* circulation.

It is necessary, however, to state that the blood which circulates through the spleen, pancreas, stomach, small intestine, and the greater part of the large intestine is not returned directly from these organs to the heart, but is collected into a large vein, termed the *portal vein*, by which it is carried to the liver. In the liver this vein divides, after the manner of an artery, and ultimately ends in capillary vessels, from which the rootlets of a series of veins, called the *hepatic veins*, arise; these carry the blood into the inferior vena cava, which conveys it to the right auricle.

From this it will be seen that the blood contained in the portal vein passes through two sets of capillary vessels: (1) those in the spleen, pancreas, stomach, &c., and (2) those in the liver.

Speaking generally, the arteries may be said to contain pure, and the veins impure, blood. This is true of the systemic, but not of the pulmonary, vessels, since it has been seen that the impure blood is conveyed from the heart to the lungs by the pulmonary arteries, and the pure blood returned from the lungs to the heart by the pulmonary veins. Arteries, therefore, must be defined as vessels which convey blood *from* the heart, and veins as vessels which return blood *to* the heart.

Fig. 464.—Diagram to show the course of the circulation of the blood.

The heart and lungs are contained within the cavity of the thorax, the walls of which afford them protection. The heart lies between the two lungs, and is there enclosed within a membranous bag, the *pericardium*, while each lung is invested by a serous membrane, the *pleura*. The skeleton of the thorax, and the shape and boundaries of the cavity, have already been described (page 252 *et seq.*).

The cavity of the thorax.—The capacity of the cavity of the thorax does not correspond with its apparent size externally, because (1) the space enclosed by the lower ribs is occupied by some of the abdominal viscera; and (2) the cavity extends above the first rib into the neck. The size of the cavity of the thorax is constantly varying during life with the movements of the ribs and Diaphragm, and with the degree of distension of the abdominal viscera. From the collapsed state of the lungs, as seen when the thorax is opened, in the dead body, it would appear as if the viscera only partly filled the cavity of the thorax, but during life there is no vacant space, that which is seen after death being filled up by the expanded lungs.

The upper opening of the thorax.— The parts which pass through the upper opening of the thorax are, from before backwards in or near the middle line, the Sterno-hyoid and Sterno-thyroid muscles, the remains of the thymus gland, the inferior thyroid veins, the trachea, œsophagus, thoracic duct, and the Longus colli muscles; at the sides, the innominate artery, the left common carotid and left subclavian arteries, the internal mammary and superior intercostal arteries, the right and left innominate veins, the pneumogastric, cardiac, phrenic, and sympathetic nerves, the greater part of the anterior primary division of the first dorsal nerve, and the recurrent laryngeal nerve of the left side. The apex of each lung, covered by the pleura, also projects through this aperture, a little above the margin of the first rib.

The **lower opening of the thorax** is wider transversely than from before backwards. It slopes obliquely downwards and backwards, so that the cavity of the thorax is much deeper behind than in front. The Diaphragm (see page 479) closes in the opening, forming the floor of the thorax. The floor is flatter at the centre than at the sides, and is higher on the right side than on the left, corresponding in the dead body to the upper border of the fifth costal cartilage on the former, and to the corresponding part of the sixth costal cartilage on the latter. From the highest point on each side the floor slopes suddenly downwards to the attachment of the Diaphragm to the ribs ; this is more marked behind than in front, so that only a narrow space is left between it and the posterior wall of the thorax.

The Pericardium

The **Pericardium** (fig. 465) is a conical sero-membranous sac, in which the heart and the commencement of the great vessels are contained. It is placed behind the sternum, and the cartilages of the third, fourth, fifth, sixth, and seventh ribs of the left side, in the interval between the pleuræ.

Its *apex* is directed upwards, and surrounds the great vessels about two inches above their origin from the base of the heart. Its *base* is attached to the

Fig. 465.—Posterior wall of the pericardial sac, showing the lines of reflection of the serous pericardium on the great vessels.

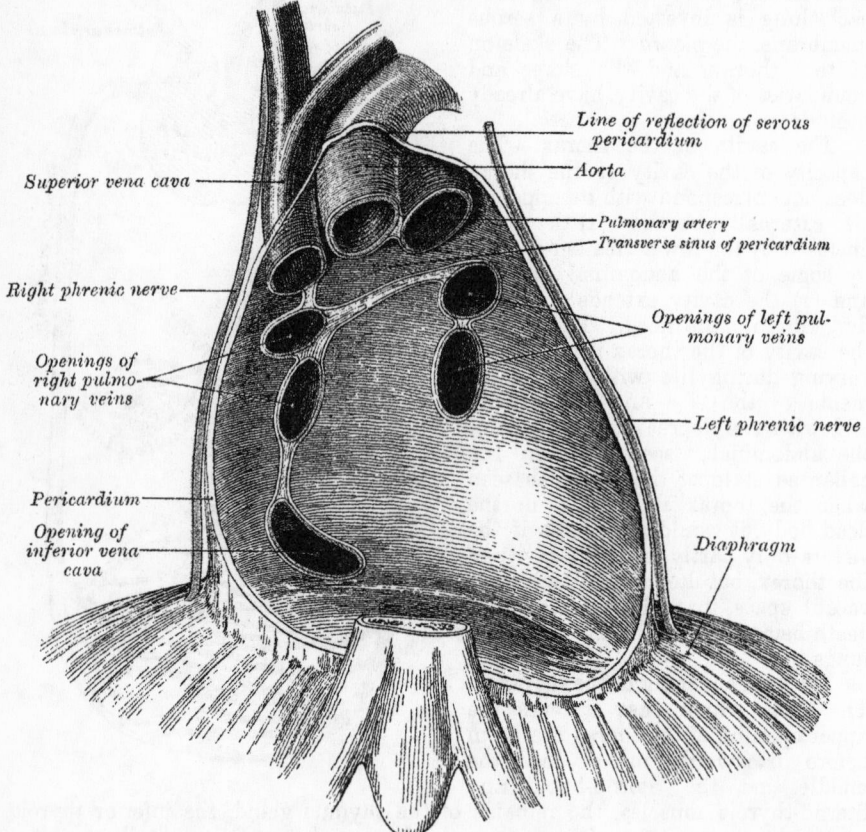

central tendon and to the left part of the adjoining muscular structure of the Diaphragm. *In front*, it is separated from the anterior wall of the thorax, in the greater part of its extent, by the lungs and pleuræ ; but a small area, some-what variable in extent, and usually corresponding with the left half of the lower portion of the gladiolus of the sternum and the inner extremities of the cartilages of the fourth and fifth ribs of the left side, comes into direct relationship with the chest-wall. The lower extremity of the thymus gland, in the child, is in contact with the front of the upper part of the pericardium. *Behind*, it rests upon the bronchi, the œsophagus, and the descending aorta. *Laterally*, it is covered by the pleuræ, and is in relation to the inner surface of the lungs ; the phrenic nerve, with its accompanying vessels, descends between the pericardium and pleura on either side.

Structure of the Pericardium.—The pericardium is a fibro-serous membrane, and consists, therefore, of two layers, an external fibrous and an internal serous.

The *fibrous layer* is a strong, dense membrane. Above, it surrounds the great vessels arising from the base of the heart, on which it is continued in the form of tubular prolongations, which are gradually lost upon their external coats ; the strongest being that which encloses the aorta. It may be traced, over these vessels, to become continuous with the deep layer of the cervical fascia. In front the pericardium is connected to the posterior surface of the sternum by two fibrous bands, the *superior* and *inferior sterno-pericardiac ligaments*; the upper passing to the manubrium, and the lower to the ensiform cartilage. On each side of the ascending aorta it sends upwards a diverticulum: the one on the left side, somewhat conical in shape, passes upwards and outwards, between the arch of the aorta and the pulmonary artery, as far as

FIG. 466.—Front view of the thorax. The ribs and sternum are represented in relation to the lungs, heart, and other internal organs.

1. Pulmonary orifice. 2. Aortic orifice. 3. Left auriculo-ventricular orifice. 4. Right auriculo-ventricular orifice.

the ductus arteriosus, where it terminates in a cæcal extremity which is attached by loose connective tissue to the obliterated duct. The one on the right side passes upwards and to the right, between the ascending aorta and vena cava superior, and also terminates in a cæcal extremity. Below, the fibrous layer is attached to the central tendon of the Diaphragm; and, on the left side, to its muscular fibres.

The vessels receiving fibrous prolongations from this membrane are, the aorta, the superior vena cava, the right and left pulmonary arteries, the four pulmonary veins, and the obliterated ductus arteriosus. As the inferior vena cava enters the pericardium through the central tendon of the Diaphragm, it receives no covering from the fibrous layer.

The *serous layer* invests the heart, and is then reflected on the inner surface of the pericardium. It consists, therefore, of a visceral and parietal portion. The former invests the surface of the heart, and the commencement of the great vessels; from these it is reflected upon the inner surface of the fibrous layer, lining, below, the upper surface of the central tendon of the Diaphragm. The serous membrane encloses the aorta and pulmonary artery in a single tube, so that a passage, termed the *transverse sinus* of the pericardium, exists between these vessels in front and the auricles behind. The membrane only partially covers the superior vena cava and the four pulmonary veins, and scarcely covers the inferior cava, as this vessel enters the heart almost directly after it has passed through the Diaphragm. Its inner surface is smooth and glistening, and secretes a serous fluid, which serves to facilitate the movements of the heart.

Arteries of the pericardium.— These are derived from the internal mammary and its musculo-phrenic branch, and from the descending thoracic aorta.

Fig. 467.—Showing relations of opened heart to front of chest.

In. Innominate artery. LCC. Left common carotid artery. LS. Left subclavian artery. Ao. Aorta. PA. Pulmonary artery. RA. Right auricle. RV. Right ventricle. VS. Ventricular septum. LV. Left ventricle. AP. Anterior papillary muscle. Inf. Infundibular segment of tricuspid valve.

Nerves of the pericardium.—These are branches from the vagus, the phrenic, and the sympathetic.

The vestigial fold of the pericardium.—Between the left pulmonary artery and subjacent pulmonary vein is a triangular fold of the serous pericardium; it is known as the *vestigial fold of Marshall*. It is formed by the duplicature of the serous layer over the remnant of the lower part of the left superior vena cava (duct of Cuvier), which, after birth, becomes obliterated, and remains as a fibrous band stretching from the left superior intercostal vein to the left auricle, where it is continuous with a small vein, the oblique vein of Marshall, which opens into the coronary sinus.

Surgical Anatomy.—Effusions of serum, pus, or blood may take place into the sac of the pericardium, and may require surgical treatment for their removal. In cases of serous effusion, and possibly also in cases of hæmorrhagic extravasation, to relieve urgent cardiac distress, paracentesis of the pericardium may be required. As it is advisable to perform this operation without transfixing the pleura, the puncture should be made either in the fifth or sixth intercostal space on the left side and close to the sternum, so as to avoid wounding the internal mammary artery, which descends about half or three-quarters of an inch from the sternal margin. Probably of the two the sixth space is to be preferred, but the operation is sometimes difficult on account of the extreme narrowness of the space. In consequence of the uncertain and varying position of the anterior reflexion of the pleura, transfixion of the pleural sac cannot always be avoided; and therefore Porter has advised that the operation should invariably be performed by open incision. *Pericardotomy* is required when the effusion is of a purulent nature. In order to do this it is necessary to excise a portion of the fifth or sixth costal cartilage. An incision is made along the left border of the sternum from the upper border of the fourth cartilage to the seventh. Transverse incisions an inch long are then made outwards from either extremity of this, and the rectangular flap thus formed reflected outwards. The fifth costal cartilage is now separated from the sternum by means of a gouge, great care being taken not to let the instrument slip and penetrate too deeply. The cartilage is then seized with lion forceps and raised, the tissues beneath it being peeled off, so as to avoid wounding the internal mammary artery or the pleura. The Triangularis sterni is now scratched through with a director or the nail of the index finger close to the sternum, and the pericardium felt for and opened, the finger guarding the pleura and left internal mammary artery.

THE HEART

The **Heart** is a hollow muscular organ of a conical form, placed between the lungs, and enclosed in the cavity of the pericardium.

Position.—The heart is placed obliquely in the chest: the broad attached end, or base, is directed upwards, backwards, and to the right, and corresponds with the dorsal vertebræ, from the fifth to the eighth inclusive; the apex is directed downwards, forwards, and to the left, and corresponds to the space between the cartilages of the fifth and sixth ribs, three-quarters of an inch to the inner side, and an inch and a half below the left nipple, or about three and a half inches from the middle line of the sternum. The heart is placed behind the lower two-thirds of the sternum, and projects farther into the left than into the right half of the cavity of the chest, extending from the median line about three inches in the former direction, and only one and a half in the latter; about one-third of the heart lies to the right and two-thirds to the left of the mesial plane. The antero-superior surface of the heart is round and convex, directed upwards and forwards, and formed chiefly by the right auricle and ventricle, together with a small part of the left ventricle. Its postero-inferior surface, which looks downwards rather than backwards, is flattened, and rests mainly upon the central tendon of the Diaphragm, but to a slight extent upon the muscular structure also. It is formed chiefly by the left ventricle. The right border of the heart is long, thin, and sharp, and is formed partly by the right auricle above and partly by the right ventricle below. The auricular portion is almost vertical, situated behind the costal cartilages of the third, fourth, and fifth ribs, about half an inch external to the right border of the sternum: the ventricular portion is nearly horizontal, and extends from the extremity of the sixth costal cartilage on the right side, behind the lower part of the body of the sternum and the sixth costal cartilage of the left side, to the apex of the heart. The left border is short, but thick and rounded; it is formed mainly by the left ventricle, and only to a slight extent by the left auricle. It passes from the second intercostal space, about an inch from the sternum, obliquely downwards and outwards, with a convexity to the left, to the apex of the heart.

Size.—The heart, in the adult, measures five inches in length, three inches and a half in breadth at the broadest part, and two inches and a half in thickness. The prevalent weight, in the male, varies from ten to twelve ounces; in the female, from eight to ten: its proportions to the body being as 1 to 169 in males; 1 to 149 in females. The heart continues to increase in weight, and also in length, breadth, and thickness, up to an advanced period of life: this increase is more marked in men than in women.

Component Parts.—As has already been stated (page 578), the heart is subdivided by a muscular septum into two lateral halves, which are named respectively right and left; and a transverse constriction subdivides each half of the organ into two cavities, the upper cavity being called the *auricle*, the lower the *ventricle*. The heart therefore consists of four chambers, viz.: a right and a left auricle, and a right and a left ventricle. The course of the blood through the heart cavities and blood-vessels has already been described (page 578).

The division of the heart into four cavities is indicated by grooves upon its surface. The groove separating the auricles from the ventricles is called the *auriculo-ventricular groove*. It is deficient, in front, where it is crossed by the root of the pulmonary artery. It contains the trunks of the nutrient vessels of the heart. The auricular portion occupies the base of the heart, and is subdivided into two cavities by a septum. The ventricles are also separated by two furrows, the *interventricular grooves*, one of which is situated on the antero-superior, the other on the postero-inferior surface; these extend from the base of the ventricular portion to near the apex of the organ; the former being situated nearer to the left border of the heart, and the latter to the right. It follows, therefore, that

FIG. 468.—The right auricle and ventricle laid open,
the anterior walls of both having been removed.

Bristle passed through right
auriculo-ventricular opening

the right ventricle forms the greater portion of the antero-superior surface of the heart, and the left ventricle more of its postero-inferior surface.

Each of these cavities must be examined.

The **Right Auricle** is a little larger than the left, its walls somewhat thinner, measuring about one line; and its cavity is capable of containing about two ounces. It consists of two parts: a principal cavity, the *atrium* or *sinus venosus*, situated posteriorly, and an anterior, smaller portion, the *appendix auriculæ*.

The *atrium* is the large quadrangular cavity placed between the two venæ cavæ; its walls are extremely thin; it is connected below with the right ventricle, and internally with the left auricle, being free in the rest of its extent.

The *appendix auriculæ*, so called from its fancied resemblance to a dog's ear, is a small conical muscular pouch, the margins of which present a dentated edge. It projects from the sinus forwards and to the left side, overlapping the root of the aorta.

To examine the interior of the right auricle, an incision should be made along its right border from the entrance of the superior vena cava to that of the inferior. A second cut is to be made from the centre of this first incision to the tip of the auricular appendix, and the flaps raised.

The internal surface of the right auricle is smooth, except in the appendix and adjacent part of the anterior wall of the sinus venosus, where the muscular wall is thrown into parallel ridges, resembling the teeth of a comb and hence named the *musculi pectinati*. These terminate behind on a vertical smooth ridge, the *crista terminalis* of His, the position of which is indicated on the surface of the distended auricle by a furrow, the *sulcus terminalis* (His) ; this represents the line of fusion of the sinus venosus of the embryo with the primitive auricle proper.

It presents the following parts for examination :

Openings
{
Superior cava.
Inferior cava.
Coronary sinus.
Foramina Thebesii.
Auriculo-ventricular.
}

Valves
{
Eustachian.
Coronary.
}

Fossa ovalis.
Annulus ovalis.
Tuberculum Loweri.
Musculi pectinati.

The *superior vena cava* returns the blood from the upper half of the body, and opens into the upper and back part of the auricle, the direction of its orifice being downwards and forwards. It has no valve.

The *inferior vena cava*, larger than the superior, returns the blood from the lower half of the body, and opens into the lowest part of the auricle, near the septum, its orifice being directed upwards and inwards, and guarded by a rudimentary valve, the *Eustachian valve*. The direction of a current of blood through the superior vena cava would consequently be towards the auriculo-ventricular orifice ; while the direction of the blood through the inferior cava would be towards the auricular septum. This is the normal direction of the two currents in fœtal life.

The *coronary sinus* opens into the auricle, between the inferior vena cava and the auriculo-ventricular opening. It returns the blood from the substance of the heart, and is protected by a semicircular fold of the lining membrane of the auricle, the *coronary valve*, or *valve of Thebesius*. The sinus, before entering the auricle, is considerably dilated—nearly to the size of the end of the little finger. Its wall is partly muscular, and, at its junction with the great coronary vein, is somewhat constricted, and furnished with a valve, consisting of two unequal segments.

The *foramina Thebesii* are depressions in the walls of the auricle : the majority of these are culs-de-sac, but about one third present the orifices of minute veins (*venæ cordis minimæ*), which return the blood directly from the muscular substance of the heart.

The *auriculo-ventricular opening* is the large oval aperture of communication between the auricle and the ventricle, to be presently described.

The *Eustachian valve* is situated between the anterior margin of the inferior vena cava and the auriculo-ventricular orifice. It is semilunar in form, its convex margin being attached to the margin of the vein ; its concave margin, which is free, terminating in two cornua, of which the left is continuous with the anterior edge of the annulus ovalis ; the right being lost on the wall of the auricle. The valve is formed by a duplicature of the lining membrane of the auricle, containing a few muscular fibres.

In the fœtus this valve is of large size, and serves to direct the blood from the inferior vena cava, through the foramen ovale, into the left auricle.

In the adult it is occasionally persistent, and may assist in preventing the reflux of blood into the inferior vena cava ; more commonly it is small, and its free margin presents a cribriform or filamentous appearance ; sometimes it is altogether wanting.

The *coronary valve* (valve of Thebesius) is a semicircular fold of the lining membrane of the auricle, protecting the orifice of the coronary sinus. It prevents the regurgitation of blood into the sinus during the contraction of the auricle. This valve may be double.

The *fossa ovalis* is an oval depression on the posterior wall of the auricle, corresponding to the situation of the foramen ovale in the fœtus. It is situated at the lower part of the auricular septum, above and to the left of the orifice of the inferior vena cava.

The *annulus ovalis* is the prominent oval margin of the foramen ovale. It is most distinct above, and at the sides ; below, it is deficient. A small slit-like valvular opening is occasionally found, at the upper margin of the fossa ovalis, which leads upwards, beneath the annulus, into the left auricle, and is the remains of the aperture between the two auricles in the fœtus.

The *tuberculum Loweri* is a small projection on the posterior wall of the auricle, above the fossa ovalis. It is most distinct in the hearts of quadrupeds ; in man it is scarcely visible. It was supposed by Lower to direct the blood from the superior cava towards the auriculo-ventricular opening.

To examine the interior of the right ventricle, its anterior wall should be turned downwards and to the right in the form of a triangular flap. This is accomplished by making two incisions : (1) from the pulmonary artery to the apex of the ventricle parallel to, but a little to the right of, the anterior interventricular furrow ; (2) another, starting from the upper extremity of the first and carried outwards parallel to, but a little below, the auriculo-ventricular furrow ; care being taken not to injure the auriculo-ventricular valve.

The **Right Ventricle** is triangular in form, and extends from the right auricle to near the apex of the heart. Its antero-superior surface is rounded and convex, and forms the larger part of the front of the heart. Its under surface is flattened, rests upon the Diaphragm, and forms only a small part of the back of the heart. Its posterior wall is formed by the partition between the two ventricles, the *septum ventriculorum*. The septum bulges into the cavity of the right ventricle, so that a transverse section of the cavity presents a semilunar outline. Its upper and inner angle is prolonged into a conical pouch, the *infundibulum*, or *conus arteriosus*, from which the pulmonary artery arises. The wall of the right ventricle is thinner than that of the left, the proportion between them being as 1 to 3. The wall is thickest at the base, and gradually becomes thinner towards the apex. The cavity equals in size that of the left ventricle, and is capable of containing about three fluid ounces.*

The following parts present themselves for examination :

Openings {Auriculo-ventricular. / Opening of the pulmonary artery.

Valves {Tricuspid. / Pulmonary.

And a muscular and tendinous apparatus connected with the tricuspid valve :

Columnæ carneæ.   Chordæ tendineæ.

The *auriculo-ventricular orifice* is the large oval aperture of communication between the auricle and ventricle. It is situated at the base of the ventricle, near the right border of the heart. It is about an inch and a half in diameter,† oval from side to side, surrounded by a fibrous ring, covered by the lining membrane of the heart ; it is considerably larger than the corresponding aperture on the left side, being sufficient to admit the ends of four fingers. It is guarded by the tricuspid valve.

The *opening of the pulmonary artery* is circular in form, and situated at the summit of the conus arteriosus, close to the septum ventriculorum. It is placed above and on the left side of the auriculo-ventricular opening, upon the

* Morrant Baker says that, ' taking the mean of various estimates, it may be inferred that each ventricle is able to contain four to six ounces of blood.'—Kirke's *Physiology*, 10th edition, p. 156.

† In the *Pathological Transactions*, vol. vi. p. 119, Dr. Peacock has given some careful researches upon the weight and dimensions of the heart in health and disease. He states, as the result of his investigations, that, in the healthy adult heart, the right auriculo-ventricular aperture has a mean circumference of 54·4 lines, or $4\frac{8}{9}$ inches ; the left auriculo-ventricular aperture a mean circumference of 44·3 lines, or $3\frac{2}{3}$ inches ; the pulmonic orifice of 40 lines, or $3\frac{1}{3}$ inches ; and the aortic orifice of 35·5 lines, or $3\frac{2}{4}$ inches ; but the dimensions of the orifices varied greatly in different cases, the right auriculo-ventricular aperture having a range of from 45 to 60 lines, and the others in the same proportion.

anterior aspect of the heart. Its orifice is guarded by the pulmonary semilunar valves.

The *tricuspid valve* consists of three segments of a triangular or trapezoidal shape, formed by a duplicature of the lining membrane of the heart, strengthened by an intervening layer of fibrous tissue. These segments are connected by their bases to the fibrous ring surrounding the auriculo-ventricular orifice, and by their sides with one another, so as to form a continuous annular membrane, which is attached round the margin of the auriculo-ventricular opening, their free margins and ventricular surfaces affording attachment to a number of delicate tendinous chords, the *chordæ tendineæ*. The largest and most movable segment is placed in front and to the left side of the auriculo-ventricular opening, interposed between that opening and the infundibulum ; hence it is called the *infundibular cusp*. Another segment is in relation to the right margin of the ventricle, the *right* or *marginal cusp* ; and a third to its posterior or septal wall, the *posterior* or *septal cusp*. The central part of each segment is thick and strong : the lateral margins are thin and translucent. The chordæ tendineæ are connected with the adjacent margins of the principal segments of the valve, and are further attached to each segment in the following manner : 1. Three or four reach the attached margin of each segment, where they are continuous with the auriculo-ventricular tendinous ring. 2. Others, four to six in number, are attached to the central thickened part of each segment. 3. The most numerous and finest are connected with the marginal portion of each segment.

The *columnæ carneæ* are the rounded muscular columns which project from nearly the whole of the inner surface of the ventricle, excepting near the opening of the pulmonary artery, where the wall is smooth. They may be classified, according to their mode of connection with the ventricle, into three sets. The first set merely form prominent ridges on the inner surface of the ventricle, being attached by their entire length on one side, as well as by their extremities. The second set are attached by their two extremities, but are free in the rest of their extent ; while the third set (*musculi papillares*) are continuous by one extremity with the wall of the heart, the opposite extremity giving attachment to the *chordæ tendineæ*. There are two papillary muscles, anterior and posterior : of these, the anterior is the larger ; its chordæ tendineæ are connected with the marginal and infundibular segments of the valve. The posterior is not always single, but sometimes consists of two or three muscular columns ; its chordæ tendineæ are connected with the septal and marginal segments. In addition to these, some few chordæ spring directly from the ventricular septum, or from small eminences on it, and passing to the septal and infundibular segments. A fleshy band, well marked in sheep and some other animals, is frequently seen passing from the base of the anterior papillary muscle to the interventricular septum. From its attachments it may assist in preventing over-distension of the ventricle, and so has been named the *moderator band*.

The right auriculo-ventricular orifice allows the blood to pass freely from the right auricle into the right ventricle, and it will be noted that the surface of the tricuspid valve next the blood-current is quite smooth. When the right ventricle contracts to force the blood into the pulmonary artery the segments of the tricuspid valve come together and close the auriculo-ventricular opening, and so prevent the blood from passing back into the auricle. The papillary muscles and chordæ tendineæ moor the segments of the valve, and prevent their being forced through into the auricle by the weight of blood behind them.

The *pulmonary valve* consists of three semilunar segments, two anterior (right and left) and one posterior, formed by a duplicature of the lining membrane, strengthened by fibrous tissue. They are attached, by their convex margins, to the wall of the artery, at its junction with the ventricle, the straight border being free, and directed upwards in the lumen of the vessel. The free margin of each is somewhat thicker than the rest of the segment,* is strengthened by a bundle of tendinous fibres, and presents, at its middle, a small projecting thickened nodule, called *corpus Arantii*, and consisting of bundles of interlacing connective-tissue

---

* The pulmonary semilunar segments have been found to be two in number instead of three (Dr. Hand, of St. Paul, Minn., in the *North-Western Med. and Surg. Journ.* July 1873), and the same variety is more frequently noticed in the aortic semilunar segments.

fibres with branched connective-tissue cells and some few elastic fibres. From this nodule tendinous fibres radiate through the segment to its attached margin, and these fibres form a constituent part of its substance throughout its whole extent, excepting two narrow lunated portions, the *lunulæ*, placed one on each side of the nodule immediately adjoining the free margin ; here the segment is thin, and formed merely by the lining membrane. During the passage of the blood along the pulmonary artery the valve is opened, and the course of the blood along the tube is uninterrupted ; but during the ventricular diastole, when the current of blood along the pulmonary artery is checked, and partly thrown back by its elastic walls, these segments come into apposition and effectually close the entrance of the tube. When the valve is closed, the lunated portions of each segment are brought into contact with one another by their opposed surfaces, the three corpora Arantii filling up the small triangular space that would be otherwise left by their approximation.

Between the semilunar segments and the commencement of the pulmonary artery are three pouches or dilatations, one behind each segment. These are the pulmonary sinuses (*sinuses of Valsalva*). Similar sinuses exist between the semilunar segments of the aortic valve and the commencement of the aorta ; they are larger than the pulmonary sinuses. The blood, in its regurgitation towards the heart, finds its way into these sinuses, and so shuts down the valve-flaps.

In order to examine the interior of the left auricle, make an incision on the posterior surface of the auricle from the pulmonary veins on one side to those on the other, the incision being carried a little way into the vessels. Make another incision from the middle of the horizontal one to the appendix.

The **Left Auricle** is rather smaller than the right, but its walls are thicker, measuring about one line and a half ; it consists, like the right, of two parts, a principal cavity, or *atrium*, and an *appendix auriculæ*.

The *atrium* is cuboidal in form, and concealed, in front, by the pulmonary artery and aorta ; internally, it is separated from the right auricle by the septum auricularum ; behind, it receives on each side the two pulmonary veins, being free in the rest of its extent.

The *appendix auriculæ* is somewhat constricted at its junction with the auricle ; it is longer, narrower, and more curved than that of the right side, and its margins more deeply indented, presenting a kind of foliated appearance. Its direction is forwards and towards the right side, overlapping the root of the pulmonary artery.

The following parts present themselves for examination:

The openings of the four pulmonary veins.
Auriculo-ventricular opening.
Musculi pectinati.

The *pulmonary veins*, four in number, open into the posterior part of the left auricle—two on either side of the middle line : they are not provided with valves. The two left veins frequently terminate by a common opening.

The *auriculo-ventricular opening* is the large oval aperture of communication between the auricle and ventricle. It is rather smaller than the corresponding opening on the opposite side (see note, page 586).

The *musculi pectinati* are fewer in number and smaller than on the right side ; they are confined to the inner surface of the appendix.

On the septum auricularum may be seen a lunated impression, bounded below by a crescentic ridge, the concavity of which is turned upwards. The depression is just above the fossa ovalis in the right auricle.

To examine the interior of the left ventricle, make an incision a little to the left of the anterior interventricular groove from the base to the apex of the heart, and carry it up thence a little to the left of the posterior interventricular groove, nearly as far as the auriculo-ventricular groove.

The **Left Ventricle** is longer and more conical in shape than the right ventricle, and on transverse section its cavity presents an oval or nearly circular outline. It forms a small part of the anterior surface of the heart, and a considerable part of its posterior surface. It also forms the apex of the heart by its projection beyond the right ventricle. Its walls are much thicker than those of the right side, the proportion being as 3 to 1. They are thickest opposite the widest part of the ventricle, becoming gradually thinner towards the base, and also towards the apex, which is the thinnest part.

The following parts present themselves for examination :

Openings { Auriculo-ventricular. }   Valves { Mitral. }
         { Aortic. }   { Aortic. }
    Chordæ tendineæ.   Columnæ carneæ.

The *auriculo-ventricular opening* is placed below and to the left of the aortic orifice. It is a little smaller than the corresponding aperture of the opposite side, admitting only two fingers ; but, like it, is broader in the transverse than in the antero-posterior diameter. It is surrounded by a dense fibrous ring, covered by the lining membrane of the heart, and guarded by the mitral valve.

Fig. 469.—The left auricle and ventricle laid open,
the posterior walls of both being removed.

*Bristle passed through left
auriculo-ventricular opening*

Passed through aortic opening

The *aortic opening* is a circular aperture, in front and to the right side of the auriculo-ventricular, from which it is separated by one of the segments of the mitral valve. Its orifice is guarded by the *aortic valve*, which consists of three semilunar segments. The portion of the ventricle immediately below the aortic orifice is often termed the *aortic vestibule* of Sibson. It possesses fibrous instead of muscular walls, and so does not collapse during the ventricular diastole ; it thus gives space for the segments of the aortic valve during its closure.

The *mitral* or *bicuspid valve* is attached to the circumference of the auriculo-ventricular orifice in the same way that the tricuspid valve is on the opposite side. It consists of two triangular cusps or segments, formed by a duplicature of the lining membrane, strengthened by fibrous tissue, and containing a few muscular fibres. The segments are of unequal size, and are larger, thicker, and altogether stronger than those of the tricuspid. The larger segment is placed in front and to the right between the auriculo-ventricular and aortic orifices, the smaller to the left and behind the opening, close to the wall of the ventricle. Two smaller segments are usually found at the angles of junction of the larger

The mitral valve-flaps are furnished with chordæ tendineæ, the mode of attachment of which is precisely similar to those on the right side; but they are thicker, stronger, and less numerous.

The *aortic valve* consists of three semilunar segments, which surround the orifice of the aorta; two are posterior (right and left) and one anterior: they are similar in structure, and in their mode of attachment, to those of the pulmonary valve. They are, however, larger, thicker, and stronger than those of the right side; the lunulæ are more distinct, and the corpora Arantii larger and more prominent. Opposite each segment the wall of the aorta presents a slight dilatation or bulging (sinus of Valsalva); they are larger than those at the commencement of the pulmonary artery.

The *columnæ carneæ* admit of a subdivision into three sets, like those upon the right side; but they are more numerous, and present a dense interlacement,

FIG. 470.—Section of the heart, showing the interventricular septum.

especially at the apex, and upon the posterior wall. The *musculi papillares* are two in number, one being connected to the anterior, the other to the posterior wall; they are of large size, and terminate by free rounded extremities, from which the chordæ tendineæ arise.

The septum between the two ventricles is thick, especially below (fig. 470). At its upper part it suddenly tapers off and becomes destitute of muscular fibres, consisting only of fibrous tissue, covered by two layers of endocardium; and on the right side also covered, during diastole, by the septal segment of the tricuspid valve. This upper portion is termed the *membranous part* of the septum, and is continued upwards and forms the septum between the aortic vestibule and the right auricle. An abnormal communication may exist between the ventricles at this part owing to defective development of the septum.

The *Endocardium* is a thin membrane which lines the internal surface of the heart; it assists in forming the valves by its reduplications, and is continuous with the lining membrane of the great blood-vessels. It is a smooth, transparent membrane, giving to the inner surface of the heart its glistening appearance. It is composed of endothelial cells resting upon a connective-tissue membrane, which contains unstriped muscle cells and elastic tissue, and is attached to the muscular structure by loose elastic tissue which contains blood-vessels and nerves. It is more opaque on the left than on the right side of the heart, thicker in the auricles than in the ventricles, and thickest in the left auricle. It is thin on the musculi pectinati, and on the columnæ carneæ; but thicker on the smooth part of the auricular and ventricular walls, and on the tips of the musculi papillares.

Structure.—The heart consists of muscular fibres, and of fibrous rings which serve for their attachment. It is closely covered by the visceral layer of the serous pericardium (*epicardium*), and its cavities are lined by the *endocardium*. Between these two membranes is the muscular wall of the heart, the *myocardium*.

The *fibrous rings* surround the auriculo-ventricular and arterial orifices: they are stronger upon the left than on the right side of the heart. The auriculo-ventricular rings serve for the attachment of the muscular fibres of the auricles and ventricles, and also for the mitral and tricuspid valves; the ring on the left side is closely connected, by its right margin, with the aortic arterial ring. Between these and the right auriculo-ventricular ring is a mass of fibrous tissue; and in some of the larger animals, as the ox and elephant, a nodule of bone, the *os cordis*.

The fibrous rings surrounding the arterial orifices serve for the attachment of the great vessels and semilunar valves. Each ring receives, by its ventricular margin, the attachment of the muscular fibres of the ventricles; its opposite margin presents three deep semicircular notches, within which the middle coat of the artery (which has three convex semicircular segments) is firmly fixed; the attachment of the artery to its fibrous ring being strengthened by the thin cellular coat and serous membrane externally, and by the endocardium within. It is opposite the margins of these semicircular notches, in the arterial rings, that the endocardium, by its reduplication, forms the semilunar valves, the fibrous structure of the ring being continued into each of the segments of the valve at this part. The middle coat of the artery in this situation is thin, and the wall of the vessel is dilated to form the sinuses of Valsalva.

The *muscular structure of the heart* consists of bands of fibres, which present an exceedingly intricate interlacement. They are of a deep red colour, and marked with transverse striæ.

The muscular fibres of the heart admit of a subdivision into two groups, those of the auricles and those of the ventricles, which are quite independent of one another.

Fibres of the Auricles.—These are disposed in two layers—a superficial layer common to both cavities, and a deep layer proper to each. The *superficial fibres* are more distinct on the anterior surface of the auricles, across the bases of which they run in a transverse direction, forming a thin and incomplete layer. Some of these fibres pass into the septum auricularum. The *internal* or *deep fibres* proper to each auricle consist of two sets, looped and annular fibres. The *looped fibres* pass upwards over each auricle, being attached by their two extremities to the corresponding auriculo-ventricular rings, in front and behind. The *annular fibres* surround the whole extent of the appendices auricularum, and are continued upon the walls of the venæ cavæ and coronary sinus on the right side, and upon the pulmonary veins on the left side, at their connection with the heart. In the appendices they interlace with the longitudinal fibres.

Fibres of the Ventricles.—These are arranged in an exceedingly complex manner, and the accounts given by various anatomists differ considerably. This is probably due partly to the fact that the various layers of muscular fibre of which the heart is said to be composed are not independent, but their fibres are interlaced to a considerable extent, and therefore any separation into layers must be in a great measure artificial; and also partly to the fact, pointed out by Henle, that there are varieties in the arrangement owing to individual differences. If the epicardium and the subjacent fat are removed from a heart which has been subjected to prolonged boiling, so as to dissolve the connective tissues, the superficial fibres of the ventricles will be exposed. They will be seen to commence at the base of the heart, where they are attached to the tendinous rings around the orifices, and to pass obliquely downwards towards the apex, with a direction from right to left. At the apex the fibres turn suddenly inwards, into the interior of the ventricle, forming what is called the *vortex*. On the back of the heart it will be seen that the fibres pass continuously from one ventricle to the other over the interventricular groove ; and the same thing will be noticed on the front of the heart at the upper and lower end of the anterior interventricular groove, but in the middle portion of this groove the fibres passing from one ventricle to the other are interrupted by fibres emerging from the septum along the groove ; many of the superficial fibres pass in also at this groove to the septum. The vortex is produced, as stated above, by the sudden turning inwards of the superficial fibres in a peculiar spiral manner into the interior of the ventricle. Those fibres which descend on the posterior surface of the heart, enter the left ventricle at the vortex, and, ascending, form the posterior part of the inner layer of muscular fibres lining this cavity and the right (posterior) musculus papillaris ; those fibres which descend on the front of the heart, to reach the apex, also pass, at the vortex, into the interior of the ventricle, where they form the remainder of the innermost layer of the ventricle and the left (anterior) musculus papillaris. The fibres forming the inner layer of the wall of the ventricle ascend to be attached to the fibrous rings around the orifices.

By dissection these superficial fibres may be removed as a thin stratum, and it will then be found that the ventricles are made up of oblique fibres, super-imposed in layers one on the top of another, and assuming gradually a less oblique direction as they pass to the middle of the thickness of the ventricular wall, so that in the centre of the wall the fibres are transverse. Internal to this central transverse layer the fibres become oblique again, but in the opposite direction to the external ones. This division into distinct layers is, however, to a great extent artificial, as fibres cross from one layer to another, and have therefore to be divided in the dissection, and the change in the direction of the fibres is very gradual. These oblique fibres commence above at the fibrous rings at the base of the heart, and descending towards the apex they enter the septum near its lower end. In the septum the fibres which form the left ventricle may be traced in three directions : 1. some pass upwards to be attached to the central mass of fibrous tissue ; 2. others pass through the septum to become continuous with the fibres of the right ventricle ; 3. and the remainder pass through the septum to encircle the ventricle as annular fibres. Of the fibres of the right ventricle, some on entering the septum pass upwards to be attached to the central mass of fibrous tissue ; some entering the septum from behind pass forwards to become continuous with the fibres on the anterior surface of the left ventricle ; and others entering in front pass backwards to join the fibres on the posterior wall of the left ventricle. The septum therefore consists of three varieties of fibres, viz. : annular fibres special to the left ventricle ; ascending fibres, derived from both ventricles and ascending through the septum to the central fibro-cartilage ; and decussating fibres derived from the anterior wall of one ventricle and passing to the posterior wall of the other ventricle, or from the posterior wall of the right ventricle and passing to the anterior wall of the left. In addition to these fibres, there are a considerable number which appear to encircle both ventricles and which pass across the septum without turning into it.

Vessels and Nerves. —The *arteries* supplying the heart are the right and left coronary from the aorta.

The *veins* terminate in the right auricle, and will be described with the general venous system.

The *lymphatics* terminate in the thoracic and right lymphatic ducts.

The *nerves* are derived from the cardiac plexuses, which are formed partly from the cranial nerves, and partly from the sympathetic. They are freely distributed both on the surface and in the substance of the heart, the separate filaments being furnished with small ganglia.

Surface Form. —In order to show the extent of the heart in relation to the front of the chest, draw a line from the lower border of the second left costal cartilage, one inch from the sternum, to the upper border of the third right costal cartilage, half an inch from the sternum. This represents the base line, or upper limit of the organ. Take a point an inch and a half below, and three-quarters of an inch internal to the left nipple—that is, about three and a half inches to the left of the median line of the body. This represents the apex of the heart. Draw a line from this apex point, with a slight convexity downwards, to the junction of the seventh right costal cartilage to the sternum. This represents the lower limit of the heart. Join the right extremity of the first line—that is, the base line—with the right extremity of this line—that is, to the seventh right chondrosternal joint—with a slight curve outwards, so that it projects about an inch and a half from the middle line of the sternum. Lastly, join the left extremity of the base line and the apex point by a line curved slightly to the left.

The position of the various orifices is as follows : viz. the pulmonary orifice is situated in the upper angle formed by the articulation of the third left costal cartilage with the sternum ; the aortic orifice is a little below and internal to this, behind the left border of the sternum, close to the articulation of the third left costal cartilage to this bone. The left auriculo-ventricular opening is behind the sternum, rather to the left of the median line and opposite the fourth costal cartilages. The right auriculo-ventricular opening is a little lower, opposite the fourth interspace and in the middle line of the body (fig. 466).

A portion of the area of the heart thus mapped out is uncovered by lung, and therefore gives a dull note on percussion ; the remainder, being overlapped by the lung, gives a more or less resonant note. The former is known as the area of superficial cardiac dulness ; the latter, as the area of deep cardiac dulness. The area of superficial cardiac dulness is included between a line drawn from the centre of the sternum, on a level with the fourth costal cartilages, to the junction of the body of the sternum with the ensiform cartilage : from the two extremities of this line, two others are to be drawn to the position of the apex of the heart in the fifth intercostal space. Below, this area merges into the dulness which corresponds to the liver. Latham lays down the following rule as a sufficient practical guide for the definition of the portion of the heart which is uncovered by lung or pleura : 'Make a circle of two inches in diameter round a point midway between the nipple and the end of the sternum.'

Surgical Anatomy.—Wounds of the heart are often immediately fatal, but not necessarily so. They may be non-penetrating, when death may occur from hæmorrhage, if one of the coronary vessels has been wounded, or subsequently from pericarditis ; or, on the other hand, the patient may recover. Even a penetrating wound is not necessarily fatal, as a considerable number of cases have now been recorded in which the wound has been sutured. A flap comprising the whole thickness of the thoracic wall must be made, the cavity of the pericardium opened, and the wound in the heart sutured.

Peculiarities in the Vascular System of the Fœtus

The chief peculiarities in the heart of the fœtus are the direct communication between the two auricles through the foramen ovale, and the large size of the Eustachian valve. There are also several minor peculiarities. Thus, the position of the heart is vertical until the fourth month, when it commences to assume an oblique direction. Its size is also very considerable as compared with that of the body, the proportion at the second month being 1 to 50 ; at birth it is as 1 to 120 ; while in the adult the average is about 1 to 160. At an early period of fœtal life the auricular portion of the heart is larger than the ventricular, the right auricle being more capacious than the left ; but towards birth the ventricular portion becomes the larger. The thickness of both ventricles is, at first, about equal, but towards birth the left becomes the thicker of the two.

The *foramen ovale* is situated at the lower and back part of the auricular septum, forming a communication between the auricles. It remains as a free oval opening until the middle period of fœtal life. About this period a fold grows up from the posterior wall of the auricle, to the left of the foramen ovale, and advances over the opening, so as to form a sort of valve, which allows the blood to pass only from the right to the left auricle, and not in the opposite direction.

The *Eustachian valve* is directed upwards on the left side of the opening of the inferior vena cava, and serves to direct the blood from this vessel through the foramen ovale into the left auricle.

The peculiarities in the arterial system of the fœtus are the communication between the pulmonary artery and the descending aorta by means of the *ductus arteriosus*, and the communication between the internal iliac arteries and the placenta by means of the *umbilical arteries*.

The *ductus arteriosus* is a short tube, about half an inch in length at birth, and of the diameter of a goose-quill. In the early condition it forms the continuation of the pulmonary artery, and opens into the descending aorta, just below the origin of the left subclavian artery ; and so conducts the chief part of the blood from the right ventricle into this vessel. When the branches of the

Q Q

pulmonary artery have become larger relatively to the ductus arteriosus, the latter is chiefly connected to the left pulmonary artery ; and the fibrous cord, which is all that remains of the ductus arteriosus in later life, will be found to be attached to the root of that vessel.

The *umbilical* or *hypogastric arteries* arise from the internal iliacs, in addition to the branches given off from those vessels in the adult. Ascending along the sides of the bladder to its apex, they pass out of the abdomen at the umbilicus and are continued along the umbilical cord to the placenta, coiling round the umbilical vein. They carry to the placenta the blood which has circulated in the system of the foetus.

The peculiarity in the venous system of the foetus is the communication established between the placenta and the liver and portal vein, through the umbilical vein ; and the inferior vena cava through the ductus venosus.

Fœtal Circulation

The blood destined for the nutrition of the foetus is returned from the placenta to the foetus by the umbilical vein. This vein enters the abdomen at the umbilicus, and passes upwards along the free margin of the suspensory ligament of the liver to the under surface of that organ, where it gives off two or three branches to the left lobe, one of which is of large size ; and others to the lobus quadratus and lobulus Spigelii. At the transverse fissure it divides into two branches : of these, the larger is joined by the portal vein, and enters the right lobe ; the smaller branch continues outwards, under the name of the ductus venosus, and joins the left hepatic vein at the point of junction of that vessel with the inferior vena cava. The blood, therefore, which traverses the umbilical vein, reaches the inferior vena cava in three different ways. The greater quantity circulates through the liver with the portal venous blood, before entering the vena cava by the hepatic veins ; some enters the liver directly, and is also returned to the inferior cava by the hepatic veins : the smaller quantity passes directly into the vena cava, by the junction of the ductus venosus with the left hepatic vein.

In the inferior cava, the blood carried by the ductus venosus and hepatic veins becomes mixed with that returning from the lower extremities and wall of the abdomen. It enters the right auricle, and, guided by the Eustachian valve, passes through the foramen ovale into the left auricle, where it becomes mixed with a small quantity of blood returned from the lungs by the pulmonary veins. From the left auricle it passes into the left ventricle ; and from the left ventricle into the aorta, by means of which it is distributed almost entirely to the head and upper extremities, a small quantity being probably carried into the descending aorta. From the head and upper extremities the blood is returned by the tributaries of the superior vena cava to the right auricle, where it becomes mixed with a small portion of the blood from the inferior cava. From the right auricle it descends over the Eustachian valve into the right ventricle ; and from the right ventricle, passes into the pulmonary artery. The lungs of the foetus being inactive, only a small quantity of the blood of the pulmonary artery is distributed to them, by the right and left pulmonary arteries, and is returned by the pulmonary veins to the left auricle : the greater part passes through the ductus arteriosus into the commencement of the descending aorta, where it becomes mixed with a small quantity of blood transmitted by the left ventricle into the aorta. Through this vessel it descends to supply the lower extremities and viscera of the abdomen and pelvis, the chief portion being, however, conveyed by the umbilical arteries to the placenta.

From the preceding account of the circulation of the blood in the foetus, it will be seen :

1. That the placenta serves the purposes of nutrition and excretion, receiving the impure blood from the foetus, and returning it charged with additional nutritive material.

2. That nearly the whole of the blood of the umbilical vein traverses the liver before entering the inferior cava ; hence the large size of this organ, especially at an early period of foetal life.

3. That the right auricle is the point of meeting of a double current, the blood in the inferior cava being guided by the Eustachian valve into the left auricle,

while that in the superior cava descends into the right ventricle. At an early period of fœtal life it is highly probable that the two streams are quite distinct; for the inferior cava opens almost directly into the left auricle, and the Eustachian valve would exclude the current along the vein from entering the

FIG. 471.—Plan of the fœtal circulation.

Ductus arteriosus

Right Lung

Left Lung

Left lobe

Right Ventricle

Left At.

Right Lobe

Bladder

Umbilical Vein

Internal iliac artery

Placenta

In this plan the figured arrows represent the kind of blood, as well as the direction which it takes in the vessels. Thus—arterial blood is figured ≫--------≫; venous blood, ≫————≫; mixed (arterial and venous) blood, ≫-------≫

right ventricle. At a later period, as the separation between the two auricles becomes more distinct, it seems probable that some mixture of the two streams must take place.

4. The pure blood carried from the placenta to the fœtus by the umbilical vein, mixed with the blood from the portal vein and inferior cava, passes almost

directly to the arch of the aorta, and is distributed by the branches of that vessel to the head and upper extremities : hence the large size and perfect development of those parts at birth.

5. The blood contained in the descending aorta, chiefly derived from that which has already circulated through the head and limbs, together with a small quantity from the left ventricle, is distributed to the lower extremities : hence the small size and imperfect development of these parts at birth.

CHANGES IN THE VASCULAR SYSTEM AT BIRTH

At birth, when respiration is established, an increased amount of blood from the pulmonary artery passes through the lungs, which now perform their office as respiratory organs, and, at the same time, the placental circulation is cut off. The foramen ovale becomes gradually closed by about the tenth day after birth : the valvular fold above mentioned becomes adherent to the margins of the foramen for the greater part of its circumference, but above a slit-like opening is left between the two auricles, and this sometimes persists.

The *ductus arteriosus* begins to contract immediately after respiration is established, becomes completely closed from the fourth to the tenth day, and ultimately degenerates into an impervious cord, which serves to connect the left pulmonary artery to the descending aorta.

Of the *umbilical* or *hypogastric arteries*, the portion continued on to the bladder from the trunk of the corresponding internal iliac remains pervious, as the superior vesical artery ; and the part extending from the side of the bladder to the umbilicus becomes obliterated between the second and fifth days after birth, and projects as a fibrous cord towards the abdominal cavity, carrying on it a fold of peritoneum and separating two of the fossæ of the peritoneum, spoken of in the section on the surgical anatomy of direct inguinal hernia.

The *umbilical vein* and *ductus venosus* become completely obliterated between the second and fifth days after birth, and ultimately dwindle to fibrous cords ; the former becoming the round ligament of the liver, the latter the fibrous cord which, in the adult, may be traced along the fissure of the ductus venosus.

THE ARTERIES

Arteries are cylindrical tubular vessels, which serve to convey blood from both ventricles of the heart to every part of the body. These vessels were named arteries (ἀήρ, *air* ; τηρεῖν, *to contain*) from the belief entertained by the ancients that they contained air. To Galen is due the honour of refuting this opinion ; he showed that these vessels, though for the most part empty after death, contain blood in the living body.

The distribution of the systemic arteries is like a highly ramified tree, the common trunk of which, formed by the aorta, commences at the left ventricle of the heart, the smallest ramifications corresponding to the circumference of the body and the contained organs. The arteries are found in nearly every part of the body, with the exception of the hair, nails, epidermis, cartilages, and cornea ; and the larger trunks usually occupy the most protected situations, running, in the limbs, along the flexor side, where they are less exposed to injury.

There is considerable variation in the mode of division of the arteries : occasionally a short trunk subdivides into several branches at the same point, as may be observed in the coeliac and thyroid axes ; or the vessel may give off several branches in succession, and still continue as the main trunk, as is seen in the arteries of the limbs ; but frequently the division is dichotomous, as, for instance, the aorta dividing into the two common iliacs ; and the common carotid into the external and internal.

The branches of arteries arise at very variable angles ; some, as the superior intercostal arteries from the aorta, arise at an obtuse angle ; others, as the lumbar arteries, at a right angle ; or, as the spermatic, at an acute angle. An artery from which a branch is given off is smaller in size, but retains a uniform diameter until a second branch is derived from it. A branch of an artery is smaller than the trunk from which it arises; but if an artery divides into two branches, the combined area of the two vessels is, in nearly every instance, somewhat greater than that of the trunk; and the combined area of all the arterial branches greatly exceeds the area of the aorta; so that the arteries collectively may be regarded as a cone, the apex of which corresponds to the aorta, the base to the capillary system.

The arteries, in their distribution, communicate with one another, forming what is called an *anastomosis* (*ἀνά*, *between*; *στόμα*, *mouth*), or inosculation: and this communication is very free between the large as well as between the smaller branches. The anastomosis between trunks of equal size is found where great activity of the circulation is requisite, as in the brain; here the two vertebral arteries unite to form the basilar, and the two anterior cerebral arteries are connected by a short communicating trunk; it is also found in the abdomen, the intestinal arteries having very ample anastomoses between their larger branches. In the limbs, the anastomoses are most numerous and of largest size around the joints; the branches of an artery above inosculating with branches from the vessels below. These anastomoses are of considerable interest to the surgeon, as it is by their enlargement that a *collateral circulation* is established after the application of a ligature to an artery for the cure of aneurism. The smaller branches of arteries anastomose more frequently than the larger; and between the smallest twigs these inosculations become so numerous as to constitute a close network that pervades nearly every tissue of the body.

Throughout the body generally the larger arterial branches pursue a straight course; but in certain situations they are tortuous; thus the facial artery in its course over the face, and the arteries of the lips, are extremely tortuous in their course, to accommodate themselves to the movements of the parts. The uterine arteries are also tortuous, to accommodate themselves to the increase of size which the organ undergoes during pregnancy. Again, the internal carotid and vertebral arteries, previous to their entering the cavity of the skull, describe a series of curves, which are probably intended to diminish the velocity of the current of blood, by increasing the extent of surface over which it moves, and adding to the amount of impediment which is produced by friction.

The arteries are dense in structure, of considerable strength, highly elastic, and, when divided, they preserve, although empty, their cylindrical form. Their structure has been described on page 51.

In the description of the arteries, the efferent trunk of the pulmonic circulation, the pulmonary artery, will be first considered; and then the efferent trunk of the systemic circulation, the aorta, and its branches.

PULMONARY ARTERY (fig. 472)

The **pulmonary artery** conveys the venous blood from the right side of the heart to the lungs. It is a short, wide vessel, about two inches in length and 1⅓ inch (30 mm.) in diameter, arising from the left side of the base (conus arteriosus) of the right ventricle. It extends obliquely upwards and backwards, passing at first in front and then to the left of the ascending aorta, as far as the under surface of the arch, but on a plane posterior to it, where it divides, about the level of the intervertebral substance between the fifth and sixth dorsal vertebræ, into two branches of nearly equal size, the *right* and *left pulmonary arteries*.

Relations.—The whole of this vessel is contained, together with the ascending aorta, in the pericardium. It is enclosed with the aorta in a single tube of the visceral layer of the serous pericardium, which is continued upwards upon them from the base of the heart and connects them together. The fibrous layer of the pericardium becomes gradually lost upon the external coat of its two branches. In front, the pulmonary artery is separated from the anterior extremity of the second left intercostal space by the pleura and left lung, in addition to the pericardium; it rests at first upon the ascending aorta, and higher up lies in front of the left auricle on a plane posterior to the ascending aorta. On each side of its origin is the appendix of the corresponding auricle and a coronary artery. the left coronary artery passing, in the first part of its course, behind the vessel. The superficial cardiac plexus lies above its bifurcation, between it and the arch of the aorta.

The **right pulmonary artery,** longer and larger than the left, runs horizontally outwards, behind the ascending aorta and superior vena cava, to the root of the right lung, where it divides into two branches, of which the lower and larger supplies the middle and lower lobes ; the upper and smaller is distributed to the upper lobe.

The **left pulmonary artery,** shorter and somewhat smaller than the right, passes horizontally in front of the descending aorta and left bronchus to the root of the left lung, where it divides into two branches for the two lobes.

FIG. 472.—The arch of the aorta, and its branches.

FIG. 473.
Plan of the branches.

The root of the left pulmonary artery is connected to the under surface of the arch of the aorta by a short fibrous cord, the *ligamentum arteriosum* ; this is the remains of a vessel peculiar to fœtal life, the *ductus arteriosus.*

The terminal branches of the pulmonary artery will be described with the anatomy of the lung.

THE AORTA

The **aorta** (ἀορτή, *arteria magna*) is the main trunk of a series of vessels which convey the oxygenated blood to the tissues of the body for their nutrition. This vessel commences at the upper part of the left ventricle, where it is about $1\frac{1}{8}$ inch in diameter, and after ascending for a short distance, arches backwards, and to the left side, over the root of the left lung, then descends within the thorax on the left side of the vertebral column, passes through the aortic opening in the Diaphragm, and entering the abdominal cavity, terminates, considerably diminished in size (about $\frac{1}{6}$ of an inch in diameter), opposite the lower border of the fourth lumbar vertebra, where it divides into the right and left common iliac arteries. Hence it is divided into the *ascending aorta*, the *arch* of the aorta, and the *descending aorta*, which last is again divided into the *thoracic* and *abdominal aorta*, from the position of these parts.

ASCENDING AORTA

The **ascending aorta** is about two inches in length. It commences at the upper part of the base of the left ventricle, on a level with the lower border of the third costal cartilage behind the left half of the sternum ; it passes obliquely upwards, forwards, and to the right, in the direction of the heart's axis, as high as the upper border of the second right costal cartilage, describing a slight curve in its course, and being situated, when distended, about a quarter of an inch behind the posterior surface of the sternum. At its commencement, opposite the segments of the aortic valve, it presents three small dilatations called the sinuses of Valsalva. These segments are placed one in front and two behind, and serve the purpose of preventing regurgitation of blood into the cavity of the ventricle. At the union of the ascending with the transverse part of the aorta the calibre of the vessel is increased, owing to a bulging outwards of its right wall. This dilatation is termed the *great sinus of the aorta*, and on section presents a somewhat oval figure. This portion of the aorta is contained in the cavity of the pericardium, and, together with the pulmonary artery, is invested by a tube of the serous pericardium, continued on to them from the surface of the heart.

Relations.—The ascending aorta is covered at its commencement by the trunk of the pulmonary artery and the right auricular appendix, and, higher up, is separated from the sternum by the pericardium, the right pleura, and the anterior margin of the right lung, some loose areolar tissue, and the remains of the thymus gland ; *behind*, it rests upon the right pulmonary artery and left auricle. On the *right side*, it is in relation with the superior vena cava and right auricle, the former lying partly behind it ; on the *left side*, with the pulmonary artery.

PLAN OF THE RELATIONS OF THE ASCENDING AORTA

In front.
Pulmonary artery.
Right auricular appendix.
Pericardium.
Right pleura and lung.
Remains of thymus gland.

Right side.
Superior vena cava.
Right auricle.

Ascending Aorta.

Left side.
Pulmonary artery.

Behind.
Right pulmonary artery.
Left auricle.

Branches.—The only branches of the ascending aorta are the coronary arteries. They supply the heart, and are two in number, right and left, arising near the commencement of the aorta immediately above the free margin of the semilunar valves.

The **Right Coronary Artery,** about the size of a crow's quill, arises from the anterior sinus of Valsalva. It passes forwards between the pulmonary artery and the right auricular appendix, then runs obliquely to the right side, in the groove between the right auricle and ventricle, and curving around the right border of the heart, runs along its posterior surface as far as the posterior interventricular groove, where it divides into two branches, one of which (*transverse*) continues onwards in the groove between the left auricle and ventricle, and anastomoses with the left coronary; the other (*descending*) courses along the posterior interventricular furrow, supplying branches to both ventricles and to the septum, and anastomosing at the apex of the heart with the descending branches of the left coronary.

This vessel sends a large branch (*marginal*) along the thin margin of the right ventricle to the apex; which in its course gives off numerous small branches to the anterior and posterior surfaces of the ventricle. It also gives off a branch close to its origin (*right auricular*), which passes upwards between the right auricle and the aorta, and distributes branches to the right auricle, the auricular septum, the aorta, and pulmonary artery.

The **Left Coronary Artery,** larger than the former, arises from the left posterior sinus of Valsalva; it passes forwards between the pulmonary artery and the left auricular appendix, and divides into two branches. Of these, one (*transverse*) passes transversely outwards in the left auriculo-ventricular groove, and winds around the left border of the heart to its posterior surface, where it anastomoses with the transverse branch of the right coronary; the other (*descending*) passes along the anterior interventricular groove to the apex of the heart, where it anastomoses with the descending branches of the right coronary. The left coronary supplies the left auricle and its appendix, gives branches to both ventricles, and numerous small branches to the pulmonary artery, and commencement of the aorta.*

Peculiarities.—These vessels occasionally arise by a common trunk, or their number may be increased to three; the additional branch being of small size. More rarely, there are two additional branches.

ARCH OF THE AORTA

The **arch,** or **transverse aorta,** commences at the upper border of the second chondro-sternal articulation of the right side, and passes at first upwards and backwards and from right to left, and then from before backwards and downwards, to the left side of the lower border of the fourth dorsal vertebra behind. It thus forms two curvatures: one with its convexity upwards, the other with its convexity forwards and to the left. Its upper border is usually about an inch below the upper margin of the sternum.

Between the origin of the left subclavian artery and the attachment of the ductus arteriosus the lumen of the fœtal aorta is considerably narrowed, forming what is termed the *aortic isthmus,* while immediately beyond the ductus arteriosus the vessel presents a fusiform dilatation which His has named the *aortic spindle* —the point of junction of the two parts being marked in the concavity of the arch by an indentation or angle. These conditions persist, to some extent, in the adult, where His found that the average diameter of the spindle exceeded that of the isthmus by 3 mm. (about one-eighth of an inch).

Relations.—The *anterior surface* of the arch of the aorta is covered by the pleuræ and lungs and the remains of the thymus gland, and crossed towards the left side by the left pneumogastric and phrenic nerves, the inferior cervical cardiac nerve of the left pneumogastric, the superior cardiac nerve from the left sympathetic, and by the left superior intercostal vein. Its *posterior surface* lies on the trachea, just above its bifurcation, and on the great, or deep, cardiac plexus. To its right side are the œsophagus, thoracic duct, and left recurrent laryngeal nerve. Its *upper border* is in relation with the left innominate vein; and from its upper part are given off the innominate, left common carotid, and left subclavian arteries. Its *lower border* is in relation with the bifurcation

---

* According to West, there is a very free and complete anastomosis between the two coronary arteries (*Lancet*, June 2, 1883, p. 945). This, however, is not the view generally held by anatomists, for, with the exception of the anastomosis mentioned above in the auriculo-ventricular and interventricular grooves, it is believed that the two arteries only communicate by very small vessels in the substance of the heart.

of the pulmonary artery, the remains of the ductus arteriosus, which is connected with the left division of that vessel, and the superficial cardiac plexus; the left recurrent laryngeal nerve winds round it from before backwards, while the left bronchus passes below it.

PLAN OF THE RELATIONS OF THE ARCH OF THE AORTA

Above.

Left innominate vein.
Innominate artery.
Left carotid.
Left subclavian.

In front.

Pleuræ and lungs.
Remains of thymus gland.
Left pneumogastric nerve.
Left phrenic nerve.
Cardiac branches from vagus
    and sympathetic.
Left superior intercostal vein.

Arch of Aorta.

Behind.

Trachea.
Deep cardiac plexus.
Œsophagus.
Thoracic duct.
Left recurrent nerve.

Below.

Bifurcation of pulmonary artery.
Remains of ductus arteriosus.
Superficial cardiac plexus.
Left recurrent nerve.
Left bronchus.

Peculiarities.—The height to which the aorta rises in the chest is usually about an inch below the upper border of the sternum; but it may ascend nearly to the top of that bone. Occasionally it is found an inch and a half, more rarely two or even three inches below this point.

In Direction.—Sometimes the aorta arches over the root of the right instead of the left lung, and passes down on the right side of the spine, a condition which is found in birds.

FIG. 474.—Relation of great vessels at base of heart, seen from above.
(From a preparation in the Museum of the Royal College of Surgeons of England.)

In such cases all the viscera of the thoracic and abdominal cavities are transposed. Less frequently, the aorta, after arching over the root of the right lung, is directed to its usual position on the left side of the spine, this peculiarity not being accompanied by any transposition of the viscera.

In Conformation.—The aorta occasionally divides, as in some quadrupeds, into an ascending and a descending trunk, the former of which is directed vertically upwards, and subdivides into three branches, to supply the head and upper extremities. Sometimes the aorta subdivides soon after its origin into two branches, which soon reunite. In one of these cases the œsophagus and trachea were found to pass through the interval left by their division; this is the normal condition of the vessel in the reptilia.

Surgical Anatomy.—Of all the vessels of the arterial system, the aorta, and more especially its arch, is most frequently the seat of disease; hence it is important to consider some of the consequences that may ensue from aneurism of this part.

It will be remembered that the ascending aorta is contained in the pericardium, just behind the sternum, being crossed at its commencement by the pulmonary artery and right auricular appendix, and having the right pulmonary artery behind, the vena cava on the right side, and the pulmonary artery and left auricle on the left side.

Aneurism of the ascending aorta, in the situation of the sinuses of Valsalva, in the great majority of cases, affects the anterior sinus; this is mainly owing to the fact that the regurgitation of blood upon the sinuses takes place chiefly on the anterior aspect of the vessel. As the aneurismal sac enlarges, it may compress any or all of the structures in immediate proximity with it, but chiefly projects towards the right anterior side; and, consequently, interferes mainly with those structures that have a corresponding relation with the vessel. In the majority of cases, it bursts into the cavity of the pericardium, the patient suddenly drops down dead, and, upon a post-mortem examination, the pericardial sac is found full of blood; or it may compress the right auricle, or the pulmonary artery, and adjoining part of the right ventricle, and open into one or the other of these parts, or it may press upon the superior vena cava.

Aneurism of the ascending aorta, originating above the sinuses, most frequently implicates the right anterior wall of the vessel, where, as has been explained, there exists a normal dilatation, the great sinus of the aorta; this is probably mainly owing to the blood being impelled against this part. The direction of the aneurism is also chiefly towards the right of the median line. If it attains a large size and projects forwards, it may absorb the sternum and the cartilages of the ribs, usually on the right side, and appear as a pulsating tumour on the front of the chest, just below the manubrium; or it may burst into the pericardium, or may compress, or open into the right lung, the trachea, bronchi, or œsophagus.

Regarding the transverse aorta, the student is reminded that the vessel lies on the trachea, the œsophagus, and thoracic duct; that the recurrent laryngeal nerve winds around it; and that from its upper part are given off three large trunks, which supply the head, neck, and upper extremities. Now, an aneurismal tumour taking origin from the posterior part of the vessel, its most usual site, may press upon the trachea, impede the breathing, or produce cough, hæmoptysis, or stridulous breathing, or it may ultimately burst into that tube, producing fatal hæmorrhage. Again, its pressure on the laryngeal nerves may give rise to symptoms which so accurately resemble those of chronic laryngitis, that the operation of tracheotomy has in some cases been resorted to, from the supposition that disease existed in the larynx; or it may press upon the thoracic duct and destroy life by inanition; or it may involve the œsophagus, producing dysphagia, and has not infrequently been mistaken for œsophageal stricture; or it may burst into the œsophagus, when fatal hæmorrhage will occur. Compression or stretching of the sympathetic filaments may, in the former case, produce dilatation of the pupil; in the latter, contraction, if the conducting power is abolished, on the affected side. This has proved to be an important diagnostic sign in this disease. Again, the innominate artery, or the subclavian, or left carotid, may be so obstructed by clots as to produce a weakness, or even a disappearance, of the pulse in one or the other wrist, or in the left temporal artery; or the tumour may present itself at or above the manubrium, generally either in the median line, or to the right of the sternum, and may simulate an aneurism of one of the arteries of the neck.

Branches (figs. 472, 473).—The branches given off from the arch of the aorta are three in number: the innominate artery, the left common carotid, and the left subclavian.

Peculiarities. Position of the Branches.—The branches, instead of arising from the highest part of the arch (their usual position), may be moved more to the right, arising from the commencement of the transverse or upper part of the ascending portion; or the distance from one another at their origin may be increased or diminished, the most frequent change in this respect being the approximation of the left carotid towards the innominate artery.

The Number of the primary branches may be reduced to a single vessel, or more commonly two: the left carotid arising from the innominate artery; or (more rarely), the carotid and subclavian arteries of the left side arising from a left innominate artery. But the number may be increased to four, from the right carotid and subclavian arteries arising directly from the aorta, the innominate being absent. In most of these latter cases the right subclavian has been found to arise from the left end of the arch ; in other cases it was the second or third branch given off instead of the first. Another common form in which there are four primary branches is that in which the left vertebral artery arises from the arch of the aorta between the left carotid and subclavian arteries. Lastly, the number of trunks from the arch may be increased to five or six ; in these instances, the external and internal carotids arise separately from the arch, the common carotid being absent on one or both sides. In some cases, where six branches have been found, it has been due to a separate origin of the vertebral on both sides.

Number usual, Arrangement different.—When the aorta arches over to the right side, the three branches have an arrangement the reverse of what is usual, the innominate supplying the left side, and the carotid and subclavian (which arise separately) the right side. In other cases, where the aorta takes its usual course, the two carotids may be joined in a common trunk, and the subclavians arise separately from the arch, the right subclavian generally arising from the left end of the arch.*

In some instances other arteries are found to arise from the arch of the aorta. Of these the most common are the bronchial, one or both, and the thyroidea ima ; but the internal mammary and the inferior thyroid have been seen to arise from this vessel.

INNOMINATE ARTERY

The **innominate artery (brachio-cephalic)** is the largest branch given off from the arch of the aorta. It arises, on a level with the upper border of the second right costal cartilage, from the commencement of the arch of the aorta in front of the left carotid, and, ascending obliquely upwards, backwards, and outwards to the upper border of the right sterno-clavicular articulation, divides into the right common carotid and right subclavian arteries. This vessel varies from an inch and a half to two inches in length.

Relations.—*In front*, it is separated from the first piece of the sternum by the Sterno-hyoid and Sterno-thyroid muscles, the remains of the thymus gland, the left innominate and right inferior thyroid veins which cross its root, and sometimes the inferior cervical cardiac branch of the right pneumogastric. *Behind*, it lies upon the trachea, which it crosses obliquely. On the *right side* are the right innominate vein, the superior vena cava, the right phrenic nerve, and the pleura ; and on the *left side*, the remains of the thymus gland, the origin of the left carotid artery, the left inferior thyroid vein, and the trachea.

Branches.—The innominate usually gives off no branches ; but occasionally a small branch, the *thyroidea ima*, arises from this vessel. It also sometimes gives off a *thymic or bronchial branch*.

The **Thyroidea ima** ascends in front of the trachea to the lower part of the thyroid body, which it supplies. It varies greatly in size, and appears to compensate for deficiency or absence of one of the other thyroid vessels. It occasionally is found to arise from the aorta or the right common carotid, the subclavian or internal mammary arteries.

PLAN OF THE RELATIONS OF THE INNOMINATE ARTERY

In front.

Sternum.
Sterno-hyoid and Sterno-thyroid muscles.
Remains of thymus gland.
Left innominate and right inferior thyroid veins.
Inferior cervical cardiac branch from right pneumogastric nerve.

Right side.		*Left side.*
Right innominate vein.		Remains of thymus.
Superior vena cava.	Innominate Artery.	Left carotid.
Right phrenic nerve.		Left inferior thyroid vein.
Pleura.		Trachea.

Behind.
Trachea.

Peculiarities in point of Division.—When the bifurcation of the innominate artery varies from the point above mentioned, it sometimes ascends a considerable distance

---

* The anomalies of the aorta and its branches are minutely described by Krause in Henle's *Anatomy* (Brunswick, 1868), vol. iii. p. 203 *et seq.*

above the sternal end of the clavicle; less frequently it divides below it. In the former class of cases, its length may exceed two inches; and, in the latter, be reduced to an inch or less. These are points of considerable interest for the surgeon to remember in connection with the operation of tying this vessel.

Position.—When the aorta arches over to the right side, the innominate is directed to the left side of the neck instead of the right.

Collateral Circulation.—Allan Burns demonstrated, on the dead subject, the possibility of the establishment of the collateral circulation after ligature of the innominate artery, by tying and dividing that artery, after which, he says, ' Even coarse injection, impelled into the aorta, passed freely by the anastomosing branches into the arteries of the right arm, filling them and all the vessels of the head completely.' * The branches by which this circulation would be carried on are very numerous; thus, all the communications across the middle-line between the branches of the carotid arteries of opposite sides would be available for the supply of blood to the right side of the head and neck; while the anastomosis between the superior intercostal of the subclavian and the first aortic intercostal (see *infra* on the collateral circulation after obliteration of the thoracic aorta) would bring the blood, by a free and direct course, into the right subclavian: the numerous connections, also, between the intercostal arteries and the branches of the axillary and internal mammary arteries would, doubtless, assist in the supply of blood to the right arm, while the deep epigastric, from the external iliac, would, by means of its anastomosis with the internal mammary, compensate for any deficiency in the vascularity of the wall of the chest.

Surgical Anatomy.—Aneurism of the innominate artery not infrequently occurs as an accompaniment to aneurism of the arch of the aorta. It causes bulging of the right sterno-clavicular articulation, pushing forwards the Sterno-mastoid muscle and filling up the suprasternal notch. It produces serious pressure symptoms: from pressure on the innominate veins it may cause œdema of the upper extremities, and of the head and neck; from pressure on the trachea it produces dyspnœa; and on the right recurrent laryngeal nerve, hoarseness and laryngeal cough.

Although the operation of tying the innominate artery has been performed by several surgeons, not many successes have been recorded. The chief danger of the operation appears to be the frequency of secondary hæmorrhage; but in the present day, with the practice of aseptic surgery and our greater knowledge of the use of the ligature, more favourable results may be anticipated. Other causes of death after operation are pleurisy, pericarditis, and suppurative cellulitis. The main obstacles to the operation are, as the student will perceive from his dissection of this vessel, the deep situation of the artery behind and beneath the sternum, and the number of important structures which surround it in every part.

In order to apply a ligature to this vessel, the patient is to be placed upon his back with the thorax slightly raised, the head bent a little backwards, and the shoulder on the side of the aneurism strongly depressed, so as to draw out the artery from behind the sternum into the neck. An incision three or more inches long is then made along the anterior border of the Sterno-mastoid muscle, terminating at the sternal end of the clavicle. From this point, a second incision is carried about the same length along the upper border of the clavicle. The skin is then dissected back, and the Platysma divided on a director: the sternal end of the Sterno-mastoid is now brought into view, and a director being passed beneath it, and close to its under surface, so as to avoid any small vessels, it is to be divided; in like manner the clavicular origin is to be divided throughout the whole or greater part of its attachment. By pressing aside any loose cellular tissue or vessels that may now appear, the Sterno-hyoid and Sterno-thyroid muscles will be exposed, and must be divided, a director being previously passed beneath them. The inferior thyroid veins may come into view, and must be carefully drawn either upwards or downwards, by means of a blunt hook, or tied with double ligatures and divided. After tearing through a strong fibro-cellular lamina, the right carotid is brought into view, and being traced downwards, the arteria innominata is arrived at. The left innominate vein should now be depressed; the right innominate vein, the internal jugular vein, and the pneumogastric nerve drawn to the right side; and a curved aneurism needle may then be passed around the vessel, close to its surface, and in a direction from below upwards and inwards; care being taken to avoid the right pleural sac, the trachea, and cardiac nerves. The ligature should be applied to the artery as high as possible, in order to allow room between it and the aorta for the formation of the coagulum. The importance of avoiding the thyroid plexus of veins during the primary steps of the operation, and the pleural sac while including the vessel in the ligature, should be most carefully borne in mind. After the artery has been secured, the common carotid should be tied about half an inch above its origin, and also the thyroidea ima if the vessel is of any size. The severed muscles are united by buried sutures.

---

* *Surgical Anatomy of the Head and Neck*, p. 62.

ARTERIES OF THE HEAD AND NECK

The artery which supplies the head and neck is the Common Carotid; it ascends in the neck and divides into two branches, viz.: (1) the External Carotid, supplying the superficial parts of the head and face and the greater part of the neck; (2) the Internal Carotid, supplying to a great extent the parts within the cranial cavity.

COMMON CAROTID ARTERIES

The **common carotid arteries,** although occupying a nearly similar position in the neck, differ in position, and in their relations at their origin. The right carotid arises from the innominate artery, behind the right sterno-clavicular articulation; the left from the highest part of the arch of the aorta. The left carotid is therefore longer, and at its origin is contained within the thorax. The course and relation of that portion of the left carotid which intervenes between the arch of the aorta and the left sterno-clavicular articulation, will first be described (see fig. 472).

The *left carotid within the thorax* ascends obliquely outwards from the arch of the aorta to the root of the neck. *In front*, it is separated from the first piece of the sternum by the Sterno-hyoid and Sterno-thyroid muscles, the anterior portions of the left pleura and lung, the left innominate vein, and the remains of the thymus gland; *behind*, it lies on the trachea, œsophagus, and thoracic duct. *Internally*, it is in relation with the innominate artery, the inferior thyroid veins, and the remains of the thymus gland; *externally*, with the left pneumogastric and phrenic nerves, left pleura, and lung. The left subclavian artery is posterior and slightly external to it.

PLAN OF THE RELATIONS OF THE LEFT COMMON CAROTID.
THORACIC PORTION

In front.

Sternum.
Sterno-hyoid and Sterno-thyroid muscles.
Anterior parts of left lung and pleura.
Left innominate vein.
Remains of thymus gland.

<table>
<tr><td>Internally.</td><td rowspan="2">
Left Common Carotid. Thoracic Portion.</td><td>Externally.</td></tr>
<tr><td>Innominate artery.
Inferior thyroid veins.
Remains of thymus gland.</td><td>Left pneumogastric nerve.
Left phrenic nerve.
Left pleura and lung.
Left subclavian artery.</td></tr>
</table>

Behind.

Trachea.
Œsophagus.
Thoracic duct.
Left subclavian artery.

In the neck, the two common carotids resemble each other so closely, that one description will apply to both. Each vessel passes obliquely upwards, from behind the sterno-clavicular articulation, to a level with the upper border of the thyroid cartilage, opposite the lower border of the third cervical vertebra, where it divides into the external and internal carotids; these names being derived from the distribution of the arteries to the external parts of the head and face, and to the internal parts of the cranium and orbit respectively.

At the lower part of the neck the two common carotid arteries are separated from each other by a very small interval, which contains the trachea; but at the upper part, the thyroid body, the larynx and pharynx project forwards between the two vessels, and give them the appearance of being placed farther back in this situation. The common carotid artery is contained in a sheath, derived from the deep cervical fascia, which also encloses the internal jugular vein and pneumo-gastric nerve, the vein lying on the outer side of the artery, and the nerve between the artery and vein, on a plane posterior to both. On opening the sheath, these three structures are seen to be separated from one another, each being enclosed in a separate fibrous investment.

Relations.—At the lower part of the neck the common carotid artery is very deeply seated, being covered by the integument, superficial fascia, Platysma, and

FIG. 475.—Superficial dissection of the right side of the neck, showing the carotid and subclavian arteries.

deep cervical fascia, the Sterno-mastoid, Sterno-hyoid, and Sterno-thyroid muscles, and by the Omo-hyoid, opposite the cricoid cartilage ; but in the upper part of its course, near its termination, it is more superficial, being covered merely by the integument, the superficial fascia, Platysma, deep cervical fascia, and inner margin of the Sterno-mastoid, and, when the latter is drawn back-wards, it is seen to be contained in a triangular space, bounded behind by the

Sterno-mastoid, above by the posterior belly of the Digastric, and below by the anterior belly of the Omo-hyoid. This part of the artery is crossed obliquely, from within outwards, by the sterno-mastoid artery; it is crossed also by the superior and middle thyroid veins, which terminate in the internal jugular; and descending on its sheath in front is seen the descendens hypoglossi nerve, this filament being joined by one or two branches from the cervical nerves, which cross the vessel from without inwards. Sometimes the descendens hypoglossi is contained within the sheath. The superior thyroid vein crosses the artery near its termination; the middle thyroid a little below the level of the cricoid cartilage, and the anterior jugular vein just above the clavicle; the latter, however, is separated from the artery by the Sterno-hyoid and Sterno-thyroid muscles. *Behind*, the artery is separated from the transverse processes of the vertebræ by the Longus colli and Rectus capitis anticus major, the sympathetic cord being interposed between it and the muscles. The recurrent laryngeal nerve and inferior thyroid artery cross behind the vessel at its lower part. *Internally*, it is in relation with the trachea and thyroid gland, the latter overlapping it, the inferior thyroid artery and recurrent laryngeal nerve being interposed: higher up, with the larynx and pharynx. On its *outer side* are placed the internal jugular vein and pneumogastric nerve.

At the lower part of the neck, the internal jugular vein on the right side diverges from the artery, but on the left side it approaches it, and often overlaps its lower part. This is an important fact to bear in mind during the performance of any operation on the lower part of the left common carotid artery.

PLAN OF THE RELATIONS OF THE COMMON CAROTID ARTERY

In front.

Integument, and superficial fascia.
Deep cervical fascia.
Platysma.
Sterno-mastoid.
Sterno-hyoid.
Sterno-thyroid.

Omo-hyoid.
Descendens and Communicans hypoglossi nerves.
Sterno-mastoid artery.
Superior and middle thyroid veins.
Anterior jugular vein.

Externally.

Internal jugular vein.
Pneumogastric nerve.

Common Carotid.

Internally.

Trachea.
Thyroid gland.
Recurrent laryngeal nerve.
Inferior thyroid artery.
Larynx.
Pharynx.

Behind.

Longus colli.
Rectus capitis anticus major.
Recurrent laryngeal nerve.

Sympathetic cord.
Inferior thyroid artery.

At the angle of bifurcation of the common carotid artery on its posterior aspect is a reddish-brown oval body, known as the *carotid gland*. It is similar in structure to the coccygeal gland, which is situated on the middle sacral artery.

Peculiarities as to Origin.—The *right common carotid* may arise above or below the upper border of the sterno-clavicular articulation. This variation occurs in one out of about eight cases and a half, and the origin is more frequently below than above: or the artery may arise as a separate branch from the arch of the aorta, or in conjunction with the left carotid. The *left common carotid* varies in its origin more than the right. In the majority of abnormal cases it arises with the innominate artery, or, if the innominate artery is absent, the two carotids arise usually by a single trunk. It rarely joins with the left subclavian, except in cases of transposition of the arch.

Peculiarities as to point of Division.—The most important peculiarities of this vessel, from a surgical point of view, relate to its place of division in the neck. In the majority of abnormal cases, this occurs higher than usual, the artery dividing into two branches opposite the hyoid bone, or even higher; more rarely, it occurs below, opposite the middle of the larynx, or the lower border of the cricoid cartilage; and one case is related by Morgagni, where the common carotid, only an inch and a half in length, divided at the root of the neck. Very rarely, the common carotid ascends in the neck without any subdivision, either the external or the internal carotid being wanting;

and in a few cases the common carotid has been found to be absent, the external and internal carotids arising directly from the arch of the aorta. This peculiarity existed on both sides in some instances, on one side in others.

Occasional Branches.—The common carotid usually gives off no branch previous to its bifurcation ; but it occasionally gives origin to the superior thyroid, or its laryngeal branch, the ascending pharyngeal, the inferior thyroid, or, more rarely, the vertebral artery.

Surface Marking.—The carotid arteries are covered throughout their entire extent by the Sterno-mastoid muscle, but their course does not correspond to the anterior border of the muscle, which passes in a somewhat curved direction from the mastoid process to the sterno-clavicular joint. The course of the artery is indicated more exactly by a line drawn from the upper part of the sternal end of the clavicle below, to a point midway between the angle of the jaw and the mastoid process above. The portion of this line below the level of the upper border of the thyroid cartilage would represent the course of the vessel.

Surgical Anatomy.—*Aneurisms* are not commonly met with on the common carotid ; when they do occur they are usually situated low down at the root of the neck, or just below the point of bifurcation of the vessel. They do not frequently assume a large size, and are more commonly found on the right side. As they increase in size they displace the trachea and larynx, and therefore dyspnœa becomes a prominent symptom. Dysphagia also may be present from pressure on the œsophagus, especially if the aneurism is on the left side ; and pressure on the recurrent laryngeal nerve may produce hoarseness and laryngeal cough. It is important to bear in mind that an enlarged gland in the superior carotid triangle, receiving a transmitted pulsation from the carotid artery, may simulate aneurism of that vessel, but may be distinguished from it by the character of the pulsation, which is not distensile.

Digital compression of the common carotid is sometimes required, and is best effected by compressing the vessel with the thumb against the anterior tubercle of the transverse process of the sixth cervical vertebra (see page 162). The operation of tying the common carotid artery may be necessary in a case of wound of that vessel or its branches, in aneurism, or in a case of pulsating tumour of the orbit or skull. If the wound involves the trunk of the common carotid, it will be necessary to tie the artery above and below the wounded part. But in cases of aneurism, or where one of the branches of the common carotid is wounded in an inaccessible situation, it may be judged necessary to tie the trunk. In such cases, the whole of the artery is accessible, and any part may be tied, except close to either end. When the case is such as to allow of a choice being made, the lower part of the carotid should never be selected as the spot upon which to place a ligature, for not only is the artery in this situation placed very deeply in the neck, but it is covered by three layers of muscles, and, on the left side, the internal jugular vein, in the great majority of cases, passes obliquely in front of it. Neither should the upper end be selected, for here the superior thyroid vein and its tributaries would give rise to very considerable difficulty in the application of a ligature. The point most favourable for the operation is that part of the vessel which is at the level of the cricoid cartilage. It occasionally happens that the carotid artery bifurcates below its usual position : if the artery be exposed at its point of bifurcation, both divisons of the vessel should be tied near their origin, in preference to tying the trunk of the artery near its termination ; and if, in consequence of the entire absence of the common carotid, or from its early division, two arteries, the external and internal carotids, are met with, the ligature should be placed on that vessel which is found on compression to be connected with the disease.

In this operation, the direction of the vessel and the inner margin of the Sterno-mastoid are the chief guides to its performance. The patient should be placed on his back with the head thrown back and turned slightly to the opposite side : an incision is to be made, three inches long, in the direction of the anterior border of the Sterno-mastoid, so that the centre corresponds to the level of the cricoid cartilage : after dividing the integument, superficial fascia, and Platysma, the deep fascia must be cut through on a director, so as to avoid wounding numerous small veins that are usually found beneath. The head may now be brought forwards so as to relax the parts somewhat, and the margins of the wound held asunder by retractors. The descendens hypoglossi nerve may now be exposed, and must be avoided, and the sheath of the vessel having been raised by forceps, is to be opened to a small extent over the artery at its inner side. The internal jugular vein may present itself alternately distended and relaxed ; this should be compressed both above and below, and drawn outwards, in order to facilitate the operation. The aneurism needle is passed from the outside, care being taken to keep the needle in close contact with the artery, and thus avoid the risk of injuring the internal jugular vein, or including the vagus nerve. Before the ligature is tied, it should be ascertained that nothing but the artery is included in it.

Ligature of the Common Carotid at the Lower Part of the Neck.—This operation is sometimes required in cases of aneurism of the upper part of the carotid, especially if the sac is of large size. It is best performed by dividing the sternal origin of the Sterno-mastoid muscle, but may be done in some cases, if the aneurism is not of very large size, by an incision along the anterior border of the Sterno-mastoid, extending down to the sterno-clavicular articulation, and by then retracting the muscle. The easiest and best plan, however, is to make an incision two or three inches long down the lower part of the anterior border of the Sterno-mastoid muscle to the sterno-clavicular joint, and a second incision, starting from the termination of the first, along the upper border of the clavicle for about two inches. This incision is made through the superficial and deep fascia and the sternal origin of the muscle exposed. This is to be divided on a director and turned up, with the superficial structures, as a triangular flap. Some loose connective tissue is to be divided or torn through, and the outer border of the Sterno-hyoid muscle exposed. In doing this, care must be taken not to wound the anterior jugular vein, which crosses the muscle to reach the external jugular or subclavian vein. The Sterno-hyoid, and with it the Sterno-thyroid, are to be drawn inwards by means of a retractor, and the sheath of the vessel is exposed. This must be opened with great care on its inner or tracheal side, so as to avoid the internal jugular vein. This is especially necessary on the left side, where the artery is commonly overlapped by the vein. On the right side there is usually an interval between the artery and the vein, and not the same risk of wounding the latter.

The common carotid artery, being a long vessel without any branches, is particularly suitable for the performance of Brasdor's operation for the cure of an aneurism of the lower part of the vessel. Brasdor's procedure consists in ligaturing the artery on the distal side of the aneurism, and in the case of the common carotid there are no branches given off from the vessel between the aneurism and the site of the ligature, hence the flow of blood through the sac of the aneurism is diminished and cure takes place in the usual way, by the deposit of laminated fibrin.

Collateral Circulation.—After ligature of the common carotid, the collateral circulation can be perfectly established, by the free communication which exists between the carotid arteries of opposite sides, both without and within the cranium, and by enlargement of the branches of the subclavian artery on the side corresponding to that on which the vessel has been tied—the chief communication outside the skull taking place between the superior and inferior thyroid arteries, and the profunda cervicis and arteria princeps cervicis of the occipital; the vertebral taking the place of the internal carotid within the cranium.

EXTERNAL CAROTID ARTERY

The **external carotid artery** (fig. 475) commences opposite the upper border of the thyroid cartilage, and, taking a slightly curved course, passes upwards and forwards, and then inclines backwards to the space behind the neck of the condyle of the lower jaw, where it divides into the temporal and internal maxillary arteries. It rapidly diminishes in size in its course up the neck, owing to the number and large size of the branches given off from it. In the child, it is somewhat smaller than the internal carotid; but in the adult, the two vessels are of nearly equal size. At its commencement, this artery is more superficial, and placed nearer the middle line than the internal carotid, and is contained in the triangular space bounded by the Sterno-mastoid behind, the Omo-hyoid below, and the posterior belly of the Digastric and Stylo-hyoid above.

Relations.—The external carotid artery is covered by the skin, superficial fascia, Platysma, deep fascia, and anterior margin of the Sterno-mastoid, crossed by the hypoglossal nerve, and by the lingual and facial veins; it is afterwards crossed by the Digastric and Stylo-hyoid muscles, and higher up passes deeply into the substance of the parotid gland, where it lies beneath the facial nerve and the junction of the temporal and internal maxillary veins. *Internally* is the hyoid bone, wall of the pharynx, the superior laryngeal nerve, and the ramus of the jaw, from which it is separated by a portion of the parotid gland. *Externally*, in the lower part of its course, is the internal carotid artery. *Behind* it, near its origin, is the superior laryngeal nerve; and higher up, it is separated from the internal carotid by the Stylo-glossus and Stylo-pharyngeus muscles, the glosso-pharyngeal nerve, and part of the parotid gland.

R R

PLAN OF THE RELATIONS OF THE EXTERNAL CAROTID

In front.

Skin, superficial fascia.
Platysma and deep fascia.
Anterior border of Sterno-mastoid.
Hypoglossal nerve.
Lingual and facial veins.
Digastric and Stylo-hyoid muscles.   [substance.
Parotid gland with facial nerve and temporo-maxillary vein in its

Internally.

Hyoid bone.
Pharynx.
Superior laryngeal nerve.
Parotid gland.
Ramus of jaw.

External
Carotid.

Externally.

Internal carotid artery.

Behind.

Superior laryngeal nerve.
Stylo-glossus.
Stylo-pharyngeus.
Glosso-pharyngeal nerve.
Parotid gland.

Surface Marking.—The position of the external carotid artery may be marked out with sufficient accuracy by a line drawn from the front of the meatus of the external ear to the side of the cricoid cartilage, slightly arching the line forwards.

Surgical Anatomy.—The application of a ligature to the external carotid may be required in cases of wound of this vessel, or of its branches when these cannot be tied, and in some cases of pulsating tumours of the scalp or face. The operation has not received the attention which it deserves, owing to the fear which surgeons have entertained of secondary hæmorrhage, on account of the number of branches given off from the vessel. Cripps, however, has shown that this fear is not well founded.[*] The operation is to be preferred to ligature of the common carotid, as it does not interfere with the cerebral circulation. The seat of election for ligature is between the origin of the superior thyroid and lingual branches, about a finger's breadth below the tip of the great cornu of the hyoid bone. To tie the vessel, an incision is to be made from the angle of the jaw to the upper border of the thyroid cartilage, and the superficial tissues and the deep fascia divided. The anterior border of the Sterno-mastoid must be retracted and the lower border of the parotid gland raised, so as to expose the tendon of the Digastric muscle and the hypoglossal nerve, which cross the artery. The great difficulty in doing this is due to the plexus of veins derived from the superior thyroid and lingual veins, which overlie the artery. If necessary, these must be ligatured and divided. Care must be taken not to mistake the lingual and facial, when they arise by a common trunk, as they sometimes do, for the external carotid. When the vessel is exposed, the needle is to be passed from without inwards, carefully avoiding the superior laryngeal nerve, which lies in close proximity to the artery. The circulation is at once re-established by the free communication between most of the large branches of the artery (facial, lingual, superior thyroid, occipital) and the corresponding arteries of the opposite side, and by the anastomosis of its branches with those of the internal carotid, and of the occipital with branches of the subclavian, &c.

Branches.—The external carotid artery gives off eight branches, which, for convenience of description, may be divided into four sets. (See fig. 476.)

Anterior.	*Posterior.*	*Ascending.*	*Terminal.*
Superior Thyroid.	Occipital.	Ascending	Superficial Temporal.
Lingual.	Posterior Auricular.	Pharyngeal.	Internal Maxillary.
Facial.			

The student is here reminded that many variations are met with in the number, origin, and course of these branches in different subjects; but the above arrangement is that which is found in the great majority of cases.

The **Superior Thyroid Artery** (figs. 475 and 478) is the first branch given off from the external carotid, being derived from that vessel just below the great cornu of the hyoid bone. At its commencement it is quite superficial, being covered by the integument, fascia, and Platysma, and is contained in the

[*] *Med.-Chir. Trans.* vol. lxi. p. 229.

triangular space bounded by the Sterno-mastoid, Digastric, and Omo-hyoid muscles. After running upwards and inwards for a short distance, it curves downwards and forwards, in an arched and tortuous manner, to the upper part of the thyroid gland, passing beneath the Omo-hyoid, Sterno-hyoid, and Sterno-thyroid muscles and supplying them. It distributes numerous branches to the upper part of the gland, anastomosing with its fellow of the opposite side, and with the inferior thyroid arteries. The branches supplying the gland are generally three in number: one, the largest, supplies principally the anterior surface of the gland; it courses along the inner border of the lobe as far as the upper border of the isthmus, and then passes in the substance of the isthmus to the middle line of the neck, where it anastomoses with the corresponding artery of the opposite side: a second branch courses along the external border of the lobe and supplies this portion of the gland, and the third passes to the posterior surface, the upper part of which it supplies. Besides the arteries distributed to the muscles by which it is covered and to the substance of the gland, the branches of the superior thyroid are the following:

FIG. 476.—Plan of the branches of the external carotid.

Infrahyoid. Superior laryngeal.
Sterno-mastoid. Crico-thyroid.

The **infrahyoid** is a small branch which runs along the lower border of the hyoid bone beneath the Thyro-hyoid muscle; after supplying the muscles connected to that bone, it forms an arch, by anastomosing with the vessel of the opposite side.

The **sterno-mastoid** runs downwards and outwards across the sheath of the common carotid artery, and supplies the Sterno-mastoid and neighbouring muscles and integument. There is frequently a separate branch from the external carotid distributed to the Sterno-mastoid muscle.

The **superior laryngeal**, larger than either of the preceding, accompanies the internal laryngeal nerve, beneath the Thyro-hyoid muscle; it pierces the thyro-hyoid membrane, and supplies the muscles, mucous membrane, and glands of the larynx, anastomosing with the branch from the opposite side.

The **crico-thyroid** is a small branch which runs transversely across the crico-thyroid membrane, communicating with the artery of the opposite side.

Surgical Anatomy.—The superior thyroid, or one of its branches, is often divided in cases of cut throat, giving rise to considerable hæmorrhage. In such cases, the artery should be secured, the wound being enlarged for that purpose, if necessary. The operation may be easily performed, the position of the artery being very superficial, and the only structures of importance covering it being a few small veins. The operation of tying the superior thyroid artery, in bronchocele, has been performed in numerous instances with partial or temporary success. When, however, the collateral circulation between this vessel and the artery of the opposite side, and the inferior thyroid, is completely re-established, the tumour usually regains its former size, and hence the operation has been given up, especially as better results are obtained by other means. Both thyroid arteries on the same side, and indeed all the four thyroid arteries, have been tied in enlarged thyroid.

The position of the sterno-mastoid branch is of importance in connection with the operation of ligature of the common carotid artery. It crosses and lies on the sheath of this vessel and may chance to be wounded in opening the sheath. The position of the crico-thyroid branch should be remembered, as it may prove the source of troublesome hæmorrhage during the operation of laryngotomy.

The **Lingual Artery** (figs. 478, 481) arises from the external carotid between the superior thyroid and facial; it first runs obliquely upwards and inwards to the great cornu of the hyoid bone; it then curves downwards and forwards, forming a loop which is crossed by the hypoglossal nerve, and passing beneath the Digastric and Stylo-hyoid muscles, it runs horizontally forwards, beneath the Hyo-glossus, and finally, ascending almost perpendicularly to the tongue, turns forwards on its lower surface as far as the tip, under the name of the *ranine artery*.

Relations.—Its first, or oblique, portion is superficial, being contained in the same triangular space as the superior thyroid artery, resting upon the Middle constrictor of the pharynx, and covered by the Platysma, and fascia of the neck. Its second, or curved, portion also lies upon the Middle constrictor, being covered at first by the tendon of the Digastric and the Stylo-hyoid muscle, and afterwards by the Hyo-glossus, the latter muscle separating it from the hypoglossal nerve. Its third, or horizontal, portion lies between the Hyo-glossus and Genio-hyo-glossus muscles. The fourth, or terminal, part, under the name of the *ranine*, runs along the under surface of the tongue to its tip : here it is very superficial, being covered only by the mucous membrane ; above it is the Lingualis inferior, and on the outer side the Genio-hyo-glossus. The hypoglossal nerve crosses the lingual artery, and then becomes separated from it, in the second part of its course, by the Hyo-glossus muscle.

The branches of the lingual artery are, the

Suprahyoid.	Sublingual.
Dorsalis Linguæ.	Ranine.

The **suprahyoid** runs along the upper border of the hyoid bone, supplying the muscles attached to it and anastomosing with its fellow of the opposite side.

The **dorsalis linguæ** (fig. 481) arises from the lingual artery beneath the Hyo-glossus muscle (which, in the figure, has been partly cut away, to show the vessel) ; it ascends to the back part of the dorsum of the tongue, and supplies the mucous membrane in this situation, the tonsil, soft palate, and epiglottis ; anastomosing with its fellow from the opposite side. This artery is frequently represented by two or three small branches.

The **sublingual** arises at the anterior margin of the Hyo-glossus muscle, and runs forward between the Genio-hyo-glossus and Mylo-hyoid to the sublingual gland. It supplies the substance of the gland, giving branches to the Mylo-hyoid and neighbouring muscles, the mucous membrane of the mouth and gums. One branch runs behind the alveolar process of the lower jaw in the substance of the gum to anastomose with a similar artery from the other side.

The **ranine** is the terminal portion of the lingual artery ; it pursues a tortuous course and runs along the under surface of the tongue, below the Inferior lingualis, and above the mucous membrane ; it lies on the outer side of the Genio-hyo-glossus, accompanied by the lingual nerve. On arriving at the tip of the tongue, it has been said to anastomose with the artery of the opposite side ; but this is denied by Hyrtl. In the mouth, these vessels are placed one on each side of the frænum.

Surgical Anatomy.—The lingual artery may be divided near its origin in cases of cut throat, a complication that not infrequently happens in this class of wounds ; or severe hæmorrhage, which cannot be restrained by ordinary means, may ensue from a wound, or deep ulcer, of the tongue. In the former case, the primary wound may be enlarged if necessary, and the bleeding vessel secured. In the latter case, it has been suggested that the lingual artery should be tied near its origin. Ligature of the lingual artery has been also occasionally practised, as a palliative measure, in cases of cancer of the tongue, in order to check the progress of the disease by starving the growth, and it is sometimes tied, as a preliminary measure to removal of the tongue. The operation is a difficult one, on account of the depth of the artery, the number of important parts by which it is surrounded, the loose and yielding nature of the parts upon which it is supported, and its occasional irregularity of origin. An incision is to be made in a curved direction from a finger's breadth external to the symphysis of the jaw downwards to the cornu of the hyoid bone, and then upwards to near the angle of the jaw. Care must be taken not to carry this incision too far backwards, for fear of endangering the facial vein. In the first incision the skin, superficial fascia, and Platysma will be divided, and the deep fascia exposed. This is then to be incised and the submaxillary gland exposed and pulled upwards by retractors. A triangular space is now seen, bounded internally by the posterior border of the Mylo-hyoid muscle ; below and externally, by the tendon of the Digastric ; and above, by the hypoglossal nerve. The floor of the space is formed by the Hyo-glossus muscle, beneath which the artery lies. The parts are to be drawn forwards by a blunt hook inserted beneath the tendon of the Digastric muscle, and the fibres of the Hyo-glossus cut through horizontally just above the Digastric. The vessel will then be exposed ; and in passing the aneurism needle, care must be taken not to open the pharynx. The hypoglossal nerve must also be avoided.

Troublesome hæmorrhage may occur in the division of the frænum in children, if the ranine artery, which lies on each side of it, is wounded. The student should remember that the operation is always to be performed with a pair of blunt-pointed scissors, and the mucous membrane only is to be divided by a very superficial cut, which cannot endanger any vessel. The scissors, also, should be directed away from the tongue. Any further liberation of the tongue which may be necessary can be effected by tearing.

The **Facial Artery (external maxillary)** (fig. 477) arises a little above the lingual, and passes obliquely upwards, beneath the Digastric and Stylo-hyoid muscles, and frequently beneath the hypoglossal nerve; it now runs forwards under cover of the body of the lower jaw, lodged in a groove on the posterior surface of the submaxillary gland; this may be called the cervical part of the artery. It then curves upwards over the body of the jaw at the anterior inferior angle of the Masseter muscle; passes forwards and upwards across the cheek to the angle of the mouth, then upwards along the side of the nose, and terminates at the inner canthus of the eye, under the name of the *angular artery*. This vessel, both in the neck and on the face, is remarkably tortuous: in the former situation, to accommodate itself to the movements of the pharynx in deglutition; and in the latter, to the movements of the jaw, and the lips and cheeks.

Relations.—*In the neck*, its origin is superficial, being covered by the integument, Platysma, and fascia; it then passes beneath the Digastric and Stylo-hyoid muscles, and part of the submaxillary gland. It lies upon the middle constrictor of the pharynx, and is separated from the Stylo-glossus and Hyo-glossus muscles by a portion of the submaxillary gland. *On the face*, where it passes over the body of the lower jaw, it is comparatively superficial, lying immediately beneath the Platysma. In this situation its pulsation may be distinctly felt, and compression of the vessel against the bone can be effectually made. In its course over the face, it is covered by the integument, the fat of the cheek, and, near the angle of the mouth, by the Platysma, Risorius, and Zygomatici muscles. It rests on the Buccinator, the Levator anguli oris, and the Levator labii superioris (sometimes piercing or else passing under this last muscle). The facial vein lies to the outer side of the artery, and takes a more direct course across the face, where it is separated from the artery by a considerable interval. In the neck it lies superficial to the artery. The branches of the facial nerve cross the artery, and the infra-orbital nerve lies beneath it.

The branches of this vessel may be divided into two sets: those given off below the jaw (cervical), and those on the face (facial).

Cervical Branches.	*Facial Branches.*
Inferior or Ascending palatine.	Inferior labial.
Tonsillar.	Inferior coronary.
Submaxillary.	Superior coronary.
Submental.	Lateral nasal.
Muscular.	Angular.

The **inferior** or **ascending palatine** (fig. 481) passes up between the Stylo-glossus and Stylo-pharyngeus to the outer side of the pharynx, along which it is continued between the Superior constrictor and the Internal pterygoid to near the base of the skull. It divides, near the Levator palati, into two branches: one follows the course of the Levator palati, and, winding over the upper border of the Superior constrictor, supplies the soft palate and the palatine glands, anastomosing with its fellow of the opposite side and with the posterior palatine branch of the internal maxillary artery; the other pierces the Superior constrictor and supplies the tonsil and Eustachian tube, anastomosing with the tonsillar and ascending pharyngeal arteries.

The **tonsillar branch** (fig. 481) passes up between the Internal pterygoid and Stylo-glossus, and then ascends along the side of the pharynx, perforating the Superior constrictor, to ramify in the substance of the tonsil and root of the tongue.

The **submaxillary** or **glandular branches** consist of three or four large vessels, which supply the submaxillary gland, some being prolonged to the neighbouring muscles, lymphatic glands, and integument.

The **submental**, the largest of the cervical branches, is given off from the facial artery just as that vessel quits the submaxillary gland: it runs forwards upon the Mylo-hyoid muscle, just below the body of the jaw, and beneath the Digastric;

after supplying the surrounding muscles, and anastomosing with the sublingual artery by branches which perforate the Mylo-hyoid muscle, it arrives at the symphysis of the chin, where it turns over the border of the jaw and divides into a superficial and a deep branch ; the former passes between the integument and Depressor labii inferioris, supplies both, and anastomoses with the inferior labial artery. The deep branch passes between the latter muscle and the bone, supplies the lip, and anastomoses with the inferior labial and mental arteries.

The **muscular branches** are distributed to the Internal pterygoid and Stylo-hyoid in the neck, and to the Masseter and Buccinator on the face.

The **inferior labial** passes beneath the Depressor anguli oris, to supply the muscles and integument of the lower lip, anastomosing with the inferior coronary and submental branches of the facial, and with the mental branch of the inferior dental artery.

FIG. 477.—The arteries of the face and scalp.*

The **inferior coronary** is derived from the facial artery, near the angle of the mouth ; it passes upwards and inwards beneath the Depressor anguli oris, and, penetrating the Orbicularis oris muscle, runs in a tortuous course along the edge of the lower lip between this muscle and the mucous membrane, inosculating with the artery of the opposite side. This artery supplies the labial glands, the mucous membrane, and muscles of the lower lip ; and anastomoses with the inferior labial and the mental branch of the inferior dental artery.

The **superior coronary** is larger and more tortuous than the preceding. It follows the same course along the edge of the upper lip, lying between the mucous membrane and the Orbicularis oris, and anastomoses with the artery of the opposite side. It supplies the textures of the upper lip, and gives

* The muscular tissue of the lips must be supposed to have been cut away, in order to show the course of the coronary arteries.

off in its course two or three vessels which ascend to the nose. One, named the *inferior artery of the septum*, ramifies on the septum of the nares as far as the point of the nose; another, the *artery of the ala*, supplies the ala of the nose.

The **lateral nasal** is derived from the facial, as that vessel ascends along the side of the nose; it supplies the ala and dorsum of the nose, anastomosing with its fellow, the nasal branch of the ophthalmic, the inferior artery of the septum, the artery of the ala, and the infra-orbital.

The **angular artery** is the termination of the trunk of the facial; it ascends to the inner angle of the orbit, embedded in the fibres of the Levator labii superioris alæque nasi, and accompanied by a large vein, the *angular*; it distributes some branches on the cheek which anastomose with the infra-orbital, and, after supplying the lachrymal sac and Orbicularis palpebrarum muscle, terminates by anastomosing with the nasal branch of the ophthalmic artery.

The anastomoses of the facial artery are very numerous, not only with the vessel of the opposite side, but, in the neck, with the sublingual branch of the lingual; with the ascending pharyngeal; with the posterior palatine, a branch of the internal maxillary, by its inferior or ascending palatine and tonsillar branches; on the face, with the mental branch of the inferior dental as it emerges from the mental foramen; with the transverse facial, a branch of the temporal; with the infra-orbital, a branch of the internal maxillary; and with the nasal branch of the ophthalmic.

Peculiarities.—The facial artery not infrequently arises by a common trunk with the lingual. This vessel is also subject to some variations in its size, and in the extent to which it supplies the face. It occasionally terminates as the submental, and not infrequently extends only as high as the angle of the mouth or nose. The deficiency is then compensated for by enlargement of one of the neighbouring arteries.

Surgical Anatomy.—The passage of the facial artery over the body of the jaw would appear to afford a favourable position for the application of pressure in cases of hæmorrhage from the lips, the result either of an accidental wound or during an operation; but its application is useless, except for a very short time, on account of the free communication of this vessel with its fellow, and with numerous branches from different sources. In a wound involving the lip, it is better to seize the part between the fingers, and evert it, when the bleeding vessel may be at once secured with pressure-forceps. In order to prevent hæmorrhage in cases of removal of diseased growths from the part, the lip should be compressed on each side between the fingers and thumb, or by a pair of specially devised clamp-forceps, while the surgeon excises the diseased part. In order to stop hæmorrhage where the lip has been divided in an operation, it is necessary, in uniting the edges of the wound, to pass the sutures through the cut edges, almost as deep as its mucous surface; by these means, not only are the cut surfaces more neatly and securely adapted to each other, but the possibility of hæmorrhage is prevented by including in the suture the divided artery. If the suture is, on the contrary, passed through merely the cutaneous portion of the wound, hæmorrhage occurs into the cavity of the mouth. The student should, lastly, observe the relation of the angular artery to the lachrymal sac, and it will be seen that, as the vessel passes up along the inner margin of the orbit, it ascends on its nasal side. In operating for fistula lachrymalis, the sac should always be opened on its outer side, in order that this vessel may be avoided.

The **Occipital Artery** (figs. 477, 478) arises from the posterior part of the external carotid, opposite the facial, near the lower margin of the Digastric muscle. At its origin, it is covered by the posterior belly of the Digastric and Stylo-hyoid muscles, and the hypoglossal nerve winds around it from behind forwards; higher up, it passes across the internal carotid artery, the internal jugular vein, and the pneumogastric and spinal accessory nerves; it next ascends to the interval between the transverse process of the atlas and the mastoid process of the temporal bone, and passes horizontally backwards, grooving the surface of the latter bone, being covered by the Sterno-mastoid, Splenius, Trachelo-mastoid, and Digastric muscles, and resting upon the Rectus lateralis, the Superior oblique, and Complexus muscles; it then changes its course and passes vertically upwards, pierces the fascia which connects the cranial attachment of the Trapezius with the Sterno-mastoid, and ascends in a tortuous course over the occiput, as high as the vertex, where it divides into numerous branches. It is accompanied in the latter part of its course by the great occipital nerve.

The branches given off from this vessel are:

Muscular.	Sterno-mastoid.	Auricular.
Meningeal.	Arteria princeps cervicis.	

The **muscular branches** supply the Digastric, Stylo-hyoid, Splenius, and Trachelo-mastoid muscles.

The **sterno-mastoid** is a large and constant branch, generally arising from the artery close to its commencement, but sometimes springing directly from the external carotid. It passes downwards and backwards over the hypoglossal nerve, and enters the substance of the muscle, in company with the spinal accessory nerve.

The **auricular branch** supplies the back part of the concha. It frequently gives off a branch, which enters the skull through the mastoid foramen and

FIG. 478.—The occipital artery and its relations.
(From a dissection by Gerald S. Hughes.)

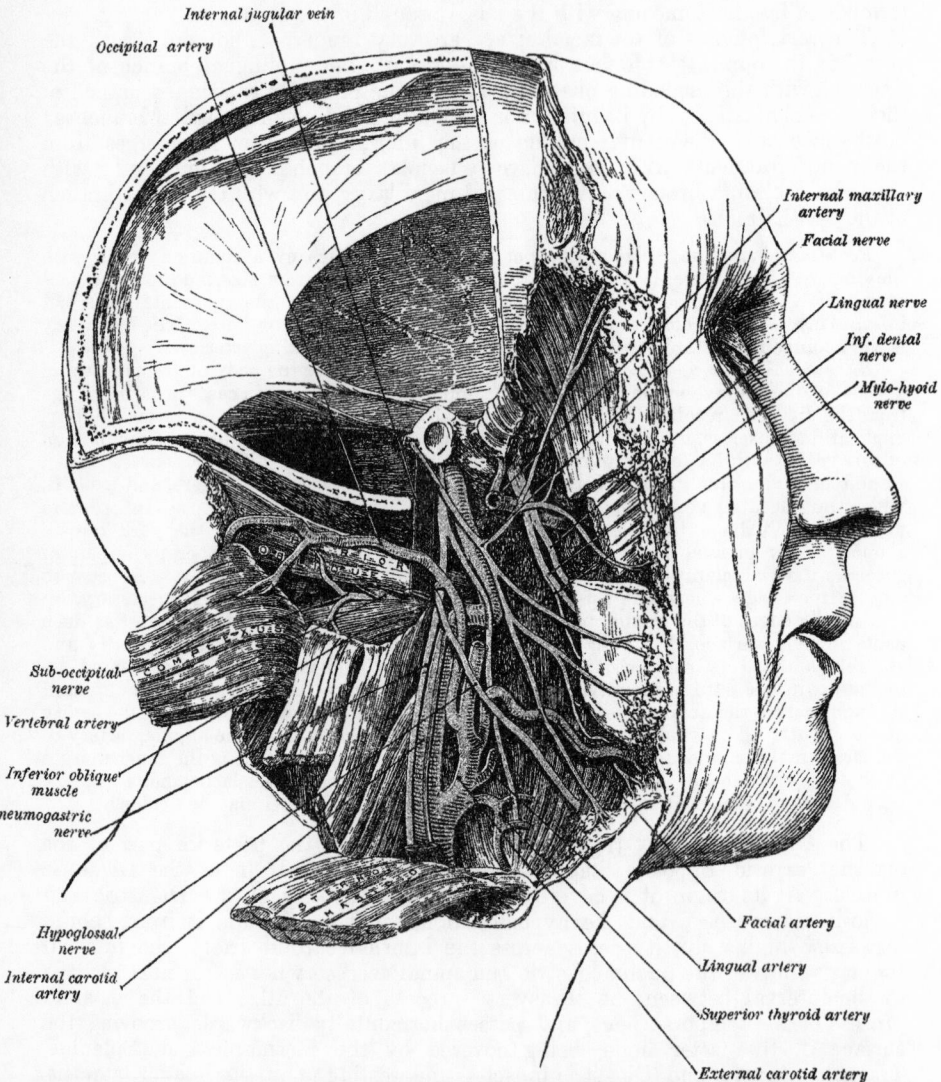

supplies the dura mater, the diploë, and the mastoid cells. This branch is sometimes given off directly from the occipital artery, and is then known as the *mastoid* branch.

The **meningeal branch** ascends with the internal jugular vein, and enters the skull through the foramen lacerum posterius, to supply the dura mater in the posterior fossa.

The **arteria princeps cervicis** (fig. 481), the largest branch of the occipital, descends along the back part of the neck, and divides into a superficial and deep portion. The former runs beneath the Splenius, giving off branches which perforate that muscle to supply the Trapezius, anastomosing with the superficial cervical artery, a branch of the Transversalis colli : the latter passes beneath the Complexus, between it and the Semispinalis colli, and anastomoses with branches from the vertebral and with the deep cervical artery, a branch of the superior intercostal. The anastomosis between these vessels assists in establishing the collateral circulation after ligature of the common carotid or subclavian artery.

The cranial branches of the occipital artery are distributed upon the occiput : they are very tortuous, and lie between the integument and Occipito-frontalis, anastomosing with the artery of the opposite side and with the posterior auricular and temporal arteries. They supply the back part of the Occipito-frontalis muscle, the integument, and pericranium.

The **Posterior Auricular Artery** (fig. 477) is a small vessel which arises from the external carotid, above the Digastric and Stylo-hyoid muscles, opposite the apex of the styloid process. It ascends, under cover of the parotid gland, on the styloid process of the temporal bone, to the groove between the cartilage of the ear and the mastoid process, immediately above which it divides into its two terminal branches, the auricular and mastoid.

Besides several small branches to the Digastric, Stylo-hyoid, and Sterno-mastoid muscles, and to the parotid gland, this vessel gives off three branches :

| | Stylo-mastoid. | Auricular. | Mastoid. |

The **stylo-mastoid branch** enters the stylo-mastoid foramen, and supplies the tympanum, mastoid cells, and semicircular canals. In the young subject a branch from this vessel forms, with the tympanic branch from the internal maxillary, a vascular circle, which surrounds the membrana tympani, and from which delicate vessels ramify on that membrane. It anastomoses with the petrosal branch of the middle meningeal artery by a twig, which enters the hiatus Fallopii.

The **auricular branch,** one of the terminal branches, ascends behind the ear, beneath the Retrahens auriculam muscle, and is distributed to the back part of the cartilage of the ear, upon which it ramifies minutely, some branches curving round the margin of the fibro-cartilage, others perforating it, to supply its anterior surface. It anastomoses with the posterior branch of the superficial temporal and also with its anterior auricular branches.

The **mastoid branch** passes backwards, over the Sterno-mastoid muscle, to the scalp above and behind the ear. It supplies the posterior belly of the Occipito-frontalis muscle and the scalp in this situation. It anastomoses with the occipital artery.

The **Ascending Pharyngeal Artery** (fig. 481), the smallest branch of the external carotid, is a long, slender vessel, deeply seated in the neck, beneath the other branches of the external carotid and the Stylo-pharyngeus muscle. It arises from the back part of the external carotid, near the commencement of that vessel, and ascends vertically between the internal carotid and the side of the pharynx, to the under surface of the base of the skull, lying on the Rectus capitis anticus major. It terminates by dividing into branches which supply the pharynx and soft palate. Its branches may be subdivided into five sets :

Pharyngeal.	Prevertebral.
Palatine.	Tympanic.
Meningeal.	

The **pharyngeal branches** are three or four in number. Two of these descend to supply the middle and inferior Constrictors and the Stylo-pharyngeus, ramifying in their substance and in the mucous membrane lining them.

The **palatine branch** is a vessel of variable size, which sometimes takes the place of the ascending palatine branch of the facial artery, when that vessel is of small size. It passes inwards, running upon the Superior constrictor, and sends ramifications to the soft palate and tonsil. A twig from this branch supplies the Eustachian tube.

The **prevertebral branches** are numerous small vessels, which supply the Recti capitis antici and Longus colli muscles, the sympathetic, hypoglossal, and pneumogastric nerves, and the lymphatic glands, anastomosing with the ascending cervical artery.

The **tympanic branch** is a small artery which passes through a minute foramen in the petrous portion of the temporal bone, in company with the tympanic branch of the glosso-pharyngeal nerve to supply the inner wall of the tympanum and anastomose with the other tympanic arteries.

The **meningeal branches** consist of several small vessels, which pass through foramina in the base of the skull, to supply the dura mater. One, the *posterior meningeal*, enters the cranium through the foramen lacerum posterius; a second passes through the foramen lacerum medium; and occasionally a third through the anterior condyloid foramen. They are all distributed to the dura mater.

Surgical Anatomy.—The ascending pharyngeal artery has been wounded from the throat; as in the case in which the stem of a tobacco-pipe was driven into the vessel, causing fatal hæmorrhage.

The **Superficial Temporal Artery** (fig. 477), the smaller of the two terminal branches of the external carotid, appears, from its direction, to be the continuation of that vessel. It commences in the substance of the parotid gland, behind the neck of the lower jaw, crosses over the posterior root of the zygoma, passes beneath the Attrahens auriculam muscle, lying on the temporal fascia, and divides, about two inches above the zygomatic arch, into two branches, an anterior and a posterior.

The **anterior temporal** runs tortuously upwards and forwards to the forehead, supplying the muscles, integument, and pericranium in this region, and anastomoses with the supra-orbital and frontal arteries.

The **posterior temporal**, larger than the anterior, curves upwards and backwards along the side of the head, lying superficial to the temporal fascia, and inosculates with its fellow of the opposite side, and with the posterior auricular and occipital arteries.

The superficial temporal artery, as it crosses the zygoma, is covered by the Attrahens auriculam muscle, and by a dense fascia given off from the parotid gland: it is crossed by the temporo-facial division of the facial nerve and one or two veins, and is accompanied by the auriculo-temporal nerve, which lies behind it. Besides some twigs to the parotid gland, the articulation of the jaw, and the Masseter muscle, its branches are, the

Transverse facial.   Middle temporal.
Anterior auricular.

The **transverse facial** is given off from the temporal before that vessel quits the parotid gland; running forwards through its substance, it passes transversely across the face, between Stenson's duct and the lower border of the zygoma, and divides on the side of the face into numerous branches, which supply the parotid gland, the Masseter muscle, and the integument, anastomosing with the facial, masseteric, and infra-orbital arteries. This vessel rests on the Masseter, and is accompanied by one or two branches of the facial nerve. It is sometimes a branch of the external carotid.

The **middle temporal artery** arises immediately above the zygomatic arch, and, perforating the temporal fascia, gives branches to the Temporal muscle, anastomosing with the deep temporal branches of the internal maxillary. It occasionally gives off an **orbital** branch, which runs along the upper border of the zygoma, between the two layers of the temporal fascia, to the outer angle of the orbit. This branch, which may arise directly from the superficial temporal artery, supplies the Orbicularis palpebrarum, and anastomoses with the lachrymal and palpebral branches of the ophthalmic artery.

The **anterior auricular branches** are distributed to the anterior portion of the pinna, the lobule, and part of the external meatus, anastomosing with branches of the posterior auricular.

Surgical Anatomy.—The temporal artery, as it crosses the zygoma, lies immediately beneath the skin, and its pulsations may be readily felt during the administration of an anæsthetic, or under circumstances where the radial pulse is not available; or it may be easily compressed against the bone in order to check bleeding from the temporal region of the scalp. When flaps are raised from this part of the head, as in the operation of trephining, the incisions should be shaped like a horse-shoe, with its convexity upwards, so that it shall contain the temporal artery and ensure a sufficient supply of blood. The same principle is applied, as far as possible, in making incisions to raise flaps in other parts of the scalp. Formerly the operation of arteriotomy was performed upon this vessel in cases of inflammation of the eye or brain, but now the operation is probably never performed. If the student will consider the relations of the trunk of the vessel, as it crosses the zygomatic arch, with the surrounding structures, he will observe that it is covered by a thick and dense fascia, crossed by one of the main divisions of the facial nerve and one or two veins, and accompanied by the auriculo-temporal nerve. Bleeding should not be performed in this situation, as much difficulty may arise from the dense fascia over the vessel preventing a free flow of blood, and considerable pressure is requisite afterwards to arrest the hæmorrhage. Again, a varicose aneurism may be formed by the accidental opening of one of the veins in front of the artery; or severe neuralgic pain may arise from the operation implicating one of the nervous filaments in the neighbourhood. The anterior branch, on the contrary, is subcutaneous, is a large vessel, and is readily compressed; it is consequently more suitable for the operation.

The **Internal Maxillary Artery** (fig. 479), the larger of the two terminal branches of the external carotid, arises from that vessel behind the neck of the

FIG. 479.—The internal maxillary artery, and its branches.

FIG. 480.—Plan of the branches.

lower jaw, and is at first embedded in the substance of the parotid gland; it passes inwards between the ramus of the jaw and the internal lateral ligament, and then upon the outer surface of the External pterygoid muscle to the sphenomaxillary fossa to supply the deep structures of the face. For convenience of description, it is divided into three portions: a maxillary, a pterygoid, and a spheno-maxillary.

In the first part of its course (*maxillary portion*), the artery passes horizontally forwards and inwards, between the ramus of the jaw and the internal lateral ligament. The artery here lies parallel to and a little below the auriculotemporal nerve; it crosses the inferior dental nerve, and lies along the lower border of the External pterygoid muscle.

In the second part of its course (*pterygoid portion*), it runs obliquely forwards and upwards upon the outer surface of the External pterygoid muscle, being covered by the ramus of the lower jaw, and lower part of the Temporal muscle; or it may pass on the inner surface of the External pterygoid muscle to reach the interval between its two heads, between which it passes to reach the sphenomaxillary fossa.

In the third part of its course (*spheno-maxillary portion*), it approaches the superior maxillary bone, and enters the spheno-maxillary fossa in the interval between the two heads of the External pterygoid, where it lies in relation with Meckel's ganglion, and gives off its terminal branches.

The branches of this vessel may be divided into three groups, corresponding with its three divisions.

Branches of the First or Maxillary Portion

Tympanic (anterior)	Middle meningeal.
Deep auricular.	Small meningeal.
Inferior dental.	

The **tympanic branch** passes upwards behind the articulation of the lower jaw, enters the tympanum through the Glaserian fissure, and ramifies upon the membrana tympani, forming a vascular circle around the membrane with the stylo-mastoid branch of the posterior auricular artery, and anastomosing with the Vidian and the tympanic branch from the internal carotid.

The **deep auricular branch** often arises in common with the preceding. It passes upwards in the substance of the parotid gland, behind the temporomaxillary articulation, pierces the cartilaginous or bony wall of the external auditory meatus, and supplies its cuticular lining and the outer surface of the membrana tympani. In its course it supplies a branch to the temporomandibular joint.

The **middle meningeal** is the largest of the branches which supply the dura mater. It arises from the internal maxillary, ascends between the internal lateral ligament and the External pterygoid muscle, and passes between the two roots of the auriculo-temporal nerve to the foramen spinosum of the sphenoid bone, through which it enters the cranium, and passing upwards and forwards in a groove on the greater wing of the sphenoid bone, it divides into two branches, anterior and posterior. The *anterior branch*, the larger, crosses the great ala of the sphenoid, and reaches the groove, or canal, in the anterior inferior angle of the parietal bone: it then divides into branches which spread out between the dura mater and internal surface of the cranium, some passing upwards over the parietal bone as far as the vertex, and others backwards to the occipital bone. The *posterior branch* crosses the squamous portion of the temporal, and on the inner surface of the parietal bone divides into branches which supply the posterior part of the dura mater and cranium. The branches of this vessel are distributed partly to the dura mater, but chiefly to the bones; they anastomose with the arteries of the opposite side, and with the anterior and posterior meningeal.

The middle meningeal on entering the cranium gives off the following branches: 1. Numerous small vessels to the Gasserian ganglion, and to the dura mater in this situation. 2. A *petrosal* branch, which enters the hiatus Fallopii, supplies the facial nerve, and anastomoses with the stylo-mastoid branch of the posterior auricular artery. 3. A minute *tympanic* branch, which runs in the canal for the Tensor tympani muscle, and supplies this muscle and the lining membrane of the canal. 4. *Orbital* branches, which pass through the sphenoidal fissure, or through separate canals in the great wing of the sphenoid to anastomose with the lachrymal or other branches of the ophthalmic artery. 5. *Temporal* or *anastomotic* branches, which pass through foramina in the great wing of the sphenoid, and anastomose in the temporal fossa with the deep temporal arteries.

Surgical Anatomy.—The middle meningeal is an artery of considerable surgical importance, as it may be torn in fractures of the temporal region of the skull, or, indeed, by injuries causing separation of the dura mater from the bone, without fracture, and the injury may be followed by considerable hæmorrhage between the bone and dura mater, which may produce compression of the brain, and require the operation of trephining for its relief. As the compression implicates the motor region of the cortex, paralysis on the opposite side of the body forms the prominent symptom of the lesion. The anterior branch of this artery crosses the anterior inferior angle of the parietal bone at a point 1½ inch behind the external angular process of the frontal bone, and 1¾ inch above the zygoma. From this point it passes upwards and slightly backwards to the sagittal suture, lying about ½ inch to ¾ inch behind the coronal suture. The posterior branch passes backwards over the squamous portion of the temporal bone. In order to expose the anterior branch of the artery, a point is taken 1½ inch above the zygoma and the same distance behind the external angular process of the frontal bone. This is the point where the pin of the trephine is to be applied. A horseshoe-shaped flap of skin, measuring three inches in length and transversely, is now to be made, with its base just above the zygoma, consisting of all the structures of the scalp down to the pericranium. This flap is reflected, and a crucial incision is made in the pericranium, the point of meeting of the four arms being at the spot indicated above. The pericranium is turned back and an inch trephine applied. After the crown of bone has been removed, the blood-clot is exposed, and must be gently got rid of with the scoop of a director.

The **small meningeal** is sometimes derived from the preceding. It enters the skull through the foramen ovale, and supplies the Gasserian ganglion and dura mater.

The **inferior dental** descends with the inferior dental nerve to the foramen on the inner side of the ramus of the jaw. It runs along the dental canal in the substance of the bone, accompanied by the nerve, and opposite the first bicuspid tooth divides into two branches, incisor and mental; the *former* is continued forwards beneath the incisor teeth as far as the symphysis, where it anastomoses with the artery of the opposite side; the *mental* branch escapes with the nerve at the mental foramen, supplies the structures composing the chin, and anastomoses with the submental, inferior labial, and inferior coronary arteries. Near its origin the inferior dental artery gives off a *lingual* branch, which descends with the lingual (gustatory) nerve and supplies the mucous membrane of the mouth. As the inferior dental artery enters the foramen, it gives off a *mylo-hyoid* branch, which runs in the mylo-hyoid groove, and ramifies on the under surface of the Mylo-hyoid muscle. The inferior dental artery and its incisor branch during their course through the substance of the bone give off a few twigs which are lost in the cancellous tissue, and a series of branches which correspond in number to the roots of the teeth: these enter the minute apertures at the extremities of the fangs, and supply the pulp of the teeth.

BRANCHES OF THE SECOND OR PTERYGOID PORTION

Deep temporal.	Masseteric.
Pterygoid.	Buccal.

These branches are distributed, as their names imply, to the muscles in the maxillary region.

The **deep temporal branches** are two in number, anterior and posterior, and each occupies that part of the temporal fossa indicated by its name. Ascending between the Temporal muscle and pericranium, they supply that muscle, and anastomose with the middle temporal artery; the anterior branch communicating with the lachrymal through small branches which perforate the malar bone and great wing of the sphenoid.

The **pterygoid branches,** irregular in their number and origin, supply the Pterygoid muscles.

The **masseteric** is a small branch which passes outwards, above the sigmoid notch of the lower jaw, to the deep surface of the Masseter. It supplies that muscle, and anastomoses with the masseteric branches of the facial and with the transverse facial artery.

The **buccal** is a small branch which runs obliquely forwards, between the Internal pterygoid and the ramus of the jaw, to the outer surface of the Buccinator, to which it is distributed, anastomosing with branches of the facial artery.

Branches of the Third or Spheno-maxillary Portion

Alveolar.	Vidian.
Infra-orbital.	Pterygo-palatine.
Posterior or Descending palatine.	Naso- or Spheno-palatine.

The **alveolar** or **posterior dental branch** is given off from the internal maxillary, frequently by a common branch with the infra-orbital, just as the trunk of the vessel is passing into the spheno-maxillary fossa. Descending upon the tuberosity of the superior maxillary bone, it divides into numerous branches, some of which enter the posterior dental canals, to supply the molar and bicuspid teeth and the lining of the antrum, while others are continued forwards on the alveolar process to supply the gums.

The **infra-orbital** appears, from its direction, to be the continuation of the trunk of the internal maxillary. It often arises from that vessel by a common trunk with the preceding branch, and runs along the infra-orbital canal with the superior maxillary nerve, emerging upon the face at the infra-orbital foramen, beneath the Levator labii superioris. While contained in the canal, it gives off branches which ascend into the orbit, and assist in supplying the Inferior rectus and Inferior oblique muscles and the lachrymal gland. Other branches (*anterior dental*) descend through the anterior dental canals in the bone, to supply the mucous membrane of the antrum, and the front teeth of the upper jaw. On the face, some branches pass upwards to the inner angle of the orbit and the lachrymal sac, anastomosing with the angular branch of the facial artery; other branches pass inwards towards the nose, anastomosing with the nasal branch of the ophthalmic; and other branches descend beneath the Levator labii superioris, and anastomose with the transverse facial and buccal arteries.

The four remaining branches arise from that portion of the internal maxillary which is contained in the spheno-maxillary fossa.

The **descending palatine** descends through the posterior palatine canal with the anterior palatine branch of Meckel's ganglion, and, emerging from the posterior palatine foramen, runs forwards in a groove on the inner side of the alveolar border of the hard palate to the anterior palatine canal, where the terminal branch of the artery passes upwards through the foramen of Stenson to anastomose with the naso-palatine artery. Its branches are distributed to the gums, the mucous membrane of the hard palate, and the palatine glands. While it is contained in the palatine canal, it gives off branches, which descend in the accessory palatine canals to supply the soft palate and tonsil, anastomosing with the ascending palatine artery.

Surgical Anatomy.—The position of the descending palatine artery on the hard palate should be borne in mind in performing an operation for the closure of a cleft in the hard palate, as it is in danger of being wounded, and may give rise to formidable hæmorrhage. In one case, in which it was wounded, it was necessary to plug the posterior palatine canal in order to arrest the bleeding.

The **Vidian branch** passes backwards along the Vidian canal with the Vidian nerve. It is distributed to the upper part of the pharynx and Eustachian tube, sending a small branch into the tympanum, which anastomoses with the other tympanic arteries.

The **pterygo-palatine** is a very small branch, which passes backwards through the pterygo-palatine canal with the pharyngeal nerve, and is distributed to the upper part of the pharynx and Eustachian tube.

The **spheno-palatine** passes through the spheno-palatine foramen into the cavity of the nose, at the back part of the superior meatus, and divides into several branches : one internal, the *naso-palatine* or *artery of the septum*, passes obliquely downwards and forwards along the septum nasi, supplies the mucous membrane, and anastomoses in front with the terminal branch of the descending palatine. The other branches, two or three in number, supply the mucous membrane covering the lateral wall of the nose, the antrum, and the ethmoid and sphenoid cells.

The Triangles of the Neck (fig. 475)

The student having considered the relative anatomy of the large arteries of the neck and their branches, and the relations they bear to the veins and nerves, should now examine these structures collectively, as they present themselves in certain regions of the neck, in each of which important operations are constantly being performed.

The side of the neck presents a somewhat quadrilateral outline, limited, above, by the lower border of the body of the jaw, and an imaginary line extending from the angle of the jaw to the mastoid process; below, by the prominent upper border of the clavicle; in front, by the median line of the neck; behind, by the anterior margin of the Trapezius muscle. This space is subdivided into two large triangles by the Sterno-mastoid muscle, which passes obliquely across the neck, from the sternum and clavicle below, to the mastoid process above. The triangular space in front of this muscle is called the *anterior triangle*; and that behind it, the *posterior triangle*.

Anterior Triangle of the Neck

The **anterior triangle** is bounded, in front, by a line extending from the chin to the sternum; behind, by the anterior margin of the Sterno-mastoid; its base, directed upwards, is formed by the lower border of the body of the jaw, and a line extending from the angle of the jaw to the mastoid process; its apex is below, at the sternum. This space is subdivided into three smaller triangles by the Digastric muscle above, and the anterior belly of the Omo-hyoid below. These smaller triangles are named, from below upwards, the inferior carotid, the superior carotid, and the submaxillary triangle.

The **Inferior Carotid,** or **Muscular Triangle,** is bounded, in front, by the median line of the neck from the hyoid bone to the sternum; behind, by the anterior margin of the Sterno-mastoid; above, by the anterior belly of the Omo-hyoid; and is covered by the integument, superficial fascia, Platysma, and deep fascia; ramifying between which are some of the descending branches of the superficial cervical plexus. Beneath these superficial structures are the Sterno-hyoid and Sterno-thyroid muscles, which, together with the anterior margin of the Sterno-mastoid, conceal the lower part of the common carotid artery.*

This vessel is enclosed within its sheath, together with the internal jugular vein and pneumogastric nerve; the vein lying on the outer side of the artery on the right side of the neck, but overlapping it below on the left side; the nerve lying between the artery and vein, on a plane posterior to both. In front of the sheath are a few filaments descending from the loop of communication between the descendens and communicantes hypoglossi; behind the sheath are seen the inferior thyroid artery, the recurrent laryngeal nerve, and the sympathetic nerve; and on its inner side, the trachea, the thyroid gland—much more prominent in the female than in the male—and the lower part of the larynx. By cutting into the upper part of this space, and slightly displacing the Sterno-mastoid muscle, the common carotid artery may be tied below the Omo-hyoid muscle.

The **Superior Carotid,** or **Carotid Triangle,** is bounded, behind, by the Sterno-mastoid; below, by the anterior belly of the Omo-hyoid; and above, by the Stylohyoid muscle and the posterior belly of the Digastric. It is covered by the integument, superficial fascia, Platysma, and deep fascia; ramifying between which are branches of the facial and superficial cervical nerves. Its floor is formed by parts of the Thyro-hyoid, Hyo-glossus, and the Inferior and Middle constrictor muscles of the pharynx. This space when dissected is seen to contain the upper part of the common carotid artery, which bifurcates opposite the upper border of the thyroid cartilage into the external and internal carotid. These vessels are somewhat concealed from view by the anterior margin of the Sterno-mastoid

---

* Therefore the common carotid artery and internal jugular vein are not, strictly speaking, contained in this triangle, since they are covered by the Sterno-mastoid muscle; that is to say, they lie under that muscle, which forms the posterior border of the triangle. But as they lie very close to the structures which are really contained in the triangle, and whose position it is essential to remember in operating on this part of the artery, it is expedient to study the relations of all these parts together.

muscle, which overlaps them. The external and internal carotids lie side by side, the external being the more anterior of the two. The following branches of the external carotid are also met with in this space: the superior thyroid, running forwards and downwards; the lingual, directly forwards; the facial, forwards and upwards; the occipital, backwards; and the ascending pharyngeal, directly upwards on the inner side of the internal carotid. The veins met with are: the internal jugular, which lies on the outer side of the common and internal carotid arteries; and veins corresponding to the above-mentioned branches of the external carotid—viz. the superior thyroid, the lingual, facial, ascending pharyngeal, and sometimes the occipital—all of which accompany their corresponding arteries, and terminate in the internal jugular. The nerves in this space are the following: In front of the sheath of the common carotid is the descendens hypoglossi. The hypoglossal nerve crosses both the internal and external carotids above, curving round the origin of the occipital artery. Within the sheath, between the artery and vein, and behind both, is the pneumogastric nerve; behind the sheath, the sympathetic. On the outer side of the vessels, the spinal accessory nerve runs for a short distance before it pierces the Sterno-mastoid muscle; and on the inner side of the external carotid, just below the hyoid bone, may be seen the internal laryngeal nerve; and, still more inferiorly, the external laryngeal nerve. The upper portion of the larynx and lower portion of the pharynx are also found in the front part of this space.

The **Submaxillary Triangle,** or **Digastric,** corresponds to the region of the neck immediately beneath the body of the jaw. It is bounded, above, by the lower border of the body of the jaw, and a line drawn from its angle to the mastoid process; below, by the posterior belly of the Digastric and Stylo-hyoid muscles; in front, by the anterior belly of the Digastric. It is covered by the integument, superficial fascia, Platysma, and deep fascia; ramifying between which are branches of the facial nerve and ascending filaments of the superficial cervical nerve. Its floor is formed by the Mylo-hyoid, Hyo-glossus, and Superior constrictor of the pharynx. It is divided into an anterior and posterior part by the stylo-mandibular ligament. The anterior part contains the submaxillary gland, superficial to which is the facial vein, while embedded in it is the facial artery and its glandular branches; beneath this gland, on the surface of the Mylo-hyoid muscle, are the submental artery and the mylo-hyoid artery and nerve. The posterior part of this triangle contains the external carotid artery, ascending deeply in the substance of the parotid gland; this vessel lies here in front of, and superficial to, the internal carotid, being crossed by the facial nerve, and gives off in its course the posterior auricular, temporal, and internal maxillary branches: more deeply are the internal carotid, the internal jugular vein, and the pneumogastric nerve, separated from the external carotid by the Stylo-glossus and Stylo-pharyngeus muscles, and the glosso-pharyngeal nerve.[*]

POSTERIOR TRIANGLE OF THE NECK

The **posterior triangle** is bounded, in front, by the Sterno-mastoid muscle; behind, by the anterior margin of the Trapezius; its base corresponds to the middle third of the clavicle; its apex, to the occiput. The space is crossed, about an inch above the clavicle, by the posterior belly of the Omo-hyoid, which divides it into two triangles, an upper or occipital, and a lower or subclavian.

The **Occipital,** the larger division of the posterior triangle, is bounded, in front, by the Sterno-mastoid; behind, by the Trapezius; below, by the Omo-hyoid. Its floor is formed from above downwards by the Splenius, Levator anguli scapulæ, and the Middle and Posterior scaleni muscles. It is covered by the integument, the Platysma below, the superficial and deep fasciæ. The spinal accessory nerve is directed obliquely across the space from the Sterno-mastoid, which it pierces, to the under surface of the Trapezius; below, the

[*] The same remark will apply to this triangle as was made about the inferior carotid triangle. The structures enumerated as contained in the back part of the space lie, strictly speaking, beneath the muscles which form the posterior boundary of the triangle; but as it is very important to bear in mind their close relation to the parotid gland and its boundaries (on account of the frequency of surgical operations on this gland), all these parts are spoken of together.

descending branches of the cervical plexus and the transversalis colli artery and vein cross the space. A chain of lymphatic glands is also found running along the posterior border of the Sterno-mastoid, from the mastoid process to the root of the neck.

The **Subclavian,** the smaller division of the posterior triangle, is bounded, above, by the posterior belly of the Omo-hyoid; below, by the clavicle; its base, directed forwards, being formed by the posterior border of the Sterno-mastoid. The size of the subclavian triangle varies according to the extent of attachment of the clavicular portion of the Sterno-mastoid and Trapezius muscles, and also according to the height at which the Omo-hyoid crosses the neck above the clavicle. Its height also varies according to the position of the arm, being diminished by raising the limb, on account of the ascent of the clavicle, and increased by drawing the arm downwards, when that bone is depressed. This space is covered by the integument, the Platysma, the superficial and deep fasciæ; and crossed by the descending branches of the cervical plexus. Just above the level of the clavicle, the third portion of the subclavian artery curves outwards and downwards from the outer margin of the Scalenus anticus, across the first rib, to the axilla. Sometimes this vessel rises as high as an inch and a half above the clavicle, or to any point intermediate between this and its usual level. Occasionally, it passes in front of the Scalenus anticus, or pierces the fibres of that muscle. The subclavian vein lies behind the clavicle, and is usually not seen in this space; but in some cases it rises as high as the artery, and has even been seen to pass with that vessel behind the Scalenus anticus. The brachial plexus of nerves lies above the artery, and in close contact with it. Passing transversely behind the clavicle are the suprascapular vessels; and traversing its upper angle in the same direction, the transversalis colli artery and vein. The external jugular vein runs vertically downwards behind the posterior border of the Sterno-mastoid, to terminate in the subclavian vein; it receives the transverse cervical and suprascapular veins, which occasionally form a plexus in front of the artery, and a small vein which crosses the clavicle from the cephalic. The small nerve to the Subclavius muscle also crosses this triangle about its middle, and a lymphatic gland is sometimes found in the space. Its floor is formed by the first rib with the first digitation of the Serratus magnus.

INTERNAL CAROTID ARTERY

The **internal carotid artery** supplies the anterior part of the brain, the eye and its appendages, and sends branches to the forehead and nose. Its size, in the adult, is equal to that of the external carotid, though, in the child, it is larger than that vessel. It is remarkable for the number of curvatures that it presents in different parts of its course. It occasionally has one or two flexures near the base of the skull, while in its passage through the carotid canal and along the side of the body of the sphenoid bone it describes a double curvature which resembles the italic letter *S* placed horizontally. These curvatures most probably diminish the velocity of the current of blood, by increasing the extent of surface over which it moves, and adding to the amount of impediment produced from friction.

In considering the course and relations of this vessel, it may be conveniently divided into four portions: a cervical, petrous, cavernous, and cerebral.

Cervical Portion.—This portion of the internal carotid commences at the bifurcation of the common carotid, opposite the upper border of the thyroid cartilage, and runs perpendicularly upwards, in front of the transverse processes of the three upper cervical vertebræ, to the carotid canal in the petrous portion of the temporal bone. It is superficial at its commencement, being contained in the superior carotid triangle, and lying behind and to the outer side of the external carotid, overlapped by the Sterno-mastoid, and covered by the deep fascia, Platysma, and integument: it then passes beneath the parotid gland, being crossed by the hypoglossal nerve, the Digastric and Stylo-hyoid muscles, and the occipital and posterior auricular arteries. Higher up, it is separated from the external carotid by the Stylo-glossus and Stylo-pharyngeus muscles, the glosso-pharyngeal nerve, and the pharyngeal branch of the pneumogastric. It is in relation, *behind*, with the Rectus capitis anticus major, the superior cervical ganglion of the sympathetic, and superior laryngeal nerve; *externally*, with the internal jugular vein and pneumogastric nerve, the nerve lying on a plane posterior to the artery; *internally*, with the pharynx, tonsil, the superior laryngeal nerve, and ascending pharyngeal artery. At the base of the skull the glosso-pharyngeal, vagus, spinal accessory, and hypoglossal nerves lie between the artery and the internal jugular vein.

PLAN OF THE RELATIONS OF THE INTERNAL CAROTID ARTERY IN THE NECK

In front.

Skin, superficial and deep fasciæ.
Platysma.
Sterno-mastoid.
Occipital and posterior auricular arteries.
Hypoglossal nerve.
Parotid gland.
Stylo-glossus and Stylo-pharyngeus muscles.
Glosso-pharyngeal nerve.
Pharyngeal branch of the pneumogastric.

Externally.

Internal jugular vein.
Pneumogastric nerve.

Internally.

Pharynx.
Superior laryngeal nerve.
Ascending pharyngeal artery.
Tonsil.

Behind.

Rectus capitis anticus major.
Sympathetic.
Superior laryngeal nerve.

Petrous Portion.—When the internal carotid artery enters the canal in the petrous portion of the temporal bone, it first ascends a short distance, then curves forwards and inwards, and again ascends as it leaves the canal to enter the cavity of the skull between the lingula and petrosal process. In this canal, the artery lies at first in front of the cochlea and tympanum; from the latter cavity it is separated by a thin, bony lamella, which is cribriform in the young subject, and often absorbed in old age. Farther forwards it is separated from the Gasserian ganglion by a thin plate of bone, which forms the floor of the fossa for the ganglion and the roof of the horizontal portion of the canal. Frequently this bony plate is more or less deficient, and then the ganglion is separated from the artery by fibrous membrane. The artery is separated from the bony wall of the carotid canal by a prolongation of dura mater, and is surrounded by filaments of the carotid plexus, derived from the ascending branch of the superior cervical ganglion of the sympathetic, and a number of small veins.

Cavernous Portion.—The internal carotid artery, in this part of its course, is situated between the layers of the dura mater forming the cavernous sinus, but covered by the lining membrane of the sinus. It at first ascends towards the posterior clinoid process, then passes forwards by the side of the body of the sphenoid bone, and again curves upwards on the inner side of the anterior clinoid process, and perforates the dura mater forming the roof of the sinus. In this part of its course it is surrounded by filaments of the sympathetic nerve, and has in relation with it externally the sixth nerve.

Cerebral Portion.—Having perforated the dura mater, on the inner side of the anterior clinoid process, the internal carotid passes between the second and third cranial nerves to the anterior perforated space at the inner extremity of the fissure of Sylvius, where it gives off its terminal or cerebral branches. This portion of the artery has the optic nerve on its inner side, and the third nerve externally.

Peculiarities.—The length of the internal carotid varies according to the length of the neck, and also according to the point of bifurcation of the common carotid. Its origin sometimes takes place from the arch of the aorta; in such rare instances, this vessel has been found to be placed nearer the middle line of the neck than the external carotid, as far upwards as the larynx, when the latter vessel crossed the internal carotid. The course of the vessel, instead of being straight, may be very tortuous. A few instances are recorded in which this vessel was altogether absent: in one of these the common carotid passed up the neck, and gave off the usual branches of the external carotid: the cranial

FIG. 481.—The internal carotid and vertebral arteries. Right side.

First aortic intercostal

portion of the internal carotid being replaced by two branches of the internal maxillary, which entered the skull through the foramen rotundum and ovale, and joined to form a single vessel.

Surgical Anatomy.—The cervical part of the internal carotid is very rarely wounded. Cripps, in an interesting paper in the 'Medico-Chirurgical Transactions,' compares the rareness of a wound of the internal carotid with one of the external carotid or its branches. It is, however, sometimes injured by a stab or gunshot wound in the neck, or even occasionally by a stab from within the mouth, as when a person receives a thrust from the

S S 2

end of a parasol, or falls down with a tobacco-pipe in his mouth. The relation of the internal carotid with the tonsil should be especially remembered, as instances have occurred in which the artery has been wounded during the operation of scarifying the tonsil, and fatal hæmorrhage has supervened. The indications for ligature are wounds, when the vessel should be exposed by a careful dissection and tied above and below the bleeding-point; and aneurism, which if non-traumatic may be treated by ligature of the common carotid; but if traumatic in origin, by exposing the sac and tying the vessel above and below. The incision for ligature of the cervical portion of the internal carotid should be made along the anterior border of the Sterno-mastoid, from the angle of the jaw to the upper border of the thyroid cartilage. The superficial structures being divided, and the Sterno-mastoid defined and drawn outwards, the cellular tissue must be carefully separated and the posterior belly of the Digastric and hypoglossal nerve sought for as guides to the vessel. When the artery is found, the external carotid should be drawn inwards and the Digastric muscle upwards, and the aneurism needle passed from without inwards.

Branches.—The branches given off from the internal carotid are:

From the Petrous portion	Tympanic (internal or deep). Vidian.
From the Cavernous portion	Arteriæ receptaculi. Pituitary. Ganglionic. Anterior meningeal. Ophthalmic.
From the Cerebral portion	Anterior cerebral. Middle cerebral. Posterior communicating. Anterior choroid.

The cervical portion of the internal carotid gives off no branches.

Fig. 482.—The ophthalmic artery and its branches, the roof of the orbit having been removed.

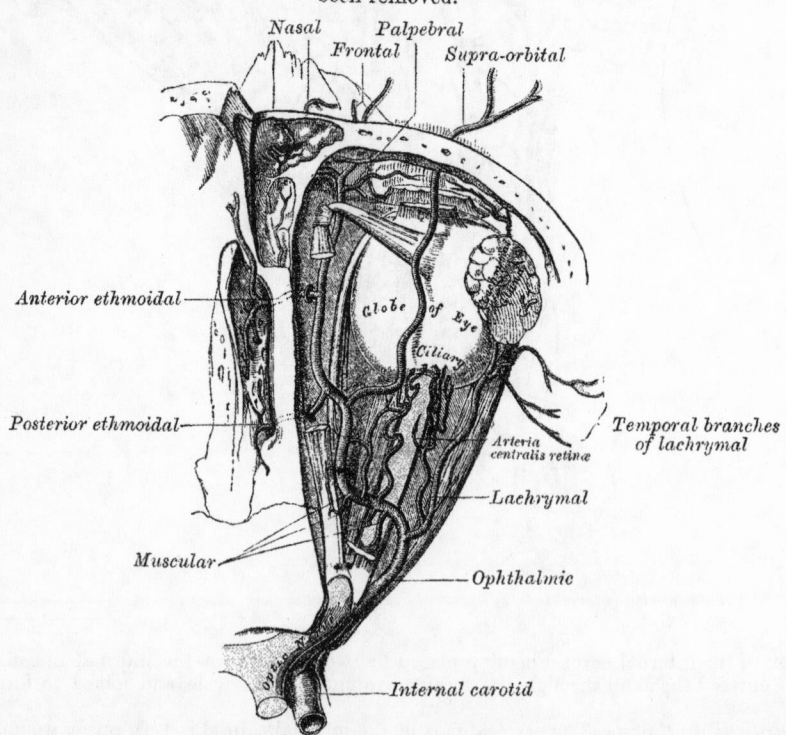

The **tympanic** is a small branch which enters the cavity of the tympanum, through a minute foramen in the carotid canal, and anastomoses with the tympanic branch of the internal maxillary, and with the stylo-mastoid artery.

The **Vidian** is a small branch, which is by no means constant. It passes through the Vidian canal and anastomoses with the Vidian branch of the internal maxillary artery.

The **arteriæ receptaculi** are numerous small vessels, derived from the internal carotid in the cavernous sinus; they supply the pituitary body, the Gasserian ganglion, and the walls of the cavernous and inferior petrosal sinuses. Some of these branches anastomose with branches of the middle meningeal.

The **pituitary** branches are one or two minute vessels which supply the pituitary body.

The **ganglionic** branches are small vessels to the Gasserian ganglion.

The **anterior meningeal** is a small branch which passes over the lesser wing of the sphenoid to supply the dura mater of the anterior fossa; it anastomoses with the meningeal branch from the posterior ethmoidal artery.

The **ophthalmic artery** arises from the internal carotid, just as that vessel is emerging from the cavernous sinus, on the inner side of the anterior clinoid process, and enters the orbit through the optic foramen, below and on the outer side of the optic nerve. It then passes over the nerve to the inner wall of the orbit, and thence horizontally forwards, beneath the lower border of the Superior oblique muscle to a point behind the internal angular process of the frontal bone, where it divides into two terminal branches, the *frontal* and *nasal*. As the artery crosses the optic nerve it is accompanied by the nasal nerve, and is separated from the frontal nerve by the Rectus superior and Levator palpebræ superioris muscles.

The branches of this vessel may be divided into an *orbital group*, which is distributed to the orbit and surrounding parts; and an *ocular group*, which supplies the muscles and globe of the eye.

Orbital Group.	*Ocular Group.*
Lachrymal.	Short ciliary.
Supra-orbital.	Long ciliary.
Posterior ethmoidal.	Anterior ciliary.
Anterior ethmoidal.	Central artery of the retina.
Internal palpebral.	Muscular.
Frontal.	
Nasal.	

The *lachrymal* is one of the largest branches derived from the ophthalmic, arising close to the optic foramen : not infrequently it is given off from the artery before it enters the orbit. It accompanies the lachrymal nerve along the upper border of the External rectus muscle, and is distributed to the lachrymal gland. Its terminal branches, escaping from the gland, are distributed to the eyelids and conjunctiva : of those supplying the eyelids, two are of considerable size and are named the *external palpebral*; they run inwards in the upper and lower lids respectively and anastomose with the internal palpebral arteries, forming an arterial circle in this situation. The lachrymal artery gives off one or two *malar branches*, one of which passes through a foramen in the malar bone, to reach the temporal fossa, and anastomoses with the deep temporal arteries; the other appears on the cheek through the malar foramen, and anastomoses with the transverse facial. A branch is also sent backwards, through the sphenoidal fissure, to the dura mater, which anastomoses with a branch of the middle meningeal artery.

Peculiarities.—The lachrymal artery is sometimes derived from one of the anterior branches of the middle meningeal artery.

The *supra-orbital artery* arises from the ophthalmic as that vessel is crossing over the optic nerve. Ascending so as to rise above all the muscles of the orbit, it runs forwards, with the supra-orbital nerve, between the periosteum and Levator palpebræ; and, passing through the supra-orbital foramen, divides into a superficial and a deep branch, which supply the integument, the muscles, and the pericranium of the forehead, anastomosing with the frontal, the anterior branch of the temporal, and the artery of the opposite side. This artery in the orbit supplies the Superior rectus and the Levator palpebræ, and sends a branch inwards, across the pulley of the Superior oblique muscle, to supply the parts at the inner canthus. At the supra-orbital foramen it frequently transmits a branch to the diploë.

The *ethmoidal branches* are two in number: posterior and anterior. The former, which is the smaller, passes through the posterior ethmoidal foramen,

FIG. 483.—The arteries of the base of the brain. The right half of the cerebellum and pons have been removed.

N.B.—It will be noticed that in the illustration the two anterior cerebral arteries have been drawn at a considerable distance from each other; this makes the anterior communicating artery appear very much longer than it really is.

supplies the posterior ethmoidal cells, and, entering the cranium, gives off a meningeal branch, which supplies the adjacent dura mater; and nasal branches which descend into the nose through apertures in the cribriform plate, anastomosing with branches of the spheno-palatine. The anterior ethmoidal artery accompanies the nasal nerve through the anterior ethmoidal foramen, supplies the anterior ethmoidal cells and frontal sinuses, and, entering the cranium, gives off a meningeal branch, which supplies the adjacent dura mater; and nasal branches, which descend into the nose, through the slit by the side of the crista galli, and running along the groove on the under surface of the nasal bone supply the skin of the nose.

The *palpebral arteries*, two in number, superior and inferior, arise from the ophthalmic, opposite the pulley of the Superior oblique muscle; they leave the orbit to encircle the eyelids near their free margin, forming a superior and an inferior arch, which lie between the Orbicularis muscle and tarsal plates; the superior palpebral inosculating, at the outer angle of the orbit, with the orbital

FIG. 484.—Vascular area of the upper surface of the cerebrum. (After Duret.)

Blue indicates the distribution of the anterior cerebral; yellow, the middle cerebral; red, the posterior cerebral. I. The part supplied by the external and inferior frontal artery. II. The part supplied by the ascending frontal. III. The part supplied by the ascending parietal. IV. The part supplied by the parieto-sphenoidal artery.

branch of the temporal artery, and with the upper of the two external palpebral branches from the lachrymal artery: the inferior palpebral inosculating, at the outer angle of the orbit, with the lower of the two external palpebral branches from the lachrymal and with the transverse facial artery, and, at the inner side of the lid, with a branch from the angular artery. From this last anastomosis a branch passes to the nasal duct, ramifying in its mucous membrane, as far as the inferior meatus.

The *frontal artery*, one of the terminal branches of the ophthalmic, passes from the orbit at its inner angle, and, ascending on the forehead, supplies the integument, muscles, and pericranium, anastomosing with the supra-orbital artery, and with the artery of the opposite side.

The *nasal artery*, the other terminal branch of the ophthalmic, emerges from the orbit above the tendo oculi, and, after giving a branch to the upper part of the lachrymal sac, divides into two branches, one of which crosses the root of the nose, the *transverse nasal*, and anastomoses with the angular artery; the other, the *dorsalis nasi*, runs along the dorsum of the nose, supplies its outer surface, and anastomoses with the artery of the opposite side, and with the lateral nasal branch of the facial.

The *ciliary arteries* are divisible into three groups, the short, long, and anterior. The *short ciliary arteries*, from six to twelve in number, arise from the ophthalmic, or some of its branches; they surround the optic nerve as they pass forwards to the posterior part of the eyeball, pierce the sclerotic coat around the entrance of the nerve, and supply the choroid coat and ciliary processes. The *long ciliary arteries*, two in number, pierce the posterior part of the sclerotic at some little distance from the optic nerve, and run forwards, along each side of the eyeball, between the sclerotic and choroid, to the ciliary muscle, where they divide into two branches; these form an arterial circle, the *circulus major*, around the circumference of the iris, from which numerous radiating branches pass forwards, in its substance, to its free margin, where they form a second arterial circle, the *circulus minor*, around its pupillary margin. The *anterior ciliary*

FIG. 485.—Vascular area of the internal surface of the cerebrum. (After Duret.)

Blue indicates the distribution of the anterior cerebral artery; red, the posterior cerebral. I. The part supplied by the anterior and internal frontal. II. The part supplied by the middle and internal frontal. III. The part supplied by the posterior and internal frontal. IV. The part supplied by the posterior temporal; and V. the part supplied by the occipital, both terminal branches of the posterior cerebral.

arteries are derived from the muscular branches; they pass to the front of the eyeball in company with the tendons of the Recti muscles, form a vascular zone beneath the conjunctiva, and then pierce the sclerotic a short distance from the cornea and terminate in the circulus major of the iris.

The *central artery of the retina* is the first and one of the smallest branches of the ophthalmic artery. It runs for a short distance within the dural sheath of the nerve, but about half an inch behind the eyeball it pierces the optic nerve obliquely, and runs forward in the centre of its substance, and enters the globe of the eye through the porus opticus. Its mode of distribution will be described in the account of the anatomy of the eye.

The *muscular branches*, two in number, superior and inferior, frequently spring from a common trunk. The superior, the smaller, often wanting, supplies the Levator palpebræ, Superior rectus, and Superior oblique. The inferior, more constant in its existence, passes forwards, between the optic nerve and Inferior rectus, and is distributed to the External, Internal, and Inferior recti, and Inferior oblique. This vessel gives off most of the anterior ciliary arteries. Additional muscular branches are given off from the lachrymal and supra-orbital arteries or from the ophthalmic itself.

The **anterior cerebral artery** arises from the internal carotid, at the inner extremity of the fissure of Sylvius. It passes forwards and inwards across the anterior perforated space, above the optic nerve, to the commencement of the great longitudinal fissure. Here it comes into close relationship with the artery of the opposite side, and the two vessels are connected together by a short anastomosing trunk, about two lines in length, the *anterior communicating artery*. From this point, the two vessels run side by side in the longitudinal fissure, curve round the genu of the corpus callosum, and turning backwards continue along its upper surface to its posterior part, where they terminate by anastomosing with the posterior cerebral arteries. In their course they give off the following branches :

Antero-median ganglionic.	Anterior internal frontal.
Inferior internal frontal.	Middle internal frontal.
Posterior internal frontal.	

The *antero-median ganglionic* is a group of small arteries which arise at the commencement of the anterior cerebral artery ; they pierce the anterior perforated space and lamina terminalis, and supply the rostrum of the sphenoid, the septum lucidum, and the head of the caudate nucleus. The *inferior internal frontal branches*, two or three in number, are distributed to the orbital surface of the frontal lobe, where they supply the olfactory lobe, gyrus rectus, and internal orbital convolution. The *anterior internal frontal branches* supply a part of the marginal convolution, and send branches over the edge of the hemisphere to the superior and middle frontal convolutions and upper part of the ascending frontal convolution. The *middle internal frontal branches* supply the corpus callosum, the convolution of the corpus callosum, the inner surface of the first frontal convolution, and the upper part of the ascending frontal convolution. The *posterior internal frontal branches* supply the lobus quadratus and adjacent outer surface of the hemisphere.

The *anterior communicating artery* is a short branch, about two lines in length, but of moderate size, connecting together the two anterior cerebral arteries across the longitudinal fissure. Sometimes this vessel is wanting, the two arteries joining together to form a single trunk, which afterwards divides. Or the vessel may be wholly, or partially, divided into two ; frequently it is longer and smaller than usual. It gives off some of the antero-median ganglionic group of vessels, which are, however, principally derived from the anterior cerebral.

The **middle cerebral artery** (fig. 487), the largest branch of the internal carotid, passes obliquely outwards along the fissure of Sylvius, and, opposite the island of Reil, divides into its terminal branches. The branches of the middle cerebral artery are :

Antero-lateral ganglionic.	Ascending parietal.
Inferior external frontal.	Parieto-temporal.
Ascending frontal.	Temporal.

The *antero-lateral ganglionic branches*, a group of small arteries which arise at the commencement of the middle cerebral artery, are arranged in two sets : one, the *internal striate*, passes upwards through the inner segments of the lenticular nucleus, and supplies it, the caudate nucleus and the internal capsule ; the other, the *external striate*, passes upwards through the outer segment of the lenticular nucleus, and supplies the caudate nucleus and the optic thalamus. One artery of this group is of larger size than the rest, and is of special importance, as being the artery in the brain most frequently ruptured ; it has been termed by Charcot, the '*artery of cerebral hæmorrhage.*' It passes up between the lenticular nucleus and the external capsule, and ultimately ends in the caudate nucleus. The *inferior external frontal* supplies the third or inferior frontal convolution (Broca's convolution) and the outer part of the orbital surface of the frontal lobe. The *ascending frontal* supplies the ascending frontal convolution. The *ascending parietal* supplies the ascending parietal convolution and the lower part of the superior parietal convolution. The *parieto-temporal* supplies the supra-marginal and angular gyri, and the posterior parts of the superior and middle temporal convolutions. The *temporal branches*, two or three in number, are distributed to the outer surface of the temporal lobe.

The **posterior communicating artery** arises from the back part of the internal carotid, runs directly backwards, and anastomoses with the posterior cerebral, a branch of the basilar. This artery varies considerably in size, being sometimes small, and occasionally so large that the posterior cerebral may be considered as arising from the internal carotid rather than from the basilar. It is frequently larger on one side than on the other side. From the posterior half of this vessel are given off a number of small branches, the *postero-median ganglionic branches*,

Fig. 486.—Vascular area of the inferior surface of the cerebrum. (After Duret.)

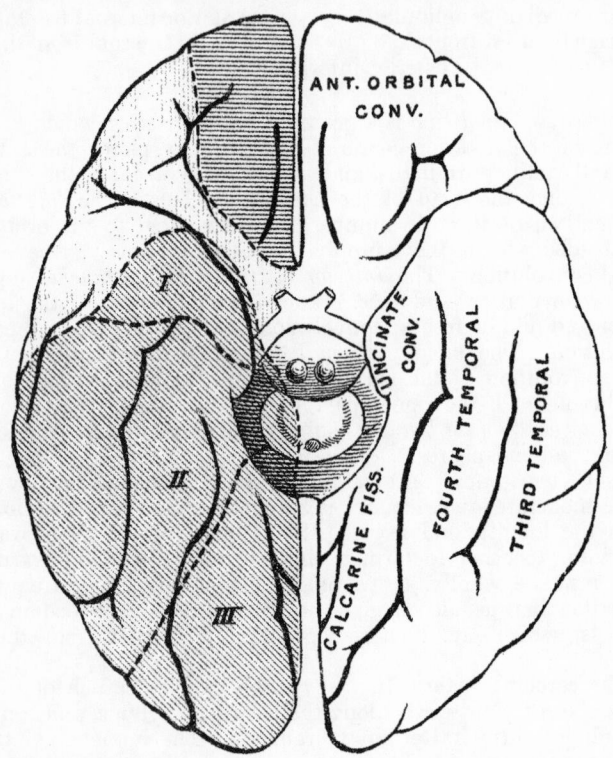

Blue indicates the distribution of the anterior cerebral; yellow, the middle cerebral; red, the posterior cerebral. I. The part supplied by the anterior temporal artery. II. The part supplied by the posterior temporal artery. III. The part supplied by the occipital artery.

which, with similar vessels from the posterior cerebral, pierce the posterior perforated space and supply the internal surfaces of the optic thalami and the walls of the third ventricle.

The **anterior choroid** is a small but constant branch, which arises from the back part of the internal carotid, near the posterior communicating artery. Passing backwards and outwards between the temporal lobe and the crus cerebri, it enters the descending horn of the lateral ventricle through the choroidal fissure and ends in the choroid plexus. It is distributed to the hippocampus major, corpus fimbriatum, velum interpositum, and choroid plexus.

THE BLOOD-VESSELS OF THE BRAIN

Investigations tend to show that the mode of distribution of the vessels of the brain has an important bearing upon a considerable number of the anatomical lesions of which this part of the nervous system may be the seat; it therefore becomes important to consider a little more in detail the manner in which the cerebral vessels are distributed.

The cerebral arteries are derived from the internal carotid and the vertebral, which at the base of the brain form a remarkable anastomosis known as the *circle of Willis*. It is formed in front by the anterior cerebral arteries, branches of the internal carotid, which are connected together by the anterior communicating; behind by the two posterior cerebrals, branches of the basilar, which are connected on each side with the internal carotid by the posterior communicating (fig. 483, page 630). The parts of the brain included within this arterial circle are the lamina terminalis, the commissure of the optic nerves, the infundibulum, the tuber cinereum, the corpora albicantia, and the posterior perforated space.

The three trunks which together supply each cerebral hemisphere arise from the circle of Willis. From its anterior part proceed the two anterior cerebrals; from its antero-lateral part the middle cerebrals, and from its posterior part the posterior cerebrals. Each of these principal arteries gives origin to two very different systems of secondary vessels. One of these systems has been named the *central ganglionic system*, and the vessels belonging to it supply the central ganglia of the brain; the other has been named the *cortical system*, and its vessels ramify in the pia mater and supply the cortex and subjacent medullary matter. These two systems, though they have a common origin, do not communicate at any point of their peripheral distribution, and are entirely independent

FIG. 487.—The distribution of the middle cerebral artery. (After Charcot.)

Perforating branches

Middle cerebral artery

of each other. Though some of the arteries of the cortical system approach, at their terminations, the regions supplied by the central ganglionic system, no communication between the two sets of vessels takes place, and there is between the parts supplied by the two systems a borderland of diminished nutritive activity, where, it is said, softening is especially liable to occur in the brains of old people.

The central ganglionic system.—All the vessels belonging to this system are given off from the circle of Willis, or from the vessels immediately after their origin from it. So that if a circle is drawn at a distance of about an inch from the circle of Willis, it will include the origin of all the arteries belonging to this system (fig. 488). The vessels of this system form six principal groups: (I) the *antero-median group*, derived from the anterior cerebrals and anterior communicating; (II) the *postero-median group*, from the posterior cerebrals and posterior communicating; (III) the right and left *antero-lateral groups*, from the middle cerebrals; and (IV) the right and left *postero-lateral groups*, from the posterior cerebrals, after they have wound round the crura cerebri. The vessels belonging to this system are larger than those of the cortical system, and are what Cohnheim has designated 'terminal' arteries—that is to say, vessels which from their origin to their termination neither supply nor receive any

FIG. 488.—Diagram of the arterial circulation at the base of the brain.
(After Charcot.)

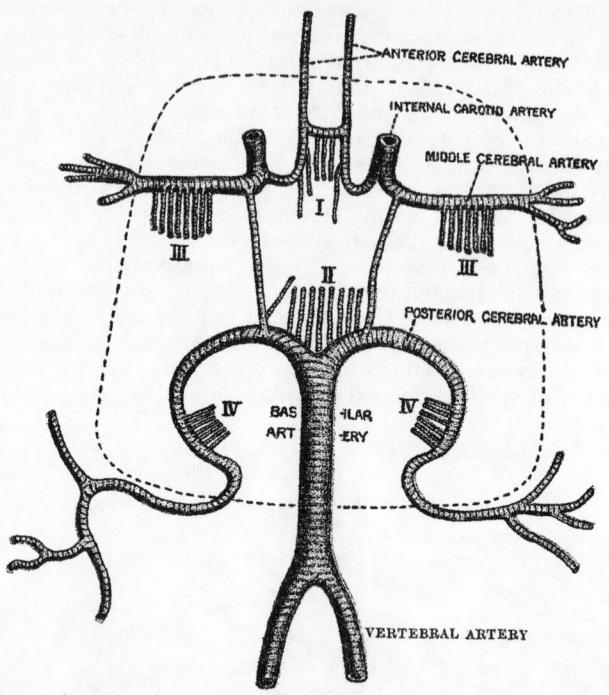

I. Antero-median group of ganglionic branches.   II. Postero-median group.   III. Right and left antero-lateral group.
IV. Right and left postero-lateral group.   The dotted line shows the limit of the ganglionic circle.

FIG. 489.—Distribution of the cortical arteries.   (After Charcot.)

I. Medullary arteries.   I'. Group of medullary arteries in the sulcus between two adjacent convolutions.   I''. Arteries
situated among the short association fibres.   2, 2. Cortical arteries.   a. Capillary network with fairly wide meshes,
situated beneath the pia mater.   b. Network with more compact, polygonal meshes, situated in the cortex.   c. Tran
sitional network with wider meshes.   d. Capillary network in the white matter.

anastomotic branch, so that, by one of the small vessels, only a limited area of the central ganglia can be injected, and the injection cannot be driven beyond the area of the part supplied by the particular vessel which is the subject of the experiment.

The cortical arterial system.—The vessels forming this system are the terminal branches of the anterior, middle, and posterior cerebral arteries, described above. These vessels divide and ramify in the substance of the pia mater, and give off nutrient arteries which penetrate the cortex perpendicularly. These nutrient vessels are divisible into two classes, the long and short. The *long*, or, as they are sometimes called, the *medullary arteries*, pass through the grey matter to penetrate the centrum ovale to the depth of about an inch and a half, without intercommunicating otherwise than by very fine capillaries, and thus constitute so many independent small systems. The *short vessels* are confined to the cortex, where they form with the long vessels a compact network in the middle zone of the grey matter, the outer and inner zones being sparingly supplied with blood (fig. 489). The vessels of the cortical arterial system are not so strictly 'terminal' as those of the central ganglionic system, but they approach this type very closely, so that injection of one area from the vessel of another area, though it may be possible, is frequently very difficult, and is only effected through vessels of small calibre. As a result of this, obstruction of one of the main branches, or its divisions, may have the effect of producing softening in a very limited area of the cortex.*

ARTERIES OF THE UPPER EXTREMITY

The artery which supplies the upper extremity continues as a single trunk from its commencement down to the elbow; but different portions of it have received different names, according to the regions through which it passes. That part of the vessel which extends from its origin to the outer border of the first rib is termed the *subclavian*; beyond this point to the lower border of the axilla, it is termed the *axillary*; and from the lower margin of the axillary space to the bend of the elbow, it is termed *brachial*; here, the single trunk terminates by dividing into two branches, the *radial* and *ulnar*, an arrangement precisely similar to what occurs in the lower limb.

Subclavian Arteries (fig. 490)

The **subclavian artery** on the right side arises from the innominate artery opposite the right sterno-clavicular articulation; on the left side it arises from the arch of the aorta. It follows, therefore, that these two vessels must, in the first part of their course, differ in their length, their direction, and their relation with neighbouring parts.

In order to facilitate the description of these vessels, more especially from a surgical point of view, each subclavian artery has been divided into three parts. The first portion, on the right side, passes upwards and outwards from the origin of the vessel to the inner border of the Scalenus anticus. On the left side it ascends nearly vertically, to gain the inner border of that muscle. The second part passes outwards, behind the Scalenus anticus; and the third part extends from the outer margin of that muscle, beneath the clavicle, to the outer border of the first rib, where it becomes the axillary artery. The first portions of the two vessels differ so much in their course, and in their relation with neighbouring parts, that they will be described separately. The second and third parts are alike on the two sides.

First Part of the Right Subclavian Artery (figs. 472, 475, 490)

The **right subclavian artery** arises from the arteria innominata, opposite the upper part of the right sterno-clavicular articulation, and passes upwards and

---

* The student who desires further information on this subject is referred to Charcot's *Localisation of Cerebral and Spinal Diseases*, p. 42 *et seq.*, whence the facts above given have been principally derived.

outwards to the inner margin of the Scalenus anticus muscle. In this part of its course it ascends a little above the clavicle, the extent to which it does so varying in different cases. It is covered, *in front*, by the integument, superficial fascia, Platysma, deep fascia, the clavicular origin of the Sterno-mastoid, the Sterno-hyoid, and Sterno-thyroid muscles, and another layer of the deep fascia. It is crossed by the internal jugular and vertebral veins, and by the pneumo-gastric nerve and the cardiac branches of the pneumogastric and sympathetic. A loop of the sympathetic nerve itself also crosses the artery, forming a ring (*annulus Vieusseni*) around the vessel. The anterior jugular vein passes outwards in front of the artery but is not in contact with it, being separated from it by the Sterno-hyoid and Sterno-thyroid muscles. Below and behind the artery is the pleura, which separates it from the apex of the lung; *behind* is the gangliated cord of the sympathetic, the Longus colli muscle and first dorsal vertebra; the recurrent laryngeal nerve winds round the lower and back part of the vessel.

PLAN OF RELATIONS OF FIRST PORTION OF THE RIGHT SUBCLAVIAN ARTERY

In front.

Skin, superficial fascia.
Platysma, deep fascia.
Clavicular origin of Sterno-mastoid.
Sterno-hyoid, and Sterno-thyroid.
Anterior jugular, internal jugular, and vertebral veins.
Pneumogastric and cardiac nerves.
Loop from the sympathetic.

Right Subclavian Artery. First Portion.

Beneath.

Pleura.
Recurrent laryngeal nerve.

Behind.

Recurrent laryngeal nerve.
Sympathetic.
Pleura and apex of lung.
Longus colli.
First dorsal vertebra.

FIRST PART OF THE LEFT SUBCLAVIAN ARTERY (fig. 472)

The **left subclavian artery** arises from the end of the arch of the aorta, opposite the fourth dorsal vertebra, and ascends nearly vertically to the inner margin of the Scalenus anticus muscle. This part of the vessel is, therefore, longer than the right, situated deeply in the cavity of the chest, and directed nearly vertically upwards instead of arching outwards like the vessel of the opposite side.

The first part of the left subclavian artery is in relation, *in front*, with the pneumogastric, cardiac, and phrenic nerves, which lie parallel with it, the left common carotid artery, left internal jugular and vertebral veins, and the com-mencement of the left innominate vein, and is covered by the Sterno-thyroid, Sterno-hyoid, and Sterno-mastoid muscles; *behind*, it is in relation with the œsophagus, thoracic duct, inferior cervical ganglion of the sympathetic, and Longus colli; higher up, however, the œsophagus and thoracic duct lie to its right side; the latter ultimately arching over the vessel to join the angle of union between the subclavian and internal jugular veins. To its *inner side* are the œsophagus, trachea, thoracic duct, and left recurrent laryngeal nerve; to its *outer side*, the left pleura and lung.

PLAN OF RELATIONS OF FIRST PORTION OF THE LEFT SUBCLAVIAN ARTERY

In front.

Pneumogastric, cardiac, and phrenic nerves.
Left common carotid artery. Thoracic duct.
Left internal jugular, vertebral, and innominate veins.
Sterno-thyroid, Sterno-hyoid, and Sterno-mastoid muscles.

Inner side.

Trachea.
Œsophagus.
Thoracic duct.
Left recurrent laryngeal nerve.

Left Subclavian Artery.

Outer side.

Pleura and left lung.

Behind.

Œsophagus and thoracic duct.
Inferior cervical ganglion of sympathetic.
Longus colli.

SECOND AND THIRD PARTS OF THE SUBCLAVIAN ARTERY (figs. 475, 490)

The **Second Portion of the Subclavian Artery** lies behind the Scalenus anticus muscle; it is very short, and forms the highest part of the arch described by that vessel.

Relations.—It is covered, *in front*, by the skin, superficial fascia, Platysma, deep cervical fascia, Sterno-mastoid, and the Scalenus anticus muscle. On the right side the phrenic nerve is separated from the second part of the artery by the Anterior scalene muscle, while on the left side the nerve crosses the first part of the artery immediately to the inner edge of the muscle. *Behind*, it is in relation with the pleura and the Middle scalene. *Above*, with the brachial plexus of nerves. *Below*, with the pleura. The subclavian vein lies below and in front of the artery, separated from it by the Scalenus anticus.

PLAN OF RELATIONS OF SECOND PORTION OF THE SUBCLAVIAN ARTERY

In front.

Skin and superficial fascia.
Platysma and deep cervical fascia.
Sterno-mastoid.
Phrenic nerve.
Scalenus anticus.
Subclavian vein.

Above.

Brachial plexus.

Subclavian Artery. Second Portion.

Below.

Pleura.

Behind.
Pleura and Middle Scalenus.

The **Third Portion of the Subclavian Artery** passes downwards and outwards from the outer margin of the Scalenus anticus to the outer border of the first rib, where it becomes the axillary artery. This portion of the vessel is the most superficial, and is contained in the Subclavian triangle (see page 625).

Relations.—It is covered, *in front*, by the skin, the superficial fascia, the Platysma, the descending clavicular branches of the cervical plexus, and the deep cervical fascia; by the clavicle, the Subclavius muscle, and the suprascapular artery and vein, and the transverse cervical vein; the nerve to the Subclavius muscle passes vertically downwards in front of the artery. The external jugular vein crosses it at its inner side, and receives the suprascapular and transverse cervical veins, which frequently form a plexus in front of it. The subclavian vein is below and in front of the artery, lying close behind the clavicle. *Behind*, it lies on the Middle scalene muscle and the trunk of the brachial plexus, which is formed by the union of the last cervical and first dorsal nerves. *Above* it, and to its outer side, are the upper trunks of the brachial plexus, and Omo-hyoid muscle. *Below*, it rests on the upper surface of the first rib.

PLAN OF RELATIONS OF THIRD PORTION OF THE SUBCLAVIAN ARTERY

In front.

Skin and superficial fascia.
Platysma and deep cervical fascia.
Descending branches of cervical plexus. Nerve to Subclavius muscle.
Subclavius muscle, suprascapular artery, and vein.
External jugular and transverse cervical veins.
Clavicle.

Above.
Brachial plexus.
Omo-hyoid.

Subclavian Artery. Third Portion.

Below.
First rib.

Behind.
Scalenus medius.
Lower cord of brachial plexus.

Peculiarities.—The subclavian arteries vary in their origin, their course, and the height to which they rise in the neck.

The origin of the right subclavian from the innominate takes place, in some cases, above the sterno-clavicular articulation ; and occasionally, but less frequently, in the cavity of the thorax, below that joint ; or the artery may arise as a separate trunk from the arch of the aorta. In such cases it may be either the first, second, third, or even the last branch derived from that vessel ; in the majority of cases, it is the first or last, rarely the second or third. When it is the first branch, it occupies the ordinary position of the innominate artery ; when the second or third, it gains its usual position by passing behind the right carotid ; and when the last branch, it arises from the left extremity of the arch, at its upper or back part, and passes obliquely towards the right side, usually behind the trachea, œsophagus, and right carotid, sometimes between the œsophagus and trachea, to the upper border of the first rib, whence it follows its ordinary course. In very rare instances, this vessel arises from the thoracic aorta, as low down as the fourth dorsal vertebra. Occasionally, it perforates the Anterior scalene ; more rarely it passes in front of that muscle. Sometimes the subclavian vein passes with the artery behind the Anterior scalene. The artery may ascend as high as an inch and a half above the clavicle, or any intermediate point between this and the upper border of the bone, the right subclavian usually ascending higher than the left.

The left subclavian is occasionally joined at its origin with the left carotid.

Surface Marking.—The course of the subclavian artery in the neck may be mapped out, by describing a curve, with its convexity upwards, at the base of the posterior triangle. The inner end of this curve corresponds to the sterno-clavicular joint, the outer end to the centre of the lower border of the clavicle. The curve is to be drawn with such an amount of convexity that its mid-point reaches half an inch above the upper border of the clavicle. The left subclavian artery is more deeply placed than the right in the first part of its course, and, as a rule, does not reach quite as high a level in the neck. It should be borne in mind that the posterior border of the Sterno-mastoid muscle corresponds to the outer border of the Scalenus anticus, so that the third portion of the artery, that part most accessible for operation, lies immediately external to the posterior border of the Sterno-mastoid.

Surgical Anatomy.—An *aneurism* may form on any part of the subclavian artery, except the intrathoracic portion of the left vessel, which is said never to be the seat of aneurism. The most common site is, however, the third portion, especially on the right side, on account of the greater exposure to injury and the greater amount of use of the right upper extremity. In this situation it may cause pressure on the brachial plexus, producing pain and numbness in the arm and fingers, with loss of power or paralysis of the muscles of these parts. Œdema of the arm may result from pressure of the subclavian vein. The external jugular vein may become distended and varicose. The treatment is unsatisfactory, since proximal ligature cannot be undertaken with much chance of success. If constitutional treatment with direct pressure on the aneurismal sac fails, the best treatment is excision of the sac, if it is still small; if it has become diffused, as it has a great tendency to, amputation at the shoulder-joint may be necessary. In aneurisms of the first portion of this artery there is œdema of the head and face, with lividity, congestion of the brain, and semi-consciousness from pressure on the internal jugular vein, and spasmodic action of the Diaphragm from pressure on the phrenic nerve.

Compression of the subclavian artery is required in cases of operations about the shoulder, in the axilla, or at the upper part of the arm; and the student will observe that

FIG. 490.—The subclavian artery, showing its relations.
(From a preparation in the Museum of the Royal College of Surgeons of England.)

there is only one situation in which it can be effectually applied, viz. where the artery passes across the upper surface of the first rib. In order to compress the vessel in this situation, the shoulder should be depressed, and the surgeon grasping the side of the neck should press with his thumb in the angle formed by the posterior border of the Sterno-mastoid with the upper border of the clavicle, downwards, backwards, and inwards against the rib; if from any cause the shoulder cannot be sufficiently depressed, pressure may be made from before backwards, so as to compress the artery against the **Middle** scalenus and transverse process of the seventh cervical vertebra. In appropriate cases, a preliminary incision may be made through the cervical fascia, and the finger may be pressed down directly upon the artery.

Ligature of the subclavian artery may be required in cases of wounds, or of aneurism in the axilla, or in cases of aneurism on the cardiac side of the point of ligature ; and the third part of the artery is that which is most favourable for an operation, on account of its being comparatively superficial, and most remote from the origin of the large branches. In those cases where the clavicle is not displaced, this operation may be performed with comparative facility; but where the clavicle is pushed up by a large aneurismal tumour in the axilla, the artery is placed at a great depth from the surface, which materially increases the difficulty of the operation. Under these circumstances, it becomes a matter of importance to consider the height to which this vessel reaches above the bone. In ordinary cases, its arch is about half an inch above the clavicle, occasionally as high as an inch and a half, and sometimes so low as to be on a level with its upper border. If the clavicle is displaced, these variations will necessarily make the operation more or less difficult, according as the vessel is more or less accessible.

The chief points in the operation of tying the third portion of the subclavian artery are as follows : The patient being placed on a table in the supine position, with the head drawn over to the opposite side, and the shoulder depressed as much as possible, the integument should be drawn downwards over the clavicle, and an incision made through it, upon that bone, from the anterior border of the Trapezius to the posterior border of the Sterno-mastoid, to which may be added a short vertical incision meeting the inner end of the preceding. The object in drawing the skin downwards is to avoid any risk of wounding the external jugular vein, for as it perforates the deep fascia above the clavicle, it cannot be drawn downwards with the skin. The soft parts should now be allowed to glide up, and the cervical fascia divided upon a director, and if the interval between the Trapezius and Sterno-mastoid muscles be insufficient for the performance of the opera-tion, a portion of one or both may be divided. The external jugular vein will now be seen towards the inner side of the wound : this and the suprascapular and transverse cervical veins which terminate in it should be held aside. If the external jugular vein is at all in the way and exposed to injury, it should be tied in two places and divided. The suprascapular artery should be avoided, and the Omo-hyoid muscle held aside if necessary. In the space beneath this muscle, careful search must be made for the vessel ; a deep layer of fascia and some connective tissue having been divided carefully, the outer margin of the Scalenus anticus muscle must be felt for, and the finger being guided by it to the first rib, the pulsation of the subclavian artery will be felt as it passes over the rib. The sheath of the vessels having been opened, the aneurism needle may then be passed around the artery from above downwards and inwards so as to avoid including any of the branches of the brachial plexus. If the clavicle is so raised by the tumour that the application of the ligature cannot be effected in this situation, the artery may be tied above the first rib, or even behind the Scalenus anticus muscle ; the difficulties of the operation in such a case will be materially increased, on account of the greater depth of the artery, and the alteration in position of the surrounding parts.

The second part of the subclavian artery, from being that portion which rises highest in the neck, has been considered favourable for the application of the ligature, when it is difficult to tie the artery in the third part of its course. There are, however, many objections to the operation in this situation. It is necessary to divide the Scalenus anticus muscle, upon which lies the phrenic nerve, and at the inner side of which is situated the internal jugular vein ; and a wound of either of these structures might lead to the most dangerous consequences. Again, the artery is in contact, below, with the pleura, which must also be avoided ; and, lastly, the proximity of so many of its large branches arising internal to this point must be a still further objection to the operation. In cases, however, where the sac of an axillary aneurism encroaches on the neck, it may be necessary to divide the outer half or two-thirds of the Scalenus anticus muscle, so as to place the ligature on the vessel at a greater distance from the sac. The operation is performed exactly in the same way as ligature of the third portion, until the Scalenus anticus is exposed, when it is to be divided on a director (never to a greater extent than its outer two-thirds), and it immediately retracts. The operation is therefore merely an extension of ligature of the third portion of the vessel.

In those cases of aneurism of the axillary or subclavian artery which encroach upon the outer portion of the Scalenus muscle to such an extent that a ligature cannot be applied in that situation, it may be deemed advisable, as a last resource, to tie the first portion of the subclavian artery. On the left side, this operation is almost impracticable ; the great depth of the artery from the surface, its intimate relation with the pleura, and its close proximity to the thoracic duct and to so many important veins and nerves, present a series of difficulties which it is next to impossible to overcome.* On the right side, the operation is practicable, and has been performed on several occasions. The main objection to the operation in this situation is the smallness of the interval which usually exists between the commencement of the vessel and the origin of the nearest branch. The operation may be performed in the following manner : The patient being placed on

---

* The operation has, however, been performed by J. K. Rodgers, by Halsted, and by Schumpert.

the table in the supine position, with the neck extended, an incision should be made along the upper border of the inner part of the clavicle, and a second along the inner border of the Sterno-mastoid, meeting the former at an angle. The attachment of both heads of the Sterno-mastoid must be divided on a director, and turned outwards; a few small arteries and veins, and occasionally the anterior jugular, must be avoided, or, if necessary, ligatured in two places and divided, and the Sterno-hyoid and Sterno-thyroid muscles divided in the same manner as the preceding muscle. After tearing through the deep fascia with the finger nail, the internal jugular vein will be seen crossing the subclavian artery; this should be pressed aside, and the artery secured by passing the needle from below upwards, by which the pleura is more effectually avoided. The exact position of the vagus, the recurrent laryngeal, the phrenic and sympathetic nerves should be borne in mind, and the ligature should be applied near the origin of the vertebral, in order to afford as much room as possible for the formation of a coagulum between the ligature and the origin of the vessel. It should be remembered, that the right subclavian artery is occasionally deeply placed in the first part of its course, when it arises from the left side of the aortic arch, and passes in such cases behind the œsophagus, or between it and the trachea.

Collateral Circulation.—After ligature of the third part of the subclavian artery, the collateral circulation is mainly established by three sets of vessels, thus described in a dissection:

' 1. A posterior set, consisting of the suprascapular and posterior scapular branches of the subclavian, anastomosing with the subscapular from the axillary.

' 2. An internal set, produced by the connection of the internal mammary on the one hand, with the superior and long thoracic arteries, and the branches from the subscapular on the other.

' 3. A middle or axillary set, consisting of a number of small vessels derived from branches of the subclavian, above; and, passing through the axilla, terminated either in the main trunk, or some of the branches of the axillary below. This last set presented most conspicuously the peculiar character of newly formed or, rather, dilated arteries, being excessively tortuous, and forming a complete plexus.

' The chief agent in the restoration of the axillary artery below the tumour was the subscapular artery, which communicated most freely with the internal mammary, suprascapular, and posterior scapular branches of the subclavian, from all of which it received so great an influx of blood as to dilate it to three times its natural size.' *

When a ligature is applied to the first part of the subclavian artery, the collateral circulation is carried on by : 1, the anastomosis between the superior and inferior thyroid ; 2, the anastomosis of the two vertebrals ; 3, the anastomosis of the internal mammary with the deep epigastric and the aortic intercostals ; 4, the superior intercostal anastomosing with the aortic intercostals ; 5, the profunda cervicis anastomosing with the princeps cervicis ; 6, the scapular branches of the thyroid axis anastomosing with the branches of the axillary; and 7, the thoracic branches of the axillary anastomosing with the aortic intercostals.

FIG. 491.—Plan of the branches of the right subclavian artery.

Branches.—The branches given off from the subclavian artery are :

Vertebral.
Internal mammary.
Thyroid axis.
Superior intercostal.

On the left side all four branches generally arise from the first portion of the vessel ; but on the right side, the superior intercostal usually springs from the second portion of the vessel. On both sides of the body, the first three branches

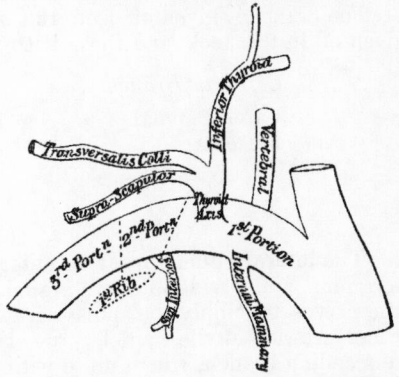

arise close together at the inner margin of the Scalenus anticus ; in the majority of cases, a free interval of from half an inch to an inch exists between the commencement of the artery and the origin of the nearest branch ; in a smaller number of cases, an interval of more than an inch exists, but it never exceeds an inch and three-quarters. In a very few instances, the interval has been found to be less than half an inch. The vertebral artery arises from the upper

* *Guy's Hospital Reports*, vol. i. 1836. Case of axillary aneurism, in which Aston Key had tied the subclavian artery on the outer edge of the Scalenus muscle, twelve years previously.

and posterior part of the artery, the internal mammary from the lower part of the artery; the thyroid axis from in front, and the superior intercostal from behind.

The **Vertebral artery** (fig. 481) is generally the first and largest branch of the subclavian; it arises from the upper and back part of the first portion of the vessel, and, passing upwards, enters the foramen in the transverse process of the sixth cervical vertebra,* and ascends through the foramina in the transverse processes of all the vertebræ above this. Above the upper border of the axis, it inclines outwards and upwards to the foramen in the transverse process of the atlas, through which it passes; it then winds backwards behind its articular process, runs in a deep groove on the upper surface of the posterior arch of this bone, and, passing beneath the posterior occipito-atlantal ligament, pierces the dura mater and arachnoid, and enters the skull through the foramen magnum. It then passes forwards and upwards, inclining from the lateral aspect to the front of the medulla oblongata. It unites, in the middle line, with the vessel of the opposite side at the lower border of the pons Varolii to form the *basilar artery*.

Relations.—For convenience of description the vertebral artery may be divided into four parts. The *first part* (*cervical*) lies between the Longus colli and Scalenus anticus muscles. It has in front of it the internal jugular and vertebral veins, and is crossed by the inferior thyroid artery, and on the left side by the thoracic duct. It rests on the transverse process of the seventh cervical vertebra, and the sympathetic cord. The *second part* (*vertebral*) runs upwards through the foramina in the transverse processes, and is surrounded by a network of nerves from the inferior cervical ganglion of the sympathetic, and a dense plexus of veins which unite to form the vertebral vein at the lower part of the neck. It is situated in front of the cervical nerves as they issue from the intervertebral foramina. The *third part* (*atlantic*) begins where the artery passes through the transverse process of the atlas, and, winding round its articular process, passes beneath the posterior occipito-atlantal ligament. In this stage it is contained in a triangular space (*suboccipital triangle*) formed by the Rectus capitis posticus major, the Superior and the Inferior oblique muscles; and at this point is covered by the Complexus muscle. The suboccipital nerve here lies between the artery and the bone. The *fourth part* (*intracranial*) winds round the medulla oblongata, and is placed between the hypoglossal nerve and the anterior root of the suboccipital nerve, beneath the first digitation of the ligamentum denticulatum, and finally ascends between the basilar process of the occipital bone and the anterior surface of the medulla oblongata.

The branches given off from this vessel may be divided into two sets—those given off in the neck, and those within the cranium.

Cervical Branches.	*Cranial Branches.*
Lateral spinal.	Posterior meningeal.
Muscular.	Anterior spinal.
	Posterior spinal.
	Posterior inferior cerebellar.
	Bulbar.

The **lateral spinal branches** enter the spinal canal through the intervertebral foramina, and divide into two branches. Of these, one passes along the roots of the nerves to supply the spinal cord and its membranes, anastomosing with the other arteries of the spinal cord; the other divides into an ascending and a descending branch, which unite with similar branches from the artery above and below, so that two lateral anastomotic chains are formed on the posterior surface of the bodies of the vertebræ, near the attachment of the pedicles. From these anastomotic chains branches are given off, to supply the periosteum and the bodies of the vertebræ, and to communicate with similar branches from the opposite side; from these latter small branches are given off which join similar branches above and below, so that a central anastomotic chain is formed on the posterior surface of the bodies of the vertebræ.

* The vertebral artery sometimes enters the foramen in the transverse process of the fifth vertebra. Smyth, who tied this artery in the living subject, found it, in one of his dissections, passing into the foramen in the seventh vertebra.

Muscular branches are given off to the deep muscles of the neck, where the vertebral artery curves round the articular process of the atlas. They anastomose with the occipital, and with the ascending and deep cervical arteries.

The **posterior meningeal** are one or two small branches given off from the vertebral opposite the foramen magnum. They ramify between the bone and dura mater in the cerebellar fossæ, and supply the falx cerebelli.

The **anterior spinal** (fig. 492) is a small branch, which arises near the termination of the vertebral, and, descending in front of the medulla oblongata, unites with its fellow of the opposite side at about the level of the foramen magnum. One of these vessels is usually larger than the other, but occasionally they are about equal in size. The single trunk, thus formed, descends on the front of the spinal cord, and is reinforced by a succession of small branches which enter the spinal canal through the intervertebral foramina ; these branches are derived from the vertebral and ascending cervical of the inferior thyroid in the neck ; from the intercostals in the dorsal region ; and from the lumbar, ilio-lumbar, and lateral

FIG. 492.—Arteries and veins of the spinal cord. Schematic arrangement of vessels as seen in a transverse section of the dorsal region of the cord. (Poirier.)

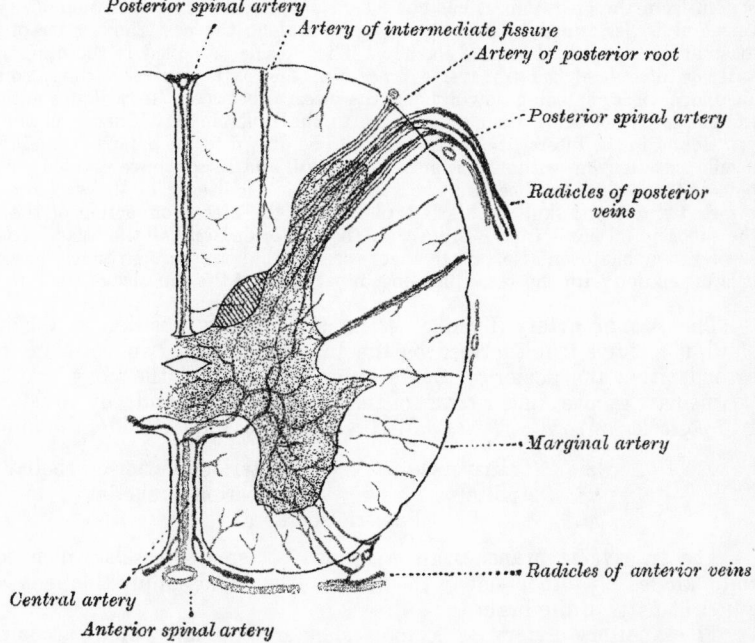

Posterior spinal artery
Artery of intermediate fissure
Artery of posterior root
Posterior spinal artery
Radicles of posterior veins
Marginal artery
Radicles of anterior veins
Central artery
Anterior spinal artery

sacral arteries in the lower part of the spine. They unite, by means of ascending and descending branches, to form a single anterior median artery, which extends as far as the lower part of the spinal cord, and is continued as a slender twig on the filum terminale. This vessel is placed in the pia mater along the anterior median fissure ; it supplies that membrane, and the substance of the cord, and sends off branches at its lower part to be distributed to the cauda equina, and ends on the central fibrous prolongation of the cord.

The **posterior spinal** (fig. 492) arises from the vertebral, at the side of the medulla oblongata ; passing backwards to the posterior aspect of the spinal cord, it descends on the spinal cord, lying behind the posterior roots of the spinal nerves ; and is reinforced by a succession of small branches, which enter the spinal canal through the intervertebral foramina, and by which it is continued to the lower part of the cord, and to the cauda equina. Branches from right and left posterior spinal arteries form a free anastomosis round the posterior roots of the spinal nerves, and communicate, by means of very tortuous transverse branches, with the vessel of the opposite side. At its commencement it gives off an ascending branch, which terminates at the side of the fourth ventricle.

The **posterior inferior cerebellar artery** (fig. 483), the largest branch of the vertebral, winds backwards round the upper part of the medulla oblongata, passing between the origin of the pneumogastric and spinal accessory nerves, over the restiform body to the under surface of the cerebellum, where it divides into two branches : an internal one, which is continued backwards to the notch between the two hemispheres of the cerebellum ; and an external one, which supplies the under surface of the cerebellum, as far as its outer border, where it anastomoses with the anterior inferior cerebellar and the superior cerebellar branches of the basilar artery. Branches from this artery supply the choroid plexus of the fourth ventricle.

The **bulbar arteries** comprise several minute vessels which spring from the vertebral and its branches and are distributed to the medulla oblongata.

Surgical Anatomy.—The vertebral artery has been tied in several instances : 1, for wounds or traumatic aneurism ; 2, in ligature of the innominate, either at the same time to prevent hæmorrhage, or later on to arrest bleeding where it has occurred at the seat of ligature ; and 3, in epilepsy. In these latter cases the treatment was recommended by Alexander, of Liverpool, in the hope that by diminishing the supply of blood to the posterior part of the brain and the spinal cord, a diminution or cessation of the epileptic fits would result. But, on account of the uncertainty as to what cases, if any, derived benefit from the operation, it has now been abandoned. The operation of ligature of the vertebral is performed by making an incision along the posterior border of the Sterno-mastoid muscle, just above the clavicle. The muscle is pulled to the inner side and the anterior tubercle of the transverse process of the sixth cervical vertebra sought for. A deep layer of fascia being now divided, the interval between the Scalenus anticus and the Longus colli, just below their attachment to the tubercle, is defined, and the artery and vein found in the interspace. The vein is to be drawn to the outer side, and the aneurism needle passed from without inwards. Ramskill and Bright have pointed out that severe pain at the back of the head may be symptomatic of disease of the vertebral artery just before it enters the skull. This is explained by the close connection of the artery with the suboccipital nerve in the groove on the posterior arch of the atlas. Disease of the same artery has been also said to affect speech, from pressure on the hypoglossal, where it is in relation with the vessel, leading to paralysis of the muscles of the tongue.

The **Basilar artery** (fig. 483), so named from its position at the base of the skull, is a single trunk formed by the junction of the two vertebral arteries ; it extends from the posterior to the anterior border of the pons Varolii, lying in its median groove, under cover of the arachnoid. It ends by dividing into the two *posterior cerebral arteries*. Its branches are, on each side, the following :

Transverse.   Anterior inferior cerebellar.
Auditory.   Superior cerebellar.
          Posterior cerebral.

The **transverse branches** are a number of small vessels which come off at right angles on either side of the basilar artery and supply the pons Varolii and adjacent parts of the brain.

The **auditory artery** is a long slender branch, which accompanies the auditory nerve into the internal auditory meatus, where it lies between the seventh and eighth nerves, and at the bottom of the meatus passes into the internal ear.

The **anterior inferior cerebellar artery** passes backwards, to be distributed to the anterior part of the under surface of the cerebellum, anastomosing with the posterior inferior cerebellar branch of the vertebral.

The **superior cerebellar arteries** arise near the termination of the basilar. They pass outwards, immediately behind the third nerves, which separate them from the posterior cerebral, wind round the crura cerebri, close to the fourth nerve, and arriving at the upper surface of the cerebellum, divide into branches which ramify in the pia mater and, reaching the circumference of the cerebellum, anastomose with the branches of the inferior cerebellar arteries. Several branches are given to the pineal gland, the valve of Vieussens, and the velum interpositum.

The **posterior cerebral arteries,** the two terminal branches of the basilar, are larger than the preceding, from which they are separated near their origin by the third nerves. Passing outwards, parallel to the superior cerebellar artery, and receiving the posterior communicating from the internal carotid, they wind round the crura cerebri, and pass to the under surface of the occipital lobes of the cerebrum, and break up into branches for the supply of the temporal and occipital lobes. The branches of the posterior cerebral artery are divided into two sets, ganglionic and cortical :

Ganglionic {Postero-median.
Posterior choroid.
Postero-lateral.

Cortical {Anterior temporal.
Posterior temporal.
Occipital.

Ganglionic.—The *postero-median ganglionic branches* (fig. 488) are a group of small arteries which arise at the commencement of the posterior cerebral artery ; these, with similar branches from the posterior communicating, pierce the posterior perforated space, and supply the internal surfaces of the optic thalami and the walls of the third ventricle. The *posterior choroid* enters the interior of the brain beneath the splenium of the corpus callosum, and supplies the velum interpositum and the choroid plexus. The *postero-lateral ganglionic branches* are a group of small arteries which arise from the posterior cerebral artery, after it has turned round the crus cerebri ; they supply a considerable portion of the optic thalamus.

Cortical.—The *cortical branches* are distributed as follows : the first (*anterior temporal*) to the uncinate gyrus ; the second (*posterior temporal*) to the temporo-occipital and the third temporal convolutions ; and the third (*occipital*) to the inner and outer surfaces of the occipital lobe.

The **Thyroid axis** (fig. 475) is a short thick trunk, which arises from the front of the first portion of the subclavian artery, close to the inner border of the Scalenus anticus muscle, and divides, almost immediately after its origin, into three branches, the *inferior thyroid, suprascapular,* and *transverse cervical.*

The **Inferior thyroid artery** passes upwards, in front of the vertebral artery and Longus colli muscle ; then turns inwards behind the sheath of the common carotid artery and internal jugular vein, and also behind the sympathetic nerve, the middle cervical ganglion resting upon the vessel, and reaching the lower border of the lateral lobe of the thyroid gland it divides into two branches, which supply the posterior and under part of the organ, and anastomose in its substance with the superior thyroid, and with the corresponding artery of the opposite side. The recurrent laryngeal nerve passes upwards, generally behind but occasionally in front of the artery. Its branches are, the

Inferior laryngeal.   Œsophageal.
Tracheal.   Ascending cervical.
                 Muscular.

The **inferior laryngeal branch** ascends upon the trachea to the back part of the larynx, in company with the recurrent laryngeal nerve, and supplies the muscles and mucous membrane of this part, anastomosing with the branch from the opposite side, and with the laryngeal branch from the superior thyroid artery.

The **tracheal branches** are distributed upon the trachea, anastomosing below with the bronchial arteries.

The **œsophageal branches** are distributed to the œsophagus, and anastomose with the œsophageal branches of the aorta.

The **ascending cervical** is a small branch which arises from the inferior thyroid, just where that vessel is passing behind the common carotid artery, and runs up on the anterior tubercles of the transverse processes of the cervical vertebræ in the interval between the Scalenus anticus and Rectus capitis anticus major. It gives branches to the muscles of the neck, which anastomose with branches of the vertebral, and sends one or two branches into the spinal canal through the intervertebral foramina to be distributed to the spinal cord and its membranes, and to the bodies of the vertebræ in the same manner as the lateral spinal branches from the vertebral. It anastomoses with the ascending pharyngeal and occipital arteries.

The **muscular branches** supply the depressors of the hyoid bone, the Longus colli, the Scalenus anticus, and the Inferior constrictor of the pharynx.

Surgical Anatomy.—The inferior thyroid artery has been tied, in conjunction with the superior thyroid, in cases of bronchocele. An incision is made along the anterior border of the Sterno-mastoid down to the clavicle. After the deep fascia has been divided, the Sterno-mastoid and carotid vessels are drawn outwards, and the carotid (Chassaignac's) tubercle sought for. The vessel will be found just below this tubercle, between the carotid sheath on the outer side and the trachea and œsophagus on the inner side. In passing the ligature, great care must be exercised to avoid including the recurrent laryngeal nerve, which is occasionally found crossing in *front* of the vessel.

The **Suprascapular artery (transversalis humeri)**, smaller than the transversalis colli, passes obliquely from within outwards, across the root of the neck. It at first passes downwards and outwards across the Scalenus anticus and phrenic nerve, being covered by the Sterno-mastoid; it then crosses the subclavian artery and the cords of the brachial plexus, and runs outwards, behind and parallel with the clavicle and Subclavius muscle, and beneath the posterior belly of the Omo-hyoid, to the superior border of the scapula, where it passes over the transverse ligament of the scapula, which separates it from the suprascapular nerve, to the supraspinous fossa. In this situation it lies close to the bone, and ramifies between it and the Supraspinatus muscle, to which it supplies branches. It then passes downwards behind the neck of the scapula, to reach the infraspinous fossa, where it anastomoses with the dorsalis scapulæ and posterior scapular arteries. Besides distributing branches to the Sterno-mastoid, Subclavius, and neighbouring muscles, it gives off a *suprasternal branch*, which crosses over the sternal end of the clavicle to the skin of the upper part of the chest; and a *supra-acromial branch*, which, piercing the Trapezius muscle, supplies the skin over the acromion, anastomosing with the acromial thoracic artery. As the artery passes over the transverse ligament of the scapula, a branch descends into the subscapular fossa, ramifies beneath that muscle, and anastomoses with the posterior and subscapular arteries. It also sends branches to the acromio-clavicular and shoulder joints, and a nutrient artery to the clavicle.

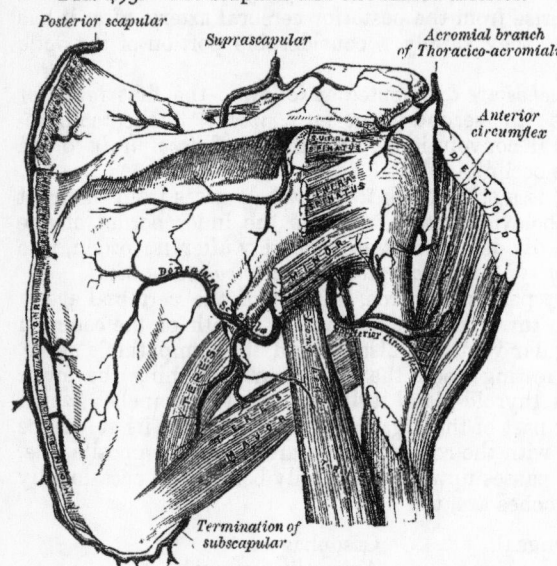

FIG. 493.—The scapular and circumflex arteries.

Posterior scapular

Suprascapular

Acromial branch of Thoracico-acromialis

Anterior circumflex

Termination of subscapular

The **Transverse cervical (transversalis colli)** passes transversely outwards, across the upper part of the subclavian triangle, to the anterior margin of the Trapezius muscle, beneath which it divides into two branches, the *superficial cervical* and the *posterior scapular*. In its passage across the neck, it crosses in front of the phrenic nerve, Scaleni muscles, and the brachial plexus, between the divisions of which it sometimes passes, and is covered by the Platysma, Sterno-mastoid, Omo-hyoid, and Trapezius muscles.

The **superficial cervical** ascends beneath the anterior margin of the Trapezius, distributing branches to it, and to the neighbouring muscles and glands in the neck, and anastomoses with the superficial branch of the arteria princeps cervicis.

The **posterior scapular** passes beneath the Levator anguli scapulæ to the superior angle of the scapula, and then descends along the posterior border of that bone as far as the inferior angle. In its course it is covered by the Rhomboid muscles, supplying them and the Latissimus dorsi and Trapezius, and anastomosing with the suprascapular and subscapular arteries, and with the posterior branches of some of the intercostal arteries.

Peculiarities.—The *superficial cervical* frequently arises as a separate branch from the thyroid axis; and the posterior scapular, from the third, more rarely from the second, part of the subclavian.

The **Internal mammary** (fig. 494) arises from the under surface of the first portion of the subclavian artery, opposite the thyroid axis. It passes downwards and inwards behind the costal cartilage of the first rib to the inner surface of the

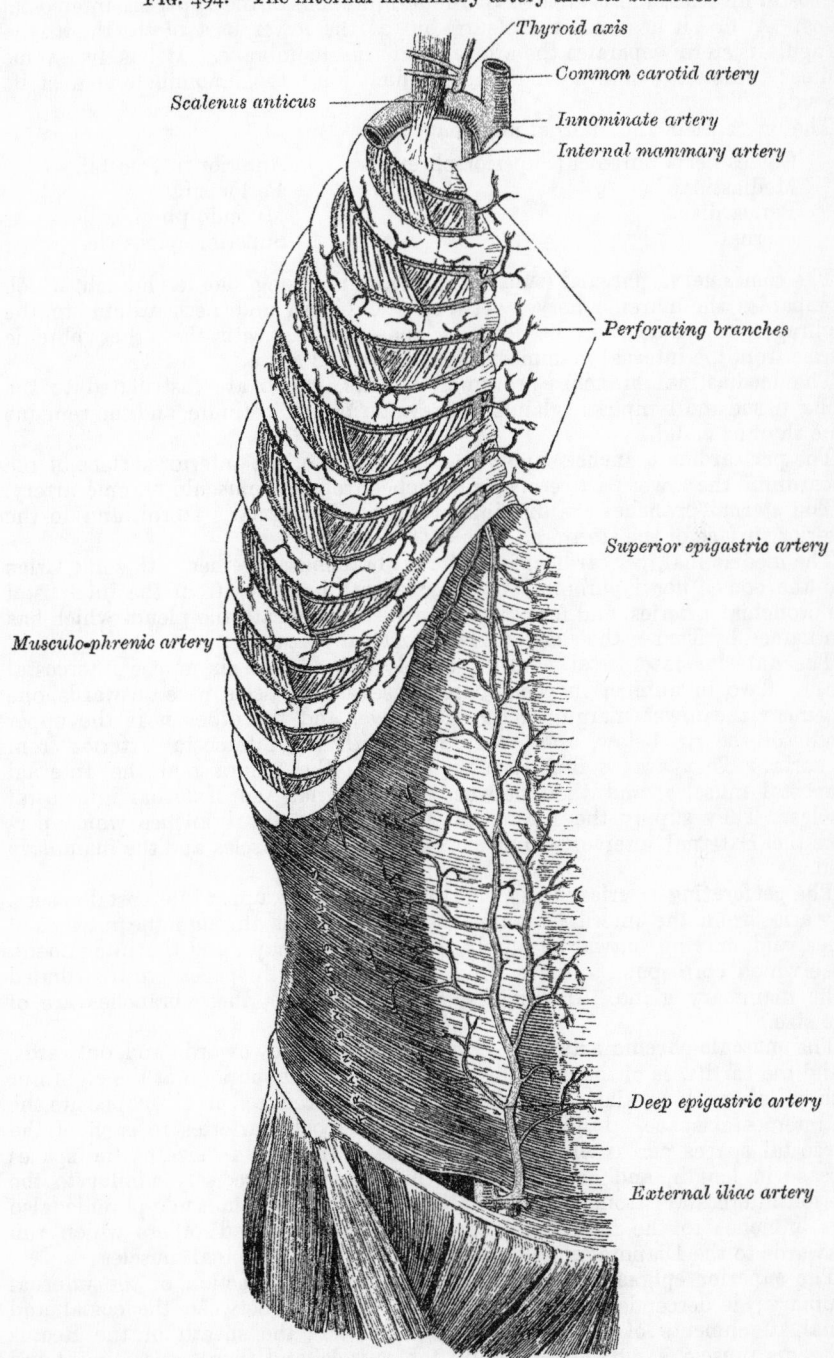

FIG. 494.—The internal mammary artery and its branches.

Thyroid axis

Common carotid artery

Scalenus anticus

Innominate artery

Internal mammary artery

Perforating branches

Superior epigastric artery

Musculo-phrenic artery

Deep epigastric artery

External iliac artery

anterior wall of the chest, resting against the costal cartilages about half an inch from the margin of the sternum ; and, at the interval between the sixth and seventh cartilages, divides into two branches, the *musculo-phrenic* and *superior epigastric*.

Relations.—At its origin it is covered by the internal jugular and subclavian veins, and as it enters the thorax is crossed from without inwards by the phrenic nerve, and then passes forwards close to the outer side of the innominate vein. In the upper part of the thorax, it lies behind the costal cartilages and Internal intercostal muscles, and is crossed by the terminations of the upper six intercostal nerves. At first it lies upon the pleura, but at the lower part of the thorax the Triangularis sterni separates the artery from this membrane. It has two venæ comites ; these unite into a single vein, which joins the innominate vein of its own side.

The branches of the internal mammary are :

Comes nervi phrenici (Superior phrenic).	Anterior intercostal.
Mediastinal.	Perforating.
Pericardiac.	Musculo-phrenic.
Sternal.	Superior epigastric.

The **comes nervi phrenici (superior phrenic)** is a long slender branch, which accompanies the phrenic nerve, between the pleura and pericardium, to the Diaphragm, to which it is distributed ; anastomosing with the other phrenic arteries from the internal mammary, and abdominal aorta.

The **mediastinal branches** are small vessels, which are distributed to the areolar tissue and lymphatic glands in the anterior mediastinum, and the remains of the thymus gland.

The **pericardiac branches** supply the upper part of the anterior surface of the pericardium, the lower part receiving branches from the musculo-phrenic artery.

The **sternal branches** are distributed to the Triangularis sterni, and to the posterior surface of the sternum.

The mediastinal, pericardiac, and sternal branches, together with some twigs from the comes nervi phrenici, anastomose with branches from the intercostal and bronchial arteries, and form a minute plexus beneath the pleura which has been named by Turner the *subpleural mediastinal plexus*.

The **anterior intercostal arteries** supply the five or six upper intercostal spaces. Two in number in each space, these small vessels pass outwards, one lying near the lower margin of the rib above, and the other near the upper margin of the rib below, and anastomose with the intercostal arteries from the aorta. They are at first situated between the pleura and the Internal intercostal muscles, and then between the Internal and External intercostal muscles. They supply the Intercostal muscles and, by branches which perforate the External intercostal muscle, the Pectoral muscles and the mammary gland.

The **perforating arteries** correspond to the five or six upper intercostal spaces. They arise from the internal mammary, pass forwards through the intercostal spaces, and, curving outwards, supply the Pectoralis major and the integument. Those which correspond to the second, third, and fourth spaces are distributed to the mammary gland. In females, during lactation, these branches are of large size.

The **musculo-phrenic artery** is directed obliquely downwards and outwards, behind the cartilages of the false ribs, perforates the Diaphragm at the eighth or ninth costal cartilage, and terminates, considerably reduced in size, opposite the last intercostal space. It gives off anterior intercostal arteries to each of the intercostal spaces across which it passes ; these diminish in size as the spaces decrease in length, and are distributed in a manner precisely similar to the anterior intercostals from the internal mammary. The musculo-phrenic also gives branches to the lower part of the pericardium, and others which run backwards to the Diaphragm, and downwards to the abdominal muscles.

The **superior epigastric** continues in the original direction of the internal mammary ; it descends through the cellular interval between the costal and sternal attachments of the Diaphragm, and enters the sheath of the Rectus abdominis muscle, at first lying behind the muscle, and then perforating it and supplying it, and anastomosing with the deep epigastric artery from the external iliac. Some vessels perforate the sheath of the Rectus, and supply the muscles of the abdomen and the integument, and a small branch, which passes inwards upon the side of the ensiform appendix, anastomoses in front of that cartilage with the artery of the opposite side. It also gives some twigs to the Diaphragm, while from the artery of the right side small branches extend into the falciform ligament of the liver and anastomose with the hepatic artery.

Surgical Anatomy.—The course of the internal mammary artery may be defined by drawing a line across the six upper intercostal spaces, half an inch from and parallel with the sternum. The position of the vessel must be remembered, as it is liable to be wounded in stabs of the chest-wall. It is most easily reached by a transverse incision in the second intercostal space.

The **Superior intercostal** (fig. 481) arises from the upper and back part of the subclavian artery, behind the Anterior scalene muscle on the right side, and to the inner side of that muscle on the left side. Passing backwards, it gives off the *deep cervical branch*, and then descends behind the pleura in front of the necks of the first two ribs, and inosculates with the first aortic intercostal. As it crosses the neck of the first rib it lies to the inner side of the anterior division of the first dorsal nerve and to the outer side of the first thoracic ganglion of the sympathetic. In the first intercostal space, it gives off a branch which is distributed in a manner similar to the distribution of the aortic intercostals. The branch for the second intercostal space usually joins with one from the highest aortic intercostal. This branch is not constant, but more commonly found on the right side; when absent, its place is supplied by an intercostal branch from the aorta. Each intercostal gives off a branch to the posterior spinal muscles, and a small one which passes through the corresponding intervertebral foramen to the spinal cord and its membranes.

The *deep cervical branch* (*profunda cervicis*) arises, in most cases, from the superior intercostal, and is analogous to the posterior branch of an aortic intercostal artery: occasionally it arises as a separate branch from the subclavian artery. Passing backwards, above the eighth cervical nerve and between the transverse process of the seventh cervical vertebra and the neck of the first rib, it runs up the back part of the neck, between the Complexus and Semispinalis colli muscles, as high as the axis, supplying these and adjacent muscles, and anastomosing with the deep branch of the arteria princeps cervicis of the occipital, and with branches which pass outwards from the vertebral. It gives off a special branch which enters the spinal canal through the intervertebral foramen between the seventh cervical and first dorsal vertebra.

ANATOMY OF THE AXILLA

The **Axilla** is a pyramidal space, situated between the upper and lateral part of the chest and the inner side of the arm.

Boundaries.—Its *apex*, which is directed upwards towards the root of the neck, corresponds to the interval between the first rib, the upper edge of the scapula, and the clavicle, through which the axillary vessels and nerves pass. The *base*, directed downwards, is formed by the integument, and a thick layer of fascia, the *axillary fascia*, extending between the lower border of the Pectoralis major in front, and the lower border of the Latissimus dorsi behind; it is broad internally, at the chest, but narrow and pointed externally, at the arm. The *anterior boundary* is formed by the Pectoralis major and minor muscles, the former covering the whole of the anterior wall of the axilla, the latter covering only its central part. The space between the inner border of the Pectoralis minor and the clavicle is occupied by the costo-coracoid membrane. The *posterior boundary*, which extends somewhat lower than the anterior, is formed by the Subscapularis above, the Teres major and Latissimus dorsi below. On the *inner side* are the first four ribs with their corresponding Intercostal muscles, and part of the Serratus magnus. On the *outer side*, where the anterior and posterior boundaries converge, the space is narrow, and bounded by the humerus, the Coraco-brachialis and Biceps muscles.

Contents.—This space contains the axillary vessels, and brachial plexus of nerves, with their branches; some branches of the intercostal nerves, and a large number of lymphatic glands, all connected together by a quantity of fat and loose areolar tissue.

Their position.—The axillary artery and vein, with the brachial plexus of nerves, extend obliquely along the outer boundary of the axillary space, from its apex to its base, and are placed much nearer the anterior than the posterior wall, the vein lying to the inner or thoracic side of the artery, and partially concealing it. At the fore part of the axillary space, in contact with the Pectoral muscles, are the thoracic branches of the axillary artery, and along the lower margin of the Pectoralis minor the long thoracic artery extends to the side of the chest. At the back part, in contact with the lower margin of the Subscapularis muscle, are the subscapular vessels and nerves; winding around the outer border of this muscle are the dorsalis scapulæ artery and veins; and, close to the neck of the humerus, the posterior circumflex vessels and the circumflex nerve are seen curving backwards to the shoulder.

Along the inner or thoracic side no vessel of any importance exists, the upper part of the space being crossed merely by a few small branches from the superior thoracic artery. There are some important nerves, however, in this situation,

Fig. 495.—The axillary artery and its branches.

Anterior circumflex

viz. the posterior thoracic or external respiratory nerve, descending on the surface of the Serratus magnus, to which it is distributed; and perforating the upper and anterior part of this wall, the intercosto-humeral nerve or nerves, passing across the axilla to the inner side of the arm.

The cavity of the axilla is filled by a quantity of loose areolar tissue, and a large number of small arteries and veins, all of which are, however, of inconsiderable size, and numerous lymphatic glands, the position and arrangement of which are described on a subsequent page.

Surgical Anatomy.—The axilla is a space of considerable surgical importance. It transmits the large vessels and nerves to the upper extremity, and these may be the seat of injury or disease: it contains numerous lymphatic glands which may require removal; in it is a quantity of loose connective and adipose tissue which may be readily infiltrated with blood or inflammatory exudation, and it may be the seat of rapidly growing tumours. Moreover, it is covered at its base by thin skin, largely supplied with sebaceous and sweat glands, which is frequently the seat of small cutaneous abscesses and boils, and of eruptions due to irritation.

Penetrating wounds in the axilla are sometimes accompanied by extensive hæmorrhage, either from wound of the main vessels, or of one of the large branches of the axillary artery, e.g. the long thoracic or the subscapular. Where the blood cannot find an easy exit externally, it collects in the axillary space and forms a large swelling which projects in the floor of the axilla and also bulges forwards the Pectoralis major. The treatment consists in freely opening up the cavity and searching for and securing the bleeding vessel.

In *suppuration in the axilla*, the arrangement of the fasciæ plays a very important part in the direction which the pus takes. As described on page 505, the costo-coracoid membrane, after covering in the space between the clavicle and the upper border of the Pectoralis minor, splits to enclose this muscle, and, reblending at its lower border, becomes incorporated with the axillary fascia at the anterior fold of the axilla. This is known as the *clavi-pectoral fascia*. Suppuration may take place either superficial to or beneath this layer of fascia ; that is, either between the Pectorals or below the Pectoralis minor : in the former case, it would point either at the anterior border of the axillary fold, or in the groove between the Deltoid and the Pectoralis major ; in the latter, the pus would have a tendency to surround the vessels and nerves, and ascend into the neck, that being the direction in which there is least resistance. Its progress towards the surface is prevented by the axillary fascia ; its progress backwards, by the Serratus magnus ; forwards, by the clavi-pectoral fascia ; inwards, by the wall of the thorax ; and outwards, by the upper limb. The pus in these cases, after extending into the neck, has been known to spread through the superior opening of the thorax into the mediastinum. In some instances which have been recorded, the pus has found its way along the course of the vessels into the upper arm.

In opening an axillary abscess, the knife should be entered in the floor of the axilla, midway between the anterior and posterior margins and near the thoracic side of the space. It is well to use a director and dressing forceps, after an incision has been made through the skin and fascia, in the manner directed by Hilton.

The student should attentively consider the relation of the vessels and nerves in the several parts of the axilla, for it is the almost universal plan, in the present day, to remove the glands from the axilla in operating for cancer of the breast. In performing such an operation, it is necessary to proceed with much caution in the direction of the outer wall and apex of the space, as here the axillary vessels are in danger of being wounded. Towards the posterior wall, it is necessary to avoid the subscapular, dorsalis scapulæ, and posterior circumflex vessels. Along the anterior wall, it is necessary to avoid the thoracic branches. In clearing out the axilla, the axillary vein should be first defined and traced up to the apex of the space by means of an elevator or other blunt instrument. The Pectoralis major is retracted by an assistant ; or, as is more commonly the practice in the present day, the sternal origin of this muscle is first removed. This proceeding not only lessens the chance of recurrence of the disease, but also enables the surgeon to clear out the axillary cavity more thoroughly. When the apex of the space is reached all fat and glands must be carefully removed and the whole axilla cleared by separating the tissues along the inner and posterior walls, so that when the proceeding is completed the axilla is cleared of all its contents except the main vessels and nerves.

AXILLARY ARTERY

The **Axillary artery**, the continuation of the subclavian, commences at the outer border of the first rib, and terminates at the lower border of the tendon of the Teres major muscle, where it takes the name of brachial. Its direction varies with the position of the limb : when the arm lies by the side of the chest, the vessel forms a gentle curve, the convexity being upwards and outwards ; when it is directed at right angles with the trunk, the vessel is nearly straight ; and when it is elevated still higher, the artery describes a curve, the concavity of which is directed upwards. At its commencement the artery is very deeply situated, but near its termination is superficial, being covered only by the skin and fascia. The description of the relations of this vessel is facilitated by its division into three portions : the first portion being that above the Pectoralis minor ; the second portion, behind ; and the third below, that muscle.

The **first portion of the axillary artery** is in relation, *in front*, with the clavicular portion of the Pectoralis major, the costo-coracoid membrane, the external anterior thoracic nerve, and the acromio-thoracic and cephalic veins ; *behind*, with the first intercostal space, the corresponding Intercostal muscle, the first and second digitations of the Serratus magnus, and the posterior thoracic and internal anterior thoracic nerves ; on its *outer side*, with the brachial plexus, from which it is separated by a little cellular interval ; on its *inner*, or thoracic side, with the axillary vein which overlaps the artery.

RELATIONS OF THE FIRST PORTION OF THE AXILLARY ARTERY

In front.

Pectoràlis major.
Costo-coracoid membrane.
External anterior thoracic nerve.
Acromio-thoracic and cephalic veins.

Outer side.		*Inner side.*
Brachial plexus.	Axillary Artery. First Portion.	Axillary vein.

Behind.

First intercostal space, and Intercostal muscle.
First and second digitations of Serratus magnus.
Posterior thoracic and internal anterior thoracic nerves.

The **second portion of the axillary artery** lies behind the Pectoralis minor. It is covered, *in front*, by the Pectoralis major and minor muscles; *behind* is the posterior cord of the brachial plexus, and it is separated from the Subscapularis by a cellular interval; on the *inner side* is the axillary vein, separated from the artery by the inner cord of the plexus and the internal anterior thoracic nerve; on the *outer side* is the outer cord of the brachial plexus. The brachial plexus of nerves surrounds the artery on three sides, and separates it from direct contact with the vein and adjacent muscles.

RELATIONS OF THE SECOND PORTION OF THE AXILLARY ARTERY

In front.

Pectoralis major and minor.

Outer side.		*Inner side.*
Outer cord of plexus.	Axillary Artery. Second Portion.	Axillary vein. Inner cord of plexus. Internal anterior thoracic nerve.

Behind.

Subscapularis.
Posterior cord of plexus.

The **third portion of the axillary artery** lies below the Pectoralis minor. It is in relation, *in front*, with the lower part of the Pectoralis major above, but covered only by the integument and fascia below, where it is crossed by the inner head of the median nerve; *behind*, with the lower part of the Subscapularis, and the tendons of the Latissimus dorsi and Teres major; on its *outer side*, with the Coraco-brachialis; on its *inner*, or thoracic side, with the axillary vein. The nerves of the brachial plexus bear the following relation to the artery in this part of its course: on the *outer side* is the median nerve, and the musculo-cutaneous for a short distance; on the *inner side*, the ulnar (between the vein and artery) and lesser internal cutaneous nerves (to the inner side of the vein); in *front* is the internal cutaneous nerve, and *behind*, the musculo-spiral and circumflex, the latter extending only to the lower border of the Subscapularis muscle.

RELATIONS OF THE THIRD PORTION OF THE AXILLARY ARTERY

In front.

Integument and fascia.
Pectoralis major.
Inner head of median nerve.
Internal cutaneous nerve.

Outer side.

Coraco-brachialis.
Median nerve.
Musculo-cutaneous nerve.

Axillary
Artery.
Third Portion.

Inner side.

Ulnar nerve.
Axillary vein.
Lesser internal cutaneous
nerve.

Behind.

Subscapularis.
Tendons of Latissimus dorsi and Teres major.
Musculo-spiral, and circumflex nerves.

Peculiarities.—The axillary artery, in about one case out of every ten, gives off a large branch, which forms either one of the arteries of the forearm, or a large muscular trunk. In the first set of cases, this artery is most frequently the radial (1 in 33), sometimes the ulnar (1 in 72), and, very rarely, the interosseous (1 in 506). In the second set of cases, the trunk has been found to give origin to the subscapular, circumflex, and profunda arteries of the arm. Sometimes, only one of the circumflex, or one of the profunda arteries, arose from the trunk. In these cases the brachial plexus surrounded the trunk of the branches, and not the main vessel.

Surface Marking.—The course of the axillary artery may be marked out by raising the arm to a right angle and drawing a line from the middle of the clavicle to the point where the tendon of the Pectoralis major crosses the prominence caused by the Coraco-brachialis as it emerges from under cover of the anterior fold of the axilla. The third portion of the artery can be felt pulsating beneath the skin and fascia, at the junction of the anterior with the middle third of the space between the anterior and posterior folds of the axilla, close to the inner border of the Coraco-brachialis.

Surgical Anatomy.—The student, having carefully examined the relations of the axillary artery in its various parts, should now consider in what situation compression of this vessel may be most easily effected, and the best position for the application of a ligature to it when necessary.

Compression of the vessel may be required in the removal of tumours, or in amputation of the upper part of the arm; and the only situation in which this can be effectually made is in the lower part of its course; by pressing on it in this situation from within outwards against the humerus, the circulation may be effectually arrested.

The axillary artery is perhaps more frequently lacerated than any other artery in the body, with the exception of the popliteal, by violent movements of the upper extremity, especially in those cases where its coats are diseased. It has occasionally been ruptured in attempts to reduce old dislocations of the shoulder-joint. This lesion is most likely to occur during the preliminary breaking down of adhesions, in consequence of the artery having become fixed to the capsule of the joint. Aneurism of the axillary artery is of frequent occurrence: a large percentage of the cases being traumatic in their origin, due to the violence to which it is exposed in the varied, extensive, and often violent movements of the limb.

The *application of a ligature to the axillary artery* may be required in cases of aneurism of the upper part of the brachial, or as a distal operation for aneurism of the subclavian; and there are only two situations in which it can be secured, viz. in the first and in the third parts of its course; for the axillary artery at its central part is so deeply seated, and, at the same time, so closely surrounded with large nervous trunks, that the application of a ligature to it in that situation would be almost impracticable.

In the *third part* of its course the operation is most simple, and may be performed in the following manner: The patient being placed on a bed, and the arm separated from the side, with the hand supinated, an incision is made through the integument forming the floor of the axilla, about two inches in length, a little nearer to the anterior than the posterior fold of the axilla. After carefully dissecting through the areolar tissue and fascia, the median nerve and axillary vein are exposed; the former having been displaced to the outer, and the latter to the inner side of the arm, the elbow being at the same time bent, so as to relax the structures and facilitate their separation, the ligature may be passed round the artery from the ulnar to the radial side. This portion of the artery is occasionally crossed by a muscular slip, the *axillary arch*, derived from the Latissimus dorsi, which may mislead the surgeon during an operation. The occasional existence of this muscular fasciculus was spoken of in the description of the muscles. It may easily be recognised by the transverse direction of its fibres.

The *first portion* of the axillary artery may be tied in cases of aneurism encroaching so far upwards that a ligature cannot be applied in the lower part of its course. Notwithstanding that this operation has been performed in some few cases, and with success, its performance is attended with much difficulty and danger. The student will remark that, in this situation, it would be necessary to divide a thick muscle, and, after incising the costo-coracoid membrane, the artery would be exposed at the bottom of a more or less deep space, with the cephalic and axillary veins in such relation with it as must render the application of a ligature to this part of the vessel particularly hazardous. Under such circumstances it is an easier, and, at the same time, more advisable operation, to tie the subclavian artery in the third part of its course.

The vessel can be best secured by a curved incision with the convexity downwards from a point half an inch external to the Sterno-clavicular joint to a point half an inch internal to the coracoid process. The limb is to be well abducted and the head inclined to the opposite side, and the incision carried through the superficial structures, care being taken of the cephalic vein at the outer angle of the incision. The clavicular origin of the Pectoralis major is then divided in the whole extent of the wound. The arm is now to be brought to the side, and the upper edge of the Pectoralis minor defined and drawn downwards. The costo-coracoid membrane is to be carefully divided on a director, close to the coracoid process, and the axillary sheath exposed; this is to be opened with especial care on account of the vein overlapping the artery. The needle should be passed from below, so as to avoid wounding the vein.

In a case of wound of the vessel, the general practice of cutting down upon, and tying it above and below the wounded point, should be adopted in all cases.

Collateral Circulation after Ligature of the Axillary Artery.—If the artery be tied above the origin of the acromial thoracic, the collateral circulation will be carried on by the same branches as after the ligature of the subclavian; if at a lower point, between the acromial thoracic and subscapular arteries, the latter vessel, by its free anastomoses with the other scapular arteries, branches of the subclavian, will become the chief agent in carrying on the circulation, to which the long thoracic, if it be below the ligature, will materially contribute, by its anastomoses with the intercostal and internal mammary arteries. If the point included in the ligature is below the origin of the subscapular artery, it will most probably also be below the origins of the two circumflex arteries. The chief agents in restoring the circulation will then be the subscapular and the two circumflex arteries anastomosing with the superior profunda from the brachial, which will be afterwards referred to as performing the same office after ligature of the brachial. The cases in which the operation has been performed are few in number, and no published account of dissections of the collateral circulation appears to exist.

The branches of the axillary artery are—

From first part, Superior thoracic.

From second part { Acromio-thoracic. / Long thoracic. / Alar thoracic.

From third part { Subscapular. / Posterior circumflex. / Anterior circumflex.

The **superior thoracic** is a small artery, which arises from the axillary separately, or by a common trunk with the acromio-thoracic. Running forwards and inwards along the upper border of the Pectoralis minor, it passes between it and the Pectoralis major to the side of the chest. It supplies these muscles, and the parietes of the thorax, anastomosing with the internal mammary and intercostal arteries.

The **acromio-thoracic** is a short trunk, which arises from the fore part of the axillary artery, its origin being generally overlapped by the upper edge of the Pectoralis minor. Projecting forwards to the upper border of the Pectoralis minor, it pierces the costo-coracoid membrane and divides into four branches— *thoracic, acromial, clavicular,* and *humeral.* The *thoracic branch* runs forwards and inwards between the two Pectoral muscles, and is distributed to them and to the mammary gland, anastomosing with the intercostal branches of the internal mammary. The *acromial branch* is directed outwards towards the acromion, supplying the Deltoid muscle, and anastomosing, on the surface of the acromion, with the suprascapular, and posterior circumflex arteries. The *clavicular branch* runs upwards and inwards to the sterno-clavicular joint, supplying this articulation, the structures around, and the Subclavius muscle.

The *humeral* or *descending branch* passes in the space between the Pectoralis major and Deltoid, in the same groove as the cephalic vein, and gives branches to both muscles.

The **long thoracic** passes downwards and inwards along the lower border of the Pectoralis minor to the side of the chest, supplying the Serratus magnus, the Pectoral muscles, and mammary gland, and sending branches across the axilla to the axillary glands and Subscapularis; it anastomoses with the internal mammary and intercostal arteries.

The **alar thoracic** is a small branch, which supplies the glands and areolar tissue of the axilla. Its place is frequently taken by branches from some of the other thoracic arteries.

The **subscapular,** the largest branch of the axillary artery, arises opposite the lower border of the Subscapularis muscle, and passes downwards and backwards along its lower margin to the inferior angle of the scapula, where it anastomoses with the long thoracic and intercostal arteries and with the posterior scapular, a branch of the transversalis colli, from the thyroid axis of the subclavian. About an inch and a half from its origin it gives off a large branch, the *dorsalis scapulæ*, and terminates by supplying branches to the muscles in the neighbourhood.

The *dorsalis scapulæ* is given off from the subscapular about an inch and a half from its origin, and is generally larger than the continuation of the vessel. It curves round the axillary border of the scapula, leaving the axilla through the space between the Subscapularis above, the Teres major below, and the long head of the Triceps externally (fig. 493), and enters the infraspinous fossa by passing under cover of the Teres minor, where it anastomoses with the posterior scapular and suprascapular arteries. In its course it gives off two sets of branches: one enters the subscapular fossa beneath the Subscapularis, which it supplies, anastomosing with the posterior scapular and suprascapular arteries; the other is continued along the axillary border of the scapula, between the Teres major and minor, and, at the dorsal surface of the inferior angle of the bone, anastomoses with the posterior scapular. In addition to these, small branches are distributed to the back part of the Deltoid muscle and the long head of the Triceps, anastomosing with an ascending branch of the superior profunda of the brachial.

The **circumflex arteries** wind round the neck of the humerus. The *posterior circumflex* (fig. 493), the larger of the two, arises from the back part of the axillary artery opposite the lower border of the Subscapularis muscle, and, passing backwards with the circumflex veins and nerve through the quadrangular space bounded by the Teres major and minor, the scapular head of the Triceps and the humerus, winds round the neck of that bone and is distributed to the Deltoid muscle and shoulder-joint, anastomosing with the anterior circumflex, and acromial thoracic arteries, and with the superior profunda branch of the brachial artery. The *anterior circumflex* (figs. 493, 495), considerably smaller than the preceding, arises nearly opposite that vessel, from the outer side of the axillary artery. It passes horizontally outwards, beneath the Coraco-brachialis and short head of the Biceps, lying upon the fore part of the neck of the humerus, and, on reaching the bicipital groove, gives off an ascending branch which passes upwards along the groove to supply the head of the bone and the shoulder-joint. The trunk of the vessel is then continued outwards beneath the Deltoid, which it supplies, and anastomoses with the posterior circumflex artery.

BRACHIAL ARTERY (fig. 496)

The **Brachial artery** commences at the lower margin of the tendon of the Teres major, and, passing down the inner and anterior aspect of the arm, terminates about half an inch below the bend of the elbow, where it divides into the *radial* and *ulnar* arteries. At first the brachial artery lies internal to the humerus; but as it passes down the arm it gradually gets in front of the bone, and at the bend of the elbow it lies midway between the two condyles.

Relations.—This artery is superficial throughout its entire extent, being covered, *in front,* by the integument, the superficial and deep fascia; the bicipital fascia separates it opposite the elbow from the median basilic vein; the median nerve crosses it from without inwards opposite the insertion of the Coraco-brachialis. *Behind,* it is separated from the long head of the Triceps by the musculo-spiral nerve and superior profunda artery. It then lies upon the inner

U U

head of the Triceps, next upon the insertion of the Coraco-brachialis, and lastly on the Brachialis anticus. By its *outer side*, it is in relation with the commencement of the median nerve, and the Coraco-brachialis and Biceps muscles, the two muscles overlapping the artery to a considerable extent. By its *inner side*, its upper half is in relation with the internal cutaneous and ulnar nerves, its lower half with the median nerve. The basilic vein lies on the inner side of the artery, but is separated from it in the lower part of the arm by the deep fascia. It is accompanied by two venæ comites, which lie in close contact with the artery, being connected together at intervals by short transverse communicating branches.

PLAN OF THE RELATIONS OF THE BRACHIAL ARTERY

In front.

Integument and fasciæ.
Bicipital fascia, median basilic vein.
Median nerve.
Overlapped by Coraco-brachialis and Biceps.

Outer side.		*Inner side.*
Median nerve (above).	Brachial Artery.	Internal cutaneous and ulnar nerve.
Coraco-brachialis.		Median nerve (below).
		Basilic vein.

Behind.

Triceps (long and inner heads).
Musculo-spiral nerve.
Superior profunda artery.
Coraco-brachialis.
Brachialis anticus.

ANATOMY OF THE BEND OF THE ELBOW

At the bend of the elbow the brachial artery sinks deeply into a triangular interval, the base of which is directed upwards, and may be represented by a line connecting the two condyles of the humerus; the sides are bounded, externally, by the inner edge of the Brachio-radialis; internally, by the outer margin of the Pronator radii teres; its floor is formed by the Brachialis anticus and Supinator brevis. This space contains the brachial artery, with its accompanying veins; the radial and ulnar arteries; the median and musculo-spiral nerves; and the tendon of the Biceps. The brachial artery occupies the middle line of this space, and divides opposite the neck of the radius into the radial and ulnar arteries; it is covered, *in front*, by the integument, the superficial fascia, and the median basilic vein, the vein being separated from direct contact with the artery by the bicipital fascia. *Behind*, it lies on the Brachialis anticus, which separates it from the elbow-joint. The median nerve lies on the inner side of the artery, close to it above; but separated from it below by the coronoid origin of the Pronator radii teres. The tendon of the Biceps lies to the outer side of the space, and the musculo-spiral nerve still more externally, situated upon the Supinator brevis, and concealed by the Brachio-radialis.

Peculiarities of the Brachial Artery as regards its Course.—The brachial artery, accompanied by the median nerve, may leave the inner border of the Biceps, and descend towards the inner epicondyle of the humerus; about two inches above the epicondyle, it usually curves round a prominence of bone, the *supracondylar process*, from which a fibrous arch is in most cases thrown over the artery; it then inclines outwards, beneath or through the substance of the Pronator radii teres muscle, to the bend of the elbow. The variation bears considerable analogy with the normal condition of the artery in some of the carnivora: it has been referred to in the description of the humerus (page 276).

As regards its Division.—Occasionally, the artery is divided for a short distance at its upper part into two trunks, which are united below. A similar peculiarity occurs in the main vessel of the lower limb.

The vessels concerned in the high division of the brachial artery are three : viz. radial, ulnar, and interosseous. Most frequently the radial is given off high up ; the other limb of the bifurcation consisting of ulnar and interosseous. In some instances the ulnar arises from the brachial above the ordinary level, and the radial and interosseous form the other limb of the division ; and occasionally the interosseous arises high up, the radial and ulnar continuing down together.

Sometimes, long slender vessels, *vasa aberrantia*, connect the brachial or axillary arteries with one of the arteries of the forearm, or a branch from them. These vessels usually join the radial.

Varieties in Muscular Relations.*—The brachial artery is occasionally concealed, in some part of its course, by muscular or tendinous slips derived from the Coraco-brachialis, Biceps, Brachialis anticus, and Pronator radii teres muscles.

Surface Marking.—The direction of the brachial artery is marked by a line drawn along the inner edge of the Biceps from the insertion of the Teres major muscle to the point midway between the epicondyles of the humerus.

Surgical Anatomy.—In spite of the fact that the brachial artery is very superficial and but little protected by surrounding tissues, it is seldom wounded. This, no doubt, is due to its situation on the inner side of the arm, which is little exposed to injury. Compression of the brachial artery is required in cases of amputation and some other operations in the arm and forearm ; and it will be observed that it may be effected in almost any part of the course of the artery. If pressure is made in the upper part of the limb, it should be directed from within outwards ; and if in the lower part, from before backwards, as the artery lies on the inner side of the humerus above, and in front of it below. The most favourable situation is about the middle of the arm, where it lies on the tendon of the Coraco-brachialis on the inner flat side of the humerus.

The application of a ligature to the brachial artery may be required in cases of wound of the vessel, and in some cases of wound of the palmar arch. It is also sometimes necessary in cases of aneurism of the brachial, the radial, ulnar, or interosseous arteries. The artery may be secured in any part of its course. The chief guides in determining its position are the surface markings produced by the inner margin of the Coraco-brachialis and Biceps, the known course of the vessel, and its pulsation, which should be carefully felt for before any operation is performed, as the

FIG. 496.—The brachial artery.

Internal cutaneous nerve

Musculo-spiral nerve

Superior profunda

Inferior profunda

Anastomotica magna

vessel occasionally deviates from its usual position in the arm. In whatever situation the operation is performed, great care is necessary, on account of the extreme thinness of the parts covering the artery, and the intimate connection which the vessel has throughout its whole course with important nerves and veins. Sometimes a thin layer of muscular fibre is met with concealing the artery ; if such is the case, it must be cut across in order to expose the vessel.

---

* See Struthers's *Anatomical and Physiological Observations.*

In the upper third of the arm the artery may be exposed in the following manner The patient being placed supine upon a table, the affected limb should be raised from the side, and the hand supinated. An incision about two inches in length should be made on the inner side of the Coraco-brachialis muscle, and the subjacent fascia cautiously divided, so as to avoid wounding the internal cutaneous nerve or basilic vein, which sometimes runs on the surface of the artery as high as the axilla. The fascia having been divided, it should be remembered that the ulnar and internal cutaneous nerves lie on the inner side of the artery, the median on the outer side, the latter nerve being occasionally superficial to the artery in this situation, and that the venæ comites are also in relation with the vessel, one on either side. These being carefully separated, the aneurism needle should be passed round the artery from the inner to the outer side.

If two arteries are present in the arm, in consequence of a high division, they are usually placed side by side; and if they are exposed in an operation, the surgeon should endeavour to ascertain, by alternately pressing on each vessel, which of the two communicates with the wound or aneurism, when a ligature may be applied accordingly; or if pulsation or hæmorrhage ceases only when both vessels are compressed, both vessels may be tied, as it may be concluded that the two communicate above the seat of disease, or are reunited.

It should also be remembered that two arteries may be present in the arm in a case of high division, and that one of these may be found along the inner intermuscular septum, in a line towards the inner epicondyle of the humerus; or in the usual position of the brachial, but deeply placed beneath the common trunk; a knowledge of these facts will suggest the precautions necessary in every case, and indicate the measures to be adopted when anomalies are met with.

In the middle of the arm the brachial artery may be exposed by making an incision along the inner margin of the Biceps muscle. The forearm being bent so as to relax the muscle, it should be drawn slightly aside, and the fascia carefully divided, when the median nerve will be exposed lying upon the artery (sometimes beneath); this being drawn inwards and the muscle outwards, the artery should be separated from its accompanying veins and secured. In this situation the inferior profunda may be mistaken for the main trunk, especially if enlarged, from the collateral circulation having become established; this may be avoided by directing the incision externally towards the Biceps, rather than inwards or backwards towards the Triceps.

The lower part of the brachial artery is of interest in a surgical point of view, on account of the relation which it bears to the veins most commonly opened in vene-section. Of these vessels, the median basilic is the largest and most prominent, and, consequently, the one usually selected for the operation. It should be remembered that this vein runs parallel with the brachial artery, from which it is separated by the bicipital fascia, and that care should be taken, in opening the vein, not to carry the incision too deep, so as to endanger the artery.

Collateral Circulation.—After the application of a ligature to the brachial artery in the upper third of the arm, the circulation is carried on by branches from the circumflex and subscapular arteries, anastomosing with ascending branches from the superior profunda. If the brachial is tied *below* the origin of the profunda arteries, the circulation is maintained by the branches of the profundæ, anastomosing with the recurrent radial, ulnar, and interosseous arteries. Two cases are described by South,[*] in which the brachial artery had been tied some time previously: in one 'a long portion of the artery had been obliterated, and sets of vessels are descending on either side from above the obliteration, to be received into others which ascend in a similar manner from below it. In the other, the obliteration is less extensive, and a single curved artery about as big as a crow-quill passes from the upper to the lower open part of the artery.'

The branches of the brachial artery are, the

Superior profunda.	Inferior profunda.
Nutrient.	Anastomotica magna.

Muscular.

The **superior profunda** arises from the inner and back part of the brachial, just below the lower border of the Teres major, and passes backwards to the interval between the outer and inner heads of the Triceps muscle, accompanied by the musculo-spiral nerve; it winds round the back of the shaft of the humerus in the spiral groove, between the outer head of the Triceps and the bone, to the outer side of the humerus, where it reaches the external intermuscular septum and divides into two terminal branches. One of these pierces

---

[*] Chelius's *Surgery*, vol. ii. p. 254. See also White's engraving, referred to by South, of the anastomosing branches after ligature of the brachial, in White's *Cases in Surgery*. Porta also gives a case (with drawings) of the circulation after ligature of both brachial and radial. (*Alterazioni Patologiche delle Arterie.*)

the external intermuscular septum, and descends, in company with the musculo-spiral nerve, to the space between the Brachialis anticus and Brachio-radialis, where it anastomoses with the recurrent branch of the radial artery; while the other, much the larger of the two, descends behind the external intermuscular septum to the back of the elbow-joint, where it anastomoses with the posterior interosseous recurrent, and across the back of the humerus with the posterior ulnar recurrent, the anastomotica magna, and inferior profunda (fig. 499). The superior profunda supplies the Triceps muscle and gives off a nutrient artery which enters the bone at the upper end of the musculo-spiral groove. Near its commencement it sends off a branch which passes upwards between the external and long heads of the Triceps muscle to anastomose with the posterior circumflex artery; and, while in the groove, a small branch which accompanies a branch of the musculo-spiral nerve through the substance of the Triceps muscle and ends in the Anconeus below the outer epicondyle of the humerus.

The **nutrient artery** of the shaft of the humerus arises from the brachial, about the middle of the arm. Passing downwards, it enters the nutrient canal of that bone, near the insertion of the Coraco-brachialis muscle.

The **inferior profunda,** of small size, arises from the brachial, a little below the middle of the arm; piercing the internal intermuscular septum, it descends on the surface of the inner head of the Triceps muscle, to the space between the inner epicondyle and olecranon, accompanied by the ulnar nerve, and terminates by anastomosing with the posterior ulnar recurrent and anastomotica magna. It sometimes supplies a branch to the front of the internal epicondyle, which anastomoses with the anterior ulnar recurrent.

The **anastomotica magna** arises from the brachial, about two inches above the elbow-joint. It passes transversely inwards upon the Brachialis anticus, and piercing the internal intermuscular septum, winds round the back part of the humerus between the Triceps and the bone, forming an arch above the olecranon fossa, by its junction with the posterior articular branch of the superior profunda. As this vessel lies on the Brachialis anticus, branches ascend to join the inferior profunda; and others descend in front of the inner epicondyle, to anastomose with the anterior ulnar recurrent. Behind the internal epicondyle a branch is given off which anastomoses with the inferior profunda and posterior ulnar recurrent arteries and supplies the Triceps.

The **muscular** are three or four large branches, which are distributed to the muscles in the course of the artery. They supply the Coraco-brachialis, Biceps, and Brachialis anticus muscles.

The **Anastomosis around the Elbow-joint** (fig. 499).—The vessels engaged in this anastomosis may be conveniently divided into those situated *in front* of and *behind* the internal and external epicondyles. The branches anastomosing *in front* of the internal epicondyle are: the anastomotica magna, the anterior ulnar recurrent, and the anterior branch of the inferior profunda. Those *behind* the internal epicondyle are: the anterior branch of the anastomotica magna, the posterior ulnar recurrent, and the posterior terminal branch of the inferior profunda. The branches anastomosing *in front* of the external epicondyle are: the radial recurrent and the anterior terminal branch of the superior profunda. Those *behind* the external epicondyle (perhaps more properly described as being situated between the external epicondyle and the olecranon) are: the anastomotica magna, the interosseous recurrent, and the posterior terminal branch of the superior profunda. There is also a large arch of anastomosis above the olecranon, formed by the interosseous recurrent joining with the anastomotica magna and posterior ulnar recurrent (fig. 499).

From this description it will be observed that the anastomotica magna is the vessel most engaged, the only part of the anastomosis in which it is not employed being that *in front* of the external epicondyle.

RADIAL ARTERY (fig. 497)

The **Radial Artery** appears, from its direction, to be the continuation of the brachial, but, in size, it is smaller than the ulnar. It commences at the bifurcation of the brachial, just below the bend of the elbow, and passes along the radial side of the forearm to the wrist; it then winds backwards, round the outer side of the carpus, beneath the extensor tendons of the thumb to the upper end of the space between the metacarpal bones of the thumb and index finger, and, finally, passes forwards between the two heads of the First dorsal interosseous muscle, into the palm of the hand, where it crosses the metacarpal bones to the ulnar border of the hand, to form the deep palmar arch. At its termination, it inosculates with the deep branch of the ulnar artery. The relations of this vessel may thus be conveniently divided into three parts, viz. in the forearm, at the back of the wrist, and in the hand.

FIG. 497.—The radial and ulnar arteries.

Radial recurrent

Extensor brevis pollicis

Superficialis volæ

Deep branch of ulnar

Princeps pollicis

Radialis indicis

Relations. — *In the forearm,* this vessel extends from opposite the neck of the radius to the fore part of the styloid process, being placed to the inner side of the shaft of the bone above, and in front of it below. It is overlapped in the upper part of its course by the fleshy belly of the Brachio-radialis muscle; throughout the rest of its course it is superficial, being covered by the integument, the superficial and deep fasciæ. In its course downwards, it lies upon the tendon of the Biceps, the Supinator brevis, the Pronator radii teres, the radial origin of the Flexor sublimis digitorum, the Flexor longus pollicis, the Pronator quadratus, and the lower extremity of the radius. In the upper third of its course, it lies between the Brachio-radialis and the Pronator radii teres; in its lower two-thirds, between the tendons of the Brachio-radialis and the Flexor carpi radialis. The radial nerve lies close to the outer side of the artery in the middle third of its course; and some filaments of the musculo-cutaneous nerve, after piercing the deep fascia, run along the lower part of the artery as it winds round the wrist. The vessel is accompanied by venæ comites throughout its whole course.

PLAN OF THE RELATIONS OF THE RADIAL ARTERY IN THE FOREARM

In front.

Skin, superficial and deep fasciæ.
Brachio-radialis.

Inner side.

Pronator radii teres.
Flexor carpi radialis.

Radial Artery
in Forearm.

Outer side.

Brachio-radialis.
Radial nerve (middle third).

Behind.

Tendon of Biceps.
Supinator brevis.
Pronator radii teres.
Flexor sublimis digitorum.
Flexor longus pollicis.
Pronator quadratus.
Radius.

At the wrist, as it winds round the outer side of the carpus, from the styloid process to the first interosseous space, it lies upon the external lateral ligament, and then upon the scaphoid bone and trapezium, being covered by the extensor tendons of the thumb, subcutaneous veins, some filaments of the radial nerve, and the integument. It is accompanied by two veins, and a filament of the musculo-cutaneous nerve.

In the hand, it passes from the upper end of the first interosseous space, between the heads of the Abductor indicis or First dorsal interosseous muscle, transversely across the palm, to the base of the metacarpal bone of the little finger, where it inosculates with the communicating branch from the ulnar artery, forming the *deep palmar arch.* It lies upon the carpal extremities of the metacarpal bones and the Interossei muscles, being covered by the Adductor obliquus pollicis, the flexor tendons of the fingers, and the Lumbricales. Alongside of it is the deep branch of the ulnar nerve, but running in the opposite direction—that is to say, from within outwards.

Peculiarities.—The origin of the radial artery, according to Quain, is, in nearly one case in eight, higher than usual ; more often arising from the axillary or upper part of the brachial, than from the lower part of this vessel. In the forearm it deviates less frequently from its position than the ulnar. It has been found lying over the fascia instead of beneath it. It has also been observed on the surface of the Brachio-radialis, instead of under its inner border ; and in turning round the wrist, it has been seen lying over, instead of beneath, the extensor tendons of the thumb.

Surface Marking.—The position of the radial artery in the forearm is represented by a line drawn from the outer border of the tendon of the Biceps in the centre of the hollow in front of the elbow-joint with a straight course to the inner side of the fore part of the styloid process of the radius, with the limb in a position of supination.

Surgical Anatomy.—The radial artery is much exposed to injury in its lower third, and is frequently wounded by the hand being driven through a pane of glass, by the slipping of a knife or chisel held in the other hand, and such-like accidents. The injury is often followed by a traumatic aneurism, for which the old operation of laying open the sac and securing the vessel above and below is required.

The operation of tying the radial artery is required in cases of wounds either of its trunk, or of some of its branches, or for aneurism : and it will be observed, that the vessel may be exposed in any part of its course through the forearm without the division of any muscular fibres. The operation in the middle or inferior third of the forearm is easily performed ; but in the upper third, near the elbow, it is attended with some difficulty, from the greater depth of the vessel, and from its being overlapped by the Brachio-radialis muscle.

To tie the artery in the upper third, an incision three inches in length should be made through the integument, in a line drawn from the centre of the bend of the elbow to the front of the styloid process of the radius, avoiding the branches of the median vein ; the fascia of the arm being divided, and the Brachio-radialis drawn a little outwards, the artery will be exposed. The venæ comites should be carefully separated from the vessel, and the ligature passed from the radial to the ulnar side.

In the middle third of the forearm the artery may be exposed by making an incision of similar length on the inner margin of the Brachio-radialis. In this situation, the radial nerve lies in close relation with the outer side of the artery, and should, as well as the veins, be carefully avoided.

In the lower third, the artery is easily secured by dividing the integument and fascia in the interval between the tendons of the Brachio-radialis and Flexor carpi radialis muscles.

The branches of the radial artery may be divided into three groups, corresponding with the three regions in which the vessel is situated.

In the forearm.	*At the wrist.*	*In the hand.*
Radial recurrent.	Posterior carpal.	Princeps pollicis.
Muscular.	Metacarpal.	Radialis indicis.
Anterior carpal.	Dorsales pollicis.	Perforating.
Superficialis volæ.	Dorsalis indicis.	Palmar interosseous.
		Palmar recurrent.

The **radial recurrent** is given off immediately below the elbow. It ascends between the branches of the musculo-spiral nerve lying on the Supinator brevis, and then between the Brachio-radialis and Brachialis anticus, supplying these muscles and the elbow-joint, and anastomosing with the anterior terminal branch of the superior profunda.

The **muscular branches** are distributed to the muscles on the radial side of the forearm.

The **anterior carpal** is a small vessel which arises from the radial artery near the lower border of the Pronator quadratus, and, running inwards in front of the carpus, anastomoses with the anterior carpal branch of the ulnar artery. In this way an arterial anastomosis, *anterior carpal arch*, is formed in front of the wrist: it is joined by branches from the anterior interosseous above, and by recurrent branches from the deep palmar arch below, and gives off branches which descend to supply the articulations of the wrist and carpus.

The **superficialis volæ** arises from the radial artery, just where this vessel is about to wind round the outer side of the wrist. Running forwards, it passes through, occasionally over, the muscles of the thumb, which it supplies, and sometimes anastomoses with the palmar portion of the ulnar artery, completing the superficial palmar arch. This vessel varies considerably in size: usually it is very small, and terminates in the muscles of the thumb; sometimes it is as large as the continuation of the radial.

The **posterior carpal** is a small vessel which arises from the radial artery beneath the extensor tendons of the thumb; crossing the carpus transversely towards the inner border of the hand, it anastomoses with the posterior carpal branch of the ulnar, forming the *posterior carpal arch*, which is joined by the termination of the anterior interosseous artery. From this arch are given off descending branches, the *dorsal interosseous arteries* for the third and fourth interosseous spaces, which run forwards on the Third and Fourth dorsal interossei muscles and divide into dorsal digital branches which supply the adjacent sides of the middle, ring, and little fingers respectively, communicating with the digital arteries of the superficial palmar arch. At their origin they anastomose with the perforating branches from the deep palmar arch.

The **metacarpal (first dorsal interosseous branch)** arises beneath the extensor tendons of the thumb, sometimes with the posterior carpal artery; running forwards on the Second dorsal interosseous muscle, it communicates, behind, with the corresponding perforating branch of the deep palmar arch; and, in front, it divides into two *dorsal digital branches*, which supply the adjoining sides of the index and middle fingers, inosculating with the digital branch of the superficial palmar arch.

The **dorsales pollicis** are two small vessels which run along the sides of the dorsal aspect of the thumb. They arise separately, or occasionally by a common trunk, near the base of the first metacarpal bone.

The **dorsalis indicis**, also a small branch, runs along the radial side of the back of the index finger, sending a few branches to the Abductor indicis.

The **princeps pollicis** arises from the radial just as it turns inwards to the deep part of the hand; it descends between the Abductor indicis and

Adductor obliquus pollicis, along the ulnar side of the metacarpal bone of the thumb, to the base of the first phalanx, where it lies beneath the tendon of the Flexor longus pollicis and divides into two branches. These make their appearance between the inner and outer insertions of the Adductor obliquus pollicis and run along the sides of the palmar aspect of the thumb, and form an arch on the palmar surface of the last phalanx, from which branches are distributed to the integument and pulp of the thumb.

The **radialis indicis** arises close to the preceding, descends between the Abductor indicis and Adductor transversus pollicis, and runs along the radial side of the index finger to its extremity, where it anastomoses with the collateral digital artery from the superficial palmar arch. At the lower border of the Adductor transversus pollicis, this vessel anastomoses with the princeps pollicis, and gives a communicating branch to the superficial palmar arch.

The **perforating arteries,** three in number, pass backwards from the deep palmar arch, through the second, third, and fourth interosseous spaces and between the heads of the corresponding Interossei muscles, to inosculate with the dorsal interosseous arteries.

The **palmar interosseous,** three or four in number, arise from the convexity of the deep palmar arch; they run downwards upon the Interossei muscles, and anastomose at the clefts of the fingers with the digital branches of the superficial arch.

The **palmar recurrent branches** arise from the concavity of the deep palmar arch. They pass upwards in front of the wrist, supplying the carpal articulations and anastomosing with the anterior carpal arch.

ULNAR ARTERY (fig. 498)

The **Ulnar Artery,** the larger of the two terminal branches of the brachial, commences a little below the bend of the elbow,

FIG. 498.—Ulnar and radial arteries. Deep view.

Anastomotica magna

Anterior ulnar recurrent

Posterior ulnar recurrent

Radial recurrent

Posterior interosseous

Muscular

Muscular

Anterior carpal Superficialis volæ

Anterior carpal

Deep branch of ulnar

and, passing obliquely downwards and inwards, reaches the inner side of the forearm at a point about midway between the elbow and the wrist; it then runs along its ulnar border to the wrist, crosses the annular ligament on the radial side of the pisiform bone, and immediately beyond this bone divides into two branches, which enter into the formation of the superficial and deep palmar arches.

Relations in the Forearm.—In its *upper half,* it is deeply seated, being covered by all the superficial Flexor muscles, excepting the Flexor carpi ulnaris; the median nerve is in relation with the inner side of the artery for about an inch and then crosses the vessel, being separated from it by the deep head of the Pronator radii teres; it lies upon the Brachialis anticus and Flexor profundus digitorum muscles. In the *lower half* of the forearm, it lies upon the Flexor profundus, being covered by the integument, the superficial and deep fasciæ, and is placed between the Flexor carpi ulnaris and Flexor sublimis digitorum muscles. It is accompanied by two venæ comites, and is overlapped in its middle third by the Flexor carpi ulnaris; the ulnar nerve lies on its inner side for the lower two-thirds of its extent, and a small branch from the nerve descends on the lower part of the vessel to the palm of the hand.

PLAN OF THE RELATIONS OF THE ULNAR ARTERY IN THE FOREARM

In front.

Superficial layer of Flexor muscles. } Upper half.
Median nerve.
Superficial and deep fasciæ.   Lower half.

Inner side.   (Ulnar Artery in Forearm.)   *Outer side.*
Flexor carpi ulnaris.
Ulnar nerve (lower two-thirds).   Flexor sublimis digitorum.

Behind.
Brachialis anticus.
Flexor profundus digitorum.

At the wrist (fig. 497) the ulnar artery is covered by the integument and fascia, and lies upon the anterior annular ligament. On its inner side is the pisiform bone. The ulnar nerve lies at the inner side, and somewhat behind the artery; here the nerve and artery are crossed by a band of fibres, which extends from the pisiform bone to the anterior annular ligament.

Peculiarities.—The ulnar artery has been found to vary in its origin nearly in the proportion of one in thirteen cases, in one case arising lower than usual, about two or three inches below the elbow, and in all other cases much higher, the brachial being more often the source of origin than the axillary.

Variations in the position of this vessel are more common than in the radial. When its origin is normal, the course of the vessel is rarely changed. When it arises high up, it is almost invariably superficial to the Flexor muscles in the forearm, lying commonly beneath the fascia, more rarely between the fascia and integument. In a few cases, its position was subcutaneous in the upper part of the forearm, and subaponeurotic in the lower part.

Surface Marking.—On account of the curved direction of the ulnar artery, the line on the surface of the body which indicates its course is somewhat complicated. First, draw a line from the front of the internal epicondyle of the humerus to the radial side of the pisiform bone; the lower two-thirds of this line represent the course of the middle and lower thirds of the ulnar artery. Secondly, draw a line from the centre of the hollow in front of the elbow-joint to the junction of the upper and middle thirds of the first line; this represents the course of the upper third of the artery.

Surgical Anatomy.—The application of a ligature to this vessel is required in cases of wound of the artery, or of its branches, or in consequence of aneurism. In the upper half of the forearm, the artery is deeply seated beneath the superficial Flexor muscles, and the application of a ligature in this situation is attended with some difficulty. An incision is to be made in the course of a line drawn from the front of the internal epicondyle of the humerus to the outer side of the pisiform bone, so that the centre of the incision is three fingers' breadth below the internal epicondyle. The skin and superficial fascia having been divided, and the deep fascia exposed, the white line which separates the Flexor carpi ulnaris from the other Flexor muscles is to be sought for, and the fascia incised in this line. The Flexor carpi ulnaris is now to be carefully separated from the other muscles, when the ulnar nerve will be exposed and must be drawn aside. Some little distance below the nerve, the artery will be found accompanied by its venæ comites, and may be ligatured by passing the needle from within outwards. In the middle and lower third of the forearm, this vessel may be easily secured by making an incision on the radial side of the tendon of the Flexor carpi ulnaris: the deep fascia being divided, and the Flexor carpi ulnaris and its companion muscle, the Flexor sublimis, being separated from each other, the vessel will be exposed, accompanied by its venæ comites, the ulnar nerve lying on its inner side. The veins being separated from the artery, the ligature should be passed from the ulnar to the radial side, taking care to avoid the ulnar nerve.

The branches of the ulnar artery may be arranged in the following groups :

In the forearm
- Anterior ulnar recurrent.
- Posterior ulnar recurrent.
- Interosseous { Anterior interosseous.
                 Posterior interosseous.
- Muscular.

At the wrist
- Anterior carpal.
- Posterior carpal.

In the hand
- Deep palmar or communicating.
- Superficial palmar arch.

The **anterior ulnar recurrent** (fig. 498) arises immediately below the elbow-joint, passes upwards and inwards between the Brachialis anticus and Pronator radii teres, supplies twigs to those muscles, and, in front of the inner epicondyle, anastomoses with the anastomotica magna and inferior profunda.

The **posterior ulnar recurrent** is much larger, and arises somewhat lower than the preceding. It passes backwards and inwards, beneath the Flexor sublimis, and ascends behind the inner epicondyle of the humerus. In the interval between this process and the olecranon, it lies beneath the Flexor carpi ulnaris, and ascending between the heads of that muscle, in relation with the ulnar nerve, it supplies the neighbouring muscles and the elbow-joint, and anastomoses with the inferior profunda, anastomotica magna, and interosseous recurrent arteries (fig. 499).

The **interosseous artery** (fig. 498) is a short trunk about half an inch in length, and of considerable size, which arises immediately below the tuberosity of the radius, and, passing backwards to the upper border of the interosseous membrane, divides into two branches, the *anterior* and *posterior interosseous*.

The **anterior interosseous** passes down the forearm on the anterior surface of the interosseous membrane, to which it is connected by a thin aponeurotic arch. It is accompanied by the interosseous branch of the median nerve, and over-lapped by the contiguous margins of the Flexor profundus digitorum and Flexor longus pollicis muscles, giving off in this situation *muscular* branches, and the *nutrient* arteries of the radius and ulna. At the upper border of the Pronator quadratus, a branch descends beneath the muscle, to anastomose in front of the carpus with the anterior carpal arch. The continuation of the artery passes behind the Pronator quadratus, and, piercing the interosseous membrane, reaches the back of the forearm, and anastomoses with the posterior interosseous artery (fig. 499). It then descends, in company with the terminal portion of the posterior interosseous nerve, to the back of the wrist to join the posterior carpal arch. The anterior interosseous gives off a long, slender branch, the *median artery* (*comes nervi mediani*), which accompanies the median nerve, and gives offsets to its substance. This artery is sometimes much enlarged, and runs with the nerve into the palm of the hand.

The **posterior interosseous artery** passes backwards through the interval between the oblique ligament and the upper border of the interosseous membrane. It appears between the contiguous borders of the Supinator brevis and the Extensor ossis metacarpi pollicis, and runs down the back of the forearm, between the superficial and deep layer of muscles, to both of which it distributes branches. Where it lies upon the Extensor ossis metacarpi pollicis and the Extensor brevis pollicis, it is accompanied by the posterior interosseous nerve. At the lower part of the forearm it anastomoses with the termination of the anterior interosseous artery, and with the posterior carpal arch. This artery gives off, near its origin, the *interosseous recurrent branch*.

The **interosseous recurrent artery** is a large vessel which ascends to the interval between the external epicondyle and olecranon, on or through the fibres of the Supinator brevis, but beneath the Anconeus, anastomosing with the posterior branch of the superior profunda, and with the posterior ulnar recurrent and anastomotica magna.

The **muscular branches** are distributed to the muscles along the ulnar side of the forearm.

FIG. 499.—Arteries of the back of the forearm and hand.

The **anterior carpal** is a small vessel which crosses the front of the carpus beneath the tendons of the Flexor profundus, and inosculates with a corresponding branch of the radial artery.

The **posterior carpal** arises immediately above the pisiform bone, and winds backwards beneath the tendon of the Flexor carpi ulnaris; it passes across the dorsal surface of the carpus beneath the extensor tendons, anastomosing with a corresponding branch of the radial artery, and forming the *posterior carpal arch*. Immediately after its origin, it gives off a small branch, which runs along the ulnar side of the fifth metacarpal bone, and supplies the ulnar side of the dorsal surface of the little finger.

The **branch to the deep palmar arch** (**deep** or **communicating branch**) (fig. 498) passes between the Abductor minimi digiti and Flexor brevis minimi digiti, near their origins; it anastomoses with the termination of the radial artery, completing the deep palmar arch.

The continuation of the trunk of the ulnar artery in the hand forms the greater part of the superficial palmar arch.

The Superficial Palmar Arch (fig. 497)

The **Superficial Palmar Arch** is formed by the ulnar artery in the hand, and is completed on the outer side by this vessel anastomosing with a branch from the radialis indicis, though sometimes the arch is completed by the ulnar anastomosing with the superficialis volæ or princeps pollicis of the radial artery. The arch passes across the palm, describing a curve, with its convexity downwards, to the space between the ball of the thumb and the index finger, where the above-mentioned anastomosis takes place.

Relations.—The superficial palmar arch is covered by the skin, the Palmaris brevis, and the palmar fascia. It lies upon the annular ligament, the Flexor brevis and Opponens of the little finger, the tendons of the superficial flexor of the fingers, the lumbrical muscles, and the divisions of the median and ulnar nerves.

Plan of the Relations of the Superficial Palmar Arch

In front.

Skin.
Palmaris brevis.
Palmar fascia.

Superficial
Palmar Arch.

Behind.

Annular ligament.
Flexor brevis and Opponens of little finger.
Superficial flexor tendons.
Divisions of median and ulnar nerves.
Lumbrical muscles.

The superficial palmar arch gives off four *digital* branches.

The **digital branches** (fig. 497) are given off from the convexity of the superficial palmar arch. They supply the ulnar side of the little finger, and the adjoining sides of the little, ring, middle, and index fingers; the radial side of the index finger and thumb being supplied from the radial artery. The digital arteries at first lie superficial to the flexor tendons, but as they pass forwards with the digital nerves to the clefts between the fingers, they lie between them, and are there joined by the interosseous branches from the deep palmar arch. The digital arteries on the sides of the fingers lie beneath the digital nerves; and, about the middle of the last phalanx, the two branches for each finger form an arch, from the convexity of which branches pass to supply the pulp of the finger.

Surface Marking.—The superficial palmar arch is represented by a curved line, starting from the outer side of the pisiform bone, and carried downwards as far as the middle third of the palm, and then curved outwards on a level with the upper end of the cleft between the thumb and index finger.

The deep palmar arch is situated about half an inch nearer to the carpus.

Surgical Anatomy.—Wounds of the palmar arches are of special interest, and are always difficult to deal with. When the superficial arch is wounded it is generally possible, by enlarging the wound if necessary, to secure the vessel and tie it; or in cases where it is found impossible to encircle the vessel with a ligature, a pair of Wells's artery clips may be applied and left on for twenty-four or forty-eight hours. Wounds of the deep arch are not so easily dealt with. It may be possible to secure the vessel by forcipressure forceps, which may be left on; or, failing this, the wound may be plugged with gauze and an outside dressing carefully bandaged on. The plug should be allowed to remain untouched for three or four days. In wounds of the deep palmar arch, a ligature may be applied to the bleeding points, from the dorsum of the hand, by resection of the upper part of the third metacarpal bone. It is useless in these cases to ligature one of the arteries of the forearm alone, and indeed simultaneous ligature of both radial and ulnar arteries above the wrist is often unsuccessful, on account of the anastomosis carried on by the carpal arches. Therefore, upon the failure of pressure to arrest hæmorrhage, it is expedient to apply a ligature to the brachial artery.

ARTERIES OF THE TRUNK

THE DESCENDING AORTA

The **Descending Aorta** is divided into two portions, the *thoracic* and *abdominal*, in correspondence with the two great cavities of the trunk in which it is situated.

THE THORACIC AORTA

The **Thoracic Aorta** commences at the lower border of the fourth dorsal vertebra, on the left side, and terminates at the aortic opening in the Diaphragm, in front of the lower border of the last dorsal vertebra. At its commencement, it is situated on the left side of the spine; it approaches the median line as it descends; and, at its termination, lies directly in front of the column. The direction of this vessel being influenced by the spine, upon which it rests, it describes a curve which is concave forwards in the dorsal region. As the branches given off from it are small, the diminution in the size of the vessel is inconsiderable. It is contained in the back part of the posterior mediastinum.

Relations.—It is in relation, *in front*, from above downwards, with the root of the left lung, the pericardium, the œsophagus, and the Diaphragm; *behind*, with the vertebral column, and the azygos minor veins; on the *right side*, with the vena azygos major, and thoracic duct; on the *left side*, with the left pleura and lung. The œsophagus, with its accompanying nerves, lies on the right side of the aorta *above*; but at the lower part of the thorax it gets in front of the aorta, and close to the Diaphragm is situated to its left side.

PLAN OF THE RELATIONS OF THE THORACIC AORTA

In front.

Root of left lung.
Pericardium.
Œsophagus.
Diaphragm.

Right side.	Thoracic Aorta.	Left side.
Œsophagus (above).		Pleura.
Vena azygos major.		Left lung.
Thoracic duct.		Œsophagus (below).

Behind.

Vertebral column.
Superior and inferior azygos minor veins.

The aorta is occasionally found to be obliterated at a particular spot, viz. at the junction of the arch with the thoracic aorta, just below the ductus arteriosus. Whether this is the result of disease, or of congenital malformation, is immaterial to our present purpose; it affords an interesting opportunity of observing the resources of the collateral circulation. The course of the anastomosing vessels, by which the blood is brought from the upper to the lower part of the artery, will be found well described in an account of two cases in the 'Pathological Transactions,' vols. viii. and x. In the former (p. 162), Sydney Jones thus sums up the detailed description of the anastomosing vessels: 'The principal communications by which the circulation was carried on, were—Firstly, the internal mammary, anastomosing with the intercostal arteries, with the phrenic of the abdominal aorta by means of the musculo-phrenic and comes nervi phrenici, and largely with the deep epigastric. Secondly, the superior intercostal, anastomosing anteriorly by means of a large branch with the first aortic intercostal, and posteriorly with the posterior branch of the same artery. Thirdly, the inferior thyroid, by means of a branch about the size of an ordinary radial, forming a communication with the first aortic intercostal. Fourthly, the transversalis colli, by means of very large communications with the posterior branches of the intercostals. Fifthly, the branches (of the subclavian and axillary) going to the side of the chest were large, and anastomosed freely with the lateral branches of the intercostals.' In the second case also (vol. x. p. 97), Wood describes the anastomoses in a somewhat similar manner, adding the remark, that ' the blood which was brought into the aorta through the anastomoses of the intercostal arteries appeared to be expended principally in supplying the abdomen and pelvis; while the supply to the lower extremities had passed through the internal mammary and epigastrics.'

Surgical Anatomy.—The student should now consider the effects likely to be produced by aneurism of the thoracic aorta, a disease of common occurrence. When we remember the great depth of the vessel from the surface, and the number of important structures which surround it on every side, it may easily be conceived what a variety of obscure symptoms are likely to arise from disease of this part of the arterial system, and how they may be mistaken for those of other affections. Aneurism of the thoracic aorta most usually extends backwards, along the left side of the spine, producing absorption of the bodies of the vertebræ, with curvature of the spine; while the irritation or pressure on the cord will give rise to pain, either in the chest, back, or loins, with radiating pain in the left upper intercostal spaces, from pressure on the intercostal nerves; at the same time the tumour may project backwards on each side of the spine, beneath the integument, as a pulsating swelling, simulating abscess connected with diseased bone; or it may displace the œsophagus, and compress the lung on one or the other side. If the tumour extend forward, it may press upon and displace the heart, giving rise to palpitation and other symptoms of disease of that organ; or it may displace, or even compress, the œsophagus, causing pain and difficulty of swallowing, as in stricture of that tube; and ultimately even open into it by ulceration, producing fatal hæmorrhage. If the disease extends to the right side, it may press upon the thoracic duct; or it may burst into the pleural cavity, or into the trachea or lung; and lastly, it may open into the posterior mediastinum. Of late years, the diagnosis of thoracic aneurism has been much facilitated by the employment of the x-rays, by means of which the outline of the sac may be demonstrated.

BRANCHES OF THE THORACIC AORTA

Pericardiac. Œsophageal.
Bronchial. Posterior mediastinal.
Intercostal.

The **pericardiac** are a few small vessels, irregular in their origin, distributed to the pericardium.

The **bronchial arteries** are the nutrient vessels of the lungs, and vary in number, size, and origin. That of the right side arises from the first aortic intercostal, or by a common trunk with the left bronchial, from the front of the thoracic aorta. Those of the left side, usually two in number, arise from the thoracic aorta, one a little lower than the other. Each vessel is directed to the back part of the corresponding bronchus along which it runs, dividing and subdividing along the bronchial tubes, supplying them, the cellular tissue of the lungs, the bronchial glands, and the œsophagus.

The **œsophageal arteries,** usually four or five in number, arise from the front of the aorta, and pass obliquely downwards to the œsophagus, forming a chain of anastomoses along that tube, anastomosing with the œsophageal branches of the inferior thyroid arteries above, and with ascending branches from the phrenic and gastric arteries below.

The **posterior mediastinal arteries** are numerous small vessels which supply the glands and loose areolar tissue in the mediastinum.

The **intercostal arteries** arise from the back of the aorta. They are generally nine in number on each side, the two superior intercostal spaces being supplied by the superior intercostal, a branch of the subclavian. The second space usually receives a considerable branch from the first aortic intercostal, which joins with the branch from the superior intercostal of the subclavian. The branch which runs along the lower border of the last rib is named the *subcostal artery*. The right intercostals are longer than the left, on account of the position of the aorta on the left side of the spine : they pass outwards, across the bodies of the vertebræ, to the intercostal spaces, being covered by the pleura, the œsophagus, thoracic duct, sympathetic nerve, and the vena azygos major ; the left passing outwards are crossed by the sympathetic ; the upper two are also crossed by the superior intercostal vein, the lower by the azygos minor veins. In each intercostal space the artery passes outwards, at first lying upon the External intercostal muscle, covered in front by the pleura and a thin fascia. It then passes between the two layers of Intercostal muscles, and having ascended obliquely to the lower border of the rib above it is continued forwards in the groove on its lower border and anastomoses with the anterior intercostal branches of the internal mammary. The first aortic intercostal anastomoses with the superior intercostal, and the last three pass between the abdominal muscles, inosculating with the epigastric in front, and with the phrenic and lumbar arteries. Each intercostal artery is accompanied by a vein and nerve, the former being above, and the latter below, except in the upper intercostal spaces, where the nerve is at first above the artery. The arteries are protected from pressure during the action of the Intercostal muscles by fibrous arches thrown across, and attached by each extremity to the bone. The lower intercostal arteries are continued anteriorly from the intercostal spaces into the abdominal wall, except the *subcostal*, which lies throughout its whole course in the abdominal wall, since it is placed below the last rib. They pass behind the costal cartilages between the Internal oblique and Transversalis muscle to the sheath of the Rectus, where they anastomose with the internal mammary and the deep epigastric arteries. Behind, the subcostal artery anastomoses with the first lumbar artery.

Each intercostal artery gives off the following branches :

> Posterior or dorsal branch.   Spinal.
>   Collateral intercostal.

The *posterior or dorsal branch* of each intercostal artery passes backwards to the inner side of the anterior costo-transverse ligament, and divides into an external and internal branch, which are distributed to the muscles and integument of the back.

The *spinal branch*, which enters the spinal canal through the intervertebral foramen, is distributed to the spinal cord and its membranes, and to the bodies of the vertebræ in the same manner as the lateral spinal branches from the vertebral.

The *collateral intercostal branch* comes off from the intercostal artery near the angle of the rib, and descends to the upper border of the rib below, along which it courses to anastomose with the anterior intercostal branch of the internal mammary.

Mammary branches are given off by the intercostal arteries in the third, fourth, and fifth spaces. They supply the mammary gland, and increase considerably in size during the period of lactation.

Surgical Anatomy.—The position of the intercostal vessels should be borne in mind in performing the operation of paracentesis thoracis. The puncture should never be made nearer the middle line posteriorly than the angle of the rib, as the artery crosses the space internal to this point. In the lateral portion of the chest, where the puncture is usually made, the artery lies at the upper part of the intercostal space, and therefore the puncture should be made just above the upper border of the rib forming the lower boundary of the space.

THE ABDOMINAL AORTA (fig. 500)

The **Abdominal Aorta** commences at the aortic opening of the Diaphragm, in front of the lower border of the body of the last dorsal vertebra, and, descending a little to the left side of the vertebral column, terminates on the body of the fourth lumbar vertebra, commonly a little to the left of the middle line,* where it divides into the two common iliac arteries. It diminishes rapidly in size, in consequence of the many large branches which it gives off. As it lies upon the bodies of the vertebræ, the curve which it describes is convex forwards, the greatest convexity corresponding to the third lumbar vertebra.

Relations.—The abdominal aorta is covered, *in front*, by the lesser omentum and stomach, behind which are the branches of the cœliac axis, and the solar

FIG. 500.—The abdominal aorta and its branches.

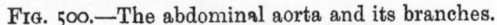

plexus: below these, by the splenic vein, the pancreas, the left renal vein, the transverse portion of the duodenum, the mesentery, and aortic plexus. *Behind*, it is separated from the lumbar vertebræ and intervening discs by the anterior

* Lord Lister, having accurately examined 30 bodies in order to ascertain the exact point of termination of this vessel, found it 'either absolutely, or almost absolutely, mesial in 15, while in 13 it deviated more or less to the left, and in 2 was slightly to the right.'— *System of Surgery*, edited by T. Holmes, 2nd ed. vol. v. p. 652.

X X.

common ligament and left lumbar veins. On the *right side* it is in relation with the inferior vena cava (the right crus of the Diaphragm being interposed above), the vena azygos major, thoracic duct, and right semilunar ganglion ; on the *left side*, with the sympathetic nerve, the left semilunar ganglion, and the fourth or ascending portion of the duodenum.

PLAN OF THE RELATIONS OF THE ABDOMINAL AORTA

In front.

Lesser omentum and stomach.
Branches of the cœliac axis and solar plexus.
Splenic vein.
Pancreas.
Left renal vein.
Transverse duodenum.
Mesentery.
Aortic plexus.

Right side.

Right crus of Diaphragm.
Inferior vena cava.
Vena azygos major.
Thoracic duct.
Right semilunar ganglion.

Abdominal Aorta.

Left side.

Sympathetic nerve.
Left semilunar ganglion.
Ascending duodenum.

Behind.

Left lumbar veins.
Vertebral column.

Surface Marking.—In order to map out the abdominal aorta on the surface of the abdomen, a line must be drawn from the middle line of the body, on a level with the distal extremity of the seventh costal cartilages, downwards and slightly to the left, so that it just skirts the umbilicus, to a zone drawn round the body opposite the highest point of the crest of the ilium. This point is generally half an inch below and to the left of the umbilicus, but as the position of this structure varies with the obesity of the individual, it is not a reliable landmark as to the situation of the bifurcation of the aorta.

Surgical Anatomy.—The abdominal aorta may be the seat of an aneurism either at its upper part, close to and often involving the cœliac axis, or at its lower part, near the bifurcation. Occasionally aneurisms are met with on some of the branches of the aorta, the mesenteric or splenic, quite independent of the main trunk. Aneurisms of the abdominal aorta near the cœliac axis communicate in nearly equal proportion with the anterior and posterior parts of the artery.

When an aneurismal sac is connected with the back part of the abdominal aorta, it usually produces absorption of the bodies of the vertebræ, and forms a pulsating tumour, that presents itself in the left hypochondriac or epigastric regions, and is accompanied by symptoms of disturbance in the alimentary canal. Pain is invariably present, and is usually of two kinds—a fixed and constant pain in the back, caused by the tumour pressing on or displacing the branches of the solar plexus and splanchnic nerves ; and a sharp lancinating pain, radiating along those branches of the lumbar nerves which are pressed on by the tumour ; hence the pain in the loins, the testes, the hypogastrium, and in the lower limb (generally of the left side). This form of aneurism usually bursts into the peritoneal cavity, or behind the peritoneum, in the left hypochondriac region ; or it may form a large aneurismal sac, extending down as low as Poupart's ligament ; hæmorrhage in these cases being generally very extensive, but slowly produced, and not rapidly fatal.

When an aneurismal sac is connected with the front of the aorta near the cœliac axis, it forms a pulsating tumour in the left hypochondriac or epigastric regions, usually attended with symptoms of disturbance of the alimentary canal, as sickness, dyspepsia, or constipation, and accompanied by pain, which is constant, but nearly always fixed, in the loins, epigastrium, or some part of the abdomen ; the radiating pain being rare, as the lumbar nerves are seldom implicated. This form of aneurism may burst into the peritoneal cavity, or behind the peritoneum, between the layers of the mesentery, or, more rarely, into the duodenum ; it rarely extends backwards so as to affect the spine.

The abdominal aorta has been tied several times, and although none of the patients permanently recovered, still, as Keen's case lived forty-eight days, the possibility of the re-establishment of the circulation may be considered to be proved. In the lower animals this artery has been often successfully tied. The vessel may be reached in several ways. In the original operation, performed by Sir A. Cooper, an incision was made in the linea alba, the peritoneum opened in front, the finger carried down among the intestines, towards the spine, the peritoneum again opened behind, by scratching through the mesentery, and the vessel thus reached. Or either of the operations, described below, for securing the common iliac artery, may, by extending the dissection a sufficient distance upwards, be made use of to expose the aorta. The chief difficulty in the dead subject consists in isolating the artery, in consequence of its great depth; but in the living subject, the embarrassment resulting from the proximity of the aneurismal tumour, and the great probability of disease in the vessel itself, add to the dangers and difficulties of this formidable operation so greatly, that it is very doubtful whether it ought ever to be performed.

Collateral Circulation.—The collateral circulation would be carried on by the anastomosis between the internal mammary and the deep epigastric; by the free communication between the superior and inferior mesenterics, if the ligature were placed above the latter vessel; or by the anastomosis between the inferior mesenteric and the internal pudic, when (as is more common) the point of ligature is below the origin of the inferior mesenteric; and possibly by the anastomoses of the lumbar arteries with the branches of the internal iliac.

The abdominal aorta may be compressed by digital or instrumental pressure. In emaciated patients the circulation may be controlled by the pressure of the fingers; but in any patient, the following plan, which has been strongly advocated by Macewen, will be found effectual. An assistant stands on the left side of the patient, on a high stool, so that his knee-joint is about on a level with the operating-table, and with his right side against the table. He then places the closed fist of his right hand on the patient's abdomen, so that the knuckle of the index finger is just above and to the left of the umbilicus, and, standing on his left foot, with the right leg crossing in front of the left, he leans by means of the closed fist on the patient's abdomen, and can exert sufficient pressure to entirely arrest the flow of blood through the aorta, without any great amount of fatigue, for any length of time that may be required. Instrumental pressure is carried out by means of a tourniquet invented by Lord Lister. It is reliable, but involves a danger of bruising the intestine or mesentery, and sometimes interferes with the respiration.

BRANCHES OF THE ABDOMINAL AORTA

Inferior phrenic.	Renal.
Cœliac axis { Gastric. / Hepatic. / Splenic.	Spermatic in male. Ovarian in female. Inferior mesenteric.
Superior mesenteric.	Lumbar.
Suprarenal.	Sacra media.

Common iliac.

The branches may be divided into two sets: 1. Those supplying the viscera. 2. Those distributed to the walls of the abdomen.

Visceral Branches.

Cœliac axis { Gastric / Hepatic. / Splenic.
Superior mesenteric.
Inferior mesenteric.
Suprarenal.
Renal.
Spermatic or Ovarian.

Parietal Branches.

Inferior phrenic.
Lumbar.
Sacra media.
Common iliac.

Of the visceral branches, the cœliac axis, the superior and inferior mesenteric are single, while the suprarenal, renal, and spermatic or ovarian are paired. The inferior phrenic, lumbar, and common iliac are paired parietal branches; the middle sacral unpaired.

CŒLIAC AXIS (fig. 501)

To expose this artery, raise the liver, draw down the stomach, and then tear through the layers of the lesser omentum and ascending layer of the transverse mesocolon.

The **Cœliac axis** is a short thick trunk, about half an inch in length, which arises from the aorta, close to the margin of the opening in the Diaphragm, and passing nearly horizontally forwards, divides into three large branches, the *gastric, hepatic,* and *splenic,* occasionally giving off one of the phrenic arteries.

Relations.—The cœliac axis is covered by the lesser omentum. On the *right side*, it is in relation with the right semilunar ganglion, and the lobus Spigelii ; on the *left side*, with the left semilunar ganglion and cardiac end of the stomach. *Below*, it is in relation to the upper border of the pancreas, and the splenic vein.

The **Gastric** or **Coronary artery**, the smallest of the three branches of the cœliac axis, passes upwards and to the left, behind the lesser sac of the peritoneum, to the cardiac orifice of the stomach, distributing branches to the œsophagus, which anastomose with the aortic œsophageal arteries ; others supply the cardiac end of the stomach, inosculating with branches of the splenic artery : it then passes from left to right, along the lesser curvature'of the stomach

FIG. 501.—The cœliac axis and its branches, the liver having been raised, and the lesser omentum and ascending layer of the transverse mesocolon removed.

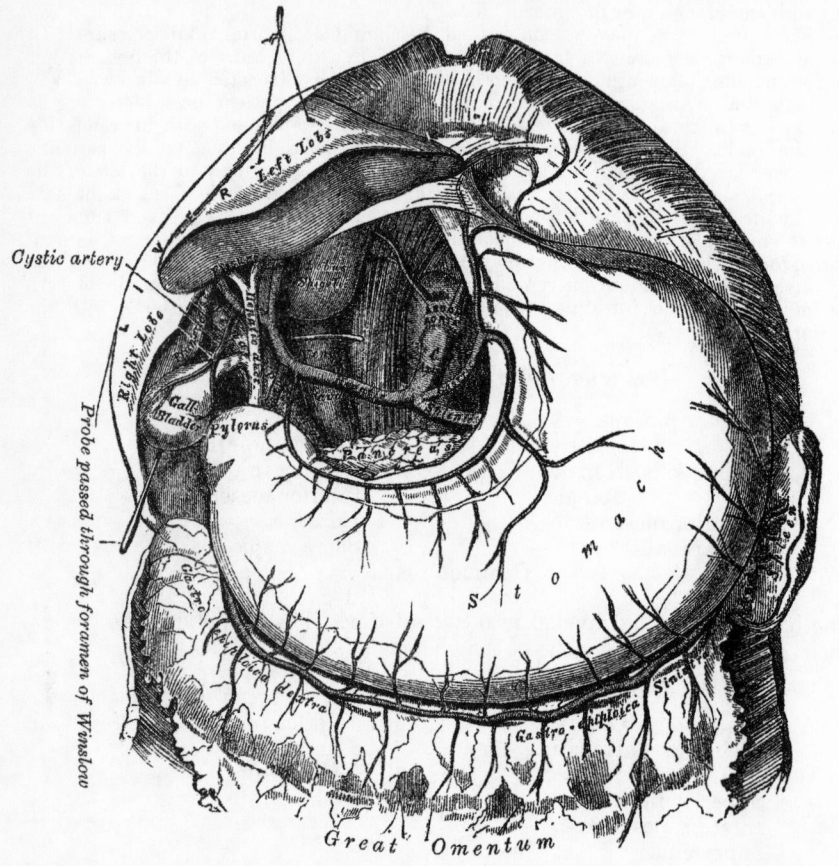

to the pylorus, lying in its course between the layers of the lesser omentum, and giving branches to both surfaces of the organ : at its termination it anastomoses with the pyloric branch of the hepatic.

The **Hepatic artery**, in the adult, is intermediate in size between the gastric and splenic ; in the fœtus, it is the largest of the three branches of the cœliac axis. It is first directed forwards and to the right, to the upper margin of the pyloric end of the stomach, forming the lower boundary of the foramen of Winslow. It then passes upwards between the layers of the lesser omentum, and in front of the foramen of Winslow, to the transverse fissure of the liver, where it divides into two branches, right and left, which supply the corresponding lobes of that organ, accompanying the ramifications of the vena portæ and hepatic ducts. The hepatic artery, in its course along the right border of the lesser omentum, is in relation with the common bile-duct and portal vein, the duct lying to the right of the artery, and the vein behind.

Its branches are, the

> Pyloric.
> Gastro-duodenal {Gastro-epiploica dextra.
> Superior pancreatico-duodenal.
> Cystic.

The **pyloric branch** arises from the hepatic, above the pylorus, descends to the pyloric end of the stomach, and passes from right to left along its lesser curvature, supplying it with branches, and inosculating with the gastric branches of the coronary artery.

The **gastro-duodenal** (fig. 502) is a short but large branch, which descends, near the pylorus, behind the first portion of the duodenum, and divides at the

FIG. 502.—The cœliac axis and its branches, the stomach having been raised and the transverse mesocolon removed.

lower border of this viscus into two branches, the *gastro-epiploica dextra* and the *pancreatico-duodenalis superior*. Previous to its division, it gives off two or three small inferior pyloric branches to the pyloric end of the stomach and pancreas.

The **gastro-epiploica dextra** runs from right to left along the greater curvature of the stomach, between the layers of the great omentum, anastomosing about the middle of the lower border of the stomach with the gastro-epiploica sinistra from the splenic artery. This vessel gives off numerous branches, some of which ascend to supply both surfaces of the stomach, while others descend to supply the great omentum.

The **pancreatico-duodenalis superior** descends between the contiguous margins of the duodenum and pancreas. It supplies both these organs, and anastomoses with the inferior pancreatico-duodenal branch of the superior mesenteric artery, and with the pancreatic branches of the splenic.

The **cystic artery** (fig. 501), usually a branch of the right hepatic, passes downwards and forwards along the neck of the gall-bladder, and divides into two branches, one of which ramifies on its free surface, the other between it and the substance of the liver.

The **Splenic artery,** in the adult, is the largest of the three branches of the cœliac axis, and is remarkable for the extreme tortuosity of its course. It passes horizontally to the left side, behind the peritoneum and along the upper border of the pancreas, accompanied by the splenic vein, which lies below it; it crosses in front of the upper part of the left kidney, and, on arriving near the spleen, divides into branches, some of which enter the hilum of that organ between the two layers of the lieno-renal ligament to be distributed to its structure, while others are distributed to the pancreas and great end of the stomach. Its branches are, the

Pancreaticæ parvæ.	Gastric (Vasa brevia).
Pancreatica magna.	Gastro-epiploica sinistra.

The **pancreatic** are numerous small branches derived from the splenic as it runs behind the upper border of the pancreas, supplying its middle and left parts. One of these, larger than the rest, is sometimes given off from the splenic near the left extremity of the pancreas; it runs from left to right near the posterior surface of the gland, following the course of the pancreatic duct, and is called the **pancreatica magna.** These vessels anastomose with the pancreatic branches of the pancreatico-duodenal arteries, derived from the hepatic on the one hand and the superior mesenteric on the other.

The **gastric (vasa brevia)** consist of from five to seven small branches, which arise either from the termination of the splenic artery, or from its terminal branches; and, passing from left to right, between the layers of the gastro-splenic omentum, are distributed to the great curvature of the stomach; anastomosing with branches of the gastric and gastro-epiploica sinistra arteries.

The **gastro-epiploica sinistra,** the largest branch of the splenic, runs from left to right along the great curvature of the stomach, between the layers of the great omentum, and anastomoses with the gastro-epiploica dextra. In its course it distributes several branches to the stomach, which ascend upon both surfaces; others descend to supply the omentum.

SUPERIOR MESENTERIC ARTERY (fig. 503)

In order to expose this vessel, raise the great omentum and transverse colon, draw down the small intestines, and cut through the peritoneum where the transverse meso-colon and mesentery join: the artery will then be exposed, just as it issues from beneath the lower border of the pancreas.

The **Superior Mesenteric artery** supplies the whole length of the small intestine, except the first part of the duodenum; it also supplies the cæcum, ascending and transverse colon; it is a vessel of large size, arising from the fore part of the aorta, about half an inch below the cœliac axis; being covered at its origin by the splenic vein and pancreas. It passes forwards, between the pancreas and transverse portion of the duodenum, crosses in front of this portion of the intestine, and descends between the layers of the mesentery to the right iliac fossa, where, considerably diminished in size, it anastomoses with one of its own branches, viz. the ileo-colic. In its course it forms an arch, the convexity of which is directed forwards and downwards to the left side, the concavity backwards and upwards to the right. It is accompanied by the superior mesenteric vein, which lies to its right side, and it is surrounded by the superior mesenteric plexus of nerves. Its branches are, the

Inferior pancreatico-duodenal.	Ileo-colic.
Vasa intestini tenuis.	Colica dextra.
Colica media.	

The **inferior pancreatico-duodenal** is given off from the superior mesenteric, or from its first intestinal branch behind the pancreas. It courses to the right between the head of the pancreas and duodenum, and then ascends to anastomose with the superior pancreatico-duodenal artery. It distributes branches to the head of the pancreas and to the transverse and descending portions of the duodenum.

The **vasa intestini tenuis** arise from the convex side of the superior mesenteric artery. They are usually from twelve to fifteen in number, and are distributed to the jejunum and ileum. They run parallel with one another between the layers of the mesentery; each vessel dividing into two branches, which unite with adjacent branches, forming a series of arches, the convexities of which are directed towards the intestine. From this first set of arches branches arise, which again unite with similar branches from either side, and thus a second

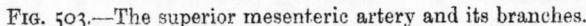

FIG. 503.—The superior mesenteric artery and its branches.

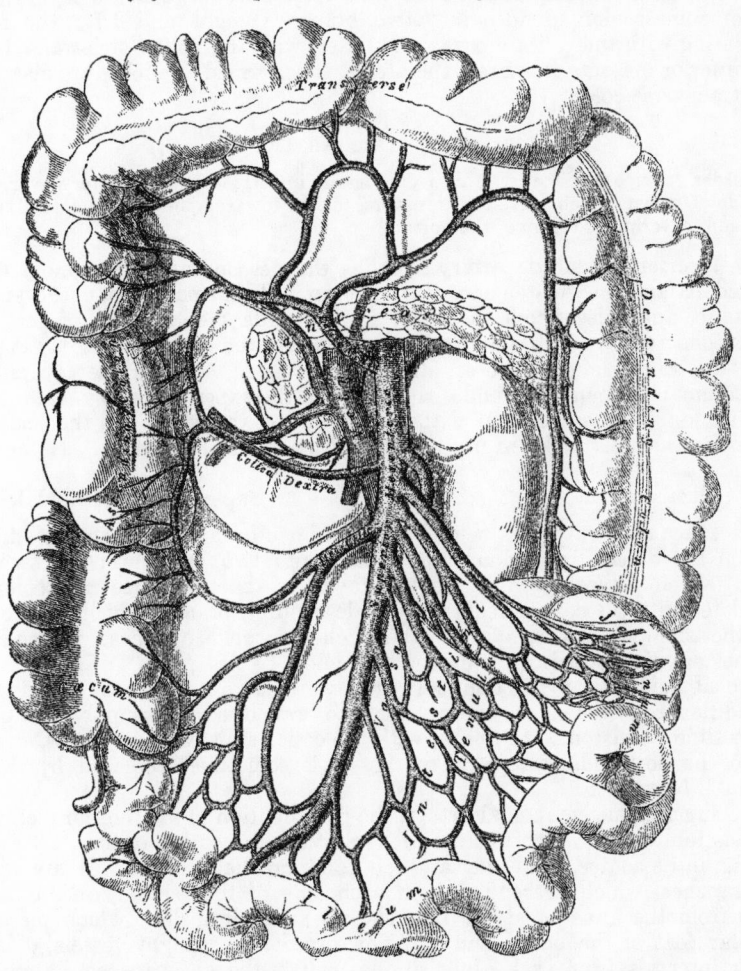

series of arches is formed; and from these latter, a third, and a fourth, or even fifth series of arches are constituted, diminishing in size the nearer they approach the intestine. From the terminal arches numerous small straight vessels arise which encircle the intestine, upon which they are distributed, ramifying between its coats. Throughout their course small branches are given off to the glands and other structures between the layers of the mesentery.

The **ileo-colic artery** is the lowest branch given off from the concavity of the superior mesenteric artery. It passes downwards and to the right behind the peritoneum towards the right iliac fossa, where it divides into two branches. Of these, the inferior division inosculates with the termination of the superior mesenteric artery, forming with it an arch, from the convexity of which branches proceed to supply the termination of the ileum, the cæcum and vermiform appendix, and the ileo-cæcal valve. The superior division inosculates with the colica dextra, and supplies the commencement of the colon. The branch which goes to the appendix is called the *appendicular artery*. It passes along the free edge of the meso-appendix, and is a vessel of considerable size.

The **colica dextra** arises from about the middle of the concavity of the superior mesenteric artery, and, passing behind the peritoneum to the middle of the ascending colon, divides into two branches : a descending branch, which inosculates with the ileo-colic ; and an ascending branch, which anastomoses with the colica media. These branches form arches, from the convexity of which vessels are distributed to the ascending colon.

The **colica media** arises from the upper part of the concavity of the superior mesenteric, and, passing downwards and forwards between the layers of the transverse mesocolon, divides into two branches, right and left : the former inosculating with the colica dextra ; the latter, with the colica sinistra, a branch of the inferior mesenteric. From the arches thus formed, branches are distributed to the transverse colon.

INFERIOR MESENTERIC ARTERY (fig. 504)

In order to expose this vessel, draw the small intestines and mesentery over to the right side of the abdomen, raise the transverse colon towards the thorax, and divide the peritoneum covering the front of the aorta.

The **Inferior Mesenteric artery** supplies the descending and sigmoid flexure of the colon, and the greater part of the rectum. It is smaller than the superior mesenteric, and arises from the front and towards the left side of the aorta, between one and two inches above its division into the common iliacs. It passes downwards to the left iliac fossa, and then descends, between the layers of the mesorectum, into the pelvis, under the name of the *superior hæmorrhoidal artery*. It lies at first in close relation with the left side of the aorta, and then passes as the superior hæmorrhoidal in front of the left common iliac artery. Its branches are, the

Colica sinistra.　　　　Sigmoid.　　　　Superior hæmorrhoidal.

The **colica sinistra** passes behind the peritoneum, in front of the left kidney, to reach the descending colon, and divides into two branches : an ascending branch, which passes between the two layers of the transverse mesocolon and inosculates with the colica media ; and a descending branch, which anastomoses with the sigmoid artery. From the arches formed by these inosculations, branches are distributed to the descending colon.

The **sigmoid artery** runs obliquely downwards across the Psoas muscle to the sigmoid flexure of the colon, and divides into branches which supply that part of the intestine ; anastomosing above, with the colica sinistra ; and below, with the superior hæmorrhoidal artery. This vessel is sometimes replaced by three or four small branches.

The **superior hæmorrhoidal artery,** the continuation of the inferior mesenteric, descends into the pelvis between the layers of the mesentery of the pelvic colon, crossing, in its course, the ureter and left common iliac vessels. It divides into two branches, which descend one on each side of the rectum, and about five inches from the anus break up into several small branches, which pierce the muscular coat of the bowel and run downwards, as straight vessels, placed at regular intervals from each other in the wall of the gut between its muscular and mucous coat, to the level of the internal sphincter ; here they form a series of loops around the lower end of the rectum, and communicate with the middle hæmorrhoidal arteries, branches of the internal iliac, and with the inferior hæmorrhoidal branches of the internal pudic.

SUPRARENAL ARTERIES (fig. 500)

The **Suprarenal arteries** are two small vessels which arise, one on each side of the aorta, opposite the superior mesenteric artery. They pass obliquely upwards and outwards, over the crura of the Diaphragm, to the under surface of the suprarenal capsules, to which they are distributed, anastomosing with suprarenal branches from the phrenic and renal arteries. In the adult these arteries are of small size ; in the fœtus they are as large as the renal arteries.

RENAL ARTERIES (fig. 500)

The **Renal arteries** are two large trunks, which arise from the sides of the aorta, immediately below the superior mesenteric artery. Each is directed outwards, across the crus of the Diaphragm, so as to form nearly a right angle with the aorta. The right is longer than the left, on account of the position of the aorta ; it passes behind the inferior vena cava. The left is somewhat higher than the right. Before reaching the hilum of the kidney, each artery divides into four or five branches ; the greater number of which lie between the renal vein and ureter, the vein being in front, the ureter behind : one branch

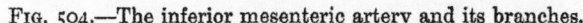

FIG. 504.—The inferior mesenteric artery and its branches.

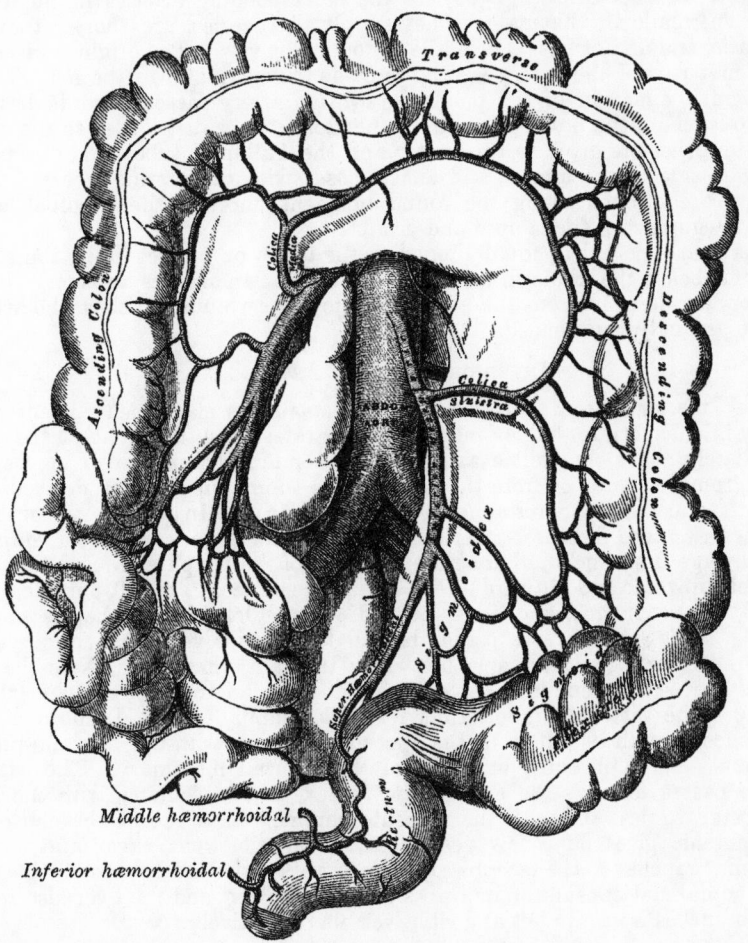

is usually situated behind the ureter. Each vessel gives off some small branches to the suprarenal capsule, the ureter, and the surrounding cellular tissue and muscles. Frequently there is a second renal artery, which is given off from the abdominal aorta either above or below the renal artery proper, the former being the more common position. Instead of entering the kidney at the hilum, these accessory renal arteries usually pierce the upper or lower part of the gland.

SPERMATIC ARTERIES

The **Spermatic arteries** are distributed to the testes. They are two slender vessels, of considerable length, which arise from the front of the aorta, a little below the renal arteries. Each artery passes obliquely outwards and downwards, behind the peritoneum, resting on the Psoas muscle, the right spermatic lying in front of the inferior vena cava and behind the terminal part of the ileum, the left behind the sigmoid flexure of the colon. It crosses obliquely over the ureter and the lower part of the external iliac artery to reach the internal abdominal ring, through which it passes, and accompanies the other constituents of the spermatic cord along the inguinal canal to the scrotum, where it becomes tortuous, and divides into several branches, two or three of which accompany the vas deferens, and supply the epididymis, anastomosing with the artery of the vas deferens; others pierce the back part of the tunica albuginea, and supply the substance of the testis. The spermatic artery supplies one or two small branches to the ureter, and in the inguinal canal gives one or two twigs to the Cremaster muscle.

The **Ovarian arteries** (fig. 506) are the corresponding arteries in the female to the spermatic in the male. They supply the ovaries, are shorter than the spermatic, and do not pass out of the abdominal cavity. The origin and course of the first part of the artery are the same as the spermatic in the male, but on arriving at the margin of the pelvis the ovarian artery passes inwards, between the two layers of the broad ligament of the uterus, to be distributed to the ovary. Small branches are given to the ureter and the Fallopian tube; and one passes on to the side of the uterus, and anastomoses with the uterine artery. Other offsets are continued along the round ligament, through the inguinal canal, to the integument of the labium and groin.

At an early period of fœtal life, when the testes or ovaries lie by the side of the spine, below the kidneys, the spermatic or ovarian arteries are short; but as these organs descend from the abdomen into the scrotum or pelvis, the arteries become gradually lengthened.

INFERIOR PHRENIC ARTERIES

The **Inferior phrenic arteries** are two small vessels, which present much variety in their origin. They may arise separately from the front of the aorta, immediately above the cœliac axis, or by a common trunk, which may spring either from the aorta or from the cœliac axis. Sometimes one is derived from the aorta, and the other from one of the renal arteries. In only one out of thirty-six cases examined did these arteries arise as two separate vessels from the aorta. They diverge from one another across the crura of the Diaphragm, and then pass obliquely upwards and outwards upon its under surface. The left phrenic passes behind the œsophagus, and runs forwards on the left side of the œsophageal opening. The right phrenic passes behind the inferior vena cava, and ascends along the right side of the aperture which transmits that vein. Near the back part of the central tendon each vessel divides into an internal and an external branch. The internal branch runs forwards, supplying the Diaphragm, and anastomosing with its fellow of the opposite side, and with the musculo-phrenic and comes nervi phrenici, branches of the internal mammary. The external branch passes towards the side of the thorax, and inosculates with the lower intercostal arteries, and with the musculo-phrenic. The internal branch of the right phrenic gives off a few vessels to the inferior vena cava; and the left one some branches to the œsophagus. Each vessel also sends capsular branches to the suprarenal capsule of its own side. The spleen and the liver also receive a few branches from the left and right vessels respectively.

LUMBAR ARTERIES

The **Lumbar arteries** are analogous to the intercostal. They are usually four in number on each side, and arise from the back part of the aorta, nearly at right angles with that vessel. They pass outwards and backwards, around the sides of the bodies of the lumbar vertebræ, behind the sympathetic nerve, the Psoas magnus muscle, and the lumbar plexus; those on the right side being covered by the inferior vena cava, and the two upper ones on each side by the corresponding crus of the Diaphragm. In the interval between the transverse processes of the vertebræ each artery divides into a *dorsal* and an *abdominal branch*.

The *dorsal branch* gives off, immediately after its origin, a *spinal branch*, which enters the spinal canal; it then continues its course backwards, between the transverse processes, and is distributed to the muscles and integument of the back, anastomosing with the similar branches of the adjacent lumbar arteries, and with the posterior branches of the intercostal arteries.   The *spinal branch* enters the spinal canal through the intervertebral foramen, to be distributed to the spinal cord and its membranes and to the bodies of the vertebræ in the same manner as the lateral spinal branches from the vertebral (see page 644).

The *abdominal branches* pass outwards, having a variable relation to the Quadratus lumborum muscle.   Most frequently the last branch passes in front of the muscle and the others behind it ; sometimes the order is reversed and the first branch passes in front of the muscle.   At the outer border of the Quadratus they are continued between the Transversalis and Internal oblique muscles of the abdomen, and anastomose with branches of the epigastric and internal mammary *in front*, the intercostals *above*, and those of the ilio-lumbar and circumflex iliac *below*.

MIDDLE SACRAL ARTERY

The **Middle Sacral artery** is a small vessel, which arises from the back part of the aorta, just at its bifurcation.   It descends upon the last lumbar vertebra, and along the middle line of the front of the sacrum, to the upper part of the coccyx ; it anastomoses with the lateral sacral arteries, and terminates in the coccygeal gland.   From it, minute branches pass to the posterior surface of the rectum.   Other branches are given off on each side, which anastomose with the lateral sacral arteries, and send off small offsets, which enter the anterior sacral foramina.   It is crossed by the left common iliac vein, and is accompanied by a pair of venæ comites ; these unite to form a single vessel, which opens into the left common iliac vein.

Coccygeal gland, or Luschka's gland.—Lying near the tip of the coccyx in a small tendinous interval formed by the union of the Levator ani muscle of each side, and just above the coccygeal attachment of the Sphincter ani, is a small conglobate body, about as large as a lentil or a pea, first described by Luschka,[*] and named by him the *coccygeal gland*.   Its most obvious connections are with the arteries of the part.

Structure.—The coccygeal gland is generally classed with the ductless glands of the body, and consists of a congeries of small arteries, with little aneurismal dilatations, derived from the middle sacral, and freely communicating with each other.   These vessels are enclosed in one or more layers of polyhedral granular cells, and the whole structure is invested in a capsule of connective tissue which sends in trabeculæ, dividing the interior into a number of spaces in which the vessels and cells are contained.   Nerves pass into this little body from the sympathetic, but their mode of termination is unknown.

COMMON ILIAC ARTERIES

The abdominal aorta divides, on the left side of the body of the fourth lumbar vertebra, into the two **common iliac arteries**.   They are about two inches in length ; diverging from the termination of the aorta, they pass downwards and outwards to the margin of the pelvis, and divide opposite the intervertebral substance, between the last lumbar vertebra and the sacrum, into two branches, the *external* and *internal iliac arteries* : the former supplying the lower extremity ; the latter, the viscera and parietes of the pelvis.

The **right common iliac** is somewhat longer than the left, and passes more obliquely across the body of the last lumbar vertebra.   In front of it are the peritoneum, the small intestines, branches of the sympathetic nerve, and, at its point of division, the ureter.   *Behind*, it is separated from the fourth and fifth lumbar vertebræ, with the intervening intervertebral disc, by the two common iliac veins.   On its *outer side*, it is in relation with the inferior vena cava, and the right common iliac vein, above ; and the Psoas magnus muscle below.   On its *inner side* is the left common iliac vein above.

The **left common iliac** is in relation, in front, with the peritoneum, the small intestines, branches of the sympathetic nerve, and the superior hæmorrhoidal artery ; and is crossed at its point of bifurcation by the ureter.   It rests on the bodies of the fourth and fifth lumbar vertebræ, with the intervening intervertebral

* *Der Hirnanhang und die Steissdrüse des Menschen*, Berlin, 1860 ; *Anatomie des Menschen*, Tübingen, 1864, vol. ii. pt. 2, p. 187.

disc. The left common iliac vein lies partly on the inner side, and partly beneath the artery; on its outer side, the artery is in relation with the Psoas magnus muscle.

PLAN OF THE RELATIONS OF THE COMMON ILIAC ARTERIES

In front.

Peritoneum.
Small intestines.
Sympathetic nerves.
Ureter.

Outer side.

Vena cava.
Right common iliac vein.
Psoas muscle.

Right
Common
Iliac.

Inner side.

Left common iliac vein.

Behind.

Fourth and fifth lumbar vertebræ.
Right and left common iliac veins.

In front.

Peritoneum.
Small intestines.
Sympathetic nerves.
Superior hæmorrhoidal artery.
Ureter.

Inner side.

Left common iliac vein.

Left
Common
Iliac.

Outer side.

Psoas magnus muscle.

Behind.

Fourth and fifth lumbar vertebræ.
Left common iliac vein.

Branches.—The common iliac arteries give off small branches to the peritoneum, Psoas magnus, ureters, and the surrounding cellular tissue, and occasionally give origin to the ilio-lumbar, or renal arteries.

Peculiarities.—The *point of origin* varies according to the bifurcation of the aorta. In three-fourths of a large number of cases, the aorta bifurcated either upon the fourth lumbar vertebra, or upon the intervertebral disc between it and the fifth; the bifurcation being, in one case out of nine below, and in one out of eleven above this point. In ten out of every thirteen cases, the vessel bifurcated within half an inch above or below the level of the crest of the ilium: more frequently below than above.

The *point of division* is subject to great variety. In two-thirds of a large number of cases it was between the last lumbar vertebra and the upper border of the sacrum; being above that point in one case out of eight, and below it in one case out of six. The left common iliac artery divides lower down more frequently than the right.

The *relative length*, also, of the two common iliac arteries varies. The right common iliac was the longer in sixty-three cases; the left in fifty-two; while they were both equal in fifty-three. The length of the arteries varied in five-sevenths of the cases examined, from an inch and a half to three inches; in about half of the remaining cases the artery was longer, and in the other half, shorter: the minimum length being less than half an inch, the maximum four and a half inches. In rare instances, the right common iliac has been found wanting, the external and internal iliacs arising directly from the aorta.

Surface Marking.—Draw a zone round the body opposite the highest part of the crest of the ilium; in this line take a point half an inch to the left of the middle line. From this draw two lines to points midway between the anterior superior spines of the ilium and the symphysis pubis. These two diverging lines will represent the course of the common and external iliac arteries. Draw a second zone round the body corresponding to the level of the anterior superior spines of the ilium: the portion of the diverging lines between the two zones will represent the course of the common iliac artery; the portion below the lower zone, that of the external iliac artery.

Surgical Anatomy.—The application of a ligature to the common iliac artery may be required on account of aneurism or hæmorrhage, implicating the external or internal iliacs. Now that the surgeon no longer dreads opening the peritoneal cavity, there can be no question that the easiest and best method of tying the artery is by a transperitoneal route. The abdomen is opened; the intestines are drawn to one side and the peritoneum covering the artery divided. The sheath is then opened, and the needle passed from within outwards. On the right side great care must be exercised in passing the needle, since both the common iliac veins lie behind the artery. After the vessel has been tied the incision in the peritoneum over the artery should be sutured. Formerly there were two different methods by which the common iliac artery was tied, without opening the peritoneal cavity: 1, an anterior or iliac incision, by which the vessel is approached more directly from the front; and 2, a posterior abdominal or lumbar incision, by which the vessel is reached from behind.

Collateral Circulation.—The principal agents in carrying on the collateral circulation after the application of a ligature to the common iliac are: the anastomoses of the hæmorrhoidal branches of the internal iliac with the superior hæmorrhoidal from the inferior mesenteric; the anastomoses of the uterine and ovarian arteries, and of the vesical arteries of opposite sides; of the lateral sacral with the middle sacral artery; of the epigastric with the internal mammary, inferior intercostal, and lumbar arteries; of the circumflex iliac with the lumbar arteries; of the ilio-lumbar with the last lumbar artery; of the obturator artery, by means of its pubic branch, with the vessel of the opposite side, and with the deep epigastric.

Compression of the Common Iliac Arteries.—The common iliac arteries are most efficiently compressed by Davy's lever. The instrument consists of a gum-elastic tube, about two feet long, in which fits a round wooden 'lever' considerably longer than the tube. A small quantity of olive oil having been injected into the rectum, the gum-elastic tube, softened in hot water, is passed into the bowel sufficiently far to permit its pressing upon the common iliac artery as it lies in the groove between the last lumbar vertebra and the Psoas muscle. The wooden lever is then inserted into the tube, and the projecting end carried towards the opposite thigh and raised, when it acts as a lever of the first order, the anus being the fulcrum. In cases where the mesorectum is abnormally short it may be impossible, without unjustifiable force, to compress the artery on the right side.

INTERNAL ILIAC ARTERY (fig. 505)

The **Internal Iliac artery** supplies the walls and viscera of the pelvis, the buttock, the generative organs, and inner side of the thigh. It is a short, thick vessel, smaller in the adult than the external iliac, and about an inch and a half in length. It arises at the bifurcation of the common iliac, opposite the lumbo-sacral articulation, and, passing downwards to the upper margin of the great sacro-sciatic foramen, divides into two large trunks, an *anterior* and *posterior*; from its anterior division a partially obliterated cord, the *hypogastric artery*, extends forwards to the bladder.

Relations.—*In front* with the ureter, which separates it from the peritoneum. *Behind*, with the internal iliac vein, the lumbo-sacral cord, and Pyriformis muscle. By its *outer side*, near its origin, with the external iliac vein, which lies between it and the Psoas magnus muscle; lower down, with the obturator nerve.

PLAN OF THE RELATIONS OF THE INTERNAL ILIAC ARTERY

In front.

Peritoneum.
Ureter.

Outer side.
Psoas magnus.

Internal Iliac.

Behind.

Internal iliac vein.
Lumbo-sacral cord.
Pyriformis muscle.

In the fœtus, the internal iliac artery (*hypogastric*) is twice as large as the external iliac, and appears to be the continuation of the common iliac. Instead of dipping into the pelvis, it passes forwards to the bladder, and ascends along the sides of that viscus to its summit, to which it gives branches; it then passes upwards along the back part of the anterior wall of the abdomen to the umbilicus, converging towards its fellow of the opposite side. Having passed through the umbilical opening, the two arteries twine round the umbilical vein, forming with it the umbilical cord; and, ultimately, ramify in the placenta. The portion of the vessel within the abdomen is called the *hypogastric artery*; and that external to that cavity, the *umbilical artery*.

At birth, when the placental circulation ceases, the upper portion of the hypogastric artery, extending from the summit of the bladder to the umbilicus,

FIG. 505.—Arteries of the pelvis.

contracts, and ultimately dwindles to a solid fibrous cord; but the lower portion, extending from its origin (in what is now the internal iliac artery) for about an inch and a half to the wall of the bladder, and thence to the summit of that organ, is not totally impervious, though it becomes considerably reduced in size, and serves to convey blood to the bladder, under the name of the *superior vesical artery.*

Peculiarities as regards Length.—In two-thirds of a large number of cases, the length of the internal iliac varied between an inch and an inch and a half; in the remaining third it was more frequently longer than shorter, the maximum length being three inches, the minimum half an inch.

The lengths of the common and internal iliac arteries bear an inverse proportion to each other, the internal iliac artery being long when the common iliac is short, and *vice versa*.

As regards its Place of Division.—The place of division of the internal iliac varies between the upper margin of the sacrum and the upper border of the sacro-sciatic foramen.

The arteries of the two sides in a series of cases often differed in length, but neither seemed constantly to exceed the other.

Surgical Anatomy.—The application of a ligature to the internal iliac artery may be required in cases of aneurism or hæmorrhage affecting one of its branches. The vessel may be best secured by an abdominal section in the median line and reaching the vessel through the peritoneal cavity. This plan has been advocated by Dennis, of New York, on the following grounds: (1) It no way increases the danger of the operation; (2) it prevents a series of accidents which have occurred during ligature of the artery by the older methods; (3) it enables the surgeon to ascertain the exact extent of disease in the main arterial trunk, and select his spot for the application of the ligature ; and (4) it occupies much less time. It should be remembered that the vein lies behind, and on the right side, a little external to the artery, and in close contact with it ; the ureter, which lies in front, must also be avoided. The degree of facility in applying a ligature to this vessel will mainly depend upon its length. It has been seen that, in the great majority of the cases examined, the artery was short, varying from an inch to an inch and a half; in these cases, the artery is deeply seated in the pelvis ; when, on the contrary, the vessel is longer, it is found partly above that cavity. If the artery is very short, as occasionally happens, it would be preferable to apply a ligature to the common iliac, or upon the external and internal iliacs at their origin.

Collateral Circulation.—The circulation after ligature of the internal iliac artery is carried on by the anastomoses of the uterine and ovarian arteries ; of the opposite vesical arteries ; of the hæmorrhoidal branches of the internal iliac with those from the inferior mesenteric ; of the obturator artery, by means of its pubic branch, with the vessel of the opposite side, and with the epigastric and internal circumflex ; of the circumflex and perforating branches of the profunda femoris with the sciatic ; of the gluteal with the posterior branches of the sacral arteries ; of the ilio-lumbar with the last lumbar ; of the lateral sacral with the middle sacral ; and of the circumflex iliac with the ilio-lumbar and gluteal.*

Branches of the Internal Iliac

From the Anterior Trunk.	*From the Posterior Trunk.*
Superior vesical.	Ilio-lumbar.
Middle vesical.	Lateral sacral.
Inferior vesical.	Gluteal.
Middle hæmorrhoidal.	
Obturator.	
Internal pudic.	
Sciatic.	
In female { Uterine. / Vaginal.	

The **superior vesical** is that part of the fœtal hypogastric artery which remains pervious after birth. It extends to the side of the bladder, distributing numerous branches to the apex and body of the organ. From one of these a slender vessel is derived, which accompanies the vas deferens in its course to the testis, where it anastomoses with the spermatic artery. This is the *artery of the vas deferens.* Other branches supply the ureter.

The **middle vesical,** usually a branch of the superior, is distributed to the base of the bladder and under surface of the vesiculæ seminales.

The **inferior vesical** arises from the anterior division of the internal iliac, frequently in common with the middle hæmorrhoidal, and is distributed to the base of the bladder, the prostate gland, and vesiculæ seminales. The branches distributed to the prostate communicate with the corresponding vessels of the opposite side.

The **middle hæmorrhoidal artery** usually arises together with the preceding vessel. It is distributed to the rectum, anastomosing with the inferior vesical

---

* For a description of a case in which Owen made a dissection ten years after ligature of the internal iliac artery, see *Med.-Chir. Trans.* vol. xvi.

and with the superior and inferior hæmorrhoidal arteries. It gives offsets to the seminal vesicle and prostate gland.

The **uterine artery** (fig. 506) passes inwards from the anterior trunk of the internal iliac to the neck of the uterus. Ascending, in a tortuous course on the side of this viscus, between the layers of the broad ligament, it distributes branches to its substance, anastomosing, near its termination, with the ovarian

FIG. 506.—The arteries of the internal organs of generation of the female,
seen from behind. (After Hyrtl.)

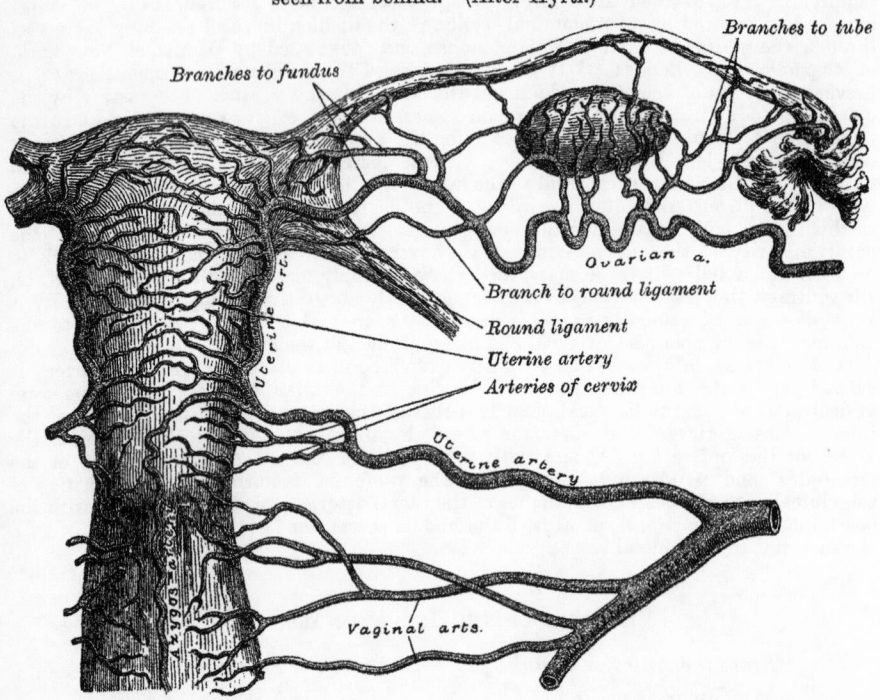

artery. It gives off branches to the cervix uteri (*cervical*), and branches which descend on the vagina, and, joining with branches from the vaginal arteries, form a median longitudinal vessel both in front and behind ; these descend on the anterior and posterior surfaces of the vagina, and are named the *azygos arteries of the vagina*.

The **vaginal artery** is analogous to the inferior vesical in the male ; it descends upon the vagina, supplying its mucous membrane, and sending branches

FIG. 507.—Variations in origin and course of obturator artery.

to the bulb of the vestibule, the neck of the bladder, and contiguous part of the rectum. It assists in forming the azygos arteries of the vagina. The vaginal artery is frequently represented by two or three branches.

The **obturator artery** usually arises from the anterior trunk of the internal iliac, sometimes from the posterior. It passes forwards, below the brim of the pelvis, to the upper part of the obturator foramen, and escaping from the pelvic cavity through a short canal, formed by a groove on the under surface of the ascending ramus of the os pubis, and the arched border of the obturator membrane, it divides into an internal and an external branch. In the pelvic cavity this vessel is in relation, externally, with the parietal pelvic fascia; internally, with the ureter, vas deferens, and peritoneum; while a little below it, is the obturator nerve.

Branches.—*Within the pelvis*, the obturator artery gives off an *iliac branch* to the iliac fossa, which supplies the bone and the Iliacus muscle, and anastomoses with the ilio-lumbar artery; a *vesical branch*, which runs backwards to supply the bladder; and a *pubic branch*, which is given off from the vessel just before it leaves the pelvic cavity. This branch ascends upon the back of the os pubis, communicating with offsets from the deep epigastric artery, and with the corresponding vessel of the opposite side. It is frequently placed on the inner side of the femoral ring. *External to the pelvis*, the obturator artery divides into an *internal* and an *external branch*, which are deeply situated beneath the Obturator externus muscle.

The *internal branch* curves downwards along the inner margin of the obturator foramen, lying beneath the Obturator externus muscle; it distributes branches to the Obturator externus, Pectineus, Adductors, and Gracilis, and anastomoses with the external branch, and with the internal circumflex artery.

The *external branch* curves round the outer margin of the foramen, also lying beneath the Obturator externus muscle, to the space between the Gemellus inferior and Quadratus femoris, where it divides into two branches: one, the smaller, courses inwards around the lower margin of the foramen and anastomoses with the internal branch and with the internal circumflex; the other inclines outwards in the groove below the acetabulum, and supplies the muscles attached to the tuberosity of the ischium and anastomoses with the sciatic artery. It sends a branch to the hip-joint through the cotyloid notch, which ramifies on the round ligament as far as the head of the femur.

Peculiarities.—In two out of every three cases the obturator arises from the internal iliac; in two out of seven from the deep epigastric; and in about one in seventy-two cases by two roots, one from either vessel. It arises in about the same proportion from the external iliac artery. The origin of the obturator from the epigastric is not commonly found on both sides of the same body.

When the obturator artery arises at the front of the pelvis from the deep epigastric, it descends almost vertically to the upper part of the obturator foramen. The artery in this course usually lies in contact with the external iliac vein, and on the outer side of the femoral ring (fig. 507, A); in such cases it would not be endangered in the operation for femoral hernia. Occasionally, however, it curves inwards along the free margin of Gimbernat's ligament (fig. 507, B), and under such circumstances would almost completely encircle the neck of a hernial sac (supposing a hernia to exist in such a case), and would be in great danger of being wounded if an operation were performed.

The **internal pudic** is the smaller of the two terminal branches of the anterior trunk of the internal iliac, and supplies the external organs of generation. Though the course of the artery is the same in the two sexes, the vessel is much smaller in the female than in the male, and the distribution of its branches somewhat different. The description of its arrangement in the male will first be given, and subsequently the differences which it presents in the female will be mentioned.

The **internal pudic artery in the male** passes downwards and outwards to the lower border of the great sacro-sciatic foramen, and emerges from the pelvis between the Pyriformis and Coccygeus muscles: it then crosses the spine of the ischium, and enters the perinæum through the lesser sacro-sciatic foramen. The artery now crosses the Obturator internus muscle, along the outer wall of the ischio-rectal fossa, being situated about an inch and a half above the lower margin of the ischial tuberosity. It is here contained in a sheath of the obturator fascia, and gradually approaches the margin of the ramus of the ischium, along which it passes forwards and upwards; it pierces the base of the superficial layer of the triangular ligament of the urethra, runs forwards along the inner margin of the ramus of the os pubis, and divides into its two terminal branches, the *dorsal artery of the penis* and the *artery of the corpus cavernosum*.

Y Y

Relations.—Within the pelvis, it lies in front of the Pyriformis muscle, the sacral plexus of nerves, and the sciatic artery, and on the outer side of the rectum (on the left side). As it crosses the spine of the ischium, it is covered by the Gluteus maximus and overlapped by the great sacro-sciatic ligament. Here the pudic nerve lies to the inner side and the nerve to the Obturator internus to the outer side of the vessel. In the perinæum it lies on the outer side of the ischio-rectal fossa, in a canal (*Alcock's canal*) formed by the splitting of the obturator fascia. It is accompanied by the pudic veins and the pudic nerve.

Peculiarities.—The internal pudic is sometimes smaller than usual, or fails to give off one or two of its usual branches; in such cases the deficiency is supplied by branches derived from an additional vessel, the *accessory pudic*, which generally arises from the internal pudic artery before its exit from the great sacro-sciatic foramen. It passes forwards along the lower part of the bladder and across the side of the prostate gland to the root of the penis, where it perforates the triangular ligament, and gives off the branches usually derived from the pudic artery. The deficiency most frequently met with is that in which the internal pudic ends as the artery of the bulb; the artery of the corpus cavernosum and the dorsal artery of the penis being derived from the accessory pudic. Or the pudic may terminate as the superficial perineal, the artery of the bulb being derived, with the other two branches, from the accessory vessel. Occasionally the accessory pudic artery is derived from one of the other branches of the internal iliac, most frequently the inferior vesical or the obturator.

Branches.—The branches of the internal pudic artery are:

Muscular.	Transverse perineal.
Inferior hæmorrhoidal.	Artery of the bulb.
Superficial perineal.	Artery of the corpus cavernosum.
Dorsal artery of the penis.	

The *muscular branches* consist of two sets: one given off in the pelvis; the other, as the vessel crosses the ischial spine. The former consists of several small offsets which supply the Levator ani, the Obturator internus, the Pyriformis, and the Coccygeus muscles. The branches given off outside the pelvis are distributed to the adjacent part of the Gluteus maximus and External rotator muscles. They anastomose with branches of the sciatic artery.

The *inferior hæmorrhoidal branches* are two or three in number, and arise from the internal pudic as it passes above the tuberosity of the ischium. Piercing the wall of Alcock's canal and crossing the ischio-rectal fossa, they are distributed to the muscles and integument of the anal region, and send offshoots round the lower edge of the Gluteus maximus to the skin of the buttock. They anastomose with the corresponding vessels of the opposite side, with the superior and middle hæmorrhoidal arteries, and with the transverse and superficial perineal vessels.

The *superficial perineal artery* supplies the scrotum, muscles and integument of the perinæum. It arises from the internal pudic, in front of the preceding branches, and turns upwards, crossing either over or under the Transversus perinæi muscle, and runs forwards, parallel to the pubic arch, in the interspace between the Accelerator urinæ and Erector penis muscles, both of which it supplies, and is finally distributed to the skin and dartos of the scrotum. In its passage through the perinæum, it lies beneath the superficial perineal fascia.

The *transverse perineal* is a small branch which arises either from the internal pudic, or from the superficial perineal artery as it crosses the Transversus perinæi muscle. It runs transversely inwards along the cutaneous surface of the Transversus perinæi muscle, which it supplies, as well as the structures between the anus and bulb of the urethra, and anastomoses with the corresponding vessel of the opposite side.

The *artery of the bulb* is a large but very short vessel, which arises from the internal pudic between the two layers of the triangular ligament, and, passing nearly transversely inwards through the fibres of the Compressor urethræ muscle, it pierces the superficial layer of the triangular ligament and ramifies in the bulb of the urethra. It gives off a small branch to Cowper's gland.

The *artery of the corpus cavernosum*, one of the terminal branches of the internal pudic, arises from that vessel while it is situated between the two layers of the triangular ligament; it pierces the superficial layer, and entering the crus penis obliquely, it runs forwards in the centre of the corpus cavernosum, to which its branches are distributed.

The *dorsal artery of the penis* ascends between the crus and pubic symphysis, and, piercing the triangular ligament, passes between the two layers of the suspensory ligament of the penis, and runs forwards on the dorsum of the penis to the glans, where it divides into two branches, which supply the glans and prepuce. On the dorsum of the penis, it lies immediately beneath the integument between the dorsal nerve and vein, the former being on its outer side. It supplies the integument and fibrous sheath of the corpus cavernosum, sending branches through the sheath to anastomose with the preceding vessel.

The **internal pudic artery in the female** is smaller than in the male. Its origin and course are similar, and there is considerable analogy in the distribution of its branches. The superficial perineal artery supplies the labia pudendi; the artery of the bulb supplies the bulbus vestibuli and the erectile tissue of the vagina; the artery of the corpus cavernosum supplies the cavernous body of the clitoris; and the arteria dorsalis clitoridis supplies the dorsum of that organ, and terminates in the glans and in the membranous fold corresponding to the prepuce of the male.

Surgical Anatomy.—The relation of the accessory pudic to the prostate gland and urethra is of the greatest interest from a surgical point of view, as this vessel is in danger of being wounded in the operation of lateral lithotomy. The student should also study the position of the internal pudic artery and its branches, when running a normal course, with regard to the same operation. The superficial and the transverse perineal arteries are, of necessity, divided in this operation, but the hæmorrhage from these vessels is seldom excessive; should a ligature be required, it can readily be applied on account of their superficial position. The artery of the bulb may be divided if the incision be carried too far forwards, and injury of this vessel may be attended with serious or even fatal consequences. The main trunk of the internal pudic artery may be wounded if the incision be carried too far outwards; but, being bound down by the strong obturator fascia, and under cover of the ramus of the ischium, the accident is not very likely to occur unless the vessel runs an anomalous course.

The artery to the bulb is of considerable importance from a surgical point of view, as it is in danger of being wounded in the operation of lateral lithotomy, an accident usually attended in the adult with alarming hæmorrhage. The vessel is sometimes very small, occasionally wanting, or even double. In some instances it arises from the internal pudic earlier than usual, and crosses the perinæum to reach the back part of the bulb. In such a case the vessel could hardly fail to be wounded in the performance of the operation of lateral lithotomy. If, on the contrary, it should arise from an accessory pudic, it lies more forward than usual, and is out of danger.

The **sciatic artery** (fig. 508), the larger of the two terminal branches of the anterior trunk of the internal iliac, is distributed chiefly on the buttock and back of the thigh. It passes down to the lower part of the great sacro-sciatic foramen, behind the internal pudic artery, resting on the sacral plexus of nerves and Pyriformis muscle, and escapes from the pelvis through this foramen between the Pyriformis and Coccygeus. It then descends in the interval between the trochanter major and tuberosity of the ischium, accompanied by the sciatic nerves, and covered by the Gluteus maximus, and is continued down the back of the thigh supplying the skin, and anastomosing with branches of the perforating arteries.

Within the pelvis, it distributes branches to the Pyriformis, Coccygeus, and Levator ani muscles; some hæmorrhoidal branches, which supply the rectum, and occasionally take the place of the middle hæmorrhoidal artery; and vesical branches to the base and neck of the bladder, vesiculæ seminales, and prostate gland. *External to the pelvis*, it gives off the following branches:

Muscular.	Anastomótic.
Coccygeal.	Articular.
Comes nervi ischiadici.	Cutaneous

The *muscular branches* supply the Gluteus maximus, anastomosing with the gluteal artery in the substance of the muscle: the external rotators, anastomosing with the internal pudic artery; and the muscles attached to the tuberosity of the ischium, anastomosing with the external branch of the obturator and the internal circumflex arteries.

The *coccygeal branch* runs inwards, pierces the great sacro-sciatic ligament, and supplies the Gluteus maximus, the integument, and other structures on the back of the coccyx.

The *comes nervi ischiadici* is a long, slender vessel, which accompanies the great sciatic nerve for a short distance; it then penetrates it, and runs in its substance to the lower part of the thigh.

The *anastomotic artery* is directed downwards across the external rotators, and assists in forming the so-called *crucial anastomosis* by anastomosing with the superior perforating and the internal and external circumflex.

FIG. 508.—The arteries of the gluteal and posterior femoral regions.

The *articular branch*, generally derived from the anastomotic, is distributed to the capsule of the hip-joint.

The *cutaneous branches* are distributed to the skin of the buttock and back of the thigh.

The **ilio-lumbar artery**, given off from the posterior trunk of the internal iliac, turns upwards and outwards behind the obturator nerve and the external iliac vessels, to the inner margin of the Psoas muscle, behind which it divides into a lumbar and an iliac branch.

The *lumbar branch* supplies the Psoas and Quadratus lumborum muscles, anastomosing with the last lumbar artery, and sends a small spinal branch through the intervertebral foramen between the last lumbar vertebra and the sacrum, into the spinal canal, to supply the cauda equina.

The *iliac branch* descends to supply the Iliacus muscle; some offsets, running between the muscle and the bone, anastomose with the iliac branch of the obturator; one of these enters an oblique canal to supply the diploë, while others run along the crest of the ilium, distributing branches to the gluteal and abdominal muscles, and anastomosing in their course with the gluteal, circumflex iliac, and external circumflex arteries.

The **lateral sacral arteries** (fig. 505) are usually two in number on each side, superior and inferior.

The *superior*, which is of large size, passes inwards, and, after anastomosing with branches from the middle sacral, enters the first or second anterior sacral foramen, supplies branches to the contents of the sacral canal, and escaping by the corresponding posterior sacral foramen, is distributed to the skin and muscles on the dorsum of the sacrum, anastomosing with the gluteal.

The *inferior* passes obliquely across the front of the Pyriformis muscle and sacral nerves to the inner side of the anterior sacral foramina, descends on the front of the sacrum, and anastomoses over the coccyx with the sacra media and opposite lateral sacral artery. In its course it gives off branches, which enter the anterior sacral foramina ; these, after supplying the contents of the sacral canal, escape by the posterior sacral foramina, and are distributed to the muscles and skin on the dorsal surface of the sacrum, anastomosing with the gluteal.

The **gluteal artery** is the largest branch of the internal iliac, and appears to be the continuation of the posterior division of that vessel. It is a short, thick trunk, which runs backwards between the lumbo-sacral cord and the first sacral nerve, and, passing out of the pelvis above the upper border of the Pyriformis muscle, immediately divides into a *superficial* and a *deep branch*. Within the pelvis, it gives off a few muscular branches to the Iliacus, Pyriformis, and Obturator internus, and just previous to quitting that cavity, a nutrient artery, which enters the ilium.

The *superficial branch* enters the deep surface of the Gluteus maximus, and divides into numerous branches, some of which supply that muscle, while others perforate its tendinous origin, and supply the integument covering the posterior surface of the sacrum, anastomosing with the posterior branches of the sacral arteries.

The *deep branch* lies under the Gluteus medius and almost immediately sub-divides into two. Of these, the *superior division*, continuing the original course of the vessel, passes along the upper border of the Gluteus minimus to the anterior superior spine of the ilium, anastomosing with the circumflex iliac and ascending branches of the external circumflex artery. The *inferior division* crosses the Gluteus minimus obliquely to the trochanter major, distributing branches to the Glutei muscles, and inosculates with the external circumflex artery. Some branches pierce the Gluteus minimus to supply the hip-joint.

Surface Marking.—The three main branches of the internal iliac, the sciatic, internal pudic, and gluteal, may occasionally be the object of surgical interference, and their positions are indicated on the surface in the following way : draw a line from the posterior superior iliac spine to the posterior superior angle of the great trochanter, with the limb slightly flexed and rotated inwards : the point of emergence of the *gluteal artery* from the upper part of the sciatic notch will correspond with the junction of the upper with the middle third of this line. Draw a second line from the posterior superior iliac spine to the outer part of the tuberosity of the ischium ; the junction of the lower with the middle third marks the point of emergence of the *sciatic* and *pudic arteries* from the great sciatic notch.

Surgical Anatomy.—Any of these three vessels may require ligaturing for a wound or for aneurism, which is generally traumatic. The *gluteal* artery is ligatured by turning the patient two-thirds over on to his face and making an incision from the posterior superior spine of the ilium to the upper and posterior angle of the great trochanter. This must expose the Gluteus maximus muscle, and its fibres are to be separated through the whole thickness of the muscle and pulled apart with retractors. The contiguous margins of the Gluteus medius and Pyriformis are now to be separated from each other, and the artery will be exposed emerging from the sciatic notch. In ligature of the *sciatic* artery, the incision should be made parallel with that for ligature of the gluteal but an inch and a half lower down. After the fibres of the Gluteus maximus have been separated, the vessel is to be sought for at the lower border of the Pyriformis ; the great sciatic nerve, which lies just above it, forming the chief guide to the artery.

EXTERNAL ILIAC ARTERY (fig. 505)

The **External Iliac artery** is larger in the adult than the internal iliac, and passes obliquely downwards and outwards along the inner border of the Psoas muscle, from the bifurcation of the common iliac to a point beneath Poupart's ligament, where it enters the thigh and becomes the femoral artery.

Relations.—*In front*, with the peritoneum, subperitoneal areolar tissue, the termination of the ileum on the right side, and the sigmoid flexure on the left, and a thin layer of fascia, derived from the iliac fascia, which surrounds the artery and vein. At its origin it is occasionally crossed by the ureter. The spermatic vessels descend for some distance upon it near its termination, and it is crossed in this situation by the genital branch of the genito-crural nerve and the deep circumflex iliac vein ; the vas deferens in the male, and the round liga-ment in the female, curve down along its inner side. *Behind*, it is in relation with the inner border of the Psoas muscle, from which it is separated by the iliac fascia. At the upper part of its course, the external iliac vein lies partly behind it, but lower down lies entirely to its inner side. *Externally,* it rests against the Psoas muscle, from which it is separated by the iliac fascia. The artery rests upon this muscle, near Poupart's ligament. Numerous lymphatic vessels and glands are found lying on the front and on the inner side of the vessel.

PLAN OF THE RELATIONS OF THE EXTERNAL ILIAC ARTERY

In front.

Peritoneum, intestines, and fascia.

Near Poupart's Ligament {
Lymphatic vessels and glands.
Spermatic vessels.
The genital branch of the genito-crural nerve.
Deep circumflex iliac vein.
}

Outer side.
Psoas magnus.
Iliac fascia.

External Iliac.

Inner side.
External iliac vein and vas deferens near Poupart's ligament.

Behind.
External iliac vein.
Psoas magnus.
Iliac fascia.

Surface Marking.—The surface line indicating the course of the external iliac artery has been already given (see page 684).

Surgical Anatomy.—The application of a ligature to the external iliac may be required in cases of aneurism of the femoral artery, ilio-femoral aneurism, or for a wound of the artery. This vessel may be secured in any part of its course, excepting near its upper end, which is to be avoided on account of the proximity of the great stream of blood in the internal iliac, and near its lower end, which should also be avoided, on account of the proximity of the deep epigastric and circumflex iliac vessels. The operation may be performed by opening the abdomen and incising the peritoneum over the artery (*transperitoneal*); or by an incision in the iliac region, division of all the structures down to, but not through, the peritoneum, and the reflection inwards of the peritoneum from the iliac fossa until the artery is reached (*retroperitoneal*).

Transperitoneal ligature.—An incision four inches in length is made in the semi-lunar line, commencing about an inch below the umbilicus, and carried through the abdominal wall into the peritoneal cavity. The intestines are then pushed upwards and held out of the way by a broad abdominal retractor, and an incision made through the peritoneum at the margin of the pelvis in the course of the artery; the vessel is secured in any part of its course which may seem desirable to the operator. The advantages of this operation appear to be, that if it is found necessary the common iliac artery can be ligatured instead of the external iliac without extension or modification of the incision; and secondly, that the vessel can be ligatured without in any way interfering with the sac of the aneurism.

Retroperitoneal ligature may be performed either by the modified Abernethy's method, which consists in making an incision from an inch above and internal to the anterior superior spine of the ilium in a curved direction, with its convexity outwards and downwards to a point an inch and a half above the middle of Poupart's ligament; or by Astley Cooper's method, in which an incision is made in a curved direction from a little above and outside the external abdominal ring to an inch from the inner side of the anterior superior spinous process of the ilium. In both, the abdominal muscles and transversalis fascia having been cautiously divided, the peritoneum should be separated from the iliac fossa and raised towards the pelvis; and on introducing the finger to the bottom of the wound, the artery may be felt pulsating along the inner border of the Psoas muscle. The external iliac vein is generally found on the inner side of the artery, and must be cautiously separated from it by the finger-nail, or handle of the knife, and the aneurism needle should be introduced on the inner side, between the artery and vein.

Collateral Circulation.—The principal anastomoses in carrying on the collateral circulation, after the application of a ligature to the external iliac, are : the ilio-lumbar with the circumflex iliac; the gluteal with the external circumflex; the obturator with the internal circumflex; the sciatic with the superior perforating and circumflex branches of the profunda artery; and the internal pudic with the external pudic. When the obturator arises from the epigastric, it is supplied with blood by branches, either from the internal iliac, the lateral sacral, or the internal pudic. The epigastric receives its supply from the internal mammary and inferior intercostal arteries, and from the internal iliac by the anastomoses of its branches with the obturator.*

Branches.—Besides several small branches to the Psoas muscle and the neighbouring lymphatic glands, the external iliac gives off two branches of considerable size, the

Deep epigastric and Deep circumflex iliac.

The **deep epigastric artery** arises from the external iliac, a few lines above Poupart's ligament. It at first descends to reach this ligament, and then ascends obliquely along the inner margin of the internal abdominal ring, between the transversalis fascia and peritoneum; continuing its course upwards, it pierces the transversalis fascia, and, passing over the semilunar fold of Douglas, enters the sheath of the Rectus muscle. It then ascends on the posterior surface of the muscle, and finally divides into numerous branches, which anastomose, above the umbilicus, with the superior epigastric branch of the internal mammary and with the inferior intercostal arteries (fig. 494). The deep epigastric artery bears a very important relation to the internal abdominal ring as it passes obliquely upwards and inwards from its origin from the external iliac. In this part of its course it lies along the lower and inner margin of the ring, and beneath the commencement of the spermatic cord. As it passes to the inner side of the internal abdominal ring, it is crossed by the vas deferens in the male and the round ligament in the female.

The branches of this vessel are the following : the *cremasteric*, which accompanies the spermatic cord, and supplies the Cremaster muscle and other coverings of the cord, anastomosing with the spermatic artery (in the female it is very small and accompanies the round ligament) ; a *pubic branch*, which runs along Poupart's ligament, and then descends behind the os pubis to the inner side of the femoral ring, and anastomoses with offsets from the obturator artery ; *muscular branches*, some of which are distributed to the abdominal muscles and peritoneum, anastomosing with the lumbar and circumflex iliac arteries : others perforate the tendon of the External oblique, and supply the integument, anastomosing with branches of the superficial epigastric.

Peculiarities.—The origin of the deep epigastric may take place from any part of the external iliac between Poupart's ligament and a point two inches and a half above it; or it may arise below this ligament, from the common femoral, or from the deep femoral.

Union with branches.—The deep epigastric frequently arises from the external iliac, by a common trunk with the obturator. Sometimes it arises from the obturator, the latter vessel being furnished by the internal iliac, or it may be formed of two branches, one derived from the external iliac, the other from the internal iliac.

Surgical Anatomy.—The deep epigastric artery follows a line drawn from the middle of Poupart's ligament towards the umbilicus; but shortly after this line crosses the linea semilunaris, the direction changes, and the course of the vessel is upwards along the line of junction of the inner with the outer two-thirds of the Rectus muscle. It has important surgical relations, and is one of the principal means, through its anastomosis with the internal mammary, of establishing the collateral circulation after ligature of either the common or external iliac arteries. It lies close to the internal abdominal ring, and is therefore *internal* to an oblique inguinal hernia, but *external* to a direct inguinal hernia, as these emerge from the abdomen. It forms the outer boundary of Hesselbach's triangle, and is in close relationship with the spermatic cord, which lies in front of it in the inguinal canal, separated only by the transversalis fascia. The vas deferens hooks round its outer side.

The **deep circumflex iliac artery** arises from the outer side of the external iliac nearly opposite the deep epigastric artery. It ascends obliquely outwards behind Poupart's ligament, contained in a fibrous sheath formed by the junction of the transversalis and iliac fasciæ, to the anterior superior spinous process of the ilium, where it anastomoses with the ascending branch of the external circumflex artery. It then passes along the inner surface of the crest of the ilium to about its middle, where it pierces the Transversalis, and runs backwards

* Sir Astley Cooper describes in vol. i. of the *Guy's Hospital Reports* the dissection of a limb eighteen years after successful ligature of the external iliac artery.

between that muscle and the Internal oblique, to anastomose with the ilio-lumbar and gluteal arteries. Opposite the anterior superior spine of the ilium it gives off a large branch, which ascends between the Internal oblique and Transversalis muscles, supplying them, and anastomosing with the lumbar and epigastric arteries.

ARTERIES OF THE LOWER EXTREMITY

The artery which supplies the greater part of the lower extremity is the direct continuation of the external iliac. It continues as a single trunk from Poupart's ligament to the lower border of the Popliteus muscle, where it divides into two branches, the anterior and posterior tibial, an arrangement exactly similar to that which occurs in the upper limb. For convenience of description, the upper part of the main trunk is named femoral, the lower part popliteal.

Femoral Artery (fig. 509)

The **Femoral artery** commences immediately behind Poupart's ligament, midway between the anterior superior spine of the ilium and the symphysis pubis, and passing down the front and inner side of the thigh, terminates at the opening in the Adductor magnus, at the junction of the middle with the lower third of the thigh, where it becomes the popliteal artery. The vessel, at the upper part of the thigh, lies in front of the hip-joint, just on a line with the innermost part of the head of the femur; in the lower part of its course it lies to the inner side of the shaft of the bone, and between these two parts, the vessel is some distance from the bone. In the upper third of the thigh it is contained in a triangular space, called *Scarpa's triangle*; and in the middle third of the thigh, in an aponeurotic canal, called *Hunter's canal*.

Scarpa's Triangle.—Scarpa's triangle corresponds to the depression seen immediately below the fold of the groin. Its apex is directed downwards, and the sides are formed externally by the Sartorius, internally by the inner margin of the Adductor longus, and above by Poupart's ligament. The floor of the space is formed from without inwards by the Iliacus, Psoas, Pectineus (in some cases a small part of the Adductor brevis), and the Adductor longus muscles; and it is divided into two nearly equal parts by the femoral vessels, which extend from near the middle of its base to its apex: the artery giving off in this situation its cutaneous and profunda branches, the vein receiving the deep femoral and internal saphenous. On the outer side of the femoral artery is the anterior crural nerve dividing into its branches. Besides the vessels and nerves, this space contains some fat and lymphatics.

Hunter's Canal.—This is an aponeurotic tunnel in the middle third of the thigh, extending from the apex of Scarpa's triangle to the femoral opening in the Adductor magnus muscle. It is bounded, in front and externally, by the Vastus internus; behind, by the Adductores longus and magnus muscles; and covered in by a strong aponeurosis which extends from the Vastus internus, across the femoral vessels to the Adductores longus and magnus; lying on the aponeurosis is the Sartorius muscle. It contains the femoral artery and vein enclosed in their own sheath of areolar tissue, the vein being behind and on the outer side of the artery, and the internal or long saphenous nerve lying at first on the outer side and then in front of the vessels.

For convenience of description and also in reference to its surgical anatomy, the femoral artery is divided into a short trunk, about an inch and a half or two inches in length, which is known as the *common femoral artery*, while the remainder of the vessel is termed the *superficial femoral*, to distinguish it from the *deep femoral* (profunda femoris), a large branch given off from the common femoral at its termination, and which by its derivation from the parent trunk marks the commencement of the superficial femoral artery.

The **common femoral artery** is very superficial, being covered by the skin and superficial fascia, superficial inguinal lymphatic glands, and the iliac portion of the fascia lata. It has, in front, filaments from the crural branch of the genito-crural nerve, the superficial circumflex iliac vein, and occasionally the superficial epigastric vein. Above, it rests on the inner margin of the Psoas muscle, which separates it from the capsular ligament of the hip-joint, and below on the Pectineus muscle; crossing behind it is the branch to the Pectineus from the anterior crural nerve. Separating the artery from the Pectineus muscle is the pubic portion of the fascia lata. The anterior crural nerve lies about half an inch to the outer side of the common femoral artery, being separated from

FIG. 509.—The femoral artery.

Scrotum

Long saphenous nerve

Anastomotica magna

Superior external articular

Anastomotica magna
Superior internal articular

Inferior internal articular

Inferior internal articular

Anterior tibial recurrent

it by a small part of the Psoas muscle. To the inner side of the artery is the femoral vein, between the margins of the Pectineus and Psoas muscles. The two vessels are enclosed in a strong fibrous sheath, the *femoral sheath*, the anterior part of which is formed by a prolongation downwards of the fascia transversalis, the posterior part by a similar prolongation of the fascia iliaca; the artery and vein are separated, however, from one another by a thin fibrous partition.

PLAN OF THE RELATIONS OF THE COMMON FEMORAL ARTERY

In front.

Skin and superficial fascia.
Superficial inguinal glands.
Iliac portion of fascia lata.
Anterior part of femoral sheath.
Crural branch of genito-crural nerve.
Superficial circumflex iliac vein.
Superficial epigastric vein.

Inner side.
Femoral vein.

Common Femoral Artery.

Outer side.
Small part of Psoas muscle, separating the artery from the anterior crural nerve.

Behind.

Posterior part of femoral sheath.
Pubic portion of fascia lata.
Nerve to Pectineus.
Psoas muscle.
Pectineus muscle.
Capsule of hip-joint.

The **superficial femoral artery** is only superficial where it lies in Scarpa's triangle. Here it is covered by the skin, superficial and deep fascia, and is crossed at the apex of the triangle by the internal cutaneous branch of the anterior crural nerve. In Hunter's canal it is more deeply seated, being covered by the integument, the superficial and deep fascia, the Sartorius and the aponeurotic covering of the canal; the long saphenous nerve crosses the artery from without inwards. In Scarpa's triangle the artery lies on the femoral vein and the profunda artery and vein, which separate it from the Pectineus muscle; in Hunter's canal it rests on the Adductor longus and Adductor magnus muscles. To the outer side, above, are the long saphenous nerve and the Vastus internus muscle with its nerve-supply, and, below, the femoral vein. To the inner side is the Adductor longus above, and the Adductor magnus and Sartorius below.

PLAN OF THE RELATIONS OF THE SUPERFICIAL FEMORAL ARTERY

In front.

Skin, superficial and deep fasciæ.
Internal cutaneous nerve.
Sartorius.
Aponeurotic covering of Hunter's canal.
Long saphenous nerve.

Inner side.
Adductor longus.
Adductor magnus.
Sartorius.

Superficial Femoral Artery.

Outer side.
Long saphenous nerve.
Nerve to Vastus internus.
Vastus internus.
Femoral vein (below).

Behind.

Femoral vein (above).
Profunda artery and vein.
Pectineus muscle.
Adductor longus.
Adductor magnus.

The *femoral vein* at Poupart's ligament lies close to the inner side of the artery, separated from it by a thin fibrous partition; at the apex of Scarpa's triangle, the vein is behind the artery, and at the lower part of Hunter's canal, to its outer side.

The *long saphenous nerve* is situated on the outer side of the artery, in the middle third of the thigh, beneath the aponeurotic covering of Hunter's canal; but not usually within the sheath of the vessels. The internal cutaneous nerve passes from without inwards across the sheath of the femoral artery at the apex of Scarpa's triangle.

Peculiarities.—Double femoral reunited.—Several cases are recorded in which the femoral artery divided into two trunks below the origin of the profunda, and became reunited near the opening in the Adductor magnus, so as to form a single popliteal artery. One of them occurred in a patient operated upon for popliteal aneurism.

Change of Position.—A few cases have been recorded in which the femoral artery was situated at the back of the thigh, the vessel being continuous above with the internal iliac, escaping from the pelvis through the great sacro-sciatic foramen, and accompanying the great sciatic nerve to the popliteal space, where its division occurred in the usual manner. The external iliac in these cases was small, and terminated in the profunda.

Position of the Vein.—The femoral vein is occasionally placed along the inner side of the artery, throughout the entire extent of Scarpa's triangle; or it may be split so that a large vein is placed on each side of the artery for a greater or less distance.

Origin of the Profunda.—This vessel sometimes arises from the inner side, and, more rarely, from the back of the common trunk; but a more important peculiarity, from a surgical point of view, is that which relates to the height at which the vessel arises from the femoral. In three-fourths of a large number of cases it arose between one and two inches below Poupart's ligament; in a few cases the distance was less than an inch; more rarely, opposite the ligament; and in one case above Poupart's ligament, from the external iliac. Occasionally the distance between the origin of the vessel and Poupart's ligament exceeds two inches, and in one case it was found to be as much as four inches.

Surface Marking.—The upper two-thirds of a line drawn from a point midway between the anterior superior spine of the ilium and the symphysis pubis to the Adductor tubercle on the inner condyle of the femur, with the thigh abducted and rotated outwards, will indicate the course of the femoral artery.

Surgical Anatomy.—*Compression* of the femoral artery, which is constantly requisite in amputations and other operations on the lower limb, and also for the cure of popliteal aneurism, is most effectually made immediately below Poupart's ligament. In this situation the artery is very superficial, and is merely separated from the ascending ramus of the os pubis by the Psoas muscle; so that the surgeon, by means of his thumb or a compressor, may effectually control the circulation through it. This vessel may also be controlled in the middle third of the thigh by placing a compress over the artery, beneath the tourniquet, and directing the pressure from within outwards, so as to press the vessel against the inner side of the shaft of the femur.

The superficial position of the femoral artery in Scarpa's triangle renders it particularly liable to be injured in wounds, stabs, or gunshot injuries in the groin. On account of the close relationship between the artery and vein, the latter vessel is also liable to be wounded in these injuries, and thus, if the patient survives, lead to arterio-venous aneurism. In wounds of the femoral artery, the question of the mode of treatment is of considerable importance. If the wound in the superficial structures is a large one, the injured vessel must be exposed and tied; but if the wound is a punctured one and the bleeding has ceased, the question will arise whether to cut down upon the artery or to trust to pressure. Cripps[*] advises, that if the wound is in the 'upper part of the thigh—that is to say, in a position where the femoral artery is comparatively superficial—the surgeon may enlarge the opening with a good prospect of finding the wounded vessel without an extensive or prolonged operation. If the wound be in the lower half of the thigh, owing to the greater depth of the artery, and the possibility of its being the popliteal that is wounded, the search is rendered a far more severe and hazardous operation, and it should not be undertaken until a thorough trial of pressure has proved ineffectual.'

Great care and attention are necessary for the successful application of pressure. The limb should be carefully bandaged from the foot upwards to the wound, which is not covered, and then onwards to the groin. The wound is then dusted with iodoform or boracic powder, and a conical pad applied over it. Rollers the thickness of the index finger are then placed along the course of the vessel above and below the wound, and the whole carefully bandaged to a back splint with a foot-piece.

The *application of a ligature* to the femoral artery may be required in cases of wound or aneurism of the popliteal or femoral, or arteries of the leg;[†] and the vessel may be exposed and tied in any part of its course. The great depth of this vessel at its

---

[*] Heath's *Dictionary of Practical Surgery*, vol. i. p. 525.

[†] Ligature of the femoral artery has been also recommended and performed for elephantiasis of the leg and acute inflammation of the knee-joint.—Maunder, *Clin. Soc. Trans.* vol. ii. p. 37.

lower part, its close connection with important structures, and the density of its sheath, render the operation in this situation one of much greater difficulty than the application of a ligature at its upper part, where it is more superficial.

Ligature of the common femoral artery is not regarded with much favour, on account of the near connection of large branches with it, viz.: the deep epigastric and the deep circumflex iliac arising just above Poupart's ligament; on account of the number of small branches which arise from it, in its short course, and on account of the uncertainty of the origin of the profunda femoris, which, if it arise high up, would be too close to the ligature for the formation of a firm coagulum. It would appear, therefore, that the most favourable situation for the application of a ligature to the femoral is at the apex of Scarpa's triangle. In order to expose the artery in this situation, an incision, between three and four inches long, should be made in the course of the vessel, the patient lying in the recumbent position, with the limb slightly flexed and abducted, and rotated outwards. A large vein is frequently met with, passing in the course of the artery to join the internal saphenous vein; this must be avoided, and the fascia lata having been cautiously divided, and the Sartorius exposed, that muscle must be drawn outwards, in order to expose fully the sheath of the vessels. The finger having been introduced into the wound, and the pulsation of the artery felt, the sheath is opened on the outer side of the vessel to a sufficient extent to allow of the introduction of the aneurism needle, but no farther; otherwise the nutrition of the coats of the vessel may be interfered with, or muscular branches which arise from the vessel at irregular intervals may be divided. In this part of the operation the long saphenous nerve and the nerve to the Vastus internus, which is in close relation with the sheath, should be avoided. The aneurism needle must be carefully introduced and kept close to the artery, to avoid the femoral vein, which lies behind the vessel in this part of its course.

To expose the artery, in Hunter's canal, an incision between three and four inches in length should be made through the integument, a finger's breadth internal to the line of the artery, so that the centre of the incision is in the middle of the thigh—i.e. midway between the groin and the knee. The fascia lata having been divided, and the outer border of the Sartorius exposed, this muscle should be drawn inwards, when the strong fascia which is stretched across from the Adductors to the Vastus internus will be exposed, and must be freely divided; the sheath of the vessels is now seen, and must be opened, and the artery secured by passing the aneurism needle between the vein and artery, in the direction from without inwards. The femoral vein in this situation lies on the outer side of the artery, the long saphenous nerve on its anterior and outer side.

Ligature of the common femoral artery is a comparatively simple proceeding. With the limb slightly flexed, abducted and rotated outwards, an incision about three inches in length is made in the course of the artery; the upper end of the incision being slightly above Poupart's ligament. The skin and superficial fascia having been divided, the iliac portion of the fascia lata is exposed and must be incised. This exposes the sheath of the vessels. A small opening must be made in this structure, and the aneurism needle passed from within outwards.

It has been seen that the femoral artery occasionally divides into two trunks below the origin of the profunda. If, in the operation for tying the femoral, two vessels are met with, the surgeon should alternately compress each, in order to ascertain which vessel is connected with the aneurismal tumour, or with the bleeding from the wound, and that one only should be tied which controls the pulsation or hæmorrhage. If, however, it is necessary to compress both vessels before the circulation in the tumour is controlled, both should be tied, as it would be probable that they became reunited, as in the instances referred to above.

Collateral Circulation.—After ligature of the femoral artery, the main channels for carrying on the circulation are the anastomoses between—(1) the gluteal and sciatic branches of the internal iliac with the internal and external circumflex and superior perforating branches of the profunda femoris; (2) the obturator branch of the internal iliac with the internal circumflex of the profunda; (3) the internal pudic of the internal iliac with the superficial and deep external pudic of the common femoral; (4) the deep circumflex iliac of the external iliac with the external circumflex of the profunda and the superficial circumflex iliac of the common femoral; and (5) the sciatic and comes nervi ischiadici of the internal iliac with the perforating branches of the profunda.[*]

Branches.—The branches of the femoral artery are, the

Superficial epigastric.	⎧ External circumflex.
Superficial circumflex iliac.	Profunda ⎨ Internal circumflex.
Superficial external pudic.	⎩ Three perforating.
Deep external pudic.	Muscular.

Anastomotica magna.

---

[*] In Porta's work (*Alterazioni Patologiche delle Arterie*, tab. xii. xiii.) there is a good representation of the collateral circulation after ligature of the femoral artery.

The **superficial epigastric** arises from the femoral, about half an inch below Poupart's ligament, and, passing through the saphenous opening in the fascia lata, by piercing the cribriform fascia, ascends on to the abdomen, in the superficial fascia covering the External oblique muscle, nearly as high as the umbilicus. It distributes branches to the superficial inguinal glands, the superficial fascia, and the integument, anastomosing with branches of the deep epigastric.

FIG. 510.—Femoral artery and its branches.
(From a preparation in the Museum of the Royal College of Surgeons of England.)

The **superficial circumflex iliac**, the smallest of the cutaneous branches, arises close to the preceding, and, piercing the fascia lata, runs outwards, parallel with Poupart's ligament, as far as the crest of the ilium; it divides into branches which supply the integument of the groin, the superficial fascia, and the superficial inguinal lymphatic glands, anastomosing with the deep circumflex iliac, and with the gluteal and external circumflex arteries.

The **superficial external pudic** (superior) arises from the inner side of the femoral artery, close to the preceding vessels, and, after passing through the saphenous opening, courses inwards, across the spermatic cord or round ligament, to be distributed to the integument on the lower part of the abdomen, the penis and scrotum in the male, and the labium in the female, anastomosing with branches of the internal pudic.

The **deep external pudic** (inferior), more deeply seated than the preceding, passes inwards across the Pectineus and Adductor longus muscles, covered by the fascia lata, which it pierces at the inner border of the thigh, its branches being distributed, in the male, to the integument of the scrotum and perinæum ; and in the female to the labium, anastomosing with branches of the superficial perineal artery.

The **profunda femoris** (*deep femoral artery*) (fig. 510) nearly equals the size of the superficial femoral. It arises from the outer and back part of the common femoral artery, from one to two inches below Poupart's ligament. At first it lies on the outer side of the superficial femoral ; it then runs behind it and the femoral vein to the inner side of the femur, and, passing downwards beneath the Adductor longus, terminates at the lower third of the thigh in a small branch, which pierces the Adductor magnus (and from this circumstance is sometimes called the fourth perforating artery), and is distributed to the Flexor muscles on the back of the thigh, anastomosing with branches of the popliteal and third perforating arteries.

Relations.—*Behind*, it lies first upon the Iliacus, and then on the Pectineus, Adductor brevis, and Adductor magnus muscles. *In front*, it is separated from the superficial femoral artery by the femoral and profunda veins above, and by the Adductor longus below. On its *outer side*, the origin of the Vastus internus intervenes between it and the femur.

PLAN OF THE RELATIONS OF THE PROFUNDA ARTERY

In front.
Superficial femoral artery.
Femoral and profunda veins.
Adductor longus.

Outer side.
Vastus internus.

Behind.
Iliacus.
Pectineus.
Adductor brevis.
Adductor magnus.

The profunda gives off the following named branches :

External circumflex.　　Internal circumflex.　　Four perforating.

The **external circumflex artery** supplies the muscles on the front of the thigh. It arises from the outer side of the profunda, passes horizontally outwards, between the divisions of the anterior crural nerve, and behind the Sartorius and Rectus muscles, and divides into ascending, transverse, and descending branches.

The *ascending branch* passes upwards, beneath the Tensor fasciæ femoris muscle, to the outer side of the hip, and anastomoses with the terminal branches of the gluteal and deep circumflex iliac arteries.

The *descending branch* runs downwards, behind the Rectus, upon the Vasti muscles, to which it gives offsets ; one long branch passes downwards in the Vastus externus as far as the knee, and anastomoses with the superior external articular branch of the popliteal artery. It is accompanied by the branch of the anterior crural nerve to the Vastus externus.

The *transverse branch*, the smallest, passes outwards over the Crureus, pierces the Vastus externus, and winds round the femur, just below the great trochanter, anastomosing at the back of the thigh with the internal circumflex, sciatic, and superior perforating arteries.

The **internal circumflex artery**, smaller than the external, arises from the inner and posterior aspect of the profunda, and winds round the inner side of the femur, passing first between the Pectineus and Psoas muscles, and then between the Obturator externus and Adductor brevis. On reaching the upper border of the Adductor brevis, it gives off two branches, one of which passes inwards to be distributed to the Adductor muscles, the Gracilis, and Obturator externus, anastomosing with the obturator artery; the other descends beneath the Adductor brevis, to supply it and the great Adductor; while the continuation of the vessel passes backwards and divides into an ascending and a transverse branch (fig. 449). The *ascending branch* runs obliquely upwards upon the tendon of the Obturator externus and under cover of the Quadratus femoris towards the digital fossa, where it anastomoses with twigs from the gluteal and sciatic arteries. The *transverse branch*, larger than the ascending, appears between the Quadratus femoris and upper border of the Adductor magnus, anastomosing with the sciatic, external circumflex, and superior perforating arteries ('the *crucial anastomosis*'). Opposite the hip-joint, the artery gives off an articular vessel, which enters the joint beneath the transverse ligament; and, after supplying the adipose tissue, passes along the round ligament to the head of the bone.

The **perforating arteries** (fig. 508), usually four in number, are so named because they perforate the tendon of the Adductor magnus muscle to reach the back of the thigh. They pass backwards close to the linea aspera of the femur under cover of small tendinous arches in the Adductor magnus. The first is given off above the Adductor brevis, the second in front of that muscle, and the third immediately below it.

The *first perforating artery* passes backwards between the Pectineus and Adductor brevis (sometimes it perforates the latter); it then pierces the Adductor magnus close to the linea aspera. It gives off branches which supply the Adductor brevis, the Adductor magnus, the Biceps, and Gluteus maximus muscles, and anastomoses with the sciatic, internal and external circumflex, and middle perforating arteries.

The *second perforating artery*, larger than the first, pierces the tendons of the Adductor brevis and Adductor magnus muscles, and divides into ascending and descending branches, which supply the flexor muscles of the thigh, anastomosing with the first and third perforating. The second artery frequently arises in common with the first. The nutrient artery of the femur is usually given off from this branch.

The *third perforating artery* is given off below the Adductor brevis; it pierces the Adductor magnus, and divides into branches which supply the flexor muscles of the thigh; anastomosing above with the higher perforating arteries, and below with the terminal branches of the profunda and the muscular branches of the popliteal.

The *fourth perforating artery* is represented by the termination of the profunda femoris artery, already described.

Muscular branches are given off from the superficial femoral throughout its entire course. They vary from two to seven in number, and supply chiefly the Sartorius and Vastus internus.

The **anastomotica magna** (fig. 511) arises from the femoral artery just before it passes through the tendinous opening in the Adductor magnus muscle, and immediately divides into a superficial and deep branch.

The *superficial branch* pierces the aponeurotic covering of Hunter's canal, and accompanies the long saphenous nerve to the inner side of the knee. It passes between the Sartorius and Gracilis muscles, and, piercing the fascia lata, is distributed to the integument of the upper and inner part of the leg, anastomosing with the inferior internal articular.

The *deep branch* descends in the substance of the Vastus internus, lying in front of the tendon of the Adductor magnus, to the inner side of the knee, where it anastomoses with the superior internal articular artery and anterior recurrent branch of the anterior tibial. A branch from this vessel crosses outwards above the articular surface of the femur, forming an anastomotic arch with the superior external articular artery, and supplying branches to the knee-joint.

POPLITEAL ARTERY

The **Popliteal artery** commences at the termination of the femoral at the opening in the Adductor magnus, and, passing obliquely downwards and outwards

FIG. 511.—Side view of the popliteal artery.
(From a preparation in the Museum of the Royal College of Surgeons of England.)

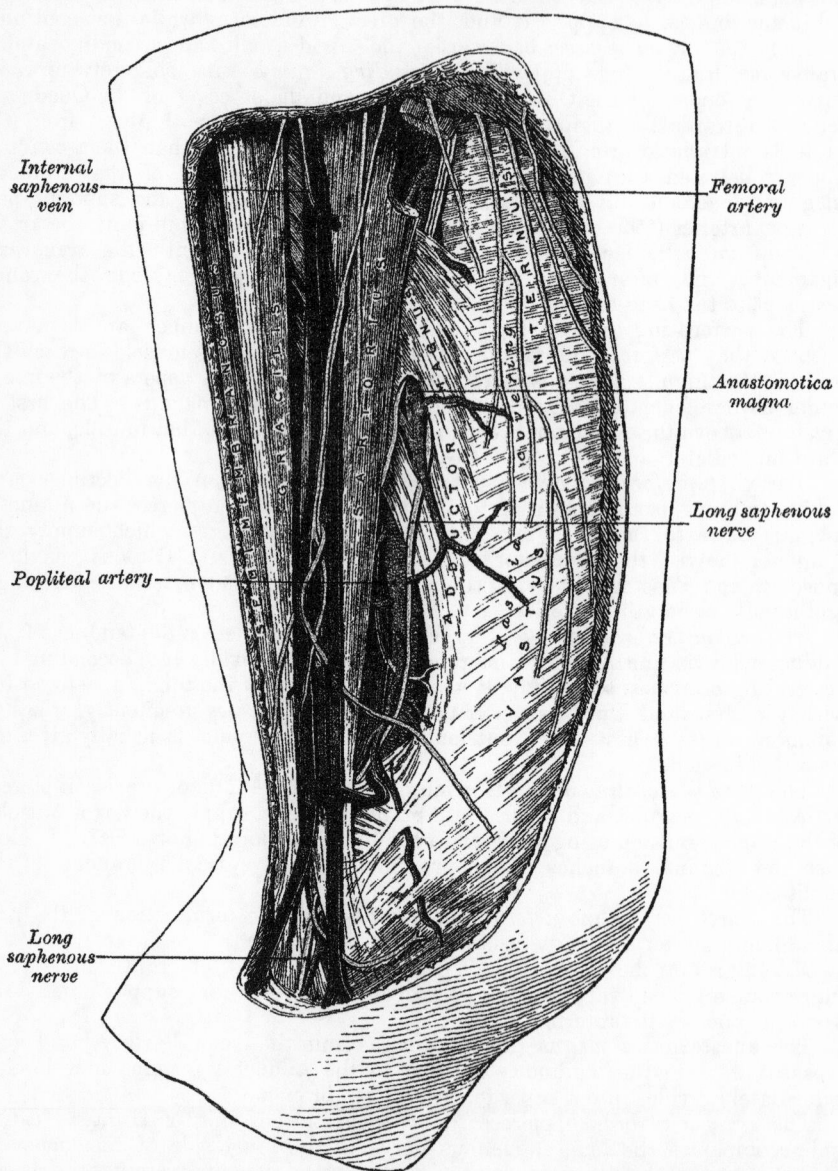

behind the knee-joint to the lower border of the Popliteus muscle, divides into the *anterior* and *posterior tibial arteries*. A portion of the artery lies in the popliteal space; but it is covered above and below, to a considerable extent, by the muscles which form the boundaries of the space, and is therefore beyond the confines of the hollow.

The Popliteal Space (fig. 512)

Dissection.—A vertical incision about eight inches in length should be made along the back part of the knee-joint; at its two extremities, transverse incisions should be carried from the inner to the outer side of the limb. The flaps of integument should be reflected in the direction shown in fig. 447, page 552.

Fig. 512.—The popliteal, posterior tibial, and peroneal arteries.

Boundaries. — The **popliteal space**, or the **ham**, is a lozenge-shaped space, widest at the back part of the knee-joint, and deepest above the articular part of the femur. It is bounded externally, above the joint, by the Biceps, and, below the joint, by the Plantaris and external head of the Gastrocnemius; internally, above the joint, by the Semimembranosus, Semitendinosus, Gracilis, Sartorius, and the tendon of the Adductor magnus; below the joint, by the inner head of the Gastrocnemius.

Above, it is limited by the apposition of the inner and outer hamstring muscles; below, by the junction of the two heads of the Gastrocnemius. The floor is formed by the lower part of the posterior surface of the shaft of the femur, the posterior ligament of the knee-joint, the upper end of the tibia, and the fascia covering the Popliteus muscle; the space is covered in by the fascia lata.

Contents.—The popliteal space contains the popliteal vessels and nerves, together with the termination of the external saphenous vein, the lower part of the small sciatic nerve, the articular branch from the obturator nerve, a few small lymphatic glands, and a considerable quantity of loose adipose tissue.

Position of contained parts. — The internal popliteal nerve descends in the middle line of the space, lying under the deep fascia and crossing the artery from without inwards. The external popliteal nerve descends on the outer side of the upper part of the space, lying close to the tendon of the Biceps muscle. More deeply at the bottom of the space are the popliteal vessels, the vein lying superficial to the artery, to which it is closely united by dense areolar tissue; the vein is a thick-walled vessel, and lies at first to the outer side of the artery, and then crosses it to gain the inner side below; sometimes it is double, the artery lying between two venæ comites, which are usually connected by short transverse branches. More deeply, and, at the upper part of the space, close to the surface of the bone, is the popliteal artery, and passing off

Z Z

from it at right angles are its articular branches. The articular branch from the obturator nerve descends upon the popliteal artery to supply the knee; and occasionally there is found deep in the space an articular filament from the great sciatic nerve. The popliteal lymphatic glands, four or five in number, surround the artery: one usually lies superficial to the vessel; another is situated between it and the bone; and the rest are placed on either side of it.

The **popliteal artery,** in its course downwards from the aperture in the Adductor magnus to the lower border of the Popliteus muscle, rests first on the inner surface of the femur, and is then separated by a little fat from the popliteal surface of the bone; in the middle of its course, it rests on the posterior ligament of the knee-joint; and below, on the fascia covering the Popliteus muscle. *Superficially*, it is covered above by the Semimembranosus; in the middle of its course, by a quantity of fat, which separates it from the deep fascia and integument; and below, it is overlapped by the Gastrocnemius, Plantaris, and Soleus muscles, the popliteal vein, and the internal popliteal nerve. The popliteal vein, which is intimately attached to the artery, lies superficial and external to it above; it then crosses it and lies to its inner side. The internal popliteal nerve is still more superficial and external above, but below the joint it crosses the artery and lies on its inner side. *Laterally*, the artery is bounded by the muscles which are situated on either side of the popliteal space.

PLAN OF THE RELATIONS OF THE POPLITEAL ARTERY

In front.

Femur.
Posterior ligament of knee.
Fascia covering the Popliteus.

Inner side.

Semimembranosus.
Internal condyle.
Gastrocnemius (inner head).

Popliteal Artery.

Outer side.

Biceps.
Outer condyle.
Gastrocnemius (outer head).
Plantaris.

Behind.

Semimembranosus.
Fascia.
Popliteal vein.
Internal popliteal nerve.
Gastrocnemius.
Plantaris.
Soleus.

Peculiarities in point of Division.—Occasionally the popliteal artery divides prematurely into its terminal branches; this unusual division occurs most frequently opposite the knee-joint. The anterior tibial under these circumstances may pass in front of the Popliteus muscle.

Unusual Branches.—The artery sometimes divides into the anterior tibial and peroneal, the posterior tibial being wanting, or very small. Occasionally the popliteal is found to divide into three branches, the anterior and posterior tibial, and peroneal.

Surface Marking.—The course of the upper part of the popliteal artery is indicated by a line drawn from the outer border of the Semimembranosus muscle at the junction of the middle and lower third of the thigh—that is to say, from a point a little internal to the upper angle of the popliteal space—obliquely downwards to the middle of the popliteal space exactly behind the knee-joint. From this point it passes vertically downwards to the level of a line drawn through the lower part of the tubercle of the tibia.

Surgical Anatomy.—The popliteal artery is not infrequently the seat of injury. It may be torn by direct violence, as by the passage of a cart-wheel over the knee, or by hyperextension of the knee; and in the dead body, at all events, the middle and internal coats may be ruptured by extreme flexion. It may also be lacerated by fracture of the lower part of the shaft of the femur, or by antero-posterior dislocation of the knee-joint. It has been torn in breaking down adhesions in cases of fibrous ankylosis of the knee, and is in danger of being wounded, and in fact has been wounded, in performing Macewen's operation of osteotomy of the lower end of the femur for genu valgum. In addition, Spencer records a case in which the popliteal artery was wounded from in front by a stab just below the knee, the knife passing through the interosseous space. The popliteal artery is more frequently the seat of aneurism than any other artery in the body, with the exception of the thoracic aorta. No doubt this is due in a great measure to the amount of movement to which it is subjected, and to the fact that it is supported by loose and lax tissue only, and not by muscles as is the case with most arteries. When the knee is acutely flexed, the popliteal artery becomes bent on itself to such an extent as to entirely arrest the circulation through it.

Ligature of the popliteal artery is required in cases of wound of that vessel, but for aneurism of the posterior tibial, it is preferable to tie the superficial femoral. The popliteal may be tied in the upper or lower part of its course; but in the middle of the ham the operation is attended with considerable difficulty, from the great depth of the artery, and from the extreme degree of tension of the lateral boundaries of the space.

In order to expose the vessel in the upper part of its course, the patient should be placed in the supine position, with the knee flexed and the thigh abducted and rotated outwards, so that it rests on its outer surface; an incision three inches in length, beginning at the junction of the middle and lower third of the thigh, is to be made parallel to and immediately behind the tendon of the Adductor magnus, and the skin, superficial and deep fascia divided. The tendon of the muscle is thus exposed, and is to be drawn forwards and the hamstring tendons backwards. A quantity of fatty tissue will now be opened up, in which the artery will be felt pulsating. This is to be separated with the point of a director until the artery is exposed. The vein and nerve will not be seen, as they lie to the outer side of the artery. The sheath is to be opened and the aneurism needle passed from before backwards, keeping its point close to the artery for fear of injuring the vein. The only structure to avoid is the long saphenous vein in the superficial incision. The upper part of the popliteal artery may also be tied by an incision on the back of the limb, along the outer margin of the Semimembranosus, but the operation is a more difficult one, as the internal popliteal nerve and the popliteal vein are first exposed, and great care has to be exercised in separating the vein from the artery.

To expose the vessel in the lower part of its course, where the artery lies between the two heads of the Gastrocnemius, the patient should be placed in the prone position with the limb extended. An incision should then be made through the integument in the middle line, commencing opposite the bend of the knee-joint, care being taken to avoid the external saphenous vein and nerve. After dividing the deep fascia, and separating some dense cellular membrane, the artery, vein, and nerve will be exposed, descending between the two heads of the Gastrocnemius. Some muscular branches of the popliteal should be avoided if possible, or, if divided, tied immediately. The leg being now flexed, in order the more effectually to separate the two heads of the Gastrocnemius, the nerve should be drawn inwards and the vein outwards, and the aneurism needle passed between the artery and vein from without inwards.

Branches.—The branches of the popliteal artery are, the

Muscular { Superior.	Superior external articular.	
{ Inferior or Sural.	Azygos articular.	
Cutaneous.	Inferior internal articular.	
Superior internal articular.	Inferior external articular.	

The **superior muscular branches,** two or three in number, arise from the upper part of the popliteal artery, and are distributed to the lower parts of the Adductor magnus and hamstring muscles; anastomosing with the fourth perforating branch of the profunda.

The **inferior muscular (sural)** are two large branches, which are distributed to the two heads of the Gastrocnemius and to the Plantaris muscle. They arise from the popliteal artery opposite the knee-joint.

The **cutaneous branches** arise either from the popliteal artery or from some of its branches; they descend between the two heads of the Gastrocnemius muscle, and, piercing the deep fascia, are distributed to the integument of the calf. One branch usually accompanies the short, or external, saphenous vein.

The **superior articular arteries,** two in number, arise one on each side of the popliteal, and wind round the femur immediately above its condyles to the front of the knee-joint. The *internal branch* runs inwards beneath the inner hamstring muscles, to which it supplies branches, above the inner head of the Gastrocnemius, and passing beneath the tendon of the Adductor magnus, divides into two branches, one of which supplies the Vastus internus, inosculating with the anastomotica magna and inferior internal articular; the other ramifies close to the surface of the femur, supplying it and the knee-joint, and anastomosing with the superior external articular artery. The internal articular artery is frequently of small size, a condition which is associated with an increase in the size of the anastomotica magna. The *external branch* passes above the outer condyle, beneath the tendon of the Biceps, and divides into a superficial and deep branch: the superficial branch supplies the Vastus externus, and anastomoses with the descending branch of the external circumflex, and the inferior external articular arteries; the deep branch supplies the lower part of the femur and knee-joint, and forms an anastomotic arch across the bone with the anastomotica magna and the inferior internal articular arteries.

FIG. 513.—Anterior tibial and dorsalis pedis arteries.

Inferior external articular

Inferior internal articular

Anterior tibial recurrent

Anterior peroneal

External malleolar

Internal malleolar

Communicating

The **azygos articular** is a small branch, arising from the popliteal artery opposite the bend of the knee-joint. It pierces the posterior ligament, and supplies the ligaments and synovial membrane in the interior of the articulation.

The **inferior articular arteries**, two in number, arise from the popliteal beneath the Gastrocnemius. The *internal* one first descends along the upper margin of the Popliteus muscle, to which it gives branches; it then passes below the inner tuberosity of the tibia, beneath the internal lateral ligament, at the anterior border of which it ascends to the front and inner side of the joint, to supply the head of the tibia and the articulation of the knee, anastomosing with the inferior external articular and superior internal articular arteries. The *external* one runs outwards above the head of the fibula, to the front of the knee-joint, passing in its course beneath the outer head of the Gastrocnemius, the external lateral ligament, and the tendon of the Biceps muscle, and divides into branches, which anastomose with the inferior internal articular artery, the superior external articular artery, and the anterior recurrent branch of the anterior tibial.

Circumpatellar anastomosis.—Around and above the patella, and on the contiguous ends of the femur and tibia, is a large network of vessels, forming a superficial and deep plexus. The *superficial plexus* is situated between the fascia and skin round about the patella, and forms three well-defined arches: one, above the upper border of the patella, in the loose connective tissue over the Quadriceps extensor muscle; the other two, below the level of the patella, are situated in the fat behind the ligamentum patellæ. The *deep plexus*, which forms a close network of vessels, lies on the lower end of the femur and upper end of the tibia around their articular surfaces, and sends numerous offsets into the interior of the joint. The arteries from which this plexus is formed are the two internal and two external articular branches of the popliteal: the anastomotica magna; the deep branch of the profunda; the descending branch of the external circumflex, and the anterior recurrent branch of the anterior tibial.

Anterior Tibial Artery (fig. 513)

The **Anterior Tibial artery** commences at the bifurcation of the popliteal, at the lower border of the Popliteus muscle, passes forwards between the two heads of the Tibialis posticus, and through the large oval aperture above the upper border of the interosseous membrane, to the deep part of the front of the leg: it here lies close to the inner side of the neck of the fibula; it then descends on the anterior surface of the interosseous membrane, gradually approaching the tibia; and, at the lower part of the leg, lies on this bone, and then on the anterior ligament of the ankle, to the bend of the ankle-joint, where it lies more superficially, and becomes the *dorsalis pedis*.

Relations.—In the upper two-thirds of its extent, the anterior tibial artery rests upon the interosseous membrane, to which it is connected by delicate fibrous arches thrown across it; in the lower third, upon the front of the tibia, and the anterior ligament of the ankle-joint. In the upper third of its course, it lies between the Tibialis anticus and Extensor longus digitorum; in the middle third, between the Tibialis anticus and Extensor proprius hallucis. At the bend of the ankle, it is crossed by the tendon of the Extensor proprius hallucis, and lies between it and the innermost tendon of the Extensor longus digitorum. It is covered, in the upper two-thirds of its course, by the muscles which lie on either side of it, and by the deep fascia; in the lower third, by the integument, anterior annular ligament, and fascia.

The anterior tibial artery is accompanied by two veins (venæ comites) which lie one on either side of the artery; the anterior tibial nerve, coursing round the outer side of the neck of the fibula, comes into relation with the outer side of the artery shortly after it has reached the front of the leg; about the middle of the leg it is placed superficial to it; at the lower part of the artery the nerve is generally again on the outer side.

Plan of the Relations of the Anterior Tibial Artery

In front.

Integument, superficial and deep fasciæ.
Anterior tibial nerve.
Tibialis anticus
Extensor longus digitorum } (overlap it in the upper part of the leg).
Extensor proprius hallucis.
Anterior annular ligament.

Inner side.		*Outer side.*
Tibialis anticus. Extensor proprius hallucis (crosses its lower part from without inwards).	Anterior Tibial.	Anterior tibial nerve. Extensor longus digitorum. Extensor proprius hallucis.

Behind.

Interosseous membrane.
Tibia.
Anterior ligament of ankle-joint.

Peculiarities in Size.—This vessel may be diminished in size, may be deficient to a greater or less extent, or may be entirely wanting, its place being supplied by perforating branches from the posterior tibial, or by the anterior division of the peroneal artery.

Course.—The artery occasionally deviates in its course towards the fibular side of the leg, regaining its usual position beneath the annular ligament at the front of the ankle. In two instances the vessel has been found to approach the surface in the middle of the leg, being covered merely by the integument and fascia below that point.

Surface Marking.—Draw a line from the inner side of the head of the fibula to a point midway between the two malleoli. In this line take a point an inch and a quarter below the head of the fibula, and the portion of the line below this point will mark the course of the artery.

Surgical Anatomy.—The anterior tibial artery is liable to be injured in fractures of the lower third of the tibia, on account of its close proximity to the bone. The application of a ligature to the vessel is rarely required, except in cases of wound or for traumatic aneurism. The operation in the upper third of the leg is attended with great difficulty, on account of the depth of the vessel from the surface. An incision about four inches in length is made in the line of the artery, from the inner side of the head of the fibula to midway between the two malleoli; the incision commencing about a hand's breadth below the level of the knee-joint. The skin and superficial structures having been divided and the deep fascia exposed, the wound must be carefully dried, its edges retracted, and the white line separating the Tibialis anticus from the Extensor longus digitorum sought for. When this has been clearly defined, the deep fascia is to be divided in this line, and the Tibialis anticus separated from adjacent muscles with the handle of the scalpel or a director until the interosseous membrane is reached. The foot is to be flexed in order to relax the muscles, and upon drawing them apart the artery will be found lying on the interosseous membrane with the nerve on its outer side or on the top of the artery. The nerve should be drawn outwards, and the venæ comites separated from the artery and the needle passed around it.

To tie the vessel in the lower third of the leg above the ankle-joint, an incision about three inches in length should be made through the integument between the tendons of the Tibialis anticus and Extensor proprius hallucis muscles, the deep fascia being divided to the same extent. The tendon on either side should be retracted, when the vessel will be seen lying upon the tibia, accompanied by the venæ comites, and with the nerve on the outer side.

Branches.—The branches of the anterior tibial artery are, the

Posterior recurrent tibial.	Muscular.
Superior fibular.	Internal malleolar.
Anterior recurrent tibial.	External malleolar.

The **posterior recurrent tibial** is not a constant branch, and is given off from the anterior tibial before that vessel passes through the interosseous space. It ascends in front of the Popliteus muscle, which it supplies, and anastomoses with the lower articular branches of the popliteal artery, giving off an offset to the superior tibio-fibular joint.

The **superior fibular** is sometimes given off from the anterior tibial, sometimes from the posterior tibial. It passes outwards, round the neck of the fibula, through the Soleus, which it supplies, and ends in the substance of the Peroneus longus muscle.

The **anterior recurrent tibial** arises from the anterior tibial, as soon as that vessel has passed through the interosseous space; it ascends in the Tibialis anticus muscle, and ramifies on the front and sides of the knee-joint, anastomosing with the articular branches of the popliteal, and with the anastomotica magna assisting in the formation of the circumpatellar plexus.

The **muscular branches** are numerous: they are distributed to the muscles which lie on either side of the vessel, some piercing the deep fascia to supply the integument, others passing through the interosseous membrane, and anastomosing with branches of the posterior tibial and peroneal arteries.

The **malleolar arteries** supply the ankle-joint. The *internal* arises about two inches above the articulation, and passes beneath the tendons of the Extensor proprius hallucis and Tibialis anticus, to the inner ankle, upon which it ramifies, anastomosing with branches of the posterior tibial and internal plantar arteries and with the internal calcanean from the posterior tibial. The *external* passes beneath the tendons of the Extensor longus digitorum and Peroneus tertius, and supplies the outer ankle, anastomosing with the anterior peroneal artery, and with ascending branches from the tarsal branch of the dorsalis pedis.

DORSALIS PEDIS ARTERY (fig. 513)

The **Dorsalis pedis**, the continuation of the anterior tibial, passes forwards from the bend of the ankle along the tibial side of the dorsum of the foot to the back part of the first intermetatarsal space, where it divides into two branches, the *dorsalis hallucis* and *communicating.*

Relations.—This vessel, in its course forwards, rests upon the anterior ligament of the ankle-joint, the astragalus, navicular, and middle cuneiform bones, and the ligaments connecting them, being covered by the integument and fascia, anterior annular ligament, and crossed near its termination by the innermost tendon of the Extensor brevis digitorum. On its *tibial side* is the tendon of the Extensor proprius hallucis ; on its *fibular side*, the innermost tendon of the Extensor longus digitorum, and the termination of the anterior tibial nerve. It is accompanied by two veins.

Plan of the Relations of the Dorsalis Pedis Artery

In front.
Integument and fascia.
Anterior annular ligament.
Innermost tendon of Extensor brevis digitorum.

Tibial side.
Extensor proprius hallucis.

Dorsalis Pedis.

Fibular side.
Extensor longus digitorum.
Anterior tibial nerve.

Behind.
Anterior ligament of ankle.
Astragalus,
Navicular,   } and their ligaments.
Middle cuneiform,

Peculiarities in Size.—The dorsal artery of the foot may be larger than usual, to compensate for a deficient plantar artery ; or it may be deficient in its terminal branches to the toes, which are then derived from the internal plantar ; or its place may be supplied altogether by a large anterior peroneal artery.

Position.—This artery frequently curves outwards, lying external to the line between the middle of the ankle and the back part of the first interosseous space.

Surface Marking.—The dorsalis pedis artery is indicated on the surface of the dorsum of the foot by a line drawn from the centre of the space between the two malleoli to the back of the first intermetatarsal space.

Surgical Anatomy.—This artery may be tied, by making an incision through the integument, between two and three inches in length, on the fibular side of the tendon of the Extensor proprius hallucis, in the interval between it and the inner border of the short Extensor muscle. The incision should not extend farther forwards than the back part of the first intermetatarsal space, as the artery divides in that situation. The deep fascia being divided to the same extent, the artery will be exposed, the nerve lying upon its outer side.

Branches.—The branches of the dorsalis pedis are, the

Tarsal.   Dorsalis hallucis.
Metatarsal—Interosseous.   Communicating.

The **tarsal artery** arises from the dorsalis pedis, as that vessel crosses the navicular bone ; it passes in an arched direction outwards, lying upon the tarsal bones, and covered by the Extensor brevis digitorum ; it supplies that muscle and the articulations of the tarsus, and anastomoses with branches from the metatarsal, external malleolar, peroneal, and external plantar arteries.

The **metatarsal artery** arises a little anterior to the preceding ; it passes outwards, over the bases of the metatarsal bones, beneath the tendons of the short Extensor, its direction being influenced by its point of origin ; and it anastomoses with the tarsal and external plantar arteries. This vessel gives off three branches, the *interosseous arteries*, which run forwards upon the three outer Dorsal interossei muscles, and, in the clefts between the toes, divide into two dorsal collateral branches for the adjoining toes. At the back part of each interosseous space these vessels receive the posterior perforating branches from the plantar arch ; and at the fore part of each interosseous space, they are joined by the anterior perforating branches, from the plantar digital arteries. The outermost interosseous artery gives off a branch which supplies the outer side of the little toe.

The **dorsalis hallucis, or first dorsal interosseous artery,** runs forwards along the outer border of the first metatarsal bone, and at the cleft between the first and second toes divides into two branches, one of which passes inwards, beneath the tendon of the Extensor proprius hallucis, and is distributed to the inner border of the great toe ; the other bifurcates to supply the adjoining sides of the great and second toes.

The **communicating artery** dips down into the sole of the foot, between the two heads of the First dorsal interosseous muscle, and inosculates with the termination of the external plantar artery, to complete the plantar arch. It here gives off its plantar digital branch, which is named the *arteria magna hallucis.* This artery passes forwards along the first interosseous space, and, after sending a branch along the inner side of the great toe, bifurcates for the supply of the adjacent sides of the great and second toes.

Posterior Tibial Artery

The **posterior tibial** is an artery of large size, which extends obliquely downwards from the lower border of the Popliteus muscle, along the tibial side of the leg, to the fossa between the inner ankle and the heel, where it divides beneath the origin of the Abductor hallucis, on a level with a line drawn from the point of the internal malleolus to the centre of the convexity of the heel, into the *internal* and *external plantar arteries.* At its origin it lies opposite the interval between the tibia and fibula ; as it descends, it approaches the inner side of the leg, lying behind the tibia, and, in the lower part of its course, is situated midway between the inner malleolus and the tuberosity of the os calcis.

Relations.—The posterior tibial artery lies successively upon the Tibialis posticus, the Flexor longus digitorum, the tibia, and the back part of the ankle-joint. It is *covered* by the deep transverse fascia, which separates it above from the Gastrocnemius and Soleus muscles : at its termination it is covered by the Abductor hallucis muscle. In the lower third, where it is more superficial, it is covered only by the integument and fascia, and runs parallel with the inner border of the tendo Achillis. It is accompanied by two veins, and by the posterior tibial nerve, which lies at first to the inner side of the artery, but soon crosses it, and is, in the greater part of its course, on its outer side.

Plan of the Relations of the Posterior Tibial Artery

In front.

Tibialis posticus.
Flexor longus digitorum.
Tibia.
Ankle-joint.

Inner side.		*Outer side.*
Posterior tibial nerve, upper third.	Posterior Tibial.	Posterior tibial nerve, lower two-thirds.

Behind.

Integument and fascia.
Gastrocnemius.
Soleus.
Deep transverse fascia.
Posterior tibial nerve.
Abductor hallucis.

Behind the Inner Ankle, the tendons and blood-vessels are arranged, under cover of the internal annular ligament, in the following order, from within outwards : First, the tendons of the Tibialis posticus and Flexor longus digitorum, lying in the same groove, behind the inner malleolus, the former being the most internal. External to these is the posterior tibial artery, having a vein on either side; and, still more externally, the posterior tibial nerve. About half an inch nearer the heel is the tendon of the Flexor longus hallucis.

Peculiarities in Size.—The posterior tibial is not infrequently smaller than usual, or absent, its place being supplied by a large peroneal artery, which passes inwards at the lower end of the tibia, and either joins the small tibial artery, or continues alone to the sole of the foot.

Surface Marking.—The course of the posterior tibial artery is indicated by a line drawn from a point an inch below the centre of the popliteal space to midway between the tip of the internal malleolus and the centre of the convexity of the heel.

Surgical Anatomy.—The *application of a ligature* to the posterior tibial may be required in cases of wound of the sole of the foot, attended with great hæmorrhage, when the vessel should be tied at the inner ankle. In cases of wound of the posterior tibial, it will be necessary to enlarge the opening so as to expose the vessel at the wounded point, excepting where the vessel is injured by a punctured wound from the front of the leg. In cases of aneurism from wound of the artery low down, the vessel should be tied in the middle of the leg. But in aneurism of the posterior tibial high up, it would be better to tie the femoral artery.

To tie the posterior tibial artery at the ankle, a semilunar incision, convex backwards, should be made through the integument, about two inches and a half in length, midway between the heel and inner ankle, or a little nearer the latter. The subcutaneous cellular tissue having been divided, a strong and dense fascia, the internal annular ligament, is exposed. This ligament is continuous above with the deep fascia of the leg, covers the vessels and nerves, and is intimately adherent to the sheaths of the tendons. This having been cautiously divided upon a director, the sheath of the vessels is exposed, and, being opened, the artery is seen with one of the venæ comites on either side. The aneurism needle should be passed round the vessel from the heel towards the ankle, in order to avoid the posterior tibial nerve, care at the same time being taken not to include the venæ comites.

The vessel may also be tied in the lower third of the leg by making an incision about three inches in length, parallel with the inner margin of the tendo Achillis. The internal saphenous vein being carefully avoided, the two layers of fascia must be divided upon a director, when the artery is exposed along the outer margin of the Flexor longus digitorum, with one of its venæ comites on either side, and the nerve lying external to it.

Ligature of the posterior tibial in the middle of the leg is a very difficult operation, on account of the great depth of the vessel from the surface. The patient being placed in the recumbent position, the injured limb should rest on its outer side, the knee being partially bent, and the foot extended, so as to relax the muscles of the calf. An incision about four inches in length should then be made through the integument, a finger's breadth behind the inner margin of the tibia, care being taken to avoid the internal saphenous vein. The deep fascia having been divided, the margin of the Gastrocnemius is exposed, and must be drawn aside, and the tibial attachment of the Soleus divided, a director being previously passed beneath it. The artery may now be felt pulsating beneath the deep fascia, about an inch from the margin of the tibia. The fascia having been divided, and the limb placed in such a position as to relax the muscles of the calf as much as possible, the veins should be separated from the artery, and the aneurism needle passed round the vessel from without inwards, so as to avoid wounding the posterior tibial nerve.

Branches.—The branches of the posterior tibial artery are, the

Peroneal.	Muscular.
Nutrient.	Communicating.
Internal calcanean.	

The **Peroneal artery** lies, deeply seated, along the back part of the fibular side of the leg. It arises from the posterior tibial, about an inch below the lower border of the Popliteus muscle, passes obliquely outwards to the fibula, and then descends along the inner border of that bone, contained in a fibrous canal between the Tibialis posticus and the Flexor longus hallucis, or in the substance of the latter muscle to the lower third of the leg, where it gives off the *anterior peroneal*. It then passes across the articulation between the tibia and fibula to the outer side of the os calcis, where it gives off its terminal branches, the *external calcanean*.

Relations.—This vessel rests at first upon the Tibialis posticus, and then, for the greater part of its course, in a fibrous canal between the origins of the Flexor longus hallucis and Tibialis posticus, covered or surrounded by the fibres of the Flexor longus hallucis. It is *covered*, in the upper part of its course, by the Soleus and deep transverse fascia; *below*, by the Flexor longus hallucis.

PLAN OF THE RELATIONS OF THE PERONEAL ARTERY

In front.

Tibialis posticus.
Flexor longus hallucis.

Outer side.

Fibula.
Flexor longus hallucis.

Peroneal Artery.

Inner side.

Flexor longus hallucis.

Behind.

Soleus.
Deep transverse fascia.
Flexor longus hallucis.

Peculiarities in Origin.—The peroneal artery may arise three inches below the Popliteus, or from the posterior tibial high up, or even from the popliteal.

Its size is more frequently increased than diminished; and then it either reinforces the posterior tibial by its junction with it, or altogether takes the place of the posterior tibial in the lower part of the leg and foot, the latter vessel only existing as a short muscular branch. In those rare cases where the peroneal artery is smaller than usual, a branch from the posterior tibial supplies its place; and a branch from the anterior tibial compensates for the diminished anterior peroneal artery. In one case the peroneal artery was entirely wanting.

The anterior peroneal is sometimes enlarged, and takes the place of the dorsal artery of the foot.

The branches of the peroneal are, the

Muscular.	Communicating.
Nutrient.	Posterior peroneal.
Anterior peroneal.	External calcanean.

Muscular branches.—The peroneal artery, in its course, gives off branches to the Soleus, Tibialis posticus, Flexor longus hallucis, and Peronei muscles.

The *nutrient artery* supplies the fibula.

The *anterior peroneal* pierces the interosseous membrane, about two inches above the outer malleolus, to reach the front of the leg, and, passing down beneath the Peroneus tertius, to the outer ankle, ramifies on the front and outer side of the tarsus, anastomosing with the external malleolar and tarsal arteries.

The *communicating* is given off from the peroneal about an inch from its lower end, and, passing inwards, joins the communicating branch of the posterior tibial.

The *posterior peroneal* passes down behind the outer ankle to the back of the external malleolus, to terminate in branches which ramify on the outer surface and back of the os calcis, and anastomose with the external malleolar and tarsal arteries.

The *external calcanean* are the terminal branches of the peroneal artery; they pass to the outer side of the heel, and communicate with the external malleolar and, on the back of the heel, with the internal calcanean arteries.

The **nutrient artery** of the tibia arises from the posterior tibial, near its origin, and after supplying a few muscular branches, enters the nutrient canal of that bone, which it traverses obliquely from above downwards. This is the largest nutrient artery of bone in the body.

The **muscular branches** of the posterior tibial are distributed to the Soleus and deep muscles along the back of the leg.

The **communicating branch** runs transversely across the back of the tibia, about two inches above its lower end, beneath the Flexor longus hallucis, to join a similar branch of the peroneal.

The **internal calcanean** are several large arteries, which arise from the posterior tibial just before its division; they are distributed to the fat and integument behind the tendo Achillis and about the heel, and to the muscles on the inner side of the sole, anastomosing with the peroneal and internal malleolar and, on the back of the heel, with the external calcanean arteries.

The **Internal Plantar artery** (figs. 514, 515), much smaller than the external, passes forwards along the inner side of the foot. It is at first situated above * the Abductor hallucis, and then between it and the Flexor brevis digitorum, both of which it supplies. At the base of the first metatarsal bone, where it is much diminished in size, it passes along the inner border of the great toe, inosculating with its digital branch. Small superficial digital branches accompany the digital branches of the internal plantar nerve and join the plantar digital arteries of the three inner spaces.

The **External Plantar artery,** much larger than the internal, passes obliquely outwards and forwards to the base of the fifth metatarsal bone. It then turns inwards to the interval between the bases of the first and second metatarsal bones, where it anastomoses with the communicating branch from the dorsalis

FIG. 514.—The plantar arteries. Superficial view.

FIG. 515.—The plantar arteries. Deep view.

Communicating branch of dorsalis pedis

Its digital branches

pedis artery, thus completing the *plantar arch*. As this artery passes outwards, it is first placed between the os calcis and Abductor hallucis, and then between the Flexor brevis digitorum and Flexor accessorius; as it passes forwards to the base of the little toe, it lies more superficially between the Flexor brevis digitorum and Abductor minimi digiti, covered by the deep fascia and integument. The remaining portion of the vessel is deeply situated; it extends from the base of the metatarsal bone of the little toe to the back part of the first interosseous space, and forms the plantar arch; it is convex forwards, lies upon the Interossei muscles, opposite the tarsal ends of the metatarsal bones, and is covered by the Adductor obliquus hallucis, the flexor tendons of the toes, and the Lumbricales.

Surface Marking.—The course of the internal plantar artery is represented by a line drawn from the mid-point between the tip of the internal malleolus and the centre of the convexity of the heel to the middle of the under surface of the great toe. The external

---

* This refers to the erect position of the body. In the ordinary position for dissection, the artery is deeper than the muscle.

plantar by a line from the same point to within a finger's breadth of the tuberosity of the fifth metatarsal bone. The plantar arch is indicated by a line drawn from this point : i.e. a finger's breadth internal to the tuberosity of the fifth metatarsal bone transversely across the foot to the back of the first interosseous space.

Surgical Anatomy.—Wounds of the plantar arch are always serious, on account of the depth of the vessel and the important structures which must be interfered with in an attempt to ligature it. They must be treated on similar lines to those of wounds of the palmar arches (see page 670). Delorme has shown that it may be ligatured from the dorsum of the foot in almost any part of its course by removing a portion of one of the three middle metatarsal bones.

Branches.—The plantar arch, besides distributing numerous branches to the muscles, integument, and fasciæ in the sole, gives off the following branches :

Posterior perforating.   Digital—Anterior perforating.

The **posterior perforating** are three small branches, which ascend through the back part of the three outer interosseous spaces, between the heads of the Dorsal interossei muscles, and anastomose with the interosseous branches from the metatarsal artery.

The **digital branches** are four in number, and supply the three outer toes and half the second toe. The *first* passes from the outer side of the plantar arch, and is distributed to the outer side of the little toe, passing in its course beneath the Abductor and short Flexor muscles. The *second*, *third*, and *fourth* run forwards along the interosseous spaces, and on arriving at the clefts between the toes divide into collateral branches, which supply the adjacent sides of the three outer toes and the outer side of the second. Near to its point of bifurcation, each digital artery sends upwards, through the fore part of the corresponding interosseous space, a small branch, the *anterior perforating artery*, which inosculates with the corresponding interosseous branch of the metatarsal artery.

From the arrangement already described of the distribution of the vessels to the toes, it will be seen that both sides of the three outer toes, and the outer side of the second toe, are supplied by branches from the plantar arch ; both sides of the great toe, and the inner side of the second, are supplied by the communicating branch of the dorsalis pedis.

THE VEINS

The veins are the vessels which serve to return the blood from the capillaries of the different parts of the body to the heart. They consist of two distinct sets of vessels, the *pulmonary* and *systemic*.

The **Pulmonary Veins** are concerned in the circulation in the lungs. Unlike other vessels of this kind, they contain arterial blood, which they return from the lungs to the left auricle of the heart.

The **Systemic Veins** are concerned in the general circulation ; they return the venous blood from the body generally to the right auricle of the heart.

The **Portal Vein,** an appendage to the systemic venous system, is confined to the abdominal cavity, and returns the venous blood from the spleen and the viscera of digestion to the liver. This vessel ramifies in the substance of the liver and breaks up into a minute network of capillaries. From these capillaries the blood is conveyed by the hepatic veins to the inferior vena cava.

The veins, like the arteries, are found in nearly every tissue of the body. They commence by minute plexuses which receive the blood from the capillaries. The branches which have their commencement in these plexuses unite together into trunks, and these, in their passage towards the heart, constantly increase in size as they receive tributaries, or join other veins. The veins are larger and altogether more numerous than the arteries ; hence, the entire capacity of the venous system is much greater than that of the arterial ; the pulmonary veins, however, only slightly exceed in capacity the pulmonary arteries. Since the combined area of the smaller venous branches is greater than the main trunks, the venous system may be compared to a cone, the summit of which corresponds to the heart, its base to the circumference of the body. In form, the veins are perfectly cylindrical like the arteries, their walls being collapsed when empty, and the uniformity of their surface being interrupted at intervals by slight constrictions, which indicate the existence of valves in their interior. They usually retain, however, the same calibre as long as they receive no branches.

The veins communicate very freely with one another, especially in certain regions of the body ; and this communication exists between the larger trunks as well as between the smaller branches. Thus, in the cavity of the cranium, and between the veins of the neck, where obstruction would be attended with imminent danger to the cerebral venous system, we find that the sinuses and larger veins have large and very frequent anastomoses. The same free communication exists between the veins throughout the whole extent of the spinal canal, and between the veins composing the various venous plexuses in the abdomen and pelvis, as the spermatic, uterine, vesical, and prostatic.

Veins have thinner walls than arteries, the difference in thickness being due to the smaller amount of elastic and muscular tissues. The superficial veins usually have thicker coats than the deep veins, and the walls of the veins of the lower limb are thicker than those of the upper.

The minute structure of these vessels has been described in the section on General Anatomy.

The systemic veins are subdivided into three sets, viz. : superficial and deep veins, and venous sinuses.

The **Superficial** or **Cutaneous Veins** are found between the layers of the superficial fascia, immediately beneath the integument ; they return the blood from these structures, and communicate with the deep veins by perforating the deep fascia.

The **Deep Veins** accompany the arteries, and are usually enclosed in the same sheath with those vessels. With the smaller arteries—as the radial, ulnar, brachial, tibial, peroneal—they exist generally in pairs, one lying on each side of the vessel, and are called *venæ comites*. The larger arteries—such as the axillary, subclavian, popliteal, and femoral—have usually only one accompanying vein. In certain organs of the body, however, the deep veins do not accompany the arteries ; for instance, the veins in the skull and spinal canal, the hepatic veins in the liver, and the larger veins returning blood from the osseous tissue.

Venous Sinuses are channels, which, in their structure and mode of distribution, differ altogether from the veins. They are found only in the interior of the skull, and consist of canals formed by a separation of the two layers of the dura mater ; their outer coat consisting of fibrous tissue, their inner of an endothelial layer continuous with the lining membrane of the veins.

THE PULMONARY VEINS

The *Pulmonary Veins* return the arterial blood from the lungs to the left auricle of the heart. They are four in number, two for each lung. The pulmonary differ from other veins in several respects : 1. They carry arterial instead of venous blood. 2. They are destitute of valves. 3. They are only slightly larger than the arteries they accompany. 4. A single vein accompanies each artery. They commence in a capillary network, upon the walls of the air-cells, where they are continuous with the capillary ramifications of the pulmonary artery, and, joining together, form one vessel for each lobule. These vessels, uniting successively, form a single trunk for each lobe, three for the right, and two for the left lung. The vein from the middle lobe of the right lung generally unites with that from the upper lobe, forming two trunks from each lung, which perforate the pericardium and open separately into the upper and back part of the left auricle. Occasionally the three veins on the right side remain separate. Not infrequently, the two left pulmonary veins terminate by a common opening.

Within the lung, the branches of the pulmonary artery are *in front*, the veins *behind*, and the bronchi *between* the two.

At the root of the lung, the upper pulmonary vein lies in front and a little below the pulmonary artery ; the lower is situated below the other structures in the lung root, and on a plane posterior to the upper vein. Behind the pulmonary artery is the bronchus.

Within the pericardium, their anterior surface is invested by the serous layer of this membrane.

The right pulmonary veins pass behind the right auricle and superior vena cava ; the left in front of the thoracic aorta.

THE SYSTEMIC VEINS

The Systemic Veins may be arranged into three groups : 1. Those of the head and neck, upper extremity and thorax, which terminate in the superior vena cava. 2. Those of the lower extremity, abdomen, and pelvis, which terminate in the inferior vena cava. 3. The cardiac veins, which open into the right auricle of the heart.

VEINS OF THE HEAD AND NECK

The Veins of the Head and Neck may be subdivided into three groups : 1. The veins of the exterior of the head and face. 2. The veins of the neck. 3. The veins of the diploë and the interior of the cranium.

VEINS OF THE EXTERIOR OF THE HEAD AND FACE

The veins of the exterior of the head and face are, the

Frontal.	Temporal.
Supra-orbital.	Internal maxillary.
Angular.	Temporo-maxillary.
Facial.	Posterior auricular.
Occipital.	

The **Frontal vein** commences on the anterior part of the skull by a venous plexus which communicates with the anterior tributaries of the temporal vein. The veins converge to form a single trunk, which runs downwards near the middle line of the forehead parallel with the vein of the opposite side, the two veins being joined at the root of the nose, by a transverse branch, called the *nasal arch*. Occasionally the frontal veins join to form a single trunk, which bifurcates at the root of the nose into the two angular veins. At the root of the nose the veins diverge, and join the *supra-orbital vein*, at the inner angle of the orbit, to form the *angular vein*.

The **Supra-orbital vein** commences on the forehead, communicating with the anterior temporal vein, and runs downwards and inwards, superficial to the Occipito-frontalis muscle, receiving tributaries from the neighbouring structures, and joins the frontal vein at the inner angle of the orbit to form the *angular vein*. Previous to its junction with the frontal vein, it sends a branch through the supra-orbital notch into the orbit, which communicates with the ophthalmic vein. As this vessel passes through the notch, it receives a diploic vein from the diploë of the frontal bone, which emerges through a foramen at the bottom of the notch.

The **Angular vein** formed by the junction of the frontal and supra-orbital veins runs obliquely downwards and outwards, on the side of the root of the nose, to the level of the lower margin of the orbit, where it becomes the facial vein. It receives the veins of the ala nasi on its inner side, and the superior palpebral veins on its outer side ; it moreover communicates with the ophthalmic vein, thus establishing an important anastomosis between this vessel and the cavernous sinus. Some small veins from the dorsum of the nose terminate in the nasal arch.

The **Facial vein** commences at the side of the root of the nose, being a direct continuation of the angular vein. It lies behind and follows a less tortuous course than the facial artery. It passes obliquely downwards and outwards, beneath the Zygomaticus major and minor muscles, descends along the anterior border of the Masseter, crosses over the body of the lower jaw, with the facial artery, and, passing obliquely outwards and backwards, beneath the Platysma and cervical fascia superficial to the submaxillary gland, unites with the anterior division of the temporo-maxillary vein to form a trunk of large size (*common facial vein*) which enters the internal jugular. From near its termination a communicating branch often runs down the anterior border of the Sterno-mastoid to join the lower part of the anterior jugular.

Tributaries.—The facial vein receives, near the angle of the mouth, a communicating tributary of considerable size (the *deep facial* or *anterior internal*

FIG. 516.—Veins of the head and neck.

maxillary vein) from the pterygoid plexus. It is also joined by the inferior palpebral, the superior and inferior labial veins, the buccal veins from the cheek, and the masseteric veins. Below the jaw it receives the submental; the inferior palatine, which returns the blood from the plexus around the tonsil and soft palate; the submaxillary vein, which commences in the submaxillary gland; and, generally, the ranine vein.

Surgical Anatomy.—There are some points about the facial vein which render it of great importance in surgery. It is not so flaccid as are most superficial veins, and, in consequence of this, remains more patent when divided. It has, moreover, no valves. It communicates freely with the intracranial circulation, not only at its commencement by its tributaries, the angular and supra-orbital veins, communicating with the ophthalmic vein, a tributary of the cavernous sinus, but also by its deep branch, which communicates through the pterygoid plexus with the cavernous sinus by branches which pass through the foramen ovale and foramen lacerum medium (see page 729). These facts have an important bearing upon the surgery of some diseases of the face ; for on account of its patency the facial vein favours septic absorption, and therefore any phlegmonous inflammation of the face following a poisoned wound is liable to set up thrombosis in the facial vein, and detached portions of the clot may give rise to purulent foci in other parts of the body. And on account of its communications with the cerebral sinuses, these thrombi are apt to extend upwards into them, and so induce a fatal issue.

The **Temporal vein** commences by a minute plexus on the side and vertex of the skull, which communicates with the frontal and supra-orbital veins in front, the corresponding vein of the opposite side, and the posterior auricular and occipital veins behind. From this network, anterior and posterior branches arise and unite above the zygoma, to form the trunk of the vein. This trunk is joined in this situation by a large vein, the *middle temporal*, which receives the blood from the substance of the Temporal muscle, and crossing the posterior root of the zygoma, enters the substance of the parotid gland, and unites with the internal maxillary vein to form the temporo-maxillary vein.

Tributaries.—The temporal vein receives in its course some parotid veins, an articular branch from the temporo-mandibular joint, anterior auricular veins from the external ear, and a vein of large size, the *transverse facial*, from the side of the face. The middle temporal vein, previous to its junction with the temporal vein, receives a branch, the *orbital vein*, which is formed by some external palpebral branches, and passes backwards between the layers of the temporal fascia.

The **Pterygoid plexus** is of considerable size, and is situated between the Temporal and External pterygoid, and partly between the two Pterygoid muscles. It receives branches which correspond with those of the internal maxillary artery. Thus it receives the middle meningeal veins, the deep temporal, the pterygoid, masseteric, buccal, alveolar, some palatine veins, and the inferior dental, and a branch which communicates with the ophthalmic vein through the spheno-maxillary fossa. This plexus communicates very freely with the facial vein ; it also communicates with the cavernous sinus, by branches through the foramen Vesalii, foramen ovale, and foramen lacerum medium, at the base of the skull.

The **Internal maxillary vein** is a short trunk which accompanies the first part of the internal maxillary artery. It is formed by a confluence of the veins of the pterygoid plexus, and passes backwards between the internal lateral ligament and the neck of the lower jaw, and unites with the temporal vein, forming the temporo-maxillary vein.

The **Temporo-maxillary vein,** formed by the union of the temporal and internal maxillary veins, descends in the substance of the parotid gland superficial to the external carotid artery but beneath the facial nerve, between the ramus of the jaw and the Sterno-mastoid muscle, and divides into two branches, an anterior, which passes inwards to join the facial vein, and a posterior, which is joined by the posterior auricular vein and becomes the external jugular.

The **Posterior auricular vein** commences upon the side of the head, by a plexus which communicates with the tributaries of the temporal and occipital veins. It descends behind the external ear and joins the posterior division of the temporo-maxillary vein, to form the external jugular. This vessel receives the stylo-mastoid vein, and some tributaries from the back part of the external ear.

The **Occipital vein** commences at the back part of the vertex of the skull, by a plexus, in a similar manner to the other veins. These unite and form a single vessel, which pierces the cranial attachment of the Trapezius and, dipping into the sub-occipital triangle, joins the deep cervical vein. Occasionally, instead of dipping into the sub-occipital triangle, it follows the course of the occipital artery and terminates in the internal jugular ; in other instances, it joins the posterior auricular and passes into the external jugular. As this vein passes across the mastoid portion of the temporal bone, it receives the mastoid vein, and thus establishes a communication with the lateral sinus.

VEINS OF THE NECK

The veins of the neck, which return the blood from the head and face, are the

External jugular.
Posterior external jugular.
Vertebral.

Anterior jugular.
Internal jugular.

The **External jugular vein** receives the greater part of the blood from the exterior of the cranium and deep parts of the face, being formed by the junction of the posterior division of the temporo-maxillary with the posterior auricular vein. It commences in the substance of the parotid gland, on a level with the angle of the lower jaw, and runs perpendicularly down the neck, in the direction of a line drawn from the angle of the jaw to the middle of the clavicle. In its course it crosses the Sterno-mastoid muscle, and in the subclavian triangle perforates the deep fascia, and terminates in the subclavian vein, on the outer side of or in front of the Scalenus anticus muscle. In the neck it is separated from the Sterno-mastoid by the investing layer of the deep cervical fascia, and is covered by the Platysma, the superficial fascia, and the integument. This vein crosses the superficial cervical nerve, and its upper half runs parallel with the great auricular nerve. The external jugular vein varies in size, bearing an inverse proportion to that of the other veins of the neck; it is occasionally double. It is provided with two pairs of valves, the lower pair being placed at its entrance into the subclavian vein, the upper pair in most cases about an inch and a half above the clavicle. The portion of vein between the two sets of valves is often dilated, and is termed the *sinus*. These valves do not prevent the regurgitation of the blood, or the passage of injection from below upwards.*

Surgical Anatomy.—Venesection used formerly to be performed on the external jugular vein, but is now probably never resorted to. The anatomical point to be remembered in performing this operation is to cut across the fibres of the Platysma myoides in opening the vein, so that by their contraction they will expose the orifice in the vein and so allow the flow of blood.

Tributaries.—This vein receives the occipital occasionally, the posterior external jugular, and, near its termination, the suprascapular, transverse cervical, and anterior jugular veins, and, in the substance of the parotid, a large branch of communication from the internal jugular.

The **Posterior external jugular vein** commences in the occipital region and returns the blood from the integument and superficial muscles in the upper and back part of the neck, lying between the Splenius and Trapezius muscles. It runs down the back part of the neck, and opens into the external jugular just below the middle of its course.

The **Anterior jugular vein** commences near the hyoid bone from the confluence of several superficial veins from the submaxillary region. It descends between the median line and the anterior border of the Sterno-mastoid, and, at the lower part of the neck, passes beneath that muscle to open into the termination of the external jugular, or, in some instances, into the subclavian vein (fig. 523). This vein varies considerably in size, bearing almost always an inverse proportion to the external jugular. Most frequently there are two anterior jugulars, a right and left; but sometimes only one. This vein receives some laryngeal veins, and occasionally a small thyroid vein. Just above the sternum, the two anterior jugular veins communicate by a transverse trunk, which receives tributaries from the inferior thyroid veins. It also communicates with the internal jugular. There are no valves in this vein.

The **Internal jugular vein** collects the blood from the interior of the cranium, from the superficial parts of the face, and from the neck. It is directly continuous with the lateral sinus, and commences in the posterior compartment of the jugular foramen, at the base of the skull (fig. 521). At its origin it is somewhat dilated, and this dilatation is called the *sinus*, or *bulb*, of the internal jugular vein. It runs down the side of the neck in a vertical direction, lying at first on the outer

* The student may refer to an interesting paper by Struthers, 'On Jugular Venesection in Asphyxia, anatomically and experimentally considered, including the Demonstration of Valves in the Veins of the Neck,' in the *Edinburgh Medical Journal* for November 1856.

side of the internal carotid, and then on the outer side of the common carotid, and at the root of the neck unites with the subclavian vein to form the innominate vein. The internal jugular vein, at its commencement, lies upon the Rectus capitis lateralis, and behind the internal carotid artery and the nerves passing through the jugular foramen; lower down, the vein and artery lie upon the same plane, the glosso-pharyngeal and hypoglossal nerves passing forwards between them; the pneumogastric descends between and behind them in the same sheath, and the spinal accessory passes obliquely outwards, behind or in front of the vein. At the root of the neck the vein of the right side is placed at a little distance from the artery; on the left side, it usually lies over the artery at its lower part. The right internal jugular vein crosses the first part of the subclavian artery. The vein is of considerable size, but varies in different individuals, the left one being usually the smaller. It is provided with a pair of valves, which are placed at its point of termination, or from half to three-quarters of an inch above it.

Tributaries.—The vein receives in its course the inferior petrosal sinus, the common facial, lingual, pharyngeal, superior and middle thyroid veins, and sometimes the occipital. At its point of junction with the common facial vein it becomes greatly increased in size.

The **inferior petrosal sinus** leaves the skull through the anterior compartment of the jugular foramen, and joins the vein near its commencement.

The **lingual veins** commence on the dorsum, sides, and under surface of the tongue, and, passing backwards, following the course of the lingual artery and its branches, terminate in the internal jugular. The ranine vein, a branch of considerable size, commencing below the tip of the tongue, may join the lingual; generally, however, it passes backwards on the Hyo-glossus muscle in company with the hypoglossal nerve, and joins the facial.

The **pharyngeal vein** begins in a minute plexus, the *pharyngeal*, at the back part and sides of the pharynx, and, after receiving meningeal tributaries, and the Vidian and spheno-palatine veins, terminates in the internal jugular. It occasionally opens into the facial, lingual, or superior thyroid vein.

The **superior thyroid vein** commences in the substance and on the surface of the thyroid gland, by tributaries corresponding with the branches of the superior thyroid artery, and terminates in the upper part of the internal jugular vein. It receives the superior laryngeal and crico-thyroid veins.

The **middle thyroid vein** collects the blood from the lower part of the lateral lobe of the thyroid gland, and, being joined by some veins from the larynx and trachea, terminates in the lower part of the internal jugular vein.

The **facial** and **occipital veins** have been described above.

Surgical Anatomy.—The internal jugular vein occasionally requires ligature in cases of septic thrombosis of the lateral sinus from suppuration in the middle ear, in order to prevent septic emboli being carried into the general circulation. This operation has been performed recently in many cases, with the most satisfactory results. The cases are generally those of chronic disease of the middle ear, with discharge of pus which perhaps has existed for many years. The patient is seized with acute septic inflammation, spreading to the mastoid cells, and, consequent on this, septic thrombosis of the lateral sinus extending to the internal jugular vein. Such cases are always extremely grave, for there is a danger of a portion of the septic clot being detached and causing septic embolism in the thoracic viscera. If the condition is suspected, the sinus should be at once explored by trephining at a point an inch behind the centre of the external auditory meatus and a quarter of an inch above Reid's base line. The condition of the sinus is then investigated, and if it is found to be thrombosed, the surgeon should at once proceed to ligature the internal jugular vein, by an incision along the anterior border of the sterno-mastoid, the centre of which is on a level with the greater cornu of the hyoid bone. The vein should be ligatured in two places and divided between. After the vessel has been secured and divided, the lateral sinus is to be thoroughly cleared out, and by removing the ligature from the upper end of the divided vein, all septic clots removed by syringing from the sinus through the vein. If hæmorrhage occurs from the distal end of the sinus, it can be arrested by careful plugging with antiseptic gauze. The internal jugular vein is also surgically important, because it is surrounded by a large number of the deep chain of cervical lymphatic glands; and when these are enlarged in tuberculous or malignant disease, they are apt to become adherent to the vessel, rendering their removal difficult and often dangerous. The proper course to pursue in these cases is to ligature the vessel above and below the glandular mass, and resect the included portion together with the glands.

The **Vertebral vein** commences in the occipital region, by numerous small tributaries, from the intraspinal venous plexus, issuing from the spinal canal above the posterior arch of the atlas ; these unite with small veins from the deep muscles at the upper and back part of the neck, and form a vessel which passes outwards and enters the foramen in the transverse process of the atlas, and descends, forming a dense plexus around the vertebral artery, in the canal formed by the foramina in the transverse processes of the cervical vertebræ. This plexus unites at the lower part of the neck into one main trunk, which emerges from the foramen in the transverse process of the sixth cervical vertebra, and terminates at the root of the neck in the back part of the innominate vein near its origin, its mouth being guarded by a pair of valves.   On the right side, it crosses the first part of the subclavian artery.

Tributaries.—The vertebral vein receives in its course a vein from the inside of the skull through the posterior condyloid foramen ; muscular veins, from the muscles in the prevertebral region : dorsi-spinal veins, from the back part of the cervical portion of the spine ; meningo-rachidian veins, from the interior of

FIG. 517.—Veins of the diploë as displayed by the removal of the outer table of the skull.

the spinal canal ; the anterior and posterior vertebral veins ; and close to its termination it is sometimes joined by a small vein from the first intercostal space which accompanies the superior intercostal artery.

The **anterior vertebral vein** commences in a plexus around the transverse processes of the upper cervical vertebræ, descends in company with the ascending cervical artery between the Scalenus anticus and Rectus capitis anticus major muscles, and opens into the vertebral vein just before its termination.

The **posterior vertebral** or **deep cervical vein** accompanies the profunda cervicis artery, lying between the Complexus and Semispinalis colli. It commences in the suboccipital region by communicating branches from the occipital vein and tributaries from the deep muscles at the back of the neck. It receives tributaries from the plexuses around the spinous processes of the cervical vertebræ, and terminates in the lower end of the vertebral vein.

VEINS OF THE DIPLOË

The diploë of the cranial bones is channelled in the adult by a number of tortuous canals, which are lined by a more or less complete layer of compact tissue.

The veins they contain are large and capacious, their walls being thin, and formed only of endothelium resting upon a layer of elastic tissue, and they present, at irregular intervals, pouch-like dilatations, or *culs-de-sac*, which serve as reservoirs for the blood. These are the veins of the diploë; they can only be displayed by removing the outer table of the skull.

In adult life, as long as the cranial bones are distinct and separable, these veins are confined to the particular bones; but in old age, when the sutures are united, they communicate with each other, and increase in size. These vessels communicate, in the interior of the cranium, with the meningeal veins, and with the sinuses of the dura mater; and, on the exterior of the skull, with the veins of the pericranium. They are divided into the *frontal*, which opens into the supra-orbital vein by an aperture in the supra-orbital notch; the *anterior temporal*, which is confined chiefly to the frontal bone, and opens into one of the deep temporal veins, through an aperture in the great wing of the sphenoid; the *posterior temporal*, which is situated in the parietal bone, and terminates in the lateral sinus by an aperture at the posterior inferior angle of the parietal bone; and the *occipital*, the largest of the four, which is confined to the occipital bone, and opens either into the occipital vein, or internally into the lateral sinus or torcular Herophili.

Cerebral Veins

The **Cerebral veins** are remarkable for the extreme thinness of their coats in consequence of the muscular tissue in them being wanting, and for the absence of valves. They pierce the arachnoid membrane, and the inner or meningeal layer of the dura mater, and open into the cranial venous sinuses. They may be divided into two sets: the superficial which are placed on the surface, and the deep veins which occupy the interior of the organ.

The **Superficial cerebral veins** ramify upon the surface of the brain, being lodged in the sulci, between the convolutions, a few running across the convolutions. They receive branches from the substance of the brain, and are named, from the position they occupy, superior, median, and inferior cerebral veins.

The **superior cerebral veins,** eight to twelve in number on each side, return the blood from the convolutions on the superior surface of the hemisphere; they pass forwards and inwards towards the great longitudinal fissure, where they receive the *median cerebral veins*; near their termination, they become invested with a tubular sheath of the arachnoid membrane, and open into the superior longitudinal sinus, in the opposite direction to the course of the current of the blood.

The **median cerebral veins** return the blood from the convolutions of the mesial surface of the corresponding hemisphere; they open into the superior cerebral veins; or occasionally into the inferior longitudinal sinus.

The **inferior cerebral veins** ramify on the lower part of the outer and on the under surface of the cerebral hemisphere. Some, collecting tributaries from the under surface of the anterior lobes of the brain, terminate in the cavernous sinus. One vein of large size, the *middle cerebral vein*, commences on the surface of the temporal lobe, and, running along the fissure of Sylvius, opens into the cavernous sinus. Another large vein, the *great anastomotic vein of Trolard*, connects the superior longitudinal and cavernous sinuses, by becoming continuous above with one of the superior cerebral veins opening into the sinus, and below by joining the middle cerebral vein which opens into the cavernous sinus. A third vein, the *posterior anastomotic vein of Labbé*, connects the middle cerebral vein, and so the cavernous sinus, with the lateral sinus by coursing over the temporal lobe. Other inferior cerebral veins commence on the under surface of the base of the brain, and unite to form from three to five veins, which open into the superior petrosal and lateral sinuses from before backwards.

The **Deep cerebral,** or **ventricular veins** (*venæ Galeni*), are two in number. Each is formed by the union of two veins, the *vena corporis striati*, and the *choroid vein*. They run backwards, parallel with one another, between the layers of the velum interpositum, and pass between the posterior extremity, or splenium, of the corpus callosum and the corpora quadrigemina, to enter the straight sinus. The two veins usually unite to form one, the *vena magna Galeni*, before opening into the straight sinus. Just before their union they receive the basilar vein.

The **vena corporis striati** commences in the groove between the corpus striatum and optic thalamus, receives numerous veins from both of these parts, and unites behind the anterior pillar of the fornix with the choroid vein, to form one of the venæ Galeni.

The **choroid vein** runs along the whole length of the outer border of the choroid plexus, receiving veins from the hippocampus major, the fornix and corpus callosum, and unites, at the anterior extremity of the choroid plexus, with the vein of the corpus striatum.

The **Basilar vein** commences at the anterior perforated space at the base of the brain by the union of a small *anterior cerebral vein*, which accompanies the anterior cerebral artery, with the *deep Sylvian vein*, which descends through the lower part of the Sylvian fissure. It passes backwards round the crus cerebri, receiving the inferior striate vein from the corpus striatum, interpeduncular veins from the interpeduncular space, ventricular veins from the middle cornu of the lateral ventricles, and tributaries from the uncinate convolution, and enters the vein of Galen just before its junction with the vein of the opposite side.

The **Cerebellar veins** occupy the surface of the cerebellum, and are disposed in two sets, superior and inferior. The *superior* pass partly forwards and inwards, across the superior vermiform process, to terminate in the straight sinus and the venæ Galeni, partly outwards to the lateral and superior petrosal sinuses. The *inferior cerebellar veins*, of large size, terminate in the lateral, superior petrosal, and occipital sinuses.

The perivascular lymphatics alluded to in the section on General Anatomy are especially found in connection with the vessels of the brain. These vessels are enclosed in a sheath, which acts as a lymphatic channel, through which the lymph is carried to the subarachnoid and subdural spaces, from which it is returned into the general circulation.

Sinuses of the Dura Mater

The sinuses of the dura mater are venous channels, analogous to the veins, situated between the two layers of the dura mater, and lined by endothelium continuous with that which lines the veins. They are fourteen in number, of which six are single and situated in the middle line; the other eight are disposed in four pairs, one sinus of each pair being placed on either side of the middle line. They may be divided into two sets: 1. Those situated at the upper and back part of the skull; 2. Those at the base of the skull. The former are, the

Superior longitudinal sinus.	Straight sinus.
Inferior longitudinal sinus.	Two lateral sinuses.
Occipital sinus.	

The **Superior longitudinal sinus** occupies the attached margin of the falx cerebri. Commencing at the foramen cæcum, through which it communicates by a small branch with the veins of the nasal fossæ, it runs from before backwards, grooving the inner surface of the frontal, the adjacent margins of the two parietal, and the superior division of the crucial ridge of the occipital bone; near the internal occipital protuberance it deviates to one or other side (usually the right), and is continued as the corresponding lateral sinus. The sinus is triangular in form, narrow in front, and gradually increases in size as it passes backwards. On examining its inner surface, it presents the internal openings of the superior cerebral veins, which run, for the most part, from behind forwards, and open chiefly at the back part of the sinus, their orifices being concealed by fibrous folds; numerous fibrous bands (*chordæ Willisii*) are also seen, extending transversely across the inferior angle of the sinus; and, lastly, small openings which communicate with *venous lacunæ* in the dura mater, into which small white bodies, the *glandulæ Pacchioni*, project. This sinus receives the superior cerebral veins, numerous veins from the diploë and dura mater, and, at the posterior extremity of the sagittal suture, veins from the pericranium, which pass through the parietal foramina.

The *torcular Herophili*, or *confluence of the sinuses*, is the term applied to the dilated extremity of the superior longitudinal sinus. It is of irregular form,

and is lodged on one side (generally the right) of the internal occipital protuberance. From it the lateral sinus of the side to which it is deflected is
derived. It receives also the blood from the occipital sinus, and is connected
across the middle line with the commencement of the lateral sinus of the
opposite side.

Surgical Anatomy.—The numerous communications which take place between this
sinus and the veins of the nose, scalp and diploë, cause it to be frequently the seat of
infective thrombosis from suppurative processes in these parts. It is also sometimes
the seat of a non-infective thrombosis in weak and debilitated individuals.

The **Inferior longitudinal sinus**, more correctly described as the *inferior
longitudinal vein*, is contained in the posterior half or two-thirds of the free
margin of the falx cerebri. It is of a cylindrical form, increases in size as it
passes backwards, and terminates in the straight sinus. It receives several
veins from the falx cerebri, and occasionally a few from the mesial surface of the
hemispheres.

The **Straight sinus** is situated at the line of junction of the falx cerebri
with the tentorium. It is triangular in form, increases in size as it proceeds
backwards, and runs obliquely downwards and backwards from the termination

Fig. 518.—Vertical section of the skull, showing the sinuses of the dura mater.

of the inferior longitudinal sinus to the lateral sinus of the opposite side to
that into which the superior longitudinal sinus is prolonged. Its terminal part
communicates by a cross branch with the torcular Herophili. Besides the
inferior longitudinal sinus, it receives the vena magna Galeni and the superior
cerebellar veins. A few transverse bands cross its interior.

The **Lateral sinuses** are of large size, and commence at the internal occipital
protuberance; one, generally the right, being the direct continuation of the
superior longitudinal sinus, the other of the straight sinus. They pass outwards
and forwards, describing a slight curve with its convexity upwards, to the base
of the petrous portion of the temporal bone, being situated, in this part of their
course, in the attached margin of the tentorium cerebelli; they then leave the
tentorium and curve downwards and inwards on each side to reach the jugular
foramen, where they terminate in the internal jugular vein. Each sinus rests,
in its course, upon the inner surface of the occipital, the posterior inferior angle
of the parietal, the mastoid portion of the temporal, and on the occipital, again,
just before its termination. These sinuses are frequently of unequal size, that
formed by the superior longitudinal sinus being the larger, and they increase
as they proceed from behind forwards. The horizontal portion is of a triangular
form, the curved portion semicylindrical. Their inner surface is smooth, and

not crossed by the fibrous bands found in the other sinuses. They receive the blood from the superior petrosal sinuses at the base of the petrous portion of the temporal bone ; they communicate with the veins of the pericranium by means of the mastoid and posterior condyloid veins, and they receive some of the inferior cerebral and inferior cerebellar veins, and some veins from the diploë. The *petro-squamous sinus,* when present, runs backwards along the junction of the petrous and squamous-temporal, and opens into the lateral sinus.

The **Occipital** is the smallest of the cranial sinuses. It is generally single, but occasionally there are two. It is situated in the attached margin of the falx cerebelli. It commences by several small veins around the margin of the foramen magnum, one of which joins the termination of the lateral sinus ; it communicates with the posterior spinal veins, and terminates in the torcular Herophili.

The sinuses at the base of the skull are, the

Two cavernous sinuses.   Two superior petrosal sinuses.
Circular sinus.   Two inferior petrosal sinuses.
                    Transverse sinus.

The **Cavernous sinuses** are named from their presenting a reticulated structure, due to their being traversed by numerous interlacing filaments. They are two in number, of irregular form, larger behind than in front, and are placed one on each side of the body of the sphenoid, extending from the sphenoidal fissure to the apex of the petrous portion of the temporal bone ; they receive anteriorly the ophthalmic veins through the sphenoidal fissure, and open behind into the

FIG. 519.—Plan showing the relative position of the structures in the right cavernous sinus, viewed from behind.

petrosal sinuses. On the inner wall of each sinus is found the internal carotid artery, accompanied by filaments of the carotid plexus and by the sixth nerve ; and on its outer wall, the third and fourth nerves, and the ophthalmic and superior maxillary divisions of the fifth nerve. These parts are separated from the blood flowing along the sinus by the lining membrane, which is continuous with the inner coat of the veins. The cavernous sinus receives some of the cerebral veins, and also a small sinus, the *spheno-parietal,* which extends inwards on the under aspect of the lesser wing of the sphenoid ; it communicates with the lateral sinus by means of the superior petrosal sinus, with the internal jugular vein through the inferior petrosal sinus and through a plexus of veins on the petrous portion of the internal carotid artery, with the pterygoid venous plexus through the foramen Vesalii or foramen ovale, and with the angular vein through the ophthalmic vein. The two sinuses also communicate with each other by means of the circular sinus.

Surgical Anatomy.—An arterio-venous communication may be established between the cavernous sinus and the internal carotid artery, as it lies in it, giving rise to a pulsating tumour in the orbit. These communications may be the result of injury, such as a bullet wound, a stab, or a blow or fall sufficiently severe to cause a fracture of the base of the skull in this situation ; or they may occur idiopathically, from the rupture of an aneurism or a diseased condition of the internal carotid artery. The disease begins with sudden noise and pain in the head, followed by exophthalmos, swelling and congestion of the lids and conjunctivæ, and development of a pulsating tumour at the margin of the orbit, with thrill and the characteristic bruit; accompanying these symptoms there may be impairment of sight, paralysis of the iris and orbital muscles, and pain of varying intensity

In some cases the opposite orbit becomes affected by the passage of the arterial blood into the opposite sinus by means of the circular sinus. Or the arterial blood may find its way through the emissary veins (see page 729) into the pterygoid plexus, and thence into the veins of the face. Pulsating tumours of the orbit may also be due to traumatic aneurism of one of the orbital arteries, and symptoms resembling those of pulsating tumour may be produced by pressure on the ophthalmic vein, as it enters the sinus, by an aneurism of the internal carotid artery.

The **Ophthalmic Veins** are two in number, superior and inferior.

The *superior ophthalmic vein* connects the angular vein at the inner angle of the orbit with the cavernous sinus; it pursues the same course as the ophthalmic artery, and receives tributaries corresponding to the branches derived from that vessel. Forming a short single trunk, it passes through the inner extremity of the sphenoidal fissure, and terminates in the cavernous sinus.

The *inferior ophthalmic vein* receives the veins from the floor of the orbit, and either passes out of the orbit through the spheno-maxillary fissure to join the

Fig. 520.—The sinuses at the base of the skull.

Torcular Herophili

pterygoid plexus of veins; or else, passing backwards through the sphenoidal fissure, it enters the cavernous sinus, either by a separate opening, or, more frequently, in common with the superior ophthalmic vein.

The **Circular sinus** is formed by two transverse vessels, the *anterior* and *posterior intercavernous sinuses*, which connect together the two cavernous sinuses; one passing in front and the other behind the pituitary body, and thus forming with the cavernous sinuses a venous circle around that body. The anterior one is usually the larger of the two, and one or other is occasionally absent.

The **Superior petrosal sinus** is situated along the superior border of the petrous portion of the temporal bone, in the front part of the attached margin of the tentorium. It is small and narrow, and connects together the cavernous and lateral sinus, opening into the latter as it curves downwards on the inner surface of the mastoid part of the temporal bone. It receives some cerebellar and inferior cerebral veins, and veins from the tympanic cavity.

The **Inferior petrosal sinus** is situated in the groove formed by the junction of the posterior border of the petrous portion of the temporal with the basilar process of the occipital. It commences in front at the termination of the cavernous sinus, and, passing through the anterior compartment of the jugular foramen, joins the commencement of the internal jugular vein. The inferior petrosal sinus receives the veins from the internal ear and also veins from the medulla, pons, and under surface of the cerebellum.

The exact relation of the parts to one another in the jugular foramen is as follows : the inferior petrosal sinus is in front, with the meningeal branch of the ascending pharyngeal artery, and is directed obliquely downwards and backwards ; the lateral sinus is situated at the back part of the foramen with a meningeal branch of the occipital artery, and between the two are the glosso-pharyngeal, pneumogastric, and spinal accessory nerves. These three sets of structures are divided from each other by two processes of fibrous tissue. The junction of the inferior petrosal sinus with the internal jugular vein takes place superficial to the nerves, so that these latter lie a little internal to the venous channels in the foramen (see fig. 521).

FIG. 521.—Relation of structures in jugular foramen.

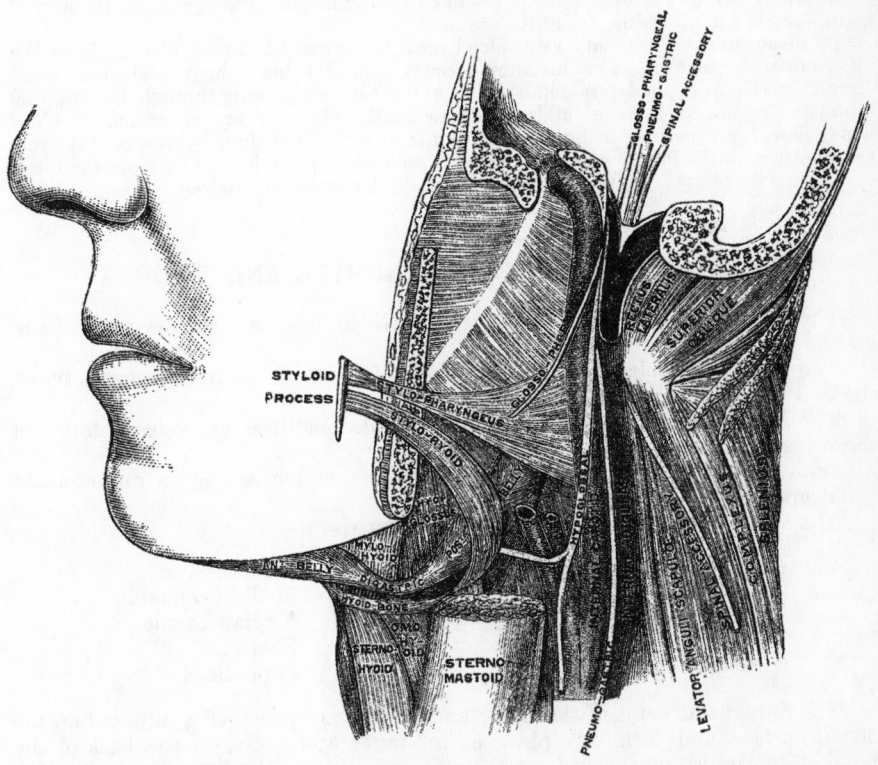

The **Transverse sinus**, or **basilar sinus**, consists of several interlacing veins between the layers of the dura mater over the basilar process of the occipital bone, which serve to connect the two inferior petrosal sinuses. They communicate with the anterior spinal veins.

Emissary veins.—The emissary veins are vessels which pass through apertures in the cranial wall and establish communications between the sinuses inside the skull and the veins external to it. Some of these are always present, others only occasionally so. They vary much in size in different individuals. The principal emissary veins are the following: 1. A vein, almost always present, which passes through the mastoid foramen and connects the lateral sinus with the posterior auricular or with an occipital vein. 2. A vein which passes through the parietal foramen and connects the superior longitudinal sinus with the veins of the scalp. 3. A plexus of minute veins which passes through the anterior condyloid foramen and connects the lateral sinus with the vertebral vein and deep veins of the neck. 4. An inconstant vein, which passes through the posterior condyloid foramen and connects the lateral sinus with the deep veins of the neck. 5. One or two veins of considerable size, which pass through the foramen ovale and connect the cavernous sinus with the pterygoid and pharyngeal plexuses. 6. Two or three small veins which pass through the foramen lacerum medium and connect the cavernous sinus with the pterygoid and pharyngeal plexuses. 7. There is sometimes a small vein passing through the foramen of Vesalius connecting the same parts. 8. A plexus of veins passing through the carotid canal and connecting the cavernous sinus with the internal jugular vein. 9. A vein passing through the foramen cæcum which connects the longitudinal sinus with the veins of the mucous membrane of the nose.

Surgical Anatomy.—These emissary veins are of great importance in surgery. In addition there are, however, other communications between the intra- and extra-cranial circulation. As, for instance, the communication of the angular and supra-orbital veins with the ophthalmic vein at the inner angle of the orbit (page 718), and the communication of the veins of the scalp with the diploic veins (page 724). Through these communications inflammatory processes commencing on the outside of the skull may travel inwards, leading to osteo-phlebitis of the diploë and inflammation of the membranes of the brain. To this in former days was to be attributed one of the principal dangers of scalp wounds and other injuries of the scalp.

By means of these emissary veins blood may be abstracted almost directly from the intracranial circulation. For instance, leeches applied behind the ear abstract blood almost directly from the lateral sinus, by means of the vein passing through the mastoid foramen. Again, epistaxis in children will frequently relieve severe headache, the blood which flows from the nose being derived from the longitudinal sinus by means of the vein passing through the foramen cæcum, which is another communication between the intra-cranial and extra-cranial circulation, and is constantly found in children.

VEINS OF THE UPPER EXTREMITY AND THORAX

The Veins of the Upper Extremity are divided into two sets, *superficial* and *deep*.

The **Superficial veins** are placed immediately beneath the integument between the two layers of superficial fascia.

The **Deep veins** accompany the arteries, and constitute the venæ comites of those vessels.

Both sets of vessels are provided with valves, which are more numerous in the deep than in the superficial.

The superficial veins of the upper extremity are, the

Superficial veins of the hand.	Median.
Anterior ulnar.	Median cephalic.
Posterior ulnar.	Median basilic.
Common ulnar.	Basilic.
Radial.	Cephalic.

The **Superficial veins of the hand and fingers** are principally situated on the dorsal surface, and form two plexuses, an inner and outer, on the back of the hand. The inner plexus is formed by the veins from the little finger, the ring finger, and the ulnar side of the middle finger; from it the anterior and posterior ulnar veins are derived. The outer plexus is formed by veins from the thumb, the index finger, and radial side of the middle finger; from it the radial vein is derived. These two plexuses communicate on the back of the hand, forming a superficial arch of veins in this situation. The superficial veins from the palm of the hand form a plexus in front of the wrist.

The **Anterior ulnar vein** commences on the anterior surface of the ulnar side of the hand and wrist, and ascends along the anterior surface of the ulnar side of the forearm to the bend of the elbow, where it joins with the posterior ulnar vein to form the common ulnar. Occasionally it opens separately into the median basilic vein. It communicates with branches of the median vein in front, and with the posterior ulnar vein behind.

The **Posterior ulnar vein** commences on the posterior surface of the ulnar side of the wrist. It runs on the posterior surface of the ulnar side of the forearm, and just below the elbow unites with the anterior ulnar vein to form the common ulnar, or else joins the median basilic to form the basilic. It communicates with the deep veins of the palm by a branch which emerges from beneath the Abductor minimi digiti muscle.

The **Common ulnar** is a short trunk which is not constant. When it exists it is formed by the junction of the two preceding veins, and, passing upwards and outwards, joins the median basilic to form the basilic vein. When it does not exist, the anterior and posterior ulnar veins open separately into the median basilic vein.

The **Radial vein** commences on the dorsal surface of the wrist, communicating with the deep veins of the palm by a branch which passes through the first interosseous space. It is a large vessel, which ascends along the radial side of the forearm, and receives numerous veins from both its surfaces. At the bend of the elbow it unites with the median cephalic to form the cephalic vein.

The **Median vein** usually arises from the outer part of the dorsal plexus, near the base of the thumb. It turns round the radial border of the wrist, ascends on the front of the forearm, and communicates with the anterior ulnar and radial veins. At the bend of the elbow it receives a branch of communication from the deep veins, and divides into two branches, the median cephalic and median basilic, which diverge from each other as they ascend.

The **Median cephalic,** usually the smaller of the two, passes upwards and outwards in the groove between the Brachio-radialis and Biceps muscles; it crosses superficial to the musculo-cutaneous nerve, and joins with the radial to form the cephalic vein. The branches of the external cutaneous nerve pass beneath this vessel.

The **Median basilic vein** passes obliquely inwards, in the groove between the Biceps and Pronator radii teres, and joins the common ulnar to form the basilic vein. This vein passes in front of the brachial artery, from which it is separated by a fibrous expansion (the *bicipital fascia*) which is given off from the tendon of the Biceps to the fascia covering the Flexor muscles of the forearm. Filaments of the internal cutaneous nerve pass in front as well as behind this vessel.*

* Cruveilhier says: 'Numerous varieties are observed in the disposition of the veins of the elbow; sometimes the common median vein is wanting; but in those cases, its two branches are furnished by the radial vein, and the cephalic is almost always in a rudimentary condition. In other cases, only two veins are found at the bend of the elbow, the radial and ulnar, which are continuous, without any demarcation, with the cephalic and basilic.'

FIG. 522.—The superficial veins of the upper extremity.

Median cephalic

External cutaneous nerve

Internal cutaneous nerve
Median basilic

Venesection is generally performed at the bend of the elbow, and as a matter of practice the largest vein in this situation is commonly selected. This is usually the median basilic, and there are anatomical advantages and disadvantages in selecting this vein. The advantages are, that in addition to its being the largest vessel, and therefore yielding a greater supply of blood, it is the least movable and can be easily steadied on the bicipital fascia on which it rests. The disadvantages are, that it is in close relationship with the brachial artery, separated only by the bicipital fascia; and formerly, when venesection was frequently practised, arterio-venous aneurism was no uncommon result of this practice. Another disadvantage is, that the median basilic is crossed by some of the branches of the internal cutaneous nerve, and these may be divided in the operation, giving rise to ' traumatic neuralgia of extreme intensity ' (Tillaux).

The **Basilic vein** is of considerable size, formed by the coalescence of the common ulnar vein with the median basilic. It runs upwards along the inner side of the Biceps muscle, passes through an opening in the deep fascia a little below the middle of the arm, and, ascending in the course of the brachial artery to the lower border of the Teres major muscle, is continued onwards as the axillary vein.

The **Cephalic vein** courses along the outer border of the Biceps muscle, lying in the same groove with the upper external cutaneous branch of the musculo-spiral nerve, to the upper third of the arm; it then passes between the Pectoralis major and Deltoid muscles, lying in the same groove with the descending or humeral branch of the acromio-thoracic artery. It pierces the costo-coracoid membrane, and crossing the axillary artery, terminates. in the axillary vein just below the clavicle. This vein is occasionally connected with the external jugular by a branch which passes upwards in front of the clavicle.

The **Deep veins of the upper extremity** follow the course of the arteries, forming their venæ comites. They are generally arranged in pairs, and are situated on either side of the corresponding artery, and connected at intervals by short transverse branches.

Two digital veins accompany each artery along the sides of the fingers: these, uniting at the base of the fingers, pass along the interosseous spaces in the palm, and terminate in the two venæ comites which accompany the superficial palmar arch. Branches from these vessels on the radial side of the hand accompany the superficialis volæ, and on the ulnar side terminate in the deep ulnar veins. The deep ulnar veins, as they pass in front of the wrist, communicate with the interosseous and superficial veins, and, at the elbow, unite with the deep radial veins to form the venæ comites of the brachial artery.

The **Interosseous veins** accompany the anterior and posterior interosseous arteries. The anterior interosseous veins commence in front of the wrist, where they communicate with the deep radial and ulnar veins; at the upper part of the forearm they receive the posterior interosseous veins, and terminate in the venæ comites of the ulnar artery.

The **Deep palmar veins** accompany the deep palmar arch, being formed by tributaries which follow the ramifications of that vessel. They communicate with the deep ulnar veins at the inner side of the hand, and on the outer side terminate in the venæ comites of the radial artery. At the wrist, they receive a dorsal and a palmar tributary from the thumb, and unite with the deep radial veins. Accompanying the radial artery, these vessels terminate in the venæ comites of the brachial artery.

The **Brachial veins** are placed one on each side of the brachial artery, receiving tributaries corresponding with the branches given off from that vessel; at the lower margin of the Subscapularis, they join the axillary vein.

These deep veins have numerous anastomoses, not only with each other, but also with the superficial veins.

The **Axillary vein** is of large size, and is the continuation upwards of the basilic vein. It commences at the lower border of the Teres major, increases in size as it ascends, by receiving tributaries corresponding with the branches of the axillary artery, and terminates immediately beneath the clavicle at the outer border of the first rib, where it becomes the subclavian vein. This vessel is covered in front by the Pectoral muscles and costo-coracoid membrane, and lies on the thoracic side of the axillary artery, which it partially overlaps. Near its termination it receives the cephalic vein. This vein is provided with a pair of valves, opposite the lower border of the Subscapularis muscle ; valves are also found at the termination of the cephalic and subscapular veins.

Surgical Anatomy.—There are several points of surgical interest in connection with the axillary vein. Being more superficial and larger than the axillary artery, which it overlaps, it is more liable to be wounded than the artery in the operation of extirpation of the axillary glands, especially as these glands, when diseased, are apt to become adherent to the vessel. When wounded, there is always a danger of air being drawn into its interior, and death resulting. This is due not only to the fact that it is near the thorax and therefore liable to be influenced by the respiratory movements, but also because it is adherent by its anterior surface to the costo-coracoid membrane, and therefore if wounded is likely to remain patulous and favour the chance of air being sucked in. This adhesion of the vein to the fascia prevents its collapsing, and therefore favours the furious bleeding which takes place in these cases.

To avoid wounding the axillary vein in the extirpation of glands from the axilla, it is always advisable to expose the vein as soon as possible ; no sharp cutting instruments should be used after the axillary cavity has been freely exposed ; and care should be taken to use no undue force in isolating the glands (see page 653). Should the vein be so embedded in the malignant deposit that the latter cannot be removed without taking away a part of the vein, this must be done, the vessel having been first ligatured above and below.

The **Subclavian vein,** the continuation of the axillary, extends from the outer border of the first rib to the inner end of the clavicle, where it unites with the internal jugular to form the innominate vein. It is in relation, in front, with the clavicle and Subclavius muscle ; behind and above, with the subclavian artery, from which it is separated internally by the Scalenus anticus muscle and phrenic nerve. Below, it rests in a depression on the first rib and upon the pleura. Above, it is covered by the cervical fascia and integument.

The subclavian vein occasionally rises in the neck to a level with the third part of the subclavian artery, and in one or two instances has been seen passing with this vessel behind the Scalenus anticus. This vessel is usually provided with valves about an inch from its termination in the innominate, just external to its junction with the external jugular vein.

Tributaries.—This vein receives the external jugular vein, sometimes the anterior jugular vein, and occasionally a small branch from the cephalic, outside the Scalenus ; and on the inner side of that muscle it joins with the internal jugular vein to form the innominate vein. At the angle of junction with the internal jugular, the left subclavian vein receives the thoracic duct ; while the right subclavian vein receives the right lymphatic duct.

The **Innominate** or **Brachio-cephalic veins** (fig. 523) are two large trunks, placed one on either side of the root of the neck, and formed by the union of the internal jugular and subclavian veins of the corresponding side.

The **Right innominate vein** is a short vessel, an inch in length, which commences at the inner end of the clavicle, and, passing almost vertically downwards, joins with the left innominate vein just below the cartilage of the first rib, close to the right border of the sternum, to form the superior vena cava. It lies superficial and external to the innominate artery ; on its right side is the phrenic nerve, and the pleura is here interposed between it and the apex of the lung. This vein, at the angle of junction of the internal jugular with the subclavian, receives the right vertebral vein ; and, lower down, the right internal mammary, right inferior thyroid, and sometimes the right superior intercostal veins.

The **Left innominate vein,** about two and a half inches in length, and larger than the right, passes from left to right across the upper and front part of the chest, at the same time inclining downwards, to unite with its fellow of the opposite side, forming the *superior vena cava.* It is in relation, in front, with the first piece of the sternum, from which it is separated by the Sterno-hyoid and Sterno-thyroid muscles, the thymus gland or its remains, and some loose areolar tissue. Behind it are the three large arteries arising from the arch of the aorta, together with the vagus and phrenic nerves. This vessel is joined by the left vertebral, left internal mammary, left inferior thyroid, and the left superior intercostal veins, and occasionally some thymic and pericardiac veins. There are no valves in the innominate veins.

Peculiarities.—Sometimes the innominate veins open separately into the right auricle ; in such cases the right vein takes the ordinary course of the superior vena cava ; but the left vein—*left superior vena cava,* as it is termed—after communicating by a small branch with the right one, passes in front of the root of the left lung, and, turning to the back of the heart, receives the cardiac veins, and terminates in the back of the right auricle.   This occasional condition in the adult is due to the persistence of the early fœtal condition, and is the normal state of things in birds and some mammalia.

The **internal mammary veins,** two in number to each artery, follow the course of that vessel, and receive branches corresponding with those derived from it. The two veins unite into a single trunk, which terminates in the corresponding innominate vein.

The **inferior thyroid veins,** two, frequently three or four, in number, arise in the venous plexus on the thyroid body, communicating with the middle and superior thyroid veins.   They form a plexus in front of the trachea, behind the Sterno-thyroid muscles.   From this plexus, a left vein descends and joins the left innominate trunk, and a right vein passes obliquely downwards and outwards across the innominate artery to open into the right innominate vein, just at its junction with the superior vena cava.   These veins receive œsophageal, tracheal, and inferior laryngeal veins, and are provided with valves at their termination in the innominate veins.

The **Superior intercostal veins** (right and left) drain the blood from two or three of the intercostal spaces below the first.   The *right* vein passes downwards and inwards and opens into the vena azygos major ; the *left* runs across the transverse aorta and opens into the left innominate vein.   It usually receives the left bronchial and left superior phrenic vein, and communicates below with the vena azygos minor superior.   The *highest intercostal vein,* i.e. the vein from the first intercostal space, opens directly into the corresponding vertebral or innominate vein.

The **Superior vena cava** receives the blood which is conveyed to the heart from the whole of the upper half of the body.   It is a short trunk, varying from two inches and a half to three inches in length, formed by the junction of the two innominate veins.   It commences immediately below the cartilage of the first rib close to the sternum on the right side, and, descending vertically, enters the pericardium about an inch and a half above the heart, and terminates in the upper part of the right auricle opposite the upper border of the third right costal cartilage.   In its course it describes a slight curve, the convexity of which is turned to the right side.

Relations.—*In front,* with the pericardium and process of cervical fascia which is continuous with it ; this separates it from the thymus gland, and from the second and third costal cartilages ; *behind,* with the root of the right lung. On its *right side,* with the phrenic nerve and right pleura ; on its *left side,* with the commencement of the innominate artery and the ascending part of the aorta, the latter overlapping it.   The portion contained within the pericardium is covered, in front and laterally, by the serous layer of that membrane.   It receives the vena azygos major just before it enters the pericardium, and several small veins from the pericardium and parts in the mediastinum.   The superior vena cava has no valves.

The **Azygos veins** are three in number ; they collect the blood from the majority of the intercostal spaces, and connect together the superior and inferior venæ cavæ.

The *vena azygos major* (vena azygos) commences opposite the first or second lumbar vertebra, by a branch connecting the right lumbar veins (the *ascending lumbar*) ; sometimes by a branch from the right renal vein, or from the inferior vena cava.   It enters the thorax through the aortic opening in the Diaphragm, and passes along the right side of the vertebral column to the fourth dorsal vertebra, where it arches forward over the root of the right lung, and terminates in the superior vena cava, just before that vessel enters the pericardium.   While passing through the aortic opening of the Diaphragm, it lies with the thoracic duct on the right side of the aorta ; and in the thorax, it lies upon the intercostal arteries, on the right side of the aorta and thoracic duct, and is partly covered by pleura.

Tributaries.—It receives the lower ten intercostal veins of the right side, the upper two or three of these opening first of all into the right superior intercostal vein.   It receives the azygos minor veins, several œsophageal, mediastinal, and pericardial veins ; near its termination, the right bronchial vein ; and

generally the right superior intercostal vein. A few imperfect valves are found
in this vein; but its tributaries are provided with complete valves.

FIG. 523.—The venæ cavæ and azygos veins, with their formative branches.

The intercostal veins on the left side, below the three upper intercostal spaces,
usually form two trunks, named the left lower, and the left upper, azygos veins.

The *vena azygos minor inferior* (vena hemiazygos) commences in the lumbar region, by a branch connecting the lumbar veins (*ascending lumbar*), or from the left renal. It passes into the thorax, through the left crus of the Diaphragm, and, ascending on the left side of the spine, as high as the ninth dorsal vertebra, passes across the column, behind the aorta, œsophagus, and thoracic duct, to terminate in the right azygos vein. It receives the four or five lower intercostal veins of the left side, and some œsophageal and mediastinal veins.

The *vena azygos minor superior* (vena hemiazygos accessoria) varies inversely with the size of the left superior intercostal. It receives veins from the intercostal spaces between the left superior intercostal vein and highest tributary of the left lower azygos. They are usually three or four in number, and join to form a trunk which ends in the right azygos vein, or in the left lower azygos. It sometimes receives the left bronchial vein. When this vein is small, or altogether wanting, the left superior intercostal vein will extend as low as the fifth or sixth intercostal space.

Surgical Anatomy.—In obstruction of the inferior vena cava, the azygos veins are one of the principal means by which the venous circulation is carried on, connecting as they do the superior and inferior venæ cavæ and communicating with the common iliac veins by the ascending lumbar veins and with many of the tributaries of the inferior vena cava.

The *bronchial veins* return the blood from the larger bronchi, and from the structures at the roots of the lungs ; that of the right side opens into the vena azygos major, near its termination ; that of the left side, into the left superior intercostal vein or left upper azygos vein. A considerable quantity of the blood which is carried to the lungs through the bronchial arteries is returned to the left side of the heart through the pulmonary veins.

THE SPINAL VEINS

The numerous venous plexuses placed upon and within the spine may be arranged into four sets :

1. Those placed on the exterior of the spinal column (the *dorsal spinal veins*).
2. Those situated in the interior of the spinal canal, between the vertebræ and the theca vertebralis (*meningo-rachidian veins*).
3. The veins of the bodies of the vertebræ (*venæ basis vertebræ*).
4. The veins of the spinal cord.

1. The **Dorsal spinal veins** commence by small branches, which receive their blood from the integument of the back of the spine, and from the muscles in the vertebral grooves. They form a complicated network, which surrounds the spinous processes, the laminæ, and the transverse and articular processes of all the vertebræ. At the bases of the transverse processes, they communicate, by means of ascending and descending branches, with the veins surrounding the contiguous vertebræ, and they join with the veins in the spinal canal by branches which perforate the ligamenta subflava. Other branches pass obliquely forwards between the transverse processes, and communicate with the intraspinal veins through the intervertebral foramina. They terminate by joining the vertebral veins in the neck, the intercostal veins in the thorax, and the lumbar and sacral veins in the loins and pelvis. In addition to these, another extra-spinal plexus of relatively small vessels is formed in front of the bodies of the vertebræ. This is termed the *anterior* spinal plexus, and communicates with deep cervical, intercostal, lumbar, and lateral sacral veins in the respective regions of the spine.

2. The **Meningo-rachidian veins.**—The principal veins contained in the spinal canal are situated between the theca vertebralis and the vertebræ. They consist of two longitudinal plexuses, one of which runs along the posterior surface of the bodies of the vertebræ (*anterior longitudinal spinal veins*). The other plexus (*posterior longitudinal spinal veins*) is placed on the inner or anterior surface of the laminæ of the vertebræ.

The *Anterior longitudinal spinal veins* consist of two large, tortuous veins, which extend along the whole length of the vertebral column, from the foramen magnum, where they communicate by a venous ring around that opening, from which tributaries are given off to form the commencement of the vertebral vein, to the base of the coccyx, being placed one on each side of the posterior surface of the bodies of the vertebræ, along the margin of the posterior common ligament.

These veins communicate together opposite each vertebra by transverse trunks, which pass beneath the ligament, and receive the large *venæ basis vertebræ* from the interior of the vertebral body. The anterior longitudinal spinal veins are least developed in the cervical and sacral regions. They are not of uniform size throughout, being alternately enlarged and constricted. At the intervertebral foramina they communicate with the dorsal spinal veins, and with the vertebral veins in the neck, with the intercostal veins in the dorsal region, and with the lumbar and sacral veins in the corresponding regions.

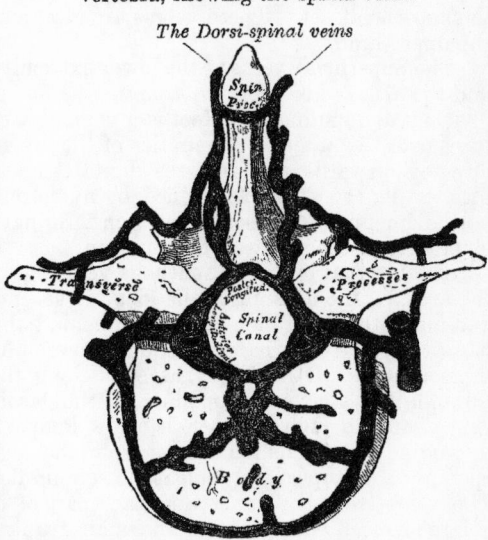

FIG. 524.—Transverse section of a dorsal vertebra, showing the spinal veins.

The *Posterior longitudinal spinal veins*, smaller than the anterior, are situated one on each side, between the inner surface of the laminæ and the theca vertebralis. They communicate (like the anterior), opposite each vertebra, by transverse trunks; and with the anterior longitudinal veins, by lateral transverse branches, which pass from behind forwards. These veins, by branches which perforate the ligamenta subflava, join with the dorsal spinal veins. From them branches are given off, which pass through the intervertebral foramina and join the vertebral, intercostal, lumbar, and sacral veins.

3. The **Veins of the bodies of the vertebræ** (*venæ basis vertebræ*) emerge from the foramina on their posterior surface, and join the transverse trunk connecting the anterior longitudinal spinal veins. They are contained in large, tortuous channels in the substance of the bones, similar in every respect to those found in the diploë of the cranial bones. These canals lie parallel to the upper and lower surfaces of the bones. They commence by small openings on the front and sides of the bodies of the vertebræ, through which communicating branches from the veins external to the bone pass into its substance, and converge to the principal canal, which is sometimes double towards its posterior part, and open into the corresponding transverse branch uniting the anterior longitudinal veins. They become greatly developed in advanced age.

FIG. 525.—Vertical section of two dorsal vertebræ, showing the spinal veins.

4. The **Veins of the spinal cord** consist of a minute, tortuous, venous plexus which covers the entire surface of the cord, being situated in the pia mater. These vessels emerge chiefly from the median furrows, and are largest in the lumbar region. In this plexus there are: (1) two median longitudinal veins, one in front of the anterior fissure, and the other behind the posterior fissure of the cord; and (2) four lateral longitudinal veins which run behind the nerve-roots. Near the base of the skull they unite, and form two or three small trunks, which communicate with the vertebral veins, and then terminate in the inferior cerebellar veins, or in the inferior petrosal sinuses. Each of the spinal nerves is accompanied by a branch as far as the intervertebral foramina, where it joins the other veins from the spinal canal.

There are no valves in the spinal veins.

3 B

VEINS OF THE LOWER EXTREMITY, ABDOMEN, AND PELVIS

The Veins of the Lower Extremity are subdivided, like those of the upper, into two sets, superficial and deep : the superficial veins being placed beneath the integument, between the two layers of superficial fascia ; the deep veins accompanying the arteries, and forming the venæ comites of those vessels. Both sets of veins are provided with valves, which are more numerous in the deep than in the superficial set. These valves are also more numerous in the lower than in the upper limb.

The **Superficial veins of the lower extremity** are the *internal* or *long saphenous*, and the *external* or *short saphenous*.

On the dorsum of the foot is a venous arch, situated in the superficial structures over the anterior extremities of the metatarsal bones. It has its convexity directed forwards, and receives digital tributaries from the upper surface of the toes ; at its concavity it is joined by numerous small veins, which form a plexus on the dorsum of the foot. The arch terminates internally in the long saphenous, externally in the short saphenous vein.

The **internal** or **long saphenous vein** (fig. 526) commences at the inner side of the arch on the dorsum of the foot ; it ascends in front of the inner malleolus, and along the inner side of the leg, behind the inner margin of the tibia, accompanied by the internal saphenous nerve. At the knee, it passes behind the inner condyle of the femur, ascends along the inside of the thigh, and, passing through the saphenous opening in the fascia lata, terminates in the femoral vein about an inch and a half below Poupart's ligament. This vein receives in its course cutaneous tributaries from the leg and thigh, and at the saphenous opening the superficial epigastric, circumflex iliac, and external pudic veins. The veins from the inner and back part of the thigh frequently unite to form a large vessel, which enters the main trunk near the saphenous opening ; and sometimes those on the outer side of the thigh join to form another large vessel ; so that occasionally three large veins are seen converging from different parts of the thigh towards the saphenous opening. The internal saphenous vein communicates in the foot with the internal plantar vein ; in the leg, with the posterior tibial veins, by branches which perforate the tibial origin of the Soleus muscle, and also with the anterior tibial veins ; at the knee, with the articular veins ; in the thigh, with the femoral vein by one or more branches. The valves in this vein vary from ten to twenty in number ; they are more numerous in the thigh than in the leg.

The **external** or **short saphenous vein** (fig. 527) commences at the outer side of the arch on the dorsum of the foot ; it ascends behind the outer malleolus, and along the outer border of the tendo Achillis, across which it passes at an acute angle to reach the middle line of the posterior aspect of the leg. Passing directly upwards, it perforates the deep fascia in the lower part of the popliteal space, and terminates in the popliteal vein, between the heads of the Gastrocnemius muscle. It receives numerous large tributaries from the back part of the leg, and communicates with the deep veins on the dorsum of the foot, and behind the outer malleolus. Before it perforates the deep fascia, it gives off a communicating branch, which passes upwards and inwards to join the internal saphenous vein. This vein contains from nine to twelve valves, one of which is always found near its termination in the popliteal vein. The external saphenous nerve lies close beside this vein.

Surgical Anatomy.—The saphenous veins are of considerable surgical importance, since a varicose condition of these vessels is more frequently met with than those in other parts of the body, except perhaps the spermatic and hæmorrhoidal veins. The main cause of this is the high blood-pressure, determined chiefly by the erect position and the length of the column of blood, which has to be propelled in an uphill direction. In normal vessels, there is only just sufficient force to perform this task ; and in those cases where there is diminished resistance of the walls of the vein, these vessels are liable to dilate, and a varicose condition is set up. This diminished resistance may be due to heredity, the vein-walls being congenitally weak ; or it may be due to inflammatory conditions, leading to softening or to absorption of surrounding tissues, and thus depriving the vessel of its accustomed support. Increased blood pressure in the veins, caused by any obstacle to the return of the venous blood, such as the pressure of a tumour, or the gravid uterus or tight gartering, may also produce varix. In the normal condition of the veins, the valves in their interior break up the column of blood into a number of smaller columns, and so to a considerable extent mitigate the ill effects of the erect position; but when the dilatation of the veins has reached a certain limit, the valves

FIG. 526.—The internal or long saphenous vein and its branches.

FIG. 527.
External or short saphenous vein.

become incapable of supporting the overlying column of blood, and the pressure is increased, tending to emphasise also the varicose condition. Both the saphenous veins in the leg are accompanied by nerves, the internal saphenous being joined by its companion nerve just below the level of the knee-joint. No doubt much of the pain of varicose veins in the leg is due to this fact. On the Continent the internal saphenous vein as it rests on the tibia just above the malleolus is sometimes selected for venesection.

The **Deep veins of the lower extremity** accompany the arteries and their branches, and are called the *venæ comites* of those vessels.

The **external** and **internal plantar veins** unite to form the **posterior tibial veins**, which accompany the posterior tibial artery, and are joined by the *peroneal veins.*

3 B 2

The **anterior tibial veins** are formed by a continuation upwards of the venæ comites of the dorsalis pedis artery. They pass between the tibia and fibula, through the large oval aperture above the interosseous membrane, and form, by their junction with the posterior tibial, the *popliteal* vein.

The valves in the deep veins are very numerous.

The **popliteal vein** is formed by the junction of the venæ comites of the anterior and posterior tibial vessels at the lower border of the Popliteus muscle ; it ascends through the popliteal space to the tendinous aperture in the Adductor magnus, where it becomes the femoral vein. In the lower part of its course, it is placed internal to the artery ; between the heads of the Gastrocnemius, it is superficial to that vessel ; but above the knee-joint, it is close to its outer side. It receives the sural veins from the Gastrocnemius muscle, the articular veins, and the external saphenous. The valves in this vein are usually four in number.

The **Femoral vein** accompanies the femoral artery through the upper two-thirds of the thigh. In the lower part of its course it lies external to the artery ; higher up, it is behind it ; and at Poupart's ligament, it lies to its inner side, and on the same plane. It receives numerous muscular tributaries, and about an inch and a half below Poupart's ligament it is joined by the profunda femoris : near its termination it is joined by the internal saphenous vein. The valves in this vein are three in number.

The **External iliac vein** commences at the termination of the femoral, beneath Poupart's ligament, and passing upwards along the brim of the pelvis, terminates opposite the sacro-iliac articulation, by uniting with the internal iliac to form the common iliac vein. On the right side, it lies at first along the inner side of the external iliac artery ; but as it passes upwards, gradually inclines behind it. On the left side, it lies altogether on the inner side of the artery. It receives, immediately above Poupart's ligament, the deep epigastric and deep circumflex iliac veins and a small pubic vein, corresponding to the pubic branch of the obturator artery. According to Friedreich, it frequently contains one, sometimes two, valves.

The deep epigastric veins.—Two veins accompany the deep epigastric artery ; they usually unite into a single trunk before their termination in the external iliac vein.

The deep circumflex iliac veins.—Two veins accompany the deep circumflex iliac artery. These unite into a single trunk which crosses the external iliac artery just above Poupart's ligament, and terminates in the external iliac vein.

The **Internal iliac vein** is formed by the venæ comites of the branches of the internal iliac artery, the umbilical arteries of the fœtus excepted. It receives the blood from the exterior of the pelvis by the gluteal, sciatic, internal pudic, and obturator veins ; and from the organs in the cavity of the pelvis by the hæmorrhoidal and vesico-prostatic plexuses in the male, and the uterine and vaginal plexuses in the female. The vessels forming these plexuses are remarkable for their large size, their frequent anastomoses, and the number of valves which they contain. The internal iliac vein lies at first on the inner side, and then behind the internal iliac artery, and terminates opposite the sacro-iliac articulation, by uniting with the external iliac to form the common iliac vein. This vessel has no valves.

The *internal pudic veins* (*venæ comites*) have the same course as the internal pudic artery. They receive tributaries corresponding to the branches of the artery except that corresponding to the dorsal artery of the penis, viz. the dorsal vein of the penis, which opens into the prostatic plexus.

The *hæmorrhoidal plexus* surrounds the lower end of the rectum, being formed by the superior hæmorrhoidal veins, tributaries of the inferior mesenteric. It commences as a series of dilated pouches, about twelve in number, which are arranged circularly at the verge of the anus and are connected by transverse branches. From these pouches veins, about six in number, pass upwards in a straight direction in the submucous tissue for about three inches ; they then pierce the muscular coat and become arranged in a circular manner at right angles to the long axis of the gut, and eventually unite to form the superior hæmorrhoidal vein.

Surgical Anatomy.—The veins of this plexus are apt to become dilated and varicose, and form piles. This is due to several anatomical reasons : the vessels are contained in very loose, lax connective tissue, so that they get less support from surrounding structures than most other veins, and are less capable of resisting increased blood pressure; the condition is favoured by gravitation, being influenced by the erect posture, either sitting or standing, and by the fact that the superior hæmorrhoidal and portal veins have no valves; the veins pass through muscular tissue and are liable to be compressed by its contraction, especially during the act of defæcation; they are affected by every form of portal obstruction.

The *vesico-prostatic plexus* surrounds the neck and base of the bladder and prostate gland. It communicates with the hæmorrhoidal plexus behind, and receives the deep dorsal vein of the penis, which enters the pelvis beneath the subpubic ligament. This plexus is supported upon the sides of the bladder by a reflection of the pelvic fascia. The veins composing it are very liable to become varicose, and often contain hard, earthy concretions, called *phleboliths*.

Surgical Anatomy.—This plexus is wounded in the operation of lateral lithotomy, and it is through it that septic matter finds its way into the general circulation after this operation.

The *dorsal veins of the penis* are two in number, a superficial and a deep. The *superficial* vein drains the prepuce and skin of the penis, and, running backwards in the subcutaneous tissue, inclines to the right or left, and opens into the corresponding superficial external pudic vein, a tributary of the internal or long saphenous vein. The *deep* vein receives the blood from the glans penis and corpora cavernosa : it courses backwards in the middle line between the dorsal arteries, underneath the deep fascia, and near the root of the penis passes between the two parts of the suspensory ligament and then through an aperture between the subpubic ligament and the triangular ligament of the urethra, and divides into two branches, which enter the prostatic plexus.

The *vaginal plexus* surrounds the vagina, being especially developed at the orifice of the canal; it communicates with the vesical plexus in front, and with the hæmorrhoidal plexus behind.

The *uterine plexus* is situated along the sides and superior angles of the uterus, between the two layers of the broad ligament, receiving, during pregnancy, large venous canals from the blood-sinuses of the placenta. The veins composing this plexus anastomose frequently with each other and with the ovarian veins. They are not tortuous like the arteries.

The **Common iliac veins** are formed by the union of the external and internal iliac veins in front of the sacro-iliac articulation; passing obliquely upwards towards the right side, they terminate upon the fifth lumbar vertebra, by uniting with each other at an acute angle to form the inferior vena cava. The *right common iliac* is shorter than the left, nearly vertical in its direction, and ascends behind and then to the outer side of its corresponding artery. The *left common iliac*, longer and more oblique in its course, is at first situated on the inner side of the corresponding artery, and then behind the right common iliac. Each common iliac receives the ilio-lumbar, and sometimes the lateral sacral veins. The left receives, in addition, the middle sacral vein. No valves are found in these veins.

The **middle sacral veins** accompany the corresponding artery along the front of the sacrum, and join to form a single vein which terminates in the left common iliac vein; sometimes in the angle of junction of the two iliac veins.

Peculiarities.—The left common iliac vein, instead of joining with the right in its usual position, occasionally ascends on the left side of the aorta as high as the kidney, where, after receiving the left renal vein, it crosses over the aorta, and then joins with the right vein to form the vena cava. In these cases, the two common iliacs are connected by a small communicating branch at the spot where they are usually united.*

The **Inferior vena cava** returns to the heart the blood from all the parts below the Diaphragm. It is formed by the junction of the two common iliac veins, on the right side of the fifth lumbar vertebra. It passes upwards along the front of the spine, on the right side of the aorta, and, having reached the liver, is contained in a groove on its posterior surface. It then perforates the Diaphragm between the mesial and right portions of its central tendon; it subsequently

* See two cases which have been described by Walsham in the *St. Bartholomew's Hospital Reports*, vols. xvi. and xvii.

inclines forwards and inwards for about an inch, and, piercing the pericardium, opens into the lower and back part of the right auricle. In front of its auricular orifice is a semilunar valve, termed the *valve of Eustachius*: this is rudimentary in the adult, but is of large size and exercises an important function in the fœtus.

Relations.—The *abdominal portion* of the inferior vena cava is in relation *in front*, from below upwards, with the mesentery, right spermatic artery, transverse portion of the duodenum, the pancreas, portal vein, and the posterior surface of the liver, which partly and occasionally completely surrounds it; *behind*, with the vertebral column, the right Psoas muscle, the right crus of the Diaphragm, the right renal and lumbar arteries, right semilunar ganglion, and the inner part of the right suprarenal body; on the *right side*, with the right kidney and ureter; on the *left side*, with the aorta.

The *thoracic portion* is only about an inch in length, and is situated partly inside and partly outside the pericardial sac. The *extra-pericardial part* is separated from the right pleura and lung by a fibrous band, named the *right phrenico-pericardiac ligament* of Teutleben. This ligament, often feebly marked, is attached below to the margin of the vena-caval opening in the Diaphragm, and above to the pericardium in front of and behind the root of the right lung. The *intrapericardiac part* is very short, and is covered antero-laterally by the serous layer of the pericardium.

Peculiarities.—In Position.—This vessel is sometimes placed on the left side of the aorta, as high as the left renal vein, after receiving which it crosses over to its usual position on the right side; or it may be placed altogether on the left side of the aorta, as far upwards as its termination in the heart: in such cases, the abdominal and thoracic viscera, together with the great vessels, are all transposed.

Point of Termination.—Occasionally the inferior vena cava joins the right azygos vein, which is then of large size. In such cases, the superior cava receives the whole of the blood from the body before transmitting it to the right auricle, except the blood from the hepatic veins, which passes directly into the right auricle.

Tributaries.—It receives in its course the following veins :

Lumbar.	Renal.	Phrenic.
Right spermatic, or ovarian.	Suprarenal.	Hepatic.

The **lumbar veins,** four in number on each side, collect the blood by dorsal tributaries from the muscles and integument of the loins, and by abdominal tributaries from the walls of the abdomen, where they communicate with the epigastric veins. At the spine, they receive veins from the spinal plexuses, and then pass forwards, round the sides of the bodies of the vertebræ beneath the Psoas magnus, and terminate in the back part of the inferior cava. The left lumbar veins are longer than the right, and pass behind the aorta. The lumbar veins are connected together by a longitudinal vein which passes in front of the transverse processes of the lumbar vertebræ, and is called the *ascending lumbar*. It forms the most frequent origin of the corresponding vena azygos, and serves to connect the common iliac, ilio-lumbar, lumbar, and azygos veins of the corresponding side of the body.

The **spermatic veins** emerge from the back of the testis, and receive tributaries from the epididymis; they unite and form a convoluted plexus, called the *spermatic plexus* (*plexus pampiniformis*), which constitutes the chief mass of the cord; the vessels composing this plexus are very numerous, and ascend along the cord in front of the vas deferens; below the external abdominal ring they unite to form three or four veins, which pass along the inguinal canal, and, entering the abdomen through the internal abdominal ring, coalesce to form two veins, which ascend on the Psoas muscle, behind the peritoneum, lying one on each side of the spermatic artery; these unite to form a single vein, which opens on the right side into the inferior vena cava, at an acute angle; on the left side into the left renal vein, at a right angle. The spermatic veins are provided with valves.*

* Rivington has pointed out that a valve is usually found at the orifices of both the right and left spermatic veins. When no valves exist at the opening of the left spermatic vein into the left renal vein, valves are generally present in the left renal vein within a quarter of an inch from the orifice of the spermatic vein.—*Journal of Anatomy and Physiology*, vol. vii. p. 163.

The left spermatic vein passes behind the sigmoid flexure of the colon, and is thus exposed to pressure from the contents of that bowel.

Surgical Anatomy.—The spermatic veins are very frequently varicose, constituting the disease known as *varicocele*. Though it is quite possible that the originating cause of this affection may be a congenital weakness of the walls of the veins of the pampiniform plexus, still it must be admitted that there are many anatomical reasons why these veins should become varicose : viz. the imperfect support afforded to them by the loose tissue of the scrotum ; their great length ; their vertical course ; their dependent position ; their plexiform arrangement in the scrotum, with their termination in one small vein in the abdomen ; their few and imperfect valves ; and the fact that they may be subjected to pressure in their passage through the abdominal wall. Varicocele is more common on the left than on the right side, and this has been accounted for by the fact that the left spermatic vein joins the left renal at a right angle ; that it is overlaid by the sigmoid flexure of the colon, and that when this portion of the gut is full of fæcal matter, in cases of constipation, its weight impedes the return of the venous blood ; and that the left spermatic veins are somewhat longer than the right.

The **ovarian veins** are analogous to the spermatic in the male ; they form a plexus near the ovary in the broad ligament and Fallopian tube, and communicate with the uterine plexus. They terminate in the same way as the spermatic veins in the male. Valves are occasionally found in these veins. These vessels, like the uterine veins, become much enlarged during pregnancy.

The **renal veins** are of large size, and placed in front of the renal arteries.* The left is longer than the right, and passes in front of the aorta, just below the origin of the superior mesenteric artery. It receives the left spermatic, the left inferior phrenic, and, generally, the left suprarenal veins. It opens into the vena cava, a little higher than the right.

The **suprarenal veins** are two in number : that on the right side terminates in the vena cava ; that on the left side, in the left renal or phrenic vein.

The **phrenic veins** follow the course of the phrenic arteries. The *two superior*, of small size, accompany the phrenic nerve and comes nervi phrenici artery, and join the internal mammary ; the left sometimes opens into the left innominate vein, and the right into the point of union of the two innominate veins. The *two inferior phrenic veins* follow the course of the phrenic arteries, and terminate, the right in the inferior vena cava, the left in the left renal vein.

The **hepatic veins** commence in the substance of the liver, in the capillary terminations of the portal vein and hepatic artery : these tributaries, gradually uniting, usually form three large veins, which constitute the *upper* group of hepatic veins ; they converge towards the posterior surface of the liver, and open into the inferior vena cava, while that vessel is situated in the groove at the back part of this organ. Of these three veins, one from the right, and another from the left lobe, open obliquely into the inferior vena cava ; that from the middle of the organ and lobulus Spigelii having a straight course. The *veins* of the lower group vary in number, and are of small size ; they come from the right and Spigelian lobes. The hepatic veins run singly, and are in direct contact with the hepatic tissue. They are destitute of valves.

Portal System of Veins

The portal venous system is composed of four large veins which collect the venous blood from the viscera of digestion (stomach, intestine, and pancreas) and from the spleen. The trunk formed by their union (*vena portæ*) enters the liver and ramifies throughout its substance after the manner of an artery and ends in capillaries, from which the blood is collected into the hepatic veins, which terminate in the inferior vena cava. The tributaries of this vein are in all cases single, and destitute of valves.

---

* The student may observe that all veins above the Diaphragm, which do not lie on the same plane as the arteries which they accompany, lie in front of them : and that all veins below the Diaphragm, which do not lie on the same plane as the arteries which they accompany, lie behind them, except the renal and profunda femoris vein.

The veins forming the portal system are, the

Superior mesenteric.   Inferior mesenteric.
Splenic.   Gastric.

Cystic.

The **superior mesenteric vein** returns the blood from the small intestines, and from the cæcum and ascending and transverse portions of the colon, corresponding with the distribution of the branches of the superior mesenteric artery. The

FIG. 528.—Portal vein and its branches.

NOTE.—In this diagram the right gastro-epiploic vein opens into the splenic vein ; generally it empties itself into the superior mesenteric, close to its termination.

large trunk formed by the union of these branches ascends along the right side and in front of the corresponding artery, passes in front of the transverse portion of the duodenum, and unites, behind the neck of the pancreas, with the splenic vein to form the vena portæ. It receives the right gastro-epiploic vein.

The **splenic vein** commences by five or six large branches, which return the blood from the substance of the spleen. These unite to form a single vessel, which passes from left to right, grooving the upper and back part of the pancreas, below the splenic artery, and terminates behind the neck of the pancreas by uniting at a right angle with the superior mesenteric to form the vena portæ. The splenic vein is of large size, but is not tortuous like the artery. It receives the vasa brevia from the left extremity of the stomach, the left gastro-epiploic vein, pancreatic branches from the pancreas, the pancreatico-duodenal vein, and the inferior mesenteric vein.

The **inferior mesenteric vein** returns the blood from the rectum, sigmoid flexure, and descending colon, corresponding with the ramifications of the branches of the inferior mesenteric artery. It lies to the left of the artery, and ascends beneath the peritoneum in the lumbar region; it passes behind the transverse portion of the duodenum and pancreas, and terminates in the splenic vein. Its hæmorrhoidal branches inosculate with those of the internal iliac, and thus establish a communication between the portal and the general venous systems.*

The **gastric veins** are two in number: one, a small vein, corresponds to the pyloric branch of the hepatic artery; the other, considerably larger, corresponds to the gastric artery. The former (*pyloric*, Walsham) runs along the lesser curvature of the stomach towards the pyloric end, receives branches from the pylorus and duodenum, and ends in the vena portæ. The latter (*coronary*, Walsham) begins near the pylorus, runs along the lesser curvature of the stomach, towards the œsophageal opening, and then passes across the front of the spine from left to right to end in the vena portæ, at a point a little above the junction of the pyloric vein.

The **Portal Vein** is formed by the junction of the superior mesenteric and splenic veins, their union taking place in front of the vena cava, and behind the neck of the pancreas. Passing upwards in the right border of the lesser omentum to the under surface of the liver, it enters the transverse fissure, where it is somewhat enlarged, forming the *sinus* of the portal vein, and divides into two branches, which accompany the ramifications of the hepatic artery and hepatic duct throughout the substance of the liver. Of these two branches the right is the larger, but the shorter, of the two. The portal vein is about three or four inches in length, and, while contained in the lesser omentum, lies behind and between the common bile duct and the hepatic artery, the former being to the right, the latter to the left. These structures are accompanied by filaments of the hepatic plexus of nerves, and numerous lymphatics, are surrounded by a quantity of loose areolar tissue (*capsule of Glisson*), and are placed between the layers of the lesser omentum.

The **Cystic Vein.**—The trunk of the vena portæ generally receives the cystic vein, although it sometimes terminates in its right branch.

The portal vein divides, in the substance of the liver, like an artery, and its minute ramifications end in capillaries, from which the blood is carried to the inferior vena cava by the hepatic veins; these veins also collect the blood which has been brought to the liver by the hepatic artery. It will therefore be seen that the blood which is carried to the liver by the portal vein passes through two sets of capillary vessels, viz.: (1) the capillaries in the stomach, intestine, pancreas, and spleen, and (2) the capillaries of the portal vein in the liver.

Surgical Anatomy.—Obstruction to the portal vein produce ascites, and may arise from many causes: as (1) the pressure of tumour on the portal vein, such as cancer, hydatid cyst, or abscess of the liver itself, enlarged lymphatic glands in the lesser omentum, or cancer of the head of the pancreas; (2) from cirrhosis of the liver, when the radicles of the portal vein are pressed upon by the contracting fibrous tissue in the portal canals; (3) from valvular disease of the heart, and back pressure on the hepatic veins, and so on the whole of the circulation through the liver.

---

* Besides this anastomosis between the portal vein and the branches of the vena cava other anastomoses between the portal and systemic veins are formed by the communication between the gastric veins and the œsophageal veins which empty themselves into the vena azygos minor; between the left renal vein and the veins of the intestines, especially of the colon and duodenum; between the veins of the round ligament of the liver and the portal veins; between the veins of the intestinal canal and the veins of the abdominal wall; and between the superficial branches of the portal veins of the liver and the phrenic veins, as pointed out by Kiernan. (See *Physiological Anatomy*, by Todd and Bowman, 1859, vol. ii. p. 348.)

CARDIAC VEINS

The veins which return the blood from the substance of the heart are, the

Great cardiac vein.
Posterior cardiac vein.
Left cardiac veins.
Right or small coronary vein.

Coronary sinus.
Oblique vein of Marshall.
Anterior cardiac veins.
Venæ Thebesii.

The **Great cardiac vein** (sometimes called the *left coronary vein*) is a vessel of considerable size, which commences at the apex of the heart, and ascends along the anterior interventricular groove to the base of the ventricles. It then curves to the left side, around the auriculo-ventricular groove, between the left auricle and ventricle, to the back part of the heart, and opens into the left extremity of the coronary sinus, its aperture being guarded by two valves. It receives tributaries from the left auricle and from both ventricles, but especially the left; one of these, ascending along the thick margin of the left ventricle, is of considerable size. The vessels joining it are provided with valves.

The **Posterior cardiac vein** (sometimes called the *middle cardiac vein*) commences by small tributaries, at the apex of the heart, communicating with those of the preceding vein. It ascends along the posterior interventricular groove to the base of the heart, and terminates in the coronary sinus, its orifice being guarded by a valve. It receives the veins from the posterior surface of both ventricles.

The **Left cardiac veins** are three or four small vessels, which collect the blood from the posterior surface of the left ventricle, and open into the lower border of the coronary sinus.

The **Right** or **small coronary vein** runs along the groove between the right auricle and ventricle, to open into the right extremity of the coronary sinus. It receives blood from the back part of the right auricle and ventricle.

The **Coronary sinus** is situated in the posterior part of the left auriculo-ventricular groove. It is about an inch in length, is of considerable size, and covered by the muscular fibres of the left auricle. It receives the veins enumerated above, and the **oblique vein of Marshall** from the back part of the left auricle; this vein is the remnant of the obliterated left Cuvierian duct of the fœtus. The great coronary sinus terminates in the right auricle, between the inferior vena cava and the auriculo-ventricular aperture, its orifice being guarded by a semilunar fold of the lining membrane of the heart, the *Thebesian valve*. All the veins joining this vessel, excepting the oblique vein of Marshall, are provided with valves.

The **Anterior cardiac veins** are three or four small vessels, which collect the blood from the anterior surface of the right ventricle. One of these (the *vein of Galen*), larger than the rest, runs along the right border of the heart. They open separately into the lower part of the right auricle.

The **Venæ Thebesii** (*venæ cordis minimæ*) are numerous minute veins, which return the blood directly from the muscular substance, without entering the venous current. They open by minute orifices (*foramina Thebesii*) chiefly into the auricles, but a few terminate in the ventricles.

THE LYMPH VASCULAR SYSTEM

THE lymph vascular system includes not only the lymphatic vessels and glands, but also the *lacteal* or *chyliferous vessels*. The lacteals are the lymphatic vessels of the small intestine, and differ in no respect from the lymphatics generally, excepting that they contain a milk-white fluid, the *chyle*, during the process of digestion, and convey it into the blood through the thoracic duct.

The **lymphatics** have derived their name from the appearance of the fluid contained in their interior (*lympha*, water). They are also called *absorbents*, from the property they possess of absorbing certain materials from the tissues and conveying them into the circulation.

The lymphatics are exceedingly delicate vessels, the coats of which are so transparent that the fluid they contain is readily seen through them. They retain a nearly uniform size, being interrupted at intervals by constrictions, which give them a knotted or beaded appearance. These constrictions are due to the presence of valves in their interior. Lymphatics have been found in nearly every texture and organ of the body which contains blood-vessels. Such non-vascular structures as cartilage, the nails, cuticle, and hair have none, but with these exceptions it is probable that eventually all parts will be found to be permeated by these vessels.

The lymphatics are arranged into a superficial and a deep set. The *superficial* lymphatics, on the surface of the body, are placed immediately beneath the integument, accompanying the superficial veins ; they join the deep lymphatics in certain situations by perforating the deep fascia. In the interior of the body they lie in the submucous areolar tissue, throughout the whole length of the gastro-pulmonary and genito-urinary tracts ; and in the subserous tissue in the cranial, thoracic, and abdominal cavities. The method of their origin has been described along with the other details of their minute anatomy. Here it will be sufficient to say that a plexiform network of minute lymphatics may be found interspersed among the proper elements and blood-vessels of the several tissues ; the vessels composing which, as well as the meshes between them, are much larger than those of the capillary plexus. From these networks small vessels emerge, which pass, either to a neighbouring gland, or to join some larger lymphatic trunk. The *deep* lymphatics, fewer in number, and larger than the superficial, accompany the deep blood-vessels. Their mode of origin is probably similar to that of the superficial vessels. The lymphatics of any part or organ exceed the veins in number, but in size they are much smaller. Their anastomoses also, especially those of the large trunks, are more frequent, and are effected by vessels equal in diameter to those which they connect, the continuous trunks retaining the same diameter.

The **lymphatic** or **absorbent glands**, named also *conglobate glands*, are small, solid, glandular bodies, situated in the course of the lymphatic and lacteal vessels. In size they vary from a hemp-seed to an almond, and their colour, on section, is of a pinkish-grey tint, excepting the bronchial glands, which in the adult are mottled with black. Each gland has a layer or capsule of cellular tissue investing it, from which prolongations dip into its substance, forming partitions. The lymphatic and lacteal vessels pass through these bodies in their passage to the thoracic and lymphatic ducts. A lymphatic or lacteal vessel, previous to entering a gland, divides into several small branches, which are named *afferent vessels*. As they enter, their external coat becomes continuous with the capsule of the gland, and the vessels, much thinned, and consisting only of their internal or endothelial coat, pass into the gland, and branch out upon and in the tissue of the capsule ; these branches opening into the lymph sinuses of the gland. From these sinuses fine branches proceed to form a plexus, the vessels of which unite to form a single *efferent vessel*, which, on emerging from the gland, is again invested with an external coat. Further details on the minute anatomy of the lymphatic vessels and glands will be found in the section on General Anatomy.

FIG. 529.—The thoracic and right lymphatic duct.

Right lymphatic duct

Intercostal glands

Lumbar glands

THORACIC DUCT

The **thoracic duct** (fig. 529) conveys the great mass of the lymph and chyle into the blood. It is the common trunk of all the lymphatic vessels of the body, excepting those of the right side of the head, neck, and thorax, and right upper extremity, the right lung, right side of the heart, and the convex surface of the liver. In the adult it varies in length from fifteen to eighteen inches, and extends from the second lumbar vertebra to the root of the neck. It commences in the abdomen by a triangular dilatation, the *receptaculum chyli* (*reservoir* or *cistern of Pecquet*), which is situated on the front of the body of the second lumbar vertebra, to the right side of and behind the aorta, by the side of the right crus of the Diaphragm. It ascends into the thorax through the aortic opening in the Diaphragm, and is then placed in the posterior mediastinum between the aorta and vena azygos major. Here it lies in front of the vertebral column, from which it is separated by the right intercostal arteries and by the azygos minor veins as the latter cross the middle line to open into the vena azygos major. Opposite the fifth dorsal vertebra, it inclines towards the left side and ascends behind the arch of the aorta, on the left side of the œsophagus, and behind the first portion of the left subclavian artery to the upper orifice of the thorax. Opposite the seventh cervical vertebra, it turns outwards in front of the vertebral vein and artery, behind the left internal carotid artery and vagus nerve, and then curves downwards over the subclavian artery, and in front of the Scalenus anticus muscle and the phrenic nerve, so as to form an arch; it terminates in the left subclavian vein, at its angle of junction with the left internal jugular vein. The thoracic duct, at its commencement, is about equal in size to the diameter of a goose-quill, but diminishes considerably in its calibre in the middle of the thorax, and is again dilated just before its termination. It is generally flexuous in its course, and constricted at intervals so as to present a varicose appearance. Not infrequently it divides in the middle of its course into two branches of unequal size, which soon reunite, or into several branches which form a plexiform interlacement. It occasionally divides, at its upper part, into two branches, right and left; the left terminates in the usual manner, while the right opens into the right subclavian vein, in connection with the right lymphatic duct. The thoracic duct has several valves throughout its course : at its termination it is provided with a pair of valves, the free borders of which are turned towards the vein, so as to prevent the passage of venous blood into the duct.

The **receptaculum chyli** receives the two lumbar lymphatic trunks, right and left, and the intestinal lymphatic trunk. The *lumbar lymphatic trunks* are formed by the union of the efferent vessels from the lymphatic glands which are situated on either side of the aorta. They drain the lymph from the lower limbs, from the walls and viscera of the pelvis, from the kidneys and suprarenal bodies, together with the deep lymphatics of the greater part of the abdominal wall. The *intestinal lymphatic trunk* drains the lymph from the stomach and small intestine, from the pancreas and spleen, and from the lower and front part of the liver.

Tributaries.—Opening into the commencement of the thoracic duct is a descending trunk from the lower posterior intercostal glands. This trunk may be duplicated, and may terminate in the receptaculum chyli. Within the thorax, the thoracic duct is joined by the lymphatic vessels from the left half of the wall of the thoracic cavity, the lymphatics from the sternal and intercostal glands, those of the left lung and pleura, left side of the heart, trachea, and œsophagus; and, just before its termination, it receives the lymphatics of the left side of the head and neck through the *left jugular trunk*, and of the left upper extremity through the *left subclavian trunk*. The last frequently opens separately from the thoracic duct.

Structure.—The thoracic duct is composed of three coats, which differ in some respects from those of the lymphatic vessels. The *internal coat* consists of a single layer of flattened lanceolate-shaped endothelial cells, with serrated borders; of a subendothelial layer similar to that found in the arteries; and an elastic fibrous coat, the fibres of which run in a longitudinal direction. The *middle coat* consists of a longitudinal layer of white connective tissue with elastic fibres, external to which are several laminæ of muscular tissue, the fibres of which are for the most part disposed transversely, but some are oblique or longitudinal, and intermixed with elastic fibres. The *external coat* is composed of areolar tissue, with elastic fibres and isolated fasciculi of muscular fibres.

The **Right Lymphatic Duct** is a short trunk, about half an inch in length, and a line or a line and a half in diameter. It is formed by the junction of the lymphatic vessels of the right side of the head and neck with those of the right upper extremity, and courses along the inner border of the Scalenus anticus at the root of the neck. It terminates in the right subclavian vein, at its angle of junction with the right internal jugular vein. Its orifice is guarded by two semilunar valves, which prevent the passage of venous blood into the duct.

Tributaries.—It receives the lymph from the right side of the head and neck through the *right jugular trunk*; from the right upper extremity, through the *right subclavian trunk*; from the right side of the thorax, the right lung and right side of the heart, and from part of the convex surface of the liver, through what is sometimes termed the *right broncho-mediastinal trunk*. These three collecting trunks frequently open separately in the angle of union of the two veins.

LYMPHATICS OF THE HEAD, FACE, AND NECK

The **lymphatic glands of the head** (fig. 530) are arranged in the following groups : (1) The *occipital*, one or two in number, placed at the back of the head, on the insertion of the Complexus, close to the occipital artery. (2) The *posterior* *auricular* or *mastoid*, usually two in number, situated on the insertion of the Sterno-mastoid to the mastoid process. Both these sets of glands are affected in cutaneous eruptions and other diseases of the scalp. (3) The *parotid* or *pre-auricular*, some of which are superficial to, and others are embedded in, the substance of the parotid gland; the latter are mostly grouped in relation to the external carotid artery and the external jugular vein. (4) The *buccal*, one or more, placed on the surface of the Buccinator muscle. (5) The *supra-maxillary*, on the outer surface of the mandible, in front of the Masseter, in contact with the facial vessels. (6) The *internal maxillary*, beneath the ramus of the jaw, in company with the internal maxillary artery. (7) The *lingual*, two or

FIG. 530.—The superficial lymphatics and glands of the head, face, and neck.

three small nodules, lying on the Hyo-glossus and on and under the Genio-hyo-glossus. (8) The *retro-pharyngeal*, lying one on each side of the middle line in front of the Rectus capitis anticus major, which intervenes between them and the lateral mass of the atlas.

The **lymphatic vessels of the scalp** are divisible into (*a*) those of the *frontal* region, which terminate in the parotid glands; (*b*) of the *temporal* region, which end in the parotid and mastoid glands; and (*c*) those of the *occipital* region, which terminate partly in the occipital glands and partly in a trunk which runs down behind the Sterno-mastoid to join the glands in the lower part of the neck.

The **lymphatic vessels of the pinna and external auditory meatus** are collected into three sets of vessels: an *anterior*, which joins the parotid glands; a *posterior*, which passes to the mastoid glands; and an *inferior*, which terminates in the cervical glands.

The **lymphatic vessels of the face** are divided into two sets, *superficial* and *deep*.

The **superficial lymphatic vessels of the face** are more numerous than those of the scalp, and commence over its entire surface. Those from the nose terminate partly in the parotid, but chiefly in those of the submaxillary group of glands. The latter receive most of the lymphatic vessels from the lips, and are often found enlarged in cases of malignant disease of those parts. Those which drain the middle part of the lower lip end in the suprahyoid glands.

The **lymphatic vessels of the cranium** consist of two sets, the *meningeal* and *cerebral*. The *meningeal lymphatics* accompany the meningeal vessels, escape through foramina at the base of the skull, and join the internal maxillary and deep cervical lymphatic glands. The *cerebral lymphatics* are described by Eshmann as being situated between the arachnoid and pia mater, as well as in the choroid plexuses of the lateral ventricles; they accompany the trunks of the carotid and vertebral arteries, and pass through foramina at the base of the skull, to terminate in the deep cervical glands. In the brain, they assume the form of perivascular lymphatic spaces around the branches of the arteries. These spaces are said, by some observers, to be lined by endothelium; but others dispute this.

The **lymphatics of the orbit** and of the temporal and zygomatic fossæ run with the branches of the internal maxillary artery to the internal maxillary glands, and afterwards to the deep cervical.

The **lymphatics of the nasal fossæ** can be injected from the subdural and subarachnoid spaces. Those from the anterior part of the fossæ terminate in the submaxillary glands; those from the posterior two-thirds of the fossæ pass partly to the retro-pharyngeal and partly to the deep cervical glands.

The **lymphatics of the tongue** are chiefly drained into the deep cervical glands which lie between the posterior belly of the Digastric and the posterior belly of the Omo-hyoid. To reach their destination the vessels take different directions : some pierce the wall of the pharynx; others run over or under the Hyo-glossus, and in the course of these small lymphatic nodules (lingual glands) are found; while others pass between the Genio-hyo-glossi muscles. Some of the vessels which drain the margins and lateral portions of the dorsum of the tongue pierce the Mylo-hyoid, and end in the submaxillary glands; while from the tip of the tongue the lymph is carried partly into the suprahyoid glands, and partly into one of the lower of the deep cervical glands.

From the upper part of the *pharynx* the lymphatics pass to the retro-pharyngeal glands, from the lower part to the deep cervical glands. From the *larynx* two sets of vessels arise, an upper and a lower. The vessels of the upper set pierce the thyro-hyoid membrane, and join the superior group of deep cervical glands. Of the lower set, some pierce the crico-thyroid membrane and join the pretracheal and deep cervical glands; while others run between the cricoid and first tracheal ring, and enter the deep glands situated above the clavicle. The lymphatics of the *thyroid body* consist of two sets of vessels : an upper, which accompanies the superior thyroid artery, and enters the superior group of deep cervical glands; and a lower, which runs partly to the pretracheal glands, and partly to the small glands which accompany the laryngeal branch of the inferior thyroid artery.

The **Lymphatic glands of the neck** are divided into two sets, *superficial* and *deep*.

The **superficial cervical glands** may be arranged in three sets: (1) the *submaxillary*, eight to ten in number, situated beneath the body of the lower jaw in the submaxillary triangle; (2) *suprahyoid* or *submental*, one or two in number, situated in the middle line of the neck, between the anterior bellies of the two Digastric muscles; (3) *pretracheal*, one or two in number, in front of the trachea. Sometimes one or two small glands are found in front of the crico-thyroid membrane (*prelaryngeal*); and (4) *cervical*, four or five in number, placed in the course of the external jugular vein between the Platysma and deep fascia. A few small glands are sometimes found accompanying the anterior jugular vein.

The **deep cervical glands** (fig. 531) are numerous and of large size; they form a chain along the sheath of the carotid artery and internal jugular vein, lying by the side of the pharynx, œsophagus, and trachea, and extending from the base of the skull to the root of the neck. They are subdivided into two sets : an *upper*, ten to twenty in number, situated about the bifurcation of the common carotid and along the upper part of the internal jugular vein; and a *lower*, ten to fifteen in number, clustered around the lower part of the internal jugular vein, and extending outwards into the supraclavicular fossa, where they are continuous with the axillary glands. Internally, this set is continuous with the mediastinal glands. A few minute glands are situated alongside of the recurrent laryngeal nerves, in relation to the lateral aspect of the trachea and œsophagus.

The **superficial** and **deep cervical lymphatic vessels** are a continuation of those already described on the cranium and face. After traversing the glands in those regions, they pass through the chain of glands which lies along the

FIG. 531.—The deep lymphatics and glands of the neck and thorax.

sheath of the carotid vessels, being joined by the lymphatics from the pharynx, œsophagus, larynx, trachea, and thyroid gland. At the lower part of the neck, after receiving some lymphatics from the thorax, they unite into a single trunk, which terminates, on the left side, in the thoracic duct; on the right side, in the right lymphatic duct.

Surgical Anatomy.—The cervical glands are very frequently the seat of tuberculous disease. This condition is most usually set up by some lesion in those parts from which they receive their lymph. This excites inflammation, which subsequently takes on a tuberculous character. It is very desirable therefore for the surgeon, in dealing with these cases, to possess a knowledge of the relation of the respective groups of glands to the periphery. The following table is extracted from Sir Frederick Treves's work on 'Scrofula and its Gland Diseases.'

Scalp.—Posterior part = suboccipital and mastoid glands. Frontal and parietal portions = parotid glands.

Lymphatic vessels from the scalp also enter the superficial cervical set of glands.

Skin of face and neck = submaxillary, parotid, and superficial cervical glands.

External ear = superficial cervical glands.

Lower lip = submaxillary and suprahyoid glands.

Buccal cavity = submaxillary and upper set of deep cervical glands.

Gums of lower jaw = submaxillary glands.

Tongue.—Anterior portion = suprahyoid and submaxillary glands. Posterior portion = upper set of deep cervical glands.

Tonsils and palate = upper set of deep cervical glands.

Pharynx.—Upper part = parotid and retro-pharyngeal glands. Lower part = upper set of deep cervical glands.

Larynx, orbit, and roof of mouth = upper set of deep cervical glands.

Nasal fossæ = retro-pharyngeal glands, upper set of deep cervical glands. Some lymphatic vessels from posterior part of the fossæ enter the parotid glands.

Lymphatics of the Upper Extremity

The **Lymphatic Glands of the Upper Extremity** (fig. 532) are divided into two sets, *superficial* and *deep*.

The **superficial lymphatic glands** are few and of small size. There are occasionally two or three in front of the elbow, and usually one or two above the internal epicondyle of the humerus, near the basilic vein, while one or two may be found lying beside the cephalic vein between the Pectoralis major and Deltoid muscles, immediately below the clavicle.

The **deep lymphatic glands** are subdivided into those in the forearm, the arm, and the axilla. In the *forearm*, a few small ones are occasionally found in the course of the radial and ulnar vessels. In the *arm*, there is a chain of small glands along the inner side of the brachial artery. One, sometimes two, fairly constant glands are situated a little above and in front of the inner epicondyle of the humerus. In the *axilla* they are of large size, vary from twenty to thirty in number, and are arranged in the following groups: (1) An *external* chain of from four to six glands lies in relation to the front and inner aspects of the axillary vessels; they receive the lymphatic vessels from the arm; (2) a *pectoral* chain, consisting of four or five glands, runs along the lower border of the Pectoralis major, and receives the lymphatics from the front of the chest and from the outer two-thirds of the mammary gland; (3) a *subscapular* chain, of six or seven glands, is placed along the lower margin of the posterior wall of the axilla, and receives the lymphatics from the lower part of the back of the neck and from the lateral and dorsal parts of the upper part of the trunk; (4) two or three *subclavian* or *infraclavicular* lymphatic glands are placed immediately beneath the clavicle; (5) a *central* group of three or four glands is embedded in the adipose tissue near the base of the axilla. The efferent vessels from the axillary glands vary from one to three or four in number. They accompany the subclavian vein into the neck, and end, on the right side, by joining the right lymphatic duct, on the left side by opening into the thoracic duct.

Surgical Anatomy.—In malignant diseases, tumours, or other affections implicating the upper part of the back and shoulder, the front of the chest and mamma, the upper part of the front and side of the abdomen, or the hand, forearm, and arm, the axillary glands are liable to be found enlarged.

The **Lymphatic vessels of the upper extremity** are divided into two sets, *superficial* and *deep*.

The **superficial lymphatic vessels of the upper extremity** commence in the lymphatic plexus, which everywhere pervades the skin, the meshes of which are, however, much finer in the palm and on the flexor aspect of the digits than elsewhere. The digital plexuses are drained by a pair of vessels on the lateral aspect of each digit, which incline backwards to reach the dorsum of the hand. From the dense plexus of the palm the vessels pass in different directions, some upwards towards the wrist, others downwards to join the digital vessels, others inwards to join the vessels on the ulnar border of the hand, and others outwards to those on the thumb. From the central part of the plexus several vessels unite

3 C

to form a trunk, which passes round the metacarpal bone of the index finger to join the vessels on the back of that digit and on the back of the thumb. Running upwards in front of and behind the wrist, the vessels are collected into radial, median, and ulnar groups, which accompany the corresponding superficial veins. A few of the ulnar lymphatics terminate in the supratrochlear glands: some of the radial vessels are collected into a trunk, which ascends with the cephalic vein to the glands between the Pectoralis major and Deltoid;

Fig. 532.—The superficial lymphatics and glands of the upper extremity.

while a few of the median vessels terminate in the glands in front of the elbow, when these are present. All the remaining vessels are continued upwards to the axilla, where they terminate in the external group of axillary glands.

The **deep lymphatic vessels of the upper extremity** accompany the deep blood-vessels. In the forearm, they consist of four sets, corresponding with the radial, ulnar, and interosseous arteries; they communicate at intervals with the superficial lymphatics, and some of them end in the glands which are occasionally found beside the arteries. In their course upwards, a few end in the glands which lie upon the brachial artery; but most of them pass to the external set of axillary glands.

LYMPHATICS OF THE LOWER EXTREMITY

The **Lymphatic Glands of the Lower Extremity** are divided into two sets, *superficial* and *deep*. The superficial are confined to the inguinal region, forming the *superficial inguinal lymphatic glands*.

The **superficial inguinal lymphatic glands,** placed immediately beneath the integument, are of large size, and vary from eight to twenty in number. They are divisible into two groups : an upper *oblique* set, disposed irregularly along Poupart's ligament, which receives the lymphatic vessels from the integument of the scrotum, penis, parietes of the abdomen, perineal and gluteal regions, and the mucous membrane of the urethra in the male and the lower part of the vagina in the female ; and an inferior *vertical* set, which surrounds the saphenous opening in the fascia lata ; a few being sometimes continued along the saphenous vein for a variable distance. This latter group receives the superficial lymphatic vessels from the lower extremity. Some of the efferent vessels of these glands terminate in the deep inguinal glands, but most of them pass to the external iliac glands.*

Surgical Anatomy.—The superficial inguinal glands frequently become enlarged in diseases implicating the parts from which their lymphatics originate. Thus in malignant or syphilitic affections of the prepuce and penis, or of the labia majora in the female, in cancer scroti, in abscess in the perinæum, or in any other diseases affecting the integument and superficial structures in those parts, or the subumbilical part of the abdominal wall, or the gluteal region, the upper chain of glands is almost invariably enlarged, the lower chain being implicated in diseases affecting the lower limb.

The following table, showing the relation of the respective groups of glands to the periphery, is extracted from Sir Frederick Treves's 'Surgical Applied Anatomy ':

Superficial vessels of lower limb = vertical set of superficial glands.

Superficial vessels of lower half of abdomen = middle glands of horizontal set.

Superficial vessels from outer surface of buttock = external glands of horizontal set.

From inner surface of buttock = internal glands of horizontal set. (A few of these vessels go to the vertical glands.)

Superficial vessels from external genitals = horizontal glands, some few going to vertical set.

Superficial vessels of perinæum = vertical set.

Deep lymphatics of lower limb = deep set of glands.

---

* These glands are variously grouped by different anatomists, and such grouping is to a large extent artificial, since vessels from the same region (e.g. perinæum and scrotum) may terminate in both of the groups above-mentioned.

FIG. 533.—The superficial lymphatics and glands of the lower extremity.

Superficial inguinal glands

The **deep lymphatic glands** are : the anterior tibial, popliteal, deep inguinal, gluteal, and ischiatic.

The **anterior tibial gland** is not constant in its existence. It is generally found by the side of the anterior tibial artery, upon the interosseous membrane at the upper part of the leg. Occasionally, two glands are found in this situation.

The **popliteal glands,** four or five in number, are of small size ; they surround the popliteal vessels, and are embedded in the cellular tissue and fat of the popliteal space.

The **deep inguinal glands,** two or three in number, are placed beneath the deep fascia on the inner side of the femoral vein, the highest being placed in the crural canal. They are of small size, and communicate with the superficial inguinal glands through the saphenous opening. Their efferent vessels end in the external iliac glands.

The **gluteal** and **ischiatic glands** are placed, the former above, the latter below, the Pyriformis muscle, resting on their corresponding vessels as they pass through the great sacro-sciatic foramen.

The **Lymphatic vessels of the lower extremity,** like the veins, may be divided into two sets, *superficial* and *deep.*

The **superficial lymphatic vessels** are placed beneath the integument in the superficial fascia, and are divisible into two groups : an internal group, which follows the course of the internal saphenous vein ; and an external group, which accompanies the external saphenous. The vessels of the *internal group* are larger and more numerous than those of the external group, and commence on the inner side and dorsum of the foot ; they pass, some in front and some behind the inner ankle, run up the leg with the internal saphenous vein, pass with it behind the inner condyle of the femur, and accompany it to the groin, where they terminate in the inferior group of superficial inguinal lymphatic glands. Some of the efferent vessels from these glands pierce the cribriform fascia and sheath of the femoral vessels, and terminate in the gland contained in the crural canal. The vessels of the *external group* arise from the outer side of the foot, ascend in front of the leg, and, just below the knee, cross the tibia from without inwards, to join the lymphatics on the inner side of the thigh. Others commence on the outer side of the foot, pass behind the outer malleolus, and accompany the external saphenous vein along the back of the leg, where they enter the popliteal glands.

The **deep lymphatic vessels of the lower extremity** are few in number, and accompany the deep blood-vessels. In the leg, they consist of three sets, the anterior tibial, peroneal, and posterior tibial, which accompany the corresponding blood-vessels, two or three with each artery ; they ascend with the blood-vessels, and enter the lymphatic glands in the popliteal space ; the efferent vessels from these glands accompany the femoral vein, and join the deep inguinal glands ; from these, the vessels pass beneath Poupart's ligament, and communicate with the external iliac glands.

The deep lymphatic vessels of the gluteal and ischiatic regions follow the course of the blood-vessels, and join the gluteal and ischiatic glands at the great sacro-sciatic foramen.

LYMPHATICS OF THE ABDOMEN AND PELVIS

The lymphatic glands of the abdomen and pelvis may be divided into (a) *parietal* and (b) *visceral* : the former are situated behind the peritoneum, in close association with the larger blood-vessels ; the latter are found in relation to the walls and blood-vessels of the viscera.

The **parietal glands** are grouped as follows :

The **external iliac glands** are from eight to ten in number, and lie along the front and lateral aspects of the external iliac vessels. Three of them are placed immediately above Poupart's ligament—one on the outer, another on the inner, and a third on the anterior aspect of the vessels ; the last, however, is sometimes absent. The principal efferents of the external iliac glands are derived from (a) the superficial and deep inguinal glands, (b) the deep lymphatics of the abdominal wall, below the level of the umbilicus, the glans penis, membranous urethra, prostate, and base of bladder. Small chains of glands are sometimes found in the course of the deep epigastric and deep circumflex iliac arteries, while an additional gland (*obturator*) may be occasionally seen on the inner aspect of the obturator foramen.

The **internal iliac glands** surround the internal iliac vessels, and receive the lymphatics corresponding to the distribution of the branches of the internal iliac artery : i.e. they receive lymphatics from all the pelvic viscera, from the prostatic and membranous parts of the urethra, from the deeper parts of the perinæum, buttock, and back of thigh

FIG. 534.—The deep lymphatic vessels and glands of the abdomen and pelvis.

The **sacral glands** are placed in the concavity of the sacrum, some being situated in the mesorectal fold (*rectal glands*) ; they receive lymphatics from the rectum and posterior wall of the pelvis.

The **common iliac glands,** four to six in number, are grouped in relation to the anterior and lateral aspects of the common iliac artery ; one or two being placed below the bifurcation of the aorta, in front of the fifth lumbar vertebra. They chiefly drain the internal and external iliac glands.

The **lumbar glands** are very numerous, and consist of lateral and mesial groups. The *lateral lumbar glands* are small, and lie behind the Psoas between the transverse processes of the vertebræ; they receive the lymphatics which accompany the lumbar arteries. The *mesial lumbar glands* are partly grouped in relation to the lateral aspects of the aorta and vena cava; but a large number are placed in front of the vessels (pre-aortic glands), while others (retro-aortic) lie behind them. Those in relation to the lateral aspects of the vessels receive the lymphatics from the common, internal, and external iliac glands, from the upper part of the uterus, from the ovary or testicle, from the kidneys, suprarenal capsules, and posterior part of the Diaphragm. Their efferents unite to form the *right and left lumbar lymphatic trunks*, which terminate in the receptaculum chyli. The pre-aortic glands are especially aggregated in relation to the cœliac axis (cœliac glands) and the origin of the superior mesenteric artery. They receive the lymphatics from the alimentary canal, liver, spleen, and pancreas; and the efferent vessels of the highest of these glands unite to form the *intestinal lymphatic trunk*, which terminates in the receptaculum chyli.

The principal groups of visceral glands will be referred to in connection with the description of the lymphatics of the abdominal and pelvic organs.

The **Lymphatic vessels of the abdomen and pelvis** may be divided into two sets, *superficial* and *deep*.

The **superficial lymphatic vessels of the walls of the abdomen and pelvis** follow the course of the superficial blood-vessels. Those derived from the integument of the lower part of the abdomen below the umbilicus follow the course of the superficial epigastric vessels and converge to the superior group of the superficial inguinal glands; a deeper set accompanies the deep epigastric and deep circumflex iliac vessels, and communicates with the external iliac glands. The superficial lymphatics from the sides of the lumbar part of the abdominal wall wind round the crest of the ilium, accompanying the superficial circumflex iliac vessels, to join the superior group of the superficial inguinal glands; the greater number, however, run backwards along with the ilio-lumbar and lumbar vessels, to join the lateral lumbar glands.

The **superficial lymphatic vessels of the gluteal region** turn horizontally round the outer side of the nates, and join the superficial inguinal glands.

The **superficial lymphatic vessels of the scrotum and perinæum** follow the course of the external pudic vessels, and terminate in the superficial inguinal glands.

The **superficial lymphatic vessels of the penis** occupy the sides and dorsum of the organ, the latter receiving the lymphatics from the skin covering the glans penis; they terminate chiefly in the upper chain of the superficial inguinal glands. The **lymphatics of the glans penis** accompany the deep dorsal vein, and form a plexus near the root of the penis. From this plexus two or three vessels pass outwards with the deep external pudic artery to the gland in the crural canal, and to the lower of the external iliac glands; one vessel runs with the spermatic cord, and ends in the outer of the three lower external iliac glands. The **deep lymphatic vessels of the penis** follow the course of the internal pudic vessels, and join the internal iliac glands.

In the female, the lymphatic vessels of the mucous membrane of the labia, nymphæ, and clitoris terminate in the upper chain of the inguinal glands.

The **deep lymphatic vessels of the abdomen and pelvis** take the course of the principal blood-vessels. Those of the parietes of the pelvis, which accompany the gluteal, ischiatic, and obturator vessels, follow the course of the internal iliac artery, and ultimately join the lumbar lymphatics.

The efferent vessels from the inguinal glands enter the pelvis beneath Poupart's ligament, where they lie in close relation with the femoral vein; they then pass through the chain of glands surrounding the external iliac vessels, and finally terminate in the lumbar glands. They receive the deep epigastric and circumflex iliac lymphatics.

The **lymphatic vessels of the bladder** arise from the entire surface of the organ. Those from the front of the bladder are directed outwards to the external iliac glands; those from the posterior surface terminate in the external, internal, and common iliac glands.

The **lymphatics of the prostate** terminate chiefly in the internal iliac and sacral glands; but one vessel is described as passing to one of the external iliac glands.

The **lymphatics of the vagina** are carried in three directions : those of the upper part, to the external iliac glands ; those of the middle part, to the internal iliac glands ; and those of the lower part, to the common iliac glands. Some lymphatics from the lower part of the vagina join those of the external genitals, and pass to the superficial inguinal glands.

The **lymphatic vessels of the uterus** consist of two sets, *superficial* and *deep* ; the former being placed beneath the peritoneum, the latter in the substance of the organ. The lymphatics of the cervix uteri enter the internal iliac and sacral glands ; those from the body and fundus of the uterus pass outwards in the broad ligaments, and, being joined by the lymphatics from the broad ligaments and Fallopian tubes, ascend with the ovarian vessels to open into the lumbar glands. In the unimpregnated uterus they are small ; but during gestation they become very greatly enlarged.

The **lymphatic vessels of the testicle** consist of two sets, *superficial* and *deep* ; the former commence on the surface of the tunica vaginalis, the latter in the epididymis and body of the testis. They form several large trunks, which ascend with the spermatic cord, and, accompanying the spermatic vessels into the abdomen, terminate in the lumbar glands ; hence the enlargement of these glands in malignant disease of the testis. The **lymphatic vessels of the ovary** have a similar termination.

The **lymphatic vessels of the kidney** arise on the surface, and also in the interior of the organ ; they join at the hilum, and, after receiving the lymphatic vessels from the ureter and suprarenal capsules, open into the lumbar glands.

The **hepatic glands** are four or five in number, and lie between the two layers of the small omentum near the transverse fissure of the liver ; their efferent vessels end in the cœliac glands.

The **lymphatic vessels of the liver** are divisible into two sets, *superficial* and *deep*. The former arise in the subperitoneal areolar tissue over the entire surface of the organ, and may be grouped into (*a*) those on the convex surface, and (*b*) those on the under surface.

(*a*) On the *convex* surface. From the back part of this surface the vessels take three courses. From its middle portion, some four or five vessels pass through the caval opening in the Diaphragm, and end in one or two glands which are situated around the terminal part of the inferior vena cava. From the right lobe, one or two vessels run on the ventral surface of the Diaphragm, and, after crossing the right crus, terminate in one of the cœliac glands. From the left lobe, a few vessels pass backwards towards the œsophageal opening, and end in the gastric or coronary glands. From the portions of the right and left lobes of the liver which are adjacent to the falciform ligament, several vessels arise which converge to form a large trunk : this pierces the Diaphragm, and ends in the glands of the anterior mediastinum. From the anterior surface, a few vessels turn round the anterior sharp margin of the liver, and end in the glands of the hepatic chain ; while others pierce the liver substance and join the deep lymphatics.

(*b*) On the *under* surface. The vessels on this surface mostly converge to the transverse fissure, and accompany the deep lymphatics which emerge from this fissure ; but one or two from the right and Spigelian lobes accompany the vena cava through the Diaphragm, and end in the glands around the terminal part of this vessel.

The *deep lymphatics* consist of ascending and descending trunks. The ascending trunks accompany the hepatic veins, and pass through the Diaphragm to end in the glands around the terminal part of the inferior vena cava. The descending trunks emerge from the transverse fissure, and end in the glands of the hepatic chain.

The **gastric glands** are found along the greater and lesser curvatures, accompanying the epiploic and coronary arteries respectively. A considerable number of glands are found in the neighbourhood of the pylorus.

The **lymphatic vessels of the stomach** consist of two sets, *superficial* and *deep* ; the former originating in the subserous, and the latter in the submucous coat. They follow the course of the blood-vessels, and may consequently be arranged into three groups. Those of the *first group* accompany the gastric vessels along the lesser curvature to the cardiac orifice, receiving branches from both surfaces of the organ, and pass ultimately to the cœliac glands. Those of the *second group* pass from the great end of the stomach, accompanying the vasa brevia, and enter the splenic lymphatic glands. The vessels of the *third group* run along the greater curvature with the right gastro-epiploic vessels towards the pylorus, and, receiving the lymphatics from the upper part of the duodenum, terminate in the cœliac glands.

The **lymphatic glands of the spleen** accompany the splenic vessels, and lie in relation to the upper part of the posterior surface of the pancreas. Its *lymphatic vessels* consist of two sets, superficial and deep : the former being placed beneath its peritoneal covering, the latter in the substance of the organ ; they accompany the blood-vessels, receive the lymphatics from the pancreas, and ultimately pass into the cœliac glands.

The *lymphatics of the pancreas* also enter the cœliac glands.

The Lymphatic System of the Intestines

The **mesenteric lymphatic glands** (*lymphatic glands of the small intestine*) are situated between the two layers of the mesentery. They vary in number from one hundred to one hundred and fifty,* and may be divided into three sets : one lying near the intestine, a second in relation to the primary branches and loops of the superior mesenteric artery, and a third along the trunk of the artery. The last group receives not only the vessels from the small intestine, but the efferents from the ileo-cæcal and colic glands.

The **ileo-cæcal glands,** some half a dozen in number, occupy the angle between the terminal part of the ileum and the commencement of the ascending colon. One or two small *appendicular* glands are usually seen in the base of the meso-appendix.

The **colic glands** are found in relation to the ascending, transverse, descending, and ileo-pelvic parts of the colon, and are associated with the colic branches of the superior and inferior mesenteric arteries.

The **rectal glands** lie between the two layers of the mesorectum, and have already been referred to in connection with the sacral glands.

The **lymphatic vessels of the small intestine** are called **lacteals** from the milk-white fluid they usually contain ; they consist of two sets, superficial and deep : the former lie between the layers of the muscular coat and between the muscular and peritoneal coats, taking a longitudinal course along the outer side of the intestine ; the latter occupy the submucous tissue, and course transversely round the intestine, accompanied by the branches of the mesenteric vessels ; they pass between the layers of the mesentery, enter the mesenteric glands, the efferents of which terminate in the pre-aortic glands.

The **lymphatic vessels of the cæcum and vermiform appendix** are numerous, since the lymphoid tissue in the appendix is extremely well developed. Those from the cæcum consist of anterior and posterior groups ; they pass to the ileo-cæcal glands, the efferents of which terminate in the mesenteric glands. Four or five vessels drain the appendix, and, after traversing the appendicular and ileo-cæcal glands, finally end in the mesenteric glands.

Lymphatic vessels of the colon.—The vessels from the ascending and transverse parts of the colon finally terminate in the mesenteric glands ; those from the descending and ileo-pelvic parts of the colon ultimately end in the lumbar and pre-aortic glands.

Lymphatic vessels of the rectum.—Those from the anus end in the inguinal glands ; those from the region of the anal canal, in the internal iliac glands ; while the vessels from the remainder of the tube terminate in the glands which are situated in the mesorectum.

The Lymphatics of the Thorax

The **lymphatics of the thorax** may be divided into parietal and visceral—the former being situated in the thoracic wall, the latter in relation to the viscera.

The **parietal lymphatic glands** include the following :

1. The **internal mammary glands** are placed at the anterior extremity of each intercostal space, by the side of the internal mammary artery. They derive

---

* Leaf says (*The Surgical Anatomy of the Lymphatics*) that frequently there are not more than forty or fifty.

afferents from the mammary gland, from the deeper structures of the anterior abdominal wall above the level of the umbilicus, from the upper surface of the liver through a small group of glands which lie behind the ensiform cartilage, and from the deeper parts of the anterior portion of the thoracic wall. The efferents of the right chain open into the right lymphatic duct, those of the left chain into the thoracic duct. Sometimes the efferents join to form a single trunk on each side, and this may open independently at the junction of the internal jugular and subclavian veins.

2. The **intercostal glands** occupy the posterior parts of the intercostal spaces, where they lie in relation to the intercostal vessels. They receive the deep lymphatics from the postero-lateral aspect of the chest. The efferents of the glands in the lower four or five spaces unite to form a trunk, which descends and opens either into the receptaculum chyli or the commencement of the thoracic duct. The efferents of the glands in the upper spaces of the left side terminate in the thoracic duct; those of the corresponding right spaces, in the right lymphatic duct.

3. The **diaphragmatic glands** lie on the thoracic aspect of the Diaphragm, and consist of: (a) two or three small glands behind the ensiform cartilage, which receive afferents from the convex surface of the liver, while their efferents join the internal mammary glands; (b) one or two glands placed near the junction of the seventh rib with its cartilage, which receive some of the lymphatics from the Diaphragm; (c) two or three small glands on each side, which are situated close to where the phrenic nerves enter the Diaphragm, and receive some of the lymphatics of the Diaphragm: those of the right side also receive afferents from the convex surface of the liver; (d) a few glands situated on the back of the diaphragmatic crura, and connected on the one hand with the lumbar glands, and on the other with the posterior mediastinal glands.

The **superficial lymphatic vessels of the thoracic wall** ramify beneath the skin. They converge to the axillary glands, those over the pectoral region running backwards, those over the Serratus magnus upwards, and those over the Trapezius and Latissimus dorsi forwards. A few lymphatic vessels from the upper part of the pectoral region pass over the clavicle to the cervical glands. Further, it must be stated that the vessels of opposite sides anastomose across the front of the sternum.

The **lymphatics of the mammary gland** mostly end in the pectoral group of axillary glands, but some from the inner part of the mamma pierce the Pectoralis major and Internal intercostals and join the internal mammary glands. A vessel has occasionally been found running upwards under the Pectoralis major to the subclavian glands.

The **deep lymphatics of the thoracic wall** consist of:

1. The lymphatics of the muscles which lie on the ribs: most of these terminate in the axillary glands.

2. The intercostal lymphatic vessels drain the Intercostal muscles and parietal pleura. Those in the posterior part of the spaces run backwards and, after receiving the vessels which accompany the posterior branches of the intercostal arteries, terminate in the posterior intercostal glands. Those in the front part of the spaces run forwards and end in the internal mammary glands. The vessels which run backwards are said to drain the External, and those running forward the Internal intercostal muscles.

3. The lymphatic vessels of the Diaphragm follow more or less closely the course of the corresponding blood-vessels, and end, behind, in the posterior mediastinal and lower intercostal glands; in front, in the internal mammary and anterior mediastinal glands, and also in those diaphragmatic glands which lie behind the junction of the seventh rib with its cartilage.

The **visceral lymphatic glands** include the following:

1. The **anterior mediastinal glands** lie in front of the pericardium, in the areolar tissue of the anterior mediastinum. They receive some afferents from the liver, and are drained partly into the superior mediastinal glands and partly into the thoracic and right lymphatic ducts.

2. The **superior mediastinal glands** are situated partly at the sides of the trachea and in the angle of its bifurcation, and partly above the aortic arch, where they are especially grouped in relation to the right innominate vein and left subclavian artery. They receive afferents from the anterior mediastinal

glands, from the thymus, heart, and pericardium; their efferents end in the thoracic and right lymphatic ducts.

3. The **posterior mediastinal glands** lie behind the pericardium in relation to the œsophagus and aorta. Their afferents are derived from the œsophagus, the posterior part of the pericardium, the Diaphragm and liver. Their efferents mostly terminate in the thoracic and right lymphatic ducts, but some join the bronchial glands.

4. The **bronchial glands** are large and numerous, and lie in relation to the extra-pulmonary bronchi and the larger branches of the intra-pulmonary bronchi. They receive the lymphatics of the lungs, and become brown or black in old age; while they are frequently the seat of tuberculous deposits.

The **lymphatic vessels of the lung** consist of two sets, *superficial* and *deep*. Those of the former are placed beneath the pleura, forming a minute plexus, which covers the outer surface of the lung; those of the latter accompany the blood-vessels, and run along the bronchi: they both terminate at the root of the lung in the bronchial glands. The efferent vessels from these glands, two or three in number, ascend upon the trachea to the root of the neck, traverse the tracheal and œsophageal glands, and terminate on the left side in the thoracic duct, and on the right side in the right lymphatic duct, or they may join with the efferent vessels from the internal mammary and superior mediastinal glands to form the *broncho-mediastinal trunk*. The left broncho-mediastinal trunk may join the thoracic duct, and the right, the right lymphatic duct; but more frequently they open independently into the junction of the internal jugular and subclavian veins.

The **cardiac lymphatic vessels** consist of two sets, *superficial* and *deep*: the vessels of the former set arise in the subserous areolar tissue of the surface, and those of the latter in the deeper tissues of the heart. They follow the course of the coronary vessels: those of the right side unite into a trunk at the root of the aorta, which, ascending across the arch of that vessel, communicates with one or more of the cardiac glands, and passes backwards to the trachea, upon which it ascends, to terminate at the root of the neck in the right lymphatic duct. Those of the left side unite into a single vessel at the base of the heart, which, passing along the pulmonary artery, and traversing some glands at the root of the aorta, ascends on the trachea to terminate in the thoracic duct.

The **thymic lymphatic vessels** arise from the under surface of the thymus gland, and enter the superior mediastinal glands, from which they emerge as two vessels: these terminate, one on each side, in the corresponding internal jugular vein.

The **lymphatic vessels of the œsophagus** form a plexus round that tube, traverse the glands in the posterior mediastinum, and after communicating with the pulmonary lymphatic vessels near the roots of the lungs, terminate in the thoracic duct.

THE NERVOUS SYSTEM

THE nervous system is the most complicated and the most highly organised of the various systems which make up the human body. It may be divided into two parts, viz. *cerebro-spinal* and *sympathetic*—a subdivision which is convenient from a descriptive point of view, but is nevertheless quite arbitrary, since the two parts are not only intimately connected, but have a common ectodermal origin.

The **cerebro-spinal nervous system** is associated with the functions of the special senses and with the voluntary movements of the body, and comprises the brain and spinal cord, together with the cranial and spinal nerves and their ganglia. The brain and spinal cord constitute the *cerebro-spinal axis*, and are continuous with each other through the foramen magnum, the former being situated within the cranium, the latter within the vertebral canal. The **sympathetic nervous system** presides over the movements of the viscera, determines the calibre of the blood-vessels, and controls the phenomena of secretion. It consists of two rows of central ganglia, situated one on either side of the middle line, in front of the vertebral column : these ganglia are intimately connected with the spinal cord and its nerves, and are also joined to each other by vertical strands of nerve-fibres so as to constitute a pair of knotted cords, the *gangliated cords of the sympathetic*, which reach from the base of the skull to the coccyx. The sympathetic system also comprises three great prevertebral plexuses, which supply the thoracic, abdominal, and pelvic viscera : in relation to the walls of these viscera, intricate nerve-plexuses and numerous peripheral ganglia are found.

The nervous system is built up of nervous and non-nervous tissues—the former consisting of nerve-cells and nerve-fibres ; the latter, of neuroglia and blood-vessels, together with certain enveloping membranes.

The minute structure of the nervous elements, and of the neuroglia, has been described in the chapter on Histology ; and an outline of the development of the nervous system furnished in that on Embryology. It may be stated here, however, that, in its earliest condition, the nervous system consists of cells only, and that the nerve-fibres arise as outgrowths or processes from the nerve-cells. Nerve-cells give off, as a rule, two kinds of processes, viz. a delicate filament termed an *axon*, or *axis-cylinder process* ; and several thicker branched offshoots, named *dendrites*. The nerve-cell, with its axon and dendrites, constitutes what is termed a *neurone* ; and it is generally maintained that each neurone possesses an individuality of its own, and that its various processes only come into contact, but are never directly continuous, with those of the other neurones : there are, however, some who hold that direct continuity is established between neighbouring neurones by means of their dendrites. Dendrites and axons are alike conductors of nervous impulses : the former, however, convey them to, the latter from, the nerve-cell ; in other words, the dendrites form the paths of reception, the axon the path of transmission.

A section across the brain or spinal cord exhibits two kinds of nervous tissue, white and grey. The white nervous tissue consists of axons which have acquired their myelin sheaths, and so constitute medullated nerve-fibres ; the grey nervous tissue is chiefly composed of nerve-cells.

THE SPINAL CORD

The spinal cord (fig. 535) is the elongated portion of the cerebro-spinal axis which occupies the upper two-thirds of the vertebral canal. It measures from seventeen to eighteen inches in length, and weighs, when divested of its membranes and nerves, a little over an ounce ; its specific gravity is about 1·387. It follows the curvatures of that part of the vertebral canal in which it lies, but does not nearly fill the cavity, its investing membranes being separated from the surrounding walls by areolar tissue and a plexus of veins. It extends from the upper

Fig. 535.—Posterior view of the spinal cord *in situ*.

border of the atlas to the lower border of the body of the first, or upper border of the body of the second, lumbar vertebra. Its position varies, however, with the degree of curvature of the spinal column, being drawn slightly upwards when the column is bent forwards. Above, it is continuous with the medulla oblongata of the brain ; below, it ends in a conical extremity, the *conus medullaris*, the apex of which is prolonged downwards as a thread-like filament, the *filum terminale*, which is fixed to the back of the upper part of the coccyx.

Thirty-one pairs of spinal nerves are attached to the lateral aspects of the cord, each nerve arising by two roots, an anterior or ventral, and a posterior or dorsal root. The anterior root consists of motor, and the posterior root of sensory fibres ; and the posterior root is further distinguished by the presence of an oval swelling, the *spinal ganglion*, which contains numerous nerve-cells. The pairs of spinal nerves are grouped as follows : Cervical 8, Dorsal 12, Lumbar 5, Sacral 5, Coccygeal 1 ; and the cord is divided for descriptive purposes into similar regions (cervical, dorsal, &c.), each of which corresponds with the extent of attachment of a group of nerves. Further, it is convenient to regard the spinal cord as being built up of a series of superimposed segments, termed *spinal segments* or *neuromeres*, each of which corresponds with the attachment of a pair of spinal nerves. It must be remembered, however, that these so-called segments are intimately linked to each other by numerous association or intersegmental fibres, and by the fibres which pass to or from the brain.

In the fœtus up to the third month, the spinal cord extends to the lower end of the spinal canal, but after this date the spinal column elongates more rapidly than the spinal cord ; and the latter, being fixed above, through its continuity with the brain, recedes upwards within the canal, so that by the fifth month of fœtal life its lower end reaches only to the base of the sacrum, and at birth to the level of the third lumbar vertebra. As a consequence of this, the spinal nerve-roots, which at first passed transversely outwards from the cord to reach their respective intervertebral foramina, become more and more oblique in direction from above downwards ; and the lower part of the canal is occupied by the lumbar and sacral nerves, which descend vertically to reach their points of exit. From their great length and the appearance they present at their attachment to the cord, the lumbar and sacral nerves are collectively termed the *cauda equina*. The filum terminale lies in the centre of this bundle of nerves, and may readily be distinguished from them by its bluish-white appearance.

The spinal cord is not quite cylindrical, being slightly flattened from before backwards ; nor is it of uniform circumference throughout, but presents two swellings or enlargements, an upper or cervical, and a lower or lumbar.

The *cervical enlargement* is the more pronounced of the two, and corresponds with the origin of the large nerves which supply the upper limbs. It extends from about the third cervical to the second dorsal vertebra, its maximum circumference (about thirty-eight millimetres) being on a level with the origin of the sixth pair of cervical nerves.

The *lumbar enlargement* gives attachment to the nerves which supply the lower limbs. It commences about the level of the ninth dorsal vertebra, and reaches its maximum circumference, of about thirty-three millimetres, opposite the last dorsal vertebra, below which it tapers gradually into the conus medullaris.

Fissures and sulci of the spinal cord.—A pair of median fissures, anterior and posterior, dip into the substance of the cord, and incompletely divide it into two symmetrical halves, which are joined across the middle line by a commissural band of nervous matter.

The anterior median fissure is wider and shallower than the posterior : it has an average depth of three millimetres, but this is increased in the lower part of the cord. It contains a double fold of pia mater, and its floor is formed by a transverse band of white substance, the *white commissure*, which is perforated by blood-vessels on their way to or from the central part of the cord.

The posterior median fissure is not an actual fissure like the anterior ; it does not contain a fold of pia mater, but merely a septum of neuroglia which is intimately united with the neuroglia in the adjacent parts of the cord, and for this reason it would be more correct to name it the *posterior median septum*. It reaches rather more than halfway into the substance of the cord, and its depth varies from four to six millimetres, but diminishes in the lower part of the cord.

On either side of the posterior median fissure, and at a short distance from it, the posterior nerve-roots are attached to the cord along a vertical furrow named the *postero-lateral sulcus*. The portion of the cord which lies between this sulcus and the posterior median fissure is named the *posterior column*. In the cervical and upper dorsal regions this column presents a longitudinal furrow, the *postero-intermediate sulcus* : this marks the position of a septum which extends into the posterior column and subdivides it into two fasciculi—an inner, named the *tract of Goll*; and an outer, the *tract of Burdach* (see fig. 541). The portion of the cord which lies in front of the postero-lateral sulcus is termed the *antero-lateral column*. The anterior nerve-roots, unlike the posterior, are not attached in linear series, and their position of exit is not marked by a sulcus. They arise by separate bundles which spring from the anterior horn of grey matter and, passing forward through the white matter, emerge over an area of some slight width. The outermost of these bundles is generally taken as a dividing line which separates the antero-lateral column into two parts, viz. an *anterior column*, between the anterior median fissure and the outermost of the anterior nerve-roots; and a *lateral column*, between the exit of these roots and the postero-lateral sulcus.

INTERNAL STRUCTURE OF THE SPINAL CORD

On examining a transverse section of the cord, it is seen to consist of grey and white nervous matter, the former being enclosed within the latter.

Grey matter of the spinal cord.—The **grey matter** of the spinal cord consists of two symmetrical portions, one in each half of the cord : these are joined across the middle line by a transverse band, termed the *grey commissure*, through which runs a minute canal, the *central canal* of the cord, just visible with the naked eye. Each half of the grey substance is shaped in the form of a comma or crescent, the concavity of which is directed outwards; and these, together with the intervening grey commissure, present on transverse section the appearance of the letter H. An imaginary line drawn transversely through the central canal serves to divide each crescent into an anterior and a posterior cornu.

The *anterior cornu* is directed forwards, and is broad and of a rounded or quadrangular shape. Its posterior part is termed the *base*, and its anterior part the *head*, but these are not differentiated from each other by any well-defined constriction. It is separated from the surface of the cord by a layer of white matter which is traversed by the bundles of the anterior nerve-roots. In the dorsal region, the postero-external part of the anterior cornu projects outwards as a triangular field, which is named the *lateral cornu*.

The *posterior cornu* is long and slender, and is directed backwards and outwards: it reaches almost as far as the postero-lateral sulcus, from which it is separated only by a thin layer of white substance, the *tract of Lissauer*. It consists of a *base*, which is directly continuous with the corresponding part of the anterior horn; a *neck* (cervix cornu) or slightly constricted portion, which is succeeded by an oval or fusiform area, termed the *head* (caput cornu), and of which the summit (apex cornu) approaches the postero-lateral sulcus. The head is capped by a **V**-shaped or crescentic mass of translucent, gelatinous neuroglia, termed the *substantia gelatinosa of Rolando*, which contains not only neuroglia-cells, but numerous small nerve-cells. Between the anterior and posterior cornua the grey matter extends as a series of processes for some distance into the lateral column, where they form a network called the *formatio reticularis*.

The quantity of grey matter, as well as the form which it assumes on transverse section, varies markedly at different levels. It is small, not only in amount but relatively to the surrounding white substance, in the dorsal region. Its amount is greatly increased in the cervical and lumbar enlargements: in the latter, and especially in the conus medullaris, its proportion to the white matter is greatest (fig. 536). In the cervical region, its posterior cornu remains comparatively narrow, while its anterior is broad and expanded; in the dorsal region, both cornua are attenuated, and the lateral horn is evident; in the lumbar enlargement, both cornua are expanded; while in the conus medullaris the grey matter assumes the form of two oval masses, one in each half of the cord, connected together by a broad grey commissure.

The **central canal** runs in the grey commissure throughout the entire length of the cord. The part of the grey commissure in front of the canal is named the *anterior grey commissure*; that behind it, the *posterior grey commissure*. The former is thin, and is in contact anteriorly with the white commissure: it contains a couple of longitudinal veins, one on either side of the middle line. The posterior grey commissure reaches from the central canal to the posterior median septum, and is thinnest in the dorsal region, and thickest in the conus medullaris. The central canal is continued upwards through the lower part of the medulla oblongata, and opens into the fourth ventricle of the brain; below, it reaches for a short distance into the filum terminale. In the lower part of the conus medullaris it exhibits a fusiform dilatation, the *terminal sinus* (*terminal ventricle of Krause* *). This has a vertical measurement of from eight to ten millimetres, is triangular on cross section with its base directed forwards, and tends to undergo obliteration after the age of forty.

Throughout the cervical and dorsal regions the central canal is situated in the anterior third of the cord: in the lumbar enlargement it is near the middle, and in the conus medullaris it approaches the posterior surface. It is filled with cerebro-spinal fluid, and is lined by ciliated, columnar epithelium, outside which is an encircling band of gelatinous neuroglia, the *substantia gelatinosa centralis*.

* *Archiv für micro. Anat.* 1875.

FIG. 536.—Shape of the grey matter in different regions of the spinal cord. (Poirier.)

I

2

3

4

5

6

Transverse sections, passing from above downwards, through: 1, the upper cervical region; 2, the cervical enlargement; 3, the dorsal region; 4, the lumbar enlargement; 5, the sacral region; 6, the filum terminale.

Structure of the Grey Matter.—The grey matter consists of neuroglia, together with numerous nerve-cells and nerve-fibres. Throughout the greater part of the grey matter the neuroglia presents the appearance of a sponge-like network, but around the central canal and on the heads of the posterior cornua it consists of the gelatinous substance already referred to (fig. 537). The nerve-cells in the grey matter are multipolar, and vary greatly in size and shape (fig. 538). They consist of : (1) motor cells, whose axons leave the cord in the spinal nerve-roots. These cells are of large size, and are situated in the anterior horn of grey matter ; the axons of nearly all of them pass out to form the anterior nerve-roots.* (2) Nerve-cells of medium or small size, whose axons do not emerge from the cord but pass transversely outwards into the white matter,

Fig. 537.—From a transverse section through the spinal cord of a calf. Magnified about 180 diameters, showing part of the central canal and the tissue immediately around it, viz. the central grey matter. (Klein and Noble Smith.)

Fig. 538.—Cells of the spinal cord. (Poirier.)

The canal is lined with epithelium, composed of ciliated, more or less conical, cells ; in most instances a filamentous process passes from the cell into the tissue underneath. This tissue contains, in a hyaline matrix, a network of fibrils ; most of these run horizontally : others have a longitudinal course, and appear therefore here cut transversely, i.e. as small dots. The nuclei correspond to the cells of the neuroglia, the cell-substance not being shown. Both the nuclei of the neuroglia cells and those of the epithelium contain three or more large disc-shaped particles.

— Collateral

-- Ascending

-- Descending

Arborisation

Diagram showing in longitudinal section the intersegmental neurones of the spinal cord. The grey and white parts correspond respectively to the grey and white substance of the spinal cord.

where some assume an ascending, and others a descending course, but where most of them divide in a T-shaped manner into a descending and an ascending process. These axons give off collaterals which enter and ramify in the grey matter, while the terminations of the axons behave in a similar manner. The length of these axons varies greatly : some are short and pass only between adjoining segments of the cord, while others are longer and connect segments which are more widely separated. These cells and their processes constitute a series of *association* or *intersegmental neurones*, which link together the different parts of the cord. The axons of most of these cells are confined to that side of the cord in which the nerve-cells are situated, but some cross to the opposite side through the anterior commissure, and are termed *crossed commissural fibres*. These cells are especially grouped into three regions, viz. in the inner part of the anterior cornu, in the lateral cornu, and in the base of the posterior cornu : in the last situation they form a well-defined column, termed *Clarke's column*. (3) Cells of the Golgi type, whose

* Lenhossek and Cajal found that in the chick embryo the axons of a few of these nerve-cells passed backwards through the posterior cornu, and emerged as the *motor fibres* of the *posterior nerve-roots*; but the presence of these in man has not yet been determined.

axons are short and entirely confined to the grey matter, in which they break up into numerous fine filaments.

Distribution of the nerve-cells (fig. 539).—Most of the nerve-cells are arranged in longitudinal columns, which appear as groups on transverse section.

Nerve-cells in the anterior horn.—These are of large size, and form two principal columns—a *mesial*, in the inner, and a *lateral*, in the outer part of the horn. In the cervical and lumbar enlargements each of these is divided into anterior and posterior parts. The anterior cells of the mesial column are large, and their axons pass into the anterior nerve-roots; the posterior cells of this column are smaller, and their axons pass across the middle line in the anterior commissure, and, ascending in the anterior column of the opposite side, terminate in the grey matter, thus forming crossed commissural fibres. The cells of the lateral column are all of large size, and their axons pass into the anterior nerve-roots, of which they form the main portion. Solitary

FIG. 539.—Mode of distribution of the nerve-cells in the grey matter. (Schematic.)
(Testut.)

1 2. Mesial and lateral groups of nerve-cells in anterior horn. 3. Nerve-cells in lateral horn. 4. 4. Column of Clarke. 5. Group of nerve-cells in substantia gelatinosa of Rolando. 6. Nerve-cell of anterior horn, the axon of which is passing into the posterior nerve-root. 7. Cells of substantia gelatinosa centralis. 8, 8′. Solitary cells. 9. Cells of Golgi. 10. Cells of origin of the tract of Gowers. 11. Anterior root. 12. Posterior root. 13. Spinal ganglion.

cells are scattered throughout the base of the anterior horn, the axons of some of which form crossed commissural fibres, while others constitute the motor fibres of the posterior nerve-root. (See footnote, page 767.)

Nerve-cells in the lateral horn.—These form a column which is best marked where the lateral horn of grey matter is differentiated, viz. in the dorsal region; but it can be traced throughout the entire length of the cord, in the form of groups of small cells which are situated in the anterior part of the formatio reticularis. The cells of this column are fusiform or star-shaped, and of a medium size: the axons of some of these pass into the anterior nerve-roots, by which they are carried to the sympathetic; while those of others pass into the anterior and lateral column, where they become longitudinal.

Nerve-cells in the posterior horn.—I. The *Column of Clarke* (*posterior vesicular column*). This occupies the inner part of the base of the posterior horn, and appears as a well-defined oval area on transverse section. It commences below at the level of the second or third lumbar nerve, and reaches its maximum size opposite the twelfth dorsal nerve. Above the level of the ninth dorsal nerve its size diminishes, and the column terminates opposite the last cervical or first dorsal nerve. It is represented, however, in the other regions of the cord by scattered cells, which become aggregated to form a *cervical nucleus* opposite the third cervical nerve, and a *sacral nucleus* in the middle and lower part of the sacral region. Its cells are of medium size, and of an oval or pyriform shape ; their axons pass into the peripheral part of the lateral column of the same side, and there ascend, under the name of the *direct* or *ascending cerebellar tract.* II. *Nerve-cells in the substantia gelatinosa of Rolando.*—These are arranged in three zones : a posterior or marginal zone, composed of large triangular or fusiform cells ; an intermediate zone of small fusiform cells, and an anterior zone of star-shaped cells. The axons of these cells pass into the lateral and posterior columns, and there assume a vertical course. In the anterior zone, some Golgi cells are found whose short axons ramify in the grey matter. III. *Solitary cells* of varying form and size are scattered throughout the posterior horn. Some of these are grouped to form the *posterior basal column* in the base of the posterior horn on the outer side of Clarke's column. The posterior basal column is well marked in the gorilla (Waldeyer), but is ill defined in man. The axons of these cells pass partly to the posterior and lateral columns of the same side, and partly through the anterior commissure to the lateral column of the opposite side. Before leaving the grey matter, a considerable number run longitudinally for a varying distance in the head of the posterior horn, forming what is termed the *longitudinal fasciculus* of the posterior horn.

A few star-shaped or fusiform nerve-cells of varying size are found in the substantia gelatinosa centralis. Their axons pass into the lateral column of the same, or of the opposite side.

The nerve-fibres in the grey matter form a dense interlacement of minute fibrils among the nerve-cells. This interlacement is formed partly of axons which pass from the cells in the grey matter to enter the white columns or nerve-roots ; partly of the axons of Golgi's cells which ramify only in the grey substance ; and partly of collaterals from the nerve-fibres in the white columns, which, as already stated, enter the grey matter and ramify within it.

White Matter of the Spinal Cord.—The white matter of the spinal cord consists of medullated nerve-fibres embedded in a sponge-like network of neuroglia, and is arranged in three columns : anterior, lateral, and posterior. The anterior column lies between the anterior median fissure and the outermost of the anterior nerve-roots ; the middle between the outermost of the anterior nerve-roots and the posterior horn ; and the posterior between the posterior horn and the posterior median fissure. On examining a transverse section which has been stained with picrocarmine, the nerve-fibres appear as small circles,

FIG. 540.—Transverse section through the white matter of the spinal cord of a calf. Magnified about 300 diameters. (Klein and Noble Smith.)

In the upper part are shown two isolated flattened nucleated cells of the neuroglia, under a somewhat higher power than the rest. In the bulk of the figure the nerve-fibres are seen in transverse section. They are of different sizes, and possess a laminated medullary sheath surrounding the axis cylinder, which was deeply stained in the preparation, and is here represented by a black dot. The nerve-fibres are embedded in the neuroglia. Among the neuroglia are also seen two branched connective-tissue cells—neuroglia cells.

in the centres of which the axis-cylinders form minute crimson dots (fig. 540). The fibres vary greatly in thickness, the smallest being found in the tracts of Goll and Lissauer, and in the inner part of the lateral column ; while the largest are situated in the anterior column, and in the peripheral part of the lateral column.

Nerve-tracts in the spinal cord.—The nerve-fibres in the white columns of

3 D

the cord are grouped into more or less definite bundles or fasciculi. These are not recognisable from each other in the normal state of the cord, and their existence has been determined by the following methods. All nerve-fibres are outgrowths from nerve-cells, and if a bundle of nerve-fibres be cut, the portions of the fibres which are separated from their cells rapidly degenerate and become atrophied, while the cells and the parts of the fibres connected with them undergo little alteration.* Similarly, if a group of nerve-cells be destroyed, the fibres arising from them undergo degeneration. Thus, if the cells of the cerebral cortex which give origin to the motor impulses be destroyed, or if the fibres arising from these cells be severed, a *downward* degeneration from the seat of injury takes place in the fibres. In the same manner, if a spinal ganglion be destroyed, or the fibres which pass from it into the spinal cord be cut, an *ascending* degeneration will extend along these fibres into the spinal cord. By adopting proper methods, such degenerated fibres can be followed to their termination.

Further, by tracing the development of the nervous system, it has been observed that at first the nerve-fibres are merely naked axis-cylinders, and that they do not all acquire their medullary sheaths at the same time ; hence the fibres can be grouped into different bundles according to the dates at which they receive their medullary sheaths.

Lastly, mention must be made of the silver method of staining nervous tissues, which has been of the greatest value in following the course and mode of termination of the axis-cylinder processes.

Nerve-tracts in the anterior column (fig. 541).—There are two principal tracts in this column, viz. the direct pyramidal, and the anterior basis bundle.

The *direct pyramidal tract* is usually of small size, but varies inversely with that of the crossed pyramidal tract in the lateral column. It lies next the anterior median fissure, and is present only in the upper part of the cord ; gradually lessening as it is traced downwards, it disappears about the middle of the dorsal region. It consists of descending fibres which arise from cells in the motor area of the cerebral hemisphere of the same side, and which, on being traced downwards through the spinal cord, cross in succession through the white commissure to the opposite side, where they end by arborising around the motor cells in the anterior horn.

The *anterior basis bundle* constitutes nearly the whole of the remainder of the anterior column. It consists of—(a) longitudinal commissural fibres which arise from cells in the grey matter, more especially from those of the mesial group in the anterior horn, and which, after a longer or shorter course, re-enter the grey matter ; (b) fibres which cross the middle line in the white commissure ; and (c) fibres of the anterior nerve-roots which pass horizontally forward to reach the surface of the cord.

Tracts in the lateral column.—The principal tracts in the lateral column are five in number, viz. the direct or ascending cerebellar, the crossed pyramidal, the tract of Gowers, the lateral basis bundle, and the tract of Lissauer.

The *direct or ascending cerebellar tract* is situated at the periphery of the posterior part of the lateral column, where it forms a flattened band which extends forwards as far as a line drawn transversely through the central canal. Internally it is in contact with the crossed pyramidal tract, in front with the tract of Gowers, behind with that of Lissauer. It commences below at the level of the third lumbar nerve, and, increasing in size as it ascends, passes to the cerebellum, mainly through its inferior, but partly through its superior, peduncle. Its fibres are formed by the axons of the cells of Clarke's column, which pass outwards through the inner part of the lateral column of the cord.

---

* It was formerly believed that the cell underwent no change under such circumstances. This, however, is not the case ; for if a nerve—the sciatic, for instance—be divided in an animal, and, after an interval of some weeks, the animal be injected with methylene blue and killed, it will be seen, on examining sections of the lumbar region of the spinal cord, that the cells are imperfectly stained, or not at all, owing to a diminution, or it may be entire disappearance, of their chromatin, a substance which in a normal cell shows marked affinity for staining reagents. Further, the body of the cell is swollen, the nucleus displaced towards the periphery, and the part of the axon still attached to the altered cell is diminished in size and somewhat atrophied. Under favourable conditions the cell is capable of reassuming its normal appearance, and its axon may commence to grow.

The *crossed pyramidal tract* extends throughout the entire length of the cord, and on transverse section appears as an oval or triangular area, in front of the posterior horn, and on the mesial aspect of the direct cerebellar tract. It consists of descending fibres which arise from the cells in the motor area of the cerebral hemisphere of the opposite side. These fibres pass downwards in company with those of the direct pyramidal tract, on the same side of the brain as that from which they take origin; but, unlike those of the direct pyramidal tract, they cross to the opposite side in the medulla oblongata (hence the name of 'crossed pyramidal'), and enter the lateral column. They end by arborising around the motor cells in the anterior horn.

The crossed and direct pyramidal tracts constitute the motor fasciculi of the spinal cord, and are formed by the axis-cylinders of the motor cells in the cerebral cortex. They descend through the internal capsule of the cerebrum, traverse the pons Varolii, and enter the anterior pyramid of the medulla oblongata on the same side of the brain as that in which they arise. In the lower part of the medulla oblongata, some two-thirds of the fibres decussate in the middle line with the corresponding fibres of the opposite side, and run down in the lateral column of the cord as the crossed pyramidal tract, while the remaining fibres do

FIG. 541.—Tracts of the spinal cord. (Testut.)

a. Anterior median fissure. *b.* Posterior median fissure. 1. Direct pyramidal tract. 2. Crossed pyramidal tract. 3. Anterior basis bundle. 3'. Lateral basis bundle. 4. Direct cerebellar tract. 5. Tract of Gowers. 6, 6'. Lateral limiting zone. 7. Tract of Burdach. 8. Tract of Goll. 9. Cornu-commissural tract. 10. Lissauer's tract.

not cross the middle line, but run down as the direct pyramidal tract in the anterior column of the same side. The fibres of this tract, however, have been seen to cross the middle line in the anterior commissure of the cord, and thus the whole of the motor fibres from one side of the brain ultimately reach the opposite side of the cord. The number of fibres which decussate in the medulla oblongata is not constant, and hence the direct and crossed pyramidal tracts vary inversely in size. The fibres of the crossed and direct pyramidal tracts do not acquire their medullary sheaths until after birth.

The *tract of Gowers (antero-lateral ascending tract)* is an extensive comma-shaped strand, which skirts the circumference of the lateral column in front of the direct cerebellar tract. Its expanded extremity, or head, lies in the angle between the direct cerebellar and crossed pyramidal tracts, while its tail reaches forward into the anterior column. It extends along the entire length of the cord, and consists of ascending fibres which are derived from cells situated at the base of the posterior horn, and which cross to the opposite side of the cord in the anterior commissure. They can be traced upwards through the medulla oblongata and pons to the cerebellum, reaching the latter through its superior peduncles, while others pass to the cerebrum. This tract—the fibres of which do not acquire their medullary sheaths until the latter half of the eighth month—is sometimes named the *lateral sensory fasciculus*, in contradistinction to the posterior sensory fasciculi which are situated in the posterior column.

The *lateral basis bundle* constitutes the remainder of the lateral column, and the portion of it which lies next the grey matter is sometimes named the *lateral limiting zone*. It is continuous in front with the anterior basis bundle, and the two together constitute the *antero-lateral ground bundle*. It consists chiefly of longitudinal commissural fibres which arise from cells throughout the grey matter, and after a longer or shorter vertical course, re-enter and ramify in the grey matter. Some of its fibres are, however, continued upwards into the brain, under the name of the *posterior longitudinal fasciculus*.

FIG. 542. — Descending fibres in the posterior column, shown at different levels. (Testut.)

Tract of Lissauer.—The fibres of this tract form a small strand which lies in relation to the tip of the posterior horn, close to where the posterior nerve-roots enter the cord. It extends throughout the entire length of the cord, and is formed by some fibres of the posterior nerve-root, which ascend for a short distance in the tract and then enter the posterior horn of grey matter.

Descending cerebellar fibres.—In addition to the fasciculi already described in the anterior and lateral columns, fibres are found in these columns which arise from cells in the cerebellum and undergo descending degeneration when this part of the brain is extirpated. They are named the *descending cerebellar fibres*, and are disseminated throughout the peripheral part of the anterior and lateral columns, where they intermingle with the fibres of the direct pyramidal tract and peripheral part of the anterior basis bundle, as well as with the fibres of the direct cerebellar tract and tract of Gowers.

Tracts in the posterior column.—This column comprises two main tracts, viz. the tract of Goll, and the tract of Burdach. These are separated from each other in the cervical and upper dorsal regions by the postero-intermediate septum, and consist mainly of ascending fibres derived from the posterior nerve-roots.

The *tract of Goll* is wedge-shaped on transverse section, and lies next the posterior median fissure, its base being at the surface of the cord, and its apex directed towards the posterior grey commissure. It increases in size from below upwards, and consists of long, thin fibres derived from the posterior nerve-roots, which ascend as far as the medulla oblongata, where they end in the gracile nucleus.

The *tract of Burdach* is triangular on transverse section, and lies between that of Goll and the posterior cornu, its base corresponding with the surface of the cord. Its fibres, larger than those of Goll's tract, are mostly derived from the same source, viz. the posterior nerve-roots. Some ascend only for a short distance in the tract, and entering the grey matter, come into close relationship with the cells of Clarke's column; while others can be traced as far as the medulla oblongata, where they end in the gracile and cuneate nuclei.

A In the conus medullaris. B. In the lumbar region. C. In the lower dorsal region. D. In the upper dorsal region. 1. Posterior horn. 2 Posterior nerve-roots. 3. Posterior median fissure. 4. Triangular strand. 5. Oval area of Flechsig. 6. Dorsal peripheral band. 7. Descending comma tract.

Occupying the ventral part of the posterior column is an ascending strand of fibres, termed the *cornu-commissural tract*. It is somewhat triangular on transverse section, and occupies the angle between the posterior grey commissure and the posterior cornu. It is best marked in the lumbar region, but can be traced into the dorsal and cervical regions, and probably consists of commissural fibres. It has been found to preserve its integrity in certain cases of locomotor ataxia.

Descending fibres in the posterior column.—The posterior column does not consist entirely of ascending fibres, but contains some which have a descending course, and these occupy different parts of the column at different levels. In the cervical and upper dorsal regions, they appear as a comma-shaped strand (*descending comma tract*) in the outer part of the tract of Burdach, the blunt end of the comma being directed towards the posterior grey commissure; in the lower dorsal region, they form a thin band (*dorsal peripheral band*) at the posterior surface of the column; in the lumbar region, they are situated by the side of the posterior median fissure, and here appear on section as a semi-elliptical bundle, which, together with the corresponding bundle of the opposite side, forms the *oval area of Flechsig*; while in the conus medullaris they assume the form of a *triangular strand* in the postero-internal part of Goll's tract. These descending fibres partly consist of the descending branches of the posterior nerve-roots, and are partly derived from nerve-cells in the posterior cornu (fig. 542).

Roots of the spinal nerves.—As already stated, each spinal nerve arises by two roots, an anterior and a posterior, which are attached to the surface of the

FIG. 543.—A spinal nerve with its anterior and posterior roots. (Testut.)

FIG. 544.—Section of the white and grey substance of the spinal cord, through the middle of the lumbar enlargement. Magnified 14 diameters.

1. A portion of the spinal cord viewed from the left side. 2. Anterior median fissure. 3. Anterior horn. 4. Posterior horn. 5. Lateral horn, or tractus intermedio-lateralis. 6. Formatio reticularis. 7. Anterior root. 8. Posterior root, with 8′, its ganglion. 9. Spinal nerve; 9′, its posterior division.

cord opposite the corresponding horn of grey matter.

Anterior nerve-roots.—The anterior nerve-roots consist of *motor* fibres, which are the axons of the nerve-cells in the ventral part of the anterior horn. Soon after their origin from the nerve-cells, these axons acquire medullary sheaths and, passing forwards and slightly outwards, emerge on the surface of the cord in two or three irregular rows over an area which measures about three millimetres in width (fig. 543).

Posterior nerve-roots.—The posterior root of each spinal nerve comprises some six or eight fasciculi which are attached in linear series along the postero-lateral sulcus. It consists of *sensory* fibres which arise from the nerve-cells in the spinal ganglia. Each ganglion cell, at first round or oval, is elongated into two processes, an internal (axon) and an external (dendrite), and so becomes a bipolar nerve-cell. These two processes gradually undergo approximation, and finally arise from a single stem in a T-shaped manner. The internal processes of the ganglion cells grow into the cord as the posterior roots of the spinal nerves, while the external are directed towards the periphery.

The posterior nerve-roots enter the cord in two chief bundles, mesial and lateral. The *mesial* strand passes directly into the tract of Burdach: it consists of coarse fibres, which acquire their medullary sheaths by the fifth month of intra-uterine life. The *lateral* strand is sometimes divided into an intermediate and an external bundle. The intermediate bundle consists of coarse fibres, which enter the gelatinous substance of Rolando; while the external is composed of fine fibres, which assume a longitudinal direction in the tract of Lissauer: the latter do not acquire their medullary sheaths until after birth.

Having entered the cord, all the posterior nerve-roots divide into ascending and descending branches, and these in their turn give off collaterals which enter

FIG. 545.—Dorsal roots entering cord and dividing into ascending and descending branches. (Van Gehuchten.)

FIG. 546.—Formation of column of Goll. (Poirier.)

a. Stem-fibre.   *b, b.* Ascending and decending limbs of bifurcation.
c. Collateral arising from stem-fibre.

First dorsal nerve

Lumbar nerves

Sacral nerves

the grey matter (fig. 545). The descending fibres are short, and soon enter the grey matter. The ascending fibres are grouped into long, short, and intermediate: the long fibres ascend in the tracts of Goll and Burdach as far as the medulla, where they end by arborising around the cells of the gracile and cuneate nuclei; the short fibres run upwards for a distance of only five or six millimetres, and enter the grey matter; while the intermediate fibres, after a somewhat longer course, have a similar destination. All the fibres which enter the grey matter end by arborising around its nerve-cells, those of intermediate length being especially associated with the cells of Clarke's column.

The course taken by the fibres of the posterior nerve-roots has been arrived at by dividing the

Spinal cord viewed from behind. To the left, the tract of Goll is shaded. To the right, the drawing shows that the column of Goll is formed by the long fibres of the posterior roots, and that in this tract the sacral nerves lie next the mesial plane, the lumbar to their outer side, and the dorsal still more laterally.

nerve-roots between their ganglia and their entrance into the spinal cord, and by subsequently examining the degenerated areas. It has been determined that the fibres pursue an oblique course upwards, being situated at first in the outer part of Burdach's tract: higher up, they occupy the middle of this column, having been displaced inwards by the accession of other entering fibres; while still higher, they pass into, and are continued upwards in, the tract of

Goll. The upper cervical fibres do not reach Goll's tract, but are entirely confined to that of Burdach. The degeneration method proves that the localisation of these fibres is very precise : the sacral nerves lying in the inner part of Goll's tract and near its periphery ; the lumbar nerves to their outer side ; the dorsal nerves still more laterally ; while the cervical nerves are confined to Burdach's tract (fig. 546).

The development of the spinal cord is described in the section on Embryology (page 107).

THE BRAIN

GENERAL CONSIDERATIONS AND DIVISIONS

The brain, or encephalon, is contained within the cranium, and constitutes the upper, greatly expanded part of the cerebro-spinal axis. In its early embryonic condition it consists of three hollow vesicles, termed the *fore-brain*, the *mid-brain*, and the *hind-brain* ; and the parts derived from each of these can be recognised without difficulty in the adult. Thus, in the process of development the wall of the hind-brain (rhombencephalon) undergoes modification to form the medulla oblongata, the pons Varolii, and cerebellum, while its cavity is expanded to form the fourth ventricle. The mid-brain (mesencephalon) forms but a small part of the adult brain : its cavity becomes the aqueduct of Sylvius, which serves as a tubular communication between the third and fourth ventricles ; while its walls are thickened to form the corpora quadrigemina and crura cerebri, which constitute the bond of union of the fore-brain with the hind-brain. The fore-brain undergoes great modification : its anterior part (telencephalon) expands laterally in the form of two hollow vesicles which become the lateral ventricles, while their surrounding walls form the cerebral hemispheres and the commissures which connect these across the middle line ; the posterior part of the fore-brain (thalamencephalon) forms the greater part of the third ventricle, and most of the structures which bound that cavity. Further details regarding these important changes are given in the chapter on Embryology (page 109).

THE HIND-BRAIN

The hind-brain comprises the parts which occupy the posterior fossa of the cranial cavity, and lie below the tentorium cerebelli, viz. the medulla oblongata, the pons Varolii, the cerebellum, and the cavity of the fourth ventricle.

THE MEDULLA OBLONGATA

The **medulla oblongata, or bulb,** forms the lowest and smallest division of the brain ; its structure, however, is extremely complex, since it gives attachment to many of the cranial nerves, and forms the connecting link between the spinal cord below and the cerebrum and cerebellum above.

It extends from the lower margin of the pons Varolii to a plane passing transversely below the decussation of the pyramids and above the first pair of cervical nerves. This plane corresponds with the upper border of the atlas behind, and the middle of the odontoid process of the axis in front, and at this level the medulla oblongata is continuous with the spinal cord. In front of it are the basi-occiput and the upper part of the odontoid process of the axis ; from these it is separated by the membranes of the brain and certain ligaments ; in front also of its upper part are the two vertebral arteries. Its posterior surface is received into the fossa between the hemispheres of the cerebellum, and forms the lower part of the floor of the fourth ventricle. The vertebral arteries pass upwards and forwards in relation to its lateral aspects, and, reaching its anterior surface, they unite at the lower border of the pons Varolii to form the basilar artery.

The medulla oblongata (fig. 547) is pyramidal in shape, its broad extremity being directed upwards towards the pons Varolii, while its narrow, lower end is continuous with the spinal cord. It measures rather over an inch in length, rather less than an inch in breadth at its widest part, and about half an inch in thickness ;

while it weighs about a quarter of an ounce. The central canal of the spinal cord is prolonged into its lower half, and then opens into the cavity of the fourth ventricle; and the medulla may therefore be divided into a lower *closed part* containing the central canal, and an upper *open part* corresponding with the lower portion of the fourth ventricle. Its surface is marked in the median line in front and behind by an anterior and a posterior fissure, which are continuous with the corresponding fissures of the spinal cord.

The **anterior fissure** contains a fold of pia mater, and extends along the entire length of the medulla: it terminates at the lower border of the pons Varolii in a small triangular expansion, the *foramen cæcum of Vicq d'Azyr*. On separating the lips of the fissure, it is seen to be interrupted below by bundles of fibres which cross obliquely from one side to the other, and constitute the *decussation of the pyramids*. Some fibres, termed the *anterior external arcuate fibres*, emerge from the anterior fissure above this decussation and curve outwards and upwards over the lateral aspect of the medulla.

The **posterior fissure** is a narrow groove which exists only in the closed part of the medulla; it becomes gradually shallower from below upwards, and finally terminates about the middle of the medulla, where the central canal of the cord expands into the cavity of the fourth ventricle.

Fig. 547.—Medulla oblongata and pons Varolii. Anterior surface.

Middle peduncle of cerebellum

These two fissures divide the closed part of the medulla into symmetrical halves: each half presenting elongated eminences which, on surface view, are continuous with the columns of the cord. In the open part of the medulla the halves are separated by the anterior median fissure, and by a median *raphé* which extends from the bottom of the fissure to the floor of the fourth ventricle. Further, certain of the cranial nerves pass through the substance of the medulla, and are attached to its surface in series with the roots of the spinal nerves; thus, the fibres of the hypoglossal nerve represent the upward continuation of the anterior nerve-roots, and emerge in linear series from a furrow termed the *pre-olivary sulcus*, or *antero-lateral furrow*. Similarly, the spinal accessory, vagus, and glosso-pharyngeal nerves correspond with the posterior nerve-roots, and are attached to the bottom of a sulcus named the *postero-lateral furrow*. Advantage is taken of this arrangement to subdivide each half of the medulla into three areas, anterior, middle, and posterior. Although these three areas appear to be directly continuous with the corresponding columns of the cord, it must be pointed out that they do not necessarily contain the same fibres, since some of the nerve-tracts of the cord terminate in the medulla, while others alter their course in passing through it.

The **anterior area** is named the **pyramid,** and lies between the anterior median fissure and the pre-olivary sulcus, from the latter of which the roots of the hypoglossal nerve issue. Its upper extremity is slightly constricted, and between it and the pons the fibres of the sixth nerve emerge; a little below the pons it becomes enlarged and prominent, and finally tapers into the anterior column of the cord, with which, at first sight, it appears to be directly continuous.

The two pyramids constitute the great motor strands of the medulla, since they contain the motor fibres which pass from the brain to the spinal cord. When these pyramidal fibres are traced downwards, it is found that some two-thirds, or more, of them leave the anterior pyramid in successive bundles, and decussate in the anterior median fissure with corresponding bundles derived from the opposite pyramid, forming what is termed the *decussation of the pyramids*, or *motor decussation*. Having crossed the middle line, they pass down in the posterior part of the lateral column as the crossed pyramidal tract. The remaining fibres—i.e. those which occupy the outer part of the pyramid—do not cross the middle line, but are carried downwards as the direct pyramidal tract (fig. 548) into the anterior column of the same side.

The number of fibres which undergo decussation may be increased or diminished, while in some cases the decussation is complete, and in others it is totally in abeyance; and hence the direct and crossed pyramidal tracts vary inversely in size, but in this connection it must be remembered that the fibres which do not cross in the medulla undergo decussation in the cord. The mode of termination of the pyramidal tracts in the cord has already been referred to (pages 770 and 771).

The greater part of the basis bundle of the anterior column of the cord is continued upwards through the medulla oblongata as a strand, which is termed the *posterior longitudinal fasciculus*.

The **lateral area** is limited in front by the antero-lateral sulcus and the roots of the hypoglossal nerve, and behind by the roots of the spinal accessory, vagus, and glosso-pharyngeal nerves. Its upper part consists of a prominent oval mass which is named the *olivary body*, while its lower part is of the same width as the lateral column of the cord, and appears on the surface to be a direct continuation of it. As a matter of fact, only a portion of the lateral column is continued upwards into this area, for the crossed pyramidal tract passes into the pyramid of the opposite side, and the direct or ascending cerebellar tract is carried into the restiform body. The remainder of the lateral column, which consists chiefly of the basis bundle and the tract of Gowers, can be traced into this area. Most of these fibres, however, dip beneath the olivary body and disappear from the surface; but a small strand remains superficial, and passes up between the olivary body and the postero-lateral sulcus. At the upper end of this strand is a depression or fossa, in which the auditory nerve is seen.

The **olivary body** is situated on the outer side of the pyramid, from which it is separated by the antero-lateral or pre-olivary sulcus, and the fibres of the hypoglossal nerve. Behind, it is separated from the postero-lateral sulcus by the small superficial strand of the lateral column already referred to. It measures about half an inch in length, and between its upper end and the pons there is a slight depression in which the roots of the seventh nerve are attached. The external arcuate fibres wind across the

FIG. 548.—Decussation of the pyramids. Scheme showing the passage of the various tracts from cord to the medulla. (Testut.)

a. Pons Varolii. *b*. Medulla from the front. *c*. Decussation of the pyramids. *d*. Section of cervical part of cord. 1. Direct pyramidal tract (in red). 2. Crossed pyramidal tract (in red). 3. Sensory tract (tracts of Goll and Burdach) (in blue). 3'. Gracile and cuneate nuclei. 4. Antero-lateral ground bundle, in dotted line. 5. Anterior pyramid. 6. Fillet. 7. Posterior longitudinal fasciculus. 8. Gowers' tract (in blue). 9. Direct cerebellar fibres (in yellow).

lower part of the pyramid and olivary body to enter the restiform body.

The **posterior area** (fig. 549) lies behind the postero-lateral sulcus and the roots of the spinal accessory, vagus, and the glosso-pharyngeal nerves, and, like the lateral area, is divisible into a lower and an upper portion. The *lower part* is limited behind by the postero-median fissure, and consists of the tracts of Goll and Burdach: the former is now known as the *funiculus gracilis*, and the latter as the *funiculus cuneatus*. The funiculus gracilis is a narrow white band placed parallel to and along the side of the posterior median fissure, and separated from the funiculus cuneatus by the postero-intermediate furrow and septum.

The gracile and cuneate funiculi are at first vertical in direction; but when traced upwards to the lower part of the floor of the fourth ventricle, they diverge from the middle line in a V-shaped manner, and each presents an elongated swelling. That on the funiculus gracilis is named the *clava*, and is produced by a subjacent nucleus of grey matter, the *gracile nucleus* ; that on the funiculus cuneatus is termed the *cuneate tubercle*, and is likewise caused by a grey nucleus, named the *cuneate nucleus*. The fibres of the tracts of Goll and Burdach terminate by arborising around the cells in those nuclei.

A third elevation, termed the **funiculus of Rolando,** is seen in the lower part of the posterior area of the medulla. It lies on the lateral aspect of the funiculus cuneatus, and is not represented by a corresponding elevation in the cord. It is produced by the substantia gelatinosa of Rolando, which here approaches the surface. Narrow below, this funiculus gradually expands above, and terminates, about half an inch below the pons, in a tubercle, the *tubercle of Rolando*. Intervening between it and the surface of the medulla is a band of nerve-fibres formed by the lower sensory or spinal root of the fifth nerve.

Fig. 549.—Posterior surface of the medulla oblongata.

The restiform body.—The upper part of the posterior area of the medulla is occupied by the restiform body, a thick rope-like strand, which is situated between the lower part of the floor of the fourth ventricle and the roots of the ninth and tenth nerves. The restiform bodies connect the cord and medulla with the cerebellum, and are named the *inferior peduncles of the cerebellum*. As they pass upwards, they diverge from each other, and assist in forming the lower parts of the lateral boundaries of the fourth ventricle ; higher up, they are directed backwards, each passing to the corresponding cerebellar hemisphere. Near where they enter the cerebellum they are crossed by several strands of fibres, which extend inwards over the floor of the fourth ventricle, and are named the *striæ acusticæ*. At first sight, the restiform body appears to be formed by the upward continuation of the funiculus gracilis and funiculus cuneatus ; this, however, is not so, as it is probable that all the fibres of these funiculi end in the gracile and cuneate nuclei. The constitution of the restiform body will be subsequently discussed.

Internal structure of the medulla oblongata.—Although the external form of the medulla oblongata bears a certain resemblance to that of the upper part of the cord, its internal structure differs widely from that of the latter, and this for the following principal reasons : (1) certain tracts of fibres which extend from the cord to the brain, and *vice versa*, undergo a rearrangement in their passage through the medulla ; (2) other tracts which exist in the cord (e.g. those of Goll and Burdach) terminate in the medulla, whereas new strands originate in the grey matter of the medulla and pass to different parts of the brain ; (3) the grey matter, which in the cord forms a continuous H-shaped column, becomes greatly modified and subdivided in the medulla, in which also new masses of grey matter are superadded ; (4) on account of the opening-out of the central canal of the cord in the cavity of the fourth ventricle, certain parts of the grey matter, which in the cord were more or less centrally situated, are displayed in the floor of that cavity ; (5) lastly, the medulla is intimately associated with many of the cranial nerves, some arising from, and others terminating in, nuclei within its substance.

The internal structure of the medulla is best studied by examining a series of transverse and of longitudinal sections (fig. 550, 551). A short description of the course taken by the principal tracts, and of the arrangement of the grey matter, will now be given.

Pyramidal tracts.—The division of the pyramids of the medulla into direct and crossed pyramidal fibres, and the course of these tracts in the cord, have already been described. In passing to reach the lateral column of the opposite side, the fibres of the crossed pyramidal tracts extend backwards through the

FIG. 550.—Section of the medulla at the level of the decussation of the pyramids. (Testut.)

1. Anterior median fissure. 2 Posterior median fissure. 3. Motor roots. 4. Sensory roots. 5. Base of the anterior horn, from which the head (5') has been detached by the crossed pyramidal tract. 6. Decussation of the crossed pyramidal tracts. 7. Posterior horns (in blue). 8. Gracile nucleus.

FIG. 551.—Section of the medulla through the lower part of the decussation of the pyramids. (Testut.)

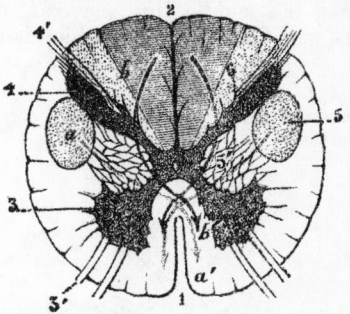

1. Anterior median fissure. 2. Posterior median fissure. 3. Anterior horn (in red), with 3', anterior root. 4. Posterior horn (in blue), with 4', posterior roots. 5. Crossed pyramidal tract. 6. Posterior column. The red arrow, a a', indicates the course the crossed pyramidal tract takes at the level of the decussation of the pyramids; the blue arrow, b b', indicates the course which the sensory fibres take.

anterior cornua, and the head of each of these horns is separated from its base. The base retains its position in relation to the ventral aspect of the central canal, and, when the latter opens into the fourth ventricle, appears in the floor of that cavity close to the middle line, where it forms the nuclei of the twelfth and sixth nerves; while above the level of the ventricle it exists as the nuclei of the third and fourth nerves in relation to the floor of the aqueduct of Sylvius. The head of the cornu is pushed outwards and forms an elongated column, the *nucleus ambiguus*, which gives origin from below upwards to the bulbar part of the spinal accessory and the motor fibres of the vagus and glossopharyngeal, and still higher to the motor fibres of the seventh and fifth nerves.

The tracts of Goll and Burdach constitute the posterior sensory fasciculi of the spinal cord. The tract of Goll is prolonged upwards as the funiculus gracilis, and expands to form the eminence of the clava: that of Burdach is carried up as the funiculus cuneatus, which enlarges to form the cuneate tubercle. The clava corresponds with a subjacent mass of grey matter, the *gracile nucleus*, and the cuneate tubercle to a second mass, the *cuneate nucleus*: these two nuclei are continuous, in front, with the central grey matter, of

FIG. 552.—Transverse section passing through the sensory decussation. (Schematic.) (Testut.)

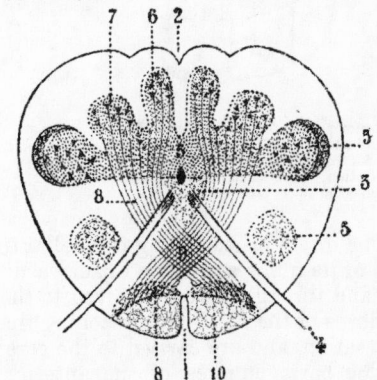

1. Anterior median fissure. 2. Posterior median fissure. 3, 3'. Head and base of anterior horn (in red). 4. Hypoglossal nerve. 6. Gracile nucleus. 7. Cuneate nucleus. 8, 8. Fillet, or sensory tract. 9. Sensory decussation. 10. Pyramidal tract.

which they may be regarded as backward projections, each being covered superficially by the fibres of the corresponding funiculus. On transverse section (fig. 552), the gracile nucleus appears as a single, more or less quadrangular mass, while the cuneate nucleus consists of two parts: a larger, somewhat triangular, *internal nucleus*, composed of small or medium sized cells, and a smaller *external nucleus* containing large cells.

The fibres of the tract of Goll and Burdach end by arborising around the cells of these nuclei, which therefore may be regarded as the nuclei of termination of the posterior sensory fasciculi (fig. 553). From the cells of these nuclei new fibres take origin : some of these are continued as the posterior external arcuate fibres into the restiform body, and through it to the cerebellum ; but most of them pass forwards through the neck of the posterior horn, thus cutting off the head from the base of the horn. Curving forwards and inwards, they reach the middle line, and there decussate with the corresponding fibres of the opposite side. Having crossed the middle line, they run upwards immediately behind the pyramidal fibres, as a flattened band, which is named the *fillet*. The decussation of these sensory fibres is situated above that of the motor fibres, and is named the *superior pyramidal* or *sensory decussation*, or *the decussation of the fillet*.

FIG. 553.—Superior termination of the posterior tracts of the spinal cord. (Testut.)

1. Posterior median fissure. 2. Tract of Goll. 3. Tract of Burdach. 4. Gracile nucleus. 5. Cuneate nucleus. 6, 6', 6''. Sensory fibres forming the fillet. 7. Sensory decussation. 8. Cerebellar fibres uncrossed (in black). 9. Cerebellar fibres crossed (in black).

The base of the posterior horn at first lies on the dorsal aspect of the central canal ; but when the latter opens into the fourth ventricle, it appears in the lateral part of the floor of that cavity. It forms the nuclei of termination of the sensory fibres of the vagus and glosso-pharyngeal, and is associated with the vestibular part of the auditory nerve and the sensory root of the seventh nerve (pars intermedia of Wrisberg). Still higher, it forms a mass of pigmented cells —the *locus cæruleus*—in which some of the sensory fibres of the fifth nerve appear to terminate. The head of the posterior horn forms a long continuous column, in which the fibres of the spinal or lower sensory root of the fifth nerve largely terminate.

The tract of Gowers, or lateral sensory fasciculus, consists of fibres which have already decussated in the anterior commissure of the cord. Some of its lower fibres appear to end in the cord, but the majority ascend in the lateral column to the medulla, where they join the fillet. The fillet, by the addition of this tract, now contains all the sensory fibres from the spinal cord ; most of these have crossed in the medulla in the sensory decussation, while the remainder (tract of Gowers) have crossed in the cord.

The direct or ascending cerebellar tract leaves the lateral area of the medulla ; most of its fibres are carried backwards to join the restiform body of the same side, and through it are conveyed to the cerebellum ; but some run upwards with the fibres of the fillet, and, reaching the posterior quadrigeminal bodies, undergo decussation, and are carried to the cerebellum through its superior peduncle.

The basis bundles of the anterior and lateral columns largely consist of intersegmental fibres, which link together the different segments of the cord ; but their upper fibres can be traced into the medulla, where, without undergoing decussation, many of them are accumulated into a strand which runs up close to the median raphé between the fillet and the floor of the fourth ventricle. This strand is named the *posterior longitudinal fasciculus*, and will be again referred to.

Grey matter of the medulla oblongata.—In addition to the gracile and cuneate nuclei, which may be regarded as backward prolongations of the central grey matter of the cord, there are several other nuclei which require consideration.

Some of these are traceable from the grey matter of the spinal cord, while others are unrepresented in the cord.

1. The **hypoglossal nucleus** is derived from the base of the anterior horn : in the lower closed part of the medulla, this is situated on the ventro-lateral aspect of the central canal ; but in the upper part it approaches the floor of the fourth ventricle, where it lies close to the middle line, under an eminence named the *trigonum hypoglossi* (fig. 554). It measures about three-quarters of an inch in length, and consists of large multipolar nerve-cells, the axons of which constitute the roots of the hypoglossal nerve. The fibres of this nerve pass forward between the anterior and lateral areas of the medulla, and emerge from the pre-olivary sulcus.

2. The **motor nucleus,** common to the **glosso-pharyngeal, vagus,** and medullary portion of the **spinal accessory** nerves, is named the *nucleus ambiguus.* It

FIG. 554.--Section of the medulla oblongata at about the middle of the olivary body.   (Schwalbe.)

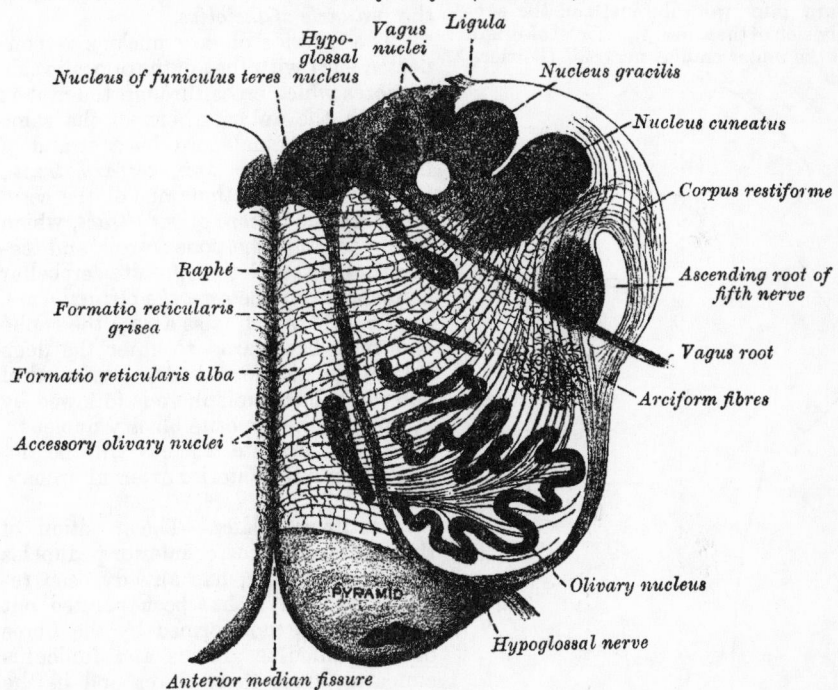

represents the head of the anterior horn, lies deeply in the formatio reticularis grisea, and extends throughout nearly the whole length of the medulla (fig. 555).

3. The **sensory nucleus,** or nucleus of termination of the sensory fibres of the **glosso-pharyngeal** and **vagus,** represents the base of the posterior horn. It measures nearly three-quarters of an inch in length, and in the lower, closed part of the medulla is situated behind the hypoglossal nucleus ; whereas in the upper, open part it lies to the outer side of that nucleus, and corresponds to an eminence, named the *trigonum vagi,* in the floor of the fourth ventricle.

4. **Nuclei of the auditory nerve.**—Towards the upper part of the medulla a considerable tract of grey matter is found immediately below that portion of the floor of the fourth ventricle which is known as the *trigonum acustici* : this forms the *inner* and *outer dorsal auditory nuclei,* in which the vestibular division of the auditory nerve terminates. Further, there are two collections of nerve-cells associated with the termination of the cochlear division of the auditory nerve, one of which is named the *ventral* or *accessory auditory nucleus,* and is situated on the ventral surface of the restiform body ; while the other lies on the outer aspect of the same body, and is termed the *tuberculum acusticum.* The auditory nuclei will be again referred to in association with the description of the pons Varolii.

5. The **olivary nuclei** are three in number on either side of the middle line, viz. the inferior olivary nucleus, and the mesial and dorsal accessory olivary nuclei. (*a*) The *inferior olivary nucleus* is the largest of these, and is situated within the olivary body. It consists of a grey folded lamina arranged in the form of an incomplete capsule, which opens internally by an aperture called the *hilum*. (*b*) The *mesial accessory olivary nucleus* lies between the inferior olivary nucleus and the anterior pyramid, and forms a curved lamina, the concavity of which is directed outwards. The fibres of the hypoglossal nerve, as they traverse the medulla, pass between the mesial accessory and the inferior olivary nuclei. (*c*) The *dorsal accessory olivary nucleus* is the smallest of the three, and appears on transverse section as a curved lamina on the dorsal aspect of the inferior olivary nucleus.

Fig. 555.—Continuation of the anterior horn of grey matter of the spinal cord into the medulla oblongata, pons, and mesencephalon. Separation of anterior horn into two columns, and the subdivision of these into the nuclei of origin of the motor cranial nerves. (Poirier.)

1. Motor oculi. 2. Pathetic. 3. Lower nucleus of fifth. 4. Abducens. 5. Facial. 6. Glosso-pharyngeal and Pneumogastric. 7. Hypoglossal. 8. Spinal accessory. 9. First cervical nerve.

These three nuclei consist of small round yellowish cells and numerous fine nerve-fibres; while entering and leaving the hilum of the inferior olivary nucleus are numerous fibres which constitute the *peduncle of the olive.*

The inferior olivary nucleus is connected—(1) with that of the opposite side by fibres which cross through the raphé; (2) with the anterior horn of the same side of the spinal cord by a strand of fibres termed the *spino-olivary tract*; (3) with the optic thalamus of the cerebrum by the *cerebro-olivary tract*, which passes through the pons Varolii and tegmentum; (4) with the opposite cerebellar hemisphere by the *cerebello-olivary tract*, the fibres of which pass across the raphé and turn backwards to enter the deep part of the restiform body. Removal of one cerebellar hemisphere is followed by atrophy of the opposite olivary nucleus.

6. The **arcuate nucleus** will be described with the anterior external arcuate fibres.

Restiform bodies.—The position of the restiform bodies, or inferior peduncles of the cerebellum, has already been referred to, and it has been pointed out that they are not formed by the fibres of the funiculus gracilis and funiculus cuneatus, since these fibres end in the gracile and cuneate nuclei. The restiform body comprises: (1) the direct or ascending cerebellar tract from the lateral column of the cord; (2) descending cerebellar fibres, many of which are disseminated throughout the peripheral part of the anterior and lateral columns of the cord, while others are conducted to the motor nuclei of the cranial nerves; and (3) the arcuate fibres, which are arranged in three sets, viz. the internal arcuate fibres, and the anterior and posterior external arcuate fibres.

The *internal arcuate fibres* form the deeper and larger part of the restiform body. They decussate in the middle line of the medulla, and having reached the opposite side, terminate partly in the gracile and cuneate nuclei, while many of them enter the hilum of the inferior olivary nucleus, and constitute the cerebello-olivary tract already described (fig. 556).

The *anterior external arcuate fibres* vary as to their prominence in different cases: in some they form an almost continuous layer covering the pyramid and olivary body, while in others they are barely visible on the surface. They arise from the cells of the gracile and cuneate nuclei, and passing forwards through the formatio reticularis, decussate in the middle line. Most of them reach the

FIG. 556.—Diagram showing the course of the arciform fibres. (Testut.)

1. Medulla, anterior surface. 2. Anterior median fissure. 3. Fourth ventricle. 4. Olivary body, with the accessory olivary nuclei. 5. Gracile nucleus. 6. Cuneate nucleus. 7. Trigeminal. 8. Inferior cerebellar peduncles, seen from in front. 9. Posterior external arcuate fibres. 10. Anterior external arcuate fibres. 11. Internal arcuate fibres. 12. Peduncle of olivary body. 13. Arcuate nucleus. 14. Pneumogastric. 15. Hypoglossal.

surface by way of the anterior median fissure, and arch outwards and backwards over the pyramid. Reinforced by others which emerge between the pyramid and olivary body, they pass backwards over the olivary body and lateral area of the medulla, and enter the outer part of the restiform body. They thus connect the cerebellum with the gracile and cuneate nuclei of the opposite side. As the fibres arch across the pyramid, they enclose a small nucleus which lies in front and to the inner side of the pyramid. This is named the *arcuate* or *pre-pyramidal nucleus*, and is directly continuous above with the nuclei pontis in the pons Varolii : it contains small fusiform cells, around which some of the arcuate fibres terminate, and from which others arise.

The *posterior external arcuate fibres* also take origin in the gracile and cuneate nuclei ; they do not undergo decussation, but pass to the restiform body of the same side.

Formatio reticularis (fig. 557). This term is applied to the coarse reticulum which occupies the anterior and lateral areas of the medulla, and is seen when transverse sections are examined. It is situated behind the pyramid and

FIG. 557.—The formatio reticularis of the medulla, shown by a horizontal section passing through the middle of the olivary body. (Testut.)

1. Anterior median fissure. 2. Fourth ventricle. 3. Formatio reticularis, with 3', its internal part (reticularis alba), and 3'', its external part (reticularis grisea). 4. Raphé. 5. Pyramid. 6. Fillet. 7. Inferior olivary nucleus with the two accessory olivary nuclei. 8. Hypoglossal nerve, with 8', its nucleus of origin. 9. Vagus nerve, with 9', its nucleus of termination. 10. External dorsal auditory nucleus. 11. Nucleus ambiguus (nucleus of origin of motor fibres of glosso-pharyngeal, vagus, and bulbar portion of spinal accessory). 12. Gracile nucleus. 13. Cuneate nucleus. 14. Head of posterior cornu, with 14', the lower sensory root of fifth nerve. 15. Fasciculus solitarius. 16. Anterior external arcuate fibres, with 16', the arcuate nucleus. 17. Nucleus lateralis. 18. Nucleus of fasciculus teres. 19. Ligula.

olivary body, extending laterally as far as the restiform bodies, and dorsally to within a short distance of the floor of the fourth ventricle. The reticulum is caused by the intersection of bundles of fibres running at right angles to each other, some being longitudinal, others more or less transverse in direction. The formatio reticularis presents a different appearance in the anterior area from what it does in the lateral area : in the former, there is almost an entire absence of nerve-cells, and hence this part is known as the *reticularis alba*; whereas nerve-cells are numerous in the lateral area, and as a consequence it presents a grey appearance, and is termed the *reticularis grisea*.

In the substance of the formatio reticularis are two small nuclei of grey matter: one is situated near the dorsal aspect of the hilum of the inferior olivary nucleus, and is named the *inferior central nucleus*, or *nucleus of Roller*; the other lies between the olivary body and the spinal root of the fifth nerve, and is termed the *nucleus lateralis*.

In the reticularis alba the longitudinal fibres form two well-defined strands, viz. : (1) the *fillet*, which lies close to the raphé immediately behind the fibres of the pyramid ; and (2) the *posterior longitudinal fasciculus*, continued upwards from the antero-lateral ground bundle of the spinal cord, and which, in the upper part of the medulla, lies between the fillet and the grey matter in the floor of the fourth ventricle. The longitudinal fibres in the reticularis grisea are derived from the lateral column of the cord after the crossed pyramidal tract has passed over to the opposite side, and the direct cerebellar tract has entered the restiform body. They form indeterminate fibres, with the exception of a bundle named the *fasciculus solitarius*, which is made up of fibres of the vagus and glosso-pharyngeal nerves. The transverse fibres of the formatio reticularis are the arcuate fibres already described (page 782).

THE PONS VAROLII

The **pons Varolii,** or antero-superior part of the hind-brain, is situated above the medulla oblongata, below the crura cerebri, and in front of the cerebellum. Somewhat cuboidal in shape, its vertical and antero-posterior diameters each measure about an inch, while its transverse diameter is about an inch and a half. From its superior surface the crura cerebri emerge, one on either side of the middle line, while the central part of this surface forms the posterior boundary of the interpeduncular space at the base of the cerebrum. The anterior part of its inferior surface is separated from the medulla oblongata by a furrow in which the sixth, seventh, and eighth cranial nerves make their appearance.

Its *ventral* or *anterior surface* is very prominent, markedly convex from side to side, less so from above downwards. It consists of transverse fibres which arch like a bridge across the middle line, and on either side are gathered into a compact mass forming the middle peduncle of the cerebellum. It rests upon the posterior surface of the basi-sphenoid, and is limited above and below by well-defined borders. It presents, in the middle line, a furrow for the lodgment of the basilar artery : this furrow is bounded on either side by an eminence—the *pyramidal eminence*—caused by the passage of the pyramidal fibres downwards through the substance of the pons. Outside these eminences, near the upper border of the pons, the fifth nerves make their exit, each consisting of a smaller, internal, motor root, and a larger, external, sensory root ; a vertical line, drawn on either side immediately outside the fifth nerve, may be taken as the boundary between the ventral surface of the pons and the middle peduncle of the cerebellum.

The posterior surface is hidden by the cerebellum ; its lateral portions are continuous with the superior cerebellar peduncles, while its middle part is free, triangular in shape, and forms the upper part of the floor of the fourth ventricle.

Structure of the pons (fig. 558).—Transverse sections of the pons Varolii show that it is composed of two parts which differ from each other in appearance and structure : thus, the anterior or ventral portion consists for the most part of fibres arranged in transverse and longitudinal bundles, together with a small amount of grey matter ; while the posterior or dorsal portion is a continuation of the reticular formation of the medulla oblongata, and is called the **tegmental**

portion, as most of its constituents are continued into the tegmentum of the crus cerebri.

The *anterior* or *ventral part* of the pons consists of—(*a*) superficial and deep transverse fibres, (*b*) longitudinal fibres, and (*c*) some small nuclei of grey matter, termed the nuclei pontis.

The *superficial transverse fibres* consist of a rather thick layer on the ventral surface of the pons, and are collected into a large rounded bundle on each side

FIG. 558.—Vertical transverse section of the pons, at its upper part. (Testut.)

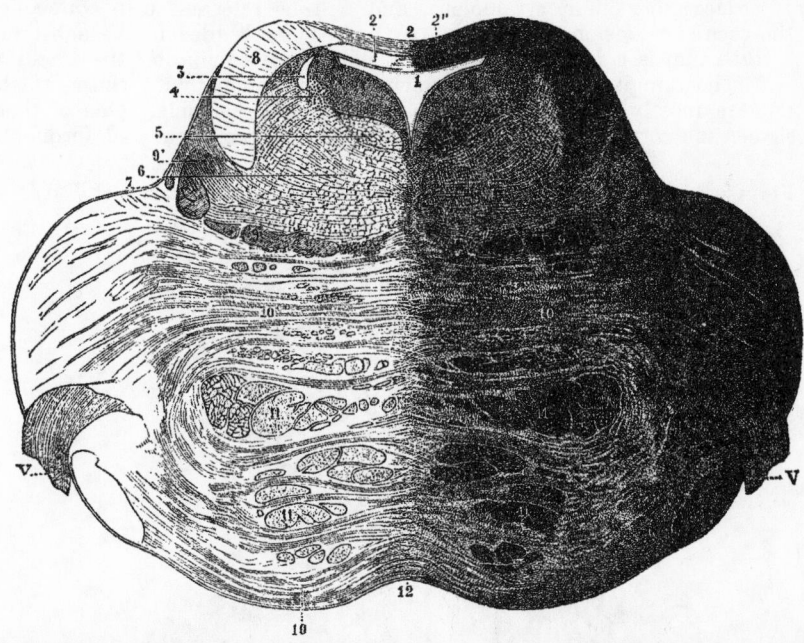

1. Fourth ventricle; its ependyma in yellow. 2. Valve of Vieussens, with 2′, its white stratum (anterior medullary velum), and 2″, its grey stratum (lingula). 3. Superior or Sylvian root of trigeminal. 4. Nerve-cells associated with this root. 5. Posterior longitudinal fasciculus. 6. Formatio reticularis. 7. Lateral sulcus. 8. Section of superior cerebellar peduncle. 9. Mesial fillet. 9′. Lateral fillet. 10, 10. Transverse fibres of pons. 11, 11. Pyramidal fibres. 12. Raphé. V. Trigeminal.

of the middle line. This bundle, with the addition of some transverse fibres from the deeper part of the pons, forms the middle peduncle of the corresponding half of the cerebellum (fig. 559).

The *deep transverse fibres* partly intersect and partly lie on the dorsal aspect of the pyramidal fibres. They course to the lateral border of the pons, and assist the superficial transverse fibres in forming the middle peduncle of the cerebellum. The further connections of the transverse fibres will be discussed with the anatomy of the cerebellum.

The *longitudinal fibres* are derived from the crura cerebri, and enter the upper surface of the pons. They stream downwards on either side of the middle line in larger or smaller bundles, separated from each other by the deeper transverse fibres of the pons; these longitudinal bundles cause a forward projection of the overlying superficial transverse fibres, and thus give rise to the pyramidal eminences on the ventral surface. Many of these fibres end in the pons, some in the nuclei pontis, and others, after decussating, in the motor nuclei of the fifth, sixth, seventh, and twelfth nerves; but most of them are carried through the pons, and at its lower surface are collected into the pyramids of the medulla. The fibres which decussate and end in the motor nuclei of the cranial nerves are derived from what is termed the geniculate bundle of the internal capsule, and they bear the same relation to the motor cells of the cranial nerves that the pyramidal fibres bear to the motor cells in the anterior horn of the cord.

The *nuclei pontis* are continuous with the arcuate nucleus in the medulla, and consist of small groups of multipolar nerve-cells which are scattered between the bundles of transverse fibres.

3 E

The *tegmental* or *dorsal portion* of the pons is chiefly composed of an upward continuation of the reticular formation and grey matter of the medulla. It is subdivided into lateral halves by a median raphé; but this does not extend into the ventral part of the pons, being there obliterated by the transverse fibres.

The tegmental portion of the pons consists of transverse and longitudinal fibres, and also contains important grey nuclei. The transverse fibres in the lower part of the pons are collected into a distinct strand, named the *trapezium* or *corpus trapezoides*. This consists of fibres which arise from the cells of the ventral or accessory auditory nucleus, and will be referred to in connection with the cochlear division of the auditory nerve. Embedded in the substance of the trapezium is a collection of nerve-cells, which constitutes the *trapezoid nucleus*. The longitudinal fibres, which are continuous with those of the medulla, are mostly collected into two bundles on either side. One of these lies between the corpus trapezoides and the reticular formation, and forms the

FIG. 559.—Superficial dissection of the medulla oblongata and pons. (Ellis.)

Superficial fibres of pons reflected

upward prolongation of the *fillet*; the second is situated near the floor of the fourth ventricle, and is the *posterior longitudinal fasciculus*. Other longitudinal fibres, which are more diffusely distributed, arise from the cells of the grey matter of the pons.

The greater part of the dorsal portion of the pons is a continuation upward of the formatio reticularis of the medulla, and, like it, presents the appearance of a network, in the meshes of which are numerous nerve-cells. Besides these scattered nerve-cells, there are some important masses of grey matter which require mention.

1. The *superior olivary nucleus* is a small isolated collection of grey matter situated on the dorsal surface of the outer part of the trapezium. Rudimentary in man, but well developed in certain animals, it exhibits the same structure as the inferior olivary nucleus, and is situated immediately above it. Some of the fibres of the corpus trapezoides end by arborising around the cells of this nucleus, while others arise from these cells. The other collection of nerve-cells

in the formatio reticularis of the pons form nuclei of origin or termination of certain cranial nerves.

2. *Nuclei of the fifth nerve.*—The nuclei of the fifth nerve in the pons are two in number : a motor and a sensory. The *motor nucleus* is situated in the upper part of the pons, close under its dorsal surface and along the line of the lateral margin of the fourth ventricle. The axis-cylinder processes of its cells form a portion of the motor root of the fifth nerve : the remaining fibres of the motor root of this nerve are formed by a tract which arises from the grey matter of the floor of the Sylvian aqueduct, and hence is named the *Sylvian* or *mesencephalic root.* The *sensory nucleus* lies external to the motor one, and beneath the superior peduncle of the cerebellum, which forms the lateral boundary of the upper half of the fourth ventricle. Some of the sensory fibres of the fifth nerve terminate in this nucleus ; but the greater number descend, under the name of the lower sensory or spinal root, to end in the substantia gelatinosa of Rolando. The roots, motor and sensory, of the fifth nerve pass through the substance of the pons and emerge near the upper margin of its ventral surface.

3. The *nucleus of the sixth nerve* is a circular mass of grey matter situated close to the floor of the fourth ventricle, above the striæ acusticæ and subjacent to the eminentia teres : it lies a little external to the ascending part of the seventh nerve. The fibres of the sixth nerve pass forward through the entire thickness of the pons on the mesial side of the superior olivary nucleus, and between the outer bundles of the pyramidal fibres, and emerge in the furrow between the lower border of the pons and the pyramid of the medulla.

4. The *nucleus of the facial nerve* is situated deeply in the reticular formation of the pons on the dorsal aspect of the superior olivary nucleus, and the roots of the nerve derived from it pursue a remarkably tortuous course in the substance of the pons. At first they pass backwards and inwards until they reach the floor of the fourth ventricle, close to the median groove, where they are collected into a round bundle. This passes upwards and forwards, producing an elevation (fasciculus teres) in the floor of the ventricle, and then takes a sharp bend, and arches outwards through the substance of the pons to emerge at its lower border in the interval between the olivary and restiform bodies of the medulla.

5. The *nuclei of the auditory nerve* are : (*a*) the *inner dorsal nucleus*, which corresponds to the trigonum acustici in the floor of the fourth ventricle : it is principally situated in the medulla, but extends above the level of the striæ acusticæ into the upper half of the floor of the fourth ventricle ; (*b*) the *outer dorsal nucleus* (*nucleus of Deiters*), which lies in the floor of the lateral angle of the fourth ventricle, and of which the external part is sometimes named the *nucleus of Bechterew* ; (*c*) the *ventral* or *accessory nucleus*, which is situated on the ventral aspect of the restiform body, between the cochlear and vestibular divisions of the auditory nerve, the latter being to its inner side ; and (*d*) the *tuberculum acusticum*, which lies on the outer side of the restiform body. The two divisions of the auditory nerve appear in the groove at the lower border of the pons, and in front of the restiform body.

THE CEREBELLUM

The **cerebellum** constitutes the largest part of the hind-brain, and is contained in the inferior occipital fossæ. It lies behind the pons Varolii and medulla oblongata, while between its central portion and these structures is the cavity of the fourth ventricle. Above it, is a fold of dura mater, named the tentorium cerebelli, which separates it from the tentorial surface of the cerebrum. It is somewhat oval in form, but constricted mesially and flattened from above downwards, its greatest diameter being from side to side. It measures from three and a half to four inches transversely, two to two and a half inches from before backwards, and is about two inches thick in the centre. Its surface is not convoluted like that of the cerebrum, but is traversed by numerous curved furrows or sulci, which vary in depth at different parts, and separate the laminæ of which it is composed.

Lobes of the cerebellum.—The cerebellum consists of three parts, a median and two lateral, which are continuous with each other, and are substantially the same in structure. The median portion is constricted, and is called the *worm* or *vermis*, from the annulated appearance which it presents owing to transverse ridges and furrows upon it ; the lateral expanded portions are named the *hemispheres*. On the upper surface of the cerebellum the worm is elevated above the level of the hemispheres, but on the under surface it is sunk almost out of sight in the bottom of a deep depression between the hemispheres ; this depression is called the *vallecula cerebelli*, and lodges the medulla oblongata. The part of the worm which lies on the upper surface of the cerebellum is named the *superior vermis* ; that on the lower surface, the *inferior vermis*. Below and behind, the hemispheres are separated by a deep notch (*posterior cerebellar notch* or *incisura marsupialis*), and in front by a broader shallower notch (*anterior cerebellar notch* or *incisura semilunaris*). The anterior notch lies close to the pons and upper part of the medulla, and its upper edge encircles the lower pair of corpora quadrigemina and the superior peduncles of the cerebellum. The posterior notch contains the upper part of the falx cerebelli. The cerebellum is characterised by its laminated or foliated appearance ; it is marked by deep, somewhat curved fissures, which lie close together, and extend for a considerable distance into its substance, dividing it into a series of layers or leaves. Upon making sections across the laminæ, it will be seen that

FIG. 560.—Upper surface of the cerebellum.   (Schäfer.)

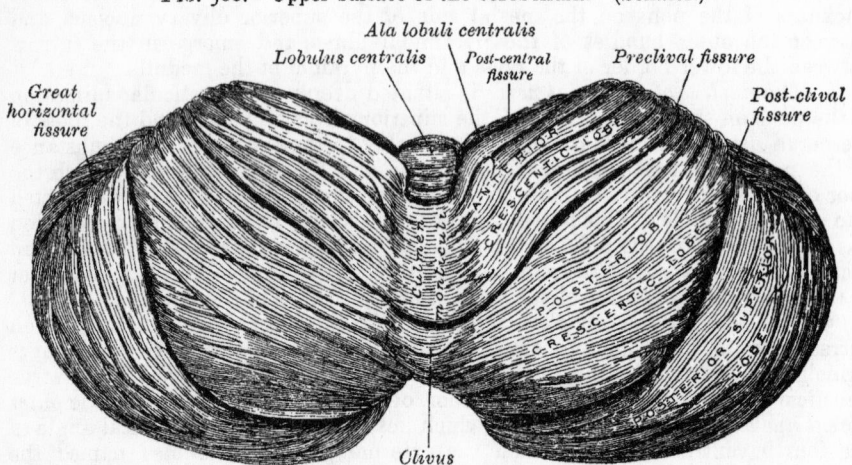

the folia, though differing in appearance from the convolutions of the cerebrum, are homologous with them, inasmuch as they consist of a central white substance with a covering or cortex of grey matter.

The largest and deepest fissure is named the *great horizontal fissure*. It commences in front at the pons, and passes horizontally round the free margin of the hemisphere to the middle line behind, and divides the cerebellum into an upper and lower portion. Several secondary but deep fissures separate the cerebellum into lobes, and these are further subdivided by shallower sulci, which separate the individual folia or laminæ from each other.

The cerebellum is connected to the cerebrum, pons, and medulla by three pairs of peduncles : a superior pair connect it with the cerebrum ; a middle pair with the pons ; and an inferior pair with the medulla.

Upper Surface of Cerebellum (fig. 560).—The superior surface of the cerebellum is elevated in the middle line and sloped towards its circumference, its hemispheres being connected together by the superior vermis ; this assumes the form of a raised median ridge, which is most prominent in front, but is not sharply defined from the hemispheres. This surface is traversed by four curved fissures, which are named, from their relations to two prominent lobes (central lobe and clivus) of the worm, (1) the precentral fissure, (2) the post-central fissure, (3) the preclival fissure, and (4) the post-clival fissure. These four fissures extend into the hemispheres and divide the entire upper surface of the cerebellum into five lobes ; but the portion of the lobe in the worm has received a different name from that in the hemispheres, although they are continuous with each other. The five lobes in the worm are named, from before backwards : (1) the *lingula*, (2) the *lobus centralis*, (3) the *culmen monticuli*, (4) the *clivus monticuli*, and (5) the *folium cacuminis*; and the corresponding lobes in the hemisphere are termed: (1) the *frænulum*, (2) the *ala lobuli centralis*, (3) *anterior crescentic*, (4) *posterior crescentic*, and (5) *posterior superior*. The arrangement of these lobes and fissures will be understood by a reference to the accompanying scheme, in which they are named in order from before backwards.

UPPER SURFACE OF CEREBELLUM

Hemisphere	Superior vermis	Hemisphere
Frænulum.	Lingula.	Frænulum.
	Precentral fissure	
Ala lobuli centralis.	Lobus centralis.	Ala lobuli centralis.
	Post-central fissure	
Anterior crescentic lobe.	Culmen monticuli.	Anterior crescentic lobe.
	Preclival fissure	
Posterior crescentic lobe.	Clivus monticuli.	Posterior crescentic lobe.
	Post-clival fissure	
Postero-superior lobe.	Folium cacuminis.	Postero-superior lobe.

The **lingula** is a small tongue-shaped process, consisting of four or five folia ; it lies in front of the lobus centralis, and is concealed by it. It is in relation, anteriorly, with the valve of Vieussens, on the dorsal surface of which it rests and with which it is connected, its white matter being continuous with that of the valve. On either side, the lingula gradually shades off, and is prolonged only for a short distance into the hemispheres, where it forms the **frænulum**. This does not stretch beyond the superior penduncle of the cerebellum, over which it lies.

The lobus centralis.—The lobus centralis is a small square lobe, situated in the anterior notch. It overlaps the lingula, and is in turn partially concealed by the culmen monticuli ; laterally, it extends along the upper and anterior part of each hemisphere, where it forms a wing-like prolongation, the **ala lobuli centralis.**

The **culmen monticuli** is much larger than the two lobes just described, and constitutes, with the succeeding lobe, the clivus, the bulk of the upper worm. In front, it partially overlaps and obscures the lobus centralis ; and behind, it is separated from the clivus by the *preclival fissure*. It forms the most prominent part of the upper worm, and is marked on its surface by three or four secondary fissures, which divide it into smaller lobules. Laterally, it is continuous with the **anterior crescentic lobes** of the hemispheres, which are distinctly differentiated from the posterior crescentic lobes by the preclival fissure. The culmen monticuli and the two anterior crescentic lobes form the **lobus culminis.**

The **clivus monticuli** is of considerable size, and consists of a group of laminæ which are separated in front from the culmen by the preclival fissure, but appear behind to be almost continuous with the folium cacuminis ; it will be found, however, on careful examination, to be separated from it by a well-defined fissure, the *post-clival fissure*. Laterally, this lobe is continued into the hemispheres as the **posterior crescentic lobes**, which are somewhat semilunar in shape, and form, with the anterior crescentic lobes, the greater part of the upper surface of the hemispheres. The two posterior crescentic lobes and the intervening clivus monticuli constitute the **lobus clivi.**

The **folium cacuminis** is a short, narrow, concealed band at the posterior extremity of the worm, consisting apparently of a single folium, but in reality marked on its upper and under surfaces by secondary fissures. Laterally, it expands in either hemisphere into a considerable lobe, which is semilunar in shape, and is situated at the postero-superior part of the hemisphere, and bounded below by the great horizontal fissure. It is named the **postero-superior lobe**, and occupies the posterior third of the upper surface of the hemisphere, forming its rounded postero-lateral border. The postero-superior lobes and the folium cacuminis form the **lobus cacuminis**.

The **Under Surface of the Cerebellum** (fig. 561) presents, in the middle line, the *inferior vermis*, buried in the vallecula, and separated from the hemisphere

FIG. 561.—Under surface of the cerebellum. (Schäfer.)

on either side by a deep groove, the *sulcus valleculæ*. Here, as on the upper surface, there are deep fissures, dividing it into separate segments or lobes ; but the arrangement is more complicated, and the relation of the segments of the worm to those of the hemisphere is less clearly marked. The fissures are three

FIG. 562.—Diagram showing fissures on under surface of the cerebellum.

F. Flocculus. N. Nodule. U. Uvula. PY. Pyramid. AM. Amygdala. BIVENT. Biventral lobe.

in number, but are not so regularly disposed as those on the upper surface (fig. 562). They are named, from their relation to the pyramid and nodule, two of the lobes of the inferior worm: (1) *post-nodular*, (2) *prepyramidal*, and (3) *post-pyramidal* fissures. The part of the worm in front of the post-nodular fissure is termed the *nodule* ; and the lobule in the hemisphere corresponding with this is the *flocculus*. The next lobe is situated between the post-nodular and prepyramidal fissures. In the vermiform process, it is known as the *uvula*, and its lateral expansion in the hemisphere is named the *amygdala* or *tonsil*. The lobule of the worm between the pre- and post-pyramidal fissures is the *pyramid*, and its corresponding part in the hemisphere is the *biventral* lobe. Finally, behind the post-pyramidal fissure in the worm is a small lobe, the *tuber valvulæ* or *tuber posticum*; this, in the hemisphere, expands into a large lobe, which occupies at least two-thirds of the inferior surface of the cerebellum, and is subdivided into two by a secondary fissure, named the *post-gracile* fissure. The anterior of the two subdivisions is named the *slender* lobe (lobus gracilis); and the posterior, the *inferior semilunar* or *postero-inferior* lobe. These fissures and lobes are here arranged, from before backwards, in a schematic form.

UNDER SURFACE OF CEREBELLUM

Hemisphere	*Inferior vermis*	*Hemisphere*
Flocculus.	Nodule.	Flocculus.
	Post-nodular fissure	
Amygdala.	Uvula.	Amygdala.
	Prepyramidal fissure	
Biventral lobe.	Pyramid.	Biventral lobe.
	Post-pyramidal fissure	
Lobus gracilis.		Lobus gracilis.
Post-gracile fissure. }	Tuber valvulæ.	{ *Post-gracile fissure.*
Inferior semilunar lobe.		Inferior semilunar lobe.

The chief fissures of the under surface, as stated above, are three in number, but are not so regularly disposed as on the upper surface. (1) The *post-nodular fissure* courses transversely across the worm, separating the nodule in front from the uvula behind. When this fissure reaches the hemispheres, it passes in front of the amygdala, and then crosses between the flocculus in front and the biventral lobe behind, and joins the anterior end of the great horizontal fissure. (2) The *prepyramidal fissure* crosses the worm between the uvula in front and the pyramid behind, then curves laterally behind the amygdala, and passes forwards along the outer border of this lobe, between it and the biventral lobe, to join the post-nodular sulcus. (3) The *post-pyramidal fissure* passes across the worm behind the pyramid and in front of the tuber valvulæ, and, in the hemispheres, courses behind the amygdala and biventral lobes, and then along the outer border of the biventral lobe to the post-nodular sulcus. It cuts off at least two-thirds of the inferior surface of the hemisphere. From it a secondary sulcus springs, which, coursing forwards and outwards, divides this surface into two parts and falls into the great horizontal fissure. This sulcus is termed the *post-gracile fissure*.

THE LOBES OF THE INFERIOR SURFACE OF THE CEREBELLUM

The Nodule and Flocculus.—The nodule is a distinct prominence, forming the anterior extremity of the inferior worm. It abuts against the roof of the fourth ventricle, and can only be distinctly seen after the cerebellum has been separated from the medulla and pons. On each side of the nodule is a thin layer of white substance, named the *inferior medullary velum*. It is semilunar in form, its convex border being continuous with the white substance of the cerebellum; it extends on either side as far as the flocculus, which it connects with the nodule. The flocculus is a prominent, irregular lobule, situated just in front of the biventral lobe, between it and the middle peduncle of the cerebellum. It is subdivided into a few small laminæ, and is connected to the inferior medullary velum by its central white core. The flocculi, together with the inferior medullary velum and nodule, constitute the **lobus noduli**.

The **Uvula** and **Amygdalæ.**—The uvula forms a considerable portion of the inferior worm; it is separated on either side from the amygdala by the *sulcus valleculæ,* at the bottom of which it is connected to the amygdala by a ridge of grey matter, indented on its surface by shallow furrows, and hence called the *furrowed band.* It is marked on its surface by three or four transverse fissures. The amygdalæ, or tonsils, are rounded masses, situated in the lateral hemispheres. Each lies in a deep fossa, termed the *bird's nest* (nidus avis), between the uvula and the biventral lobe. The uvula and tonsils form the **lobus uvulæ.**

The **Pyramid** and **Biventral Lobes** constitute the **lobus pyramidis.** The pyramid is a conical projection, forming the largest prominence of the lower worm. It is separated from the hemispheres by the sulcus valleculæ, across which it is connected to the biventral lobe by an indistinct grey band, analogous to the furrowed band already described. The biventral lobe is triangular in shape; its apex points inwards and backwards, and is joined by the connecting band to the pyramid. The external border is separated from the slender lobe by the post-pyramidal fissure. The base is directed forwards, and is on a line with the anterior border of the amygdala, and is separated from the flocculus by the post-nodular fissure.

The **Tuber Valvulæ** and the **Postero-inferior Lobes** collectively form the **lobus tuberis.** The tuber valvulæ is the most posterior division of the inferior worm. It is of small size, and laterally spreads out into the large postero-inferior lobes of the hemispheres. These lobes, as stated above, comprise at least two-thirds of the inferior surface of the hemisphere, and are divided into two by the *post-gracile fissure.* The anterior part is named the *slender lobe* (lobus gracilis); and the posterior, the *inferior semilunar lobe.* Each of them is traversed by a curved fissure; that in the slender lobe being well marked, and termed the *intragracile fissure.*

INTERNAL STRUCTURE OF THE CEREBELLUM

The cerebellum consists of white and grey matter.

The White Matter.—If a sagittal section (fig. 563) is made through either hemisphere, the interior will be found to consist of a central stem of white matter, in the interior of which is a grey mass, the *corpus dentatum.* From the surface of this central stem a series of plates of medullary matter are detached, which, covered with grey matter, form the laminæ. In consequence of the main branches from the central stem dividing and subdividing, the section presents a characteristic appearance, which is named the *arbor vitæ.* If a vertical section is made in the median plane of the cerebellum, it will be found that the central stem divides into a vertical and a horizontal branch. The *vertical* branch passes upwards to the culmen, where it subdivides freely, one of its ramifications passing forwards and upwards to the central lobe. The *horizontal* branch passes backwards to the folium cacuminis, greatly diminished in size, in consequence of having given off large secondary branches: one, from its upper surface, ascends to the clivus; the others descend, and enter the lobes in the inferior vermiform process, viz. the tuber valvulæ, the pyramid, the uvula, and the nodule. It is not necessary to describe in detail the various divisions of the white matter, as they correspond to the lobes on the surface.

The white matter of the cerebellum includes two sets of nerve-fibres: (1) the *peduncular fibres,* continuous with those of the peduncles of the cerebellum; (2) the fibres proper (*fibræ propriæ*) of the cerebellum itself.

The Peduncles of the Cerebellum.—From the anterior part of each hemisphere arise three large processes or peduncles—superior, middle, and inferior—by which the cerebellum is connected with the rest of the encephalon.

The **superior peduncles** form at first the upper lateral boundaries of the fourth ventricle, but as they extend forwards and upwards they converge on the dorsal aspect of the ventricle, and thus assist to roof it in. They may be traced as far as the corpora quadrigemina, under which they pass. They enter the upper and mesial part of the medullary substance of the hemispheres, beneath the ala lobuli centralis and the frænulum, and pass, to a great extent, into the interior of the corpus dentatum, though some of their fibres wind round it and reach the grey cortical matter, especially that of the inferior surface.

The fibres of the superior peduncles mainly emerge from the hilum of the corpus dentatum; others come from the cortex, and probably also from the smaller nuclei in the central white substance. The majority of the fibres decussate with those of the opposite peduncle below the corpora quadrigemina, and pass to the red nucleus of the tegmentum; they will be again referred to in connection with the description of the mid-brain. Fibres also connect the spinal cord with the cerebellum through its superior peduncles; these are chiefly derived from the antero-lateral ascending cerebellar tract of Gowers.

The **middle peduncles** are the largest of the three pairs. They consist of curved fibres, which, as already described, comprise most of the transverse fibres of the pons. They enter the cerebellum between the margins of the great horizontal fissure at the anterior notch, and the fibres spread out in all directions; some passing to the upper, and some to the lower part of the hemisphere, while others enter its middle region. The fibres of the middle peduncle are of two kinds: centrifugal and centripetal. The *centrifugal fibres* are the axons of the cells of Purkinje in the cerebellar cortex, and are characterised by their large size, and by the fact that they acquire their medullary sheaths early.

FIG. 563.—Sagittal section of the cerebellum, near the point of junction of the worm with the hemisphere. (Schäfer.)

They decussate in the pons, and having reached the opposite side, divide into ascending and descending branches which end in the nuclei pontis and in the motor nuclei of the cranial nerves. The *centripetal fibres* are very numerous, and are the axons of the cells of the nuclei pontis. Some pass into the middle peduncle of the same side, but the majority into that of the opposite side; they are conducted to the cerebellar cortex. It is probable that all the fibres of the middle peduncle are interrupted in the pons, and that none pass continuously from one cerebellar hemisphere to the other.

The **inferior peduncles** connect the cerebellum with the medulla oblongata. As the restiform bodies of the latter, they are directed upwards and outwards, forming part of the lateral wall of the fourth ventricle, and enter the cerebellum between the middle and superior peduncles; passing upwards, they end in the grey cortex of the upper surface of the hemisphere, some being prolonged into the white matter of the superior vermiform process. The inferior peduncles contain fibres derived from: (1) the direct cerebellar tract of the spinal cord; (2) the gracile and cuneate nuclei (crossed and uncrossed fibres); (3) the opposite olivary body of the medulla; (4) fibres which pass to join the posterior longitudinal bundle; (5) the descending cerebellar fibres, which pass down the restiform body and antero-lateral column of the cord, to terminate around the cells in the anterior horn of the cord.

The **valve of Vieussens, or superior medullary velum.**—Stretched across from one superior peduncle to the other, is a thin, transparent lamina of white matter, the *valve of Vieussens*; on to the dorsal surface of its lower half the folia of the lingula are prolonged. It forms, together with the superior peduncles, the roof of the upper part of the fourth ventricle, and is continuous with the central white stem of the cerebellum. It is narrow above, where it passes beneath the corpora quadrigemina, and broader below, at its connection with the white substance of the superior worm. A slightly elevated ridge, the *frœnulum veli*, descends upon the upper part of the valve from between the lower corpora quadrigemina, and on either side of this the fourth nerve emerges.

The **inferior medullary velum** is a thin layer of white substance, prolonged from the white centre of the medulla above and on either side of the nodule, which assists in forming a part of the roof of the fourth ventricle. Somewhat semilunar in shape, it is continuous with the white substance of the cerebellum

FIG. 564.—Diagrammatic representation of the cells of the cerebellum. (Modified from Foster's ' Physiology.')

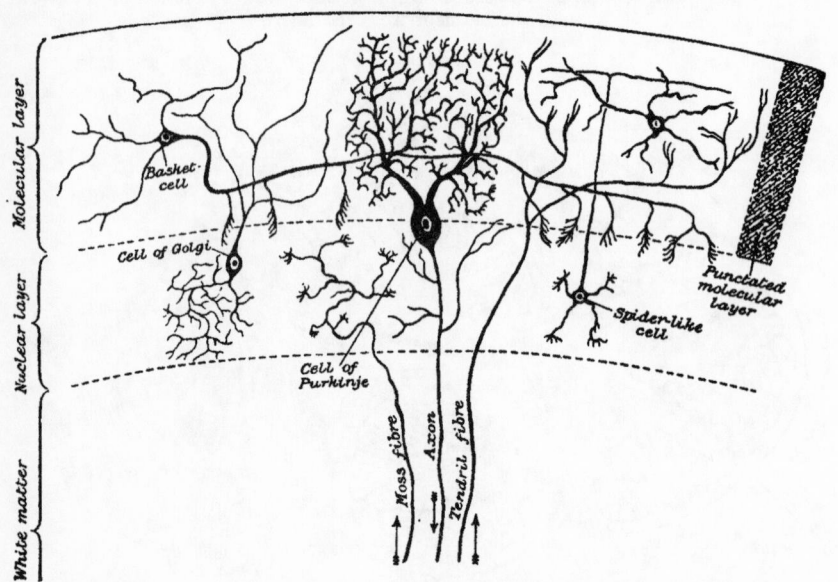

by its convex edge, while its thin concave margin is apparently free. In reality, however, it is continuous with the epithelium of the ventricle, which is prolonged downwards from the inferior medullary velum to the ligulæ.

The two medullary vela are in contact with each other along their line of emergence from the white substance of the cerebellum; and this line of contact forms the summit of the roof of the fourth ventricle, which, in a vertical section through the cavity, appears as a pointed angle.

The **fibræ propriæ of the cerebellum** are of two kinds : (1) *commissural fibres*, which cross the middle line to connect the opposite halves of the cerebellum, some at the anterior part, and others at the posterior part of the vermiform process; (2) *arcuate* or *association fibres*, which connect adjacent laminæ with each other.

The **grey matter** of the cerebellum is found in two situations : (1) on the surface, forming the cortex; (2) as independent masses in the interior.

(1) The **grey matter of the cortex** presents a characteristic foliated appearance, due to the series of laminæ which are given off from the central white matter; these in their turn give off secondary laminæ, which are covered with grey matter. This arrangement gives to the cut surface of the organ a foliated appearance (fig. 563). Externally, the cortex is covered by pia mater; internally is the medullary centre, consisting mainly of nerve-fibres.

Microscopic appearance of the cortex.—The cortex consists of two distinct
layers, viz. an external grey molecular layer, and an internal rust-coloured
granular layer. Between the two layers is an incomplete stratum of cells
which are characteristic of the cerebellum, viz. the *cells of Purkinje*.

The *external grey* or *molecular layer* (figs. 564 and 565) consists of fibres and
cells. The nerve-fibres are delicate fibrillæ, and are derived from the following
sources : (*a*) the dendrites and axon-collaterals of Purkinje's cells ; (*b*) fibres
from cells in the granular layer ; (*c*) fibres from the central white substance of
the cerebellum ; (*d*) fibres derived from cells in the molecular layer itself. In
addition to these are other fibres, which have a vertical direction. These are

FIG. 565.—Vertical section through the grey matter of the human cerebellum.
Magnified about 100 diameters.   (Klein and Noble Smith.)

*External grey or
molecular layer*

Corpuscles of Purkinje

*Internal rust-coloured
or granular layer*

White substance

the processes of large neuroglia-cells, situated in the granular layer. They pass
outwards to the periphery of the grey matter, where they expand into little
conical enlargements which form a sort of limiting membrane beneath the pia
mater, analogous to the membrana limitans interna in the retina, formed by
the fibres of Müller.

The *cells* of the molecular layer are small, and are arranged in two strata, an
outer and an inner. They all possess branched axis-cylinder processes ; those
of the inner layer run for some distance horizontally—i.e. parallel with the
surface of the folia—giving off collaterals, which pass in a vertical direction
towards the cell-bodies of Purkinje's corpuscles, around which they become

enlarged, and form a basket-like network. Hence these cells of the inner layer are named *basket-cells*.

The *cells of Purkinje* form a single stratum of large, flask-shaped cells situated at the junction of the molecular and granular layers, their bases resting against the latter; in fishes and reptiles they are arranged in several layers. The cells are flattened in a direction transverse to the long axis of the folium, and thus appear broad in sections carried across the folium, and fusiform in sections parallel to the long axis of the folium. From the neck of the flask one or more dendrites arise and pass into the molecular layer, where they subdivide and form an extremely rich arborescence, the various subdivisions of the dendrites being covered by lateral spine-like processes. This arborescence is not circular, but, like the cell, is flattened at right angles to the long axis of the folium; in other words, it does not resemble a round bush, but has been aptly compared by Obersteiner to the branches of a fruit tree trained against a trellis or a wall. Hence, in sections carried across the folium the arborescence is broad and expanded; whereas in those which are parallel to the long axis of the folium, the arborescence, like the cell itself, is seen in profile, and is limited to a narrow area.

From the bottom of the flask-shaped cell the axon arises; this passes through the granular layer, and, becoming medullated, is continued as a nerve-fibre in the subjacent white substance. This axon gives off fine collaterals as it passes through the granular layer, some of which run back into the molecular layer.

The *internal rust-coloured, granular* or *nuclear, layer* (figs. 564, 565) is characterised by containing numerous small nerve-cells or granules of a reddish-brown colour, together with many nerve-fibrils. Most of the cells are nearly spherical and provided with short dendrites, which spread out in a spider-like manner in the granular layer. Their axons pass outwards into the molecular layer, and, bifurcating at right angles, run horizontally for some distance. In the outer part of the granular layer are some larger cells, of the type termed *Golgi cells*. Their axons undergo frequent division as soon as they leave the nerve-cells, and pass into the granular layer; while their dendrites ramify chiefly in the molecular layer.

Finally, in the grey matter of the cerebellar cortex, there are fibres which come from the white centre and penetrate the cortex. The cell-origin of these fibres is unknown, though it is believed that it is probably in the grey matter of the spinal cord. Some of these fibres end in the granular layer, by dividing into numerous branches, on which are to be seen peculiar moss-like appendages: hence they have been termed by Ramón y Cajal the *moss-fibres*; they form an arborescence around the cells of the granular layer. Other fibres derived from the medullary centre can be traced into the molecular layer, where their branches cling around the dendrites of Purkinje's cells, and hence they have been named the *clinging* or *tendril fibres*.

(2) The **independent centres of grey matter in the cerebellum** are four in number on each side: one is of large size, and is known as the corpus dentatum; the other three, much smaller, are situated near the middle of the cerebellum, and are known as the nucleus emboliformis, nucleus globosus, and nucleus fastigii.

The *corpus dentatum*, or *ganglion of the cerebellum*, is situated a little to the inner side of the centre of the stem of the white matter of the hemisphere. It consists of an irregularly folded lamina, of a greyish-yellow colour, containing white fibres, and presenting on its antero-internal aspect an opening, the hilum, from which most of the fibres of the superior cerebellar peduncle emerge.

The *nucleus emboliformis* is placed immediately to the inner side of the corpus dentatum, and partly covering its hilum. The *nucleus globosus* is an elongated mass, directed antero-posteriorly, and placed to the inner side of the preceding. The *nucleus fastigii* is somewhat larger than the other two, and is situated close to the middle line at the anterior end of the superior vermiform process, and immediately over the roof of the fourth ventricle, from which it is separated by a thin layer of white matter. It is known as the *roof nucleus of Stilling*.

Weight of the Cerebellum.—Its average weight in the male is about 5 oz. 4 drs. It attains its maximum between the twenty-fifth and fortieth years; its increase after the fourteenth year being relatively greater in the female than in the male. The proportion between the cerebellum and cerebrum is, in the male, as 1 to 8·2, and in the female as 1 to 8. In the infant the cerebellum is proportionately much smaller than in the adult, the relation between it and the cerebrum being, according to Chaussier, between 1 to 13, and 1 to 26; by Cruveilhier the proportion was found to be 1 to 20.

THE FOURTH VENTRICLE (fig. 566)

The **fourth ventricle**, or cavity of the hind-brain, is a diamond-shaped space situated in front of the cerebellum and behind the pons Varolii and medulla oblongata. It is lined by ciliated epithelium, and is continuous below with the central canal of the spinal cord; above, it communicates, by means of a passage termed the aqueduct of Sylvius, with the cavity of the third ventricle. It presents four angles, and possesses a roof, a floor, and lateral boundaries.

Angles.—The *superior angle* is on a level with the upper border of the pons Varolii, and is continuous with the lower end of the aqueduct of Sylvius. The *inferior angle* is on a level with the lower end of the olivary body, and opens into the central canal of the spinal cord. The *lateral angles* correspond with the point of meeting of the three cerebellar peduncles. A little below the lateral angles, on a level with the striæ acusticæ, the ventricular cavity is prolonged outwards in the form of two narrow passages, one on either side. These are named the *lateral recesses* of the ventricle, and are situated between the restiform bodies and the flocculi, reaching as far as the origin of the glosso-pharyngeal and vagi nerves.

Lateral boundaries.—The lower part of each lateral boundary is constituted by the clava, the funiculus cuneatus, and the restiform body; the upper part, by the superior cerebellar peduncle.

Roof.—The upper portion of the roof is formed by the superior cerebellar peduncles and the valve of Vieussens; the lower portion, by the inferior medullary velum, the epithelial lining of the ventricle covered by the tela choroidea inferior, the obex and ligulæ.

The *superior peduncles*, on emerging from the central white substance of the cerebellum, pass upwards and forwards, forming at first the lateral boundaries of the upper part of the cavity; but on approaching the inferior quadrigeminal bodies, they converge, and their mesial portions overlap the cavity and form part of its roof.

The *superior medullary velum* (page 794) fills in the angular interval between the superior peduncles, and is continuous behind with the central white substance of the cerebellum; it is covered on its dorsal aspect by the lingula of the superior vermis.

The *inferior medullary velum* (page 794) is continued downwards and forwards from the central white substance of the cerebellum in front of the nodule and amygdalæ, and ends inferiorly in a thin, concave, somewhat ragged margin. Between this margin above, and the obex and ligulæ below, there is an entire absence of nervous matter; and this part of the roof is completed by the *epithelial lining of the ventricle*, which is prolonged downwards as a thin membrane (*membrana tectoria*), from the deep surface of the inferior medullary velum to the corresponding surface of the obex and ligulæ, and thence on to the floor of the ventricular cavity. This part of the roof is covered and strengthened by a portion of the pia mater, which is named the *tela choroidea inferior*.

The *obex* is a thin, triangular, grey lamina, which roofs in the lower angle of the ventricle and is attached by its lateral margins to the clavæ: it is continuous below with the posterior grey commissure of the closed part of the medulla.

The *ligulæ*, or tæniæ, are two narrow bands of white matter, one on either side, which complete the lower part of the roof of the ventricle. Each consists of an inner vertical and an outer horizontal part. The vertical part is continuous below with the obex, and is adherent by its outer border to the clava; the horizontal portion extends transversely outwards across the restiform body, below the striæ acusticæ, and roofs in the lower and posterior part of the lateral recess. It is attached by its lower margin to the restiform body, and partly encloses the choroid plexus, which, however, projects beyond it like a cluster of grapes; and hence this part of the ligula has been termed the *cornucopia* (Bochdalek).

The *tela choroidea inferior* is the name applied to the triangular fold of pia mater which is carried upwards between the cerebellum and the medulla oblongata. It consists of two layers, which are continuous with each other in front, and are more or less adherent throughout. The posterior layer covers the antero-inferior surface of the cerebellum; while the anterior is applied to the structures which form the lower part of the roof of the ventricle, and is continuous inferiorly with the pia mater on the restiform bodies and closed part of the medulla.

Openings in the roof of the fourth ventricle.—In the roof of the fourth ventricle there are three openings in the pia mater and subjacent epithelium : one of these, the *foramen of Majendie*, is situated in the middle line immediately above the inferior angle of the ventricle ; the other two (*foramina of Luschka*, or *foramina of Key and Retzius*) are found at the extremities of the lateral recesses. By means of these three foramina the cavity of the ventricle communicates with the subarachnoid space, and the cerebro-spinal fluid can pass from the ventricle into this space, or *vice versa*.

FIG. 566.—Floor of the fourth ventricle and corpora quadrigemina. (Testut.)

Choroid plexuses of the fourth ventricle.—These consist of two highly vascular inflexions of the tela choroidea inferior, which invaginate the epithelial part of the roof of the ventricle. Each consists of a vertical and a horizontal portion : the former lies close to the middle line, and extends upwards from the foramen of Majendie ; the latter passes into the lateral recess and projects beyond its apex ; they are everywhere covered by the epithelial lining of the ventricle. The vertical parts of the plexuses are distinct from each other, but the horizontal portions are joined in the middle line ; and hence the entire structure presents the form of the letter **T**, the vertical limb of which, however, is double.

Floor of the fourth ventricle (fig. 566).—This is rhomboidal in shape, its upper portion being formed by the dorsal surface of the pons Varolii, and its lower by the corresponding surface of the open part of the medulla oblongata. It is covered by a layer of grey matter continuous with the central grey matter of the spinal cord ;

1. Middle peduncle of cerebellum. 2. Superior peduncle of cerebellum. 3. Inferior peduncle of cerebellum. 1′, 2′, 3′ Sections through same. 4. Section of the valve of Vieussens. 5. Frænulum veli. 6. Posterior median fissure. 7. Postero-intermediate sulcus. 8. Postero-lateral sulcus. 9. Funiculus gracilis. 10. Funiculus cuneatus. 11. Lateral area of medulla. 12. Clava. 13. Restiform body. 14. Placed above the calamus scriptorius. 15. Obex. 16. Trigonum hypoglossi. 17. Eminentia acustici. 18. Trigonum vagi. 19. Eminentia teres. 20. Inferior fovea. 21. Superior fovea. 22. Locus cœruleus. 23. Striæ acusticæ. 24. Conductor sonorus of Bergmann. 25 Posterior quadrigeminal bodies. 26. Anterior quadrigeminal bodies. 27. Third ventricle. 28. Optic thalamus. 28′. Trigonum habenulæ. 29. Pineal body, hooked forward. 30. Sulcus lateralis. 31. Lateral fillet. 31′. Fibres passing to the valve of Vieussens. 32. Crus cerebri. IV. Fourth nerve. VIII. Eighth nerve.

superficial to this is a thin lamina of neuroglia, which constitutes the ependyma of the ventricle and is lined by ciliated epithelium. It is traversed by a median sulcus, which divides it into symmetrical halves, and it is crossed at the level of the lateral recesses by a number of white strands, named the *striæ acusticæ*. These form a portion of the cochlear division of the auditory nerve, and, sweeping round the outer aspect of the restiform body, extend inwards on the floor of the ventricle, where they disappear by passing into the median sulcus. They divide the floor into two triangular areas, an upper and a lower, which correspond, approximately, to the portions of the floor which are formed by the back of the pons and medulla respectively.

Below the striæ acusticæ, at a short distance from the median sulcus, is a small triangular depression, the *inferior fovea*, the apex of which is directed upwards, while its sides are prolonged downwards as divergent furrows. The inner of these furrows is carried towards the lower angle of the ventricle, the outer towards its lateral wall; and in this manner three small triangular areas are marked off on either side of the middle line. That which lies between the diverging furrows of the fovea is darker in colour than the other two, and is named the *ala cinerea* or *trigonum vagi*; it corresponds with the position of the sensory nuclei of the vagus and glosso-pharyngeal nerves. The area which lies between the inner limb of the fovea and the median sulcus is termed the *trigonum hypoglossi*, and contains the nucleus of the hypoglossal nerve. Its base is directed upwards, and is continuous with an elevation, the *eminentia teres*, which lies above the striæ acusticæ; its apex is directed downwards, and forms with that of the opposite side a pointed elevation, the *calamus scriptorius*. The third area, that on the outer side of the fovea inferior, is named the *trigonum acustici*, and corresponds with one of the chief nuclei of the auditory nerve. Its base is directed upwards, and is continuous with a larger eminence, termed the *eminentia* or *area acustici*, which is crossed by the striæ acustici.

In each half of that part of the floor of the ventricle which lies above the striæ acustici, a small depression, the *superior fovea*, is seen. Between it and the median sulcus is an elongated eminence, the *eminentia teres*, which overlies the nucleus of the sixth nerve, and is, in part at least, produced by the ascending portion of the root of the seventh nerve. This eminence becomes less distinct above, while below it is continuous with the trigonum hypoglossi. Outside the superior fovea is the prominence of the *area acustici* (already referred to); and above it is a bluish, depressed spot, the *locus cœruleus*, which owes its colour to a patch of deeply pigmented nerve-cells, termed the *substantia ferruginosa*, in which a part of the sensory root of the fifth nerve terminates.

THE MID-BRAIN

The **mid-brain,** or **mesencephalon,** is the short, constricted portion which connects the pons Varolii and cerebellum with the thalamencephalon and cerebral hemispheres. It is directed upwards and forwards, and consists of: (1) a ventro-lateral portion, composed of a pair of cylindrical bodies, named the *crura cerebri*; (2) a dorsal portion, consisting of four rounded eminences, named the *corpora quadrigemina*; and (3) an intervening passage or tunnel, the *aqueduct of Sylvius*, which represents the original cavity of the mid-brain and connects the third with the fourth ventricle.

The **crura cerebri** are two cylindrical masses situated at the base of the brain, and largely hidden by the temporal lobes of the cerebrum, which must be drawn aside or removed in order to expose them. They emerge from the upper surface of the pons Varolii, one on either side of the middle line, and diverging as they pass upwards and forwards, disappear into the substance of the cerebral hemispheres. Each crus is crossed from within outwards by the posterior cerebral artery; its lateral surface is in relation to the uncinate convolution of the cerebrum, and is crossed from behind forwards by the fourth nerve. Close to its point of disappearance into the cerebral hemisphere, the optic tract winds forwards around its ventro-lateral aspect. The inner surface of the crus forms the lateral boundary of the posterior part of the interpeduncular space, and is marked by a longitudinal furrow, the *oculo-motor sulcus*, from which the roots of the third or oculo-motor nerve emerge. On the outer surface of the mid-brain there is a second longitudinal furrow, termed the *sulcus lateralis*, which is prolonged downwards between the pons Varolii and the superior cerebellar peduncle. The fibres of the lateral fillet come to the surface in this sulcus, and pass backwards and upwards, to disappear under the lower quadrigeminal body.

Structure of the crura cerebri (fig. 567).—On transverse section, each crus is seen to consist of a dorsal and a ventral part, separated by a deeply pigmented lamina of grey matter, named the *substantia nigra*. The dorsal part is named the *tegmentum*; the ventral, the *crusta* or *pes*: the two crustæ are separated from each other, but the tegmenta are joined in the mesial plane by a forward prolongation of the raphé of the pons Varolii. Laterally, the tegmenta are free, and are constituted by the fibres of the lateral fillet; dorsally, they blend with the corpora quadrigemina.

The **crusta**, or **pes**, is semilunar on transverse section, and consists of longitudinal bundles of efferent fibres, which arise from the cells of the cerebral cortex and are grouped into three principal sets, viz. pyramidal, geniculate, and cortico-pontine. The *pyramidal* fibres occupy the middle three-fifths of the crusta, and are continued downwards through the pons into the pyramid of the medulla. The *geniculate* fibres—so named because they are situated in the knee-shaped bend of the internal capsule—occupy the inner fifth of the crusta, and terminate, after decussating with the corresponding fibres of the other side, in the motor nuclei of the cranial nerves. The *cortico-pontine* or *cortico-protuberantial* fibres terminate below in the nuclei pontis, and consist of anterior and posterior groups. The fibres of the posterior group arise in the temporal and occipital lobes, and occupy the outer fifth of the crusta; while those of the anterior group take origin in the frontal lobe, and are disseminated among the pyramidal and geniculate fibres.

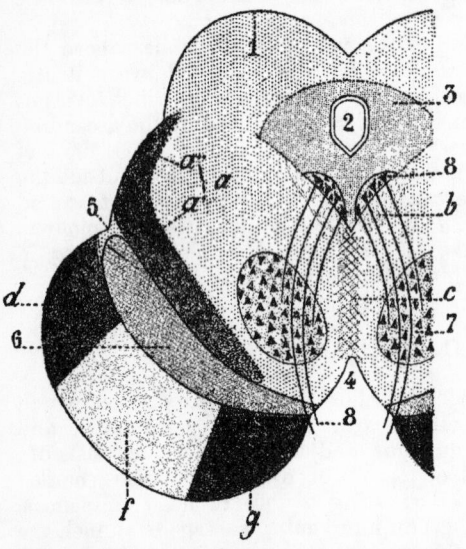

FIG. 567.—Vertical transverse section through the mid-brain. (Schematic.) (Testut.)

1. Corpora quadrigemina. 2. Aqueduct of Sylvius. 3. Central grey matter of aqueduct. 4. Interpeduncular space. 5. Sulcus lateralis. 6. Substantia nigra. 7. Red nucleus of tegmentum. 8. Third nerve, with 8', its nucleus of origin. *a.* Fillet (in blue), with *a'*, the mesial or sensory fillet, and *a''*, the lateral or auditory fillet. *b.* Posterior longitudinal fasciculus. *c.* Raphé. *d.* Posterior cortico-protuberantial tract. *e.* Portion of mesial fillet which passes to the lenticular nucleus and island of Reil. *f.* Pyramidal tract (in red). *g.* Geniculate bundle (in green).

The **substantia nigra** is a layer of grey matter containing numerous deeply pigmented, multipolar nerve-cells. Like the crusta, it is semilunar on transverse section, its concavity being directed towards the tegmentum; from its convex aspect, prolongations extend downwards between the fibres of the crusta. Thicker internally than externally, it reaches from the oculo-motor sulcus to the lateral sulcus, and extends from the upper surface of the pons to the subthalamic region; its inner part is traversed by the fibres of the third nerve as these stream forwards to reach the oculo-motor sulcus. The connections of the cells of the substantia nigra have not been definitely established.

The **tegmenta.**—The tegmental portions of the crura cerebri are continuous below with the reticular formation of the pons Varolii, and, like it, consist of longitudinal and transverse fibres, together with a considerable amount of grey matter. Three well-defined longitudinal strands can be recognised in each tegmentum, viz.: (1) the superior cerebellar peduncle; (2) the posterior longitudinal fasciculus; and (3) the fillet.

Superior cerebellar peduncle.—The fibres of the superior cerebellar peduncle chiefly emerge from the hilum of the dentate nucleus of the cerebellum; but some are connected with the cerebellar cortex, while others are probably derived from the smaller nuclei in the cerebellar white substance. They are continued upwards under cover of the quadrigeminal bodies into the tegmentum, where the fibres of the two peduncles undergo an almost complete decussation on the ventral aspect of the Sylvian aqueduct, and subsequently pass to the red nuclei. Here most of them terminate; and from the cells of the red nuclei a relay of fibres is prolonged to the optic thalamus, and from the latter a second relay is continued through the posterior part of the internal capsule to the cerebral cortex. Some of the fibres of the superior cerebellar peduncle do not terminate in the red nucleus, but are carried past it into the optic thalamus, and probably through the latter into the cortex. The fibres of the superior cerebellar peduncles are both afferent and efferent, since lesions of the cerebral cortex or of the optic thalamus cause atrophy of the red nucleus, and of the peduncle; while lesions of the cerebellum are followed by atrophy of the fibres of the peduncle, and of the opposite red nucleus.

The *posterior longitudinal fasciculus* is continuous below with the antero-lateral ground-bundle of the spinal cord; and has been traced by Edinger as far as a nucleus which is situated in the subthalamic region, immediately in front of the aqueduct of Sylvius. In the medulla oblongata and pons, it runs close to the middle line, near the floor of the fourth ventricle; and in the mid-brain, it is situated on the ventral aspect of the Sylvian aqueduct, below the nuclei of the third and fourth nerves. Its connections are imperfectly known, but it consists largely of ascending and descending intersegmental or association fibres, which connect the various nuclei of the mid- and hind-brain to each other. Many of the descending fibres arise in the superior quadrigeminal body, and, after decussating in the middle line, terminate in the motor nuclei of the pons and medulla. The ascending fibres arise from the cells of the grey matter of the upper part of the cord, and from the nuclei in the medulla and pons, and pass, without undergoing decussation, to the higher nuclei. Fibres are also carried through the posterior longitudinal fasciculus from the nucleus of the sixth nerve into the third nerve of the opposite side, and through this nerve to the Internal rectus muscle of the eyeball. Again, fibres are prolonged through this fasciculus from the nucleus of the third nerve into the seventh nerve, and are distributed to the Orbicularis palpebrarum, the Corrugator supercilii, and the Occipito-frontalis muscles.

The fillet.—The fibres of the fillet have been seen to take origin in the gracile and cuneate nuclei of the medulla oblongata, and to cross to the opposite side in the sensory decussation (page 780). They then pass upwards through the medulla, in which they are situated behind the pyramidal fibres, and between the olivary bodies. Here they are joined by the fibres of Gowers' ascending tract, these having already undergone decussation in the spinal cord. As the fillet is traced upwards, it receives additional fibres from the terminal nuclei of the sensory nerves of the opposite side. In the pons, it assumes a flattened, ribbon-like appearance, and is placed on the dorsal aspect of the trapezium. In the mid-brain, its outer part is folded backwards and forms nearly a right angle with its mesial portion; and hence it is customary to speak of the fillet as consisting of lateral and mesial parts.

The *lateral fillet* has been seen to come to the surface of the mid-brain along its lateral sulcus, and to disappear under the inferior quadrigeminal body. It consists of fibres which are derived from the terminal nuclei of the cochlear division of the auditory nerve, together with others which arise within the superior olive and the trapezoid nucleus. Most of these fibres are crossed, but some are uncrossed. Many of them pass to the inferior quadrigeminal body of the same or opposite side; but others are prolonged to the optic thalamus, and thence through the posterior part of the internal capsule to the middle and superior temporal convolutions.

The *mesial fillet* comprises that portion of the fillet which commences in the gracile and cuneate nuclei of the opposite side, and which is joined by Gowers' tract and by fibres from the terminal nuclei of the sensory nerves of the opposite side, excepting the cochlear division of the auditory. In the crus cerebri, a few of its fibres pass upwards in the outer part of the pes or crusta, on the dorsal aspect of the cortico-pontine fibres, and reach the lenticular nucleus and the island of Reil. The greater part of the mesial fillet, on the other hand, is prolonged through the tegmentum, and most of its fibres end in the optic thalamus; probably some are continued directly through the posterior part of the internal capsule to the cerebral cortex. From the cells of the optic thalamus a relay of fibres is prolonged to the cerebral cortex.

The *red nucleus*, or *nucleus of the tegmentum*, is a collection of grey matter which is situated in the anterior part of the tegmentum, and is prolonged into the posterior part of the subthalamic region. In sections at the level of the upper pair of quadrigeminal bodies, it appears as a circular mass, which is traversed by the fibres of the third nerve. It consists of multipolar nerve-cells of different sizes, and its relations to the fibres of the superior cerebellar peduncles have already been referred to (page 800).

The **corpora quadrigemina** are four rounded eminences which form the dorsal part of the mid-brain. They are situated above and in front of the valve of

3 F

Vieussens and superior peduncles of the cerebellum, and below and behind the third ventricle and posterior commissure. They are covered by the splenium of the corpus callosum, and are partly overlapped on each side by the inner angle, or pulvinar, of the posterior end of the optic thalamus; on their lateral aspect, under cover of the pulvinar, is an oval eminence, named the *internal geniculate body*. The corpora quadrigemina are arranged in pairs (upper and lower), and are separated from one another by a crucial sulcus. The longitudinal part of this sulcus expands superiorly to form a slight depression which supports the *pineal body*, a cone-like structure which projects backwards from the thalamencephalon and partly obscures the upper quadrigeminal bodies. From the lower end of the longitudinal sulcus, a white band, termed the *frænulum veli*, is prolonged downwards to the valve of Vieussens; on either side of this band the fourth cranial nerve emerges, and passes forwards on the lateral aspect of the crus to reach the base of the brain. The *upper pair*, sometimes called the *nates*, are larger and darker in colour than the lower, and are oval in shape. The *lower pair*, called the *testes*, are hemispherical, and somewhat more prominent than the upper. The upper quadrigeminal bodies are associated with the sense of sight, the lower with that of hearing. From the lateral aspect of each of the four bodies, a white band, termed the *brachium*, is prolonged upwards and forwards. The *superior brachium* extends outwards from the upper quadrigeminal body, and, passing between the pulvinar and internal geniculate body, is partly continued into an eminence called the external geniculate body, and partly into the outer portion of the optic tract. The *inferior brachium* passes forwards and upwards from the lower quadrigeminal body, and disappears under cover of the internal geniculate body.

Structure of the Corpora quadrigemina.—The *lower* quadrigeminal body consists of a compact nucleus of grey matter which contains large and small multipolar nerve-cells, and is more or less completely surrounded by white fibres derived from the lateral fillet. Most of these fibres end in the grey nucleus of the same side, but some cross the middle line and terminate in that of the opposite side. From the cells of the grey nucleus, fibres are prolonged through the inferior brachium into the tegmentum of the crus cerebri, and are carried to the optic thalamus and the cortex of the temporal lobe; other fibres cross the middle line and end in the opposite quadrigeminal body. The *upper* quadrigeminal body is covered by a thin stratum of white fibres, termed the *stratum zonale*, the majority of which are derived from the optic tract. Beneath this is the *stratum cinereum*, a layer of grey matter which resembles a cap: it is semilunar in shape, thicker in the centre than at the margins, and consists of numerous multipolar nerve-cells, for the most part of small size, embedded in a fine network of nerve-fibres. Still deeper is the *stratum opticum*, which contains large multipolar nerve-cells, separated by numerous fine nerve-fibres. Finally, there is the *stratum lemnisci*, which consists of fibres partly derived from the fillet and partly from the cells of the stratum opticum; interspersed among these fibres are many large multipolar nerve-cells. The two last-named strata are sometimes termed the *grey-white layers*, from the fact that they consist of both grey and white matter. Of the afferent fibres which reach the superior quadrigeminal body, some are derived from the fillet, but the majority have their origin in the retina and are conveyed to it through the superior brachium; all of them terminate by arborising around the cells of the grey matter. Of the fibres which arise from the cells of the grey matter, some cross the middle line to the opposite quadrigeminal body; many ascend through the superior brachium, and finally reach the cortex of the occipital lobe of the cerebrum; while others, after undergoing decussation, descend through the posterior longitudinal fasciculus to the motor nuclei in the pons and medulla. The corpora quadrigemina are larger in the lower animals than in man. In fishes, reptiles, and birds, they are hollow, and only two in number (corpora bigemina); they represent the superior quadrigeminals of mammals, and are frequently termed the optic lobes, because of their intimate connection with the optic tracts.

In close relationship with the corpora quadrigemina are the *superior peduncles of the cerebellum*, which emerge from the upper and mesial part of the cerebellar hemispheres. They run upwards and forwards, and, passing under the corpora quadrigemina, enter the tegmenta as already described (page 800).

The **aqueduct of Sylvius,** or **iter a tertio ad quartum ventriculum,** is a narrow canal, about fifteen millimetres in length, situated between the corpora quadrigemina and tegmenta, and connecting the third with the fourth ventricle. Its shape on transverse section varies at different levels, being T-shaped below, triangular above, and oval or heart-shaped about the middle of its course. It is lined by ciliated columnar epithelium, and is surrounded by a layer of grey matter, named the *central grey matter* of the aqueduct : this is continuous below with the grey substance in the floor of the fourth ventricle, and above with that of the third ventricle. Dorsally, it is partly separated from the grey matter of the quadrigeminal bodies by the fibres of the lemniscus ; ventral to it are the posterior longitudinal fasciculus, and the formatio reticularis of the tegmentum. Scattered throughout its grey matter are numerous nerve-cells of various sizes, interlaced by a network of fine fibres. Besides these scattered cells it contains three groups which constitute the nuclei of the third and fourth nerves, and the nucleus of the Sylvian or mesencephalic root of the fifth nerve. The nucleus of the fifth extends along the entire length of the aqueduct, and occupies the outer part of the grey substance, while those of the third and fourth are situated in its ventral part. The nucleus of the third measures about ten millimetres in length, and lies under the upper quadrigeminal body, beyond which, however, it extends for a short distance into the grey matter of the third ventricle ; that of the fourth is small and nearly circular, and is on a level with a plane carried transversely through the upper part of the lower quadrigeminal body.

The **posterior commissure** is a rounded band of white fibres, which stretches across the middle line behind the upper end of the Sylvian aqueduct, and in relation to the ventral aspect of the stalk of the pineal body, with which it is connected ; it forms a part of the posterior boundary of the third ventricle, but its connections have not been definitely determined. The fibres which occupy its ventral part acquire their medullary sheaths early, and are said to arise in the ganglion habenulæ and pineal body, and pass to the posterior longitudinal bundle and oculo-motor nucleus of the opposite side. The more dorsally situated fibres connect the posterior part of the optic thalamus with the tegmentum of the crus cerebri.

THE FORE-BRAIN

The **fore-brain** consists of : (1) the *thalamencephalon,* or *inter-brain* (diencephalon), which corresponds in a large measure to the third ventricle and the structures which bound it ; and (2) the *telencephalon,* which comprises the largest part of the brain, viz. the cerebral hemispheres ; these hemispheres are intimately connected with each other across the middle line, and each contains a large cavity, named the lateral ventricle. The lateral ventricles communicate, through the foramen of Monro, with the third ventricle ; but are separated from each other by a mesial septum which contains a slit-like cavity, termed the fifth ventricle, which, however, has no communication with the other brain ventricles.

THE THALAMENCEPHALON

The **thalamencephalon,** or **inter-brain,** comprises the greater part of the third ventricle, together with the structures which are in immediate relation with it. It is connected above and in front with the cerebral hemispheres ; behind, with the mid-brain. Its upper surface is entirely concealed by the parts which join the cerebral hemispheres together, and is also covered by a fold of pia mater, named the velum interpositum. Inferiorly, it reaches to the base of the brain, where it forms the structures contained in the interpeduncular space.

The **interpeduncular space** consists of a somewhat lozenge-shaped area, limited in front by the optic commissure ; behind, by the upper surface of the pons Varolii ; antero-laterally, by the converging optic tracts, which meet in front in the optic commissure ; postero-laterally, by the diverging crura cerebri. The structures seen in this space form the greater part of the floor of the third ventricle, and are named, from before backwards, the tuber cinereum, with its infundibulum and pituitary body, the corpora albicantia, and the posterior perforated space.

The **tuber cinereum** is a hollow eminence of grey matter, situated behind the optic commissure and in front of the corpora albicantia. Laterally it is continuous with the grey substance of the anterior perforated spaces, and anteriorly with a thin lamina, termed the *lamina terminalis*.* From the under surface of the tuber cinereum a conical process of grey matter projects downwards and forwards, and is attached to the posterior lobe of the pituitary body. This is named the *infundibulum* or *stalk of the pituitary body*, and, like the tuber cinereum, is formed by a portion of the grey matter of the floor of the third ventricle. Its upper part is hollow, and contains a funnel-shaped diverticulum of the ventricular cavity.

The **pituitary body** (*hypophysis cerebri*) (fig. 568) is a small reddish-grey vascular mass, weighing from five to ten grains, situated in the pituitary fossa, where it is retained by a circular fold of dura mater, named the *diaphragma sellæ*. This fold almost completely roofs in the pituitary fossa, but presents a small, central aperture, through which the infundibulum passes. The pituitary body is oval, and measures about half an inch from side to side, and about one-third of an inch from before backwards. It is very vascular, and consists of two lobes, separated from one another by a fibrous lamina; of these the anterior is the larger, of an oblong form, and somewhat concave behind, where it receives the posterior lobe, which is round. The two lobes differ in their development and in their structure. The *anterior* lobe, of a dark, reddish-brown colour, is developed from the ectoderm of the buccal cavity, and resembles to a considerable extent the thyroid body. It consists of a number of isolated vesicles and slightly convoluted tubules, lined by columnar epithelium and united together by a very vascular connective tissue The vesicles sometimes contain a colloid material similar to that found in the thyroid body. The *posterior* lobe is developed as a downgrowth from the floor of the embryonic brain, and during fœtal life contains a cavity continuous with that of the third

FIG. 568.—The pituitary body, in position. Shown in vertical section. (Testut.)

1, 1'. Anterior and posterior lobes of pituitary body. 2. Infundibulum. 3. Optic commissure. 4. Lamina terminalis. 5. Optic recess. 6. Anterior commissure. 7, 7'. Circular sinus. 8. Anterior cerebral artery. 9. Basilar artery. 10. Posterior cerebral artery. 11. Corpus albicans. 12. Crus cerebri. 13. Pons Varolii.

ventricle. This cavity becomes obliterated; and, in the adult, the lobe consists of a reticulum of connective tissue with branched cells, some of which contain pigment. In man, this lobe possesses no nervous elements; but in certain of the lower vertebrates (e.g. the fishes) nervous structures are plentiful, and the lobe is of large size.

The **corpora albicantia**, or **corpora mammillaria**, are two round white masses, each about the size of a small pea, which lie side by side below the grey matter of the ventricular floor, immediately behind the tuber cinereum. They are composed externally of white substance, and internally of grey matter: the nerve-cells of the latter are arranged in two sets, an inner of smaller, and an outer of larger cells. The white substance is mainly formed by the fibres of the anterior pillars of the fornix, which descend to the base of the brain and end in the corpora albicantia. From the cells of the grey matter of each corpus albicans

---

* In relation to the tuber cinereum there are two commissural strands which pass across from side to side. One is named the *fasciculus of the tuber cinereum*, and crosses that eminence immediately below the floor of the ventricle; it is said to end partly in the anterior pillar of the fornix, and partly in the lenticular nucleus. The other is termed the *tract* or *commissure of Meynert*; it lies above and behind the optic commissure and optic tract, and passes into the crus cerebri. The signification of these commissural strands is unknown. In some animals the commissure of Meynert is seen on the surface of the base of the brain; but in man it is buried in the grey substance.

two fasciculi arise. One, named the *bundle of Vicq d'Azyr*, passes upwards into the anterior nucleus of the optic thalamus ; the other, termed the *peduncle of the corpus albicans*, passes backwards and downwards into the tegmental region, where its mode of ending is not known. At an early period of fœtal life, there is only a single mammillary eminence ; but this becomes divided about the seventh month. In most vertebrates there is only one median corpus albicans.

The **posterior perforated space,** or **pons Tarini,** is a layer of grey matter which forms the floor of a deep fossa between the crura cerebri and immediately in front of the pons Varolii. It is pierced by numerous orifices for the passage of the postero-mesial group of central cerebral arteries.

FIG. 569.—The third and fourth ventricles.

An arrow has been placed in the position of the foramen of Monro.

The **third ventricle** (fig. 569) is the cavity of the thalamencephalon, and consists of a median cleft between the two optic thalami. Behind, it communicates with the fourth ventricle through the aqueduct of Sylvius, and in front with the lateral ventricles through the foramen of Monro. Somewhat triangular in shape, with the apex directed backwards, it presents a roof, a floor, an anterior and a posterior boundary, and a pair of lateral walls.

The *roof* of the third ventricle is formed by a layer of epithelium which stretches between the upper edges of the lateral walls of the cavity, and is continuous with the epithelial lining of the ventricle. It is covered by, and adherent to, a fold of pia mater, named the velum interpositum, from the under surface of which a pair of vascular fringed processes, the *choroid plexuses of the third ventricle*, project downwards, one on either side of the middle line, and invaginate the epithelial roof into the ventricular cavity. When the velum interpositum is reflected, the epithelial roof is torn from its lateral attachments, and is removed with it, and the cavity of the ventricle is exposed.

The *floor* of the third ventricle slopes downwards and forwards, and is formed by the structures which occupy the interpeduncular space, and behind these by the tegmenta of the crura cerebri.

The *anterior boundary* of the third ventricle is constituted below by the lamina terminalis, and above by the anterior pillars of the fornix and the anterior commissure. The *lamina terminalis* is a thin layer of grey matter, which stretches from the upper surface of the optic commissure to the rostrum of the corpus callosum. It is continuous behind with the grey matter of the tuber cinereum, and laterally with that of the anterior perforated spaces. Immediately above the optic commissure the ventricle exhibits a small angular recess or diverticulum, named the *optic* or *supra-optic recess*. The anterior pillars of the fornix are two thick strands of white substance, which curve downwards in front of the optic thalami and subsequently pass into the side-wall of the third ventricle ; while the anterior commissure is a transverse band lying directly in front of these pillars. Between the anterior pillars of the fornix and above the anterior commissure, the ventricular cavity is continued as a small recess, termed the *vulva*. At the junction of the roof with the anterior wall of the ventricle, and situated between the optic thalami behind and the anterior pillars of the fornix in front, is the *foramen of Monro*, through which the third communicates with the lateral ventricles.

The *posterior boundary* of the third ventricle consists of the pineal body, the posterior commissure, and the upper end of the aqueduct of Sylvius.

The **pineal body** (epiphysis cerebri) is a small reddish-grey body, conical in shape (hence its synonym, *conarium*), placed immediately above and behind the posterior commissure, and projecting backwards and downwards between the upper pair of quadrigeminal bodies. It lies beneath the splenium of the corpus callosum, but is separated from it by the velum interpositum, the lower layer of which envelops it. It measures about one-third of an inch in length, and its base, directed forwards, is attached by a stalk or peduncle of white matter. The *stalk* of the pineal body is divided anteriorly into two laminæ, upper and lower ; and these are separated by a small triangular recess, the *recessus pinealis* of the third ventricle. Above the pineal body is another recess, named the *recessus suprapinealis* ; this consists of a diverticulum of the epithelium which forms the ventricular roof. The ventral lamina of the pineal stalk is prolonged into the posterior commissure, which lies immediately below it. The dorsal lamina divides into two lateral strands, named the *striæ pinealis*, which extend forward, one on either side, along the optic thalamus at the junction of its mesial and upper surfaces, and blend in front with the anterior pillars of the fornix.

Structure of the pineal body.—The pineal body is destitute of nervous elements, and consists of a number of follicles, lined by epithelium and enveloped by connective tissue. These follicles contain a variable quantity of gritty material, named *acervulus cerebri* or *brain-sand*, composed of phosphate and carbonate of lime, phosphate of magnesia and ammonia, and a little animal matter.

The pineal body is generally believed to be the homologue of the pineal eye of lizards. In these animals the epiphysis is attached by an elongated stalk, and projects through an aperture in the roof of the cranium. Its extremity lies immediately under the epidermis, and, on microscopic examination, presents, in a rudimentary fashion, structures similar to those found in the eyeball. Recent observations tend to the conclusion that the pineal body arises as a paired structure, probably serially homologous with the paired eyes.

The posterior commissure and the aqueduct of Sylvius have already been described (page 803).

Lateral walls of the third ventricle.—Each lateral wall consists of an upper portion formed by the inner surface of the anterior two-thirds of the optic thalamus, and a lower part consisting of an upward continuation of the grey matter of the ventricular floor. These two parts correspond to the alar and basal laminæ respectively of the lateral walls of the fore-brain vesicle, and are separated from each by a furrow which reaches from the foramen of Monro to the aqueduct of Sylvius ; this furrow is named the *sulcus of Monro* (see page 113). The lateral wall is limited above by a delicate band of white fibres, the stria pinealis, which runs forwards along the junction of the mesial and upper surfaces of the thalamus to join the corresponding anterior pillar of the fornix. The anterior pillars of the fornix curve downwards in front of the foramen of Monro,.and then run in the lateral walls of the ventricle, where, at first, they form distinct prominences, but are subsequently lost to sight below.

The **optic thalamus** is a large ovoid mass of grey matter, situated on the lateral aspect of the third ventricle, and reaching for some distance behind that cavity. It measures about an inch and a half in length, and presents two extremities, anterior and posterior, and four surfaces, superior, inferior, internal, and external.

The *anterior extremity* is narrow, directed forwards and inwards, and lies close to the middle line, where it forms the posterior boundary of the foramen of Monro.

The *posterior extremity* is expanded, directed backwards and outwards, and overlaps the superior quadrigeminal body. Internally, it presents a well-marked, angular prominence, the *posterior tubercle*, or *pulvinar*: this is continued externally into an oval swelling, the *external geniculate body*; while beneath the pulvinar, but separated from it by the superior brachium, is a second oval eminence, the *internal geniculate body*.

The *superior surface* is free, slightly convex, and covered by a layer of white matter, termed the *stratum zonale*. It is separated externally from the caudate nucleus, a large pear-shaped mass of grey matter, by a white band, named the *tænia semicircularis*, and by the *vein of the corpus striatum*. It is divided into an inner and an outer portion by an oblique, but shallow, furrow which runs from behind, forwards and inwards. The outer part forms a portion of the floor of the lateral ventricle, and is covered by the epithelial lining of that cavity: it terminates anteriorly in a tubercle, named the *anterior tubercle* of the optic thalamus. The inner part is covered by the velum interpositum, and is excluded from the lateral and third ventricles, and is therefore destitute of an epithelial covering.

The *inferior surface* rests upon and is continuous with the upward prolongation of the tegmentum (subthalamic tegmental region), in front of which it is related to the *substantia innominata of Meynert*.

The *internal surface* constitutes the upper part of the lateral wall of the third ventricle, and is connected to that of the opposite side by the *middle* or *grey commissure*. This consists of a flattened grey band, measuring about one-third of an inch antero-posteriorly; it is not, however, a true commissure, since it contains no nervous tissues, but is composed only of neuroglial elements which pass between the inner aspects of the optic thalami.

The *external surface* is in contact with a thick band of white matter which forms the posterior limb of the internal capsule and separates the optic thalamus from the lenticular nucleus of the corpus striatum.

Structure.—The optic thalamus consists chiefly of grey matter; but its upper surface is covered by a layer of white matter, the *stratum zonale*, and its outer surface by a similar layer, termed the *external medullary lamina*. Further, the grey mass is subdivided into three chief nuclei—anterior, inner, and outer—by a white layer, named the *internal medullary lamina*. The *anterior nucleus* is the smallest, and is situated in the anterior tubercle; the fibres of the bundle of Vicq d'Azyr arise from the cells of the corpus albicans and end in this nucleus. The *inner nucleus* lies between the internal medullary lamina and the lateral wall of the ventricle, but does not reach as far back as the pulvinar. The *outer nucleus* is the largest, and is placed between the internal and external medullary laminæ, and extends backwards into the pulvinar.

Connections of the optic thalamus.—Most of the fibres of the tegmentum end in the optic thalamus, but some are probably carried through it to the cerebral cortex. Its anterior tubercle receives fibres (the bundle of Vicq d'Azyr) from the corpus albicans. Many fibres enter its posterior extremity from the optic tract. Numerous fibres arise from the cells of the optic thalamus and radiate to the different parts of the cerebral cortex. These are described under the four following groups: (*a*) the fibres of the *anterior stalk* of the optic thalamus, which pass through the anterior limb of the internal capsule to the frontal lobe; (*b*) the *posterior stalk*, or *optic radiation*, which consists of fibres that arise in the pulvinar, and pass through the extreme posterior part of the internal capsule to the occipital lobe ; (c) the *inferior stalk* is composed of fibres which leave the mesial and under aspects of the thalamus, and pass outwards beneath the lenticular nucleus to end in the cortex of the temporal lobe and island of Reil ; and (d) the *parietal stalk*, made up of fibres which pass from the outer surface of the thalamus to the parietal lobe. Fibres also pass from the optic thalamus into the corpus striatum—those destined for the caudate nucleus leave the outer surface, and those for the lenticular nucleus, the inferior surface of the thalamus.

Corpora geniculata.—These are two in number—internal and external—on each side, and have already been referred to in connection with the posterior extremity of the optic thalamus.

The *internal geniculate body* lies under the pulvinar of the optic thalamus, and on the lateral aspect of the corpora quadrigemina, and is sometimes described as a part of the mid-brain. Oval in shape, with its long axis directed forwards and outwards, it is lighter in colour and smaller in size than the external. The inferior brachium from the lower quadrigeminal body disappears under cover of it, while from its outer extremity a strand of fibres passes to join the optic tract. The internal geniculate bodies are connected with each other through the posterior part of the optic commissure by a bundle of fibres, named the *commissure of Gudden*.

The *external geniculate body* forms an oval elevation on the outer part of the posterior extremity of the optic thalamus, and is connected internally to the upper quadrigeminal body by the superior brachium. It receives numerous fibres from the optic tract, many of which end in it, while others pass over or through it into the pulvinar. It is of a dark colour, and presents a laminated arrangement consisting of alternate layers of grey and white matter. Its cells are large, multipolar, and pigmented ; their axons pass to the visual area in the occipital part of the cortex.

The upper quadrigeminal body, the pulvinar, and external geniculate body are intimately connected with sight, and constitute the *lower visual centres*. Extirpation of the eyes in newly born animals entails an arrest of their development, but has no effect on the lower quadrigeminal or internal geniculate bodies. Moreover, the latter are well developed in the mole, where the superior quadrigeminal body is rudimentary.

Trigonum habenulæ (fig. 566).—This name is given to a small, depressed, triangular area which is situated in front of the upper quadrigeminal body and between the pulvinar and the posterior part of the stria pinealis. It contains a group of multipolar nerve-cells, termed the *ganglion habenulæ*. Fibres enter it from the stria pinealis, and others pass across the middle line in the upper part of the pineal stalk to the corresponding ganglion on the opposite side. The majority of its fibres are, however, directed downwards, and form a bundle, the *fasciculus retroflexus of Meynert*, which passes on the mesial side of the red nucleus, and after decussating with the corresponding fasciculus of the opposite side, ends in a small mass of grey matter situated between the diverging crura cerebri, and named the *ganglion interpedunculare*.

Optic commissure.—The optic commissure consists of a flattened, somewhat quadrilateral band of fibres, which is situated at the junction of the floor and anterior wall of the third ventricle. Most of its fibres have their origin in the retina, and reach it through the optic nerves, which are continuous with its antero-lateral angles. In the commissure, they undergo a partial decussation : those fibres which come from the nasal or inner half of the retina decussate and enter the optic tract of the opposite side, while the fibres from the temporal or outer half of the retina do not undergo decussation, but pass back into the optic tract of the same side. Occupying the posterior part of the commissure, however, is a strand of fibres which is not derived from the optic nerves : this constitutes the *commissure of Gudden*, and has already been referred to as forming a connecting link between the internal geniculate bodies.

Optic tracts.—The optic tracts are continued backwards and outwards from the postero-lateral angles of the optic commissure. Each passes between the anterior perforated space and the tuber cinereum, and, winding round the ventro-lateral aspect of the crus cerebri, divides into a mesial and a lateral root. The former comprises the fibres of Gudden's commissure. The lateral root consists mainly of afferent fibres which arise in the retina and undergo partial decussation in the optic commissure, as described ; but it also contains a few fine efferent fibres which have their origin in the brain and their termination in the retina. When traced backwards, the fibres of the lateral root are found to end in the external geniculate body and pulvinar of the optic thalamus, and in the upper quadrigeminal body; and these three structures constitute the *lower visual centres*. Fibres arise from the nerve-cells in these centres, and pass through the hindmost part of the internal capsule, under the name of the *optic radiations*, to the cortex of the occipital lobe of the cerebrum, where the *higher visual centre* is situated. Some of the fibres of the optic radiations take an opposite course, arising from the cells of the occipital cortex and passing to the lower visual centres. Some fibres are detached from the optic tract, and pass through the crus cerebri to the nucleus of the third nerve. These may be regarded as the afferent branches for the Sphincter pupillæ and Ciliary muscles. Other fibres have been described as reaching the cerebellum through its superior peduncles; while others, again, are lost in the pons Varolii.

Subthalamic region.—Towards the anterior part of the crus cerebri the tegmentum becomes thinned out and blended with the lower surface of the optic thalamus. To this region the name of *subthalamic tegmental region* has been given. In front, it is continuous with the substantia innominata and the grey matter of the floor of the third ventricle. This subthalamic region contains a forward prolongation of the red nucleus and substantia nigra, and consists, from above downwards, of three strata: (1) *stratum dorsale*, which is directly applied to the under surface of the optic thalamus, and consists of fine longitudinal fibres; (2) *zona incerta*, which is a continuation forwards of the formatio reticularis of the tegmentum; and (3) the *corpus subthalamicum*, or *nucleus of Luys*, which comprises a greyish-yellow mass situated on the dorsal aspect of the substantia nigra. It consists of a dense reticulum of neuroglia and small nerve-cells, but its signification and connections are very obscure.

THE CEREBRAL HEMISPHERES

The **cerebral hemispheres** constitute the largest part of the encephalon, and, when viewed together from above, assume the form of a large ovoid mass which is broader behind than in front, the greatest transverse diameter corresponding with a line connecting the two parietal eminences. They are separated mesially by a deep cleft, named the *great longitudinal fissure*, and each possesses a central cavity, named the lateral ventricle.

The **great longitudinal fissure** separates the cerebral hemispheres, and contains a sickle-shaped process of dura mater, the falx cerebri. In front and behind, the fissure extends from the top to the bottom of the hemispheres and completely severs them, but its middle portion only separates them for about one-half of their vertical extent; for if the hemispheres be drawn apart it will be seen that the floor of this portion of the fissure is formed by a great central white commissure, the *corpus callosum*, which connects the hemispheres across the middle line.

Surfaces of the cerebral hemispheres.—Each hemisphere presents three surfaces: an outer, a mesial, and a lower.

The *outer surface* is convex in adaptation to the concavity of the corresponding half of the vault of the cranium. The *mesial surface* is flat and vertical, and is separated from that of the opposite hemisphere by the great longitudinal fissure and the falx cerebri. The *lower surface* is of an irregular form, and may be divided into three areas: anterior, middle, and posterior. The anterior area is concave, formed by the orbital surface of the frontal lobe, and rests on the roof of the orbit and nose; the middle area is convex, and consists of the under surface of the temporal lobe: it is adapted to the corresponding half of the middle cranial fossa. The posterior area is concave, directed inwards as well as downwards, and is named the *tentorial surface*, since it rests upon the tentorium cerebelli, which intervenes between it and the upper surface of the cerebellum.

These three surfaces are separated from each other by the following borders: (a) *supero-mesial*, between the outer and mesial surfaces; (b) *infero-lateral*, between the outer and tentorial surfaces; (c) *internal occipital*, separating the mesial and tentorial surfaces; (d) *superciliary*, between the outer surface and the orbital part of the under surface; and (e) *internal orbital*, separating the orbital from the mesial surface. The anterior extremity of the hemisphere is named the *frontal pole*; the posterior, the *occipital pole*; and the anterior end of the temporal lobe, the *temporal pole*. About two inches in front of the occipital pole on the infero-lateral border is an indentation or notch, named the *pre-occipital notch*.

The surfaces of the hemispheres are moulded into a number of irregular eminences, named *convolutions* or *gyri*, and these are separated by clefts or furrows, termed *fissures* or *sulci*. The fissures are of two kinds, *complete* and *incomplete*. The former appear early in foetal life, are few in number, and are produced by infoldings of the entire thickness of the brain-wall, and give rise to corresponding elevations in the interior of the ventricle. They comprise the hippocampal or dentate fissure, and parts of the calcarine and collateral fissures.

Fig. 570.—Upper surface of the brain, the arachnoid having been removed.

The incomplete fissures are very numerous, and only indent the central white substance, without producing any corresponding elevations in the ventricular cavity.

The convolutions and their intervening fissures are fairly constant in their arrangement; at the same time, they vary within certain limits, not only in different individuals, but on the two hemispheres of the same brain. The convoluted condition of the surface permits of a great increase of the grey matter without sacrificing much additional space. The number and extent of the convolutions, as well as the depth of the intervening sulci, appear to bear a direct relati n to the intellectual powers of the individual.

Certain of the fissures are utilised for the purpose of dividing the hemisphere into lobes: included under this category are the fissure of Sylvius, the fissure of Rolando, the parieto-occipital and calloso-marginal fissures.

The **fissure of Sylvius** constitutes a well-marked cleft on the under and outer surfaces of the hemisphere, and consists of a short stem which divides into three limbs. The *stem* is situated on the base of the brain, and commences at the outer angle of the anterior perforated spot, in a depression named the vallecula Sylvii. From this point it extends outwards between the anterior part of the temporal lobe and the orbital surface of the frontal lobe, and reaches the outer surface of the hemisphere. Here it divides into three limbs: an anterior, an ascending, and a posterior. The *anterior limb* passes forwards for nearly an inch into the inferior frontal convolution, while the *ascending limb* extends upwards into the same convolution for about an equal distance. The *posterior limb* is the longest; it is carried backwards and slightly upwards for about three inches, and terminates by an upward inflexion in the parietal lobe.

The **fissure of Rolando,** or **sulcus centralis,** is situated about the middle of the outer surface of the hemisphere, and commences in or near the great longitudinal fissure, a little behind its mid-point. It runs sinuously downwards and forwards, and terminates a little above the posterior limb of the fissure of Sylvius, and about an inch behind the ascending limb of the same fissure. It describes two chief curves: an upper or *superior genu* with its concavity directed forwards, and a lower or *inferior genu* with its concavity directed backwards. The fissure of Rolando forms an angle opening forwards, of about seventy degrees with the mesial plane.

The **Parieto-occipital fissure.**—Only a small part of this fissure is seen on the outer surface of the hemisphere, its chief part being situated on the mesial surface. That on the outer surface is usually named the external parieto-occipital fissure, and that on the mesial aspect the internal parieto-occipital fissure.

The *external parieto-occipital* fissure is situated about two inches from the posterior extremity or occipital pole of the hemisphere, and extends on to the outer surface for about half an inch.

The *internal parieto-occipital* fissure runs downwards and forwards as a deep cleft on the mesial surface of the hemisphere, and joins the calcarine fissure below and behind the posterior end of the corpus callosum. On separating the lips of this fissure, it will be seen in most cases to contain a submerged convolution.

The **calloso-marginal fissure** is situated on the mesial surface of the hemisphere; it commences below the anterior end of the corpus callosum and runs upwards and forwards nearly parallel to the rostrum of this body, and, curving in front of the genu, is continued backwards above the corpus callosum, and finally ascends to the upper margin of the hemisphere, a short distance behind the upper extremity of the fissure of Rolando. It separates the marginal convolution of the frontal lobe from the callosal convolution of the limbic lobe.

The **collateral fissure** is situated on the tentorial surface of the hemisphere, and extends from near the occipital to within a short distance of the temporal pole. Behind, it lies below and to the outer side of the calcarine fissure, from which it is separated by the gyrus lingualis; in front, it is situated between the hippocampal convolution of the limbic lobe and the anterior part of the temporo-occipital convolution.

Lobes of the hemispheres.—By means of these fissures, assisted by certain arbitrary lines, the hemisphere is subdivided into the following convolutions: the *frontal*, the *parietal*, the *temporal*, the *occipital*, and the *limbic*; to these must be added the *central lobe*, or *island of Reil*, which is buried in the fissure of Sylvius, and the *olfactory lobe*, which lies below the orbital surface of the frontal lobe and was formerly described as the olfactory nerve. The first four of these lobes are named after the bones of the skull with which they are chiefly in relation, but it must be borne in mind that their limits do not correspond with those of the bones after which they are named.

The **frontal lobe.**—On the outer surface of the hemisphere, this lobe extends from the frontal pole to the fissure of Rolando, the latter separating it from the parietal lobe. Below, it is limited by the posterior limb of the fissure of Sylvius, which intervenes between it and the temporal lobe. On the mesial surface, it is separated from the limbic lobe by the calloso-marginal fissure; and on the under surface, it is bounded behind by the stem of the Sylvian fissure. The outer surface of the frontal lobe presents three sulci, which divide it into four convolutions (fig. 573). These are named the precentral, and the superior and inferior

frontal sulci.   The *precentral sulcus* runs upwards through this lobe, parallel to
the fissure of Rolando, and is usually divided into upper and lower parts.   It

FIG. 571.—Fissures and lobes on the external surface of the cerebral hemispheres.

forms the anterior limit of a convolution, which lies between it and the fissure
of Rolando, and which is called the *ascending frontal* or *precentral convolution*.

FIG. 572.—Fissures and lobes on the internal surface of the cerebral hemispheres.

From it two sulci, the *superior* and *inferior frontal*, run forwards and downwards,
and divide the remainder of the outer surface of the lobe into three parallel

convolutions, named respectively the *superior, middle,* and *inferior frontal convolutions.*

The *ascending frontal convolution* is a simple convolution, bounded in front by the precentral sulcus, behind by the fissure of Rolando, and extending from the supero-mesial border of the hemisphere to the posterior limb of the fissure of Sylvius.

The *superior frontal convolution* is situated between the margin of the longitudinal fissure and the superior frontal sulcus. It is continuous on the inner aspect of the hemisphere with the marginal convolution, and on the orbital surface with the internal orbital convolution. It is usually more or less completely subdivided into an upper and a lower part by an antero-posterior sulcus, the *sulcus paramesialis,* which, however, is frequently interrupted by bridging convolutions.

The *middle frontal convolution* is situated between the superior and inferior frontal sulci, and is continuous with the anterior orbital convolution on the

FIG. 573.—Convolutions and sulci on the external surface of the cerebral hemisphere.

lower surface of the hemisphere. The middle frontal convolution is frequently subdivided into two by a sagittally directed sulcus, the *sulcus frontalis medius* of Eberstaller.

The *inferior frontal convolution* is situated below the inferior frontal sulcus, and extends forwards from the lower part of the precentral sulcus; it is continuous with the external and posterior orbital convolutions on the under surface of the lobe. The inferior frontal convolution is subdivided by the anterior and ascending limbs of the fissure of Sylvius into three parts, viz.: (1) the *pars orbitalis,* below the anterior limb of the fissure; (2) the *pars triangularis* ('cap' of Broca), between the two limbs; and (3) the *pars basilaris,* behind the ascending limb.

The left inferior frontal convolution is, as a rule, more highly developed than the right, and is named the *convolution of Broca,* from the fact that in 1861 Broca discovered that it was the centre for speech.

The *under surface* of the frontal lobe is concave, and rests on the orbital plate of the frontal bone (fig. 574). It is divided into four convolutions by a well-marked **H**-shaped sulcus, the *orbital sulcus.* These are named, from their position, the *internal, anterior, external,* and *posterior orbital convolutions.* The internal orbital convolution presents a well-marked antero-posterior groove or sulcus, the *olfactory sulcus,* for the olfactory tract; and the portion internal to this is named the *gyrus rectus*, and is continuous with the marginal gyrus on the mesial surface.

The *mesial* or *internal surface* of the frontal lobe is occupied by a single curved convolution, which from its situation is termed the *marginal gyrus* (fig. 575). It commences in front of the anterior perforated space, runs along the margin of the longitudinal fissure on the mesial surface of the orbital lobe, where it is continuous with the internal orbital convolution; it then ascends, and runs backwards to the point where the calloso-marginal fissure turns upwards to reach the superior border of the hemisphere. The posterior part of this convolution is sometimes marked off by a vertical fissure, and is dis-

FIG. 574.—Convolutions and sulci on the under surface of the anterior lobe.

tinguished as the *paracentral convolution*, because it is continuous with those convolutions which lie immediately in front of and behind the central fissure or fissure of Rolando.

The **parietal lobe** forms a part of both the outer and mesial surfaces of the hemisphere. It is separated from the frontal lobe by the fissure of Rolando, but its boundaries below and behind are not so definite. Posteriorly, it is limited by the external parieto-occipital fissure, and by a line carried across the hemisphere from the outer end of this fissure to the pre-occipital notch. Below, it is separated from the temporal lobe by the posterior limb of the fissure of Sylvius, and by a line carried backwards from the horizontal part of this fissure to meet the one passing downwards to the pre-occipital notch.

The outer surface of the parietal lobe is cleft by a well-marked furrow, the *intraparietal sulcus* of Turner, which consists of an oblique and a horizontal portion. The oblique part is named the *post-central sulcus*, and commences below, about midway between the lower end of the fissure of Rolando and the upturned end of the fissure of Sylvius. It runs upwards and backwards, parallel to the fissure of Rolando, and is sometimes divided into an *upper* and a *lower* ramus. It forms the posterior limit of the ascending parietal or post-central convolution.

From about the middle of the post-central sulcus, or from the upper end of its inferior ramus, the *horizontal portion* of the intraparietal sulcus is carried backwards and slightly upwards through the parietal lobe, and is prolonged, under the name of the *ramus occipitalis*, into the occipital lobe, where it divides into two parts, which form nearly a right angle with the main stem and constitute the *transverse occipital fissure*. The part of the parietal lobe above the horizontal ramus is named the superior parietal convolution ; the part below, the inferior parietal convolution.

The *ascending parietal* or *post-central convolution* extends from the great longitudinal fissure above to the horizontal limb of the fissure of Sylvius below. It lies parallel with the ascending frontal or *precentral* convolution, with which it is connected below, and also, sometimes, above the fissure of Rolando.

The *superior parietal convolution* is bounded in front by the upper part of the post-central sulcus, which lies between it and the previous convolution, but with

FIG. 575.—Convolutions and sulci on the internal surface of the cerebral hemispheres.

which it is usually connected above the upper extremity of the sulcus ; behind, it is bounded by the external parieto-occipital fissure, outside the termination of which it is joined to the occipital lobe by a narrow convolution, the *first annectant gyrus*, or *arcus parieto-occipitalis* ; below, it is separated from the inferior parietal convolution by the horizontal portion of the intraparietal sulcus ; and above, it is continuous on the inner surface of the hemisphere with the precuneus or quadrate lobe.

The *inferior parietal convolution* lies below the horizontal ramus, and behind the lower part of the post-central sulcus. It is divided from before backwards into three convolutions. One, the *supramarginal*, arches over the upturned end of the fissure of Sylvius ; it is continuous in front with the ascending parietal convolution, and behind with the superior temporal convolution. The second, the *angular*, arches over the posterior end of the superior temporal or parallel sulcus, behind which it is continuous with the middle temporal convolution. The third, the *post-parietal*, curves round the end of the second temporal sulcus, and is continuous with the third temporal convolution.

The *internal* or *mesial* surface of the parietal lobe is continuous with the external surface, over the supero-mesial border of the hemisphere. It is bounded behind by the internal parieto-occipital fissure; in front, by the upturned end of the calloso-marginal fissure; and below, it is separated from the limbic lobe by the *post-limbic fissure*. It is of small size, and consists of a square-shaped convolution, which is termed the *quadrate lobe* or *precuneus*.

The **occipital lobe** is small and pyramidal in shape; it presents three surfaces: an outer, a mesial, and a tentorial.

Its *outer* surface is limited in front by the external parieto-occipital fissure, and by a line carried from the outer end of this fissure to the pre-occipital notch. This surface is traversed by the transverse occipital and the lateral occipital sulci. The *transverse occipital sulcus* is continuous with the posterior end of the ramus occipitalis of the intraparietal sulcus, and runs across the upper part of the lobe, a short distance behind the external parieto-occipital fissure. The *lateral occipital sulcus* extends from behind forwards, and divides the outer surface of the occipital lobe into an *upper* and a *lower* convolution, which are continuous in front with the parietal and temporal lobes. Elliot Smith has named the lateral occipital sulcus the *sulcus lunatus*; he regards it as the representative, in the human brain, of the 'Affenspalte' of the brain of the ape.

The *mesial* surface of the occipital lobe is bounded in front by the internal parieto-occipital fissure, and is traversed by the calcarine fissure, which subdivides it into the cuneus and the lingual lobule. The *calcarine fissure* commences near the occipital pole in a bifid extremity; it runs almost horizontally forwards, and ends in the substance of the limbic lobe, a little below the posterior extremity of the corpus callosum. It is joined at an acute angle by the *internal parieto-occipital fissure*, and the wedge-shaped area between the two fissures is named the *cuneus*. The *lingual lobule* lies between the calcarine fissure and the posterior part of the collateral fissure, and extends, therefore, on to the tentorial surface. Behind, it reaches the occipital pole; in front, it is continued on to the tentorial surface of the temporal lobe, and joins the hippocampal convolution of the limbic lobe. The anterior portion of the calcarine fissure gives rise to the prominence of the *hippocampus minor*, or *calcar avis*, in the interior of the lateral ventricle.

The *tentorial* surface of the occipital lobe is limited in front by an imaginary line carried inwards from the pre-occipital notch, and consists of the posterior part of the *occipito-temporal convolution* and the lower part of the *lingual lobule*, which are separated from each other by the posterior segment of the *collateral fissure*.

The **temporal lobe** presents upper, lower, and tentorial surfaces.

The *upper* surface forms the lower limit of the fissure of Sylvius, and overlaps the island of Reil. On opening out the fissure of Sylvius, three or four gyri will be seen springing from the depth of the hinder end of the fissure, and running obliquely forwards and outwards on the posterior part of the upper surface of the first temporal convolution; these are named the *transverse temporal gyri*, or *gyri of Heschl*.

The *outer* surface is bounded above by the posterior limb of the fissure of Sylvius, and by the imaginary line continued backwards from it; below, it is limited by the infero-lateral border of the hemisphere. It is divided by two sulci, which are termed the first and second temporal sulci. The *first temporal sulcus* is well marked, and runs from before backwards through the temporal lobe, some little distance below, but parallel with, the posterior limb of the fissure of Sylvius; and hence it is often termed the *parallel sulcus*. The *second temporal sulcus* takes the same direction as the first, but is situated at a lower level, and is usually subdivided into two or more parts. These two sulci subdivide this surface of the temporal lobe into three convolutions. The *first* or *superior temporal convolution* lies between the posterior limb of the fissure of Sylvius and the parallel sulcus, and is continuous behind with the supramarginal and angular convolutions. The *second* or *middle temporal convolution* is placed between the first and second temporal sulci, and is joined posteriorly with the angular and post-parietal convolutions. The *third* or *inferior temporal convolution* is placed below the second temporal sulcus, and is connected behind with the lower occipital convolution; it also extends round the infero-lateral border on to the tentorial surface, where it is limited by the occipito-temporal sulcus about to be described.

The *tentorial* surface is concave, looks downwards and inwards, and is directly continuous posteriorly with the tentorial surface of the occipital lobe. It is traversed by the *occipito-temporal sulcus*, which extends from near the occipital pole behind to within a short distance of the temporal pole in front; but is frequently subdivided by bridging gyri. To the outer side of this fissure is the narrow tentorial part of the third temporal convolution, and to its inner side the *occipito-temporal convolution*, which extends from the occipital to the temporal pole. This convolution is limited internally by the collateral fissure, which separates it from the lingual lobule behind and from the hippocampal convolution of the limbic lobe in front.

The **central lobe**, or **island of Reil** (fig. 576), lies deeply in the Sylvian fissure, and can only be seen when the lips of that fissure are widely separated, since it is overlapped and hidden by the convolutions which bound the fissure. These convolutions are termed the *opercula of the insula*; they are separated from each other by the three limbs of the Sylvian fissure, and named the orbital, frontal, fronto-parietal, and temporal opercula. The *orbital operculum* lies below the anterior limb of the fissure, the *frontal* between the anterior and ascending limbs, the *fronto-parietal* between the ascending limb and the upturned end of the

FIG. 576.—The island of Reil. Left side. The overlapping parts of the hemisphere have been removed.

1, 2, 3. Gyri breves. 4, 5. Gyrus longus, bifurcated at its upper extremity. Between the gyri breves and the gyrus longus is seen the sulcus centralis.

posterior limb, and the *temporal* below the posterior limb. The frontal operculum may be of small size; and when this is the case, the anterior and ascending limbs of the fissure of Sylvius arise from a common stem which lies between the orbital and fronto-parietal opercula. The island of Reil is surrounded by a deep *limiting sulcus*, which separates it from the frontal, parietal, and temporal lobes. When the opercula have been removed it presents the form of a triangular eminence, the apex of which is directed downwards and inwards towards the anterior perforated space. It is divided into a *precentral* and a *post-central* lobe by the *sulcus centralis*, which runs backwards and upwards from the apex of the insula. The precentral lobe is further subdivided by shallow sulci into three or four short convolutions, the *gyri breves*, while the post-central lobe is named the *gyrus longus* and is often bifurcated at its upper extremity. The grey matter of the insula is continuous with that of the different opercula, while its mesial surface corresponds with the lenticular nucleus of the corpus striatum.

The Limbic lobe.—The term limbic lobe (*grande lobe limbique*) was introduced by Broca in 1878, and under it he included two convolutions, viz. the callosal and hippocampal, which together arch round the corpus callosum and the dentate or hippocampal fissure. These he separated on the morphological

3 G

ground that they are well developed in animals possessing a keen sense of smell (osmatic animals), such as the dog and fox. To the lobe thus defined the following parts must be added, viz.: (1) the laminæ of the septum lucidum, the fornix and its fimbriæ, and the infracallosal gyrus, which may be regarded as forming an inner or deep arch; (2) the peduncles and longitudinal striæ of the corpus callosum, together with the fasciola cinerea and gyrus dentatus, which form a middle arch. The outer arch is constituted by the callosal and hippo-campal convolutions: the inner and middle arches are separated from each other by the corpus callosum.

The **callosal convolution** is an arch-shaped convolution, lying in close relation to the superficial surface of the corpus callosum, from which it is separated by a slit-like fissure, the *callosal fissure*. It commences below the rostrum of the corpus callosum, curves round in front of the genu, extends along the upper surface of the body, and finally turns downwards behind the splenium, where it is connected by a narrow *isthmus* with the gyrus hippocampi. It is separated from the marginal convolution by the calloso-marginal sulcus, and from the quadrate lobe by the post-limbic sulcus.

The **hippocampal convolution** (gyrus hippocampi) is bounded above by the hippocampal or dentate fissure, and below by the anterior part of the collateral fissure. Behind, it is continuous superiorly, through the isthmus, with the callosal convolution, and inferiorly with the lingual lobule. Its anterior extremity is recurved in the form of a hook, and is named the *uncus*. Running in the substance of the callosal and hippocampal convolutions, and connecting them together, is a tract of arched fibres, named the *cingulum*. The outer root of the olfactory tract passes into the anterior extremity of the hippocampal convolution, and the inner root into the commencement of the callosal con-volution, so that these two convolutions, with the addition of the olfactory tract, present a racquet-like appearance—the olfactory tract constituting the handle and the two convolutions the circumference of the blade.

The **dentate convolution** (formerly named the *dentate fascia*) is situated above the gyrus hippocampi, from which it is separated by the hippocampal or dentate fissure. It is overlapped by the fimbria, from which it is separated by the fimbrio-dentate fissure, and is a narrow, elongated convolution, the free surface of which presents a notched or toothed appearance, hence its name. Posteriorly it is prolonged as a delicate lamina, the *fasciola cinerea*, around the splenium of the corpus callosum, and becomes continuous on the upper surface of that body with its mesial and lateral longitudinal striæ. Anteriorly it is prolonged into the notch produced by the recurving of the uncus, where it forms a sharp bend; from here it can be traced as a delicate band (*band of Giacomini*) over the uncus, on the outer surface of which it is lost.

The remaining structures which contribute to the formation of the limbic lobe will be subsequently described.

The **dentate** or **hippocampal fissure** commences immediately behind the posterior extremity of the corpus callosum, and runs forwards between the hippocampal and dentate convolutions to terminate in the uncus. It is a complete fissure (see page 810), and gives rise to the prominence of the hippo-campus major in the descending horn of the lateral ventricle. It separates the hippocampal and dentate convolutions from each other.

The **olfactory lobe** is situated on the under surface of the frontal lobe. It is rudimentary in man and some other mammals, but in most vertebrates it is well developed and consists of a distinct extension of the cerebral hemisphere, enclosing a portion of the anterior horn of the lateral ventricle. In man it is long and slender, and may be described as consisting of two parts, the *anterior* and *posterior olfactory lobules*.

The *anterior olfactory lobule* is made up of: (1) the olfactory bulb; (2) the olfactory tract; (3) the trigonum olfactorium; and (4) the area of Broca.

(1) The *olfactory bulb* is an oval mass of a reddish-grey colour, which rests on the cribriform plate of the ethmoid bone, and forms the anterior expanded extremity of the olfactory tract. Its under surface receives the olfactory nerves, which pass upwards through the cribriform plate from the olfactory region of the nose. Its minute structure will be subsequently described.

(2) The *olfactory tract* is a band of white matter, triangular on section, the apex being directed upwards. It lies in the olfactory sulcus on the under

surface of the frontal lobe. Traced backwards, it is seen to divide into two roots, an outer and an inner. The *outer root* passes across the outer part of the anterior perforated space to the nucleus amygdalæ and the anterior part of the gyrus hippocampi. The *inner root* turns sharply inwards behind Broca's area, and ends in the callosal convolution; in other words, the inner root is continuous with one extremity and the outer root with the other extremity of the limbic lobe.

(3) The *trigonum olfactorium* is a small triangular area of grey matter, situated between the diverging roots of the olfactory tract in front of the anterior perforated space, and is sometimes described as the *middle* or *grey root* of the tract.

(4) The *area of Broca* is a small triangular field, situated in front of the inner root of the olfactory tract, and separated from the peduncle of the corpus

FIG. 577.—Base of the brain.

callosum and the posterior olfactory lobule by a fissure (*fissura prima*). It is continuous with the commencement of the callosal convolution, and is cut off from the gyrus rectus by an oblique fissure (*fissura serotina*).

The *posterior olfactory lobule*, or *anterior perforated space*, is marked off from the anterior lobule by the fissura prima, and is situated at the commencement of the fissure of Sylvius. Internally it is bounded by the peduncle of the corpus callosum, and is continuous with the lamina cinerea. Posteriorly it is bounded by the optic tract, and it is partially concealed by the temporal lobe which overlaps it. It has received the name of anterior perforated space from its being perforated by numerous openings, which transmit blood-vessels to the interior of the brain, and it corresponds to the under surface of the lenticular nucleus and part of the claustrum.

Interior of the cerebrum.—If the upper part of either hemisphere be removed with a knife, at a level about half an inch above the corpus callosum, the internal white matter will be exposed. It is an oval-shaped centre, of white substance, surrounded by a narrow convoluted margin of grey matter, which presents nearly an equal thickness in every part. This white central mass has been called the *centrum ovale minus*. Its surface is studded with numerous minute red dots (*puncta vasculosa*), produced by the escape of blood from divided blood-vessels. In inflammation or great congestion of the brain these are very numerous, and of a dark colour. If the remaining portions of the hemispheres be separated from each other, a broad band of white substance will be observed, connecting them at the bottom of the longitudinal fissure ; this is the

Fig. 578.—Section of the brain. Made on a level with the corpus callosum.

corpus callosum. The margins of the hemispheres which overlap this portion of the brain are called the *labia cerebri*. Each labium is part of the callosal convolution already described ; and the slit-like interval between it and the upper surface of the corpus callosum is termed the *callosal fissure* (fig. 575). The hemispheres should now be sliced off to a level with the upper surface of the corpus callosum, when the white substance of that structure will be seen connecting the two hemispheres. The large expanse of medullary matter now exposed, surrounded by the convoluted margin of grey substance, is called the *centrum ovale majus of Vieussens* (fig. 578).

The **corpus callosum** is the great transverse commissure which unites the cerebral hemispheres and roofs in the lateral ventricles. A good conception of its position and size is obtained by examining a mesial section of the brain, when it is seen to form an arched structure about four inches in length. Its anterior extremity reaches to within about an inch and a half of the frontal pole, and its posterior extremity about two and a half inches from the occipital pole of the hemisphere.

The *anterior* extremity is named the *genu*, and is bent downwards and backwards in front of the septum lucidum; diminishing rapidly in thickness, it is prolonged backwards, under the name of the *rostrum*, which is connected below with the lamina terminalis. The anterior cerebral arteries are in contact with the under aspect of the rostrum; they then arch over the front of the genu, and are carried backwards above the body of the corpus callosum.

FIG. 579.—Vertical median section of the encephalon, showing the parts in the middle line.

1. Callosal convolution. Above it is the calloso-marginal fissure.
2. Fissure of Rolando.
3. The parieto-occipital fissure.
4. 4 point to the calcarine fissure, which is just above the numbers. Between 2 and 3 is the quadrate lobe. Between 3 and 4 is the cuneate lobe.
5. The corpus callosum.
6. The septum lucidum.
7. The fornix.
8. Anterior crus of the fornix, descending to the base of the brain, and turning on itself to form the corpus albicans. The bundle of Vicq d'Azyr is indicated by a dotted line.
9. The optic thalamus. Behind the anterior crus of the fornix, a shaded part indicates the foramen of Monro; in front of the number an oval mark shows the position of the grey or middle commissure.
10. The velum interpositum.
11. The pineal body.
12. The corpora quadrigemina.
13. The crus cerebri.
14. The valve of Vieussens (to the right of the number).
15. The pons Varolii.
16. The third nerve.
17. The pituitary body.
18. The optic nerve.
19 points to the anterior commissure, indicated by the oval outline behind the number.

The *posterior* extremity is termed the *splenium*, and constitutes the thickest part of the corpus callosum. It overlaps the velum interpositum and the midbrain, and terminates in a thick, convex, free border. When a mesial section of the splenium is examined, it is seen that the posterior end of the corpus callosum is acutely bent forwards, the upper and lower parts being applied to each other.

The *upper* surface is convex from before backwards, and is nearly an inch in width. Its mesial part forms the bottom of the great longitudinal fissure, and is in contact posteriorly with the lower border of the falx cerebri. Laterally it is overlapped by the callosal convolution, but is separated from it by the slit-like callosal fissure. It is traversed by numerous ridges and furrows, and is covered by a thin layer of grey matter continuous on either side with the grey matter of the callosal convolution, and better marked in animals which possess a keen sense of smell. On either side of the middle line is a slightly elevated band, named the *mesial longitudinal stria*; and still more externally, under the callosal convolution, is a second band, termed the *lateral longitudinal stria* or *tænia tecta*. Traced backwards, the longitudinal striæ are found to be continuous with the fasciola cinerea and the dentate convolution (page 818). Traced forwards, they pass round the genu and extend backwards on the under surface of the rostrum, where they form the *peduncles of the corpus callosum* or *gyrus subcallosus*. The peduncle of the corpus callosum passes backwards across the anterior perforated space, and ends in the apex of the temporal lobe: further, it is connected with the anterior pillar of the fornix, and with the olfactory tract.

The *lower* surface is concave, and forms on either side of the middle line the roof of the lateral ventricle. Mesially, this surface is attached in

FIG. 580.—The lateral ventricles of the brain.

front to the septum lucidum; behind this it is fused with the upper surface of the body of the fornix, while the splenium is in contact with the velum interpositum.

On each side, the fibres of the corpus callosum radiate in the white substance and pass to the various parts of the cerebral cortex. The greater thickness of the two extremities of this commissure is explained by the fact that the fibres from the anterior and posterior parts of each hemisphere cannot pass directly across, but have to take a curved direction. The part of the corpus callosum which curves forwards on each side from the genu into the frontal lobe is called the *forceps minor*. The part which curves backwards from each side of the splenium into the occipital lobe is known as the *forceps major*. Between these two parts on each side is the main body of the fibres : these constitute the *tapetum* or *mat*, and extend laterally into the temporal lobe, and cover in the body of the lateral ventricle.

An incision should now be made through the corpus callosum, on either side of the middle line, when two large irregular cavities will be exposed, which extend through a great part of the length of each hemisphere. These are the lateral ventricles.

The lateral ventricles (figs. 580 and 581).—The lateral ventricles, two in number, right and left, are irregular cavities situated in the lower and inner parts of the cerebral hemisphere, one on either side of the middle line. They are separated from each other by a mesial vertical partition, the *septum lucidum*, but communicate with the third ventricle and indirectly with each other through the *foramen of Monro*. They are lined by a thin, diaphanous membrane, the *ependyma*, which is covered by ciliated epithelium, and are moistened by cerebrospinal fluid, which, even in health, may be secreted in considerable amount. Each lateral ventricle consists of a central cavity or *body*, and three prolongations from it, termed *cornua*. The *anterior* cornu curves forwards and outwards into the frontal lobe; the *posterior* backwards and inwards into the occipital lobe; and the *middle* descends into the temporal lobe.

The **body** of the lateral ventricle extends from the foramen of Monro to the splenium of the corpus callosum. It is an irregularly curved cavity, triangular in shape on transverse section, and presents a roof, a floor, and an inner wall. Its *roof* is formed by the under surface of the corpus callosum; its *floor* is formed by the following parts, enumerated in their order of position, from before backwards : the *caudate nucleus of the corpus striatum*, *tænia semicircularis* and *vein of the corpus striatum*, the outer portion of the upper surface of the *optic thalamus*, the *choroid plexus*, and the lateral part of the *fornix*; its *inner wall* is the posterior part of the septum lucidum, which separates it from the opposite ventricle, and connects the under surface of the corpus callosum with the fornix.

The **anterior cornu** passes forwards and outwards, with a slight inclination downwards, from the foramen of Monro into the frontal lobe, curving round the anterior extremity of the caudate nucleus. It is bounded above by the corpus callosum, and below by the upper surface of its reflected portion, the rostrum. It is bounded internally by the anterior portion of the septum lucidum, and externally by the head of the caudate nucleus of the corpus striatum. Its apex reaches the posterior surface of the genu of the corpus callosum.

The **posterior cornu** passes into the occipital lobe, its direction being backwards and outwards, and then inwards; its concavity is therefore directed inwards. Its roof is formed by the fibres of the corpus callosum passing to the temporal and occipital lobes. On its inner wall is seen a longitudinal eminence, which is an involution of the ventricular wall produced by the calcarine sulcus; this is called the *hippocampus minor*, or *calcar avis*. Just above this the

FIG. 581.—Drawing of a cast of the ventricular cavities viewed from above. (Testut.)

a, a', a''. The three horns—anterior, posterior, and middle—of the left lateral ventricle. *b.* Third ventricle. *c.* Fourth ventricle. *d.* Commencement of central canal of cord. 1. Inferior angle of fourth ventricle. 2. Superior angle. 3. Lateral angle. 4. Sylvian aqueduct. 5. Recessus suprapinealis. 6. Vulva. 7. Junction of descending and posterior horns.

forceps major of the corpus callosum, sweeping round to enter the occipital
lobe, causes another projection, which is known as the *bulb of the posterior
horn*. The hippocampus minor and bulb of the posterior horn are extremely
variable in their degree of development, being in some cases ill defined, while in
others they are unusually prominent.

The **middle** or **descending cornu,** the largest of the three, traverses the
temporal lobe of the brain, forming in its course a remarkable curve round the
back of the optic thalamus. It passes at first backwards, outwards, and downwards, and then curves forwards and inwards, to within an inch of the apex of
the temporal lobe, its direction being fairly well indicated on the surface of the
brain by that of the parallel sulcus. Its upper boundary, or roof, is formed
chiefly by the under surface of the tapetum of the corpus callosum, but the tail
of the nucleus caudatus of the corpus striatum and the tænia semicircularis are
also prolonged into it, and extend forwards in the roof of the descending horn
to its extremity, where they end in a mass of grey matter, the *amygdaloid nucleus.*
Its lower boundary, or floor, presents for examination the following parts : the

Fig. 582.—Transverse vertical section of the brain, through the fore part of the foramen
    magnum, looked at from the front. (After Hirschfeld and Leveillé.)

hippocampus major, the *fimbria* or *tænia hippocampi,* the *eminentia collateralis,*
and the *choroid plexus.* When the choroid plexus is removed, a cleft-like
opening is formed along the mesial wall of the descending cornu. This cleft
constitutes the lower part of the *choroidal fissure,* and through it the ventricular
cavity opens on to the surface of the hemisphere.

The **hippocampus major,** or **cornu Ammonis,** is a curved eminence, about two
inches long, which extends throughout the entire length of the floor of the
descending horn. Its lower extremity is enlarged, and presents two or three
rounded elevations with intervening depressions, which give it a paw-like
appearance, and hence it is named the *pes hippocampi.* If a transverse section
be made through the hippocampus major, it will be seen that this eminence is
produced by the folding of the wall of the hemisphere to form the dentate or
hippocampal fissure. The main mass of the hippocampus major consists of
grey matter ; but on its ventricular surface is a thin layer of white matter, known
as the *alveus,* which is continuous with the fimbria of the fornix and is covered

by the ependyma of the ventricle. Macarthy of Montreal has shown* that if the alveus and superficial strata of grey matter be reflected from the surface of the hippocampus by an incision carried along its convexity, the 'core' of the hippocampus, as he terms it, presents, in many cases, a corrugated or crimped appearance.

The **eminentia collateralis** is an elongated eminence lying to the outer side and parallel with the hippocampus major. It corresponds with the middle part of the collateral fissure, and its size depends on the depth and direction of that furrow. It is continuous behind with a flattened triangular area which is situated between the posterior and descending cornua, and named the *trigonum ventriculi*.

The fimbria is a continuation of the posterior pillar of the fornix, and will be discussed with that body; while a description of the choroid plexus will be found on page 830.

The **grey nuclei**, or **basal ganglia of the hemisphere,** consist of the corpus striatum, the claustrum, and the amygdaloid nucleus.

The **corpus striatum** has received its name from the striped appearance which its section presents, in consequence of diverging white fibres being mixed

FIG. 583.—Transverse section of the middle horn of the lateral ventricle.
(From a drawing by F. A. Barton.)

EMINENTIA COLLATERALIS

HIPPOCAMPUS MAJOR

GREY MATTER OF THE HIPPOCAMPUS

DENTATE CONVOLUTION

UNCINATE CONVOLUTION

UNCINATE PROCESS

with the grey matter which forms the greater part of its substance. A part of this body is embedded in the white substance of the hemisphere, and is therefore external to the ventricle. It is termed the *extra-ventricular portion*, or the *nucleus lenticularis*; a part, however, is visible in the ventricle, and is named the *intraventricular portion*, or the *nucleus caudatus*.

The **nucleus caudatus** (fig. 584) is a pear-shaped, highly arched mass of grey matter; its broad extremity, or *head*, is directed forwards into the fore part of the body and anterior cornu of the lateral ventricle, and is continuous with the grey matter of the anterior perforated space and with the anterior end of the lenticular nucleus; its narrow end, or *tail*, is directed outwards and backwards on the outer side of the optic thalamus, from which it is separated by the tænia semicircularis. It is then continued downwards into the roof of the descending cornu, where it terminates in the *nucleus amygdalæ*, at the apex of the temporal lobe. It is covered by the lining of the ventricle, and crossed by some veins of considerable size. It is separated from the extra-ventricular portion, in

* *Journal of Anatomy and Physiology*, vol. xxxiii.

the greater part of its extent, by a lamina of white matter, which is called the *internal capsule*, but the two portions of the corpus striatum are united in front.

The **nucleus lenticularis,** or extra-ventricular portion of the corpus striatum, is placed outside the caudate nucleus and optic thalamus, and is only seen in sections of the hemisphere. When divided horizontally, it presents, to some extent, the appearance of a biconvex lens, while a vertical transverse section of its central part presents a somewhat triangular outline. It does not extend as far forwards or backwards as the nucleus caudatus. It is bounded externally by a lamina of white matter called the *external capsule,* on the outer surface of which is a thin layer of grey matter termed the *claustrum.* Its anterior extremity is continuous with the lower part of the head of the caudate nucleus and with the grey matter of the anterior perforated space.

Fig. 584.—Middle part of a horizontal section through the cerebrum at the level of the dotted line in the small figure of one hemisphere.   (From Ellis, after Dalton.)

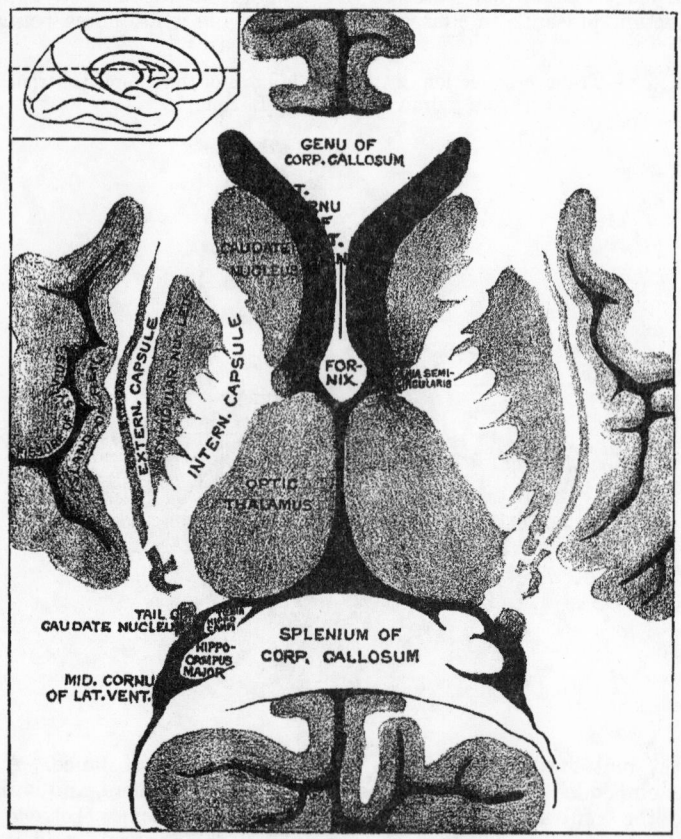

Upon making a transverse vertical section through the middle of the nucleus lenticularis it is seen to present two white laminæ, the *medullary laminæ,* parallel with its lateral border, which divide it up into three zones. The outer and largest zone is of a reddish colour and is known as the *putamen,* while the two inner are paler and of a yellowish tint and together constitute the *globus pallidus.* All three zones are marked by fine radiating white fibres, which are most distinct in the putamen.

The grey matter of the corpus striatum is traversed by nerve-fibres, some of which are believed to originate in it. The cells are multipolar, both large and small; those of the lenticular nucleus contain yellow pigment. The caudate and lenticular nuclei are not only directly continuous with each other anteriorly, but are connected to each other by numerous fibres. The corpus striatum is also connected: (1) to the cerebral cortex, by what are termed the *cortico-striate* fibres; (2) to the optic thalamus, by fibres which pass through the internal capsule, and by a strand named the *ansa lenticularis*; (3) to the crus cerebri, by fibres which leave the lower aspect of the caudate and lenticular nuclei.

The **claustrum** is a thin layer of grey matter, situated on the outer surface of the external capsule. On transverse section it is seen to be triangular, with its apex directed upwards and its base downwards. Its inner surface, which is contiguous to the external capsule, is smooth, but its outer surface presents ridges and furrows which correspond with the convolutions and sulci of the island of Reil, with which it is in close relationship. The claustrum is regarded as a detached portion of the grey matter of the island of Reil, from which it is separated by a layer of white fibres, the *capsula extrema* or *band of Baillarger* Its cells are small and spindle-shaped, and contain yellow pigment; they are similar to those found in the deepest layer of the cortex.

The **amygdaloid nucleus** is an ovoid mass of grey matter, situated in the roof of the descending horn at its lower extremity. It is merely a localised thickening of the grey cortex, and is continuous with that of the uncus; the tænia semi-circularis and the tail of the caudate nucleus end in it.

The **internal capsule** is a flattened band of white fibres, which lies between the lenticular nucleus on the outer side and the caudate nucleus and optic thalamus on the inner side. In horizontal section (figs. 584, 585) it is seen to be somewhat abruptly curved, with its convexity inwards; the prominence of the curve is called the *genu*, and projects between the caudate nucleus and the optic thalamus. The portion in front of the genu is termed the *anterior limb*, and separates the lenticular from the caudate nucleus; the portion behind the genu is the *posterior limb*, and separates the lenticular nucleus from the optic thalamus.

The anterior limb of the internal capsule contains: (1) fibres which pass from the optic thalamus to the frontal lobe (cortico-thalamic); (2) fibres connecting the lenticular and caudate nuclei (lenticulo-caudate); (3) fibres connecting the cortex with the corpus striatum (cortico-striate); and (4) fibres passing from the frontal lobe to the nuclei pontis (cortico-protu-berantial). The fibres which occupy the region of the genu are named the *geniculate fibres*. They originate in the motor part of the cerebral cortex, and, after passing downwards in the inner fifth of the crusta, undergo decussation and end in the motor nuclei of the cranial nerves of the opposite side. The anterior two-thirds of the posterior limb contains the *pyramidal fibres*, which arise in the motor area of the cortex and, passing downwards through the middle three-fifths of the crusta, are continued into the pyramids of the medulla. The

Fig. 585.—Horizontal section of the internal capsule. (Schematic.) (Testut.) Below the horizontal section, there is shown a transverse section of the corresponding crus cerebri.

1. Optic thalamus. 2, 2'. Caudate nucleus. 3. Lenticular nucleus. 4. Claustrum. 5. Island of Reil. 6. Lateral ventricle. 7. Anterior stalk of optic thalamus. 8. Posterior stalk of optic thalamus, or optic radiations. *a*. Anterior segment of internal capsule. *b*. Geniculate bundle (in green). *c*. Pyramidal fibres (in red). *d*. Posterior cortico-pontine fibres. *e*. Fillet.

posterior third of the posterior limb contains: (1) sensory fibres, which are largely derived from the optic thalamus, but some of which may be continued upwards from the mesial fillet; (2) the fibres of optic radiation, which pass from the lower visual centres to the cortex of the occipital lobe; (3) auditory fibres, from the lateral fillet to the temporal lobe ; and (4) cortico-protuberantial fibres, which pass from the occipital and temporal lobes to the nuclei pontis.

The fibres of the internal capsule radiate widely as they pass to the various parts of the cerebral cortex, forming the *corona radiata* and intermingling with the fibres of the corpus callosum.

The **external capsule** is a lamina of white matter, situated on the outer side of the lenticular nucleus, between it and the claustrum, and is continuous with the internal capsule below and behind the lenticular nucleus. It probably contains fibres derived from the optic thalamus, the anterior white commissure, and the subthalamic region.

The **substantia innominata of Meynert** is a stratum consisting partly of grey and partly of white matter, which lies below the anterior part of the optic thalamus and lenticular nucleus. It consists of three layers, superior, middle, and inferior. The *superior* layer is named the *ansa lenticularis*, and its fibres, derived from the medullary lamina of the lenticular nucleus, pass inwards to end in the optic thalamus and subthalamic region, while others are said to terminate in the tegmentum and red nucleus. The *middle* layer consists of nerve-cells and nerve-fibres : fibres enter it from the parietal lobe through the external capsule, while others are said to connect it with the posterior longitudinal fasciculus. The *lower* layer forms the main part of the inferior stalk of the optic thalamus, and connects this body with the temporal lobe and the island of Reil.

The **tænia semicircularis** is a narrow, whitish band of medullary substance, situated in the depression between the caudate nucleus and the optic thalamus. Anteriorly, its fibres are partly continued into the anterior pillar of the fornix ; some, however, pass over the anterior commissure to the grey matter between the caudate nucleus and septum lucidum, while others are said to penetrate the caudate nucleus. Posteriorly, it is continued into the roof of the middle or descending horn of the lateral ventricle, at the extremity of which it enters the *nucleus amygdalæ*. Superficial to it is a large vein, *vena corporis striati*, which receives numerous small veins from the surface of the corpus striatum and optic thalamus ; it runs forwards to the foramen of Monro and joins with the vein which drains the choroid plexus, to form the corresponding *vena Galeni*. On the surface of the vein of the corpus striatum is a narrow band of white fibres, named the *lamina cornea*.

The remains of the corpus callosum should now be removed in order to expose the fornix.

The **fornix** (figs. 579, 580) is a longitudinal, arch-shaped lamella of white matter, situated beneath the corpus callosum, with which it is continuous behind, but from which it is separated in front by the septum lucidum. It may be described as consisting of two symmetrical halves, one for either hemisphere. The two portions are not united to each other in front and behind, but their central parts are joined together in the middle line. The two anterior, separated parts are called the *anterior pillars* (*columnæ fornicis*) ; the intermediate united portions constitute the *body of the fornix* ; and the posterior parts, which are also separated from each other, are called the *posterior pillars* (*crura fornicis*).

The *body of the fornix* is triangular, narrow in front and broad behind. The mesial part of its upper surface is connected to the septum lucidum in front and to the corpus callosum behind. The lateral portion of this surface forms part of the floor of the lateral ventricle, and is covered by the ventricular epithelium. The under surface rests upon the velum interpositum, which separates it from the epithelial roof of the third ventricle, and from the inner portions of the upper surfaces of the optic thalami. Its outer edge overlaps the choroid plexuses, and is continuous with the epithelial covering of these structures.

The *anterior pillars* arch downwards in front of the foramen of Monro and behind the anterior commissure. They become separated by a narrow interval, and each descends through the grey matter in the lateral wall of the third ventricle to the base of the brain, where it terminates in the corpus albicans. From the grey matter of the corpus albicans a fasciculus of fibres, termed the *bundle of Vicq d'Azyr*, takes origin and is prolonged into the anterior nucleus of the optic thalamus. The anterior pillar of the fornix and the bundle of Vicq d'Azyr together form a loop resembling the figure of 8, but the continuity of the loop is broken in the corpus albicans. The anterior pillar of the fornix is joined by the peduncle of the pineal body and by the superficial fibres of the tænia semicircularis, and is also said to receive fibres from the septum lucidum. Zuckerkandl describes an *olfactory fasciculus* which becomes detached from the main portion of the anterior pillar of the fornix, and passes downwards in front of the anterior commissure to the base of the brain, where it divides into two bundles, one joining the inner root of the olfactory tract; the other, the peduncle of the corpus callosum, and, through the latter, reaching the hippocampal convolution.

The *posterior pillars* are prolonged backwards from the body of the fornix. They are flattened bands, and at their commencement are intimately connected by their upper surfaces with the under aspect of the corpus callosum. Diverging from one another, each curves round the posterior extremity of the optic thalamus, and passes downwards and forwards into the descending horn of the lateral ventricle. Here it lies along the concavity of the hippocampus major, on the surface of which some of its fibres are spread out to form the *alveus*, while the remainder are continued as a narrow white band, the *fimbria* or *tænia hippocampi*, which is prolonged into the uncus or hook of the gyrus hippocampi. Its inner edge overlaps the dentate convolution (page 818), from which it is separated by the *fimbrio-dentate fissure*; from its outer margin, which is thin and ragged, the ventricular epithelium is reflected over the choroid plexus as the latter projects into the choroidal fissure.

The posterior pillars of the fornix are connected to each other across the middle line by a thin lamina which contains transverse fibres, and has been named the *psalterium*, or *lyra*, from its fancied resemblance to a harp. In the new-born child a space exists between the fornix and the corpus callosum, which usually becomes obliterated shortly after birth. It is compressed from above downwards, and presents a triangular shape; its apex is directed forwards, and is said to be continuous with the cavity of the fifth ventricle. This space is named the *ventricle of the fornix*, or, after its discoverer, the *ventricle of Verga*.

Foramen of Monro.—Between the anterior pillars of the fornix and the anterior extremities of the optic thalami, an oval aperture is seen on each side: this is the foramen of Monro. The two openings descend towards the middle line and lead into the upper part of the third ventricle. Through this foramen the lateral ventricles communicate with the third ventricle, and consequently with each other. Behind its epithelial lining, the two choroid plexuses are joined to each other across the middle line.

The **anterior commissure** is a bundle of white fibres, which connects the two cerebral hemispheres across the middle line, and is placed in front of the anterior pillars of the fornix. On transverse section it is seen to be oval in shape, its long diameter being vertical in direction and measuring about one-fifth of an inch. Its fibres can be traced outwards and backwards on each side beneath the corpus striatum into the substance of the temporal lobe. It serves in this way to connect the two temporal lobes, but it also contains fibres from the olfactory tract of the opposite side, the decussation of which in the anterior commissure may serve to explain the condition of crossed anosmia, e.g. where there is a lesion in one temporal lobe with a loss of smell in the olfactory area of the opposite side of the nose.

The **septum lucidum** is a thin, double, vertically placed partition, which forms the internal boundary of the body and anterior horn of the lateral ventricle. It consists of two distinct laminæ, separated in part of their extent by a narrow chink or interval, called the *fifth ventricle*. It is a thin, semitransparent septum, attached, above, to the under surface of the corpus callosum; below, to the anterior part of the fornix behind, and the reflected portion of the corpus callosum in front. It is triangular in form, broad in front and narrow behind; its inferior angle corresponds with the upper part of the anterior commissure. The outer surface of each lamina is directed towards the lateral ventricle, and is covered by the ependyma of that cavity, while its mesial surface bounds the cavity of the fifth ventricle.

Fifth ventricle.—The fifth ventricle is generally regarded as a part of the great longitudinal fissure, which has become shut off by the union of the hemispheres in the formation of the corpus callosum above and the fornix below. Each half of the septum is therefore formed by the median wall of the hemisphere, and consists of an internal layer of grey matter, derived from the grey matter of the cortex, and an external layer of white substance continuous with the white matter of the cerebral hemispheres. The fifth ventricle differs from the other ventricles of the brain, inasmuch as it is not developed from the cavity of the cerebral vesicles, is not lined by ciliated epithelium but by altered pia mater, and does not communicate with the general ventricular cavity; further, the fluid it contains is of the nature of lymph.

The **choroid plexus of the lateral ventricle** is a highly vascular fringe-like process of pia mater, which appears as if it were contained within the ventricular cavity. The plexus, however, is not actually within the cavity, as it is everywhere covered by a layer of epithelium continuous with the epithelial lining of

FIG. 586.—The fornix, velum interpositum, and middle or descending cornu of the lateral ventricle.

the ventricle, which therefore shuts it out of the ventricular cavity. It extends from the foramen of Monro, where it is joined with the plexus of the opposite ventricle, to the extremity of the descending horn. The part in relation to the body of the ventricle forms the vascular fringed margin of a triangular process of pia mater, named the *velum interpositum*, and projects from under cover of the outer edge of the fornix. It lies upon the upper surface of the optic thalamus, from which the epithelium is reflected over the plexus on to the edge of the fornix. The portion in relation to the descending horn lies in the concavity of the hippocampus major and overlaps the fimbria: from the outer edge of the latter the epithelium is reflected over the plexus on to the roof of the cornu. It consists of minute and highly vascular villous processes, each with an afferent and an efferent vessel. The cells of the epithelium which covers it often contain yellowish fat molecules. The *arteries* of the plexus are: (a) the anterior choroidal, a branch of the internal carotid, which enters the plexus at the extremity of the descending horn ; and (b) the posterior choroidal, one or two small branches of the posterior cerebral, which pass forwards under the splenium. The *veins* of the choroid plexus unite to form a prominent vein, which courses from behind forwards to the foramen of Monro and joins with the vein of the corpus striatum to form the corresponding vein of Galen.

When the choroid plexus is pulled away, the continuity of the epithelium which covers it, with that which lines the ventricle, is severed, and a cleft-like space is produced. This is named the **choroidal fissure** ; like the plexus, it extends from the foramen of Monro to the extremity of the descending horn. The upper part of this fissure, i.e. the part nearest the foramen of Monro, is situated between the lateral edge of the fornix and the upper surface of the optic thalamus ; further back, at the beginning of the descending horn, it is between the commencement of the fimbria and the posterior end of the optic thalamus, while in the descending horn it lies between the fimbria in the floor and the tænia semicircularis in the roof of the cornu : through this part of the fissure the descending horn opens on to the tentorial surface of the hemisphere.

The **velum interpositum,** or **tela choroidea superior** (fig. 586), is a double fold of pia mater, triangular in shape, which lies beneath the fornix. The lateral portions of its lower surface rest upon the optic thalami, while its mesial portion is in contact with the epithelial roof of the third ventricle. Its apex is situated at the foramen of Monro ; its base corresponds with the splenium of the corpus callosum, and occupies the interval between that structure above and the corpora quadrigemina and pineal body below. This interval, together with the lower portions of the choroidal fissures, is sometimes spoken of as the *great transverse fissure* of the brain. At its base the two layers of the velum separate from each other, and are continuous with the pia mater investing the brain in this region. Each of its lateral margins is modified to form the highly vascular fringed structure which constitutes the portion of the choroid plexus seen in the body of the lateral ventricle, and which is continued round the posterior end of the optic thalamus into the descending horn. The two plexuses are connected with each other at the foramen of Monro, and are prolonged backwards on the under surface of the velum as the *choroid plexuses of the third ventricle,* which invaginate the epithelial roof of this cavity. It is supplied by the anterior and posterior choroidal arteries, already described. The veins of the velum interpositum are named the *venæ Galeni* ; they are two in number, and run backwards· between its layers, each being formed at the foramen of Monro by the union of the vein of the corpus striatum with the choroid vein. The venæ Galeni unite posteriorly into a single trunk, the *vena magna Galeni,* which passes out beneath the splenium and terminates in the straight sinus (fig. 518).

Structure of the Cerebral Hemispheres

The **cerebral hemispheres,** like the other parts of the great nerve-centre, are composed of grey and white matter : the former covers their surfaces, and is termed the *cortex* ; the latter occupies the interior of the hemispheres, and is named the *medullary centre.*

The **white matter of the cerebral hemispheres** consists of medullated fibres, varying in size and arranged in bundles, separated by neuroglia. They may be divided into three distinct systems, according to the course they take. 1. Projection fibres, which connect the hemisphere with the lower parts of the brain and with the spinal cord. 2. Transverse or commissural fibres, which unite together the two hemispheres. 3. Association fibres, which connect different structures in the same hemisphere. These are, in many instances, collateral branches of the projection fibres, but others are the axons of independent cells.

1. The **projection fibres** consist of efferent and afferent fibres, which connect the cortex to the lower parts of the brain and to the cord. The principal efferent strands are : (1) the motor tract, which occupies the genu and anterior two-thirds of the posterior limb of the internal capsule, and which consists of (a) the geniculate fibres, which decussate and end in the cranial motor nuclei of the opposite side ; and (b) the pyramidal fibres, which are prolonged through the pyramid of the medulla into the spinal cord : (2) the cortico-protuberantial

fibres, which end in the nuclei pontis. The chief afferent fibres are: (1) those fibres of the fillet which are not interrupted in the optic thalamus; (2) those fibres of the superior cerebellar peduncles which are not interrupted in the red nucleus and optic thalamus; (3) the numerous fibres which arise within the optic thalamus, and pass through its stalks to the different parts of the cortex (page 807); (4) the optic and acoustic fibres, the former passing to the occipital, the latter to the temporal lobe.

2. The **transverse** or **commissural fibres** connect the two hemispheres. They include: (a) the transverse fibres of the corpus callosum; (b) the anterior commissure; (c) the posterior commissure; (d) the lyra, and have already been described.

3. **Association fibres.**—These connect different parts of the same hemisphere, and are of two kinds : (1) those which unite adjacent convolutions, *short association fibres*; (2) those which pass between more distant parts, *long association fibres*.

The *short association fibres* are situated immediately beneath the grey substance of the cortex of the hemispheres, and connect together adjacent convolutions.

The *long association fibres* include the following : (a) the uncinate fasciculus ; (b) the cingulum ; (c) the superior longitudinal fasciculus ; (d) the inferior longitudinal fasciculus ; (e) the perpendicular fasciculus ; (f) the occipito-frontal fasciculus ; and (g) the fornix.

(a) The *uncinate fasciculus* passes across the bottom of the Sylvian fissure, and connects the convolutions of the frontal lobe with the anterior end of the temporal lobe.

(b) The *cingulum* is a band of white matter which is contained within the callosal convolution of the limbic lobe. Commencing in front at the anterior perforated space, it passes forwards and upwards parallel with the rostrum, winds round the genu, runs in the convolution from before backwards, immediately above the corpus callosum, turns round its posterior extremity, and passes into the hippocampus major, through which it courses to its anterior extremity.

(c) The *superior longitudinal fasciculus* consists of fibres which pass backwards from the frontal lobe above the lenticular nucleus and island of Reil ; some of these terminate in the occipital lobe, while others pass downwards and forwards into the temporal lobe.

(d) The *inferior longitudinal fasciculus* is a collection of fibres which connects the temporal and occipital lobes, running along the outer wall of the descending and posterior cornua of the lateral ventricle.

(e) The *perpendicular fasciculus* runs vertically through the front part of the occipital lobe, and connects the inferior parietal lobule with the fourth temporal convolution.

(f) The *occipito-frontal fasciculus* passes backwards from the frontal lobe, along the outer border of the caudate nucleus, and on the mesial aspect of the corona radiata, and its fibres radiate in a fan-like manner and pass into the occipital and temporal lobes on the outer aspect of the posterior and descending cornua. Déjérine regards the fibres of the tapetum as being derived from this fasciculus, and not from the corpus callosum.

(g) The *fornix* connects the hippocampal convolution with the corpus albicans, and, by means of the bundle of Vicq d'Azyr, with the optic thalamus (see page 828). Through the fibres of the lyra it probably also unites the opposite hippocampal convolutions.

The **grey matter of the hemisphere** is divided into : (1) That of the cerebral cortex. (2) That of the basal ganglia, viz. the nucleus caudatus and the nucleus lenticularis of the corpus striatum ; the claustrum and the amygdaloid nucleus. They are, with the exception of the amygdaloid nucleus, situated to the inner side of the island of Reil, and form with this convolution the oldest part of the hemisphere, for they are the first parts of the encephalon to be differentiated in the development of the individual. They are simply semi-detached local thickenings of the grey cortex, and have already been described (page 825). The optic thalamus is not reckoned as a basal ganglion of the cerebral hemisphere, since it forms part of the thalamencephalon.

STRUCTURE OF THE CEREBRAL CORTEX

The cerebral cortex differs in its thickness and in its minute structure in different parts of the hemisphere. For instance, it is thinner in the occipital region than in the pre- and post-central gyri, and it is also much thinner at the bottom of the sulci than on the top of the convolutions. Again, the minute structure of the precentral differs from that of the post-central gyrus, and areas possessing a specialised type of cortex can be mapped out in the occipital lobe.

On examining a section through the cortex with a lens, it is seen to consist of alternating white and grey layers thus disposed from the surface inwards : (1) a thin layer of white substance; (2) a layer of grey substance ; (3) a second layer of white substance (outer band of Baillarger or band of Gennari); (4) a second grey layer ; (5) a third white layer (inner band of Baillarger) ; (6) a third grey layer, which rests on the medullary substance of the convolution.

The cortex is made up of nerve-cells which vary in size and shape, and of nerve-fibres, which are either medullated or naked axis-cylinders, embedded in a matrix of neuroglia.

Nerve-cells.—According to Cajal, whose description is now generally accepted, the nerve-cells are arranged in four layers, named from the surface inwards as follows : (1) the molecular layer, (2) the layer of small pyramidal cells, (3) the layer of large pyramidal cells, (4) the layer of polymorphous cells.

The molecular layer.—In this layer the cells are polygonal, triangular, or fusiform in shape. Each polygonal cell gives off some four or five dendrites, while its axon may arise directly from the cell or from one of its dendrites. The axons and dendrites of these cells ramify in the molecular layer. Each triangular cell gives off two or three dendrites, from one of which the axon arises, the dendrites and the axon ramifying in the molecular layer. The fusiform cells are placed with their long axes parallel to the surface and are mostly bipolar, each pole being prolonged into a dendrite, which runs horizontally for some distance and furnishes ascending branches. Their axons, two or three in number, arise from the dendrites, and, like them, take a horizontal course, giving off numerous ascending collaterals. The distribution of the axons and dendrites of all three sets of cells is limited to the molecular layer.

The **layer of small** and the **layer of large pyramidal cells.**—The cells in these two layers may be studied together, since, with the exception of the difference in size and the more superficial position of the smaller cells, they resemble each other. The average length of the small cells is from 10 to 15 μ ; that of the large cells from 20 to 30 μ. The body of each cell is pyramidal in shape, its base being directed to the deeper parts and its apex towards the surface. It contains granular pigment, and stains deeply with ordinary reagents. The nucleus is nucleolated, of large size, and round or oval in shape. The base of the cell gives off the axis-cylinder, and this passes into the central white substance, giving off collaterals in its course, and is distributed as a projection, commissural, or association fibre. Both the apical and basal parts of the cell give off dendrites. The apical dendrite is directed towards the surface, and ends in the molecular layer by dividing into numerous branches, all of which may be seen, when prepared by the silver or methylene-blue method, to be studded with projecting bristle-like processes. The largest pyramidal cells are found in the upper part of the precentral gyrus and in the paracentral lobule. These, which are often arranged in groups or nests of from three to five, are named the *giant-cells* of Betz. In the former situation they may exceed 50 μ in length and 40 μ in breadth, while in the paracentral lobule they may attain a length of 65 μ.

Layer of polymorphous cells.—The cells in this layer, as their name implies, are very irregular in contour ; they may be fusiform, oval, triangular, or star-shaped. Their dendrites are directed outwards, towards, but do not reach, the molecular layer; their axons pass into the subjacent white matter.

There are two other kinds of cells in the cerebral cortex, but their axons pass in a direction opposite to that of the pyramidal and polymorphous cells, among which they lie. They are: (a) the *cells of Golgi*, the axons of which do not become medullated, but divide immediately after their origin into a large number of branches, which are directed towards the surface of the cortex ; (*b*) the *cells of Martinotti*, which are chiefly found in the polymorphous layer. Their dendrites are short, and may have an ascending or descending course, while their axons pass out into the molecular layer and form an extensive horizontal arborisation.

Nerve-fibres.—These fill up a large part of the intervals between the cells, and may be medullated or non-medullated—the latter comprising the axons of

Fig. 587.—Cerebral cortex. (Poirier.) To the left, the groups of cells ; to the right, the systems of fibres. Quite to the left of the figure a sensory nerve-fibre is shown.

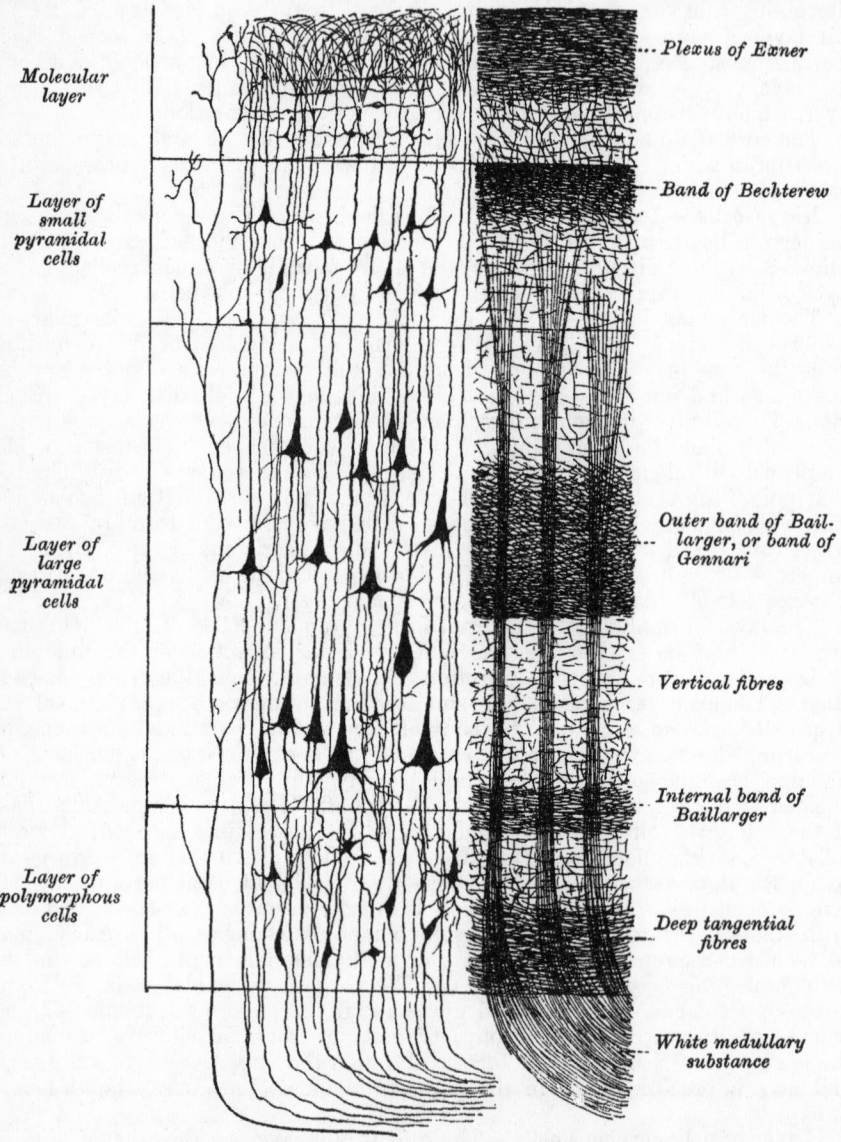

Molecular layer

Layer of small pyramidal cells

Layer of large pyramidal cells

Layer of polymorphous cells

Plexus of Exner

Band of Bechterew

Outer band of Baillarger, or band of Gennari

Vertical fibres

Internal band of Baillarger

Deep tangential fibres

White medullary substance

the smallest pyramidal cells and the cells of Golgi. In their direction the fibres may be either transverse (tangential or horizontal) or vertical (radial). The *tangential fibres* run parallel to the surface of the hemisphere, intersecting the vertical fibres at a right angle. They consist of several strata, of which the following are the most important : (1) a stratum of white fibres covering the superficial aspect of the molecular layer (plexus of Exner) ; (2) the band of

Bechterew, which is situated in the outer part of the layer of small pyramidal cells; (3) the external band of Baillarger (band of Gennari or band of Vicq d'Azyr), which runs through the layer of large pyramidal cells; (4) the internal band of Baillarger, which intervenes between the layer of large pyramidal cells and the polymorphous layer; (5) the deep tangential fibres which lie in the lower part of the polymorphous layer. According to Cajal, the transverse fibres consist of (a) the collaterals of the pyramidal and polymorphous cells and of the cells of Martinotti; (b) the arborisations of the axons of Golgi's cells; (c) the collaterals and terminal arborisations of the projection, commissural, or association fibres. *The vertical fibres.*—Some of these, viz. the axons of the pyramidal and polymorphous cells, are directed towards the central white matter, while others, the terminations of the projection, commissural, or association fibres, pass outwards to end in the cortex. The axons of the cells of Martinotti are also ascending fibres.

SPECIAL TYPES OF CEREBRAL CORTEX

It has been already pointed out that the minute structure of the cortex differs in different regions of the hemisphere; and A. W. Campbell * has endeavoured to prove, as the result of an exhaustive examination of a series of human and anthropoid brains, ' that there exists a direct correlation between physiological function and histological structure.' The principal regions where the 'typical' structure is departed from will now be referred to.

1. In the calcarine fissure, and the convolutions bounding it, the inner band of Baillarger is absent, while the band of Gennari is of considerable thickness and forms a characteristic feature of this region of the cortex. If a section be examined microscopically, an additional layer is seen to be interpolated between the molecular layer and the layer of small pyramidal cells. This extra layer consists of two or three strata of fusiform cells, the long axes of which are at right angles to the surface. Each cell gives off two dendrites, external and internal, from the latter of which the axon arises and passes into the white central substance. In the layer of small pyramidal cells, fusiform cells, identical with the above, are seen, as well as ovoid or star-like cells with ascending axons (cells of Martinotti). This is the *visual area* of the cortex, and it has been shown by J. S. Bolton † that in old-standing cases of optic atrophy the thickness of Gennari's band is reduced by nearly 50 per cent.

A. W. Campbell says: 'Histologically, two distinct types of cortex can be made out in the occipital lobe. The first of these coats the walls and bounding convolutions of the calcarine fissure, and is distinguished by the well-known line of Gennari or Vicq d'Azyr; the second area forms an investing zone a centimetre or more broad around the first, and is characterised by a remarkable wealth of fibres, as well as by curious pyriform cells of large size richly stocked with chromophilic elements—cells which seem to have escaped the observation of Ramón y Cajal, Bolton, and others who have worked at this region. As to the functions of these two regions there is abundant evidence, anatomical, embryological, and pathological, to show that the first or calcarine area is that to which visual sensations primarily pass, and we are gradually obtaining proof to the effect that the second investing area is constituted for the interpretation and further elaboration of these sensations. These areas therefore deserve the names *visuo-sensory* and *visuo-psychic*.'

2. The precentral gyrus is characterised by the presence of the giant-cells of Betz and by ' a wealth of nerve-fibres immeasurably superior to that of any other part' (Campbell), and in these respects differs from the post-central gyrus. These two gyri, together with the paracentral lobule, have long been regarded as containing the 'motor areas' of the hemisphere; but Sherrington and Grünbaum have shown ‡ that in the chimpanzee the motor area never extends on to the free

---

* *Proceedings of the Anatomical Society of Great Britain and Ireland,* June 1903; and *Proceedings of the Royal Society,* vols. lxxii. and lxxiv. A full account of Campbell's valuable work on the minute anatomy of the cerebral cortex will shortly be published by the Cambridge University Press, under the title *Histological Studies on the Localisation of Cerebral Function.*

† *Phil. Trans. of Royal Society,* Series B, vol. cxciii. p. 165.

‡ *Transactions of the Pathological Society of London,* vol. liii.

face of the post-central convolution, but 'occupies unbrokenly the whole length of the precentral convolution, and in most cases the greater part or the whole of its width. It extends into the depth of the Rolandic fissure, occupying the anterior wall, and in some places the floor, and in some extending even into the deeper part of the posterior wall of the fissure.'

3. In the hippocampus major the molecular layer is very thick and contains a large number of Golgi cells. It has been divided into three strata : (a) S. convolutum or S. granulosum, containing many tangential fibres ; (b) S. lacunosum, presenting numerous lymphatic or vascular spaces ; (c) S. radiatum, exhibiting a rich plexus of fibrils. The two layers of pyramidal cells are condensed into one, and the cells are mostly of large size. The axons of the cells in the polymorphous layer may run in an ascending, descending, or horizontal direction. Between the polymorphous layer and the ventricular ependyma is the white substance of the alveus.

4. In the rudimentary dentate convolution the molecular layer contains some pyramidal cells, while the layer of pyramidal cells is almost entirely represented by small ovoid cells.

5. **The olfactory bulb.**—In many of the lower animals this contains a cavity which communicates through the hollow olfactory stalk with the lateral ventricle. In man the original cavity is filled up by neuroglia and its wall becomes thickened, but much more so on its ventral than on its dorsal aspect. Its dorsal part contains a small amount of grey and white matter, but it is scanty and ill defined. A section through the ventral part shows it to consist of the following layers from without inwards : (1) A layer of olfactory nerve-fibres, which are the non-medullated axons prolonged from the olfactory cells of the nose, and which reach the bulb by passing through the cribriform plate of the ethmoid bone. At first they cover the bulb, and then penetrate it to end by forming synapses with the dendrites of the mitral cells, presently to be described. (2) *Glomerular layer.*—This contains numerous spheroidal reticulated enlargements, termed *glomeruli*, which are produced by the branching and arborisation of the processes of the olfactory nerve-fibres with the descending dendrite of the mitral cells. (3) *Molecular layer.*—This is formed of a matrix of neuroglia, embedded in which are the *mitral cells*. These cells are pyramidal in shape, and the basal part of each gives off a thick dendrite, which descends into the glomerular layer, where it arborises as indicated above, and others which interlace with similar dendrites of neighbouring mitral cells. The axons pass through the next layer into the white matter of the bulb, from which, after becoming bent on themselves at a right angle, they are continued into the olfactory tract. (4) *Nerve-fibre layer.*—This lies next the central core of neuroglia, and its fibres consist of the axons or afferent processes of the mitral cells which are passing to the brain ; some efferent fibres are, however, also present, and terminate in the molecular layer, but nothing [is known as to their exact origin.

Weight of the Encephalon.—The average weight of the brain, in the adult male, is 49½ oz., or a little more than 3 lb. avoirdupois ; that of the female, 44 oz. ; the average difference between the two being from 5 to 6 oz. The prevailing weight of the brain, in the male, ranges between 46 oz. and 53 oz. ; and, in the female, between 41 oz. and 47 oz. In the male, the maximum weight out of 278 cases was 65 oz. and the minimum weight 34 oz. The maximum weight of the adult female brain, out of 191 cases, was 56 oz., and the minimum weight 31 oz. According to Luschka, the average weight of a man's brain is 1,424 grammes (about 45 oz.), of a woman's 1,272 grammes (about 41 oz.) ; and according to Krause, 1,570 grammes (about 48½ oz.) for the male, and 1,350 grammes (about 43 oz.) for the female. It appears that the weight of the brain increases rapidly up to the seventh year, more slowly to between sixteen and twenty, and still more slowly to between thirty and forty, when it reaches its maximum. As age advances and the mental faculties decline, the brain diminishes slowly in weight, to the extent of about an ounce for each subsequent decennial period. These results apply alike to both sexes.

The size of the brain was formerly said to bear a general relation to the intellectual capacity of the individual. Cuvier's brain weighed rather more than 64 oz., that of the late Dr. Abercrombie 63 oz., and that of Dupuytren 62½ oz. On the other hand, the brain of an idiot seldom weighs more than 23 oz. But these facts are by no means conclusive, and it is well known that these weights have been equalled by the brains of persons who never displayed any remarkable intellect. Haldennan, of Cincinnati, has recorded the case of a mulatto, aged 45, whose brain weighed 68⅜ oz.; he had been a slave, and was never regarded as particularly intelligent; he was illiterate, but is said to have been reserved, meditative, and economical. Ensor, district medical officer at Port Elizabeth, reports that the brain of Carey, the Irish informer, weighed 61 oz. M. Nikiforoff has published an article on the subject of the weight of brains in the 'Novosti.' According to him, the weight of the brain has no influence whatever on the mental faculties. It ought to be remembered that the significance of the weight of the brain should depend upon the proportion it bears to

FIG. 588.—Side view of the brain of man, showing the localisation of various functions. (After Ferrier.)

1. Centre for movements of opposite leg and foot. 2, 3, 4. Centres for complex movements of the arms and legs, as in swimming. 5. Extension forwards of the arm and hand. 6. Supination of the hand and flexion of the forearm. 7, 8. Elevators and depressors of the angle of the mouth. 9, 10. Movements of the lips and tongue. 11. Retraction of the angle of the mouth. 12. Movements of the eyes. 13, 13'. Vision. 14. Hearing. a, b, c, d. Movements of the wrists and fingers.

the dimensions of the whole body and to the age of the individual. It is equally important to know what was the cause of death, for long illness or old age exhausts the brain. To define the real degree of development of the brain, it is therefore necessary to have a knowledge of the condition of the whole body; and, as this is usually lacking, the mere record of weight possesses little significance.

The human brain is heavier than that of any of the lower animals, except the elephant and whale. The brain of the former weighs from eight to ten pounds; and that of a whale, in a specimen seventy-five feet long, weighed rather more than five pounds.

Cerebral Localisation.—Physiological and pathological research have now gone far to prove that a considerable part of the surface of the brain may be mapped out into a series of more or less definite areas, each of which is intimately connected with some well-defined function.

This is not the place, nor can space be given here, to describe these localities. But the two accompanying drawings from Ferrier (figs. 588, 590), and that of the brain of the chimpanzee by Sherrington and Grünbaum (fig. 589), have been introduced, and will serve to indicate the position of some of the more important areas. On comparing these figures, it will be seen that whereas Ferrier devotes both the precentral and post-central convolutions to the motor areas, Sherrington and Grünbaum limit them practically to the former gyrus.

Cerebral Topography.—The relation of the principal fissures and convolutions of the cerebrum to the outer surface of the scalp has been the subject of much investigation, and many systems have been devised by which one may localise these parts from an examination of the external surface of the head.

These plans can only be regarded as approximately correct for several reasons : in the first place, because the relations of the convolutions and sulci to the surface vary in different individuals; secondly, because the surface area of the scalp is greater than the surface area of the brain, so that lines drawn on the one cannot correspond exactly to

FIG. 589.—Side view of the brain of the chimpanzee, showing the localisation of various functions. (Sherrington and Grünbaum.)

sulci or convolutions on the other; and thirdly, because the sulci and convolutions in two individuals are never precisely alike. Nevertheless, the principal fissures and convolutions can be mapped out with sufficient accuracy for all practical purposes, so that any particular convolution can be exposed by removing with the trephine a certain portion of the skull's area.

The various landmarks on the outside of the skull, which can be easily felt, and which serve as indications of the position of the parts beneath, have been already referred to (see page 245), but there are certain other points or landmarks which require alluding to, in order to facilitate the description of the relation of the fissures and convolutions of the brain to the external surface of the skull.

A line drawn horizontally backwards from the middle of the infra-orbital margin, through the centre of the outlet of the external auditory meatus, will represent what is known as *Reid's base-line.* A spot on this base-line in the hollow between the tragus of the ear and the condyle of the lower jaw is known as the *pre-auricular point.*

The longitudinal fissure.—This corresponds to a line drawn from the nasion to the inion.

The Sylvian fissure.—In order to mark out this fissure, a point must be defined by carrying horizontally backwards a line for $1\frac{2}{8}$ of an inch (thirty-five millimetres) from the external angular process of the frontal bone, and from the posterior extremity of this line a vertical line upwards for half an inch (twelve millimetres); the upper end of this second line is the point in question, and is known as the *Sylvian point*. It marks the spot where the Sylvian fissure divides. Another and simple plan of defining the Sylvian point has been devised by E. H. Taylor. He divides the distance between the nasion and inion into four equal parts, and draws one line from the junction of the third and fourth parts, reckoning from before backwards, to the external angular process of the frontal bone; and a second line, from the junction of the first and second segments to the centre of the external auditory meatus. The Sylvian point will be the spot where these two lines cross one another, and the first line from this point onwards will lie over the

FIG. 590.—Top view of the brain of man, showing the localisation of various functions. (After Ferrier.)

1. Centre for movements of opposite leg and foot.   2, 3, 4. Centres for complex movements of the arms and legs, as in swimming.   5. Extension forwards of the arm and hand.   6. Supination of the hand and flexion of the forearm
7, 8. Elevators and depressors of the angle of the mouth.   12. Movements of the eyes.   13, 13'. Vision.   14. Hearing.
a, b, c, d. Movements of the wrists and fingers.

posterior limb of the fissure of Sylvius. The *Sylvian line*—that is to say, the line on the surface of the skull which lies over the posterior limb of the fissure of Sylvius—is usually marked out by drawing a line from the Sylvian point to the lower part of the parietal eminence. The ascending limb of the fissure of Sylvius may be marked out by drawing a line upwards, at right angles to the Sylvian line, for nearly an inch (two centimetres); and the horizontal limb by a line of the same length, drawn horizontally forwards from the same point.

The fissure of Rolando.—In order to define this fissure, its upper extremity, *superior Rolandic point*, must be first found. This is situated in the sagittal suture, at a point 55·6 per cent. of the distance between the nasion and the inion. For all practical purposes the plan suggested by Thane is sufficient. He takes the middle point of a line between the nasion and the inion, and fixes the superior Rolandic point at half an inch behind this point. The inferior Rolandic point is defined by drawing a line at right angles to the base-line of Reid, from the pre-auricular point to the Sylvian line; this it meets about an inch from the Sylvian point. By joining these two points, the Rolandic line, which overlies the fissure of Rolando, is mapped out. It forms an angle, opening forwards, of about seventy degrees with the median line. The Rolandic line does not, how-ever, actually define the limits of the Rolandic fissure; for this fissure does not quite reach the middle line of the head above, and it terminates below, a short distance above the Sylvian line. Reid has devised another plan for mapping out this fissure. He draws two perpendicular lines from the base-line to the top of the head: one (fig. 592, D E) from the pre-auricular point, and the other (F G) from the posterior border of the mastoid process at its root. A line drawn from the upper extremity of the posterior line (F) to the point where the Sylvian line crosses the anterior one would indicate the position of the fissure of Rolando.

FIG. 591.—Drawing to illustrate cranio-cerebral topography. (Taken from a cast in the Museum of the Royal College of Surgeons of England, prepared by Cunningham.)

The *external parieto-occipital fissure* runs outwards at right angles to the great longitudinal fissure for about an inch, from a point one-fifth of an inch in front of the lambda (posterior fontanelle). Reid states that if the posterior limb of the fissure of Sylvius be continued backwards to the sagittal suture, the last inch of this line will indicate the position of the sulcus.

The *pre-central* and *post-central sulci* are situated three-fifths of an inch in front of and behind the Rolandic fissure respectively: they are nearly parallel with this fissure, and extend as low as the Sylvian line.

The *superior frontal fissure* may be mapped out by drawing a line from the junction of the upper and middle third of the precentral sulcus, in a direction parallel with the longitudinal fissure, to a point midway between the middle line of the forehead and the temporal crest, an inch and a half above the supra-orbital notch.

The *inferior frontal fissure* follows the course of the superior temporal ridge, commencing at the junction of the middle and lower third of the precentral sulcus.

The *intraparietal fissure* begins on a level with the junction of the middle and lower third of the fissure of Rolando, on a line carried across the head from the back of the root of one auricle to that of the other. After passing upwards, it curves backwards, lying parallel to the longitudinal fissure, midway between it and the parietal eminence; it then curves downwards to end midway between the lambda or junction of the sagittal and lambdoidal sutures and the parietal eminence.

The *lateral ventricles* may be circumscribed, according to Poirier, by describing a quadrilateral figure on the side of the head. The upper limit is a horizontal line drawn two inches above and parallel with the zygoma: this defines the roof of the ventricular cavity. The lower limit is a second horizontal line drawn half an inch above the zygoma: this indicates the level of the extremity of the descending horn of the ventricle. Two vertical lines—one drawn through the junction of the anterior and middle thirds of the zygomatic arch, and the other two inches behind the tip of the mastoid process—indicate the extent of the anterior horn in front and the posterior horn behind.

Fig. 592.—Relation of the principal fissures and convolutions of the cerebrum to the outer surface of the scalp. (Reid.)

Surgical Anatomy.—In cases of distension of the ventricles with cerebro-spinal fluid, which causes symptoms very closely resembling cerebral tumour, recourse has been had to trephining for the purpose of draining these cavities. Different routes have been adopted by surgeons. Poirier advocates tapping the descending horn. He trephines the skull rather more than an inch and a half (four centimetres) above the external auditory meatus, and thus exposes the second temporal convolution. By introducing a trocar, the middle horn is reached at the depth of about an inch. Kocher recommends the puncture to be made an inch and a quarter in front of the precentral sulcus, and the same distance from the middle line of the head. This corresponds to the position of the superior frontal fissure; and the brain substance requires perforating to the extent of about two inches in a direction downwards and backwards, in order to reach the ventricle.

Keen, of Philadelphia, who was one of the first to draw attention to this subject, trephined at a spot an inch and a quarter above and behind the external auditory meatus, and directed his trocar inwards and slightly forwards and upwards, aiming for a point two and a half inches above the opposite meatus, and perforated the brain substance to the depth of an inch and three-quarters before the ventricle was reached.

NERVE-TRACTS

The anatomy of the various parts of the central nervous system having been described, a short account will now be given of the course taken by its more

FIG. 593.—The motor tract. (Poirier.)

Pyramidal cells of cortex

Geniculate bundle

Pyramidal or motor decussation

Direct pyramidal tract

Motor cell in anterior horn of cord

Muscle

Crossed pyramidal tract

important nerve-tracts, and of the direction in which impulses pass along them. The methods employed in elucidating this complex subject have already been referred to (page 770).

MOTOR, EFFERENT, OR DESCENDING TRACT

The constituent fibres of this tract are the axis-cylinder processes of cells situated in the motor area of the cortex. At first they are somewhat widely diffused, but as they descend through the corona radiata they gradually approach each other and pass between the lenticular nucleus and optic thalamus in the genu and anterior two-thirds of the posterior limb of the internal capsule, those which occupy the genu being named the geniculate fibres, while the remainder constitute the pyramidal fibres. Proceeding downwards they enter the crusta or pes of the crus cerebri, the pyramidal fibres occupying the middle three-fifths, and the geniculate fibres the innermost fifth of this structure. The geniculate fibres then decussate in the middle line with the corresponding fibres of the opposite side, and end by arborising around the cells of the motor nuclei of the cranial nerves. The pyramidal fibres are continued downwards into the anterior pyramids of the medulla, and the transit of the fibres from the medulla is effected by two paths. The fibres nearest to the anterior median fissure cross the middle line, forming the *decussation of the pyramids*, and descend in the opposite side of the cord as the indirect or crossed pyramidal tract. Throughout the length of the spinal cord fibres from this column pass into the grey matter, to terminate by ramifying around the cells of the anterior horn. The more laterally placed portion of the motor tract does not decussate in the medulla, but descends as the direct or uncrossed pyramidal tract; these fibres, however, end in the anterior grey horn of the opposite side of the spinal cord by passing across in the anterior white commissure. There is considerable variation in the extent to which decussation takes place in the medulla, the commonest condition being that in which about two-thirds or three-fourths of the fibres decussate in the medulla and the remainder in the cord.

The axons of the motor cells in the anterior horn pass out as the fibres of the anterior roots of the spinal nerves, along which the impulses are conducted to the muscles of the trunk and limbs.

From this it will be seen that all the fibres of the motor tract pass to the nuclei of the motor nerves on the opposite side of the brain or cord, a fact which explains why a lesion involving the motor area of one side causes paralysis of the muscles of the opposite side of the body. Further, it will be seen that there is a break in the continuity of the motor chain : in the case of the cranial nerves this break occurs in the nuclei of these nerves; and in the case of the spinal nerves, in the anterior horn of grey matter.

Other Descending Tracts

1. The cortico-protuberantial fibres arise in the cerebral cortex and descend to end in the nuclei pontis. They consist of two groups : (1) those from the frontal lobe pass through the anterior limb of the internal capsule, and in the crusta are disseminated among the pyramidal and geniculate fibres ; (2) those from the temporal and occipital lobes proceed through the posterior limb of the internal capsule and occupy the outer fifth of the crusta. From the nuclei pontis the fibres of the second link in the chain take origin, and, after crossing to the opposite side, are carried to the cortex of the cerebellum through the middle peduncles. The descending cerebellar fibres in the anterior and lateral columns of the spinal cord are by some regarded as a third link of this chain.

2. A small tract arises from the cells of the caudate and lenticular nuclei and descends to end in the crus cerebri.

Sensory, Afferent, Ascending Tract

Sensory impulses are conveyed to the spinal cord through the posterior roots of the spinal nerves. On entering the cord these root-fibres divide into descending and ascending branches ; the former soon enter the grey matter : of the latter some end in the grey matter after a longer or shorter course, while others are continued directly into the posterior columns of the cord, where they form the tracts of Goll and Burdach (posterior sensory fasciculi). From the cells of the posterior horn, fibres arise which cross the middle line and ascend in the peripheral part of the lateral column as the tract of Gowers (lateral sensory fasciculus). Some observers maintain that part of the sensory fibres ascend in the anterior column. The tracts of Goll and Burdach end by arborising around the cells of the gracile and cuneate nuclei, and from these cells the fibres of the mesial fillet take origin and cross to the opposite side in the *sensory decussation*.

The mesial fillet is then joined by the fibres of Gowers' tract which have already crossed in the cord, and in its further course receives fibres from the cranial sensory nuclei of the opposite side, with the exception of the cochlear division of the auditory nerve. Ascending through the crus, the fillet gives off some fibres to the lenticular nucleus and island of Reil, but the greater part of it is

FIG. 594.—The sensory tract. (Poirier.)

Fillet

Sensory cranial nerve

Sensory decussation

Gracile and cuneate nuclei

Tract of Goll

Tract of anterior column

Tract of lateral column

Anterior median fissure

Posterior horn

Posterior nerve-root

carried into the optic thalamus, where most of its fibres terminate—only a small proportion being continued directly into the cerebral cortex. From the grey matter of the optic thalamus the fibres of the third link in the chain arise and pass to the cerebral cortex. The fibres from the terminal nuclei of the cochlear nerve pass upwards in the lateral fillet, and are carried through the posterior part of the internal capsule to the temporal lobe. Further, Gowers' tract gives off a fasciculus which reaches the cerebellum through its superior peduncles.

OTHER ASCENDING TRACTS

The direct cerebellar tract begins about the level of the second lumbar vertebra, and is the continuation upwards of the axis cylinders of Clarke's column. At the upper end of the cord most of its fibres pass into the restiform body and through this reach the cerebellum ; some, however, pass through the pons and are carried to the cerebellum through its superior peduncle. This tract seems to lose some of its fibres in the cord, since the area of its degeneration resulting from a section of the lower part of the cord diminishes from below upwards ; only some of its fibres therefore pass directly to the cerebellum. On the other hand, the tract is reinforced by an accession of fibres from the cord itself, so that its transverse area is greater above than below.

Most of the fibres of the superior cerebellar peduncle arise within the dentate nucleus of the cerebellum or from the cerebellar cortex, and after crossing to the opposite side, are continued to the red nucleus and optic thalamus, from the cells of which a relay of fibres is prolonged to the cerebral cortex.

MENINGES OF THE BRAIN AND SPINAL CORD

The brain and spinal cord are enclosed within three membranes. These are named from without inwards : the dura mater, the arachnoid membrane, and the pia mater.

THE DURA MATER

The **Dura Mater** is a thick and dense inelastic membrane, which forms an external covering for the brain and spinal cord. The portion which encloses the brain differs in several essential particulars from that which surrounds the spinal cord, and therefore it is necessary to describe them separately ; but at the same time it must be distinctly understood that the two form one complete membrane, and are continuous with each other at the foramen magnum.

The **cranial dura mater** lines the interior of the skull, and serves the twofold purpose of an internal periosteum to the bones and a membrane for the protection of the brain. Its outer surface is rough and fibrillated, and adheres closely to the inner surface of the bones ; the adhesion being most marked opposite the sutures and at the base of the skull. Its inner surface is smooth and lined by a layer of endothelium. It sends four processes into the cavity of the skull, which divide it into a number of freely communicating compartments, for the lodgment and protection of the different parts of the brain ; and it is prolonged to the outer surface of the skull, through the various foramina which exist at the base, and thus becomes continuous with the pericranium ; its fibrous layer forms sheaths for the nerves which pass through these apertures. At the base of the skull, it sends a fibrous prolongation into the foramen cæcum ; it sends a series of tubular prolongations round the filaments of the olfactory nerves as they pass through the cribriform plate, and another round the nasal nerve as it passes through the nasal slit ; a prolongation is also continued through the sphenoidal fissure into the orbit, and another is carried into the same cavity through the optic foramen, forming a sheath for the optic nerve, which is continued as far as the eyeball. In the posterior fossa it sends a process into the internal auditory meatus, ensheathing the facial and auditory nerves ; another through the jugular foramen, forming a sheath for the structures which pass through this opening ; and a third through the anterior condyloid foramen. Around the margin of the foramen magnum it is closely adherent to the bone, and is continuous with the spinal dura mater. It is composed of two layers closely connected together, except in certain situations, where, as already described (page 717), they separate to form sinuses for the passage of venous blood. Upon the outer surface of the cranial dura mater, in the situation of the longitudinal sinus, may be seen numerous small whitish bodies, the *glandulæ Pacchioni*.

Processes of the Cranial Dura Mater.—The processes of the cranial dura mater, which project into the cavity of the skull, are formed by a reduplication of the inner or meningeal layer of the membrane, and are four in number: the falx cerebri, the tentorium cerebelli, the falx cerebelli, and the diaphragma sellæ.

The *falx cerebri*, so named from its sickle-like form, is a strong, arched process which descends vertically in the longitudinal fissure between the hemispheres of the brain. It is narrow in front, where it is attached to the crista galli of the ethmoid bone; and broad behind, where it is connected with the upper surface of the tentorium cerebelli. Its upper margin is convex and attached to the inner surface of the skull, in the middle line, as far back as the internal occipital protuberance; it contains the superior longitudinal sinus. Its lower margin is free, concave, and presents a sharp curved edge, which contains the inferior longitudinal sinus.

The *tentorium cerebelli* is an arched lamina of dura mater, elevated in the middle, and inclining downwards towards the circumference. It covers the upper surface of the cerebellum, and supports the occipital lobes of the brain and prevents them pressing upon the cerebellum. It is attached, behind, by its convex border to the transverse ridges upon the inner surface of the occipital bone, and there encloses the lateral sinuses; in front, to the superior margin of the petrous portion of the temporal bone on either side, enclosing the superior petrosal sinuses, and at the apex of this bone the free or anterior border and the attached or external border meet, and, crossing one another, are continued forwards to be fixed to the anterior and posterior clinoid processes respectively. To the middle line of its upper surface the posterior border of the falx cerebri is attached, the straight sinus being placed at their line of junction. Its anterior border is free and concave, and bounds a large oval opening, the *incisura tentorii*, for the transmission of the crura cerebri.

The *falx cerebelli* is a small triangular process of dura mater, received into the indentation between the two lateral lobes of the cerebellum behind. Its base is attached, above, to the under and back part of the tentorium; its posterior margin, to the lower division of the vertical crest on the inner surface of the occipital bone. As it descends, it sometimes divides into two smaller folds, which are lost on the sides of the foramen magnum.

The *diaphragma sellæ* is a horizontal process, in the form of a small circular fold, which constitutes a roof for the sella turcica. This almost completely covers the pituitary body, presenting merely a small central opening for the infundibulum to pass through.

Structure.—The cranial dura mater consists of white fibrous tissue, with connective-tissue cells and elastic fibres arranged in flattened laminæ, which are imperfectly separated by lacunar spaces and blood-vessels into two layers, *endosteal* and *meningeal*. The *endosteal layer* is the internal periosteum for the cranial bones, and contains the blood-vessels for their supply. At the margin of the foramen magnum it becomes continuous with the periosteum lining the spinal canal. The *meningeal* or *supporting layer* is lined on its inner surface by a layer of nucleated endothelium, similar to that found on serous membranes: these cells were formerly regarded as belonging to the arachnoid membrane. By its reduplication the meningeal layer forms the falx cerebri, the tentorium and falx cerebelli, and the diaphragma sellæ. The two layers are connected by fibres which intersect each other obliquely.

The *arteries* of the dura mater are very numerous, but are chiefly distributed to the bones. Those in the anterior fossa are the anterior meningeal branches of the anterior and posterior ethmoidal and internal carotid, and a branch from the middle meningeal. Those in the middle fossa are the middle and small meningeal of the internal maxillary, a branch from the ascending pharyngeal, which enters the skull through the foramen lacerum medium basis cranii; branches from the internal carotid, and a recurrent branch from the lachrymal. Those in the posterior fossa are meningeal branches from the occipital: one of which enters the skull through the jugular foramen, and the other through the mastoid foramen; the posterior meningeal from the vertebral; occasionally meningeal branches from the ascending pharyngeal, which enter the skull through the jugular and condyloid foramina; and a branch from the middle meningeal.

The *veins* which return the blood from the cranial dura mater, and partly from the bones, anastomose with the diploic veins. They terminate in the various sinuses, with the exception of the two which accompany the middle meningeal artery and pass out of the skull at the foramen spinosum to join the pterygoid plexus through which their contents are drained into the internal maxillary vein ; above, they communicate with the superior longitudinal sinus. Many of the meningeal veins do not open directly into the sinuses, but indirectly through a series of ampullæ, termed *venous lacunæ*. These are found on each side of the superior longitudinal sinus, especially near its middle portion, and are often invaginated by Pacchionian bodies ; they also exist near the lateral and straight sinuses. They communicate with the underlying cerebral veins, and also with the diploic and emissary veins.

The nerves of the cranial dura mater are filaments from the Gasserian ganglion, from the ophthalmic, superior maxillary, inferior maxillary, vagus, and hypoglossal nerves, and from the sympathetic.

The **spinal dura mater** forms a loose sheath around the cord, and represents only the inner or meningeal layer of the cranial dura mater. The outer or en-dosteal layer ceases at the foramen magnum, where it becomes continuous with the periosteum lining the spinal canal. The dura mater is separated from the bony walls of the spinal canal by a space, the *epidural space*, in which is contained a quantity of loose areolar tissue and a plexus of veins. The situation of these veins between the dura mater of the cord and the periosteum of the vertebræ corresponds therefore to that of the cranial sinuses between the meningeal and endosteal layers of the cranial dura mater. It is attached to the circumference of the foramen magnum, and to the second and third cervical vertebræ ; it is also connected to the posterior common ligament, especially near the lower end of the spinal canal, by fibrous slips ; it extends below as far as the second or third piece of the sacrum : here it becomes impervious, and, ensheathing the filum terminale, descends to the back of the coccyx, where it blends with the periosteum. The sheath of the dura mater is much larger than is necessary for the accommodation of its contents, and its size is greater in the cervical and lumbar regions than in the dorsal. Its inner surface is smooth. On each side may be seen the double openings which transmit the two roots of the corresponding spinal nerve, the dura mater being continued in the form of a tubular prolongation on them as they pass through the intervertebral foramina. These prolongations are short in the upper part of the spine, but gradually become longer below, forming a number of tubes of fibrous membrane, which enclose the lower spinal nerves and are contained in the spinal canal.

FIG. 595.—The spinal cord and its membranes.

The chief peculiarities of the dura mater of the cord, as compared with that investing the brain, are the following : The dura mater of the cord is not adherent to the bones of the spinal canal, which have an independent periosteum. It does not send partitions into the fissures of the cord, as in the brain. Its fibrous laminæ do not separate to form venous sinuses, as in the cranial dura mater.

Structure.—The spinal dura mater resembles in structure the meningeal or supporting layer of the cranial dura mater, consisting of white fibrous and elastic tissue, arranged in bands or lamellæ, which, for the most part, are parallel with one another and have a longitudinal arrangement. Its internal surface is covered by a layer of endothelial cells, which gives this surface its smooth appearance. It is sparingly supplied with blood-vessels ; and some few nerves have been traced into it.

Subdural space.—The dura mater is separated from the arachnoid by a potential space, the *subdural space*. The two membranes are, in fact, in contact with each other, except where they are separated by a minute quantity of fluid, which just serves to keep the two opposing surfaces moist.

The Arachnoid Membrane

The arachnoid (ἀράχνη, εἶδος, *like a spider's web*), so named from its extreme thinness, is a delicate membrane which envelops both the brain and cord, lying between the pia mater internally and the dura mater externally.

The *cranial part of the arachnoid* invests the brain loosely, and does not dip into the sulci between the convolutions, nor into the fissures, with the exception of the longitudinal fissure. On the upper surface of the cerebrum the arachnoid is thin and transparent, and may be easily demonstrated by injecting a stream of air beneath it by means of a blowpipe. At the base of the brain the arachnoid is thicker and slightly opaque towards the central part, where it extends across between the two temporal lobes in front of the pons Varolii, so as to leave a considerable interval between it and the brain.

The *spinal part of the arachnoid* is a thin, delicate tubular membrane, which loosely invests the cord. Above, it is continuous with the cerebral arachnoid; below, it widens out and invests the cauda equina and the nerves

Fig. 596.—Diagrammatic representation of a section across the top of the skull, showing the membranes of the brain, &c. (Modified from Testut.)

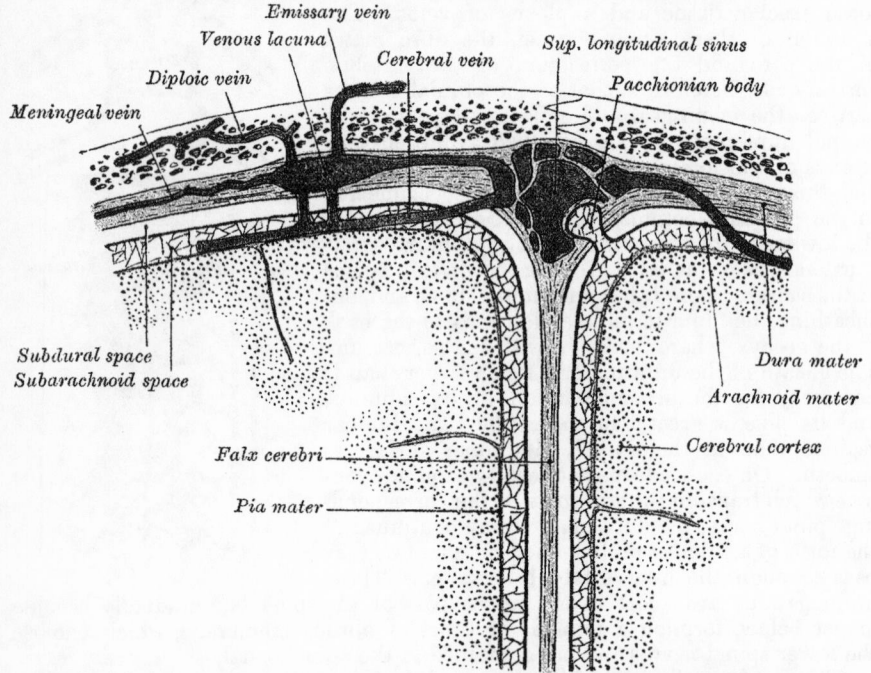

proceeding from it. Its outer surface is in contact with the inner surface of the dura mater, but for the most part the membranes are not connected together, though here and there they may be joined together by isolated connective-tissue trabeculæ, which are most numerous on the posterior surface of the cord. The space between the two membranes is the *subdural space*.

The arachnoid membrane surrounds the nerves which arise from the brain and spinal cord and encloses them in loose sheaths as far as their point of exit from the skull and spinal canal.

Structure.—The arachnoid consists of bundles of white fibrous and elastic tissue intimately blended together. Its outer surface is covered with a layer of endothelium. Vessels of considerable size, but few in number, and, according to Bochdalek, a rich plexus of nerves derived from the motor division of the fifth, the facial, and the spinal accessory nerves, are found in the arachnoid.

The **subarachnoid space** is the interval between the arachnoid and pia mater. It is not, properly speaking, a *space*, for it is occupied everywhere by a spongy tissue consisting of trabeculæ of delicate connective tissue, and intercommunicating channels in which the subarachnoid fluid is contained. This so-called space is small on the surface of the hemispheres of the brain; on the summit of each convolution the pia mater and arachnoid membrane are in close contact; but in the sulci between the convolutions, triangular spaces are left, in which the subarachnoid trabecular tissue is found, for the pia mater dips into the sulci, whereas the arachnoid bridges across them from convolution to convolution. At the base of the brain, in certain situations, the arachnoid is separated by wider intervals from the pia mater, forming larger spaces, which have received the name of *cisternæ*, and in these the subarachnoid tissue is less abundant and the communicating channels larger than in those regions where the two membranes are more closely approximated.

The three principal of these spaces have been named the *cisterna magna*, the *cisterna pontis*, and the *cisterna basalis*; but it should be clearly understood that these spaces communicate freely with each other. The *cisterna magna* is a space, triangular on sagittal section, caused by the arachnoid bridging over the interval between the under surfaces of the hemispheres of the cerebellum and the medulla. It is continuous with the posterior part of the subarachnoid space of the spinal cord through the foramen magnum. The *cisterna pontis* is a considerable interval between the pia mater and the arachnoid on the ventral aspect of the pons Varolii. It contains the basilar artery, and is continuous behind with the subarachnoid space of the spinal cord, and with the cisterna magna; and in front of the pons with the cisterna basalis. The *cisterna basalis* is a wide interval left between the pia mater and the arachnoid, in consequence of the latter membrane extending across between the two temporal lobes. It encloses the crura cerebri and the structures contained in the interpeduncular space, and contains the circle of Willis. In front, the cisterna basalis extends forwards on to the upper surface of the corpus callosum, for the arachnoid stretches across from one cerebral hemisphere to the other, immediately beneath the free border of the falx cerebri, and thus leaves a space in which the anterior cerebral arteries are contained. Again, another space is formed in front of either temporal lobe by the arachnoid bridging across the fissure of Sylvius without dipping down to the bottom of the fissure. This space is a prolongation from the cisterna basalis, and contains the middle cerebral artery. The subarachnoid space communicates with the general ventricular cavity of the brain by three openings: one of these is in the middle line at the inferior boundary of the fourth ventricle, where an opening in the pia-matral covering of this cavity, the *foramen of Majendie*, exists, and permits the passage of fluid from the one space to the other. The other two communications are at the extremities of the lateral recesses of the fourth ventricle, behind the upper roots of the glosso-pharyngeal nerves; they are named the *foramina of Key and Retzius*. It is stated by Meckel that the lateral ventricles also communicate with the subarachnoid space at the apices of their descending horns.

The spinal part of the subarachnoid space is a very wide interval between the arachnoid membrane and the pia mater, and is largest at the lower part of the spinal canal, where the arachnoid membrane encloses the nerves which form the cauda equina. Superiorly, it is continuous with the cranial subarachnoid space, through which it communicates with the general ventricular cavity of the brain, by means of the openings, in the pia mater, in the roof of the fourth ventricle (*foramen of Majendie* and *foramina of Key and Retzius*). It is partially divided by a longitudinal membranous septum, the *septum posticum*, which serves to connect the arachnoid with the pia mater, opposite the posterior median fissure of the spinal cord, and forms a partition, which is incomplete and cribriform above, but more perfect in the dorsal region of the spine. It consists of bundles of white fibrous tissue interlacing with each other. Each of these divisions of the spinal subarachnoid space is further subdivided by the *ligamenta denticulata*, which will be described with the pia mater.

The cerebro-spinal fluid fills up the subarachnoid space. In the spine it is so abundant as to completely fill up the whole of the space included in the dura mater. It is a clear limpid fluid, having a saltish taste and a slightly alkaline reaction. According to Lassaigne, it consists of 98·5 parts of water, the remaining 1·5 per cent. being solid matters, animal and saline. It varies in quantity, being most abundant in old persons, and is quickly reproduced. Its chief use

3 I

is probably to afford mechanical protection to the nervous centres, and to prevent the effects of concussions communicated from without.

The **glandulæ Pacchioni**, or **arachnoid villi**, are small, fleshy-looking elevations, usually collected into clusters of variable size, which may be seen upon the outer surface of the dura mater, in the vicinity of the superior longitudinal sinus, and in some other situations. Little pits or depressions will be found on the corresponding parts of the calvarium, into which these excrescences are received. Upon laying open the superior longitudinal sinus, others of these bodies will be found protruding into its interior. These bodies are not glandular in structure, but, according to Luschka, are enlarged normal villi of the arachnoid. On each side of the sinus, and communicating with it, are large venous spaces, named *lacunæ laterales*, situated in the dura mater, and into these the villi project. As they grow they push the thinned dura mater before them, and cause absorption of the bone from pressure, and so produce the pits or depressions on the inner wall of the calvarium. A Pacchionian body consists of the following parts : 1. In the interior is a core of subarachnoid tissue, which is continuous with the meshwork of the general subarachnoid tissue through a narrow pedicle, by which the Pacchionian body is attached to the arachnoid. 2. Around this tissue is a layer of arachnoid membrane, which limits and encloses the subarachnoid tissue. 3. Outside this, again, is the thinned wall of the lacuna, which is separated from the arachnoid covering the body by a space which corresponds to and is continuous with the subdural space. 4. And finally, if the body projects into the longitudinal sinus, it will be covered by the greatly thinned upper wall of the sinus. It will be seen, therefore, that fluid injected into the subarachnoid space will find its way into the Pacchionian bodies, and it has been found experimentally that it passes from these bodies into the venous sinuses into which these bodies project. The Pacchionian bodies are supposed to be the means by which excess of cerebro-spinal fluid is got rid of, when its quantity is increased above normal.

These bodies are not seen in infancy, and very rarely until the third year. They are usually found after the seventh year ; and from this period they increase in number as age advances.

The Pia Mater

The **pia mater** is a vascular membrane, consisting of a minute plexus of blood-vessels, held together by an extremely fine areolar tissue. The *cerebral pia mater* invests the entire surface of the brain, dipping down between the convolutions and laminæ, and is prolonged into the interior, as an invagination forming the velum interpositum or tela choroidea superior, and the choroid plexuses of the lateral and third ventricles. As it passes over the roof of the fourth ventricle, it forms the tela choroidea inferior and the choroid plexuses of this ventricle. Upon the surfaces of the hemispheres, where it covers the grey matter of the convolutions, it gives off from its inner surface a multitude of sheaths, which surround minute vessels, that extend perpendicularly for some distance into the cerebral substance (see fig. 489, page 636). On the cerebellum the membrane is more delicate ; the vessels from its inner surface are shorter, and its relations to the cortex are not so intimate. The *spinal pia mater* is thicker, firmer, and less vascular than that of the brain : this is due to the fact that it consists of two layers, the outer or additional one being composed of bundles of connective-tissue fibres, arranged for the most part longitudinally. Between the layers are cleft-like lymphatic spaces which communicate with the subarachnoid space, and a number of blood-vessels which are enclosed in perivascular lymphatic sheaths, into which the lymphatic spaces open. The spinal pia mater covers the entire surface of the cord, to which it is very intimately adherent, and in front it sends a process backwards into the anterior fissure. A longitudinal fibrous band extends along the middle line on its anterior surface, called by Haller the *linea splendens* ; and a somewhat similar band, the *ligamentum denticulatum*, is situated on each side. At the point where the cord terminates, the pia mater becomes contracted and is continued down as a long, slender filament (*filum terminale*), which descends through the centre of the mass of nerves forming the cauda equina. It perforates the dura mater about the level of the second or third lumbar vertebra, receiving a sheath from it, and extends downwards as far as the base of the coccyx, where it blends with the periosteum. It assists in maintaining the cord in its position during the movements of the trunk, and is, from this circumstance, called the *central ligament of the cord*.

The pia mater of both the brain and the spinal cord forms a sheath for the nerves as they emerge from the central nervous matter. This is closely connected with the nerve, and blends with its common membranous investment.

The **Ligamentum Denticulatum** (fig. 597) is a narrow fibrous band situated on each side of the spinal cord throughout its entire length, and separating the anterior from the posterior roots of the spinal nerves. It has received its name from the serrated appearance which it presents. Its inner border is continuous with the pia mater, at the side of the cord. Its outer border presents a series of triangular tooth-like processes, the points of which are fixed at intervals to the dura mater. These processes are twenty-one in number, on each side, the first being attached to the dura mater, opposite the margin of the foramen magnum between the vertebral artery and the hypoglossal nerve ; and the last near the lower end of the cord. Its function is to support the cord in the fluid by which it is surrounded.

FIG. 597.—Transverse section of the spinal cord and its membranes.

Surgical Anatomy.—Evidence of great value in the diagnosis of meningitis may sometimes be obtained by puncturing the theca of the cord and withdrawing some of the cerebro-spinal fluid ; and the operation is regarded by some as curative, under the supposition that the draining of some of the cerebro-spinal fluid relieves the patient by diminishing the intracranial pressure. The operation is performed by inserting a trocar, of the smallest size, between the lamina of the third and fourth, or the fourth and fifth lumbar vertebræ, through the ligamenta subflava. The spinal cord, even of a child at birth, does not reach below the third lumbar vertebra, and therefore the canal may be punctured between the third and fourth lumbar vertebræ without any risk of injuring this structure. The point of puncture is indicated by laying the child on its side and dropping a perpendicular line from the highest point of the crest of the ilium ; this will cross the upper border of the spine of the fourth lumbar vertebra, and will indicate the level at which the trocar should be inserted a little to one side of the middle line.

CRANIAL NERVES

There are twelve pairs of cranial nerves ; they arise from some part of the cerebro-spinal centre, and are transmitted through foramina in the base of the cranium. They have been named numerically, according to the order in which they pass through the dura mater lining the base of the skull. Other names are also given to them derived from the parts to which they are distributed, or from their functions. Taken in their order, from before backwards, the different pairs are named as follows :

1st. Olfactory.	7th. Facial.
2nd. Optic.	8th. Auditory.
3rd. Motor oculi.	9th. Glosso-pharyngeal.
4th. Trochlear (Pathetic).	10th. Pneumogastric or Vagus.
5th. Trifacial (Trigeminus).	11th. Spinal accessory.
6th. Abducent.	12th. Hypoglossal.

All the cranial nerves are connected to some part of the surface of the brain. This is termed their *superficial* or *apparent origin*. But their fibres can, in all cases, be traced into the substance of the brain to special centres of grey matter, termed *nuclei*. As already stated in the chapter on Embryology, the cranial nerves, with the exception of the first and second, are developed in a similar manner to the spinal nerves. The motor or efferent cranial nerves arise from groups of nerve-cells situated within the brain, and such groups of cells constitute the *nuclei of origin of the motor nerves*. The sensory or afferent cranial nerves arise outside the brain from groups of nerve-cells or ganglia derived from the neural crest or ganglion ridge, and situated on the trunks of the nerves ; these ganglia must therefore be looked upon as the *nuclei of origin of the sensory nerves.* The central processes of these ganglion cells grow into the brain, and there end by arborising around the cells of certain nuclei or collections of nerve-cells, which are termed the *nuclei of termination of the sensory nerves.* The nuclei of origin of the motor nerves and the nuclei of termination of the sensory nerves are brought into relationship with the cerebral cortex, the former through the geniculate bundle of the internal capsule, the latter through the fillet. The geniculate fibres arise from the cells of the motor area of the cortex, and after crossing the middle line, end by arborising around the cells of the nuclei of origin of the motor nerves. On the other hand, fibres arise from the cells of the nuclei of termination of the sensory nerves, and after crossing the opposite side, join the fillet, and thus connect these nuclei, directly or indirectly, with the cerebral cortex. The nerves pass through foramina or tubular prolongations in the dura mater, leave the skull through foramina in its base, and are prolonged to their final distribution.

OLFACTORY NERVES

The **Olfactory nerves** (*nn. olfactorii*) are the nerves of smell, and are distributed to the mucous membrane of the olfactory region of the nose: this region comprises the superior turbinated bone, and the corresponding part of the nasal septum. Usually described as arising from the under surface of the olfactory bulb which rests upon the cribiform plate of the ethmoid bone, the real origin of these nerves is to be looked for in the olfactory cells of the nasal mucous membrane, the central or deep processes of these cells being prolonged upwards as the nerves. The nerve filaments form a plexiform network in the mucous membrane, and are then collected into about twenty branches, which pierce the cribiform plate of the ethmoid bone in two groups, an outer and an inner, and terminate in the glomeruli of the olfactory bulb (see page 818). Each branch receives tubular sheaths from the dura mater and pia mater, the former being lost in the periosteum of the nose, the latter in the neurilemma of the nerve.

The olfactory nerves differ in structure from other nerves in being composed exclusively of non-medullated fibres. They consist of axis-cylinders with a distinct, nucleated sheath, in which there are, however, fewer nuclei than in ordinary non-medullated nerve-fibres.

The olfactory centre in the cortex is not definitely known. It is generally associated with the uncus of the hippocampal convolution, the trigonum olfactorium, the part of the callosal convolution which lies below the genu and rostrum of the corpus callosum, and also the posterior olfactory lobule.

Surgical Anatomy.—In severe injuries to the head, the olfactory bulb may become separated from the olfactory nerves, thus producing loss of smell (*anosmia*), and with this a considerable loss in the sense of taste, since much of the perfection of the sense of taste is due to the sapid substances being also odorous, and simultaneously exciting the sense of smell.

SECOND NERVE (fig. 598)

The **Second** or **Optic nerve** (*n. opticus*), the nerve of sight, is distributed exclusively to the eyeball. The nerves of opposite sides are connected together at the commissure, and from the back of the commissure they may be traced to the brain, under the name of the *optic tracts.*

The *optic tract*, at its connection with the brain, is divided into two bands, external and internal. The *external* band is the larger; it arises from the external geniculate body and from the pulvinar of the optic thalamus, and is partly continuous with the brachium of the upper quadrigeminal body. The *internal* band curves round the crusta, and ends in the internal geniculate body ; its fibres are merely commissural, forming Gudden's commissure. From this origin the tract winds obliquely across the under surface of the crus cerebri, in the form of a flattened band, and is attached to the crus by its anterior margin. It then assumes a cylindrical form, and, as it passes forwards, is connected with the tuber cinereum and lamina terminalis. It finally joins with the tract of the opposite side to form the *optic commissure.*

The *commissure* or *chiasma*, somewhat quadrilateral in form, rests upon the olivary eminence and on the anterior part of the diaphragma sellæ, being bounded, above, by the lamina terminalis; behind, by the tuber cinereum; on either side, by the anterior perforated space. Within the commissure, the optic nerves of the two sides undergo a partial decussation.

FIG. 598.—The left optic nerve and optic tracts.

The fibres which form the inner margin of each tract and posterior part of the commissure have no connection with the optic nerves. They simply pass across the commissure from one hemisphere of the brain to the other, and connect the internal geniculate bodies of the two sides. They are known as the *commissure of Gudden*. The remaining and principal part of the commissure consists of two sets of fibres, crossed and uncrossed. The *crossed*, which are the more numerous, occupy the central part of the chiasma, and pass from the optic nerve of one side to the optic tract of the other, decussating in the commissure with similar fibres of the opposite optic nerve. The *uncrossed* fibres occupy the outer part of the chiasma, and pass from the nerve of one side to the tract of the same side.*

The great majority of the fibres of the optic nerve consist of the afferent axons of nerve-cells in the retina. Some few, however, are efferent fibres, and grow out from the brain. The afferent fibres end in arborisations around the cells in the external geniculate body, pulvinar, and upper quadrigeminal body, which constitute the *lower visual centres*. From these nuclei other fibres are prolonged to the *cortical visual centre*, which, according to most observers, is situated in the cuneus, and probably also in the lingual lobule of the occipital lobe (see page 808).

FIG. 599.—Course of the fibres in the optic commissure.

Uncrossed fibres

Crossed fibres

Commissure of Gudden

It should be stated that some fibres are detached from the optic tract, and pass through the crus cerebri to the nucleus of the third nerve. These fibres are small, and may be regarded as afferent branches for the Sphincter pupillæ and Ciliary muscles. Other fibres pass to the cerebellum through its superior peduncles, while others, again, are lost in the pons.

The *optic nerves* arise from the fore part of the commissure, and, diverging from one another, become rounded in form and firm in texture, and are enclosed in a sheath derived from the pia mater and arachnoid. As each nerve passes through the corresponding optic foramen, it receives a sheath from the dura mater; and as it enters the orbit this sheath divides into two layers, one of which becomes continuous with the periosteum of the orbit; the other forms the proper sheath of the nerve, and surrounds it as far as the sclerotic. The nerve passes forwards and outwards through the cavity of the orbit, pierces the sclerotic and choroid coats at the back part of the eyeball, about one-eighth of an inch to the nasal side of its centre, and expands into the internal layer of the retina. A small artery, the *arteria centralis retinæ*, perforates the optic nerve a little behind the globe, and runs along its interior in a tubular canal of fibrous tissue. It supplies the inner surface of the retina, and is accompanied by corresponding veins. The retina is described with the anatomy of the eyeball.

Surgical Anatomy.—The optic nerve is peculiarly liable to become the seat of neuritis or undergo atrophy in affections of the central nervous system, and as a rule the pathological relationship between the two affections is exceedingly difficult to trace.

---

* A specimen of congenital absence of the optic commissure is to be found in the Museum of the Westminster Hospital. See also Henle, *Nervenlehre*, p. 393, ed. 2.

There are, however, certain points in connection with the anatomy of this nerve which tend to throw light upon the frequent association of these affections with intracranial disease : (1) From its mode of development, and from its structure, the optic nerve must be regarded as a prolongation of the brain-substance, rather than as an ordinary cerebro-spinal nerve. (2) As it passes from the brain, it receives sheaths from the three cerebral membranes, a perineural sheath from the pia mater ; an intermediate sheath from the arachnoid ; and an outer sheath from the dura mater, which is also connected with the periosteum, as it passes through the optic foramen. These sheaths are separated from each other by spaces, which communicate with the subdural and subarachnoid spaces respectively. The innermost or perineural sheath sends a process around the arteria centralis retinæ into the interior of the nerve, and enters intimately into its structure. Thus inflammatory affection of the meninges or of the brain may readily extend themselves along these spaces, or along the interstitial connective tissue in the nerve.

Fig. 600.—Figure showing the course of the optic nerve-fibres from the retina to the cerebral cortex. (Testut.)

o.s. Left eye.   o.d. Right eye.   t. Temporal zone of the retina. n. Nasal zone.   T, N. Temporal portion and nasal portion of the field of vision for the left eye.   T', N'. The same for the right eye.   D. Right half, and s, left half of the field of vision.   1. Optic nerve, with 1', its direct fibres (in red) ; 1″, its crossed fibres (in yellow on the left side, in blue on the right).   2. Optic commissure.   3. The optic tract, with 3', direct fibres ; 3″, crossed fibres ; 3‴, commissure of Gudden. 4. Lower visual centres.   5. Optic radiations.   5'. Direct fibres. 5″. Crossed fibres.   6. Cerebral cortex (mesial surface of occipital lobe).   7. Posterior horn of lateral ventricle.

The course of the fibres in the optic commissure has an important pathological bearing, and has been the subject of much controversy. Microscopic examination, experiments, and pathology all seem to point to the fact that there is a partial decussation of the fibres, each optic tract supplying the corresponding half of each eye, so that the right tract supplies the right half of each eye, and the left tract the left half of each eye. At the same time Charcot believes, and his view has met with general acceptation, that the fibres which do not decussate at the optic commissure have already decussated in the corpora quadrigemina, so that lesion of the cerebral centre of one side causes complete blindness of the opposite eye, because both sets of decussating fibres are destroyed ; whereas if one tract, say the right,

be destroyed by disease, there will be blindness of the right half of both retinæ.

An antero-posterior section through the commissure would divide the decussating fibres, and would therefore produce blindness of the inner half of each eye ; while a section at the margin of the side of the optic commissure would produce blindness of the external half of the retina of the same side.

The optic nerve may also be affected in injuries or diseases involving the orbit ; in fractures of the anterior fossa of the base of the skull ; in tumours of the orbit itself, or those invading this cavity from neighbouring parts.

THIRD NERVE (figs. 601, 602, 603, 604)

The **Third** or **Motor oculi** nerve (*n. oculo-motorius*) supplies all the muscles of the orbit, except the Superior oblique and External rectus ; it also supplies, through its connection with the ciliary ganglion, the Sphincter muscle of the iris and the Ciliary muscle. It is rather a large nerve, of rounded form and firm texture.

The fibres of the third nerve arise from a nucleus of cells which lies in the grey matter of the floor of the aqueduct of Sylvius on the level of the upper quadrigeminal body. The nucleus also extends in front of the aqueduct and reaches a short distance into the floor of the third ventricle. From their nucleus of origin the fibres pass forward through the tegmentum, the red nucleus and the inner part of the substantia nigra, forming a series of curves with their convexity outwards, and emerge in the oculo-motor sulcus on the inner side of the crus cerebri.

The nucleus of the oculo-motor nerve does not consist of a continuous column of cells, but is broken up into a number of smaller nuclei, which may be

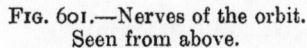

FIG. 601.—Nerves of the orbit. Seen from above.

FIG. 602.—Figure showing the different groups of cells, which constitute, according to Perlia, the nucleus of origin of the motor oculi nerve. (Testut.)

1. Posterior dorsal nucleus. 1'. Posterior ventral nucleus. 2. Anterior dorsal nucleus. 2'. Anterior ventral nucleus. 3. Central nucleus. 4. Nucleus of Edinger and Westphal. 5. Antero-internal nucleus. 6. Antero-external nucleus. 7. Trunk of the motor oculi nerve. 8. Crossed fibres. 9. Trochlear nerve, with 9', its nucleus of origin, and 9", its decussation. 10. Third ventricle. M, M. Median line.

arranged in two groups, anterior and posterior. Those of the posterior group are six in number, five of which are symmetrical on the two sides of the middle line, while the sixth is centrally placed and is common to the nerves of both sides. The anterior group consists of two nuclei, an antero-internal and an antero-external (fig. 602).

The nucleus of the third nerve gives fibres to the seventh nerve, which probably supply the Orbicularis palpebrarum, Corrugator supercilii, and anterior belly of the Occipito-frontalis. It is also connected with the nuclei of the fourth and sixth nerves, with the cerebellum, upper quadrigeminal body and cortex of the occipital lobe of the cerebrum.

The nucleus of the third nerve, considered from a physiological standpoint, can be subdivided into several smaller groups of cells, each group controlling a particular muscle. The nerves to the different muscles appear to take their origin from behind forwards, as follows : Inferior oblique, Inferior rectus, Superior rectus, Levator palpebræ, and Internal rectus ; while from the anterior end of the nucleus the fibres for accommodation and for the Sphincter pupillæ take their origin.

On emerging from the brain, the nerve is invested with a sheath of pia mater, and enclosed in a prolongation from the arachnoid. It passes between the superior cerebellar and posterior cerebral arteries, and then pierces the dura mater in front of and external to the posterior clinoid process, passing between the free and attached borders of the tentorium, which are prolonged forwards to be connected with the anterior and posterior clinoid processes of the sphenoid bone. It passes along the outer wall of the cavernous sinus, above the other orbital nerves, receiving in its course one or two filaments from the cavernous plexus of the sympathetic, and a communicating branch from

FIG. 603.—Plan of the motor oculi nerve. (After Flower.)

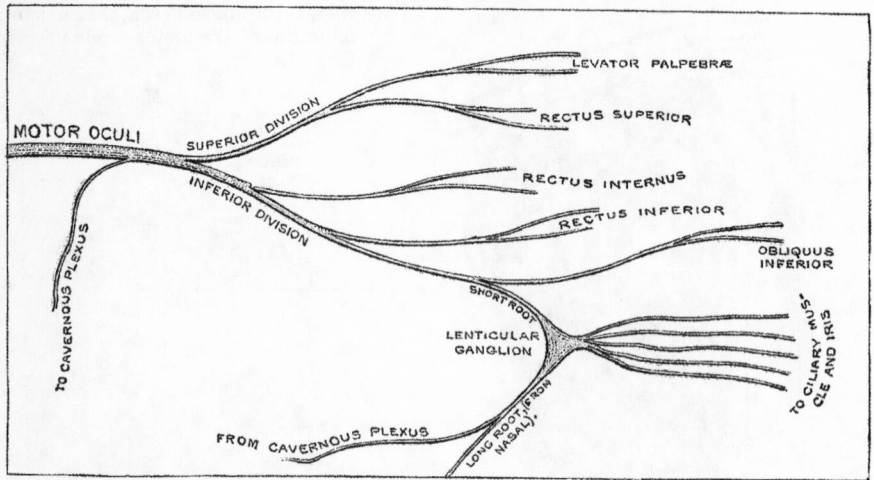

the first division of the fifth. It then divides into two branches, which enter the orbit through the sphenoidal fissure, between the two heads of the External rectus muscle. On passing through the fissure, the nerve is placed below the fourth and the frontal and lachrymal branches of the ophthalmic nerve, while the nasal nerve is placed between its two divisions.

The *superior division*, the smaller, passes inwards over the optic nerve, and supplies the Superior rectus and Levator palpebræ. The *inferior division*, the larger, divides into three branches. One passes beneath the optic nerve to the Internal rectus ; another, to the Inferior rectus ; and the third, the longest of the three, runs forwards between the Inferior and External recti to the Inferior oblique. From this latter a short, thick branch is given off to the lower part of the lenticular ganglion, and forms its inferior root. All these branches enter the muscles on their ocular surface, with the exception of the nerve to the Inferior oblique, which enters the muscle at its posterior border.

Surgical Anatomy.—Paralysis of the third nerve may be the result of many causes: such as cerebral disease ; or conditions causing pressure on the cavernous sinus ; or periostitis of the bones entering into the formation of the sphenoidal fissure. It results, when complete, in (1) ptosis, or drooping of the upper eyelid, in consequence of the Levator palpebræ being paralysed ; (2) external strabismus, on account of the unopposed action of the External rectus muscle, which is not supplied by the third nerve, and is not therefore paralysed ; (3) dilatation of the pupil, because the sphincter fibres of the iris are paralysed ; (4) loss of power of accommodation, as the Sphincter pupillæ, the Ciliary muscle, and the Internal rectus are paralysed ; (5) slight prominence of the eyeball, owing to most of its muscles being relaxed. Occasionally paralysis may affect only a part of the nerve—that is to say, there may be, for example, a dilated and fixed pupil, with ptosis, but no other signs. Irritation of the nerve causes spasm of one or other of the muscles supplied by it ; thus, there may be internal strabismus from spasm of the Internal rectus ; accommodation for near objects only from spasm of the Ciliary muscle ; or myosis, contraction of the pupil, from irritation of the Sphincter of the pupil.

FOURTH NERVE (fig. 601)

The **Fourth** or **Trochlear nerve** (*n. trochlearis*), the smallest of the cranial nerves, supplies the Superior oblique muscle of the eyeball.

It arises from a nucleus of cells situated in the floor of the Sylvian aqueduct, opposite the upper part of the inferior quadrigeminal body. From its origin the nerve runs outwards and downwards through the tegmentum, and then turns backwards and inwards into the upper part of the valve of Vieussens. Here it decussates with the corresponding nerve of the opposite side and emerges from the surface of the velum at the side of the frænulum veli, immediately behind the lower quadrigeminal body.

Emerging from the valve of Vieussens, the nerve is directed outwards across the superior peduncle of the cerebellum, and then winds forwards round the outer side of the crus cerebri, immediately above the pons Varolii, pierces the dura mater in the free border of the tentorium cerebelli, just behind, and external to, the posterior clinoid process, and passes forwards in the outer wall of the cavernous sinus, between the third nerve and the ophthalmic division of the fifth. It crosses the third nerve and enters the orbit, through the sphenoidal fissure. It now becomes the highest of all the nerves, lying at the inner extremity of the fissure internal to the frontal nerve. In the orbit it passes inwards, above the origin of the Levator palpebræ, and finally enters the orbital surface of the Superior oblique muscle.

Branches of communication.—In the outer wall of the cavernous sinus the fourth nerve forms communications with the ophthalmic division of the fifth and with the cavernous plexus of the sympathetic. In the sphenoidal fissure it occasionally gives off a branch to the lachrymal nerve. *Branches of distribution.*—It gives off a recurrent branch, which passes backwards between the layers of the tentorium, dividing into two or three filaments which may be traced as far back as the wall of the lateral sinus.

Surgical Anatomy.—The fourth nerve when paralysed causes loss of function in the Superior oblique, so that the patient is unable to turn his eye downwards and outwards. Should the patient attempt to do this, the eye of the affected side is twisted inwards, producing diplopia or double vision. Accordingly, it is said that the first symptom of loss of function in this muscle is giddiness when going down hill or in descending stairs, owing to the double vision induced by the patient looking at his steps while descending.

FIFTH NERVE

The **Fifth** or **Trifacial nerve** (*n. trigeminus*) is the largest cranial nerve. It resembles a spinal nerve : (1) in arising by two roots, a motor and a sensory ; and (2) in having a ganglion developed on its sensory root. It is the great sensory nerve of the head and face, and the motor nerve of the muscles of mastication. It divides into three divisions : the first and second are entirely sensory, the third is partly sensory and partly motor.

It emerges from the side of the pons Varolii, near its upper border, by a small *motor* and a large *sensory* root—the former being situated in front and to the inner side of the latter.

The fibres of the *motor root* arise from two nuclei, an upper and a lower. The *upper nucleus* consists of a strand of cells which occupies the whole length of the lateral portion of the grey matter of the Sylvian aqueduct. The *lower* or *chief nucleus* is situated in the upper part of the pons Varolii, close to its dorsal surface, and along the line of the lateral margin of the fourth ventricle. The fibres from the upper nucleus constitute the *mesencephalic* or *Sylvian root* : they descend through the mid-brain, and, entering the pons, join with the fibres from the lower nucleus ; and the motor root, thus formed, passes forward through the pons to its point of emergence.

The fibres of the *sensory root* arise from the cells of the Gasserian ganglion, which lies in a cavity of the dura mater near the apex of the petrous part of the temporal bone. They pass backwards and inwards below the superior petrosal sinus and tentorium cerebelli, and entering the pons, divide into upper and lower roots. The upper root terminates partly in a nucleus which is situated in the pons on the outer side of the lower motor nucleus, and partly in the locus cæruleus ; the lower root descends through the pons and medulla, and ends in the upper part of the substantia gelatinosa of Rolando. This lower root is sometimes named the *ascending root* of the fifth nerve.

The **Gasserian ganglion** occupies a cavity (*cavum Meckelii*) in the dura mater which is situated in a depression near the apex of the petrous part of the temporal bone. It is somewhat crescentic in shape, with its convexity directed forwards ; internally it is in relation with the internal carotid artery and the posterior part of the cavernous sinus. The motor root runs forwards and outwards in front and to the inner side of the sensory root, and then passes below the ganglion without having any connection with it. It leaves the skull through the foramen ovale, and, immediately below this foramen, joins the inferior maxillary nerve. Besides the motor root, the large superficial petrosal nerve lies underneath the ganglion.

Branches of communication.—This ganglion receives, on its *inner side*, filaments from the carotid plexus of the sympathetic. *Branches of distribution.*—It gives off minute branches to the tentorium cerebelli and the dura mater, in the middle fossa of the cranium. From its convex border, which is directed forwards and outwards, three large branches proceed, viz. : the *ophthalmic, superior maxillary*, and *inferior maxillary*. The ophthalmic and superior maxillary consist exclusively of fibres derived from the ganglion, and are solely nerves of common sensation. As already stated, the third division, or inferior maxillary, is joined outside the cranium by the motor root. This, therefore, strictly speaking, is the only portion of the fifth nerve which can be said to resemble a spinal nerve.

OPHTHALMIC NERVE (figs. 601, 604, 605)

The **Ophthalmic** (*n. ophthalmicus*), or first division of the fifth, is a sensory nerve. It supplies sensory branches to the cornea, ciliary muscle, and iris ; the lachrymal gland, the conjunctiva ; a part of the mucous membrane of the nasal fossæ ; and the integument of the eyebrow, forehead, and nose. It is the smallest of the three divisions of the fifth, and arises from the upper part of the Gasserian ganglion. It is a short, flattened band, about an inch in length, which passes forwards along the outer wall of the cavernous sinus, below the third and fourth nerves ; just before entering the orbit, through the sphenoidal fissure, it divides into three branches, *lachrymal, frontal*, and *nasal*.

Branches of communication.—The ophthalmic nerve is joined by filaments from the cavernous plexus of the sympathetic, and communicates with the third, fourth, and sixth nerves.

Branches of distribution.—The ophthalmic nerve gives off a recurrent filament which passes between the layers of the tentorium, and the nerve then divides into

Lachrymal.	Frontal.	Nasal.

The **lachrymal** is the smallest of the three branches of the ophthalmic. It sometimes receives a filament from the fourth nerve, but this is possibly derived from the branch of communication which passes from the ophthalmic to the fourth. It passes forwards in a separate tube of dura mater, and enters the orbit through the narrowest part of the sphenoidal fissure. In the orbit it runs along the upper border of the External rectus muscle, with the lachrymal artery, and communicates with the temporo-malar branch of the superior maxillary. It enters the lachrymal gland and gives off several filaments, which supply the gland and the conjunctiva. Finally, it pierces the superior palpebral ligament, and terminates in the integument of the upper eyelid, joining with filaments of the facial nerve. The lachrymal nerve is occasionally absent, when its place is taken by the temporal branch of the superior maxillary. Sometimes the latter branch is absent, and a continuation of the lachrymal is substituted for it.

The **frontal** is the largest division of the ophthalmic, and may be regarded both from its size and direction, as the continuation of the nerve. It enters the orbit above the muscles, through the sphenoidal fissure, and runs forwards along the middle line, between the Levator palpebræ and the periosteum. Midway between the apex and base of the orbit it divides into two branches, supratrochlear and supra-orbital.

The *supratrochlear branch*, the smaller of the two, passes inwards, above the pulley of the Superior oblique muscle, and gives off a descending filament, which joins with the infratrochlear branch of the nasal nerve. It then escapes from the orbit between the pulley of the Superior oblique and the supra-orbital foramen, curves up on to the forehead close to the bone, ascends beneath the Corrugator supercilii and Occipito-frontalis muscles, and dividing into branches, which pierce these muscles, it supplies the integument of the lower part of the forehead on either side of the middle line and sends filaments to the conjunctiva and skin of the upper eyelid.

The *supra-orbital branch* passes forwards through the supra-orbital foramen, and gives off, in this situation, palpebral filaments to the upper eyelid. It then ascends upon the forehead, and terminates in two branches, an inner and an outer, which supply the integument of the cranium, reaching nearly as far back as the parieto-occipital suture. They are at first situated beneath the Occipito-frontalis, the inner branch perforating the frontal portion of the muscle, the outer branch its tendinous aponeurosis. From its two branches, small twigs pass to the pericranium.

The **nasal nerve** is intermediate in size between the frontal and lachrymal, and is more deeply placed than the other branches of the ophthalmic. It enters the orbit between the two heads of the External rectus, and extends obliquely

FIG. 604.—Nerves of the orbit and ophthalmic ganglion. Side view.

inwards across the optic nerve, beneath the Superior rectus and Superior oblique muscles, to the inner wall of the orbit. Here it passes through the anterior ethmoidal foramen, and entering the cavity of the cranium, traverses a shallow groove on the front part of the cribriform plate of the ethmoid bone, and runs down, through the slit by the side of the crista galli, into the nose, where it divides into two branches, an internal and an external. The *internal branch* supplies the mucous membrane near the fore part of the septum of the nose. The *external branch* descends in a groove on the inner surface of the nasal bone, and supplies a few filaments to the mucous membrane covering the fore part of the outer wall of the nares as far as the inferior spongy bone; it then leaves the cavity of the nose, between the lower border of the nasal bone and the upper lateral cartilage, and, passing down beneath the Compressor nasi, supplies the integument of the ala and the tip of the nose, joining with the facial nerve.

The branches of the nasal nerve are, the *ganglionic, ciliary,* and *infra-trochlear.*

The *ganglionic* is a slender branch, about half an inch in length, which usually arises from the nasal between the two heads of the External rectus. It passes forwards on the outer side of the optic nerve, and enters the postero-superior angle of the ciliary ganglion, forming its superior or long root. It is sometimes joined by a filament from the cavernous plexus of the sympathetic, or from the superior division of the third nerve.

The *long ciliary nerves*, two or three in number, are given off from the nasal as it crosses the optic nerve. They join the short ciliary nerves from the ciliary ganglion, pierce the posterior part of the sclerotic, and running forwards between it and the choroid, are distributed to the Ciliary muscle, iris, and cornea.

The *infratrochlear branch* is given off just before the nasal nerve enters the anterior ethmoidal foramen. It runs forwards along the upper border of the Internal rectus, and is joined, near the pulley of the Superior oblique, by a filament from the supratrochlear nerve. It then passes to the inner angle of the eye, and supplies the integument of the eyelids and side of the nose, the conjunctiva, lachrymal sac, and caruncula lachrymalis.

Ophthalmic Ganglion (figs. 604, 605)

Associated with the three divisions of the fifth nerve are four small ganglia. The *ophthalmic ganglion* is connected with the first division ; the *spheno-palatine* or *Meckel's ganglion* with the second ; and the *otic* and *submaxillary ganglia* with the third. All the four receive sensory filaments from the fifth, and motor and sympathetic filaments from various sources ; these filaments are called the *roots of the ganglia*.

The **Ophthalmic, Lenticular,** or **Ciliary ganglion** is a small, quadrangular, flattened ganglion, of a reddish-grey colour, and about the size of a pin's head, situated at the back part of the orbit between the optic nerve and the External rectus muscle, lying generally on the outer side of the ophthalmic artery. It is enclosed in a quantity of loose fat, which makes its dissection somewhat difficult.

Its *branches of communication*, or *roots*, are three, all of which enter its posterior border. One, the long or sensory root, is derived from the nasal branch of the ophthalmic, and joins its postero-superior angle. The second, the short or motor root, is a short, thick nerve, occasionally divided into two parts, which is derived from the branch of the third nerve to the Inferior oblique muscle, and is connected with the postero-inferior angle of the ganglion. The third, the sympathetic root, is a slender filament from the cavernous plexus of the sympathetic. This is frequently blended with the long root, though it sometimes passes to the ganglion separately. According to Tiedemann, this ganglion receives a filament of communication from the spheno-palatine ganglion.

Its *branches of distribution* are the short ciliary nerves. These are delicate filaments, from six to ten in number, which arise from the fore part of the ganglion in two bundles, connected with its superior and inferior angles ; the lower bundle is the larger. They run forwards with the ciliary arteries in a wavy course, one set above and the other below the optic nerve, and are accompanied by the long ciliary nerves from the nasal. They pierce the sclerotic at the back part of the globe, pass forwards in delicate grooves on its inner surface, and are distributed to the Ciliary muscle, iris, and cornea. Tiedemann has described one small branch as penetrating the optic nerve with the arteria centralis retinæ.

Superior Maxillary Nerve (fig. 606)

The **Superior maxillary** (*n. maxillaris*), or second division of the fifth, is a sensory nerve. It is intermediate, both in position and size, between the ophthalmic and inferior maxillary. It commences at the middle of the Gasserian ganglion as a flattened plexiform band, and passing horizontally forwards, it leaves the skull through the foramen rotundum, where it becomes more cylindrical in form, and firmer in texture. It then crosses the spheno-maxillary fossa, and inclines outwards on the back of the superior maxilla ; it then enters the orbit through the spheno-maxillary fissure, traverses the infra-orbital canal in the floor of the orbit, and appears upon the face at the infra-orbital foramen.*
At its termination, the nerve lies beneath the Levator labii superioris muscle,

---

* After it enters the infra-orbital canal, the nerve is frequently called the *infra-orbital*.

FIG. 605.—Plan of the fifth cranial nerve. (After Flower.)

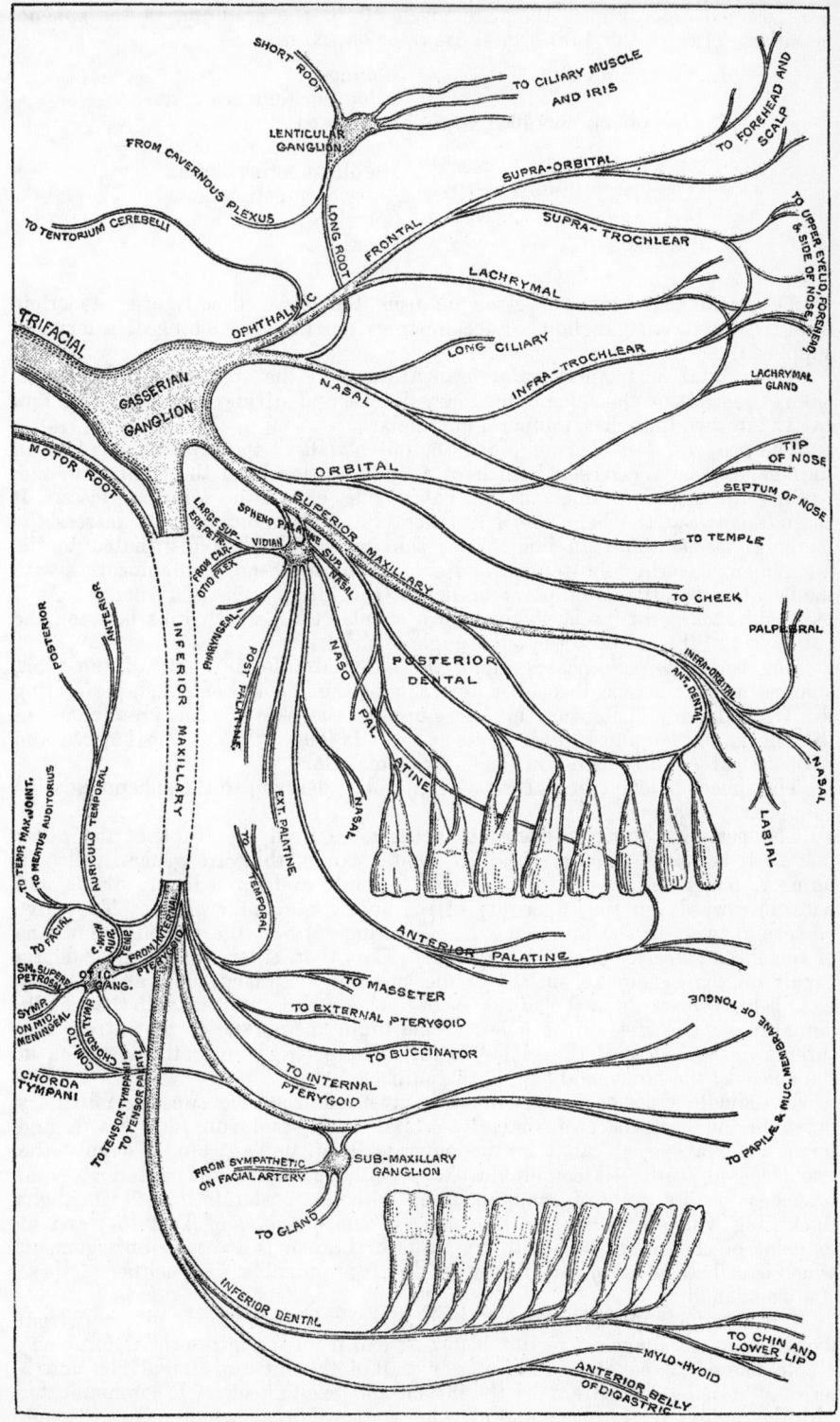

and divides into a leash of branches, which spread out upon the side of the nose, the lower eyelid, and upper lip, joining with filaments of the facial nerve.

Branches of distribution.—The branches of this nerve may be divided into four groups, according as they are given off in the cranium, in the spheno-maxillary fossa, in the infra-orbital canal, or on the face.

In the cranium . . .	Meningeal.
In the spheno-maxillary fossa	{ Orbital or temporo-malar. Spheno-palatine. Posterior superior dental.
In the infra-orbital canal .	{ Middle superior dental. Anterior superior dental.
On the face . . .	{ Palpebral. Nasal. Labial.

The **meningeal branch** is given off from the nerve directly after its origin from the Gasserian ganglion ; it accompanies the middle meningeal artery and supplies the dura mater.

The **orbital** or **temporo-malar branch** arises in the spheno-maxillary fossa, enters the orbit by the spheno-maxillary fissure, and divides at the back of that cavity into two branches, temporal and malar.

The *temporal branch* runs along the outer wall of the orbit in a groove in the malar bone, receives a branch of communication from the lachrymal, and, passing through a foramen in the malar bone, enters the temporal fossa. It ascends between the bone and substance of the Temporal muscle, pierces the temporal fascia about an inch above the zygoma, and is distributed to the integument covering the temple and side of the forehead, communicating with the facial and auriculo-temporal branch of the inferior maxillary nerve. As it pierces the temporal fascia, it gives off a slender twig, which runs between the two layers of the fascia to the outer angle of the orbit.

The *malar branch* passes along the external inferior angle of the orbit, emerges upon the face through a foramen in the malar bone, and, perforating the Orbicularis palpebrarum muscle, supplies the skin on the prominence of the cheek, and is named *subcutaneus malæ*. It joins with the facial nerve and with the palpebral branches of the superior maxillary.

The **spheno-palatine branches,** two in number, descend to the spheno-palatine ganglion.

The **posterior superior dental branches** arise from the trunk of the nerve just as it is about to enter the infra-orbital canal ; they are generally two in number, but sometimes arise by a single trunk, and immediately divide and pass downwards on the tuberosity of the superior maxillary bone. They give off several twigs to the gums and neighbouring parts of the mucous membrane of the cheek (*superior gingival branches*). They then enter the posterior dental canals on the zygomatic surface of the superior maxillary bone, and, passing from behind forwards in the substance of the bone, communicate with the middle dental nerve, and give off branches to the lining membrane of the antrum and three twigs to each of the molar teeth. These twigs enter the foramina at the apices of the fangs, and supply the pulp.

The **middle superior dental branch** is given off from the superior maxillary nerve in the back part of the infra-orbital canal, and runs downwards and forwards in a special canal in the outer wall of the antrum to supply the two bicuspid teeth. It communicates with the posterior and anterior dental branches. At its point of communication with the posterior branch is a slight thickening which has received the name of the *ganglion of Valentin* ; and at its point of communication with the anterior branch is a second enlargement, which is called the *ganglion of Bochdalek*. It is probable that neither of these is a true ganglion.

The **anterior superior dental branch,** of considerable size, is given off from the superior maxillary nerve just before its exit from the infra-orbital foramen ; it enters a special canal in the anterior wall of the antrum, and divides into a series of branches which supply the incisor and canine teeth. It communicates with the middle dental nerve, and gives off a *nasal branch*, which passes through a minute canal into the nasal fossa, and supplies the mucous membrane of the fore part of the inferior meatus and the floor of this cavity, communicating with the nasal branches from Meckel's ganglion.

The **palpebral branches** pass upwards beneath the Orbicularis palpebrarum. They supply the integument and conjunctiva of the lower eyelid, joining at the outer angle of the orbit with the facial nerve and malar branch of the orbital.

FIG. 606.—Distribution of the second and third divisions of the fifth nerve and submaxillary ganglion.

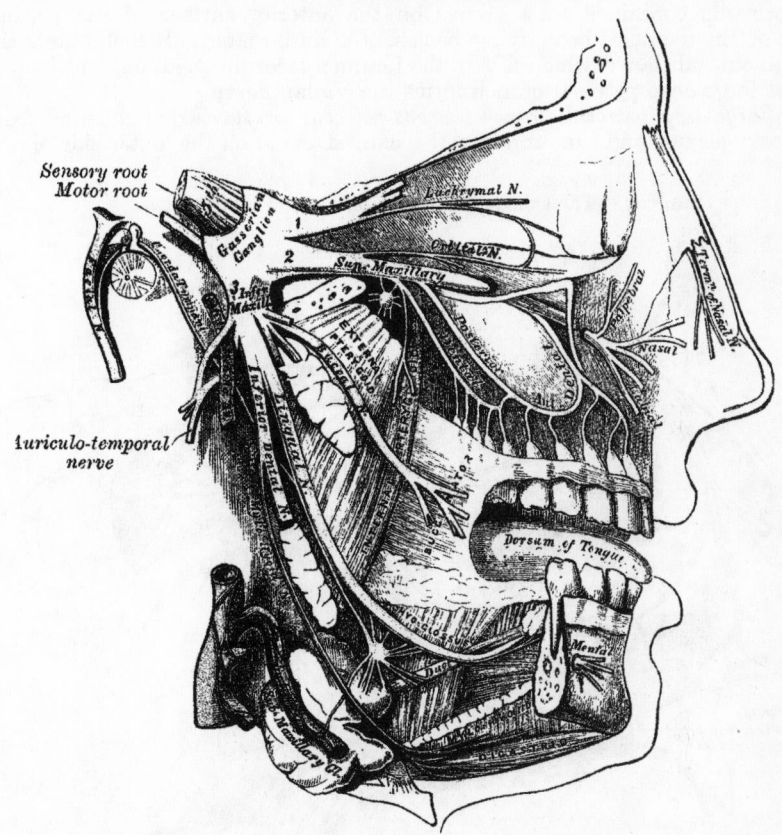

The **nasal branches** pass inwards ; they supply the integument of the side of the nose, and join with the nasal branch of the ophthalmic.

The **labial branches,** the largest and most numerous, descend beneath the Levator labii superioris, and are distributed to the integument of the upper lip, the mucous membrane of the mouth, and labial glands.

All these branches are joined, immediately beneath the orbit, by filaments from the facial nerve, forming an intricate plexus, the *infra-orbital.*

SPHENO-PALATINE GANGLION (fig. 607)

The **spheno-palatine ganglion** (*Meckel's*), the largest of the cranial ganglia, is deeply placed in the spheno-maxillary fossa, close to the spheno-palatine foramen. It is triangular or heart-shaped, of a reddish-grey colour, and is situated just below the superior maxillary nerve as it crosses the fossa.

Branches of communication.—Like the other ganglia of the fifth nerve, the spheno-palatine possesses a motor, a sensory, and a sympathetic root. Its *sensory root* is derived from the superior maxillary nerve through its two spheno-palatine branches. These branches of the nerve, given off in the spheno-maxillary fossa, descend to the ganglion. Their fibres, for the most part, pass in front of the ganglion, as they proceed to their destination, in the palate and nasal fossa, and are not incorporated in the ganglionic mass; some few of the fibres, however, enter the ganglion, constituting its sensory root. Its *motor root* is probably derived from the facial nerve through the large superficial petrosal nerve, but this nerve consists chiefly of sensory fibres; and its *sympathetic root*, from the carotid plexus, through the large deep petrosal nerve. These two nerves join together to form a single nerve, the *Vidian*, before their entrance into the ganglion.

The *large superficial petrosal branch (nervus petrosus superficialis major)* is given off from the geniculate ganglion of the facial nerve in the aqueductus Fallopii; it passes through the hiatus Fallopii; enters the cranial cavity, and runs forwards contained in a groove on the anterior surface of the petrous portion of the temporal bone, lying beneath the dura mater. It then enters the cartilaginous substance which fills in the foramen lacerum medium, and joining with the large deep petrosal branch forms the Vidian nerve.

The *large deep petrosal branch (nervus petrosus profundus)* is given off from the carotid plexus, and runs through the carotid canal on the outer side of the

FIG. 607.—The spheno-palatine ganglion and its branches.

internal carotid artery. It then enters the cartilaginous substance which fills in the foramen lacerum medium, and joins with the large superficial petrosal nerve to form the Vidian.

The *Vidian nerve*, formed in the cartilaginous substance which fills in the middle lacerated foramen by the junction of the two preceding nerves, passes forwards, through the Vidian canal, with the artery of the same name, and is joined by a small ascending branch, the *sphenoidal branch*, from the otic ganglion. Finally, it enters the spheno-maxillary fossa, and joins the posterior angle of Meckel's ganglion.

Branches of distribution.—These are divisible into four groups, viz. : *ascending* branches, which pass to the orbit; *descending*, to the palate; *internal*, to the nose; and *posterior*, to the pharynx and nasal fossæ.

The *ascending branches* are two or three delicate filaments, which enter the orbit by the spheno-maxillary fissure, and supply the periosteum. According to Luschka, some filaments pass through foramina in the suture between the os planum of the ethmoid and frontal bones to supply the mucous membrane of the posterior ethmoidal and sphenoidal sinuses.

The *descending* or *palatine branches* are distributed to the roof of the mouth, the soft palate, tonsil, and lining membrane of the nose. They are almost a direct continuation of the spheno-palatine branches of the superior maxillary nerve, and are three in number : anterior, middle, and posterior.

The *anterior* or *large palatine nerve* descends through the posterior palatine canal, emerges upon the hard palate at the posterior palatine foramen, and passes forwards in a groove in the hard palate, nearly as far as the incisor teeth. It supplies the gums, the mucous membrane and glands of the hard palate, and communicates in front with the termination of the naso-palatine nerve. While in the posterior palatine canal, it gives off *inferior nasal branches*, which enter the nose through openings in the palate bone, and ramify over the inferior turbinated bone and middle and inferior meatuses ; and, at its exit from the canal, a palatine branch is distributed to both surfaces of the soft palate.

The *middle* or *external palatine nerve* descends, through one of the accessory palatine canals, distributing branches to the uvula, tonsil, and soft palate. It is occasionally wanting.

The *posterior* or *small palatine nerve* descends with a minute artery through the small posterior palatine canal, emerging by a separate opening behind the posterior palatine foramen. It supplies the soft palate, tonsil, and uvula, and was formerly believed to supply the Levator palati and Azygos uvulæ muscles, but these are probably supplied by the spinal accessory through the pharyngeal plexus. The middle and posterior palatine join with the tonsillar branches of the glosso-pharyngeal to form a plexus around the tonsil (*circulus tonsillaris*).

The *internal branches* are distributed to the septum and outer wall of the nasal fossæ. They are the superior nasal and the naso-palatine.

The *superior nasal branches*, four or five in number, enter the back part of the nasal fossa by the spheno-palatine foramen. They supply the mucous membrane covering the superior and middle spongy bones, and the lining of the posterior ethmoidal cells, a few being prolonged to the upper and back part of the septum.

The *naso-palatine nerve* (*Cotunnius*) also enters the nasal fossa through the spheno-palatine foramen, and passes inwards across the roof of the nose, below the orifice of the sphenoidal sinus, to reach the septum ; it then runs obliquely downwards and forwards along the lower part of the septum, to the anterior palatine foramen, lying between the periosteum and mucous membrane. It descends to the roof of the mouth through the anterior palatine canal. The two nerves are here contained in separate and distinct canals, situated in the inter-maxillary suture, and termed the *foramina of Scarpa*, the left nerve being anterior to the right one. In the mouth, they become united, supply the mucous membrane behind the incisor teeth, and join with the anterior palatine nerves. The naso-palatine nerve furnishes a few small filaments to the mucous membrane of the septum.

Posterior branch.—The *pharyngeal nerve* is a small branch arising from the back part of the ganglion. It passes through the pterygo-palatine canal with the pterygo-palatine artery, and is distributed to the mucous membrane of the upper part of the pharynx, behind the Eustachian tube.

INFERIOR MAXILLARY NERVE (fig. 606)

The **Inferior maxillary nerve** (*n. mandibularis*) distributes branches to the teeth and gums of the lower jaw, the integument of the temple and external ear, the lower part of the face and lower lip, and the muscles of mastication; it also supplies a large branch to the tongue. It is the largest of the three divisions of the fifth, and is made up of two roots : a large or sensory root proceeding from the inferior angle of the Gasserian ganglion ; and a small or motor root, which passes beneath the ganglion, and unites with the sensory root, just after its exit from the skull through the foramen ovale. Immediately beneath the base of the skull, this nerve divides into two trunks, anterior and posterior. Previous to its division, the primary trunk gives off from its inner side a recurrent (meningeal) branch, and the nerve to the Internal pterygoid muscle.

The **recurrent branch** is given off directly after its exit from the foramen ovale. It passes backwards into the skull through the foramen spinosum with the middle meningeal artery. It divides into two branches, anterior and posterior, which accompany the main divisions of the artery and supply the dura mater.

3 K

The posterior branch also supplies the mucous lining of the mastoid cells. The anterior branch communicates with the meningeal branch of the superior maxillary nerve.

The **internal pterygoid nerve,** given off from the inferior maxillary, previous to its division, is intimately connected at its origin with the otic ganglion. It is a slender branch, which passes inwards to enter the deep surface of the Internal pterygoid muscle.

The *anterior* and smaller division, which receives nearly the whole of the motor root, divides into branches which supply the muscles of mastication and the skin and mucous membrane of the cheek. They are, the masseteric, deep temporal, long buccal, and external pterygoid.

The **masseteric branch** passes outwards, above the External pterygoid muscle, in front of the temporo-mandibular articulation and behind the tendon of the Temporal muscle; it crosses the sigmoid notch with the masseteric artery, to the deep surface of the Masseter muscle, in which it ramifies nearly as far as its anterior border. It gives a filament to the temporo-mandibular joint.

The **deep temporal branches** are two in number, anterior and posterior. They pass above the upper border of the External pterygoid muscle and enter the deep surface of the Temporal muscle. The *posterior branch,* of small size, is placed at the back of the temporal fossa. It sometimes arises from the masseteric branch. The *anterior branch* is frequently given off from the buccal nerve, and then passes upwards over the upper head of the External pterygoid muscle. Frequently a third branch (*middle deep temporal*) is present.

The **long buccal branch** passes forwards between the two heads of the External pterygoid, and] downwards beneath or through the fibres of the Temporal muscle, and emerges from under the anterior border of the Masseter to reach the surface of the Buccinator, upon which it unites with the buccal branches of the facial nerve. It gives a branch to the External pterygoid during its passage through that muscle, and may give off the anterior deep temporal nerve. The long buccal branch supplies the integument over the Buccinator muscle, and the mucous membrane lining its inner surface.

The **external pterygoid nerve** frequently arises in conjunction with the long buccal, but it may be given off separately from the anterior trunk of the nerve. It enters the muscle on its inner surface.

The *posterior* and larger division of the inferior maxillary nerve is for the most part sensory, but receives a few filaments from the motor root. It divides into three branches: auriculo-temporal, lingual (gustatory), and inferior dental.

The **auriculo-temporal nerve** generally arises by two roots, between which the middle meningeal artery passes. It runs backwards beneath the External pterygoid muscle to the inner side of the neck of the lower jaw. It then turns upwards with the temporal artery, between the external ear and condyle of the jaw, under cover of the parotid gland, and, escaping from beneath this structure, ascends over the zygoma, and divides into two temporal branches.

The auriculo-temporal nerve communicates with the facial and with the otic ganglion. The branches of communication with the facial, usually two in number, pass forwards, from behind the neck of the condyle of the jaw, to join this nerve at the posterior border of the Masseter muscle. They form one of the principal branches of communication between the facial and the fifth nerve. The filaments of communication with the otic ganglion are derived from the commencement of the auriculo-temporal nerve.

The *branches of distribution* of the auriculo-temporal nerve are:

Anterior auricular.	Articular.
Branches to the meatus auditorius.	Parotid.
Superficial temporal.	

The *anterior auricular branches* are usually two in number. They supply the front of the upper part of the pinna, being distributed principally to the skin covering the front of the helix and tragus.

Branches to the meatus auditorius, two in number, enter the canal between the bony and cartilaginous portion of the meatus. They supply the skin lining the meatus; the upper one sending a filament to the membrana tympani.

A *branch to the temporo-mandibular articulation* is usually derived from the auriculo-temporal nerve.

The *parotid branches* supply the parotid gland.

The *superficial temporal* accompanies the temporal artery to the vertex of the skull, and supplies the integument of the temporal region, communicating with the facial nerve, and the temporal branch of the temporo-malar, from the superior maxillary.

The **lingual nerve (gustatory)** supplies the mucous membrane of the anterior two-thirds of the tongue, and is deeply placed throughout the whole of its course. It lies at first beneath the External pterygoid muscle, together with the inferior dental nerve, being placed to the inner side and in front of this nerve, and is occasionally joined to it by a branch which may cross the internal maxillary artery. The chorda tympani also joins it at an acute angle in this situation. The nerve then passes between the Internal pterygoid muscle and the inner side of the ramus of the jaw, and crosses obliquely to the side of the tongue over the Superior constrictor and Stylo-glossus muscles, and then between the Hyo-glossus muscle and deep part of the submaxillary gland; it finally runs across Wharton's duct, and along the tongue to its apex, lying immediately beneath the mucous membrane.

The *branches of communication* are with the facial (through the chorda tympani), the inferior dental and hypoglossal nerves, and the submaxillary ganglion. The branches to the submaxillary ganglion are two or three in number; those connected with the hypoglossal nerve form a plexus at the anterior margin of the Hyo-glossus muscle.

The *branches of distribution* supply the mucous membrane of the mouth, the gums, the sublingual gland, and the mucous membrane of the anterior two-thirds of the tongue; the terminal filaments communicate, at the tip of the tongue, with the hypoglossal nerve.

The **inferior dental** is the largest of the three branches of the inferior maxillary nerve. It passes downwards with the inferior dental artery, at first beneath the External pterygoid muscle, and then between the internal lateral ligament and the ramus of the jaw to the dental foramen. It then passes forwards in the dental canal of the inferior maxillary bone, lying beneath the teeth, as far as the mental foramen, where it divides into two terminal branches, incisor and mental.

The branches of the inferior dental are, the mylo-hyoid, dental, incisive, and mental.

The *mylo-hyoid* is derived from the inferior dental just as that nerve is about to enter the dental foramen. It descends in a groove on the inner surface of the ramus of the jaw, in which it is retained by a process of fibrous membrane. It reaches the under surface of the Mylo-hyoid muscle, which it supplies together with the anterior belly of the Digastric.

The *dental branches* supply the molar and bicuspid teeth. They correspond in number to the fangs of those teeth: each nerve entering the orifice at the point of the fang, and supplying the pulp of the tooth.

The *incisive branch* is continued onwards within the bone to the middle line, and supplies the canine and incisor teeth.

The *mental branch* emerges from the bone at the mental foramen, and divides beneath the Depressor anguli oris into three branches; one descends to supply the skin of the chin, and sometimes two ascend to supply the skin and mucous membrane of the lower lip. These branches communicate freely with the facial nerve.

Two small ganglia are connected with the inferior maxillary nerve: the otic with the trunk of the nerve; and the submaxillary with its lingual branch.

Otic Ganglion (fig. 608)

The **Otic ganglion** (*Arnold's*) is a small, oval-shaped, flattened ganglion of a reddish-grey colour, situated immediately below the foramen ovale, on the inner surface of the inferior maxillary nerve, surrounding the origin of the internal pterygoid nerve. It is in relation, *externally*, with the trunk of the inferior maxillary nerve, at the point where the motor root joins the sensory portion; *internally*, with the cartilaginous part of the Eustachian tube, and the origin of the Tensor palati muscle; *behind* it is the middle meningeal artery.

Branches of communication.—This ganglion is connected with the internal

3 K 2

pterygoid branch of the inferior maxillary nerve by two or three short, delicate filaments. From this it may obtain a motor root, and possibly also a sensory root, as these filaments from the nerve to the Internal pterygoid may contain sensory fibres. It communicates with the glosso-pharyngeal and facial nerves, through the small superficial petrosal nerve continued from the tympanic plexus, and through this communication it probably receives its sensory root from the glosso-pharyngeal and its motor root from the facial; its communication with the sympathetic is effected by a filament from the plexus surrounding the middle meningeal artery. The ganglion also communicates with the auriculo-temporal

FIG. 608.—The otic ganglion and its branches.

nerve by a branch which is probably derived from the glosso-pharyngeal, and which passes to the ganglion, and through it and the auriculo-temporal nerve to the parotid gland. A slender filament (*sphenoidal*) ascends from it to the Vidian nerve, and a small branch communicates with the chorda tympani.

Its *branches of distribution* are, a filament to the Tensor tympani, and one to the Tensor palati. The former passes backwards, on the outer side of the Eustachian tube; the latter arises from the ganglion, near the origin of the internal pterygoid nerve, and passes forwards. The fibres of these nerves are, however, mainly derived from the nerve to the Internal pterygoid muscle.

SUBMAXILLARY GANGLION (fig. 606)

The **Submaxillary ganglion** is of small size, fusiform in shape, and situated above the deep portion of the submaxillary gland, on the Hyo-glossus muscle, near the posterior border of the Mylo-hyoid; it is connected by filaments with the lower border of the lingual (gustatory) nerve.

Branches of communication.—This ganglion is suspended from the lingual (gustatory) nerve by two filaments which join the front and back parts of the ganglion. It also receives a branch from the chorda tympani of the facial, and communicates with the sympathetic by filaments from the sympathetic plexus around the facial artery.

Branches of distribution.—These are five or six in number; they arise from the lower part of the ganglion, and supply the mucous membrane of the mouth and Wharton's duct, some being lost in the submaxillary gland. The branch of communication from the lingual to the fore part of the ganglion is by some regarded as a branch of distribution, by which filaments of the chorda tympani pass from the ganglion to the nerve, and by it are conveyed to the sublingual gland and the tongue.

Surface Marking.—It will be seen from the above description that the terminal branches of the three divisions of the fifth nerve emerge from foramina on to the face: the terminal branch of the first division emerging through the supra-orbital foramen; that of the second through the infra-orbital foramen; and the third through the mental foramen. The supra-orbital foramen is situated at the junction of the internal and middle thirds of the supra-orbital arch. If a straight line is drawn from this point to the lower border of the inferior maxillary bone, so that it passes between the two bicuspid teeth of the lower jaw, it will pass over the infra-orbital and mental foramina; the former being situated about one centimetre (two-fifths of an inch) below the margin of the orbit, and the latter varying in position according to the age of the individual. In the adult it is midway between the upper and lower borders of the inferior maxillary bone; in the child it is nearer the lower border, and in the edentulous jaw of old age it is close to the upper margin.

Surgical Anatomy.—The fifth nerve may be injured in its entirety; or its sensory or motor root or one of its primary main divisions may be alone implicated. In injury to the sensory root there is anæsthesia of the half of the face on the side of the lesion, with the exception of the skin over the parotid gland; insensibility of the conjunctiva, followed by destructive inflammation of the cornea, partly from loss of trophic influence, and partly from the irritation produced by the presence of foreign bodies on it, which are not perceived by the patient, and therefore not expelled by the act of winking; dryness of the nose, loss to a considerable extent of the sense of taste, and diminished secretion of the lachrymal and salivary glands. In injury to the motor root, there is impaired action of the lower jaw, from paralysis of the muscles of mastication on the affected side.

The fifth nerve is often the seat of neuralgia, and each of the three divisions has been divided, or a portion of the nerve excised, for this affection. The supra-orbital nerve may be exposed by making an incision an inch and a half in length along the supra-orbital margin below the eyebrow, which is to be drawn upwards, the centre of the incision corresponding to the supra-orbital notch. The skin and Orbicularis palpebrarum having been divided, the nerve can be easily found emerging from the notch, and lying in some loose cellular tissue. It should be drawn up by a blunt hook and divided, or, what is better, a portion of it removed. The infra-orbital nerve has been divided at its exit by an incision on the cheek; or the floor of the orbit has been exposed, the infra-orbital canal opened up, and the anterior part of the nerve resected; or the whole nerve, together with Meckel's ganglion as far back as the foramen rotundum, has been removed. This latter operation, though undoubtedly a severe proceeding, appears to have been followed by the best results. The operation is performed as follows: the superior maxillary bone is first exposed by a T-shaped incision, one limb passing along the lower margin of the orbit, the other from the centre of this vertically down the cheek to the angle of the mouth. The nerve is to be found, divided, and a piece of silk tied to it as a guide. A small trephine (half-inch) is applied to the bone, below, but including the infra-orbital foramen, and the antrum opened. The trephine is now applied to the posterior wall of the antrum, and the spheno-maxillary fossa exposed. The infra-orbital canal is opened up from below by fine cutting-pliers or a chisel, and the nerve drawn down into the trephine hole, it being held on the stretch by means of the piece of silk; it is severed with fine curved scissors as near the foramen rotundum as possible, any branches coming off from the ganglion being also divided.* The mental branch of the inferior dental nerve has been divided at its exit from the foramen by an incision made through the mucous membrane where it is reflected from the alveolar process on to the lower lip; or a portion of the trunk of the inferior dental nerve has been resected by an incision on the cheek through the Masseter muscle, exposing the outer surface of the ramus of the jaw. A trephine was then applied over the position of the inferior dental foramen and the outer table removed, so as to expose the inferior dental canal. The nerve was dissected out of the portion of the canal exposed, and, having been divided after its exit from the mental foramen, it was, by traction on the end exposed in the trephine hole, drawn out entire, and cut off as high up as possible.† The inferior dental nerve has also been divided by an incision within the mouth, the bony point guarding the inferior dental foramen forming the guide to the nerve. The buccal nerve may be divided by an incision through the mucous membrane of the mouth and the Buccinator just in front of the anterior border of the ramus of the lower jaw (Stimson).

The lingual (gustatory) nerve is occasionally divided with the view of relieving the pain in cancerous disease of the tongue. This may be done in that part of its course where it lies below and behind the last molar tooth. If a line is drawn from the middle of the crown of the last molar tooth to the angle of the jaw it will cross the nerve, which lies about half an inch behind the tooth, parallel to the bulging alveolar ridge on the inner side of the body of the bone. If the knife is entered three-quarters of an inch behind and below the last molar tooth, and carried down to the bone, the nerve will be divided. Hilton divided it opposite the second molar tooth, where it is covered only by

---

* Camochan, *Amer. Journ. Med. Science*, 1858, p. 136.
† Mears, *Trans. Amer. Surg. Assoc.* vol. ii. p. 469.

the mucous membrane, and Lucas pulls the tongue forwards and over to the opposite side, when the nerve can be seen standing out as a firm cord under the mucous membrane by the side of the tongue, and can be easily seized with a sharp hook and divided, or a portion excised. This is a simple enough operation on the cadaver, but when the disease is extensive and has extended to the floor of the mouth, as is generally the case when division of the nerve is required, the operation is not practicable. Kocher, in order to avoid a wound in the mouth, advises that division of the lingual nerve should be performed from without. He makes an incision from just below the tip of the mastoid process to the middle of the hyoid bone, so that it passes an inch below and behind the angle of the jaw. He divides the skin, superficial fascia, Platysma, and deep fascia, and thus exposes the submaxillary gland. This structure is then turned upwards, and the lingual nerve will be found lying on the Hyo-glossus muscle.

In some cases of neuralgia of the fifth nerve, when all other operative measures have failed, the Gasserian ganglion has been removed in whole or in part with a certain measure of success. Rose was the first to perform this operation; and he reached the ganglion by trephining the base of the skull in the position of the foramen ovale, after dividing the zygomatic arch, in front and behind, and turning it and the Masseter muscle downwards, and cutting through the coronoid process of the lower jaw, and turning it and the Temporal muscle upwards. A more efficient method appears to be that known as the Krause-Hartley method. The bone forming the temporal fossa having been removed to a sufficient extent, the temporal lobe of the brain, with the dura mater, is gradually raised from the middle fossa, until the foramen spinosum, with the middle meningeal artery passing through it, is exposed. This vessel is to be ligatured in two places and divided between the ligatures; and then by further raising the temporal lobe, the foramina ovale and rotundum will be exposed, with the second and third divisions of the fifth nerve passing through them. These nerves are to be clearly defined and divided. The dura mater is then to be raised from the ganglion, when the ophthalmic nerve will be exposed and must be divided, and the ganglion, by means of a little careful dissection, raised from its bed and removed.

Sixth Nerve (fig. 604)

The **Sixth** or **Abducent nerve** supplies the External rectus muscle of the eyeball.

Its fibres arise from a small nucleus situated in the upper part of the floor of the fourth ventricle, close to the middle line and beneath the eminentia teres. They pass downwards and forwards through the pons, and emerge in the furrow between the lower border of the pons and the upper end of the pyramid of the medulla oblongata.

From the nucleus of the sixth nerve fibres pass through the posterior longitudinal bundle to the oculo-motor nerve of the opposite side, along which they are carried to the Internal rectus muscle. The External rectus of one eye and the Internal rectus of the other may therefore be said to receive their nerves from the same nucleus—a factor of great importance in connection with the conjugate movements of the eyeball, and one that may explain certain paralytic phenomena of the Recti muscles, which are often associated with lesions in the pons.

The nerve pierces the dura mater on the basilar surface of the sphenoid bone, runs through a notch immediately below the posterior clinoid process, and enters the cavernous sinus. It passes forwards through the sinus, lying on the outer side of the internal carotid artery. It enters the orbit through the sphenoidal fissure, and lies above the ophthalmic vein, from which it is separated by a lamina of dura mater. It then passes between the two heads of the External rectus, and is distributed to that muscle on its ocular surface.

Branches of communication.—The sixth nerve is joined by several filaments from the carotid and cavernous plexuses, and by one from the ophthalmic nerve.

The sixth nerve, together with the third, fourth, and the ophthalmic division of the fifth, as they pass to the orbit, bear a certain relation to each other in the cavernous sinus, at the sphenoidal fissure, and in the cavity of the orbit, which will now be described.

In the *cavernous sinus* (fig. 519), the third, fourth, and ophthalmic division of the fifth are placed on the outer wall of the sinus, in their numerical order, both from above downwards, and from within outwards. The sixth nerve lies at the outer side of the internal carotid artery. As these nerves pass forwards to the sphenoidal fissure, the third and fifth nerves become divided into branches, and the sixth approaches the rest; so that their relative position becomes considerably changed.

In the *sphenoidal fissure* (fig. 610), the fourth, and the frontal and lachrymal divisions of the ophthalmic lie upon the same plane, the first being most internal, the last external; and they enter the cavity of the orbit above the

FIG. 609.—Figure showing the mode of innervation of the Internal and External recti muscles of the eye, after Duval and Laborde. (Testut.)

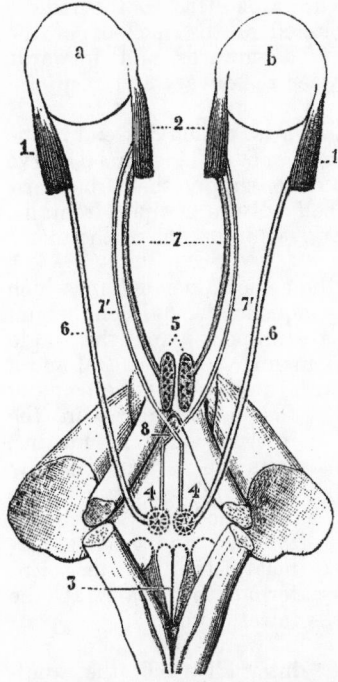

a. Left eyeball. b. Right eyeball. 1, 1. External rectus muscle. 2. Internal rectus muscle. 3. Floor of fourth ventricle. 4. Nucleus of sixth nerve. 5. Nucleus of third nerve. 6. Sixth nerve. 7. Nerve to Internal rectus arising from the nucleus of the third nerve of the same side. 7'. The nerve to the Internal rectus of the other side arising from the opposite nucleus. 8. Decussation of the fibres of sixth nerve to Internal recti.

FIG. 610.—Relations of structures passing through the sphenoidal fissure.

muscles. The remaining nerves enter the orbit between the two heads of the External rectus. The superior division of the third is the highest of these; beneath this lies the nasal branch of the ophthalmic; then the inferior division of the third; and the sixth lowest of all.

In the *orbit*, the fourth, and the frontal and lachrymal divisions of the ophthalmic lie on the same plane immediately beneath the periosteum, the fourth nerve being internal and resting on the Superior oblique, the frontal resting on the Levator palpebræ, and the lachrymal on the External rectus. Next in order comes the superior division of the third nerve lying immediately beneath the Superior rectus, and then the nasal branch of the ophthalmic, crossing the optic nerve to reach the inner side of the orbit. Beneath these is found the optic nerve, surrounded in front by the ciliary nerves, and having the lenticular ganglion on its outer side, between it and the External rectus. Below the optic are the inferior division of the third, and the sixth, the latter lying on the outer side of the orbit.

Surgical Anatomy.—The sixth nerve is more frequently involved in fractures of the base of the skull than any of the other cranial nerves. The result of paralysis of this nerve is internal or convergent squint. Diplopia is also present. When injured so that its function is destroyed, there is, in addition to the paralysis of the External rectus muscle, often a certain amount of contraction of the pupil, because some of the sympathetic fibres to the radiating muscle of the iris are conveyed through this nerve.

SEVENTH NERVE (figs. 611, 612, 613)

The **Seventh** or **Facial nerve,** like the fifth, consists of a motor root, which is usually described as the *facial nerve,* and a sensory root, which is named the *pars intermedia.* The two roots emerge separately at the lower border of the pons Varolii in the recess between the olivary and restiform bodies, the motor root being the more internal; immediately to the outer side of the sensory root is the auditory nerve.

The motor root supplies the muscles of the face, scalp and pinna, the Buccinator and Platysma, the Stylo-hyoid and posterior belly of the Digastric, and the Stapedius muscle of the tympanic cavity.   The sensory root is the nerve of taste for the anterior two-thirds of the tongue, and the vaso-dilator nerve of the submaxillary and sublingual glands.

The *motor* root takes origin from a nucleus of cells which lies deeply in the reticular formation of the lower part of the pons Varolii.   This nucleus is situated above the nucleus ambiguus, behind the superior olive and internal to the lower sensory root of the fifth nerve.   From this origin the fibres pursue a curved course in the substance of the pons.   They first pass backwards and inwards towards the floor of the fourth ventricle, and, reaching the posterior extremity of the nucleus of the sixth nerve, run upwards close to the middle line beneath the eminentia teres.   At the anterior end of the nucleus of the sixth nerve they make a second bend, and run downwards and forwards through the pons to their point of emergence between the olivary and restiform bodies.

From the nucleus of the third nerve some fibres arise which descend in the posterior longitudinal fasciculus and join the motor root of the facial nerve before it leaves the pons.   These fibres are said to supply the Orbicularis palpebrarum, Corrugator supercilii, and anterior belly of the Occipito-frontalis, as these muscles have been observed to escape paralysis in lesions of the motor nucleus of the facial nerve.

The *sensory* root (pars intermedia) arises from the geniculate ganglion, which is situated on the genu of the facial nerve in the aqueductus Fallopii, behind the hiatus Fallopii.   The cells of this ganglion are unipolar, and the single process given off from each divides in a T-shaped manner into a central and a peripheral branch.   The central branches run inwards, and, leaving the trunk of the facial nerve in the internal auditory meatus, form the pars intermedia; the peripheral branches are continued into the chorda tympani and great superficial petrosal nerves.   Entering the brain at the lower border of the pons between the motor root internally and the auditory nerve externally, the fibres of the sensory root

Fig. 611.—The course and connections of the facial nerve in the temporal bone.

pass into the substance of the medulla and terminate in the upper part of the nucleus of the glosso-pharyngeal and in the fasciculus solitarius.

From their superficial attachment to the brain, the two roots of the facial nerve pass outwards and forwards with the auditory nerve to the internal auditory meatus.   In the meatus the motor root lies in a groove on the upper and anterior surface of the auditory nerve, the sensory root (pars intermedia) being placed between the two; here the sensory root joins the motor root to form the **facial nerve.**

At the bottom of the meatus, the facial nerve enters the aqueductus Fallopii, and follows the course of that canal through the petrous portion of the temporal bone, from its commencement at the bottom of the internal meatus, to its termination at the stylo-mastoid foramen.   It is at first directed outwards between the cochlea and vestibule towards the inner wall of the tympanum; it then bends suddenly backwards and arches downwards behind the tympanum to the stylo-mastoid foramen.   The point where it changes its direction is named the *genu*; it presents a reddish gangliform swelling, the *geniculate ganglion*, or nucleus of the sensory root of the nerve.   On emerging from the stylo-mastoid foramen, it runs forwards in the substance of the parotid gland, crosses the external carotid artery, and divides behind the ramus of the lower jaw into two primary branches, *temporo-facial* and *cervico-facial*, from which numerous offsets are distributed over the side of the head, face, and upper part of the neck, supplying the superficial muscles in these regions.   As the primary **branches** and their offsets diverge from each other, they present somewhat the appearance of a bird's foot or claws; hence the name of *pes anserinus* is given to the divisions of the facial nerve in and near the parotid gland.

The communications of the facial nerve may be arranged as follows :

In the internal auditory meatus . With the auditory nerve.

From the geniculate ganglion .
{ With Meckel's ganglion by the large superficial petrosal nerve.
With the otic ganglion by a branch which joins the small superficial petrosal nerve.
With the sympathetic on the middle meningeal by the external superficial petrosal nerve.

In the Fallopian aqueduct . . { With the auricular branch of the pneumo-gastric.

At its exit from the stylo-mastoid foramen . . . .
{ With the glosso-pharyngeal.
„ pneumogastric.
„ great auricular.
„ auriculo-temporal.

Behind the ear . . . . With the small occipital.
On the face . . . . . With the three divisions of the fifth.
In the neck . . . . . With the superficial cervical.

In the internal auditory meatus some minute filaments pass from the pars intermedia and from the facial to the auditory nerve.

The large superficial petrosal nerve arises from the geniculate ganglion, and consists chiefly of sensory branches which are distributed to the mucous membrane of the soft palate ; but it probably also contains a few motor fibres which form the motor root of Meckel's ganglion. It passes forwards through the hiatus Fallopii, and runs in a groove on the anterior surface of the petrous portion of the temporal bone beneath the Gasserian ganglion, to the foramen lacerum medium. It receives a twig from the tympanic plexus, and in the foramen is joined by the great deep petrosal, from the sympathetic plexus on the internal carotid artery, to form the Vidian nerve. This nerve passes forward through the Vidian canal and ends in Meckel's ganglion. The geniculate ganglion is connected with the otic ganglion by a branch which joins the small superficial petrosal nerve, and also with the sympathetic filaments accompanying the middle meningeal artery, by the external petrosal (Bidder). From the ganglion, according to Arnold, a twig is sent back to the auditory nerve. Just before the facial nerve emerges from the stylo-mastoid foramen, it generally receives a twig of communication from the auricular branch of the pneumo-gastric.

After its exit from the stylo-mastoid foramen, it sends a twig to the glosso-pharyngeal, another to the pneumogastric nerve, and communicates with the great auricular branch of the cervical plexus, and with the auriculo-temporal branch of the inferior maxillary nerve in the parotid gland, with the small occipital behind the ear ; on the face with the terminal branches of the three divisions of the fifth, and in the neck with the superficial or transverse cervical.

Branches of Distribution

Within the aqueductus Fallopii . { Tympanic, to the Stapedius muscle.
Chorda tympani.

At its exit from the stylo-mastoid foramen . . . . .
{ Posterior auricular.
Digastric.
Stylo-hyoid.

On the face . . . . . .
{ Temporo-facial { Temporal.
Malar.
Infra-orbital.
Cervico-facial { Buccal.
Supramandibular.
Inframandibular.

The **tympanic branch** arises from the nerve opposite the pyramid ; it passes through a small canal in the pyramid, and supplies the Stapedius muscle.

The **chorda tympani** is given off from the facial as it passes vertically downwards at the back of the tympanum, about a quarter of an inch before its exit from the stylo-mastoid foramen. It passes from below upwards and forwards in a distinct canal, and enters the cavity of the tympanum, through an aperture (*iter chordæ posterius*) on its posterior wall, close to the inner aspect of the posterior border of the membrana tympani and on a level with the upper end of the handle of the malleus, and becomes invested with mucous membrane. It passes forwards through the cavity of the tympanum, between the fibrous and mucous layers of the membrana tympani, and over the handle of the malleus, emerging from that cavity through a foramen which is situated at the inner end of the Glaserian fissure, and named the *iter chordæ anterius*, or *canal of Huguier*. It then descends between the two Pterygoid muscles on the inner aspect of the spine of the sphenoid bone, which it sometimes grooves, and joins, at an acute angle, the posterior border of the lingual nerve. Some of its fibres

Fig. 612.—Plan of the facial nerve.

enter the submaxillary ganglion, and through it are distributed to the submaxillary and sublingual glands; the remaining fibres are continued onwards through the muscular substance of the tongue to the mucous membrane covering its anterior two-thirds. Before joining the lingual nerve it receives a small communicating branch from the otic ganglion. As already stated, the chorda tympani nerve is derived from the sensory root (pars intermedia) of the lingual nerve.

The **posterior auricular nerve** arises close to the stylo-mastoid foramen, and passes upwards in front of the mastoid process, where it is joined by a filament from the auricular branch of the pneumogastric, and communicates with the mastoid branch of the great auricular and with the small occipital. As it ascends between the meatus and mastoid process it divides into auricular and occipital branches. The *auricular branch* supplies the Retrahens auriculam and the small intrinsic muscles on the cranial surface of the pinna. The *occipital branch*, the larger, passes backwards along the superior curved line of the occipital bone, and supplies the occipital portion of the Occipito-frontalis.

The **digastric branch** arises close to the stylo-mastoid foramen; it divides into several filaments, which supply the posterior belly of the Digastric; one of these perforates that muscle to join the glosso-pharyngeal nerve.

The **stylo-hyoid branch** frequently arises in conjunction with that to the posterior belly of the Digastric; it is long and slender, and passes inwards to enter the Stylo-hyoid muscle about its middle.

The **temporo-facial,** the larger of the two terminal branches, passes upwards and forwards through the parotid gland, crosses the external carotid artery and temporo-maxillary vein, and passes over the neck of the condyle of the jaw, being connected in this situation with the auriculo-temporal branch of the inferior

FIG. 613.—The nerves of the scalp, face, and side of the neck.

Termination
of supratrochlear
of infratrochlear
of nasal

maxillary nerve, and divides into branches, which are distributed over the temple and upper part of the face; these are divided into three sets: temporal, malar, and infra-orbital.

The *temporal branches* cross the zygoma to the temporal region, supplying the Attrahens and Attollens auriculam muscles, and join with the temporal branch of the temporo-malar, a branch of the superior maxillary, and with the auriculo-temporal branch of the inferior maxillary. The more anterior branches supply the frontal portion of the Occipito-frontalis, the Orbicularis palpebrarum, and Corrugator supercilii muscles, and join with the supra-orbital and lachrymal branches of the ophthalmic.

The *malar branches* pass across the malar bone to the outer angle of the orbit, where they supply the Orbicularis palpebrarum muscle, and join with filaments from the lachrymal nerve and the malar branch of the superior maxillary nerve.

The *infra-orbital branches*, of larger size than the rest, pass horizontally forwards to be distributed between the lower margin of the orbit and the mouth. The *superficial branches* run beneath the skin and above the superficial muscles of the face which they supply : some branches are distributed to the Pyramidalis nasi, joining at the inner angle of the orbit with the infratrochlear and nasal branches of the ophthalmic. The *deep branches* pass beneath the Zygomatici and the Levator labii superioris, supplying them and the Levator anguli oris, and form a plexus (*infra-orbital*) by joining with the infra-orbital branch of the superior maxillary nerve and the buccal branches of the cervico-facial. These branches also supply the Levator labii superioris alæque nasi and the small muscles of the nose.

The **cervico-facial division of the facial nerve** passes obliquely downwards and forwards through the parotid gland, crossing the external carotid artery. In this situation it is joined by branches from the great auricular nerve. Opposite the angle of the lower jaw it divides into branches which are distributed on the lower half of the face and upper part of the neck. These may be divided into three sets : buccal, supramandibular, and inframandibular.

The *buccal branches* cross the Masseter muscle. They supply the Buccinator and Orbicularis oris, and join with the infra-orbital branches of the temporo-facial division of the nerve, and with filaments of the long buccal branch of the inferior maxillary nerve.

The *supramandibular branches* pass forwards beneath the Platysma and Depressor anguli oris, supplying the muscles of the lower lip and chin, and communicating with the mental branch of the inferior dental nerve.

The *inframandibular (cervical) branches* run forward beneath the Platysma, and form a series of arches across the side of the neck over the suprahyoid region. One of these branches descends vertically to join with the superficial cervical nerve from the cervical plexus ; others supply the Platysma.

Surgical Anatomy.—The facial nerve is more frequently paralysed than any of the other cranial nerves. The paralysis may depend upon either (1) central causes, i.e. blood clots or intracranial tumours pressing on the nerve before its entrance into the internal auditory meatus. It is also one of the nerves involved in 'bulbar paralysis.' Or (2) it may be paralysed in its passage through the petrous bone, by damage due to middle ear disease, or by fractures of the base. Or (3) it may be affected at or after its exit from the stylo-mastoid foramen. This is commonly known as 'Bell's paralysis.' It may be due to exposure to cold or to injury of the nerve, either from accidental wounds of the face, or during some surgical operation, as removal of parotid tumours, opening of abscesses, or operations on the lower jaw.

When the cause is central, the sixth nerve is usually paralysed as well, and there is also hemiplegia on the opposite side. In these cases the electrical reactions are the same as in health ; whereas, when the paralysis is in the course of the nerve, the reaction is usually lost. When the nerve is paralysed in the petrous bone, in addition to the paralysis of the muscles of expression, there is loss of taste in the anterior part of the tongue, and the patient is unable to recognise the difference between bitters and sweets, acids and salines, from involvement of the chorda tympani. The mouth is dry, because the salivary glands are not secreting ; and the sense of hearing is affected from paralysis of the Stapedius. When the cause of the paralysis is from fracture of the base of the skull, the auditory nerve and the petrosal nerves, which are connected with the geniculate ganglion, are also involved. When the injury to the nerve is after its exit from the stylo-mastoid foramen, all the muscles of expression, except the Levator palpebræ, together with the posterior belly of the Digastric and Stylo-hyoid, are paralysed. There is smoothness of the forehead, and the patient is unable to frown ; the eyelids cannot be closed and the lower lid droops, so that the punctum is no longer in contact with the globe, and the tears run down the cheek ; there is smoothness of the cheek and loss of the naso-labial furrow ; the nostril cannot be dilated ; the mouth is drawn to the sound side, and there is inability to whistle ; food collects between the cheek and gum from paralysis of the Buccinator. The facial nerve is at fault in cases of so-called 'histrionic spasm,' which consists in an almost constant and uncontrollable twitching of some or all of the muscles of the face. This twitching is sometimes so severe as to cause great discomfort and annoyance to the patient, and to interfere with sleep, and for its relief the facial nerve has been stretched. The operation is performed by making an incision behind the ear, from the root of the mastoid process to the angle of the jaw. The parotid is turned forwards and the dissection carried along the anterior border of the Sterno-mastoid muscle and mastoid process, until the upper border of the posterior belly of the Digastric is found. The nerve is parallel to this on about a level of the middle of the mastoid process. When found, the nerve must be stretched by passing a blunt hook beneath it and pulling it forwards and outwards. Too great force must not be used, for fear of permanent injury to the nerve.

EIGHTH NERVE

The **Eighth** or **Auditory nerve** is the special nerve of the sense of hearing, being distributed exclusively to the internal ear.

Origin of the Eighth Nerve.—The eighth nerve consists of two sets of fibres, which, although differing in their central connections, are both concerned in the transmission of afferent impulses from the internal ear to the medulla and pons, and from there, by means of new fibres which arise from collections of grey matter in these structures, to the cerebrum and cerebellum. One set of fibres forms the vestibular root of the nerve, and arises from the cells in the ganglion of Scarpa, which is situated in the internal auditory meatus; the other set constitutes the cochlear root, and takes origin from the cells in the ganglion spirale or ganglion of Corti, which occupies the spiral canal of the cochlea.

Both of these ganglia consist of bipolar nerve-cells; one process from each of the cells passes inwards to the brain, the other outwards to the internal ear. At its connection with the brain the eighth nerve occupies the groove between the pons and medulla, where it is situated between the restiform body, which is behind, and the seventh nerve, which is in front.

Vestibular or ventral root. The fibres of this root enter the medulla to the inner side of those of the cochlear root, and pass between the restiform body, which is to their outer side, and the inferior sensory root of the fifth, which lies to their inner side. They then divide into an ascending and a descending set. The fibres of the latter end by arborising round the cells of the *internal nucleus*, which is situated in the *trigonum acustici* in the floor of

Fig. 614.—Terminal nuclei of the vestibular root of the auditory nerve, with their upper connections. (Schematic.) (Testut.)

1. Posterior or cochlear root, with its two nuclei; 2. Accessory nucleus. 3. Tuberculum acusticum. 4. Anterior or vestibular root. 5. Internal nucleus. 6. Nucleus of Deiters. 7. Nucleus of Bechterew. 8. Inferior or descending root of auditory. 9. Ascending cerebellar fibres. 10. Fibres going to raphé. 11. Fibres taking an oblique course. 12. Fillet. 13. Inferior sensory root of trigeminal. 14. Pyramidal tract. 15. Raphé. 16. Fourth ventricle. 17. Restiform body. 18. Origin of striæ acusticæ.

the fourth ventricle. The ascending fibres either end in the same manner or in the *external nucleus*, which is situated to the outer side of the trigonum acustici and farther from the ventricular floor. It is described as consisting of two parts, an inner, the *nucleus of Deiters*, and an outer, the *nucleus of Bechterew*. Some of the axons of the cells of the external nucleus, and possibly also of the internal nucleus, are continued upwards through the restiform body to the roof nuclei of the opposite side of the cerebellum, to which also are prolonged other fibres of the vestibular root without undergoing a relay in the nuclei of the medulla. A second set of fibres from the internal and external nuclei end partly in the tegmentum, while the remainder ascend in the posterior longitudinal bundle to arborise around the cells of the nuclei of the oculo-motor nerve.

Cochlear or dorsal root.—This part of the nerve is placed externally to the vestibular root. Its fibres end in two nuclei, one of which, the *accessory nucleus*, lies immediately in front of the restiform body; the other, the *tuberculum acusticum*, somewhat to its outer side.

The striæ acusticæ, or medullary striæ, are the axons of the cells of the tuberculum acusticum. They pass backwards and inwards over the restiform body, and across the floor of the fourth ventricle towards the middle line. Here they dip into the substance of the pons, to end around the cells of the *superior olive* of the same or opposite side. There are, however, other fibres, and these are both direct and crossed, which do not arborise around the tegmental nuclei but pass into the lateral fillet. The cells of the accessory nucleus give origin to fibres which pass transversely in the pons and constitute the trapezium. The description given as to the mode of ending of the striæ acusticæ is applicable to that of the trapezoid fibres, viz. around the cells of the superior olive or of the *trapezoid nucleus* (which lies ventral to the olive) of the same or opposite side, while others, crossed or uncrossed, pass directly into the lateral fillet.

If the further connections of the cochlear nerve of one side, say the left, be considered, it is found that they lie to the outer side of the main sensory tract, the fillet, and are therefore termed the *lateral fillet*. The fibres comprising the

FIG. 615.—Terminal nuclei of the cochlear nerve, with their upper connections. (Schematic.)   (Testut.)

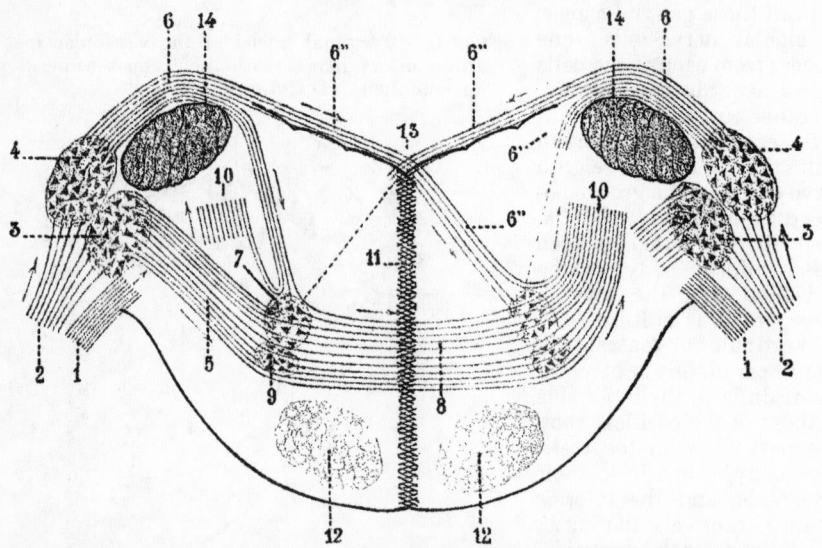

The vestibular root with its terminal nuclei and thin efferent fibres have been suppressed. On the other hand, in order not to obscure the trapezoid body, the efferent fibres of the terminal nuclei on the right side have been resected in a considerable portion of their extent. The trapezoid body, therefore, shows only one half of its fibres, viz. those which come from the left.
1. Anterior or vestibular root of the auditory, divided at its entrance into the bulb. 2. Posterior or cochlear root. 3. Accessory nucleus of auditory nerve. 4. Tuberculum acusticum. 5. Efferent fibres of accessory nucleus. 6. Efferent fibres of tuberculum acusticum, forming the striæ acusticæ, with 6', their direct bundle going to the superior olivary body of the same side: 6'', their decussating bundles going to the superior olivary body of the opposite side. 7. Superior olivary body. 8. Trapezoid body. 9. Trapezoid nucleus. 10. Central acoustic tract (lateral fillet). 11. Raphé. 12. Pyramidal tract. 13. Fourth ventricle. 14. Restiform body.

left lateral fillet arise in the superior olive or trapezoid nucleus of the same or opposite side, while others are the uninterrupted fibres already alluded to, and these are either crossed or uncrossed, the former being the axons of the cells of the right accessory nucleus or of the cells of the right tuberculum acusticum, while the latter are derived from the same cells of the left side. In the upper part of the lateral fillet there is a collection of nerve-cells, the *nucleus of the lateral fillet*, around the cells of which some of the fibres arborise and from the cells of which axons originate to continue upwards the tract of the lateral fillet. The ultimate ending of the left lateral fillet is partly in the opposite internal geniculate body, and partly in the quadrigeminal bodies of the same or opposite side. From the cells of these bodies new fibres arise which ascend in the posterior limb of the internal capsule to reach the posterior three-fifths of the first left temporal convolution and the transverse temporal gyri of Heschl.

The auditory nerve contains a few efferent fibres which arise in the quadrigeminal bodies, the nucleus of the lateral fillet, trapezoid nucleus, and superior clive.

The auditory nerve after leaving the medulla passes forwards across the posterior border of the middle peduncle of the cerebellum, in company with the facial nerve, from which it is partially separated by the auditory artery. It then enters the internal auditory meatus with the facial nerve. At the bottom of the meatus it receives one or two filaments from the facial nerve, and then divides into its two branches, *cochlear* and *vestibular*. The auditory nerve is soft in texture, and is destitute of neurilemma ; its distribution will be described with the anatomy of the ear.

Surgical Anatomy.—The auditory nerve is frequently injured, together with the facial nerve, in fractures of the middle fossa of the base of the skull implicating the internal auditory meatus. The nerve may be either torn across, producing permanent deafness, or it may be bruised or pressed upon by extravasated blood or inflammatory exudation, when the deafness will in all probability be temporary. The nerve may also be injured by violent blows on the head without any fracture of the bones of the skull taking place, and deafness may arise from loud explosions from dynamite, &c., probably from some lesion of this nerve, which is more liable to be injured than the other cranial nerves on account of its structure. The test that the nerve is destroyed, and that the deafness is not due to some lesion of the auditory apparatus, is obtained by placing a vibrating tuning-fork on the head. The vibrations will be heard in cases where the auditory apparatus is at fault, but not in cases of destruction of the auditory nerve.

NINTH PAIR (figs. 616, 617, 618)

The **Ninth** or **Glosso-pharyngeal nerve** contains both motor and sensory fibres, and is distributed, as its name implies, to the tongue and pharynx, being the nerve of ordinary sensation to the mucous membrane of the pharynx, fauces, and tonsil ; and the nerve of taste to the parts of the tongue to which it is distributed.

FIG. 616.—Origin, ganglia, and communications of the ninth, tenth, and eleventh cranial nerves.

Its *superficial origin* is by three or four filaments, closely connected together, from the upper part of the medulla oblongata, in the groove between the olivary and the restiform body.

The *sensory* fibres arise from the cells of the jugular and petrous ganglia, which are situated on the trunk of the nerve, and will be presently described. When traced into the medulla, some of the sensory fibres terminate by arborising around the cells of the upper part of a nucleus which lies beneath the ala cinerea in the lower part of the floor of the fourth ventricle. Many of the fibres, however, contribute to form a strand, named the *fasciculus solitarius*, which descends in the medulla. Associated with this strand are numerous nerve-cells, around which the fibres of the fasciculus terminate.

The *motor* fibres take origin from the cells of the *nucleus ambiguus*, which lies some distance from the lower part of the floor of the fourth ventricle in the lateral area of the medulla, and which is continuous below with the anterior grey cornu of the spinal cord. From this nucleus the fibres are first directed backwards, and then they bend forwards and outwards to join the fibres of the sensory root. The nucleus ambiguus gives origin to the motor branches of the glosso-pharyngeal and vagus, and to the bulbar part of the spinal accessory.

From its superficial origin, it passes outwards across the flocculus, and leaves the skull at the central part of the jugular foramen, in a separate sheath of the dura mater, external to and in front of the pneumogastric and spinal accessory nerves (fig. 521). In its passage through the jugular foramen, it grooves the lower border of the petrous portion of the temporal bone ; and, at its exit from the skull, passes forwards between the jugular vein and internal carotid artery, and descends in front of the latter vessel, and beneath the styloid process and the muscles connected with it, to the lower border of the Stylo-pharyngeus. The nerve now curves inwards, forming an arch on the side of the neck and lying upon the Stylo-pharyngeus and Middle constrictor of the pharynx. It then passes beneath the Hyo-glossus, and is finally distributed to the mucous membrane of the fauces and base of the tongue, and the mucous glands of the mouth and tonsil.

In passing through the jugular foramen, the nerve presents, in succession, two gangliform enlargements. The superior, the smaller, is called the *jugular ganglion*; the inferior and larger, the *petrous ganglion*, or the *ganglion of Andersch*.

The **superior** or **jugular ganglion** is situated in the upper part of the groove in which the nerve is lodged during its passage through the jugular foramen. It is of very small size, and involves only the lower part of the trunk of the nerve. It is usually regarded as a detached portion of the lower ganglion.

The **inferior** or **petrous ganglion** is situated in a depression in the lower border of the petrous portion of the temporal bone; it is larger than the superior, and involves the whole of the fibres of the nerve. From this ganglion arise those filaments which connect the glosso-pharyngeal with the pneumogastric and sympathetic nerves.

The glosso-pharyngeal nerve communicates with the pneumogastric, sympathetic, and facial.

The branches to the pneumogastric are two filaments, which arise from the petrous ganglion, one passing to the auricular branch, and the other to the upper ganglion of the pneumogastric.

The petrous ganglion is connected by a filament with the superior cervical ganglion of the sympathetic.

The branch of communication with the facial perforates the posterior belly of the Digastric. It arises from the trunk of the nerve below the petrous ganglion, and joins the facial just after its exit from the stylo-mastoid foramen.

The *branches of distribution* of the glosso-pharyngeal are, the tympanic, carotid, pharyngeal, muscular, tonsillar, and lingual.

The **tympanic branch** (*Jacobson's nerve*) arises from the petrous ganglion, and enters a small bony canal on the under surface of the petrous portion of the temporal bone; the lower opening of which is situated on the bony ridge which separates the carotid canal from the jugular fossa. It ascends to the tympanum, enters that cavity by an aperture in its floor close to the inner wall, and divides into branches, which are contained in grooves upon the surface of the promontory, forming the tympanic plexus. This plexus gives off: (1) the small superficial petrosal nerve; (2) a branch to join the great superficial petrosal nerve; and (3) branches to the tympanic cavity, all which will be described in connection with the anatomy of the ear.

The **carotid branches** descend along the trunk of the internal carotid artery as far as its commencement, communicating with the pharyngeal branch of the pneumogastric, and with branches of the sympathetic.

The **pharyngeal branches** are three or four filaments which unite, opposite the Middle constrictor of the pharynx, with the pharyngeal branches of the pneumogastric and sympathetic nerves, to form the pharyngeal plexus; branches from this plexus perforate the muscular coat of the pharynx and supply its muscles and mucous membrane.

The **muscular branch** is distributed to the Stylo-pharyngeus.

The **tonsillar branches** supply the tonsil, forming a plexus (*circulus tonsillaris*) around this body, from which filaments are distributed to the soft palate and fauces, where they communicate with the palatine nerves.

The **lingual branches** are two in number: one supplies the circumvallate papillæ and the mucous membrane covering the surface of the base of the tongue; the other perforates its substance, and supplies the mucous membrane and follicular glands of the posterior part of the tongue and communicates with the lingual nerve.

TENTH PAIR (figs. 617, 618)

The **Tenth** or **Pneumogastric nerve** (*nervus vagus*) has a more extensive distribution than any of the other cranial nerves, passing through the neck and thorax to the upper part of the abdomen. It is composed of both motor and sensory fibres. It supplies the organs of voice and respiration with motor and sensory fibres; and the pharynx, œsophagus, stomach, and heart with motor fibres.

FIG. 617.—Plan of the glosso-pharyngeal, pneumogastric, and spinal accessory nerves.
(After Flower.)

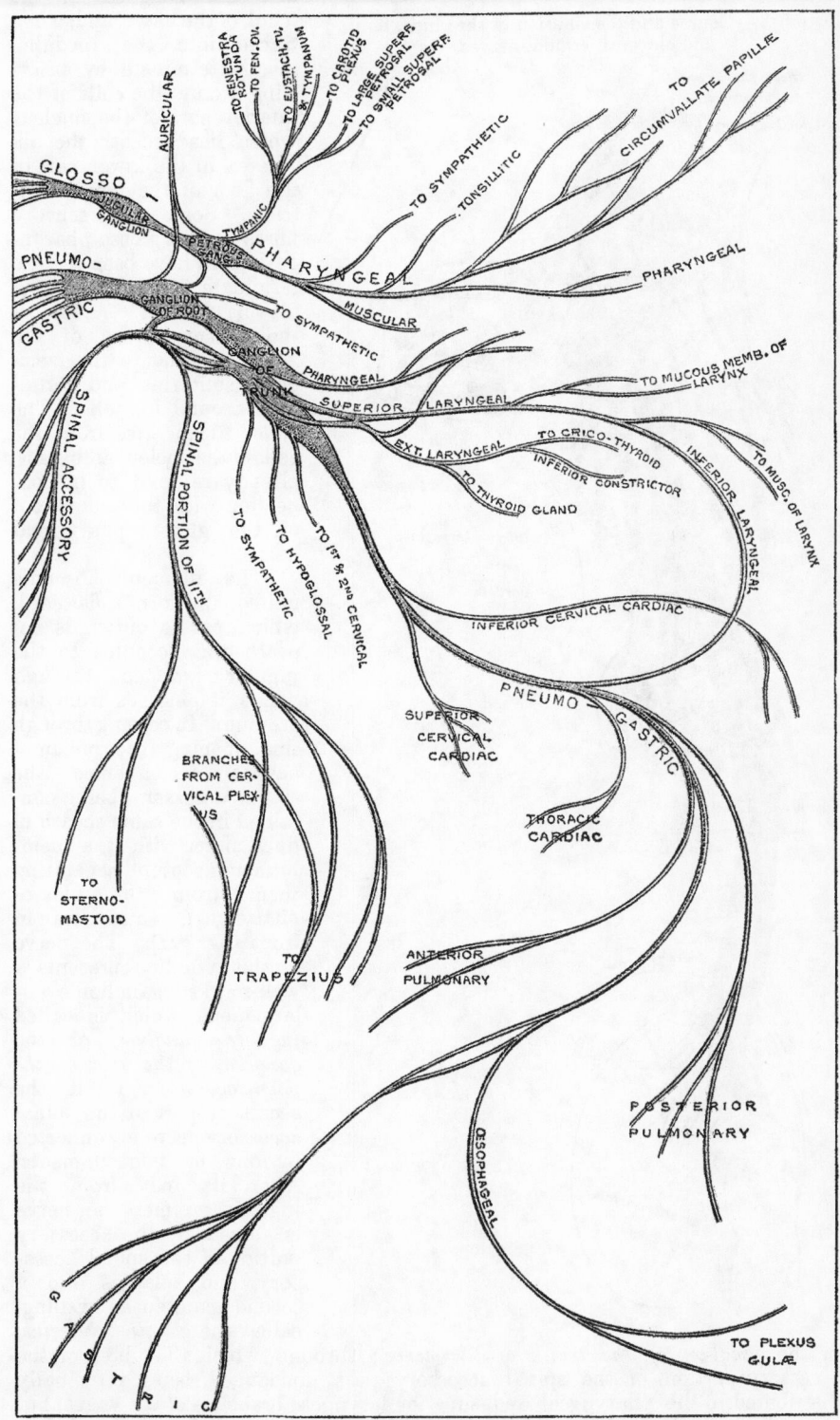

The *superficial origin* of the pneumogastric nerve is by eight or ten filaments from the groove between the olivary and the restiform bodies below the glosso-pharyngeal. The *sensory* fibres arise from the cells of the ganglion of the root

FIG. 618.—Course and distribution of the ninth, tenth, and eleventh cranial nerves.

Glosso-pharyngeal
Pneumogastric
Spinal accessory

TRAPEZIUS

Internal laryngeal nerve

External laryngeal

Inferior laryngeal

Cardiac

Pulmonary brs.

DIAPHRAGM

Gastric

Anterior surface of stomach

and the ganglion of the trunk of the nerve, and, when traced into the medulla, mostly terminate by arborising around the cells of the inferior part of the nucleus which lies beneath the ala cinerea in the lower part of the floor of the fourth ventricle. Some of the sensory fibres of the glosso-pharyngeal nerve have been seen to terminate in the upper part of this nucleus. A few of the sensory fibres of the vagus descend in the fasciculus solitarius and terminate around its cells. The *motor* fibres arise from the cells of the nucleus ambiguus, already referred to in connection with the motor root of the glosso-pharyngeal nerve.

The filaments become united, and form a flat cord, which passes outwards beneath the flocculus to the jugular foramen, through which it emerges from the cranium. In passing through this opening, the pneumogastric accompanies the spinal accessory, being contained in the same sheath of dura mater with it, a membranous septum separating them from the glosso-pharyngeal, which lies in front (fig. 521). The nerve in this situation presents a well-marked ganglionic enlargement, which is called *jugular ganglion*, or the *ganglion of the root of the pneumogastric*: to it the accessory part of the spinal accessory nerve is connected by one or two filaments. After its exit from the jugular foramen the nerve is joined by the accessory portion of the spinal accessory, and enlarges into a second gangliform swelling, called the *ganglion inferius*, or the *ganglion of the trunk of the nerve*; through which the fibres of the accessory portion of the spinal accessory pass unchanged, being principally distributed to the pharyngeal and superior laryngeal branches of the vagus, but some of the filaments from it are continued into the trunk of the vagus below the ganglion, to be distributed with the recurrent laryngeal nerve and probably also with the cardiac nerves. The nerve passes vertically down the neck within the sheath of the carotid vessels lying between the internal carotid artery and internal jugular vein as far as the thyroid cartilage, and then between the same vein and the common carotid artery to the root of the neck. From this downwards, the course of the nerve differs on the two sides of the body.

On the *right side*, the nerve passes across the subclavian artery between it and the right innominate vein, and descends by the side of the trachea to the back part of the root of the lung, where it spreads out in a plexiform network (*posterior pulmonary plexus*), from the lower part of which two cords descend on the œsophagus, upon which they divide, forming, with branches from the opposite nerve, the œsophageal plexus (*plexus gulæ*) ; below, these branches are collected into a single cord, which runs along the back part of the œsophagus, enters the abdomen, and is distributed to the posterior surface of the stomach, joining the left side of the solar plexus, and sending filaments to the splenic plexus and a considerable branch to the cœliac plexus.

On the *left side*, the pneumogastric nerve enters the chest between the left carotid and subclavian arteries, behind the left innominate vein. It crosses the arch of the aorta, and descends behind the root of the left lung, forming the *posterior pulmonary plexus*, and along the anterior surface of the œsophagus, where it unites with the nerve of the right side in forming the *plexus gulæ*, to the stomach, distributing branches over its anterior surface, some extending over the great *cul-de-sac*, and others along the lesser curvature. Filaments from these branches enter the gastro-hepatic omentum, and join the hepatic plexus.

The *ganglion of the root* is of a greyish colour, circular in form, about two lines in diameter.

Connecting branches.—To this ganglion the accessory portion of the spinal accessory nerve is connected by several delicate filaments ; it also has a communicating twig with the petrous ganglion of the glosso-pharyngeal, with the facial nerve by means of its auricular branch, and with the sympathetic by means of an ascending filament from the superior cervical ganglion.

The *ganglion of the trunk* (inferior) is a plexiform cord, cylindrical in form, of a reddish colour, and about an inch in length ; it involves the whole of the fibres of the nerve, and passing through it is the accessory portion of the spinal accessory nerve, which blends with the pneumogastric below the ganglion, and is then principally continued into its pharyngeal and superior laryngeal branches.

Connecting branches.—This ganglion is connected with the hypoglossal, the superior cervical ganglion of the sympathetic, and the loop between the first and second cervical nerves.

The *branches of the pneumogastric* are :

In the jugular fossa . . .	{ Meningeal. { Auricular.
In the neck	(Pharyngeal. ⎧ Superior laryngeal. ⎨ Recurrent laryngeal. (Cervical cardiac.
In the thorax	(Thoracic cardiac. ⎧ Anterior pulmonary. ⎨ Posterior pulmonary. (Œsophageal.
In the abdomen	Gastric.

The **meningeal branch** is a recurrent filament given off from the ganglion of the root in the jugular foramen. It passes backwards through the jugular foramen, and is distributed to the dura mater covering the posterior fossa of the base of the skull.

The **auricular branch** (*Arnold's*) arises from the ganglion of the root, and is joined soon after its origin by a filament from the petrous ganglion of the glosso-pharyngeal ; it passes outwards behind the jugular vein, and enters a small canal on the outer wall of the jugular fossa. Traversing the substance of the temporal bone, it crosses the aqueductus Fallopii about two lines above its termination at the stylo-mastoid foramen ; here it gives off an ascending branch, which joins the facial : the continuation of the nerve reaches the surface by passing through the auricular fissure between the mastoid process and the tympanic plate, and divides into two branches, one of which communicates with the posterior auricular nerve, while the other supplies the integument at the back part of the pinna and the posterior part of the external auditory meatus.

The **pharyngeal branch,** the principal motor nerve of the pharynx, arises from the upper part of the inferior ganglion of the pneumogastric. It consists principally of filaments from the accessory portion of the spinal accessory ; it passes across the internal carotid artery to the upper border of the Middle constrictor, where it divides into numerous filaments, which join with those from the glosso-pharyngeal, superior laryngeal (its external branch), and sympathetic, to form the pharyngeal plexus, from which branches are distributed to the muscles and mucous membrane of the pharynx and the muscles of the soft palate, except the Tensor palati and possibly the Azygos uvulæ and Levator palati (see page 460). From the pharyngeal plexus a minute filament is given off, which descends and joins the hypoglossal nerve as it winds round the occipital artery.

The **superior laryngeal** is larger than the preceding, and arises from the middle of the inferior ganglion of the pneumogastric. In its course it receives a branch from the superior cervical ganglion of the sympathetic. It descends, by the side of the pharynx, behind the internal carotid, where it divides into two branches, the external and internal laryngeal.

The *external laryngeal branch,* the smaller, descends by the side of the larynx, beneath the Sterno-thyroid, to supply the Crico-thyroid muscle. It gives branches to the pharyngeal plexus and the Inferior constrictor, and communicates with the superior cardiac nerve, behind the common carotid artery.

The *internal laryngeal branch* descends to the opening in the thyro-hyoid membrane, through which it passes with the superior laryngeal artery, and is distributed to the mucous membrane of the larynx. A small branch communicates with the recurrent laryngeal nerve. The branches to the mucous membrane are distributed, some in front to the epiglottis, the base of the tongue, and the epiglottidean glands ; while others pass backwards, in the aryteno-epiglottidean fold, to supply the mucous membrane surrounding the superior orifice of the larynx, as well as the membrane which lines the cavity of the larynx as low down as the vocal cord. The filament which joins with the recurrent laryngeal descends beneath the mucous membrane on the inner surface of the thyroid cartilage, where the two nerves become united.

The **inferior** or **recurrent laryngeal,** so called from its reflected course, is the motor nerve of the larynx. It arises on the right side, in front of the subclavian artery ; winds from before backwards round that vessel, and ascends obliquely to the side of the trachea behind the common carotid, and either in front of or behind the inferior thyroid artery. On the left side, it arises in front of the arch of the aorta, and winds from before backwards round the aorta at the point where the remains of the ductus arteriosus are connected with it, and then ascends to the side of the trachea. The nerves on both sides ascend in the groove between the trachea and œsophagus, and, passing under the lower border of the Inferior constrictor muscle, enter the larynx behind the articulation of the inferior cornu of the thyroid cartilage with the cricoid, being distributed to all the muscles of the larynx, excepting the Crico-thyroid. It communicates with the superior laryngeal nerve, and gives off a few filaments to the mucous membrane of the lower part of the larynx.

The recurrent laryngeal, as it winds round the subclavian artery and aorta, gives off several cardiac filaments, which unite with the cardiac branches from the pneumogastric and sympathetic. As it ascends in the neck it gives off œsophageal branches, more numerous on the left than on the right side, which supply the mucous membrane and muscular coat of the œsophagus ; tracheal branches to the mucous membrane and muscular fibres of the trachea ; and some pharyngeal filaments to the Inferior constrictor of the pharynx.

The **cervical cardiac branches,** two or three in number, arise from the pneumogastric, at the upper and lower part of the neck.

The *superior branches* are small, and communicate with the cardiac branches of the sympathetic. They can be traced to the great or deep cardiac plexus.

The *inferior branch* arises at the lower part of the neck, just above the first rib. That from the right vagus passes in front or by the side of the arteria innominata, and communicates with one of the cardiac nerves proceeding to the great or deep cardiac plexus; that from the left runs in front of the arch of the aorta, and joins the superficial cardiac plexus.

The **thoracic cardiac branches,** on the right side, arise from the trunk of the pneumogastric, as it lies by the side of the trachea, and from its recurrent laryngeal branch; but on the left side from the recurrent nerve only; passing inwards, they terminate in the deep cardiac plexus.

The **anterior pulmonary branches,** two or three in number, and of small size, are distributed on the anterior aspect of the root of the lungs. They join with filaments from the sympathetic, and form the *anterior pulmonary plexus.*

The **posterior pulmonary branches,** more numerous and larger than the anterior, are distributed on the posterior aspect of the root of the lung: they are joined by filaments from the third and fourth (sometimes also first and second) thoracic ganglia of the sympathetic, and form the *posterior pulmonary plexus.* Branches from both plexuses accompany the ramifications of the air-tubes through the substance of the lungs.

The **œsophageal branches** are given off from the pneumogastric both above and below the pulmonary branches. The lower are more numerous and larger than the upper. They form, together with branches from the opposite nerve, the *œsophageal plexus* or *plexus gulæ.* From this plexus branches are distributed to the back of the pericardium.

The **gastric branches** are the terminal filaments of the pneumogastric nerve. The nerve on the right side is distributed to the posterior surface of the stomach, and joins the left side of the cœliac plexus and the splenic plexus. The nerve on the left side is distributed over the anterior surface of the stomach, some filaments passing across the great *cul-de-sac,* and others along the lesser curvature. They unite with branches of the right nerve and with the sympathetic, some filaments passing through the lesser omentum to the hepatic plexus.

Surgical Anatomy.—The functions of the pneumogastric nerve may be interfered with by damage to its nucleus of origin in the medulla; by thickening or growth from the meninges or bones, or aneurism of the basilar artery before its exit from the skull; injuries such as gunshot or punctured wounds in the neck, or injuries during such operations as ligature of the carotid artery, removal of tuberculous glands or other deep-seated tumours. The pneumogastric may also be compressed by aneurisms of the carotid artery. The symptoms produced by paralysis of the nerve are palpitation, with increased frequency of the pulse, constant vomiting, slowing of the respiration, and a sensation of suffocation. The laryngeal nerves are of considerable importance in considering some of the morbid conditions of the larynx. When the peripheral terminations of the superior laryngeal nerve are irritated by some foreign body passing over them, reflex spasm of the glottis is the result. When the trunk of this same nerve is pressed upon, by, for instance, a goitre or an aneurism of the upper part of the carotid, there is a peculiar dry, brassy cough. When the nerve is paralysed, there is anæsthesia of the mucous membrane of the larynx, so that foreign bodies can readily enter the cavity, and, in consequence of its supplying the Crico-thyroid muscle, the vocal cords cannot be made tense, and the voice is deep and hoarse. Paralysis of the superior laryngeal nerves may be the result of bulbar paralysis; may be a sequel to diphtheria when both nerves are usually involved; or it may, though less commonly, be caused by the pressure of tumours or aneurisms, when the paralysis is generally unilateral. Irritation of the inferior laryngeal nerves produces spasm of the muscles of the larynx. When both these recurrent nerves are paralysed, the vocal cords are motionless, in the so-called ' cadaveric position '—that is to say, in the position in which they are found in ordinary tranquil respiration; neither closed as in phonation, nor open as in deep inspiratory efforts. When one recurrent nerve is paralysed, the cord of the same side is motionless, while the opposite one crosses the middle line to accommodate itself to the affected one; hence phonation is present, but the voice is altered and weak in timbre. The recurrent laryngeal nerves may be paralysed in bulbar paralysis or after diphtheria, when the paralysis usually affects both sides; or they may be affected by the pressure of aneurisms of the aorta, innominate or subclavian arteries; by mediastinal tumours; by bronchocele; or by cancer of the upper part of the œsophagus, when the paralysis is often unilateral.

The **Eleventh** or **Spinal Accessory nerve** consists of two parts : one, the accessory part to the vagus, and the other the spinal portion.

The **bulbar** or **accessory part** is the smaller of the two. Its fibres arise from the cells of the *nucleus ambiguus* and emerge as four or five delicate filaments from the side of the medulla, below the roots of the vagus. It passes outwards to the jugular foramen, where it interchanges fibres with the spinal portion or becomes united to it for a short distance ; it is also connected, in the foramen, with the upper ganglion of the vagus by one or two filaments. It then passes through the foramen, and becoming again separated from the spinal portion it is continued over the surface of the ganglion of the trunk of the vagus, being adherent to its surface, and is distributed principally to the pharyngeal and superior laryngeal branches of the pneumogastric. Through the pharyngeal branch it probably supplies the Azygos uvulæ and Levator palati muscles (see page 460). Some few filaments from it are continued into the trunk of the vagus below the ganglion, to be distributed with the recurrent laryngeal nerve and probably also with the cardiac nerves.

The **spinal portion** is firm in texture, and its fibres arise from the motor cells in the outer part of the anterior horn of the grey matter of the spinal cord as low as the fifth cervical nerve. Passing outwards and backwards through the lateral white column of the cord, they emerge on its surface and unite to form a single trunk, which ascends between the ligamentum denticulatum and the posterior roots of the spinal nerves, enters the skull through the foramen magnum, and is then directed outwards to the jugular foramen, through which it passes, lying in the same sheath as the pneumogastric, but separated from it by a fold of the arachnoid. In the jugular foramen, it receives one or two filaments from the accessory portion, or else joins it for a short distance and then separates from it again. At its exit from the jugular foramen, it passes backwards, either in front of or behind the internal jugular vein, and descends obliquely behind the Digastric and Stylo-hyoid muscles to the upper part of the Sterno-mastoid. It pierces that muscle, and passes obliquely across the posterior triangle, to terminate in the deep surface of the Trapezius. This nerve gives several branches to the Sterno-mastoid during its passage through it, and joins in its substance with branches from the second cervical, which supply the muscle. In the posterior triangle it joins with the second and third cervical nerves, while beneath the Trapezius it forms a sort of plexus with the third and fourth cervical nerves, and from this plexus fibres are distributed to the muscle.

Surgical Anatomy.—The functions of the spinal accessory nerve may be interfered with either by central changes ; or at its exit from the skull, by fractures running across the jugular foramen ; or in the neck, by inflamed lymphatic glands, &c. When irritated, it causes clonic spasm of the Sterno-mastoid and Trapezius muscles, or, as it is termed, spasmodic torticollis. In cases of this affection in which all previous palliative treatment has failed, and the spasms are so severe as to undermine the patient's health, division or excision of a portion of the spinal accessory nerve has been resorted to. This may be done along either the anterior or posterior border of the Sterno-mastoid muscle. The former operation consists in making an incision, three inches in length, from the apex of the mastoid process along the anterior border of the Sterno-mastoid muscle. The anterior border of the muscle is defined and pulled backwards, so as to stretch the nerve, which is then to be sought for, beneath the Digastric muscle, about two inches below the apex of the mastoid process. The other operation is performed by making an incision along the posterior border of the muscle, so that the centre of the incision corresponds to the middle of this border of the muscle. The superficial structures having been divided and the border of the muscle defined, the nerve is to be sought for as it emerges from the muscle to cross the occipital triangle. When found, it is to be traced upwards through the muscle, and a portion of it excised above the point where it gives off its branches to the Sterno-mastoid. In this operation one of the descending branches of the superficial cervical plexus is liable to be mistaken for the nerve.

The **Twelfth** or **Hypoglossal nerve** is the motor nerve of the tongue.

Its fibres arise from the cells of the *hypoglossal nucleus*, which is an upward prolongation of the base of the anterior horn of grey matter of the cord. This nucleus is about three-quarters of an inch in length, and its upper part corresponds with the *trigonum hypoglossi*, which is situated close to the middle line in the lower half of the floor of the fourth ventricle. The lower part of the nucleus extends downwards into the closed part of the medulla, and there lies in relation to the ventro-lateral aspect of the central canal. The fibres run forwards through the entire thickness of the medulla, between its anterior and lateral areas, and emerge in the pre-olivary sulcus between the pyramid and the olivary body.

The filaments of this nerve are collected into two bundles, which perforate the dura mater separately, opposite the anterior condyloid foramen, and unite together after their passage through it. In those cases in which the anterior condyloid foramen in the occipital bone is double, these two portions of the nerve

FIG. 619.—Plan of the hypoglossal nerve.

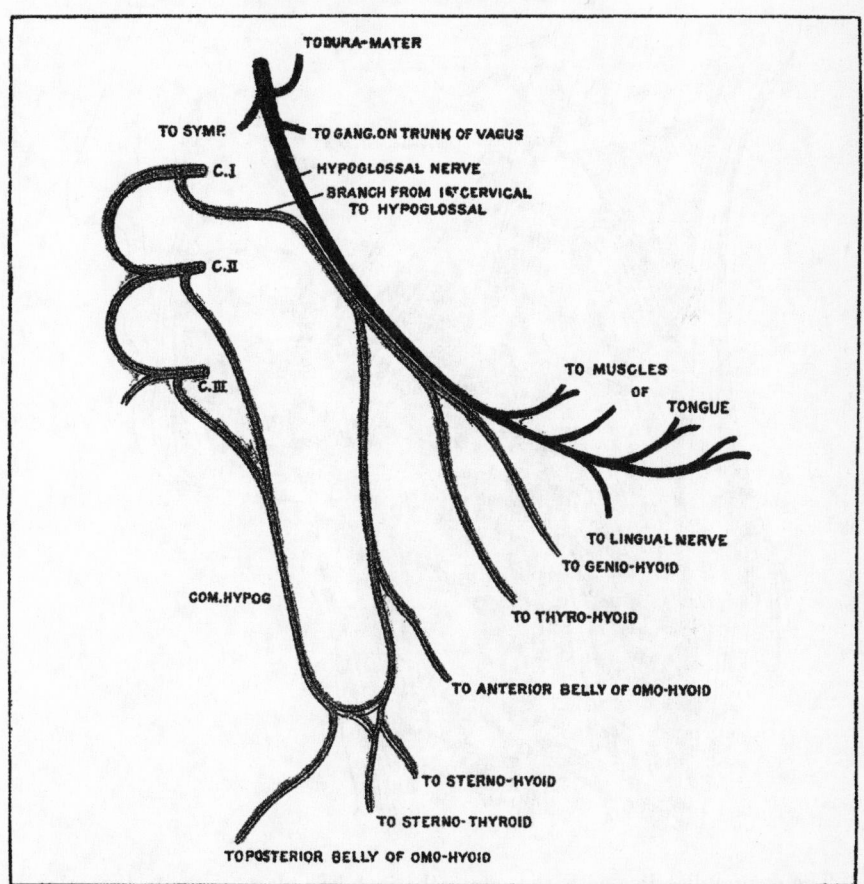

are separated by the small piece of bone which divides the foramen. The nerve descends almost vertically to a point corresponding with the angle of the jaw. It is at first deeply seated beneath the internal carotid artery and internal jugular vein, and intimately connected with the pneumogastric nerve; it then passes forwards between the vein and artery, and lower down in the neck becomes superficial below the Digastric muscle. The nerve then loops round the occipital artery, and crosses the external carotid and its lingual branch below the tendon. of the Digastric muscle. It passes beneath the tendon of the Digastric, the Stylo-hyoid, and the Mylo-hyoid muscles, lying between the last-named muscle and the Hyo-glossus, and communicates at the anterior border of the Hyo-glossus with the lingual (gustatory) nerve; it is then continued forwards in the fibres of the Genio-hyo-glossus muscle as far as the tip of the tongue, distributing branches to its muscular substance.

The *branches of communication* are, with the

Pneumogastric.	First and second cervical nerves.
Sympathetic.	Lingual (gustatory).

The communication with the pneumogastric takes place close to the exit of the nerve from the skull, numerous filaments passing between the hypoglossal and lower ganglion of the pneumogastric through the mass of connective tissue which unites the two nerves. It also communicates with the pharyngeal plexus by a minute filament as it winds round the occipital artery.

FIG. 620.—Hypoglossal nerve, cervical plexus, and their branches.

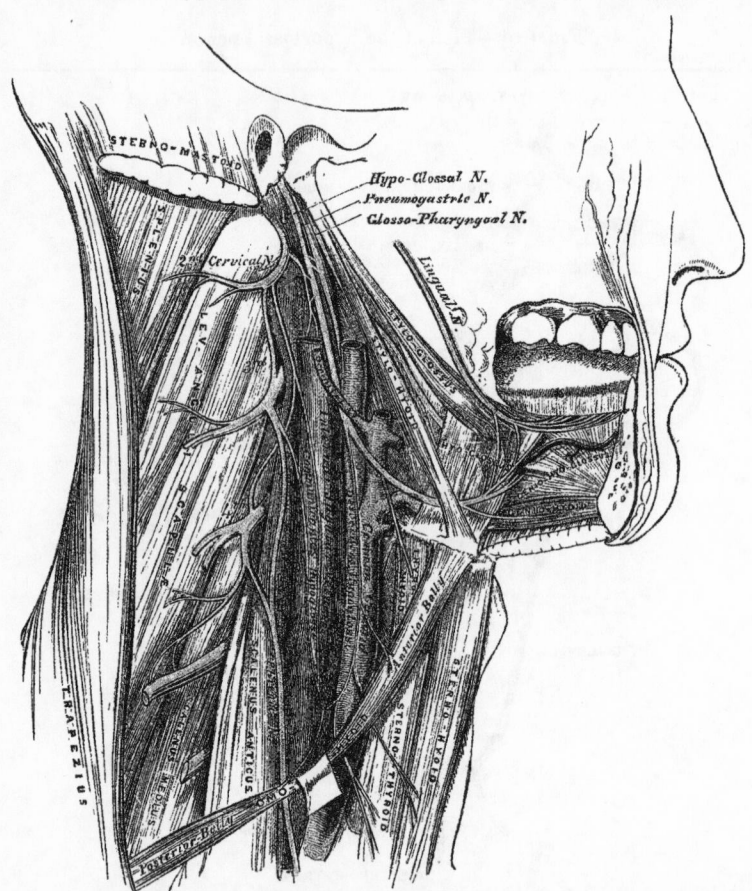

The communication with the sympathetic takes place opposite the atlas by branches derived from the superior cervical ganglion, and in the same situation the nerve is joined by a filament derived from the loop connecting the first two cervical nerves.

The communication with the lingual (gustatory) takes place near the anterior border of the Hyo-glossus muscle by numerous filaments which ascend upon it.

The *branches of distribution* are, the

Meningeal.	Thyro-hyoid.
Descendens hypoglossi.	Muscular.

Meningeal branches.—As the hypoglossal nerve passes through the anterior condyloid foramen it gives off, according to Luschka, several filaments to the dura mater in the posterior fossa of the base of the skull; these filaments are probably derived from a branch which passes from the first cervical nerve to the hypoglossal nerve.

The **descendens hypoglossi** is a long slender branch, which quits the hypoglossal where it turns round the occipital artery. It consists mainly of fibres which pass to the hypoglossal from the loop between the first and second cervical nerves in the above-mentioned communication. It descends in front of the sheath of the carotid vessels, giving off a branch to the anterior belly of the Omohyoid, and then joins the communicating branches from the second and third cervical nerves, just below the middle of the neck, to form a loop, the *ansa hypoglossi*. From the convexity of this loop branches pass to supply the Sterno-hyoid, Sterno-thyroid, and the posterior belly of the Omo-hyoid. According to Arnold, another filament descends in front of the vessels into the chest, and joins the cardiac and phrenic nerves.

The **thyro-hyoid** is a small branch, arising from the hypoglossal near the posterior border of the Hyo-glossus ; it passes obliquely across the great cornu of the hyoid bone, and supplies the Thyro-hyoid muscle.

The **muscular branches** are distributed to the Stylo-glossus, Hyo-glossus, Genio-hyoid, and Genio-hyo-glossus muscles. At the under surface of the tongue numerous slender branches pass upwards into the substance of the organ to supply its intrinsic muscles.

Surgical Anatomy.—The hypoglossal nerve is an important guide in the operation of ligature of the lingual artery (see page 612). It runs forwards on the Hyo-glossus just above the great cornu of the hyoid bone, and forms the upper boundary of the triangular space in which the artery is to be sought for by cutting through the fibres of the Hyo-glossus. In cases where it has been injured on one side of the neck, unilateral paralysis of the tongue is the result ; the tongue, when protruded, being directed to the side on which the injury has taken place.

THE SPINAL NERVES

The **spinal nerves** take their origin from the spinal cord, and are transmitted through the intervertebral foramina on either side of the spinal column. There are thirty-one pairs of spinal nerves, which are arranged into the following groups, corresponding to the region of the spine through which they pass :

Cervical	8 pairs
Dorsal	12 „
Lumbar	5 „
Sacral	5 „
Coccygeal	1 pair

It will be observed that each group of nerves corresponds in number with the vertebræ in that region, except the cervical and coccygeal.

Each spinal nerve arises by two roots, an anterior or motor root, and a posterior or sensory root, the latter being distinguished by a ganglion, termed the *spinal* ganglion.

Roots of the Spinal Nerves

The **Anterior** or **Ventral Roots** are *motor* in function, and arise from the motor cells in the anterior horn of grey matter ; and, passing through the overlying white substance, emerge on the anterior surface of the cord in three or four irregular rows.

The **Posterior** or **Dorsal Roots** are *sensory* in function, and arise from the nerve-cells of the spinal ganglia, from which they can be traced into the cord in two main bundles, the course of which has already been studied (page 774).

The *anterior roots* are smaller than the posterior, devoid of ganglionic enlargements, and their component fibrils are collected into two bundles, near the intervertebral foramina.

The *posterior roots* of the nerves are larger, but the individual filaments are finer and more delicate than those of the anterior. As their component fibrils pass outwards, towards the aperture in the dura mater, they coalesce into two bundles, receive a tubular sheath from that membrane, and enter the ganglion which is developed upon each root.

The posterior root of the first cervical nerve forms an exception to the rule, being smaller than the anterior.

Ganglia of the Spinal Nerves

A **ganglion** is found upon the posterior root of each of the spinal nerves. These ganglia are of an oval form, and of a reddish colour; they bear a proportion in size to that of the nerves upon which they are situated, and are placed in the intervertebral foramina, external to the point where the nerves perforate the dura mater. Each ganglion is bifid internally, where it is joined by the two bundles

Fig. 621.—Diagram to show the composition of a peripheral nerve-trunk. (Böhm and Davidoff.)

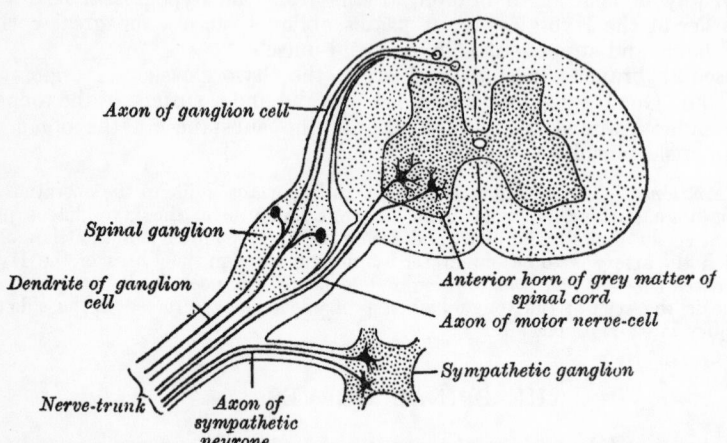

Axon of ganglion cell

Spinal ganglion

Dendrite of ganglion cell

Anterior horn of grey matter of spinal cord

Axon of motor nerve-cell

Sympathetic ganglion

Nerve-trunk

Axon of sympathetic neurone

of the posterior root, the two portions being united into a single mass externally. The ganglia upon the first and second cervical nerves form exceptions to this rule, being placed on the arches of the vertebræ over which the nerves pass. The ganglia of the sacral nerves are placed within the spinal canal; and that on the coccygeal nerve, also in the canal, is situated at some distance from the origin of the posterior root.

Distribution of the Spinal Nerves

Immediately beyond the ganglion the two roots coalesce, their fibres intermingle, and the trunk thus formed constitutes the *spinal nerve*; it passes out of the intervertebral foramen, and divides into a posterior division for the supply of the posterior part of the body, and an anterior division for the supply of the anterior part of the body; each division contains fibres from both nerve-roots.

Each spinal nerve receives a branch (grey ramus) from the sympathetic; while in the case of the dorsal nerves, the first and second lumbar, and the second and third sacral spinal nerves, a branch (white ramus) passes from the spinal nerve into the corresponding ganglion of the sympathetic.

Before dividing, each spinal nerve gives off a small *recurrent* or *meningeal* branch, which is joined by a filament from the communicating branch of the sympathetic. It passes inwards through the intervertebral foramen and supplies the dura mater, sending branches to the bones and ligaments.

The **posterior divisions of the spinal nerves** are generally smaller than the anterior; they arise from the trunk resulting from the union of the nerve-roots, in the intervertebral foramina; and, passing backwards, divide into internal and external branches, which are distributed to the muscles and integument on the posterior part of the trunk. The first cervical, the fourth and fifth sacral, and the coccygeal, do not divide into external and internal branches.

The **anterior divisions of the spinal nerves** supply antero-lateral parts of the trunk, and the limbs. They are for the most part larger than the posterior divisions. In the dorsal region the anterior divisions of the spinal nerves are quite separate from each other, and are uniform in their distribution; but in the cervical, lumbar, and sacral regions they form intricate plexuses previous to their distribution.

POINTS OF EMERGENCE OF THE SPINAL NERVES

The roots of the first and second spinal nerves run almost transversely outwards from their origin in the cord; all the others pass obliquely downwards and outwards to their point of exit from the intervertebral foramina: the amount of obliquity increasing from above downwards, and the level of their emergence from the cord is within certain limits variable, and of course does not correspond to that of their emergence from the intervertebral foramina. The accompanying table, from Macalister, shows the relation of these points of origin from the spinal cord to the bodies and spinous processes of the vertebræ.

Level of Body of	No. of Nerve	Level of tip of Spine of	Level of Body of	No. of Nerve	Level of tip of Spine of
C. 1	C. 1	—	D. 8	D. 9	7 d.
2	2	—	9	10	8 d.
	3	1 c.	10	11	9 d.
3	4	2 c.	—	12	10 d.
4	5	3 c.	11	L. 1	11 d.
5	6	4 c.	—	{ 2	—
6	7	5 c.	12	{ 3 }	12 d.
—	8	6 c.		{ 4 }	
7	D. 1	7 c.		{ 5 }	—
D. 1	2	1 d.		{ S. 1 }	
2	3	—	L. 1	2	
3	4	2 d.		{ 3 }	
4	5	3 d.		4	1 L.
5	6	4 d.	—	5	
6	7	5 d.	—	C. 1	
7	8	6 d.	L. 2	—	

CERVICAL NERVES

The **roots of the cervical nerves** increase in size from the first to the fifth, and then remain the same size to the eighth. The posterior roots bear a proportion to the anterior of 3 to 1, which is much greater than in any other region, the individual filaments being also much larger than those of the anterior roots. The posterior root of the first cervical is an exception to this rule: it is smaller than the anterior root. In direction, the roots of the cervical are less oblique than those of the other spinal nerves. The first cervical nerve is directed a little upwards and outwards; the second is horizontal; the others are directed obliquely downwards and outwards, the lowest being the most oblique, and consequently longer than the upper, the distance between their place of origin and their point of exit from the spinal canal never exceeding the depth of one vertebra.

The *trunk of the first cervical nerve* leaves the spinal canal, between the occipital bone and the posterior arch of the atlas; the second, between the posterior arch of the atlas and the lamina of the axis; and the eighth (the last), between the last cervical and first dorsal vertebræ.

Each nerve, at its exit from the intervertebral foramen, divides into a posterior and an anterior division. The anterior divisions of the four upper cervical nerves form the cervical plexus. The anterior divisions of the four lower cervical nerves, together with the greater part of the anterior division of the first dorsal nerve, form the brachial plexus.

POSTERIOR DIVISIONS OF THE CERVICAL NERVES (fig. 622)

The **posterior division of the first cervical**, or **sub-occipital**, **nerve** differs from the posterior divisions of the other cervical nerves in not dividing into an internal and external branch. It is larger than the anterior division, and escapes from the spinal canal between the occipital bone and the posterior arch of the atlas, lying beneath the vertebral artery. It enters the sub-occipital triangle formed by the Rectus capitis posticus major, the Obliquus superior, and Obliquus inferior, and supplies the Recti and Obliqui muscles, and the Complexus. From the branch which supplies the Inferior oblique a filament is given off which joins the second cervical nerve. This nerve also occasionally gives off a cutaneous filament, which accompanies the occipital artery, and communicates with the great and small occipital nerves.

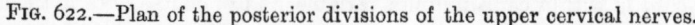

FIG. 622.—Plan of the posterior divisions of the upper cervical nerves.

The **posterior division of the second cervical nerve** is three or four times greater than the anterior division, and the largest of all the posterior cervical divisions. It emerges from the spinal canal between the posterior arch of the atlas and lamina of the axis, below the Inferior oblique. It supplies a twig to this muscle, and receives a communicating filament from the first cervical. It then divides into a large internal and a small external branch.

The *internal branch*, called, from its size and distribution, the *great occipital*, ascends obliquely inwards between the Obliquus inferior and Complexus, and pierces the latter muscle and the Trapezius near their attachments to the cranium. It is now joined by a filament from the posterior division of the third cervical nerve, and, ascending on the back part of the head with the occipital artery, divides into two branches, which supply the integument of the scalp as far forwards as the vertex, communicating with the occipitalis minor. It gives off muscular branches to the Complexus, and occasionally an auricular branch to the back part of the ear. The *external branch* is often joined by the external branch of the posterior division of the third, and supplies the Splenius, Trachelo-mastoid, and Complexus.

The **posterior division of the third cervical** is smaller than the preceding, but larger than the fourth; it differs from the posterior divisions of the remaining cervical nerves in supplying an additional filament, the *third occipital nerve*, to the integument of the occiput. The posterior division of the third nerve, like the others, divides into an internal and external branch. The *internal branch* passes between the Complexus and Semispinalis, and, piercing the Splenius and Trapezius, supplies the skin over the latter muscle; the *external branch* joins with that of the posterior division of the second to supply the Splenius, Trachelo-mastoid, and Complexus.

The *third occipital nerve* arises from the internal or cutaneous branch beneath the Trapezius; it then pierces that muscle, and supplies the skin on the lower and back part of the head. It lies to the inner side of the great occipital, with which it is connected.

The posterior division of the sub-occipital nerve and the internal branches of the posterior divisions of the second and third cervical nerves are occasionally joined beneath the Complexus by communicating branches. This communication is described by Cruveilhier as the *posterior cervical plexus*.

The **posterior divisions of the fourth, fifth, sixth, seventh, and eighth cervical nerves** (fig. 629) pass backwards, and divide, behind the Intertransversales muscles, into internal and external branches. The *internal branches*, the larger, are distributed differently in the upper and lower part of the neck. Those derived from the fourth and fifth nerves pass between the Complexus and Semispinalis muscles, and, having reached the spinous processes, perforate the aponeurosis of the Splenius and Trapezius, and are continued outwards to the integument over the Trapezius, while those derived from the three lowest cervical nerves are small and placed beneath the Semispinalis colli, which they supply, and then pass into the Interspinales, Multifidus spinæ and Complexus, and send twigs through this latter muscle to supply the integument near the spinous processes (Hirschfeld). The *external branches* supply the muscles at the side of the neck, viz. the Cervicalis ascendens, Transversalis colli, and Trachelo-mastoid.

Anterior Divisions of the Cervical Nerves

The **anterior division of the first cervical nerve** is of small size. It escapes from the spinal canal through a groove upon the posterior arch of the atlas. In this groove it lies beneath the vertebral artery, to the inner side of the Rectus capitis lateralis. As it crosses the foramen in the transverse process of the atlas, it receives a filament from the sympathetic. It then descends, in front of this process, to communicate with an ascending branch from the second cervical nerve.

Communicating filaments connect the loop between this nerve and the second with the pneumogastric, the hypoglossal, and sympathetic, and some branches are distributed to the Rectus lateralis and the two Anterior recti. The fibres which communicate with the hypoglossal, simply pass through the latter nerve to become for the most part the descendens hypoglossi. According to Valentin, the anterior division of the sub-occipital distributes filaments to the occipito-atlantal articulation, and mastoid process of the temporal bone.

The **anterior division of the second cervical nerve** escapes from the spinal canal, between the posterior arch of the atlas and the lamina of the axis, and, passing forwards on the outer side of the vertebral artery, divides, in front of the Intertransverse muscle, into an ascending branch, which joins the first cervical; and one or two descending branches which join the third. It gives off the small occipital; a branch to assist in forming the great auricular; another to assist in forming the superficial cervical; one of the communicantes hypoglossi, and a filament to the Sterno-mastoid, which communicates in the substance of the muscle with the spinal accessory.

The **anterior division of the third cervical nerve** is double the size of the preceding. At its exit from the intervertebral foramen, it passes downwards and outwards beneath the Sterno-mastoid, and divides into two branches. The ascending branch joins the anterior division of the second cervical; the descending branch passes down in front of the Scalenus anticus, and communicates with the fourth. It gives off the larger part of the great auricular and superficial cervical nerves; one of the communicantes hypoglossi; a branch to the supraclavicular nerves; a filament to assist in forming the phrenic; and muscular branches to the Levator anguli scapulæ and Trapezius; that to the Trapezius communicates beneath the muscle with the spinal accessory. Sometimes the nerve to the Scalenus medius is derived from this source.

The **anterior division of the fourth cervical** is of the same size as the preceding. It receives a branch from the third, sends a communicating branch to the fifth cervical, and, passing downwards and outwards, divides into numerous filaments, which cross the posterior triangle of the neck, forming the supraclavicular nerves. It gives origin to the main part of the phrenic nerve, while it is contained in the intertransverse space, and sometimes supplies a branch to the Scalenus medius muscle. It gives a branch to the Levator anguli scapulæ, and also one to the Trapezius, which unites with the branch given off from the third nerve, and communicates beneath the muscle with the spinal accessory.

The **anterior divisions of the fifth, sixth, seventh, and eighth cervical nerves** are remarkable for their size. They are much larger than the preceding nerves, and are all of equal dimensions. They assist in the formation of the brachial plexus.

CERVICAL PLEXUS

The cervical plexus (fig. 623) is formed by the anterior divisions of the four upper cervical nerves. It is situated opposite the four upper cervical vertebræ, resting upon the Levator anguli scapulæ and Scalenus medius muscles, and covered by the Sterno-mastoid.

Its branches are divided into two groups, *superficial* and *deep*, which may be thus arranged:

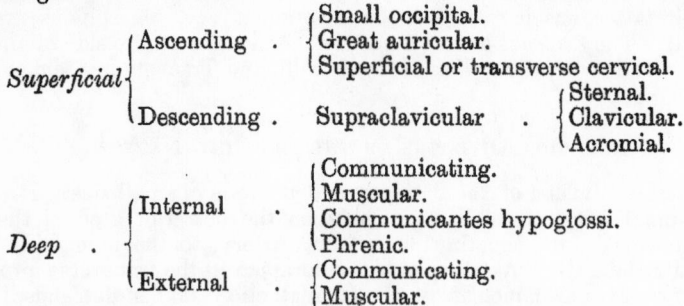

SUPERFICIAL BRANCHES OF THE CERVICAL PLEXUS

The **Small Occipital** (fig. 623) arises from the second cervical nerve, sometimes also from the third; it curves round the posterior border of the Sterno-mastoid, and ascends, running parallel to the posterior border of the muscle, to the back part of the side of the head. Near the cranium it perforates the deep fascia, and is continued upwards along the side of the head behind the ear, supplying the integument, and communicating with the great occipital, the great auricular, and with the posterior auricular branch of the facial.

This nerve gives off an *auricular branch*, which supplies the integument of the upper and back part of the auricle, communicating with the mastoid branch of the great auricular. This branch is occasionally derived from the great occipital nerve. The small occipital varies in size; it is sometimes duplicated.

The **Great Auricular** is the largest of the ascending branches. It arises from the second and third cervical nerves, winds round the posterior border of the Sterno-mastoid, and, after perforating the deep fascia, ascends upon that muscle beneath the Platysma to the parotid gland, where it divides into facial, auricular, and mastoid branches.

The *facial branches* are distributed to the integument of the face over the parotid gland; others penetrate the substance of the gland, and communicate with the facial nerve.

FIG. 623.—Plan of the cervical plexus.

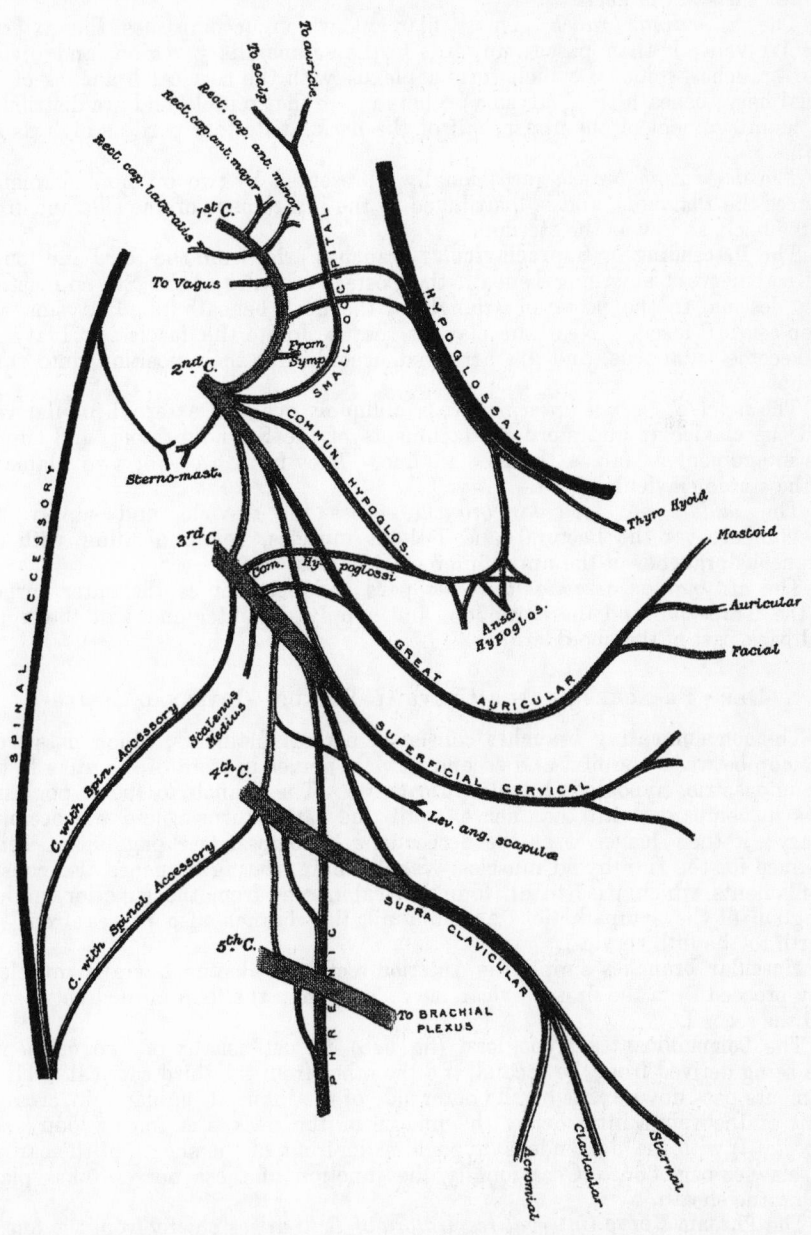

The *auricular branches* ascend to supply the integument of the back of the pinna, except at its upper part, communicating with the auricular branches of the facial and pneumogastric nerves. A filament pierces the pinna to reach its outer surface, where it is distributed to the lobule and lower part of the concha.

The *mastoid branch* communicates with the small occipital and with the posterior auricular branch of the facial, and is distributed to the integument behind the ear.

The **Superficial** or **Transverse Cervical** arises from the second and third cervical nerves, turns round the posterior border of the Sterno-mastoid about its middle, and, passing obliquely forwards beneath the external jugular vein to the anterior border of the muscle, perforates the deep cervical fascia, and divides beneath the Platysma into two branches, which are distributed to the antero-lateral parts of the neck.

The *ascending branch* gives a filament which accompanies the external jugular vein; it then passes upwards to the submaxillary region, and divides into branches, some of which form a plexus with the cervical branches of the facial nerve beneath the Platysma; others pierce that muscle, and are distributed to the integument of the upper half of the neck, at its fore part, as high as the chin.

The *descending branch* (occasionally represented by two or more filaments) pierces the Platysma, and is distributed to the integument of the side and front of the neck, as low as the sternum.

The **Descending** or **Supraclavicular** branches arise from the third and fourth cervical nerves: emerging beneath the posterior border of the Sterno-mastoid, they descend in the posterior triangle of the neck beneath the Platysma and deep cervical fascia. Near the clavicle they perforate the fascia and Platysma to become cutaneous, and are arranged, according to their position, into three groups.

The *inner* or *sternal branches* cross obliquely over the external jugular vein and the clavicular and sternal attachments of the Sterno-mastoid, and supply the integument as far as the median line. They furnish one or two filaments to the sterno-clavicular joint.

The *middle* or *clavicular branches* cross the clavicle, and supply the integument over the Pectoral and Deltoid muscles, communicating with the cutaneous branches of the upper intercostal nerves.

The *external* or *acromial branches* pass obliquely across the outer surface of the Trapezius and the acromion, and supply the integument of the upper and back part of the shoulder.

DEEP BRANCHES OF THE CERVICAL PLEXUS. INTERNAL SERIES

The **communicating branches** consist of several filaments, which pass from the loop between the first and second cervical nerves in front of the atlas to the pneumogastric, hypoglossal, and sympathetic. The branch to the hypoglossal runs in company with that nerve, until the latter turns round the occipital artery; it then leaves it as the descendens hypoglossi (see page 889), being destined for the Infrahyoid muscles. The communicating branches also consist of filaments, which pass to all four cervical nerves from the superior cervical ganglion of the sympathetic. A communicating branch also passes from the fourth to the fifth cervical.

Muscular branches supply the Anterior recti and Rectus lateralis muscles; they proceed from the first cervical nerve, and from the loop formed between it and the second.

The **Communicantes Hypoglossi** (fig. 620) consist usually of two filaments, one being derived from the second, and the other from the third cervical. These filaments pass downwards on the outer side of the internal jugular vein, cross in front of the vein a little below the middle of the neck, and form a loop (*ansa hypoglossi*) with the descendens hypoglossi in front of the sheath of the carotid vessels (see page 889). Occasionally, the junction of these nerves takes place within the sheath.

The **Phrenic Nerve** (*internal respiratory of Bell*) arises chiefly from the fourth cervical nerve with a few filaments from the third and a communicating branch from the fifth. It descends to the root of the neck, running obliquely across the front of the Scalenus anticus and beneath the Sterno-mastoid, the posterior belly of the Omo-hyoid, and the Transversalis colli and suprascapular vessels. It next passes over the first part of the subclavian artery, between it and the subclavian vein, and, as it enters the chest, crosses the internal mammary artery near its origin. Within the chest, it descends nearly vertically in front of the root of the lung, and by the side of the pericardium, between it and the mediastinal portion of the pleura, to the Diaphragm, where it divides into branches, which separately pierce that muscle, and are distributed to its under surface.

The two phrenic nerves differ in their length, and also in their relations at the upper part of the thorax.

The *right nerve* is situated more deeply, and is shorter and more vertical in direction than the left; it lies on the outer side of the right vena innominata and superior vena cava.

The *left nerve* is rather longer than the right, from the inclination of the heart to the left side, and from the Diaphragm being lower on this than on the opposite side. At the root of the neck, it is crossed by the thoracic duct; in the superior mediastinum, it is placed between the left common carotid and left subclavian arteries, and crosses in front of the vagus and the arch of the aorta.

In the thorax each phrenic nerve is accompanied by a branch of the internal mammary artery, the comes nervi phrenici.

Each nerve supplies filaments to the pericardium and pleura, and near the chest is joined by a filament from the sympathetic, and, occasionally, by one from the union of the descendens hypoglossi with the spinal nerves; it frequently receives a filament from the nerve to the Subclavius muscle. Branches have been described as passing to the peritoneum.

From the *right nerve*, one or two filaments pass to join in a small ganglion with phrenic branches of the solar plexus: and branches from this ganglion are distributed to the hepatic plexus, the suprarenal capsule, and inferior vena cava. From the *left nerve*, filaments pass to join the phrenic plexus of the sympathetic, but without any ganglionic enlargement.

DEEP BRANCHES OF THE CERVICAL PLEXUS. EXTERNAL SERIES

Communicating branches.—The deep branches of the external series of the cervical plexus communicate with the spinal accessory nerve, in the substance of the Sterno-mastoid muscle, in the posterior triangle, and beneath the Trapezius.

Muscular branches are distributed to the Sterno-mastoid, Trapezius, Levator anguli scapulæ, and Scalenus medius.

The branch for the Sterno-mastoid is derived from the second cervical; the Trapezius and Levator anguli scapulæ receive branches from the third and fourth. The Scalenus medius is supplied sometimes by the third, sometimes by the fourth, and occasionally by both nerves.

THE BRACHIAL PLEXUS (fig. 624)

The **Brachial Plexus** is formed by the union of the anterior divisions of the four lower cervical nerves and the greater part of the anterior primary division of the first dorsal nerve, receiving usually a fasciculus from the fourth cervical nerve, and frequently one from the second dorsal nerve. It extends from the lower part of the side of the neck to the axilla. It is very broad and presents little of a plexiform arrangement at its commencement. It is narrow opposite the clavicle, becomes broad, and forms a more dense interlacement in the axilla, and divides opposite the coracoid process into numerous branches for the supply of the upper limb. The nerves which form the plexus are all similar in size, but their mode of communication is subject to some variation, so that no one plan can be given as applying to every case. The following is, however, the most constant arrangement: The fifth and sixth cervical unite together soon after their exit from the intervertebral foramina to form a common trunk. The eighth cervical and first dorsal also unite to form one trunk. So that the nerves forming the plexus, as they lie on the Scalenus medius, external to the outer border of the Scalenus anticus, are blended into three trunks: an upper one, formed by the junction of the fifth and sixth cervical nerves; a middle one, consisting of the seventh cervical nerve; and a lower one, formed by the junction of the eighth cervical and first dorsal nerves. As they pass beneath the clavicle, each of these three trunks divides into two divisions, an *anterior* and a *posterior*.*

* The posterior division of the lower trunk is very much smaller than the others, and is frequently derived entirely from the eighth cervical nerve.

3 M

The anterior divisions of the upper and middle trunks then unite to form a cord, which is situated on the outer side of the middle part of the axillary artery, and is called the *outer cord* of the brachial plexus. The anterior division of the lower trunk passes down on the inner side of the axillary artery in the middle of the axilla, and forms the *inner cord* of the brachial plexus. The posterior divisions of all three trunks unite to form the *posterior cord* of the brachial plexus, which is situated behind the second portion of the axillary artery. From this posterior cord are given off the three subscapular nerves, before it divides into the circumflex and musculo-spiral nerves.

The brachial plexus communicates with the cervical plexus by a branch from the fourth to the fifth nerve, and with the phrenic nerve by a branch from the fifth cervical, which joins that nerve on the Anterior scalenus muscle : the fifth and sixth cervical nerves are joined by filaments from the middle cervical

FIG. 624.—Plan of the brachial plexus.

ganglion of the sympathetic, the seventh and eighth cervical from its inferior ganglion, and the first dorsal nerve from its first thoracic ganglion, close to their exit from the intervertebral foramina.

Relations.—*In the neck*, the brachial plexus lies in the posterior triangle, being covered by the skin, Platysma, and deep fascia : it is crossed by the descending superficial cervical nerves, the posterior belly of the Omo-hyoid, the external jugular vein, and the Transversalis colli artery. When the posterior scapular artery arises from the third part of the subclavian, it usually passes between the roots of the plexus. The plexus emerges from between the Anterior and Middle scaleni muscles ; its upper part lies above and to the outer side of the third part of the subclavian artery, while the trunk formed by the union of the eighth cervical and first dorsal is placed behind the artery ; it next passes behind the clavicle, the Subclavius muscle and suprascapular vessels, and lies upon the first serration of the Serratus magnus, and the Subscapularis muscles. *In the axilla* it is placed on the outer side of the first portion of the axillary artery ; it surrounds the artery in the second part of its course, one cord lying upon the outer side of that vessel, one on the inner side, and one behind it ; and at the lower part of the axillary space gives off its terminal branches to the upper extremity.

Branches.—The branches of the brachial plexus are arranged into two groups, viz. those given off above the clavicle, and those below that bone.

BRANCHES ABOVE THE CLAVICLE

Communicating.   Posterior thoracic.
Muscular.   Suprascapular.

The **communicating branch** with the phrenic is derived from the fifth cervical nerve or from the loop between the fifth and sixth; it joins the phrenic on the Anterior scalenus muscle. The communications with the sympathetic have already been referred to.

The **muscular branches** supply the Longus colli, Scaleni, Rhomboidei, and Subclavius muscles. Those for the Longus colli and Scaleni arise from the four lower cervical nerves at their exit from the intervertebral foramina. The rhomboid branch (*posterior scapular*) arises from the fifth cervical, pierces the Scalenus medius, and passes beneath the Levator anguli scapulæ, which it occasionally supplies, to the Rhomboid muscles. The nerve to the Subclavius is a small filament, which arises from the fifth cervical at its point of junction with the sixth nerve; it descends in front of the third part of the subclavian artery and the lower cords of the plexus to the Subclavius muscle, and is usually connected by a filament with the phrenic nerve.

The **posterior thoracic nerve** (*long thoracic, external respiratory of Bell*) (fig. 627) supplies the Serratus magnus, and is remarkable for the length of its course. It usually arises by three roots from the fifth, sixth, and seventh cervical nerves; but the root from the seventh nerve may be absent. The roots from the fifth and sixth nerves pierce the Middle scalene muscle, while that from the seventh nerve emerges from in front of the muscle. The nerve passes down behind the brachial plexus and the axillary vessels, resting on the outer surface of the Serratus magnus. It extends along the side of the chest to the lower border of that muscle, supplying filaments to each of its digitations.

The **suprascapular nerve** (fig. 628) arises from the cord formed by the fifth and sixth cervical nerves; passing obliquely outwards beneath the Trapezius and the Omo-hyoid, it enters the supraspinous fossa through the suprascapular notch, below the transverse or suprascapular ligament, and, passing beneath the Supraspinatus muscle, curves round the external border of the spine of the scapula to the infraspinous fossa. In the supraspinous fossa it gives off two branches to the Supraspinatus muscle, and an articular filament to the shoulder-joint; and in the infraspinous fossa it gives off two branches to the Infraspinatus muscle, besides some filaments to the shoulder-joint and scapula.

BRANCHES BELOW THE CLAVICLE

The branches given off below the clavicle are derived from the three cords of the brachial plexus, in the following manner:

From the outer cord arise the external anterior thoracic nerve, the musculo-cutaneous, and the outer head of the median.

From the inner cord arise the internal anterior thoracic nerve, the internal cutaneous, the lesser internal cutaneous (nerve of Wrisberg), the ulnar and the inner head of the median.

From the posterior cord arise the three subscapular nerves; the cord then divides into the musculo-spiral and circumflex nerves.

These may be arranged according to the regions they supply:

To the chest	Anterior thoracics.
To the shoulder	{ Subscapulars. { Circumflex.
To the arm, forearm, and hand .	{ Musculo-cutaneous. { Internal cutaneous. { Lesser internal cutaneous. { Median. { Ulnar. { Musculo-spiral.

3 M 2

The fasciculi of which these nerves are composed may be traced through the plexus to the spinal nerves from which they originate. They are as follows:

External anterior thoracic from 5th, 6th, and 7th cervical.
Internal anterior thoracic   „   8th cervical and 1st dorsal.
Subscapular   „   5th, 6th, 7th, and 8th cervical.
Circumflex   „   5th and 6th cervical.
Musculo-cutaneous   „   5th and 6th cervical.
Internal cutaneous   „   8th cervical and 1st dorsal.
Lesser internal cutaneous   „   1st dorsal; sometimes also from the
                                    8th cervical.
Median   „   6th, 7th, and 8th cervical, and 1st dorsal.
Ulnar   „   8th cervical and 1st dorsal.
Musculo-spiral   „   6th, 7th, and 8th cervical; sometimes also
                                    from the 5th cervical and 1st dorsal.

The **Anterior Thoracic Nerves** (fig. 627), two in number, supply the Pectoral muscles.

The *external anterior thoracic nerve*, the larger of the two, arises from the outer cord of the brachial plexus, through which its fibres may be traced to the fifth, sixth, and seventh cervical nerves. It passes inwards, across the axillary artery and vein, pierces the costo-coracoid membrane, and is distributed to the under surface of the Pectoralis major. It sends down a communicating filament to join the internal nerve, which forms a loop in front of the first part of the axillary artery.

The *internal anterior thoracic nerve* arises from the inner cord, and through it from the eighth cervical and first dorsal. It passes behind the first part of the axillary artery, curves forwards between the axillary artery and vein, and joins with the filament from the external nerve. It then passes to the under surface of the Pectoralis minor muscle, where it divides into a number of branches, which supply the muscle on its under surface. Some two or three branches pass through the muscle to supply the Pectoralis major.

The **Subscapular Nerves**, three in number, supply the Subscapularis, Teres major, and Latissimus dorsi muscles. The fasciculi of which they are composed may be traced to the fifth, sixth, seventh, and eighth cervical nerves.

The *upper* or *short subscapular nerve*, the smallest, enters the upper part of the Subscapularis muscle; this nerve is frequently represented by two branches.

The *lower subscapular nerve* enters the axillary border of the Subscapularis, and terminates in the Teres major. The latter muscle is sometimes supplied by a separate branch.

The *middle* or *long subscapular*, the largest of the three, follows the course of the subscapular artery, along the posterior wall of the axilla to the Latissimus dorsi, in which it may be traced as far as its lower border.

The **Circumflex Nerve** (fig. 628) supplies some of the muscles, and part of the integument of the shoulder, and gives a branch to the shoulder-joint. It arises from the posterior cord of the brachial plexus, in common with the musculo-spiral nerve, and its fibres may be traced through the posterior cord to the fifth and sixth cervical nerves. It is at first placed behind the axillary artery, between it and the Subscapularis muscle, and passes downwards and outwards to the lower border of that muscle. It then winds backwards, in company with the posterior circumflex artery, through a quadrilateral space, bounded above by the Subscapularis, below by the Teres major, internally by the long head of the Triceps, and externally by the surgical neck of the humerus, and divides into two branches.

The *upper branch* winds round the surgical neck of the humerus, beneath the Deltoid, with the posterior circumflex vessels, as far as the anterior border of that muscle, supplying it, and giving off cutaneous branches, which pierce the muscle and ramify in the integument covering its lower part.

The *lower branch*, at its origin, distributes filaments to the Teres minor and back part of the Deltoid muscles. Upon the filament to the former muscle an oval enlargement usually exists. The nerve then pierces the deep fascia, and supplies the integument over the lower two-thirds of the posterior surface of the Deltoid, as well as that covering the long head of the Triceps.

The circumflex nerve, before its division, gives off an articular filament, which enters the shoulder-joint below the Subscapularis.

The **Musculo-cutaneous Nerve** (fig. 627) arises from the outer cord of the brachial plexus, opposite the lower border of the Pectoralis minor, receiving

FIG. 625.—Cutaneous nerves of right upper extremity. Anterior view.

FIG. 626.—Cutaneous nerves of right upper extremity. Posterior view.

filaments from the fifth and sixth cervical nerves. It perforates the Coraco-brachialis muscle, passes obliquely between the Biceps and Brachialis anticus, to the outer side of the arm, and, a little above the elbow, becomes cutaneous by perforating the deep fascia on the outer side of the tendon of the Biceps. This nerve, in its course through the arm, supplies the Coraco-brachialis, Biceps, and the greater part of the Brachialis anticus muscles. The branch to the Coraco-brachialis is given off from the nerve close to its origin, and in some instances, especially in early life, as a separate filament from the outer cord of the plexus. It is derived from the seventh nerve, and is by some anatomists regarded as a separate nerve, more or less closely incorporated with the musculo-cutaneous. The branches to the Biceps and Brachialis anticus are given off after the nerve has pierced the Coraco-brachialis. The nerve also sends a small branch to the bone, which enters the nutrient foramen with the accompanying artery, and a filament, from the branch supplying the Brachialis anticus, to the elbow-joint.

The cutaneous portion of the nerve passes behind the median cephalic vein, and divides, opposite the elbow-joint, into an anterior and a posterior branch.

The *anterior branch* descends along the radial border of the forearm to the wrist, and supplies the integument over the outer half of its anterior surface. At the wrist-joint it is placed in front of the radial artery, and some filaments, piercing the deep fascia, accompany that vessel to the back of the wrist, and supply the carpus. The nerve then passes downwards to the ball of the thumb, where it terminates in cutaneous filaments. It communicates with a branch from the radial nerve, and with the palmar cutaneous branch of the median.

The *posterior branch* passes downwards, along the back part of the radial side of the forearm, to the wrist. It supplies the integument of the lower two-thirds of the forearm, communicating with the radial nerve and the external cutaneous branch of the musculo-spiral.

The musculo-cutaneous nerve presents frequent irregularities. It may adhere for some distance to the median and then pass outwards, beneath the Biceps, instead of through the Coraco-brachialis. Frequently some of the fibres of the median run for some distance in the musculo-cutaneous and then leave it to join their proper trunk. Less frequently the reverse is the case, and the median sends a branch to join the musculo-cutaneous. Instead of piercing the Coraco-brachialis the nerve may pass under it or through the Biceps. Occasionally it gives a filament to the Pronator teres, and it has been seen to supply the back of the thumb when the radial nerve was absent.

The **Internal Cutaneous Nerve** (fig. 627) is one of the smallest branches of the brachial plexus. It arises from the inner cord in common with the ulnar and internal head of the median, and, at its commencement, is placed on the inner side of the axillary, and afterwards of the brachial artery. It derives its fibres from the eighth cervical and first dorsal nerves. It passes down the inner side of the arm, pierces the deep fascia with the basilic vein, about the middle of the upper arm, and, becoming cutaneous, divides into two branches, anterior and posterior.

This nerve gives off, near the axilla, a filament, which pierces the fascia and supplies the integument covering the Biceps muscle, nearly as far as the elbow. This filament lies a little external to the common trunk, from which it arises.

The *anterior branch*, the larger of the two, passes usually in front of, but occasionally behind, the median basilic vein. It then descends on the anterior surface of the ulnar side of the forearm, distributing filaments to the integument as far as the wrist, and communicating with a cutaneous branch of the ulnar nerve.

The *posterior branch* passes obliquely downwards on the inner side of the basilic vein, in front of the internal epicondyle of the humerus to the back of the forearm, and descends on the posterior surface of its ulnar side as far as the wrist, distributing filaments to the integument. It communicates with the lesser internal cutaneous, the lower external cutaneous branch of the musculo-spiral, and the dorsal branch of the ulnar nerve.

The **Lesser Internal Cutaneous Nerve** (*nerve of Wrisberg*) is distributed to the integument on the inner side of the arm (fig. 627). It is the smallest of the branches of the brachial plexus, and arising from the inner cord, with the internal cutaneous and ulnar nerves, receives its fibres from the first dorsal nerve, and sometimes from the eighth cervical. It passes through the axillary space, at first lying behind, and then on the inner side of the axillary vein, and communicates with the intercosto-humeral nerve. It descends along the inner side of the brachial artery to the middle of the arm, where it pierces the deep fascia, and is distributed to the integument of the back part of the lower third of

FIG. 627.—Nerves of the left upper extremity.

External anterior thoracic

Internal anterior thoracic

Musculo-cutaneous

Median

Ulnar

Internal cutaneous

Musculo-spiral

Posterior interosseous

Radial

Anterior interosseous

Ulnar

Dorsal branch

Deep branch

the arm, extending as far as the elbow, where some filaments are lost in the integument in front of the inner epicondyle, and others over the olecranon. It communicates with the posterior branch of the internal cutaneous nerve.

In some cases the nerve of Wrisberg and intercosto-humeral are connected by two or three filaments, which form a plexus at the back part of the axilla. In other cases, the intercosto-humeral is of large size, and takes the place of the nerve of Wrisberg, receiving merely a filament of communication from the brachial plexus, which represents the latter nerve. In other cases, this filament is wanting, the place of the nerve of Wrisberg being supplied entirely from the intercosto-humeral.

The **Median Nerve** (fig. 627) has received its name from the course it takes along the middle of the arm and forearm to the hand, lying between the ulnar and the musculo-spiral and radial nerves. It arises by two roots, one from the outer, and one from the inner cord of the brachial plexus; these embrace the lower part of the axillary artery, uniting either in front or on the outer side of that vessel. It receives filaments from the sixth, seventh, and eighth cervical and the first dorsal nerves. As it descends through the arm, it lies at first on the outer side of the brachial artery, crosses that vessel in the middle of its course, usually in front of, but occasionally behind it, and lies on its inner side to the bend of the elbow, where it is placed beneath the bicipital fascia, and is separated from the elbow-joint by the Brachialis anticus. *In the forearm* it passes between the two heads of the Pronator radii teres and crosses the ulnar artery, from which it is separated by the deep head of the Pronator radii teres. It descends beneath the Flexor sublimis, lying on the Flexor profundus, to within two inches above the annular ligament, where it becomes more superficial, lying between the tendons of the Flexor sublimis and Flexor carpi radialis, beneath, and rather to the radial side of, the tendon of the Palmaris longus, covered by the integument and fascia. It then passes beneath the anterior annular ligament into the hand. In its course through the forearm it is accompanied by a branch of the anterior interosseous artery.

Branches.—With the exception of the nerve to the Pronator teres, which sometimes arises above the elbow-joint, the median nerve gives off no branches in the arm. As it passes in front of the elbow, it supplies one or two articular twigs to the joint. *In the forearm* its branches are, muscular, anterior interosseous, and palmar cutaneous.

The *muscular branches* supply all the superficial muscles on the front of the forearm, except the Flexor carpi ulnaris. These branches are derived from the nerve near the elbow.

The *anterior interosseous* supplies the deep muscles on the front of the forearm, except the inner half of the Flexor profundus digitorum. It accompanies the anterior interosseous artery along the interosseous membrane, in the interval between the Flexor longus pollicis and Flexor profundus digitorum muscles, supplying the whole of the former and the outer half of the latter, and terminates below in the Pronator quadratus and wrist-joint.

The *palmar cutaneous branch* arises from the median nerve at the lower part of the forearm. It pierces the fascia above the annular ligament, and descending over that ligament, divides into two branches: of which the *outer* supplies the skin over the ball of the thumb, and communicates with the anterior cutaneous branch of the musculo-cutaneous nerve; and the *inner* supplies the integument of the palm of the hand, communicating with the cutaneous branch of the ulnar.

In the palm of the hand, the median nerve is covered by the integument and palmar fascia, and crossed by the superficial palmar arch. It rests upon the tendons of the flexor muscles. In this situation it becomes enlarged, somewhat flattened, of a reddish colour, and divides into two branches. Of these, the *external* supplies a muscular branch to some of the muscles of the thumb, and digital branches to the thumb and radial side of the index finger; the *internal* supplies digital branches to the contiguous sides of the index and middle, and of the middle and ring fingers.

The *branch to the muscles of the thumb* is a short nerve, which divides to supply the Abductor, Opponens, and the superficial head of the Flexor brevis pollicis muscles; the remaining muscles of this group being supplied by the ulnar nerve.

The *digital branches* are five in number. The *first* and *second* pass along the borders of the thumb, the external branch communicating with branches of the radial nerve. The *third* passes along the radial side of the index finger, and supplies the First lumbrical muscle. The *fourth* subdivides to supply the adjacent sides of the index and middle fingers, and sends a branch to the Second lumbrical muscle. The *fifth* supplies the adjacent sides of the middle and ring fingers, and communicates with a branch from the ulnar nerve.

Each digital nerve, opposite the base of the first phalanx, gives off a dorsal branch, which joins the dorsal digital nerve from the radial, and runs along the side of the dorsum of the finger, to end in the integument over the last phalanx. At the end of the finger, the digital nerve divides into a palmar and a dorsal branch : the former of which supplies the extremity of the finger, and the latter ramifies round and beneath the nail. The digital nerves, as they run along the fingers, are placed superficially to the digital arteries.

The **Ulnar Nerve** (fig. 627) is placed along the inner or ulnar side of the upper limb, and is distributed to the muscles and integument of the forearm and hand. It is smaller than the median, behind which it is placed, diverging from it in its course down the arm. It arises from the inner cord of the brachial plexus, in common with the inner head of the median and the internal cutaneous nerve, and derives its fibres from the eighth cervical and first dorsal nerves. At its commencement it lies to the inner side of the axillary artery, and holds the same relation with the brachial artery to the middle of the arm. Here it pierces the internal intermuscular septum, runs obliquely across the internal head of the Triceps, and descends to the groove between the internal epicondyle and the olecranon, accompanied by the inferior profunda artery. *At the elbow*, it rests upon the back of the inner epicondyle, and passes into the forearm between the two heads of the Flexor carpi ulnaris. *In the forearm*, it descends in a perfectly straight course along its ulnar side, lying upon the Flexor profundus digitorum, its upper half being covered by the Flexor carpi ulnaris, its lower half lying on the outer side of the muscle, covered by the integument and fascia. The ulnar artery, in the upper third of its course, is separated from the ulnar nerve by a considerable interval ; but in the rest of its extent the nerve lies to its inner side. *At the wrist* the ulnar nerve crosses the annular ligament on the outer side of the pisiform bone, to the inner side of, and a little behind, the ulnar artery, and immediately beyond this bone divides into two branches, superficial and deep palmar.

The branches of the ulnar nerve are :

In the forearm
{
Articular (elbow).
Muscular.
Palmar cutaneous.
Dorsal cutaneous.
Articular (wrist).
}

In the hand
{
Superficial palmar.
Deep palmar.
}

The *articular branches* to the elbow-joint consist of several small filaments. They arise from the nerve as it lies in the groove between the inner epicondyle and olecranon.

The *muscular branches* are two in number : one supplying the Flexor carpi ulnaris ; the other, the inner half of the Flexor profundus digitorum. They arise from the trunk of the nerve near the elbow.

The *palmar cutaneous* arises from the ulnar nerve about the middle of the forearm, and runs downwards, lying on the ulnar artery, giving off some filaments to the vessel. Just above the annular ligament, it perforates the deep fascia and ends in the integument of the palm, communicating with the palmar branch of the median nerve.

The *dorsal cutaneous branch* arises about two inches above the wrist ; it passes backwards beneath the Flexor carpi ulnaris, perforates the deep fascia, and, running along the ulnar side of the back of the wrist and hand, divides into branches : one of these supplies the inner side of the little finger ; a second supplies the adjacent sides of the little and ring fingers ; a third joins the branch of the radial nerve which supplies the adjoining sides of the middle and ring fingers, and assists in supplying them ; a fourth is distributed to the metacarpal region of the hand, communicating with a branch of the radial nerve.

On the little finger the dorsal digital branches only extend as far as the base of the terminal phalanx, and on the ring finger as far as the base of the second phalanx; the more distal parts of these digits are supplied by dorsal branches derived from the palmar digital branches of the ulnar.

The *superficial palmar branch* supplies the Palmaris brevis, and the integument on the inner side of the hand, and terminates in two digital branches, which are distributed, one to the ulnar side of the little finger, the other to the adjoining sides of the little and ring fingers, the latter communicating with a branch from the median. The digital branches are distributed to the fingers in the same manner as the digital branches of the median.

FIG. 628.—The suprascapular, circumflex, and musculo-spiral nerves.

Suprascapular

Circumflex

Musculo-spiral

Posterior interosseous

The *deep palmar branch*, accompanied by the deep branch of the ulnar artery, passes between the Abductor and Flexor brevis minimi digiti muscles; it then perforates the Opponens minimi digiti and follows the course of the deep palmar arch beneath the flexor tendons. At its origin it supplies the muscles of the little finger. As it crosses the deep part of the hand, it supplies all the Interossei muscles and the two inner Lumbricales; it ends by supplying the Adductores transversus et obliquus pollicis and the inner head of the Flexor brevis pollicis. It also sends articular filaments to the wrist-joint.

It will be remembered that the inner part of the Flexor profundus digitorum is supplied by the ulnar nerve; the two inner Lumbricales, which are connected with the tendons of this part of the muscle, are therefore supplied by the same nerve. The outer part of the Flexor profundus is supplied by the median nerve: the two outer Lumbricales, which are connected with the tendons of this part of the muscle, are therefore supplied by the same nerve. Brooks stated that in twelve instances out of twenty-one he found that the third lumbrical received a twig from the median nerve, in addition to its branch from the ulnar.

The **Musculo-spiral nerve** (fig. 628), the largest branch of the brachial plexus, supplies the muscles of the back part of the arm and forearm, and the integument of the same parts, as well as that of the back of the hand. It arises from the posterior cord of the brachial plexus, of which it may be regarded as the continuation. It receives filaments from the sixth, seventh, and eighth, and sometimes also from the fifth cervical and first dorsal nerves. At its commencement it is placed behind the axillary and upper part of the brachial arteries, passing down in front of the tendons of the Latissimus dorsi and Teres major. It winds round the humerus in the musculo-spiral groove with the superior profunda artery, passing from the inner to the outer side of the bone, between the internal and external heads of the Triceps muscle. It pierces the external intermuscular septum, and descends between the Brachialis anticus and Brachio-radialis to the front of the external epicondyle, where it divides into the radial and posterior interosseous nerves.

The branches of the musculo-spiral nerve are:

Muscular.	Radial.
Cutaneous.	Posterior interosseous.

The *muscular branches* are divided into internal, posterior, and external; they supply the Triceps, Anconeus, Brachio-radialis, Extensor carpi radialis longior, and Brachialis anticus. These branches are derived from the nerve, at the inner side, back part, and outer side of the arm respectively.

The internal muscular branches supply the inner and middle heads of the Triceps muscle. That to the inner head of the Triceps is a long, slender filament, which lies close to the ulnar nerve, as far as the lower third of the arm, and is therefore frequently spoken of as the *ulnar collateral*.

The posterior muscular branch, of large size, arises from the nerve in the groove between the Triceps and the humerus. It divides into branches, which supply the inner and outer heads of the Triceps and the Anconeus muscles. The branch for the latter muscle is a long, slender filament, which descends in the substance of the inner head of the Triceps to the Anconeus.

The external muscular branches supply the Brachio-radialis, Extensor carpi radialis longior, and (usually) the outer part of the Brachialis anticus.

The *cutaneous branches* are three in number, one internal and two external.

The internal cutaneous branch arises in the axillary space, with the inner muscular branch. It is of small size, and passes through the axilla to the inner side of the arm, supplying the integument on its posterior aspect nearly as far as the olecranon. In its course it crosses beneath the intercosto-humeral, with which it communicates.

The two external cutaneous branches perforate the outer head of the Triceps, at its attachment to the humerus. The *upper* and smaller one passes to the front of the elbow, lying close to the cephalic vein, and supplies the integument of the lower half of the arm on its anterior aspect. The *lower* branch pierces the deep fascia below the insertion of the Deltoid, and passes down along the outer side of the arm and elbow, and then along the back part of the radial side of the forearm to the wrist, supplying the integument in its course, and joining, near its termination, with the posterior cutaneous branch of the musculo-cutaneous nerve.

The **radial nerve** passes along the front of the radial side of the forearm to the commencement of its lower third. It lies at first a little to the outer side of the radial artery, concealed beneath the Brachio-radialis. In the middle third of the forearm, it lies beneath the same muscle, in close relation with the outer side of the artery. It quits the artery about three inches above the wrist, passes beneath the tendon of the Brachio-radialis, and, piercing the deep fascia at the outer border of the forearm, divides into two branches.

The external branch, the smaller of the two, supplies the integument of the radial side and ball of the thumb, joining with the anterior branch of the musculo-cutaneous nerve.

The internal branch communicates, above the wrist, with the posterior cutaneous branch from the musculo-cutaneous, and, on the back of the hand, with the dorsal cutaneous branch of the ulnar nerve. It then divides into four digital nerves, which are distributed as follows: the first supplies the ulnar side of the thumb; the second, the radial side of the index finger; the third, the adjoining sides of the index and middle fingers; and the fourth, the adjacent borders of the middle and ring fingers.* The latter nerve communicates with a filament from the dorsal branch of the ulnar nerve.

* According to Hutchinson, the digital nerve to the thumb reaches only as high as the root of the nail: the one to the forefinger as high as the middle of the second phalanx: and the one to the middle and ring fingers not higher than the first phalangeal joint.—*London Hos. Gaz.* vol. iii. p. 319.

The **posterior interosseous nerve** winds to the back of the forearm round the outer side of the radius, passes between the two planes of fibres of the Supinator brevis, and is prolonged downwards between the superficial and deep layer of muscles, to the middle of the forearm. Considerably diminished in size, it descends on the interosseous membrane, beneath the Extensor longus pollicis, to the back of the carpus, where it presents a gangliform enlargement from which filaments are distributed to the ligaments and articulations of the carpus. It supplies all the muscles on the radial side and posterior aspect of the forearm, excepting the Anconeus, Brachio-radialis, and Extensor carpi radialis longior.

Surgical Anatomy.—The brachial plexus may be injured by falls from a height on to the side of the head and shoulder, whereby the nerves of the plexus are violently stretched, and under these circumstances the fifth cervical nerve sustains the greatest amount of injury, and the subsequent paralysis may be confined to the muscles supplied by this nerve. These are the Deltoid, Biceps, Brachialis anticus, and the Brachio-radialis, with sometimes the Supra- and Infra-spinatus and the Supinator brevis. The position of the limb, under such conditions, is characteristic: the arm hangs by the side and is rotated inwards; the forearm is extended and pronated. The arm cannot be raised from the side; all power of flexion of the elbow is lost, as is also supination of the forearm. This is known as Erb's paralysis, and a very similar condition is occasionally met with in new-born children, either from injury to the fifth nerve from the pressure of forceps used in effecting delivery, or from traction of the head in breech presentations. The brachial plexus may also be injured by violent traction on the arm, or by efforts at reducing a dislocation of the shoulder-joint; and the amount of paralysis will depend upon the amount of injury to the constituent nerves. When the entire plexus is involved, the whole of the upper extremity will be paralysed and anæsthetic. In these cases the injury appears to be rather a tearing away of the roots of the nerves from their origin in the spinal cord, than a solution of continuity in the nerves themselves. The nerves of the brachial plexus may also be lacerated by direct violence, generally in gunshot injury. In these cases, the lower nerves—that is to say, the last cervical and the first dorsal—are generally the ones involved. When these trunks are injured, the muscles of the hand are the ones principally implicated, those of the shoulder and arm escaping. The fingers become clawed, and there is inability to abduct and adduct them. The brachial plexus in the axilla is often damaged from the pressure of a crutch, producing the condition known as 'crutch paralysis.' In these cases the musculo-spiral appears most frequently to be the nerve which is chiefly implicated; the ulnar nerve being the one that appears to suffer next in frequency.

The *circumflex nerve* is of particular surgical interest. On account of its course round the surgical neck of the humerus, it is liable to be torn in fractures of this part of the bone, and in dislocations of the shoulder-joint, leading to paralysis of the Deltoid, and, according to Erb, inflammation of the shoulder-joint is liable to be followed by a neuritis of this nerve from extension of the inflammation to it.

Hilton takes the circumflex nerve as an illustration of a law which he lays down, that 'the same trunks of nerves whose branches supply the groups of muscles moving a joint, furnish also a distribution of nerves to the skin over the insertions of the same muscles, and the interior of the joint receives its nerves from the same source.' In this way he explains the fact that an inflamed joint becomes rigid, because the same nerves which supply the interior of the joint supply the muscles also which move that joint.

The *median nerve* is liable to injury in wounds of the forearm. When paralysed, there is loss of flexion of the second phalanges of all the fingers, and of the terminal phalanges of the index and middle fingers. Flexion of the terminal phalanges of the ring and little fingers is effected by that portion of the Flexor profundus digitorum which is supplied by the ulnar nerve. There is power to flex the proximal phalanges through the Interossei. The thumb cannot be flexed or opposed, and is maintained in a position of extension and adduction. There is loss in the power of pronating the forearm; the Brachio-radialis has the power of bringing the forearm into a position of mid-pronation, but beyond this no further pronation can be effected. The wrist can be flexed, if the hand is first adducted, by the action of the Flexor carpi ulnaris. There is loss or impairment of sensation on the palmar surface of the thumb, index, middle, and outer half of ring fingers, and on the dorsal surface of the same fingers over the last two phalanges; except in the thumb, where the loss of sensation would be limited to the back of the last phalanx. In order to expose the median nerve, for the purpose of stretching, an incision should be made along the radial side of the tendon of the Palmaris longus, which serves as a guide to the nerve.

The *ulnar nerve* is also liable to be injured in wounds of the forearm. When paralysed, there is impaired power of ulnar flexion and extension, and upon an attempt being made to flex the wrist, the hand is drawn to the radial side from paralysis of the Flexor carpi ulnaris: there is inability to spread out the fingers from paralysis of the

Interossei, and from the same cause the fingers, especially the ring and little fingers, cannot be flexed at the metacarpo-phalangeal joints or extended at the interphalangeal joints, and the hand assumes a claw shape from the action of the opposing muscles; there is loss of power of flexion in the little and ring fingers; and there is inability to adduct the thumb. The muscles of the hypothenar eminence become wasted. Sensation is lost, or impaired, in the skin supplied by the nerve. In order to expose the nerve in the lower part of the forearm, an incision should be made along the outer border of the tendon of the Flexor carpi ulnaris, and the nerve will be found lying on the ulnar side of the ulnar artery.

The *musculo-spiral nerve* is probably more frequently injured than any other nerve of the upper extremity. In consequence of its close relationship to the humerus, as it lies in the musculo-spiral groove, it is often torn or injured in fractures of this bone, or subsequently involved in the callus that may be thrown out around a fracture and thus pressed upon and its functions interfered with. It is also liable to be contused against the bone by kicks or blows, or to be divided by wounds of the arm. When paralysed, the hand is flexed at the wrist and lies flaccid. This is known as 'drop-wrist.' The fingers are also flexed, and on an attempt being made to extend them, the last two phalanges only will be extended, through the action of the Interossei; the first phalanges remaining flexed. There is no power of extending the wrist. Supination is completely lost when the forearm is extended on the arm, but is possible to a certain extent if the forearm is flexed so as to allow of the action of the Biceps. The power of extension of the forearm is lost on account of paralysis of the Triceps, if the injury to the nerve has taken place near its origin. The best position in which to expose the nerve, for the purpose of stretching, is to make an incision along the inner border of the Brachio-radialis, just above the level of the elbow-joint. The skin and superficial structures are to be divided and the deep fascia exposed. The white line in this structure indicating the border of the muscle is to be defined, and the deep fascia divided in this line. By now raising the Brachio-radialis, the nerve will be found lying beneath it, on the Brachialis anticus.

DORSAL NERVES (fig. 629)

The **dorsal** or **thoracic nerves** are twelve in number on each side. The first appears between the first and second dorsal vertebræ, and the twelfth between the last dorsal and first lumbar.

The *roots of the dorsal nerves* are of small size, and vary very little from the second to the last. Both roots are very slender; the posterior roots only slightly exceeding the anterior in thickness. They gradually increase in length from above downwards, and, in the lower part of the dorsal region, pass down in contact with the spinal cord for a distance equal to the height of, at least, two vertebræ, before they emerge from the spinal canal. They then join in the intervertebral foramen, and, at their exit, divide into two primary divisions, a posterior (dorsal) and an anterior (intercostal).

The first, the second, and the last dorsal nerves are peculiar in some respects.

Posterior Divisions of the Dorsal Nerves

The **posterior divisions of the dorsal nerves,** which are smaller than the anterior, pass backwards between the transverse processes, and divide into internal and external branches.

The *internal branches of the six upper nerves* pass inwards between the Semi-spinalis dorsi and Multifidus spinæ muscles, which they supply; and then, piercing the origins of the Rhomboidei and Trapezius muscles, become cutaneous by the side of the spinous processes and ramify in the integument. The internal branches of the *six lower nerves* are distributed to the Multifidus spinæ, without giving off any cutaneous filaments.

The *external branches* increase in size from above downwards. They pass through the Longissimus dorsi, to the cellular interval between it and the Ilio-costalis, and supply those muscles, as well as their continuations upwards to the head, and the Levatores costarum; the five or six lower nerves also give c cutaneous filaments, which pierce the Serratus posticus inferior and Latissimu. dorsi, in a line with the angles of the ribs, and then ramify in the integument.

The *cutaneous branches of the posterior primary divisions of the dorsal nerves* are twelve in number. The six upper cutaneous nerves are derived from the internal branches of the posterior divisions of the corresponding dorsal nerves.

They pierce the origins of the Rhomboidei and Trapezius muscles, and become cutaneous by the side of the spinous processes, and then ramify in the integument. They are frequently furnished with gangliform enlargements. The six lower cutaneous nerves are derived from the external branches of the posterior divisions of the corresponding dorsal nerves. They pierce the Serratus posticus inferior and Latissimus dorsi, in a line with the angles of the ribs, and then ramify in the integument.

Anterior Divisions of the Dorsal Nerves

The **anterior divisions of the dorsal nerves** are twelve in number on each side. Eleven of them are situated between the ribs, and are therefore termed *intercostal*; the twelfth lies below the last rib. They are, for the most part, distributed to the parietes of the thorax and abdomen, separately from each other, i.e. without being joined in a plexus ; in which respect they differ from the other spinal nerves. Each nerve is connected with the adjoining ganglia of the sympathetic by one or two filaments. The intercostal nerves may be divided into two sets, from the difference they present in their distribution. The six upper, with the exception of the first and the intercosto-humeral branch of the second, are limited in their distribution to the parietes of the chest. The five lower supply the parietes of the chest and abdomen ; the twelfth dorsal is distributed to the abdominal wall and the skin of the buttock.

The first dorsal nerve.—The anterior division of the first dorsal nerve divides into two branches : one, the larger, leaves the thorax in front of the neck of the first rib, and enters into the formation of the brachial plexus ; the other and smaller branch runs along the first intercostal space, forming the *first intercostal nerve*, and terminates on the front of the chest, by forming the first anterior cutaneous nerve of the thorax. Occasionally this anterior cutaneous branch is wanting. The first intercostal nerve as a rule gives off no lateral cutaneous branch ; but sometimes a small branch is given off, which communicates with the intercosto-humeral. It frequently receives a connecting twig from the second dorsal nerve, which passes upwards over the neck of the second rib.

The upper dorsal nerves.—The anterior divisions of the second, third, fourth, fifth, and sixth dorsal nerves and the small branch from the first dorsal are confined to the parietes of the thorax, and are named *thoracic intercostal nerves*. They pass forwards in the intercostal spaces with the intercostal vessels, being situated below the latter. At the back of the chest they lie between the pleura and the External intercostal muscles, but are soon placed between the two planes of Intercostal muscles as far as the middle of the rib. They then enter the substance of the Internal intercostal muscles, and, running amidst their fibres as far as the costal cartilages, they gain the inner surface of the muscles and lie between them and the pleura. Near the sternum, they cross in front of the internal mammary artery and Triangularis sterni muscle, pierce the Internal intercostal muscles, the anterior intercostal membranes, and Pectoralis major muscle, and supply the integument of the front of the chest and over the mammary gland, forming the anterior cutaneous nerves of the thorax ; the branch from the second nerve is joined with the supraclavicular nerves of the cervical plexus.

Branches.—Numerous slender muscular filaments supply the Intercostals, the Infracostales, the Levatores costarum, Serratus posticus superior, and Triangularis sterni muscles. Some of these branches, at the front of the chest, cross the costal cartilages from one to another intercostal space.

Lateral cutaneous nerves.—These are derived from the intercostal nerves, midway between the vertebræ and sternum ; they pierce the External intercostal and Serratus magnus muscles, and divide into two branches, anterior and posterior.

The *anterior branches* are reflected forwards to the side and the fore part of the chest, supplying the integument of the chest and mamma ; those of the fifth and sixth nerves supply the upper digitations of the External oblique.

The *posterior branches* are reflected backwards, to supply the integument over the scapula and Latissimus dorsi.

The lateral cutaneous branch of the second intercostal nerve is of large size, and does not divide, like the other nerves, into an anterior and a posterior

FIG. 629.—Superficial and deep distribution of the posterior divisions of the spinal nerves (after Hirschfeld and Leveillé). On the left side the cutaneous branches are represented lying on the superficial layer of muscles. On the right side the superficial muscles have been removed, the Splenius capitis and Complexus divided in the neck, and the Erector spinæ divided and partly removed in the back, so as to expose the posterior divisions of the spinal nerves near their origin.

a, a. Lesser occipital nerve from the cervical plexus. 1. External muscular branches of the first cervical nerve, and union by a loop with the second. 2. Placed on the Rectus capitis posticus major muscle, marks the great occipital nerve (2′), passing round the short muscles and piercing the Complexus : the external branch is seen to the outside. 3. External branch from the posterior division of the third nerve. 3′. Its internal branch, sometimes called the third occipital. 4′ to 8′. The internal branches of the several corresponding nerves on the left side. The external branches of these nerves, proceeding to muscles, are displayed on the right side. *d* 1 to *d* 6 and thence to *d* 12. External muscular branches of the posterior divisions of the 12 dorsal nerves on the right side. *d* 1′ to *d* 6′. The internal cutaneous branches of the six upper dorsal nerves on the left side. *d* 7′ to *d* 12′. Cutaneous twigs from the external branches of the six lower dorsal nerves. *l, l.* External branches from the posterior divisions of several lumbar nerves on the right side, piercing the muscles, the lower descending over the gluteal region. *l′ l′.* The same, more superficially, on the left side. *s, s.* The issue and union by loops of the posterior divisions of four sacral nerves on the right side. *s′, s′.* Some of those distributed to the skin on the left side.

branch. It is named, from its origin and distribution, the *intercosto-humeral nerve* (fig. 627). It pierces the External intercostal muscle and the Serratus magnus, crosses the axilla to the inner side of the arm, and joins with a filament from the nerve of Wrisberg. It then pierces the fascia, and supplies the skin of the upper half of the inner and back part of the arm, communicating with the internal cutaneous branch of the musculo-spiral nerve. The size of this nerve is in inverse proportion to the size of the other cutaneous nerves, especially the nerve of Wrisberg. A second intercosto-humeral nerve is frequently given off from the third intercostal. It supplies filaments to the armpit and inner side of the arm.

The lower dorsal nerves.—The anterior divisions of the seventh, eighth, ninth, tenth, and eleventh dorsal nerves are continued anteriorly from the intercostal spaces into the abdominal wall; hence these nerves are named *thoracico-abdominal intercostal nerves*. They have the same arrangement as the upper ones as far as the anterior extremities of the intercostal spaces, where they pass behind the costal cartilages, and between the Internal oblique and Transversalis muscles, to the sheath of the Rectus, which they perforate. They supply the Rectus muscle, and terminate in branches which become subcutaneous near the linea alba. These branches are named the anterior cutaneous nerves of the abdomen. They are directed outwards as far as the lateral cutaneous nerves, supplying the integument of the front of the belly. The lower intercostal nerves supply the Intercostal and Abdominal muscles—the last three giving branches to the Serratus posticus inferior—and, about the middle of their course, give off lateral cutaneous branches, which pierce the External intercostal and External oblique muscles, in the same line as the lateral cutaneous nerves of the thorax, and divide into anterior and posterior branches, which are distributed to the integument of the abdomen and back; the anterior branches supply the digitations of the External oblique muscle, and extend downwards and forwards nearly as far as the margin of the Rectus : the posterior branches pass backwards to supply the skin over the Latissimus dorsi.

The anterior division of the **last dorsal** is larger than the others ; it runs along the lower border of the last rib, often gives a communicating branch to the first lumbar nerve, and passes under the external arcuate ligament of the Diaphragm. It then runs in front of the Quadratus lumborum, perforates the Transversalis, and passes forwards between it and the Internal oblique, to be distributed in the same manner as the lower intercostal nerves. It communicates with the ilio-hypogastric branch of the lumbar plexus, and gives a branch to the Pyramidalis muscle.

The *lateral cutaneous branch of the last dorsal* is remarkable for its large size. It does not divide into an anterior and posterior branch like the lateral cutaneous branches of the intercostal nerves, but perforates the Internal and External oblique muscles, passes downwards over the crest of the ilium in front of the iliac branch of the ilio-hypogastric (fig. 636), and is distributed to the integument of the front part of the gluteal region, some of its filaments extending as low down as the trochanter major.

Surgical Anatomy.—The lower seven intercostal nerves and the ilio-hypogastric from the first lumbar nerve, supply the skin of the abdominal wall. They run downwards and inwards fairly equidistant from each other. The sixth and seventh supply the skin over the 'pit of the stomach'; the eighth corresponds to about the position of the middle linea transversa; the tenth to the umbilicus; and the ilio-hypogastric supplies the skin over the pubes and external abdominal ring. There are several points of surgical interest about the distribution of these nerves, and it is important to remember their origin and course, for in many diseases affecting the nerve-trunks at or near the origin, the pain is referred to their peripheral terminations. Thus, in Pott's disease of the spine, children will often be brought to the surgeon suffering from pain in the belly. This is due to the fact that the nerves are irritated at the seat of disease as they issue from the spinal canal. When the irritation is confined to a single pair of nerves, the sensation complained of is often a feeling of constriction, as if a cord were tied round the abdomen, and in these cases the situation of the sense of constriction may serve to localise the disease in the spinal column. In other cases where the bone disease is more extensive, and two or more nerves are involved, a more general, diffused pain in the abdomen is complained of. A similar condition is sometimes present in affections of the cord itself, as in tabes dorsalis.

Again, it must be borne in mind that the same nerves which supply the skin of the abdomen supply also the planes of muscle, which constitute the greater part of the abdominal wall. Hence, it follows that any irritation applied to the peripheral terminations of the cutaneous branches in the skin of the abdomen is immediately followed by reflex contraction of the abdominal muscles. A good practical illustration of this may sometimes be seen in watching two surgeons examine the abdomen of the same patient. One, whose hand is cold, causes an immediate reflex contraction of the abdominal muscles, so that the belly wall becomes rigid and not nearly so suitable for examination; the other, who has taken the precaution to warm his hand, examines the abdomen without exciting any reflex contraction. The supply of both muscles and skin from the same source is of importance in protecting the abdominal viscera from injury. A blow on the abdomen, even of a severe character, will do no injury to the viscera if the muscles are in a condition of firm contraction; whereas in cases where the muscles have been taken unawares, and the blow has been struck while they were in a state of rest, an injury insufficient to produce any lesion of the abdominal wall has been attended with rupture of some of the abdominal contents. The importance, therefore, of immediate reflex contraction upon the receipt of an injury cannot be over-estimated, and the intimate association of the cutaneous and muscular fibres in the same nerve produces a much more immediate response on the part of the muscles to any peripheral stimulation of the cutaneous filaments than would be the case if the two sets of fibres were derived from independent sources.

Again, the nerves supplying the abdominal muscles and skin derived from the lower intercostal nerves are intimately connected with the sympathetic supplying the abdominal viscera, through the lower thoracic ganglia from which the splanchnic nerves are derived. In consequence of this, in laceration of the abdominal viscera and in acute peritonitis, the muscles of the belly wall become firmly contracted, and thus as far as possible preserve the abdominal contents in a condition of rest.

LUMBAR NERVES

The **lumbar nerves** are five in number on each side. The first appears between the first and second lumbar vertebræ, and the last between the fifth lumbar vertebra and the base of the sacrum.

The **roots of the lumbar nerves** are the largest, and their filaments the most numerous, of all the spinal nerves, and they are closely aggregated together upon the lower end of the cord. The anterior roots are smaller than the posterior; but there is not the same disproportion between the roots as in the cervical nerves. The roots of these nerves have a vertical direction, and are of considerable length, more especially the lower ones, since the spinal cord does not extend beyond the lower border of the first, or upper border of the second lumbar vertebra. The roots become joined in the intervertebral foramina; and the nerves, so formed, divide at their exit into two divisions, posterior and anterior.

Posterior Divisions of the Lumbar Nerves

The **posterior divisions of the lumbar nerves** (fig. 629) diminish in size from above downwards; they pass backwards between the transverse processes, and divide into internal and external branches.

The *internal branches*, the smaller, pass inwards close to the articular processes of the vertebræ, and supply the Multifidus spinæ and Interspinales muscles.

The *external branches* supply the Erector spinæ and Intertransverse muscles. From the three upper branches, cutaneous nerves are derived which pierce the aponeurosis of the Latissimus dorsi muscle at the outer border of the Erector spinæ muscle, and descend over the back part of the crest of the ilium, to be distributed to the integument of the gluteal region; some of the filaments passing as far as the trochanter major.

Anterior Divisions of the Lumbar Nerves

The **anterior divisions of the lumbar nerves** increase in size from above downwards. At their origin, they communicate with the lumbar ganglia of the sympathetic by long, slender filaments, which accompany the lumbar arteries round the sides of the bodies of the vertebræ, beneath the Psoas muscle. The nerves pass obliquely outwards behind the Psoas magnus, or between its

3 N

fasciculi, distributing filaments to it and the Quadratus lumborum. The first three and the greater part of the fourth are connected together in this situation by anastomotic loops, and form the *lumbar plexus*. The smaller part of the fourth joins with the fifth to form the *lumbo-sacral cord*, which assists in the formation of the sacral plexus. The fourth nerve is named the *nervus furcalis*, from the fact that it is subdivided between the two plexuses.

Lumbar Plexus (fig. 630)

The **Lumbar plexus** is formed by the loops of communication between the anterior divisions of the first three and the greater part of the fourth lumbar nerves. The plexus is narrow above, and the first lumbar often receives a

FIG. 630.—Plan of the lumbar and sacral plexuses.

branch from the last dorsal nerve; it is broad below, where it is joined to the sacral plexus by the lumbo-sacral cord. It is situated in the substance of the Psoas muscle near its posterior part, in front of the transverse processes of the lumbar vertebræ.

The mode in which the plexus is arranged varies in different subjects. It differs from the brachial plexus in not forming an intricate interlacement, but the several nerves of distribution arise from one or more of the spinal nerves, somewhat in the following manner: the first lumbar nerve frequently receives a branch from the last dorsal, and divides into an upper and lower branch; the upper and larger branch subdivides into the ilio-hypogastric and ilio-inguinal;

the lower and smaller branch unites with a branch of the second lumbar to form the genito-crural nerve. The remainder of the second nerve, and the third and fourth nerves, divide into anterior and posterior divisions. The anterior division of the second unites with the anterior divisions of the third and fourth nerves to form the obturator nerve. The posterior divisions of the second and third nerves divide into two branches, a smaller branch from each uniting to form the external cutaneous nerve, and a larger branch from each joining with the posterior division of the fourth lumbar nerve to form the anterior crural. The accessory obturator, when it exists, is formed by the union of two small branches given off from the third and fourth nerves.*

From this arrangement it follows that the ilio-hypogastric and ilio-inguinal are derived entirely from the first lumbar nerve; the genito-crural from the first

Fig. 631.—The lumbar plexus and its branches.

and second nerves; the external cutaneous from the second and third; the anterior crural and obturator by fibres derived from the second, third, and fourth; and the accessory obturator, when present, from the third and fourth.

* In most cases the fourth lumbar is the *nervus furcalis*; but this arrangement is frequently departed from. The third is occasionally the lowest nerve which enters the lumbar plexus, giving at the same time some fibres to the sacral plexus, and thus forming the nervus furcalis; or both the third and fourth may be furcal nerves. When this occurs, the plexus is termed *high* or *prefixed*. More frequently the fifth nerve is divided between the lumbar and sacral plexuses, and constitutes the nervus furcalis; and when this takes place, the plexus is distinguished as a *low* or *post-fixed* plexus. These variations necessarily produce corresponding modifications in the sacral plexus.

The branches of the lumbar plexus are, the

Ilio-hypogastric.

Ilio-inguinal.

Genito-crural.

Anterior crural.

External cutaneous.

Obturator.

Accessory obturator.

The **Ilio-hypogastric nerve** arises from the first lumbar nerve. It emerges from the outer border of the Psoas muscle at its upper part, and crosses obliquely in front of the Quadratus lumborum to the crest of the ilium. It then perforates the Transversalis muscle at its posterior part, near the crest of the ilium, and divides between it and the Internal oblique into two branches, iliac and hypogastric.

The *iliac branch* pierces the Internal and External oblique muscles immediately above the crest of the ilium, and is distributed to the integument of the gluteal region, behind the lateral cutaneous branch of the last dorsal nerve (fig. 636). The size of this nerve bears an inverse proportion to that of the lateral cutaneous branch of the last dorsal nerve.

The *hypogastric branch* (fig. 632) continues onwards between the Internal oblique and Transversalis muscles. It then pierces the Internal oblique, and becomes cutaneous by perforating the aponeurosis of the External oblique, about an inch above, and a little to the outer side of the external abdominal ring, and is distributed to the integument of the hypogastric region.

The ilio-hypogastric nerve communicates with the last dorsal and ilio-inguinal nerves.

The **Ilio-inguinal nerve,** smaller than the preceding, arises with it from the first lumbar nerve. It emerges from the outer border of the Psoas just below the ilio-hypogastric, and, passing obliquely across the Quadratus lumborum and Iliacus muscles, perforates the Transversalis, near the fore part of the crest of the ilium, and communicates with the ilio-hypogastric nerve between that muscle and the Internal oblique. The nerve then pierces the Internal oblique, distributing filaments to it, and accompanying the spermatic cord through the external abdominal ring, and is distributed to the integument of the upper and inner part of the thigh, to the skin over the root of the penis and upper part of the scrotum in the male, and to the skin covering the mons Veneris and labium majus in the female. The size of this nerve is in inverse proportion to that of the ilio-hypogastric. Occasionally it is very small, and ends by joining the ilio-hypogastric; in such cases, a branch from the ilio-hypogastric takes the place of the ilio-inguinal, or the latter nerve may be altogether absent.

The **Genito-crural nerve** arises from the first and second lumbar nerves. It passes obliquely through the substance of the Psoas, and emerges from its inner border, close to the vertebral column, opposite the disc between the third and fourth lumbar vertebræ; it then descends on the surface of the Psoas muscle, under cover of the peritoneum, and divides into a genital and crural branch.

The *genital branch* passes outwards on the Psoas magnus, and pierces the fascia transversalis, or passes through the internal abdominal ring; it then descends along the back part of the spermatic cord to the scrotum, and supplies, in the male, the Cremaster muscle. In the female, it accompanies the round ligament, and is lost upon it.

The *crural branch* descends on the external iliac artery, sending a few filaments round it, and, passing beneath Poupart's ligament to the thigh, enters the sheath of the femoral vessels, lying superficial and a little external to the femoral artery. It pierces the anterior layer of the sheath of the vessels, and, becoming superficial by passing through the fascia lata, it supplies the skin of the anterior aspect of the thigh as far as midway between the pelvis and knee. On the front of the thigh it communicates with the outer branch of the middle cutaneous nerve, derived from the anterior crural.

A few filaments from this nerve may be traced on to the femoral artery; they are derived from the nerve as it passes beneath Poupart's ligament.

The **External cutaneous nerve** arises from the second and third lumbar nerves. It emerges from the outer border of the Psoas muscle about its middle, and crosses the Iliacus muscle obliquely, towards the anterior superior spine of the ilium. It then passes under Poupart's ligament and over the Sartorius

FIG. 632.—Cutaneous nerves of lower extremity. Front view.

FIG. 633.—Nerves of the lower extremity. Front view.

muscle into the thigh, where it divides into two branches, anterior and posterior.

The *anterior branch* descends in an aponeurotic canal formed in the fascia lata, becomes superficial about four inches below Poupart's ligament, and divides into branches which are distributed to the integument along the anterior and outer part of the thigh, as far down as the knee. The terminal filaments of this nerve frequently communicate with the middle and internal cutaneous, and with the patellar branch of the long saphenous, forming with them the *patellar plexus*.

The *posterior branch* pierces the fascia lata, and subdivides into filaments which pass backwards across the outer and posterior surface of the thigh, supplying the integument from the level of the great trochanter as far as the middle of the thigh.

The **Obturator nerve** supplies the Obturator externus and Adductor muscles of the thigh, the articulations of the hip and knee, and occasionally part of the integument of the thigh and leg. It arises from the second, third, and fourth lumbar nerves. Of these, the branch from the third is the largest, while that from the second is often very small. It descends through the inner fibres of the Psoas muscle, and emerges from its inner border near the brim of the pelvis; it then passes behind the commencement of the external iliac vessels, which separates it from the ureter, and runs along the lateral wall of the pelvis, above the obturator vessels, to the upper part of the obturator foramen, where it enters the thigh, and divides into an anterior and a posterior branch, separated by some of the fibres of the Obturator externus (fig. 449), and lower down by the Adductor brevis muscle.

The *anterior branch* (fig. 633) passes down in front of the Adductor brevis, being covered by the Pectineus and Adductor longus; and at the lower border of the latter muscle communicates with the internal cutaneous and internal saphenous nerves, branches of the anterior crural, forming a kind of plexus. It then descends upon the femoral artery, to which it is finally distributed. The nerve, near the obturator foramen, gives off an articular branch to the hip-joint. Behind the Pectineus, it distributes muscular branches to the Adductor longus and Gracilis, and usually to the Adductor brevis, and in rare cases to the Pectineus, and receives a communicating branch from the accessory obturator nerve when that nerve is present.

Occasionally the communicating branch to the internal cutaneous and internal saphenous nerves is continued down, as a cutaneous branch, to the thigh and leg. When this is so, it emerges from beneath the lower border of the Adductor longus, descends along the posterior margin of the Sartorius to the inner side of the knee, where it pierces the deep fascia, communicates with the long saphenous nerve, and is distributed to the integument of the inner side of the leg as low down as its middle. When this communicating branch is small, its place is supplied by the internal cutaneous nerve.

The *posterior branch of the obturator nerve* pierces the anterior part of the Obturator externus, sending branches to supply this muscle; it then passes behind the Adductor brevis on the front of the Adductor magnus, where it divides into numerous muscular branches, which supply the Adductor magnus, and the Adductor brevis when the latter does not receive a branch from the anterior division of the nerve. It also gives off an articular filament to the knee-joint.

The *articular branch for the knee-joint* is sometimes absent; it either perforates the lower part of the Adductor magnus, or passes through the opening which transmits the femoral artery, and enters the popliteal space; it then descends upon the popliteal artery, as far as the back part of the knee-joint, where it perforates the posterior ligament, and is distributed to the synovial membrane. It gives filaments to the artery in its course.

The **Accessory obturator nerve** (fig. 631) is present in about 29 per cent. of cases. It is of small size, and arises by separate filaments from the third and fourth lumbar nerves. It descends along the inner border of the Psoas muscle, crosses the ascending ramus of the os pubis, and passes under the outer border of the Pectineus muscle, where it divides into numerous branches. One of these supplies the Pectineus, penetrating its under surface; another is distributed to the hip-joint; while a third communicates with the anterior branch of the obturator nerve. When this nerve is absent, the hip-joint receives two branches from the obturator nerve. Occasionally it is very small and becomes lost in the capsule of the hip-joint.

The **Anterior crural nerve** (figs. 631, 633) is the largest branch of the lumbar plexus. It supplies muscular branches to the Iliacus, Pectineus, and all the muscles on the front of the thigh, excepting the Tensor fasciæ femoris; cutaneous filaments to the front and inner side of the thigh, and to the leg and foot; and articular branches to the hip and knee. It arises from the second, third, and fourth lumbar nerves. It descends through the fibres of the Psoas muscle, emerging from it at the lower part of its outer border; and passes down between it and the Iliacus behind the fascia iliaca; it then runs beneath Poupart's ligament, into the thigh, where it becomes somewhat flattened, and divides into an anterior and a posterior part. Under Poupart's ligament, it is separated from the femoral artery by a portion of the Psoas muscle.

Within the abdomen the anterior crural nerve gives off from its outer side some small branches to the Iliacus, and a branch to the femoral artery, which is distributed upon the upper part of that vessel. The origin of this branch varies: it occasionally arises higher than usual, or it may arise lower down in the thigh.

In the thigh the following branches are given off:

From the Anterior Division	From the Posterior Division
Middle cutaneous.	Long saphenous.
Internal cutaneous.	Muscular.
Muscular.	Articular.

The *middle cutaneous nerve* (fig. 632) pierces the fascia lata (and generally the Sartorius) about three inches below Poupart's ligament, and divides into two branches, which descend in immediate proximity along the fore part of the thigh, to supply the integument as low as the front of the knee, where they communicate with the internal cutaneous and the patellar branch of the internal saphenous nerve, to form the patellar plexus. In the upper part of the thigh the outer division of the middle cutaneous communicates with the crural branch of the genito-crural nerve.

The *internal cutaneous nerve* passes obliquely across the upper part of the sheath of the femoral artery, and divides in front, or at the inner side of that vessel, into two branches, an anterior and a posterior or internal.

The *anterior branch* runs downwards on the Sartorius, perforates the fascia lata at the lower third of the thigh, and divides into two branches: one supplies the integument as low down as the inner side of the knee; the other crosses to the outer side of the patella, communicating in its course with the patellar branch of the long saphenous nerve.

The *posterior* or *internal branch* descends along the inner border of the Sartorius muscle to the knee, where it pierces the fascia lata, communicates with the long saphenous nerve, and gives off several cutaneous branches. The nerve then passes down the inner side of the leg, to the integument of which it is distributed. This nerve, beneath the fascia lata, at the lower border of the Adductor longus, joins in a plexiform network (*subsartorial plexus*) by uniting with branches of the long saphenous and obturator nerves (fig. 633). When the communicating branch from the obturator nerve is large and continued to the integument of the leg, the internal branch of the internal cutaneous is small, and terminates in the plexus, occasionally giving off a few cutaneous filaments.

The internal cutaneous nerve, before dividing, gives off a few filaments, which pierce the fascia lata, to supply the integument of the inner side of the thigh, accompanying the long saphenous vein. One of these filaments passes through the saphenous opening; a second becomes subcutaneous about the middle of the thigh; and a third pierces the fascia at its lower third.

Muscular branches of the anterior division.—The *nerve to the Pectineus* arises from the anterior crural immediately below Poupart's ligament, and passes inwards behind the femoral sheath to enter the anterior surface of the muscle; it is often duplicated. The *nerve to the Sartorius* arises in common with the middle cutaneous.

The *long* or *internal saphenous nerve* is the largest of the cutaneous branches of the anterior crural. It approaches the femoral artery where this vessel passes beneath the Sartorius, and lies in front of it, beneath the aponeurotic covering of Hunter's canal, as far as the opening in the lower part of the Adductor magnus. It then quits the artery, and descends vertically along the inner side of the knee beneath the Sartorius, pierces the fascia lata, opposite the interval between the tendons of the Sartorius and Gracilis, and becomes subcutaneous. The nerve then passes along the inner side of the leg, accompanied by the internal saphenous vein, descends behind the internal border of the tibia, and, at the lower third of the leg, divides into two branches : one continues its course along the margin of the tibia, terminating at the inner ankle ; the other passes in front of the ankle, and is distributed to the integument along the inner side of the foot, as far as the ball of the great toe, communicating with the internal branch of the musculo-cutaneous nerve.

Branches.—The long saphenous nerve, about the middle of the thigh, gives off a communicating branch which joins the subsartorial plexus.

At the inner side of the knee it gives off a large *patellar branch* (*nervus cutaneus patellæ*) which pierces the Sartorius and fascia lata, and is distributed to the integument in front of the patella. This nerve communicates above the knee with the anterior branch of the internal cutaneous and with the middle cutaneous ; below the knee, with other branches of the long saphenous ; and, on the outer side of the joint, with branches of the external cutaneous nerve, forming a plexiform network, the *plexus patellæ*. The patellar branch is occasionally small, and terminates by joining the internal cutaneous, which supplies its place in front of the knee.

Below the knee, the branches of the long saphenous nerve are distributed to the integument of the front and inner side of the leg, communicating with the cutaneous branches from the internal cutaneous, or from the obturator nerve.

The *muscular branches of the posterior division* supply the four parts of the Quadriceps extensor muscle.

The *branch to the Rectus muscle* enters its under surface high up, sending off a small filament to the hip-joint.

The *branch to the Vastus externus,* of large size, follows the course of the descending branch of the external circumflex artery to the lower part of the muscle. It gives off an articular filament to the knee-joint.

The *branch to the Vastus internus* is a long branch which runs down on the outer side of the femoral vessels in company with the internal saphenous nerve. It enters the muscle about its middle, and gives off a filament, which can usually be traced downwards on the surface of the muscle to the knee-joint.

The *branches to the Crureus* are two or three in number, and enter the muscle on its anterior surface about the middle of the thigh ; a filament from one of these descends through the muscle to the Subcrureus and the knee-joint.

The *articular branch to the hip-joint* is derived from the nerve to the Rectus.

The *articular branches to the knee-joint* are three in number. One, a long, slender filament, is derived from the nerve to the Vastus externus ; it penetrates the capsular ligament of the joint on its anterior aspect. Another is derived from the nerve to the Vastus internus. It can usually be traced downwards on the surface of this muscle to near the joint ; it then penetrates the muscular fibres, and accompanies the deep branch of the anastomotica magna artery, pierces the capsular ligament of the joint on its inner side, and supplies the synovial membrane. The third branch is derived from the nerve to the Crureus.

THE SACRAL AND COCCYGEAL NERVES

The *sacral nerves* are five in number on each side. The four upper ones pass from the sacral canal, through the sacral foramina ; the fifth through the foramen between the sacrum and coccyx.

The *roots of the upper sacral nerves* are the largest of all the spinal nerves ; while those of the lowest sacral and coccygeal nerve are the smallest. They are longer than those of any of the other spinal nerves, on account of the spinal cord not extending beyond the first or second lumbar vertebra. From their great length, and the appearance they present in connection with their attachment to the spinal cord, the roots of origin of these nerves are called collectively the *cauda equina*.

Each sacral and coccygeal nerve separates into two divisions, posterior and anterior.

The **posterior divisions of the sacral nerves** (fig. 634) are small, diminish in size from above downwards, and emerge, except the last, from the sacral canal by the posterior sacral foramina.

The *three upper ones* are covered, at their exit from the sacral canal, by the Multifidus spinæ, and divide into internal and external branches.

The *internal branches* are small, and supply the Multifidus spinæ.

The *external branches* join with one another, and with the last lumbar and fourth sacral nerves, in the form of loops on the posterior surface of the sacrum. From these loops branches pass to the outer surface of the great sacro-sciatic ligament, where they form a second series of loops beneath the Gluteus maximus.

FIG. 634.—The posterior sacral nerves.

ANTERIOR BRANCHES
OF LOWER SACRAL NERVES

Cutaneous branches from this second series of loops, usually two or three in number, pierce the Gluteus maximus along a line drawn from the posterior superior spine of the ilium to the tip of the coccyx. They supply the integument over the posterior part of the gluteal region.

The *posterior divisions of the two lower sacral nerves* are situated below the Multifidus spinæ. They are of small size, and do not divide into internal and external branches, but join with each other, and with the coccygeal nerve, so as to form loops on the back of the sacrum, filaments from which supply the integument over the coccyx.

The **coccygeal nerve** divides into its anterior and posterior division in the spinal canal. The *posterior division* is the smaller. It does not divide, but receives, as already mentioned, a communicating branch from the last sacral, and is lost in the integument over the back of the coccyx.

The **anterior divisions of the sacral nerves** diminish in size from above downwards. The four upper ones emerge from the anterior sacral foramina: the anterior division of the fifth, after emerging from the spinal canal through its terminal opening, curves forwards between the sacrum and the coccyx. All the anterior sacral nerves communicate with the sacral ganglia of the sympathetic, at their exit from the sacral foramina. The *first nerve*, of large size, unites with the lumbo-sacral cord, formed by the fifth lumbar and a branch from the fourth. The *second*, equal in size to the preceding, and the *third*, about one-fourth the size of the second, unite with this trunk, and form, with a small fasciculus from the fourth, the *sacral plexus*; a visceral branch is given off from the third nerve to the bladder.

The **anterior primary division of the fourth sacral nerve** sends a branch to join the pudic nerve, and a communicating filament to join the fifth sacral nerve. The remaining portion of the nerve divides into visceral and muscular branches. The *visceral branches* are distributed to the viscera of the pelvis, communicating with the sympathetic nerve. These branches ascend upon the rectum and bladder, and in the female upon the vagina, communicating with branches of the sympathetic from the pelvic plexus. The *muscular branches* are distributed to the Levator ani, Coccygeus, and Sphincter ani externus. The branch to the Sphincter ani (*perineal branch of fourth sacral*) pierces the Coccygeus muscle, or passes between this muscle and the Levator ani, so as to reach the ischio-rectal fossa, where it is found lying close to the coccyx. Cutaneous filaments arise from this branch, which supply the integument between the anus and coccyx. Another cutaneous branch is frequently given off from this nerve, though sometimes from the pudic (Schwalbe). It perforates the great sacro-sciatic ligament, and, winding round the lower border of the Gluteus maximus, supplies the skin over the lower and inner part of this muscle.

The **anterior primary division of the fifth sacral nerve,** after passing from the lower end of the sacral canal, curves forwards through the fifth sacral foramen, formed between the lower part of the sacrum and the transverse process of the first piece of the coccyx. It pierces the Coccygeus muscle, and descends upon its anterior surface to near the tip of the coccyx, where it again perforates the muscle, to be distributed to the integument over the back part and side of the coccyx. This nerve communicates above with the fourth sacral, and below with the coccygeal nerve, and supplies the Coccygeus muscle.

The **anterior division of the coccygeal nerve** is a delicate filament which escapes at the termination of the sacral canal; it passes downwards behind the rudimentary transverse process of the first piece of the coccyx, and curves forwards, through the notch between the first and second pieces, piercing the Coccygeus muscle and descending on its anterior surface to near the tip of the coccyx, where it again pierces the muscle, to be distributed to the integument over the back part and side of the coccyx. It is joined by a branch from the fifth sacral as it descends on the surface of the Coccygeus muscle.

SACRAL PLEXUS (figs. 630, 635)

The **Sacral plexus** is formed by the lumbo-sacral cord, the anterior divisions of the three upper sacral nerves, and part of that of the fourth. These nerves proceed in different directions: the upper ones obliquely downwards and outwards, the lower one nearly horizontally, and they all unite into two bands: an *upper* and larger which is formed by the lumbo-sacral cord with the first, and the greater parts of the second and third sacral nerves, and is termed the *sciatic band*; and a *lower* and smaller formed by branches from the second, third, and fourth sacral nerves, and is named the *pudic band*. The upper band is prolonged into the great sciatic nerve, and the lower into the pudic. Each nerve which enters the plexus divides into an anterior or ventral, and a posterior or dorsal division, and the branches which arise from the plexus may be arranged into two groups, according to their origin from these divisions.

The sacral plexus is triangular in form, its base corresponding with the exit of the nerves from the sacrum, its apex with the lower part of the great sacro-sciatic foramen. It rests upon the anterior surface of the Pyriformis, and is covered in front by the pelvic fascia, which separates it from the sciatic and pudic branches of the internal iliac artery, and from the viscera of the pelvis.

The branches of the sacral plexus are:

Muscular.	Small sciatic.
Superior gluteal.	Perforating cutaneous.
Inferior gluteal.	Pudic.

Great sciatic.

The **Muscular branches** supply the Pyriformis, Obturator internus, the two Gemelli, and the Quadratus femoris. The branch to the Pyriformis arises from the posterior divisions of the first and second sacral nerves; the branch to the Obturator internus arises from the anterior branches of the fifth lumbar and the first and second sacral nerves: it passes out of the pelvis through the great sacro-sciatic foramen below the Pyriformis, crosses the spine of the ischium, and re-enters the pelvis through the lesser sacro-sciatic foramen to enter the inner surface of the Obturator internus; the branch to the Gemellus

Fig. 635.—Side view of pelvis, showing sacral nerves.

superior arises in common with the nerve to the Obturator internus: it enters the muscle at the upper part of its posterior surface; the small branch to the Gemellus inferior and Quadratus femoris also arises from the anterior branches of the fifth lumbar and the first and second sacral nerves: it passes through the great sacro-sciatic foramen below the Pyriformis, and courses down beneath the great sciatic nerve, the Gemelli and tendon of the Obturator internus, and supplies the muscles on their deep or anterior surface. It gives off an articular branch to the hip-joint. A second articular branch is occasionally derived from the upper part of the sacral plexus.

The **Superior gluteal nerve** (fig. 637) arises from the posterior branches of the fourth and fifth lumbar and first sacral nerves: it passes from the pelvis through the great sacro-sciatic foramen above the Pyriformis muscle, accompanied by the gluteal vessels, and divides into a superior and an inferior branch.

The *superior branch* follows the line of origin of the Gluteus minimus, in company with the superior branch of the deep division of the gluteal artery, and supplies the Gluteus medius.

The *inferior branch* crosses obliquely between the Gluteus minimus and medius, distributing filaments to both these muscles, and terminates in the Tensor fasciæ femoris. It accompanies the inferior branch of the deep division of the gluteal artery.

The **Inferior gluteal** arises from the posterior branch of the fifth lumbar and first two sacral nerves, and is intimately connected with the small sciatic at its origin. It passes out of the pelvis through the great sciatic foramen, beneath the Pyriformis muscle, and divides into a number of branches which enter the Gluteus maximus muscle, on its deep surface.

The **Small sciatic nerve** (fig. 637) supplies the integument of the perinæum and back part of the thigh and leg. It is usually formed by the union of two branches, which arise from the second and third nerves of the sacral plexus. It issues from the pelvis through the great sacro-sciatic foramen below the Pyriformis muscle, descends beneath the Gluteus maximus with the sciatic artery, and at the lower border of that muscle passes along the back part of the thigh, beneath the fascia lata and over the long head of the Biceps, to the lower part of the popliteal region, where it pierces the fascia and becomes cutaneous. It then accompanies the external saphenous vein to about the middle of the leg, its terminal filaments communicating with the external saphenous nerve.

The branches of the small sciatic nerve are all cutaneous, and are grouped as follows : gluteal, perineal, and femoral.

The *gluteal cutaneous branches* (*ascending*) consist of two or three filaments, which turn upwards round the lower border of the Gluteus maximus to supply the integument covering the lower and outer part of that muscle.

The *perineal cutaneous branches* are distributed to the skin at the upper and inner side of the thigh, on its posterior aspect. One branch, longer than the rest, the *inferior pudendal*, curves forwards below the tuber ischii, pierces the fascia lata, and passes forwards beneath the superficial fascia of the perinæum to be distributed to the integument of the scrotum in the male and the labium in the female, communicating with the superficial perineal and inferior hæmorrhoidal nerves.

The *femoral cutaneous branches* (*descending*) are numerous filaments, derived from both sides of the nerves, which are distributed to the back and inner and outer sides of the thigh, to the skin covering the popliteal space, and to the upper part of the leg.

The **Perforating cutaneous nerve** usually arises from the posterior aspect of the second and third sacral nerves, and is of small size. It is continued backwards through the great sacro-sciatic ligament, and, winding round the lower border of the Gluteus maximus, supplies the integument covering the inner and lower part of that muscle.

The **Pudic nerve** is the direct continuation of the pudic band of the sacral plexus, and derives its fibres from the anterior branches of the second, third, and fourth sacral nerves. It leaves the pelvis, through the great sacro-sciatic foramen, below the Pyriformis. It then crosses the spine of the ischium, and re-enters the pelvis through the lesser sacro-sciatic foramen. It accompanies the pudic vessels upwards and forwards along the outer wall of the ischio-rectal fossa, being contained in a sheath of the obturator fascia, termed *Alcock's canal*, and divides into two terminal branches, viz. the perineal nerve, and the dorsal nerve of the penis or clitoris. Before its division, it gives off the inferior hæmorrhoidal nerve.

The *inferior hæmorrhoidal nerve* is occasionally derived separately from the sacral plexus. It passes across the ischio-rectal fossa, with its accompanying vessels, towards the lower end of the rectum, and is distributed to the Sphincter ani externus and to the integument round the anus. Branches of this nerve communicate with the inferior pudendal and superficial perineal nerves at the fore part of the perinæum.

The *perineal nerve*, the inferior and larger of the two terminal branches of the pudic, is situated below the pudic artery. It accompanies the superficial perineal artery in the perinæum, dividing into cutaneous and muscular branches.

The cutaneous branches (superficial perineal) are two in number, posterior

FIG. 636.—Cutaneous nerves of lower extremity. Posterior view.

FIG. 637.—Nerves of the lower extremity.* Posterior view.

* N.B.—In this diagram the communicans tibialis and communicans peronei are not in their normal position. They have been displaced by the removal of the superficial muscles.

and anterior. The *posterior* or *external branch* pierces the base of the triangular ligament of the urethra, and passes forwards along the outer side of the urethral triangle in company with the superficial perineal artery; it is distributed to the skin of the scrotum. It communicates with the inferior hæmorrhoidal, the inferior pudendal, and the other superficial perineal nerve. The *anterior* or *internal branch* also pierces the base of the triangular ligament, and passes forwards nearer to the middle line, to be distributed to the inner and back part of the scrotum. Both these nerves supply the labia majora in the female.

The muscular branches are distributed to the Transversus perinæi, Accelerator urinæ, Erector penis, and Compressor urethræ. A distinct branch is given off from the nerve to the Accelerator urinæ, which pierces this muscle, and supplies the corpus spongiosum, ending in the mucous membrane of the urethra. This is the *nerve to the bulb.*

The *dorsal nerve of the penis* is the deepest division of the pudic nerve; it accompanies the pudic artery along the ramus of the ischium; it then runs forwards along the inner margin of the ramus of the os pubis, between the superficial and deep layers of the triangular ligament. Piercing the superficial layer it gives a branch to the corpus cavernosum, and passes forwards, in company with the dorsal artery of the penis, between the layers of the suspensory ligament on to the dorsum of the penis, along which it is carried as far as the glans to which it is distributed.

In the female the dorsal nerve is very small, and supplies the clitoris.

The **Great sciatic nerve** (fig. 637) supplies nearly the whole of the integument of the leg, the muscles of the back of the thigh, and those of the leg and foot. It is the largest nervous cord in the body, measuring three-quarters of an inch in breadth, and is the continuation of the upper or sciatic band of the sacral plexus. It passes out of the pelvis through the great sacro-sciatic foramen, below the Pyriformis muscle. It descends between the trochanter major and tuberosity of the ischium, along the back part of the thigh to about its lower third, where it divides into two large branches, the *internal* and *external popliteal nerves.*

This division may take place at any point between the sacral plexus and the lower third of the thigh. When the division occurs at the plexus, the external popliteal nerve usually pierces the Pyriformis muscle. As the nerve descends along the back of the thigh, it rests upon the posterior surface of the ischium, the nerve to the Quadratus femoris, and the External rotator muscles of the thigh, in company with the small sciatic nerve and artery, being covered by the Gluteus maximus; lower down, it lies upon the Adductor magnus, and is covered by the long head of the Biceps.

The *branches* of the nerve, before its division, are articular and muscular.

The *articular branches* arise from the upper part of the nerve; they supply the hip-joint, perforating the posterior part of its fibrous capsule. These branches are sometimes derived from the sacral plexus.

The *muscular branches* are distributed to the flexors of the leg: viz. the Biceps, Semitendinosus, and Semimembranosus, and to the Adductor magnus. The nerve to the short head of the Biceps comes from the external popliteal part of the great sciatic, while the other muscular branches arise from the internal popliteal portion, as may be seen in those cases where the two popliteal nerves emerge separately on the buttock.

The **Internal popliteal nerve,** the larger of the two terminal branches of the great sciatic, arises from the anterior branches of the last two lumbar and first three sacral nerves. It descends along the back part of the thigh, through the middle of the popliteal space to the lower part of the Popliteus muscle, where it passes with the popliteal artery beneath the arch of the Soleus, and becomes the posterior tibial. It is overlapped by the hamstring muscles above, and then becomes more superficial, and lies to the outer side of, and some distance from, the popliteal vessels; opposite the knee-joint, it is in close relation with the vessels, and crosses to the inner side of the artery. Below, it is overlapped by the Gastrocnemius.

The *branches* of this nerve are, articular, muscular, and a cutaneous branch, the *communicans tibialis nerve.*

The *articular branches,* usually three in number, supply the knee-joint; two of these accompany the superior and inferior internal articular arteries; and a third, the azygos articular artery.

The *muscular branches*, four or five in number, arise from the nerve as it lies between the two heads of the Gastrocnemius muscle ; they supply that muscle, the Plantaris, Soleus, and Popliteus. The branch which supplies the Popliteus turns round its lower border and is distributed to its deep surface.

The *communicans tibialis* descends between the two heads of the Gastrocnemius muscle, and, about the middle of the back of the leg, pierces the deep fascia, and joins a communicating branch (*communicans peronei*) from the external popliteal nerve to form the external, or short, saphenous (fig. 636).

The *external saphenous nerve*, formed by the communicating branches of the internal and external popliteal nerves, passes downwards and outwards near the outer margin of the tendo Achillis, lying close to the external saphenous vein, to the interval between the external malleolus and the os calcis. It winds round the outer malleolus, and is distributed to the integument along the outer side of the foot and little toe, communicating on the dorsum of the foot with the musculo-cutaneous nerve. In the leg, its branches communicate with those of the small sciatic.

The **Posterior tibial nerve** (fig. 637) commences at the lower border of the Popliteus muscle, and passes along the back part of the leg with the posterior tibial vessels to the interval between the inner malleolus and the heel, where it divides beneath the internal annular ligament of the ankle into the *external* and *internal plantar nerves*. It lies upon the deep muscles of the leg, and is covered in the upper part by the muscles of the calf, lower down by the skin, the superficial and deep fasciæ. In the upper part of its course, it lies to the inner side of the posterior tibial artery; but it soon crosses that vessel, and lies to its outer side as far as the ankle. In the lower third of the leg, it is placed parallel with the inner margin of the tendo Achillis.

FIG. 638.—The plantar nerves.

The *branches of the posterior tibial nerve* are muscular, calcaneo-plantar, and articular.

The *muscular branches* arise either separately or by a common trunk from the upper part of the nerve. They supply the Soleus, Tibialis posticus, Flexor longus digitorum, and Flexor longus hallucis muscles ; the branch to the latter muscle accompanying the peroneal artery. The branch from the posterior tibial nerve to the Soleus enters its deep surface, while the branch which this muscle receives from the internal popliteal enters its superficial aspect.

The *calcaneo-plantar* (*internal calcanean*) *branch* perforates the internal annular ligament, and supplies the integument of the heel and inner side of the sole of the foot.

The *articular branch* is given off just above the bifurcation of the nerve, and supplies the ankle-joint.

The **internal plantar nerve** (fig. 638), the larger of the two terminal branches of the posterior tibial, accompanies the internal plantar artery along the inner side of the foot. From its origin at the inner ankle it passes beneath the Abductor hallucis, and then forwards between this muscle and the Flexor brevis digitorum ; it divides opposite the bases of the metatarsal bones into four digital branches, and communicates with the external plantar nerve.

Branches.—In its course, the internal plantar nerve gives off *cutaneous branches*, which pierce the plantar fascia, and supply the integument of the sole of the foot ; *muscular branches*, which supply the Abductor hallucis and

Flexor brevis digitorum; *articular branches* to the articulations of the tarsus and metatarsus; and *four digital branches.* The three outer branches pass between the divisions of the plantar fascia in the clefts between the toes : the first (innermost) branch becomes cutaneous about the middle of the sole, between the Abductor hallucis and Flexor brevis digitorum. They are distributed in the following manner : the *first* supplies the inner border of the great toe, and sends a filament to the Flexor brevis hallucis muscle ; the *second* bifurcates, to supply the adjacent sides of the great and second toes, sending a filament to the first Lumbricalis muscle ; * the *third digital branch* supplies the adjacent sides of the second and third toes ; the *fourth* supplies the corresponding sides of the third and fourth toes, and receives a communicating branch from the external plantar nerve. Each digital nerve gives off cutaneous and articular filaments ; and opposite the last phalanx sends a dorsal branch, which supplies the structures around the nail, the continuation of the nerve being distributed to the ball of the toe. It will be observed, that the distribution of these digital branches is precisely similar to those of the median nerve in the hand.

The **external plantar nerve,** the smaller of the two, completes the nervous supply to the structures of the sole of the foot, being distributed to the little toe and one-half of the fourth, as well as to most of the deep muscles, its distribution being similar to that of the ulnar in the hand. It passes obliquely forwards with the external plantar artery to the outer side of the foot, lying between the Flexor brevis digitorum and Flexor accessorius ; and, in the interval between the former muscle and Abductor minimi digiti, divides into a superficial and a deep branch. Before its division, it supplies the Flexor accessorius and Abductor minimi digiti.

The *superficial branch* separates into two digital nerves : one, the smaller of the two, supplies the outer side of the little toe, the Flexor brevis minimi digiti, and the two Interossei muscles of the fourth metatarsal space ; the other and larger digital branch supplies the adjoining sides of the fourth and fifth toes, and communicates with the internal plantar nerve.

The *deep* or *muscular branch* accompanies the external plantar artery into the deep part of the sole of the foot, beneath the tendons of the Flexor muscles and Adductor obliquus hallucis, and supplies all the Interossei (except those in the fourth metatarsal space), the three outer Lumbricales, the Adductor obliquus hallucis, and the Adductor transversus hallucis.

The **External popliteal** or **peroneal nerve** (fig. 637), about one-half the size of the internal popliteal, arises from the dorsal branches of the last two lumbar and first two sacral nerves. It descends obliquely along the outer side of the popliteal space to the head of the fibula, close to the inner margin of the Biceps muscle. It is easily felt beneath the skin behind the head of the fibula, at the inner side of the tendon of the Biceps. It lies between the tendon of the Biceps and outer head of the Gastrocnemius muscle, winds round the neck of the fibula, between the Peroneus longus and the bone, and divides beneath the muscle into the anterior tibial and musculo-cutaneous nerves.

The *branches of the external popliteal nerve,* previous to its division, are articular and cutaneous.

The *articular branches* are three in number ; two of these accompany the superior and inferior external articular arteries to the outer side of the knee. The upper one occasionally arises from the great sciatic nerve before its bifurcation. The third (*recurrent*) articular nerve is given off at the point of division of the external popliteal nerve ; it ascends with the anterior recurrent tibial artery through the Tibialis anticus muscle to the front of the knee, which it supplies.

The *cutaneous branches,* two or three in number, supply the integument along the back part and outer side of the leg ; one of these, larger than the rest, the *communicans peronei,* arises near the head of the fibula, crosses the external head of the Gastrocnemius to the middle of the leg, and joins with the communicans tibialis to form the external saphenous. This nerve occasionally exists as a separate branch, which is continued down as far as the heel.

The **Anterior tibial nerve** (fig. 633) commences at the bifurcation of the peroneal nerve, between the fibula and upper part of the Peroneus longus, passes obliquely forwards beneath the Extensor longus digitorum to the fore part of the

---

* See footnote, page 573.

interosseous membrane, and comes into relation with the anterior tibial artery above the middle of the leg ; it then descends with the artery to the front of the ankle-joint, where it divides into an external and an internal branch. This nerve lies at first on the outer side of the anterior tibial artery, then in front of it, and again at its outer side at the ankle-joint.

The *branches of the anterior tibial nerve,* in its course through the leg, are muscular to the Tibialis anticus, Extensor longus digitorum, Peroneus tertius, and Extensor proprius hallucis muscles, and an *articular branch* to the ankle-joint.

The *external terminal branch* passes outwards across the tarsus, beneath the Extensor brevis digitorum, and, having become enlarged like the posterior interosseous nerve at the wrist, supplies the Extensor brevis digitorum. From the enlargement three minute *interosseous branches* are given off, which supply the tarsal joints and the metatarso-phalangeal joints of the second, third, and fourth toes. The first of these sends a filament to the second dorsal interosseous muscle.

The *internal branch,* the continuation of the nerve, accompanies the dorsalis pedis artery along the inner side of the dorsum of the foot, and, at the first interosseous space, divides into two branches, which supply the adjacent sides of the great and second toes, communicating with the innermost branch of the musculo-cutaneous nerve. Before it divides it gives off an *interosseous branch* to the first space, which supplies the metatarso-phalangeal joint of the great toe and sends a filament to the First dorsal interosseous muscle.

The **Musculo-cutaneous nerve** (fig. 633) supplies the muscles on the fibular side of the leg, and the integument over the greater part of the dorsum of the foot. It passes forwards between the Peronei muscles and the Extensor longus digitorum, pierces the deep fascia at the lower third of the leg, on its front and outer side, and divides into two branches. This nerve, in its course between the muscles, gives off muscular branches to the Peroneus longus and brevis, and cutaneous filaments to the integument of the lower part of the leg.

The *internal branch of the musculo-cutaneous nerve* passes in front of the ankle-joint, and divides into two branches, one of which supplies the inner side of the great toe, the other, the adjacent sides of the second and third toes. It also supplies the integument of the inner ankle and inner side of the foot, communicating with the internal saphenous nerve, and joining with the anterior tibial nerve, between the great and second toes.

The *external branch,* the smaller, passes along the outer side of the dorsum of the foot, and divides into two branches, the inner being distributed to the contiguous sides of the third and fourth toes, the outer to the opposed sides of the fourth and fifth toes. It also supplies the integument of the outer ankle and outer side of the foot, communicating with the short saphenous nerve.

The branches of the musculo-cutaneous nerve supply all the toes excepting the outer side of the little toe, and the adjoining sides of the great and second toes, the former being supplied by the external saphenous, and the latter by the internal branch of the anterior tibial. It frequently happens, however, that some of the outer branches of the musculo-cutaneous are absent, their place being then taken by branches of the external saphenous nerve.

Surgical Anatomy.—The lumbar plexus passes through the Psoas muscle, and therefore in psoas abscess, any or all of its branches may be irritated, causing severe pain in the part to which the irritated nerves are distributed. The genito-crural nerve is the one which is most frequently implicated. This nerve is also of importance as it is concerned in one of the principal reflexes employed in the investigation of diseases of the spine. If the skin over the inner side of the thigh just below Poupart's ligament, the part supplied by the crural branch of the genito-crural nerve, be gently tickled in a male child, the testicle will be noticed to be drawn upwards, through the action of the Cremaster muscle, supplied by the genital branch of the same nerve. The same result may sometimes be noticed in adults, and can almost always be produced by severe stimulation. This reflex, when present, shows that the portion of the cord from which the first and second lumbar nerves are derived is in a normal condition.

The anterior crural nerve is in danger of being injured in fractures of the true pelvis, since the fracture most commonly takes place through the ascending ramus of the os pubis, at or near the point where this nerve crosses the bone. It is also liable to be injured in fractures and dislocations of the femur, and is likely to be pressed upon, and its functions impaired, in some tumours growing in the pelvis. Moreover, on account of

3 O

its superficial position, it is exposed to injury in wounds and stabs in the groin. When this nerve is paralysed, the patient is unable to flex his hip completely, on account of the loss of motion in the Iliacus; or to extend the knee on the thigh, on account of paralysis of the Quadriceps extensor cruris; there is complete paralysis of the Sartorius, and partial paralysis of the Pectineus. There is loss of sensation down the front and inner side of the thigh, except in that part supplied by the crural branch of the genito-crural, and by the ilio-inguinal. There is also loss of sensation down the inner side of the leg and foot as far as the ball of the great toe.

The obturator nerve is of special surgical interest. It is rarely paralysed alone, but occasionally in association with the anterior crural. The principal interest attached to it is in connection with its supply to the knee; pain in the knee being symptomatic of many diseases in which the trunk of this nerve, or one of its branches, is irritated. Thus it is well known that in the earlier stages of hip-joint disease the patient does not complain of pain in that articulation, but on the inner side of the knee, or in the knee-joint itself, both these articulations being supplied by the obturator nerve, the final distribution of the nerve being to the knee-joint. Again, the same thing occurs in sacro-iliac disease: pain is complained of in the knee-joint, or on its inner side. The obturator nerve is in close relationship with the sacro-iliac articulation, passing over it, and, according to some anatomists, distributing filaments to it. Further, in cancer of the sigmoid flexure, and even in cases where masses of hardened fæces are impacted in this portion of the gut, pain is complained of in the knee. The left obturator nerve lies beneath the sigmoid flexure, and is readily pressed upon and irritated when disease exists in this part of the intestine. Finally, pain in the knee forms an important diagnostic sign in obturator hernia. The hernial protrusion as it passes out through the opening in the obturator membrane presses upon the nerve and causes pain in the parts supplied by its peripheral filaments. When the obturator nerve is paralysed, the patient is unable to press his knees together or to cross one leg over the other, on account of paralysis of the Adductor muscles. Rotation outwards of the thigh is impaired from paralysis of the Obturator externus. Sometimes there is loss of sensation in the upper half of the inner side of the thigh.

The great sciatic nerve is liable to be pressed upon by various forms of pelvic tumours, giving rise to pain along its trunk, to which the term sciatica is applied. Tumours growing from the pelvic viscera, or bones, aneurisms of some of the branches of the internal iliac artery, calculus in the bladder when of large size, accumulation of fæces in the rectum, may all cause pressure on the nerve inside the pelvis, and give rise to sciatica. Outside the pelvis exposure to cold, violent movements of the hip-joint, exostoses or other tumours growing from the margin of the sacro-sciatic foramen, may also give rise to the same condition. When paralysed there is loss of motion in all the muscles below the knee, and loss of sensation in the same situation, except the upper half of the back of the leg, which is supplied by the small sciatic, and in the upper half of the inner side of the leg, when the communicating branch of the obturator is large (see page 918).

The great sciatic nerve has been frequently cut down upon and stretched, or has been acupunctured for the relief of sciatica. The nerve has also been stretched in cases of locomotor ataxy, the anæsthesia of leprosy, &c. In order to define it on the surface, a point is taken at the junction of the middle and lower third of a line stretching from the posterior superior spine of the ilium to the outer part of the tuber ischii, and a line drawn from this to the middle of the upper part of the popliteal space. The line must be slightly curved with its convexity outwards, and as it passes downwards to the lower border of the Gluteus maximus is slightly nearer the tuber ischii than the great trochanter, as it crosses a line drawn between these two points. The operation of stretching the sciatic nerve is performed by making an incision over the course of the nerve about the centre of the thigh. The skin, superficial structures, and deep fascia having been divided, the interval between the inner and outer hamstrings is to be defined, and these muscles pulled inwards and outwards with retractors. The nerve will be found a little to the inner side of the Biceps. It is to be separated from the surrounding structures, hooked up with the finger, and stretched by steady and continuous traction for two or three minutes. The sciatic nerve may also be stretched by what is known as the 'dry' plan. The patient is laid on his back, the foot is extended, the leg flexed on the thigh, and the thigh strongly flexed on the abdomen. While the thigh is maintained in this position, the leg is forcibly extended to its full extent, and the foot as fully flexed on the leg.

The position of the external popliteal, close behind the tendon of the Biceps on the outer side of the ham, should be remembered in subcutaneous division of the tendon. After the tendon is divided, a cord often rises up close beside it, which might be mistaken for a small undivided portion of the tendon, and the surgeon might be tempted to reintroduce his knife and divide it. This must never be done, as the cord is the external popliteal nerve, which becomes prominent as soon as the tendon is divided.

THE SYMPATHETIC NERVOUS SYSTEM

The **Sympathetic Nervous System** consists of (1) a series of ganglia, connected together by intervening trunks, extending from the base of the skull to the coccyx, one on each side of the middle line of the body, which are connected with at least one spinal nerve by one or more *rami communicantes*, and give off numerous nerve-fibres, *rami efferentes*, to the blood-vessels and viscera; (2) three great gangliated plexuses, or aggregations of nerves and ganglia, situated in front of the spine in the thoracic, abdominal, and pelvic cavities respectively; and (3) smaller secondary gangliated plexuses to certain viscera, as the heart, the intestines, the suprarenal capsules, and the arteries.*

Each **gangliated cord** may be traced upwards from the base of the skull into its cavity by an ascending branch, which passes through the carotid canal, forms a plexus on the internal carotid artery, and is connected with certain cranial nerves. Below, they converge and unite together in a single ganglion (*ganglion impar*) placed in front of the coccyx. The ganglia of these cords are distinguished as cervical, dorsal, lumbar, and sacral, and except in the neck they nearly correspond in number to the vertebræ against which they lie. They may be thus arranged:

Cervical portion	. .	3 pairs of ganglia.
Dorsal ,,	. .	12 ,, ,,
Lumbar ,,	. .	4 ,, ,,
Sacral ,,	. .	4 or 5 ,, ,,

In the neck the ganglia are situated in front of the transverse processes of the vertebræ; in the dorsal region, in front of the heads of the ribs; in the lumbar region, on the sides of the bodies of the vertebræ; and in the sacral region, in front of the sacrum. Each ganglion consists of cells and fibres; the latter being of two kinds, medullated and non-medullated. The *medullated nerve-fibres* arise from the anterior primary divisions of the spinal nerves and are derived from both nerve-roots; they may terminate in the ganglion by forming arborisations around the cells; or they may pass through the ganglion to end in relation with the cells of the great gangliated plexuses; or thirdly, they may pass through the ganglion to the one above or below through the trunk of communication. They are termed *white rami communicantes*, and exist only in connection with the dorsal and upper two lumbar nerves, and with the second and third sacral nerves.

The *non-medullated nerve-fibres* are the axis-cylinder processes of the ganglion cells. Like the medullated fibres, they terminate in three ways: (1) some pass through the trunk of communication to the ganglion above or below, where they end in arborisations around its nerve-cells; (2) others form rami efferentes to the vessels or viscera, or pass to the great gangliated plexuses; and (3) the remainder (grey rami communicantes) are conveyed to the spinal nerves, passing from *all* the sympathetic ganglia to *all* the spinal nerves.

The **three great gangliated plexuses** are situated in front of the spine in the thoracic, abdominal, and pelvic regions, and are named respectively, the *cardiac*, the *solar* or *epigastric*, and the *hypogastric plexus*. They consist of collections of nerves and ganglia; the nerves being derived from the gangliated cords, and from the cerebro-spinal nerves. They distribute branches to the viscera.

The **branches of distribution** derived from the gangliated cords, from the prevertebral plexuses, and also from the smaller gangliated plexuses, are principally destined for the blood-vessels and thoracic and abdominal viscera, supplying the involuntary muscular fibres of the coats of the vessels and the hollow viscera, and the secreting cells, as well as the muscular coats of the vessels in the glandular viscera.

CERVICAL PORTION OF THE GANGLIATED CORD

The cervical portion of the gangliated cord consists of three ganglia on each side, which are distinguished, according to their position, as the superior, middle,

---

* Certain ganglia, connected with the fifth cranial nerve, the ciliary, Meckel's, otic, and submaxillary, and already described (pages 860 et seq.), may be regarded as part of the sympathetic system.

and inferior cervical. This portion of the sympathetic cord receives no white rami communicantes from the cervical spinal nerves, its spinal fibres being derived from the white communicating rami of the upper dorsal nerves, which enter the corresponding dorsal ganglia of the sympathetic, and through these ascend into the neck.

The **Superior cervical ganglion,** the largest of the three, is placed opposite the second and third cervical vertebræ. It is of a reddish-grey colour, and usually fusiform in shape; sometimes broad and flattened, and occasionally constricted at intervals; it is believed to be formed by the coalescence of four ganglia, corresponding to the four upper cervical nerves. It is in relation, in front, with the sheath of the internal carotid artery, and internal jugular vein; behind, it lies on the Rectus capitis anticus major muscle.

Its branches may be divided into superior, inferior, external, internal, and anterior.

The *superior branch* appears to be a direct prolongation of the ganglion. It is soft in texture, and of a reddish colour. It ascends by the side of the internal carotid artery, and, entering the carotid canal in the temporal bone, divides into two branches, which lie, one on the outer, and the other on the inner side of that vessel.

The *outer branch*, the larger of the two, distributes filaments to the internal carotid artery, and forms the *carotid plexus.*

The *inner branch* also distributes filaments to the internal carotid, and, continuing onwards, forms the *cavernous plexus.*

The **carotid plexus** is situated on the outer side of the internal carotid. Filaments from this plexus occasionally form a small gangliform swelling on the under surface of the artery, which is called the *carotid ganglion.* The carotid plexus communicates with the Gasserian ganglion, with the sixth nerve, and the spheno-palatine ganglion, and distributes filaments to the wall of the carotid artery, and to the dura mater (Valentin); it also communicates with Jacobson's nerve (tympanic branch of the glosso-pharyngeal).

The *communicating branches with the sixth nerve* consist of one or two filaments which join that nerve as it lies upon the outer side of the internal carotid. Other filaments are also connected with the Gasserian ganglion. The communication with the spheno-palatine ganglion is effected by a branch, the *large deep petrosal,* which is given off from the plexus on the outer side of the artery, and which passes through the cartilage filling up the foramen lacerum medium, and joins the great superficial petrosal to form the Vidian nerve. The Vidian nerve then proceeds through the Vidian canal to the spheno-palatine ganglion. The communication with Jacobson's nerve is effected by two branches, one of which is called the *small deep petrosal nerve,* and the other the *carotico-tympanic;* the latter may consist of two or three delicate filaments.

The **cavernous plexus** is situated below, and internal to that part of the internal carotid which is placed by the side of the sella turcica, in the cavernous sinus, and is formed chiefly by the internal division of the ascending branch from the superior cervical ganglion. It communicates with the third, the fourth, the ophthalmic division of the fifth, and the sixth nerves, and with the ciliary ganglion, and distributes filaments to the wall of the internal carotid. The branch of communication with the third nerve joins that nerve at its point of division; the branch to the fourth nerve joins it as it lies on the outer wall of the cavernous sinus; other filaments are connected with the under surface of the trunk of the ophthalmic nerve; and a second filament of communication joins the sixth nerve.

The filament of connection with the ciliary ganglion arises from the anterior part of the cavernous plexus and enters the orbit through the sphenoidal fissure; it may join the nasal branch of the ophthalmic nerve, or be continued forwards as a separate branch.

The terminal filaments from the carotid and cavernous plexuses are prolonged along the internal carotid, forming plexuses which entwine round the anterior and middle cerebral arteries and the ophthalmic artery: along the former vessels they may be traced to the pia mater; along the latter, into the orbit, where they accompany each of the subdivisions of the vessel, a separate plexus passing, with the arteria centralis retinæ, into the interior of the eyeball.

FIG. 639.—The sympathetic nerve.

Carotid plexus

Superior cervical ganglion

Middle cervical ganglion

Inferior cervical ganglion

Dorsal ganglia

Lumbar ganglia

Sacral ganglia

Ganglion impar

Pharyngeal branches

Cardiac branches

Deep cardiac plexus

Superficial cardiac plexus

Solar plexus

Aortic plexus

Hypogastric plexus

The filaments prolonged on to the anterior communicating artery connect the sympathetic nerves of the right and left sides.

The **inferior** or **descending branch of the superior cervical ganglion** communicates with the middle cervical ganglion.

The **external branches** are communicating, and consist of grey rami communicantes to the four upper cervical nerves and to certain of the cranial nerves. Sometimes the branch to the fourth cervical nerve may come from the cord connecting the upper and middle cervical ganglia. The branches of communication with the cranial nerves consist of delicate filaments, which pass from the superior cervical ganglion to the ganglion of the trunk of the pneumogastric, and to the hypoglossal nerve. A separate filament (*nervus jugularis*) from the ganglion passes upwards to the base of the skull, and subdivides to join the petrosal ganglion of the glosso-pharyngeal, and the ganglion of the root of the pneumogastric in the jugular foramen.

The **internal branches** are peripheral, and are the *pharyngeal*, and *superior cardiac nerve*. The *pharyngeal branches* pass inwards to the side of the pharynx, where they join with branches from the glosso-pharyngeal, pneumogastric, and external laryngeal nerves to form the *pharyngeal plexus*.

The *superior cardiac nerve* (*nervus superficialis cordis*) arises by two or more branches from the superior cervical ganglion, and occasionally receives a filament from the cord of communication between the first and second cervical ganglia. It runs down the neck behind the common carotid artery, lying upon the Longus colli muscle ; and crosses in front of the inferior thyroid artery, and recurrent laryngeal nerve. The course of the nerves on the two sides then differs.

The *right superior cardiac nerve*, at the root of the neck, passes either in front of or behind the subclavian artery, and along the arteria innominata, to the back part of the arch of the aorta, where it joins the deep cardiac plexus. This nerve, in its course, is connected with other branches of the sympathetic ; about the middle of the neck it receives filaments from the external laryngeal nerve ; lower down, one or two twigs from the pneumogastric ; and as it enters the thorax it is joined by a filament from the recurrent laryngeal. Filaments from this nerve communicate with the thyroid branches from the middle cervical ganglion.

The *left superior cardiac nerve*, in the chest, runs in front of the left common carotid artery and the arch of the aorta, to the superficial cardiac plexus.

The **anterior branches** ramify upon the external carotid artery and its branches, forming round each a delicate plexus, on the nerves composing which small ganglia are occasionally found. The plexuses accompanying some of these arteries have important communications with other nerves. That surrounding the facial communicates with the submaxillary ganglion by one or two filaments ; and that accompanying the middle meningeal artery sends an offset to the otic ganglion, and another, the *external superficial petrosal nerve*,' to the geniculate ganglion of the facial nerve.

¶ The **Middle cervical ganglion** (*thyroid ganglion*) is the smallest of the three cervical ganglia, and is occasionally altogether wanting. It is placed opposite the sixth cervical vertebra, usually upon, or close to, the inferior thyroid artery ; hence the name 'thyroid ganglion,' assigned to it by Haller. It is probably formed by the coalescence of two ganglia corresponding to the fifth and sixth cervical nerves.

It is joined by grey rami communicantes to the fifth and sixth cervical nerves.

It gives off the thyroid and the middle cardiac nerves.

The *thyroid branches* are small filaments, which accompany the inferior thyroid artery to the thyroid gland ; they communicate, on the artery, with the superior cardiac nerve, and, in the gland, with branches from the recurrent and external laryngeal nerves.

The *middle cardiac nerve* (*nervus cardiacus magnus*), the largest of the three cardiac nerves, arises from the middle cervical ganglion, or from the cord between the middle and inferior ganglia. On the right side it descends behind the common carotid artery ; and at the root of the neck passes either in front of or behind the subclavian artery ; it then descends on the trachea, receives a few filaments from the recurrent laryngeal nerve, and joins the right half of the deep cardiac plexus. In the neck, it communicates with the superior cardiac and recurrent laryngeal nerves. On the left side, the middle cardiac nerve

FIG. 640.—Plan of the cervical portion of the sympathetic. (After Flower).

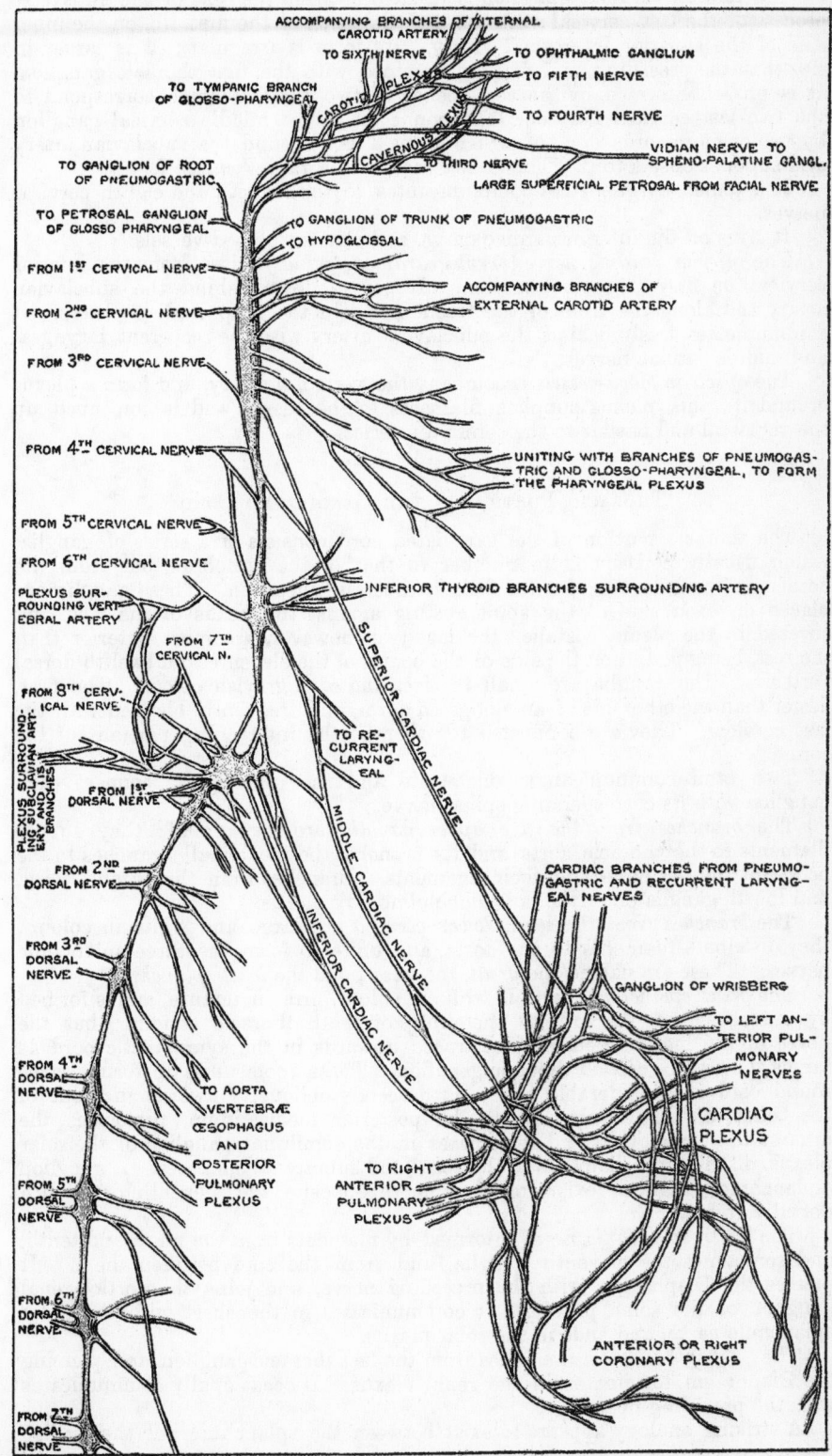

enters the chest between the left carotid and subclavian arteries, and joins the left half of the deep cardiac plexus.

The **Inferior cervical ganglion** is situated between the base of the transverse process of the last cervical vertebra and the neck of the first rib, on the inner side of the superior intercostal artery. Its form is irregular; it is larger in size than the preceding, and frequently joined with the first thoracic ganglion. It is probably formed by the coalescence of two ganglia which correspond to the two last cervical nerves. It is connected to the middle cervical ganglion by two or more cords : one of which forms a loop around the subclavian artery and supplies offsets to it. This loop is named the *ansa Vieussenii*.

It is joined by grey rami communicantes to the seventh and eighth cervical nerves.

It gives off the inferior cardiac nerve, and offsets to blood-vessels.

The *inferior cardiac nerve (nervus cardiacus minor)* arises from the inferior cervical or first thoracic ganglion. It passes down behind the subclavian artery and along the front of the trachea, to join the deep cardiac plexus. It communicates freely behind the subclavian artery with the recurrent laryngeal and middle cardiac nerves.

The *offsets to blood-vessels* accompany the vertebral artery, and form a plexus around it; this plexus supplies filaments to the vessel, and is continued up the vertebral and basilar to the cerebral arteries.

THORACIC PORTION OF THE GANGLIATED CORD

The thoracic portion of the gangliated cord consists of a series of ganglia, which usually correspond in number to that of the vertebræ; but, from the occasional coalescence of two, their number is uncertain. These ganglia are placed on each side of the spine, resting against the heads of the ribs, and covered by the pleura costalis; the last two, however, are more anterior than the rest, being placed on the side of the bodies of the eleventh and twelfth dorsal vertebræ. The ganglia are small in size, and of a greyish colour. The first, larger than the others, is of an elongated form, and frequently blended with the last cervical. They are connected together by the intervening portions of the cord.

Two rami communicantes, one white and the other grey, connect each ganglion with its corresponding spinal nerve.

The *branches from the five upper ganglia* are very small; they supply filaments to the thoracic aorta and its branches, besides small branches to the bodies of the vertebræ and their ligaments. Branches from the second, third, and fourth ganglia enter the posterior pulmonary plexus.

The *branches from the seven lower ganglia* are large, and white in colour; they distribute filaments to the aorta, and unite to form the three splanchnic nerves. These are named the *great*, the *lesser*, and the *least splanchnic*.

The *great splanchnic nerve* is white in colour, firm in texture, and is formed by branches from the fifth to the ninth or tenth thoracic ganglia; but the fibres in the higher roots may be traced upwards in the sympathetic cord as far as the first or second thoracic ganglion. These roots unite to form a large round cord of considerable size. It descends obliquely inwards in front of the bodies of the vertebræ along the posterior mediastinum, perforates the crus of the Diaphragm, and terminates in the semilunar ganglion of the solar plexus, distributing filaments to the renal and suprarenal plexuses. A ganglion (splanchnic ganglion) exists on this nerve opposite the eleventh or twelfth dorsal vertebra.

The *lesser splanchnic nerve* is formed by filaments from the ninth and tenth, and sometimes the eleventh ganglia, and from the cord between them. It pierces the Diaphragm with the preceding nerve, and joins the aortico-renal ganglion of the solar plexus. It communicates in the chest with the great splanchnic nerve, and ends in the solar plexus.

The *least splanchnic nerve* arises from the last thoracic ganglion, and, piercing the Diaphragm, terminates in the renal plexus. It occasionally communicates with the preceding nerve.

A striking analogy appears to exist between the splanchnic and the cardiac nerves. The cardiac nerves are three in number; they arise from the three cervical ganglia, and are distributed to a large and important organ in the thoracic cavity. The splanchnic nerves, also three in number, are connected probably with all the dorsal ganglia, and are distributed to important organs in the abdominal cavity.

LUMBAR PORTION OF THE GANGLIATED CORD

The lumbar portion of the gangliated cord is situated in front of the vertebral column, along the inner margin of the Psoas muscle. It consists usually of four ganglia, connected together by interganglionic cords. It is continuous above with the thoracic portion beneath the internal arcuate ligament of the Diaphragm, and below with the sacral portion behind the common iliac artery. The·ganglia are of small size, and placed much nearer the median line than the thoracic ganglia.

Grey rami communicantes connect all the ganglia with the lumbar spinal nerves. There may be two from each ganglion, but the arrangement is not so uniform as in other regions. The first and second, and sometimes the third, lumbar nerves send white rami communicantes to the upper two or three ganglia. From the situation of the lumbar ganglia, these branches are longer than in the other regions. They accompany the lumbar arteries around the sides of the bodies of the vertebræ, passing beneath the fibrous arches from which some of the fibres of the Psoas muscle arise.

Of the *branches of distribution*, some pass inwards, in front of the aorta, and help to form the aortic plexus. Other branches descend in front of the common iliac arteries, and, joining over the promontory of the sacrum, assist in forming the hypogastric plexus. Numerous delicate filaments are also distributed to the bodies of the vertebræ, and the ligaments connecting them.

PELVIC PORTION OF THE GANGLIATED CORD

The pelvic portion of the gangliated cord is situated in front of the sacrum, along the inner side of the anterior sacral foramina. It consists of four or five small ganglia on each side, connected together by interganglionic cords, and continuous above with the lumbar portion. Below, these cords converge and unite on the front of the coccyx, by means of a small ganglion (the *coccygeal ganglion*, or *ganglion impar*).

Grey rami communicantes pass from the ganglia to the sacral and coccygeal nerves, but no white rami communicantes are derived from the spinal nerves, and in this respect the sacral ganglia resemble the cervical and lower lumbar ganglia.

The *branches of distribution* communicate, on the front of the sacrum, with the corresponding branches from the opposite side; some, from the first two ganglia, pass to join the pelvic plexus, and others form a plexus, which accompanies the middle sacral artery and sends filaments to the coccygeal gland.

THE GREAT PLEXUSES OF THE SYMPATHETIC

The great plexuses of the sympathetic are the large aggregations of nerves and ganglia, above alluded to, situated in the thoracic, abdominal, and pelvic cavities, and named the cardiac, solar, and hypogastric plexuses. They consist not only of sympathetic fibres derived from the ganglia, but of fibres from the central nervous axis, which are conveyed through the white rami communicantes. From the plexuses branches are given to the thoracic, abdominal, and pelvic viscera.

CARDIAC PLEXUS

The cardiac plexus is situated at the base of the heart, and is divided into a *superficial part*, which lies in the concavity of the arch of the aorta, and a *deep part*, which lies between the trachea and aorta. The two plexuses are, however, closely connected.

The **superficial cardiac plexus** lies beneath the arch of the aorta, in front of the right pulmonary artery. It is formed by the superior cardiac branch of the left sympathetic and the inferior cervical cardiac branch of the left pneumogastric A small ganglion (*cardiac ganglion of Wrisberg*) is occasionally found connected with these nerves at their point of junction. This ganglion, when present, is situated immediately beneath the arch of the aorta, on the right side of the ductus arteriosus. The superficial cardiac plexus gives branches (*a*) to the deep cardiac plexus beneath the arch of the aorta; (*b*) to the right or anterior coronary plexus; and (*c*) to the left anterior pulmonary plexus.

The **deep cardiac plexus** is situated in front of the trachea at its bifurcation, above the point of division of the pulmonary artery, and behind the arch of the aorta. It is formed by the cardiac nerves derived from the cervical ganglia of the sympathetic, and the cardiac branches of the recurrent laryngeal and pneumogastric. The only cardiac nerves which do not enter into the formation of this plexus are the superior cardiac branch of the left sympathetic, and the inferior cervical cardiac branch from the left pneumogastric, which pass to the superficial plexus.

The branches from the *right side* of this plexus pass, some in front of, and others behind, the right pulmonary artery; the former, the more numerous, transmit a few filaments to the anterior pulmonary plexus, and are then continued onwards to form part of the right coronary plexus; those behind the pulmonary artery distribute a few filaments to the right auricle, and are then continued onwards to form part of the left coronary plexus.

The *left side* of the plexus is connected with the superficial cardiac plexus, and gives filaments to the left auricle of the heart, and to the anterior pulmonary plexus, and is then continued to form the greater part of the left coronary plexus.

The **left coronary plexus** is larger than the right, and accompanies the left coronary artery: it is chiefly formed by filaments prolonged from the left side of the deep cardiac plexus, and by a few from the right side. It gives branches to the left auricle and ventricle.

The **right coronary plexus** is formed partly from the superficial and partly from the deep cardiac plexus. It accompanies the right coronary artery, and gives branches to the right auricle and ventricle.

EPIGASTRIC OR SOLAR PLEXUS (figs. 639, 641)

The **Epigastric** or **Solar plexus** supplies all the viscera in the abdominal cavity. It consists of a great network of nerves and ganglia situated behind the stomach and lesser sac of the peritoneum, and in front of the aorta and crura of the Diaphragm. It surrounds the cœliac axis and root of the superior mesenteric artery, extending downwards as low as the pancreas, and outwards to the supra-renal capsules. This plexus, and the ganglia connected with it, receive the great and small splanchnic nerves of both sides, and some filaments from the right pneumogastric. It distributes filaments, which accompany, under the name of plexuses, all the branches from the front of the abdominal aorta.

Of the ganglia of which the solar plexus is partly composed the principal are the two **semilunar ganglia,** which are situated one on each side of the plexus, and are the largest peripheral ganglia in the body. They are large irregular gangliform masses, formed by the aggregation of smaller ganglia, having inter-spaces between them. They are situated in front of the crura of the Diaphragm, close to the suprarenal capsules: the one on the right side lies beneath the inferior vena cava; the upper part of each ganglion is joined by the great splanchnic nerve, and to the inner side of each the branches of the solar plexus are connected. A portion of the semilunar ganglion at its lower end is segmented off, and is named the *aortico-renal ganglion*. This receives the lesser splanchnic nerve, and gives off the greater part of the renal plexus.

From the epigastric or solar plexus are derived the following:

Phrenic or Diaphragmatic plexus.
Suprarenal plexus.
Renal plexus.   Cœliac plexus { Gastric plexus.
Spermatic plexus.   Hepatic plexus.
                                                 Splenic plexus.
                    Superior mesenteric plexus.
              Aortic plexus.

The **phrenic plexus** accompanies the phrenic artery to the Diaphragm, some filaments passing to the suprarenal capsule. It arises from the upper part of the semilunar ganglion, and is larger on the right than on the left side. It receives

FIG. 641.—Lumbar portion of the gangliated cord, with the solar and hypogastric plexuses. (After Henle.)

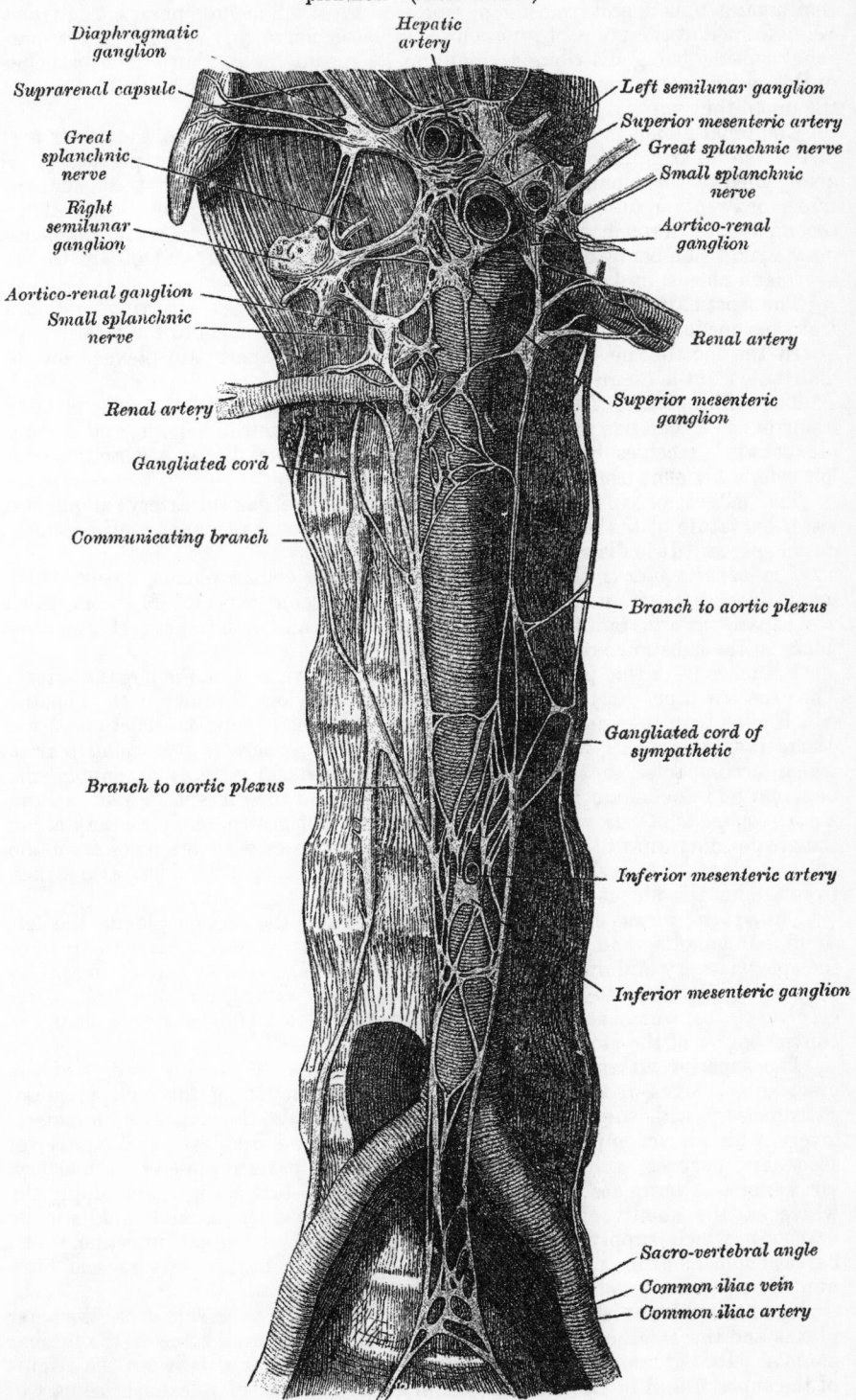

one or two branches from the phrenic nerve. At the point of junction of the right phrenic plexus with the phrenic nerve, is a small ganglion (*ganglion diaphragmaticum*). This ganglion distributes branches to the inferior vena cava, suprarenal capsule, and the hepatic plexus. There is no ganglion on the left side.

The **suprarenal plexus** is formed by branches from the solar plexus, from the semilunar ganglion, and from the phrenic and great splanchnic nerves, a ganglion being formed at the point of junction of the latter nerve. It supplies the suprarenal capsule, being distributed chiefly to its medullary portion. The branches of this plexus are remarkable for their large size, in comparison with the size of the organ they supply.

The **renal plexus** is formed by filaments from the solar plexus, the lower part of the semilunar ganglion (aortico-renal ganglion), and the aortic plexus. It is also joined by the smallest splanchnic nerve. The nerves from these sources, fifteen or twenty in number, have numerous ganglia developed upon them. They accompany the branches of the renal artery into the kidney; some filaments on the right side being distributed to the inferior vena cava, and others to the spermatic plexus, on both sides.

The **spermatic plexus** is derived from the renal plexus, receiving branches from the aortic plexus. It accompanies the spermatic vessels to the testes.

In the female, the **ovarian plexus** arises like the spermatic plexus, and is distributed to the ovaries and fundus of the uterus.

The **cœliac plexus**, of large size, is a direct continuation from the solar plexus: it surrounds the cœliac axis, and subdivides into the gastric, hepatic, and splenic plexuses. It receives branches from the lesser splanchnic nerves, and, on the left side, a filament from the right pneumogastric.

The *gastric* or *coronary plexus* accompanies the gastric artery along the lesser curvature of the stomach, and joins with branches from the left pneumogastric nerve. It is distributed to the stomach.

The *hepatic plexus*, the largest offset from the cœliac plexus, receives filaments from the left pneumogastric and right phrenic nerves. It accompanies the hepatic artery, ramifying upon its branches, and upon those of the vena portæ in the substance of the liver.

Branches from this plexus accompany all the divisions of the hepatic artery. Thus there is a *pyloric plexus* accompanying the pyloric branch of the hepatic, which joins with the gastric plexus and pneumogastric nerves. There is also a *gastro-duodenal plexus*, which subdivides into the *pancreatico-duodenal plexus*, which accompanies the superior pancreatico-duodenal artery, to supply the pancreas and duodenum, joining with branches from the mesenteric plexus; and a *gastro-epiploic plexus*, which accompanies the right gastro-epiploic artery along the greater curvature of the stomach, and anastomoses with branches from the splenic plexus. A *cystic plexus*, which supplies the gall-bladder, also arises from the hepatic plexus, near the liver.

The *splenic plexus* is formed by branches from the cœliac plexus, the left semilunar ganglia, and from the right pneumogastric nerve. It accompanies the splenic artery and its branches to the substance of the spleen, giving off, in its course, filaments to the pancreas (*pancreatic plexus*), and the *left gastro-epiploic plexus*, which accompanies the gastro-epiploica sinistra artery along the convex border of the stomach.

The **superior mesenteric plexus** is a continuation of the lower part of the great solar plexus, receiving a branch from the junction of the right pneumogastric nerve with the cœliac plexus. It surrounds the superior mesenteric artery, which it accompanies into the mesentery, and divides into a number of secondary plexuses, which are distributed to all the parts supplied by the artery, viz. pancreatic branches to the pancreas; intestinal branches, which supply the whole of the small intestine; and ileo-colic, right colic, and middle colic branches, which supply the corresponding parts of the great intestine. The nerves composing this plexus are white in colour and firm in texture, and have numerous ganglia developed upon them near their origin.

The **aortic plexus** is formed by branches derived, on each side, from the solar plexus and the semilunar ganglia, receiving filaments from some of the lumbar ganglia. It is situated upon the sides and front of the aorta, between the origins of the superior and inferior mesenteric arteries. From this plexus arise part of the spermatic, the inferior mesenteric, and the hypogastric plexuses; and it distributes filaments to the inferior vena cava.

The *inferior mesenteric plexus* is derived chiefly from the left side of the aortic plexus. It surrounds the inferior mesenteric artery, and divides into a number of secondary plexuses, which are distributed to all the parts supplied by the artery, viz. the left colic and sigmoid plexuses, which supply the descending and ilio-pelvic parts of the colon: and the superior hæmorrhoidal plexus, which supplies the rectum, and joins in the pelvis with branches from the pelvic plexuses.

HYPOGASTRIC PLEXUS

The **Hypogastric plexus** supplies the viscera of the pelvic cavity. It is situated in front of the promontory of the sacrum, between the two common iliac arteries, and is formed by the union of numerous filaments, which descend on each side from the aortic plexus, and from the lumbar ganglia. This plexus contains no evident ganglia; it bifurcates, below, into two lateral portions, which form the *pelvic plexuses*.

PELVIC PLEXUSES

The **Pelvic plexuses** supply the viscera of the pelvic cavity, and are situated at the sides of the rectum in the male, and at the sides of the rectum and vagina in the female. They are formed by a continuation of the hypogastric plexus, by branches from the second, third, and fourth sacral nerves, and by a few filaments from the first two sacral ganglia. At the point of junction of these nerves, small ganglia are found. From these plexuses numerous branches are distributed to all the viscera of the pelvis. They accompany the branches of the internal iliac artery.

The **inferior hæmorrhoidal plexus** arises from the upper part of the pelvic plexus. It supplies the rectum, joining with branches of the superior hæmorrhoidal plexus.

The **vesical plexus** arises from the fore part of the pelvic plexus. The nerves composing it are numerous, and contain a large proportion of spinal nerve-fibres. They accompany the vesical arteries, and are distributed to the side and base of the bladder. Numerous filaments also pass to the vesiculæ seminales and vas deferens; those accompanying the vas deferens join, on the spermatic cord, with branches from the spermatic plexus.

The **prostatic plexus** is continued from the lower part of the pelvic plexus. The nerves composing it are of large size. They are distributed to the prostate gland, vesiculæ seminales, and erectile structure of the penis. The nerves supplying the erectile structure of the penis consist of two sets, the small and large cavernous nerves. They are slender filaments, which arise from the fore part of the prostatic plexus, and, after joining with branches from the internal pudic nerve, pass forwards beneath the pubic arch.

The *small cavernous nerves* perforate the fibrous covering of the penis, near its root.

The *large cavernous nerve* passes forwards along the dorsum of the penis, joins with the dorsal nerve of the penis, and is distributed to the corpora cavernosa and corpus spongiosum.

The **vaginal plexus** arises from the lower part of the pelvic plexus. It is distributed to the walls of the vagina, to the erectile tissue of the vestibule, and to the clitoris. The nerves composing this plexus contain, like the vesical, a large proportion of spinal nerve-fibres.

The **uterine plexus** accompanies the uterine artery to the side of the organ between the layers of the broad ligament; it is distributed to the uterus, and communicates with the ovarian plexus.

ORGANS OF SPECIAL SENSE

THE Organs of the Senses are five in number : viz. those of Touch, of Taste, of Smell, of Sight, and of Hearing. The skin, which is the principal seat of the sense of touch, has been described in the section on General Anatomy. The remaining four are the Organs of Special Sense.

THE TONGUE

The tongue is the organ of the special sense of taste. It is situated in the floor of the mouth, in the interval between the two lateral portions of the body of the lower jaw.

Its base, or root, is directed backwards, and connected with the os hyoides by the Hyo-glossi and Genio-hyo-glossi muscles and the hyo-glossal membrane ; with the epiglottis by three folds (*glosso-epiglottic*) of mucous membrane ; with the soft palate by means of the anterior pillars of the fauces ; and with the pharynx by the Superior constrictors and the mucous membrane. Its apex, or tip, thin and narrow, is directed forwards against the inner surface of the lower incisor teeth. The under surface of the tongue is connected with the lower jaw by the Genio-hyo-glossi muscles ; from its sides, the mucous membrane is reflected to the inner surface of the gums ; and from its under surface on to the floor of the mouth, where, in the middle line, it is elevated into a distinct vertical fold, the *frænum linguæ*. To the outer side of the frænum is a slight fold of the mucous membrane, the *plica fimbriata*, the free edge of which exhibits a series of fringe-like processes.

The tip of the tongue, part of the under surface, its sides, and dorsum are free.

The dorsum of the tongue is convex, marked along the middle line by a raphé, which divides it into symmetrical halves ; this raphé terminates behind, about an inch from the base of the organ, in a depression, the *foramen cæcum*, from which a shallow groove, the *sulcus terminalis* of His, runs outwards and forwards on each side to the lateral margin of the tongue. The part of the dorsum of the tongue in front of this groove, forming about two-thirds of its upper surface, is rough and covered with papillæ ; the posterior third is smoother, and contains numerous muciparous glands and lymphoid follicles.

Structure of the tongue.—The tongue is partly invested by mucous membrane and a submucous fibrous layer. It consists of symmetrical halves, separated from each other, in the middle line, by a fibrous septum. Each half is composed of muscular fibres arranged in various directions (page 454), containing much interposed fat, and supplied by vessels and nerves.

The **mucous membrane** invests the entire extent of the free surface of the tongue. On the dorsum it is thicker behind than in front, and is continuous with the sheaths of the muscles attached to it, through the submucous fibrous layer. On the under surface of the organ, where it is thin and smooth, it can be traced on each side of the frænum, through the ducts of the submaxillary and the sublingual glands. As it passes over the borders of the organ, it gradually assumes its papillary character.

The **structure of the mucous membrane of the tongue** differs in different parts. That covering the under surface of the organ is thin, smooth, and identical in structure with that lining the rest of the oral cavity. The mucous membrane covering the tongue behind the foramen cæcum and sulcus terminalis is thick and freely movable over the subjacent parts. It contains a large number of lymphoid follicles, which together constitute what is sometimes termed the *lingual tonsil*. Each follicle forms a rounded eminence, the centre of which is perforated by a minute orifice leading into a funnel-shaped cavity or recess ; around this recess are grouped numerous oval or rounded nodules of lymphoid tissue, each enveloped by a capsule derived from the submucosa, while opening into the bottom of the recesses are also seen the ducts of mucous glands. The mucous membrane on the anterior part of the dorsum of the tongue is thin and intimately adherent to the muscular tissue, and covered with

Fig. 642.—Upper surface of the tongue.

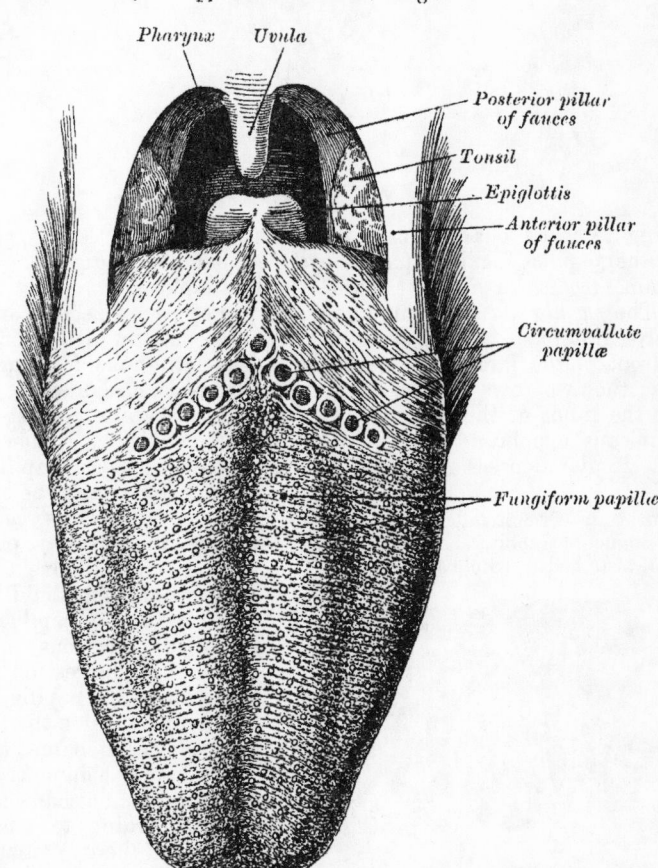

minute eminences, the *papillæ* of the tongue. It consists of a layer of connective tissue, the *corium* or *mucosa*, supporting numerous papillæ, and covered, as well as the papillæ, with epithelium.

The epithelium is of the scaly variety, like that of the epidermis. It covers the free surface of the tongue, as may be demonstrated by maceration or boiling, when it can be easily detached entire : it is much thinner than on the skin : the intervals between the large papillæ are not filled up by it, but each papilla has a separate investment from root to summit. The deepest cells may sometimes be detached as a separate layer, corresponding to the rete mucosum, but they never contain colouring matter.

The *corium* consists of a dense felt-work of fibrous connective tissue, with numerous elastic fibres, firmly connected with the fibrous tissue forming the septa between the muscular bundles of the tongue. It contains the ramifications of the numerous vessels and nerves from which the papillæ are supplied, large plexuses of lymphatic vessels, and the glands of the tongue.

The papillæ of the tongue.—These are papillary projections of the corium. They are thickly distributed over the anterior two-thirds of its upper surface,

FIG. 643.—The three kinds of papillæ, magnified.

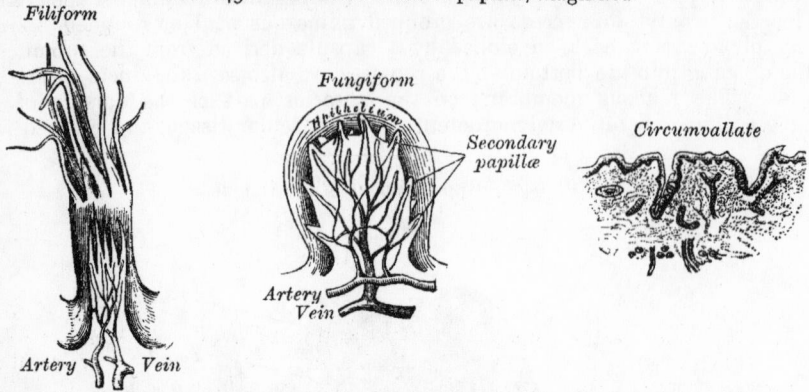

giving to it its characteristic roughness. The varieties of papillæ met with are the papillæ maximæ (*circumvallatæ*), papillæ mediæ (*fungiformes*), papillæ minimæ (*conicæ* or *filiformes*), and papillæ simplices.

The *papillæ maximæ* (circumvallatæ) are of large-size, and vary from eight to twelve in number. They are situated on the dorsum of the tongue immediately in front of the foramen cæcum and sulcus terminalis, forming a row on each side; the two rows run backwards and inwards, and meet in the middle line, like the limbs of the letter V inverted. Each papilla consists of a projection of mucous membrane from $\frac{1}{20}$ to $\frac{1}{12}$ of an inch wide, attached to the bottom of a circular depression of the mucous membrane; the papilla is shaped like a truncated cone; the smaller end being directed downwards and attached to the tongue, the broader part or base projecting a little above the surface of the tongue and being studded with numerous small secondary papillæ and covered by stratified squamous epithelium. The cup-shaped depression forms a kind of fossa round the papilla, and the mucous membrane outside the fossa forms a circular elevation, named the wall (vallum). Immediately behind the apex of the V is the foramen cæcum, mentioned above. This, according to His, represents the remains of the invagination which forms the median rudiment of the thyroid body, and for a time opens by a duct, the *thyro-glossal duct*, on to the dorsum of the tongue. It may extend downwards towards the hyoid bone, beyond which the lower part of the duct is represented by the pyramid of the thyroid body. Kanthack, however, disputed this view.[*]

FIG. 644.—Circumvallate papillæ of tongue of rabbit, showing position of taste-buds. (Stöhr.)

a. Duct of gland. *d.* Serous gland. *g.* Taste-buds. *l.* Primary septa, and *l'*, secondary septa, of papillæ. *n.* Medullated nerve. *M.* Muscular fibres.

The *papillæ mediæ* (fungiformes), more numerous than the preceding, are scattered irregularly and sparingly over the dorsum of the tongue; but are found chiefly at its sides and apex. They are easily recognised, among the other papillæ, by their large size, rounded eminences, and deep red colour.

[*] *Journal of Anat. and Physiol.* 1891.

They are narrow at their attachment to the tongue, but broad and rounded at their free extremities, and covered with secondary papillæ.

The *papillæ minimæ* (conicæ or filiformes) cover the anterior two-thirds of the dorsum of the tongue. They are very minute, more or less conical or filiform in shape, and arranged in lines corresponding in direction with the two rows of the papillæ circumvallatæ; excepting at the apex of the organ, where their direction is transverse. Projecting from their apices are numerous filiform processes, or secondary papillæ; these are of a whitish tint, owing to the thickness and density of the epithelium of which they are composed, and which has here undergone a peculiar modification, the cells having become cornified and elongated into dense, imbricated, brush-like processes. They contain also a number of elastic fibres, which render them firmer and more elastic than the papillæ of mucous membrane generally.

Simple papillæ, similar to those of the skin, cover the whole of the mucous membrane of the tongue, as well as the larger papillæ. They consist of closely set, microscopic elevations of the corium, containing a papillary loop, covered by a layer of epithelium.

Structure of the papillæ.—The papillæ apparently resemble in structure those of the cutis, consisting of a cone-shaped projection of connective tissue, covered with a thick layer of squamous epithelium, and contain one or more capillary loops, among which nerves are distributed in great abundance. If the epithelium is removed, it will be found that they are not simple elevations like the papillæ of the skin, for the surface of each is studded with minute conical processes of the mucous membrane, which form secondary papillæ (Todd and Bowman). In the papillæ circumvallatæ, the nerves are numerous and of large size; in the papillæ fungiformes, they are also numerous, and terminate in a plexiform network, from which brush-like branches proceed; in the papillæ filiformes, their mode of termination is uncertain.

Taste-buds.—This name is applied to certain flask-shaped groups of modified epithelial cells which are found on the tongue and adjacent parts. They occupy nests in the stratified epithelium, and are present in large numbers on the sides of the circumvallate papillæ, and to a less extent on their opposed walls. They are also found on the fungiform papillæ over the back part and sides of the tongue, and in the general epithelial covering in the same areas. They are very plentiful over the fimbriæ linguæ, and are also present on the under aspect of the soft palate, and on the posterior surface of the epiglottis. They are flask-like in shape, their broad base resting on the corium, and their neck opening by an orifice, the *gustatory pore*, between the cells of the epithelium. They are formed by two kinds of cells: supporting cells and gustatory cells. The *supporting cells* are mostly arranged like the staves of a cask, and form an outer envelope for the bud. Some, however, are found in the interior of the bud between the gustatory cells. The *gustatory cells* occupy the central portion of the bud; they are spindle-shaped, and each possesses a large spherical nucleus near the middle of the cell. The peripheral end of the cell terminates at the gustatory pore in a fine hair-like filament, the *gustatory hair*. The central process passes towards the deep extremity of the bud, and there ends in a single or bifurcated varicose filament, which was formerly supposed to be continuous with the terminal fibril of a nerve; the investigations of Lenhossék and others would seem to prove, however, that this is not so, but that the nerve-fibrils after losing their medullary sheaths enter the taste-bud, and terminate in fine extremities between the gustatory cells. Other nerve-fibrils may be seen ramifying between the supporting cells and terminating in fine extremities; these, however, are believed to be nerves of ordinary sensation and not gustatory.

FIG. 645.—Taste-buds.

a. Supporting cell. *b.* Gustatory cell.

Glands of the tongue.—The tongue is provided with mucous and serous glands.

The *mucous glands* are similar in structure to the labial and buccal glands. They are found especially at the back part behind the circumvallate papillæ, but are also present at the apex and marginal parts. In this connection the glands of Blandin or Nuhn require special notice. They are situated on the under surface of the apex of the tongue, one on either side of the frænum, where they are covered by a fasciculus of muscular fibres derived from the Stylo-glossus and Inferior lingualis. They are from half an inch to nearly an inch long, and about the third of an inch broad, and each opens by three or four ducts on the under surface of the apex.

The *serous glands* occur only at the back of the tongue in the neighbourhood of the taste-buds, their ducts opening for the most part into the fossæ of the circumvallate papillæ. These glands are racemose, the duct branching into several minute ducts, which terminate in alveoli, lined by a single layer of more

Fig. 646.—Under surface of tongue, showing position and relations of gland of Blandin or Nuhn. (From a preparation in the Museum of the Royal College of Surgeons of England.)

Bristles in ducts of glands

Gland of Blandin or Nuhn

Lingual nerve   Ranine artery

or less columnar epithelium. Their secretion is of a watery nature, and probably assists in the distribution of the substance to be tasted over the taste area. (Ebner.)

The *fibrous septum* consists of a vertical layer of fibrous tissue, extending throughout the entire length of the middle line of the tongue, from the base to the apex, though not quite reaching the dorsum. It is thicker behind than in front, and occasionally contains a small fibro-cartilage, about a quarter of an inch in length. It is well displayed by making a vertical section across the organ.

The **hyo-glossal membrane** is a strong fibrous lamina, which connects the under surface of the base of the tongue to the body of the hyoid bone. This membrane receives, in front, some of the fibres of the Genio-hyo-glossi muscles.

Vessels of the tongue.—The main artery of the tongue is the lingual branch of the external carotid,. but the facial and ascending pharyngeal also give branches to it. The *veins* open into the internal jugular.

Muscles of the tongue.—The muscular fibres of the tongue run in various directions. These fibres are divided into two sets, Extrinsic and Intrinsic, which have already been described (page 454).

The *lymphatics of the tongue* have also been described on page 751.

The *nerves of the tongue* are: (1) the lingual branch of the third division of the fifth, which is distributed to the papillæ at the fore part and sides of the tongue, and forms the nerve of ordinary sensibility for its anterior two-thirds ; (2) the chorda tympani branch of the facial nerve, which runs in the sheath of the lingual, is generally regarded as the nerve of taste for the anterior two-thirds : this nerve is a continuation of the sensory root of the facial (pars intermedia of Wrisberg) ; (3) the lingual branch of the glosso-pharyngeal, which is distributed to the mucous membrane at the base and sides of the tongue, and to the papillæ circumvallatæ, and which supplies both sensory and gustatory filaments to this region ; (4) the hypoglossal nerve, which is the motor nerve to the muscular substance of the tongue ; (5) the superior laryngeal, which sends some fine branches to the root near the epiglottis. Sympathetic filaments also pass to the tongue from the nervi molles on the lingual and other arteries supplying it.

Surgical Anatomy.—The diseases to which the tongue is liable are numerous, and its surgical anatomy of importance, since any or all of the structures of which it is composed — muscles, connective tissue, mucous membrane, glands, vessels, nerves, and lymphatics — may be the seat of morbid changes. It is not often the seat of congenital defects, though a few cases of vertical cleft have been recorded, and it is occasionally, though much more rarely than is commonly supposed, the seat of 'tongue tie,' from shortness of the frænum (see page 613).

There is, however, one not uncommon condition, which must be regarded as congenital, though sometimes it does not evidence itself until a year or two after birth. This is an enlargement of the tongue which is due primarily to a dilatation of the lymph-channels and a greatly increased development of the lymphatic tissue throughout the organ. This is often aggravated by inflammatory changes induced by injury or exposure, and the tongue may assume enormous dimensions and hang out of the mouth, giving the child an imbecile expression. The treatment consists in excising a V-shaped portion and bringing the cut surfaces together with deeply placed silver sutures. Compression has been resorted to in some cases and with success, but it is difficult to apply.

Acute inflammation of the tongue, which may be caused by injury and the introduction of some septic or irritating matter, is attended by great swelling from infiltration of its connective tissue, which is in considerable quantity. This renders the patient incapable of swallowing or speaking, and may seriously impede respiration. It may run on to suppuration, and the formation of an acute abscess. Chronic abscess, which has been mistaken for cancer, may also occur in the substance of the tongue.

The mucous membrane of the tongue may become chronically inflamed, and presents different appearances in the various stages of the disease, to which the terms leucoplakia, psoriasis, and ichthyosis have been given.

The tongue, being very vascular, is often the seat of nævoid growths, and these have a tendency to increase rapidly.

The tongue is frequently the seat of ulceration, which may arise from many causes, as from the irritation of jagged teeth, dyspepsia, tuberculosis, syphilis, and cancer. Of these the cancerous ulcer is the most important and probably also the most common. The variety is the squamous epithelioma, which soon develops into an ulcer with an indurated base. It produces great pain, which speedily extends to all parts supplied with sensation by the fifth nerve, especially to the region of the ear. The pain in these cases is conducted to the ear and temporal region by the lingual nerve, and from it to the other branches of the inferior maxillary nerve, especially the auriculo-temporal. Possibly pain in the ear itself may be due to implication of the fibres of the glosso-pharyngeal nerve, which by its tympanic branch is conducted to the tympanic plexus.

Cancer of the tongue may necessitate removal of a part or the whole of the organ, and many different methods have been adopted for its excision. It may be removed from the mouth by the écraseur or the scissors. Probably the better method is by the scissors, usually known as Whitehead's method. The mouth is widely opened with a gag ; the tongue transfixed with a stout silk ligature, by which to hold and make traction on it, and the reflection of mucous membrane from the tongue to the jaw, and the insertion of the Genio-hyo-glossus first divided with a pair of curved, blunt-pointed scissors. The Palato-glossus is also divided. The tongue can now be pulled well out of the mouth. The base of the tongue is cut through by a series of short snips, each bleeding vessel being dealt with as soon as divided, until the situation of the main artery is reached. The remaining undivided portion of tissue is to be seized with a pair of Wells' forceps; the tongue removed, and the vessel secured. In the event of the artery being accidentally injured, hæmorrhage can be at once controlled, by passing the forefinger over the tongue till it touches the epiglottis, and then turning it towards the side on which the artery is to be compressed, and pushing it forcibly against the jaw (Heath). In cases where the disease is confined to one side of the tongue, this operation may be modified by splitting the tongue down the centre and removing only the affected half.

In cases where the submaxillary glands are involved, Kocher's operation should be resorted to. Having performed a preliminary tracheotomy, Kocher removes the tongue from the neck by an incision from near the lobule of the ear, down the anterior border of the Sterno-mastoid to the level of the great cornu of the hyoid bone, then forwards to the body of the hyoid bone, and upwards to near the symphysis of the jaw. The lingual artery is now secured, and by a careful dissection the submaxillary lymphatic glands and the tongue removed. Regnoli advocated the removal of the tongue by a semilunar incision in the submaxillary triangle, along the line of the lower jaw, and a vertical incision from the centre of the semilunar one backwards to the hyoid bone. Care must be taken not to carry the first incision too far backwards, so as to wound the facial arteries. The tongue is thus reached through the floor of the mouth, pulled out through the external incision, and removed with the écraseur or knife. The great objection to this operation is that all the muscles which raise the hyoid bone and larynx are divided, and that therefore the movements of deglutition and respiration are interfered with.

Finally, where both sides of the floor of the mouth are involved in the disease, or where very free access is required on account of the extension backwards of the disease to the pillars of the fauces and the tonsil, or where the lower jaw is involved, the operation devised by Sedillot must be performed. This is done by an incision through the central line of the lip, across the chin, and down as far as the hyoid bone. The lower jaw is sawn through at the symphysis, and the two halves of the bone forcibly separated from each other. The mucous membrane is separated from the bone, and the Genio-hyo-glossi detached from the bone, and the Hyo-glossi divided. The tongue is then drawn forwards, and removed close to its attachment to the hyoid bone. Any glands which are enlarged can be removed, and if the bone is implicated in the disease, it can also be removed by freeing it from the soft parts externally and internally, and making a second section with the saw beyond the diseased part.

Formerly many surgeons before removing the tongue performed a preliminary tracheotomy: (1) to prevent blood entering the air passages; and (2) to allow the patient to breathe through the tube and not inspire air which had passed over a sloughy wound, and which was loaded with septic organisms and likely to induce septic pneumonia. By the judicious use of iodoform, this secondary evil may be obviated, and the preliminary tracheotomy is now usually dispensed with.

THE NOSE

The nose is the peripheral organ of the sense of smell : by means of the peculiar properties of its nerves, it protects the lungs from the inhalation of deleterious gases, and assists the organ of taste in discriminating the properties of food.

The organ of smell consists of two parts : one external, the *outer nose*; the other internal, the *nasal fossæ*.

The *outer nose* (nasus externus) is the more anterior and prominent part of the organ of smell. Of a triangular form, it is directed downwards, and projects from the centre of the face, immediately above the upper lip. Its summit, or *root*, is connected directly with the forehead. Its inferior part, or *base*, presents two elliptical orifices, the nostrils or anterior nares, separated from each other by an antero-posterior septum, the *columna*. The margins of these orifices are provided with a number of stiff hairs, or *vibrissæ*, which arrest the passage of foreign substances carried with the current of air intended for respiration. The lateral surfaces of the nose form, by their union in the middle line, the *dorsum*, the direction of which varies considerably in different individuals ; the upper part of the dorsum is supported by the nasal bones, and is named the *bridge*. The lateral surface terminates below in a rounded eminence, the *ala nasi*.

The nose is composed of a framework of bones and cartilages, the latter being slightly acted upon by certain muscles. It is covered by the integument, and lined by mucous membrane, and supplied with vessels and nerves.

The *bony framework* occupies the upper part of the organ: it consists of the nasal bones, and the nasal processes of the superior maxillary.

The *cartilaginous framework* consists of five pieces, the two upper and the two lower lateral cartilages, and the cartilage of the septum.

The *upper lateral cartilages* are situated below the free margin of the nasal bones : each cartilage is flattened, and triangular in shape. Its anterior margin is thicker than the posterior, and is continuous above with the cartilage of the septum, but separated from it below by a narrow fissure. Its posterior margin is attached to the nasal bone and the nasal process of the superior maxilla. Its inferior margin is connected by fibrous tissue with the lower lateral cartilage; one surface is turned outwards, the other inwards towards the nasal cavity.

The *lower lateral cartilages* are two thin, flexible plates, situated immediately below the preceding, and bent upon themselves in such a manner as to form the

FIG. 647.—Cartilages of the nose, seen from below.

Lower lateral cartilage

Sesamoid cartilages

FIG. 648.—Cartilages of the nose. Side view.

inner and outer walls of each orifice of the nostril. The portion which forms the inner wall (*crus mediale*), thicker than the rest, is loosely connected with the same part of the opposite cartilage, and forms a part of the columna. Its inferior border, free, rounded, and projecting, forms, with the thickened integument and subjacent tissue, and the corresponding parts of the opposite side, the septum mobile nasi. The part which forms the outer wall (*crus laterale*) is curved to correspond with the ala of the nose; it is oval and flattened, narrow behind, where it is connected with the nasal process of the superior maxilla by a tough fibrous membrane, in which are found three or four small cartilaginous plates termed *sesamoid cartilages* or *cartilagines minores*. Above, it is connected by fibrous tissue to the upper lateral cartilage and front part of the cartilage of the septum; below, it falls short of the margin of the nostril, the ala being formed by fatty and fibrous tissue covered by skin. In front, the lower lateral cartilages are separated by a notch which corresponds with the point of the nose.

The *cartilage of the septum* is somewhat quadrilateral in form, thicker at its margins than at its centre, and completes the separation between the nasal fossæ in front. Its anterior margin, thickest above, is connected with the nasal bones, and is continuous with the anterior margins of the two upper lateral cartilages. Below, it is connected to the inner portions of the lower lateral cartilages by fibrous tissue. Its posterior margin is connected with the perpendicular lamella of the ethmoid; its inferior margin with the vomer and the palate processes of the superior maxillary bones.

FIG. 649.—Bones and cartilages of septum of nose. Right side.

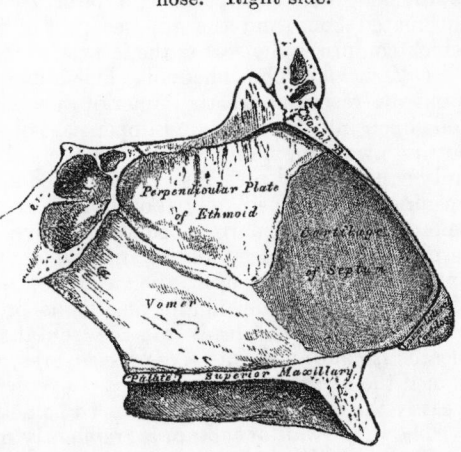

It may be prolonged backwards (especially in children) for some distance between the vomer and perpendicular plate of the ethmoid, forming what is termed the *processus sphenoidalis*. The septal cartilage does not reach as far as the lowest part of the nasal septum. This is formed by the inner portions of the lower lateral cartilages and by the skin; it is freely movable, and hence is termed the *septum mobile nasi*.

These various cartilages are connected to each other, and to the bones by a tough fibrous membrane, which allows the utmost facility of movement between them.

The *muscles of the nose* have been described on page 435.

The *integument* covering the dorsum and sides of the nose is thin, and loosely connected with the subjacent parts; but where it forms the tip and the alæ of the nose it is thicker and more firmly adherent, and is furnished with a large number of sebaceous follicles, the orifices of which are usually very distinct.

The *arteries of the outer nose* are the lateralis nasi from the facial, and the inferior artery of the septum from the superior coronary, which supply the alæ and septum; the sides and dorsum being supplied from the nasal branch of the ophthalmic and the infra-orbital. The *veins* terminate in the facial and ophthalmic.

The *nerves* for the muscles of the nose are derived from the facial, while the skin receives its branches from the infra-orbital, infratrochlear, and nasal branches of the ophthalmic.

NASAL FOSSÆ

The nasal fossæ are two irregular cavities situated in the middle of the face, and extending from before backwards. They open in front by the two anterior nares, and terminate, behind, by the posterior nares in the naso-pharynx. The *anterior nares* are somewhat pear-shaped apertures, each measuring about one inch antero-posteriorly and half an inch transversely at their widest part. The *posterior nares*, or *choanæ*, are two oval openings, which are smaller in the living or recent subject than in the skeleton, because they are narrowed by the mucous membrane. Each measures an inch in the vertical, and half an inch in the transverse direction in a well-developed adult skull.

For the description of the bony boundaries of the nasal fossæ, see section on Osteology (page 242).

Inside the aperture of the nostril is a slight dilatation, the *vestibule*, which is bounded externally by the ala and outer plate of the lower lateral cartilage, and internally by the mesial plate of the same cartilage. It is lined by skin, and contains hairs and sebaceous glands, and extends as a small pouch, the *ventricle*, towards the point of the nose. The fossa, above and behind the vestibule, is divided into two parts: an *olfactory* portion, consisting of the superior turbinated bone and the opposed part of the septum, and a *respiratory* portion, which comprises the rest of the fossa.

Outer wall.—The sphenoidal air-sinus opens into the *spheno-ethmoidal recess*, a narrow recess above the superior turbinated bone. The posterior ethmoidal cells open into the front and upper part of the superior meatus. On raising or cutting away the middle turbinated bone the outer wall of the middle meatus is fully exposed, and presents (1) a rounded elevation, termed the *bulla ethmoidalis*, opening on or immediately above which are the orifices of the middle ethmoidal cells; (2) a deep, narrow, curved groove, in front of the bulla ethmoidalis, termed the *hiatus semilunaris*, into which the anterior ethmoidal cells and the antrum of Highmore open, the orifice of the latter being placed near the roof of the antrum. The middle meatus is prolonged, above and in front, into the *infundibulum*, which leads into the frontal sinus. The anterior extremity of the meatus is continued into a depressed area, which lies above the vestibule and is named the *atrium*. The *nasal duct* opens into the anterior part of the inferior meatus, the opening being overlapped sometimes by a fold of mucous membrane.

The inner wall or septum is frequently more or less deflected from the mesial plane, thus limiting the size of one fossa and increasing that of the other. Ridges or spurs of bone growing outwards from the septum are also sometimes present. Immediately over the incisive foramen at the lower edge of the cartilage of the septum a depression, the *naso-palatine recess*, may be seen. In the septum close to this recess a minute orifice may be discerned: it leads backwards into a blind pouch, the rudimentary *organ of Jacobson*, which is supported by a strip of cartilage, the *cartilage of Jacobson*. This organ is well developed in many of the lower animals, where it apparently plays a part in the sense of smell, since it is supplied by twigs of the olfactory nerve and lined by epithelium corresponding to that which lines the olfactory region of the nose.

The *mucous membrane* lining the nasal fossæ is called *pituitary*, from the nature of its secretions; or *Schneiderian*, from Schneider, the anatomist who first showed that the secretion proceeded from the mucous membrane, and not, as was formerly imagined, from the brain. It is intimately adherent to the periosteum or perichondrium, over which it lies. It is continuous with the skin through the anterior nares, and with the mucous membrane of the naso-pharynx through the posterior nares. From the nasal fossæ its continuity with the con-junctiva may be traced, through the nasal duct and lachrymal canals; with the lining membrane of the tympanum and mastoid cells, through the Eustachian tube; and with the frontal, ethmoidal, and sphenoidal sinuses, and the antrum of Highmore, through the several openings in the meatuses. The mucous mem-brane is thickest, and most vascular, over the turbinated bones. It is also thick

FIG. 650.—Transverse vertical section of the nasal fossæ. (Henle.)

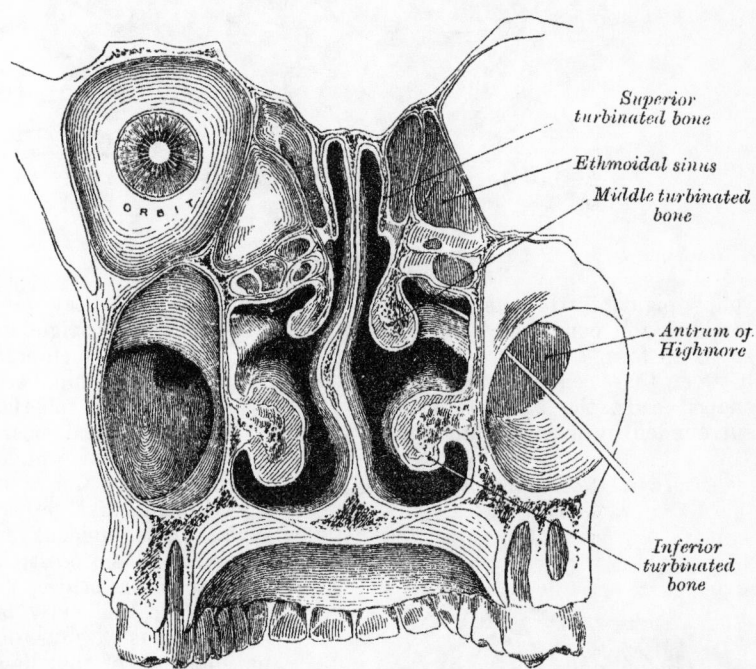

Superior
turbinated bone

Ethmoidal sinus

Middle turbinated
bone

Antrum of
Highmore

Inferior
turbinated
bone

over the septum; but, in the intervals between the spongy bones, and on the floor of the nasal fossæ, it is very thin. Where it lines the various sinuses and the antrum of Highmore, it is thin and pale.

Owing to the great thickness of this membrane, the nasal fossæ are much narrower, and the turbinated bones, especially the middle and inferior, appear larger and more prominent than in the skeleton. From the same circumstance, also, the various apertures communicating with the meatuses are considerably narrowed.

The vestibule, as already stated, is lined by modified skin, and contains hairs or vibrissæ which guard the entrance of the nostril.

Structure of the mucous membrane.—The epithelium covering the mucous membrane differs in its character according to the functions of the part of the nose in which it is found. In the respiratory portion of the nasal cavity the epithelium is columnar and ciliated. Interspersed among the columnar ciliated cells are goblet or mucin cells, while between their bases are found smaller pyramidal cells. In this region, beneath the epithelium and its basement mem-brane, is a fibrous layer infiltrated with lymph-corpuscles, so as to form in many parts a diffuse adenoid tissue, and beneath this a nearly continuous layer of

smaller and larger glands, some mucous and some serous, the ducts of which open
upon the surface.   In the olfactory region the mucous membrane is yellowish in
colour and the epithelial cells' are columnar and non-ciliated; they are of two
kinds, supporting cells and olfactory cells.   The *supporting cells* contain oval
nuclei, which are situated in the deeper parts of the cells and constitute the zone
of oval nuclei; the outer part of the cell is columnar, and contains granules of

FIG. 651.—Section of the olfactory mucous membrane.   (Cadiat.)

a. Epithelium.   *b.* Glands of Bowman.   *c.* Nerve bundles.

yellow pigment, while its deeper portion is prolonged as a delicate process which
ramifies and communicates with similar processes from neighbouring cells,
so as to form a network in the deep part of the mucous membrane.   Lying
between these central processes of the supporting cells are a number of spindle-
shaped cells, the *olfactory cells*, which consist of a large spherical nucleus
surrounded by a small amount of granular protoplasm, and possessing two
processes, one of which runs out-
wards between the columnar epithelial cells, and projects on the
surface of the mucous membrane as
a fine, hair-like process, the *olfactory
hair*; the other or deep process runs
inwards, is frequently beaded, and
is continuous with one of the terminal filaments of the olfactory nerve.
Beneath the epithelium, extending
through the thickness of the mucous
membrane, is a layer of tubular,
often branched, glands, the *glands
of Bowman*, identical in structure
with serous glands.

FIG. 652.—Nerves of septum of nose.
Right side.

The *arteries of the nasal fossæ*
are the anterior and posterior ethmoidal, from the ophthalmic, which
supply the ethmoidal cells, frontal
sinuses, and roof of the nose; the
spheno-palatine, from the internal
maxillary, which supplies the mucous membrane covering the spongy bones,
the meatuses and septum; the inferior artery of the septum, from the superior
coronary of the facial; the infra-orbital and alveolar branches of the internal
maxillary, which supply the lining membrane of the antrum; and the pterygo-
palatine branch of the same artery, which supplies the sphenoidal sinus.   The
ramifications of these vessels form a close, plexiform network, beneath and in
the substance of the mucous membrane.

The *veins of the nasal fossæ* form a close cavernous-like network beneath the mucous membrane. This cavernous appearance is especially well marked over the lower part of the septum and over the middle and inferior turbinated bones. Some of the veins open into the spheno-palatine vein ; others join the facial vein ; some accompany the ethmoidal arteries, and terminate in the ophthalmic vein ; and, lastly, a few communicate with the veins in the interior of the skull, through the foramina in the cribriform plate of the ethmoid bone, and the foramen cæcum.

The *lymphatics* can be injected from the subdural and subarachnoid spaces, and form a plexus in the superficial portion of the mucous membrane. The lymph is drained partly into one or two glands which lie near the great cornu of the hyoid bone, and partly into a gland situated in front of the axis.

The *nerves* of ordinary sensation are, the nasal branch of the ophthalmic, filaments from the anterior dental branch of the superior maxillary, the Vidian, the naso-palatine, the large or anterior palatine, and nasal branches of Meckel's ganglion.

The *olfactory*, the special nerve of the sense of smell, is distributed to the olfactory region, already referred to (page 852).

The *nasal branch of the ophthalmic* distributes filaments to the fore part of the septum, and outer wall of the nasal fossæ.

Filaments from the anterior dental branch of the superior maxillary supply the inferior meatus and inferior turbinated bone.

The *Vidian nerve* supplies the upper and back part of the septum, and superior spongy bone ; and the *upper nasal branches* from the spheno-palatine ganglion have a similar distribution.

The *naso-palatine nerve* supplies the middle of the septum.

The *large*, or *anterior palatine nerve*, supplies the *lower nasal branches* to the middle and lower spongy bones.

Surgical Anatomy.—Instances of congenital deformity of the nose are occasionally met with, such as complete absence of the nose, an aperture only being present ; or perfect development on one side, and suppression or malformation on the other; or there may be imperfect apposition of the nasal bones, so that the nose presents a median cleft or furrow. Deformities which have been acquired are much more common, such as flattening of the nose, the result of syphilitic necrosis ; or imperfect development of the nasal bones in cases of congenital syphilis ; or a lateral deviation of the nose may result from fracture.

The skin over the alæ and tip of the nose is thick and closely adherent to subjacent parts. Inflammation of this part is therefore very painful, on account of the tension. It is largely supplied with blood, and, the circulation here being terminal, vascular engorgement is liable to occur, especially in women at the menopause, and in both sexes from disorders of digestion, exposure to cold, &c. The skin of the nose also contains a large number of sebaceous follicles, and these, as the result of intemperance, are apt to become affected and the nose reddened, congested, and irregularly swollen. To this the term ' grog-blossom ' is popularly applied. In some of these cases there is enormous hypertrophy of the skin and subcutaneous tissues, producing pendulous masses, termed lipomata nasi. Epithelioma and rodent ulcer may attack the nose, the latter being the more common of the two. Lupus and syphilitic ulceration frequently attack the nose, and may destroy the whole of the cartilaginous portion. In fact, lupus vulgaris begins more frequently on the ala of the nose than in any other situation.

Cases of congenital occlusion of one or both nostrils, or adhesion between the ala and septum, may occur and may require immediate operation, since the obstruction much interferes with sucking. Bony closure of the posterior nares may also occur.

To examine the nasal cavities, the head should be thrown back and the nose drawn upwards, the parts being dilated by some form of speculum. It can also be examined with the little finger or a probe, and in this way foreign bodies detected. A still more extensive examination can be made by Rouge's operation, which was introduced for the cure of ozæna, by the removal of any dead bone which may be present in this disease. The whole framework of the nose is lifted up by an incision made inside the mouth, through the junction of the upper lip with the bone ; the septum nasi and the lateral cartilages are divided with strong scissors till the anterior nares are completely exposed. The posterior nares can be explored by reflected light from the mouth, through which they can be illuminated. The examination is very difficult to carry out, and, as a rule, sufficient information regarding the presence of foreign bodies or tumours in the naso-pharynx can be obtained by the introduction of the finger behind the soft palate through the mouth. The septum of the nose may be displaced or may deviate from the middle line : this may be the result of an injury or of some congenital defect in its development; in the latter case the deviation usually occurs along the line of union of the vomer and mesethmoid and rarely occurs before the seventh year. Sometimes the deviation may be so great that the septum may come into contact with the outer wall of the nasal fossæ, and may even become adherent to it, thus producing complete obstruction. Perforation of the septum is not an uncommon affection, and may arise from several causes : syphilitic or tuberculous ulceration, blood tumour or abscess of the septum, and especially in workmen exposed to the vapour of bichromate of potash, from the irritating and corrosive action of the fumes. When small, the perforation may cause a peculiar whistling sound during respiration. When large, it may lead to the falling in of the bridge of the nose.

Epistaxis is a very common affection in children. It is rarely of much consequence, and will almost always subside ; but in the more violent hæmorrhages of later life it may be necessary to plug the posterior nares. In performing this operation it is desirable to remember the size of the posterior nares. A ready method of regulating the bulk of the plug to fit the opening is to make it of the same size as the terminal phalanx of the thumb of the patient to be operated on.

Nasal polypus is a very common disease, and presents itself in three forms : the gelatinous, the fibrous, and the malignant. The first is by far the most common. It grows from the mucous membrane of the outer wall of the nasal fossa, where there is an abundant layer of highly vascular submucous tissue ; rarely from the septum, where the mucous membrane is closely adherent to the cartilage and bone, without the intervention of much, if any, submucous tissue. Their most common seat is probably the middle turbinated bone. The fibrous polypus generally grows from the base of the skull behind the posterior nares, or from the roof of the nasal fossæ. The malignant polypi, both sarcomatous or carcinomatous, may arise in the nasal cavities and the nasopharynx ; or they may originate in the antrum, and protrude through its inner wall into the nasal fossa.

Rhinoliths, or nose-stones, may sometimes be found in the nasal cavities, from the formation of phosphate of lime, either upon a foreign body or a piece of inspissated secretion.

THE EYE

The eyeball is contained in the cavity of the orbit. In this situation it is securely protected from injury, while its position is such as to ensure the most extensive range of sight. It is acted upon by numerous muscles, by which it is capable of being directed to different parts ; it is supplied by vessels and nerves, and is additionally protected in front by several appendages, such as the eyebrow, eyelids, &c.

The eyeball is embedded in the fat of the orbit, but is surrounded by a thin membranous sac, the *capsule of Tenon*, which isolates it, so as to allow of free movement.

The **capsule of Tenon** consists of a thin membrane which envelops the eyeball from the optic nerve to the ciliary region, separating it from the orbital fat and forming a socket in which it plays. Its inner surface is smooth, and is in contact with the outer surface of the sclerotic, the *perisclerotic lymph-space* only intervening. This lymph-space is continuous with the subdural and subarachnoid spaces, and is traversed by delicate bands of connective tissue which extend between the capsule and the sclerotic. The capsule is perforated behind by the ciliary vessels and nerves and by the optic nerve, being continuous with the sheath of the latter. In front it blends with the ocular conjunctiva, and with it is attached to the ciliary region of the eyeball. It is perforated by the muscles which move the eyeball, and on each it sends a tubular sheath. The sheath of the Superior oblique is carried as far as the fibrous pulley of that muscle ; that on the Inferior oblique reaches as far as the floor of the orbit, to which it gives off a slip. The sheaths on the Recti are gradually lost in the perimysium, but they give off important expansions. The expansion from the Superior rectus blends with the tendon of the Levator palpebræ ; that of the Inferior rectus is attached to the inferior tarsal plate. These two Recti, therefore, will exercise some influence on the movements of the eyelids. The expansions from the sheaths of the Internal and External recti are strong, especially the one from the latter muscle, and are attached to the lachrymal and malar bones respectively. As they probably check the action of these two Recti they have been named the *internal* and *external check ligaments*.

Lockwood has also described a thickening of the lower part of the capsule of Tenon, which he has named the *suspensory ligament of the eye*. It is slung like a hammock below the eyeball, being expanded in the centre and narrow at its extremities, which are attached to the malar and lachrymal bones respectively.*

The eyeball is composed of segments of two spheres of different sizes. The anterior segment is one of a small sphere; it is transparent, and forms about one-sixth of the eyeball. It is more prominent than the posterior segment, which is one of a much larger sphere, and is opaque, and forms about five-sixths of the globe. The term *anterior pole* is applied to the central point of the anterior curvature of the eyeball, and that of *posterior pole* to the central point of its posterior curvature; a line joining the two poles forms its *sagittal axis*. The axes of the eyeballs are nearly parallel, and therefore do not correspond to the axes of the orbits, which are directed outwards. The optic nerves follow the direction of the axes of the orbits, and are therefore not parallel; each enters its eyeball about 1 mm. below and 3 mm. to the inner or nasal side of the posterior pole. The eyeball measures rather more in its transverse and antero-posterior diameters than in its vertical diameter, the former amounting to about 24 mm., the latter to about 23·5 mm.; in the female all three diameters are rather less than in the male. At birth the eyeball has a diameter of about 17·5 mm., while at puberty it measures from 20 to 21 mm.

The eyeball is composed of three investing tunics and of three refracting media.

Tunics of the Eye

From without inwards the three tunics are:

    I. Sclerotic and Cornea.
    II. Choroid, Ciliary Body, and Iris.
    III. Retina.

I. The Sclerotic and Cornea

The sclerotic and cornea (fig. 653) form the external tunic of the eyeball; they are essentially fibrous in structure, the sclerotic being opaque, and forming the posterior five-sixths of the globe; the cornea, which forms the remaining sixth, being transparent.

The **Sclerotic** (σκληρός, *hard*) has received its name from its extreme density and hardness; it is a firm, unyielding, fibrous membrane, serving to maintain the form of the globe. It is much thicker behind than in front. Its *external surface* is of a white colour, and is in contact with the inner surface of the capsule of Tenon; it is quite smooth, except at the points where the Recti and Obliqui muscles are inserted into it, and its anterior part is covered by the conjunctival membrane: hence the whiteness and brilliancy of the front of the eyeball. Its *inner surface* is stained of a brown colour and marked by grooves, in which are lodged the ciliary nerves and vessels; it is loosely connected by an exceedingly fine cellular tissue (*lamina fusca*) with the outer surface of the choroid, an extensive lymph-space (*perichoroidal*) intervening between the sclerotic and choroid. Behind, it is pierced by the optic nerve, and is continuous with its fibrous sheath, which is derived from the dura mater. At the point where the optic nerve passes through the sclerotic, this tunic forms a thin cribriform lamina, the *lamina cribrosa*; the minute orifices in this lamina serve for the transmission of the nervous filaments, and the fibrous septa dividing them from one another are continuous with the membranous processes which separate the bundles of nerve-fibres. One of these openings, larger than the rest, occupies the centre of the lamella; it transmits the central artery of the retina to the interior of the eyeball. Around the cribriform lamella are numerous small apertures for the transmission of the ciliary vessels and nerves, and about midway between the margin of the cornea and the entrance of the optic nerve are four or five large apertures, for the transmission of veins (*venæ vorticosæ*). In front, the fibrous tissue of the sclerotic is directly continuous with that of the cornea, but the opaque sclerotic slightly overlaps the outer surface of the transparent cornea.

* See a paper by C. B. Lockwood, *Journal of Anatomy and Physiology*, vol. xx. part i. p. 1.

Structure.—The sclerotic is formed of white fibrous tissue intermixed with fine elastic fibres, and of flattened connective-tissue corpuscles, some of which are pigmented, contained in cell-spaces between the fibres. These fibres are aggregated into bundles, which are arranged chiefly in a longitudinal direction. It yields gelatin on boiling. Its vessels are not numerous, the capillaries being of small size, uniting at long and wide intervals. Its nerves are derived from the ciliary nerves, but their exact mode of ending is not known.

The **Cornea** is the projecting transparent part of the external tunic of the eyeball, and forms the anterior sixth of the globe. It is almost circular in shape, occasionally a little broader in the transverse than in the vertical direction. It is convex anteriorly, and projects forwards from the sclerotic in the same manner that a watch-glass does from its case. Its degree of curvature varies in different individuals, and in the same individual at different periods of life; it is more prominent in youth than in advanced life, when it becomes flattened

FIG. 653.—Horizontal section of the eyeball.

The cornea is dense and of uniform thickness throughout; its posterior surface is perfectly circular in outline, and exceeds the anterior surface slightly in extent, from the latter being overlapped by the sclerotic.

Structure (fig. 654).—The cornea consists of four layers, namely: (1) a layer of stratified epithelium, continuous with that of the conjunctiva; (2) a thick central fibrous structure, the *substantia propria*; (3) a homogeneous elastic lamina; and (4) a single layer of endothelial cells, forming part of the lining membrane of the anterior chamber of the eyeball.

The *conjunctival epithelium*, which covers the front of the cornea proper, consists of several strata of epithelial cells. The cells of the deepest layer are columnar: then follow two or three layers of polyhedral cells, the majority of which present finger-like processes (i.e. prickle-cells) similar to those found in the cuticle. Lastly, there are three or four layers of scaly epithelium, with flattened nuclei.

The *substantia propria* is fibrous, tough, unyielding; perfectly transparent, and continuous with the sclerotic. It is composed of about sixty flattened lamellæ, superimposed one on another. These lamellæ are made up of bundles of modified connective tissue, the fibres of which are directly continuous with the fibres of the sclerotic. The fibres of each lamella are for the most part parallel with each other; those of alternating lamellæ at right angles to each other. Fibres, however, frequently pass from one lamella to the next.

The lamellæ are connected with each other by an interstitial cement-substance, in which are spaces, the *corneal spaces*. These are stellate in shape and have numerous offsets, by which they communicate with each other. Each contains a cell, the *corneal corpuscle*, which resembles in form the space in which it is lodged; it does not, however, entirely fill it.

FIG. 654.—Vertical section of human cornea from near the margin. (Waldeyer.) Magnified.

1. Epithelium.   2. Anterior homogeneous lamina.   3. Substantia propria.   4. Posterior homogeneous (elastic) lamina.   5. Endothelium of the anterior chamber.   *a.* Oblique fibres in the anterior layer of the substantia propria. *b.* Lamellæ the fibres of which are cut across, producing a dotted appearance. *c.* Corneal corpuscles appearing fusiform in section. *d.* Lamellæ the fibres of which are cut longitudinally.   *e.* Transition to the sclerotic, with more distinct fibrillation, and surmounted by a thicker epithelium. *f.* Small blood-vessels cut across near the margin of the cornea.

Immediately beneath the conjunctival epithelium, the cornea proper presents certain characteristics which have led some anatomists to regard it as a distinct membrane, and it has been named by Bowman the *anterior elastic lamina*. It differs, however, from the posterior elastic lamina or membrane of Descemet in many essential particulars, presenting evidence of fibrillar structure, and not having the same tendency to curl inwards, or to undergo fracture, when detached from the other layers of the cornea. It consists of extremely closely interwoven fibrils, similar to those found in the rest of the cornea proper, but contains no corneal corpuscles. It may be regarded as a part of the proper tissue of the cornea.*

The *posterior elastic lamina* (*membrane of Descemet or Demours*), which covers the posterior surface of the substantia propria, presents no structure recognisable under the microscope. It consists of an elastic, and perfectly transparent homogeneous membrane, of extreme thinness, which is not rendered opaque by either water, alcohol, or acids. It is very brittle, but its most remarkable property is its extreme elasticity, and the tendency which it presents to curl up, or roll upon itself, with the attached surface innermost, when separated from the proper substance of the cornea. Its use appears to be (as suggested by Jacob) 'to preserve the requisite permanent correct curvature of the flaccid cornea proper.'

At the margin of the cornea this posterior elastic membrane breaks up into fibres to form a reticular structure at the outer angle of the anterior chamber, the intervals between the fibres forming small cavernous spaces, the *spaces of Fontana*. These little recesses communicate with a circular canal in the substance of the sclerotic close to its junction with the cornea. This is the *canal of Schlemm*, or *sinus venosus sclerœ*; it communicates internally with the anterior chamber through the spaces of Fontana, and externally with the scleral veins. Some of the fibres of this reticulated structure are continued into the front of the iris, forming the *ligamentum pectinatum iridis*; while others are connected with the fore part of the sclerotic and choroid.

The *endothelial lining of the aqueous chamber* covers the posterior surface of the elastic lamina, is reflected on to the front of the iris, and also lines the spaces of Fontana. It consists of a single layer of polygonal, flattened, nucleated cells, similar to those lining other serous cavities.

Arteries and Nerves.—The cornea is a non-vascular structure, the capillary vessels terminating in loops at its circumference. Lymphatic vessels have not as yet been demonstrated in it, but are represented by the channels in which the bundles of nerves run; these are lined by an endothelium, and are continuous with the cell-spaces. The nerves are numerous, twenty-four to thirty-six in number (Kölliker); forty to forty-five (Waldeyer and Sümisch): they are derived from the ciliary nerves. They form an *annular plexus* around the periphery of the cornea, from which fibres enter the substantia propria. They lose their medullary sheaths and ramify throughout its substance in a delicate network, and their terminal filaments form a firm and closer plexus on the surface of the cornea proper beneath the epithelium. This is termed the *subepithelial plexus*, and from it fibrils are given off which ramify between the epithelial cells, forming a network which is termed the *intra-epithelial plexus*.

Dissection.—In order to separate the sclerotic and cornea, so as to expose the second tunic, the eyeball should be immersed in a small vessel of water, and held between the finger and thumb. The sclerotic is then carefully incised, in the equator of the globe, till the choroid is exposed. One blade of a pair of probe-pointed scissors is now introduced through the opening thus made, and the sclerotic divided around its entire circumference, and removed in separate portions. The front segment being then drawn forwards, the handle of the scalpel should be pressed gently against it at its connection with the periphery of the iris, and these being separated, a quantity of perfectly transparent fluid will escape; this is the aqueous humour. In the course of the dissection, the ciliary nerves may be seen lying in the loose cellular tissue between the choroid and sclerotic, or contained in delicate grooves on the inner surface of the latter membrane.

II. The Choroid, Ciliary Body, and Iris

The Second Tunic of the Eye (*tunica vasculosa oculi*) is formed from behind forwards by the choroid, the ciliary body, and the iris.

The choroid is the vascular and pigmentary tunic of the eyeball, investing the posterior five-sixths of the globe, and extending as far forwards as the ora serrata of the retina; the ciliary body connects the choroid to the circumference of the iris. The iris is the circular diaphragm behind the cornea, and presents in its centre a large rounded aperture, the *pupil*.

---

* This layer has been called by Reichert the 'anterior limiting layer,' a name which appears more applicable to it than that of 'anterior elastic lamina.'

The **Choroid** is a thin, highly vascular membrane, of a dark brown or chocolate colour, which invests the posterior five-sixths of the globe, and is pierced behind by the optic nerve, and in this situation is firmly adherent to the sclerotic. It is thicker behind than in front. Externally, it is loosely connected by the lamina fusca with the inner surface of the sclerotic. Its inner surface is attached to the pigmented layer of the retina.

Structure.—The choroid consists mainly of a dense capillary plexus and of small arteries and veins, carrying the blood to and returning it from this plexus. On its external surface, i.e. the surface next the sclerotic, is a thin membrane, the *lamina suprachoroidea*, composed of delicate non-vascular lamellæ— each lamella consisting of a network of fine elastic fibres among which are branched pigment-cells. The spaces between the lamellæ are lined by endothelium, and open freely into the perichoroidal lymph-space, which, in its turn, communicates with the perisclerotic space by the perforations in the sclerotic through which the vessels and nerves are transmitted.

Internal to this lamina is the *choroid proper*, and in consequence of the small arteries and veins being arranged on the outer surface of the capillary network, it is customary to describe this as consisting of two layers: the outermost, composed of small arteries and veins, with pigment-cells interspersed between them; and the inner, consisting of a capillary plexus. The *external layer* or *lamina vasculosa* consists, in part, of the larger branches of the short ciliary arteries which run forwards between the veins, before they bend inwards to terminate in the capillaries; but is formed principally of veins, which are named, from their arrangement, *venæ vorticosæ*. They converge to four or five equidistant trunks, which pierce the sclerotic midway between the margin of the cornea and the entrance of the optic nerve. Interspersed between the vessels are dark star-shaped pigment-cells, the offsets from which, communicating with similar branchings from neighbouring cells, form a delicate network or stroma, which, towards the inner surface of the choroid, loses its pigmentary character. The *internal layer* consists of an exceedingly fine capillary plexus, formed by the short ciliary vessels, and is known as the *lamina chorio-capillaris* or *tunica Ruyschiana*. The network is close, and finer at the hinder part of the choroid than in front. About half an inch behind the cornea its meshes become larger, and are continuous with those of the

FIG. 655.—The choroid and iris. (Enlarged.)

FIG. 656.—The veins of the choroid. (Enlarged.)

ciliary processes. These two laminæ are connected by an *intermediate stratum*, which is destitute of pigment-cells and consists of fine elastic fibres. On the inner surface of the lamina chorio-capillaris is a very thin, structureless, or, according to Kölliker, faintly fibrous membrane, called the *lamina basalis* or *membrane of Bruch*; it is closely connected with the stroma of the choroid, and separates it from the pigmentary layer of the retina.

Tapetum.—This name is applied to the iridescent appearance which is seen in the outer and posterior part of the choroid of many animals.

The ciliary body should now be examined. It may be exposed, either by detaching the iris from its connection with the Ciliary muscle, or by making a transverse section of the globe, and examining it from behind.

The **ciliary body** comprises the orbiculus ciliaris, the ciliary processes, and the Ciliary muscle.

The *orbiculus ciliaris* is a zone of about one-sixth of an inch in width, directly continuous with the anterior part of the choroid; it presents numerous ridges arranged in a radial manner.

The *ciliary processes* are formed by the plaiting and folding inwards of the various layers of the choroid (i.e. the choroid proper and the lamina basalis) at

FIG. 657.—The arteries of the choroid and iris.
The sclerotic has been mostly removed. (Enlarged.)

its anterior margin, and are received between corresponding foldings of the sus-pensory ligament of the lens, thus establishing a connection between the choroid and inner tunic of the eye. They are arranged in a circle, and form a sort of plaited frill behind the iris, round the margin of the lens. They vary from sixty to eighty in number, lie side by side, and may be divided into large and small; the former are about one-tenth of an inch in length, and the latter, consisting of about one-third of the entire number, are situated in the spaces between them, but without regular alternation. They are attached by their periphery to three or four of the ridges of the orbiculus ciliaris, and are continuous with the layers of the choroid: their opposite extremities are free and rounded, and are directed towards the posterior chamber of the eyeball and circumference of the lens. In front, they are continuous with the circumference of the iris. Their posterior surfaces are connected with the suspensory ligament of the lens.

Structure.—The ciliary processes are similar in structure to the choroid, but the vessels are larger, and have chiefly a longitudinal direction. They are covered on their inner surface by two strata of black pigment-cells, which are continued forwards from the retina, and are named the *pars ciliaris retinæ*. In the stroma of the ciliary processes there are also stellate pigment-cells, which, however, are not so numerous as in the choroid itself.

The *Ciliary muscle* (Bowman) consists of unstriped fibres: it forms a greyish, semitransparent, circular band, about one-eighth of an inch broad, on the outer surface of the fore part of the choroid. It is thickest in front, and gradually becomes thinner behind. It consists of two sets of fibres, *radial* and *circular*. The former, much the more numerous, arise at the point of junction of the cornea and sclerotic, and partly also from the ligamentum pectinatum iridis, and, passing backwards, are attached to the ciliary processes and orbiculus ciliaris. One bundle, according to Waldeyer, is continued backwards to be inserted into the sclerotic. The circular fibres are internal to the radiating ones and to some extent unconnected with them, and have a circular course around the attachment of the iris. They are sometimes called the 'ring muscle' of Müller, and were formerly described as the ciliary ligament. They are well developed in hypermetropic, but are rudimentary or absent in myopic eyes. The Ciliary muscle is admitted to be the chief agent in accommodation, i.e. in adjusting the eye to the vision of near objects. When it contracts, it draws forwards the ciliary processes,

FIG. 658.—Section of the eye, showing the relations of the cornea, sclerotic, and iris, together with the Ciliary muscle and the spaces of Fontana near the angle of the anterior chamber. (Waldeyer.)

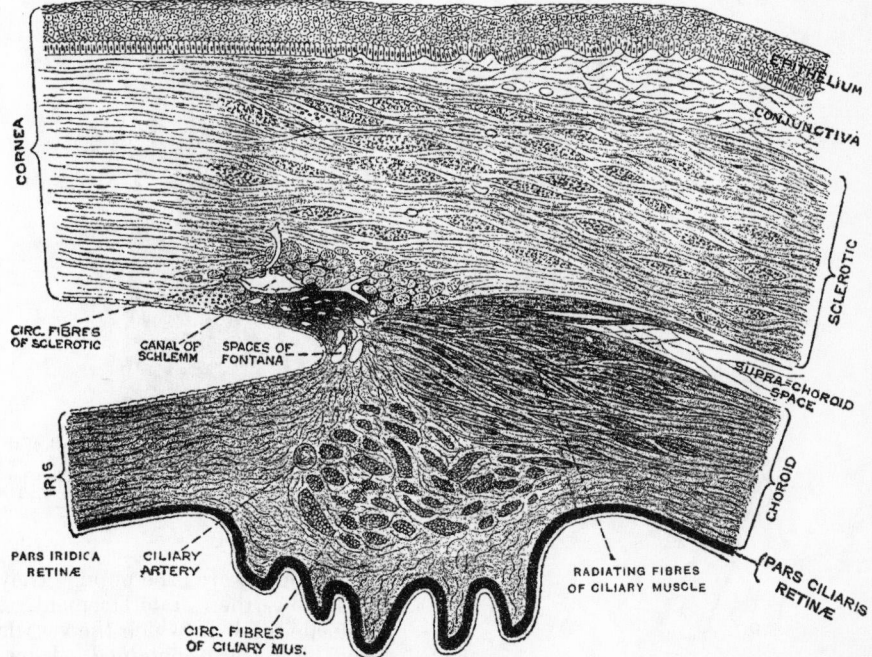

relaxes the suspensory ligament of the lens, and thus allows the anterior surface of the lens to become more convex. The pupil is at the same time slightly contracted.

The **Iris** (*iris*, a rainbow) has received its name from its various colours in different individuals. It is a thin, circular-shaped, contractile curtain, suspended in the aqueous humour behind the cornea, and in front of the lens, being perforated a little to the nasal side of its centre by a circular aperture, the *pupil*, for the transmission of light. By its circumference it is continuous with the ciliary body, and is also connected with the posterior elastic lamina of the cornea by means of the pectinate ligament; its inner or free edge forms the margin of the pupil; its surfaces are flattened, and look forwards and backwards, the anterior towards the cornea, the posterior towards the ciliary processes and lens. The anterior surface of the iris is variously coloured in different individuals, and marked by lines which converge towards the pupil. The posterior surface is of a deep purple tint, from being covered by two layers of pigmented, columnar epithelium, which are continuous posteriorly with the pars ciliaris retinæ. This pigmented epithelium is termed the *pars iridica retinæ*, though it is sometimes named *uvea*, from its resemblance in colour to a ripe grape.

3 Q

Structure.—The iris is composed of the following structures :

1. In front is a layer of flattened endothelial cells placed on a delicate hyaline basement-membrane. This layer is continuous with the epithelial layer covering the membrane of Descemet, and in men with dark-coloured irides the cells contain pigment-granules.

2. *Stroma.*—The stroma consists of fibres and cells. The former are made up of delicate bundles of fibrous tissue, of which some few fibres have a circular direction at the circumference of the iris ; but the chief mass consists of fibres

Fig. 659.—Diagrammatic representation of the course of the vessels in the eye. Horizontal section. (Leber.) Arteries and capillaries red ; veins blue.

Fig. 660.—Vessels of the choroid, ciliary processes, and iris of a child. (Arnold.) Magnified 10 times.

a. Capillary network of the posterior part of the choroid, ending at *b,* the ora serrata. *c.* Arteries of the corona ciliaris, supplying the ciliary processes, *d,* and passing into the iris, *e. f.* The capillary network close to the pupillary margin of the iris.

O. Entrance of optic nerve. *a.* Short posterior ciliary arteries. *b.* Long posterior ciliary arteries. *c.* Anterior ciliary vessels. *d.* Posterior conjunctival vessels. *d'.* Anterior conjunctival vessels. *e.* Central vessels of the retina. *f.* Vessels of the inner sheath of the optic nerve. *g.* Vessels of the outer sheath. *h.* Vorticose vein. *i.* Short posterior ciliary vein. *k.* Branches of the short posterior ciliary arteries to the optic nerve. *l.* Anastomosis of choroidal vessels with those of optic nerve. *m.* Chorio-capillaris. *n.* Episcleral vessels. *o.* Recurrent artery of the choroid. *p.* Circulus iridis major (in section). *q.* Vessels of iris. *r.* Vessels of ciliary process. *s.* Branch from ciliary muscle to vorticose vein. *t.* Branch from ciliary muscle to anterior ciliary vein. *u.* Canal of Schlemm. *v.* Capillary loop at margin of cornea.

radiating towards the pupil. They form, by their interlacement, a delicate mesh, in which the vessels and nerves are contained. Interspersed between the bundles of connective tissue are numerous branched cells with fine processes. In dark eyes many of them contain pigment-granules, but in blue eyes and the pink eyes of albinos they are unpigmented.

3. The *muscular fibre* is involuntary, and consists of circular and radiating fibres. The *circular fibres* (sphincter pupillæ) surround the margin of the pupil on the posterior surface of the iris, like a sphincter, forming a narrow band about one-thirtieth of an inch in width ; those near the free margin being closely aggregated ; those more external somewhat separated, and forming less complete circles. The *radiating fibres* (dilator pupillæ) converge from the circumference towards the centre, and blend with the circular fibres near the margin of the pupil. These fibres are regarded by some as elastic, not muscular.

4. *Pigment.*—The situation of the pigment-cells differs in different irides. In the various shades of blue eyes, the only pigment-cells are several layers of small

round or polyhedral cells, filled with dark pigment, situated on the posterior surface of the iris, and continuous with the pigmentary lining of the ciliary processes. The colour of the eye in these individuals is due to this colouring matter showing more or less through the texture of the iris. In the albino, even this pigment is absent. In the grey, brown, and black eye there are, as mentioned above, pigment-granules to be found in the cells of the stroma and in the epithelial layer on the front of the iris; to these the dark colour of the eye is due.

The *arteries of the iris* are derived from the long and anterior ciliary, and from the vessels of the ciliary processes (see page 632). The long ciliary arteries, two in number, having reached the attached margin of the iris, divide into an upper and lower branch, and encircling the iris, anastomose with corresponding branches from the opposite side; into this vascular zone (circulus major) the anterior ciliary pour their blood. From this zone vessels converge to the free

FIG. 661.—The iris, viewed from in front, with its great and small arterial circles.
(Testut.)

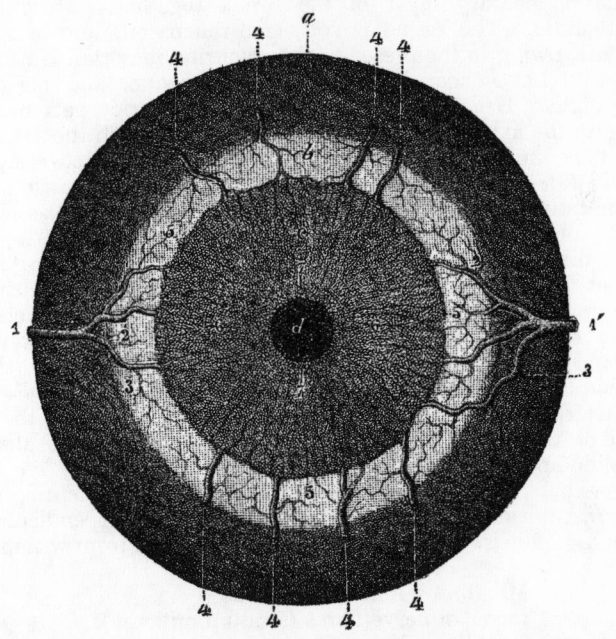

a. Choroid. b. Ciliary muscle. c. Iris. d. Pupil. 1 and 1'. The two long ciliary arteries, with 2, their ascending branch of bifurcation; 3, their descending branch of bifurcation. 4. The anterior ciliary arteries. 5. Circulus major; 6, its branches radiating through the iris. 7. Circulus minor around the pupil.

margin of the iris, and these communicate by branches from one to another and thus form a second zone (circulus minor) in this situation (figs. 659 and 661).

The *nerves of the choroid and iris* are derived from the ciliary branches of the lenticular ganglion, and the long ciliary from the nasal branch of the ophthalmic division of the fifth. They pierce the sclerotic around the entrance of the optic nerve, and run forwards in the perichoroidal space, and supply the blood-vessels of the choroid. After reaching the iris they form a plexus around its attached margin; from this are derived non-medullated fibres which terminate in the circular and radiating muscular fibres. Their exact mode of termination has not been ascertained. Other fibres from the plexus terminate in a network on the anterior surface of the iris. The fibres derived from the motor root of the lenticular ganglion (third nerve) supply the circular fibres, while those derived from the sympathetic supply the radiating fibres.

Membrana pupillaris.—In the fœtus, the pupil is closed by a delicate transparent vascular membrane, the *membrana pupillaris*, which divides the space in which the iris is suspended into two distinct chambers. This membrane contains numerous minute vessels, continued from the margin of the iris to those on the front part of the capsule of the lens. These vessels have a looped arrangement, and converge towards each other without anastomosing. Between the seventh and eighth month the membrane begins to disappear by its gradual absorption, from the centre towards the circumference, and at birth only a few fragments are present. It is said sometimes to remain permanent and produce blindness.

III. The Retina

The **Retina** is a delicate nervous membrane, upon the surface of which the images of external objects are received. Its outer surface is in contact with the choroid; its inner with the vitreous body. Behind, it is continuous with the optic nerve; it gradually diminishes in thickness from behind forwards; and, in front, extends nearly as far as the ciliary body, where it appears to terminate in a jagged margin, the *ora serrata*. Here the nervous tissues of the retina end, but a thin prolongation of the membrane extends forwards over the back of the ciliary processes and iris, forming the *pars ciliaris retinæ* and *pars iridica retinæ* already referred to. This forward prolongation consists of the pigmentary layer of the retina together with a stratum of columnar epithelium. The retina is soft, semitransparent, and of a purple tint in the fresh state, owing to the presence of a colouring material named *rhodopsin* or *visual purple*; but it soon becomes clouded, opaque, and bleached when exposed to sunlight. Exactly in the centre of the posterior part of the retina, corresponding to the axis of the eye, and at a point in which the sense of vision is most perfect, is an oval yellowish spot, called, after its discoverer, the *yellow spot* or *macula lutea* of Sömmerring; having a central depression, the *fovea centralis*. The retina in the situation of the fovea centralis is exceedingly thin, and the dark colour of the choroid is distinctly seen through it; so that it presents more the appearance of a foramen, and hence the name 'foramen of Sömmerring' at first given to it. It exists only in man, the quadrumana, and some saurian reptiles. About one-eighth of an inch (three millimetres) to the inner side, and about one millimetre below the level of the yellow spot, is the point of entrance of the optic nerve (*porus opticus* or *optic disc*), the circumference of which is slightly raised so as to form an eminence (*colliculus nervi optici*); the arteria centralis retinæ pierces its centre. This is the only part of the surface of the retina from which the power of vision is absent, and is termed the 'blind spot.'

Structure.—The retina is an exceedingly complex structure, and, when examined microscopically by means of sections made perpendicularly to its surface, is found to consist of ten layers, which are named from within outwards, as follows:

1. Membrana limitans interna.
2. Layer of nerve-fibres (stratum opticum).
3. Ganglionic layer, consisting of nerve-cells.
4. Inner molecular, or plexiform, layer.
5. Inner nuclear layer, or layer of inner granules.
6. Outer molecular, or plexiform, layer.
7. Outer nuclear layer, or layer of outer granules.
8. Membrana limitans externa.
9. Jacob's membrane (layer of rods and cones).
10. Pigmentary layer (tapetum nigrum).

1. The *membrana limitans interna* is the most internal layer of the retina, and is in contact with the hyaloid membrane of the vitreous body. It is derived from the supporting framework of the retina, with which tissue it will be described.

2. The *layer of nerve-fibres* is formed by the expansion of the optic nerve. This nerve passes through all the other layers of the retina, except the membrana limitans interna, to reach its destination. As the nerve passes through the lamina cribrosa of the sclerotic coat, the fibres of which it is composed lose their medullary sheaths and are continued onwards, through the choroid and retina, as simple axis cylinders. When these non-medullated fibres reach the internal surface of the retina they radiate from their point of entrance over the surface of the retina, grouped in bundles, and in many places, according to

Michel, arranged in plexuses. Most of the fibres in this layer are centripetal, and are the direct continuations of the axis-cylinder processes of the cells of the next layer, but a few of them (centrifugal fibres) pass through it and the next succeeding layer to ramify in the inner molecular and inner nuclear layers,

FIG. 662.—The arteria centralis retinæ, yellow spot, &c., the anterior half of the eyeball being removed. (Enlarged.)

where they terminate in enlarged extremities (fig. 665, 1, *m*). The layer is thickest at the optic nerve entrance, and gradually diminishes in thickness towards the ora serrata.

3. The *ganglionic layer* consists of a single layer of large ganglion-cells, except in the macula lutea, where there are several strata. The cells are some-

FIG. 663.   FIG. 664.

Vertical sections of the human retina : Fig. 663, half an inch from the entrance of the optic nerve. Fig. 664, close to the latter. 1. Layer of rods and cones, Jacob's membrane, bounded underneath by the *membrana limitans externa*. 2. Outer nuclear layer. 3. Outer molecular layer. 4. Inner nuclear layer. 5. Inner molecular layer. 6. Ganglionic layer. 7. Layer of nerve-fibres. 8. Sustentacular fibres of Müller. 9. Their attachment to the *membrana limitans interna*.

what flask-shaped ; their rounded internal surface resting on the preceding layer and sending off an axon which is prolonged as a nerve-fibre into the fibrous layer. From the opposite extremity numerous thicker processes (dendrites) extend into the inner molecular layer, where they branch out into flattened arborisations at different levels (fig. 665, VII). The ganglion-cells vary much in size, and the dendrites of the smaller ones as a rule arborise in the inner molecular layer as soon as they enter it; while the processes of the larger cells ramify close to the inner nuclear layer.

4. The *inner molecular layer* is made up of a dense reticulum of minute fibrils, formed by the interlacement of the dendrites of the ganglion-cells with those of the cells of the inner nuclear layer, immediately to be described. Within the reticulum formed by these fibrils a few branched spongioblasts are sometimes embedded.

5. The *inner nuclear layer* is made up of a number of closely packed cells, of which there are three different kinds. (1) A large number of oval cells, which are commonly regarded as bipolar nerve-cells, and are much more numerous than either of the other kind. They each consist of a large oval body placed vertically to the surface, and containing a distinct nucleus: they are surrounded by a small amount of protoplasm, which is prolonged into two processes; one of these passes inwards into the inner molecular layer, is varicose in appearance, and ends in a terminal ramification, which is often in close proximity to the ganglion-cells (fig. 665, I, c). The other process passes outwards, into the outer molecular layer, and there breaks up into a number of branches. According to Cajal there are two varieties of these bipolar cells: one in which the outer process arborises around the knobbed ends of the rod-fibres, and the inner around the cells of the ganglionic layer; these he calls *rod-bipolars* (fig. 665, I, c d); the others are those in which the outer process breaks up in a horizontal ramification, in contact with the end of a cone-fibre; these are the *cone-bipolars*, and their inner process breaks up into its terminal ramifications in the inner molecular layer (fig. 665, I, e). (2) At the innermost part of this inner nuclear layer is a stratum of cells, which are named by Cajal *amacrine* cells, from the fact that they have no axis-cylinder process, but they give a number of short protoplasmic processes, which extend into the inner molecular layer and there ramify (fig. 665, I, h). There are also at the outermost part of this layer some cells, the processes of which extend into and ramify in the outer molecular layer. These are the *horizontal* cells of Cajal. (3) Some few cells are also found in this layer, connected with the fibres of Müller, and will be described with those structures.

6. The *outer molecular layer* is much thinner than the inner molecular layer; but, like it, consists of a dense network of minute fibrils, derived from the processes of the horizontal cells of the preceding layer and the outer processes of the bipolar cells, which ramify in it, forming arborisations around the ends of the rod-fibres and with the branched foot-plates of the cone-fibres.

DESCRIPTION OF FIGURE 665.

I. Section of the dog's retina. A. Outer molecular layer. B. Inner molecular layer. a. Cone-fibre. b. Rod-fibre and nucleus. c d. Bipolar cells (inner granules) with vertical ramification of outer processes destined to receive the enlarged ends of rod-fibres. e. Bipolars with flattened ramification for ends of cone-fibres. f. Giant bipolar with flattened ramification. g. Cell sending a neuron or nerve-fibre process to the outer molecular layer. h. Amacrine cell with diffuse arborisation in inner molecular layer. i. Nerve-fibrils passing to outer molecular layer. j. Centrifugal fibres passing from nerve-fibre layer to inner molecular layer. m. Nerve-fibril passing into inner molecular layer. n. Ganglionic cells.

II. Horizontal or basal cells of the outer molecular layer of the dog's retina. A. Small cell with dense arborisation. B. Large cell, lying in inner nuclear layer but with its processes branching in the outer molecular. a. Its horizontal neuron. c. Medium-sized cell of the same character.

III. Cells from the retina of the ox. a. Rod-bipolars with vertical arborisations. b c d e. Cone-bipolars with horizontal ramification of outer process. h. Cells lying on the outer surface of the outer molecular layer, and ramifying within it. i j m. Amacrine cells within the substance of the inner molecular layer.

IV. Neurons or axis-cylinder processes belonging to horizontal cells of the outer molecular layer, one of them, b, ending in a close ramification at a.

V. Nervous elements connected with the inner molecular layer of the ox's retina. A. Amacrine cell, with long processes ramifying in the outermost stratum. B. Large amacrine with thick processes ramifying in second stratum. c. Flattened amacrine with long and fine processes ramifying mainly in the first and fifth strata. D. Amacrine with radiating tuft of fibrils destined for third stratum. E. Large amacrine, with processes ramifying in fifth stratum. F. Small amacrine, branching in second stratum. G H. Other amacrines destined for fourth stratum. a. Small ganglion-cell sending its processes to fourth stratum. b. A small ganglion-cell with ramifications in three strata. c. A small cell ramifying ultimately in first stratum. d. A medium-sized ganglion-cell ramifying in fourth stratum. e. Giant-cell, branching in third stratum. f. A bi-stratified cell ramifying in second and fourth strata.

VI. Amacrines and ganglion-cells from the dog. A. Amacrine with radiating tuft. B. Large amacrine passing to third stratum. c and G. Small amacrines with radiations in second stratum. F. Small amacrine passing to third stratum. D. Amacrine with diffuse arborisation. E. Amacrine belonging to fourth stratum. a d e g. Small ganglion-cells, ramifying in various strata. b f. Large ganglion-cells showing two different characters of arborisation. i. Bi-stratified cell.

VII. Amacrines and ganglion-cells from the dog. A B C. Small amacrines ramifying in middle of molecular layer. b d g h i. Small ganglion-cells showing various kinds of arborisation. f. A larger cell, similar in character to g, but with longer branch. a c e. Giant-cells with thick branches ramifying in the first, second, and third layers. l l. Ends of bipolars branching over ganglion-cells.

FIG. 665.—Elements of the retina of mammals, displayed by the chromate of silver method of Golgi. (Cajal.) (Copied from 'Quain's Anatomy.')

7. *The outer nuclear layer.*—Like the inner nuclear layer, this layer contains several strata of clear oval nuclear bodies ; they are of two kinds, and on account of their being respectively connected with the rods and cones of Jacob's membrane, are named rod-granules and cone-granules. The *rod-granules* are much the more numerous, and are placed at different levels throughout the layer. Their nuclei present a peculiar cross-striped appearance, and prolonged from either extremity of the granule is a fine process : the outer process is continuous with a single rod of Jacob's membrane ; the inner passes inwards towards the outer molecular layer and terminates in an enlarged extremity, and is embedded in the tuft into which the outer process of the rod-bipolars break up. In its course it presents numerous varicosities. The *cone-granules*, fewer in number than the rod-granules, are placed close to the membrana limitans externa, through which they are continuous with the cones of Jacob's membrane. They do not present any cross-striping, but contain a pyriform nucleus, which almost completely fills the cell. From their inner extremity a thick process passes inwards to the outer molecular layer, where it expands into a pyramidal enlargement or foot-plate, from which are given off numerous fine fibrils, which come in contact with the outer processes of the cone-bipolars.

8. *The membrana limitans externa.*—This layer, like the membrana limitans interna, is derived from the fibres of Müller, with which structures it will be described.

9. *Jacob's membrane (layer of rods and cones).*—The elements which compose this layer are of two kinds, *rods* and *cones*, the former being much more numerous than the latter. The *rods* are of nearly uniform size, and arranged perpendicularly to the surface. Each rod consists of two portions, an outer and inner, which are of about equal length. The segments differ from each other as regards refraction and in their behaviour with colouring reagents, the inner portion becoming stained by carmine, iodine, &c., the outer portion remaining unstained with these reagents, but staining yellowish-brown with osmic acid. The outer portion of each rod is marked by transverse striæ, and tends to break up into a number of thin discs superimposed on one another. It also exhibits faint longitudinal markings. The inner portion of each rod, at its deeper part where it is joined to the outer process of the rod-granule, is indistinctly granular ; its more superficial part presents a longitudinal striation, being composed of fine, bright, highly refracting fibrils. The visual purple or rhodopsin is only found in the outer segments of the rods.

The *cones* are conical or flask-shaped, their broad ends resting upon the membrana limitans externa, the narrow pointed extremity being turned to the choroid. Like the rods, they are made up of two portions, outer and inner ; the outer portion is a short conical process, which, like the outer segment of the rods, exhibits transverse striæ. The inner portion resembles the inner portion of the rods in structure, presenting a superficial striated and deeper granular part ; but differs from it in size, being bulged out laterally and presenting a flask shape. The chemical and optical characters of the two portions are identical with those of the rods.

10. *The pigmentary layer, or tapetum nigrum.*—The most external layer of the retina, formerly regarded as a part of the choroid, consists of a single layer of hexagonal epithelial cells, loaded with pigment-granules. They are smooth externally, where they are in contact with the choroid, but internally they are prolonged into fine, straight processes, which extend between the rods, this being especially the case when the eye is exposed to light. In the eyes of albinos, the cells of the pigmentary layer are present, but they contain no colouring matter.

Supporting framework of the retina.—Almost all these layers of the retina are connected together by a supporting framework, formed by the *fibres of Müller*, or *radiating fibres*, from which the membrana limitans interna et externa are derived. These fibres are found stretched between the two limiting layers, ' as columns between a floor and a ceiling,' and passing through all the nervous layers, except Jacob's membrane. Each commences on the inner surface of the retina by an expanded, often forked, base, which sometimes contains a spheroidal body, staining deeply with hæmatoxylin, the edges of the bases of adjoining fibres being united and thus forming a boundary line, which is the membrana limitans interna. As they pass through the nerve-fibre and ganglionic layers they give off few lateral branches; in the inner nuclear layer they give off numerous lateral processes for the support of the inner granules, while in the outer nuclear layer they form a network around the rod- and cone-fibrils, and unite to form the external limiting membrane at the bases of the rods and cones. In the inner nuclear layer each fibre of Müller presents a clear oval nucleus, which is sometimes situated at the side of, sometimes altogether within, the fibre.

Macula lutea and fovea centralis.—The structure of the retina at the yellow spot presents some modifications. In the macula lutea (1) the nerve-fibres are wanting as a continuous layer; (2) the ganglionic layer consists of several strata of cells, instead of a single layer; (3) in Jacob's membrane there are no rods, but only cones, and these are longer and narrower than in other parts; and

Fig. 666.—The layers of the retina (diagrammatic). (After Schultze.)

Pigmentary layer

Jacob's membrane

Membrana limitans externa

Outer nuclear layer

Outer molecular layer

Inner nuclear layer

Fibre of Müller

Inner molecular layer

Ganglionic layer

Layer of nerve-fibres
Membrana limitans interna

(4) in the outer nuclear layer there are only cone-fibres, which are very long and arranged in curved lines. At the fovea centralis the only parts which exist are (1) the cones of Jacob's membrane; (2) the outer nuclear layer, the cone-fibres of which are almost horizontal in direction; (3) an exceedingly thin inner molecular layer; (4) the pigmentary layer, which is thicker and its pigment more pronounced than elsewhere. The colour of the macula seems to imbue all the layers except Jacob's membrane; it is of a rich yellow, deepest towards the centre, and does not appear to consist of pigment-cells, but simply a staining of the constituent parts.

At the ora serrata the nervous layers of the retina terminate abruptly, and the retina is continued onwards as a single layer of elongated columnar cells covered by the pigmentary layer. This prolongation is known as the *pars ciliaris retinæ*, and can be traced forwards from the ciliary processes on to the back of the iris, where it is termed the *pars iridica retinæ* or *uvea*.

From the description given of the nervous elements of the retina it will be seen that there is no direct continuity between the structures which form its different layers except between the ganglionic and nerve-fibre layers, the majority of the nerve-fibres being formed of the axons of the ganglionic cells. In the inner molecular layer the dendrites of the ganglionic layer interlace with those of the cells of the inner nuclear layer, while in the outer molecular layer a like synapsis occurs between the processes of the inner granules and the rod and cone elements.

The *arteria centralis retinæ* and its accompanying vein pierce the optic nerve, and enter the globe of the eye through the porus opticus. The artery immediately bifurcates into an upper and a lower branch, and each of these again divides into an inner or nasal and an outer or temporal branch, which at first run between the hyaloid membrane and the nervous layer; but they soon enter the latter, and pass forwards, dividing dichotomously. From these branches a minute capillary plexus is given off, which does not extend beyond the inner nuclear layer. The macula receives small twigs from the temporal branches and others directly from the central artery; these do not, however, reach as far as the fovea centralis, which has no blood-vessels. The branches of the arteria centralis retinæ do not anastomose with each other—in other words, they are 'terminal arteries.' In the fœtus, a small vessel passes forwards, through the vitreous humour, to the posterior surface of the capsule of the lens.

REFRACTING MEDIA

The Refracting media are three, viz. :

Aqueous humour.   Vitreous body.   Crystalline lens.

I. AQUEOUS HUMOUR

The **aqueous humour** completely fills the anterior and posterior chambers of the eyeball. It is small in quantity (scarcely exceeding, according to Petit, four or five grains in weight), has an alkaline reaction, in composition is little more than water, less than one-fiftieth of its weight being solid matter, chiefly chloride of sodium.

The *anterior chamber* is the space bounded in front by the cornea; behind, by the front of the iris and the central part of the lens. The *posterior chamber* is a narrow chink between the peripheral part of the iris, the suspensory ligament of the lens, and the ciliary processes.

In the adult, these two chambers communicate through the pupil; but in the fœtus of the seventh month, when the pupil is closed by the membrana pupillaris, the two chambers are quite separate.

II. VITREOUS BODY

The **vitreous body** forms about four-fifths of the entire globe. It fills the concavity of the retina, and is hollowed in front, forming a deep concavity, the *fossa patellaris*, for the reception of the lens. It is perfectly transparent, of the consistence of thin jelly, and is composed of an albuminous fluid enclosed in a delicate transparent membrane, the *membrana hyaloidea*. It has been supposed, by Hannover, that from its inner surface numerous thin lamellæ are prolonged inwards in a radiating manner, forming spaces in which the fluid is contained. In the adult, these lamellæ cannot be detected even after careful microscopic examination in the fresh state, but in preparations hardened in weak chromic acid, it is possible to make out a distinct lamellation at the periphery of the body; and in the fœtus a peculiar fibrous texture pervades the mass, the fibres joining at numerous points, and presenting minute nuclear granules at their point of junction. In the centre of the vitreous humour, running from the entrance of the optic nerve to the posterior surface of the lens, is a canal, filled with fluid and lined by a prolongation of the hyaloid membrane. This is the *canal of Stilling*, which in the embryonic vitreous humour conveyed the minute vessel from the central artery of the retina to the back of the lens.

The fluid from the vitreous body resembles nearly pure water ; it contains, however, some salts, and a little albumen.

The hyaloid membrane encloses the whole of the vitreous humour. In front of the ora serrata it is thickened by the accession of radial fibres and is termed the *zonule of Zinn* or *zonula ciliaris*. It presents a series of radially arranged furrows, in which the ciliary processes are accommodated and to which they are adherent, as evidenced by the fact that when removed some of their pigment remains attached to the zonule. The zonule of Zinn splits into two layers, one of which is thin and lines the fossa patellaris, the other is named the *suspensory ligament of the lens*; it is thicker, and passes over the ciliary body to be attached to the capsule of the lens a short distance in front of its equator. Scattered and delicate fibres are also attached to the region of the equator itself. This ligament retains the lens in position, and is relaxed by the contraction of the radial fibres of the Ciliary muscle, so that the lens is allowed to become more convex. Behind the suspensory ligament there is a sacculated canal, the *canal of Petit*, which encircles the equator of the lens and which can be easily inflated through a fine blowpipe inserted through the suspensory ligament.

In the fœtus, the centre of the vitreous humour presents the canal of Stilling, already referred to, which transmits a minute artery to the capsule of the lens. *In the adult*, no vessels penetrate its substance; so that its nutrition must be carried on by the vessels of the retina and ciliary processes, situated upon its exterior.

III. Crystalline Lens

The **crystalline lens,** enclosed in its capsule, is situated immediately behind the pupil, in front of the vitreous body, and encircled by the ciliary processes, which slightly overlap its margin.

The *capsule of the lens* is a transparent, highly elastic, and brittle membrane, which closely surrounds the lens. It rests, behind, in the fossa patellaris in the fore part of the vitreous body; in front, it is in contact with the free border of the iris, this latter receding from it at the circumference, thus forming the posterior chamber of the eye; and it is retained in its position chiefly by the suspensory ligament of the lens, already described. The capsule is much thicker in front than behind, and structureless in texture ; when ruptured, the edges roll up with the outer surface innermost.

Fig. 667.—The crystalline lens, hardened and divided. (Enlarged.)

The anterior surface of the lens is covered by a single layer of transparent, columnar, nucleated cells. At the circumference of the lens, these cells undergo a change in form : they become elongated, and Babuchin states that he can trace the gradual transition of the cells into proper lens-fibres, with which they are directly continuous. There is no epithelium on the posterior surface.

In the fœtus, a small branch from the arteria centralis retinæ runs forwards, as already mentioned, through the vitreous humour to the posterior part of the capsule of the lens, where its branches radiate and form a plexiform network, which covers its surface, and they are continuous round the margin of the capsule with the vessels of the pupillary membrane, and with those of the iris. In the adult no vessels enter its substance.

The *lens* is a transparent, biconvex body, the convexity being greater on the posterior than on the anterior surface. The central points of its anterior and posterior surfaces are known as its *anterior* and *posterior poles*. It measures from nine to ten millimetres in the transverse diameter, and about four millimetres in the antero-posterior. It consists of concentric layers, of which the external in the fresh state are soft and easily detached (*substantia corticalis*) ; those beneath are firmer, the central ones forming a hardened nucleus (*nucleus lentis*). These laminæ are best demonstrated by boiling, or immersion in alcohol, and consist of minute parallel fibres, which are hexagonal prisms, the edges being dentated, and the dentations fitting accurately into each other ; their breadth is about $\frac{1}{5000}$th of an inch. Faint lines radiate from the anterior and posterior poles to the circumference of the lens. In the adult there may be six or more of these, but in the foetus they are only three in number and diverge from each other at angles of 120 degrees (fig. 668). On the anterior surface one line ascends vertically and the other two diverge downwards and outwards. On the posterior surface one ray descends vertically and the other two diverge upwards. They correspond with the free edges of an equal number of septa in the lens, along which the ends of the lens fibres come into apposition and are joined by transparent amorphous substance. The fibres run in a curved manner from

FIG. 668.—Diagram to show the direction and arrangement of the radiating lines on the front and back of the foetal lens. A. From the front. B. From the back.

A.　　　　　　　　　　B.

the septa on the anterior surface to those on the posterior surface. No fibres pass from pole to pole, but they are arranged in such a way that fibres which commence near the pole on the one aspect of the lens terminate near the peripheral extremity of the plane on the other, and *vice versa*. The fibres of the outer layers of the lens are nucleated, and together form a layer (nuclear layer) on the surface of the lens, most distinct towards its circumference.

The *changes produced in the lens by age* are the following :

In the foetus, its form is nearly spherical, its colour of a slightly reddish tint ; it is not perfectly transparent, and is so soft as to break down readily on the slightest pressure.

In the adult, the posterior surface is more convex than the anterior ; it is colourless, transparent, and firm in texture.

In old age it becomes flattened on both surfaces, slightly opaque, of an amber tint, and it also increases in density.

The *arteries of the globe of the eye* are the short, long, and anterior ciliary arteries, and the arteria centralis retinæ. They have been already described (see page 632).

The *ciliary veins* are seen on the outer surface of the choroid, and are named, from their arrangement, the *venæ vorticosæ*. They converge to four or five equidistant trunks which pierce the sclerotic midway between the margin of the cornea and the entrance of the optic nerve. Another set of veins accompanies the anterior ciliary arteries and opens into the ophthalmic vein.

The *ciliary nerves* are derived from the nasal branch of the ophthalmic nerve and from the ciliary or ophthalmic ganglion.

Surgical Anatomy.—From a surgical point of view the cornea may be regarded as consisting of three layers: (1) of an external epithelial layer, developed from the epiblast, and continuous with the external epithelial covering of the rest of the body, and therefore in its lesions resembling those of the epidermis; (2) of the cornea proper, derived from the mesoblast, and associated in its diseases with the fibro-vascular structures of the body; and (3) the posterior elastic layer with its endothelium, also derived from the mesoblast and having the characters of a serous membrane, so that inflammation of it resembles inflammation of the other serous and synovial membranes of the body.

The cornea contains no blood-vessels except at its periphery, where numerous delicate loops, derived from the anterior ciliary arteries, may be demonstrated on the anterior surface of the cornea. The rest of the cornea is nourished by lymph, which gains access to the proper substance of the cornea and the posterior layer through the spaces of Fontana. This lack of a direct blood-supply renders the cornea very apt to inflame in the cachectic and ill-nourished. In cases of granular lids, there is a peculiar affection of the cornea, called *pannus*, in which the anterior layers of the cornea become vascularised, and a rich network of blood-vessels may be seen upon it; and in interstitial keratitis new vessels extend into the cornea, giving it a pinkish hue to which the term ' salmon patch ' is applied. The cornea is richly supplied with nerves, derived from the ciliary, which enter the cornea through the fore part of the sclerotic and form plexuses in the stroma, terminating between the epithelial cells by free ends or in corpuscles. In cases of glaucoma the ciliary nerves may be pressed upon as they course between the choroid and sclerotic, and the cornea becomes anæsthetic. The sclerotic has very few blood-vessels and nerves. The blood-vessels are derived from the anterior ciliary, and form an open plexus in its substance. As they approach the corneal margin this arrangement is peculiar. Some branches pass through the sclerotic to the ciliary body ; others become superficial and lie in the epi-scleral tissue, and form arches, by anastomosing with each other, some little distance behind the corneal margin. From these arches numerous straight vessels are given off, which run forwards to the cornea, forming its marginal plexus. In inflamma-tion of the sclerotic and episcleral tissue these vessels become conspicuous, and form a pinkish zone of straight vessels radiating from the corneal margin, commonly known as the *zone of ciliary injection*. In inflammation of the iris and ciliary body this zone is present, since the sclerotic speedily becomes involved when these structures are inflamed. But in inflammation of the cornea the sclerotic is sel-dom much affected, though the cornea and sclerotic are structurally continuous. This would appear to be due to the fact, that the nutrition of the cornea is derived from a different source than that of the sclerotic. The sclerotic may be ruptured subcu-taneously without any laceration of the conjunctiva, and the rupture usually occurs near the corneal margin, where the tunic is thinnest. It may be complicated with lesions of adjacent parts—lacera-tion of the choroid, retina, iris, or suspensory liga-ment of the lens—and is then often attended with hæmorrhage into the anterior chamber, which masks the nature of the injury. In some cases the lens has escaped through the rent in the sclerotic and has been found under the conjunctiva. Wounds of the sclerotic are always dangerous, and are often followed by inflammation, suppuration, and by sympathetic ophthalmia.

One of the functions of the choroid is to provide nutrition for the retina, and to convey vessels and nerves to the ciliary body and iris. Inflammation of the choroid is therefore followed by grave disturb-ance in the nutrition of the retina, and is attended with early interference with vision. In its diseases it bears a considerable analogy to those which affect the skin, and, like it, is one of the places from which melanotic sarcomata may grow. These tumours contain a large amount of pigment in their cells, and grow only from those parts where pigment is naturally present. The choroid may be ruptured, without injury to the other tunics, as well as partici-pating in general injuries of the eyeball. In cases of uncomplicated rupture, the injury is usually at its posterior part, and is the result of a blow on the front of the eye. It is attended by considerable hæmorrhage, which for a time may obscure vision, but, in most cases, this is restored, as soon as the blood is absorbed.

The iris may be absent, either in part or alto-gether as a congenital condition, and in some instances the pupillary membrane may remain persistent, though it is rarely complete. And, again, the iris may be the seat of a malformation, termed *coloboma*, which consists in a deficiency or cleft, which in a great number of cases is clearly due to an arrest in development. In these cases it is found at the lower aspect, extending directly downwards from the pupil, and the gap frequently extends through the choroid to the

Fig. 669. — Section through the margin of the lens, showing the transition of the epithe-lium into the lens - fibres. (Babuchin.)

entrance of the optic nerve. In some rarer cases the gap is found in other parts of the iris, and is then not associated with any deficiency of the choroid. Wounds of the iris, especially if complicated with injury to the ciliary body, may be followed by serious consequences. If septic matter is introduced, and a suppurative inflammation is set up, complete loss of vision may result; and, what is perhaps of greater consequence, similar inflammatory changes may be set up in the sound eye, from spreading of the infective process through the connective tissue surrounding the optic nerve to the commissure, and thence down the opposite nerve to the sound eye. The iris is abundantly supplied with blood-vessels and nerves, and is therefore very prone to become inflamed. And when inflamed, in consequence of the intimate relationship which exists between the vessels of the iris and choroid, this latter tunic is very liable to participate in the inflammation. And, in addition, inflammation of adjacent structures, the cornea and sclerotic, is apt to spread into the iris. The iris is covered with epithelium, and partakes of the character of a serous membrane, and, like these structures, is liable to pour out a plastic exudation, when inflamed, and contract adhesions, either to the cornea in front (*synechia anterior*), or to the capsule of the lens behind (*synechia posterior*). In iritis the lens may become involved, and the condition known as secondary cataract may be set up. Tumours occasionally commence in the iris; of these, cysts, which are usually congenital, and sarcomatous tumours, are the most common and require removal. Gummata are not infrequently found in this situation. In some forms of injury of the eyeball, as the impact of a spent shot, the rebound of a twig, or a blow with a whip, the iris may be detached from the Ciliary muscle, the amount of detachment varying from the slightest degree to separation of the whole iris from its ciliary connection.

The retina, with the exception of its pigment layer and its vessels, is perfectly transparent when examined by the ophthalmoscope, so that its diseased conditions are recognised by its loss of transparency. In retinitis, for instance, there is more or less dense and extensive opacity of its structure, and not infrequently extravasations of blood into its substance. Hæmorrhages may also take place into the retina, from rupture of a blood-vessel without inflammation.

The retina may become displaced from effusion of serum between it and the choroid, or by blows on the eyeball, or may occur without apparent cause in progressive myopia, and in this case the ophthalmoscope shows an opaque, tremulous cloud. Glioma, a form of sarcoma, and essentially a disease of early life, is occasionally met with in connection with the retina.

The lens has no blood-vessels, nerves, or connective tissue in its structure, and therefore is not subject to those morbid changes to which tissues containing these structures are liable. It does, however, present certain morbid or abnormal conditions of various kinds. Thus, variations in shape, absence of the whole or a part of the lens, and displacements are among its congenital defects. Opacities may occur from injury, senile changes, malnutrition, or errors in growth or development. These opacities give rise to *cataract*, of which the senile variety is the most common. They vary as to the part of the lens in which the opacity commences, and are classified accordingly, as nuclear, cortical, lamellar, anterior and posterior polar. Senile changes may take place in the lens, impairing its elasticity and rendering it harder than in youth, so that it loses its power of altering its curvature to suit the requirements of near vision. This condition is known as *presbyopia*. And, finally, the lens may be dislocated or displaced by blows upon the eyeball; and its relations to surrounding structures altered by adhesions or the pressure of new growths.

There are two particular regions of the eye which require special notice: one of these is known as the 'filtration area,' and the other as the 'dangerous area.' The *filtration area* is the circumcorneal zone immediately in front of the iris. Here are situated the cavernous spaces of Fontana, which communicate with the canal of Schlemm, through which the chief transudation of fluid from the eye is now believed to take place. If any obstruction to this transudation occurs, increased intra-ocular tension is set up, and the disease known as *glaucoma* results. The *dangerous area of the eye* is the region in the neighbourhood of the ciliary body, and wounds or injuries in this situation are peculiarly dangerous; for inflammation of the ciliary body is liable to spread to many of the other structures of the eye, especially to the iris and choroid, which are intimately connected by nervous and vascular supplies.

APPENDAGES OF THE EYE

The appendages of the eye (*tutamina oculi*) include the eyebrows, the eyelids, the conjunctiva, and the lachrymal apparatus, viz. the lachrymal gland, the lachrymal sac, and the nasal duct.

The **eyebrows** (*supercilia*) are two arched eminences of integument, which surmount the upper circumference of the orbit on each side, and support numerous short, thick hairs, directed obliquely on the surface. In structure, the eyebrows consist of thickened integument, connected beneath with the Orbicularis palpebrarum, Corrugator supercilii, and Occipito-frontalis muscles. These muscles serve, by their action on this part, to control to a certain extent the amount of light admitted into the eye.

The **eyelids** (*palpebræ*) are two thin, movable folds, placed in front of the eye, protecting it from injury by their closure. The upper lid is the larger, and the more movable of the two, and is furnished with a separate elevator muscle, the *Levator palpebræ superioris*. When the eyelids are open, an elliptical space (*fissura palpebrarum*) is left between their margins, the angles of which correspond to the junction of the upper and lower lids, and are called *canthi*.

The *outer canthus* is more acute than the inner, and the lids here lie in close contact with the globe : but the *inner canthus* is prolonged for a short distance inwards towards the nose, and the two lids are separated by a triangular space, the *lacus lacrimalis*. At the commencement of the lacus lacrimalis, on the margin of each eyelid, is a small conical elevation, the *lachrymal papilla*, the apex of which is pierced by a small orifice, the *punctum lacrimale*, the commencement of the lachrymal canal.

The *eyelashes* (*cilia*) are attached to the free edges of the eyelids ; they are short, thick, curved hairs, arranged in a double or triple row at the margin of the lids : those of the upper lid, more numerous and longer than the lower, curve upwards ; those of the lower lid curve downwards, so that they do not interlace in closing the lids. Near the attachment of the eyelashes are the openings of a number of glands, *glands of Moll*, arranged in several rows close to the free margin of the lid. They are regarded as enlarged and modified sweat-glands.

Structure of the eyelids. — The eyelids are composed of the following structures taken in their order from without inwards :

Integument, areolar tissue, fibres of the Orbicularis muscle, tarsal plate and its ligament, Meibomian glands and conjunctiva. The upper lid has, in addition, the aponeurosis of the Levator palpebræ.

The *integument* is extremely thin, and continuous at the margin of the lids with the conjunctiva.

The *subcutaneous areolar tissue* is very lax and delicate, seldom contains any fat, and is extremely liable to serous infiltration.

Fig. 670.—Vertical section through the upper eyelid. (After Waldeyer.)

a. Skin. *b*. Orbicularis palpebrarum. *b'*. Marginal fasciculus of orbicularis (ciliary bundle). *c*. Levator palpebræ. *d*. Conjunctiva. *e*. Tarsal plate. *f*. Meibomian gland. *g*. Sebaceous gland. *h*. Eyelashes. *i*. Small hairs of skin. *j*. Sweat-glands. *k*. Posterior tarsal glands.

The *fibres of the Orbicularis muscle*, where they cover the palpebræ, are thin, pale in colour, and possess an involuntary action.

The *tarsal plates* are two thin elongated plates of dense connective tissue, about an inch in length. They are placed one in each lid, contributing to their form and support.

The *superior* tarsal plate, the larger, is of a semilunar form, about one-third of an inch in breadth at the centre, and becoming gradually narrowed at each extremity. To the anterior surface of this plate the aponeurosis of the Levator palpebræ is attached.

The *inferior* tarsal plate, the smaller, is thinner, and of an elliptical form.

The *free* or *ciliary margin* of these plates is thick, and presents a perfectly straight edge. The *attached* or *orbital margin* is connected to the circumference of the orbit by the fibrous membrane of the lids with which it is continuous. The outer angle of each plate is attached to the malar bone by the external tarsal ligament. The inner angles of the two plates terminate at the commencement of the lacus lacrimalis; they are attached to the nasal process of the superior maxilla by the internal tarsal ligament or tendo oculi.

The *palpebral* ligaments are membranous expansions situated one in each lid, and are attached marginally to the edge of the orbit, where they are continuous with the periosteum. The superior ligament blends with the tendon of the Levator palpebræ, the inferior with the inferior tarsal plate. Externally the two ligaments fuse to form the external tarsal ligament, just referred to; internally they are much thinner and, becoming separated from the internal tarsal ligament, are fixed to the lachrymal bone immediately behind the lachrymal sac. Together, the ligaments form an incomplete septum, the *septum orbitale*, which is perforated by the vessels and nerves which pass from the orbital cavity to the face and scalp.

The *Meibomian glands* (fig. 671) are situated upon the inner surface of the eyelids, between the tarsal plates and conjunctiva, and may be distinctly seen

FIG. 671.—The Meibomian glands &c. seen from the inner surface of the eyelids.

Puncta lacrimalia

through the mucous membrane on everting the eyelids, presenting the appearance of parallel strings of pearls. They are about thirty in number in the upper eyelid, and somewhat fewer in the lower. They are embedded in grooves in the inner surface of the tarsal plates, and correspond in length with the breadth of each plate; they are, consequently, longer in the upper than in the lower eyelid. Their ducts open on the free margin of the lids by minute foramina, which correspond in number to the follicles. The use of their secretion is to prevent adhesions of the lids.

Structure of the Meibomian glands.—These glands are a variety of the cutaneous sebaceous glands, each consisting of a single straight tube or follicle, having a cæcal termination, and with numerous small lateral diverticula opening into it. The tubes consist of a basement-membrane, lined at the mouths of the tubes by stratified epithelium; the deeper parts of the tubes and the lateral offshoots are lined by a layer of polyhedral cells. They are thus identical in structure with the sebaceous glands.

The **conjunctiva** is the mucous membrane of the eye. It lines the inner surface of the eyelids, and is reflected over the fore part of the sclerotic and cornea. In each of these situations, its structure presents some peculiarities.

The *palpebral portion of the conjunctiva* is thick, opaque, highly vascular, and covered with numerous papillæ, its deeper parts presenting a considerable amount of lymphoid tissue. At the margin of the lids it becomes continuous with the lining membrane of the ducts of the Meibomian glands, and, through the lachrymal canals, with the lining membrane of the lachrymal sac and nasal duct. At the outer angle of the upper lid the lachrymal ducts open on its free surface ; and at the inner angle of the eye it forms a semilunar fold, the *plica semilunaris*. The folds formed by the reflection of the conjunctiva from the lids on to the eye are called the *superior* and *inferior palpebral folds*, the former being the deeper of the two. Upon the *sclerotic*, the conjunctiva is loosely connected to the globe ; it becomes thinner, loses its papillary structure, is transparent, and only slightly vascular in health. Upon the *cornea*, the conjunctiva consists only of epithelium, constituting the stratified epithelium of the cornea, already described (see page 956). *Lymphatics* arise in the conjunctiva in a delicate zone around the cornea, from which the vessels run to the ocular conjunctiva.

At the point of reflection of the conjunctiva from the lid on to the globe of the eye, termed the *fornix conjunctivæ*, are a number of mucous glands, which are much convoluted. They are chiefly found in the upper lid. Other glands, analogous to lymphoid follicles, and called by Henle *trachoma glands*, are found in the conjunctiva, and, according to Strohmeyer, are chiefly situated near the inner canthus of the eye. They were first described by Brush, in his description of Peyer's patches of the small intestine, as ' identical structures existing in the under eyelid of the ox.'

The nerves in the conjunctiva are numerous and form rich plexuses. According to Krause, they terminate in a peculiar form of tactile corpuscle, which he terms ' terminal bulb.'

The *caruncula lacrimalis* is a small, reddish, conical-shaped body, situated at the inner canthus of the eye, and filling up the small triangular space in this situation, the *lacus lacrimalis*. It consists of a small island of skin containing sebaceous and sweat glands, and is the source of the whitish secretion which constantly collects at the inner angle of the eye. A few slender hairs are attached to its surface. On the outer side of the caruncula is a slight semilunar fold of mucous membrane, the concavity of which is directed towards the cornea ; it is called the *plica semilunaris*. Müller found smooth muscular fibres in this fold, and in some of the domesticated animals a thin plate of cartilage has been discovered. This structure is considered to be the rudiment of the third eyelid in birds, the *membrana nictitans*.

Lachrymal Apparatus (fig. 672)

The lachrymal apparatus consists of the lachrymal gland, which secretes the tears, and its excretory ducts, which convey the fluid to the surface of the eye. This fluid is carried away by the lachrymal canals into the lachrymal sac, and along the nasal duct into the cavity of the nose.

The **lachrymal gland** is lodged in a depression at the outer angle of the orbit, on the inner side of the external angular process of the frontal bone. It is of an oval form, about the size and shape of an almond. Its upper convex surface is in contact with the periosteum of the orbit, to which it is connected by a few fibrous bands. Its under concave surface rests upon the convexity of the eyeball, and upon the Superior and External recti muscles. Its vessels and nerves enter its posterior border, while its anterior margin is closely adherent to the back part of the upper eyelid, where it is covered to a slight extent by the reflection of the conjunctiva. The fore part of the gland is separated from the rest by a fibrous septum ; hence it is sometimes described as a separate lobe, called the *palpebral portion of the gland*

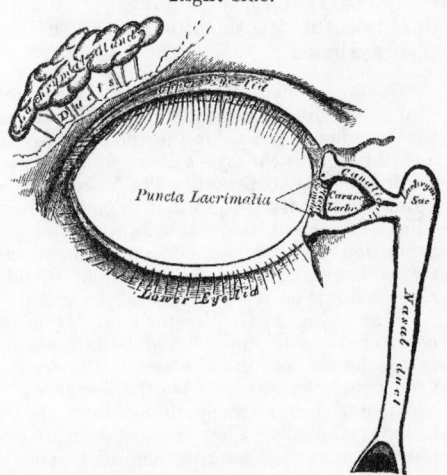

Fig. 672.—The lachrymal apparatus. Right side.

Puncta Lacrimalia

(*accessory gland of Rosenmüller*). Its ducts, from six to twelve in number, run obliquely beneath the mucous membrane for a short distance, and, separating from each other, open by a series of minute orifices on the upper and outer half of the conjunctiva, near its reflection on to the globe. These orifices are arranged in a row, so as to disperse the secretion over the surface of the membrane.

Structure of the lachrymal gland.—In structure and general appearance the lachrymal resembles the serous salivary glands (page 1047). In the recent state the cells are so crowded with granules that their limits can hardly be defined. They contain oval nuclei, and the cell protoplasm is finely fibrillated.

The **lachrymal canals** commence at the minute orifices, *puncta lacrimalia*, on the summit of a small conical elevation, the *lachrymal papilla*, seen on the margin of the lids, at the outer extremity of the lacus lacrimalis. The *superior canal*, the smaller and shorter of the two, at first ascends, and then bends at an acute angle, and passes inwards and downwards to the lachrymal sac. The *inferior canal* at first descends, and then, abruptly changing its course, passes almost horizontally inwards to the lachrymal sac. They are dense and elastic in structure and somewhat dilated at their angle. The mucous membrane is covered with stratified scaly epithelium, placed on a basement membrane. Outside the latter is a layer of striped muscular fibres, continuous with the Tensor tarsi muscle, which at the bases of the lachrymal papillæ are circularly arranged and form a kind of sphincter.

The **lachrymal sac** is the upper dilated extremity of the nasal duct, and is lodged in a deep groove formed by the lachrymal bone and nasal process of the superior maxillary. It is oval in form, its upper extremity being closed in and rounded, while below it is continued into the nasal duct. It is covered by a fibrous expansion derived from the tendo oculi, and on its deep surface it is crossed by the Tensor tarsi muscle (Horner's muscle, page 431), which is attached to the ridge on the lachrymal bone.

Structure.—The lachrymal sac consists of a fibrous elastic coat, lined internally by mucous membrane : the latter being continuous, through the lachrymal canals, with the mucous lining of the conjunctiva, and through the nasal duct with the pituitary membrane of the nose.

The **nasal duct** is a membranous canal, about three-quarters of an inch in length, which extends from the lower part of the lachrymal sac to the inferior meatus of the nose, where it terminates by a somewhat expanded orifice, provided with an imperfect valve, the *valve of Hasner*, formed by a fold of the mucous membrane. It is contained in an osseous canal, formed by the superior maxillary, the lachrymal, and the inferior turbinated bones, is narrower in the middle than at either extremity, and takes a direction downwards, backwards, and a little outwards. It is lined by mucous membrane, which is continuous below with the pituitary lining of the nose. This membrane in the lachrymal sac and nasal duct is covered with columnar epithelium as in the nose. This epithelium is in places ciliated.

Surface Form.—The palpebral fissure, or opening between the eyelids, is elliptical in shape, and differs in size in different individuals and in different races of mankind. In the Mongolian races, for instance, the opening is very small, merely a narrow fissure, and this makes the eye appear small in these races, whereas the size of the eyeball is relatively very constant. The normal form of the fissure is oblique, in a direction upwards and outwards, so that the outer angle is on a slightly higher level than the inner. This is especially noticeable in the Mongolian races, in whom, owing to the upward projection of the malar bone and the shortness of the external angular process of the frontal bone, the tarsal plate of the upper lid is raised at its outer part, and gives a very oblique direction to the palpebral fissure.

When the eyes are directed forwards, as in ordinary vision, the upper part of the cornea is covered by the upper lid, and the lower margin of the cornea corresponds to the level of the lower lid, so that about the lower three-fourths of the cornea are exposed under ordinary circumstances. On the margins of the lids, about a quarter of an inch from the inner canthus, are two small openings, the *puncta lacrimalia*, the commencement of the lachrymal canals. They are best seen by everting the eyelids. In the natural condition they are in contact with the conjunctiva of the eyeball, and are maintained in this position by the Tensor tarsi muscle, so that the tears running over the surface of the globe easily find their way into the lachrymal canals. The position of the lachrymal sac into which the canals open is indicated by a little tubercle (page 247), which is plainly to be felt on the lower margin of the orbit. The lachrymal sac lies immediately above and to the inner side of this tubercle, and a knife passed through the skin in this situation would open the cavity. The position of the lachrymal sac may also be indicated by the tendo oculi, or internal tarsal ligament. If both lids be drawn outwards so as to tighten the skin at the inner angle, a prominent cord will be seen beneath the tightened skin. This is the *tendo oculi*, which lies directly over the lachrymal sac, bisecting it, and thus forming a useful guide to its situation. A knife entered immediately beneath the tense cord would open the lower part of the sac. A probe introduced through this opening can be readily passed downwards through the duct into the inferior meatus of the nose. The direction of the duct is downwards, outwards, and backwards, and this course should be borne in mind in passing the probe, otherwise the point may be driven through the thin bony walls of the canal. A convenient plan is to direct the probe in such a manner, that if it were pushed onwards it would strike the first molar tooth of the lower jaw on the same side of the body. In other words, the surgeon standing in front of his patient should carry in his mind the position of the first molar tooth, and should push his probe onwards in such a way as if he desired to reach this structure.

Beneath the internal angular process of the frontal bone, the pulley of the Superior oblique muscle of the eye can be plainly felt by pushing the finger backwards between the upper and inner angle of the eye and the roof of the orbit; passing backwards and outwards from this pulley the tendon can be felt for a short distance.

Surgical Anatomy.—The eyelids are composed of various tissues, and consequently are liable to a variety of diseases. The skin which covers them is exceedingly thin and delicate, and is supported on a quantity of loose and lax subcutaneous tissue, which contains no fat. In consequence of this it is very freely movable, and is liable to be drawn down by the contraction of neighbouring cicatrices, and thus produce an eversion of the lid, known as *ectropion*. Inversion of the lids (*entropion*) from spasm of the Orbicularis palpebrarum or from chronic inflammation of the palpebral conjunctiva may also occur. The eyelids are richly supplied with blood, and are often the seat of vascular growths, such as nævi. Rodent ulcer also frequently commences in this situation. The loose cellular tissue beneath the skin is liable to become extensively infiltrated either with blood or inflammatory products, producing very great swelling. Even from very slight injuries to this tissue, the extravasation of blood may be so great as to produce considerable swelling of the lids and complete closure of the eye, and the same is the case when inflammatory products are poured out. The follicles of the eyelashes or the sebaceous glands associated with these follicles may be the seat of inflammation, constituting the ordinary 'sty.' The Meibomian glands are affected in the so-called 'tarsal tumour:' the tumour, according to some, being caused by the retained secretion of these glands; by others it is believed to be a neoplasm connected with the gland. The ciliary follicles are liable to become inflamed, constituting the disease known as *blepharitis ciliaris* or 'blear eye.' Irregular or disorderly growth of the eyelashes not infrequently occurs; some of them being turned towards the eyeball and producing inflammation and ulceration of the cornea, and possibly eventually complete destruction of the eye. The Orbicularis palpebrarum may be the seat of spasm, either in the form of slight quivering of the lids; or repeated twitchings, most commonly due to errors of refraction in children; or more continuous spasm, due to some irritation of the fifth or seventh cranial nerve. The Orbicularis may be paralysed, generally associated with paralysis of the other facial muscles. Under these circumstances, the patient is unable to close the lids, and, if he attempts to do so, rolls the eyeball upwards under the upper lid. The tears overflow from displacement of the lower lid, and the conjunctiva and cornea, being constantly exposed and the patient being unable to wink, become irritated from dust and foreign bodies. In paralysis of the Levator palpebræ superioris there is drooping of the upper eyelid and other symptoms of implication of the third nerve. The eyelids may be the seat of bruises, wounds, or burns. In these latter injuries, adhesions of the margins of the lids to each other, or adhesion of the lids to the globe may take place. The eyelids are sometimes the seat of emphysema, after fracture of some of the thin bones forming the inner wall of the orbit. If shortly after such an injury the patient blows his nose, air is forced from the nostril through the lacerated structures into the connective tissue of the eyelids, which suddenly swell up and present the peculiar crackling characteristic of this affection.

Foreign bodies frequently get into the conjunctival sac and cause great pain, especially if they come in contact with the corneal surface, during the movements of the lid and the eye on each other. The conjunctiva is often involved in severe injuries of the eyeball, but is seldom ruptured alone; the most common form of injury to the conjunctiva alone is from a burn, either from fire, strong acids, or lime. In these cases union is liable to take place between the eyelid and the eyeball. The conjunctiva is often the seat of inflammation arising from many different causes, and the arrangement of the conjunctival vessels should be remembered as affording a means of diagnosis between this condition and injection of the sclerotic, which is present in inflammation of the deeper structures of the globe. The inflamed conjunctiva is bright red; the vessels are large and tortuous, and greatest at the circumference, shading off towards the corneal margin ; they anastomose freely and form a dense network, and they can be emptied or displaced by gentle pressure.

The lachrymal gland is occasionally, though rarely, the seat of inflammation, either acute or chronic ; it is also sometimes the seat of tumours, benign or malignant, and for these may require removal. This may be done by an incision through the skin, just below the eyebrow ; and the gland, being invested with a special capsule of its own, may be isolated and removed, without opening the general cavity of the orbit. The canaliculi may be obstructed, either as a congenital defect, or by some foreign body, as an eyelash or a dacryolith, causing the tears to run over the cheek. The canaliculi may also become occluded as a result of burns or injury ; overflow of the tears may in addition result from deviation of the puncta, or from chronic inflammation of the lachrymal sac. This latter condition is set up by some obstruction to the nasal duct, frequently occurring in tuberculous subjects. In consequence of this the tears and mucus accumulate in the lachrymal sac, distending it. Suppuration in the lachrymal sac is sometimes met with ; this may be the sequel of a chronic inflammation ; or may occur after some of the eruptive fevers, in cases where the lachrymal passages were previously quite healthy. It may lead to lachrymal fistula.

THE EAR

The organ of hearing is divisible into three parts : the external ear, the middle ear or tympanum, and the internal ear or labyrinth.

THE EXTERNAL EAR

The **external ear** consists of an expanded portion named *pinna* or *auricle*, and the auditory canal, or *meatus*. The former serves to collect the vibrations of the air by which sound is produced ; the latter conducts those vibrations to the tympanum.

The **pinna** or **auricle** (fig. 673) is of an ovoid form, with its larger end directed upwards. Its outer surface is irregularly concave, directed slightly forwards,

FIG. 673.—The pinna, or auricle. Outer surface.

and presents numerous eminences and depressions which result from the foldings of its fibro-cartilaginous element. To each of these, names have been assigned. Thus the external prominent rim of the auricle is called the *helix*. Another curved prominence, parallel with and in front of the helix, is called the *antihelix* ; this divides above into two crura, which enclose a triangular depression, the *fossa of the antihelix* (*fossa triangularis*). The narrow curved depression between the helix and the antihelix is called the *fossa of the helix* (*scapha*) ; the antihelix describes a curve round a deep, capacious cavity, the *concha*, which is partially divided into two parts by the *crus helicis* or the commencement of the helix ; the upper part is termed the *cymba conchæ*, the lower part the *cavum conchæ*. In front of the concha, and projecting backwards over the meatus, is a small pointed eminence, the *tragus* : so called from its being generally covered on its under surface with a tuft of hair, resembling a goat's beard. Opposite the tragus, and separated from it by a deep notch (*incisura intertragica*), is a small tubercle, the *antitragus*. Below this is the *lobule*, composed of tough areolar and adipose tissues, and wanting the firmness and elasticity of the rest of the pinna. Where the helix turns downwards a small tubercle, the *tubercle of Darwin*, is frequently seen. This tubercle is very evident about the sixth month of fœtal life ; at this stage the human pinna has a close resemblance to that of some of the adult monkeys.

The cranial surface of the pinna presents elevations which correspond to the depressions on its outer surface and after which they are named, e.g. eminentia conchæ, eminentia triangularis, &c.

Structure of the pinna.—The pinna is composed of a thin plate of yellow fibro-cartilage, covered with integument, and connected to the surrounding parts by the extrinsic ligaments and muscles ; and to the commencement of the external auditory canal by fibrous tissue.

The *integument* is thin, closely adherent to the cartilage, and covered with hairs furnished with sebaceous glands, which are most numerous in the concha and scaphoid fossa. The hairs are most numerous and largest on the tragus and antitragus.

The *cartilage of the pinna* consists of one single piece : it gives form to this part of the ear, and upon its surface are found all the eminences and depressions above described. It does not enter into the construction of all parts of the auricle ; thus it does not form a constituent part of the lobule ; it is deficient, also, between the tragus and beginning of the helix, the notch between them being filled up by dense fibrous tissue. At the front part of the pinna, where the helix bends upwards, is a small projection of cartilage, called the *spina helicis*, while the lower part of the helix is prolonged downwards as a tail-like process, the *cauda helicis*; this is separated from the antihelix by a fissure, the *fissura anti-tragohelicina*. The cranial aspect of the cartilage exhibits a transverse furrow, the *sulcus antihelicis transversus*, which corresponds with the inferior crus of the antihelix and separates the prominence produced by the concha from that caused by the fossa triangularis. The eminentia conchæ is crossed by a vertical ridge (*ponticulus*) which gives attachment to the Retrahens auriculam muscle. The cartilage of the pinna presents several intervals or fissures in its substance, which partially separate the different parts. The fissure of the helix is a short vertical slit, situated at the fore part of the pinna. Another fissure, the fissure of the tragus, is seen upon the anterior surface of the tragus. The cartilage of the pinna is very pliable, elastic, of a yellowish colour, and belongs to that form of cartilage which is known under the name of yellow fibro-cartilage.

FIG. 674.—The muscles of the pinna.

The *ligaments of the pinna* consist of two sets : 1. The extrinsic set, or those connecting it to the side of the head. 2. The intrinsic set, or those connecting various parts of its cartilage together.

The *extrinsic ligaments* are two in number, anterior and posterior. The *anterior ligament* extends from the spina helicis and tragus to the root of the zygoma. The *posterior ligament* passes from the posterior surface of the concha to the outer surface of the mastoid process of the temporal bone.

The chief *intrinsic ligaments* are: (1) a strong fibrous band, stretching across from the tragus to the commencement of the helix, completing the meatus in front, and partly encircling the boundary of the concha; and (2) a band which extends between the antihelix and the cauda helicis. Other less important bands are found on the cranial surface of the pinna.

The *muscles of the pinna* (fig. 674) consist of two sets : 1, the *extrinsic*, which connect it with the side of the head, moving the pinna as a whole, viz. the Attollens, Attrahens, and Retrahens auriculam (page 430); and 2, the *intrinsic*, which extend from one part of the auricle to another, viz. :

Helicis major.	Antitragicus.
Helicis minor.	Transversus auriculæ.
Tragicus.	Obliquus auriculæ.

The *Musculus helicis major* is a narrow vertical band of muscular fibres, situated upon the anterior margin of the helix. It arises, below, from the crus helicis, and is inserted into the anterior border of the helix, just where it is about to curve backwards.

The *Musculus helicis minor* is an oblique fasciculus, which covers the crus helicis.

The *Tragicus* is a short, flattened band of muscular fibres situated upon the outer surface of the tragus, the direction of its fibres being vertical.

The *Antitragicus* arises from the outer part of the antitragus: its fibres are inserted into the cauda helicis and antihelix. This muscle is usually very distinct.

The *Transversus auriculæ* is placed on the cranial surface of the pinna. It consists of scattered fibres, partly tendinous and partly muscular, extending from the convexity of the concha to the prominence corresponding with the groove of the helix.

The *Obliquus auriculæ* (Tod) consists of a few fibres extending from the upper and back part of the concha to the convexity immediately above it.

FIG. 675.—A front view of the organ of hearing. Right side.

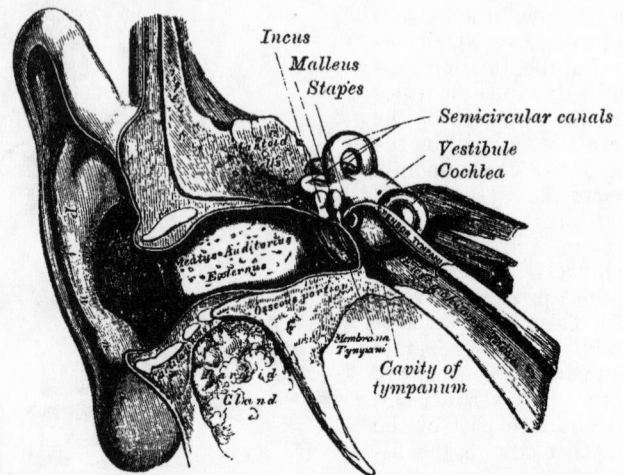

The *arteries of the pinna* are the posterior auricular from the external carotid, the anterior auricular from the temporal, and an auricular branch from the occipital artery.

The *veins* accompany the corresponding arteries.

The *nerves* are : the great auricular, from the cervical plexus ; the auricular branch of the pneumogastric ; the auriculo-temporal branch of the inferior maxillary nerve; the small occipital from the cervical plexus, and the great occipital or internal branch of the posterior division of the second cervical nerve. The muscles of the pinna are supplied by the facial nerve.

The **Auditory Canal** (*meatus auditorius externus*) extends from the bottom of the concha to the membrana tympani (fig. 675). It is about an inch and a half in length if measured from the tragus ; from the bottom of the concha its length is about an inch. It forms a sort of **S**-shaped curve, and is directed at first inwards, forwards, and slightly upwards (*pars externa*); it then passes inwards and backwards (*pars media*), and lastly is carried inwards, forwards, and slightly downwards (*pars interna*). It forms an oval cylindrical canal, the greatest diameter being in the vertical direction at the external orifice, but nearly in the transverse direction at the tympanic end. It presents two constrictions, one near the inner end of the cartilaginous portion, and another, the *isthmus*, in the osseous portion, about three-quarters of an inch from the bottom of the concha. The membrana tympani, which occupies the termination of the meatus, is obliquely directed, in consequence of which the floor of the canal is longer than the roof, and the anterior wall longer than the posterior. The auditory canal is formed partly by cartilage and membrane, and partly by bone, and is lined by skin.

The *cartilaginous portion* is about one-third of an inch (eight millimetres) in length, it is formed by the cartilage of the pinna, prolonged inwards, and firmly attached to the circumference of the auditory process of the temporal bone. The cartilage is deficient at its upper and back part, its place being supplied by fibrous membrane. This part of the canal is rendered extremely movable by two or three deep fissures (*incisuræ Santorini*), which extend through the cartilage in a vertical direction.

The *osseous portion* is about two-thirds of an inch (sixteen millimetres) in length, and narrower than the cartilaginous portion. It is directed inwards and a little forwards, forming a slight curve in its course, the convexity of which is upwards and backwards. Its inner end, which communicates, in the dry bone, with the cavity of the tympanum, is smaller than the outer and sloped, the anterior wall projecting beyond the posterior about two lines ; it is marked, except at its upper part, by a narrow groove, the *sulcus tympanicus*, for the insertion of the membrana tympani. Its outer end is dilated and rough in the greater part of its circumference, for the attachment of the cartilage of the pinna. Its vertical transverse section is oval, the greatest diameter being from above downwards and backwards. The front and lower parts of this canal are formed by a curved plate of bone, the tympanic plate, which, in the fœtus, exists as a separate ring (*annulus tympanicus*), incomplete at its upper part. See section on Osteology (page 197).

The *skin* lining the meatus is very thin, adheres closely to the cartilaginous and osseous portions of the tube, and covers the surface of the membrana tympani, forming its outer layer. After maceration, the thin pouch of epidermis, when withdrawn, preserves the form of the meatus. In the thick subcutaneous tissue of the cartilaginous part of the meatus are numerous ceruminous glands, which secrete the ear-wax. They resemble in structure sweat-glands, and their ducts open on the surface of the skin.

Relations of the meatus.—In front of the osseous part is the condyle of the mandible, which, however, is separated from the cartilaginous part by the retromandibular part of the parotid gland. The movements of the jaw influence to some extent the lumen of this latter portion. Behind the osseous part are the mastoid air-cells, separated from it by a thin layer of bone.

The *arteries* supplying the meatus are branches from the posterior auricular, internal maxillary, and temporal.

The *nerves* are chiefly derived from the auriculo-temporal branch of the inferior maxillary nerve and the auricular branch of the pneumogastric.

Surface Form.—At the point of junction of the osseous and cartilaginous portions, the tube forms an obtuse angle, which projects into the tube at its antero-inferior wall. This produces a sort of constriction in this situation, and renders it a narrow portion of the canal—an important point to be borne in mind in connection with the presence of foreign bodies in the ears. The cartilaginous is connected to the bony part by fibrous tissue which renders the outer part of the tube very movable, and therefore by drawing the pinna upwards and backwards the canal is rendered almost straight. At the external orifice are a few short, crisp hairs, which serve to prevent the entrance of small particles of dust, or flies and other insects. In the external auditory meatus the secretion of the ceruminous glands serves to catch any small particles which may find their way into the canal, and prevent their reaching the membrana tympani, where their presence might excite irritation. In young children the meatus is very short, the osseous part consisting merely of a bony ring (the *annulus tympanicus*), which supports the membrana tympani. In the fœtus, the osseous part is entirely absent. The shortness of the canal in children should be borne in mind in introducing the aural speculum, so that it be not pushed in too far, at the risk of injuring the membrana tympani ; indeed, even in the adult the speculum should never be introduced beyond the constriction which marks the junction of the osseous and cartilaginous portions. In using this instrument it is advisable that the pinna should be drawn upwards, backwards, and a little outwards, so as to render the canal as straight as possible, and thus assist the operator in obtaining, by the aid of reflected light, a good view of the membrana tympani. Just in front of the membrane is a well-marked depression, situated on the floor of the canal, and bounded by a somewhat prominent ridge; in this foreign bodies may become lodged. By aid of the speculum, combined with traction of the auricle upwards and backwards, the whole of the membrana tympani is rendered visible. It is a pearly-grey membrane, slightly glistening in the adult, placed obliquely, so as to form with the floor of the meatus a very acute angle (about fifty-five degrees), while with the roof it forms an obtuse angle. At birth it is more horizontal, situated in almost the same plane as the base of the skull. About midway between the anterior and posterior margins of the membrane, and extending from the centre obliquely upwards, is a reddish-yellow streak; this is the handle of the malleus, which is inserted into the membrane. At the upper part of this streak, close to the roof of the meatus, a little white, rounded prominence is plainly to be seen; this is the processus brevis of the malleus, projecting against the membrane. The membrana tympani does not present a plane surface; on the contrary, its centre is drawn inwards, on account of its connection with the handle of the malleus, and thus the external surface is rendered concave.

THE MIDDLE EAR, OR TYMPANUM

The **middle ear,** or **tympanum,** is an irregular cavity, compressed from without inwards, and situated within the petrous bone. It is placed above the jugular fossa; the carotid canal lying in front, the mastoid cells behind, the meatus auditorius externally, and the labyrinth internally. It is filled with air, and communicates with the naso-pharynx by the Eustachian tube. The tympanum is traversed by a chain of movable bones, which connect the membrana tympani with the labyrinth, and serve to convey the vibrations communicated to the membrana tympani across the cavity of the tympanum to the internal ear.

The tympanic cavity consists of two parts: the *atrium* or *tympanic cavity* proper, opposite the tympanic membrane, and the *attic* or *epitympanic* recess, above the level of the upper part of the membrane; the latter contains the upper half of the malleus and the greater part of the incus. Its diameter, including the attic, measures about fifteen millimetres vertically and transversely. From without inwards it measures about six millimetres above and four millimetres below; opposite the centre of the tympanic membrane it is only about two millimetres. It is bounded externally by the membrana tympani and meatus; internally, by the outer surface of the internal ear; and communicates, behind, with the mastoid antrum and through it with the mastoid cells; and in front with the Eustachian tube and canal for the Tensor tympani. Its roof and floor are formed by thin osseous laminæ, the one forming the roof being a thin plate situated on the anterior surface of the petrous portion of the temporal bone, close to its angle of junction with the squamous portion of the same bone.

The **roof** (*paries tegmentalis*) is broad, flattened, and formed of a thin plate of bone (*tegmen tympani*), which separates the cranial and tympanic cavities. It is prolonged backwards so as to roof in the mastoid antrum; it is also carried forwards to cover in the canal for the Tensor tympani muscle. Its outer edge corresponds with the remains of the petro-squamous suture.

The **floor** (*paries jugularis*) is narrow, and is separated by a thin plate of bone (*fundus tympani*) from the jugular fossa. It presents, near the inner wall, a small aperture for the passage of Jacobson's nerve.

The **outer wall** is formed mainly by the membrana tympani, partly by the ring of bone into which this membrane is inserted. This ring of bone is incomplete at its upper part, forming a notch (*incisura Rivini*). Close to it are three small apertures: the iter chordæ posterius, the Glaserian fissure, and the iter chordæ anterius.

The *iter chordæ posterius* is in the angle of junction between the posterior and external walls of the tympanum, immediately behind the membrana tympani and on a level with the upper end of the handle of the malleus; it leads into a minute canal, which descends in front of the aquæductus Fallopii, and terminates in that canal near the stylo-mastoid foramen. Through it the chorda tympani nerve enters the tympanum.

The *Glaserian fissure* opens just above and in front of the ring of bone into which the membrana tympani is inserted; in this situation it is a mere slit about a line in length. It lodges the long process and anterior ligament of the malleus, and gives passage to the tympanic branch of the internal maxillary artery.

The *iter chordæ anterius* is seen at the inner end of the preceding fissure; it leads into a canal (*canal of Huguier*), which runs parallel with the Glaserian fissure. Through it the chorda tympani nerve leaves the tympanum.

The **internal wall of the tympanum** (*paries labyrinthica*) (fig. 676) is vertical in direction, and looks directly outwards. It presents for examination the following parts:

Fenestra ovalis.   Promontory.
Fenestra rotunda.   Ridge of the aquæductus Fallopii.

The *fenestra ovalis* is a reniform opening leading from the tympanum into the vestibule; its long diameter is directed horizontally, and its convex border is upwards. The opening in the recent state is occupied by the base of the stapes, which is connected to the margin of the foramen by its annular ligament.

The *fenestra rotunda* is an aperture placed at the bottom of a funnel-shaped depression, leading into the cochlea. It is situated below and rather behind the fenestra ovalis, from which it is separated by a rounded elevation, the *promontory*; it is closed in the recent state by a membrane (*membrana tympani secundaria*, Scarpa). This membrane is concave towards the tympanum, convex towards the

FIG. 676.—View of inner wall of tympanum. (Enlarged.)

Chorda tympani

cochlea. It consists of three layers: the external, or mucous, derived from the mucous lining of the tympanum; the internal from the lining membrane of the cochlea; and an intermediate, or fibrous layer.

The *promontory* is a rounded hollow prominence, formed by the projection outwards of the first turn of the cochlea; it is placed between the fenestræ, and is furrowed on its surface by three small grooves, which lodge branches of the tympanic plexus. A minute spicule of bone frequently connects the promontory to the pyramid.

The *rounded eminence of the aquæductus Fallopii*, the prominence of the bony canal in which the facial nerve is contained, traverses the inner wall of the tympanum above the fenestra ovalis, and behind that opening curves nearly vertically downwards along the posterior wall.

The **posterior wall of the tympanum** (*paries mastoidea*) is wider above than below, and presents for examination the

Opening of the antrum.   Pyramid.   Fossa incudis.

The *opening of the antrum* is a large irregular aperture, which extends backwards from the epitympanic recess and leads into a considerable air space, the *antrum mastoideum* (see page 193). The antrum communicates with large irregular cavities contained in the interior of the mastoid process, the *mastoid air-cells*. These cavities vary considerably in number, size, and form; they are lined by mucous membrane, continuous with that lining the cavity of the tympanum. On the inner wall of the opening into the antrum is a rounded eminence, situated above and behind the eminence of the aquæductus Fallopii, which corresponds with the position of the ampullated extremities of the superior and external semicircular canals.

The *pyramid* is a conical eminence, situated immediately behind the fenestra ovalis, and in front of the vertical portion of the eminence above described; it is hollow in the interior, and contains the Stapedius muscle; its summit projects forwards towards the fenestra ovalis, and presents a small aperture, which transmits the tendon of the muscle. The cavity in the pyramid is prolonged downwards and backwards in front of the aquæductus Fallopii, with which it communicates by a minute canal which transmits a twig from the facial nerve to the Stapedius muscle.

The *fossa incudis* is a small depression which is situated in the lower and back part of the epitympanic recess, and lodges the short process of the incus.

The **anterior wall of the tympanum** (*paries carotica*) is wider above than below; it corresponds with the carotid canal, from which it is separated by a thin plate of bone, perforated by the tympanic branch of the internal carotid artery, and by a communicating branch which connects the sympathetic plexus on the internal carotid artery with the tympanic plexus on the promontory It presents for examination the

Canal for the Tensor tympani.　　　Orifice of the Eustachian tube.
The processus cochleariformis.

The orifice of the canal for the Tensor tympani and the orifice of the Eustachian tube are situated at the upper part of the anterior wall, being separated from each other by a thin, delicate, horizontal plate of bone, the *processus cochleariformis*. These canals run from the tympanum forwards, inwards, and a little downwards, to the retiring angle between the squamous and petrous portions of the temporal bone.

The *canal for the Tensor tympani* is the superior and the smaller of the two; it is rounded and lies beneath the forward prolongation of the tegmen tympani. It extends on to the inner wall of the tympanum and ends immediately above the fenestra ovalis. The processus cochleariformis passes backwards below this part of the canal, forming its outer wall and floor; it expands above the anterior extremity of the fenestra ovalis and terminates by curving outwards so as to form a pulley over which the tendon passes.

The *Eustachian tube* is the channel through which the tympanum communicates with the naso-pharynx. Its length is an inch and a half (thirty-six millimetres), and its direction downwards, forwards, and inwards, forming an angle of about forty-five degrees with the sagittal plane and one of from thirty to forty degrees with the horizontal plane. It is formed partly of bone, partly of cartilage and fibrous tissue.

The *osseous portion* is about an inch in length. It commences in the anterior wall of the tympanum, below the processus cochleariformis, and, gradually narrowing, terminates at the angle of junction of the petrous and squamous portions, its extremity presenting a jagged margin which serves for the attachment of the cartilaginous portion.

The *cartilaginous portion*, about an inch in length, is formed of a triangular plate of elastic fibro-cartilage, the apex of which is attached to the margin of the inner extremity of the osseous canal, while its base lies directly under the mucous membrane of the naso-pharynx, where it forms an elevation or cushion behind the pharyngeal orifice of the tube. The upper edge of the cartilage is curled upon itself, being bent outwards so as to present on transverse section the appearance of a hook; a groove or furrow is thus produced, which opens below and externally, and this part of the canal is completed by fibrous membrane. The cartilage is fixed to the base of the skull, and lies in a groove between the petrous-temporal and the greater wing of the sphenoid; this groove ends opposite the middle of the internal pterygoid plate. The cartilaginous and bony portions of the tube are not in the same plane, the former inclining downwards a little more than the latter. The diameter of the canal is not uniform throughout, being greatest at the pharyngeal orifice and least at the junction of the bony and cartilaginous portions, where it is named the *isthmus*; it again expands somewhat as it approaches the tympanic cavity. The position and relations of the pharyngeal orifice are described with the anatomy of the naso-pharynx. Through this canal the mucous membrane of the pharynx is continuous with that which lines the tympanum. The mucous membrane is covered with ciliated epithelium and is thin in the osseous portion, while in the cartilaginous portion it contains many mucous glands and near the pharyngeal orifice a considerable amount of adenoid tissue, which has been named by Gerlach the *tube-tonsil*. The tube is opened during deglutition by the Salpingo-pharyngeus and Dilator tubæ muscles. The latter arises from the hook of the cartilage and from the membranous part of the tube, and blends below with the Tensor palati.

The **membrana tympani** separates the cavity of the tympanum from the bottom of the external meatus. It is a thin, semi-transparent membrane, nearly oval in form, somewhat broader above than below, and directed very obliquely downwards and inwards so as to form an angle of about fifty-five degrees with the floor of the meatus. Its longest diameter is directed from above and behind, downwards and forwards, and measures from nine to ten millimetres; its shortest diameter measuring from eight to nine millimetres. The greater part of its circumference is thickened to form an annular ring which is fixed in a groove, the *sulcus tympanicus*, at the inner extremity of the meatus. This sulcus is deficient superiorly at the incisure or notch of Rivinus. From the extremities of this notch two bands, the *anterior* and *posterior malleolar folds*, are prolonged to the short process of the malleus. The small, somewhat triangular part of the membrane situated above these folds is lax and thin, and is named the *membrana flaccida* of Shrapnell; in it a small orifice is sometimes seen. The handle of the malleus is firmly attached to the inner aspect of the membrane as far as its centre, which it draws inwards towards the cavity of the tympanum. The outer surface of the membrane is thus concave, and the most depressed part of this concavity is named the *umbo* or *navel*.

Structure.—This membrane is composed of three layers, an external (cuticular), a middle (fibrous), and an internal (mucous). The *cuticular lining* is derived from the integument lining the meatus. The fibrous layer consists of two strata, an external of *radial fibres*, which diverge from the handle of the malleus, and an internal of *circular fibres*, which are plentiful around the circumference but sparse and scattered near the centre of the membrane. Branched or *dendritic* fibres, as pointed out by Grüber, are also present, especially in the posterior half of the membrane.

FIG. 677.—The membrana tympani viewed from its internal surface. (Testut.)

The malleus has been resected immediately beyond its processus brevis, in order to show the tympano-malleolar folds and the membrane of Shrapnell.
1. Membrana tympani. 2. Umbo or navel. 3. Handle of the malleus. 4. Processus brevis. 5. Anterior tympano-malleolar fold. 6. Posterior tympano-malleolar fold. 7. Membrane of Shrapnell. 8. Anterior pouch of Tröltsch. 9. Posterior pouch of Tröltsch. 10. Fibro-cartilaginous ring. 11. Glaserian fissure. 12. Eustachian tube. 13. Iter chordæ posterius. 14. Iter chordæ anterius. 15. Facet for short process of the incus. 16. Prominentia styloidea.

The arteries of the membrana tympani are derived from the deep auricular branch of the internal maxillary, which ramifies beneath the cuticular layer and from the stylo-mastoid branch of the posterior auricular and tympanic branch of the internal maxillary, which are distributed on the mucous surface. The superficial veins open into the external jugular; those on the mucous surface drain themselves partly into the lateral sinus and veins of the dura mater and partly into a plexus on the Eustachian tube. The membrane receives its nervous supply from the auriculo-temporal branch of the inferior maxillary, the auricular branch of the vagus, and the tympanic branch of the glosso-pharyngeal.

OSSICLES OF THE TYMPANUM (fig. 678)

The tympanum is traversed by a chain of movable bones, three in number, the *malleus*, *incus*, and *stapes*. The first is attached to the membrana tympani, the last to the fenestra ovalis, the incus being placed between the two, and connected to both by delicate articulations.

The **Malleus,** so named from its fancied resemblance to a hammer, consists of a head, neck, and three processes, viz.: the handle or manubrium, the processus gracilis, and the processus brevis.

The head is the large upper extremity of the bone; it is oval in shape, and articulates posteriorly with the incus, being free in the rest of its extent. The facet for articulation with the incus is constricted near the middle, and is divided by a ridge into an upper, larger, and lower, lesser part, which form nearly a right angle with each other. Opposite the constriction the lower margin of the facet projects in the form of a process, the *cog-tooth* or *spur* of the malleus.

FIG. 678.—The small bones of the ear seen from the outside. (Enlarged.)

The *neck* is the narrow contracted part just beneath the head; and below this is a prominence, to which the various processes are attached.

The *manubrium* is a vertical process of bone, which is connected by its outer margin with the membrana tympani. It is directed downwards, inwards, and backwards; it decreases in size towards its extremity, where it is curved slightly forwards, and flattened from within outwards. On the inner side, near its upper end, is a slight projection, into which the tendon of the Tensor tympani is inserted.

The *processus gracilis* is a long and very delicate process, which passes from the eminence below the neck forwards and outwards to the Glaserian fissure, to which it is connected by ligamentous fibres. In the fœtus this is the longest process of the malleus, and is in direct continuity with the cartilage of Meckel.

The *processus brevis* is a slight conical projection, which springs from the root of the manubrium; it is directed outwards, and is attached to the upper part of the tympanic membrane and, by means of the malleolar folds, to the extremities of the notch of Rivinus.

The **Incus** has received its name from its supposed resemblance to an anvil, but it is more like a bicuspid tooth, with two roots, which differ in length, and are widely separated from each other. It consists of a body and two processes.

The *body* is somewhat quadrilateral but compressed laterally. On its anterior surface is a deeply concavo-convex facet, which articulates with the head of the malleus; in the fresh state it is covered with cartilage and the joint lined with synovial membrane.

The two processes diverge from one another nearly at right angles.

The *short process,* somewhat conical in shape, projects almost horizontally backwards, and articulates with a depression, the *fossa incudis,* in the lower and back part of the epitympanic recess.

The *long process,* longer and more slender than the preceding, descends nearly vertically behind and parallel to the handle of the malleus, and, bending inwards, terminates in a rounded globular projection, the *os orbiculare* or *lenticular process,* which is tipped with cartilage, and articulates with the head of the stapes. In the fœtus the os orbiculare exists as a separate bone.

The **Stapes,** so called from its close resemblance to a stirrup, consists of a head, neck, two crura, and a base.

The *head* presents a depression, tipped with cartilage, which articulates with the os orbiculare.

The *neck,* the constricted part of the bone succeeding the head, receives the insertion of the Stapedius muscle.

The *two crura* diverge from the neck and are connected at their extremities by a flattened oval-shaped plate (the *base*), which forms the foot-plate of the stirrup and is fixed to the margin of the fenestra ovalis by ligamentous fibres. Of the two crura the anterior is shorter and less curved than the posterior.

Ligaments of the Ossicula.—These small bones are connected with each other, and with the walls of the tympanum, by ligaments, and moved by small muscles. The articular surfaces of the malleus and incus, and the orbicular process of the incus and head of the stapes, are covered with cartilage, connected together by delicate capsular ligaments, and lined by synovial membrane. The ligaments connecting the ossicula with the walls of the tympanum are five in number : three for the malleus, one for the incus, and one for the stapes.

The *anterior ligament of the malleus* was formerly described by Sömmerring as a muscle (*Laxator tympani*). It is now, however, believed by most observers to consist of ligamentous fibres only. It is attached by one extremity to the neck of the malleus, just above the processus gracilis, and by the other to the anterior wall of the tympanum, close to the Glaserian fissure, some of its fibres being prolonged through the fissure to reach the spine of the sphenoid.

The *superior ligament of the malleus* is a delicate, round bundle of fibres which descends perpendicularly from the roof of the epitympanic recess to the head of the malleus.

The *external ligament of the malleus* is a triangular plane of fibres passing from the posterior part of the notch in the tympanic ring (*incisura Rivini*) to the head of the malleus.

The *posterior ligament of the incus* is a short, thick, ligamentous band which connects the extremity of the short process of the incus to the fossa incudis in the epitympanic recess.

The inner surface and the circumference of the base of the stapes are covered with hyaline cartilage, and the *annular ligament of the stapes* connects the circumference of the base to the margin of the fenestra ovalis.

A *superior ligament of the incus* has been described by Arnold, but it is little more than a fold of mucous membrane.

The **muscles of the tympanum** are two :

Tensor tympani.   Stapedius.

The *Tensor tympani*, the larger, is contained in the bony canal, above the osseous portion of the Eustachian tube, from which it is separated by the processus cochleariformis. It arises from the cartilaginous portion of the Eustachian tube and the adjoining part of the great wing of the sphenoid, as well as from the osseous canal in which it is contained. Passing backwards through the canal, it terminates in a slender tendon which enters the tympanum and makes a sharp bend outward round the extremity of the processus cochleariformis, and is inserted into the handle of the malleus, near its root. It is supplied by a branch from the otic ganglion.

The *Stapedius* arises from the side of a conical cavity, hollowed out of the interior of the pyramid ; its tendon emerges from the orifice at the apex of the pyramid, and, passing forwards, is inserted into the posterior aspect of the neck of the stapes. Its surface is aponeurotic, its interior fleshy ; and its tendon occasionally contains a slender bony spine, which is constant in some mammalia. It is supplied by a branch of the facial nerve.

Actions.—The Tensor tympani draws the membrana tympani inwards, and thus heightens its tension. The Stapedius draws the head of the stapes backwards, and thus causes the base of the bone to rotate on a vertical axis drawn through its own centre : in doing this the back part of the base would be pressed inwards towards the vestibule, while the fore part would be drawn from it. It probably compresses the contents of the vestibule.

The **mucous membrane of the tympanum** is continuous with the mucous membrane of the pharynx, through the Eustachian tube. It invests the ossicles, and the muscles and nerves contained in the tympanic cavity ; forms the internal layer of the membrana tympani, and the outer layer of the membrana tympani secundaria, and is reflected into the mastoid antrum and cells, which it lines throughout. It forms several vascular folds, which extend from the walls of the tympanum to the ossicles ; of these one descends from the roof of the tympanum to the head of the malleus and upper margin of the body of the incus, a second invests the Stapedius muscle : other folds invest the chorda tympani nerve and the Tensor tympani muscle. These folds separate off pouch-like cavities, and give the interior of the tympanum a somewhat honeycomb appearance. One of these pouches is well marked, viz. the *pouch of Prussak*, which lies between the neck of the malleus and the membrana flaccida. Two other folds may be mentioned : they are formed by the mucous membrane which envelops the chorda tympani nerve and are situated, one in front of, and the other behind the handle of the malleus ; they are named the *anterior* and *posterior recesses of Tröltsch*. In the tympanum this membrane is pale, thin, slightly vascular and covered for the most part with columnar ciliated epithelium, but over the pyramid, ossicula, and membrana tympani it possesses a flattened non-ciliated epithelium. In the antrum and mastoid cells its epithelium is also non-ciliated. In the osseous portion of the Eustachian tube the membrane is thin ; but in the cartilaginous portion it is very thick, highly vascular, and provided with numerous mucous glands ; the epithelium which lines the tube is columnar and ciliated.

The **arteries supplying the tympanum** are six in number. Two of them are larger than the rest, viz. the tympanic branch of the internal maxillary, which supplies the membrana tympani ; and the stylo-mastoid branch of the posterior auricular, which supplies the back part of the tympanum and mastoid cells. The smaller branches are—the petrosal branch of the middle meningeal, which enters through the hiatus Fallopii ; a branch from the ascending pharyngeal and another from the Vidian, which accompany the Eustachian tube ; and the tympanic branch from the internal carotid, given off in the carotid canal and perforating the thin anterior wall of the tympanum.

The **veins of the tympanum** terminate in the pterygoid plexus, the superior petrosal sinus, and the middle meningeal vein.

The **nerves of the tympanum** constitute the tympanic plexus, which ramifies upon the surface of the promontory. The plexus is formed by (1) the tympanic branch of the glosso-pharyngeal ; (2) the small deep petrosal nerve ; (3) the small superficial petrosal nerve ; and (4) a branch which joins the great superficial petrosal.

The *tympanic branch of the glosso-pharyngeal* (Jacobson's nerve) enters the tympanum by an aperture in its floor close to the inner wall and divides into branches, which ramify on the promontory and enter into the formation of the plexus. The *small deep petrosal nerve* from the carotid plexus of the sympathetic passes through the wall of the carotid canal, and joins the branches of Jacobson's nerve. The branch to the great superficial petrosal passes through an opening on the inner wall of the tympanum in front of the fenestra ovalis. The *small superficial petrosal nerve*, derived from the otic ganglion, passes through a foramen in the middle fossa of the base of the skull (sometimes the foramen ovale), passes backwards and enters the petrous bone through a small aperture, situated external to the hiatus Fallopii on the anterior surface of this bone ; it then courses downwards through the bone, and, passing by the gangliform enlargement of the facial nerve, receives a connecting filament from it and enters the tympanic cavity, where it communicates with Jacobson's nerve, and assists in forming the tympanic plexus.

The *branches of distribution* of the tympanic plexus are supplied to the mucous membrane of the tympanum ; one special branch passing to the fenestra ovalis, another to the fenestra rotunda, and a third to the Eustachian tube. The small superficial petrosal may be looked upon as the continuation of the nerve of Jacobson through the plexus to the otic ganglion.

In addition to the tympanic plexus there are the nerves supplying the muscles. The Tensor tympani is supplied by a branch from the third division of the fifth through the otic ganglion, and the Stapedius by the tympanic branch of the facial.

The *chorda tympani* nerve crosses the tympanic cavity. It is given off from the facial, as it passes vertically downwards at the back of the tympanum, about a quarter of an inch before its exit from the stylo-mastoid foramen. It passes from below upwards and forwards in a distinct canal, and enters the cavity of the tympanum through an aperture, *iter chordæ posterius*, already described (page 984), and becomes invested with mucous membrane. It passes forwards, through the cavity of the tympanum, crossing internal to the membrana tympani and over the upper part of the handle of the malleus to the anterior wall of the tympanum, and emerges from that cavity through the *iter chordæ anterius*, or *canal of Huguier*. It is invested by the fold of mucous membrane already mentioned.

Internal Ear, or Labyrinth

The **internal ear** is the essential part of the organ of hearing, receiving the ultimate distribution of the auditory nerve. It is called the **labyrinth,** from the complexity of its shape, and consists of two parts : the *osseous labyrinth*, a series of cavities channelled out of the substance of the petrous bone, and the *membranous labyrinth*, the latter being contained within the former.

The Osseous Labyrinth

The **osseous labyrinth** consists of three parts : the *vestibule, semicircular canals*, and *cochlea*. These are cavities hollowed out of the substance of the bone, and lined by periosteum ; they contain a clear fluid, perilymph, or liquor Cotunnii, in which the membranous labyrinth is situated.

The **Vestibule** (fig. 679) is the common central cavity of communication between the parts of the internal ear. It is situated on the inner side of the tympanum, behind the cochlea, and in front of the semicircular canals. It is

FIG. 679.—The osseous labyrinth laid open. (Enlarged.)

Opening of
Aquæductus vestibuli

Bristle passed through
Foramen rotundum

Opening of
Aquæductus cochleæ

somewhat ovoid in shape from before backwards, flattened from within outwards, and measures about one-fifth of an inch from before backwards, as well as from above downwards, and about one-eighth of an inch from without inwards. On its *outer* or *tympanic wall* is the fenestra ovalis, closed, in the recent state, by the base of the stapes, and its annular ligament. On its *inner wall*, at the fore part, is a small circular depression, the *recessus sphæricus*, which is perforated, at its anterior and inferior part, by several minute holes (*macula cribrosa media*) for the passage of filaments of the auditory nerve to the saccule ; and behind this depression is a vertical ridge, the *crista vestibuli*. This ridge bifurcates below to enclose a small depression, the *fossa cochlearis*, which is perforated by a number of holes for the passage of filaments of the auditory nerve which supply the posterior end of the ductus cochlearis. At the hinder part of the inner wall is the orifice of the *aquæductus vestibuli*, which extends to the posterior surface of the petrous portion of the temporal bone. It transmits a small vein, and contains a tubular prolongation of the lining membrane of the vestibule, the *ductus endolymphaticus*, which ends in a *cul-de-sac* between the layers of the dura mater within the cranial cavity. On the *upper wall* or *roof* is a transversely oval depression, the *recessus ellipticus*, separated from the recessus sphæricus by the crista vestibuli already mentioned. *Behind*, the semicircular canals open into the vestibule by five orifices. In *front* is an elliptical opening, which communicates with the scala vestibuli of the cochlea by an orifice, *apertura scalæ vestibuli cochleæ.*

The **semicircular canals** are three bony canals, situated above and behind the vestibule. They are of unequal length, compressed from side to side, and describe the greater part of a circle. They measure about one-twentieth of an inch in diameter, and each presents a dilatation at one end, called the *ampulla,* which measures more than twice the diameter of the tube. These canals open into the vestibule by five orifices, one of the apertures being common to two of the canals.

The *superior semicircular canal,* fifteen to twenty millimetres in length, is vertical in direction, and is placed transversely to the long axis of the petrous portion of the temporal bone; on the anterior surface of which its arch forms a round projection. It describes about two-thirds of a circle. Its outer extremity, which is ampullated, communicates by a distinct orifice with the upper part of the vestibule; the opposite end of the canal, which is not dilated, joins with the upper part of the posterior canal to form the *crus commune,* which opens into the upper and inner part of the vestibule.

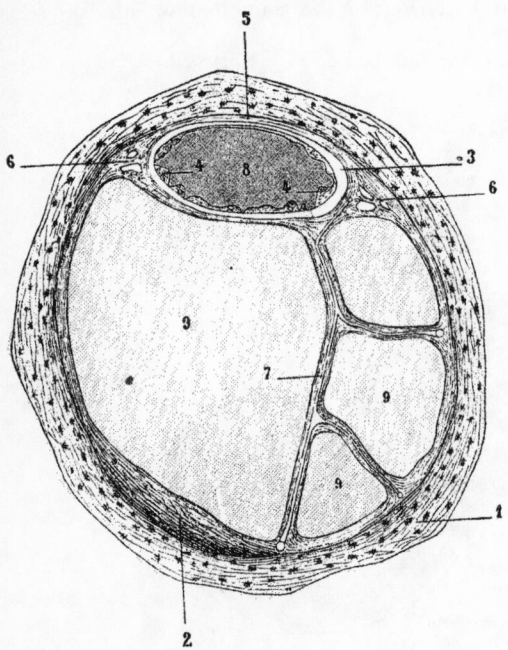

Fig. 680.—Transverse section of a human semicircular canal, after Rüdinger. (Testut.)

1. Bony semicircular canal. 2. Periosteum. 3. Membranous semicircular canal, with 4, papilliform processes on its internal surface. 5. Connective tissue binding the membranous canal to the periosteum. 6, 6. Fibrous bands uniting the free surface of the membranous canal to the periosteum. 7. Vessels. 8. Endolymphatic space. 9, 9. Perilymphatic space.

The *posterior semicircular canal,* also vertical in direction, is directed backwards, nearly parallel to the posterior surface of the petrous bone; it is the longest of the three, measuring from eighteen to twenty-two millimetres; its lower or ampullated end opens into the lower and back part of the vestibule, its opposite end joining to form the common canal already mentioned.

The *external* or *horizontal canal* is the shortest of the three. It measures from twelve to fifteen millimetres, and its arch is directed outwards and backwards; thus each semicircular canal stands at right angles to the other two. Its ampullated end corresponds to the upper and outer angle of the vestibule, just above the fenestra ovalis, where it opens close to the ampullary end of the superior canal; its opposite end opens by a distinct orifice at the upper and back part of the vestibule. 'The external canal of one ear is very nearly in the same plane as that of the other; while the superior canal of one ear is nearly parallel to the posterior canal of the other.' *

The **Cochlea** bears some resemblance to a common snail-shell: it forms the anterior part of the labyrinth, is conical in form, and placed almost horizontally in front of the vestibule; its apex is directed forwards and outwards, with a slight inclination downwards, towards the upper and front part of the inner wall of the tympanum; its base corresponds with the bottom of the internal auditory meatus, and is perforated by numerous apertures for the passage of the cochlear division of the auditory nerve. It measures nearly a quarter of an inch (five millimetres) from base to apex, and its breadth across the base is somewhat greater (about nine millimetres). It consists of a conical-shaped central axis,

* Crum Brown, *Journal of Anatomy and Physiology,* vol. viii.

the *modiolus* or *columella* ; of a canal, the inner wall of which is formed by the central axis, wound spirally around it for two turns and three-quarters, from the base to the apex, and of a delicate lamina (the *lamina spiralis ossea*) which projects from the modiolus, and, following the windings of the canal, partially subdivides into two. In the recent state a membrane, the *membrana basilaris*, stretches from the free border of this lamina to the outer wall of the bony cochlea and completely separates the canal into two passages, which, however, communicate with each other at the apex of the modiolus by a small opening, named the *helicotrema*.

The *modiolus* or *columella* is the central axis or pillar of the cochlea. It is conical in form, and extends from the base to the apex of the cochlea. Its base is broad, and appears at the bottom of the internal auditory meatus, where it corresponds with the area cochleæ ; it is perforated by numerous orifices, which transmit filaments of the cochlear division of the auditory nerve ; the nerves for the first turn and a half pass through the foramina of the tractus spiralis foraminosus ; those for the apical turn through the foramen centrale.

FIG. 681.—The cochlea and vestibule, viewed from above. (Testut.)

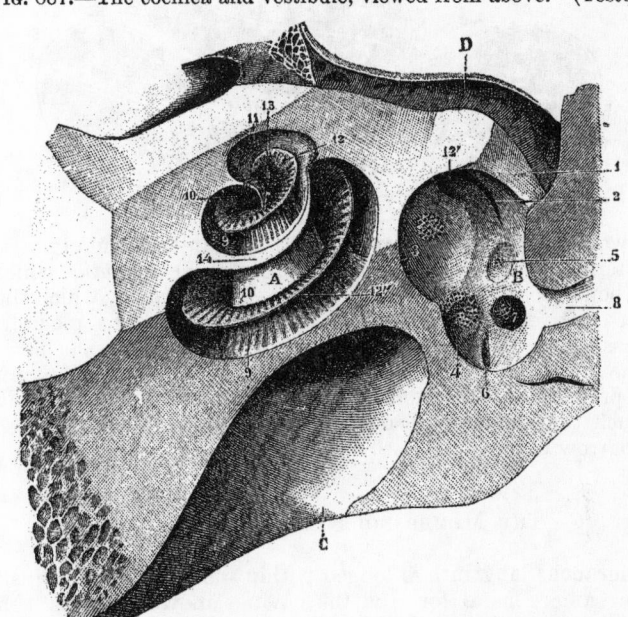

All the hard parts which form the roof of the internal ear have been removed with the saw. A. Cochlea. B. Vestibule. C. Internal auditory meatus. D. Tympanic cavity. 1. Lower border of fenestra ovalis. 2. Fissura vestibuli. 3. Recessus sphæricus. 4. Recessus ellipticus. 5. Fossa cochlearis. 6. Orifice of the aquæductus vestibuli. 7. Inferior opening of the posterior semicircular canal. 8. Non-ampullated end of external semicircular canal. 9. Scala tympani of the cochlea. 10. Scala vestibuli. 11. Cupola. 12. Lamina spiralis ossea, with 12', its vestibular origin ; 12'', its external border. 13. Helicotrema. 14. Bony wall of cochlea.

The foramina of the tractus spiralis foraminosus pass up through the modiolus and successively bend outwards to reach the attached margin of the lamina spiralis ossea. Here they become enlarged, and by their apposition form a spiral canal (*canalis spiralis modioli*), which follows the course of the attached margin of the lamina spiralis ossea and lodges the ganglion spirale (*ganglion of Corti*). The foramen centrale is continued as a canal up the middle of the modiolus to its apex. The modiolus diminishes rapidly in size in the second and succeeding coil.

The bony canal of the cochlea (fig. 680) takes two turns and three-quarters round the modiolus. It is a little over an inch in length (about thirty millimetres) and diminishes gradually in size from the base to the summit, where it terminates in a *cul-de-sac*, the *cupola*, which forms the apex of the cochlea. The commencement of this canal is about the tenth of an inch in diameter ; it diverges from the modiolus towards the tympanum and vestibule, and presents three openings. One, the *fenestra rotunda*, communicates with the tympanum ; in the recent

3 S

state this aperture is closed by a membrane, the *membrana tympani secundaria*. Another aperture, of an elliptical form, enters the vestibule. The third is the aperture of the aqueductus cochleæ, leading to a minute funnel-shaped canal, which opens on the basilar surface of the petrous bone and transmits a small vein, and also forms a communication between the subarachnoid space of the skull and the scala tympani.

The *lamina spiralis ossea* is a bony shelf or ledge which projects outwards from the modiolus into the interior of the spiral canal, and, like the canal, takes two and three-quarter turns round the modiolus. It reaches about halfway towards the outer wall of the spiral tube, and partially divides its cavity into two passages or scalæ, of which the upper is named the *scala vestibuli*,

Fig. 682.—The cochlea laid open. (Enlarged.)

while the lower is termed the *scala tympani*. Near the summit of the cochlea the lamina terminates in a hook-shaped process, the *hamulus*, which assists in forming the boundary of a small opening, the *helicotrema*, by which the two scalæ communicate with each other. From the canalis spiralis modioli numerous foramina pass outwards through the osseous spiral lamina as far as its free edge. In the lower part of the first turn a second bony lamina (*lamina spiralis secundaria*) projects inwards from the outer wall of the bony tube; it does not, however, reach the primary osseous spiral lamina, so that if viewed from the vestibule a narrow fissure, the *fissura vestibuli*, is seen between them.

THE MEMBRANOUS LABYRINTH (fig. 683)

The **membranous labyrinth** is lodged within the bony cavities just described, and has the same general form as the cavities in which it is contained; it is, however, considerably smaller, and is separated from the bony walls by a quantity of fluid, the *perilymph*. It does not float loosely in this fluid, but in certain places is fixed to the walls of the cavity. The membranous labyrinth contains fluid, the *endolymph*, and on its walls the ramifications of the auditory nerve are distributed.

Within the osseous vestibule the membranous labyrinth does not quite preserve the form of the bony cavity, but presents two membranous sacs, the *utricle* and the *saccule*. The *utricle* is the larger of the two, of an oblong form, compressed laterally, and occupies the upper and back part of the vestibule, lying in contact with the recessus ellipticus and the part below it. That portion which is lodged in the recess forms a sort of pouch or *cul-de-sac*, the floor and anterior wall of which are much thicker than elsewhere, and form the *macula acustica utriculi*, which receives the utricular filaments of the auditory nerve and has attached to its internal surface a layer of calcareous particles (otoliths). The cavity of the utricle communicates behind with the membranous semicircular canals by five orifices. From its anterior wall is given off a small canal, the *ductus utriculo-saccularis*, which opens into the ductus endolymphaticus from the saccule.

The *saccule* is the smaller of the two vestibular sacs; it is globular in form, and lies in the recessus sphæricus near the opening of the scala vestibuli of the cochlea. Its anterior part exhibits an oval thickening, the *macula acustica sacculi*, to which are distributed the saccular filaments of the auditory nerve. Its cavity does not directly communicate with that of the utricle. From the posterior wall is given off a canal, the *ductus endolymphaticus*; this duct is joined by the ductus utriculo-saccularis, and then passes along the aquæductus vestibuli and ends in a blind pouch on the posterior surface of the petrous portion of the temporal bone, where it is in contact with the dura mater. From the lower part of the saccule a short tube, the *canalis reuniens* of Hensen, passes downwards and outwards to open into the ductus cochlearis near its vestibular extremity (fig. 683).

The *membranous semicircular canals* are about one-fourth of the diameter of the osseous canals, but in number, shape, and general form they are precisely similar, and each presents at one end an ampullary enlargement. They open by five orifices into the utricle, one opening being common to the inner end of the superior and the upper end of the posterior canal. In the ampullæ the wall is thickened, and projects into the cavity as a fiddle-shaped, transversely placed elevation, the *septum transversum*, in which the nerves end.

The utricle, saccule, and membranous canals are held in position by numerous fibrous bands which stretch across the space between them and the bony walls.

Structure.—The walls of the utricle, saccule, and semicircular canals consist of three layers. The *outer layer* is a loose and flocculent structure, apparently

Fig. 683.—The membranous labyrinth. (Enlarged.)

composed of ordinary fibrous tissue, containing blood-vessels and some pigment-cells. The *middle layer*, thicker and more transparent, forms a homogeneous membrana propria, and presents on its internal surface, especially in the semicircular canals, numerous papilliform projections, which, on the addition of acetic acid, exhibits an appearance of longitudinal fibrillation and elongated nuclei. The *inner layer* is formed of polygonal nucleated epithelial cells. In the maculæ of the utricle and saccule, and in the transverse septa of the ampullæ of the canals, the middle coat is thickened and the epithelium is columnar, and consists of *supporting cells* and *hair-cells*. The former are fusiform, and their deep ends are attached to the membrana propria, while their free extremities are united to form a thin cuticle. The hair-cells are flask-shaped, and their deep, rounded ends do not reach the membrana propria, but lie between the supporting cells. The deep part of each contains a large nucleus, while its more superficial part is granular and pigmented. The free end is surmounted by a long, tapering, hair-like filament, which projects into the cavity. The filaments of the auditory nerve enter these parts, and having pierced the outer and the thickened middle layers, they lose their medullary sheath, and their axis cylinders ramify between the hair-cells.

Two small rounded bodies termed *otoliths*, and consisting of a mass of minute crystalline grains of carbonate of lime, held together in a mesh of delicate

FIG. 684.—Right human membranous labyrinth, removed from its bony enclosure and viewed from the antero-lateral aspect. (G. Retzius.)

FIG. 685.—The same from the postero-mesial aspect. (G. Retzius.)

1. External semicircular canal ; 1′, its ampulla. 2. Posterior canal ; 2′, its ampulla. 3. Superior canal ; 3′, its ampulla. 4. Conjoined limb of superior and posterior canals (sinus utriculi superior). 5. Utricle. 5′. Recessus utriculi. 5″. Sinus utriculi posterior. 6. Ductus endolymphaticus. 7. Canalis utriculo-saccularis. 8. Nerve to ampulla of superior canal. 9. Nerve to ampulla of external canal. 10. Nerve to recessus utriculi (in fig. 684 the three branches appear conjoined). 10′. Ending of nerve in recessus utriculi. 11. Facial nerve. 12. Lagena cochleæ. 13. Nerve of cochlea within spiral lamina. 14. Basilar membrane. 15. Nerve-fibres to macula of saccule. 16. Nerve to ampulla of posterior canal. 17. Saccule. 18. Secondary membrane of tympanum. 19. Canalis reuniens. 20. Blind ending of cochlear canal in vestibule. 21. Outer wall of cochlea. 22. Spiral ligament. 23 Section of the seventh and eighth nerves within internal auditory meatus (the separation between them is not apparent in the section).

fibrous tissue, are contained in the walls of the utricle and saccule opposite the distribution of the nerves. According to Bowman, a calcareous material is also sparingly scattered in the cells lining the ampullæ of the semicircular canals.

The **membranous cochlea, ductus cochlearis,** or **scala media** consists of a spirally arranged tube enclosed in the bony canal of the cochlea and lying along its outer wall. The manner in which it is formed will now be described.

FIG. 686.—Floor of scala media, showing the organ of Corti, &c.

As already stated, the osseous spiral lamina extends only part of the distance between the modiolus and the outer bony wall of the cochlea, while a membrane, the *membrana basilaris*, stretches from its free edge to the outer wall of the cochlea, and completes the roof of the scala tympani. A second and more delicate membrane, the *membrane of Reissner*, extends from the thickened periosteum

FIG. 687.—The lamina reticularis and subjacent structures. (Schematic.)
(Testut.)

A. Internal pillar of Corti, with *a*, its plate. B. External pillar (in yellow). C. Tunnel of Corti. D. Membrana basilaris. E. Inner hair-cells. 1, 1′. Internal and external-borders of the membrana reticularis. 2, 2′, 2″. The three rows of circular holes (in blue). 3. First row of phalanges (in yellow). 4, 4′, 4″. Second, third, and fourth rows of phalanges (in red). 6, 6′, 6″. The three rows of outer hair-cells (in blue). 7, 7′, 7″. Cells of Deiters. 8. Cells of Claudius.

covering the lamina spiralis ossea to the outer wall of the cochlea, to which it is attached at some little distance from the outer edge of the membrana basilaris. A canal is thus shut off between the scala tympani below and the scala vestibuli above; this is the *membranous canal of the cochlea, ductus cochlearis,* or *scala media*. It is triangular on transverse section, its roof being formed by the membrane of Reissner, its outer wall by the periosteum which lines the bony canal, and its floor by the membrana basilaris, and the outer part of the lamina spiralis ossea, on the former of which is placed the organ of Corti. Reissner's membrane is thin and homogeneous, and is covered on its upper and under surfaces by a layer of epithelium. The periosteum, which forms the outer wall of the ductus cochlearis, is greatly thickened and altered in character, forming what is called the *ligamentum spirale*. It projects inwards below as a triangular prominence, the *crista basilaris*, which gives attachment to the outer edge of the membrana basilaris, and immediately above which is a concavity, the *sulcus spiralis externus*. The upper portion of the ligamentum spirale contains numerous capillary loops and small blood-vessels, and forms what is termed the *stria vascularis*.

The lamina spiralis ossea consists of two plates of bone extending outwards; between these are the canals for the transmission of the filaments of the auditory nerve. On the upper plate of that part of the osseous spiral lamina which is outside Reissner's membrane the periosteum is thickened to form the *limbus laminæ spiralis*, and this terminates externally in a concavity, the *sulcus spiralis*

FIG. 688.—Limbus laminæ spiralis and membrana basilaris. (Schematic.) (Testut.)

1, 1'. Upper and lower lamellæ of the lamina spiralis ossea. 2. Limbus laminæ spiralis, with *a*, the teeth of the first row; *b*, *b'*, the auditory teeth of the other rows; *c*, *c'*, the interdental grooves and the cells which are lodged in them. 3. Sulcus spiralis internus, with 3', its labium vestibulare, and 3'', its labium tympanicum. 4. Foramina nervosa, giving passage to the nerves from the ganglion spirale or ganglion of Corti. 5. Vas spirale. 6. Zona arcuata, and 6', zona pectinata of the basilar membrane, with *a*, its hyaline layer; *β*, its connective-tissue layer. 7. Arch of Corti, with 7', its inner rod, and 7'', its outer rod. 8. Feet of the internal rods, from which the cells are removed. 9. Feet of the external rods. 10. Membrane of Reissner, at its origin.

internus, which presents, on section, the form of the letter **C**; the upper part of the letter, formed by the overhanging extremity of the limbus, is named the *labium vestibulare*; the lower part, prolonged and tapering, is called the *labium tympanicum*, and is perforated by numerous foramina (*foramina nervosa*) for the passage of the cochlear nerves. Externally, the labium tympanicum is continuous with the membrana basilaris. The upper surface of the labium vestibulare is intersected at right angles by a number of furrows, between which are numerous elevations; these present the appearance of teeth along the free surface and margin of the labium, and have been named by Huschke the *auditory teeth*. The limbus is covered by a layer of what at first sight appears to be squamous epithelium, since the deeper parts of the cells occupy the intervals between the elevations and between the auditory teeth. This layer is continuous on the one hand with that which lines the sulcus spiralis internus, and on the other with that which covers the under aspect of Reissner's membrane. The basilar membrane may be divided into two areas, inner and outer. The inner is thin, and is named the *zona arcuata*: it supports the organ of Corti. The outer is thicker and striated, and is termed the *zona pectinata*. The under surface of the membrane is covered by a layer of vascular connective tissue.

One of these vessels in this tissue is somewhat larger than the rest, and is named the *vas spirale*; it lies below Corti's tunnel.

Organ of Corti.*—This organ (fig. 689) is situated upon the inner part of the membrana basilaris, and appears at first sight as a papilla, winding spirally throughout the whole length of the ductus cochlearis, from which circumstance it has been designated the *papilla spiralis*. More accurately viewed, it is seen to be composed of a remarkable arrangement of cells, which may be likened to the keyboard of a pianoforte. Of these cells, the two central ones are rodlike bodies, and are called the inner and outer *rods of Corti*. Their bases are expanded and placed on the basilar membrane, at some little distance from each other, while their intermediate portions are inclined towards each other, so that the rods meet at their opposite extremities, and form a series of arches roofing over a minute tunnel, the *tunnel of Corti*, between them and the basilar membrane; this tunnel ascends spirally through the whole length of the cochlea.

The *inner rods*, some 6,000 in number, rest by means of expanded footplates on the basilar membrane, close to the labium tympanicum; they project obliquely upwards and outwards, and terminate above in expanded extremities, each of which resembles in shape the upper end of the ulna, with its sigmoid

Fig. 689.—Section through the organ of Corti. Magnified. (G. Retzius.)

cavity, coronoid and olecranon processes. On the outer side of the rod, in the angle formed between it and the basilar membrane, is a nucleated mass of protoplasm; while on the inner side is a row of epithelial cells (*inner hair-cells*), surmounted by a brush of fine, stiff, hairlike processes. On the inner side of these cells are two or three rows of columnar supporting cells, which are continuous with the cubical cells lining the sulcus spiralis internus.

The *outer rods*, numbering about 4,000, also rest by broad foot-plates on the basilar membrane; they incline upwards and inwards, and their upper extremity resembles the head and bill of a swan: the back of the head fitting into the concavity—the analogue of the sigmoid cavity—of one or more of the internal rods, and the bill projecting outwards as a phalangeal process of the membrana reticularis, presently to be described.

In the head of these outer rods is an oval portion, where the fibres of which the rod appears to be composed are deficient, and which stains more deeply with carmine than the rest of the rod. At the base of the rod, on its internal side—that is to say, in the angle formed by the rod with the basilar membrane—is a similar protoplasmic mass to that found on the outer side of the base of the inner rod; these masses of protoplasm are probably the undifferentiated portions of the cells from which the rods are developed. External to the outer rod are

* Corti's original paper is in the *Zeitschrift f. Wissen. Zool.* iii. 109.

three or four successive rows of epithelial cells, more elongated than those found on the internal side of the inner rod, but, like them, furnished with minute hairs or cilia. These are termed the *outer hair-cells*, in contradistinction to the *inner hair-cells* above referred to. The outer hair-cells number about 12,000, the inner about 3,500.

The *hair-cells* are somewhat oval in shape; their free extremities are on a level with the heads of Corti's rods, and from each some twenty fine hairlets project and are arranged in the form of a crescent, the concavity of which opens inwards. The deep ends of the cells are rounded and contain large nuclei : they only reach as far as the middle of Corti's rods, and are in contact with the ramifications of the nervous filaments. Between the rows of the outer hair-cells are rows of supporting cells, called the *cells of Deiters*; their expanded bases are planted on the basilar membrane, while the opposite end of each presents a clubbed extremity or *phalangeal* process. Immediately to the outer side of Deiters's cells are some five or six rows of columnar cells, the *supporting cells of Hensen*. Their bases are narrow, while their upper parts are expanded and form a rounded elevation on the floor of the ductus cochlearis. The columnar

FIG. 690.—Longitudinal section of the cochlea, showing the relations of the scalæ, the ganglion spirale, &c.

s.v. Scala vestibuli.   s.t. Scala tympani.   s.m. Scala media.   l.s. Ligamentum spirale.
G.S. Ganglion spirale.

cells lying outside Hensen's cells are termed the *cells of Claudius*. A space is seen between the outer rods of Corti and the adjacent hair-cells; this is called the *space of Nuel*.

The *lamina reticularis* or *membrane of Kölliker* is a delicate framework perforated by rounded holes. It extends from the heads of the outer rods of Corti to the external row of the outer hair-cells, and is formed by several rows of 'minute fiddle-shaped cuticular structures,' called *phalanges*, between which are circular apertures containing the free ends of the hair-cells. The innermost row of phalanges consists of the phalangeal processes of the outer rods of Corti ; the outer rows are formed by the modified free ends of Deiters's cells.

Covering over these structures, but not touching them, is the *membrana tectoria*, or membrane of Corti, which is attached to the limbus laminæ spiralis close to the inner edge of the membrane of Reissner. It is thin near its inner margin, and overlies the auditory teeth of Huschke. Its outer half is thick, and along its lower edge, opposite the inner hair-cells, is a clear band, named *Hensen's stripe*. Externally, the membrane becomes much thinner, and is attached to the outer row of Deiters's cells (Retzius).

The *inner surface of the osseous labyrinth* is lined by an exceedingly thin fibro-serous membrane, analogous to a periosteum, from its close adhesion to the inner surfaces of these cavities, and performing the office of a serous membrane by its free surface. It lines the vestibule, and from this cavity is continued into the semicircular canals and the scala vestibuli of the cochlea, and through the helicotrema into the scala tympani. A delicate tubular process is prolonged along the aqueduct of the vestibule to the inner surface of the dura mater. This membrane is continued across the fenestra ovalis and rotunda, and consequently has no communication with the lining membrane of the tympanum. Its attached surface is rough and fibrous, and closely adherent to the bone; its free surface is smooth and pale, covered with a layer of epithelium, and secretes a thin, limpid fluid, the *perilymph*.

The ductus cochlearis or scala media is closed above and below. The upper blind extremity is termed the *lagena*, and is attached to the cupola at the upper part of the helicotrema; the lower end is lodged in the recessus cochlearis of the vestibule. Near this blind extremity, the scala media receives the *canalis reuniens of Hensen* (fig. 683), a very delicate canal, by which the ductus cochlearis is brought into continuity with the saccule.

The **arteries of the labyrinth** are the internal auditory, from the basilar, and the stylo-mastoid, from the posterior auricular. The internal auditory divides at the bottom of the internal meatus into two branches: cochlear and vestibular.

The cochlear branch subdivides into twelve or fourteen twigs, which traverse the canals in the modiolus, and are distributed, in the form of a capillary network, in the lamina spiralis and basilar membrane.

The vestibular branches accompany the nerves, and are distributed, in the form of a minute capillary network, in the substance of the membranous labyrinth.

The **veins** (auditory) of the vestibule and semicircular canals accompany the arteries, and, receiving those of the cochlea at the base of the modiolus, terminate in the posterior part of the superior petrosal sinus or in the lateral sinus.

The **auditory nerve,** the special nerve of the sense of hearing, divides, at the bottom of the internal auditory meatus, into two branches, the cochlear and vestibular.

The *vestibular nerve*, the posterior of the two, presents, as it lies in the internal auditory meatus, a ganglion, the *ganglion of Scarpa*, the cells of which constitute the real origin of the nerve; it divides into three branches, which pass through minute openings at the upper and back part of the bottom of the meatus (*area vestibularis superior*), and, entering the vestibule, are distributed to the utricle and to the ampullæ of the external and superior semicircular canals.

The nervous filaments enter the ampullary enlargements opposite the septum transversum, and arborise around the hair-cells. In the utricle and saccule the nerve-fibres pierce the membrana propria of the maculæ, and end in arborisations round the hair-cells.

The *cochlear nerve* gives off the branch to the saccule, the filaments of which are transmitted from the internal auditory meatus through the foramina of the *area vestibularis inferior*, which lies at the lower and back part of the floor of the meatus. It also gives off the branch for the ampulla of the posterior semicircular canal, which leaves the meatus through the *foramen singulare*.

The rest of the cochlear nerve divides into numerous filaments at the base of the modiolus; those for the basal and middle coils pass through the foramina in the tractus foraminosus, those for the apical coil through the canalis centralis, and the nerves bend outwards to pass between the lamellæ of the osseous spiral lamina. Occupying the spiral canal of the modiolus is the *ganglion spirale*, consisting of bipolar nerve-cells, which really constitute the true cells of origin of this nerve, one pole being prolonged centrally to the brain and the other peripherally to the hair-cells of Corti's organ. Reaching the outer edge of the osseous spiral lamina, they pass through the foramina in the labium tympanicum; some end by arborising around the bases of the inner hair-cells, while others pass between Corti's rods and through the tunnel, to terminate in a similar manner in relation to the outer hair-cells.

Surgical Anatomy.—Malformations, such as imperfect development of the external parts, absence of the meatus, or supernumerary auricles, are occasionally met with. Or the pinna may present a congenital fistula, which is due to defective closure of the first visceral cleft, or rather of that portion of it which is not concerned in the formation of the Eustachian tube, tympanum, and meatus. The skin of the auricle is thin and richly supplied with blood, but in spite of this it is often the seat of frost-bite, due to the fact that it is much exposed to cold, and lacks the usual underlying subcutaneous fat found in most other parts of the body. A collection of blood is sometimes found between the cartilage and perichondrium (*hæmatoma auris*), usually the result of traumatism, but not necessarily due to the cause. It is said to occur most frequently in the ears of the insane. Keloid sometimes grows in the auricle around the puncture made for earrings, and epithelioma occasionally affects this part. Deposits of urate of soda are often met with in the pinna in gouty subjects.

The external auditory meatus can be most satisfactorily examined by light reflected down a funnel-shaped speculum; by gently moving the latter in different directions the whole of the canal and membrana tympani can be brought into view. The points to be noted are, the presence of wax or foreign bodies; the size of the canal; and the condition of the membrana tympani. Accumulation of wax is often the cause of deafness, and may give rise to very serious consequences, causing ulceration of the membrane and even absorption of the bony wall of the canal. Foreign bodies are not infrequently introduced into the ear by children, and, when situated in the first portion of the canal, may be removed with tolerable facility by means of a minute hook or loop of fine wire, aided by reflected light; but when they have slipped beyond the narrow middle part of the meatus, their removal is in no wise easy, and attempts to effect it, in inexperienced hands, may be followed by destruction of the membrana tympani and possibly the contents of the tympanum. The calibre of the external auditory canal may be narrowed by inflammation of its lining membrane, running on to suppuration; by periostitis; by polypi, sebaceous tumours, and exostoses. The membrana tympani, when seen in a healthy ear, 'reflects light strongly, and, owing to its peculiar curvature, presents a bright area of triangular shape at its lower and anterior portion.' From the apex of this, proceeding upwards and slightly forwards, is a white streak formed by the handle of the malleus, while near the upper part of the membrane may be seen a slight projection, caused by the short process of the malleus. In disease, alterations in colour, lustre, curvature or inclination, and perforation must be noted. Such perforations may be caused by a blow, a loud report, a wound, or as the result of suppuration in the middle ear.

The upper wall of the meatus is separated from the cranial cavity by a thin plate of bone; the anterior wall is separated from the temporo-mandibular joint and parotid gland by the bone forming the glenoid fossa; and the posterior wall is in relation with the mastoid cells, hence inflammation of the external auditory meatus may readily extend to the membranes of the brain, to the temporo-mandibular joint, or to the mastoid cells; and, in addition to this, blows on the chin may cause fracture of the wall of the meatus.

The nerves supplying the meatus are, the auricular branch of the pneumogastric, the auriculo-temporal, and the great auricular. The connections of these nerves explain the fact of the occurrence, in cases of any irritation of the meatus, of constant coughing and sneezing, from implication of the pneumogastric, or of yawning from implication of the auriculo-temporal. No doubt also the association of earache with toothache in cancer of the tongue is due to implication of the same nerve, a branch of the fifth, which supplies also the teeth and the tongue. The vessels of the meatus and membrana tympani are derived from the posterior auricular, temporal, and internal maxillary arteries. The upper half of the membrana tympani is much more richly supplied with blood than the lower half. For this reason, and also to avoid the chorda tympani nerve and ossicles, incisions through the membrane should be made at the lower and posterior part.

The principal point in connection with the surgical anatomy of the tympanum is its relations to other parts. Its roof is formed by a thin plate of bone, which, with the dura mater, is all that separates it from the temporal lobe of the brain. Its floor is immediately above the jugular fossa behind and the carotid canal in front. Its posterior wall presents the openings of the mastoid cells. On its anterior wall is the opening of the Eustachian tube. Thus it follows that in disease of the middle ear we may get subdural abscess, septic meningitis, or abscess of the cerebrum or cerebellum, from extension of the inflammation through the bony roof; thrombosis of the lateral sinus, with or without pyæmia, by extension through the floor; or mastoid abscess, by extension backwards. In addition to this, fatal hæmorrhage may occur from the internal carotid in destructive changes of the middle ear. And in throat disease the inflammation may extend up the Eustachian tube to the middle ear. The Eustachian tube is accessible from the nose. If the nose and mouth be closed, and an attempt made to expire air, a sense of pressure with dulness of hearing is produced in both ears, from the air finding its way up the Eustachian tube and bulging out the membrana tympani. During the act of swallowing, the pharyngeal orifice of the tube, which is normally closed, is opened, probably by the action of the Dilator tubæ muscle. This fact was employed by Politzer in devising an easy method of inflating the tube. The nozzle of an indiarubber syringe is inserted into the nostril; the patient takes a mouthful of water and holds it in his mouth; both nostrils are closed with the finger and thumb to prevent the escape of air, and the patient is then requested to swallow ; as he does so, the air is forced out of the syringe into his nose, and is driven into the Eustachian tube, which is now open. The impact of the air against the membrana tympani can be heard, if the membrane is sound, by means of a piece of indiarubber tubing, one end of which is inserted into the meatus of the patient's ear, the other into that of the surgeon. The direct examination of the Eustachian tube is made by the Eustachian catheter. This is passed along the floor of the nostril, close to the septum, with the point touching the floor, to the posterior wall of the pharynx. When this is felt, the catheter is to be withdrawn about half an inch, and the point rotated outwards through a quarter of a circle and pushed again slightly backwards, when it will enter the orifice of the tube, and will be found to be caught, and air forced into the catheter will be heard impinging on the tympanic membrane, if the ears of the patient and surgeon are connected by an indiarubber tube.

ORGANS OF VOICE AND
RESPIRATION

THE LARYNX

THE Larynx is the organ of voice, and is placed at the upper part of the air-passage. It is situated between the trachea and base of the tongue, at the upper and fore part of the neck, where it forms a considerable projection in the middle line. On either side of it lie the great vessels of the neck; behind, it forms part of the boundary of the pharynx, and is covered by the mucous membrane lining that cavity. Its vertical extent corresponds to the fourth, fifth, and sixth cervical vertebræ, but it is placed somewhat higher in the female and also during childhood. In infants between six and twelve months of age Symington found that the tip of the epiglottis was a little above the level of the cartilage between the odontoid process and body of the axis, and that between infancy and adult life the larynx descends for a distance equal to two vertebral bodies and two intervertebral discs. According to Sappey the average measurements of the adult larynx are as follows:

	In males	In females
Vertical diameter . . .	44 mm.	36 mm.
Transverse diameter . . .	43 ,,	41 ,,
Antero-posterior diameter . .	36 ,,	26 ,,
Circumference	136 ,,	112 ,,

Until puberty the larynx of the male and that of the female differ little in size. In the female its further increase at puberty is only slight, whereas in the male it is great; all the cartilages are enlarged and the thyroid becomes prominent as the *pomum Adami* in the middle line of the neck, while the length of the glottis is nearly doubled.

The larynx is broad above, where it presents the form of a triangular box, flattened behind and at the sides, and bounded in front by a prominent vertical ridge. Below, it is narrow and cylindrical. It is composed of cartilages, which are connected together by ligaments and moved by numerous muscles. It is lined by mucous membrane, which is continuous above with that of the pharynx and below with that of the trachea.

The **Cartilages of the Larynx** are nine in number, three single, and three paired, as follows:

Thyroid.	Two Arytenoid.
Cricoid.	Two Cornicula laryngis.
Epiglottis.	Two Cuneiform.

The **thyroid** (θυρεός, *a shield*) is the largest cartilage of the larynx. It consists of two lateral lamellæ or alæ, united at an acute angle in front, forming a vertical projection in the middle line which is prominent above, and called the *pomum Adami*. This projection is subcutaneous, more distinct in the male than in the female, and occasionally separated from the integument by a bursa mucosa.

Each lamella is quadrilateral in form. Its *outer surface* presents an *oblique ridge*, which passes downwards and forwards from a tubercle, situated near the root of the superior cornu, to another on the lower border. This ridge gives attachment to the Sterno-thyroid and Thyro-hyoid muscles, and the portion of cartilage included between it and the posterior border to part of the Inferior constrictor muscle.

The *inner surface* of each ala is smooth, slightly concave, and covered by mucous membrane above and behind; but in front, in the receding angle formed by the junction of the alæ, are attached the epiglottis, the true and false vocal cords, the Thyro-arytenoid and Thyro-epiglottidean muscles, and the thyro-epiglottidean ligament.

The *upper border of the thyroid cartilage* is sinuously curved, being concave at its posterior part, just in front of the superior cornu, then rising into a convex outline, which dips in front to form the sides of a notch, the *thyroid notch*, in the middle line, immediately above the pomum Adami. This border gives attachment throughout its whole extent to the thyro-hyoid membrane.

The *lower border* is nearly straight in front, but behind, close to the cornu, is concave. It is connected to the cricoid cartilage, in and near the median line, by the middle portion of the crico-thyroid membrane; and, on each side, by the Crico-thyroid muscle.

FIG. 691.—Side view of the thyroid and cricoid cartilages.

The *posterior borders*, thick and rounded, terminate, above, in the *superior cornua*, and below, in the *inferior cornua*. The two superior cornua are long and narrow, directed upwards, backwards, and inwards, and terminate in conical extremities, which give attachment to the lateral thyro-hyoid ligaments. The two inferior cornua are short and thick; they pass downwards, with a slight inclination forwards and inwards, and each presents, on its inner surface, a small oval articular facet for articulation with the side of the cricoid cartilage. The posterior border receives the insertion of the Stylo-pharyngeus and Palato-pharyngeus muscles on each side.

During infancy the alæ of the thyroid cartilage are joined to each other by a narrow, lozenge-shaped strip, named the *intrathyroid cartilage*. This strip extends from the upper to the lower border of the cartilage in the middle line, and is distinguished from the alæ by being more transparent and more flexible.

The **cricoid cartilage** is so called from its resemblance to a signet ring (κρίκος, *a ring*). It is smaller, but thicker and stronger than the thyroid cartilage, and forms the lower and back parts of the cavity of the larynx. It consists of two parts: a quadrate portion, situated behind, and a narrow ring or arch, one-fourth or one-fifth of the depth of the posterior part, situated in front. The posterior square portion rapidly narrows at the sides of the cartilage, at the expense of the upper border, into the anterior portion.

Its *posterior portion* is very deep and broad, and measures from above downwards about an inch (two to three centimetres); it presents, on its posterior surface, in the middle line, a vertical ridge for the attachment of the longitudinal fibres of the œsophagus; and on either side a broad depression for the Crico-arytenoideus posticus muscle.

Its *anterior portion* is narrow and convex, and measures vertically about one-fourth or one-fifth of an inch (·7 to ·5 centimetre); it affords attachment externally in front and at the sides to the Crico-thyroid muscles, and behind, to part of the Inferior constrictor.

At the point of junction of the posterior quadrate portion with the rest of the cartilage is a small round elevation on each side, for articulation with the inferior cornua of the thyroid cartilage.

The *lower border* of the cricoid cartilage is horizontal, and connected to the upper ring of the trachea by fibrous membrane.

Its *upper border* is directed obliquely upwards and backwards, owing to the great depth of the posterior surface. It gives attachment, in front, to the middle portion of the crico-thyroid membrane; at the sides, to the lateral portions of the same membrane and to the lateral Crico-arytenoid muscles; behind, it presents, in the middle, a shallow notch, and on each side of this is a smooth, oval surface, directed upwards and outwards, for articulation with the base of the arytenoid cartilage.

FIG. 692.—The cartilages of the larynx. Posterior view.

The *inner surface* of the cricoid cartilage is smooth, and lined by mucous membrane.

The **arytenoid cartilages** are two in number, and situated at the upper border of the cricoid cartilage, at the back of the larynx. Each cartilage is pyramidal in form, and presents for examination three surfaces, a base, and an apex.

The *posterior surface* is triangular, smooth, concave, and gives attachment to the Arytenoid muscle.

The *antero-external surface* is somewhat convex and rough. It presents rather below its centre a transverse ridge, to the inner extremity of which is attached the false vocal cord; to the outer part, as well as to the surfaces above and below it, the Thyro-arytenoid muscle is inserted.

The *internal surface* is narrow, smooth, and flattened, covered by mucous membrane, and forms the lateral boundary of the respiratory part of the glottis.

Arytenoid cartilages, base

The *base* of each cartilage is broad, and presents a concave smooth surface, for articulation with the cricoid cartilage. Two of its angles require special mention: the *external*, which is short, rounded, and prominent, projects backwards and outwards, and is termed the *muscular process*; it gives insertion to the Posterior crico-arytenoid muscle behind and to the Lateral crico-arytenoid in front. The *anterior angle*, also prominent, but more pointed, projects horizontally forwards, and gives attachment to the true vocal cord. This angle is called the *vocal process*.

The *apex* of each cartilage is pointed, curved backwards and inwards, and surmounted by a small conical, cartilaginous nodule, the *corniculum laryngis.*

The **cornicula laryngis** (*cartilages of Santorini*) are two small conical nodules, consisting of yellow elastic cartilage, which articulate with the summit of the arytenoid cartilages and serve to prolong them backwards and inwards. They are situated in the posterior part of the *aryteno-epiglottidean folds* of mucous membrane, and are sometimes united to the arytenoid cartilages.

The **cuneiform cartilages** (*cartilages of Wrisberg*) are two small, elongated pieces of yellow elastic cartilage, placed one on each side, in the aryteno-epiglottidean fold, where they give rise to small whitish elevations on the inner surface of the mucous membrane, just in front of the arytenoid cartilages.

The **epiglottis** is a thin lamella of fibro-cartilage, of a yellowish colour, shaped like a leaf, and placed behind the tongue in front of the superior opening of the larynx. Its free extremity is broad and rounded; its attached part is long, narrow, and connected to the receding angle between the two alæ of the thyroid cartilage, just below the median notch, by a ligamentous band, the *thyro-epiglottic ligament*. Its anterior surface is connected to the upper border of the body of the hyoid bone by an elastic ligamentous band, the *hyo-epiglottic ligament*.

Its *anterior* or *lingual surface* is curved forwards towards the tongue, and covered at its upper, free part by mucous membrane, which is reflected on to the sides and base of the organ, forming a median and two lateral folds, the *glosso-epiglottidean folds*; the lateral folds are partly attached to the wall of the pharynx. The depressions between the epiglottis and the base of the tongue, on either side of the median fold, are named the *valleculæ*. The lower part of its anterior surface lies behind the hyoid bone, the thyro-hyoid membrane, and upper part of the thyroid cartilage, but is separated from these structures by a mass of fatty tissue.

Its *posterior* or *laryngeal surface* is smooth, concave from side to side, concavo-convex from above downwards; its lower part projects backwards as an elevation, the *tubercle* or *cushion*; when the mucous membrane is removed, the surface of the cartilage is seen to be studded with a number of small mucous glands, which are lodged in little pits upon its surface. To its sides the aryteno-epiglottidean folds are attached.

Structure.—The cornicula laryngis and cuneiform cartilages, the epiglottis, and the apices of the arytenoids are composed of yellow fibro-cartilage, which shows little tendency to calcification; on the other hand the thyroid, cricoid, and the greater part of the arytenoids consist of hyaline cartilage, and become more or less ossified as age advances. Ossification commences about the twenty-fifth year in the thyroid cartilage, somewhat later in the cricoid and arytenoids; by the sixty-fifth year these cartilages may be completely converted into bone.

Ligaments.—The ligaments of the larynx are *extrinsic*, i.e. those connecting the thyroid cartilage and epiglottis with the hyoid bone, and the cricoid cartilage with the trachea; and *intrinsic*, those which connect the several cartilages of the larynx to each other.

The ligaments connecting the thyroid cartilage with the hyoid bone are three in number—the thyro-hyoid membrane, and the two lateral thyro-hyoid ligaments.

The *thyro-hyoid membrane*, or *middle thyro-hyoid ligament*, is a broad, fibro-elastic, membranous layer, attached below to the upper border of the thyroid cartilage, and above to the upper margin of the posterior surface of the body and greater cornua of the hyoid bone, thus passing behind the posterior surface of the body of the hyoid, and being separated from it by a synovial bursa, which facilitates the upward movement of the larynx during deglutition. It is thicker in the middle line than at either side, and is pierced, in the latter situation, by the superior laryngeal vessels and the internal laryngeal nerve. Its anterior surface is in relation with the Thyro-hyoid, Sterno-hyoid, and Omo-hyoid muscles, and with the body of the hyoid bone.

The *two lateral thyro-hyoid ligaments* are round, elastic cords, which pass between the superior cornua of the thyroid cartilage and the extremities of the greater cornua of the hyoid bone. A small cartilaginous nodule (*cartilago triticea*), sometimes bony, is frequently found in each.

The ligament connecting the epiglottis with the hyoid bone is the *hyo-epiglottic*. In addition to this extrinsic ligament, the epiglottis is connected to the tongue by the three glosso-epiglottidean folds of mucous membrane, which may also be considered as extrinsic ligaments of the epiglottis.

The *hyo-epiglottic ligament* is an elastic band, which extends from the anterior surface of the epiglottis, near its apex, to the upper border of the body of the hyoid bone.

The ligaments connecting the thyroid cartilage to the cricoid are also three in number—the crico-thyroid membrane, and the capsular ligaments.

The *crico-thyroid membrane* is composed mainly of yellow elastic tissue. It consists of three parts, a central, triangular portion and two lateral portions. The *central* part is thick and strong, narrow above and broad below. It connects together the contiguous margins of the thyroid and cricoid cartilages. It is convex, concealed on each side by the Crico-thyroid muscle, but subcutaneous in the middle line; it is crossed horizontally by a small anastomotic arterial arch, formed by the junction of the two crico-thyroid arteries. The *lateral* portions are thinner and lie close under the mucous membrane of the larynx. They extend from the superior border of the cricoid cartilage to the inferior margin of the true vocal cords, with which they are continuous, so that these cords may be regarded as the free borders of the lateral portions of the crico-thyroid membrane. They extend from the vocal processes of the arytenoid cartilages to the receding angle of the thyroid cartilage near its centre. The lateral portions are lined internally by mucous membrane, and covered by the lateral Crico-arytenoid and Thyro-arytenoid muscles.

A *capsular ligament* encloses the articulation of the inferior cornu of the thyroid with the cricoid cartilage on each side, strengthened posteriorly by a well-marked fibrous band. The articulation is lined by synovial membrane.

The ligaments connecting the arytenoid cartilages to the cricoid are two *capsular ligaments*, and two *posterior crico-arytenoid ligaments*. The *capsular ligaments* are thin and loose capsules attached to the margin of the articular surfaces; they are lined internally by synovial membrane. The *posterior crico-arytenoid ligaments* extend from the cricoid to the inner and back part of the base of the arytenoid cartilage.

The ligament connecting the epiglottis with the thyroid cartilage is the thyro-epiglottic.

The *thyro-epiglottic ligament* is a long, slender, elastic cord which connects the apex of the epiglottis with the receding angle of the thyroid cartilage, immediately beneath the median notch, above the attachment of the vocal cords.

The *crico-tracheal ligament* connects the cricoid cartilage with the first ring of the trachea. It resembles the fibrous membrane which connects the cartilaginous rings of the trachea to each other.

The Joints of the Larynx.—The articulation between the inferior cornu of the thyroid cartilage and the cricoid cartilage on each side is a diarthrodial one, and permits of a rotatory and gliding movement. The rotatory movement is one in which the inferior cornua of the thyroid cartilage rotate upon the cricoid cartilage around an axis passing transversely through both joints. The gliding movement consists in a limited shifting of the cricoid on the thyroid in different directions.

The articulation between the arytenoid cartilages and the cricoid is also a diarthrodial one, and permits of two varieties of movement: one a rotation of the arytenoid on a vertical axis, whereby the vocal process is moved outwards or inwards, and the opening of the rima glottidis increased or diminished; the other is a gliding movement, and allows the arytenoid cartilages to approach or recede from each other. The two movements are associated, the gliding inwards being connected with inward rotation, and the gliding outwards with outward rotation. The posterior crico-arytenoid ligaments prevent any forward movement of the arytenoid cartilages on the cricoid.

Interior of the Larynx.—The *cavity of the larynx* extends from its superior aperture to the lower border of the cricoid cartilage. It is divided into two parts by the projection inwards of the true vocal cords, between which is a narrow triangular fissure or chink, the *rima glottidis*. The portion of the cavity of the larynx above the true vocal cords, sometimes called the *vestibule*, is broad and triangular in shape, and corresponds to the interval between the alæ of the thyroid cartilage; it contains the false vocal cords, and between these and the true vocal cords are the ventricles of the larynx. The portion below the true vocal cords widens out, and is at first of an elliptical and lower down of a circular form, and is continuous with the tube of the trachea.

The *superior aperture of the larynx* (fig. 693) is a triangular or cordiform opening, wide in front, narrow behind, and sloping obliquely downwards and backwards. It is bounded, in front, by the epiglottis; behind, by the apices of the arytenoid cartilages and the cornicula laryngis; and laterally, by folds of mucous membrane, enclosing ligamentous and muscular fibres, stretched between the sides of the epiglottis and the apices of the arytenoid cartilages: these are the *aryteno-epiglottidean folds*, on the margins of which the cuneiform cartilages form more or less distinct whitish prominences.

The *rima glottidis* is the elongated fissure or chink between the inferior or true vocal cords in front, and between the bases and vocal processes of the arytenoid cartilages behind. It is therefore frequently subdivided into an anterior interligamentous or *vocal* portion (*glottis vocalis*) and a posterior intercartilaginous or *respiratory* part (*glottis respiratoria*). Posteriorly it is limited by the mucous membrane passing between the arytenoid cartilages. The vocal portion averages about three-fifths of the length of the entire aperture. The rima glottidis is the narrowest part of the cavity of the larynx, and its level corresponds with the bases of the arytenoid cartilages. Its length, in the male, measures rather less than an inch (twenty-three millimetres); in the female it is shorter by five or six millimetres, or three lines. The width and shape of the rima glottidis vary with the movements of the vocal cords and arytenoid

FIG. 693.—Larynx, viewed from above. (Testut.)

cartilages during respiration and phonation. In the condition of rest, i.e. when these structures are uninfluenced by muscular action, as in quiet respiration, the glottis vocalis is triangular, with its apex in front and its base behind—the latter being represented by a line, about eight millimetres long, connecting the anterior extremities of the vocal processes, while the inner surfaces of the arytenoids are parallel to each other, and hence the glottis respiratoria is rectangular. During extreme adduction of the cords, as in the emission of a high note, the glottis vocalis is reduced to a linear slit by the apposition of the cords, while the glottis respiratoria is triangular, its apex corresponding to the anterior extremities of the vocal processes of the arytenoids, which are approximated by the inward rotation of the cartilages. Conversely in extreme abduction of the cords, as in forced inspiration, the arytenoids and their vocal processes are rotated outwards, and the glottis respiratoria is triangular in shape but with its apex directed backwards. In this condition the entire glottis is somewhat lozenge-shaped, the sides of the glottis vocalis diverging from before backwards, those of the glottis respiratoria diverging from behind forwards—the widest part of the aperture corresponding with the attachment of the cords to the vocal processes.

The *superior* or *false vocal cords*, so called because they are not directly concerned in the production of the voice, are two thick folds of mucous membrane, enclosing a narrow band of fibrous tissue, the *superior thyro-arytenoid ligament*, which is attached in front to the angle of the thyroid cartilage immediately below

3 T

the attachment of the epiglottis, and behind to the antero-external surface of the arytenoid cartilage, a short distance above the vocal process. The lower border of this ligament, enclosed in mucous membrane, forms a free crescentic margin, which constitutes the upper boundary of the ventricle of the larynx.

The *inferior* or *true vocal cords*, so called from their being concerned in the production of sound, are two strong bands, named the *inferior thyro-arytenoid ligaments*. Each ligament consists of a band of yellow elastic tissue, attached in front to the depression between the two alæ of the thyroid cartilage, and behind to the vocal process of the base of the arytenoid. Its lower border is continuous with the thin lateral part of the crico-thyroid membrane. Its upper border forms the lower boundary of the ventricle of the larynx. Externally, the Thyro-arytenoideus muscle lies parallel with it. It is covered internally by mucous membrane, which is extremely thin, and closely adherent to its surface.

FIG. 694.—Vertical section of the larynx and upper part of the trachea.

Glosso-epiglottic fold

Epiglottis

Hyoid bone

Aryteno-epiglottic fold

Arytenoid cartilage

Vocal cords, ventricle between them

Arytenoideus muscle

Thyroid cartilage

Crico-thyroid membrane

Cricoid cart.

Cricoid cartilage

Trachea

The *ventricle of the larynx* is an oblong fossa, situated between the superior and inferior vocal cords on each side, and extending nearly their entire length. This fossa is bounded, above, by the free crescentic edge of the false vocal cord; below, by the straight margin of the true vocal cord; externally, by the mucous membrane covering the corresponding Thyro-arytenoideus muscle. The anterior part of the ventricle leads up by a narrow opening into a cæcal pouch of mucous membrane of variable size, called the sacculus laryngis.

The *sacculus laryngis*, or laryngeal pouch, is a membranous sac, placed between the superior vocal cord and the inner surface of the thyroid cartilage, occasionally extending as far as its upper border or even higher: it is conical in form, and curved slightly backwards. On the surface of its mucous membrane are the openings of sixty or seventy mucous glands, which are lodged in the submucous areolar tissue. This sac is enclosed in a fibrous capsule, continuous below with the superior thyro-arytenoid ligament: its laryngeal surface is covered by a few delicate muscular fasciculi, which arise from the apex of the arytenoid cartilage and become lost in the fold of mucous membrane extending between the arytenoid cartilage and the side of the epiglottis (they were named by Hilton the *compressor sacculi laryngis*); while its exterior is covered by the Thyro-arytenoideus and Thyro-epiglottideus muscles. These muscles compress the sacculus laryngis, and discharge the secretion it contains upon the chordæ vocales, the surfaces of which it is intended to lubricate.

Muscles.—The muscles of the larynx are *extrinsic*, passing between the larynx and parts around: these have been described elsewhere; and *intrinsic*, confined entirely to the larynx itself.

The intrinsic muscles are, the

Crico-thyroid.　　　　　　　　　　Crico-arytenoideus lateralis.
Crico-arytenoideus posticus.　　　　Arytenoideus.
Thyro-arytenoideus.

The *Crico-thyroid* is triangular in form, and situated at the fore part and side of the cricoid cartilage. It arises from the front and lateral part of the cricoid cartilage; its fibres diverge, passing obliquely upwards and outwards, to be inserted into the lower border of the thyroid cartilage, and into the anterior border of the lower cornua.

The inner borders of these two muscles are separated in the middle line by a triangular interval, occupied by the central part of the crico-thyroid membrane.

The *Crico-arytenoideus posticus* arises from the broad depression occupying each lateral half of the posterior surface of the cricoid cartilage; its fibres pass upwards and outwards, converging to be inserted into the posterior surface of the muscular process at the base of the arytenoid cartilage. The upper fibres are nearly horizontal, the middle oblique, and the lower almost vertical.

The *Crico-arytenoideus lateralis* is smaller than the preceding, and of an oblong form. It arises from the upper border of the side of the cricoid cartilage, and, passing obliquely upwards and backwards, is inserted into the front of the muscular process of the arytenoid cartilage.

The *Arytenoideus* is a single muscle, filling up the posterior concave surfaces of the arytenoid cartilages. It arises from the posterior surface and outer border of one arytenoid cartilage, and is inserted into the corresponding parts of the opposite cartilage. It consists of three planes of fibres, two oblique and one transverse. The *oblique fibres*, the more superficial, form two fasciculi, which pass from the base of one cartilage to the apex of the opposite one, and which therefore cross each other like the limbs of the letter **X**. The *transverse fibres*, the deeper and more numerous, pass transversely across between the two cartilages; hence the Arytenoideus was formerly considered as three muscles, the transverse and the two oblique. A few of the oblique fibres are continued round the outer margin of the cartilage, and are prolonged into the aryteno-epiglottidean fold. They are sometimes described as a separate muscle, the *Aryteno-epiglottideus*.

The *Thyro-arytenoideus* is a broad, flat muscle, which lies parallel with the outer side of the true vocal cord. It arises in front from the lower half of the receding angle of the thyroid cartilage, and from the crico-thyroid membrane. Its fibres pass backwards and out-

Fig. 695.—Muscles of larynx. Side view. Right ala of thyroid cartilage removed.

Cornicula laryngis

Articular facet for inferior cornu of thyroid cartilage

wards, to be inserted into the base and anterior surface of the arytenoid cartilage. This muscle consists of two fasciculi, an inner and an outer.* The *inner portion* is a triangular band which is inserted into the vocal process of the arytenoid cartilage, and into the adjacent portion of its anterior surface; it lies parallel with the true vocal cord, to which it is adherent. The *outer fasciculus*, the thinner, is inserted into the anterior surface and outer border of the arytenoid cartilage above the preceding fibres; it lies on the outer side of the sacculus laryngis, immediately beneath the mucous membrane.

A considerable number of the fibres of the Thyro-arytenoideus are prolonged into the aryteno-epiglottidean fold, where some of them become lost, while others are continued forwards to the margin of the epiglottis. They have received a distinctive name, *Thyro-epiglottideus*, and are sometimes described as a separate muscle.

* Henle describes these two portions as separate muscles, under the names of External and Internal thyro-arytenoid.

Actions.—In considering the actions of the muscles of the larynx, they may be conveniently divided into two groups, viz.: 1. Those which open and close the glottis. 2. Those which regulate the degree of tension of the vocal cords.

1. The muscles which open the glottis are the Crico-arytenoidei postici; and those which close it are the Crico-arytenoidei laterales and the Arytenoideus. 2. The muscles which regulate the tension of the vocal cords are the Crico-thyroidei, which elongate and render them tense; and the Thyro-arytenoidei, which relax and shorten them.

The *Crico-arytenoidei postici* separate the vocal cords, and, consequently, open the glottis, by rotating the arytenoid cartilages outwards around a vertical axis passing through the crico-arytenoid joints; so that their vocal processes and the vocal cords attached to them become widely separated.

The *Crico-arytenoidei laterales* close the glottis, by rotating the arytenoid cartilages inwards, so as to approximate their vocal processes.

The *Arytenoideus muscle* approximates the arytenoid cartilages, and thus closes the opening of the glottis, especially at its back part.

The *Crico-thyroid muscles* produce tension and elongation of the vocal cords. This is effected as follows: when the Crico-thyroid muscles contract they draw down the lower border of the thyroid cartilage and slightly advance its inferior cornua, and thus the distance between the vocal processes and the angle of the thyroid is increased, and the cords are consequently elongated.

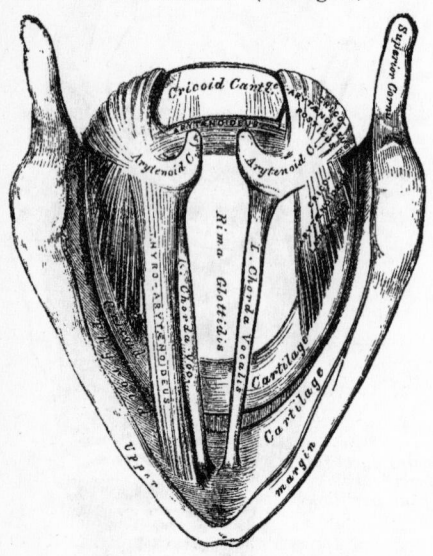

FIG. 696.—Interior of the larynx, seen from above. (Enlarged.)

The *Thyro-arytenoidei muscles*, consisting of two parts having different attachments and different directions, are rather complicated as regards their action. Their main use is to draw the arytenoid cartilages forwards towards the thyroid, and thus shorten and relax the vocal cords. But, owing to the connection of the inner portion with the vocal cord, this part, if acting separately, is supposed to modify its elasticity and tension, and the outer portion, being inserted into the outer part of the anterior surface of the arytenoid cartilage, may rotate it inwards, and thus narrow the rima glottidis by bringing the two cords together.

The manner in which the superior aperture of the larynx is closed during deglutition is referred to on page 460.

The **Mucous Membrane of the Larynx** is continuous above with that lining the mouth and pharynx, and is prolonged through the trachea and bronchi into the lungs. It lines the posterior surface and the upper part of the anterior surface of the epiglottis, to which it is closely adherent, and forms the aryteno-epiglottidean folds which form the lateral boundaries of the superior aperture of the larynx. It lines the whole of the cavity of the larynx; forms, by its reduplication, the chief part of the superior, or false, vocal cord; and, from the ventricle, is continued into the sacculus laryngis. It is then reflected over the true vocal cords, where it is thin, and very intimately adherent; covers the inner surface of the crico-thyroid membrane and cricoid cartilage; and is ultimately continuous with the lining membrane of the trachea. The fore part of the anterior surface and the upper half of the posterior surface of the epiglottis, the upper part of the aryteno-epiglottidean folds, and the true vocal cords are covered by stratified squamous epithelium; all the rest of the laryngeal mucous membrane is covered by columnar ciliated cells.

Glands.—The mucous membrane of the larynx is furnished with numerous muciparous glands, the orifices of which are found in nearly every part; they are very plentiful upon the epiglottis, being lodged in little pits in its substance; they are also found in large numbers along the posterior margin of the aryteno-epiglottidean fold, in front of the arytenoid cartilages, where they are termed the *arytenoid glands*. They exist also in large numbers upon the inner surface of the sacculus laryngis.   None are found on the free edges of the vocal cords.

Vessels and Nerves.—The *arteries of the larynx* are the laryngeal branches derived from the superior and inferior thyroid. The *veins* accompany the arteries : those accompanying the superior laryngeal artery join the superior thyroid vein, which opens into the internal jugular vein ; while those accompanying the inferior laryngeal artery join the inferior thyroid vein, which opens into the innominate vein. The *lymphatics* consist of two sets, superior and inferior. The former accompany the superior laryngeal artery and pierce the thyro-hyoid membrane, to terminate in the glands situated near the bifurcation of the common carotid artery. Of the latter, some pass through the crico-thyroid membrane, and open into a gland lying in front of that membrane or in front of the upper part of the trachea, while others pass to the deep cervical glands and to the glands which accompany the inferior thyroid artery. The *nerves* are derived from the internal and external laryngeal branches of the superior laryngeal nerve, from the inferior or recurrent laryngeal, and from the sympathetic. The internal laryngeal nerve is almost entirely sensory, but some motor filaments are said to be carried by it to the Arytenoideus muscle. It divides into a branch which is distributed to both surfaces of the epiglottis, a second to the aryteno-epiglottidean folds, and a third, the largest, which supplies the mucous membrane over the back of the larynx and communicates with the recurrent laryngeal. The external laryngeal nerve supplies the Crico-thyroid muscle. The recurrent laryngeal passes upwards under the lower border of the Inferior constrictor, and enters the larynx between the cricoid and thyroid cartilages.   It supplies all the muscles of the larynx except the Crico-thyroid and perhaps a part of the Arytenoideus. The sensory branches of the laryngeal nerves form subepithelial plexuses, from which fibres pass to end between the cells covering the mucous membrane.

Over the posterior surface of the epiglottis, in the aryteno-epiglottidean folds, and less regularly in some other parts, taste-buds, similar to those in the tongue, are found.

The Trachea (fig. 697)

The **trachea,** or **windpipe,** is a cartilaginous and membranous tube, which extends from the lower part of the larynx, on a level with the sixth cervical vertebra, to the upper border of the fifth dorsal vertebra, where it divides into the two bronchi, one for each lung. The trachea is nearly but not quite cylindrical, being flattened posteriorly ; it measures about four inches and a half in length ; its diameter, from side to side, is from three-quarters of an inch to an inch, being always greater in the male than in the female.

Relations.—The anterior surface of the trachea is convex, and covered, *in the neck,* from above downwards, by the isthmus of the thyroid gland, the inferior thyroid veins, the arteria thyroidea ima (when that vessel exists), the Sterno-thyroid and Sterno-hyoid muscles, the cervical fascia, and, more superficially, by the anastomosing branches between the anterior jugular veins ; *in the thorax,* it is covered from before backwards by the first piece of the sternum, the remains of the thymus gland, the left innominate vein, the arch of the aorta, the innominate and left carotid arteries, and the deep cardiac plexus. Posteriorly it is in relation with the œsophagus ; laterally, *in the neck,* it is in relation with the common carotid arteries, the lateral lobes of the thyroid gland, the inferior thyroid arteries, and recurrent laryngeal nerves ; and, *in the thorax,* it lies in the upper part of the interpleural space (superior mediastinum), and is in relation on the right side to the pleura and right vagus, and near the root of the neck to the innominate artery ; on its left side are the recurrent laryngeal nerve, the aortic arch, the left common carotid and subclavian arteries.

The **Right Bronchus,** wider, shorter, and more vertical in direction than the left, is about an inch in length, and enters the right lung nearly opposite the fifth dorsal vertebra. The vena azygos major arches over it from behind ; and the right pulmonary artery lies below and then in front of it. About three-quarters of an inch from its commencement it gives off a branch to the upper lobe of the right lung. This is termed the *eparterial* bronchus, because it is given off above the right pulmonary artery. The bronchus now passes below the artery, and is known as the *hyparterial* bronchus. It divides into two branches for the middle and lower lobes.

The **Left Bronchus** is smaller and longer than the right, being nearly two inches in length. It enters the root of the left lung, opposite the sixth dorsal vertebra, about an inch lower than the right bronchus. It passes beneath the arch of the aorta, crosses in front of the œsophagus, the thoracic duct, and the descending aorta, and has the left pulmonary artery lying at first above, and then in front of it. The left bronchus has no branch corresponding to the

FIG. 697.—Front view of cartilages of larynx; the trachea and bronchi.

eparterial branch of the right bronchus, and therefore it has been supposed by some that there is no upper lobe to the left lung, but that the so-called upper lobe corresponds to the middle lobe of the right lung.

The further subdivision of the bronchi will be considered with the anatomy of the lung.

If a transverse section is made across the trachea, a short distance above its point of bifurcation, and a bird's-eye view taken of its interior (fig. 698), the septum placed at the bottom of the trachea and separating the two bronchi will be seen to occupy the left of the median line, and the right bronchus appears to be a more direct continuation of the trachea than the left, so that any solid body dropping into the trachea would naturally be directed towards the right bronchus. This tendency is aided by the larger size of the right tube as compared with its fellow. This fact serves to explain why a foreign body in the trachea more frequently falls into the right bronchus.*

Structure.—The trachea is composed of imperfect cartilaginous rings, fibrous membrane, muscular fibres, mucous membrane, and glands.

The **cartilages** vary from sixteen to twenty in number : each forms an imperfect ring, which surrounds about two-thirds of the cylinder of the trachea, being deficient behind, where the tube is completed by fibrous membrane. The cartilages are placed horizontally above each other, separated by narrow membranous intervals. They measure about two lines in depth, and half a line in thickness. Their outer surfaces are flattened, but internally they are convex, from being thicker in the middle than at the margins. Two or more of the cartilages often unite, partially or completely, and they are sometimes bifurcated at their extremities. They are highly elastic, but may become calcified in advanced life. In the right bronchus the cartilages vary in number from six to eight ; in the left, from nine to twelve. They are shorter and narrower than those of the trachea. The peculiar cartilages are the first and the last.

The *first cartilage* is broader than the rest, and often divided at one end ; it is connected by fibrous membrane with the lower border of the cricoid cartilage, with which, or with the succeeding cartilage, it is sometimes blended.

FIG. 698.—Transverse section of the trachea, just above its bifurcation, with a bird's-eye view of the interior.

Left   *Right*

The *last cartilage* is thick and broad in the middle, in consequence of its lower border being prolonged into a triangular hook-shaped process, which curves downwards and backwards between the two bronchi. It terminates on each side in an imperfect ring, which encloses the commencement of the bronchi. The cartilage above the last is somewhat broader than the rest at its centre.

The fibrous membrane.—The cartilages are enclosed in an elastic fibrous membrane, which consists of two layers ; one, the thicker of the two, passing over the outer surface of the ring, the other over the inner surface : at the upper and lower margins of the cartilages the two layers blend together to form a single membrane, which connects the rings one with another. They are thus, as it were, embedded in the membrane. In the space behind, between the extremities of the rings, the membrane forms a single distinct layer.

The **muscular fibres** are disposed in two layers, longitudinal and transverse. The *longitudinal fibres* are external, and consist merely of a few scattered longitudinal bundles.

The *transverse fibres* (Trachealis muscle, Todd and Bowman) are internal and form a thin layer, which extends transversely between the ends of the cartilages and the intervals between them at the posterior part of the trachea. The muscular fibres are of the unstriped variety.

The **mucous membrane** is continuous above with that of the larynx, and below with that of the bronchi. Microscopically, it consists of areolar and lymphoid tissue, and presents a well-marked basement-membrane, supporting a layer of columnar, ciliated epithelium ; between the deeper ends of the columnar cells are smaller triangular cells, the bases of which, often branched, are attached to the basement-membrane. These triangular cells are mucus-secreting, and may be seen as goblet or chalice cells when their contents have been discharged. In the deepest part of the mucous membrane, and especially between the mucous and submucous layers, longitudinally arranged fibres are very abundant and form a distinct layer.

The **tracheal glands** are found in great abundance at the posterior part of

* Reigel asserts that the entry of a foreign body into the *left* bronchus is by no means so infrequent as is generally supposed. See also *Med.-Chir. Trans.* vol. lxxi. p. 121.

the trachea. They are racemose glands, and consist of a basement-membrane lined by columnar mucus-secreting cells. They are situated at the back of the trachea, outside the layer of muscular tissue, between it and the outer fibrous layer. Their excretory ducts pierce the muscular and inner fibrous layers, and pass through the submucous and mucous layers to open on the surface of the mucous membrane. Some glands of smaller size are also found at the sides of the trachea, between the layers of fibrous tissue connecting the rings, and others immediately beneath the mucous coat. The secretion from these glands serves to lubricate the inner surface of the trachea.

Vessels and Nerves.—The trachea is supplied with blood by the inferior thyroid arteries. The *veins* terminate in the thyroid venous plexus. The *nerves* are derived from the pneumogastric and its recurrent branches, and from the sympathetic.

Surface Form.—In the middle line of the neck, some of the cartilages of the larynx can be readily distinguished. In the receding angle below the chin, the hyoid bone can easily be made out (see page 228), and a finger's breadth below it is the pomum Adami, the prominence between the upper borders of the two alæ of the thyroid cartilage. About an inch below this, in the middle line, is a depression corresponding to the crico-thyroid space, in which the operation of laryngotomy is performed. This depression is bounded below by a prominent arch, the anterior part of the cricoid cartilage, below which the trachea can be felt, though it is only in the emaciated that the separate rings can be distinguished. The lower part of the trachea is not easily made out, for as it descends in the neck it takes a deeper position, and is farther removed from the surface. The level of the vocal cords corresponds to the middle of the anterior margin of the thyroid cartilage.

With the laryngoscope the following structures can be seen: the base of the tongue, and the upper surface of the epiglottis, with the glosso-epiglottic ligaments; the superior aperture of the larynx; bounded on either side by the aryteno-epiglottidean folds, in which may be seen two rounded eminences, corresponding to the cornicula and cuneiform cartilages. Beneath these the true and false vocal cords, with the ventricle between them. Still deeper, the cricoid cartilage, and some of the anterior parts of the rings of the trachea, and sometimes, in deep inspiration, the bifurcation of the trachea.

Surgical Anatomy.—*Foreign bodies* often find their way into the air-passages. These may consist of large soft substances, as pieces of meat, which may become lodged in the upper aperture of the larynx, or in the rima glottidis, and cause speedy suffocation unless rapidly got rid of, or unless an opening is made into the air-passages below, so as to enable the patient to breathe. Smaller bodies, frequently of a hard nature, such as cherry or plum stones, small pieces of bone, buttons, &c., may find their way through the rima glottidis into the trachea or bronchus, or may become lodged in the ventricle of the larynx. The dangers then depend not so much upon the mechanical obstruction as upon the spasm of the glottis which they excite from reflex irritation. When lodged in the ventricle of the larynx, they may produce very few symptoms beyond sudden loss of voice or alteration in the voice sounds, immediately following the inhalation of the foreign body. When, however, they are situated in the trachea, they are constantly striking against the vocal cords during expiratory efforts, and produce attacks of dyspnœa from spasm of the glottis. When lodged in the bronchus, they usually become fixed there, and, occluding the lumen of the tube, cause a loss of the respiratory murmur on the affected side, which is, as stated above, more often the right.

Beneath the mucous membrane of the upper part of the air-passages, there is a considerable amount of submucous tissue, which is liable to become much swollen from effusion in inflammatory affections, constituting the disease known as 'œdema of the glottis.' This effusion does not extend below the level of the vocal cords, on account of the fact that the mucous membrane is closely adherent to these structures, without the intervention of any submucous tissue. So that, in cases of this disease, in which it is necessary to open the air-passages to prevent suffocation, the operation of laryngotomy is sufficient.

Chronic laryngitis is an inflammation of the mucous glands of the larynx, which occurs in those who speak much in public, and is known as 'Clergyman's sore throat.' It is due to the dryness induced by the large amount of cold air drawn into the air-passages during prolonged speaking, which excites increased activity of the mucous glands to keep the parts moist, and this eventually terminates in inflammation of these structures.

Ulceration of the larynx may occur from syphilis, either as a superficial ulceration, or from the softening of a gumma; from tuberculous disease (laryngeal phthisis), or from malignant disease (epithelioma).

The air-passages may be opened in three different situations: by a vertical incision through the centre of the thyroid cartilage (*thyrotomy*); through the crico-thyroid membrane (*laryngotomy*), or in some part of the trachea (*tracheotomy*); and to these some surgeons have added a fourth method, by opening the crico-thyroid membrane and dividing the cricoid cartilage with the upper ring of the trachea (*laryngo-tracheotomy*).

Thyrotomy is usually performed for the purpose of removing growths from the vocal cords, or for extracting foreign bodies from the ventricle of the larynx. A median incision is made from the upper border of the body of the hyoid bone to the lower border of the cricoid cartilage, and is carried through the subcutaneous tissues and deep fascia between the margins of the Sterno-hyoid muscles. An incision is then made in the crico-thyroid membrane, and one blade of a stout, sharp-pointed pair of scissors is introduced beneath the lower border of the thyroid cartilage, and this structure is divided from below upwards. Great care must be taken to cut exactly in the middle line to avoid wounding the vocal cords. If the two halves of the cartilage are now drawn apart, a very good view of the interior of the larynx will be obtained.

Laryngotomy is anatomically a simple operation: it can readily be performed, and should be employed in those cases where the air-passages require opening in an emergency for the relief of some sudden obstruction to respiration. The crico-thyroid membrane is very superficial, being covered only in the middle line by the skin, superficial fascia, and the deep fascia. On either side of the middle line it is also covered by the Sterno-hyoid and Sterno-thyroid muscles, which diverge from each other at their upper parts, leaving a slight interval between them. On these muscles rest the anterior jugular veins. The only vessel of any importance in connection with this operation is the crico-thyroid artery, which crosses the crico-thyroid membrane, and which may be wounded, but rarely gives rise to any trouble. The operation is performed thus: the head being thrown back and steadied by an assistant, the finger is passed over the front of the neck, and the crico-thyroid depression felt for. A vertical incision is then made through the skin, in the middle line over this spot, and carried down through the fascia until the crico-thyroid membrane is exposed. A cross cut is then made through the membrane, close to the upper border of the cricoid cartilage, so as to avoid, if possible, the crico-thyroid artery, and a tracheotomy tube introduced. It has been recommended, as a more rapid way of performing the operation, to make a transverse instead of a longitudinal cut through the superficial structures, and thus to open at once the air-passages. It will be seen, however, that in operating in this way the anterior jugular veins are in danger of being wounded.

Tracheotomy may be performed either above or below the isthmus of the thyroid body: or this structure may be divided and the trachea opened behind it.

The isthmus of the thyroid gland usually crosses the second and third rings of the trachea; along its upper border is frequently to be found a large transverse communicating branch between the superior thyroid veins; and the isthmus itself is covered by a venous plexus, formed between the thyroid veins of the opposite sides. Theoretically, therefore, it is advisable to avoid dividing this structure in opening the trachea.

Above the isthmus the trachea is comparatively superficial, being covered by the skin, superficial fascia, deep fascia, Sterno-hyoid and Sterno-thyroid muscles, and a second layer of the deep fascia, which is attached above to the lower border of the hyoid bone and descends beneath the muscles to the thyroid body, where it divides into two layers and encloses the isthmus.

Below the isthmus the trachea lies much more deeply, and is covered by the Sterno-hyoid and the Sterno-thyroid muscles, and a quantity of loose areolar tissue, in which is a plexus of veins, some of which are of large size; they converge to two trunks, the inferior thyroid veins, which descend on either side of the median line on the front of the trachea, and open into the innominate vein. In the infant the thymus gland ascends a variable distance along the front of the trachea; and opposite the episternal notch the windpipe is crossed by the left innominate vein. Occasionally also in young subjects the innominate artery crosses the tube obliquely above the level of the sternum. The thyroidea ima artery, when that vessel exists, passes from below upwards, along the front of the trachea.

From these observations it must be evident that the trachea can be more readily opened above than below the isthmus of the thyroid body.

Tracheotomy above the isthmus is performed thus: the patient should, if possible, be laid on his back on a table in a good light. A pillow is to be placed under the shoulders and the head thrown back and steadied by an assistant. The surgeon standing on the right side of his patient makes an incision from an inch and a half to two inches in length in the median line of the neck from the top of the cricoid cartilage. The incision must be made exactly in the middle line so as to avoid the anterior jugular veins, and after the superficial structures have been divided, the interval between the Sterno-hyoid muscles must be found, the raphé divided, and the muscles drawn apart. The lower border of the cricoid cartilage must now be felt for, and the upper part of the trachea exposed from this point downwards in the middle line. Bose has recommended that the layer of fascia in front of the trachea should be divided transversely at the level of the lower border of the cricoid cartilage, and, having been seized with a pair of forceps, pressed downwards with the handle of the scalpel. By this means the isthmus of the thyroid gland is depressed and is saved from all danger of being wounded, and the trachea cleanly exposed. The trachea is now transfixed with a sharp hook and drawn forwards in order to steady it, and is then opened by inserting the knife into it and dividing the two or three upper rings by cutting upwards. If the trachea is to be opened beneath the isthmus, the incision must be made from a little below the cricoid cartilage to the top of the sternum.

In the child the trachea is smaller, more deeply placed, and more movable than in the adult. In fat or short-necked people, or in those in whom the muscles of the neck are prominently developed, the trachea is more deeply placed than in the opposite conditions.

A portion of the larynx or the whole of it has been removed for malignant disease. The results which have been obtained from the removal of the whole of it have not been very satisfactory, and the cases in which the operation is justifiable are very few. It may be removed by a median incision through the soft parts; freeing the cartilages from the muscles and other structures in front; separating the larynx from the trachea below, and dissecting off the deeper structure from below upwards.

The Pleuræ

Each lung is invested by an exceedingly delicate serous membrane, the **pleura**, which encloses the organ as far as its root, and is then reflected on to the pericardium, chest-wall, and Diaphragm. The portion of the serous membrane investing the surface of the lung and dipping into the fissures between its lobes,

Fig. 699.—Front view of chest, showing relations of pleuræ and lungs to the chest-wall. The blue lines indicate the lines of the reflection of the pleuræ; the red, the outlines of the lungs and their fissures.

Line of pleural reflection
on to Diaphragm

Lower edge of lung

Incisura cardiaca

is called the *visceral layer of the pleura,* or *pleura pulmonalis;* while that which lines the inner surface of the chest and covers the Diaphragm and pericardium, is called the *parietal layer of pleura.* The space between these two layers is called the *cavity of the pleura,* but it must be borne in mind that in the healthy condition the two layers are in contact and that there is no real cavity, until the lung becomes collapsed and a separation of it from the wall of the chest takes place. Each pleura is therefore a shut sac, one occupying the right, the other the left half of the thorax; and they are perfectly separate from each other. The two pleuræ do not meet in the middle line of the chest, excepting anteriorly opposite the second and third pieces of the sternum: a space being left between them, which contains all the viscera of the thorax excepting the lungs; this space is named the *mediastinum*.

Different portions of the parietal pleura have received special names which indicate their position: thus, that portion which lines the inner surface of the ribs and Intercostal muscles is the *costal pleura*; that which coats the convex surface of the Diaphragm is the *diaphragmatic pleura*; that which rises into the neck, over the summit of the lung, is the *cervical pleura*; and that which is applied to the adjacent structures in the mediastinum, or space intervening between the two pleural cavities, is the *mediastinal pleura*.

Reflections of the Pleura (figs. 699, 700).—Commencing at the sternum, the pleura passes outwards, lines the costal cartilages, the inner surface of the ribs and Intercostal muscles, and at the back part of the thorax passes over the gangliated cord of the sympathetic and its branches, and is reflected upon the sides of the bodies of the vertebræ, where it is separated by a narrow interval, the *posterior mediastinum*, from the opposite pleura. From the vertebral column the pleura passes to the side of the pericardium, which it covers to a slight extent; it then covers the back part of the root of the lung, from the lower border of which a triangular fold descends vertically by the side of the posterior mediastinum to the Diaphragm. This fold is the broad ligament of the lung (*ligamentum latum pulmonis*). From the posterior aspect of the lung root, the pleura may be traced over the convex surface of the lung, the summit and base, and also over the sides of the fissures between the lobes, on to its inner surface and the front part of its root; from this it is reflected on to the pericardium, and from it to the back of the sternum.

FIG. 700.—Lateral view of chest, showing relations of right pleura and lung to the chest-wall. The blue line indicates the line of pleural reflection; the red lines, the outline and fissures of the lung.

Subclavian artery

Below, it covers the upper surface of the Diaphragm, and extends, in front, as low as the costal cartilage of the seventh rib; at the side of the chest, to the lower border of the tenth rib on the left side and to the upper border of the same rib on the right side; and behind, it reaches as low as the twelfth rib, and sometimes even as low as the transverse process of the first lumbar vertebra. *Above*, its apex projects, in the form of a *cul-de-sac*, through the superior opening of the thorax into the neck, extending from one to two inches above the anterior extremity of the first rib, and receives the summit of the corresponding lùng; this sac is strengthened, according to Sibson, by a dome-like expansion of fascia, attached in front to the posterior border of the first rib, and behind to the anterior border of the transverse process of the seventh cervical vertebra. This is covered and strengthened by a few spreading muscular fibres derived from the Scaleni muscles.

In the front of the chest, where the parietal layer of the pleura is reflected backwards to the pericardium, the two pleural sacs are in contact for a short distance. At the upper part of the chest, behind the manubrium, they are not in contact; the point of reflection being represented by a line drawn from the sterno-clavicular articulation to the mid-point of the junction of the manubrium to the body of the sternum. From this point the two pleuræ descend in close contact to the level of the fourth costal cartilages. Here the line of reflection on the right side is continued onwards in nearly a straight line to the lower end of the gladiolus and then turns outwards, while on the left side the line of reflection diverges outwards and is continued downwards, close to the left border of the sternum, as far as the sixth costal cartilage. The inferior limit of the pleura is on a considerably lower level than the corresponding limit of the lung, but does not extend to the attachment of the Diaphragm, so that below the line

FIG. 701.—A transverse section of the thorax, showing the relative position of the viscera, and the reflections of the pleuræ.

of reflection of the pleura from the chest-wall on to the Diaphragm the latter is in direct contact with the rib cartilages and the Internal intercostal muscles. Moreover, in ordinary inspiration the thin margin of the base of the lung does not extend as low as the line of the pleural reflection, with the result that the costal and diaphragmatic pleuræ are here in contact, the narrow slit between the two being termed the *phrenico-costal sinus*. A similar condition exists behind the sternum and rib cartilages, where the anterior thin margin of the lung falls short of the line of pleural reflection, and where the slit-like cavity between the two layers of pleura forms what is sometimes called the *costo-mediastinal sinus*.

The line along which the right pleura is reflected from the chest-wall to the Diaphragm starts in front, immediately below the seventh costo-sternal joint, and runs downwards and backwards behind the seventh costal cartilage so as to cross the tenth rib in the mid-axillary line, from which it is prolonged to the twelfth dorsal spine. The reflection of the left pleura is slightly higher in front, and follows the ascending part of the sixth costal cartilage.

The free surface of the pleura is smooth, polished, and moistened by a serous fluid; its attached surface is intimately adherent to the surface of the lung, and to the pulmonary vessels as they emerge from the pericardium; it is also adherent to the upper surface of the Diaphragm: throughout the rest of its extent it is somewhat thicker, and may be separated from the adjacent parts with extreme facility.

The right pleural sac is shorter, wider, and reaches higher in the neck than the left.

Ligamentum latum pulmonis.—From the above description it will be seen that the root of the lung is covered in front and behind by pleura, and that at its lower border the investing layers come into contact. Here they form a sort of mesenteric fold, the *ligamentum latum pulmonis*, which extends as far as the Diaphragm between the pericardium and the lower part of the inner surface of the lung, having a free falciform border below, between the lung and the Diaphragm. It serves to retain the lower part of the lung in position.

Vessels and Nerves.—The *arteries of the pleura* are derived from the intercostal, the internal mammary, the musculo-phrenic, thymic, pericardiac, and bronchial vessels. The *veins* correspond to the arteries. The *lymphatics* are very numerous. The *nerves* are derived from the phrenic and sympathetic (Luschka). Kölliker states that nerves accompany the ramification of the bronchial arteries in the pleura pulmonalis.

Surgical Anatomy.—In operations upon the kidney, it must be borne in mind that the pleura usually extends below the level of the inner portion of the last rib, and may therefore be opened in these operations, especially when the last rib is removed in order to give more room.

The Mediastinum

The **Mediastinum** is the space left in and near the median line of the chest by the non-approximation of the two pleuræ. It extends from the sternum in front to the spine behind, and contains all the viscera in the thorax excepting the lungs. The mediastinum may be divided for purposes of description into two parts: an upper portion, above the upper level of the pericardium, which is named the *superior mediastinum* (Struthers); and a lower portion, below the upper level of the pericardium. This lower portion is again subdivided into three parts, viz.: that which is in front of the pericardium, the *anterior mediastinum*; that which contains the pericardium and its contents, the *middle mediastinum*; and that which is behind the pericardium, the *posterior mediastinum*.

The **superior mediastinum** is that portion of the interpleural space which lies between the manubrium sterni in front, and the upper dorsal vertebræ behind. It is bounded below by a plane passing backwards from the junction of the manubrium and gladiolus sterni to the lower part of the body of the fourth dorsal vertebra, and laterally by the pleuræ. It contains the origins of the Sterno-hyoid and Sterno-thyroid muscles and the lower ends of the Longi colli muscles; the arch of the aorta; the innominate artery and the thoracic portions of the left carotid and subclavian arteries; the innominate veins and the upper half of the superior vena cava; the left superior intercostal vein; the pneumogastric, cardiac, phrenic, and left recurrent laryngeal nerves; the trachea, œsophagus, and thoracic duct; the remains of the thymus gland and some lymphatic glands.

The **anterior mediastinum** is bounded in front by the sternum, laterally by the pleuræ, and behind by the pericardium. It is narrow above, but widens out a little below, and, owing to the oblique course taken by the left pleura, it is directed from above downwards and to the left. Its anterior wall is formed by the left Triangularis sterni muscle and the fifth, sixth, and seventh left costal cartilages. It contains a quantity of loose areolar tissue, some lymphatic vessels which ascend from the convex surface of the liver, two or three lymphatic glands (anterior mediastinal glands), and the small mediastinal branches of the internal mammary artery.

The **middle mediastinum** is the broadest part of the interpleural space. It contains the heart, enclosed in the pericardium, the ascending aorta, the lower half of the superior vena cava, with the vena azygos major opening into it, the bifurcation of the trachea and the two bronchi, the pulmonary artery dividing into its two branches and the right and left pulmonary veins, the phrenic nerves, and some bronchial lymphatic glands.

The **posterior mediastinum** (fig. 702) is an irregular triangular space running parallel with the vertebral column; it is bounded in front by the pericardium above and by the posterior aspect of the Diaphragm below, behind by the vertebral

FIG. 702.—The posterior mediastinum.

column from the lower border of the fourth dorsal vertebra, and on either side by the pleura. It contains the descending thoracic aorta, the greater and lesser azygos veins, the pneumogastric and splanchnic nerves, the œsophagus, thoracic duct, and some lymphatic glands.

THE LUNGS

The **Lungs** are the essential organs of respiration; they are two in number, placed one on each side of the chest, separated from each other by the heart and other contents of the mediastinum. Each lung is conical in shape, and presents for examination an apex, a base, two borders, and two surfaces (see fig. 704).

The *apex* is rounded, and extends into the root of the neck, reaching from an inch to an inch and a half above the level of the anterior end of the first rib. A furrow runs upwards and outwards immediately below the apex, and is produced by the subclavian artery as it curves outwards in front of the pleura.

The *base* is broad, concave, and rests upon the convex surface of the Diaphragm, which separates the right lung from the right lobe of the liver and the left lung from the left lobe of the liver, the stomach, and spleen. Since the Diaphragm extends higher on the right than on the left side, it follows that the concavity on the base of the right lung is deeper than that on the left. Laterally and behind, the base is bounded by a thin, sharp margin which projects for some distance into the phrenico-costal sinus of the pleura, between the lower

FIG. 703.—Transverse section through the upper margin of the third dorsal vertebra. (Braune.)

ribs and the costal attachment of the Diaphragm. The base of the lung descends during inspiration and ascends during expiration, and its relation to the chest-wall is indicated in figs. 699 and 700.

The *external* or *costal surface* is smooth, convex, of considerable extent, and corresponds to the form of the cavity of the chest, being deeper behind than in front. It is in contact with the costal pleura, and presents slight grooves corresponding with the overlying ribs.

The *inner* or *mediastinal surface* is in contact with that portion of the pleura which forms the lateral boundary of the mediastinal space. It presents a deep concavity which accommodates the pericardial sac; this is larger and deeper in the left than in the right lung on account of the heart projecting farther to the left than to the right side of the mesial plane. Above and behind this concavity is a triangular depression named the *hilus*, where the structures which form the root of the lung enter and leave the viscus. These structures are invested by pleural membrane, which, below the hilus, forms the ligamentum latum pulmonis. On the *right* lung, immediately above the hilus, is an arched furrow which accommodates the vena azygos major; while running upwards, and then arching outwards some little distance below the apex, is a wide groove for the superior vena cava and right innominate vein, and behind this, nearer the apex, is a second furrow for the innominate artery. Along the back part of the inner surface is a vertical groove for the œsophagus ; this groove becomes less distinct below, owing to the inclination of the lower part of the œsophagus to the left of the middle line.   In front and to the right of the lower part of the œsophageal groove, the inner surface is applied to the pleural covering of the right and posterior aspects of the thoracic part of the inferior vena cava—this vessel being accommodated in a deep concavity.   On the *left* lung, immediately above the hilus, is a well-marked curved furrow produced by the aortic arch, and passing upwards from this towards the apex are two grooves, the anterior accommodating the left common carotid artery, the posterior the left subclavian artery.   Behind the hilus and pericardial depression is a vertical furrow produced by the descending thoracic aorta, and in front of this, near the base of the lung, the lower part of the œsophagus causes a shallow depression.

The *posterior border* is broad and rounded, and is received into the deep concavity on either side of the spinal column.   It is much longer than the anterior border, and projects, below, into the upper part of the phrenico-costal sinus.

The *anterior border* is thin and sharp, and overlaps the front of the pericardium. The anterior border of the right lung is almost vertical, and projects into the costo-mediastinal sinus of the pleura ; that of the left presents, below, an angular notch, the *incisura cardiaca,* in which the pericardium is exposed. Opposite this incisure the anterior margin of the left lung is situated some little distance to the outer side of the line of reflection of the corresponding part of the pleura.

Fissures and lobes of the lungs.—The *left lung* is divided into two lobes, an upper and a lower, by an *oblique fissure,* which extends from the outer to the inner surface of the lung both above and below the hilus.   As seen on the surface, this fissure commences on the inner aspect of the lung at the upper and posterior part of the hilus, and runs backwards and upwards to the posterior border, which it crosses at a point about two and a half inches below the apex. It then extends downwards and forwards over the outer surface, and reaches the lower border a little behind its anterior extremity, and its further course can be followed upwards and backwards across the inner surface as far as the lower part of the hilus.   The *upper lobe* lies above and in front of this fissure, and includes the apex, the anterior border, and a considerable part of the outer surface and the greater part of the inner surface of the lung.   The *lower lobe,* the larger of the two, is situated below and behind the fissure, and comprises almost the whole of the base, a large portion of the outer surface, and the greater part of the posterior border.

The *right lung* is divided into three lobes, upper, middle, and lower, by an oblique and a horizontal fissure.   The *oblique fissure* separates the lower from the middle and upper lobes, and corresponds closely with the fissure in the left lung.   Its direction is, however, more vertical, and it cuts the lower border about three inches behind its anterior extremity.   The *horizontal fissure* separates the upper from the middle lobe.   It begins in the oblique fissure near the posterior border of the lung, and, running horizontally forwards, cuts the anterior border on a level with the sternal end of the fourth costal cartilage ; on the inner surface it may be traced backwards to the hilus.   The *middle lobe,* the smallest of the lobes of the right lung, lies between the horizontal fissure and the lower part of the oblique fissure ; it is wedge-shaped, and includes the lower part of the anterior border and the anterior part of the base of the lung.

The *right lung* is the larger and heavier ; it is broader than the left, owing to the inclination of the heart to the left side ; it is also shorter by an inch, in consequence of the Diaphragm rising higher on the right side to accommodate the liver.

The root of the lung.—A little above the middle of the inner surface of each lung, and nearer its posterior than its anterior border, is its root, by which the lung is connected to the heart and the trachea.   The root is formed by the bronchial tube, the pulmonary artery, the pulmonary veins, the bronchial arteries and veins, the pulmonary plexuses of nerves, lymphatics, bronchial glands, and areolar tissue, all of which are enclosed by a reflection of the pleura.   The root of the right lung lies behind the superior vena cava and part of the right auricle, and below the vena azygos major.   That of the left lung passes beneath the arch of the aorta and in front of the descending aorta ; the phrenic nerve with its accompanying artery and vein, and the anterior pulmonary plexus lie in front of each, and the pneumogastric and posterior pulmonary plexus behind each; below each is the ligamentum latum pulmonis.

The chief structures composing the root of each lung are arranged in a similar manner from before backwards on both sides, viz.: the upper of the two pulmonary veins in front; the pulmonary artery in the middle; and the bronchus together with the bronchial vessels, behind. From above downwards, on the two sides, their arrangement differs, thus:

On the right side their position is—bronchus, pulmonary artery, pulmonary veins; but on the left side their position is—pulmonary artery, bronchus, pulmonary veins. The lower of the two pulmonary veins is situated below the bronchus at the apex or lowest part of the hilus. It should be noted that the entire right bronchus does not lie above the right pulmonary artery, but only

FIG. 704.—Front view of the heart and lungs.

its eparterial branch (see below), which passes to the upper lobe of the right lung; the divisions of the bronchus for the middle and lower lobes lie below the artery.

Division of the bronchi.—Just as the lungs differ from each other in the number of their lobes, so the bronchi differ in their mode of subdivision.

The *right* bronchus gives off, about an inch from the bifurcation of the trachea, a branch for the upper lobe. This branch arises above the level of the pulmonary artery, and is therefore named the *eparterial bronchus*. All the other divisions of the main stem come off below the pulmonary artery, and consequently are termed *hyparterial bronchi*. The first of these is distributed to the middle lobe, and the main tube then passes downwards and backwards into the lower lobe, giving off in its course a series of large ventral and small dorsal branches. The ventral and dorsal branches arise alternately, and are usually eight in number—four of each kind. The branch to the middle lobe is regarded as the first of the ventral series.

3 U

The *left* bronchus passes below the level of the pulmonary artery before it divides, and hence all its branches are *hyparterial*; it may therefore be looked upon as equivalent to that portion of the right bronchus which lies on the distal side of its eparterial branch. The first branch of the left bronchus arises about two inches from the bifurcation of the trachea, and is distributed to the upper lobe. The main stem then enters the lower lobe, where it divides into ventral and dorsal branches similar to those in the right lung. The branch to the upper lobe of the left lung is regarded as the first of the ventral series.

Aeby regarded the absence of a left eparterial bronchus as indicating the absence of the corresponding lobe of the lung, and considered the middle lobe of the right lung the homologue of the upper lobe of the left. His conclusions, however, are not universally accepted.

The *weight* of both lungs together is about forty-two ounces, the right lung being two ounces heavier than the left; but much variation is met with according to the amount of blood or serous fluid they may contain. The lungs are heavier in the male than in the female, their proportion to the body being, in the former, as 1 to 37, in the latter as 1 to 43. The specific gravity of the lung-tissue varies from 0·345 to 0·746.

The *colour* of the lungs at birth is a pinkish-white; in adult life it is a dark slate-colour, mottled in patches; and as age advances, this mottling assumes a black colour. The colouring matter consists of granules of a carbonaceous substance deposited in the areolar tissue near the surface of the organ. It increases in quantity as age advances, and is more abundant in males than in females. The posterior border of the lung is usually darker than the anterior.

The *surface* of the lung is smooth, shining, and marked out into numerous polyhedral areas, indicating the lobules of the organ: each of these areas is crossed by numerous lighter lines.

The *substance* of the lung is of a light, porous, spongy texture; it floats in water, and crepitates when handled, owing to the presence of air in the air-cells; it is also highly elastic; hence the collapsed state of these organs when they are removed from the closed cavity of the thorax.

Structure.—The lungs are composed of an external serous coat, a subserous areolar tissue, and the pulmonary substance or parenchyma.

The *serous coat* is derived from the pleura; it is thin, transparent, and invests the entire organ as far as the root.

The *subserous areolar tissue* contains a large proportion of elastic fibres; it invests the entire surface of the lung, and extends inwards between the lobules.

The *parenchyma* is composed of lobules, which, although closely connected together by an interlobular areolar tissue, are quite distinct from one another, and may be teased asunder without much difficulty in the fœtus. The lobules vary in size: those on the surface are large, of pyramidal form, the base turned towards the surface; those in the interior smaller, and of various forms. Each lobule is composed of one of the ramifications of a bronchial tube and its terminal air-cells, and of the ramifications of the pulmonary and bronchial vessels, lymphatics, and nerves; all of these structures being connected together by areolar tissue.

The primary branches of the *bronchus*, upon entering the substance of the lung, divide and subdivide successively in a bipinnate manner into smaller tubes throughout the entire organ. Each of the smaller subdivisions of the bronchus enters a pulmonary lobule, and is termed a *lobular bronchial tube* or *bronchiole*. Its wall now begins to present irregular dilatations, *air-cells* or *alveoli*, at first sparingly and on one side of the tube only, but as it proceeds onwards these dilatations become more numerous and surround the tube on all sides, so that it loses its cylindrical character. The bronchiole now becomes enlarged, and is termed the *atrium* or *alveolar passage*; from it are given off, on all sides, ramifications, called *infundibula*, which are closely beset in all directions by *alveoli* or *air-cells*. Within the lungs the bronchial tubes are circular, and present certain peculiarities of structure.

Changes in the structure of the bronchi in the lungs.—1. *In the lobes of the lungs.*—In the lobes of the lungs the following changes take place. The *cartilages* are not imperfect rings, but consist of thin laminæ, of varied form and size, scattered irregularly along the sides of the tube, being most distinct at the points of division of the bronchi. They may be traced into tubes, the diameter of which is only one-fourth of a line. Beyond this point the tubes are wholly membranous. The fibrous coat is continued into the smallest ramifications of the bronchi. The muscular coat is disposed in the form of a continuous layer of annular fibres, which may be traced upon the smallest bronchial tubes, and consists of the unstriped variety of muscular tissue. The mucous membrane lines the bronchi and its ramifications throughout, and is covered with columnar ciliated epithelium.

2. *In the lobules of the lung.*—In the lobular bronchial tubes and in the infundibula the following changes take place. The muscular tissue begins to disappear, but can be traced as far as the infundibula, where irregular fasciculi are still to be found. The fibrous coat becomes thinner and degenerates into areolar tissue. The epithelium becomes non-ciliated and flattened. This occurs gradually; thus, in the lobular bronchial tubes, patches of non-ciliated, flattened epithelium may be found scattered among the columnar ciliated cells ; then these patches become more and more numerous, until in the infundibula and air-cells all the epithelium is of the non-ciliated pavement variety. In addition to these flattened cells, there are small polygonal granular cells in the air-sacs in clusters of two or three, between the others.

The air-cells are small polyhedral recesses, composed of a fibrillated connective tissue, and surrounded by a few involuntary muscular and elastic fibres. Under the microscope, granular, rounded, amœboid cells (eosinophile leucocytes), often containing carbonaceous particles, are seen lying free in their cavities. The air-cells are well seen on the surface of the lung, and vary from $\frac{1}{200}$th to $\frac{1}{70}$th of an inch in diameter ; being largest on the surface at the thin borders and at the apex, and smallest in the interior.

The *pulmonary artery* conveys the venous blood to the lungs ; it divides into branches which accompany the bronchial tubes, and terminates in a dense capillary network, upon the walls of the intercellular passages and air-cells. In the lung, the branches of the pulmonary artery are usually above and in front of a bronchial tube, the vein below.

The *pulmonary capillaries* form plexuses which lie immediately beneath the mucous membrane, in the walls and septa of the air-cells, and of the infundibula. In the septa between the air-cells the capillary network forms a single layer. The capillaries form a very minute network, the meshes of which are smaller than the vessels themselves ; * their walls are also exceedingly thin. The arteries of neighbouring lobules are independent of each other, but the veins freely anastomose together.

The *pulmonary veins* commence in the pulmonary capillaries, the radicles coalescing into larger branches which run through the substance of the lung, independently from the minute arteries and bronchi. After freely communicating with other branches they form large vessels, which ultimately come into relation with the arteries and bronchial tubes, and accompany them to the hilum of the organ. Finally they open into the left auricle of the heart, conveying oxygenated blood to be eventually distributed to all parts of the body by the aorta.

The *bronchial arteries* supply blood for the nutrition of the lung ; they are derived from the thoracic aorta or from the upper aortic intercostal arteries, and, accompanying the bronchial tubes, are distributed to the bronchial glands, and upon the walls of the larger bronchial tubes and pulmonary vessels. Those supplying the bronchial tubes form a capillary plexus in the muscular coat, from which branches are given off to form a second plexus in the mucous coat. This plexus communicates with branches of the pulmonary artery, and empties itself into the pulmonary vein. Others are distributed in the interlobular areolar tissue, and terminate partly in the deep, partly in the superficial, bronchial veins. Lastly, some ramify upon the surface of the lung, beneath the pleura, where they form a capillary network.

The *bronchial vein* is formed at the root of the lung, receiving superficial and deep veins corresponding to branches of the bronchial artery. It does not, however, receive all the blood supplied by the artery, as some of it passes into

---

* The meshes are only 0·002''' to 0·008''' in width, while the vessels are 0·003''' to 0·005'''.—Kölliker, *Human Microscopic Anatomy.*

the pulmonary veins. It terminates on the right side in the vena azygos major, and on the left side in the superior intercostal or left upper azygos vein.

The **lymphatics** consist of a superficial and deep set; they terminate at the root of the lung, in the bronchial glands.

Nerves.—The lungs are supplied from the anterior and posterior pulmonary plexuses, formed chiefly by branches from the sympathetic and pneumogastric. The filaments from these plexuses accompany the bronchial tubes, upon which they are lost. Small ganglia are found upon these nerves.

Surface Form.—The apex of the lung is situated in the neck, behind the interval between the two heads of origin of the Sterno-mastoid. The height to which it rises above the clavicle varies very considerably, but is generally about an inch. It may, however, extend as much as an inch and a half or an inch and three-quarters, or, on the other hand, it may scarcely project above the level of this bone. In order to mark out the anterior margin of the lung, a line is to be drawn from the apex point, an inch above the level of the clavicle, and rather nearer the posterior than the anterior border of the Sterno-mastoid muscle, downwards and inwards across the sterno-clavicular articulation and first piece of the sternum until it meets, or almost meets, its fellow of the other side at the level of the articulation of the manubrium and gladiolus. From this point the two lines are to be drawn downwards, rather to the left of the mesial line but close to it, as far as the level of the articulation of the fourth costal cartilages to the sternum. From here the two lines diverge, the left at first passing outwards with a slight inclination downwards, and then taking a bend downwards with a slight inclination outwards to the apex of the heart, and thence to the sixth costo-chondral articulation. The direction of the anterior border of this part of the left lung is denoted with sufficient accuracy by a curved line, with its convexity directed upwards and outwards from the articulation of the fourth right costal cartilage of the sternum to the fifth intercostal space, an inch and a half below, and three-quarters of an inch internal to the left nipple. The continuation of the anterior border of the right lung is marked by a prolongation of its line from the level of the fourth costal cartilages vertically downwards as far as the sixth, when it slopes off along the line of the sixth costal cartilage to its articulation with the rib.

After expiration the lower border of the lung may be marked out by a slightly curved line, with its convexity downwards, from the articulation of the sixth costal cartilage with its rib to the spinous process of the tenth dorsal vertebra. If vertical lines are drawn downwards from the nipple, the mid-axillary line, and the apex of the scapula, while the arms are raised from the sides, they should intersect this convex line, the first at the sixth, the second at the eighth, and the third at the tenth rib. It will thus be seen that the pleura (see page 1019) extends farther down than the lung, so that it may be wounded, and a wound pass through its cavity into the Diaphragm, and even injure the abdominal viscera, without the lung being involved.

The posterior border of the lung is indicated by a line drawn from the level of the spinous process of the seventh cervical vertebra, down either side of the spine, corresponding to the costo-vertebral joints as low as the spinous process of the tenth dorsal vertebra. The trachea bifurcates opposite the spinous process of the fourth dorsal vertebra, and from this point the two bronchi are directed outwards.

The position of the great fissure in each lung may be indicated by a line drawn from the second dorsal spine round the side of the chest to the sixth rib in the nipple line. The smaller or secondary fissure in the right lung is indicated by a line drawn from the preceding, where it bisects the mid-axillary line, to the junction of the fourth costal cartilage to the sternum.

Surgical Anatomy.—The lungs may be wounded or torn in three ways: (1) By compression of the chest, without any injury to the ribs. (2) By a fractured rib penetrating the lung. (3) By stabs, gunshot wounds, &c.

The first form, where the lung is ruptured by external compression without any fracture of the ribs, is very rare and usually occurs in young children, and affects the root of the lung, i.e. the most fixed part, and thus, implicating the great vessels, is frequently fatal. It would seem *a priori* a most unusual injury, and its exact mode of causation is difficult to interpret. The probable explanation is that immediately before the compression is applied a deep inspiration is taken and the lungs are fully inflated; owing then to spasm of the glottis at the moment of compression, the air is unable to escape from the lung, which is not able to recede, and consequently gives way.

In the second variety, when the wound in the lung is produced by the penetration of a broken rib, both the pleura costalis and pulmonalis must necessarily be injured, and consequently the air taken into the wounded air-cells may find its way through these wounds into the cellular tissue of the parietes of the chest, producing emphysema. This it may do without collecting in the pleural cavity; the two layers of the pleura are so intimately in contact that the air passes straight through from the wounded lung into the subcutaneous tissue. Emphysema constitutes therefore the most important sign of injury to the lung in cases of fracture of the ribs. Pneumothorax, or air in the pleural cavity, is much more likely to occur in injuries to the lung of the third variety—that is to say, from external wounds, from stabs, gunshot injuries, and such like—in which case air passes either from the wound of the lung or from the external wound into the cavity of the pleura during the respiratory movements. In these cases there is generally no emphysema of the subcutaneous tissue unless the external wound is small and valvular, so that the air is drawn into the wound during inspiration, and then forced into the cellular tissue around during expiration because it cannot escape from the external wound. Occasionally in wounds of the parietes of the chest no air finds its way into the cavity of the pleura, because the lung at the time of the accident protrudes through the wound and blocks the opening. This takes place where the wound is large, and constitutes one form of *hernia* of the lung. Another form of hernia of the lung occurs, though very rarely, after wounds of the chest wall, when the wound has healed and the cicatrix subsequently yields from the pressure of the viscus behind. It forms a globular, elastic, crepitating swelling, which enlarges during expiratory efforts, falls in during inspiration, and disappears on holding the breath.

An incision into the lung is occasionally required in cases of abscess the result of pneumonia or the presence of a foreign body, and from an abscess in the liver which has made its way through the Diaphragm into the lung substance, and also in cases of hydatid disease. In these cases there is always risk of hæmorrhage, and it has been recommended that the lung tissue should be penetrated by the actual cautery, rather than with the knife. Unless adhesions have formed between the two layers of the pleura, the pleural cavity must necessarily be opened, and there is the further risk of pneumothorax, and possibly septic infection. It is therefore advisable to suture the lung to the opening in the thoracic wall, and wait for adhesions to form before perforating the lung.

Excision of a portion of the lung for tumour and other morbid conditions has been performed, but the difficulties and risks of the operation are considerable.

ORGANS OF DIGESTION

THE Apparatus for the Digestion of the Food consists of the alimentary canal and of certain accessory organs.

The **alimentary canal** is a musculo-membranous tube, about thirty feet in length, extending from the mouth to the anus, and lined throughout its entire extent by mucous membrane. It has received different names in the various parts of its course: at its commencement is the *mouth*, where provision is made for the mechanical division of the food (*mastication*), and for its admixture with a fluid secreted by the salivary glands (*insalivation*); beyond this are the organs of deglutition, the *pharynx* and the *œsophagus*, which convey the food into that part of the alimentary canal (the *stomach*) in which the principal chemical changes occur, and in which the reduction and solution of the food take place; the stomach is followed by the *small intestine*, which is divided for purposes of description into three parts, the *duodenum*, the *jejunum*, and *ileum*, and in which the nutritive principles of the food are separated and absorbed; finally the small intestine terminates in the *large intestine*, which is made up of *cæcum*, *colon*, and *rectum*, the last terminating on the surface of the body at the *anus*. The *accessory organs* are the *teeth*, for purposes of mastication; the three pairs of *salivary glands*—the *parotid*, *submaxillary*, and *sublingual*—the secretion from which mixes with the food in the mouth and converts it into a bolus and acts chemically on one of its constituents; the *liver* and *pancreas*, two large glands in the abdomen, the secretions of which, in addition to that of numerous minute glands in the walls of the alimentary canal, assist in the process of digestion.

Alimentary Canal

Mouth.	Small intestine	Duodenum. Jejunum. Ileum.
Pharynx.		
Œsophagus.	Large intestine	Cæcum. Colon. Rectum.
Stomach.		

Accessory Organs

Teeth

Salivary glands	Parotid. Submaxillary. Sublingual.	Liver. Pancreas.

The **mouth** (*oral* or *buccal cavity*) is placed at the commencement of the alimentary canal (fig. 705); it is a nearly oval-shaped cavity, in which the mastication of the food takes place. It consists of two parts: an outer, smaller portion, the vestibule (vestibulum oris), and an inner, larger part, the cavity proper of the mouth (cavum oris proprium).

The *vestibulum oris* is a slit-like space, bounded in front and laterally by the lips and cheeks; behind and internally by the gums and teeth. It communicates with the surface of the body by the *aperture of the mouth*. Above and below, it is limited by the reflection of the mucous membrane from the lips and cheeks to the gum covering the upper and lower alveolar arch respectively. It receives the secretion from the parotid glands, and communicates, when the jaws are closed, with the cavum oris by an aperture on each side behind the wisdom teeth, and by narrow clefts between opposing teeth.

The *cavum oris proprium* is bounded laterally and in front by the alveolar arches with their contained teeth; behind, it communicates with the pharynx by a constricted aperture termed the *isthmus faucium*. It is roofed in by the hard and soft palate, while the greater part of the floor is formed by the tongue, the remainder being completed by the reflection of the mucous membrane from the sides and under surface of the tongue to the gum lining the inner aspect of the mandible. It receives the secretion from the submaxillary and sublingual glands.

The *mucous membrane* lining the mouth is continuous with the integument at the free margin of the lips, and with the mucous lining of the pharynx behind; it is of a rose-pink tinge during life, and very thick where it covers the hard parts bounding the cavity. It is covered by stratified epithelium.

The **lips** are two fleshy folds which surround the orifice of the mouth, formed externally of integument and internally of mucous membrane, between which

FIG. 705.—Sectional view of the nose, mouth, pharynx, &c.

are found the Orbicularis oris muscle, the coronary vessels, some nerves, areolar tissue, and fat, and numerous small labial glands. The inner surface of each lip is connected in the middle line to the gum of the corresponding jaw by a fold of mucous membrane, the *frænum labii superioris* and *inferioris*—the former being the larger of the two.

The *labial glands* are situated between the mucous membrane and the Orbicularis oris, round the orifice of the mouth. They are circular in form, about the size of small peas, their ducts opening by minute orifices upon the mucous membrane. In structure they resemble the salivary glands.

The **cheeks** form the sides of the face, and are continuous in front with the lips. They are composed externally of integument; internally of mucous membrane; and between the two of a muscular stratum, besides a large quantity of fat, areolar tissue, vessels, nerves, and buccal glands.

The *mucous membrane* lining the cheek is reflected above and below upon the gums, and is continuous behind with the lining membrane of the soft palate. Opposite the second molar tooth of the upper jaw is a papilla, the summit of which presents the aperture of the duct of the parotid gland. The principal muscle of the cheek is the Buccinator; but numerous other muscles enter into its formation, viz.: the Zygomatici, Risorius, and Platysma myoides.

The *buccal glands* are placed between the mucous membrane and Buccinator muscle: they are similar in structure to the labial glands, but smaller. About five of larger size than the rest are placed between the Masseter and Buccinator muscles around the distal extremity of Stenson's duct; their ducts open in the mouth opposite the last molar tooth. They are called *molar glands*.

The **gums** are composed of a dense fibrous tissue, closely connected to the periosteum of the alveolar processes, and surrounding the necks of the teeth. They are covered by smooth and vascular mucous membrane, which is remarkable for its limited sensibility. Around the necks of the teeth this membrane presents numerous fine papillæ; and from this point it is reflected into the alveolus, where it is continuous with the periosteal membrane lining that cavity.

Surgical Anatomy.—The gums are occasionally the seat of considerable hypertrophy, forming a lobulated, vascular fold growing up in front and behind the teeth, so as almost to bury them. They may also become swollen and congested, bleeding freely and often becoming ulcerated. The condition is known as *spongy gums*, and may occur in scurvy, from the administration of mercury, in stomatitis and dyspepsia, and in ill-fed tuberculous children. They are very tender, mastication is painful, and there is often considerable fœtor. The margin of the gums presents a blue line in cases of lead-poisoning. The collection of tartar, which consists of the secretion from the gums, mixed with fragments of food and salivary salts, may give rise to a condition known as *pyorrhœa alveolaris*, which is an inflammatory condition of the gums, followed by the gradual absorption of the alveolus and the falling out of the teeth. The condition, however, may occur independently of tartar, and by some is believed to be constitutional. Fibrous tumours (epulis), myeloid growths, and epithelioma are met with in the gums.

THE TEETH

The human subject is provided with two sets of teeth, which make their appearance at different periods of life. Those of the first set appear in childhood, and are called the *temporary, deciduous,* or *milk teeth*. Those of the second set, which also appear at an early period, continue until old age, and are named *permanent*.

The *temporary teeth* are twenty in number: four incisors, two canines, and four molars, in each jaw.

The *permanent teeth* are thirty-two in number: four incisors (two central and two lateral), two canines, four bicuspids, and six molars, in each jaw.

The dental formulæ may be represented as follows:

Temporary Teeth

	mol.	can.	in.	in.	can.	mol.	
Upper jaw . .	2	I	2	2	I	2	} Total 20
Lower jaw . .	2	I	2	2	I	2	

Permanent Teeth

	mol.	bic.	can.	in.	in.	can.	bic.	mol.	
Upper jaw .	3	2	I	2	2	I	2	3	} Total 32
Lower jaw .	3	2	I	2	2	I	2	3	

General Characters.—Each tooth consists of three portions: the *crown*, or *body*, projecting above the gum; the *root*, consisting of one or more *fangs*, entirely concealed within the alveolus; and the *neck*, the constricted portion, between the crown and root.

The *roots of the teeth* are firmly implanted within the alveoli; these depressions are lined with periosteum, which is reflected on to the tooth at the point of the fang, and covers it as far as the neck. At the margin of the alveolus, the periosteum becomes continuous with the fibrous structure of the gums.

In consequence of the curve of the dental arch, such terms as anterior, posterior, internal and external, as applied to the teeth, are misleading and confusing. Special terms are therefore applied to the different surfaces of a tooth: that surface which is directed towards the lips or cheek is known as the *labial* surface; that which is directed towards the tongue is described as the *lingual* surface; that surface which is directed towards the mesial line, supposing the teeth were arranged in a straight line outwards from the central incisor, is known as the *proximal* surface; while that which is directed away from the mesial line is called the *distal* surface.

The teeth in the upper jaw form a larger arch than those in the lower jaw, so that they slightly overlap those of the mandible both in front and at the sides in the normal condition. In consequence of the greater width of the upper central incisors over those of the lower, the other teeth in the upper jaw are thrown somewhat distally and the two sets do not quite correspond to each other when the mouth is closed: thus the canine tooth of the upper jaw rests partly on the canine of the lower jaw and partly on the first premolar, and in the molar teeth the cusps of the teeth of the upper jaw lie behind the corresponding cusps of the teeth of the lower jaw. The two series, however, terminate nearly at the same point behind; this is mainly due to the smaller size of the molars in the upper jaw.

PERMANENT TEETH

The **incisors,** or cutting teeth, are so named from their presenting a sharp cutting edge, adapted for biting the food. They are eight in number, and form the four front teeth in each jaw.

The *crown* is directed vertically, and is chisel-shaped, being bevelled at the expense of its lingual surface, so as to present a sharp horizontal cutting edge, which, before being subjected to attrition, presents three small prominent points separated by two slight notches. It is convex, smooth, and highly

FIG. 706.—Permanent teeth. Right side. (Burchard.)

polished on its labial surface; concave on its lingual surface, where, in the teeth of the upper jaw, it is frequently marked by a V-shaped eminence, situated near the gum, the apex, where the two arms of the eminence meet, being directed upwards. This is known as the *basal ridge* or *cingulum.*

The *neck* is constricted.

The *fang* is long, single, conical, transversely flattened, thicker in front than behind, and slightly grooved on each side in the longitudinal direction.

The *incisors of the upper jaw* are altogether larger and stronger than those of the lower jaw. They are directed obliquely downwards and forwards. The two central ones are larger than the two lateral, and the root is more rounded.

The *incisors of the lower jaw* are smaller than those of the upper jaw: the two central ones are smaller than the two lateral, and are the smallest of all the incisor teeth. They are placed vertically in the jaw and are somewhat bevelled in front, where they have been worn down by contact with the overlapping edge of the upper teeth. The cingulum is absent.

The **canine teeth** are four in number: two in the upper, and two in the lower jaw; one being placed distally to each lateral incisor. They are larger and stronger than the incisors, especially the fangs, which sink deeply into the jaws, and cause well-marked prominences upon its surface.

Fig. 707.—Temporary teeth. Left side.

The *crown* is large and conical, very convex on its labial surface, a little hollowed and uneven on its lingual surface, and tapering to a blunted point or cusp, which projects beyond the level of the other teeth.

The *root* is single, but longer and thicker than that of the incisors, conical in form, compressed laterally, and marked by a slight groove on each side.

The *upper canine teeth* (popularly called *eye-teeth*) are larger and longer than the lower, and situated a little distally to them. They usually present a distinct basal ridge.

The *lower canine teeth* are placed mesially to the upper, so that their summits correspond to the interval between the upper canine tooth and the neighbouring incisors on each side.

The **bicuspid teeth** (*premolars*, or false molars) are eight in number: four in each jaw, two being placed distally to each of the canine teeth. They are smaller and shorter than the canines.

The *crown* is compressed proximo-distally, and surmounted by two pyramidal eminences, or cusps, separated by a groove; hence their name, *bicuspid*. Of the two cusps the labial is larger and more prominent than the lingual.

Fig. 708.—Front and side views of the teeth and jaws. (Cryer.)

The *neck* is oval.

The *root* is generally single, compressed, and presents a deep groove on each side, which indicates a tendency in the root to become double. The apex is generally bifid.

The *upper bicuspids* are larger, and present a greater tendency to the division of their roots than the lower; this is especially marked in the first upper bicuspid.

The **molar teeth** (*multicuspidati*, or true molars) are the largest of the permanent set and are adapted from the great breadth of their crowns for grinding and pounding the food. They are twelve in number: six in each jaw, three being placed distally to the second bicuspids.

The *crown* is nearly cubical in form, convex on its labial and lingual surfaces, flattened on its proximal and distal aspects; the upper surface being surmounted by four or five tubercles, or cusps (four in the upper, five in the lower molars), separated from each other by a crucial depression; hence their name, *multicuspid*.

The *neck* is distinct, large, and rounded.

The *root* is subdivided into two or three fangs: three in the teeth of the upper jaw, and two in those of the lower. Each of these fangs presents an aperture at its summit.

The crown of the *first molar tooth* in the *upper jaw* has usually four cusps; the root consists of three fangs, widely separated from one another, two being labial, the other lingual.

The crown of the *first molar tooth* in the *lower jaw* is larger than that of the upper; it has five cusps, and its root consists of two fangs, one being placed proximally, the other distally: they are both compressed from before backwards, and grooved on their contiguous faces, indicating a tendency to division.

The *second molar* is a little smaller than the first.

The crown has three or four cusps in the upper, and usually five in the lower jaw.

The root has three fangs in the upper jaw, and two in the lower, the characters of which are similar to those of the preceding tooth.

The *third molar tooth* is called the *wisdom-tooth* (*dens sapientiæ*), from its late appearance through the gum.

Its crown is nearly as large as that of the second molar, but is smaller than that of the first. In the upper jaw it. is usually furnished with three cusps, the two lingual ones being blended; in the lower jaw there are five cusps as in the other molars.

The root is generally single, short, conical, slightly curved, and grooved so as to present traces of a subdivision into three fangs in the upper, and two in the lower jaw.

TEMPORARY TEETH

The **temporary** or **milk teeth** are smaller, but, generally speaking, resemble in form the teeth which bear the same names in the permanent set. The hinder of the two temporary molars is the largest of all the milk teeth, and is succeeded by the second permanent bicuspid. The first upper molar has only three cusps —two labial, one lingual; the second upper molar has four cusps. The first lower molar has four cusps; the second lower molar has five. The fangs of the temporary molar teeth are smaller and more divergent than those of the permanent set, but in other respects bear a strong resemblance to them.

STRUCTURE OF THE TEETH

On making a vertical section of a tooth (fig. 709), a cavity will be found in the interior. This cavity is situated in the interior of the crown and the centre of each fang, and opens by a minute orifice at the extremity of the latter. The shape of the cavity corresponds somewhat with that of the tooth; it forms what is called the *pulp cavity*, and contains a soft, highly vascular, and sensitive substance, the *dental pulp*. The pulp consists of a loose connective tissue consisting of fine fibres and cells; it is richly supplied with vessels and nerves, which enter the cavity through the small aperture at the point of each fang. The cells of the pulp are partly found permeating the matrix, and partly arranged as a layer on the wall of the pulp cavity. These latter cells are named the *odontoblasts of Waldeyer*. These cells, during the development of the tooth, are columnar in shape, but later on, after the dentine is fully formed, they become flattened and resemble the osteoblasts found in the osteogenetic layer of the periosteum of bone. They have two fine processes, the outer or distal one passing into a dental tubule, the inner being continuous with the processes of the connective-tissue cells of the pulp matrix.

The solid portion of the tooth consists of three distinct structures, viz. the proper dental substance, which forms the larger portion of the tooth, the *ivory* or *dentine*; a layer which covers the exposed part of the crown, the *enamel*; and a thin layer, which is disposed on the surface of the fang, the *cement* or *crusta petrosa*.

The **ivory,** or **dentine** (fig. 711), forms the principal mass of a tooth; in its central part is the cavity enclosing the pulp. It is a modification of osseous tissue, from which it differs, however, in structure. On microscopic examination

FIG. 709.—Vertical section of a tooth *in situ* (15 diameters).

FIG. 710.—Vertical section of a molar tooth.

Crown

Neck

Fang

FIG. 711.—Vertical section of a bicuspid tooth.   (Magnified.)

Enamel worn away

Crown

Neck

Fang

is placed in the pulp cavity, opposite the cervix or neck of the tooth; the part above it is the crown, that below is the root (fang). 1. Enamel with radial and concentric markings. 2. Dentine with tubules and incremental lines. 3. Cement or crusta petrosa, with bone corpuscles. 4. Dental periosteum. 5. Bone of lower jaw.

it is seen to consist of a number of minute wavy and branching tubes, having distinct parietes. They are called the *dentinal tubules*, and are embedded in a dense homogeneous substance, the *intertubular tissue*.

The *dentinal tubules* (fig. 709) are placed parallel with one another, and open at their inner ends into the pulp cavity. In their course to the periphery they present two or three curves, and are twisted on themselves in a spiral direction. These tubes vary in direction : they are vertical in the upper portion of the crown, becoming oblique and then horizontal in the neck and upper part of the root, while towards the lower part of the root they are inclined downwards. The tubules, at their commencement, are about $\frac{1}{4500}$ of an inch in diameter; in their course they divide and subdivide dichotomously, so as to give to the cut surface of the dentine a striated appearance. From the sides of the tubes, especially in the fang, ramifications of extreme minuteness are given off, which join together in loops in the intertubular substance, or terminate in small dilatations, from which branches are given off. Near the periphery of the dentine, the finer ramifications of the tubules terminate imperceptibly by free ends. The dentinal tubules have comparatively thick walls, consisting, in addition to the intertubular tissue, of an elastic homogeneous membrane, the *dentinal sheath* of Neumann, which resists the action of acids; they contain slender cylindrical prolongations, first described by Tomes, and named *Tomes's fibres* or *dentinal fibres*. These dentinal fibres are analogous to the soft contents of the canaliculi of bone.

The *intertubular substance* is translucent, and contains the chief part of the earthy matter of the dentine. In it are a number of fine fibrils, which are continuous with the fibrils of the dental pulp (Mummery). After the earthy matter has been removed by steeping a tooth in weak acid, the animal basis remaining may be torn into laminæ which run parallel with the pulp cavity, across the direction of the tubes. A section of dry-dentine often displays a series of somewhat parallel lines— the *incremental lines of Salter*. These lines are composed of imperfectly calcified dentine arranged in layers. In consequence of the imperfection in the calcifying process, little irregular cavities are left, termed *interglobular spaces*. These spaces are found especially towards the outer surface of the dentine, where they form a layer, which is sometimes known as the *granular layer* (fig. 712). They have received their name from the fact that they are surrounded by minute nodules or globules of dentine. Other curved lines may be seen parallel to the surface. These are the *lines of Schreger*, and are due to the optical effect of simultaneous curvature of the dentinal fibres.

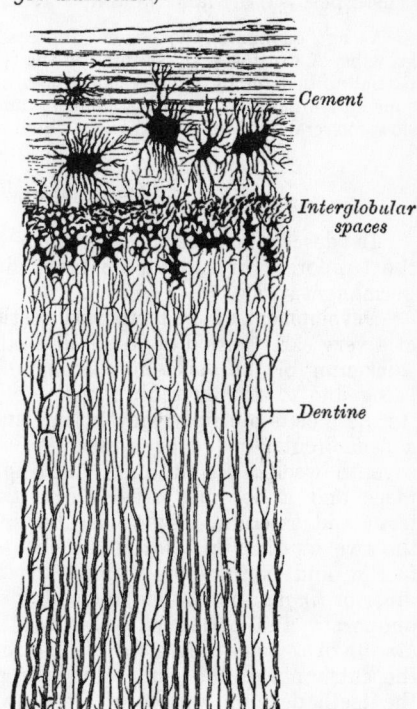

Fig. 712.—Transverse section of a portion of the root of a canine tooth. (Magnified 300 diameters.)

Cement

Interglobular spaces

Dentine

Chemical Composition.—According to Berzelius and Bibra, dentine consists of 28 parts of animal and 72 of earthy matter. The animal matter is resolvable by boiling into gelatin. The earthy matter consists of phosphate of lime, carbonate of lime, a trace of fluoride of calcium, phosphate of magnesia, and other salts.

The **enamel** is the hardest and most compact part of a tooth, and forms a thin crust over the exposed part of the crown, as far as the commencement of the fang. It is thickest on the grinding surface of the crown, until worn away by attrition, and becomes thinner towards the neck. It consists of minute hexagonal rods or columns. They lie parallel with one another, resting by one extremity upon the dentine, which presents a number of minute depressions for their reception; and forming the free surface of the crown by the other extremity. The columns are directed vertically on the summit of the crown, horizontally at the sides; they are about $\frac{1}{5500}$ of an inch in diameter, and pursue a more or less wavy course. Each column is a six-sided prism and presents numerous dark transverse shadings; these shadings are probably due to the manner in which the columns are developed in successive stages, producing shallow constrictions, as will be subsequently explained. Another series of lines, having a brown appearance, and denominated the *parallel striæ* or *coloured lines* *of Retzius*, are seen on a section of the enamel. According to Ebner, they are produced by air in the interprismatic spaces; others believe that they are the result of true pigmentation.

Numerous minute interstices intervene between the enamel fibres near their dentinal ends, a provision calculated to allow of the permeation of fluids from the dentinal tubule into the substance of the enamel. It is a disputed point whether the dentinal fibres penetrate a certain distance between the rods of the enamel or not. No nutritive canals exist in the enamel.

Chemical Composition.—According to Bibra, enamel consists of 96·5 per cent. of earthy matter, and 3·5 per cent. of animal matter.* The earthy matter consists of phosphate of lime, with traces of fluoride of calcium, carbonate of lime, phosphate of magnesia, and other salts.

The **crusta petrosa**, or **cement**, is disposed as a thin layer on the roots of the teeth, from the termination of the enamel, as far as the apex of the fang, where it is usually very thick. In structure and chemical composition it resembles bone. It contains, sparingly, the lacunæ and canaliculi which characterise true bone ; the lacunæ placed near the surface have the canaliculi radiating from the side of the lacunæ towards the periodontal membrane ; and those more deeply placed join with the adjacent dental tubules. In the thicker portions of the crusta petrosa, the lamellæ and Haversian canals peculiar to bone are also found.

As age advances, the cement increases in thickness, and gives rise to those bony growths, or exostoses, so common in the teeth of the aged; the pulp cavity also becomes partially filled up by a hard substance, intermediate in structure between dentine and bone (*osteo-dentine*, Owen ; *secondary dentine*, Tomes). It appears to be formed by a slow conversion of the dental pulp, which shrinks, or even disappears.

DEVELOPMENT OF THE TEETH

In describing the development of the teeth, the mode of formation of the temporary or milk teeth must first be considered, and then that of the permanent series.

Development of the temporary teeth.—The development of these teeth begins at a very early period of fœtal life—about the sixth week. It commences as a thickening of the epithelium along the line of the future jaw ; the thickening being due to a rapid multiplication of the more deeply situated epithelial cells. As these cells multiply they extend into the subjacent mesoblast, and thus form a semicircular ridge or strand of cells, enclosed by mesoblast. About the seventh week a longitudinal splitting or cleavage of this strand of cells takes place, and it becomes divided into two strands ; the separation beginning in front and extending laterally : the process occupying four or five weeks. Of the two strands thus formed, the *outer* or *labial* forms the future labio-dental furrow, and is therefore termed the *labio-dental strand* : while the other, the *inner* or *lingual*, is the ridge of cells in connection with which the teeth, both temporary and permanent, are developed. Hence it is known as the *dental lamina* or *common dental germ*. It forms a flat band of cells, which grows into the substance of the embryonic jaw, at first horizontally inwards, and then, as the teeth develop, vertically, i.e. upwards in the upper jaw, and downwards in the lower jaw. While still maintaining a horizontal direction, it has two edges : one, the *attached edge*, which is continuous with the epithelium lining the mouth ; the other, the *free edge*, projecting inwards, and embedded in the mesoblastic tissue of the embryonic jaw. Along its line of attachment to the buccal epithelium is a shallow groove, the *dental furrow*.

About the ninth week this dental lamina begins to develop enlargements along its free border. These are ten in number in each jaw, and each corresponds to a future milk tooth. They consist of masses of epithelial cells ; and the cells of the deeper part—that is, the part farthest from the margin of the jaw—increase rapidly and spread out in all directions. Each mass thus comes to assume a flask shape, connected with the general epithelial lining of the mouth by a narrow neck, embraced by mesoblast. They are now known as *special dental*

---

* Tomes disputes this, and says that enamel is an inorganic substance, and that what has been regarded as organic matter is in reality merely water in combination with the salts.

germs. After a time the lower expanded portion, or body of the flask, inclines outwards, so as to form an angle with the superficial constricted portion, which is sometimes known as the *neck* of the special dental germ. About the tenth week the mesoblastic tissue beneath these special dental germs becomes differentiated into papillæ; these grow upwards, and come in contact with the epithelial cells of the special dental germs, which become folded over them like a hood or cap. There is, then, at this stage a papilla or papillæ, which have already begun to assume somewhat the shape of the crown of the future tooth, and from which the dentine and pulp of the tooth are formed, surmounted by a dome or cap of epithelial cells, from which the enamel is derived.

In the meantime, while these changes have been going on, the dental lamina has been extending backwards behind the special dental germ corresponding to the second molar tooth of the temporary set, and at about the seventeenth week it presents an enlargement, the special dental germ for the first permanent molar, soon followed by the formation of a papilla in the mesoblastic tissue for the same tooth. This is followed by a further extension backwards of the dental lamina, with the formation of another enlargement and its corresponding papilla

FIG. 713.—Vertical section of the inferior maxilla of an early human fœtus.
(Magnified 25 diameters)

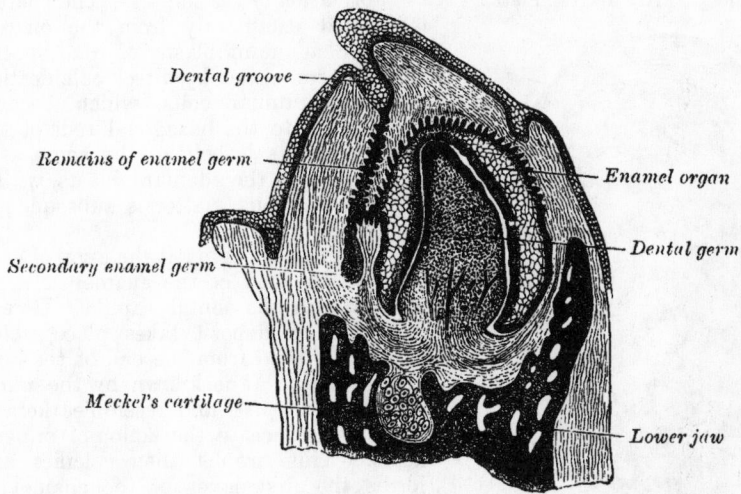

Dental groove

Remains of enamel germ

Enamel organ

Secondary enamel germ

Dental germ

Meckel's cartilage

Lower jaw

about the sixth month after birth for the second molar. And finally the process is repeated for the third molar, its papilla appearing about the fifth year of life.

After the formation of the special dental germs, the dental lamina undergoes atrophic changes and becomes cribriform, except on the lingual and lateral aspects of each of the special germs of the temporary teeth, where it undergoes a local thickening, forming the special dental germ of each of the successional permanent teeth—i.e. the ten anterior ones in each jaw. Here the same process goes on as has been described in connection with those of the milk teeth : that is, they recede into the substance of the gum behind the germs of the temporary teeth. As they recede they become flask-shaped, form an expansion at their distal extremity, and finally meet a papilla, which has been formed in the mesoblast, just in the same manner as was the case in the temporary teeth. The apex of the papilla indents the dental germ, which encloses it, and forming a cap for it, becomes converted into the enamel, while the papilla forms the dentine and pulp of the permanent tooth.

The special dental germs consist at first of rounded or polyhedral epithelial cells; after the formation of the papillæ, these cells undergo a differentiation into three classes. Those which are in immediate contact with the papilla become elongated, and form a layer of well-marked columnar epithelium coating the papilla. They are the cells which form the enamel fibres, and are therefore termed *enamel cells* or *adamantoblasts*. The cells of the outer layer of the special dental germ, which are in contact with the inner surface of the dentinal sac, presently to be described, are much shorter, cubical in form, and are named the *external enamel epithelium*. All the intermediate round cells of the dental germ between these two layers undergo a peculiar change. They become stellate in shape and develop processes, which unite to form a network into which fluid is secreted, which has the appearance of a jelly, and to which the name of enamel pulp is given. This transformed special dental germ is now known under the name of *enamel organ*.

While these changes are going on, a sac is formed around each enamel organ from the surrounding mesoblastic tissue. This is known as the *dental sac*, and is a vascular membrane of connective tissue. It grows up from below, and thus encloses the whole tooth germ; as it grows it causes the neck of the enamel organ to atrophy and disappear; so that all communication between the enamel organ and the superficial epithelium is cut off. At this stage there are vascular papillæ surmounted by inverted caps of epithelial cells, the whole being surrounded by membranous sacs. The cap consists of an internal layer of cells— the enamel cells or adamantoblasts—in contact with the papilla; of an external layer of cells—the external enamel epithelium—lining the interior of the dentinal sac; and of an intermediate mass of stellate cells, with anastomosing processes— the enamel pulp (fig. 714).

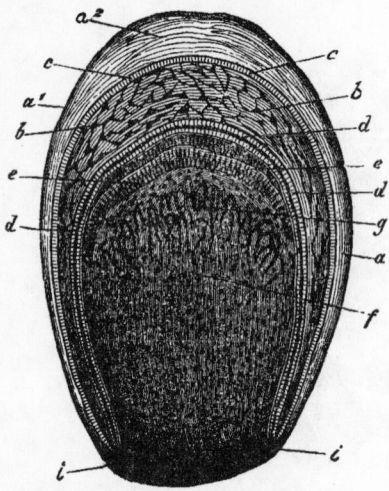

Fig. 714.—Dental sac of a human embryo at an advanced stage of development. Partly diagrammatic.

a. Wall of the sac, formed of connective tissue, with its outer stratum, *a¹*, and its inner, *a²*. *b.* Enamel organ. *c.* The external enamel epithelium. *d.* The enamel cells. *e.* Dentine cells. *f.* Dental papilla. *g, i.* Transition of the wall of the follicle into the tissue of the dental germ.

Formation of the enamel.—The enamel is formed exclusively from the enamel cells or adamantoblasts of the special dental germ, either by direct calcification of the columnar cells, which become elongated into the hexagonal rods of the enamel; or, as is believed by some, as a secretion from the adamantoblasts, within which calcareous matter is subsequently deposited.

The process begins at the apex of each cusp, at the end of the enamel cells, in contact with the dental papilla. Here a fine globular deposit takes place, being apparently shed from the end of the adamantoblasts. It is known by the name of *enamel droplet*, and resembles keratin in its resistance to the action of mineral acids. This droplet then calcifies and forms the first layer of the enamel; a second droplet now appears and calcifies, and so on; successive droplets of keratin-like material are shed from the adamantoblasts and form successive layers of enamel, the adamantoblasts gradually receding as each layer is produced, until at the termination of the process they have almost disappeared. The intermediate cells of the enamel pulp atrophy and disappear, so that the newly formed calcified material and the external enamel epithelium come into apposition. This latter layer, however, soon disappears on the emergence of the tooth beyond the gum. After its disappearance the crown of the tooth is still covered by a distinct membrane, which remains persistent for some time. This is known as the *cuticula dentis*, or *Nasmyth's membrane*, and is believed to be the last-formed layer of enamel derived from the *adamantoblast*, which has not become calcified. It forms a horny layer, which may be separated from the subjacent calcified mass by the action of strong acids. It is marked by the hexagonal impressions of the enamel prisms, and when stained by nitrate of silver, shows the characteristic appearance of epithelium.

Formation of the dentine.—While these changes are taking place in the epithelium to form the enamel, contemporaneous changes are occurring in the differentiated mesoblast of the dental papillæ which result in the formation of the dentine. As before stated, the first germ of the dentine consists in the formation of papillæ, corresponding in number to the teeth, from the soft mesoblastic tissue which bounds the depressions containing the special enamel germ. The papillæ grow upwards into the enamel germ and become covered by it, both being enclosed in a vascular connective tissue, the dentinal sac, in the manner above described. Each papilla then constitutes the formative pulp from which the dentine and permanent pulp are developed; it consists of rounded cells, and is very vascular, and soon begins to assume the shape of the future tooth. The next step is the appearance of the *odontoblasts*, which have a relation to the development of the teeth similar to that of the osteoblasts in the formation of bone. They are formed from the cells of the periphery of the papilla—that is to say, from the cells in immediate contact with the adamantoblasts of the special dental germ. These cells become elongated; one end of the elongated cell resting against the epithelium of the special dental germs, the other being tapered and often branched. By the direct transformation of the peripheral ends of these cells, or by a secretion from them, a layer of uncalcified matrix is formed which caps the cusp or cusps, if there are more than one, of the papillæ. In this matrix islets of calcification make their appearance, and coalescing give rise to a continuous layer of calcified material, which caps each cusp and constitutes the first layer of dentine. The odontoblasts having thus formed the first layer, retire towards the centre of the papilla, and as they do so produce successive layers of dentine from their peripheral extremities—that is to say, they form the dentinal matrix in which calcification subsequently takes place. As they thus retire from the periphery of the papilla, they leave behind them filamentous processes of cell protoplasm, provided with finer side processes; these are surrounded by calcified material, and thus form the dentinal tubules, and, by their side branches, the anastomosing tubules, whereby the dentinal tubules communicate: the processes of protoplasm contained within them, constituting the dentinal fibres (Tomes' fibres) which, as mentioned above, are found within the tubules. In this way the entire thickness of the dentine is developed, each tubule being completed throughout its whole length by a single odontoblast. The central part of the papilla does not undergo calcification, but persists as the pulp of the tooth. In this process of formation of dentine it has been shown that an uncalcified matrix is first developed, and that in this matrix islets of calcification appear, which subsequently blend together to form a cap to each cusp: in like manner successive layers are produced, which ultimately become blended with each other. In certain places this blending is not complete, portions of the matrix remaining uncalcified between the successive layers; this gives rise, in the macerated tooth, to little spaces, which are the interglobular spaces alluded to above.

Formation of the cement.—The root of the tooth begins to be formed shortly before the crown emerges through the gum, but is not completed until some time afterwards. It is produced by a downgrowth of the epithelium of the dental germ, which extends almost as far as the situation of the apex of the future fang, and determines the form of this portion of the tooth. This fold of epithelium is known as the *epithelial sheath*, and on its papillary surface odontoblasts appear, which in turn form dentine, so that the dentine formation is identical in the crown and root of the tooth. After the dentine of the root has been developed, the vascular tissues of the dental sac begin to break through the epithelial sheath, and spread over the surface of the fang as a layer of bone-forming material. In this osteoblasts make their appearance, and the process of ossification goes on in identically the same manner as in the ordinary intra-membranous ossification of bone. In this way the cement is formed, and consists of ordinary bone, containing canaliculi and lacunæ.

Formation of the alveoli.—About the fourteenth week of embryonic life the dental lamina becomes enclosed in a trough or groove of mesoblastic tissue, which at first is common to all the dental germs, but subsequently becomes divided, by bony septa, into loculi, each loculus containing the special dental germ of a temporary tooth and its corresponding permanent tooth. After birth each cavity becomes subdivided, so as to form separate loculi (the future alveoli) for the milk tooth and its corresponding permanent tooth. Although at one time the whole of the growing tooth is contained in the cavity of the alveolus, the latter never completely encloses it, since there is always an aperture over the top of the crown filled by soft tissue, by which the dental sac is connected with the surface of the gum, and which in the permanent teeth is called the *gubernaculum dentis*.

Development of the permanent teeth.—The permanent teeth as regards their development may be divided into two sets: (1) those which replace the temporary teeth, and which, like them, are ten in number in each jaw; these are the *successional permanent teeth*; and (2) those which have no temporary predecessors, but are superadded at the back of the temporary dental series. These are three in number on either side in each jaw, and are termed *superadded permanent teeth*. They are the three molars of the permanent set, the molars of the temporary set being replaced by the premolars or bicuspids of the permanent set. The development of the successional permanent teeth—the ten anterior ones in either jaw—will first be considered. As already stated, the original dental lamina, after the formation of the special dental germs of the temporary teeth, undergoes atrophic changes, except at one spot behind and lateral to each of the special germs of the milk teeth; here a local thickening takes place, and forms the special dental germ of each successional permanent tooth. In each of these, identically the same changes go on as took place in the special germs of the temporary teeth: they elongate and recede into the gum behind the germs of the milk teeth; a papilla springs up from the mesoblastic tissue to meet them, and the two become enclosed in a dental sac; the formation of the dentine and enamel takes place in the same way as in the temporary teeth. During their development the permanent teeth, enclosed in their sacs, come to be placed on the lingual side of the temporary teeth and more distant from the margin of the future gum, and, as already stated, are separated from them by bony partitions. As the crown of the permanent tooth grows, absorption of these bony partitions and of the fang of the temporary tooth takes place, through the agency of *osteoclasts*, which appear at this time, and finally nothing but the crown of the temporary tooth remains. This is shed or removed, and the permanent tooth takes its place.

The superadded permanent teeth are developed in the manner already described, by extensions backward of the posterior part of the dental lamina of the immediately preceding tooth. About the seventeenth week, that portion of the common dental germ of the last temporary tooth which lies behind the tooth, and which has remained unaltered, is prolonged backwards, and forms the special dental germ of the first permanent molar, and into this a papilla projects. In a similar manner about the fourth month after birth, a further extension backwards having taken place, a second enlargement occurs, into which a papilla projects about the sixth month, and thus the rudiment of the second molar is formed. Finally a third enlargement takes place, posterior to the other two, for the third molar, and its papilla becomes visible about the fifth year.

Eruption.—When the calcification of the different tissues of the tooth is sufficiently advanced to enable it to bear the pressure to which it will be afterwards subjected, its eruption takes place; the tooth making its way through the gum. The gum is absorbed by the pressure of the crown of the tooth against it, which is itself pressed up by the increasing size of the fang. At the same time the septa between the dental sacs, at first fibrous in structure, ossify, and constitute the alveoli; these firmly embrace the necks of the teeth, and afford them a solid basis of support.

The eruption of the temporary teeth commences at the seventh month, and is completed about the end of the second year, those of the lower jaw preceding those of the upper.

The following, according to C. S. Tomes, are the most usual times of eruption:

Lower central incisors	6 to 9 months.
Upper incisors	8 to 10 months.
Lower lateral incisors and first molars	15 to 21 months.
Canines	16 to 20 months.
Second molars	20 to 24 months.

Calcification of the permanent teeth proceeds in the following order in the lower jaw; in the upper jaw it takes place a little later: the first molar, soon after birth; the central and lateral incisors, and the canine, about six months after birth ; the bicuspids, at the second year, or a little later ; the second molar, about the end of the second year ; third molar, about the twelfth year.

The eruption of the permanent teeth takes place at the following periods, the teeth of the lower jaw preceding those of the upper by a short interval :

First molars	6th year.
Two central incisors	7th year.
Two lateral incisors	8th year.
First bicuspids	9th year.
Second bicuspids	10th year.
Canines	11th to 12th year.
Second molars	12th to 13th year.
'Wisdom' teeth	17th to 25th year.

Towards the sixth year, before the shedding of the temporary teeth begins, there are twenty-four teeth in each jaw, viz. the ten temporary teeth and the crowns of all the permanent teeth except those of the third molars.

THE PALATE

The **palate** forms the roof of the mouth : it consists of two portions, the hard palate in front, the soft palate behind.

The **hard palate** is bounded in front and at the sides by the alveolar arches and gums; behind, it is continuous with the soft palate. It is covered by a dense structure formed by the periosteum and mucous membrane of the mouth, which are intimately adherent. Along the middle line is a linear ridge or raphé, which terminates anteriorly in a small papilla, corresponding with the inferior opening of the anterior palatine fossa. On either side and in front of the raphé the mucous membrane is thick, pale in colour, and corrugated ; behind, it is thin, smooth, and of a deeper colour : it is covered with squamous epithelium, and furnished with numerous glands (palatal glands), which lie between the mucous membrane and the surface of the bone.

The **soft palate** (*velum pendulum palati*) is a movable fold, suspended from the posterior border of the hard palate, and forming an incomplete septum between the mouth and pharynx. It consists of a fold of mucous membrane enclosing muscular fibres, an aponeurosis, vessels, nerves, adenoid tissue, and mucous glands. When occupying its usual position (i.e. relaxed and pendent), its anterior surface is concave, continuous with the roof of the mouth, and marked by a medium ridge or raphé, which indicates its original separation into two lateral halves. Its posterior surface is convex, and continuous with the mucous membrane covering the floor of the posterior nares. Its upper border is attached to the posterior margin of the hard palate, and its sides are blended with the pharynx. Its lower border is free.

Hanging from the middle of its lower border is a small, conical-shaped, pendulous process, the *uvula* ; and arching outwards and downwards from the base of the uvula on each side are two curved folds of mucous membrane, containing muscular fibres, called the *arches* or *pillars of the soft palate* or *pillars of the fauces.*

The *anterior pillars* run downwards, outwards, and forwards to the sides of the base of the tongue, and are formed by the projection of the Palato-glossi muscles, covered by mucous membrane.

The *posterior pillars* are nearer to each other, and larger than the anterior ; they run downwards, outwards, and backwards to the sides of the pharynx, and are formed by the projection of the Palato-pharyngei muscles, covered by mucous membrane. The anterior and posterior pillars are separated below by a triangular interval, in which the tonsil is lodged.

The aperture by means of which the mouth communicates with the pharynx is called the *isthmus of the fauces.* It is bounded, above, by the soft palate ; below, by the dorsum of the tongue ; and on each side, by the anterior pillar of the fauces.

The *mucous membrane of the soft palate* is thin, and covered with squamous epithelium on both surfaces, excepting near the orifice of the Eustachian tube, where it is columnar and ciliated.*   Beneath the mucous membrane on the oral surface of the soft palate is a considerable amount of adenoid tissue.   The palatine glands form a continuous layer on its posterior surface and round the uvula.

The *aponeurosis of the soft palate* is a thin but firm fibrous layer attached above to the posterior border of the hard palate, and becoming thinner towards the free margin of the velum.   Laterally, it is continuous with the pharyngeal aponeurosis.   It forms the framework of the soft palate, and is joined by the tendon of the Tensor palati muscle.

The *muscles of the soft palate* are five on each side : the Levator palati, Tensor palati, Azygos uvulæ, Palato-glossus, and Palato-pharyngeus (see page 458). The following is the relative position of the structures in a dissection of the soft palate from the posterior or pharyngeal to the anterior or oral surface. Immediately beneath the mucous membrane is a thin stratum of muscular fibres, the posterior fasciculus of the Palato-pharyngeus muscle, joining with its fellow of the opposite side in the middle line.   Beneath this is the Azygos uvulæ, consisting of two rounded fleshy fasciculi which are in contact with each other in the median line of the soft palate.   Next comes the aponeurosis of the Levator palati joining with the muscle of the opposite side in the middle line.   Fourthly, the anterior fasciculus of the Palato-pharyngeus, thicker than the posterior, and separating the Levator palati from the next muscle, the Tensor palati.   This muscle terminates in a tendon which, after winding round the hamular process, expands into a broad aponeurosis in the soft palate, anterior to the other muscles which have been enumerated.   Finally we have a thin muscular stratum, the Palato-glossus muscle, placed in front of the aponeurosis of the Tensor palati, and separated from the oral mucous membrane by glands and adenoid tissue.

THE SALIVARY GLANDS (fig. 715)

The principal salivary glands communicating with the mouth, and pouring their secretion into its cavity, are the parotid, submaxillary, and sublingual.

The **parotid gland,** so called from being placed near the ear (παρά, near ; οὖς, ὠτός, the ear), is the largest of the three salivary glands, varying in weight from half an ounce to an ounce.   It lies upon the side of the face, immediately below and in front of the external ear.   It is limited above by the zygoma : below, by the angle of the jaw, and by a line drawn between it and the mastoid process ; anteriorly, it extends to a variable extent over the Masseter muscle ; posteriorly, it is bounded by the external meatus, the mastoid process, and the Sterno-mastoid and Digastric muscles, slightly overlapping the two muscles.

The *anterior surface* of the parotid gland is grooved to embrace the posterior margin of the ramus of the lower jaw, and advances beneath the ramus, between the two Pterygoid muscles and superficial to the ramus over the Masseter muscle. Its *outer surface*, slightly lobulated, is covered by the integument and parotid fascia, and has one or two lymphatic glands resting on it.   Its *inner surface* extends deeply into the neck, by means of two large processes, one of which dips behind the styloid process, and projects beneath the mastoid process and the Sterno-mastoid muscle ; the other is situated in front of the styloid process, and passes into the back part of the glenoid fossa, behind the articulation of the lower jaw.   The structures passing through the parotid gland are, the external carotid artery, giving off its three terminal branches : the posterior auricular artery emerges from the gland behind ; the temporal artery above ; the transverse facial, a branch of the temporal, in front ; and the internal maxillary winds through it as it passes inwards, behind the neck of the jaw.   Superficial to the external carotid artery is the trunk formed by the union of the temporal and internal maxillary veins : a branch, connecting this trunk with the internal jugular, passes through the gland.   It is also traversed by the facial nerve and its branches, which emerge at its anterior border ; branches of the great auricular nerve pierce the gland to join the facial, and the auriculo-temporal branch of the

---

* According to Klein, the mucous membrane on the nasal surface of the soft palate in the fœtus is covered throughout by columnar ciliated epithelium, which subsequently becomes squamous ; and some anatomists state that it is covered with columnar ciliated epithelium, except at its free margin, throughout life.

inferior maxillary nerve emerges from the upper part of the gland. The internal carotid artery and internal jugular vein lie close to its deep surface.

The **duct of the parotid gland** (*Stenson's*) is about two inches and a half in length. It commences by numerous branches from the anterior part of the gland, crosses the Masseter muscle, and at its anterior border it turns inwards nearly at a right angle and passes into the substance of the Buccinator muscle, which it pierces ; it then runs for a short distance obliquely forwards between the Buccinator and mucous membrane of the mouth, and opens upon the inner surface of the cheek by a small orifice, opposite the second molar tooth of the upper jaw. While crossing the Masseter it receives the duct of a small detached portion of the gland, *socia parotidis*, which occasionally exists as a separate lobe, just beneath the zygomatic arch. In this position it has the transverse facial artery above it and some branches of the facial nerve below it.

Structure.—The parotid duct is dense, its wall being of considerable thickness; its canal is about the size of a crow-quill, but at its orifice on the inner

FIG. 715.—The salivary glands.

aspect of the cheek its lumen is greatly reduced in size ; it consists of a thick external fibrous coat which contains contractile fibres, and of an internal or mucous coat lined with short columnar epithelium.

Surface Form.—The direction of the duct corresponds to a line drawn across the face about a finger's breadth below the zygoma—that is, from the lower margin of the concha to midway between the red margin of the upper lip and the ala of the nose.

Vessels and Nerves.—The *arteries* supplying the parotid gland are derived from the external carotid, and from the branches given off by that vessel in or near its substance. The *veins* empty themselves into the external jugular, through some of its tributaries. The *lymphatics* terminate in the superficial and deep cervical glands, passing in their course through two or three lymphatic glands, placed on the surface and in the substance of the parotid. The *nerves* are derived from the plexus of the sympathetic on the external carotid artery, the facial, the auriculo-temporal, and great auricular nerves.

It is probable that the branch from the auriculo-temporal nerve is derived from the glosso-pharyngeal through the otic ganglion (see page 868). At all events, in some of the lower animals this has been proved experimentally to be the case.

The **submaxillary gland** is situated below the jaw, in the anterior part of the submaxillary triangle of the neck. It is irregular in form, and weighs about two drachms. It is covered by the integument, Platysma, deep cervical fascia, and the body of the lower jaw, corresponding to a depression on the inner surface of that bone; and lies upon the Mylo-hyoid, Hyo-glossus, and Stylo-glossus muscles, a portion of the gland passing beneath the posterior border of the Mylo-hyoid. In front of it is the anterior belly of the Digastric; behind, it is separated from the parotid gland by the stylo-mandibular ligament, and in front from the sublingual gland by the Mylo-hyoid muscle. The facial artery lies embedded in a groove in its posterior and upper border.

The **duct of the submaxillary gland** (*Wharton's*) is about two inches in length, and its walls are much thinner than those of the parotid duct. It commences by numerous branches from the deep portion of the gland which lies on the upper surface of the Mylo-hyoid muscle, and passes forwards and inwards between the Mylo-hyoid and the Hyo-glossus and Genio-hyo-glossus muscles, then between the sublingual gland and the Genio-hyo-glossus, and opens by a narrow orifice on the summit of a small papilla, at the side of the frænum linguæ. On the Hyo-glossus muscle it lies between the lingual and hypoglossal nerves, but at the anterior border of the muscle it crosses under the lingual nerve, and is then placed above it.

Vessels and Nerves.—The *arteries* supplying the submaxillary gland are branches of the facial and lingual. Its *veins* follow the course of the arteries. The *nerves* are derived from the submaxillary ganglion, through which it receives filaments from the chorda tympani of the facial and lingual branch of the inferior maxillary, sometimes from the mylo-hyoid branch of the inferior dental, and from the sympathetic.

The **sublingual gland** is the smallest of the salivary glands. It is situated beneath the mucous membrane of the floor of the mouth, at the side of the frænum linguæ, in contact with the inner surface of the lower jaw, close to the symphysis. It is narrow, flattened, shaped somewhat like an almond, and weighs about a drachm. It is in relation, *above*, with the mucous membrane; *below*, with the Mylo-hyoid muscle; *in front*, with the lower jaw, and its fellow of the opposite side; *behind*, with the deep part of the submaxillary gland; and *internally*, with the Genio-hyo-glossus, from which it is separated by the lingual nerve and Wharton's duct. Its excretory ducts (*ducts of Rivinus*) are from eight to twenty in number; some join Wharton's duct; others open separately into the mouth, on the elevated crest of mucous membrane, caused by the projection of the gland, on either side of the frænum linguæ. One or more join to form a tube, which opens into the Whartonian duct: this is called the *duct of Bartholin*.

Vessels and Nerves.—The sublingual gland is supplied with blood from the sublingual and submental arteries. Its nerves are derived from the lingual, the chorda tympani, and sympathetic.

Structure of Salivary Glands.—The salivary are compound racemose glands, consisting of numerous lobes, which are made up of smaller lobules, connected together by dense areolar tissue, vessels, and ducts. Each lobule consists of the ramifications of a single duct, 'branching frequently in a tree-like manner,' the branches terminating in dilated ends or alveoli on which the capillaries are distributed. These alveoli, however, as Pflüger points out, are not necessarily spherical, though they may assume that form; sometimes they are perfectly cylindrical, and very often they are mutually compressed. The alveoli are enclosed by a basement-membrane, which is continuous with the membrana propria of the duct. It presents a peculiar reticulated structure, having the appearance of a basket with open meshes, and consisting of a network of branched and flattened nucleated cells.

The alveoli of the salivary glands are of two kinds, which differ in the appearance of their secreting cells, in their size, and in the nature of their secretion. The one variety secretes a ropy fluid, which contains mucin, and has therefore been named the *mucous*; while the other secretes a thinner and more watery fluid, which contains serum-albumin, and has been named *serous* or *albuminous.* The sublingual gland consists of mucous, the parotid of serous alveoli. The submaxillary contains both mucous and serous alveoli, the latter, however, preponderating.

Both kinds of alveoli are lined by cells, and it is by the character of these cells that the nature of the gland is chiefly to be determined. In addition, however, the alveoli of the serous glands are smaller than those of the mucous ones.

The **cells in the mucous alveoli** are spheroidal in shape, glassy, transparent, and dimly striated in appearance. The nucleus is usually situated in the part of the cell which is next the basement-membrane, against which it is sometimes flattened. The most remarkable peculiarity presented by these cells is, that each gives off an extremely fine process, which is curved in a direction parallel to the surface of the alveolus, lies in contact with the membrana propria, and overlaps the processes of neighbouring cells. The cells contain a quantity of mucin, to which their clear, transparent appearance is due.

Here and there in the alveoli are seen peculiar half-moon-shaped bodies, lying between the cells and the membrana propria of the alveolus. They are termed the *crescents of Gianuzzi*, or the *demilunes of Heidenhain* (fig. 716), and are composed of polyhedral granular cells, which Heidenhain regards as young epithelial cells destined to supply the place of those salivary cells which have undergone disintegration. This view, however, is not accepted by Klein.

Fig. 716.—A highly magnified section of the submaxillary gland of the dog, stained with carmine. (Kölliker.)

Serous alveoli.—In the serous alveoli the cells almost completely fill the cavity, so that there is hardly any lumen perceptible. Instead of presenting the clear transparent appearance of the cells of the mucous alveoli, they exhibit a granular appearance, due to distinct granules, of an albuminous nature, embedded in a closely reticulated protoplasm.

The ducts which originate out of the alveoli are lined at their commencement by epithelium which differs little from the pavement type. As the ducts enlarge, the epithelial cells change to the columnar type, and the part of the cell next the basement-membrane is finely striated. The lobules of the salivary glands are richly supplied with blood-vessels which form a dense network in the inter-alveolar spaces. Fine plexuses of nerves are also found in the interlobular tissue. The nerve-fibrils pierce the basement-membrane of the alveoli, and end in branched varicose filaments between the secreting cells. There is no doubt that ganglia are to be found in some salivary glands in connection with the nerve-plexuses in the interlobular tissue; they are present in the submaxillary, but not in the parotid.

In the submaxillary and sublingual glands the lobes are larger and more loosely united than in the parotid.

Mucous Glands.—Besides the salivary glands proper, numerous other glands are found in the mouth. In addition to keeping the mouth moist during the intervals when food is not being masticated, they assist the larger salivary glands in the insalivation of the food. Many of these glands are found at the posterior part of the dorsum of the tongue, behind the circumvallate papillæ, and also along its margins as far forwards as the apex. Others lie around and in the tonsil between its crypts, and large numbers are present in the soft palate, the lips and cheeks. These glands are of the same structure as the larger salivary glands, and are of the mucous or mixed type.

Surface Form.—The orifice of the mouth is bounded by the lips: two thick, fleshy folds covered externally by integument and internally by mucous membrane, and consisting of muscles, vessels, 'nerves, areolar tissue, and numerous small glands. The size of the orifice of the mouth varies considerably in different individuals, but seems to bear a close relation to the size and prominence of the teeth. Its corners usually correspond to the outer border of the canine teeth. In the Mongolian tribes, where the front teeth are large and inclined forward, the mouth is large; and this, combined with the thick and everted lips, which appear to be associated with prominent teeth, gives to the Negro's face much of the peculiarity by which it is characterised. The smaller teeth, and the slighter prominence of the alveolar arch of the more highly civilised races, render the orifice of the mouth much smaller, and thus a small mouth is an indication of intelligence, and is regarded as an evidence of the higher civilisation of the individual.

Upon looking into the mouth, the first thing to be noted is the tongue, the upper surface of which will be seen occupying the floor of the cavity. This surface is convex, and is marked along the middle line by a raphé, which divides it into two symmetrical portions. The anterior two thirds are rough and studded with papillæ; the posterior third, smooth and tuberculated, is covered by numerous glands which project from the surface. Upon raising the tongue, the mucous membrane which invests its upper surface may be traced over its sides on to its under surface, from which it is reflected over the floor of the mouth on to the inner surface of the lower jaw, a part of which it covers. As it passes over the borders of the tongue it changes its character, becoming thin and smooth, and losing the papillæ which are to be seen on the upper surface. In the middle line the mucous membrane on the under surface of the tip of the tongue forms a distinct fold, the *frænum linguæ*, by which this organ is connected to the symphysis of the jaw. Occasionally it is found that this frænum is rather shorter than natural, and, acting as a bridle, prevents the complete protrusion of the tongue. When this condition exists and an attempt is made to protrude the organ, the tip will be seen to remain buried in the floor of the mouth, and the dorsum of the tongue is rendered very convex, and more or less extruded from the mouth: at the same time a deep furrow will be noticed to appear in the middle line of the anterior part of the dorsum. Sometimes, a little external to the frænum, the ranine vein may be seen immediately beneath the mucous membrane. The corresponding artery, being more deeply placed, does not come into view, nor can its pulsation be felt with the finger. On either side of the frænum, in the floor of the mouth, is a longitudinal elevation or ridge, produced by the projection of the sublingual gland, which lies immediately beneath the mucous membrane. Close to the attachment of the frænum to the tip of the tongue may be seen on either side the slit-like orifice of Wharton's ducts, into which a fine probe may be passed without much difficulty. By everting the lips the smooth mucous membrane lining them may be examined, and may be traced from them on to the outer surface of the alveolar arch. In the middle line, both of the upper and lower lip, a small fold of mucous membrane passes from the lip to the bone, constituting the *fræna*; these are not so large as the frænum linguæ. By pulling outwards the angle of the mouth the mucous membrane lining the cheeks can be seen, and on it may be perceived a little papilla which marks the position of the orifice of Stenson's duct—the duct of the parotid gland. The exact position of the orifice of the duct is opposite the second molar tooth of the upper jaw. The introduction of a probe into this duct is attended with considerable difficulty. The teeth are the next objects which claim our attention upon looking into the mouth. There are, as stated above, ten in either jaw in the temporary set, and sixteen in the permanent set. The gums, in which they are implanted, are dense, firm, and vascular.

At the back of the mouth is seen the *isthmus of the fauces*, or, as it is popularly called, 'the throat:' this is the space between the pillars of the fauces on either side, and is the means by which the mouth communicates with the pharynx. Above, it is bounded by the soft palate, the anterior surface of which is concave and covered with mucous membrane, which is continuous with that lining the roof of the mouth. Projecting downwards from the middle of its lower border is a conical-shaped projection, the *uvula*. On either side of the isthmus of the fauces are the anterior and posterior pillars, formed by the Palato-glossus and Palato-pharyngeus muscles respectively, covered over by mucous membrane. Between the two pillars on either side is situated the tonsil.

When the mouth is wide open a prominent tense fold of mucous membrane may be seen and felt extending upwards and backwards from the position of the fang of the last molar tooth to the posterior part of the hard palate. This is caused by the Pterygo-mandibular ligament which is attached by one extremity to the apex of the internal pterygoid plate, and by the other to the posterior extremity of the mylo-hyoid ridge of the lower jaw. It connects the Buccinator with the Superior constrictor of the pharynx. The fang of the last molar tooth indicates the position of the lingual (gustatory) nerve, where it is easily accessible, and can with readiness be divided in cases of cancer of the tongue (see page 869). On the inner side of the last molar tooth we can feel the hamular process of the internal pterygoid plate of the sphenoid bone, around which the tendon of the Tensor palati plays. About one-third of an inch in front of the hamular process and the same distance directly inwards from the last molar tooth is the situation of the opening of the posterior palatine canal, through which emerges the posterior or descending palatine branch of the internal maxillary artery, and one of the descending palatine nerves from Meckel's ganglion. The exact position of the opening on the subject may be ascertained by driving a needle through the tissues of the palate in this situation, when it will be at once felt to enter the canal. The artery emerging from the opening runs forwards in a groove in the bone, just internal to the alveolar border of the hard palate, and may be wounded in the operation for the cure of cleft palate. Under these circumstances the palatine canal may require plugging. By introducing the finger into the mouth the anterior border of the coronoid process of the jaw can be felt, and is especially prominent when the jaw is dislocated. By throwing the head well back a considerable portion of the posterior wall of the pharynx may be seen through the isthmus faucium, and on introducing the finger the anterior surfaces of the bodies of the upper cervical vertebræ may be felt immediately beneath the thin muscular stratum forming the wall of the pharynx. The finger can be hooked round the posterior border of the soft palate, and by turning it forwards, the posterior nares, separated by the septum, can be felt, or the presence of any adenoid or other growths in the naso-pharynx ascertained.

The Pharynx

The **pharynx** is that part of the alimentary canal which is placed behind the nose, mouth, and larynx. It is a musculo-membranous tube, somewhat conical in form, with the base upwards, and the apex downwards, extending from the under surface of the skull to the level of the cricoid cartilage in front, and that of the sixth cervical vertebra behind.

The pharynx is about five inches in length, and broader in the transverse than in the antero-posterior diameter. Its greatest breadth is immediately below the base of the skull, where it projects on either side, behind the orifice of the Eustachian tube, as a recess termed the fossa of Rosenmüller ; its narrowest point is at its termination in the œsophagus. It is limited, *above*, by the body of the sphenoid and basilar process of the occipital bone ; *below*, it is continuous with the œsophagus ; *posteriorly*, it is connected by loose areolar tissue with the cervical portion of the vertebral column, and the Longi colli and Recti capitis antici muscles ; *anteriorly*, it is incomplete, and is attached in succession to the internal pterygoid plate, the pterygo-mandibular ligament, the lower jaw, the tongue, hyoid bone, and thyroid and cricoid cartilages ; *laterally*, it is connected to the styloid processes and their muscles, and is in contact with the common and internal carotid arteries, the internal jugular veins, and the glosso-pharyngeal, pneumogastric, hypoglossal, and sympathetic nerves, and above with a small part of the Internal pterygoid muscles. Seven openings communicate with it, viz. : the two posterior nares, the two Eustachian tubes, the mouth, larynx, and œsophagus.

The pharynx may be subdivided from above downwards into three parts : nasal, oral, and laryngeal. The nasal part of the pharynx (*pars nasalis*), or naso-pharynx, lies behind the nose and above the level of the soft palate : it differs from the two lower parts of the tube in that its cavity always remains patent. In front it communicates through the choanæ with the nasal fossæ. On its lateral wall is the pharyngeal orifice of the Eustachian tube, which presents the appearance of a vertical cleft bounded behind by a firm prominence, the *cushion*, caused by the inner extremity of the cartilage of the tube impinging on the deep surface of the mucous membrane. A vertical fold of mucous membrane, the *plica salpingo-pharyngea*, stretches from the lower part of the cushion to the pharynx ; it contains the Salpingo-pharyngeus muscle. A second and smaller mucous fold may be seen stretching from the upper part of the cushion to the palate, the *plica salpingo-palatina*. Behind the orifice of the Eustachian tube is a deep recess, the *fossa of Rosenmüller*, which represents the remains of the upper part of the second branchial cleft.

The oral part of the pharynx (*pars oralis*) reaches from the soft palate to the level of the hyoid bone. It opens anteriorly, through the isthmus faucium, into the mouth, while in its lateral wall, between the two pillars of the fauces, is the tonsil.

The **Tonsils** (*amygdalæ*) are two prominent bodies situated one on each side of the fauces, between the anterior and posterior pillars of the soft palate. They are of a rounded form, and vary considerably in size in different individuals. A recess, the *fossa supra-tonsillaris*, may be seen, directed upwards and backwards, above the tonsil. His regards this as the remains of the lower part of the second visceral cleft. It is covered by a fold of mucous membrane termed the *plica triangularis*. Externally the tonsil is in relation with the inner surface of the Superior constrictor, to the outer side of which are the ascending palatine and tonsillar arteries and the Internal pterygoid muscle. The internal carotid artery lies behind and to the outer side of the tonsil, and nearly an inch (twenty to twenty-five millimetres) distant from it. The *outer surface* of the tonsil corresponds in position to the angle of the lower jaw. Its *inner surface* presents from twelve to fifteen orifices, leading into small crypts or recesses, from which numerous follicles branch out into the substance of the gland. These follicles are lined by a continuation of the mucous membrane of the pharynx, covered with epithelium; around each follicle is a layer of closed capsules embedded in the submucous tissue. These capsules are analogous to those of Peyer's glands, consisting of adenoid tissue. No openings from the capsules into the follicles can be recognised. They contain a thick greyish secretion. Surrounding each follicle is a close plexus of lymphatics. From these plexuses the lymphatic vessels pass to the deep cervical glands in the neighbourhood of the greater cornu of the hyoid bone, behind and below the angle of the jaw; these glands frequently become enlarged in affections of the tonsils.

The *arteries* supplying the tonsil are the dorsalis linguæ from the lingual, the ascending palatine and tonsillar from the facial, the ascending pharyngeal from the external carotid, the descending palatine branch of the internal maxillary, and a twig from the small meningeal.

The *veins* terminate in the tonsillar plexus, on the outer side of the tonsil.

The *nerves* are derived from Meckel's ganglion, and from the glosso-pharyngeal.

Surgical Anatomy of the Tonsils.—The tonsils can be easily inspected by instructing the patient to throw the head back and open his mouth widely; the tongue at the same time being depressed by a spatula or tongue-depressor. The normal tonsil should not project beyond the plane of the anterior pillar of the fauces. They are prone to become enlarged, especially in tuberculous children; and when much increased in size they cause great trouble, owing to obstruction to respiration and deglutition. The tonsils may be the seat of acute inflammation, which may run on to suppuration, requiring evacuation of the pus. The incision into the tonsil should always be made from in front backwards and inwards. Another form of acute inflammation of the tonsil is follicular tonsillitis, due to the lodgment of micro-organisms in the crypts of the tonsil. It may be the seat of a sarcomatous growth; or of an epithelioma, generally spreading to it from neighbouring parts, rather than commencing in the tonsil itself. The removal of the tonsil is, as a rule, a very simple operation, and is not usually attended with much hæmorrhage, unless the patient is suffering from hæmophilia.

The laryngeal part of the pharynx (*pars laryngea*) reaches from the hyoid bone to the lower border of the cricoid cartilage, where it is continuous with the œsophagus. In front it presents the triangular aperture of the larynx, the base of which is directed forwards and is formed by the epiglottis, while its lateral boundaries are constituted by the aryteno-epiglottidean folds. On either side of the laryngeal orifice is a recess, termed the *sinus pyriformis*; it is bounded internally by the aryteno-epiglottidean fold, externally by the thyroid cartilage and thyro-hyoid membrane.

Structure.—The pharynx is composed of three coats: mucous, fibrous, and muscular.

The *pharyngeal aponeurosis*, or *fibrous coat*, is situated between the mucous and muscular layers. It is thick above where the muscular fibres are wanting, and is firmly connected to the basilar process of the occipital and petrous portion of the temporal bones. As it descends it diminishes in thickness, and is gradually lost. It is strengthened posteriorly by a strong fibrous band, which is attached above to the pharyngeal spine on the under surface of the basilar portion of the occipital bone, and passes downwards, forming a median raphé, which gives attachment to the Constrictor muscles of the pharynx.

The *mucous coat* is continuous with that lining the Eustachian tubes, the nares, the mouth, and the larynx. In the naso-pharynx it is covered by columnar ciliated epithelium ; in the buccal and laryngeal portions the epithelium is of the squamous variety. Beneath the mucous membrane are found racemose mucous glands ; they are especially numerous at the upper part of the pharynx around the orifices of the Eustachian tubes. Throughout the pharynx are also numerous crypts or recesses, the walls of which are surrounded by lymphoid tissue, similar to what is found in the tonsils. Across the back part of the pharyngeal cavity, between the two Eustachian tubes, a considerable mass of this tissue exists, and has been named the *pharyngeal tonsil.* Above this, in the middle line, an irregular, flask-shaped depression of the mucous membrane is sometimes seen extending up as far as the basilar process of the occipital bone. It is known as the *bursa pharyngea,* and was regarded by Luschka as the remains of the diverticulum, which is concerned in the development of the anterior lobe of the pituitary body. Other anatomists believe that it is connected with the formation of the pharyngeal tonsil.

The *muscular coat* has been already described (page 456).

Surgical Anatomy of the Pharynx.—The pharynx is sometimes the seat of a pouch-like dilatation of its walls, in which the food collects when the patient swallows. A cure is effected by removing the diverticulum and accurately suturing the opening in the pharynx which has been made. The internal carotid artery is in close relation with the pharynx, so that its pulsations can be felt through the mouth. It has been occasionally wounded by sharp-pointed instruments, introduced into the mouth and thrust through the wall of the pharynx. In aneurism of this vessel in the neck, the tumour necessarily bulges into the pharynx, as this is the direction in which it meets with the least resistance, nothing lying between the vessel and the mucous membrane except the thin Constrictor muscle, whereas on the outer side there is the dense cervical fascia, the muscles descending from the styloid process and the margin of the Sterno-mastoid.

The mucous membrane of the pharynx is very vascular, and is often the seat of inflammation, frequently of a septic character, since the numerous recesses are prone to lodge micro-organisms. And, in addition, owing to its exposed situation, the mucous membrane is liable to be irritated by agents introduced during inspiration. The inflammation may be attended with serious consequences : it may extend up the Eustachian tube and involve the middle ear ; it may spread to the entrance of the larynx, causing œdema and seriously interfering with respiration, or, invading the lymphatics, it may spread to the loose and lax tissue surrounding the pharyngeal wall, and may extend far and wide ; sometimes into the posterior mediastinum along the œsophagus. Abscess may form in the connective tissue behind the pharynx, between it and the vertebral column, constituting what is known as retro-pharyngeal abscess. This is most commonly due to caries of the cervical vertebræ ; but may also be caused by suppuration of a lymphatic gland, which is situated in this position opposite the axis, and which receives lymphatics from the nares ; by a gumma ; or by acute pharyngitis. In these cases the pus may be easily evacuated by an incision, with a guarded bistoury, through the mouth, but, for aseptic reasons, it is desirable that the abscess should be opened from the neck. In some instances this is perfectly easy ; the abscess can be felt bulging at the side of the neck and merely requires an incision for its relief, but this is not always so, and then an incision should be made along the posterior border of the Sterno-mastoid and the deep fascia divided. A director is now to be inserted into the wound, the fore finger of the left hand being introduced into the mouth and pressure made upon the swelling. This acts as a guide, and the director is to be pushed onwards until pus appears in the groove. A pair of sinus forceps is now inserted along the director and the opening into the cavity dilated.

Tumours of the pharynx may be innocent fibromata or papillomata, or malignant, most commonly epithelioma, secondary to disease of adjacent parts. For this condition pharyngotomy may be required, either lateral by means of a free incision along the anterior border of the Sterno-mastoid, or anterior by making a transverse incision immediately below the hyoid bone, cutting through the thyro-hyoid membrane and detaching the epiglottis from the back of the tongue.

Foreign bodies not infrequently become lodged in the pharynx, and most usually at its termination at about the level of the cricoid cartilage, just beyond the reach of the finger, as the distance from the arch of the teeth to the commencement of the œsophagus is about six inches.

THE ŒSOPHAGUS

The **œsophagus,** or **gullet,** is a muscular canal, about nine or ten inches in length, extending from the pharynx to the stomach. It commences at the upper border of the cricoid cartilage, opposite the sixth cervical vertebræ, descends along the front of the spine, through the posterior mediastinum, passes through the Diaphragm, and, entering the abdomen, terminates at the cardiac orifice of the stomach, opposite the eleventh dorsal vertebra. The general direction of the œsophagus is vertical; but it presents two slight curves in its course. At its commencement it is placed in the median line; but it inclines to the left side as far as the root of the neck, gradually passes to the middle line again, and finally deviates to the left as it passes forwards to the œsophageal opening of the Diaphragm. The œsophagus also presents an antero-posterior flexure, corresponding to the curvature of the cervical and thoracic portions of the spine. It is the narrowest part of the alimentary canal, being most contracted at its commencement, and at the point where it passes through the Diaphragm.

FIG. 717. — Accessory muscular fibres between the œsophagus and pleura, and œsophagus and trachea. (From a preparation in the Museum of the Royal College of Surgeons of England.)

Relations.—*In the neck*, the œsophagus is in relation, *in front*, with the trachea; and at the lower part of the neck, where it projects to the left side, with the thyroid gland; *behind*, it rests upon the vertebral column and Longi colli muscles; *on each side* it is in relation with the common carotid artery (especially the left, as it inclines to that side), and part of the lateral lobes of the thyroid gland; the recurrent laryngeal nerves ascend between it and the trachea; to its left side is the thoracic duct.

In the thorax, the œsophagus is at first situated a little to the left of the median line; it then passes behind the aortic arch, separated from it by the trachea, and descends in the posterior mediastinum, along the right side of the aorta, nearly to the Diaphragm, where it passes in front and a little to the left of the aorta, previous to entering the abdomen. It is in relation, *in front*, with the trachea, the left bronchus, the pericardium, and the Diaphragm; *behind*, it rests upon the vertebral column, the Longi colli muscles, the right intercostal arteries, the thoracic duct, and azygos minor veins; and below, near the Diaphragm, upon the front of the aorta. On its *left* side, in the superior mediastinum, are the terminal part of the arch of the aorta, the left subclavian artery, the thoracic duct, and left pleura, while running upwards in the angle between it and the trachea is the left recurrent laryngeal nerve; below, it is in relation with the descending thoracic aorta. On its *right* side are the right pleura and the vena azygos major, the latter of which it overlaps. The pneumogastric nerves descend in close contact with it, the right nerve passing down behind, and the left nerve in front of it; the two nerves uniting to form a plexus (the *plexus gulæ*) around the tube.

In the lower part of the posterior mediastinum the thoracic duct lies to the right side of the œsophagus; higher up, it is placed behind it, and, crossing about the level of the fourth dorsal vertebra, is continued upwards on its left side.

The *abdominal portion* of the œsophagus lies in the œsophageal groove on the posterior surface of the left lobe of the liver. It measures about half an inch in length, and its front and left aspects only are covered by peritoneal membrane.

Structure.—The œsophagus has three coats: an external or muscular; a middle or areolar; and an internal or mucous coat.

The *muscular coat* is composed of two planes of fibres of considerable thickness: an external longitudinal, and an internal circular.

The *longitudinal fibres* are arranged, at the commencement of the tube, in three fasciculi: one in front, which is attached to the vertical ridge on the posterior surface of the cricoid cartilage ; and one at each side, which is continuous with the muscular fibres of the pharynx: as they descend they blend together, and form a uniform layer, which covers the outer surface of the tube.

Accessory slips of muscular fibres are described by Cunningham as passing between the œsophagus and the left pleura, where the latter covers the thoracic aorta, or the root of the left bronchus, or the back of the pericardium, as well as other rarer accessory fibres. In fig. 717, taken from a dissection in the Museum of the Royal College of Surgeons of England, several of these accessory slips may be seen passing from the œsophagus to the pleura, and two slips to the back of the trachea just above its bifurcation.

The *circular fibres* are continuous above with the Inferior constrictor; their direction is transverse at the upper and lower parts of the tube, but oblique in the central part.

The muscular fibres in the upper part of the œsophagus are of a red colour, and consist chiefly of the striped variety; but below, they consist for the most part of involuntary fibres.

The *areolar coat* connects loosely the mucous and muscular coats.

The *mucous coat* is thick, of a reddish colour above, and pale below. It is disposed in longitudinal folds, which disappear on distension of the tube. Its surface is studded with minute papillæ, and it is covered throughout with a thick layer of stratified pavement epithelium. Beneath the mucous membrane, between it and the areolar coat, is a layer of longitudinally arranged non-striped muscular fibres. This is the *muscularis mucosæ.* At the commencement of the œsophagus it is absent, or only represented by a few scattered bundles ; lower down it forms a considerable stratum.

The *œsophageal glands* are numerous small compound racemose glands scattered throughout the tube: they are lodged in the submucous tissue, and each opens upon the surface by a long excretory duct. They are most numerous at the lower part of the tube, and form a ring round the cardiac orifice.

Vessels of the œsophagus.—The arteries supplying the œsophagus are derived from the inferior thyroid branch of the thyroid axis of the subclavian ; from the descending thoracic aorta, from the gastric branch of the cœliac axis, and from the left inferior phrenic of the abdominal aorta. They have for the most part a longitudinal direction.

Nerves of the œsophagus.—The nerves are derived from the pneumogastric and from the sympathetic; they form a plexus, in which are groups of ganglion-cells, between the two layers of the muscular coats, and also a second plexus in the submucous tissue.

Surgical Anatomy.—The relations of the œsophagus are of considerable practical interest to the surgeon, as he is frequently required, in cases of stricture of this tube, to dilate the canal by a bougie, when it is of importance that the direction of the œsophagus, and its relations to surrounding parts, should be remembered. In cases of malignant disease of the œsophagus, where its tissues have become softened from infiltration of the morbid deposit, the greatest care is requisite in directing the bougie through the strictured part, as a false passage may easily be made, and the instrument may pass into the mediastinum, or into one or the other pleural cavity, or even into the pericardium.

The student should also remember that obstruction of the œsophagus, and consequent symptoms of stricture, are occasionally produced by an aneurism of some part of the aorta pressing upon this tube. In such a case, the passage of a bougie could only hasten the fatal issue.

In passing a bougie the left forefinger should be introduced into the mouth, and the epiglottis felt for, care being taken not to throw the head too far backwards. The bougie is then to be passed beyond the finger until it touches the posterior wall of the pharynx. The patient is now asked to swallow, and at the moment of swallowing the bougie is passed gently onwards, all violence being carefully avoided.

It occasionally happens that a foreign body becomes impacted in the œsophagus, which can neither be brought upwards nor moved downwards. When all ordinary means for its removal have failed, excision is the only resource. This, of course, can only be performed when it is not very low down. If the foreign body is allowed to remain, extensive inflammation and ulceration of the œsophagus may ensue. In one case the foreign body ultimately penetrated the intervertebral substance, and destroyed life by inflammation of the membranes and substance of the cord.

The operation of œsophagotomy is thus performed. The patient being placed upon his back, with the head and shoulders slightly elevated, an incision, about four inches in length, should be made on the left side of the trachea, from the thyroid cartilage downwards, dividing the skin, Platysma, and deep fascia.　The edges of the wound being separated, the Omo-hyoid muscle should, if necessary, be divided, and the fibres of the Sterno-hyoid and Sterno-thyroid muscles drawn inwards.　The pretracheal layer of the cervical fascia must now be divided, and when this is done, the left lobe of the thyroid body will be exposed and must be drawn upwards and inwards, and the carotid vessels, being brought into view, must be drawn outwards, and retained in that position by retractors: the œsophagus will now be exposed, and should be divided over the foreign body, which can then be removed.　Great care is necessary to avoid wounding the thyroid vessels, the thyroid gland, and the laryngeal nerves.

The œsophagus may be obstructed not only by foreign bodies, but also by changes in its coats, producing stricture, or by pressure on it from without of new growths or aneurism, &c.

The different forms of stricture are: (1) the spasmodic, usually occurring in nervous women, and intermittent in character, so that the dysphagia is not constant; (2) fibrous, due to cicatrisation after injuries, such as swallowing corrosive fluids or boiling water; and (3) malignant, usually epitheliomatous in its nature.　This is situated generally either at the upper end of the tube, opposite the cricoid cartilage, or at its lower end at the cardiac orifice, but is also occasionally found at that part of the tube where it is crossed by the left bronchus.

The operation of œsophagostomy has occasionally been performed in cases where the stricture in the œsophagus is at the upper part, with a view to making a permanent opening below the stricture through which to feed the patient; but the operation has been far from successful, and the risk of setting up diffuse inflammation in the loose planes of connective tissue deep in the neck is so great that it would appear to be better, if any operative interference is undertaken, to perform gastrostomy.　The operation of œsophagostomy is performed in the same manner as œsophagotomy, but the edges of the opening in the œsophagus are stitched to the skin incision.

THE ABDOMEN

The **Abdomen** is the largest cavity in the body.　It is of an oval form, the extremities of the oval being directed upwards and downwards: the upper being formed by the under surface of the Diaphragm, the lower by the upper concave surface of the Levatores ani.　The cavity is wider above than below, and measures more in the vertical than in the transverse diameter.　In order to facilitate description, it is artificially divided into two parts: an upper and larger part, the *abdomen proper*; and a lower and smaller part, the *pelvis*.　These two cavities are not separated from each other, but the limit between them is marked by the brim of the true pelvis.

The abdomen proper differs from the other great cavities of the body in being bounded for the most part by muscles and fasciæ, so that it can vary in capacity and shape according to the condition of the viscera which it contains; but, in addition to this, the abdomen varies in form and extent with age and sex.　In the adult male, with moderate distension of the viscera, it is oval or barrel-shaped, but at the same time flattened from before backwards.　In the adult female, with a fully developed pelvis, it is conical with the apex above, and in young children it is conical with the apex below.

Boundaries.—The boundary between the thorax and abdomen is the Diaphragm.　This muscle forms a dome over the abdomen, and the cavity extends high into the bony thorax, reaching on the right side, in the nipple line, to the upper border of the fifth rib; on the left side it falls below this level by about an inch.　Below, the abdomen is limited by the structures which clothe the inner surface of the bony pelvis, principally the Levatores ani and Coccygei muscles on either side.　These muscles are sometimes termed the *Diaphragm of the pelvis*. The **abdomen proper** is bounded *in front*, and *at the sides*, by the lower ribs, the abdominal muscles, and the iliac fossæ; *behind*, by the vertebral column and the Psoas and Quadratus lumborum muscles; *above*, by the Diaphragm; *below*, by the brim of the pelvis.　The muscles forming the boundaries of the cavity are lined upon their inner surface by a layer of fascia, differently named according to the part which it covers.

The abdomen contains the greater part of the alimentary canal; some of the accessory organs to digestion, viz. the liver and pancreas; the spleen, the kidneys, and suprarenal capsules.　Most of these structures, as well as the wall of the cavity in which they are contained, are more or less covered by an extensive and complicated serous membrane, the *peritoneum*.

FIG. 718.—Topography of thoracic and abdominal viscera.

The *apertures* found in the walls of the abdomen, for the transmission of structures to or from it, are, the *umbilicus* (in the fœtus), for the transmission of the umbilical vessels ; the *caval opening* in the Diaphragm, for the transmission of the inferior vena cava ; the *aortic opening*, for the passage of the aorta, vena azygos major, and thoracic duct ; and the *œsophageal opening*, for the œsophagus and pneumogastric nerves. *Below*, there are two apertures on each side : one for the passage of the femoral vessels, and the other for the transmission of the spermatic cord in the male, and the round ligament in the female.

Regions.—For convenience of description of the viscera, as well as of reference to the morbid conditions of the contained parts, the abdomen is artificially divided into nine regions by imaginary planes, two horizontal and two sagittal, passing through the cavity, the edges of the planes being indicated by lines drawn on the surface of the body. Of the horizontal planes the upper or *infracostal* is indicated by a line encircling the body at the level of the lowest points of the tenth costal cartilages, the lower by a line carried round the trunk at the level of the highest points of the iliac crests as seen from the front. The latter is the *intertubercular plane* of Cunningham, who has pointed out * that its level corresponds with the prominent and easily defined tubercle on the iliac crest about two inches behind the anterior superior iliac spine. By means of these imaginary planes the abdomen is divided into three zones, which are named from above downwards the *subcostal*, *umbilical*, and *hypogastric* zones. Each of these is further subdivided into three regions by the two sagittal planes, which are indicated on the surface by lines drawn vertically through a point halfway between the anterior superior iliac spine and the symphysis pubis.†

The middle region of the upper zone is called the *epigastric* ; and the two lateral regions, the *right* and *left hypochondriac*. The central region of the middle zone is the *umbilical* ; and the two lateral regions, the *right* and *left lumbar*. The middle region of the lower zone is the *hypogastric* or *pubic region* ; and the lateral regions are the *right* and *left iliac* or *inguinal* (fig. 719). The viscera contained in these different regions are the following :

Right Hypochondriac	*Epigastric Region*	*Left Hypochondriac*
The greater part of right lobe of the liver, the hepatic flexure of the colon, and outer part of the right kidney.	Part of the stomach, including both cardiac and pyloric orifices, the left lobe and part of the right lobe of the liver and the gall-bladder, the pancreas, part of the duodenum, the suprarenal capsules, and greater parts of the kidneys.	The fundus of the stomach, the spleen and tail of the pancreas, the splenic flexure of the colon, and outer part of the left kidney.
Right Lumbar	*Umbilical Region*	*Left Lumbar*
Ascending colon, lower extremity of the right kidney, and some convolutions of the small intestines.	The transverse colon, part of the great omentum and mesentery, transverse part of the duodenum, and some convolutions of the jejunum and ileum, and lower end of right kidney.	Descending colon, part of the omentum, and some convolutions of the small intestines.

* *Journal of Anatomy and Physiology*, vol. xxvii.

† Anatomists are far from agreed as to the best method of subdividing the abdominal cavity, but that given above is the one which is generally adopted in this country. Addison,[1] in a careful analysis of the abdominal viscera in a large number of subjects, adopts the following lines: (1) a median, from the symphysis pubis to the ensiform cartilage ; (2) two lateral lines drawn vertically through a point midway between the anterior superior iliac spine and the symphysis pubis ; (3) an upper transverse line halfway between the symphysis pubis and the suprasternal notch ; and (4) a lower transverse line midway between the last and the upper border of the symphysis pubis. The upper transverse line corresponds with what he has termed the *transpyloric plane*, from the fact that in most cases it cuts through the pylorus.

Right Iliac (Inguinal)

The cæcum and termination of small intestine, and sometimes the vermiform appendix.

Hypogastric Region

Convolutions of the small intestines, part of the sigmoid flexure of the colon; sometimes the vermiform appendix; the bladder in children, and in adults if distended, and the uterus during pregnancy.

Left Iliac (Inguinal)

Part of the sigmoid flexure of the colon, and some convolutions of the small intestine.

If the anterior abdominal wall is reflected in the form of four triangular flaps by means of vertical and transverse incisions—the former from the ensiform cartilage to the symphysis pubis, the latter from flank to flank at the level of the umbilicus—the abdominal or peritoneal cavity is freely opened into and the contained viscera are in part exposed * (fig. 718).

FIG. 719.—The regions of the abdomen and their contents.
(Edge of costal cartilages in dotted outline.)

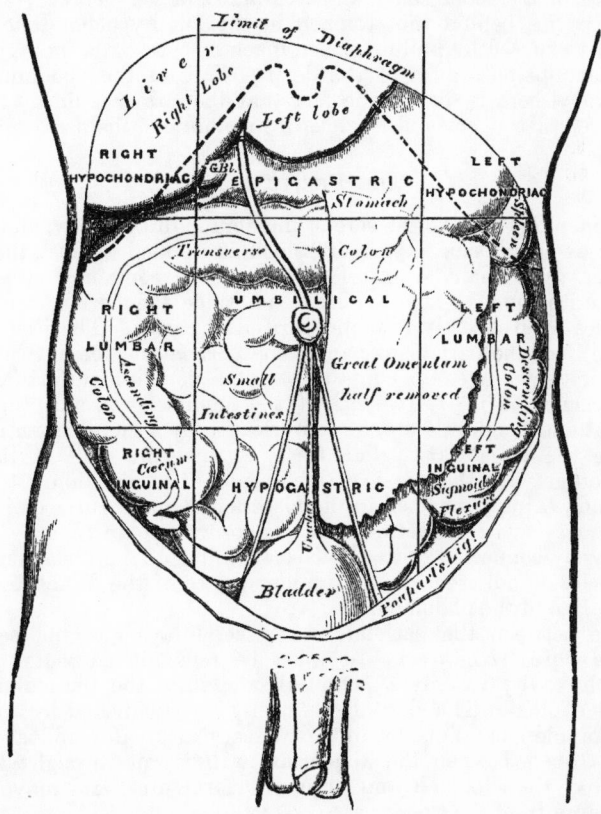

Above and to the right side is the liver, situated chiefly under the shelter of the right ribs and their cartilages, but extending across the middle line and reaching for some distance below the level of the ensiform cartilage. To the left of the liver is the stomach, from the lower border of which an apron-like fold of peritoneum, the *great omentum*, descends for a varying distance, and obscures, to a greater or lesser extent, the other viscera. Below it, however, some of the coils of the small intestine can generally be seen, while in the right and left iliac regions respectively the *cæcum* and the *sigmoid flexure* of the colon

* It must be borne in mind that, although the term abdominal cavity is used, there is, under normal conditions, only a potential cavity or lymph-space, since the viscera are everywhere in contact with the parietes.

3 Y

are partly exposed. The bladder occupies the anterior part of the pelvis, and, if distended, will project above the symphysis pubis; the rectum lies in the concavity of the sacrum, but is usually obscured by the coils of the small intestine.

If the stomach is followed from left to right it will be found to be continuous with the first part of the small intestine, or *duodenum*, the point of continuity being marked by a thickened ring which indicates the position of the pyloric valve. The duodenum passes towards the under surface of the liver, and then curving downwards, is lost to sight. If, however, the great omentum be thrown upwards over the chest, the terminal part of the duodenum will be observed passing across the spine towards the left side, where it becomes continuous with the *coils of the jejunum and ileum*. These measure some twenty feet in length, and if followed downwards will be seen to end in the right iliac fossa by opening into the *cæcum* or commencement of the *large intestine*. From the cæcum the large intestine takes an arched course, passing at first upwards on the right side, then across the middle line and downwards on the left side, and forming respectively the *ascending, transverse, and descending parts of the colon*. In the left iliac region and pelvis it assumes the form of a loop, the *sigmoid flexure* or *ilio-pelvic colon*, and terminates in the *rectum*.

The *spleen* lies behind the stomach in the left hypochondriac region, and may be in part exposed by pulling the stomach over towards the right side.

The glistening appearance of the deep surface of the abdominal wall and of the exposed viscera is due to the fact that the former is lined and the latter more or less completely covered by a serous membrane, the *peritoneum*.

THE PERITONEUM

The peritoneum is the largest serous membrane in the body, and consists, in the male, of a closed sac, a part of which is applied against the abdominal parietes, while the remainder is reflected over the contained viscera. In the female the peritoneum is not a closed sac, since the free extremities of the Fallopian tubes open directly into the peritoneal cavity. The part of the peritoneum which lines the parietes is named the *parietal* portion of the peritoneum; that which is reflected over the contained viscera constitutes the *visceral* portion of the peritoneum. The *free surface* of the membrane is smooth, covered by a layer of flattened endothelium, and lubricated by a small quantity of serous fluid. Hence the viscera can glide freely against the wall of the cavity or upon one another with the least possible amount of friction. The *attached surface* is rough, being connected to the viscera and inner surface of the parietes by means of *areolar tissue*, termed the *subserous areolar tissue*. The parietal portion is loosely connected with the fascial lining of the abdomen and pelvis, but is more closely adherent to the under surface of the Diaphragm and also in the middle line of the abdomen.

The space between the parietal and visceral layers of the peritoneum is named the *peritoneal cavity*; but it must be remembered that under normal conditions this cavity is only a potential one, since the parietal and visceral layers are in contact. The peritoneal 'cavity' is subdivided by a constriction, termed the foramen of Winslow, into two sacs, a greater and a lesser. The greater sac is opened when the abdominal wall is cut through; the lesser is situated behind the stomach and adjoining structures, and may be regarded as a diverticulum from the greater sac.

The peritoneum differs from the other serous membranes of the body in presenting a much more complex arrangement, and one which can only be clearly understood by following the changes which take place in the alimentary canal during its development; the student therefore is advised to preface his study of the peritoneum by reviewing the chapter dealing with this subject in the section on Embryology (page 135).

To trace the continuity of the membrane from one viscus to another, and from the viscera to the parietes, it is necessary to follow its reflections in the vertical and horizontal directions, and in doing so it matters little where a start is made.

If the stomach is drawn downwards a fold of peritoneum will be seen stretching from its lesser curvature to the transverse fissure of the liver (fig. 720).

This is the *gastro-hepatic*, or *small omentum*, and consists of two layers; these, on being traced downwards, split to envelop the stomach, covering respectively its antero-superior and postero-inferior surfaces. At the greater curvature of the stomach they again come into contact and are continued downwards in front of the transverse colon, forming the anterior two layers of the *great* or *gastro-colic* omentum. Reaching the free edge of this fold they are reflected upwards as its two posterior layers, and thus the great omentum consists of four layers of peritoneum. Followed upwards the two posterior layers separate to enclose the transverse colon, above which they once more come into contact and pass backwards to the abdominal wall as the *transverse mesocolon*. Reaching the abdominal wall at the upper border of the transverse part of the duodenum, the two layers of the transverse mesocolon become separated from each other

FIG. 720.—The reflections of the peritoneum, as seen in a vertical section of the abdomen.

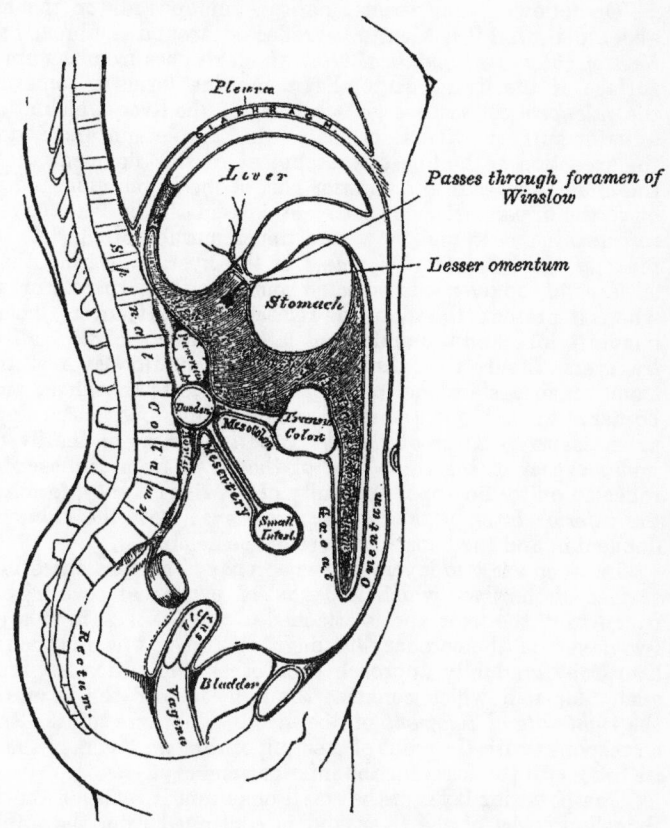

and take different directions; the upper or anterior layer (ascending layer of transverse mesocolon) ascends in front of the pancreas, and its further course will be followed presently. The lower or posterior layer is carried downwards, as the anterior layer of the *mesentery*, by the superior mesenteric vessels to the small intestine around which it may be followed and subsequently traced upwards as the posterior layer of the mesentery to the abdominal wall. From the posterior abdominal wall it sweeps downwards over the aorta into the pelvis, where it invests the first part of the rectum (pelvic part of colon) and attaches it to the front of the sacrum by a fold termed the *mesorectum* or *pelvic mesocolon*. Leaving first the sides and then the front of the rectum it is reflected on to the back of the bladder, and, after covering the posterior and upper aspects of this viscus, is carried by the urachus and obliterated hypogastric arteries on to the posterior surface of the anterior abdominal wall. Between the rectum and bladder it forms a pouch, the *recto-vesical pouch*, the bottom of which is about on a level with the middle of the vesiculæ seminales, i.e. about three inches from the orifice of the anus. When the bladder is distended the peritoneum is carried up with the expanded viscus, so that a considerable part of the anterior surface of the latter lies directly against the abdominal wall without the intervention of the peritoneal membrane.

In the female the peritoneum is reflected from the rectum on to the upper part of the posterior vaginal wall, forming the *recto-vaginal pouch* or *pouch of Douglas*, which is bounded on each side by a crescentic fold, the *fold of Douglas*. It is then carried over the posterior aspect and fundus of the uterus on to its anterior surface, which it covers as far as the junction of the body and cervix uteri, forming here a second, but shallower depression, the *utero-vesical pouch*. It is also reflected from the sides of the uterus to the lateral walls of the pelvis as two expanded folds, the *broad ligaments of the uterus*, in the free margin of each of which can be felt a thickened cord-like structure, the *Fallopian tube*.

On following the parietal peritoneum upwards on the back of the anterior abdominal wall it is seen to be reflected around a fibrous band, the *ligamentum teres* or *obliterated umbilical vein*, which reaches from the umbilicus to the under surface of the liver. Here the membrane forms a somewhat triangular fold, the *falciform* or '*suspensory*' *ligament* of the liver, which attaches the upper and anterior surfaces of that organ to the Diaphragm and abdominal wall. With the exception of the line of attachment of this ligament the peritoneum covers the under surface of the anterior part of the Diaphragm and is reflected from it on to the upper surface of the right lobe of the liver as the *superior layer of the coronary ligament*, and on to the upper surface of the left lobe as the *superior layer of the left lateral ligament* of the liver. Covering the upper and anterior surfaces of the liver it is reflected round its sharp margin on to its under surface, where it presents the following relations : (1) It covers the lower aspect of the quadrate lobe and the under and lateral aspects of the gall-bladder, and at the transverse fissure is continuous with the anterior layer of the small omentum, from which a start was made. (2) It invests the under surface and posterior border of the left lobe, and is reflected from its upper surface on to the Diaphragm as the *superior layer of the left lateral ligament* of the liver. (3) It covers the under aspect of the right lobe of the liver, from the back part of which it is reflected on to the upper extremity of the right kidney, forming, in this situation, the *inferior layer of the coronary ligament* ; from the kidney it is carried to the duodenum and the hepatic flexure of the colon.

Between the two layers of the coronary ligament there is a large triangular surface of the liver which is devoid of peritoneal covering : this is named the *bare area* of the liver, and is attached to the Diaphragm by areolar tissue. If the two layers of the coronary ligament be traced towards the right margin of the liver, they gradually approach each other and ultimately fuse to form a small triangular fold which connects the right lobe to the Diaphragm, and is termed the *right lateral ligament* of the liver. The apex of the triangular bare area corresponds with the point of meeting of the two layers of the coronary ligament ; its base with the fossa for the inferior vena cava.

The posterior layer of the small omentum is reflected on to the caudate and Spigelian lobes of the liver, and is continued from the upper extremity of the latter lobe to the Diaphragm, forming the upper limit of the lesser sac of the peritoneum. Between the two layers of the small omentum the hepatic artery and portal vein ascend to, and the bile-duct descends from, the liver. When followed to the right the small omentum is seen to form a distinct free border, around which its anterior and posterior layers are continuous with each other ; and if the finger be introduced behind this free border, it passes into the lesser sac of the peritoneum through a somewhat constricted ring termed the foramen of Winslow. At the left extremity of the transverse fissure the small omentum assumes a vertical direction, and is attached along the bottom of the fissure for the ductus venosus.

The *foramen of Winslow* forms the communication between the greater and lesser sacs of the peritoneum, and can be readily located by passing the finger upwards and towards the left along the neck of the gall-bladder. It is bounded, *in front*, by the free border of the small omentum, with the bile-duct, hepatic artery, and portal vein between its two layers ; *behind*, by the parietal layer of the peritoneum covering the inferior vena cava ; *above*, by the caudate lobe of the liver ; and *below*, by the first part of the duodenum and by the peritoneum, which covers the hepatic artery as the latter passes forwards beneath the foramen of Winslow, previous to ascending between the two layers of the small omentum.

The *lesser sac of the peritoneum* is merely a diverticulum of the greater sac— the two sacs being continuous with each other through the foramen of Winslow. *In front*, the lesser sac is bounded, from above downwards, by the Spigelian lobe of the liver, the small omentum, the stomach, and the anterior two layers of the great omentum. *Behind*, it is limited from below upwards by the two posterior layers of the great omentum, the transverse colon, and the ascending layer of the transverse mesocolon, which covers the upper surface of the pancreas, the left suprarenal capsule, and the upper end of the left kidney. To the right of the œsophageal opening of the Diaphragm it is formed by that portion of the Diaphragm which supports the Spigelian lobe of the liver. Laterally, the lesser sac extends from the foramen of Winslow to the spleen, where it is limited by the posterior layer of the gastro-splenic omentum. In the fœtus it reaches as low as the free margin of the great omentum ; but in the adult its vertical extent is usually more limited, owing to adhesions between the layers of the omentum. The extent of the lesser sac and its relations to surrounding parts can be definitely made out by tearing through the small omentum and inserting the hand through the opening thus made.

During a considerable part of fœtal life the transverse colon is suspended from the posterior abdominal wall by a mesentery of its own—the two posterior layers of the great omentum passing, at this stage, in front of the colon. This condition occasionally persists throughout life, but as a rule adhesion occurs between the mesentery of the transverse colon and the posterior layer of the great omentum, with the result that the colon appears to receive its peritoneal covering by the splitting of the two posterior layers of the latter fold.

In addition to tracing the peritoneum vertically, it is necessary to trace it horizontally. Below the transverse colon, the arrangement is extremely simple, as it includes only the greater sac (fig. 720). Above the level of the transverse colon it is more complicated, on account of the existence of the two sacs.

Starting from the linea alba, below the level of the transverse colon, and tracing the continuity in a horizontal direction to the right, the peritoneum covers the internal surface of the abdominal wall almost as far as the outer border of the Quadratus lumborum muscle ; it encloses the cæcum, and is reflected over the sides and anterior surface of the ascending colon, fixing it to the abdominal wall, from which it can be traced over the Psoas muscle and inferior vena cava towards the middle line. It then passes along the mesenteric vessels to invest the small intestine, and back again to the large vessels in front of the spine, forming the mesentery, between the layers of which are contained the mesenteric blood-vessels, nerves, lacteals, and glands. It is then carried over the left Psoas muscle ; it covers the sides and anterior surface of the descending colon, and, reaching the abdominal wall, is continued along it to the middle line of the abdomen.

Above the transverse colon (fig. 721) the peritoneum can be traced, forming the greater and lesser sacs, and their communication with each other through the foramen of Winslow can be demonstrated. Commencing in the middle line of the abdomen, the membrane may be traced lining its anterior wall, and sending a process backwards to encircle the obliterated umbilical vein or round ligament of the liver, forming the falciform ligament of the liver. Continuing its course to the right, it is reflected over the front of the upper part of the right kidney, across the vena cava inferior and aorta, and over the left kidney to the hilum of the spleen, forming the anterior layer of the *lieno-renal ligament*. From the hilum of the spleen it is reflected to the stomach, forming the posterior layer of the *gastro-splenic omentum*. It covers the posterior surface of the stomach, and from its lesser curvature it passes around the portal vein, hepatic artery, and bile-duct, and back again to the stomach, as the lesser omentum, and thus it forms the anterior boundary of the foramen of Winslow. It now covers the front of the stomach, and upon reaching the cardiac extremity it passes to the spleen, immediately in front of the hilum, forming the anterior layer of the gastro-splenic omentum. It can then be traced over the surface of the spleen, to which it gives a serous covering, and is reflected from the posterior border of the hilum on to the left kidney, forming the posterior layer of the lieno-renal ligament.

Numerous peritoneal folds extend between the various organs or connect them to the parietes. These serve to hold them in position, and, at the same time, enclose the vessels and nerves proceeding to them. Some of these folds are called *ligaments*, such as the ligaments of the liver and the false ligaments of the bladder. Others, which connect certain parts of the intestine with the abdominal wall, constitute the *mesenteries*; and lastly, those which proceed from the stomach to certain viscera in its neighbourhood are called *omenta*.

The **Ligaments,** formed by folds of the peritoneum, include those of the liver, spleen, bladder, and uterus. They will be found described with their respective organs.

The **Omenta** are: the lesser omentum, the great omentum, and the gastro-splenic omentum.

The *lesser omentum (gastro-hepatic)* is the duplicature which extends between the liver and the lesser curvature of the stomach. It is extremely thin, and

Fig. 721.—Transverse section of peritoneum.

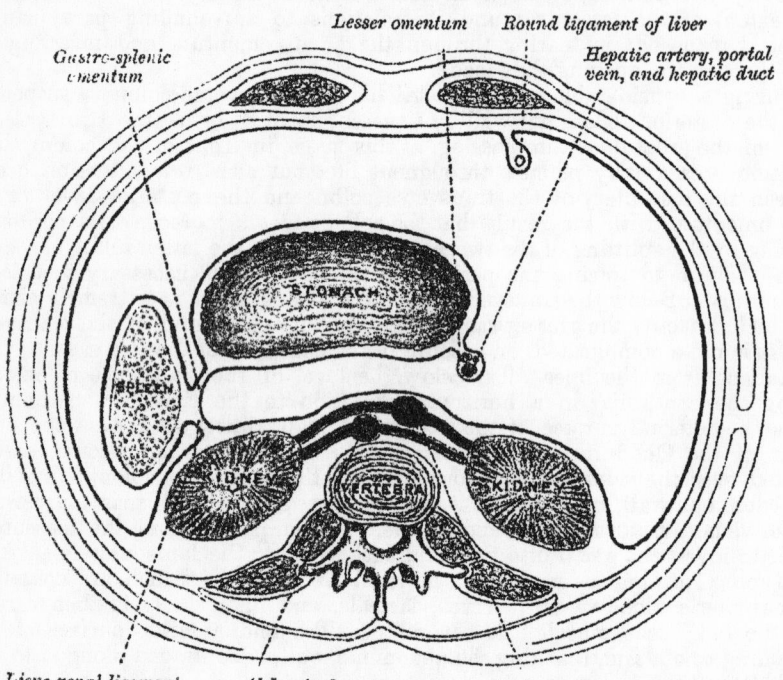

consists of the two layers of peritoneum which cover respectively the anterior and posterior surfaces of the stomach. When these two layers reach the lesser curvature of the stomach, they join together and ascend as the double fold to the transverse fissure of the liver; to the left of this fissure the double fold is attached to the bottom of the fissure of the ductus venosus, along which it is carried to the Diaphragm, where the two layers separate to embrace the end of the œsophagus. At the right border the lesser omentum is free, and the two layers of which it is composed are continuous. The anterior layer, which belongs to the greater sac, turns round the hepatic vessels to become continuous with the posterior layer belonging to the lesser one. They here form a free, rounded margin, which constitutes the anterior boundary of the foramen of Winslow. Between the two layers, close to this free margin, are the hepatic artery, the common bile-duct, the portal vein, lymphatics, and the hepatic plexus of nerves—all these structures being enclosed in loose areolar tissue, called *Glisson's capsule*. Between the layers where they are attached to the stomach run the gastric artery and the pyloric branch of the hepatic artery.

The *great omentum (gastro-colic)* is the largest peritoneal fold. It consists of four layers of peritoneum, two of which descend from the stomach, one from its anterior, the other from its posterior surface, and, uniting at its lower border, descend in front of the small intestines, sometimes as low down as the pelvis; they then turn upon themselves, and ascend again as far as the transverse colon, where they separate and enclose that part of the intestine. These individual layers may be easily demonstrated in the young subject, but in the adult they are more or less inseparably blended. The left border of the great omentum is continuous with the gastro-splenic omentum; its right border extends as far only as the duodenum. The great omentum is usually thin, presents a cribriform appearance, and always contains some adipose tissue, which in fat subjects accumulates in considerable quantity. Its use appears to be to protect the intestines from the cold, and to facilitate their movement upon each other during their vermicular action. Between its two anterior layers is the anastomosis between the right and left gastro-epiploic arteries.

The *gastro-splenic omentum* is the fold which connects the margins of the hilum of the spleen to the *cul-de-sac* of the stomach, being continuous by its lower border with the great omentum. It contains the vasa brevia vessels.

The **Mesenteries** are: the mesentery proper, the transverse mesocolon, the sigmoid mesocolon, and the mesorectum. In addition to these there are sometimes present an ascending and a descending mesocolon.

The *mesentery* is the broad, fan-shaped fold of peritoneum which connects the convolutions of the jejunum and ileum with the posterior wall of the abdomen. Its *root*—the part connected with the structures in front of the vertebral column—is narrow, about six inches in length, and is directed obliquely from the duodeno-jejunal flexure at the left side of the second lumbar vertebra to the right iliac fossa (fig. 722). Its intestinal border is about twenty feet in length; and here its two layers separate to enclose the intestine, and form its peritoneal coat. Its breadth, between its vertebral and intestinal borders, is about eight inches. Its *upper border* is continuous with the under surface of the transverse mesocolon : its *lower border*, with the peritoneum covering the cæcum and ascending colon. It serves to retain the small intestines in their position, and contains between its layers the rami intestini tenuis of the superior mesenteric artery, with their accompanying veins and plexuses of nerves, the lacteal vessels, and mesenteric glands.

In most cases the peritoneum covers only the front and sides of the ascending and descending parts of the colon. Sometimes, however, these are surrounded by the serous membrane and attached to the posterior abdominal wall by an ascending and a descending mesocolon respectively. At the place where the transverse colon turns downwards to form the descending colon, a fold of peritoneum is continued to the Diaphragm opposite the tenth and eleventh ribs. This is the *phreno-colic ligament*; it passes below the spleen, and serves to support this organ, and therefore it has received the second name of *sustentaculum lienis*.

The *transverse mesocolon* is a broad fold, which connects the transverse colon to the posterior wall of the abdomen. It is formed by the two ascending or posterior layers of the great omentum, which, after separating to surround the transverse colon, join behind it, and are continued backwards to the spine, where they diverge in front of the anterior border of the pancreas. This fold contains between its layers the vessels which supply the transverse colon.

The *sigmoid mesocolon* is the fold of peritoneum which retains the sigmoid flexure in connection with the left Psoas muscle. Between the two layers of this fold run the sigmoid arteries.

The *mesorectum* is the narrow fold which connects the upper part of the rectum with the front of the sacrum. It contains the superior hæmorrhoidal vessels.*

The *appendices epiploicæ* are small pouches of the peritoneum filled with fat and situated along the colon and upper part of the rectum. They are chiefly appended to the transverse colon.

Retro-peritoneal fossæ.—In certain parts of the abdominal cavity there are recesses of peritoneum forming *culs-de-sac* or pouches, which are of surgical

---

* The sigmoid mesocolon and mesorectum are sometimes described together under the name of the *pelvic mesocolon*.

interest in connection with the possibility of the occurrence of retro-peritoneal hernia. One of these is the lesser sac of the peritoneum, which may be regarded as a recess of peritoneum through the foramen of Winslow, in which a hernia may take place, but there are several others, of smaller size, which require mention.

These recesses or fossæ may be divided into three groups, viz.: 1, the duodenal fossæ; 2, pericæcal fossæ; and 3, the intersigmoid fossa.

1. *Duodenal fossæ.*—Moynihan has described no less than nine fossæ as occurring in the neighbourhood of the duodenum. Three of these are fairly

Fig. 722.—Diagram devised by Delépine to show the lines along which the peritoneum leaves the wall of the abdomen to invest the viscera.

constant, and are the only ones which require mention. (*a*) The *inferior duodenal fossa* is the most constant of all the peritoneal fossæ in this region, being present in from 70 to 75 per cent. of cases. It is situated opposite the third lumbar vertebra on the left side of the ascending portion of the duodenum. The opening into the fossa is directed upwards, and is bounded by a thin sharp fold of peritoneum with a concave margin, called the *inferior duodenal fold*. The tip of the index finger introduced into the fossa under the fold passes some little distance up behind the ascending or fourth portion of the duodenum.

(b) The *superior duodenal fossa* is the next most constant pouch or recess, being present in from 40 to 50 per cent. of cases. It often coexists with the inferior one, and its orifice looks downwards, in the opposite direction to the preceding fossa. It lies to the left of the ascending portion of the duodenum. It is bounded by the free edge of the *superior duodenal fold*, which presents a semilunar margin; to the right it is blended with the peritoneum covering the ascending duodenum, and to the left with the peritoneum covering the perirenal tissues. The fossa is bounded in front by the superior duodenal fold; behind by the second lumbar vertebra; to the right by the duodenum. Its depth is two centimetres, and it terminates in the angle formed by the left renal vein crossing the aorta. This fossa is of importance, as it is in relation with the inferior mesenteric vein: that is to say, the vein almost always corresponds to the line of union of the superior duodenal fold with the posterior parietal peritoneum. (c) The *duodeno-jejunal fossa* can be seen by pulling the jejunum downwards and to the right, after the transverse colon has been pulled upwards. It will appear as an almost circular opening, looking downwards and to the right, and bounded by two free borders or folds of peritoneum, the *duodeno-mesocolic ligaments*. The opening admits the little finger into the fossa, to the depth of from two to three centimetres. The fossa is bounded above by the pancreas, to the right by the aorta, and to the left by the kidney; beneath is the left renal vein. The fossa exists in from 15 to 20 per cent. of cases, and has never yet been found in conjunction with any other form of duodenal fossa.

2. *Pericæcal fossæ.*—There are at least three pouches or recesses to be found in the neighbourhood of the cæcum, which are termed *pericæcal fossæ*. (1) The *ileo-colic fossa* (superior ileo-cæcal) is formed by a fold of peritoneum, the ileo-colic fold, arching over a branch of the ileo-colic artery, which supplies the ileo-colic junction, and appears to be the direct continuation of the artery. The fossa is a narrow chink situated between the ileo-colic fold in front, and the mesentery of the small intestine, the ileum, and the small portion of the cæcum behind. (2) The *ileo-cæcal fossa* (inferior ileo-cæcal) is situated behind the angle of junction of the ileum and cæcum. It is formed by a fold of peritoneum (the ileo-cæcal fold or bloodless fold of Treves), the upper border of which is attached to the ileum, opposite its mesenteric attachment, and the lower border, passing over the ileo-cæcal junction, joins the mesentery of the appendix, and sometimes the appendix itself; hence this fold has been called the ileo-appendicular. Between this fold and the mesentery of the vermiform appendix is the ileo-cæcal fossa. It is bounded above by the posterior surface of the ileum and the mesentery; in front and below by the ileo-cæcal fold, and behind by the upper part of the mesentery of the appendix. (3) The *subcæcal fossa* (retro-cæcal) is situated immediately behind the cæcum, which has to be raised to bring it into view. It varies much in size and extent. In some cases it is sufficiently large to admit the index finger, and extends upwards behind the ascending colon in the direction of the kidney: in others it is merely a shallow depression. It is bounded and formed by two folds: one, the *parieto-colic*, which is attached by one edge to the abdominal wall from the lower border of the kidney to the iliac fossa and by the other to the postero-external aspect of the colon; and the other, *mesenterico-parietal*, which is in reality the insertion of the mesentery into the iliac fossa. In some instances the subcæcal fossa is double.

3. The *Intersigmoid fossa* is constant in the fœtus and during infancy, but disappears in a certain percentage of cases as age advances. Upon drawing the sigmoid flexure upwards, the left surface of the sigmoid mesocolon is exposed, and on it will be seen a funnel-shaped recess of the peritoneum, lying on the external iliac vessels, in the interspace between the Psoas and Iliacus muscles. This is the orifice leading to the fossa intersigmoidea, which lies behind the sigmoid mesocolon, and in front of the parietal peritoneum. The fossa varies in size; in some instances it is a mere dimple, whereas in others it will admit the whole of the index finger.

Any of these fossæ may be the site of a retro-peritoneal hernia. The pericæcal fossæ are of especial interest, because hernia of the vermiform appendix frequently takes place into one of them, and may there become strangulated. The presence of these pouches also explains the course which pus has been known to take in cases of perforation of the appendix, where it travels upwards behind the ascending colon as far as the Diaphragm.*

THE STOMACH

The **Stomach** is the principal organ of digestion. It is the most dilated part of the alimentary canal, and is situated between the termination of the œsophagus and the commencement of the small intestine. Its form is somewhat pyriform with the large end (*fundus*) directed upwards and the small end bent to the right. It is situated in the left hypochondriac and epigastric regions, and is placed, in part, immediately behind the anterior wall of the abdomen and beneath the Diaphragm. Viewing the stomach from in front it appears that the right margin of the œsophagus is continued downwards as the upper two-thirds of the lesser curvature of the stomach, the remaining third of this border bending sharply backwards and to the right, to complete the smaller curvature (fig. 723). The greater curvature begins at the left border of the termination of the œsophagus in a somewhat acute angle; it then passes upwards and to the left to the under surface of the Diaphragm, with which it lies in contact for some distance, and then sweeps downwards

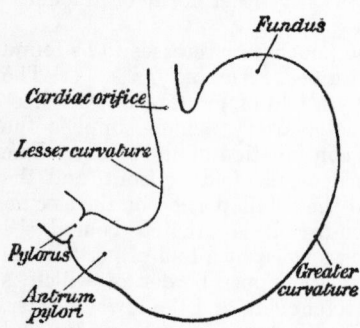

FIG. 723.—Diagrammatic outline of the stomach.

Fundus

Cardiac orifice

Lesser curvature

Pylorus

Antrum pylori

Greater curvature

with a convexity to the left, and, continued across the middle line of the body, finally turns upwards and backwards, to terminate at the commencement of the small intestine. It will thus be seen that the stomach may be divided into a main or *cardiac* portion, the long axis of which is directed downwards, with a little inclination forwards and to the right, and a smaller or *pyloric* portion, the long axis of which is horizontal with an inclination backwards. Of the two openings, the *cardiac orifice*, by which it communicates with the œsophagus, is situated slightly to the left of the middle line of the body to the right of the *fundus*, or dilated upper extremity of the stomach, and is directed downwards; the other, the *pyloric orifice*, by which it communicates with the small intestine, is on a lower plane, close to the right of the mid-line, and looks directly backwards.

The stomach has two surfaces, called anterior and posterior, and two borders, termed the greater and lesser curvatures.

Surfaces.—With regard to the so-called anterior and posterior surfaces of the stomach, it must be borne in mind that these names are not strictly correct, as the anterior surface is directed upwards as well as forwards, and the posterior downwards as well as backwards.

The *anterior* surface has a somewhat flattened appearance when the stomach is empty, but when the stomach is full the surface becomes convex. It is in relation with the Diaphragm; the thoracic wall formed by the anterior parts of the seventh, eighth, and ninth ribs of the left side; the left lobe of the liver; and the anterior abdominal wall. Between the part covered by the liver and that covered by the left ribs there is a triangular segment of the anterior wall of the stomach, which is in contact with the abdominal wall and is the only part of the stomach which is visible when the abdominal wall is removed and the viscera allowed to remain *in situ*. It measures about forty square centimetres and is of great importance to the surgeon, as the stomach can readily be reached in this situation. When the stomach is empty the transverse colon may be found lying in front of the lower part of its anterior surface. The whole of this surface of the stomach is covered by peritoneum.

The *posterior* surface of the stomach is in relation with the Diaphragm, the gastric surface of the spleen, the left suprarenal capsule, the upper part of the left kidney, the anterior surface of the pancreas, the splenic flexure of the colon,

* On the anatomy of these fossæ, see the *Arris and Gale Lectures* by Moynihan, 1899.

and the ascending layer of the transverse mesocolon. These structures form a shallow concavity or *bed* on which this surface of the stomach rests. The transverse mesocolon separates the stomach from the duodeno-jejunal junction and commencement of the jejunum. Almost the whole of this surface is covered with peritoneum, but behind the cardiac orifice there is a small portion of the stomach which is uncovered by peritoneum and is in contact with the Diaphragm and frequently with the upper portion of the left suprarenal capsule.

The *lesser curvature* of the stomach extends between the cardiac and pyloric orifices along the right border of the organ. It descends in front of the left crus of the Diaphragm, along the left side of the eleventh and twelfth dorsal vertebræ, and then turning to the right it crosses the first lumbar vertebra and ascends to the pylorus. It gives attachment to the two layers of the gastro-hepatic omentum, between which are situated the gastric and pyloric arteries.

The *greater curvature* is directed to the left, and is four or five times as long as the lesser curvature. Starting from the cardiac orifice it forms an arch to the left with its convexity upwards, the highest point of which is on a level with the costal cartilage of the sixth rib of the left side. It then passes nearly straight downwards, with a slight convexity to the left, as low as the costal cartilage of the ninth rib and then turns to the right to end at the pylorus. As it crosses the median line the lowest edge of the greater curvature is about two fingers' breadth above the umbilicus. The lower part of the greater curvature gives attachment to the two anterior layers of the great omentum, between which are situated the gastro-epiploic vessels.

The *cardiac orifice* is the opening by which the œsophagus communicates with the stomach. It is therefore sometimes termed the *œsophageal opening*. It is the most fixed part of the stomach, and is situated about two inches below the highest part of the fundus on a level with the body of the eleventh dorsal vertebra to the left and a little in front of the aorta. This would correspond on the anterior surface of the body to the left seventh costal cartilage, a little more than an inch (three centimetres) from its junction with the sternum.

The *pyloric orifice* communicates with the duodenum, the aperture being guarded by a valve. Its position varies with the movements of the stomach. When the stomach is empty the pylorus is situated just to the right of the median line of the body, on a level with the upper border of the first lumbar vertebra. On the anterior surface of the body its position would be indicated by a point an inch below the tip of the ensiform cartilage and a little to the right. As the stomach becomes filled the pylorus moves to the right, and in a fully distended stomach may be situated about two inches to the right of the median line. Near the pylorus the stomach exhibits a slight dilatation, which is named the *antrum pylori*, and when the stomach is fully distended, this part becomes elongated to the right so as to extend beyond the pylorus itself.

The size of the stomach varies considerably in different subjects. When moderately distended its greatest length, from the top of the fundus to the lowest part of the greater curvature, is from ten to twelve inches; and its diameter at the widest part from four to five inches. The distance between the two orifices is from three to six inches, and the measurement from the anterior to the posterior wall three and a half inches. Its weight is about four ounces and a half, and its average capacity in the adult male is about a quart.

Alterations in Position.—There is no organ in the body the position and size of which present such frequent alterations as the stomach. *When empty*, it lies at the back part of the abdomen, some distance from the surface. Its pyloric end is situated close to or very slightly to the right of the middle line, covered in front by the left lobe of the liver, and being on a level with the first lumbar vertebra. When empty, the stomach assumes a more or less cylindrical form, especially noticeable at its pyloric end. *When the stomach is distended*, its surfaces, which are flattened when the organ is empty, become convex, and the stomach is brought well against the anterior wall of the abdomen. Its fundus expands and rises considerably above the level of the cardiac orifice: in doing this the Diaphragm is forced upwards, contracting the cavity of the chest; hence the dyspnœa complained of, from inspiration being impeded. The apex of the heart is also tilted upwards; hence the oppression in this region and the palpitation experienced in extreme distension of the stomach. The left lobe of the liver is pushed to the right side. When the stomach becomes distended the change in the position of the pylorus is very considerable; it is shifted to the right, about two inches from the median line, and lies under cover of the liver, near the neck of the gall-bladder. In consequence of the distension of the stomach the lesser *cul-de-sac* bulges over the pylorus, concealing it from view, and causing it to undergo a rotation, so that its orifice is directed backwards. The stomach is displaced downwards by the descent of the Diaphragm during inspiration, and elevated by the pressure of the abdominal muscles during expiration. *Pressure from without,* as from tight lacing, pushes the stomach down towards the pelvis. In disease, also, the position and connection of the organ may be greatly changed, from the accumulation of fluid in the chest or abdomen, or from alteration in size of any of the surrounding viscera. *Variations according to age.*—In an early period of development the stomach is vertical, and in the new-born child it is more vertical than later on in life, as owing to the large size of the liver it is more pushed over to the left side of the abdomen, and the whole of the anterior surface is covered by the left lobe of this organ.

FIG. 724.—The mucous membrane of the stomach and duodenum with the bile-ducts.

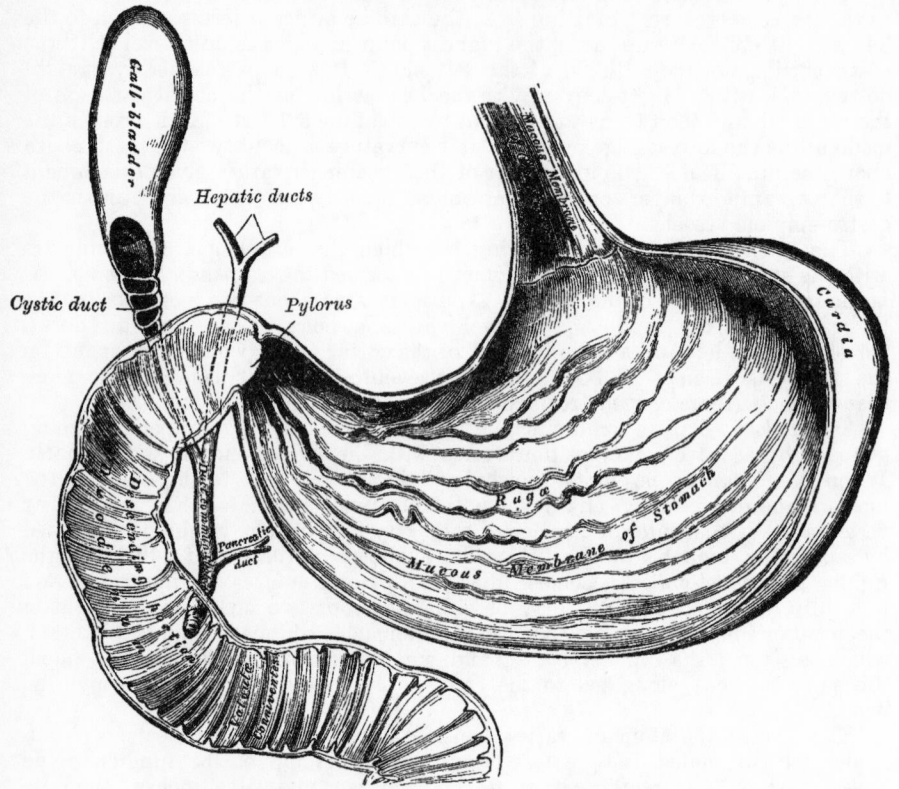

On looking into the pyloric end of the stomach, the mucous membrane is found projecting inwards in the form of a circular fold, the *pyloric valve,* leaving a narrow circular aperture, about half an inch in diameter, by which the stomach communicates with the duodenum.

The *pyloric valve* is formed by a reduplication of the mucous membrane of the stomach, containing numerous circular muscular fibres, which are aggregated into a thick ring; the longitudinal fibres and serous membrane being continued over the fold without assisting in its formation.

Structure.—The wall of the stomach consists of four coats: serous, muscular, areolar, and mucous, together with vessels and nerves.

The *serous coat* is derived from the peritoneum, and covers the entire surface of the organ, excepting along the greater and lesser curvatures, at the points of attachment of the greater and lesser omenta; here the two layers of peritoneum leave a small triangular space, along which the nutrient vessels and nerves pass. On the posterior surface of the stomach, close to the cardiac orifice, there is also a small area uncovered by peritoneum, where the organ is in contact with the under surface of the Diaphragm.

The *muscular coat* (fig. 725) is situated immediately beneath the serous covering, to which it is closely connected. It consists of three sets of fibres : longitudinal, circular, and oblique.

The *longitudinal fibres* are most superficial ; they are continuous with the longitudinal fibres of the œsophagus, and radiate in a stellate manner from the cardiac orifice. They are most distinct along the curvatures, especially the lesser, but are very thinly distributed over the surfaces. At the pyloric end they are more thickly distributed, and continuous with the longitudinal fibres of the small intestine.

The *circular fibres* form a uniform layer over the whole extent of the stomach beneath the longitudinal fibres. At the pylorus they are most abundant, and are aggregated into a circular ring, which projects into the lumen, and forms, with the fold of mucous membrane covering its surface, the *pyloric valve*. They are continuous with the circular fibres of the œsophagus.

The *oblique fibres* are limited chiefly to the cardiac end of the stomach, where they are disposed as a thick uniform layer, covering both surfaces, some passing obliquely from left to right, others from right to left, round the cardiac end.

FIG. 725.—The muscular coat of the stomach.

The *areolar* or *submucous coat* consists of a loose, filamentous, areolar tissue, connecting the mucous and muscular layers. It supports the blood-vessels previous to their distribution to the mucous membrane : hence it is sometimes called the *vascular coat*.

The *mucous membrane* is thick ; its surface smooth, soft, and velvety. In the fresh state it is of a pinkish tinge at the pyloric end, and of a red or reddish-brown colour over the rest of its surface. In infancy it is of a brighter hue, the vascular redness being more marked. It is thin at the cardiac extremity, but thicker towards the pylorus. During the contracted state of the organ it is thrown into numerous plaits or rugæ, which, for the most part, have a longitudinal direction, and are most marked towards the lesser end of the stomach, and along the greater curvature (fig. 724). These folds are entirely obliterated when the organ becomes distended.

Structure of the Mucous Membrane.—When examined with a lens, the inner surface of the mucous membrane presents a peculiar honeycomb appearance from being covered with small shallow depressions or alveoli, of a polygonal or hexagonal form, which vary from $\frac{1}{100}$ to $\frac{1}{200}$ of an inch in diameter, and are separated by slightly elevated ridges. In the bottom of the alveoli are seen the orifices of minute tubes, the *gastric glands*, which are situated perpendicularly side by side throughout the entire substance of the mucous membrane. The surface of the mucous membrane of the stomach is covered by a single layer of columnar epithelium ; it lines the alveoli, and also for a certain distance the mouths of the gastric glands. This epithelium commences very abruptly at the cardiac orifice, where the cells suddenly change in character from the stratified epithelium of the œsophagus. The cells are elongated, and consist of two parts, the inner or attached portions being granular, and the outer or free parts being clear and occupied by a muco-albuminous substance.

The gastric glands are of two kinds, which differ from each other in structure, and it is believed also in the nature of their secretion. They are named respectively *pyloric* and *cardiac* or *oxyntic glands*. They are both tubular in character, and are formed of a delicate basement-membrane, lined by epithelium. The basement-membrane consists of flattened transparent endothelial cells, with

Fig. 726.—Pyloric gland.   Fig. 727.—Cardiac gland.

processes which extend between and support the epithelium. The *pyloric glands* (fig. 726) are most numerous at the pyloric end of the stomach, and from this fact have received their name. They consist of two or three short, closed tubes opening into a common duct, the external orifice of which is situated at the bottom of an alveolus. The cæcal tubes are wavy, and are of about equal length with the duct. The tubes and duct are lined throughout with epithelium, the duct being lined by columnar cells, continuous with the epithelium lining the surface of the mucous membrane of the stomach, the tubes with shorter and more cubical cells which are finely granular. The *cardiac glands* (fig. 727) are found all over the surface of the stomach, but occur most numerously at the cardiac end. Like the pyloric glands they consist of a duct, into which open two or more cæcal tubes. The duct, however, in these glands is shorter than in the other variety, sometimes not amounting to more than one-sixth of the whole length of the gland ; it is lined throughout by columnar epithelium. At the point where the terminal tubes open into the duct, and which is termed the neck, the epithelium alters, and consists of short columnar or polyhedral, granular cells, which almost fill the tube, so that the lumen becomes suddenly constricted;

and is continued down as a very fine channel. They are known as the *chief* cells or *central* cells of the glands. Between these cells and the basement-membrane are found other darker granular-looking cells, studded throughout the tube at intervals, and giving it a beaded or varicose appearance. These are known as the *parietal* or *oxyntic* cells. Between the glands the mucous membrane consists of a connective-tissue framework, with lymphoid tissue. In places, this latter tissue, especially in early life, is collected into little masses, which to a certain extent resemble the solitary glands of the intestine, and are by some termed the *lenticular* glands of the stomach. They are not, however, so distinctly circumscribed as the solitary glands. Beneath the mucous membrane, and between it and the submucous coat, is a thin stratum of involuntary muscular fibre (*muscularis mucosæ*), which in some parts consists only of a single longitudinal layer; in others of two layers, an inner circular, and an outer longitudinal.

Vessels and Nerves.—The arteries supplying the stomach are : the gastric, the pyloric and right gastro-epiploic branches of the hepatic, the left gastro-epiploic and vasa brevia from the splenic. They supply the muscular coat, ramify in the submucous coat, and are finally distributed to the mucous membrane. The arrangement of the vessels in the mucous membrane is somewhat peculiar. The arteries break up at the base of the gastric tubules into a plexus of fine capillaries which run upwards between the tubules, anastomosing with each other, and ending in a plexus of larger capillaries, which surround the mouths of the tubes, and also form hexagonal meshes around the alveoli. From these latter the *veins* arise, and pursue a straight course downwards, between the tubules, to the submucous tissue ; they terminate either in the splenic and superior mesenteric veins, or directly in the portal vein. The *lymphatics* are numerous ; they consist of a superficial and deep set, which pass to the lymphatic glands found along the two curvatures of the organ. The *nerves* are the terminal branches of the right and left pneumogastric, the former being distributed upon the back, and the latter upon the front part of the organ. A great number of branches from the solar plexus of the sympathetic are also distributed to it.

Surface Form.—The stomach lies for the most part in the left hypochondriac region, but also slightly in the epigastric region, and is partly in contact with the abdominal wall, partly under cover of the lower ribs on the left side, and partly under the left lobe of the liver. Its cardiac orifice corresponds to the seventh left costal cartilage, about an inch from the sternum. The pyloric orifice would be pierced by a needle passed through the abdominal wall, five centimetres (two inches) below the junction of the right seventh costal cartilage with the sternum, to the disc between the last dorsal and the first lumbar vertebra (Macalister). The fundus of the stomach reaches as high as the level of the sixth costal cartilage of the left side, being a little below and behind the apex of the heart. The portion of the stomach which is in contact with the abdominal wall, and is therefore accessible for opening in the operations of gastrotomy and gastrostomy, is represented by a triangular space, the base of which is formed by a line drawn from the tip of the tenth costal cartilage on the left side to the tip of the ninth costal cartilage on the right, and the sides by two lines drawn from the extremity of the eighth costal cartilage on the left side to the ends of the base line.

Surgical Anatomy.—Operations on the stomach are frequently performed. By 'gastrotomy' is meant an incision into the stomach for the removal of a foreign body, the opening being immediately afterwards closed—in contradistinction to 'gastrostomy,' the making of a more or less permanent fistulous opening. *Gastrotomy* is probably best performed by an incision in the linea alba, especially if the foreign body is large, by a cut from the ensiform cartilage to the umbilicus ; but may be performed by an incision over the foreign body itself, where this can be felt, or by one of the incisions for gastrostomy, to be mentioned immediately. The peritoneal cavity is opened, and the point at which the stomach is to be incised decided upon. This portion is then brought out of the abdominal wound and sponges carefully packed around. The stomach is now opened by a transverse incision and the foreign body extracted. The wound in the stomach is then closed by Lembert's sutures, i.e. by sutures passed through the peritoneal and muscular coats in such a way that the peritoneal surfaces on each side of the wound are brought into apposition, and in this way the wound is closed. *Gastrostomy* was formerly done in two stages by the *direct* method. The first stage consisted in opening the abdomen, drawing up the stomach into the external wound, and fixing it there ; and the second stage, performed from two to four days afterwards, consisted in opening the stomach. The operation is now done by a *valvular* method. An incision is commenced opposite the eighth intercostal space, two inches from the median line, and carried downwards for three inches. By this incision the fibres of the Rectus muscle are exposed and these are separated from each other in the same line with a steel director. The posterior layer of the sheath, the Transversalis muscle and fascia, and the peritoneum are then divided, and the peritoneal cavity opened. The anterior wall of the stomach is now seized and drawn out of the wound and a silk suture passed through its muscular and serous coats at the point selected for opening the viscus. This is held by an assistant so that a long conical diverticulum of the stomach protrudes from the external wound, and the parietal peritoneum and the posterior layer of the sheath of the Rectus are sutured to it. A second incision is made through the skin, over the margin of the costal cartilage, above and a little to the outer side of the first incision. With a pair of dressing forceps a track is made under the skin through the subcutaneous tissue from the one opening to the other, and the diverticulum of the stomach is drawn along this track by means of the suture inserted into it, so that its apex appears at the second opening. A small perforation is now made into the stomach through this protruding apex, and its margins carefully and accurately sutured to the margin of the external wound. The remainder of this incision and the whole of the first incision are then closed in the ordinary way and the wound dressed.

In cases of gastric ulcer perforation sometimes takes place, and this was formerly regarded as an almost fatal complication. In the present day, by opening the abdomen and closing the perforation, which is generally situated on the anterior surface of the stomach, a considerable percentage of cases is cured, provided the operation is done within twelve or fifteen hours after the perforation has taken place. The opening is best closed by bringing the peritoneal surfaces on either side into apposition by means of Lembert's sutures.

Excision of the pylorus has occasionally been performed, but the results of this operation are by no means favourable, and, in cases of cancer of the pylorus, before operative proceedings are undertaken, the tumour has become so fixed and has so far implicated surrounding parts that removal of the pylorus is impossible and gastro-enterostomy has to be substituted. The object of this operation is to make a fistulous communication between the stomach, on the cardiac side of the disease, and the small intestine, as high up as is possible.

In cases of cancer of the stomach involving other parts than the pylorus, the question of removing the whole or greater part of the stomach has to be considered. This operation has been performed by Schlatter and others with success.

Digital dilatation of the pylorus for simple stricture was first performed by Loreta. He exposed the stomach and opened it by a transverse incision near the pylorus. He then inserted the forefingers of both hands and passed these through the pylorus and stretched it with some degree of force. The operation has now, however, dropped out of use and been replaced by pyloroplasty. This consists in making a longitudinal incision from the stomach through the pylorus into the duodenum, and converting this longitudinal incision into a transverse one by traction at the centre of the incision, and retaining it permanently in this position by sutures.

The stomach is seldom ruptured from external violence, on account of its protected position. If it occurs it is when the organ is distended with food. The stomach is sometimes injured in gunshot wounds. There is intense shock and severe pain, localised at first to the seat of the injury, but soon radiating over the whole abdomen. The treatment consists in opening the peritoneal cavity, clearing away all the extravasated contents of the stomach and repairing the rent.

THE SMALL INTESTINE

The small intestine is a convoluted tube, extending from the pylorus to the ileo-cæcal valve, where it terminates in the large intestine. It is about twenty feet in length,* and gradually diminishes in size from its commencement to its termination. It is contained in the central and lower parts of the abdominal cavity, and is surrounded above and at the sides by the large intestine ; a portion of it extends below the brim of the pelvis and lies in front of the rectum ; it is in relation, in front, with the great omentum and abdominal parietes ; and connected to the spine by a fold of peritoneum, the mesentery. The small intestine is divisible into three portions : the duodenum, the jejunum, and ileum.

The **duodenum** has received its name from being about equal in length to the breadth of twelve fingers (ten inches). It is the shortest, the widest, and the most fixed part of the small intestine, and has no mesentery, being only

---

* Treves states that, in one hundred cases, the average length of the small intestine in the adult male was 22 feet 6 inches, and in the adult female 23 feet 4 inches : but that it varies very much, the extremes in the male being 31 feet 10 inches in one case, and 15 feet 6 inches in another, a difference of over 15 feet. He states that he has convinced himself that the length of the bowel is independent, in the adult, of age, height, and weight.

partially covered by peritoneum. Its course presents a remarkable curve, some-what of the shape of an imperfect circle, so that its termination is not far removed from its starting-point.

In the adult the course of the duodenum is as follows : commencing at the pylorus it passes backwards, upwards, and to the right, beneath the quadrate lobe of the liver to the neck of the gall-bladder, varying slightly in direction according to the degree of distension of the stomach : it then takes a sharp curve and descends along the right margin of the head of the pancreas, for a variable distance, generally to the level of the upper border of the body of the fourth lumbar vertebra. It now takes a second bend, and passes from right to left across the vertebral column, having a slight inclination upwards ; and on the left side of the vertebral column it ascends for about an inch, and then terminates opposite the second lumbar vertebra in the jejunum. As it unites with the jejunum it turns abruptly forwards, forming the *duodeno-jejunal flexure.* From the above description it will be seen that the duodenum may be divided into four portions : superior, descending, transverse, and ascending.

The first or *superior portion* (fig. 728) is about two inches in length. Begin-ning at the pylorus it ends at the neck of the gall-bladder. It is the most movable of the four portions. It is almost completely covered by peritoneum derived from the two layers of the lesser omentum, but a small part of its posterior surface near the neck of the gall-bladder and the inferior vena cava is uncovered. It is in such close relation with the gall-bladder that it is usually found to be stained by bile after death, especially on its anterior surface. It is in relation above and in front with the quadrate lobe of the liver and the gall-bladder ; behind with the gastro-duodenal artery, the common bile-duct, and the vena porta ; and below with the head and neck of the pancreas.

The second or *descending portion* is between three and four inches in length, and extends from the neck of the gall-bladder on a level with the first lumbar vertebra along the right side of the vertebral column as low as the upper border of the body of the fourth lumbar vertebra. It is crossed in its middle third by the transverse colon, the posterior surface of which is uncovered by peritoneum and is connected to the duodenum by a small quantity of connective tissue (fig. 728). The portions of the descending part of the duodenum above and below this interspace are named the supra- and infra-colic portions, and are covered in front by peritoneum. The infra-colic part is covered by the right leaf of the mesentery. Posteriorly the descending portion of the duodenum is uncovered by peritoneum. It is in relation, in front, with the transverse colon, and above this with the liver ; behind with the inner part of the right kidney, to which it is connected by loose areolar tissue, the renal vessels, the vena cava inferior, and the Psoas magnus below ; at its inner side is the head of the pancreas, and the common bile-duct ; to its outer side is the hepatic flexure of the colon. The common bile-duct and the pancreatic duct perforate the inner side of this portion of the intestine obliquely, some three or four inches below the pylorus. The relations of the second part of the duodenum to the right kidney present considerable variations.

The third or *transverse portion* (pre-aortic portion) is from two to three inches in length. It commences at the right side of the upper border of the fourth lumbar vertebra and passes from right to left, with a slight inclination upwards, in front of the great vessels and crura of the Diaphragm, and ends in the fourth portion in front of the abdominal aorta. It is crossed by the superior mesenteric vessels and the mesentery. Its front surface is covered by peritoneum, except near the middle line, where it is crossed by the superior mesenteric vessels. Its posterior surface is uncovered by peritoneum, except towards its left extremity, where the posterior layer of the mesentery may sometimes be found covering it to a variable extent. This surface rests upon the aorta, the vena cava inferior, and the right crus of the Diaphragm. By its upper surface this portion of the duodenum is in relation with the head of the pancreas.

The fourth or *ascending portion* of the duodenum is about an inch in length. It ascends on the left side of the vertebral column and aorta, as far as the level of the upper border of the second lumbar vertebra, where it turns abruptly forwards to become the jejunum, forming the *duodeno-jejunal flexure.* It is covered entirely in front and partly at the sides by peritoneum, derived from the left portion of the mesentery. It touches the left kidney, slightly overlapping its inner margin, and rests upon the left crus of the Diaphragm.

FIG. 728.—Relations of duodenum, pancreas and spleen.
(From a cast by Birmingham.*)

The dotted lines represent the attachment of the transverse mesocolon.

The first part of the duodenum, as stated above, is somewhat movable, but the rest is practically fixed and is bound down to neighbouring viscera and the

* In the subject from which the cast was taken the left kidney was lower than normal.

posterior abdominal wall by the peritoneum. In addition to this, the fourth part of the duodenum and the duodeno-jejunal flexure are fixed by a structure to which the name of *musculus suspensorius duodeni* has been given. This structure commences in the connective tissue around the cœliac axis and left crus of the Diaphragm, and passes downwards to be inserted into the superior border of the duodeno-jejunal curve and a part of the ascending duodenum, and from this it is continued into the mesentery. It possesses, according to Treitz, plain muscular fibres mixed with the fibrous tissue, of which it is principally made up. It is of little importance as a muscle, but acts as a suspensory ligament.

Vessels and Nerves.—The *arteries* supplying the duodenum are the pyloric and pancreatico-duodenal branches of the hepatic, and the inferior pancreatico-duodenal branch of the superior mesenteric. The *veins* terminate in the splenic and superior mesenteric. The *nerves* are derived from the solar plexus.

Jejunum and Ileum.—The remainder of the small intestine from the termination of the duodenum is named *jejunum* and *ileum*; the former term being given to the upper two-fifths and the latter to the lower three-fifths. There is no morphological line of distinction between the two, and the division is arbitrary; but at the same time it must be noted that the character of the intestine gradually undergoes a change from the commencement of the jejunum to the termination of the ileum, so that a portion of the bowel taken from these two situations would present characteristic and marked differences. These are briefly as follows :

The *jejunum*, which derives its name from the Latin word *jejunus* (empty), because it was formerly supposed to be empty after death, is wider, its diameter being about an inch and a half, and is thicker, more vascular, and of a deeper colour than the ileum, so that a given length weighs more. Its valvulæ conniventes are large and thickly set, and its villi are larger than in the ileum. The glands of Peyer are almost absent in the upper part of the jejunum, and in the lower part are less frequently found than in the ileum, and are smaller and tend to assume a circular form. By grasping the jejunum between the finger and thumb the valvulæ conniventes can be felt through the walls of the gut ; these being absent in the lower part of the ileum, it is possible in this way to distinguish the upper from the lower part of the small intestine.

The *ileum*, so called from the Greek word εἰλεῖν (to twist), on account of its numerous coils and convolutions, is narrow, its diameter being an inch and a quarter, and its coats thinner and less vascular than those of the jejunum. It possesses but few valvulæ conniventes, and they are small and disappear entirely towards its lower end, but Peyer's patches are larger and more numerous. The jejunum for the most part occupies the umbilical and left iliac regions, while the ileum occupies chiefly the umbilical, hypogastric, right iliac, and pelvic regions. Its terminal part usually lies in the pelvis, from which it ascends over the right iliac vessels and Psoas muscle ; it ends in the right iliac fossa by opening into the inner side of the commencement of the large intestine. The jejunum and ileum are attached to the posterior abdominal wall by an extensive fold of peritoneum, the *mesentery*, which allows the freest motion, so that each coil can accommodate itself to changes in form and position. The mesentery is fan-shaped ; its posterior border or root, about six inches in length, is attached to the posterior abdominal wall from the left side of the body of the second lumbar vertebra to the right iliac fossa, crossing successively the third part of the duodenum, the aorta, the inferior vena cava, the ureter, and right Psoas muscle (fig. 722). Its breadth between its vertebral and intestinal borders is about eight inches, and is greater in the middle than at its upper and lower extremities. According to Lockwood it tends to increase in breadth as age advances. Between the two layers of which it is composed are contained blood-vessels, nerves, lacteals, and lymphatic glands, together with a variable amount of fat.

Meckel's diverticulum.—This consists of a pouch which projects from the lower part of the ileum in about 2 per cent. of subjects. Its average position is about three feet above the ileo-cæcal valve, and its average length about two inches. Its calibre is generally similar to that of the ileum, and its blind extremity may be free or may be connected with the abdominal wall or with some other portion of the intestine by a fibrous band. It represents the remains of the proximal part of the vitelline or omphalo-mesenteric duct, the duct of communication between the umbilical vesicle and the alimentary canal in early fœtal life.

Structure.—The wall of the small intestine is composed of four coats : serous, muscular, areolar, and mucous.

The *serous coat* is derived from the peritoneum. The first or ascending portion of the duodenum is almost completely surrounded by this membrane near its pyloric end, but is only covered in front at the other extremity; the second or descending portion is covered by it in front, except where it is carried off by the transverse colon; and the third or transverse portion lies behind the peritoneum, which passes over it, without being closely incorporated with the other coats of this part of the intestine, and is separated from it in and near the middle line by the superior mesenteric vessels. The remaining portion of the small intestine is surrounded by the peritoneum, excepting along its attached or mesenteric border; here a space is left for the vessels and nerves to pass to the gut.

The *muscular coat* consists of two layers of fibres, an external or longitudinal, and an internal or circular layer. The *longitudinal fibres* are thinly scattered over the surface of the intestine, and are more distinct along its free border. The *circular fibres* form a thick, uniform layer; they surround the cylinder of the intestine in the greater part of its circumference, and are composed of plain muscle-cells of considerable length. The muscular coat is thicker at the upper than at the lower part of the small intestine.

The *areolar* or *submucous coat* connects together the mucous and muscular layers. It consists of loose, filamentous areolar tissue, which forms a nidus for the subdivision of the nutrient vessels, previous to their distribution to the mucous surface.

The *mucous membrane* is thick and highly vascular at the upper part of the small intestine, but somewhat paler and thinner below. It consists of the following structures : next the areolar or submucous coat is a layer of unstriped muscular fibres, the *muscularis mucosæ*; internal to this is a quantity of retiform tissue, enclosing in its meshes lymph-corpuscles, and in which the blood-vessels and nerves ramify. Lastly, a basement-membrane, supporting a single layer of epithelial cells, which throughout the intestines are columnar in character. They are granular in appearance, and each possesses a clear oval nucleus. At their superficial or unattached ends they present a distinct layer of highly refracting material, marked by vertical striæ, which were formerly believed to be minute channels, by which the chyle was taken up into the interior of the cell, and by them transferred to the lacteal vessels of the mucous membrane.

The mucous membrane presents for examination the following structures, contained within it or belonging to it :

Valvulæ conniventes.		Duodenal glands.
Villi.	Glands	Solitary glands.
Simple follicles.		Peyer's or Agminated glands.

The **valvulæ conniventes** (valves of Kerkring) are large folds or valvular flaps projecting into the lumen of the bowel. They are composed of reduplications or folds of the mucous membrane, the two layers of the fold being bound together by submucous tissue; they contain no muscular fibres, and, unlike the folds in the stomach, they are permanent, and are not obliterated when the intestine is distended. The majority extend transversely across the cylinder of the intestine for about one-half or two-thirds of its circumference, but some form complete circles, and others have a spiral direction; the latter usually extend a little more than once round the bowel, but occasionally two or three times. The spiral arrangement is the characteristic one of the shark family of fishes. The larger folds are about one-third of an inch in depth at their broadest part; but the greater number are of smaller size. The larger and smaller folds alternate with each other. They are not found at the commencement of the duodenum, but begin to appear about one or two inches beyond the pylorus. In the lower part of the descending portion, below the point where the bile and pancreatic ducts enter the intestine, they are very large and closely approximated. In the transverse portion of the duodenum and upper half of the jejunum they are large and numerous, but from this point, down to the middle of the ileum, they diminish considerably in size. In the lower part of the ileum they almost entirely disappear; hence the comparative thinness of this portion of the intestine, as compared with the duodenum and jejunum. The valvulæ conniventes retard the passage of the food along the intestines, and afford a more extensive surface for absorption.

The **villi** are minute, highly vascular processes, projecting from the mucous membrane of the small intestine throughout its whole extent, and giving to its surface a velvety appearance. According to Rauber, they are short and leaf-shaped in the duodenum, tongue-shaped in the jejunum, and filiform in the ileum. They are largest and most numerous in the duodenum and jejunum, and become fewer and smaller in the ileum. Krause estimates their number in the

FIG. 729.—Diagrammatic section of a villus. (Watney.)

ep. Epithelium only partially shaded in. *l.* Central chyle-vessel; the cells forming the vessel have been less shaded to distinguish them from the cells of the parenchyma of the villus. *m.* Muscle-fibres running up by the side of the chyle-vessel. It will be noticed that each muscle-fibre is surrounded by the reticulum, and by this reticulum the muscles are attached to the cells forming the membrana propria, as at *e'*, or to the reticulum of the villus. *lc.* Lymph-corpuscles, marked by a spherical nucleus and a clear zone of protoplasm. *l'.* Upper limit of the chyle-vessel. *e, e, e'.* Cells forming the membrana propria. It will be seen that there is hardly any difference between the cells of the parenchyma, the endothelium of the *upper* part of the chyle-vessel, and the cells of the membrana propria. *v.* Blood-vessels. *z.* Dark line at the base of the epithelium formed by the reticulum. It will be seen that the reticulum penetrates between all the other elements of the villus. The reticulum contains thickenings or 'nodal points.' The diagram shows that the cells of the upper part of the villus are larger and contain a larger zone of protoplasm than those of the lower part. The cells of the upper part of the chyle-vessel differ somewhat from those of the lower part, in that they more nearly resemble the cells of the parenchyma.

upper part of the small intestine at from fifty to ninety in a square line; and in the lower part from forty to seventy; the total number for the whole length of the intestine being about four millions.

Structure of the villi (fig. 729).—The structure of the villi has been studied by many eminent anatomists. The description here followed is that of Watney,[*] whose researches have an important bearing on the physiology of the absorption of fat, which is the peculiar function of this part of the intestine.

The essential parts of a villus are: the lacteal vessel, the blood-vessels, the epithelium, the basement-membrane, and muscular tissue of the mucosa, these structures being supported and held together by retiform lymphoid tissue.

These structures are arranged in the following manner. Situated in the

* *Phil. Trans.* vol. clxvi. pt. ii.

centre of the villus is the lacteal, terminating near the summit in a blind extremity; running along this vessel are unstriped muscular fibres; surrounding it is a plexus of capillary vessels, the whole being enclosed by a basement-membrane, and covered by columnar epithelium. Those structures which are contained within the basement-membrane—namely, the lacteal, the muscular tissue, and the blood-vessels—are surrounded and enclosed by a delicate reticulum which forms the matrix of the villus, and in the meshes of which are found large flattened cells, each with an oval nucleus, and, in smaller numbers, lymph-corpuscles. These latter are to be distinguished from the larger cells of the villus by their behaviour with reagents, by their size, and by the shape of their nucleus, which is spherical. Transitional forms, however, of all kinds are met with between the lymph-corpuscle and the proper cells of the villus. Nerve-fibres are contained within the villi: they form ramifications throughout the reticulum.

The *lacteals* are in some cases double, and in some animals multiple. Situated in the axis of the villi, they commence by dilated cæcal extremities

FIG. 730.—Villi of small intestine.   (Cadiat.)

Capillaries

Lymph-trunk

Capillaries

Lymph-trunk

Small artery   Lymphatic plexus

near to, but not quite at, the summit of the villus. The walls are composed of a single layer of endothelial cells, the interstitial substance between the cells being continuous with the reticulum of the matrix.

The *muscular fibres* are derived from the muscularis mucosæ, and are arranged in longitudinal bundles around the lacteal vessel, extending from the base to the summit of the villus, and giving off, laterally, individual muscle-cells, which are enclosed by the reticulum, and by it are attached to the basement-membrane.

The *blood-vessels* form a plexus between the lacteal and the basement-membrane, and are enclosed in the reticular tissue. In the interstices of the capillary plexus, which they form, are contained the cells of the villus.

These structures are surrounded by the basement-membrane, which is made up of a stratum of endothelial cells, and upon which is placed a layer of columnar epithelium. The reticulum of the matrix is continuous through the basement-membrane (that is, through the interstitial substance between the individual endothelial cells) with the interstitial cement-substance of the columnar cells on the surface of the villus. Thus we are enabled to trace a direct continuity between the interior of the lacteal and the surface of the villus by means of the reticular tissue, and it is along this path that the chyle passes in the process of absorption by the villi. That is to say, it passes first of all into the columnar epithelial cells, and, escaping from them, is carried into the reticulum of the villus, and thence into the central lacteal.

The **simple follicles**, or *crypts of Lieberkühn* (figs. 731, 732), are found in considerable numbers over every part of the mucous membrane of the small intestine. They consist of minute tubular depressions of the mucous membrane, arranged perpendicularly to the surface, upon which they open by small circular apertures. They may be seen with the aid of a lens, their orifices appearing as minute dots, scattered between the villi. Their walls are thin, consisting of a basement-membrane lined by columnar epithelium, and covered on their exterior by capillary vessels.

FIG. 731.—Longitudinal section of crypts of Lieberkühn. Goblet-cells seen among the columnar epithelial cells. (Klein and Noble Smith.)

The **duodenal** or **Brunner's glands** are limited to the duodenum. They are small, flattened, granular bodies embedded in the submucous areolar tissue, and open upon the surface of the mucous membrane by minute excretory ducts. They are largest and most numerous near the pylorus, forming an almost complete layer in the first and upper half of the second portions of the duodenum. They then begin to diminish in number, and practically disappear at the junction of the duodenum and jejunum. They are small compound acino-tubular glands, and much resemble the small glands which are found in the mucous membrane of the mouth. They are believed by Watney to be direct continuations of the pyloric glands of the stomach. They consist of a number of tubular alveoli, lined by epithelium, and opening by a single duct on the inner surface of the intestine.

The **solitary glands** (*glandulæ solitariæ*) are found scattered throughout the mucous membrane of the small intestine, but are most numerous in the lower part of the ileum. They are small, round, whitish bodies, from half a line to a line in diameter. Their free surface is covered with villi, and each gland is surrounded by the openings of the follicles of Lieberkühn. They are now recognised as lymph-follicles, and consist of a dense interlacing retiform tissue closely packed with lymph-corpuscles, and permeated with an abundant capillary network (fig. 733). The interspaces of the retiform tissue are continuous with larger lymph-spaces which surround the gland, through which they communicate with the lacteal system. They are situated partly in the submucous tissue, partly in the mucous membrane, where they form slight projections of its epithelial layer, after having penetrated the muscularis mucosæ. The villi which are situated on them are generally absent from the very summit (or 'cupola,' as Frey calls it) of the gland.

FIG. 732.—Transverse section of crypts of Lieberkühn. (Klein and Noble Smith.)

Peyer's glands (agminated glands) (figs. 733 to 736) may be regarded as aggregations of solitary glands, forming circular or oval patches from twenty to thirty in number, and varying in length from half an inch to four inches. They are largest and most numerous in the ileum. In the lower part of the jejunum they are small, of a circular form, and few in number. They are occasionally seen in the duodenum. They are placed lengthwise in the intestine, and are situated in the portion of the tube most distant from the attachment of the mesentery. Each patch is formed of a group of the above-described solitary glands covered with mucous membrane, and in almost every respect is similar in structure to them. Each patch is surrounded by a circle of the crypts of Lieberkühn, but the patches do not, as a rule, possess villi on their free surfaces. They are best marked in the young subject, become indistinct in middle age, and sometimes disappear altogether in advanced life. They are freely supplied with blood-vessels, which form an abundant plexus around each follicle and give off fine branches which permeate the lymphoid tissue in the interior of the follicle. The lacteal plexuses which are found throughout the small intestine are especially abundant around these patches; here they form rich plexuses with sinuses around the glands (fig. 736).

Fig. 733.—Transverse section through the equatorial plane of three of Peyer's follicles from the rabbit.

... Capillary network

Large circular vessel

Vessels and Nerves.—The jejunum and ileum are supplied by the superior mesenteric artery, the branches of which, having reached the attached border of the bowel, run between the serous and muscular coats, with frequent inosculations to the free border, where they also anastomose with other branches running round the opposite surface of the gut. From these vessels numerous branches are given off, which pierce the muscular coat, supplying it and forming an intricate plexus in the submucous tissue. From this plexus minute vessels pass to the glands and villi of the mucous membrane. The veins have a similar course and arrangement to the arteries. The *lymphatics of the small-intestines* (lacteals) are arranged in two sets, those of the mucous membrane, and those of the muscular coat. The lymphatics of the villi commence in these structures in the manner described above, and form an intricate plexus in the mucous and submucous tissue, being joined by the lymphatics from the

Fig. 734.—Patch of Peyer's glands. From the lower part of the ileum.

Fig. 735.—A portion of a Peyer's patch magnified.

lymph-spaces at the bases of the solitary glands, and from this pass to larger vessels at the mesenteric border of the gut. The lymphatics of the muscular coat are situated to a great extent between the two layers of muscular fibres, where they form a close plexus, and throughout their course communicate freely with the lymphatics from the mucous membrane, and empty themselves in the same manner into the commencement of the lacteal vessels at the attached border of the gut.

The *nerves of the small intestines* are derived from the plexuses of sympathetic nerves around the superior mesenteric artery. From this source they run to a

FIG. 736.—Vertical section of one of Peyer's patches from man, injected through its lymphatic canals.

a. Villi with their chyle-passages. *b.* Follicles of Lieberkühn. *c.* Muscularis mucosæ. *d.* Cupola or apex of solitary glands. *e.* Mesial zone of glands. *f.* Base of glands. *g.* Points of exit of the chyle-passages from the villi, and entrance into the true mucous membrane. *h.* Retiform arrangement of the lymphatics in the mesial zone. *i.* Course of the latter at the base of the glands. *k.* Confluence of the lymphatics opening into the vessels of the submucous tissue. *l.* Follicular tissue of the latter.

FIG. 737.—Meissner's plexus. (Klein and Noble Smith.)

Multipolar ganglion-cells

Single ganglion-cell

plexus of nerves and ganglia situated between the circular and longitudinal muscular fibres (*Auerbach's plexus*), from which the nervous branches are distributed to the muscular coats of the intestine. From this plexus a secondary plexus (*Meissner's plexus*) is derived, and is formed by branches which have perforated the circular muscular fibres (fig. 737). This plexus lies between the muscular and mucous coats of the intestine. It is also gangliated, and from it the ultimate fibres pass to the muscularis mucosæ and to the villi and mucous membrane.

THE LARGE INTESTINE

The large intestine extends from the termination of the ileum to the anus. It is about five feet in length, being one-fifth of the whole extent of the intestinal canal. It is largest at its commencement at the cæcum, and gradually diminishes as far as the rectum, where there is a dilatation of considerable size just above the anus. It differs from the small intestine in its greater size, its more fixed position, its sacculated form, and in possessing certain appendages to its external coat, the *appendices epiploicæ*. Further, its longitudinal muscular fibres do not form a continuous layer around the gut but are arranged in three longitudinal bands or *tæniæ*. The large intestine, in its course, describes an arch, which surrounds the convolutions of the small intestine. It commences in the right iliac region, in a dilated part, the *cæcum*. It ascends through the right lumbar and hypochondriac regions to the under surface of the liver; it here takes a bend (the *hepatic flexure*) to the left, and passes transversely across the abdomen on the confines of the epigastric and umbilical regions, to the left hypochondriac region; it then bends again (the *splenic flexure*), and descends through the left lumbar region to the left iliac fossa, where it becomes convoluted, and forms the *sigmoid flexure*; finally it enters the pelvis, and descends along its posterior wall to the anus. The large intestine is divided into the cæcum, colon, and rectum.

FIG. 738.—The cæcum and colon laid open to show the ileo-cæcal valve.

The **cæcum** (*cæcus, blind*) is the large blind pouch, or *cul-de-sac*, situated below the ileo-cæcal valve, in which the large intestine commences (fig. 738). Its blind end is directed downwards, and its open end upwards, communicating directly with the colon, of which this blind pouch appears to be the beginning or head, and hence the old name of *caput cæcum coli* was applied to it. Its size is variously estimated by different authors, but on an average it may be said to be two and a half inches in length and three in breadth. It is situated in the right iliac fossa, above the outer half of Poupart's ligament: it rests on the Ilio-psoas muscle and lies immediately behind the abdominal wall. As a rule, it is entirely enveloped by peritoneum, but in a certain number of cases (5 per cent., Berry) the peritoneal covering is not complete, so that the upper part of the posterior surface is uncovered and connected to the iliac fascia by connective tissue. The cæcum lies quite free in the abdominal cavity and enjoys a considerable amount of movement, so that it often becomes herniated down the right inguinal canal, and has occasionally been found in an inguinal hernia on the left side. The cæcum varies in shape, but, according to Treves, in man it may be classified under one of four types. In early fœtal life it is short, conical, and broad at the base, with its apex turned upwards and inwards towards the ileo-cæcal junction. It then resembles the cæcum of some of the monkey tribe, e.g. Mangabey monkey. As the fœtus grows the cæcum increases in length more than in breadth, so that it forms a longer tube than in the primitive form and without the broad base, but with the same inclination inwards of the apex towards the ileo-cæcal junction. This form is seen in others of the monkey tribe : e.g. the spider monkey. As development goes on, the lower part of the tube ceases to grow and the upper part becomes greatly increased, so that at birth there is a narrow tube, the vermiform appendix, hanging from a conical projection, the cæcum. This is the infantile form, and as it may persist throughout life, in about 2 per cent. of cases, it is regarded by Treves as the *first* of his four types of human cæca. The cæcum is conical and the appendix rises from its apex. The three longitudinal bands start from the appendix and are equidistant from each other. In the *second* type, the conical cæcum has become quadrate by the growing out of a saccule on either side of the anterior longitudinal band. These saccules are of equal size, and the appendix arises from between them, instead of from the apex of a cone. This type is found in about 3 per cent. of cases. The *third* type is the normal type of man. Here the two saccules, which in the second type were uniform, have grown at unequal rates : the right with greater rapidity than the left. In consequence of this an apparently new apex has been formed by the growing downwards of the right saccule, and the original apex, with the appendix attached, is pushed over to the left towards the ileo-cæcal junction. The three longitudinal bands still start from the base of the appendix, but they are now no longer equidistant from each other, because the right saccule has grown between the anterior and postero-external bands, pushing them over to the left. This type occurs in about 90 per cent. of cases. The *fourth* type is merely an exaggerated condition of the third ; the right saccule is still larger, and at the same time the left saccule has become atrophied, so that the original apex of the cæcum, with the appendix, is close to the ileo-cæcal junction, and the anterior band courses inwards to the same situation. This type is present in about 4 per cent. of cases.

The **vermiform appendix** is a long, narrow, worm-shaped tube, which starts from what was originally the apex of the cæcum, and may pass in several directions : upwards behind the cæcum ; to the left behind the ileum and mesentery ; or downwards and inwards into the true pelvis. It varies from one to nine inches in length, its average being about three inches. It is retained in position by a fold of peritoneum, the *meso-appendix*, derived from the left leaf of the mesentery. This fold, in the majority of cases, is more or less triangular in shape, and as a rule extends along the entire length of the tube. Between its two layers lies a considerable branch of the ileo-colic artery, the appendicular artery. The canal of the appendix is small, extends throughout the whole length of the tube, and communicates with the cæcum by an orifice which is placed below and behind the ileo-cæcal opening. It is sometimes guarded, according to Gerlac by a semilunar valve formed by a fold of mucous membrane, but this is by no means constant. Its coats are the same as those of the intestine : serous, muscular, submucous, and mucous.

Structure.—The *serous* coat forms a complete investment for the tube, except along the narrow line of attachment of its mesentery in its proximal two-thirds. The *longitudinal muscular fibres* entirely surround the tube as one continuous layer, and do not form three bands as in the greater part of the large intestine. The *submucous tissue* contains an abundant supply of adenoid tissue, especially in young subjects. The *mucous membrane* is lined by columnar epithelium and resembles that of the rest of the large intestine, but the simple follicles are few in number.

It is stated that the vermiform appendix tends to undergo obliteration as an involution change of a functionless organ.

The **ileo-cæcal valve** (*valvula Bauhini*).—The lower end of the ileum terminates by opening into the inner and back part of the large intestine, at the point of junction of the cæcum with the colon. The opening is guarded by a valve, consisting of two segments, an upper or colic and lower or cæcal, which project into the lumen of the large intestine. If the intestine has been inflated and dried, the segments are of a semilunar shape. The upper one, nearly horizontal in direction, is attached by its convex border to the line of junction of the ileum with the colon ; the lower segment, which is longer and more concave, is attached to the line of junction of the ileum with the cæcum. At each end of the aperture the two segments of the valve coalesce, and are continued as a narrow membranous ridge around the canal for a short distance, forming the *fræna* or *retinacula* of the valve. The left or anterior end of the aperture is rounded; the right or posterior is narrow and pointed. In the fresh condition, or in specimens which have been hardened *in situ*, the segments project as thick cushion-like folds into the lumen of the large gut, while the opening between them may present the appearance of a slit or may be somewhat oval in shape.

Each segment of the valve is formed by a reduplication of the mucous membrane and of the circular muscular fibres of the intestine, the longitudinal fibres and peritoneum being continued uninterruptedly across from one portion of the intestine to the other. When these are divided or removed, the ileum may be drawn outwards, and all traces of the valve will be lost, the ileum appearing to open into the large intestine by a funnel-shaped orifice of large size.

The surface of each segment of the valve directed towards the ileum is covered with villi, and presents the characteristic structure of the mucous membrane of the small intestine; while that turned towards the large intestine is destitute of villi, and marked with the orifices of the numerous tubular glands peculiar to the mucous membrane of the large intestine. These differences in structure continue as far as the free margin of the valve.

When the cæcum is distended, the margins of the opening are approximated so as to prevent any reflux into the ileum. This is believed to be due to tension or stretching of the retinacula of the valve.

The **colon** is divided into four parts: the ascending, transverse, descending, and the sigmoid flexure.

The **ascending colon** is smaller than the cæcum, with which it is continuous. It passes upwards, from its commencement at the cæcum, opposite the ileo-cæcal valve, to the under surface of the right lobe of the liver, on the right of the gall-bladder, where it is lodged in a shallow depression, the impressio colica; here it bends abruptly forwards and to the left, forming the *hepatic flexure*. It is retained in contact with the posterior wall of the abdomen by the peritoneum, which covers its anterior surface and sides, its posterior surface being connected by loose areolar tissue with the Quadratus lumborum muscle, and with the front of the lower and outer part of the right kidney (fig. 739). Sometimes the peritoneum completely invests it, and forms a distinct but narrow mesocolon.* It is in relation, in front, with the convolutions of the ileum and the abdominal parietes.

The **transverse colon**, the longest part of the large intestine, passes transversely from the right hypochondriac region across the abdomen, opposite the confines of the epigastric and umbilical zones, into the left hypochondriac region, where it curves downwards beneath the lower end of the spleen, forming the *splenic flexure*. In its course it describes an arch, the concavity of which is directed backwards towards the vertebral column and a little upwards; hence the name *transverse arch of the colon*. This is the most movable part of the colon, being almost completely invested by peritoneum, and connected to the spine behind by a large and wide duplicature of that membrane, the *transverse mesocolon*. It is in relation, by its upper surface, with the liver and gall-bladder, the great curvature of the stomach, and the lower end of the spleen; by its under surface, with the small intestines; by its anterior surface, with the anterior layers of the great omentum and the abdominal parietes; its posterior surface is in relation from right to left with the second portion of the duodenum, the head of the pancreas, and some of the convolutions of the jejunum and ileum.

The **splenic flexure** is situated at the junction of the transverse and descending parts of the colon, and is in relation with the lower end of the spleen and the tail of the pancreas. It lies at a higher level than, and on a plane posterior to, the hepatic flexure, and is attached to the Diaphragm, opposite the tenth and eleventh ribs, by a peritoneal fold which is named the *phreno-colic* or

---

* Treves states that, after a careful examination of one hundred subjects, he found that in fifty-two there was neither an ascending nor a descending mesocolon. In twenty-two there was a descending mesocolon, but no trace of a corresponding fold on the other side. In fourteen subjects there was a mesocolon to both the ascending and the descending segments of the bowel; while in the remaining twelve there was an ascending mesocolon, but no corresponding fold on the left side. It follows, therefore, that in performing lumbar colotomy a mesocolon may be expected upon the left side in 36 per cent. of all cases, and on the right in 26 per cent.—*The Anatomy of the Intestinal Canal and Peritoneum in Man*, 1885, p. 55.

costo-colic ligament, and which assists in supporting the lower end of the spleen (see page 1063).

The **descending colon** passes downwards through the left hypochondriac and lumbar regions along the outer border of the left kidney. At the lower end of

FIG. 739.—Diagram of the relations of the large intestine and kidney, from behind.

the kidney it turns inwards towards the outer border of the Psoas muscle, along which it descends to the crest of the ilium, where it terminates in the sigmoid flexure. It is retained in position by the peritoneum, which covers its anterior surface and sides, its posterior surface being connected by areolar tissue with the outer border of the left kidney, and with the Quadratus lumborum muscle (fig. 739). It is smaller in calibre and more deeply placed than the ascending colon, and is more frequently covered with peritoneum on its posterior surface than the ascending colon (Treves).

The **sigmoid flexure** is the narrowest part of the colon: it is situated in the left iliac fossa, commencing from the termination of the descending colon, at the margin of the crest of the ilium, and ending in the rectum opposite the left sacro-iliac articulation. It curves in the first place forwards, downwards, and inwards for about two inches and then forms a loop, which varies in length and position and which terminates in the rectum.

Treves and subsequently Jonnesco have pointed out that this description of the sigmoid flexure is inaccurate and that the loop described above does not exist. They include together the sigmoid flexure and the first part of the rectum, which they say form a single loop, which cannot be divided into parts. This loop, which is of the shape of the Greek letter Ω (omega), commences at the margin of the crest of the ilium and terminates opposite the third piece of the sacrum in the middle line. Artificially, it may be divided for purposes of description into two parts: (1) iliac colon, and (2) pelvic colon.

The *iliac colon* commences at the crest of the ilium and terminates at the brim of the true pelvis. It curves forwards, downwards, and inwards, lying in front of the Iliacus and Psoas muscles, and is covered by peritoneum on its sides and anterior surface only.

The *pelvic colon* corresponds to the portion of the sigmoid flexure which forms the loop mentioned above, together with the first part of the rectum. It is in relation behind with the external iliac vessels, the left Pyriformis muscle, and left sacral plexus of nerves, which separate it from the anterior surface of the sacrum. In front, it is separated from the bladder in the male, and the uterus in the female, by convolutions of small intestine. It is completely surrounded by peritoneum, which forms a mesentery for it (*pelvic mesocolon*) and comprises the sigmoid mesocolon and the mesorectum of the older description. In its left layer is the intersigmoid fossa (see page 1065).

The **rectum** is the terminal part of the large intestine, and terminates at the anal orifice. As stated above, its superior limit cannot be determined precisely; but the brim of the true pelvis, opposite the left sacro-iliac joint, is arbitrarily given as its point of commencement. From this point it passes downwards, backwards, and to the right to the level of the third sacral vertebra, where it lies in the middle line. This is the *first part* of the rectum. The *second part* curves forwards and is continued downwards as far as the apex of the prostate gland, about an inch in front of, and a little below, the tip of the coccyx. From this point the bowel is directed downwards and backwards and terminates at the anal orifice. This is the third portion of the rectum, or, as described by Symington, the *anal canal*. It will be seen, therefore, that the rectum presents two antero-posterior curves: the upper one, with its convexity backwards, is due to the conformation of the sacro-coccygeal column. The lower one has its convexity forwards, and is angular. Two lateral curves are also described: one to the right, opposite the junction of the third and fourth sacral vertebræ; and the other to the left, opposite the sacro-coccygeal articulation; they are of little importance.

The length of the rectum is about eight inches. The first part measures four inches, the second three, and the third from one to one and a half, being rather longer in the male than in the female. The calibre of the first part of the rectum is similar to that of the colon, while near the termination of the second part the tube is dilated to form what is known as the *rectal ampulla*. When the tube is empty its anterior and posterior walls are in apposition with each other, forming a transverse slit. The third part of the rectum, or anal canal, is also a slit, with, however, an antero-posterior direction, so that its lateral walls are in apposition (fig. 740).

The first portion of the rectum is surrounded by peritoneum, and is connected to the anterior surface of the sacrum by a double fold, called the *mesorectum*, which is continuous above with the sigmoid mesocolon. The mesorectum is triangular in shape, its apex ending below at the third sacral vertebra; between its two layers are the superior hæmorrhoidal vessels. The second portion has no mesorectum, but is covered in front and laterally by peritoneum at its upper part; gradually the peritoneum leaves its sides, and about an inch above the prostate is reflected from the anterior surface of the bowel on to the posterior wall of the bladder in the male, and the upper fifth of the posterior wall of the vagina in the female, forming the recto-vesical and recto-vaginal pouches respectively. The third portion of the rectum has no peritoneal covering.

The level at which the peritoneum leaves the anterior wall of the rectum to be reflected on to the viscus in front of it is of considerable importance from a surgical point of view, in connection with removal of the lower part of the rectum. It is higher in the male than in the female. In the former the height of the recto-vesical pouch is about three inches: that is to say, the height to which an ordinary index finger can reach from the anus. In the female the height of the recto-vaginal pouch is about two and a quarter inches from the anal orifice.

The *first portion* of the rectum is in relation, behind, with the mesorectum and the superior hæmorrhoidal vessels, the left Pyriformis muscle, and left sacral plexus of nerves, which separate it from the anterior surface of the upper

FIG. 740.—Coronal section through the anal canal. (Symington.)

B. Cavity of bladder. V.D. Vas deferens. S.V. Seminal vesicle. R. Second part of rectum. A.C. Anal canal;
L.A. Levator ani. I.S. Internal sphincter. E.S. External sphincter.

sacral vertebræ; to its left side are the branches of the left internal iliac artery and the left ureter; in front it is separated, in the male, from the posterior surface of the bladder; in the female, from the posterior surface of the uterus and its appendages, by some convolutions of the small intestine, and frequently by the sigmoid flexure of the colon. The *second portion* of the rectum is in relation, behind, with the sacrum and coccyx, and with a fibro-muscular mass which intervenes between it and the coccyx and is named the *ano-coccygeal body* (Symington); in front, in the male, with the recto-vesical pouch, the triangular portion of the base of the bladder, the vesiculæ seminales, and vasa deferentia, and more anteriorly with the under surface of the prostate. In the female, with the posterior wall of the vagina below, and the recto-vaginal pouch above, in which are some convolutions of small intestine. The *third portion* or *anal canal* is invested by the Internal sphincter, supported by the Levatores ani muscles, and surrounded at its termination by the External sphincter; in the empty condition it presents the appearance of an antero-posterior longitudinal slit. Behind it is the ano-coccygeal body, and in front of it are, in the male, the membranous portion and bulb of the urethra and the base of the triangular ligament; and in the female it is separated from the lower end of the vagina by a mass of muscular and fibrous tissue, named the *perineal body*. Laterally is the fat which occupies the ischio-rectal fossæ.

Structure.—The large intestine has four coats : serous, muscular, areolar, and mucous.

The *serous coat* is derived from the peritoneum, and invests the different portions of the large intestine to a variable extent. The cæcum is completely covered by the serous membrane, except in about 5 per cent. of cases, where the upper part of the posterior surface is uncovered. The ascending and descending colons are usually covered only in front and at the sides ; a variable amount of the posterior surface is uncovered.* The transverse colon is almost completely invested, the parts corresponding to the attachment of the great omentum and transverse mesocolon being alone excepted. The sigmoid flexure is partially surrounded above, and entirely surrounded below. The upper part of the rectum is completely invested by the peritoneum, except along the attachment of the mesorectum ; the middle portion is covered above on its anterior surface and sides ; below, on its anterior aspect only ; and the lower portion is entirely devoid of any serous covering. In the course of the colon and upper part of the rectum, the peritoneal coat is thrown into a number of small pouches filled with fat, called *appendices epiploicæ*. They are chiefly appended to the transverse colon.

The *muscular coat* consists of an external longitudinal and an internal circular layer of non-striped muscular fibres.

The *longitudinal fibres* do not form a continuous layer over the whole surface of the large intestine. In the cæcum and colon they are especially collected into three flat longitudinal bands or tæniæ, each of about half an inch in width. The vermiform appendix is surrounded by a uniform layer of longitudinal muscular fibres, and these bands commence at the attachment of the appendix to the cæcum : one, the posterior, is placed along the attached border of the intestine ; the anterior, the largest, corresponds along the arch of the colon to the attachment of the great omentum, but is in front in the ascending and descending parts of the colon and in the sigmoid flexure ; the third, or lateral band, is found on the inner side of the ascending and descending parts of the colon, and on the under aspect of the transverse colon. These bands are shorter than the other coats of the intestine, and serve to produce the sacculi which are characteristic of the cæcum and colon ; accordingly, when they are dissected off, the tube can be lengthened, and its sacculated character becomes lost. In the sigmoid flexure the longitudinal fibres become more scattered ; but upon its lower part, and round the rectum, they spread out and form a layer, which completely encircles this portion of the gut, but is thicker on the anterior and posterior surfaces, where it forms two bands, than on the lateral surfaces. In addition, two bands of plain muscular tissue arise from the second and third coccygeal vertebræ, and pass downwards and forwards to blend with the longitudinal muscular fibres on the posterior wall of the anal canal. These are known as the *recto-coccygeal* muscles.

The *circular fibres* form a thin layer over the cæcum and colon, being especially accumulated in the intervals between the sacculi ; in the rectum they form a thick layer, especially at its lower end, where they become numerous, and constitute the Internal sphincter.

The *areolar coat* connects the muscular and mucous layers closely together.

The *mucous membrane*, in the cæcum and colon, is pale, smooth, destitute of villi, and raised into numerous crescentic folds which correspond to the intervals between the sacculi. In the rectum it is thicker, of a darker colour, more vascular, and connected loosely to the muscular coat, as in the œsophagus. When the lower part of the rectum is contracted, its mucous membrane is thrown into a number of folds, some of which, near the anus, are longitudinal in direction, and are effaced by the distension of the gut. Besides these there are certain permanent folds, of a semilunar shape, known as Houston's valves.† They are usually three in number ; sometimes a fourth is found, and occasionally only two are present. One is situated near the commencement of the rectum, on the right side ; another extends inwards from the left side of the tube, opposite the middle of the sacrum ; the largest and most constant one projects

* See footnote, page 1084.   † *Dublin Hosp. Reports*, vol. v. p. 163.

backwards from the fore part of the rectum, opposite the base of the bladder. When a fourth is present, it is situated nearly an inch above the anus on the back of the rectum. These folds are about half an inch in width, and contain some of the circular fibres of the gut. In the empty state of the intestine they overlap each other, as Houston remarks, so effectually as to require considerable manœuvring to conduct a bougie or the finger along the canal of the intestine. Their use seems to be, ' to support the weight of fæcal matter, and prevent its urging towards the anus, where its presence always excites a sensation demanding its discharge.'

The lumen of the anal canal presents, in its upper half, a number of vertical folds, produced by an infolding of the mucous membrane and some of the muscular tissue. They are known as the *columns of Morgagni*, and are separated from one another by furrows, which terminate below in small valve-like folds, which join together the lower ends of the columns of Morgagni and are situated at the point where the proctodeum and rectum meet. They are termed *anal valves*, and, according to Ball, fissure of the anus is due to one of them being caught by a scybalous mass and torn from its upper connections.

As in the small intestine, the mucous membrane consists of a muscular layer, the muscularis mucosæ ; of a quantity of retiform tissue in which the vessels ramify ; of a basement-membrane and epithelium, which is of the columnar

FIG. 741.—Minute structure of large intestine.

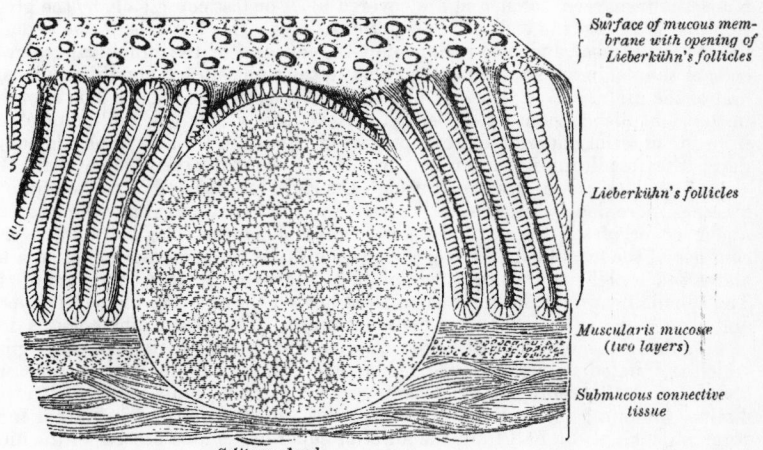

Surface of mucous membrane with opening of Lieberkühn's follicles

Lieberkühn's follicles

Muscularis mucosæ (two layers)

Submucous connective tissue

Solitary gland

variety, and exactly resembles the epithelium found in the small intestine. The mucous membrane of this portion of the bowel presents for examination simple follicles and solitary glands.

The *simple follicles* are minute tubular prolongations of the mucous membrane arranged perpendicularly, side by side, over its entire surface ; they are longer, more numerous, and placed in much closer apposition than those of the small intestine ; and they open by minute rounded orifices upon the surface, giving it a cribriform appearance.

The *solitary glands* (fig. 741) of the large intestine are most abundant in the *cæcum* and *vermiform appendix*, but are irregularly scattered also over the rest of the intestine. They are similar to those of the small intestine.

Vessels and Nerves.—The arteries supplying the large intestine give off large branches, which ramify between and supply the muscular coats, and, after dividing into small vessels in the submucous tissue, pass to the mucous membrane. The rectum is supplied mainly by the superior hæmorrhoidal branch of the inferior mesenteric, but also at its lower end by the middle hæmorrhoidal from the internal iliac, and the inferior hæmorrhoidal from the pudic artery. The superior hæmorrhoidal, the continuation of the superior mesenteric, divides into two branches, which run down either side of the rectum to within about five inches of the anus ; they here split up into about six branches, which pierce the muscular coat and descend between it and the mucous membrane in a longitudinal direction, parallel with each other as far as the Internal sphincter, where they anastomose with the other hæmorrhoidal arteries and form a series of loops around the anus. The veins of the rectum commence in a plexus of vessels which surrounds the lower extremity of the intestinal canal. In the vessels forming this plexus are small saccular dilatations just within the margin of the anus; from the plexus about six vessels of considerable size are given off. These ascend between the muscular and mucous coats for about five inches, running parallel to each other; they then pierce the muscular coat, and, by their union, form a single trunk, the superior hæmorrhoidal vein. This arrangement is termed the *hæmorrhoidal plexus*; it communicates with the tributaries of the middle and inferior hæmorrhoidal veins at its commencement, and thus a communication is established between the systemic and portal circulations. The nerves are derived from the plexuses of the sympathetic nerve around the branches of the superior and inferior mesenteric arteries which supply the large intestine. They are distributed in a similar way to those in the small intestine. The lymphatic vessels of the large intestine are found in the submucosa, where they form a wide-meshed network, and also, more deeply seated, beneath the simple follicles. Those from the colon open into the mesenteric glands; those from the sigmoid flexure into the lumbar glands; those from the rectum enter the glands which are situated in the hollow of the sacrum; and those around the anus open into the glands in the groin.

Surface Form.—The coils of the small intestine occupy the front of the abdomen, below the transverse colon, and are covered more or less completely by the great omentum, For the most part the coils of the jejunum occupy the left side of the abdominal cavity, i.e. the left lumbar and iliac regions and the left half of the umbilical region; while the coils of the ileum are situated to the right, in the right lumbar and iliac regions, in the right half of the umbilical region, and also in the hypogastric region. The cæcum is situated in the right iliac region. Its position varies slightly, but the mid-point of a line drawn from the anterior superior spinous process of the ilium to the symphysis pubis will about mark the middle of its lower border. It is comparatively superficial. From it the ascending colon passes upwards through the right lumbar and hypochondriac regions, and becomes more deeply situated as it ascends to the hepatic flexure, which is deeply placed, under cover of the liver. The transverse colon crosses the belly transversely on the confines of the umbilical and epigastric regions; its lower border being on a level slightly above the umbilicus, its upper border just below the greater curvature of the stomach. The splenic flexure of the colon is situated behind the stomach in the left hypochondrium, and is on a higher level than the hepatic flexure. The descending colon is deeply seated, passing down through the left hypochondriac and lumbar regions to the sigmoid flexure, which is situated in the left iliac region and can be felt in thin persons, with relaxed abdominal walls, rolling under the fingers when empty, and when distended forming a distinct tumour. The position of the base of the vermiform appendix is indicated by a point an inch and a half from the anterior superior spinous process of the ilium, in a line drawn from this process to the umbilicus. This is known as *McBurney's spot*. Another mode of defining the position of the base of the appendix is to draw a line between the anterior superior spines of the ilia, marking the point where this line intersects the right semilunar line.

Upon introducing the finger into the rectum, the membranous portion of the urethra can be felt exactly in the middle line if an instrument has been introduced into the bladder; above this the prostate gland can be recognised by its shape and hardness and any enlargement detected; behind the prostate the fluctuating wall of the bladder when full can be felt, and if thought desirable it can be tapped in this situation; on either side and behind the prostate the vesiculæ seminales can be readily felt, especially if enlarged by tuberculous disease. Behind, the coccyx is to be felt; and on the mucous membrane one or two of Houston's folds. The ischio-rectal fossæ can be explored on either side, with a view to ascertaining the presence of deep-seated collections of pus. Finally, it will be noted that the finger is firmly gripped by the sphincter for about an inch up the bowel. By gradual dilatation of the sphincter, the whole hand can be introduced into the rectum so as to reach the descending colon. This method of exploration is rarely, however, required for diagnostic purposes.

Surgical Anatomy.—The small intestines are much exposed to injury, but, in consequence of their elasticity and the ease with which one fold glides over another, they are not so frequently ruptured as would otherwise be the case. Any part of the small intestine may be ruptured, but probably the most common situation is the transverse duodenum, on account of its being more fixed than other portions of the bowel, and because it is situated in front of the bodies of the vertebræ, so that if this portion of the intestine is struck by a sharp blow, as from the kick of a horse, it is unable to glide out of the way, but is compressed against the bone and so lacerated. Wounds of the intestine sometimes occur. If the wound is a small puncture, under, it is said, three lines in length, no extravasation of the contents of the bowel takes place. The mucous membrane becomes everted and plugs the little opening. The bowels, therefore, may be safely punctured with a fine capillary trocar, in cases of excessive distension of the intestine with gas, without fear of extravasation. A longitudinal wound gapes more than a transverse, owing to the greater amount of circular muscular fibres. The small intestine, and most frequently the ileum, may become strangulated by internal bands, or through apertures, normal or abnormal. The bands may be formed is several different ways : they may be old peritoneal adhesions from previous attacks of peritonitis ; or an adherent omentum from the same cause ; or the band may be formed by Meckel's diverticulum, which has contracted adhesions at its distal extremity ; or it may be the result of the abnormal attachment of some normal structure, as the adhesion of two appendices epiploicæ, or an adherent vermiform appendix or Fallopian tube. Intussusception or invagination of the small intestine may take place in any part of the jejunum and ileum, but the most frequent situation is at the ileo-cæcal valve, the valve forming the apex of the entering tube. This form may attain great size, and it is not uncommon in these cases to find the valve projecting from the anus. Stricture, the impaction of foreign bodies, and twisting of the gut (*volvulus*) may lead to intestinal obstruction.

Resection of a portion of the intestine may be required in cases of gangrenous gut ; in cases of intussusception ; for the removal of new growth in the bowel ; in dealing with artificial anus ; and in cases of rupture. The operation is termed *enterectomy*, and is performed as follows : the abdomen having been opened and the amount of bowel requiring removal having been determined upon, the gut must be clamped on either side of this portion in order to prevent the escape of any of the contents of the bowel during the operation. The portion of bowel is then separated above and below by means of scissors. If the portion resected is small, it may be simply removed from the mesentery at its attachment and the bleeding vessels tied ; but if it be large it will be necessary to take away a triangular piece of the mesentery, and, having secured the vessels, suture the cut edges of this structure together. In doing this, care must be taken not to leave any intestine projecting beyond the line of the section of mesentery, as gangrene is very likely to occur in the projecting gut if this is done. The surgeon then proceeds to unite the cut ends of the bowel together by the operation of what is termed end-to-end anastomosis. There are many ways of doing this, which may be divided into two classes : one where the anastomosis is made by means of some mechanical appliance, such as Murphy's button, or one of the forms of decalcified bone bobbin ; and the other, where the operation is performed by suturing the ends of the bowel in such a manner that the peritoneum covering the free divided ends of the bowel is brought into contact, so that speedy union may ensue.

The vermiform appendix is very liable to become inflamed, because it contains a relatively large amount of lymphoid tissue, which is prone to bacterial infection. This condition may be set up by the appendix becoming twisted, owing to the shortness of its mesentery, in consequence of distension of the cæcum. As the result of this its blood supply, which is mainly through one large artery running in the mesentery, becomes interfered with. Again, in rarer cases, the inflammation is set up by the impaction in it of a solid mass of fæces or a foreign body, or by the inspissation of its mucous secretion in catarrhal conditions. The inflammation may result in ulceration and perforation, or if the torsion is very acute in gangrene of the appendix. These conditions may require operative interference, and in cases of recurrent attacks of appendicitis it is generally advisable to remove this diverticulum of the bowel. In doing this, care must be taken not to divide the fibres of the abdominal muscles more than is absolutely necessary, in order to prevent subsequent weakening of the abdominal parietes and the occurrence of a ventral hernia. The procedure recommended by Battle appears to fulfil this requirement. He divides the superficial structures an inch internal to the right semilunar line, and opens the sheath of the Rectus. He then separates the Rectus from the posterior layer of the sheath with his finger, and retracts it inwards and divides the posterior layer of the sheath and peritoneum in the same line as the external incision. After the appendix is removed, he sutures the posterior layer of the sheath and peritoneum with one row of sutures ; then allows the Rectus to return to its natural position, and sutures the anterior layer of the sheath and the superficial structures with separate rows of sutures. McBurney's incision is also deserving of notice, but possesses the disadvantage that it does not afford much room. He separates the abdominal muscles in the direction of their muscular and tendinous fibres, instead of cutting across them. He makes his incision in the direction of the fibres of the External oblique muscle, and when the muscle and aponeurosis are exposed he separates the fibres in the long axis of the wound. Widely retracting its edges, he now traverses the Internal oblique and Transversalis in the direction of their fibres, and finally divides the transversalis fascia and peritoneum.

In external hernia the ileum is the portion of bowel most frequently herniated. When a part of the large intestine is involved it is usually the cæcum, and this may occur even on the left side. In some few cases the vermiform appendix has been the part implicated in cases of strangulated hernia, and has given rise to serious symptoms of obstruction.

Chronic ulcer of the duodenum is sometimes met with, probably produced by the same causes as chronic ulcer of the stomach. It may perforate and set up a rapidly fatal peritonitis, or it may open into one of the large duodenal vessels and cause death from hæmorrhage. An acute ulcer sometimes, but rarely, follows extensive burns of the skin.

The diameter of the large intestine gradually diminishes from the cæcum, which has the greatest diameter of any part of the bowel, to the point of junction of the sigmoid flexure with the rectum, at, or a little below, which point stricture most commonly occurs, and diminishes in frequency as one proceeds upwards to the cæcum. When distended by some obstruction low down, the outline of the large intestine can be defined throughout nearly the whole of its course—all, in fact, except the hepatic and splenic flexures, which are more deeply placed; the distension is most obvious in the two flanks and on the front of the abdomen just above the umbilicus. The cæcum, however, is the portion of the bowel which becomes most distended. It may assume enormous dimensions, and has been known to give way from the distension, causing fatal peritonitis. The hepatic flexure and the right extremity of the transverse colon are in close relationship with the liver, and abscess of this viscus sometimes bursts into the gut in this situation. The gall-bladder may become adherent to the colon, and gall-stones may find their way through into the gut, where they may become impacted or may be discharged per anum. The mobility of the sigmoid flexure renders it more liable to become the seat of a volvulus or twist than any other part of the intestine. It generally occurs in patients who have been the subjects of habitual constipation, and in whom therefore the mesocolon is elongated. The gut at this part being loaded with fæces, from its weight falls over the gut below, and so gives rise to the twist.

There are several points of practical interest in connection with the mesentery which merit notice. 1. The depth of the mesentery—that is to say, the distance from its parietal to its intestinal attachment—is never more than eight inches, generally nearer six or seven; but under certain abnormal conditions it may become elongated, and this would appear to favour the occurrence of hernia of the intestine. 2. Not only may the depth of the mesentery be increased, but its point of attachment to the posterior abdominal wall may yield, and descend over the lumbar spine. This condition, which is known under the name of *enteroptosis*, usually occurs in women who have borne many children, and is attended with general relaxation of the abdominal parietes. It produces a characteristic appearance, the abdomen being prominent and pendulous below, while above it is flattened and constricted. 3. Holes are sometimes present in the mesentery, and these may be congenital, or may be the result of injury. They are of practical importance, since a knuckle of intestine may become herniated into one of them, causing acute strangulation. 4. The lymphatic glands contained between the two layers of the mesentery are frequently the seat of tuberculous deposit, especially in children, constituting the disease known as *tabes mesenterica*.

The surgical anatomy of the rectum is of considerable importance. There may be congenital malformations due to arrest of, or imperfection in, development. Thus, there may be no inflection of the ectoderm (see page 141), and consequently a complete absence of the anus; or the hind-gut may be imperfectly developed, and there may be an absence of the rectum, though the anus is developed; or the inflection of the ectoderm may not communicate with the termination of the hind-gut from want of solution of continuity in the septum which in early fœtal life exists between the two. The mucous membrane is thick and but loosely connected to the muscular coat beneath, and thus favours prolapse, especially in children. The vessels of the rectum are arranged, as mentioned above, longitudinally, and are contained in the loose cellular tissue between the mucous and muscular coats, and receive no support from surrounding tissues, and this favours varicosity. Moreover, the veins, after running upwards in a longitudinal direction for about five inches in the submucous tissue, pierce the muscular coats, and are liable to become constricted at this spot by the contraction of the muscular wall of the gut. In addition to this there are no valves in the superior hæmorrhoidal veins, and the vessels of the rectum are placed in a dependent position, and are liable to be pressed upon and obstructed by hardened fæces. The anatomical arrangement, therefore, of the hæmorrhoidal vessels explains the great tendency to the occurrence of piles. The presence of the Sphincter ani is of surgical importance, since it is the constant contraction of this muscle which prevents an ischio-rectal abscess from healing, and causes it to become a fistula. Also the reflex contraction of this muscle is the cause of the severe pain complained of in fissure of the anus. The relations of the peritoneum to the bowel are of importance in connection with the operation of removal of the lower end of the rectum for malignant disease. This membrane gradually leaves the rectum as it descends into the pelvis; first leaving its posterior surface, then the sides, and then the anterior surface, to become reflected, in the male on to the posterior wall of the bladder, forming the recto-vesical pouch, and in the female on to the posterior wall of the vagina, forming Douglas's pouch. The recto-vesical pouch of peritoneum extends to within three inches from the anus, so that it is not desirable to remove more than two and a half inches of the entire circumference of the bowel for fear of the risk of opening the peritoneum. When, however, the disease is confined to the posterior surface of the rectum, or extends farther in this direction, a greater amount of the posterior wall of the gut may be removed, as the peritoneum does not extend on this surface to a lower level than five inches from the margin of the anus. The recto-vaginal or Douglas's pouch in the female extends somewhat lower than the recto-vesical pouch of the male, and therefore it is necessary to remove a less length of the tube in this sex. Of recent years, however, much more extensive operations have been done for the removal of cancer of the rectum, and in these the peritoneal cavity has necessarily been opened. If, in these cases, the opening is plugged with antiseptic wool until the operation is completed and then the edges of the wound in the peritoneum accurately brought together with sutures, no evil result appears to follow. For cases of cancer of the rectum which are too low to be reached by abdominal section, and too high to be removed by the ordinary operation from below, Kraske has devised an operation which goes by his name. The patient is placed on his right side and an incision is made from the second sacral spine to the anus. The soft parts are now separated from the back of the left side of the sacrum as far as its left margin, and the greater and lesser sacro-sciatic ligaments are divided. A portion of the lateral mass of the sacrum, commencing on the left border at the level of the third posterior sacral foramen, and running downwards and inwards through the fourth foramen to the cornu, is now cut away with a chisel. The left side of the wound being now forcibly drawn outwards, the whole of the rectum is brought into view, and the diseased portion can be removed, leaving the anal portion of the gut, if healthy. The two divided ends of the gut can then be approximated and sutured together in front, the posterior part being left open for drainage.

The colon frequently requires opening in cases of intestinal obstruction, and by some surgeons this operation is performed in cases of cancer of the rectum, as soon as the disease is recognised, in the hope that the rate of growth may be retarded by removing the irritation produced by the passage of fæcal matter over the diseased surface. The operation of colotomy may be performed either in the inguinal or lumbar region; but inguinal colotomy has in the present day almost superseded the lumbar operation. The main reason for preferring this operation is that a spur-shaped process of the mesocolon can be formed which prevents any fæcal matter finding its way past the artificial anus, and becoming lodged on the diseased structures below. The sigmoid flexure being almost entirely surrounded by peritoneum, a coil can be drawn out of the wound and the greater part of its calibre removed, leaving the remainder attached to the mesocolon, which forms a spur, much the same as in an artificial anus caused by sloughing of the gut after a strangulated hernia, and this prevents any fæcal matter finding its way from the gut above the opening into that below. The operation is performed by making an incision two or three inches in length from a point one inch internal to the anterior superior spinous process of the ilium, parallel to Poupart's ligament. The various layers of abdominal muscles are cut through, and the peritoneum opened and sewn to the external skin. The sigmoid flexure is now sought for, and pulled out of the wound and fixed by passing a needle threaded with carbolised silk through the mesocolon close to the gut, and then through the abdominal wall. The intestine is now sewn to the skin all round, the sutures passing only through the serous and muscular coats. The wound is dressed, and on the second to the fourth day, according to the requirements of the case, the protruded coil of intestine is opened and removed with scissors.

Lumbar colotomy is performed by placing the patient on the side opposite to the one to be operated on, with a firm pillow under the loin. A line is then drawn from the anterior superior to the posterior superior spine of the ilium, and the mid-point of this line (Heath) or half an inch behind the mid-point (Allingham) is taken, and a line drawn vertically upwards from it to the last rib. This line represents, with sufficient correctness, the position of the normal colon. An oblique incision four inches in length is now made midway between the last rib and the crest of the ilium, so that its centre bisects the vertical line, and the following parts successively divided: (1) The skin, superficial fascia, with cutaneous vessels and nerves and deep fascia. (2) The posterior fibres of the External oblique and anterior fibres of the Latissimus dorsi. (3) The Internal oblique. (4) The lumbar fascia and the external border of the Quadratus lumborum. The edges of the wound are now to be held apart with retractors, and the transversalis fascia will be exposed. This is to be opened with care, commencing at the posterior angle of the incision. If the bowel is distended, it will bulge into the wound, and no difficulty will be found in dealing with it. If, however, the gut is empty, this bulging will not take place, and the colon will have to be sought for. The guides to it are the lower end of the kidney, which will be plainly felt, and the outer edge of the Quadratus lumborum. The bowel having been found, is to be drawn well up into the wound, and it may be opened at once and the margins of the opening stitched to the skin at the edge of the wound; or, if the case is not an urgent one, it may be retained in this position by two harelip pins passed through the muscular coat, the rest of the wound closed, and the bowel opened in three or four days, when adhesion of the bowel to the edges of the wound has taken place.

THE LIVER

The **Liver** is the largest gland in the body, and is situated in the upper and right parts of the abdominal cavity, occupying almost the whole of the right hypochondrium, the greater part of the epigastrium, and not uncommonly extending into the left hypochondrium as far as the mammary line. In the male it weighs from fifty to sixty ounces, in the female from forty to fifty. It is relatively much larger in the fœtus than in the adult, constituting, in the former, about one-eighteenth, and in the latter, about one thirty-sixth of the entire body weight. Its greatest transverse measurement is from eight to nine inches. Vertically, near its lateral or right surface, it measures about six or seven inches, while its greatest antero-posterior diameter is on a level with the upper end of the right kidney and is from four to five inches. Opposite the vertebral column its measurement from before backwards is reduced to about three inches. Its consistence is that of a soft solid ; it is, however, friable and easily lacerated ; its colour is a dark reddish-brown, and its specific gravity is 1·05.

To obtain a correct idea of its shape it must be hardened *in situ*, and it will then be seen to present the appearance of a wedge, the base of which is directed

FIG. 742.—The liver. Upper surface. (Slightly modified from His's model.)

Left lateral ligament

Right lateral ligament

to the right and the thin edge towards the left. Symington describes its shape as that ' of a right-angled triangular prism with the right angles rounded off.' It possesses five surfaces, viz. : superior, inferior, anterior, posterior, and lateral.

The superior and anterior surfaces are separated from each other by a thick rounded border, and are attached to the Diaphragm and anterior abdominal wall by a triangular or falciform fold of peritoneum, the *suspensory* or *falciform ligament*, which divides the liver into two unequal parts, termed the right and left lobes. Except along the line of attachment of this ligament to the liver, these two surfaces are covered by peritoneum.

The *superior surface* (fig. 742) comprises a part of both lobes, and, as a whole, is convex, and fits under the vault of the Diaphragm ; its central part, however, presents a shallow depression, which corresponds with the position of the heart on the upper surface of the Diaphragm. It is separated from the anterior, posterior, and lateral surfaces by thick, rounded borders. Its left extremity is separated from the under surface by a prominent sharp margin.

The *anterior surface* is large, triangular in shape, and comprises also a part of both lobes. It is directed forwards, and the greater part of it is in contact with the Diaphragm, which separates it from the right lower ribs and their cartilages. In the middle line it lies behind the ensiform cartilage, to the left of which it is protected by the seventh and eighth left costal cartilages. In the angle between the diverging rib cartilages of opposite sides the anterior surface is in contact with the abdominal wall. It is separated from the inferior surface by a sharp margin, and from the superior and lateral surfaces by thick rounded borders.

The *lateral* or *right surface* is convex from before backwards and slightly so from above downwards. It is directed towards the right side, forming the base of the wedge, and lies against the lateral portion of the Diaphragm, which separates it from the lower part of the left pleura and lung, outside which are the right costal arches from the seventh to the eleventh inclusive.

The *under* or *visceral surface* (figs. 743, 744) is uneven, concave, directed downwards, backwards and to the left, and is in relation with the stomach and duodenum, the hepatic flexure of the colon, and the right kidney and suprarenal capsule. The surface is divided by a longitudinal fissure into a right and left lobe, and is almost completely invested by peritoneum ; the only parts where this covering is absent are where the gall-bladder is attached to the liver and at the transverse fissure, where the two layers of the lesser omentum are separated

FIG. 743.—The liver. Posterior and inferior surfaces. (Drawn from His's model.)

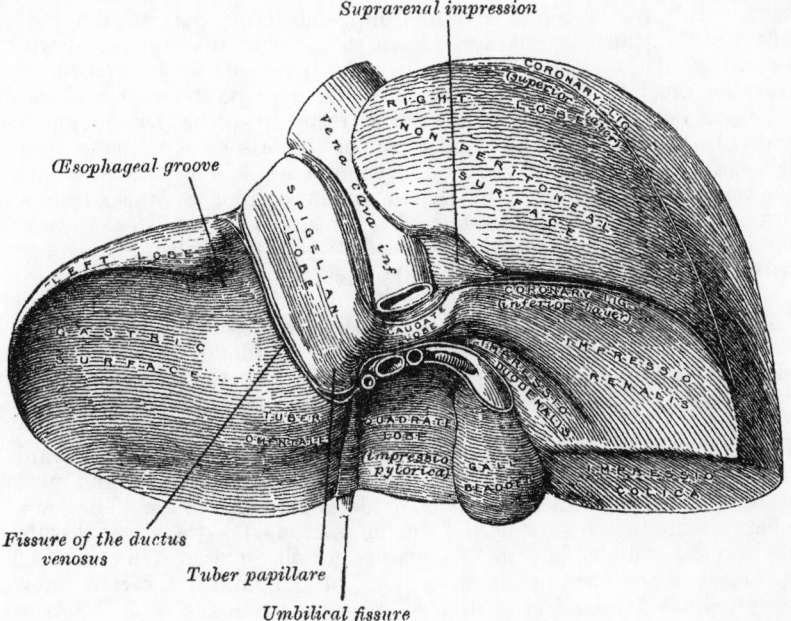

from each other by the blood-vessels and duct of the viscus. The under surface of the left lobe presents behind and to the left a depression where it is moulded over the cardiac part of the stomach, and to the right and near the centre a rounded eminence, the *tuber omentale*, which fits into the concavity of the lesser curvature, lying in front of the anterior layer of the lesser omentum. The under surface of the right lobe is divided into two unequal portions by a fossa, which lodges the gall-bladder, the *fossa vesicalis* ; the portion to the left, the smaller of the two, is somewhat oblong in shape, its antero-posterior diameter being greater than its transverse. It is known as the *quadrate lobe*, and is in relation with the pyloric end of the stomach and the first portion of the duodenum. The portion of the under surface of the right lobe to the right of the fossa vesicalis presents two shallow concave impressions, one situated behind the other, the two being separated by a ridge. The anterior of these two impressions, the *impressio colica*, is produced by the hepatic flexure of the colon ; the posterior, the *impressio renalis*, is occupied by the upper part of the right kidney and lower part of the right suprarenal capsule. To the inner side of the renal impression is a third and slightly marked impression, lying between it and the neck of the gall-bladder. This is caused by the second portion of the duodenum, and is known as the *impressio duodenalis*. Just in front of the vena cava is a narrow strip of liver tissue, the *caudate lobe*, which connects the right inferior angle of the Spigelian lobe to the under surface of the right lobe. It forms the upper boundary of the foramen of Winslow.

The *posterior surface* (fig. 743) is rounded and broad behind the right lobe, but narrow on the left. Over a large part of its extent it is not covered by peritoneum ; this uncovered portion is about three inches broad, and is in direct contact with the Diaphragm. It is marked off from the upper surface by the line of reflection of the upper or anterior layer of the coronary ligament. It is in the same way marked off from the under surface of the liver by the line of reflection of the lower layer of the coronary ligament. In its centre this posterior surface is bent on itself around the vertebral column and crura of the Diaphragm, the two portions of the surface, having different directions, forming an angle in this situation. To the right of this the inferior vena cava is lodged in an indentation in the liver substance, lying between the uncovered area and the Spigelian lobe. Close to the right of this indentation and immediately above the renal impression is a small triangular depressed area (*impressio supra-renalis*), the greater part of which is devoid of peritoneum ; it lodges the right suprarenal capsule. To the left of the inferior vena cava is the *Spigelian lobe*, which lies between the fissure for the vena cava and the fissure for the ductus venosus. It projects below and in front, and forms part of the posterior boundary of the transverse fissure. Here, to the right, it is connected with the under surface of the right lobe of the liver by the caudate lobe, and to the left it presents a tubercle, the *tuber papillare*. It is opposite the tenth and eleventh dorsal vertebræ, and its posterior surface rests upon the Diaphragm, being separated from it merely by the upper part of the lesser sac of the peritoneum. Its left surface is free, and is separated from the left lobe by the fissure for the ductus venosus. This lobe is nearly vertical in position, is longer from above downwards than from side to side, and is somewhat concave in the transverse direction. On the posterior surface to the left of the Spigelian lobe is a groove indicating the position of the œsophageal orifice of the stomach.

The *inferior border* is thin and sharp, and marked opposite the attachment of the falciform ligament by a deep notch, the *umbilical notch*, and opposite the cartilage of the ninth rib by a second notch for the fundus of the gall-bladder. In adult males this border generally corresponds with the lower margin of the thorax in the right nipple line ; but in women and children it usually projects below the ribs.

The *left extremity of the liver* is thin and flattened from above downwards.

Fissures (fig. 743).—Five fissures are seen upon the under and posterior surfaces of the liver, which serve to divide it into five lobes. They are, the umbilical fissure, the fissure of the ductus venosus, the transverse fissure, the fissure for the gall-bladder, and the fissure for the inferior vena cava. They are arranged in the form of the letter **H**. The left limb of the **H** is known as the *longitudinal fissure*, and is divided into two parts, the *umbilical fissure* in front, and the *fissure of the ductus venosus* behind. The right limb is formed in front by the *fissure for the gall-bladder*, and behind by the *fissure for the inferior vena cava* ; these two fissures are separated from each other by the caudate lobe. The connecting bar of the **H** is the *transverse* or *portal fissure*. It separates the quadrate lobe in front from the caudate and Spigelian lobes behind.

The *longitudinal fissure* is a deep groove, which extends from the notch on the anterior margin of the liver to the upper border of the posterior surface of the organ. It separates the right and left lobes ; the transverse fissure joins it, at right angles, and divides it into two parts. The anterior part is called the *umbilical fissure* ; it lodges the umbilical vein in the fœtus, and its remains (the round ligament) in the adult ; it lies between the quadrate lobe and the left lobe of the liver, and is often partially bridged over by a prolongation of the hepatic substance, the *pons hepatis*. The posterior part of the longitudinal fissure contains the ductus venosus, and is known as the *fissure of the ductus venosus*.

The *fissure of the ductus venosus* is the back part of the longitudinal fissure, and is situated mainly on the posterior surface of the liver. It lies between the left lobe and the lobe of Spigelius. It lodges in the fœtus the ductus venosus, and in the adult a slender fibrous cord, the obliterated remains of that vessel.

The *transverse* or *portal fissure* is a short but deep fissure, about two inches in length, extending transversely across the under surface of the left portion of the right lobe, nearer to its posterior surface than its anterior border. It joins, nearly at right angles, with the longitudinal fissure, and separates the quadrate lobe in front from the caudate and Spigelian lobes behind. By the older anatomists this fissure was considered the gateway (*porta*) of the liver; hence the large vein which enters at this fissure was called the *portal vein*. Besides this vein, the fissure transmits the hepatic artery and nerves, and the hepatic duct and lymphatics. At their entrance into the fissure, the hepatic duct lies in front and to the right, the hepatic artery to the left, and the portal vein behind and between the duct and artery.

The *fissure for the gall-bladder* (*fossa vesicalis*) is a shallow, oblong fossa, placed on the under surface of the right lobe, parallel with the longitudinal fissure. It extends from the anterior free margin of the liver, which is notched for its reception, to the right extremity of the transverse fissure.

FIG. 744.—Posterior and under surfaces of the liver. (His.)

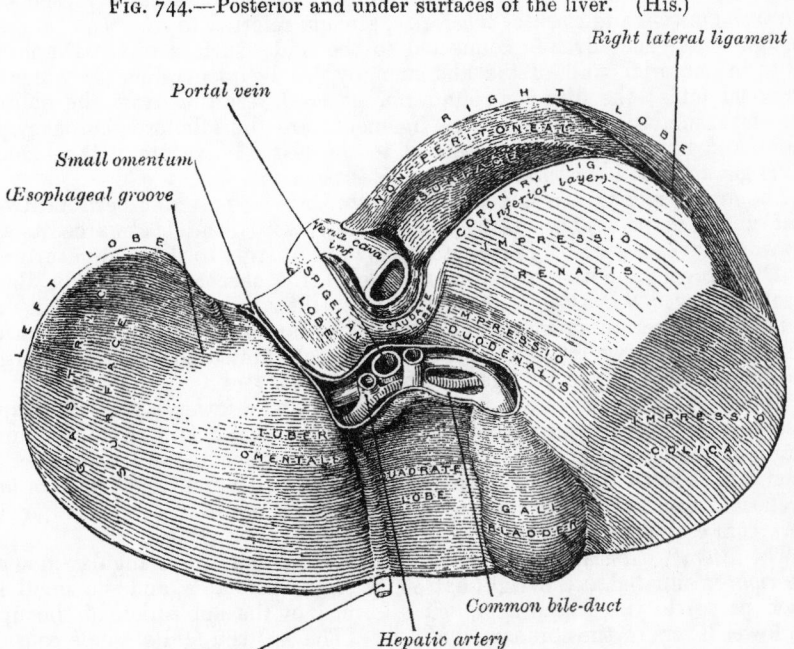

The *fissure for the inferior vena cava* is a short deep fissure, occasionally a complete canal, in consequence of the substance of the liver surrounding the vena cava. It extends obliquely upwards from the lobus caudatus, which separates it from the transverse fissure, on the posterior surface of the liver, and separates the Spigelian from the right lobe. On slitting open the inferior vena cava the orifices of the hepatic veins will be seen opening into this vessel at its upper part, after perforating the floor of this fissure.

Lobes.—The lobes of the liver, like the ligaments and fissures, are five in number—the right lobe, the left lobe, the lobus quadratus, the lobus Spigelii, and the lobus caudatus, the last three being merely parts of the right lobe.

The *right lobe* is much larger than the left; the proportion between them being as six to one. It occupies the right hypochondrium, and is separated from the left lobe, on its upper and anterior surfaces by the falciform ligament; on its under and posterior surfaces by the longitudinal fissure; and in front by the umbilical notch. It is of a somewhat quadrilateral form, its under and posterior surfaces being marked by three fissures: the transverse fissure, the fissure for the gall-bladder, and the fissure for the inferior vena cava, which separate its left part into three smaller lobes: lobus Spigelii, lobus quadratus, and lobus caudatus. The impressions on the right lobe have already been described.

The *lobus quadratus* is situated on the under surface of the right lobe, bounded in front by the inferior margin of the liver; behind by the transverse fissure; on the right, by the fossa for the gall-bladder; and on the left, by the umbilical fissure.

The *lobus Spigelii* is situated upon the posterior surface of the right lobe of the liver. It looks directly backwards, and is nearly vertical in direction. It is bounded, below, by the transverse fissure; on the right, by the fissure for the vena cava; and, on the left, by the fissure for the ductus venosus. Its left upper angle forms part of the groove for the œsophagus.

The *lobus caudatus*, or tailed lobe, is a small elevation of the hepatic substance extending obliquely outwards, from the lower extremity of the lobus Spigelii to the under surface of the right lobe. It is situated behind the transverse fissure, and separates the fissure for the gall-bladder from the commencement of the fissure for the inferior vena cava.

The *left lobe* is smaller and more flattened than the right. It is situated in the epigastric and left hypochondriac regions. Its upper surface is slightly convex and is moulded on to the Diaphragm; its under surface presents the *gastric impression* and *omental tuberosity*, already referred to.

Ligaments.—The liver is connected to the under surface of the Diaphragm and to the anterior wall of the abdomen by five ligaments, four of which are peritoneal folds; the fifth is a round, fibrous cord, resulting from the obliteration of the umbilical vein. These ligaments are the falciform, coronary, two lateral, and round. It is also attached to the lesser curvature of the stomach by the gastro-hepatic or small omentum (see page 1062).

The *falciform ligament* is a broad and thin antero-posterior peritoneal fold, falciform in shape, its base being directed downwards and backwards, its apex upwards and backwards. It is attached by one margin to the under surface of the Diaphragm, and the posterior surface of the sheath of the right Rectus muscle as low down as the umbilicus; by its hepatic margin it extends from the notch on the anterior margin of the liver, as far back as its posterior surface. It is composed of two layers of peritoneum closely united together. Its base or free edge contains the round ligament between its layers.

The *coronary ligament* consists of an upper and a lower layer. The *upper layer* is formed by the line of reflection of the peritoneum from the upper margin of the bare area of the liver to the under surface of the Diaphragm, and is continuous with the right layer of the falciform ligament. The *lower layer* is reflected from the lower margin of the bare area on to the right kidney and suprarenal capsule.

The *lateral ligaments*, two in number, right and left, are triangular in shape. The *right* is situated at the right extremity of the bare area, and is a small fold which passes to the Diaphragm, being formed by the apposition of the upper and lower layers of the coronary ligament. The *left* is a fold of some considerable size, which connects the posterior part of the upper surface of the left lobe to the Diaphragm; its upper layer is continuous with the left layer of the falciform ligament.

The *round ligament* (ligamentum teres) is a fibrous cord resulting from the obliteration of the umbilical vein. It ascends from the umbilicus, in the free margin of the falciform ligament, to the notch in the anterior border of the liver, from which it may be traced along the longitudinal fissure on the under surface of the liver; on the posterior surface it is continued upwards as the obliterated ductus venosus as far as the inferior vena cava.

Vessels.—The vessels connected with the liver are, the hepatic artery, the portal vein, and the hepatic veins.

The *hepatic artery* and *portal vein*, accompanied by numerous lymphatics and nerves, ascend to the transverse fissure, between the layers of the gastro-hepatic omentum. The *bile-duct*, lying in company with them, descends from the transverse fissure between the layers of the same omentum. The relative position of the three structures is as follows: the bile-duct lies to the right, the hepatic artery to the left, and the portal vein behind and between the other two. These are enveloped in a loose areolar tissue, the *capsule of Glisson*, which accompanies the vessels in their course through the *portal canals*, in the interior of the organ.

The *hepatic veins* convey the blood from the liver, and are described on page 743. They have very little cellular investment, and what there is binds their parietes closely to the walls of the canals through which they run; so that, on section of the organ, they remain widely open and are solitary, and may be easily distinguished from the branches of the portal vein, which are more or less collapsed, and always accompanied by an artery and duct.

Structure of the Liver.—The substance of the liver is composed of lobules, held together by an extremely fine areolar tissue, and of the ramifications of the portal vein, hepatic ducts, hepatic artery, hepatic veins, lymphatics, and nerves; the whole being invested by a serous and a fibrous coat.

The *serous coat* is derived from the peritoneum, and invests the greater part of the surface of the organ. It is intimately adherent to the fibrous coat.

The *fibrous coat* lies beneath the serous investment, and covers the entire surface of the organ. It is difficult of demonstration, excepting where the serous coat is deficient. At the transverse fissure it is continuous with the capsule of Glisson, and, on the surface of the organ, with the areolar tissue separating the lobules.

The *lobules* form the chief mass of the hepatic substance; they may be seen either on the surface of the organ, or by making a section through the gland. They are small granular bodies, about the size of a millet seed, measuring from one-twentieth to one-tenth of an inch in diameter. In the human subject their outline is very irregular; but in some of the lower animals (for example, the pig) they are well defined, and, when divided transversely, have a polygonal outline. If divided longitudinally they are more or less foliated or oblong. The bases of the lobules are clustered round the smallest radicles (*sublobular*) of the hepatic veins, to which each is connected by means of a small branch which issues from the centre of the lobule (*intralobular*). The remaining part of the surface of each lobule is imperfectly isolated from the surrounding lobules by a thin stratum of areolar tissue, in which is contained a plexus of vessels (the *interlobular plexus*) and ducts. In some animals, as the pig, the lobules are completely isolated from one another by the interlobular areolar tissue.

If one of the sublobular veins be laid open, the bases of the lobules may be seen

FIG. 745.—Longitudinal section of an hepatic vein. (After Kiernan.)

Orifices of intralobular veins

through the thin wall of the vein on which they rest, arranged in the form of a tesselated pavement, the centre of each polygonal space presenting a minute aperture, the mouth of an intralobular vein (fig. 745).

Microscopic appearance.—Each lobule consists of a mass of cells (*hepatic cells*), surrounded by a dense capillary plexus, composed of vessels which penetrate from the circumference to the centre of the lobule, and terminate in the *intralobular vein*, which runs through its centre, to open at its base into one of the *sublobular veins*. Between the cells are also the minute commencements of the bile-ducts. Therefore, in the lobule we have all the essentials of a secreting gland; that is to say: (1) *cells*, by which the secretion is formed; (2) *blood-vessels*, in close relation with the cells, containing the blood from which the secretion is derived; (3) *ducts*, by which the secretion, when formed, is carried away. Each of these structures must be further considered.

(1) The *hepatic cells* are more or less spheroidal in form; but may be rounded, flattened, or many-sided from mutual compression. They vary in size from $\frac{1}{1000}$ to $\frac{1}{2000}$ of an inch in diameter. They consist of a honeycomb network, and contain one or sometimes two distinct nuclei. The nucleus contains an intranuclear network and one or two refractile nucleoli. Embedded in the honeycomb network are numerous yellow particles, the colouring-matter

of the bile, and oil-globules.   The cells adhere together by their surfaces so as to form rows, which radiate from the centre to the circumference of the lobules.* As stated above, they are the chief agents in the secretion of the bile.

(2) *The blood-vessels.*—The blood in the capillary plexus, around the liver cells, is brought to the liver principally by the portal vein, but also to a certain extent by the hepatic artery.   For the sake of clearness, the distribution of the blood derived from the hepatic artery may be considered first.

The *hepatic artery*, entering the liver at the transverse fissure with the portal vein and hepatic duct, ramifies with these vessels through the portal canals.   It gives off *vaginal branches*, which ramify in the capsule of Glisson, and appear to be destined chiefly for the nutrition of the coats of the vessels and ducts.   It also gives off *capsular branches*, which reach the surface of the organ, terminating in its fibrous coat in stellate plexuses.   Finally it gives off *interlobular branches*, which form a plexus on the outer side of each lobule, to supply the walls of the interlobular veins and the accompanying bile-ducts.   From this plexus lobular branches enter the lobule and end in the capillary network between the cells.   Some anatomists, however, doubt whether it transmits any blood directly to the capillary network.

The *portal vein* also enters at the transverse fissure and runs through the portal canals, enclosed in Glisson's capsule, dividing into branches in its course, which finally break up into a plexus (the *interlobular plexus*) in the interlobular spaces.   In their course these branches receive the vaginal and capsular veins, corresponding to the vaginal and capsular branches of the hepatic artery (fig. 746).   Thus it will be seen that all the blood carried to the liver by the portal vein and hepatic artery, except perhaps that derived from the interlobular branches of the hepatic artery, directly or indirectly, finds its way into the interlobular plexus.   From this plexus the blood is carried into the lobule by fine branches which converge from the circumference to the centre of the lobule, and are connected by transverse branches (fig. 747).   In the interstices of the network of vessels thus formed are situated, as before said, the liver-cells ; and here it is that, the blood being brought into intimate connection with the liver-cells, the bile is secreted.   Arrived at the centre of the lobule, all these minute vessels empty themselves into one vein, of considerable size, which runs down the centre of the lobule from apex to base, and is called the *intralobular vein.* At the base of the lobule this vein opens directly into the *sublobular vein*, with which the lobule is connected, and which, as before mentioned, is a radicle of the hepatic vein.   The sublobular veins, uniting into larger and larger trunks, end at last in the hepatic veins, which do not receive any intralobular veins. Finally, the hepatic veins, as mentioned on page 743, converge to form three large trunks which open into the inferior vena cava, while that vessel is situated in the fissure appropriated to it at the back of the liver.

(3) *The bile-ducts.*—Having shown how the blood is brought into intimate relation with the hepatic cells in order that the bile may be secreted, it remains now to consider the means by which the secretion is carried away.   Several views have prevailed as to the mode of origin of the hepatic ducts ; it seems,

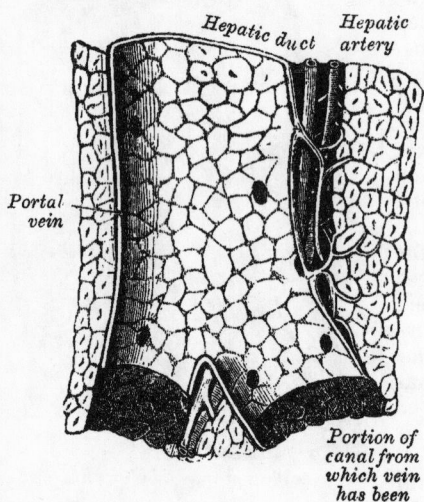

FIG. 746.—Longitudinal section of a small portal vein and canal. (After Kiernan.)

Hepatic duct   *Hepatic artery*

Portal vein

Portion of canal from which vein has been removed

* Delépine states that there are evidences of the arrangement of these cells in the form of columns, which form tubes with narrow lumina branching from terminal bile-ducts.   This branching is evidenced by a divergence of the columns from lines extending between adjacent portal vessels.   The columns of cells group round terminal bile-ducts and not round the so-called intralobular veins.—*Lancet*, vol. i. 1895, p. 1254.

however, to be clear that they commence by little passages which are formed between the cells, and which have been termed *intercellular biliary passages* or *bile capillaries*. These passages are merely little channels or spaces left between the contiguous surfaces of two cells, or in the angle where three or

FIG. 747.—Horizontal section of liver (dog).

Trunk of intralobular vein

Interlobular vein

Intralobular vein

more liver-cells meet (fig. 748), and it seems doubtful whether there is any delicate membrane forming the wall of the channel. The channels thus formed radiate to the circumference of the lobule, and form a plexus (*interlobular*) between the lobules. From this plexus ducts are derived which pass into the portal canals, become enclosed in Glisson's capsule, and, accompanying the

FIG. 748.—Section of liver.

FIG. 749.—A transverse section of a small portal canal and its vessels. (After Kiernan.)

Biliary duct

Hepatic cells

Capillary

Biliary duct

1. Portal vein. 2. Interlobular branches. 3. Vaginal branches. 4. Hepatic duct. 5. Hepatic artery.

portal vein and hepatic artery (fig. 749), join with other ducts to form two main trunks, which leave the liver at the transverse fissure, and by their union form the *hepatic duct*.

Structure of the ducts.—The walls of the smallest biliary ducts, which lie in the interlobular spaces, consist of a connective-tissue coat, in which are muscle-cells, arranged both circularly and longitudinally, and an epithelial layer, consisting of short columnar cells. In the larger ducts, which lie in the portal canals, there are a number of orifices disposed in two longitudinal rows, which were formerly regarded as the openings of mucous glands, but which are merely the orifices of tubular recesses. They occasionally anastomose, and from the sides of them saccular dilatations are given off.

Lymphatics of the liver.—The lymphatics in the substance of the liver commence in lymphatic spaces around the capillaries of the lobules; they accompany the vessels of the interlobular plexus, often enclosing and surrounding them. These unite and form larger vessels, which run in the portal canals, enclosed in Glisson's capsule, and emerge at the portal fissure to be distributed in the manner described. Other superficial lymphatics form a close plexus, under the peritoneum, where this membrane covers the liver, and pass in various directions through the ligaments of the liver (page 759).

Nerves of the liver.—The nerves of the liver, derived from the left pneumogastric and sympathetic, enter at the transverse fissure and accompany the vessels and ducts to the interlobular spaces. Here, according to Korolkow, the medullated fibres are distributed almost exclusively to the coats of the blood-vessels; while the non-medullated enter the lobules and ramify between the cells.

EXCRETORY APPARATUS OF THE LIVER

The excretory apparatus of the liver consists of (1) the *hepatic duct*, which, as we have seen, is formed by the junction of the two main ducts, which pass out of the liver at the transverse fissure, and are formed by the union of the bile capillaries; (2) the *gall-bladder*, which serves as a reservoir for the bile; (3) the *cystic duct*, which is the duct of the gall-bladder; and (4) the *common bile-duct*, formed by the junction of the hepatic and cystic ducts.

The hepatic duct.—Two main trunks of nearly equal size issue from the liver at the transverse fissure, one from the right, the other from the left lobe; these unite to form the hepatic duct, which then passes downwards and to the right for about an inch and a half, between the layers of the lesser omentum, where it is joined at an acute angle by the cystic duct, and so forms the common bile-duct (ductus communis choledochus). The hepatic duct is accompanied by the hepatic artery and portal vein.

The **gall-bladder** is a conical or pear-shaped musculo-membranous sac, lodged in a fossa on the under surface of the right lobe of the liver, and extending from near the right extremity of the transverse fissure to the anterior border of the organ. It is from three to four inches in length, one inch in breadth at its widest part, and holds from eight to ten drachms. It is divided into a fundus body, and neck. The *fundus*, or broad extremity, is directed downwards, forwards, and to the right, and projects beyond the anterior border of the liver; the *body* and *neck* are directed upwards and backwards to the left. The upper surface of the gall-bladder is attached to the liver by connective tissue and vessels. The under surface is covered by peritoneum, which is reflected on to it from the surface of the liver. Occasionally the whole of the organ is invested by the serous membrane, and is then connected to the liver by a kind of mesentery.

Relations.—The *body of the gall-bladder* is in relation, by its upper surface, with the liver, to which it is connected by areolar tissue and vessels; by its under surface, with the commencement of the transverse colon; and farther back, with the upper end of the descending portion of the duodenum or sometimes with the pyloric end of the stomach, or first portion of the duodenum. The *fundus* is completely invested by peritoneum; it is in relation, in front, with the abdominal parietes, immediately below the ninth costal cartilage; behind with the transverse arch of the colon. The *neck* is narrow, and curves upon itself like the letter **S**; at its point of connection with the cystic duct it presents a well-marked constriction.

When the gall-bladder is distended with bile or calculi, the fundus may be felt through the abdominal parietes, especially in an emaciated subject: its relations will also serve to explain the occasional occurrence of abdominal biliary fistulæ, through which biliary calculi may pass out, and of the passage of calculi from the gall-bladder into the stomach, duodenum, or colon, which sometimes happens.

Structure.—The gall-bladder consists of three coats: serous, fibrous and muscular, and mucous.

The *external* or *serous coat* is derived from the peritoneum; it completely invests the fundus, but covers the body and neck only on their under surface.

The *fibro-muscular coat* is a thin but strong layer, which forms the framework of the sac, consisting of dense fibrous tissue, which interlaces in all directions, and is mixed with plain muscular fibres, which are disposed chiefly in a longitudinal direction, a few running transversely.

The *internal* or *mucous coat* is loosely connected with the fibrous layer. It is generally of a yellowish-brown colour, and is everywhere elevated into minute rugæ, by the union of which numerous meshes are formed; the depressed intervening spaces having a polygonal outline. The meshes are small in the fundus and neck, and large near the centre of the sac. Opposite the neck of the gallbladder the mucous membrane projects inwards in the form of oblique ridges or folds, forming a sort of screw-like valve.

The mucous membrane is covered with columnar epithelium, and secretes an abundance of thick viscid mucus; it is continuous through the hepatic duct with the mucous membrane lining the ducts of the liver, and through the common bile-duct with the mucous membrane of the alimentary canal.

The **cystic duct,** the smallest of the three biliary ducts, is about an inch and a half in length. It passes backwards, downwards and to the left from the neck of the gall-bladder, and joins the hepatic duct to form the common bile-duct. The mucous membrane lining its interior is thrown into a series of crescentic folds, from five to twelve in number, similar to those found in the neck of the gall-bladder. They project into the duct in regular succession, and are directed obliquely round the tube, presenting much the appearance of a continuous spiral valve. When the duct is distended, the spaces between the folds are dilated, so as to give to its exterior a twisted appearance.

The **common bile-duct** (ductus communis choledochus) is about three inches in length, of the diameter of a goose-quill, and formed by the junction of the cystic and hepatic ducts.

It descends along the right border of the lesser omentum behind the first portion of the duodenum, in front of the vena portæ, and to the right of the hepatic artery; it then passes between the head of the pancreas and descending portion of the duodenum, and, running for a short distance along the right side of the pancreatic duct, near its termination, passes, with it, obliquely between the mucous and muscular coats. The two ducts open by a common orifice upon the summit of a papilla, situated at the inner side of the descending portion of the duodenum, a little below its middle and about three or four inches from the pylorus.

Structure.—The coats of the large biliary ducts are, an external or fibrous, and an internal or mucous. The fibrous coat is composed of strong fibro-areolar tissue, with a certain amount of muscular tissue, arranged, for the most part, in a circular manner around the duct. The mucous coat is continuous with the lining membrane of the hepatic ducts and gall-bladder, and also with that of the duodenum; and, like the mucous membrane of these structures, its epithelium is of the columnar variety. It is provided with numerous mucous glands, which are lobulated and open by minute orifices, scattered irregularly in the larger ducts. The coats of the smallest biliary ducts, which lie in the interlobular spaces, are a connective-tissue coat, in which, according to Heidenhain, are muscle-cells, arranged both circularly and longitudinally, and an epithelial layer, consisting of short columnar cells.

Surface Relations.—The liver is situated mainly in the right hypochondriac and the epigastric regions, and is moulded to the arch of the Diaphragm. In the greater part of its extent it lies under cover of the lower ribs and their cartilages, but in the epigastric region it comes in contact with the abdominal wall, in the subcostal angle. The *upper limit of the right lobe of the liver* may be defined in the middle line by the junction of the mesosternum with the ensiform cartilage; on the right side the line must be carried upwards as far as the fifth rib cartilage in the line of the nipple and then downwards to reach the seventh rib at the side of the chest. The *upper limit of the left lobe* may be defined by continuing this line to the left with an inclination downwards to a point about two inches to the left of the sternum on a level with the sixth left costal cartilage. The *lower limit of the liver* may be indicated by a line drawn half an inch below the lower border of the thorax on the right side as far as the ninth right costal cartilage, and thence obliquely upwards across the subcostal angle to the eighth left costal cartilage. A slightly curved line with its convexity to the left from this point, i.e. the eighth left costal cartilage, to the termination of the line indicating the upper limit, will denote the left margin of the liver. Birmingham teaches that the limits of the normal liver may be marked out on the surface of the body in the following manner. Take three points : 1, half an inch below the right nipple; 2, half an inch below the tip of the tenth rib; and 3, one inch below the left nipple. Join 1 and 3·by a line slightly convex upwards; join 1 and 2 by a line slightly convex outwards, and 2 and 3 by a line slightly convex downwards. The fundus of the gall-bladder approaches the surface behind the anterior extremity of the ninth costal cartilage, close to the outer margin of the right Rectus muscle.

It must be remembered that the liver is subject to considerable alterations in position, and the student should make himself acquainted with the different circumstances under which this occurs, as they are of importance in determining the existence of enlargement or other diseases of the organ.

The position of the liver varies according to the posture of the body. In the erect position in the adult male, the edge of the liver projects about half an inch below the lower edge of the right costal cartilages, and its anterior border can be often felt in this situation if the abdominal wall is thin. In the supine position the liver gravitates backwards, and recedes above the lower margin of the ribs, and cannot then be detected by the finger. In the prone position it falls forward, and can then generally be felt in a patient with loose and lax abdominal walls. Its position varies also with the ascent or descent of the Diaphragm. In a deep inspiration the liver descends below the ribs; in expiration it is raised behind them. Again, in emphysema, where the lungs are distended, and the Diaphragm descends very low, the liver is pushed down : in some other diseases, as phthisis, where the Diaphragm is much arched, the liver rises very high up. Pressure from without, as in tight-lacing, by compressing the lower part of the chest, displaces the liver considerably : its anterior edge frequently extending as low as the crest of the ilium ; and its convex surface is often at the same time deeply indented from the pressure of the ribs. Again, its position varies greatly according to the greater or less distension of the stomach and intestines. When the intestines are empty, the liver descends in the abdomen; but when they are distended, it is pushed upwards. Its relations to surrounding organs may also be changed by the growth of tumours, or by collections of fluid in the thoracic or abdominal cavities.

Surgical Anatomy.—On account of its large size, its fixed position, and its friability, the liver is more frequently ruptured than any of the abdominal viscera. The rupture may vary from a slight scratch to an extensive and complete laceration of its substance, dividing it into two parts. Sometimes an internal rupture, without laceration of the peritoneal covering, takes place, and such injuries are most susceptible of repair; but small tears of the surface may also heal ; when, however, the laceration is extensive, death usually takes place from hæmorrhage, on account of the fact that the hepatic veins are contained in rigid canals in the liver-substance and are unable to contract, and are moreover unprovided with valves. The liver may also be torn by the end of a broken rib perforating the Diaphragm. It may be injured by stabs or other punctured wounds, and when these are inflicted through the chest wall both pleural and peritoneal cavities may be opened up, and both lung and liver wounded. In cases of wound of the liver from the front, hernia of a part of this viscus may take place, but generally can be easily replaced. In cases of laceration of the liver, when there is evidence that bleeding is going on, the abdomen must be opened, the laceration sought for, and the bleeding arrested. This may be done temporarily by introducing the forefinger into the foramen of Winslow and placing the thumb on the gastro-hepatic omentum and compressing the hepatic artery and portal vein between the two. Any bleeding points can then be seen and tied and the margins of the laceration, if small, brought together and sutured by means of a blunt curved needle passed from one side of the wound to the other. All sutures must be passed before any are tied, and this must be done with the greatest gentleness as the liver substance is very friable. When the laceration is extensive it must be packed with iodoform gauze, the end of which is allowed to hang out of the external wound. Abcess of the liver is of not infrequent occurrence, and may open in many different ways on account of the relations of this viscus to other organs. Thus it has been known to burst into the lungs and the pus coughed up, or into the stomach and the pus vomited ; it may burst into the colon, or into the duodenum ; or, by perforating the Diaphragm, it may empty itself into the pleural cavity. It often makes its way forwards, and points on the anterior abdominal wall, and finally it may burst into the peritoneal or pericardiac cavities. Abscesses of the liver frequently require opening, and this must be done by an incision in the abdominal wall, in the thoracic wall, or in the lumbar region, according to the direction in which the abscess is tracking. The incision through the abdominal wall is to be preferred when possible. The abdominal wall is incised over the swelling, and unless the peritoneum is adherent, sponges are packed all round the exposed liver surface and the abscess opened, if deeply seated preferably by the thermo-cautery. Hydatid cysts are more often found in the liver than in any of the other viscera. The reason of this is not far to seek. The embryo of the egg of the tænia echinococcus, being liberated in the stomach by the disintegration of its shell, bores its way through the gastric walls and usually enters a blood-vessel, and is carried by the blood-stream to the hepatic capillaries, where its onward course is arrested, and where it undergoes development into the fully formed hydatid. Tumours of the liver have recently been subjected to surgical treatment by removal of a portion of the organ. The abdomen is opened and the diseased portion of liver exposed; the circulation is controlled by compressing the portal vein and the hepatic artery in the gastro-hepatic omentum and a wedge-shaped portion of liver containing the tumour removed; the divided vessels are ligatured and the cut surfaces brought together and sutured in the manner directed above.

When the *gall-bladder* or one of its main ducts is ruptured, which may occur independently of laceration of the liver, death usually results from peritonitis. If the symptoms have led to the performance of a laparotomy and a rent is found, it should be sutured if small, or the gall-bladder removed if it is extensive. If the cystic duct is torn, its intestinal end must be closed and the gall-bladder removed. In rupture of either of the other ducts, the only thing which can be done is to provide for free drainage, in the hope that a biliary fistula may form.

The gall-bladder may become distended in cases of obstruction of its duct or the common bile-duct, or from a collection of gall-stones in its interior, thus forming a large tumour. The swelling is pear-shaped, and projects downwards and forwards to the umbilicus. It moves with respiration, since it is attached to the liver. To relieve this condition, the gall-bladder must be opened and the gall-stones removed. The operation is performed by an incision, two or three inches long in the right semilunar line, commencing at the costal margin. The peritoneal cavity is opened, and the tumour having been found, sponges are packed round it to protect the peritoneal cavity, and it is aspirated. When the contained fluid has been evacuated the flaccid bladder is drawn out of the abdominal wound and its wall incised to the extent of an inch; any gall-stones in the bladder are now removed and the interior of the sac sponged dry. If the case is one of obstruction of the duct, an attempt must be made to dislodge the stone by manipulation through the wall of the duct; or it may be crushed from without by the fingers or carefully padded forceps. If this does not succeed, the safest plan is to incise the duct, extract the stone, and close the incision by fine sutures in two layers. After all obstruction has been removed, four courses are open to the surgeon: 1. The wound in the gall-bladder may be at once sewn up, the organ returned into the abdominal cavity, and the external incision closed. 2. The edges of the incision in the gall-bladder may be sutured to the external wound, and a fistulous communication established between the gall-bladder and the exterior; this fistulous opening usually closes in the course of a few weeks. 3. The gall-bladder may be connected with the intestinal canal, preferably the duodenum, by means of a lateral anastomosis; this is known as cholecystenterostomy. 4. The gall-bladder may be completely removed.

THE PANCREAS

Dissection.—The pancreas may be exposed for dissection in three different ways: 1. By raising the liver, drawing down the stomach, and tearing through the gastro-hepatic omentum and the ascending layer of the transverse mesocolon. 2. By raising the stomach, the arch of the colon, and great omentum, and then dividing the inferior layer of the transverse mesocolon and raising its ascending layer. 3. By dividing the two layers of peritoneum, which descend from the great curvature of the stomach to form the great omentum; turning the stomach upwards, and then cutting through the ascending layer of the transverse mesocolon (see fig. 720, page 1059).

The **Pancreas** (παν-κρέας, *all flesh*) is a compound racemose gland, analogous in its structure to the salivary glands, though softer and less compactly arranged than those organs. It is long and irregularly prismatic in shape, and has been compared to a human or a dog's tongue: its right extremity being broad, is called the *head*—this is connected to the main portion of the organ, the *body*, by a slight constriction, the *neck*; while its left extremity gradually tapers to form the *tail*. It is situated transversely across the posterior wall of the abdomen, at the back of the epigastric and left hypochondriac regions. Its length varies from five to six inches, its breadth is an inch and a half, and its thickness from half an inch to an inch, being greater at its right extremity and along its upper border. Its weight varies from two to three and a half ounces, but it may reach six ounces.

The *right extremity* or *head of the pancreas* (fig. 750) is shaped like a crook, the convexity of which conforms to the concavity of the duodenum, which is slightly overlapped by it. It is flattened from before backwards. Behind, the head of the pancreas is in relation with the inferior vena cava, the left renal vein, the right crus of the Diaphragm, and the aorta. The superior mesenteric vessels, passing forwards between the pancreas and duodenum, are lodged in

4 B

a notch in the head of the viscus to the right of the neck; they then cross in front of the head, near its left border. The anterior surface is in contact with the commencement of the transverse colon, and below this is covered by peritoneum, which is continuous with the transverse mesocolon. The common bile-duct descends behind, between the descending portion of the duodenum and the head of the pancreas; and the superior pancreatico-duodenal artery descends in front between the same parts.

The *neck of the pancreas* is about an inch long, and passes upwards and to the left, having the first part of the duodenum above it, and the termination of the fourth portion below. It lies in front of the commencement of the portal vein, and is grooved on the right by the gastro-duodenal and superior pancreatico-duodenal arteries. The pylorus lies just above it.

The *body and tail of the pancreas* are somewhat prismatic in shape, and have three surfaces: anterior, posterior, and inferior.

The *anterior surface* is somewhat concave, and is directed forwards and upwards: it is covered by the posterior surface of the stomach which rests upon

Fig. 750.—The pancreas and its relations.

it, the two organs being separated by the lesser sac of the peritoneum. At its right extremity there is a well-marked prominence, the *omental tuberosity*, which abuts against the posterior surface of the small omentum.

The *posterior surface* is devoid of peritoneum and separated from the vertebral column by the aorta, the splenic vein, the left kidney and its vessels, the left suprarenal capsule, the pillars of the Diaphragm, and the origin of the superior mesenteric artery.

The *inferior surface* is narrow and covered by peritoneum; it lies upon the duodeno-jejunal flexure and on some coils of the jejunum; its left extremity rests on the splenic flexure of the colon.

The *superior border* of the body is blunt and flat to the right; narrow and sharp to the left, near the tail. It commences to the right in the omental tuberosity, and is in relation with the cœliac axis, from which the hepatic artery courses to the right just above the gland, while the splenic branch runs in a groove along this border to the left.

The *anterior border* separates the anterior from the inferior surface, and along this border the two layers of the transverse mesocolon diverge from one another: one passing upwards over the anterior surface, the other backwards over the inferior surface.

The *lesser end* or *tail of the pancreas* is narrow; it extends to the left as far as the lower part of the gastric surface of the spleen, and it is in contact with the splenic flexure of the colon.

Birmingham describes the body of the pancreas as projecting forwards as a prominent ridge into the abdominal cavity and forming a sort of shelf on which the stomach lies. He says: 'The portion of the pancreas to the left of the middle line has a very considerable antero-posterior thickness; as a result the anterior surface is of considerable extent, it looks strongly upwards and forms a large and important part of the shelf. As the pancreas extends to the left towards the spleen it crosses the upper part of the kidney, and is so moulded on to it that the top of the kidney forms an extension inwards and backwards of the upper surface of the pancreas and extends the bed in this direction. On the

Fig. 751.—The duodenum and pancreas.
The liver has been lifted up and the greater part of the stomach removed. (Testut.)

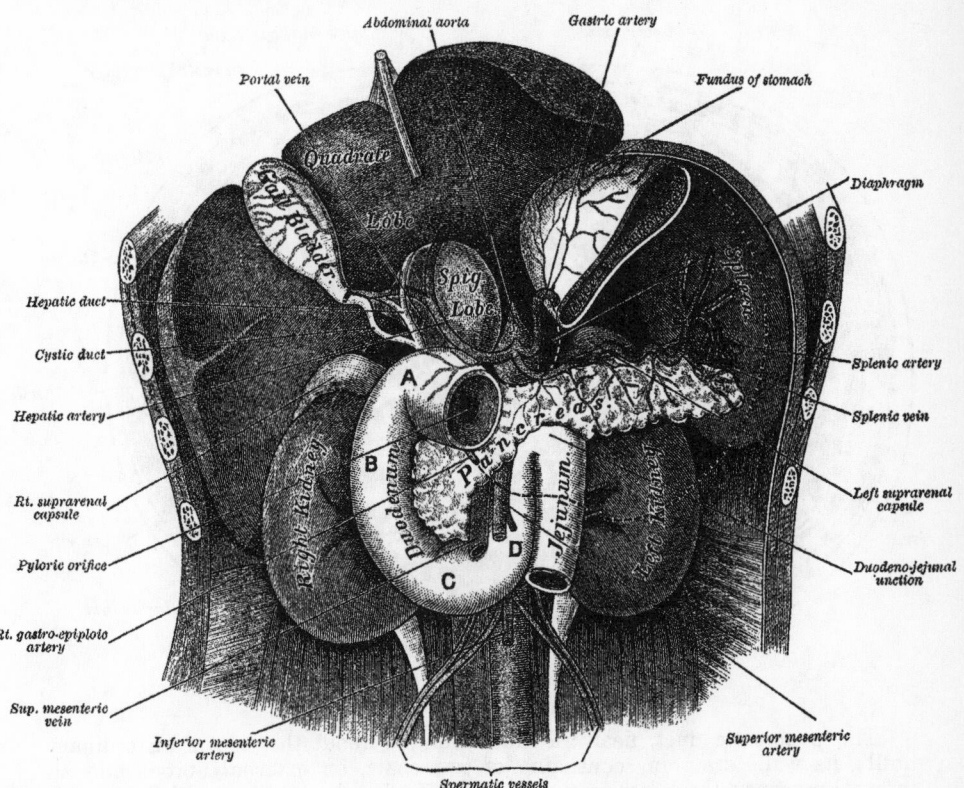

A, B, C, D. The four portions of the duodenum.

other hand, the extremity of the pancreas comes in contact with the spleen in such a way that the plane of its upper surface runs with little interruption upwards and backwards into the concave gastric surface of the spleen, which completes the bed behind and to the left, and running upwards, forms a partial cap for the wide end of the stomach ' * (see fig. 728).

The principal excretory duct of the pancreas, called the **pancreatic duct** or **canal of Wirsung,** from its discoverer, extends transversely from left to right through the substance of the pancreas. In order to expose it, the superficial portion of the gland must be removed. It commences by the junction of the small ducts of the lobules situated in the tail of the pancreas, and, running from right to left through the body, it receives the ducts of the various lobules composing the gland. Considerably augmented in size, it reaches the neck, and

* *Journal of Anatomy and Physiology,* vol. xxxi. pt. I, p. 102.

turning downwards, backwards and to the right, it comes into relation with the common bile-duct, lying to its left side ; leaving the head of the gland it passes very obliquely through the mucous and muscular coats of the duodenum, and terminates by an orifice common to it and the common bile-duct upon the summit of an elevated papilla, situated at the inner side of the descending portion of the duodenum, three or four inches below the pylorus.

Sometimes the pancreatic duct and the common bile-duct open separately into the duodenum. Frequently there is an accessory duct, which is given off from the canal of Wirsung in the neck of the pancreas and passes horizontally to the right to open into the duodenum about an inch above the orifice of the main duct. It receives the ducts from the lower part of the head, and is known as the *ductus pancreaticus accessorius* or *ductus Santorini*.

FIG. 752.—Transverse section through the middle of the first lumbar vertebra, showing the relations of the pancreas. (Braune.)

The pancreatic duct, near the duodenum, is about the size of an ordinary quill : its walls are thin, consisting of two coats, an external fibrous and an internal mucous ; the latter is smooth, and furnished near its termination with a few scattered follicles.

In **structure,** the pancreas resembles the salivary glands. It differs from them, however, in certain particulars, and is looser and softer in its texture. It is not enclosed in a distinct capsule, but is surrounded by areolar tissue, which dips into its interior, and connects together the various lobules of which it is composed. Each lobule, like the lobules of the salivary glands, consists of one of the ultimate ramifications of the main duct, terminating in a number of cæcal pouches or alveoli, which are tubular and somewhat convoluted. The minute ducts connected with the alveoli are narrow and lined with flattened cells. The alveoli are almost completely filled with secreting cells, so that scarcely any lumen is visible. In some animals those cells which occupy the centre of the alveolus are spindle-shaped, and are known as the *centro-acinar cells of Langerhans.* The true secreting cells which line the wall of the alveolus are very characteristic. They are columnar in shape and present two zones : an outer one clear and finely striated next the basement-membrane, and an inner granular one next the lumen. During activity the granular zone occupies the greater part of the cell : before the cells are called into action, while in a condition of rest, the outer or clear zone is the larger. In some of the secreting cells of the pancreas is a spherical mass, staining more easily than the rest of the cell ; this is termed the *paranucleus*, and is believed to be an extension from the nucleus. The connective tissue between the alveoli presents in certain parts collections of cells, which are termed *inter-alveolar cell-islets*.

Vessels and Nerves.—The *arteries of the pancreas* are derived from the splenic and the pancreatico-duodenal branches of the hepatic and the superior mesenteric. Its *veins* open into the splenic and superior mesenteric veins. Its *lymphatics* terminate in the lumbar glands. Its *nerves* are filaments from the splenic plexus.

Surface Form.—The pancreas lies in front of the second lumbar vertebra, and can sometimes be felt, in emaciated subjects, when the stomach and colon are empty, by making deep pressure in the middle line about three inches above the umbilicus.

Surgical Anatomy. — Inflammation of the pancreas has of late years received considerable attention. It appears to be due to infection of the pancreatic ducts by micro-organisms from the duodenum in cases of gastro-duodenal catarrh, or from the biliary passages in which a gall-stone is lodged. Acute cases usually terminate fatally from peritonitis, but subacute and chronic cases generally run on to suppuration, and an operation for the evacuation of the pus becomes necessary. The best mode of reaching the site of the disease is by an incision in the middle line of the abdomen above the umbilicus. When the cavity of the abdomen has been opened, the two layers of peritoneum passing from the great curvature of the stomach to form the great omentum must be divided, the stomach turned upwards, and the ascending layer of the transverse mesocolon cut through. Careful drainage will be necessary, and this is best effected by an opening in the left loin just below the last rib. Cysts of the pancreas are sometimes met with. They may be the result of traumatism, when they generally contain blood, or they may be due to retention from obstruction of a duct, or by pressure on the main duct by a gall-stone. They may attain a large size, and cause symptoms by pressing on the stomach, Diaphragm, or common bile-duct. They generally push their way forwards between the stomach and transverse colon, and may then be felt as a definite tumour in the middle line of the upper part of the abdomen. The tumour is fixed and does not move with respiration. The treatment consists in opening the abdomen in the middle line, incising the cyst, evacuating its contents, and fixing its walls to the deeper layers of the abdominal wall. When they are situated in the tail of the pancreas they may be removed. The pancreas is occasionally the seat of cancer. It usually affects the head, and therefore often speedily involves the common bile-duct, leading to persistent jaundice ; or it may press upon the portal vein, causing ascites, or upon the stomach, causing pyloric obstruction. It has been said that the pancreas is the only abdominal viscus which has never been found in a hernial protrusion ; but even this organ has been found, in company with other viscera, in rare cases of diaphragmatic hernia. The pancreas has been known to become invaginated into the intestine, and portions of the organ have sloughed off. In cases of excision of the pylorus great care must be exercised to avoid wounding the pancreas, as the escape of the pancreatic fluid may be attended with serious results. According to Billroth, it is likely, in consequence of its peptonising qualities, to dissolve the cicatrix of the stomach.

THE URINARY ORGANS

THE KIDNEYS

THE Kidneys, two in number, secrete the urine and are situated in the back part of the abdomen.

They are placed one on either side of the vertebral column, behind the peritoneum, and surrounded by a mass of fat and loose areolar tissue. Their upper extremity is on a level with the upper border of the twelfth dorsal vertebra, their lower extremity on a level with the third lumbar. The right kidney is usually on a slightly lower level than the left, probably on account of the vicinity of the liver.

Each kidney is about four and a half inches in length, two to two and a half in breadth, and rather more than one inch in thickness. The left is somewhat longer, though narrower, than the right. The weight of the kidney in the adult male varies from four and a half ounces to six ounces, in the adult female from four to five and a half ounces. The combined weight of the two kidneys in proportion to the body is about 1 in 240.

The kidney has a characteristic form, and presents for examination two surfaces, two borders, and an upper and lower extremity.

Its *anterior surface* is convex, looks forwards and outwards, and is partially covered by peritoneum. In both kidneys a small portion of the upper part of the anterior surface is in relation with the suprarenal capsule. The *right* kidney, in its upper three-fourths, with the exception of the suprarenal area, is in contact with the posterior part of the under surface of the right lobe of the liver, on which it produces a concave impression, the *impressio renalis* (page 1095). Towards its inner border it is covered by the second part of the duodenum, while its lower and outer part is in relation with the hepatic flexure of the colon. The relation of the second part of the duodenum to the front of the right kidney is a varying one. The *left* kidney is covered above by the posterior surface of the stomach, below the stomach by the pancreas, behind which are the splenic vessels. Its lower half is in contact with some of the coils of the small intestine and sometimes with the third part of the duodenum. Near its outer border the anterior surface lies behind the spleen and the splenic flexure of the colon.

The kidneys are partly covered in front by peritoneum and partly uncovered. On the right kidney, the *hepatic area*—that is to say, that portion of the kidney which produces the renal impression on the liver—is covered by peritoneum, which therefore separates the kidney from the liver: the *duodenal* and *colic areas* are not peritoneal, and these structures are connected to the kidney by loose connective tissue; at the lower and inner extremity is a small area, which is in contact with some of the coils of the small intestine, and is covered by peritoneum and is in relation with branches of the right colic vessels. On the left kidney, the *gastric area* is covered by the peritoneum of the lesser sac; the *pancreatic* and *colic areas* are non-peritoneal: while as on the right side, at the lower and inner extremity, is an area, *mesocolic*, which is covered by the peritoneum of the greater sac and is in relation with the left colic vessels.

The *posterior surface* of the kidney is directed backwards and inwards. It is entirely devoid of peritoneal covering, being embedded in areolar and fatty tissue. It lies upon the Diaphragm, the anterior layer of the lumbar aponeurosis (which separates it from the Quadratus lumborum), the external and internal arcuate ligaments, the Psoas muscle and the tendon of the Transversalis muscle, one or two of the upper lumbar arteries, the last dorsal, ilio-hypogastric, and ilio-inguinal nerves. The right kidney rests upon the twelfth rib, the left usually on the eleventh and twelfth. The Diaphragm separates the kidney from the pleura as it dips down to form the phrenico-costal sinus, but frequently the muscular fibres of the Diaphragm are defective or absent over a triangular area immediately above the external arcuate ligament, and when this is the case the perirenal areolar tissue is in immediate apposition with the diaphragmatic pleura.

The *external border* is convex, and is directed outwards and backwards, towards the postero-lateral wall of the abdomen. On the left side it is in contact, at its upper part, with the spleen.

The *internal border* is concave, and is directed forwards and a little downwards. It presents a deep longitudinal fissure, bounded by prominent

FIG. 753.—The relations of the kidneys, from behind.

1. Spleen. 2. Descending colon. 3. Line of pleural reflection. 4. Liver. 5. Subcostal artery. 6. Last dorsal nerve.
7. Line indicating outer edge of Quadratus lumborum. 8. Ilio-inguinal nerve. 9. Ilio-hypogastric nerve.
10. Ascending colon. 11. Psoas.

overhanging anterior and posterior lips. This fissure is named the *hilus*, and allows of the passage of the vessels, nerves, and ureter into and out of the kidney.

The *superior extremity*, directed slightly inwards as well as upwards, is thick and rounded, and is surmounted by the suprarenal capsule, which covers also a small portion of the anterior surface.

The *inferior extremity*, directed a little outwards as well as downwards, is smaller and thinner than the superior. It extends to within two inches of the crest of the ilium.

At the hilus of the kidney the relative position of the main structures passing into and out of the kidney is as follows: the vein is in front, the artery in the

middle, and the duct or ureter behind and directed downwards. By a knowledge
of these relations, the student may distinguish between the right and left kidney.
The kidney is to be laid on the table, before the student, on its posterior surface,
with its lower extremity towards the observer—that is to say, with the ureter
behind and *below* the vessels; the hilus will then be directed to the side to
which the kidney belongs.

General Structure of the Kidney.—The kidney is surrounded by a distinct
investment of fibrous tissue, which forms a firm, smooth covering to the organ.
It closely invests it, but can be easily stripped off, in doing which, however,
numerous fine processes of connective tissue and small blood-vessels are torn
through. Beneath this coat, a thin wide-meshed network of unstriped muscular
fibre forms an incomplete covering to the organ. When the fibrous coat is
stripped off, the surface of the kidney is found to be smooth and even, and
of a deep red colour.

In infants, fissures extending for some depth may be seen on the surface of
the organ, a remnant of the lobular construction of the gland. The kidney
is dense in texture, but is easily lacerable by mechanical force. In order to
obtain a knowledge of the structure of the gland, a vertical section must be made
from its convex to its concave border, and the loose tissue and fat removed
from around the vessels and the excretory duct (fig. 754). It will be then seen
that the kidney consists of a central cavity, surrounded at all parts but one by
the proper kidney-substance. This central cavity is called the *sinus*, and is
lined by a prolongation of the fibrous coat of the kidney, which is continued
round the lips of the hilus. Through this hilus the blood-vessels of the kidney
and its excretory duct pass, and therefore these structures, upon entering or
leaving the kidney, are contained within the sinus. The excretory duct or
ureter, after entering, dilates into a wide, funnel-shaped expansion, named the
pelvis of the kidney. This divides into two or three tubular divisions, which
subdivide into several short, truncated branches, named *calyces* or *infundibula*,
all of which are contained in the central cavity of the kidney. The blood-vessels
of the kidney, after passing through the hilus, are contained in the sinus or
central cavity, lying between its lining membrane and the excretory apparatus
before entering the kidney-substance.

This central cavity, as before mentioned, is surrounded on all sides, except
at the hilus, by the substance of the kidney, which consists of two parts, viz.
an external, granular, investing part, which is called the *cortical portion*, and
an internal part, the *medullary portion*, made up of a number of dark-coloured
pyramidal masses, the pyramids of Malpighi, the bases of which rest on the
cortical part, while their apices converge towards the sinus, where they form
prominent papillæ, which project into the interior of the calices.

The *cortical substance* is of a bright reddish-brown colour, soft, granular, and
easily lacerable. It is found everywhere immediately beneath the capsule, and
is seen to extend in an arched form over the base of each medullary pyramid.
The parts dipping in between the pyramids, towards the sinus through which
the arteries and nerves enter, and the veins and lymphatics emerge from the
kidney, are called the *cortical columns* or *columnæ Bertini* (A A', fig. 754); while
that portion which stretches from one cortical column to the next, and intervenes between the base of the pyramid and the capsule (which is marked by
the dotted line, extending from A to A' in fig. 754), is called a *cortical arch*,
the depth of which varies from a third to half an inch.

The *medullary substance*, as before said, is seen to consist of red-coloured,
striated, conical masses, the *pyramids of Malpighi*; the number of which,
varying from eight to eighteen, corresponds to the number of lobes of which
the fœtal kidney is composed. The bases of the pyramids are surrounded by
the cortical arches and directed towards the circumference of the kidney; the
sides are contiguous with the cortical columns; while the apices, known as
the *renal papillæ*, project into the calyces of the ureter, each calyx receiving
one, two, or perhaps three papillæ.

These two parts, *cortical* and *medullary*, so dissimilar in appearance, are
very similar in structure, being made up of urinary tubes and blood-vessels,
united and bound together by a connecting matrix or stroma.

Minute Anatomy.—The *tubuli uriniferi*, of which the kidney is for the most
part made up, commence in the cortical portion of the kidney, and, after pursuing

a very circuitous course through the cortical and medullary parts, finally terminate at the apices of the Malpighian pyramids by open mouths (fig. 755), so that the fluid which they contain is emptied, through the calices, into the dilated extremity of the ureter contained in the sinus of the kidney. If the surface of one of the papillæ is examined with a lens, it will be seen to be studded over with small depressions, from sixteen to twenty in number, and if pressure is made on a fresh kidney, fluid will be seen to exude from these depressions. They are the orifices of the tubuli uriniferi, which terminate in this situation. The tubuli uriniferi commence in the cortical portion of the kidney as the *Malpighian bodies*, which are small rounded masses, varying in size, but of an average of about $\frac{1}{120}$ of an inch in diameter. They are of a deep red colour, and are found only in the cortical portion of the kidney. Each of these little bodies is composed of two parts : a central glomerulus of

FIG. 754.—Vertical section of kidney.   FIG. 755.—Plan of uriniferous tubes.

vessels, called a *Malpighian tuft*; and a membranous envelope, the *Malpighian capsule*, or *capsule of Bowman*, which is a small pouch-like commencement of a uriniferous tubule.

A A. Malpighian bodies. B B. Margin of medullary structure. C C C. Loops of Henle. D D D. Straight tubes cut of. E. Commencing straight tubes. F. Termination of straight tube.

The *Malpighian tuft*, or vascular glomerulus, is a network of convoluted capillary blood-vessels, held together by scanty connective tissue and grouped into from two to five lobules. This capillary network is derived from a small arterial twig, the *afferent vessel*, which pierces the wall of the capsule, generally at a point opposite that at which the latter is connected with the tube; and the resulting vein, the *efferent vessel*, emerges from the capsule at the same point. The afferent vessel is usually the larger of the two (fig. 756). The *Malpighian*, or *Bowman's capsule*, which surrounds the glomerulus, is formed of a hyaline membrane, supported by a small amount of connective tissue, which is continuous with the connective tissue of the tube. It is lined on its inner surface by a layer of squamous epithelial cells, which are reflected from the lining membrane on to the glomerulus, at the point of entrance or exit of the afferent and efferent vessels. The whole surface of the glomerulus is covered with a continuous layer of the same cells, on a delicate supporting membrane, which with the cells dips in between the lobules of the glomerulus, closely surrounding them (fig. 757). Thus between the glomerulus and the capsule a space is left, forming a cavity lined by a continuous layer of cells; this cavity varies in size according to the state of secretion and the amount of fluid present in it. The cells, as above stated, are squamous in the adult, but in the fœtus and young subject they are polyhedral or even columnar.

The *tubuli uriniferi*, commencing in the Malpighian bodies, present, during their course, many changes in shape and direction, and are contained partly in the medullary and partly in the cortical portions of the organ. At their junction with the Malpighian capsule they exhibit a somewhat constricted portion, which is termed the *neck*. Beyond this the tube becomes convoluted, and pursues a considerable course in the cortical structure, constituting the *proximal convoluted tube*. After a time the convolutions disappear, and the tube approaches the medullary portion of the kidney in a more or less spiral manner. This section of the tube has been called the *spiral tube of Schachowa*. Throughout this portion of their course the tubuli uriniferi are contained entirely in the cortical structure, and present a fairly uniform calibre. They now enter the medullary portion, suddenly become much smaller, quite straight in direction, and dip down for a variable depth into the pyramids, constituting the *descending limb of Henle's loop*. Bending on themselves, they form what is termed the *loop of Henle*, and re-ascending, they become suddenly enlarged and again spiral in direction, forming the *ascending limb of Henle's loop* and re-enter the cortical structure. This portion of the tube does not present

FIG. 756.—Minute structure of kidney.

FIG. 757.—Malpighian body.

a uniform calibre, but becomes narrower as it ascends, and is irregular in outline. As a narrow tube it enters the cortex and ascends for a short distance, when it again becomes dilated, irregular, and angular. This section is termed the *irregular tubule*; it terminates in a convoluted tube, which exactly resembles the proximal convoluted tubule, and is called the *distal convoluted tubule*. This again terminates in a narrow *curved tube*, which enters the straight or collecting tube.

Each *straight* or *collecting tube* commences by a small orifice on the summit of one of the papillæ, thus opening and discharging its contents into the interior of one of the calices. These tubes run from the apex to the base of the pyramid, dividing dichotomously in their course and slightly diverging from each other. Thus dividing and subdividing, they reach the base of the pyramid, and enter the cortical structure greatly increased in number. Upon entering the cortical portion they continue a straight course for a variable distance, and are arranged in groups, called *medullary rays*, several of these groups corresponding to a single pyramid. The tubes in the centre of the group are the longest, and reach almost to the surface of the kidney, while the external ones are shorter, and advance only a short distance into the cortex. In consequence of this arrangement the cortical portion presents a number of conical masses, the apices of which reach the periphery of the organ, and the bases are applied to the medullary portion. These are termed the *pyramids of Ferrein*. As they run through the cortical portion, the straight tubes receive on either side the curved extremity of the distal convoluted tubules.

It will be seen from the above description that the tubes are continuous from their commencement in the Malpighian bodies to their termination at the orifices on the apices of the pyramids of Malpighi; and that the urine, the secretion of which commences in the capsule, will find its way through these tubes into the calices of the kidney, and so into the ureter. Commencing at the capsule, the tube first presents a narrow constricted portion, (1) the *neck*. (2) It forms a wide convoluted tube, the *proximal convoluted tube*. (3) It becomes spiral, the *spiral tubule of Schachowa*. (4) It enters the medullary structure as a narrow, straight tube, the *descending limb of Henle's loop*. (5) It forms Henle's loop and becoming dilated, it ascends somewhat spirally, and, gradually diminishing in calibre, again enters the cortical structure, the *ascending limb of Henle's loop*. (6) It now becomes irregular and angular in outline, the *irregular tubule*. (7) It then becomes convoluted, the *distal convoluted tubule*. (8) Diminishing in size,

FIG. 758.—Uriniferous tube.

N.B.—For the sake of clearness the epithelial cells have been represented more highly magnified than the tubes in which they are contained.

it forms a curve, the *curved tubule*. (9) Finally it joins a straight tube, the *straight collecting tube*, which is continued downwards through the medullary substance to open at the apex of a pyramid.

The Tubuli Uriniferi : their Structure.—The tubuli uriniferi consist of basement-membrane lined with epithelium. The epithelium varies considerably in different sections of the uriniferous tubes. In the neck the epithelium is continuous with that lining the Malpighian capsule, and like it consists of flattened cells each containing an oval nucleus (fig. 758, A). In the proximal convoluted tubule and the spiral tubule of Schachowa the epithelium is polyhedral in shape, the sides of the cells not being straight, but fitting into each other, and in some animals so fused together that it is impossible to make out the lines of junction. In the human kidney the cells often present an angular projection of the surface next the basement-membrane. These cells are made up of more or less rod-like fibres, which rest by one extremity on the basement-membrane, while the other projects towards the lumen of the tube. This gives to the cells the appearance of distinct striation (Heidenhain) (fig. 758, B). In the descending limb of Henle's loop the epithelium resembles that found in the Malpighian capsule and the commencement of the tube, consisting of flat transparent epithelial plates with an oval nucleus (figs. 758, A; 759). In the ascending limb, on the other hand, the cells partake more of the character of those described as existing in the proximal convoluted tubule, being polyhedral in shape, and presenting the same appearance of striation. The nucleus, however, is not situated in the centre of the cell, but near the lumen (fig. 758, C). After the ascending limb of Henle's loop becomes narrower upon entering the cortical structure, the striation appears to be confined to the outer part of the cell; at all events it is much more distinct in this situation; the nucleus, which appears flattened and angular, being still situated near the lumen (fig. 758, D). In the irregular tubule, the cells undergo a still further change, becoming very angular, and presenting thick bright rods or markings, which render the striation much more distinct than in any other section of the urinary tubules (fig. 758, H). In the distal convoluted tubule the epithelium appears to be somewhat similar to that which has been described as existing in the proximal convoluted tubule, but presents a peculiar refractive appearance (fig. 758, B). In the curved tubule, just before its entrance into the straight collecting tube, the epithelium varies greatly as regards the shape of the cells, some being angular with short processes, others spindle-shaped, others polyhedral (fig. 758, E).

FIG. 759.*—Longitudinal section of Henle's descending limb.

FIG. 760.—Longitudinal section of straight tube.

a. Membrana propria.
b. Epithelium.

a. Cylindrical or cubical epithelium.
b. Membrana propria.

In the straight tubes the epithelium is more or less columnar: in its papillary portion the cells are distinctly columnar and transparent (figs. 760, 761); but as the tube approaches the cortex the cells are less uniform in shape: some are polyhedral, and others angular with short processes (fig. 758, F and G).

The Renal Blood-vessels.—The kidney is plentifully supplied with blood by the renal artery, a large offset of the abdominal aorta. Previously to entering the kidney, each artery divides into four or five branches, which are distributed to its substance. At the hilus these branches lie between the renal vein and ureter, the vein being in front, the ureter behind; one branch usually lies behind the ureter. Each vessel gives off some small branches to the suprarenal capsules, the ureter, and the surrounding cellular tissue and muscles. Frequently there is a second renal artery, which is given off from the abdominal aorta at a lower level, and supplies the lower portion of the kidney. It is termed the *inferior renal artery.* Sometimes an additional artery enters the upper part of the kidney. The branches of the renal artery, while in the sinus, give off a few twigs for the nutrition of the surrounding tissues, and terminate in the *arteriæ propriæ renales*, which enter the kidney proper in the columns of Bertini. Two of these pass to each pyramid of Malpighi, and run along its sides for its entire

* From the *Handbook for the Physiological Laboratory.*

length, giving off as they advance the afferent vessels of the Malpighian bodies in the columns. Having arrived at the bases of the pyramids, they make a bend in their course, so as to lie between the bases of the pyramids and the cortical arches, where they break up into two distinct sets of branches devoted to the supply of the remaining portions of the kidney.

FIG. 761.—Transverse section of pyramidal substance of kidney of pig, the blood-vessels of which are injected.

a. Large collecting tube, cut across, lined with cylindrical epithelium. *b*. Branch of collect-ing tube, cut across, lined with epithelium with shorter cylinders. *c, d*. Henle's loops cut across. *e*. Blood-vessels cut across. D. Connective-tissue ground-substance.

The *first set*, the *interlobular arteries* (figs. 762, 763, B), are given off at right angles from the side of the arteriæ propriæ renales looking towards the cortical substance, and passing directly outwards between the pyramids of Ferrein they

FIG. 762.—Diagrammatic sketch of the blood-vessels of the kidney.

FIG. 763.—A portion of fig. 762 enlarged. (The references are the same.)

A *a*. Proper renal artery and vein, the former giving off the renal afferents, the latter receiving the renal efferents. B *b*. Interlobular artery and vein, the latter commencing from the stellate veins, and receiving branches from the plexus around the tubuli contorti, the former giving off renal afferents. c. Straight tube, surrounded by tubuli contorti, with which it communicates, as more fully shown in fig. 755. D. Margin of medullary substance. E E E. Receiving tubes, cut off. F *f*. Arteriolæ et venæ rectæ, the latter arising from (G) the plexus at the medullary apex.

reach the capsule, where they terminate in the capillary network of this part. In their outward course they give off lateral branches; these are the *afferent vessels* for the Malpighian bodies (see page 1113), and, having pierced the capsule, end in the Malpighian tufts. From each tuft the corresponding *efferent vessel* arises, and, having made its egress from the capsule near to the point where the afferent vessel entered, breaks up into a number of branches, which form a dense *plexus* around the adjacent urinary tubes (fig. 764).

The *second set of branches* from the arteriæ propriæ renales supply the medullary pyramids, which they enter at their bases; and, passing straight through their substance to their apices, terminate in the venous plexuses found in that situation. They are called the *arteriæ rectæ* (figs. 762, 763, F).

The *renal veins* arise from *three sources*, viz.: the veins beneath the capsule, the plexuses around the convoluted tubules in the cortical arches, and the plexuses situated at the apices of the pyramids of Malpighi. The veins beneath the capsule are stellate in arrangement, and are derived from the capillary network of the capsule, into which the terminal branches of the interlobular arteries break up. These join to form the *venæ interlobulares*, which pass inwards between the pyramids of Ferrein, receive branches from the plexuses around the convoluted tubules, and, having arrived at the bases of the Malpighian pyramids, join with the venæ rectæ, next to be described (figs. 762, 763, *b*).

The *venæ rectæ* are branches from the plexuses at the apices of the medullary pyramids, formed by the terminations of the arteriæ rectæ. They pass outwards in a straight course between the tubes of the medullary structure, and joining, as above stated, the venæ interlobulares, form the proper renal veins (figs. 762, 763, *f*).

FIG. 764.—Diagrammatic representation of the blood-vessels in the substance of the cortex of the kidney.

m. Region of the medullary ray. *b*. Region of the tortuous portion of the tubules. *ai*. Arteria interlobularis. *vi*. Vena interlobularis. *va*. Vas afferens. *gl*. Glomerulus. *ve*. Vas efferens. *v*. Venous twig of the interlobularis. (From Ludwig, in Stricker's 'Handbook.')

These vessels, *venæ propriæ renales*, accompany the arteries of the same name, running along the entire length of the sides of the pyramids; and, having received in their course the efferent vessels from the Malpighian bodies in the adjacent cortical structure, quit the kidney substance to enter the sinus. In this cavity they join the corresponding veins from the other pyramids to form the *renal vein*, which emerges from the kidney at the hilus and opens into the inferior vena cava; the left being longer than the right and crossing in front of the abdominal aorta.

Nerves of the Kidney.—The nerves of the kidney, although small, are about fifteen in number. They have small ganglia developed upon them, and are derived from the renal plexus, which is formed by branches from the solar plexus, the lower and outer part of the semilunar ganglion and aortic plexus, and from the lesser and smallest splanchnic nerves. They communicate with the spermatic plexus, a circumstance which may explain the occurrence of pain in the testicle in affections of the kidney. So far as they have been traced, they seem to accompany the renal artery and its branches, but their exact mode of termination is not known.

The *lymphatics* consist of a superficial and deep set, which terminate in the lumbar glands.

Connective tissue, or *intertubular stroma.*—Although the tubules and vessels are closely packed, a small amount of connective tissue, continuous with the capsule, binds them firmly together. This tissue was first described by Goodsir, and subsequently by Bowman. Ludwig and Zawarykin have observed distinct fibres passing around the Malpighian bodies; and Henle has seen them between the straight tubes composing the medullary structure.

Surface Form.—The kidneys, being situated at the back part of the abdominal cavity and deeply placed, cannot be felt unless enlarged or misplaced. The greater part of each kidney lies in the epigastric region, i.e. internal to the mid-Poupart plane, but a small part is situated outside this plane, viz. in the hypochondriac region. The lower end of the left kidney is usually on a level with the subcostal plane: that of the right extends for about half an inch below this plane. The hilus is on a level with Addison's transpyloric plane. The left is somewhat higher than the right. According to Morris, the position of the kidney may be thus defined: *Anteriorly*. ' 1. A horizontal line through the umbilicus is below the lower edge of each kidney. 2. A vertical line carried upwards to the costal arch from the middle of Poupart's ligament has one third of the kidney to its outer side, and two thirds to its inner side, i.e. between this line and the median line of the body.' In adopting these lines it must be borne in mind that the axes of the kidneys are not vertical, but oblique, and if continued upwards would meet about the ninth dorsal vertebra. *Posteriorly.* The upper end of the left kidney would be defined by a line drawn horizontally outwards from the spinous process of the eleventh dorsal vertebra, and its lower end by a point two inches above the iliac crest. The right kidney would be half to three-quarters of an inch lower. Morris lays down the following rules for indicating the position of the kidney on the posterior surface of the body : ' 1. A line parallel with, and one inch from, the spine between the lower edge of the tip of the spinous process of the eleventh dorsal vertebra, and the lower edge of the spinous process of the third lumbar vertebra. 2. A line from the top of this first line outwards at right angles to it for two and three-quarter inches. 3. A line from the lower end of the first transversely outwards for two and three-quarter inches. 4. A line parallel to the first, and connecting the outer extremities of the second and third lines just described.'

The hilus of the kidney lies about two inches from the middle line of the back at the level of the spinous process of the first lumbar vertebra.

Surgical Anatomy.—Malformations of the kidney are not uncommon. There may be an entire absence of one kidney, but, according to Morris, the number of these cases is ' excessively small : ' or there may be congenital atrophy of one kidney, when the kidney is very small, but usually healthy in structure. These cases are of great importance, and must be duly taken into account, when nephrectomy is contemplated. A more common malformation is where the two kidneys are fused together. They may be only joined together at their lower ends by means of a thick mass of renal tissue, so as to form a horseshoe-shaped body, or they may be completely united, forming a disc-like kidney, from which two ureters descend into the bladder. These fused kidneys are generally situated in the middle line of the abdomen, but may be misplaced as well. In some mammals (e.g. ox and bear) the kidney consists of a number of distinct lobules ; this lobulated condition is characteristic of the kidney of the human fœtus, and traces of it may persist in the adult.

One or both kidneys may be misplaced as a congenital condition, and remain fixed in this abnormal position. They are then very often misshapen. They may be situated higher, though this is very uncommon, or lower than normal or removed farther from the spine than usual or they may be displaced into the iliac fossa, over the sacro-iliac joint, on to the promontory of the sacrum, or into the pelvis between the rectum and bladder or by the side of the uterus. In these latter cases they may give rise to very serious trouble. The kidney may also be misplaced as a congenital condition, but may not be fixed. It is then known as a *floating kidney.* It is believed to be due to the fact that the kidney is completely enveloped by peritoneum which then passes backwards to the spine as a double layer, forming a mesonephron, which permits of movement taking place. The kidney may also be misplaced as an acquired condition ; in these cases the kidney is mobile in the tissues by which it is surrounded, either moving in its capsule or else moving with the capsule in the perinephritic tissues. This condition is known as *movable kidney,* and is more common in the female than in the male. Sometimes the pelvis is duplicated, while a double ureter is not very uncommon. In some rare instances a third kidney may be present.

Injuries of the kidney are generally due to some severe crushing force, as from being run over by a heavy waggon or cart, or from the abdomen being compressed between the buffers of two railway carriages. When a laceration occurs on the posterior surface of the organ, infiltration of blood and urine takes place into the retro-peritoneal connective tissue ; this is often followed by suppuration, and death may ensue from septic poisoning. When the laceration is in front, the peritoneum may be torn and extravasation of blood and urine take place into the peritoneal cavity. Death may occur from hæmorrhage or peritonitis. Occasionally, when rupture involves the pelvis of the kidney or the commencement of the ureter, this duct may become blocked, and hydronephrosis follow. Sometimes the kidney may be bruised by blows in the loin, or by being compressed between the lower ribs and the ilium when the body is violently bent forwards. This is followed by a little transient hæmaturia, which, however, speedily passes off.

The loose cellular tissue around the kidney may be the seat of suppuration, constituting *perinephritic abscess.* This may be due to injury, to disease of the kidney itself, or to extension of inflammation from neighbouring parts. The abscess may burst into the pleura, constituting empyema ; into the colon or bladder ; or may point externally in the groin or loin.

Tumours of the kidney, of which perhaps sarcoma, in children, is the most common, may be recognised by their position and fixity ; by the resonant colon lying in front of them ; by their not moving with respiration ; and by their rounded outline not presenting a notched anterior margin like the spleen, with which they are most likely to be confounded. The examination of the kidney should be bimanual ; that is to say, one hand should be placed in the flank and firm pressure made forwards, while the other hand is buried in the abdominal wall, over the situation of the organ. Manipulation of the kidney frequently produces a peculiar sickening sensation, with sometimes faintness.

The kidney is mainly held in position by the mass of fatty matter in which it is embedded. If this fatty matter is loose or lax or is absorbed, the kidney may become movable and may give rise to great pain. This condition occurs, therefore, in badly nourished people, or in those who have become emaciated from any cause, and is more common in women than in men. It must not be confounded with the *floating kidney* : this is a congenital condition due to the development of a mesonephron, which permits the organ to move more or·less freely. The two conditions cannot, however, be distinguished until the abdomen is opened or the kidney explored from the loin.

The kidney has, of late years, been frequently the seat of surgical interference. It may be exposed for exploration or the evacuation of pus (nephrotomy) ; it may be incised for the removal of stone (nephro-lithotomy) ; it may be sutured when movable or floating (nephrorraphy) ; or it may be removed (nephrectomy).

The kidney may be exposed either by a lumbar or abdominal incision. The operation is best performed by a lumbar incision, except in cases of very large tumours, or of wandering kidneys, with a loose mesonephron, on account of the advantages which it possesses of not opening the peritoneum, and affording admirable drainage. It may be performed either by an oblique, a vertical, or a transverse incision. Perhaps the most desirable, as affording the best means for exploring the whole surface of the kidney, is an incision from the tip of the last rib, backwards to the edge of the Erector spinæ. This incision must not be quite parallel to the rib, but its posterior end must be at least three-quarters of an inch below it, lest the pleura be wounded. This cut is quite sufficient for an exploration of the organ. Should it require removal, a vertical incision can be made downwards to the crest of the ilium, along the outer border of the Quadratus lumborum. The structures divided are, the skin, the superficial fascia with the cutaneous nerves, the deep fascia, the posterior border of the External oblique muscle of the abdomen, and the outer border of the Latissimus dorsi ; the Internal oblique and the posterior aponeurosis of the Transversalis muscle ; the outer border of the Quadratus lumborum ; the deep layer of the lumbar fascia, and the transversalis fascia. The fatty tissue around the kidney is now exposed to view and must be separated by the fingers, or a director, in order to reach the kidney.

The abdominal operation is best performed by an incision in the linea semilunaris on the side of the kidney to be removed, as recommended by Langenbuch ; the kidney is then reached from the outer side of the colon, ascending or descending, as the case may be, and the vessels of the colon are not interfered with. If the incision is made in the linea alba, the kidney is reached from the inner side of the colon, and the vessels running to supply it must necessarily be interfered with. The incision is made of varying length according to the size of the kidney, commencing just below the costal arch. The abdominal cavity is opened. The intestines are held aside, and the outer layer of the mesocolon incised, so that the fingers can be introduced behind the peritoneum and the renal vessels sought for. These are then to be ligatured : if tied separately, care must be taken to ligature the artery first. The kidney must now be enucleated, and the vessels and ureter divided, and the latter tied, or if thought necessary stitched to the edge of the wound. The particular advantage of the abdominal operation is that the condition of the other kidney can be ascertained by manual examination, before the removal of the diseased kidney is finally decided upon. Kocher, however, states that this exploration can be effected through the lumbar incision by incising the peritoneum at the outer margin of the wound and inserting the hand across the middle line.

The operations of *nephro-lithotomy*, for the removal of calculi from the kidney, and *nephrotomy*, or incision of the kidney for abscess, &c., are generally performed by the lumbar incision.

Nephrorrhaphy is the name given to the operation for fixing a movable kidney. The kidney is reached by the lumbar incision, and its posterior surface denuded of its fatty capsule. Three stitches of medium thickness are passed through the transversalis fascia and muscles and through the cortical portion of the kidney, securing a good hold of it. When these sutures are tied, the kidney is tightly anchored in position.

THE URETERS

The **Ureters** are the two tubes which convey the urine from the kidneys to the bladder. Each commences within the sinus of the corresponding kidney as a number of short cup-shaped tubes, termed *calyces* or *infundibula*, which encircle the *renal papillæ*. Since a single calyx may enclose more than one papilla the calyces are generally fewer in number than the pyramids—the former varying from seven to thirteen, the latter from eight to eighteen. The calyces join to form two or three short tubes, and these unite to form a funnel-shaped dilatation, wide above and narrow below, named the **pelvis of the kidney** which is situated behind the renal vessels and lies partly inside and partly outside the renal sinus. It is usually placed on a level with the spinous process of the first lumbar vertebra, in which situation it is accessible behind the peritoneum.

The **ureter proper** measures from ten to twelve inches in length, and is a thick-walled narrow cylindrical tube which is directly continuous near the lower end of the kidney with the tapering extremity of the pelvis. It runs downwards and inwards in front of the Psoas muscle, and, entering the pelvic cavity, finally opens into the base of the bladder.

In the *abdominal part* of its course it lies behind the peritoneum on the inner part of the Psoas muscle, and is crossed obliquely by the spermatic vessels. It enters the pelvic cavity by crossing either the termination of the common, or the commencement of the external, iliac vessels.

At its origin the *right* ureter is usually covered by the second part of the duodenum, and in its course downwards lies to the right of the inferior vena cava and is crossed by the right colic artery, while near the pelvic brim it passes behind the lower part of the mesentery and the terminal part of the ileum. The *left* ureter is crossed by the left colic artery, and near the brim of the pelvis passes behind the pelvic colon and its mesentery.

Within the pelvis the ureter runs at first downwards on the lateral wall of the pelvic cavity under cover of the peritoneum, lying in front of the internal iliac vessels and on the inner side of the obliterated hypogastric artery and the obturator nerve and vessels. Opposite the lower part of the great sacro-sciatic foramen it inclines inwards behind the vas deferens (which crosses to its inner side) and reaches the base of the bladder, where it is situated in front of the upper end of the seminal vesicle and at a distance of about two inches from the opposite ureter. Finally, the ureters run obliquely for about three-quarters of an inch through the wall of the bladder and open by slit-like apertures into the cavity of the viscus at the lateral angles of the trigone. When the bladder is distended the openings of the ureters are about two inches apart, but when it is empty and contracted the distance between them is diminished by one half. Owing to their oblique course through the coats of the bladder, their upper and lower walls become closely applied to each other when the viscus is distended, and, acting as a valve, prevent regurgitation of urine from the bladder.

In the *female*, the ureter forms, as it lies in relation to the wall of the pelvis, the posterior boundary of a shallow depression named the *fossa ovarii*, in which the ovary is situated. It then runs inwards and forwards on the lateral aspect of the cervix uteri and upper part of the vagina to reach the base of the bladder. In this part of its course it is accompanied for about an inch by the uterine artery, which then crosses in front of the ureter and ascends between the two layers of the broad ligament. The ureter is distant about three-quarters of an inch from the lateral aspect of the neck of the uterus.

The ureter is sometimes duplicated, and the two tubes may remain distinct as far as the base of the bladder. On rare occasions they open separately into the bladder cavity.

Structure.—The *ureter* is composed of three coats: fibrous, muscular, and mucous.

The *fibrous coat* is continuous at one end with the capsule of the kidney at the floor of the sinus; while at the other it is lost in the fibrous structure of the bladder.

In the pelvis of the kidney the *muscular coat* consists of two layers, longitudinal and circular: the longitudinal fibres become lost upon the sides of the papillæ at the extremities of the calyces; the circular fibres may be traced, surrounding the medullary structure in the same situation. In the ureter proper the muscular fibres are very distinct, and are arranged in three layers: an external longitudinal, a middle circular, and an internal layer, less distinct than the other two, but having a general longitudinal direction. According to Kölliker this internal layer is only found in the neighbourhood of the bladder.

The *mucous coat* is smooth, and presents a few longitudinal folds which become effaced by distension. It is continuous with the mucous membrane of the bladder below, while it is prolonged over the papillæ of the kidney above. Its epithelium is of a peculiar character, and resembles that found in the bladder. It is known by the name of 'transitional' epithelium (see fig. 15, page 13). It consists of several layers of cells, of which the innermost—that is to say, the

4 C

cells in contact with the urine—are quadrilateral in shape, with a concave margin on their outer surface, into which fits the rounded end of the cells of the second layer. These, the intermediate cells, more or less resemble columnar epithelium, and are pear-shaped, with a rounded internal extremity which fits into the concavity of the cells of the first layer, and a narrow external extremity which is wedged in between the cells of the third layer. The external or third layer consists of conical or oval cells varying in number in different parts, and presenting processes which extend down into the basement-membrane.

The *arteries* supplying the ureter are branches from the renal, spermatic, internal iliac, and inferior vesical.

The *nerves* are derived from the inferior mesenteric, spermatic, and pelvic plexuses.

Surgical Anatomy.—Subcutaneous rupture of the ureter is not a common accident, but occasionally occurs from a sharp, direct blow on the abdomen, as from the kick of a horse. It may be either torn completely across or only partially divided, and, as a rule, the peritoneum escapes injury. If torn completely across, the urine collects in the retro-peritoneal tissues; if it is not completely divided, the lumen of the tube may become obstructed and hydronephrosis or pyonephrosis result. If the injury is diagnosed, the divided ureter should be sutured. The lower end is closed by means of a Lembert's suture; and the upper end is implanted into a longitudinal incision made in the lower portion, and carefully sutured in that position. The ureter may be accidentally wounded in some pelvic operations, such as removal of the uterus; if this should happen the divided ends must be sutured together, or, failing to accomplish this, the upper end must be implanted into the bladder or the intestine.

THE PELVIS

The **cavity of the pelvis** is that part of the general abdominal cavity which is below the level of the linea ilio-pectinea, the promontory of the sacrum behind, and the pubic crests in front.

Boundaries.—The cavity is bounded, behind, by the sacrum, the coccyx, the Pyriformis muscles, and the great sacro-sciatic ligaments; in front and at the sides, by the ossa pubis and ischia, covered by the Obturator muscles; above, it communicates with the cavity of the abdomen; and below, the outlet is closed by the triangular ligament, the Levatores ani and Coccygei muscles, and the visceral layer of the pelvic fascia, which is reflected from the wall of the pelvis on to the viscera.

Contents.—The viscera contained in this cavity are the urinary bladder, the rectum, some of the generative organs peculiar to each sex, and some convolutions of the small intestines: they are partially covered by the peritoneum, and supplied with blood-vessels, lymphatics, and nerves.

THE BLADDER

The **Bladder** is a musculo-membranous sac which acts as a reservoir for the urine; and as its size, position, and relations vary according to the amount of fluid it contains, it is necessary to study it as it appears: (*a*) in the empty condition, and (*b*) in the condition of distension.

The empty bladder.—When hardened *in situ*, the empty bladder has the form of 'a flattened inverted tetrahedron.'* It presents a base, an apex, a superior and an inferior surface. The *base* is triangular in shape, and is directed downwards and backwards towards the rectum, from which it is separated by the recto-vesical fascia, the vesiculæ seminales, and the terminal portions of the vasa deferentia. The *apex* is directed forwards towards the upper part of the symphysis pubis, and from it a fibrous cord is continued upwards on the back of the anterior abdominal wall to the umbilicus. This cord is named the urachus, and represents the fibrous remains of the intra-abdominal part of the fœtal allantois (see page 151). The peritoneum is carried by it from the apex of the bladder on to the abdominal wall to form what is termed the anterior or superior false ligament of the bladder. The *superior surface* is triangular,

---

* Dixon, *Journal of Anatomy and Physiology*, vol. xxxiv.

bounded on either side by a lateral border which separates it from the inferior surface, and behind by a posterior border which intervenes between it and the base, and is represented by a line joining the two ureters. The lateral borders extend from the ureters to the bladder apex, and from them the peritoneum is carried to the walls of the pelvis as the lateral false ligaments of the bladder. On either side of the bladder the peritoneum shows a depression, which is named the *paravesical fossa*. The superior surface is directed upwards, is covered by peritoneum, and is in relation with the pelvic colon and some of the coils of the small intestine. When the bladder is empty and firmly contracted, this surface is convex and the lateral and posterior borders are rounded; whereas if the bladder be relaxed it is concave, and the interior of the viscus, as seen in a vertical mesial section, presents the appearance of a **V**-shaped slit with a shorter

FIG. 765.—Vertical section of bladder, penis, and urethra.

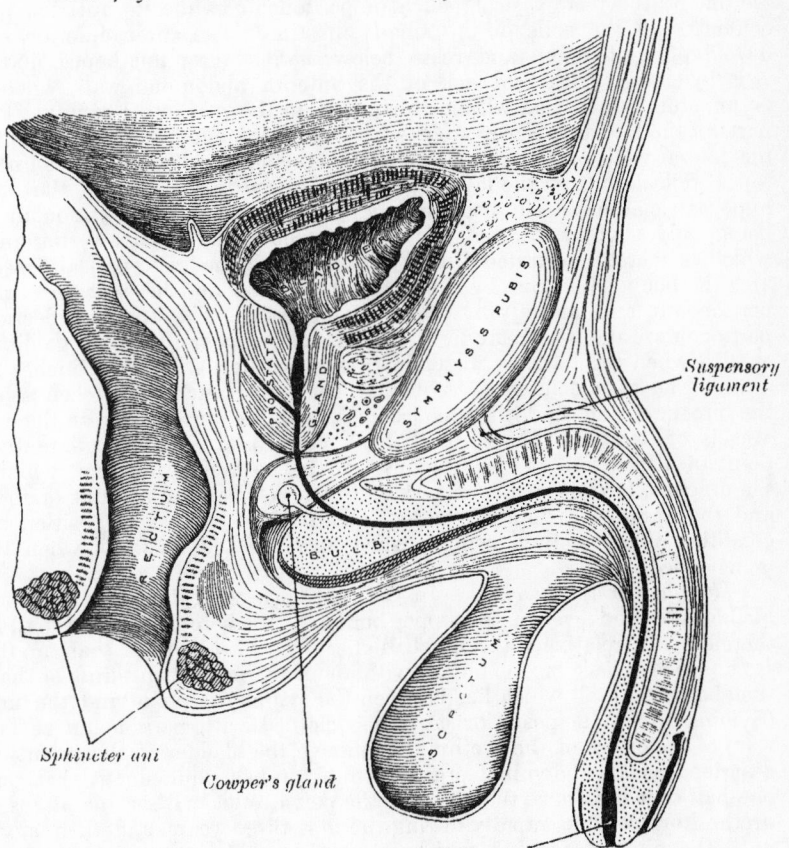

Suspensory ligament

Sphincter ani

Cowper's gland

Fossa navicularis

posterior and a longer anterior limb—the apex of the **V** corresponding with the orifice of the urethra. The *inferior* surface is directed downwards and is uncovered by peritoneum. It may be divided into a posterior or prostatic area and two antero-lateral surfaces. The prostatic area is somewhat triangular: it rests upon and is in direct continuity with the base of the prostate gland; this area is usually named the *neck* of the bladder, and from it the urethra emerges. The so-called *neck* or *cervix of the bladder* corresponds with the commencement of the urethra. There is, however, no tapering part which would constitute a true neck, but the bladder suddenly contracts to the opening of the urethra. The infero-lateral portions of the inferior surface are directed downwards and outwards: in front, they are separated from the symphysis pubis by a mass of fatty tissue which is named the *retro-pubic pad*; behind, they are in contact with the fascia which covers the Levatores ani muscles.

When the bladder is empty it is placed entirely within the pelvis, below the level of the obliterated hypogastric arteries, and below the level of those portions of the vasa deferentia which are in contact with the lateral wall of the pelvis; after they cross the ureters the vasa deferentia come into contact with the base of the bladder. As the viscus becomes filled, its base, being more or less fixed, is only slightly depressed; while its superior surface gradually rises into the abdominal cavity, carrying with it its peritoneal covering, and at the same time rounding off the posterior and lateral borders.

The distended bladder.—When the bladder is moderately full it contains about a pint and assumes an oval form; the long diameter of the oval measures about five inches and is directed upwards and forwards. In this condition it presents a postero-superior, an antero-inferior, and two lateral surfaces, a summit and a base. The postero-superior surface is directed upwards and backwards, and is covered by peritoneum: behind, it is separated from the rectum by the recto-vesical pouch of peritoneum, while its anterior part is in contact with the coils of the small intestine. The antero-inferior surface is devoid of peritoneum, and rests, below, against the pubic bones, above which it is in contact with the back of the anterior abdominal wall, where it may be punctured or incised without opening into the peritoneal cavity. The lower parts of the lateral surfaces are destitute of peritoneum, and are in contact with the lateral walls of the pelvis. When the bladder is distended, the line of peritoneal reflection from the lateral surface is raised to the level of the obliterated hypogastric artery. The base or fundus undergoes little alteration in position, being only slightly lowered. It exhibits, however, a narrow triangular area, which is merely separated from the rectum by the recto-vesical fascia. This area is bounded below by the prostate, above by the recto-vesical fold of peritoneum, and laterally by the vasa deferentia. The line of reflection of the peritoneum from the rectum to the bladder appears to undergo little or no change when the latter is distended; it is situated about four inches from the anus. The vasa deferentia frequently come in contact with each other above the prostate, and under such circumstances the lower part of the triangular area is obliterated. The summit is directed upwards and forwards above the point of attachment of the urachus, and hence the peritoneum, which follows the urachus, forms a pouch of varying depth between the summit of the bladder and the anterior abdominal wall. The bladder varies in position with the condition of the rectum, being pushed upwards and forwards when the latter is distended.

The bladder in the child.—In the new-born child the urethral orifice of the bladder is at the level of the upper border of the symphysis pubis; the bladder therefore lies relatively at a much higher level in the infant than in the adult. Its anterior surface is 'in contact with about the lower two-thirds of that part of the abdominal wall which lies between the symphysis pubis and the umbilicus' (Symington*). Its posterior surface is clothed with peritoneum as far as the level of the orifice of the urethra. Although the bladder of the infant is usually described as an abdominal organ, Symington has pointed out that only about one-half of it lies above the plane of the pelvic inlet. Disse maintains that the urethral orifice sinks rapidly during the first three years, and then more slowly until the ninth year, after which it remains stationary until puberty, when it again slowly descends and reaches its adult position.

The female bladder.—In the female, the bladder is in relation behind with the uterus and the upper part of the vagina. It is separated from the anterior surface of the body of the uterus by the recto-vesical pouch of peritoneum, but below the level of this pouch it is connected to the front of the cervix uteri and the upper part of the anterior wall of the vagina by areolar tissue. When the bladder is empty the uterus rests upon its superior surface. The female bladder is said by some to be more capacious than that of the male, but probably the opposite is the case.

Ligaments.—The bladder is retained in its place by ligaments, which are divided into true and false. The true ligaments are five in number: two anterior, two lateral, and the urachus. The false ligaments, also five in number, are formed by folds of the peritoneum.

* *The Anatomy of the Child.*

The *anterior true ligaments (pubo-prostatic)* extend from the back of the pubic bones, one on either side of the symphysis, to the front of the neck of the bladder, over the anterior surface of the prostate gland. These ligaments are formed by the recto-vesical fascia, and contain a few muscular fibres prolonged from the bladder.

The *lateral true ligaments*, also formed by the recto-vesical fascia, are broader and thinner than the preceding. They are attached to the lateral parts of the prostate, and to the sides of the base of the bladder.

The *urachus* is the fibro-muscular cord already mentioned, extending between the summit of the bladder and the umbilicus. It is broad below, at its attachment to the bladder, and becomes narrower as it ascends.

Fig. 766.—Vertical median section of the male pelvis. (Henle.)

The *false ligaments of the bladder* are two posterior, two lateral, and one anterior or superior.

The *two posterior* pass forwards from the sides of the rectum to the posterior and lateral aspect of the bladder, and form the lateral boundaries of the recto-vesical pouch of the peritoneum.

The *two lateral ligaments* are reflections of the peritoneum, from the lateral walls of the pelvis to the sides of the bladder.

The *anterior* or *superior ligament (ligamentum suspensorium)* is the fold of peritoneum extending from the summit of the bladder to the abdominal wall. It is carried off from the bladder by the urachus.

Structure.—The bladder is composed of four coats: serous, muscular submucous, and mucous.

The *serous coat* is a partial one, and is derived from the peritoneum. It invests the superior surface and the upper part of the lateral surfaces, and is reflected from these parts on to the abdominal and pelvic walls.

The *muscular coat* consists of three layers of unstriped muscular fibres: an external layer, composed of fibres having for the most part a longitudinal arrangement; a middle layer, in which the fibres are arranged, more or less, in a circular manner; and an internal layer, in which the fibres have a general longitudinal arrangement.

The *fibres of the external longitudinal layer* arise from the posterior surface of the body of the os pubis in both sexes (*musculi pubo-vesicalis*), and in the male from the adjacent part of the prostate gland and its capsule. They pass, in a more or less longitudinal manner, up the anterior surface of the bladder, over its apex, and then descend along its posterior surface to its base, where they become attached to the prostate in the male, and to the front of the vagina in the female. At the sides of the bladder the fibres are arranged obliquely and intersect one another. This layer has been named the *Detrusor urinæ muscle*.

The *fibres of the middle circular layer* are very thinly and irregularly scattered on the body of the organ, and though to some extent placed transversely to the long axis of the bladder, are for the most part arranged obliquely. Towards the lower part of the bladder, round the cervix and commencement of the urethra, they are disposed in a thick circular layer, forming the *sphincter vesicæ*, which is continuous with the muscular fibres of the prostate gland.

FIG. 767.—Superficial layer of the epithelium of the bladder. Composed of polyhedral cells of various sizes, each with one, two, or three nuclei. (Klein and Noble Smith.)

FIG. 768.—Deep layers of epithelium of bladder, showing large club-shaped cells above, and smaller, more spindle-shaped cells below—each with an oval nucleus. (Klein and Noble Smith.)

The *internal longitudinal layer* is thin, and its fasciculi have a reticular arrangement, but with a tendency to assume for the most part a longitudinal direction.

Two bands of oblique fibres, originating behind the orifices of the two ureters, converge to the back part of the prostate gland, and are inserted, by means of a fibrous process, into the middle lobe of that organ. They are the *muscles of the ureters*, described by Sir C. Bell, who supposed that during the contraction of the bladder they serve to retain the oblique direction of the ureters, and so prevent the reflux of the urine into them.

The *submucous coat* consists of a layer of areolar tissue, connecting together the muscular and mucous coats, and intimately united to the latter.

The *mucous coat* is thin, smooth, and of a pale rose colour. It is continuous above through the ureters with the lining membrane of the uriniferous tubes, and below with that of the urethra. It is connected loosely to the muscular coat by a layer of areolar tissue, and is therefore thrown into folds or *rugæ* when the bladder is empty. Over the trigonum vesicæ the mucous membrane is closely attached to the muscular coat, and is not thrown into folds, but is smooth and flat. The epithelium covering it is of the transitional variety, consisting of a superficial layer of polyhedral flattened cells, each with one, two, or three nuclei; beneath these, a stratum of large club-shaped cells, with their narrow extremities directed downwards and wedged in between smaller spindle-shaped cells, containing an oval nucleus (figs. 767, 768). There are no true glands in the mucous membrane of the bladder, though certain mucous follicles which exist, especially near the neck of the bladder, have been regarded as such.

Objects seen on the inner surface.—Upon the inner surface of the bladder are seen the orifices of the ureters, the trigone, and the commencement of the urethra.

The orifices of the ureters.—These are situated at the base of the trigone, and are about two inches apart when the bladder is distended ; they are about an inch and a half from the base of the prostate and the commencement of the urethra.

The *trigone of the bladder* (*trigonum vesicæ*) is a triangular smooth surface, with the apex directed forwards, situated at the base of the bladder, immediately behind the urethral orifice. It is paler in colour than the rest of the interior, and never presents any rugæ, even in the collapsed condition of the organ, owing to the intimate adhesion of its mucous membrane to the subjacent tissue. It is bounded at each posterior angle by the orifice of the ureter, and in front by the orifice of the urethra. A band of non-striped muscular fibres passes across between the terminations of the ureters, and forms what is sometimes named the *ligamentum uretericum*. When the bladder is illuminated, this presents the appearance of a pale ridge leading to the opening of the ureter on either side ; it therefore constitutes an important guide in the operation of introducing a catheter into the ureter. Projecting from the lower and anterior part of the bladder, and reaching to the orifice of the urethra, is a slight elevation of mucous membrane, called the *uvula vesicæ*. It is formed by a thickening of the submucous tissue.

The *arteries* supplying the bladder are the superior, middle, and inferior vesical in the male, with additional branches from the uterine and vaginal in the female. They are all derived from the anterior trunk of the internal iliac. The obturator and sciatic arteries also supply small visceral branches to the bladder.

The *veins* form a complicated plexus round the neck, sides, and base of the bladder, and terminate in the internal iliac veins.

The *lymphatics* form two plexuses, one in the muscular and another in the submucous coat ; they are most numerous in the neighbourhood of the trigone. They accompany the blood-vessels, and ultimately terminate in the internal iliac glands.

The *nerves* are derived from the pelvic plexuses of the sympathetic and from the third and the fourth sacral nerves ; the former supplying the upper part of the organ, the latter its base and neck. According to F. Darwin the sympathetic fibres have ganglia connected with them, which send branches to the vessels and muscular coat.

Surface Form.—The surface form of the bladder varies with its degree of distension and under other circumstances. In the young child it is represented by a conical figure, the apex of which, even when the viscus is empty, is situated in the hypogastric region, above the level of the symphysis pubis. In the adult, when the bladder is empty, its apex does not reach above the level of the upper border of the symphysis pubis, and the whole organ is situated in the pelvis ; the neck, in the male, corresponding to a line drawn horizontally backwards, through the symphysis a little below its middle. As the bladder becomes distended it gradually rises out of the pelvis into the abdomen and forms a swelling in the hypogastric region, which is perceptible to the hand, as well as to percussion. In extreme distension it reaches into the umbilical region. Under these circumstances the lower part of its anterior surface, for a distance of about two inches above the symphysis pubis, is closely applied to the abdominal wall, without the intervention of peritoneum, so that it can be tapped by an opening in the middle line just above the symphysis pubis, without any fear of wounding the serous membrane. When the rectum is distended, the prostatic portion of the urethra is elongated and the bladder lifted out of the pelvis and the peritoneum pushed upwards. Advantage is taken of this by some surgeons in performing the operation of suprapubic cystotomy. The rectum is distended by an indiarubber bag, which is introduced into this cavity empty, and then filled with ten or twelve ounces of water. If now the bladder is injected with about half a pint of some antiseptic fluid, it will appear above the pubes plainly perceptible to the sight and touch. The peritoneum will be pushed out of the way, and an incision three inches long may be made in the linea alba, from the symphysis pubis upwards, without any great risk of wounding the peritoneum. Other surgeons object to the employment of this bag, as its use is not unattended with risk, and because it causes pressure on the prostatic sinuses and produces congestion of the vessels over the bladder and a good deal of venous hæmorrhage.

When distended the bladder can be felt in the male, from the rectum, behind the prostate, and fluctuation can be perceived by a bimanual examination, one finger being introduced into the rectum and the distended bladder tapped on the front of the abdomen with the finger of the other hand. This portion of the bladder—that is, the portion felt in the rectum by the finger—is also uncovered by peritoneum, and the bladder may here be punctured from the rectum, in the middle line, without risk of wounding the serous membrane.

Surgical Anatomy.—A defect of development, in which the bladder is implicated, is known under the name of *extroversion of the bladder*. In this condition the lower part of the abdominal wall and the anterior wall of the bladder are wanting, so that the posterior surface of the bladder presents on the abdominal surface, and is pushed forwards by the pressure of the viscera within the abdomen, forming a red vascular tumour, on which the openings of the ureters are visible. The penis, except the glans, is rudimentary and is cleft on its dorsal surface, exposing the floor of the urethra, a condition known as *epispadias*. The pelvic bones are also arrested in development (see page 312).

The bladder may be ruptured by violence applied to the abdominal wall, when the viscus is distended, without any injury to the bony pelvis, or it may be torn in cases of fracture of the pelvis. The rupture may be either intraperitoneal or extraperitoneal: that is, may implicate the superior surface of the bladder in the former case, or one of the other surfaces in the latter. Rupture of the antero-inferior surface alone is, however, very rare. Until recently intraperitoneal rupture was uniformly fatal, but now abdominal section and suturing the rent with Lembert's suture is resorted to, with a very considerable amount of success. The sutures are inserted only through the peritoneal and muscular coats in such a way as to bring the serous surfaces at the margins of the wound into apposition, and one is inserted just beyond each end of the wound. The bladder should be tested as to whether it is water-tight before closing the external incision.

The muscular coat of the bladder undergoes hypertrophy in cases in which there is any obstruction to the flow of urine. Under these circumstances the bundles of which the muscular coat consists become much increased in size, and, interlacing in all directions, give rise to what is known as the *fasciculated bladder*. Between these muscular bundles the mucous membrane may bulge out, forming sacculi, constituting the *sacculated bladder*, and in these little pouches phosphatic concretions may collect, forming *encysted calculi*. The mucous membrane is very loose and lax, except over the trigone, to allow of the distension of the viscus.

Various forms of tumour have been found springing from the wall of the bladder. The innocent tumours are the papilloma and the mucous polypus arising from the mucous membrane; the fibrous, from the submucous tissue; and the myoma, originating in the muscular tissue; and, very rarely, dermoid tumours, the exact origin of which it is difficult to explain. Of the malignant tumours, epithelioma is the most common, but sarcoma is occasionally found in the bladder of children.

In doubtful cases, the cystoscope may prove to be an aid in diagnosis. This instrument consists of a tube in which is fixed a small electric light, the wires of which run through the shaft of the instrument. Upon introducing this down the urethra, the bladder can be examined with the eye, and a villous growth or other tumour, a calculus, or an ulcer can be detected; or the orifices of the ureters can be examined, and renal hæmaturia diagnosed, and the kidney from which the blood comes definitely settled. Again, the presence of minute tuberculous ulceration near the mouth of the ureter on the affected side may establish the diagnosis, not only of tuberculous kidney, but also of the side in which the disease is located. Caspar has utilised the cystoscope in catheterising the ureter, by causing a groove to be made on one side of the shaft of the instrument, along which a fine bougie can be passed. The mouth of the ureter is first found by the surgeon, and the bougie then projected into the field of vision and guided directly into the opening.

Puncture of the bladder may be performed either above the symphysis pubis or through the rectum, in both cases without wounding the peritoneum. The former plan is generally to be preferred, since in puncture by the rectum a permanent fistula may be left from abscess forming between the rectum and the bladder; or pelvic cellulitis may be set up; moreover, it is exceedingly inconvenient to keep a cannula in the rectum. In some cases, in performing this operation the recto-vesical pouch of peritoneum has been wounded, inducing fatal peritonitis. The operation, therefore, has been almost completely abandoned.

Access to the bladder, for the purpose of removing calculi, tumours, and portions of the prostate in enlargement of that organ, is almost always effected by the suprapubic route in the present day, the old perineal operation being now rarely resorted to. In the female, owing to the shortness of the urethra and ready dilatability, calculi and foreign bodies and new growths, when of small size, may be removed by the urethral route.

Suprapubic cystotomy is performed by first injecting ten or twelve ounces of some weak antiseptic fluid into the bladder. Then, with or without distending the rectum (*vide supra*), a vertical median incision is made in the hypogastric region, from three to four inches in length, immediately above the symphysis and extended between the Pyramidales and Recti muscles until the transversalis fascia is reached. This is divided and some fatty tissue exposed (space of Retzius). Upon separating this, the anterior surface of the bladder will be exposed and will be recognised by its muscular fibres. A needle on a handle should be passed through its coats on either side of the spot selected for the opening, and two long pieces of silk inserted. The bladder is incised between these stays, which are held by an assistant and form a useful guide to the opening in the bladder when the fluid has escaped.

THE MALE URETHRA

The **urethra in the male** extends from the neck of the bladder to the meatus urinarius at the end of the penis. It presents a double curve in the ordinary relaxed state of the penis (fig. 765). Its length varies from seven to eight inches; and it is divided into three portions, the *prostatic, membranous,* and *spongy,* the structure and relations of which are essentially different. Except during the passage of the urine or semen, the urethra is a mere transverse cleft or slit, with its upper and under surfaces in contact. At the meatus urinarius the slit is vertical, and in the prostatic portion somewhat arched.

The **prostatic portion** is the widest and most dilatable part of the canal. It passes through the prostate gland, from its base to its apex, lying nearer its anterior than its posterior surface. It is about an inch and a quarter in length; the form of the canal is spindle-shaped, being wider in the middle than at either extremity, and narrowest below, where it joins the membranous portion. A transverse section of the canal as it lies in the prostate is horse-shoe in shape, the convexity being directed forwards (fig. 770), since the direction of the canal is nearly vertical.

Upon the posterior wall or floor of the canal is a narrow longitudinal ridge, the *verumontanum,* or *caput gallinaginis,* formed by an elevation of the mucous membrane and its subjacent tissue. It is eight or nine lines in length, and a line and a half in height; and contains, according to Kobelt, muscular and erectile tissues. When distended, it may serve to prevent the passage of the semen backwards into the bladder. On each side of the verumontanum is a slightly depressed fossa, the *prostatic sinus,* the floor of which is perforated by numerous apertures, the *orifices of the prostatic ducts* from the lateral lobes of the gland; the ducts of the middle lobe open behind the verumontanum. At the fore part of the verumontanum, in the middle line, is a depression, the *sinus pocularis* (*vesicula prostatica*); and upon or within its margins are the slit-like openings of the ejaculatory ducts. The sinus pocularis forms a *cul-de-sac* about a quarter of an inch in length, which runs upwards and backwards in the substance of the prostate behind the middle lobe; its prominent anterior wall partly forms the verumontanum. Its walls are composed of fibrous tissue, muscular fibres, and mucous membrane, and numerous small glands open on its inner surface. It has been called by Weber, who discovered it, the *uterus masculinus,* from its being developed from the united lower ends of the atrophied Müllerian ducts, and therefore homologous with the uterus and vagina in the female.

The **membranous portion of the urethra** extends between the apex of the prostate and the bulb of the corpus spongiosum. It is the narrowest part of the canal (excepting the meatus), and measures three-quarters of an inch along its upper, and half an inch along its lower surface, in consequence of the bulb projecting backwards beneath it. Its anterior concave surface is placed about an inch below and behind the pubic arch, from which it is separated by the dorsal vessels and nerves of the penis, and some muscular fibres. Its posterior convex surface is separated from the rectum by a triangular space, which constitutes the perinæum. The membranous portion of the urethra lies between the inferior and superior layers of the triangular ligament. As it pierces the inferior layer, the fibres around the opening are prolonged over the tube. It is also surrounded by the Compressor urethræ muscle.

The **spongy portion** is the longest part of the urethra, and is contained in the corpus spongiosum. It is about six inches in length, and extends from the termination of the membranous portion to the meatus urinarius. Commencing just below the triangular ligament, it passes forwards for a short distance; and then, in the flaccid condition of the penis, it bends downwards and forwards. It is narrow, and of uniform size in the body of the penis, measuring about a quarter of an inch in diameter; being dilated behind, within the bulb; and again anteriorly within the glans penis, where it forms the *fossa navicularis*.

The *bulbous portion* is a name given, in some descriptions of the urethra, to the posterior part of the spongy portion contained within the bulb.

The *meatus urinarius* is the most contracted part of the urethra; it is a vertical slit, about three lines in length, bounded on each side by two small labia.

The inner surface of the lining membrane of the urethra, especially on the floor of the spongy portion, presents the orifices of numerous mucous glands and

Fig. 769.—The male urethra, laid open on its anterior (upper) surface. (Testut.)

Veru-montanum

Prostate

Cowper's gland
Bulb

Crus

Corpus spongiosum

Glans

Meatus

Prostatic portion

Membranous portion

Spongy portion

Fossa navicularis

follicles situated in the submucous tissue, and named the *glands of Littré*. Besides these there are a number of small pit-like recesses, or *lacunæ*, of varying sizes. Their orifices are directed forwards, so that they may easily intercept the point of a catheter in its passage along the canal. One of these lacunæ, larger than the rest, is situated on the upper surface of the fossa navicularis, about an inch and a half from the orifice; it is called the *lacuna magna*. Into the bulbous portion are found opening the ducts of Cowper's glands.

Structure.—The urethra is composed of a continuous mucous membrane, supported by a submucous tissue which connects it with the various structures through which it passes.

The *mucous coat* forms part of the genito-urinary mucous membrane. It is continuous with the mucous membrane of the bladder, ureters, and kidneys; externally, with the integument covering the glans penis; and is prolonged into the ducts of the glands which open into the urethra, viz. Cowper's glands and the prostate gland; and into the vasa deferentia and vesiculæ seminales, through the ejaculatory ducts. In the spongy and membranous portions the mucous membrane is arranged in longitudinal folds when the tube is empty. Small papillæ are found upon it, near the orifice; and its epithelial lining is of the columnar variety, excepting near the meatus, where it is squamous.

The *submucous tissue* consists of a vascular erectile layer; outside which is a layer of unstriped muscular fibres, arranged in a circular direction, which separates the mucous membrane and submucous tissue from the tissue of the corpus spongiosum.

Surgical Anatomy.—The urethra may be ruptured by the patient falling astride of any hard substance and striking his perinæum, so that the urethra is crushed against the pubic arch. Bleeding will at once take place from the urethra, and this, together with the bruising in the perinæum and the history of the accident, will point to the nature of the injury.

The surgical anatomy of the urethra is of considerable importance in connection with the passage of instruments into the bladder. Otis was the first to point out that the urethra is capable of great dilatability, so that, excepting through the external meatus, an instrument corresponding to 18 English gauge (29 French) can usually be passed without damage. The orifice of the urethra is not so dilatable, and therefore frequently requires slitting. A recognition of this dilatability caused Bigelow to very considerably modify the operation of lithotrity and introduce that of litholapaxy. In passing catheters, especially fine ones, the point of the instrument should be kept as far as possible along the upper wall of the canal, as it is otherwise very liable to enter one of the lacunæ. Stricture of the urethra is a disease of very common occurrence, and is generally situated in the spongy part of the urethra, most commonly in the bulbous portion, just in front of the membranous urethra, but in a very considerable number of cases in the penile or ante-scrotal part of the canal.

Congenital defects of the urethra occur occasionally. The one most frequently met with is where there is a cleft on the floor of the urethra owing to an arrest of union in the middle line. This is known as *hypospadias*, and the cleft may vary in extent. The simplest, and by far the most common form, is where the deficiency is confined to the glans penis. The urethra ends at the point where the extremity of the prepuce joins the body of the penis in a small valve-like opening. The prepuce is also cleft on its under surface and forms a sort of hood over the glans. There is a depression on the glans in the position of the normal meatus. This condition produces no disability and requires no treatment. In more severe cases the penile portion of the urethra is cleft throughout its entire length, and the opening of the urethra is at the point of junction of the penis and scrotum. The under surface of the penis in the middle line presents a furrow lined by a moist mucous membrane, on either side of which is often more or less dense fibrous tissue stretching from the glans to the opening of the urethra, which prevents complete erection taking place. Great discomfort is induced during micturition, and connection is impossible. The condition may be remedied by a series of plastic operations, but these should not be attempted until the patient has reached the age of puberty.

The worst form of this condition is where the urethra is deficient as far back as the perinæum, and the scrotum is cleft. The penis is small and bound down between the two halves of the scrotum, so as to resemble an hypertrophied clitoris. The testicles are often retained. The condition of parts, therefore, very much resembles the external organs of generation of the female, and male children the victims of this malformation have been brought up as girls. The halves of the scrotum, deficient of testicles, resemble the labia, the cleft between them looks like the orifice of the vagina, and the diminutive penis is taken for an enlarged clitoris. There is no remedy for this condition.

A much more uncommon form of malformation is where there is an apparent deficiency of the upper wall of the urethra. This is named *epispadias*, and would appear to be really a deficiency of the lower wall of the urethra, with torsion of the penis. The deficiency may vary in extent; when it is complete the condition is associated with extroversion of the bladder (see page 1128). In less extensive cases, where there is no extroversion, there is an infundibuliform opening into the bladder. The penis is usually dwarfed and turned upwards, so that the glans lies over the opening. For the relief of this condition, Cantwell has introduced a most ingenious operation.

THE FEMALE URETHRA

The **female urethra** is a narrow membranous canal, about an inch and a half in length, extending from the neck of the bladder to the meatus urinarius. It is placed beneath the symphysis pubis, embedded in the anterior wall of the vagina and its direction is obliquely downwards and forwards, its course being slightly curved, the concavity directed forwards and upwards. Its diameter when undilated is about a quarter of an inch. The urethra perforates the triangular ligament, and its external orifice is situated directly in front of the vaginal opening and about an inch behind the glans clitoridis.

Structure.—The urethra consists of three coats : muscular, erectile, and mucous.

The *muscular coat* is continuous with that of the bladder; it extends the whole length of the tube, and consists of a circular stratum of muscular fibres. In addition to this, between the two layers of the triangular ligament, the female urethra is surrounded by the Compressor urethræ, as in the male.

A *thin layer of spongy erectile tissue*, containing a plexus of large veins, intermixed with bundles of unstriped muscular fibre, lies immediately beneath the mucous coat.

The *mucous coat* is pale, continuous externally with that of the vulva, and internally with that of the bladder. It is thrown into longitudinal folds, one of which, placed along the floor of the canal, resembles the verumontanum in the male urethra. It is lined by laminated epithelium, which becomes transitional near the bladder. Its external orifice is surrounded by a few mucous follicles.

The urethra, from not being surrounded by dense resisting structures, as in the male, admits of marked dilatation, which enables the surgeon to remove, with considerable facility, calculi, or other foreign bodies, from the cavity of the bladder.

MALE ORGANS OF GENERATION

The Prostate Gland

THE **Prostate Gland** (προΐστημι, *to stand before*) is a firm, partly glandular and partly muscular body, which is placed immediately below the neck of the bladder and around the commencement of the urethra. It is situated in the pelvic cavity, behind the lower part of the symphysis pubis, above the deep layer of the triangular ligament, and in front of the rectum, through which it may be distinctly felt, especially when enlarged. It is about the size of a

FIG. 770.—Transverse section of the prostate gland; showing the urethra, with the eminence of the caput gallinaginis; beneath it the sinus pocularis and ejaculatory ducts. (Enlarged.)

chestnut and somewhat conical in shape, and presents for examination a base, an apex, an anterior, a posterior and two lateral surfaces.

The *base* is directed upwards, and is applied to the under surface of the bladder. The greater part of this surface is directly continuous with the bladder wall: the urethra penetrates it nearer its anterior than its posterior border.

The *apex* is directed downwards, and is in contact with the deep layer of the triangular ligament, which it touches.

The *posterior surface* is flattened from side to side and convex from above downwards; it rests on the rectum, and is distant about an inch and a half from the anus. Near its upper border there is a depression through which the two common ejaculatory ducts enter the prostate. This depression serves to divide the posterior surface into a lower larger and an upper smaller part. The upper smaller part constitutes the *middle lobe* of the prostate and intervenes between the ejaculatory ducts and the urethra; it varies greatly in size, and in some cases is destitute of glandular tissue. The lower larger portion sometimes presents a shallow median furrow, which imperfectly separates it into two *lateral* lobes: these form the main mass of the gland and are directly continuous with each other behind the urethra. In front of the urethra they are connected by a band which is named the *anterior commissure*: this consists of the same tissues as the capsule and is devoid of glandular substance.

The *anterior surface* measures about an inch from above downwards, but is narrow and convex from side to side. It is placed about three-quarters of an inch behind the pubic symphysis, from which it is separated by a plexus of veins and a quantity of loose fat. It is connected to the pubic bone on either side by the pubo-prostatic ligaments. The urethra emerges from this surface a little above and in front of the apex of the gland.

The *lateral surfaces* are prominent, and are covered by the anterior portions of the Levatores ani muscles, which are, however, separated from the gland by a plexus of veins.

The prostate measures about an inch and a half transversely at the base, three-quarters of an inch in its antero-posterior diameter, and an inch and a quarter in its vertical diameter. Its weight is about four and a half drachms. It is held in its position by the anterior ligaments of the bladder (*pubo-prostatic*); by the deep layer of the triangular ligament, which invests the commencement of the membranous portion of the urethra and prostate gland; and by the anterior portions of the Levatores ani muscles, which pass backwards from the os pubis and embrace the sides of the prostate. These portions of the Levatores ani, from the support they afford to the prostate, are named the *Levator prostatæ*.

The prostate gland is perforated by the urethra and the ejaculatory ducts. The urethra usually lies along the junction of its anterior with its middle third. The ejaculatory ducts pass obliquely downwards and forwards through the posterior part of the prostate, and open into the prostatic portion of the urethra.

Structure.—The prostate is immediately enveloped by a thin but firm fibrous capsule, distinct from that derived from the recto-vesical fascia, and separated from it by a plexus of veins. This capsule is firmly adherent to the prostate and is structurally continuous with the stroma of the gland, being composed of the same tissues, viz. non-striped muscle and fibrous tissue. The substance of the prostate is of a pale reddish-grey colour, of great density, and not easily torn. It consists of glandular substance and muscular tissue.

The *muscular tissue*, according to Kölliker, constitutes the proper stroma of the prostate; the connective tissue being very scanty, and simply forming thin trabeculæ between the muscular fibres, in which the vessels and nerves of the gland ramify. The muscular tissue is arranged as follows: immediately beneath the fibrous capsule is a dense layer, which forms an investing sheath for the gland; secondly, around the urethra, as it lies in the prostate, is another dense layer of circular fibres, continuous above with the internal layer of the muscular coat of the bladder, and below blending with the fibres surrounding the membranous portion of the urethra. Between these two layers, strong bands of muscular tissue, which decussate freely, form meshes in which the glandular structure of the organ is embedded. In that part of the gland which is situated in front of the urethra the muscular tissue is especially dense, and there is here little or no gland tissue; while in that part which is behind the urethra the muscular tissue presents a wide-meshed structure, which is densest at the base of the gland—that is, near the bladder—becoming looser and more sponge-like towards the apex of the organ.

The *glandular substance* is composed of numerous follicular pouches, opening into elongated canals, which join to form from twelve to twenty small excretory ducts. The follicles are connected together by areolar tissue, supported by prolongations from the fibrous capsule and muscular stroma, and enclosed in a delicate capillary plexus. The epithelium which lines the canals and the terminal vesicles is of the columnar variety. The prostatic ducts open into the floor of the prostatic portion of the urethra.

Vessels and Nerves.—The *arteries* supplying the prostate are derived from the internal pudic, vesical, and hæmorrhoidal. Its *veins* form a plexus around the sides and base of the gland ; they receive in front the dorsal vein of the penis, and terminate in the internal iliac veins. The *nerves* are derived from the pelvic plexus.

Surgical Anatomy.—The relation of the prostate to the rectum should be noted : by means of the finger introduced into the gut, the surgeon detects enlargement or other disease of this organ ; he can feel the apex of the gland, which is the guide to Cock's operation for stricture ; he is enabled also by the same means to direct the point of a catheter, when its introduction is attended with difficulty either from injury or disease of the membranous or prostatic portions of the urethra. When the finger is introduced into the bowel, the surgeon may, in some cases, especially in boys, learn the position, as well as the size, of a calculus in the bladder ; and in the operation for its removal, if, as is not infrequently the case, it should be lodged behind an enlarged prostate, it may be displaced from its position by pressing upwards the base of the bladder from the rectum. The prostate gland is occasionally the seat of suppuration, due either to injury, gonorrhœa, or tuberculous disease. The gland is enveloped in a dense unyielding capsule, which determines the course of the abscess, and also explains the great pain which is present in the acute form of the disease. The abscess most frequently bursts into the urethra, the direction in which there is least resistance, but may occasionally burst into the rectum, or more rarely in the perinæum. In advanced life the prostate sometimes becomes considerably enlarged and projects into the bladder so as to impede the passage of the urine. According to Messer's researches, conducted at Greenwich Hospital, it would seem that such obstruction exists in 20 per cent. of all men over sixty years of age. In some cases the condition affects principally the lateral lobes, which may undergo considerable enlargement without causing much inconvenience. In other cases it would seem that the middle lobe enlarges most, and even a small enlargement of this lobe may act injuriously, by forming a sort of valve over the urethral orifice, preventing the passage of the urine ; and the more the patient strains, the more completely will it block the opening into the urethra. In consequence of the enlargement of the prostate, a pouch is formed at the base of the bladder behind the projection, in which urine collects, and cannot be entirely expelled. It becomes decomposed and ammoniacal, and leads to cystitis. For this condition 'prostatectomy' is sometimes done. The bladder is opened by an incision above the symphysis pubis, the mucous membrane incised, and in those cases where the enlargement is due to encapsuled masses—adenomata or fibro-myomata—within the prostate, these can be shelled out without much difficulty. In other cases where there is a general hypertrophy of all the tissues of the prostate, it becomes necessary to enucleate the entire organ by the use of a suitable scoop, aided by the finger. Recently, most satisfactory results have followed this operation.

COWPER'S GLANDS

Cowper's Glands are two small, rounded, and somewhat lobulated bodies, of a yellow colour, about the size of peas, placed behind the membranous portion of the urethra, between the two layers of the triangular ligament. They lie close above the bulb, and are enclosed by the transverse fibres of the Compressor urethræ muscle. Their existence is said to be constant : they gradually diminish in size as age advances.

Structure.—Each gland is made up of several lobules, held together by a fibrous investment. Each lobule consists of a number of acini, lined by columnar epithelial cells, opening into one duct, which, joining with the ducts of other lobules outside the gland, form a single excretory duct. The excretory duct of each gland, nearly an inch in length, passes obliquely forwards beneath the mucous membrane, and opens by a minute orifice on the floor of the bulbous portion of the urethra.

THE PENIS

The **Penis** consists of a root, body, and extremity or *glans penis*.

The *root* is firmly connected to the rami of the os pubis and ischium by two strong tapering fibrous processes, the *crura* ; and to the front of the symphysis pubis by the *suspensory ligament,* a strong band of fibrous tissue which passes downwards from the front of the symphysis pubis to the upper surface of the root of the penis, where it splits into two portions and blends with the fascial sheath of the organ.

The *extremity*, or *glans penis*, presents the form of an obtuse cone, flattened from above downwards. At its summit is a vertical fissure, the orifice of the urethra (*meatus urinarius*). The base of the glans forms a rounded projecting border, the *corona glandis*; and behind the corona is a deep constriction, the *cervix*. Upon both of these parts, numerous small sebaceous glands are found, the *glandulæ Tysonii odoriferæ*.* They secrete a sebaceous matter of very peculiar odour, which probably contains caseine, and becomes easily decomposed.

The *body of the penis* is the part between the root and extremity. In the flaccid condition of the organ it is cylindrical, but when erect has a triangular prismatic form with rounded angles, the broadest side being turned upwards, and called the *dorsum*. The body is covered by integument, and contains in its interior a large portion of the urethra. The integument covering the penis is remarkable for its thinness, its dark colour, its looseness of connection with the deeper parts of the organ, and its containing no adipose tissue. At the root of the penis, the integument is continuous with that upon the pubes and scrotum : and at the neck of the glans it leaves the surface, and becomes folded upon itself to form the *prepuce* or *foreskin*.

The internal layer of the prepuce is attached behind to the cervix, and resembles in appearance a mucous membrane ; from the cervix it is reflected over the glans penis, and at the meatus urinarius is continuous with the mucous lining of the urethra.

The integument covering the glans penis contains no sebaceous glands ; but projecting from its free surface are a number of small, highly sensitive papillæ. At the back part of the meatus urinarius a fold of integument passes backwards to the bottom of a depressed raphé, where it is continuous with the prepuce ; this fold is termed the *frænum præputii*.

Structure of the Penis.—The penis is composed of a mass of erectile tissue, enclosed in three cylindrical fibrous compartments. Of these, two, the *corpora cavernosa*, are placed side by side along the upper part of the organ ; the third, or *corpus spongiosum*, encloses the urethra, and is placed below.

The **Corpora Cavernosa** form the chief part of the body of the penis. They consist of two fibrous cylindrical tubes, placed side by side, and intimately connected along the median line for their anterior three-fourths, while at their back part they separate from each other to form the *crura*, which are two strong tapering processes firmly connected to the rami of the os pubis and ischium. Each crus commences by a blunt-pointed process in front of the tuberosity of the ischium ; and, before its junction with its fellow to form the body of the penis, it presents a slight enlargement, named by Kobelt the *bulb of the corpus cavernosum*. Just beyond this point they become constricted, and retain an equal diameter to their anterior extremities, where each forms a single rounded end, which is received into a fossa in the base of the glans penis. A median groove on the upper surface lodges the dorsal vein of the penis, and the groove on the under surface receives the corpus spongiosum. The root of the penis is connected to the symphysis pubis by the suspensory ligament.

Structure.—The corpora cavernosa are surrounded by a strong fibrous envelope, consisting of two sets of fibres : the one, longitudinal in direction, being common to the two corpora cavernosa, and investing them in a common covering ; the other, internal, circular in direction, being proper to each corpus cavernosum. The internal circular fibres of the two corpora cavernosa form, by their junction in the mesial plane, an incomplete partition or septum between the two bodies.

The *septum* between the two corpora cavernosa is thick and complete behind, but in front it is incomplete, and consists of a number of vertical bands, which are arranged like the teeth of a comb, and hence it has received the name of *septum pectiniforme*. These bands extend between the dorsal and the urethral surfaces of the corpora cavernosa. This fibrous investment is extremely dense, of considerable thickness, and consists of bundles of shining white fibres, with

---

* Stieda (*Comptes-rendus du XII Congrès International de Médecine*, Moscow, 1897) asserts that Tyson's glands are never found on the corona glandis, and that what have hitherto been mistaken for glands are really large papillæ.

an admixture of well-developed elastic fibres, so that it is possessed of great elasticity.

From the internal surface of the fibrous envelope, as well as from the sides of the septum, numerous bands or cords are given off, which cross the interior of the corpora cavernosa in all directions, subdividing them into a number of separate compartments, and giving the entire structure a spongy appearance. These bands and cords are called *trabeculæ*, and consist of white fibrous tissue, elastic fibres, and plain muscular fibres. In them are contained numerous arteries and nerves.

The component fibres of which the trabeculæ are composed are larger and stronger round the circumference than at the centre of the corpora cavernosa; they are also thicker behind than in front. The interspaces, on the contrary, are larger at the centre than at the circumference, their long diameter being directed transversely; they are largest anteriorly. They are occupied by venous blood, and are lined by a layer of flattened cells similar to the endothelial lining of veins.

The whole of the structure of the corpora cavernosa, contained within the fibrous sheath, consists therefore of a sponge-like tissue of areolar spaces, freely communicating with each other and filled with venous blood. The spaces may therefore be regarded as large cavernous veins.

FIG. 771.—From the peripheral portion of the corpus cavernosum penis under a low magnifying power. (Copied from Langer.)

1. *a.* Capillary network. *b.* Cavernous spaces. 2. Connection of the arterial twigs (*a*) with the cavernous spaces.

The arteries bringing the blood to these spaces are the arteries of the corpora cavernosa and branches from the dorsal arteries of the penis, which perforate the fibrous capsule, along the upper surface, especially near the fore part of the organ.

These arteries on entering the cavernous structure divide into branches, which are supported and enclosed by the trabeculæ. Some of these terminate in a capillary network, the branches of which open directly into the cavernous spaces; others assume a tendril-like appearance, and form convoluted and somewhat dilated vessels, which were named by Müller *helicine arteries.* They project into the spaces, and from them are given off small capillary branches to supply the trabecular structure. They are bound down in the spaces by fine fibrous processes, and are more abundant in the back part of the corpora cavernosa (fig. 771).

The blood from the cavernous spaces is returned by a series of vessels, some of which emerge in considerable numbers from the base of the glans penis and converge on the dorsum of the organ to form the dorsal vein; others pass out on the upper surface of the corpora cavernosa and join the dorsal vein; some emerge from the under surface of the corpora cavernosa and, receiving branches from the corpus spongiosum, wind round the sides of the penis to terminate in the dorsal vein; but the greater number pass out at the root of the penis and join the prostatic plexus.

The **Corpus Spongiosum** encloses the urethra, and is situated in the groove on the under surface of the corpora cavernosa. It commences posteriorly below the superficial layer of the triangular ligament of the urethra, between the diverging crura of the corpora cavernosa, where it forms a rounded enlargement, the *bulb*; and terminates, anteriorly, in another expansion, the *glans penis*, which overlaps the anterior rounded extremity of the corpora cavernosa. The central portion, or body of the corpus spongiosum, is cylindrical, and tapers slightly from behind forwards.

The *bulb* varies in size in different subjects; it receives a fibrous investment from the superficial layer of the triangular ligament, and is surrounded by the Accelerator urinæ muscle. The urethra enters the bulb nearer its upper than its lower surface, being enveloped by a layer of erectile tissue, a thin prolongation of which is continued backwards round the membranous and prostatic portions of the canal to the neck of the bladder, lying between the two layers of muscular tissue. The portion of the bulb below the urethra presents a partial division into two lobes, being marked externally by a linear raphé, while internally there projects, for a short distance, a thin fibrous septum, which is more distinct in early life.

Structure.—The corpus spongiosum consists of a strong fibrous envelope, enclosing a trabecular structure, which contains in its meshes erectile tissue. The fibrous envelope is thinner, whiter in colour, and more elastic than that of the corpora cavernosa. The trabeculæ are more delicate, nearly uniform in size, and the meshes between them smaller than in the corpora cavernosa: their long diameter, for the most part, corresponding with that of the penis. The external envelope or outer coat of the corpus spongiosum is formed partly of unstriped muscular fibre, and a layer of the same tissue immediately surrounds the canal of the urethra.

The *lymphatics of the penis* consist of a superficial and a deep set: the superficial vessels are derived from a dense network on the skin of the glans and prepuce and from the mucous membrane of the urethra, and terminate in the superficial inguinal glands; the deep vessels emerge from the corpora cavernosa and corpus spongiosum, and, passing beneath the pubic arch, join the deep lymphatics of the pelvis.

The *nerves* are derived from the internal pudic nerve and the pelvic plexuses. On the glans and bulb some filaments of the cutaneous nerves have Pacinian bodies connected with them, and, according to Krause, many of them terminate in a peculiar form of end-bulb (see page 46).

Surgical Anatomy.—The penis occasionally requires removal for malignant disease. Usually, removal of the ante-scrotal portion is all that is necessary, but sometimes it is requisite to remove the whole organ from its attachment to the rami of the ossa pubis and ischia. The former operation is performed either by cutting off the whole of the anterior part of the penis with one sweep of the knife; or, what is better, cutting through the corpora cavernosa from the dorsum, and then separating the corpus spongiosum from them, dividing it at a level nearer the glans penis. The mucous membrane of the urethra is then slit up, and the edges of the flap attached to the external skin, in order to prevent contraction of the orifice, which might otherwise take place. The vessels which require ligature are the two dorsal arteries of the penis, the arteries of the corpora cavernosa, and the artery of the septum. When the entire organ requires removal, the patient is placed in the lithotomy position, and an incision is made through the skin and subcutaneous tissue round the root of the penis, and carried down the median line of the scrotum as far as the perinæum. The two halves of the scrotum are then separated from each other, and a catheter having been introduced into the bladder as a guide, the spongy portion of the urethra below the triangular ligament is separated from the corpora cavernosa and divided, the catheter having been withdrawn just behind the bulb. The suspensory ligament is now severed, and the crura separated from the bone with a periosteum scraper, and the whole penis removed. The membranous portion of the urethra, which has not been removed, is now to be attached to the skin at the posterior extremity of the incision in the perinæum. The remainder of the wound is to be brought together, free drainage being provided for.

THE TESTES AND THEIR COVERINGS (fig. 772)

The **Testes** are two glandular organs, which secrete the semen; they are situated in the scrotum, being suspended by the spermatic cords. At an early period of fœtal life the testes are contained in the abdominal cavity, behind the peritoneum. Before birth they descend to the inguinal canal, along which they pass with the spermatic cord, and, emerging at the external abdominal ring, they descend into the scrotum, becoming invested in their course by numerous coverings derived from the serous, muscular, and fibrous layers of the abdominal parietes, as well as by the scrotum. The coverings of the testes are, the

> Skin ⎫ Scrotum.
> Dartos ⎰
> Intercolumnar, or External spermatic fascia.
> Cremasteric fascia.
> Infundibuliform fascia (Internal spermatic fascia).
> Tunica vaginalis.

The **scrotum** is a cutaneous pouch which contains the testes and part of the spermatic cords. It is divided on its surface into two lateral portions by a median line, or *raphé*, which is continued forwards to the under surface of the penis, and backwards along the middle line of the perinæum to the anus. Of these two lateral portions the left is longer than the right, and corresponds with the greater length of the spermatic cord on the left side. Its external aspect varies under different circumstances: thus, under the influence of warmth, and

FIG. 772.—Transverse section through the left side of the scrotum and the left testicle. The sac of the tunica vaginalis represented in a distended condition. (Diagrammatic.) (Delépine.)

in old and debilitated persons, it becomes elongated and flaccid; but, under the influence of cold, and in the young and robust, it is short, corrugated, and closely applied to the testes.

The scrotum consists of two layers, the integument and the dartos.

The **integument** is very thin, of a brownish colour, and generally thrown into folds or rugæ. It is provided with sebaceous follicles, the secretion of which has a peculiar odour, and is beset with thinly scattered, crisp hairs, the roots of which are seen through the skin.

The **dartos** is a thin layer of loose reddish tissue, endowed with contractility: it forms the proper tunic of the scrotum, is continuous, around the base of the scrotum, with the two layers of the superficial fascia of the groin and perinæum, and sends inwards a distinct septum, *septum scroti*, which divides it into two cavities for the testes, the septum extending between the raphé and the under surface of the penis, as far as its root.

The dartos is closely united to the skin externally, but connected with the subjacent parts by delicate areolar tissue, upon which it glides with the greatest facility. The dartos is very vascular, and consists of a loose areolar tissue, containing unstriped muscular fibre, but no fat. Its contractility is slow, and excited by cold and mechanical stimuli, but not by electricity.

The **intercolumnar fascia** is a thin membrane, which is derived from the pillars of the external abdominal ring during the descent of the testis in the foetus, and is prolonged downwards around the surface of the cord and testis. It is separated from the dartos by loose areolar tissue, which allows of considerable movement of the latter upon it, but is intimately connected with the succeeding layers.

The **cremasteric fascia** consists of scattered bundles of muscular fibres (*Cremaster muscle*), connected together into a continuous covering by intermediate areolar tissue. The muscular fibres are continuous with the lower border of the Internal oblique muscle (see page 487); they are supplied by the genital branch of the genito-crural nerve.

The **infundibuliform fascia** is a thin membranous layer, which loosely invests the surface of the cord. It is a continuation downwards of the fascia transversalis. Beneath it is a quantity of loose connective tissue which connects this layer of fascia with the spermatic cord and posterior part of the testicle. This connective tissue is continuous above with the subserous areolar tissue of the abdomen. These two layers, the infundibuliform fascia and the tissue beneath it, are known collectively as the *fascia propria*.

The **tunica vaginalis** is described with the testis.

Vessels and Nerves.—The *arteries* supplying the coverings of the testis are: the superficial and deep external pudic, from the femoral; the superficial perineal branch of the internal pudic; and the cremasteric branch from the epigastric. The *veins* follow the course of the corresponding arteries. The *lymphatics* terminate in the inguinal glands. The *nerves* are the ilio-inguinal branch of the lumbar plexus, the two superficial perineal branches of the internal pudic nerve, and the inferior pudendal branch of the small sciatic nerve.

The **spermatic cord** extends from the internal or deep abdominal ring, where the structures of which it is composed converge, to the back part of the testicle. In the abdominal wall the cord passes obliquely along the inguinal canal, lying at first beneath the Internal oblique, and upon the fascia transversalis; but nearer the pubes, it rests upon Poupart's ligament, having the aponeurosis of the External oblique in front of it, and the conjoined tendon behind it. It then escapes at the external ring, and descends nearly vertically into the scrotum. The left cord is rather longer than the right, consequently the left testis hangs somewhat lower than its fellow.

Structure of the spermatic cord.—The spermatic cord is composed of arteries, veins, lymphatics, nerves, and the excretory duct of the testicle. These structures are connected together by areolar tissue, and invested by the layers brought down by the testicle in its descent.

The *arteries of the cord* are: the spermatic, from the aorta; the artery of the vas deferens, from the superior vesical; and the cremasteric, from the deep epigastric.

The *spermatic artery*, a branch of the abdominal aorta, escapes from the abdomen at the internal or deep abdominal ring, and accompanies the other constituents of the spermatic cord along the inguinal canal and through the external abdominal ring into the scrotum. It then descends to the testicle, and, becoming tortuous, divides into several branches, two or three of which accompany the vas deferens and supply the epididymis, anastomosing with the artery of the vas deferens: others pierce the back of the tunica albuginea and supply the substance of the testis.

The *cremasteric artery* is a branch of the deep epigastric artery. It accompanies the spermatic cord and supplies the Cremaster muscle and other coverings of the cord, anastomosing with the spermatic artery.

The *artery of the vas deferens*, a branch of the superior vesical, is a long slender vessel, which accompanies the vas deferens, ramifying upon the coats of that duct, and anastomosing with the spermatic artery near the testis.

The *spermatic veins* emerge from the back of the testis, and receive tributaries from the epididymis: they unite and form a convoluted plexus (*plexus pampiniformis*), which forms the chief mass of the cord; the vessels composing this plexus are very numerous, and ascend along the cord in front of the vas deferens; below the external or superficial abdominal ring they unite to form three or four veins, which pass along the inguinal canal, and, entering the abdomen through the internal or deep abdominal ring, coalesce to form two veins. These again unite to form a single vein, which opens on the right side into the inferior vena cava, at an acute angle, and on the left side into the renal vein at a right angle.

The *lymphatic vessels* terminate in the lumbar glands.

The *nerves* are the spermatic plexus from the sympathetic, joined by filaments from the pelvic plexus which accompany the artery of the vas deferens.

Surgical Anatomy.—The scrotum forms an admirable covering for the protection of the testicle. This body, lying suspended and loose in the cavity of the scrotum and surrounded by a serous membrane, is capable of great mobility, and can therefore easily slip about within the scrotum, and thus avoid injuries from blows or squeezes. The skin of the scrotum is very elastic and capable of great distension, and on account of the looseness and amount of subcutaneous tissue, the scrotum becomes greatly enlarged in cases of œdema, to which this part is especially liable as a result of its dependent position. The scrotum is occasionally the seat of epithelioma; this is no doubt due to the rugæ on its surface, which favour the lodgment of dirt, and this, producing irritation, is the exciting cause of the disease, which is especially common in chimney-sweeps from the lodgment of soot. The disease is very much less common than it used to be ; this is probably due to the better hygienic conditions of the working classes. The scrotum is also the part most frequently affected by elephantiasis.

On account of the looseness of the subcutaneous tissue, large extravasations of blood may take place from very slight injuries. It is therefore generally recommended never to apply leeches to the scrotum, since they may lead to ecchymosis, but rather to puncture one or more of the superficial veins of the scrotum in cases where local blood-letting from this part is judged to be desirable. The muscular fibre in the dartos causes contraction and considerable diminution in the size of a wound of the scrotum, as after the operation of castration, and is of assistance in keeping the edges together, and covering the exposed parts.

The Testes

The **testes** are suspended in the scrotum by the spermatic cords. As the left spermatic cord is rather longer than the right one, the left testicle hangs somewhat lower than its fellow. Each gland is of an oval form, compressed laterally, and having an oblique position in the scrotum ; the upper extremity being directed forwards and a little outwards ; the lower, backwards and a little inwards ; the anterior convex border looks forwards and downwards ; the posterior or straight border, to which the cord is attached, backwards and upwards.

The anterior border and lateral surfaces, as well as both extremities of the organ, are convex, free, smooth, and invested by the visceral layer of the tunica vaginalis. The posterior border, to which the cord is attached, receives only a partial investment from that membrane. Lying upon the outer edge of this posterior border is a long, narrow, flattened body, named, from its relation to the testis, the *epididymis* (δίδυμος, testis). It consists of a central portion, or *body* ; an upper enlarged extremity, the head, or *globus major* ; and a lower pointed extremity, the tail, or *globus minor*. The globus major is intimately connected with the upper end of the testicle by means of its efferent ducts ; and the globus minor is connected with its lower end by cellular tissue, and a reflection of the tunica vaginalis. The outer surface and upper and lower ends of the epididymis are free and covered by serous membrane ; the body is also completely invested by it, excepting along its posterior border, and between the body and the testicle is a pouch or *cul-de-sac*, named the *digital fossa*. The epididymis is connected to the back of the testis by a fold of the serous membrane. Attached to the upper end of the testis, close to the globus major, are two small pedunculated bodies. One of them is pear-shaped, and attached by its narrow stalk, the other is small and sessile ; they are believed to be the remains of the upper extremity of the Müllerian duct (page 150), and are termed the *hydatids of Morgagni*; some observers, however, regard the stalked hydatid as being a rudiment of the pronephros. When the testicle is removed from the body, the position of the vas deferens, on the posterior surface of the testicle and inner side of the epididymis, marks the side to which the gland has belonged.

Size and Weight.—The average dimensions of this gland are from one and a half to two inches in length, an inch in breadth, and an inch and a quarter in the antero-posterior diameter; and the weight varies from six to eight drachms, the left testicle being a little the larger.

The testis is invested by three tunics : the tunica vaginalis, tunica albuginea, and tunica vasculosa.

The **tunica vaginalis** is the serous covering of the testis. It is a pouch of serous membrane, derived from the peritoneum during the descent of the testis in the fœtus from the abdomen into the scrotum. After its descent, that portion of the pouch which extends from the internal ring to near the upper part of the gland becomes obliterated, the lower portion remains as a shut sac, which invests the outer surface of the testis, and is reflected on to the internal surface of the scrotum; hence it may be described as consisting of a visceral and parietal portion.

The *visceral portion of the tunica vaginalis* covers the greater part of the testis and epididymis, connecting the latter to the testis by means of a distinct fold. From the posterior border of the gland it is reflected on to the internal surface of the scrotum.

FIG. 773.—The testis *in situ*, the tunica vaginalis having been laid open.

The *parietal portion of the tunica vaginalis* is far more extensive than the visceral portion, extending upwards for some distance in front, and on the inner side of the cord, and reaching below the testis. The inner surface of the tunica vaginalis is free, smooth, and covered by a layer of endothelial cells. The interval between the visceral and parietal layers of this membrane constitutes the cavity of the tunica vaginalis.

The obliterated portion of the pouch may generally be seen as a fibro-cellular thread lying in the loose areolar tissue around the spermatic cord ; sometimes this may be traced as a distinct band from the upper end of the inguinal canal, where it is connected with the peritoneum, down to the tunica vaginalis ; sometimes it gradually becomes lost on the spermatic cord. Occasionally no trace of it can be detected. In some cases it happens that the pouch of peritoneum does not become obliterated, but the sac of the peritoneum communicates with the tunica vaginalis. This may give rise to one of the varieties of oblique inguinal hernia (page 1182). Or in other cases the pouch may contract, but not become entirely obliterated ; it then forms a minute canal leading from the peritoneum to the tunica vaginalis.*

The **tunica albuginea** is the fibrous covering of the testis. It is a dense membrane, of a bluish-white colour, composed of bundles of white fibrous tissue, which interlace in every direction. Its outer surface is covered by the tunica vaginalis, except at the points of attachment of the epididymis to the testicle, and along its posterior border, where the spermatic vessels enter the gland. This membrane surrounds the glandular structure of the testicle, and, at its posterior border, is reflected into the interior of the gland, forming an incomplete vertical septum, called the *mediastinum testis (corpus Highmori).*

The *mediastinum testis* extends from the upper, nearly to the lower extremity of the gland, and is wider above than below. From the front and sides of this septum numerous slender fibrous cords and imperfect septa (*trabeculæ*) are given off, which radiate towards the surface of the organ, and are attached to the

---

* It is recorded that in the post-mortem examination of Sir Astley Cooper, this minute canal was found on both sides of the body. Sir Astley Cooper states that when a student he suffered from inguinal hernia ; probably this was of the congenital variety, and the canal found after death was the remains of the one down which the hernia travelled.—*Lancet*, vol. ii. 1824, p. 116.

inner surface of the tunica albuginea. They therefore divide the interior of the organ into a number of incomplete spaces, which are somewhat cone-shaped, being broad at their bases at the surface of the gland, and becoming narrower as they converge to the mediastinum. The mediastinum supports the vessels and ducts of the testis in their passage to and from the substance of the gland.

The **tunica vasculosa** is the vascular layer of the testis, consisting of a plexus of blood-vessels, held together by delicate areolar tissue. It covers the inner surface of the tunica albuginea and the different septa in the interior of the gland, and therefore forms an internal investment to all the spaces of which the gland is composed.

Structure.—The glandular structure of the testis consists of numerous lobules (*lobuli testis*). Their number, in a single testis, is estimated by Berres at 250, and by Krause at 400. They differ in size according to their position, those in the middle of the gland being larger and longer. The lobules are conical in shape, the base being directed towards the circumference of the organ, the apex towards the mediastinum. Each lobule is contained in one of the intervals between the fibrous cords and vascular processes which extend between the mediastinum testis and the tunica albuginea, and consists of from one to three, or more, minute convoluted tubes, the *tubuli seminiferi*. The tubes may be separately unravelled, by careful dissection under water, and may be seen to commence either by free cæcal ends or by anastomotic loops. The total number of tubes is estimated by Lauth at 840, and their average length two feet and a quarter. Their diameter varies from $\frac{1}{200}$ to $\frac{1}{150}$ of an inch. The tubuli are pale in colour in early life, but in old age they acquire a deep yellow tinge, from containing much fatty matter. Each tube consists of a basement layer, formed of epithelioid cells united edge to edge, outside which are other layers of flattened cells arranged in interrupted laminæ, which give to the tube an appearance of striation in cross section. The cells of the outer layers gradually pass into the interstitial tissue. Within the basement-membrane are epithelial cells arranged in several irregular layers, which are not always clearly separated, but which may be arranged in three different groups. Among these cells may be seen the *spermatozoa* in different stages of development. 1. Lining the basement-membrane and forming the outer zone is a layer of cubical cells, with small nuclei ; these are known as the *lining cells* or *spermatogonia*. The nucleus of some of them may be seen to be in the process of indirect division (*karyokinesis*, page 3), and in consequence of this daughter cells are formed, which constitute the second zone. 2. Within this first layer is to be seen a number of larger cells with clear nuclei, arranged in two or three layers ; these are the *intermediate cells* or *spermatocytes*. Most of these cells are in a condition of karyokinetic division, and the cells which result from this division form those of the next layer, the *spermatoblasts* or *spermatids*. 3. The third layer of cells therefore consists of the spermatoblasts or spermatids, and each of these, without further subdivision, becomes a *spermatozoon*. They are ill-defined granular masses of protoplasm, of an elongated form, with a nucleus, which becomes the head of the future spermatozoon. In addition to these three layers of cells others are seen, which are termed the *supporting cells*, or *cells of Sertoli*. They are elongated and columnar, and project inwards from the basement-membrane towards the lumen of the tube. They give off numerous lateral branches, which form a reticulum for the support of the three groups of cells just described. As development of the spermatozoa proceeds the latter group themselves around the inner extremities of the supporting cells. The nuclear portion of the spermatozoon, which is partly embedded in the supporting cell, is differentiated to form the head of the spermatozoon, while the cell protoplasm becomes lengthened out to form the middle piece and tail, the latter projecting into the lumen of the tube. Ultimately the heads are liberated and the spermatozoa are set free. The structure of the spermatozoa is described on page 76.

The tubules are enclosed in a delicate plexus of capillary vessels, and are held together by an intertubular connective tissue, which presents large inter-stitial spaces lined by endothelium, which are believed to be the rootlets of the lymphatic vessels of the testis.

In the apices of the lobules, the tubuli become less convoluted, assume a nearly straight course, and unite together to form from twenty to thirty larger ducts, of about $\frac{1}{50}$ of an inch in diameter, and these, from their straight course, are called *vasa recta*.

The *vasa recta* enter the fibrous tissue of the mediastinum, and pass upwards and backwards, forming, in their ascent, a close network of anastomosing tubes which are merely channels in the fibrous stroma, lined by flattened epithelium and having no proper walls; this constitutes the *rete testis*. At the upper end of the mediastinum, the vessels of the rete testis terminate in from twelve to fifteen or twenty ducts, the *vasa efferentia*: they perforate the tunica albuginea, and carry the seminal fluid from the testis to the epididymis. Their course is at first straight; they then become enlarged, and exceedingly convoluted, and form a series of conical masses, the *coni vasculosi*, which together constitute the globus major of the epididymis. Each cone consists of a single convoluted duct, from six to eight inches in length, the diameter of which gradually decreases from the testis to the epididymis. Opposite the bases of the cones the efferent vessels open at narrow intervals into a single duct, which constitutes, by its complex convolutions, the body and globus minor of the epididymis.

When the convolutions of this tube are unravelled, it measures upwards of twenty feet in length; it increases in diameter and thickness as it approaches the vas deferens. The convolutions are held together by fine areolar tissue, and by bands of fibrous tissue.

FIG. 774.—Vertical section of the testicle, to show the arrangement of the ducts.

The vasa recta are of smaller diameter than the seminal tubes, and have very thin parietes. They, like the channels of the rete testis, are lined by a single layer of flattened epithelium. The vasa efferentia and the tube of the epididymis have walls of considerable thickness, on account of the presence in them of muscular tissue, which is principally arranged in a circular manner. These tubes are lined by columnar ciliated epithelium.

The **vas deferens,** the excretory duct of the testis, is the continuation of the canal of the epididymis. Commencing at the lower part of the globus minor, it is at first very tortuous, but gradually becoming less twisted it ascends along the posterior border of the testis and inner side of the epididymis, and along the back part of the spermatic cord, through the inguinal canal to the internal or deep abdominal ring. From the ring it curves round the outer side of the epigastric artery, and ascends for about an inch in front of the external iliac artery. It is next directed backwards and slightly downwards, and, crossing the external iliac vessels obliquely, enters the pelvic cavity, where it lies between the peritoneal membrane and the lateral wall of the pelvis, and passes on the inner side of the obliterated hypogastric artery and the obturator nerve and vessels. It then crosses in front of the ureter, and, reaching the inner side of this tube, bends to form an acute angle, and runs inwards and slightly forwards between the base of the bladder and the upper end of the seminal vesicle. Reaching the inner side of the seminal vesicle, it is directed downwards and inwards in contact with it, gradually approaching the vas of the opposite side. Here it lies between the base of the bladder and the rectum, where it is enclosed, together with the seminal vesicle, in a sheath derived from the recto-vesical layer of the pelvic fascia. Lastly, it is directed downwards to the base of the prostate, where it becomes greatly narrowed, and is joined at an acute angle by the duct of the seminal vesicle to form the common ejaculatory duct, which traverses the prostate gland behind its middle lobe and opens into the urethra, close to the sinus pocularis. The vas deferens presents a hard and cord-like sensation to the fingers; it is of cylindrical form, and about a line and a quarter in diameter. Its walls are dense, measuring one-third of a line ; and its canal is extremely small, measuring about half a line. At the base of the bladder it becomes enlarged and sacculated, and this portion is termed the *ampulla.*

Structure.—The vas deferens consists of three coats : 1. An external or areolar coat. 2. A muscular coat, which in the greater part of the tube consists of two layers of unstriped muscular fibre : an outer, longitudinal in direction, and an inner, circular ; but in addition to these, at the commencement of the vas deferens, there is a third layer, consisting of longitudinal fibres, placed internal to the circular stratum, between it and the mucous membrane. 3. An internal, or mucous coat, which is pale, and arranged in longitudinal folds ; its epithelial lining is of the columnar variety.

A long narrow tube, the *vas aberrans of Haller,* is occasionally found connected with the lower part of the canal of the epididymis, or with the commencement of the vas deferens. It extends up into the cord for about two or three inches, where it terminates by a blind extremity, which is sometimes bifurcated. Its length varies from an inch and a half to fourteen inches, and it may become dilated towards its extremity ; more commonly it retains the same diameter throughout. Its structure is similar to that of the vas deferens. Occasionally it is found unconnected with the epididymis.

Organ of Giraldès.—This term is applied to a small collection of convoluted tubules, situated in front of the lower part of the cord above the globus major of the epididymis. These tubes are lined with columnar ciliated epithelium, and probably represent the remains of a part of the Wolffian body.

The testicle, developed in the lumbar region, may be arrested or delayed in its transit to the scrotum. It may be retained in the abdomen ; or it may be arrested at the internal abdominal ring, or in the inguinal canal ; or it may just pass out of the external ring without finding its way to the bottom of the scrotum. When retained in the abdomen it gives rise to no symptoms, other than the absence of the testicle from the scrotum ; but when it is retained in the inguinal canal it is subjected to pressure and may become inflamed and painful, and the inflammation may extend to the peritoneum, producing general peritonitis. The testicle when first formed is believed to be normal, but if retained it undergoes degenerative changes and becomes functionally useless ; so that a man in whom both testicles are retained (*anorchism*) is sterile, though he may not be impotent. The absence of one testicle is termed *monorchism.* When a testicle is retained in the inguinal canal it is often complicated with a congenital hernia, the funicular process of the peritoneum not being obliterated. In addition to the cases above described, where there is some arrest in the descent of the testicle, this organ may descend through the inguinal canal, but may miss the scrotum and assume some abnormal position. The most common form is where the testicle, emerging at the external abdominal ring, slips down between the scrotum and thigh and comes to rest in the perinæum. This is known as *perineal ectopia.* Again, in other cases, the testicle, upon emerging at the external ring, slips downwards and outwards over the spine of the os pubis and becomes located over the situation of the saphenous opening. It has been stated that occasionally the testicle may leave the abdomen by the femoral ring and pass down the crural canal to present on the front of the thigh. This form has been termed *ectopia cruralis.* It is very rare, and it is doubtful if it ever really exists.

Finally, the testicle may reach the scrotum, but may occupy an abnormal position in it. It may be inverted, so that its posterior or attached border is directed forwards and the tunica vaginalis is situated behind. Should a hydrocele occur, and tapping be resorted to, the trochar may be thrust into the testicle, if the operation is performed in the ordinary way, and care is not taken beforehand to ascertain the position of the gland.

Recently, several cases of torsion of the spermatic cord, resulting in acute strangulation of the testicle, have been recorded. In some it has been attributed to a strain or twist, and in several instances the condition has been associated with a late descent of the organ.

Surgical Anatomy.—Fluid collections of a serous character are very frequently found in the scrotum. To these the term *hydrocele* is applied. The most common form is the ordinary *vaginal hydrocele,* in which the fluid is contained in the sac of the tunica vaginalis, which is separated, in its normal condition, from the peritoneal cavity by the whole extent of the inguinal canal. Another form, the *congenital hydrocele,* is where the fluid is in the sac of the tunica vaginalis, but this cavity communicates with the general peritoneal cavity, its tubular process remaining pervious. A third variety, known as an *infantile* hydrocele, occurs in those cases where the tubular process becomes obliterated only at its upper part, at or near the internal abdominal ring. It resembles the vaginal hydrocele, except as regards its shape, the collection of fluid extending up the cord into the inguinal canal. Fourthly, a very rare form of hydrocele is known as *hydrocele of the funicular process.* In this variety the processus vaginalis has become closed at its lower end, but has remained unobliterated throughout the rest of its extent. This may become distended with fluid, producing a sausage-shaped tumour in the inguinal canal, which is reducible and may be mistaken for an inguinal hernia.

Encysted hydrocele, or *spermatocele,* is the name given to a cyst usually found in connection with the globus major of the epididymis. Among its contents are found, in many instances, a varying number of spermatozoa. In addition to these different forms of hydrocele, an *encysted hydrocele of the cord* is sometimes met with. This may be due to a small portion of the funicular process of the peritoneum remaining unobliterated and becoming filled with fluid, or to the formation of an independent cyst in the connective tissue of the cord. A *diffused hydrocele of the cord,* which would appear to be merely a dropsy of the cellular tissue of the cord, has been described.

The testicle frequently requires removal for malignant disease; in tuberculous disease; in cystic disease; in cases of large hernia testis, and in some instances of incompletely descended or misplaced testicles. The operation of castration has also been performed for enlargement of the prostate; since it has been found that removal of the testicles is followed by very rapid and often considerable diminution in the size of the prostate. The operation is, however, one of severity, and is frequently followed by death, as the patients on whom it is performed are advanced in life. Reginald Harrison has proposed to substitute for it excision of a portion of the vasa deferentia. The operation of castration is a comparatively simple one. An incision is made into the tunica vaginalis from the external ring to the bottom of the scrotum. The coverings are shelled off the organ, and the mesorchium, stretching between the back of the testicle and the scrotum, divided. The cord is then isolated, and an aneurism needle, armed with a double ligature, passed under it, as high as is thought necessary, and the cord tied in two places, and divided between the ligatures. Sometimes, in cases of malignant disease, it is desirable to open the inguinal canal and tie the cord as near the internal abdominal ring as possible.

VESICULÆ SEMINALES

The **seminal vesicles** are two lobulated membranous pouches, placed between the base of the bladder and the rectum, serving as reservoirs for the semen, and secreting a fluid to be added to the secretion of the testicles. Each sac is somewhat pyramidal in form, the broad end being directed backwards, upwards and

FIG. 775.—Base of the bladder, with the vasa deferentia and vesiculæ seminales.

Left ejaculatory duct

outwards. They measure about two and a half inches in length, about five lines in breadth, and two or three lines in thickness. They vary, however, in size, not only in different individuals, but also in the same individual on the two sides. Their *anterior surface* is in contact with the base of the bladder, extending from near the termination of the ureters to the base of the prostate gland. Their *posterior surface* rests upon the rectum, from which they are separated by the recto-vesical fascia. Their *upper extremities* diverge from each other, and are separated from the bladder by the vas deferens and the lower end of the ureter, and are partly covered by peritoneum. Their *lower extremities* are pointed, and converge towards the base of the prostate gland, where each joins with the corresponding vas deferens to form the ejaculatory duct. Along the inner margin of each vesicle runs the enlarged and convoluted vas deferens. The inner borders of the vesicles, and the corresponding vas deferens, form the lateral boundaries of a narrow triangular space, limited behind by the recto-vesical peritoneal fold ; the portion of the bladder included in this space rests on the rectum.

Each vesicle consists of a single tube, coiled upon itself, and giving off several irregular cæcal diverticula ; the separate coils, as well as the diverticula, being connected together by fibrous tissue. When uncoiled, this tube is about the diameter of a quill, and varies in length from four to six inches ; it terminates posteriorly in a *cul-de-sac* ; its anterior extremity becomes constricted into a narrow straight duct, which joins with the corresponding vas deferens, and forms the ejaculatory duct.

The **ejaculatory ducts** are two in number, one on either side of the middle line. Each is formed by the union of the duct from the vesicula seminalis with the vas deferens, and is about three-quarters of an inch in length. They commence at the base of the prostate, and run forwards and downwards between its middle and lateral lobes, and along the sides of the sinus pocularis, to terminate by separate slit-like orifices close to or just within the margins of the sinus. The ducts diminish in size, and also converge towards their termination.

Structure.—The vesiculæ seminales are composed of three coats : an *external* or *areolar coat* ; a *middle* or *muscular coat*, which is thinner than in the vas deferens, arranged in two layers, an outer longitudinal, and inner circular ; an *internal* or *mucous coat*, which is pale, of a whitish-brown colour, and presents a delicate reticular structure, like that seen in the gall-bladder, but the meshes are finer. The epithelium is columnar.

The coats of the ejaculatory ducts are extremely thin. They are : an *outer fibrous layer*, which is almost entirely lost after their entrance into the prostate ; a *layer of muscular fibres*, consisting of an outer thin circular and an inner longitudinal layer ; and *mucous membrane*.

Vessels and Nerves.—The *arteries* supplying the vesiculæ seminales are derived from the middle and inferior vesical and middle hæmorrhoidal. The veins and lymphatics accompany the arteries. The nerves are derived from the pelvic plexuses.

Surgical Anatomy.—The vesiculæ seminales are often the seat of an extension of the disease in cases of tuberculosis of the testicle, and should always be examined from the rectum before coming to a decision with regard to castration in this affection.

FEMALE ORGANS OF GENERATION

EXTERNAL ORGANS

THE **external organs of generation in the female** are : the mons Veneris, the labia majora and minora, the clitoris, the meatus urinarius, and the orifice of the vagina. The term 'vulva' or 'pudendum,' as generally applied, includes all these parts.

The **mons Veneris** is the rounded eminence in front of the pubic symphysis formed by a collection of fatty tissue beneath the integument. It becomes covered with hair at the time of puberty.

The **labia majora** are two prominent longitudinal cutaneous folds, extending downwards and backwards from the mons Veneris to the anterior boundary of

FIG. 776.—The vulva. External female organs of generation.

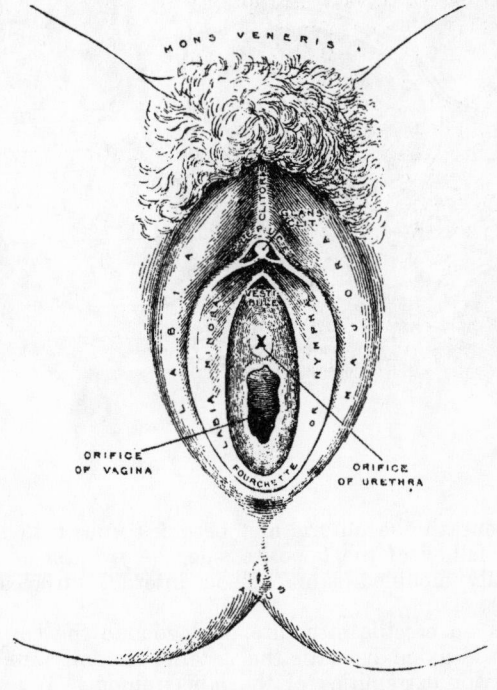

the perinæum, and enclosing the common urino-sexual opening. Each labium has two surfaces, an outer, which is pigmented and covered with strong, crisp hairs ; and an inner, which is smooth and is beset with large sebaceous follicles and is continuous with the genito-urinary mucous tract ; between the two there is a considerable quantity of areolar tissue, fat, and a tissue resembling the dartos of the scrotum, besides vessels, nerves, and glands. The labia are thicker in front, where they form by their meeting the *anterior commissure*. Posteriorly they are not really joined, but appear to become lost in the neighbouring integument, terminating close to, and nearly parallel with, each other. Together with the connecting skin between them, they form the *posterior commissure*, or posterior boundary of the vulval orifice. The interval between the posterior commissure and the anus, from an inch to an inch and a quarter in length, constitutes the perinæum. The *fourchette* is the anterior edge of the perinæum, and between it and the hymen is a depression, the *fossa navicularis*. The labia correspond to the scrotum in the male.

The **labia minora**, or **nymphæ**, are two small cutaneous folds, situated within the labia majora, and extending from the clitoris obliquely downwards, outwards, and backwards for about an inch and a half on each side of the orifice of the vagina, between which and the labia majora they are lost. Anteriorly, each labium minus divides into two portions: the upper division passes above the clitoris to meet its fellow of the opposite side, forming a fold which overhangs the glans clitoridis and is named the *prepuce* (præputium clitoridis); the lower

FIG. 777.—Vertical median section of the female pelvis.   (Henle.)

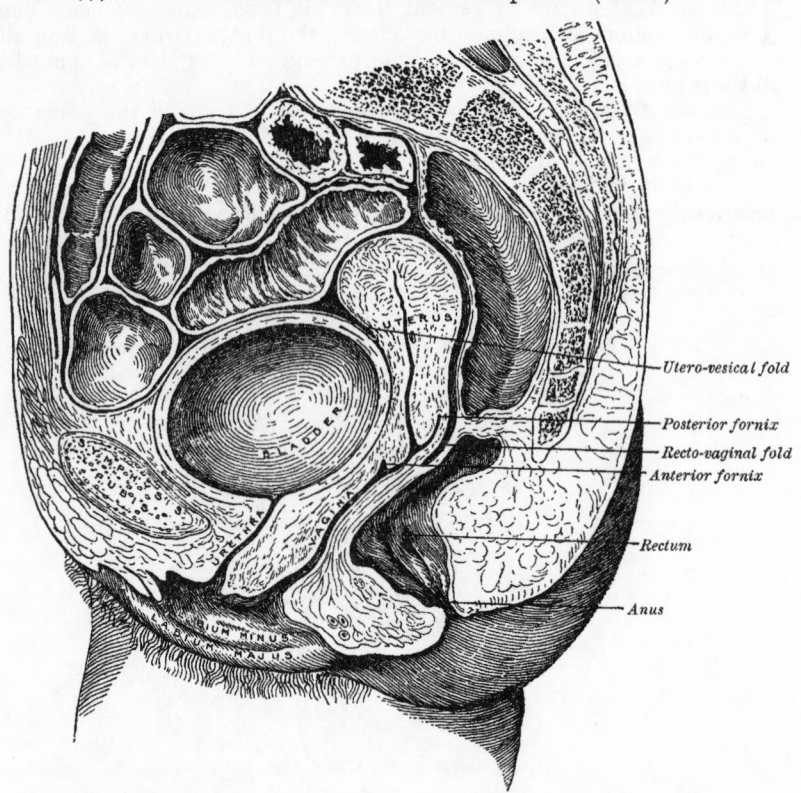

Utero-vesical fold

Posterior fornix
Recto-vaginal fold
Anterior fornix

Rectum

Anus

division passes beneath the clitoris and becomes united to its under surface, forming, with its fellow of the opposite side, the *frænum of the clitoris*. The nymphæ are really modified skin. Their internal surfaces have numerous sebaceous follicles.

The **clitoris** is an erectile structure, analogous to the corpora cavernosa of the penis. It is situated beneath the anterior commissure, partially hidden between the anterior extremities of the labia minora. It is connected to the rami of the os pubis and ischium on each side by a crus; the body is short and concealed beneath the labia; the free extremity, or *glans clitoridis*, is a small rounded tubercle, consisting of spongy erectile tissue, and highly sensitive. It is provided, like the penis, with a suspensory ligament, and with two small muscles, the Erectores clitoridis, which are inserted into the crura of the clitoris. The clitoris consists of two corpora cavernosa, composed of erectile tissue enclosed in a dense layer of fibrous membrane, united together along their inner surfaces by an incomplete fibrous pectiniform septum.

Between the clitoris and the entrance of the vagina is a triangular smooth surface, bounded on each side by the nymphæ; this is the **vestibule**.

The orifice of the urethra (**meatus urinarius**) is situated at the back part of the vestibule, about an inch below the clitoris, and near the margin of the vagina, surrounded by an elevation of the mucous membrane. Below the meatus urinarius is the orifice of the vagina, more or less closed in the virgin by a membranous fold, the *hymen*.

The **hymen** varies much in shape. Its commonest form is that of a ring, generally broadest posteriorly; sometimes it is represented by a semilunar fold, with its concave margin turned towards the pubes. A complete septum stretched across the lower part of the vaginal orifice is called 'imperforate hymen.' Occasionally it is cribriform, or its free margin forms a membranous fringe, or it may be entirely absent. It may persist after copulation, so that it cannot be considered as a test of virginity. When the hymen has been ruptured, small rounded elevations known as the *carunculæ myrtiformes* are found as the remains of this structure.

Glands of Bartholin.—On each side of the commencement of the vagina, and behind the hymen, is a round or oblong body, of a reddish-yellow colour, and of the size of a horse-bean, analogous to Cowper's gland in the male. It is called the *gland of Bartholin*. Each gland opens by means of a long single duct, immediately external to the hymen, in the angle or groove between it and the nympha.

Bulbi vestibuli.—Extending from the clitoris, along either side of the vestibule, and lying a little above the nymphæ, are two oblong masses, about an inch in length, consisting of a plexus of veins, enclosed in a thin layer of fibrous membrane. These bodies are narrow in front and rounded below, and lie on the superficial aspect of the triangular ligament of the urethra: they are termed the *bulbi vestibuli*, and are analogous to the bulb of the corpus spongiosum in the male. Immediately in front of these bodies is a smaller venous plexus, continuous with the bulbi vestibuli behind and the glans clitoridis in front: it is called the *pars intermedia*, and is considered as analogous to that part of the body of the corpus spongiosum which lies in front of the bulb.

INTERNAL ORGANS

The **internal organs of generation** are: the vagina, the uterus and its appendages, the Fallopian tubes, the ovaries and their ligaments.

The **vagina** extends from the vulva to the uterus. It is situated in the cavity of the pelvis, behind the bladder, and in front of the rectum. Its direction is curved upwards and backwards, at first in the line of the outlet, and afterwards in that of the axis of the cavity of the pelvis. Its walls are ordinarily in contact, and the usual shape of its lower part on transverse section is that of an **H**, the transverse limb being slightly curved forwards or backwards, while the lateral limbs are somewhat convex towards the median line; its middle part has the appearance of a transverse slit. Its length is two and a half to three inches along its anterior wall, and three and a half inches along its posterior wall. It is constricted at its commencement, dilated in the middle, and narrowed near its uterine extremity; it surrounds the vaginal portion of the cervix uteri, a short distance from the os, its attachment extending higher up on the posterior than on the anterior wall of the uterus. To the recess behind the cervix the term *posterior fornix* is applied, while the smaller recess in front is termed *anterior fornix*.

Relations.—The *anterior surface* of the vagina is in relation with the base of the bladder, and with the urethra. Its *posterior surface* is connected for the lower three-fourths of its extent to the anterior wall of the rectum, the upper fourth being separated from that tube by the recto-vaginal pouch of peritoneum, or pouch of Douglas. The orifice of the vagina is separated from the anus by the perinæum. Its sides are enclosed between the Levatores ani muscles. As the terminal portions of the ureters pass forward and inwards to reach the base of the bladder, they run one on either side of the lateral aspect of the upper part of the vagina.

Structure.—The vagina consists of an internal mucous lining, of a muscular coat, and between the two of a layer of erectile tissue.

The *mucous membrane* is continuous above with that lining the uterus. Its inner surface presents a longitudinal ridge or raphé on its anterior and on its posterior wall. These ridges are called the *columns of the vagina*, and from them numerous transverse ridges or rugæ extend outwards on either side. These rugæ are divided by furrows of variable depth, giving to the mucous membrane the appearance of being studded over with conical projections or papillæ; they are most numerous near the orifice of the vagina, especially in females before parturition. The epithelium covering the mucous membrane is of the squamous variety. The submucous tissue is very loose, and contains numerous large veins, which by their anastomoses form a plexus, together with smooth muscular fibres derived from the muscular coat; it is regarded by Gussenbauer as an erectile tissue. It contains a number of mucous crypts, but no true glands.

The *muscular coat* consists of two layers: an external longitudinal, which is by far the stronger, and an internal circular layer. The longitudinal fibres are continuous with the superficial muscular fibres of the uterus. The strongest fasciculi are those attached to the recto-vesical fascia on each side. The two layers are not distinctly separable from each other, but are connected by oblique decussating fasciculi, which pass from the one layer to the other. In addition to this, the vagina at its lower end is surrounded by a band of striped muscular fibres, the *sphincter vaginæ* (see page 501).

External to the muscular coat is a layer of connective tissue, containing a large plexus of blood-vessels.

The *erectile tissue* consists of a layer of loose connective tissue, situated between the mucous membrane and the muscular coat; embedded in it is a plexus of large veins, and numerous bundles of unstriped muscular fibres, derived from the circular muscular layer. The arrangement of the veins is similar to that found in other erectile tissues.

THE UTERUS

The **uterus,** or **womb,** is a hollow muscular organ, with thick walls situated within the pelvis between the bladder and the rectum. The ova, when discharged from the ovaries, are conducted to the uterine cavity by the Fallopian tubes, which open into it at its upper lateral angles. When impregnation takes place, the uterus retains and supports the fertilised ovum during the development of the fœtus, and effects its expulsion at the time of parturition.

In the virgin state the uterus is pear-shaped, flattened from before backwards, and situated in the cavity of the pelvis, between the bladder and the rectum; it is retained in its position by the round and broad ligaments on each side, and projects into the upper end of the vagina below. Its upper end, or base, is directed upwards and forwards; its lower end, or apex, downwards and backwards, in the line of the axis of the inlet of the pelvis. It therefore forms an angle with the vagina, since the direction of the vagina corresponds to the axis of the cavity and outlet of the pelvis. The uterus measures about three inches in length, two in breadth at its upper part, and nearly an inch in thickness, and it weighs from an ounce to an ounce and a half.

It consists of two parts: (1) the *body*, with its upper broad extremity, the *fundus*; and (2) the *cervix*, or *neck*, which is situated partly above and partly in the vagina. The fundus is placed below the level of the brim of the pelvis, and its direction varies with the condition of the bladder.

The division between the body and cervix is indicated externally by a slight constriction, and by the reflection of the peritoneum from the anterior surface of the uterus on to the bladder, and internally by a narrowing of the canal, called the *internal os*.

The *body* gradually narrows from the fundus to the neck. Its anterior surface is flattened, covered by peritoneum, which becomes separated from it at its union with the cervix, in order to form the utero-vesical pouch, which lies between the uterus and bladder. Its *posterior surface* is convex transversely, covered by peritoneum throughout, and separated from the rectum by some convolutions of the intestine. Its *lateral margins* are concave, and give attachment on each side to the Fallopian tube above, the round ligament below and in front of this, and the ligament of the ovary behind it. These three structures are enclosed within a duplicature of the peritoneal membrane, which extends from the lateral margin of the uterus to the wall of the pelvis, and is named the *broad ligament* of the uterus.

The *cervix* is the lower constricted segment of the uterus; around its circumference is attached the upper end of the vagina, which extends upwards a greater distance behind than in front.

The *supravaginal portion* of the cervix is not covered by peritoneum in front; a pad of cellular tissue is interposed between it and the bladder. Behind, the peritoneum is extended over it. The *vaginal portion* is the rounded lower end projecting into the vagina. On its surface is a small aperture, the *os uteri*, generally circular in shape, but sometimes oval or almost linear. The anterior

FIG. 778.—Douglas's pouch.
(From a preparation in the Museum of the Royal College of Surgeons of England.)

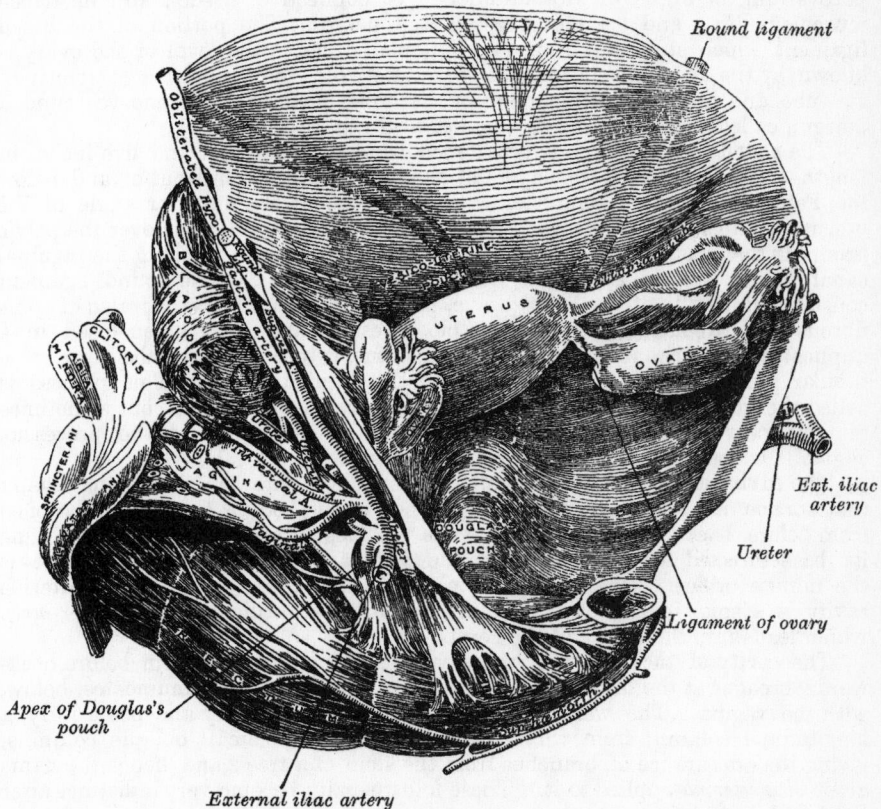

lip of the os is thicker and shorter than the posterior lip: under ordinary circumstances, both are in contact with the posterior wall of the vagina. In nulliparous women these lips are smooth, but after parturition they are usually fissured and irregular. The terminal portions of the ureters run downwards and inwards, one on either side of the cervix but at a distance from it of about three-quarters of an inch.

Ligaments.—The ligaments of the uterus are eight in number: one anterior; one posterior; two lateral or broad; two utero-sacral; and, lastly, two round ligaments.

The *anterior ligament* consists of the utero-vesical fold of peritoneum, which is reflected on to the bladder from the front of the uterus, at the junction of the cervix and body.

The *posterior ligament* consists of the recto-vaginal fold of peritoneum, which is reflected from the back of the upper fourth of the vagina on to the front of the rectum. It forms the bottom of a deep pouch called *Douglas's pouch* (fig. 778), which is bounded in front by the posterior wall of the uterus, the supravaginal cervix, and the upper fourth of the vagina; behind, by the rectum; and laterally by two crescentic folds of peritoneum which pass backwards from the cervix uteri on either side of the rectum to the posterior wall of the pelvis. These folds are named the *folds of Douglas*, or *recto-uterine folds*. They contain a considerable amount of fibrous tissue and non-striped muscular fibres which are attached to the front of the sacrum and constitute the *utero-sacral ligaments*.

The *two lateral* or *broad ligaments* pass from the sides of the uterus to the lateral walls of the pelvis. Together with the uterus they form a septum across the female pelvis, which divides that cavity into two portions. In the anterior part is contained the bladder; in the posterior part, the rectum, and in certain conditions some coils of the small intestine and a part of the pelvic colon. Between the two layers of each broad ligament are contained: (1) the Fallopian tube superiorly; (2) the round ligament; (3) the ovary and its ligament; (4) the parovarium, or organ of Rosenmüller; (5) connective tissue; (6) unstriped muscular fibre; and (7) blood-vessels and nerves. The portion of the broad ligament which stretches from the Fallopian tube to the level of the ovary is known by the name of the *mesosalpinx*. Between the fimbriated extremity of the tube and the lower attachment of the broad ligament is a concave rounded margin, called the *infundibulo-pelvic ligament* (fig. 780).

The **round ligaments** are two flattened bands between four and five inches in length, situated between the layers of the broad ligament in front of and below the Fallopian tubes. Commencing on each side at the superior angle of the uterus, this ligament is directed forwards, upwards, and outwards over the pelvic brim. It then passes through the internal abdominal ring and along the inguinal canal to the labium majus, in which it becomes lost. The round ligament consists principally of muscular tissue, prolonged from the uterus; also of some fibrous and areolar tissue, besides blood-vessels and nerves, enclosed in a duplicature of peritoneum, which, in the fœtus, is prolonged in the form of a tubular process for a short distance into the inguinal canal. This process is called the *canal of Nuck*. It is generally obliterated in the adult, but sometimes remains pervious even in advanced life. It is analogous to the processus vaginalis which precedes the descent of the testis.

The **cavity of the uterus** is small in comparison with the size of the organ : that portion of the cavity which corresponds to the body is triangular, flattened from before backwards, so that its walls are closely approximated, and having its base directed upwards towards the fundus. At each superior angle is the minute orifice of the Fallopian tube. At the inferior angle of the uterine cavity is a small constricted opening, the internal orifice (*ostium internum*), which leads into the cavity of the cervix.

The **cavity of the cervix** is somewhat fusiform, flattened from before backwards, broader at the middle than at either extremity, and communicates, below, with the vagina. The wall of the canal presents, anteriorly and posteriorly, a longitudinal column, from which proceed a number of small oblique columns, giving the appearance of branches from the stem of a tree; and hence the name *arbor vitæ uterina* applied to it. These folds usually become very indistinct after the first labour.

Structure.—The uterus is composed of three coats: an external or serous coat, a middle or muscular coat, and an internal or mucous coat.

The *serous coat* is derived from the peritoneum; it invests the fundus and the whole of the posterior surface of the uterus; but covers the anterior surface only as far as the junction of the body and cervix. In the lower fourth of the posterior surface the peritoneum, though covering the uterus, is not closely connected with it, being separated from it by a layer of loose cellular tissue and some large veins.

The *muscular coat* forms the chief bulk of the substance of the uterus. In the unimpregnated state it is dense, firm, of a greyish colour, and cuts almost like cartilage. It is thick opposite the middle of the body and fundus, and thin at the orifices of the Fallopian tubes. It consists of bundles of unstriped muscular fibres, disposed in layers, intermixed with areolar tissue, blood-vessels, lymphatic vessels, and nerves. In the impregnated state the muscular tissue becomes more prominently developed, and is disposed in three layers : external, middle, and internal.

The external layer is placed beneath the peritoneum, disposed as a thin plane on the anterior and posterior surfaces. It consists of fibres, which pass transversely across the fundus, and, converging at each superior angle of the uterus, are continued on to the Fallopian tube, the round ligament, and the ligament of the ovary : some passing at each side into the broad ligament, and others running backwards from the cervix into the sacro-uterine ligaments.

The middle layer of fibres, which is thickest, presents no regularity in its arrangement, being disposed longitudinally, obliquely, and transversely. It contains most blood-vessels.

The internal or deep layer consists of circular fibres arranged in the form of two hollow cones, the apices of which surround the orifices of the Fallopian tubes, their bases intermingling with one another on the middle of the body of the uterus. At the internal os these circular fibres form a distinct sphincter.

FIG. 779.—The arteries of the internal organs of generation of the female, seen from behind. (After Hyrtl.)

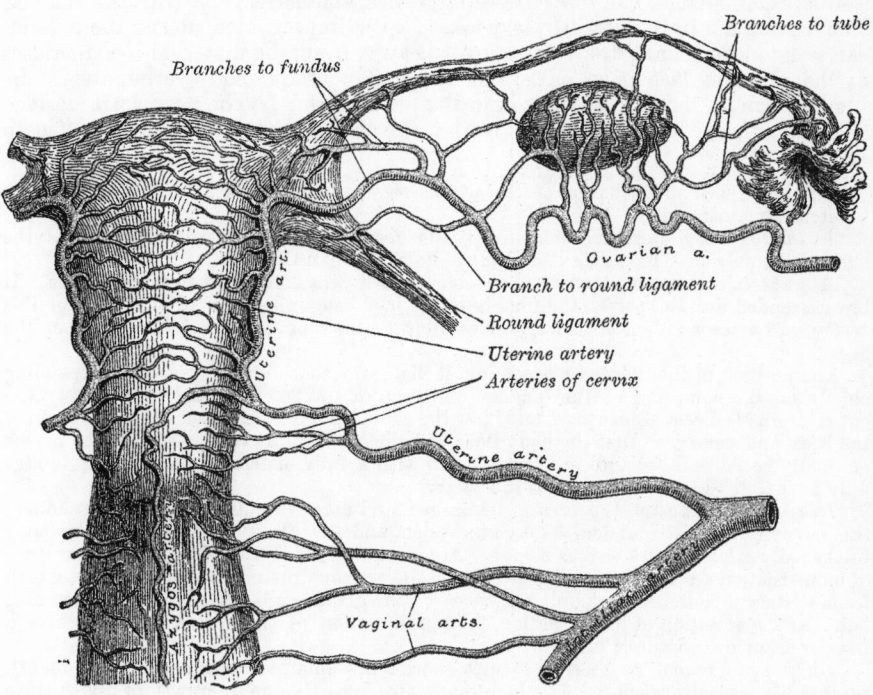

The *mucous membrane* is thin, smooth, and closely adherent to the subjacent tissue. It is continuous, through the fimbriated extremity of the Fallopian tubes, with the peritoneum ; and, through the os uteri, with the lining of the vagina.

In the body of the uterus the mucous membrane is smooth, soft, of a pale red colour, lined by columnar ciliated epithelium, and presents, when viewed with a lens, the orifices of numerous tubular follicles, arranged perpendicularly to the surface. It is unprovided with any submucosa, but is intimately connected with the innermost layer of the muscular coat, which by some anatomists is regarded as the muscularis mucosæ. In structure the corium differs from ordinary mucous membrane, consisting of an embryonic nucleated and highly cellular form of connective tissue in which run numerous large lymphatics. In it are the tube-like *uterine glands*, which are of small size in the unimpregnated uterus, but shortly after impregnation become enlarged and elongated, presenting a contorted or waved appearance (see page 92). They consist of a delicate membrane, lined by an epithelium, which becomes ciliated towards the orifices.

4 E

In the cervix the mucous membrane is sharply differentiated from that of the uterine cavity. It is thrown into numerous oblique ridges, which diverge from an anterior and posterior longitudinal raphé, presenting an appearance which has received the name of *arbor vitæ*. In the upper two-thirds of the canal, the mucous membrane is provided with numerous deep glandular follicles, which secrete a clear viscid alkaline mucus; and, in addition, extending through the whole length of the canal, are a variable number of little cysts, presumably follicles, which have become occluded and distended with retained secretion. They are called the *ovula Nabothi*. The mucous membrane covering the lower half of the cervical canal presents numerous papillæ. The epithelium of the upper two-thirds is cylindrical and ciliated, but below this it loses its cilia, and gradually changes to squamous epithelium close to the external os.

Vessels and Nerves.—The *arteries of the uterus* are the uterine, from the internal iliac; and the ovarian, from the abdominal aorta. They are remarkable for their tortuous course in the substance of the organ, and for their frequent anastomoses. The termination of the ovarian artery meets the termination of the uterine artery, and forms an anastomotic trunk from which branches are given off to supply the uterus, their disposition being, as shown by Sir John Williams, circular. The *veins* are of large size, and correspond with the arteries. They terminate in the uterine plexuses. In the impregnated uterus the arteries carry the blood to, and the veins convey it away from, the maternal blood-sinuses of the placenta (see page 94). The *lymphatics* of the body terminate in the lumbar glands, those of the cervix in the pelvic glands. The *nerves* are derived from the inferior hypogastric and ovarian plexuses, and from the third and fourth sacral nerves.

The form, size, and situation of the uterus vary at different periods of life and under different circumstances.

In the fœtus the uterus is contained in the abdominal cavity, projecting beyond the brim of the pelvis. The cervix is considerably larger than the body.

At puberty the uterus is pyriform in shape, and weighs from eight to ten drachms. It has descended into the pelvis, the fundus being just below the level of the brim of this cavity. The arbor vitæ is distinct, and extends to the upper part of the cavity of the organ.

The position of the uterus in the adult is liable to considerable variation, depending chiefly on the condition of the bladder and rectum. When the bladder is empty the entire uterus is directed forwards, and is at the same time bent on itself at the junction of the body and cervix, so that the body lies upon the bladder. As the latter fills, the uterus gradually becomes more and more erect, until with a fully distended bladder the fundus may be directed backwards towards the sacrum.

During menstruation the organ is enlarged, and more vascular, its surfaces rounder; the os externum is rounded, its labia swollen, and the lining membrane of the body thickened, softer, and of a darker colour. According to Sir J. Williams, at each recurrence of menstruation a molecular disintegration of the mucous membrane takes place, which leads to its complete removal, only the bases of the glands embedded in the muscle being left. At the cessation of menstruation, by a proliferation of the remaining structures a fresh mucous membrane is formed.

During pregnancy the uterus becomes enormously enlarged, and in the ninth month reaches the epigastric region. The increase in size is partly due to growth of pre-existing muscle, and partly to development of new fibres.

After parturition the uterus nearly regains its usual size, weighing about an ounce and a half; but its cavity is larger than in the virgin state, its vessels are tortuous, and its muscular layers are more defined; the external orifice is more marked, and its edges present a fissured surface.

In old age the uterus becomes atrophied, and paler and denser in texture; a more distinct constriction separates the body and cervix. The ostium internum is frequently, and the ostium externum occasionally, obliterated, while the labia almost entirely disappear.

Surgical Anatomy.—From what has been said above, it will be evident that the uterus is held in position mainly by the broad ligaments, the utero-sacral folds, and the blood-vessels. The cervix is the most fixed part, being retained in position by the vagina and supported by the Levatores ani muscles. As the cervix is more or less fixed and the body comparatively free, there is usually a certain degree of flexion of the latter at the point where the two parts meet. The normal position of the body with regard to the cervix is one of anteversion, combined with slight anteflexion; its anterior surface resting on the upper surface of the bladder, when this viscus is empty or contains only a small quantity of urine. As the bladder fills, the uterus is displaced backwards, so that when complete distension of the bladder takes place it may become retroverted; but returns to its normal position when the bladder is emptied. The body of the uterus is maintained in its natural position by the pressure of the coils of intestine resting on its posterior surface, but is capable of undergoing a considerable range of movement. A certain amount of anteversion and retroversion can take place without the condition being regarded as pathological, but when the degree of flexion becomes considerable it must be regarded as a morbid condition. This is especially true of retroversion and retroflexion. The former is a falling back of the whole uterus, so that the cervix points upwards towards the pubes, and the latter is a bending backwards of the body, the cervix remaining in its normal position. The two conditions are usually combined. Prolapse of the uterus is another common infirmity. The organ sinks to an abnormally low level in the pelvis, and sometimes protrudes beyond the vulva. This is due to the supporting mechanism of the uterus being insufficient to support the strain thrown upon it, on account of its relaxation.

The uterus may require removal in cases of malignant disease or for fibroid tumours. Carcinoma is the most common form of malignant disease of the uterus, though cases of sarcoma do occur. It may show itself either as a columnar carcinoma or as a squamous carcinoma; the former commencing either in the cervix or body of the uterus, the latter always commencing in the epithelial cells of the mucous covering of the vaginal surface of the cervix. The columnar form may be treated in the early stage, before fixation has taken place, by removal of the uterus, either through the vagina or by means of abdominal section. The former operation is the better of the two, and is attended by a much smaller death rate. Vaginal hysterectomy is performed by placing the patient in the lithotomy position and introducing a large duckbill speculum. The cervix is then seized with a volsellum and pulled down as far as possible and the mucous membrane of the vagina incised around the cervix and as near to it as the disease will allow, especially in front, where the ureters are in danger of being wounded. A pair of dressing forceps are then pushed through into Douglas's pouch and opened sufficiently to allow of the introduction of the two forefingers, by means of which the opening is dilated laterally as far as the sacro-uterine ligaments. A somewhat similar proceeding is adopted in front, but here the bladder has to be separated from the anterior wall of the uterus for about an inch before the vesico-uterine fold of peritoneum can be reached. This is done by carefully burrowing upwards with a director and stripping the tissues off the anterior uterine wall. When the vesico-uterine pouch has been opened and the opening dilated laterally, the uterus remains attached only by the broad ligaments, in which are contained the vessels that supply the uterus. Before division of the ligaments, these vessels have to be dealt with. The forefinger of the left hand is introduced into Douglas's pouch, and an aneurism needle, armed with a long silk ligature, is inserted into the vesico-uterine pouch, and is pushed through the broad ligament about an inch above its lower level and at some distance from the uterus. One end of the ligature is now pulled through the anterior opening, and in this way we have the lowest inch of the broad ligament, in which is contained the uterine artery (fig. 779), enclosed in a ligature. This is tied tightly, and the operation is repeated on the other side. The broad ligament is then divided on either side, between the ligature and the uterus, to the extent to which it has been constricted. By traction on the volsellum which grasps the cervix, the uterus can be pulled considerably further down in the vagina, and a second inch of the broad ligament is treated in a similar way. This second ligature will embrace the pampiniform plexus of veins, and, when the broad ligament has been divided on either side, it will be found that a third ligature can be made to pass over the Fallopian tube and top of the broad ligament, after the uterus has been dragged down as far as possible. After the third ligature has been tied and the structures between it and the uterus divided, this organ will be freed from all its connections and can be removed from the vagina. This canal is then sponged out and lightly dressed with gauze; no sutures being used. The gauze may be removed at the end of the second day. In squamous epithelioma, amputation of the cervix is all that is necessary in those cases where the disease is recognised before it has invaded the walls of the vagina or the neighbouring broad ligaments. The operation consists in removing a wedge-shaped piece of the uterus, including the cervix, through the vagina and attaching the cut surfaces of the stump to the anterior and posterior vaginal walls, so as to prevent retraction. In the treatment of uterine fibroids which require operative interference, removal of the whole of the uterus together with the tumours through an abdominal incision gives the most satisfactory results; for, if the tumour is large, its size acts as a barrier to its safe delivery through the pelvis and genital passages. After the abdomen has been opened the uterine vessels are secured and the broad ligaments divided in a similar manner to that employed in vaginal hysterectomy, except that the proceeding is commenced from above. When the first two ligatures have been tied, and the broad ligament divided, it will be found that the uterus can be raised out of the pelvis. A transverse incision is now made through the peritoneum, where it is reflected from the anterior surface of the uterus on to the back of the bladder and the serous membrane peeled from the surface of the uterus until the vagina is reached. The anterior wall of this canal is then cut across. The uterus is now turned forwards and the peritoneum at the bottom of Douglas's pouch incised transversely, and the posterior wall of the vagina cut across, until it meets the incision on the anterior wall. The uterus is now almost free, and is held only by the lower part of the broad ligament on either side, containing the uterine artery. A third ligature is made to encircle this, and, after having been tied, the structures are divided between the ligature and the uterus. The organ can now be removed. The vagina is plugged with gauze, and the external wound closed in the usual way. The vagina acts as a drain, and therefore the opening into it is not sutured.

THE FALLOPIAN TUBES

The **Fallopian tubes,** or **oviducts,** convey the ova from the ovaries to the cavity of the uterus. They are two in number, one on each side, situated in the upper margin of the broad ligament, extending from each superior angle of the uterus to the side of the pelvis. Each tube is about four inches in length; and is described as consisting of three portions: (1) the *isthmus,* or inner constricted third; (2) the *ampulla,* or outer dilated portion, which curves over the ovary; and (3) the *infundibulum,* with its *ostium abdominale,* surrounded by fimbriæ, one of which is attached to the ovary, the *fimbria ovarica.* The Fallopian tube is directed outwards as far as the lower or uterine pole of the ovary, and then

FIG. 780.—Broad ligament seen from behind. (Henle.)

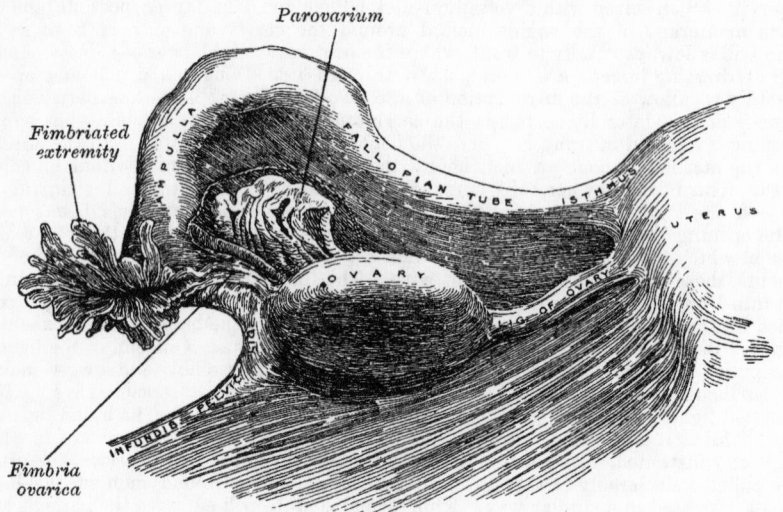

ascends along the anterior border of the ovary as far as the upper or tubal pole, over which it arches; finally it turns downwards and ends in relation to the posterior border and inner surface of the ovary. The uterine opening is minute, and will only admit a fine bristle; the abdominal opening is somewhat larger. In connection with the fimbriæ of the Fallopian tube, or with the broad ligament close to them, there are frequently one or more small pedunculated vesicles. These are termed the *hydatids of Morgagni.*

Structure.—The Fallopian tube consists of three coats: serous, muscular, and mucous.

The *external* or *serous coat* is peritoneal.

The *middle* or *muscular coat* consists of an external longitudinal and an internal circular layer of muscular fibres continuous with those of the uterus.

The *internal* or *mucous coat* is continuous with the mucous lining of the uterus, and, at the free extremity of the tube, with the peritoneum. It is thrown into longitudinal folds, which in the outer, larger part of the tube, or ampulla, are much more extensive than in the narrow canal of the isthmus. The lining epithelium is columnar ciliated. This form of epithelium is also found on the inner surface of the fimbriæ; while on the outer or serous surfaces of these processes the epithelium gradually merges into the endothelium of the peritoneum.

The Ovaries

The **ovaries** (*testes muliebres*, Galen) are analogous to the testes in the male. They are two oval-shaped bodies, situated one on each side of the uterus in relation to the lateral wall of the pelvis, and attached to the back of the broad ligament of the uterus behind and below the Fallopian tubes. Each ovary presents an outer and an inner surface, an upper and a lower extremity, and an anterior and a posterior border. It lies in a shallow depression, named the *fossa ovarii*, on the lateral wall of the pelvis; this fossa is bounded above by the external iliac vessels, in front by the obliterated hypogastric artery, and behind by the ureter. The long axis of the ovary is vertical when the woman is standing erect. The *upper* or *tubal* extremity is near the external iliac vein, while the *lower* or *uterine* end is directed downwards towards the pelvic floor. The *outer* surface is in contact with the parietal peritoneum, which lines the fossa ovarii; the *inner* surface is to a large extent covered by the fimbriated extremity of the Fallopian tube. The *anterior* or *straight* border is directed towards the obliterated hypogastric artery, and is attached to the back of the broad ligament by a short fold named the *mesovarium*. Between the two layers of this fold the blood-vessels and nerves pass to reach the hilus of the ovary. The *posterior* or *convex* border is free, and is directed towards the ureter. The Fallopian tube arches over the ovary, running upwards in relation to its anterior border, then curving over its upper or tubal pole, and finally passing downwards on its posterior border and inner surface. To its upper end is attached the ovarian fimbria of the Fallopian tube and a fold of peritoneum, the *suspensory ligament of the ovary*, which is directed upwards over the iliac vessels and contains the ovarian vessels. The lower end is usually narrower than the upper, and is attached to the lateral angle of the uterus immediately behind the Fallopian tube by a rounded cord termed the *ligament of the ovary*, which lies within the broad ligament and contains some non-striped muscular fibres.

The exact position of the ovary has been the subject of considerable difference of opinion, and the description here given applies to the ovary of the nulliparous woman. The ovary becomes displaced during the first pregnancy, and probably never again returns to its original position.

The ovaries are of a greyish-pink colour, and present either a smooth or a puckered, uneven surface. They are each about an inch and a half in length, three-quarters of an inch in width, and about a third of an inch in thickness, and weigh from one to two drachms.

Structure.—The ovary consists of a number of Graafian vesicles, embedded in the meshes of a stroma or framework, the surface of the ovary being covered by a layer of columnar cells which constitute the *germinal epithelium of Waldeyer*. This gives to the ovary a dull grey colour as compared with the shining smoothness of the peritoneum; and the transition between the pavement epithelium of the peritoneum and the columnar cells which cover the ovary is usually marked by a line around the anterior border of the ovary.

Stroma.—The stroma is a peculiar soft tissue, abundantly supplied with blood-vessels, consisting for the most part of spindle-shaped cells with a small amount of ordinary connective tissue. These cells have been regarded by some anatomists as unstriped muscle-cells, which, indeed, they most resemble (His); by others as connective-tissue cells (Waldeyer, Henle, and Kölliker). On the surface of the organ this tissue is much condensed, and forms a layer composed of short connective-tissue fibres, with fusiform cells between them. This was formerly regarded as a distinct fibrous covering, and was termed the *tunica albuginea*, but is nothing more than a condensed layer of the stroma of the ovary.

Graafian follicles.—Upon making a section of an ovary, numerous round transparent vesicles of various sizes are to be seen; they are the *Graafian follicles*, or ovisacs containing the ova. Immediately beneath the superficial covering is a layer of stroma, in which are a large number of minute vesicles, of uniform size, about $\frac{1}{100}$ of an inch in diameter. These are the Graafian follicles in their earliest condition, and the layer where they are found has been termed the *cortical layer*. They are especially numerous in the ovary of the young child. After puberty, and during the whole of the child-bearing period, large and mature, or almost mature, Graafian follicles are also found in the cortical layer in small numbers, and also 'corpora lutea,' the remains of follicles which have burst and are undergoing atrophy and absorption. Beneath this superficial stratum, other large and more mature Graafian follicles are found embedded in the ovarian stroma. These increase in size as they recede from the surface towards a highly vascular stroma in the centre of the organ, termed the *medullary substance* (*zona vasculosa*, Waldeyer). This stroma forms the tissue of the hilus by which the ovary is attached, and through which the blood-vessels enter : it does not contain any Graafian follicles.

The larger *Graafian follicles* consist of an external fibro-vascular coat, connected with the surrounding stroma of the ovary by a network of blood-vessels ; and an internal coat, named *ovicapsule*, which is lined by a layer of nucleated cells, called the *membrana granulosa*. The fluid contained in the interior of the vesicles is transparent and albuminous, and in it is suspended the ovum. In that part of the mature Graafian follicle which is nearest the surface of the ovary, the cells of the membrana granulosa are connected into a mass which projects into the cavity of the follicle. This is termed the *discus proligerus*, and in this the ovum is embedded.*

The ova are usually regarded as being formed from the germinal epithelium on the surface of the ovary. This becomes thickened, and in it are seen some

FIG. 781.—Section of the ovary. (After Schrön.)

FIG. 782.—Section of the Graafian follicle. (After Von Baer.)

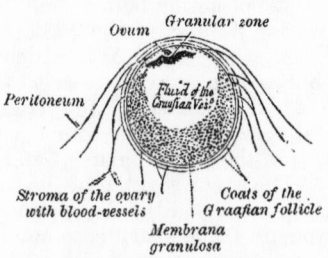

1. Outer covering. 1'. Attached border. 2. Central stroma. 3. Peripheral stroma. 4. Blood-vessels. 5. Graafian follicles in their earliest stage. 6, 7, 8. More advanced follicles. 9. An almost mature follicle. 9'. Follicle from which the ovum has escaped. 10. Corpus luteum.

cells which are larger and more rounded than the rest : these are termed the *primordial ova*. The germinal epithelium grows downwards in the form of tubes or columns, termed the *egg tubes* of Pflüger, into the ovarian stroma, which grows outwards between the tubes, and ultimately cuts them off from the germinal epithelium. These tubes are further subdivided into rounded *nests* or groups, each containing a primordial ovum which undergoes further development and growth while the surrounding cells of the nest form the epithelium of the Graafian follicle.†

The development and maturation of the Graafian follicles and ova continue uninterruptedly from puberty to the end of the fruitful period of woman's life, while their formation commences before birth. Before puberty the ovaries are small, the Graafian follicles contained in them are disposed in a comparatively thick layer in the cortical substance ; here they present the appearance of a large number of minute closed vesicles, constituting the early condition of the Graafian follicles ; many, however, never attain full development, but shrink and disappear. At puberty the ovaries enlarge, and become more vascular, the Graafian follicles are developed in greater abundance, and their ova are capable of fecundation.

Discharge of the ovum.—The Graafian follicles, after gradually approaching the surface of the ovary, burst : the ovum and fluid contents of the follicles are liberated, and escape on the exterior of the ovary, passing thence into the Fallopian tube.‡

* For a description of the ovum, see page 73.   † See footnote on page 152.

‡ This is effected either by application of the tube to the ovary, or by a curling upwards of the fimbriated extremity, so that the ovum is caught as it falls.

In the fœtus, the ovaries are situated, like the testes, in the lumbar region, near the kidneys. They may be distinguished from those bodies at an early period by their elongated and flattened form, and by their position, which is at first oblique and then nearly transverse. They gradually descend into the pelvis.

Lying above the ovary in the broad ligament between it and the Fallopian tube is the *organ of Rosenmüller*, called also the *parovarium* or *epoöphoron*. This is the remnant of a fœtal structure, the development of which is described on page 148. In the adult it consists of a few closed convoluted tubes, lined with epithelium, which converge towards the ovary at one end and at the other are united by a longitudinal tube, which is the homologue of the *duct of Gärtner* in the cow. This duct terminates in a bulbous enlargement (see fig. 190). The parovarium is connected at its uterine extremity with the remains of the Wolffian duct. A few scattered rudimentary tubules, best seen in the child, are situated in the broad ligament between the parovarium and the uterus. These constitute the *paroöphoron of Waldeyer*.

Vessels and Nerves.—The *arteries of the ovaries and Fallopian tubes* are the ovarian from the aorta. Each enters the attached border, or hilus, of the corresponding ovary. The *veins* follow the course of the arteries; they form a plexus near the ovary, the *pampiniform plexus*. The *nerves* are derived from the inferior hypogastric or pelvic plexus, and from the ovarian plexus, the Fallopian tube receiving a branch from one of the uterine nerves.

MAMMARY GLANDS

The **mammæ,** or breasts, secrete the milk, and are accessory glands of the generative system. They exist in the male as well as in the female; but in the former only in the rudimentary state, unless their growth is excited by peculiar circumstances. In the female they are two large hemispherical eminences situated towards the lateral aspect of the pectoral region, corresponding to the intervals between the second and sixth ribs, and extending from the side of the sternum to near the mid-axillary line, their outer and lower parts resting on the Serratus magnus muscles. Their weight and dimensions differ at different periods of life, and in different individuals. Before puberty they are of small size, but enlarge as the generative organs become more completely developed. They increase during pregnancy, and especially after delivery, and become atrophied in old age. The left mamma is generally a little larger than the right. Their bases are nearly circular, flattened or slightly concave, and have their long diameter directed upwards and outwards towards the axilla; they are separated from the Pectoralis major and Serratus magnus muscles by a layer of fascia. The outer surface of the mamma is convex, and presents, just below the centre, a small conical prominence, the nipple (*mammilla*). The surface of the nipple is dark-coloured, and surrounded by an *areola* having a coloured tint. In the virgin the areola is of a delicate rosy hue; about the second month after impregnation it enlarges and acquires a darker tinge, which increases as pregnancy advances, becoming in some cases of a dark brown, or even black colour. This colour diminishes as soon as lactation is over, but is never entirely lost throughout life. These changes in the colour of the areola are of importance in forming a conclusion in a case of suspected first pregnancy.

The **nipple** is a cylindrical or conical eminence, capable of undergoing a sort of erection from mechanical excitement, a change mainly due to the contraction of its muscular fibres. It is of a pink or brownish hue, its surface wrinkled and provided with papillæ; and it is perforated by from fifteen to twenty orifices, the apertures of the lactiferous ducts. Near the base of the nipple, and upon the surface of the areola, are numerous large sebaceous glands (*glands of Montgomery*), which become much enlarged during lactation, and present the appearance of small tubercles beneath the skin. These glands secrete a peculiar fatty substance, which serves as a protection to the integument of the nipple during the act of sucking. The nipple consists of numerous vessels, intermixed with plain muscular fibres, which are principally arranged in a circular manner around the base: some few fibres radiating from base to apex.

Structure.—The mamma consists of gland-tissue; of fibrous tissue, connecting its lobes; and of fatty tissue in the intervals between the lobes. The gland-tissue, when freed from fibrous tissue and fat, is of a pale reddish colour, firm in texture, circular in form, flattened from before backwards, thicker in the centre than at the circumference, and presenting several inequalities on its surface, especially in front.   It consists of numerous lobes, and these are composed of lobules, connected together by areolar tissue, blood-vessels, and ducts.   The smallest lobules consist of a cluster of rounded alveoli, which open into the smallest branches of the lactiferous ducts : these ducts unite to form larger ducts, which terminate in a single canal, corresponding with one of the chief subdivisions of the gland.   The number of excretory ducts varies from fifteen to twenty ; they are termed the *tubuli lactiferi*, or *galactophori*.   They converge towards the areola, beneath which they form dilatations, or *ampullæ*, which serve as reservoirs for the milk, and, at the base of the nipple, become contracted, and pursue a straight course to its summit, perforating it by separate orifices considerably narrower than the ducts themselves.   The ducts are composed of areolar tissue, with longitudinal and transverse elastic fibres ; muscular fibres are entirely absent ; their mucous lining is continuous, at the point of the nipple, with the integument.   The epithelium of the mammary gland differs according to

FIG. 783.—Dissection of the lower half of the female breast during the period of lactation.   (Luschka.)

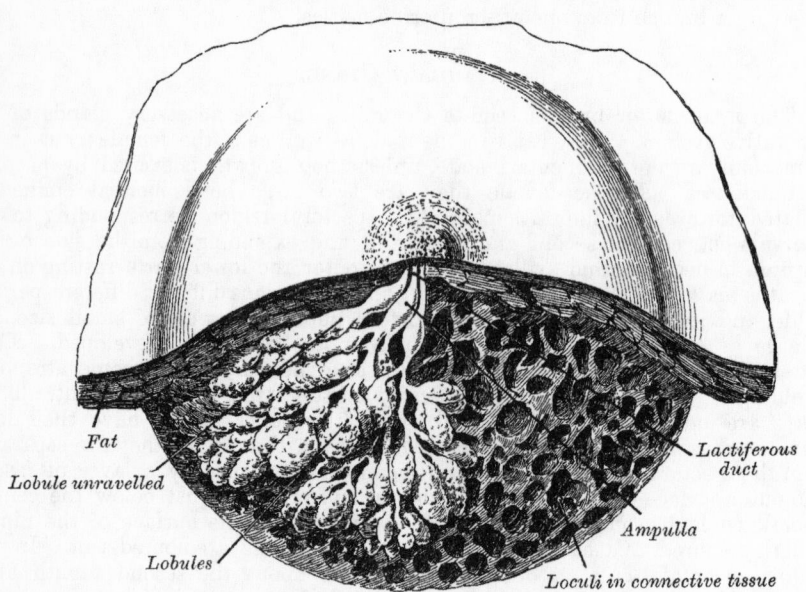

the state of activity of the organ.   In the gland of a woman who is not pregnant or suckling, the alveoli are very small and solid, being filled with a mass of granular polyhedral cells.   During pregnancy the alveoli enlarge, and the cells undergo rapid multiplication.   At the commencement of lactation, the cells in the centre of the alveolus undergo fatty degeneration, and are eliminated in the first milk, as *colostrum corpuscles*.   The peripheral cells of the alveolus remain, and form a single layer of granular, short columnar cells, with a spherical nucleus, lining the limiting membrana propria.   The cells, during the state of activity of the gland, are capable of forming, in their interior, oil-globules, which are then ejected into the lumen of the alveolus, and constitute the milk-globules.

The *fibrous tissue* invests the entire surface of the breast, and sends down septa between its lobes, connecting them together.

The *fatty tissue* surrounds the surface of the gland, and occupies the interval between its lobes.   It usually exists in considerable abundance, and determines the form and size of the gland.   There is no fat immediately beneath the areola and nipple.

Vessels and Nerves.—The *arteries* supplying the mammæ are derived from the thoracic branches of the axillary, the intercostals, and internal mammary. The *veins* describe an anastomotic circle round the base of the nipple, called by Haller the *circulus venosus*. From this, large branches transmit the blood to the circumference of the gland, and end in the axillary and internal mammary veins. The *lymphatics*, for the most part, run along the lower border of the Pectoralis major to the axillary glands; some few, from the inner side of the breast, perforate the intercostal spaces and empty themselves into the anterior mediastinal glands. The *nerves* are derived from the anterior and lateral cutaneous nerves of the thorax.

THE DUCTLESS GLANDS

THERE are certain organs which are very similar to secreting glands, but differ from them in one essential particular, viz. they do not possess any ducts by which their secretion is discharged. These organs are known as *ductless glands*. They are capable of *internal secretion*—that is to say, of forming substances, out of materials brought to them in the blood, which have a certain influence upon the nutritive changes going on in the body. This secretion is carried away, either directly by the veins into the blood-stream, or indirectly through the medium of the lymphatics.

These glands include the thyroid and the parathyroids, the thymus, the spleen, and the suprarenal capsules, which will be described in this section. They also include the lymphatic glands, which have already been described in the section on the Vascular System; and the small coccygeal and carotid bodies, which have been alluded to in the same section.

THE THYROID BODY

The **Thyroid Body** is a highly vascular organ, situated at the front and sides of the neck, and consists of two lateral lobes connected across the middle line by a narrow transverse portion, the *isthmus*.

The weight of the thyroid body is somewhat variable but is usually about one ounce. It is slightly heavier in the female, in whom it becomes enlarged during menstruation and pregnancy.

The lobes are conical in shape, the apex of each being directed upwards and outwards as far as the junction of the middle with the lower third of the thyroid cartilage; the base looks downwards, and is on a level with the fifth or sixth tracheal ring.

The *external* or *superficial surface* is convex, and covered by the skin, the superficial and deep fasciæ, the Sterno-mastoid, the anterior belly of the Omohyoid, the Sterno-hyoid and Sterno-thyroid muscles, and beneath the last muscle by the pre-tracheal layer of the deep fascia, which forms a capsule for the gland.

The *deep* or *internal surface* is moulded over the underlying structures, viz. the thyroid and cricoid cartilages, the trachea, the Inferior constrictor and posterior part of the Crico-thyroid muscles, the œsophagus (particularly on the left side of the neck), the superior and inferior thyroid arteries, and the recurrent laryngeal nerves.

The *anterior border* is thin, and inclines obliquely from above downwards and inwards towards the middle line of the neck, while the *posterior border* is thick and overlaps the common carotid artery. Each lobe is about two inches in length, its greatest width is about an inch and a quarter, and its thickness about three-quarters of an inch.

The *isthmus* connects together the lower thirds of the two lateral lobes; it measures about half an inch in breadth, and the same in depth, and usually covers the second and third rings of the trachea. Its situation and size present, however, many variations—facts of importance in the operation of tracheotomy. In the middle line of the neck it is covered by the skin and fascia, and close to the middle line, on either side, by the Sterno-hyoid. Across its upper border runs a branch of the superior thyroid artery; at its lower border are the inferior thyroid veins. Sometimes the isthmus is altogether wanting.

A third lobe, of conical shape, called the *pyramid*, frequently arises from the upper part of the isthmus, or from the adjacent portion of either lobe, but most commonly the left, and ascends as high as the hyoid bone. It is occasionally quite detached, or may be divided into two or more parts.

A fibrous or muscular band is sometimes found attached, above, to the body of the hyoid bone, and below to the isthmus of the gland, or its pyramidal process. When muscular, it is termed the *Levator glandulæ thyroideæ.*

Small detached portions of thyroid tissue are sometimes found in the vicinity of the lateral lobes or above the isthmus. These are termed *accessory thyroids.*

Structure.—The thyroid body is invested by a thin capsule of connective tissue, which projects into its substance and imperfectly divides it into masses of irregular form and size. When the organ is cut into, it is of a brownish-red colour, and is seen to be made up of a number of closed vesicles, containing a yellow glairy fluid, and separated from each other by intermediate connective tissue.

According to Baber, the vesicles of the thyroid of the adult animal are generally closed cavities; but in some young animals (e.g. young dogs) the vesicles are more or less tubular and branched. This appearance he supposes to be due to the mode of growth of the gland, and merely indicating that an

Fig. 784.—Minute structure of thyroid. From a transverse section of the thyroid of a dog. (Semi-diagrammatic.) (Baber.)

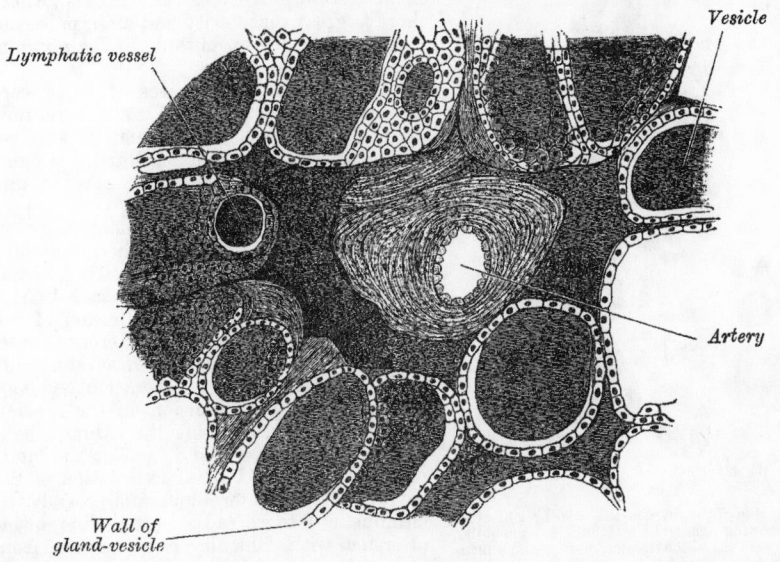

increase in the number of vesicles is taking place. Each vesicle is lined by a single layer of epithelium, the cells of which, though differing somewhat in shape in different animals, have always a tendency to assume a columnar form. Between the epithelial cells exists a delicate reticulum. The vesicles are of various sizes and shapes, and contain as a normal product a viscid, homogeneous, semi-fluid, slightly yellowish material, which frequently contains blood; red corpuscles are found in it in various stages of disintegration and decolorisation, the yellow tinge being probably due to the hæmoglobin, which is thus set free from the coloured corpuscles. Baber has also described in the thyroid gland of the dog large round cells ('parenchymatous cells'), each provided with a single oval-shaped nucleus, which migrate into the interior of the gland-vesicles.

The capillary blood-vessels form a dense plexus in the connective tissue around the vesicles, between the epithelium of the vesicles and the endothelium of the lymph-spaces, which latter surround a greater or smaller part of the circumference of the vesicle. These lymph-spaces empty themselves into lymphatic vessels which run in the interlobular connective tissue, not uncommonly surrounding the arteries which they accompany, and communicate with a network in the capsule of the gland. Baber has found in the lymphatics of the thyroid a viscid material which is morphologically identical with the normal constituent of the vesicle.

Vessels and Nerves.—The *arteries* supplying the thyroid are the superior and inferior thyroid, and sometimes an additional branch (thyroidea media, or ima) from the innominate artery, or the arch of the aorta, which ascends upon the front of the trachea. The arteries are remarkable for their large size and frequent anastomoses. The *veins* form a plexus on the surface of the gland, and on the front of the trachea, from which arise the superior, middle, and inferior thyroid veins; the superior and middle terminate in the internal jugular, the inferior in the innominate vein. The *lymphatics* are numerous, of large size, and end in the thoracic and right lymphatic ducts. The *nerves* are derived from the middle and inferior cervical ganglia of the sympathetic.

Surgical Anatomy.—The thyroid gland is subject to enlargement, which is called goitre. This may be due to hypertrophy of any of the constituents of the gland. The simplest (*parenchymatous goitre*) is due to an enlargement of the follicles. The *fibroid* is due to increase of the interstitial connective tissue. The *cystic* is that in which one or more large cysts are formed from dilatation and possibly coalescence of adjacent follicles. The *pulsating goitre* is where the vascular changes predominate over the parenchymatous, and the vessels of the gland are especially enlarged. Finally, there is the *exophthalmic goitre* (Graves's disease), where there is great vascularity and often pulsation accompanied by exophthalmos, palpitation, and rapid pulse.

FIG. 785.—Minute structure of thymus gland.

1. Upper portion of the thymus of a fœtal pig of 2″ in length, showing the bud-like lobuli and glandular elements. 2. Cells of the thymus, mostly from a man. *a.* Free nuclei. *b.* Small cells. *c.* Larger. *d.* Larger, with oil-globules, from the ox. *e,f.* Cells completely filled with fat, at *f* without a nucleus. *g, h.* Concentric bodies: *g.* An encapsulated nucleated cell. *h.* A composite structure of a similar nature.

A large number of cases of what were formerly supposed to be goitre are now known to be cases of adenomatous enlargement, where an adenoma, starting in one part of the gland, gradually spreads and involves the whole organ.

Where, in spite of treatment, a goitre continues to grow, especially when symptoms of tracheal pressure are commencing, operative interference becomes necessary. In a certain percentage of cases, division of the isthmus is attended with satisfactory results, particularly in relieving the respiration. It is a comparatively simple operation, and consists in making an incision in the median line of the neck, exposing the isthmus, ligaturing it at either end and excising the intermediate portion. Partial extirpation of the thyroid, viz. one lateral lobe, and possibly the isthmus, is a more radical proceeding, but is attended with difficulty and danger: from hæmorrhage, which may be almost uncontrollable if the capsule is accidentally opened; from risk of wounding the internal jugular vein or the recurrent laryngeal nerve. It must be borne in mind that the removal of the whole of the thyroid body is followed by myxœdema. In performing the operation, the particular form of external incision may be varied according to circumstances; and the points to bear in mind are that care must be taken to avoid tearing the capsule, and that the thyroid arteries should be ligatured before an attempt is made to remove the body. In ligaturing the inferior thyroid artery, the position of the recurrent laryngeal nerve must be borne in mind, so as not to include it in the ligature.

Parathyroids.—These are small brownish-red bodies, with an average diameter of about a quarter of an inch, situated near the thyroid gland, from which, however, they differ in structure, being composed of masses of cells arranged in a more or less columnar fashion with numerous intervening capillaries. They measure on an average about a quarter of an inch in length, and from a sixth to an eighth of an inch in breadth, and usually present the appearance of flattened oval discs. They are divided, according to their situation, into *postero-superior* and *antero-inferior*. The postero-superior, usually two in number, are the more constant in position, and are situated, one on each side, at the level of the lower border of the cricoid cartilage, behind the junction of the pharynx and œsophagus and in front of the prevertebral fascia. The antero-inferior, also usually two in number, may be applied to the lower edge of the lateral lobe, or placed at some little distance below the thyroid body, or found in relation to one of the inferior thyroid veins.*

The Thymus Gland

The **Thymus Gland** is a temporary organ, attaining its full size at the end of the second year, when it ceases to grow, and gradually dwindles, until at puberty it has almost disappeared. If examined when its growth is most active, it will be found to consist of two lateral lobes placed in close contact along the

Fig. 786.—Minute structure of thymus gland. Follicle of injected thymus from calf, four days old, slightly diagrammatic, magnified about 50 diameters. The large vessels are disposed in two rings, one of which surrounds the follicle, the other lies just within the margin of the medulla. (Watney.)

A and B. From thymus of camel, examined without addition of any reagent. Magnified about 400 diameters. A. Large colourless cell, containing small oval masses of hæmoglobin. Similar cells are found in the lymph-glands, spleen, and medulla of bone. B. Coloured blood-corpuscles.

middle line, situated partly in the superior mediastinum, partly in the neck, and extending from the fourth costal cartilage upwards, as high as the lower border of the thyroid gland. It is covered by the sternum, and by the origins of the Sterno-hyoid and Sterno-thyroid muscles. Below, it rests upon the pericardium, being separated from the arch of the aorta and great vessels by a layer of fascia. In the neck it lies on the front and sides of the trachea, behind the Sterno-hyoid and Sterno-thyroid muscles. The two lobes generally differ in size; they are occasionally united, so as to form a single mass; and sometimes separated

* See article concerning the parathyroid glands, by Welsh, *Journal of Anatomy and Physiology*, vol. xxxii.

by an intermediate lobe. The thymus is of a pinkish-grey colour, soft and lobulated on its surfaces. It is about two inches in length, one and a half in breadth below, and about three or four lines in thickness. At birth it weighs about half an ounce.

Structure.—Each lateral lobe is composed of numerous lobules, held together by delicate areolar tissue; the entire gland being enclosed in an investing capsule of a similar but denser structure. The primary lobules vary in size from a pin's head to a small pea, and are made up of a number of small nodules or follicles, which are irregular in shape and are more or less fused together, especially towards the interior of the gland. Each follicle consists of a medullary and a cortical portion, which differ in many essential particulars from each other. The *cortical portion* is mainly composed of lymphoid cells, supported by a delicate reticulum. In addition to this reticulum, of which traces only are found in the medullary portion, there is also a network of finely branched cells, which is continuous with a similar network in the medullary portion. This network forms an adventitia to the blood-vessels. In the *medullary portion* there are but few lymphoid cells, but there are, especially towards the centre, granular cells and concentric corpuscles. The granular cells are rounded or flask-shaped masses, attached (often by fibrillated extremities) to blood-vessels and to newly formed connective tissue. The concentric corpuscles are composed of a central mass, consisting of one or more granular cells, and of a capsule which is formed of epithelioid cells, which are continuous with the branched cells forming the network mentioned above.

Each follicle is surrounded by a capillary plexus, from which vessels pass into the interior, and radiate from the periphery towards the centre, and form a second zone just within the margin of the medullary portion. In the centre of the medulla there are very few vessels, and they are of minute size.

Watney has made the important observation that haemoglobin is found in the thymus, either in cysts or in cells situated near to, or forming part of, the concentric corpuscles. This haemogoblin varies from granules to masses exactly resembling coloured blood-corpuscles, oval in the bird, reptile, and fish; circular in all mammals, except in the camel. He has also discovered, in the lymph issuing from the thymus, similar cells to those found in the gland, and, like them, containing haemoglobin, in the form of either granules or masses. From these facts he arrives at the physiological conclusion that the thymus is one source of the coloured blood-corpuscles.

Vessels and Nerves.—The *arteries* supplying the thymus are derived from the internal mammary, and from the superior and inferior thyroid. The *veins* terminate in the left innominate vein, and in the thyroid veins. The *lymphatics* are of large size, arise in the substance of the gland, and are said to terminate in the internal jugular vein. The *nerves* are exceedingly minute; they are derived from the pneumogastric and sympathetic. Branches from the descendens hypoglossi and phrenic reach the investing capsule, but do not penetrate into the substance of the gland.

The Spleen

The **Spleen** is situated principally in the left hypochondriac region, its upper and inner extremity extending into the epigastric region; lying between the fundus of the stomach and the Diaphragm. It is the largest of the ductless glands, and measures about five inches in length. It is of an oblong, flattened form, soft, of very brittle consistence, highly vascular, and of a dark purplish colour.

Surfaces.—The *external* or *diaphragmatic surface* is convex, smooth, and is directed upwards, backwards, and to the left, except at its upper end, where it is directed slightly inwards. It is in relation with the under surface of the Diaphragm, which separates it from the ninth, tenth, and eleventh ribs of the left side, and the intervening lower border of the left lung and pleura.

The *internal surface* is divided by a ridge into an anterior or gastric, and a posterior or renal portion.

The *gastric surface*, which is directed forwards and inwards, is broad and concave, and is in contact with the posterior wall of the great end of the stomach;

and below this with the tail of the pancreas. It presents near its inner border a long fissure, termed the *hilus*. This is pierced by several irregular apertures, for the entrance and exit of vessels and nerves.

The *renal surface* is directed inwards and downwards. It is somewhat flattened, is considerably narrower than the gastric surface, and is in relation with the upper part of the outer surface of the left kidney and occasionally with the left suprarenal capsule.

The *upper end* is directed inwards, towards the vertebral column, where it lies on a level with the eleventh dorsal vertebra. The *lower end*, sometimes termed the *basal surface*, is flat, triangular in shape, and rests upon the splenic flexure of the colon and the phreno-colic ligament, and is generally in contact with the tail of the pancreas. The *anterior* border is free, sharp, and thin, and is often notched, especially below. It separates the phrenic from the gastric surface. The *posterior* border is more rounded and blunter than the anterior. It separates the renal surface from the diaphragmatic surface. It corresponds to the lower border of the eleventh rib and lies between the Diaphragm and

FIG. 787.—The spleen, showing its gastric and renal surfaces.

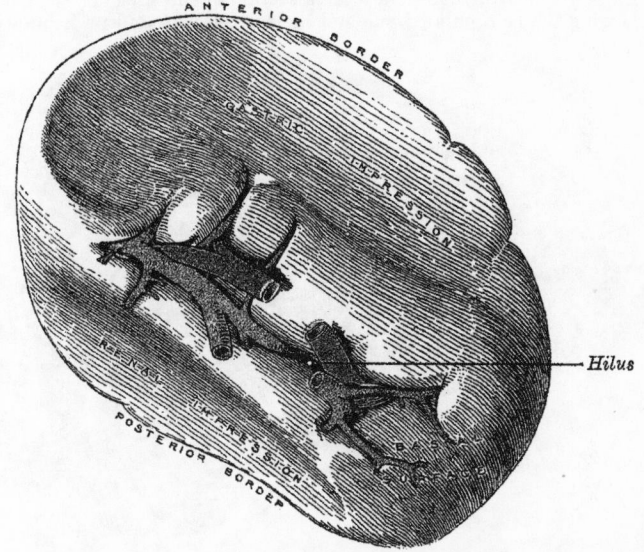

left kidney. The *internal* border or intermediate margin is the ridge which seperates the renal and gastric surfaces. The *inferior* border separates the diaphragmatic from the basal surface.

The spleen is almost entirely surrounded by peritoneum, which is firmly adherent to its capsule. It is held in position by two folds of this membrane : one, the *lieno-renal ligament*, is derived from the layers of peritoneum forming the greater and lesser sacs, where they come into contact between the left kidney and the spleen ; the splenic vessels pass between its two layers (fig. 721) ; the second, the *gastro-splenic omentum*, also formed of two layers, derived from the greater and lesser sacs respectively, where they meet between the spleen and stomach (fig. 721). The vasa brevia of the splenic artery and vein run between the two layers of the gastro-splenic omentum. Its lower end or basal surface is supported by the phreno-colic ligament (see page 1063).

The size and weight of the spleen are liable to very extreme variations at different periods of life, in different individuals, and in the same individual under different conditions. *In the adult*, it is usually about five inches in length, three inches in breadth, and an inch or an inch and a half in thickness, and weighs about seven ounces. *At birth*, its weight, in proportion to the entire body, is almost equal to what is observed in the adult, being as 1 to 350 : while in the adult it varies from 1 to 320 and 400. *In old age*, the organ not only diminishes in weight, but decreases considerably in proportion to the entire body, being as 1 to 700.   The size of the spleen is increased during and after digestion, and varies according to the state of nutrition of the body, being large in highly fed, and small in starved animals.   In malarial fever it becomes much enlarged, weighing occasionally from eighteen to twenty pounds.

Frequently in the neighbourhood of the spleen, and especially in the gastro-splenic and great omenta, small nodules of splenic tissue may be found, either isolated or connected to the spleen by thin bands of splenic tissue.   They are known as *supernumerary* or *accessory spleens*.   They vary in size from a pea to a plum.

Structure.—The spleen is invested by two coats : an external serous, and an internal fibro-elastic coat.

The *external* or *serous coat* is derived from the peritoneum ; it is thin, smooth, and in the human subject intimately adherent to the fibro-elastic coat. It invests the entire organ, except at the hilus and along the lines of reflection of the lieno-renal ligament and gastro-splenic omentum.

The *fibro-elastic coat* forms the framework of the spleen.   It invests the organ, and at the hilus is reflected inwards upon the vessels in the form

Fig. 788.—Transverse section of the spleen,
showing the trabecular tissue and the splenic vein and its tributaries.

of sheaths.   From these sheaths, as well as from the inner surface of the fibro-elastic coat, numerous small fibrous bands, *trabeculæ* (fig. 789), are given off in all directions ; these uniting, constitute the framework of the spleen.   This resembles a sponge-like material, consisting of a number of small spaces or *areolæ*, formed by the trabeculæ, which are given off from the inner surface of the capsule, or from the sheaths prolonged inwards on the blood-vessels.   In these spaces or areolæ is contained the *splenic pulp*.

The proper coat, the sheaths of the vessels and the trabeculæ, consist of a dense mesh of white and yellow elastic fibrous tissues, the latter predominating. It is owing to the presence of this tissue that the spleen possesses a considerable amount of elasticity, which allows of the very great variations in size that it presents under certain circumstances.   In addition to these constituents of this tunic, there is found in man a small amount of non-striped muscular fibre ; and in some mammalia (e.g. dog, pig, and cat) a large amount, so that the trabeculæ appear to consist chiefly of muscular tissue.   It is probably owing to this structure that the spleen exhibits, when acted upon by the galvanic current, faint traces of contractility.

The *proper substance of the spleen*, or *spleen pulp*, is a soft mass of a dark reddish-brown colour, resembling grumous blood.   When examined, by means of a thin section, under the microscope, it is found to consist of a number of branching cells, and of an intercellular substance. The cells are connective-tissue corpuscles, and have been named the *sustentacular* or *supporting cells of the pulp.* The processes of these branching cells communicate with each other, thus forming a delicate reticulated tissue in the interior of the areolæ formed by the trabeculæ of the capsule, so that each primary space may be considered to be divided into a number of smaller spaces by the junction of the processes of the branching corpuscles. These secondary spaces are full of blood, in which, however, the white corpuscles are found to be in larger proportion than they are in ordinary blood. Large, rounded cells, termed *splenic cells*, are also seen ; these are capable of amœboid movement, and often contain pigment and red blood-corpuscles in their interior. The sustentacular cells are either small, uni-nucleated, or larger, multi-nucleated cells ; they do not stain deeply with carmine, and in this respect differ from the cells of the Malpighian bodies, presently to be described (W. Müller), but like these cells they possess amœboid move-ments (Cohnheim). In many of them may be seen deep red or reddish-yellow granules of various sizes, which present the characters of the hæmatin of the blood. Unchanged blood-discs may be seen in these cells, but more frequently blood-discs are found which are altered both in form and colour. In fact, blood-

FIG. 789.—Transverse section of the human spleen,
showing the distribution of the splenic artery and its branches.

corpuscles in all stages of disintegration may be noticed to occur within them. Klein has pointed out that sometimes these cells, in the young spleen, contain a proliferating nucleus ; that is to say, the nucleus is of large size, and presents a number of knob-like projections, as if small nuclei were budding from it by a process of gemmation. This observation is of importance, as it may explain one possible source of the colourless blood-corpuscles.

The interspaces or areolæ formed by the framework of the spleen are thus filled by a delicate reticulum of branched connective-tissue corpuscles, the interstices of which are occupied by blood, and in which the blood-vessels terminate in the manner now to be described.

Blood-vessels of the Spleen.—The splenic artery is remarkable for its large size in proportion to the size of the organ, and also for its tortuous course. It divides into six or more branches, which enter the hilus of the spleen and ramify throughout its substance (fig. 789), receiving sheaths from an involution of the external fibrous tissue. Similar sheaths also invest the nerves and veins.

Each branch runs in the transverse axis of the organ, from within outwards, diminishing in size during its transit, and giving off in its passage smaller branches, some of which pass to the anterior, others to the posterior part. These ultimately leave the trabecular sheaths, and terminate in the proper substance of the spleen in small tufts or pencils of minute arterioles, which open into the

4 F

interstices of the reticulum formed by the branched sustentacular cells. Each of the larger branches of the artery supplies chiefly that region of the organ in which the branch ramifies, having no anastomosis with the majority of the other branches.

The *arterioles*, supported by the minute trabeculæ, traverse the pulp in all directions in bundles or pencilli of straight vessels. Their external coat, on leaving the trabecular sheaths, consists of ordinary connective tissue, but it gradually undergoes a transformation, becomes much thickened, and converted into adenoid material.* This change is effected by the conversion of the connective tissue into adenoid tissue; the bundles of connective tissue becoming looser and laxer, their fibrils more delicate, and containing in their interstices an abundance of lymph-corpuscles (W. Müller).

The altered coat of the arterioles, consisting of adenoid tissue, presents here and there thickenings of a spheroidal shape, the *Malpighian bodies of the spleen.* These bodies vary in size from about $\frac{1}{100}$ of an inch to $\frac{1}{25}$ of an inch in diameter. They are merely local expansions or hyperplasiæ of the adenoid tissue, of which the external coat of the smaller arteries of the spleen is formed.

FIG. 790.—Part of a Malpighian capsule of the spleen of man. (Klein and Noble Smith.)

a. Arterial branch in longitudinal section.
b. Adenoid tissue, still containing the lymph-corpuscles; only their nuclei are shown.
c. Adenoid reticulum, the lymph-corpuscles accidentally removed.

They are most frequently found surrounding the arteriole, which thus seems to tunnel them, but occasionally they grow from one side of the vessel only, and present the appearance of a sessile bud growing from the arterial wall. Klein, however, denies this, and says it is incorrect to describe the Malpighian bodies as isolated masses of adenoid tissue, but that they are always formed around an artery, though there is generally a greater amount on one side than on the other, and that, therefore, in transverse sections, the artery, in the majority of cases, is found in an eccentric position. These bodies are visible to the naked eye on the surface of a fresh section of the organ, appearing as minute dots of a semi-opaque whitish colour in the dark substance of the pulp. In minute structure they resemble the adenoid tissue of lymphatic glands, consisting of a delicate reticulum, in the meshes of which lie ordinary lymphoid cells.

The reticulum of the tissue is made up of extremely fine fibrils, and is comparatively open in the centre of the corpuscle, becoming closer at its periphery. The cells which it encloses, like the supporting cells of the pulp, are possessed of amœboid movements, but when treated with carmine become deeply stained, and can be easily recognised from those of the pulp.

The arterioles terminate in capillaries, which traverse the pulp in all directions; their walls become much attenuated, lose their tubular character, and the cells of the adenoid tissue of which they are composed become altered, presenting a branched appearance, and acquiring processes which are directly connected with the processes of the sustentacular cells of the pulp (fig. 791). In this manner the capillary vessels terminate, and the blood flowing through them finds its way into the interstices of the reticulated tissue formed by the branched connective-tissue corpuscles of the splenic pulp. Thus the blood passing through the spleen is brought into intimate relation with the elements of the pulp, and no doubt undergoes important changes.

After these changes have taken place the blood is collected from the interstices of the tissue by the rootlets of the veins, which commence much in the same way as the arteries terminate. Where a vein is about to commence the connective-tissue corpuscles of the pulp arrange themselves in rows, in such a

* According to Klein, it is the sheath of the small vessel which undergoes this transformation, and forms a 'solid mass of adenoid tissue which surrounds the vessel like a cylindrical sheath.'—*Atlas of Histology*, p. 424.

way as to form an elongated space or sinus. They become changed in shape, being elongated and spindle-shaped, and overlap each other at their extremities. They thus form a sort of endothelial lining of the path or sinus, which is the radicle of a vein. On the outer surface of these cells are seen delicate transverse lines or markings, which are due to minute elastic fibrillæ arranged in a circular manner around the sinus. Thus the channel obtains an external investment, and gradually becomes converted into a small vein, which after a time presents a coat of ordinary connective tissue, lined by a layer of fusiform epithelial cells, which are continuous with the supporting cells of the pulp. The smaller veins unite to form larger ones, which do not accompany the arteries, but soon enter the trabecular sheaths of the capsule, and by their junction form six or more branches, which emerge from the hilus, and, uniting, constitute the splenic vein, the largest radicle of the vena portæ.

The veins are remarkable for their numerous anastomoses, while the arteries hardly anastomose at all.

The lymphatics originate in two ways, i.e. from the sheaths of the arteries and in the trabeculæ. The former accompany the blood-vessels, the latter pass

FIG. 791.—Section of spleen, showing the termination of the small blood-vessels.

Supporting cells   *Vessel undergoing lymphoid change*

Vessel continuous with processes of supporting cells

Small artery

Supporting cell

Supporting cell

to the superficial lymphatic plexus which may be seen on the surface of the organ. The two sets communicate at the hilus, and ultimately terminate in the thoracic duct.

The *nerves* are derived from branches of the right and left semilunar ganglia, and from the right pneumogastric nerve.

Surface Form.—The spleen is situated under cover of the ribs of the left side, being separated from them by the Diaphragm, and above by a small portion of the lower margin of the left lung and pleura. Its position corresponds to the ninth, tenth, and eleventh ribs. It is placed very obliquely. 'It is oblique in two directions, viz. from above downwards and outwards, and also from above downwards and forwards' (Cunningham). 'Its highest and lowest points are on a level respectively with the ninth dorsal and first lumbar spines ; its inner end is distant about an inch and a half from the median plane of the body, and its outer end about reaches the mid-axillary line' (Quain).

Surgical Anatomy.—Injury of the spleen is less common than that of the liver, on account of its protected situation and connections. It may be ruptured by direct or indirect violence ; torn by a broken rib ; or injured by a punctured or gunshot wound. When the organ is enlarged, the chance of rupture is increased. The great risk is hæmorrhage, owing to the vascularity of the organ, and the absence of a proper system of capillaries. The injury is not, however, necessarily fatal, and this would appear to be due, in a great measure, to the contractile power of its capsule, which narrows the wound and prevents the escape of blood. In cases where the diagnosis is clear and the symptoms indicate danger to life, laparotomy must be performed, and if the hæmorrhage cannot be stayed by ordinary surgical methods, the spleen must be removed. The spleen may become displaced, producing| great pain from stretching of the vessels and nerves, and this may require removal of the organ. The spleen may become enormously enlarged in certain diseased conditions, such as ague, leukæmia, syphilis, valvular disease of the heart,

4 F 2

or without any obtainable history of previous disease. It may also become enlarged in lymphadenoma, as a part of a general blood-disease. In these cases the tumour may fill a considerable part of the abdomen and extend into the pelvis, and may be mistaken for ovarian or uterine disease.

The spleen is sometimes the seat of cystic tumours, especially hydatids, and of abscess. These cases require treatment by incision and drainage ; and in abscess great care must be taken, if there are no adhesions between the spleen and abdominal wall, to prevent the escape of any of the pus into the peritoneal cavity. If possible the operation should be performed in two stages. Sarcoma and carcinoma are occasionally found in the spleen, but very rarely as a primary disease.

Extirpation of the spleen has been performed for wounds or injuries, in floating spleen, in simple hypertrophy, and in leukæmic enlargement ; but in the last condition the operation is now regarded as unjustifiable, as every case in which it has been performed has terminated fatally. The incision is best made in the left semilunar line ; the spleen is isolated from its surroundings, and the pedicle transfixed and ligatured in two portions, before the tumour is turned out of the abdominal cavity, if this is possible, so as to avoid any traction on the pedicle, which may cause tearing of the splenic vein. In applying the ligature care must be taken not to include the tail of the pancreas, and in lifting out the organ to avoid rupturing the capsule.

SUPRARENAL CAPSULES

The **Suprarenal Capsules** are two small flattened bodies, of a yellowish colour, situated at the back part of the abdomen, behind the peritoneum, and immediately above and in front of the upper end of each kidney ; hence their name. The right one is somewhat triangular in shape, bearing a resemblance to a cocked hat ; the left is more semilunar, usually larger and placed at a higher level than the right. They vary in size in different individuals, being sometimes so small as to be scarcely detected : their usual size is from an inch and a quarter to nearly two inches in length, rather less in width, and from two to three lines in thickness. Their average weight is from one to one and a half drachms each.

Relations.—The relations of the suprarenal capsules differ on the two sides of the body. The *right suprarenal* is situated behind the inferior vena cava and right lobe of the liver, and in front of the Diaphragm and upper end of the right kidney. It is roughly triangular in shape ; its base, directed downwards, is in contact with the inner and anterior aspects of the upper end of the right kidney. It presents two surfaces for examination, an anterior and posterior. The *anterior surface* looks forwards and outwards, and has two areas : an inner, narrow and non-peritoneal, which lies behind the inferior vena cava ; and an outer, somewhat triangular, in contact with the liver. The upper part of this latter surface is devoid of peritoneum, and is in relation with the bare area of the liver near its lower and inner angles, while its inferior portion is covered by peritoneum, reflected on to it from the inferior layer of the coronary ligament. A little below the apex, and near the anterior border of the capsule, is a short furrow termed the hilus, from which the suprarenal vein emerges to join the inferior vena cava. The *posterior surface* is divided into upper and lower parts by a curved ridge : the upper, slightly convex, rests upon the Diaphragm ; the lower, concave, is in contact with the upper end and the adjacent part of the anterior surface of the kidney. The *left suprarenal*, slightly larger than the right, is crescentic in shape, its concavity being adapted to the inner border of the upper part of the left kidney. It presents an inner border which is convex, and an outer which is concave ; its upper border is narrow, and its lower rounded. Its *anterior surface* has two areas : an upper one, covered by the peritoneum forming the lesser sac, which separates it from the cardiac end of the stomach and sometimes from the superior extremity of the spleen ; and a lower one, which is in contact with the pancreas and splenic artery, and is therefore not covered by the peritoneum. On the anterior surface, near its lower end, is a furrow or hilus which is directed downwards and forwards, and from which the suprarenal vein emerges. Its *posterior surface* presents a vertical ridge, which divides it into two areas. The ridge lies in the sulcus between the kidney and crus of the Diaphragm, while the outer area, which is thin, rests on the kidney, and the inner and smaller area on the left crus of the Diaphragm.

The surface of the suprarenal capsule is surrounded by areolar tissue containing much fat, and closely invested by a thin fibrous coat, which is difficult to remove, on account of the numerous fibrous processes and vessels which enter the organ through the furrows on its anterior surface and base.

Small accessory suprarenals are often to be found in the connective tissue around the suprarenals. The smaller of these, on section, show a uniform surface, but in some of the larger a distinct medulla can be made out.

Structure.—On making a perpendicular section, the suprarenal capsule is seen to consist of two substances : external or cortical, and internal or medullary. The former, which constitutes the chief part of the organ, is of a deep yellow

Fig. 792.—Vertical section of the suprarenal capsule. (From Elberth, in Stricker's 'Manual.')

Fig. 793.—Minute structure of suprarenal capsule.

Fig. 794.—Minute structure of suprarenal capsule.

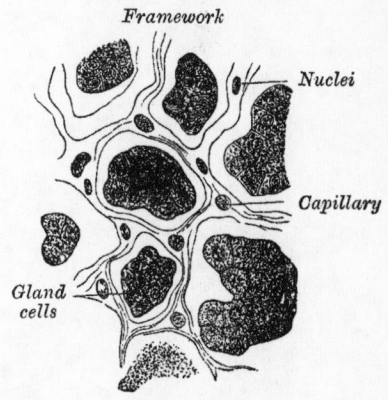

colour. The medullary substance is soft, pulpy, and of a dark brown or black colour, whence the name *atrabiliary capsules* formerly given to these organs.

The **cortical portion** consists chiefly of narrow columnar masses, placed perpendicularly to the surface. This arrangement is due to the disposition of the capsule, which sends processes into the interior of the gland ; these are connected with each other by transverse bands, so as to form a series of intercommunicating spaces. These spaces are of slight depth near the surface of the organ, so that there the section somewhat resembles a net : this is termed the *zona glomerulosa* ; but they become much deeper or longer farther in, so as to resemble pipes or tubes placed endwise, the *zona fasciculata*. Still deeper down, near the medullary part, the spaces become again of small extent : this is named the *zona reticularis*. These processes or trabeculæ, derived from the capsule and forming the framework of the spaces, are composed of fibrous connective tissue, with longitudinal bundles of unstriped muscular fibres. Within the interior of the spaces are found groups of polyhedral cells, which are finely granular in appearance, and contain a spherical nucleus, and not infrequently fat-globules. These groups of cells do not entirely fill the spaces, and thus between them and the trabeculæ of the framework is a channel, which is believed to be a lymph path or sinus, and which communicates with certain passages between the cells composing the group. The lymph path is supposed to open into a plexus of efferent lymphatic vessels which are contained in the capsule.

In the **medullary portion,** the fibrous stroma seems to be collected together into a much closer arrangement, and forms bundles of connective tissue which are loosely applied to the large plexus of veins of which this part of the organ mainly consists. In the interstices lie a number of cells compared by Frey to those of columnar epithelium. They are coarsely granular, do not contain any fat, and some of them are branched. Luschka has affirmed that these branches are connected with the nerve-fibres of a very intricate plexus which is found in the medulla; this statement has not been verified by other observers, for the tissue of the medullary substance is less easy to make out than that of the cortical, owing to its rapid decomposition.

The numerous arteries which enter the suprarenal bodies, from the sources mentioned below, penetrate the cortical part of the gland, where they break up into capillaries in the fibrous septa, and these converge to the very numerous veins of the medullary portion, which are collected together into the suprarenal vein, which emerges from the hilus of the gland.

The *arteries* supplying the suprarenal capsules are numerous and of comparatively large size; they are derived from the aorta, the phrenic, and the renal; they subdivide into numerous minute branches previous to entering the substance of the gland.

The *suprarenal vein* returns the blood from the medullary venous plexus and receives several branches from the cortical substance; on the right side it opens into the inferior vena cava, on the left into the renal vein.

The *lymphatics* terminate in the lumbar glands.

The *nerves* are exceedingly numerous, and are derived from the solar and renal plexuses, and, according to Bergmann, from the phrenic and pneumogastric nerves. They enter the lower and inner part of the capsule, traverse the cortex, and terminate around the cells of the medulla. They have numerous small ganglia developed upon them, from which circumstance the organ has been conjectured to have some function in connection with the sympathetic nervous system.

THE SURGICAL ANATOMY OF INGUINAL HERNIA

Dissection (fig. 410).—For dissection of the parts concerned in inguinal hernia, a male subject, free from fat, should always be selected. The body should be placed in the supine position, the abdomen and pelvis raised by means of blocks, and the lower extremities rotated outwards, so as to make the parts as tense as possible. If the abdominal walls are flaccid, the cavity of the abdomen should be inflated through an aperture made at the umbilicus. An incision should be made along the middle line, from a little below the umbilicus to the symphysis pubis, and continued along the front of the scrotum ; and a second incision, from the anterior superior spine of the ilium to just below the umbilicus. These incisions should divide the integument ; and the triangular-shaped flap included between them should be reflected downwards and outwards, when the superficial fascia will be exposed.

The superficial fascia of the abdomen.—This fascia has already been described (page 482). In the neighbourhood of the groin it is divisible into two layers, between which are found the superficial vessels and nerves and the superficial inguinal lymphatic glands. The superficial layer (fascia of Camper) forms part of the general subcutaneous fatty covering of the body ; the deeper layer (fascia of Scarpa) is thinner and more membranous in character. In the middle line it is intimately adherent to the linea alba ; below, it blends with the inner half of Poupart's ligament and with the fascia lata of the thigh below the ligament ; below and internally, in the male, it can be traced as a thin layer along the cord over the scrotum and penis and onwards into the perinæum, where it blends with the deep layer of the superficial fascia. In the female, it is continued into the labia majora.

In the inguinal region, three small branches of the femoral artery—the *superficial epigastric*, the *superficial circumflex iliac*, and the *superficial external pudic*—ramify in this fascia, and distribute branches to the superficial inguinal lymphatic glands and the integument (see page 701). The veins accompanying these vessels are usually much larger than the arteries : they terminate in the internal saphenous vein.

The integument in this situation is supplied by the *hypogastric branch of the ilio-hypogastric nerve*, which is distributed to the hypogastric region ; and the *ilio-inguinal*, which is distributed to the integument of the upper and inner part of the thigh, and to the scrotum in the male and the labium in the female.

The superficial inguinal lymphatic glands are situated between the two layers of the superficial fascia ; they are of large size, and vary from eight to twenty in number. They are divisible into two groups : an upper, disposed irregularly along Poupart's ligament, which receives the lymphatic vessels from the integument of the scrotum, penis, and the mucous membrane of the urethra in the male, and the lower part of the vagina and urethra in the female ; also from the parietes of the abdomen, below the level of the umbilicus, and from the perineal and gluteal regions : and an inferior group, which surrounds the saphenous opening in the fascia lata, a few lymphatics being sometimes continued along the saphenous vein to a variable extent. The inferior group receives the superficial lymphatic vessels from the lower extremity.

Upon removing these superficial structures, the aponeurosis of the External oblique muscle of the abdomen is exposed. This has already been described (page 483). Just above and to the outer side of the crest of the os pubis, an interval may be seen in this aponeurosis, called the *external abdominal ring*.

This is formed by a splitting of the fibres of the aponeurosis, which diverge from each other, to be inserted into the symphysis pubis and the spine of the os pubis respectively, forming a triangular opening, the base of which is the crest of the os pubis.   These diverging fibres are called the *columns* or *pillars of the ring* (see page 485).

The **external abdominal ring** gives passage to the spermatic cord in the male and the round ligament in the female, and through it an inguinal hernia descends. It is much larger in men than in women, on account of the greater size of the spermatic cord as compared with the round ligament of the uterus; and hence the great frequency of inguinal hernia in men.

Stretching across between the two pillars of the external abdominal ring and the lower part of the aponeurosis of the External oblique is a series of curved

FIG. 795.—Inguinal hernia.   Superficial dissection.

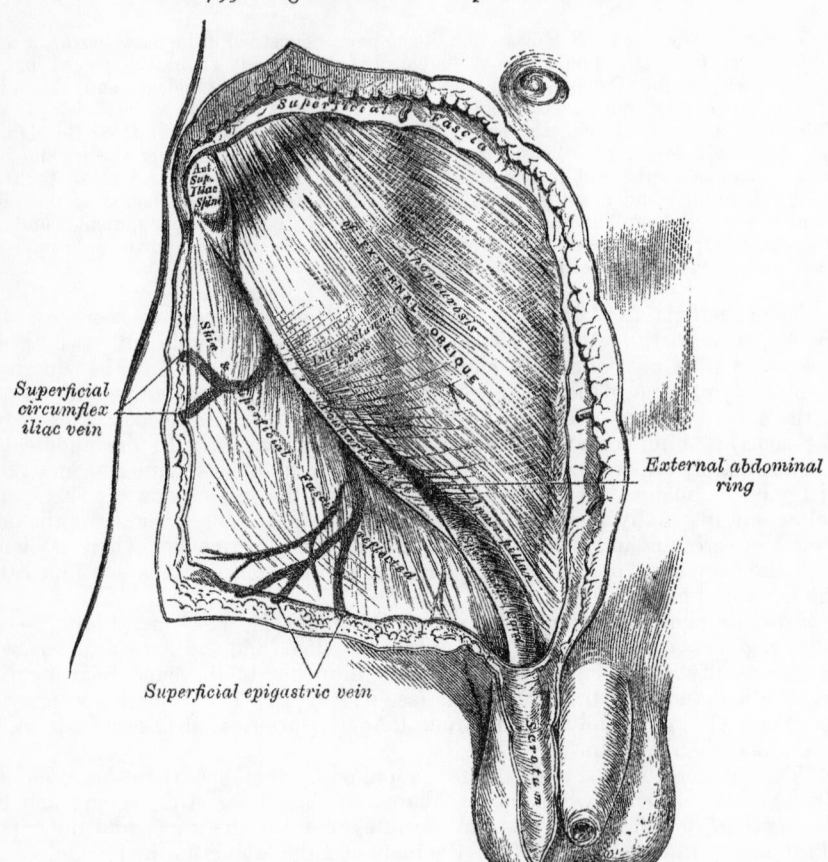

Superficial
circumflex
iliac vein

External abdominal
ring

Superficial epigastric vein

tendinous fibres (*intercolumnar fibres*).   They are thickest below, where they arise from Poupart's ligament; and are inserted into the linea alba, describing a curve, with the convexity downwards.   The fibres are connected together by delicate fibrous tissue, and thus constitute a fascia which is called the *inter-columnar fascia*, which forms an envelope for the spermatic cord or round ligament, after they have passed through the external abdominal ring.   It constitutes one of the coverings of an inguinal hernia, when it descends into the scrotum.

If the finger be introduced a short distance into the external ring, and the limb is extended and rotated outwards, the aponeurosis of the External oblique, together with the iliac portion of the fascia lata, will be felt to become tense, and the external ring much contracted; if the limb, on the contrary, be flexed upon the pelvis and rotated inwards, this aponeurosis is relaxed, and the external ring sufficiently enlarged to admit the finger with comparative ease; hence the patient should always be put in the latter position when the taxis is applied for the reduction of an inguinal hernia, in order that the abdominal wall may be relaxed as much as possible.

Dissection.—The aponeurosis of the External oblique should be removed by dividing it across in the same direction as the external incisions, and reflecting it downwards and outwards; great care is requisite in separating it from the aponeurosis of the muscle beneath. The lower part of the Internal oblique and the Cremaster are then exposed, together with the inguinal canal, which contains the spermatic cord (fig. 796). The mode of insertion of Poupart's and Gimbernat's ligaments and the triangular fascia of the abdomen into the os pubis should also be examined. (See page 485.)

The **Internal oblique muscle** has been previously described (page 486). The portion which is now exposed is partly muscular and partly tendinous in structure. Those fibres which arise from Poupart's ligament, few in number, and paler in

FIG. 796.—Inguinal hernia.
Dissection showing the Internal oblique and Cremaster.

colour than the rest, arch downwards and inwards across the spermatic cord, and, becoming tendinous, are inserted conjointly with those of the Transversalis into the crest of the os pubis and pectineal line, forming what is known as the conjoined tendon of the Internal oblique and Transversalis. This tendon is inserted immediately behind the inguinal canal and external abdominal ring, serving to protect what would otherwise be a weak point in the abdominal wall. Sometimes this tendon is insufficient to resist the pressure from within, and is carried forwards in front of the protrusion through the external ring, forming one of the coverings of direct inguinal hernia; or the hernia forces its way through the fibres of the conjoined tendon.

The **Cremaster muscle** has been previously described (page 487). The loops which it forms over the cord and testicle are united together by areolar tissue, and constitute a thin covering to these structures, the *cremasteric fascia.*

In the descent of an oblique inguinal hernia, which takes the same course as the spermatic cord, the Cremaster muscle forms one of its coverings. This muscle becomes largely developed in cases of hydrocele and large, old scrotal herniæ. No such muscle exists in the female, but an analogous structure is developed in those cases where an oblique inguinal hernia descends beneath the margin of the Internal oblique.

Dissection.—The Internal oblique should be detached from Poupart's ligament, separated from the Transversalis to the same extent as in the previous incisions, and reflected inwards on to the sheath of the Rectus (fig. 797). The deep circumflex iliac vessels, which lie between these two muscles, form a valuable guide to their separation.

The **Transversalis muscle** has been previously described (page 488). The portion which is now exposed is partly muscular and partly tendinous in structure; it arises from the outer third of Poupart's ligament, its fibres curve downwards and inwards, and are inserted together with those of the Internal oblique into the lower part of the linea alba, into the crest of the os pubis and the pectineal line, forming what is known as the conjoined tendon of the

FIG. 797.—Inguinal hernia. Dissection showing the Transversalis muscle, the transversalis fascia, and the internal abdominal ring.

Internal oblique and Transversalis. Between the lower border of this muscle and Poupart's ligament, a space is left, in which the transversalis fascia is seen.

The **inguinal** or **spermatic canal** contains the spermatic cord in the male, and the round ligament in the female. It is an oblique canal, about an inch and a half in length, directed downwards and inwards, and placed parallel with, and a little above, Poupart's ligament. It commences, above, at the internal or deep abdominal ring, which is the point where the cord enters the inguinal canal; and terminates, below, at the external or superficial ring. It is bounded, *in front*, by the integument and superficial fascia, by the aponeurosis of the External oblique throughout its whole length, and by the Internal oblique for its outer third; *behind*, by the triangular fascia, the conjoined tendon of the Internal oblique and Transversalis, the transversalis fascia, and the subperitoneal fat and peritoneum; *above*, by the arched fibres of the Internal oblique and Transversalis; *below*, by the union of the transversalis fascia with Poupart's ligament. The deep epigastric artery passes upwards and inwards behind the canal, lying close to the inner side of the internal abdominal ring. The interval between this artery and the outer head of the Rectus abdominis is named

Hesselbach's triangle, the base of which is formed by Poupart's ligament. That form of hernia in which the intestine follows the course of the spermatic cord along the inguinal canal is called *oblique inguinal hernia*.

The **transversalis fascia** has already been described (page 491). Throughout the greater part of its extent it is a thin aponeurotic membrane, but in the inguinal region it is thick and dense in structure. Here the spermatic cord in the male, and the round ligament in the female, pass through it, and the point where they pass through is called the *internal* or *deep abdominal ring*. This opening is not visible externally owing to a prolongation of the transversalis fascia on these structures, forming the infundibuliform fascia.

The **internal** or **deep abdominal ring** is situated in the transversalis fascia, midway between the anterior superior spine of the ilium and the symphysis pubis, and about half an inch above Poupart's ligament (see page 492). From its circumference a thin, funnel-shaped membrane, the *infundibuliform fascia*, is continued round the cord and testis, enclosing them in a distinct pouch. When the sac of an oblique inguinal hernia passes through the internal or deep abdominal ring, the infundibuliform fascia constitutes one of its coverings.

The extra-peritoneal connective tissue.—Between the transversalis fascia and the peritoneum is a quantity of loose areolar tissue. It has been described on page 492.

The *deep epigastric artery* arises from the external iliac artery a few lines above Poupart's ligament. It at first descends to reach this ligament, and then ascends obliquely along the inner margin of the internal or deep abdominal ring, lying between the transversalis fascia and the peritoneum; and, running upwards, pierces the transversalis fascia and enters the sheath of the Rectus muscle by passing over the semilunar fold of Douglas. Consequently the deep epigastric artery bears a very important relation to the internal abdominal ring, as it extends obliquely upwards and inwards, from its origin from the external iliac. In this part of its course it lies along the lower and inner margin of the internal ring, and beneath the commencement of the spermatic cord. Close to its origin it is crossed by the vas deferens in the male, and by the round ligament in the female.

The **peritoneum**, corresponding to the inner surface of the internal ring, presents a well-marked depression, the depth of which varies in different subjects. A thin fibrous band is continued from it along the front of the cord for a variable distance, and becomes ultimately lost. This is the remains of the pouch of peritoneum, which in the fœtus precedes the cord and testis into the scrotum and the obliteration of which commences soon after birth. In some cases the fibrous band can only be traced a short distance; but occasionally it may be followed, as a fine cord, as far as the upper end of the tunica vaginalis. Sometimes the tube of peritoneum is only closed at intervals, and presents a sacculated appearance; or a single pouch may extend along the whole length of the cord, which may be closed above; or the pouch may be directly continuous with the peritoneum by an opening at its upper part.

In the female fœtus the peritoneum is also prolonged in the form of a tubular process for a short distance into the inguinal canal. This process is called the *canal of Nuck*. It is generally obliterated in the adult, but sometimes it remains pervious even in advanced life.

In order to understand the relation of the peritoneum to inguinal hernia, it is necessary to view the anterior abdominal wall from its internal aspect, when it will be seen as shown in fig. 798. Between the upper margin of the front of the pelvis and the umbilicus, the peritoneum, when viewed from behind, will be seen to be raised into five folds, with intervening depressions, by more or less prominent bands which converge to the umbilicus. One of these is situated in the median line, and is caused by the urachus, the remnant of the allantois; it extends from the summit of the bladder to the umbilicus. The fold of peritoneum covering it is known as the *plica urachi*. On either side of this is a prominent band, caused by the obliterated hypogastric artery, which extends obliquely upwards and inwards to the umbilicus. This is covered by a fold of peritoneum, which is known as the *plica hypogastrica*. To either side of these three cords is the deep epigastric artery, which ascends obliquely upwards and inwards from a point midway between the symphysis pubis and the anterior superior spine of the ilium to the semilunar fold of

Douglas, in front of which it disappears. It is covered by a fold of peritoneum, which is known as the *plica epigastrica.* Between these raised folds are depressions of the peritoneum, constituting so-called fossæ. The most internal, between the plica urachi and the plica hypogastrica, is known as the *internal inguinal fossa* (fovea supravesicalis). The middle one is situated between the plica hypogastrica and the plica epigastrica, and is termed the *middle inguinal fossa* (fovea inguinalis mesialis). The external one is external to the plica epigastrica, and is known as the *external inguinal fossa* (fovea inguinalis lateralis). Occasionally the deep epigastric artery corresponds in position to the obliterated hypogastric artery, and then there is but one fold on each side of the middle line, and the two external fossæ are merged into one. In the usual condition of the parts

FIG. 798.—Posterior view of the anterior abdominal wall in its lower half. The peritoneum is in place, and the various cords are shining through. (After Joessel.)

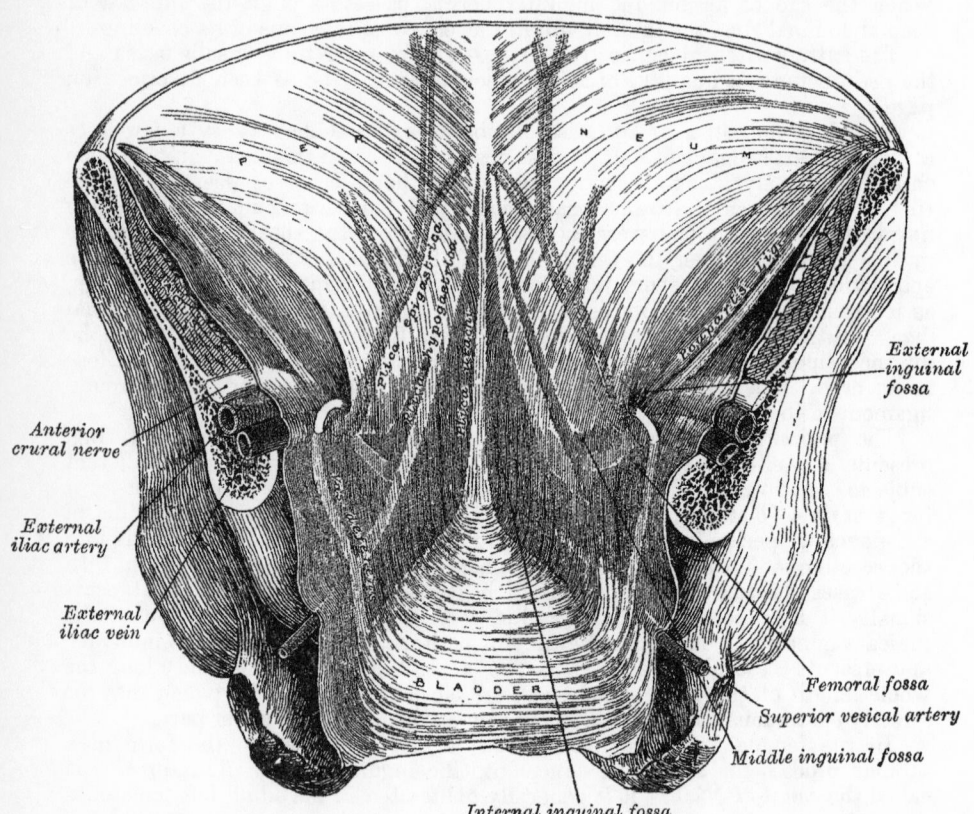

Anterior
crural nerve

External
iliac artery

External
iliac vein

External
inguinal
fossa

Femoral fossa

Superior vesical artery

Middle inguinal fossa

Internal inguinal fossa

the floor of the external inguinal fossa corresponds to the internal abdominal ring, and into this fossa an oblique inguinal hernia descends. To the inner side of the plica epigastrica are the two internal fossæ, and through either of these a direct hernia may descend, as will be subsequently explained (page 1184). The whole of this space—that is to say, the space between the deep epigastric artery, the margin of the Rectus and Poupart's ligament—is commonly known as *Hesselbach's triängle.* These three depressions or fossæ are situated above the level of Poupart's ligament, and in addition to them is another below the ligament, corresponding to the position of the femoral ring, and into which a femoral hernia descends. This is known as the *femoral fossa.*

INGUINAL HERNIA

Inguinal hernia is that form of protrusion which makes its way through the abdomen in the inguinal region.

There are two principal varieties of inguinal hernia: external or oblique, and internal or direct.

External or *oblique inguinal hernia*, the more frequent of the two, takes the same course as the spermatic cord. It is called *external*, from the neck of the sac being on the outer or iliac side of the deep epigastric artery.

Internal or *direct inguinal hernia* does not follow the same course as the cord, but protrudes through the abdominal wall, on the inner or pubic side of the deep epigastric artery.

OBLIQUE INGUINAL HERNIA

In **oblique inguinal hernia** the intestine escapes from the abdominal cavity at the internal ring, pushing before it a pouch of peritoneum, which forms the hernial sac (fig. 800, A). As it enters the inguinal canal it receives an

FIG. 799.—Oblique inguinal hernia, showing its various coverings.
(From a preparation in the Museum of the Royal College of Surgeons of England.)

investment from the extra-peritoneal tissue, and is enclosed in the infundibuliform process of the transversalis fascia. In passing along the inguinal canal it displaces upwards the arched fibres of the Transversalis and Internal oblique muscles, and is surrounded by the fibres of the Cremaster. It then passes along the front of the cord, and escapes from the inguinal canal at the external ring, receiving an investment from the intercolumnar fascia. Lastly, it descends into the scrotum, receiving coverings from the superficial fascia and the integument.

The coverings of this form of hernia, after it has passed through the external ring, are, from without inwards, the integument, superficial fascia, intercolumnar fascia, cremasteric fascia, infundibuliform fascia, extra-peritoneal tissue, and peritoneum.

This form of hernia lies in front of the vessels of the spermatic cord, and seldom extends below the testis, on account of the intimate adhesion of the coverings of the cord to the tunica vaginalis.

The *seat of stricture* in oblique inguinal hernia is either at the external ring; in the inguinal canal, caused by the fibres of the Internal oblique or Transversalis; or at the internal ring; most frequently in the last situation. If it is situated at the external ring, the division of a few fibres at one point of its circumference is all that is necessary for the replacement of the hernia. If in the inguinal canal, or at the internal ring, it may be necessary to divide the aponeurosis of the External oblique, so as to lay open the inguinal canal. In dividing the stricture, the direction of the incision should be upwards.

When the intestine passes along the inguinal canal, and escapes from the external ring into the scrotum, it is called *complete oblique inguinal*, or *scrotal hernia*. If the intestine does not escape from the external ring, but is retained in the inguinal canal, it is called *incomplete inguinal hernia*, or *bubonocele*. In each of these cases, the coverings which invest it will depend upon the extent to which it descends in the inguinal canal.

There are some other varieties of oblique inguinal hernia depending upon congenital defects in the processus vaginalis. The testicle in its descent from the abdomen into the scrotum is preceded by a pouch of peritoneum, the *processus vaginalis*, which about the period of birth becomes shut off from the general peritoneal cavity by a closure of that portion of the pouch which extends from the internal abdominal ring to near the upper part of the testicle; the lower portion of the pouch remaining persistent as the tunica vaginalis. It would appear that this closure commences at two points, viz. at the internal abdominal ring, and at the top of the epididymis, and gradually extends until, in the normal condition, the whole of· the intervening portion is converted into a fibrous cord. From failure in the completion of this process, variations in the relation of the hernial protrusion to the testicle and tunica vaginalis are produced, which constitute distinct varieties of inguinal hernia, and which have received separate names, and are of surgical importance. These are congenital, infantile, encysted, and hernia of the funicular process.

Congenital hernia (fig. 800, B).—Where the pouch of peritoneum which precedes the cord and testis in its descent remains patent throughout, and is unclosed at any point, the cavity of the tunica vaginalis communicates directly with that of the peritoneum. The intestine descends along this pouch into the cavity of the tunica vaginalis, which constitutes the sac of the hernia, and the gut lies in contact with the testicle. Though this form of hernia is termed 'congenital,' it must be borne in mind that this term does not imply that the hernia existed at birth, but merely a condition of things which may allow of the descent of the hernia at any moment. As a matter of fact, congenital herniæ frequently do not appear till adult life.

Infantile and encysted herniæ.—Where the pouch of peritoneum is occluded at the internal ring only, and remains patent throughout the rest of its extent, two varieties of oblique inguinal hernia may be produced, which have received the names of infantile and encysted herniæ. In the *infantile* form (fig. 800, c) the bowel pressing upon the septum and the peritoneum in its immediate neighbourhood causes it to yield and form a sac, which descends behind the tunica vaginalis; so that, in front of the bowel, there are three layers of peritoneum, the two layers of the tunica vaginalis and its own sac. In the *encysted* form (fig. 800, D) pressure in the same position—namely, at the occluded spot in the pouch—causes the septum to yield and form a sac which projects *into* and not *behind* the tunica vaginalis, as in the infantile form, and thus it constitutes a sac within a sac, so that in front of the bowel there are two layers of peritoneum, one layer of the tunica vaginalis and its own sac.

Hernia into the funicular process (fig. 800, E).—Where the pouch of peritoneum is occluded at the lower point only—that is, just above the testicle—the intestine descends into the pouch of peritoneum as far as the testicle, but is prevented from entering the sac of the tunica vaginalis by the septum which

FIG. 800.—Varieties of oblique inguinal hernia.

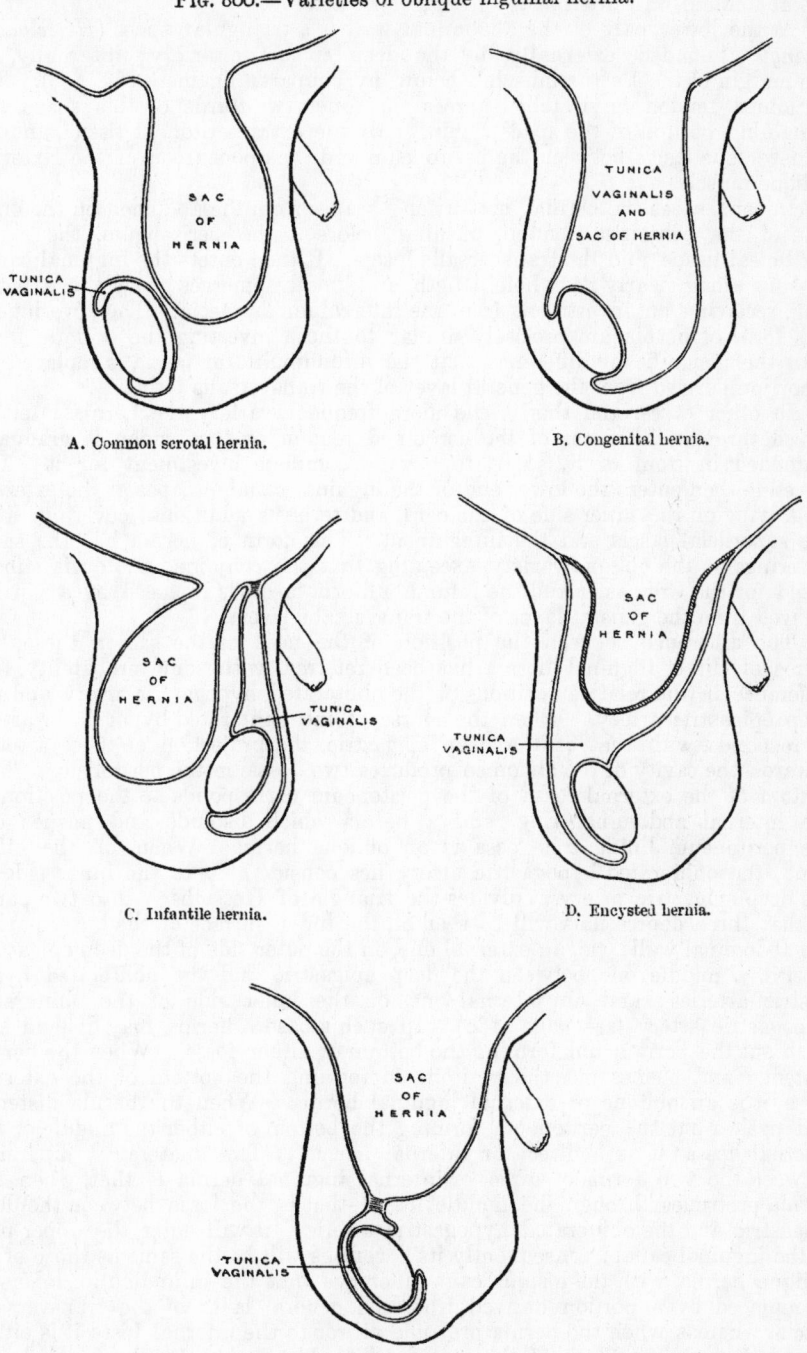

A. Common scrotal hernia.

B. Congenital hernia.

C. Infantile hernia.

D. Encysted hernia.

E. Hernia into the funicular process.

has formed between it and the pouch, so that it resembles the congenital form in all respects, except that, instead of enveloping the testicle, that body can be felt below the hernia.

DIRECT INGUINAL HERNIA

In **direct inguinal hernia** the protrusion makes its way through some part of the abdominal wall internal to the epigastric artery.

At the lower part of the abdominal wall is a triangular space (*Hesselbach's triangle*), bounded, externally, by the deep epigastric artery; internally, by the margin of the Rectus muscle; below, by Poupart's ligament (fig. 798). The conjoined tendon is stretched across the inner two-thirds of this space, the remaining portion of the space having only the extra-peritoneal tissue, and the transversalis fascia between the peritoneum and the aponeurosis of the External oblique muscle.

In some cases the hernial protrusion escapes from the abdomen on the outer side of the conjoined tendon, pushing before it the peritoneum, the extra-peritoneal tissue, and the transversalis fascia. It then enters the inguinal canal, passing along nearly its whole length, and finally emerges from the external ring, receiving an investment from the intercolumnar fascia. The coverings of this form of hernia are precisely similar to those investing the oblique form, with the insignificant difference that the infundibuliform fascia is replaced by a portion derived from the general layer of the transversalis fascia.

In other cases, and this is the more frequent variety, the hernia is either forced through the fibres of the conjoined tendon, or the tendon is gradually distended in front of it, so as to form a complete investment for it. The intestine then enters the lower end of the inguinal canal, escapes at the external ring, lying on the inner side of the cord, and receives additional coverings from the superficial fascia and the integument. This form of hernia has the same coverings as the oblique variety, excepting that the conjoined tendon is substituted for the Cremaster and the infundibuliform fascia is replaced by a portion derived from the general layer of the transversalis fascia.

The difference between the position of the neck of the sac in these two forms of direct inguinal hernia has been referred, with some probability, to a difference in the relative positions of the obliterated hypogastric artery and the deep epigastric artery. When the course of the obliterated hypogastric artery corresponds with that of the deep epigastric, the projection of these arteries towards the cavity of the abdomen produces two fossæ in the peritoneum. The bottom of the external fossa of the peritoneum corresponds to the position of the internal abdominal ring, and a hernia which distends and pushes out the peritoneum lining this fossa is an oblique hernia. When, on the other hand, the obliterated hypogastric artery lies considerably to the inner side of the deep epigastric artery, it divides the triangle of Hesselbach into two parts, so that three depressions will be seen on the inner surface of the lower part of the abdominal wall: viz. an external one, on the outer side of the deep epigastric artery; a middle one, between the deep epigastric and the obliterated hypogastric arteries; and an internal one, on the inner side of the obliterated hypogastric artery (see page 1180). In such a case a hernia may distend and push out the peritoneum forming the bottom of either fossa. When the hernia distends and pushes out the peritoneum forming the bottom of the external fossa, it is an oblique or external inguinal hernia. When the hernia distends and pushes out the peritoneum forming the bottom of either the middle or the internal fossa, it is a direct or internal hernia. The anatomical difference between the two forms of direct or internal inguinal hernia is that, when the hernia protrudes through the middle fossa—that is, the fossa between the deep epigastric and the obliterated hypogastric arteries—it will enter the upper part of the inguinal canal; consequently its coverings will be the same as those of an oblique hernia, with the insignificant difference that the infundibuliform fascia is replaced by a portion derived from the general layer of the transversalis fascia, whereas when the hernia protrudes through the internal fossa it is either forced through the fibres of the conjoined tendon, or the tendon is gradually distended in front of it, so as to form a complete investment for it. The intestine then enters the lower part of the inguinal canal, and escapes from the external abdominal ring lying on the inner side of the cord.

This form of hernia has the same coverings as the oblique variety, excepting that the conjoined tendon is substituted for the Cremaster, and the infundibuliform fascia is replaced by a portion derived from the general layer of the transversalis fascia.

The *seat of stricture* in both varieties of direct hernia is usually found either at the neck of the sac, or at the external ring. In that form of hernia which perforates the conjoined tendon, it not infrequently occurs at the edges of the fissure through which the ,gut passes. In dividing the stricture, the incision should in all cases be directed upwards.*

If the hernial protrusion passes into the inguinal canal, but does not escape from the external abdominal ring, it forms what is called *incomplete direct hernia*. This form of hernia is usually of small size, and in corpulent persons very difficult of detection.

Direct inguinal hernia is of much less frequent occurrence than the oblique, their proportion being, according to Cloquet, as one to five. It occurs far more frequently in men than in women, on account of the larger size of the external ring in the former sex. It differs from the oblique in its smaller size and globular form, dependent most probably on the resistance offered to its progress by the transversalis fascia and conjoined tendon. It differs also in its position, being placed over the os pubis, and not in the course of the inguinal canal. The deep epigastric artery runs on the outer or iliac side of the neck of the sac, and the spermatic cord along its external and posterior side, not directly behind it, as in oblique inguinal hernia.

---

* In all cases of inguinal hernia, whether oblique or direct, it is proper to divide the stricture directly upwards; the reason of this is obvious, for by cutting in this direction the incision is made parallel to the deep epigastric artery—either external to it, in the oblique variety; or internal to it, in the direct form of hernia; and thus all chance of wounding the vessel is avoided. If the incision was made outwards, the artery might be divided if the hernia was direct; and if made inwards it would stand an equal chance of injury if the case was one of oblique inguinal hernia.

THE SURGICAL ANATOMY OF FEMORAL HERNIA

Dissection.—The dissection of the parts comprised in the anatomy of femoral hernia should be performed, if possible, upon a female subject free from fat. The subject should lie upon its back; a block is first placed under the pelvis, the thigh everted, and the knee slightly bent, and retained in this position. An incision should then be made from the anterior superior spinous process of the ilium along Poupart's ligament to the symphysis pubis; a second incision should be carried transversely across the thigh about six inches beneath the preceding; and these are to be connected together by a vertical one carried along the inner side of the thigh. These several incisions should divide merely the integument; this is to be reflected outwards, when the superficial fascia will be exposed.

The **superficial fascia** forms a continuous layer over the whole of the thigh, consisting of areolar tissue, containing in its meshes much fat, and capable of being separated into two or more layers, between which are found the superficial vessels and nerves. It varies in thickness in different parts of the limb. In the groin it is thick, and the two layers are separated from one another by the superficial inguinal lymphatic glands, the internal saphenous vein, and several smaller vessels. One of these layers, the superficial, is continuous with the superficial fascia of the abdomen.

The superficial layer should be detached by dividing it across in the same direction as the external incisions; its removal will be facilitated by commencing at the lower and inner angle of the space, detaching it at first from the front of the internal saphenous vein, and dissecting it off from the anterior surface of that vessel and its tributaries; it should be reflected outwards, in the same manner as the integument. The cutaneous vessels and nerves, and superficial inguinal glands, are then exposed, lying upon the deep layer of the superficial fascia. These are the internal saphenous vein, and the superficial epigastric, superficial circumflex iliac, and superficial external pudic vessels, as well as numerous lymphatics ascending with the saphenous vein to the inguinal glands.

The *internal* or *long saphenous vein* ascends along the inner side of the thigh, and, passing through the saphenous opening in the fascia lata, terminates in the femoral vein about an inch and a half below Poupart's ligament. This vein receives, at the saphenous opening, the superficial epigastric, the superficial circumflex iliac, and the superficial external pudic veins.

The three small arteries in this situation are branches of the femoral, and have already been described (page 701). The superficial external pudic and the superficial epigastric arteries pass through the saphenous opening; the superficial circumflex iliac pierces the fascia lata in the immediate neighbourhood of the saphenous opening.

The superficial veins.—The veins accompanying these superficial arteries are usually much larger than the arteries: they terminate in the internal or long saphenous vein at the saphenous opening.

The *superficial inguinal lymphatic glands* have been described in the section on the lymph vascular system (see page 755). The lower group surrounds the saphenous opening, and receives the superficial lymphatic vessels from the lower extremity.

Two small nerves are found in this situation: (1) the *ilio-inguinal*, from the first lumbar nerve, which supplies the skin of the upper and inner part of the thigh, and the scrotum in the male and the labium in the female; and (2) the *crural branch of the genito-crural*, from the first and second lumbar nerves, which supplies the skin of the anterior aspect of the thigh as far as midway between the pelvis and the knee (see page 916).

The *deep layer of the superficial fascia* is a very thin fibrous layer placed beneath the subcutaneous nerves and upon the surface of the fascia lata. It covers the saphenous opening in this fascia, and in consequence of being perforated here by the internal saphenous vein and by numerous lymphatics and blood-vessels, this portion of the deep layer of the superficial fascia has received the name of cribriform fascia (see page 543). The cribriform fascia is frequently described as a part of the fascia lata. A femoral hernia, in passing through the saphenous opening, receives the cribriform fascia as one of its coverings.

The deep layer of superficial fascia having been removed, the fascia lata is exposed.

The **fascia lata** has been already described with the muscles of the front of the thigh (page 544). At the upper and inner part of the thigh, a little below

FIG. 801.—Femoral hernia. Superficial dissection.

Poupart's ligament, a large oval-shaped aperture is observed after the superficial and cribriform fasciæ have been cleared away; it transmits the internal saphenous vein and other smaller vessels, and is called the *saphenous opening*.

The **saphenous opening** is an oval-shaped aperture, measuring about an inch and a half in length, and half an inch in width. It is situated at the upper and inner part of the front of the thigh, below Poupart's ligament, and is directed obliquely downwards and outwards.

The *outer margin* of the saphenous opening is of a semilunar form, thin, strong, sharply defined, and lies on a plane considerably anterior to the inner margin. If this edge is traced upwards, it will be seen to form a curved elongated process, the *falciform process*, or *superior cornu*, which ascends in front of the femoral vessels, and, curving inwards, is attached to Poupart's ligament and to the spine of the os pubis and pectineal line, where it is continuous with the pubic portion. If traced downwards, it is found to be continuous with another curved margin, the concavity of which is directed upwards and inwards: this is the inferior cornu of the saphenous opening, and is blended with the pubic portion of the fascia lata covering the Pectineus muscle.

The *inner boundary of the opening* is on a plane posterior to the outer margin and behind the level of the femoral vessels; it is much less prominent and defined than the outer, from being stretched over the subjacent Pectineus muscle. It is through the saphenous opening that a femoral hernia passes after descending along the crural canal.

If the finger is introduced into the saphenous opening while the limb is moved in different directions, the aperture will be found to be greatly constricted

FIG. 802.—Femoral hernia, showing fascia lata and saphenous opening.

on extending the limb or rotating it outwards, and to be relaxed on flexing and inverting it: hence the necessity for placing the limb in the latter position in employing the taxis for the reduction of a femoral hernia.

Dissection.—The iliac portion of the fascia lata, but not its falciform process should now be removed by detaching it from the lower margin of Poupart's ligament, carefully dissecting it from the subjacent structures, and turning it inwards, when the sheath of the femoral vessel is exposed descending beneath Poupart's ligament (fig. 803).

Poupart's ligament, or the **crural arch,** is the lower border of the aponeurosis of the External oblique muscle, which extends from the anterior superior spine of the ilium to the spine of the os pubis. From this latter point it is reflected backwards, to be attached to the pectineal line for about half an inch, forming Gimbernat's ligament. Its general direction is curved downwards towards the thigh, where it is continuous with the fascia lata. Its outer half is rounded and oblique in direction. Its inner half gradually widens at its attachment to the os pubis, is more horizontal in direction, and lies beneath the spermatic cord. Nearly the whole of the space included between the crural arch and the innominate bone is filled in by the parts which descend from the abdomen into the thigh (fig. 804). The outer half of the space is occupied by the Iliacus and Psoas muscles, together with the external cutaneous and anterior crural nerves. The pubic half of the space is occupied by the femoral vessels included in their sheath, and by the origin of the Pectineus muscle; a small oval-shaped interval existing between the femoral vein and the inner wall of the sheath, which is

FIG. 803.—Femoral hernia. Iliac portion of fascia lata removed, and sheath of femoral vessels and femoral canal exposed.

occupied merely by a little loose areolar tissue, a few lymphatic vessels, and occasionally by a small lymphatic gland; this is the crural ring, through which the gut descends in femoral hernia.

Gimbernat's ligament (figs. 804, 805) is that part of the aponeurosis of the External oblique muscle which is reflected backwards and outwards, from the spine of the os pubis, to be inserted into the pectineal line. It is about half an inch in length, larger in the male than in the female, almost horizontal in direction in the erect posture, and of a triangular form, with the base directed outwards. Its *base*, or outer margin, is concave, thin, and sharp, and lies in contact with the femoral sheath. Its *apex* corresponds to the spine of the os pubis. Its *posterior margin* is attached to the pectineal line, and is continuous with the pubic portion of the fascia lata. Its *anterior margin* is continuous with Poupart's ligament.

Femoral sheath.—The femoral or crural sheath is a prolongation downwards of the fasciæ that line the abdomen, the transversalis fascia passing down in front of the femoral vessels, and the iliac fascia descending behind them; these fasciæ are directly continuous on the iliac side of the femoral artery, but a small space exists between the femoral vein and the point where they are continuous on the pubic side of that vessel, which constitutes the femoral or crural canal. The femoral sheath, funnel-shaped, wide above and narrow below, is closely adherent to the contained vessels about an inch below the saphenous opening, being blended with the areolar sheath of the vessels. The outer border of the sheath is perforated by the genito-crural nerve. Its inner border is pierced by the internal saphenous vein and numerous lymphatic vessels. In front, it is covered by the iliac portion of the fascia lata; and behind it is the pubic portion of the same fascia.

If the anterior wall of the sheath is removed, the femoral artery and vein are seen lying side by side, a thin septum separating the two vessels, while another

FIG. 804.—Structures which pass beneath the crural arch.

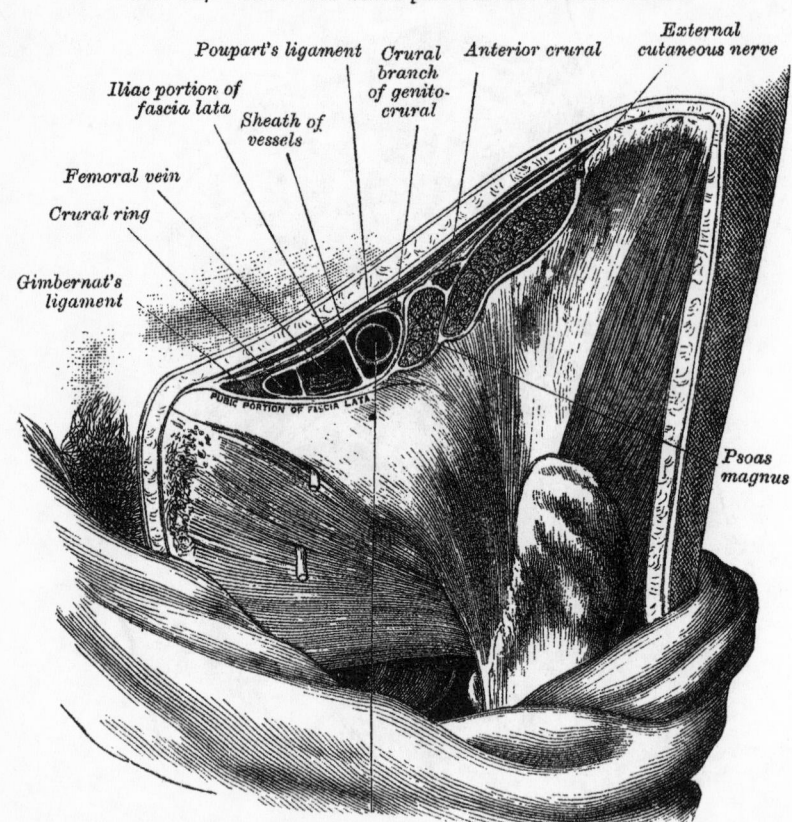

septum may be seen lying just internal to the vein, and cutting off a small space between the vein and the inner wall of the sheath. The septa are stretched between the anterior and posterior walls of the sheath, so that the two vessels are enclosed in separate compartments. The interval left between the vein and the inner wall of the sheath contains a little loose areolar tissue, a few lymphatic vessels, and occasionally a small lymphatic gland; this is the *femoral* or *crural canal*, through which the intestine descends in femoral hernia.

Deep crural arch.—Passing across the front of the femoral sheath on the abdominal side of Poupart's ligament, and closely connected with it, is a thickened band of fibres, called the *deep crural arch*. It is apparently a thickening of the transversalis fascia, which is joined externally to the centre of Poupart's ligament, and arches across the front of the femoral sheath, to be inserted by a broad attachment into the pectineal line, behind the conjoined tendon. In some subjects this structure is not very prominently marked, and not infrequently it is altogether wanting.

The **crural canal** is the narrow interval between the femoral vein and the inner wall of the femoral sheath. It exists as a distinct canal only when the sheath has been separated from the vein by dissection, or by the pressure of a hernia or tumour. Its length is from a quarter to half an inch, and it extends from Gimbernat's ligament to the upper part of the saphenous opening.

The *anterior wall* of the crural canal is very narrow, and formed by a continuation downwards of the transversalis fascia, under Poupart's ligament, covered by the falciform process of the fascia lata.

The *posterior wall* is formed by a continuation downwards of the iliac fascia covering the pubic portion of the fascia lata.

The *outer wall* is formed by the fibrous septum separating it from the inner side of the femoral vein.

The *inner wall* is formed by the junction of the processes of the transversalis and iliac fasciæ which form the inner side of the femoral sheath, and lies in contact, at its commencement, with the outer edge of Gimbernat's ligament.

FIG. 805.—The relations of the femoral and internal abdominal rings, seen from within the abdomen. Right side.

This canal has two orifices : an upper one, the *crural ring*, closed by the septum crurale ; and a lower one, the *saphenous opening*, closed by the cribriform fascia.

The **crural ring** (figs. 804, 805) is the upper opening of the crural canal. It is bounded, in front, by Poupart's ligament and the deep crural arch ; behind, by the os pubis, covered by the Pectineus muscle, and the pubic portion of the fascia lata ; internally, by the base of Gimbernat's ligament, the conjoined tendon, the transversalis fascia, and the deep crural arch ; externally, by the fibrous septum lying on the inner side of the femoral vein. The crural ring is of an oval form ; its long diameter, directed transversely, measures about half an inch ; it is larger in the female than in the male, and this is one of the reasons of the greater frequency of femoral hernia in the former sex.

Position of parts around the ring.—The spermatic cord in the male, and round ligament in the female, lie immediately above the anterior margin of the femoral ring, and may be divided in an operation for femoral hernia if the incision for the relief of the stricture is not of limited extent. In the female this is of little importance, but in the male the spermatic artery and vas deferens may be divided.

The *femoral vein* lies on the outer side of the ring.

The *deep epigastric artery*, in its passage upwards and inwards from the external iliac artery, passes across the upper and outer angle of the crural ring, and is consequently in danger of being wounded if the stricture is divided in a direction upwards and outwards.

The *communicating branch* between the deep epigastric and obturator arteries lies in front of the ring.

The circumference of the ring is thus seen to be bounded by vessels in every part, excepting internally and behind. It is in the former position that the stricture is divided in cases of strangulated femoral hernia.

When the *obturator artery* arises from the deep epigastric, a condition which occurs twice in every seven subjects, the vessel bears a very important relation to the crural ring. In most cases it descends on the inner side of the external iliac vein to the obturator foramen, and will consequently lie on the outer side of the crural ring, where there is no danger of its being wounded in the operation for dividing the stricture in femoral hernia (see fig. 507, A, page 688). Occasionally, however, the obturator artery curves along the free margin of Gimbernat's ligament in its passage to the obturator foramen; it would, consequently, skirt along the greater part of the circumference of the crural ring, and could hardly avoid being wounded in the operation (see fig. 507, B, page 688).

Septum crurale.—The portion of the extra-peritoneal tissue which closes the crural ring is named the *septum crurale*. This serves as a barrier to the protrusion of a hernia through this part. Its upper surface is slightly concave, and supports a small lymphatic gland. Its under surface is turned towards the femoral canal. The septum crurale is perforated by numerous apertures for the passage of lymphatic vessels, connecting the deep inguinal lymphatic glands with those surrounding the external iliac artery. On the abdominal aspect of the septum crurale is the parietal layer of the peritoneum, which, when viewed from above, presents a slight depression, termed the *femoral fossa*.

The size of the crural canal, the degree of tension of its orifices, and, consequently, the degree of constriction of a hernia, vary according to the position of the limb. If the leg and thigh are extended, abducted, or everted, the femoral canal and its orifices are rendered tense, from the traction on these parts by Poupart's ligament and the fascia lata, as may be ascertained by passing the finger along the canal. If, on the contrary, the thigh is flexed upon the pelvis, and at the same time adducted and rotated inwards, the femoral canal and its orifices become considerably relaxed; for this reason the limb should always be placed in the latter position when the application of the taxis is made in attempting the reduction of a femoral hernia.

Descent of the hernia.—From the preceding description it follows that the crural ring must be a weak point in the abdominal wall: hence it is that, when violent or long-continued pressure is made upon the abdominal viscera, a portion of intestine may be forced into it, constituting a femoral hernia; and the changes in the tissues of the abdomen which are produced by pregnancy, together with the larger size of the crural ring in the female, serve to explain the frequency of this form of hernia in women.

When a portion of the intestine is forced through the crural ring, it carries before it a pouch of peritoneum, which forms what is called the *hernial sac*; it receives an investment from the extra-peritoneal tissue or septum crurale, and descends along the crural canal in the inner compartment of the sheath of the femoral vessels as far as the saphenous opening; at this point it changes its course, being prevented from extending farther down the sheath, on account of the narrowing of the latter and its close contact with the vessels, and also from the close attachment of the superficial fascia and femoral sheath to the lower part of the circumference of the saphenous opening; the tumour is, consequently, directed forwards, pushing before it the cribriform fascia, and then curves upwards over Poupart's ligament and the lower part of the tendon of the External oblique, being covered by the superficial fascia and integument. While the hernia is contained in the crural canal, it is usually of small size, owing to the resisting nature of the surrounding parts; but when it has escaped from the saphenous opening into the loose areolar tissue of the groin it becomes considerably enlarged. The direction taken by a femoral hernia in its descent is at first downwards, then forwards and upwards; this should be borne in mind, as in the application of the taxis for the reduction of a femoral hernia pressure should be directed in the reverse order.

Coverings of the hernia.—The coverings of a femoral hernia, from within outwards, are : peritoneum, the septum crurale, femoral sheath, cribriform fascia, superficial fascia, and integument.*

Varieties of femoral hernia.—If the intestine descends along the femoral canal only as far as the saphenous opening, and does not escape from this aperture, it is called *incomplete femoral hernia.* The small size of the protrusion in this form of hernia, on account of the firm and resisting nature of the canal in which it is contained, renders it an exceedingly dangerous variety of the disease, from the extreme difficulty of detecting the existence of the swelling, especially in corpulent subjects. The coverings of an incomplete femoral hernia would be, from without inwards : integument, superficial fascia, falciform process of fascia lata, femoral sheath, septum crurale, and peritoneum. When, however, the hernial tumour protrudes through the saphenous opening, and directs itself forwards and upwards, it forms a *complete femoral hernia.* Occasionally the hernial sac descends on the iliac side of the femoral vessels, or in front, or even sometimes behind them.

The *seat of stricture of a femoral hernia* varies : it may be in the peritoneum at the neck of the hernial sac ; in the greater number of cases it would appear to be at the point of junction of the falciform process of the fascia lata with the lunated edge of Gimbernat's ligament; or at the margin of the saphenous opening in the thigh. The stricture should in every case be divided in a direction upwards and inwards ; and the extent necessary in the majority of cases is about two or three lines. By these means, all vessels or other structures of importance, in relation with the neck of the hernial sac, will be avoided.

In connection with the subject of hernia, it should be mentioned that the spine of the os pubis forms an important landmark in serving to differentiate the inguinal from the femoral variety. The position and mode of defining this process of bone has already been alluded to (page 312). If the finger is placed upon it, and the hernial protrusion is above and to the inner side of the finger, the hernia is of the inguinal variety ; if, on the other hand, the protrusion is below and to the outer side of the finger, the hernia is a femoral one.

---

* Sir Astley Cooper has described an investment for femoral hernia, under the name of ' fascia propria,' lying immediately external to the peritoneal sac, but frequently separated from it by more or less adipose tissue. Surgically it is important to remember the existence (at any rate, the occasional existence) of this layer, on account of the ease with which an inexperienced operator may mistake the fascia for the peritoneal sac, and the contained fat for omentum. Anatomically, this fascia appears identical with what is called in the text ' the extra-peritoneal tissue,' the areolar tissue being thickened and caused to assume a membranous appearance by the pressure of the hernia.

THE SURGICAL ANATOMY OF
THE PERINÆUM

Dissection.—The student should select a well-developed muscular subject free from fat, and the dissection should be commenced early, in order that the parts may be examined in as recent a state as possible. A staff having been introduced into the bladder, and the subject placed in the position shown in fig. 806, the scrotum should be raised upwards, and retained in that position, and the rectum moderately distended with tow.

The **perinæum** corresponds to the inferior aperture or outlet of the pelvis. Its deep boundaries are, in front, the pubic arch and subpubic ligament; behind, the tip of the coccyx; and on each side, the rami of the os pubis and ischium, the tuberosities of the ischium, and great sacro-sciatic ligaments. The space included by these boundaries is somewhat lozenge-shaped, and is limited on the surface of the body by the scrotum in front, by the buttocks behind, and on each side by the inner side of the thighs. A line drawn transversely between the anterior part of the tuberosity of the ischium, on each side, in front of the anus, divides this space into two portions. The anterior portion contains the penis and urethra, and is called the *urogenital triangle* or *perinæum proper*. The posterior portion contains the termination of the rectum, and is called the *ischio-rectal* or *anal region*.

The Ischio-rectal Region

The **ischio-rectal region** contains the termination of the rectum and a deep fossa, filled with fat, on each side of the intestine, between it and the tuberosity of the ischium : this is called the *ischio-rectal fossa.*

The ischio-rectal region presents in the middle line the *aperture of the anus* ; around this orifice the integument is thrown into numerous folds, which are obliterated on distension of the intestine. The integument is of a dark colour, continuous with the mucous membrane of the rectum, and provided with numerous follicles, which may inflame and suppurate and be mistaken for fistulæ. The veins round the margin of the anus are occasionally much dilated, forming a number of hard pendent masses, of a dark bluish colour, covered partly by mucous membrane and partly by the integument. These tumours constitute the disease called *external piles.*

Dissection (fig. 806).—Make an incision through the integument, along the median line, from the base of the scrotum to the anterior extremity of the anus : carry it round the margins of this aperture to its posterior extremity, and continue it backwards to about an inch behind the tip of the coccyx. A transverse incision should now be carried across the base of the scrotum, joining the anterior extremity of the preceding ; a second, carried in the same direction, should be made in front of the anus ; and a third at the posterior extremity of the first incision. These incisions should be sufficiently extensive to enable the dissector to raise the integument from the inner side of the thighs. The flaps of skin corresponding to the ischio-rectal region should now be removed. In dissecting the integument from this region great care is required, otherwise the Corrugator cutis ani and External sphincter will be removed, as they are intimately adherent to the skin.

The **superficial fascia** is exposed on the removal of the skin : it is very thick, areolar in texture, and contains much fat in its meshes. In it are found ramifying two or three branches of the perforating cutaneous nerve; these turn round the inferior border of the Gluteus maximus, and are distributed to the integument around the anus.

In this region, and connected with the lower end of the rectum, are four muscles: the Corrugator cutis ani; the two Sphincters, External and Internal; and the Levator ani.

These muscles have been already described (see page 494).

The **ischio-rectal fossa** is situated between the lower end of the rectum and the tuberosity of the ischium. It is triangular in shape; its base, directed to the surface of the body, is formed by the integument of the ischio-rectal region; its *apex*, directed upwards, corresponds to the point of division of the obturator fascia, and the thin membrane given off from it, which covers the outer surface of the Levator ani (ischio-rectal or anal fascia). Its dimensions are about an inch in breadth at the base, and about two inches in depth, being deeper behind than in front. It is bounded, *internally*, by the Sphincter ani, Levator ani, and Coccygeus muscles; *externally*, by the tuberosity of the ischium and the obturator fascia, which covers the inner surface of the Obturator internus muscle; *in front*, it is limited by the line of junction of the superficial fascia with the base of the triangular ligament; and *behind*, by the margin of the Gluteus maximus, and the great sacro-sciatic ligament. This space is filled with a large mass of adipose tissue, which explains the frequency with which abscesses in the neighbourhood of the rectum burrow to a considerable depth.

FIG. 806.—Dissection of perinæum and ischio-rectal region.

If the subject has been injected, on placing the finger on the outer wall of this fossa, the internal pudic artery, with its accompanying veins and the two divisions of the nerve, will be felt about an inch and a half above the margin of the ischiatic tuberosity, but approaching nearer the surface as they pass forwards along the inner margin of the pubic arch. These structures are enclosed in a sheath (canal of Alcock) formed by the obturator fascia, the perineal nerve lying below the artery and the dorsal nerve of the penis above it. Crossing the space transversely about its centre, are the inferior hæmorrhoidal vessels and nerves, which are distributed to the integument of the anus, and to the muscles of the lower end of the rectum. These vessels are occasionally of large size, and may give rise to troublesome hæmorrhage when divided in the operation of lithotomy, or in that for fistula in ano. At the back part of this space, near the coccyx, may be seen a branch of the fourth sacral nerve; and at the fore part of the space, the superficial perineal vessels and nerves can be seen for a short distance.

THE UROGENITAL TRIANGLE IN THE MALE

The deep boundaries of the urogenital triangle are, laterally, the rami of the pubic bones and ischia, meeting, in front, at the pubic arch: behind, an imaginary transverse line, extending between the anterior parts of the tuberosities of the ischia. The lateral boundaries are, in the adult, from three inches to three inches and a half in length; and the base from two inches to three inches and a half in breadth; the average extent of the space being two inches and three-quarters.

The variations in the diameter of this space are of extreme interest in connection with the operation of lithotomy, and the extraction of a stone from the cavity of the bladder. In those cases where the tuberosities of the ischia are near together it would be necessary to make the incision in the operation of lateral lithotomy less oblique than if the tuberosities were widely separated, and the perineal space, consequently, wider. The perinæum is subdivided by the median raphé into two equal parts. Of these, the left is the one in which the operation of lateral lithotomy is performed.

In the middle line the triangle is convex, and corresponds to the bulb of the urethra. The skin is of a dark colour, thin, freely movable upon the subjacent parts, and covered with sharp crisp hairs, which should be removed before the dissection of the part is commenced. In front of the anus a prominent ridge or *raphé* commences, continuous with the raphé of the scrotum.

Fig. 807.—The superficial muscles and vessels of the perinæum.

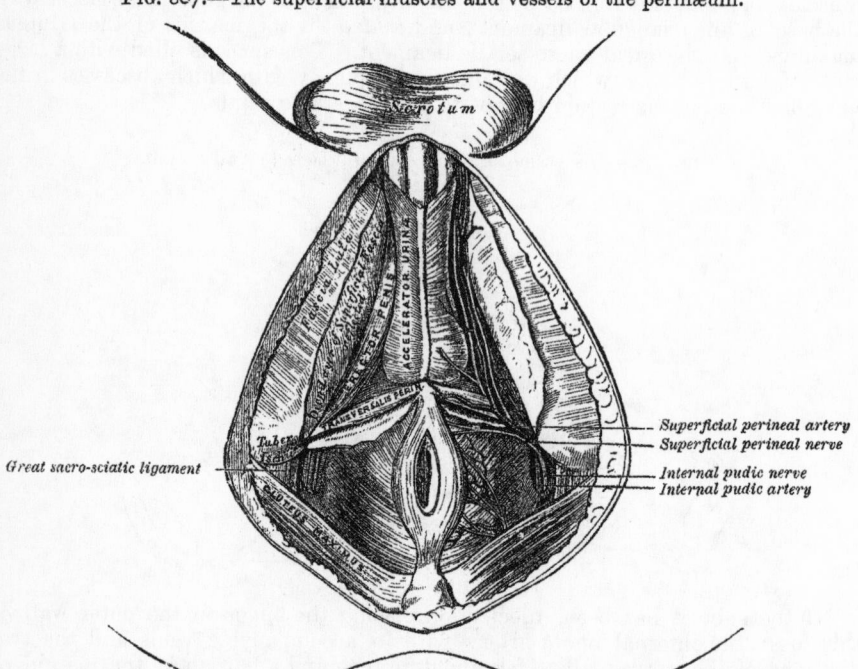

Upon removing the skin and superficial structures from this region, in the manner shown in fig. 806, a plane of fascia will be exposed, covering in the triangular space and stretching across from one ischio-pubic ramus to the other. This is the *deep layer of the superficial fascia*, or *fascia of Colles*. It has already been described (page 496). It is a layer of considerable strength, and encloses and covers a space in which are contained muscles, vessels, and nerves. It is continuous, in front, with the dartos of the scrotum; on each side, it is firmly attached to the margin of the ischio-pubic rami and to the tuberosity of the ischium; and posteriorly, it curves behind the Transversi perinæi muscles to join the base of the triangular ligament.

It is between this layer of fascia and the triangular ligament of the urethra that extravasation of urine most frequently takes place in cases of rupture of the urethra. The triangular ligament of the urethra (see page 498) is attached to the ischio-pubic rami, and in front to the subpubic ligament. It is clear, therefore, that when extravasation of fluid takes place between these two layers, it cannot pass backwards, because the two layers are continuous with each other around the Transversi perinæi muscles; it cannot extend laterally, on account of the connection of both these layers to the rami of the os pubis and ischium; it cannot find its way into the pelvis, because the opening into this cavity is closed by the triangular ligament, and, therefore, so long as these two layers remain intact, the only direction in which the fluid can make its way is forwards into the areolar tissue of the scrotum and penis, and thence on to the anterior wall of the abdomen.

When the deep layer of the superficial fascia is removed, a space is exposed, between this fascia and the triangular ligament, in which are contained the superficial perineal vessels and nerves and some of the muscles connected with the penis and urethra, viz.: in the middle line, the Accelerator urinæ; on each side, the Erector penis; and behind, the Transversus perinæi; together with the crura of the corpora cavernosa and the bulb of the corpus spongiosum. Here also is seen the *central tendinous point of the perinæum*. This is a fibrous point in the middle line of the perinæum between the bulb of the urethra and the anus, being about half an inch in front of the latter. At this point four muscles converge and are attached, viz. the External sphincter ani, the Accelerator

FIG. 808.—Deep perineal fascia. On the left side the inferior layer has been removed.

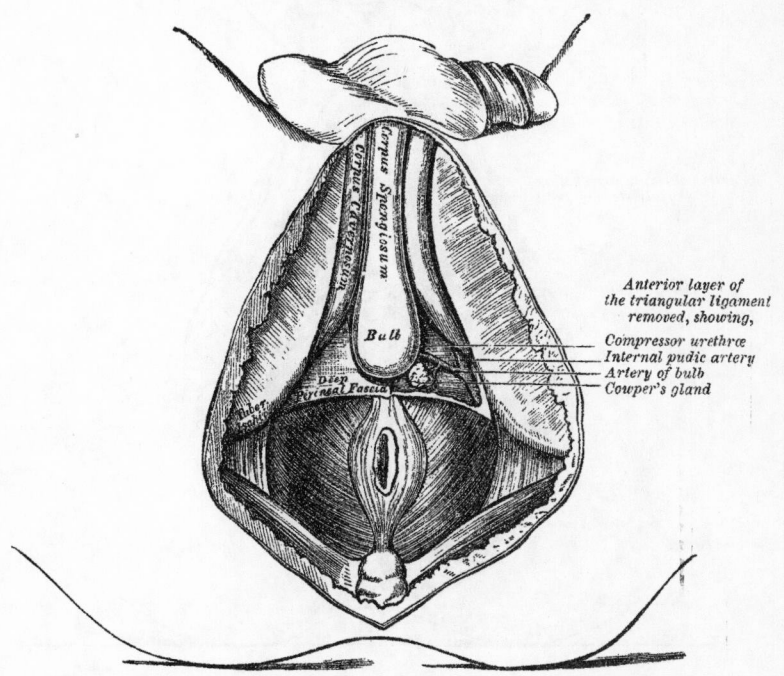

urinæ, and the two Transversi perinæi muscles; so that by the contraction of these muscles, which extend in opposite directions, it serves as a fixed point of support.

The Accelerator urinæ, the Erector penis, and the Transversus perinæi muscles have been already described (page 497). They form a triangular space, bounded, internally, by the Accelerator urinæ; externally, by the Erector penis; and behind, by the Transversus perinæi. The floor of this space is formed by the triangular ligament of the urethra; and running from behind forwards in it are the superficial perineal vessels and nerves, and the transverse perineal artery coursing along the posterior boundary of the space, on the Transversus perinæi muscle.

The Accelerator urinæ and Erector penis should now be removed, when the triangular ligament of the urethra will be exposed, stretching across the front of the outlet of the pelvis. The urethra is seen perforating its centre, just behind the bulb; and on each side is the crus penis, connecting the corpus cavernosum with the rami of the ischium and os pubis.

The **triangular ligament**, which has already been described (see page 498), consists of two layers, the inferior or superficial layer of which is now exposed. It is united to the superior or deep layer behind, but is separated in front by an interval in which are contained certain structures.

The *inferior layer of the triangular ligament* consists of a strong fibrous membrane, the fibres of which are disposed transversely, which stretches across from one ischio-pubic ramus to the other, and completely fills in the pubic arch; it is attached in front to the subpubic ligament, except in the middle line, where a small interspace is left for the dorsal vein of the penis. In the erect position of the body it is almost horizontal. It is perforated by the urethra in the middle line, and on each side of the urethral opening by the ducts of Cowper's glands and by the arteries of the bulb; in front, and external to this, by the artery of the corpus cavernosum, immediately before this vessel enters the crus penis. Near its apex the ligament is perforated by the termination of the pudic artery and by the dorsal nerve of the penis. The *crura penis* are exposed, lying

FIG. 809.—A view of the position of the viscera at the outlet of the pelvis.

superficial to this ligament. They are attached by blunt-pointed processes to the rami of the os pubis and ischium, in front of the tuberosities, and passing forwards and inwards, are continued as the corpora cavernosa. In the middle line the bulb and corpus spongiosum are exposed by the removal of the Accelerator urinæ muscle.

If the superficial layer of the triangular ligament is detached on either side, the deep perineal interspace will be exposed and the following parts will be seen, between it and the deep layer of the ligament, viz.: the subpubic ligament in front, close to the symphysis pubis; the dorsal vein of the penis; the membranous portion of the urethra and the Compressor urethræ muscle; Cowper's glands and their ducts; the pudic vessels and the dorsal nerve of the penis; the artery and nerve of the bulb, and a plexus of veins.

The *superior layer of the triangular ligament* is derived from the obturator fascia, and is continuous with it along the pubic arch. Behind, it joins with the superficial layer of the triangular ligament, and is continuous with the anal fascia. Above it is the prostate gland, supported by the anterior fibres of the Levator ani, which act as a sling for the gland and form the Levator prostatæ muscle. The superior layer of the triangular ligament is continuous round the anterior free edge of this muscle with the recto-vesical layer covering the prostate gland. The superior layer of the triangular ligament is perforated by the urethra. Between the two layers of the triangular ligament are situated the membranous part of the urethra, enveloped by the Compressor urethræ muscle; the ducts of Cowper's glands; the arteries to the bulb; the pudic vessels and the dorsal nerve of the penis. The membranous part of the urethra is about three-quarters of an inch in length, and passes downwards and forwards behind the symphysis pubis, from which it is distant about an inch. It is the narrowest part of the tube, and is enveloped, as has already been stated, by the Compressor urethræ muscle.

The **Compressor urethræ** has already been described (page 500). In addition to this muscle and immediately beneath it, *circular muscular fibres* surround the

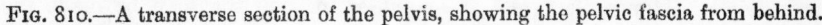

FIG. 810.—A transverse section of the pelvis, showing the pelvic fascia from behind.

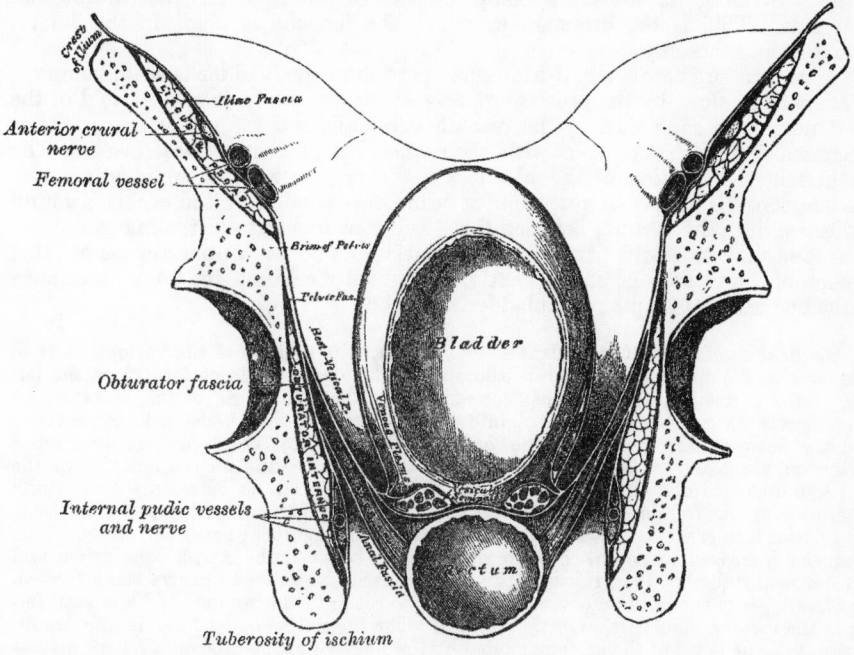

membranous portion of the urethra from the bulb in front to the prostate behind, and are continuous with the muscular fibres of the bladder. These fibres are involuntary.

Cowper's glands are situated immediately below the membranous portion of the urethra, close behind the bulb and below the artery of the bulb.

The **pudic vessels** and **dorsal nerve of the penis** are placed along the inner margin of the pudic arch (pages 689 and 926).

The **artery of the bulb** passes transversely inwards, from the internal pudic along the base of the triangular ligament, between its two layers, accompanied by a branch of the pudic nerve (page 690). If the deep layer of the triangular ligament is removed and the crus penis of one side detached from the bone, the under or perineal surface of the Levator ani is brought fully into view. This muscle, with the triangular ligament in front and the Coccygeus and Pyriformis behind, closes the outlet of the pelvis.

The **Levator ani and Coccygeus** muscles have already been described (pages 494–496).

Position of the Viscera at the Outlet of the Pelvis.—Divide the central tendinous point of the perinæum, separate the rectum from its connections by dividing the fibres of the Levator ani, which descend upon the sides of the prostate gland, and draw the gut backwards towards the coccyx, when the under surface of the prostate gland, the neck and base of the bladder, the vesiculæ seminales, and the vasa deferentia will be exposed.

The **prostate gland** is a pale, firm, glandular body which is placed immediately below the neck of the bladder, around the commencement of the urethra. It is situated in the pelvic cavity, behind the lower part of the symphysis pubis, above the deep layer of the triangular ligament, and rests upon the rectum, through which it may be distinctly felt, especially when enlarged. In shape and size it resembles a chestnut. Its base is directed upwards towards the neck of the bladder. Its apex is directed downwards to the deep layer of the triangular ligament, which it touches. Its posterior surface is smooth, marked by a slight longitudinal furrow, and rests on the second part of the rectum, to which it is connected by areolar tissue. Its anterior surface is flattened, marked by a slight longitudinal furrow, and placed about three-quarters of an inch behind the pubic symphysis.

The prostate measures about an inch and a half in its transverse diameter at the base, an inch in its antero-posterior diameter, and three-quarters of an inch in depth. Hence the greatest extent of incision that can be made in it without dividing its substance completely across, is obliquely backwards and outwards. This is the direction in which the incision is made in the lateral operation of lithotomy.

Above the prostate a small triangular portion of the bladder is seen, bounded, in front and below, by the prostate gland; above, by the recto-vesical fold of the peritoneum; on each side, by the vesicula seminalis and the vas deferens. It is separated from direct contact with the rectum by the recto-vesical fascia. The relation of this portion of the bladder to the rectum is of extreme interest to the surgeon. In cases of retention of urine this portion of the organ is found projecting into the rectum, between three and four inches from the margin of the anus, and may be easily perforated, without injury to any important parts; this portion of the bladder is, consequently, occasionally selected for the performance of the operation of tapping the bladder.

Surgical Anatomy.—The student should consider the position of the various parts in reference to the operation of lateral lithotomy. This operation is performed on the left side of the perinæum, as it is most convenient for the right hand of the operator. A grooved staff having been introduced into the bladder, the first incision is commenced midway between the anus and the back of the scrotum (i.e. in an ordinary adult perinæum, about an inch and a half in front of the anus), a little on the left side of the raphé, and carried obliquely backwards and outwards to midway between the anus and tuberosity of the ischium. The incision divides the integument and superficial fascia, the inferior hæmorrhoidal vessels and nerves, and the superficial perineal vessels. It is gradually increased in depth, through the interval between the Accelerator urinæ and Erector penis muscles, the Transversus perinæi muscles, vessels and nerves being divided. The forefinger of the left hand is now thrust upwards into the wound, and the staff felt for in the membranous portion of the urethra. The finger-nail being fixed in the groove of the staff, the point of the knife is guided by the finger into the groove. When this has been entered, the knife is turned outwards and pushed along the groove in the staff into the bladder. This thrust will divide the membranous portion of the urethra and part of the left lobe of the prostate gland, to the extent of about an inch. The knife is then withdrawn, and the forefinger of the left hand passed along the staff into the bladder. The position of the stone having been ascertained, the staff is removed, and the forceps introduced over the finger into the bladder. If the stone is very large, the opposite side of the prostate may be notched before the forceps is introduced: the finger is now withdrawn, and the blades of the forceps opened and made to grasp the stone, which must be extracted by slow and cautious undulating movements.

Parts divided in the operation.—The various structures divided in this operation are as follows : the integument, superficial fascia, inferior hæmorrhoidal vessels and nerves and probably the superficial perineal vessels and nerves, the posterior fibres of the Accelerator urinæ, the Transversus perinæi muscle and artery, the triangular ligament, the anterior fibres of the Levator ani, part of the Compressor urethræ, the membranous and prostatic portions of the urethra, and part of the prostate gland.

Parts to be avoided in the operation.—In making the necessary incisions in the perinæum for the extraction of a calculus, the following parts should be avoided. The primary incision should not be made too near the middle line, for fear of wounding the bulb of the corpus spongiosum or the rectum; nor too far externally, otherwise the pudic artery may be implicated as it ascends along the inner border of the pubic arch. If the incisions are carried too far forwards, the artery of the bulb may be divided; if carried too far backwards, the entire breadth of the prostate and neck of the bladder may be cut through, which allows the urine to become infiltrated behind the pelvic fascia into the loose areolar tissue between the bladder and rectum, instead of escaping externally :

diffuse inflammation is consequently set up, and peritonitis, from the close proximity of the recto-vesical peritoneal fold, is the result. If, on the contrary, only the anterior part of the prostate is divided, the urine makes its way externally, and there is less danger of infiltration taking place.

During the operation it is of great importance that the finger should be passed into the bladder *before* the staff is removed ; if this is neglected, and if the incision made in the prostate and neck of the bladder is too small, great difficulty may be experienced in introducing the finger afterwards ; and in the child, where the connections of the bladder to the surrounding parts are very loose, the force made in the attempt is sufficient to displace the bladder upwards into the abdomen, out of the reach of the operator. Such a proceeding has not infrequently occurred, producing the most embarrassing results, and total failure of the operation.

It is necessary to bear in mind that the arteries in the perinæum occasionally take an abnormal course. Thus the artery of the bulb, when it arises, as sometimes happens, from the pudic opposite the tuber ischii, is liable to be wounded in the operation for lithotomy, in its passage forwards to the bulb. The accessory pudic may be divided near the posterior border of the prostate gland, if this is completely cut across ; and the prostatic veins, especially in people advanced in life, are of large size, and give rise, when divided, to troublesome hæmorrhage.

THE FEMALE PERINÆUM

The **female perinæum** presents certain differences from that of the male, in consequence of the whole of the structures which constitute it being perforated in the middle line by the vulvo-vaginal passage.

The **superficial fascia,** as in the male, consists of two layers : of which the superficial one is continuous with the superficial fascia over the rest of the body ; and the deep layer, corresponding to the fascia of Colles in the male, is like it attached to the ischio-pubic ramus, and in front is continued forwards through the labia majora to the inguinal region. It is of less extent than the male, in consequence of being perforated by the aperture of the vulva.

On removing this fascia the muscles of the female perinæum, which have already been described (page 500), are exposed. The Sphincter vaginæ, corresponding to the Accelerator urinæ in the male, consists of an attenuated plane of fibres, forming an orbicular muscle around the orifice of the vagina, instead of being united in a median raphé, as in the male. The Erector clitoridis is proportionately reduced in size, but differs in no other respect; and the Transversus perinæi is similar to the muscle of the same name in the male.

The triangular ligament of the urethra is not so strongly marked as in the male. It transmits the urethra and the tube of the vagina.

The **Compressor urethræ** (*Transversus perinæi profundus*) corresponds with the Compressor urethræ in the male. It arises from the ischio-pubic ramus, and, passing inwards, its anterior fibres blend with the muscle of the opposite side in front of the urethra ; its middle fibres, the most numerous, are inserted into the side of the vagina, and the posterior fibres join the central point of the perinæum.

The distribution of the internal pudic artery is the same as in the male (see page 691), and the pudic nerve has also a similar arrangement, the dorsal nerve being, however, very small and supplying the clitoris.

The corpus spongiosum is divided into two lateral halves, which are represented by the *bulbi vestibuli* and *pars intermedia* (see page 1149).

The **perineal body** fills up the interval between the lower part of the vagina and the rectum. Its base is covered by the skin lying between the anus and vagina on what is called the 'perinæum.' Its anterior surface lies behind the posterior vaginal wall, and its posterior surface lies in front of the anterior rectal wall and the anus. It measures about an inch and a quarter from before backwards, and laterally extends from one tuberosity of the ischium to the other. In it are situated the muscles belonging to the external organs of generation. Through its centre runs the transverse perineal septum, which is of great strength in women, and forms on either side, behind the posterior commissure, a hard, ill-defined body, consisting of connective tissue, with much yellow elastic tissue and interlacing bundles of involuntary muscular fibres, in which the voluntary muscles of the perinæum are inserted.

4 H

PELVIC FASCIA

The **pelvic fascia** (fig. 811) is a thin membrane which lines the cavity of the pelvis, and is continuous over the back part of the ilio-pectineal line with the iliac fascia. It is attached to the brim of the pelvis, for a short distance, at the side of the cavity, and to the inner surface of the bone round the attachment of the Obturator internus. At the posterior border of this muscle, it is continued backwards as a very thin membrane in front of the Pyriformis muscle and sacral nerves to the front of the sacrum. In front it follows the attachment of the Obturator internus to the bone, arches beneath the obturator vessels, completing the orifice of the obturator canal, and at the front of the pelvis is attached to the lower part of the symphysis pubis. At the level of a line extending from the lower part of the symphysis pubis to the spine of the ischium, is a thickened

FIG. 811.—Side view of the pelvic viscera of the male subject, showing the pelvic and perineal fascia.

whitish band, termed the *white line* ; this marks the attachment of the Levator ani muscle to the pelvic fascia; and corresponds to its point of division into two layers, the *obturator* and *recto-vesical*.

The **obturator fascia** descends and covers the Obturator internus muscle. It is a direct continuation of the parietal pelvic fascia below the white line above mentioned, and is attached to the pubic arch, the ischial tuberosities, and to the margin of the great sacro-sciatic ligaments. This fascia forms a canal for the pudic vessels and nerve in their passage forwards to the perinæum, and gives off a thin membrane which covers the perineal aspect of the Levator ani muscle, called the *ischio-rectal (anal) fascia*. From its attachment to the rami of the os pubis and ischium a process is given off which is continuous with a similar process from the opposite side, so as to close the front part of the outlet of the pelvis, forming the deep layer of the triangular ligament.

The **recto-vesical fascia** (*visceral layer of the pelvic fascia*) descends into the pelvis upon the upper surface of the Levator ani muscle, and invests the prostate, bladder, and rectum. From the inner surface of the symphysis pubis a short rounded band is continued, on each side of the middle line, to the upper surface of the prostate and neck of the bladder, forming the pubo-prostatic **or** anterior true ligaments of the bladder. At the side, this fascia is connected to the prostate, enclosing this gland and the vesico-prostatic plexus of veins, and is continued on to the side of the bladder, forming the lateral true ligaments of the organ. Another prolongation invests the vesiculæ seminales, and passes across between the bladder and rectum, being continuous with the same fascia of the opposite side. Another thin prolongation is reflected round the surface of the lower end of the rectum. The Levator ani muscle arises from the point of division of the pelvic fascia; the visceral layer of the fascia descending upon and being intimately adherent to the upper surface of the muscle, while the under surface of the muscle is covered by a thin layer derived from the obturator fascia, called the *ischio-rectal* or *anal fascia*. In the female, the vagina perforates the recto-vesical fascia, and receives a prolongation from it.

INDEX